Anne E. Hochwalt

BEST AND TAYLOR'S
Physiological Basis
of Medical Practice

ELEVENTH EDITION

BEST AND TAYLOR'S
Physiological Basis
of Medical Practice

ELEVENTH EDITION

EDITED BY

John B. West, M.D., Ph.D.

Professor of Physiology and Medicine
University of California, San Diego
School of Medicine
La Jolla, California

WILLIAMS & WILKINS
Baltimore • Hong Kong • London • Sydney

Editor: Maureen K. Vardoulakis
Associate Editor: Carol-Lynn Brown
Copy Editors: Andrea Clemente, Stephen Siegforth
Design: Joanne Janowiak
Illustration Planning: Wayne Hubble
Production: Raymond E. Reter

Accurate indications, adverse reactions, and dosage schedules for drugs are provided
in this book, but it is possible that they may change. The reader is urged to review
the package information data of the manufacturers of the medications mentioned.

Made in the United States of America

First Edition, 1937
Second Edition, 1939
Third Edition, 1943
Fourth Edition, 1945
Fifth Edition, 1950
Sixth Edition, 1955
Seventh Edition, 1961
Eighth Edition, 1966
Ninth Edition, 1973
Tenth Edition, 1979

Spanish Editions, 1939, 1941, 1943, 1947, 1954, 1982
Portuguese Editions, 1940, 1945, 1976
Rumanian Editions, 1958, 1966
Italian Editions, 1959, 1979
Polish Editions, 1960, 1971
Taiwan Edition, 1980
Indian Edition, 1967
Asian Edition, 1967
Philippine Edition (English), 1981

Library of Congress Cataloging in Publication Data

Best, Charles Herbert, 1899–
    Best and Taylor's Physiological basis of medical practice.

    Includes bibliographies and index.
    1. Human physiology. I. Taylor, Norman Burke, 1885–   . II. West, John
Burnard. III. Title. IV. Title: Physiological basis of medical practice.
[DNLM: 1. Physiology. QT 104 B561p]
QP34.5.B4713 1984   612   83-6613
ISBN 0-683-08944-7

88 89 90 91   10 9 8 7 6 5 4 3

Preface to the First Edition

Physiology is a science in its own right and the laboratory worker who pursues his research quite detached from medical problems need offer no apology for his academic outlook. Indeed some of the most valuable contributions to medical science have been the outcome of laboratory studies whose applications could not have been foreseen. Nevertheless, we feel that the teacher of physiology in a medical school owes it to his students, whose ultimate interest it must be conceded is in the diagnosis and treatment of disease, to emphasize those aspects of the subject which will throw light upon disorders of function. The physiologist can in this way play a part in giving the student and practitioner a vantage point from which he may gain a rational view of pathological processes.

We have endeavored to write a book which will serve to link the laboratory and the clinic, and which will therefore promote continuity of physiological teaching throughout the pre-clinical and clinical years of the undergraduate course. It is hoped that when the principles underlying diseased states are pointed out to the medical student, and he is shown how a knowledge of such principles aids in the interpretation of symptoms or in directing treatment, he will take a keener interest in physiological studies. When such studies are restricted to the classical aspects of the subject, apparently remote from clinical application, the student is likely to regard them only as a task which his teachers in their inscrutable wisdom have condemned him to perform. Too often he gains the idea, from such a course, that physiology is of very limited utility and comes to believe that, having once passed into the clinical years, most of what he has "crammed" for examination purposes may be forgotten without detriment to his more purely medical studies. Unfortunately, he does not always realize at this stage in his education how great has been the part which physiological discoveries have played in the progress of medicine, and that the practice of today has evolved from the "theories" of yesterday.

Many physiological problems can be approached only through animal experimentation. Advances in many fields, most notably in those of carbohydrate metabolism, nutrition, and endocrinology, bear witness to the fertility of this method of research. On the other hand, many problems can be elucidated only by observations upon man, and physiology has gained much from clinical research. The normal human subject as an experimental animal possesses unique advantages for many types of investigation; and in disease, nature produces abnormalities of structure and function which the physiological laboratory can imitate only in the crudest way. Within recent years the clinical physiologist, fully realizing these advantages and the opportunities afforded by the hospital wards, has contributed very largely to physiological knowledge. In many instances, clinical research has not only revealed the true nature of the underlying process in disease, but has cast a light into some dark corner of physiology as well; several examples of clinical investigation which have pointed the way to the physiologist could be cited. In the last century, knowledge of the processes of disease was sought mainly in studies of morbid *anatomy*; biochemistry was in its infancy and many of the procedures now commonly employed for the investigation of the human subject had not been devised. Today, the student of scientific medicine is directing his attention more and more to the study of morbid *physiology* in his efforts to solve clinical problems. This newer outlook has borne fruit in many fields. It has had the beneficent result of drawing the clinic and the physiological and biochemical laboratories onto common ground from which it has often been possible to launch a joint attack upon disease. We feel that this modern trend in the field of research should be reflected in the teaching of medical students, and have therefore given greater prominence to clinical aspects of the subject than is usual in physiological texts.

In order to understand the function of an organ it is usually essential to have a knowledge of its structure. For this reason we have followed the plan of preceding the account of the physiology of a part by a short description of its morphology and, in many instances, of its nerve and blood supply. The architecture and functions of the central nervous system are so intimately related that some space has been devoted to a description of the more important fiber tracts and grey masses of the cerebrum, cerebellum and spinal cord.

We wish to thank our colleagues in physiology, biochemistry and anatomy whom we have drawn upon on so many occasions for information and advice; without their generous help the undertaking would have been an almost impossible one. We are also deeply grateful for the unstinted assistance which we have received from our friends on the clinical staff, several of whom have read parts of the text in manuscript or in proof.

October 15, 1936

C.H.B.
N.B.T.

Preface to the Eleventh Edition

We agree. Indeed this new edition returns this book closer to its original laudable objectives. The revisions are very extensive. Of the nine sections, seven have been completely rewritten or thoroughly revised by new section editors. We welcome the new section editors, Frank P. Brooks, M.D., Darrell D. Fanestil, M.D., Gordon N. Gill, M.D., Nora Laiken, Ph.D., Robert B. Livingston, M.D., Samuel I. Rapaport, M.D., John Ross, Jr., M.D., and Daniel Steinberg, M.D., Ph.D.

Of the two remaining sections, the first (General Physiological Processes) has been brought up to date, and much of the physiological chemistry has been incorporated into a new section on Metabolism. The only other section remaining from the 10th edition (Respiration) has been updated, and the chapter on Control of Respiration completely rewritten. Thus, this new edition represents a major change.

The guiding principle of this edition has been to produce a book that emphasizes the clinical relevance of physiology, yet is authoritative with appropriate scientific rigor. In general, the section editors are people who teach first year medical students and who have firsthand knowledge of the issues which are important in clinical medicine. We have tried to produce a text that is clearly written, didactic, not necessarily encyclopedic, and which recognizes the limited time available to medical students.

We wish to thank Michelle Lambert for nearly all the new illustrations, and the editors and staff of Williams & Wilkins for their cooperation.

November 1984

For the editors
J.B.W.

Section Editors

General Physiological Processes
Irving L. Schwartz, M.D.

Cardiovascular System
John Ross, Jr., M.D.

Blood and Lymph
Samuel I. Rapaport, M.D.

Body Fluids and Renal Function
Nora Laiken, Ph.D., and Darrell D. Fanestil, M.D.

Respiration
John B. West, M.D., Ph.D.

Gastrointestinal System
Frank P. Brooks, M.D.

Metabolism
Daniel Steinberg, M.D., Ph.D.

Endocrine System
Gordon N. Gill, M.D.

Neurophysiology
Robert B. Livingston, M.D.

Contributors

SECTION 1

Irving L. Schwartz, M.D.
Harold and Golden Lamport Distinguished Service
    Professor
Mount Sinai School of Medicine
New York, New York

Madeleine A. Kirchberger, Ph.D.
Associate Professor of Physiology and Biophysics
Mount Sinai School of Medicine
New York, New York

Sandra K. Masur, Ph.D.
Assistant Professor of Physiology and Biophysics
Mount Sinai School of Medicine
New York, New York

George J. Siegel, M.D.
Professor of Neurology
The University of Michigan
School of Medicine
Ann Arbor, Michigan

Herman R. Wyssbrod, Ph.D.
Associate Professor of Physiology and Biophysics
Mount Sinai School of Medicine
New York, New York

SECTION 2

John Ross, Jr., M.D.
Professor of Medicine
University of California, San Diego
School of Medicine
La Jolla, California

James W. Covell, M.D.
Professor of Medicine
University of California, San Diego
School of Medicine
La Jolla, California

SECTION 3

Samuel I. Rapaport, M.D.
Professor of Medicine
University of California, San Diego
School of Medicine
La Jolla, California

Helen M. Ranney, M.D.
Professor of Medicine
University of California, San Diego
School of Medicine
La Jolla, California

Raymond Taetle, M.D.
Associate Professor of Pathology and Medicine
University of California, San Diego
School of Medicine
La Jolla, California

SECTION 4

Nora D. Laiken, Ph.D.
Assistant Dean
University of California, San Diego
School of Medicine
La Jolla, California

Darrell D. Fanestil, M.D.
Professor of Medicine
University of California, San Diego
School of Medicine
La Jolla, California

SECTION 5

John B. West, M.D., Ph.D.
Professor of Physiology and Medicine
University of California, San Diego
School of Medicine
La Jolla, California

SECTION 6

Frank P. Brooks, M.D.
Professor of Medicine and Physiology
University of Pennsylvania
School of Medicine
Philadelphia, Pennsylvania

SECTION 7
Daniel Steinberg, M.D., Ph.D.
Professor of Medicine
University of California, San Diego
School of Medicine
La Jolla, California

C. Wayne Bardin, M.D.
Director, Population Council
Center for Biomedical Research
Rockefeller University
New York, New York

Rosemarie B. Thau, Ph.D.
Scientist, Population Council
Center for Biomedical Research
Rockefeller University
New York, New York

SECTION 8
Gordon N. Gill, M.D.
Professor of Medicine
University of California, San Diego
School of Medicine
La Jolla, California

SECTION 9
Robert B. Livingston, M.D.
Professor of Neurosciences
University of California, San Diego
School of Medicine
La Jolla, California

Contents

SECTION ONE: GENERAL PHYSIOLOGICAL PROCESSES

edited by Irving L. Schwartz

SECTION TWO: CARDIOVASCULAR SYSTEM

edited by John Ross, Jr.

SECTION THREE: BLOOD AND LYMPH

edited by Samuel I. Rapaport

SECTION FOUR: BODY FLUIDS AND RENAL FUNCTION

edited by Nora D. Laiken and Darrell D. Fanestil

SECTION FIVE: RESPIRATION

edited by John B. West

SECTION SIX: GASTROINTESTINAL SYSTEM

edited by Frank P. Brooks

SECTION SEVEN: METABOLISM

edited by Daniel Steinberg

SECTION EIGHT: ENDOCRINE SYSTEM

edited by Gordon N. Gill

SECTION NINE: NEUROPHYSIOLOGY

edited by Robert B. Livingston

General Physiological Processes

CHAPTER 1

By Way of Introduction: The Cell

The science of mammalian physiology involves the study of dynamic interrelationships that exist among cells, tissues, and organs, and reaches ultimately to the level of the organism as a whole. In this chapter we shall consider the structure and function of the organelles which comprise the cell. One group of organelles is bounded by a limiting membrane; a second group is not so delimited. The former group encompasses the nucleus, endoplasmic reticulum, Golgi apparatus, lysosome, peroxisome, and mitochondrion—and in this category we also include the plasma (outer) membrane of the cell as a whole. The group of organelles that are not bounded by a membrane includes the chromosomes, nucleoli, microtubules, ribosomes, microfilaments, and centrioles.

Although there was some knowledge of the cell boundary and subcellular organelles prior to the advent of the electron microscope, the complicated ultrastructure of intracellular membrane systems was unforeseen (Fig. 1.1).

The following two membrane properties are particularly significant in accounting for differences of function among the various membrane-delimited subcellular organelles: (a) the property of selective permeability and transport which maintains compartments of specific character within the cell; and (b) the ability to provide a matrix which lends its surface as well as its interior for the arrangement of an ordered sequence of molecules (enzymes, cofactors, carriers) in a functionally meaningful pattern.

CELL MEMBRANE

The cell membrane is a permeability barrier. If a cell is placed in hypotonic solution and if it contains molecules which cannot penetrate its outer mem-

brane, it will swell; conversely, it will shrink if placed in a hypertonic medium. In both instances water moves down its concentration gradient (Chapter 2). Thus, the cell behaves as an osmometer. Nonpolar molecules (gases, lipids) move freely across the membrane; polar molecules penetrate the membrane much less readily and indeed it is the selective permeability of the plasma membrane to certain ions which determines the excitability characteristics of nerve and muscle cells.

Chemical Composition

Although the general chemical nature of the membrane found at the cell boundary was predicted on the basis of physiological data, the detailed molecular structure of the membrane is not yet known. Models of cell membranes prepared by combining their lipid and protein constituents (partly known from chemical analysis of purified cell membrane preparations) exhibit some physiological characteristics similar to those of natural membranes. The role of lipid-lipid interactions in membrane structure has been at the center of attention because such interactions can explain much of the presently known phenomena of membrane transport. Quantitative studies of isolated cell membranes revealed that enough lipid is present to be arranged as a bilayer coating the cell. Artificial mixtures of extracted cellular polar lipids (lecithin, phospholipid, and steroids) under appropriate conditions will form a bimolecular layer spontaneously. Presumably the polar (hydrophilic) ends of the lipids form the two outer borders, making them available for interaction with other polar molecules such as proteins (Fig. 1.2). On a weight basis membranes contain a significantly larger amount of protein than lipid (ratio up to 4:1); however, due to

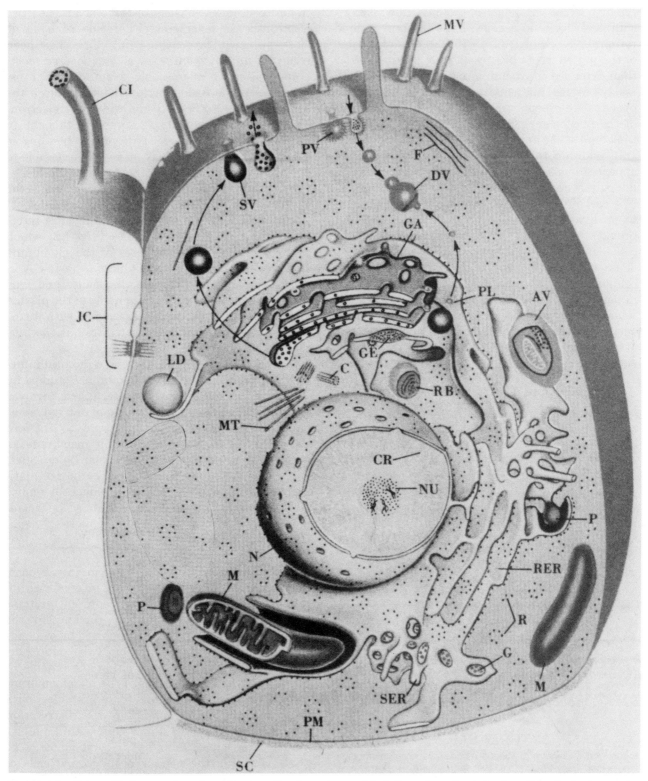

Figure 1.1. Schematic diagram of a cell and its organelles drawn to reveal their three-dimensional structure. *AV*, autophagic vacuole; *C*, centriole; *CI*, cilium; *CR*, chromatin; *DV*, digestion vacuole; *F*, microfilaments; *G*, glycogen; *GA*, Golgi apparatus; *JC*, junctional complex; *LD*, lipid droplet; *M*, mitochondrion; *MT*, microtubules; *MV*, microvillus; *N*, nucleus; *NU*, nucleolus; *P*, peroxisome; *PL*, primary lysosome; *PM*, plasma membrane; *PV*, pinocytic vesicle; *R*, ribosomes and polysomes; *RB*, residual body; *RER*, rough endoplasmic reticulum; *SER*, smooth endoplasmic reticulum; *SV*, secretion vacuole. The organelles have been drawn only roughly to scale. The sizes and relative amounts of different organelles can vary considerably from one cell type to another. [Adapted frm Novikoff and Holtzman (1976).]

the high molecular weight of proteins, this relationship is reversed on a molar basis (protein to lipid ratio ranging from 1:100 to 10:100). At least some of the protein must be present in a globular rather than extended form; unfolding of enzymic protein results in loss of activity. In fact, some of the proteins are probably not restricted to the surfaces of the plasma membranes but rather extend into the bimolecular lipid layer (Fig. 1.2*B*).

In addition to lipid and protein, carbohydrate appears to be associated with the cell membrane as lipopolysaccharide and as protein-polysaccharide. The carbohydrate moieties of the membrane serve to modify the electrical charge at its surface and provide specific surface-binding sites. Cytochemically demonstrable polysaccharide-protein complex is associated with many cell surfaces as an extracellular layer. In some places, particularly at luminal surfaces, this layer forms a fuzzy coat—often referred to as a glycocalyx (Fig. 1.3)—which may act as a crude filter and/or facilitate the attachment of molecules for endocytic (see below) transport across the cell membrane.

Structure of the Unit Membrane

A pattern generally found in almost all cellular membranes prepared for microscopy by conven-

A

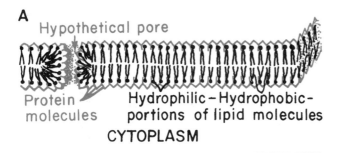

B.

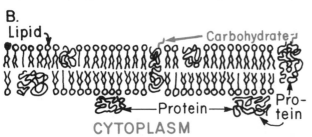

Figure 1.2. Diagram of various models of the cell membrane (after H. A. Davson, J. F. Danielli, S. J. Singer, and others). *A* represents the classic bimolecular lipid leaflet coated by protein; note inclusion of a hypothetical hydrophilic pore. *B* represents the involvement of integral and peripheral membrane proteins with the lipid components. Presumably the hydrophobic portions of certain proteins allow their integration within the hydrophobic lipid regions. Both lipid and protein components and associated groups of either surface, cytoplasmic or extracellular, may differ from one another. [From Novikoff and Holtzman (1976).]

tional techniques consists of three layers, i.e., two electron-dense layers on either side of a single electron-lucent layer (Fig. 1.3). This has been termed a *unit membrane*, or a *three-layered membrane*. Electron microscopic examination of osmium tetroxide-fixed, sectioned tissue shows the cell membrane to be 7–10 nm wide. The electron-lucent line in the unit membrane is thought to represent the lipid layer. Hydrophobic bonding in the lipid bilayer region may make it inaccessible to osmium deposition. Thus, the two electron-dense lines would result from deposition of osmium at the surfaces of this bilayer. There is much physiological evidence to suggest that the lipid bilayer is interrupted by hydrophilic molecules (Fig. 1.2) which connect the two outer surfaces of the membrane and provide transmembrane channels (a few Å in width) for transfers of water molecules and ions (Chapter 2). The protein components of the plasma membrane are either tightly associated with it (*integral*) or more readily dissociated from it (*peripheral*).

Additional electron microscopic structural information comes from unfixed membranes studied by the freeze-etch technique. This technique involves the quick freezing and sublimation of unfixed membranes. Membranes so treated tend to fracture along the middle layer of the unit membrane as would be expected from a bimolecular lipid leaflet. The evaporation of a thin layer of carbon or platinum onto the exposed surface produces a replica which is viewed in the electron microscope. Intramembrane particles are seen in replicas of membranes having integral proteins and also in various lipid-cholesterol mixtures.

In most metabolically active membranes, repeating structures are visible at various intervals within the layer. These structures may represent proteins that extend through much of the thickness of the membrane (Fig. 1.4). This is supported by evidence from other techniques which label the outer portions of integral proteins at either the outside or inside surface. The positions of the intramembranous particles may be more or less stable. With varying physiological conditions the distribution pattern of these particles may shift within a membrane. This lateral mobility of components, e.g., proteins within the membrane is seen if cells of different origins are caused to fuse. Rather than a patchwork of the two original membrane particle patterns the result is an intermixture of both characteristics.

On the other hand, certain membrane junctions are characterized by very ordered and stable patterns in freeze fracture (Fig. 1.4).

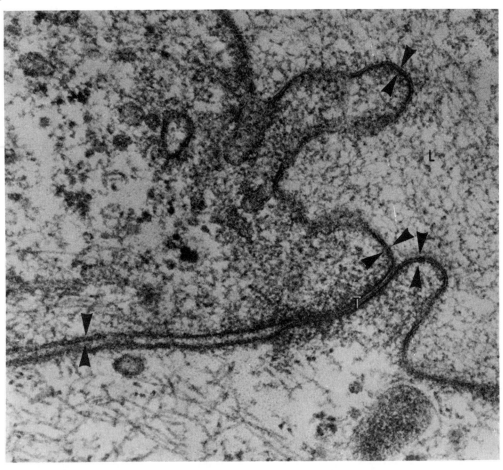

Figure 1.3. Electron micrograph of portions of epithelial cells (*E*) bordering the lumen (*L*) of the toad urinary bladder, as seen in a thin section. The three-layered membrane structure of two-cell membranes is seen between the *arrowheads*. In the tight junction (*T*) between the cells the external space is obliterated between the two cell membranes so that a five-layered image is seen (compare with Figure 1.4). The "fuzz" in the lumen is associated with the cell surface (see text) (×58,000).

Features of Membrane Structure Related to Transport

Although the physiology of the movement of molecules across membranes is dealt with in Chapter 2, we will here indicate a few selected examples of transport phenomena correlated with structural features of the cell membrane. The lipid bilayer serves as a transport medium for lipid-soluble molecules to gain entry into the cell, whereas protein-lined hydrophilic "pores" probably provide channels for diffusion of polar entities such as water and ions. Surprisingly, certain lipid bilayers show rates of movement sufficient to account for osmosis. The permeability to water, however, seems to be inversely related to the proportion of cholesterol in the phospholipid membranes. Specialized membrane-bound proteins seem to be involved in enzymatically controlled transport processes. For example, a particular type of adenosine triphosphatase (ATPase) is involved in the active transport of Na^+ and K^+ ions, a variety of "permeases" are involved in the initiation of transmembrane sugar transport, and adenylcyclase, a membrane-bound enzyme that converts adenosine triphosphate (ATP) to $3'$-$5'$-cyclic adenosine monophosphate, serves to mediate the actions of several hormones which attach to "receptor sites" on the exoplasmic surface of their target cells. A model for how transport might be achieved through a hypothetical pore comes from studies in which antibiotics, e.g., valinomycin, are added to artificial membranes and selectively increase the rates at which the Na^+ and K^+ can diffuse through the membrane. The hydrophobic outside of the circular valinomycin molecule could insert itself within the lipid bilayer, and its hydrophilic interior could transport the ions through the membrane.

Endocytosis and Exocytosis

Membrane components are subject to continual turnover. In certain cell types, portions of the membrane invaginate into the cell and pinch off to form

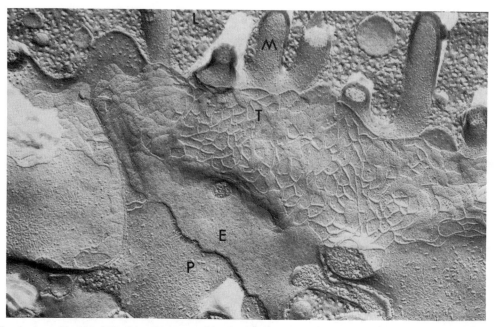

Figure 1.4. Toad urinary bladder epithelial cells prepared by the freeze fracture technique. The fracture has resulted in exposure of the inner or protoplasmic (*P*) leaflet of the membrane in which numerous particles are visible. Fewer particles are seen in the extracellular (*E*) fracture face. The fused region (*central dark line* in Fig. 1.3) of the tight junction (*T*) appears as a long cylindrical protuberance and complementary groove. The weblike nature of the extensive tight junction presumably prevents the intercellular movement of most materials out of the bladder lumen (*L*). A microvillus is seen at *M*. Greater understanding of fine structure is achieved by putting together the different sorts of information from images such as Figures 1.3 and 1.4 (×50,000). [Courtesy of J. B. Wade.]

the boundary of an intracellular vesicle, vacuole, or tubule. External material is carried into the cell by this process, referred to as *endocytosis* (Fig. 1.1; see also Chapter 2). This material and its enclosing membrane may fuse with lysosomes, or after delivering endocytosed material to an intracellular endosome (see Fig. 1.7), the specialized endocytic vesicle membrane may return to the plasma membrane (see below). *Exocytosis*, on the other hand, involves the fusion of membranes of intracellular origin, such as those delimiting secretory granules, with the cell membrane. These granules release their internal material to the outside of the cell but their membranes may return as endocytic vesicles to the Golgi area. As a result there is a flow of membrane and of material enclosed in membrane-delimited spaces between the surface and intracellular compartments. Similar exchanges seem to occur between certain intracellular organelles, notably the endoplasmic reticulum and the Golgi apparatus. In the various processes involving formation of vesicles (e.g., pinocytic endocytosis) there is often a "bristle" coating of fuzz on the cytoplasmic side of the vesicle membrane (Figs. 1.1 and 1.7). Exocytosis and incorporation into the surface of intracellular membrane-containing transport units (e.g., channels) may provide for rapid changes of cell permeability in response to hormonal stimulation.

Structural Aspects of Increasing the Cell Surface

Stable evaginations and invaginations of the cell membrane are important elements in providing a dramatic increase in surface area contact between cell and environment. Microvilli, finger-like evaginations, are generally associated with cell surfaces involved in absorption processes, such as in the intestine and kidney (Figs. 1.3 and 1.4). Conversely, in striated muscle, one finds an invagination of the cell membrane, the transverse tubule, associated with each sarcomere. Since the transverse tubules are continuous with the surface membrane they provide a direct route for the communication of alterations at the cell surface to the contractile system deep within the muscle fiber (Chapter 4).

THE NUCLEUS

Function of Nucleus

The nucleus has two principal functions: replication of deoxyribonucleic acid (DNA) and synthesis of ribosomal, messenger and transfer ribonucleic acids (RNAs). Because it is best understood, we shall discuss in some detail ribosomal RNA synthesis which occurs in the nucleolus.

Each nucleus posesses one or more *nucleoli*, not delimited by membranes. Each nucleolus consists of a roughly spherical, dense array of fibrils and granules rich in RNA. Often the nucleolus is found in intimate association with special regions of DNA (known as nucleolar organizer regions) which are presumed to carry the information for *ribosomal RNA*. Nucleolar RNA (45S predominantly) is almost certainly a precursor form of ribosomal RNA found in the cytoplasm; if one "labels" RNA synthesized in the nucleolus with radioactive nucleotides, labeled RNA molecules are subsequently detected in the cytoplasm. These RNA molecules complex with protein and form, respectively, a 30S and a 60S subunit. One small and one large subunit combine in the cytoplasm to form a *ribosome* (15–25 nm) in diameter. There may be several million ribosomes in a given cell.

The specific function of ribosomal RNA is not well understood; generally speaking, however, all three types of RNA are involved in the translation of genetic information contained in the DNA molecule into specific proteins that are synthesized in the cytoplasm. The ribosomes interact in the process of protein synthesis with two other types of RNA: large *messenger RNA* molecules (mRNAs) determine the sequence of amino acids in proteins by specifying the order of attachment of the small *transfer RNAs* carrying the appropriate amino acids. Our belief that DNA is the template for these RNAs is derived largely from experiments with procaryotic (bacterial) cells.

As the information in a messenger RNA molecule is being "read," several ribosomes attach via their smaller subunits to the mRNA. The combination of an mRNA and its attached ribosomes is referred to as a *polysome*. Each ribosome of a polysome synthesizes a polypeptide chain, so that several chains will be produced simultaneously by a polysome. The nascent peptide seems to be attached to the larger ribosomal subunit; completed protein is released to the cytoplasm. An average polypeptide may be synthesized in 10 to 20 s.

Morphology of the Nucleus

The interphase nucleus is readily seen in the light microscope as a spheroidal body with a "suggestion" of internal organization (Fig. 1.1). The DNA-containing material can be specifically stained. The nuclear chromatin can be resolved into two types: euchromatin (loosely coiled) and heterochromatin (compact). It seems likely that the euchromatic regions are more active in the transcription process than the heterochromatic regions, i.e., there is little demonstrable RNA synthesis in chromosomes that are largely or entirely composed of heterochromatin, as in the case of sperm cells, polymorphonuclear leukocytes, and the Barr body (one of the X chromosomes of female cells). The association of euchromatin with active transcription may account for some of the selective genetic expression associated with characteristic chromosomal uncoiling patterns found in different tissues within the same organism or at different developmental stages in the same tissue.

Chemical analysis has shown that the chromosome consists of DNA associated with basic proteins (histones) and with other (nonhistone) protein. It has been speculated that the complexing of histone with DNA may have a protective or structural function (preventing alteration or denaturation of the DNA, controlling coiling, etc.), or the histone may have a repressor function (interfering with the template activity of DNA). The amounts of RNA and nonbasic nuclear protein seem to vary in parallel with the metabolic activity of the cell—e.g., sperm cell nuclei have essentially neither RNA nor nonbasic protein.

Isolated chromosomes studied by electron microscopy appear as masses of fibers around 25 nm in diameter, or may have a beaded look with periodic DNA coiling around histone groups (nucleosomes). An individual DNA double helix coated with protein measures less than 5 nm, and while it is known that the fibers of chromosomes are coiled, the nature of the packaging of nucleic acid and proteins is yet to be described. Nevertheless, there are theories, consistent with current evidence, suggesting that a single chromosome contains one, or at most a very few, extremely elongated DNA molecules complexed with protein and coiled into a fiber structure which is seen in the electron microscope.

INTRACELLULAR MEMBRANE SYSTEMS

Nuclear Envelope

The boundary of the nucleus, the nuclear envelope, is a double membrane complex (Fig.1.5). Each membrane is approximately 7–8 nm thick. The envelope, a flattened sac with an enclosed perinuclear space, resembles the rough endoplasmic reticulum (ER) (see below): (1) the cytoplasmic surface of the outer (cytoplasmic) nuclear membrane has granules which appear to be ribosomes; (2) direct continuities are seen between the cytoplasmic portion of the nuclear membrane and the ER (Fig. 1.5); and (3) the presence of certain enzymes can be demonstrated cytochemically in both the perinuclear space and the cisternae of the ER.

The inner surface of the nuclear membrane is often associated with chromatin and an "internal dense lamella"; the latter may provide some rigidity

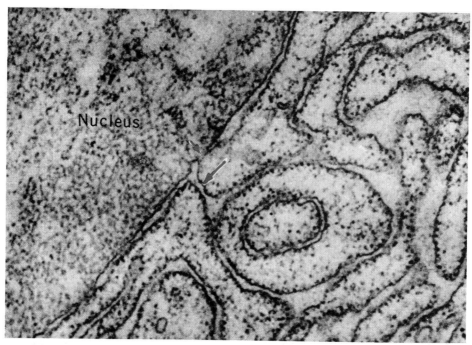

Figure 1.5. Electron micrograph of a portion of a serous cell of mouse salivary gland. The rough endoplasmic reticulum (*ER*) (*large arrow*) of the cell is continuous with the outer membrane of the nuclear envelope. A nuclear pore is indicated by the *small arrow* (osmium fixation; ×84,000). [From Fawcett (1966).]

to the structure. The inner and outer membranes of the envelope join at intervals to form "pores" tens of nanometers in diameter (Fig. 1.5).

How does a "directive" of the nucleus reach the cytoplasm or, conversely, how do cytoplasmic and other external feedback messages reach the nucleus? Nonnuclear substances can act as inducers or repressors of the synthesis of specific proteins in the cytoplasm. This almost certainly requires interaction with genes. Furthermore, most gene products (e.g., mRNA) must leave the nucleus and enter the cytoplasm to express their effects. Permeability properties of the nucleus are too complex to be explained by simple holes. The morphology of the nuclear boundary provides for two alternative routes for the transfer of information—either across membranes of the perinuclear sac or through "pores." The pores are often referred to as "annuli" to emphasize that they are not simple holes but rather organized regions: often pores are seen which contain a diaphragm or plug. In addition, the membrane adjacent to the pore may show morphological traces of special organization.

To date, the morphological evidence in support of the physiological and biochemical data on transnuclear transport through pores rests mainly on a few observations, such as the movement of electrondense material (thought to be RNA-containing granules) through the nuclear pores in the insect salivary gland and some other tissues, and the movement of a marker (colloidal gold) into the nucleus when it is injected into the cytoplasm of ameba. Clear morphological evidence on passage of material across the nuclear membranes, as distinct from the pores, is not available.

Endoplasmic Reticulum

Often an integrated biochemical and ultrastructural investigation (involving cell disruption, isolation, and analysis of a homogeneous organelle population) leads to the clearest understanding of organelle function in situ. From such studies, in a variety of cell types, a fraction of membrane-delimited vesicles, referred to as *microsomes*, is recovered. The microsomes perform several functions, including the provision of a base for the attachment of ribosomes, the biosynthesis of lipids, and, in the case of striated muscle, the accumulation and release of calcium. In the intact cell, microsomal vesicles are not found as such; rather, one observes a tubular network known as the endoplasmic reticulum (ER) (Figs. 1.1 and 1.5). It is assumed that the majority of microsomal vesicles represent a reproducible, preparative artifact arising during cell disruption by the shearing into fragments and closing up of the tubules and sacs of the endoplasmic reticulum.

ROUGH ER

As noted above, the endoplasmic reticulum (and/ or the microsomes derived from the ER) can provide a base for the attachment of ribosomes (Fig. 1.6). Such ribosome-carrying endoplasmic reticulum is referred to as *rough ER*. Microsomal vesicles derived from this rough ER were found to be capable of protein synthesis, the newly synthesized protein appearing in the vesicle lumen. In the rough ER in situ the nascent protein likewise can be demonstrated within the reticulum lumen. This unidirectional passage into the lumen is thought to result from the folding of the original .5 to 1-nm wide protein into a three-dimensional structure which is large enough to be retained.

The rough ER seems to grow by synthesizing more of itself. The newly made rough ER may lose

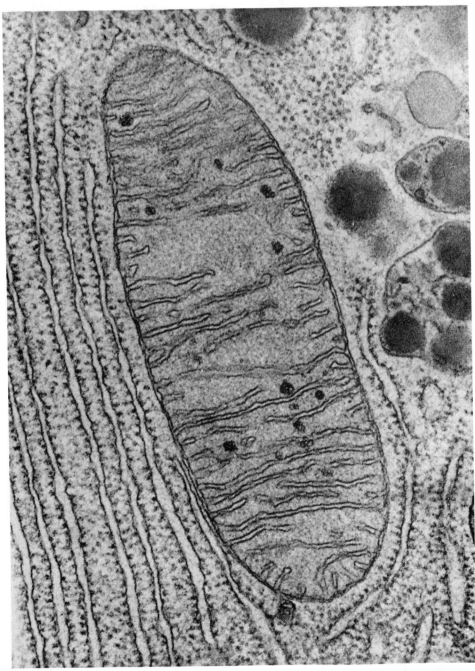

Figure 1.6. Electron micrograph of a portion of a pancreatic cell showing the mitochondrial structures labeled in the diagrammatic representation in Figure 1.8. A parallel array of cisternae of rough endoplasmic reticulum (*ER*) is also evident (×95,000). [From Fawcett (1966).]

its ribosomes and thus become converted to *smooth ER*. The relative proportions of rough and smooth ER vary within different cells; for example, the rough ER is extensive in cells which specialize in synthesizing protein for export, while the smooth ER is extensive in steroid-secreting cells.

SMOOTH ER

As noted above, the ER which lacks ribosomes is referred to as the smooth ER. The membrane of the endoplasmic reticulum carries enzymes which are important in several biosynthetic pathways. For example, the enzymes required for the synthesis of steroid are found in microsomal fractions of steroid-secreting cells. Enzymes involved in triglyceride synthesis as well as phospholipid synthesis are also found in this fraction, the phospholipid sometimes appearing in the ER as small fat droplets. In liver cells, important drug-degrading enzymes are associated with the smooth ER. Also, in the hepatocytes the close spatial relationship of the smooth ER with glycogen, the major form of glucose, suggests that the smooth ER may function in glycogen metabolism. In muscle, the smooth ER (sarcoplasmic reticulum) controls the local concentration of calcium ions near the contractile machinery and thereby influences the contraction and relaxation process (Chapter 4).

Golgi Apparatus

The Golgi apparatus is believed to be a site for the concentration of protein and polysaccharide. It is also a site for completion of the synthesis of the carbohydrate moiety of glycoprotein, e.g., the synthesis of the carbohydrate moieties of thyroglobulin and immunoglobulin begins in the ER, but the terminal sugars are added in the Golgi apparatus. In the case of synthesis of polysaccharides destined for secretion, the precursors are first seen in the Golgi apparatus. Therefore, the apparatus is believed to be the site of synthesis and packaging of polysaccharides for secretion. These products are usually packaged as "granules" within Golgi-derived vacuoles or vesicles which then migrate away from the Golgi apparatus. The enzymes involved in the polymerization of polysaccharide or addition of carbohydrate to protein, glycosyl transferases, have recently been used as marker enzymes for the biochemical isolation of Golgi apparatus fractions.

The Golgi apparatus consists of a stack of several membranous saccules with associated vacuoles and vesicles (Fig. 1.7). The ER in some cell types is assumed to contribute to the "forming face" or "outer" surface of the Golgi apparatus. Within the stacked membranes of the Golgi apparatus materials are concentrated as they pass from the saccules on the "outer" surface to those forming the "inner" surface of the apparatus (Fig. 1.1). In exocrine and endocrine cells the mature secretory granules are generally found in association with the "inner" saccules.

Lysosomes

Lysosomes have been found in virtually all animal cells which have been studied. As organelles they are best defined by biochemical and cytochemical criteria: a lysosome is a membrane-delimited body containing demonstrable acid-hydrolase activity. Over 30 acid hydrolases are known to occur in the lysosomes; these enzymes can digest essentially all macromolecules. Material to be digested becomes enclosed within lysosomal membranes permitting isolated, controlled degradation. The means of development of an acid pH within the vacuole is not known. There are numerous findings suggesting that release of hydrolases from the lysosome into the cell may be important in various pathological states. In silicosis (miner's disease) it is believed that macrophages of the lung take up silica into phagocytic vacuoles which, upon fusion with lysosomes (Fig. 1.7), make the lysosomal membranes leaky. In some inflammations, hydrolases may be released at the surface of the phagocytic cells and affect the adjacent tissues. In many cases these findings are yet to be fully evaluated.

Microscopic identification of lysosomes often consists of demonstration of acid phosphatase activity within a membrane-delimited body; the assumption is that other lysosomal hydrolases are also present. Morphologically, the lysosomes are a motley group of subcellular bodies (Fig. 1.7). Their appearance depends largely on the origin of the enclosed material which is destined for intracellular digestion by the lysosomal hydrolases. In polymorphonuclear white blood cells Golgi-derived lysosomal granules fuse with phagocytic vacuoles formed as a result of endocytosis of foreign material. *Autophagic* vacuoles are lysosomes which contain bits of the cell's own substance which have been separated along with the hydrolases from the rest of the cytoplasm within a membrane-delimited space. Autophagia, which may be enhanced by stress, is hypothesized to be important for the turnover of some cell constituents.

The degraded soluble products of lysosomal hydrolysis can either enter the anabolic pool to be reused in biosynthesis or to be secreted. An example of the latter is the secretion of thyroid hormone: thyroid colloid travels from the follicle lumen

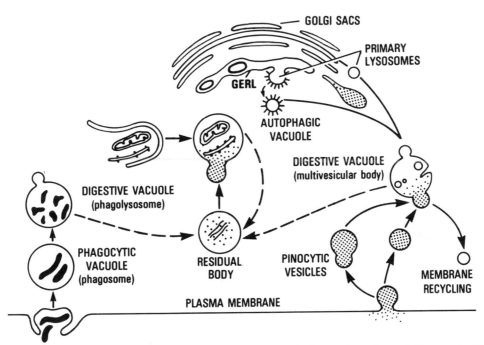

Figure 1.7. Diagram suggesting probable interrelations of some organelles in the membrane-bounded transport of lysosomal hydrolases and the formation of lysosomes. Illustrated are: (1) Fusion of primary lysosomes (Golgi vesicles) with phagocytic and pinocytic vesicles resulting in the formation of digestive vacuoles (multivesicular bodies). Some of the vesicles are "coated" (see text). (2) Vesicles and autophagic vacuoles are among the structures that may form from closely interrelated components of the ER and Golgi apparatus (GERL). GERL is defined as Golgi-associated endoplasmic reticulum from which lysosomes form. (3) Formation of residual bodies by the accumulation of indigestible residues within lysosomes. [From Holtzman and Novikoff (1983).]

within endocytic vacuoles which then fuse with the lysosomes; the colloid is hydrolyzed and thyroxine is released. It is also noteworthy that indigestible residues accumulate within lysosomes, a phenomenon which accompanies the aging process in neural and other cells. The accumulation of lipid deposits in blood vessels may be one factor contributing to the development of atherosclerosis.

Peroxisomes

The peroxisomes constitute another group of membrane-delimited bodies. They are often associated with the ER from which they are thought to derive (see Fig. 1.1). They are concerned with the metabolism of peroxide: peroxisomes contain enzymes which destroy hydrogen peroxide (such as catalase) and other enzymes which produce hydrogen peroxide (such as D-amino acid oxidase and, in some species, urate oxidase). Peroxisomes have, thus far, been found in essentially all cell types, of which liver and kidney are among the best established examples. The function of peroxisomes is currently under active investigation, and there is evidence that, in some species, they are involved in carbohydrate synthesis from fat and in the degradation of purines and fats, as well as in the detoxification of hydrogen peroxide.

Mitochondria

The early cytologists noted that mitochondria were closely associated with motile processes and were situated in regions of intense metabolic activity such as sites of active transport. Biochemists have since shown that two major metabolic pathways, the tricarboxylic acid cycle and the electron transport chain, are situated within the mitochondrion; thus, this organelle is involved in the metabolism of lipids, amino acids, and carbohydrates.

The "typical" mitochondrion (Fig. 1.6) has a length of 5 to 10 μm and a diameter of 0.5 to 1.0 μm. It is bounded externally by two lipoprotein membranes (each about 7 nm thick), the inner one of which is thrown into folds termed cristae or tubules (Figs. 1.6 and 1.8). Within the inner membrane is a matrix containing granules, RNA and DNA. The DNA is a small circular molecule and is thought to provide some, but not all, of the necessary information for replication of the mitochondria. The RNA is responsible for synthesis of a few of the proteins of the mitochondrion; the majority of the mitochondrial enzymes are synthesized on cytoplasmic ribosomes under the direction of the nuclear DNA. From physiological and biochemical evidence, it is generally believed that the respiratory enzymes and the components of oxidative

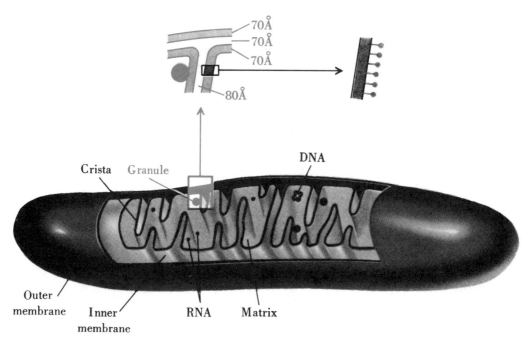

Figure 1.8. Diagrammatic representation of a mitochondrion. [Data from Ornstein, Palade, Sjostrand, Fernandez-Moran, and others. From Novikoff and Holtzman (1976).]

phosphorylation are associated in an ordered array on the inner membrane. The order is thought to promote the sequential interaction of substrates and enzymes in these multienzyme systems with concomitant conservation of energy. Therefore, much effort has been directed toward isolation of modular physiological units from inner membrane fractions. The fact that multienzyme assemblies can be obtained from mitochondria has been encouraging. Moreover, morphological studies support the multienzyme assembly hypothesis. First, the respiratory activity of mitochondria is roughly proportional to the amount of inner membrane—a finding which could be explained by presuming an increased number of repeating units or respiratory assemblies associated with the increased membrane area. Second, a repetitive array of particles found in certain preparations suggests a similarly repetitive assemblage of inner membrane enzyme systems. These regularly spaced particles are found to be attached by small stalks to the inner membrane of isolated mitochondria, unfixed and negatively stained. (In negative staining, the surface details of the material under study appear as light objects against a dark background). Surprisingly, after isolation the stalked particles contained an ATPase predominantly; however, it is suspected that the equilibrium of the reaction catalyzed by the ATPase in the stalked particles is reversed in the intact mitochondrial oxidative phosphorylation system, and that it couples phosphorylation of adenosine

diphosphate (ADP) with electron transport. This suggestion is supported by the fact that dissociation of the particle from submitochondrial preparations abolishes the ability of the preparation to carry out oxidative phosphorylation; when the particles are added back, oxidative phosphorylation returns too. These particles are closely associated with the assemblage of enzymes and cytochromes that are distributed in the inner membrane and that form the electron transport system of the cell. Normally these two essential processes, namely, ATP formation and electron transport, are tightly coupled. The morphology of the inner membrane enzyme system remains a matter of current dispute.

The outer membrane of the mitochondrion has a different enzyme content, a larger percentage of lipids, and is more permeable to simple sugars than the inner membrane. In addition to the demonstrably membrane-anchored enzymes of both inner and outer membranes, certain enzymes (such as those of the tricarboxylic acid cycle) are solubilized after disruption of the mitochondria. These enzymes are presumed to be situated in the matrix or possibly loosely attached to a mitochondrial membrane.

Physiologists can clearly define the metabolic state of the mitochondria in terms of electron transport and oxidative phosphorylation. Careful electron microscopic study of isolated mitochondria, and in a few cases of mitochondria in situ, has resulted in the identification of two functionally related states: (1) the condensed state in which the matrix appears dense and the space between the inner and outer membrane is enlarged; this state is seen when oxidative phosphorylation is proceeding at a rapid rate under conditions of excess ADP and inorganic phosphate, and (2) the orthodox state (see Fig. 1.6) when ADP and P_i are rate-limiting in oxidative phosphorylation. It is hoped that this kind of structural alteration can be explained eventually in terms of the interaction of mitochondrial macromolecules.

MICROTUBULES AND MICROFILAMENTS

The asymmetry observed in certain cell types is sharply at variance with the picture of an idealized cell in which all the organelles surround a central nucleus symmetrically. A nerve cell in which the axon runs several feet as an extension of the perikaryon, an elongated muscle cell and a squamous (or columnar) epithelial cell, all exemplify such asymmetry. Likewise, the nonrandom (asymmetric) movement of subcellular elements is exemplified by transport of neurosecretory products (axonal flow), sliding myofilaments in muscle contraction, and chromosomal movement in cell division. Electron microscopy of aldehyde-fixed tissue has revealed a morphological basis for asymmetric structure and movement in the form of entities referred to as *microtubules* and *microfilaments*; these structures are not membrane-delimited.

Microtubules and Centrioles

In cross-section microtubules are 20–30 nm in diameter and may be followed for several microns in longitudinal sections. They are found in many regions in which phase-contrast and polarizing microscopy had previously demonstrated the presence of formed oriented, elongated elements. Microtubules are often associated with oriented movement, e.g., axonal transport of neurotransmitter from the nerve cell body (where synthesis occurs) to the synaptic terminal where transmitter is released (see Chapter 3). The microtubule subunits are also synthesized in the cell body but assembled in the axon. One of the best established examples of microtubule association with movement is chromosomal movement in the mitotic spindle. The mitotic poles—towards which the microtubules of the spindle orient and towards which the chromosomes move—have centrioles, usually two per cell.

In the interphase cell the pair of centrioles is generally found with long axes at right angles to each other. Each centriole is a cylinder 0.15 μm in

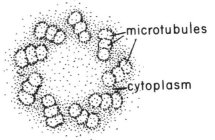
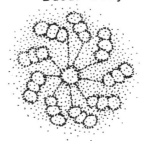
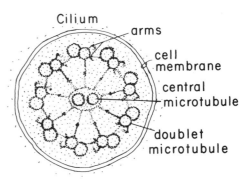

Figure 1.9. Diagrammatic representations of organelles composed of microtubules. [Adapted from Fawcett (1966), Gibbons (1967), and Satir (1974).]

diameter and 0.5 μm in length, composed of nine sets of microtubule-like elements (Fig. 1.9).

The organization of microtubules can be disrupted by physical means (freezing or high pressure) or by chemical treatment, especially with colchicine. When this is done, motion is inhibited and some of the structure collapses. Therefore, the affinity of microtubular protein for colchicine serves as a means of identifying microtubular protein in a cell fraction. Isolated and disrupted microtubules yield protein subunits of approximately 6×10^4 daltons. These appear to be globular subunits which may be arranged in a helical fashion to form the microtubules. In normal cells the microtubular protein appears to be present in a form which is assembled into tubules (e.g., for the formation of the mitotic spindle) under appropriate, but as yet not understood, stimulation.

Cilia and Flagella

Microtubule-like structures may also be organized into organelles as diagrammed in Fig. 1.9. Cilia and flagella are rapidly beating cell processes which extend 10 to 200 μm from the cell and are surrounded by a membrane which is continuous with the plasma membrane. The intracellular basal bodies of cilia and flagella are also composed of microtubular structures arranged in the pattern of nine basic units (often referred to as "9 + 0"); they are widely assumed to be an alternate form of the centriole.

Cilia and flagella generally have, in addition to the basic nine outer sets, a central pair of microtubules ("9 + 2"). It is not known how these elements interact in the matrix to produce beats; one possibility is that the sliding of tubules within a doublet is the motile force. The process of beating requires cellular energy as indicated by the findings that exogenous ATP can cause beating in isolated cilia and flagella and that the "arms" of the nine sets of microtubules contain an ATPase. The tubules of the cilia are composed of molecules similar to those of the other cellular microtubules.

Microfilaments

Microfilaments are a heterogeneous class of long, thin nontubular structures of various diameters. Among the most commonly seen microfilaments are those which appear to serve as the structural core of microvilli. Also frequently encountered are tonofilaments on the intracellular side of desmosomes (see below). The best example of association of microfilaments with motion is the extensively developed myofilament system which forms the basis of muscle contraction (Chapter 4). The myofilament proteins actin and myosin, have now been localized in many other cell types as a result of improved techniques for the cellular localization of specific proteins. Results of the application of antibody to actin have implicated some form of this protein as a constant component of thin (5 nm

diameter) microfilaments. Also cellular thick microfilaments (10 nm diameter) are myosin-like, as shown by antibody and other cytochemical techniques.

THE JUNCTIONAL COMPLEX

We have thus far limited our view of the cell to those entities circumscribed by and including its boundary, the plasma membrane. Cells rarely are continuous with one another, usually a space of 10–20 nm separates them. The cells are associated in tissues by various means; the best described is the *junctional complex* in epithelial cells (Fig. 1.1). In this complex the plasma membranes of two adjacent cells contribute to specialized attachment sites: a tight junction (*zonula occludens*), a desmosome (*macula adherens*), and between these two usually a less well-defined *zonula adherens* where the two membranes are separated by a constant 20-nm space. The desmosome may contain organized extracellular material between the two cell membranes, which may be seen as an additional dense line in parallel with the membranes.

In the region of the tight junction, the outermost layers of the two cellular unit membranes appear to be very closely associated or fused with one another (Fig. 1.1); externally applied tracer molecules (such as ferritin, an iron-containing electron-dense protein) cannot penetrate between the cells at the tight junction. Movement across epithelial cell layers with tight junctions requires a pathway through cells rather than around them. Certain cells are considered to be electronically coupled in that the usual insulation effect on passage of an applied electric current is greatly reduced at a specialized juntional area. This area has been called a *gap junction* and appears to have many small regions of contact between the plasma membranes with obliteration of the adjacent dense lines. Communication between and coordination of the individual cells of cardiac muscle may be effected via such gap junctions.

CHAPTER 2

Properties and Functions of Cell Membranes

Several aspects of structure and function of the cell membrane—often referred to as the plasma membrane—were reviewed in Chapter 1. In this chapter we shall further consider the physicochemical properties of this important boundary structure and examine the processes by which it regulates the molecular traffic between the extracellular and intracellular fluids.

The two sides of the cell membrane appear to have different properties. One may therefore provisionally label the two sides of the membrane as the *outside* (extracellular) and *inside* (cytoplasmic) surfaces. In the case of the cell membrane the "outside" surface is in contact with the cell environment, and the "inside" surface is in contact with the cytoplasm.

PHYSICOCHEMICAL PROPERTIES

Electrical Properties

Even before the nature of the cell membrane had been demonstrated by electron microscopy, electrical measurements gave rise to the interesting generalization that the boundaries of all kinds of cells shared a common property: the capacitance per unit area of the cell boundary material was about 1 μf/cm^2. This figure can be combined with the 7.5-nm thickness of the unit membrane in an interesting manner. The cell membrane can be pictured, to the first approximation, to be the dielectric material of a charged condenser. In general the charge is negative on the inside surface and positive on the outside surface. There is a difference of potential across the membrane, for example in resting nerve, on the order of 0.06 v. The intensity of an electric field is usually expressed in dimensions of volts per meter. Thus, the electric field within the cell membrane of resting nerve is on the order of

$$(0.06 \text{ v})/(7.5 \times 10^{-9} \text{ m}) = 8,000,000 \text{ v/m}$$

The dielectric strength of a commercial insulator is the maximum electric field (volts/meter) to which the material can be subjected before breaking down. The highest dielectric strength for a good commercial insulator such as rubber is about 1,000,000 v/m. Thus, the cell membrane material, viewed as an electrical insulator, must have a very high dielectric strength, i.e., 8,000,000 v/m.

A second number of interest which we are able to estimate is the dielectric constant, k, of the cell membrane material. The capacitance of a condenser is a measure of the amount of charge which can be stored in the plates of the condenser when a given voltage is impressed across these plates. The capacitance, C, is directly proportional to the cross-sectional area, A, of the condenser, and to the dielectric constant, k, of the insulator between the condenser plates; the capacitance is inversely proportional to the thickness of the dielectric, x, that is, the distance between the condenser plates. In mks units $C = \epsilon_0 kA/x$, where ϵ_0 is the constant 9×10^{-12} farads(f)/m. Solving for k and substituting we find that:

$$\begin{aligned} k &= Cx/\epsilon_0 A \\ &= \frac{(10^{-6} \text{ f})(75 \times 10^{-10} \text{ m})}{(9 \times 10^{-12} \text{ f/m})(10^{-4} \text{ m}^2)} \\ &\cong 8 \end{aligned}$$

Again compared to commercial electrical insulators a dielectric constant of 8 is relatively high. We conclude that the nerve (cell) membrane has a relatively high dielectric constant and dielectric strength.

Molecular Properties

The postulated molecular structure of the cell membrane is best understood in terms of the hydrophilic and hydrophobic nature of certain chemical groups. Hydrophilic groups are polar groups which may or may not bear a formal electrical charge. Substances containing these groups tend to dissolve in water by interacting with the dipolar water molecules; such interactions are enhanced as a consequence of the high dielectric constant of water. Accordingly, there is a reduction in the strength of the interactions of polar solutes (dipolar molecules, ions) with each other. Amino acids and sugar molecules, for example, are hydrophilic. The long chain hydrocarbons (e.g., long chain fatty acids) on the other hand, are hydrophobic; such molecules are relatively insoluble in water. The aggregation of hydrophobic molecules in the presence of water has been explained as follows. The structure of water is in an ordered, ice-like state in the vicinity of nonpolar molecules. When nonpolar molecules leave the water phase and form an aggregate, the degree of order of the water phase is decreased; the entropy (degree of disorder) of the whole system is thereby increased. Since systems tend to change spontaneously towards a state in which the entropy is maximized, nonpolar (hydrophobic) molecules tend to separate from the water phase and to aggregate, i.e., they do not "dissolve" in water. This complex set of ideas for explaining the association of nonpolar molecules is represented in the current literature by the term "hydrophobic bond."

A long chain fatty acid is a hydrophobic molecule for the most part, but it bears a hydrophilic (carboxyl) group at one end. When hydrophobicity exceeds hydrophilicity, such a substance does not appreciably dissolve molecule by molecule in water. If enough molecules of this kind are present, however, they can enter the water phase and simultaneously satisfy their hydrophobic and hydrophilic tendencies by forming spherical or lamellar *micelles*. A micelle is a multimolecular structure in which the hydrophilic parts of its constituent molecules are at the surface in contact with water, while the hydrophobic parts are in the interior in a milieu of other hydrophobic groups.

A considerably simplified model of the cell membrane pictures it to be a lamellar micelle consisting of two rows of phospholipid molecules (Fig. 1.10). The hydrophobic hydrocarbon chains are on the interior. The two surfaces of the micelle are covered by a macromolecular material. The asymmetry of this membrane is attributed to differences in the macromolecular materials which line the two surfaces of the micelle. During the past decade, evidence has accumulated to suggest that there is considerable lateral mobility within the lipid moiety of the plasma membrane ("fluid mosaic model") (Hubbell and McConnell, 1968, 1971; Singer and Nicolson, 1972; Singer, 1974).

The biological role played by the cell membrane

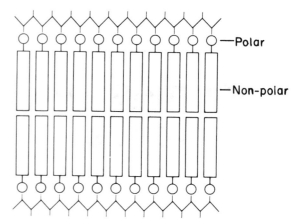

Figure 1.10.   Model of cell membrane as laminar micelle.

can best be interpreted from the point of view of evolution. It is unlikely that at the time of the origin of life the first organism was already equipped with a membrane boundary. It is more likely that this early organism was a relatively simple, self-replicating molecule. With the development of biological complexity, however, an "organelle" arose which isolated the organism from its environment. This organelle, the cell membrane, had selective value, since it tended to reduce the influx of injurious molecules from the environment while tending to preserve useful molecules for the exclusive use of the organism.

From this point of view the cell membrane should be thought of as a barrier to diffusion. Basically, the cell membrane should be regarded as a relatively impermeable rather than a permeable structure.

We have considered the cell membrane to be basically a lipid micelle. To the *first* approximation, it is convenient to treat the cell membrane as if it were a thin oil film (Fig. 1.11*A*). This crude approximation suffices to explain a number of important permeability and electrical properties of the cell membrane. For example, the movement of a substance across the cell membrane may be viewed as follows: Each individual molecule of the substance leaves the aqueous phase on one side of the membrane and dissolves in the oil film. The molecule then diffuses across the oil film and enters the aqueous phase on the other side.

A slightly more sophisticated model of the cell membrane is required to explain a number of experimental findings which can best be interpreted by assuming that some molecules can pass through the membrane without leaving the aqueous phase. In other words, to the *second* approximation, the cell membrane behaves as if it were an oil film pierced by pores (Fig. 1.11*B*). The aqueous phase

A.   **B.**

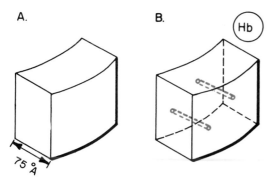

Figure 1.11. Simple models of cell membrane. *A*, First approximation. Oil film (7.5 nm thick). *B*, Second approximation. Oil film (7.5 nm thick) is perforated by pores. Assuming 0.7 nm for the average diameter of a pore, each pore is about 10 pore diameters long. The diameter of a hemoglobin molecule (*Hb*) is shown to the same scale.

is continuous across the membrane through the pores. These pores may be statistical rather than permanent in nature, that is, they may be temporary holes in the oil film created by the thermal agitation. From osmotic studies on the human erythrocyte, it has been inferred that the effective diameter of individual pores is about 0.7 nm and that the total pore area is only about 0.0001 of the total surface area of the cell. From the size of the pores it is evident that only small particles (e.g., water, urea) can cross the cell membrane via the pores. Larger molecules will have to cross the membrane in the region which we have referred to as the "oil film." Thus, for the purposes of our survey, it will be useful to view the cell membrane as a perforated oil film. To this simple model more complex features (e.g., carrier molecules, enzymes, virus attachment sites) can be added as needed.

We may now consider the manner and the relative rates at which substances of biological significance cross the cell membrane. The following modes of transport will be discussed. (1) *Bulk Flow*. In this mode water molecules do not pass through the membrane by dissolving in it and diffusing across it, one by one; rather, the motion of each molecule is correlated with that of neighboring water molecules, that is, water flows through the pores in the membrane in bulk as it does through a pipe. (2) *Diffusion*. Solute and water molecules cross the membrane individually as a consequence of their thermal motion. (3) *Carrier-Mediated Diffusion*. Solute molecules combine with membrane-bound carrier molecules. The carrier-solute complex diffuses or otherwise moves across the membrane. (4) *Active Transport*. Molecules move across the membrane by a carrier-mediated process which requires a coupled input of metabolic energy.

TRANSPORT

Physical Basis of Diffusion

Consider a tube which is divided into two parts by a thin, removable partition. The tube contains an aqueous solution of some substance. The concentration of this substance is higher to the right of the partition than to the left. After the partition is carefully withdrawn it is found that with time the concentration on the left increases and that on the right decreases. Ultimately the concentration becomes the same throughout the tube. The physical explanation of this phenomenon is quite simple. Every solute molecule has motion of its own and in addition is under constant bombardment from adjacent solute and solvent molecules. As a result a given solute molecule follows an erratic path through the liquid. Because there is initially a greater concentration of solute at the right of our tube than at the left, however, more solute molecules move from right to left than from left to right until the solute concentration throughout the whole tube is equalized. Thus, although each individual solute molecule moves at random, the average motion of a large collection of such molecules results in a net flow of solute from a region of high concentration into a region of low concentration.

DIFFUSION THROUGH A MEMBRANE

In view of our interest in the diffusion of molecules through the cell membrane we shall define a number, the *permeability coefficient*, which is descriptive of the ease with which a *given solute* can diffuse through a *given membrane*. Clearly this number will be a function both of the nature of the membrane and of the nature of the solute molecule.

Consider a tube which is divided into two parts by a fixed membrane of thickness, x (Fig. 1.12).

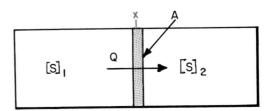

Figure 1.12. Permeability. Chamber divided by a thin membrane of area A and thickness x. The concentration of a given substance is $[S]_1$ on the left and $[S]_2$ on the right. The rate, Q, in moles/sec, at which the substance diffuses from left to right is:

$$Q = pA([S]_1 - [S]_2)$$

where p is the permeability coefficient of the membrane for the given substance. Note that p may include such factors as the solute diffusion coefficient, the membrane thickness, the geometry of aqueous channels in the case of polar solutes, etc.

The tube is filled with a solution containing a solute, S, which is at concentration $[S]_1$ to the left of the membrane and at concentration $[S]_2$ to the right. Assuming that $[S]_1$ is greater than $[S]_2$ there will be a net transfer of solute across the membrane from left to right. To simplify the mathematics we assume that over the period of observation the concentrations $[S]_1$ and $[S]_2$ (in moles/cm^3) remain constant with time. These restrictions can be approximated in a real system by taking measurements over a sufficiently short period so that the concentrations do not change appreciably, and by stirring the solutions on each side of the membrane. Under these conditions there will be a net *rate of flow*, Q (moles/sec), of solute across the membrane. The flow rate, Q, is proportional to the *concentration gradient*, $([S]_1 - [S]_2)/x$. (When the concentration is equal on both sides, that is, when $[S]_1 = [S]_2$, then Q is zero.) Q is also proportional to the area, A (cm^2), of the membrane, i.e., the greater the area the greater the flow rate. Introducing the proportionality constant, p, we obtain the equation:

$$Q = pA([S]_1 - [S]_2). \qquad (1)$$

It can be seen that p includes the membrane thickness, x, and has the dimensions of centimeters per second:

$$p = \frac{Q}{A([S]_1 - [S]_2)} = \frac{\text{moles/sec}}{(\text{cm}^2)(\text{moles/cm}^3)} = \frac{\text{cm}}{\text{sec}} \qquad (2)$$

p is the *permeability coefficient* for the given solute and membrane—the greater p, the greater the rate at which the solute diffuses across the membrane. From *Eq. 2* it can be observed that p defines the flow rate of solute per unit membrane area per unit concentration gradient maintained across the membrane.

Any rate of transfer of mass (particles) or energy across a boundary may be referred to as a *flux*. Thus, the term Q in *Eq. 1* above may be spoken of as the net flux of solute across the membrane area, A. In recent years the letter J has been used as a general designation for flux whether the moving particles are molecules, ions, electrons, or packets of energy, such as calories or quanta, etc. In this chapter we are using J to represent the flux, Q, per unit membrane area, A. In other words,

$$J = Q/A.$$

Therefore, *Eq. 1* for simple diffusion can be rewritten as:

$$J \equiv Q/A = p([S]_1 - [S]_2). \qquad (3)$$

If the solute particles are ions, the movement or flux can be measured in terms of specific ionic electrical current, I, instead of flux, J. Current and flux are related as follows:

$$I = zFJ$$

where z is the valence of the ion and F is Faraday's constant (96,500 coul/eq). Typical units of current are μamp or μamp/cm^2 if normalized. A handy equation for converting electrical current to ionic flux is:

$$I \; (\mu\text{amp/cm}^2) = 26.8 \cdot z \cdot J \; (\mu\text{mole/hr-cm}^2)$$

PHYSICOCHEMICAL CRITERIA FOR RATES OF DIFFUSION ACROSS ALL MEMBRANES

Lipid Solubility

Let us consider a vessel in which two aqueous phases are separated by an oil phase (analogous to a lipoidal cell membrane). A solute, S, is introduced into aqueous phases 1 and 2 to establish concentrations $[S]_1$ and $[S]_2$, respectively. If equilibrium is established across each interface, the concentrations in the oil phase (membrane) and the adjoining aqueous phases can be related to B, the oil-water partition coefficient, which is defined as follows:

$$B = [S]_{1 \text{ (oil)}}/[S]_{1 \text{ (water)}} = [S]_{2 \text{ (oil)}}/[S]_{2 \text{ (water)}}$$

where $[S]_{1 \text{ (oil)}}$ and $[S]_{2 \text{ (oil)}}$ are the concentrations of the substance in the oil phases at oil-water interfaces 1 and 2, respectively, and $[S]_{1 \text{ (water)}}$ and $[S]_{2 \text{ (water)}}$ are the concentrations of the substance in the water phases 1 and 2, respectively. There will be a gradient of concentration in the oil phase (membrane) when the concentrations established in the aqueous phases are unequal. A gradient across the oil phase gives rise to diffusion of the substance through this phase. The net flux per unit area, $J \equiv Q/A$, from aqueous phase 1 to aqueous phase 2 is given by:

$$J \equiv Q/A = p'([S]_{1 \text{ (oil)}} - [S]_{2 \text{ (oil)}})$$

where p' is a proportionality coefficient. Knowledge of the oil-water partition coefficient, B, makes it possible to substitute the concentration in the aqueous phases for the concentration in the oil phase at the interfaces (where an equilibrium distribution is assumed to prevail); thus:

$$J \equiv Q/A = p'B([S]_{1 \text{ (water)}} - [S]_{2 \text{ (water)}}) \qquad (4)$$

The apparent permeability coefficient, $p'B$, corresponds to p in *Eq. 3*. It is seen that flux through the oil phase (membrane) is proportional to the oil-water partition coefficient, B, all other factors being equal. For many substances of biological interest B is very much less than one; for substances which can enter the cell membrane only to a limited extent, Q is small. In general, although there are exceptions, the greater the oil-water partition coefficient, B, for a substance, the greater its rate of transfer through the cell membrane (Overton's rule).

Hydrophilicity

Chemical groups are said to be polar when the positive nuclei and negative electrons are so unevenly arrayed that the local net electrical charge does not average to zero. Let us consider some examples. Ionized groups, that is, groups which have lost or gained electrons, are polar. Groups in which there has been no net gain or loss of charge, but where a displacement of charge within the group has led to a concentration of negative charge in one region, and positive charge in another, are also polar. The alcoholic hydroxyl group ($-OH$) is polar because the two valence electrons shared by the oxygen atom with a hydrogen atom and a carbon atom (as in CH_3OH or CH_3CH_2OH) are attracted toward the more electronegative oxygen atom. This is also why water is a highly polar substance.

The greater the number of hydroxyl groups on an organic molecule, for example, the greater the hydrophilicity. As one would expect from the lipid (hydrophobic) constitution of biological membranes, the permeability of the cell membrane to a substance is inversely related to the hydrophilic nature of that substance. Thus, hydrophilic substances, such as sodium ions, potassium ions, sugars, and amino acids, will penetrate the cell membrane only very slowly by simple diffusion. In Table 1.3 it is seen that the permeability constant of sodium ion, Na^+, is only about one hundred-millionth that of water. It is equally evident that, from the point of view of the economy of the cell, certain hydrophilic substances must be brought into the cell in appreciable quantities. As we shall see, such substances are transported into the cell by a process which involves membrane-bound carrier molecules. While it is generally true that electrically charged particles penetrate the cell membrane only very slowly by simple diffusion, it is to be noted that in the case of the mammalian red blood cell, the anions OH^-, Cl^-, and HCO_3^- as well as the cation H^+ have permeabilities which are very high compared to that of the cations Na^+ and K^+. In general, the electrical charge distribution in the walls of the pores favors the diffusion of anions over cations.

Molecular Size

In general it may be said for a series of homologous substances that the greater the molecular weight, the smaller the permeability coefficient. This follows from the fact that the larger a particle the slower its thermal motion and, consequently, the smaller its rate of diffusion. For polar molecules, such as proteins, nucleic acids, and viruses, the permeability coefficient is exceedingly small not only because of low diffusibility but also because of molecular sieving, i.e., large polar molecules tend to be sieved by the cell membrane owing to the geometry (size and shape) and the electrical characteristics (charge distribution) of its aqueous channels. However, since such particles do indeed cross the cell membrane, we must seek an explanation for their transport in terms of special mechanisms other than simple diffusion.

Carrier-Mediated Transport

THE CARRIER MODEL

We have noted that many substances, such as amino acids and sugars, which are vital to the economy of the cell can be expected to cross the cell membrane by simple diffusion only at low rates. A considerable body of experimental evidence has accumulated to suggest the following mode of transport: The substrate to be transported combines with specific, membrane-bound *carrier* molecules and crosses the cell membrane as part of a substrate-carrier molecular complex. The carrier molecules are assumed to be confined to the membrane and to shuttle back and forth from one side of the membrane to the other. Thus the carrier has been visualized as a molecular ferryboat which increases the rate of transport across a barrier. In another view the carrier is thought to rotate within the membrane exposing binding sites first to the compartment of origin of the transport process and then to the compartment of destination for the transported entity. More generally, the term carrier mediation has been used to refer to any transport system that is more complicated than simple diffusion. Presumably the carrier reduces the thermal energy needed by the substrate molecule to leave the aqueous phase and to enter the membrane phase. Consider a sugar molecule which is strongly bound to the surrounding water by hydrogen bonds. For this molecule to enter the nonaqueous phase, i.e., the membrane, it is necessary to await a thermal collision of sufficient magnitude to break its hydrogen bonding to the water. If, however, there is a carrier molecule in the membrane to which the sugar molecules can readily attach, then the thermal energy needed by the sugar molecules to enter the membrane is decreased because the hydrogen bonds between the sugar and the water molecules can be replaced by bonds between the sugar and the carrier molecules. Thus, the entry of the sugar molecules into the membrane is greatly facilitated. At the present time many laboratories are engaged in the isolation and characterization of membrane components possessing the required specificity to serve as carriers for the substrate molecules that are believed to undergo carrier-mediated transport in biological systems.

Carrier-mediated transport can be distinguished from simple diffusion by the following criteria.

Saturation

When there is no carrier, *Eq. 1* tells us that for simple diffusion the rate of transport is proportional to the concentration difference, $[S]_1 - [S]_2$. If this difference doubles, the rate of transport doubles. This is not the case for carrier-mediated transport. Here, concentrations can be reached at which virtually all the carrier molecules are in constant use; then further increases in concentration will not appreciably increase the rate of transport, since unoccupied carriers are not available. The carrier system is saturated and the carrier-mediated transport operates at the maximum possible rate.

Competitive Inhibition

Two substances *A* and *B* which attach to the same carrier *compete* for the carrier system. If *B* is added to the extracellular fluid, the cellular uptake of *A* is decreased because molecules of *B* now occupy carrier molecules which were previously available for the transport of *A*. The competition, of course, is mutual, so that addition of *A* reduces the rate of uptake of *B*.

Specificity

One carrier molecule may transport one amino acid or group of amino acids, while another transports one sugar or group of sugars. If the addition of a substance *X* does not decrease the carrier-mediated uptake of a substance *A*, then *A* and *X* do not share the same carrier system and the chemical specificity of the substance-carrier interaction is thereby demonstrated. This specificity is never perfect and a given carrier may carry, with varying efficiencies, a number of substances of related chemical structure.

Noncompetitive Inhibition

A substance *I* may interact with a carrier in such a manner as to prevent the carrier from combining with its usual substrate, *A*. In this case, even though *I* is not being transported, the transport of *A* is inhibited.

Finally, it should be noted that carrier-mediated transport generally shows the high temperature coefficients characteristic of chemical reactions.

The carrier-mediated transport of a substance along a concentration gradient (i.e., in the direction of simple diffusion) is called *facilitated diffusion*. On the level of the organism an analogy to facilitated diffusion may be found in the transport of oxygen from the lungs to the tissues. The *carrier* is the hemoglobin molecule; the substrate being transported is oxygen. The carrier manifests *specificity*—it combines with oxygen but not with nitrogen; *saturation*—an increase in alveolar oxygen tension beyond certain levels does not yield a proportional increase in oxygen transport; and *competition*—carbon monoxide competes with oxygen for the same carrier and prevents oxygen uptake.

ELECTROCHEMICAL POTENTIAL ($\tilde{\mu}$)

The transport processes which we have discussed so far have dealt with the movement of molecules from regions of high concentration to regions of low concentration. For generality and scientific precision we should be referring to the movement of molecules from regions of high *electrochemical potential* to regions of low electrochemical potential. In the interest of simplicity and clarity for students who have not been exposed to physical chemistry, however, we shall continue to use the term, concentration, interchangeably with the more rigorous term, electrochemical potential. Nonetheless, it is appropriate at this time to consider briefly the meaning of the electrochemical potential of a substance, *S*, in solution ($\tilde{\mu}_S$).

$\tilde{\mu}_S$ is a measure of the "useful" energy content of *S*. Typical units are joule/mole or kcal/mole. Electrochemical potential is composed of a chemical part and an electrical part:

$$\tilde{\mu}_S = \underset{\substack{\uparrow \\ \text{chemical} \\ \text{part}}}{\mu_S} + \underset{\substack{\uparrow \\ \text{electrical} \\ \text{part}}}{z_S F \psi} \qquad (5)$$

where z_S is the valence of *S*, *F* is Faraday's constant, and Ψ is the electrical potential (voltage) measured with respect to any reference point. The chemical part may be separated into component parts:

$$\mu_S = \mu_S^0(T, P) + RT \ln a \qquad (6)$$

where *a* is the activity, *R* is the gas constant, *T* is the absolute temperature, and *P* is the pressure. μ_S^0, as indicated, is a function only of *T* and *P*, i.e., it is constant when *T* and *P* are constant. The activity *a* may be broken down as follows:

$$a = \gamma \cdot [S] \qquad (7)$$

where γ is the activity coefficient (usually in the neighborhood of unity) and $[S]$ is the concentration of S. Therefore, the electrochemical potential on side i (i being 1 or 2) is:

$$\mu_{S_{(i)}} = \mu_S^0(T_i, P_i) + RT_i \ln (\gamma_i[S]_i) + z_s F\psi_i \qquad (8)$$

One of the "forces" driving S across the membrane is the difference in electrochemical potential across the membrane. In biological systems, there is usually neither a significant temperature difference ($T_1 = T_2$) nor a significant pressure difference ($P_1 = P_2$) across the membrane, and the activity coefficients are treated as equal ($\gamma_1 = \gamma_2$). In this case,

$$\Delta\tilde{\mu}_S = \tilde{\mu}_{S(1)} - \tilde{\mu}_{S(2)} \qquad (9)$$

$$= RT \ln ([S]_1/[S]_2) + z_s F\Delta\psi \qquad (10)$$

$$\uparrow \qquad\qquad \uparrow$$

chemical part   electrical part

where $\Delta\psi = \psi_1 - \psi_2$.

Active Transport

When a solute moves downhill from a region of high electrochemical potential to a region of low electrochemical potential, the transport process is generally thermodynamically spontaneous or dissipative in character in that the free energy of the matter under observation decreases during transport (Fig. 1.13*A*); thus, no expenditure of energy is required on the part of the cell. A transport process in which the net flow of substance is from a region of high to a region of low electrochemical potential, however, need not be exclusively spontaneous; for example, a metabolic energy source ($\Delta\mu_M$) might be coupled to such a downhill process so as to increase the net rate of flow of substance above that which would prevail in a purely dissipative process. A situation of this type is sketched in Fig. 1.14 to show the net flux of a substance, J_S^{NET}, being driven by a metabolic energy source, $\Delta\mu_M$, as well as by the electrochemical potential gradient, $\Delta\tilde{\mu}_S$. This is an illustration of *active downhill* transport. Another case of active downhill transport is observed when the metabolic component of transport (i.e., that component which is linked to $\Delta\mu_M$) is oriented in the reverse direction from but overwhelmed by the electrochemical potential gradient $\Delta\tilde{\mu}_S$ (Fig. 1.15). It is also possible for active transport to be manifest when the electrochemical potential gradient has been eliminated ($\Delta\tilde{\mu}_S = 0$). In this situation, referred to as *active level* flow, $\Delta\mu_M$ alone drives the transport process (Fig. 1.16).

In all cells there exist mechanisms whereby substances can be transported uphill across membranes from regions of low to regions of high elec-

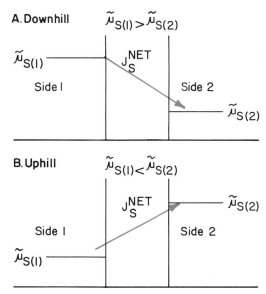

Figure 1.13. If the net flow of a substance is in a direction from higher to lower electrochemical potential, it is moving downhill; if net flow is from lower to higher electrochemical potential, it is moving uphill.

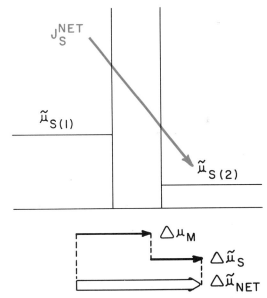

Figure 1.14. Active downhill transport. $\Delta\mu_M$ represents the effective input of metabolic energy from the cell for the transport of S. This effective input will be less than the free energy available from the coupled metabolic reaction to the extent that the efficiency of coupling between metabolism and transport is less than 100%. The metabolic coupling in this scheme is oriented to drive S from side 1 to side 2. $\Delta\tilde{\mu}_S$ is the electrochemical potential difference of S. $\Delta\tilde{\mu}_{NET}$ represents the net driving "force" due to $\Delta\mu_M$ plus $\Delta\tilde{\mu}_S$. In this example of active downhill transport, both $\Delta\tilde{\mu}_M$ and $\tilde{\mu}_S$ act to drive S from side 1 to side 2.

trochemical potential (Fig. 1.13*B*). These mechanisms are in most cases examples of *active uphill* transport in which a metabolic process (off-equilibrium breakdown of a metabolite) is coupled to and

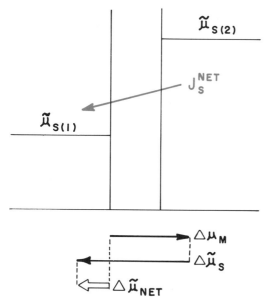

Figure 1.15. Reversed active downhill transport. $\Delta\mu_M$ and $\Delta\tilde{\mu}_S$ are oriented to drive S in opposite directions. In this example, $\Delta\mu_M$ is "overwhelmed" by $\Delta\tilde{\mu}_S$ so that $\Delta\tilde{\mu}_{NET}$ is oriented to drive S from side 2 to 1 in the reverse of its "normal" direction of active transport. If coupling between the metabolic and transport processes is highly efficient, then in principle, the reversed downhill transport of S can serve to generate metabolites (e.g., adenosine 5'-triphosphate) from breakdown products (e.g., adenosine 5'-diphosphate *plus* inorganic phosphate).

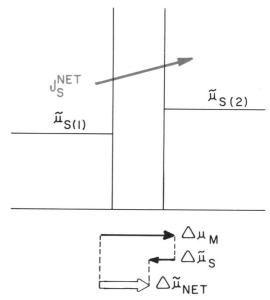

Figure 1.17. Active uphill transport, $\Delta\mu_M$ and $\Delta\tilde{\mu}_S$ are oriented to drive S in opposite directions. In this example, $\Delta\tilde{\mu}_S$ is "overwhelmed" by $\Delta\mu_M$ so that $\Delta\tilde{\mu}_{NET}$ is oriented to drive S from side 1 to side 2. S gains "useful" energy during its net transport from side 1 to side 2. In effect, the active transport mechanism is responsible for the transfer of energy from metabolites to S.

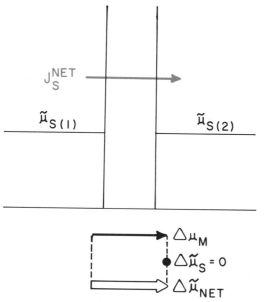

Figure 1.16. Active level flow. Only $\Delta\mu_M$ provides the net driving "force" for the transport of S. In undergoing net transport from side 1 to side 2, S neither gains nor loses "useful" energy ($\tilde{\mu}_S$). The direction of net transport during level flow can be used to define the "normal" direction of active transport.

thus provides the energy for the transport process (Fig. 1.17). The movement of every mole of substance against its electrochemical potential gradient must be paid for by the expenditure of an amount of energy which depends upon the efficiency of the coupling between the energy-providing process and the transport work accomplished. The energy required for active transport is derived from the hydrolysis of substances such as adenosine triphosphate (ATP). If there is no coupling with a metabolic process, transport is passive. For example, a substance may move uphill passively if its transport is coupled with the downhill movement of another substance rather than with a metabolic process. On the other hand, as we have seen, an actively transported substance may move uphill, downhill, or remain at a given level with respect to its electrochemical potential, $\tilde{\mu}_S$. Thus, an active transport process is defined, not by demonstrating that flux is thermodynamically uphill, but *only* by demonstrating that flux is coupled to metabolism.

In mammalian cells the existence of active transport processes for sodium and potassium ions can be readily inferred from the concentrations of these ions in the intracellular and extracellular fluids (Table 1.1). In general, though with exceptions, the concentration of K^+ is greater in the cell than it is in the extracellular fluid, while the concentration of Na^+ is lower in the cell than it is in the extracellular fluid. From the magnitudes of the permeabilities of cell membranes to these ions (Table 1.3) it can be shown that the intra- and extracellular concentrations would equalize by diffusion over a period of several hours. It follows that concentra-

Table 1.1

Concentrations (meq/l of H_2O) of certain ions in human plasma and erythrocytes

	K^+	Na^+	Ca^{++}	Mg^{++}	Cl^-	HCO_3^-
Plasma	5	140	3	1	110	28
Erythrocytes	150	15	Trace	3	70	27
Ratio (P/E)	0.03	9		0.3	1.6	~1

Adapted from Davson (1964).

tion gradients are maintained in vivo across the cell membrane by active processes which pump the ions against the gradients at rates sufficient to offset passive back diffusion (leakage down the gradients).

Active transport is carrier-mediated. Therefore the criteria which define it include: (1) saturation; (2) competitive inhibition; (3) specificity; and (4) noncompetitive inhibition; as noted above, however, the new criterion which must be added is the requirement for metabolic energy. As already noted, a substance can be made to move against a concentration gradient only by the expenditure of energy. If the energy supply is blocked, active transport ceases. As an example we may consider the human erythrocyte, which under normal circumstances has a potassium concentration which is higher in the cell than it is in the serum. When erythrocytes are refrigerated, active transport ceases and potassium ions slowly leak out of the cells until the intracellular and extracellular potassium concentrations are equal. When the blood temperature is returned to 37°C, providing there is an energy source such as glucose in the serum, active transport resumes and the intracellular potassium concentration increases toward normal levels. Similar observations have been made in vitro in experiments with muscle tissue, kidney slices, salivary gland slices, etc.

A model for active transport is shown in Figure 1.18. In addition to the carrier, there is an enzyme, ϵ, and an energy source, W. The substance to be transported combines reversibly with a carrier molecule on the extracellular surface of the membrane. The substance-carrier complex diffuses to the intracellular side of the cell membrane where the complex dissociates and the substance discharges into the cell. In this model the enzyme, ϵ, which is localized on the intracellular side of the membrane, alters the carrier molecule to an inactive form, that is a form which has reduced affinity for the substance. The alteration of the carrier is an enzymatic process which does not require an input of energy. The inactivated (unloaded) carrier passes to the extracellular side of the membrane where it is converted to its active form by an energy-requiring process. Now it can again combine with the substance to be transported and the cycle then repeats.

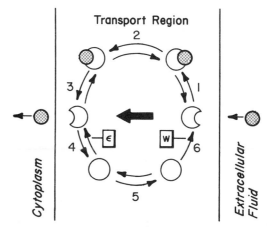

Figure 1.18. Model for active transport. The carrier molecules (*large circles*) extend to the membrane region. There is a net transport of substance (*small circles*) from the extracellular fluid to the cytoplasm. The *large arrow* indicates the direction of transport. *1*, Substance combines with carrier. *2*, Substance-carrier complex passes to cytoplasmic side. *3*, Substance-carrier complex dissociates. Substance is discharged into cytoplasm. *4*, Carrier is enzymatically degraded into an inactive form, i.e., a form which exhibits low affinity for the substance. *5*, Inactive carrier passes to extracellular side. *6*, Carrier is activated by an energy-requiring process. (Adapted from Heinz and Walsh, 1958.)

It should be pointed out that there exists no direct evidence for this model; in fact recent studies of the structure and function of carrier systems suggest that the binding site for the transported substrate and the enzyme complex with the function designated in Figure 1.18 as ϵ as well as the function which facilitates the energy input, W, are all part of a single molecule. In the case of the system responsible for the active transport of sodium and potassium ions, the relevant carrier-enzyme molecule (namely, Na^+-K^+-adenosine triphosphatase) has two subunits (Kyte, 1971; Guidotti, 1972; Dahl and Hokin, 1974; Sweadner and Goldin, 1980), the larger subunit, designated α, having a molecular weight of approximately 95,000 daltons and the smaller subunit, a glycoprotein designated β, having a molecular weight of approximately 40,000 daltons.

Despite the above-noted recent findings the model shown in Figure 1.18 is very helpful in considering a wide range of experimental findings related to active transport. It might be thought that active transport should give rise to a continuously increasing concentration gradient. This is not the case for the following reasons:

1. As the intracellular concentration of the transported substance increases, the dissociation of the substance-carrier complex is opposed by mass action, and, therefore, the probability of a loaded carrier molecule returning from the intracellular to the extracellular side increases.

2. As the concentration gradient builds up there

will be an increase in the passive back diffusion (leak down the gradient in the direction opposite to that of the active transport) so that a steady state is attained in which the rate of active transport equals the rate of back leak.

Thus, the ratio of the intracellular concentration to the extracellular concentration of the transported substance becomes constant with time.

The adaptive value and evolutionary significance of active transport is clear: The total amount of an amino acid in the environment of a microorganism, for example, may be very large, though its concentration and rate of entry into the cell may be too small to sustain an adequate level of protein synthesis. In the presence of an energy-coupled carrier system the amino acid can be concentrated within the cell. Such a system should not be thought of as operating only between the cell and its outside. It can operate at the level of any membrane system— making it possible, for example, for a cell organelle to concentrate a needed constituent. This is the case for the mitochondrion, in which many ions are accumulated. The general features of active transport and facilitated diffusion hold true for a wide range of transport phenomena. Active transport processes for amino acids, sugars, and ions (Na^+, K^+, Cl^-, Mg^{2+}, etc.) have been demonstrated and referred to by various terms, such as the sodium and potassium "pump" of nerve, muscle, and erythrocytes, and the active secretion and active reabsorptive processes of the kidney, gastrointestinal tract, and exocrine glands.

UNIDIRECTIONAL FLUXES AND NET FLUX

Clearly the net flux across a membrane will be resultant of the two unidirectional fluxes. The unidirectional flux of solute, S, from side 1 to 2 ($J_S^{1\to2}$) can be determined by introducing an isotope, S^*, to serve as a tracer for S and then measuring the rate of appearance of S^* on side 2. The unidirectional flux from side 2 to 1 ($J_S^{2\to1}$) can be determined by introducing a second isotope of S, S^{**}, into side 2 and measuring its rate of appearance on side 1. These operations are depicted schematically in Figure 1.19. *Net* flux is given by:

$$J_S^{1\to2(net)} = J_S^{1\to2} - J_S^{2\to1}$$

If $J_S^{i\to j(net)} > 0$, material S disappears from compartment i and appears in compartment j.

Unidirectional flux is of interest because the net flux is often most easily and usefully determined as the difference between the two unidirectional fluxes.

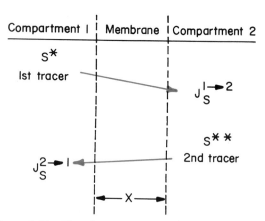

Figure 1.19. Use of tracer isotopes to measure net flux.

CIS- AND TRANS-

Cis is a prefix referring to the side from which a unidirectional flux originates and *trans-* is a prefix referring to the side to which a unidirectional flux moves.

	cis-side	*trans*-side
$J_S^{1\to2}$	1	2
FLUX		
$J_S^{2\to1}$	2	1

In general we may represent unidirectional fluxes as follows:

$$\underset{J_S^{i\to j}}{\overset{\textit{cis}\text{-side } \textit{trans}\text{-side}}{\searrow \downarrow}} \quad \text{or} \quad J_S^{cis\to trans}$$

It is often important to know whether biologically active agents (drugs, hormones), which modify the transport of a substance, act on one or both unidirectional fluxes and whether such action is from the *cis*-side, the *trans*-side, or both.

Pinocytosis

In discussing the relationship between permeability and molecular size it was pointed out that with respect to diffusion the permeability of the cell membrane to large molecules, such as proteins, is negligible. Yet the passage of macromolecules into cells is well documented, for example, (1) the genetic transformation of bacteria by transforming principle (deoxyribonucleic acid); (2) the infection of cells by viruses; and (3) the penetration of cells by antibodies. One mechanism whereby such macromolecules enter cells is called pinocytosis, a phenomenon related to phagocytosis. This phenomenon was noted under the general heading of endocytosis in Chapter 1. In pinocytosis the cell membrane binds and engulfs extracellular material (receptor-mediated endocytosis) forming membrane-bounded vacuoles and vesicles (endosomes) (Fig.

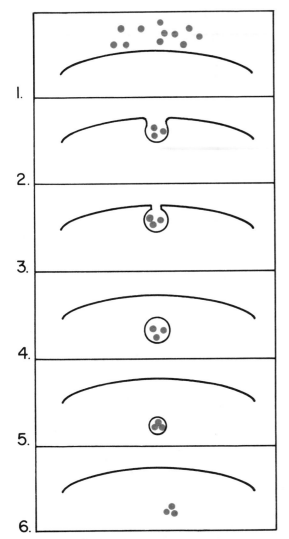

Figure 1.20. Pinocytosis. Molecules which stimulate pinocytosis are indicated by *dots*. *1,* Molecules bind to the outer surface of the cell membrane. *2,* The membrane invaginates in the region of this binding, and vesicles and vacuoles form (receptor-mediated endocytosis). *3,* The vesicles and vacuoles (endosomes) travel inwards and fuse with primary lysosomes, thus initiating the lysosomal enzymatic breakdown of the endosomal membrane and contents. (Adapted from Bennett, 1956.)

1.20) which travel into the cell interior. The content of the vacuole faces the "outside" surface of the former cell membrane. Thus, from the point of view of the asymmetry of the membrane the vacuolar content lies outside the cytoplasm, just as the content of the gastrointestinal tract lies, in reality, outside of an organism. In pinocytosis, however, the cell membrane appears to have properties in the cell interior which are different from those which it had on the cell surface. The cell membrane of the amoeba, for example, is highly impermeable to glucose. When pinocytosis is stimulated, how-

ever, glucose uptake from the vacuolar fluid becomes considerable. It is believed that macromolecules pass from the interior of the pinocytic vacuole into the cytoplasm as a consequence of the fusion of the vacuole (or vesicle) with a primary lysosome and the resultant breakdown of its membrane and contents.

Pinocytosis is not simply the entrapment of a volume of extracellular fluid with the subsequent uptake of solutes from the fluid. In the amoeba the concentration of protein in the pinocytic vacuole exceeds that in the extracellular fluid. This anomaly was explained when it was shown that the protein was adsorbed to the cell membrane prior to the formation of the pinocytic vacuole. Thus, it appears likely that the transport process involved in pinocytosis is primarily one of plasma membrane movement into the interior of the cell and, therefore, that the ingestion of the extracellular fluid which is trapped in the vacuole may be a by-product of the process. Thus, pinocytosis might be regarded as a transport process in which the carrier is the cell membrane.

Phylogenetic Variability

Thus far we have emphasized the similarity of various biological membranes. It is appropriate now to take note of the great diversity of properties manifested by biological membranes. In Table 1.2 we note a 100-fold difference in permeability to water of the cell membranes of *Amoeba proteus* and the human erythrocyte. Even comparisons within the class of erythrocytes show marked differences. The permeabilities to glycerol of the cell membranes of erythrocytes of two evolutionary cousins, the rat and the ox, differ by a factor of more than 100. We have already considered several generalizations relating the chemical structure of a solute molecule to its permeability coefficient. These generalizations were derived largely from studies on plant cells but, with many exceptions, apply also to mammalian cells.

Table 1.2
Permeability to water of some cell membranes

Species	Extracellular Fluid	Permeability
		(μ^3 water/μ^2 surface area/atm)
Amoeba proteus	Fresh water	0.03
Arbacia egg	Sea water	0.4
Human erythrocyte	Serum	3.0

Adapted from Prosser and Brown (1961).

MOVEMENT OF WATER ACROSS THE CELL MEMBRANE

Because of the special role which water plays in biological systems and because of the unique physical characteristics of this molecule we shall treat the diffusion of water across the cell membrane as a special case. Compared to other molecules water diffuses across the cell membrane at a very rapid rate. The permeability coefficient of water is 400 times greater than that of urea (Table 1.3), although the latter molecule itself has a very high permeability constant compared to other molecules of biological significance. As a result, the water within an erythrocyte exchanges with water from outside the erythrocyte in a small fraction of one second. Nevertheless, it is important to recognize that the cell membrane is a rather impermeable structure as demonstrated by the fact that a water layer of equal thickness would have a permeability to water molecules 10,000 to 100,000 times larger. Thus, when we say that the cell membrane is highly permeable to water we are speaking only in relative terms, that is, we are noting the permeability of the membrane to water as compared with its permeability to other molecules.

Osmosis

Osmotic phenomena arise as a consequence of the high rate at which water, compared to other cell constituents (solute), crosses the cell membrane. Insight into osmotic phenomena can be most easily obtained by studying a comparable system of ideal gases.

Consider a container (Fig. 1.21*a*) divided into two equal parts by a partition. There are an equal number of ideal gas molecules on each side of the partition. What distinguishes the two sides is that on the left there are 12 big molecules (*marbles*) and 4 small molecules (*dots*), a total of 16 molecules, while on the right there are 16 small molecules. (In the biological system the dots correspond to water molecules, the marbles to solute molecules which penetrate the cell membrane very slowly, if at all.)

Table 1.3
Orders of magnitude of the permeability coefficients of erythrocyte membranes

Substance	Oil/Water Partition Coefficient	Permeability Coefficient, cm/s	Species
Water	10^{-3}	10^{-2}	Ox
Urea	10^{-4}	10^{-4}	Ox
Cl$^-$		10^{-4}	Man
K$^+$		10^{-8}	Man
Na$^+$		10^{-10}	Man

For references see Davson (1964).

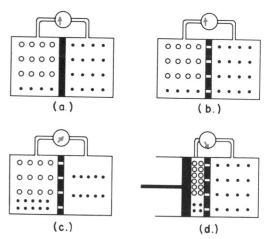

Figure 1.21. Ideal gas model of osmosis. *a*, An equal number of particles on each side of the partition leads to a zero pressure difference between the two compartments. *b*, Holes are drilled through the partition. These are of such size that the dots, but not the marbles, can pass from one compartment to the next. *c*, At equilibrium the concentration of dots is the same on both sides of the partition. Due to the presence of the marbles, the pressure is higher on the left than it is on the right. *d*, The net flow of dots from right to left is averted by decreasing the volume of the left compartment and thereby increasing the concentration of dots on the left. The pressure difference between the two compartments when net flow of dots is zero is analogous to the stopping pressure that can be used to counteract the osmotic pressure, π.

Now it is a characteristic of the ideal gas that its pressure depends only on the concentration of particles and *not* on the *nature* (large or small molecules) of the gas. Hence, the pressure is the same on each side of the partition. Let us now drill holes through the partition of a size that the dots can pass through but the marbles cannot (Fig. 1.21*b*). What happens next is a consequence of the fact that in a mixture of ideal gases each component behaves as if the other components did not exist. Thus, dots migrate through the partition until the number of dots on the left equals the number of dots on the right (Fig. 1.21*c*). More generally, we can say that the migration will occur in a direction to equalize the number of dots per unit volume (concentration on both sides of the partition). The migration can be prevented by decreasing the volume on the left side of the chamber so as to equalize the initial concentration of dots on both sides of the *semipermeable* partition (Fig. 1.21*d*). The decrease in volume is accompanied by an increase in the pressure of the gas in the left chamber. This increase in pressure required to prevent the new flow of "solvent" (dots) between the two chambers is analogous to the stopping pressure that can be used to counteract the *osmotic pressure, π*, of a "solution" (marbles dissolved in dots) with respect to a pure "solvent" (dots).

To transpose this problem back to the cell: the

perforated partition corresponds (in its osmotic properties) to the semipermeable cell membrane. (A semipermeable membrane is one which cannot be penetrated by solute molecules or which is very much more readily penetrated by solvent than by solute molecules.) The small particles (dots) correspond to the solvent (water), the large particles (marbles), the solute. As compared to water the marbles represent solute molecules which penetrate the cell membrane slowly, if at all. For an aqueous solution the osmotic pressure, π (in atmospheres), can be estimated from the equation:

$$\pi V = n_0 R T$$

In this equation V is the volume (in liters) of the solution, n_0 is the number of moles of solute particles in this volume, T is the absolute temperature (°K) and R is the universal gas constant 8.2×10^{-2} (liter·atm)/(mole·°K). If the extracellular and intracellular fluids have the same osmotic pressure and no solute penetrates the cell membrane, then there is no *net* flow of water across the cell membrane and the cell does not swell or shrink. Mammalian cells do not swell or shrink when placed in a 0.9% NaCl solution. The molecular weight of NaCl is 58.5; therefore, the molarity of this solution is 0.15. However, the NaCl is almost completely ionized so that there are approximately 2 osmotically active particles (Na^+ and Cl^-) for every NaCl molecule. Hence, $n_0 = 2 \times 0.15 = 0.3$ moles (per liter), $V = 1$ liter, $T = 310°$ (at body temperature), and the osmotic pressure of the interior of the mammalian cell is approximately:

$$\pi = n_0 R T / V$$
$$= (0.3)(8.2 \times 10^{-2})(3.1 \times 10^2)/1 = 7.5 \text{ atm}$$

Any system, such as the cell or the mitochondrion, which is bounded by a semipermeable membrane, manifests osmotic properties. In the mammalian organism the osmotic pressure of the blood and extracellular fluids is regulated very precisely, and water equilibrates rapidly across the cell membranes. In the amoeba, a fresh water organism, there is a constant influx of water into the organism, and a continuing energy expenditure acts indirectly (via ion pumping) to drive the water out again. Thus, the relatively low permeability to water of the amoeba as compared with the human erythrocyte (Table 1.2) is an important adaptation for this organism.

Chapters 1 and 2 are based in part on material prepared for edition 8 by Joseph Engelberg.

CHAPTER 3

Excitation, Conduction, and Transmission of the Nerve Impulse

The nervous system is a complex array of specialized structures which serve to receive, store, and transmit information—thereby integrating the activities of spatially separated cells, tissues, and organs and making it possible for a multicellular organism to function as a coordinated unit in terms of growth, development, and the ability to do work and to adapt to changes in the environment. Our efforts to understand the function of the nervous system involve the concepts and languages of many disciplines, ranging from mathematics and physics to the physiology and behavior of human beings in all of their complexity. In this chapter, however, we will be concerned primarily with the generation and propagation of the impulses which constitute the currency of all neural transactions.

STRUCTURE OF NERVOUS TISSUE

The Neuron

The neuron consists of a body (soma, perikaryon) and two types of processes—the dendrite and the axon (Fig. 1.22a). In vertebrates, the bodies of the nerve cells lie within the grey matter of the central nervous system or in outlying ganglia, e.g., posterior spinal root, cranial, or autonomic ganglia. The white matter of the brain and spinal cord and of the peripheral nerves is composed of bundles of nerve fibers. The core of each nerve fiber is formed by a process of a nerve cell, and many of them are surrounded by a sheath of myelin which gives them a white appearance. The grey matter receives a rich blood supply from the vessels of the pia mater; the blood supply to the white substance is much less profuse.

There are a number of different types of nerve cell; those in which axon and dendrite arise by a common stem are called unipolar, and those in which the axon and the dendrite or dendrites spring from opposite or at least different parts of the soma are called bipolar or multipolar. The cell bodies or somata are of various sizes and forms—stellate, round, pyramidal, fusiform, etc.

After fixation and staining by special techniques, various structures are seen in the cytoplasm or perikaryon of the nerve cell body: (1) neurofibrils; (2) Nissl bodies or tigroid substance; (3) Golgi apparatus; (4) mitochondria; (5) ribosomes; and (6) the endoplasmic reticulum (see also Chapter 1). Electron microscopy has revealed much of the detailed structure of these intracellular entities. Their structure and function appear to be the same in all cells studied so far. Mitochondria, found along the entire length of the axon, contain all the enzymes required for the respiratory activity of the cell, and are, therefore, responsible for those functions dependent upon aerobic metabolism. The neurofibrils appear as fine filaments which stream through the cytoplasm from dendrites to axon (Fig. 1.22b); they enter the latter process and extend to its terminations. The Nissl bodies composed of ribonucleic acid and polysomes are granular masses stainable with basic dyes and occur in the perikaryon and dendrites but not in the axon. They give a striped or tigroid appearance to the cell. They are absent from the region of origin of the axon (axon hillock) and vary in size and number with the state of the neuron; they undergo disintegration (chromatolysis) in a fatigued or injured cell or in one whose axon has been sectioned. This means that the synthetic machinery of the perikaryon is at least partially regulated by events in the peripheral proc-

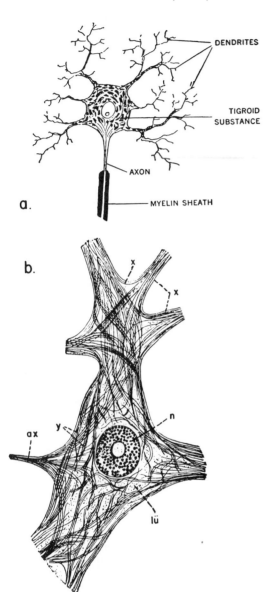

Figure 1.22. *a,* Different parts of the neuron. *b,* Neurofibrils in a cell from the anterior gray column of the human spinal cord. *ax,* axon; *lū,* interfibrillar spaces; *n,* nucleus; *x,* neurofibrils passing from one dendrite to another; *y,* neurofibrils passing through the body of the cell.

esses of the cell. The nature of this feedback control is not known. The internal reticular apparatus of Golgi is a coarse network seen within the cells when special methods—e.g., impregnation with silver chromate—are employed which leave the Nissl bodies and the neurofibrils invisible.

The nucleus of the nerve cell contains one and sometimes two nucleoli but, as a rule, no centrosome. The absence of a centrosome indicates that the highly specialized nerve cell has lost its power of division. Nerve cells once destroyed are replaced merely by neuroglia.

The linking together of neurons to form conducting pathways is effected by the contact (but

not union) of the axon terminal of one nerve cell with the body, dendritic process, or in some instances, the axon of another. Such a junction, without anatomical continuity, is called a *synapse.*

Though the nerve cell frequently possesses more than one dendrite, the axon is single. The axon may be long and contribute to one of the tracts of the central nervous system forming the white matter, or it may terminate as a peripheral nerve fiber. Such cells are referred to as Golgi I type. In the Golgi II type cell, the axon is short and ends within the grey matter by making contact with another neuron. The axon arises from a small elevation on the surface of the cell body, the axon hillock. It may give off short collateral branches or run as an unbranched fiber, not dividing until it has reached its destination. The dendrite is the receptive process of the neuron; the axon is the discharging process, i.e., the former transmits the impulse toward the cell body and the latter away from the cell body. Within the central nervous system the dendrite is usually short and possesses many branches, but in the peripheral sensory nerves it is comparable in length with an axon. Nerve fibers which carry impulses to the central nervous system are termed *afferent*; those conveying impulses from the central nervous system to the periphery are called *efferent*. A mixed nerve contains fibers of both types. The fact that conduction is unidirectional is a consequence of the properties of synapses, for nerve fibers per se can be made to conduct in either direction.

As with other cells, the cytoplasm of the soma is enveloped by a plasma or unit membrane composed of a bimolecular leaflet of lipid material covered and penetrated by protein. This extends over the processes of the nerve cell. The electrical properties of nerve depend upon the plasma membrane.

The Nerve Fiber

The white matter of the central nervous system and the peripheral nerves are composed of thousands of individual nerve fibers. Within the grey matter the axons are enclosed only by the plasma membrane; but upon leaving the gray substance they acquire a sheath of lipid material called myelin. The myelin sheath is associated with somatic nerves of large diameter and generally not with nerve fibers of small diameter. Hence, we speak of myelinated (medullated) or unmyelinated (unmedullated) nerves. Myelinated fibers, usually larger than 1 μm in diameter, conduct faster than the smaller unmyelinated fibers.

The myelin sheath of the somatic nerves consists of compressed layers of Schwann cells spiraling concentrically to form a wrapping around the axons. The outer layer of the myelin contains the flattened nuclei of the Schwann cells and is termed the neurilemma or sheath of Schwann. Myelinated nerves appear as if constricted at regular intervals along their course. This appearance is due to the absence of myelin at these points and the dipping inwards of the neurilemma; these points are known as the nodes of Ranvier. The segments between the nodes vary in length in different nerves, but the usual internodal distance is about 1 mm. Each internodal segment of the neurilemma consists of a single Schwann cell. The fibers forming the white matter of the central nervous system and optic nerves have no neurilemma; this membrane is replaced by glial cells which produce the myelin sheaths as the Schwann cells do peripherally.

Compared with the myelin sheath the neurilemma in the region of the node of Ranvier under appropriate conditions becomes highly permeable to sodium and potassium ions. There is good evidence that the conduction of the impulse along a myelinated nerve is a "leaping" from node to node over the intersegmental regions rather than a continuous process. This type of conduction is called saltatory (from the Latin word *saltus*, a leaping).

MYELINATION OF FIBER TRACTS IN THE CENTRAL NERVOUS SYSTEM

The nerve fibers in the various conducting pathways receive their myelin sheaths at different ages, and it is generally believed that the myelination of a given tract and the time at which it commences to function coincide. The sensory tracts become myelinated first, those of the posterior columns of the spinal cord between the 4th and 5th months of fetal life (human). The spinocerebellar tracts are myelinated later, and the motor paths, e.g., corticospinal (pyramidal) tracts, do not begin to be invested with myelin sheaths until the 2nd month of life and are not completely myelinated until about the 2nd year, about the time when the child has learned to walk. The fibers of association paths, for the most part, myelinate at still later dates. The high insulating property of myelin, in addition to confining the nerve impulse to individual fibers and thus preventing cross-stimulation of adjacent axons, serves to increase the rate of conduction.

The neuroglia or "glia" (Greek, "glue") is a special type of interstitial tissue. Its cells are of three kinds, astrocytes, oligodendrocytes, and microglia. The microglia appear to have a phagocytic role, since they wander into the central nervous system from the meninges and blood vessels and increase in number during inflammatory processes. Astrocytic

processes are found abutting blood vessels and investing synaptic structures, neuronal bodies, and neuronal processes. Although the functions of astrocytes are not precisely known, putative roles include involvement in support, transport mechanisms, inflammatory and reparative reactions, and isolation of neuronal elements. The oligodendroglia are responsible for the formation of the myelin layers around axons within the central nervous system.

Degeneration and Regeneration of Nerve

When a peripheral nerve is cut, the part of the nerve separated from the cell body shows a series of chemical and physical degenerative changes. At the same time the fibers of the proximal stump of the nerve, those still attached to their cell bodies, grow distally toward the separated part of the nerve; these changes constitute the process of regeneration (Fig. 1.23).

DEGENERATION

The degenerative period may be divided into an early and late phase. Shortly after the nerve has been sectioned, the axon swells, the myelin sheath begins to form bead-like structures, and a series of enlargements (round, fatty fragments) appear. The axon breaks up, and its parts are devoured by macrophages. Up to 3 days after section, the distal nerve will continue to conduct an impulse. Changes in the action potential can be observed as early as

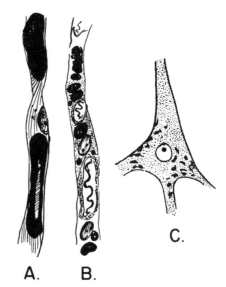

Figure 1.23. Degenerating nerve stained with osmic acid. *A* shows appearance of distal segment of nerve fiber 2 days after section; note large masses of myelin derived from medullary sheath. *B*, 5 days after section, smaller myelin particles together with droplets of fatty acids and fragmented neurofibrils. *C*, Retrograde degeneration in cell body, disintegration of Nissl bodies.

2 days after section. After the 3rd day, the ability of the nerve to conduct has seriously deteriorated and after the 5th day, an impulse can no longer be evoked. The first period (3–5 days) is one in which the obvious changes in nerve are largely structural. Changes in the ultrastructure of the myelin sheath can be shown to occur during this period, as well as changes in the endoplasmic reticulum, mitochondria, neurofibrils, and the plasma membrane.

Up to the 8th day, little or no changes in lipid histochemistry may be detected, although changes in nerve cholinesterase and failure in the ability to synthesize acetylcholine have been demonstrated about the 3rd day. The lipids appear to maintain their original chemical structure. From the 8th to the 32nd day after section, the myelin gradually disappears, while Schwann cells and macrophages have both increased greatly in number. Why the myelin breaks down is not clear. It has been suggested that the macrophages and Schwann cells secrete enzymes which aid in the destruction of myelin. The principal myelin lipids are free cholesterol, and two lipids containing sphingosine: cerebroside and sphingomyelin. At the time of the disappearance of the myelin, cholesterol esters appear in large quantities, and free cholesterol disappears.

The changes just described are generally known as Wallerian degeneration. In addition there are other changes that occur in the neuron on the proximal side of the section (retrograde degeneration). The nerve fiber as far centrally as the first node of Ranvier shows changes similar in nature to those just described. In the cell body itself, swelling of the cytoplasm and nucleus occurs and the Nissl granules undergo disintegration (chromatolysis). Atrophy of the cell body may ultimately result.

REGENERATION

Following section of a nerve, the fibers in the central stump begin to send out branches consisting of outgrowths of the axon near the cut tip. Up to 50 branches may sprout from a cut axon. At the same time, there is a rapid proliferation of the Schwann cells. If a gap larger than 3 mm exists between the central and peripheral stumps, the fibers tend to intermesh and form a tumor-like swelling called a neuroma. In such an event regeneration will probably never occur. For this reason it is necessary to join accurately by suture the proximal and peripheral stumps. If the neuroma is composed of sensory fibers it may be very painful to pressure, often a troublesome complication following amputation. When peripheral degeneration has proceeded for enough time so that the peripheral stump contains only empty neurilemmal tubes,

one of the outgrowing sprouts enters a tube and grows to form a new axon. Only one of the many sprouts enters an empty tube. The ability of axons to grow into a peripheral stump has led some observers to invoke a special chemical attraction between the nerve fiber and the terminal organ. This is the doctrine of neurotropism.

The rate of growth of regenerating nerve is from 1 to 4 mm/day. It has been established, in a variety of experiments, that there is a flow of axoplasm down the normal nerve at about the same rate.

PRINCIPLES OF BIOELECTRICITY

It will be recalled that the practical unit for quantitation of charge is the coulomb which equals 6.2×10^{17} electrons. A positive charge results from the removal of electrons from a neutral body, as when one electron is removed from the neutral sodium atom to form a sodium ion.

An electric field exists in the space around a charge and extends to infinity. The strength of this field is measured by the force which is exerted on a unit charge placed at any point in the field. The intensity of the field decreases inversely as the square of the distance from the charge (Coulomb's law). In order to move a charge against the field, work must be performed. Thus, it takes work to move a positive charge up to a region where another positive charge is situated, and the potential at any point in the field is defined as the work that must be done on a unit positive charge to move it up to that point from an infinite distance. This is the absolute definition of potential which is seldom necessary in practice. The significant quantity in experimental work is the difference in potential between two points which is defined as the work necessary to move a unit charge from one point to the other. The work is measured in joules. The practical unit of potential difference is the volt, i.e., the potential difference against which 1 joule of work is done in the transfer of 1 coulomb.

Origin of Potentials

A difference in potential is associated with a flow of current in a medium (e.g., electrons in metal conductors and charge-carrying ions in solutions of electrolytes). The current flow requires energy since it encounters resistance in the medium. The relationship between E, the potential difference (in volts), I, the current (in amperes which is the same as coulombs per second) and R, the resistance (in ohms), is given by Ohm's law as follows, $E = IR$. In order to produce electrical currents, special sources of electrical energy are available, such as the battery and generator. Such sources of energy are said to produce an electromotive force or emf. In solutions, electrical energy is produced by chemical reactions which result in the separation of charge at electrodes. Other sources of energy produce separation of charge by the flow of ions from solutions of high concentration, such separation of charge itself constituting a form of stored energy which is manifest as a difference in electrical potential.

Two sources of energy are important in understanding the origin of biological potentials; these sources give rise to emfs called *electrode* or *concentration* potentials and *diffusion* or *membrane* potentials. Concentration potentials arise in every measurement of potential difference in solution and are generally to be avoided in biological measurements. They arise whenever an electrode is dipped into a solution containing an ion in common with the electrode. Thus, an electrode made of silver and coated with silver chloride inserted into a solution of sodium chloride will produce a potential difference between the electrode and the solution. The silver dissolves in the solution as silver ion, Ag^+, leaving the electrode negative. The positive Ag^+ ion remains at the electrode, forming a double layer of charge with the negative electrode. Charge is therefore separated at the electrodes. If two such electrodes are set up, each dipping into a different concentration of sodium chloride (a concentration cell), double layers will be set up at each electrode, but the double layer at one electrode will be more charged. When two such electrodes are connected by a wire, a current will flow. The emf produced between the two electrodes can be calculated from the Nernst equation for the concentration cell,

$$E = 2.303(RT/F)\log(C_1/C_2)$$

where the electrolyte is univalent. R is the gas constant (equal to 1.99 cal/mole/°C, or 8.3 joules/mole/°C), T absolute temperature (°K), F the Faraday (23,050 cal/v/mole, 96,494 coul/mole), and C_1 and C_2 are the concentrations of electrolyte in each solution. Note, that according to the formula if the concentrations are equal in each solution, the emf is zero. The method of producing a concentration cell is shown in Fig. 1.24.

DIFFUSION POTENTIALS

Let us place two solutions containing different amounts of NaCl in contact by means of membrane freely permeable to the electrolyte. An emf whose origin is quite different from that of the electrode potentials we have discussed will arise at the membrane. The potential difference arises from the diffusion of ions across the membrane or at any junction between such solutions. They are therefore

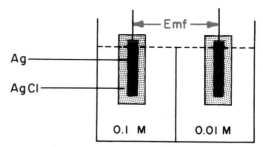

Figure 1.24. A concentration cell. Both solutions contain sodium chloride. The partition between the compartments is freely permeable to water and salt.

called *diffusion, membrane,* or *junction* potentials. The sodium chloride in the solution of higher concentration tends to diffuse across the membrane into the other solution, the sodium ion diffusing faster than the chloride ion. The emf at the membrane depends upon the relative rates (mobilities) at which the sodium and chloride ions move and the relative concentrations of sodium chloride. Since Na^+ diffuses faster than Cl^-, a separation of charge is produced; the solution into which a new flow of Na^+ occurs becoming positive. At biological membranes, somewhat special hindrances to the movement of ions are present. Potassium and chloride usually diffuse at the same rate in solution; however, the ability of chloride to move across the cell membrane is more limited than that of potassium. Under these circumstances, potassium diffuses ahead of chloride, and an excess of potassium ions appears on one side of the membrane. Although this excess is very small, it is sufficient to set up an emf or potential difference, so that the solution containing the lesser concentration of potassium chloride becomes positive to that containing the higher concentration. There is, in effect, a greater loss of potassium from the more concentrated to the less concentrated solution. The more concentrated solution becomes negative, since it has lost positive charge. The concentration of K^+ increases until the emf has become sufficiently negative to keep any further K^+ from diffusing across the membrane, this is the equilibrium potential for K^+. The diffusion potential at a membrane is given by a formula similar to the Nernst equation in which the latter has been modified to take into account the mobilities of the ions. The equation is:

$$E = [(u - v)/(u + v)]2.303(RT/F)\log(C_1/C_2)$$

where the electrolyte is univalent, u and v are the mobilities of the cation and anion, respectively, i.e., the rates at which the ions move under unit field strength in solution. The remaining symbols are the same as those in the Nernst equation. Note that when the anion and cation mobilities are equal, $u = v$, and the emf is zero.

TRANSMEMBRANE POTENTIAL

Methods are available for measuring the potential difference across the cell membrane at rest and during the activity associated with impulse generation and propagation. In some cells, such as the giant axon of the squid, it is possible to insert an electrode into the body of the axon down its length and, by placing another electrode outside the cell, to measure the transmembrane potential directly. In other cells, where this technique is not feasible, another method is used. A glass capillary is drawn out to a fine tip less than 1 μm in diameter. The capillary is filled with saturated KCl or NaCl and serves as a microelectrode. It is inserted directly into the soma, muscle fiber, or even peripheral nerve fiber. The potential difference is measured between the microelectrode and a large, nonpolarizable electrode (connected via a KCl bridge) located outside the cell; this is the transmembrane potential. Such a method always involves the possibility that the injury caused by the puncture will result in a gradual fall in the potential and death of the cell. The membrane potentials of a variety of cells are shown in Table 1.4; they range from 61 to 94 mv, with the cell interior negative in relation to the external medium.

In recent years two approaches have been developed for the assessment of the contributions to transmembrane current flow of individual ionic channels. The first of these methods involves analysis of the minute variations (noise analysis) in the conductance of a large segment of cell membrane (Begenisich and Stevens, 1975; Conti et al., 1976); the second of these methods involves the measurement of the ionic current flow through very small patches of cell membrane (Neher et al., 1978; Hamill et al., 1981).

Theory of the Membrane Potential

The original theory of the membrane potential, largely valid today in broad outline, was first enunciated by Bernstein (1912) at the turn of this century. It is well known that the concentration of potassium is much higher inside cells than outside. Table 1.5 gives the value of the ratio of inside to outside concentrations for potassium and other important ions in various tissues. The ratios for K^+ vary from 23 to 68. Bernstein maintained that the

Table 1.4
Resting and action potentials

Tissue	Resting Potential, mv	Action Potential, mv
Loligo axon	61	96
Sepia axon	62	122
Carcinus axon	82	134
Frog myelinated nerve fiber	71	116
Frog striated muscle fiber	88	119
Frog cardiac muscle fiber	70	90
Dog cardiac muscle fiber	90	121
Kid cardiac muscle fiber	94	135

Table 1.5
Ionic content of nerve and muscle cells

Tissue	Sodium In, mM	Sodium Out, mM	Sodium Ratio	Potassium In, mM	Potassium Out, mM	Potassium Ratio	Chloride In, mM	Chloride Out, mM	Chloride Ratio
Carcinus nerve		460		380	10	38		540	
Carcinus nerve		460		230	10	23		540	
Frog nerve (Nov.)	37	120	0.31	110	2.5	44		120	
Frog nerve (Mar.)				170		68			
Frog sartorius muscle	15	120	0.12	125	2.5	50	1.2	120	0.01
Frog sartorius muscle	26	120	0.22	115	2.5	46	11	120	0.092
Rat cardiac muscle	13	150	0.087	140	2.7	52		140	
Dog skeletal muscle	12	150	0.08	140	2.7	48		140	

Ionic fluxes of resting membrane

			pmol/cm^{-1}/sec^{-1}					
Sepia axon	61	31	17	58				
			(11)	(33)				
Carcinus axon			19	22				
Frog sartorius muscle	13	16	7	5				
Frog sartorius muscle		5–10		20				
Frog abdominal muscle		5	10	10				
Frog ext. long. dig. IV muscle[a]			4	5				

[a] Extensor longus digitorum.

membrane potential was the result of the outward diffusion of potassium ions from the cells. Since the resting potential is a diffusion potential in which the mobility of the chloride anion was taken as zero and sodium was also presumed to be unable to penetrate the cell, the magnitude of the membrane potential can be calculated from the formula for the diffusion potential in which $v = 0$. The formula reduces to that of Nernst,

$$E = 2.3(RT/F)\log(K_{outside}/K_{inside})$$

Partial verification for this theory has been obtained by Hodgkin and Huxley (1952b) in several experiments in which the external potassium concentration was varied and resting potential of the squid axon measured. It was shown that the resting potential varied directly with the logarithm of the external concentration of potassium over a wide range. Table 1.5 shows that the sodium is largely present outside cells. The resting potential, V_r, was shown to be independent of the external sodium concentration [$Na_{outside}$] in a series of experiments on the squid axon in which [$Na_{outside}$] was varied over a wide range with no effect on V_r. Similar findings have been encountered in experiments on frog cardiac and skeletal muscle.

Ionic Distributions and the Membrane Potential

If no other ion but potassium could penetrate the membrane, the membrane potential would, as noted above, be given by the equation for a diffusion potential in which the anion mobility is zero. The resulting equation is then equal to the Nernst equation for a concentration cell. Actually, however, potassium does not pass through the membrane alone; there is always some transmembrane flux of the sodium and chloride ions. Two schemes for the membrane potential have been proposed to take into account the flow through the membrane of ions in addition to potassium. These are the Goldman equation and the Hodgkin-Huxley equivalent circuit for the resting membrane. Tests of these equations are usually made by plotting the membrane potential against the external potassium concentration and observing how closely the equations agree with the experimentally derived curve. All the equations, the Nernst relation included, are satisfactory over some range of potassium concentrations.

In order to understand the derivation and significance of the Nernst equation, let us consider the problem of moving an ion in a solution in which a potential difference is present. Work must be done with or against two forces, the electrical field, and any concentration difference in the ions. Thus, both "electrical" and "concentration" work must be performed. The electrical work (in joules) necessary to move 1 mole of a univalent ion (i.e., one equivalent) against a potential difference, E (in volts), is given by the expression: work = FE, where F is the Faraday (96,494 coul/mole). The concentration work required to move a mole of ion from a concentration C_1 to a higher concentration C_2 is given by the expression: work = 2.303 $RT \log C_2/C_1$, where R is the gas constant and T the absolute temperature. The total work is given by the sum of the electrical and the concentration work, or work = $FE + 2.303\ RT \log C_2/C_1$. When the system is at equilibrium, the work required is zero (this is the definition of thermodynamic equilibrium), and the potential difference is given by the Nernst equation,

$$E = 2.303(RT/F)\log(C_1/C_2)$$

The physical interpretation of this equation is that the tendency of an ion to diffuse down its concentration gradient is countered by the buildup of an electric field at the junction of the two solutions. The direction of the field is such as to hold back the ion from further movement. Note also that it is necessary to move only a very minute number of ions in order to produce the restraint required to obtain equilibrium of electric and concentration gradients. As an example of an application of the Nernst equation, we calculate the transmembrane potential difference which might exist in frog muscle on the assumption that the muscle cell membrane is permeable only to K^+. The internal concentration of potassium is 155 meq/l; the external concentration is 4 meq/l. Substituting these values into the Nernst equation at a temperature of 27°C, we obtain for the potential difference,

$$E = 2.303(8.3 \times 300/96,500)\log(4/155)$$
$$= -95\ mv$$

The Goldman equation will not be derived here. However, in the derivation, the assumptions are made that the total flow of current through the membrane is zero, that is, that the flow of negative charges is equal to the flow of positive charges, and also that the drop in potential across the membrane is linear. Using these assumptions it can be shown that the resting potential will be given by the relation

$$E = 2.303 \frac{RT}{F} \log \frac{P_K C_K^o + P_{Na} C_{Na}^o + P_{Cl} C_{Cl}}{P_K C_K + P_{Na} C_{Na} + P_{Cl} C_{Cl}^o}.$$

In this equation R, T, and F have their usual significance. P represents the permeability of the membrane to the ion and the superscript "O" above the concentration, i.e., $C°$, represents the concentration outside the cell, the unlabeled C, the concentration within the cell. When the permeabilities to Na^+ and Cl^- are taken to be zero, the equation reduces to the Nernst equation. If permeabilities are assumed to have the ratio,

$$P_K : P_{Na} : P_{CL} :: 1 : 0.04 : 0.45,$$

then the emf of the membrane as given by the equation agrees rather well with the value measured in the squid axon, and the resting membrane potential agrees well with the calculated value over a more than 50-fold variation of concentration of external potassium.

The Hodgkin-Huxley equivalent circuit constitutes a third approach to the analysis of the nerve cell membrane potential. The membrane is assumed to contain separate channels through which each of the ions passes without interference from the others. The total electrical current flow through the membrane is again assumed to be zero. Each ion, in passing through its channel, encounters resistance to its movement through the membrane. Three such channels are shown, one for Na^+, K^+, and Cl^- (Fig. 1.25); each is represented by a battery (equilibrium potential) whose emf is calculated from the Nernst relation and whose opposition to current flow is shown as an electrical resistance. Often the reciprocal of resistance, the conductance (g), is used to represent the channel permeability. If the total current from the membrane is zero, it can be shown that the membrane potential, E, resulting from the three emfs in parallel is given by the equation

$$E = \frac{E_K g_K + E_{Na} g_{Na} + E_{Cl} g_{Cl}}{g_{Na} + g_K + g_{Cl}}.$$

The gs represent the conductance of the membrane; they correspond approximately to the permeabilities to each ion. The emfs are the equilibrium potentials for each ion. Thus, in frog muscle the equilibrium potential for potassium is -95 mv, for sodium $+65$ mv, and for chloride -90 mv, values which have been calculated for each ion from the Nernst equation and the concentration of ions inside and outside the cell. Assuming further that the potassium conductance is 100 times greater than the Na and Cl conductances (i.e., that $g_K = 100 \, g_{Na}$

$= 100 \, g_{Cl}$), which is approximately true for frog muscle, we obtain for the emf across the membrane,

$$\frac{-95(g) + 65(g/100) + (-90)(g/100)}{g + g/100 + g/100}$$

or -95 mv, a value very close to the actual membrane potential. The Hodgkin-Huxley equivalent circuit will prove very important in discussing the mechanism of the action potential in later sections.

EXCITABILITY AND CONDUCTIVITY OF THE NERVE FIBER

The generation and transmission of a nerve impulse involves two conceptually independent but operationally related processes, excitation and conduction.

When a nerve is stimulated, electrical events which are not propagated occur in the membrane in the vicinity of the electrodes. If the local events have particular characteristics (described below), the membrane potential undergoes an abrupt change, termed the *action potential*, which is self-propagated along the axon. Excitation refers to the events leading to the generation of an action potential; conduction refers to the propagation of the action potential which proceeds away from the site of excitation, much as a wave travels in a taut string. It is fundamental that nerve may be excited anywhere along its length and propagation is away from the point of stimulation in both directions. However, when a junction (synapse) intervenes between nerve fibers in the mammalian nervous system, then propagation can only continue in one direction. Experimentally many different kinds of stimuli (electrical, thermal, mechanical, chemical) may be utilized to excite nerve or muscle; however, all of those stimuli operate by depolarizing the nerve fiber. Since electrical stimuli of any intensity, shape (wave form), and duration may be easily produced both accurately and repetitively, this form of stimulation is used universally to study the phenomena of excitation and propagation. Moreover, the nerve impulse is electrical in nature, and many of its effects as an excitatory agent can be simulated by the electrical stimulus.

Characteristics of the Stimulus

Nerve responds to electrical stimulation provided that the electrical stimulus fulfills certain specific criteria. It must be of sufficient *intensity* and *duration* to reduce the transmembrane potential from its resting value to a critical voltage which, when achieved, results in the development of a propagated impulse. This reduction in membrane potential is termed *depolarization*, and the critical voltage

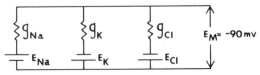

Figure 1.25. The Hodgkin-Huxley equivalent circuit for the membrane. The symbol E represents the "equilibrium" potential as determined for each ion from the Nernst equation. The symbol g represents the conductance of each ion.

required for impulse propagation is termed the *threshold voltage* or *critical firing potential*. The relationship between the intensity and duration of the initial stimulus is further discussed below. A current just adequate to cause an impulse is called a *threshold* stimulus. Intensities below threshold are referred to as *subliminal*. Thresholds vary only slightly if the temperature and external ionic composition are maintained constant.

Another characteristic of the stimulus is its rate of rise. If the current is increased too slowly the nerve will not respond. Figure 1.26 shows two linearly rising currents, one of which is able to reach threshold. The other current rises too slowly, and the nerve is able to accommodate to the passage of the current. *Accommodation*, therefore, consists of a rise in threshold of the tissue during stimulation. To minimize accommodations, it is convenient to employ stimulus currents which rise extremely rapidly. Two such stimuli are shown in Figure 1.27, the square wave and the exponential pulse.

Even the rapidly rising pulses of Figure 1.27, if too short in duration, would not result in an impulse. The two properties, intensity and duration, obviously interact, and it is therefore important to be aware of the relationship between the threshold stimulus intensity and the duration of the stimulus—the *strength-duration* relationship.

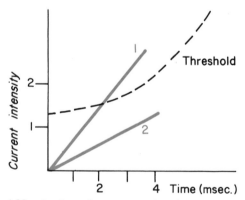

Figure 1.26. A schematic representation of the effect of rate of rise of current. Curves *1* and *2* represent two stimuli with different rates of rise. Stimulus 2 never reaches threshold (*dashed line*). Stimulus 1 attains threshold.

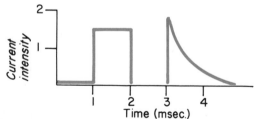

Figure 1.27. Two commonly used stimuli: a rectangular pulse (*left*) and an exponential pulse (*right*).

For this purpose, the following kind of experiment is usually performed. A stimulus of a fixed duration, e.g., 1 ms, is applied to a nerve through two electrodes, one of which is the cathode (−), the other the anode (+). The threshold stimulus is determined by increasing the current until a response is obtained—at the particular duration selected. A series of other durations are selected and the thresholds determined. The thresholds obtained at each duration are plotted as a function of the duration. The curve so obtained is called the strength-duration relationship for nerve (Fig. 1.28).

The curve is accurately described over most of its course by the empirical relationship $I = I_0 (1 - e^{-kt})^{-1}$, where I_0 and k are constants, t is the stimulus duration, and I the threshold current. For short durations, in which accommodation is presumably slight, the relation is approximated by the equation, $It =$ constant. This relationship may be interpreted as follows: The current I is the charge per unit time which is placed on the membrane in a time t. The product of the current and time is, therefore, equivalent to a constant charge. The equation implies that a critical amount of charge must be placed on the membrane, whatever the current or duration.

What is the meaning of this critical amount of charge? Placing a charge on the membrane by an appropriately oriented stimulus current reduces the net charge on the membrane, i.e., part of the charge on the membrane is neutralized. Such a partial neutralization of charge is equivalent to a *depolarizing* of the membrane, i.e., to a reduction in the membrane potential from its resting negative value toward 0. Increased polarization or *hyperpolarizing* results in an increase in the membrane potential, i.e., increasing negativity.

A feature of excitation discovered quite early was the fact that the nerve impulse originated at the cathodal stimulating electrode, i.e., at the negative stimulating electrode. Figure 1.29 illustrates the

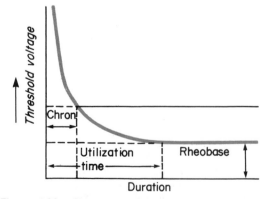

Figure 1.28. Strength-duration curve; *chron*, chronaxie.

lines of current that flow into and out of the electrodes. At the cathode, the current flows outward through the membrane; at the anode, where current is flowing inward, excitation is hindered. Continuous current flowing through the nerve depolarizes the membrane in the vicinity of the cathode and hyperpolarizes it in the vicinity of the anode. Thus, it is easier to stimulate a nerve in the vicinity of the cathode because the membrane potential has been lowered; conversely, at the anode, the membrane potential is increased and higher currents are required to excite the underlying tissue. These effects of direct current on the excitability of nerve and muscle have been referred to as electrotonic phenomena; the depolarization at the cathode as *catelectrotonus*, and the hyperpolarization at the anode *anelectrotonus*.

Summation of the Local Excitatory State

The application of a brief subthreshold stimulus has a residual effect on nerve, even though an impulse is not elicited. This residual effect is revealed by the fact that the application of a second subthreshold stimulus within a millisecond can elicit a response. The longer the interval between the two stimuli, the more intense the second stimulus must be to yield a response. The important point, however, is the fact that two subthreshold stimuli can sum their effect on the nerve membrane so that a response is evoked. The interpretation of this experiment is that the first subthreshold stimulus produces a change in the membrane potential which lasts for a millisecond or more and that this change facilitates the effect of a second stimulus. The change in the membrane caused by the first stimulus is referred to as the "local excitatory state" in the vicinity of the electrode. It is a nonpropagated response of the membrane which is associated with the depolarization of the membrane in the region of the cathode (see below).

THE LOCAL EXCITATORY STATE

The experiments on subthreshold summation

and the electrotonic properties of nerve led to the expectation that the membrane potential should be decreased in the vicinity of the cathode, and indeed, this expectation has been affirmed by direct experiments in which the changes in membrane potential were measured in the vicinity of the cathode and anode.

It is necessary first to consider the purely passive changes in the membrane potential which result from the fact that nerve has the properties of an electric cable, namely electrical resistance and capacitance (the ability to store charge). The capacitance, C, is defined as the charge, Q, which must be placed on two surfaces in order to produce a unit potential difference, V, between them ($C = Q/V$). The units of C are farads when Q is expressed in coulombs and V in volts. The capacitance of the nerve membrane is 1 $\mu f/cm^2$.

For each unit length of nerve, there is an external electrical resistance to current flow through the extracellular medium and an intracellular resistance to flow through the cytoplasm. The transmembrane resistance constitutes a third electrical resistance. Each unit length of nerve may be considered as an electrical cable (Fig. 1.30) consisting of these resistances and a condenser (the capacitance). If a stimulating current is passed through a section of nerve, its condensers will charge, but the most distant ones will be least charged since more external and internal resistance is included between them and the source of current. If one were to measure the charging process at any moment, the voltage would be highest at the electrodes and would decrease exponentially as one proceeded away from the stimulating electrodes (Fig. 1.31). This charging process is almost instantaneous and represents a nonpropagated buildup of charge along the nerve. It extends for some distance along the nerve, but it is never propagated as a wave.

When nerve is stimulated, two electrical events take place. The membrane charges passively as a

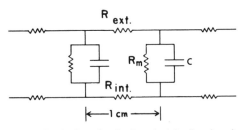

Figure 1.30. Equivalent circuit of a nerve (radius 1 μm) considered as an electric cable. $R_{ext.}$ is the resistance of 1 cm of the external medium; generally low. $R_{int.}$ is the resistance of 1 cm of the internal axoplasm (10^7 Ω) and R_m is the resistance of 1 cm of the membrane to radial currents—of the order of 10^{10} Ω. The capacitance, C, is 1 $\mu f/cm^2$ of nerve surface.

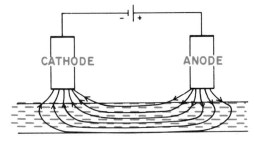

Figure 1.29. Illustrating the passage of direct current through tissue. Anode is defined as the electrode which sends current *into* tissue.

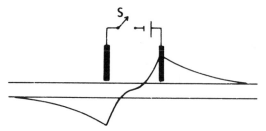

Figure 1.31. Illustration of the passive, electrotonic potential along the nerve resulting from the passage of direct current into the nerve.

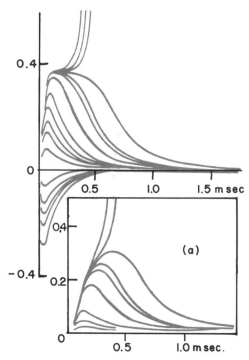

Figure 1.32. Electrical changes at stimulating electrode produced by shocks with relative strengths, successively from above, 1.00 (upper 6 curves), 0.96, 0.85, 0.71, 0.57, 0.43, 0.21, −0.21, −0.43, −0.57, −0.71, −1.00. The ordinate scale gives the potential as a fraction of the propagated spike, which was about 40 mv in amplitude. The 0.96 curve is thicker than the others, because the local response has begun to fluctuate very slightly at this strength. The width of the line indicates the extent of fluctuation. *Inset* (a), Responses produced by shocks with strengths, successively from above, 1.00 (upper 5 curves), 0.96, 0.85, 0.71, 0.57; obtained from curves in upper figure by subtracting anodic changes from corresponding cathodic curves. Ordinate as above. [From Hodgkin (1939).]

cable and, at the same time, the nerve begins to react physiologically at the cathode. Events at the anode are passive, and since excitation does not occur there, the time course of charging at the anode may be taken as that for the passive or physical charging of nerve. At the cathode, passive charging occurs and, in addition, an active process which we have called the local excitatory process. If the passive process at the cathode is subtracted graphically from the overall recorded response, the local excitatory process should be obtained. The passive physical process at the anode and cathode are proportional to the stimulating current. In order to measure the local responses, an experiment is performed in which the potential difference between electrodes placed at the anode and cathode is measured with respect to a distant electrode. Figure 1.32 shows the results of the potential measurements and also the result of subtracting the passive response from the overall response to obtain the local potential which corresponds to an active, nonlinear process at the electrode.

The significance of the local potential is that it demonstrates the time course of the depolarization at the cathode. When the depolarization at this electrode reaches a critical value an action potential will be initiated which will propagate away from the electrode. Such local responses can be found in many tissues. For example, receptors must be depolarized before they give rise to an action potential; the potential representing the local excitatory state of the receptor is called the generator potential. An active process also occurs at the neuromuscular end-plate which is called the end-plate potential; this depolarization initiates the propagated action potential of muscle. At the synapses of neurons, a local, nonpropagated potential called the excitatory postsynaptic potential (EPSP) may be recorded which gives rise to the nerve impulse of the neuron. The critical event in the excitation of all these cells is the local, nonpropagated depolarization. When this depolarization reaches threshold magnitude, the explosive, propagated change in

membrane potential called the nerve impulse or action potential is set off.

THE NERVE IMPULSE

Recording the Action Potential

The externally recorded action potential may be obtained by stimulating a nerve at one end and picking up the responses of the nerve some distance away with two recording electrodes and a recording device. Figure 1.33 shows the arrangement by which the action potential of nerve is usually recorded with a cathode ray oscilloscope. Unlike any instrument previously employed for this purpose (e.g., the string galvanometer or the capillary electrometer), the moving part of the oscilloscope is a stream of electrons and consequently has virtually no mass, and thus, no inertia. It is, therefore, capable of recording very rapid changes in electrical potential. The instrument consists of an evacuated tube; an electron stream from a hot cathode strikes a flu-

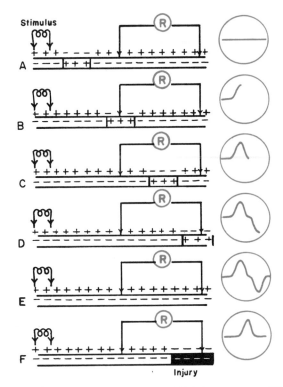

Figure 1.33. Passage of action potential down a nerve. Stimulus is applied at left. Cathode ray oscilloscope is indicated at *R*. The recorded potential is shown at the face of the cathode ray oscilloscope at the extreme right of each figure. A monophasic potential is shown in figure *F*.

The response on the cathode ray oscilloscope shows up as a diphasic variation of potential. The cathode ray oscilloscope records the potential under one recording electrode as a function of time with respect to the other recording electrode. When the nerve is stimulated, an electrical impulse is generated which travels along the fiber. The arrival of this impulse at the first electrode causes the oscilloscope beam to deflect in one direction. When the impulse passes between the electrodes, no recorded potential is observed. But when the impulse reaches the second electrode, a potential is recorded which is opposite in polarity to the first and accordingly, causes the oscilloscope beam to deflect in the opposite direction. We, therefore, say that there are two phases present in the recorded action potential. It is quite clear that one can eliminate the second phase from the recording by preventing the negativity of the impulse from reaching the second electrode. This can be done by crushing the nerve either in the region between the two electrodes or directly under the second electrode. Of course, if the crush is at the first electrode, no potential will be recorded at all, since the action potential cannot pass a dead region of nerve. The potential recorded when the nerve is crushed at the second electrode is called a monophasic potential and is shown also in the accompanying figure (Fig. 1.33*F*).

THE COMPOUND NATURE OF THE ACTION CURRENT RECORDED FROM A NERVE TRUNK

Erlanger and Gasser (1937) studied the action potential of mixed nerve trunks by means of the cathode ray oscillograph. They showed that the recorded "spike" is actually compounded of the individual spikes of many axons which were classified into three main types of nerve fiber—referred to as the A, B, and C groups (Fig. 1.34). Several properties of nerve are correlated with the diameters of the fibers: the larger the fiber diameter, the greater the conduction velocity; the greater the magnitude of electrical response, the lower the threshold of excitation and the shorter the duration of response and the refractory period. The relationship of conduction velocity to diameter of the nerve fiber is a linear one (Fig. 1.35). The amplitude of the externally recorded potential is also linearly related to the fiber diameter.

The A group is composed of the largest fibers, 1–20 μm in diameter, with conduction rates from 5 m/s or less for the smallest fiber to 100 m/s for the largest. The fibers of the A group are all myelinated, are both sensory and motor in function, and are found in such somatic nerves as the sciatic and saphenous nerves.

orescent screen upon which it produces a spot of light. On either side of the electron stream is placed a vertical plate. A potential difference is created between the pair of plates; the electric field set up across the path of the stream deflects it horizontally and sweeps it across the screen. The spot of light is converted into a horizontal streak. By means of a sweep oscillator the horizontal deflections are repeated many times per second. A second pair of horizontal plates is placed one above, the other below, the electron stream. It is this latter set of plates into which the nerve action potential is fed so that a vertical deflection of the electron stream results with production of a standing wave which is photographed and a permanent record thus obtained. The speed of the horizontal movement of the spot of light enables the time scale to be calculated; this can be varied by altering the potential applied to the vertical pair of plates. The magnitude of the action potential is determined from the height of the wave. Before reaching the recording system the action current is amplified several thousand times by passing it through an amplifier. This is necessary because the electron stream requires about 50 v to cause a deflection of 1 cm on the face of the tube.

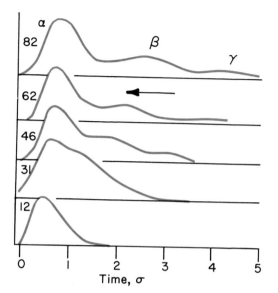

Figure 1.34. Cathode ray oscillograph records of the action currents in the sciatic nerve of the bullfrog after conduction from the point of stimulation through the distances (in millimeters) shown at the left. The action potentials might be compared to runners in a race who become separated along the course as the faster contestants outstrip the slower; thus, in a record at 82 mm from the point of stimulation three waves are shown, whereas at 12 mm the potentials are fused, and only one large wave appears. [Adapted from Erlanger and Gasser (1937).]

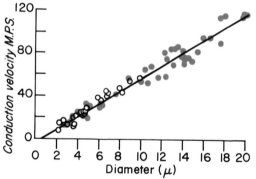

Figure 1.35. Linear relation between diameter and conduction velocity of mammalian nerve fibers. Each point represents a determination of the maximum conduction velocity in meters per second and of the diameter in micra of the largest fiber of an individual nerve. *Dots*, adult nerves; *circles*, immature nerves. [Adopted from Hursh (1939).]

The B fibers are myelinated and have diameters from 1–3 μm and conduction velocities from about 3 m/s to 14 m/s. The B fibers are found solely in preganglionic autonomic nerves. The C group, composed of the smallest fibers (less than 1 μm in diameter), are unmyelinated and have a conduction rate of around 2 m/s or less; many are found in cutaneous and visceral nerves. They have a high threshold, 30-fold that of the A group. The A group of fibers makes by far the greatest quantitative contribution to the compound action potential, and

the C group the least. The electrical potentials recorded from both A and C fibers exhibit slow variations in potential following the action potential, negative and positive after-potentials, but the B group does not exhibit a negative after-potential with a single response (though a negative after-potential does appear upon repetitive stimulation). The B fibers are the most susceptible to asphyxia, the C fibers, the least so.

The linear relationship between fiber diameter and conduction velocity holds also for growing nerves of young animals. During growth the diameters of the nerve fibers enlarge. Conduction velocity increases proportionately so that the time taken for an impulse to travel from the toes of a kitten a few days old to the spinal cord is the same as for a full-grown cat. Thus, the kitten and the cat react to stimulation with about equal promptness.

The diameters of the regenerating fibers in a sectioned or crushed nerve also enlarge gradually, and conduction velocities increase accordingly, the relationship again being a linear one. The maximum conduction velocity is not reached until maximum diameter of the fiber is attained. If the axons of the nerve alone are interrupted, the sheaths of the nerve fibers remaining intact, the diameters and conduction velocities may reach those of the normal nerve; this rarely occurs if the nerve has been completely severed.

Conduction Rates

The velocity of the nerve impulse varies in different nerve fibers in accordance with their diameters, the thicker fibers conducting more rapidly than the fibers of smaller diameter. In the large afferents from muscle spindles of the mammals, the rate is from 80 to 120 m/s. Sensory nerves of the skin being of smaller diameter have slower conduction rates. Nonmedullated fibers conduct more slowly than medullated fibers. Some of the fibers subserving pain sensation and those of the sympathetic nervous system have a very slow conduction rate.

The following table gives approximate conduction rates in nerves of several different animals:

Medullated nerve, mammal, 37°C: 120 m/s
Medullated nerve, dogfish, 20°C: 35 m/s
Medullated nerve, frog, 20°C: 30 m/s
Nonmedullated nerve, crab, 22°C: 1.5 m/s
Nonmedullated nerve, mammal, 37°C: 1 m/s
Nonmedullated nerve, olfactory of pike: 20°C: 0.2 m/s
Nonmedullated nerve, in fishing filament of physalia, 26°C: 0.12 m/s
Nonmedullated nerve, in *Anadon*: 0.05 m/s
Compare the velocity of sound in air at 0°C: 331 m/s

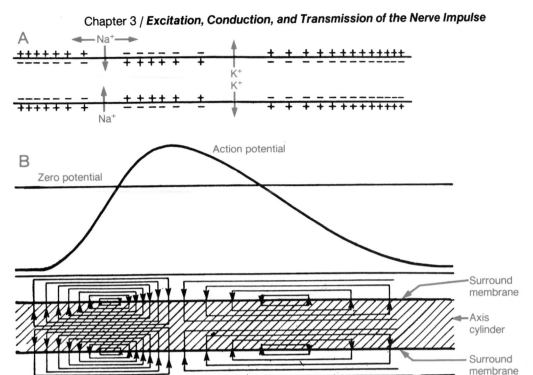

Figure 1.36. *A*, The movement of sodium and potassium during the action potential, traveling in the direction of the *arrow*. The charge in each region is also shown. *B*, Upper curve shows the potential distribution of the impulse along the nerve. The lower part shows the flow of currents in the external medium and within the fiber. There is a reversal of potential during the spike. [Adapted from Eccles (1953).]

By an indirect method of measurement the rates of conduction in various human postganglionic sympathetic nerves have been found to be from 0.85 to 2.30 m/s. The lower figures were obtained for the nerves of the leg, the higher ones for the nerves of the chest.

ACTION CURRENTS AND EXCITATION

The externally recorded action potential represents a variation in electrical potential along the nerve. An alternative way of looking at the nerve impulse is as a set of currents flowing out of the membrane ahead of the area of greatest depolarization or, more particularly, as the movements of these currents along the nerve (Fig. 1.36). The term action current is as appropriate as action potential. The advantage of understanding the currents of nerve as well as the potential becomes apparent when one considers how the self-exciting properties of the nerve impulse arise; for it is the currents themselves which act to depolarize the nerve. The analogy is often made that the nerve acts like a fuse along which the ignition progresses. Action currents leave the nerve ahead of the region of depolarization, acting as a virtual cathode since they have the direction of currents flowing into a cathodal electrode. The hypothesis that action currents act to depolarize the region ahead of them was substantiated by experiments in which it was

demonstrated that if action currents were allowed to enter but not to excite a region beyond a narcotized stretch of the nerve, then the action currents gave rise to two phenomena beyond the blocked region: an increase in excitability and a depolarization of the fiber. Neither the increase in excitability nor the depolarization were large enough to set up an impulse; the block diminished the intensity of the currents, but it was clear from this experiment that the local currents generated by the action potential (as distinct from the action potential itself) could cause depolarization and, therefore, a change in excitability.

Another hypothesis which is a part of the general theory of self-excitation holds that in order for a propagated nerve impulse to be generated, not only is a depolarization necessary, but there must also be an increase in permeability to ions at the depolarized region. Thus, depolarization brings about an increased ion flow through the "not yet excited" membrane bordering on the depolarized region. In other words, during the action potential a path is opened in the membrane through which the charge from resting membrane can flow. Thus, by discharging resting membrane ahead of itself, the action potential becomes self-propagating. Evidence for this characteristic of self-propagating activity was obtained by Curtis and Cole (1938) who showed that there was an increased permeability to ions during activity. This was demonstrated as a decrease in transverse resistance of the membrane from a resting value of 1000 Ω/cm^2 to 25 Ω/cm^2 during the rise of the action potential in the squid axon. Decreases in membrane resistance have also been demonstrated in other tissues, such as muscle, during propagated activity.

"All-or-None" Principle

A stimulus which is just capable of exciting a nerve fiber (threshold stimulus) sets up an impulse which is no different from one set up by a much stronger stimulus. The impulse generated by the weak stimulus is conducted just as rapidly and is equal in magnitude to that generated by the strong stimulus when judged by the action current developed or the mechanical response of the muscle it innervates. Thus, the propagated disturbance established in a single nerve fiber cannot be varied by grading the intensity or duration of the stimulus, i.e., the nerve fiber under a given set of conditions gives a maximal response or no response at all. To make use again of the train of gunpowder analogy—the flame of a match applied to the powder fuse will start a traveling spark no less intense than one started by the flame of a torch. The restoration of the strength of the impulse to its original value after passing from a narcotized region into normal nerve also shows the "all or none" nature of nervous conduction. The well-known fact that a strong stimulus applied to a nerve *trunk* causes a compound action current of greater amplitude, and a greater muscular response than a weaker stimulus is due to the fact that the nerve trunk is composed of many fibers, each of which supplies a group of muscle fibers. The weak stimulus excites only a proportion of the units of the nerve trunk, whereas a maximal stimulus excites them all. For example, the cutaneous dorsi muscle of the frog is supplied by a nerve which contains only 8 or 9 fibers; each of these innervates about 20 muscle fibers. When the nerve was stimulated by shocks of gradually increasing intensity, the muscular responses did not show a similar continuous rise in amplitude; on the contrary, the responses of the muscle increased in a series of well-defined steps. In other words, increasing the stimulus intensity produced no effect for a time upon the amplitude of the muscular response, but then a slight increase in strength of stimulus produced a sudden rise in amplitude. The steps which were never greater in number than the number of fibers in the nerve were due to additional fibers becoming excited as the strength of stimulus reached a certain value.

It must also be remembered that the "all or none" principle applies only for the condition of the nerve at the point where, and the moment when, the impulse arises. A stimulus which will give rise to a response of a certain magnitude under one condition of the nerve may give a much smaller response under other conditions, e.g., during the relative refractory period (see below), narcosis, oxygen lack, etc.

Absolute and Relative Refractory Periods of Nerve

For a brief interval following the passage of an impulse along the nerve fiber, a second stimulus, however strong, is unable to evoke a response. This interval is called the absolute refractory period. In a frog's sciatic nerve at a temperature of about 15°C the absolute refractory period has a duration of between 2 and 3 ms. Its duration is roughly the same as the action potential "spike." It is much shorter in mammalian nerve (0.4–1.0 ms in large medullated nerve fibers).

The period during which the nerve is absolutely refractory is succeeded by one in which the nerve, though it will not respond to a stimulus of the same strength as it did before the passage of the impulse, will respond to a somewhat stronger one. The excitability of the nerve gradually increases and the strength of stimulus necessary for excitation becomes progressively less (Fig. 1.37). In the end, the restoration of excitability is complete, and the nerve responds to a stimulus of no greater strength than that which is capable of exciting a resting nerve. The period following the absolute refractory phase and during which the excitability gradually rises to normal is called the relative refractory period. The time required for the excitability of the nerve to return to about 95% of its resting level ranged from 10 to 30 ms (full recovery may not be attained until the lapse of 100 ms). It should be pointed out that the failure of the nerve to conduct a second impulse is not due simply to lowered excitability at the point in the nerve where the original stimulus was applied, for during the absolute refractory period a stimulus applied to any other point upon the nerve likewise fails to set up an impulse. The passage of the impulse along the nerve leaves in its wake a change of state in the membrane organization. A certain time is required for the changes associated with the passage of the impulse to become reversed and the nerve restored to its resting condition.

The refractory period renders a continuous excitatory state of the nerve impossible. Fusion or summation of impulses does not occur. The refractory period obviously must also set an upper limit on the frequency of the impulses. In the mammal the absolute refractory period is about 0.5 ms. The intervals between impulses cannot be shorter than

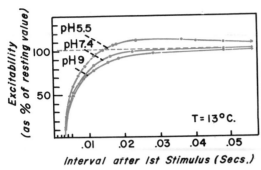

Figure 1.37. Recovery of excitability in nerve perfused with fluids of different pH. [Adapted from Adrian (1935).]

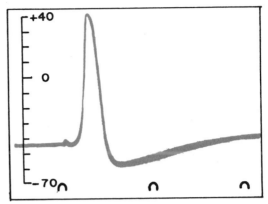

Figure 1.38. Action potential recorded with an internal electrode from a squid giant axon. The scale shows the internal potential in millivolts relative to the outside bath. Time marks are in 2-ms intervals. [Adapted from Hodgkin and Huxley (1945).]

the absolute refractory period; the maximum impulse frequency is around 1000/s. At this rate the impulses are travelling in the *relative* refractory period of their predecessors and are weaker and more slowly conducted. In frog nerve with its refractory period of from 2 to 3 ms, the maximal impulse frequency is between 250 and 300/s.

Classical Mechanism of the Action Potential

The mechanism by which the action potential is produced has not been examined, but the role of the currents in exciting a region ahead of the approaching nerve impulse was noted. It is now appropriate to consider the origin of the action potential in terms of the prevalent ions, namely, sodium and potassium. The resting potential was shown to arise from a diffusion of potassium ions across the cell membrane. The action potential has a more complex origin. Recent investigations have demonstrated that the action potential results chiefly from the movement of sodium ions into the nerve fiber during the early, rising part of the nerve impulse, and that there is a movement outward of potassium ions later during the fall of the action potential.

At the turn of the century, the action potential was attributed to a complete depolarization of the membrane, creating an area more negative than other parts of the nerve into which the ionic currents flowed. This concept, first enunciated by Bernstein (1912), could not be tested with the methods and tissues available until 1940. At that time two groups, one in England and the other in America, performed the critical experiments which led to a reexamination of the Bernstein concept. These experiments utilized the giant axon of the squid to make a direct measurement of the transmembrane potential. This axon is 0.1 mm or more in diameter, so that it is possible to insert one electrode directly into it along its length and to place another electrode outside. The resting potential was measured and the nerve stimulated to obtain an action poten-

tial. The results of such an experiment are illustrated in Figure 1.38. According to the classical concept, the action potential arises from the fall of the membrane potential toward zero, the nerve being completely depolarized when the membrane potential reaches zero. In the case of the squid axon instead of a simple depolarization of the membrane, an "overshoot" of the potential past the zero baseline was noted. Such an overshoot represents a reversal of the potential from the resting state so that during the peak of the action potential, the inside of the cell becomes about 50 to 60 mv positive to the outside (Fig. 1.38). This is an astonishing result for, on first consideration, there does not appear to be any mechanism by which the potential difference could invert. Bernstein had considered the resting potential a diffusion potential resulting from the permeability of nerve to potassium and the impermeability to sodium and chloride, and the electrochemical events during the action potential were attributed to an equal flow of anion and cation other than potassium into the nerve fibers, so that the diffusion potential across the membrane disappeared when the sodium and chloride flows were permitted to take place. (In a diffusion cell the emf is proportional to the difference of the mobilities of cations and anions, and if these are equal the emf is zero.) This hypothesis, while showing how the membrane potential could become zero, did not appear to contain an explanation of how the reversal of potential (which has been referred to as overshoot) might arise.

THE MODERN CONCEPT OF THE ACTION POTENTIAL

Closer consideration of the details of ionic flow pointed to sodium as the ion which might give rise to the reversal of potential. Since the concentration of sodium is higher outside than inside the nerve, a flow of sodium inward would tend to make the inside positive with respect to the outside. The experimental result establishing that sodium is the most important cation involved in both the production of the action potential and of the overshoot is simple. The squid axon is normally surrounded by sea water osmotically equivalent to a 0.3 M sodium chloride solution. If the concentration of the sodium in the sea water is changed so that less sodium is present, the amplitude of the action potential will diminish. In such experiments, the resting potential is unaffected by altering the sodium concentration. Accordingly it was concluded that sodium is essential to the production of the action potential including the overshoot. Fig. 1.39 illustrates the type of result obtained from such substitution experiments. Note that in applying half sea water, the osmotic lack of sodium is balanced by substituting an osmotically equivalent amount of a nonelectrolyte such as sucrose or an electrolyte such as choline chloride.

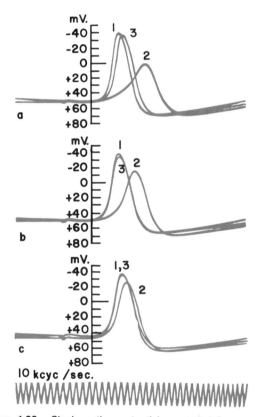

Figure 1.39. Single action potentials recorded from a squid giant axon. *1,* Normal action potential in sea water. *2,* Action potential after equilibration in medium with altered sodium. *a* is 0.33 times normal, *b* is 0.5 times normal, and *c* is 0.71 times normal in concentration. *3,* Action potential after return to normal sodium solution.

Voltage "Clamping"

Various experiments were carried out to determine the dependence of the sodium and potassium ionic currents on the membrane potential. Unfortunately, it is not possible to measure one ionic current flow during a depolarization of nerve directly and independently of another. The changes in current flow affect the membrane potential, which in turn affects the current flow. This concept of a mutual influence of current and potential is important for the understanding of the mechanism of the action potential, but it is necessary to eliminate the interaction operationally if one wishes to study the effect of a change in membrane potential on the ionic flexes. Such independence of current and potential is achieved by depolarizing the membrane to a given value and then subsequently maintaining the potential at that value by provision of current via an external circuit. The method is referred to as the *voltage clamp method* because the membrane potential is "clamped" (fixed) at an invariant value. Thus, it became possible to investigate the ionic flows resulting from a given step of depolarization. Note that once the potential has been clamped there will be no flow of current into the capacitive element of the membrane, since to put charge on a condenser requires a changing potential. The current, therefore, flows only through the external, internal, and transmembrane resistances noted above.

In such voltage clamp experiments it has been observed that the current may change while the depolarization is being maintained. This appears to violate Ohm's law, which states that current flow in a system depends upon the voltage ($I = E/R$), and the voltage across the membrane is held constant in these experiments. The explanation of these findings rests in the fact that the resistance or permeability of the membrane changes during a voltage clamp experiment, and thus, although the potential is constant, extensive changes in current occur.

The effect of a small clamp on the current flow through the membrane is shown in Figure 1.40. The depolarizing clamp causes a diphasic current which is first directed inward (into the nerve fiber) and then outward. If a solution which lacks sodium is substituted for the outside bath, the initial inward current disappears, but the outward current remains unaffected. The early, inward current is, therefore, attributed to sodium inflow. The dependence of the sodium current on the amount of depolarization is revealed by the experiments in which progressively greater amounts of depolarization are used (Fig. 1.40). As the membrane potential is decreased (greater clamp voltage), the inward current increases. With further increases in the depolarization, however, the sodium current decreases in magnitude and ultimately disappears.

The results of a brief depolarization or clamp may be outlined as follows: (1) There occurs, initially, a brief inward current resulting from an inward movement of sodium ions, followed by an outward flow of potassium ions. (2) The amount of sodium flowing depends upon the extent of depolarization; over a rather broad range it is seen that the greater the depolarization, the larger the sodium current. (3) The potassium current is relatively unaffected by the depolarization but is greatly affected by a hyperpolarization of the membrane. Increasing the membrane potential from, for example, −60 to −80 mv increases the potassium current.

Hodgkin-Huxley Theory

To understand the mechanism of the action potential, it is necessary to examine the results of the voltage clamp experiments in terms of a detailed theory—evolved by A. Hodgkin (1964) and A. F. Huxley (1964)—which explains many of the phenomena of excitation and conduction in nerve and muscle. According to this theory, the nerve membrane is represented as containing three channels through which sodium, potassium, and chloride ions may move independently. The individual ions are forced through the membrane by their electrochemical potential gradients. A schematic circuit representing the three-channel hypothesis is shown in

Figure 1.41. The value of each emf is given by the Nernst equation, the three emfs being equivalent to the three "equilibrium" potentials (as discussed above in the section on the membrane potential). It is important to understand that the network is only an electrical "equivalent" and that the channels are not physically located one next to the other. Each of the channels represents a group of many identical channels in a unit area of the membrane lumped together schematically as one channel. The resistance signifies the opposition encountered by each ion in the unit area of membrane. The reciprocal of the resistance, the conductance (g), is proportional to the permeability of the membrane to the ion.

The sequence of ionic events in an action potential may now be described as follows: A depolarizing voltage (stimulus) is applied to the membrane and the sodium influx increases in the direction of its concentration gradient. If this Na^+ influx is greater than the flux of potassium and chloride (both of which constitute current flowing in the opposite direction), then the net sodium entry causes a change in the membrane potential. But, as noted above, the change in potential across the membrane brings about an increased sodium influx, which in turn leads to a further decrease in the membrane potential. Therefore, the influx of sodium builds up quite rapidly. This mutual effect of sodium influx and membrane potential constitutes the *regenerative* factor in impulse transmission and is an example of a positive feedback process. In terms of the equivalent circuit of Figure 1.41, the action potential develops in the following way. The increase in permeability to sodium is essentially a

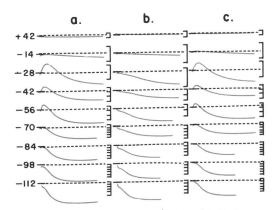

Figure 1.40. Currents flowing through squid giant axon membrane during a steady voltage clamp. Figures on left are clamped values of the membrane potential in millivolts relative to resting potential (outside minus inside). Minus values represent depolarizations. Columns *a* and *c*, axon in sea water. Column *b*, axon in choline sea water without sodium. Vertical scale, 1 div = 0.5 ma/cm³. Time dots are 1 ms apart. [Adapted from Hodgkin and Huxley (1952d).]

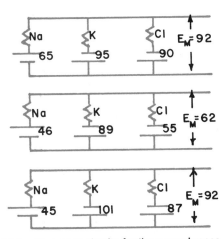

Figure 1.41. Equivalent circuits for three membranes. *Upper* is for mammalian skeletal muscle fiber; *middle* is for squid giant axon; and *lower* is for frog sartorius muscle fiber. E_m is resting potential; all values are in millivolts.

decrease in resistance. The potential difference across the membrane, which is the resultant of the three emfs in parallel, will approach that emf which has the lowest resistance in series with it. The resistance to potassium and chloride does not change. The membrane potential therefore tends to approach the equilibrium potential of the sodium ion which, in volts, is

$$E_{Na} = 0.058 \log(Na_{outside}/Na_{inside}).$$

The equilibrium potential of the sodium ion is approximated but never quite reached by the peak of the overshoot. It is possible that an increased efflux of potassium and an active process of sodium extrusion accounts for the difference between the experimentally observed peak of the overshoot and the Na equilibrium potential, i.e., the theoretically expected peak.

The rapid rise in sodium current does not continue for very long and, as shown by the clamp experiments, the sodium currents are rapidly followed by a potassium efflux from the axon. The sodium influx is terminated by a process called *sodium inactivation* which develops during the increase in sodium flux. Evidence for the existence of such an inactivation process has been adduced in clamp experiments of another type. For example, if one applies a clamp of the order of 10 mv, this voltage change is insufficient to cause a sodium influx, yet it can nevertheless set into action the process of inactivation—as demonstrated by the fact that, following the small depolarization, a larger depolarization of about 40 mv does not evoke the magnitude of sodium current that would have developed if the smaller voltage had not been applied initially. In other words the small initial voltage clamp rendered the larger subsequent voltage clamp less effective in changing the permeability of the membrane to Na^+. Thus, by following small clamps by larger clamps at various intervals, the time course of development of the inactivation can be ascertained (Fig. 1.42). Thus, the inactivation process turns off the sodium current and thereby contributes toward restoration of the resting state.

A second process also aids in the restoration of the resting membrane potential. It was noted that a potassium efflux follows the sodium influx. The time course of efflux of potassium ions corresponds to the fall of the action potential. The increase in conductance to potassium means that the change in potential across the membrane is now being driven by the electrochemical potential due to the potassium gradient. This net efflux of cations, the consequence of the increased permeability to potassium ions (sometimes referred to as rectification)

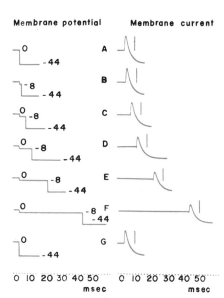

Figure 1.42. Development of "inactivation" during constant depolarization of 8 mv. *Left column,* displacement of membrane potential from the resting value in mv. *Right column,* membrane currents as a function of time for the displacement in the left column. The vertical lines show the sodium current expected in the absence of the conditioning step. [Adapted from Hodgkin and Huxley (1952d).]

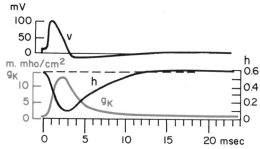

Figure 1.43. Time course of inactivation (*h*) and potassium conductance (*g*$_k$) during a nonpropagated action potential. *Upper curve* is action potential in response to a 15 mv initial depolarization. [Adapted from Hodgkin and Huxley (1952d).]

tends to restore the membrane potential to its original state of internal negativity. Figure 1.43 shows the time course of the increase in conductance of potassium in relation to the sodium inactivation process.

Both the sodium and potassium flows during the phase of increased permeability are in the direction of their concentration gradients. The flow of ions during the action potential is passive and depends upon the concentration gradient and on interactions among ions and with the channel (Hille, 1978). The production of a nerve impulse would be expected therefore to be independent of metabolism. This deduction is substantiated by experiments in which both aerobic and anaerobic processes are blocked by inhibitors. Nerves may continue to conduct for hours, even when stimulated at high rates under the influence of metabolic inhibitors. The increased permeability to sodium and potassium persists, although the means for pumping sodium and potassium have been cut off. This is not to say that transport processes are not important for maintaining the ionic inequalities. If nerve is continually stimulated while being treated with an inhibitor, it fails to pump out the sodium which accumulates inside the nerve both during the resting state and during excitation—and it also fails to reaccumulate potassium to restore its resting intracellular K^+ concentration. Ultimately, therefore, the nerve will depolarize as the result of potassium loss and/or sodium influx. Thus, the ion pumps and associated metabolic activity which maintain the steady state ionic gradients across the nerve cell membrane are necessary in the long run, but neither the resting potential nor the action potential depends primarily upon metabolism for their production.

How does this theory of the action potential explain other important phenomena of nerve, such as refractoriness and accommodation? The absolute refractory period is the result of the inability of nerve to respond to a stimulus no matter how intense. This inability corresponds to the period during the action potential when sodium influx is completely inactivated. It is impossible to turn on the sodium influx machinery in any degree whatever at the height of the inactivation process. During the relative refractory period, the nerve is under the influence of partial sodium inactivation and increased potassium efflux. The potassium efflux tends to maintain the membrane potential at the original unexcited level. The restorative processes of sodium inactivation and potassium efflux (rectification) must be opposed by very large depolarizations in order to get sufficient sodium influx to start regeneration. Similarly, accommodation, which is a rise in threshold during the persistent application of a linearly rising current, may be explained as the result of growth of inactivation at the cathode of an exciting current, much as inactivation developed in the clamp experiments which demonstrated the existence of the process.

Little is known concerning the molecular events within the nerve membrane that are triggered by depolarization and then lead to increased sodium conductance. It is clear, however, that these events are reversible.

As noted above, the excitatory process of axon and muscle membranes involves voltage-dependent transient increases in permeability to Na^+ and K^+ ions in specific channels for each. In mammalian myelinated nerve, the relative importance of potassium channels diminishes; for example, rabbit sciatic nerve nodal membrane lacks K^+ channels and depends on inactivation of Na^+ channels and leakage conductance for repolarization.

Ca^{++} channels also exist to account for Ca^{++} action potentials in certain locations (see below). While the steady-state membrane conductance to specific cations is a function of the membrane potential, the average conductance of single channels does not depend on the membrane potential. The simplest interpretation of this apparent paradox is that the single channels have two conductance states, open or closed, and the number of open channels depends on the membrane potential.

The opening of a channel is itself associated with a small current called the "gating current" (Armstrong, 1975; Armstrong and Bezanilla, 1977). This gating current has been measured for Na^+ channels in squid axon and frog nerve node of Ranvier by blocking the Na^+ channel with tetrodotoxin, the K^+ channel with tetraethylammonium, and then determining the current change associated with one or more step increments in the clamped membrane voltage. Tetrodotoxin, while blocking the entry into the Na^+ channel, does not prevent the opening and closing of the gate and, under this condition, the gating current is a large fraction of the total charge displacement. The gating mechanism behaves as if three or four charged particles cooperate in the opening of each channel (Keynes, 1979). The energy source for the gate opening seems to be the electric field; no chemical sources have been identified. The closing of the channels, it is now believed, involves a further rearrangement of the gating particles, rather than a separate and independent set of blocking components. Hille (1978) has estimated that the cross-sectional dimensions of the Na^+ channel are 3×5 Å. In frog node of Ranvier there are about 3000 sodium channels per μm^2. The single channel conductance is 3–8 pmho. A value of 5 pmho corresponds to 3×10^6 ions per second for a 100-mv driving force (because of this high rate the channel is considered to be an aqueous pathway); this ion flux is far in excess of turnover rates for carriers. However, this is less than the theoretical maximum estimated for free diffusion, and there is saturation at high concentrations of sodium ion. Therefore, there appear to be interactions between ions and the channels.

In some manner the electric field "acts on components attached to or within the membrane, and the electric work done on them is the energy injected into the gating process. The location of the field-sensing components of the channel is unknown, but one might expect them to be immersed in a fluid region of the insulating dielectric of the membrane rather than in the very polar, ion-permeable part" (Hille, 1978). The development of useful models of voltage-dependent regulation of channel conductances must await more understanding than we now have of ion channel chemistry.

Increasing calcium ions at the extracellular surface of the membrane raises the threshold for excitation (i.e., it shifts the critical firing potential to a more depolarized level) while decreasing the calcium ion concentration lowers the threshold for excitation by shifting the critical firing potential closer to the resting potential. Ca^{++} may exert these effects by combining with membrane surface-negative charges near the voltage-sensing component of the channel gate.

Recently, attention has been drawn to voltage-dependent calcium currents, i.e., calcium action potentials (Llinas and Walton, 1980) which have been identified in neuronal cell bodies, dendrites, and presynaptic terminals. Calcium action potentials are much slower than sodium action potentials; they are not blocked by tetrodotoxin or removal of extracellular sodium. They are blocked by removal of extracellular calcium and by calcium channel blockers. Also they are associated in many cases with a calcium-activated increase in potassium conductance. The calcium action potential and the calcium-activated potassium conductance are components of a negative feedback mechanism which influences both membrane excitability and the level of intracellular calcium. The calcium-activated increase in potassium permeability operates to terminate the calcium action potential and thereby reduces the influx of calcium ions (Llinas and Walton, 1980; Thompson and Aldrich, 1980). The role of intracellular Ca^{++} in transmitter release will be considered below.

NERVE METABOLISM
Intermediary Metabolism

Nerve, like other tissues, contains the enzymatic apparatus for glycolysis, the citric acid (Krebs) cycle, and the electron transport system and, thus, can generate and store energy in the form of adenosine triphosphate (ATP). The pathway of glycolysis is concerned with the breakdown of glucose to pyruvate and/or lactate. Only a small amount of ATP is produced in this pathway which is anaerobic. The major part of the ATP is produced aerobically in the Krebs cycle and the electron transport system (for details see discussion of the energy-generating systems of cells in Chapter 4).

Under anaerobic conditions, e.g., in an atmosphere of nitrogen, resting nerve converts glucose to lactate quantitatively. However, the nerve at rest produces very little lactic acid aerobically, although it has been shown that there is a clear-cut increase in lactic acid production in an oxygen atmosphere (aerobic glycolysis) when the nerve is stimulated at 100/s.

Nerve conduction continues under anaerobic conditions (nitrogen atmosphere); for, as noted previously, the action potential does not depend immediately upon metabolism. Ultimately, of course, conduction will fail under anaerobic conditions because of the gradual accumulation of sodium and consequent depolarization of the nerve fiber.

Further evidence for the existence of glycolytic systems in nerve comes from the effect of metabolic inhibitors on nerve metabolism. Iodoacetic acid, which interferes with glycolysis, causes a decrease in the oxygen consumption of nerve. At the same time the effect of the inhibitor may be slowed by the administration of sodium lactate, the latter probably forming pyruvate which is then oxidized. The cofactor necessary for the formation of pyruvic acid from lactic acid, nicotinamide adenine dinucleotide, is found in frog nerve. It has also been shown that the administration of glucose partially prevents the loss of potassium and gain of sodium resulting from anoxia of nerve.

Evidence for the presence of the Krebs cycle in nerve comes from the effect of inhibitors on the oxygen consumption of frog nerve. Methylfluoracetate, an inhibitor which acts in the Krebs cycle, inhibits the rate of oxygen uptake and ultimately causes a failure of conduction. If the nerve is preequilibrated with sodium fumarate or sodium succinate, the inhibition is overcome, and the nerve regains its ability to conduct and to consume oxygen. Fumarate and succinate are intermediates in the Krebs cycle and serve to keep going the portion of the cycle devoted to converting succinate or fumarate to carbon dioxide and water (see energy-generating systems, Chapter 4).

The importance of the electron transport scheme has been demonstrated by experiments in which the cytochrome oxidase of nerve has been inactivated by carbon monoxide. When nerve is poisoned with carbon monoxide, the cytochrome oxidase fails to function, and molecular oxygen cannot be reduced for its ultimate reaction with metabolically derived hydrogen to form water. The nerve behaves as though it were in nitrogen. However, it is possible to dissociate the carbon monoxide from the cytochrome oxidase by shining visible light on the nerve. The nerve then regains its ability to consume oxygen and will conduct an action potential.

Resting and Active Metabolism

In the resting state, the oxygen consumption of frog sciatic nerve is quite low, 30–40 mm^3/g wet weight/h in contrast to that of mammalian nerve which ranges from 200–300 mm^3/g/h. These values may be compared with those of brain which is of the order of 2,000 mm^3/g/h. Corresponding to the consumption of oxygen, there is a resting release of heat by nerve which amounts to 0.15 cal/g/h for frog sciatic nerve.

The oxygen consumption of nerve in vitro is accompanied by a release of carbon dioxide such that the respiratory quotient (moles CO_2/moles O_2) is 0.8. This value indicates that the fuel of nerve is not exclusively glucose which would yield a respiratory quotient (RQ) of 1, as indeed proves to be the case when the RQ of mammalian brain is measured in vivo. During excitation the respiratory quotient of the extra oxygen consumption of nerve changes to 0.9, suggesting that the recovery processes following excitation use a different substrate than is used during the resting state.

How much of the resting heat production (metabolism) of nerve can be attributed to the heat production of the process pumping Na^+ out of nerve and K^+ into the cell during the resting state? One can calculate the energy required to extrude sodium against both the electrical force in the membrane and the concentration gradient. A similar calculation may be made of potassium uptake. If the process is assumed to be inefficient and to produce heat equal to the work done (50% mechanical efficiency) an estimate of the heat production can be obtained. In the case of sodium, it is necessary to pump the ion against both the electric field and its concentration gradient, since the ion is positive and it must be forced out against the negative attraction of the interior. The electric work required is given by EF (ionic flux). The concentration work is obtained from the relationship 2.303 RT log C_2/C_1, where the symbols have their usual significance. Substituting the values for frog nerve of $E = 0.070$ v, $F = 23,050$ cal/v/mole, $C_2/C_1 = 3$ (sodium), and a sodium flux of 10^{-5} mole/g/h, one obtains an electrical work of 0.015 cal/g/h and a concentration work of 0.007 cal/g/h. A similar calculation for potassium shows that the energy necessary to pump this ion into nerve is 0.019 cal/g/h. The work necessary to pump both sodium and potassium is 0.041 cal/g/h. If the pump has a mechanical efficiency of 50% then its operation—complete with inefficiency—accounts for 0.082 cal or about 50% of the total energy consumption of 0.15 cal/g/h noted above. The value of 0.041 cal/g/h corresponds to 4.1 kcal/mole of Na^+ and K^+ exchanged. Thus,

if 8–10 kcal/mole is taken as the free energy of ATP hydrolysis, it is seen that approximately 2–3 moles of Na^+ can be exchanged for 2–3 moles of K^+ for each mole of ATP converted to adenosine diphosphate plus inorganic phosphate.

When nerve is stimulated, there is an increase in the oxygen consumption and an increased loss of heat. The extra oxygen consumption may double. The increase in heat production and oxygen consumption parallel one another and depend upon the frequency of stimulation. The increase in oxygen consumption rises gradually with increase in frequency of stimulation, leveling off at about 100 impulses/s. The extra heat which accompanies short tetanic stimulation has been shown by Hill (1938, 1959) and co-workers to consist of two phases: an early one lasting only 2–3 s is called the initial heat, which amounts to a few percent of the total extra heat; this is followed by the remainder of the heat production which is called the delayed heat. The latter is presumably concerned with the recovery process in nerve involving the pumping of ions. The extra oxygen consumption of nerve may be eliminated by inhibitors (azide and methylfluoracetate) which have no effect on the action potential.

Abbott et al. (1958), using improved thermal methods, have been able to measure the heat output resulting from a single shock to a crab nerve. They found that the initial heat accompanying an action potential occurred in two phases, an early, rapid portion, 1 ms in duration, which was positive (exothermic), and a delayed, negative (endothermic) heat of absorption. The actual heats during an impulse were $+14 \times 10^{-16}$ and -12×10^{-16} cal/g/impulse. The difference gives the net initial heat production. Calculations by Hodgkin indicate that the initial heat can be accounted for by the heat production of the action currents flowing within and without the nerve. But other processes such as heat produced by the mixing of sodium entering nerve may also account for the initial heat.

Thus, active cation transport is the major energy-consuming process in specialized tissues such as nerve. McIlwain (1966) has shown that electrical pulsation of brain slices in vitro results in cation-dependent respiratory stimulation accompanied by acceleration of active cation fluxes. About 30% of the increased respiration may be accounted for by the increased cation flux. The maintenance or restoration of cation gradients in nerve as well as other tissues has been correlated with a sodium-and-potassium-stimulated adenosine triphosphatase present in membranes. This enzyme utilizes the chemical bond energy in ATP to translocate Na^+

and K^+ across the membrane against their respective concentration gradients.

SYNAPTIC TRANSMISSION

In the nervous system of mammals, the formation of conducting pathways does not involve direct anatomical continuity between one neuron and another or between a neuron and a cell of an effector organ. An axon gives rise to many expanded terminal branches (presynaptic terminal boutons). The postsynaptic component of the synapse may be formed by any part of the surface of the second neuron, with the exception generally of the axon hillocks; it usually is a dendrite (axodendritic synapse), but it may be part of the cell body (axosomatic synapse) or part of the membrane of another axon (axoaxonic synapse). A single neuron may be involved in many thousands of synaptic connections, but in every case the impulse transmission can occur only in one direction. In the mammalian central nervous system there is a cleft of about 20 nm (the *synaptic cleft*) separating the presynaptic axon terminal (the *presynaptic membrane*) and the surface of the postsynaptic component (the *postsynaptic membrane*).

At the sites of apposition, both the presynaptic and postsynaptic membranes exhibit a plaque of electron-dense material on their cytoplasmic surfaces. Beneath this plaque on the presynaptic membrane and within the presynaptic bouton, there are vesicles ranging in diameter from 10 to 50 nm and often numerous mitochondria.

The action potential does not cross the synaptic cleft but instead causes the release of a transmitter substance which is stored in the vesicles of the presynaptic terminals. The release of transmitter requires the entry of calcium ions into the presynaptic terminal from the extracellular milieu upon arrival of the action potential. The transmitter molecule then crosses the gap, attaches to "*receptors*" on the postsynaptic membrane, which is then either *excited* or *inhibited* depending on which ionic channels are opened by the binding of the transmitter to the postsynaptic receptor (see below).

Figure 1.44 shows by autoradiography the distribution of three different receptors in human cerebrum.

Chemical transmission was first demonstrated by Otto Loewi in a classical experiment in which the heart of a frog was slowed by perfusing it with blood taken from the heart of a second frog that had been slowed as a result of vagal stimulation. Loewi (1921) concluded that the vagal stimulation had caused the release of a substance identified as acetylcholine—which altered the excitability of cardiac muscle.

Excitatory Postsynaptic Potential (EPSP)

Eccles (1953) and co-workers have inserted finely drawn-out microelectrodes made of glass and filled with KCl or NaCl into the cell body of motoneurons of the spinal cord. When the microelectrode is inserted, a resting potential of about 70 mv is observed, with the interior of the cell negative to the outside. If the motoneuron is antidromically excited, i.e., by stimulating the axon of the motoneuron, the motoneuron is excited without the intervention of the synapse. An action potential with an overshoot of about +20 mv appears. Thus, antidromic excitation gives rise to an action potential in the postsynaptic cell, the motoneuron, which is no different in its general characteristics than that obtained from squid axon or the muscle fiber.

The events of interest in synaptic excitation, however, occur when excitation is delivered orthodromically, i.e., to the afferent neuron of the dorsal root, so that the presynaptic endings of this neuron are excited. In such experiments in which a synapse is present, an additional potential appears in recording from the postsynaptic cell which is a sign of postsynaptic depolarization. Fig. 1.45 shows the response of the motoneuron to a presynaptic volley. The slow potential change which appears just before the action potential is called the excitatory postsynaptic potential or EPSP. It consists of a depolarization of the membrane which may last for 20 ms and can best be seen when the extent of depolarization is insufficient to give rise to an action potential. When the membrane potential falls to a critical value of about −60 mv, the motoneuron fires, and an action potential appears superimposed on the EPSP. Temporal summation of such EPSPs is also possible. If two subliminal volleys are sent in over the same nerve, each volley produces an effect which is manifested by an EPSP. The EPSPs will then sum, and if the critical level of depolarization is reached, an impulse will be set off.

The EPSP is monophasic and nonpropagating; it represents a depolarization which is localized to the soma of the motoneuron. Unlike the action potential, the EPSP is a potential which is not all or none in character, since it can be augmented simply by increasing the intensity of the input volley. Moreover, the EPSPs of different inputs can sum on a postsynaptic cell to produce a greater depolarization. These characteristics show that the EPSP is produced in a process which is fundamentally different from that of the action potential.

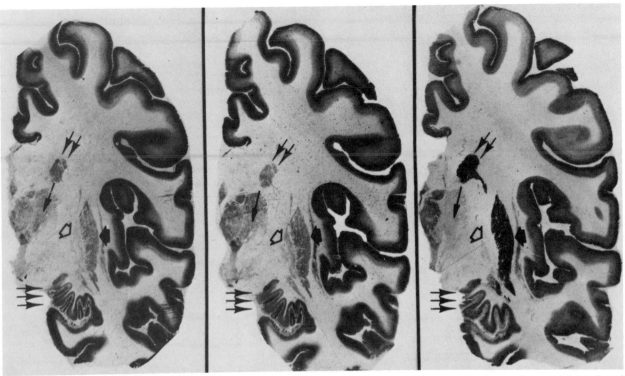

Figure 1.44.   Autoradiograms of the regional localization of GABA (γ-aminobutyric acid), benzodiazepine, and muscarinic cholinergic receptors in 30-μm coronal frozen sections of normal human brain. Images were made onto tritium-sensitive film as described by Penney and Young (1982). Images reflect the binding of tritiated ligands to membrane receptor molecules in the tissue sections. The darker regions of the image reflect higher amounts of receptor binding; the lighter regions, lower amounts. *Left,* [³H]muscimol binding to GABA receptors. Muscimol is a potent GABA agonist which binds to GABA receptors with high affinity. Binding is highest in cerebral cortex and hippocampus, intermediate in putamen and thalamus, much less in globus pallidus, and negligible in white matter. *Middle,* [³H]-flunitrazepam binding to benzodiazepine receptors. Flunitrazepam is a potent benzodiazepine agonist. Benzodiazepines are drugs used as antianxiety and muscle relaxant agents and have been shown to interact with specific brain receptors closely linked to GABA receptors in the membranes. Binding is very similar in distribution to that seen with [³H]muscimol. *Right,* [³H]quinuclidinylbenzilate (QNB) binding to muscarinic cholinergic receptors. QNB is a potent muscarinic antagonist, similar to but more potent than atropine. Binding is highest in caudate and putamen, intermedite in cerebral cortex and hippocampus, less in thalamus, little in globus pallidus, and negligible in white matter. *Narrow dark arrow,* thalamus; *open arrow,* globus pallidus; *closed arrow,* putamen; *double arrow,* caudate; *triple arrow,* hippocampus. The claustrum is the narrow band just lateral to the putamen in all images. [Courtesy of J. B. Penney and A. B. Young.]

Several lines of evidence indicate that the EPSP arises from an influx of all ions through the postsynaptic membrane into the cell. One group of experiments concerns setting the membrane at a given membrane potential by passing current across the membrane. A double-barreled microelectrode was inserted into a motoneuron. One electrode was used to depolarize or hyperpolarize the neuron with a direct current, the other to record the EPSP. When the membrane potential was increased, the amplitude and rate of rise of the EPSP increased. Conversely, when the membrane potential was decreased, the EPSP decreased. Reversal of the direction of the EPSP occurred when the membrane potential was set up at zero volts, which is therefore the equilibrium value for the EPSP. The explanation of this effect goes back to the experiments on the squid axon. It will be recalled that when the

axon was clamped at the equilibrium potential of an ion, a reversal of current flow occurred through the membrane above and below this value of clamp voltage. We may imagine that the motoneuron postsynaptic membrane contains channels through which ions can flow to cause the development of an EPSP. These channels are additional to those serving as producers of the action potential. A diagram of the equivalent circuit for such channels is shown in Fig. 1.46. When the membrane potential is set at zero, the equilibrium potential of the EPSP, a change in resistance of this channel can have no effect on the recorded potential. No EPSP will develop. Under normal circumstances the membrane potential is determined by the equilibrium potential for potassium, but when the EPSP channels decrease in resistance, the potential recorded tends to approach the equilibrium potential for the

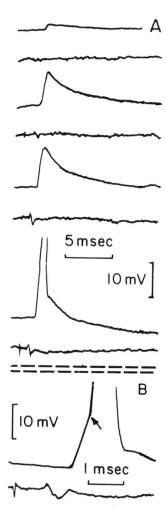

Figure 1.45. Intracellular potentials set up in a biceps-semiten-dinosus neuron by various sizes of volleys in the afferent nerve (lower records). *A* shows synaptic potentials (upper records) of graded size, the largest setting up an action potential. In *B* a faster record of this response is shown. Note spike arising at *arrow* from more slowly rising synaptic potential. [From Brock et al. (1952).]

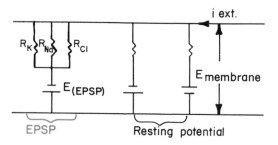

Figure 1.46. Equivalent circuit of motoneuron membrane for excitatory postsynaptic potential (*EPSP*). Left circuit represents channels giving rise to EPSP. Right circuits represent circuit of polarized membrane.

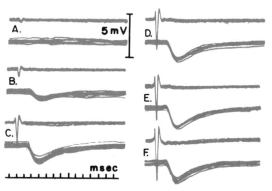

Figure 1.47. Lower records give intracellular responses of biceps-semitendinosus motoneuron to a quadriceps volley of progressively increasing size, as is shown by the upper records which are recorded from the L6 dorsal root by a surface electrode (downward deflection signaling negativity). Note three gradations in the size of the IPSP; from *A* to *B*, from *B* to *C*, and from *D* to *E*. Voltage scale gives 5 mv for intracellular records, downward deflection indicating membrane hyperpolarization. [From Coombs et al. (1955).]

EPSP. Now the only process which can have an equilibrium potential of zero is one in which the membrane is freely permeable to all ions. This process will have an equilibrium potential of zero because it is a diffusion potential, and the value of the emf developed in a diffusion potential depends directly upon the difference of mobilities of the cations and anions. During the production of the EPSP it is presumed that the mobilities are equal so that the process has an equilibrium potential of zero.

Inhibitory Potentials

If a monosynaptic reflex is evoked by ortho-dromic excitation, an inhibition of the reflex can be produced by stimulating the afferent nerve from an ipsilateral antagonistic muscle as described more fully in the section on inhibition. The effect of the inhibitory stimulus is to cause a hyperpolarization of the motoneuronal postsynaptic membrane in a direction opposite to that of the EPSP (Fig. 1.47). This hyperpolarization consequent upon an inhibitory stimulus is called the inhibitory postsynaptic potential or IPSP. Its time course is the same as that of EPSP, of the order of 20 ms, and corresponds to the curve of inhibition obtained from reflex studies. Indeed, the IPSP is a mirror image of the EPSP in the same motoneuron. The IPSP, being a hyperpolarization of the membrane, renders it less excitable. Any EPSP which occurs during an IPSP will generate currents which will be less effective in causing excitation of the cell.

The problem of determining the nature of the current responsible for the IPSP has been attacked by the same procedures described for the EPSP. Using double-barreled microelectrodes, it is found that setting the membrane potential at −80 mv will cause the disappearance of the IPSP. The potential −80 mv is the average of that determined by potassium (−70) and chloride (−90), suggesting that the IPSP results from an increased flow of K^+ and Cl^- making the inside more negative. Further evidence, however, points to chloride as the principal ion involved in producing the IPSP. The method of electrophoretic injection has been used to demonstrate this. One barrel of a double microelectrode is filled with KCl and made negative so that this electrode effectively drives chloride ions into the cell; it is a hyperpolarizing current since it is internally negative. The other barrel electrode records the IPSP. It is found that the injection of chloride into the cell converts the IPSP, elicited by a group I_a afferent volley, from a hyperpolarization to a depolarization (EPSP). The injection of chloride caused an increased internal concentration of chloride. The transmitter causing inhibition results in an increased permeability to chloride, but the chloride ionic gradient is now opposite to its normal direction so that a depolarization will occur instead of a hyperpolarization when the membrane becomes especially permeable to chloride. Chloride now flows out and makes the inside of the cell less negative. A number of anions were injected internally, and all those with less than a certain hydrated ion diameter caused a reversal of the IPSP to an EPSP. Apparently, the inhibitory substance opens pores which allow ions below a certain size to pass. Since chloride is the only anion which under in vivo conditions exists in a high enough concentration to flow in during such an inhibition, it is presumed that chloride is the main contributor to the inhibitory currents.

TYPES OF INHIBITION

Some restraint must be placed upon the ability of muscle and neural circuits to respond. In reaching for an object, for example, the muscle must be controlled if the movement is to be accurate and the muscle is not to overshoot the mark. Certainly reciprocal innervation of some type must often be employed so that when a muscle is activated its antagonist will be inhibited; otherwise, persistent opposition of an undesired sort will be encountered. These two reasons give some indication of the usefulness of inhibition in motor movement. A third less obvious necessity for inhibition arises from the very complexity of neural activity in which some form of "negative feedback" or inhibition is necessary in order to keep the complex neural networks from overactivity. An abnormal form of such activity is observed, for example, in strychnine poisoning in which inhibitory neurons have been shown to be inactivated. In a strychninized animal any stimulus leads to persistent neural activity and convulsion. Again, in the disorder known as parkinsonism, periodic activity in the form of a tremor of a limb manifests itself, presumably as a result of injury to inhibitory systems which restrain normal motor activity.

Three kinds of inhibition have been extensively studied: (1) direct; (2) presynaptic; and (3) Renshaw cell or recurrent inhibition. A fourth type of inhibition, called indirect inhibition and extensively discussed by Sherrington, is not now considered to be inhibition per se but a form of occlusion.

Direct Inhibition

The direct form of inhibition can be observed by evoking a monosynaptic reflex and then depressing the amplitude of the reflex by stimulating an ipsilateral skin nerve (type II fibers). It is necessary to stimulate the inhibitory nerve shortly before eliciting the reflex. By varying the interval between the inhibitory and excitatory stimulus, the time course of direct inhibition can be obtained. The inhibitory nerve is stimulated first. A curve of the form of Figure 1.48 is obtained in which maximum inhibition is obtained at an interval of 0.5 ms between the two stimuli and with a delay lasting for about 10 ms. The interval 0.5 ms is equivalent to one synaptic delay so that at least one inhibitory interneuron is interposed between the afferent neuron and the motoneuron. It is believed at present that this interneuron in the inhibitory pathway possesses the ability to secrete a substance at its terminals which hyperpolarizes the membrane and leads to an IPSP. Thus, a neuron which is normally excitatory to other neurons may exert an inhibitory action by the interposition of an inhibitory neuron.

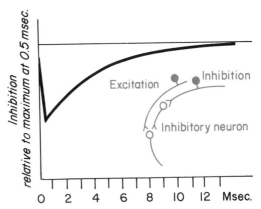

Figure 1.48. Time course of direct inhibition. *Insert* shows circuit containing inhibitory neuron. [Adapted from Lloyd (1946).]

Direct inhibition is very susceptible to strychnine injection which can completely abolish it. This effect of strychnine is the basis for the explanation of its convulsant activity—all inhibition of the direct type having been removed, the slightest stimulus causes a tremendous extensor response.

Presynaptic Inhibition

Early in the studies of the electrical responses evoked by reflex action, electrodes were placed on the surface of the spinal cord or on dorsal roots, and large potentials which were positive to distant regions, called P waves, were recorded. It was suspected that this positivity might be the sign of some inhibitory presynaptic activity going on in the cord. To demonstrate presynaptic inhibition, a reflex response is elicited in extensor motoneurons. This reflex will be inhibited if it is preceded by stimuli from any nerve entering the cord. Group I_a and I_b fibers are most effective in causing the inhibition, but groups II and III will also serve. The time course of such an inhibition is shown in Figure 1.48. Unlike direct inhibition, maximum inhibition is observed when the inhibitory stimulus precedes the excitatory by about 20 ms. The inhibition may endure for as long as 200 to 300 ms and is extremely resistant to strychnine. Several lines of evidence indicate that this inhibition is presynaptic. Recording within the motoneuron at the time that inhibition is produced shows no changes in the membrane potential. Neither a hyperpolarization nor a depolarization is produced by an inhibitory volley. In-

stead, it is noted that the magnitude of the EPSP which is set up by the reflex volley is diminished by the inhibitory stimulus and that the magnitude of the EPSP parallels the curve of inhibition (Fig. 1.49). It has also been shown that there is a depolarization of the presynaptic fine terminals entering the dorsal region of the cord. Microelectrodes are inserted into and just outside a nerve fiber. The undesirable potential of neighboring neurons which is recorded inside a nerve fiber together with the membrane potential, may be obviated by subtracting the externally recorded potential from the internally recorded to obtain the true membrane potential. It is found that the time course of the depolarization produced by inhibitory volleys parallels the time course of inhibition.

Although the mechanism is not clear, depolarization in axoaxonal synapses on presynaptic terminals may reduce the amount of excitatory transmitter released from the presynaptic terminals, thus decreasing the EPSP amplitude. Since long delays are involved in presynaptic inhibition, many interneurons are interposed between the first and last neurons of the chain.

Renshaw Cell Inhibition

In 1946, Renshaw discovered that a volley of impulses delivered to motor axons causes an inhibition of all types of motoneurons at the segmental level. This type of inhibition has therefore been called antidromic because the inhibition may be evoked by firing backwards over the motor roots

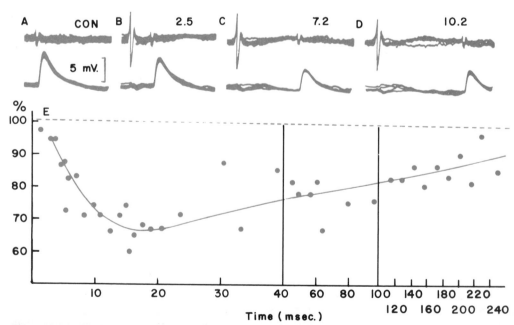

Figure 1.49. Time course of presynaptic inhibition (*E*). Inhibitory stimulus is maximal group I volley in biceps-semitendinosus nerve. Monosynaptic EPSPs were evoked at various intervals after the inhibitory stimulus by stimulating the gastrocnemius-soleus nerve maximally. *A* shows control EPSP; *B*, *C*, and *D* are the EPSP at intervals marked above the records.

into the spinal cord. When such antidromic excitation was used, an afterdischarge of quite high frequency was observed in microelectrode recordings from the ventral horn. It was also shown that the discharge did not occur in the motoneuron but in neighboring cells near the motoneuron—in cells which discharge with a high frequency when the antidromic excitation occurs. At the same time the motoneuron is inhibited. These neighboring cells are therefore believed to cause the inhibition. The motoneuron displays a hyperpolarization which has all the characteristics of an inhibitory postsynaptic potential, and the IPSP lasts for the period of time corresponding to the discharge in the neighboring cells which are called Renshaw cells. An anatomical pathway has been suggested to explain these results (Fig. 1.50). The motoneuron gives off a collateral to the Renshaw cells; the Renshaw cell axon returns to the motoneuron and inhibits it.

The repetitive discharge of the Renshaw cell is presumed to be caused by an accumulation of acetylcholine at the junction between the motoneuron collateral fiber and the Renshaw cell. It has been shown, for example, that the discharge of the Renshaw call may be prolonged by anticholinesterases which allow acetylcholine to accumulate. It is also possible to inhibit the discharge with β-dihydroerythroidine, a substance which blocks the action of acetylcholine. The functional significance of this pathway is not clear. It appears to serve as a general synaptic inhibitor and may act to limit the frequencies of impulses going to the motor end-plate.

ELECTRICAL AND CHEMICAL TRANSMISSION

The events at a synapse may be interpreted according to either a chemical or electrical theory of synaptic transmission. The chemical theory of transmission attributes the response of the postsynaptic cell to a chemical substance (transmitter) released from the presynaptic nerve terminals. The electrical mechanism of transmission presupposes the activation of postsynaptic neurons by electrical currents flowing out from the presynaptic terminals into the postsynaptic cell. Examples of each of these types of transmission may be found in various kinds of synapses among vertebrates and invertebrates. It may be stated, however, that no firm evidence of electrical transmission at mammalian synapses has as yet been obtained.

The present day understanding of chemical transmission is based primarily on studies of the neuromuscular junction because of its accessibility to investigation. The chemical mechanism of transmitter release and action may be schematically outlined as in Figure 1.51.

Cholinergic Transmission

The action potential, upon invasion of the presynaptic terminal, causes release of transmitter material. In the case of neuromuscular junction in which the transmitter is acetylcholine, this substance is stored in clear vesicles in the presynaptic nerve terminal. In their classical experiments with frog muscle fibers, Fatt and Katz (1952) observed that under certain conditions the muscle end-plate (postsynaptic membrane) exhibited intermittent,

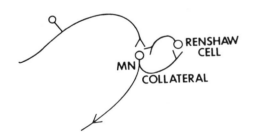

Figure 1.50. Schematic drawing of recurrent inhibition.

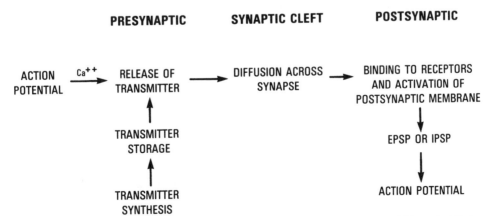

Figure 1.51. Outline of processes involved in storage and release of chemical transmitter. It should be noted that some transmitters have receptors in the presynaptic membrane as well as in the postsynaptic membrane; occupation and activation of these presynaptic receptors serve to self-inhibit and thereby modulate release.

spontaneous miniature potentials in a random manner. These miniature end-plate potentials (MEPPs), of insufficient amplitude to depolarize the membrane to threshold, were related to random release of acetylcholine packets from the presynaptic vesicles. When the external Ca^{++} concentration was reduced, the amplitude of the end-plate potentials decreased in a stepwise manner; under various conditions the amplitudes were integral multiples of the miniature components. However, graded ionophoretic application of acetylcholine directly onto the end-plate produced a continuously graded potential change, thus indicating that the stepwise electrical changes were not due to the nature of the receptor response, but, rather, to the nature of the transmitter release. Their conclusion was that the transmitter is released in quanta or packets of uniform size. Depolarization of the presynaptic terminal increased the frequency of quanta release, thus allowing the MEPPs to summate and reach threshold for firing of an action potential in the postsynaptic cell. Other studies have shown that entry of Ca^{++} is essential for the increased quantal release during the action potential and that external Mg^{++} antagonizes the Ca^{++} action; however, the detailed mechanism for the coupling of excitation to secretion of transmitter is not known.

The released acetylcholine diffuses across the synaptic cleft in a very short time (<0.6 ms) to attach to receptor sites on the postsynaptic membrane. Specific acetylcholine receptor molecules can be extracted and purified from muscle end-plates (Changeux, 1981). The mechanism by which attachment of acetylcholine to its receptor alters the channel permeability in producing the end-plate potential probably involves rearrangement of the receptor structure. The acetylcholine receptor complex, consisting of five polypeptide subunits, contains the acetylcholine binding sites and the ionic channels. The action of acetylcholine is terminated by its hydrolysis catalyzed by acetylcholinesterase present in the postsynaptic membrane. Numerous observations indicate that the quantal nature of transmission is applicable to interneuronal synapses in many portions of the vertebrate nervous system and also to other transmitters, for example, norepinephrine. In the nervous system, combination of the transmitter with receptor gives rise to the EPSP or IPSP. However, direct demonstrations of specific transmitters and their mode of release in the mammalian central nervous system have not yet been achieved.

The basis for establishing that chemical transmission occurs involves the following purely chemical criteria. A transmitter substance must be found (1) which is produced presynaptically, that is, the pertinent enzyme systems for its production are present in the presynaptic terminals; (2) which is released and detected when the presynaptic terminals are stimulated; (3) the action of which is dissipated by a specific enzyme (as in the case of acetylcholinesterase which destroys acetylcholine) or by reuptake (as in the case of norepinephrine) or by simple diffusion away from the postsynaptic membrane; and (4) the action of which, when applied directly onto the postsynaptic membrane, mimics the effects of presynaptic nerve stimulation. In the case of acetylcholine, the transmitter at the neuromuscular junction, such criteria have been met.

Additionally, acetylcholine has been shown to be a transmitter in autonomic ganglia, postganglionic parasympathetic terminals, certain postganglionic sympathetic terminals (sweat glands and vasodilator fibers), and in motoneuron collateral-Renshaw cell synapses in the spinal cord. Examples of some of the evidence bearing on the role of acetylcholine as a transmitter in the central nervous system are: (1) choline acetylase and vesicles containing acetylcholine are present in nerve endings from many nuclei of the nervous system; (2) acetylcholine release has been detected in the cerebral cortex after excitation of a sensory nerve, but only in eserinized preparations in which acetylcholine cannot be destroyed; (3) acetylcholinesterase is abundantly present in neural tissue; and (4) injection of acetylcholine either directly or electrophoretically will activate many cells in the nervous system.

Other Transmitters

There is substantial evidence that norepinephrine is the transmitter released at postganglionic sympathetic nerve terminals. The inactivation of norepinephrine action involves reuptake into the presynaptic terminal. Within the central nervous system norepinephrine is found abundantly in the hypothalamus and brainstem reticular formation where it may have a transmitter role. On the other hand, dopamine, a precursor of norepinephrine, is abundant in the corpus striatum, where it is thought to act as a transmitter. Histochemical fluorescent studies have shown that dopamine is present in nerve terminals within the corpus striatum that derive from cell bodies located in the substantia nigra. In Parkinson's disease, which is associated with symptoms of basal ganglia dysfunction, there is degeneration of this biochemically defined nigra-striatal path and depletion of dopamine in the corpus striatum. Administration of L-dopa, which is decarboxylated to form dopamine in vivo, ameliorates the symptoms of parkinsonism. Both catecholamines, dopamine and norepinephrine, are stored in granular vesicles in presynaptic nerve terminals.

A number of other substances have been suggested as possible synaptic transmitters and inhibitors. Serotonergic axons extend from median raphe nuclei in the brainstem to many areas of the brain, but particularly to the hypothalamus. Evidence is accumulating that γ-aminobutyric acid is an inhibitory transmitter in many portions of the central nervous system and that glycine acts as an inhibitory transmitter in the spinal cord. In Huntington's disease, another type of disorder of the basal ganglia, neurons containing γ-aminobutyrate are affected (Perry et al., 1973). Glutamic acid and aspartic acid have been suggested as important excitatory transmitters. These are as effective as acetylcholine in activating neurons when injected into the cells electrophoretically. These amino acids, however, have not been found associated with vesicles, and mechanisms of their release are not known. Postulated inactivation mechanisms involve reuptake into synaptic and possibly glial membranes. A number of peptides (e.g., substance P, neurotensin, endorphins, enkephalins, thyrotropin-releasing hormone, luteinizing hormone-releasing hormone, angiotensin, bradykinin, bombesin, oxytocin, vasopressin, melanocyte-stimulating hormone (MSH), MSH release-inhibiting factor, adrenocorticotrophic hormone, somatostatin) and other substances (e.g., histamine, prostaglandins, and purines) found in nerve tissues may also play roles in transmission or modulation of synaptic function, but their precise roles remain to be elaborated (Hedquist, 1974; Hughes et al., 1975; Takahashi and Otsuka, 1975). The interested student is referred to section III of a recent text on neurochemistry (Siegel et al., 1981) for additional reading on synaptic function.

Greengard (1976) has suggested that cyclic nucleotides (adenosine 3′,5′-phosphate, guanosine 3′,5′-phosphate), which serve in many tissues to translate hormonal messages into intracellular responses, may also mediate certain postsynaptic cell responses to neurotransmitters.

There are other indications that chemical transmission takes place at mammalian synapses. Some synaptic delay would be anticipated if the release, diffusion, and attachment of a substance were factors important in transmission. Thus, the finding of a synaptic delay at vertebrate synapses of 0.5 ms is consistent with a chemical theory. A second phenomenon which is consistent with the chemical theory is the fatigability of synapses when rates of stimulation are increased. At 40 or 50 impulses/s, many synapses within the nervous system fail to transmit an impulse. According to the chemical theory, such a failure would be in accord with the expectation that there will be a frequency of stimulation above which the synthesis of transmitter is unable to keep pace with its release from the terminals. At such high frequencies, the amount of transmitter released per impulse would decrease and fail to activate the postsynaptic membrane. A third, well-known observation is the susceptibility of synapses to anoxia and metabolic inhibitors. Thus, cutting off the supply of oxygen interferes with the synthesis of ATP and other chemical materials necessary for the manufacture of the transmitter substance. The supply of transmitter diminishes and the amount released per impulse again becomes inadequate to activate the postsynaptic membrane. A fourth phenomenon is the repetitive discharge of neurons as noted in the case of the Renshaw cell where a repetitive discharge ensues from a single shock. This repetitive discharge is the result of a prolonged depolarization lasting more than 100 ms. The prolonged depolarization is attributable to the accumulation of the transmitter substance at the synapse.

Electrical Transmission

Criteria for the presence of electrical transmission are essentially the opposite of those for chemical transmission. Thus a system which transmits impulses across junctions at high frequencies with no delay and in which anoxia and metabolic inhibitors have little or no effect is presumed to be an electrical synapse. Such systems behave, indeed, as though no junction were present, although there may be an anatomical junction with little or no space between the presynaptic and the postsynaptic membranes. Such electrical junctions have not yet been discovered in the mammalian nervous system, but artificial synapses called "ephapses" have been created at junctions between nerve fibers so that an incoming electrical impulse will either modify the excitability in a contiguous nerve or actually set off an action potential. Among the invertebrates, several examples of purely electrical junctions have been discovered. The giant motor synapses of the crayfish are designed to transmit electrically in one direction only. Simultaneous recording from both presynaptic and postsynaptic cells has demonstrated that practically no delay (about 0.1 ms) occurs. The postsynaptic response has the shape of an EPSP, and spikes may be generated from sufficiently strong EPSPs. Such synapses appear to act as rectifiers, i.e., they pass current more easily in one direction than the other.

CHAPTER 4

Excitation and Contraction of Skeletal Muscle

GENERAL CHARACTERISTICS OF MUSCLE

In this chapter only skeletal muscle will be discussed. A thorough understanding of skeletal muscle physiology will provide a basis for studying cardiac and smooth muscle, both of which differ significantly from skeletal muscle in certain respects and will be discussed in subsequent chapters. Much of our knowledge of mammalian skeletal muscle stems from studies of frog skeletal muscle, which is similar to mammalian skeletal muscle and to which frequent reference will be made.

The functions of muscle tissue are development of tension and shortening. The nervous system coordinates the activity of different parts of muscle tissue to produce useful movements and postures. The effect of muscle activity is transferred to the skeleton by means of tendons; the translational displacements between various parts of the muscle mass, displacements that are associated with movement, are facilitated by the interposition of connective tissue septa. The latter structures are present at places where the translational displacements are most pronounced.

The tension developed by a muscle in the body is graded and adjusted to the load. If contraction is associated with shortening, the tension is adjusted both to the load and to the velocity of the shortening. The graded response of muscle is due to variation in the degree of activation of the tissue by the motor nerves.

STRUCTURE

Myofibers, Myofibrils, and Myofilaments

Skeletal muscle is composed of numerous parallel elongated cells referred to as *muscle fibers* or *myofibers*. These are about 10 to 100 μm in diameter and vary with the length of the muscle, often extending its entire length. Under the electron microscope, the subcellular structure of the skeletal muscle fibers is seen to be composed of smaller fibrous structures 1 μm in diameter, *myofibrils*, which are separated by cytoplasm and arranged in parallel fashion along the long axis of the cell (Fig. 1.52). Each myofibril is further subdivided into thick and thin *myofilaments* which are referred to simply as *thick* and *thin filaments*. Thin filaments are about 7 nm wide and 1.0 μm long, and thick filaments are about 10 to 14 nm wide (in mammals) and 1.6 μm long. The arrangement of the thick and thin filaments is responsible for the cross-striated appearance of the muscle, which results from a regular repetition of dense cross-bands (1.6 μm in length) separated by less dense bands. The dense cross-bands, referred to as A-bands because they are strongly anisotropic,* contain the thick filaments arranged neatly in parallel. The less dense segments, the *I bands*, contain the thin filaments, which extend symmetrically in opposite directions from a dense thin line, the *Z line*. The Z line appears in embryonic muscle before the thin filaments develop and is comprised of a lattice-like protein, intertwined with the thin filaments which extend perpendicularly from the Z line for 1 μm on either side. The term I band is based on the fact that this zone was originally considered to be isotropic. Although it is now recognized to be weakly birefringent, it is very much less birefringent than the A band.

The width of the I band varies with the degree of stretch or shortening of the muscle fiber (Fig. 1.53). Since the length of the individual thin filaments is 1 μm and there are two sets of such filaments extending from the Z line, which is 0.05 μm wide, the total width of the Z line-thin filament complex is 2.05 μm. The Z-line structure contributes toward keeping the thin filaments arranged in register and with a regular spacing. The gap between the ter-

---

* Optically anisotropic substances have different refractive indices in two different directions and hence are birefringent. Optically isotropic substances have only one refractive index in all directions.

a.

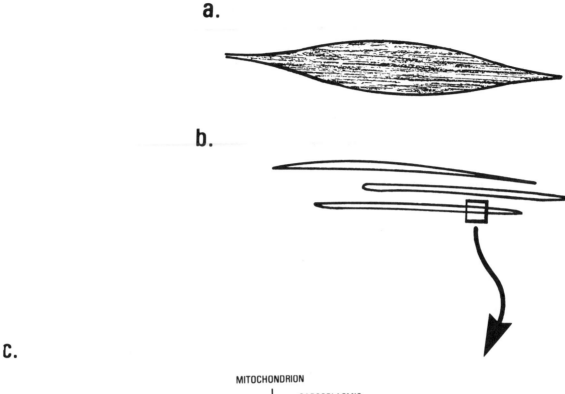

b.

c.

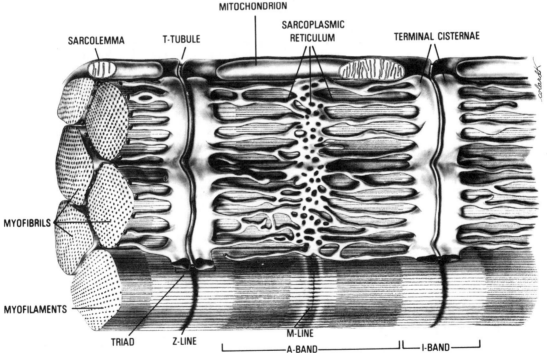

Figure 1.52. Structure of skeletal muscle. *a*, Whole muscle. *b*, Muscle fibers. *c*, Schematic representation of a segment of a skeletal muscle fiber. [Adapted from Fawcett and McNutt (1969).]

minations of the thin filaments is called the *H zone*, and the darker area in the center of the H zone, the *M line*.

Sarcomere

The molecular basis for the difference in isotropicity between A and I bands described above be-

comes evident from an understanding of the structure of the *sarcomere*, the fundamental contractile unit of muscle. A sarcomere consists of the region between two consecutive Z lines; thus, this unit consists of one A band and one-half I band at each end of the A band. Its length at rest varies between 2.0 and 2.6 μm in frog and mammalian muscle.

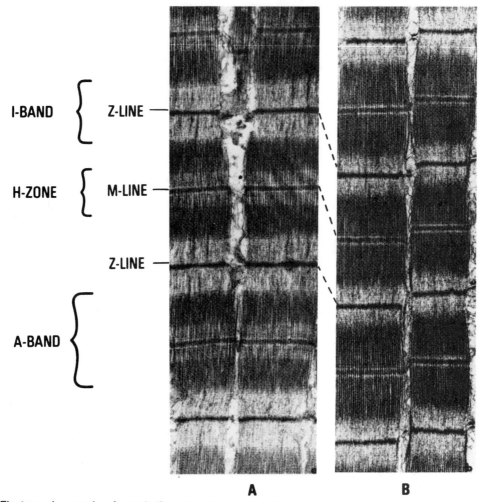

I-BAND { **Z-LINE** —

H-ZONE { **M-LINE** —

Z-LINE —

A-BAND {

A   **B**

Figure 1.53. Electron micrographs of muscle (frog sartorius) fixed at different stages of shortening. *A,* Muscle had shortened only very slightly and a wide H-zone is still visible (×15,500). *B,* As in *A* but allowed to shorten further so that a very short H-zone (and a shorter I-band) is visible (× approximately 15,500). [From Huxley, H. E. (1964)].

The length of the sarcomere depends on the degree of muscle contraction. At the *rest length* of muscle, defined as the normal length of the muscle in situ, the two sets of filaments in most muscles interdigitate with a rather extensive zone of overlap at each end of the A band (Fig. 1.54*A*). If a muscle is allowed to shorten, the relationship between the half-width of the I bands, the width of the overlap zone, and the width of the H band is the same as described above (Fig. 1.53). At sarcomere lengths less than 2 μm, that is, less than the sum of the lengths of the two sets of thin filaments in the sarcomere, a somewhat more dense zone appears in the middle of the A band. This phenomenon is due to a double overlap of the thin filaments (Fig. 1.54*D*). An increased density seen in the region of the M line is called a contraction band, specifically a CM band. A second contraction band (CZ band), due to the penetration of the Z line by the ends of thick filaments, is seen at sarcomere lengths of less

than 1.5μm. Alternatively, it could be due to ends of the thin filaments reversing direction at the Z line.

If the muscle is stretched, the zone of overlap decreases in width in proportion to the increase in half-width of the I band. The zone in the middle of the A band bounded by the two zones of overlap, the H zone, also increases in width, this increase corresponding to the decrease in width of both zones of overlap. This means that when a muscle is stretched, the thin filaments are being pulled out of the A band, while the thick filaments remain constant (Fig. 1.54*B* and *C*). At a sarcomere length of about 3.6 μm, the thin filaments have been pulled out completely, and the two sets of filaments appear now in an end-to-end arrangement.

In accord with the above considerations, the microscopic patterns seen in cross-sections of muscle fibers depend on the level of the section (Fig. 1.55). For instance, near the Z band, only thin filaments

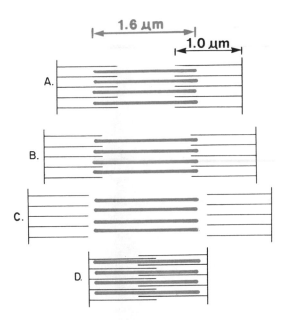

Figure 1.54. Schematic illustration of the arrangement of thick filaments and thin filaments in the sarcomere at various degrees of stretch and shortening. *A* and *B* show different degrees of overlap of the two types of filaments, *C* illustrates the case when the thin filaments have been completely pulled out from the A band with a gap between A band and the thin filaments. In the extensively shortened sarcomere in *D* a zone of double overlap of thin filaments has developed in the center of the A band.

are observed, whereas at the site of overlap of thick and thin filaments, each thick filament is surrounded by six thin filaments and each thin filament by three thick filaments. A cross-section through the M line shows the thin connections between adjoining thick filaments. A representative electron micrograph of a cross-section through several myofibrils is seen in Figure 1.56.

Sarcolemma

The sarcolemma is the outer membrane surrounding each muscle fiber. It is comprised of two components, the *plasma membrane* proper (plasmalemma) and a *basement membrane*. The main function of the plasma membrane in muscle contraction is to spread the wave of depolarization originating at the motor end plate over the entire cell surface to initiate contraction. The plasma membrane has the general double membrane structure of most biological membranes. The basement membrane on the outer surface of the plasma membrane consists of a thin mucopolysaccharide-rich coating containing fine collagen fibrils, which fuse with the tendons at the ends of the muscle. It may affect membrane permeability in some as yet unknown manner. Besides shielding the muscle fiber, the sar-

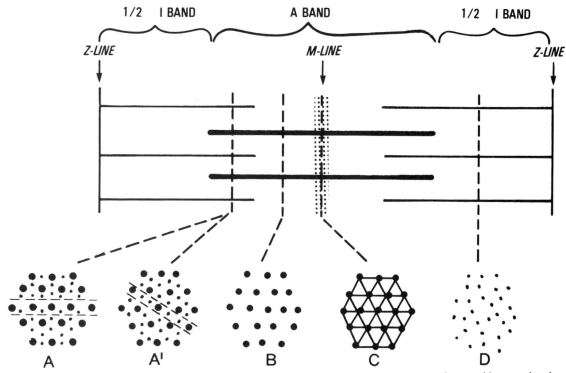

Figure 1.55. Diagram illustrating the arrangement of the thick and thin filaments in a sarcomere as they would appear in a longitudinal section of a muscle fiber. Immediately below are diagrams of transverse sections taken through the A band where the thick and thin filaments overlap (*A* and *A′*); the A band where there is no overlap, i.e., H zone, (*B*); the M line (*C*); and the I band (*D*). If a longitudinal section were cut in the plane which is indicated by the dotted lines in (*A*), it would show one (apparently) thin filament between each two thick ones. If it were cut in the other plane shown by dotted lines in (*A′*), it would show two thin filaments between each two thick ones. [Adapted from Huxley and Hanson (1960).]

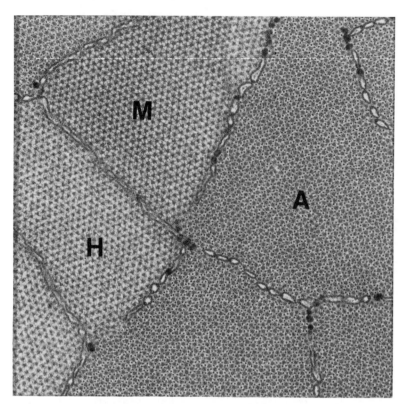

Figure 1.56. Cross-section through several myofibrils, at the level of the A band, illustrating the relative disposition of thin and thick filaments in the sarcomere. At right, thin filaments occupy a trigonal position in the hexagonal lattice of thick filaments (*A*). At left, in the H zone (*H*), only thick filaments are present. Cross-links join thick filaments to each other at the M line (*M*), located in the center of the sarcomere. (×33,000). [From Franzini-Armstrong and Peachey (1981).]

colemma imparts a characteristic resistance to stretch to the muscle as a whole. Because this resistance is anatomically disposed in parallel with the contractile elements, it is usually lumped together with the resistance contributed by the connective tissue surrounding the muscle fibers, and both are referred to as *parallel elastic elements.*

Tubular extensions of the sarcolemma, called *T-tubules* or *sarcotubules*, extend deep into the fiber at the level of either the Z line or at the junction of the A-I bands, depending on the type of muscle fiber or the animal species. The T-tubules, about 300 μm in diameter, allow a wave of depolarization to pass rapidly into the fiber so that deep lying myofibrils may be activated.

Sarcoplasm

The sarcoplasm of a muscle fiber consists of the contents of the sarcolemma, excluding the proteins of the contractile elements and nuclei. Specifically it consists of the cytoplasm and the usual cytoplasmic organelles: mitochondria (sarcosomes), sarcoplasmic reticulum, Golgi apparatus, and liposomes.

Sarcoplasmic Reticulum

The sarcoplasmic reticulum is an elaborately anastomosing tubular network running parallel to the myofilaments. The diameter of the tubules is approximately 1 μm. The slender tubules often extend the full length of the sarcomere and are closed at each end. The dilated ends of the tubules are called *terminal cisternae.* The sarcoplasmic reticulum can thus be divided into the longitudinal sarcoplasmic reticulum and the terminal cisternae. A group of one T-tubule and two terminal cisternae is called a *triad* because of the triple spaces seen in cross-sections of these structures (Fig. 1.57). In the narrow space between the sarcoplasmic reticulum membrane and the T-tubule (about 12 to 14 nm) and connecting the two membranes at periodic distances, there is present amorphous material called "feet." These structures may play a role in excitation-contraction coupling, but their precise function is unknown.

The functions of the sarcoplasmic reticulum are the release of calcium during muscle contraction and the sequestration and storage of calcium during muscle relaxation. Some of the earliest evidence for the function of the sarcoplasmic reticulum came from the studies in frog skeletal muscle by Huxley and Taylor, who found that following exploratory stimulation with minute electrical currents applied with a microelectrode at intervals of 1 μm along the length of the sarcolemmal membrane, a local con-

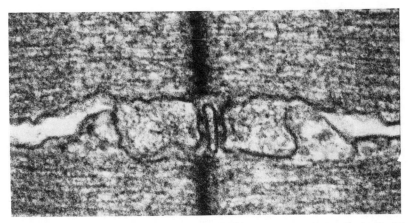

Figure 1.57. Electron micrograph of a longitudinal section of a muscle fiber showing a triad at the level of the Z-line. Two "feet" are visible on each side of the T-tubule (×120,000). [From Franzini-Armstrong (1970).]

traction was elicited *only* when the stimulatory microelectrode was placed at the level of the Z line. These findings revealed an important relationship between the T tubules and the adjoining cisternae of the triad and also between the triad as a whole and the contractile elements (myofilaments) of the muscle fiber. Somewhat later, in an autoradiographic study by Winegrad which allowed localization of radiolabeled calcium during muscle contraction and relaxation, evidence was obtained to indicate that during muscle excitation, calcium stored in the terminal cisternae is released to allow binding to troponin (see below) and interaction of the thick and thin filaments. Muscle relaxation was correlated with the sequestration of the calcium at the level of the A band by the longitudinal sarcoplasmic reticulum.

Nuclei

Skeletal muscle fibers are multinucleated. In embryonic tissue the nuclei are located in the center of the fiber; however, in the course of differentiation, they come to occupy a peripheral position.

PROTEINS OF THE CONTRACTILE ELEMENTS

Thin Filaments

Thin filaments are composed primarily of three types of protein: actin, tropomyosin, and troponin in a ratio of 7:1:1. The so-called "functional unit" necessary for relaxation consists of 7 actin monomers, 1 tropomyosin, and 1 troponin complex.

ACTIN

In vitro, actin molecules exist in two states: G-actin and F-actin. G-actin is a monomeric globular protein with a molecular weight of 42,000 and a diameter of 4 to 5 nm. F-actin, a fibrous polymer consisting of approximately 300 or more monomers of

G-actin, is formed, in vitro, when salts at physiologic ionic strength are added to G-actin monomers. In mammalian skeletal muscle, only F-actin is found. The basic structure of the thin filament consists of two strands of F-actin polymers intertwined in the conformation of a double stranded helix (Fig. 1.58).

The amino acid composition of actin obtained from muscles of different animals and from different cell types including fibroblasts and platelets is very similar. Each monomer of skeletal muscle actin contains binding sites for other actin monomers, myosin, tropomyosin, troponin I, ATP, cations. The link between actin monomers in G-actin is noncovalent, probably due to hydrophobic interaction.

TROPOMYOSIN

Tropomyosin, an elongated protein about 41 nm long and 20 nm in diameter, consists of two α-helical chains, each with a molecular weight of approximately 35,000, wound around each other to form a coiled-coil. It lies in the two grooves, 180° apart, formed by the double stranded helix consisting of F-actin (Fig. 1.59). One molecule of tropomyosin extends over seven actin monomers.

TROPONIN

Troponin consists of a complex of three separate proteins: troponin-T (abbreviated TN-T), troponin-C (TN-C), and troponin-I (TN-I). Each troponin complex is bound to a tropomyosin molecule. While only actin and myosin are directly involved in tension generation, the tropomyosin and troponin complex are known to regulate the actin-myosin interaction, hence are called regulatory proteins (Fig. 1.59).

TN-T (molecular weight, 30,000) binds the other two troponin subunits to tropomyosin. TN-C (molecular weight, 18,000) is the calcium acceptor pro-a.

b.

Figure 1.58. Structure of F-actin. *a*, Model showing double stranded helix of F-actin. F-actin is a polymer of an indeterminate number of G-actin monomers. *b*, Electron micrograph of a segment of an F-actin filament. (Approximately ×230,000). [Courtesy of James Spudich.]

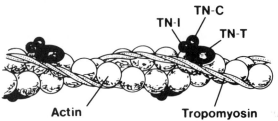

Figure 1.59. Model of thin filament of muscle. Thin filament consists of a double stranded helix of F-actin, two double stranded helices of tropomyosin, and a troponin complex, consisting of TN-T, TN-C, and TN-I, located at each group of seven actin monomers. [Reprinted with permission from McCubbin and Kay (1980). Copyright 1980 American Chemical Society; based on model by Ebashi et al. (1969).]

tein of the troponin complex. It binds to TN-T and to TN-I (molecular weight, 22,000), which is believed to induce the inhibitory conformation of the actin-tropomyosin filament. If indeed the actin is masked during relaxation (this is the traditional viewpoint), then tropomyosin masks the actin when tropomyosin is in its inhibitory conformation. TN-I probably participates in shifting tropomyosin to this position. Strictly speaking, TN-I is not the prime mover, since there is 1 TN-I molecule per 7 actin monomers, and TN-I is too small to "mask" 7 actin monomers. In vitro, it is possible to design an experiment (as Hartshorne (1970) did) to demonstrate that a TN-C–TN-I complex (troponin-B) at a 1:1 ratio with actin can mask actin and cause relaxation. However, these are not physiological conditions.

OTHER PROTEINS ASSOCIATED WITH THIN FILAMENTS

α-Actinin is a protein localized at the level of the Z-line. Its physiologic function is unclear, but it may play a role in the formation or attachment of F-actin to the Z-line, where it may serve as an attachment plate. *β-Actinin*, on the other hand, is a protein located presumably at the free ends of the thin filaments. It may serve as an ending factor during the polymerization of G-actin.

Thick Filaments

Thick filaments consist primarily of myosin. Various other proteins, mentioned below, are associated with myosin; their function remains unclear.

MYOSIN

Myosin, a dimer with a molecular weight of about 480,000 measured under nondenaturing conditions, consists of two globular heads, which hydrolyze ATP and interact with actin, and a rod-like region, which confers stability to the molecule. In in vitro studies, purified myosin may be treated with proteolytic enzymes in order to study the properties of different segments of the molecule (Fig. 1.60). Treatment of myosin with trypsin splits the protein into two components: light meromyosin (LMM) (molecular weight, 140,000 to 160,000), which consists of the tail (rod-like) part of the native molecule, and heavy meromyosin (HMM) (molecular weight, 340,000 to 380,000), the primarily globular end of the native molecule, to which is attached a segment of the rod-like portion. Papain cleaves HMM or myosin into a globular protein, S_1, and a rod-like protein, S_2, or myosin rod consisting of LMM and S_2. Both HMM-S_2 (molecular weight 60,000) and LMM are insoluble, whereas HMM-S_1 is soluble in solutions of low ionic strength. Myosin adenosine triphosphatase (ATPase) activity, i.e., the ability to hydrolyze ATP, is retained in HMM and in HMM-S_1 obtained with papain; LMM and HMM-S_2 are totally inert in this respect. The ATPase activity of myosin obtained from different types of skeletal muscle correlates with the shortening velocity of the particular muscle type.

When column-purified myosin obtained from fast skeletal muscle is examined by sodium dodecyl sulfate-polyacrylamide gel electrophoresis (denaturing conditions), it is found to consist of heavy chains (molecular weight, 200,000) and three light chains called L_1 (molecular weight, 25,000), L_2 (molecular weight, 18,000), and L_3 (molecular weight, 16,000). The designations g_1, g_2, and g_3 are used synonymously. These light chains are bound noncovalently to the globular head, but their precise location is unknown, although L_2 may be located at the hinge region between S_1 and S_2 and can be lost during enzymatic digeston of myosin. Light chains can also be dissociated from the heavy chains by treatment wth 12 M urea or 5 M guanidine.

The L_2 light chains, two per myosin molecule,

a.

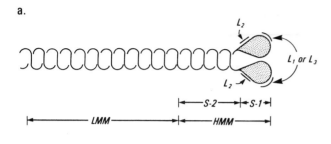

b.

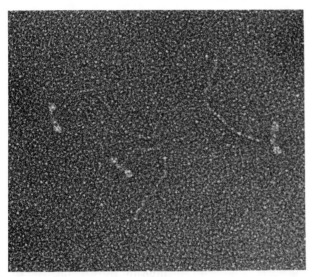

Figure 1.60. Structure of the myosin molecule. *a*, Schematic representation: The molecule is a dimer consisting of two globular heads joined to a rod-like tail region. Hinge-regions occur at the head-tail junction and in the tail near the site of proteolytic cleavage of the molecule into LMM and HMM. The exact location of the light chains (L_1, L_2, and L_3) on the myosin heads is not known. *b*, Electron micrograph of shadow cast myosin molecules magnified approximately 271,000 times. [Courtesy of S. S. Margossian and H. S. Slater.]

one on each head, are also called DTNB light chains because they can be dissociated with 5,5'-dithiobis-(2-nitrobenzoic acid) (DTNB). The L_2 light chain is not essential for either the myosin or actin-activated ATPase activity, but may modulate the latter (Pemrick, S. M., 1980). The other two light chains, L_1 and L_3, are called the alkali light chains because they may be removed with alkali treatment. It is thought that the L_3 light chain is derived from the L_1 light chain. Traditionally, removal of these light chains has resulted in loss of ATPase activity, hence they have been called essential light chains. Recent technology, however, has made it possible to demonstrate ATPase activity in the absence of alkali light chains (Silvaramakrishnam and Burke, 1982).

Myosin molecules will aggregate in a particular pattern to form filaments, similar to the thick filaments seen in intact muscle, when the ionic strength of the medium is reduced, e.g., from 0.5 M to 0.1 M KCl (Fig. 1.61). The morphology of the thick filaments agrees with the assumption that the myosin molecules aggregate with their globular ends directed towards the ends of the filaments. The middle smooth shaft corresponds to a close packing of the rod-shaped part of the molecules, and the series of projections correspond to their globular ends.

Skeletal muscle myosin is heterogenic or polymorphic with respect to both the heavy and light chains. Various isozymes can be identified in different muscle types by certain types of gel electrophoresis or by immunocytochemical methods. These methods have been used to distinguish three isomyosins in fast skeletal muscle fibers and two isomyosins in slow fibers, all with the same molecular weight. The presence of these isomyosins correlates with the ATPase activity. In fast fibers, there are three light chains, LC_1^f, LC_2^f, and LC_3^f, whereas in slow fibers, there are three different forms of light chains designated LC_{1a}^s, LC_{1b}^s, and LC_2^s. The molecular weights of these light chains range from 14,000 to 31,000. The possibility of fast and slow myosins existing in the same fiber cannot be excluded particularly during states of transition such as occur in cross-innervation experiments, during maturation, in response to exercise or changes in hormonal status. For example, changes in circulating levels of thyroid hormone are known to affect muscle phenotype.

OTHER PROTEINS ASSOCIATED WITH THICK FILAMENTS

When myosin preparations obtained by conventional methods were originally examined by SDS-gel electrophoresis, they were found to contain low amounts of other proteins which appeared as faintly staining bands on the gels in addition to the heavy and light chains of myosin. The protein bands, starting at the top of the gel, were designated A, B, C, D, E, F, G, H, I, and J, and L_1, L_2 and L_3, which referred to the light chains. Band A corresponds to the heavy chains of myosin; J-band turned out to be actin. C-protein (monomer of 140,000 daltons), the best studied of the rest of the proteins, is associated with the A band on either side of the M line and binds to myosin or light meromyosin at physiologic ionic strength, indicating interaction with the thick-filament backbone. In vitro, C-protein can be shown to inhibit actomyosin ATPase activity; whether a similar regulatory effect exists in intact muscle is unknown.

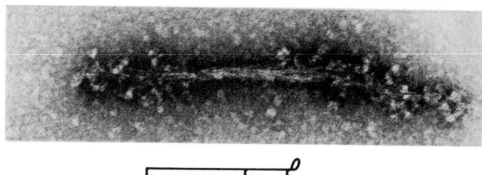

LMM   HMM

Figure 1.61.   Electron micrograph of a synthetic thick filament formed by aggregation of purified myosin in 0.1 M KCl. Projecting cross-bridges can be seen. Filaments containing light meromyosin alone (not shown) have no projections. (Approximately ×162,000). [From Huxley (1969).] Shown immediately below in order to illustrate the pattern of aggregation are diagrams of a single myosin molecule and a synthetic thick filament. *LMM,* light meromyosin; *HMM,* heavy meromyosin.

Additional proteins have been associated with the thick filaments. There are two principal components of the M-band region. One is a dimer of molecular weight about 88,000 identified as the native form of creatine kinase. The second component of the M-band is a 165,000 molecular weight protein.

In molluscan muscle, the thick filaments consist of a core of rod-like proteins called paramyosin and a surface layer of myosin molecules. The presence of paramyosin in vertebrate muscle is not established.

In Vitro Studies of the Contractile Process

The properties of the contractile proteins can be studied in a number of in vitro systems. Purified myosin can be used to study its ATPase activity. Myosin ATPase can be activated by Mg^{++}, K^+ (plus EDTA), NH_4^+, or Ca^{++}. All of these ions can be utilized as all bind to ATP; however, the physiologic substrate of myosin in MgATP. Greatest activation is obtained with the physiologic activator actin in the presence of MgATP.

Actin and myosin combine to form actomyosin, a highly viscous complex or thick gel. If the actin filament contains troponin and tropomyosin, it is called *"regulated actin,"* which means that the interaction of actin and myosin in the presence of ATP is Ca^{2+} sensitive. In the absence of Ca^{2+}, troponin is calcium free and tropomyosin is in its inhibitory position on actin (see above). As a result, little or no interaction occurs between actin and myosin. This in vitro situation is analogous to muscle relaxation. In the presence of Ca^{2+}, regulated actin is no longer exerting an inhibitory influence on the system, actin-myosin interaction is optimal, and the in vitro situation is similar to muscle contraction. This cycle of relaxation and contraction can be followed in a spectrophotometer due to changes in light scattering associated with ATP mediated changes in the actomyosin gel. For example, the addition of ATP to an actomyosin gel first dissociates the actin and myosin resulting in a clearing phase (minimal light scattering), which is very long (time scale of hours) in the absence of Ca^{2+} and reasonably short in the presence of Ca^{2+} (3 to 5 minutes) since ATP is rapidly hydrolyzed only in the latter case. When the ATP level falls below a critical value, a phenomenon known as *superprecipitation* (increase in light scattering) occurs due to shrinking and loss of water content in the actomyosin gel. Superprecipitation is believed to be analogous to an elementary contraction process.

The contraction process can also be studied in glycerinated fibers which have been subjected to prolonged treatment with 40 to 50% glycerol. Glycerol treatment of this type results in loss of cellular membranes, including the sarcoplasmic reticulum plus the regulatory proteins of the thin filament while leaving the contractile proteins (actin and myosin) intact. Addition of ATP causes contraction of the fibers.

THEORIES OF CONTRACTION

Historical Development

From the 1840s until the 1920s, the *viscoelastic theory*, also called the *new elastic body theory*, was postulated to explain muscle contraction. According to this theory, muscle acts like a stretched spring (or new elastic body) contained in a viscous medium. The amount of energy to be released upon contraction would depend on how much the muscle or spring was stretched; when released, the muscle or spring would liberate all of its energy in an all or none fashion irrespective of the work done. Thus it had been believed that a preset amount of additional energy is fed into the contractile machinery when a muscle contracts (in association with the assumption by the muscle of new elastic characteristics), and that this present increment of energy could be proportioned between work and heat, the total amount of energy (work + heat) remaining constant. That this was not the case was demonstrated in the 1920s by Fenn, who observed that the total energy released by a muscle (work + heat) increases as muscle work increases. This finding became known as the *Fenn Effect*.

Fenn's finding indicated that the energy release is determined not only by the activation process but also by the work load imposed on the muscle. Thus the mechanical function was shown to control chemical reactions in muscle cells, the molecular mechanism being finely attuned to the work demand, i.e., the performance of a given increment of muscular work calls for the generation and delivery of a commensurate increment of energy. While these findings add little insight into the molecular mechanism of muscle contraction, they are significant because they showed that the viscoelastic theory of muscle contraction, a view held for 80 years, was incorrect.

Since the identification of actin and myosin as the contractile proteins, various other theories of contraction and relaxation have been postulated. According to the *continuous filament theory,* during contraction actin and myosin combined to form one continuous filament, which underwent folding and thereby shortening. The folding was postulated to be due either to thermal agitation or to loss of water molecules from the intrinsic chemical structure of fibers.

Electron microscopic observations did not support the continuous filament theory, because actin and myosin appears in regularly spaced arrays. After contraction, the lengths of thick and thin filaments were observed to be altered and only their relative position changed. Therefore a new theory, known as the *sliding filament theory,* was proposed by H. E. Huxley (1969) on the basis of electron microscopic observations and X-ray diffraction patterns of muscle fibers both at rest and in the contracted state. As biochemical data on the contractile proteins became available, the sliding filament theory was transformed into the *cross-bridge theory* of muscle contraction. These theories are described below.

Sliding Filament and Cross-Bridge Theories

The sliding filament theory states that muscle contraction is the result of two overlapping sets of filaments sliding past each other. Specifically, the thin filaments at each end of the sarcomere move in opposite directions toward the center and between the thick filaments to which they are linked by cross-bridges. Polarity of the thick filaments, as described below, is thus a requisite for contraction.

The molecular basis for the sliding motion of the filaments became evident with the elucidation of the structure of actin and myosin. The globular heads of myosin form cross-bridges with the actin monomers, hence Huxley's new designation cross-bridge theory.

Huxley suggested that the cross-bridges move to and fro, first attaching to the thin filaments and pulling them toward the center of the A band during the cycle and then detaching again prior to their return stroke—similar to the action of a ratchet. The cross-bridge theory is thus also known as the *ratchet theory* of muscle contraction. The cross-bridges are attached to thin filaments while force is being developed and serve as the agents through which the mechanical force is transmitted.

The cross-bridges consist of the globular heads of the myosin and α-helical tail by which the cross-bridges attach to the backbone of the thick filament. Myosin molecules are arranged in the thick filaments with a definite structural polarity so that the heads of the molecules are always directed away from the midpoint of the filament. Thus, all cross-bridges in one half of an A band have the same orientation (polarity). This polarity is reversed in the opposite half of the A band. The actin monomers are also oriented oppositely on either side of their attachments to the Z line. During contraction the sets of thin filaments in each half sarcomere are drawn toward the center of the A band and subjected to sliding forces oriented in opposite directions.

Using X-ray diffraction Huxley had initially predicted that the contractile components (later identified as actin and myosin) must have different structures. Further study of X-ray diffraction patterns indicated that in striated muscle of vertebrates the cross-bridge projections on the thick filaments are arranged so that there are three pairs of cross-bridges per 360° turn (Fig. 1.62). Each pair of bridges projects out from the myosin backbone in opposite directions, so that the two bridges in any pair form an angle of 180° with each other, i.e., any pair of cross-bridges form a straight line at any given level. Taking any pair of cross-bridges as a reference, the nearest neighbor pair of bridges on either side occurs at a distance of 14.3 nm and is rotated relative to the reference pair by 120°. This arrangement continues so that the full pattern repeats itself at intervals of 3 × 14.3 nm or 42.9 nm.

The X-ray diffraction patterns of the thin fila-

ments show a double helical structure, with subunits (the G-actin monomers) repeating at 5.46 nm intervals along each strand. The position of any subunit on one strand is staggered relative to the position of its nearest neighbor subunit on the other strand by half a subunit period, i.e., 2.73 nm. The chains twist around each other with cross-over points at 36 to 37 nm intervals, so that the pitch of the helix formed by either of the two chains is 72 to 74 nm (Fig. 1.63).

On the basis of these observations, a model (Fig. 1.64) was suggested in which the cross-bridges are attached to the backbone of the thick filaments and extend outward beyond the myosin backbone by a short subunit tail. This tail is attached to the thick filament only at one end. This type of attachment makes it possible for the cross-bridges to move away from the backbone toward the thin filament. As noted previously, the attachment of the heavy meromyosin tail to the light meromyosin of the thick filament backbone is susceptible to trypsin digestion and, therefore, probably represents a nonhelical flexible region of the molecule. Similarly, the junction between the linear part of the HMM tail

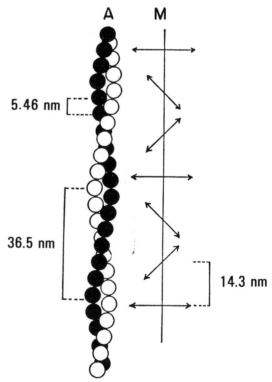

Figure 1.63. Diagram of the arrangement of G-actin monomers in A filaments derived from X-ray diffraction and electron microscope observations. Both the pitch of the helix and the subunit repeat differ from those of the M filaments, indicated schematically alongside. Thus, cross-bridges between filaments would act asynchronously, and a sequence of them would develop a fairly steady force as the filaments moved. [Adapted from Huxley, H.E. (1969).]

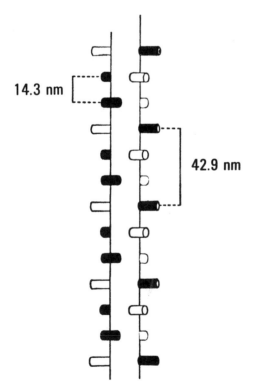

Figure 1.62. Diagram of cross-bridge arrangement on thick myosin-containing filaments of frog sartorius muscle which would account for the observed X-ray pattern. [Adapted from Huxley (1969).]

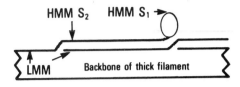

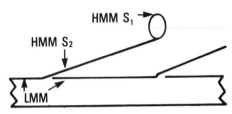

Figure 1.64. Suggested behavior of myosin molecules in the thick filaments. The light meromyosin (*LMM*) part of the molecule is bonded into the backbone of the filament, while the linear portion of the heavy meromyosin (*HMM*) component can tilt further out from the filament (by bending at the HMM-LMM junction), allowing the globular part of HMM (that is, the S_1 fragment) to attach to actin over a range of different side spacings, while maintaining the same orientation. [From Huxley, H. E. (1969).]

with the globular head is also susceptible to enzymatic digestion, and this site of attachment probably also constitutes a flexible site. Therefore, the presence of two flexible points on the tail of the cross-bridge would give the HMM head the mobility needed for it to become properly oriented toward the thin filament during contraction. Because of these flexible points or hinge regions, the cross-bridge projections may change either in orientation or length at different sarcomere lengths when the lateral distance between thick and thin filaments changes (Fig. 1.65). Because the lateral distance between thick and thin filaments increases as the fiber shortens, the volume of the muscle remains constant during contraction.

At rest, cross-bridges do not reach the thin filaments, whereas in rigor 50 to 100% of the cross-bridges are attached to the thin filaments. In an active contraction, the number of cross-bridges attached to the thin filament is proportional to the force of contraction but is only a small percentage (<50%) of the total number of available cross-bridges.

On the basis of both mechanical and biochemical considerations, the flexible tail of the cross-bridge cannot itself account for the development of tension during contraction, inasmuch as the ATPase activity of HMM remains with the globular head after the linear tail is removed by enzymatic treatment. However, the orientation of the linear tail to the thick filament backbone suggests that the tail serves to sustain the tension which is developed by the globular end of the HMM cross-bridge. Obser-

vations of interactions between cross-bridges and thin filaments indicate that globular HMM or S_1 binds tightly to actin at least during part of the contraction cycle. According to the Huxley-Simmons model of muscle contraction (see Harrington, 1979), tilting of the globular HMM head of the cross-bridge is associated with a simultaneous stretching of the spring-like elastic component in S_2; retraction of the elastic component produces the power stroke or movement. This model of muscle contraction is depicted in Figure 1.66.

During the cyclic attachment and detachment of the cross-bridges, ATP is hydrolyzed by a series of biochemical reactions, which have not yet been firmly established in vivo (see Harrington, 1979). Figure 1.67 shows a scheme of reactions which can be formulated based on in vitro studies. In this scheme, "charged" myosin-nucleotide intermediates are indicated by asterisks. The charged intermediates are detected by their ability to fluoresce.

Actin and myosin in the absence of ATP are bound (AM) due to the very high affinity of myosin for actin. This is called a rigor complex. If ATP is then introduced into a system which contains a large number of rigor complexes, ATP binds to the actomyosin and dissociates the complex (step 2 to form M*·ATP (steps 7 and 8). Normally, when ATP is not limiting and Ca^{++} is present, ATP binds

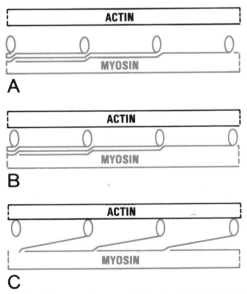

Figure 1.65. Diagram showing relative positions of filaments and cross-bridges at two different interfilament spacings [(A) 25 nm and (B) 20 nm] corresponding, in frog sartorius muscle, to sarcomere lengths of ~2.0 and ~3.1 μm. The X-ray diagram (not shown) suggests that in a relaxed muscle the cross-bridges do not project very far toward the thin (actin) filaments. During contraction or in rigor, the cross-bridges could attach to the actin filaments by bending at two flexible junctions, as shown in (C). [From Huxley, H. E. (1969).]

to AM (step 1) and dissociates the complex to form M*·ATP (step 2). Regardless of the pathway by which M*·ATP is generated, a second myosin-ATP intermediate is formed (step 3) in which the ATP is already hydrolyzed but the energy is stored. The second charged intermediate will now react with actin (step 4), and the stored energy is transformed to movement as the products of ATP hydrolysis are released (steps 5 and 6). It is thus clear that the ATP not only provides the energy for movement but also allows the dissociation of AM in order for the myosin head to attach to another actin monomer to repeat the cycle. Thus, the normal sequence of steps is steps 1, 2, 3, 4, 5, and 6. Each of these reactions can be described in terms of rate constants for the forward and backward direction. This se-

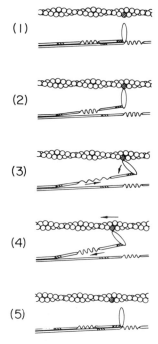

Figure 1.66. Huxley-Simmons model of muscle contraction. (*1*) resting state, (*2*) attachment of S-1 to actin (thin) filament, (*3*) rotation of S-1 while it is attached to the actin filament and simultaneous stretching of the spring-like elastic component in S-2, (*4*) power stroke resulting from retraction of elastic component, (*5*) return of cross-bridge to resting state. [From Harrington (1979).]

quence of reactions may be repeated all along the thick filament where there is overlap of thick and thin filaments. In the absence of actin, ATP is hydrolyzed at step 3 in the sequence $7 \rightarrow 8 \rightarrow 3 \rightarrow 9 \rightarrow 10$. This situation does not change much in the presence of actin and excess ATP, i.e., 80 to 90% of the myosin "heads" execute the cycle in the following direction: $1 \rightarrow 2 \rightarrow 3 \rightarrow 4 \rightarrow 5 \rightarrow 6 \rightarrow 1$. In the absence of actin, the rate-limiting step is 9; in the presence of actin, it is step 4. Actin is termed a cofactor of myosin ATPase, not only because it changes the rate-limiting step, but also because actin increases, by a factor of 100 to 200 fold, the rate of release of ADP and P_i from actomyosin (at steps 5 and 6). Under normal conditions ATP is present in the cell in millimolar concentrations and therefore does not limit this cyclic process, which is instead controlled by Ca^{2+} at step 4.

As described above, in the absence of Ca^{2+}, the inhibitory conformation of regulated actin prevents formation of the AM**·ADP·P_i complex. This has been the traditional view until recently when it became possible to demonstrate that 10 to 20% of the "heads" of myosin hydrolyze ATP at step 11. This introduces the concept that relaxation need not be associated with dissociation of actin and myosin but rather with a particular conformation of myosin bound to actin and with ability of actin to respond to this conformation of myosin (Stein et al., 1979; Murray et al., 1981).

From the known free energy of hydrolysis of ATP, it seems possible that splitting of one molecule of ATP per cross-bridge can provide enough energy for the thick and thin filaments to slide past each other for a distance of 5 to 10 nm during a maximal shortening. This is consistent with a model in which one molecule of ATP is split at each cross-bridge during one tension-generating cycle and in which each cross-bridge can go through its cycle only once during a relative movement of the filaments over a distance of 5 to 10 nm. The extent of sliding during a twitch may be about 1 μm.

In this context, there will be neither tension development nor significant energy release (ATP

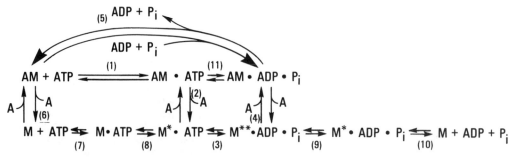

Figure 1.67. Scheme of reactions of hydrolysis of ATP by myosin based on in vitro studies. modified Lymn-Taylor model. [Courtesy of S. Pemrick.]

cleavage) when this "make and break" filament cycle is not completed. Thus, of chemical energy release in the muscle (in the presence of Ca^{++}) can be controlled (1) by the tension developed, which is proportional to the number of cross-bridges that had time to attach themselves to the thin filaments at any given shortening velocity and (2) by the distance shortened, which would be proportional to the number of "make and break" cycles. This kind of system would release energy efficiently and in proportion to the work done.

When all the ATP in the muscle is lost and the muscle goes into rigor, as is seen in *rigor mortis*, it is believed that a large portion of the cross-bridges become attached to thin filaments. Thus, the effect of ATP on actin-myosin interaction is two fold: (1) to provide energy for movement, and (2) to reduce the affinity between actin and myosin to allow cyclic interaction of the two proteins. The latter function of ATP is often called its *plasticizing effect*.

MECHANICAL CHARACTERISTICS OF MUSCLE

The Single Muscle Twitch

The contraction-relaxation of a skeletal muscle in response to a single stimulus is called a *twitch* (Fig. 1.68). When a muscle is activated by a single stimulus while extended by a moderate load, there is a brief lag between the arrival of the stimulus and the initiation of tension development. This lag is referred to as the *latent period,* which lasts approximately 2 to 4 ms. The initial part of the latent period is due to the spread of the action potential along the plasma membrane and T-tubules and transmission of the signal to the sarcoplasmic reticulum to cause calcium release. When the tension

has reached a value exceeding the stretching force exerted by the load, the muscle shortens rapidly. This development of contractile force is transient, and as the contractile force decreases, the muscle lengthens and returns to its relaxed condition.

During the latent period, the muscle may lengthen slightly, a phenomenon called *latency relaxation.* The subcellular events during this period are unknown, although it has been suggested that latency relaxation reflects the disruption of long-lasting cross-bridge bonding between thick and thin filaments just prior to the initiation of the more rapid "make and break" cross-bridge bonding associated with the sliding process. More recently it was proposed that latency relaxation is due to a change in Ca^{++} permeability of the membranes of the terminal cisternae which causes the tension-bearing ability of the sarcoplasmic reticulum to decline; tension is developed as soon as some of the Ca^{++} has diffused to the troponin. Latency relaxation is most marked at longer lengths of muscle and is not at all evident when the muscle length is substantially (10% or more) below the normal body length.

Summation and Tetanus

If a skeletal muscle is stimulated and a second stimulus is applied before relaxation is complete, a second contraction, which develops a greater tension, is fused to the first contraction. This is called *summation.* A possible explanation for the increase in tension may be that Ca^{++}, which would normally have been sequestered by the sarcoplasmic reticulum, remains in the sarcoplasm from the previous contraction and together with additional Ca^{++} released with the second stimulus constitutes more

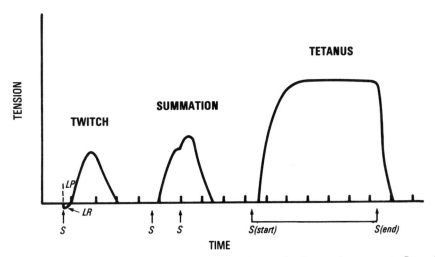

Figure 1.68.   Single muscle twitch, summation, and tetanus. Each division on the time scale represents 5 ms. *S*, stimulus; *LP*, latent period; *LR*, latency relaxation. Latency relaxation may or may not be observed, i.e., it is not routinely recorded under normal experimental conditions. High frequency stimulation is necessary to produce tetanic tension, which although not shown above, may begin to fall despite continued stimulation due to muscle fatigue.

activator Ca^{++} than would be available if relaxation had been complete.

If the stimulus is repeated at a sufficiently high rate, the muscle will not relax between each stimulus but rather will remain in a contracted state which is referred to as *tetanus*. The plateau of such a tetanic contraction exceeds the peak of a single twitch (Fig. 1.68). If the stimulus is of an intensity high enough to activate all of the fibers within the muscle, the maximum contraction of which the muscle is capable is produced. This occurs only rarely physiologically as most routine daily tasks do not require a maximum voluntary contraction of skeletal muscle. Tetanus, whether of a single fiber or an intact muscle, cannot be maintained for prolonged periods due to fatigue, the exact source of which remains unknown.

Changes in the Excitability of Muscle during the Contraction-Relaxation Cycle

The action potential of a skeletal muscle fiber is similar to that of a nerve except that it is about 5 msec. in duration instead of 2 msec. Like the nerve action potential, it has an *absolute refractory period* (also called effective refractory period) and a *relative refractory period* (see Chap. 3). The action potential of a skeletal muscle fiber typically has a long negative afterpotential during which time the fiber is hyperexcitable. This is called the *supernormal period*.

Refractoriness of the plasma membrane plays little role in the physiologic properties of skeletal muscle. This is because the action potential is very brief and the refractory period is over before the mechanical response even begins (Fig. 1.69). Therefore it is possible to tetanize a skeletal muscle fiber. That is, a second stimulus is applied soon after an initial stimulus and its action potential will produce a sec-

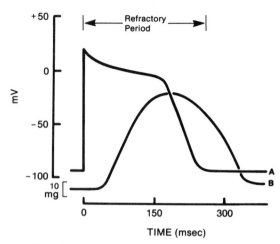

Figure 1.70. Electrical (A) and mechanical (B) responses of a cardiac ventricular muscle cell. Because the refractory period of the action potential extends well into the relaxation phase of the mechanical response, complete fusion or tetanus cannot occur.

ond action potential before the mechanical response to the first stimulus is complete resulting in fusion of twitches.

In contrast to skeletal muscle, cardiac muscle cannot be tetanized. In skeletal muscle tetanus may be useful as during a sustained maximum voluntary contraction (although the muscle soon fatigues). In cardiac muscle, a sustained contraction would result in a cessation of the pumping of blood. The reason that cardiac muscle cannot be tetanized is that the action potential is long relative to the mechanical response (Fig. 1.70). The plasma membrane therefore is in a refractory state until the muscle is well into the relaxation phase so that complete fusion or tetanus cannot occur.

The Motor Unit

In normal skeletal muscles, fibers probably never contract as isolated individuals. Instead several of them contract at almost the same moment, all being supplied by branches of the axon of one spinal motor neuron. In other words, when the motor neuron is excited by a stimulus at or above threshold, all muscle fibers innervated by the neuron will be activated. The single motor neuron and the group of muscle fibers which it innervates are called a *motor unit* (Fig. 1.71); this is the smallest part of the muscle that can be made to contract independently. The number of muscle fibers in this unit varies in different muscles, from 2 to 3 to more than 1,000; the size of the unit is correlated with the precision with which the tension developed by the muscle is graded.

The muscle fibers of a motor unit normally contract sharply upon the arrival of impulses at frequencies ranging up to 50/s. This frequency seems to be the upper physiological limit for axonal prop-

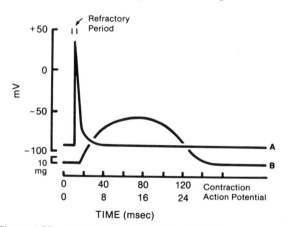

Figure 1.69. Electrical (A) and mechanical (B) responses of a fast twitch skeletal muscle fiber. Because the refractory period of the action potential is over before the mechanical response of the fiber even begins, skeletal muscle can be tetanized by rapid stimulation above a certain critical frequency.

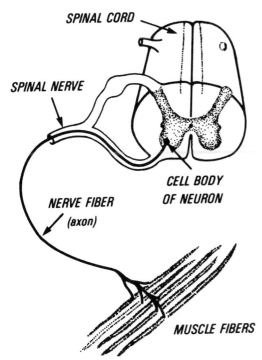

Figure 1.71. Diagram of a motor unit. [From Basmajian (1978).]

agation of motor neurons in most mammals including man. However, motor units may be deliberately fired at much slower frequencies, even to as low as single isolated contractions at will. Normal human beings can learn easily to provide various frequencies of impulses—usually below 16/s—from their spinal motor neurons.

The number of striated muscle fibers that are served by one axon, i.e., the number in a motor unit, varies widely. Generally, muscles controlling fine movements and adjustments (such as those attached to the ossicles of the ear and of the eyeball, larynx, and pharynx) have the smallest number of muscle fibers per motor unit. The muscles that move the eye have less than 10 fibers per unit; the muscles of the middle ear, 10 to 125; the laryngeal muscles 2 to 3; and the pharyngeal muscles have 2 to 6. These are all rather small and delicate muscles and they control fine or delicate movements. On the other hand, large, coarse-acting muscles have motor units with many muscle fibers, e.g., the human gastrocnemius has 2,000 or more.

Even the largest bundles of muscle fibers are quite small, and so a strong contraction of a skeletal muscle requires the participation of many motor units. Further, there is a complete asynchrony of the motor unit contractions imposed by asynchronous volleys of impulses coming down the many axons. Thus, with motor units contracting and relaxing at differing rates, a smooth pull results. (In certain disturbances the contractions become synchronized, resulting in a visible tremor.)

The fibers in a motor unit of the rabbit's sartorius may be scattered and intermingled with fibers of other units. Thus, the individual muscle bundles seen in routine histology do not correspond to individual motor units as such. In rat muscle, the fibers of a motor unit are widely scattered. In man, a similar condition is probable; the spike potentials of each motor unit in the biceps brachii are localized to an approximately circular region of 5 mm diameter in which the fibers of the unit are confined. However, the potentials can be traced in their spread to over 20 mm distance; thus, the area of 5 mm diameter includes many overlapping motor units.

Gradation of Muscle Tension: All or None Response

The gradation of the tension developed by skeletal muscle depends on variation in the number of motor units activated and on the frequency of stimulation. Stimulation at *threshold intensity* (minimum effective stimulation) evokes a minimal twitch tension. There is one response to one stimulus. If the intensity of the stimulus is increased stepwise, the amplitude of the twitch tension will likewise increase. However, with continued increase in the intensity of stimulation, a maximal contractile response is attained, after which a further increase in the intensity of stimulation no longer results in a change in the amplitude of the twitch tension (Fig. 1.72).

The explanation for these responses lies in the fact that with low intensity stimuli, only the most excitable nerve fibers respond and activate their respective motor units, but as the intensity of the stimuli is increased, more motor units are recruited, until finally all motor units are responding, at which time the twitch tension is maximal at a particular frequency of stimulation, and therefore an increase in intensity of the stimuli can no longer elicit an increased response. The individual muscle fibers have long been believed to respond to activation in an *"all or none"* manner, that is, to exhibit contractions of uniform intensity once their particular threshold of excitability had been reached. In seeming conflict with the classical "all or none" concept of muscle fiber contraction are the findings of Costantin and Taylor (1973) which indicate that the

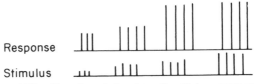

Response

Stimulus

Figure 1.72. Recruitment of motor units with increasing stimulus intensity. The response of the individual unit is "all or none," but the threshold of excitability differs among the units.

contraction of single muscle fibers can be graded by varying the magnitude of membrane depolarization, resulting in activation of variable numbers of myofibrils within the fibers. However, in the in situ muscle it is not possible to vary the magnitude of action potential, which remains all or none thus accounting for the all-or-none property of the contractile response of a skeletal muscle fiber to neural stimulation.

Elastic Elements and Force Generation

Skeletal muscle was originally thought of as a two-component assembly consisting of an elastic element in series with a viscous contractile element, and a second elastic element in parallel with the two-component assembly (Fig. 1.73). The *parallel elastic element* has already been identified as consisting largely of the sarcolemma; the *series elastic element* is now thought to reside in the hinge regions of myosin.

If a muscle is rapidly stretched during the early part of a contraction, it is capable of developing a considerably greater tension than that recorded under normal conditions (i.e., without stretching). This property has been interpreted to reveal the presence in the muscle of an elastic element coupled in series with the contractile elements. The *quick stretch experiments* also show that the contractile machinery is activated already during the latent period and that this activation, which Hill referred to as the "active state" (see below), very rapidly reaches a maximum which is maintained at a constant level for only a short period of time.

The existence of a series elastic element can also be demonstrated by *quickly releasing* one point of fixation of an isometrically contracting muscle and allowing it to shorten before again fixing it at a new (shorter) length. Immediately after release (even if the muscle is shortened by as little as one mm) the tension falls abruptly to zero and then redevelops gradually. This observation suggests that the initial loss of tension is due to elastic recoil of a series elastic element in the sarcomere and that the subsequent slower redevelopment of tension is the result of shortening of the contractile apparatus per se.

Hill (1938) believed that much of the elasticity resided in the tendons and in the connective tissues in series with the contractile elements. Later, however, A. F. Huxley and Simmons (1973) demonstrated that even in an isolated single skeletal muscle fiber, devoid of any tendinous attachment, quick release results in an initial abrupt loss of tension followed by a slow redevelopment of tension. Since the degree of the slow tension development depends on the precise instant of the quick release in the contraction-relaxation cycle, A. F. Huxley (1974) has suggested that the site of series elasticity as well as the site of tension generation are both in the sarcomere. These observations support the idea that the cross-bridges themselves, probably the hinge regions of the myosin, are the sites of tension generation in muscle—indeed, that they are independent force generators—and, therefore, that the degree of overlap between the thick and thin filaments is directly proportional to the level of tension generated in a twitch (see below).

Resting Tension

It has been observed that muscle which is neither stimulated nor stretched beyond the normal body length still maintains some degree of tension, a phenomenon which is revealed by the simple fact that if a tendon is cut in vivo its associated muscle shortens. D. K. Hill (1970) has obtained evidence that resting tension is due to residual long-lasting interactions between thick and thin filaments in the absence of Ca^{++}. This factor, however, cannot explain the sharply rising resting tension at long sarcomere lengths where cross-bridge formation is not possible (see text below and Fig. 1.75). The nature of the increased stiffness of muscle at long sarcomere lengths is unknown.

Isometric Contraction

When the two ends of a muscle are held at fixed points, stimulation causes the development of force without change in muscle length called isometric tension. Isometric tension can be recorded in either a single muscle fiber or a whole muscle, both of which can also be tetanized isometrically. In the whole muscle, if a stimulus of an intensity sufficient to activate all of the fibers within the muscle is given at a high enough (critical) frequency, a sustained maximal isometric contraction called maximal isometric tetanus is developed.

ISOMETRIC LENGTH-TENSION CURVE

Isometric tension depends not only on the frequency of fiber or muscle stimulation but also on the initial length of the fiber or muscle. The length-tension relationship for isometric contraction (Fig. 1.74) shows a maximum tension at a length slightly

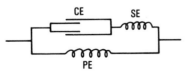

Figure 1.73. Model of skeletal muscle fiber to show contractile and elastic elements. The contractile element (*CE*) is in series with a series elastic element (*SE*). The CE and SE are in parallel with the parallel elastic element (*PE*).

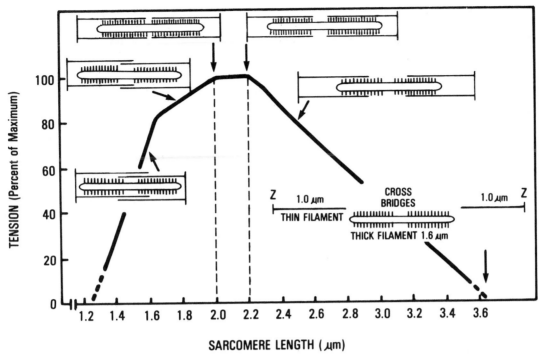

Figure 1.74. Length-tension relation in skeletal muscle fibers. Ordinate, Isometric tetanic tension developed (total tension during tetanus minus resting tension). Abscissa, Striation spacing in micrometers. Diagrams alongside the length-tension curve show critical stages in the degree of overlap of the thick and thin filaments. Optimum overlap is between 2.0 and 2.2 μm. L_{max} is at 3.65 μm. [Adapted from Gordon et al. (1966).]

less than the resting length of the muscle in the body. This amounts to about 3.5 kg/cm^2. Active tension declines as the muscle is stretched beyond normal body length. These length-tension relationships can be explained in terms of the sliding filament theory of muscle contraction. When tension is minimal, the length of the muscle is referred to as *initial length* (L_i). With L_i at considerably less than body length, the thin filaments overlap each other in the center of the sarcomere (Fig. 1.74) to such an extent as to reduce the number of active sites available for interaction with myosin. As the sarcomere is stretched, the "abnoraml" overlap of the thin filaments is progressively reduced, and accordingly, more active sites become available for linkage to the myosin cross-bridges. The length at which tension becomes maximal is defined as *optimal length* (Lo), which is between 2.0 and 2.2 um and is the normal rest length of muscle in the body (normal body length). In this condition, the position of the thick and thin filaments is such as to provide a maximum number of active sites for interaction (Fig. 1.74). As the muscle is further stretched beyond the optimal length the overlap of thick and thin filaments is again progressively reduced, and contractile tension diminishes until a point is reached at which there is no overlap (Fig. 1.74), and, consequently, no active tension can be developed. At this point the muscle is at *maximal length* (L_{max}).

It should be appreciated that Figure 1.74 shows the level of *active isometric tetanic tension* development at sarcomere lengths ranging from 1.2 to 3.6 μm. The change in *resting tension* at various sarcomere lengths is represented in a general sense by the curve shown in Figure 1.75. The *total tension* developed by the muscle is the sum of the resting tension and the active tension as the sarcomere length is increased. The plot of total tension against sarcomere length generates an N-shaped curve rising steeply as the sarcomere length increases beyond 3.6 μm (reflecting the dominance of resting tension when the sarcomere is stretched so that it approaches (L_{max}).

For many years the interaction of the thick and thin filaments has been the basis for an explanation of the length-tension curve. More recent studies, some of which were done with the calcium-sensitive protein aequorin, indicate that the release of calcium from the sarcoplasmic reticulum may also be length-dependent and may be the dominant factor in the ascending limb of the length-tension curve.

Isotonic Contraction

When one end of the muscle is free, stimulation produces shortening while exerting a constant force. This is called an isotonic contraction. The degree of shortening of a muscle bears a direct relationship to

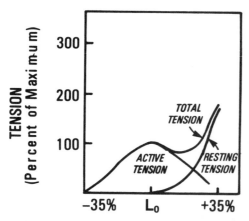

Figure 1.75. Relationship of resting, active, and total tensions in a skeletal muscle. Total tension is the sum of the active tension developed during contraction plus the resting tension in the muscle prior to stimulation. Lo, optimal muscle length [Adapted from Spiro and Sonnenblick, 1964.]

the load upon which it exerts force. A progressive increase in the frequency of stimulation at suprathreshold levels will evoke a partial and then a complete state of tetanus which, under conditions of isotonic contraction, is characterized by sustained maximal shortening. The additional shortening observed during tetanus as compared with the shortening observed during a single twitch indicates that in the latter circumstance the muscle is not fully activated.

FORCE-VELOCITY RELATIONSHIP

If a muscle is contracting isotonically (i.e., shortening against a load), the velocity of contraction is inversely related to the load which the muscle is lifting. If the muscle is unloaded, it shortens with maximum velocity, and as the load increases, the velocity of shortening declines. When the load reaches a maximum (which is equal to the maximum isometric tension the muscle can develop), the muscle no longer shortens. The relationship between load and the velocity of shortening is hyperbolic as shown in Figure 1.76. Hill (1938) adduced the following empirical equation on the basis of data obtained from frog sartorius muscle tested at 0°C:

$$(P + a)(V + b) = \text{constant} \qquad (1)$$

where P is the load applied to the muscle, V is the velocity of shortening, and a and b are constants with dimensions of force and velocity, respectively. This equation states that force is inversely related to velocity. It can be seen from Figure 1.76 that when $P = P_0$, $V = 0$, P_0 being defined as the

maximum isometric tension that the muscle can develop; therefore

$$(P_0 + a)b = \text{constant} \qquad (2)$$

Therefore

$$(P + a)(V + b) = (P_0 + a)b \qquad (3)$$

Eq. 3 can be rearranged to take the following form, which is known as the Hill equation:

$$(P + a)V = (P_0 - P)b. \qquad (4)$$

The constant a is independent of temperature and is similar to α, the coefficient of the heat of shortening, discussed below. Unlike a, b is temperature-dependent, exhibiting a Q_{10} ranging from 2.0 to 2.5. Thus, b appears to be related to chemical processes underlying the mechanism of shortening. The value of b varies from muscle to muscle but is similar for the same type of muscle, i.e., fast or slow twitch skeletal or cardiac or smooth muscle. Force-velocity curves are useful for comparing different muscles with respect to V_{max}, a, and b. The constants a and b may be derived from the linearized form of the Hill equation (see inset to Fig. 1.76) in which the slope is equal to $1/b$ and the y intercept is a/b. V_{max}, which correlates with the rate of myosin ATPase activity, may be derived from *Eq. 3*. From *Eq. 3* it follows that when

$$P = 0 \quad \text{and} \quad V = V_{max},$$

then

$$a \cdot V_{max} = b \cdot P_0$$

and, rearranging,

$$V_{max} = (b/a) \cdot P_0. \qquad (5)$$

Long before the advent of the sliding filament theory the significance of this observation was recognized by Hill (1938), who postulated "that the inverse dependence of energy release upon load could be explained if each active point in the molecular machinery (i.e., those sites which generate tension and effect shortening) could exist in one of the two following states: (1) in combination with each other and thereby developing tension or (2) uncombined or free, and thereby able to take part in chemical transformations. At any instant the distribution of the active points between those two states would be determined by the load which the muscle bears. If the ends of the muscle are fixed, so that no external shortening occurs and the tension developed is at its maximum (isometric contraction), all active points would tend to be in the combined state. Thus, P_0 (maximal tension, P at zero velocity) should be determined by the total *number* of active points. With lighter loads fewer

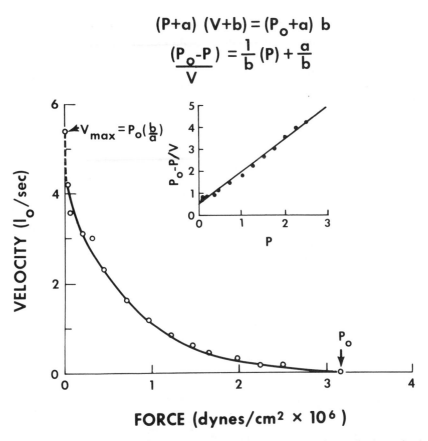

$$(P+a)\ (V+b) = (P_o+a)\ b$$

$$\frac{(P_o-P)}{V} = \frac{1}{b}\ (P) + \frac{a}{b}$$

Figure 1.76. Force-velocity data from a typical soleus muscle. *Open circles* are experimentally determined points. *Solid line* was obtained as regression fitted to data plotted (*inset*) using linearized form of Hill equation. *Dashed line* is extrapolation to value of V_{max} computed from equation for $P = 0$. Velocity is given in fiber lengths (l_0)/s. P, force; V, velocity; a and b, constants. Note that P is force normalized with respect to cross-sectional area—i.e., P is actually a stress. [From Murphy and Beardsly (1974).]

points would be in the "combined" state and more in the "free" state, until both shortening velocity and the rate of energy release are maximal when the muscle shortens freely at zero load. Here, because essentially all active points are in the free state, the maximal shortening velocity (V_{max}) will be determined by the intrinsic rate at which the active points are able to undergo cyclic position changes which effect shortening. As a result, the limiting *rate* at which the active points of the contractile machinery couple and uncouple should determine both V_{max} and the maximal rate of energy release."

If the term "active point" is replaced by the phrase "actin-myosin interaction," Hill's entire analysis can be translated into present day concepts of the contractile process, including the correlation of V_{max} and the rate of myosin ATPase activity (Bárány, 1967).

Heat Production during the Contraction-Relaxation Cycle

Muscle contraction is associated with heat production, a sign of the chemical and physical events associated with contraction. The generation of heat by muscle can be measured by an elaborate system of thermocouples connected to an ultrasensitive galvanometer. The temperature change in muscle can be measured to within about $10^{-5}°C$, and the time course of heat liberation can be resolved within a few milliseconds.

The heat production of unstimulated muscle amounts to 0.0002 cal/g·min and is referred to as "*resting heat.*" The heat produced during a single muscle twitch is about 0.0003 cal/g and appears as (1) an initial very rapid heat production which starts before any shortening has occurred; and (2) heat liberated during shortening. If the muscle performs work, additional heat is liberated in proportion to the work done (Fenn effect).

ACTIVATION HEAT

The initial heat liberation during a twitch, called activation heat, is associated with the development of the active state and is derived from initial processes other than cross-bridge cycling. These initial processes may include release of calcium from the terminal cisternae, binding of calcium to troponin, reorganization of the thick and thin filaments, e.g., movement of cross-bridges away from the backbone of the thick filament, and the active uptake of calcium by the sarcoplasmic reticulum as soon as Ca^{++} is outside the sarcoplasmic reticulum in which the Ca^{++} pump is activated, i.e., immediately after calcium release. The latter process, which requires ATP hydrolysis, accounts for about 50% of the activation heat. Activation heat may be measured at L_{max} where there is no interference from cross-bridge interaction. It has been presumed to be independent of the extent of shortening, of length, of the velocity of shortening, of oxygen supply, and of work done. However, recent studies with aequorin, a protein which emits light in the presence of Ca^{++}, indicate that the amount of calcium released during muscle activation may depend both on the frequency of stimulation and on muscle length (Rüdel and Taylor, 1973, and Taylor et al., 1975). Therefore, activation heat measured at L_{max} may differ from that measured at other muscle lengths, a finding which must be considered in future studies.

SHORTENING HEAT

Heat liberated during shortening other than the activation heat is called shortening heat. It may be estimated by obtaining the difference in heat produced during shortening and during a comparable period of isometric contraction, measured at either the same muscle length and with the same degree of stimulation or at the same level of active tension development as is obtained during shortening. The origin of the shortening heat is not established. While shortening heat depends on cross-bridge cycling, some of it appears to be attributable to cross-bridge reactions which do not split ATP.

The shortening heat, H_s, is proportional to the degree of shortening:

$$H_s = \alpha x$$

where α is the coefficient of the heat of shortening and x is the distance representing the decrease in length of the shortened muscle.

MAINTENANCE HEAT

Heat produced during an isometric tetanus is called maintenance heat. It consists of a labile heat with a duration of a few seconds and a stable heat, persisting during the duration of the tetanus. The labile and stable heats consist largely of activation heat and actin-myosin interactions, respectively, although the exact source of these heats is unknown.

RECOVERY HEAT

Further heat is liberated during recovery. This "recovery heat" is associated with oxidative chemical synthetic processes. The recovery heat is about equal to the total of the initial energy consisting of activation heat, shortening heat, and work. Since about half this initial energy is derived from chemical reactions that are reversed during recovery, it appears that about half the energy associated with the recovery reactions is wasted heat.

THERMOELASTIC HEAT

Resting muscle has rubber-like elasticity, that is, it shortens when warmed and lengthens when cooled. In activated muscle, the elasticity is no longer rubber-like but spring-like. The change in elastic properties as the muscle contracts is associated with the absorption of heat. An equivalent amount of heat, called thermoelastic heat, is released during muscle relaxation. Thus no net loss or gain of heat occurs as a result of changes in muscle elasticity during a full contraction-relaxation cycle. These changes in thermoelastic heat must nevertheless be corrected for in other heat measurements.

The Active State

The active state of muscle which develops following an adequate stimulus commences with the onset of latency relaxation. The maximal intensity of the active state (I_{as}) was originally defined as the maximal isometric tension which the contractile elements of muscle can develop independently of the effects of the elastic elements. Only a part of this maximal tension is actually manifest in a twitch because of the damping effect of series elastic elements; i.e., the "springlike" quality of the series elastic elements initially absorbs the energy developed during the onset of the active state in the sarcomere.

I_{as} can be determined at various times after stimulation by measuring the maximum load that a muscle can sustain after a small quick stretch. Shortly after a muscle starts to contract (about 20 ms after stimulation for frog muscle), a quick stretch demonstrates that, in fact, the load-bearing capacity (i.e., the capacity to develop tension) of the contractile element has reached its peak. Thereafter I_{as} declines exponentially.

Observations obtained from "quick stretch" experiments provided the basis for the formulation of the concept of the "active state" by Hill (1927). Because maximal tension (in a quick stretch experiment) is always observed almost immediately after stimulation, Hill postulated that in the simple twitch response all the contractile elements undergo a rapid transition from a fully resting state to a fully activated state which persists for a variable period of time, and then declines before peak contractile tension is reached.

The quick stretch methods for mapping the time course of the active state interrupt the instantaneous changes and provide only discontinuous "steady-state" glimpses of the total time course. Therefore, a new technique was developed which does not require any manipulation of the muscle and which provides a continuous map of the active state contour. In this technique, tension is transduced into an electrical signal which, in turn, is differentiated so that the change in muscle tension as a function of time, dP/dt, can be recorded for a muscle twitching isometrically (Fig. 1.77). Inasmuch as the stretching of the elastic elements (which develops the isometric tension) is a reflection of the state of activity of the contractile elements, the dP/dt curve in effect indicates the instantaneous changes in the contour of the active state. Observations recorded by this "continuous" technique have confirmed (1) that the maximum intensity of the active state is not developed instantaneously; (2) that after attaining maximum intensity, the active state declines immediately; (3) that the intensity of the active state depends on the initial length of the muscle prior to stimulation; and (4) that various conditions which increase contractile tension affect either the peak or the duration of the active state.

Findings based on measurements of the active state of muscle can be readily understood in terms of the availability of Ca^{++} to troponin. Thus we know now that (1) intracellular Ca^{++} concentrations are highest when maximum I_{as} is attained; (2) Ca^{++} is rapidly sequestered by the Ca^{++}-activated calcium pump of the sarcoplasmic reticulum, resulting in a decline in the active state; (3) Ca^{++}

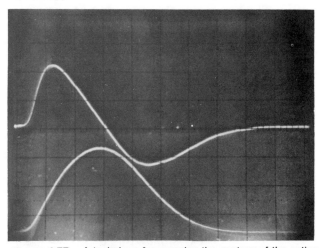

Figure 1.77. A technique for mapping the contour of the active state. The monophasic curve is a recording of isometric tension in rat tibialis anterior muscle measured at 22°C. The biphasic curve is a recording of dP/dt, the rate of change in the development and release of tension. 20 ms per division. [Courtesy of Z. Penefsky.]

release from the sarcoplasmic reticulum is dependent on the initial length of the muscle; and (4) factors which affect the amount and duration of Ca^{++} availability affect the intensity and duration of the active state.

Muscle Relaxation

For many years, a physiologic factor which reverses the contractile process was sought. In the early 1950s, Marsh observed that a shrunken actomyosin gel could be made to take up water by the addition of a supernatant fluid which had been recovered from the centrifugation of ATP-treated muscle homogenates. This muscle extract could relax contracted glycerinated muscle fibers. The unknown component responsible for these effects was called "relaxing factor." It was later found to sediment in a moderate gravitational field, and it therefore appeared to be particulate in nature.

Many investigators attempted to find a small molecule which, on release from the particulate, "relaxed" the ATP-contracted actomyosin system. However, several years elapsed before it became apparent that the relaxing effect could be due to the removal of an element from the contracted actomyosin system, and therefore that the relaxing factor could be the system involved in the removal and sequestration of this element.

In the mid-1950s it was observed that the addition of ethylenediaminetetraacetic acid (a Ca^{++} chelator) in the presence of ATP mimicked the physiological action of the relaxing factor; but it was not until 1959 that Weber demonstrated that the ATP-actomyosin system depended on minute amounts of Ca^{++} for ATPase activity.

The effect of the relaxing factor was then conceived to be a chelating system which was responsible for the removal of an agent, Ca^{++}, which regulated the actin-myosin interaction. The relaxing factor was identified on further study as the fragmented sarcoplasmic reticulum recovered from a muscle homogenate. It was shown that these vesicular fragments could indeed concentrate Ca^{++} against very high gradients.

Much of the protein of sarcoplasmic reticular membranes consists of the *calcium pump protein*, a 100,000-dalton protein that has Ca^{++}-activated ATPase activity. The calcium pump protein transports Ca^{++} present in the sarcoplasm during contraction into the sarcoplasmic reticular tubules causing muscle relaxation. For every two molecules of Ca^{++} transported, one molecule of ATP is hydrolyzed. The rapid uptake of calcium by the sarcoplasmic reticulum lowers the sarcoplasmic Ca^{++} from approximately 10–20 μM during the height of contraction to 0.1 μM or less, allowing the muscle to relax. The calcium is sequestered within the tubules where the total calcium concentration is in the millimolar range. The calcium pump is able to pump Ca^{++} against gradients of several thousand fold (micromolar concentrations of Ca^{++} in the sarcoplasm and millimolar concentrations within the tubules). The sequestered calcium is presumed to move to the terminal cisternae, which are known to contain a calcium-binding protein called *calsequestrin*, which can bind large quantities of calcium. This calcium becomes available for release from the sarcoplasmic reticulum for the next contraction.

Events leading to muscle relaxation in the intact muscle may be summarized as follows. The sarcoplasmic reticular calcium pump is activated by the Ca^{++} present in the sarcoplasm during contraction. Since the free Ca^{++} is in equilibrium with the calcium bound to TN-C, the latter is dissociated from the TN-C-Ca complex as Ca^{++} ions and removed from the sarcoplasm. Actin-myosin interaction is now inhibited, and the contractile elements return to their resting state. At this point all of the Ca^{++} that had been released from the sarcoplasmic reticulum is presumed to be resequestered. Recent studies suggest that a protein called *parvalbumin* may serve as an intermediate calcium-binding protein in the movement of Ca^{++} from the myofibrils to the sarcoplasmic reticulum and thus facilitates muscle relaxation.

Staircase

In 1871 Bowditch, working with the frog heart, observed that if the heart had been inactive for a period of time a sudden series of stimulations induced a corresponding series of contractions that increased in amplitude until a steady-state was reached. This phenomenon was called "staircase" or "Treppe."

Contrariwise, if the rate of stimulation of the heart was suddenly decreased, a series of consecutively declining contractions was obtained, until a second steady-state was reached (negative staircase or negative Treppe). The same phenomena are observed in skeletal muscle, but because of the shorter refractory period in this tissue the rate of stimulation must be much higher to elicit staircase effects.

Staircase phenomena are believed to be the result of changes in the intracellular distribution of Ca^{++} with changes in frequency. A positive staircase may be the result of more Ca^{++} being available to the contractile proteins. A negative staircase could be the result of a lag in calcium movement within the sarcoplasmic reticulum, and therefore less calcium would be available for subsequent contractions.

The physiological significance of staircase phenomena, demonstrated in isolated muscle, is not known. It is conceivable that a positive staircase phenomenon may account in part for the improved performance of athletes after "warm-up" exercises.

Posttetanic Potentiation

For a brief period of time after cessation of tetanic stimulation of a fast muscle (such as the extensor digitorum longus) a single stimulus induces a twitch tension of higher amplitude than that elicited in a muscle not subjected to previous tetanization. This "posttetanic potentiation" (PTP) declines exponentially and is no longer manifest after the fifth and sixth posttetanic twitch.

The mechanism of PTP is not known; the phenomenon, however, may simply reflect a transient increase in the level of calcium in the cytosol during the tetanization.

Contracture

Agents which lead to a sustained elevation of cytosolic calcium (by promoting release of calcium or by inhibiting reaccumulation of calcium by the SR) can induce a state of prolonged contraction *in the absence of action potentials*. This phenomenon is referred to as contracture. If muscle cell membranes are depolarized by high concentrations of K^+, a state of contracture will ensue because the depolarization triggers calcium release, increases calcium conductance, and inhibits calcium sequestration.

Clinically, a contracture analogous to pharmacologic contracture is produced in McArdle's phosphorylase deficiency, a condition characterized by an inability of muscle to relax after strenuous exercise. The reason for this appears to be that the muscle cannot utilize glycogen and produce ATP, which is necessary for calcium uptake by the sarcoplasmic reticulum and in turn for dissociation of actin and myosin and thus for relaxation.

ACTIVATION OF MUSCLE

Neuromuscular Transmission

The activity of muscle is controlled by the central nervous system through the motor innervation of the myofibers. Each motor nerve fiber splits up into a number of branches that make contact with the surface of the individual muscle fibers via several bulb-shaped endings. These endings are arranged in a group, and with a specialized structure of the surface of the muscle fiber, they form an entity which is referred to as a *neuromuscular junction, myoneural junction,* or *motor end-plate* (Fig. 1.78).

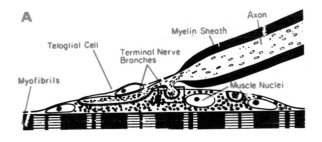

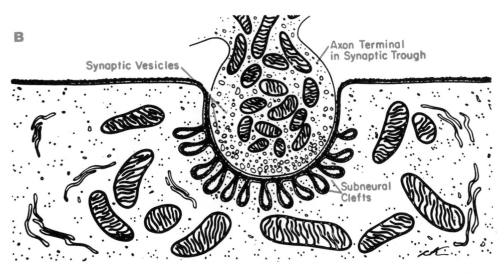

Figure 1.78. Motor end-plate for neuromuscular transmission in striated muscle. *A*, Longitudinal section. *B*, Electron microscopic appearance of the rectangular area indicated in *A* showing the ultrastructure of the junction between an axon terminal and the muscle fiber membrane. The invagination of the muscle fiber plasma membrane (sarcolemma) forms the synaptic trough into which the axon terminal protrudes. The space between the plasma membrane of the axon terminal and the invaginated sarcolemma is referred to as the neuromuscular cleft or synaptic gap. The invaginated sarcolemma (postjunctional or postsynaptic membrane) has many folds (subneural clefts) which greatly increase its surface area. Acetylcholine is stored in the synaptic vesicles located in the axon terminal. Acetylcholine receptor protein and acetylcholinesterase are both associated with the postjunctional membrane [From Bloom and Fawcett (1975).]

Development of End-Plate Potentials and Activation of Skeletal Muscle Fibers

The action potential conducted along the nerve fiber enhances the endocytic release into the neuromuscular "cleft" of acetylcholine (ACh) from packets (vesicles) located in the nerve endings. A small number of vesicles release their contents intermittently from unstimulated nerve endings, thereby accounting for the quantized subthreshold *miniature end-plate potentials* (MEPP) described below and in Chapter 3. When an action potential reaches the nerve terminals in the region of a motor end-plate, there is an enhanced permeability to Ca^{++} ions which increases the exocytic release of ACh from several hundred vesicles at the presynaptic (prejunctional) membrane—so that the number of ACh molecules that diffuses across the junctional gap to react with specific ACh receptor pro-

tein in the postjunctional membrane equals or exceeds the threshold amount needed for induction of an action potential in the muscle fiber. Excess ACh is rapidly inactivated via hydrolysis by acetylcholinesterase, which is present on the surface of the postjunctional membrane.

The receptor density is many times greater at the neuromuscular junction than at any other site on the sarcolemmal membrane. Thus the structural and chemical specificity of the ACh receptor explains why ACh is a very much more effective agonist when applied to the myoneural junction than when applied elsewhere on the sarcolemma. It is of interest that the innervation of skeletal muscle plays an important role in the differential sensitivity of various regions of the sarcolemma to ACh. For example, following section of a motor nerve the total number of receptors is increased more than 10-fold, and the ACh receptor material, normally concentrated at the end-plate, becomes dispersed along the sarcolemma and therefore the entire surface of the sarcolemma, including the neuromuscular junctional area, becomes equally sensitive to ACh. After the motor nerve has regenerated, however, and functional innervation of the muscle is reestablished, the accumulation of receptor decreases throughout the sarcolemma and the sensitivity of the nonjunctional portion of the sarcolemma diminishes.

There is evidence that the ACh receptor molecule is composed of five subunits with a molecular weight in the range of 40,000 to 70,000 and that one or two molecules of ACh interact with one molecule of receptor. At rest the electrical potential difference across the muscle cell membrane (i.e., the resting potential) amounts to approximately -90 mv (inside negative). Because of the difference in concentration of Na^+ ions (high outside the muscle cell) and K^+ ions (high inside the muscle cell) and because the formation of an ACh-receptor complex results in an increase in the ionic permeability of the postjunctional membrane, the release of threshold amounts of ACh into the neuromuscular junctional cleft gives rise to a sudden influx of Na^+ ions and an efflux of K^+ ions through the plasma membrane. This short-circuits the adjacent parts of the plasma membrane and causes a drop in the membrane potential to the level where the membrane becomes electrically excited (i.e., the membrane potential rises to the critical firing level, approximately -50 mv), and a wave of depolarization spreads along the muscle fiber (in the same manner as the nerve impulse is propagated along an axon). The electrical response developed as a consequence of the acetylcholine effect on membrane permeability can be recorded as an action potential of approximately 130 mv with the inside of the membrane briefly becoming positive. The resting potential of about -90 mv is, therefore, not only abolished but, as in the nerve cell, the distribution of positive and negative charges is reversed. The conduction of one impulse involves only a minute amount of Na^+ ions (about 4×10^{-12} moles/cm^2) entering the muscle fiber and an equivalent leakage of K^+ ions to the extracellular space. Through a recovery mechanism, involving active transport processes, the ion concentration differences between the muscle cell interior and the extracellular space is reestablished by Na^+ extrusion and K^+ uptake.

The activity of acetylcholinesterase rapidly abolishes the effect of acetylcholine. The electrical response of the plasma membrane of the muscle fiber is therefore efficiently controlled by the nerve im-

pulses reaching the motor end-plate. Even without any such impulses, there are, as noted above, minute spike potentials developed at the motor end-plate region of the muscle plasma membrane (MEPP), which are due to a spontaneous release of ACh-containing vesicles in the nerve terminals of the neuromuscular junction. The concentration of ACh that results from this "leakage" is generally not high enough to produce a sufficient permeability change in the plasma membrane of the muscle fiber to raise the membrane potential to -50 mv, the threshold for the development of a propagated electric response. The minute spike potentials therefore remain localized to the end-plate region.

Actions of Drugs on the Neuromuscular Junction

DRUGS THAT MIMIC ACH

Drugs, such as carbamylcholine and nicotine mimic the effect of ACh at the motor end-plate. However, because these drugs are either not inactivated or are very slowly inactivated by cholinesterase, their action is much more prolonged than that of ACh. In moderate dosage these drugs will cause a muscle spasm: in high dosage, however, they induce a state of partial or complete paralysis (depolarizing block) similar to the "desensitization" phenomenon observed when the neuromuscular junction is exposed to ACh for an extended period. This type of diminution or ablation of response has been observed in other receptor systems and is believed to be due to a gradual conversion of the transmitter-receptor complex from an active to an inactive form. The inactive complex then dissociates to generate an inactive or "desensitized" free receptor which only slowly reverts to the normal, active receptor.

DRUGS THAT BLOCK TRANSMISSION

Drugs such as *d*-tubocurarine and its congeners bind to the acetylcholine receptor with high affinity, but the curare-receptor complex does not alter the ionic permeability of the postjunctional membrane but rather prevents ACh from binding. Thus curare or *d*-tubocurarine diminishes or blocks neuromuscular transmission by acting as a competitive antagonist of ACh at the motor end-plate, an effect which can cause a fatal paralysis of the muscles needed for respiration.

DRUGS THAT INACTIVATE CHOLINESTERASE AND THEREFORE STIMULATE

Yet certain other drugs—e.g., physostigmine, neostigmine, and diisopropyl fluorophosphate

(DFP)—inactivate acetylcholinesterase and thus permit ACh to accumulate within the neuromuscular cleft and in turn to stimulate neuromuscular transmission excessively. These drugs initially induce a persistent contraction (spasm) but also can cause depolarizing block at higher doses. DFP is particularly dangerous in this respect because it inactivates cholinesterase virtually irreversibly (i.e., for several weeks), whereas physostigmine and neostigmine remain bound to cholinesterase only for a few hours—after which time they dissociate from the enzyme which then becomes fully capable of hydrolyzing ACh. These latter drugs are used for the treatment of myasthenia gravis, a disease in which neuromuscular transmission is defective because of impairment of receptor function by anti-receptor antibodies which react with the postjunctional membrane. The rationale of this treatment, therefore, is simply to inhibit the enzymatic action of ACh esterase so that sufficient amounts of ACh can accumulate to restore a normal level of neuromuscular transmission despite the inadequate numbers of functional receptors at myasthenic neuromuscular junctions.

Propagation of Excitation across and within a Muscle Fiber

The impulse or action potential conducted along the muscle fiber from the end-plate region accounts for a spreading of a state of excitation *along* the fiber. The activation of the contractile elements, which are distributed over the whole cross-section of the fiber, depends on spreading of the excitatory state from the plasma membrane transversally *across* the muscle fiber via the T-tubules.

The currents flowing during the action potential are not directly responsible for activating the contractile mechanism. It is not possible for any substance liberated at the surface of the muscle fiber in connection with the electric impulse propagation along the plasma membrane to activate the contractile elements by diffusion of the substance through the fiber, because such a mechanism would be too slow. The activation is, furthermore, highly temperature-dependent, which also militates against a mechanism involving simple diffusion of an activating substance.

The activation of the contractile machinery ultimately depends on the effect of an activating factor, namely Ca^{++}, capable of generating the active state. Ca^{++} is the final mediator of the process linking excitation of the plasma membrane to the initiation of contraction (excitation-contraction coupling), the action of Ca^{++} resulting from its binding to troponin. An impulse transmitted by the T system to the SR brings about the release of Ca^{++} from the SR. The released Ca^{++} then initiates activation of the contractile machinery (see below).

Electrical Stimulation Via the Sarcolemma

Although the normal path of the electrical activity is via the neuromuscular junction, it is possible to stimulate the sarcolemma directly. If the intensity of the stimulus is subthreshold, a brief local depolarization can be induced, which will reverse upon the removal of the stimulus. However, if the stimulus depolarizes the membrane to the critical firing potential (threshold), an action potential will be observed. Whereas the subthreshold response is decremental and does not propagate far from the site of stimulation, the action potential is self-regenerative and will be conducted in both directions away from the site of stimulation.

Action potentials recorded from fast skeletal muscles depolarize the membrane rapidly, overshoot, and then repolarize rapidly to a level of approximately −90 mv. At this level of membrane potential the repolarization process develops a relatively long time constant. The membrane voltage during this latter period of slower repolarization is referred to as a *negative afterpotential,* which gradually returns towards the normal resting potential, also similar to the response of nerve fibers. The contribution of various sites on the sarcolemma (surface membrane, transverse tubules) to the ionic fluxes which determine the shape of the action potential has been investigated extensively, and it has been shown that depolarization of the muscle cell membrane is largely due to Na^+ ion influx in most species. Although the transverse tubules appear to be an extension of the sarcolemma, the membranes may have different conductance properties, as suggested by the analysis of the action potential recorded after the T-tubules have been destroyed with hypertonic Ringer's glycerol solution and the analysis of membrane currents after voltage clamping.

The usual rapid repolarization that follows a spike generated by the muscle membrane is due to a delayed rectification of the membrane. Inactivation of the conductance of Na^+ ions and increase in the conductance of K^+ ions are responsible for the quick phase of repolarization, following which there is an early and late negative afterpotential, with the membrane remaining slightly depolarized for about 0.25 s.

The origin of the fast phase of repolarization

(delayed rectification) seems to be on the surface membrane, whereas the afterpotential seems to originate mostly in the transverse tubules. It has been shown that if the transverse tubules are destroyed by exposure to hypertonic solutions, muscle will no longer contract in response to stimulation (excitation-contraction uncoupling); in addition, the total membrane capacitance is reduced from 6.5 to 2.6 $\mu F/cm$ and the afterpotentials will disappear, although the earlier features of the action potential (spike and the rapid phase of repolarization) persist.

The afterpotentials may be due to K^+ ion accumulation in the tubules. Using rubidium in place of potassium, it has been observed that the afterpotentials can be suppressed.

The contribution of Cl^- ions to the ionic currents of the muscle cell membrane has long been controversial. It is now known that chloride conductance (G_{Cl}) is pH-sensitive. At values of pH of about 5.6, G_{Cl} is inhibited. Under such conditions it has been shown that G_K in the tubules changes in direction during the afterpotential; this anomalous rectification seems to be a property of the surface membrane; whereas the slow rectification (which accounts for the return from maximal hyperpolarization to the resting potential at the end of the action potential cycle) seems to take place within the tubules (which occupy a total of 0.002 to 0.003% of the fiber volume).

Many of the observations noted above have been verified by the observations of Adrian et al. (1970) to the effect that the potassium current can be divided into three separate components: (1) A current in a delayed rectifier channel of the muscle cell membrane which reaches a maximum in about 100 ms at a membrane potential of -30 mv and then declines with a time constant of about 4 ms until the fiber is repolarized to -100 mv. This current inactivates completely. (2) A channel in which repolarization is associated with an increase in G_K, which is one or two orders of magnitude slower than the K current in the delayed rectifier channel. (3) A very slow inward K current which has a time constant of 0.25 s at -150 mv. In addition, a path for active uptake of K^+ ion is also postulated. The data on current flow obtained with the voltage clamping technique also suggest that (1) the delayed rectifier channel must be in the surface membrane, whereas the two other channels must be in the T tubules; (2) that G_K rather than tubular K concentration changes during the positive and negative afterpotentials; and (3) that the afterpotentials can be explained if one supposes that potassium ions do not carry all the current or that some variable other than membrane potential is the factor which controls potassium permeability.

Recapitulation of the Role of Mono- and Divalent Ions

1. K^+ ion, the major component of the intracellular fluid, functions chiefly to maintain the resting potential, which approximates the K^+ equilibrium potential. If the concentration of K^+ in the extracellular fluid is increased, resting membrane potential will decline according to the Nernst equation $E = RT/F \ln [K_I]/[K_E]$, where K_I is the intracellular K^+ concentration and K_E is the extracellular K^+ concentration.

2. Na^+ ion, the major component of the extracellular fluid, functions to maintain the osmotic pressure (osmolality) of the extracellular compartment. The permeability of the muscle cell membrane to Na^+ is very low at rest but increases markedly following stimulation. Thus, Na^+ ions "rush" into the cell, and these positive charges depolarize the membrane. Because Na^+ permeability is voltage-dependent, more Na^+ ions "rush" into the cell as depolarization progresses in what amounts to an autocatalytic, positive feedback process which is propagated along the muscle membrane. This self-regenerative process is qualitatively identical to the action potential of nerve. For a brief instant the inside is positive with respect to the outside. At this point (the peak of the overshoot) the membrane potential approaches the Na^+ equilibrium potential. However, the increased Na^+ permeability quickly reverses, permeability to K^+ increases, and the repolarization process is initiated.

The monovalent ion, Li^+, can replace Na^+ as a carrier of positive charge into the muscle fiber; however, Li^+ ions tend to accumulate inside the fiber because there is no active transport mechanism to remove Li^+ from the intracellular space.

3. The role of Ca^{++} ions in muscle physiology will be discussed below in connection with the relaxation of muscle and under the heading of "Excitation-Contraction Coupling."

4. Mg^{++} ions are important in many enzyme reactions, including those of glucose metabolism and myosin ATPase activity. The cytoplasmic concentration of Mg^{++} has been estimated to be in the millimolar range.

5. Cl^-, the major anion in the extracellular fluid, passively follows the cations moving across the muscle cell membrane distributing according to the Gibbs-Donnan equilibrium.

Regulatory Role of Calcium

Ca^{++} is needed to release the inhibition of ATP hydrolysis by myosin in the presence of regulated actin. In the absence of Ca^{++} (intracellular concentration less than 0.1 μM), tropomyosin sterically blocks the attachment of the myosin head to the actin according to one hypothesis (Fig. 1.79). This steric hindrance by the elongate tropomyosin molecule occurs along the entire length of the F-actin. In addition, at each seven actin monomers of F-actin, the TN-I is bound to both actin and tropomyosin in the absence of Ca^{++}. In the presence of Ca^{++}, when Ca^{++} binds to the TN-C, there is a conformational change in the troponin complex whereby the TN-I dissociates from the actin and the tropomyosin. This allows the tropomyosin to move closer to the groove of the two-stranded actin helix. The tropomyosin now no longer blocks the interaction of myosin with actin, and muscle contraction occurs.

Excitation-Contraction Coupling

From the foregoing it is clear that the stimulation of the muscle, either through its nerve or directly, sets up a process which starts on the sarcolemma and ends in the contractile machinery. It should be recalled that minute electrical currents at the level of the Z band in frog skeletal muscle induce local contractions and also that, in the frog, the transverse tubular system is located at the level of the Z lines—whereas, in crab muscle, local stimulation is effective only at the A-I junction area and the transverse tubules are also found in this area. It should also be recalled that when the T tubules are destroyed by exposure of muscle fibers to glycerin, electrical stimuli no longer elicit a contractile response, even though the action potentials persist. Thus, it seems clear that the depolarization process initiated at the myoneural junction in the postjunctional membrane must then travel to the core of the myofibril through the T-tubule system.

Electron micrographs have shown that the closed end of the T tubules are in close proximity (5 nm) to the terminal cisternae. At the site of junction of the T tubules and the terminal cisternae (triad), densities ("feet") have been observed which suggest connections between the elements of the cisternae and the T tubules. By the freeze-etching method, Franzini-Armstrong (1973) has demonstrated channels between the T tubules and the lateral cisternae. A new line of evidence by Chandler and Schneider (1976) suggests that during rest these channels are closed, but when an action potential reaches the T tubules, a charge transfer occurs which in some manner leads to a conformational change in the membrane of the cisternae and thereby to an opening of the channels. Ionic currents, thus generated, are presumed to induce the release of Ca^{++} ions from the cisternae into the sarcoplasm.

Release of the Ca^{++} into the sarcoplasm has been demonstrated by injecting aequorin into muscle. This substance (which can be extracted from jelly fish) luminesces in the presence of Ca^{++} ions with an intensity that is proportional to the Ca^{++} ion concentration. After stimulation, luminescence increases, reaches a maximum during the course of tension development, and decreases to a minimum by the time peak contractile tension is developed.

As noted above, calcium ions that are released into the sarcoplasm bind to troponin C, and the

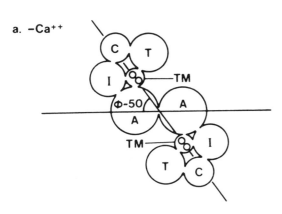

a. $-Ca^{++}$

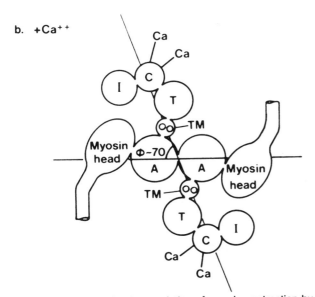

b. $+Ca^{++}$

Figure 1.79. Model for the regulation of muscle contraction by Ca^{++}. Abbreviations are as follows: *A*, actin; *TM*, tropomyosin; *I*, troponin I; *C*, troponin C; *T*, troponin T. [Reprinted from McCubbin and Kay (1980). Copyright 1980 American Chemical Society]; adapted from Potter and Gergely (1974).]

TNC-Ca complex serves to release the inhibitory effect of troponin I on the actomyosin interaction. Then by alternate binding and unbinding of the cross-bridges to successive sites on the actin chain, the thin filaments slide towards the center of the sarcomere, thus shortening the muscle.

The entire process can be abbreviated as follows: Stimulus → Depolarization of the sarcolemma → Action potential initiated and propagated along the T tubules → Calcium released from the SR system → Calcium ions diffuse and attach to the active site on troponin C → Inhibitory effect of troponin I on the interaction of actin and myosin is removed → thin filaments slide along thick filaments shortening the sarcomere. This series of events is fully reversible under normal conditions. Following repolarization there is an increase in the uptake of calcium ions by the sarcoplasmic reticulum. As this sequestration proceeds, more Ca^{++} ions dissociate from their binding sites on troponin C resulting in a restoration of actomyosin-inhibiting property of troponin I and, accordingly, in a return of the thick and thin filaments to the resting state (relaxation). Indeed, there is evidence that the rate of muscle relaxation is proportional to the removal of calcium from the contractile protein.

ENERGY-GENERATING SYSTEMS

The energy for muscle contraction is derived from chemical reactions in the muscle fiber. The main energy source is glycogen. This energy is made available without any oxygen being consumed, even when oxygen is present. The oxidative (aerobic) chemical reactions are therefore not directly associated with muscle contraction but with the recovery processes which operate to provide energy in a form that is readily available to the contractile machinery, thereby securing a quick response to a stimulus.

Adenosine Phosphates

The compound that appears to be ultimately involved in providing energy to the contractile machinery is adenosine triphosphate. In this process, energy is utilized in a reaction coupled with the hydrolytic splitting of the terminal phosphate group from ATP:

$$ATP \rightarrow ADP + P_i$$

This reaction involves a liberation of about 7,300 cal/mole of ATP. The bond between the terminal phosphate group and the neighboring phosphate group is often referred to as a "high energy" or "energy-rich" bond designated by the symbol ~. It must be realized when using these terms that the energy is contained in the molecule as a whole in the case of so-called "high energy" compounds, such as ATP. Much of the free energy is released when these compounds are hydrolyzed. The structure of adenosine triphosphate can be written as follows:

$$A-O-\overset{\overset{\displaystyle O}{\|}}{\underset{\underset{\displaystyle OH}{|}}{P}}-O-\overset{\overset{\displaystyle O}{\|}}{\underset{\underset{\displaystyle OH}{|}}{P}}-O-\overset{\overset{\displaystyle O}{\|}}{\underset{\underset{\displaystyle OH}{|}}{P}}-OH$$

The ATPase activity of myosin permits the hydrolysis of ATP to take place at the contractile machinery itself where energy is transferred from ATP to the myosin-actin system to allow muscle contraction.

The restoration of ATP involves phosphorylation of ADP. Prompt resynthesis of ATP is achieved by phosphate group transfer from creatine phosphate to ADP, catalyzed by creatine kinase (Fig. 1.80). Creatine phosphate, depleted in this process, is ultimately regenerated by the transfer of phosphate from ATP to creatine. In this way a second high-energy compound, creatine phosphate, is built up in the muscle during recovery. The ATP content of resting muscle is about 5×10^{-6} moles/g of muscle tissue, which is four to six times less than the content of creatine phosphate. The ATP and the creatine phosphate represent energy stored in a form that can be utilized rapidly. Recharging of ATP through oxidative phosphorylation is a much slower process. Since the high energy phosphate storage in the muscle is sufficient to allow the muscle to contract only 50–100 times, the energy reservoirs are rapidly depleted. For maintenance of muscle contractility, a steady supply of energy is required to recharge ADP.

Since the reactions associated with the contraction, dephosphorylation of ATP, as well as the phosphate transfer from creatine phosphate to ADP, are nonoxidative processes, it becomes ob-

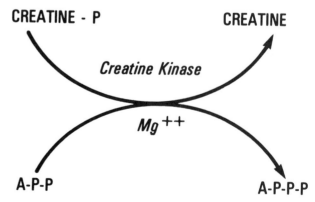

Figure 1.80. Resynthesis of ATP with phosphate from phosphocreatine.

vious that no oxygen is consumed during contraction provided that no recovery metabolism is maintained in parallel. Oxygen is consumed during the recovery phase in connection with oxidative phosphorylation of ADP.

The energy required for the formation of the high energy ATP and creatine phosphate is supplied by glycolysis and by the oxidation of acetyl CoA (derived from pyruvic acid) in the citric acid cycle with the associated oxidative phosphorylation by the respiratory chain. Glucose metabolism via the pentose shunt pathway is negligible in skeletal muscle. In addition to glycogen, fatty acids and amino acids can furnish fuel for the muscle, and in resting muscle, glycogen breakdown is not required to maintain metabolism. Muscle metabolism can be maintained by the uptake of carbohydrates and other compounds from the blood. In resting muscle, carbohydrates are responsible for about 60% of the energy requirements of the tissue.

Glycogenesis

The synthesis of glycogen from glucose is called glycogenesis. Muscle as well as liver cells are able to store large amounts of glucose in the form of glycogen. Many other cells store lesser amounts. The first step in glycogenesis involves the phosphorylation of glucose to glucose 6-phosphate, a reaction catalyzed by *hexokinase* and associated with utilization of energy-rich phosphate in the form of ATP.

$$\text{Glucose} + \text{ATP} \xrightarrow{\text{Mg}^{++}} \text{glucose 6-phosphate} + \text{ADP}$$

The product, glucose 6-phosphate, is trapped within the cell as cell membranes are poorly permeable to the compound. It is converted to glucose 1-phosphate by the enzyme *phosphoglucomutase* (Fig. 1.81). This reaction requires Mg^{++} and glucose 1,6-diphosphate as cofactors. Glucose 1-phosphate is then activated by reacting with uridine triphosphate (UTP) to form UDP-glucose in a reaction catalyzed by *UDP-glucose pyrophosphorylase*. The glucose moiety of UDP-glucose is then transferred to a preexisting glycogen molecule (primer) to which it is attached in an 1,4-linkage involving the 1 carbon and 4 carbon of adjoining glucose molecules. This reaction is catalyzed by the enzyme *glycogen synthase* (UDP-glucose glycogen-transglucosylase). The formation of such bonds gives rise to linear chains of glucose residues. At intervals, chains of 8 to 12 glucose residues are picked up and transferred to make a branch which is attached to the main chain of glucose residues by means of a 1,6 linkage. This reaction is catalyzed by the

"branching enzyme" (amylo-1,4-1,6 transglucosylase). In this way, a branch point is established, and the glycogen molecule grows to a large, highly branched structure (Fig. 1.82), with a molecular weight that ranges from 240,000 to 10^7.

The rate of glycogen synthesis is controlled by a regulatory control mechanism. The rate-limiting enzyme in this process is glycogen synthase, which exists in two forms called glycogen synthase D (D, dependent) and glycogen synthase I (I, independent). Glycogen synthase D, the less active form, is dependent on glucose 6-phosphate for activity, whereas glycogen synthase I, the more active form, is independent. In order for glycogen synthase D to be converted to glycogen synthase I, it may be dephosphorylated by a *glycogen synthase phosphatase*. Conversely, for conversion of glycogen synthase I to glycogen synthase D, the latter must be phosphorylated by cyclic AMP-dependent protein kinase. The enzyme is activated when cellular cyclic AMP concentrations are increased as a result of the stimulation of adenylate cyclase by epinephrine. Epinephrine is released from the adrenals in response to stress when energy demands by the muscle may be increased, as during the "fight or flight" response. At such a time it would be inappropriate

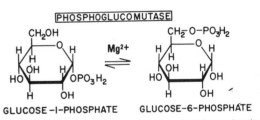

Figure 1.81. Interconversion of glucose phosphates by the enzyme phosphoglucomutase.

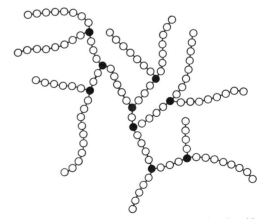

Figure 1.82. Part of a branched glycogen molecule with each circle representing one glucose residue. *Solid circles* represent branch points of glucose residues whose 6-carbon is joined in 1,6-linkages of glucose residues, whereas *open circles* represent linear chains of glucose residues consisting of 1,4-linkages.

for the muscle to store available glucose in the form of glycogen when it is needed for energy.

Glycogenolysis

Glycogenolysis is the breakdown of glycogen to glucose. During glycogenolysis (Fig. 1.83), glycogen is first partially degraded to glucose 1-phosphate; this degradation is catalyzed by *glycogen phosphorylase*, an enzyme whose activity is confined to the 1,4 linkages. All of the residues on the branches except for the proximal residue are transferred to the main chain by the debranching enzyme. The 1,6 linkage involving the proximal residue is then split by *amylo-(1,6)-glucosidase*, releasing this branch point residue as free glucose. Thereafter the remaining molecule can be degraded completely by glycogen phosphorylase.

Glucose 1-phosphate is converted to glucose 6-phosphate in a reaction catalyzed by *phosphoglucomutase*, the same enzyme utilized in glycogenesis (see above). The glucose 6-phosphate enters the main pathway of glycolysis to form pyruvic acid followed by either complete oxidation via the citric acid cycle and its associated oxidative phosphorylation system or conversion to lactic acid under anaerobic conditions. The lactic acid is then reconverted to pyruvic acid and metabolized further when O_2 is again available to the muscle or is metabolized in the liver. Glucose 6-phosphate cannot be converted to free glucose in muscle tissue because it, unlike the liver, kidney, and small intestine, lacks the enzyme glucose 6-phosphatase necessary for this conversion.

The rate-limiting step in glycogenolysis is at the level of glycogen phosphorylase, and an elaborate control mechanism exists. Glycogen phosphorylase exists in an active form called *phosphorylase a* and a less active form called *phosphorylase b*. These two forms of phosphorylase differ in several respects. Phosphorylase *b* has a requirement for AMP, whereas phosphorylase *a* does not. Phosphorylase *a* exists as a tetramer which has been phosphorylated on serine residues by another enzyme called

phosphorylase b kinase (see below). When these phosphate groups are removed by the action of an enzyme called *phosphorylase phosphatase*, there is a rearrangement of the enzyme to a dimeric form. Phosphorylase *b* is found in resting muscle.

Phosphorylase *b* kinase requires Ca^{++} for activation. The Ca^{++} requirement reflects the fact that calmodulin, a calcium-binding protein, forms one of the subunits of this enzyme. During muscle activation when cytosolic Ca^{++} rises, phosphorylase *b* kinase becomes activated and thereby increases glycogenolysis, which supplies energy for contraction. Additional activation of phosphorylase *b* kinase may occur when it too, like phosphorylase *b*, is phosphorylated, in this case by cyclic AMP-dependent protein kinase, which can also be called phosphorylase *b* kinase kinase when phosphorylase *b* kinase is its substrate. When the phosphorylase system is activated by phosphorylation, the synthase is inactivated by the very same process. This biochemical switch permits glycogen breakdown to occur without the concurrent competition of synthesis.

The scheme in Figure 1.84 summarizes the pathways of glycogenesis and glycogenolysis.

Glycolysis

Glycolysis is the degradation of glucose to pyruvic acid. It occurs in the sarcoplasm, and the enzymes involved are part of the easily extracted soluble proteins of muscle. In glycolysis, also known as the *Embden-Meyerhof pathway* of glucose metabolism, two molecules of ATP are formed per molecule of

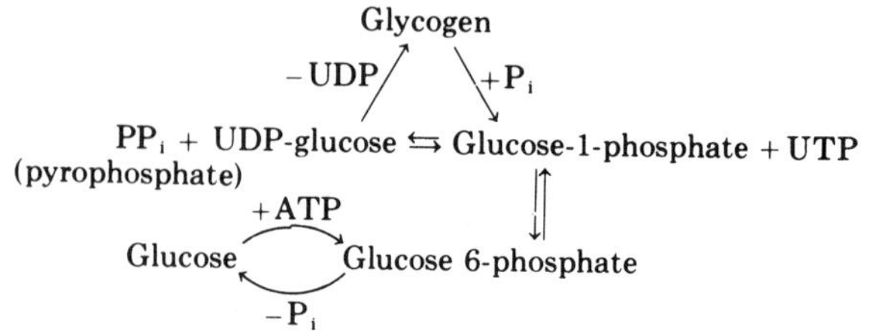

$$\text{PP}_i + \text{UDP-glucose} \rightleftharpoons \text{Glucose-1-phosphate} + \text{UTP}$$

Figure 1.84. Summary of pathways of glycogenesis and glycogenolysis.

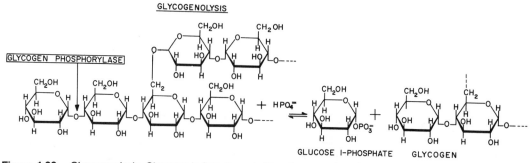

Figure 1.83. Glycogenolysis. Glycogen is first degraded by glycogen phosphorylase, which acts at 1,4-linkages.

glucose degraded.

$$\text{Glucose} + 2\,\text{ADP} + 2\,\text{P}_i \rightarrow 2\,\text{pyruvic acid} + 2\,\text{ATP}$$

The conversion of free glucose to glucose 6-phosphate is bypassed when glycolysis follows glycogenolysis because glucose 6-phosphate is then formed directly from glucose 1-phosphate in a reaction not involving utilization of phosphate-bond energy. The net gain of glycolysis will, therefore, in this case, be 3 moles of ATP per mole of glucose residues in glycogen. When free glucose is derived from the bloodstream rather than by glycogenolysis within the muscle, it is converted to glucose 6-phosphate by hexokinase in the reaction described under Glycogenesis.

Hexokinase and numerous other enzymes in glucose metabolism require Mg^{++} or other metals as cofactors for activity. These cofactors are indicated in Figure 1.85, which summarizes the reaction steps in glycolysis.

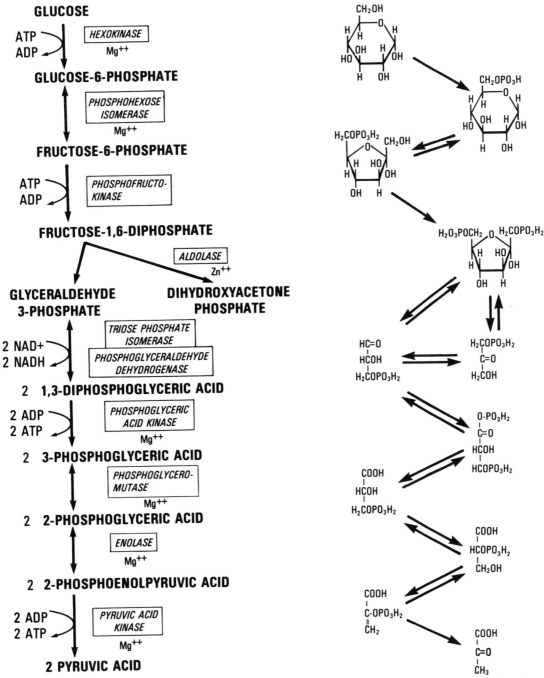

Figure 1.85. Glycolysis. All steps are reversible except the conversions of glucose to glucose 6-phosphate, of 2-phosphoenol pyruvic acid to pyruvic acid, and of fructose-6-phosphate to fructose-1,6-diphosphate.

Glucose 6-phosphate is converted to fructose 6-phosphate in a reaction catalyzed by *phosphoglucose isomerase*. A second phosphate group is introduced at the 1-carbon of fructose 6-phosphate to yield fructose 1,6-diphosphate. The phosphate group is transferred from ATP, and the transfer is catalyzed by the enzyme *phosphofructokinase*. Since 1 mole of ATP is dephosphorylated per mole of fructose 1,6-diphosphate formed, this reaction involves expenditure of energy.

In the next step, the 6-carbon compound fructose 1,6-diphosphate is split between carbons 3 and 4 to yield two 3-carbon compounds, dihydroxyacetone phosphate and D-glyceraldehyde 3-phosphate. The enzyme *aldolase* catalyzes this cleavage.

The two triosephosphates formed are interconvertible by isomerization, and dihydroxyacetone phosphate is converted to D-glyceraldehyde 3-phosphate in a reaction catalyzed by *triosephosphate isomerase*.

D-Glyceraldehyde 3-phosphate is then converted to 3-phosphoglyceric acid. The formation of 3-phosphoglyceric acid consists of an oxidation of an aldehyde to an acid and takes place in the following steps:

(1) 3-Phosphoglyceraldehyde + NAD$^+$ + P$_i$ → 1,3-diphosphoglyceric acid + NADH, catalyzed by *phosphoglyceraldehyde dehydrogenase*, and
(2) 1,3-Diphosphoglyceric acid + ADP → 3-phosphoglyceric acid + ATP, catalyzed by *phosphoglyceric acid kinase*.

It is seen that the energy of the oxidation of the aldehyde of 3-phosphoglyceraldehyde to the carboxylic acid of 3-phosphoglyceric acid has been preserved in the formation of an anhydride bond in the 1-position of 1,3-diphosphoglyceric acid and utilized to form ATP. Intact sulfydryl groups in phosphoglyceraldehyde dehydrogenase are critical for enzyme activity. The inhibition of glycolysis by iodoacetate is due to its reaction with these sulfhydryl groups. This is an example of the coupling of an energy-releasing oxidation with phosphorylation of ADP in order to retain energy in a chemically available form, thus preventing it from being dissipated as heat. Oxidation of NADH is also coupled to ADP phosphorylation and yields 3 moles of ATP per mole of NADH (see below).

The 3-phosphoglyceric acid is transformed into 2-phosphoglyceric acid in an intermediate step in which 2,3-diphosphoglyceric acid is utilized as a cofactor. The overall reaction is catalyzed by *phosphoglyceromutase* and is similar in mechanism to that involving glucose 1-phosphate and glucose 6-phosphate.

2-Phosphoglyceric acid is converted by dehydra-

tion to phosphoenolpyruvic acid by *enolase*. Fluoride, a potent inhibitor of glycolysis, acts on this enzyme. Phosphoenolpyruvic acid is a high-energy compound. A transfer of its phosphate group to ADP yields one molecule of ATP plus pyruvic acid in a reaction catalyzed by pyruvic acid or *pyruvate kinase*. This is the last reaction during glycolysis that can furnish energy for the synthesis of ATP.

NADH formed as a result of conversion of 3-phosphoglyceraldehyde to 3-phosphoglyceric acid can be oxidized to NAD$^+$ in one of two ways. In the presence of O_2, the NADH is oxidized by a transfer of electrons to the cytochrome chain in the mitochondria. This process is described further under Oxidative Phosphorylation. In the absence of O_2, NADH oxidation is coupled to lactic acid formation. The coupling occurs in contracting muscle with normal blood supply when muscle activity exceeds a certain intensity and O_2 becomes limiting relative to demand. Lactic acid can thus be formed under both aerobic and anaerobic conditions. The amount of lactic acid formed represents an *oxygen debt* which can be repaid during the recovery period of the muscle. Alternatively, if the muscle contraction continues, lactic acid is released into the bloodstream and carried to the liver where it is converted to glucose 6-phosphate via gluconeogenesis. The glucose is then released by the action of the liver's glucose 6-phosphatase to be used as energy by the muscle. The conversion of lactic acid to glucose by liver which is returned to the muscle is called the *Cori cycle*.

The coupling of the reactions leading to restoration of NAD$^+$ can be summarized as shown in Figure 1.86. No oxygen, however, is consumed during glycolysis, and the presence of oxygen is not required.

The main control point in glycolysis occurs at the level of phosphofructokinase, which is a rate-limiting enzyme. The activity of this enzyme is increased by fructose 6-phosphate, ADP, AMP, and inorganic phosphate. Recently, fructose 2,6-diphosphate has been found to be a powerful physiological regulator, producing stimulation. Epinephrine also activates phosphofructokinase because it stimulates the production of cyclic AMP which, in turn, helps activate an enzyme catalyzing the production of the potent activator fructose 2,6-diphosphate. On the other hand, ATP at high concentrations and phosphocreatine inhibit phosphofructokinase activity. A particularly important inhibitor is citrate, which is produced by aerobic metabolism via the citric acid cycle. Therefore, when oxygen is present, glycolysis is slowed reflecting the more efficient production of ATP by oxidative metabolism.

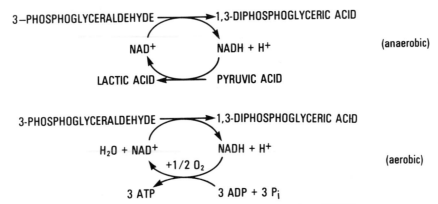

Figure 1.86. Anaerobic and aerobic restoration of NAD^+

Citric Acid Cycle

The citric acid cycle (tricarboxylic cycle or Krebs cycle), occurring in the mitochondria, represents a cyclic sequence of chemical reactions by which acetic acid bound to coenzyme A (CoA) as acetyl coenzyme A is oxidized to CO_2 and H_2O and CoA is re-formed (Fig. 1.87 and 1.88). Pyruvic acid derived from glycolysis furnishes one source of acetyl CoA. Pyruvic acid is oxidized in a series of reactions involving several enzymes and four cofactors which form a complex constituting *pyruvic acid dehydrogenase.* The net effect of these reactions is summarized in Figure 1.89.

The acetyl CoA is an energy-rich compound and can also be formed from fatty acids and certain amino acids. The citric acid cycle, therefore, represents a common pathway in the oxidative degradation of carbohydrates, fats, and proteins. Glucose and fatty acid metabolism represent, however, the main sources of acetyl CoA.

In the citric acid cycle, acetyl CoA reacts with the 4-carbon oxaloacetic acid to form the 6-carbon citric acid, a reaction catalyzed by citrate synthase, formerly referred to as the *"condensing enzyme."* Citric acid is converted to its isomer, isocitric acid. Oxidative decarboxylation of isocitric acid results in the formation of α-ketoglutarate (plus CO_2) in a series of reactions analogous to the pyruvate dehydrogenase reactions (see above). Succinyl CoA (plus CO_2) is then formed from α-ketoglutarate and CoA. The energy of the thioester bond in succinyl CoA is used to form guanosine 5′-triphosphate (GTP) from guanosine 5′-diphosphate (GDP) in a reaction producing free succinate and free CoA. Succinate is oxidized to form fumarate and reduced flavin adenine dinucleotide ($FADH_2$). Hydration of fumarate to malate and finally oxidation of malate to form oxaloacetate completes the cycle. One molecule of the 2-carbon acetyl fragment is oxidized during the cycle, with two molecules of CO_2 being produced

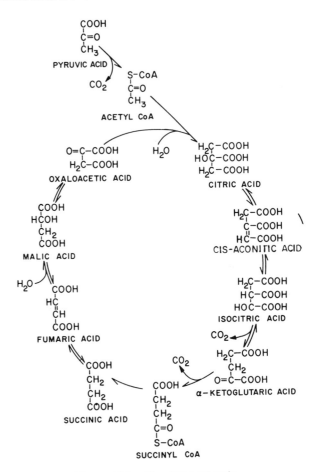

Figure 1.87. The citric acid cycle.

and four hydrogen atoms removed. The net exchange during one cycle is shown in the following scheme:

$$CH_3COOH + 2\,O_2 \rightarrow 2\,CO_2 + 2\,H_2O + energy$$

The five energy-yielding reactions involve three steps in which NADH is produced, one step in which $FADH_2$ is produced and one substrate-level phosphorylation, where GTP is produced and subsequently converted to ATP by a kinase.

(1) Isocitrate + NAD^+ → α-ketoglutarate + CO_2 + NADH + H^+

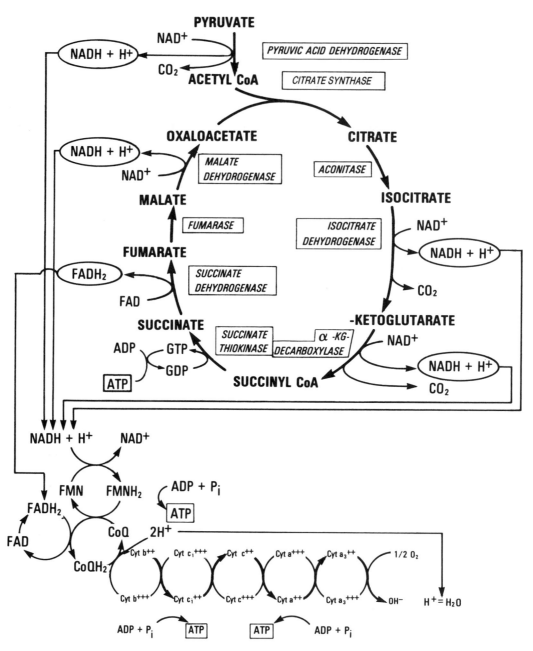

Figure 1.88.   The citric acid cycle and the electron transport chain.

(2) α-Ketoglutarate + NAD$^+$ + CoA $\rightarrow$ succinyl CoA + CO$_2$ + NADH + H$^+$

(3) Succinyl CoA + GDP + P$_i$ $\rightarrow$ succinate + CoA + GTP

(4) Succinate + FAD $\rightarrow$ fumarate + FADH$_2$

(5) Malate + NAD$^+$ $\rightarrow$ oxaloacetate + NADH + H$^+$

Oxidative Phosphorylation

The reactions in the citric acid cycle are coupled to the respiratory chain, consisting of a sequence of hydrogen and electron carriers (Fig. 1.88) also located within the mitochondria.

The electron carriers are iron-containing hemoproteins or cytochromes, which transfer electrons by alternating between reduced and oxidized states, the iron switching between ferrous and ferric forms

(Fig. 1.90). An intermediate step involving the removal of hydrogen from NADH by NADH dehydrogenase (which is a flavoprotein) precedes the passage of electrons along the series of cytochromes in the carrier chain. The flavoprotein is oxidized by ubiquinone (coenzyme Q, CoQ), a lipid-soluble quinone acting as a coenzyme. Ubiquinone, which can drift laterally in the membrane, shuttles redox equivalents between FMNH$_2$ and cytochrome b (Cyt b) and between FADH$_2$ and Cyt b, all of which are membrane bound. In the step involving oxidation of ubiquinone, the electron from hydrogen is transferred from ubiquinone to the first cytochrome in the cytochrome series, Cyt b, and the proton is donated from the membrane to the medium. Elec-

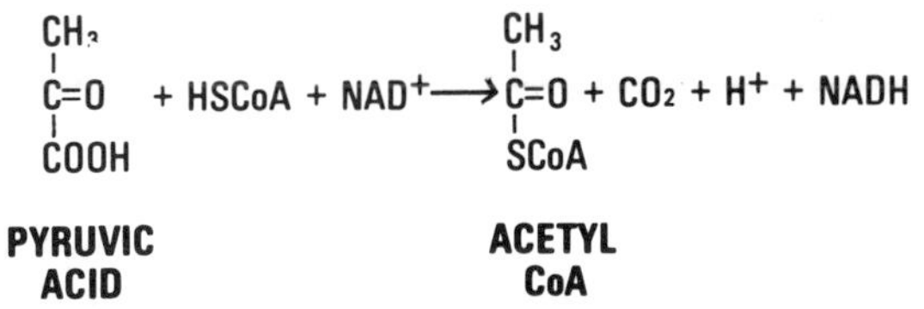

$$CH_3 \\ | \\ C=O \ + HSCoA + NAD^+ \longrightarrow \begin{matrix} CH_3 \\ | \\ C=O \\ | \\ SCoA \end{matrix} + CO_2 + H^+ + NADH \\ | \\ COOH$$

PYRUVIC   **ACETYL**
ACID   **CoA**

Figure 1.89. Summary of Reaction by which acetyl CoA is produced as pyruvic acid is oxidized.

A. $CoQH_2 + 2\ Fe^{+++} \longrightarrow CoQ + 2\ H^+ + Fe^{++}$

B. $\frac{1}{2}\ O_2 + H_2O + 2\ Fe^{++} \longrightarrow 2\ OH^- + 2\ Fe^{+++}$

Net. $CoQH_2 + \frac{1}{2}\ O_2 \longrightarrow CoQ + H_2O$

Figure 1.90. Summary of reactions in which ubiquinone (CoQH$_2$), oxygen, and cytochrome oxidase interact to form water.

trons are now passed along the chain of cytochromes, involving in addition to cytochrome b, cytochrome c_1 (Cyt c_1), cytochrome c (Cyt c), cytochrome a (Cyt a), and cytochrome a_3 (Cyt a_3). The complex consisting of Cyt a plus Cyt a_3 is called cytochrome oxidase, which can react directly with oxygen. When oxygen is reduced by cytochrome oxidase, hydroxyl ions are formed that combine with the protons generated at the oxidation of ubiquinone to form water. The net effects of the electron transport along the respiratory chain are illustrated in Figure 1.88.

During the electron transfer along the respiratory chain, starting with NADH, three ADP molecules are phosphorylated to yield three ATP molecules per atom of oxygen. This is expressed as a P/O ratio of oxidative phosphorylation of 3. The energy for the formation of ATP is derived at three points along the respiratory chain; these three points are considered to be in the segment between NADH and cytochrome b, between cytochromes b and c, and at the oxidation of cytochrome c by *cytochrome oxidase*. The minimum difference in oxidation-reduction potential which is required to provide about 7,300 cal/mole for phosphorylation of ADP is 0.2 V. At the fourth energy-yielding reaction in the citric-acid cycle described above, succinate is oxidized to fumarate by succinate dehydrogenase. Since the oxidation of NADH is bypassed, the P/O ratio is 2 instead of 3.

The exact mechanism by which the phosphorylation of ADP is coupled to the electron transport along the respiratory chain is unknown. However, an abundance of evidence has been obtained in favor of the *chemiosmotic hypothesis* first postulated by Peter Mitchell in 1961. According to this hypothesis, energy is transferred to ATP from a proton-motive gradient consisting of a charge gra-

dient and a pH gradient across the mitochondrial membrane by an ATPase complex—i.e., by a H$^+$ pump that is being driven "backwards" by the gradient. If this coupling mechanism is inhibited by drugs, respiration can, under experimental conditions, proceed without the generation of ATP, indicating that the electron transport system can remain intact when the phosphorylation is uncoupled. Under physiological conditions, however, respiration and phosphorylation are tightly coupled, and the rate-limiting factors in these processes are the concentrations of ADP and inorganic phosphate. The ADP concentration, on the other hand, is determined by the rate at which ATP is utilized. This represents a self-regulatory mechanism by which the rate of energy generation is adjusted to the requirements of the cell.

The mitochondrial matrix contains the enzymes for the oxidative decarboxylation of pyruvate and the reactions of the citric acid cycle, whereas the components of the respiratory chain are arranged in proper sequence in the inner membranes of the mitochondria. As noted in Chapter 1, the mitochondrion has a double membrane, consisting of an outer and an inner membrane. The inner membrane forms plates called cristae, which are arranged more or less in parallel and usually extend perpendicular to the long axis of the mitochondrion almost across its whole diameter. The cristae are separated by the mitochondrial matrix (Fig. 1.91). Another type of constraint which may be imposed upon the mitochondrial membrane is the orientation of the binding sites of the electron carriers. For example, the binding sites for NADH dehydrogenase and succinic acid dehydrogenase are assumed to face the

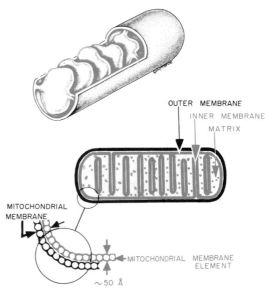

OUTER MEMBRANE
INNER MEMBRANE
MATRIX

MITOCHONDRIAL
MEMBRANE

MITOCHONDRIAL MEMBRANE ELEMENT

$\sim 50\ \text{Å}$

Figure 1.91. The structure of mitochondria.

matrix at the surface of the membrane elements. It has been speculated that the electron transfer might be facilitated by the close arrangement of the components of the respiratory chain, which would allow transfer with small changes in orientation of the individual electron carriers.

Finally it should be noted that the mitochondria are impermeable to NADH. Therefore special energy-requiring mechanisms, called substrate shuttles, exist to transfer the reducing equivalents of NADH produced during glycolysis from the cytoplasm into the mitochondria.

Energy Yield in Glucose Oxidation

Table 1.6 summarizes the energy-yielding and energy-requiring reactions during glucose degradation. In reactions 4, 5, and 9, ATP is formed from ADP participating in the reaction. In the other reactions, except for the reaction 10, ATP generation is a consequence of NADH oxidation in the respiratory system of mitochondria.

$$NADH + H^+ + \tfrac{1}{2} O_2 + 3\ ADP + 3\ P_i \rightarrow NAD^+ + H_2O + 3\ ATP$$

NAD is bypassed in reaction 10 and hydrogen is transferred to FAD, and ATP is formed in connection with a coupling of $FADH_2$ oxidation and ADP phosphorylation according to the following net reaction:

$$FADH_2 + \tfrac{1}{2} O_2 + 2\ ADP + 2\ P_i \rightarrow FAD + H_2O + 2\ ATP$$

In reactions 6, 7, and 8 1 mole of CO_2 is formed for each mole of NAD formed. In glycolytic degradation of glucose equivalents derived from glycogenolysis reaction 1 is not involved, and the total number of moles of ATP formed per mole glucose equivalents is, therefore, 39.

Total oxidation of glucose to CO_2 and H_2O yields

686,000 cal/mole, but only 56,000 cal/mole are produced when the glucose degradation does not proceed further than to lactic acid. The efficiency of glycolysis, that is, of the energy-generating reactions under anaerobic conditions, can be calculated by assuming that the phosphorylation of three ADP molecules represents about 24,000 cal/mole, as greater than 8,000 cal/mole are released when phosphate is transferred from high-energy compounds, e.g. phosphoenolypyruvate, to ADP. This is 24,000/56,000 × 100%, or about 43% of the potentially available energy during glycolysis following glycogenolysis. In comparison with the energy released during complete oxidation of glucose—686,000 cal/mole—the energy stored in three molecules of ATP will represent 24,000/686,000 × 100% or about 3% of the total energy available as chemical energy in glucose. This shows the low energy yields of anaerobic energy metabolism. In comparison, 39 molecules of ADP are phosphorylated in the aerobic degradation of a glucose equivalent derived from glycogenolysis, representing a free energy storage of about 312,000 cal. The efficiency of the oxidative degradation of glucose is therefore 312,000/686,000 × 100% or about 45%, which is in good agreement with the efficiency estimated from measurements of the recovery heat.

Electromyography

The structural basis of electromyography is the motor unit (for definition, see section entitled The Motor Unit).

MOTOR UNIT POTENTIAL

During the normal twitch of a muscle fiber, a minute electrical potential is generated, which is dissipated into the surrounding tissue. The dura-

Table 1.6
Energy yield in glucose degradation

Reaction	ATP yield
(1) Glucose + ATP → glucose 6-phosphate + ADP	−1
(2) Fructose 6-phosphate + ATP → fructose 1,6-diphosphate + ADP	−1
(3) 2 D-Glyceraldehyde 3-phosphate + 2 NAD$^+$ + 2 P$_i$ ⇌ 2 1,3-diphosphoglyceric acid + 2 NADH + 2 H$^+$	+6[a]
(4) 2 1,3-diphosphoglyceric acid + 2 ADP ⇌ 2,3-phosphoglyceric acid + 2 ATP	+2
(5) 2 Phosphoenolpyruvic acid + 2 ADP → 2 pyruvic acid + 2 ATP	+2
(6) 2 Pyruvic acid + 2 CoA + 2 NAD$^+$ → 2 acetyl CoA + 2 CO$_2$ + 2 NADH + 2 H$^+$	+6
(7) 2 Isocitric acid + 2 NAD$^+$ ⇌ 2 α-ketoglutaric acid + 2 CO$_2$ + 2 NADH + 2 H$^+$	+6
(8) 2 α-Ketoglutaric acid + 2 CoA + NAD$^+$ → 2 succinyl CoA + 2 NADH + 2 H$^+$ + 2 CO$_2$	+6
(9) 2 Succinyl CoA + 2 GDP + 2 P$_i$ → 2 succinic acid + 2 CoA + 2 GTP 2 GTP + 2 ADP ⇌ 2 GDP + 2 ATP	+2
(10) 2 Succinic acid + 2 FAD ⇌ 2 fumaric acid + 2 FADH$_2$	+4
(11) 2 Malic acid + 2 NAD$^+$ ⇌ 2 oxaloacetate acid + 2 NADH + 2 H$^+$	+6
Total	+38

[a] Substrate shuttles transferring the reducing equivalents of NADH from the cytoplasm into the mitochondria require energy. Therefore the actual ATP yield at step 3 is only +4 or +5.

tion of the action potential associated with this twitch is about 1 to 2 ms or even 4 ms. Since all the muscle fibers of a motor unit do not contract at exactly the same time—some being delayed for several milliseconds—the electrical potential developed by the single twitch of all the fibers in the motor unit is prolonged. The electrical result of the motor-unit twitch, then, is an electrical discharge lasting about 5 to 8 ms (and often as long as 12 ms). The majority of these motor-unit potentials have an amplitude of around 0.5 mv. When displayed on a cathode ray oscilloscope, the result is a sharp spike that is most often biphasic (Fig. 1.92), but it may also have a more complex form, depending on physical and other factors.

Generally, the larger the motor unit potential registered, the larger is the motor unit producing it. However, complicating factors, such as distance of the unit from the electrodes and the types of electrodes and equipment used, determine the final size and pattern of the individual motor unit potential that is recorded. There are, furthermore, characteristic variations with different normal muscles, e.g., the smallness of potentials in facial muscles as compared with those in muscles of the extremity. Advancing age may result in a slight prolongation of potentials.

ELECTROMYOGRAPHIC TECHNIQUE

The types and construction of electrodes used in electromyography vary widely. The two main types of electrodes used for the study of muscle dynamics are surface (or "skin") electrodes and inserted (usually wire or needle) electrodes.

Basically, an electromyograph is a high-gain amplifier with selectivity for frequencies in the range of 10 to several 1,000 Hz. It has been suggested that

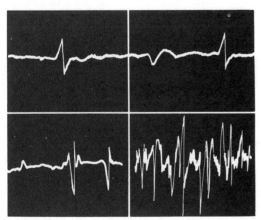

Figure 1.92. Normal electromyograms. The single potential in the upper corner has a measured amplitude of 0.8 mv and duration of 7 ms.

the sharply peaked spectra of motor unit potentials derived with surface electrodes render the use of amplifiers with limited frequency response quite practical, and it has been found that 20–200 Hz is the most desirable frequency range for the elimination of amplifier "noise," general nonmuscular "tissue noise," and movement artifact without significant loss of motor-unit potentials.

CLINICAL APPLICATION

Electromyography is most useful for distinguishing nerve from muscle disease and for differentiating weakness and abnormalities of sensation due to peripheral nerve and/or muscle disease from abnormalities of the central nervous system. Moreover, because one is able to measure conduction velocity in nerve, it is possible to diagnose damage to the myelin sheath, in which case conduction velocity is slowed. Typically, electromyograms are obtained (a) at rest, when normally there is no spontaneous muscle activity, (b) during slight muscle contraction in order to assess the size and duration of activity of motor units, thus enabling differentiation between myopathic and neuropathic diseases, and (c) during maximal muscle activity in order to determine if there is abnormal recruitment of motor units, which, again, can differentiate whether the problem is myopathic or neuropathic.

According to the Henneman principle (Henneman et al., 1974), under normal conditions, the smaller potentials appear first with a slight contraction. As the force is increased, larger and larger potentials are recruited, this being the normal pattern of recruitment. This normal pattern is absent in cases of partial paralysis due to injuries or lesions of the lower motor neuron, i.e., the small potentials never appear, apparently because only the larger motor units have survived.

Two common abnormalities that may be detected at rest when normal muscle is quiescent are fibrillation and fasciculation. *Fibrillation* is the contraction of single muscle cells which have become completely dissociated from nervous control due to destruction of a motor nerve and subsequent degeneration of the distal part of the axon by disease or wilting of the nerve. Fibrillation occurs for several weeks and then ceases as the muscle cells atrophy. Because the motor unit has been disrupted, there can be no summation of muscle fibers. Therefore, the potentials must be measured from single muscle cells with the aid of bipolar needle electrodes inserted into the muscle.

Fasciculation is the contraction of groups of muscle cells integrated by a single axon into a motor unit. Fasciculation occurs when an anterior motor neuron is destroyed, as in motor neuron disease or the axon is severed. Spontaneous impulses will arise as the distal axon atrophies. Until atrophy is complete, the muscle cells comprising the motor unit will contract in response to these spontaneous impulses. These contractions are therefore strong enough to allow measurement of potentials with skin electrodes.

FUNCTION OF WHOLE SKELETAL MUSCLE

Gross Organization

Approximately 600 skeletal muscles, composed of millions of muscle fibers, account for 40 to 45% of the human body weight. Surrounding individual muscle fibers is connective tissue called the *endomysium*, which contains small blood vessels and nerves. Groups of muscle fibers, collected into bundles of fascicles, are bound together by a more dense layer of collagenous and elastic fibers called the *perimysium*. Finally the connective tissue which binds the fascicles into definitive muscle is called the *epimysium*. The connective tissues of all these three layers are actually in continuity with each other. At the ends of the elongated muscle, the connective tissue forms a common bundle of fibers called the *tendon*.

The arrangement of the muscle fibers in a muscle is variable (see Fig. 1.93) and has been classified as follows:

1) Muscle fibers parallel to the long axis of the muscle (e.g., the straplike sartorius, the pectineus, and the fusiform biceps brachii and flexor carpi radialis).

2) Muscle fibers oblique to the long axis of the muscle. Muscles whose fibers are arranged so that they insert obliquely into a tendon include several varieties: (1) unipennate (e.g., flexor pollicis longus); (2) bipennate (e.g., rectus femoris); (3) circumpennate (e.g., tibialis anterior); and (4) multipennate (e.g., deltoid).

3) Muscle fibers arranged in a radial or triangular shape (e.g., adductor longus, pectoralis minor).

Muscles whose component fibers are in parallel array are able to shorten over greater distances than muscles whose fibers are organized in the pennate manner. The penniform arrangement of muscle fibers is associated with those muscles which exert great force rapidly over a short distance.

In order for muscles to perform their function, they are attached, for the most part, from one bone across a joint or joints to another bone by *tendons* or flattened sheets of connective tissue called *aponeuroses*. The tendons and aponeuroses are formed

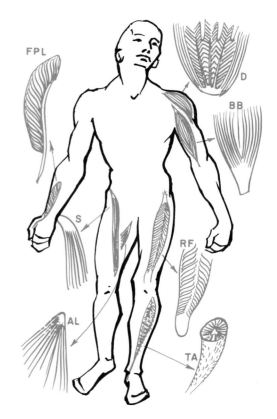

Figure 1.93. The form and distribution of several varieties of striated skeletal muscle: the pennate, flexor pollicis longus (*FPL*); multipennate, deltoid (*D*); fusiform, biceps brachii (*BB*), bipennate, rectus femoris (*RF*); circumpennate, tibialis anterior (*TA*), radial, adductor longus (*AL*) and straplike sartorius (*S*). (The male figure redrawn from the First Muscle Tabule in Fabrica by Vesalius).

of collagen, which is flexible but practically inextensible.

The points of muscular attachment are called the origin and the insertion. The *origin* of a muscle is usually the fixed or proximal attachment, while the *insertion* is the distal attachment to the bone which is moved. There are exceptions and in many cases the designation of the origin and insertion of a muscle has become a matter of anatomical convention. The force developed by an uncontrolled maximal contraction of a whole muscle is of little practical use in the movements of man and animal. Therefore, the force and movement generated in a given circumstance by a given muscle depends upon the time-integrated individual activities of its component motor units. For fine control of movement one or more motor units are employed. As more force is required the number of stimuli carried by each motor unit nerve is increased, along with the recruitment of additional motor units. The grading of muscular activity is the result, therefore, of asynchronous firing of the motor units of a whole muscle.

In considering the function of muscles in man, it

should be noted that no muscle acts alone even in the simplest movement, i.e., a variety of muscles which are described as agonists, synergists, and antagonists are involved in each action. The muscle which produces the movement is termed the *agonist*, while the muscle which opposes the movement is termed the *antagonist*. The *synergists* are muscles which act together to produce a movement which no muscle could produce alone. An example of synergistic function can be seen in the adduction of the hand at the wrist. The flexor carpi ulnaris will both flex and adduct the hand. To produce adduction alone, the extensor carpi ulnaris must be brought into action to offset flexion and yet permit adduction of the hand by the flexor carpi ulnaris. Examples of antagonistic muscles are the biceps brachii and triceps brachii, which are mutually antagonistic with respect to movement at the elbow joint.

The antagonist is usually reflexly inhibited when the agonist is brought into action, e.g., if one holds a weight on the palm of the hand with the elbow flexed perpendicular to the body, the biceps brachii (in this case the agonist) develops tension while the triceps brachii (the antagonist) is flaccid. This action is reversed—the triceps being contracted and the biceps relaxed—when the hand presses down onto a table with the arms flexed. When however, the arm is extended and the elbow joint is fixed both the biceps and the triceps are contracted. Large, translational movements may occur between muscles that exhibit antagonistic effects, such as flexors and extensors, and such muscles are also frequently separated by extensive connective tissue septa.

Not all striated muscles are associated with the movement of skeletal parts at joints or with the fixation or maintenance of joint stability as related to posture. Muscles serve as *sphincters* when they encircle an orifice (e.g., eyelid, lips, and anus). Striated muscle is found in the upper two-thirds of the esophagus in a tubular arrangement which aids in the swallowing mechanism. The diaphragm is a thin muscular sheet which upon contraction serves to enlarge the thorax and compress the abdomen, effects which are associated with inspiration. This action is reversed during expiration when the muscle relaxes. In addition, there are certain cutaneous muscles which have at least one of their attachments to the skin, e.g., the mimetic facial muscles and the panniculus carnosus of the trunk.

Force Transmission to the Skeleton

ROLE OF TENDONS

The force generated by a muscle is transmitted to the skeleton by means of its tendons. The sub-microscopic collagen filaments of tendons extend along the surface of the individual muscle fibers at the tapering ends of the muscle fibers. The surface area of the muscle fibers is fairly large here due to longitudinal folds of the plasma membrane. There is close contact between collagen filaments and the muscle fiber surface.

The shape of a tendon varies with the shape of the muscle to which it is attached, but the muscle fibers are always oriented at an angle to the main direction of the tendon. Muscles in which the angle is relatively large are called pennate muscles. This angle, which never exceeds 10–20°, serves to prevent fraying of the tendon when the diameter of the fibers increases in connection with contraction and shortening.

A tendon may form one or several flat sheets at the region of contact with the muscle, and the muscle fibers can impinge on one or both sides of such a sheet. The length of the fibers extending between tendons is usually considerably shorter than the overall length of the muscle. The distance bridged by the muscle fibers is related to the range of shortening under physiological conditions and to the fact that under these conditions the maximal shortening of individual muscle fibers is kept within a certain limit, viz., about 30% of the maximal length of the muscle in the body. This is presumably an adjustment to the relationship between the maximal tension developed by a muscle as a function of its length.

In places where a large force is required but where the range of shortening is limited, muscles are characterized by multipennate structure, with the tendons forming several septa arranged in two planes. In this case the effective cross-section of the muscle is considerably larger than the anatomical cross-section. The septa of the tendons in the human deltoid muscle were found to develop gradually during embryologic growth and to be reduced in "old age." It is therefore likely that the gross structure of a muscle is dynamically adjusted to its functional requirements as is its cross-sectional area.

EFFECT OF EXERCISE

The cross-sectional area of a muscle increases when the muscle is forced repeatedly to develop maximal or close-to-maximal tension, as in regular exercise or work that requires forceful muscle action. This increase is due to an increase of the cross-sectional area of the individual muscle fibers (*hypertrophy*) but not to a formation of new fibers (*hyperplasia*). Hypertrophy will persist and increase only if the work load is continuously increased during a training period. If the load is kept constant, the growth of the muscle stops when its strength has been adjusted to the load.

Muscles exposed to sustained rhythmic activity over fairly long periods of time at a load below maximum show an increase in the density of the network of blood capillaries extending between the muscle fibers. This appears as an adjustment to the requirements for increased oxygenation associated with this form of activity. In animals, experiments have shown the effect to be more pronounced in the young than in adults.

Types of Skeletal Muscle Fibers

It has been recognized for many years that muscles vary in color, not only from species to species but also within the same animal. Striking examples of this are the crimson red pectoralis of the pigeon as contrasted with the stark white chicken pectoralis and the red soleus lying near the white gatrocnemius in the lower limb of animals and man. Using morphological, histochemical, biochemical, and physiologic properties as criteria, one is able to distinguish several distinct muscle types in the animal kingdom. The voluntary skeletal muscles of mammals consist wholly of twitch-type fibers, which can be characterized by their ability to produce (on nerve stimulation) a propagated action potential and to respond with fast shortening or tension development. Other types of muscle fibers have evolved for specific purposes in other species and include insect flight muscle (which will not be described further) and slow fibers, which produce tonic (as opposed to phasic) contractions. Slow fibers are widely distributed in the animal kingdom, but in mammals they are found only in the extrinsic muscles of the ear and in the extraocular muscles, which have both slow and twitch types.

SLOW-TYPE FIBERS

Although students of human physiology will be concerned primarily with the twitch-type muscle of the skeletal musculature (see below), various properties of slow-type fibers will be given in order to compare them with twitch-type fibers. The slow fibers differ fundamentally from the twitch fibers because they have lower levels of resting potential (50 to 70 mv as compared to 80 to 90 mv for fast fibers), they do not produce propagated action potentials, they exhibit a longer latency of response (6 to 8 ms as compared to 2 to 4 ms for fast fibers), and they do not respond to stimulation by rapid shortening or rapid tension development. Only repeated stimulation results in a significant rise in tension by these slow muscle fibers. In addition, the slow muscle fibers are innervated by numerous nerve endings (the *en grappe* type), distributed widely over the whole fiber length, whereas the twitch fibers are innervated by a single discrete, localized nerve ending (the *en plaque* type) with a distinct neuromuscular junction. The numerous areas of nerve contact, which locally activate the contractile apparatus of the fiber, compensate for the inability of a slow fiber to produce propagated muscle impulses.

The slow-type muscle system has been studied in greatest detail in the frog. It produces the prolonged contraction during the male's amplexus, which may last for days during the mating season. In this animal tension development by slow fibers following reflex activation is similar to tension development after direct nerve stimulation (provided that the direct stimulation is of sufficient intensity to activate most of the small nerve fibers serving the muscle). The motor units of the slow fiber system may be normally activated, therefore, by a synchronous high frequency discharge from the spinal cord. In the frog, the nerve fibers to the slow muscle fibers are small and slow conducting (2 to 8 m/s), while the nerve fibers to the twitch fibers are large and fast conducting (8 to 40 m/s). No overlap of innervation between the two muscle groups has been found. The slow muscle fibers are reflexly activated even during periods when the twitch fibers are at rest. The slow fibers can also be activated during twitch fiber activity and can maintain large tensions originally produced by the twitch fibers over very long periods by virtue of low frequency stimulation. A functionally useful synergism has been demonstrated by the twitch and slow fibers in the frog iliofibularis muscle. First, small nerve stimulation at 30/s results in a smooth rise in tension. If a burst of twitch activity is added, via the large nerve system, the final tension developed exceeds that which would have been developed without the twitch activity. Second, when stimulation ceases there is a very gradual relaxation of the generated tension. If a burst of twitch activity is initiated during the slow relaxation phase, the residual tension in the slow fibers collapses rapidly.

Slow fibers and twitch fibers have similarities and differences in their fine structure. In both, the myofilaments of the myofibrils are of the same length, even though sarcomere lengths may differ. However, the slow fibers have no discrete M lines, the H zones are poorly defined, and the Z lines are thick. Although the myofibrils tend to be large and vaguely defined in cross-section, this cannot be considered as the sole criterion for slow fibers.

A striking difference between twitch and slow fibers is the absence or poor development of the transverse tubular or T system in the slow fiber: In the twitch fiber the T tubule is in continuity with the sarcolemma; but, in those slow fibers in which a T system can be recognized, the small delicate tubules have no specific localization with respect to the sacrolemma or the contractile apparatus.

The motor nerve endings on slow muscle fibers, as revealed by electron microscopic studies, are readily distinguishable from those of twitch muscle fibers. The specific region of nerve-muscle contact is small compared with that on twitch fibers. In addition, there are no junctional folds in the muscle membrane or sarcolemma in the region of the nerve ending. Such junctional folds are a constant feature of the motor end-plate in twitch fibers.

TWITCH-TYPE FIBERS

Classification

Twitch-type fibers can be further classified into three distinct subtypes which have been variously designated through the years and now frequently are referred to as fast fatigable, slow fatigable, and fast fatigue resistant (Table 1.7). Fast fatigable fibers, also called white fibers, have high myofibrillar ATPase activity, high glycolytic enzyme activities, an intermediate glycogen content, a low myoglobin content, and a small mitochondrial content. They produce a fast twitch and fatigue rapidly. The

latter property is related to their limited glycogen content and low capacity for oxidative metabolism. Fast fatigue-resistant fibers also have high myofibrillar ATPase activity, a high glycogen content, and an intermediate level of glycolytic enzymes but have a large mitochondrial content and a high myoglobin content, related to their resistance to fatigue. Slow fatigue-resistant fibers have low myofibrillar ATPase activity, low glycogen content but a high myoglobin content, high mitochondrial oxidative enzyme activities, and an intermediate mitochondrial content. The latter two fiber types are red in color. Thus red fibers can be either fast or slow contracting.

The red color of muscle is due to *myoglobin* which is related to the hemoglobin of red blood cells. Myoglobin from skeletal muscles has a molecular weight of 16,700 and contains 1 heme (iron-porphyrin complex) per molecule. The hemoglobin of the blood is composed of 4 hemes/molecule and differs also in the protein globin. Myoglobin has a greater affinity for oxygen than does hemoglobin and may serve as an oxygen reservoir and facilitate the transport of oxygen within the muscle fiber. In addition to its high affinity for oxygen, myoglobin loads and unloads its oxygen with great speed. Myoglobin is probably synthesized within the muscle fiber much as hemoglobin is synthesized within the red blood cell. A direct relationship has been

Table 1.7
Morphological and histochemical types of twitch fibers in mammalian skeletal muscle

Classifications			
Burke et al. (1973)	Fast fatigable	Slow fatigue-resistant	Fast fatigue-resistant
Other designations in the literature:			
(a)	Fast-twitch white	Slow-twitch intermediate	Fast-twitch red
(b)	White	Medium	Red
(c)	A	B	C
(d)	II	I	II
(e)	I	III	II
(f)	White	Intermediate	Red
(g)	A	C	B
Morphological properties			
Mitochondrial content	Small	Intermediate	Large
Z line	Narrow	Intermediate	Broad
Fiber diameter	Large	Intermediate	Small
Neuromuscular junction	Large and complex	Intermediate	Small and simple
Histochemical properties			
Distribution of succinic dehydrogenase	Even network	Even network	Predominantly subsarcolemmal
Oxidative enzyme activities	Low	High or intermediate	Intermediate or high
Mitochondrial ATPase	Low	Intermediate	High
Glycolytic activities	High	Low or variable	Intermediate or low
Myoglobin content	Low	High	High
Glycogen content	Intermediate	Low	High
Myofibrillar ATPase at pH 9.4	High	Low or variable	High
pH sensitivity of myofibrillar ATPase	Acid labile, alkali stable	Acid stable, alkali labile	Acid labile, alkali stable
Formaldehyde sensitivity of myofibrillar ATPase	Sensitive		Stable

Table, in modified form, and pertinent references (not shown) are from Close, 1972.

found between the myoglobin content of muscle and the rate of blood flowing through a given muscle mass, usually referred to as the fractional blood flow and expressed as the percentage of cardiac output perfusing 100 g of muscle.

The *density of the capillary network* (as measured by the alkaline phosphatase reaction) is highest in the slow, fatigue-resistant fibers, lowest in fast fatigable fibers, and intermediate in fast fatigue-resistant fibers. Myocardial fibers and the external oculomotor fibers have a denser capillary network than other muscles. Twitch time is proportional to the fractional blood flow (thus, red muscles twitch more slowly than white muscles).

Metabolism of the red fibers is mostly aerobic, while white muscles can also function anaerobically as illustrated by the increase in the lactic acid content in the veins draining an exercising muscle. Such anaerobic metabolism leads to "oxygen debt," from which the muscle must ultimately recover by oxidative replenishment of its energy stores. White fast muscles undergo only short bursts of activity and possess little myoglobin. It should be recalled, however, that most muscles are a mixture of both red (slow and fast fatigue-resistant and white fast fatigable) fibers.

Both white and red fast-twitch muscles not only are able to contract faster than red slow-twitch muscle, but they also relax faster. The shorter relaxation time of either white or red fast-twitch muscle correlates with a faster rate of calcium transport by the sarcoplasmic reticulum compared to slow-twitch muscle, as determined in in vitro studies with sarcoplasmic reticulum vesicles.

Heterogeneity of Muscle

The above fiber types have been well characterized in the cat, where some muscles consist almost exclusively of one fiber type. In the human, most skeletal muscles are mixed, that is, they contain different fiber types in various percentages, although a single fiber type may predominate in some muscles. For example, the human soleus consists largely of slow, fatigue-resistant fibers. The anatomically equivalent muscle may consist of different percentages of fibers in various subspecies of the same animal. Thus the leg muscles are primarily red in the rabbit but primarily white in the hare.

Whole muscles provide a spectrum of speeds of contraction and relaxation as shown in Figure 1.94. It can be seen that the contraction time of an extraocular muscle (7.5 ms at 37°C) is five times faster than that of the gastrocnemius muscle (40 ms) and 12 times faster than that of the soleus muscle (90 ms). The extraocular muscles and the

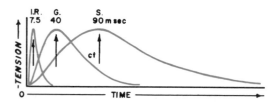

Figure 1.94. Illustration of the isometric contraction-relaxation (twitch) curves for three mammalian skeletal muscles. The lateral rectus (*l.r.*), gastrocnemius (*G*), and soleus (*S*) represent the extremes (fast and slow) and midrange in a spectrum of contraction-relaxation times for mammalian muscles. The *arrows* indicate peak tension development which corresponds to the contraction time for each muscle (lateral rectus, 7.5 ms; gastrocnemius, 40 ms; and soleus, 90 ms).

soleus represent the fast and slow extremes in most mammals. It is important to remember that whole muscles are not normally activated synchronously to yield the maximal twitch response of the kind recorded in Figure 1.94. Furthermore, the twitch response of isolated whole muscles does not reveal the heterogeneous nature of the tissue either in functional or structural terms. The gastrocnemius illustrates the heterogeneity of muscle. The white gastrocnemius muscle of the cat is in the midrange of muscle speed (contraction time is about 40 ms and contraction-relaxation time about 160 ms). The contraction time of the motor units which comprise the gastrocnemius vary from 17.8 to 129 ms, with a distribution peak between 30 and 40 ms. The peak distribution corresponds to the contraction time for the twitch of the isolated whole muscle.

Effects of Hormones on Muscle

Key effects of hormones on skeletal muscle are outlined only briefly below, as these are discussed in greater detail in chapters on the endocrine system.

Insulin is necessary for the uptake of glucose by the muscle fiber and most but not all other cells. It acts either on the facilitated diffusion of glucose across the cell membrane or on the phosphorylation of glucose by hexokinase, or on both steps of glucose uptake.

Glucocorticoids, as exemplified by cortisol, mobilize amino acids from extrahepatic tissues, primarily from muscle, by producing a decrease in protein synthesis. The amino acids are then carried via the bloodstream to the liver where they participate in gluconeogenesis, which is also stimulated by cortisol. In the normal state, glucocorticoids maintain blood sugar and glycogen content of liver and muscle. In Cushing's syndrome and other conditions such as clinical treatment with steroids where there is prolonged elevation of circulating levels of glucocorticoids, glucocorticoids produce a diabetes-like state because of their ability to raise blood sugar and to decrease the sensitivity of muscle and other insulin-sensitive tissues to insulin. Glucocorticoids thus are said to have a diabetogenic effect. Mobilization of amino acids from muscle may result in a generalized decrease in muscle mass and may contribute to the muscle wasting which is commonly seen in Cushing's syndrome (cortisol excess).

Testosterone has a marked anabolic effect on muscle due to its ability to stimulate protein synthesis.

Epinephrine has already been described as an important regulator of glycogen synthesis and glycogenolysis because of its ability to stimulate cyclic AMP formation in the cell. In white fast-contracting skeletal muscle, it produces a slight inotropic effect, i.e., it enhances twitch tension. This effect requires pharmacologic doses of epinephrine and therefore probably does not occur under normal physiologic conditions. In red slow-contracting skeletal muscle, epinephrine in physiologic doses causes a decrease in twitch tension and an increase in the rate of muscle relaxation. The latter effect appears to be on the sarcoplasmic reticulum of slow-contracting skeletal muscle, which like cardiac sarcoplasmic reticulum, contains a low molecular weight protein called phospholamban, which, when phosphorylated by cyclic AMP-dependent protein kinase, increases the activity of the calcium pump associated with these membranes (Kirchberger and Tada, 1976).

Thyroid hormone, as mentioned above, is known to affect the myosin phenotype or isozyme pattern in a particular muscle. Clinically, hyperthyroidism produces a myopathy which is characterized by muscle weakness and fatigability. The symptoms may be related to a change in skeletal muscle phenotype either as a result of direct effect of thyroid hormone or its lack on the myosin isozyme composition of the muscle or an indirect effect due to a change in the pattern of neural activation of the muscle with a resultant loss of fast-twitch fibers (Johnson et al., 1980). In the latter case, a change in myosin light chains would be expected to occur based on our knowledge gained from cross-innervation experiments. In cardiac muscle, thyroid hormone produces isozymes of myosin which differ with respect to the heavy chain.

BIBLIOGRAPHY

ABBOTT, B. C., A. V. HILL, AND J. V. HOWARTH. The positive and negative heat production associated with a nerve impulse. *Proc. Roy. Soc. (London) Ser. B.* 148: 149–187, 1958.

ADRIAN, E. D. *The Mechanism of Nervous Action.* Philadelphia: Univ. Pennsylvania Press, 1935 (Reprinted, 1959).

ADRIAN, R. H. The effect of internal and external potassium concentration on the membrane potential of frog muscle. *J. Physiol. (London)* 133: 631–658, 1956.

ADRIAN, R. H., L. L. COSTANTIN, AND L. D. PEACHEY. Radial spread of contraction in frog muscle fibres. *J Physiol. (London)* 204: 231–257, 1969.

ADRIAN, R. H., W. K. CHANDLER, AND A. L. HODGKIN. Slow changes in potassium permeability in skeletal muscle. *J. Physiol. (London)* 208: 645–668, 1970.

ANFINSEN, C. *The Molecular Basis of Evolution.* New York: John Wiley and Sons, 1960.

ARMSTRONG, C. M. Ionic pores, gates, and gating currents. *Qtr. Rev. Biophys.* 7: 179–210, 1975.

ARMSTRONG, C. M., AND F. BEZANILLA. Inactivation of the sodium channel. II. Gating current experiments. *J. Gen. Physiol.* 70: 567–570, 1977.

BÁRÁNY, M. ATPase activity of myosin correlated with speed of muscle shortening. *J. Gen. Physiol.* 208: 197–218, 1967.

BASMAJIAN, J. V. *Muscles Alive: Their Functions Revealed by Electromyography,* Ed. 4. Baltimore: Williams & Wilkins, 1978.

BEGENISICH, T., AND C. F. STEVENS. How many conductance states do potassium channels have? *Biophys. J.* 15: 843–846, 1975.

BENNETT, H. S. The concepts of membrane flow and membrane vesiculation as mechanisms for active transport and ion pumping. *J. Biophys. Biochem. Cytol.* 2(Suppl.):99–103, 1956.

BERNSTEIN, J. *Elektrobiologie.* Braunschweig, Germany: Friedr. Vieweg Sohn, 1912.

BLAND, J. H. (ed.). Seminars on the Lesch-Nyhan syndrome. *Fed. Proc.* 27: 1027–1112, 1968.

BLOOM, W., AND D. W. FAWCETT. *A Textbook of Histology,* Ed. 2. Philadelphia: W. B. Saunders, 1975.

BOWDITCH, H. P; Über die Eigenthumlichkeiten der Reizbarkeit, welche die Muskelfasern des Herzens zeigen. *Arb. Physiol. Anstalt, Leipzig* 6: 139–176, 1871.

BROCK, L. G., J. S. COOMBS, AND J. C. ECCLES. The recording of potentials from motoneurons with an intracellular electrode. *J. Physiol. (London)* 117: 431–460, 1952.

BRODSKY, W. A., A. E. SHAMOO, AND I. L. SCHWARTZ. Dissipative transport processes. In: *Handbook of Neurochemistry,* edited by A. Lajtha. New York: Plenum Press, 1971, vol. V, pt. B, chap. 20, p. 645–681.

BRONNER, F., AND A. KLEINZELLER (eds.). *Current Topics in Membranes and Transport.* New York: Academic Press, vols. 1–16, 1970–1982.

BROOKHAVEN NATIONAL LABORATORY. *Structure, Function and Evolution in Proteins,* report of a symposium. Brookhaven Symposia in Biology, 21, vol. I and II, 1969.

BURKE, R. E., D. N. LEVINE, P. TSAIRIS, AND F. E. ZAJAC, III. Physiological types and histochemical profiles in motor units of the cat gastrocnemius. *J. Physiol. (London)* 234: 723–748, 1973.

BURKE, R. E., AND P. TSAIRIS. The correlation of physiological properties with histochemical characteristics in single muscle units. *Ann. N.Y. Acad. Sci.* 228: 145–159, 1974.

CALDWELL, P. C., A. L. HODGKIN, R. D. KEYNES, AND T. L. SHAW. The effects of injecting "energy-rich" phosphate compounds on the active tranport of ions in the giant axons of *Loligo. J. Physiol. (London)* 152: 561–590, 1960.

CHANDLER, W. K., AND M. F. SCHNEIDER. Time-course of potential spread along a skeletal muscle fiber under voltage clamp. *J. Gen. Physiol.* 67: 165–184, 1976.

CHANGEUX, J.-P. The acetylcholine receptor: An "allosteric" membrane protein. *Harvey Lectures* 75: 85–255, 1981.

CHAPMAN, D., AND D. F. H. WALLACH (eds.). *Biological Membranes.* London: Academic Press, 1968–1976, vols. 1–3.

CLOSE, R. I. Dynamic properties of skeletal muscle. *Physiol. Rev.* 52: 129–197, 1972.

COLD SPRING HARBOR SYMPOSIA ON QUANTITATIVE BIOLOGY. *The Mechanism of Muscle Contraction.* Cold Spring Harbor, New York: Cold Spring Harbor Laboratory Press, 1972, vol. 37.

COLE, K. S., AND J. W. MOORE. Ionic current measurements in the squid giant axon membrane. *J. Gen. Physiol.* 44: 123–167, 1960.

COLLANDER, R. Cell membranes: their resistance to penetration and their capacity for transport. In: *Plant Physiology,* edited by F. C. Steward. New York: Academic Press, 1959, vol. II.

CONNELLY, C. M. Recovery processes and metabolism of nerve. In: *Biophysical Science,* edited by J. L. Oncley. New York: John Wiley and Sons, chap. 51, p. 475–484, 1959.

CONTI, F., B. HILLE, B. NEUMCKE, W. NONNER, AND R. STÄMPFLI. Measurement of the conductance of the sodium channel from current fluctuations at the node of Ranvier. *J. Physiol. (London)* 262: 699–727, 1976.

COOMBS, J. S., J. C. ECCLES, AND P. FATT. The inhibitory suppression of reflex discharges from motoneurones. *J. Physiol. (London)* 130: 396–413, 1955.

CONSTANTIN, L. L., AND S. R. TAYLOR. Graded activation in frog muscle fibres. *J. Gen. Physiol.* 61: 424–443, 1973.

CURTIN, N. A., AND R. C. WOLEDGE. Energy changes and muscular contraction. *Physiol. Rev.* 58: 690–761, 1978.

CURTIS, H. J., AND K. S. COLE. Transverse electric impedance of the squid giant axon. *J. Gen. Physiol.* 21: 757–765, 1938.

DAHL, J. L., AND L. E. HOKIN. The sodium-potassium adenosinetriphosphatase. *Ann. Rev. Biochem.* 43: 327–356, 1974.

DINNO, M. A., AND A. B. CALLAHAN (eds.). *Membrane Biophysics: Structure and Function in Epithelia.* New York: Alan R. Liss, 1981.

DOWBEN, R. M. *General Physiology.* New York: Harper & Row. 1969.

DuPRAW, E. J. *Cell and Molecular Biology.* New York: Academic Press, 1968.

EBASHI, S., AND M. ENDO. Calcium ion and muscle contraction. *Progr. Biophys. Mol. Biol.* 18: 123–183, 1968.

EBASHI, S., M. ENDO, AND I. OHTSUKI. Control of muscle contraction. *Qtr. Rev. Biophys.* 2: 351–385, 1969.

ECCLES, J. C. *The Neurophysiological Basis of Mind.* London: Oxford Univ. Press, 1953.

EDMAN, K. A. The rising phase of the active state in single skeletal muscle fibers of the frog. *Acta Physiol. Scand.* 79: 167–173, 1970.

EINSTEIN, A. *Investigations on the Theory of the Brownian Movement.* New York: Dover Publications, 1926.

ELLIOT, A., AND G. OFFER. Shape and flexibility of the myosin molecule. *J. Mol. Biol.* 123: 505–519, 1978.

ERLANGER, J., AND H. S. GASSER. *Electrical Signs of Nervous Activity.* Philadelphia: Univ. Pennsylvania Press, 1937.

FATT, P., AND B. KATZ. Spontaneous subthreshold activity of motor nerve endings. *J. Physiol (London)* 117: 109–128, 1952.

FAWCETT, D. *The Cell, an Atlas of Fine Structure.* Philadelphia: W.B. Saunders, 1966.

FAWCETT, D. W., AND N. S. McNUTT. The ultrastructure of the cat myocardium. I. Ventricular papillary muscle. *J. Cell Biol.* 42: 1–45, 1969.

FENN, W. O. A quantitative comparison between the energy liberated and the work performed by the isolated sartorius muscle of the frog. *J. Physiol. (London)* 58: 175–203, 1923.

FENN, W. O., AND B. S. MARSH. Muscular force at different speeds of shortening. *J. Physiol. (London)* 85: 277–297, 1935.

FINEAN, J. B., R. COLEMAN, AND R. H. MITCHELL. *Membranes and Their Cellular Functions,* Ed. 2. New York: Halsted Press, John Wiley and Sons, 1978.

FORRESTER, T., AND H. SCHMIDT. An electrophysiological investigation of the slow fibre system in the frog rectus abdominis muscle. *J. Physiol. (London)* 207: 477–491, 1970.

FRANKENHAEUSER, B., AND A. L. HODGKIN. The aftereffects of impulses of the giant nerve fibers of *Loligo. J. Physiol. (London)* 131:341–376, 1956.

FRANZINI-ARMSTRONG, C. Studies of the triad. I. Structure of the junction in frog twitch fibers. *J. Cell Biol.* 47: 488–499, 1970.

FRANZINI-ARMSTRONG, C. Membranous system in muscle

fibers. In: *The Structure and Function of Muscle*, edited by G. H. Bourne. New York: Academic Press, 1973, vol. 2, pt. 2, p. 531–619.

FRANZINI-ARMSTRONG, C., AND L. D. PEACHEY. Striated muscle—contraction and control mechanisms. *J. Cell Biol.* 91: 166s–186s, 1981.

FRYE, L. D., AND M. EDIDIN. The rapid intermixing of cell surface antigens after formation of mouse-human heterokaryons. *J. Cell Sci.* 7: 319–335, 1970.

GALL, J. G., K. R. PORTER, AND P. SIEKEVITZ (eds.). Discovery in cell biology. *J. Cell Biol.* 91: 1s–306s, 1981.

GASSER, H. S., AND HILL, A. V. The dynamics of muscular contraction. *Proc. Roy. Soc. (London) Ser. B.* 96: 398–437, 1924.

GEREN, B. B. The formation from the Schwann cell surface of myelin in the peripheral nerves of chick embryos. *Exp. Cell. Res.* 7: 558–562, 1954.

GIBBONS, I. R. The organization of cilia and flagella. In: *Molecular Organization and Biological Function*, edited by J. M. Allen. New York: Harper and Row, 1967, Chap. 8, p. 211–237.

GORDON, A. M., AND E. B. RIDGWAY. Length-dependent electromechanical coupling in single muscle fibers. *J. Gen. Physiol.* 68: 653–669, 1976.

GORDON, A. M., A. F. HUXLEY, AND F. J. JULIAN. The variation in isometric tension with sarcomere length in vertebrate muscle fibres. *J. Physiol. (London)* 184: 170–192, 1966.

GREASER, M. L., AND J. GERGELY. Reconstituion of troponin activity from three protein components. *J. Biol. Chem.* 246: 4226–4233, 1971.

GREENGARD, P. Possible role for cyclic nucleotides and phosphorylated membrane proteins in postsynaptic actions of neurotransmitters. *Nature (London)* 260: 101–108, 1976.

GUIDOTTI, G. Membrane proteins. *Ann. Rev. Biochem.* 41: 731–752, 1972.

HAMILL, O. P., A. MARTY, E. HEHER, B. SAKMANN, AND F. J. SIGWORTH. Improved patch-clamp techniques for high-resolution current recordings from cells and cell-free membrane patches. *Pflügers Arch.* 391: 85–100, 1981.

HARRINGTON, W. F. Contractile proteins of muscle. In: *The Proteins*, Ed. 3, edited by H. Neurath and R. L. Hill. New York: Academic Press, 1979, vol. IV, chap. 3, p. 245–409.

HARTSHORNE, D. J. Interactions of desensitized actomyosin with tropomyosin, troponin A, troponin B, and polyanions. *J. Gen. Physiol.* 55: 585–601, 1970.

HEDQVISTK, P. Role of the α-receptor in the control of noradrenaline release from sympathetic nerves. *Acta Physiol. Scand.* 90: 158–165, 1974.

HEINZ, E. *Mechanics and Energetics of Biological Transport*. Berlin: Springer-Verlag, 1978.

HEINZ, E., AND P. M. WALSH. Exchange diffusion, transport and intracellular level of amino acids in Ehrlich carcinoma cells. *J. Biol. Chem.* 233: 1488–1493, 1958.

HENNEMAN, E., H. P. CLAMANN, J. D. GILLIES, AND R. D. SKINNER. Rank order of motoneurons within a pool: law of combination. *J. Neurophysiol.* 37: 1338–1349, 1974.

HILL, A. V. Muscular Movement in Man. New York: McGraw-Hill, 1927.

HILL, A. V. The heat of shortening and the dynamic constants of muscle. *Proc. R. Soc. (London) Ser. B* 126: 136–195, 1938.

HILL, A. V. The abrupt transition from rest to activity in muscle. Proc. Roy. *Soc. B,* 136: 399–419, 1949.

HILL, A. V. The heat production of muscle and nerve. *Ann. Rev. Physiol.* 21: 1–18, 1959.

HILL, A. V. *First and Last Experiments in Muscle Mechanics*. London: Cambridge Univ. Press, 1970.

HILL, D. K. The effect of temperature on the resting tension of frog's muscle in hypertonic solutions. *J. Physiol. (London)* 208: 741–756, 1970.

HILLE, B. Pharmacological modifications of the sodium channels of frog nerve. *J. Gen. Physiol.* 51: 199–219, 1968.

HILLE, B. Ionic channels in excitable membranes: current problems and biophysical approaches. *Biophys. J.* 22: 283–294, 1978.

HODGKIN, A. L. The subthreshold potentials in a crustacean nerve fiber. *Proc. Roy. Soc. (London) Ser. B* 126: 87–121, 1939.

HODGKIN, A. L. Ionic movements and electrical ativity in giant nerve fibres. *Proc. Roy. Soc. (London) Ser. B* 148: 1–37, 1957.

HODGKIN, A. L. The ionic basis of nervous conduction. *Science* 145: 1148–1154, 1964.

HODGKIN, A. L. *The Conduction of the Nerve Impulse*. Springfield, IL: Charles C Thomas, 1964.

HODGKIN, A. L., AND A. F. HUXLEY. Resting and action potentials in single nerve fibres. *J. Physiol. (London)* 104: 176–195, 1945.

HODGKIN, A. L., AND A. F. HUXLEY. Currents carried by sodium and potassium ions through the membrane of the giant axon of *Loligo. J. Physiol. (London)* 116: 449–472, 1952a.

HODGKIN, A. L., AND A. F. HUXLEY. The components of membrane conductance in the giant axon of *Loligo. J. Physiol. (London)* 116: 473–496, 1952b.

HODGKIN, A. L., AND A. F. HUXLEY. The dual effect of membrane potential on sodium conductance in the giant axon of *Loligo. J. Physiol. (London)* 116: 497–506, 1952c.

HODGKIN, A. L., AND A. F. HUXLEY. A quantitative description of membrane current and its application to conduction and excitation in nerve. *J. Physiol. (London)* 117: 500–544, 1952d.

HODGKIN, A. L., AND R. D. KEYNES. The mobility and diffusion coefficient of potassium in giant axons from *Sepia. J. Physiol. (London)* 119: 513–528, 1953.

HODGKIN, A. L., AND R. D. KEYNES. Active transport of cations in giant axons from *Sepia* and *Loligo. J. Physiol. (London)* 128: 28–60, 1955.

HODGKIN, A. L., AND R. D. KEYNES. Experiments on the injection of substances into squid giant axons by means of a microsyringe. *J. Physiol. (London)* 131: 592–616, 1956.

HODGKIN, A. L., A. F. HUXLEY, AND B. KATZ. Measurement of current-voltage relations in the membrane of the giant axon of *Loligo. J. Physiol. (London)* 116: 424–448, 1952.

HOFFMAN, J. F. (ed.). *Membrane Transport Processes*. New York: Raven Press, 1977, vol. 1.

HOLTER, H. Pinocytosis. *Int. Rev. Cytol.* 8: 481–504, 1959.

HOLTZMAN, E., AND A. B. NOVIKOFF. *Cells and Organelles*, Ed. 3. New York: Holt, Rinehart and Winston, 1982.

HOMSHER, E., AND C. J. KEAN. Skeletal muscle energetics and metabolism. *Ann. Rev. Physiol.* 40: 93–131, 1978.

HUBBELL, W. L., AND H. M. McCONNELL. Spin-label studies of the excitable membranes of nerve and muscle. *Proc. Natl. Acad. Sci. U. S. A.* 61: 12–16, 1968.

HUBBELL, W. L., AND H. M. McCONNELL. Molecular motion in spin-labeled phospholipids and membranes. *J. Am. Chem. Soc.* 93: 314–326, 1971.

HUGHES, J., T. W. SMITH, H. W. KOSTERLITZ, L. A. FOTHERGILL, B. A. MORGAN, AND H. R. MORRIS. Identification of two related pentapeptides from the brain with potent opiate agonist activity. *Nature (London)* 258: 577–579, 1975.

HURSH, J. B. Conduction velocity and diameter of nerve fibers. *Am. J. Physiol.* 127: 131–139, 1939.

HUXLEY, A. F. Electrical processes in nereve conduction. In: *Ion Transport across Membranes.* edited by H. T. Clarke and D. Nachmansohn. New York: Academic Press, 1954, p. 23–24.

HUXLEY, A. F. Muscle structure and theories of contraction. *Progr. Biophys.* 7: 255–318, 1957.

HUXLEY, A. F. Excitation and conduction in nerve: quantitative analysis. *Science* 145: 1154–1159, 1964.

HUXLEY, A. F. Muscular contraction. *J. Physiol. (London)* 243: 1–43, 1974.

HUXLEY, A. F., AND R. M. SIMMONS. Mechanical transients and the origin of muscular force. *Cold Spring Harbor Symp. Quant. Biol.* 37: 669–686, 1973.

HUXLEY, A. F., AND R. STÄMPFLI. Evidence for saltatory conduction in peripheral myelinated nerve fibres. *J. Physiol. (London)* 108: 315–339, 1949.

HUXLEY, A. F., AND R. STÄMPFLI. Direct determination of membrane resting potential and action potential in single myelinated nerve fibres. *J. Physiol. (London)* 112: 476–495, 1951.

HUXLEY, A. F., AND R. W. STRAUB. Local activation and interfibrillar structures in striated muscle. *J. Physiol. (London)* 143: 40p–41p, 1958.

HUXLEY, A. F., AND R. E. TAYLOR. Local activation of striated muscle fibres. *J. Physiol. (London)* 144: 426–441, 1958.

HUXLEY, H. E. Electron microscope studies on the structure of natural and synthetic protein filaments from striated muscle. *J. Mol. Biol.* 7: 281–308, 1963.

HUXLEY, H. E. Structural evidence concerning the mechanism of contraction in striated muscle. In: *Muscle: Proceedings of a Symposium Held at the Faculty of Medicine, University of Alberta*, edited by W. M. Paul, E. E. Daniel, C. M. Kay, and G. Monckton. Oxford: Pergamon, p. 3–28, 1964.

HUXLEY, H. E. The mechanism of muscular contraction. *Science* 164: 1356–1365, 1969.

HUXLEY, H. E., AND J. HANSON. The molecular basis of contraction in cross-striated muscles. In: *The Structure and Function of Muscle*, edited by G. H. Bourne. New York: Academic Press, 1960, vol. 1, chap. 7, p. 183–227.

JOHNSON, M. A., F. L. MASTAGALIA, AND A. G. MONTGOMERY. Changes in myosin light chains in the rat soleus after thyroidectomy. *FEBS Lett.* 110: 230–235, 1980.

KAPLAN, D. M., AND R. S. CRIDDLE. Membrane structural proteins. *Physiol. Rev.* 51: 249–273, 1971.

KATZ, B. The relation between force and speed in muscular contraction. *J. Physiol. (London)* 96: 45–64, 1939.

KATZ, B. *Nerve, Muscle and Synapse*. New York: McGraw-Hill, 1966.

KATZ, B. *The Release of Neural Transmitter Substances*. Springfield, IL: Charles C Thomas, 1969.

KATZ, A. M., AND A. J. BRADY. Mechanical and biochemical correlates of cardiac contraction (I). *Mod. Concepts Cardiovasc. Dis.* 40: 39–43, 1971.

KENDREW, J. C. Side-chain interactions in myoglobin. *Brookhaven Symp. Biol.* 15: 216–228, 1962.

KEYNES, R. D. The ionic movements during nervous activity. *J. Physiol. (London)* 114: 119–150, 1951.

KEYNES, R. D. Ion channels in the nerve-cell membrane. *Sci. Am.* 240: 126–135, 1979.

KINNE, R., AND I. L. SCHWARTZ. Resolution of the epithelial cell envelope into luminal and contraluminal plasma membranese as a tool for the analysis of transport processes and hormone action. In: *Membranes and Disease*, edited by L. Bolis, J. F. Hoffman, and A. Leaf. New York: Raven Press, 1976.

KINNE, R., AND I. L. SCHWARTZ. Isolated membrane vesicles in the evaluation of the nature, localization, and regulation of renal transport processes. *Kidney International* 14: 547–556, 1978.

KIRCHBERGER, M. A., AND M. TADA. Effects of adenosine 3′:5′-monophosphate-dependent protein kinase on sarcoplasmic reticulum isolated from cardiac and slow and fast contracting skeletal muscles. *J. Biol. Chem.* 251: 725–729, 1976.

KNAPPEIS, G. G., AND F. CARLSEN. The ultrastructure of the M line in skeletal muscle. *J. Cell. Biol.* 38: 202–211, 1968.

KUSHMERICK, M. J. Energy balance in muscle contraction: a biochemical approach. *Curr. Topics Bioenergetics* 6: 1–37, 1977.

KUSHNER, D. Self-assembly of biological structure. *Bact. Rev.* 33: 302–345, 1969.

KYTE, J. Purification of the sodium- and potassium-dependent adenosine triphosphatase from canine and renal medulla. *J. Biol. Chem.* 246: 4157–4165, 1971.

LEACH, S. J. (ed.). *Physical Principles and Techniques of Protein Chemistry*. New York: Academic Press, 1969, vol. 7, part A.

LEHNINGER, A. L. *The Mitochondrion*. New York: W. A. Benjamin, 1964.

LIMA-DE-FARIA, A. (ed.) *Handbook of Molecular Cytology*. Amsterdam: North Holland, 1969.

LLINÁS, R., AND K. WALTON. In: *The Cell Surface and Neuronal Function*, edited by C. W. Cotman, G. Poste, and G. L. Nicolson. New York: Elsevier/North-Holland Biomedical Press, 1980, chap. 3, p. 87–118.

LLOYD, D. P. C. Facilitation and inhibition of spinal motoneurones. *J. Neurophysiol.* 9: 421–438, 1946.

LOEWI, O. Über humoral Übertagbarkeit der Herznervenwirkung. I. Mitteilung. *Pflügers Arch.* 189: 239–242, 1921.

LOEWY, A. G., AND P. SIEKEVITZ. *Cell Structure and Function*, Ed. 2. New York: Holt, Rinehart and Winston, 1969.

LYMN, R. W., AND W. W. TAYLOR. Mechanisms of adenosinetriphosphate hydrolysis by actomyosin. *Biochemistry* 10: 4617–4624, 1971.

MACPHERSON, L., AND D. R. WILKIE. The duration of the active state in a muscle twitch. *J. Physiol. (London)* 124: 292–299, 1954.

McCUBBIN, W. D., AND C. M. KAY. Calcium-induced conformational changes in the troponin-tropomyosin complexes of skeletal and cardiac muscles and their roles in the regulation of contraction-relaxation. *Acct. Chem. Res.* 13: 185–192, 1980.

McILWAIN, H. *Biochemistry and the Central Nervous System*, Ed. 3. Boston: Little, Brown and Co., 1966.

McPHEDRAN, A. M., R. B. WUERKER, AND E. HENNEMAN. Properties of motor units in a homogeneous red muscle (soleus) of the cat. *J. Neurophysiol.* 28: 71–84, 1965.

MARGOULIES, M. (ed.). *Protein and Polypeptide Hormones*. Amsterdam: Excerpta Medica Foundation, 1968.

MITCHELL, P. Vectorial chemistry and the molecular mechanisms of chemiosmotic coupling: power transmission by proticity. *Trans. Biochem. Soc.* 4: 662–663, 1976.

MITCHELL, P. Keilin's respiratory chain concept and its chemiosmotic consequences. *Science* 206: 1148–1159, 1979.

MORGAN, D. L., AND U. PROSKE. Vertebrate slow muscle: its structure, pattern of innervation, and mechanical properties. *Physiol. Rev.* 64: 103–169, 1984.

MULIERI, L. A., AND N. R. ALPERT. Activation heat and latency relaxation in relation to calcium movement in skeletal and cardiac muscle. *Can. J. Physiol. Pharmacol.* 60: 415–588, 1982.

MURER, H., AND R. KINNE. The use of membrane vesicles to study epithelial transport porcesses. *J. Membrane Biol.* 55: 81–95, 1980.

MURPHY, R. A., AND A. C. BEARDSLEY. Mechanical properties of the cat soleus muscle in situ. *Am. J. Physiol.* 227: 1008–1013, 1974.

MURRAY, J. M., AND A. WEBER. The cooperative action of muscle proteins. *Sci. Am.* 230: 58–71, 1974.

MURRAY, J. M., A. WEBER, AND M. K. KNOX. Myosin sulfragment 1 binding to relaxed actin filaments and steric model of relaxation. *Biochemistry* 20: 641–649, 1981.

NEHER, E., B. SAKMANN, AND J. H. STEINBACH. The extracellular patch clamp: a method for resolving currents through individual open channels in biological membranes. *Pflügers Arch.* 375: 219–228, 1978.

NERNST, W. Zür Theorie des elektrischen Reizes. *Pflügers Arch.* 122: 275–314, 1908.

NEW YORK HEART ASSOCIATION. The contractile process. Proceedings of a Symposium. *J. Gen. Physiol.* 50(Suppl. 6): 5–292, 1967.

NORRIS, F. H., JR., AND R. L. IRWIN. Motor unit area in a rat muscle. *Am. J. Physiol.* 200: 944–946, 1961.

NOVIKOFF, A. B., AND E. HOLTZMAN. *Cells and Organelles*, 2nd ed. New York: Holt, Rinehart and Winston, 1976.

PAGE, S. G. Comparison of the fine structures of frog slow and twitch muscle fibres. *J. Cell. Biol.* 26: 477–497, 1965.

PARSONS, C., AND K. R. PORTER. Muscle relaxation: evidence for an intrafibrillar restoring force in vertebrate striated muscle. *Science* 153: 426–427, 1966.

PEACHEY, L. D. The sarcoplasmic reticulum and transverse tubules of the frogs's sartorius. *J. Cell Biol.* 25: 209–231, 1965.

PEMRICK, S. M. (1980). The phosphorylated L_2 light chain of skeletal myosin is a modifier of the actomyosin ATPase. *J. Biol. Chem.* 255: 8836–8841, 1980.

PENEFSKY, Z. P. A model active state in cardiac muscle: A study of the first derivative of isometric tension. In: *Research in Physiology*, edited by F. F. Kao, K. Koizumi, and M. Vassalle. Bologna: Aulo Gaggi Publisher, 1971, p. 239–251.

PENNEY, J. B., AND A. B. YOUNG. Quantitative autoradiography of neurotransmitter receptors in Huntington disease. *Neurology* 32: 1391–1395, 1982.

PERRY, S. V., H. A. COLE, J. F. HEAD, AND F. J. WILSON. Localization and mode of action of the inhibitory protein component of the troponin complex. *Cold Spring Harbor Symp. Quant. Biol.* 37: 251–262, 1972.

PERRY, T. L., S. HANSEN, AND M. KLOSTER. Huntington's chorea: deficiency of γ-amino-butyric acid in brain. *N. Engl. J. Med.* 288: 337–342, 1973.

PETERS, A., S. L. PALAY, AND H. DEF. WEBSTER. *The Fine Structure of the Nervous System: The Neurons and Supporting Cells.* Philadelphia: W. B. Saunders, 1976.

PETERS, J. P., AND D. D. VAN SLYKE. *Quantitative Clinical Chemistry*, Ed. 2 Baltimore: Williams & Wilkins, 1946, vol. 1, p. 937.

PETERSON, I., AND E. KUGELBERG. Duration and form of action potential in the normal human muscle. *J. Neurol. Neurosurg. Psychiatry* 12: 124–128, 1949.

PINTO DA SILVA, P., AND D. BRANTON. Membrane splitting in freeze-etching. *J. Cell Biol.* 45: 598–605, 1970.

PORTER, K. R., AND A. B. NOVIKOFF. The 1974 Nobel Prize for Physiology or Medicine. *Science* 186: 516–520, 1974.

PORTER, K. R., AND G. E. PALADE. Studies on the endoplasmic reticulum. III. Its form and distribution in striated muscle cells. *J. Biophys. Biochem. Cytol.* 3: 269–300, 1957.

POST, R. L., C. R. MERRITT, C. R. KINSOLVING, AND C. D. ALBRIGHT. Membrane adenosine triphosphatase as a participant in the active transport of sodium and potassium in the human erythrocyte. *J. Biol. Chem.* 235: 1796–1802, 1960.

POTTER, J. D., AND J. GERGELY. Troponin, tropomyosin, and actin interactions in the Ca^{2+} regulation of muscle contraction. *Biochemistry* 13: 2697–2703, 1974.

PREWITT, M. A., AND B. SAFALSKY. Enzymic and histochemical changes in fast and slow muscles after cross innervation. *Am. J. Physiol.* 218: 69–74, 1970.

PROSSER, C. L., AND F. A. BROWN, JR. *Comparative Animal Physiology*, Ed. 3. Philadelphia: W. B. Saunders, 1961.

RALL, J. A. Sense and nonsense about the Fenn effect. *Am. J. Physiol.* 242: H1–H6, 1982.

REIS, D. J., AND G. F. WOOTEN. The relationship of blood flow to myoglobin, capillary density, and twitch characteristics in red and white skeletal muscle in cat. *J. Physiol. (London)* 210: 121–135, 1970.

RENSHAW, B. Central effects of centripetal impulses in axons of spinal ventral roots. *J. Neurophysiol.* 9: 191–204, 1946.

RIDGEWAY, E. B., AND C. C. ASHLEY. Calcium transients in single muscle fibers. *Biochem. Biophys. Res. Commun.* 29: 229–234, 1967.

ROBERTSON, J. D. New observations on the ultrastructure of the membranes of frog peripheral nerve fibers. *J. Biophys. Biochem. Cytol.* 3: 1043–1047, 1957.

ROBERTSON, J. D. Unit membranes: a review with recent new studies of experimental alterations and a new subunit structure in synaptic membranes. In: *Cellular Membranes in Development*, edited by M. Locke. New York: Academic Press, 1964, p. 1–81.

RÜDEL, R., AND S. R. TAYLOR. Aequorin luminescence during contraction of amphibian skeletal muscle. *J. Physiol. (London)* 233: 5P–6P, 1973.

SANDOW, A., S. R. TAYLOR, A. ISAACSON, AND J. J. SEGUIN. Electromechanical coupling in potentiation of muscular contraction. *Science* 143: 577–579, 1964.

SATIR, P. How cilia move. *Sci. Am.* 231: 44–52, 1974.

SCHNEIDER, M. F. Linear electrical properties of the transverse tubules and surface membrane of skeletal muscle fibers. *J. Gen. Physiol.* 56: 640–671, 1970.

SCHNEIDER, M. F., AND W. K. CHANDLER. Effect of membrane potential on the capacitance of skeletal muscle fibers. *J. Gen. Physiol.* 67: 125–163, 1976.

SCHOENHEIMER, R., AND D. RITTENBERG. Study of intermediary metabolism of animals with aid of isotopes. *Physiol. Rev.* 20: 218–248, 1940.

SCHRÖDER, E., AND K. LÜBKE. *The Peptides*. New York: Academic Press, 1966, vol. II.

SEMENZA, G., AND E. CARIFOLI (eds.). *Biochemistry of Membrane Transport, FEBS Symposium No. 42.* Berlin: Springer-Verlag, 1977.

SIEGEL, G. J., AND R. W. ALBERS. Nucleosidotriphosphate phosphohydrolases. In: *Handbook of Neurochemistry*, edited by A. Lajtha. New York: Plenum Press, 1970 vol. IV, chap. 2, p. 13–44.

SIEGEL, G., R. W. ALBERS, B. W. AGRANOFF, AND R. KATZMAN (eds.). *Basic Neurochemistry*, Ed. 3. Boston: Little, Brown and Co., 1981.

SINGER, S. J. The molecular organization of membranes. *Ann. Rev. Biochem.* 43: 805–833, 1974.

SINGER, S. J., AND G. L. NICOLSON. The fluid mosaic model of the structure of cell membranes. *Science* 175: 720–731, 1972.

SIVARAMAKRISHNAN, M., AND M. BURKE. The free heavy chain of vertebrate skeletal myosin subfragment 1 shows full enzymatic activity. *J. Biol. Chem.* 257: 1102–1105, 1982.

SKOU, J. C. The influence of some cations on an adenosine triphosphatase from peripheral nerve. *Biochim. Biophys. Acta* 23: 394–401, 1957.

SPIRO, D., AND E. H. SONNENBLICK. Comparison of the ultrastructural basis of the contractile process in heart and skeletal muscle. *Circ. Res.* 15 (Suppl. II) : 14–36, 1964.

SQUIRE, J. M. Muscle filament structure and muscle contraction. *Ann. Rev. Biophys. Bioeng.* 4: 137–163, 1975.

STARR, R., AND G. OFFER. Polypeptide chains of intermediate molecular weight in myosin preparations. *FEBS Lett.* 15: 40–44, 1971.

STEIN, W. D. *The Movement of Molecules Across Cell Membranes.* New York: Academic Press, 1967.

STEIN, J. M., AND H. A. PADYKULA. Histochemical classification of individual skeletal muscle fibers of the rat. *Am. J. Anat.* 110: 103–123, 1962.

STEIN, L. A., R. P. SCHWARZ, JR., P. B. CHOCK, AND E. EISENBERG. Mechanism of actomyosin adenosine triphosphatase. Evidence that adenosine 5′-triphosphate hydrolysis can occur without disruption of the actomyosin complex. *Biochemistry* 18: 3895–3909, 1979.

SWEADNER, K. J., AND S. M. GOLDIN. Active transport of sodium and potassium ions: Mechanism, function and regulation. *N. Engl. J. Med.* 302: 777–783, 1980.

TAKAHASHI, T., AND M. OTSUKA. Regional distribution of Substance P in the spinal cord and nerve roots of the cat, and the effects of dorsal root section. *Brain Research* 87: 1–11, 1975.

TANFORD, C. *The Physical Chemistry of Macromolecules.* New York: John Wiley and Sons, 1961.

TAYLOR, J. H. (ed.) *Molecular Genetics.* New York: Academic Press, 1963 and 1967, pts. I and II.

TAYLOR, S. R., R. RÜDEL, AND J. R. BLINKS. Calcium transients in amphibian muscle. *Fed. Proc.* 34: 1379–1381, 1975.

THOMPSON, S. H., AND R. W. ALDRICH. In: *The Cell Surface and Neuronal Function*, edited by C. W. Cotman, G. Poste, and G. L. Nicolson. New York: Elsevier/North-Holland Biomedical Press, 1980, chap. 2, p. 49–85.

TOSTESON, D. C. (ed.). *Membrane Transport in Biology Series.* Berlin: Springer-Verlag, 1978 and 1979, vols. I and II.

TOSTESON, D. C., Y. A. OVCHINNIKOV, AND R. LATORRE (eds.). *Membrane Transport Processes.* New York: Raven Press, 1977, vol. 2.

TOWER, D. B. Molecular transport across neural and nonneural membranes. In: *Properties of Membranes*, edited by D. B. Tower, S. A. Luse, and H. Grundfest. New York: Springer Publishing Co., 1962, p. 1–42.

TREGEAR, R. T., AND S. B. MARSTON. The crossbridge theory. *Ann. Rev. Physiol.* 41: 723–736, 1979.

VAN HARREVELD, A. On the force and size of motor units in the rabbit sartorius muscle. *Am. J. Physiol.* 151: 96–106, 1947.

VERATTI, E. Ricerche sulla fine struttura della fibra muscolare striata. *Arch. Ital. Biol.* 37: 449–454, 1902.

WEBER, A. On the role of calcium in the activity of adenosine 5′-triphosphate hydrolysis of actomyosin. *J. Biol. Chem.* 234: 2764–2769, 1959.

WEISSMANN, G., AND R. CLAIBORNE (eds). *Cell Membranes: Biochemistry, Cell Biology and Pathology.* New York: Hospital Practice Publishing Co., 1975.

WILKIE, D. H. Thermodynamics and the interpretation of biological heat measurements. *Progr. Biophys.* 10: 259–298, 1960.

WINEGRAD, S. Autoradiographic studies of intracellular calcium in frog skeletal muscle. *J. Gen. Physiol.* 48: 455–479, 1965.

WOLFE, S. L. *Biology of the Cell,* Ed. 2. Belmont, CA: Wadsworth Publishing Co., 1981.

WUERKER, R. B., A. M. McPHEDRAN, AND E. HENNEMAN. Properties of motor units in a heterogeneous pale muscle (M. gastrocneumius) of cat. *J. Neurophysiol.* 28: 85–99, 1965.

WYSSBROD, H. R., W. N. SCOTT, W. A. BRODSKY, AND I. L. SCHWARTZ. Carrier-mediated transport processes. In: *Handbook of Neurochemistry,* edited by A. Lajtha. New York: Plenum Press, 1971, vol. V, pt. B, chap. 21, p. 683–819.

YOUNG, J. Z. The functional repair of nervous tissue. *Physiol. Rev.* 22: 318–374, 1942.

Cardiovascular System

CHAPTER 5

Introduction to the Cardiovascular System

The heart and circulation constitute primarily a transport and exchange system. Large blood vessels in the circulatory loop provide the pathways for distributing and collecting blood-borne materials. The heart generates the energy for moving the blood through the circuit, and in the various organs and tissues, exchange of oxygen, carbon dioxide, and other metabolites occurs across the extremely thin walls of the capillaries. (For the historical development of concepts about the heart and circulation, see Fishman and Richards, 1964.)

An understanding of certain principles concerned with heart and circulatory function is fundamental to an appreciation of disease states, as well as of the action of various pharmacologic agents. Therefore, as we examine the physiology of the normal heart and circulation it will be useful to reinforce these physiologic principles with some examples of the derangements that can occur in human disease. Physicians encounter a great deal of heart and circulatory disease; for example, nearly 30 million people in the United States have cardiovascular disease, and about ¾ of a million people die each year from heart attack alone. Cardiovascular disease is responsible for approximately ½ of all male deaths between the ages of 35 and 64 yr, far more than accounted for by cancer. Such figures emphasize not only the importance to the physician of obtaining an appreciation of basic cardiocirculatory phenomena, but also the need for additional research in this area.

In this chapter the general components of the heart and circulation will be briefly and simply described. The details of blood vessel and cardiac structure and function will be considered in the ensuing chapters, and the regulation of certain special circulations of the body (such as those supplying the brain, the kidney, the lung, and the gastrointestinal tract) are presented in other sections of this book.

GENERAL COMPONENTS OF THE CIRCULATORY SYSTEM

Each type of blood vessel has a different function, and whereas some organs have a highly specialized circulatory supply (such as the kidney and liver), the basic components supplying most tissues and organs consist of: artery→arteriole→capillary→venule→vein.

Arteries

The arteries serve to deliver oxygenated blood to the various organs. They have a relatively thick muscular wall and are sufficiently distensible to smooth the pulsations generated by the heart (Chapter 7). The channel of the arteries ranges from several cm in diameter (the aorta, the largest artery, which directly leaves the heart) through progressive branchings down to tiny arteries 1 mm or less in diameter.

The arteries are particularly susceptible to atherosclerosis, a buildup of cholesterol and other lipids in the artery walls, which can produce narrowed regions in the channel. When such a diseased artery supplies the brain, for example, a stroke can ensue.

Arterioles

The smallest branches of the arteries are the arterioles, small thick-walled muscular vessels which regulate the resistance to flow through the various organs and tissues of the body, thereby controlling the distribution and rate of blood flow to these regions. The ratio of the thickness of the vessel wall to the diameter of the vessel is high, and contraction or relaxation of the smooth muscle in the wall thereby allows them to serve as "stopcocks" which can readily control the vessel caliber (Fig. 2.1). The state of contraction of this "vascular smooth muscle" is regulated by several factors, including nerve impulses which reach nerve terminals lying within the smooth muscle, circulating vasoactive hormones, and metabolites carried in the blood; it can also be affected by a variety of pharmacologic agents. The degree of contraction of these sphincter-like arterioles not only regulates the amount of blood flow to an organ, but when a sufficiently large number of arterioles are involved the pressure is affected in the large arteries behind them (Fig. 2.1). This effect can be likened to constriction or release of a screw-clamp placed on a garden hose; a change in resistance will raise or lower the pressure in the hose between the open faucet and the screw-clamp.

A disease which involves the arterioles is high blood pressure, or hypertension (see Chapter 16). The "tone" or state of contraction of the smooth muscle of the arterioles is increased in this disorder, and since the heart tends to maintain its output of blood to meet the needs of the body, the increased

resistance to flow provided by the arterioles throughout the body elevates the pressure in the large arteries, with a number of unfavorable consequences.

Capillaries

The blood vessels continue to divide into smaller and smaller branches to reach the smallest unit, the capillary, a tiny thin-walled vessel through which exchange of materials occurs by diffusion and ultrafiltration. The distribution of blood flow through the capillary bed of an organ is further controlled in part by another set of tiny sphincter vessels, the precapillary sphincters, which contain some smooth muscle (Chapter 6). The thin wall and small cross-sectional area of the individual capillaries are ideally suited for exchange. The capillary channel (lumen) is approximately 5–7 μm in diameter (just sufficient to accommodate a red blood cell (see Fig. 2.17). The volume of the capillary bed is small (4% of the total blood volume), but the vast number of capillaries provides a large total cross-sectional area, which leads to a very slow rate of blood flow in that segment of the circulation, thereby further favoring transcapillary exchange.

Alterations in the capillaries may occur during disease states. For example, an increase in the permeability of the capillary wall can lead to leakage of intravascular proteins, fluid, and other substances out of the vascular bed (see Shock, Chapter 18).

Venules

These small vessels have a relatively large channel and a thinner muscular wall than the arteries, and they form the low pressure collecting system for the venous (relatively deoxygenated) blood as it leaves the capillaries. The venules, like the arterioles, are innervated by the autonomic nervous system, and the smooth muscle in their walls can contract or relax, thereby contributing to the pressure in the capillary bed as well as to the overall size (capacitance) of the venous compartment.

Veins

The venules progressively join together to form larger veins which eventually merge into the two largest veins in the body, the venae cavae (superior and inferior vena cava). These two large, thin-walled vessels return venous blood directly to the heart. The veins contain the largest proportion of blood within the circulation (about 60%) and thereby serve as a blood reservoir; they can accommodate large changes in blood volume with rather

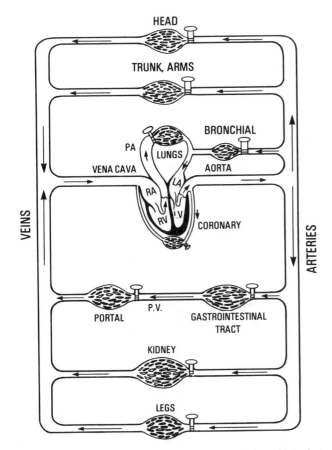

Figure 2.1. The pulmonary and systemic circulations. Note that the pulmonary circulation is connected in series to the remainder of the circulation (blood flows first through the chambers of the right heart and lungs, and then through chambers of the left heart). The organ systems operate in parallel from the systemic circulation. The small bronchial arteries supply oxygenated blood to the tissue of the lungs. *PA*, pulmonary artery; *RA*, right atrium; *LA*, left atrium; *LV*, left ventricle; *RV*, right ventricle; *PV*, portal vein.

small changes in pressure and are much more distensible than the arteries (Chapter 7).

Certain disease processes can primarily affect the veins. For example, inflammation of the wall of a vein can lead to the formation of a blood clot on the wall; eventually, it may dislodge and travel to the lungs. Such "pulmonary emboli" are unable to pass through the capillaries in the lung and can cause serious obstruction to blood flow.

Lymphatics

The lymphatic vessels are tiny, thin-walled channels that are not connected in sequence with the main blood vessels of the body, but rather serve as an additional drainage and transport system. They form a branching network throughout the tissues and organs of the body, and the fluid contained within them (lymph) also passes through the lymph nodes and is eventually channeled into a large vein. The ends of the lymphatic vessels are in direct communication with the extravascular fluid of the various organs, including the gastrointestinal tract, and they serve to return proteins, fats, and other substances to the circulation (Chapter 19).

Many diseases can involve the lymphatics; infection or scarring with obstruction of lymphatics can lead to swelling of the tissues (edema), and cancer cells are often seeded via the lymphatics into the lymph nodes.

THE PULMONARY AND SYSTEMIC CIRCULATIONS

There are two separate major circulations of the body which are linked in sequence, so that blood flows through the circulation to the body and then to the lungs, before returning again to the body.

The path of blood flow through these two circulations can be traced in Figure 2.1, the venous blood returning from the body being pumped by the right side of the heart into the pulmonary artery to the lungs (an unusual "artery" in that it contains venous blood) and, from there, into the pulmonary capillaries within the lung where uptake of oxygen and release of carbon dioxide occur. The blood then flows to the pulmonary veins (again, unusual "veins" in that they contain oxygenated blood) and back to the left-sided chambers of the heart, where it is pumped to the arteries and capillaries of the rest of the body.

The pulmonary circulation (sometimes called the "central" or "lesser" circulation) carries a relatively low pressure (about one-fifth that in the systemic arteries, with a mean pressure of 15–20 mmHg in the pulmonary artery) and serves a special function in providing a low resistance pathway for the entire output of blood from the body to traverse the lungs (Section 6, Chapter 36). The pulmonary circulation, in addition to gas exchange, serves a secondary function as a blood volume reservoir for the left heart (it contains about 10% of the total blood volume), and its high distensibility allows it to readily accommodate the large increases in blood flow that can occur (for example, during physical exercise) without manifesting a substantial change in pressure.

The systemic circulation (sometimes called the "peripheral" circulation) also has special properties (Chapters 6 and 7) and provides a high pressure source of oxygenated blood to supply the organs of the body (Fig. 2.1). It also drains the venous blood back to the right side of the heart; as mentioned, the veins serve an important volume reserve function and contain most of the blood in the circulation.

The mean pressures within the components of the systemic circulation drop progressively as blood flows from the arteries to the veins, with the highest pressure existing in the systemic arteries and the lowest in the veins. In the capillary bed, the pressure has fallen sufficiently to give a mean hydrostatic pressure that is in relative equilibrium with the plasma osmotic pressure (Chapter 6), thereby allowing fluid movement out of the thin-walled capillaries to be balanced by fluid gain due to the osmotic effect. Blood flow is slow in the capillaries, and an analogy to the marked slowing effect of the very large cross-sectional area of the capillary bed, compared to that of the arteries, is to allow water from a rapidly running hose to flow over a large screen; the high velocity of flow through the hose abruptly slows as filming over the screen takes place (see relation between volume flow and flow velocity, Chapter 6).

FEATURES OF OVERALL CIRCULATORY ORGANIZATION AND CONTROL

There are numerous miniature circulations to each of the organs and tissues of the body. These circulations operate *in parallel* off the systemic arteries, with the larger arteries serving as a "pressure reservoir" for oxygenated blood (Fig. 2.1). The regulation of blood flow through each organ is dependent upon its moment-to-moment function and metabolic needs. For example, the blood flow through a skeletal muscle is largely determined by local factors related to the varying metabolic requirements of the muscle as it alters its contractile work. If the leg muscle contracts repeatedly while running, the smooth muscle in the walls of the arterioles supplying the leg will relax to allow an increase in blood flow. In this manner, a series of "variable stopcocks" (the arteriolar resistance vessels) can regulate flow to each organ, operating semi-independently and in parallel off the large arteries (Fig. 2.1).

The general resistance equation: a simplified equation (discussed in more detail in Chapter 7) which describes these pressure-flow relations can be termed the "general resistance equation":

$$\text{Flow} = \frac{\text{driving pressure or pressure difference across a vascular bed}}{\text{vascular resistance}}$$

This equation, like Ohm's law of electrical circuits (current flow = voltage/resistance), makes intuitive sense. Again considering the garden hose analogy, the rate of flow will be directly proportional to the driving pressure from the faucet and inversely proportional to the amount of resistance provided by the screw clamp on the hose; as pressure increases with resistance constant, the flow will increase. As vascular resistance increases, in order to hold the pressure constant the flow from the faucet must be decreased; or, if flow is maintained constant as the screwclamp is tightened, the driving pressure must increase.

Regulation of the Systemic Arterial Pressure

The systemic arterial pressure (SAP) is held *relatively constant* by means of reflexes from the autonomic nervous system (Chapter 16) which control both the function of the heart and, hence, the cardiac output (Q) as well as the *total* peripheral vascular resistance (TPVR). In this context, the general resistance equation may be arranged as follows:

$$SAP = Q \times TPVR$$

The TPVR, of course, represents the net effect of the resistance in *all* of the organ beds of the systemic circulation. When sympathetic nerve impulse traffic increases (as during a change from supine to upright posture) norepinephrine stored in the nerve terminals around the arterioles is released. It then acts on smooth muscle cell receptors (α-receptors), causing them to contract (vasoconstriction), thereby increasing vascular resistance. Alternatively, when the blood pressure rises, sympathetic impulses decrease (through a feedback reflex), and vascular resistance decreases, tending to restore the blood pressure toward normal. Thus, through feedback mechanisms the blood pressure in the systemic arteries tends to be protected at a relatively constant level, so that an adequate pressure head is provided for perfusion of the various organs of the body.

Using the above resistance equation as applied to the regulation of arterial pressure, one can understand why the blood pressure does not drop markedly and remain low when we assume the erect posture from a lying position. Since it is known that a large volume of blood is transiently sequestered in the lower extremities and abdomen, such storage of blood should lead to a decrease in the return of blood to the heart and thereby to a drop in pressure. However, the pressure drops only mildly as the cardiac output (Q) initially falls because there is a prompt reflex increase in vascular resistance (TPVR) produced by the autonomic nervous system. Also, reflex neural stimulation of the heart leads to an increased heart rate and augmented force of the heartbeat, and reflex contraction of the veins further tends to restore the cardiac output toward normal.

Local Organ Blood Flow

The blood flow to an organ (local Q), as mentioned earlier, is largely controlled by organ metabolism, the products of which affect local vascular resistance (local VR). The general resistance equation may be restated as follows:

$$\text{Local Q} = SAP/\text{Local } VR$$

As mentioned, the systemic arterial pressure is held relatively constant by overall circulatory reflexes. Therefore, in effect, each organ bed takes the flow of blood that it requires by regulating its local VR; as organ flow requirements increase at a constant SAP, local VR decreases, and local Q must increase.

GENERAL FEATURES OF THE HEART

The heart is a 4-chambered organ composed of a special type of muscle called "myocardium" (see James *et al.* (1982) and Netter (1969) for details of cardiac anatomy). There are two separate pumps, the right ventricle and the left ventricle, each of which is supplied with a "booster pump"—the right atrium and the left atrium (the atria have ear-like appendages and are sometimes called "auricles") (Fig. 2.2*A*). As we shall see in Chapter 8, the electrical excitation of the heart muscle is organized so that the atrial booster pumps are timed to contract just prior to the contraction of the ventricles; thus, the two atria contract nearly together, helping to fill the main pumping chambers just before the ventricles contract to eject blood into the pulmonary artery and the aorta.

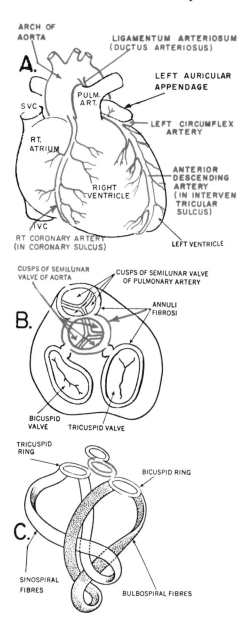

Figure 2.2. (*A*) Front view of the human heart. (*B*) Fibrous skeleton and valves of the heart viewed from above. (*C*) Spiral muscles of the ventricle. [Redrawn from R. C. Truex (1967).]

The Pericardium

A thin, membranous sac envelops the heart, and the space between this sac and the heart contains a small amount of fluid which serves to lubricate the heart surface as it moves during contraction. The pericardium is composed primarily of inextensible fibrous tissue which, under normal conditions, is relatively flaccid, but which, if it is acutely overstretched, becomes quite stiff. Thus, during acute overtransfusion of blood or other forms of rapid fluid overload, the pericardium can limit overdistension of the heart (Chapter 18). The pericardial

sac also can become inflamed, a condition called "pericarditis." If this condition becomes chronic, the pericardium may become sufficiently thickened to limit normal filling of the heart.

The Five Cardiac Subsystems

The heart may be conveniently considered as five subsystems, each of which will be considered in more detail in subsequent chapters. Cardioactive drugs and various disease processes tend to affect each of these subsystems somewhat differently.

The Electrical Pacemaker and the Conducting System of the Heart

As shown in Figure 2.3, the self-firing pacemaker of the heart (sinoatrial node) is situated in the right atrium; the electrical impulse it generates travels into the two atria, which are stimulated to contract, and then into a specialized conduction system, consisting of the atrioventricular node or junction which delays the impulse; it then travels down a rapid conduction system (the bundle branches) to the two ventricles, causing them to contract.

The pacemaker and specialized conduction tissues are influenced by the intrinsic innervation to the heart (the autonomic nervous system), which controls the heart rate and the velocity with which impulses are conducted through the specialized system.

Disorders of the heart's rhythm are common; disease of the pacemaker can result in rapid or irregular heart rhythms, whereas damage to the atrioventricular junction or specialized conduction system to the ventricles can result in faulty conduction of the electrical impulse to the ventricles from the atrium, so-called "heart block" (Chapter 9).

The Heart Muscle

The muscle of the heart (myocardium) is composed of branched interconnected muscle fibers (cells) which are structured to rapidly transmit the electrical impulse from cell to cell. The walls of the two atria are thin (appproximately 2 mm thick in the human heart) (Fig. 2.4). The wall of the right ventricle in the normal human heart is about 3–4mm in thickness, and the left ventricle measures about 8–9mm (Fig. 2.4), the difference in wall thickness relating to the higher pressure that the left ventricle is required to generate. The muscle bundles of the ventricles tend to wind spirally around the pumping chambers, originating and ending their insertion onto the fibrous portion (annuli fibrosi) of the heart surrounding the heart valves

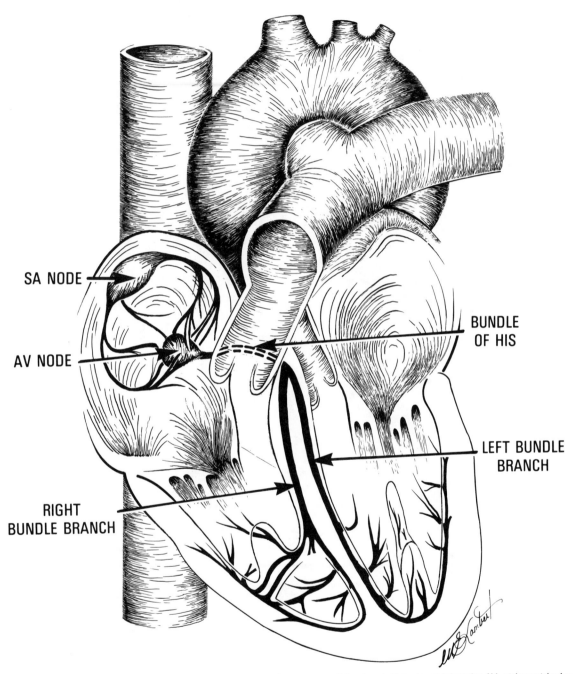

SA NODE

AV NODE

RIGHT
BUNDLE BRANCH

BUNDLE
OF HIS

LEFT BUNDLE
BRANCH

Figure 2.3. Diagram of the electrical pacemaking and conducting system of the heart. *SA*, sinoatrial node; *AV*, atrioventricular node.

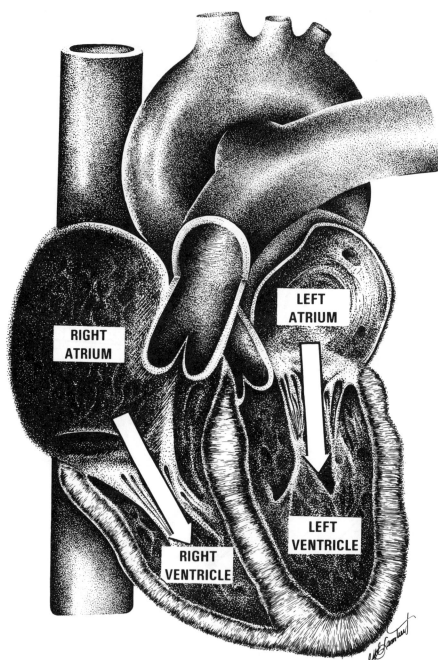

Figure 2.4.   Diagram showing the heart muscle. Notice the thin muscular walls of the two atria. The wall of the right ventricle is considerably thinner than that of the left ventricle. The valves of the heart are shown here diagrammatically during diastole. Note that the two sets of valves (semilunar and atrioventricular) operate out of phase (see Fig. 2.6), and here the mitral and tricuspid valves are open, whereas the aortic and pulmonic valves are closed.

(Fig. 2.2*B* and *C*). In the left ventricle, the muscle is arranged across the wall in a particular manner, with the fibers winding in a counterclockwise direction at the inner wall, in a horizontal (circumferential) direction in the midwall, and in a clockwise direction at the outer wall. This is shown in Figure 2.5 as the deviation of fiber angles from the ventricular long axis. It can be seen that most of the fibers lying in the center of the left ventricular wall run at right angles to the long axis, whereas those near the inner and outer surfaces run longitudinally, more parallel with the long axis (Fig. 2.5). (See Streeter *et al.* (1969) and Streeter (1979).)

The force and speed of contraction of the heart muscle is variable and is affected by such factors as the length of the fibers, and by stimulation of the nerve supply to the heart (adrenergic nerve terminals release norepinephrine, a powerful stimulant of contraction, into the myocardium).

Many diseases directly attack the heart muscle

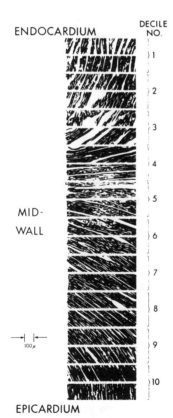

Figure 2.5. Myocardial fiber directions across the wall of the dog's left ventricle from endocardium (the inner wall, decile 1) to epicardium (the outer wall, decile 10). The fibers are viewed as if facing the front of the heart, and it can be seen that the bulk of the fibers in the midwall tend to run in a circumferential (horizontal) direction, whereas fibers in the inner (endocardial) and outer (epicardial) layers tend to run more vertically, along the long axis of the heart. [From Streeter et al. (1969).]

(for example, viral infestation is termed "myocarditis"), and other chronic forms of myocardial inflammation may accompany diseases such as rheumatic fever. Extensive damage to the heart muscle results in heart failure (Chapter 18).

The Valves of the Heart

There are four one-way valves within the heart (Figs. 2.4 and 2.6). Two of them lie between the ventricles and the great arteries and are called semilunar valves (Fig. 2.6); the aortic valve prevents blood from leaking backwards from the aorta into the left ventricle, and the pulmonic valve performs a similar function between the pulmonary artery and the right ventricle (Fig. 2.4). The two other valves, called atrioventricular valves, prevent leakage of blood backward from the ventricles into the atria when the ventricles are contracting to eject blood into the great vessels (Fig. 2.6); the two-leaflet mitral valve serves this function between the left ventricle and left atrium, and the three-leaflet tricuspid valve prevents backward leakage of blood from the right ventricle into the right atrium (Fig. 2.2*B* and 2.4).

The two sets of valves, semilunar and atrioventricular, function out of sequence, that is, when one set is open, the other is closed. As shown in Figure 2.4, the aortic and pulmonic valves are closed when the ventricles are in the resting phase of the heart beat, called *diastole*, whereas the atrioventricular valves are open during diastole, allowing the two pumping chambers to fill from the atria (Fig. 2.4). During the contraction phase of the heartbeat, called *systole*, the ventricles develop pressure and eject blood, and the atrioventricular (mitral and tricuspid) valves snap shut, while the semilunar (aortic and pulmonic) valves open (Fig. 2.6).

Understanding the function of the valves becomes important in considering the significance of normal and abnormal heart sounds, as they are heard over the chest wall through the stethoscope (Chapter 13).

A number of disease states can affect one or more valves of the heart. Such disorder can be acquired, as with rheumatic fever (which leads to valve scarring), or due to a congenital deformity. Either can lead to valve leakage, or to valve narrowing with partial obstruction to blood flow.

The Coronary Circulation

The coronary arteries are the first branches to arise from the aorta, and flow through these vessels supplies oxygen and nutrients to the heart itself (Fig. 2.2*A* and 2.7). Since the heart is an active

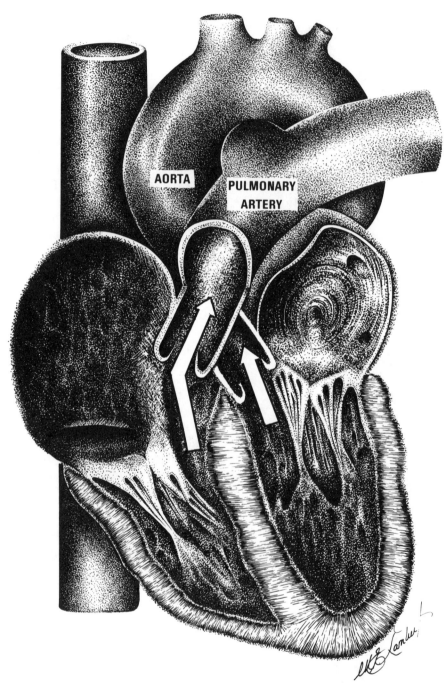

Figure 2.6. As the ventricular wall moves inward to eject blood during systole, the atrioventricular valves (mitral and tricuspid) are shut and the semilunar valves (aortic and pulmonic) open, whereas during ventricular filling, opening and closure are reversed (see Fig. 2.4). For further discussion see text.

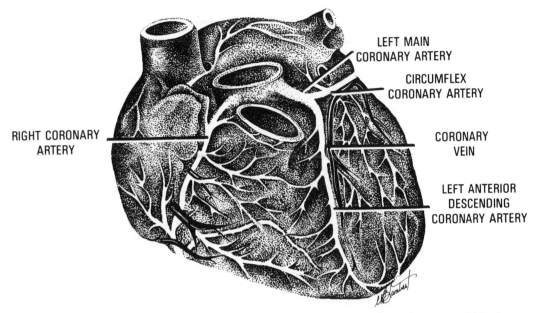

Figure 2.7.　Simplified scheme of the coronary arteries. In most human subjects, the left coronary artery divides into two branches supplying the front wall of the heart (anterior descending coronary artery) and the lateral wall of the left ventricle (circumflex coronary artery), whereas the single right coronary artery provides branches to the right ventricle and then circles backward to supply the posterior wall of the left ventricle (*dashed lines*).

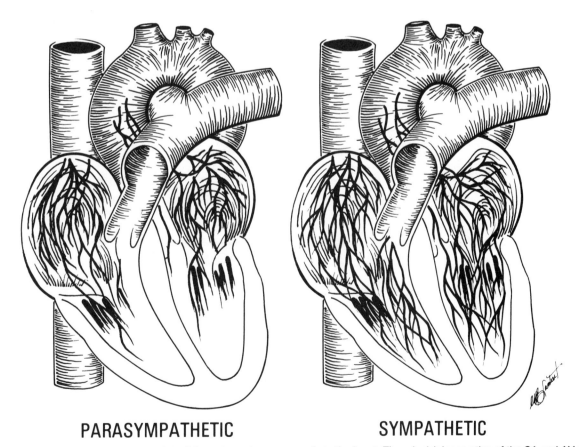

Figure 2.8.　Diagrammatic representation of the autonomic nerve supply to the heart. There is rich innervation of the SA and AV nodes and of both atrial and ventricular myocardium by the sympathetic nervous system. The parasympathetic system predominantly supplies the SA and AV nodes and the atrial myocardium with only a few branches to the ventricles.

organ, the coronary circulation provides a rich blood supply to the myocardium.

Disease of the coronary arteries, most often due to atherosclerosis, can cause impairment of the blood supply to the heart. When a coronary artery closes off completely, a heart attack (death of the heart muscle) usually occurs rapidly, often with fatal consequences (Chapter 15).

The Autonomic Innervation of the Heart

Sympathetic nerve fibers (adrenergic nerve endings containing norepinephrine) ramify in the heart muscle of the atria and ventricles, as well as in the sinus node and the atrioventricular junction (Fig. 2.8). Stimulation of this system speeds up the electrical pacemaker, facilitates conduction through the specialized conduction system, and increases the force of heart muscle contraction. Parasympathetic fibers (cholinergic nerve endings containing acetylcholine) ramify in the pacemaker, the atrioventricular junction, and in the muscle of the two atria (there is little cholinergic innervation of the ventricles) (Fig. 2.8). Stimulation of the parasympathetic nervous system slows the pacemaker, delays conduction through the specialized conduction system, and reduces the force of the contraction of the atria (Chapter 16).

Under normal resting conditions, the parasympathetic and sympathetic systems interact to regulate the heartbeat at a rate of about 70/min. Activation of the sympathetic nervous system by emotional or physical stress (see Section 9, Chapter 3) increases the heart rate and the force of the heartbeat, thereby augmenting the output of blood from the heart. Simultaneously, sympathetic stimulation of the peripheral arteries and veins also causes release of norepinephrine into the walls of those vessels and into the circulation from the adrenal glands, thereby raising the blood pressure and increasing the return of venous blood to the heart (Chapter 16).

CHAPTER 6

Structure-Function Relations in the Peripheral Circulation

In order to understand the function of the circulation and the dynamics of pressure and flow throughout the peripheral vascular bed, it is necessary to consider in some detail the structural characteristics of various vascular components in relation to their function. In general, the system of conduits that constitute the circulation can be diagrammed as shown in Figure 2.9. On the left, large phasic pressure changes are generated by the left ventricle as it delivers the stroke volume into the aorta. The elastic properties of the large arteries and aorta allow them to rapidly expand as the volume of blood enters the arterial bed. This expansion of the larger arteries during systole allows them to transiently store blood, which is then passed onward by recoil during diastole (the arteries acting much like an accessory pump, thereby maintaining blood flow during diastole); this process has been termed the "Windkessel Effect" (Fig. 2.9, *dashed lines, solid lines*). Thus, the sharply phasic character of blood flow during systole as it crosses the aortic valve is progressively modified as it moves further down the arterial system, to yield a more and more continuous flow pattern.

The resistance vessels, the small arteries and arterioles, primarily determine the systemic vascular resistance and arterial pressure (Fig. 2.9), whereas the most distal vessels in the arterial distribution system (the tiny terminal arterioles and metarterioles) contribute a relatively small component to the vascular resistance but open and close to regulate blood flow through the capillaries; these "sphincter vessels" (diagrammed as stopcocks in Fig. 2.9) therefore determine the area of the capillary bed perfused. The postcapillary resistance vessels (small venules) help to regulate capillary hydrostatic pressure and thereby affect fluid exchange in the capillaries (Fig. 2.9). Finally, the larger venules and veins of the systemic circulation contain most of the blood volume, and relatively minor changes in the diameter of these vessels can modify the return of blood to the heart by changing their capacity to store blood. (The veins are therefore termed the "capacitance vessels" (Fig. 2.9).)

The progressive fall of intravascular pressure in the normal circulation is also shown in Figure 2.9, beginning with the pump (left ventricle). The precapillary resistance vessels (arterioles and the precapillary sphincter region) provide the largest pressure drop, from a mean pressure of about 85 mmHg in the aorta to 38 mmHg (Fig. 2.9). Across the capillary vessels, the pressure drops from about 38 mmHg to about 25 mmHg (in the postcapillary venules), and the remainder of the pressure fall occurs in the postcapillary resistance venules to reach the normal venous pressure of 5 mmHg in the large capacitance veins. In terms of vascular resistance, about 70% lies in the precapillary area, 10% in the capillaries, and 20% in the postcapillary region (Mellander and Johansson, 1968).

The mean velocities of blood flow through the various segments can be deduced from the geometry of the systemic circulation, since at any point the mean linear velocity equals total volume flow divided by the cross-sectional area (*CSA*) of the vascular bed:

$$\text{Velocity cm/s} = \frac{\text{Volume flow cm}^3/\text{s}}{CSA \text{ cm}^2}$$

The *smallest* cross-sectional area which receives the entire output of blood from the heart is in the aorta (about 3.2 cm² in man) and, accordingly, the mean flow velocity is *highest* in that vessel. For example, if the output of blood from the normal resting heart (cardiac output) is about 6 l/min (100 cm3/s) divided by 3.2 cm² equals a mean velocity of 31 cm/s. The peak aortic flow velocity and other vessel diameters and velocities are shown in Table 2.1. As the arteries progressively divide into smaller and smaller vessels the total cross-sectional area increases, and flow ve-

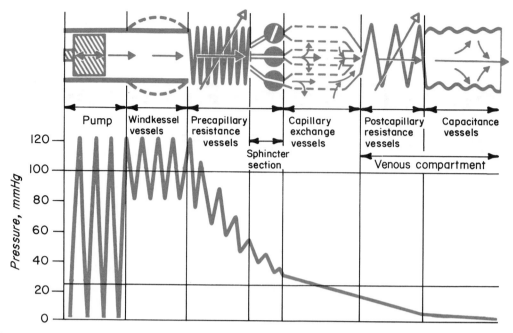

Figure 2.9. Functionally differentiated segments of the vascular bed. [From Mellander and Johansson (1968).]

Table 2.1

Vessel diameter, blood velocity, and Reynolds number for the systemic circulation of man

Structure	Diameter, cm	Blood Ve-locity, cm/s	Tube Reynolds[a] Number
Ascending aorta	2.0–3.2	63[b]	3,600–5,800
Descending aorta	1.6–2.0	27[b]	1,200–1,500
Large arteries	0.2–0.6	20–50[b]	110–850
Capillaries	0.0005–0.001	0.05–0.1[c]	0.0007–0.003
Large veins	0.5–1.0	15–20[c]	210–570
Venae cavae	2.0	11–16[c]	630–900

From Whitmore (1968). [a] Assuming viscosity of blood is 0.035 poise. [b] Mean peak value. [c] Mean velocity over indefinite period of time.

locity correspondingly falls (Fig. 2.10A and B). Each capillary is tiny (Table 2.1), but since the overall capillary bed contains many millions of vessels, it has a total cross-sectional area several hundred times that of the aorta (cross-sectional area = 2000 cm²), and hence the mean blood flow velocity falls several hundredfold, yielding the *slowest* mean flow velocity in capillaries (approximately 0.05 cm/s). This slowing of flow velocity in the capillaries (Fig. 2.10B) provides a highly favorable setting for exchange of gases and metabolites across the short length of the capillaries. Moreover, by the time the blood reaches the capillaries, the flow is relatively constant due to the damping effect of the elastic walls of the arteries (discussed below). The two venae cavae exhibit a mean velocity about one-half that of the descending aorta (Table 2.1).

The distribution of the blood volume within the various segments of the circulation is shown in

Table 2.2 and Figure 2.10C. The total volume of blood contained in a normal human subject is about 7% of body weight (5 l in a 75-kg man), so that the total blood volume must circulate about once per minute. Despite a very large cross-sectional area, the numerous capillaries are short and of small caliber, and contain only about 4–6% of the total blood volume; the systemic arteries contain about 12–14% of the blood volume, whereas the veins (mainly the small veins) contain nearly two-thirds (64%) of the total blood volume, emphasizing their role as a reservoir for blood. About 9–10% of the blood volume is contained in the pulmonary circulation, and 6–7% is within the heart itself during diastole (Table 2.2) (Milnor, 1968).

The distribution of the cardiac output to the various major organ systems of the body is shown schematically in Figure 2.11. The arrangement of these circulations "in parallel" off the main arterial bed is evident. In the resting state, approximately 15% of the cardiac output goes to the brain, 15% to the muscles, 30% to the gastrointestinal tract, and 20% to the kidneys; the coronary circulation takes about 5%, and the skin and bones 10%. During exercise, of course, the proportion going to the coronary circulation and the working skeletal muscles increases markedly. Notice that the arterial blood supply which nourishes the lung tissue (bronchial arteries, a circulation distinct from the blood flow through the pulmonary artery) recirculates directly back to the left atrium and left ventricle, so that the left ventricle pumps slightly more blood than the right ventricle (less than 3% of the cardiac output) (Fig. 2.11).

Volume Flow Through the Systemic Circulation

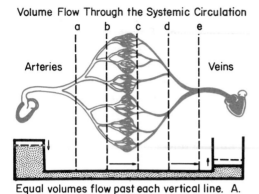

Equal volumes flow past each vertical line. A.

The Relation Between Cross-sectional Area and the Velocity of Flow in the Systemic Circulation

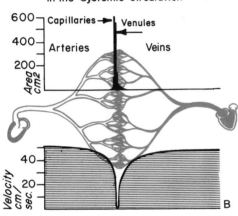

C. Pressures and Volumes of Blood in the Systemic Circulation

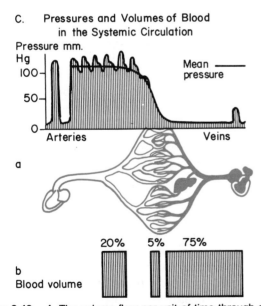

Figure 2.10. *A,* The volume flow per unit of time through each segment of the circulation must equal that entering and leaving the system despite great differences in cross-sectional area. *B,* The cross-sectional areas of various segments of the circulation computed from a 13-kg dog are graphed (*bars*) above velocity of blood flow, to show their inverse relationship. Flow through the capillaries averages about 0.07 cm/s. *C,* Approximate pressures

Table 2.2

Estimated distribution of blood in vascular system of hypothetical adult man[a]

Region	Volume			
	ml		%	
Heart (diastole)	360		7.2	
Pulmonary				
Arteries	130		2.6	
Capillaries	110	440	2.2	8.8
Veins	200		4.0	
Systemic				
Aorta and large arteries	300		6.0	
Small arteries	400		8.0	
Capillaries	300	4,200	6.0	84.0
Small veins	2,300		46.0	
Large veins	900		18.0	
	5,000		100	

From Milnor (1968). [a] Age, 40 years; weight, 75 kg; surface area, 1.85 m.[2]

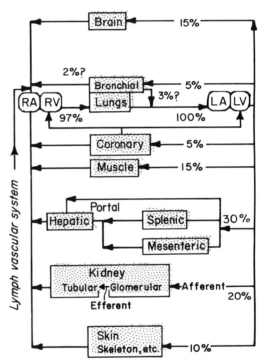

Figure 2.11. Stylized block diagrams of the circulatory system. A scheme of the circulatory system including the major vascular components (heart, arteries, microcirculation, veins) which distribute to and collect blood from the tissues. The special circulatory beds to be discussed separately are indicated and the relationship to the lymphatic collecting duct system represented. *Solid lines* depict arteries and veins, *arrowheads* show the direction of flow, and *shaded blocks* indicate the microcirculation (arterioles, capillaries, and venules). Approximate percentages of cardiac output going to the various circulatory beds are indicated near each line.

and volumes throughout the circulation are depicted schematically. Pressure is high and pulsatile in the arteries, falls rapidly across the resistance vessels, and the pulsation is damped out in the capillaries and veins. Pressure is low and nonpulsatile in most capillaries and veins. The blood volumes depicted are approximations. [Redrawn from Rushmer (1961).]

VASCULAR SMOOTH MUSCLE

The presence of smooth muscle cells in the walls of the arteries, arterioles, and veins give these vessels the capability of changing their caliber. The smooth muscle cells are in general arranged circumferentially around the vessel, and are found in varying numbers in different components of the circulation (for general reviews on vascular smooth muscle see Bülbring et al., 1970; Bohr, 1978, and Johansson, 1981). The smooth muscle occupies the middle portion of the blood vessel wall, tending to be arranged in a spiral or a helical manner in larger vessels, with a dominantly circumferential direction in the small arterioles. Smooth muscle cells vary widely in size among species (in the pig, for example, they average 110 μm long and 3 μm in diameter). They are generally narrow and tend to be arranged in muscle bundles, which can respond as an "effector unit." Nexus junctions (similar to "gap junctions" of striated muscle) have been identified by electron microscopy and appear to represent sites for electrical coupling between cells; other areas of the cell-to-cell junctions appear to serve for tension transmission. The cells are packed with myofilaments, which lie largely parallel to the long axis; their arrangement is less uniform than in skeletal muscle, and the myofilaments appear to attach to the plasma membrane and to "dark bodies" internally (which may be somewhat analogous to the Z-lines of skeletal muscle, see Section 1, Chapter 4). Sarcoplasmic reticulum is present in smooth muscle, although it is less well developed than in skeletal or cardiac tissue. Smooth muscle cells appear to have a rather low resting transmembrane potential (-40 to -60 mV), and excitation spikes occur due to induced alterations in membrane potential (Somlyo and Somlyo, 1980). In some types of vascular smooth muscle, spontaneous action potentials are observed as a consequence of changing resting membrane potential.

Arteries down to the precapillary vessels and veins are innervated by adrenergic nerve fibers, which tend to be located in the outer muscular wall or at the boundary with the outer connective tissue coat. The innervation of the veins is somewhat less dense than that of the arteries. Arterioles of about 50 μm in diameter have seven or eight nerve fiber bundles giving rise to a network that surrounds the vessel. The discharge rate of the autonomic nerves to the arterioles appears to be sufficient to maintain a resting level of contraction or tone. There appears to be close contact of nerve terminals with the muscle cells in small arteries, while in larger arteries the neurotransmitter norepinephrine released from neural storage vesicles may encounter large diffusion distances. There is some evidence that "key cells" once activated by neurotransmitter may activate cells toward the inner wall of larger arteries by initiating propagated action potentials, which travel via the nexus junctions. In different parts of the vascular tree removal of autonomic vasoconstrictor fibers has different effects; certain regions of the skin are under powerful sympathetic control and show little intrinsic muscle tone after denervation, whereas regions such as the brain, heart, and skeletal muscle exhibit autoregulation (see below), tend to metabolically override neural influences, and exhibit a strong degree of intrinsic tone after sympathectomy (Johnson, 1980).

The sarcoplasmic reticulum appears to be the major calcium storage site in vascular smooth muscle, with contraction and relaxation under cytoplasmic calcium control (Johansson, 1981). Thus, the levels of intracellular calcium appear to determine the contractile state of vascular smooth muscle. (This does not appear to operate through the troponin-tropomyosin mechanism, and there is some evidence that calcium control of myosin-light chain phosphorylation is involved through activation of light chain kinase; this may be the site of cyclic-AMP action.) Current evidence indicates that many vasoactive substances operate to alter the force of smooth muscle contraction through changes in intracellular calcium. Norepinephrine and some other agents that stimulate contraction also may produce greater depolarization and increase the frequency of spike potentials. However, pharmacologic effects clearly can occur without changes in membrane depolarization, and it is considered likely that alterations in membrane permeability (such as to calcium) are important. Locally produced metabolites also affect vasomotor tone including CO_2, lactic acid, ATP breakdown products (such as adenosine), and other potent vasoactive substances such as histamine, prostaglandin, and serotonin (Somlyo and Somlyo, 1981). However, their role in regulating *basal* vascular tone is uncertain.

Smooth muscle exhibits a length-active tension relation (enhanced stretch increasing the force of contraction). In this connection, transmural distending pressure within the vessel is important, either increased pressure or pulsatile pressure producing a myogenic vascular contraction (increased tone); the mechanism of this response is not clearly established. As mentioned, some smooth muscles have inherent myogenic tone as a result of electric spike discharge or nonelectrogenic mechanisms.

With long-standing increases in transmural pressure, vascular smooth muscle (like other muscles) undergoes hypertrophy (increased cell size), with thickening of the muscular layer. In such thickened arterioles, a given degree of smooth muscle shortening produces a greater reduction in luminal diameter, leading to hyperreactivity of the arterioles in hypertension.

TENSION IN VASCULAR WALLS

Calculation of tension in blood vessels is based on the law of Laplace used for calculation of tension in the walls of soap bubbles. This law states that at any point on the curvature of the membrane, if a slit is made there will be a force T (in dynes) tending to pull the slit apart; the force is directly proportional to the pressure across the membrane (inside minus outside pressure) and to the principal radius of curvature (r) of the membrane at that point. Since in a cylinder one radius of curvature is infinite (or zero) the law can be stated as:

$$T = P \cdot r$$

This law indicates that the smaller the radius the greater the mechanical advantage in terms of the tension that must be borne by the vessel wall. Thus,

a capillary with its very small diameter (7 μm) can support an intravascular pressure of 25 mmHg with a tension of only 16 dynes, whereas the aorta with its large radius carries a tension in the wall of 170,000 dynes (Fig. 2.12*A*). This point is also illustrated using a balloon of nonuniform shape (Fig. 2.12*C*); the tension in the wall is much lower at the small end of the balloon than in the central, dilated portion.

In thicker-walled structures, such as the arteries, this equation is modified to include the wall thickness so that

$$T = P \cdot r/h$$

where h = wall thickness.

Thus, when the wall is relatively thick, as in the arterioles, a relatively lower wall tension exists, and the smooth muscle bears a smaller load despite a relatively high intravascular pressure; this provides a mechanical advantage for changing blood vessel diameter in these small vessels. This equation also has importance in considering aortic aneurysm (a large saccular bulging area, usually commencing in an area of disease in the aortic wall). In an aneurysm, any rise in blood pressure produces an in-

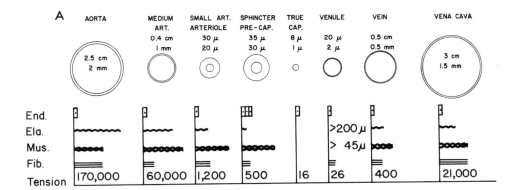

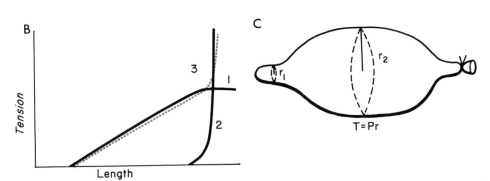

Figure 2.12. Illustrated are the lumen diameters, wall thickness, relative amounts of different tissues, and wall tensions (dynes/cm²) in the various blood vessels. *Lower left*: tension-length diagrams of elastic (*1*) and collagen fibers (*2*) presented alone and in combination (*3*). *Lower right*: balloon demonstrating pressure-tension relation. The pressure (*P*) within the balloon is equally transmitted to all parts of the contained air (Pascal's law), and the tension (*T*) in the balloon varies with the radius (*r*) in that portion, Laplace's law. [Adapted from Burton (1954) and Wolfe (1952).]

crease in radius, with further stretching (and thinning) of the aortic wall, thereby resulting in a further rise in wall tension; this sequence leads to further stretching, etc. Thus, an aneurysm can become unstable, and expansion eventually leads to rupture of the aorta in some individuals (see also influence of lateral pressure in locally dilated vessels, Chapter 7).

THE LARGE ARTERIES AND VEINS

The relative sizes and structures of various components of the vascular system are summarized in Figure 2.12*A* (Schwartz et al., 1981). The upper numbers represent the diameter of the lumen and the wall thickness (Burton, 1954). It can be seen that the channel of the aorta is very large (2.5 cm), with a relatively thin wall (2 mm). The ratio of wall thickness to lumen diameter rapidly increases as the smaller resistance arterioles are reached. Notice also that the proportion of muscle increases as the arteries become smaller, whereas the relative proportion of elastic tissue decreases. The outer layer of connective tissue around a blood vessel is termed the *adventitia*; the middle muscular layer is called the *media*, and in most arteries there are one or more *elastic layers*, one of which lies inside the muscular wall.

All blood vessels are lined by a single layer of cells, the *endothelium*. In the arteries and the veins the cells of the endothelium are thicker than in capillaries (see below), and there tend to be few fenestrations between cells. The important role of the endothelium in preventing intravascular blood clotting is discussed in Section 3, Chapter 25.

The blood vessels, like many biological tissues, do not obey Hooke's law when passively stretched. (Hooke's law states that when simple homogeneous materials are stretched, elastic tension is proportional to the degree of elongation.) The more blood vessels are stretched the more they resist stretch; thus, blood vessels develop a higher resting tension the more they are distended, becoming progressively stiffer or less compliant. The pressure-volume curve of a large artery such as the aorta is shown in Figure 2.13. Notice the relatively steep rise in pressure as volume is changed, giving a nearly exponential relation. This shape of the curve reflects, in large part, initial stretching of the elastic fibers, which are partially folded and tend to obey Hooke's law up to their point of yield (Fig. 2.12*B*, curve 1). At that point, the collagen fibers (fibrous tissue) come into play as the vessel reaches its "elastic limit"; these fibers are very stiff and have a steep length-tension relation (small change in length with large change in tension) (Fig. 2.12*B*,

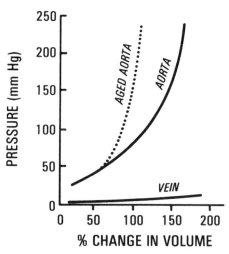

Figure 2.13. Pressure-volume characteristics of a large artery such as the aorta in a normal middle-aged human subject (*middle curve*) and in an aged human subject (*dotted curve*); notice the steeper (less compliant or stiffer) relationship in the older individual. The vein exhibits a much higher compliance over the normal operating volume (compliance equals the slope of the relationship at any given volume, dV/dP).

curve 2). The resultant of these two curves is a mixed relation, the lower flatter portion being determined largely by elastic fibers (as well as muscle) and the upper steep portion being related to the nondistensible collagen (Fig. 2.12*B*, curve 3) (Burton, 1959, 1972). The pressure-volume characteristics of larger arteries such as the aorta tend to become stiffer with age (*dotted line*, Fig. 2.13). This alteration would explain how a larger pulse pressure results from a normal stroke volume in older subjects (Chapter 7).

As shown in Figure 2.12*A*, the larger veins and venae cavae tend to have a large lumen and a relatively thin wall. The pressure-volume curve of veins tends to be quite flat over the physiologic range of pressures, that is, large changes in the volume of blood contained in the reservoirs can occur with minimal change in the pressure, a property valuable to the capacitance function of the venous bed (Fig. 2.13); eventually, at very high distending volumes, the venous bed will also reach its elastic limit.

The walls of the veins contain smooth muscle, and the veins are innervated down to the level of the venules. Thus, reflex changes in sympathetic tone affect the caliber of the veins and thereby the size of the storage compartment (Chapter 16), as well as to some degree the compliance of their walls.

THE ARTERIOLES

The arterioles have a high muscle content, and the ratio of wall thickness to channel diameter is high, thereby favoring their role in regulating vascular resistance (Fig. 2.12*A*). The arterioles are innervated by the autonomic nerves. Several phenomena, which occur predominantly in arteriolar beds, deserve discussion.

Critical Closure

The vessel radius and the wall thickness operate in a dynamic, changing relationship as vascular resistance is modulated in small blood vessels, and (in addition to aneurysm formation, mentioned above) there is another situation in which a blood vessel may become unstable. Such instability tends to occur in vessels with active basal tone (the arterioles) and takes place when the intravascular or transmural pressure falls below a critical value (normally the tissue pressure is low and the transmural pressure can be equated to the intravascular pressure). When this pressure (the "distending pressure" in the arteriole) drops below a certain value, the blood vessel may close altogether, a phenomenon called *critical closure*. The intravascular pressure at which such closure occurs is affected by the amount of active smooth muscle tone in the vessel wall and is termed the "critical closing pressure" (Burton, 1972). For example, a series of different zero-flow intercepts can be produced as the critical closing pressure of a given vascular bed is changed by varying the vasomotor tone with different levels of sympathetic nerve stimulation (Fig. 2.14). It can also be seen from this diagram, that

critical closure can occur at a transmural pressure of 30 mmHg, for example, when there is a relatively high degree of sympathetic nerve stimulation, whereas with low sympathetic tone substantial flow remains at this pressure level (Girling, 1952).

Autoregulation

Some vascular beds, particularly those with relatively low neurogenic control (such as the coronary, skeletal muscle, and cerebral circulations), exhibit "autoregulation" (for general discussion see Johnson 1964, 1980; Haddy and Scott, 1977); that is, regardless of changes in perfusion pressure over a substantial range, the local vascular resistance will change to maintain the blood flow relatively constant (see also discussion of this phenomenon in the coronary circulation, Chapter 15). Thus, as long as the metabolic requirements of the tissue supplied by that vascular bed remain constant, the blood flow will be regulated by local mechanisms. This type of autoregulatory response is diagrammed in Figure 2.15 (*solid line*); the curve has a sigmoid shape with a central relatively flat portion where flow is regulated, a steep portion at the lower end of the pressure-flow curve where autoregulation fails, and another region at the high end of the pressure-flow curve where autoregulation is incomplete (Fig. 2.15). The shape of the curve and the position of the lower "knee" (the point at which flow starts to fall) differ in various tissues, and the speed of the autoregulatory response also may differ (for example, in the heart it takes 8–10 s for auto-

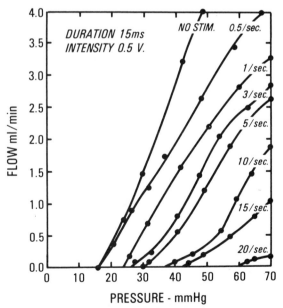

Figure 2.14. Pressure-flow relations in the rabbit ear at various levels of stimulation of the sympathetic nerve supply. Note that increased frequency of sympathetic stimulation produces vasoconstriction, with a shift of the curves to the right and a decreased slope, together with a higher pressure intercept (higher critical closing pressure). [From Girling (1952).]

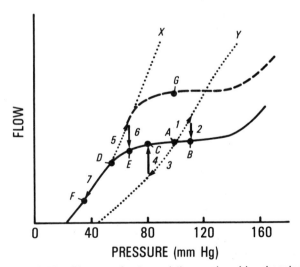

Figure 2.15. Diagram of autoregulation produced by changing the perfusion pressure in a vascular bed, while keeping the metabolic activity constant (*solid line*). Autoregulation at a higher level of metabolic activity is shown by the *dashed line*. The *dotted lines* show pressure-flow relations prior to the autoregulatory response. For further discussion see text.

regulation to occur after the pressure is suddenly changed).

In Figure 2.15, two pressure-flow curves in the steady state at different levels of vasomotor tone in the absence of autoregulation are shown by the *dotted lines* (X and Y), and the autoregulated curve is shown by the *solid line*. For example, at a perfusion pressure of about 95 mmHg, at a given level of resting metabolic function of the muscle, blood flow is regulated at point A (if the tissue were metabolically active so that its oxygen consumption were increased, flow at that pressure would be metabolically regulated to a higher level, so-called "active hyperemia," point G, the *dashed line* indicating that the curve is shifted upward) (Stainsby, 1962). If from point A the perfusion pressure is suddenly increased to 110 mmHg, flow will initially rise along curve Y (arrow 1), but active constriction of the muscle of the arterioles will then occur, increasing the vascular resistance, so that flow is regulated (arrow 2) back to nearly the same level but at the higher perfusion pressure (point B). Conversely, if the perfusion pressure is suddenly dropped from 95 mmHg to 80 mmHg flow will initially fall along curve Y (arrow 3), but then smooth muscle relaxation occurs, lowering the vascular resistance and returning flow to near the normal level (arrow 4, point C) but at the lower perfusion pressure. Curve X (Fig. 2.13) represents the unregulated pressure-flow curve when the arteriolar bed is *maximally* dilated (its lowest vascular tone). Point D is situated on the lower knee of the autoregulated curve; if perfusion pressure is raised to about 65 mmHg, the flow will initially follow the curve of maximum dilatation (arrow 5) and then return to a slightly increased flow at point E (arrow 6). On the other hand, if at that point perfusion pressure is lowered *below point D*, e.g., to 35 mmHg, since the autoregulatory reserve is exhausted, flow will *fall* as perfusion pressure is lowered (arrow 7 to point F), i.e., the pressure-flow relation follows curve X.

The mechanism of autoregulation remains controversial (Johnson, 1980). Three mechanisms have been proposed: *myogenic, metabolic,* and *tissue pressure.* Some investigators believe that a myogenic response related to the degree of stretch of the smooth muscle is the predominant factor producing alterations in vascular resistance during autoregulation. In fact, an isolated vascular muscle strip will, when stretched, contract to a length shorter than that existing prior to the stretch. Others believe that a chemical mediator is likely to be the dominant factor; thus, under conditions of constant oxygen consumption if a vasodilator metabolite were being produced at a constant rate, transiently increasing the flow would wash out the metabolite at a more rapid rate and reduce the concentration of the vasodilator, thereby increasing vascular resistance; conversely, if perfusion pressure were dropped, the vasodilator would wash out at a reduced rate, increase its concentration, and decrease vascular resistance. Such chemical mediators undoubtedly function during active hyperemia (muscle contraction), but their nature remains unknown in skeletal muscle and in most other tissues; much information suggests that in the coronary circulation adenosine is important (see Chapter 15). A third potential mechanism in autoregulation is a change in tissue pressure, at least in some tissues (such as the kidney) in which the vascular bed lies within an encapsulated organ. Thus, increased perfusion pressure might result in filtration of fluid and an increase in tissue pressure, which could then tend to compress the arterioles and reduce flow back towards its previous level (and vice versa). It is possible that during autoregulation, all three mechanisms may participate, their relative importance depending on the tissue involved.

Some of the most significant experiments indicating a role for myogenic phenomena are those showing an increase in precapillary resistance in response to elevation of the *venous* pressure which decreases flow. This response of the arteriolar smooth muscle to stretch appears to contradict the metabolic theory which would predict that vasodilation should accompany decreased flow (Johnson, 1980). Direct observations of arterioles in the microcirculation have also supported the presence of myogenic responses, showing decreases in diameter in response to increased arterial pressure, and arteriolar dilation in response to a lowered pulse pressure. In fact, studies on pulse pressure effects have suggested that about one-third of the basal tone of arterioles in skeletal muscle of cats may be of myogenic origin stimulated through pulsatile effects on the arterioles (Grande et al., 1979).

THE MICROCIRCULATION

For practical purposes we shall consider the microcirculation as comprised of the smallest arterioles and the precapillary sphincter region, the capillaries, and the venules (for general reviews see Landis, 1963; Zweifach, 1973; Fung, 1977; and Weideman et al., 1982).

The Capillaries

The structure of capillary endothelium varies widely in different organs, being relatively continuous in the brain and heavily fenestrated in the liver (for review see Karnovsksy, 1968). The special structure of capillaries in the kidney and lung are described in Sections 4 and 5. The capillaries vary in diameter from 3 to 8 μm (depending on the species; ordinarily, they are about the size of a red blood cell), and they are lined by a single layer of endothelial cells (Fig. 2.16). The cells of the capillary endothelium are thinner than those of the arteries and veins and, in addition to fenestrations, in some tissues they may exhibit clefts between cells (Fig. 2.17); they also contain vesicles (Fig. 2.17) which, in some instances, form tubular channels entirely across the endothelial cells. The capillary endothelium in the heart and skeletal muscle is about 1 μm thick (except at the nucleus) and shows intercellular clefts and vesicles, but no fenestrations. These clefts or slits (sometimes called aqueous pores) average about 2.5 nm in width, and show narrowed areas (tight junctions)

Figure 2.16. Section of capillary wall stained with silver nitrate to demonstrate the intercellular cement. Nuclei of the cells stained with hematoxylin.

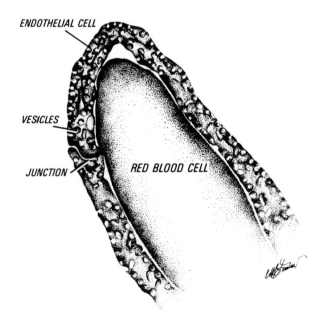

Figure 2.17. Diagrammatic representation of a portion of a capillary, such as might be found in skeletal muscle. A portion of a red blood cell contained within the capillary is also shown. The junction between two endothelial cells is illustrated, and multiple vesicles some of which open on the interior and exterior of the endothelial cell are shown.

which may limit large molecule movement (Fig. 2.17). The clefts have been calculated to be relatively few (less than 0.2% of the capillary surface area) and allow bulk flow of water and ready diffusion of smaller, water-soluble substances such as NaCl (except in the brain). As molecular size approaches pore size, however, diffusion is increasingly limited (Landis and Pappenheimer, 1963) (see subsequent section on transcapillary exchange). In the transport of larger molecules there is controversy about the relative importance of fenestrations and vesicles (the latter process is termed *pinocytosis* or vesicular transfer).

The basement membrane which lies outside the endothelial cells may be particularly important in limiting permeability in some tissues, but in many capillary beds overall permeability is affected both by the endothelium and the basement membrane. Some rigidity to the single layer of endothelial cells of the capillaries is also provided by the basement membrane. Outside the basement membrane is the "ground substance," composed largely of mucopolysaccharides, which also may affect diffusion properties across the capillary wall. The capillaries contain no smooth muscle and are not subject to active control.

The precapillary region is occupied by small blood vessels that contain some smooth muscle. This region functionally controls blood flow distribution and hence the area of the capillary bed perfused. It encompasses a relatively wide range of vessel sizes, including small arterioles (metarterioles) (Fig. 2.18) and precapillary arterioles (9–16 μm) which contain only a single layer of smooth muscle cells and give rise directly to several lateral capillary branches in many tissues. The precapillary sphincters are smaller and usually serve only one or two capillaries. Some of the metarterioles give rise to so-called "thoroughfare channels" which in some tissues allow bypass of the capillary bed (Zweifach, 1973). Thoroughfare or preferential channels in some

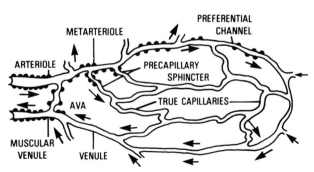

Figure 2.18. Terminal vascular bed, with preferential channels. *AVA*, arteriovenous anastamosis. [Adapted from Zweifach (1949), as redrawn by Burton (1972).]

beds, such as the mesentery, offer lower resistance arteriovenous connections and also exhibit lateral sphincters serving capillaries (Fig. 2.18). However, in other beds (particularly skeletal muscle) thoroughfare channels are sparse or absent, and the arterioles branch into terminal arterioles (about 15 μm in diameter) which then subdivide irregularly into many capillaries (Zweifach, 1973). There is a tremendous variety of origins, lengths, diameters, and flow velocities between (and within) capillary beds (for example, see Klitzman and Johnson, 1982).

The postcapillary venules (15–20 μm) contain smooth muscle and are capable of regulating their caliber; they constitute the main postcapillary resistance region, although it is probable that postcapillary resistance lies in venules up to 200 μm in diameter. The collecting venules (35–45 μm) contain no smooth muscle and they, in turn, join the large venules (up to 1000 μm), which do have smooth muscle and begin to serve the capacitance function performed by larger venules and veins.

Control of the Microcirculation

A variety of methods has been used to study the microcirculation, including electron microscopy, high power study of living tissue (vital microscopy), high speed cinemicrophotography or television microscopy, (by which red cell velocities and vessel diameters can be measured, red cell velocities on the order of a 1 mm/s having been measured in a number of tissues in animals as well as in man), and microtechniques for determining intravascular pressures (Goodman, 1974; Wiedman, 1963, Wiederhielm and Weston, 1973). Such techniques have allowed measurement of capillary density as well as the dynamic behavior of the microcirculation; for example, as mentioned local contractile responses to passive stretch have been demonstrated in some tissues. Microscopic studies using fluorescent staining of nerve terminals have shown adrenergic innervation down to the level of the precapillary sphincter in some tissues, and direct stimulation of the perivascular nerves has allowed measurements of contractile responses in both the small arteries and veins. β-adrenergically mediated dilatation of precapillary sphincters has also been demonstrated directly by injection of β-agonists, and a variety of potent naturally occurring vasoactive substances has been shown to have direct or indirect effects on the microcirculation. These include CO_2 (a particularly potent regulator of the cerebral circulation), P_{O_2} (a potent vasodilator in skeletal muscle), pH, and polypeptides such as bradykinin which are released from lysosomes or by activation of the

blood components during tissue injury. Certain prostaglandins and histamine are also potent vasodilators.

There is no agreement as to which local factors regulate flow through the capillary bed in most tissues. Many of the chemical mediators mentioned above, as well as adenosine, have been considered. Some factor related to oxygen metabolism undoubtedly is involved, and increasing local tissue metabolism produces increased flow. In addition, decreasing the P_{O_2} in the arterial blood enhances local perfusion, whereas increasing P_{O_2} is associated with diminished capillary density and a decrease in perfusion (Duling and Klitzman, 1980). Thus, capillary density is sensitive to tissue P_{O_2} in most tissues, although the related direct chemical mediators remain unknown (Duling, 1980). The tissue P_{O_2}, in turn, is regulated by the capillary P_{O_2} and the rate of tissue O_2 consumption ($\dot{V}O_2$) (these factors together with the intercapillary distance determine the gradient for O_2 diffusion, see next section). Thus, regulation of flow and capillary density (surface area) are highly important control mechanisms in the microcirculation. The regulation of resistance vessel caliber in the pre- and postcapillary regions also affects capillary pressure, thereby influencing transcapillary filtration, as further discussed in the next section.

TRANSCAPILLARY EXCHANGE

The capillary endothelium is relatively impermeable to proteins and large molecules. Substances such as crystalloids and gases in solution are relatively freely diffusible, and delivery or uptake of some diffusion-limited substances are importantly related to the capillary transit time. Hydrostatic pressure and colloid osmotic pressure largely govern the movement of fluid in and out of the capillaries, whereas larger molecules move by several processes. It is convenient to consider transcapillary transport in terms of three processes: diffusion, filtration, and large molecule movement. (For reviews see Landis, 1964; Landis and Pappenheimer, 1963).

Diffusion

This is by far the major process by which nutrients and metabolic products cross the capillary membrane. The rate of movement of solutes, and the fluid carrying them, is far greater than that transferred by filtration. Small, lipid-soluble molecules (such as CO_2 and O_2) move freely by diffusion across the endothelial cells, as well as through the capillary pores. Water and water-soluble materials are believed to cross mainly through the microscopic clefts (pores or slits) between endothelial cells. For small, water-soluble molecules such as NaCl and glucose pore size is not an important consideration, and the rate of delivery of the substance to the capillary bed (blood flow rate) primarily determines their transport; movement of these substances is *flow-limited*. Larger molecules (such as sucrose, 358 daltons) begin to be limited by pore size. In general, such lipid-insoluble *diffusion-limited* substances show diffusion rates that are inversely related to their molecular size.

There are five main factors which affect diffusion between the intravascular and tissue compartments. These include: (1) the *intercapillary distance*; (2) the *blood flow*; (3) the *concentration gradient* for the solute; (4) the *capillary permeability*; (5) the *capillary surface area*.

The effects of intercapillary distance and blood flow rate on O_2 transport serve as a useful example for other readily diffusible substances (although O_2 delivery is complicated by the equilibration of O_2 in solution with hemoglobin*; see discussion in Section 5, Chapter 36). The diffusion of O_2 out of a capillary is dependent on two types of gradients, the longitudinal and radial gradients.

The longitudinal gradient from inside to outside the capillary exhibits a nearly exponential fall from the arterial to the venous end of the capillary. It is affected by capillary length, by the flow rate (an increased flow at a fixed tissue $\dot{V}O_2$ will result in a higher venous P_{O_2}, and vice versa), and by the tissue $\dot{V}O_2$ (an increased $\dot{V}O_2$ at a fixed flow rate will produce a lower venous P_{O_2}, and vice versa).

The radial gradient has usually been analyzed using Krogh's tissue cylinder concept (Krogh, 1919, 1959), which considers the factors responsible for radial diffusion from a capillary. The Krogh equation employs the radius of the capillary and the radius of the tissue cylinder around it, together with the capillary concentration or partial pressure of a substance and its diffusion coefficient, to calculate the concentration or partial pressure of the substance at a given distance from the capillary† (see Fig. 5.20, Section 5, Chapter 36).

---

* The *level* of hemoglobin (Hgb) is an important determinant of O_2 delivery, as well as the characteristics of O_2-Hgb *dissociation*. O_2 begins to diffuse through the arterial wall before it reaches the capillary, but most of the O_2 is unloaded from Hgb when the capillary P_{O_2} falls below 60 mmHg, as the dissociation curve steepens, and small further decreases in P_{O_2} then result in large reductions in Hgb saturation (see Fig. 5.12, Section 5, Chapter 36).

† For oxygen:

$$\text{mean } P_{T_{O_2}} = P_{cap_{O_2}} - A \left| \frac{\dot{V}O_2 R^2}{4D} \right|$$

where P_T and P_{cap} = partial pressures in tissue and capillary, A

The radial gradient can be calculated at any point along the capillary. Obviously, if two capillaries are far apart, tissue P_{O_2} can fall to a very low value between them, whereas if the intercapillary distance is small (capillary density and surface area high), the tissue P_{O_2} can be well maintained. Under conditions such as exercise in skeletal muscle, the high $\dot{V}O_2$ may produce a low tissue P_{O_2}, despite marked capillary recruitment. It has been emphasized that at high $\dot{V}O_2$ in heart muscle, the great variability of capillary lengths and flow velocities in capillaries can result in inhomogeneity of tissue P_{O_2}; under some conditions this can lead to "islands" where tissue P_{O_2} is insufficient to maintain mitochondrial respiration, so-called "tissue hypoxia," and decreased tissue function or cell damage can result (mitochondrial function decreases at tissue P_{O_2} below 0.1 mmHg) (Honig and Bourdeau-Martini, 1973).

Fick's law of diffusion considers the remaining three variables (concentration gradient, capillary permeability, and surface area) as follows:

$$\dot{M} = D \cdot \frac{A}{T}(C_i - C_o)$$

where $\dot{M}$ = the diffusion rate of the material, D = the free diffusion coefficient for that substance (which is related to its molecular weight), A = capillary surface area, T is the thickness of the membrane (to which diffusion rate is inversely proportional), and $C_i - C_o$ = the concentration gradient for the substance (see further discussion relative to the lung, Section 5, Chapter 37).

For a thin-walled membrane, the values for T and A are usually replaced by PS (in which P = permeability and S = surface area of the capillary), and the equation then becomes:

$$\dot{M} = PS(C_i - C_o)$$

Since the permeability rarely changes (except under pathologic conditions), PS becomes proportional to the capillary surface area under many conditions. Some substances (such as O_2) begin to diffuse out of the arteries well before the capillary membrane is reached, but all demonstrate a progressive fall of the concentration gradient as they pass from the arterial to the venous end of the capillary bed. Of course, CO_2 concentration is higher outside the capillaries than within, and this substance moves *into* the capillary in accordance with its concentration difference.

Most metabolites, nutrients, and other common

= area of the tissue cylinder outside the capillary, R = radius of the cylinder from the center of the capillary (related directly to intercapillary distance), and D = O_2 diffusion coefficient.

substances (such as urea and glucose) are quite readily diffusible, and only small concentration differences are required for their transport. For most of these *flow-limited* (not diffusion-limited) substances, their transport can then be determined from the standard Fick equation:

$$\dot{M} = Q(C_A - C_V)$$

where Q = the blood flow rate and $C_A - C_V$ = arteriovenous concentration difference.

Filtration

Fluid movement results from a hydrostatic or osmotic pressure difference across a membrane (actually causing bulk flow of fluid, a process distinct from diffusion). Such bulk fluid movement occurs through the membrane or through pores. The plasma proteins, primarily albumin and globulin, are in high concentration within the blood compartment (6.0–8.0 g/100 ml or g/dl), whereas protein concentrations in the interstitial fluid are relatively low (0.7–2.0 g/100 ml). The proteins (with hyaluronic acid in solution in the interstitial space) produce *colloid osmotic pressure* (oncotic pressure) which is determined primarily by the number of molecules in solution. Thus, about 75% of the total oncotic pressure of the plasma results from albumin which has a relatively small molecular weight (69,000), and the remainder is due largely to globulins. It should be pointed out that some proteins can and do leak through pores in the capillary wall or are moved by vesicular transport, and these proteins and other substances that filter out of the capillary wall are returned to the circulation by the lymphatic system (as discussed in Chapter 19).

Several factors determine the rate at which fluid is filtered across the capillary membrane. These include:

1. The *filtration coefficient* for a given capillary wall. This depends on the tissue (for example, this coefficient is much higher for water in the kidney than in skeletal muscle), and the coefficient also may vary under different physiologic conditions.

2. The *capillary hydrostatic pressure*. There is, of course, a gradient of pressure from the arterial to the venous end of the capillary (Fig. 2.19), and the intracapillary pressure can be modified by changes in the caliber of the precapillary as well as the postcapillary resistance vessels.

3. The *interstitial fluid hydrostatic pressure*. The level of hydrostatic pressure in the interstitial fluid is controversial, both small positive and small negative values having been reported under normal conditions (the average value is probably close to zero) (Guyton, 1963). Under abnormal conditions,

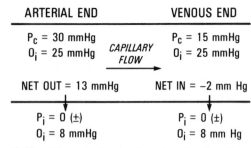

Figure 2.19. Diagram of a capillary showing the effects of hypothetical plasma and tissue hydrostatic and oncotic pressures on transcapillary fluid movement. For further discussion, see text. *P*, pressure; *O*, oncotic pressure in the capillary (*c*) and interstitial (*i*) regions.

however, this pressure may become substantially positive (such as with severe swelling [edema] of the legs or other tissues).

4. The *colloid osmotic pressure of the plasma*. This is determined by the plasma protein concentration. The colloid osmotic pressure tends to draw *water* into the capillary in accordance with its concentration difference. Of course, in disease states where the serum albumin is low, this pressure can become abnormally reduced.

5. The *colloid osmotic pressure of the interstitial fluid* (discussed earlier).

An equation developed by E. H. Starling, modified to include interstitial colloidal pressure, has been applied to describe ultrafiltration across a single blood capillary. It should be understood, however, that the capillary beds of various organs differ markedly in their filtration properties (for example, the kidney *filters* throughout its entire capillary length, whereas *absorption* may occur predominantly in some capillary beds in the intestinal wall).‡ In the general scheme shown in Figure 2.19, filtration predominantly occurs at the arterial end of the capillary and absorption at the venous end. *Starling's law of ultrafiltration* states that:

$$FM = K(P_c + \pi_i - P_i - \pi_c)$$

where FM = fluid movement (+ outward, − inward), K = filtration coefficient of capillary wall, P_c = hydrostatic pressure in the capillaries, P_i = hydrostatic pressure in the interstitial fluid, π_c =

---

‡ This simplified Starling scheme for a single capillary is further greatly complicated by the finding that permeability coefficients may vary in different portions of the same capillary bed and also may be significantly higher at the venous end of capillary beds. The view that venous vessels have a higher filtration coefficient (which may tend to counterbalance the less marked inward pressure at the venous end of the capillary) has been supported by studies showing movements of various dyes, as well as by electron microscopic studies suggesting more fenestrations in venular endothelium (Zweifach, 1973).

oncotic pressure in the plasma in the capillary, and π_i = oncotic pressure in the interstitial fluid.

Thus, net filtration rate will be determined by the difference between the forces tending to move fluid *outward* (P_c and π_i) and forces tending to move fluid *inward* (P_i and π_c). If the coefficient is ignored, it can be seen from Figure 2.19 that at the arterial end of this particular capillary, the net driving pressure outward is about $38 - 25$ mmHg $= 13$ mmHg (causing bulk flow of water toward the interstitial fluid, (*arrows*)), whereas toward the venous end of the capillary the net driving pressure is inward, $23 - 25$ mmHg $= -2$ mmHg (causing fluid movement into the capillary (arrows)). In skin, the outward pressure at the arterial end has been calculated at $+7$ mmHg and the inward pressure at the venous end at -9 mmHg (Landis and Pappenheimer, 1963).

A more important controlling mechanism than this delicate hydrostatic-osmotic balance in the individual capillaries of many tissues, however, is the presence of vasomotion in the precapillary beds. Thus, direct measurements in the microcirculation have shown spontaneous, relatively prolonged periods of contraction or even closure of terminal arterioles or precapillary sphincters, producing a very low capillary hydrostatic pressure (5–10 mmHg); this type of behavior dominates in some tissues, such as skin and muscle, so that the oncotic pressure in the plasma favors fluid reabsorption. Occasionally, the precapillary vessels spontaneously open widely, and capillary pressures as high as 40–50 mmHg (favoring net filtration) are transiently recorded (Wiederhielm and Weston, 1973).

As shown in Figure 2.20, an increase in either arterial or venous pressure will increase capillary hydrostatic pressure, and vice versa. Capillary pressure (P_c) is controlled both by the precapillary resistance (R_1) and the postcapillary resistance (R_2). An increase in arteriolar (precapillary) resistance (R_1) will *lower* capillary pressure, whereas an increase in venous (postcapillary) resistance (R_2) will *increase* capillary pressure. The normal ratio

of $R_1/R_2 = 4/1$ (that is, 70–80% of the resistance is in the precapillary region). Since the venous resistance is much lower than the arterial resistance, an increase in venous pressure has a substantially larger effect on capillary hydrostatic pressure than does a change in arterial pressure. If R_1 over R_2 increases it will favor net reabsorption, whereas if it decreases net filtration will be favored. Both of these resistances can operate relatively independently to maintain the R_1/R_2 ratio relatively constant. For example, when the venous pressure rises, as upon assuming the standing position, capillary pressure will tend to rise; however, the increased transmural pressure results in myogenic constriction of the precapillary vessels (arterioles and sphincters), which may actually close capillary beds and reduce the capillary surface available for filtration; therefore, fluid accumulation (edema) usually does not occur under these conditions (see discussion of edema, Chapter 19).

It should be pointed out that the coefficient in the Starling equation, which importantly influences the movement of fluid, is affected by the capillary surface area as well as the permeability of the membrane; the surface area may be greatly changed during capillary "recruitment" or by closure of the precapillary sphincters. If the other values in the Starling equation are known, and membrane permeability is assumed not to change, the coefficient can be calculated and will provide some measure of the relative capillary surface area (the "capillary filtration coefficient," Landis and Pappenheimer, 1963).

Large Molecule Movement

Such movement can occur by vesicular transport or directly through capillary fenestrations. In vesicular transport, enclosure of plasma within vesicles occurs and the vesicle then moves from one side of the cell membrane to the other, thereby transporting large molecules (such as proteins) from the inner to the outer membrane surface. This process has also been called micropinocytosis; its relative importance in many tissues remains controversial. Fenestrae ("windows") which allow large molecule movements exist in many capillary membranes, although in certain tissues such as skeletal muscle and nervous tissue fenestrae are few. Fenestrated endothelial cells are found in the intestine, kidney, endocrine organs, and very large vascular openings are present in the liver and bone marrow. As mentioned, even in capillaries without apparent fenestrae, some protein "leak" occurs, and this has been theoretically accounted for by a very few large clefts (Landis and Pappenheimer, 1963).

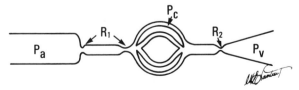

Figure 2.20. Diagram of an artery, vein, and a microcirculatory unit. P_a equals pressure in the artery. P_v equals pressure in the vein. Resistance (R_1) in the arteriole and precapillary resistance vessel is shown together with that of the postcapillary venule (R_2). PC, capillary pressure.

CHAPTER 7

Dynamics of the Peripheral Circulation

William Harvey (English physician, 1578–1657) largely dispelled the idea that blood ebbed and flowed through the arteries and veins. After studying in Padua under Fabricius, whose study of the venous valves apparently first generated the idea that blood flowed through the veins toward the heart, Harvey arrived at the conclusion that the blood must circulate around and around. In a series of experiments in animals involving vessel ligations, and measurements of the volume of blood flowing from severed vessels, he concluded that "the movement of the blood is constantly in a circle, and is brought about by the beat of the heart." He also measured the volume of blood contained in the ventricles and calculated that the amount ejected by the ventricles in ½ hour was greatly in excess of the volume of blood contained in the entire body, thereby reinforcing his conclusion that the blood must circulate (despite the fact that microscope was just coming into use and capillaries had not yet been demonstrated). These discoveries were summarized in a classic book published by Harvey in 1628 (see translation, 1928).

This chapter will consider primarily those features of the blood itself and of the peripheral circulation which are responsible for the patterns of pressure and blood flow through the arteries and veins consequent to the pumping action of the heart. Such properties of the pulmonary circulation are considered in Section 5, Chapter 36.

Certain physical properties of the blood are important in understanding the pressure and flow characteristics of the circulation. (The science of rheology is concerned with investigation of the deformation and flow of matter.) A number of different types of cells and formed elements are carried in the blood plasma (see Section 3, Chapter 20). With a normal red blood cell (erythrocyte) count of 5 million/mm^3 and a normal white blood cell count of approximately 7000/mm^3, the red cells constitute about 99.9% of the circulating cell population. Indeed, the red cells are so numerous that normally they occupy about 40% of the volume of the blood, a value expressed as a percent termed the *hematocrit*. Blood plasma is a viscous fluid, and the number of red cells and their characteristics also importantly influence this property (for review, see Dintenfass, 1976).

BLOOD VISCOSITY

Throughout most of the circulation, blood flow is *laminar* (except at the valves and certain other areas, as well as in abnormally narrowed regions produced by atherosclerosis); that is, blood flows through the vessels in an "orderly" manner, with flow velocity zero at the vessel wall (where friction is sufficiently high to prevent slippage between the fluid and wall) and progressively increasing toward the center of the vessel. There is resistance to movement between the layers of blood across the channel of the vessel, and it is this "inner friction," or resistance to slippage, that gives the blood its *viscosity* (for reviews see Rosenblum, 1977, and Cokelet et al., 1980).

A variety of instruments (viscometers) can determine the viscosity of fluids, and one simple approach involves the passage of fluid at a known pressure through a tube of precisely known dimensions; the viscosity of the fluid can then be calculated from the flow rate using the Poiseuille equation (see below). A *Newtonian fluid*, such as water, is defined as one which maintains a *constant* viscosity at any flow velocity whereas a non-Newtonian fluid changes viscosity at different velocities. In a complex non-Newtonian suspension such as blood, the viscosity is most importantly determined by the red blood cells but also to a small degree by the proteins in solution in the plasma. In such a suspension, the apparent viscosity of blood *changes* considerably, depending upon the velocity of blood flow, the hematocrit, and the size of the blood vessel.

Viscosity was defined by Newton as the ratio of sheer stress (force resisting or pushing flow) to the shear strain rate of a moving fluid. Thus:

$$F = A \cdot \eta \cdot \Delta V / \Delta X$$

where F represents the force, as drag or resistance, of one layer opposing the movement of an adjacent layer (or, conversely, the tendency of a moving layer traveling at a faster velocity to pull a slower layer along), A = area between the two layers, η = coefficient of viscosity; and $\Delta V / \Delta X$ = the velocity gradient between the two layers (the change of velocity per unit distance, ΔX, at right angles to the flow movement). Rearranging the equation:

$$\eta = \frac{F/A}{\Delta V / \Delta X} = \frac{\text{shear stress}}{\text{shear strain rate}}$$

For a Newtonian fluid flowing through a cylindrical tube and exhibiting laminar flow, solution of this equation for each of multiple layers yields a *parabolic velocity profile* across the channel, with the fastest velocity in the center and a velocity assumed to be 0 at the boundary with the wall (Fig. 2.21). With normal laminar flow of blood in arteries and veins, the profile is not strictly parabolic, although in large vessels with diameters much greater than those of the red cell, blood behaves as a relatively Newtonian fluid, and the profile resembles that shown in Figure 2.21.

The dimensions of viscosity are dynes/cm² divided by (cm/s)/cm (or dyne s/cm²), a unit termed the "poise" (after Poiseuille). The practical unit 0.01 poise is termed the "centipoise," and water at room temperature has a viscosity of about 1 centipoise. For convenience, the viscosity of various

fluids is often related to that of water at the same temperature and is termed the "relative viscosity." At a normal hematocrit, blood has a viscosity 3 or 4 times higher than that of water (relative viscosity of 3 to 4).

Plasma behaves like a Newtonian fluid, but because of the suspended red cells changes in whole blood viscosity are markedly nonlinear with alterations in flow velocity. Thus, as flow *velocity decreases, viscosity increases,* and vice versa (Fig. 2.22). The shape of this viscosity-velocity relationship is determined entirely by behavior of the red blood cells (Chien, 1966 and 1975). At low velocities (below shear rates of 1, Fig. 2.22), "rouleaux" formation occurs, with increasing aggregates of stacked red cells causing increasing viscosity (the blood proteins fibrinogen and globulin promote this, and the effect is lost in their absence). At increasing velocities (shear rates over 1, Fig. 2.22) less marked changes (decreases) in viscosity occur, which appear to relate to deformation of the red cells (elongation) with improvement of their flow characteristics; this progressive decrease in viscosity is prevented by making red cells nondeformable (see Chien, 1966). It can be seen from Fig. 2.22, that at the high velocities occurring in large blood vessels (20–60 cm/s, see Chapter 6), the curve is relatively flat, with viscosity changing little with flow velocity changes (relatively Newtonian behavior). At slower flow velocities the velocity effect on viscosity is marked (viscosity effects may become important in the veins, particularly when the cardiac output is very low). In certain disease states, such as sickle-cell anemia, the red blood cells become less deformable, and this can lead to an increase in blood viscosity. The viscosity is also increased when abnormally high concentrations of proteins are present in the plasma, as occurs in some disease states (Dintenfass, 1976).

The viscosity of blood is also markedly dependent

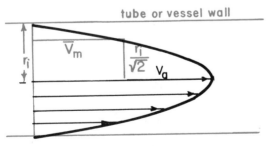

Figure 2.21. Parabolic flow velocity profile plot. The plot represents calculation of flow velocity (*V*) at various radial distances (*r*) from the center of the tube to the wall (where flow is assumed to be stationary, *V* = 0). Flow is fastest in the central axial stream (*V*ₐ) and decreases to 0 at the wall. At a distance $r/\sqrt{2}$ from the center of the tube, the flow velocity is half that along the central axis and equals the mean flow velocity (*V*ₘ) along the tube. The arrows indicate relative flow velocities at various radial distances (*r*) from the central axis.

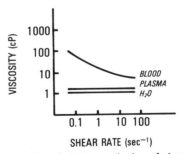

Figure 2.22. Relationship between the log of viscosity and the shear rate (related to blood flow velocity). Notice the nonlinear relation in whole blood and the essentially flat (Newtonian) relationship in plasma and water (the viscosity of plasma being slightly higher than that of water). [Adapted from Chien (1966).]

upon the hematocrit, and this effect, as determined *in vitro*, is diagrammed in Figure 2.23. The effect of hematocrit on blood viscosity increases sharply above a hematocrit of about 50%, such as in diseases associated with overproduction of red blood cells (i.e., polycythemia vera where hematocrits of 60–70% can occur), or in individuals living at high altitudes. In such conditions, substantial increases in the blood viscosity can occur with unfavorable effects on blood flow and the work of the heart. It has been found that relative blood viscosity in the intact circulation (as in the hind limb of a dog) is actually considerably lower over a range of hematocrits than that measured in glass tubes, although the shape of the curves is somewhat similar.

It might be expected that slow blood flow velocity in the capillaries would result in extremely high viscosity. However, an important effect of vessel size on viscosity prevents this occurrence, a *decrease* in viscosity occurring at tube radii below 0.5–1.0 mm (Fig. 2.24). This phenomenon is known as the *Fahraeus-Lindqvist effect* (1931), and it tends to favor flow through the capillaries, as well as through the small arterioles. This effect may, in large part, explain the lower blood viscosity as measured *in vivo* that is mentioned above. The reason for the effect is complex (see Fung, 1977 and 1981), but it has been ascribed at least in part to a phenomenon causing the layers of blood next to the vessel wall to be relatively cell-free and leading to lower hematocrits in small blood vessels. Thus, there is a tendency for red cells to move toward the higher velocity channel in the center of a tube, increasing the hematocrit there while the hematocrit decreases near the walls; the cell-poor layer near the wall becomes proportionally larger as the vessel diameter becomes smaller. In microvessels, side branches can therefore have a much lower

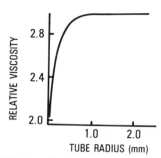

Figure 2.24. Relationship between relative viscosity and the radius of the tube in which fluid is flowing. Notice the sharp drop in viscosity at radii below 1.0. [Adapted from Fahraeus and Lindqvist (1931).]

hematocrit, a phenomenon sometimes called "plasma skimming." Indeed, in the capillary bed of skeletal muscle, for example, the hematocrit is well below 20%, compared to 40–45% in the large vessels (see Klitzman and Johnson, 1982). The lowering of hematocrit results in a blood viscosity closer to that of plasma, and undoubtedly it is an important component of the fall of viscosity (Fahraeus-Lindqvist effect) in small vessels (Fig. 2.24). Without this effect, the overall slowing of blood velocity in the capillaries and adherence of red cells into aggregates at slow velocities would markedly enhance capillary blood viscosity.

The red blood cell, which is shaped like a biconcave disc, is able to deform and to squeeze through openings much smaller than its normal diameter (about 8.5 μm). As we have seen, this immense *deformability of the red blood cell* membrane is an extremely important property and allows the red cells to readily crowd through complex, narrow capillary networks (Schmid-Schonbein, 1976, and Cokelet et al. 1980). The red blood cell membrane will not tolerate *stretch,* however, and if the volume of the red blood cell is increased, the cell tends to become more spherical, and it can break easily (a property that becomes important in certain types of anemia). (For general discussion of red cell structure and function see Section 3, Chapter 24).

The rate of flow of the plasma through blood vessels has been found to be substantially slower than the rate of movement of the red blood cells. This phenomenon results from the fact, mentioned above, that the central (axial) portion of the blood stream contains a greater proportion of red cells moving at a greater velocity. Therefore, the mean transit time of the plasma, which moves more uniformly across the wall, is slower.

VASCULAR RESISTANCE

The simplest expression of resistance to flow through a hydraulic circuit is like Ohm's law for

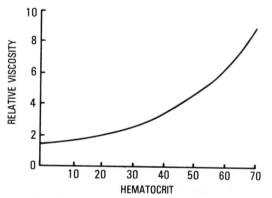

Figure 2.23. Relationship between relative viscosity and the hematocrit in whole blood. Notice that at a normal hematocrit, blood is 3 to 4 times as viscous as water and, at higher hematocrits, the viscosity increases sharply.

electrical circuits, in which $I = E/R$ (where $I =$ current flow, $E =$ voltage, and $R =$ resistance of the circuit). The analogy in the vascular bed is the general resistance equation mentioned in Chapter 5:

$$Q = \frac{P_A - P_V}{R}$$

where $Q =$ blood flow, P_A and $P_V =$ pressure (P) difference across the vascular bed between artery (A) and vein (V) and $R =$ resistance in arbitrary units. The application of this equation for defining the control of blood flow to an organ (also discussed in Chapter 5) is:

$$Q = \frac{P_A - P_V}{\text{local } VR}$$

where local $VR =$ organ or tissue vascular resistance.

This equation is useful for describing the total vascular resistance across all the systemic organ beds of the body, which maintains the blood pressure, rearranged as:

$$SAP = CO \times TPVR$$

where $SAP =$ systemic arterial pressure, $CO =$ cardiac output, and $TPVR =$ total peripheral vascular resistance.

When resistances are arranged *in series* (as in flow across an organ bed, or when vascular lesions occur in two portions of the same vessel), according to the laws of electrical circuits the total resistance to flow is the *sum* of the resistances (Fig. 2.25, *left*).

When resistances occur *in parallel* preferential flow occurs through the pathways of least resistance and the total resistance is less than the resistance of any individual vessel, being the reciprocal of the sum of the reciprocals of individual resistances (Fig. 2.25, *right panel*). *Conductance*, the reciprocal of resistance, is used as a measure of the ability of a blood vessel to accept blood flow at a given pressure difference, expressed as milliliters per second per mmHg:

$$C = \frac{1}{R}$$

where $C =$ conductance and $R =$ resistance. The total conductance (C_T, which is equal to $1/TPVR$) is therefore equal to the *sum* of the individual *conductances* in each segment for a given pressure drop:

$$C_T = C_1 + C_2 + C_3$$

In a parallel circuit, adding a resistance pathway even if it is higher than any of the individual resistances in the circuit will add another (small)

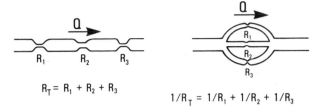

$$R_T = R_1 + R_2 + R_3$$

$$1/R_T = 1/R_1 + 1/R_2 + 1/R_3$$

Figure 2.25. (*Left panel*) Diagram showing the calculation of total resistance (R_T) and various resistances aligned in *series* as blood flows through a vascular circuit. (*Right panel*) Diagram showing the calculation of total resistance (R_T) as blood flows through various resistances aligned in a *parallel* circuit.

conductance and so will raise the total conductance, while lowering total resistance. (Such calculations also explain how a very large number of high resistance units, such as capillary beds, despite their smaller radii, can have a lower total resistance than the arteriolar bed.) If one resistance in a parallel circuit drops substantially below the others, the total conductance will rise and the total resistance will fall since, from the equation in Figure 2.25, total resistance in a parallel circuit must always be less than that of any individual resistance.*

In a circuit such as the peripheral circulation where a great many organs are arranged in parallel, modest increases or decreases of resistance in a given organ will not have a major impact on total vascular resistance or blood pressure. However, tremendous differences in flow rates exist through the various organs at rest (and, therefore, there are large differences in regional resistances since perfusion pressure is the same). The flow differences that would exist between organs during theoretical maximum flow rates to each organ are shown in Figure 2.26, and the potential maximal flows are shown in the accompanying table. Note that with maximal dilation of all beds, total flow would be extremely high (38 L, exceeding the limit for cardiac output), so that some beds must undergo compensatory vasoconstriction (by reflex control) when others dilate.

While these equations are highly useful for clarifying interrelationships among pressure, flow, and resistance (and for making calculations in the clinical setting, see Chapter 13), it is important to understand in more detail the factors that are responsible for vascular resistance in the circulation, since disturbance of these factors can occur in

---

* For example, if a parallel circuit has 2 pathways, with each having a resistance of 4 units, the total resistance is 2 units (¼ + ¼ = ½). If the resistance in one limb falls markedly to 1 unit (a 4-fold drop), the total conductance will be 0.25 + 1 = 1.25, and the total resistance (reciprocal of total conductance) becomes 0.8 units.

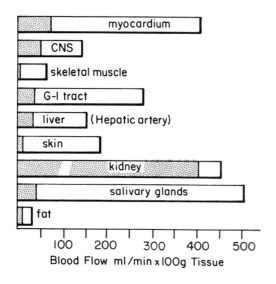

	Rest. blood flow (1/min.)	Max. blood flow (1/min.)	Organ weight (kg)
Myocardium	0.21	1.2	0.3
CNS	0.75	2.1	1.5
Skeletal muscle	0.75	18.0	30.0
G-I tract	0.7	5.5	2.0
Liver (hep. artery)	0.5	3.0	1.7
Skin	0.2	3.8	2.1
Kidney	1.2	1.4	0.3
Salivary glands	0.02	0.25	0.05
Fat	0.8	3.0	10.0
Total ≈	5.1	38.0	48.0

Figure 2.26. (*Above*) Bar graph representing approximate blood flows in various organs at maximal vasodilatation (*total bar*) and at "rest" (*hatched areas*) in ml/min × 100 g of tissue at perfusion pressure of 100 mmHg. (*Below*) Approximate values for regional blood flows in a 70-kg man at "rest" and at maximal dilatation, deduced from organ weights. The organs included comprise about 70% of total body weight. [Redrawn from Mellander and Johansson (1968).]

disease states. Thus, resistance to flow may be altered by a number of factors, including the characteristics of blood flow, the diameter of the vessels in question, the distending pressure in the vessels, and the viscosity of the blood.

Poiseuille's Law

The various factors influencing resistance emerge from an analysis of the Poiseuille equation, developed from a study of steady laminar flow by a Newtonian liquid in narrow rigid glass tubes.

The law, developed by J. L. M. Poiseuille, French physicist and physician (Poiseuille, 1846) states that:

$$Q = \pi \frac{(P_1 - P_2)r^4}{8\,\eta L}$$

where Q = flow rate, $P_1 - P_2$ = the pressure differ-

ence across the circuit, r = radius of the tube, η = viscosity of the liquid, and L = length of the tube.

The Poiseuille equation illustrates the important relationship between vascular resistance and the length of the tube and blood viscosity; thus, *if either one doubles, the flow will be halved* (or if flow stays constant, the pressure gradient will double). The equation also emphasizes the enormous influence of vessel radius; *if the radius doubles*, the *flow will increase 16-fold.*

Since the experiments were originally performed using a Newtonian fluid, if the caliber and length of the tube are kept constant, the viscosity will not vary as the flow is changed and, hence, the resistance remains constant. However, if those factors which influence resistance are considered separately, the resistance is seen to be directly proportional to the viscosity and length of the tube and inversely proportional to π and the 4th power of the radius:

$$R = \frac{8L\eta}{\pi r^4}$$

Note that this equation contains a constant $\left(\dfrac{8}{\pi}\right)$, a geometric component, $\left(\dfrac{L}{r^4}\right)$ and the viscosity component (η).

Most of these factors cannot be measured in the intact circulation. Therefore, the simpler resistance equation is usually employed, which can be obtained by substituting R for the above factors into Poiseuille's equation to obtain:

$$Q = P_1 - P_2/R \text{ or } R = \frac{P_1 - P_2}{Q}$$

For calculation of vascular resistances in the intact circulation, and the normal range of values in man, see Chapter 13.

Resistance in Distensible Tubes

An important difference between Poiseuille's experimental preparation and the circulation is the fact that blood vessels are distensible. In nondistensible tubes using Newtonian fluids, as pressure increases flow increases linearly, and the resistance (the slope of the pressure-flow plot) remains constant (Fig. 2.27, *dashed lines*). On the other hand, in a vascular bed (with no intrinsic vasomotor responses) the pressure-flow relation is nonlinear; as the pressure difference increases the flow is augmented in a nearly exponential manner as the blood vessel caliber is augmented (Fig. 2.27, *right-hand solid line*), and as the pressure rises the resistance falls (Fig. 2.27, *lefthand solid line*). In beds

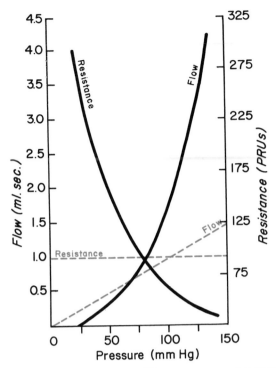

Figure 2.27. The relationships of flow, pressure, and resistance in nondistensible tubes using Newtonian fluids (*broken lines*) at constant temperature (Poiseuille's law) and blood in vessels (*solid lines*). See text for details.

with higher degrees of vasomotor tone (but without autoregulation), the pressure-flow curves are progressively shifted to the right and flattened, and the intercept at zero flow (critical closing pressure) becomes progressively higher, as discussed in a previous section (see Fig. 2.14).

TURBULENT BLOOD FLOW

When the blood flow rate exceeds a certain velocity, laminar flow no longer occurs, the velocity profile becomes blunted, frictional losses within the blood become high, and the resistance to flow rises. (This may occur, for example, in blood traversing a narrowed area in an artery produced by atherosclerosis.) The appearance of such turbulent flow is related to the blood velocity, but also to the diameter of the vessel and the density and the viscosity of the fluid. The *Reynolds number* (a dimensionless value) can be calculated from these variables:

$$R = \frac{VD\rho}{\eta}$$

where R = Reynolds number, V = mean velocity, D = tube diameter, ρ = fluid density, and η = fluid viscosity.

Turbulent flow usually occurs when the Reynolds number exceeds about 3000. Since the viscosity of blood is relatively high, the Reynolds number for

turbulent flow is not exceeded in most parts of the circulation (see Table 2.1, Chapter 6). Laminar flow is silent, but with the onset of turbulent flow sounds occur. In the vascular bed, these sounds may be heard with a stethoscope under certain circumstances (such as beyond a constricting vascular lesion) as vascular bruits, or tapping sounds beyond the cuff of a sphygmomanometer (Korotkoff sounds) (see Fig. 2.124, Chapter 13). Also, at very high cardiac outputs, the Reynolds number may be high enough in the aorta and pulmonary artery to produce systolic heart murmurs, even in the normal circulation. It is also evident that at large vessel diameters and high flow velocities when there is a low blood viscosity, as in anemia, the Reynolds number is particularly likely to be exceeded.

It should be recognized that as flow increases, if the critical Reynolds number is exceeded, the kinetic losses of energy due to the resulting turbulent flow will produce a *decrease* in the flow at the same pressure difference, a factor which can affect the blood flow through diseased blood vessels. Turbulent flow with increased shear stress and strain acting at the vascular wall are important in promoting atherosclerotic lesions at certain points in the arteries (see Fry, 1968).

DETERMINANTS OF THE MEAN ARTERIAL PRESSURE

The static pressure in a closed container refers to the force exerted against a unit area of the container's wall, although typically when pressure is measured in the vascular system we refer to the height of a column of fluid supported by the force within the vascular bed. For example, when Stephen Hales (Hales, 1733) measured the arterial blood pressure in a horse, he ligated a cannula into an artery, affixed it to a glass tube 9 feet in length, and found that the column of blood from the living horse rose about 8 feet above the level of the heart. Thus, the weight of the column of blood provided a measure of the blood pressure. A mercury manometer is more convenient for measuring pressure, and it is customary to report the height to which a column of mercury (Hg) is pushed by the intravascular pressure, a column 100 mm high reflecting a blood pressure of 100 mmHg. Sometimes, pressure in the veins is measured in centimeters of water: 1 mmHg = 1.36 cm H_2O. Commonly, for experimental purposes and during cardiac catheterization in man, electronic gauges are used (see Chapter 13) which measure the pressure waveforms and the mean pressures by means of conductance coils or by a strain-gauge attached to a rigid membrane (Fig. 2.123, Chapter 13), and these gauges are calibrated against a mercury manometer.

Kinetic and Potential Energy

When blood is flowing into the peripheral circulation, the total fluid energy (*E*) per unit volume of blood equals the sum of the pressure, plus a factor related to the influence of gravity, plus the kinetic energy:

$$E = P + \rho g h + \tfrac{1}{2}\rho V^2$$

where *P* is static pressure expressed as force (dynes/cm^2), ρ = fluid density in gm/cm^3, *g* = the gravity acceleration constant of 980 cm/s^2, *h* = the height of the fluid column above or below a reference level, and *V* = flow velocity in cm/s.

If all parts of the fluid system are at the same level, the gravitational factor can be ignored (this is the case in the supine, but not the erect posture). The kinetic energy, which is proportional to the square of the velocity, is often ignored but can become important under certain circumstances; for example, although blood normally flows from a higher pressure to a lower pressure, if the kinetic energy component (velocity) is sufficiently high, blood occasionally can flow from a lower to a higher pressure. The driving forces for forward flow are not directly affected by the different heights of various portions of the circulation in the erect body position; thus, as pointed out by A. C. Burton (Burton, 1972), like the principle of the syphon, it is only the *difference* in the total pressure and energy at one point in the circulation from that in another that determines flow, not whether the flow is uphill or downhill. However, if the static pressure in the arterial bed is measured with a manometer in the erect posture, the pressure will differ by the height of the fluid column above or below the heart; thus, if the pressure at the heart level is 100 mmHg one would predict a pressure in the toes of about

185 mmHg and in the head of about 50 to 60 mmHg. This factor is even more important in the distensible veins, as discussed subsequently.

As described in Chapter 6, the linear velocity of blood flow (*V*) is inversely proportional to the cross-sectional area (*CSA*) of the vascular stream (*V* = *Q/CSA*), and the kinetic energy component of the flowing blood (which is a function of *V*^{2}) can importantly affect the pressure, depending upon how pressure is measured. For example, if the orifice of the pressure-measuring tube faces upstream toward the oncoming flow (Fig. 2.28*b*), the pressure will be higher than if the orifice faces downstream because of the kinetic energy component. The pressure measured facing the flow, sometimes called the "end pressure," therefore measures the total fluid energy. On the other hand, when pressure is measured by a *laterally* positioned tube, it will vary depending on the velocity of flow. The *Bernoulli principle* applies to the conservation of energy in flowing fluids and indicates, in simplest terms, that the lateral pressure in a stream flowing at high velocity (Fig. 2.28) is less than when it flows at low velocity. This occurs because more of the total energy is dissipated as kinetic energy ($\tfrac{1}{2}\rho V^2$) in the faster moving stream. Thus, as shown Fig. 2.28*f*, when the diameter of a vessel widens and velocity slows, the kinetic energy term is decreased, and the pressure is higher than at *e*. Increased lateral pressure in regions of arteries that are abnormally enlarged may promote aneurysm formation, in addition to those factors related to tension in the wall due to the Laplace effect already discussed (Chapter 6).

Because of velocity changes and progressive frictional energy losses, the intravascular pressure tends to be different at every point in the circula-

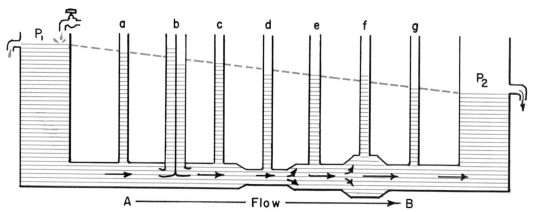

Figure 2.28. Schema illustrating, in principle, pressure changes along a tube AB through which liquid flows steadily owing to the pressure difference $P_1 - P_2$. Pressure is measured by the fluid levels in the side tubes *a–g*. Tube diameter changes at two locations, *d* and *f*. The *dashed line* indicates a theoretical pressure drop along such a tube if there were no changes in tube diameter. The Bernoulli principle is depicted by the pressure levels in *d*, as compared with *c* and *e*, and in *f*, as compared with *e* and *g*.

tion. In the aorta, the lateral pressure is quite dependent on kinetic energy losses. However, as the circulation branches into smaller and smaller arteries and velocity slows, the kinetic energy component becomes relatively unimportant, and the pressure drop in the smaller vessels is mainly due to frictional loss (as between *c* and *e*, Figure 2.28).

At a normal cardiac output, the kinetic energy has been calculated to be less than 5% of the total fluid energy at the peak systolic pressure in the aorta; however, at high cardiac outputs, since kinetic energy is proportional to the square of velocity, it may comprise nearly 15% of the total energy. Under such conditions this factor could significantly affect the total energy output and total energy requirements of the heart, but it is often neglected in calculations of total cardiac work.

PHASIC ARTERIAL PULSE WAVES AND THE ARTERIAL PRESSURE PULSE

The stroke volume is delivered into the aorta at high velocity, and after the completion of left ventricular ejection, there is a brief period of slight reverse flow which ceases as the aortic valve closes; the flow then remains at zero just above the aortic valve during diastole until the next contraction (Fig. 2.29). The distensibility of the arterial system "smooths" the flow pulse as it traverses the arterial bed, and it also effectively lowers the systolic pressure that would be developed if the ventricle were to eject the same volume of blood into a rigid tube; this tends to decrease the workload of the heart during systole, and the energy stored in the stretched wall of the arteries during ejection is then returned as elastic recoil with further flow and pressure generation during diastole by this "passive" mechanism (see Windkessel effect, Chapter 6). Thus, beyond the initial part of the aorta, flow

is maintained above zero between heart contractions.

The static pressure-volume relation of the aorta has been discussed in Chapter 6 (see Fig. 2.13), and in a young human subject a 100% change in volume produces approximately a 50 mmHg change in pressure; in older subjects the aorta is less compliant (stiffer), and the same change in volume results in a larger increase in pressure. However, the pressure developed in the arterial bed in response to delivery of the stroke volume by the heart is dependent not only on the static elastic properties (compliance or stiffness) of the aorta and arterial bed, but also upon their dynamic stiffness, that is, the elastic properties are also dependent upon the *rate* of delivery of the stroke volume, and such frequency-dependent responses imply that the aorta also has viscoelastic properties, discussed further below.

The *pulse pressure* is defined as the difference between the peak systolic and minimum diastolic pressure, measured at any point in the systemic arterial circulation (Fig. 2.30). The *mean pressure* in the arterial circulation can be determined by electronic integration of the pressure signal obtained with a pressure transducer, or it can be determined from the recorded pressure waveform by planimetric integration of the area under the systolic and diastolic pressure waves (Fig. 2.30). If only the systolic and diastolic pressures are known (as by measurement with a sphygmomanometer), the mean pressure can be estimated as the diastolic pressure plus ⅓ of the pulse pressure (Fig. 2.30).

Under steady-state conditions, the *mean* arterial pressure (MAP) is determined *only* by the cardiac output and the total peripheral vascular resistance ($MAP = CO \times TPVR$), and the dynamic properties of the vessels have little influence. However, with sudden changes in cardiac output, the *time* required

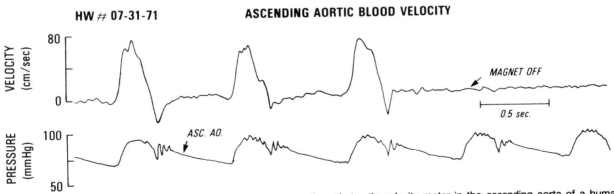

Figure 2.29. Aortic flow velocity recorded with an electromagnetic catheter tip velocity meter in the ascending aorta of a human subject. High fidelity pressure is also recorded in the ascending aorta. Zero flow velocity is indicated with the magnet off. Tracing was obtained by Gabe *et al.* (1969).

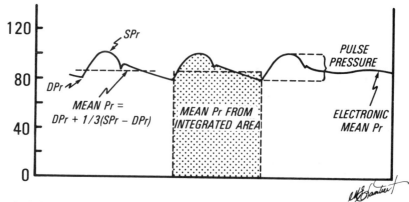

Figure 2.30. Various methods for determining the mean arterial pressure from the arterial pressure pulse. *SPr*, systolic pressure; *DPr*, diastolic pressure. For further discussion, see text.

for attaining a new level of steady-state arterial pressure will depend upon the compliance (or capacitance) of the arterial bed. For example, at a fixed peripheral vascular resistance if the cardiac output is suddenly increased in the presence of a very compliant (high-capacitance) arterial bed, time will be required to fill the arterial bed and to reach a new steady-state level of arterial pressure; conversely, if the system is noncompliant, the new level of pressure will be reached almost immediately.

Characterizing the Arterial Pressure Wave Forms

The shape of the arterial pressure wave forms changes substantially when it is measured in different portions of the arterial bed; for example, the peak systolic pressure is higher, and the pulse pressure is larger in the femoral artery of the leg than in the thoracic aorta (within the chest or thorax).

Analysis of pressure wave forms has been clarified considerably by describing them as a set of sine waves (at a fundamental frequency, together with its harmonics, called a Fourier series) (see Taylor, 1966; O'Rourke and Taylor, 1966; Milnor, 1975). Thus, the repetitive heartbeat produces a steady state of pressure oscillation, with resonation and amplification occurring at certain harmonics, while at other harmonics there is damping.

A pressure-pulse recorded from a peripheral artery is shown in Figure 2.31 (*top tracing*) together with its first five harmonics; the harmonic of lowest frequency (the fundamental harmonic) is labeled I, the second harmonic, occurring at twice the fundamental frequency, is labeled II, the third, occurring at three times the fundamental frequency, is marked III, etc. As shown in Figure 2.32, sine waves oscillate evenly around the mean value, and a complete wave, including positive and negative portions of the cycle, encompasses 360°. The maximum or

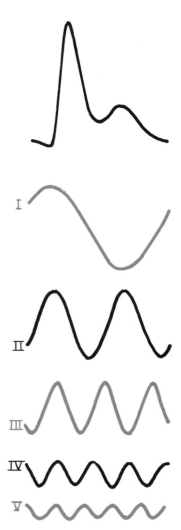

Figure 2.31. Pressure pulse from a peripheral artery and its first five harmonics. [From Taylor (1966).]

minimum amplitude of the sine wave is called the "modulus," and relationships between different harmonics can be described in terms of the "phase angle," the number of degrees by which the wave either lags behind or leads another harmonic (Fig.

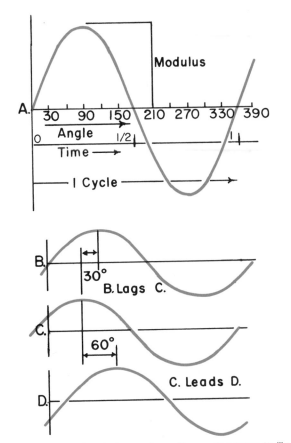

Figure 2.32. (*A*) Simple harmonic motion: a sine wave to illustrate the definition of modulus. (*B*, *C*, and *D*) Sine waves to illustrate the definition of phase lag and phase lead. [From Taylor (1966).]

2.32). For example, the phase angle between waves C and D at their peak amplitude is 60°. As discussed subsequently, moduli can be determined for both pressure and flow waves at a given harmonic, and their phase angles can be compared.

Pressure Pulse Wave Transmission

In the intact circulation, from points of branching and from the tapering of vessels into small resistance vessels, pulse waves are reflected backwards toward the central aorta, and these returning pressure waves interact with the oncoming waves. For example, if the aorta were cross-clamped at some distance from the heart, the reflected wave could double the amplitude of the oncoming wave (assuming no energy loss). On the other hand, if sufficient lag occurred that the reflected waves were ½ cycle (180°) out of phase, the reflected wave would subtract from the oncoming wave.

Another feature of pulse wave transmission is related to the compliance of the arterial bed. The smaller blood vessels tend to be less compliant (stiffer) than the aorta, and this reduced compliance results in an increased velocity of pulse-wave

transmission toward the peripheral vessels. Also, as pressure increases, arterial compliance decreases (see Fig. 2.13, Chapter 6); therefore, the peaks of the pressure waves tend to travel faster than the lowest (diastolic) points of pressure, so as the wave form moves peripherally there is a progressive movement of the peak toward the onset of the pressure pulse (a reduction in time to peak pressure).

Without such phenomena, the viscous forces in the blood vessel walls and the blood would tend to produce a *decrease* in the amplitude of the traveling pressure pulse wave, rather than the observed increase (Fig. 2.33). In fact, high-frequency oscillations, such as the incisura and positive oscillation (dicrotic wave) that follow aortic valve closure (Fig. 2.29), *do* tend to disappear as they are transmitted down the aorta. The contour of the pressure pulse as it moves toward the periphery is shown in Figure 2.33. Note that from the arch of the aorta (above the aortic valve) down to the point where the aorta branches into the femoral arteries, there is a progressive increase in the amplitude of the systolic pressure and the pulse pressure, the time to the peak pressure diminishes, the systolic component narrows, and a secondary positive wave appears. Note the increasing amplitude of several of the harmonics of the pulse (particularly harmonics I and II) as the distance from the arch of the aorta toward the periphery increases (Fig. 2.33). It seems likely that these changes in the pulse wave are due to the several factors discussed above, including amplification by reflected waves, tapering of ves-

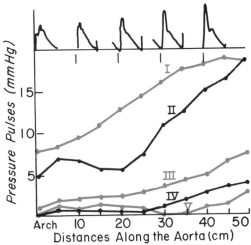

Figure 2.33. Pressure pulses recorded along the aorta of a dog from the arch to the femoral artery, with distance in centimeters. The pulse pressure increases, the dicrotic notch disappears, and a secondary wave develops on the descending limb. In the panel below are the first five harmonic components of these waves. This is the amplification referred to in the text. [From Taylor (1966).]

sels, decreased compliance from central to peripheral vessels, and effects due to the more rapid travel of the wave at peak pressure. The secondary wave does not appear to be due to the dicrotic wave (which, as mentioned, is poorly transmitted), but rather to an initial central and second peripheral reflection of the primary systolic wave (with amplification of harmonic III; for example, see Fig. 2.31).

Determinants of the Arterial Pulse Pressure

Several factors importantly affect the magnitude of the *pulse pressure* measured in a given circulatory state (in addition to the location at which the pressure wave is measured, discussed above). These are the stroke volume, the aortic compliance (or capacitance, the inverse slope of the pressure-volume relation, dV/dP), and the diastolic flow or "runoff" from the aorta. The stroke volume clearly can affect the pulse pressure and, as shown in the aorta of a normal middle-aged individual, a larger stroke volume will increase the pulse pressure (Fig. 2.34, *panel A,* beat 1 to beat 2). If the mean arterial pressure in the same individual is acutely and markedly elevated, the artery will be stretched to a stiffer portion of its pressure-volume curve (reduced compliance), and under these conditions the same stroke volume will cause a larger pulse pressure with a proportionally greater increase in systolic pressure than diastolic pressure (Fig. 2.34, *panel B,* beat 3). When aortic compliance decreases, as with age, the entire pressure-volume relationship is steeper (reduced compliance), and under these conditions the same (normal) stroke volume as in beat 1 *(panel A)* can result in an increased pulse pressure (Fig. 2.34, *panel C,* beat 4).

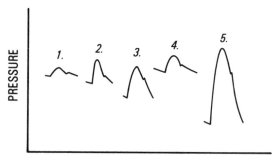

Figure 2.35. Factors affecting the arterial pulse pressure and the arterial wave form. Beat 1 is normal. Beat 2 shows the effects of an increased stroke volume. Beat 3 illustrates the effects of decreased aortic compliance. Beat 4 shows the effect of elevating the mean arterial pressure. Beat 5 indicates the consequences of aortic valve regurgitation. For further discussion see text.

If the heart rate is slowed while the cardiac output remains relatively constant (as may occur with complete heart block, Chapter 9), the stroke volume rises and the pulse pressure increases (Figure 2.35, beat 2); conversely, if the heart rate is increased (as by electrical pacing) and the cardiac output is unchanged (which is the usual response in the normal circulation), the stroke volume and pulse pressure diminish, although the mean arterial pressure remains constant. An isolated decrease in aortic compliance produced experimentally causes an increase in peak pressure and an increased pulse pressure (Fig. 2.35, beat 3). The effects of chronic hypertension are also shown in Figure 2.35 (beat 4). Since in most cases of hypertension the cardiac output remains normal, the rate of fall of pressure (diastolic runoff) may not change appreciably, although the increased peripheral vascular resistance at an unchanged cardiac output produces higher

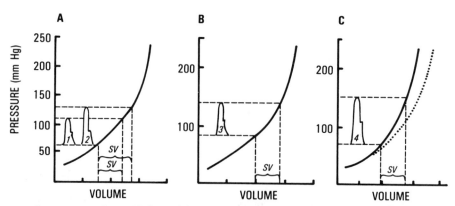

Figure 2.34. Diagrammatic pressure-volume relations of the aorta in human subjects (*curved lines* in each panel). *SV,* stroke volume. In *panel A,* increasing the stroke volume delivered into the aorta increases the arterial pulse pressure (beat 1 and beat 2). In *panel B,* when the arterial pressure is elevated, the same stroke volume as shown in *beat 1* of panel A results in a larger pulse pressure (beat 3) because the aorta is operating on a steeper portion of its pressure-volume curve. In *panel C,* the normal pressure-volume relation shown in *panels A* and *B* is indicated by the *curved dotted line,* and the pressure-volume relationship seen in some elderly subjects, which exhibits decreased aortic compliance (increased stiffness), is indicated by the *solid curved line.* Under these conditions the same stroke volume as in *beat 3* of panel *B* causes an increased pulse pressure (beat 4).

mean and diastolic pressures. If the hypertension is severe enough to place the aorta on a steeper portion of its pressure-volume relation, the pulse-pressure will rise as well (Fig. 2.35, beat 4). Finally, the pulse pressure can be importantly affected by abnormally rapid diastolic runoff. This may occur with aortic regurgitation, in which the pulse pressure increases for two reasons: first, the stroke volume is greatly increased (the ventricle must eject the amount of blood that filled the ventricle from the left atrium, as well as that which regurgitated backward across the aortic valve during the previous diastole) and, secondly, the diastolic pressure falls rapidly in the aorta because of the combination of normal runoff into the peripheral circulation and rapid retrograde filling of the left ventricle from the aorta (Fig. 2.35, beat 5). A similar change in the pulse-pressure occurs in arteriovenous fistula, in which the fistula provides a "short circuit" connecting an artery and a vein, resulting in a high cardiac output (increased stroke volume) and abnormally rapid diastolic runoff through the fistula.

PULSATILE PRESSURE-FLOW RELATIONS

With pulsatile flow in a stiff tube, the flow velocity wave lags behind the driving pressure wave, and as the frequency of the pressure pulse increases, the delay between the two pulses decreases while the amplitude of the flow velocity wave falls. In addition to frequency, a number of other factors influence pressure-flow relations in the vascular bed (including the blood viscosity and vessel characteristics) (see Attinger, 1968; McDonald, 1974).

The driving pressure generated by the ventricle accelerates the column of blood in the aorta (flow leads pressure at the origin of the aorta), and once the blood starts to move through the arteries it will continue to move as long as there is a sufficient hydrostatic pressure difference (the term "pressure gradient" is more properly used to refer to a rate of change of pressure per unit distance). The blood will cease to move when the pressure and inertial forces are counterbalanced by resistance forces in the blood vessels. The traveling pressure wave (which precedes the flow velocity wave in the peripheral arteries) can be palpated in the arteries, and its transmission time can be estimated by noting the more rapid arrival of the wave in the carotid artery than in the more distant femoral artery. Transmission of the pressure pulse produces an oscillating pressure difference (Fig. 2.36), and if pressure is recorded at two different sites, one upstream and the other a short distance downstream, the pressure difference is positive during the upstroke of the upstream pressure pulse, and when

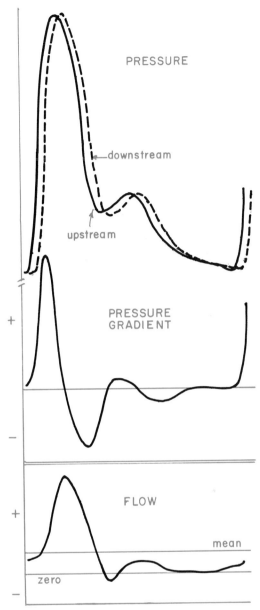

Figure 2.36. Generation of a pressure gradient by a traveling pulse wave and resultant oscillatory flow. Note small reversal of flow (below zero). Example from the femoral artery of a dog. [From Taylor (1966).]

the delayed pressure wave arrives downstream the pressure difference becomes negative (Fig. 2.36). The flow generated by the positive pressure difference lags behind to some degree (Fig. 2.36), as illustrated here by pressure and flow in a peripheral artery. In this case, the negative pressure wave may actually produce a small reversal of flow after the initial pulse.

Impedance

Since blood flow (like pressure) is repetitive and pulsatile, it can be subjected to harmonic (Fourier) analysis like that discussed above for the pressure wave. When this is done, relations between pressure and flow can then be determined by comparing the two Fourier series; of course, since the pressure pulse and the flow pulse waves occur at the same rate, their fundamental and other harmonics are at the same frequency (see Taylor and O'Rourke, 1966; Milnor, 1975).

The idea of impedance in pulsatile circuits is concerned with the ratio of pressure to flow, and therefore it represents the "resistance" to pulsatile flow. When this ratio is calculated in a nonpulsatile circuit (analogous to a direct (DC) current flow), the ratio equals the resistance. The analogy to hydraulic impedance in electrical terms is also available for alternating (AC) current flow; instead of Ohm's law, when dealing with a capacitive circuit the impedance (Z) is determined by the ratio of the sinusoidal voltage to the sinusoidal current flow, and the magnitude of the impedance will vary as a function of the frequency of the voltage wave, the capacitance, and other factors in the circuit. Similarly, in a hydraulic circuit when the flow is pulsatile the impedance includes not only the opposition to flow afforded by friction but also that due to the vascular compliance and the blood mass. In practice, the arterial impedance is determined at zero frequency (steady flow), and over a range of frequencies (analysis of pressure and flow tracings is in the time domain, whereas analysis of impedance is in the frequency domain). When measurements are made just above the aortic valve in the ascending aorta, the "aortic input impedance" is determined (Fig. 2.37). At zero frequency (steady flow), the input impedance is maximal, and the peripheral vascular resistance is usually calculated. The ratios of the sinusoidal pressure to the corresponding sinusoidal flow term are then compared over a limited range of frequencies and presented both as an *amplitude* (or modulus) of the impedance (ratio of the magnitude or modulus of the pressure term to the modulus of the flow term, Fig. 2.37, *upper panel*) and as a *phase angle* of the impedance, i.e., the number of degrees that flow lags or precedes the pressure wave by subtracting the velocity phase angle from the pressure phase angle (Fig. 2.37, *lower panel*). Notice that the moduli of impedance fall rapidly at frequencies above 0 and level off at about 2 cycles/s; note also that the phase angles at the origin of the aorta are negative at low frequencies (flow leading pressure), whereas above 7 cycles/s they became positive (pressure leading flow).

The average impedance at higher frequencies (2–12 cycles/s) constitutes the "characteristic impedance." Characteristic impedance is determined by

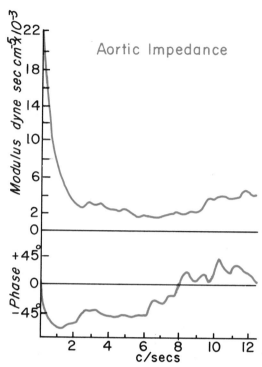

Figure 2.37. Modulus and phase of the input impedance of the arterial system of a dog measured at the ascending aorta. Note the steep fall in the modulus from the value of the peripheral resistance to an almost constant value for frequencies above 1.5 to 2.0 cps. [From Taylor (1966).]

the properties of the aorta itself and its geometry and is considered to be independent of reflected waves. Therefore, changing the compliance of the aorta would be expected to alter the characteristic impedance (for example, clamping the aorta increases impedance). However, simultaneous alterations in the tone of small arteries in the intact circulation can result in calculation errors due to reflected waves.

Impedance Effects on Ventricular Function

It is of interest to compare the effects on left ventricular function of a change in peripheral vascular resistance vs. those of a change in aortic compliance. In the time domain, if pressures and flows are analyzed when peripheral vascular resistance is increased by a vasopressor drug while the left ventricular end-diastolic pressure is held constant, a marked increase in peak systolic and mean aortic and left ventricular pressure and a decrease in the stroke volume occur. If the capacitance or compliance against which the left ventricle ejects is now decreased without changing the mean aortic pressure (an artificial system with an air-buffered reservoir can be used), peak systolic pressure rises slightly, diastolic pressure falls markedly and, again, the stroke volume is markedly reduced. The contractility (inotropic state) of heart muscle is unchanged during these conditions, but the percentage of volume ejected by the ventricle falls in both situations because of the altered loading conditions (see Chapter 12). When such events are analyzed by impedance spectra, the vasopressor drug results in an increase in resistance during steady flow (zero frequency), but there is little or no change in impedance at higher frequencies or in the phase angles, suggesting little change in aortic properties. On the other hand, with decreased aortic compliance the steady flow term (resistance) does not change, but impedance is increased over the higher frequencies, indicating an increase in the characteristic impedance of the aorta. Such studies, while complex and open to considerable theoretical and experimental difficulties, serve to illustrate the complex interaction between the pump and the peripheral circulation (Milnor, 1975; Nordergraff, 1975) (see further discussion in Chapter 12 of impedance effects on left ventricular function).

DETERMINANTS OF THE VENOUS PRESSURE

A number of properties of the veins directly or indirectly influence the venous pressure (for reviews, see Wood, 1965, and Shepherd and Vanhoutte, 1975). These properties include their high compliance (see Fig. 2.13, Chapter 6) and therefore the large volume of blood that they store (64% of the total blood volume and about 80% of the systemic blood volume, Table 2.2), their tone, as affected by adrenergic nerve impulses, and their potential collapse, which can occur in larger veins at several locations (depending on their distending pressure and the external pressure). The abdominal viscera may compress the veins, for example, causing partial collapse; also during normal inspiration, as the blood is drawn into the chest by the negative intrathoracic pressure, the neck veins tend to collapse as vena cava pressure falls (Fig. 2.38).

In the supine position the "central" mean venous pressure in normal subjects is 0–4 mmHg (up to about 6.0 cm H_2O) using the right atrium as the reference point (this reference point is largely independent of body position because the right ventricle, as a demand-pump, tends to maintain the atrial pressure constant) (Guyton and Jones, 1973). This level of pressure in the right atrium affects the pressure in the peripheral veins and, normally, in the small veins the pressure is 8–9 mmHg higher than in the right atrium; therefore, if the right atrial pressure rises, the peripheral venous pressure also rises, maintaining the pressure difference favoring forward flow. The right atrial pressure, in turn, is

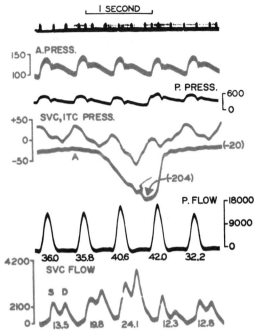

Figure 2.38. Effect of spontaneous respiration on venous return and cardiac output (closed-chest animal). Tracings (*top to bottom*): time and base line aortic pressure in millimeters of mercury, pulmonary artery, superior vena cava (*SVC*), and intrathoracic pressure *(ITC)* in millimeters of water, pulmonary arterial (*P.*) and superior vena cava flow (lower tracing) in mL per minute. (*A*) Beginning of inspiration; (*S*) acceleration of superior vena cava flow during ventricular systole; (*D*) initial acceleration of superior vena cava flow during ventricular diastole. Stroke volume (in mL) under pulmonary arterial flow curve. Flow (mL/min) through superior vena cava during each cardiac cycle at bottom of record. Electrical frequency response of both flow meters reduced from 400 to 40 cycles/s. Superior vena cava pressure curve damped. [From Brecher (1956).]

affected by the function of the right ventricle. If the right ventricle fails, right atrial pressure will rise and tend to impede venous return, whereas good ventricular function will maintain the right atrial pressure at low values.

Other factors can affect the right atrial pressure, including the blood volume, which influences the venous return (increased blood volume increases venous return, and vice versa; see Chapter 17). The *venous tone* and, hence, the capacitance of the peripheral venous bed are under autonomic control (Chapter 16), and sympathetic stimulation increases peripheral venous tone, displaces blood centrally, and can thereby raise the right atrial and venous pressures; withdrawal of sympathetic tone, on the other hand, leads to increased venous capacitance and can lower the venous pressure.

The *resistance* to venous return also can be altered, for example, by forcible abdominal compression, which will collapse the large abdominal veins, lower the right atrial pressure, and increase the peripheral venous pressure. The right atrial pressure can also be markedly elevated by forcible expiration against a closed glottis (the "Valsalva maneuver") to increase *intrathoracic pressure*; this will markedly elevate the right atrial and peripheral venous pressures. Conversely, an effort to inspire against a closed glottis (the "Mueller maneuver") will produce a negative intrathoracic pressure and lower the right atrial and peripheral venous pressures and increase vena caval flow (a magnification of the effect of normal inspiration, Fig. 2.38).

The Hydrostatic Effects of Posture

If a normal adult individual is placed on a tilt table which is then abruptly moved from the supine to the upright position, and if the leg muscles are not contracted, the pressure in a large leg vein near the foot will promptly rise from a few mmHg to that of the weight of the blood column in the veins; the column is about 4 feet or 120 cm in height (the column below the heart) and, therefore, will yield a pressure in the foot of almost 90 mmHg (1 cm of water or blood = 1.36 mmHg) (Fig. 2.39). Conversely, if the right atrial pressure is 1 or 2 mmHg, the 50-cm distance above the right atrium of a vein of the head in the upright tilt will result in a negative pressure in the head vein of over −30 mmHg (Fig. 2.39) (the veins of the cerebral circulation are normally supported externally and do not collapse under such pressures). The pressures in neck and arm veins will collapse under upright conditions, although the position of the arms is important (if the arm is held downward, the hand will be below the right atrium and will exhibit a positive venous pressure). As described earlier, the high positive pressure in the feet as a subject is tilted upward produces the same increment of pressure in the arteries as in the veins, and the same

difference in pressure remains between these two components of the circulation; therefore, since the total energy for forward flow is unchanged, flow will continue to the foot. However, as the venous pressure in the legs rises, the compliant veins are stretched, and a substantial volume of blood becomes sequestered in the legs and the abdomen, so that the venous return to the heart transiently decreases and the cardiac output drops markedly before compensatory reflex responses occur.

If the individual remains in the upright posture without moving the legs, the sustained high pressure in the veins will lead to a high capillary pressure, and edema (swelling) of the legs will occur after many minutes (indeed, soldiers on drill who stand for long periods of time at attention, without movement, can develop leg swelling). Ordinarily, however, assumption of the erect posture is associated with tensing of the muscles in the legs and is followed by walking. Since the peripheral veins have sets of *one-way valves* at frequent intervals along their course, as soon as the muscles contract the vein is emptied, and the valve then prevents backflow of blood into the vein from above. Thus, the venous valves interrupt the continuous column of blood in the veins. This mechanism acts as a "muscle pump," which results in a prompt marked lowering of the venous pressure and effective filtration pressure in the lower extremities to 8–10 mmHg upon walking (Stegall, 1966).

Sometimes, inflammatory processes in the veins or formation of blood clots leads to damage or malfunction of the venous valves (thrombophlebitis). Under these circumstances the venous pressure may remain high, resulting in further stretching of the venous valves, enlargement of the veins (varicose veins), and constantly elevated pressures in the leg veins in the erect posture as well as during walking. This, in turn, causes chronic edema of the ankles and lower legs.

Flow and Pressure Waves in the Veins

The characteristics of the pressure pulses in the right atrium and large systemic veins are described in Chapter 13; the *a*, *c*, and *v* waves are transmitted into visible venous pulsations in the neck from the right atrium, with some delay. When blood flow is recorded in the vena cava, there is a large positive flow wave during right ventricular systole, as the ejecting right ventricle moves away from the right atrium to produce a relative increase of atrial volume, and increased inflow of blood into the right atrium occurs (the tricuspid valve is closed) to produce the *v* wave in the pressure tracing (Fig. 2.133). Flow again increases early in diastole as the

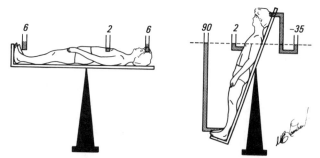

Figure 2.39. (*Left panel*) Diagram of the venous pressures (mmHg) in a normal, supine human subject, the venous pressure in the legs and head is slightly higher than in the central chest. (*Right panel*) Following sudden upward rotation on a tilt table, the venous pressure rises to 90 mmHg in the legs but remains at 2 mmHg near the heart; it becomes 35 mmHg negative in the head. For further discussion, see text.

tricuspid valve opens and blood fills the relaxing right ventricle (Fig. 2.38). There is some evidence that increased inflow at this time is also related to restoring forces in the right ventricle, giving a suction effect (see Brecher and Galletti, 1963, for review of potential suction effects). The increase of flow in the vena cava during inspiration produced by the negative intrathoracic pressure (Fig. 2.38) is also illustrated in Figure 2.125.

METHODS FOR ASSESSING PERIPHERAL BLOOD VESSELS

In diagnosing peripheral vascular disease in man (such as of the cerebral arteries, or the arteries and veins to a leg) a number of methods can be employed (see Bernstein, 1978). Angiography, with injection of an x-ray opaque liquid, allows x-ray visualization of regions of blockage in arteries or veins (see Chapter 13). Indirect measurements of blood flow in arteries can be made. For example, using transcutaneous Doppler flow probes (reflected ultrasound waves from the moving blood) abnormal flow wave patterns, or loss of reactive hyperemia, can be detected in limbs with atherosclerotic lesions (Fronek et al, 1976). Venous occlusion plethysmography, in which the change in volume or circumference of a limb is measured while a cuff is inflated to obstruct venous outflow (not arterial inflow), allows measurement of the blood flow rate; release of the cuff can provide information on whether or not there is obstruction to venous outflow by thrombophlebitis.

CHAPTER 8

Electrical Impulse Formation and Conduction in the Heart

The electrical impulses that drive the heart originate in a group of pacemaker cells lying in the sinoatrial (SA) node, and these action potentials then spread rapidly to both the right and left atrium (depolarizing them to contract). The wave of impulses then enters a conduction system made up of specialized cells, which gathers the impulses at the atrioventricular (AV) junction. There they are delayed (which allows atrial contraction to boost the filling of the ventricles prior to the onset of their contraction), and the wave then enters rapidly conducting fibers of the specialized conduction system. These carry the depolarizing wave to the left and right ventricles, causing them to contract (slightly out of sequence, the left side before the right) (see Fig. 2.4, Chapter 5).

In this chapter we will first discuss the genesis of electrical activity and electrical properties of several types of cardiac cells, since these properties are important in understanding the sequence of electrical activation, as well as how disorders of impulse formation and conduction can occur. Then, the functional anatomy and physiology of the various components of the normal cardiac conduction system will be described, and the normal sequence of cardiac excitation will be described. This sequence is due largely to differences in the ability of cells to conduct impulses and is responsible for the form of the normal electrocardiogram (Chapter 9).

BASIS OF THE RESTING MEMBRANE POTENTIAL OF HEART CELLS AND THE CARDIAC PACEMAKER

The genesis of resting cellular transmembrane potentials is complex, and its theoretical basis is detailed in Section 1, Chapter 3. It will be useful here to consider briefly the major ionic movements that produce the resting membrane potential in heart muscle cells, specialized conducting cells, and pacemaker cells (For more detailed reviews of cur-

rent concepts concerning this topic see Brady, 1979; Fozzard, 1979; Noble, 1979; and Wallace, 1982).

Resting Membrane Potential of Cardiac Muscle Cells

If the cell membrane or sarcolemma of a resting muscle cell in the ventricle is punctured by means of a microelectrode, a potential difference will be recorded across the cell wall, with the inside of the cell carrying a negative potential relative to the outside. The level of this negative potential differs somewhat in various regions of the heart; in resting ventricular muscle (which is not receiving action potentials from a pacemaker) it is about -90 mV.

The basis for the resting transmembrane potential rests primarily on the selective permeability of the cell membrane to various ions. There are well known differences in concentrations of anions and cations between the intracellular and extracellular fluid compartments, Na^+ and Cl^- forming the major cation and anion in the extracellular fluid, whereas in the intracellular fluid K^+ is the major cation and the anion charge is carried by Cl^-, organic acids, proteins, phosphates, and sulfates. These ionic compositions can be mimicked in an experimental chamber in which an impermeable membrane has been placed to represent the cell boundary; under these conditions, the sum of the positive and negative charges on either side of the membrane will be equal, maintaining electrical balance, and no potential difference will be measured with a voltmeter placed across the membrane. However, if the membrane is made permeable to one ion only, that ion will attempt to cross the membrane down its concentration gradient. If the ion is K^+, for example, ions will commence to move across the membrane from the region of high K^+ concentration (the side of the chamber representing the intracellular fluid). As a few K^+ ions cross the membrane, the intracellular side will become electrically negative in the region close to the membrane (since positive charges are leaving). The buildup of positive charge on the other side of the membrane (representing the extracellular fluid) will counteract the tendency for K^+ ions to move down their concentration gradient, however, eventually causing the ion movement to cease. This sequence of events involves only a very *few* ions close to the membrane and will not result in measurable changes in the *concentration* of ions on either side. The potential difference existing across a semipermeable membrane when the movement of ions has ceased can be called the equilibrium potential for that ion (in this case for K^+).

In cardiac muscle, the resting transmembrane potential is, in fact, predominantly determined by selective permeability to K^+, with relatively slight permeability to Na^+ and less to Cl^-. The equilibrium potential for any ion can be calculated, provided its intra and extracellular concentrations are known, using the Nernst equation.

$$V_{eq} = -\frac{61.5}{Z} \log \frac{[C]_i}{[C]_o}$$

where V_{eq} = equilibrium potential for a given ion, Z is the charge (valence) of an ion, and $[C]_o$ and $[C]_i$ represent its extracellular and intracellular concentrations, respectively (see Section 1, Chapter 3 for the more general form of the 61.5 Nernst equation and the derivation of the constant from the gas constant (R), Faraday's constant (F) and T, the absolute temperature). The equilibrium potential for a given ion can be either positive or negative, depending on its concentration difference, and the logarithmic relation means that relatively large charges in the ratio of ion concentrations inside and outside the cell result in relatively small changes in the resting transmembrane potential.

When the Nernst equation is employed to calculate the resting transmembrane potential for K^+ ($Z = 1$) it is written as:

$$V_{eq} \text{ K} = -61.5 \log \frac{[K+]_i}{[K+]_o}$$

Table 2.3 shows the relative concentrations for cations across the cell membrane in heart cells, and by substituting 150 mM and 4 mM as the normal

concentrations for K^+ inside and outside the cell, a resting membrane potential of -97 mV can be calculated. If the Nernst equation is used to predict equilibrium potentials for K^+ in an ideal system in which the $[K^+]_i$ is held constant and the $[K^+]_o$ is varied, the Nernst equation predicts an inverse relation between the level of transmembrane potential and increasing extracellular $[K^+]$; that is, the cell becomes relatively more depolarized, or less negative, as the $[K^+]_o$ rises (Fig. 2.40). If the transmembrane potential is actually recorded in a resting muscle cell using an intracellular microelectrode, when the external $[K^+]$ is varied, a relation close to that predicted by the Nernst equation is observed; however, the measured curve (*solid line*) lies closer to the theoretical relation (*dashed line*) at high $[K^+]_o$ concentrations and deviates to some degree at lower concentrations, with a resting value at a normal $[K^+]_o$ of 4 mM close to -90 mV (*dotted lines*, Fig. 2.40).

The cell membrane, as mentioned, is permeable not only to K^+ but also to Na^+, and the latter ion can also influence the resting membrane potential. Even though wide fluctuations of extracellular Na^+ alone have only a small effect on the resting membrane potential (because of the dominance of the K^+ permeability effect), the deviation from the theoretical line can nevertheless be explained in part by some degree of membrane permeability to Na^+. Also at lower K^+ concentrations, selective permeability to K^+ appears to decrease. When the Nernst equation is applied to calculate the equilibrium potential for Na^+, considering the membrane as selectively permeable to that cation alone, a positive potential of $+42$ mV can be calculated

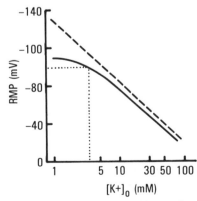

Figure 2.40. Relations between the resting membrane potential (*RMP*) and the extracellular concentration of potassium $[K^+]_o$ in cardiac cells. The *dashed straight line* indicates the theoretical relation predicted by the Nernst equation, and the *solid line* the relation as actually measured. The *dotted lines* show that at a normal potassium concentration of 4.0 mM the RMP is approximately -90 mV.

Table 2.3
Concentrations of selected cations in mammalian cardiac muscle (mEq/l or mM)

	Extracellular	Intracellular
K^+	4	150–160
Na^+	145	30–32
Ca^{++}	5	0.00007

(using intracellular and extracellular concentrations of 30 and 145 mM, respectively). Thus, Na^+ tends to move from its high concentration outside the cell toward the inside of the cell along its concentration gradient, leaving the inner side of the membrane relatively positive compared to the extracellular fluid. As mentioned, the small leakage of Na^+ into the cell is primarily responsible for the shift downward of the actual curve (Fig. 2.40), causing the interior of the cell to be less negative than predicted by the Nernst equation. This small inward sodium movement is continuously balanced by a small outward K^+ movement, so that there is no change in the transmembrane potential with time when the cardiac muscle cell is at rest (see Fig. 2.41, phase 4). (This is not the case in pacemaker cells.)

In order to calculate the *combined* effects of various ions on the transmembrane potential when the membrane is permeable to more than one ion, the relative permeability of the membrane to each ion must be considered and the *Goldman—Hodgkin—Katz equation* can be employed (for the general form of this equation see Section 1, Chapter 3). For two or more ions:

$$V_m = -61.5 \log \frac{P_K [K^+]_i + P_{Na} [Na^+]_i + P_{Cl} [Cl^-]_i}{P_K [K^+]_o + P_{Na} [Na^+]_o + P_{Cl} [Cl^-]_o}$$

Or, for two ions:

$$V_m = -61.5 \log \frac{[K^+]_i}{[K^+]_o} + \frac{[Na^+]_i (P_{Na}/P_K)}{[Na^+]_o (P_{Na}/P_K)}$$

where V_m = transmembrane potential, and P = the permeability for each ion (defined as the quantity of the ion that diffuses across each unit area of membrane per unit of concentration gradient per unit membrane thickness).

This constant-field equation makes clear the importance of the *ratio* of permeabilities of ions in determining the transmembrane potential. When the selective permeability to K^+ is much higher than that to Na^+ (as is the case with the normal *resting* cell), the V_m approaches that calculated by the Nernst equation for K^+.

Resting Membrane Potential of Pacemaker Cells

Cells in the cardiac pacemaker region (SA node) as well as in the AV junction and the His-Purkinje network (the specialized conducting system) do not exhibit constant resting potentials, but are capable of *spontaneous diastolic depolarization*. This phenomenon has been most fully studied in the large cells which comprise the rapid conducting pathway (Purkinje cells, see below), and in these cells the mechanism for this gradual fall in the negative

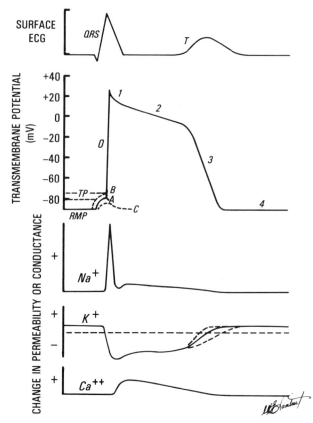

Figure 2.41. Relations between the surface electrocardiogram (*ECG*), the transmembrane potential, and diagrams of the relative changes in ion permeability (or conductance) that are responsible for the action potential. The (0–4) phases of a typical fast action potential are indicated, the resting membrane potential (*RMP*) being phase 4. The threshold potential (*TP*) is also shown; *C* represents a subthreshold stimulus (nonpropagated); *A* a threshold stimulus, and *B* is a threshold stimulus at a lower (less negative) threshold potential *(upper dashed line)*. The large rapid increase in Na^+ conductance during phase 0 is shown, together with the sustained small increase in Na^+ conductance (slow Na^+ current) during phase 2 of the action potential. The large, less rapid drop in K^+ conductance during phase 0 of the action potential is indicated and the *dashed lines* of altered K^+ conductance during phase 3 indicate induced changes in P_K. The *dashed horizontal line* indicates that K^+ conductance is above zero at rest. The increased Ca^{++} conductance during phase 2 of the action potential is shown. For further discussion, see text.

resting potential toward zero (it becomes progressively less negative, see Fig. 2.42C and D) is a gradual *decrease* in K^+ permeability, while the permeability to other ions remains constant. As seen from the Goldman equation, a decreasing value of P_K will increase the ratio of P_{Na}/P_K, thereby enhancing the effect of the Na^+ gradient on V_m and producing a relative increase of positive charge inside the cell due to the Na^+ leak (hence a progressively less negative transmembrane potential). Also, because of the decreasing K^+ permeability

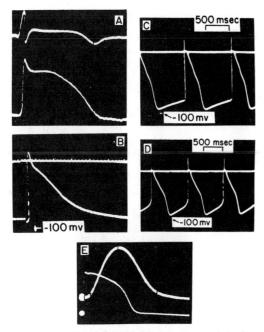

Figure 2.42. *A,* Transmembrane action potential of a single ventricular fiber (*lower curve*) and simultaneous electrogram of the intact heart in situ. The upstroke of the action potential coincides with the R wave of the electrogram, the plateau with the ST segment, and the phase 3 repolarization with the T wave. [Adapted from Brooks and associates.] *B,* Normal transmembrane potential recorded from a single fiber of the intact atrium. [Adapted from Brooks and associates.] *C,* Transmembrane action potential of a single Purkinje fiber in a nonpacemaker activity (paced from above). Note slow loss of resting potential during diastole. [Adapted from Brooks and associates.] *D,* Spontaneous activity recorded from Purkinje fiber in a pacemaker area. Note slow depolarization during diastole and then abrupt onset of spontaneous depolarization. [Adapted from Brooks and associates.] *E,* Temporal relation between action potential and mechanical contraction recorded from a papillary muscle. [Adapted from Dudel and Trautwein (1954).]

there is less K^+ movement outward, and the movement of Na^+ inward is therefore unbalanced.

This phenomenon may also be responsible, at least in part, for the gradual depolarization of resting cells in the SA node and AV junction, although the complex anatomy and small size of cells in these regions have hindered elucidation of mechanisms (Vassalle, 1977). These pacemaker cells may well have a larger permeability to Na^+ than muscle cells, as evidenced by lower (less negative) levels of resting transmembrane potential, and it has also been suggested that Ca^{++} permeability may be significantly higher than in muscle cells. It has been further postulated that gradually increasing Ca^{++} permeability may play an important role in the diastolic depolarization of cells in the SA node and AV junction (for example, Ca^{++} antagonist drugs markedly reduce progressive diastolic depolarization in those regions).

ACTION POTENTIALS IN CARDIAC TISSUE

When pacemaker cells spontaneously depolarize, they eventually reach their threshold potential and produce a spontaneous *action potential*, which completely depolarizes the cell; in such cells, action potentials typically occur at a regular rate, depending on the rate of diastolic depolarization. As in nerve fibers (see Section 1, Chapter 3), once an action potential is initiated in a given cell it will travel to adjacent cells, and if the magnitude of the action potential is sufficient to exceed the *threshold potential* of those cells a response will be triggered, which rapidly results in the complete depolarization of the adjacent cells (the triggered response is called a "propagated" action potential). Such action potentials result from rapid *changes in membrane permeability* to several ions and, of course, they provide the stimulus which carries the pacemaker potential through the specialized conduction system and initiates the contraction of heart muscle.

In pacemaker cells, the property of spontaneous diastolic depolarization followed by action potentials is referred to as *automaticity*. The threshold potential is about −70 mV in ventricular muscle cells (Fig. 2.41), although it varies somewhat depending on physiologic conditions, or on the presence of certain drugs. In atrial muscle, the threshold potential is 30–60 mV. An example of action potentials actually recorded from a spontaneously firing cell in the ventricular conducting system is shown in Figure 2.42*D* (for full discussion of automaticity see Vassalle, 1977).

Fast-Response Action Potentials

Fast-response action potentials are characterized by relatively high (more negative) and relatively constant resting membrane potentials, and by a very rapid onset of the action potential. Cells in the atrial and ventricular muscle, specialized tracts between the two atria, and parts of the conduction system exhibit this type of action potential (Fig. 2.41 and 2.42*A* and *B*).

Such tissues exhibit rapid (muscle) or very rapid (specialized conduction tissue) conduction velocity, in contrast to the slow conduction in the atrioventricular junction. The rapid upstroke is termed phase 0, the recovery of the initial overshoot to a positive membrane potential phase 1, the plateau period phase 2, repolarization phase 3, and the resting membrane potential phase 4 (Fig. 2.41). These phases of the action potential are related to marked and rapid changes of ion permeability through separate ionic channels in the cell membrane. The passage of ions across these channels is considered to be largely controlled by "gates" which open or close in response to changes in membrane potential. Experimentally, it is possible to measure, as electrical currents, the ion fluxes across the membrane by maintaining the membrane voltage constant (so-called "voltage clamping," see Section 1, Chapter 3). The *conductance* of the cell (g)* is generally measured in such experiments; it provides an electrophysiologic measure that is related to membrane permeability (as permeability increases, conductance increases). (For detailed discussion or methods and current concepts of the action potential see Noble (1979) and Fozzard (1979).)

Phase 0, the rapid upstroke which follows sudden depolarization of the cell to its threshold potential, results from a voltage-dependent very rapid increase in sodium permeability (P_{Na}) and conductance (g_{Na}). Movement through these fast sodium channels, which is thought to represent sudden opening of pores or gates in the membrane, is favored both by the electrostatic charge across the membrane (positive outside, negative inside) and by the large concentration gradient from outside to inside for the positively charged sodium ion. A sharp drop in P_K also begins during phase 0 of the action potential (Fig. 2.41).

One model for the action of these pores (developed in nerve tissue) includes two electrically charged gates on each pore that can open and close (Hodgkin, 1964; Huxley, 1964). In such a model, the change in membrane potential from −90 to about −60 mV somehow affects the conformation of one of the Na⁺ gates, and opens the fast Na⁺ channel, lowering the transmembrane potential; this, in turn, opens more channels and/or further widens the opening in each gate; this sequence has been termed a *regenerative process*, a positive feedback mechanism which rapidly drives the action potential to a positive value at the end of phase 0 (Fig. 2.41). The inward Na⁺ current is rapidly slowed as the electrical gradient across the cell membrane reverses, the inward current continuing at a positive transmembrane potential only because of the large concentration gradient for Na⁺ which persists (recall that only a small number of ions actually traverse the membrane relative to the total number of ions available in the extracellular space). The "positive overshoot" of the action potential ceases at about +20 mV, and in the gating model this "turnoff" of phase 0 is considered to result from closure of the second set of gates on each channel,

which respond to a change in membrane potential in an opposite manner to the first set, i.e., they progressively close as the transmembrane potential becomes less negative; this is a slower process than the opening of the first set of gates, however, so that phase 0 depolarization is rapid. As predicted by the Goldman equation, during the overshoot the transmembrane potential tends toward, but does not reach, the Na⁺ equilibrium potential, because these gates close before the membrane potential can reach the Na⁺ equilibrium potential (+42 mV). As might be anticipated, when the Na⁺ channels are open the transmembrane potential becomes very sensitive to the external Na⁺ concentration (see Goldman equation); V_m is then closely related to the log Na⁺ (just as the resting transmembrane potential is related to the log $[K^+_o]$ when the K⁺ channels are partially open in phase 4).

Phase 0 of the fast action potential can be specifically blocked by tetrodotoxin, an agent which blocks only the Na⁺ fast channels, and under these conditions, the rapid upstroke is replaced by a slow-response action potential (see below). The rapid spike during phase 0 of a fast-response action potential usually results in a propagated depolarization of the entire cell (an all-or-nothing response), as well as cell-to-cell propagation as discussed subsequently.

Phase 1 of the action potential starts the repolarization process of the membrane (Fig. 2.41). Its mechanism is not fully elucidated, although closing of the sodium channels is an important factor. Phase 1 is particularly prominent in Purkinje cells.

Phase 2, the so-called "plateau" of the action potential, results from several mechanisms. It is an important feature of fibers in the rapid conduction system to the ventricles as well as of the ventricular muscle itself, and it prolongs the duration of the action potential to 200–300 ms. During that period it is not possible to initiate another action potential because the fast Na⁺ channels are inactivated, and therefore the heart is not subject to tetanic contracture. (It is "refractory," see below.) The action potential in cardiac muscle contrasts with the brief action potential of skeletal muscle, which readily permits summated contractions (Section 1, Chapter 4).

Two mechanisms appear to be largely responsible for maintaining the plateau of the action potential: a slow inward Ca⁺⁺ current and/or a slow inward Na⁺ current. Voltage-dependent activation of the slow Ca⁺⁺ channel with increased Ca⁺⁺ permeability and conductance appears to be most important for maintenance of the plateau (for review, see Reuter, 1974). The Ca⁺⁺ channel opens when the

---

* Conductance is the reciprocal of the electrical resistance of a membrane $(g = 1/R)$ and expresses the ease with which an ion driven by an electrical potential penetrates a membrane.

transmembrane potential depolarizes to about $-35mV$, and some increase in Na^+ permeability (apparently through Na^+ slow channels) also occurs at this point. The slow inward movement of Ca^{++} and Na^+ ions is counterbalanced by movement of K^+ outward along its concentration gradient; following the initial drop in P_K it gradually increases, and this outward K^+ movement appears to remain sufficiently large to balance the inward currents (Fig. 2.41), thereby maintaining a relatively steady plateau.

The inward Ca^{++} current is important in influencing heart muscle contraction (Chapter 10). It can be blocked by Ca^{++} antagonists such as verapamil (which appears to close Ca^{++} slow channels) and by Mn^{++}. Movement of Ca^{++} through the slow channels may be enhanced by catecholamines, and it is increased by elevated $[Ca^{++}]_o$ (increased gradient). Such channels have also been considered to be activated and inactivated by the movement of two gates, as with the fast Na^+ channel. When the calcium slow channels are blocked, some plateau of the transmembrane potential is still observed due to the inward slow Na^+ current and the outward movement of K^+, but the plateau is abbreviated and occurs at a more negative transmembrane potential.

Phase 3 of the action potential involves relatively rapid repolarization of the cell membrane. It commences with inactivation of the Ca^{++} and Na^+ slow channels, and a relatively rapid increase in P_K (or g_K), which produces a rapid outward movement of K^+ and also contributes to restoration of the relatively negative cell interior (Fig. 2.41). This change in the K^+ channel also appears to be voltage-dependent, and it progressively increases as the transmembrane potential becomes more negative, leading to a regenerative recovery process.

The P_K during phase 3 is sensitive to a number of factors, which can alter the duration of the action potential (Fig. 2.41). For example, the drug lidocaine, or an increase $[K^+]_o$ will increase P_K; also, acetylcholine increases P_K in atrial muscle (which has acetylcholine receptors), and all of these stimuli shorten the action potential duration. A number of other influences also can decrease action potential duration, including epinephrine and norepinephrine, digitalis, and increased frequency of contraction. Other drugs, such as quinidine, as well as decreased heart rate or lowered $[K]_o$, increase the action potential duration (Fig. 2.41).

Phase 4, the rest phase, was discussed previously. It is clear that active membrane pumps must remove the Na^+ that enters the cell during the action potential, as well as that which enters by diffusion during diastole. The energy-dependent membrane

Na^+-K^+ ATPase removes these Na^+ ions from within the cell, and the K^+ ions which left the cell during repolarization are also returned by this pump, both ions being moved against their concentration gradients (see Schwartz, 1974). Ca^{++} entering the cell during the action potential is exchanged for Na^+ during the rest phase of the cycle by a non-energy-dependent electrogenic mechanism (Chapter 10).

A number of factors can modify the diastolic threshold potential, making a larger or smaller triggering potential necessary (Fig. 2.41). Certain drugs, such as quinidine, lower the threshold potential (make it less negative) and therefore the cells are more difficult to depolarize (less excitable) requiring a larger triggering potential (Fig. 2.41B). Catecholamines make the threshold potential more negative, i.e., less change in the diastolic potential is required to trigger an action potential.

Fast response cells tend to exhibit inactivation of the fast channels at less negative membrane potentials (an effect that can be produced by increased $[K^+]_o$), so that very slow depolarization of the cell towards its threshold may cause inactivation of impulse propagation.

Slow-Response Action Potentials

Cells with action potentials of this type are found in the SA and AV nodes, and in the AV junctional region where they may be either of pacemaker or nonpacemaker type. Also, fast-response cells have the capability of being converted to slow-response fibers when they are partially depolarized to a membrane potential well below the threshold for the fast response (occasionally this occurs in injured cardiac tissue, as during ischemia, see Chapter 14). The general form of the slow-response action potential is illustrated in Figure 2.43. It shows a lower resting membrane potential and a lower (less negative) threshold potential than the fast fibers. Also, there is a slower upstroke, with loss of the spike of phase 0, and the amplitude of the action potential is less with only a small positive phase. Although a plateau is present, repolarization is fairly rapid, leading to a shorter action potential duration than in fast fibers (Fig. 2.43).

Slow-response action potentials are thought to be mediated entirely by slow inward Ca^{++} and Na^+ currents. Like the gates in fast-response cells, cells in the SA and AV nodes appear to have Ca^{++} and Na^+ gates that exhibit voltage-dependent inactivation in the later phases of the action potential. The Ca^{++} slow channels may be mainly responsible for phase 0 in these cells, since it is inhibited by Ca^{++} antagonists such as verapamil, and by Mn^{++}, and

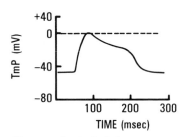

Figure 2.43. Diagram of a typical slow response of action potential of a cardiac cell. Notice the slow upstroke, lack of a phase 1, and the shorter action potential duration than the fast response shown in Figure 2.41.

is insensitive to tetrodotoxin (Wit and Cranefield, 1974). When fast-response Purkinje cells are converted to the slow response, either the slow Na⁺ channel or the Ca⁺⁺ channel may be operative (Wallace, 1982).

MEMBRANE PROPERTIES

In addition to changes in the threshold potential (which can change the excitability of the cell), a number of other properties of the cell membrane contribute to the ability of the heart's pacemaker and conduction system to effectively conduct to the muscle and to depolarize it. These properties relate primarily to the time of recovery of excitability of the cell membrane (that is, how long it is refractory to an outside stimulus), the type of response it exhibits (fast or slow), and the conducting properties of cells and cell networks.

Refractoriness

In single cells, and in groups of cells, time is required for the cell to recover partial and full excitability during the repolarization process. These periods have been divided into several segments (Hoffman, 1969).

The absolute refractory period can be determined in single cells and constitutes that period during which the membrane cannot be reexcited by an outside stimulus, regardless of the level of external voltage applied (Fig. 2.44). In networks of cells, the absolute refractory period cannot be accurately determined because of different recovery times of various cells in the network, and the *effective* refractory period is usually determined for such cell networks.

The effective refractory period constitutes that period during which only a local response can be produced by a *larger than normal* depolarizing stimulus (Fig. 2.44). Thus, during the effective refractory period the membrane can respond, but a propagated action potential that will carry the impulse throughout the cell network cannot be generated.

The relative refractory period commences at the end of the effective refractory period and constitutes that time interval late in the action potential during which a *propagated* action potential can be generated but with a depolarizing stimulus that is *larger* than normal (Fig. 2.44).

The supernormal period is a short interval during which the cell is *more* excitable than normal; that is, a weaker than usual depolarizing stimulus can initiate a propagated action potential (Fig. 2.44).

The full recovery time constitutes the period from the onset of the action potential to the end of the supernormal period.

In cells of this type (generally rapidly conducting) the recovery of excitability (or refractoriness) is mainly *voltage-dependent*, as discussed further below, whereas in slow-response cells (generally associated with slow conduction velocity) it is mainly *time-dependent*. Thus, in cells of the *slow-response* type, repolarization of the cell to its resting potential does not necessarily coincide with the recovery of excitability. In slow fibers such as in the SA node and AV junction the refractory period can

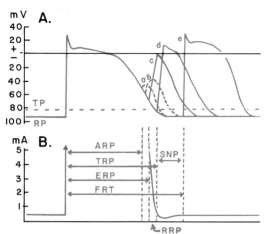

Figure 2.44. *A,* Normal transmembrane action potential and responses to series of stimuli applied during and after end of repolarization. *B,* Approximate durations of absolute refractory period (*ARP*), total refractory period (*TRP*), effective refractory period (*ERP*), full recovery time (*FRT*), supernormal period (*SNP*), and relative refractory period (*RRP*). In *A,* responses shown as *dotted lines* (*a* and *b*) are graded responses which do not propagate. Response *c* is earliest propagated response and defines end of effective refractory period. Response *d* is elicited at the time when transmembrane potential is close to level of threshold potential (*TP*) and occurs during supernormal period. Response *e* is elicited after end of repolarization and is normal in terms of rising velocity and amplitude, defining end of full recovery time. Changes in threshold shown in *B* are related to an arbitrary scale of current strength required. Curve shows onset of inexcitability coincident with phase 0 of transmembrane action potential, gradual decrease in threshold during phase 3 of repolarization, and restoration of full excitability after end of phase 3 of transmembrane action potential. [Redrawn from Hoffman (1969).]

extend *well beyond* the action potential into the rest period (phase 4), causing a marked delay in the cell's ability to respond to an early stimulus.

Different groups of cells in the cardiac conduction system have different refractory periods, and a method for studying these refractory periods in the whole heart will be discussed in a subsequent section. The refractory periods of various muscle cell types can be altered by physiologic stimuli (such as the catecholamines) and by various pharmacologic agents. The relatively long refractory periods of the cardiac conduction system and of heart muscle (which are due to the long plateau of these action potentials) are responsible for inability to produce a sustained contraction of the heart by a series of rapid impulses (which sometimes can occur during abnormal cardiac rhythms) and also prevent single extra depolarizations (so called premature depolarizations) which occur spontaneously from occurring so early that one contraction superimposes on another.

A number of drugs that are used to treat cardiac rhythm disturbances affect the refractory periods of heart cells. In particular, quinidine and like compounds increase the action potential duration and increase even more the duration of the effective refractory period of cells in the rapid conduction tracts and myocardium (thereby prolonging the period during the action potential when extra responses cannot be propagated). Lidocaine, on the other hand, shortens the action potential duration, but it has less shortening effect on the effective refractory period, so that, again, the ratio ERP/APD (effective refractory period/action potential duration) is increased, and a similar result on extra impulses or rapid rhythms ensues.

Membrane Responsiveness

During the latter part of the effective refractory period in fast fibers, a large depolarizing stimulus can produce a nonpropagated response which has a slow initial rate of rise, and during the relative refractory period propagated action potentials can be generated (by larger depolarizing stimuli than normal) which, again, have a slower than normal upstroke (phase 0) (Fig. 2.44 *a–d*).

In cells exhibiting voltage-dependent refractoriness, the maximum rate of rise of the action potential (dV/dt or $\dot{V}_{max}$) is related to the membrane potential from which it is initiated, so that during the late downslope of the action potential, as the transmembrane potential becomes progressively more negative, $\dot{V}_{max}$ increases along a sigmoid curve (Fig. 2.45). This so-called *"membrane response curve"* is important in that ability of an impulse to

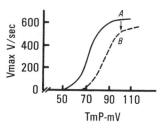

Figure 2.45. Membrane response curves for cardiac cells. *Vmax*, the maximum upstroke velocity of phase 0 of the action potential in volts (V) per second. *TmP* is the transmembrane potential (−mV). Notice the sigmoid relation in normal cardiac cells (*curve A*), with upstroke velocity increasing with more negative transmembrane potentials. This relationship is shifted to the right by antiarrhythmic agents such as quinidine (*dashed curve B*), so that at any level of transmembrane potential Vmax is reduced and conduction velocity is slowed.

be conducted depends on its initial upstroke velocity (the faster $\dot{V}_{max}$ the more readily it is conducted, see below) and, in addition, the response curve can be shifted by certain drugs. For example, the antiarrhythmic agent quinidine decreases membrane responsiveness by shifting the entire membrane response curve to the right and downward, so that $\dot{V}_{max}$ is reduced at any level of transmembrane potential (Fig. 2.45, *A* to *B*).

Impulse Conduction

The conduction of action potentials has been studied in detail in nerve fibers (Section 1, Chapter 3), and similar processes appear to occur in cardiac cells. Local current flow occurs through the electrically conductive solutions on either side of the cell membrane, and the local current flowing at the "front" produced by a local zone of depolarization can spread progressively over the entire cell. If the impulse is propagated, it will also spread to adjacent cells.

Several factors determine the speed and effectiveness with which an action potential is conducted. The amplitude and rate of change of the action potential ($\dot{V}_{max}$) is important, a more rapid phase 0 increasing conduction velocity. The magnitude of the potential and its rate of rise will also determine whether or not it reaches the threshold of adjacent cells. The anatomy of the conducting cells is also important, increased diameter of the cells resulting in increased conduction velocity. When a cell network is organized so that cell size progressively decreases and multiple interconnections exist, conduction velocity is slowed.

Since the resting membrane potential influences $\dot{V}_{max}$, this potential has an indirect effect on conduction velocity. Increases in $[K^+]_o$, by depolarizing the membrane, can result in a marked reduction in

$\dot{V}_{max}$. Drugs such as quinidine affect Na^+ conductance and shift the membrane responsiveness relationship, thereby also slowing $\dot{V}_{max}$. If the conducting cells are of the slow-response type (as in the SA or AV nodes), the upstroke velocity is slow and conduction velocity is low.

The so-called "cable properties" of the conduction system and cardiac muscle cells also determine how well the impulse is carried, and these properties in nerve fibers are discussed in detail in Section 1, Chapter 3. As with flow through a cable or wire, conduction velocity is directly related to fiber radius. The other passive cable properties can be determined by applying a current at a given point and determining its spread along the cell (Fig. 2.46). Although because of the complex cell structure and cell networks of the heart such analysis is difficult, in simpler models an equivalent electrical circuit can be used to represent the membrane resistance and capacitance, the extracellular resistance, and the intracellular (myoplasmic) resistance for each segment of the cell (Fig. 2.46). With application of

a subthreshold square wave current, the induced local potential increases slowly as the capacitance is charged, and the magnitude of the induced (subthreshold) potential then decreases as its distance from the stimulation point increases (Fig. 2.46). These changes are determined by the input resistance of the tissue and its space and time constants, (consideration of which is beyond the scope of this discussion). A subthreshold current flow precedes a propagated action potential, and the membrane capacitive effects as well as the cell properties and the resistance of the extracellular fluid determine the decay of a subthreshold stimulus as it moves down a cell or fiber. (For further details of cable theory see Brady (1974) and Wallace (1982).

ANATOMY AND PHYSIOLOGY OF THE CARDIAC PACEMAKING AND CONDUCTION SYSTEMS

The characteristics of action potentials (amplitude and duration) and fiber diameters in several

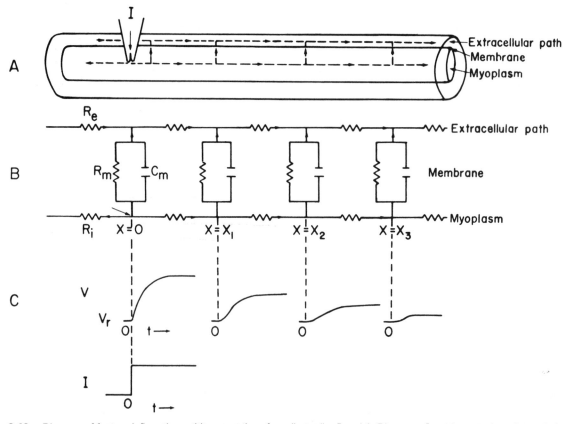

Figure 2.46.   Diagram of factors influencing cable properties of cardiac cells. *Panel A*, Diagram of an elongated cardiac cell showing its cell membrane and intracellular contents (myoplasm), surrounded by a cylinder of extracellular fluid (extracellular path). A stimulating electrode supplies current flow from a current source (*I*). *Panel B*, Equivalent circuit of the cell showing repeating resistance-capacitance units along the cell membrane (R_m and C_m, respectively) at distances X = 0, X_1, X_2, X_3, together with resistances in the extracellular path (R_e) and resistances in the intracellular path or myoplasm (R_i). *Panel C*, Membrane voltage responses (superimposed on the resting potential, V_R), as a function of time at the various distances along the membrane in response to a subthreshold square wave current (*I*) supplied at time 0 (lower tracing). [From Brady (1974).]

Table 2.4
Cardiac resting and action potentials[a]

Site	Species	Fiber Diameter, μm	MRP, mv	AP, mv	D_{AP},[b] msec
Sinus node	Frog	5	40	53	200
	Rabbit		60	66	200
AV node	Dog		53	58	350
Atrium	Frog	5	70	90	400
	Rabbit	10	78	92	150
	Dog	10	85	105	200
	Man	10	70	75	
Ventricle	Frog	10	80	95	600
	Dog	16	80	100	400
	Man	16	87	115	200
Purkinje fiber	Dog	30	90	120	300
	Sheep	75	94	130	400
Skeletal muscle[c]	Frog	130	88	120	1.5
Unmyelinated nerve fiber[c]	Squid	400	65	85	0.4

From Hecht (1968). [a] MRP, membrane resting potential; AP, action potential; D_{AP}, duration of action potential. [b] The values for D_{AP} depend on temperature and heart rate—some average values are given which show the difference in fiber size and in duration of the action potential in comparison to cardiac tissue. [c] Values for skeletal muscle and nerve fiber are given for comparison: note particularly the differences in fiber size and in duration of the action potential in comparison to cardiac tissue.

Table 2.5
Fiber diameter and conduction velocity in human hearts

Site	Fiber Diameter, μm	Conduction Velocity m/s
Sinus node	2–7	
Atrium	3–17	0.8–1.0
AV fibers (about 5 mm long)	3–11	0.05
His bundle branch	9–18	2.0
Ventricular myocardium	10–12	
Subendocardium		1
Subepicardium		0.4–1

From Bauereisen (1970).

species are shown in Table 2.4. Cell diameters and conduction velocities in man are shown in Table 2.5.

The SA Node

The sinoatrial (SA) node is situated near the junction of the superior vena cava and the wall of the right atrium (Fig. 2.47). In man, it is about 1.5 cm long and 0.5 cm wide. Its cells contain a few myofilaments, and the central stellate cells (so-called P cells) that probably constitute the pacemaker cells contain fewer myofilaments than the adjacent cells, which may serve to conduct the impulse out of the node (James et al., 1982).

Sinus node action potentials are difficult to record, but appear to exhibit a resting membrane potential of about −60 mV, and the pacemaker cells show diastolic depolarization (gradual increase in slope during phase 4). The action potential reaches threshold at about −40 mV; the upstroke velocity

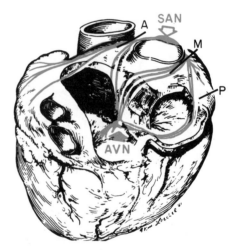

Figure 2.47. Diagram of the internodal pathways of the human heart. *A,* Anterior internodal pathway which sends fibers to the atrioventricular node (*AVN*) and interatrial fibers to the left atrium (Bachmann's bundle); *M,* middle internodal tract; *P,* posterior internodal tract. From James (1970a).]

is slower than in fast-response cells and does not exhibit a sustained plateau (Fig. 2.48). As mentioned earlier, SA-nodal pacemaker cells exhibit spontaneous diastolic depolarization probably through a progressive reduction in K^+ permeability, but it has also been postulated that a change in Ca^{++} permeability may produce an inward Ca^{++} current. Thus, Ca^{++} antagonists such as verapamil and to a lesser extent, nifedipine markedly slow or abolish the rate of diastolic depolarization (Wit and Cranefield, 1974).

A number of factors in addition to verapamil can

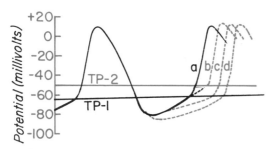

Figure 2.48. Diagram of transmembrane potential from a sino-atrial nodal pacemaker cell. On the right a series of succeeding pacemaker potentials depict the mechanisms by which the discharge rate of an automatic cell may be altered. *a*, The basic rate. *b*, The delaying effect of increasing threshold potential from TP-1 to TP-2. *c*, The slope of diastolic depolarization is decreased without a change in threshold and delays firing. *d*, The delaying effect of increasing the maximum level of diastolic depolarization (hyperpolarization) with no changes in threshold potential or diastolic depolarization slope. [From Hoffman et al. (1966).]

affect the rate of firing of the SA node. These include atropine which speeds phase 4 depolarization by blocking acetylcholine effects (acetylcholine slows phase 4), and catecholamines, which increase the firing rate (an effect counteracted by β-adrenergic blocking drugs). The various mechanisms by which the frequency of the cardiac pacemaker can theoretically be affected are summarized in Figure 2.48 and include changing the level of the initial diastolic potential (a more negative initial potential slows the sinus rate, Fig. 2.48*a* to *d*, and vice versa), altering the slope of diastolic depolarization (*a* to *c*) and a change in the threshold potential (*a* to *b*) (Hoffman and Cranefield, 1960, and Vassalle, 1977).

Under normal conditions, the rate of firing of the SA node is under the control of the autonomic nervous system through the interplay of reflexly induced release of acetylcholine from the vagus nerve terminals (parasympathetic system) and release of norepinephrine from the sympathetic nerve terminals (both of these autonomic divisions heavily innervate the SA and AV nodes). Acetylcholine produces both increased polarization (hyperpolarization) of the membrane as well as a reduction in the slope of phase 4 by increasing membrane permeability to K^+, thereby slowing the rate of the SA node. (Increased P_K results in a larger K^+ leak and a more negative cell interior, thereby partially offsetting the depolarizing inward Na^+ leak). Increased catecholamine levels can result either from norepinephrine released from nerve endings, or circulating epinephrine and norepinephrine released from the adrenal glands. The mechanism of their action on phase 4 is uncertain, but it has been postulated that these hormones may increase Ca^{++} permeability, thereby increasing the

rate of Ca^{++} influx and the slope of phase 4 depolarization.

If pacemaker cells are driven at a more rapid rate than their natural frequency, they become hyperpolarized, and if the driving frequency is then abruptly slowed the pacemaker cells will exhibit a long interval before the next spontaneous depolarization. This mechanism is called *overdrive suppression*, and it is important in maintaining control of the normal heart by the sinus node, since slower potential pacemakers elsewhere in the heart are suppressed. If the heart is artificially driven by an external electrical pacemaker, hyperpolarization of the SA node tends to occur. If the external pacemaking is then stopped, the next natural SA node cycle length will therefore be substantially prolonged. This mechanism is called "overdrive suppression" of the SA node, and it also applies to other potential pacemaker sites within the heart. Sometimes such electrical pacing is used to suppress abnormal rhythms originating in the ventricles or other regions of the heart.

The Interatrial or Internodal Tracts

These constitute specialized pathways of preferential conduction from the sinus node to both atria and to the AV junction (James et al., 1982). Three major routes can be identified (Fig. 2.47). The anterior internodal pathway sends fibers both to the AV node and to the left atrium (the latter branch is called Bachmann's bundle), the middle internodal tract goes only to the AV node, and the posterior internodal tract goes behind the superior vena cava to send a few branches to the left atrium but then continues to the AV node (Fig. 2.47). The cells in these tracts appear to constitute a mixture of ordinary muscle cells and cells which resemble the Purkinje fibers of the specialized conduction system (see below). Purkinje fibers and specialized tracts can conduct impulses at about 2.0 m/s compared with conduction velocities of 0.8–1.0 m/s in atrial muscle (Table 2.5). There is some evidence that cells in the internodal tracts can exhibit automaticity under some circumstances and that, rarely, they can conduct to the AV junction without depolarizing the atria.

The AV Junction

The cells in the atrioventricular node resemble those in the SA node. The AV node is positioned at the lower part of the septum dividing the two atria (Fig. 2.49). Three functional regions have been identified within the region called the *AV junction*: the nodal region proper, the atrionodal region which lies above the AV node, and the nodal-His region

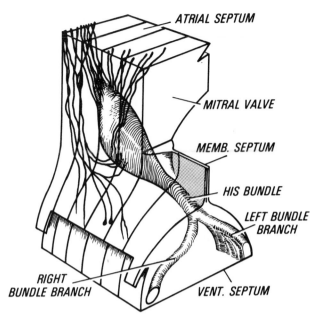

Figure 2.49. Schema of the atrioventricular node and His bundle. Atrial septum, interatrial septum; Vent. septum, interventricular septum; Memb. Septum; membranous portion of the interventricular septum. Fibers from internodal tracts are shown entering the AV node. The bundle of His runs along the lower margin of the membranous interventricular septum. It gives rise to a single right bundle branch, and then multiple left branches form a vertical sheet down the left septal subendocardium to form the origin of the left bundle branch. [From James (1970a).]

which is situated below the node toward the ventricles; from the latter region the bundle of His (see below) passes toward the interventricular septum (Fig. 2.49) (James et al., 1982). It is in the AV junction that impulses traversing the atria are collected and delayed for a period of time (normally up to 200 ms), so that contraction of the atria precedes the activation of the ventricles.

In the AV node, the fiber diameters are small, and there are multiple sub-branches, which contribute to the slow rate of conduction and the tendency for an impulse to die out in this region; moreover, the action potentials are of the slow-response type. The slow recovery of such cells contributes to their inability to be depolarized again following a second rapid stimulus, making this region subject to complete block of impulse transmission. Also, recovery of excitability may initially occur in one direction only, leading to so-called "retrograde block" at a time when forward conduction has recovered. The ability of the AV junction to slow and to block rapid impulses is called *decremental conduction*, which reflects the high degree of refractoriness of the AV node and its very slow conduction velocity (0.01–0.10 m/s). It is in the atrionodal region, as well as in the node itself, that the major conduction delay occurs in the passage of the impulse through the

AV junction. Action potentials in the nodal-His region are transitional between those in the AV-node and the bundle of His.

There are pacemaker cells in the junctional tissue (but not in the AV node itself), which are subject to acceleration and deceleration by sympathetic and parasympathetic stimulation, respectively.

A number of factors affect conduction through the AV junction. Acetylcholine (ACh) released from vagal nerve endings slows conduction through the AV node by decreasing conduction velocity and increasing the refractory period (increasing decremental conduction). Drugs such as digitalis (which has an indirect effect through central vagal stimulation), drugs which inhibit acetylcholinesterase and thereby block ACh breakdown (such as edrophonium), and the Ca^{++} antagonist verapamil, all slow conduction. Atropine (which blocks ACh) and quinidine (which inhibits vagal effects) speed AV conduction, as do the catecholamines. Thus, norepinephrine released from nerve terminals in the AV junction, or circulating catecholamines, increase the amplitude and upstroke velocity of AV nodal action potentials.

When an increased number of impulses arrives at the AV junction, its refractoriness is increased, and many of the rapid impulses are blocked. This means that when the atria are firing at an extremely rapid rate due to an abnormal rhythm (such as atrial fibrillation), the ventricles respond at a much slower frequency. Usually conduction from the ventricles backwards to the atria (so-called retrograde conduction) is blocked at a lower frequency rate than antegrade conduction.

Increased refractoriness in response to an increased number of stimuli received by the AV junction is an important property, and it has significant effects in the presence of certain rapid rhythm disorders (Chapter 9). Decremental conduction implies that early impulses may *partially* penetrate the junction (so-called "concealed conduction"), so that the next impulse finds the cells of the junction only partially recovered, i.e., they remain in a more refractory state than previously. In fact, delivery of a very early electrical stimulus by an electrical pacemaker, or increasingly rapid electrical pacing of the atrium up to the point of AV block, provide a means of determining the refractory period of the AV junction (see subsequent section). It should be recognized, however, in contrast to *electrical* pacing, that when the SA nodal rate increases under *normal* conditions (as with exercise or excitement) the refractoriness of the AV junction does *not* increase. In fact, it decreases because the increase in heart rate under these circumstances is due to activation of the sympathetic nervous system and withdrawal of parasympathetic influence (see Chapter 16) which simultaneously affect the AV junction to improve conduction.

The His-Purkinje System and the Bundle Branches

As shown in Figure 2.49, a large bundle of specialized fibers originates from the lower end of the AV junction called the bundle of His (W. His, Jr., Leipzig, 1893). These fibers course on the right side of the upper portion of the interventricular septum for a distance of over 1.0 cm and then divide into the *right bundle branch* and *left bundle branch* (Fig. 2.49) which supply the right and left ventricles, respectively; these bundle branches reach the ventricular walls and then ramify over the inner walls, branching into smaller bundles termed *Purkinje fibers* (see Fig. 2.4, Chapter 5). This network of small bundles is present throughout the subendocardial regions of both chambers. The cells in the His-Purkinje system (so-called *Purkinje cells*) are the largest cells in the heart and contain only a few muscle fibers; they also exhibit longer nexus junctions (low impedance cell-to-cell connections) than working myocardial cells, thereby favoring rapid impulse conduction (Legato, 1973). In the human heart, the right bundle branch is longer and considerably thinner than the left bundle branch (Figs. 2.4 and 2.49). The latter begins as a thick bundle of fibers, which then divides into two divisions, a thinner anterior division coursing to the front wall of the left ventricle and a thicker posterior division which supplies the posterolateral wall of the left ventricle. Each of these regions of the specialized conduction system can be damaged in certain disease states, leading to block in the transmission of the impulse to various regions of the heart.

The action potentials recorded in the His-Purkinje system resemble those in ventricular muscle cells but are of longer duration; they are fast-response action potentials, with a dip during phase 1 and a prominent plateau. The characteristics of these action potentials, coupled with a large cell (fiber) diameter result in very rapid conduction velocity from the AV node through this system (2.0–4.0 m/s). Some cells in the His-Purkinje system can exhibit spontaneous diastolic depolarization (Fig. 2.42*D*), and there are also variations in the duration of the action potentials, cells in the Purkinje fiber network near the ventricles exhibiting longer action potentials than cells higher in the His-Purkinje system (see Fig. 2.50). These so-called "gate cells" therefore can slow or block rapid sequences of action potentials before they reach the ventricles; their effective refractory period tends to

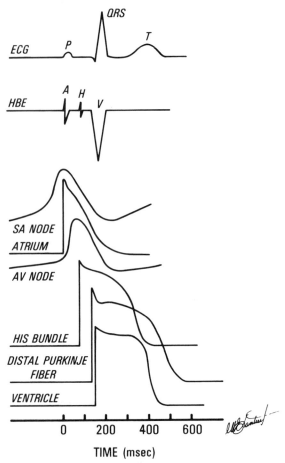

Figure 2.50. Diagram of simultaneous recordings of the surface electrocardiogram (*ECG*), the His bundle electrogram (*HBE*) (which shows activity recorded from the atrium (*A*), the His spike (*H*) and the ventricle (*V*)), and corresponding intracellular action potentials at various regions in the pacemaking and conducting system, and myocardium. (For further discussion see text).

prolong with increased heart rate, and early premature depolarizations originating in the ventricles may also be extinguished by the gate regions.

As might be anticipated, the conduction velocity of action potentials and the refractory periods of cells in the His-Purkinje system can be affected by a number of drugs and other stimuli.

Atrial and Ventricular Muscle

The muscle of both the atrium and the ventricle consists of relatively small diameter cells (10–15 μm) arranged in a branching network (see Fig. 2.78*A*, Chapter 10). There are special regions in the intercalated discs which separate the cells called nexus or gap junctions (mentioned above), which allow rapid electrical transmission of the impulse from cell to cell (other regions of the intercalated disc are thick and serve to transmit tension between cells) (see Fig. 2.79*A*, Chapter 10). Thus, the myocardium is a *functional syncytium*, through which the depolarizing wave front can spread relatively rapidly throughout the muscle. However, conduction in atrial and ventricular muscle is much slower than in the His-Purkinje system (Table 2.5).

Action potentials in the atrium tend to exhibit a somewhat shorter and steeper plateau (phase 2) than those in ventricular myocardium, and repolarization is slower (Figs. 2.42 and 2.50). Since the atrium is heavily innervated by the vagus nerve (in contrast to the ventricles), vagal stimulation with release of ACh can increase K^+ permeability and hence shorten the duration of the refractory period (see Fig. 2.41); it may also decrease the Ca^{++} inward current and shorten the duration of phase 2 of the action potential. In addition, the increased P_K produced by ACh hyperpolarizes the cells, so that vagal stimulation results in a faster phase 0 and improved conduction through cells that are less refractory.

Digitalis, which indirectly stimulates the vagus, can increase the rate of an atrial rhythm abnormality, or prevent an abnormality by such effects on the atrial muscle. Quinidine, on the other hand, decreases conduction velocity in atrial and ventricular muscle by shifting the membrane response curve, and it also tends to extinguish premature impulses primarily by increasing the effective refractory period in muscle as well as in Purkinje cells.

THE SEQUENCE OF NORMAL CARDIAC EXCITATION

A special catheter having multiple electrodes can be passed into the venous system and positioned so that one or more electrodes are located in the right atrium, another is placed on the relatively exposed portion of the bundle of His, and an electrode at the catheter tip is situated in the right ventricle. Depolarization of the atrium and ventricles can then be easily recorded and, with electronic amplification, a small electrical spike can also be recorded as the impulse travels through the His bundle (Fig. 2.50). (The latter event is not detectable on the ordinary electrocardiogram but can sometimes be recorded on high gain computer-processed, averaged beats.) The relation of these events to the electrocardiogram (ECG) recorded on the body surface (to be discussed in the next chapter), is shown in Figure 2.50. The P wave (atrial depolarization) is followed by a pause and then by electrical depolarization of the ventricles (the QRS complex); this pause, between the onset of the P wave and the onset of the QRS complex (the P-R interval), constitutes the delay in the AV junction, and it is during the latter part of this interval that the *His bundle electrogram* is normally recorded (Fig. 2.50). The T wave (Fig. 2.50) is ventricular repolarization.

Representative action potentials from the various regions of the heart during the entire sequence of its electrical excitation are summarized diagrammatically in Figure 2.50. Because of its very small size, action potentials from the SA node cannot be recorded on the ECG. (Like the His bundle, such small groups of cells are electrically silent on the body surface ECG, although direct recordings have now been made in experimental animals.) The impulse from the SA node spreads relatively slowly through the atrial muscle to produce the P wave, is delayed in the AV junction, and then rapidly traverses the His-Purkinje system to reach first the left ventricle, followed shortly by the right ventricle (conduction through the right bundle branch is slightly slower). The impulse subsequently travels through and depolarizes the ventricular muscle at a slower velocity, the duration of the QRS complex being much longer than the His bundle spike. The T wave of ventricular repolarization corresponds to repolarization of ventricular muscle cells (phase 3).

Testing the Refractory Period of the Conduction System

Using the special catheter described above for recording atrial, ventricular, and His bundle potentials, it is possible to determine the refractory periods of the various regions of the normal heart by electrically pacing the heart (see Wit et al., 1970, for details). The use of this technique to determine the refractory periods of the AV node and His-Purkinje system is illustrated in Figure 2.51. The atrium is electrically paced and a second depolarization A_2 is applied at increasingly shorter intervals; as this is done, the interval between the first and second beats A_1, A_2 can be plotted on the abscissa, and the corresponding response of the ventricles (the V_1, V_2 interval) can be plotted on the ordinate. Since the refractoriness of the AV node increases with increasing frequency of stimulation, as $A_1 A_2$ decreases a point is reached where the ventricular response deviates from the line of identity, and $V_1 V_2$ fails to decrease as $A_1 A_2$ is further diminished (Fig. 2.51A). This response indicates that the PR interval is progressively prolonging at shorter stimulus intervals. As the $A_1 A_2$ interval is decreased even further the interval between V_1 and V_2 starts to *increase* and, with further shortening of $A_1 A_2$, conduction to the ventricle fails entirely (Fig. 2.51A). The longest $A_1 A_2$ interval at which ventricular contraction still occurs constitutes the *effective refractory period* of the AV junction.

The interval between the onset of the atrial impulse and the His bundle spike (the A-H interval) and the interval from the His bundle spike to the

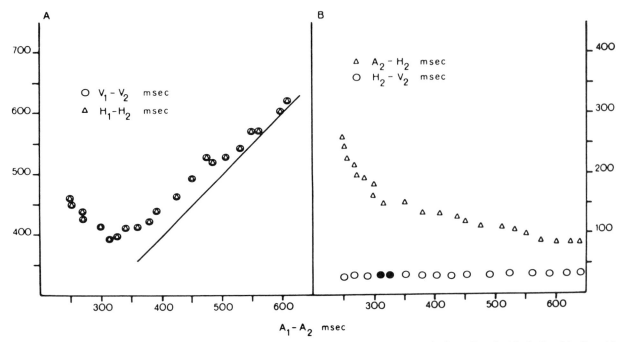

Figure 2.51.   Atrioventricular conduction in the human heart studied by applying progressively earlier electrical stimuli to the atrium. *Panel A*, The intervals between an initial and a paced, premature atrial stimulus (A_1–A_2) are shown on the abscissa, and on the ordinate are shown the intervals between the initial and premature ventricular complexes (V_1–V_2, *open circles*) and the initial and premature His spikes (H_1–H_2, *open triangles*); the latter two are identical and deviate from the line of identity at progressively shorter A_1–A_2 intervals, until conduction ceases at an interval of about 250 ms (for further discussion see text). *Panel B*, The corresponding responses of the A_2–H_2 interval is shown by the *open triangles*, and indicates progressive prolongation of conduction in the AV junction as A_1–A_2 shortens. The constant H_2V_2 interval (*open circles*) indicates lack of any conduction delay in the HIs bundle as A_1–A_2 is shortened. [Wit et al., (1970).]

onset of ventricular depolarization (the H-V interval) can also be measured. In normal man, the A-H interval is 60–130 ms, and the H-V interval is 35-55 ms. As A_1 A_2 is decreased by the pacemaker virtually all of the increasing delay occurs in the A-H interval (the AV junction), whereas the H-V interval (H_2,V_2) stays quite constant (Fig. 2.51*B*). Thus, the His-Purkinje system normally does not contribute to the increasing delay between the P wave and the QRS (increasing PR interval) produced by pacing at a rapid rate. However, in disease states the His-Purkinje system can function abnormally, and progressive lengthening of the H-V interval as well as the A-H interval can be observed with the above test; sometimes, H-V conduction even fails before A-H conduction. Therefore, complete heart block (Chapter 9) can sometimes be due to failure to conduct through the His-Purkinje system, rather than the AV junction.

THE HIERARCHY OF CARDIAC PACEMAKERS

A number of potential pacemakers exists in the cardiac conduction system. Normally, the SA node is the fastest pacemaker (70/min at rest) and it dominates the other pacemakers by producing overdrive suppression. However, if the sinus node is markedly slowed (as by intense vagal stimulation or by disease), a pacemaker lower in the heart (often in the AV junction) will commence to fire at its natural rate, which is slower than the SA node (50-60/min). If there is failure of both SA node and the AV junctional pacemaker (or if complete heart block is present), cells in the His-Purkinje system will usually drive the normal heart at a much slower rate (30-40/min). Sometimes, however, particularly in patients with heart disease, this escape frequency may be insufficient to maintain the cardiac output and blood pressure at levels compatible with life

CHAPTER 9

Electrocardiography and Disorders of Cardiac Rhythm

The electrical potentials generated by the heart can be detected by means of an electrode placed directly on the surface of the heart and, when amplified, such a recording is called an "electrogram." Since the tissues of the body contain electrolytes and are conductive, these electrical potentials spread throughout the body, and even though the changes in electrical potential become greatly attenuated they can be detected at the skin. A modern recording instrument (electrocardiograph) amplifies and records signals on moving paper, transcribing a voltage-time signal called the *electrocardiogram*. In clinical medicine, the electrocardiogram is useful for many purposes, including identification of abnormal spread of excitation through the conduction system and heart muscle, detection of changes in the size of the cardiac chambers (which alters the amplitude of the electrical signals), identification of damage to the heart (such as after a heart attack), and assessment of abnormally rapid, slow, or irregular heart rhythms.

THE CARDIAC DIPOLE

As electrical depolarization spreads across a muscle, there is an abrupt reversal of charge at the wave front as the outside of the cell changes from relatively positive to relatively negative. This wave front may be thought of as a moving dipole, consisting of positive and negative poles closely spaced together and situated in a volume conductor (Fig. 2.52). When such a dipole exists, current flow occurs, by convention, from the positive to the negative pole (actually cations flow toward the negative pole and anions and electrons toward the positive pole); the most dense current flow is close to the poles, with current density decreasing as distance from the poles increases (Fig. 2.52, *solid lines*). Each pole also generates a potential field in three dimensions, with isopotential lines arranged as shown in Figure 2.52 (*dashed lines* show isopotential lines in two dimensions). The *central vertical line* midway between the two poles is at zero potential, and the magnitude of the voltage at any distance from the dipole can be calculated: along the axis of the dipole (the *horizontal line* traversing both poles), the voltage is inversely related to the square of the distance from the center of the dipole, and when off the axis of the dipole, the potential varies with the cosine of the angle from the axis of the dipole. Also, the potential at *any* location varies with the strength of the dipole.

If a piece of isolated cardiac muscle is placed in a volume conductor and arranged so that it can be electrically stimulated (depolarized) at one end, while the positive pole of a recording galvanometer is placed at the other end, a characteristic tracing is obtained as the dipole of the depolarizing wave front moves across the muscle toward the recording electrode (Fig. 2.53). By convention, a positive deflection is recorded when the positive recording pole faces the positive pole of an approaching dipole (in this arrangement, the other electrode of the galvanometer is placed at a distant ground site where it is near zero potential). As shown in Figure 2.53 (*panel A*), no potential difference is recorded

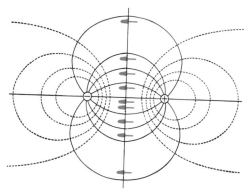

Figure 2.52. Electrical field of a dipole in a volume conductor. *Solid lines* are current flow. *Dashed lines* are isopotential lines.

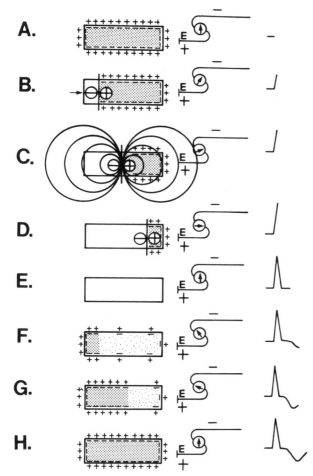

process, which is rapid and displays a sharp positive spike (which corresponds to the QRS of the electrocardiogram and the rapid upstroke (phase 0) of the action potential).

During phase 2 (the plateau) of the action potential, the muscle remains completely depolarized, and the recording line remains isoelectric (Fig. 2.53E). Repolarization proceeds more slowly, and recovery begins simultaneously but is completed earliest at the end where stimulation began. Therefore, the regaining of positive charge is most marked at the end away from the electrode, with movement of the repolarization wave front in the manner of a reverse dipole. Thus, the relatively negative wave front of the dipole approaches the electrode, leading to a progressively increasing negative deflection (*panel F*) which then declines as the inhomogeneous recovery process is slowly completed (*panel G* and *H*).

If the positive recording electrode is placed at the midpoint of the muscle (electrode E_2, Fig. 2.54) it

Figure 2.53.   Depolarization and repolarization of a strip of myocardium immersed in a volume conductor. Electrode *E* is connected to the positive pole of a recording galvanometer which is paired with a distant electrode at near zero potential. The advancing dipole at the border between activated and inactivated muscle produces an electrical field which moves toward the electrode (*panel B*). The electrode is intersected by isopotential lines of increasing force (*panel C*) until all difference of potential is extinguished by completion of depolarization (*panel E*). Recovery occurs in all parts of the muscle nearly simultaneously but is completed at the stimulated end first (*panel G*), thereby producing relative negativity at the side facing the electrode and giving rise to a negative repolarization wave. For further discussion see text.

between the electrodes when the muscle is at rest. Following stimulation of the end of the muscle (*panels B* and *C*), as the potential on the outside of the muscle is reversed and the dipole is propagated across the muscle, a progressively larger positive deflection is recorded as the isopotential lines of progressively larger potential approach the recording electrode (*E*). In panels *D* and *E* (Fig. 2.53), the maximum potential is reached just as the muscle is entirely depolarized, and then as the zero potential line at the center of the dipole reaches the end of the muscle, the recorded potential difference returns to zero. This completes the depolarization

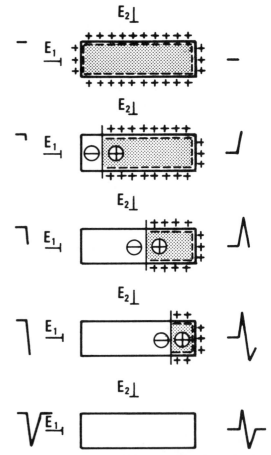

Figure 2.54.   Form of complexes inscribed from an electrode placed at position E_1 or E_2 during depolarization of a strip of myocardium suspended in a volume conductor. Each electrode paired with a distant indifferent electrode. Tracings on the *left* side are related to E_1 while tracings on the *right* relate to E_2 electrode.

can be surmised that tracings of the configuration shown in the righthand column of recordings will be obtained; the depolarization tracing initially will be positive, then will reach zero potential midway in the depolarization spike (as the zero isopotential point of the dipole passes under electrode E_2), and, finally, will become negative as the negative end of the dipole moves away to the opposite end of the muscle. The potential differences recorded when the electrode is placed at the opposite end of the muscle (compared to Fig. 2.53) are also shown in Fig. 2.54 (electrode E_1); a negative complex is evident (Fig. 2.54, *lefthand column* of recordings).

ELECTRICAL DEPOLARIZATION AND REPOLARIZATION OF THE HEART

The chambers of the entire heart contain millions of cells that are depolarized nearly simultaneously. First the atria (P wave) and then the ventricles (QRS complex) are rapidly depolarized, with both sets of chambers depolarizing slightly out of sequence, and then they repolarize more slowly with greater dispersion of the recovery of excitability (see Fig. 2.50). At any instant in time, the entire electromotive force generated by the heart can be thought of as a dipole-centered in the middle of a large volume conductor (the body), and it is the sequence of these instantaneous dipoles recorded from the body surface throughout the cardiac cycle that makes up the electrocardiogram. Because of the shape of the heart and the sequence of its activation, the net voltage measured at the body surface at any moment is the sum of forces acting in *many* directions. Thus, there are large *cancellation* effects; for example, when the wave front is heading toward the posterior wall of the left ventricle it is simultaneously spreading toward its anterior wall. The *net* magnitude and direction of the voltage will then depend on the relative numbers of

cells being depolarized in various regions of the heart at any moment in time. This process results in an electromotive force that has both magnitude and direction at any instant, that is, it constitutes a *vector*. The *magnitude* of the vector will depend largely on the *number* of cells that are being depolarized at any instant (and the distance of the heart from the recording electrode), whereas the *direction* of the vector will depend on which anatomic regions of the heart (having different locations and muscle masses) are dominant at that instant.

Atrial Depolarization

As shown in Figure 2.55, we can show the sequence of atrial depolarization (the P wave) by a series of instantaneous vectors presented by arrows that have both magnitude and direction. By convention, the *point* of the arrow represents the *positive* end of the dipole. The arrows change in amplitude and direction (right to left, up (superior) or down (inferior), and front (anterior) or back (posterior), with all directions being given as if the observer is *facing the heart*, with right being to the observer's left. For convenience, the origin of each vector arrow is placed at a central point. As the impulse sweeps over the right atrium from the sinoatrial node, its amplitude initially is small and then increases as the wave spreads initially anteriorly and inferiorly and then from right to left (Fig. 2.55) during right atrial depolarization. The vector then moves to the left and posteriorly as the left atrium is depolarized (Fig. 2.55). The vector arrows are small because the muscle of the atrial walls is thin (correspondingly, the P wave on the electrocardiogram is normally a low amplitude wave).

These instantaneous vectors can be conveniently summarized as three vectors: the initial P vector which points to the right and anteriorly (Fig. 2.55); the average or mean P vector, which is directed to

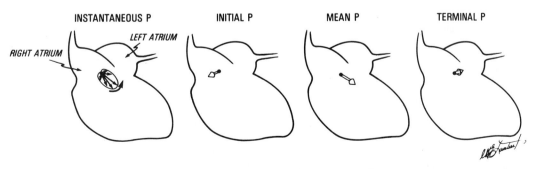

Figure 2.55. Diagram showing the heart and vectors of atrial depolarization originating from a point at the center of the heart. The left diagram shows a series of instantaneous, sequential P wave vectors, initially directed vertically then from left to right and finally superiorly (the tips of the arrows are connected to form a loop). The right three diagrams show that the initial average vector points toward the right, inferiorly and anteriorly; the mean P wave vector points toward the left and anteriorly, and the terminal P wave vector is directed toward the left and posteriorly.

the left and anteriorly (Fig. 2.55); and the terminal P vector, which is small and directed to the left and posteriorly (Fig. 2.55). Repolarization of the atria is not well seen on the electrocardiogram, usually being obscured by the onset of the much larger voltage of *ventricular depolarization.*

Ventricular Depolarization

Passage of the electrical wave front into the AV node and subsequent entry into the His bundle are electrically silent events in the standard electrocardiogram. Then, as the muscle of the ventricles depolarizes, the process can be represented by a sequence of instantaneous vectors which comprise the QRS complex (Fig. 2.56). The first impulses that leave the His bundle reach the left side of the muscle of the septum (partition) between the two ventricles, leading to a small initial vector that is directed anteriorly and commonly toward the right (Fig. 2.56). Since the Purkinje fibers lie on the inner wall of the ventricles, the wave front crosses the interventricular septum from the inside surface of the left ventricle outward; this direction of depolarization from the inner (endocardial) surface to the outer (epicardial) surface subsequently occurs elsewhere in the ventricular walls. As the wave front spreads over the two ventricles, the left ventricle, because of its much larger mass, dominates the magnitude and direction of the vectors. Since the left ventricle is situated *behind* (posterior to) the right ventricle, the instantaneous vectors which are initially to the right quickly become posterior and show increasing magnitude in the leftward direction (Fig. 2.56). Thus, the simultaneous depolarization of the thin-walled right ventricle is not seen, with its anteriorly directed forces being cancelled by the much larger forces originating from the thick-walled left ventricle. The instantaneous vectors remain leftward and posterior for a period, and then as depolarization of the heart is completed the normal vectors often point somewhat to the right (and either posteriorly or anteriorly), since the last region in the heart to be depolarized lies centrally in the region of the mitral and aortic valves (Fig. 2.56).

This series of instantaneous vectors representing ventricular depolarization can be summarized as three vectors: the initial QRS forces, which represent mainly depolarization of the interventricular septum and are directed to the right and anteriorly (Fig. 2.56); the mean QRS vector, which is large and directed to the left and posteriorly (Fig. 2.56); and the terminal QRS vector, which is often directed posteriorly and superiorly (Fig. 2.56) (there can be considerable variability, however, in the superior or interior direction of both the initial and terminal forces of the normal heart).

Ventricular Repolarization

If ventricular repolarization began at the same location and traveled in the same direction as the depolarization wave, that is, from the endocardium (where the Purkinje fibers are situated) across the wall to the epicardium, we would expect to see a slow negative wave occurring in the direction opposite to the depolarization wave. Thus, it would be directed superiorly to the right and anteriorly (Fig. 2.57, *far left*), with the mean T wave vector being opposite to the mean QRS vector. However, this is *not* the case in the whole heart, because repolarization occurs in the direction opposite as the QRS, dominated by the left ventricle. Thus, repolarization occurs from epicardium to endocardium, leading to a T wave having the *same* direction as the depolarization wave (Fig. 2.57, *center* and *right*). However, repolarization is a slower process, fewer cells are being repolarized at any moment, and the repolarization wave (T wave) is longer (which could be displayed as more vectors per unit time) and is of lower amplitude than depolarization.

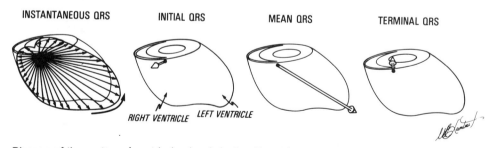

INSTANTANEOUS QRS   INITIAL QRS   MEAN QRS   TERMINAL QRS

RIGHT VENTRICLE   LEFT VENTRICLE

Figure 2.56.   Diagram of the vectors of ventricular depolarization (the atria are removed). The instantaneous, sequential QRS vectors are represented in the left diagram as a series of arrows originating from a central point in the heart, their tips being connected by a line to form a loop. In the right three diagrams the initial average QRS vector constituting septal depolarization is directed rightward and anteriorly; the mean QRS vector is directed leftward and posteriorly; and the terminal QRS vector is directed superiorly and posteriorly.

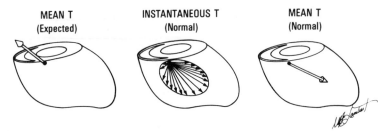

Figure 2.57. Diagram of the vectors of ventricular repolarization showing, in the *center panel*, instantaneous, sequential T wave vectors. The *left diagram* shows that the expected mean T wave vector would be 180° opposite to the mean QRS vector (see Fig. 2.56), if repolarization followed the same pattern as depolarization (see also Fig. 2.53). However, the normal mean T wave vector (*right panel*) is directed toward the left and posteriorly, in the same direction as the mean QRS vector (for further discussion see text).

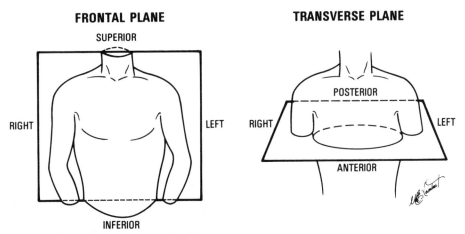

Figure 2.58. Planes transecting the body which are represented on the standard electrocardiogram. The frontal plane (*left*) passes vertically through the body, and the transverse (horizontal) plane transects the body through the chest. Note that the left and right sides of the planes are labelled as if the observer is facing the subject, that is, the subject's left is to the observer's right.

One theory for the opposite direction of repolarization holds that the buildup of pressure by contraction of the left ventricle compresses the subendocardial (inner wall) tissue, thereby delaying repolarization in that area.

RECORDING THE ELECTROCARDIOGRAM

For recording the electrocardiogram (ECG), metallic leads attached to wires are placed on the skin using electrolyte paste, and by arranging the electronic recording circuits appropriately it is possible to "look" at the cardiac vectors from different vantage points. A recording stylus moves up and down as the amplitude of the instantaneous vectors changes while the recording paper moves at a constant speed, allowing measurement of the duration of various events. In practice, pairs of recording electrodes or a single exploring electrode (which are termed electrocardiographic "leads") are arranged to view the heart both in the *frontal plane* (a plane passed vertically through the body as if directly facing it) (Fig. 2.58) and in the horizontal or *transverse plane* (a plane transsecting the body) (Fig. 2.58).

Frontal Plane Electrocardiogram

In the early part of this century the development of an improved string galvanometer led Einthoven (W. Einthoven, Dutch physiologist, 1860–1927) to develop a three-lead system which is still in use for recording the electrocardiogram. In this system, electrodes are placed on the right arm, the left arm, and left leg. An additional electrode is attached to the right leg and is grounded. These leads are connected in such a way that, for practical purposes, they form an equilateral triangle surrounding the heart in the frontal plane at the body surface (Fig. 2.59A). ("Einthoven's triangle"). (The leads are shown as if they are positioned on the two shoulders and the lower chest, although they are actually placed on the arms and legs; the effect is the same). *Lead I* measures the potential difference between the right and left side of the chest, with the *positive pole* of the recording system *toward the left arm*

FRONTAL PLANE

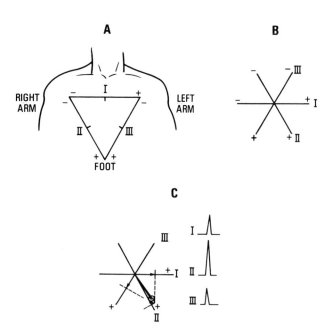

Figure 2.59.   The triaxial reference system. Einthoven's triangle is represented in panel *A*, and in panel *B* the three lead axes are rearranged to cross a central point. In panel *C*, a QRS vector is drawn together with its projections on the standard limb leads I, II, and III.

(Fig. 2.59). *Lead II* records between the right arm and the left foot with its *positive pole* being toward the *left leg* (Fig. 2.59), and *lead III* is arranged to record between the left arm and the foot with its *positive pole* also being toward the *left leg* (Fig. 2.59). These leads form a simplified equivalent circuit based on the (not strictly true) assumptions that these anatomic locations are situated at a great distance from the heart, equidistant from the central cardiac dipole and each other and in the same plane, and that the fluids and body tissues act as a homogeneous conductive medium.

These three leads may then be centered into a *triaxial reference system* in which the bisected lines cross a central point and are positive or negative on either side of this point (Fig. 2.59*B*). Left and right always refers to the left and right sides of the body as viewed by an observer facing the torso (Fig. 2.59*A*). If we now place a vector into this triaxial reference system (Fig. 2.59*C*), the direction and magnitude of the *projection* of the vector on any lead will determine its electrocardiographic voltage in that lead. The large vector arrow has a posterior direction, but this cannot be detected in the frontal plane, since only the magnitude of the vector projected on the frontal plane leads is being recorded. Thus, if a light were thrown from below straight upward toward the *large arrow*, the *shadow* of the

arrow will be projected on lead I as an arrow (Fig. 2.59*C*). A perpendicular dropped from the tip of the arrow to the lead I axis, since the light rays are perpendicular to the lead axis, would give the magnitude of the vector in lead I. To determine the projection of the vector on lead II, a light would be thrown from the left shoulder perpendicular to that lead, and an arrow will be projected on lead II (Fig. 2.59*C*). Similarly, if a light were thrown from below toward the right shoulder so that its rays were perpendicular to the lead III axis, the shadow would then be projected on lead III as an arrow (Fig. 2.59*C*).

At any instant (such as the peak of the depolarization wave), "Einthoven's law" states that voltages will conform to the following equation:

$$\text{lead II} = \text{lead I} + \text{lead III*}$$

This equation provides a means of checking whether or not the leads are properly attached to the arms and legs.

As discussed earlier, by convention any dipole that is moving towards the *positive* recording pole creates an *upright* or *positive* deflection on the electrocardiographic tracing, and the arrow tip of a vector represents its positive end. Therefore, the deflection of the pen recording the mean vectors shown in Figure 2.59*C* will yield a positive wave in each lead (notice the position of the positive poles of leads I, II, and III), and the magnitude of the waves will reflect the lengths of the arrow projected on each lead, with the largest magnitude being in lead II (Fig. 2.59*C*).

The standard "limb leads" are bipolar, measuring the potential differences between two sites. The electrocardiogram in the frontal plane has three leads added to these three standard leads. These extra leads allow other vantage points from which to view the cardiac vectors, and they are called the *augmented unipolar limb leads*. The augmented limb leads are more amplified than the standard limb leads, and to equate them the amplitude of the standard leads must be multiplied by 1.15. They are unipolar because the potential is compared to zero, not to the potential at another site. They are recorded by connecting the three standard limb terminals together through resistances and using

---

* According to Kirchoff's second law of circuits, the potentials measured sequentially in any three bipolar leads completing a closed circuit would be: I + II + III = 0. However, Einthoven arranged lead II so that in the normal electrocardiogram all 3 leads would exhibit upright QRS deflections, so that with Einthoven's system Kirchoff's law becomes I − II + III = 0, or II = I + III. Thus at any instant, this equation holds for the potential differences recorded simultaneously by these 3 leads.

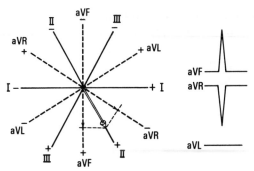

Figure 2.60. The hexaxial reference system showing the addition of the augmented limb leads (aVR, aVL, and aVF) to the standard limb leads I, II, and III. A QRS vector is shown projected onto the augmented limb leads. For further discussion see text.

this central point as a zero reference electrode; the other (positive) pole then becomes equivalent to an "exploring electrode." The augmented limb leads are arranged so that the positive electrode is positioned at three locations, equidistant between leads I, II, and III. One has its positive pole at the right upper chest (lead aVR), one at the left upper chest (lead aVL), and one at the foot (lead aVF), as shown in Figure 2.60. Thus, they lie between the standard leads giving a *hexaxial reference system*. Notice that lead aVR is positive superiorly and to the right and negative inferiorly and to the left, lead aVL is positive superiorly and to the left and negative inferiorly and to the right, and aVF is positive toward the feet and negative superiorly (Fig. 2.60). If we now place a vector in this system (Fig. 2.60) and project its shadows on leads aVF and aVR, the deflections will be equal, but the arrow on aVR is now projecting toward the negative pole, so that the ECG deflection is negative in that lead (Fig. 2.60), while it points toward the positive pole of aVF and is upright in that lead. Note that the vector is perpendicular to lead aVL, so that when a light is projected perpendicular to lead aVL, no shadow is cast and no deflection is recorded (*i.e.*, the tracing is "isoelectric"). Keep in mind that aVR is negative in the midst of the positive poles of several leads, and positive in the midst of negative poles (Fig. 2.60).

The Transverse (Horizontal) Plane

There are six leads recorded in the transverse plane with the usual electrocardiogram (total of 12 leads), and they provide information on how the instantaneous cardiac vectors are directed anteriorly and posteriorly (Fig. 2.58). These are termed the *precordial leads* (because the electrodes are arranged on the front wall of the chest, over the heart or precordium). Like the augmented limb leads, these electrodes also act as unipolar exploring elec-

trodes because they are connected to a central ground, which remains near zero potential. As shown in Figure 2.61, if the chest is represented as a cylinder and we look down from above, the six leads, called V_1, V_2, V_3, V_4, V_5, and V_6, are shown diagrammatically, and all of their positive terminals are positioned on the front of the chest. Therefore, vectors which project onto the positive one-half of each lead axis will yield a positive ECG deflection and vectors which project onto the negative one-half will produce a negative deflection. The *initial* and *mean* vectors of ventricular depolarization previously described (see Fig. 2.56) are also drawn in Figure 2.61, and the approach employed with the frontal plane lead system can be used to determine how these vectors are projected on the transverse plane leads. Note that in V_1, the normal initial vector is directed anteriorly and to the right (septal depolarization), whereas the mean vector of ventricular depolarization is directed posteriorly and to the left (Fig. 2.61). Thus, when these vectors are projected on Lead V_1 a small upward (positive) wave is first seen, followed by large negative deflection (Fig. 2.61). On the other hand, when the same vectors are projected onto Lead V_6, the initial force is projected on the negative side of V_6, so that an initial small downward (negative) deflection is seen, followed by a large positive deflection as the arrow is projected on the positive one-half of that lead (Fig. 2.61). Lead V_4 is also shown, and the two arrows are nearly perpendicular to that lead axis,

TRANSVERSE PLANE

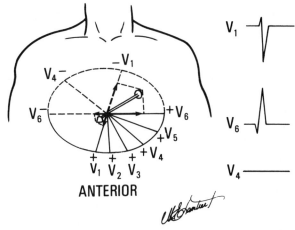

ANTERIOR

Figure 2.61. The six precordial leads, V_{1-6}, of the transverse plane lead system. Initial and mean QRS vectors are projected on selected leads of this system, giving a negative mean vector in V_1, a positive mean vector in V_6, and an isoelectric tracing in V_4 (since the mean vector is perpendicular to this lead). For further discussion see text.

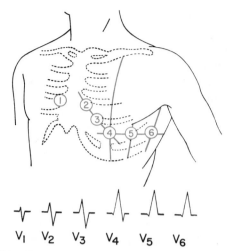

Figure 2.62. Positions of the chest electrodes for six standard precordial leads. V_1, fourth intercostal space to right of sternum. V_2, fourth intercostal space at left of sternum. V_4, fifth interspace at midclavicular line. V_3, midway between V_2 and V_4. V_5, anterior axillary line, same level as V_4. V_6, midaxillary line, same level as V_4 and V_5. Below is shown general form of QRS in normal precordial leads. V_3 represents transition from predominant negativity on the right to predominant positivity on left. [From Lipman and Massie (1959).]

so that no projection occurs on this lead and zero deflection is recorded (Fig. 2.61).

In practice, the precordial leads are arranged in accordance with the spaces between the ribs, with the electrode being sequentially applied at the locations shown in Figure 2.62. Notice that V_6 lies in the frontal plane, so that deflections in this lead will resemble those recorded from standard lead I.

THE FORM OF THE ELECTROCARDIOGRAM

In analyzing a standard ECG tracing, it is necessary to know the voltage calibration, the significance of the "time lines" on the recording paper (which moves at 25 mm/s), as well as some terminology that has been devised to identify the various ECG waves. As shown in Figure 2.63 (a recording of lead I in a normal individual), each horizontal small box represents 0.04 s (40 ms), and the time lines of the larger boxes (made up of 5 small boxes) indicate 0.2 s. The atrial or ventricular rates can be measured by counting the number of small boxes between P waves or QRS complexes and dividing the number of small boxes into 1500, or the number of large boxes into 300. In the vertical direction, the ECG is calibrated so that 1 mV causes a deflection of 10 small boxes (10 mm).

The wave of atrial depolarization is called the *P wave*, the wave of ventricular depolarization the *QRS complex* (Fig. 2.63). When the initial wave of the QRS complex is negative, it is called a Q wave.

The Q wave is usually followed by a positive deflection, the R wave, and then by a small negative deflection, the S wave, to form the *QRS complex*, although not all of these waves need be present in any single lead to apply this term. If a second positive deflection follows the S wave, it is labeled R′, and if a second S wave follows the R′, it is called S′. The first positive deflection is called the R wave, whether or not it is preceded by a Q wave. The wave of repolarization is called the *T wave*, and sometimes this is followed by a small afterpotential called the *U wave* (see Fig. 2.63). The electrocardiograph is operated through an alternating current circuit. Therefore, when no voltage *change* is occurring the baseline is flat, as between the P wave and the QRS complex, and during the plateau of the ventricular action potential (the *ST segment*) (Fig. 2.63) (see also Fig. 2.42A, Chapter 8, showing a simultaneous recording of an action potential and an electrogram).

Various other measurements made on the electrocardiogram have particular importance in abnormal states; these are indicated in Figure 2.63. The height and width of the P wave are measured because either or both may be abnormal when the atria are enlarged. The interval from the onset of the P wave to the onset of the QRS complex, the *PR interval*, is important in that it reflects conduction from the atria to the ventricles through the AV junction and His-Purkinje system (it also includes the time of atrial depolarization; the potential from the SA node is not recorded on the surface electrocardiogram, nor is the potential from the His bundle). The PR interval normally is about 0.14 s but may range between 0.12 and 0.2 s in normal individuals, depending largely on the heart rate and the tone of the autonomic nervous system. It can

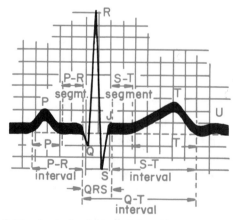

Figure 2.63. A single electrocadiographic complex giving nomenclature of the deflections and the intervals. [From Burch and Winsor (1968).]

be lengthened by functional or anatomic block at any point along the specialized conduction system. The *duration of the QRS* complex normally measures about 0.08 s (normal is up to 0.09 s). The *QT interval* is measured from the onset of the QRS complex to the end of the T wave (Fig. 2.63). The QT interval is normally 0.26 to 0.45 s, and it may be prolonged or shortened by changes in heart rate, by alterations in electrolyte concentrations, or by various drugs that affect the rate of ventricular repolarization. Often, the QT interval is corrected for heart rate (QT_c) by dividing it by the square root of the interval between R waves.

THE NORMAL 12-LEAD ELECTROCARDIOGRAM

In considering the form of the normal electrocardiogram, it is useful to examine the initial, terminal, and mean vectors for the atria and ventricles (shown in Figs. 2.55 and 2.56), and to project the *loop* connecting the tips of these arrows onto the frontal plane leads (the hexaxial reference system), as well as onto the transverse plane leads (the unipolar precordial leads).

The P wave

The mean vector of the P loop in the frontal plane generally points to the left parallel to lead II, giving an upright deflection in that lead (Fig. 2.55).

In the transverse plane the P wave is mostly anterior, giving an initial positive deflection in lead V_1, followed by a small negative (posterior) deflection due to left atrial depolarization (Fig. 2.55) (the left atrium is behind the right). If the atrium is depolarized from an abnormal location (other than the SA node), such as near the inferior vena cava, the depolarization wave can spread upward and to the right, instead of inferiorly and to the left, giving negative P waves in leads I and II. If the right atrium is enlarged, the amplitude of the P wave in the frontal plane leads is increased, but if the left atrium is enlarged there may be a prominent negative deflection in the second half of the P wave seen in lead V_1 (since the left atrium is situated posteriorly and depolarizes after the right), with some prolongation of the total duration of the P wave.

The QRS Complex

Using the initial and terminal forces together with the mean QRS vector, the normal QRS loop is plotted in the frontal plane in Figure 2.64 (*left*). The projections on lead I show a small negative wave (Q wave) followed by a large R wave. The projections on lead II show a small Q wave, a large R wave, and a small negative S wave (toward the negative pole of lead II). The projections on lead III show a relatively low amplitude QRS complex,

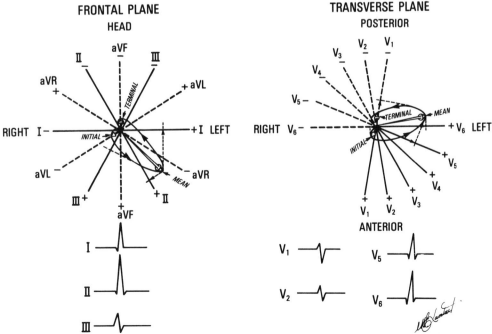

Figure 2.64. The normal QRS complex. Projection of initial, terminal, and mean normal QRS vectors, and the vector loop connecting the tips of these vectors, projected on the hexaxial lead system in the frontal plane (*left panel*), and the precordial lead system in the transverse plane (*right panel*). Notice the counterclockwise direction of the heart-shaped vector loop in both the frontal and transverse planes. For further discussion see text.

with a small R wave followed by an S wave (as the loop goes first toward the positive pole of lead III and then toward the negative pole). As the series of instantaneous vectors occur throughout the cardiac cycle, connection of their arrow points gives a time sequence and a direction to the inscription of the loop, as shown in Figure 2.64 (see also Fig. 2.56). In this example, the loop travels counterclockwise in the frontal plane. Note that the normal QRS loop tends to be "heart shaped."

The projection of these three QRS vectors on the transverse plane, together with the QRS loop, is shown in Figure 2.64 (*right*). The projection on V_1 shows a small initial R wave, as the initial septal forces approach the positive pole of V_1, followed by a deep S wave as the dominant left ventricular forces of the mean and terminal vectors project on the negative pole of V_1 (as mentioned, the positive and negative regions are determined by transsecting each lead axis by a perpendicular line at its midpoint). In Lead V_2 there is a slightly larger R wave as the loop moves toward the positive pole of that lead and an equal S wave as the loop moves counterclockwise to project on the negative half of lead V_2. In this example, lead V_2 is approximately at the "transition point" between a negative and a positive QRS complex, i.e., that lead has the most nearly *equal* positive and negative deflections. The R wave then becomes progressively larger (and the S wave smaller) in the more lateral V leads, so that in lead V_6 a small initial Q wave (septal depolarization) now projects on the negative pole; there is a large

R wave as the mean vector projects on the positive pole of that lead and a small terminal negative S wave.

The Mean Electrical Axis of the QRS Complex in the Frontal Plane

This is an important descriptor of the electrocardiogram, since it can be abnormal in a variety of cardiac disorders. It is determined by the anatomical position of the heart (vertical or horizontal, with the latter more likely in obese individuals) and by the direction of electrical depolarization of the ventricles. The *mean QRS axis* is determined from the mean QRS vector and is reported in degrees, based on the hexaxial reference system (Fig. 2.65). The normal mean QRS frontal plane axis lies between −30° and +100°. If the mean QRS axis is −30° or more negative, there is *left axis deviation*. If the mean QRS axis lies at or to the right of 100°, there is *right axis deviation* (Fig. 2.65A).

Two simple approaches can be used to determine the mean QRS axis. First, study the six limb leads to determine in which single lead the QRS complex is most nearly isoelectric (i.e., the negative and positive areas are most equal), and in which lead the QRS deflection is largest. For example, in Figure 2.65A, lead III shows a large upright deflection because the loop is wholly on its positive side, whereas lead aVR shows equal positive and negative deflections, the positive deflection being first upright (toward) the positive pole of aVR and then

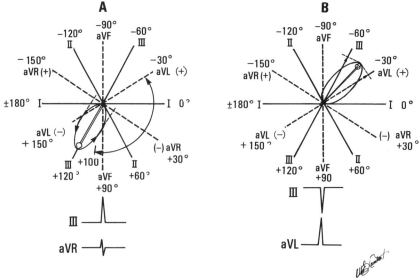

Figure 2.65. The mean QRS axis. Two methods for determining the mean electrical axis of the QRS complex are shown. *A* shows abnormal right axis deviation; the QRS loop is clockwise, and the lead to which the axis is most perpendicular is identified (in lead aVR the positive and negative areas of the QRS complex are most nearly equal). *B*, There is left axis deviation, and the two leads on which the QRS complex projects equally are identified (here leads III and aVL).

downward, since there is a clockwise loop passing first on the positive side and then on the negative side of aVR. Since the loop is perpendicular to aVR, the mean electrical axis is +120°, indicating right axis deviation. Therefore, with this approach the lead to which the QRS is most perpendicular (isoelectric) is determined, and the lead to which it is most parallel (largest deflection area) aids in the identification.

A second method for determining the mean QRS axis is to find two leads on which the QRS projects equally; the mean QRS axis then lies equidistant between the two leads. In Figure 2.65*B*, the mean electrical axis of the loop projects equally large negative and positive deflections on leads III and lead aVL, respectively. In this instance, the axis is −45°, indicating left axis deviation.

THE VECTORCARDIOGRAM

Given the standard electrocardiogram, it is evident that if the initial, mean, and terminal vector arrows are determined from the 12 scalar leads, a vector loop can be constructed in the frontal and horizontal planes. However, there is a type of electrocardiography called *vectorcardiography* which directly records vector loops using leads to record in the frontal, transverse, and sagittal planes (the latter plane transsects the body vertically from front to back, at right angles to the frontal plane; see Fig. 2.58). Vectorcardiography is sometimes used in the clinical setting to assist in the analysis of complex QRS patterns. A variety of lead systems has been developed (using numerous electrodes placed on the body torso, such as the Frank system involving 8 leads) in an effort to obtain three leads that are as close as possible to being orthogonal (at right angles) and equidistant from the heart. However, because of the complex shape of the body, no perfect lead system has yet been devised.

As the vector loops are inscribed (P, QRS, and T), a series of drop-shaped marks can be inscribed on a persistence oscilloscope, and the image can then be photographed. Each mark occurs at a given time interval (for example, .025 s), and the trailing edge of the drop is narrow, allowing ready determination of the duration and direction of various portions of the vector loops simply by counting the number of marks and noting the direction of the marks (for details, see Winsor, 1972). The vectorcardiogram is not useful for analyzing the heart's rhythm.

ABNORMALITIES OF THE QRS AMPLITUDE

Differences in amplitude of the QRS complexes from individual to individual can represent a gain or a loss of muscle mass, differences in the distance of the heart from the chestwall, or an accumulation of fluid in the pericardial sac (this produces low voltage or low amplitude QRS complexes). The amplitude of QRS complexes also can be increased by slowed conduction through the ventricles, which can lead to an abnormal direction and larger than normal amplitude of the depolarization wave due primarily to loss of simultaneous opposing forces when the affected force is late (see subsequent section). When increased QRS amplitude is due to increased muscle mass, it is called *ventricular hypertrophy* (see also Chapter 18). Hypertrophy of the left ventricle can occur, for example, with long-standing high blood pressure or with valvular heart disease (the left ventricular wall thickens and the chamber size may also enlarge). Hypertrophy of the right ventricle can occur when the pressure in the pulmonary artery is elevated due to lung disease, or when there is a congenital heart defect.

Right Ventricular Hypertrophy

Since the right ventricle lies anterior and to the right of the left ventricle, hypertrophy of that chamber would be expected to produce deviation of the mean electrical axis toward the right with an increase in the anterior forces. An example of right ventricular hypertrophy is shown in Figure 2.66 (*upper panels*). Notice that the mean electrical axis is approximately isoelectric to lead aVR and negative in Lead I, giving a mean QRS axis of +120° (right axis deviation). Notice also the tall R wave in lead V_1, which is abnormal and indicates a prominent anterior vector in the transverse plane. A valuable exercise is to plot the approximate frontal and horizontal plane QRS loops based on the initial, mean, and terminal forces of this electrocardiogram.

Left Ventricular Hypertrophy

This abnormality tends to produce a magnification of forces along the direction of the usual mean QRS vector generated primarily by the left ventricle. An electrocardiogram showing left ventricular hypertrophy is reproduced in Figure 2.66 (*lower panels*). Notice that the mean QRS axis points to the left, with the largest projection negatively on lead III, while aVL and aVF show equal positive and negative amplitudes; therefore, the mean QRS axis is approximately −60° (left axis deviation). The QRS voltage is larger than normal, being largest in V_2 in the transverse plane (large S wave). Again, a useful exercise is to plot the initial, terminal, and mean vectors and to draw the QRS loops in the frontal and horizontal planes.

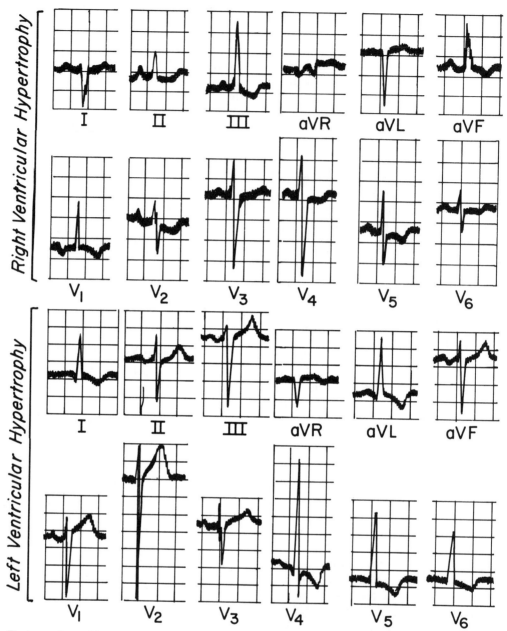

Figure 2.66. Electrocardiographic patterns of ventricular hypertrophy. Right ventricular hypertrophy (*upper panels*). The mean and terminal forces are shifted to the right and anteriorly. Left ventricular hypertrophy (*lower panels*). The initial and mean forces are shifted to the left and posteriorly.

Notice that the T waves are abnormally directed in this tracing, being opposite to the QRS in lead I and V_4, V_5, and V_6 as well as aVL. This indicates that in this condition, repolarization is following an abnormal direction, from endocardium to epicardium. These T wave changes are secondary to the hypertrophy, since the QRS is of relatively normal width and shows no evidence of damage to the muscle (see below).

MYOCARDIAL INFARCTION

When an area of heart muscle is damaged sec-

ondary to occlusion of a coronary artery, within a few hours the muscle cells in the involved region tend to die and depolarize completely (see Chapter 15). This leaves an electrical "window", with unopposed voltages generated by other regions of the heart producing abnormal deviation of the instantaneous vectors of the heart. Typically, death of the heart muscle ("myocardial infarction") leads to an initial vector pointing *away* from the location of the infarct, due to depolarization of normal muscle on the opposite side of the heart (away from the damaged area). There is also loss of QRS voltage due to

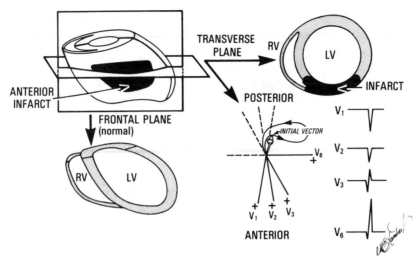

Figure 2.67. Diagrammatic representation of anterior myocardial infarction (infarction of the anterior wall of the left ventricle). As shown in the *upper left* and *lower left* diagrams the frontal plane does not pass through the infarct, and therefore the frontal plane leads are not greatly affected (a larger infarct would affect these leads). However, the transverse plane passes through the infarct region (*upper left* and *upper right* diagrams). This results in an initial vector in the precordial leads (in the transverse plane) which is directed posteriorly, in a direction opposite to normal, resulting in Q-waves in leads V_1, V_2, and V_3.

the death of tissue. These changes often persist in patients who recover from the heart attack because of the formation of a scar in the involved region.

Anterior Myocardial Infarction

The QRS complex is deformed by an anterior myocardial infarction, as shown in the simplified diagram in Figure 2.67. The location on the damaged area is indicated by the *black zone* on the frontal view of the left ventricle and on the transverse section through the left ventricle. Note that in the *frontal* plane, a small anterior infarct does not greatly affect the balance of the QRS forces since it is centered in the middle of the heart; therefore, the initial and mean QRS vectors, if plotted on the triaxial lead system, would result in relatively normal QRS complexes in leads I, II, and III. However, in the *transverse plane* (Fig. 2.67), the infarct can be seen to involve the interventricular septum and the front wall of the left ventricle and to greatly disturb the QRS. The initial vector on the transverse lead system points directly posteriorly, as the normal anterior septal force disappears and the unopposed normal forces on the posterior wall of the heart dominate. This results in the loss of the normal R wave in V_1 to V_3, with abnormal Q waves in those leads, whereas lead V_6 remains relatively normal (Fig. 2.67).

Inferior Myocardial Infarction

Damage to the inferior wall of the left ventricle by coronary occlusion is shown by the black zone on the frontal plane, so that the frontal plane

vectors are importantly affected (Fig. 2.68), but a transverse plane through the middle of the heart does not reveal the infarct (Fig. 2.68). The initial vector in the frontal plane points upward and to the left (away from the infarct), producing a Q wave in leads II and III and in lead aVF, since it projects on the negative poles of these three leads (Fig. 2.68). Thus, unopposed normal forces on the high left lateral wall of the heart dominate the early portion of the QRS complex in inferior myocardial infarction. In contrast, if the infarct is small, the QRS in leads recorded from the transverse plane can appear relatively normal.

A variety of other infarct locations resulting from the obstruction of various coronary arteries, or their branches, produce different locations of Q-waves in specific leads.

ABNORMAL CARDIAC REPOLARIZATION

As discussed in Chapter 8, a number of factors affect the rate of repolarization of individual cardiac muscle cells (phase 3 of the action potential), and such effects in the whole heart are reflected by changes in the T waves and ST segments of the ECG. For example, the T wave can be altered by an abnormal pathway of repolarization secondary to an abnormal pathway of depolarization (inverted T waves in Fig. 2.66), the QT interval is affected by an abnormal duration of repolarization, and the T wave and ST segment are influenced by electrolyte abnormalities, drugs, or tissue damage. These changes may affect repolarization directly, or through their effect on depolarization.

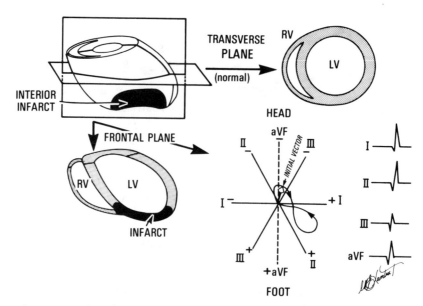

Figure 2.68. Diagrammatic representation of inferior myocardial infarction (infarction of the inferior or diaphragmatic wall of the left ventricle). The *left upper* and *right upper* diagrams show that the transverse plane does not pass through the infarct, and therefore the precordial leads remain normal. The *left upper* and *left lower* diagrams show that the frontal plane transects the infarct. Therefore in the hexaxial lead system, the initial QRS vector is abnormally directed superiorly and to the left, giving Q waves in leads II, III, and aVF.

Depression and Elevation of the ST Segments

Depression of the ST segments can occur transiently during angina pectoris (chest pain), due to inadequate coronary blood flow to the inner (subendocardial) wall of the left ventricle, causing partial depolarization of the deep layers of myocardium and leading to a potential difference between them and the uninjured superficial layers (see Chapter 15). During the acute phase of myocardial infarction, *elevation* of the ST segments occurs in those leads showing the area of damage (Q waves), which generally involves nearly the full thickness of the left ventricular wall. The explanation of ST segment elevation is based on the supposition that a boundary exists between a region of normal and damaged cells and that an abnormal current could flow between the two regions.

According to one theory (the theory of diastolic current of injury), the injured region is considered to be partially or completely depolarized at rest (with the damaged area therefore appearing electronegative with respect to the normal regions during diastole). A "current of injury" therefore flows during electrical diastole (Fig. 2.69, *upper left panel*). This current disappears when the entire heart is fully depolarized during electrical systole (Fig. 2.69, *upper right panel*). Partial diastolic depolarization has been documented by direct intracellular recordings from cells within a zone of ischemia, at a time when ST segment elevations were evident on an epicardial electrogram. In the ECG recorded from the body surface, the AC electronic circuitry com-

pensates for this base-line shift, which otherwise would be recorded as a depression of the segment between the end of the T wave and onset of the P wave (Figure 2.69, *upper left*). Therefore, with this explanation, the ST segment elevation during cardiac depolarization on the surface ECG is only apparent.

Another theory (systolic current of injury hypothesis) holds that a so-called "primary" or "true" ST segment elevation occurs during electrical systole due to failure of the injured area to depolarize adequately (or as a consequence of early repolarization) (Fig. 2.69, *lower panels*). According to this theory, the injury area is considered to be normally polarized at rest, and it remains polarized during electrical systole (phase 2 of the action potential), so that there is current flow during electrical systole with the injured region electropositive with respect to the depolarized normal zones (Fig. 2.69, *lower right panel*). This mechanism may play some role, but most of the evidence suggests that a diastolic current of injury is the dominant factor (for review, see Ross, 1976a). The effect most probably relates to loss of intracellular potassium causing local elevation of extracellular [K$^+$], which partially depolarizes the involved membranes during diastole.

Abnormal T Waves

T waves opposite in direction to the QRS often occur with ventricular hypertrophy due to abnormal spread of the repolarization process (see Fig. 2.66), and such T waves also can be seen when there is

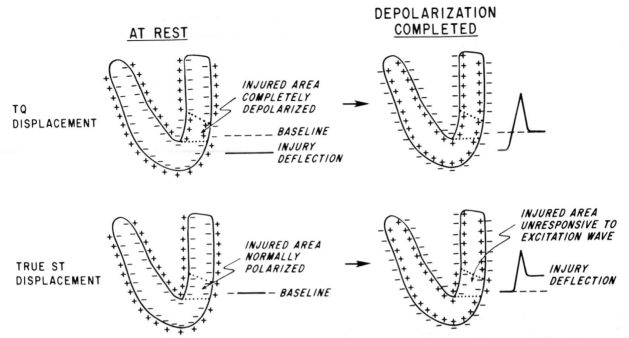

Figure 2.69. Diagram of potential mechanisms for ST segment displacement. In the upper two diagrams, at rest an injury current is produced in a lead facing the damaged area of the left ventricle *(dotted lines)* caused by partial depolarization of the cells (negative outer wall) during electrical diastole. This causes a displacement of the baseline at rest between the end of the T wave and the onset of the QRS (TQ displacement). During ventricular depolarization (the plateau phase of the electrogram), the transmembrane potential becomes zero (all cells are depolarized), and it appears as if the ST segment is elevated. Another theory *(lower two panels)* holds that the injured area is normally polarized during the rest phase of the cycle, so that the baseline is not displaced. However, during ventricular depolarization the injured area is not depolarized, so that a relatively positive area faces the recording electrode during the action potential *(right lower panel)*, resulting in a "true" ST segment elevation. (Abnormally early repolarization of the injured area, while other cells are still depolarized, could also produce such an effect). [From Ross (1976).]

an abnormal duration of repolarization, as with block in one of the bundle branches of the left ventricle or right ventricle (see below). In the bundle branch block, the depolarization wave spreads slowly through the muscle of the chamber supplied by the involved bundle branch, rather than through the specialized conduction system, and the wave of repolarization is similarly delayed and dispersed, spreading abnormally from endocardium to epicardium. Under these circumstances, the T wave often is very large as well, as it spreads slowly over the involved ventricle; this relates to the lack of normal canceling forces, which occurred earlier as the opposite ventricle was normally depolarized and repolarized. A variety of other T wave changes occur in disease of the heart muscle, the pericardium, and in myocardial infarction, but discussion of these is beyond the scope of this text.

Drugs and Electrolytes

Digitalis can alter the ST segments and T waves (Fig. 2.70), and antiarrhythmic drugs of the quinidine type reduce P_k, thereby prolonging the action potential duration and lengthening the QT interval (Fig. 2.70). High serum Ca^{++} concentration shortens phase 2 of the action potential and the QT interval; low Ca^{++} decreases P_k, lengthens phase 2,

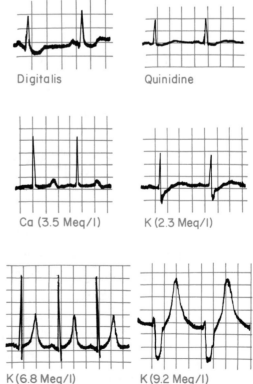

Figure 2.70. Patterns on the electrocardiogram of digitalis and quinidine effect, low serum calcium, low serum potassium, and high serum potassium.

and slows the rate of repolarization, thereby prolonging the QT interval (Fig. 2.70). Low serum K^+ reduces P_k and reduces the amplitude of the T wave, with prolongation of the QT interval and appearance of a prominent U wave which may resemble the T wave (Fig. 2.70). High serum K^+ shortens action potential duration as P_k is increased, leading to a short QT interval with peaked, tall T waves (Fig. 2.70). With a very high concentration of serum K^+, sufficient membrane depolarization takes place to impair conduction; the PR interval prolongs and the QRS widens to merge with the tall peaked T wave (Fig. 2.70).

ABNORMALITIES OF CARDIAC RHYTHM AND CONDUCTION

When the P wave is of normal configuration, each P wave is conducted to the ventricles, and the resting heart rate is within the range of 60–100 beats/min, *normal sinus rhythm* is said to exist (Fig. 2.71). Regardless of the mechanism of rhythm disturbance, when the heart rate exceeds 100 beats/min *tachycardia* is present, and when it is below 60 beats/min *bradycardia* is present. There is normally a fluctuation of heart rate that is synchronous with the respiratory cycle, and on taking the pulse during an inspiration, the pulse accelerates slightly whereas, upon breathing out, it slows. This normal response is called *sinus arrhythmia* (Fig. 2.71). (The

mechanism is related to stimulation of stretch receptors in the aorta and carotid arteries as the blood pressure changes, and perhaps to stimulation of receptors in the heart itself as well, which reflexly affect the heart rate as described in Chapter 16.)

Abnormalities of Impulse Formation and Cardiac Rhythm

Abnormal cardiac rhythms (called "arrhythmias" or "dysrhythmias") can be produced by a number of mechanisms. Such mechanisms are not well understood in the whole heart in many instances, but certain phenomena appear important in causing some of them. These mechanisms include: (1) reentry; (2) firing of an abnormal (ectopic) focus, which can occur intermittently as a single impulse to disturb the normal cardiac rhythm or, sometimes, can discharge continuously; and (3) "triggered automaticity," which sometimes occurs when an action potential is followed by an after potential that is of sufficient amplitude to reach threshold.

Reentry can be responsible for single early (premature) depolarizations, often referred to as "premature beats," of either the ventricles or the atria. It can also be the mechanism for a sustained rapid rhythm in either the atria or the ventricles. Reentry can occur at a number of sites in the heart, wherever adjacent tissues having different properties are available to set up a circular pathway (such as

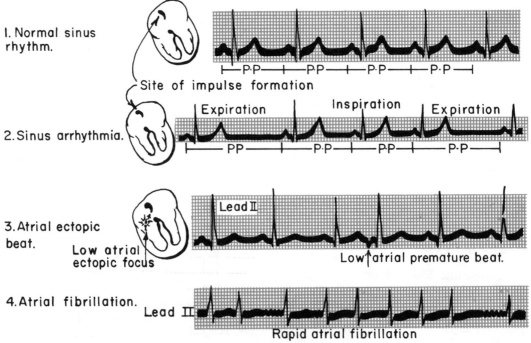

Figure 2.71. Normal and abnormal electrocardiograms. *1,* Normal sinus rhythm. *2,* Sinus arrhythmia. *3,* Atrial ectopic beat. *4,* Atrial fibrillation. The intervals between P waves (*PP*) are marked. [From Burch and Winsor (1966).]

between the atrium and the AV node, or between a zone of acute myocardial infarction and a normal region). One commonly used model for reentry has been a branching Purkinje fiber attached to the ventricular myocardium (Fig. 2.72). Conditions required for some types of reentry are: *slowed conduction in one region* and *unidirectional block.* In this diagram, the impulse normally would travel rapidly through both Purkinje fiber bundles (A and B) from the main bundle (MB) to rapidly and simultaneously depolarize the ventricular muscle (VM). As shown in Figure 2.50 (Chapter 8), action potentials are of longer duration in the His-Purkinje system than in the ventricular muscle, so that the specialized fibers are normally refractory to reentry from the muscle. The *stippled area* in Figure 2.72 represents a zone of abnormal conduction in branch B, which is refractory to the antegrade impulse and markedly slows or blocks it. The impulse now travels only through branch A to depolarize the ventricle, but as the impulse travels through the muscle it finds branch B not refractory, and it enters this branch retrogradely (Fig. 2.72) (branch B exhibits unidirectional block). The impulse is then able to travel through the *stippled area* in a retrograde manner to provide early re-excitation (reentry) of the main bundle, and it also may reenter branch A in a circular movement. The impulses recorded from cells directly in such branches are shown in the *upper right* tracings of Figure 2.72, and the electrocardiogram (recorded below) shows a wide, early QRS complex following each normal QRS (the latter are preceded by P waves). The broad QRS indicates a ventricular ectopic beat (premature ventricular depolarization), with the early impulse widened because it must travel slowly by an abnormal route. Such reentry mechanisms can also be rapid and repetitive (a circus movement) and can take over the ventricular rhythm.

Another region where reentry can occur under certain conditions is in the upper region of the AV junction. Here, an abnormal degree of decremental conduction can allow impulses to be delayed for a sufficient period of time for the atrial muscle to recover its excitability, and the impulse can then reenter the atrium, causing a premature atrial beat or a sustained tachycardia. Another example occurs when a congenital accessory bundle connects the atria and ventricles, bypassing the AV junction. This bundle has a much shorter refractory period than the AV junction, so that normally the impulse goes antegradely from the atrium to the ventricle through *both* the AV junction and the accessory

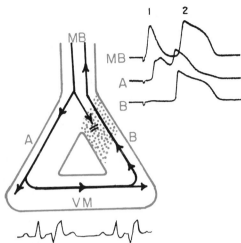

Figure 2.72. Reentry phenomenon as recorded in a loop of Purkinje fiber bundles (*A* and *B*) and ventricular muscle (*VM*). The shaded area in branch B represents an area of unidirectional block. Conduction through this area along B toward VM is blocked, but an impulse arriving later from VM to B can be conducted retrogradely. Conduction throughout the loop is slowed. At *top right*, transmembrane action potentials from a similar arrangemnt of canine Purkinje fibers are shown. Conduction was delayed with elevated K⁺ and catecholamines. *MB,* action potential from the main bundle leading into the loop. Action potential *1* in trace MB results from antegrade conduction in this bundle. *A,* action potential recorded as the impulse propagates through branch A. *B,* action potential recorded as the impulse is conduced retrogradely through branch B. Action potential *2* in MB is the retrogradely conducted reentrant impulse which is recorded further up MB and appears later than the second (reentry) impulse in branch A. The *bottom trace* shows coupled ventricular ectopic systoles following each normally conducted complex which might result from this type of reentry. [From Wit et al. (1974).]

pathway. However, if a premature beat occurs for any reason, since the accessory pathway recovers its excitability early, it may conduct the premature beat in a retrograde fashion to the atrium. This will set up a rapid circular movement, forward through the AV junction and retrograde through the accessory pathway, leading to a tachycardia.†

Atrial dysrhythmias can be of many types, but only four will be discussed briefly. *Atrial premature contractions* are identified as an earlier than normal P wave (often of unusual shape; see Fig. 2.71) which is usually followed by a *normal* QRS. The next (normal) atrial impulse then arrives at about the normal interval after the premature P wave, so that there is no additional pause after the premature beat. (Compare with ventricular premature contractions.) *Atrial tachycardia* (called "supraventricular tachycardia" if P waves cannot clearly be iden-

---

† This condition is called the Wolff-Parkinson-White syndrome.

tified) often begins and ends suddenly ("paroxysmal tachycardia"), and in many instances the mechanism of this dysrhythmia has been found to be reentry through the AV junction (see AV junctional premature beat, Fig. 2.73). The ventricular rate in such a dysrhythmia is frequently 150–200 beats/min. *Atrial flutter* is an even more rapid dysrhythmia, which is most likely due to a circular movement of the depolarizing wave around a damaged central zone. This dysrhythmia tends to produce a "sawtooth" base-line depicting a regular atrial rate of about 300/min. The many impulses produce increased refractoriness of the AV junction so that the ventricles often beat at a slower rate than the atria (often at 150/min). Therapy of the above three dysrhythmias ordinarily involves either slowing of conduction in the normal pathway, improving or blocking conduction through the abnormal pathway, or both, using drugs such as digitalis or quinidine.

Atrial fibrillation (Fig. 2.71) is a common dysrhythmia and, provided the ventricular response is not too fast, it is quite compatible with a long life. Usually it occurs in the presence of heart disease, often with enlargement of one or both atria, but sometimes it is seen in "paroxysmal" form in otherwise normal subjects. Fibrillation of the atria is probably related to multiple small reentry loops (sometimes called "microreentry") producing completely disordered, very rapid depolarizations of the atria; this causes them to simply quiver, rather than to contract effectively (with loss of the atrial booster pump). These depolarizations of atrial fibrillation occur at rates of 500–800/min, sometimes being difficult to discern on the ECG base-line (Fig. 2.71) and, being totally, irregular, they bombard the AV junction to cause varying degrees of decremental conduction, with most impulses failing to penetrate to the ventricles (see also discussion of AV node refractoriness in Chapter 8). The result is a totally irregular ventricular response which can actually be slower than that in atrial flutter because the more rapid bombardment of the AV junction in atrial fibrillation increases AV node refractoriness (Chapter 8). Once through the AV junction, the impulses travel by the normal conduction system; the QRS complexes therefore appear normal. The arterial *pulse* is irregular and of varying amplitude, not only because of the different times available between beats for ventricular filling (varying end-diastolic volume causes a variable stroke volume) but also because of the effect of changing rate on the force of contraction (force-frequency relation; see Chapter 12). Atrial fibrillation is often treated

by attempting to slow the ventricular rate to the normal range with digitalis (which exerts a central vagal effect on the AV junction, increasing its refractoriness) or with β-adrenergic blocking drugs (which antagonize sympathetic effects on the AV junction). It can also be terminated by applying a brief electrical shock across the chest (this approach can also be used to treat other arrhythmias). This electrical "countershock" momentarily depolarizes all of the cells in the heart, allowing the sinus node to fire and conduct normally, thereby abolishing the dysrhythmia. Recurrence may then be prevented in many (not all) cases by the use of an antiarrhythmic drug such as quinidine.

Ventricular dysrhythmias are also of many types, and only three important abnormalities will be discussed. *Ventricular premature beats* can be single or multiple, can originate in one or more different locations (hence, giving different configurations to the QRS), or they may occur regularly after every normal beat (bigeminy). The most common type of ventricular premature beat occurs at a fixed period after the prior normal QRS (suggesting either triggered automaticity or a reentry mechanism). Such a beat is shown in Figure 2.73; the impulse arises from the ectopic focus or reentry site in the ventricles or Purkinje system and spreads slowly through the ventricular muscle giving a wide and bizarre QRS configuration. Figure 2.74 illustrates normal conduction from the SA node through the atrium (A), corresponding to the P waves, conduction through the AV node, and ventricular (V) depolarization occurring rapidly to give normal QRS complexes (first three normal cycles). A ventricular premature depolarization then occurs (PVD, often called also a ventricular premature contraction, or VPC), and spreads in retrograde fashion, giving the widened QRS which is closely "coupled" to the third normal QRS. Notice that the *next* sinus impulse occurs on schedule (P) but that its P wave is obscured by the premature beat (*), and the normal QRS complex does not occur because the impulse enters the AV junction at a time when it is refractory because of previous retrograde penetration of the AV junction by the premature ventricular impulse (Fig. 2.74). The following P wave again occurs on schedule and is followed by a normal QRS. Notice that in contrast to a premature atrial beat, which depolarizes the atrium and "resets" the SA node (Fig. 2.71), the sinoatrial node usually is *not* reset after a ventricular premature beat. Therefore, the interval between the two normal QRS cycles bracketing the premature beat is equal to exactly two normal cycle lengths, leading to a so-

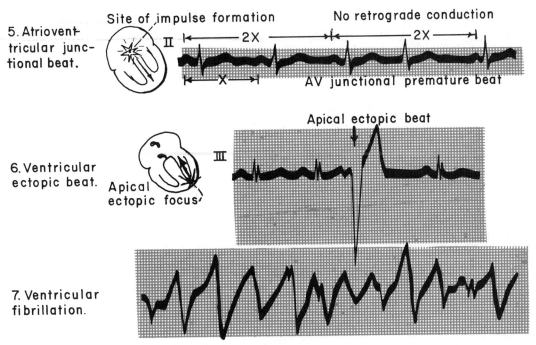

Figure 2.73. Abnormal electrocardiograms. *5,* Atrioventricular junctional beat, *6,* Ventricular ectopic beat originating at the apical region of the left ventricle. *7,* Ventricular fibrillation. [From Burch and Winsor (1966).]

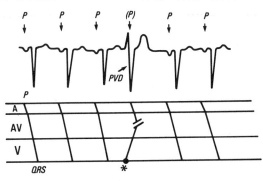

Figure 2.74. Ladder diagram showing a premature ventricular depolarization (*PVD*) (or premature ventricular contraction, shown by the *asterisk* and the wide QRS on the electrocardiogram, *top tracing*). The ladder diagram shows conduction through the atrium (*A*), the AV junction (*AV*), and the ventricular myocardium (*V*). Notice that the PVD originates in the ventricle and spreads slowly in a retrograde manner, partially penetrating the AV junction. There it collides with the impulse that originated from the normal P wave (which is obscured in the abnormal ventricular complex), thereby blocking antegrade conduction of the normal P wave through the AV junction. The next P wave then occurs on schedule exactly two cycles after the last normally conducted P wave, yielding a "compensatory pause." For further discussion see text.

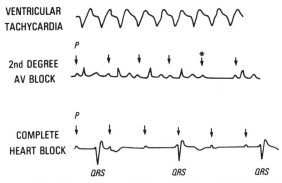

Figure 2.75. Diagrammatic tracings showing ventricular tachycardia (*upper panel*), second degree (2°) AV block of the Wenkebach type (*middle panel; asterisk* shows nonconducted P wave), and third degree (3°) AV block (complete heart block, *lower panel*).

called "compensatory pause" after the premature ventricular contraction (Fig. 2.73 and 2.74). (A similar phenomenon is seen with the AV junctional premature beat which fails to depolarize the atrium in Fig. 2.73). In some settings, such as after a heart attack, ventricular premature beats can lead to more serious ventricular rhythm disturbances (see below), and they can be suppressed by chronic oral administration of an antiarrhythmic drug.

Ventricular tachycardia is a rapid series of regular ventricular ectopic beats showing widened bizarre QRS complexes (that is, a series of rapid premature ventricular beats) (Fig. 2.75). Ventricular tachycardia often occurs at a rate of 140–170/min, and it may be sustained or it may be paroxysmal. Because of the abnormal sequence of contraction of the ventricles, and the lack of a synchronized atrial impulse, the blood pressure and cardiac output often drop sharply during this dysrhythmia; the dysrhythmia is also dangerous, because it tends to degenerate into ventricular fibrillation. Episodes of ventricular tachycardia (occurring, for example, after a heart attack) are often treated by drugs which shorten the ratio of the relative refractory period to the action potential duration, and decrease membrane responsiveness (such as lidocaine, quinidine, or procaine amide). If this approach is not effective, electrical countershock is applied to the chest wall. *Ventricular fibrillation* is a totally disorganized, very rapid rhythm of the ventricles similar to fibrillation of the atria (Fig. 2.73). It leads to totally ineffective pumping of the ventricles (which quiver at a rate several hundred times/ minute), and it is rapidly fatal, if not treated promptly, since brain death will occur within a few minutes. Ventricular fibrillation (as well as complete cessation of ventricular activity) is termed "cardiac arrest." When this occurs cardiopulmonary resuscitation almost always will be required, using mouth-to-mouth breathing and chest compression to support the circulation (see Section 5, Chapter 40), followed by *electrical countershock* with an electrical defibrillator.

Abnormalities of Conduction through the Specialized System

Partial or complete "block" can occur at any location in the specialized conduction system, and only the most important types will be mentioned briefly.

Disease of the SA node may involve "sinus arrest," or occasionally "exit block" from the sinus node in which the impulse is unable to penetrate outside the nodal region. These two conditions lead to absence of P waves on the electrocardiogram and, generally, the AV junction takes over as the heart's pacemaker.

Partial and complete heart block are due to disease of the AV junction or His-Purkinje system which impairs conduction between the atria and ventricles. "First degree AV block" refers to simple prolongation of the PR interval beyond 0.2 s. Such block is usually in the AV junction, but sometimes it is due to delayed conduction through the His-Purkinje system. "Second degree AV block" occurs when *some* of the impulses from the atria do not pass through the AV junction of His-Purkinje System to the ventricles. The PR interval is usually prolonged and, sometimes, every other QRS complex is "dropped" to produce 2:1 AV block; sometimes, there is progressive prolongation of the PR interval, followed by a dropped QRS, the so-called "Wenckebach type" of second degree AV block (Fig. 2.75). In *complete heart block* or "third degree AV block" there is *no* conduction from the atria to the ventricles (Fig. 2.75). In this condition, the ventric-

ular rate is usually slow (about 40/min) and the QRS is widened, since it originates from a pacemaker in the Purkinje fibers, while the nonconducted P waves proceed at a normal rate. Patients who survive with this abnormality may be quite limited and may be subject to fainting spells ("Stokes-Adams attacks") (see Chapter 8, p. 161, for the differentiation of impaired conduction due to block in the AV junction from that due to block in the His-Purkinje system, using His bundle electrocardiography).

The usual treatment for serious disease of the SA node, for permanent severe second degree AV block, or for complete heart block, is the implantation of an electronic cardiac pacemaker with its tip positioned in the right ventricle. The pacemaker is set to fire impulses at a normal rate. For patients in whom complete heart block is intermittent, pacemakers are available which fire impulses at a regular rate only when the patient's own pacemaker slows below a preset rate (so-called "demand pacemakers").

Right bundle branch block occurs when there is damage to the right branch of the specialized conduction system. It results in a wide QRS complex (0.12 s or more) because the right ventricular depolarization wave must pass slowly through the muscle, rather than through the conduction system. The vectors and representative scalar leads for this abnormality are shown in Figure 2.76. Since the

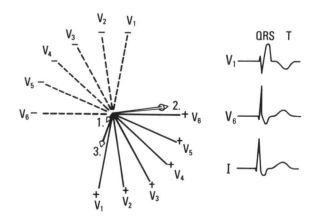

Figure 2.76. Right bundle branch block. This conduction disturbance is diagrammed in the transverse plane, showing 3 ECG leads. In the initial vector (*1*) septal depolarization is normally directed anteriorly to the right and the mean QRS vector (*2*) is also directed normally. However, the terminal vector (*3*) is delayed as the right ventricle is depolarized by spreading through the muscle rather than through the right bundle branch. It is unopposed and therefore a large terminal force directed anteriorly and to the right, resulting in a large late positive wave in lead V_1 (rSR′) and a late negative S wave in lead V_6. Since both lead V_6 and lead I are in the frontal plane, there is also a broad QRS in lead I due to a wide S wave representing the delayed rightward depolarization of the right ventricle.

initial septal forces in the normal ECG originate prior to depolarization of the right bundle, the initial septal depolarization is normal (vector 1, Fig. 2.76), and the QRS forces are normally directed initially toward the left ventricle (the left bundle branch is not involved) (vector 2, Fig. 2.76). However, the impulse then spreads relatively slowly through the muscle to reach the right ventricle, circumventing the damaged right bundle branch, and this causes an abnormal rightward and anterior force which is large and unopposed late in the QRS complex, giving a wide S wave (vector 3, Fig. 2.76). The loop in the transverse plane shows a normal initial vector and a large, late anterior and rightward force leading to an rSR′ in lead V_1 (conventionally, as mentioned earlier, the second of two waves is marked "prime"; the smaller and larger of two waves are written with small and large letters, respectively) (Fig. 2.76). Lead V_6, like the frontal plane loop in lead I, shows a terminal vector which is slowed and directed to the right, giving a large, late S wave (Fig. 2.76).

Left bundle branch block results from damage to the left branch of the conduction system and also results in a broadened QRS complex (0.12 s or more). As shown in Figure 2.77 the initial vector is abnormal, since septal and left ventricular depolarization must now originate from the right ventricle (the right bundle branch is intact), and the impulse then spreads from right to left and posteriorly (vector 1, Fig. 2.77). The prolonged late conduction through the muscle of the left ventricle leads to

TRANSVERSE PLANE

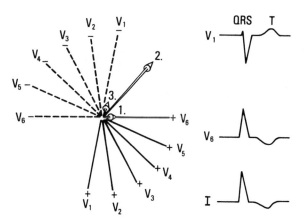

Figure 2.77. Left bundle branch block. This conduction disturbance is diagrammed in the transverse plane. The abnormal initial vector (*1*) is directed to the left, and delayed, unopposed depolarization of the left ventricle then causes a larger than normal mean and terminal QRS vector (*2* and *3*) directed to the left and posteriorly. This results in a broad QRS with an abnormally small (or absent) R wave in lead V_1 and an absent septal Q wave in lead V_6. Lead V_6 resembles the tracing in lead I on the frontal plane.

unopposed forces and an enlarged mean vector (vector 2) directed more posteriorly than normal. The result is no R wave (or only a tiny R wave) in lead V_1 with a deep wide S wave, and the absence of a septal Q wave in lead V_6 with a tall wide R wave (Fig. 2.77). In the frontal plane, the tracing in lead I resembles that of lead V_6 (Fig. 2.77).

CHAPTER 10

Cardiac Muscle: Cardiac Structure-Function Relations and Excitation-Contraction Coupling

The structure of heart muscle (myocardium) allows its function to be regulated by several intrinsic mechanisms (in addition to the autonomic nervous system). Before considering how various forms of physiologic stress or drugs can affect the whole heart, it is important to understand certain of these basic structure-function relations in *isolated cardiac muscle*, as well as how various external factors affect its performance (Chapter 11). The heartbeat is largely controlled by "involuntary" mechanisms, some of which are intrinsic to the muscle itself and independent of neurohumoral influences. Thus, the force of contraction can be modified by such stimuli as the resting length of the muscle, the frequency of electrical stimulation, and the concentration of various ions. These intrinsic mechanisms that govern cardiac muscle contraction are the subject of this chapter.

THE ANATOMY OF MYOCARDIUM

The muscular walls of the mammalian heart are composed of a branching network of muscle fibers or cells. The working muscle cells in the human heart are approximately 10–20 μm wide and 50–100 μm long (Fig. 2.78A). Each muscle cell is dominantly occupied by bundles of myofilaments, each bundle being organized into a series of repeating subunits called sarcomeres. These band-like patterns are responsible for the striated nature of both skeletal and cardiac muscle, as observed under the light microscope. By electron microscopy, the sarcomere can be seen to be composed of thick and thin myofilaments, the thick myosin filaments forming the dark A band and thin actin filaments constituting the light I bands. The I bands of each adjacent sarcomere are anchored at the Z-line (Fig. 2.78B and C). On cross-section through the A band,

it can be seen that there is an overlap of thick and thin filaments, the thin filaments forming a hexagonal array around each myosin filament, whereas a cross-section through the I band shows only thin filaments (Fig. 2.78D). From the ends of the two Z-lines bounding each sarcomere, the thin filaments protrude toward the center of the sarcomere, each one measuring about 1 μm in length, while the myosin filament is approximately 1.5 μm in length; the center of the myosin filament shows a widened area which connects adjacent thick filaments to form the M-line (Fig. 2.78C). (For review of ultrastructure see Legato, 1969; McNutt and Fawcett, 1974; Summer and Johnson, 1979; James et al., 1982.) The thick and thin filaments slide back and forth during contraction and relaxation, much as two hair brushes interdigitate when pushed together. The molecular structure of cardiac actin and myosin resembles that of skeletal muscle (for details, see Section 1, Chapter 4, and also Hanson and Long (1965) and Katz (1977)). Other cardiac cells, including the cells in the nodal regions and the Purkinje cells of the conduction system, contain relatively few myofibrils (James et al., 1982).

The myocardial cells are connected to one another by special junctions called intercalated disks (Fig. 2.79). The intercalated disk exhibits a differing structure during its course between cells, the thick transverse portions (fascia adherens) being composed of Z-substance, and undoubtedly serving to transmit force from cell to cell, whereas other (longitudinally placed) regions are often much thinner and in some regions exhibit so-called "nexus junctions" (Fig. 2.79). The latter regions have a particular structure shown on freeze cleaving with scanning electron microscopy to consist of a series of interdigitating pits and projections, each of

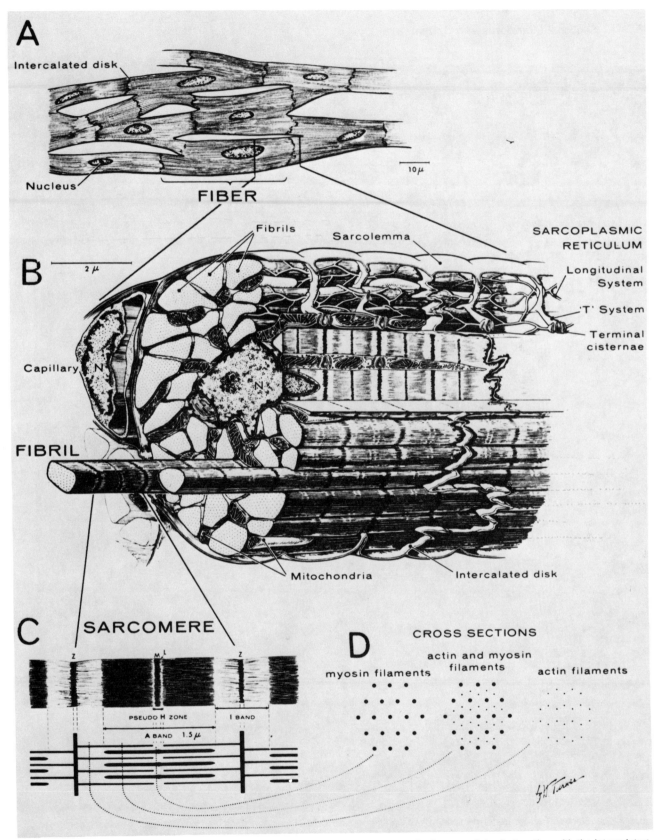

Figure 2.78. Diagram of myocardial structure. *Panel A* shows the branching network of myocardial cells together with the intercalated discs separating individual cells (fibers). The two vertical bars show the section of the cell with the tip of its nucleus magnified in *panel B* to show the myofibrils, transverse and longitudinal tubules. In *panel C* a sarcomeric unit of a myofibril is further magnified to show that the dark and light striations constituting the A and I bands, respectively, result from the arrangement of the actin and myosin filaments. In *Panel D*, three cross-sections through the sarcomere are shown. One through the light I band shows only actin filaments in a hexagonal array, one through the A band (center) shows thick myosin filaments in the central regions of the hexagonal actin filament array, and a section through the light zone adjacent to the central M line (*panel C*) shows only myosin filaments. [From Braunwald et al. (1976).]

Figure 2.79.   Electron micrograph of portions of two cardiac muscle cells showing in particular the intercalated disc. *S*, sarcomere; *ECS*, extracellular space (*large vertical arrowhead* shows origin of the boundary between the two adjacent cells); *Pm*, plasma membrane; *Pl*, internal plasma membrane; *M*, mitochondrion; *G*, Golgi apparatus. The features of the intercalated disc, which runs both in transverse and longitudinal directions relative to the long axis of the cell, are as follows: *FA*, fascia adherens (the thickened region of the intercalated disc into which the myofilaments insert and which serves to transmit force from cell to cell); *NSP*, a widened portion of the intercalated disk; *D*, desmosome; *N*, the nexus junction (the so-called "gap injunction" which appears to serve for electrical conduction from cell to cell). [From Legato (1973).]

which contains a hole; it is thought that these areas may allow communication of cytoplasm from cell to cell and serve as low impedance pathways for rapid transmission of the electrical impulse throughout the network of heart muscle cells (McNutt and Weinstein, 1973). The desmosomes (Fig. 2.79) may serve a similar function. The branched structure of cardiac muscle facilitates this process, in contrast to skeletal muscle which is composed of nonbranched muscle cell (motor) units which are activated by voluntary nerve endings (Section 1, Chapter 4).

The myocardium is rich in mitochondria which comprise 25–30% of the myocardium (many more than in skeletal muscle), and glycogen stores are plentiful (Figs. 2.78 and 2.79).

THE SARCOMERE LENGTH-TENSION RELATION

The sliding filament model for muscle was first developed in 1958 for frog skeletal muscle by Huxley and Huxley, and Huxley and Niedergierke (for review see Huxley, 1974). Much experimental evidence supports this model, which states that during contraction and relaxation of muscle the myofilaments remain at *constant length* as they slide back and forth between one another.

Relation between Active Tension and Length

It has been observed in single frog skeletal muscle fibers that there is a relation between the length to which the sarcomeres within the fiber are stretched under resting conditions and the tension developed by the fiber when it is then stimulated electrically to contract. The relation shows that as sarcomere length (the distance from Z line to Z line) is increased from about 1.7 to 2.2 μm the developed tension increases (the so-called "ascending limb") (Fig. 2.80); between 2 and 2.2 μm there is a plateau of the curve, and between 2.2 and 3.5 μm tension falls as sarcomere length increases (the so-called "descending limb") (Fig. 2.80). As in skeletal muscle, a sarcomere length-tension relationship can also be demonstrated in intact heart muscle (Fig. 2.81), but a descending limb is not evident in living muscle (Julian, 1975). This finding is related primarily to the greater stiffness of heart muscle in the relaxed, unstimulated state, as discussed further below.

The shape of the sarcomere length-tension relation in skeletal muscle has been attributed to the degree of overlap of active sites or cross-bridges on the thick and thin filaments; thus, it is known that the central 0.2 μm region of the thick filament is

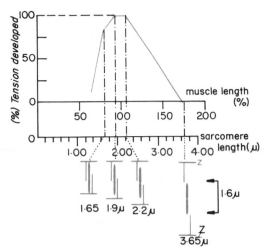

Figure 2.80. Sarcomere length in relation to tension development for skeletal muscle. Each thin (actin) filament is 1.0 μ in length, and in frog skeletal muscle the thick (myosin) filaments are 1.65 μ in length. Developed tension is constant between sarcomere lengths of 2.0 and 2.2 μ, and falls at sarcomere lengths below 2.0 μ or above 2.2 μ. [Redrawn from Hanson and Lowy (1965).]

free of active sites, thereby explaining the plateau of the length-tension relation in single fibers (Fig. 2.80). The mechanism for the descending limb observed in skeletal muscle can be explained by the progressive decrease in overlap of the regions of active sites on the thick and thin filaments as the actin filaments are progressively disengaged (Fig. 2.80), and tension reaches zero in single fibers at a sarcomere length of approximately 3.5 μm where no overlap exists (Fig. 2.80). On the other hand, the explanation for the ascending limb both in skeletal and cardiac muscle is much less clear. In the region of sarcomere lengths between 1.5 and 2.0 μm, there is "double overlap" of thin filaments with thick filaments, since below 2.0 μm these 1 μm long thin filaments cross the M-line and interdigitate with each other (Fig. 2.80). This double overlap can be seen as a darker region in the A band at short sarcomere lengths on electron micrographs. The rise of tension on the ascending limb has been attributed to lessening of interference with tension development by progressively decreasing extent of double overlap of the actin filaments; also, the possibility of wider separation between thick and thin filaments due to cell shape changes with increasing distance in shorter cells has been postulated to play a role. There is also evidence, particularly in isolated skeletal muscle, that as muscle length is increased along the ascending limb of function, there is increased activation of the muscle and vice versa (Jewell, 1977). That is, the muscle may be partially "deactivated" due to reduced calcium release at shorter muscle lengths (see Ciba Symposium on Starling's law, 1974). There are no data on this mechanism in the whole heart, however.

Regardless of the mechanism for the ascending limb, the phenomenon of increasing sarcomere length associated with increased force of muscle contraction forms the ultrastructural basis for the **Frank-Starling relation** or "Starling's law of the heart," one of the intrinsic mechanisms regulating the function of the whole heart.

Relations between Passive (Resting) Tension and Muscle Length

In skeletal muscle, tension does not rise appreciably as the resting muscle is stretched (Fig. 2.80) until sarcomere lengths exceed approximately 3 μm (Spiro and Sonnenblick, 1964). However, in isolated cardiac muscle (as well as in the whole heart), as the resting muscle is passively stretched, tension in the muscle initially rises slowly and then increases rapidly above sarcomere lengths of 1.8 to 2 μm (Fig. 2.81). Resting tension in heart muscle is relatively low at physiological muscle lengths (sar-

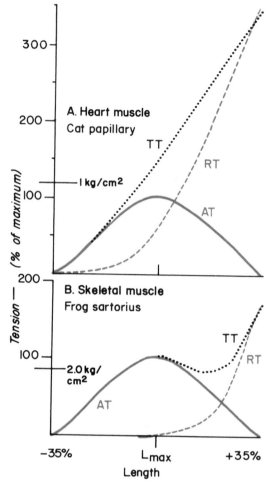

Figure 2.82. Length-tension diagrams for heart (*A*) and skeletal (*B*) muscle compared. *RT*, resting tension; *AT*, actively developed tension; *TT*, total tension (AT + RT). [Redrawn from Spiro and Sonnenblick (1964).]

comeres below 2.2 μm), the unactivated resting muscle being freely extensible, but the muscle becomes extremely stiff when it is stretched further (Spiro and Sonnenblick, 1964; Spotnitz et al., 1966). Thus, there is limit in the degree to which cardiac muscle can be stretched, since at sarcomere lengths of 2.3 to 2.4 μm any further stretch is resisted by the very high resting force.

As we shall see, this is a useful property since it prevents overdistension of the heart during changes in the return of blood to the heart (otherwise, the heart might readily be forced into a "descending limb" of function). This ability to develop high resting tension may relate to the connective tissue content of heart muscle and its branching interconnecting structure (which differ from skeletal muscle). Thus, in skeletal muscle which is attached to bone, the resting tension is still low at long sarcomere lengths, and a descending limb of active tension is readily produced (Fig. 2.82). Other structural and functional differences between skeletal and

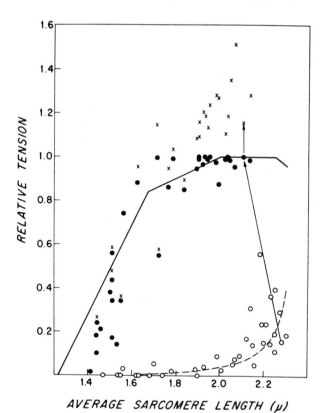

Figure 2.81. Sarcomere length-tension relation in living papillary muscles of the rat. The resting length-tension relation is shown by the *open circles*, the total active tension developed by the *xs*, and the developed tension by the *closed circles*. Note the absence of a descending limb of the length-active tension relation [From Julian and Solling (1975).]

Table 2.6.
A comparison between skeletal and cardiac muscle.

	Skeletal Muscle	Cardiac Muscle
1. Innervation and structure	a. Motor end plates (Ach) and motor units	a. Functional syncytium with propagated impulse throughout
	b. No autonomic fibers to muscle	b. Sympathetic and vagal fibers to muscle
	c. Small T tubules	c. Large T tubules and cell to cell electrical transmission
2. Alterations in performance	a. Length-tension relation	a. Length-tension relation
	b. Ability to summate contractions	b. Cannot summate contractions
	c. Force-velocity relation	c. Force-velocity relation
	d. Recruitment of more fibers	d. Cannot recruit more fibers
	e. Does not alter inotropic state (except temperature)	e. Inotropic state readily altered (can shift length-tension and force-velocity curves).
	f. Contractility not responsive to external Ca^{++} and drugs	f. Contractility responsive to external Ca^{++}, drugs, catecholamines
3. Stress-strain curve resting muscle	Compliant at normal resting length	Relatively noncompliant at normal resting length

cardiac muscle are shown in Table 2.6 (for review of skeletal muscle physiology see Section 1, Chapter 4, and Buller, 1975).

EXCITATION-CONTRACTION COUPLING

The German physiologist Ringer discovered in the 19th century that when calcium was removed from the medium bathing an isolated frog heart, electrical activity recorded by an electrogram from the heart was not affected, but contraction of the heart ceased. The calcium ion (Ca^{++}) is now known to be the key link in the coupling of electrical excitation of muscle to its mechanical contraction (see also Section 1, Chapter 4). Subsequently, it was shown that local electrical stimulation of the sarcolemma of a muscle fiber at a region near the Z-line produced local contraction of adjacent sarcomeres (Huxley, 1959), and also that insertion of a micropipette through the sarcolemma with local infusion of calcium within the cell caused local contraction of nearby myofibrils. (For general reviews on calcium and excitation-contraction coupling in muscle see Ebashi, 1976; Fozzard, 1977; and Winegrad, 1979.)

Further insight into how the concentration of calcium affects contraction is provided by studying the ATPase activity of reconstituted cardiac actomyosin made with actin containing the tropomyosin/troponin complex (Fig. 2.83). When the calcium concentration is changed from about 10^{-7} to 10^{-5} M, myosin ATPase is markedly stimulated. This contrasts with the calcium sensitivity of myosin ATPase alone (Fig. 2.83) and illustrates that troponin and tropomyosin confer "calcium sensitivity" to the ATPase (Ebashi, 1976). As described in Section 1, Chapter 4, the long rod-shaped molecule of tropomyosin is anchored to actin along with the

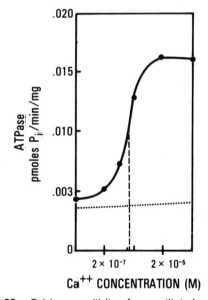

Figure 2.83. Calcium sensitivity of reconstituted cardiac actomyosin which includes the tropomyosin-troponin complex (*solid line*) expressed as ATPase activity. Note the marked increase in ATPase activity between calcium (Ca^{++}) concentrations of 2×10^{-7} and 2×10^{-5}M. The lower *dotted line* shows the lack of calcium sensitivity of myosin alone, and the *vertical dashed line* indicates the calcium concentration at half-maximum activation [Redrawn from Katz (1970).]

globular protein troponin, and when calcium is bound to troponin, the tropomyosin undergoes a conformational change which shifts it into the groove between the two actin strands, thereby uncovering active sites and allowing myosin filaments to form reversible bonds with actin filaments. The calcium binds to one component of troponin (TN-C) which releases another troponin component (TN-I, the inhibitory component) from actin, causing tropomyosin to shift to a nonblocking position. The precise mechanisms by which ATP is hydrolyzed by the ATPase bound to myosin as active sites form and break and how this interaction is transmitted into mechanical work of the muscle remain to be fully established. (For review see Hanson and Long (1965) and Noble and Pollack (1977).) However, during contraction the myosin cross-bridge changes its angle with release of ADP and P, and the cross-bridge bond to actin is broken when ATP again binds to myosin (thus, ATP is required for relaxation). For further details see Huxley (1974) and Katz (1977).

The importance of calcium concentration in the contraction of cardiac muscle is further demonstrated by use of muscle fiber preparations from which the sarcolemma is removed. If a strip of isolated cardiac muscle is placed in glycerol for many hours (alternatively, skeletal muscle fibers can be mechanically "skinned"), most of the membranes and low molecular weight compounds including ATP go into solution, whereas the contractile proteins remain in relatively normal structural relationships. If the muscle is then bathed in buffer containing magnesium and ATP, glycerinated fibers do not contract as long as the calcium remains low. However, as the calcium concentration is increased the tension is elevated (Fig. 2.84), the sigmoid-shaped curve showing the most rapid increase of tension development between 10^{-5} and 10^{-7} M calcium (resembling the sigmoid curve of ATPase activation by calcium) (Henry et al., 1972).

THE "CALCIUM CYCLE" IN CARDIAC MUSCLE

The structural basis for the transmission of the electrical impulse to the interior of the myocardial

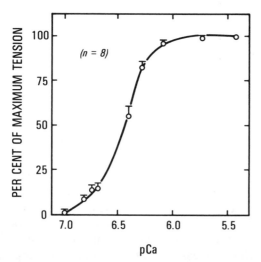

Figure 2.84. Tension developed by glycerinated cardiac muscle fibers, expressed as a percentage of maximum tension at a high calcium concentration (pCa = 5.4) Fibers held at fixed length. [Adapted from Henry et al. 1972).]

Figure 2.85. Diagrammatic concepts of triadic and diadic relationships in cardiac and skeletal muscle sarcomeres. LS = lateral sac. [Redrawn from Legato (1969).]

cell, where it promotes calcium release, is shown in Figure 2.78B. From the surface of the cell narrow, hollow tubes which are direct extensions of the sarcolemma project inward, penetrating the intracellular space. These so-called transverse tubules (T tubules) abut on the Z-bands of the sarcomeres and carry electrical depolarization inward. In heart muscle, the T tubules are much larger than in skeletal muscle (consonant with the important role for transmitted electrical activation in cardiac muscle).

The sarcoplasmic reticulum constitutes a separate system of tubules; this network of small interconnecting tubules surrounds the sarcomeres (Fig. 2.78B) and contains a high concentration of calcium due to the activity of an ATP-utilizing calcium pump. The sarcoplasmic reticulum does not connect directly with the sarcolemma or the T tubules, which are continuous with the sarcolemma, but its flattened saccular structures (cisternae) abut against the sarcolemma and T tubules. In cardiac muscle, where one or two lateral sarcotubules wrap around a transverse tubule, the appearance in cross-section is that of a "diad" or "triad" (Fig. 2.85), whereas skeletal muscle shows more dilated lateral sacs.

The movements of calcium during a single cardiac cycle are presented diagrammatically in Figure 2.86. A relatively small amount of calcium moves inward across the sarcolemma during the action potential, most of the calcium needed for muscle activation being rapidly released from the sarcoplasmic reticulum. (For details see Endo, 1977.) Following contraction the calcium is rapidly reaccumulated by the sarcoplasmic reticulum, and calcium also moves out of the cell across the sarcolemma in exchange for sodium during diastole.

Inward Transsarcolemmal Calcium Movement

Depolarization and repolarization during an action potential produce a sequence of ion fluxes involving primarily sodium, potassium, and calcium (Chapter 8). First, sodium enters (fast inward current), followed by an influx of calcium (slow inward current) (Reuter, 1974), and toward the end of the action potential, calcium influx ceases and repolarization is accompanied by potassium efflux. The ATP-dependent sodium-potassium pump (ATPase) in the sarcolemma maintains the electrical potential during diastole by carrying sodium to the outside of the cell while potassium moves back to the interior of the cell. Although calcium enters the cell during the action potential and contributes to the intracellular pool, the amount entering is much less than that needed to activate contraction (approximately 1/5). However, as this calcium moves in across the sarcolemma (including the T tubules) (Fig. 2.86) it plays a significant role in stimulating release of calcium from the sarcoplasmic reticulum (see below).

Release of Calcium from the Sarcoplasmic Reticulum and Other Sites

The sarcoplasmic reticulum contains much more calcium than is necessary for contraction, and only part of it is released with each cardiac cycle. Exactly how calcium release is triggered from the sarcoplasmic reticulum is not known (Fig. 2.86). It has been postulated that there may be electrical depolarization of the lateral sacs. In addition, stimula-

tion of calcium release from the sarcoplasmic reticulum by a small increase in calcium (so-called "regenerative" calcium release or "calcium-generated calcium release") has been shown to be operative (Fabiato and Fabiato, 1979). Negatively charged subsarcolemmal binding sites also appear to provide other intracellular rapid release sites for loosely bound calcium (Fig. 2.86).

In the regulation of the heart muscle contraction, the concept of *incomplete activation* applies; that is, much more calcium is available in the sarcoplasmic reticulum (and on rapid release sites) than is actually used during a given contraction. Moreover, the myofibrils are not fully activated to achieve maximum tension development under normal conditions. Therefore, a substantial "contractile reserve" is available which may be called upon under altered conditions.

Calcium Reuptake by the Sarcoplasmic Reticulum

Reuptake of calcium by the sarcoplasmic reticulum (Fig. 2.86) is an active ATP-dependent process (Endo, 1977) and is responsible for the termination

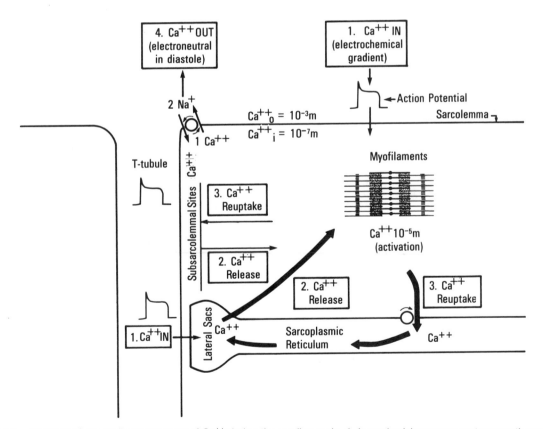

Figure 2.86. Diagram of the cyclic movements of Ca^{++} during the cardiac cycle. *1,* Inward calcium movement across the sarcolemma during phase 2 of the action potential. Such movement also appears to trigger Ca^{++} release from the sarcoplasmic reticulum. *2,* Major release of Ca^{++} occurs from the sarcoplasmic reticulum, with some release from subsarcolemmal binding sites as well. *3,* Ca^{++} reuptake is predominantly into the sarcoplasmic reticulum by an energy-dependent pump, but Ca^{++} is also rebound at subsarcolemmal binding sites. *4,* Electroneutral Ca^{++} and Na^+ exchange occurs across the sarcolemma with extrusion of Ca^{++}.

of contraction (Fig. 2.86). By this mechanism, the so-called "cardiac relaxing system," the calcium concentration is rapidly lowered to about 10^{-7} M.

There is evidence in skeletal muscle that calcium reuptake occurs primarily in the longitudinal portions of the sarcoplasmic reticulum, from which it then moves from to the lateral sacs for storage. However, the lateral sacs are much larger in skeletal muscle than in cardiac tissue, and it is uncertain whether this movement occurs in the heart.

Calcium Extrusion from the Cell (Na-Ca Exchange)

An exchange mechanism exists in the sarcolemma which *removes* calcium from the cell during the resting (diastolic) phase of the cardiac cycle (Figs. 2.86 and 2.87) (Langer, 1971; Reuter, 1974). Calcium concentration is relatively high outside the cell (10^{-3} or more), producing a concentration gradient for calcium movement inward in the resting phase of the cycle, since intracellular calcium ion concentration is maintained at a low level by the

sarcoplasmic reticulum. The sarcolemmal exchange mechanism is important for two reasons: first, for maintaining an outward movement of calcium to counterbalance the inward gradient during diastole, and secondly to remove calcium that enters during the action potential (Fig. 2.86). This exchange mechanism must be considered in detail when analyzing the effects of changes in cardiac rate and other influences on transsarcolemmal calcium movements.

The system exchanges two sodium ions for one calcium ion (Fig. 2.87), these two ions being thought to compete for a carrier site on either side of the membrane. The exchange is electroneutral, the rate of calcium movement being dependent upon the sodium gradient created by sodium-potassium ATPase, and at equilibrium the electroneutral exchange leads to the following distribution ratio:

$$\frac{[Ca]_i}{[Ca]_o} = \frac{[Na]_i^2}{[Na]_o^2}$$

where i and o indicate ion concentrations inside

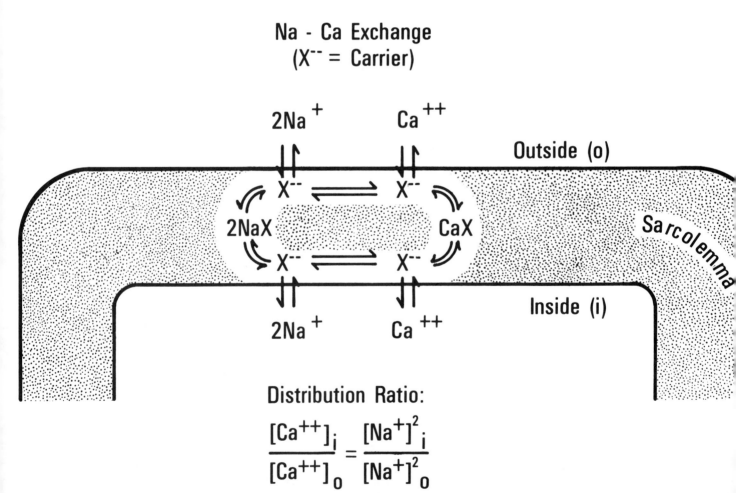

Na - Ca Exchange
(X^{--} = Carrier)

Distribution Ratio:

$$\frac{[Ca^{++}]_i}{[Ca^{++}]_o} = \frac{[Na^+]^2_i}{[Na^+]^2_o}$$

Figure 2.87.   Diagram of electroneutral Na^+-Ca^{++} exchange by means of a hypothesized carrier in the sarcolemma. Two Na^+ ions enter for each Ca^{++} ion carried to the outside of the cell (some recent work suggests this ratio may be closer to 3:1).

and outside the cell, respectively. The difference between the sodium concentrations outside and inside the cell can be termed the sodium gradient.

Any intervention which *increases the sodium gradient* will accelerate the carrier operation and *increase calcium removal* from the cell, whereas *reduction of the sodium gradient* slows the carrier and *results in calcium accumulation* inside the cell (Fig. 2.87) (Reuter, 1974). For example, when $[Na]_i$ increases (as occurs with inhibition of sodium-potassium ATPase produced by digitalis), since sodium normally is higher outside the cell the transsarcolemmal gradient is reduced, the carrier slows, and $[Ca]_i$ increases. This increased intracellular calcium increases contraction, as discussed further below. The same carrier mechanism operates in the giant squid axon, and in this large cell when sodium is placed directly in the interior of the cell with a micropipette (decreasing the transmembrane sodium gradient), a measurable decrease in calcium efflux follows. When $[Na]_o$ is lowered, the sodium gradient is also reduced, carrier operation slows and, again, $[Ca]_i$ accumulates. Increasing calcium concentration outside the cell can also increase intracellular calcium accumulation.

INOTROPIC MECHANISMS IN HEART MUSCLE

Suppose a piece of isolated cardiac muscle (such as the papillary muscle of the cat) is maintained in an oxygenated bath and held at a fixed length (isometric contraction), while active tension development following electrical stimulation is measured with each twitch. Then we can very simply define a *positive inotropic influence* as one that produces an increase in the peak force developed during contraction and a *negative inotropic influence* as a decrease in the force generated under similar conditions. As we shall learn in the next chapter, a positive inotropic stimulus also increases the rate of force development and the rate of fall of force (relaxation), whereas a negative inotropic influence produces opposite effects. As mentioned earlier, the degree of activation of mammalian cardiac muscle by calcium normally is well below maximum, giving a considerable reserve for modulation of the inotropic state of heart muscle.

A number of mechanisms can affect the level of inotropic state, thereby increasing or decreasing the force of contraction of the muscle. These include changes in the action potential itself, alterations of ion concentrations, and effects of a variety of drugs and hormones.

Action Potential Changes

When an isolated cardiac muscle is arranged so that the voltage across the membrane is held constant and the effects of various interventions on current flow across the membrane are measured (so-called "voltage clamping"), the inward calcium current during the plateau phase of the action potential can be measured. This phase of the action potential provides a "calcium gate," and it can be demonstrated that if the duration of the plateau phase is increased, or the degree of depolarization is greater, or if external calcium concentration is raised, a positive inotropic effect is produced on contraction. Conversely, shortening of the plateau phase of depolarization, reduced degree of depolarization, or lowering of external calcium has a negative inotropic effect, leading to less active force development (Wood et al., 1969; Reuter, 1974). Certain hormones and drugs may also affect the amount of calcium entering the cell during the time of the slow inward calcium current.

The occurrence of the calcium inward current during the normal action potential is now well established in isolated muscle. However, the *quantitative* role of such inward transmembrane calcium flux in the activation of contraction and in the physiologic regulation of the level of inotropic state in the whole heart has not yet been fully established.

Effects of Changing External Ion Concentrations

As discussed, increasing external calcium concentration results in increased tension development both in isolated heart muscle and in the whole heart. Reducing external calcium diminishes the force of contraction, and lowering external sodium concentration has opposite effects (reducing the sodium gradient and promoting calcium accumulation as discussed earlier); in fact, in the frog heart when external calcium is markedly reduced to produce dissociation between electrical activity and mechanical contraction, if external sodium concentration is then lowered the dissociation is corrected and contraction returns. Reduction of external potassium concentration results indirectly in an increase of intracellular calcium and a positive inotropic effect. This occurs as a consequence of the ensuing decrease in intracellular potassium with a compensatory increase in intracellular sodium, leading to a decrease in the sodium gradient and decreased efflux of calcium.

It is noteworthy that in skeletal muscle, which contains a much more extensive sarcoplasmic reticulum and depends very little on transmembrane

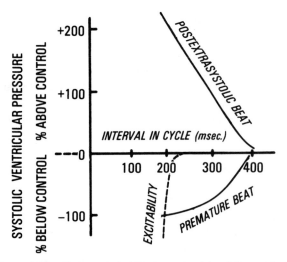

Figure 2.88. Recovery of ability to contract (systolic ventricular pressure development) in a premature beat as a function of time in the cycle after the previous beat, showing that the earlier in the cycle the weaker the premature beat. The earlier the premature beat the more forceful the postextrasystolic beat, as shown by the inverse relation between lengthening of the interval in the cycle and the systolic pressure development by the postextrasystolic beat following the premature beat (*top line*). The *dashed line* shows the recovery of electrical excitability. [Redrawn from Hoffman et al. (1956).]

calcium influx during contraction, changing the external calcium concentration has little or no effect on the force of contraction.

Time- and Frequency-dependent Changes in Cardiac Contraction

In heart muscle, the frequency and sequence with which action potentials are generated can importantly affect the force of active contraction. These phenomena are related to a number of factors already discussed that are concerned with calcium flux and storage, and the effects are readily demonstrable in the whole heart as well as in isolated muscle. (For reviews, see Koch-Weser and Blinks (1962) and Johnson (1979).)

Recovery of the ability to contract after an active twitch (after the electrical refractory period) has been observed to be progressive (Siebens et al., 1959). When electrically stimulated single beats are produced at progressively increasing intervals after the refractory period, the earlier (premature) beats are weaker (Fig. 2.88). Such reduced contractions are due, at least in part, to lack of full recovery of calcium stores in the sarcolemma and on other

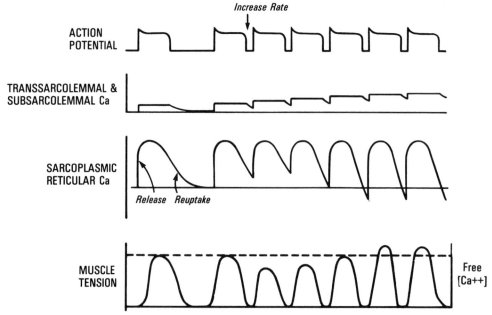

Figure 2.89. Bowditch staircase (force-frequency relation). Hypothetical diagram of how changes in calcium uptake and release might result in a positive inotropic effect due to increased frequency of contraction of isolated cardiac muscle. The action potential of the muscle is shown above and the isometric muscle tension in the lower tracing. With increased rate, an increased amount of calcium per beat transverses the sarcolemma during the action potential per unit time (increased number of action potentials) and may result in more calcium storage in the subsarcolemmal binding sites; a lag in the extrusion of calcium by the sodium-calcium exchange mechanism may also occur (second line). The release and reuptake of sarcoplasmic reticular calcium is shown in the third line; with increased rate, initially less calcium is released since reuptake is not complete and peak tension falls, but as intracellular calcium stores increase, the rate of uptake and the amount of available calcium stored in the sarcoplasmic reticulum increase. Therefore, the muscle tension and the amount of free calcium released per beat (fourth tracing) increase to a new steady state.

rapid release sites. Thus, a smaller than normal release of calcium can be presumed to occur with each of these early contractions.

Changes in the frequency of action potentials per minute in cardiac muscle produces an initial slight drop in developed force which is then followed by a progressive rise in the force of contraction over the course of several beats to a new higher level of force at steady state. This response constitutes the so-called "*Bowditch staircase*" or "force treppe" and represents a positive inotropic effect (Fig. 2.89) (Bowditch, 1871). Conversely, abruptly slowing the frequency of contraction results in an initially slightly higher force of contraction, followed over the course of several beats by a new steady state level at a lower level of force (a *negative staircase*; i.e., a negative inotropic effect of decreased frequency). These effects have also been termed "force-frequency" relations.

Changes in calcium movement during these maneuvers are complex, and several factors are undoubtedly involved as diagrammed in Figure 2.89. During the initial few contractions after frequency is increased, tension falls very transiently by the mechanism alluded to in the section on recovery of ability to contract. With increased frequency, there is less time available per minute during the rest phases of the cycle for electroneutral calcium extrusion by the sarcolemma, and it has also been postulated that a "lag" in the outward pumping of sodium by membrane pump (which is unable to keep up with the suddenly increased sodium load) produces a decrease in the sodium gradient and further enhances the accumulation of intracellular calcium (Fig. 2.89) (Langer, 1971). These factors, coupled with the increased influx of calcium due to the increased number of action potentials per minute, tend to produce enhanced loading of the sarcoplasmic reticulum and other rapid-release sites which reaches a steady state over the course of several contractions, resulting in a higher free Ca^{++} and more force development after each action potential (Fig. 2.89).

Postextrasystolic potentiation is a phenomenon which *follows* the occurrence of a premature electrical impulse early after a preceding contraction (in the whole heart, a spontaneous "extrasystole" may occur at this time). This produces a weak beat, followed by a marked augmentation of the active force in *the beat following the extra beat* (Hoffman et al., 1956). This positive inotropic effect then dissipates over the course of the next two or three cycles (Fig. 2.90). The degree of potentiation is related directly to the degree of prematurity of the extra beat (Fig. 2.88) (Seibens et al., 1959). Again, the changes in calcium movements responsible for this phenomenon are complex. In the whole heart, the early electrical impulse is followed by a *pause* because the next *normal* impulse reaches the heart when it is electrically refractory (Fig. 2.90). This in itself enhances the amount of calcium stored and released on the subsequent contraction, but the

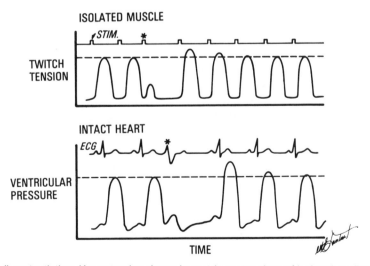

Figure 2.90. Postextrasystolic potentiation. *Upper tracing* shows isometric contractions of isolated cardiac muscle (electrical stimulus artifact *(STIM)* in which the third stimulus *(asterisk)* is premature. This produces a weak beat but is followed by a potentiated beat, and peak isometric force then returns to normal over three cardiac cycles. End-diastolic tension is unchanged udner these conditions. In the *lower tracing,* the left ventricular pressure pulse recorded in the intact left ventricle is shown together with the electrocardiogram (ECG). The third beat, a ventricular premature depolarization (asterik), produces an early beat that develops little pressure and is followed by a compensatory pause which results in an increased end-diastolic pressure on the following beat and potentiation of the systolic pressure. Two mechanisms are therefore operative in postextrasystolic potentiation under these conditions: the effect of premature electrical stimulation and the added cardiac filling of the postextrasystolic beat.

early electrical impulse also causes early release of an extra bolus of calcium which, along with the long pause before the next beat, leads to increased calcium uptake by the sarcoplasmic reticulum by the time of the next beat, resulting in increased contraction. In the whole heart an *additional* mechanism is involved in postextrasystolic potentiation: the compensatory pause following the extrasystole (see Chapter 9) results in increased filling and increased stretch on the fibers, which also results in an increased force of contraction (Fig. 2.90).

If a series of two closely spaced electrical stimuli are delivered to the heart, a positive inotropic effect is achieved. This phenomon has been called "paired electrical stimulation" (Braunwald et al., 1976) and produces a pronounced increase in the force of contraction of isolated muscle and the whole heart, even when the basic frequency of the heart is unchanged from that of preceding single stimuli.

Computer models have been developed to describe these complex force-frequency interrelations relative to the calcium cycle, and they can be used to predict various phenomena observed in the contraction of heart muscle and the whole heart (Kaufmann et al., 1974).

Pharmacologic Agents and Hormones

Digitalis glycosides, as mentioned, inhibit the sodium-potassium ATPase in the sarcolemma, slow sodium extrusion by this active pump and therefore diminish the sodium gradient (Schwartz, 1976); this should lead to increased intracellular calcium and a positive inotropic effect (increased peak isometric force). There is some evidence that catecholamines, such as epinephrine and norepinephrine, act to increase the inward calcium current of the action potential, although catecholamines and other agents which alter cyclic-AMP levels (such as the methyl xanthines which are phosphodiesterase inhibitors) may produce their positive inotropic effect by other mechanisms. There is evidence, for example, that the increased activity of c-AMP dependent protein kinase produced by the catecholamines increases the calcium sensitivity of the ATPase in the sarcoplasmic reticular calcium pump, causing increased calcium transport. This effect could speed muscle relaxation, as well as favor more calcium release (Katz, 1977; Stull and Mayer, 1979).

SUMMARY OF THE ROLE OF CALCIUM IN INOTROPIC STATE CHANGES

Based on studies in isolated cardiac muscle, Wood, Heppner and Weidmann (1969) have proposed definitions of the relationships between calcium and inotropic state changes which may be summarized as follows:

1. The amount of calcium bound to rapid release sites in the sarcoplasmic reticulum and subsarcolemmal regions just prior to the action potential determines the level of inotropic state or contractility of the following muscle contraction (amount and rate of tension development).
2. The amount of calcium bound at these sites at any moment depends on many factors including:
   a. The characteristics and shape of prior action potentials (which affect the influx of calcium ions during the slow inward calcium current). It should be emphasized, however, that in the whole heart such changes in action potential shape are relatively minor, and the major changes due to electrical phenomena result from mechanism (b) below.
   b. Prior intersystolic intervals, that is, the frequency of contraction or "force-frequency relation" and postextrasystolic potentiation.
   c. The effect of ions on the outward calcium transport system (sodium-calcium exchange). This includes such factors as the external calcium and sodium concentrations.

CHAPTER 11

Mechanical Performance of Isolated Cardiac Muscle

The performance of the whole heart is importantly affected by the loads imposed upon the heart by the level of blood pressure and the return of blood from the various organs. Such changes in loading affect the performance of the muscle in the whole heart in the same way that they influence an isolated piece of muscle subjected to varying loading conditions (Sonnenblick, 1962). In addition to the intrinsic factors discussed in the last chapter, a number of extrinsic factors (also largely involuntary) can affect the performance of the intact heart, including sympathetic (adrenergic) and parasympathetic (cholinergic) nervous stimuli and catecholamines circulating in the blood, which influence myocardial inotropic state, blood pressure, and the heart rate. All of these responses can be better appreciated if the responses of isolated muscle to altered loading conditions and inotropic state are understood (Abbott and Mommaerts, 1959; Sonnenblick, 1962).

The term "active state" is sometimes used to describe activation and the level of inotropic state and may be considered in terms of the presence of a sufficient level of free calcium to trigger and maintain contraction. The time course of the "active state" does not follow exactly the level of tension developed by the muscle (Erdman, 1968), since its onset precedes tension development. In addition, the level of tension builds up more slowly than the level of "active state," since tension development in the muscle is delayed by the stretching of elastic components within the muscle and its attachments.* A muscle can perform only a few

---

* Muscle models are beyond the scope of this discussion, but for purposes of describing the differences between active state and time course of tension development, a two-component model for muscle may be mentioned. Muscle behaves as if a passive, undamped elastic element is attached in series to the contractile elements (CE) (Fig. 2.91). This "series elastic" (SE) element

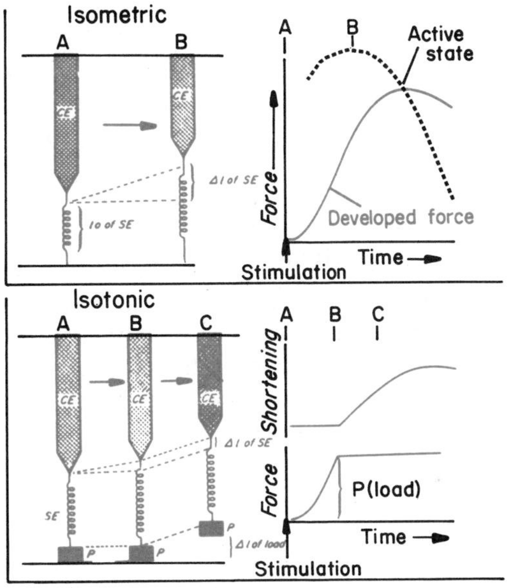

Figure 2.91. Sequence of events during contraction. *Upper panel*, Isometric contraction showing events in the Hill model (*left*) and curves of force development (*solid line*) and active state (*dashed line*). *A*, Initial resting state. *B*, An instant during active contraction. The difference between the active state curve and developed force curve is owing to the time needed for contractile element (CE) shortening to stretch series elastic element (*SE*). *Lower panel*, Afterloaded isotonic contraction (*left*) and curves of force and shortening (*right*) as functions of time. [Redrawn from Sonnenblick, (1966).]

tasks: it can develop force or tension, it can shorten with no load, or it can shorten and lift a load (perform work). (For reviews on the mechanics of muscle contraction see Sonnenblick, 1962, 1966; Mirsky et al., 1974; Braunwald et al., 1976; Brady, 1979; Fung, 1981).

---

transmits force to the attachments of the muscle. The time required to stretch the SE delays and modifies the characteristics of tension development (Fig. 2.91). In some models for muscle, a parallel elastic element which bears the resting force on the muscle is also added.

THE ISOMETRIC CONTRACTION (TWITCH) OF ISOLATED CARDIAC MUSCLE

The type of contraction of isolated muscle that is simplest to understand is the so-called *isometric contraction*, in which the muscle length is held fixed throughout contraction (Fig. 2.92). Under these conditions, when the resting muscle is electrically stimulated to contract, it develops force, but it cannot shorten (an analogy is an unsuccessful attempt to lift a heavy object with one arm, which results in force development by the biceps muscle but no movement). Thus, tension rises, peaks, and then falls as the muscle is activated and relaxes (Fig. 2.92).

In skeletal muscle a twitch can also be produced by a brief pulse of rapid electrical impulses, or (in contrast to cardiac muscle which is electrically refractory for a time following an electrical impulse) a sustained delivery of rapid pulses produces a prolonged muscle contraction, a so-called tetanic contraction (Buller, 1975).

Effects of Changing Muscle Length

When the resting length of cardiac muscle is increased by adding more preload, more active force is developed if the muscle is then stimulated electrically to contract isometrically. Conversely, if the resting muscle length is reduced (muscle shortened by reducing preload) less active tension is developed

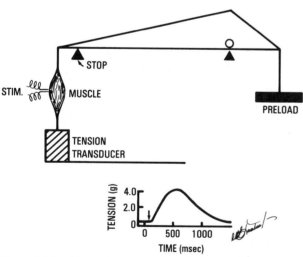

Figure 2.92. Study of isolated cardiac muscle using a lever arm arranged so that the muscle will contract isometrically. The muscle is first stretched to a given resting length by a given preload (weight), and a stop is then positioned so that when the muscle develops force it cannot shorten and move the lever arm; therefore, contraction is isometric when the muscle is then stimulated electrically to contract (STIM). Tension developed by the muscle is recorded by a transducer, and the isometric twitch following electrical stimulation (*vertical arrow*, *lower panel*) is shown as a function of time.

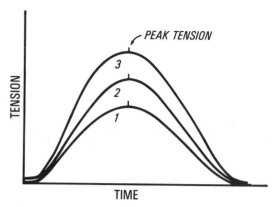

Figure 2.93. Diagram of isometric twitches from isolated cardiac muscle produced by progressively increasing preloads. Twitches obtained at different times are superimposed. Notice the increase in resting muscle tension and peak tension development in beats 1–3, and as peak tension rises the time to peak tension remains constant.

(Fig. 2.93). (See also ascending limb of sarcomere length-tension curve, Figs. 2.80 and 2.81). The time to peak tension remains the same in such contractions, and since peak tension increases, the speed of tension development (dT/dt) also increases (Fig. 2.93).

This length-tension response is presumably related primarily to a change in the degree of overlap of active sites on the myofilaments, although as discussed earlier, it is possible that changes in the level of activation by calcium release also play a role under these conditions (Taylor, 1974; Jewell, 1977).

Effects of Changing Inotropic State

Intensity of the active state or inotropic state of cardiac muscle (terms usually synonymous with "contractility," sometimes called "contractile state") is related in many situations to the availability of calcium, and perhaps thereby to the number and frequency of active bond formations between the myofilaments. Changes in the level of inotropic state are reflected by changes in the peak tension developed in the isometric twitch (Fig. 2.94). Also, very pronounced changes occur in the *rate* of tension development and the rate of relaxation (Brutsaert et al., 1978), associated with changes in the *duration* of contraction (Fig. 2.94). Such changes due to alterations in inotropic state can occur when there is *no* change in resting muscle length. It should be noted that positive inotropic stimuli (such as norepinephrine and digitalis) produce an increased in peak isometric tension, augmented velocity of both tension development and relaxation (decreased time to peak tension), and shortening of the duration of contraction (contrac-

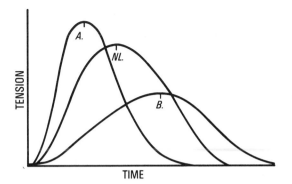

Figure 2.94. Diagram of isometric twitches of isolated cardiac muscle initiated from the same resting muscle length and tension. *NL.*, normal twitch. *A.* shows the effect of a positive inotropic stimulus, with an increased peak tension and rate of tension development and shortening of the duration of the twitch. *B.*, Effects of a negative inotropic intervention which depresses peak tension and rate of tension development, and prolongs the duration of the twitch.

tion A, Fig. 2.94). Negative inotropic stimuli (such as some anesthetic agents) produce a decreased peak tension, slowed velocity of tension development and increased time to peak tension, prolongation of the duration of isometric twitches, and delayed rate of relaxation (contraction B, Fig. 2.94).

THE ISOTONIC CONTRACTION OF ISOLATED CARDIAC MUSCLE

A piece of cardiac (or skeletal) muscle can be arranged in a muscle bath using a lever arm and weights to produce a so-called isotonic contraction (Fig. 2.95) (Sonnenblick, 1966). One end of the muscle is attached to a force transducer and the other end to the lever arm (which operates on a fulcrum). The opposite end of the lever arm is attached to a pan upon which various weights can be added. The following steps are used to produce an isotonic contraction.

1. A small weight (termed the "preload") is placed in the pan to produce a given degree of stretch of the resting muscle.
2. A stop is then positioned at the lever arm (Fig. 2.95) to prevent further stretching of the resting muscle when additional weights are added.
3. More weights are added to the pan (this added weight is not transmitted to the resting muscle). The total weight in the pan will then be encountered by the muscle only when it is stimulated to contract and develop force. If this total load is not too much for the muscle to lift at a given degree of resting stretch, the pan will be lifted, and this weight when it is lifted in the air is termed the "afterload" (the load *after* the muscle starts to shorten and lift the weights in the pan, which include the preload). The ensuing contraction (Fig. 2.95) is called isotonic

because once the weight is lifted into the air the muscle will shorten and then lengthen while bearing the same load (hence, isotonic).

It should be noted that the first portion of an isotonic contraction is isometric (since sufficient force must first be developed by the muscle to lift the afterload), and the last portion of the contraction (after the muscle has relaxed sufficiently to reposition the lever on its stop) is *also* isometric as the force falls back to the level of the preload (Fig. 2.95). Of course, if the weight is too heavy for the muscle to lift at a given preload, an isometric twitch will ensue; by definition, such a contraction is not afterloaded.

The afterloaded isotonic contraction in isolated muscle resembles to some extent the beat of the intact heart, which first must develop pressure to open the valves to the pulmonary artery and aorta before it can eject blood. However, in the whole heart the load is not isotonic, since it varies as aortic pressure changes during the course of ventricular ejection. Another difference is that the ventricle ejects its contents and then relaxes at a smaller volume or muscle length, whereas the iso-

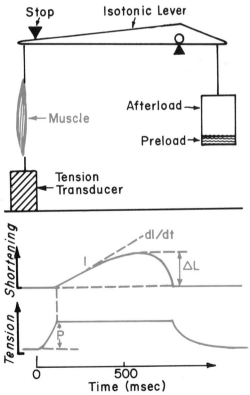

Figure 2.95. *Upper panel*, Arrangement for study of mechanics in isolated papillary muscle. *Lower panel*, Records of shortening and tension of afterloaded isotonic contraction. *P*, afterload; *dl/dt*, initial shortening velocity. [Redrawn from Sonnenblick et al. (1970).]

lated muscle lifts the weight and returns it to the previous position, lengthening as it relaxes back to its initial resting muscle length. (See subsequent discussion and Fig. 2.101.)

In a single isotonic contraction, by measuring the movement of the lever arm (such as with a photoelectric device), muscle length (L) can be continuously recorded, and the amount or extent of shortening (ΔL) and the rate of change of muscle length (dL/dt) also can be determined (Fig. 2.95). When only the *preload* is changed in a series of isotonic contractions, while the afterload remains unchanged, both the amount and velocity of shortening increase as the preload is increased, and they decrease as preload is reduced (Fig. 2.96).

When only the *afterload* of an isotonic contraction is changed while the preload is maintained constant, the performance of the muscle is also importantly affected, the extent and velocity of shortening both decreasing as the afterload is in-

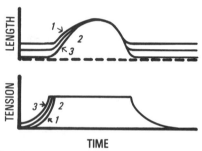

Figure 2.96.   Diagram of superimposed isotonic contractions of isolated cardiac muscle in which the preload is progressively increased (beats 1–3) while the afterload is kept constant. Notice that with higher preloads, length changes (expressed as shortening) become progressively larger, and the initial velocity of shortening also increases (isometric relaxation is also delayed at higher preloads, not shown).

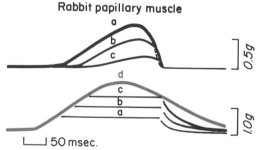

Figure 2.97.   Afterloaded contractions. Relationship between papillary muscle afterload, shortening and duration of mechanical activity. Isotonic contraction curves, *a, b, c* (*upper*) superimposed on the same time base with isometric myogram, *c, b, a* (*lower*). Curve *d* is the ordinary isometric contraction myogram. With progressively higher afterloads (*a, b, c, lower panel*) the extent and velocity of shortening decrease (*a, b, c, upper panel*). The more the muscle is allowed to shorten during the isotonic phase, the briefer is the isometric portion of the myogram. [Redrawn from Edman (1968).]

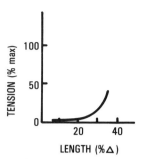

Figure 2.98.   The passive or resting length-tension relation of isolated cardiac muscle.

creased, whereas performance increases as the afterload is decreased (Fig. 2.97); thus, as might be predicted, when one attempts to lift a heavy load it cannot be lifted very far or fast, whereas if the load is lightened it is possible to lift it a greater distance at a faster speed.

THE FUNCTION CURVES OF CARDIAC MUSCLE

The mechanical loading conditions on an isolated cardiac muscle can be varied over a wide range, while the muscle's contractile responses are recorded at each level of altered preload or afterload. From these responses, "function curves" can be constructed, and such curves can also be performed after an inotropic stimulus. Function curves of this type are particularly relevant to the study of function of the whole heart, where loading conditions are often changing, and various pharmacologic or biochemical factors can also influence heart performance. We may consider one passive curve and six "active" function curves of cardiac muscle.

The Passive Length-Tension Curve

When the length of a resting muscle is progressively increased in a stepwise manner, by progressively adding weights in preload increments, the tension rises at each step to yield the passive or resting length-tension curve (Fig. 2.98). For practical purposes, we can consider that the shape or position of this curve is *not altered* appreciably by changes in mechanical loading conditions or the inotropic state of the muscle during a given experiment (see also Fig. 2.81).†

---

† This point is somewhat controversial, some investigators having reported small shifts in the passive length-tension curve under certain conditions, such as changes in muscle temperature. Resting cardiac muscle and the heart also exhibit certain time-dependent plastic or viscous properties when suddenly stretched (stress-relaxation) or suddenly loaded without a stop-mechanism (creep), which are beyond the scope of this discussion (LeWinter et al., 1979). Chronic changes in the muscle (as produced by scar or hypertrophy) can markedly alter the shape or position of the passive length-tension relation.

The Isometric Length-Tension Curve and End-Systolic Length-Tension Relation

When a muscle is stimulated to contract isometrically and each contraction (or series of contractions) is produced at a progressively increased resting muscle length (higher preload), the active length-tension curve can be obtained by plotting the peak twitch tension for each twitch vs. the associated resting muscle length (Fig. 2.99). Total active tension (*dotted line*) includes the preload (*dashed line*, Fig. 2.99); "developed" tension is sometimes used in which the preload is not included (Fig. 2.99, *solid line*). The isometric length-tension curve corresponds to the ascending limb of the sarcomere length-tension curve (Figs. 2.80 and 2.81).

The isometric length-tension curve also provides a framework for isotonic contractions. Experiments in isolated muscle have shown that the length and tension at the *end* of isotonic contractions against the same afterload tend to be independent of the initial length (preload) of the contraction (Fig. 2.100), but length and tension at the end of shortening are affected by the level of afterload (regardless of the initial length) (Fig. 2.100) (Downing and Sonnenblick, 1964). Thus, the tension developed by a muscle at the end of its shortening phase tends to be the *same* as the tension it would develop if it contracted isometrically from the same (shorter) resting muscle length as that at the end of shortening (Fig. 2.100). In other words, the entire isometric length-tension curve provides the *limit* for isotonic shortening, independent of preload and afterload. This *end-systolic length-tension relation* determined from isotonic contractions alone therefore is highly useful, since it represents points lying on the *isometric length-tension curve.*

In considering such isotonic contractions in isolated muscle, it should be understood that the muscle shortens to its minimum length and then lengthens again to its original resting length while

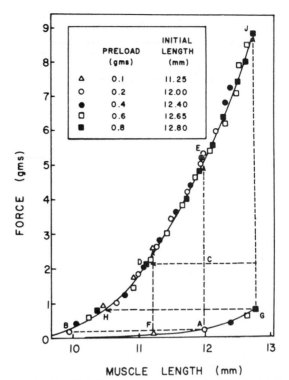

Figure 2.100. Relation between force or tension and muscle length in the papillary muscle of the cat. The resting length-tension curve is shown over a range of preloads and muscle lengths (indicated in the *inset*), and active force is plotted at increasing afterloads at a very low level (*A* to *B*), a higher level (*G* to *H*, and *I* to *D*), and for isometric contractions (*F* to *D*, *A* to *E*, and *G* to *J*). Notice that regardless of initial muscle length, the active isometric length-tension curve forms the limit of shortening for isotonic contractions. [From Downing and Sonnenblick (1964).]

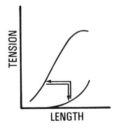

Figure 2.101. Passive and isometric length-tension curves for isolated cardiac muscle, together with the pathway of an isotonic contraction (*arrows*). The contraction first develops tension and then shortens to the length-tension curve; it then lengthens while still bearing the afterload and relaxes isometrically.

bearing the load (Fig. 2.101), in contrast to the mode of contraction of the heart, mentioned earlier.

The Length-Shortening Curve

As mentioned, the extent of shortening of the muscle depends upon the amount of stretch of the resting muscle when the afterload is kept constant (Fig. 2.102*A*). This can be represented on the length-tension diagram as progressively increased shortening at the same afterload, as increasing pre-

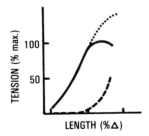

Figure 2.99. The active length-tension relation of isolated cardiac muscle. The passive length-tension relation is indicated by the lower *dashed curve*, the developed active tension by the *solid line*, and the total tension (developed tension + passive tension) is indicated by the *dotted line*.

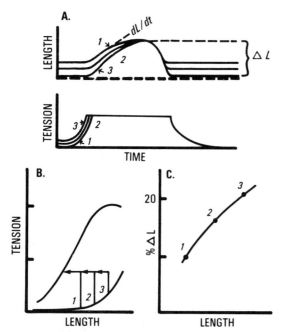

Figure 2.102. *Panel A,* Three isotonic contractions at the same afterload with increasing preloads showing progressively increasing extent of shortening (ΔL) and increased initial velocity of shortening (*dL/dt,* the slope of the initial portion of the shortening tracings). *Panel B,* Representation of contractions *1, 2,* and *3* shown in panel A on the length-tension diagram, showing that as resting muscle length increases in contractions developing the same afterload, active shortening progressively increases. *Panel C,* The relation between increasing muscle length and percent active shortening of the muscle (%ΔL) is plotted as a function curve.

loads cause increasing resting muscle lengths (Fig. 2.102*B*). From such data, a length shortening curve showing a positive relation (increased resting length causing increased shortening) can be plotted for any single level of afterload (Fig. 2.102*C*).

The Length-Velocity Curve

There is also a positive relation between resting muscle length and the velocity of shortening when afterload is held constant, velocity increasing as preload increases and vice versa (Fig. 2.102*A*).

The Force-Velocity Curve

This curve is determined by plotting the initial velocity of shortening, dl/dt, in a series of isotonic contractions produced against progressively varied levels of afterload (Fig. 2.103*A*), the force-velocity relation being determined with each contraction beginning at the *same* resting preload and length. As the afterload is progressively increased, the velocity of muscle shortening progressively declines (Fig. 2.103*A*) until, finally, the muscle is unable to lift the weight (at that particular preload) and an isometric contraction ensues. Several features of the force-velocity curve (Fig. 2.103*B*) have been described:

1. It exhibits an *inverse* curvilinear relation. In skeletal muscle, the curve is hyperbolic and has been described by an empirical equation (Hill, 1939).

2. "P_0" is the isometric tension and velocity = zero (Fig. 2.103*B*) which corresponds to one end of the force-velocity curve. This point is common to one point on the isometric active length-tension curve at that particular resting muscle length. P_0 is altered by changing the level of inotropic state.

3. "V_{max}" is the velocity of muscle shortening at zero load (Fig. 2.103*B*). It is usually determined by extrapolation, since a muscle contracting with only its preload is not operating at zero load. V_{max} in cardiac muscle is increased by positive inotropic stimuli and decreased by negative inotropic influences. It has

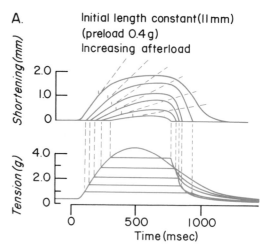

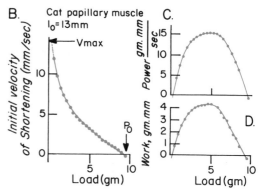

Figure 2.103. *A,* Curves of the time course of shortening and tension with increasing afterload. The initial velocity of shortening (represented by the slope of the *dashed lines* in the *top curves*) decreases with increasing afterloads. *B,* Force-velocity curve (cat papillary muscle) showing this inverse relation between afterload and initial velocity of shortening. *C,* The relation between power $\left(\text{load} \times \text{velocity} = p \cdot \dfrac{dl}{dt} \right)$ and load (*p*). *D,* The relation between work (load × distance = $\int_{r0}^{r1} p \cdot dl = p \cdot \Delta L$) and load. [Redrawn from Sonnenblick (1966).]

been suggested that such changes in the *un-loaded* (maximum) velocity may reflect a change in the intrinsic rate of "making and breaking" of bonds at the active sites on the myofilaments produced by these influences as well as an increased number of bonds.‡

The Force-Shortening Curve

When the resting muscle length is held constant and a series of contractions produced as the after-load is progressively increased the extent of shortening progressively diminishes (Fig. 2.103*A*), yielding an inverse relation between afterload and active shortening. This may also be shown within the framework of the length-tension diagram: as afterload increases from the same preload, shortening decreases and vice versa (Fig. 2.105 *left*). Also, these data can be plotted as an inverse relation between increasing afterload and decreasing ΔL (Fig. 2.105, *right*), the force-shortening curve.

There is an unusual relationship between the afterload and the *work* performed by a muscle, where work = afterload × extent of shortening (Fig. 2.103*D*). Work is zero at the left end of this curve, since a large amount of shortening against zero load produces no work; the curve then rises, as the product of work and shortening increases, and subsequently it falls as the afterload rises further, reaching zero again at P_0, when contraction is isometric and no shortening occurs (Fig. 2.103*D*). Thus, the work (and power or rate of work performance, Fig. 2.103*C*) of the muscle is highly *load-dependent*.

Three Intrinsic Mechanisms that Affect the Performance of Isolated Cardiac Muscle

Each of three major factors can affect the force, speed, and extent of shortening of cardiac muscle:

---

‡ There is evidence that $\dot{V}_{max}$ of isolated cardiac muscle is little affected by altering resting muscle length (Fig. 2.104). It has been suggested that this is due to an increased or decreased *number* of active sites available when muscle length is changed without alterations of the intrinsic rate or turnover of the active sites in the *unloaded* contracting muscle. (An analogy can be made to a group of runners pacing together at maximum running rate; adding more runners of like ability will not increase the speed of the group). On the other hand, as discussed in previous sections, a change in the preload *does* affect the extent and velocity of shortening of the *loaded* muscle and when force-velocity curves are repeated at higher levels of preload this effect results in skewing of the force-velocity curve upward and to the right with a constant $\dot{V}_{max}$ obtained by extrapolation (Fig. 2.104). This displacement could also be predicted from the fact that peak isometric force is increased by increasing preload (Fig. 2.104, *inset*). Thus, at any level of afterload (other than zero) the muscle is faster and can shorten more at a higher preload, vice versa (Fig. 2.104). (A large number of runners can jointly carry a heavy load further and faster than a few runners.)

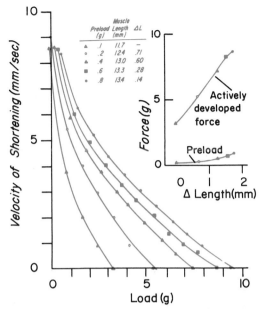

Figure 2.104. Force-velocity curves of cat papillary muscle with increasing preload which increased initial fiber length (inset represents length-tension curve at these preloads). Increasing preload (fiber length) does not alter V_{max} but increases actively developed force. [Redrawn from Sonnenblick et al. (1970).]

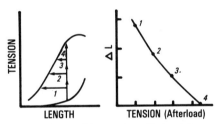

Figure 2.105. Effects of increasing afterload from a constant preload plotted on the length-tension diagram (*left panel*). Beats 1, 2, and 3 show effects of increasing afterload, and beat 4 is an isometric contraction. In the *right panel*, the extent of shortening (ΔL) in the same three beats is plotted against the tension or afterload, and an inverse relation is described.

(1) resting muscle length (preload); (2) the level of afterload; (3) the inotropic state ("contractility").

The frequency of contraction is a 4th, less important, mechanism, but it clearly can affect the amount of work done *per minute*, as the number of contractions per minute changes. In addition, it can exert a positive or negative inotropic effect through the Bowditch phenomenon (force-frequency relation) discussed earlier.

IDENTIFICATION OF CHANGES IN INOTROPIC STATE

A change in resting muscle length is not defined as a change of inotropic state, even though it alters muscle performance. Likewise, a change in the afterload also alters muscle performance but is not defined as a change in the level of inotropic state.

For practical purposes, these factors may be considered as *mechanical* determinants of muscle performance. A change in the inotropic state of the myocardium produces a change in the performance of heart muscle (force development, velocity, extent

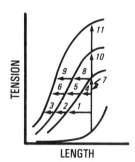

Figure 2.106. Effects of alterations of intropic state on isometric and isotonic contractions from the same resting muscle length. Beat 10 represents an isometric contraction at the normal level of contractility, which forms a point on the isometric active length-tension relation. Beat 7 is an isometric contraction at a depressed level of inotropic state, which reaches the depressed isometric length-tension relation, and beat 11 is an isometric contraction in the presence of increased inotropic state, which reaches an elevated isometric length active tension curve. At the same level of afterload beats 1 and 3 exhibit less and more active shortening, respectively, than beat 2 (the control beat), and this is the case at any given level of afterload. Notice that beat 9 shortens more than beat 8, while the depressed muscle is unable to shorten at that afterload.

of shortening) that is *independent* of alterations in the preload and the afterload. We can now *define* a change in inotropic state as a *shift* of an entire function curve of muscle. Thus, in cardiac muscle a shift or displacement of the force-velocity, force-shortening, length-active tension, length-velocity, and length-shortening curves can occur through alterations in the level of inotropic state, and the myocardium thereby can *modulate* its performance through a variety of biochemical and neurohumoral stimuli. By performing a function curve before and after an inotropic intervention, it can be determined whether the intervention had a positive or a negative inotropic effect. The most useful single curve for identifying such changes is the isometric length-tension curve and the framework that it provides for studying the extent of shortening through the length-shortening and force-shortening relations (Fig. 2.106).

A positive inotropic stimulus shifts the length-active tension curve upwards and to the left so that for any resting muscle length more peak tension is developed isometrically (Fig. 2.106). More shortening occurs in isotonic contractions from the same preloads against any given level of afterload (Fig. 2.106), leading to an upward shift of the curve relating resting length to active shortening (Fig. 2.107). Also, more shortening can occur at any level

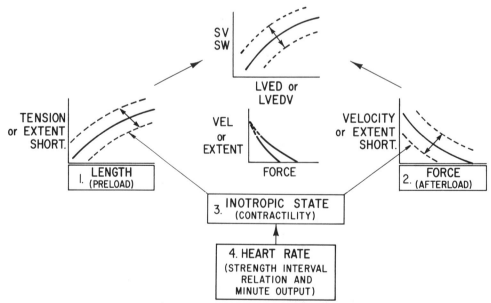

Figure 2.107. Factors influencing the performance of isolated muscle and the whole heart. The effects of the two "mechanical" determinants of performance, the resting muscle length (preload) and the tension or force (afterload) are represented by graphs 1 and 2, respectively. The preload affects isometric tension as well as the extent and velocity of shortening, and the afterload affects the extent and velocity of shortening. Upon these two factors plays the third important determinant, the inotropic state (3), which can shift the function curves of the muscle up (positive inotropic influence) or down (negative inotropic influence) (graphs 1 and 2). The frequency of contraction or heart rate (4) affects the inotropic state through the strength-interval relation, but it also affects the work per minute or the output of the heart. The force-velocity or force-shortening curve also is shifted in the characteristic manner by changes in preload (shown in the center insert; see also Fig. 2.104). The *top panel* indicates that function curves of the whole heart relating left ventricular end-diastolic pressure or volume (*LVED* or *LVEDV*) to the stroke volume and stroke work (*SV* and *SW*) are also shifted by positive and negative inotropic influences.

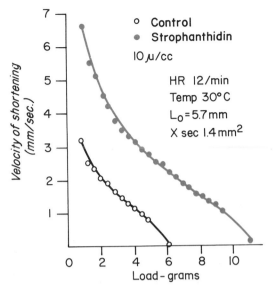

Figure 2.108. Strophanthidin effect on force velocity relation (strophanthidin is a digitalis preparation). Both V_{max} and maximal force of contraction are increased. [Redrawn from Sonnenblick et al. (1970).]

of afterload (Fig. 2.106) (leading to a shift upward of the force-shortening relation) (Fig. 2.107). Opposite shifts on these curves occur with a negative inotropic stimulus (Figs. 2.106 and 2.107). In addition, the force-velocity relation is shifted in a characteristic parallel manner so that *both* the velocity at zero load (V_{max}) and the peak isometric tension at zero velocity (P_0) are increased with a positive inotropic stimulus and decreased with a negative inotropic stimulus (graph 2, Fig. 2.107). An example of force-velocity curves actually recorded from the isolated papillary muscle of the cat before and after addition to the muscle bath of the positive inotropic agent acetyl strophanthidin (a digitalis preparation) is shown in Figure 2.108.

FACTORS AFFECTING THE LEVEL OF INOTROPIC STATE

Inotropic state, the intrinsic level of myocardial contractility, is altered by many factors. These may now be summarized as the following.

Intrinsic Factors

1. The force-frequency relation provides a positive inotropic stimulus with increased frequency of contraction and vice versa. Its effect, and the positive inotropic effect of postextrasystolic potentiation, are mediated through variations in calcium release and/or reuptake.
2. Insufficient blood flow (ischemia) produces a negative inotropic effect due to oxygen lack, acidosis, and other factors.
3. Chronic myocardial depression or "myocardial failure" reflects a negative inotropic state due to damage to the heart muscle.

Extrinsic Factors

1. Local release of norepinephrine into the myocardium due to activation of the adrenergic nerve acts as a positive inotropic stimulus.
2. Release of acetylcholine due to activation of the parasympathetic nerve acts as a negative inotropic stimulus in atrial muscle, and to a small extent in the ventricular muscle as well, although the innervation of the ventricles by the parasympathetic nervous system is sparse.
3. Pharmacologic agents of many types affect the level of inotropic state: digitalis, amrinone, and sympathomimetic agents such as isoproterenol have a positive inotropic influence, whereas many antiarrhythmic drugs, barbiturates, calcium antagonists, and anesthetic agents have a negative inotropic effect.
4. Increases in extracellular calcium concentration produce a positive inotropic effect on cardiac (not skeletal) muscle, and vice versa.
5. Certain hormones other than epinephrine and norepinephrine, such as glucagon and thyroid hormone, have positive iotropic effects.

ANALOGIES FROM ISOLATED MUSCLE TO THE WHOLE HEART

The three major mechanisms described above which influence the performance of isolated muscle also affect the performance of the whole heart. Just as the effects of these factors can be detected by examining the various function curves of muscle, by making certain analogies and assumptions the performance of the ventricles in the intact heart can be understood within a similar framework (Ross, 1966). Thus, if the physiology of isolated cardiac muscle is clearly understood, insight is gained into the function of the whole heart.

In the research setting, it is possible to measure pressure in the heart and by knowing the dimensions of the heart and its wall thickness, the force or tension developed in the myocardial wall during systole can be calculated (analogous to the afterload in an isolated muscle). Likewise, by assuming a spherical or other simplified model (such as an ellipsoid) for the shape of the left ventricle, it is possible to calculate the circumference of the chamber or its diastolic volume (analogous to resting muscle length); the end-diastolic pressure in the ventricle also can be equated to the preload.

In order to estimate the force in the myocardial wall, a simplified equation based on the law of Laplace as applied to a sphere can be employed:

$$T = \frac{P \cdot r}{2h}$$

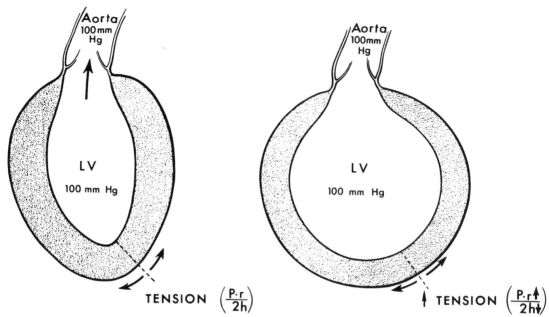

Figure 2.109. Afterload = wall tension (stress). Diagram showing how the systolic wall tension or wall stress (which represents the afterload in the myocardial fibers during left ventricular ejection) is affected by the geometry of the left ventricle (*LV*). A simplified Laplace relation relating pressure (*P*) to ventricular radius (*r*) and wall thickness (*h*) is shown. The acutely dilated ventricle on the right is developing the same systolic aortic and left ventricular pressure as the normal chamber on the left (100 mmHg). However, since the radius of the chamber on the right is increased and its wall thickness is decreased, the wall tension is elevated compared to the chamber on the left.

where T = mean wall tension (force), P = intraventricular pressure, r = ventricular radius and h = the wall thickness (Burton, 1957; Badeer, 1963; for more complex analyses, see Mirsky, 1974). As the heart enlarges, even if the systolic pressure in the chamber (during ejection) stays constant, the afterload (tension) will increase, since radius increases and the wall thickness must fall as the ventricle is stretched (muscle mass stays constant) (Fig. 2.109).

The stroke volume (the amount of volume ejected per beat) is often measured when studying the function of the whole heart, and the stroke volume can be related to the shortening of muscle of the wall by using the equation for a sphere (volume = $4/3 \, \pi r^3$) to calculate the change in radius or to calculate circumferential shortening (circumference = $2 \, \pi r$). Using calculations of this type, length-

tension, force-velocity, force-shortening, and other relations have been determined in hearts subjected to various loading conditions and shifts in the function curves described due to changes in inotropic state (Ross, 1966).

Such analogies between the function of the whole heart and that of the isolated muscle are highly useful in the research setting for understanding how the performance of the heart is altered by various physiologic responses, disease conditions, and pharmacologic agents. They are also helpful for understanding various simplified approaches currently used for examining heart function, which often employ systolic aortic or ventricular pressure (as an index of afterload), end-systolic pressure-volume relations, and analysis of the relations between end-diastolic volume and the stroke volume.

CHAPTER 12

The Cardiac Pump

We can now consider how the heart, as a hollow muscular organ, can meet its obligations as a pump and respond to changes in venous return and loading conditions by altering ventricular volume, heart rate, and the inotropic state of the myocardium. For practical purposes, in this chapter we will consider the heart as it functions in "isolated heart" and "heart lung" preparations, in which the influence of the brain and neurohumoral factors which affect the intact circulation are removed. In such a preparation, a reservoir or pump is available for rapid blood transfusion or removal of blood from the circulation, and a "Starling resistor" allows regulation of the aortic pressure and outflow resistance (Fig. 2.110) (Sarnoff et al., 1958a).

THE EVENTS OF THE VENTRICULAR CYCLE

The sequence of events during the entire cardiac cycle, including both the right and left atria and ventricles, will be described in Chapter 13. For the purposes of this discussion, we will consider the sequence of events only in a single chamber, the *left ventricle* (Wiggers, 1954). Electrical depolarization of the heart muscle is, of course, followed by the mechanical events of atrial and ventricular contraction. The electrical depolarization of atrial muscle, recorded by the electrocardiogram as the P wave, is followed by contraction of the right and the left atria almost together (slightly out of sequence), and after a delay in the AV node the QRS complex is followed by contraction of both ventricles (again, slightly out of sequence).

The pressure changes within the left ventricle, the flow of blood into and out of that chamber, and the ventricular volume change during the cardiac cycle are summarized in Figure 2.111. Commencing at the QRS, the ventricle begins to contract after a brief delay (about 50 ms) and to develop pressure as it squeezes against the blood held within it. This causes deceleration of blood flowing into the ventricle through the mitral valve leading to closure of

the valve, an event which is further assisted as the pressure continues to rise rapidly in the ventricle, while that in the left atrium stays relatively constant. Following closure of the atrioventricular (mitral) valve, as the ventricle continues to develop pressure note that ventricular volume remains constant for a time, the period of "isovolumetric contraction" (Fig. 2.111). The pressure in the ventricle then reaches that in the aorta, and when the left ventricular systolic pressure exceeds the aortic pressure, the aortic valve opens and ejection of blood into the aorta begins (Fig. 2.111). Before ejection is complete, relaxation of the ventricle has already begun, and therefore the pressures in the ventricle and aorta begin to fall together, at first

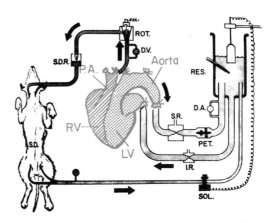

Figure 2.110. Version of isolated heart preparation. The coronary venous return is passed through a donor dog to maintain the blood at a normal biochemical environment before it is infused into the coronary system of the isolated heart. In older versions of the Starling isolated heart or heart-lung preparation in which the coronary venous blood circulating through the myocardial wall was simply reoxygenated and returned to the coronary system, progressive deterioration in performance characteristics developed within 60 to 90 min. *S.D.*, support dog; *S.D.R.*, support dog reservoir; *SOL*, solenoid valve electrically operated by microswitch at top of reservoir float; *S.R.*, air-filled Starling resistance; *PET.*, Potter electroturbinometer; flow meter *RES.*, reservoir; *I.R.*, water-filled inflow Starling resistance to regulate cardiac inflow; *D.V.* venous densitometer; *ROT.*, rotameter. [From Sarnoff et al. (1958).]

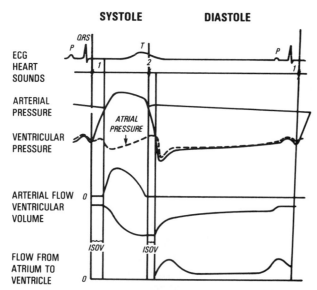

Figure 2.111. Diagram of the relations between the electrocardiogram (ECG), the heart sounds, the pressure in an artery and ventricle (either the right or left ventricle) and the corresponding atrium, flow in the artery leaving the ventricle, the volume of the ventricular chamber, and flow across the atrioventricular valve during diastole. The isovolumetric phases (*ISOV*) of ventricular contraction and relaxation are indicated. Notice at the far right at end of diastole, the P wave is followed by a presytolic (a) wave in atrial and ventricular pressure pulses, which produces an increase in ventricular volume and an accelerated flow from atrium to ventricle just prior to the first heart sound which occurs as the atrioventricular valve closes. For further discussion see text. For atrial pressure see Chapter 13.

slowly and then more rapidly; then, as blood deceleration occurs and pressure approaches that in the aorta (recall that runoff of blood from the aorta is largely determined by the peripheral factors), the aortic valve snaps shut, ventricular pressure falls below that in the aorta (Fig. 2.111) and continues to fall rapidly. This period of the cardiac cycle is called "isovolumetric relaxation," since there is no inflow to or outflow from the ventricle (Fig. 2.111). Up to this point, the entire sequence of events encompasses *ventricular systole.*

As the pressure in the ventricle continues to fall it eventually becomes lower than that in the left atrium, the atrioventricular valve (mitral valve) opens, and blood flow across that valve commences (Fig. 2.111). With the onset of flow from atrium into the ventricle the period of *ventricular diastole* commences. The flow from the atrium into the ventricle is at first rapid, then it slows somewhat (Fig. 2.111), only to speed up again as the left atrium contracts following the next P wave; the latter event leads to a rapid rise in the left ventricular end-diastolic volume and pressure (sometimes called the "atrial kick") and reflects the "booster pump" func-

tion of the atrium (Fig. 2.111). Ventricular diastole ends with the arrival of the next QRS complex and the onset of ventricular isovolumetric contraction. Ventricular end-diastolic pressure is measured after the atrium has contracted and represents the final point ·of diastole, just at the onset of ventricular isovolumetric contraction; it corresponds with the largest volume or end-diastolic volume of the ventricle (Fig. 2.111).

Ventricular systole and diastole are also reflected by events in the aorta (shown as the aortic pressure pulse in Fig. 2.111). The maximum systolic pressure in the aorta, just above the aortic valve, is normally close to the peak left ventricular systolic pressure (slightly lower). The lowest pressure reached in the aorta before the next pressure wave arrives from the ventricle is called the aortic diastolic pressure. The pressure in the aorta as it crosses the ventricular pressure (at the second heart sound) (Fig. 2.111) corresponds closely with the pressure in the ventricle at the end of left ventricular ejection which is termed the end-systolic ventricular pressure.

FUNCTIONAL GEOMETRY OF VENTRICULAR CONTRACTION

The left ventricle has a somewhat ellipsoidal shape and normally it ejects about two-thirds of its contents; that is, two-thirds of the volume present at end-diastole is ejected with each heart beat (Fig. 2.112) (Holt et al., 1968). Most of the inward movement of the left ventricular wall during ejection occurs near the minor or short axis of the ventricle, only a small reduction occurring in the long axis, giving the ventricle an even more elliptical shape at the end of systole (Fig. 2.112). As the ventricle fills during diastole it becomes somewhat more spherical, as the short axis enlarges more than the long axis.

The right ventricle has a much more irregular shape, and its free wall wraps around the interventricular septum (see Fig. 2.61, transverse plane).

THE PRESSURE-VOLUME LOOP FROM SINGLE CONTRACTIONS OF THE LEFT VENTRICLE

Referring to Figure 2.111 and plotting simultaneous points of the ventricular volume curve and the left ventricular pressure curve, it is possible to construct a loop, representing ventricular diastolic filling and contraction, that provides a highly useful way of describing the activity of the ventricle (Fig. 2.113), as discussed further below.

L.L. #G47952   Anteroposterior   Lateral

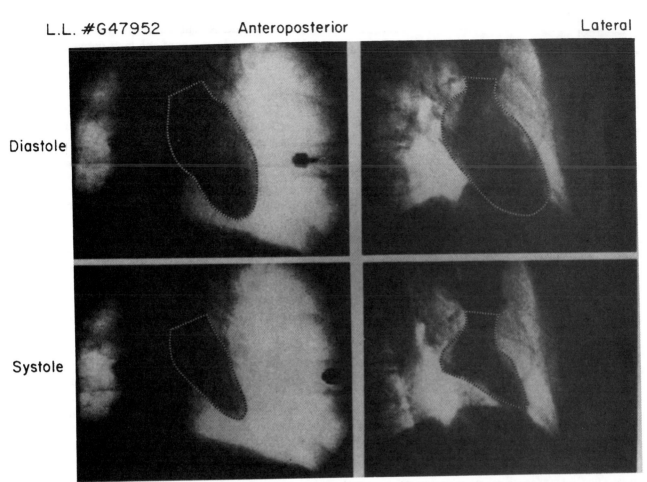

Diastole

Systole

Figure 2.112.   Representative frames from cineangiograms with injection of contrast medium into the left ventricle in the frontal (anteroposterior) and lateral projections. The upper two panels show the end of diastole in these two projections, and the lower two panels show the end of systole (outlined for clarity). Note the ellipsoidal shape of the left ventricle in diastole [From Rackley et al. (1967).]

The Passive (Resting) Pressure-Volume Curve of the Ventricle

As with isolated cardiac muscle, the myofilaments do not appear to interact appreciably in the relaxed ventricle, and they slide freely between one another. Therefore, if blood is allowed to fill a relaxed ventricle, the shape of the resulting relationship between ventricular pressure and volume is determined primarily by the properties of the tissues supporting the myofilaments, in particular the connective tissue or collagen. As the normal left ventricle fills progressively (Fig. 2.113, point 4 to point 1), a nonlinear relation between the rise in pressure resulting from each increment of volume is observed, the so-called passive pressure-volume curve. The shape of this curve is primarily due to the high compliance of relaxed supporting tissues when heart volume is small and the reduced compliance of the heart wall when these tissues begin to reach their elastic limit as the volume of the ventricle continues to increase, leading to steepen-

ing of the pressure-volume relation. (It can also be influenced by other factors, including the pericardium. For reviews see Covell and Ross, 1973; Glantz and Parmley, 1978; and Shabetai, 1981. See also Chapter 17 for pericardial effects.)

As can be seen from Figure 2.114, in the normal curve when ventricular volume is low, a large change in volume occurs with only a small change in pressure, whereas at large ventricular volumes a small volume change is accompanied by a large pressure change. The inverse slope of this relation at any point on the curve (dV/dP) can be called the chamber compliance, and the slope (dP/dV) represents the chamber stiffness (Fig. 2.114). The shape of this relationship has importance because it indicates that the heart is intrinsically difficult to overdistend; thus, on the steep portion of the pressure-volume curve, sarcomere lengths in the ventricular wall only slightly exceed 2.2 μm, and further efforts to increase volume only result in a very large rapid increase in pressure with little further over-

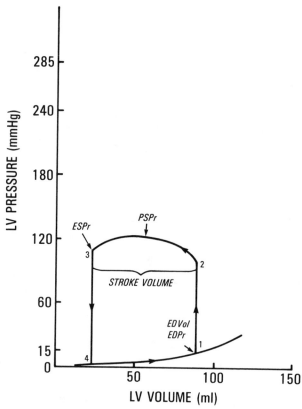

Figure 2.113. Left ventricular (*LV*) pressure-volume loop and the diastolic pressure-volume curve. During a normal contraction, point *1* shows end-diastolic (*ED*) pressure (*Pr*) and volume (*Vol*), point *2* the onset of left ventricular ejection which continues to point *3* (the distance on the volume axis representing the stroke volume). This is followed by isovolumetric relaxation (point *3* to point *4*) and then by filling of the ventricle along its diastolic pressure-volume curve (point *4* to point *1*). The end-systolic (*ES*) pressure-volume point is indicated, as is the peak systolic pressure (*PSPr*).

stretch of the sarcomeres. In the presence of heart disease, the shape and position of the diastolic pressure-volume relation can change considerably. For example, with a thickened left ventricular wall due to hypertrophy, the slope may be increased throughout its course and the pressure may be higher throughout diastole, indicating a less compliant (stiffer) ventricle (Fig. 2.114, abnormal curve).

In a single cardiac cycle, during the phase of ventricular filling the ventricle will tend to move up on a portion of its diastolic pressure-volume curve, as mentioned. Normally, the left ventricle operates on the lower, relatively flat portion of the curve (Fig. 2.113). (For application of this type of analysis to the human heart, see Grossman and McLaurin, 1976).

Systolic Left Ventricular Contraction

Beginning at point 1 on Figure 2.113 (the point of end-diastolic pressure and volume), we can trace

the remainder of the cardiac cycle by considering the moment-to-moment relation between left ventricular pressure and ventricular volume obtained from data shown in Figure 2.111. During isovolumetric contraction, pressure builds up in the ventricle with zero change in ventricular volume (point 1 to point 2, Fig. 2.113). After the onset of left ventricular ejection into the aorta (point 2, Fig. 2.113) ventricular pressure then rises and falls (together with the aortic pressure), and the volume of the ventricle progressively diminishes, reaching its minimum at the point of end-systolic pressure and volume (point 3). Ventricular pressure then falls at the smallest ventricular volume without further change in the volume (isovolumetric relaxation) until point 4, the lowest ventricular pressure which marks the onset of ventricular filling. The sequence of events gives a counter-clockwise loop (see *arrows*, Fig. 2.113), and the difference between the volume at end-diastole and that at the end of ejection is the stroke volume (Fig. 2.113) (Dodge et al., 1966).

We have previously considered the path of tension development and shortening in a piece of isolated cardiac muscle contracting isotonically. Although this type of contraction somewhat resembles the mode of contraction of the whole heart, it does not form a loop; that is, as shown in Figure 2.101, the muscle develops tension, shortens, and

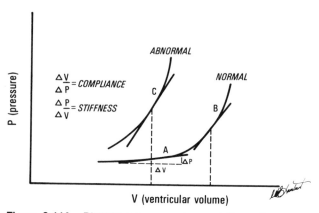

Figure 2.114. Diastolic pressure-volume relations of a normal and abnormal ventricle. The slope drawn at *A* shows a relatively flat relation between changes of volume (ΔV) and the accompanying change in pressure (ΔP), a large change of the volume producing a small change in pressure. Since the ventricle normally operates at relatively low pressures, it is relatively compliant in this range (compliance is the ratio of volume change to pressure change). Stated another way, on that portion of the curve stiffness is low (stiffness = the ratio of pressure change to volume change, the inverse of compliance). The normal ventricle becomes less compliant (more stiff) at large ventricular volumes (slope at point *B*, *dashed line*). In one type of abnormal ventricle (ventricular hypertrophy), the pressure-volume relation is steepened and shifted to the left, so that at the same ventricular volume shown as point *A* (*dashed line*) the compliance is reduced and stiffness increased (point *C*).

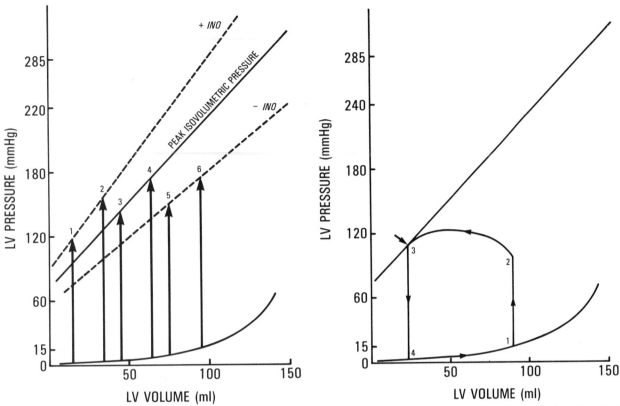

Figure 2.115. *Left panel,* Left ventricular (*LV*) pressure-volume relations, showing the relation between left ventricular diastolic volume and peak active isovolumetric pressure (*solid line*), produced by a series of isovolumetric contractions at differing end-diastolic volumes such as beats *3* and *4*. The active peak isovolumetric pressure volume curve is shifted upward and to the left by a positive inotropic stimulus (+ *INO*, isovolumetric beats *1* and *2*), and to the right by a negative inotropic stimulus (− *INO*, beats *5* and *6*). *Right panel,* Normal passive and peak active isovolumetric pressure-volume relation combined with the pressure-volume loop. Notice that the pressure-volume point at the end of left ventricular ejection (point *3, arrow*) falls near the isovolumetric pressure-volume relation.

then begins to relax with the weight still borne by the muscle, so that it lengthens while bearing the same tension before relaxing isometrically. Of course, it would be possible to keep the muscle short while allowing tension to fall isometrically and then to stretch the muscle passively to its resting end-diastolic length, thereby even more closely imitating the contraction of the whole heart (in which, ejection and aortic valve closure cause the ventricle to relax at a small ventricular volume). Recall that the shortest muscle length during active muscle shortening falls approximately on the isometric length-active tension curve of the muscle at any level of inotropic state (Fig. 2.100), a feature that also has an analogue in the whole heart (discussed below).

THE ISOVOLUMETRIC PRESSURE-VOLUME RELATION

If the isolated heart preparation is arranged with pressure at a very high level in the aorta so that the left ventricle cannot eject blood, and if the end-diastolic volume of the ventricle is then progressively lowered and raised over a wide range, an essentially linear relation between ventricular end-diastolic volume and the maximum (peak) pressure developed by isovolumetric contractions of the left ventricle is described (Fig. 2.115, *left*) (Monroe et al., 1961; Sagawa et al., 1967). (A similar linear relation can be plotted if ventricular diastolic volume is plotted against calculated peak isovolumetric tension developed by the ventricle, resembling the isometric length-active tension relation in isolated muscle) (Taylor et al., 1969.)* It has also been shown in the isolated heart that a positive inotropic stimulus, such as norepinephrine or digitalis, shifts this isovolumetric pressure-volume relation upward and to the left with an increase in its slope (with little change in the intercept on the volume axis), whereas a negative inotropic intervention can shift the relation downward and to the right with a reduced slope (Fig. 2.115, *left*).

---

* For detailed analysis of isolated hearts contracting under controlled conditions so that isovolumetric pressure-volume and tension-volume, as well as isotonic contractions, can be determined, see Covell et al., 1969, Burns et al., 1973, Suga et al., 1964, and Weber et al., 1976.

Further, it has been demonstrated that the end-systolic pressure-volume point at the end of left ventricular ejection (point 3, Fig. 2.115, *right*) falls on or close to this isovolumetric pressure-volume relation (Fig. 2.115, *right*). This point can be measured at the aortic incisura on the aortic pulse wave marking the end of ejection (or by finding the maximum ratio of left ventricular pressure to volume). This behavior of the heart allows the passive pressure-volume curve, the linear end-systolic pressure-volume relation, and the pressure-volume loop of the left ventricle to be combined in a single diagram (Fig. 2.115, *right*) (For review see Sagawa, 1978).

FOUR MAJOR FACTORS THAT INFLUENCE VENTRICULAR PERFORMANCE

The same factors that affected the performance of isolated muscle can be shown to importantly influence the performance of the whole ventricle (both the right and left ventricles), although only the left ventricle will be considered in this discussion. These factors include the *preload*, the *afterload*, and the *inotropic state* and, with the whole heart, a fourth important determinant, the *heart rate*, must be added (for review see Braunwald and Ross, 1979) (Fig. 2.107).

Preload

In the whole heart, the preload should constitute the tension in the wall at the end of diastole (which determines the resting fiber length), but for practical purposes the ventricular end-diastolic volume or the ventricular end-diastolic pressure are used to indicate the preload. The preload affects the performance of the heart through "Starling's law of the heart." As stated by E. H. Starling (1915): "The mechanical energy set free on passage from the resting to the contracted state is a function of the length of the muscle fiber, i.e., of the area of chemically active surfaces." This mechanism was also demonstrated in 1895 by Otto Frank in the isolated frog heart, and later Patterson, Piper, and Starling (1914) used the heart-lung preparation of the dog to show that as the end-diastolic volume of the ventricle (inflow from reservoir, Fig. 2.110) was increased (aortic pressure being held relatively constant) the stroke volume of the ventricle increased. Hence, the law is sometimes referred to as the "Frank-Starling mechanism," although earlier investigators had also noted the phenomenon.

Based on the principles of contraction previously described for isolated cardiac muscle, the effects of this mechanism on the potential for active isovolumetric pressure development by the ventricle and for altered shortening of the ventricular wall are evident. As shown in Figure 2.116, if the end-diastolic volume of the ejecting left ventricle is increased and the aorta is then suddenly occluded for one beat, two phenomena are seen. First, in the

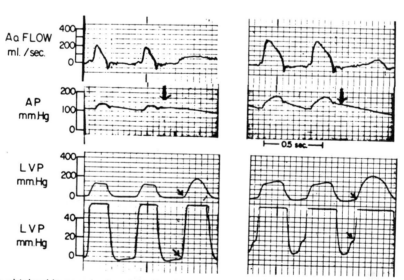

Figure 2.116.   Tracings obtained in an experimental animal in which aortic flow *(Ao Flow)* was measured in the ascending aorta with an electromagnetic flow meter, together with aortic pressure *(AP)*, left ventricular pressure *(LVP)*, and left ventricular pressure at high gain to allow visualization of the end-diastolic pressure *(small diagonal arrows, lower tracing)*. The preparation also allowed sudden occlusion of the aortic valve with a balloon *(large vertical arrow)*, so that the third contraction in each panel is isovolumetric. In the *lefthand panel*, the left ventricular end-diastolic pressure is low (1 or 2 mmHg) and in the *righthand panel* a large transfusion has markedly elevated the left ventricular end-diastolic pressure to about 23 mmHg. Note in the right panel that the two beats prior to aortic occlusion show an increased stroke volume and peak velocity of ejection despite the increased aortic pressure, compared to the left panel, and that the isovolumetric beat *(third beat, right panel)* develops a considerably higher peak pressure than the third beat in the left panel.

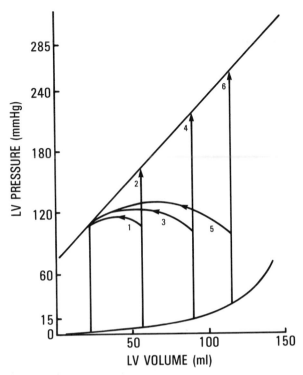

Figure 2.117. Left ventricular (*LV*) pressure-volume diagram showing effects of progressively increasing end-diastolic volume. In isovolumetric contractions (beats *2*, *4*, and *6*) the peak pressure is increased at larger end-diastolic volumes, and at nearly the same left ventricular pressure during ejection the stroke volume progressively increases (beats *1*, *3*, and *5*).

ejecting beats the stroke volume and velocity of ejection are increased at the higher end-diastolic pressure and volume, and secondly the peak pressure in the isovolumetric beat is augmented (Fig. 2.116, *right panel*).

This response can also be represented using the pressure-volume diagram, as shown in Figure 2.117. Both isovolumetric and ejecting contractions are shown at increasing end-diastolic volumes. The ejecting contractions (beats 1, 3, and 5) reach the same left ventricular end-systolic volume and pressure, and the delivery of a larger stroke volume as well as the potential for higher peak isovolumetric pressure at increasing end-diastolic volumes (in beats 2, 4, and 6) are evident (Fig. 2.117). It follows that if ventricular end-diastolic volume is reduced, peak isovolumetric pressure and stroke volume will fall.

Afterload

For practical purposes we will consider the systolic aortic pressure or systolic left ventricular pressure as the major determinant of the afterload, recognizing that the tension or force borne by the fibers of the ventricular wall (Fig. 2.109) provides a better measure of afterload (the aortic input

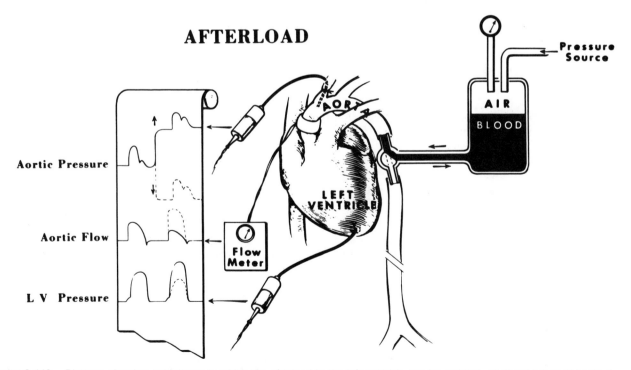

Figure 2.118. Diagram showing experiment in which the afterload in the left ventricle can be suddenly changed between beats by rapidly infusing or withdrawing blood from a reservoir attached to a compressed air source, the reservoir being connected to the aorta through a solenoid valve which is triggered from the electrocardiogram (Ross et al., 1966). The effects of suddenly increasing or decreasing aortic pressure during diastole are shown diagrammaticaly in the tracings: increasing aortic and left ventricular pressure decreases aortic flow, and decreasing aortic pressure (*dashed lines*) increases aortic flow.

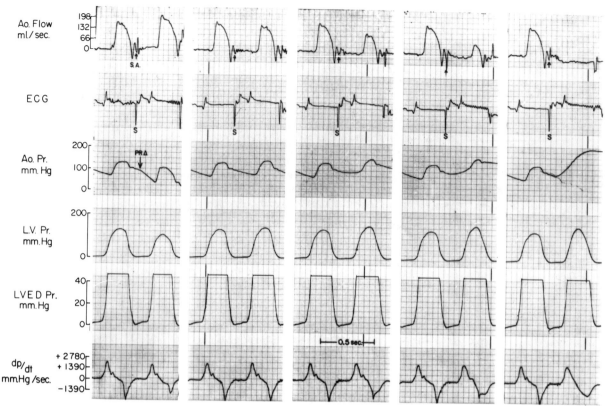

Figure 2.119. Actual tracings obtained in an experiment using the preparation shown in Figure 2.118. The first beat in each of the five panels represents a control contraction, and the second beat shows the effect of decreasing or increasing aortic pressure. *Ao* = Aortic, *Pr* = pressure, *LV* = left ventricular, *LVED Pr* = left ventricular end-diastolic pressure at high gain, and the first derivative of the left ventricular pressure (*dP/dt*) is shown in the lower tracing. *S* on the electrocardiographic tracing (and stimulus artifact *S.A.* on the flow tracing) shows the point of electrical triggering of the solenoid valve which causes withdrawal or infusion of blood into the aorta from the reservoir. In the *lefthand panel*, the aortic pressure drop (*PR∆, arrow*) resulted in a pronounced fall in diastolic aortic pressure, and the next contraction originates from the same left ventricular end-diastolic pressure but develops a lower left ventricular systolic pressure; hence, the stroke volume and peak aortic ejection velocity increased. Progressively higher aortic pressures in the middle three panels cause a progressive reduction of the aortic flow (stroke volume) and peak flow velocity, and in the *righthand panel* the pressure is raised so high that the ventricle cannot eject from that end-diastolic pressure; an isovolumetric contraction results (flow rate is zero). [From Ross et al. (1966).]

impedance also affects the afterload, see Chapter 7).† If an experimental preparation is devised so that the end-diastolic volume of the ventricle is kept constant while the aortic pressure is suddenly varied for a single beat by means of a solenoid valve

---

† Under certain circumstances, full understanding of the factors that *oppose* ventricular ejection may require actual measurement of the force on the myocardial fibers (the true afterload) (Sonnenblick and Downing, 1963). The factors which influence wall force are complex and involve the inertial properties of the blood and the elastic properties of the aorta, both of which can change continuously during systole, as well as the size of the ventricle and the thickness of its walls. Left ventricular performance can be affected not only by changes in mean aortic pressure and resistance, but also by the compliance characteristics and impedance of the peripheral circulation, as discussed in Chapter 7 (Milnor, 1975; Noordergraaf, 1975). However, such changes do not negate the force-velocity-length framework, because changes in impedance are reflected by changes in instantaneous myocardial wall stress (afterload) which, in turn, affects the shortening of the muscle (Pouleur et al., 1980). The input impedance constitutes a means for determining those external factors which contribute to the instantaneous pressure-flow relation and thereby contribute importantly to the systolic

loading conditions on the ventricle. Thus, factors affecting impedance, as well as factors relating to the geometry of the ventricle (heart size and wall thickness) determine the instantaneous afterload (force or tension) on the muscle fibers. The calculation of the input impedance remains a research procedure, since it requires a number of assumptions as well as high-fidelity measurement of pressure and flow at nearly the same site in the aorta, but it has found application, for example, in studying the affects of changes in the peripheral circulation in heart failure (Pepine, 1978). See Chapter 7, pages 144–145.

When afterload is measured as wall force or tension, we find that in the normal heart the wall force is maximum early in ejection and then rapidly falls, since ventricular size (radius) progressively diminishes and the wall thickness increases more rapidly than systolic pressure falls during later ejection (see Fig. 2.124). This importance of wall tension in heart failure, and the application of the law of Laplace under these circumstances, will be discussed in Chapter 18.

(Fig. 2.118), the effects of changing the afterload *alone* on left ventricular performance can be examined (Ross et al., 1966). The response of the

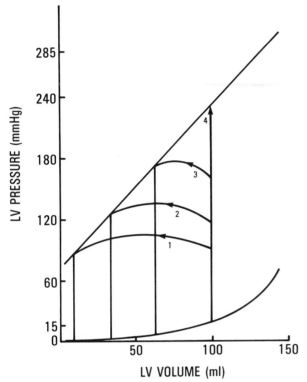

Figure 2.120. Pressure-volume diagram showing the effects of progressively increasing left ventricular systolic pressure from a constant left ventricular end-diastolic volume. There is progressive reduction of the stroke volume in beats *1*, *2*, and *3*. Beat *4* represents an isovolumetric contraction (compare to Fig. 2.119).

ventricle remarkably resembles that in isolated cardiac muscle; when aortic pressure is lowered during diastole (while the aortic valve is shut), the ensuing beat encounters a lower aortic pressure, and the left ventricular stroke volume is larger and ejected at a more rapid velocity (Figs. 2.118 and 2.119). As the aortic pressure is sequentially raised (Fig. 2.118 and on the second beat of each two-beat sequence in Fig. 2.119), the stroke volume and peak velocity of ejection progressively fall (notice that preload, as reflected by the left ventricular end-diastolic pressure, is constant) until the final panel in which aortic pressure is raised to a level higher than the amount of systolic pressure the ventricle can develop at that particular preload (which was set in this experiment at a relatively low level by means of a pump), and an isovolumetric contraction results (Fig. 2.119). Thus, an inverse relation exists between the level of afterload (systolic left ventricular and aortic pressure in this instance) and the extent of wall shortening (stroke volume) and the maximum rate of velocity of ejection from a given level of preload (Fig. 2.119), just as in isolated muscle.

This type of response is illustrated in the pressure-volume framework in Figure 2.120. Beat 1 illustrates a high stroke volume at a low level of left ventricular systolic pressure, beats 2 and 3 show progressively reduced stroke volumes at higher levels of systolic pressure, and beat 4 is an isovolumetric contraction, all beats originating from the same ventricular end-diastolic volume.

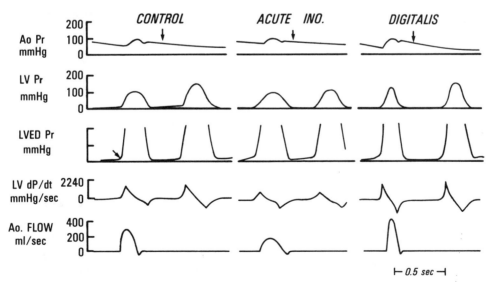

Figure 2.121. Tracings from an experiment similar to that shown in Figure 2.116 in which an isovolumetric beat could be produced by sudden inflation of the balloon in the ascending aorta. The filling of the ventricle is controlled by means of a pump in this experiment, so that the ventricular end-diastolic pressure and volume are held constant throughout. Heart rate is also constant. Control conditions are shown in the *lefthand panel* and a negative inotropic intervention (*middle panel*) produces a decrease in the stroke volume, a marked fall in the peak isovolumetric pressure, and a fall in the peak rate of pressure rise (*dP/dt*) of the left ventricle during isovolumetric systole. The positive inotropic effect of digitalis is shown on the *right* and results in an increase in stroke volume, peak isovolumetric pressure, and dP/dt.

Inotropic State

Tracings obtained in an animal in which end-diastolic pressure (preload) and systolic pressure (afterload) were held relatively constant (by means of a pump or a reservoir and a Starling resistor, Fig. 2.110) while a negative inotropic agent was given, followed by the administration of a large dose of a positive inotropic agent (digitalis), are shown in Figure 2.121. Note that following the administration of the depressant drug, stroke volume and the velocity of ejection fall, and when the aorta is occluded there is less peak pressure development during isovolumetric systole; following administration of digitalis, contractility rises above the control level with a rise in stroke volume and velocity of ejection, and there is a considerable increase in the peak rate of pressure development during isovolumetric contraction (Fig. 2.121).

These responses are diagrammed using the pressure-volume diagram in Figure 2.122. For clarity, the beats are drawn at slightly different left ventricular systolic pressures. Beat 1 shows a normal stroke volume and beat 2 an isovolumetric beat, both of which reach the normal linear relation between left ventricular end-systolic pressure and

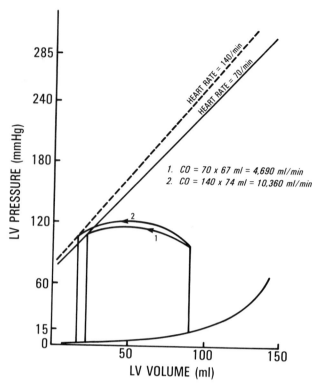

Figure 2.123. Pressure volume diagram showing the effects of increasing heart rate while maintaing the end-diastolic pressure in volume of the ventricle constant with a pump, or by transfusion. The end-systolic or isovolumetric pressure-volume relation is shifted upward and to the left by the force-frequency effect (*dashed line*), and beat 2 at the faster heart rate therefore delivers a slightly larger stroke volume. The effect of doubling the contraction frequency from 70 to 140/min at a constant end-diastolic volume is to more than double the cardiac output.

volume. A negative inotropic intervention shifts the linear end-systolic pressure-volume relation down with reduction in its slope, and beat 5 exhibits a markedly reduced stroke volume, and there is a reduced maximum peak isovolumetric pressure (beat 6, Fig. 2.122). Finally, a positive inotropic stimulus shifts the end-systolic pressure-volume relation upward and increases its slope. Beat 3 delivers a larger stroke volume than control, and beat 4 reaches a higher peak isovolumetric pressure (Fig. 2.122).

Heart Rate

As in isolated cardiac muscle, changing the frequency of cardiac contraction has an effect on the myocardial inotropic state through the force-frequency (staircase) relation, an increase in heart rate increasing myocardial contractility. Although the effects of altered contractility through the force-frequency relation are relatively small in the whole heart, the effect of changing heart rate on overall cardiac performance per minute (cardiac output) can be very large. For example, in Figure 2.123, the

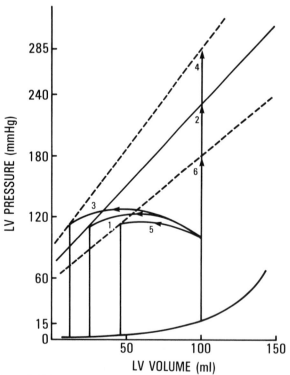

Figure 2.122. Pressure-volume diagram showing the effects of positive (*upper dashed line*) and negative inotropic stimuli (*lower dashed line*) compared to control (*solid line*); peak pressure in isovolumetric contractions is increased and decreased, respectively (beats *4* and *6*, compared to beat *2*), and the stroke volume is augmented with the positive inotropic stimulus (beat *3*) and reduced with the negative inotropic intervention (beat *5*).

filling of the heart is artificially maintained from a reservoir or by a pump to keep the end-diastolic volume constant while the heart rate is doubled from 70 to 140 beats per minute. Beat 1 represents the control condition, and in beat 2 a slightly larger stroke volume is delivered by each beat because of the leftward shift of the end-systolic pressure volume relation (due to the positive inotropic effect of increased heart rate) (Fig. 2.123). Under these conditions, however, increasing the heart rate more than doubles the cardiac output (the product of stroke volume × heart rate, Fig. 2.123), indicating that the heart rate can be a powerful determinant of cardiac performance per minute under certain circumstances.

THE PHYSIOLOGIC IMPORTANCE OF THE FOUR MAJOR DETERMINANTS OF PERFORMANCE

Under controlled conditions, we have varied each of the four major determinants of cardiac performance while holding the other three constant, in order to demonstrate their separate effects. However, it is important to realize that all of the determinants are operating together in the intact circulation, and they may change in *different* directions. Also, the ventricle responds to changes in loading conditions during the course of *each* cardiac cycle. For example, if the afterload is varied *during* ejection in single cardiac beats by suddenly increasing aortic outflow resistance, the systolic wall force (afterload) during that beat progressively rises and the instantaneous fiber length and volume do not diminish as rapidly, thereby reducing the stroke volume (Fig. 2.124). Notice, however, that the ventricle still reaches the end-systolic volume-wall tension relation at the end of ejection in each beat (Fig. 2.124) (Taylor et al., 1969).

Thus, all four factors operate simultaneously in the intact animal, or in man, to constantly modulate the heart's performance to meet different physiologic conditions (changing posture, exercise, excitement) by altering the heart rate, the level of inotropic state (through the neurohumoral control of the heart and circulation), and the loading conditions (preload and afterload), as reflected in changing volumes of the right and left ventricles and the varying aortic pressure.

In Chapter 17 we will consider the interaction of cardiac function and the peripheral circulation in integrated circulatory responses. However, it will be useful to consider here some additional features of the major determinants of ventricular performance and how they are expressed in the moment-to-moment regulation of the heart.

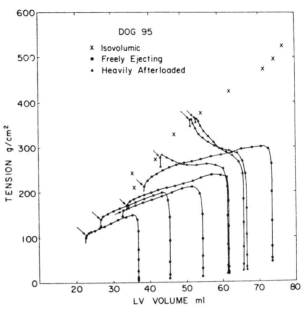

Figure 2.124. Relations between left ventricular (*LV*) volume and calculated wall tension in an experimental animal. The *x*'s represent peak tension developed by isovolumetric contractions at varying end-diastolic volumes. Portions of volume-tension loops are also shown, the *arrows* indicating the volume-tension point at the end of the left ventricular ejection. Notice that these points fall on the isovolumetric volume-tension relation both in freely ejecting beats from differing end-diastolic volumes, as well as in the heavily afterloaded beats produced by inflation of a balloon in the ascending aorta during ventricular ejection (see beats with rising wall tension during late ejection). [From Taylor et al. (1969).]

The Frank-Starling Mechanism (Preload)

Starling's law operates continuously on a beat-to-beat basis to control the output of all four chambers of the heart. This mechanism is also occasionally termed "heterometric autoregulation" of the heart (intrinsic regulation due to changing muscle length). It is expressed importantly in the following ways.

1. Any change in the return of blood to the right or left ventricle is immediately (on that same beat) expressed as a corresponding increase or decrease in performance. For example, if one very suddenly stands up, blood will be pooled in the legs, venous return to the right ventricle will drop, and its stroke volume will immediately drop. Conversely, if one is lying in a supine posture and the legs are suddenly elevated, return of venous blood to the heart from the legs will immediately increase, and the stroke volume of the right ventricle will rise.

2. Moment-to-moment changes in the output of blood from the right ventricle, after a delay of several beats (to account for the transit time of blood through the lungs) are followed by similar changes in the output of blood from the left ventricle; that is, an increased volume of blood pumped into the lungs will subsequently reach the left heart

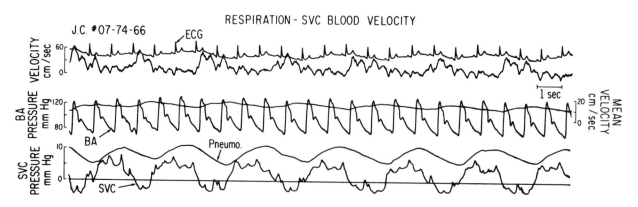

Figure 2.125. Effects of respiration on flow velocity in the superior vena cava (*upper tracing*), and pressure in the superior vena cava (*SVC, lower panel*), and on the pressure in the brachial artery (*BA, middle panel*) in a human subject. The pneumotachograph (*Pneumo*) records respiration, downward movement indicating inspiration. For further discussion see text. [Tracings obtained by I. T. Gabe et al. (1969).]

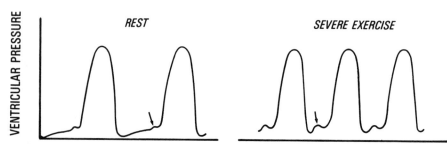

Figure 2.126. Diagrammatic tracings showing the effect of the atrial booster pump (*arrow*) on ventricular end-diastolic pressure at rest (*left panel*) and during severe exercise (*right panel*). During the short diastole of severe exercise atrial transport becomes important in maintaining ventricular filling.

causing an increased left ventricular end-diastolic volume (enhanced stretch of the fibers) and thereby produce a correspondingly enhanced stroke volume of the left ventricle. Were this mechanism *not* to operate, blood would rapidly accumulate in or be drained from the lungs, and the balance of the two ventricles could not be maintained.

This *balance* between the stroke volumes of the left and right ventricles over time is also manifested during normal respiration, and this type of response is illustrated in Figure 2.125. The velocity of flow in the superior vena cava (SVC) is recorded continuously together with the presssure in a human subject by means of calibrated catheter-tip velocity meter and a pressure transducer; the arterial pressure (BA) is also recorded. During inspiration, increased negative intrathoracic pressure lowers the SVC pressure (by creating an increased gradient for drawing blood from the peripheral veins into the right heart), and the vena caval flow velocity therefore rises transiently (Fig. 2.125). There is then a delay as this extra volume of blood is pumped through the lungs by the right ventricle. Initially, there is a transient, small reduction in the aortic pressure and stroke volume during inspiration and an increase in heart rate (the mechanisms for these

responses will be discussed in Chapter 16). Then, while expiration is occurring the increased output from the right heart reaches the left heart, and the left ventricular stroke volume and arterial pressure rise transiently (Fig. 2.125). Thus, the changes in stroke volume of the two ventricles are out of phase, but over a 1-minute period of time the integrated stroke volumes from the right and the left ventricles are equal, providing an illustration of the importance of the Frank-Starling mechanism.

3. Atrial muscle also is responsive to stretch, and as an increased volume of blood reaches the right heart or left heart it also stretches the atrium (or vice versa, if venous return decreases). This leads to variation in the force of atrial contraction with a consequent change in the "booster pump" function of the atrium (Mitchell et al., 1962; Mitchell and Shapiro, 1969). For example, when venous return to the right heart rises, increased atrial contraction helps to increase the end-diastolic volume of the right ventricle and hence its stroke volume, further assisting the heart in its role as a "demand pump" (a pump which delivers whatever volume of blood it receives).

The atrial booster pump (Mitchell et al., 1962) can become highly important in exercise (as well as

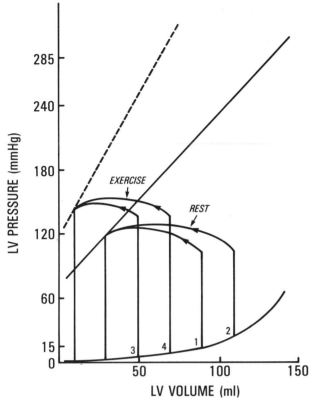

Figure 2.127. Left ventricular (*LV*) pressure-volume diagram showing the effect of increasing ventricular end-diastolic volume both at rest and during moderate exercise. At rest, increasing the end-diastolic volume (beat *1* to beat *2*) produces an increase in the stroke volume. During exercise, the end-diastolic volume is reduced (beat *3*) and the end-systolic pressure-volume relation is shifted upward and to the left (*dashed line*). If end-diastolic volume is increased (beat *4*), an increase in the stroke volume occurs even though the ventricle is smaller than under resting, control conditions (beat *1*), indicating continued operation of the Frank-Starling mechanism. Such changes might occur, for example, with deep respiration at rest and during exercise.

in disease). During normal strenuous exercise, when the heart rate is very high, the large number of systolic contractions per minute tends to markedly abbreviate the duration of diastole (Fig. 2.126) (time spent in systole is largely at the expense of diastolic time), despite the slight shortening of each beat produced by the positive inotropic effect. A forceful atrial contraction between each ventricular beat under these conditions greatly improves the transport of blood through the ventricles (Fig. 2.126).

4. Under abnormal conditions Starling's law may play an important compensatory role (see Chapter 18). For example, with complete heart block, when the cardiac rate is very slow (Chapter 9), filling of the ventricle increases during each prolonged diastole, and the increased end-diastolic volume therefore allows the ventricle to generate a much larger than normal stroke volume, so that the cardiac

output per minute may be maintained at a level compatible with life.

5. Starling's law continues to operate even when the cardiac volume is reduced below normal (as in shock, or during moderate exercise). Under exercise conditions, for example, the inotropic state is enhanced, and even though overall ventricular size is below the resting level, an increase or decrease of venous return to the ventricle will still result in a change in end-diastolic volume and hence in the stroke volume (Fig. 2.127).

Afterload

As mentioned, the peak or mean systolic aortic pressure or pulmonary artery pressure during ejection provides useful measures of the afterload. Although aortic pressure changes usually occur in the intact organism during activation of the autonomic nervous system, which also produces changes in myocardial contractility and heart rate (Chapter 16), the systolic aortic pressure alone can be varied by drug administration. Also, in some disease states rapid alterations in the afterload can suddenly oc-

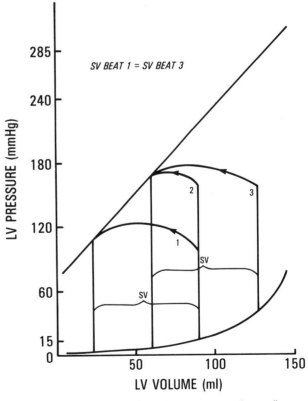

Figure 2.128. Left ventricular (*LV*) pressure-volume diagram showing the compensation for an increase in afterload. Beat *1* shows the control condition, beat *2* the initial response to an increase in afterload (the stroke volume falls), and beat *3* the compensatory increase in end-diastolic volume at the higher left ventricular systolic pressure, in which the stroke volume (*SV*) is restored to normal. (For further discussion see text).

cur; for example, pulmonary embolism (blood clot to the pulmonary artery) can cause a sudden increase in the afterload on the right ventricle.

1. A pure vasoconstrictor such as phenylephrine may be used to raise the blood pressure as, for example, when hypotension occurs during anesthesia, or to purposely produce an alteration of ventricular function in testing cardiac reserve (Chapter 17). Under the latter conditions, elevating the aortic pressure alone would *reduce* the stroke volume; however, as shown in Figure 2.128, in the intact circulation compensatory use of the Frank-Starling mechanism tends to restore the stroke volume of

the normal heart to normal. If beat 1 represents the control conditions (for simplicity, we consider a 3-beat sequence) and beat 2 encounters a higher systolic pressure and afterload as the drug is given, the stroke volume will drop; since the residual volume of beat 2 is therefore increased (the volume at the end of ejection is larger than in beat 1), and since the right ventricle is unaffected by the vasopressor effect on the systemic circulation, the right ventricular stroke volume will remain normal and augment the diastolic filling of the next beat by a normal amount. Consequently, beat 3 reaches a larger end-diastolic volume and the Frank-Starling

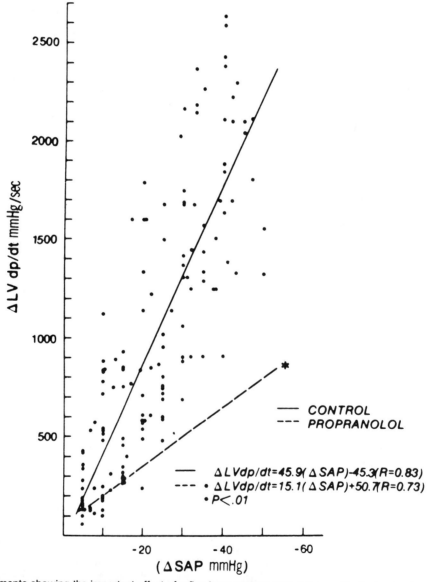

Figure 2.129. Experiments showing the important effect of reflex baroreceptor stimulation on the inotropic state of the left ventricle, as measured by the change in the rate of isovolumetric LV pressure development ($\Delta LV\ dP/dt$). The response shows the increase in dP/dt present a few seconds after correction of a *prior* decrease in systemic arterial pressure (ΔSAP). Note the positive correlation between decreasing SAP and increase of dP/dt. The increase is greatly attenuated by administration of the β-adrenergic blocking agent propranolol (*dashed line*). [From Yoran et al. (1981).]

mechanism restores the stroke volume to the control level, but at a higher systolic aortic pressure (Fig. 2.128).

2. "Homeometric autoregulation" is a term that has been applied to describe the response of the left ventricle to a very sudden increase in aortic pressure. It has also been called the "Anrep effect" (after the physiologist E. Anrep who first described it) (for review see Monroe et al., 1974). In experiments in animals in which a sudden marked increase in aortic pressure is produced, ventricular end-diastolic pressure initially rises and the stroke volume then recovers (described above) as the Frank-Starling mechanism compensates for the pressure increase. However, a gradual fall in left ventricular end-diastolic pressure then occurs in such experiments, while the stroke volume remains normal and the left ventricular and aortic systolic pressures remain elevated. This fall in end-diastolic pressure at constant level of left ventricular systolic performance has been considered by some to represent a change in myocardial contractility (which somehow results from the increased aortic pressure itself). The term homeometric autoregulation (meaning at the same muscle length) has been used to distinguish it from the "heterometric autoregulation" of the Frank-Starling mechanism.

More recent experiments indicate that this response probably represents recovery from transient insufficiency of blood flow to the inner wall of the heart (ischemia) caused by the sudden severe pressure change; that is, initial subendocardial ischemia is followed by coronary vasodilatation, with recovery from the brief insult. Whether or not homeometric autoregulation is of significance in the intact circulation remains controversial (Monroe et al., 1974).

Inotropic State (Myocardial Contractility)

The level of sympathetic or adrenergic "tone" modulates the myocardial inotropic state and the heart rate through local norepinephrine release in the myocardium. Using peak left ventricular dP/dt (the maximum rate of pressure change during isovolumetric contraction) as a measure of contractility shortly *after* stimulation of the baroreceptor reflex by lowering blood pressure (systolic and end-diastolic pressures had returned to normal), a linear change in peak dP/dt is seen following the blood pressure change (Fig. 2.129). Therefore, a decrease in the blood pressure produces a reflex positive inotropic effect on the heart muscle in the intact circulation (Chapter 16).

Stimulation of the vagus nerve has a marked negative inotropic effect on the atria (and slows

heart rate), particularly when background sympathetic tone is elevated (as discussed in Chapter 16). Under experimental conditions, a clear-cut negative inotropic effect on the ventricles of electrical stimulation of the vagus has been shown experimentally (DeGeest et al., 1965), although the effect is relatively small and its importance under normal physiologic conditions is unclear, since there is little cholinergic innervation of the ventricles.

A variety of pharmacologic agents, hormones, and metabolic factors also can profoundly affect myocardial performance by changing myocardial inotropic state. Also, the inotropic state of the myocardium is depressed in various types of heart failure. One of the goals of measuring performance of the heart in patients (Chapter 17) is to determine whether or not the function of the heart muscle is impaired compared to normal. For example, in patients with valvular heart disease, long-standing wear and tear on the muscle results in hypertrophy with some degree of scarring, and the severity of this myocardial impairment can affect whether or not the patient is a candidate for an operation to

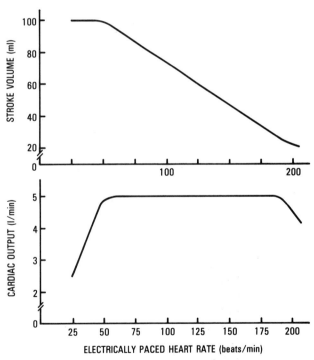

Figure 2.130. Diagram of the effects on the stroke volume and cardiac output of changing heart rate by electrical pacing of the right atrium or vagal stimulation. At very low heart rates (25–50 beats/min) the ventricle is maximally filled, and the cardiac output progressively drops as heart rate is slowed, without a change in the stroke volume. Over the range of 50–180/min the cardiac output does not change appreciably, so that the stroke volume progressively falls as heart rate is increased. Above about 180/ min, the cardiac output falls with increasing rates, probably due to impaired ventricular filling. For further discussion see text.

replace the defective heart valve. In some patients, the degree of myocardial depression (reduced inotropic state) or failure (Chapter 18) is so great that the risk of the operation becomes very high, and the benefits of valve replacement even if the patient survives operation are negligible.

Heart Rate

As mentioned earlier, changes in the heart rate can have profound importance in affecting the cardiac output, particularly under conditions such as exercise when the venous return to the heart is increased. However, when the heart rate is changed by artificial pacing (as by electrical stimulation of the right atrium with a pacemaker catheter), the cardiac output in normal subjects does not change over a wide range of paced rates (Ross et al., 1965). Under these conditions, the cardiac output is sta-

bilized by metabolic and reflex factors, and the venous return to the heart does not change, despite the altered rate and the enhanced myocardial contractility due to the force-frequency relation. (This type of response is also described as an interaction between the heart and the peripheral circulation in Chapter 17.) As shown in Figure 2.130, over a range of heart rates from about 50 to 180 beats/min the cardiac output remains relatively constant, and hence a progressive fall in the stroke volume occurs as rate is increased (cardiac output = heart rate × stroke volume) (Fig. 2.130). At higher paced heart rates, however, the cardiac output begins to fall, probably because there is inadequate time available for complete cardiac filling (Fig. 2.130); also, at low heart rates cardiac output falls since the ventricle reaches its maximum ability to deliver an increased stroke volume by use of the Frank-Starling mechanism (Fig. 2.130).

CHAPTER 13

Intracardiac and Arterial Pressures and the Cardiac Output: Cardiac Catheterization

In experimental animals and in human subjects it is possible to measure the pressure wave forms within the heart and great vessels and to determine the cardiac output by passing flexible tubes (catheters) which are only 1–3 mm in diameter and opaque to x-rays, so that their travel can be guided using a fluoroscope. Although pressures were measured in the hearts of animals by means of catheters in the 19th century, the first person to pass a catheter into the human heart was a young German intern, Werner Forssmann, who, in 1929, passed a catheter through a vein in his own arm, pushed it further into the vena cava and then placed its tip in the right atrium where the position was verified by an x-ray. The technique did not come into general use, however, until the 1940s and 1950s when two physicians in the United States, André Cournand and Dickinson Richards, developed special small catheters for use in the heart, and in patients with heart disease they passed such catheters into the right ventricle and pulmonary artery to record pressure and measure the cardiac output. For this accomplishment, Forssmann, Cournand, and Richards received the Nobel Prize in 1956.

Through a catheter, it is possible not only to inject indicators for measurement of cardiac output by the indicator dilution principle but also to withdraw blood samples from various sites within the heart and circulation, to apply the Fick principle for measuring cardiac output, to record pressures through the fluid-filled lumen of the catheter, and to inject liquids that are opaque to x-rays and therefore permit visualization of the heart and blood vessels. Using the latter technique (called angiography), high-speed motion pictures can be taken as the contrast medium is injected into a ventricle or an artery, allowing visualization of the pumping heart (Fig. 2.112), various congenital malformations of the heart, and identification of atherosclerotic lesions within blood vessels (such as in the coronary arteries, Fig. 2.153. (For details of cardiac catheterization methods of left and right sides of heart, see Grossman (1980).)

MEASUREMENT OF PRESSURE

When a catheter (or a needle, which is sometimes inserted directly into an artery) is filled with saline, the pressure wave form is transmitted from the catheter tip located within a heart chamber, pulmonary artery, or the aorta to the external end where the wave form can be recorded electronically by means of a transducer (Grossman, 1980). A typical transducer of the strain gauge type is shown in Figure 2.131 *C*, along with several other types of pressure transducers. The device has a fluid-filled chamber which covers a thin metal diaphragm; the back of the diaphragm is attached to a metal plate which is sensitive to slight displacement of the diaphragm, with a change in resistance through the metal elements being proportional to the displacement of the diaphragm. The diaphragm faithfully follows the pressure wave forms transmitted to it through a fluid-filled catheter, and a voltage proportional to the pressure is generated, amplified, and recorded on a strip chart recorder. Calibration of the transducer with a mercury manometer allows measurement of absolute pressure, and it is important to set the reference level at the stopcock (Fig. 2.131 *C*) to zero atmospheric pressure, at the level of the heart. Pressures recorded through such a catheter-transducer system are subject to some artifact from microbubbles in the catheter or transducer, and these must be carefully removed before use; in addition, movement of the catheter within

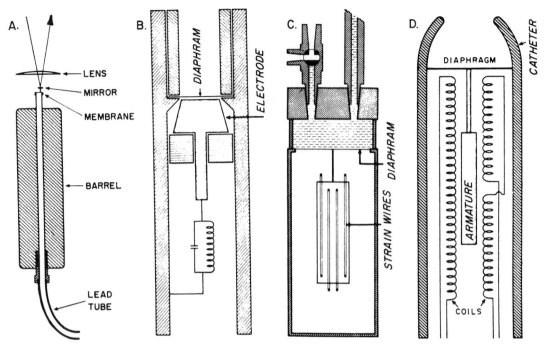

Figure 2.131. Externally recording pressure gauges. *A*), Principle of the classical optical method for determining biological pressures. A flexible, short, nondistensible lead tube offering little damping effect connects to the pressure source. *B*, Capacitance transducer; *C*, strain gauge transducer; *D*, Catheter-mounted inductance transducer.

the heart due to cardiac motion can cause artifacts by producing oscillation of the fluid column contained within the catheter. With care, the frequency response of such systems can be made adequate up to 10 or 15 cycles/second (Hz). More recently, miniature pressure transducers have been developed, of inductance, strain gauge, or piezo, electric types, that are small enough to be placed at the tip of a cardiac catheter (Fig. 2.131*D*), and these devices provide much higher fidelity pressure wave forms (up to 100 Hz), although they are expensive.

The arterial pressure can also be measured indirectly by the auscultatory method using an inflatable cuff placed on an extremity, usually the arm, with the pressure inside the cuff measured either with a mercury or aneroid manometer; the above device is called a *sphygmomanometer*. Pressure in the cuff is first inflated above the level of systolic pressure in the aorta which collapses the artery, and as one listens over the artery in the arm (brachial artery) with a stethoscope, no sound is heard. The cuff is then slowly deflated (2 to 3 mmHg/sec), and as soon as the systolic pressure is reached, a small spurt of blood passes beneath the cuff during each cardiac systole, producing a snapping sound due to turbulence and closure of the vessel (Fig. 2.132). As soon as this sound is audible, the systolic pressure is noted. As the cuff is then further deflated, the sounds change in character and intensity, and the flow becomes continuous, at which point the sound disappears and the diastolic pres-

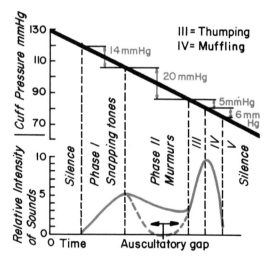

Figure 2.132. Characteristics of the auscultatory method of measuring blood pressure. [Redrawn from Geddes (1970).]

sure is recorded (Fig. 2.132). The changes in sound quality as the cuff pressure is lowered (Korotkoff sounds) are multiple,* but only the initial sound and the point of sound disappearance are now taken

---

* The Korotkoff sounds (Fig. 2.132) are: Phase I: faint tapping of increasing intensity (sound appearance up to 10–25 mmHg lower); Phase II: softer bruit for next 15–20 mmHg (sounds occasionally disappear near the "ausculatory gap," and systolic pressure should also usually be checked by palpating the pulse appearance); Phase III: sounds louder and clearer; Phase IV: muffling of sounds; Phase V: sounds disappear. The last phase corresponds clearly with the diastolic pressure at rest measured directly by intraarterial needle and transducer.

as systolic and diastolic pressures (a normal value would be recorded, for example, as 120/70 mmHg).

The Pressure Pulses in the Right and Left Sides of the Heart and the Heart Sounds

Knowledge of the sequence of electrical and mechanical events of both the right and left sides of the heart is central to understanding the shapes of the pressure waves recorded in the atria and ventricles, as well as the *timing* of heart sounds and murmurs (Wiggers, 1954). These events are shown diagrammatically in Figure 2.133, and a reference summary of terminology used to describe the ven-

tricular cycle with the average duration of these periods at an average heart rate of 75/min is shown in Table 2.7.

First, the basic pattern of ventricular pressure development and blood flow into and out of the ventricle should be recalled; during ventricular systole, flow across the atrioventricular valve is zero, and there is a period at the onset of ventricular systole (isovolumetric systole) when no flow occurs across any heart valve (see Fig. 2.111). During ventricular ejection, flow into the aorta or pulmonary artery takes place through the open semilunar valves and, following ejection during the period of

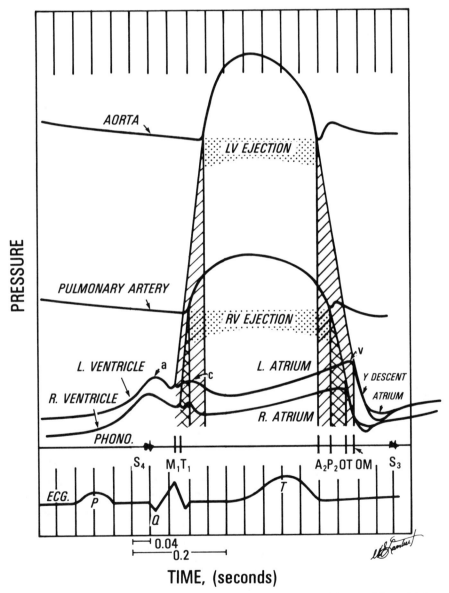

Figure 2.133. Diagram of pressure wave forms in the aorta and pulmonary artery, right ventricle and left ventricle (*RV* and *LV*), recorded together with the heart sounds on a phonocardiogram (*PHONO*) and the electrocardiogram (*ECG*). The *cross-hatched areas* show isovolumetric phases of right and left ventricular contraction and relaxation, and the *strippled areas* indicate the periods of right and left ventricular ejection. OT = opening of tricuspid valve, om = opening of mitral valve. For further discussion, see text.

Table 2.7
The ventricular contraction and relaxation cycle

360 ms (left ventricle)	**Systole**[b]
	1. Onset of ventricular contraction
	2. Atrioventricular (inlet) valve closure[a]
	3. *Isovolumetric ventricular contraction* (LV = 60 ms)
	(RV = 15 ms)
	4. Semilunar valve opening[a]
	5. *Ventricular Ejection* (LV = 200 ms)
	(RV = 270 ms)
	6. Semilunar (outlet) valve closure[a]
	7. *Isovolumetric ventricular relaxation* (LV = 100 ms)
500 ms	**Diastole**[c]
Total 860 ms (about 0.8 s)	8. Atrioventricular (inlet) valve opening[a]
	9. *Ventricular filling* (about 0.5 s or 500 ms)
	a. Rapid filling
	b. Slow filling (diastasis)
	c. Atrial systole or contraction (rapid filling again)

The 4 most important periods of the cycle are italicized. [a] Valves open and close. [b] Ventricular systole: events 1 through 7. [c] Ventricular Diastole: events 8 and 9. Event 7 is considered to be diastolic by many authors but for practical purposes may be included as part of ventricular systole. (For further details of cardiac cycle, see Wiggers (1954).)

isovolumetric relaxation, no flow takes place when all four valves are again closed. During diastole, there is early rapid filling followed by a phase of slow filling and then rapid filling again as atrial sytole occurs (Fig. 2.111).

The entire sequence of pressure events in the right and left atria and ventricles is shown diagrammatically, together with the electrocardiogram and the heart sounds, in Figure 2.133. The P wave, spreading from the sinus node in the right atrium, first depolarizes the right atrium (the first mechanical event in the cardiac cycle) and shortly thereafter the left atrium with subsequent contraction of that chamber. The slightly asynchronous contraction of the right and then the left atrium occurs late in diastole and is reflected by a buildup of pressure within the atrial chambers and in the ventricles (the mitral and tricuspid valves are still open), and the resulting positive wave is termed the *a* wave (Fig. 2.133). The *a* wave in the right atrium slightly precedes and normally is at a slightly lower pressure than that in the left atrium, and similar waves are produced in the ventricular pressure tracings as the pressure rises in those chambers due to the increase of volume provided by the atrial booster pump (see ventricular volume tracing, Fig. 2.111). In disease conditions, when the ventricles are stiff, a thudding sound may be audible just following atrial contraction. This sound is termed a "gallop" and is called an "atrial diastolic gallop" or fourth heart sound (S₄) (Fig. 2.133). Gallop sounds are usually related to rapid filling of a ventricle and may be likened to blowing up a paper bag which suddenly reaches its "elastic limit" and produces a popping sound.

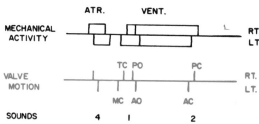

Figure 2.134. Sequence of valve motion as related to mechanical activity during the cardiac cycle. Right heart events above the line; left heart events below the line. *ATR.*, atrium; *VENT.*, ventricle; *RT.*, right; *LT.*, left; *MC.*, mitral closure; *TC.*, tricuspid closure. *PO.*, pulmonic opening; *AO.*, aortic opening; *AC.*, aortic closure; *PC.*, pulmonic closure; 4, 1, 2, fourth, first, and second heart sounds. Duration of isometric contraction for each ventricle is indicated in each ventricular rectangle by a verticle line. [Redrawn from Lewis (1962).]

After the *a* wave, the onset of isovolumetric ventricular contraction (first the left ventricle, then the right ventricle) follows the onset of the QRS complex by about 50 ms (Fig. 2.133). The left ventricular pressure then commences to rise rapidly and as the mitral valve snaps shut, the first heart sound (S₁) begins with closure of the mitral valve, followed soon thereafter by the tricuspid component of the first heart sound (M₁, T₁); Figs. 2.133 and 2.134.† (For review of heart sounds, see Craige

---

† The first heart sound is relatively long and of low frequency. At least 4 components have been identified, beginning slightly before the rise in left ventricular pressure and ending during early ejection. The second sound is shorter and of higher frequency. It results primarily from vibrations set up in the blood and blood vessel walls as the pulmonic and aortic valves tense after their closure. (For detailed discussion of the origins of heart sounds, see Bruns, (1959); Butterworth (1960); and Stein (1981).)

(1980).) Isovolumetric contraction of the two ventricles then occurs. The rapid rise of pressure in both chambers takes place at about the same rate; therefore, since the pulmonary artery pressure is only about one-fifth that in the aorta, the pulmonic valve opens sooner than the aortic valve, and the duration of isovolumetric contraction of the right ventricle is correspondingly shorter than that of the left ventricle (15 ms *vs.* 60 ms)(Table 2.7). No heart sounds accompany opening of the aortic and pulmonary valve. Thus, as shown in Figures 2.133 and 2.134, the entire duration of right ventricular isovolumetric contraction is enclosed within the isovolumetric period of the left ventricle. Isovolumetric ventricular contraction is accompanied by a positive wave in the atrial pressure tracings, which remain much lower than the ventricular pressure tracings because the tricuspid and mitral valves are shut. This small wave is termed the *c* wave. It is due in part to ballooning of the tricuspid and mitral valves backward into the atrium as higher pressure develops in the ventricular chambers (Fig. 2.133). A negative wave follows the *c* wave as the ventricles move away from the atria during early ejection (the so-called *x* descent, not labeled in Fig. 2.133). Throughout ventricular systole the tricuspid and mitral valves remain closed, but blood continues to flow into the atria from the venae cavae to the right atrium and from the pulmonary veins to the left atrium, producing a third positive wave in the atrial pressure tracings, the *v* wave (Fig. 2.133).

During ventricular ejection into the aorta and pulmonary artery, the pressure in the right ventricle is only 1 mmHg or so higher than that in the pulmonary artery and the pressure in the left ventricle exceeds that in the ascending aorta by 1 or 2 mmHg. It is during the period of ventricular ejection that the noise (or murmur) due to turbulent flow across an abnormally narrowed pulmonic or aortic valve would be heard. The ventricles then begin to relax, and the pressures also start to fall in the pulmonary artery and the aorta, as the rate of forward blood flow decreases sharply (this phase of the cardiac cycle has, in the past, sometimes been called "protodiastole"). The rapidly falling ventricular pressures then drop below those in the aorta and pulmonary artery. Deceleration of blood flow with reversal of the pressure gradient leads to closure of the aortic and pulmonary valves. The period of ventricular ejection, called the ejection time, is longer in the right ventricle than in the left (Fig. 2.133) (approximately 270 ms vs. 200 ms), so that pulmonic valve closure occurs *after* aortic valve closure. Therefore, two snapping sounds are audible as these valves close, and together they comprise the second heart sound (S_2), with its two compo-

nents termed A_2 and P_2 for aortic and pulmonic valve closure, respectively (Figs. 2.133 and 2.134). As is seen in Table 2.7, at a normal heart rate diastole is considerably longer than systole. This gives the normal heart sounds a "1–2, pause" sequence.

Isovolumetric relaxation then occurs in the two ventricles, the right ventricular pressure rapidly falls below that in the right atrium and, when the tricuspid valve opens, the *v* wave rapidly falls (the so-called *y* "descent"), marking the onset of rapid filling of the ventricle. A few msec thereafter, the falling left ventricular pressure pulse crosses the left atrial pressure pulse, and when the mitral valve opens, the *v* wave in that chamber also rapidly falls during early diastole (Fig. 2.133). As rapid ventricular filling continues, the *y* descent ends because pressures begin to increase rapidly in the ventricles and atria simultaneously. Toward the end of this rapid filling wave, another sound is sometimes audible (often in the normal heart in younger individuals, or particularly when the ventricle becomes stiff in heart failure); this sound when abnormal is sometimes termed a "ventricular diastolic gallop," or third heart sound (S_3) (Fig. 2.133). Ventricular filling then slows, and the pressures in the right atrium and the right ventricle, and those in the left atrium and the left ventricle, rise slowly together (with the atrial pressure slightly exceeding the ventricular pressure) (Fig. 2.133). This period of slow ventricular filling, which is sometimes called diastasis, is lost at rapid heart rates. Toward the end of diastole, the next P wave occurs, initiating the next *a* wave, and the cardiac cycle recurs.

The sequence of events and flow rates across the mitral or tricuspid valve make it quite easy to understand the motions of the mitral leaflet when an echocardiogram is recorded (Fig. 2.135). As the

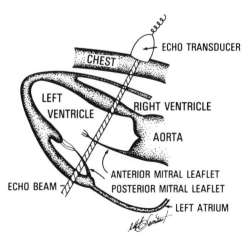

Figure 2.135. Diagram of the path of an echocardiographic beam passing through the chambers of the heart and the tips of the mitral valve leaflets.

echo beam traverses the leaflets, rapid inflow initially moves the leaflets wide apart during diastole, and as the flow slows down, the leaflets move closer together (Fig. 2.136); they open again widely with atrial systole and then abruptly close and remain opposed throughout ventricular systole (Fig. 2.136). The echodiagram is extremely useful in detecting abnormalities of valve motion and in detecting thickening of the leaflets (Feigenbaum, 1976).

Two or three of the normal atrial waves (a, c, and v) are usually visible in the neck veins. They can be recorded from the skin of the neck with an air-filled transducer and can provide considerable information about abnormal events on the *right side* of the heart. For example, if the tricuspid valve is leaking, blood fills the right atrium during ventricular systole from the right ventricle, flowing backward through the incompetent valve as well as forward from the venae cavae. The v wave visible in the neck veins is then larger and earlier than normal.

The pressure wave forms in the left atrium may also become distinctly abnormal in various disease states. For example, when the mitral valve is narrowed (with "mitral stenosis" often due to the scarring of rheumatic fever), a higher pressure is required in the left atrium to force blood across the narrowed mitral valve orifice during diastole. As one might predict, the rate of emptying of the left atrium also is delayed, resulting in a slow fall of the downslope of the v wave (slowed y descent).

THE PULMONARY ARTERY WEDGE PRESSURE

The left atrial pressure can be measured indirectly by passing a small flexible catheter, which has a small inflatable balloon at its tip, into the pulmonary artery. This is accomplished by advancing the tip of the catheter into a small pulmonary artery where the balloon is deflated (Fig. 2.137A); the pressure recorded from the tip of the catheter through the fluid-filled lumen is that in the pulmonary artery (Fig. 2.137A). When the balloon is inflated, the catheter is swept by the blood flow further out into the pulmonary artery, where it "wedges" and therefore obstructs blood flow through the small artery. This results in a fall of pressure in the segment of artery beyond the inflated balloon, and equilibrium across the pulmonary capillary bed then occurs between the pressure in the pulmonary veins and that in the left atrium; the result is a somewhat "damped" but relatively accurate recording of the left atrial pressure pulse (Fig. 2.137B), and the mean wedge pressure accurately reflects that in the left atrium. (This pressure is also sometimes called the pulmonary "capillary" wedge pressure.)

Such balloon "float" catheters are frequently used in critically ill patients at the bedside, since they can be inserted into a vein and passed without using x-ray guidance. The balloon is inflated, causing the catheter to be carried by the blood flow through the right heart chambers into the pulmonary artery. The position of the catheter tip can be determined simply by recording the different pressure waves as the catheter passes through the chambers of the right heart. Special balloon catheters are now manufactured for the purpose of measuring cardiac output. These have two lumens; one opens at the tip for measuring pressure, and the other more proximal opening is used to inject cold saline into the right atrium. There is also a thermistor at the catheter tip so that cardiac output can be measured by the "thermal dilution method" (see below).

Notice from Figure 2.133 and Table 2.8, in which the upper limits of normal for pressure values in the normal human heart are shown, that the pressure pulses in the left atrium are higher than those in the right atrium and that the left ventricular end-diastolic pressure is higher than that of the right ventricle. The differences are related primarily to two factors: first, the right ventricle has a much thinner wall than the left ventricle and is less stiff (more compliant), as shown in Figure 2.138. Therefore, when contraction of the two atria delivers an equal volume of blood into the two ventricles, a larger rise in pressure occurs in the left ventricle because its diastolic pressure-volume curve is steeper (Fig. 2.138). The second major factor contributing to higher pressures in the left atrium than in the right is that the left atrium and pulmonary veins contain a smaller volume and are less com-

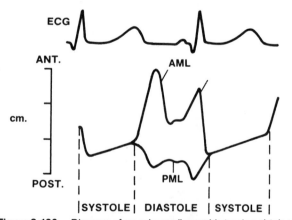

Figure 2.136. Diagram of an echocardiographic tracing obtained from the mitral valve leaflets (Fig. 2.135). During systole, both leaflets are closed, forming a single line, whereas during diastole, the anterior mitral leaflet *(AML)* moves anteriorly toward the front of the chest *(ANT.)* in a biphasic pattern; the posterior mitral leaflet *(PML)* moves similarly but in the opposite direction toward the posterior chest *(POST.)*. For further discussion, see text.

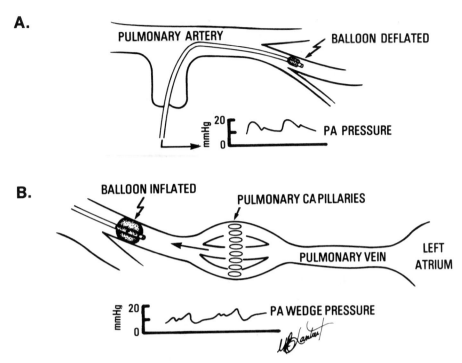

Figure 2.137. Technique for measurement of the pulmonary artery wedge pressure. *A,* Balloon at tip of catheter is deflated, and the pulmonary artery (*PA*) pressure is recorded. *B,* Magnified drawing showing the balloon inflated; pressure from the left atrium is transmitted across the pulmonary capillary bed into the obstructed segment of the pulmonary artery (*arrow*) to yield the pulmonary artery (*PA*) wedge pressure tracing.

Table 2.8
Normal intracardiac pressures in man[a]

		Adult Man[b]	
		Upper Limits of Normal, mmHg	Average Level, mmHG
Right atrium	a	7	5
	v	5	4
	Mean	6[a]	3
Right ventricle	S/D	30/5[a]	18/4
Pulmonary artery	S/D	30/15[a]	18/12
	Mean	20[a]	15
PA wedge	a	16	5
	v	21	10
	Mean	12[a]	8
Left atrium	a	16	10
	v	21	13
	Mean	12[a]	8
Left ventricle		140/12[a]	130/8
Systemic arterial	S/D	140/90[a]	130/75

S, systolic; D, diastolic (measured at end diastole). [a] Important values to remember at upper limit of normal. [b] *Other normal values* are as follows. Cardiac index: adults 2.5 to 3.5 l/min/m^2 BSA; AVO$_2$ difference, upper limits of normal at rest 5.2 vol%[a]; total body O$_2$ consumption, 110–150 ml/min/ m^2 BSA[a]; pulmonary vascular resistance (average 2 RU, upper limit 3 RU or 300 $\frac{\text{dynes-s}}{\text{cm}^5}$); systemic vascular resistance (average, 20 RU, 700–1500 $\frac{\text{dynes-s}}{\text{cm}^5}$); 1 RU = 80 $\frac{\text{dynes-s}}{\text{cm}^5}$ (1333 is the constant for conversion of mmHg to dynes/cm^2, and flow is expressed in ml/s for this calculation).

pliant than the right atrium and the venae cavae. Thus, equal volumes of blood flowing into the left and right atria during ventricular systole (when the mitral and triscuspid valve are closed) result in a larger *v* wave in the left atrium than in the right atrium. Higher *a* and *v* waves on the left side lead to a higher mean pressure under normal conditions, and the upper limit of normal for the left ventricular end-diastolic pressure is about twice that for the right ventricle (Table 2.8)

MEASUREMENT OF CARDIAC OUTPUT

Many methods are used for measuring the cardiac output in experimental animals, including placement of an electromagnetic device directly around the aorta for determining blood flow velocity, insertion of mechanical recording flowmeters by cannulating the aorta and diverting blood through the flowmeter (for example, displacement of a float is proportional to velocity of blood flow), and use of indirect techniques, including the Fick principle and the indicator dilution method, as discussed below. The latter two methods are useful in human studies, and it is also possible to place an electromagnetic catheter tip velocity meter in the aorta or pulmonary artery. (The volume flow can then be

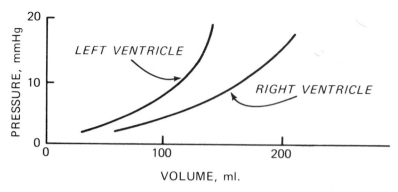

Figure 2.138.   Diagram of pressure-volume curves of the left and right ventricles, indicating that the right ventricle is more compliant.

calculated if the cross-sectional area of the vessel is known by angiography or echocardiography).

The total output of blood from the heart (the *cardiac output*) is equal to the product of the heart rate and the stroke volume. For example, in an adult human subject of medium size, the stroke volume might be 84 ml at a heart rate of 70/min, yielding a cardiac output of 5880 ml/min, or 5.88 l/min. The cardiac output is often expressed as the *cardiac index*, which is the cardiac output per square meter of body surface area (BSA). This correction is helpful because, as might be expected, large individuals have a higher cardiac output at rest than small individuals. The total body energy requirements at rest across species also correlate well with the cardiac output. In addition to body size, the amount of O_2 carried in the blood and the percent of O_2 extraction in the tissues importantly affect the cardiac output. For example, the blood of fish has a low O_2 carrying capacity, and cardiac output is proportionately higher. When anemia is present in the human subject, O_2 supply to the tissues is maintained by an increase in the cardiac output. The range of normal for the cardiac index is 2.4–3.5 l/min/m^2 body surface area (Table 2.8); it is slightly higher in males than females and considerably higher in children.

The Fick Method for Cardiac Output

The Fick principle (Adolph Fick, 1870) is an important concept, which can be applied for calculating the cardiac output, or the blood flow rate to any organ. It can also be used for calculating the oxygen consumption of the entire body or any organ, provided the flow rate and the O_2 content of blood samples are known. It did not become possible to measure cardiac output accurately in man by means of the Fick method until a sample of mixed venous blood could be obtained from the pulmonary artery through a cardiac catheter (a requirement discussed further below).

The Fick principle actually is an indicator dilution method, the indicator in this case being a physiologic one, oxygen. In the generalized form of the Fick equation the amount of any indicator accumulated or removed per minute (X) at a given site must be equal to the rate of flow per minute (Q) of the substance carrying the indicator multiplied by the difference between the concentration of indicator in the carrier before and after it passes that site ($C_1 - C_2$). The analogy may be made to a coal train passing a dumping site. Each coal car is equally filled upon arriving at the site, and each dumps the same quantity at the dumping site; by knowing the amount of coal in the cars that are arriving and in the cars that are leaving, the amount dumped by each can be determined ($C_1 - C_2$). It is readily apparent that the rate of accumulation of the coal per minute (X) will equal the product of the number of cars passing the site per minute ($\dot{Q}$) and the difference in content between a coal car arriving at and leaving the site: $X = \dot{Q}(C_1 - C_2)$ which, in order to calculate flow rate, can be arranged as:

$$\dot{Q} = \frac{X}{C_1 - C_2}$$

As usually applied to measure cardiac output in man, O_2 uptake is measured at the lungs by collecting a known volume of expired gas in a spirometer and analyzing it for residual O_2 content (the O_2 content of inspired air is known) (Fig. 2.139). The O_2 concentration in blood entering the lungs is measured from a sample of blood drawn through a catheter placed in the pulmonary artery, and the O_2 content of blood leaving the lungs is measured from a sample of blood collected from any peripheral artery, such as the brachial artery (Grossman, 1980) (the small amount of blood that normally shunts through the lungs is ignored, and the oxygen content of the systemic arterial bed is similar everywhere in the circulation). In practice (Fig. 2.139), O_2 uptake at the lungs per minute ($\dot{V}O_2$) is measured for a 3-min period, and blood samples are drawn simultaneously from the pulmonary artery (*V*) and peripheral artery (*A*) during the middle minute of the gas collection period (Fig. 2.139) to obtain the arteriovenous O_2 difference (AVO_2):

$$\dot{Q} = \frac{\dot{V}O_2}{AVO_2}$$

Example: O_2 uptake measured at the lungs was found to equal 240 ml/min, arterial blood O_2 con-

THE FICK METHOD

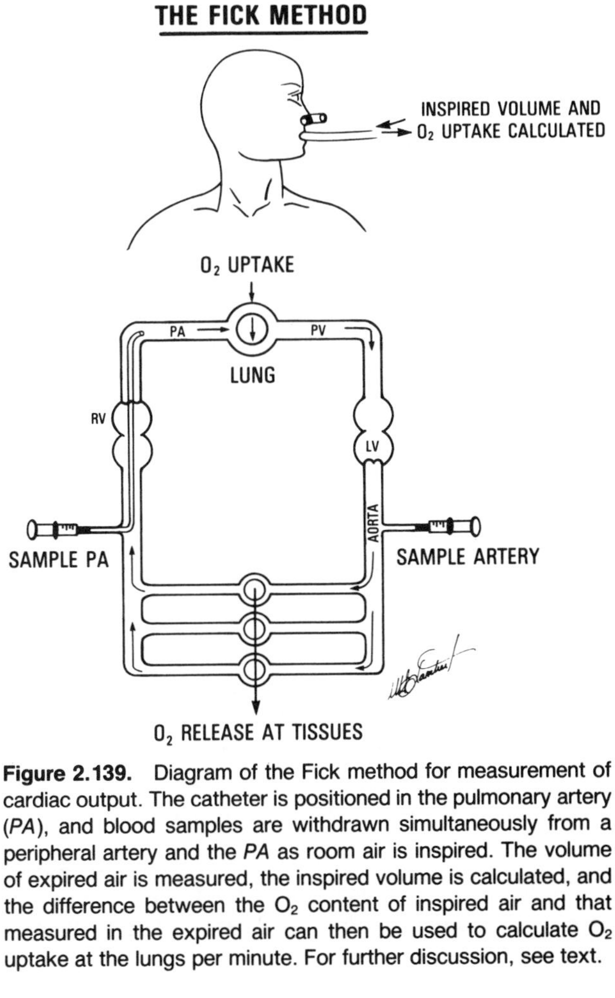

Figure 2.139. Diagram of the Fick method for measurement of cardiac output. The catheter is positioned in the pulmonary artery (*PA*), and blood samples are withdrawn simultaneously from a peripheral artery and the *PA* as room air is inspired. The volume of expired air is measured, the inspired volume is calculated, and the difference between the O_2 content of inspired air and that measured in the expired air can then be used to calculate O_2 uptake at the lungs per minute. For further discussion, see text.

tent = 18 vol% (18 ml O_2 per 100 ml blood = 0.18 ml O_2 per ml), and venous O_2 content = 14 vol%. Solving with the Fick equation, the blood flow or $\dot{Q}$ = 6000 ml/min. As measured, this value represents the pulmonary blood flow, but in the steady state it also equals the cardiac output to the body since, under basal resting conditions, the oxygen uptake of the lung is equal to the O_2 utilized by the tissues of the body (Fig. 2.139).

There are several prerequisites which must be fulfilled if the Fick equation is to be applied for measuring cardiac output.

1. A sample of *mixed* venous blood must be obtained from the pulmonary artery, since the O_2 content of the venous blood draining the various organs of the body varies widely, and some organs extract much more O_2 than others (Table 2.9). For example, the cerebral arteriovenous difference is about 6.5 vol% (and drains into the superior vena cava), while the difference in O_2 content across the heart is about 12 vol% (and drains into the right atrium); the renal circulation has an arteriovenous difference of about 2 vol% (and drains into the inferior vena cava). Therefore, the oxygen contents of blood in the venae cavae and right atrium can differ widely, usually being higher in the inferior vena cava because of the high renal contribution, whereas sampling in the right atrium near the exit of the coronary vein would result in a low O_2 content which is not representative of the entire body. It is apparent that in order to obtain a sample from the whole body it is necessary to draw the sample from a site where the venous blood is well mixed. This mixing is accomplished in the right ventricle, and a pulmonary artery sample therefore is representative of the combined venous drainage. The difference between the mixed venous O_2 content and the arteriol O_2 content for the whole body at rest is usually 3–4 vol%, the upper limit of normal at rest being 5.2 vol%, and it can rise much higher during heavy exercise.

When the Fick equation is applied to measure

Table 2.9
Distribution of cardiac output and oxygen usage

| Region | Weight | | Blood Flow | | | Oxygen Usage | | | O_2 Extraction | |
	kg	% Total	l/min	% Total	ml/100 g/min	ml/min	% Total	ml/100 g/min	Venous O_2 ml/100 ml blood	AVO_2
Total	70	100	5.4	100		250	100		14.5	4.5
Brain	1.54	2.2	0.83	15	54	63	23	3.7	12.5	6.5
Heart	0.33	0.5	0.22	4	70	23	9	7.0	7	12
Liver, intestines	2.86	4.0	1.54	29	54	55	20	1.95	15	4.0
Kidney	0.33	0.5	1.43	27	430	20	7	6.0	17	2.0
Skeletal muscle	34.0	50.0	0.92	17	2.7	55	20	0.16	12.5	6.5
Exercise			35.0	80	100	2700	90	7.9	3	16.0

blood flow to any organ, the venous sample must be representative of the entire venous drainage from that organ.

2. The *total amount* of the substance taken up or removed must be accurately known (no indicator can be lost). The Fick equation can also be applied when foreign substances are infused or cleared from the circulation, and other physiologic substances such as CO_2 production have also been used in the Fick equation to measure cardiac output. In the case of the application of the Fick method using O_2, it is necessary to make certain that all the expired gas is collected (there must be no leakage around the mouthpiece or noseclip).

3. A *steady state* must exist. There must be no marked change in the rate of ventilation during the measurement period (for example, transiently increased ventilation might result in more oxygen uptake at the lungs than is used by the peripheral tissues).

A steady circulatory state must exist during the measurement period (for example, if the subject were to become somewhat excited during the measurement period, the cardiac output might rise without a corresponding increase in ventilation, leading to inaccuracy in the measurement).

Given these precautions, the Fick principle can be widely applied in a variety of physiologic settings. For example, if blood flow to an organ is measured by an independent method (such as a flowmeter) and the arteriovenous O_2 difference across the organ is measured, the Fick equation in the form:

$$VO_2 = \dot{Q}(AVO_2)$$

can be used to calculate the oxygen consumption of the organ.

Indicator Dilution Methods

A variety of indicators has been used for measuring cardiac output or organ blood flow (Dow, 1956). The earliest were dyes, the concentration of which could be measured in the blood by light transmission (densitometer) techniques. This approach is still employed using indocyanine green dye, the concentration of which is measured as a sample of blood is withdrawn continuously through a recording densitometer. The thermal dilution method is now more widely used and involves the injection of chilled saline (of known volume and temperature) while the degree of cooling of the blood (dilution of the cold injectate) is measured by means of a calibrated thermistor placed directly in the blood stream.

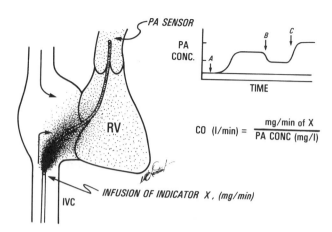

Figure 2.140. Continuous infusion method for measurement of cardiac output. The indicator is infused into the inferior vena cava (*IVC*) or right atrium, mixes in the right ventricle (*RV*), and passes into the pulmonary artery (*PA*) where the PA concentration (*CONC*) is recorded. The indicator injection is commenced at A and reaches a steady state at B; the cardiac output then increases, resulting in a lower *PA* concentration, and at C the cardiac output falls and *PA* concentration rises (less dilution).

THE CONTINUOUS INFUSION METHOD (Fig. 2.140)

If an indicator (such as a dye or cold saline) is infused at a constant rate at one location (for instance, the vena cava) and a sample is drawn during a steady circulatory state further along in the circulatory system beyond a mixing chamber (for example, in the pulmonary artery), cardiac output can be calculated provided there is no recirculation (Fig. 2.140). The cardiac output will be directly proportional to rate of infusion of the indicator (which is known and constant) and inversely proportional to concentration at the sampling site. Thus:

$$\dot{Q} = X(\text{mg/min})/PA \text{ concentration (mg/l)}$$

That is, the higher the blood flow, the more dilute the indicator (which is being delivered at a constant rate); the lower the flow, the higher the indicator concentration at the sampling site (Fig. 2.140).

This approach would be highly useful were it not for the problem of recirculation which, of course, distorts the concentration in the pulmonary artery as soon as it occurs. A second sensor can be placed in the vena cava upstream from the injection site which samples the recirculation concentration to make a correction, but this approach is cumbersome. If the indicator is a highly insoluble gas (such as krypton) that is eliminated in one passage through the lungs, flow can be calculated by this technique, provided an appropriate sensor is available.

The continuous infusion approach has had some success for measuring blood flow in individual organs by continuously infusing cold saline into a vein draining the organ and by sampling with a thermistor further downstream. Adequate mixing appears to be obtained in some veins having high flow (such as the coronary sinus), since only a small amount of cold solution is injected to measure local blood flow, and the total amount reaching the body for recirculation is negligible.

THE RAPID (BOLUS) INJECTION METHOD

Cardiac output can also be determined by rapidly injecting a small amount of an indicator (green dye or a bolus of cold saline) into the circulation and sampling at a downstream site to produce an "indicator dilution curve," which represents a graphic recording of the indicator concentration vs. time. The approach commonly employed was derived experimentally by G. N. Stewart (1896) and subsequently modified by Hamilton (1930); hence, it is commonly called the Stewart-Hamilton method (Dow, 1956).

In practice, 2 or 3 ml of the indicator is rapidly injected intravenously, with a needle for sampling being inserted into a systemic artery. The indicator injected into the vein mixes in the right heart and is diluted by the entire systemic blood flow. It is then further mixed and dispersed in the central circulation and left heart, and its concentration in the artery is sampled by withdrawing with a syringe at a constant rate through a densitometer. Alternatively, a thermistor can be placed directly in the arterial stream. It is also possible to apply this method by injecting a bolus of cold saline into the right atrium and sampling with a thermistor placed in the pulmonary artery (Fig. 2.140), or dye or cold solution can be rapidly injected into and mixed by the left ventricle and sampled from a peripheral artery.

It is important to understand that once the indicator is injected and becomes mixed with the *entire* volume of blood passing a given site at a given time, only a *sample* of blood need be taken from *any* site further along in the circulation beyond the point of mixing. This principle is illustrated in a very simplified form in Figure 2.141. In the example, 6 mg of dye are injected into a container of fluid and thoroughly mixed. A stopcock is then opened, and the fluid is allowed to run out through a tube. When a sample of the flowing stream is continuously drawn past a densitometer, a concentration of 1 mg/l is recorded (Fig. 2.141). This means that the indicator must have been *diluted* in 6 l of fluid. If the duration of the recording is 2.0 min (the duration of the indicator-dilution "curve"), it can be intuitively seen that the flow rate (cardiac output) through the tube must have been 3 l/min. The general form of the equation is:

$$\dot{Q} = \frac{X}{\bar{c}t}$$

where X = amount of indicator, $\bar{c}$ = mean concentration, and t = time (duration of the curve), and in the example above, $\dfrac{6 \text{ mg}}{1 \text{ mg/l} \times 2 \text{ min}} = 3$ l/min.

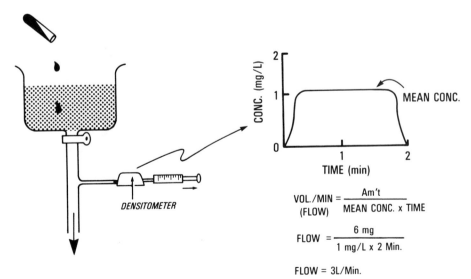

Figure 2.141. Simplified representation of an indicator dilution curve. The indicator is mixed in the container, and the stopcock is then opened to allow the container to empty as a sample is drawn by a syringe at a constant rate through a densitometer to yield a time concentration curve. For further discussion, see text.

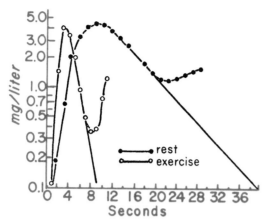

Figure 2.142. Indicator dilution determinations of cardiac output at rest and during exercise, employing the dye T-1824 (Evans blue). *Abscissa*, time in seconds; *Ordinate*, dye concentration in mg/l. Flow (F) equals the amount of dye injected (E) divided by the average concentration of the dye (Cdt), e.g., F = E/Cdt. At rest, flow in 39 s = 5 mg/dye injected/1.6 mg/l (average concentration) = 3.1 in 39 s. Converting to l/min, F = 3.1 × 60/39 = 4.7 1/min. During exercise, flow in 9 s = 5 mg/1.51 mg/l = 3.31. Flow/min = 3.3 × 60/9 = 21.9 1/min. [From Asmussen and Nielsen (1952).]

Time is often given in seconds, and the equation is written:

$$\dot{Q} = \frac{60 \cdot X}{\bar{c}t}$$

This equation resembles that shown for the continuous infusion method (note the resemblance to the Fick equation), and it can be intuitively seen that the lower the mean concentration (more dilution) and the shorter the duration of the curve, the higher the cardiac output, and vice versa (Fig. 2.142).

For the rapid injection method to be used reliably, several prerequisites must be fulfilled.

The effects of recirculation must be eliminated, that is, recirculation of indicator before the initial time-concentration curve is complete must be detected and removed. Recirculation poses a particular problem with this method, since early recirculation (as occurs through the coronary arteries, the first branches of the aorta) may not be apparent from the shape of the indicator dilution curve, although the bulk of the recirculating dye is usually detectable as a second wave on the trailing edge of the curve. Hamilton discovered that the downslope of an indicator dilution curve is a simple exponential function. This may be thought of in simplest terms as washout of the indicator from a ventricle of the heart. For example, if a dye is injected suddenly into the left ventricle and sampled in the aorta, since the ejection fraction remains constant during a sequence of heartbeats during a steady state, the concentration of the dye is diluted with each successive heartbeat by a constant fraction of undyed blood flowing in from the left atrium. This progressive reduction or concentration by a constant fraction results in an exponential fall in the concentration of the curve recorded in the aorta and is sometimes called a "washout curve."

In practice, recirculation is eliminated by plotting the curve on semilog paper (Fig. 2.142) as the log of dye concentration vs. time. Recirculation is then readily detected as a deviation from the straight line downslope, and the primary curve can be calculated by linearly extrapolating this downslope to zero. The *mean area* of the primary curve can be readily determined, and the *mean concentration* of the indicator during the curve can be computed.

No indicator can be lost during injection; it must be entirely mixed, and subsequently a representative sample must be obtained as the blood entirely mixed with indicator passes the sampling site.

The indicator must also exhibit the property called "stationarity," that is, it must be capable of being completely mixed and carried with the blood (a particulate dye must not settle out and thereby be lost in transit to the sampling site).

There must be a steady circulatory state during the entire measurement period of the curve, since sudden changes in blood flow will distort the characteristics of the curve and invalidate the principles of the method.

USE OF PRESSURE AND FLOW TO CALCULATE VASCULAR RESISTANCE

When the rate of blood flow and the mean pressure difference across a vascular bed are known, a measure of vascular resistance can be determined. As discussed in Chapter 6, calculation of true resistance involves application of the Poiseuille equation and knowledge of the length and radius of the vascular bed and the viscosity of blood, values which are not available. Nevertheless, measures of vascular resistance and the "vascular resistance unit" are highly useful in studying human disease.

For example, the pulmonary and systemic vascular resistances can be estimated in the following manner using the general resistance equation:

$$R = \frac{P_1 - P_2}{\dot{Q}}$$

where R = resistance, P_1 = mean arterial pressure, and P_2 = mean venous pressure. Thus:

$$PVR = \frac{P_{PA} - P_{LA}}{\dot{Q}}$$

where PVR = pulmonary vascular resistance, P_{PA} = pulmonary artery pressure, P_{LA} = left atrial pressure, and $\dot{Q}$ = pulmonary blood flow.

If the mean *PA* pressure is 20 mmHg, the mean *LA* pressure is 10 mmHg, and the pulmonary blood flow is 5 l/min, the *PVR* = 2 *RU* (dimensionless resistance units). Note from Table 2.8 that the upper limit of normal is 3 resistance units. Also:

$$TPVR = \frac{P_{SA} - P_{SV}}{\dot{Q}}$$

where *TPVR* = total peripheral vascular resistance, P_{SA} = mean pressure in a systemic artery, P_{SV} = mean pressure in a systemic vein (or the right atrium), and $\dot{Q}$ = the cardiac output.

If the P_{SA} is 105 mmHg, the venous pressure is 5 mmHg, and the cardiac output 5 l/min, the *TPVR* is 20 RU. This is a normal value for the systemic vascular resistance, which is about 10 times that of the normal pulmonary circulation. Vascular resistance is also sometimes expressed in dynes-s/cm^5 instead of units, and this calculation is shown in footnote b of Table 2.8.

These resistance relations may be strikingly altered in disease states. For example, the systemic vascular resistance is elevated in hypertension. In patients who have a hole between two chambers of the heart (such as a ventricular septal defect), there is a left-right shunt of blood due to the higher pressures on the left side of the heart. This short circuit of blood causes an extra volume of blood to recirculate continuously through the right ventricle and lungs so that, in addition to the forward cardiac output, the left heart must pump this extra quantity of blood around and around, resulting in an elevated pulmonary blood flow (it may be several times the output of blood into the aorta). However, sometimes in this disorder the very high pulmonary blood flow produces a high pressure in the pulmonary artery, and eventually the arterioles in the lungs become thickened so that their resistance rises, causing the left-right shunt to disappear. Under these circumstances, it is no longer possible to correct the shunt by surgically repairing the hole between the heart chambers. However, the pulmonary artery pressure may be high simply because the blood flow is very high through the lungs (in which case an operation could be performed), and the pulmonary vascular resistance is normal; the distinction between these two conditions is obviously important in managing the patient.

Application of the equations given above can solve this diagnostic problem. For example, if the mean pulmonary artery pressure is elevated to 50 mmHg, the mean left atrial pressure is 26 mmHg, and the pulmonary blood flow is elevated at 12 l/min (calculated by the Fick principle), the pulmonary vascular resistance is 2 RU, a normal value. Alternatively, if the pulmonary artery pressure is 50 mmHg, the mean left atrial pressure is 2 mmHg, and the pulmonary blood flow is reduced to only 2 l/min, the pulmonary vascular resistance is markedly elevated to 24 RU, a value at the systemic level.

Cardiac Energetics and Myocardial Oxygen Consumption

The heart is a constantly working organ, and therefore its energy use is high. Since it must develop a high pressure, the left ventricle in particular consumes a large amount of energy, and while it constitutes only about 0.3% of the body weight in man it uses about 7.0% of body's resting oxygen consumption. The heart must be continuously active, and important differences exist between the metabolism of cardiac and skeletal muscle; thus, there are many more mitochondria in heart muscle (about 30% by volume), it can utilize more substrates than skeletal muscle (which relies predominantly on glucose), and under normal conditions the energy production in heart muscle is entirely by aerobic means (skeletal muscle can readily use the anaerobic pathway). In fact, as we shall see, cardiac muscle does not develop an appreciable "oxygen debt" during each contraction, unlike skeletal muscle which develops an oxygen debt that is "repaid" during periods of rest.

There are several ways in which the energetics of muscle can be considered. The *total energy* (*TE*) *output* of the muscle can be equated with the mechanical work (*W*) it performs plus the heat (*H*) liberated:

$$TE = W + H$$

The *total chemical energy used* (TE_c) by the muscle can also be determined by measuring the loss of high energy phosphate ($\sim P$) so that

$$TE_c = \sim P \text{ depleted}$$

In an aerobic muscle, such as the myocardium, we can equate the oxygen utilized in resynthesizing $\sim P$ with the total chemical energy so that:

$$TE_c = \dot{V}O_2$$

Since the O_2 consumption of the heart, or the left ventricle, can be readily determined, whereas total heat release from the heart or its total high energy phosphate usage cannot, most of our knowledge concerning energy expenditure by the heart is based on determinations of myocardial oxygen consumption ($M\dot{V}O_2$) under varying conditions. However, it will be worthwhile to first consider briefly heat production and high energy phosphate use in skeletal and cardiac muscle.

HEAT PRODUCTION BY MUSCLE

The experimental techniques and description of heat release measurements in isolated skeletal muscle (the frog sartorius muscle is often used) are described in detail in Section 1, Chapter 4. (See also Hill (1939) and Gibbs and Chapman (1979) for a review of cardiac muscle energetics.) When such a muscle is at rest and not contracting, if it is deprived of O_2 its resting heat production is reduced by about one-half and lactic acid and other products of anaerobic metabolism accumulate. When oxygen is then resupplied a corresponding amount of extra heat production can be measured, which corresponds to repayment of the "oxygen debt." If the resting muscle is stimulated by repetitive electrical depolarizations to contract isometrically (tetanic contraction), an immediate burst of additional heat production is measured. This heat release begins even before the onset of contraction (it is termed the *initial heat*), and during maintenance of the tetanic contraction a steady output of heat above the resting level is also observed (*maintenance heat*). At the end of the tetanic contraction, heat production rapidly falls back toward, but not to, the resting level, and a small but prolonged output of additional heat occurs for several minutes (*the recovery heat*), which is approximately equal to the initial heat. If the O_2 consumption is simultaneously measured the late recovery heat follows the time course of an additional consumption of O_2 above the basal level. Finally, if the contraction takes place under anaerobic conditions, the initial heat occurs without a measurable recovery heat. Therefore, in skeletal muscle, the recovery heat after normal contractions is related to delayed oxidative "recovery" metabolism.

A component of the initial heat has been identified that is not related to the mechanical contraction of the muscle. The measurement is made by overstretching skeletal muscle to about 3.5 μm (so that no overlap of the myofilaments is present), and under these conditions electrical stimulation still produces some heat liberation (the *activation heat*). This heat is presumed to be related to the release and reuptake of activator calcium by energy-dependent processes.

Of course, contractions of the heart are normally not isometric, and measurements have also been made during isotonic contractions of skeletal muscle against varying afterloads. Under these conditions, W. O. Fenn (Fenn, 1923) showed that *more heat* is liberated by the isotonic contraction (which shortens actively) than by an isometric contraction developing the *same* tension (Fig. 2.143). (The isotonic contraction is at a longer resting muscle length so that it can develop the same active tension as the isometric contraction (Fig. 2.143, *lower panel*). It was found, moreover, that the extra heat associated with shortening was proportional to the work performed (Fig. 2.143), and this phenomenon is often called the "Fenn effect." A. V. Hill and others have developed equations for calculating the total energy output of a contracting muscle using certain constants and the various heat components discussed above plus measurement of the mechanical work performed (tension × shortening).

Heat Release in Cardiac Muscle

Measurement of the heat production by isolated cardiac muscle has proved to be more difficult than in skeletal muscle, since single twitches must be analyzed rather than tetanic contractions. However, measurements have been made in mammalian muscle during single twitches (Gibbs, 1974), and significant differences have been observed in the patterns of heat release compared to skeletal muscle. The resting heat release is higher in cardiac muscle. Also, it has been found that following single twitches of cardiac muscle, the recovery heat is negligible. This finding suggests that oxidative recovery metabolism begins *during* contraction and is nearly or entirely complete by the end of the contraction, so that an "oxygen debt" does not occur. Such a conclusion is supported by studies in which NADH fluorescence was measured during

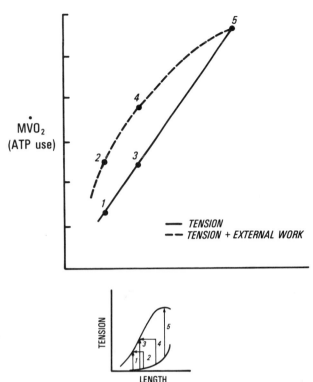

Figure 2.143. Relations between energy used by the myocardium (*MV̇O₂*, *ATP*) and isometric tension development (*solid line*) or tension development associated with shortening (*dashed line*) to perform external work (the Fenn effect). The dots indicate isometric and isotonic contractions which are diagrammed in the lower panel. Three isometric contractions (beats *1*, *3*, and *5*) are shown (*vertical arrows*), and when the resting lengths in beats *1* and *3* are increased to those of beats *2* and *4*, isotonic contractions can be produced at the same level of tension that was produced isometrically in beats *1* and *3*, and extra energy is liberated. Beat *5* is at the peak of the isometric length-tension curve, no shortening occurs, and the *solid* and *dashed lines* in the upper panel converge (point *5*).

cardiac muscle contractions (Chapman, 1972). In these experiments, a transient reduction in fluorescence occurred during the onset of contraction (NADH → NAD in the respiratory chain).

In cardiac muscle that has been shortened markedly so that no tension can be developed, heat release that may be related to the activation heat has been measured. When the muscle is then stimulated electrically the ensuing series of action potentials is accompanied by a so-called "tension independent heat" (Gibbs, 1969, 1974). The amount of this heat varies importantly in magnitude when the level of inotropic state of such noncontracting muscles is varied; for example, if catecholamines or a digitalis glycoside is administered, the tension-independent heat increases substantially (Fig. 2.144), suggesting that oxidative metabolism is importantly stimulated by alterations in contractility.

Figure 2.144. Tension-independent heat production in cardiac muscle which is markedly shortened so that no tension development can occur. The upright spikes initiate three contractions in each panel which produces an associated burst of heat release, as shown. The *left tracing* is under control conditions, and the *right tracing* was obtained after addition of a digitalis preparation. [From Gibbs and Gibson (1969).]

HIGH ENERGY PHOSPHATE USE BY HEART MUSCLE

As discussed in detail in Section 1, Chapter 4, the contractile event in both skeletal and cardiac muscle is closely linked to the hydrolysis of ATP by myosin ATPase. The ATP is rapidly replenished from the larger pool of creatine phosphate (CP) in equilibrium with ATP, and these high-energy phosphates are rapidly resynthesized by oxidative metabolism in heart muscle. If oxidative metabolism is blocked by removal of O_2, and if resynthesis of ATP by glycolysis is also blocked by use of iodoacetate, CP loss as the muscle contracts can be measured by rapid freezing and chemical analysis of the muscle after a given number of contractions. Under these conditions, the quantity of creatine phosphate hydrolyzed during isometric contractions is directly related to the amount of active tension developed and the number of contractions (Davies, 1966).

It has further been shown in cardiac muscle metabolically blocked in this manner (Pool and Sonnenblick, 1967) that if the muscle is stimulated to contract in the presence of a positive inotropic agent there is *increased* depletion of CP. Thus, under these conditions, for a given number of contractions that develop the same integral of tension as in the absence of the inotropic agent, more CP is hydrolyzed. Observations have also been made in isolated cardiac muscles while the O_2 consumption of the muscle was determined. Oxygen consumption is nearly linearly related to the level of peak tension developed in isometric contractions, and an extra O_2 cost of shortening against a load in isotonic contractions was determined (Fig. 2.143). Finally, it was shown that if a positive inotropic drug such as digitalis is administered, there is an increased oxygen consumption at all levels of isotonic tension (Fig. 2.145) (Coleman, 1967).

A COUPLED FRAMEWORK FOR CONTRACTION AND METABOLISM

The studies described above on heat release, high energy phosphate usage, and oxygen consumption in isolated muscle suggest that a close coupling exists between the initial chemical and mechanical events and the recovery metabolism. As shown in Figure 2.146, electrical activation produces calcium release, thereby triggering mechanical contraction which, in turn, determines how much ATP and CP are utilized. Thus, the initial chemical (ATP use) and mechanical (contractile) events do not involve oxidative metabolism; however, they *determine* the rate of oxygen consumption because of the close coupling with the recovery metabolism (Fig. 2.146), which in heart muscle appears to occur *during* the contraction. Uncoupling of oxidative phosphorylation could, of course, result in "slippage of the gears" (at coupling B), with more O_2 consumed than is used in high energy phosphate resynthesis. Also, there is evidence that different types of cardiac work require different energy expenditures, so that coupling A (Fig. 2.146) may be variable. Finally, as shown in Figure 2.146, events that do not directly relate to mechanical contraction, such as increased use of high energy phosphates by the calcium pump in the sarcoplasmic reticulum during inotropic stimulation of the myocardium (shown as "maintenance processes," many of which also use ATP; Fig. 2.146) could also increase O_2 consumption.

OXYGEN CONSUMPTION OF THE WHOLE HEART

Oxygen consumption ($M\dot{V}O_2$) of the whole heart can be determined using the Fick equation, provided the arteriovenous (AV) O_2 difference across the coronary bed is known, and the coronary blood

Chapter 14 / *Cardiac Energetics and Myocardial Oxygen Consumption*   **239**

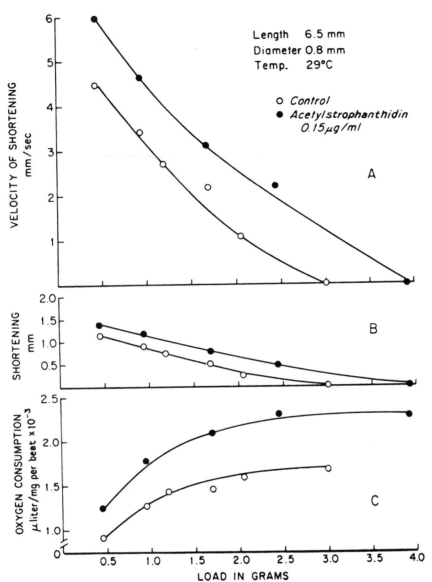

Figure 2.145. Oxygen consumption in isolated cardiac muscle before (*open circles*) and after the administration of a digitalis preparation acetyl strophanthidin (*closed circles*) into the muscle bath. Notice that at the same levels of afterload, the velocity of shortening is increased with some increase in the extent of shortening, together with stimulation of the oxygen consumption. [From Coleman (1967).]

flow can be measured:

$$M\dot{V}O_2 = CBF \times AVO_2.$$

Often, the O_2 consumption of the left ventricle alone is determined by measuring the O_2 content of blood in the large vein which drains that chamber (the coronary sinus) and in an artery, with measurements of coronary blood flow in experimental animals by means of a flowmeter on the left main coronary artery, or use of the microsphere technique, or in human subjects by using an indirect technique (Chapter 15).

The O_2 content of venous blood draining the heart is low relative to that of other organs, giving a wide arteriovenous O_2 difference. Although the percentage of O_2 extracted (AVO_2/AO_2) can in-

crease somewhat during severe exercise or other forms of stress, this reserve is small, and whenever cardiac O_2 demands rise or fall, it is primarily through adjustments in the coronary blood flow that oxygen delivery to the myocardium is varied.

THE DETERMINANTS OF MYOCARDIAL OXYGEN CONSUMPTION ($M\dot{V}O_2$)

A number of factors contribute to the O_2 consumption of the beating heart. (For reviews see Braunwald et al. (1976) and Gibbs and Chapman (1979).) For simplicity, we will focus mainly on the factors that affect the energy requirements of the major pumping chamber, the left ventricle, which consumes about 8.0 ml O_2/100 g of myocardium/

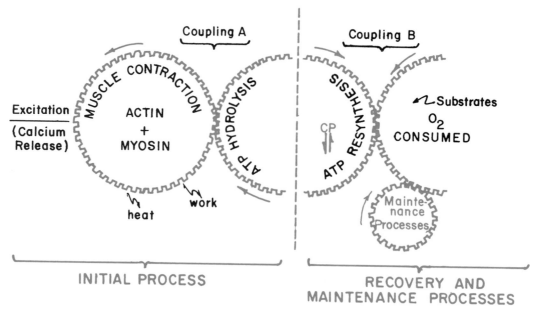

Figure 2.146. Diagram of mechanochemical coupling in cardiac muscle. The interaction of actin and myosin during contraction is initiated by Ca^{++} release (*left hand panel*, initial process) and utilizes ATP through "Coupling A." Oxidative phosphorylation provides for the formation of creatine phosphate (*CP*) and adenosine triphosphate (*ATP*), CP serving as a reservoir for high energy phosphate bonds available for ATP resynthesis. Oxygen utilization is related to CP and ATP synthesis through "Coupling B." Oxygen is also consumed in various cellular maintenance processes. [Redrawn from Braunwald et al. (1967).]

min in a normal human subject at rest. These determinants include:

Basal requirements
Activation energy
Heart rate
Tension or pressure development
Shortening against a load
Level of inotropic state

Basal O_2 Consumption

In the potassium-arrested mammalian left ventricle, the O_2 consumption is about 1.5 ml/100 g myocardium/min or almost 19% of the O_2 consumption of the normally contracting ventricle (McKeever et al., 1958). These basal energy requirements are related mainly to maintenance of chemical processes of the cell unrelated to active contraction.

Activation Energy

The amount of O_2 expended in activating contraction is unknown for the whole heart although, as discussed previously, the tension-independent heat and the related O_2 cost of changing inotropic state may be related to extra activation energy expended in operation of the sarcoplasmic reticular calcium pump. The energy required for electrical activation alone (action potential and repolarization) has been measured in the whole heart with calcium removed from the perfusing medium, and

this O_2 cost is small (less than 0.05% of the total $M\dot{V}O_2$) (Klocke et al., 1966).

The Four Major Determinants

Those factors which remain primarily determine the $M\dot{V}O_2$ of the intact, contracting heart as it changes its activity (Fig. 2.147). They are related to the performance of the myocardium and include the following.

HEART RATE

The frequency of cardiac contraction is a very important determinant of $M\dot{V}O_2$. There is a nearly linear relation between increases in heart rate and increases in cardiac O_2 requirement. In fact, under controlled experimental conditions when the heart rate is doubled the $M\dot{V}O_2$ slightly more than doubles (Fig. 2.148). Thus, as the heart rate is increased, even though tension is held constant, each beat requires slightly more oxygen (an effect which may be related to the positive inotropic effect of the force-frequency relation) (Boerth et al., 1969).

SYSTOLIC PRESSURE AND TENSION DEVELOPMENT

An almost linear increase in $M\dot{V}O_2$ also results as either the peak systolic pressure or the calculated peak systolic tension developed by the left ventricle is increased (Fig. 2.149). This occurs either in ventricles made to contract isovolumetrically or in the ejecting heart.

Under many conditions, the product of systolic blood pressure × the heart rate been found to adequately reflect changes in $M\dot{V}O_2$. For example, in the isolated supported heart preparation, the so-called "tension-time index" (TTI), which constitutes the integrated area under the left ventricular pressure curve per beat or per minute was found to correlate very highly with the O_2 consumption per beat or per minute. This index, developed by S. J. Sarnoff (Sarnoff et al., 1958b), was highly useful in an experimentally controlled setting, in which the heart is separated from autonomic connections, and spontaneous changes in inotropic state do not occur. However, as discussed subsequently, when changes take place in the level of inotropic state the TTI is no longer a useful predictor of $M\dot{V}O_2$ (see below).

EJECTION OF BLOOD (WALL SHORTENING)

This appears to be a relatively minor determinant of O_2 consumption (Fig. 2.147) and constitutes the

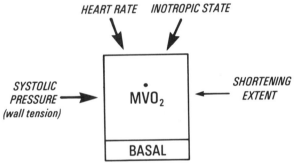

Figure 2.147. Diagram of the major determinants of myocardial oxygen consumption, above the basal level.

equivalent of the Fenn effect (muscle shortening against a load) discussed earlier in isolated muscle. This extra O_2 cost is proportional to the extra work performed (Fig. 2.143). Recent studies indicate that the volume work of the ventricle ejecting under normal conditions contributes an extra 15% to the $M\dot{V}O_2$ above that which would have been developed with the ventricle contracting isovolumetrically at the same systolic pressure (Fig. 2.143) (Burns and Covell, 1972).

It has been known for many years, that cardiac work (stroke volume × mean pressure) which is performed by ejecting a relatively large stroke volume against a low aortic pressure is much less costly in terms of O_2 consumption than the same amount of work (stroke volume × pressure) performed against a high pressure. This important point relates to the lower energy cost of shortening than of pressure development, and will be discussed further under the section on cardiac efficiency.

THE LEVEL OF INOTROPIC STATE

When the above three factors (the heart rate, mean pressure or tension, and the stroke volume) are held constant in an experimental setting another important determinant of $M\dot{V}O_2$ can be readily demonstrated (Sonnenblick et al., 1965; Ross et al., 1965). This determinant was not recognized until studies were made in relatively intact animals during stresses such as exercise, types of interventions that were not examined in the heart-lung preparation. In such an experiment, when a substance having a positive inotropic effect on the

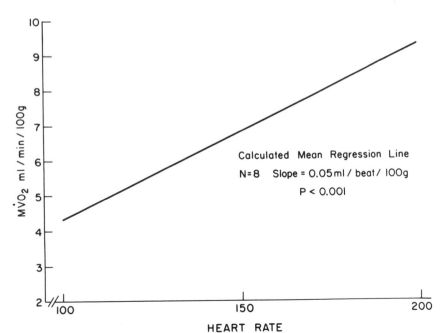

Figure 2.148. Relation between the heart rate and myocardial oxygen consumption in hearts in which the peak wall tension did not change. [From Boerth et al. (1969).]

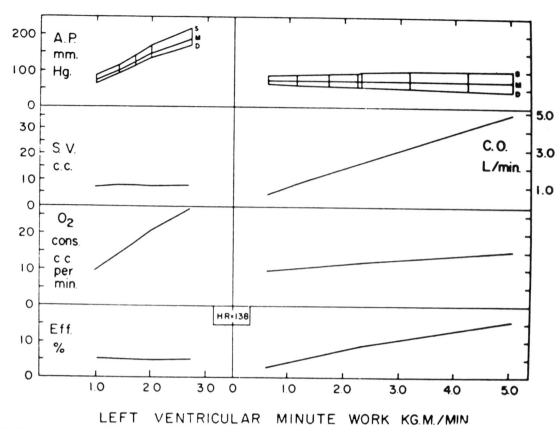

Figure 2.149. Data obtained in the isolated supported heart showing the hemodynamic determinants of cardiac O_2 consumption, left ventricular minute work, and efficiency (*Eff.*). Heart rate was constant at 138 beats/min. In the *lefthand panel*, increasing the aortic pressure (*A.P.*) without a change in the stroke volume markedly increases the O_2 consumption per minute, and the efficiency (the ratio of work to O_2 consumption) does not change. In the *righthand panel* the cardiac output (*C.O.*) is markedly increased while the aortic pressure is held constant. The cardiac O_2 consumption increases only slightly, and therefore the cardiac efficiency increases markedly. [From Sarnoff et al. (1958b).]

myocardium such as calcium, digitalis, or norepinephrine is administered, a pronounced additional stimulation of $M\dot{V}O_2$ occurs (Sonnenblick et al., 1965). As shown in Figure 2.150 the administration of norepinephrine produced a marked increase in the velocity of contraction evidenced by an increased maximum rate of pressure development (peak dP/dt), and despite a constant heart rate and stroke volume, and nearly unchanged arterial pressure, the $M\dot{V}O_2$ has almost doubled. At the same time, the positive inotropic stimulus shortens the duration of contraction, and the area under the left ventricular pressure pulse is diminished (Fig. 2.150); since the heart rate is constant in this experiment, the TTI falls while $M\dot{V}O_2$ increases, and it is clear that under such conditions directional changes in TTI do *not* follow changes in $M\dot{V}O_2$. The mechanisms for changes in $M\dot{V}O_2$ with altered inotropic state are not fully elucidated. They may be related to energy costs associated with the calcium pump, as suggested earlier, or to increased velocity of shortening (Sonnenblick et al., 1965), or

to the increased extent of wall shortening (against a load) (Rooke and Feigle, 1982), with additional kinetic energy expenditure (Chapter 7).

If a negative inotropic substance is given, such as a large dose of a β-blocking drug or a barbiturate, the $M\dot{V}O_2$ falls when the other three determinants of $M\dot{V}O_2$ are held constant in the experimental setting (Graham et al., 1967).

INTERRELATIONS BETWEEN THE FOUR DETERMINANTS OF MVO_2

In the intact animal, or in man, any of the three major $M\dot{V}O_2$ determinants (heart rate, pressure development, inotropic state) can become the most important factor influencing energy requirements of the heart (Fig. 2.151). For example, if a marked increase in the blood pressure is produced by infusing a vasoconstrictor drug such as phenylephrine (which has little inotropic effect), the oxygen consumption of the heart will rise largely due to the energy cost of increased pressure development, even though some reflex slowing of heart rate occurs

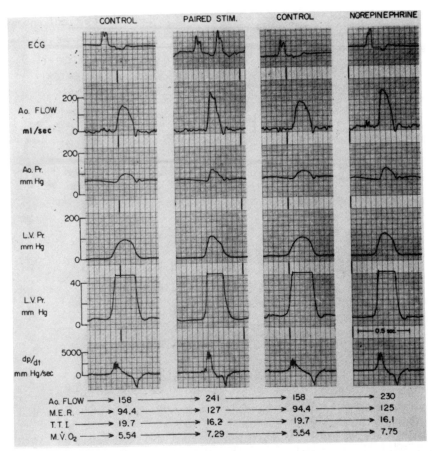

Figure 2.150. Effects of stimulating the inotropic state of the heart at a constant heart rate, stroke volume, and mean aortic pressure. In the *left two panels* is shown the positive inotropic effect of paired electrical stimulation (repeated postextrasystolic potentiation), and in the *right two panels* the effect of norepinephrine. *Ao. FLOW*, aortic flow measured with an electromagnetic flowmeter; *Ao. Pr.*, aortic pressure; *L.V. Pr.*, left ventricular pressure shown at full scale and at high gain to show the left ventricular end-diastolic pressure, and the first derivative of the left ventricular pressure (*dP/dt*). With norepinephrine, for example, the peak velocity of flow increases, the duration of contraction shortens, and the dP/dt rises. Little change occurs in the determinants of $M\dot{V}O_2$ other than inotropic state, although wall shortening must increase as well. $M\dot{V}O_2$ increases from 5.54 to 7.75 ml/min. [From Sonnenblick et al. (1965).]

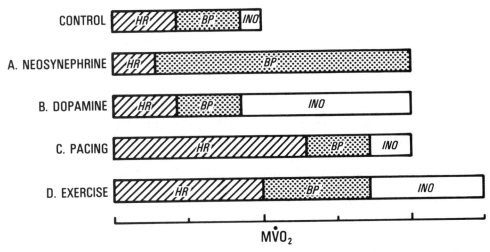

Figure 2.151. Diagram showing the manner in which various major determinants of cardiac energy utilization could influence the myocardial oxygen consumption. *HR*, heart rate; *BP*, blood pressure; *INO*, inotropic state. For further discussion, see text.

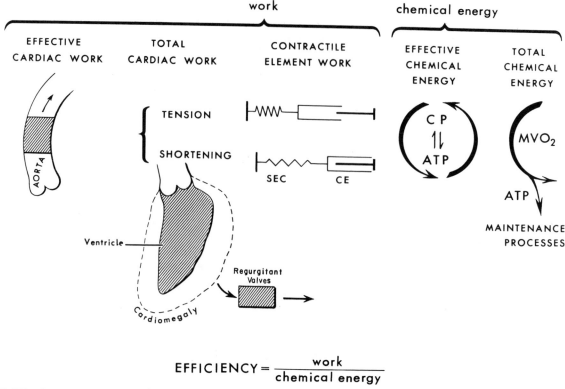

$$\text{EFFICIENCY} = \frac{\text{work}}{\text{chemical energy}}$$

Figure 2.152.   Diagram of the various factors that can be used to calculate the efficiency of the intact heart. The general efficiency equation is shown at the bottom of the figure. Chemical energy can be represented by the total (the $M\dot{V}O_2$) or by the effective energy utilized in the contractile process, represented by ATP and its equilibrium with creatine phosphate (*CP*). Work can be represented by work of the contractile elements (*CE*), in which CE must stretch the series elastic component (*SEC*) to achieve shortening and force in the total muscle fiber. Work can also be represented by the total cardiac work, which is dependent upon ventricular geometry and is measured in terms of the tension and shortening of the myocardial fibers; this work can be influenced by factors such as cardiomegaly and regurgitation of the cardiac valves. Finally, work can be represented by the effective cardiac work, represented by the stroke volume and the pressure against which it is delivered into the aorta. [From Braunwald et al. (1976).]

(Fig. 2.151*A*). If a positive inotropic agent such as dopamine is administered intravenously, stimulation of $M\dot{V}O_2$ will occur through increased myocardial contractility (Fig. 2.151*B*). Electrical pacing of the heart at a rate twice normal will double the $M\dot{V}O_2$, largely through increased number of contractions per minute, with a small contribution from enhanced inotropic state (the staircase effect) (Fig. 2.151*C*). The response to exercise is shown in Figure 2.151*D*. At rest, sympathetic (adrenergic) tone is low, and there is little stimulation of the myocardial inotropic state. However, during exercise a number of factors operate together to markedly increase $M\dot{V}O_2$. These include activation of the sympathetic nervous system with an increase in the heart rate and myocardial inotropic state, and as exercise becomes severe the systolic blood pressure increases and stroke volume rises (Fig. 2.151*D*).

CARDIAC WORK AND EFFICIENCY

Efficiency represents the fraction of the total energy of contraction that appears as mechanical work. In the whole heart, the $M\dot{V}O_2$ represents the total energy expended, and the work is usually determined as the so-called "effective" or "external work" (equal to the product of the stroke volume and the mean aortic pressure, or the mean systolic left ventricular pressure)*:

$$\text{Efficiency} = \frac{\text{cardiac work/min}}{M\dot{V}O_2}.$$

The kinetic factor of cardiac work ordinarily comprises less than 5% of the total work, and this factor is usually neglected. Often, since the resting $M\dot{V}O_2$ does not contribute directly to energy expended in contraction, it is subtracted from the total $M\dot{V}O_2$, and under these conditions the efficiency of the left ventricle in anesthetized animals

---

* To solve the efficiency equation it is necessary to employ in the equation the "work equivalent" of $M\dot{V}O_2$. One milliliter of O_2 has a caloric equivalent of 4.86 calories, and the constant for conversion of calories to kilogram meters is 0.4268. Therefore, the energy equivalent of 1 ml of O_2 is 2.06 kg m, the units in which work is generally expressed.

has been calculated at about 15%. The efficiency of isolated cardiac muscle and the whole heart can be expressed in several ways, as illustrated in Figure 2.152.

In considering the efficiency of the whole heart, several factors are of importance. As mentioned, the heart uses relatively little extra oxygen to deliver an increased cardiac output (and substantially increase the minute work) at a normal aortic pressure (low oxygen cost of the Fenn effect), making the ratio of work to $M\dot{V}O_2$ high and hence increasing efficiency (Fig. 2.149). With increasing aortic pressures at a constant cardiac output, however, $M\dot{V}O_2$ is elevated substantially as the minute work increases and efficiency may not change (Fig. 2.149); the stroke volume also can fall as aortic pressure is elevated, and work may actually decrease as $M\dot{V}O_2$ rises, leading to decreased efficiency.

Enhanced inotropic state and heart rate stimulate $M\dot{V}O_2$, and it might be expected that under conditions of mild exercise, particularly in an untrained individual (in whom the heart rate response is large and the sympathetic nervous system response is substantial), the increase in blood pressure and cardiac output (work) will be small, so that cardiac efficiency may diminish substantially. The effects of β-adrenergic blockade may be useful under these circumstances in a patient who has angina pectoris (see Chapter 15), since the heart rate response and inotropic stimulation during exercise are markedly attenuated by β-blockade; the venous return and cardiac output increase under these circumstances, but the cardiac output now rises through an increase in the stroke volume (use of the Frank-Starling mechanism), rather than by an increase in heart rate and inotropic state. This leads to an increase in cardiac efficiency, since the minute work is the same, but the $M\dot{V}O_2$ is substantially lower than prior to β blockade.

SUBSTRATE UTILIZATION BY THE MYOCARDIUM

The details of intermediary cellular metabolism, including aerobic and anaerobic glycolysis and oxidative metabolism, are presented in Section 1, Chapter 4. Important differences exist between the metabolism of cardiac and skeletal muscle, particularly in relation to the large number of substrates that heart muscle can utilize to generate high energy phosphates, predominantly by aerobic means. Fatty acids are utilized in preference to glucose when they are available, but also lactate, pyruvate, ketone bodies, and amino acids can be metabolized

(Table 2.10). Normally 60–70% of oxidative metabolism comes from fatty acids, although intermediates formed during aerobic glycolysis are used in the citric acid cycle. During cardiac work overload, glycolysis becomes a more important source of high energy phosphate production through aerobic glycolysis, although anaerobic glycolysis occurs to some extent in heart muscle during ischemia as discussed briefly below. A few key features of fatty acid and carbohydrate metabolism in the heart should be noted. (For detailed reviews see Neely and Morgan (1972) and Morgan and Neely (1982).)

Fatty Acid Metabolism

Circulating plasma triglycerides are broken down to free fatty acids at capillary and cell membranes by lipoprotein lipases, and the subsequent uptake of fatty acids by myocardial cells is not energy-dependent but depends on their plasma concentration. There is also some intracellular triglyceride storage, and these endogenous triglycerides can be mobilized to form free fatty acids when there is deprivation of glucose or of exogenous free fatty acids. Smaller activated fatty acid subunits are utilized within the mitochondria, and longer chain fatty acids employ the carnitine transport system to enter the mitochondrial matrix where β oxidation makes acetyl-CoA available for oxidation in the Kreb's cycle, with eventual production of high energy phosphates in the respiratory chain. Free fatty acid oxidation is controlled by the rate of removal of acetyl-CoA in the citric acid cycle, and a high NADH:NAD ratio inhibits β oxidation.

During increased cardiac work the rate of oxidative phosphorylation is increased; feedback regulation of the citric acid cycle occurs through the ratios of ATP:ADP and NADH:NAD. Thus, accumulation of ADP with a reduction of the ATP:ADP ratio and accumulation of NAD, with lowering of the NADH-NAD ratio, stimulate the cycle, and also accelerate high energy phosphate production in isolated mitochondria.

Table 2.10

Percentage contribution of various metabolic substrates to total myocardial oxygen usage

Carbohydrate, %		Noncarbohydrate, %	
Glucose	17.90	Fatty acids	67.0
Pyruvate	0.54	Amino Acids	5.6
Lactate	16.46	Ketones	4.3
Total	34.90[a]	Total	76.9[a]

From Bing (1955). [a] The total is more than 100% because some of the substrates removed from the blood, in amounts measured by the arteriovenous differences, were stored by the heart or not completely metabolized.

Glucose Metabolism

Glucose transport across myocardial cell membrane occurs by facilitated diffusion and is related to the blood glucose level. In the absence of insulin, there is a threshold for glucose transport of about 60 mg%, but in the presence of insulin this threshold is lowered to 10 mg%. Increased cardiac work results in increased glucose transport into the cell, whereas oxidation of free fatty acids inhibits glucose transport into the cell.

Within the myocardial cell, glucose can then either be stored as glycogen or enter the aerobic glycolytic pathway, which supplies about 30% of the oxidative production of high energy phosphates, as acetyl-CoA produced from pyruvate by the action of pyruvate dehydrogenase enters the mitochondria. Product inhibition largely regulates the rate of aerobic glycolysis, the conversion of fructose-6-phosphate to fructose-1-6-diphosphate by phosphofructokinase being inhibited by ATP, CP, and citrate, and pyruvate dehydrogenase being inhibited by ATP, NADH, and acetyl-CoA.

Hypoxemia and Ischemia

Inadequate O_2 content in the blood, or hypoxemia, as may occur on breathing low O_2 tension (or in congenital heart disease in which there is right to left shunting of blood) tends to result in high rates of blood flow through the organs of the body. Vasodilation therefore serves as the predominant means by which O_2 delivery to the tissues is maintained. *Ischemia*, on the other hand, is a condition in which inadequate O_2 delivery to the tissues occurs because of low blood flow, usually without hypoxemia of the arterial blood (see Rovetto et al.,

1973). Not only is there inadequate O_2 delivery, but the low blood flow fails to wash out accumulating metabolic products and hydrogen ions, which can lead to further impairment of metabolism.

With hypoxemia, O_2 lack slows oxidative production of ATP but removes inhibition of phosphofructokinase and stimulates the glycolytic production of high energy phosphates. Accumulation of NADH also promotes the production of lactate from pyruvate. In heart muscle, such lactate production occurs *only* during hypoxemia or ischemia (Rovetto et al., 1973).

With ischemia as may occur with insufficient coronary blood flow due to coronary atherosclerosis (see Chapter 15), glycolysis initially is stimulated, but failure to wash out lactate and other acid products because of the reduced blood flow leads to inhibition of phosphofructokinase; accumulation of NADH and other products inhibits the action of pyruvate dehydrogenase, and marked slowing of glycolysis ensues. NADH accumulation also inhibits other steps in the glycolytic pathway.

When heart muscle is completely deprived of O_2 by coronary occlusion, glycogen stores are rapidly depleted, and there is insufficient ATP production by glycolysis to maintain normal ATP concentrations. Contraction rapidly diminishes in the involved region, and within about 30 seconds it ceases; insufficient high energy phosphate is produced to maintain even a basal level of cell metabolism, and irreversible cell damage occurs in about 40 minutes. Partial ischemia can be tolerated for a much longer period, but if low blood flow persists glycogen stores become exhausted, contraction is impaired, and tissue death eventually may result.

CHAPTER 15

The Coronary Circulation

The coronary arteries are the first arterial branches to arise from the aorta just above the aortic valve. They are particularly vital arteries, since they supply blood directly to the muscle of the pumping chambers and to the conducting system of the heart, and control of their blood flow is intimately related to cardiac function. The heart must develop pressure and flow in the aorta to allow perfusion of all the other organs of the body, and it must also "autoperfuse" itself, in that it must develop sufficient pressure in the aorta to promote flow through its own vascular supply. Clearly, when the heart becomes too weak to accomplish this (as may occur, for example, during severe shock or the acute phase of a heart attack) a cycle of deterioration due to inadequate blood supply leading to further weakening of the heart ensues, and the blood pressure falls further with eventual brain death. (For reviews on the physiology of the coronary circulation see Gregg and Fisher, 1963; Berne and Rubio, 1979; and Klocke and Ellis, 1980.)

ANATOMY OF THE CORONARY CIRCULATION

The distribution of the coronary arteries varies among mammalian species. In most there are two main coronary arteries, the right and the left, which originate from the sinuses of Valsalva just above the aortic valve leaflets (see Fig. 2.8). In the dog, the right coronary artery is diminutive and provides blood to only the right ventricle and right atrium. In the pig and in primates, such as the baboon, the coronary anatomy tends to resemble that found in human subjects. In about 90% of human subjects, the right coronary artery is "dominant," being a large vessel that traverses down the groove between the right atrium and right ventricle, supplying branches to both of those chambers and then, upon reaching the posterior aspect of the heart, it turns toward the cardiac apex (the so-called posterior descending branch) and supplies blood to the lower (inferior) wall of the left ventricle (Fig. 2.8) (for details see James et al., 1982). This artery also usually supplies a branch to the atrioventricular node. The left coronary artery in man commences as a short left main branch and then divides into two branches, the anterior descending branch which supplies the septum between the right and left ventricles and the anterior surface of the left ventricle, and the circumflex branch which supplies the lateral aspect and part of the posterior surface of the left ventricle (Figs. 2.8 and 2.167). There are many variations in coronary artery anatomy, and in some subjects a so-called "left dominant" pattern exists (the right coronary artery being small and more like that found in the dog) and the circumflex coronary artery supplies the inferior wall of the left ventricle. These patterns of distribution have importance, since whenever an artery becomes obstructed by an atherosclerotic lesion, specific areas of the heart that are involved by the lack of coronary blood flow can be identified (see Chapter 9, electrocardiographic changes in myocardial infarction).

The coronary arteries divide into smaller and smaller branches as they course on the surface of the heart (the so-called epicardial arteries) and these give off tiny arteries that course at right angles through the heart muscle into the wall of the ventricle (intramural vessels). The heart has a rich blood supply, since the heart is continuously active and performs a large amount of work. Therefore, a large supply of oxygen per unit of muscle is necessary, and the potential capillary density in the myocardium is high. In the rat left ventricle, for example, when the coronary bed is maximally dilated the capillary number is as high as $3,000/mm^2$ (Honig and Bourdeau-Martini, 1973). A constant supply of oxygen is necessary for this essentially aerobic organ, and the capillaries pass through the myocardium in close proximity to the working muscular elements and the mitochondria, yielding a relatively short diffusion distance for oxygen.

The venous blood from the heart is collected in venules which course with the small arteries and then converge on the left side of heart to form the coronary sinus, a large vein running in the groove between the left atrium and left ventricle and opening into the right atrium near the tricuspid valve. This vein is sufficiently large in humans that a plastic catheter or tube can be placed under fluoroscopic control directly within it to allow withdrawal of blood samples for determining oxygen content, or the content of other substances necessary for determining coronary blood flow. The veins from the right side of the heart (right atrium, right ventricle) do not form a common channel, but drain individually directly from the myocardium at multiple sites within the right atrium and right ventricle.

Coronary Collateral Vessels

Collateral channels are blood vessels (usually small) which allow blood to flow directly from one artery to another; hence, if an artery becomes obstructed, when collateral channels exist they can allow arterial blood to enter the blocked vessel beyond the site of obstruction (Fig. 2.153). In the dog, a large number of tiny collaterals exists even in the normal heart, and they can be readily seen under magnification as vessels somewhat larger than capillaries on the surface of the heart that provide direct anastamoses (connections) between adjacent arteries. Under normal conditions such collaterals appear to have little function, but they can undergo substantial enlargement if a coronary artery is occluded.

There is evidence that the pig has few collateral channels after coronary occlusion compared to the dog (Table 2.11); collateral flow can be collected from the distal end of an occluded vessel or measured with the microsphere technique (see below). In baboons and in most human subjects few collaterals are present under normal conditions. However, such channels may develop over time and become highly important when atherosclerotic disease involves the coronary arteries (Fig. 2.153) (For reviews on the coronary collateral circulation see Schaper (1971) and Bloor (1974).)

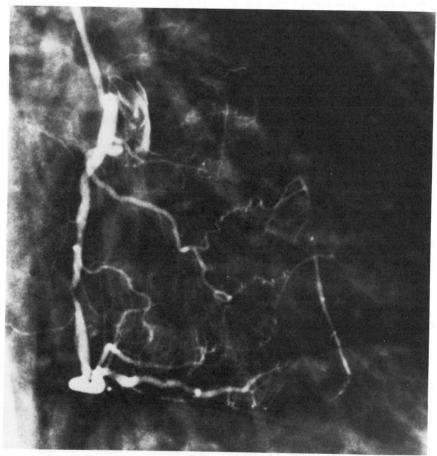

Figure 2.153. Coronary arteriogram with injection of contrast material into the right coronary artery (lateral view of the heart). The catheter is visible at the *upper left*, and several areas of narrowing, one nearly completely occluding the coronary artery, are evident. A small "bridging collateral" bypasses this very narrowed area, and other small collateral vessels can be seen reaching, and faintly opacifying, a vessel at the right which runs vertically. This is the left anterior descending coronary artery, which was found to be completely blocked at its origin from the left main coronary artery, and supplied only by these collaterals. [Courtesy of Dr. M. Judkins.]

MEASUREMENT OF CORONARY BLOOD FLOW AND RESISTANCE

When a flow-sensing device (such as an electromagnetic flowprobe) is placed directly around one of the coronary arteries supplying the left ventricle, blood flow through the artery exhibits a unique pattern. Blood flow through the coronary artery falls sharply during left ventricular systole, and with the onset of diastole it rises abruptly and then falls gradually during the remainder of the diastolic period (Fig. 2.154). A similar pattern of blood flow is seen in the right coronary artery when this vessel is "dominant" and supplies the posterior wall of the left ventricle as well as the right ventricle (as in human subjects). However, as might be anticipated, in the dog in which the right coronary artery supplies only the low-pressure right ventricle and atrium, the less forceful contraction of the right ventricle does not inhibit flow from the higher pressure aorta, and the pattern of flow therefore closely resembles the pressure pulse in the aorta, rising in systole and falling gradually during diastole. Such patterns indicate that the contraction of the *left ventricle* squeezes the small intramural arteries, so that the systolic aortic pressure cannot maintain blood flow through the increased resistance offered by the compressed vessels, producing a marked fall in coronary flow during systole in that chamber (Fig. 2.154).

Coronary blood flow was first measured in man by the *inert gas washout technique* (Kety, 1960), a method which is still employed during cardiac catheterization. The method is applicable to measure flow in any organ. The procedure usually employed for determining coronary flow involves having the subject inhale a mixture of room air and an inert gas for a number of minutes, until the concentration of inert gas in the blood is in equilibrium with that in the heart muscle, in accordance with its partition coefficient. The inhalation of the gas is then abruptly stopped, and a series of blood samples is drawn simultaneously from a needle placed in a peripheral artery and from a catheter positioned in the coronary sinus (cardiac vein). Since the gas is selected to have a low solubility, it is rapidly eliminated (blown off) as blood passes through the lungs, and therefore the arterial concentration rapidly falls toward zero (Fig. 2.155). However, the gas is more slowly eliminated from the organ under study, and the washout rate is proportional to the blood flow (Fig. 2.155). In accordance with a modified Fick equation, the arteriovenous difference can be calculated as the difference between the integrals of the arterial and coronary sinus concentrations (Fig. 2.155) and knowing the partition coefficient in ml of gas per gram of myocardium, coronary blood flow in ml/100 gm left ventricle/min can be calculated. In the original method, nitrous oxide gas was used, which involved complex manometric measurements of the volume of gas in the blood samples; more recently, less soluble gases such as hydrogen and helium have often been employed, low concentrations of which can be readily measured by chromatographic methods (see Klocke et al., 1969). Using such inert gas techniques, the level of coronary blood flow in normal human subjects at rest is 60–90 ml/100 gm left ventricle/min. During exercise and other forms of stress, coronary blood flow increases, and the coronary flow "reserve" is 4–5 times the basal level of flow.

Another technique for measuring coronary blood flow frequently used in experimental studies is the *radioactive microsphere technique*. The method is particular useful in the heart because it allows determination of *regional myocardial blood flow*,

Table 2.11
Collateral coronary blood flow after acute coronary artery occlusion as determined by 3 different methods in 3 animal species

Species	Method			
	Retrograde Flow, %	TM Distribution, %	¹³³Xe ia, %	¹³³Xe im, %
Dog	8	16	33.5	11
Pig	0.5	5.3	23.5	
Sheep			35.0	

From Schaper (1971). TM signifies tracer microsphere distribution. Xenon-133 was injected either into the coronary artery (ia) before occlusion, or directly into the heart muscle (im). Values are expressed as a percentage of the normal myocardial flow taken as 100%.

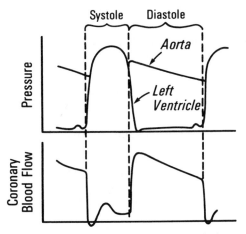

Figure 2.154. Diagram of aortic and left ventricular pressure pulses and blood flow through the left main coronary artery of a dog measured with a flowmeter. The abrupt drop in flow during systole is evident, and the diastolic flow tends to fall together with decreasing diastolic pressure in the aorta.

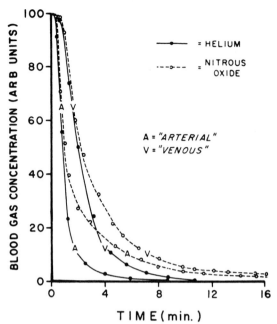

Figure 2.155. Inert gas washout for determining organ blood flow by inhaling the gas to equilibrium, followed by cessation of inhalation of the foreign gas and measurement of its concentration in serial samples of arterial and venous blood. (In the case of the heart, samples are taken from an artery and the coronary sinus). These data are from an *in vitro* model system to illustrate the principle. The more rapid elimination from the arterial blood and from the organ of the less soluble gas helium compared to nitrous oxide is shown; nevertheless, knowing the quantity of either gas taken up by the organ (such as the myocardium) per unit mass (from the partition coefficient of the gas relative to blood), and the integrated arteriovenous difference of the gas, the same blood flow is calculated using both gases. [From Klocke et al. (1969).]

including the *distribution* of blood flow across the ventricular wall. As applied for the measurement of coronary blood flow, a batch of radioactive microspheres (usually measuring 9–15 μm in diameter) is suspended in a saline-detergent solution and injected into the left atrium; it is mixed by blood flow within the left ventricle, ejected into the aorta, and then perfuses the coronary arteries. The microspheres lodge in only a few capillaries, so that there is no damage or effect on flow, but the number of spheres trapped per unit of myocardial tissue is directly proportional to the myocardial blood flow to that region. Following one or more injections, the experiment is terminated, the heart excised, and the radioactivity in samples of myocardium is directly measured in a gamma scintillation counter.

In practice, a number of isotopes are injected during experimental conditions at different coronary blood flows. Since each batch of microspheres is labelled with an isotope having a differing energy peak, the quantity of radioactivity at a given peak for a given isotope (i.e., an isotope of cesium or strontium) can be detected using a gamma spec-

trometer. In order to determine absolute myocardial blood flow, the so-called "reference withdrawal method" is used to determine a proportion between the rate of blood from a sampling site and the counts in that sample, and the (unknown) coronary blood flow and the counts in the myocardium (Domenec et al., 1969). With this approach, blood is withdrawn by a motor-driven syringe from the ascending aorta at a known rate during the injection of microspheres, and the following equation is applied:

$$Q_m/Q_r = C_m/C_r$$

where Q_m = blood flow per gm myocardium, Q_r = blood flow into the syringe (reference sample, r), C_m = counts/min in myocardium, and C_r = counts/min in reference sample. Myocardial blood is then calculated as:

$$Q_m = \frac{C_m}{C_r} \cdot Q_r.$$

In experimental animals, portions of the ventricular wall can be divided into a section from the endocardium, the midwall, and the epicardium, thereby allowing measurements of total transmural blood flow and its distribution across the wall.

Another method for determining coronary blood flow, which can be used in patients undergoing cardiac catheterization, is the *thermodilution technique* with coronary sinus sampling (Ganz et al., 1971). With this approach, a special catheter having an end hole is passed well up into the coronary vein (coronary sinus), a temperature sensor is placed further down on the catheter near the orifice of the coronary sinus, and the continuous injection method is used (see Chapter 13). Cold saline is injected continuously through the catheter and is diluted by the coronary sinus blood flow, the amount of dilution being proportional to the coronary sinus flow sampled at the thermister. Alternatively, rapid injection of the cold saline can be carried out using a spray tip catheter, and the bolus is sampled downstream in the coronary sinus from the same catheter.

Coronary Vascular Resistance

Determination of coronary vascular resistance (CVR) can be made from the ratio of mean pressure in the aorta (P), or in the coronary artery, to the mean coronary blood flow (CBF) per minute as:

$$\text{mean } CVR = P/CBF$$

If left ventricular pressure is measured, the ventricular diastolic pressure is subtracted from the aortic pressure, since this value provides the perfusion pressure difference across the coronary vascular bed. Coronary vascular resistance is high during systole, when the intramural arteries are compressed by the left ventricle, and when phasic blood flow is determined by the flowmeter technique, coronary vascular resistance is often measured at the end of diastole using the pressure and flow at that point in time. This method is free of mechanical effects on flow from left ventricular contraction and is therefore more representative of the tone of the coronary resistance vessels themselves.

MYOCARDIAL O_2 SUPPLY

Since extraction of O_2 by the coronary circulation is high, the major factor affecting O_2 delivery (O_2 supply) is the coronary blood flow (see next section). However, other factors also influence O_2 delivery. These include factors affecting O_2 availability: the PO_2 in arterial blood, the hemoglobin level, and the position of the oxyhemoglobin dissociation curve (see Section 5, Chapter 36). Also, several local factors including tissue PO_2 may directly or indirectly influence capillary density (see discussion of microcirculatory regulation in Chapter 6). Thus, while changes in blood flow through the major coronary arteries due to alterations in coronary arteriolar resistance are the most important determinant of O_2 delivery to the myocardium, other mechanisms may have an effect, particularly during stress or under abnormal conditions.

THE DETERMINANTS OF CORONARY BLOOD FLOW

The most important single factor determining the rate of coronary blood flow in the normal mammalian heart is the requirement of the heart for oxygen; thus, *metabolic* regulation of coronary vascular resistance is the major determinant. Systolic compression of the coronary arteries and coronary perfusion pressure affect coronary blood flow, particularly under abnormal conditions. Neurogenic control of coronary vascular resistance has also been demonstrated. These opposing determinants are summarized in Figure 2.156. Circulating neurohumoral factors and pharmacologic agents also affect coronary vascular tone under some circumstances.

1. Metabolic control (O_2 demands)
2. Coronary perfusion pressure
3. Systolic compression
4. Autonomic nervous system
5. Circulating catecholamines and other vasoactive substances and drugs.

Metabolic Factors

The coronary blood flow can be shown to have a nearly linear correlation with the level of myocar-

dial oxygen consumption ($M\dot{V}O_2$) in the normal heart (Fig. 2.157), coronary blood flow increasing or decreasing as the $M\dot{V}O_2$ increases or decreases (Eckenhof et al., 1947). As mentioned, the arteriovenous oxygen difference is wide in the heart under resting conditions; thus, O_2 extraction is nearly complete, the coronary sinus blood oxygen content being low (about 5 vol%) leading to an AVO_2 difference of about 12 vol% (the highest in the body). The requirements for increased oxygen by the heart are therefore met, not primarily by increased oxygen extraction, but by increased coronary blood flow.

Since the most important determinant of the $M\dot{V}O_2$ is the mechanical activity of the heart and its inotropic state, the determinants of $M\dot{V}O_2$ are also the determinants of coronary blood flow. They include the systolic ventricular pressure or wall tension, the heart rate, the amount of shortening against a load, and the level of inotropic state or contractility (see Chapter 14).

Precisely how the alterations in the $M\dot{V}O_2$ are translated into increases in coronary blood flow, and vice versa, remains controversial. Perfusion of the coronary circulation with blood having a low O_2 content is a potent stimulus to coronary vasodilatation, and increased K^+ concentration, decreased pH, and increased P_{CO_2} also have vasodilator effects. However, much evidence has been accumulated to indicate that the major physiologic vasodilator in the heart is adenosine, a metabolic product of ATP breakdown (Berne and Rubio, 1979). Thus, the concentration of adenosine (an extremely potent vasodilator) in the myocardium, interstitial space, and the coronary venous blood can be linked with level of coronary blood flow as follows:

$$ATP \rightarrow ADP \rightarrow AMP \rightarrow adenosine \rightarrow vasodilatation$$
$$\downarrow$$
$$inosine \text{ or } hypoxanthine$$

AMP is broken down by 5′-nucleotidase, located primarily in the sarcolemma and T tubules of car-

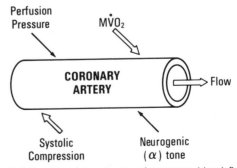

Figure 2.156. The determinants of coronary blood flow. The major determinant is metabolic, the myocardial oxygen requirement ($M\dot{V}O_2$); systolic compression limits forward flow, and a neurogenic increase in resistance is a minor determinant.

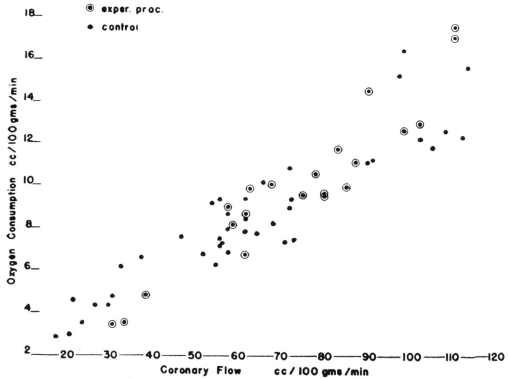

Figure 2.157.   Relationship between MV̇O₂ and coronary blood flow. Various experimental procedures were used to alter cardiac work. [From Eckenhof et al. (1947).]

diac muscle cells, and the dephosphorylation of AMP forms adenosine. Adenosine, in turn, is rapidly broken down by adenosine deaminase, an intracellular enzyme which is also found in the red blood cell, so that measurement of adenosine is complicated by its rapid disappearance. Adenosine may also reenter the myocardial cell.

The adenosine hypothesis is attractive, since it directly links the metabolic products of increased or decreased ATP usage to the regulation of coronary blood flow. Thus, as ATP usage increases, increased coronary blood flow would occur at a new steady-state of adenosine production. Also, increased washout of this metabolite by augmented coronary blood flow, as with a sudden increase in coronary perfusion pressure alone, would lower adenosine concentration and produce vasoconstriction, thereby serving to bring coronary blood flow back toward control; conversely, if coronary blood flow transiently dropped because of hypotension, for example, the slowed rate of flow would result in accumulation of this metabolite in the myocardium, with coronary vasodilatation.

When a coronary artery is briefly occluded completely and the occlusion then released, a marked increase in coronary blood flow ensues for a period of time. This response is termed *reactive hyperemia*, and the length of the hyperemia increases as the duration of the coronary occlusion is prolonged

(Fig. 2.158). (See Gregg (1950) and Coffman and Gregg (1960).) Reactive hyperemia has been attributed to the accumulation of adenosine or other metabolites during the period of obstructed coronary blood flow, leading to vasodilatation in the distribution of the artery, although lowered perfusion pressure could also produce myogenic vasodilation. The vasodilation becomes evident when the occlusion is released, leading to the marked transitory increase in coronary blood flow; as metabolic recovery of the tissue takes place, the high blood flow washes out the accumulated metabolite, and blood flow is subsequently restored to normal. As shown in Figure 2.158, the "flow debt" is markedly overpaid in the coronary circulation by the reactive hyperemia response.

Coronary Perfusion Pressure

The pressure in the coronary arteries and the aorta importantly affects the level of coronary blood flow, particularly during the diastolic interval when the aortic pressure and coronary blood flow are directly related and fall together (Fig. 2.154). However, the level of mean and systolic aortic pressures also importantly influences the MV̇O₂, and when the aortic pressure increases or decreases coronary blood flow will change as a consequence of the metabolically induced vasodilation or vasoconstriction as the work of the left ventricle changes. Thus

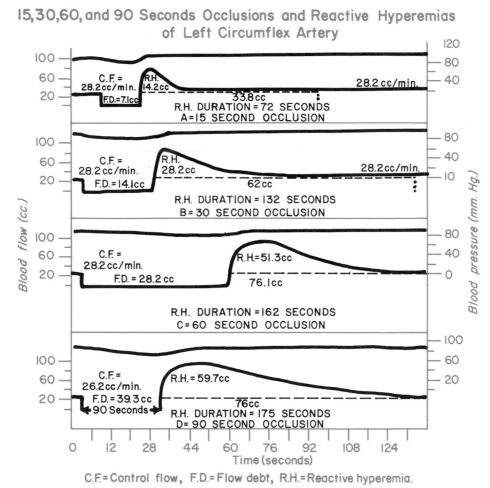

15,30,60, and 90 Seconds Occlusions and Reactive Hyperemias of Left Circumflex Artery

C.F.= Control flow, F.D.= Flow debt, R.H.= Reactive hyperemia.

Figure 2.158. Reactive hyperemic response. In each panel *top trace* is mean aortic pressure and *bottom trace* blood flow (recorded with a rotameter) into the left circumflex coronary artery perfused from a carotid artery. A, Occlusion of 15 s. Duration of reactive hyperemia (R.H., 72 s) measured from release of occlusion to return of a stable flow rate (28.2 cc/min). Flow debt (F.D., 7.1 cc) is calculated by multiplying duration of occlusion (15 s) by control flow (28.2 cc/min). Reactive hyperemic blood flow (R.H., 14.2 cc) was considered to be the total flow during reactive hyperemic period (48.0 cc) less 33.8 cc, the flow that would have occurred in this period at the control flow rate, or:

$$\left(\frac{72\ s}{60\ s} \times 28.2\ cc\ (control\ flow/min) = 33.8\ cc\right)$$

The same terminology applies to panels *B*, *C*, and *D*.

the relationship between aortic pressure and coronary blood flow is complex.

It is possible to design an experiment in which the aortic pressure and left ventricular pressure, the heart rate, and the inotropic state of the heart are held *constant* while coronary perfusion pressure *alone* is varied by means of a separate cannulation and perfusion of a coronary artery (Fig. 2.159) (Mosher et al., 1964). In such an experiment when the coronary perfusion pressure from the separate pressure-regulated blood reservoir is abruptly lowered and maintained at a new level, while the cardiac work is held constant, a transient rapid decrease in coronary blood flow ensues which is then followed over 10–12 s by return of the flow toward

the previous level (Fig. 2.160); conversely, when the coronary perfusion pressure is abruptly elevated, coronary blood flow rapidly rises initially but then returns toward the control level. Thus, the coronary vascular resistance varies *directly* with the coronary perfusion pressure under these conditions. When a series of such abrupt pressure changes is carried out over a perfusion pressure range of 10 to 170 mmHg (left ventricular and aortic systolic pressures remaining constant at a normal level of approximately 120 mmHg), a plot relating the perfusion pressure to coronary blood flow is obtained (Fig. 2.161, *solid circles*). This pressure-flow plot reveals a *plateau* between perfusion pressures of approximately 60 and 150 mmHg, over which cor-

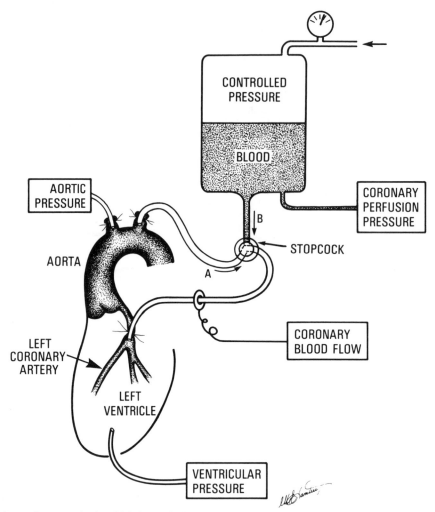

Figure 2.159.   Experimental preparation in which the perfusion pressure to a coronary artery can be abruptly changed, while the aortic pressure and function of the left ventricle remain unaffected, allowing demonstration of autoregulation. A three-way stopcock allows blood to either perfuse the coronary artery from a cannula situated in the aorta (*A*) or from the controlled pressure reservoir (*B*). Coronary blood flow is measured with a flowmeter as shown. For further discussion, see text.

onary blood flow is relatively independent of coronary perfusion pressure; this relationship is termed *autoregulation* of coronary blood flow (see additional discussion of this phenomenon in Chapter 6).

At the upper ranges of perfusion pressure, autoregulation *fails*, and coronary blood flow rises at these very high coronary perfusion pressures (Fig. 2.161). Likewise, at pressures below approximately 60 mmHg, autoregulation *fails* and coronary blood flow falls as coronary perfusion pressure drops (Fig. 2.161). The latter portion of the autoregulation curve is of great importance, because it indicates the point at which the coronary bed is maximally vasodilated, i.e., *autoregulatory reserve* has been fully utilized. For example, as coronary perfusion pressure is changed from a mean pressure of 100 to 70 mmHg, the initial drop in coronary blood flow is followed by active vasodilation, with a drop in coronary resistance (resistance = perfusion pres-

sure/coronary flow; since pressure has fallen and flow is unchanged, resistance must have fallen). However, when perfusion pressure is dropped from 60 to 40 mmHg, the blood flow falls, indicating the absence of further coronary vasodilation.

If the initial flow values *prior* to the autoregulatory responses (the lowest initial flow point after the pressure fall, see Fig. 2.160, or the initial flow increase if perfusion pressure is raised) are plotted against coronary perfusion pressure the pressure-flow relation is characteristic of a vascular bed that does not autoregulate (*exponential dashed lines*, Figs. 2.161 and 2.162).* The instantaneous pres-

---

* Notice the similarity of these "instantaneous" pressure-flow curves at difference levels of coronary vascular resistance in Figure 2.162 to those in a vascular bed that does not exhibit autoregulation at various degrees of sympathetic stimulation (Fig. 2.14); note also the change in critical closing pressure in both circumstances.

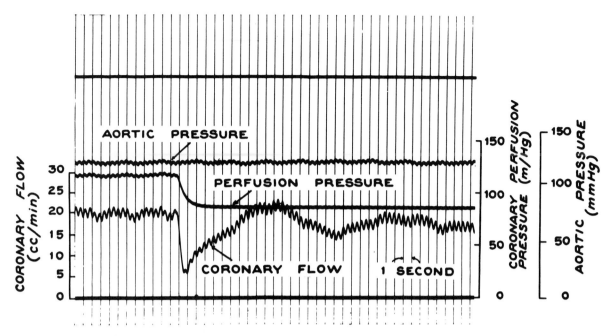

Figure 2.160. Experiment demonstrating coronary autoregulation obtained with experimental preparation shown in Figure 2.159. Note that the mean pressure in the aorta is constant, and when the mean coronary perfusion pressure is abruptly dropped, the coronary blood flow falls sharply and then recovers over 8–10 s to near the control value. [From Moser et al. (1964).]

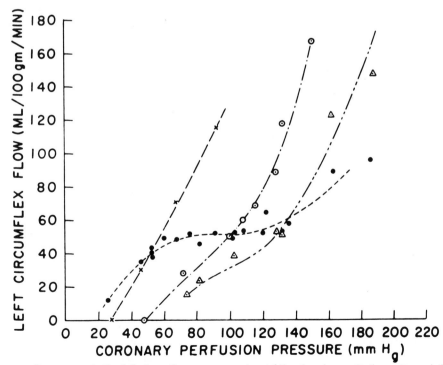

Figure 2.161. Pressure-flow curves in the left circumflex coronary artery of the dog demonstrating autoregulation of coronary blood flow (*solid circles with sigmoid dashed curve*). The three curves shown by the *x*'s, *open circles*, and *triangles* indicate instantaneous pressure-flow curves, before coronary autoregulation had occurred, at three different levels of coronary perfusion pressure and coronary tone, or vascular resistance (50 mmHg, 100 mmHg, and 125 mmHg, respectively). For further discussion, see text. [From Moser et al. (1964).]

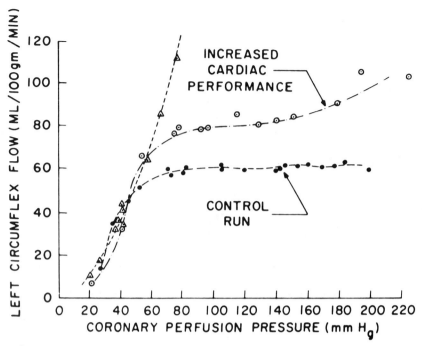

Figure 2.162. Pressure-flow curves demonstrating coronary autoregulation down to a perfusion pressure of approximately 60 mmHg (*solid circles*). When cardiac work was increased by increasing the mean aortic pressure (*open circles*), coronary blood flow was autoregulated at a higher level. The instantaneous pressure-flow relation with *maximum* coronary dilation is steep (*open triangles*), and indicates a marked dependency of coronary blood flow upon perfusion pressure below a critical pressure. For further discussion, see text. [From Moser et al. (1964).]

sure-flow curve in Figure 2.162 (*triangles*) is the same as the pressure-flow plot obtained after administration of a coronary vasodilator such as nitroglycerin or adenosine; it is the relationship of *maximum coronary vasodilation*, in which flow is highly dependent on perfusion pressure, exhibiting a nearly exponential change in coronary blood flow as pressure is changed (characteristic of a passive vascular bed).

The mechanism of pressure-induced autoregulation (which is also seen in certain other vascular beds such as the cerebral and skeletal muscle circulations) is controversial (see Chapter 6). The so-called myogenic hypothesis holds that at least a portion of the response is due to altered stretch of the smooth muscle in the walls of the coronary arteries. An important component of the autoregulatory response may also relate to metabolic control of coronary vascular tone; as discussed earlier, with the increased coronary blood flow accompanying a sudden increase in coronary perfusion pressure, washout of adenosine could rapidly occur, returning flow toward the normal level, whereas with a drop in perfusion pressure and decreased flow, diminished washout with accumulation of adenosine could result in coronary vasodilation.

It should be recognized that the *plateau* of the coronary pressure-flow curve in the autoregulating

bed can shift. That is, the *level* of autoregulated blood flow is subject to change depending upon the metabolic demands ($M\dot{V}O_2$) of the left ventricle, (Mosher et al., 1964). When $M\dot{V}O_2$ (left ventricular pressure, heart rate, or inotropic state) is augmented, the plateau is shifted upward (Fig. 2.162), so that coronary blood flow is regulated at an increased level over a given level of coronary perfusion pressures, or vice versa. It should be noted, however, that the pressure-flow curve of the maximally dilated bed (*triangles*, Fig. 2.162) continues to provide the limit for coronary blood flow autoregulation, regardless of the level of the plateau, and that maximum coronary vasodilation occurs at a *higher* perfusion pressure when the $M\dot{V}O_2$ is augmented (or at a lower pressure when cardiac energy requirements are reduced, Fig. 2.162) (see also Fig. 2.15, Chapter 6).

These responses have importance in coronary heart disease, when an atherosclerotic lesion may partially obstruct a coronary artery. Under these conditions, if the lesion is sufficiently severe a pressure difference exists across the lesion with reduced perfusion pressure beyond the narrowed area. Under normal resting conditions, as long as the perfusion pressure beyond the narrowing is above approximately 60 mmHg total coronary blood flow will be maintained, since autoregulation allows distal coronary vasodilation. However, with a drop in aortic pressure, the distal perfusion pressure will fall below this critical value, the vasodilatory reserve will be exceeded, and coronary blood flow will fall with unfavorable effects on the myocardium (see section on ischemia). It can also be seen from Figure 2.162 that when the $M\dot{V}O_2$ is increased (as during exercise), if the perfusion pressure beyond the narrowed area remains at 60 mmHg, coronary flow will be too low to meet the increased O_2 demands. The resulting insufficient O_2 supply to the myocardium in that region can cause chest pain during exertion, as discussed subsequently.

Systolic Compression

It has been pointed out that active contraction of the left ventricle causes mechanical compression with squeezing of the intramyocardial coronary blood vessels, a phenomenon that is responsible for the pattern of reduced *total* coronary blood flow observed during ventricular systole in coronary arteries supplying the left ventricle (see Fig. 2.154) (Gregg, 1950; Kirk and Honig, 1964; Downy and Kirk, 1974). This compression increases the average resistance to coronary blood flow, and if the contraction of the heart is suddenly stopped momentarily in an experimental setting, coronary blood flow rises abruptly (Fig. 2.163).

When the *transmural* distribution of blood flow is analyzed systolic compression of the coronary circulation becomes particularly important when blood flow to the *inner wall* of the left ventricle is considered, as discussed below.

TRANSMURAL DISTRIBUTION OF BLOOD FLOW

Systolic contraction of the left ventricular myocardium is responsible for an important difference between the pattern of blood flow to the inner (subendocardial) regions of the left ventricular wall and the outer (subepicardial) regions. In fact, during ventricular systole, blood flow in the subendocardial region for practical purposes is zero and then rises rapidly during diastole, whereas in the subepicardial layers blood flow rises during systole and then remains relatively high during diastole (Fig. 2.164). Blood flow to the midwall region is intermediate between these two patterns (Fig. 2.164). As mentioned the pattern of *total* coronary blood flow shows reduced flow during systole, with an abrupt rise and then a slow fall during diastole. These patterns of regional flow across the wall have been verified by injecting microspheres to perfuse the coronary arteries *only* during the phase of ventricular systole (Hess and Bache, 1976); such studies demonstrate that few microspheres reach the subendocardium under these conditions, whereas flow is roughly comparable during systole and diastole in the midwall. This means that under normal conditions coronary vascular resistance is lower in the subendocardium, in order to compensate (during *diastole*) for this systolic flow limitation (Fig. 2.164). The *mean* blood flow (ml/min) to the subendocardium is slightly higher than to the subepicardium, yielding an endocardial:epicardial blood flow ratio of about 1.1:1 under normal conditions, and it has been postulated that the explanation for this observation may lie in a slightly higher O_2 consumption of the inner myocardial fibers since they must shorten further and perform more work.

The differing flow pattern in the inner and outer wall of the left ventricle is particularly important

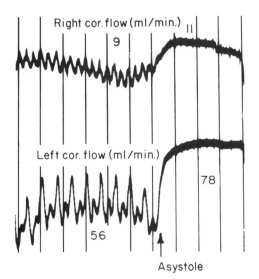

Figure 2.163. Response of left and right coronary flow to vagal induced asystole. Note the prompt increase with the induction of asystole. Time lines are 1 s apart. [From Sabiston and Gregg (1957).]

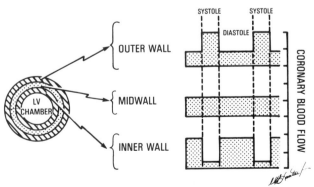

Figure 2.164. Diagrammatic representation of transmural coronary blood flow distribution in three layers of the left ventricular (*LV*) wall measured with radioactive microspheres during systole and diastole. Note that systolic flow predominates in the outer wall, whereas diastolic flow with minimal systolic flow is present in the inner wall. For further discussion, see text.

in certain disease states, and makes the subendocardium vulnerable to insufficient perfusion pressure. Thus, analysis of the *transmural* distribution of myocardial blood flow, with consideration of blood flow to the inner wall of the left ventricle, is important under some conditions (see subsequent section).

The Autonomic Nervous System

Although metabolic and mechanical factors are normally the most important determinants of the level of coronary blood flow, the autonomic nervous system, via sympathetic nerve fibers directly supplying the coronary arteries, has some influence at rest and under other circumstances. Parasympathetic stimulation, as well as circulating catecholamines and certain other substances also can have an effect.

The sympathetic nerve fibers innervating the coronary arteries contain norepinephrine, and its release stimulates α-adrenergic receptors located in the smooth muscle of the coronary vessels. Local metabolic factors largely override the influence of these nerves, but there is some evidence that the basal level of coronary blood flow is somewhat higher at a comparable level of oxygen consumption after surgical removal of the nerve tissue around the coronary arteries, and vasoconstriction due to activation of the sympathetic nerves to the coronary arteries can be demonstrated experimentally under some circumstances (see Feigle, 1967, for review). Direct electrical stimulation of the sympathetic nerves to the heart causes an increase in coronary blood flow due to increased contractility and heart rate (increased $M\dot{V}O_2$), but simultaneous activation of coronary vasoconstrictor fibers is suggested by a fall in coronary venous O_2 content (so that increased sympathetic tone to the coronary arteries may have prevented complete metabolic vasodilation). Such an effect is more clearly shown when sympathetic nerve stimulation is carried out (or norepinephrine is injected) after blockade of the myocardial β_1 receptors with the β-adrenergic blocking drug propranolol; under these conditions, the increase in $M\dot{V}O_2$ is blocked, and an *increase* in coronary vascular resistance can be demonstrated during nerve stimulation. Similar sympathetic coronary vasoconstriction during exercise can also be unmasked by β-adrenergic blockade and is prevented by α-adrenergic blockade (Aung-Din et al., 1981). There is also experimental evidence that some degree of reflex adrenergic constrictor tone exists (Vatner et al., 1970). Thus, electrical stimulation of the carotid sinus nerves (which inhibits sympathetic tone and produces reflex periph-

eral vasodilation, Chapter 16) causes a fall in the coronary vascular resistance (Vatner, 1970) despite a decrease of $M\dot{V}O_2$ (resulting from a reflex slowing of heart rate and lowered arterial pressure, which should cause increased coronary vascular resistance). Nevertheless, the level of sympathetic "tone" in the coronary bed is low relative to that in many other vascular beds, and under most circumstances metabolic control predominates. A role for coronary sympathetic vasoconstriction in patients with atherosclerotic coronary artery lesions, or in patients with spasm of large coronary arteries (discussed further below), has been suggested but remains uncertain at present.

Direct stimulation of the vagal nerve supply to the heart causes mild coronary vasodilation when bradycardia is prevented by electrical pacing of the heart (Feigle, 1965), although this effect is not likely to be of much importance in the intact circulation. Intracoronary administration of acetylcholine produces pronounced coronary vasodilation.

Reflexes which appear to originate in the coronary arteries and produce reflex peripheral vasodilation have been described, but their role is unclear.

Circulating Catecholamines and Other Substances

β-Adrenergic (β_2) receptors exist in the smooth muscle of the coronary vascular bed, and their stimulation by blood-borne epinephrine or injected sympathomimetic drugs can produce coronary vasodilation. Such direct effects can be difficult to demonstrate, however, because these substances stimulate $M\dot{V}O_2$ and thereby produce secondary, metabolic coronary vasodilatation. Under experimental conditions, a direct coronary vasodilatory effect due to the sympathomimetic drug isoproterenol can be shown in the potassium-arrested heart, where perfusion of isoproterenol into the coronary arteries produces a marked drop in coronary vascular resistance in the absence of a change in $M\dot{V}O_2$ (Klocke et al., 1965). If a "cardioselective" β blocker is administered to block only the myocardial β_1 receptors (but not β_2 receptors in the coronary arteries), administration of isoproterenol also produces direct coronary vasodilation, while $M\dot{V}O_2$ remains unaffected.

A number of other substances in the body as well as a variety of drugs can affect coronary blood flow indirectly by altering $M\dot{V}O_2$, or exhibit direct actions on the coronary vascular smooth muscle. Direct coronary vasodilators include nitroglycerin, prostacyclin, and calcium antagonist drugs such as verapamil and nifedipine. Several of these agents are useful in the treatment of chest pain due to coronary atherosclerosis, or in the therapy of coronary artery spasm (Berne and Rubio, 1979).

Coronary vasoconstrictors include vasopressin, ergonovine, and angiotensin II. Ergonovine is sometimes administered in the cardiac catheterization laboratory in order to induce coronary spasm in the diagnosis of patients suspected of having chest pain due to this phenomenon; its effects are readily reversed by the administration of nitroglycerin or a calcium antagonist.

CORONARY VASODILATOR RESERVE

Myocardial O_2 demands are met primarily by moment to moment alterations in coronary blood flow, and a substantial vasodilator reserve exists in the normal heart. This can be demonstrated by experimentally performing brief coronary occlusions, which are followed by reactive hyperemia, as discussed earlier (Fig. 2.158). Also, during stresses such as exercise, a uniform large increase of coronary blood flow occurs across the left ventricular wall (as measured with microspheres) (Fig. 2.165). In addition, intracoronary infusion of nitroglycerin or adenosine produces a marked increase in coronary blood flow (without significant change in MVO_2). When the dose of adenosine is progressively increased up to the limit of coronary vasodilator reserve, the reserve is maximally utilized first in the subendocardial regions of the left ventricle. This occurs because the increase of reserve flow in the subendocardium takes place mainly during diastole, and in diastole there is already some encroachment on subendocardial vasodilator reserve (Fig. 2.164), as discussed earlier (Bache and Cobb, 1977). Some additional reserve then remains in the subepicardial regions when the subendocardium is maximally dilated by adenosine. Under conditions of maximal coronary vasodilation by adenosine (Bache and Cobb, 1977), blood flow to the subendocardium becomes directly proportional to the so-called diastolic pressure-time index (DPTI), the

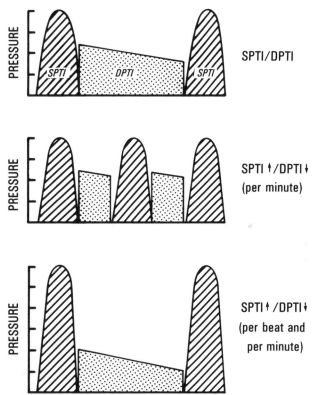

Figure 2.166 Diagram of the systolic pressure-time index (SPTI) and the diastolic pressure-time index (DPTI), which provide a guide to the adequacy of myocardial oxygen demands and coronary blood flow to the subendocardium, respectively, during maximum coronary vasodilatation. (SPTI = the area under the systolic aortic pressure pulse and DPTI = the area under the diastolic aortic pressure pulse). Normal conditions are shown in the *top panel*. During exercise in the presence of maximum coronary vasodilatation (*middle panel*), the SPTI/minute increases, whereas the time for diastolic flow to the subendocardium (DPTI) decreases; coronary blood flow may become inadequate to meet myocardial oxygen demands under these conditions. In the *bottom panel*, which illustrates severe aortic regurgitation, the SPTI is increased because of high systolic pressure, whereas the DPTI is reduced because of the low aortic pressure due to the regurgitant leak. The ratio is therefore increased, and in the presence of maximum coronary vasodilatation in the subendocardium, insufficiency of coronary flow to that region may result.

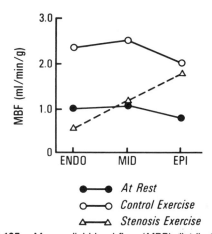

Figure 2.165. Myocardial blood flow (*MBF*) distribution across the wall of the left ventricle in three layers, *ENDO* (subendocardium), *MID* (midwall), and *EPI* (subepicardium) measured with radioactive microspheres. Data in an animal standing at rest on a treadmill (*closed circles*) show the slight predominance of subendocardial blood flow. During exercise under normal conditions, there is a uniform increase of blood flow across the wall (*open circles*, control exercise). When the left circumflex coronary artery was partially constricted, blood flow in the region supplied by that vessel showed a marked fall in subendocardial blood flow during exercise (*open triangles*), although the subepicardial flow increased normally. Also, during exercise with coronary stenosis, the region of myocardium supplied by the left circumflex coronary artery exhibited severe impairment of contraction, whereas under normal conditions contraction increased during exercise. [Courtesy of K. P. Gallagher.]

integrated area under the aortic pressure pulse during diastole (Fig. 2.166) (Buckberg et al., 1972).

Under conditions of maximal coronary vasodilation, a decrease in the time of diastolic perfusion per minute, as when the heart rate increases (an increased number of contractions is at the expense of diastolic time), would lead to a reduction in subendocardial coronary blood flow despite increasing $M\dot{V}O_2$ (Fig. 2.166). Also, if the diastolic blood pressure falls at an unchanged heart rate, coronary blood flow will also fall (Fig. 2.166), whereas if blood pressure rises or heart rate slows the DPTI (and coronary blood flow to the subendocardium will increase (Fig. 2.166). These observations have led to the idea that the *ratio* of the systolic pressure-time index (SPTI) (the same as the "tension-time index" discussed in Chapter 14), as an indicator of $M\dot{V}O_2$, to the DPTI can provide an estimate of the adequacy of subendocardial perfusion under conditions of maximal coronary dilatation (Hoffman and Buckberg, 1978). This concept has importance to regions of the heart when stenosis or narrowing of a coronary artery due to coronary atherosclerosis causes the perfusion pressure to drop, resulting in maximum dilatation of the coronary bed beyond the area of stenosis (see below). The SPTI/DPTI also can have relevance to conditions in some valvular lesions such as severe aortic regurgitation, where a high systolic pressure (increased SPTI) and low diastolic pressure (reduced DPTI) coexist and could lead to subendocardial underperfusion (Fig. 2.166) and an imbalance in O_2 supply and demand.

CORONARY HEART DISEASE AND INADEQUATE CORONARY BLOOD FLOW

Understanding of the relationships between myocardial O_2 requirements and the coronary blood supply becomes of great importance in dealing with clinical disorders of the coronary arteries. An example of the interplay between coronary perfusion and $M\dot{V}O_2$ is the setting in which a coronary artery is narrowed but not completely occluded, and partially supplied by collateral vessels (Fig. 2.154). Under these conditions, coronary blood flow may be adequate to supply the resting metabolic needs of the myocardium, but when O_2 demands are increased as during physical exercise, the blood supply to that region becomes insufficient. Thus, in the zone of myocardium supplied by the diseased coronary artery, the increased heart rate in the presence of maximal subendocardial vasodilation can result in a fall in subendocardial blood flow during exercise (Fig. 2.165), and regional contrac-

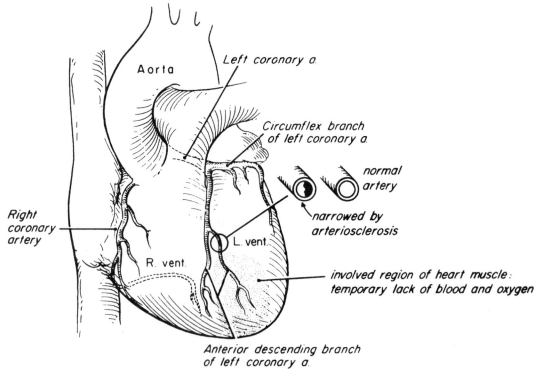

Figure 2.167. Diagram showing the mechanism of angina pectoris. A coronary atherosclerotic lesion is present in a branch of the left anterior descending coronary artery. During exercise, blood flow through the narrowed area becomes inadequate and the region of muscle supplied by that branch (*stippled area*) becomes ischemic, causing chest pain or angina pectoris. [Drawing by G. Gloege. From Ross and O'Rourke (1976).]

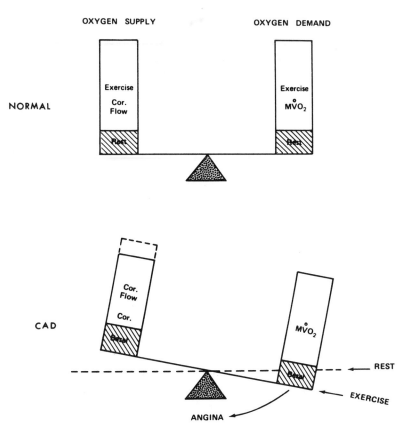

OXYGEN SUPPLY   OXYGEN DEMAND

NORMAL

CAD

Figure 2.168. Diagram representing oxygen supply and oxygen demand to the myocardium as a scale, which is normally in balance (*upper panel*) both at rest (*cross-hatched areas*) and during exercise, when the increased myocardial oxygen demands are met by increased oxygen supply (coronary flow rises). In the presence of coronary artery disease (*CAD, lower panel*), the scale may be in balance under basal conditions at rest (*cross-hatched areas*, and *dashed horizontal line*), but during exercise the myocardial oxygen demand increases while the increase in coronary blood flow is inadequate. The scale is therefore unbalanced, producing angina pectoris.

tion may rapidly deteriorate (Gallagher et al., 1982). In this setting, when the level of coronary blood supply relative to O_2 needs becomes insufficient, the condition is termed *ischemia* and it often leads to a clinical syndrome termed *angina pectoris*, or pain beneath the breastbone (Fig. 2.167). Occasionally, this syndrome is also produced by spasm of one or more large coronary arteries in the absence of an increase in $\dot{M}VO_2$ (Maseri, 1978).

Exercise-induced ischemia with angina pectoris due to the presence of an atherosclerotic coronary lesion can be treated by a variety of measures, which involve either decreasing $\dot{M}VO_2$ (reducing O_2 demands) or improving myocardial O_2 supply through enhanced coronary blood flow, or both (Fig. 2.168). One important way of decreasing myocardial O_2 demands is the administration of nitroglycerin (a smooth muscle relaxant which generally dilates arteries and veins), thereby reducing the blood pressure and heart size and diminishing $\dot{M}VO_2$. Nitroglycerin and other vasodilators also may increase O_2 supply by enhancing coronary collateral blood flow. Nitroglycerin and calcium antagonists (Schwartz, 1982) are highly useful in the treatment

of angina pectoris due to coronary artery spasm. Another important approach for diminishing O_2 demands (Fig. 2.168) is use of β-adrenergic blocking drugs (such as propranolol) which, by slowing the heart rate and diminishing the increases in blood pressure and contractility produced by exercise, allow the same exercise workload on the body to be achieved at a lower level of cardiac work and $\dot{M}VO_2$; angina pectoris may thereby be prevented.† Limitation of the heart rate increase by β blocking drugs also results in a longer time for diastolic coronary perfusion, and also should improve subendocardial blood flow during exercise by this mechanism. Another approach for treating angina pectoris is to increase O_2 supply (Fig. 2.168) by surgical means with the placement of one or more grafts to bypass the obstructions in the coronary arteries.

Abrupt obstruction of a coronary artery, usually caused by thrombosis (clotting) at the site of an

---

† If the heart rate cannot increase, cardiac output rises by use of the Frank-Starling mechanism when the venous return increases during exercise, so that β blockade does not impair the ability of the body to respond to moderate levels of exertion.

atherosclerotic lesion, produces within 1–2 min loss of contraction in the involved region with local systolic expansion of the wall (Tennant and Wiggers, 1935; Theroux et al., 1974). If sustained beyond about 20–40 min, coronary occlusion produces an area of muscle damage or necrosis, which is termed an acute myocardial infarction, or "heart attack." Research is currently underway in an effort to limit the damage by improving the coronary blood supply through use of thrombolytic (clot-lysing) drugs or by increasing collateral blood flow (use of nitroglycerin or calcium antagonists). Reducing the O_2 requirements of adjacent surviving tissue (which may have have marginal levels of perfusion) is also under study using, for example, β-adrenergic blocking drugs.

CHAPTER 16

Cardiovascular Control and Integrated Responses

The circulatory system is regulated by neural and hormonal control systems, which are superimposed on local metabolic control of vascular resistance. As we have seen in previous chapters, the flow to each organ system may be dominantly controlled by the metabolic needs of that system (autoregulation), but a complex series of neural and hormonal systems also alter the heart and vascular system in order to maintain a constant level of arterial pressure and maintain the cardiac output. These systems are described in detail in the next sections.

ANATOMY OF THE NEURAL CONTROL SYSTEMS

The neural control mechanisms have inputs to the central nervous system, which sense and relay various parameters of circulatory function, such as arterial pressure, and outputs, which regulate cardiac function, vascular volume, vascular resistance, and flow through the peripheral circulation (Peis, 1965; Korner, 1979). These are shown schematically in Figure 2.169, which diagrams central nervous system input connections (termed afferent nerves) on the left and outputs (efferent nerves) on the right. There is, of course, a bilateral distribution of both inputs and outputs.

AFFERENT CONNECTIONS (INPUTS)

The Carotid Sinuses

A dilatation exists at the base of each internal carotid artery where it joins the external carotid artery. A branch of the glossopharyngeal nerve (IXth cranial nerve) supplies that region, and reflexes from the carotid sinus are abolished when the glossopharyngeal nerve is sectioned. There is a rich sensory innervation of the carotid sinus wall. With appropriate staining, diffuse arborizing nerve networks can be demonstrated, as well as lamella-like receptors. These endings are located in the

connective tissue (adventitia) of the vascular wall and are oriented parallel the long axis of the vessel. All appear to be receptors that sense stretch of the vessel.

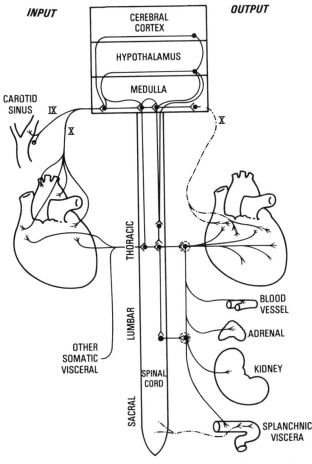

Figure 2.169. Schematic diagram of the neural connections important in circulation control. Although the system is bilaterally symmetrical, in this schematic drawing inputs (afferent fibers to the *left*) and outputs (sympathetic and parasympathetic efferent fibers, to the *right*) are shown only on one side. [Adapted from Korner (1979).]

The Carotid Bodies

These small organs, composed of rounded clumps of polyhedral cells near the carotid sinuses, are highly vascular, containing a network of vascular sinusoids in close connection with special sensory cells which respond to chemical stimuli and have afferent fibers passing to the IXth nerve. Blood flow through the carotid bodies is extremely high.

Aortic Baroreceptors

A network of branching nerves arborizes in the aortic wall in the region of the aortic arch, carotid, and subclavian arteries (Fig. 2.169). They are similar to the carotid sinus receptors, but their afferent fibers run through the vagus nerve (Xth cranial). There are both myelinated and nonmyelinated fibers in the carotid sinus, and the aortic arch.

Aortic Bodies

These chemoreceptors lie near the aortic arch and pulmonary artery, and recent studies have also identified such receptors near the coronary arteries. Structurally, they resemble the carotid bodies. Their afferent fibers pass toward the central nervous system in the vagus nerve.

Other Receptors

Mechanosensitive and chemosensitive receptors resembling those in the aortic arch and carotid sinus are also found in the lungs, the walls of the atria (Fig. 2.169), and the junctions of the venae cavae and pulmonary veins with the atria; a few are found in the ventricles, coronary arteries, and coronary sinus.

EFFERENT CONNECTIONS (OUTPUTS)

Cardiac Efferent Connections

The autonomic nerve supply to the heart consists of many mixed fibers, containing both vagal and sympathetic nerves, which course over the aorta, pulmonary artery, and vena cava to reach the cardiac chambers, conducting system, and coronary arteries.

SYMPATHETIC FIBERS

Fibers from the central nervous system transverse the spinal cord and synapse in the sympathetic chain (there are multiple cervical ganglia and paravertebral ganglia, represented by two ganglia in Fig. 2.169). Efferent fibers from these ganglia provide a rich network of postganglionic nerve end-ings, whose neurotransmitter is norepinephrine, to the atria, ventricles, sinus node, and atrioventricular (AV) node (Fig. 2.8).

VAGAL (PARASYMPATHETIC) FIBERS

The vagal control of the heart is mediated by fibers which synapse in ganglia in the heart (Fig. 2.169, shown as −·−); these ganglia provide a network of fibers which contain acetylcholine and supply the atria, sinoatrial (SA) node, and AV node (see Fig. 2.8). A few fibers also appear to supply the ventricular muscle. Other parasympathetic fibers reach various glands, the gastrointestinal tract, and the pelvic organs from the sacral outflow (Fig. 2.169).

Efferent Connections to the Peripheral Circulation

VASOCONSTRICTOR FIBERS

Distribution

These nerves were discovered in 1852 by Claude Bernard, who stimulated the cervical sympathetic nerve in the rabbit and observed constriction of the vessels of the ear. They belong to the thoracolumbar (sympathetic) division of the autonomic nervous system (Fig. 2.169). The constrictor fibers arise from groups of nerve cells situated in the intermediolateral columns of the spinal gray matter, extending in man from the first thoracic to the second or third lumbar segment, inclusive. Axons from these cells synapse with neurons lying in sympathetic ganglia outside the cerebrospinal axis. All the arterioles of the body, wherever situated, are supplied with these neurons which have their source in this relatively limited region of the central nervous system. Their adrenergic nerve endings contain norepinephrine and have been identified in all types of blood vessels, except the true capillaries (Chapter 6). Precapillary resistance vessels (small arteries and arterioles) have, in general, a rich innervation, although in the smallest precapillary vessels, the number of fibers is small. The venules have fewer adrenergic fibers than the larger veins which, themselves, are less richly innervated than the precapillary vessels. It is solely through such sympathetic fibers traveling with somatic nerve trunks that constrictor impulses are conveyed to the minute vessels of the limbs. Vasoconstrictor fibers to the head and neck are conveyed from the sympathetic chain through plexuses investing the blood vessels, and via peripheral nerve trunks (cervical and certain cranial nerves). The vessels of the abdomen and pelvis are supplied with fibers which pass along the vascular walls from plexuses surrounding the aorta and its branches. The sympathetic fibers to the heart arise chiefly in five upper dorsal segments of the cord, pass to the stellate ganglia and upper dorsal ganglia as white rami, and proceed to the heart by a complex plexus.

Site and Mode of Action

These nerve fibers constitute a group of powerful vasoconstrictor mechanisms. The physiological impulses discharge rate of the vasoconstrictor fibers is 1–2/s to maintain normal vessel tone and reaches 10 impulses/s, with maximal physiological excitation. The chemical transmitter released at the smooth muscle cell is norepinephrine, and norepinephrine is released into the blood stream on intense sympathetic stimulation. Because these constrictor fibers exert control over the resistance vessels, a marked increase in blood flow in a vasoconstricted limb immediately follows sympathetic or ganglionic blockade. During prolonged constriction accumulation of "vasodilator metabolites" can oppose neurogenic influence.

In addition, these fibers exert strong control over the capacitance vessels, mainly, the veins. They can alter greatly the venous return to the heart and, thus, markedly influence heart size and cardiac output. The significance of vasomotor fibers in controlling vascular capacitance has been verified in functionally isolated parts of the superficial and deep venous system of dog and man, by recording pressure changes evoked by various types of reflex stimulation. Veins in the intact forearm constrict during reflex sympathetic stimulation by cold, excitement, etc. Similar observations have been made on the capacity vessels of the splanchnic area. These effects are abolished by agents which block eurotransmission at any level (central, ganglionic, nerve terminals, or receptors).

Functional Significance

Vasoconstrictor fibers are of major importance in short term homeostasis of blood pressure and blood flow, including those reflex adjustments that arise from baro- and chemoreceptors. They also control blood flow through the skin and thus influence peripheral heat exchange. Vasodilation, particularly when it is reflex in nature, is predominantly an inhibition of vasoconstriction. However vasodilator mechanics as described below may play a lesser role. In fulfilling their role as the main neural modulators of the peripheral circulation, the vasoconstrictor system may exhibit generalized, segmental, or re-

gional function, depending upon the type of stimulus (Simon and Riedel, 1975).

VASODILATOR FIBERS

Vasodilator impulses emerge from the central nervous system by (1) the thoracolumbar sympathetic outflow; (2) the cranial outflow of the parasympathetic division reaching the periphery by way of the chorda tympani nerve (supplying the salivary glands), glossopharyngeal, and vagus nerves; (3) the sacral outflow of the pelvic nerve; and (4) the posterior spinal nerve roots.

Sympathetic Vasodilator Fibers

The question of vasodilator fibers in the sympathetic system has been much debated. The existence of a sympathetic vasodilator (cholinergic) innervation to skeletal muscle of the dog and cat has been established by stimulating sympathetic nerves following α-adrenergic blockade (Folkow, 1961); but it appears to be absent in rabbits and certain other mammals. It appears that these vasodilator nerves can be activated by stimulation of the hypothalamus, and also from the motor cortex. The functional significance of such sympathetic vasodilator fibers is not clear. It has been suggested that they are activated when the motor cortex becomes active, presumably in association with, or in anticipation of, skeletal muscular exercise.

The presence of vasodilator fibers has not been verified in man. The possibility that vasodilator fibers are present in humans is suggested by the observation that blood flow to the forearm increases during fainting, in spite of the generalized fall in blood pressure.

Parasympathetic Vasodilator Fibers

Parasympathetic vasodilator fibers run to restricted cranial and sacral areas such as the cerebral vessels, tongue, salivary glands, external genitalia, and to the bladder and rectum. These fibers are probably not concerned with baro- and chemoreceptor control of blood vessels, nor are they tonically active. It is generally believed that they are cholinergic, but their vasodilator effect is notably resistant to atropine. In salivary glands and in the skin the nerves that bring about cholinergic secretion of the endocrine glands may cause to be released into the circulation an enzyme that acts on plasma protein to produce the vasodilator polypeptide, bradykinin, but this conclusion is uncertain

(Kellermeyer and Graham, 1968), and the functional significance of bradykinin remains unknown.

Histaminergic Vasodilator Fibers

The presence of vasodilator fibers to the hind limb of the dog, which apparently liberate histamine, has been demonstrated (Oberg, 1976). The vasodilatation is not prevented by muscariwic or β-adrenergic receptor blockade. It is abolished by antihistamines and is associated with release of histamine into the blood. The function of these fibers is unknown.

Antidromic Vasodilator Impulses

Stimulation of the peripheral segments of the cut posterior roots of sacral nerves causes dilatation of vessels of the dog's paw. This observation is at variance with the Bell-Magendie principle that the dorsal roots contain only afferent fibers. It appears, however, that when these cut afferent fibers are stimulated electrically they conduct in the direction opposite to their usual function, i.e., antidromically.

Dorsal root vasodilation may represent a special case of activation of what is in fact an axon reflex. Peripheral parts of cutaneous afferent fibers are capable of producing what is known as an "axon reflex." The stimulus applied to the skin travels up one branch of a fiber to a bifurcation, then out along another branch of the same afferent fiber to its ending, where arrival of the impulse brings about a local vasodilatation. Inasmuch as these are afferent nerve terminals, conventional neurotransmitter arrangements may not apply here. Various substances have been proposed as the transmitter—a histamine-like compound, or acetylcholine, or adenosine triphosphate. An axon reflex may be induced by any factor that injures the surface of the skin, by trauma, cooling, heating, frostbite, etc. It is believed to contribute to local defense and repair by increasing the blood flow in response to the injury (see Chapter 19).

ADRENAL MEDULLA VS. NEURAL CONTROL OF BLOOD VESSELS

In addition to vasomotor control by way of nerves ending on vascular smooth muscle there is another type of neurohumoral control that is autonomic in nature but that acts upon vascular smooth muscle through catecholamines circulating in blood. One of the oldest known neurohumoral relationships is secretion of epinephrine (with smaller quantities of norepinephrine) from the adrenal medulla following stimulation of (1) splanchnic nerves, (2) lateral columns of the spinal cord, (3) vasomotor center in medulla oblongata, or (4) the lateral hypothalamus.

ADRENERGIC RECEPTORS IN VASCULAR SMOOTH MUSCLE

Receptors that respond to catecholamines are called adrenergic; at least two pharmacological types, α and β, have been identified. A third receptor for cholinergic effects, the muscariwic receptor, is postulated for sympathetic cholinergic vasodilation of vessels in skeletal muscle. These receptors are part of the target cells, since denervation does not prevent the response to circulating catecholamines. In vascular smooth muscle α-receptors mediate vasoconstriction and β-receptors mediate vasodilation. Norepinephrine acts primarily on α-receptors and is their natural activator when released from adrenergic sympathetic nerve endings. Following pharmacological α-receptor blockade intraarterially infused norepinephrine can produce vasodilatation, presumably the result of some β-receptor stimulation by norepinephrine. Sympathetic nerve stimulation fails to produce vasodilation under these circumstances. These observations originally led to a supposition that β-receptors are not innervated, or at least are not acted on by neurally released norepinephrine. However, in skeletal muscle when α-adrenergic vasoconstriction is prevented by an α-receptor blocking agent, and cholinergic vasodilatation is prevented by atropine, sympathetic nerve stimulation produces vasodilation (Viveros et al., 1968). Thus, whereas sympathetic nerve activity ordinarily results in a predominant α-receptor vasoconstriction, in skeletal muscle at least, a sympathetic β-receptor vasodilation can be unmasked under special experimental conditions. The dominant vasodilation produced by circulating epinephrine in some vascular beds can be changed to vasoconstriction by β-adrenergic receptor blockade, due to residual α-adrenergic stimulation by epinephrine.

It has been widely assumed, but never proved, that the hormones secreted by the adrenal medulla contribute significantly to vasomotor control of the blood vessels. But in a direct comparison of the separate actions of the hormones and of vasomotor fibers on peripheral resistance in representative vascular beds, most blood vessels have been found to be dominated by their vasomotor fibers (Mellander and Johansson, 1968). The exception is vessels in skeletal muscle, where physiological concentrations of epinephrine cause almost maximal dilation with a significant reduction in the ratio of precapillary to postcapillary resistance. Secreted epinephrine, therefore, is believed to produce vasodilatation in, e.g., exercise. If the amount of secreted epinephrine is large, as following hemorrhage, α-adrenergic effects predominate and reinforce neurally mediated α-receptor vasoconstrictor effects, especially in the renal vascular bed (Korner, 1974).

CENTRAL CONNECTIONS

In the past several years concepts concerning central nervous integration of cardiorespiratory control have been rapidly evolving. It is no longer appropriate to consider cardiorespiratory control as a series of independent reflex arcs with localized central nervous connections. There is now good evidence for modulation of afferent traffic at several different levels within the brain, and there are possibilities for interaction at various levels of efferent responses extending from the cortex to the spinal cord (Korner, 1979). Studies of reflex effects in man and in unanesthetized animals (Vatner and Boettcher, 1978; Eckberg, 1976; Van Citters and Franklin, 1969) where the central nervous system is not depressed by anesthetics emphasize the important role of the central nervous system in integrating cardiovascular responses.

The anatomy of the central connections important in circulatory control is complex and properly the subject of neurophysiology. However, there are five major areas in the central nervous system that are of major importance to circulatory control: spinal cord, medulla oblongata, hypothalamus cerebellum, and cerebral cortex. The functional interconnections between these areas are shown schematically in Figure 2.169.

As depicted in Figure 2.170 and described in more detail below, stimulation and regional ablation experiments have shown that specific regions of the brain stem may have predominant vasoconstrictor or cardioaccelerator activity. However, most recent work has shown extensive interconnections between ascending and descending pathways which emphasize the interaction between areas of the brain stem and has failed to show specific anatomic localization of functional areas (Hilton et al., 1980).

The Spinal Cord

The preganglionic sympathetic neurons may, under some circumstances, exhibit spontaneous activity that is independent of an excitatory drive from afferent fibers or from higher levels of the nervous system. This phenomenon has been observed in animals with the cervical spinal cord sectioned so that it is isolated from central inputs. Under these circumstances local changes in the tension of oxygen or carbon dioxide are believed to be responsible for the "spontaneous" activity.

Various types of afferent stimulation are able to call forth reflex vasoconstriction via the spinal cord. For example, pain or cold stimulation of the skin induces a segmental constriction of splanchnic vessels in spinal animals. Cutaneous vasodilatation occurs when the skin is moderately warmed. Effects that are more intense and more widespread are seen in humans who have suffered transverse lesions of the spinal cord, some of the more obvious of these reflexes arising from distension of hollow viscera (Guttman and Whitteridge, 1947). In patients with high spinal transection, distension of the bladder raised the blood pressure from normal levels to 300/140 mmHg. Vasoconstriction was seen in the hand, together with a fall in blood flow through the leg. These phenomena seem to represent a dissemination of the reflex process widely throughout vasoconstrictor neurons of the spinal cord following activation of stretch receptors in the bladder wall.

Medulla

Neurons subserving vasomotor function from the spinal cord are under the control of higher order neurons located particularly in the floor of the fourth ventricle of the medulla. Local electrical stimulation has revealed a lateral "pressor area" and a medial "depressor area" causing vasoconstriction and vasodilatation, respectively. The vasodilatation is caused by inhibition of vasoconstrictor tone, specific vasodilator fibers not being involved. Together these two areas are known as the vasomotor center (Fig. 2.170). In the intact animal or human this center, in turn, is controlled by still higher neurons from hypothalamus and cerebral cortex. This area receives afferent information and other neural inputs from more central areas that may become integrated into patterns of vasomotor activity.

The state of activity of the medullary vasomotor center thus depends upon afferent nerve impulses received from various organs and regions of the body, as well as from other nervous centers, respiratory centers, etc., and upon the chemical composition of the blood. The contractility of the heart, the relative peripheral resistance and, hence, the cardiac output and its distribution to the several regions or organs of the body are all mainly controlled by areas of the medulla.

The vasomotor center exhibits inherent automaticity, in that it continues to discharge and to maintain arterial blood pressure through vasoconstriction even after elimination of all incoming nerve influences (Peiss, 1965). Section of the brain stem above the medulla does not affect blood pressure; this indicates that upper levels do not domi-

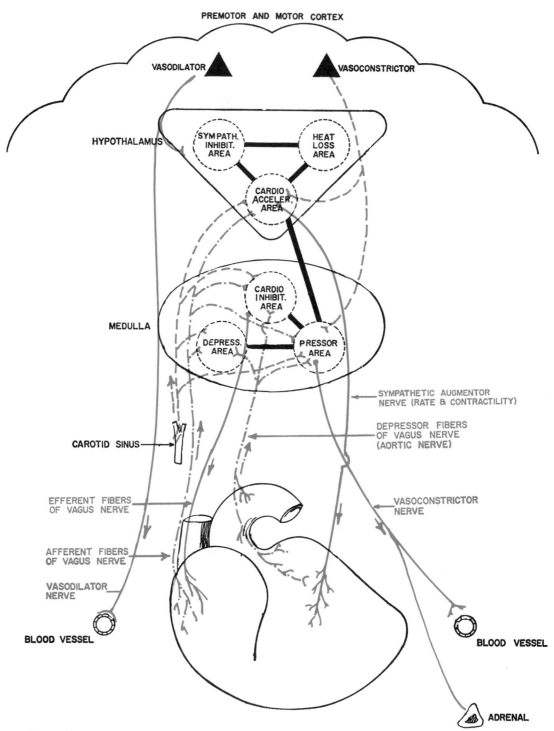

Figure 2.170. Schematic diagram of the central connections involved in cardiac and respiratory control.

nate the medullary level, even though they can modify its state of activity. Another experiment that demonstrates the nature of the medullary control and its interaction with the spinal level is section of the spinal cord in the lower cervical region in the experimental animal as described earlier. This interrupts the stream of vasoconstric-

tor impulses passing from the medulla to the spinal cord: the blood vessels dilate and arterial blood pressure falls. One additional stage of this experiment may be carried out. After a time the spinal animal will recover a near normal blood pressure; due to activity of spinal neurons, section of the splanchnic nerves once more leads to hypotension and vasodilatation, this time only in the splanchnic bed. (In normal animal section of the splanchnic nerves doubles the flow of blood in the viscera they innervate (Burton-Opitz, 1903).) But once again, in the spinal animal with section of the splanchnic nerves, a certain degree of tone ultimately returns to splanchnic blood vessels. It is intrinsic to the smooth muscle of the vessels and is known as myogenic tone. It develops still more slowly than the tone that originates in sympathetic neurons of the spinal cord.

Hypothalamus

Both increases and decreases in heart rate and arterial blood pressure have been reported by investigators using electrical currents to stimulate the hypothalamus. In most of the experiments cardiac output and regional blood flows have not been measured. Several attempts have been made to assign vasopressor effects to one part of the hypothalamus and vasodepressor changes to another. The results have been inconclusive, even though many discussions of this subject state that the rostral hypothalamus is vasodepressor and parasympathetic in function, whereas the caudal hypothalamus is sympathetic and vasopressor (Fig. 2.170).

Another and principal function of the hypothalamus is control of body temperature. The rostral hypothalamus and preoptic area contain neurons that protect the body against overheating; they also control the discharge to the vasoconstrictor fibers of the cutaneous blood vessels, and, thus, play an important role in adjusting blood pressure. Electrical stimulation or local cooling of this area brings about a rise in blood pressure (vasoconstriction), while direct warming of this region produces a fall in blood pressure (vasodilation). The cutaneous arterioles, and, especially, the arteriovenous anastomoses (shunts), are the vessels most sensitively engaged in control of heat loss (Chapter 19).

Cerebral Cortex

A number of studies of cortical control of vasoconstriction are available, but relatively little is known about the importance of circulatory adjustments originating in the cortex. Stimulation of the motor and premotor cerebral cortex results in marked elevation of blood pressure with constriction of the cutaneous, splanchnic, and renal vessels, and, at the same time, a considerable vasodilation in the skeletal muscle. It is believed that these higher centers play significant roles in the blood pressure responses to pain and anxiety, and exercise, but may, as mentioned below, have their most important effects in modulating the response of "lower centers" in the brain stem to afferent input.

Interaction between Central Areas

The major role of the central nervous system in cardiovascular control is to integrate different information ("sensory inputs") and adjust the tonic outflow of autonomic nerve traffic. In the regulation of autonomic outflow there is integration of descending traffic with inputs from all levels in the central nervous system impinging upon preganglionic vagal and sympathetic fibers. In the classic "defense" reaction which is elicited by stimulating small areas of the hypothalamus there is a rise in blood pressure, cardiac output, and heart rate, with vasoconstriction in renal, intestinal, and skin beds and marked cholinergic vasodilation in muscle (Peis, 1965). If one stimulates lower in the brain stem the responses tend to be more specific, e.g., only splanchnic or skin vasoconstriction may occur. Thus there is some evidence for anatomic integration of responses.

There are fluctuations in vagal efferent activity synchronous with respiration, which largely account for the normal respiratory variation in heart rate. These fluctuations are abolished by factors which inhibit the central areas responsible for the control of respiration and are not seen when respiration is suspended (apnea). Thus, cyclic variations in activity in areas of the brain stem not related to cardiovascular function may normally influence efferent autonomic activity to the heart and circulation. There are also phasic changes in sympathetic nerve activity that are strongly linked to the activity of respiratory motor neurons. These are most apparent as fluctuations in aortic blood pressure and are discussed in more detail later in this chapter.

Afferent input from the IXth and Xth (cranial) nerves enters the medulla and synapses in an area termed the nucleus tractus solitarius. From here there are "long latency" (presumably involving many synapses) connections to the pons, thalamus, and the cerebral hemispheres (Figs. 2.169 and 2.170). Other ascending traffic comes from "visceral" afferents (cardiac and pulmonary receptors, and somatic afferents). There is this evidence that sensory input is integrated. Thus, as described later in this section, during exercise sensory inputs from exercising muscle significantly influences the sensitivity of the baroreceptor system to changes in blood pressure.

REFLEX AND HORMONAL REGULATION OF THE HEART AND CIRCULATION

During local adjustments of blood flow to meet metabolic needs the arterial pressure driving the blood through the vessels must be maintained. According to the general resistance equation discussed in Chapter 7 (resistance = pressure/flow), if there is vasodilation with decreased resistance and increased flow, and yet the pressure gradient is found to be unchanged, some extrinsic regulatory system must have intervened to maintain arterial pressure in the vessels. In the case of a local decrease in the arterial resistance due to muscular exercise, for example, the resistance in other beds may increase, and regulatory mechanisms may also act to increase the cardiac output. These extrinsic regulatory mechanisms govern heart rate, cardiac contractility, blood vessel caliber, and the distribution of blood volume.

The control mechanisms responsible for maintaining arterial pressure may be divided into *short-term processes*, which are effective over a period of seconds to hours, and *long-term processes* that operate over days to weeks. The former are largely neural, utilizing receptors in the heart and blood vessels to sense blood pressure and the autonomic nervous system to regulate cardiac function and arteriolar diameter. The long-term controls are largely hormonal and renal. They regulate both arterial resistance and the blood volume. Both of these systems are negative feedback control systems and have many properties in common with other biological control systems which regulate a wide variety of body functions, ranging from intracellular metabolism to body posture.

The engineering theory which describes control systems was initially developed in connection with speed controls on steam engines, and now has progressed into a large and complex field utilizing modern computing equipment. The approach requires detailed knowledge of the function of the system being examined and the dynamic response of its various components. The application of this approach to cardiovascular control has recently been summarized by Sagawa (1972). The approach is extremely useful for investigation of the circulation, but it is beyond the scope of this discussion and frequently beyond the level of our current knowledge of the individual components of the cardiovascular systems. However, it is important to understand the basic principles of negative feedback control systems, since they form the basis for most cardiovascular control systems. A schematic diagram of a simplified negative feedback control system is shown in Figure 2.171. There are three

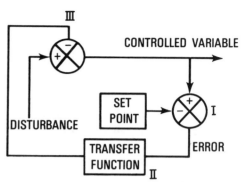

Figure 2.171. Schematic diagram of a negative feedback control system. A disturbance applied to the system at I is sensed at I and compared to some reference point (*SET POINT*). The difference between the set point and the controlled variable (*ERROR*) is used to return the controlled variable (*II* and *III*) to its original value.

essential components to this type of control system: a sensing device (I); a mechanism to change the controlled variable (III) and a function which determines which way and how much to alter the variable (II). Let us take the simple example of a gasoline engine the speed of which (RPM) we want to maintain constant. RPM is thus the controlled variable (Fig. 2.171) and we must have some way to sense the magnitude of this variable (block I), a tachometer in this case.

If we want to maintain RPM at 1000 (this value is termed the set point) we must sense the RPM and compare it to the set point. The difference between the set point and the controlled variable is the error signal. The error signal must be acted on by some function (II) to provide a signal to the plant (or engine in this example) to change the controlled variable. In our example it might adjust the throttle. Note that the change made will always be in the opposite direction from the original disturbance. Thus, the term "negative feedback." If, for example, we put a heavier load on the engine and reduce the RPM, the system would work to open the throttle to return RPM to the set point.

To further illustrate the nature of a negative feedback system let us consider the example of the arterial baroreceptor system utilizing the generalized diagram shown above. Suppose that methoxamine, an agent which constricts the arteries, is given to increase blood pressure. In the control diagram, the agent would act in block III at the *arrow* labeled disturbance to produce a 60 mmHg increase in blood pressure. Thus a 60 mmHg error would be determined at block I. The response of the transfer function is negative with respect to the initiating stimulus, and the overall effect will be to return the blood pressure towards normal. In this case a withdrawal of sympathetic arterial constrictor tone would be initiated by the central nervous system. If the initial mean arterial pressure is 100 mmHg and is increased to 160 mmHg by the drug, following "compensation" by the control system arterial blood pressure would return to perhaps 120 mmHg. The overall gain of this system can be estimated as the ratio of the amount of correction to the remaining error. For the baroreceptors, the amount of correction would be 40 mmHg, the remaining error, 20 mmHg, for an overall gain of −2 (Guyton, 1972). If the gain were increased the residual error would be less.

In the example above we have considered the steady-state values achieved by a negative feedback control system. Since the blood pressure sensory systems and the changes in sympathetic tone take some time, the transitory response to an abrupt change in blood pressure will be different. Infrequently, the control system overshoots the eventual steady-state value and may in certain circumstances oscillate for a period of time around the eventual steady-state result. Fortunately, most of the body's control systems and the arterial baroreceptor system are quite well damped (there is significant time delay in reacting to a particular stimulus), and the eventual steady-state value is approached without substantial oscillation or overshoot.

BARORECEPTOR BLOOD PRESSURE CONTROL SYSTEM

Etienne Marey, a French physician, first noted in 1859 the inverse relationship between blood pressure and heart rate, and largely through the later pioneering work of Hering (1927) and Heymans (1933), we have come to understand that the carotid sinus and aortic arch baroreceptor systems are responsible for a highly effective blood pressure control system. This system consists of stretch receptors located in vessel walls which send information reflecting the level of blood pressure to the central nervous system which, in turn, sends efferent impulses to the cardiovascular system which then alter blood pressure. These sensitive stretch receptors are diagrammed in Figure 2.172. As mentioned earlier, they are located throughout the aortic arch along the major thoracic vessels, and at the bifurcation of the carotid arteries in the neck. Fibers from the latter area course through the carotid sinus nerve (These branches of the IXth nerve are sometimes termed "Hering's nerve") and thence to the medulla. Fibers from the aortic arch sensors course through the vagus nerve (these branches going to the aortic arch and heart are sometimes termed cardiac depressor nerves) (Fig. 2.172).

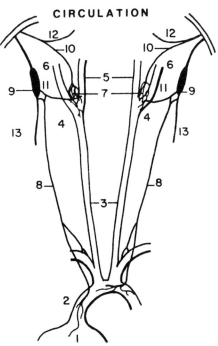

CIRCULATION

Figure 2.172. Innervation of the carotid sinus and arch of aorta: *1,* heart; *2,* arch of aorta; *3,* common carotid; *4,* carotid sinus; *5,* external carotid; *6,* internal carotid; *7,* carotid bodies; *8,* cardiac depressor nerve; *9,* ganglion of vagus; *10,* sinus nerve, branch of the glossopharyngeal nerve; *11,* nerve branch connecting the carotid sinus with the vagus ganglion; *12,* glossopharyngeal nerve; and *13,* vagus nerve. [Adapted from Heymans et al. (1933).]

The precise mechanism by which distortion activates these receptors is unknown. However, there is increasing evidence that increased stretch alters the conductance of sodium and potassium ions across the nerve (Kunze, 1978). Thus, it is possible that local alterations in sodium content of the vessel wall, as occurs during congestive heart failure, may be important in regulating receptor function (Brown, 1981). The receptors themselves respond only to distortion, and if the carotid sinus is encased in a plaster cast before raising the blood pressure, there will be no alteration in carotid sinus nerve traffic.

In Figure 2.173 panels *A* and *B* schematically represent recordings from a single carotid sinus nerve fiber at several levels of mean carotid sinus blood pressure, each vertical spike representing a single depolarization from this fiber. At a very low average aortic blood pressure (50 mmHg) there is little activity. The activity is normally correlated in time to the systolic aortic pressure increase, indicating that these receptors are sensitive to rate of change of pressure as well as average level. The average level of pressure at which a fiber first fires is termed the threshold. As mean pressure within the carotid sinus is progressively increased with the pulse pres-

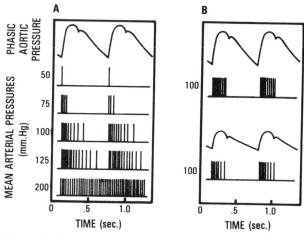

Figure 2.173. Relationship between phasic blood pressure in the carotid sinus and aorta and frequency of nerve traffic (represented by single vertical spikes). In panel *A* nerve traffic at several different mean levels of aortic pressure is shown, and in panel *B* activity at two levels of pulse pressure at the same mean pressure.

sure constant (Fig. 2.173, panel *A*) nerve traffic is increased until it becomes continuous (at 200 mmHg in this example). With a fixed level of mean pressure, the number of impulses per second increases as the pulse pressure increases (Fig. 2.173, panel *B*).

Both the thresholds for firing and the frequency of impulses per unit change in the aortic pressure (gain) may vary substantially from fiber to fiber, but the average threshold in the carotid sinus at which firing begins is normally not less than 50 mmHg, and the maximum output is achieved at approximately 170 mmHg (Fig. 2.174). Although the sensitivity of myelinated and nonmyelinated fibers is similar the threshold is usually higher in nonmyelinated fibers. The sigmoid shape of the relationship between nerve traffic and carotid sinus pressure leads to two important conclusions concerning the sensitivity of the baroreceptors. First, pressures below approximately 50 mmHg are not sensed, and therefore alterations in blood pressure below this level will result in no change in baroreceptor output. Secondly, the slope of the relationship between nerve traffic (impulse frequency) and blood pressure (gain) is maximal at the normal level of mean arterial pressure (frequently referred to as the set point); thus, the system works most effectively here. It will return blood pressure to this level with the least error and the least time lag (see discussion of Fig. 2.171). At pressures above 170 mmHg, where the gain becomes 0, changes in blood pressure will no longer result in changes in nerve activity, and hence the system will be ineffective in controlling blood pressure. Since these receptors are also sensitive to rate of change of pressure, an increase in pulse pressure at the same mean pressure

increases the frequency of carotid nerve traffic. Although the receptors in the aortic arch have a slightly higher threshold and a somewhat reduced gain, the receptors in the carotid sinus and aortic arch are in most other respects quite similar.

The increase in afferent carotid sinus nerve or vagal (aortic) nerve traffic acts through the central nervous system to return blood pressure to its control value. This is achieved by a decrease in efferent sympathetic constrictor tone to the peripheral arteries, and an increase in parasympathetic cardiac tone. As shown in Figure 2.174, panel *B*, as carotid sinus traffic increases over the physiologic range of arterial pressures, sympathetic nerve impulse frequency declines. Thus the net result of an increase in pressure in the carotid sinus will be to reduce sympathetic tone to the blood vessels and thereby to reduce blood pressure. Note that the *dashed line* in Figure 2.174 again indicates the response to increased pulse pressure at the same mean pressure. Several factors will contribute to the reduction of blood pressure. There will be withdrawal of sympathetic vasoconstriction tone to arterioles resulting in a fall of systolic and diastolic arterial pressure

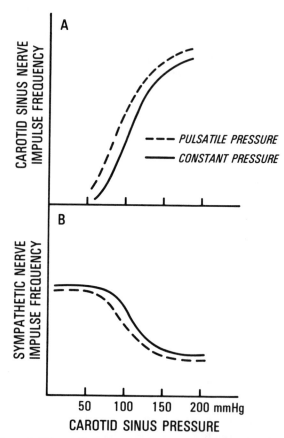

Figure 2.174. *A*, Relationship of frequency of firing of the carotid sinus nerve and carotid sinus pressure. *B*, Frequency of firing of postganglionic sympathetic fibers in response to changes in carotid sinus pressure. [Adapted from Kezdi and Geller (1968).]

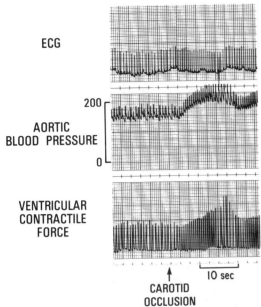

ECG

AORTIC
BLOOD PRESSURE

200

0

VENTRICULAR
CONTRACTILE
FORCE

↑
CAROTID
OCCLUSION

10 sec

Figure 2.175. Recordings of aortic blood pressure, electrocardiogram, and right ventricular contractile force (measured with a gauge sewn to the right ventricular free wall) during abrupt sustained bilateral carotid occlusion. At the *arrow* both carotid arteries are occluded below the carotid sinus. The initial increase in blood pressure, heart rate, and contractile force is due to the activation of the baroreceptor reflex. Since the vagus nerves are still intact the secondary falls in heart rate and blood pressure are due to the firing of aortic arch baroreceptors in response to the *increased* blood pressure.

and a fall in vasoconstriction tone to capacitance vessels (Alexander, 1954) resulting in decreased cardiac filling; an increase in vagal tone will reduce heart rate.

The hemodynamic effects of lowering carotid sinus pressure are demonstrated in the experiment illustrated in Figure 2.175, which shows the effect of sudden occlusion of the carotid arteries in an open chest dog in which a force gauge is sewn to the ventricle. Following bilateral carotid occlusion the drop in pressure in the carotid sinuses causes a rise in blood pressure, by increasing sympathetic stimulation to the blood vessels this occurs 10–15 s following carotid occlusion (the normal "time lag" for a sympathetic response). The effects of sympathetic nerve activation on myocardial inotropic state and the ventricular function curve are also illustrated in Figures 2.129 and 2.184. The increase in heart rate preceeds the increase in blood pressure (the time lag for vagal activity is 1 to 2 s). The subsequent stabilization and fall in blood pressure illustrates the action of the aortic receptors (which in contrast to the carotid sinuses sense the elevated blood pressure).

It is important to remember that the changes in efferent sympathetic and parasympathetic activity induced by the baroreceptor system have a variety of effects on the heart and circulation. For example, decreases in carotid sinus pressure increases sympathetic tone and will produce an increase in blood pressure; an increase in cardiac contractility, as reflected in the change in the ventricular function curve (Chapter 17); and decreased venous capacity with increased mean circulatory pressure and shift the venous return relationship (Chapter 13). The combination of these effects will produce a new equilibrium point resulting in an increase in cardiac output (Chapter 13).

Modifications of the Baroreceptor Response

Alterations in the baroreceptor blood pressure control system occur in both physiologic and pathophysiologic circumstances. These alterations may occur at all levels of the system from the interaction of receptor with the vessel wall to neuroeffector junction. The nerve endings in the carotid sinus normally adapt to a prolonged stimulus. This adaptation consists of a gradual reduction in firing frequency occurring within the carotid sinus. The mechanism of this type of adaptation is unknown, but it may have to do with altered arrangement of the coupling between the vessel wall and the receptor itself. The interaction between receptors and the vessel wall is importantly altered in several disease states. One example is the alteration occurring in essential hypertension (high blood pressure of unknown cause (Aars, 1975)). The vessel wall becomes stiffer in hypertension; thus, for a given increase in carotid sinus pressure there is less distortion of the receptor. Therefore, the gain of the system, the relationship between pressure and change in nerve traffic, is reduced (Kezdi, 1967). In addition, the threshold for firing has been reset upwards (to the new and higher level (Kezdi, 1968)). The effect would be to shift the entire relationship between carotid sinus nerve traffic and blood pressure (Fig. 2.174) to the right. The factors responsible for these changes are not well understood, but it is clear that they involve both changes in the vessel wall (stiffer) and in the receptor. Receptor sensitivity may even be increased, partially compensating for the vessel wall changes (Brown, 1981).

In congestive heart failure the gain of the baroreceptor control system is reduced without substantial change in threshold for firing (Higgins, 1972). Although the mechanism for this change is unknown, heart failure affects the baroreceptor control system at several levels. There is an increased sodium content of the arterial wall which may alter its physical properties or change the receptor response. Moreover, the efferent limb of the baroreceptor control system is altered with reduced stores of neurotransmitters in the myocardium and an altered response to sympathetic stimulation (Schmid, 1981).

Important changes in the baroreceptor response also can occur when information is integrated in the central nervous system. For example, recalling Figure 2.169 it is clear that afferent nerve traffic from other sympathetic inputs or centrally from "higher centers" may impinge upon nerve traffic coming from the baroreceptors and alter the net change of efferent sympathetic and parasympathetic traffic induced by a given change in blood pressure. The clearest example of this has been shown to occur in man during exercise, when increasing levels of exercise progressively reduce baroreceptor gain. This reduction is proportional to the extent of exercise (Korner, 1979).

OTHER REFLEXOGENIC AREAS

Two other major groups of receptors contribute importantly to circulatory regulation. One major group of receptors responds to alterations in their chemical environment (chemoreceptors). These receptors are located in specialized cells near the high pressure receptors (Heymans, 1931). The other group are stretch receptors located in the large vessels in the thorax, in the lung, in both atria, and in the ventricles. These receptors, frequently termed "low pressure receptors," are sensitive to distortion.

Chemoreceptors

Chemoreceptors may be divided conveniently into two groups, arterial and cardiopulmonary. The first group are located in the carotid and aortic bodies and respond to physiologic alterations in blood gases and pH (Biscoe, 1971). The second group are located in the heart and lungs and are sensitive to minute amounts of substances such as veratrum alkaloids, phenyldiguanide, nicotine, bradykinin, and capsaicin. Such substances stimulate a variety of afferent vagal endings in the heart, lungs, and great vessels (Bevan, 1962). Recent studies have shown that these receptors may be sensitive to some substances formed and released in their vicinity such as bradykinin, prostaglandins, or perhaps metabolic products associated with ischemia.

ARTERIAL CHEMORECEPTORS

As mentioned earlier, the carotid and aortic bodies are small, discrete structures situated at the bifurcation of each common carotid artery or along the major intrathoracic arteries. Both the carotid and aortic chemoreceptors are sensitive to changes in the partial pressures of carbon dioxide, hydrogen ion, and oxygen in the blood. Blood flow through the carotid body per gram of tissue is by far the largest in the body. Direct measurements indicate flows of 2 l/100 g/min through this organ (recall that the flow to the left ventricular muscle is approximately 100 ml/100 g/min).

The physiologic chemoreceptors play an important role in the control of respiration (see Chapter 39) and are thought to be of relatively minor importance in the control of the circulation. The mechanism by which changes in blood chemical composition effect a response in the arterial chemoreceptors is not clearly understood. The threshold for changes in afferent traffic of the chemoreceptors from the carotid body and aortic arch is fairly high. Vascular effects in dogs require a severe degree of carotid body hypoxia; when arterial P_{O_2} progressively falls below 70 mmHg afferent traffic increases and there is linear increase of sympathetic efferent traffic until a maximum is reached at a P_{O_2} in the mid 30s (Pelletier, 1972). Increases in nerve impulses from afferent nerves may be recorded with CO_2 tensions above 30 mmHg and with small decreases in pH either together or independently (Neil, 1972).

With respiration controlled (eliminating the effects of the pulmonary stretch receptors on the circulation) the major cardiac effect of stimulation (increasing afferent traffic) of the physiologic chemoreceptors is profound bradycardia, conduction defects (a common result of vagal efferent stimulation), and a diminution of cardiac contractility. Bradycardia is greatly diminished by the use of atropine but may not be abolished until sympathetic pathways to the heart are also blocked, suggesting a synergistic withdrawal of sympathetic tone (MacLeod and Scott, 1964). Carotid body reflexes not only slow the heart but also depress contractility. Downing et al. (1962), utilizing isolated heart preparations, showed that there was a depression of the ventricular function curve during carotid body hypoxia. Although the efferent neuromechanisms for this change are still controversial, they appear to involve both alterations in parasympathetic and sympathetic tone. The major effect on the peripheral circulation of hypoxic chemoreceptor stimulation is vasoconstriction (Pelletier, 1972; Biscoe, 1971; Heymans and Neil, 1958).

Recent evidence suggests that when stimulated with hypovic blood aortic and carotid bodies elicit a directionally opposite response. Although the reflex effect of stimulation of the carotid bodies is primary inhibitory (bradycardia), stimulation of aortic bodies causes an increase in heart rate and ventricular function. Perfusion of aortic bodies with hypoxic and hypercapnic blood produces the same degree of vasoconstriction as perfusion of the carotid bodies (Daly and Angell-James, 1975).

There are also direct effects of hypoxia on the brain and spinal cord, causing an increase in sympathetic efferent discharge with hypertension and tachycardia (Downing et al., 1963; DeGeest et al., 1965).

For most terrestrial mammals the peripheral arterial chemoreceptors have been shown to play a major role in the cardiovascular response to diving (Jones and Purves, 1970). On immersion of vertebrates, sensory receptors in the larynx are stimulated and trigger a prompt suppression of respiration. During the resultant hypoxia peripheral chemoreceptors are stimulated, and bradycardia and hypertension are the dominant cardiovascular response (Daly and Angell-James, 1975).

NONPHYSIOLOGIC CHEMORECEPTORS

In 1867 von Bezold discovered that the intravenous injection of crude veratrum alkaloids caused bradycardia and hypotension. Later Jarisch and his associates in 1930 to 1948 showed that these cardiovascular effects were due to stimulation of afferent vagal endings in the heart and lungs (hence the Bezold-Jarisch reflex; see Heymans and Neil, 1958). They exist in the afferent endings, lungs, heart, and great vessels and the abdominal viscera (Dawes and Comroe, 1954; Coleridge and Coleridge, 1977). In the heart, nerve endings are present in the right ventricle and atria, but there are more nerve endings in the left ventricle. They normally have very sparse resting discharges, but the endings are extremely sensitive to chemicals. Stimulation of these fibers with veratridine produces a profound cardiac vagal efferent discharge, similar to that of the physiologic chemoreceptors discussed above. The peripheral effects are opposite in direction, and stimulation of these receptors induces vasodilation and hypotension due to withdrawal of sympathetic tone.

Most current electroneurographic evidence has shown that ventricular fibers may be excited by a variety of agents, including veratradine, nicotine, phenyl diguanidine, acetylstrophanthidin, and capsaicin, all of which may evoke the chemoreceptor reflex (Coleridge et al., 1978). There is increasing evidence that these types of fibers may also be sensitive to endogenously released substances, such as prostaglandins serotonin, lactic acid, and bradykinin. These substances may be released by the myocardium during hypoxia and ischemia and may provide an important pathophysiologic role for these receptors.

In recent years it has become apparent that there are also afferent fibers from chemosensitive receptors coursing through sympathetic pathways to the spinal cord. These afferents may be sensitive to nonphysiologic substances and in vagotomized animals their activation with substances which also excite the chemoreceptors discussed above produces a tachycardia and vasoconstriction with an increase in contractility (Peterson and Brown, 1971). It is likely that these responses may be important in the excitatory reflexes occurring in myocardial ischemia in man.

In summary, the chemoreceptors are important in the control of respiration. Although they have extensive effects on the circulation they are not, however, thought to be of major importance in the control of the circulation under normal circumstances.

Intrathoracic Stretch Receptors

These receptors are located in the thorax and are sensitive to local changes in distortion. They are found along pulmonary vessels, other large intrathoracic vessels, all chambers of the heart, and the bronchi. Afferent connections of these receptors are located either in the vagus or in sympathetic tracts synapsing in the upper five thoracic sympathetic ganglia. The most important of these receptors for circulatory control are located within the four cardiac chambers. These cardiac receptors participate in reflexes which are important in the regulation of heart rate, blood pressure, and blood volume. Moreover, it is quite likely that these receptors are also important in adaptations occurring under pathologic conditions, including chronic cardiac dilatation and myocardial ischemia. Their role in the regulation of the circulation is considered in more depth under integrative responses.

The cardiac receptors are located in both atria and ventricles and may have either myelinated or nonmyelinated afferent connections. Nerve endings are found within the subendocardium and in the subepicardium distributed along the coronary vessels. Innervation of the myocardium is present but less dense. All receptors respond to distortion or stretch with increased nerve traffic. The most important atrial receptors appear to be those with vagal afferents, and the larger myelinated fibers of this group have been studied extensively. The endings of these fibers are located primarirly in the right and left veno-atrial junction. Cardiac reflexes associated with stimulation of these vagally innervated receptors, produce bradycardia and hypotension. Stimulation of the whole atrium tends to stimualte more unmylenated endings which are depressor in nature producing bradycardia and decreased peripheral resistance.

INTEGRATED CONTROL OF CIRCULATORY FUNCTIONS

The receptor systems discussed above all contribute importantly to cardiovascular homeostasis. In the next several sections we will consider their role in the integrated control of circulating blood volume, heart rate, and cardiac function. We will discuss the response of neural control systems to an acute stress, such as exercise, and a chronic circulatory stress, hypertension, and introduce the concept of longer term circulatory control.

Before considering these integrated responses it is helpful to recall the sites and mechanisms of vasomotion that are important in circulatory control (see Fig. 2.9). Arterial resistance and systemic blood pressure are primarily under the control of the resistance vessels, small arteries, and arterioles. The sympathetic (vasoconstrictor) nervous system plays an important role in regulating the diameter of these vessels through continuous levels of neural discharge (tone). The resistance across a given parallel vascular bed is primarily the result of the balance between this tone and the local metabolic and myogenic factors discussed in Chapter 7. This resistance, of course, also determines flow through the bed at a constant pressure and thus can regulate the circulating volume of a regional bed. The ability of a regional circulation to alter its resting flow varies markedly in different regional beds, with the myocardium, skeletal muscle, the splanchnic circulation, and skin all showing the capacity to increase flow by 3- to 4-fold under conditions of maximal vasodilation. The cerebral bed, however, is less sensitive to neural control (Rapela, 1967).

Referring again to Figure 2.9 there are two further sites of neurally controlled vasomotion: First on the balance between pre- and postcapillary resistances and secondly on the capacitance vessels. From the Starling equilibrium recall that the filtration of fluids across the capillary is dependent on the oncotic and hydrostatic pressure gradients. The latter is determined by both neural and local factors. Thus changes in the ratio of pre- to postcapillary resistance can produce either net absorption or filtration of circulating volume across the capillary (Chapter 7). Regulation of tone to the capacitance beds also has important effects on circulating blood volume and venous return as discussed below.

REGULATION OF CIRCULATING BLOOD VOLUME

The maintenance of adequate tissue perfusion is dependent on a sufficient level of cardiac output. The factors which influence the performance of the heart and therefore affect its output are discussed in Chapter 12. In the section below we will consider the factors which influence circulating blood volume and thereby determine venous return and cardiac output.

The kidney is the dominant organ associated with control of blood volume, and there are important reflex and hormonal systems which regulate salt and water excretion. These systems are discussed in detail in Section 5, Chapter 32 on the regulation of volume and osmolality of body fluids. Maintenance of normal volume and osmolality of the body fluids requires that the input and output of water and sodium are matched each day. Both the low pressure (atrial) stretch receptors and the arterial baroreceptors participate in this regulation. Receptors in the left atrium (and probably pulmonary veins) which respond to stretch have been implicated in the regulation of blood volume (Gauer et al., 1970; Goetz et al., 1975). Acute distention of the whole left atrium, as with a balloon, causes diuresis. Similarly, procedures which decrease central blood volume, e.g., hemorrhage, positive pressure breathing, have the opposite effect. The effects on fluid volume are mediated reflexly by changes in antidiuretic hormone, aldosterone release, and renal blood flow.

Figure 2.176 illustrates several of these effects in response to graded hemorrhage. With 10% loss of blood volume the firing of the atrial stretch receptors is reduced, and with further reduction in blood volume aortic pressure is reduced decreasing nerve traffic from the aortic and carotid sinus baroreceptors. Lowering of traffic from the atrial receptors increases the secretion of antidiuretic hormone which acts to reduce water excretion by the kidney (see Section 5). Reduction of carotid sinus and aortic baroreceptor traffic increases renal sympathetic nerve traffic constricting the afferent arteriole and reducing filtration of fluids by the kidney. Moreover, increased sympathetic nerve traffic will increase the secretion of renin by the granular cells of the kidney. Renin is a proteolytic enzyme released by the kidney which acts on a substrate protein in plasma to liberate the decapeptide angiotensin I (Fig. 2.177). This inactive prohormone is in turn converted into a potent vasoconstrictor hormone angiotensin II by a converting enzyme. Angiotensin II performs three major related roles:

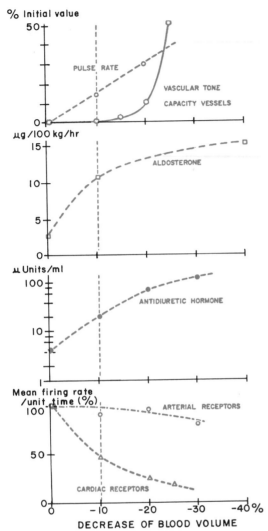

Figure 2.176. Responses to a graded hemorrhage. With loss of the first 10% of blood volume the firing rate of atrial (*cardiac*) receptors falls to one-half. As the loss exceeds 20% there is a fall in the firing rate of aortic (*arterial*) baroreceptors. The plasma aldosterone and antidiuretic hormone concentrations rise, causing sodium and water retention. The pulse rate increases steadily as blood volume falls. [Redrawn from Gauer et al., (1970).]

mone. Moreover, atrial receptors are the most important sensors reflexly altering pre- to postcapillary resistance, and can evoke sustained vasoconstriction in the intestinal vascular bed. Increased pre- to postcapillary resistance increases plasma volume, by favoring fluid uptake by the capillaries, and vasoconstriction of intestinal capacitance vessels can mobilize substantial amounts of blood into the rest of the circulation. These neurogenic vasomotor adjustments for producing rapid increases in blood volume appear to be important in maintaining cardiovascular homeostasis (Mellander and Johansson, 1968).

The reflex control of capacitance vessels by the baroreceptor systems, an important part of the response to a decrease in blood volume, is illustrated by the following experiment in which the effects of changing blood pressure in the carotid sinus is examined. The aortic nerve is cut, eliminating the response of the aortic baroreceptors. In this preparation all the blood coming from the systemic veins is diverted to a reservoir and returned at a constant rate to the right heart. Thus, at a constant cardiac output regulated by the pump, changes in the reservoir volume reflect changes in venous capacitance. As shown in Figure 2.178, progressive decreases in the pressure within the isolated carotid sinus produce an increase in aortic pressure and an increase in volume in the reservoir due to increased sympathetic vasoconstrictor tone to the capacitance vessels.

first, increasing the arterial pressure by direct vasoconstriction; second, increasing the renal retention of salt and water through stimulation of aldosterone secretion; and third, angiotensin II is thought to directly stimulate thirst. All of these responses tend to correct the loss of circulating volume induced by hemorrhage. The relative importance of these mechanisms for control of circulatory volume is discussed in Chapter 32. However, the sensitive response of the hypothalamic osmoreceptors to decreases in plasma osmolality may be the dominant normal control mechanism. These receptors stimulate the release of antidiuretic hor-

CONTROL OF HEART RATE

The cells of the cardiac conduction system exhibit spontaneous depolarization; the intrinsic cardiac rate is normally dominated by the fastest pacemaker site (usually the SA node). In the denervated heart, the rate ranges from 100 to 110 beats/min, and continuous control by the autonomic nervous system (predominately the parasympathetic nervous system) is responsible for a suppression of the

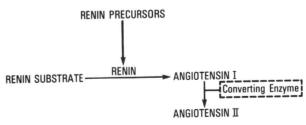

Figure 2.177. Production of angiotensin II from renin.

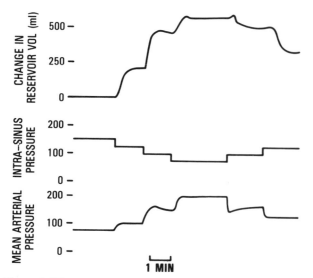

Figure 2.178. Results of an experiment in which blood pressure in the isolated carotid sinus is varied. All venous blood returning to the heart is drained into a reservoir and then pumped at a constant rate back to the pulmonary artery. Changes in reservoir volume thus represent changes in venous capacity. [Adapted from Shoukas et al. (1973).]

normal intrinsic rate of the SA node by 20 to 30 beats/min.

The sinoatrial node is richly innervated by both sympathetic and parasympathetic fibers. At rest there is continuous vagal nerve traffic to this region and relatively little resting sympathetic activity. This results in suppression of the intrinsic rate and the low resting normal heart rate. However, changes in both sympathetic and parasympathetic tone may increase or decrease heart rate. In contrast to other regions of the myocardium, combined effects of changes in sympathetic and parasympathetic efferent activity at the sinus node do not result in entirely independent effects. Thus the parasympathetic effects on the sinoatrial node predominate, and when there is high parasympathetic activity an increase in sympathetic activity will produce only a modest increase in heart rate. However, an alteration in parasympathetic activity will produce a large change in heart rate relatively independent of existing levels of sympathetic nervous activity (Levy, 1971).

Due to the high concentration of choline esterase in the sinoatrial node the effects of parasympathetic activity are very transient. The effects of a single burst of parasympathetic activity will depend to a large extent on the timing between atrial depolarization (P wave) and the arrival of the parasympathetic efferent activity. This transitory effect of parasympathetic activity was first shown in the classic studies of Brown and Eccles (1934). A burst of vagal activity which arrives just before a P wave

is more effective in reducing heart rate and prolonging the RR interval than a burst of vagal activity which arrives just after a P wave (Levy et al., 1969). Later work has shown that repetitive vagal stimulation can lead either to large fluctuations in heart rate, depending on the stimulus to P wave interval, or even "pace" the heart by maintaining a constant timing relationship between vagal stimulation and the P wave (Levy et al., 1972).

The AV node is also richly innervated by both sympathetic and parasympathetic branches of the autonomic nervous system. In contrast to SA nodal tissue the effects of simultaneous stimulation of the sympathetic and parasympathetic systems at the AV node appear to be independent. Thus, an increase in sympathetic nerve traffic will decrease AV nodal conduction time independent of the level of vagal tone (Levy, 1971).

Although there is extensive overlap between the distribution of right and left sympathetic and parasympathetic nerves to the SA and AV nodes, stimulation of the left sympathetic stellate ganglion will predominantly stimulate the left side of the ventricle and the AV node (Levy et al., 1966). Indeed stimulation of the sympathetic nerve on the left may lead to acceleration of phase 4 depolarization in an area near the His bundles to such an extent that this area then becomes the predominant cardiac pacemaker.

In addition to the possibility of altering cardiac pacemaker sites the autonomic nervous system is thought to play an important role in the genesis of arrhythmias. As discussed in Chapter 8 norepinephrine, the sympathetic neurotransmitter, not only induces changes in phase 4 depolarization but also increases phase 0 dv/dt in the "slow fibers" of the AV node or in ischemic tissue. These changes can induce alterations in ventricular excitability and refractoriness and may be an important mechanism in the generation of arrhythmias when there is increased sympathetic nerve traffic.

In 1915 Bainbridge showed that rapidly infusing saline into the right atrium resulted in an increase in heart rate. This effect has been thought to be due to stimulation of mylenated atrial stretch receptors mentioned earlier. Later studies have shown that the nature of this reflex depends upon a variety of factors, including the extent of stimulation of baroreceptors which tend to produce an opposite effect. Moreover, the magnitude of tachycardia observed is extremely sensitive to the initial cardiac rate. With a high initial cardiac rate, even a bradycardic response to volume infusion may be observed. In the conscious animal a tachycardia may be induced by volume loading or by acute hemorrhage. Presumably the Bainbridge reflex dominated the baroreceptor response when blood volume was acutely raised, and the baroreceptor system dominated the response during acute reduction in blood volume (Vatner and Boettcher, 1978).

EFFECTS OF RESPIRATION

There are important effects of respiration on the heart and circulatory system. In the central nervous system there are interactions between respiratory "pacemakers" and efferent autonomic tone which directly affect heart rate and blood pressure. These pacemakers induce variations in systemic blood pressure at the same period as the frequency of respiration. These blood pressure waves, termed Traube-Hering waves, after their discoverers, are the result of increased efferent sympathetic nerve traffic occurring with the onset of respiration. There are also variations in vagal tone that are centrally mediated and related to respiration. These result in a decrease in vagal activity occurring during inspiration and an increase in tone during expiration. These latter two effects are normally thought to be independent of a variety of reflex effects mediated by either low or high pressure baroreceptors or pulmonary stretch receptors.

There are also mechanical effects of respiration on left ventricular output. With inspiration there is a decrease in intrathoracic pressure which results in an increase in right atrial inflow (and perhaps, a secondary increase in heart rate mediated by the "Bainbridge" effect), whereas left atrial inflow is transiently reduced consequent to the increase in pulmonary vascular capacity. This results in lowering of aortic blood pressure which occurs most predominantly in early inspiration and is normally over by midinspiration. Secondly, when the increase in systemic venous return eventually reaches

the left atrium, the increased left ventricular outflow results in an increase in systemic pressure returning blood pressure to its peak level, an effect which normally occurs in mid- or end expiration. This increase in arterial pressure frequently results in reflex cardiac slowing. Thus, the normal responses of cardiac frequency to inspiration are represented by an increase in heart rate during inspiration due to central factors and possibly to a Bainbridge-like response, and a slowing of heart rate during late inspiration and early expiration when aortic pressure is increasing.

The Valsalva maneuver, a forced expiration against a closed glottis, is an excellent demonstration of the effects of changes in intrathoracic pressure on heart rate and blood pressure. As shown in Figure 2.179 with the initiation of a forced expiration against a closed glottis intrathoracic pressure rises to extremely high levels. This abruptly limits venous return, as reflected by the sharp decrease in pulmonary arterial flow velocity. Since this increase in intrathoracic pressure is transmitted directly to systemic blood vessels there is a slight increase in systemic blood pressure. Next, the decreased venous return reduces left ventricular output and decreases blood pressure. This hemodynamic response and the compensation are diagrammed in Figure 2.191, Chapter 17. Thus the major chronotropic (heart rate) response to the Valsalva maneuver is tachycardia mediated by the baroreceptor response to hypotension. Since baroreflex sensitivity and ventricular function curves are both altered in congestive heart failure, the response to the Valsalva maneuver can be used to detect the presence of congestive failure. In congestive heart failure these responses are blunted, and very frequently there will be no hypotension and tachycardia because the baroreceptor gain is reduced and the depressed cardiac function curve is flat.

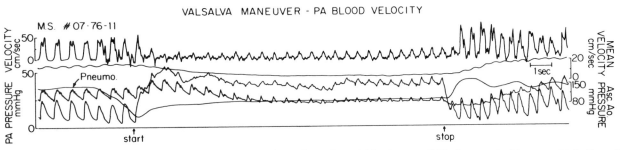

Figure 2.179. The effects of the Valsalva maneuver on hemodynamic variables in man. With the onset of increased intrathoracic pressure, pulmonary artery and aortic pressures are increased. There is an abrupt decrease in PA flow velocity due to the reduced venous return. This results in aortic hypotension and tachycardia secondary to baroreceptor activation. [Adapted from Mason et al. (1970).]

CIRCULATORY ADJUSTMENTS WITH EXERCISE

The circulatory response to exercise involves a complex series of adjustments resulting in a large increase (up to 8-fold) in cardiac output which is proportional to the increased metabolic demands placed on the body. These adjustments are designed to assure that the metabolic needs of the exercising muscles are met, that hyperthermia does not occur, and that blood flow to essential organ systems is protected. In strenuous exercise, when the cardiac output has increased 7- or 8-fold, it may attain values of 35 to 40 l/min, while total oxygen consumption of the body is 10 to 12 times greater than at rest (Khouri et al., 1965). In the next section we will discuss the cardiovascular responses necessary to achieve these large increases in cardiac output. There are two different types of exercise which may have different cardiovascular effects. In static (isometric) exercise force is generated without muscle shortening and thus no external "work" is performed. In dynamic exercise, external work (shortening against a load) is performed. Although there are some quantitative differences in the circulatory response to these two types of exercise, these differences may be due more to the amount of muscle mass involved than to a basic difference in the type of circulatory response (Asmussen, 1981; Blomquist et al., 1981). The hemodynamic responses to moderate exercise in normal man are diagrammed in Figure 2.180.

Cardiac Response to Exercise

HEART RATE

At the transition from rest to heavy work, the pulse frequency rises very rapidly, reaching levels of 160 to 180/min (Fig. 2.180). During short bouts of near maximal exercise, heart rates as high as 240 to 270/min have been recorded in normal young persons. The almost instant acceleration in cardiac rate is due to vagal withdrawal and not associated with an increase in sympathetic tone. The initial rapid increase suggests a central command, or a rapid reflex from mechanoreceptors in the active muscles (Hollander and Bouman, 1975). Later increases in heart rate stem from reflex activation of the pulmonary stretch receptors and reflexes from exercising muscles and are due to increases in sympathetic tone and vagal withdrawal as well as increases in circulating catecholamines.

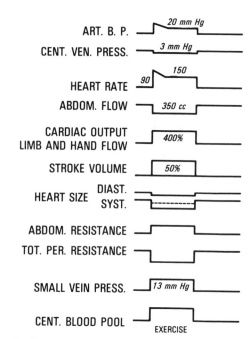

Figure 2.180. Hemodynamic responses to moderate dynamic exercise in man.

VENTRICULAR FUNCTION

Moderate exercise by healthy dogs, or by normal but untrained human subjects in the supine position, may be accompanied by an increased cardiac output that results primarily from tachycardia with little or no progressive increase in stroke volume. But if the exercise is performed in the upright position, where stroke volume at rest is less than with the subject recumbent, exercise is accompanied by increases in the stroke volume, heart rate, and cardiac output (Fig. 2.180). On the other hand, the response of the heart of trained athletes, or patients whose hearts continually pump a large volume of blood (as in the presence of valvular insufficiency), is usually different. In both of these situations the heart may be enlarged at rest, and increases in cardiac output are likely to be brought about by enhancement of the stroke volume, and tachycardia.

Taken together, the evidence indicates that in dog or in man the increase in cardiac output during moderate exercise is achieved mainly by an increase in rate and inotropic state with a constant or only slightly increased stroke volume. The explanation, however, cannot end here. If output is increased under any circumstances, the added blood expelled can come only from an increased venous return. During exercise this is achieved by increased sympathetic tone and a shift of the venous return relationship (see Fig. 2.188*H*, Chapter 17). This increased venous return must be pumped at a similar or higher end-diastolic volume, emphasizing the importance of increased inotropic state to the exercise response (Fig. 2.192, Chapter 17).

REGIONAL CIRCULATORY RESPONSES

The changes in regional blood flow with exercise are summarized in Figures 2.180 and 2.181. As expected, the increased cardiac output supplies an enormously elevated skeletal muscle blood flow. This local hyperemia is due primarily to the buildup of vasodilator metabolites in the exercising muscles. The increase in flow to skeletal muscle is directly proportional to the increase in body oxygen consumption (Fig. 2.181). Skin flow initially decreases until 50 to 60% of the maximal workload is reached; this increase favors body cooling. As exercise proceeds body fluid lost as sweat due to thermoregulation produces dehydration, and the hematocrit increases. Further increase in workload causes a progressive decrease in skin flow as cutaneous sympathetic vasoconstrictor tone continues to rise, apparently overcoming thermoregulatory vasodilator responses (Fig. 2.181). Cerebral blood flow remains unchanged, and coronary flow increases in proportion to cardiac work. These findings are, in general, based on indirect determinations of blood flow applied one at a time to various vascular beds. In dogs with implanted flow meters, renal artery flow was found to remain unchanged, while flow to the liver and stomach decreased with short-term treadmill exercise (Millard et al., 1972). Such responses appear to be characteristic of short work periods and brief bouts of exhausting exercise. On the other hand, in prolonged moderate work or sustained strenuous exercise blood flow distribution to the kidney, liver, gastrointestinal tract, and skin must be maintained at levels adequate for continued function of these organs.

BODY ARTERIOVENOUS O₂ DIFFERENCE

In addition to changes in cardiac output and local blood flow, there are other mechanisms that maintain oxygen supply to the tissues (see also Chapter 6). In exercise the O_2 extraction rises, but considerably more slowly than the cardiac output, the maximum arteriovenous O_2 difference being of the order of 13–16 ml O_2/100 ml of blood (about three times the resting value). Since most of this increase in O_2 delivery presumably occurs in the exercising muscle, the oxygen extraction in muscle must be very nearly complete whenever the average AV dif-

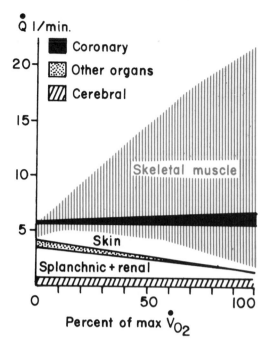

Figure 2.181. The changes of regional distribution of cardiac output as workload is increased. $\dot{Q}$, blood flow; $\dot{V}_{O_2}$, oxygen consumption as a measure of workload. (Redrawn from Clausen (1969).]

ference for the entire body is as high as 16 ml/100 ml. (A similarly high extraction has been observed in heart muscle during maximal activity.) For well trained athletes, however, the AVO_2 difference does not attain these high values.

Overall Hemodynamic Responses in Exercise

During dynamic exercise the degree of increase in heart rate, stroke volume, cardiac output, and oxygen extraction depends upon the severity of the exercise and the amount of muscle mass involved. When the exercise is strenuous and there is a very large increase in cardiac output, the work of the left ventricle may increase 4-fold. Total calculated peripheral resistance decreases significantly, because of vasodilation in active muscle, but systolic arterial pressure and the pulse pressure usually increase (Fig. 2.180). This is accompanied by decreased flow and increased resistance in the splanchnic vascular bed. One should note that the flow in the pulmonary circulation is just as great as in the systemic vessels; this flow is characterized by a mild increase in pulmonary arterial pressure during strenuous exercise, with a decrease in the resistance of the lung vasculature. Finally, the central venous, atrial, and ventricular end-diastolic pressures generally do not rise while ventricular contractility increases.

Physical conditioning or training affects the cardiopulmonary and skeletal muscular systems in a variety of ways which improve work performance (Clausen, 1969). Stroke volume, blood volume, total hemoglobin, the skeletal muscle capillary bed, and mitochondrial content of muscle are all increased. Thus, all oxygen delivery and utilization systems involved in exercise are augmented. The capacity to increase cardiac output by an increase in stroke volume is reflected in lower heart rates at rest and for given work levels. Muscle blood flow at a given workload is actually reduced (Clausen, 1969). This occurs because there is an increased oxygen extraction in trained muscle, made possible by the increase in oxidative mitochondrial enzymes and in number and size of mitochondria. Thus, "white" muscle fibers come to resemble "red" muscle fibers in histological appearance and enzymatic activity, an adaptation favoring aerobic metabolism and endurance. The net effect of these changes is that following physical training, at a given workload the myocardial and skeletal muscle blood flow can actually be reduced, and a lesser reduction in splanchnic blood flow is required for adequate muscle perfusion. These findings explain the failure to record reduction in mesenteric and renal blood flow in racing sledge dogs, as contrasted with reported decreases in splanchnic flow in exercising men not trained for endurance performances (VanCitters et al., 1969).

HYPERTENSION

Hypertension is a condition in which the systemic arterial blood pressure is elevated (Dusten, 1974; NIH Report of the Hypertension Task Force, 1979). Usually a diastolic arterial pressure (determined by the cuff method (see Chapter 13)), above 90 mmHg is considered elevated (JAMA, 1979). Long-standing hypertension without treatment (or only with sporadic therapy) results in increased morbidity and mortality associated with a variety of cardiovascular diseases such as atherosclerotic disease with its myriad of sequelae (e.g., stroke, coronary artery disease) and renal vascular disease (JAMA, 1979).

The increase in systemic arterial pressure may be due to either an increase in vascular resistance or increased cardiac output. Since cardiac output is usually normal for long periods of time after hypertension has become established, the main cause of the sustained increase in blood pressure must be increased systemic vascular resistance. This sustained increase in resistance may result from excessive vasoconstriction and/or structural abnor-

malities in the arterioles. The basis for this generalized vasoconstriction is usually not known. Indeed, hypertension is not a single disorder but may resemble an elevated body temperature, in that it signals the presence of a variety of pathogenic processes. The initial stimulus may be excessive vasoconstriction (Laragh, 1981), increased cardiac output, increased intravascular volume, or decreased vascular compliance (Guyton et al., 1971; Coleman et al., 1974). In only about 5% of the cases is it possible to find the exact cause of the elevated blood pressure. These cases are frequently due to a specific hormonal or neural abnormality and are discussed later in this section as secondary forms of hypertension. In the majority of cases it is not possible to determine the exact cause of the hypertension (frequently called essential hypertension). The initiating stimulus may be a mechanism which increases cardiac output, such as an increase in intravascular volume or one that primarily alters vascular compliance.

Let us first consider the control of intravascular volume. As discussed in more detail in Section 5, Chapter 32, the kidney is the major organ that regulates extracellular (and thus intravascular) volume. The kidney is extremely sensitive to increases in perfusion pressure, responding with increases in urinary output of water and sodium (Selkurt, 1951; Guyton, 1972). These normal responses are mediated by changes in glomerular filtration rate and suppression of renin secretion, with resultant inhibition of aldosterone secretion by the adrenal gland and increase in sodium loss. Impairment of this response due to loss of renal tissue or renal ischemia (Goldblatt et al., 1934), or, for example, to an excess of aldosterone secreted by an adrenal gland tumor will result in hypertension largely secondary to volume control abnormalities. Eventually patients whose initial hypertension is due to primary abnormalities in renal function will also exhibit increases in peripheral resistance due either to the autoregulatory response of arterioles to increased pressure or to eventual structural damage. Early increases in circulating volume and thus cardiac output are postulated by some (Guyton, 1972) to be involved in the genesis of essential hypertension. However, other factors, including those producing predominant vasoconstriction, are more likely to play a major role (Kaplan, 1980).

Another view has been that hypertension may be initiated through repeated activation of the sympathetic nervous system by emotional stress (Brody et al., 1959). However, it has been difficult to demonstrate experimentally sustained increases in blood pressure due to sympathetic activation. Baroreceptor sensitivity and threshold are known to be altered in experimental hypertension (Aars, 1975), as discussed earlier. However, these changes are not thought to be primary, and even denervation of the carotid and aortic baroreceptor response probably does not produce sustained hypertension (Cowley et al., 1973).

Support for the involvement of the sympathetic nervous system in hypertension comes primarily from the impressive therapeutic results obtained with drugs that interfere with the biosynthesis, storage, release, action on receptors, or inactivation of catecholamines. Whether this is due to the action of catecholamines in the central nervous system (as may be the case with clonidine, for example) at the level of the peripheral arteries, or by interference with the renin-induced release of angiotensin II, or aldosterone is not known, and all may play a role. It is also clear that renin-induced release of the potent vasoconstrictor angiotensin II plays an important role in some cases, since administration of drugs which block conversion of angiotensin I to angiotensin II produce an immediate fall in blood pressure and may also alter secretion of aldosterone induced by angiotensin II (Case et al., 1977).

There are several well understood causes of hypertension. These are commonly called secondary forms of hypertension. For example, the hypertension associated with a narrowing of the artery to a kidney is called renal hypertension. The narrowing reduces both blood pressure and blood flow to that kidney. Because of this reduction, the kidney produces more renin which increases the amount of circulating angiotensin. This causes a renin-dependent type of high blood pressure. Various surgical procedures have been used to relieve the obstruction and allow the kidney to receive its blood supply at a normal pressure level.

Hypertension may accompany a tumor of the outer portion of adrenal gland that produces excessive amounts of aldosterone. This leads to increased sodium and water stores in the body, increased blood volume and cardiac output. Because blood volume is increased, the kidney has turned off its production of renin so plasma renin activity is extremely low. This is a salt- and water-dependent hypertension, which may be controlled by removal of the tumor, control of salt and water intake, or drugs that block the effects of aldosterone.

When a person has impaired kidney function, sodium and water may be retained in the body. For some patients, this retention does not affect blood pressure because control mechanisms are normal. However, for others, control mechanisms do not work normally, and hypertension results. In patients with severe kidney disease, eliminating excess salt and water by kidney dialysis may be necessary to reduce blood pressure to normal.

A tumor of the adrenal medulla (pheochromocytoma) produces catecholamines (norepinephrine or epinephrine). Like increased activity of sympathetic nerves, the catecholamines that come from the tumor raise blood pressure through increasing vascular resistance and sometimes cardiac output, as well. Plasma renin activity and, consequently, aldosterone production are variably affected, and blood volume is abnormally low.

In summary, chronically elevated blood pressure may be the result of a wide spectrum of pathogenic processes involving every organ system that impinges upon blood pressure control. In particular, the kidney in controlling circulating blood volume, the central nervous system, as well as the peripheral sympathetic autonomic nervous system in regulating vasomotor tone play dominant and linked roles in chronic hypertension.

CHAPTER 17

Frameworks for Analysis of Ventricular Function and Overall Circulatory Performance

Ventricular and peripheral circulatory function, and the interaction between the two, should be understood within frameworks that are useful for measuring responses to physiologic stress, adaptations to disease, and pharmacologic effects. Several approaches for assessing the function of the ventricle were introduced in Chapter 12, including analysis of pressure-volume loops, peak ventricular dP/dt, the ejection fraction, and the end-systolic pressure-volume relation (for review see Ross, 1982). In this chapter, we will summarize such approaches, together with the analysis of the ventricular function curve, and also introduce an additional method: the interaction between cardiac output and the venous return from the peripheral circulation. These frameworks will then be applied to analyze selected physiologic perturbations.

ANALYSIS OF SINGLE CARDIAC CONTRACTIONS

The function of the right or left ventricle is often assessed in experimental animals, and in human subjects, by measuring its volume or dimensions. This can be accomplished by placing a catheter in the left or right ventricle and injecting radiopaque contrast medium while X-ray films are taken (angiocardiography). The films are calibrated, ventricular dimensions measured, and the volumes at different points in the cardiac cycle calculated, assuming an ellipsoidal shape of the left ventricle (see Fig. 2.112, Chapter 12) (Dodge et al., 1966). It is also possible to implant X-ray opaque markers on the walls of the heart in experimental animals in order to determine cardiac dimensions and calculate heart volumes, or to use implanted ultrasonic dimension gauges to continuously determine the long and short axes of the left ventricle. If pressure is measured simultaneously, it is possible to construct pressure-volume loops under a variety of conditions and to determine the linear end-systolic pressure-volume relation (Chapter 12).

In clinical studies, during cardiac catheterization the ventricular volume is determined by angiography, and the most useful and practical measure of ventricular function in the resting state with this technique is the ejection fraction. Sometimes, high fidelity pressure measurements also allow determination of the rate of change of left ventricular pressure (maximum dP/dt) and, as discussed further below, if the left ventricular end-diastolic pressure and the stroke volume (or stroke work) are known in two different circulatory states, it often is possible to determine whether or not a change in inotropic state has occurred from the analysis of only two contractions (see below).

The Ejection Fraction

The approach for determining the ejection fraction from an angiocardiogram using high-speed X-ray motion pictures (cineradiography) is shown in Figure 2.112 (Chapter 12). The volume of the ventricle is calculated both at end-diastole (the end-diastolic frame being selected just after the onset of the QRS complex) and at end-systole (the smallest ventricular volume observed in the motion picture series). The ejection fraction (EF), which is the ratio of the stroke volume to the end-diastolic volume, is then calculated from the end-diastolic volume (EDV) and the end-systolic volume (ESV) as follows:

$$EF = \frac{EDV - ESV}{EDV}$$

and expressed as the percentage of the end-diastolic volume that is ejected.

The volumes of the ventricle and the ejection

fraction can also be estimated by so-called "noninvasive" techniques. These include the use of echocardiography to determine the internal diameter of one axis of the left ventricle at end-diastole and end-systole (Fig. 2.135), from which the percent shortening of the diameter, or the volumes and the ejection fraction is estimated (using the formula for a sphere) (Feigenbaum, 1976). The radionuclide technique involves injection of a radioisotope, such as technetium-99m-labelled red blood cells, which remains for a time within the vascular space. An image of either the left or right ventricle is then obtained using a γ scintillation camera, and the end-diastolic and end-systolic counts of the image are used to compute the ejection fraction. Alternatively, a "washout" curve can be recorded as the isotope passes through the chamber on its first circulation (see Chapter 13); by knowing the ratio of counts of sequential beats on the downslope of the washout curve, the ejection fraction can be calculated (for review on these techniques, see Ashburn et al., 1978).

The ejection fraction has been shown to sensitively detect depression of myocardial contractility (reduced inotropic state) under resting basal conditions. The normal ejection fraction lies between 55 and 80% (average 67%) and is similar across various mammalian species (Holt, 1968). When myocardial depression is present the left ventricular ejection fraction is reduced *below 55%* (Dodge et al., 1966).

The ejection fraction of the right ventricle can also be determined by angiography using somewhat more complicated analysis, since the shape of that chamber is neither spherical nor elliptical (see Chapter 12). The various models used to calculate right ventricular volume have remained somewhat controversial. The lower limit of normal for the ejection fraction of the right ventricle is about 45%, a value confirmed by the radionuclide method which relies on total counts within the area of the right ventricle, and therefore is not dependent upon geometric assumptions.

More sophisticated techniques for analysis of left ventricular function by angiocardiography have also been employed, combining pressure and volume analysis to determine the force-velocity relation of the whole heart (but these approaches are beyond the scope of present discussion).

Peak dP/dt

The maximum rate of rise of left (or right) ventricular pressure measured with a high-fidelity catheter system has been used to assess the basal contractility of the heart, or to define a change in contractility produced by a drug or other intervention. As shown in Figure 2.182, the peak dP/dt during isovolumic left ventricular contraction occurs just before opening of the aortic valve, and the lower limit of normal is about 1200 mmHg/s. The maximum slope can be determined manually by drawing the tangent to the steepest point of left ventricular pressure tracing during isovolumetric contraction (Fig. 2.182), but ordinarily peak dP/dt is determined from the first derivative of the pressure by means of an electronic differentiating circuit (Fig. 2.182). (See Fig. 2.121, Chapter 12, for changes in peak dP/dt induced by positive and negative inotropic interventions.)

THE VENTRICULAR FUNCTION CURVE

Ventricular function curves can be plotted for both the right and left ventricles, and in some disease states the function of one or the other ventricle (or both) can be impaired. The shape of ventricular function curves and how they are affected by inotropic influences are determined by the three major factors which affect cardiac muscle performance (Chapter 11), that is, the preload (the positive relation between preload and tension development or muscle shortening); the afterload (the inverse relation between afterload and the extent and velocity of shortening); and the level of inotropic state (which shifts these curves upward or downward) (Fig. 2.107, Chapter 11). Direct measurements of wall force and shortening are not readily performed in laboratory studies, or in human subjects, and a simpler type of function curve is usually employed, in which the ventricular end-diastolic pressure (or sometimes the end-diastolic volume) is plotted against some measure of systolic ventricular performance, usually the stroke volume

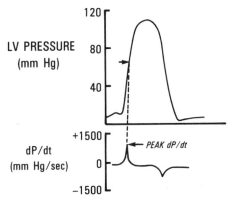

Figure 2.182. Relationship between left ventricular (*LV*) pressure and its first derivative, or rate of change of pressure (*dP/dt*) determined electronically in the lower tracing. Maximum dP/dt can also be determined manually by drawing a slope to the steepest portion of the left ventricular pressure tracing (*dashed line, arrow*).

or the stroke work. Sometimes the stroke power (work per unit time) or peak dP/dt are used.

In the isolated heart preparation (Chapter 12), this type of curve is produced by progressively infusing fluid (or blood) into the circulation and obtaining a range of points relating the end-diastolic pressure or volume to heart performance. In such an experiment the mean arterial pressure is kept constant, but this is not possible in the intact circulation where the systolic pressure tends to rise as fluid is infused. In addition, it should be recognized that the size of the heart increases, and its wall becomes thinner during volume overload of the ventricle, so that by the Laplace relation (Chapter 12) tension in the wall during ejection tends to increase. Therefore, such function curves do not represent a pure relation between preload and heart performance, but rather a composite effect, and the curve tends to flatten as the ventricular volume becomes larger. This is because the effect of increased afterload tends to limit the increase in stroke volume produced by the augmented preload (Fig. 2.107, Chapter 11. In other words, the stroke volume increase would be larger if the afterload were held truly constant.) Nevertheless, the effect on the normal heart of increasing diastolic volume is dominant as shown in the *top panel*, Figure 2.107 (Chapter 11), and as the ventricular end-diastolic pressure or volume increases, the stroke volume and the stroke work increase. Typically, a function curve is determined during one state and compared with a function curve determined during another condition (such as after a drug has been administered), and the *position* of the curve then can define whether or not a positive or negative change in inotropic state has occurred. Thus, by definition, a function curve of the ventricle is shifted upward and to the left by a positive inotropic intervention, and downward and to the right by a negative inotropic influence resulting in a "family of curves" (Sarnoff et al., 1958a) (Fig. 2.184).

The Starling Curve

A Starling curve generally indicates the relation between the ventricular end-diastolic pressure or volume and the *stroke volume*. As mentioned, experimentally such curves should ideally be compared at a constant level of mean arterial pressure. The Starling curve is the simplest to determine and, in practice, it is often plotted in intact animals or in man, even when the arterial pressure does not remain constant. Indeed, sometimes Starling curves are produced by administration of a vasopressor drug (that increases the afterload) by plotting of

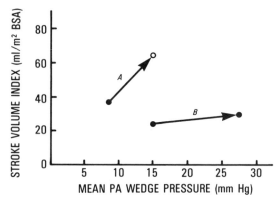

Figure 2.183. Examples of segments of ventricular function curves determined in the clinical setting as the relationship between the mean pulmonary artery (*PA*) wedge pressure (as a measure of the cardiac filling pressure or mean left atrial pressure) and the stroke volume index (stroke volume corrected for body surface area, *BSA*). Fluid is infused intravenously and *arrow A* shows the response in a subject with a relatively normal left ventricle, a large rise in stroke volume occurring with a modest increase in PA wedge pressure. *Arrow B* indicates the response in a patient who has had a large heart attack, in which a large rise in the PA wedge pressure occurs with a minimal change in the stroke volume.

the relation between the rising end-diastolic pressure and the stroke volume over a range of doses of the drug (angiotensin or phenylephrine often are used); such curves are useful for describing differences, for example, between the normal and the failing heart (Chapter 18). Alternatively, the effect of a small transfusion of saline solution may be used to test ventricular functional reserve, as in patients following a heart attack (Fig. 2.183). If the heart is failing in such patients, a large rise in the left ventricular end-diastolic pressure or the left ventricular "filling pressure" (as determined by the mean pulmonary artery wedge pressure, see Chapter 13) may be associated with little change in the stroke volume (minimal ventricular reserve), in contrast to the response of the relatively normal ventricle (Fig. 2.183).

Ventricular end-diastolic pressure often is used instead of end-diastolic volume in describing cardiac function curves because it is much easier to measure. In this setting, the nearly exponential relation between ventricular end-diastolic volume and end-diastolic pressure becomes significant (see Fig. 2.185, *lower panel*). Thus, over the range of low end-diastolic pressures (below 12 mmHg in the left ventricle) relatively large changes in ventricular diastolic volume occur with relatively small changes in diastolic pressure, whereas at high left ventricular diastolic volumes small changes in end-diastolic volume are accompanied by *large* changes in the end-diastolic pressure.

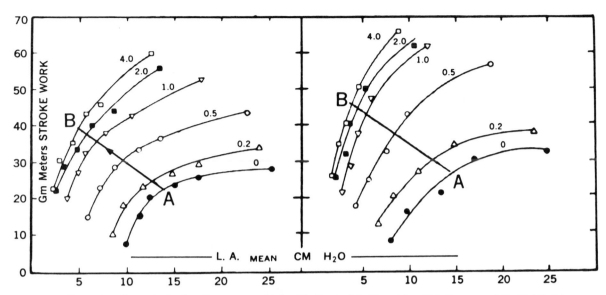

Figure 2.184. Examples of ventricular function curves obtained in the isolated heart in which the mean left atrial (*L.A.*) pressure is plotted against the stroke work of the left ventricle over a range of left atrial pressures. The progressive shift upward of the curves from *A* to *B* is due to a progressive increase of left ventricular inotropic state produced by stimulation in the cardiac sympathetic nerves at the frequencies indicated (0–4 cycles/s). The *left* and *righthand panels* indicate responses in the same heart over 1 h later. [From Sarnoff et al. (1958a).]

The Ventricular Function Curve Using Stroke Work

The term ventricular function curve sometimes in the past has been applied specifically to indicate the relation between the ventricular end-diastolic pressure (or mean atrial pressure) and the ventricular *stroke work* (the product of the mean arterial pressure or the mean ventricular systolic pressure and the stroke volume)* (Sarnoff et al., 1958a). Again, a shift in the *position* of the function curve for either ventricle is *defined* as a change in inotropic state (Figs. 2.184 and 2.185), and increasing degrees of change of inotropic state will shift the curve progressively upward (Fig. 2.184).

However, just as the work of isolated muscle is "load-dependent" (see Fig. 2.103, panel *D*, Chapter 11), the ventricular stroke work is also dependent upon the *level* of aortic pressure. As in isolated muscle, at very low levels of systolic pressure the product of mean arterial pressure and stroke volume will be low, and at very high levels of aortic pressure when the stroke volume is reduced, the stroke work will also be low. Thus, if large changes in aortic pressure occur during the performance of such curves, alterations in the stroke work are not entirely dependent upon the level of inotropic state at any given end-diastolic pressure.

In using any type of function curve of the ventricle, it is important to understand clearly the difference between *moving up and down* on a *single* function curve due to altered preloading or afterloading, and changing the *position* of the entire function curve by altering the inotropic state of the ventricle.

Using the ventricular function curve concept, and having only two points of information about heart function, often it is possible to infer whether or not an intervention has caused a change in inotropic state. As shown in Figure 2.185, if the stroke volume or stroke work *decreases* or does not change while the end-diastolic pressure shows no change or *increases* (control point to points *A*, *B* or *C*, Fig. 2.185), it can be inferred that the curve has shifted downward and that inotropic state has become depressed between the two determinations (assuming no major change in aortic pressure). On the other hand, if the stroke volume or stroke work shows no change or *increases*, while the end-diastolic pressure shows no change or *decreases*, a positive inotropic influence must have been operative (control point to points *D*, *E*, or *F*, Fig. 2.185). However, if the stroke work or stroke volume increases *and* the end-diastolic pressure or volume also increases, or if they *both* decrease, one cannot infer that a change in inotropic state has occurred without knowing the shape of the *entire* curve, since these changes could simply represent moving up or down on a single function curve. A variety of normal physiologic

---

* The stroke work is usually expressed in gram meters (g M) where pressure in centimeters H_2O × stroke volume (in cm^3) = stroke work ÷ 100. (1 mmHg = 1.36 cm H_2O).

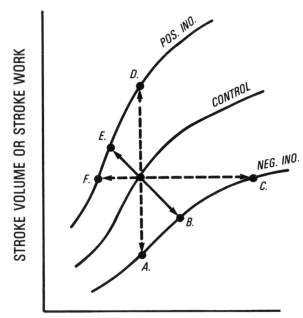

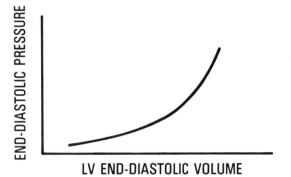

Figure 2.185. *Upper panel,* Ventricular function curves showing the relation between left ventricular stroke volume or stroke work to left ventricular end-diastolic pressure during control conditions *(center curve),* positive inotropic *(POS. INO.)* and negative inotropic *(NEG. INO.)* interventions. The arrows indicate how measurement only of a control and one other point can indicate a decrease in inotropic state *(points A, B, or C)* or an increase in inotropic state *(points D, E, or F).* For further discussion, see text. The *lower panel* shows the nonlinear relation between left ventricular *(LV)* end-diastolic volume and end-diastolic pressure.

use of the linear relation between the end-systolic volume and the end-systolic pressure of the ventricle has been studied extensively in the isolated heart preparation (Suga and Sagawa, 1973; Sagawa 1978), and it has also been extended to study intact animals by infusing a range of doses of a vasoconstrictor drug (which does not itself have appreciable inotropic effects, such as the α-adrenergic agonist phenylephrine) (Mahler et al., 1975). In man, the end-systolic ventricular volume can be determined by performing two or more angiocardiograms during infusion of the vasoconstrictor. The end-systolic volumes are then related to the corresponding ventricular pressures at the end of ejection. This approach also has been applied using noninvasive techniques for measuring ventricular dimensions or volume (echocardiography and radionuclide methods). The linear relation has been found to shift downward and to the right in the presence of myocardial disease in human subjects (Fig. 2.186), and to shift upward and to the left (with steepening of its slope) during acute positive inotropic interventions in experimental animals and in man (Fig. 2.186) (for example, Mehmel et al., 1981).

Table 2.12
Factors that affect the preload

Posture
Blood volume
Intrinsic venous tone
Sympathetic neural tone to veins and vasoactive drugs
Atrial contraction
Muscular activity
Intrapleural pressure
Intrapericardial pressure

Table 2.13
Factors that influence myocardial inotropic state

Reflex adrenergic neural stimulation of the heart
Vagal tone (primarily affects atrium)
Circulating catecholamines
Force-frequency relation (heart rate, postextrasystolic potentiation)
Myocardial failure
Myocarditis
Myocardial infarction and ischemia
Myocardial failure in late hypertrophy
Cardioactive (positive inotropic) drugs
Digitalis
Amrinone
Sympathomimetic amines
Cardiac depressant drugs
Certain anesthetic agents
Antiarrhythmic drugs
Certain calcium antagonists
External ion concentrations
Calcium
Sodium

stimuli and pathological influences can affect the preload (end-diastolic volume of the ventricle) (Table 2.12), and a number of influences can alter the level of inotropic state and thereby shift the ventricular function curves (Table 2.13).

THE END-SYSTOLIC PRESSURE-VOLUME RELATION

The pressure-volume loop of the left ventricle was discussed in Chapter 12 and related to the performance of isolated cardiac muscle (in which the isometric length-tension curve provides the limit of shortening for isotonic contractions). The

The relationship between this type of analysis and ventricular function curves before and after administration of a positive inotropic agent is shown in Figure 2.187. Ejecting beats *A*, *B*, and *C* before the inotropic intervention are shown as end-diastolic volume is increased (*left panel*) and as a ventricular function curve (*right panel*). When beats arising from the same three end-diastolic volumes are plotted after the positive inotropic agent is given, the end-systolic pressure-volume relation is shifted and the stroke volumes increase

(beats *D*, *E*, and *F*, *left panel*), displacing the function curve upward (*right panel*, Fig. 2.187). Thus, the framework using pressure-volume loops and end-systolic pressure-volume relations can be readily interchanged with ventricular function curve responses.

The linear relation between ventricular end-systolic volume and end-systolic pressure is of particular importance, because it provides a method for detecting changes in the level of inotropic state under acutely changing conditions that is *independent of* the end-diastolic volume (*preload*) and the aortic pressure (as a measure of *afterload*). Thus, a given heartbeat will arrive at end-ejection and fall on this linear relation regardless of the starting point for end-diastolic volume and the level of aortic pressure that it encounters during ejection, and the entire relationship will be shifted acutely only by a change in inotropic state.

It should be recognized, however, that under conditions where there is *chronic* change in the shape and size of the ventricle, or in the thickness of its wall, the systolic pressure is *not* indicative of the level of afterload, and under those conditions the wall force must be calculated in order to define the linear relation between end-systolic volume and wall force (Sasayama et al., 1977) (see also section on hypertrophy, Chapter 18). Also the linear end-systolic pressure-volume relation may not provide a unique descriptor of inotropic state when the transmural pressure of the ventricle is altered (as in pericardial disease).

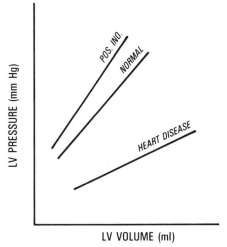

Figure 2.186. Diagrammatic linear relations between left ventricular (*LV*) end-systolic pressure and end-systolic volume in normal hearts and in patients with myocardial disease. The influence of an acute positive inotropic stimulus (*POS. INO.*) on this relation is also shown.

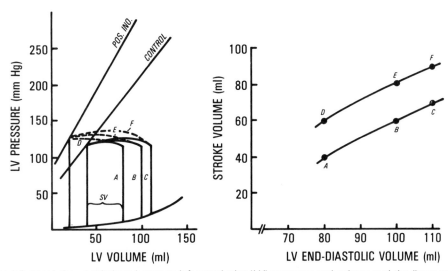

Figure 2.187. The *left panel* shows relations between left ventricular (*LV*) pressure and volume and the linear end-systolic pressure-volume relations during control conditions and a positive inotropic stimulus. The pressure-volume loops are also shown with progressive increase in end-diastolic volume and stroke volume under control conditions (*beats A, B, and C*), and during the positive inotropic stimulation at a slightly higher systolic pressure, for clarity (*beats D, E, and F*). *Right panel* shows the same beats diagrammed in left panel plotted as ventricular function curves relating left ventricular (*LV*) end-diastolic volume to the stroke volume. *A, B,* and *C*, control conditions; *D, E,* and *F*, positive inotropic stimulus.

VENOUS RETURN AND CARDIAC OUTPUT CURVES

Under normal conditions, the cardiac output is limited by the rate of venous return. Thus, as emphasized in previous chapters, the heart serves as a "demand pump," and its pumping capability far exceeds the level of cardiac output called for under normal circumstances (except, perhaps, during the most extreme exercise). There are three cardiac reserve mechanisms: (1) the heart rate (there is a reserve of about 130 beats/min, up to a maximum of about 200 beats/min in younger subjects); (2) the stroke volume (a considerable reserve is available from the Frank-Starling mechanism, if needed); (3) positive inotropic stimulation which increases the degree of systolic emptying (and the ejection fraction, thereby further augmenting the stroke volume). The total reserve for stroke volume is about 80 ml. Using all these reserve mechanisms, the cardiac output can change from a value of 5 l/min at rest up to about 25–30 l/min during maximum exercise in a human subject. Given the fact that the cardiac output at rest is far below the heart's pumping ability, it follows that peripheral factors which influence the *return* of blood to the heart must determine its output at rest.

In a steady state, the forward cardiac output must equal the venous return to the heart, and from the general resistance equation (Chapter 5):

Cardiac output (or venous return)
$$= \frac{\text{Systemic arterial pressure}}{\text{Total peripheral vascular resistance}}$$

Since the blood pressure is maintained relatively constant, changes in the total peripheral vascular resistance importantly affect the cardiac output and venous return. Local flow to the various organs of the body is affected by the local vascular resistance of that particular organ which, in turn, is determined by its metabolic needs. Therefore, the contribution of vasodilatation occurring in response to increased metabolic needs of a given organ or tissue (in the legs, for example, during brisk walking) will affect the total peripheral vascular resistance, and thereby influence the venous return to the heart. Indeed, the total cardiac output and venous return are related in a positive and nearly linear manner to the $\dot{V}O_2$ of the whole body, both in individuals as they alter their activity, and in different species having widely differing resting metabolic rates.

A number of other factors also affect the venous return. These include the volume of blood contained in the circulation, the resistance on the venous side of the circulation (including blood viscosity effects), and the compliance and capacitance of the various segments of the vascular bed including the veins. Therefore, it will be useful to assess how such peripheral circulatory factors can affect the venous return by considering a different type of function curve, one that is related to the peripheral circulation rather than the heart.

The Venous Return Curve

The venous return curve was developed to describe the influence of peripheral circulatory factors on the return of blood to the heart, when the heart is effectively removed from the circulatory loop. (For detailed monographs see Guyton et al. (1973) and Guyton (1955).) In such experiments, the heart and lungs are replaced by a pump and an artificial oxygenator, and when compensatory peripheral circulatory reflexes are absent, a function curve relating the level of right atrial pressure to the venous return can be determined. The mean right atrial pressure is plotted as the independent variable, and as the right atrial pressure is progressively elevated in abrupt, brief steps, the venous return falls (Fig. 2.188C, curve b). This occurs not only because the pressure gradient for return of blood from the periphery to the right heart is diminished but because of the important influence of the high capacitance of the veins ($\Delta V / \Delta P$) which is 18–19 times that of the arteries. Thus, as relatively small changes in right atrial pressure occur, blood is dammed back into the venous bed; this proportionately reduces the volume in the remainder of the circulation (but to only a small degree in the veins because of the brevity of each step) and causes an associated fall in arterial pressure as the venous return and cardiac output drop. At an average pressure of about +7 mmHg in the right atrium the venous return becomes zero (Fig. 2.188C, curve b). At this point, when blood flow stops, the pressure is the same everywhere in the systemic circulation, a point termed the "mean systemic pressure" (or the "mean systemic filling pressure") (Guyton, 1973). Guyton has equated this mean systemic pressure with the driving pressure for return of blood from the periphery to the right atrium, and hence when the right atrial pressure equals the mean systemic pressure, flow must cease.† The mean systemic pressure

---

† It has been argued (Levy, 1979) that it is not the right atrial pressure per se that influences the venous return to the heart by affecting the pressure gradient between the periphery and the right atrium, but that changes in right atrial pressure are rather a *consequence* of changes in cardiac output which cause redistribution of blood to the central (pulmonary) circulation and the arterial bed from the venous compartment. Thus, in a *closed* circulation (rather than using the right atrial reservoir employed by Guyton) it can be shown that as cardiac output is increased

is importantly influenced by the volume of blood contained in the circulation, as well as by changes in the tone of smooth muscle in the blood vessel walls (see below).

Notice that a knee and plateau of the venous return curve occur at a pressure of about -2 mmHg and below (Fig. 2.188C, curve *b*). At this point the large intrathoracic veins tend to collapse, so that as the right atrial pressure is lowered further, no further increase in venous return occurs; under these conditions, the pressure in the veins *outside* the negative pressure in the partially collapsed intrathoracic veins remains at about zero. The point of collapse varies, depending on the degree of filling of the circulation (see below). Notice also that under normal conditions (Fig. 2.188C, curve *b*), the *knee* of the venous return curve occurs at the level of a normal cardiac output (about 5 l/min), suggesting that changing cardiac *pumping ability alone* at this point (as by increasing inotropic state) will not increase cardiac output further.

Effects of Altered Blood Volume and Venous Tone

When blood is infused into the circulation (to produce a 20–30% increase in blood volume) the venous return curve is shifted upward in a symmetrical manner (Fig. 2.188C, curve *a*). Thus, the mean systemic pressure is increased by increased intravascular volume, and the knee and plateau are also shifted upward by increased blood volume (as expected, the plateau which is determined by venous collapse is also shifted by altered venous volume). The opposite effect occurs when blood volume is reduced (Fig. 2.188C, curve *c*), and there is a lower mean systemic pressure and reduced plateau. Thus,

---

by means of a pump the right atrial pressure will fall, and vice versa. When curves are plotted relating right atrial (or venous) pressure to the cardiac output in such a closed circulation, they are affected by various interventions in a manner identical with the venous return curve (Levy, 1979). In analyzing curves it can be shown that at a given constant value for total resistance, and at a constant blood volume, the ratio of the capacitance of the venous bed C_V ($C = \Delta_v/\Delta_c$) to that of the arterial bed, C_a, is 19:1, and assuming that these capacitances remain constant at a given blood volume, as flow is progressively increased through such a system the following equation applies: $\Delta P_a/\Delta P_v = C_v/C_a$. Thus, at any level of flow the rapid rise of pressure in the noncompliant arterial compartment can be calculated together with the fall in pressure in the more compliant venous compartment, as the blood is translocated into the arterial compartment at progressively higher flow rates (consideration of the lungs and central circulation is omitted in such a model). Thus, the *flow rate* is responsible for the pressure gradient (the difference between P_A and P_V at any level of flow) rather than the right atrial pressure *per se* (Levy, 1979).

the venous return at any level of right atrial pressure is markedly affected by the degree of distension of the circulation.

In addition, increased or decreased tone of the smooth muscle in the vascular bed shifts the venous return curve in a similar, symmetrical manner. To a large extent, this effect is due to changes in venous tone (the arterial bed volume is small), which alters the volume of the venous bed (like fluid infusion or withdrawal) and thereby also affects the mean systemic pressure. Increased tone through enhancement of sympathetic stimulation shifts the venous return curve upward (Fig. 2.188C, curve *a*), and withdrawal of sympathetic tone shifts the curve downard to curve *c*.

Changes in the Resistance to Venous Return

If a marked change in the *total* resistance to venous return is produced, such as by producing an arteriovenous fistula (a direct connection between an artery and vein), the resistance in all components of the peripheral circulation (arterial and venous) is markedly lowered. At *zero flow*, however, the effect of such a resistance change is zero. (For example, when the circulation is at rest progressive constriction of a small area of the inferior vena cava cannot affect the position of the venous return curve or alter the mean systemic pressure.) Hence, a change in vascular resistance will not affect the zero intercept of the venous return curve on the pressure axis (Fig. 2.188D). However, at all levels of blood flow, lowering the total vascular resistance by means of an arteriovenous fistula shifts the venous return curve upward and to the right, with a pronounced change in the slope of the relationship. Conversely, increasing the resistance decreases the venous return at all levels of right atrial pressure except zero, rotating the curve downward and to the left (Fig. 2.188D). Changes in arteriolar resistance have a much less pronounced effect on shifting the venous return curve than changes in the venous resistance (Guyton, 1959; Guyton et al., 1973). This difference is related primarily to the much larger capacitance ($\Delta V/\Delta P$) of venous than the arterial bed, so that with an increase in venous resistance, a large volume of blood is stored in the venous bed, whereas with increased arteriolar resistance very little blood is dammed up in the noncompliant arterial bed.

The venous return during changing vascular resistance can be derived from the following equation (Guyton et al., 1973):

$$\text{Venous return} = \frac{MSP - RAP}{R_V + R_A/19}$$

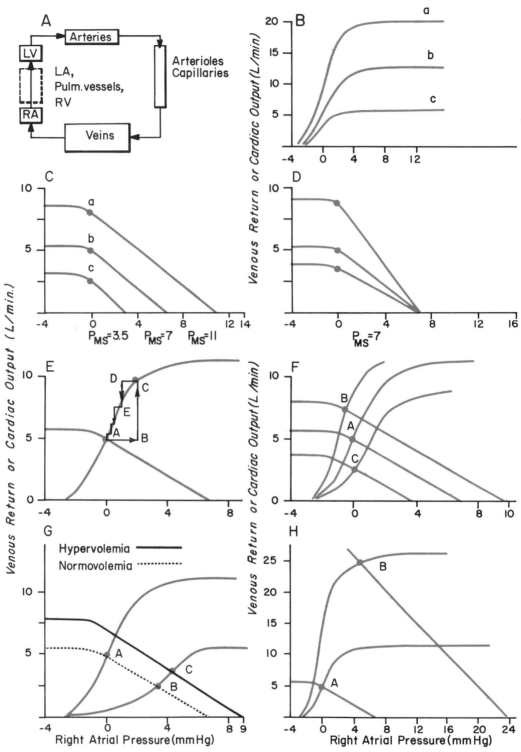

Figure 2.188. Cardiac and systemic function (venous return) curves. *A*, Simplified model of the circulation. Cardiac function curves may be plotted from determinations of right atrial pressure (*RA*) and left ventricular output (*LV*). The resultant curves characterize steady-state flow through the heart and lungs. When the only parameters measured were LV output and RA pressure, the possible effects of hemodynamic influences of the right ventricle, pulmonary vessels, and left atrium (*dotted square*) are ignored. *B*, Cardiac output curves for the heart. *a*, Hyperdynamic cardiac state. *b*, Normal state. *c*, Hypodynamic cardiac state. *C*, Systemic function curves: the effect of changing mean systemic pressure (*Pms*). *a*, Pms of 11 mmHg; *b*, Pms of 7 mmHg; *c*, Pms of 3.5 mmHg. *D*, Systemic function curves; the effect of changing resistance to venous return. *E*, Cardiac function (cardiac output) and systemic function (venous return) curves plotted on the same axes; (*A*) The equilibrium point with coordinates of 5 l/min cardiac output and venous return (ordinate) and 0 mmHg right atrial pressure (abscissa). These coordinates represent stable values at which the system is operating. When a perturbation occurs (such as a sudden shift of right atrial pressure from *A* to *B*), the coordinates for cardiac output reach a new point (*C*). In a stable

where MSP = mean systemic pressure, RAP = mean right atrial pressure, R_V = venous resistance, R_A = arteriolar resistance, and 19 = sum of the arteriolar and venous capacitances (1 + 18).

This equation indicates that if the total peripheral resistance increases by 20%, with all of the resistance increase occurring in the arterioles, venous return will be reduced by only about 6%. However, if this 20% resistance change occurs entirely on the venous side, venous return will be reduced by about 53% (a 9-fold difference). This phenomenon has been noted for many years in that a striking drop in cardiac output occurs with mild compression of the inferior vena cava in the abdomen.

The Cardiac Output Curve

In a simplified model of the circulation, such as that described above, it is possible in a steady state to equate the output of the right heart and the left heart with the venous return. (Under normal circumstances the low resistance in the lungs can be ignored). A curve relating the mean right atrial pressure to the cardiac output can then be constructed (Guyton, 1955; Guyton et al., 1973) called the "cardiac output curve" (Fig. 2.188B). Like the ventricular function curve, the cardiac output curve plots the performance of the heart (cardiac output) as a positive relation against the cardiac filling pressure (in this case the right atrial pressure represents filling of the whole heart), so that the basic shape of the curve predominantly reflects operation of the Frank-Starling mechanism.

However, in contrast to the ventricular function curve, the shifts of the entire cardiac output curve are *not* limited to inotropic state changes but can be produced by a variety of influences *in addition* to the inotropic state (by definition, ventricular function curves are shifted *only* by changes in inotropic state). Thus, *any* influence that improves the ability of the heart to pump blood will shift the cardiac output curve upward (Fig. 2.188B, curve *a*), and factors that impair the overall function of the heart will shift the curve downward (Fig. 2.188B, curve *c*). Influences that can affect cardiac performance and shift the *cardiac output curve* upward or downward are shown in Table 2.14.

Table 2.14
Factors that can shift the cardiac output curve

Shift Upward	Shift Downward
Decreased afterload (decreased vascular resistance)	Increased afterload (increased vascular resistance)
Increased inotropic state (Table 2.13)	Decreased inotropic state (Table 2.13)
Increased heart rate	Decreased heart rate
Cardiac hypertrophy	Pericardial fluid (tamponade)
Negative pressure breathing	Positive pressure breathing

The Equilibrium between Venous Return and Cardiac Output

The interaction between the peripheral circulation and the cardiac pump can be described simultaneously by plotting the venous return curve and the cardiac output curve on the same graph (Fig. 2.188E). In a steady state, under a given set of peripheral circulatory conditions and a certain level of cardiac function, it is apparent that the two curves will intersect at that point where the venous return and cardiac output are equal (termed the "equilibrium point" by Guyton, 1955). This framework allows the effects of an intervention that may alter peripheral circulatory factors, cardiac function, or both, to be plotted in a single graphic representation.

In a stable system, following a perturbation the pressures and cardiac output will tend to return toward the equilibrium point. For example, if a volume of blood were suddenly withdrawn from the arterial side and injected into the venous side, right atrial pressure would increase (Fig. 2.188E, point B). However, the increase in cardiac output would then increase venous return momentarily (point C), and over the next few cardiac cycles blood would be transferred back from the venous to the arterial compartment. Thus, given unchanged compliances and resistances in the systemic and venous beds, the venous return and cardiac output would again come into equilibrium near the previous point (Fig. 2.188E, point A).

It should be noted again that under normal conditions the venous return and cardiac output curves intersect near the knee of the venous return curve (Fig. 2.188E). Hence, unless factors also operate to

system, a series of changes is initiated (*D, E* etc.), and the original equilibrium value (*A*) is regained. *F*, The effect of altering sympathetic tone. (*A*) Control cardiac function and systemic function curves; (*B*) The effect of sympathetic nerve stimulation; (*C*) The effect of sympathetic nerve inhibition (e.g., with spinal anesthesia). *G*, Shifts in cardiac and systemic function curves in congestive heart failure. (*A*) Equilibrium point of control curves; (*B*) New equilibrium point when the heart fails (decrease in cardiac output and increase in right atrial pressure); (*C*) Equilibrium point established when the increase in blood volume which is expected in heart failure occurs. Cardiac output is substantially increased owing to this shift in the equilibrium point. *H*, The effect of exercise on cardiac and systemic function curves. (*A*) The control equilibrium point; (*B*), The new equilibrium point when cardiac output and right atrial pressure are increased during strenuous exercise.

shift the venous return curve upward, a shift upward of the cardiac output curve by any of the factors listed in Table 2.14 would *not* tend to affect the cardiac output, since the cardiac output curve would shift on the flat portion of the venous return curve. An example of a complex response in which *both* curves are shifted is shown in Figure 2.188F, in which enhancement of sympathetic tone has produced venoconstriction, increasing the mean systemic pressure and shifting the venous return curve upward, while positive inotropic stimulation of the heart has also shifted the cardiac output curve upward to produce an increased cardiac output at a slightly lower mean atrial pressure (Fig. 2.188F, point A to B). Withdrawal of sympathetic tone produces opposite effects, and a lower cardiac output occurs at a somewhat higher mean right atrial pressure (point A to point C). It is apparent from Figure 2.188F that if the venous return curve alone is shifted upward or downward (for example, where the three venous return curves intersect with the *middle* cardiac output curve), the intersections with the cardiac output curve represent the effects of moving up or down on a single cardiac output curve due to operation of the Frank-Starling mechanism. More complicated graphical analyses of cardiac output regulation are possible involving plotting of cardiac output and venous return curves for both right and left ventricles on the same graph, and computer modelling of the circulation by this technique also has been performed (Guyton et al., 1973).

SELECTED PHYSIOLOGIC RESPONSES

The purpose of the following descriptions is to illustrate that several different frameworks can be usefully employed for examining the cardiac and peripheral circulatory responses to various forms of physiologic stress.

Cardiac Pacing

When electrical stimulation of the right atrium is carried out in the normal human subject, as the heart rate is increased the mean left and right atrial pressures fall, but there is no change in the cardiac output despite the positive inotropic stimulation induced by the force-frequency relationship (Ross et al., 1965). Based on the hemodynamic principles discussed above cardiac output would not be expected to change. Thus, cardiac pacing shifts the cardiac output curve upward and to the left, but since the normal heart operates on the knee of the venous return curve, little effect or no effect on the cardiac output and venous return occurs (Fig. 2.189).

Supine to Erect Posture

A sudden change from the supine to the upright position on a tilt table produces a response somewhat resembling acute blood loss, due to pooling of blood in the veins of the lower extremities. The result is a decreased effective mean systemic pressure (relative to the level of the heart), which shifts the venous return curve downward and to the left, resulting in an immediate, severe fall in the cardiac output (Fig. 2.190, point A to point B). The fall in

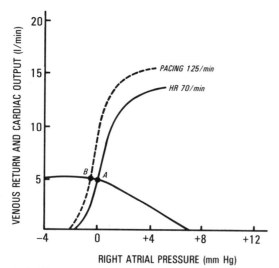

Figure 2.189. Normal venous return and cardiac output curves during electrical pacing at a normal heart rate (HR 70/min) and 125/min (*dashed line, cardiac output curve*). Despite the shift upward of the cardiac output curve, since the heart is operating on a flat portion of the venous return curve no change in the cardiac output occurs (*point A to point B*).

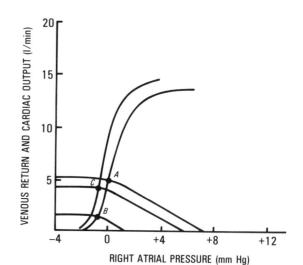

Figure 2.190. Venous return and cardiac output curves during upright body tilting. Control conditions are shown at point A, and during the initial phase of the tilt (*point B*) a marked fall of the cardiac output occurs; when reflex compensations occur the cardiac output improves (*point C*). For further discussion, see text.

arterial pressure causes reflexes which increase ventricular contractility, heart rate, and the mean systemic pressure, and these compensations then partially restore the cardiac output by shifting both the venous return and cardiac output curves upward (Fig. 2.190, point C).

A similar response occurs after acute blood loss, and in addition a further upward shift of the venous return curve would occur over several hours; thus, the acute blood loss produces a drop in capillary pressure, and osmotic forces then draw fluid into the circulation, thereby tending to further increase the mean systemic pressure.

The Valsalva Maneuver and Positive Pressure Breathing

The reflexes occurring during the Valsalva maneuver in the normal and failing heart are described in Chapter 16. The Valsalva maneuver (forced expiration against a closed glottis, or straining) produces a marked increase of the intrapleural pressure. This elevation of the intrapleural pressure shifts the entire cardiac output curve to the right (cardiac filling pressures are now high relative to peripheral pressures), and a very marked initial drop in the cardiac output occurs (Fig. 2.191, point A to B), associated with a drop in the arterial pressure. However, the associated strong contraction of the abdominal muscles markedly increases the intraabdominal pressure, thereby increasing the mean systemic pressure; this, coupled with displacement of blood from the central circulation

(lungs) to the periphery by the high intrathoracic pressure, shifts the venous return curve upward. Also, after 15–20 s a marked baroreceptor reflex results from the drop in arterial pressure, which further shifts the venous return curve upward and also displaces the cardiac output curve upward (increased heart rate and inotropic state), thereby partially restoring the cardiac output (Fig. 2.191, point C). Following release of the Valsalva maneuver, there is a marked overshoot of the blood pressure and cardiac output due to persistence of sympathetic tone to the arterioles and myocardium, and to the increased venous return as intrapleural pressure abruptly drops. Also, a sharp, reflex bradycardia occurs as the arterial pressure rises abruptly ("post-Valsalva overshoot").

Positive pressure breathing can be given during general anesthesia, and it can occur during scuba diving or while playing certain musical instruments. Similar effects to the Valsalva maneuver occur initially, with a drop in cardiac output during steady-state positive pressure breathing (Guyton et al., 1973). By the same compensatory mechanisms shown in Figure 2.191 the cardiac output can be restored to nearly normal in this setting.

Exercise

The circulatory responses to exercise are described in more detail in Chapter 16. However, it will be useful here to mention certain general circulatory responses within the frameworks under discussion. The effects of moderate exercise on the pressure-volume loops and end-systolic pressure-volume relations of the left ventricle are shown in Chapter 12 (Fig. 2.127). The positive inotropic effect of exercise due to effects of adrenergic nerve stimulation and circulating catecholamines shifts the end-systolic pressure-volume relation upward and to the left, delivering the same or a smaller stroke volume at a higher systemic arterial pressure, but from the same or an even lower left ventricular end-diastolic volume and pressure; in this setting the associated marked acceleration of heart rate is the predominant factor increasing the cardiac output (Fig. 2.123).

Exercise responses within the ventricular function curve framework are shown in Figure 2.192A. From the resting left ventricular function curve (point A) mild exercise results in enhanced inotropic state with a shift upward of the curve, which produces a fall of left ventricular end-diastolic pressure with little change in stroke work or stroke volume, the increase in the cardiac output occurring primarily through increased heart rate (Fig. 2.192A, point B and inset data). With more severe exercise,

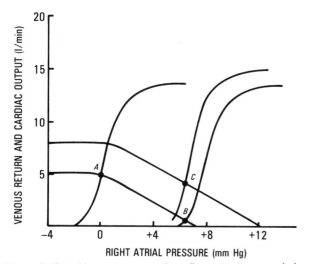

Figure 2.191. Venous return and cardiac output curves during various stages of the Valsalva maneuver. Control conditions are shown at point A, and the initial response to the Valsalva maneuver with a marked decrease in the cardiac output is at point B; reflex compensations partially restore the cardiac output at point C. For further discussion, see text.

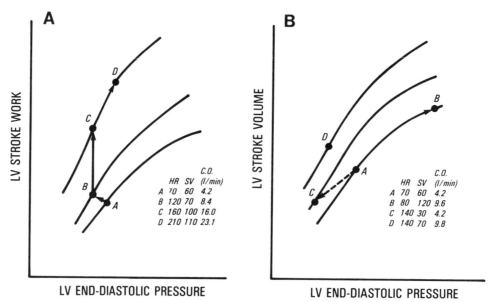

Figure 2.192. *Panel A*, Responses to exercise within the framework of left ventricular (*LV*) end-diastolic pressure and the stroke work. Cardiac outputs are calculated in the table. Point *A* shows the resting conditions; point *B*, mild exercise; point *C*, moderately severe exercise; point *D*, maximal exercise in an untrained subject. For further discussion, see text. *Panel B*, Relations between left ventricular end-diastolic pressure and stroke volume, showing diagrammatically the responses to maximal exercise after cardiac denervation and adrenalectomy, or β-adrenergic blockade (*point A* to *point B*). Point *A* is also used as the control point for a normal heart to compare the effects of cardiac pacing alone, which shifts the ventricular function curve, and produces a fall in the stroke volume (*point A* to *point C*) (although cardiac output is unchanged). If moderate exercise is performed while the pacing is continued at a rate of 140 (see table), in the absence of β blockade a further shift upward to the ventricular function curve due to sympathetic stimulation occurs, but the cardiac output increase is achieved through an increase in stroke volume and end-diastolic pressure (*points C* to *D*). For further discussion, see text.

there is a further shift of the ventricular function curve upward and to the left, stroke work increasing as both the stroke volume and the blood pressure rise, and there is little further change in the left ventricular end-diastolic pressure (circulating catecholamines and adrenergic stimulation to the myocardium are now near maximal) (Fig. 2.192*A*, point *C*). With maximal exertion, the left ventricular end-diastolic pressure and volume reach normal (same as point *A*) or even above normal levels, as the maximum heart rate is achieved and the stroke volume rises by the Frank-Starling mechanism (Fig. 192*A*, point *D*).

The exercise response also can be analyzed in the venous return-cardiac output framework as shown in Figure 2.188*H*. The resting conditions are shown at point *A*, and the responses during very strenuous exercise in the untrained normal subject are indicated at point *B*. During initial, milder exercise the mean systemic pressure is elevated by tensing of the abdominal muscles and the exercising muscles, accompanied by some increase in the resistance to venous return; however, as exercise continues, metabolic vasodilatation in the muscles coupled with a marked increase in sympathetic tone further elevates the mean systemic pressure and the venous return curve. Simultaneously, the cardiac output

curve is shifted upward by sympathetic stimulation and by lowering of total peripheral vascular resistance, allowing the cardiac output to reach 25 l/min at a near maximum heart rate, with only a slight rise in the right atrial pressure above normal (Fig. 2.188*H*). In trained athletes, such as long distance runners, the heart undergoes hypertrophy of the volume overload type (see Chapter 18) with enlargement of the end-diastolic volume of the chambers together with some thickening of the wall, and a normal level of basal inotropic state. This allows the cardiac output curve to shift even higher, and cardiac outputs of 35 l/min or more may be reached during severe exercise with a right atrial pressure at or below normal (for review of the chronic adaptations to training, see Wilmore (1977) and Chapter 16).

In experimental animals, when the sympathetic adrenergic response of the heart is blocked by surgically denervating the heart (Donald and Shepherd, 1963) and the effects of circulating catecholamines are blocked by adrenalectomy or β-adrenergic blockade, the ability to exercise maximally becomes impaired (Fig. 2.192 *panel B*). Under these conditions, the ability of the heart to shift its function curve disappears, the heart rate response is only partially impaired (direct atrial stretch may increase the heart rate, see Chapter 16), but most of the cardiac output increase is now accomplished by an increase in stroke volume via the Frank-Starling mechanism and increased ventricular end-diastolic pressure (Donald, 1974) (Fig. 2.192*B*, point *A* to point *B*). Somewhat similar patterns of responses to exercise are seen in patients who are treated with β-adrenergic-blocking drugs.

When the normal heart is paced by an electrical pacemaker under resting conditions to a rate simulating that during moderate exercise, as discussed earlier the cardiac output does not change and the stroke volume therefore falls (Fig. 2.192*B*, point *A* to point *C*, also Fig. 2.189). If pacing is maintained during moderate exercise, the heart rate fails to rise (since the heart is already paced to a level exceeding that which the heart would have reached at that level of exercise); however, there is autonomic stimulation of the myocardium, shifting the left ventricular function curve upward and to the left. Under these conditions, the increase of cardiac output is achieved by an increase in the stroke volume, through increased venous return and operation of the Frank-Starling mechanism and by decreased end-systolic volume at the enhanced level of inotropic state (Fig. 2.192*B*, point *C* to point *D*) (Ross et al., 1965).

CHAPTER 18

Heart Failure, Hypertrophy, and Other Abnormal Cardiocirculatory States

In this chapter, selected abnormal cardiac and circulatory states will be introduced, in order to illustrate the application of physiologic principles. These conditions include heart failure, certain aspects of pericardial function, cardiac hypertrophy, anemia, and hemorrhagic shock; hypertension is discussed briefly in Chapter 16.

HEART FAILURE

Failure of the whole heart can be defined in a rather simple manner if the heart is considered as a "black box" (Fig. 2.193, *upper panel*). When the heart fails it is unable to meet its obligations as a pump, and either the cardiac output falls, the venous pressures rise,* or both. In the early stages of heart failure, the mechanisms discussed below may compensate for the basic cardiac abnormality and serve to maintain the cardiac output and filling pressures within the normal range. At this stage, *mild heart failure* can be defined as: Abnormal elevation of the venous pressure (left or right) during exercise (or other stress), or failure of the cardiac output to rise normally during exercise, or both.

In later stages of heart failure compensatory mechanisms become inadequate, and *severe heart failure* can be defined as: Elevation of the venous pressures, left or right sided reduction of the cardiac output, or both, in the resting state.

In older terminology the tendency for the pressure gradient generated by the heart to drive blood toward the heart was termed *vis a tergo* (force from behind), and an elevation of venous pressure resist-

ing the return of blood to the heart was termed *vis a fronte*. Heart failure characterized by reduced cardiac output (reduced *vis a tergo*) was sometimes called "forward heart failure," whereas elevation of the venous pressure (*vis a fronte*) has been termed "backward heart failure." Often, however, the two phenomena occur simultaneously (Fig. 2.193, *upper panel*).

Causes of Heart Failure

Many factors can result in failure of the heart as a pump, and most of these can be related to the cardiac subsystems (Chapter 5). Heart failure can result from abnormal electrical conduction; for example, long-standing complete heart block with a very slow heart rate and large stroke volume can result in marked cardiac enlargement and eventual heart failure. Atherosclerotic disease of the coronary arteries can produce myocardial infarction with heart failure. Several kinds of valvular deformities can produce heart failure; for example, obstruction of the mitral valve (mitral stenosis) causes a high pressure in the left atrium and pulmonary circulation causing left heart failure and can produce secondary failure of the right ventricle; in aortic stenosis the hypertrophied left ventricle (see next section on hypertrophy) eventually can fail, and in aortic regurgitation severe left ventricular enlargement with coronary insufficiency can also lead to heart failure. Such failure of the heart muscle to contract normally results from the long-standing overload with hypertrophy, eventual loss of myofibrils from hypertrophied cells (Schwartz et al., 1981), and microscopic scarring. Mechanical constriction of the heart by a scarred and thickened pericardium can also produce all the signs of severe heart failure (Shabetai et al., 1970; Shabetai, 1981).

---

* The ventricular "filling pressures" are elevated, either the left atrial (and pulmonary venous pressure), the right atrial (and systemic venous pressure), or both.

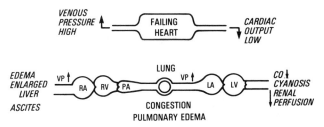

Figure 2.193. Heart failure. *Upper diagram* shows common features of heart failure, that is, either low cardiac output or high venous pressure (or both) at rest or during exercise. *Lower diagram* shows the physiologic consequences of low forward cardiac output delivered by the left side cardiac chambers (*LV, LA,* left ventricle and left atrium) with decreased cardiac output (*CO*), cyanosis, and decreased renal perfusion. The effects of increased venous pressure (*VP*) in the pulmonary veins due to an increased left atrial pressure causes pulmonary congestion and edema and can cause secondary elevation of the pulmonary artery (*PA*) pressure and right heart pressures (some lesions can cause right-sided heart failure alone). This, in turn, leads to an increased venous pressure (*VP*) in the systemic veins, which can cause tissue edema, enlargement of the liver, and fluid accumulation in the abdomen.

Finally, primary diseases of heart muscle such as infections (viral myocarditis), toxic factors (certain antitumor agents, and possibly alcohol), and "cardiomyopathies" (heart muscle diseases of unknown cause) are associated with damage and loss of heart muscle cells. Such primary and secondary causes of heart muscle damage result in *myocardial failure* (that is, depression of the myocardial inotropic state).

The cellular causes for such myocardial failure remain controversial; multiple etiologies are likely, including the possibility of abnormal excitation-contraction coupling (restoration of intracellular calcium by digitalis, for example, improves the contraction of the failing heart); abnormalities in mitochondrial function, or in the structural proteins (myosin ATPase is low in some types of experimental heart failure); and, particularly in hypertrophy, inadequate capillary blood supply may lead to undersupply of substrates to the heart. (For reviews on cellular mechanisms in heart failure see Meerson, 1969; Scheuer, 1970; Fanburg, 1974; Katz, 1976; and Braunwald et al., 1976). It does not seem likely that simple overstretch of the sarcomeres (descending limb of the sarcomere length-tension relation) is responsible for myocardial failure, since it does not occur in acute or chronic volume overload (Ross et al., 1971). Moreover, reducing the filling of the failing heart by obstructing venous return does not *improve* its function (move it backward off of a descending limb) (Ross and Braunwald, 1964b).

Physiologic Effects of Heart Failure

The right heart may fail alone (for example, in long-standing right ventricular overload in chronic lung disease). More commonly, the left side of the heart fails, causing elevation of pressures in the pulmonary circulation with secondary right heart failure. The potential sites of physiologic abnormalities are illustrated in Figure 2.193, *lower panel.* Severe heart failure with a reduced cardiac output can produce fatigue, a weakened pulse, peripheral vasoconstriction with cyanosis of the extremities, and decreased flow through the organs of the body, including the kidneys. Elevation of the left atrial pressure and the pulmonary venous pressure behind the failing left heart produces perivascular edema and "congestion" of the lungs (Fig. 2.193), which increases their stiffness thereby causing shortness of breath; if this pressure elevation is severe, leakage of fluid directly into the lung air sacs causes respiratory distress and impairment of oxygen exchange ("pulmonary edema"). With either primary or secondary failure of the right ventricle (due to elevation of pulmonary artery pressure), the venous pressure in the systemic veins eventually becomes elevated (Fig. 2.193, *lower panel*), and if the elevation is sufficiently severe leakage of fluid out of the capillaries can occur, causing fluid accumulation in the extravascular spaces (edema) in dependent locations such as the legs; enlargement of the liver with eventual leakage of fluid into the abdomen (ascites) may follow. This stage of circulatory failure is called "congestive heart failure."

Compensatory Mechanisms in Heart Failure

When the heart fails, a number of mechanisms come into play which tend to preserve the blood pressure and the cardiac output. The primary compensatory mechanisms are as follows.

1. Increased Cardiac Size. This takes place initially through acute dilation, allowing use of the Frank-Starling mechanism, but chronic heart failure then results in gradual further heart enlargement associated with some degree of hypertrophy. Hypertrophy is particularly marked when the heart failure is secondary to a long-standing increase in cardiac work (such as in hypertension or valvular heart disease). Thus, when heart failure is due to myocardial disease, the heart as seen on a chest X-ray often is enlarged.

This gradual, late ventricular enlargement helps to maintain the stroke volume by allowing greatly reduced shortening of each unit of myocardium to produce a normal or only mildly reduced stroke volume. This effect is illustrated in Figure 2.194.

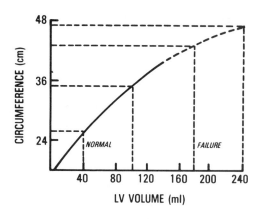

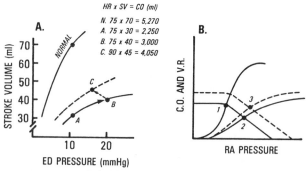

Figure 2.195. *Heart* failure as diagrammed in a framework of curves relating the stroke volume (*SV*) and left ventricular end-diastolic (*ED*) pressure (*panel A*), and in a framework of cardiac output (*C.O.*) and venous return (*V.R.*) curves (*panel B*). Reflex and other compensations are shown in panel A as points B and C, with movements up a depressed function curve and a shift of the ventricular function curve together with reflex increase in heart rate (Table). In panel B, the normal state (*point 1*) and the failure state are shown (*point 2*) with compensations indicated by the *dashed lines* and point 3. For further discussion, see text.

△ CIRC.	SV	EFx
NORMAL: 26%	60ml	60%
FAILURE: 9%	60ml	25%

Figure 2.194. Relation between left ventricular volume and the circumference of the normal and chronically failing ventricle. The *curve* shows the relation between the circumference and volume of a sphere, and its *dashed extension* shows the influence of chronic progressive left ventricular dilation. The *straight dashed lines* intersecting the normal portion of the curve indicate end-diastole and end-systole, and in failure the much larger end-diastolic and end-systolic volumes and circumferences are also indicated by *straight dashed lines*. Notice that in the normal ventricle, 26% shortening of the internal chamber circumference can produce a stroke volume of 60 ml with a normal ejection fraction (60%). In heart failure, from a much larger end-diastolic volume (120 ml) the depressed myocardium can now shorten the internal circumference by only 9%; however, the dilated ventricle could theoretically still deliver a normal stroke volume of 60 ml at a much lower ejection fraction (25%).

End-diastole and end-systole are shown in the normal heart, delivering a normal stroke volume. A 25% shortening of the circumference of the chamber produces a 60-ml stroke volume and a normal ejection fraction (60%). With chronic heart failure, the end-diastolic volume increases along the *dashed curved line* extending the normal curvilinear relationship between ventricular circumference and volume (for a sphere volume = $4/3\pi$ r^3) (Fig. 2.914, *Failure*). Under these conditions, the end-diastolic volume is markedly increased (240 ml) and the systolic shortening of the diameter is only 9%, yet the stroke volume is maintained. The increased end-diastolic volume, even though the ejection fraction is decreased (to 25%), can produce a normal stroke volume.

The responses to acute left ventricular failure are shown as ventricular function (Starling) curves of the ventricle in Figure 2.195A. Acute heart failure results in displacement of the normal left ventricular function curve downward and to the right. Without compensation by the Frank-Starling mechanism, stroke volume would drop markedly (normal to point A), and the cardiac output would

be greatly reduced. However, the stroke volume is partially restored by the higher left ventricular end-diastolic pressure. The heart rate is somewhat increased (see Table, point B), and the cardiac output therefore partially recovers; further compensations also occur (see below).

2. Autonomic and Renal Compensations. With decreased cardiac output and stroke volume, the systemic arterial compartment is relatively underfilled, a condition which activates the baroreceptors (Chapter 16). The reflex adrenergic responses elevate the peripheral vascular resistance, restoring the blood pressure and they also increase the heart rate (Zelis et al., 1973). The result is a shift of the left ventricular function curve upward and to the left (*dashed line*, Fig. 2.195A). The positive inotropic effect of adrenergic stimulation tends to increase the stroke volume and lower the left ventricular end-diastolic pressure to some degree (point C). In addition, reflex elevation of the heart rate can further restore the cardiac output (*C.O.*) toward normal (Fig. 2.195A, point C, and Table).

Mechanisms also come into play which increase the circulating blood volume, thereby sustaining the increased cardiac volume. Thus, underperfusion of the kidneys (due to the decreased cardiac output, activation of the sympathetic nervous system, and perhaps elevated venous pressure acting on the renal veins) activates the renin-angiotensin system (Chapter 32) which, in turn, stimulates aldosterone secretion (the role of the kidney in heart failure was reviewed by Barger et al., 1959). The reduced renal blood flow per se, together with an increased plasma aldosterone level (Chapter 32), promote sodium and water reabsorption by the kidney,† serving to increase the circulating blood volume and the volume of the heart. This effect is encompassed by the increased end-diastolic pressure of points *B* and *C*, Figure 2.195*A*. Within the venous return curve framework (Fig. 2.195*B*), acute depression of the cardiac output curve by heart failure (point *1* to point *2*) results initially in a marked reduction of the cardiac output. Some shift upward of the cardiac output curve occurs by reflex autonomic stimulation and the increase in heart rate (Fig. 2.195*B*, dashed line), but even more importantly, activation of the sympathetic nervous system increases the venous tone (Zelis et al., 1973) and the mean systemic pressure; together with increased blood volume brought about by renal effects, this shifts the venous return curve upward (Fig. 2.195*B*, *dashed venous return curve*). By these mechanisms the cardiac output can be restored to near-normal, despite depression of the cardiac output curve (point *3*, Fig. 195*B*). In the later phases of heart failure, cardiac autonomic compensations become impaired (Braunwald et al., 1976), and fluid retention by the kidney becomes excessive (Barger et al, 1959).

The Treatment of Heart Failure

Traditional forms of treatment for heart failure have included use of the positive inotropic agent *digitalis*, which can shift the ventricular function curve and the cardiac output curves upward (see Fig. 2.121, Chapter 12). *Diuretics* are added to rid the tissues of excessive interstitial fluid (peripheral edema and pulmonary edema), and to lower excessively elevated venous pressures. Of course, it may be possible to treat failure of the different subsystems of the heart by *specific therapies* aimed at the underlying cause of the heart failure. For example, heart block can be treated by implantation of a permanent electrical pacemaker; serious valvular lesions can be corrected by surgical replacement of the diseased valve with an artificial prosthesis; severe coronary artery narrowing can be treated with antianginal drugs or by surgical bypass-grafting of the coronary arteries; and severe hypertension is treated by appropriate medications.

In late stage hypertension or aortic stenosis, and in severe left ventricular myocardial failure of any cause, the afterload on the left ventricle (reflected either by the peak systolic pressure or the wall

† In experimental heart failure, exercise increases renal vascular resistance despite a less than normal increase in the blood pressure so that renal blood flow can drop markedly during exercise in the presence of heart failure (Millard et al., 1972).

tension) is usually elevated, leading to a state of *afterload mismatch* (Ross, 1976b). In this situation, the preload reserve of the ventricle is generally exhausted, and it is possible to show that any further increase of the afterload will result in a sharp drop in the stroke volume (Fig. 2.196). Thus, when the ventricle moves from a normal conditon (Fig. 2.196, beat *1*), to severe heart failure with a high left ventricular end-diastolic volume and pressure (the heart is now operating on a steep portion of the diastolic pressure-volume curve), the limit of preload reserve may be reached (beat *2*, Fig. 2.196); under these conditions, if an additional pressure load is applied (such as by infusing the vasopressor agent phenylephrine), the ventricle behaves *as if* the preload were fixed (compare with Fig. 2.120, Chapter 12). Thus, the response to the vasopressor will be a sharp drop in the stroke volume (beat *2* to *3*, Fig. 2.196), as demonstrated in patients with heart failure (Ross and Braunwald, 1964).

In the severely failing heart, when the peripheral vascular resistance is elevated by adrenergic compensations and the ventricle is operating at the limit of its preload reserve, the dilated ventricle can exhibit an elevated wall stress, even though the

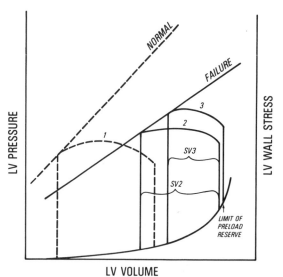

Figure 2.196. Left ventricular (*LV*) pressure-volume relation and pressure-volume loop under normal conditions (*dashed lines*) and during heart failure when the linear end-systolic pressure-volume relation is shifted downward and to the right (failure). Beat 2 to beat 3 shows the effect of acutely increasing the left ventricular systolic pressure with a vasoconstrictor when there is little or no preload reserve; the stroke volume drops (SV2 to SV3). In chronic heart failure, the left ventricle may be operating under basal conditions similar to beat 3 and left ventricular wall stress (*right-hand ordinate*) may be elevated, despite a normal left ventricular systolic pressure. Under these circumstances use of a vasodilator drug may relieve this "afterload mismatch" and allow the ventricle to improve the stroke volume by lowering the wall stress (*beat 3 to beat 2*). For further discussion, see text.

systolic pressure in the ventricle may be normal (Fig. 2.196, righthand ordinate, beat *3*). Remember, a large radius and a thin wall will yield a high wall stress even in the presence of a normal systolic pressure (Fig. 2.109). Under these conditions, it can be reasoned that *reducing the afterload* on the failing ventricle will improve its stroke volume (moving the left ventricle from beat *3*, to *2*, Fig. 2.196).

Afterload reduction (vasodilator therapy) has now become an important form of treatment for acute and chronic heart failure. (For review see Chatterjee and Parmley, 1977). Experiments on venous return curves in the normal circulation and in acute severe left ventricular failure (produced by multiple coronary artery ligations), allow understanding of how peripheral and circulatory factors interact in vasodilator treatment using nitroprusside (an agent which dilates both arterioles and veins by relaxing vascular smooth muscle) (Pouleur et al., 1980).

In the *normal* circulation, intravenous infusion of nitroprusside leads to a reduction of cardiac output despite a lower systemic vascular resistance and more favorable systolic loading conditions on the normal left ventricle. This occurs because there is concomitant dilation of the venous bed, which is only partly compensated by a small shift of blood volume from the central to the peripheral circulation. Therefore, the venous return curve is displaced downward and to the left (Fig. 2.197) with a reduced mean systemic pressure and effective systemic blood volume. The reduction in the venous return to the right heart is responsible for a *reduction* in right ventricular output (and hence left ventricular output), despite the more favorable systolic loading conditions on the normal left ventricle (Fig. 2.197).

In the presence of acute experimental left ventricular failure in which the left ventricular end-diastolic pressures were elevated to over 20 mmHg, an opposite effect occurred, with an increase in cardiac output during nitroprusside infusion (Fig. 2.198). Again, venodilatation took place in the peripheral circulation; however, this time a large shift of blood volume from the distended central circulation to the peripheral circulation occurred with vasodilator therapy (the failing heart is now able to unload by ejecting against a lower vascular resistance, and blood dammed up behind it is released). This blood volume shift exactly counterbalanced the tendency for nitroprusside to lower the effective systemic blood volume and the mean systemic pressure in these experiments (Pouleur et al., 1980). Therefore, the venous return curve was *not* shifted downward (Fig. 2.198), and the marked shift upward of the cardiac output curve, due to

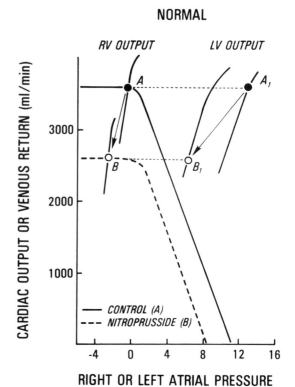

Figure 2.197. Relation between cardiac output and venous return in the normal heart (open chest, anesthetized dog). The inverse relation between right atrial pressure and venous return is shown under control conditions (*solid lines*) and during nitroprusside infusion (*dashed lines*). Segments of cardiac output curves relating right ventricular output (*RV OUTPUT*) to right atrial pressure and left ventricular output (*LV OUTPUT*) to left atrial pressure are also shown in these two conditions. Under control conditions, the cardiac output is limited by the venous return, and the equilibrium (*point A*) where the venous return and cardiac output curves intersect is on the plateau of the venous return curve. In the steady-state, the right ventricular and left ventricular outputs are in equilibrium (*dashed horizontal line* and *point A₁*). During nitroprusside infusion, venodilation produces a drop in mean systemic pressure and a shift downward of the venous return curve, the right ventricle reaching a new equilibrium point at a lower cardiac output (*point B*) which, in turn, is in equilibrium with a lower left ventricular output at a lower mean left atrial pressure (*point B₁*). Thus, despite upward shifts of the right and left ventricular function curves due to reduced afterload with lowered impedance to ejection, the cardiac output is lower. [Adapted from Pouleur et al. (1980).]

correction of the afterload mismatch on the failing ventricle, could now be expressed as an increase in the cardiac output (Fig. 2.198). Thus, is it clear that the effects of vasodilator drugs on the *peripheral circulation* as well as on the heart are highly important in determining the overall responses to acute vasodilator therapy (Pouleur et al., 1980).

A number of other drugs can decrease vascular resistance. For example, hydralazine (a vascular smooth muscle relaxant) has an effect largely limited to the arterioles, and it increases cardiac output

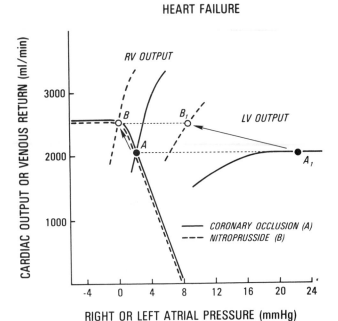

HEART FAILURE

Figure 2.198. Relations between cardiac output and venous return in acute heart failure produced by experimental coronary occlusion. Under control conditions (*solid lines*), the intersect between right ventricular output and venous return (*point A*) is on the ascending portion of the venous return curve, cardiac output being limited by the failing left ventricle which is operating on a flat and depressed cardiac output curve (*point A₁*). Following nitroprusside infusion (*dashed lines*) there is no shift of the venous return curve, a downward shift being prevented by a redistribution of the central blood volume to the periphery (see text). There is now a marked shift upward of the function curve of the left ventricle due to reduced afterload with correction of afterload mismatch, and a marked drop in left ventricular filling pressure occurs. The shift upward of the left and right ventricular output curves is now accompanied by an increase in the cardiac output (*point B*), and at equilibrium the left ventricle (*point B₁*) is now operating at a lower filling pressure with an improved cardiac output. [Adapted from Pouleur et al. (1980).]

in large measure by shifting the cardiac output curve upward (through lowered afterload), with little effect on the venous bed (in contrast to nitroprusside). In normal subjects this results in little change in the cardiac output, since the function curve is shifted to the left on the flat portion of the venous return curve. In heart failure, however, when the failing ventricle is operating below the knee of the venous return curve (the cardiac output curve is shifted downward and to the right), such an agent substantially improves the cardiac output.

Effects of the Pericardium in Overtransfusion: Possible Role in Heart Failure

The effect of the pericardium on the diastolic pressure-volume relation has been controversial. It appears to have little effect under normal conditions, but large downward shifts of the entire left

ventricular diastolic pressure-volume curve have been described in patients with heart failure during treatment with nitroprusside; these shifts were associated with little change in the left ventricular diastolic volume but a large reduction of the left ventricular diastolic pressure. That such shifts in the pressure-volume curve may reflect an effect of the pericardium, rather than an alteration of left ventricular diastolic compliance, has been indicated by laboratory studies in acutely distended hearts of conscious dogs (Shirato et al., 1978; also Glantz and Parmley, 1978; Shabetai, 1981). Following acute circulatory distension by transfusion, the entire pressure-dimension curve was displaced upward (Fig. 2.199, *upper panel*). When nitroprusside was

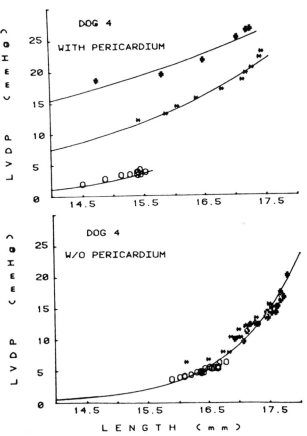

Figure 2.199. Relations between the length of a segment of the left ventricle (representing the volume of the chamber) and the left ventricular diastolic pressure (ordinate, in mmHg) in a conscious dog. Points were obtained during slow cardiac filling (diastasis). The *upper panel* shows this relation with the pericardium intact before (*open symbols*) and after the intravenous infusion of dextran to produce acute cardiac dilatation (*asterisks, upper curve*); the middle curve (*x's*) shows the effect of an intravenous infusion of nitroprusside in the presence of such acute cardiac dilatation. In the *lower panel*, the same dog is studied again without (*W/O*) the pericardium, following its surgical removal. The same interventions, volume loading and nitroprusside, are produced. The ventricle now appears to be operating on a single diastolic pressure-length relation. [From Shirato et al. 1978).]

infused under these conditions with the pericardium intact there was downward shift of the *entire curve* (Fig. 2.199, *upper panel*). When the experiment was repeated after the pericardium was removed, however, the left ventricle now moved upward (with transfusion) and downward (nitroprusside) on a *single* curve (Fig. 2.199, *lower panel*). This finding suggests that with acute cardiac distension by overtransfusion, and perhaps in acute heart failure, the limit of pericardial distension can be reached, and elevated *intrapericardial pressure* can contribute to high ventricular filling pressures. Moreover, decreases in intrapericardial pressure may contribute substantially to the large reductions in left ventricular filling pressure observed during vasodilator therapy in some patients with acute or subacute heart failure.

CARDIAC HYPERTROPHY

Several types of cardiac hypertrophy (increased size of heart muscle cells causing cardiac enlargement) can be identified. One type occurs in trained athletes, and another is left ventricular hypertrophy of unknown cause; hypertrophy also occurs in response to valvular heart disease or to hypertension, and it also takes place in normal regions of

the left ventricle after a heart attack, tending to compensate for scar formation in damaged regions.

Two basic types of hypertrophy that occur in response to stress generally are defined: that caused by pressure overload and that due to volume overload of the heart. For simplicity, we will confine the discussion primarily to the effects of these two types of overload on the left ventricle. (For reviews concerned with enhanced protein synthesis and other responses in hypertrophy, see Meerson (1969) and Zak and Rabinowitz (1979).)

Pressure Overload

Pressure overload on the left ventricle is most commonly produced by high blood pressure (hypertension, see Chapter 16), which raises the systolic and diastolic pressures in the aorta, and therefore increases the left ventricular systolic pressure. Narrowing of the aortic valve (aortic stenosis) also produces a high systolic pressure in the left ventricle, with a lower than normal pressure in the aorta. (The right ventricle can also undergo hypertrophy in response to pulmonary artery hypertension, or pulmonic valve stenosis.)

The basic response of the left ventricle to sustained pressure overload is to develop *concentric*

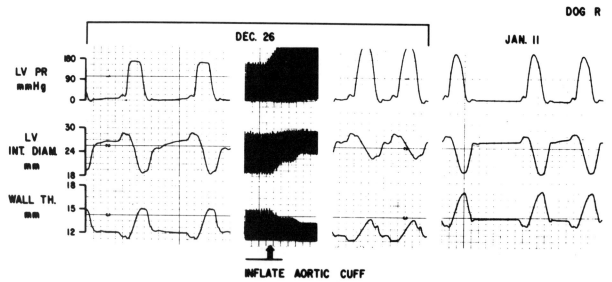

Figure 2.200. Responses of the left ventricle of a conscious dog to acute and chronic pressure overload produced by means of a cuff on the ascending aorta. Tracings on the *left* show left ventricular pressure (*LV PR*), internal diameter (*INT. DIAM.*), and the thickness of the left ventricular wall (*WALL TH.*) measured with implanted ultrasonic dimension gauges. Notice the expansion of left ventricular diameter during atrial systole (synchronous with the presystolic wave in the left ventricular pressure tracing), the shortening of the diameter during left ventricular ejection followed by rapid and slow elongation during diastolic filling; the wall thickness shows a mirror image to the internal diameter, thinning with atrial systole, and thickening as the ventricle ejects blood. Sudden inflation of the aortic cuff produces a high left ventricular systolic pressure (off scale approaching 240 mmHg) and a decreased extent of systolic shortening of the internal diameter with decreased systolic wall thickening (*middle rapid tracing*). Approximately 2 weeks later (January 11th) the left ventricular systolic pressure is still over 200 mmHg, but now the wall thickness has markedly increased, whereas the chamber diameter is normal (concentric hypertrophy). This adaptation has allowed the ventricle to develop nearly normal systolic shortening at a high left ventricular systolic pressure. [Tracings obtained by Sasayama et al. (1976).]

hypertrophy. This adaptation, which occurs over months or years, consists of increased *thickness* of the left ventricular wall (protein synthesis is stimulated, and the individual cells of the myocardium greatly enlarge with an increased number of myofilaments) but *without* an increase in the size of the left ventricular chamber (ventricular end-diastolic volume remains unchanged). Such a response in an experimental animal subjected to constriction of the ascending aorta for several weeks by means of an implanted inflatable cuff is shown in Figure 2.200 (Sasayama et al., 1976). Notice that the initial response to aortic constriction is that to an acute, severe increase in afterload: an increase in end-diastolic pressure with thinning of the left ventricular wall, and a fall in the extent of systolic wall shortening due to markedly increased systolic pressure (wall tension is increased during this acute response, see below). However, over the next few weeks the left ventricular wall thickens, and the left ventricular chamber size and the shortening of the wall return to normal (Fig. 2.200, *righthand tracings*). This chronic adaptation is diagrammed in Figure 2.201, together with the changes in factors which determine mean wall tension by the Laplace relation. The normal ventricle during systole is in the *left panel*, and it can be seen that with chronic pressure overload the high systolic pressure, with no change in ventricular radius, is compensated for by the increased wall thickness, resulting in an

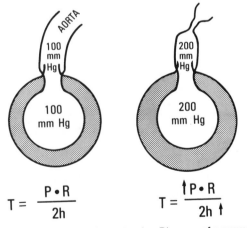

Figure 2.201. Concentric hypertrophy. Diagram of a normal left ventricle developing a systolic pressure of 100 mmHg during ventricular ejection. A simplified Laplace relation for wall tension (*T*) is shown where *P* = pressure, *R* = chamber radius and *h* = wall thickness. In the *right* figure, constriction of the aorta above the aortic valve has been produced to raise the left ventricular systolic pressure to 200 mmHg, and the chronic adaptation has occurred. Notice that the chamber has remained the same size, but the wall has thickened markedly due to hypertrophy. Thus, the increased pressure has been counterbalanced by the increased wall thickness, and wall tension (*T*) remains normal.

unchanged tension in the left ventricular wall during systole (Fig. 2.201, *right panel*). Thus, the chronic adaptation to pressure overload returns the afterload (wall tension) on the muscle fibers toward normal, despite the elevated systolic pressure, and this near-normal level of afterload allows preservation of the extent of shortening of the wall and the delivery of a normal stroke volume.

This functional response is diagrammed within the end-systolic pressure-volume framework in Figure 2.202*A*. Beat *A* shows the normal pressure-volume loop reaching the normal end-systolic pressure-volume relation, and beat *2* illustrates the chronic adaptation to pressure overload with concentric hypertrophy in which a normal stroke volume is delivered from the same end-diastolic volume but at a much higher level of left ventricular systolic pressure. Note that the linear end-systolic pressure-volume relation is shifted to the left and upward in this chronic adaptation (Fig. 2.202*A*), despite the fact that no change in the myocardial inotropic state is caused by the hypertrophy. Thus, under these chronic conditions the left ventricular end-systolic pressure-volume relation is an unreliable indicator of myocardial inotropic state; however, it does indicate hyperfunction of the left ventricle (this would result in a shift upward of the cardiac output curve). This difficulty with the use of the end-systolic pressure-volume relation for defining inotropic state in chronic adaptations can be overcome by the use of left ventricular wall tension or wall stress (instead of left ventricular pressure) on the ordinate of the graph (Fig. 2.202*B*). Under these conditions, it can be seen that both the normal loop (beat *A*) and that after the chronic adaptation (beat *B*) are now superimposable (since the systolic wall stress during ejection is normal) and fall on the same linear end-systolic volume-wall stress relation (see Sasayama et al., 1977, for further details). Eventually, with long-standing pressure overload, myocardial failure supervenes (Meerson, 1969).

Volume Overload

In this setting, the left ventricle must adapt to the requirement for a larger than normal stroke volume. Such a need may arise in conditions which place a volume overload on the left (or right) ventricle such as chronic strenuous exercise in the athlete, anemia, mitral regurgitation, aortic regurgitation, systemic arteriovenous fistula, or a left-to-right shunt due to ventricular septal defect. (The right ventricle may experience a volume overload in lesions such as tricuspid regurgitation, or a left-to-right shunt caused by an atrial septal defect).

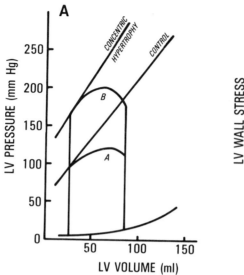
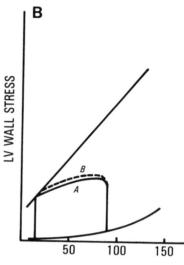

Figure 2.202. *Panel A,* Left ventricular (*LV*) pressure-volume relations and linear end-systolic pressure-volume relations before (curve *A*) and after (curve *B*) the development of concentric left ventricular hypertrophy. Left ventricular end-diastolic volume and the stroke volume are unchanged, but the left ventricular systolic pressure is much higher after hypertrophy so that the linear end-systolic pressure volume relation is shifted upward and to the left (For further discussion, see text). *Panel B,* The same responses as in panel *A* using left ventricular wall stress instead of pressure, the curves are now superimposable, since chronic thickening of the left ventricular wall lowers the systolic wall stress despite the elevated left ventricular systolic pressure.

The basic response of the left ventricle to these forms of sustained volume overload is termed *eccentric hypertrophy.* In this adaptation, the left ventricle undergoes hypertrophy in such a way that the entire chamber enlarges, with a large increase in the left ventricular volume but relatively little increase in wall thickness (Ross, 1974). This results in a marked increase in overall muscle mass and the weight of the ventricle. Because of the curvilinear relation between the volume of a spherical chamber and its radius, a *normal* percentage shortening of the circumference when the end-diastolic volume is greatly increased can deliver a much larger than normal stroke volume (see Fig. 2.194). This adaptation to chronic volume overload is shown in Figure 2.203; the normal left ventricle is on the left. With chronic ventricular dilation the stroke volume is tripled with a normal percent shortening of the circumference (and the sarcomeres) (Fig. 2.203, *right*). By this mechanism, in valvular regurgitation both the amount regurgitated and the forward stroke volume can thereby be delivered, and in other conditions of volume overload a high forward stroke volume and cardiac output can be maintained.

The response to chronic volume overload is represented within the end-systolic pressure-volume framework in Figure 2.204. The response of the normal heart to an acute volume increase is shown in beats *A* and *B.* In eccentric hypertrophy, the entire diastolic (resting) pressure-volume is shifted

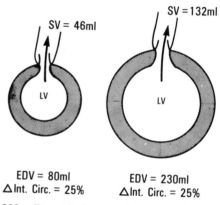

Figure 2.203. Eccentric hypertrophy. Normal ventricle shown on the *left* delivers a normal stroke volume with shortening of the chamber internal diameter of 25%. With left ventricular dilatation due to chronic volume overload, the same percentage shortening of the internal diameter now produces a nearly 3-fold increase in the stroke volume.

to the right, and with the ventricle operating at a much larger end-diastolic volume a much larger stroke volume can be delivered (Fig. 2.204, beat *C*).

ANEMIA

In anemia (Section 4, Chapter 24), there is a reduction of circulating red blood cells with a normal intravascular blood volume, leading to a low hematocrit. The reduced hematocrit causes decreased viscosity of blood (Chapter 6), which lowers arterial and venous resistances, shifting the venous return curve (Fig. 2.205). Also, the need for in-

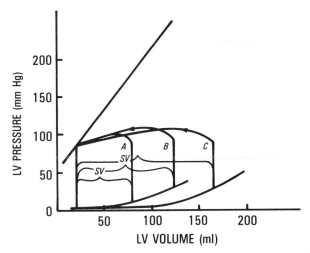

Figure 2.204. Left ventricular (*LV*) pressure-volume relations and the linear end-systolic pressure-volume relation during acute volume overloading (beat *A* to beat *B*) and after chronic eccentric hypertrophy has occurred. *SV*, stroke volume. The entire diastolic pressure-volume relation is shifted to the right, and under these circumstances the chronically dilated ventricle which has normal inotropic state can deliver a much larger stroke volume (*beat C*).

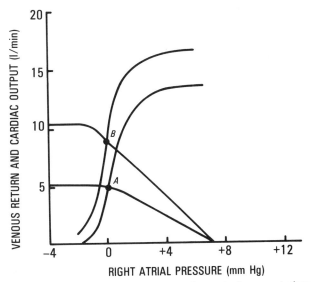

Figure 2.205. Venous return and cardiac output curves before (*A*) and after the adapations to severe anemia (*point B*). Notice the substantial increase in cardiac output. For further discussion, see text.

creased oxygen delivery to the tissues (to compensate for the low oxygen-carrying capacity of the blood) results in metabolic vasodilatation, with further lowering of peripheral vascular resistance. The lowered vascular resistance decreases the afterload on the left ventricle and shifts the cardiac output curve upward and to the left. Chronic hypertrophy of the volume overload type due to the high cardiac output, shifts the cardiac output curve further upward. Thus, a new equilibrium point is reached at

a substantially increased cardiac output during anemia (Fig. 2.205, point *B*, compared to the normal state, point *A*).

HEMORRHAGIC SHOCK: AN ABNORMAL PERIPHERAL CIRCULATORY STATE

The hemodynamic features of one form of shock will be discussed briefly in order to illustrate certain abnormalities and compensatory mechanisms that can occur in various segments of the circulation. However, detailed discussion of mechanisms and neurohumoral responses in various forms of shock is beyond the scope of this discussion (for reviews see Mills and Mayer, 1965; Zweifach and Fronek, 1975; Selkurt, 1976; and Franklin and Doelp, 1980).

In the human subject, intense bleeding due to trauma produces a marked fall in cardiac output and a secondary severe drop in the blood pressure, so-called hypotensive or "hemorrhagic shock." This condition can be mimicked in an experimental animal by attaching an artery to a blood reservoir where the pressure can be lowered to a given level (40 mmHg, for example) and maintained there for many minutes before the blood is restored to the animal. When the blood pressure is maintained at this level for a sufficiently long period of time, so-called "irreversible shock" occurs; that is, even when the blood is retransfused, the blood pressure and cardiac output fail to return to normal, and the animal eventually dies.

In the early phase of shock, the loss of blood results in a decreased venous pressure and venous return to the heart, and a diminished cardiac output; in the intact circulation (if a pressure reservoir did not maintain the low arterial pressure) the resulting hypotension would result in baroreceptor reflexes (and later to chemoreceptor and cerebral reflexes resulting from diminished cerebral blood flow, see Chapter 16) which would lead to increased peripheral vascular resistance, reflex constriction of the veins (with central displacement of blood volume), and tachycardia. All of these mechanisms would tend to restore the cardiac output and arterial pressure toward normal, and if the hemorrhage were not too severe this compensation might be adequate. In the experimental shock model the blood pressure cannot be restored by these mechanisms because of the pressure reservoir, and the blood mobilized by reflexes is initially transferred instead into the reservoir. Later in the shock state, as the pressure in the capillaries remains low, the unopposed colloid osmotic pressure tends to bring extravascular fluid into the circulation, and over time the hematocrit will fall (a manifestation of this dilution effect); this late compensatory mechanism may prevent a sustained shock state in the intact circulation when there is only a modest blood loss. However, again, in the experimental model it will only transfer more fluid into the pressure reservoir. The hypotension also causes increased release of circulating catecholamines from the adrenal glands, which serve to further stimulate the myocardium, augment heart rate, and raise the peripheral vascular resistance (Chien, 1967). Insufficient blood flow to the kidneys results in the release of angiotensin (a vasoconstrictor) as well as elevation of aldosterone levels which, along with direct effects of the low flow on the kidney, causes sodium and water retention with increased intravascular fluid volume.

The compensatory vasoconstriction initially involves primarily the skin, muscle, and splanchnic circulations (and to a much lesser degree the renal circulation), whereas the autoregulating coronary and cerebral circulations maintain relatively good flow.

After 1 or 2 h of severe hypotension, the shock state gradually becomes irreversible. If the shock remains untreated, sustained low blood flow with ischemia (insufficient O_2 and nutrient supply due to reduced flow) in the central nervous system results in decreased activity of the vasomotor centers and gradual loss of reflex compensations.

Moreover, toxic substances absorbed when there is ischemia of the bowel, as well as progressive acidosis, can reduce the reactivity of vascular smooth muscle to catecholamines. This loss of vascular reactivity tends to result in sequestering of blood in the peripheral circulation (and depletion of blood from the reservoir in the experimental model). In addition, ischemia of the capillaries causes increased permeability, with reversal of fluid transfer (loss of intravascular fluid). The intense acidosis and metabolic products from ischemic tissues may lead to thrombosis of small blood vessels.

Damage to the internal organs results in the accumulation of other toxic factors, and one of these (the so-called "myocardial depressant factor") appears to reduce myocardial contractility in the late phase of shock (Lefer and Martin, 1970). Severe acidosis also tends to decrease myocardial contractility, and the ventricular function curve shifts downward in late shock (ischemia in the subendocardial regions may also be contributory). Finally, severely reduced flow through the bronchial arteries which supply the lung tissue eventually results in pulmonary damage, with edema and a decrease in the arterial oxygen saturation. This adds to generalized tissue hypoxia and ischemia and the downward spiral of circulatory function, as compensatory mechanisms fail.

CHAPTER 19

Special Circulations, the Fetal Circulation, and the Lymphatics

This chapter will consider the circulations to the skin and skeletal muscle, the intrauterine fetal circulation and changes at birth, as well as aspects of the lymphatic circulation and edema formation. Other circulations to various organs such as the kidney, lungs, abdominal viscera, and brain are discussed in Sections 4, 5, 6, and 9, respectively.

CIRCULATION TO THE SKIN

The blood supply to the skin serves two main purposes: the relatively small nutrient flow which is primarily controlled locally, and the potentially large flow concerned with heat loss regulation which is predominantly under reflex control. Thus, most of the blood flow to the skin is neurogenically rather than metabolically controlled. (For reviews of skin functions and the skin circulation see Nicoll and Cortese (1972) and Helwick and Mostofi (1980).)

In the skin, the arterioles, upon approaching the bases of the papillae (the layer of the corium immediately underlying the epidermis), turn horizontally, and give rise to metaterioles from which originate, in turn, hairpin-shaped endothelial tubes—the *capillary loops*. The proximal or arterial limb of the capillary loop ascends in the papilla and then turns upon itself to form the venous limb. The latter on reaching the base of the papilla joins with the venous limbs of neighboring loops to form a *collecting venule*. The collecting venules anastomose with one another to form a rich plexus—the *subpapillary venous* plexus—which runs horizontally beneath the bases of the papillae and drains into deeper veins. The capillary loops can be seen readily in the living skin under the low power of the microscope. The vessels at the base of the human fingernail are shown diagrammatically in Figure 2.206.

The *color* of the skin is dependent upon blood in the capillary loops and the subpapillary venous

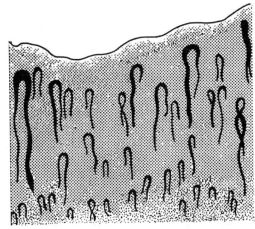

Figure 2.206. The bed of the finger nail in a healthy subject, showing the capillary loops and the summits of the skin papillae. [Adapted from Lewis (1927).]

plexus. The vessels of the plexus, though more deeply placed, present a greater area parallel to the skin, whereas the capillary loops are disposed chiefly at right angles to the skin surface. When little blood is contained in the superficial vessels, the skin is unusually pale and more transparent, and the deeper venous plexuses then contribute largely to the color of the skin, often adding a leaden tint to the pallor. When the overlying vessels are open and the skin is well supplied with blood, these deeper vessels are hidden from view.

The *hue* of the skin, i.e., the dominance of the reddish or of the bluish hue, depends upon the extent to which the oxyhemoglobin becomes reduced during the passage of the blood through the cutaneous vessels. The degree of reduction of hemoglobin will depend, as a rule, upon the rate of blood flow. With rapid flow, the blood is more arterial in character; with slow flow, more venous, and the color of the skin becomes bluish.

Sympathetic adrenergic vasoconstrictor fibers, which are tonically active, innervate the cutaneous vessels. The vascular smooth muscle of cutaneous nutrient arterioles have both α- and β-adrenergic receptors, while the arteriovenous anastomoses have only α-receptors (Korner, 1974). The physiological importance of cutaneous β-receptors is not known. In the hands and feet a maximal cutaneous blood flow results from sympathetic blockade, but in the forearm sympathectomy increases flow only if skin and core temperatures are low.

Body warming elicits a reflex active vasodilation in the hand or foot, in addition to a release of vasoconstrictor tone (a decreased number of impulses traveling over the sympathetic nerve fibers) (Shepherd, 1963). The reflex may originate either in cutaneous receptors or by central nervous system stimulation. The vessels themselves are also sensitive to warm temperature, for the dilation following blocking of vasomotor nerves can be augmented by local heating of the hand.

Arteriovenous (AV) anastomoses exist in some areas of the skin consisting of communications between smaller arteries and arterioles and the corresponding venous channels, through which the blood may be shunted to the venous plexuses and capillary areas short circuited (Fig. 2.207). These AV anastomoses do not appear to be under metabolic control but, rather, are governed chiefly by reflex influences from temperature receptors and from central nervous system centers regulating heat loss by the skin (see below).

The superficial large and small veins of the skin have a nerve supply and respond by constriction to circulating epinephrine and norepinephrine. The small veins may contract separately and independently from the large veins or arterial vessels during sympathetic nerve stimulation. When the veins constrict, an increase in pressure results. Thus, a reduced blood flow may occur through the vascular bed without any obstruction or constriction of the arterial tree. In humans, forearm cutaneous vein segments, which were isolated in situ by compression, have been demonstrated to constrict following

a variety of normal and noxious stimuli. This venoconstriction was blocked by chemical or surgical sympathectomy or infiltration of an anesthetic solution around the vein. Venous tone was decreased during vasovagal syncope and by stroking the skin over the vein. The fact that cutaneous veins constrict when cooled and are not affected by warming is evidence for the existence of only a vasoconstrictor nervous supply. The neurogenic control of the skin circulation is discussed further in Chapter 16.

Role of Skin Circulation in Heat Transfer

The blood vessels of the skin have a large capacity to alter flow and change heat loss through the outer layers of the skin, and with maximum vasodilation the skin can lose 7 or 8 times as much heat as during full vasoconstriction; the rate of blood flow to the skin at room temperature is many times the relatively small nutrient flow requirement. The nutrient circulation does exhibit both reactive hyperemia and autoregulation, and the skin circulation also participates in the usual baroreceptor reflexes. However, during changes in temperature the skin vascular resistance comes under other reflex control, particularly the heavily innervated arteriovenous anastomoses located primarily in the hands, feet, ears, and face (which, combined, present a large skin surface area) (Fig. 2.207). Total skin blood flow during marked cooling can be very low, or reach several liters per minute during extreme heating. In the latter circumstance, compensatory vasoconstriction must occur in other vascular beds if the blood pressure is to be maintained.

The arteriovenous anastomoses receive adrenergic sympathetic fibers which are largely under the control of the central nervous system, and there is also evidence (discussed below) that when sweating occurs at higher temperatures chemically induced vasodilation has an important role. The center controlling heat transfer is located in the anterior hypothalamus. Direct cooling of this center or perfusion with cool blood produces cutaneous vasoconstriction, whereas heating it or perfusing it with warm blood produces sweating and cutaneous vasodilation.

Local cooling of the skin appears to cause local reflex vasoconstriction even in the absence of a change in core temperature (possibly due to increased sensitivity of the vessels to adrenergic nerve impulses, as well as a local cord reflex). That such vasoconstriction is reflex and may involve local temperature receptors is illustrated by the response to immersion of one hand in cold water when the circulation to that arm is occluded: cutaneous vasoconstriction results. When the circulation to the

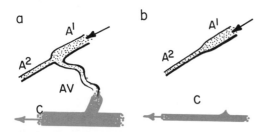

Figure 2.207.   Reaction of anastomosis and associated vessels to lowering of body temperature. A^1, artery; A^2, arteriole; AV, arteriovenous anastomosis; C, vein. a, Anastomosis open. b, Anastomosis closed. [Adapted from Grant (1930).]

immersed hand is intact and the cooling is sustained, the blood temperature is lowered and leads to direct cooling of the hypothalamic center, with further reflex vasoconstriction. With prolonged severe cold exposure local vasodilation of the skin vessels occurs despite sustained low total skin flow. (This response is responsible for the rubor or redness of face in severe cold and may serve a protective function against freezing.)

Application of local heat causes local vasodilation, with opening both of the ordinary resistance vessels as well as the arteriovenous anastomoses, and as warmed blood reaches the hypothalamus reflex vasodilation occurs elsewhere in the skin. The mechanism of vasodilation is not entirely clear, although when sweating is initiated by the hypothalamic center at higher temperatures there is evidence that a local vasodilator may be involved. One postulated mechanism includes sympathetic cholinergic activation of the sweat glands, which release an enzyme (kallikrein) which, in turn, induces the formation of the potent vasodilator bradykinin. Blockade of bradykinin production need not greatly impair this response, however, and other local products may be involved. This mechanism may also play a role in the central nervous system-mediated blushing response produced by anger or anxiety.

Vascular Responses of Skin to Stimulation by Mechanical and Other Agencies

THE WHITE REACTION

If the surface of the skin is stroked lightly with a blunt "pointed" instrument, a line of pallor appears in 15 to 20 s which traces the path taken by the instrument. The line attains its maximal intensity in ½ to 1 min and then gradually fades to disappear in 3 to 5 min. The white reaction proper is due to direct stimulation of vessel walls and has no nervous basis. It has been shown to be due to the tension exerted upon the walls of the minute venous vessels—collecting venules and especially of the subpapillary venous plexus—as they respond to the stimulus by contraction.

THE TRIPLE RESPONSE

This comprises: (1) *the red reaction,* (2) *the flare,* and (3) *the wheal.*

The Red Reaction

If the pointed instrument is drawn more firmly across the skin, especially of the forearm or back, a red instead of a white band appears after a somewhat shorter latent period (3 to 15 s), reaches its maximum in ½ to 1 min, and then gradually fades. Like the white reaction it is strictly localized to the line of stroke; it is due to *dilation* of the venular vessels. On either side of this a pale area of capillary constriction (white reaction) may appear. The red reaction can be induced in its full intensity in the skin from which the circulation has been occluded, so it is due to active dilation of the venular vessels and not merely a passive result of arteriolar dilation. The red reaction is not dependent upon nervous mechanisms since it occurs after section and degeneration of the cutaneous nerves.

The Spreading Flush or Flare

If the stimulus is a bit stronger, the reddening of the skin is not confined to the line of stroke but surrounds it for a variable distance (1–10 cm) according to the intensity of the injury inflicted. The temperature in the suffused area is definitely raised. This flare reaction appears a few seconds (15 to 30) after the local red line, and fades sooner. It also is due to dilation of the arterioles and venules, since it does not appear after the circulation of the part has been occluded by means of a tourniquet. But unlike the red reaction the flare is dependent upon local nervous mechanisms (axon reflex). It occurs after the nerves are divided but not after they have degenerated.

Local Edema or Wheal

When the stimulus is still more intense, the skin along the line of the injury becomes blanched and raised above the surrounding area to a height of 1 or 2 mm or even more. Such a wheal or welt can be produced in a normal person by the lash of a whip and other types of strong, localized stimulation. In susceptible individuals, even light stimulation, such as drawing a pencil with moderate pressure over the skin of the back, will produce linear wheals surrounded by a diffuse red halo along the pencil's track. In this way letters or other designs may be embossed upon the skin. This phenomenon is spoken of as *dermographism.* The wheal makes its appearance in 1 to 3 min from the time of injury and is at its maximum in 3 to 5 min. It is preceded by, but then replaces, the usual red reaction; it is surrounded by the flare described above. The raised patch at first is clearly demarcated, but as time passes it increases in width and decreases in height, loses its sharpness and finally, although perhaps not for some hours, disappears. The wheal is due to the transudation of fluid from the minute vessels involved previously in the red reaction: it is, therefore, a localized edema. Increased permeability of the capillary wall is judged to be the immediate cause. Wheal production does not depend upon a nervous mechanism.

The triple response has been attributed to the release of some diffusible substance by the injured cells. Injection of histamine produces a similar reaction, but whether histamine or some other vasoactive substance (e.g., ATP, bradykinin, etc.) is involved is not known.

CIRCULATION IN SKELETAL MUSCLE

Blood flow rate in resting mammalian skeletal muscle ranges from 2 to 5 ml/min/100 g in man to 10 to 20 ml/min/100 g in the cat and rabbit (Korner, 1974). One reason for this difference is the proportion of red and white fibers in the muscle groups studied, red muscle having about double the resting flow of white muscle. Skeletal muscle in general undergoes very active and widely ranging fluctuations in blood flow.

Skeletal muscle comprises the major component of the body mass, over 40% of normal body weight. The $\dot{V}O_2$ and blood flow to skeletal muscle are low at rest, but both increase very markedly during muscular exercise. The circulation to skeletal muscle is particularly suited to such a response, and exhibits a high level of basal vascular tone, considerable sympathetic tone at rest, as well as pronounced autoregulation manifesting a high degree of local control (see Chapter 6). Also, at rest, skeletal muscle has a large capillary reserve, with most of the capillaries being functionally closed by cyclic contractions of the precapillary vessels, and 4- to 5-fold increase in the number of capillaries can occur during exercise. Finally, there is considerable reserve for the extraction of oxygen during exercise, since the venous blood draining resting muscle has a high O_2 content. Further reserves for O_2 are provided by myoglobin in red skeletal muscle, which has a high aerobic capacity. In contrast, white or mixed muscle has a high anaerobic reserve, with increased content of anaerobic enzymes, relatively fewer mitochondria, and the absence of myoglobin. (For reviews on the circulation to skeletal muscle see Shepherd, 1963; Korner, 1974; and Olsson, 1981.)

Neural Control of Muscle Blood Flow

SYMPATHETIC VASOCONSTRICTOR FIBERS

Control of muscle blood flow over a wide range is effected by variations in α-adrenergic sympathetic tone (see α and β receptors, Chapter 16) acting on the resistance vessels. This forms the basis of most reflex regulation of vascular resistance in skeletal muscle. The transmitter substance is norepinephrine which stimulates α receptors in vascular smooth muscle. Maximal stimulation of these fibers doubles vascular resistance in red muscle and increases it 7-fold in white muscle (Korner, 1974). These fibers also produce venoconstriction. Following sympathectomy muscle blood flow increases transiently and then returns to control values owing to intrinsic regulation. During exercise, the effect of α-adrenergic stimulation on the arterioles is markedly reduced or abolished by the effects of intrinsic vasodilator mechanisms (see below). Venoconstriction, however, is well maintained (Mellander and Johansson, 1968). Sympathetic vasoconstrictor fibers are primarily concerned in postural and other reflex, homeostatic, hemodynamic adjustments, in regulation of pre- to postcapillary resistance, and in changes in the caliber of capacitance vessels (veins). The vasoconstriction they evoke is reflexly influenced by arterial baroreceptors, chemoreceptors, and cardiac baroreceptors (see Chapter 16).

SYMPATHETIC VASODILATOR FIBERS

Sympathetic nerve stimulation after α-adrenergic blockade produces in skeletal muscle a vasodilation which is blocked by atropine. The transmitter substance is thought to be acetylcholine acting on muscarinic receptors. This sympathetic outflow is not activated by the baro- and chemoreceptors mentioned above. It appears to originate in the cerebral cortex via fibers that have connections in the hypothalamus. Such sympathetic vasodilation, blocked by atropine, can be demonstrated in the dog, cat, fox, sheep, and goat, but not in primates, monkey, or rat. It has been postulated to occur in man during stressful mental activity and vasovagal fainting (Whelan, 1967). Such vasodilator responses are transient, and resistance returns to control levels in about a minute, despite continued nerve stimulation. This type of active vasodilatation is thought to be part of a centrally integrated neural mechanism that controls cardiovascular adjustments (increased heart rate, blood pressure, and muscle blood flow) in preparation for exercise or the defense reaction, although such a role has not been clearly demonstrated in human subjects.

A *histaminergic vasodilator* neural system has been suggested for muscle (Beck et al., 1966; Brody, 1966). Histamine is found in the venous blood during nerve stimulation, and antihistamines block the vasodilatation. Its vasodilator effects, however, are evanescent, histaminergic nerves have not been identified, and it can be demonstrated only when adrenergic effects are blocked. The possible functional significance of this postulated system is not understood.

β-ADRENERGIC RECEPTORS

The effects of circulating catecholamines on muscle blood flow are explained on the basis that both α and β receptors exist in the vascular smooth muscles. α Receptors are innervated by the sympathetic vasoconstrictor nerves and respond to circulating norepinephrine, which has little B_2 stimulating activity. β Receptors cause vasodilatation. Epinephrine stimulates both α and β receptors. Injected intravenously in man it initiates a 5- or 6-fold increase in muscle blood flow, which later returns to a value only twice the control level despite continued infusion (Whelan, 1967). Sympathectomy does not abolish this increase; thus, it is not neurogenic. Intraarterial injections also cause vasodilatation, but flow rapidly returns to or below the control level. When α-adrenergic receptors are blocked pharmacologically, epinephrine produces a sustained increase in blood flow. Thus, epinephrine stimulates α receptors to cause vasoconstriction, and β receptors to cause vasodilatation, and its overall effect is the sum of these opposing actions. Since the α receptors in muscle have the higher threshold of the two types, the usual effect of physiological amounts of epinephrine is to increase muscle blood flow, and this is one mechanism of vasodilation during exercise when circulating epinephrine is increased.

Local Control of Resistance Vessels in Skeletal Muscle

The neurally mediated changes in muscle blood flow are small in contrast to those caused by local intrinsic mechanisms. The perfused muscle bed exhibits autoregulation which minimizes changes in flow caused by alterations in perfusion pressure. Autoregulation is present in muscle both at rest and during exercise.

The hyperemia that accompanies muscle contractions is believed to be secondary to local chemical factors rather than neurally mediated for the following reasons: (1) In a curarized muscle the direct stimulation of the muscle elicits an increased blood flow, while motor nerve excitation does not. (2) Humans with sympathectomized extremities have no muscular disabilities during exercise. (3) Local or reflex warming or cooling has little effect on this intrinsic vasomotor tone in muscles. It has been suggested, therefore, that products of muscle cell metabolism act directly on the smooth muscle cells of vessels, or perhaps indirectly via a local nervous mechanism such as peripheral ganglion cells or an axon reflex. There are, however, no known ganglion cells among muscle fibers.

Thus, local metabolites override neural control during exercise, although the precise mechanism is unknown. Low P_{O2}, increased P_{CO2} tension, lactic acid, hydrogen ions, bradykinin, histamine, acetylcholine, adenosine, and potassium ions have all been suggested as local determinants of exercise vasodilation (Chapter 6). Most of the metabolites have been tested by infusion into the muscular vascular bed, or by determining their concentration in venous blood leaving the muscle following exercise hyperemia. Each one has received only a little support as the dilator agent, but a combination of two or more of these factors may provide the answer.

Bradykinin has been shown to be a very potent vasodilator in skeletal muscle by plethysmography and radioisotope disappearance rates. No change in oxygen consumption occurs despite the large increase in human calf blood flow. There is no evidence, however, for the presence of bradykinin in skeletal muscle. Also, the venous blood draining exercising muscle does not contain increased bradykinin activity, although the half-life of bradykinin in the blood is extremely short.

With rhythmic contraction of calf or forearm muscles, blood flow is rapidly increased both during and immediately following exercise. An increased flow (vasodilator metabolite mediated) alternating with decreased flow (mechanical compression of vessels) parallels the relaxation and contraction of the muscle (Fig. 2.208). Mechanical obstruction, metabolite formation, and the action of muscles on the venous flow are probably all intricately involved in the control of muscular blood flow of exercise. It has been shown that an increase in venous pressure may alter arterial inflow to muscle.

As mentioned, the sympathetic nervous system probably does not have a significant effect on the circulatory changes during exercise, for vasodilation is similar during and after exercise in normal and sympathectomized limbs. Moreover, the vascular bed of the exercising hind limb of the dog becomes less responsive to sympathetic stimulation as the strength of contraction and oxygen uptake increase, although the blood flow of exercising skeletal muscle in the dog, except during maximal exercise, is reduced by sympathetic nerve stimulation, by intraarterial norepinephrine or epinephrine infusions, or by carotid sinus stimulation. The hypothalamic vasodilator sympathetic pathway, however, does not affect exercising muscle, (see Chap-

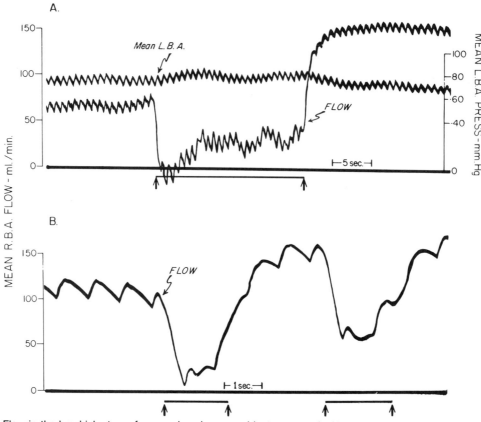

Figure 2.208. Flow in the brachial artery of a conscious human subject measured with an electromagnetic flowmeter during cardiac catheterization. *Panel A* shows mean pressure in the opposite (left) brachial artery (*L.B.A.*), together with mean blood flow during sustained handgrip using the right arm (*vertical arrows and bar below*); flow drops sharply due to mechanical squeezing of the vessels by muscle contraction, rises gradually during continued contraction, and exhibits reactive hyperemia after cessation of handgrip. *Panel B* shows tracings at rapid paper speed illustrating the effects of two brief handgrips (*vertical arrows*) on phasic brachial artery blood flow. [From Gault et al. (1966).]

ters 16 and 17 for further discussion of circulatory responses to exercise).

Reactive Hyperemia

Following a period of complete occlusion of the arterial supply to a limb or following vigorous contractions (Fig. 2.208*A*), the blood flow at first increases markedly, and then later returns to the control level. This phenomenon has been named "reactive hyperemia." The response occurs both in skin and muscle, and in the coronary circulation (Chapter 15), and can be decreased in the skin by cooling, epinephrine, or tobacco smoking. The amount and duration of reactive hyperemia correlate with the previous length of arterial occlusion. This has been shown to be true for occlusion periods up to 10 min.

The blood flow or the oxygen debt incurred during the period of occlusion can be calculated by multiplying the control blood flow or the oxygen usage before circulatory arrest by the duration of occlusion. This assumes that the blood flow and oxygen usage of the muscle would have remained

at the previous control rates if the flow had not been occluded, and that the metabolic rate of the muscle is not affected by the circulatory arrest. In the resting isolated gracilis muscle of the dog as well as in the human forearm, the oxygen debt is approximately repaid for different periods of occlusion, but the blood flow debt repayment ranges anywhere from 50 to 200% of that expected. The phenomenon of reactive hyperemia has also been demonstrated in dog's hind limb during exercise. Flow debts then are underpaid or barely repaid. Oxygen debts are entirely or partially repaid depending upon the level of muscle performance and the duration of the ischemic period.

In resting muscle and during light exercise both increased blood flow and oxygen extraction are involved in repayment of the oxygen debts, but during medium and strong exercise the increased blood flow is most important. Since during strong exercise oxygen debts are often not repaid in the presence of a *decreased* arteriovenous oxygen difference, oxygen extraction is the limiting factor.

The metabolites believed to cause exercise hyperemia have also been postulated as eliciting reactive hyperemia. Since reactive hyperemia occurs in sympathectomized and denervated limbs, nervous system control is considered to be not important. The mechanical effect of the lack of pressure in the vessels during the period of occlusion has also been demonstrated to play a role in the dilation (See Chapter 6 for further discussion of myogenic vs. metabolic theories).

Arteriovenous Anastomoses in Skeletal Muscle

The question of whether arteriovenous shunts actually exist as functional units in skeletal muscle is still unsettled. Microscopic studies on the circulation of rat skeletal muscle reveal many communications between small arteries and veins which do not enter the muscle proper. During inactivity, most of the flow is through these anastomoses and not to the muscle fibers. When the rats are bled in small amounts, the muscular arterioles close, and flow in the capillaries ceases, while it persists in the anastomotic channels. That such communications function in man, however, and possess a means of shunt regulation has not been demonstrated. Indirect evidence for the existence of functional anastomoses is the decreased oxygen consumption and increased lactic acid production during vasodilation; such findings would occur if capillary flow were reduced. The disappearance rate of a radioisotope from a depot in skeletal muscle is thought to represent only capillary blood flow and not shunt flow. During emotional stress, total forearm blood flow increases (plethysmograph), but radioisotope clearance from forearm muscle does not change. Such studies have led to the concept of a dual circulation in resting muscle; one circulatory pathway has been referred to as "nutritional" and the other as "nonnutritional."

Collateral Blood Flow

When the major artery to a vascular bed is occluded, smaller vessels arising from arteries above the occlusion supply the area with blood. This collateral flow may prove adequate for all functions. For example, some patients withstand an embolic occlusion of the femoral artery with no adverse effects to the limb. Since collateral arteries function within seconds of arterial occlusion, it is thought that these vessels already exist, but then increase in size following arterial occlusion. In patients with long-standing thrombosis of the aortic bifurcation, aortography often reveals vessels as large as the femoral artery arising from the mesenteric arteries and supplying the limbs. In normal young subjects, plethysmographic studies of calf blood flow show

that mechanical compression of the femoral artery at first decreases flow to one-sixth of the previous resting level, with recovery to normal flow within 6 min despite continued compression. The recovered flow is non-, or only slightly, pulsatile.

Reflex warming of the body and ganglionic blockade increase the collateral blood flow and thus demonstrate that it is mediated in part by the sympathetic nervous system. Exercise of the leg usually increases the collateral flow, but the increase may be due to a rise in systemic blood pressure.

When the external iliac or femoral artery is acutely occluded in dogs or cats, the flow drops to one-third of the previous resting level and then gradually rises to 50 to 100% of normal. In dogs with acute or chronic arterial occlusions, the blood pressure below the obstruction initially falls to low levels, but approaches normal in a few weeks.

FETAL AND NEONATAL CIRCULATION

The placenta serves as the "fetal lung" and access port to the maternal circulation for uptake of nutritional substances and excretion of metabolic wastes. The major circulatory features of the fetus are diagrammed in Figure 2.209. As shown, in the fetus (in contrast to the adult) the right and left ventricles do not operate in series, but largely in parallel because of the interconnections between the right and left sides of the circulation provided by the open flap between the right and the left atria (the foramen ovale) and the patent ductus arteriosus connecting the pulmonary artery and the aorta. Thus, both chambers pump blood into the aorta at different sites, and the lungs are almost entirely bypassed (Fig. 2.209). The placental circulation, supplied by the umbilical arteries, operates in parallel off the fetal circulation and has a relatively low vascular resistance, accepting about 40% of the combined output of the two fetal ventricles (Fig. 2.209).

Most studies on the physiology of the fetal and neonatal circulation have been carried out on sheep and goats. These species differ from man in having two umbilical veins rather than one. They possess cotyledonary placentas and have a relatively smaller brain. Results and interpretations reported in the literature contain unresolved differences in certain details (Dawes, 1968). Some important differences have been resolved by more recent work in the fetus, studied in the uterus some days after insertion of instrumentation (Rudolph and Heymann, 1974). Also, experiments in isolated tissues provide evidence that in fetal heart muscle about 30% of the muscle mass is constituted by myofila-

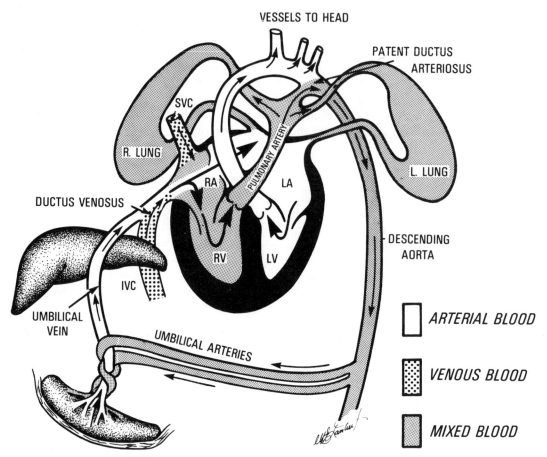

Figure 2.209.   Diagram of normal fetal circulation. For discussion, see text.

ments (in contrast to 60% in the adult), and that isolated fetal muscle can develop less tension than adult myocardium; these findings, together with responses of the intact fetal heart to pressure and volume loading indicate considerably reduced functional reserve compared to the adult heart (Rudolph and Heymann, 1974). In the fetus, the combined output of the left and right ventricle is about equally divided between the placenta and fetal membranes, and the body of the fetus itself. Studies using radioactive microspheres to measure blood flow distribution (see Chapter 14) indicate that there is a very high rate of blood flow in the descending aorta of the fetus (490 ml/kg fetal weight/min) of which nearly half goes to the placenta via the umbilical arteries (Fig. 2.209) (Rudolph, 1970). About two-thirds of the blood returning to the fetal heart enters the inferior vena cava and one-third from the superior vena cava, and about 40% of the inferior vena cava flow traverses at the foramen ovale to the left heart, whereas little flow from the superior vena cava reaches the left side. As discussed below, this pattern of flow favors the supply of more highly oxygenated blood from the placenta to

the left-sided heart chambers. About 90% of the output of the right ventricle traverses the patent ductus arteriosus to reach the descending aorta, and only about 7% traverses the lungs in the intact fetus of the sheep near term, since the lungs are collapsed and the pulmonary vascular resistance is high (Rudolph, 1970).

The blood circulating through the placenta enters with an oxygen saturation of about 58% and returns in the umbilical veins 80% saturated (contrasted with 98% oxygen saturation of maternal arterial blood). The fetal *ductus venosus* carries some of this blood, bypassing the liver, directly into the inferior vena cava, where it joins blood from the lower trunk and extremities (26% saturated) and from the liver. Mixed blood reaching the left atrium has a saturation of 60–65%.

Anatomically, the foramen ovale opens directly off the inferior vena cava so as to favor a flow of blood into the left atrium (Fig. 2.209). There is evidence that the edge of the interatrial septum, the *crista dividens,* divides the inferior vena cava blood into two unequal streams. According to this view the larger stream is composed mainly of umbilical vein blood and is shunted through the foramen ovale into the left atrium. The smaller stream mixes in the right atrium with superior vena cava blood from the upper body and with blood from the heart. This less saturated blood largely passes into the right ventricle and is pumped out the pulmonary artery. Pulmonary arterial resistance is relatively high, the pressure being about 5 mmHg above that in the aorta; consequently, most of the right ventricular output flows through the ductus arteriosus into the aorta. This blood enters the aorta at a point distal to the origins of the arteries to the head and upper limbs, and is thus directed to the posterior body and umbilical arteries (Fig. 2.209). The larger stream of more highly oxygenated blood from the foramen ovale and the blood entering the left atrium from the pulmonary veins mix and are pumped out of the left ventricle into the ascending aorta (Fig. 2.209) to reach the arteries supplying the cerebral circulation, the heart, and the upper extremities.

Fetal hemoglobin has a greater affinity for oxygen than adult hemoglobin so that at equal oxygen tension fetal blood carries more oxygen than maternal blood. Also fetal tissues are more resistant to hypoxia than are adult. These two factors make possible adequate oxygen supply to fetal tissues, despite a relatively low oxygen tension in blood leaving the placenta.

Changes at Birth

With separation from the placental circulation, the peripheral resistance rises and asphyxia develops in the infant. The pressure in the aorta rises, and gasping respiratory movements expand the lungs. Constriction of the muscular umbilical veins squeezes up to 100 ml of blood into the fetal veins ("placental transfusion"), and ventilatory movements aid venous return. With lung expansion, pulmonary blood flow increases as pulmonary vascular resistance falls below one-fifth the in utero value. The major changes that occur at birth include closure of the placental circulation, shutting of the patent foramen ovale, and closure of the patent ductus arteriosus. It has been shown that the vascular smooth muscle of the umbilical artery constricts in response to increased P_{O_2}, decreased temperature, and stretch (Rudolph, 1974), all changes which occur at birth, and hence umbilical arteries markedly contract. Spontaneous respiratory movements begin in the fetus during the last two-thirds of the gestation period (although only a small amount of fluid in the trachea moves), and at birth it appears that tactile stimulation from the skin, decreased temperature, as well as chemoreceptor drive

all stimulate spontaneous respiration. This, in turn, results in opening of the lungs with a marked increase in pulmonary blood flow as the pulmonary vascular resistance falls.

The vascular smooth muscle in the pulmonary arteries responds in an opposite manner to that of the umbilical arteries and patent ductus, a low P_{O_2} causing constriction (in utero) and the increased P_{O_2} at birth producing dilation of the pulmonary arterioles (Rudolph and Heymann, 1974). Increased P_{O_2} causes marked constriction of the smooth muscle in the wall of the *patent ductus arteriosus* by unknown mechanisms, and the ductus becomes more sensitive to this stimulus as gestation advances (Rudolph and Heymann, 1974). Although a clear physiologic role for prostaglandins in regulation of the patent ductus has not been demonstrated, there is some evidence that a vasodilator prostaglandin from the placenta (PGE) helps to maintain patency of the ductus in utero, and local prostacyclin production in the ductal wall could operate to prevent closure after birth (Heymann and Rudolph, 1981). It has also been shown that substances which inhibit prostaglandin synthesis (such as indomethacin or aspirin) can facilitate closure of a patent ductus arteriosus; this approach has been used to close persistent patency of the ductus in some prematurely born human infants (Nadas, 1976). The consequence of all the above events at birth is a marked change in the hemodynamic status of the fetal circulation. The cessation of inflow from the placental circulation together with a marked drop in the pulmonary artery and right ventricular pressures as the lung circulation opens lowers the right atrial pressure, while the increased total peripheral vascular resistance of the infant due primarily to removal of the low vascular resistance of the placental circulation raises the aortic pressure, and this, together with the increased pulmonary blood flow, elevates the left atrial pressure. This results in a higher pressure in the left than in the right atrium which closes the flap of the patent foramen ovale. Since the pulmonary artery pressure drops markedly as the pulmonary vascular resistance falls, the pressure in the aorta exceeds that in the pulmonary artery, and a left to right shunt exists through the patent ductus arteriosus until its functional closure is complete.

This is the transitional (neonatal) state of the circulatory system. Functional closure of the ductus is usually complete within 1 day after birth. Anatomical closure of the *foramen ovale* and *ductus arteriosus* may take several weeks. Normally, however, there is little flow through these channels after the first day or two. With their complete closure the adult separation of pulmonary and systemic vascular systems is accomplished. The only shunt remaining in the adult is that through the bronchial arteries and capillary system supplying the tissue of the lungs, from which a small amount of blood returns to the left heart via pulmonary veins, without having gone through the right heart.

Persistence of a patent ductus arteriosus is a relatively common congenital abnormality. This is usually treated by surgical closure of the persistent channel to eliminate the left to right shunt, although, as mentioned, use of prostaglandin inhibitors is under study. Congenital narrowing of the aorta can occur near the site of the patent ductus arteriosus (coarctation of the aorta) and eventually lead to marked elevation of the blood pressure in the upper body. Persistence of a communication between the left and right atrium causing a left to right shunt of blood (atrial septal defect) also is relatively common, as is congenital narrowing of the pulmonic valve. The most common congenital malformation is ventricular septal defect (a communication between the two pumping chambers), which also results in a left to right shunt of blood.

In summary, the fetal circulation is especially arranged to meet the needs of a rapidly growing organism living in a state of relative hypoxia. These requirements include high cardiac output, partial bypass of the lung circulation, and high placental blood flow. At birth, the fetus abruptly becomes an air-breather, hypoxia disappears, blood-gas exchange occurs in the lungs, and the large placental flow is abolished. As a result, the cardiac output is reduced, pulmonary artery pressure drastically declines, and the two sides of the heart function serially, rather than in parallel. Blood oxygen tension is the chief determinant of the pattern of fetal blood flow distribution and the changes in this distribution which occur at birth.

THE LYMPHATICS

Structure of the Lymphatic System

Lymphatics are present in most tissues in close association with the blood capillaries (Fig. 2.210). One of their functions, to return excess tissue fluid and proteins to the intravascular compartment, is intimately related to the movement of these substances across the capillary endothelium. The role of osmotic pressure and hydrostatic pressure in controlling the flow of fluids out of the vascular system is discussed elsewhere (Chapter 6). In Figure 2.210 the knoblike ends of lymphatic capillaries are shown almost in contact with blood capillaries. The direction of flow of fluids and solutes out of the

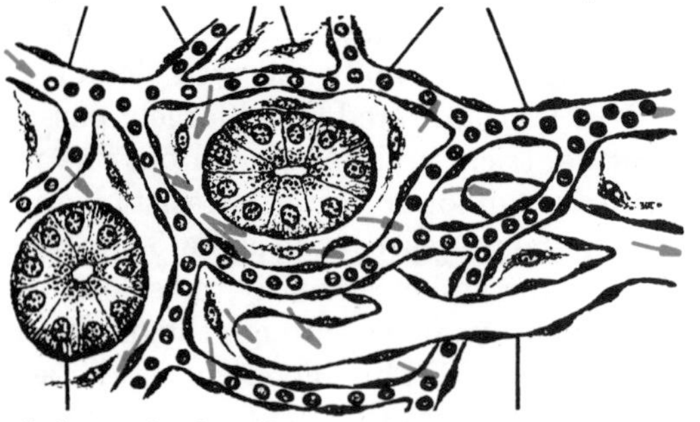

Figure 2.210. Diagram showing the relationship of lymphatic capillaries to blood capillaries and to the tissue fluids around the acini of a gland. Similar relationships exist in most of the organs of the body. *Arrows* indicate the direction of flow of fluid leaving the arterial capillaries, permeating the connective tissue spaces as tissue fluid, and re-entering the blood capillaries on the venous side. The lymphatic capillaries supplement the venous capillaries in the drainage of fluid from the tissues to the circulatory system. [From Copenhaver (1967).]

arterial capillary and into tissue spaces and back into the venous capillaries and into lymphatic capillaries is shown by the *arrows*.

By combining the injection of colloidal materials and electron microscopy, fundamental observations have been made upon the structure and function of the lymphatic capillary (Leak and Burke, 1966, 1968; Cliff and Nicoll, 1970; Leak, 1976). The endothelium of the lymphatic capillaries is similar to that of the blood capillaries. There is no basement membrane. The diameters of the lymphatic capillaries vary widely from dilated terminal capillary bulbs to structures that are either very narrow and slit-like, or dilated capillaries, depending upon their state of function. Pinocytosis has been clearly demonstrated; it is one method of transportation of material from the tissue spaces into the lymph capillaries. The electron micrographic studies of the lymphatic endothelium show that in some areas away from the nucleus, the endothelium is very attentuated, being in some places little more than the two membranes in width. There is considerable regional specialization of terminal lymphatic structure, but in most tissues the blind-ended channels tend to interconnect, and usually there are wide gaps or fenestrae between the cells of the lymphatic walls (Zweifach, 1974). In addition, there are collagenous anchoring filaments attached to lymphatic endothelium and extending out to unite with collagenous bundles in the adjoining tissue. The detail of this structure is shown in Figure 2.211.

When fluid accumulates in the interstitial spaces by transudation (or is injected), one can visualize

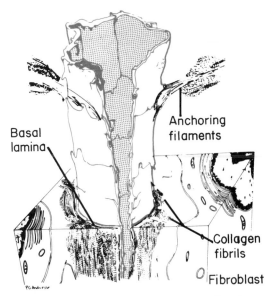

Figure 2.211. A three-dimensional, interpretative diagram of lymphatic capillary that was reconstructed from collated electron micrographs. The three-dimensional relation of the lymphatic capillary to the surrounding connective tissue area is illustrated. The lymphatic anchoring filaments appear to originate from the endothelial cells and extend among collagen bundles, elastic fibers, and cells of the adjoining tissue area, thus providing a firm connection between the lymphatic capillary wall and the surrounding connective tissue. Irregular basal lamina and collagen fibers are as marked. [From Leak and Burke (1968).]

that this fluid will occupy space between the lymphatic capillary and tissue collagenous fibers. Since the fibers attached to the lymphatic endothelium and the tissue collagen are inelastic one would expect the overlapping endothelium to be pulled apart so as to provide large pores through which the edema fluid under greater pressure can flow into the lumen of the lymphatic capillary. The combination of a mechanism to open up the spaces between endothelial cells and an active process of pinocytosis gives two mechanisms by which fluid and its solutes can be rapidly transported from blood capillaries through the interstitial spaces and then into the lymph capillaries. In the bat wing Cliff and Nicoll (1970) have shown that the terminal lymphatic capillary bulbs enter into collecting ducts that possess a contractile wall and valves. These emerge into transport lymphatic channels with very active contractility and valves. The transport lymphatics combine into larger and larger lymphatic vessels propelling the lymph to and through the lymph nodes and toward the great veins. Ultimately major lymphatic vessels are formed in the extremities and within the viscera, and then merge into the right lymphatic and thoracic ducts. From these, lymph pours into the blood stream by way of the right and left subclavian veins.

In the course of flowing from the periphery to the entrance in the blood vessels, the lymph flows through one or more lymph nodes. As the lymphatics approach the node, they break up into finer channels called the afferent lymphatics. These penetrate into the sinuses of the cortex of the lymph node, then through the cortex and the medullary sinuses which are lined by phagocytic cells. One or more large efferent lymphatics formed by confluence of the numerous medullary lymphatics emerge from the hilum of the lymph node and progress onwards to join the major lymphatic vessels.

In the walls of the abdominal cavity the lymphatics are most abundant on the undersurface of the diaphragm, where the greatest lymphatic absorption takes place from the peritoneal cavity. There is also substantial absorption into the lymphatics of the omentum. Lymphatic systems exist in all organs with the exception of the central nervous system and the cornea. Protein is not readily absorbed from the interior of the alveoli into the lymph capillaries because of the poor lymphatic supply, or possibly because it cannot readily penetrate the alveolar wall. Water passes readily into the lung capillaries, however, because of the low hydrostatic pressure within them.

The Flow of Substances between Blood and Lymph

There is a rapid and continual exchange between the intravascular and extravascular compartments by diffusion (Chapter 6). Transcapillary exchange by filtration takes place on an exceptionally large scale without any perturbation to the circulating blood volume or the electrolyte balance; this movement of fluid into and out of the blood capillaries depends on the balance of hydrostatic and osmotic pressures acting across the capillary endothelium. The osmotic force exerted by the circulating plasma protein molecules within the blood capillaries counterbalances the hydrostatic pressure in the capillaries. The capillary wall, however, is not completely impermeable to plasma proteins, and all protein molecules leak through to a certain extent. Nevertheless, they are found in the interstitial fluid and lymph in significantly lower concentration than in plasma (Yoffey and Courtice, 1956). The accumulation of proteins in the interstitial fluid, with an increasing osmotic pressure that would disrupt the balance of forces controlling the exchange of fluid across the capillary membrane, is precluded by the rapid flow of protein molecules directly into the lymphatics, presumably by a combination of pinocytosis and flow through the gaps in the lymphatic capillary endothelium.

The escape of plasma proteins into interstitial fluid varies with the tissue. It has been shown that a quantity equal to roughly 50–100% of the circulating plasma proteins escapes across the blood capillary membrane and reenters the blood through the lymphatics each day (Yoffey and Courtice, 1956).

Composition and Flow of Lymph

The *protein concentration* of lymph is always lower than that of plasma, even though there is considerable variation between lymphatics, depending upon the organ. As oral or intravenous fluid intake is increased, the protein concentration of lymph decreases.

Lymph contains all of the *coagulation factors*, and clots, although less readily than blood plasma. There is a much higher concentration of some coagulation factors in the hepatic lymph in contrast to peripheral lymph, because these factors are made within the liver. Antibodies are found in the lymph, and will be discussed in more detail in Chapter 22. Almost all of the enzymes that are found in plasma are also found in lymph to a lesser extent.

The *electrolyte* concentration in lymph is not essentially different from that of plasma. The total cation concentration is slightly lower in the lymph than in the plasma, and chloride and bicarbonate levels tend to be higher. The direction of these differences in concentration is consistent with the Gibbs-Donnan equilibrium operating on two phases whose concentrations of nondiffusible ions (proteins) differ.

Of the plasma *lipids*, cholesterol and phospholipid, being mainly associated with protein as lipoprotein, are present in the lymph in concentrations which vary with the level of protein in the lymph (Yoffey and Courtice, 1956). Neutral fat in the form of chylomicrons depends on the degree of fat absorption from the gastrointestinal tract. Immediately after meals there are large quantities of lipoproteins and fats in the lymph coming from the gastrointestinal tract. Between meals, the fat in the thoracic duct drops to low levels.

The other constituents of the plasma that are readily diffusible, nonprotein substances, and nonelectrolytes are present in the lymph in concentrations approximately the same as in the plasma.

The lymph contains *cells*. Lymphocytes of all sizes and degrees of maturity are the most numerous. A very rare monocyte and macrophage is found. Platelets are not observed. Red cells, when present, indicate the degree of bleeding into the tissues as a result of injury, or when found in the thoracic duct suggest the presence of intestinal parasites. Granulocytes are found in the lymph draining from areas of infection. Plasma cells are also found in very small numbers. The concentration of lymphocytes in efferent lymph leaving a lymph node is always much greater than that in afferent lymph.

Any condition that increases the outpouring of fluid from the capillaries into the tissues will increase the flow of lymph if there is no obstruction. The factors influencing the flow are the contractility of the lymph vessels, activity of the skeletal muscles, peristalsis, and a differential pressure between interstitial spaces and lymphatics. Presumably, the retention of fluid inside of the lymph capillary is in part due to the higher pressure of the interstitial fluid outside. Propulsion of lymph from one region to another is brought about by the extrinsic and intrinsic forces described earlier. The lymph pressure in the vessels will rise and the lymph will move centrally to a region where the pressure is lower. The numerous valves within the lymphatics prevent retrograde flow of the lymph.

Functions of the Lymphatic System

A most important function is obviously the return of protein, water, and electrolytes from tissue spaces to the blood. The lymphatics are exceptionally important in the absorption of nutrients, particularly fats, from the gastrointestinal tract.

The lymphatics serve also as a transport mechanism to remove red blood cells that have been lost into the tissues as the result of hemorrhage, or bacteria that may have invaded the tissues. When there is an infection in a distant part of the body, the regional lymph node becomes inflamed as a result of the localization of bacteria or toxins carried in the lymph to the gland. The lymph nodes contain a very efficient filtration system in the cortical and medullary sinuses; these are lined with phagocytic cells that engulf bacteria and red cells or other particulate material. The efficacy of the filtration system can be demonstrated by the direct perfusion of pathogenic bacteria into lymphatics that are afferent to a lymph node. Simultaneous culturing of the efferent lymphatic will show it to be sterile. The capability of the lymph node to filter out pathogenic bacteria can be overwhelmed, however.

Conditions That Increase the Lymph Flow

INCREASE IN CAPILLARY PRESSURE

Landis and Gibbon (1933) found that filtration from the capillaries shows a definite increase when the venous pressure rises above 12 to 15 cm of water. The rate of filtration from the capillaries is directly proportional to the increase in venous pressure (Fig. 2.212). In their experiments, when the venous pressure was increased to any given level the filtration rate increased rapidly at first, but gradually slowed and finally ceased. This falling off in the filtration rate they ascribed to a rise in extracapillary pressure due to the fluid accumulation, which opposed the hydrostatic pressure within the blood capillary. The accumulation of extracellular fluid may be expected therefore to be greater in regions that are loose in texture and where the skin is readily stretched. In persons with firm, resistant skin, edema, for the same reasons, is later in appearing and less pronounced.

Increased pressure in the portal vein or in the hepatic veins, produced for example by obstruction, causes increased filtration into the tissues of the abdominal viscera and a great increase in the volume of lymph flowing through the thoracic duct. Increase in arterial pressure, on the other hand, does not increase filtration in animals until the pressure reaches around 300 mmHg.

INCREASE IN CAPILLARY SURFACE AREA

This increases the leakage of protein and fluid. It may follow any change that causes distension of

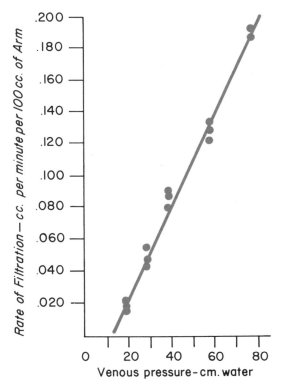

Figure 2.212. Rates of filtration produced during 30 min by venous pressures between 20 and 80 cm of water. [Adapted from Landis and Gibbon (1933).]

the capillary vessels, e.g., (1) increase in capillary pressure; (2) increase in the local temperature; or (3) infusion of fluid.

INCREASE IN CAPILLARY PERMEABILITY

(1) *Rise in temperature* locally may increase capillary permeability. (2) *Capillary poisons* such as peptone increase the flow of lymph from the thoracic duct, probably as a result of their injurious effect upon the abdominal capillaries. Other substances that increase lymph flow in this way are extracts of strawberries, crayfish, mussels, leeches, histamine, and foreign proteins. To what extent this is due to a change in permeability and to what extent it is a consequence of capillary dilation is not known. (3) *Reduced oxygen supply* to the tissues increases lymph flow, probably because of dilation of blood vessels but possibly also through damage to the capillary endothelium.

HYPERTONIC SOLUTIONS

The intravenous injection of a concentrated solution of glucose, sodium sulfate, or sodium chloride causes an increased flow of lymph from the thoracic duct. These substances in concentrated solution may increase permeability of the capillary wall but also exert an osmotic effect that alters the normal equilibrium between the extravascular and intravascular fluids. Water at first enters the plasma from the tissue spaces, particularly of the muscles and subcutaneous tissues of the limbs, which in consequence show a fall in volume; the brain shrinks. Thus, removal of fluid may actually extend to the fluids within the cells which undergo shrinkage; general desiccation of the tissues may result. The blood volume for the time is greatly augmented, the excess fluid for the most part being accommodated in the capacious capillary and venous areas of the abdomen. As the electrolyte injected moves out of the plasma and into the interstitial fluid it carries water with it. The viscera—liver, kidneys, spleen, and intestines—then increase in volume, due not only to the distension of their vascular beds, but more importantly to the great outpouring of fluid as well as of protein that occurs through the capillary walls. This fluid then swells the volume of lymph in the thoracic duct. In this way these substances produce a redistribution of fluid.

The injection of isotonic saline will also increase the passage of protein and fluid from the capillaries and increase lymph flow. This too is probably a consequence of the changes in osmotic and hydrostatic pressures in the capillaries.

INCREASED FUNCTIONAL ACTIVITY

When a gland or muscle becomes active, an increase in lymph flow occurs which starts a little after the commencement of the secretory or contractile response, but is nearly synchronous with the increased metabolism resulting from the activity. The increased flow is ascribed to (1) formation of metabolites that increase the osmotic effect of the tissue fluids and cause more fluid to leave the vessels; (2) vasodilation, with increased capillary pressure and increased fluid and protein leakage.

During rest the flow along the lymph vessels of the muscles and subcutaneous tissues is slight, and the protein content of the lymph is high. During activity the protein concentrations fall, since less transuded water undergoes reabsorption into the blood and more is carried away by the lymph channels. This may occur even though the leakage of protein from the vessels is greater. The contracting muscles exert a pumping effect upon the lymph, driving it along the lymphatic vessels.

MASSAGE AND PASSIVE MOVEMENT

These act to a certain extent like muscular activity. They augment the blood flow, capillary pressure, and capillary surface and so increase lymph formation. The manipulations and movements of the muscles serve to propel the lymph along the lymphatic channels.

EDEMA

Edema is a term applied to an excessive accumulation of fluids in the tissue spaces and is due to a disturbance in the mechanisms of fluid interchange, which has been considered in the preceding pages. Instead of there being a perfect balance between the inward and outward flow of fluid through the capillary membrane, absorption is exceeded by transudation. The particular factor or factors of the mechanism that are disordered are not always clear, and a satisfactory explanation of all forms of edema cannot be given. But from previous discussions it is evident that the following factors will tend to increase the volume of interstitial fluid: (1) reduction in the protein concentration of plasma (edema commences when the serum albumin concentration has fallen below 2.5 g/dl); (2) a general or a local rise in capillary blood pressure; (3) increased permeability of the capillary membrane; (4) increase in the filtering surface as when the capillaries dilate; and (5) obstruction of the lymph channels.

There is a tendency for the accumulation of edema fluid to progress so far and then become stationary, provided that the conditions producing it remain constant, since eventually the interstitial pressure rises to reach a critical value and limits further transudation of fluid. Several factors operate to limit edema formation even under normal conditions. In fact, the venous pressure can rise to 10 mmHg without obvious edema, and some fall in the plasma proteins also can occur without edema formation. Factors preventing edema include: (1) With slight edema, interstitial fluid hydrostatic pressure rises, and there is dilution of interstitial proteins (both effects would counteract further fluid filtration). (2) A rise in venous pressure causes myogenic arteriolar constriction (see autoregulation, Chapter 6), which will lower capillary pressure and decrease capillary surface area. (3) As tissue pressure rises with mild edema, the tethered lymphatics are widely open and, as discussed, there is a marked increase in lymph flow. (4) Finally, as the connective tissue and skin are stretched by edema toward their elastic limit, further edema formation will be restricted.

Since edema is only a symptom of some primary condition it may have a variety of causes, according to the particular disease with which it is associated.

1. Cardiac Edema. In congestive heart failure there is both an increase in extracellular fluid and in salt retention. The volume of this extracellular fluid can be reduced by restriction of the salt intake or by the administration of diuretics that remove both water and salt. Or it can be discharged by improving cardiac action by digitalis.

2. Mechanical Obstruction of Veins. When the main veins leading from a part are obstructed by any mechanism, as in cirrhosis of the liver, thrombosis, etc., an increased transudation of fluid occurs. This is due in part to the rise in intracapillary pressure and an increase in the filtering surface, but the permeability of the capillary wall may also be increased.

3. Edema Due to Renal Disease. In chronic nephritis, edema is not usually pronounced unless the heart is failing; however, in the nephrotic syndrome it is an outstanding feature. In this condition a reduction in plasma protein as a consequence of the loss of protein in the urine leads to the passage of an abnormally large volume of fluid from the capillaries throughout the body. This in turn is probably responsible in some way for renal retention of salt and as a consequence of this, retention of water.

4. Inflammatory Edema. In this type, several factors combine to produce the fluid infiltration of the tissues. Increased capillary pressure occurs, due to dilation of the vessels and local slowing of the blood stream as well as to thrombosis and obstruction of the returning veins. There is an increase in the filtering surface too, and the lymphatics for a variable distance from the inflammatory area are also seriously damaged by the bacterial toxin or other injurious agent, so that a fluid with a high protein content escapes from the vessels. The edema is localized to an area of varying extent surrounding the injured site.

5. Edema Caused by Malnutrition or Toxic Substances. Edema may occur in the anemias or in conditions in which the general nutrition of the body suffers. When the diet is deficient in vitamins, or there is too little fat or protein in the diet, edema may occur, as in beriberi, scurvy, "war edema," or in the faulty nutrition of infants. In animals, edematous conditions have actually been induced by general underfeeding, or by a diet deficient in fat and soluble vitamins, or by one deficient in protein alone. The factors responsible for the increased transudation in these cases are not always clear, but in others there is a marked lowering of plasma protein (more particularly in the albumin), which alone is sufficient to account for the edema.

Certain chemical substances such as arsenic, salts of heavy metals, and the toxins of certain infectious diseases, such as diphtheria, acute nephritis, etc., are known to act as capillary poisons and apparently cause edema in this way. Histamine causes local edema at the point of injection by inducing capillary dilation and increased permeability of the membrane.

6. Chronic Lymphatic Obstruction. Widespread obstruction of the lymph vessels may result from congenital or familial disorders of the lymph vessels. Acquired lymphedema results from obstruction of the lymph channels by cancer, scars, operative removal of lymph nodes, and fibrosis caused by X-ray therapy. It may also follow a low grade lymphangitis from parasites such as filariasis.

Severe edema may occur in the foot and ankle after fracture, dislocations, and extensive hemorrhage due to injury to lymphatic capillaries. It may be months, particularly in older persons, before lymphatic drainage is reestablished. Compression bandages enhance lymphatic drainage.

For the absorption of fluid, the pleural cavities depend upon the lymph channels; the accumulations of fluid may occur here as a result of lymphatic obstruction. The edema associated with carcinoma is due chiefly to the filling of the lymphatic channels with cords of cancer cells, as well as to venous obstruction caused by the pressure of the growth. The tissue fluid in these types of edema has a relatively high concentration of protein.

In the early stages, it cannot be distinguished from any other form of soft pitting edema. On examination of the fluid, the high protein concentration separates it from cardiac and nephritic edema. Ultimately, lymphedema causes fibrosis in the tissues, and in time the tissues become hard and brawny. The skin may be thick and folded.

BIBLIOGRAPHY

AARS, H. The baroreflex in arterial hypertension. *Scand. J. Clin. Lab. Invest.* 35: 97–102, 1975.

ABBOTT, B. C., AND W. F. MOMMAERTS. A study of inotropic mechanisms in the papillary muscle preparation. *J. Gen. Physiol.* 42: 533–551, 1959.

ALEXANDER, R. S. The participation of the venomotor system in pressor reflexes. *Circ. Res.* 2: 405–409, 1954.

ALPERT, N. R., AND M. S. GORDON. Myofibrillar adenosine triphosphatase activity in congestive heart failure. *Am. J. Physiol.* 202: 940–946, 1962.

ASHBURN, W. L., H. R. SCHELBERT, AND J. W. VERBA. Left ventricular ejection fraction—a review of several radionuclide angiographic approaches using the scintillation camera. *Prog. Cardiovasc. Dis.* 20: 267–284, 1978.

ASMUSSEN, E. Similarities and dissimilarities between static and dynamic exercise. *Circ. Res.* 48: I 3–I 10 (Suppl.), 1981.

ASMUSSEN, E., AND M. NIELSEN. The cardiac output in rest and work determined simultaneously by the acetylene and the dye dilution methods. *Acta. Physiol. Scand.* 27: 217–230, 1952.

ATTINGER, E. O. Analysis of pulsatile blood flow. *Adv. Biomed. Eng. Med. Phys.* 1: 1–59, 1968.

AUNG-DIN, R., J. H. MITCHELL, AND J. C. LONGHURST. Reflex alpha-adrenergic coronary vasoconstriction during hindlimb static exercise in dogs. *Circ. Res.* 48: 502–509, 1981.

AYERS, C. R., J. O. DAVIES, F. LIBERMAN, C. C. J. CARPENTER, AND M. BERMAN. The effects of chronic hepatic venous congestion on the metabolism of d,l-aldosterone and daldosterone. *J. Clin. Invest.* 41: 884–895, 1962.

BACHE, R. J., AND F. R. COBB. Effect of maximal coronary vasodilation on transmural myocardial perfusion during tachycardia in the awake dog. *Circ. Res.* 41: 648–653, 1977.

BADEER, H. S. Contractile tension in the myocardium. *Am. Heart J.* 66: 432–434, 1963.

BARGER, A. C., F. P. MULDOWNEY, AND M. R. LUBOWITZ. Role of the kidney in the pathogenesis of congestive heart failure. *Circulation* 20: 273–285, 1959.

BAUEREISEN, E. Herz. In: *Kurzgefasstes Lehrbuch der Physiologie*, edited by W. D. Keidel. Stuttgart: Thieme Verlag, 1970, p. 71–99.

BERNARD, C. Sur les effects de la section de la portion cephalique de grand sympathique. *Compte Rendu des Seances de la Societe de Biologie*, November, 1852, p. 168–170, 1964.

BERNE, R. M., AND R. RUBIO. Coronary circulation. In: *Handbook of Physiology*, Sect. 2, edited by R. M. Berne. Washington DC: American Physiological Society, 1979, vol. 1, p. 873–952.

BERNE, R. M., AND R. RUBIO. Adenine nucleotide metabolism in the heart. *Circ. Res.* 35: 109, 1974.

BERNSTEIN, E. F. (ed.). *Non-Invasive Diagnostic Techniques in Vascular Disease.* St. Louis: C. V. Mosby, 1978.

BEVAN, J. A. Action of lobeline and capsaicine on afferent endings in the pulmonary artery of the cat. *Circ. Res.* 10: 792–797, 1962.

BEVEGARD, B. S., AND J. T. SHEPHERD. Regulation of the circulation during exercise in man. *Physiol. Rev.* 47: 178–213, 1967.

BEZOLD, A. VON, AND L. HIRT. Ueber die physiologischen Wirkungen des essigsauren Veratrins. *Physiol. Lab. Wurzburg. Untersuchungen* 1: 75–156, 1867.

BIRCH, G. E., AND T. WINSOR. *A Primer of Electrocardiography*, 6th ed. Philadelphia: Lea & Febiger, 1972.

BING, R. J. Myocardial metabolism. *Circulation* 12: 635–647, 1955.

BISCOE, T. J. Carotid body: Structure and function. *Physiol. Rev.* 51: 437–495, 1971.

BISHOP, V. S., AND H. L. STONE. Quantitative description of ventricular output curves in conscious dogs. *Circ. Res.* 20: 581–586, 1967.

BLOMQVIST, C. G., S. F. LEWIS, W. F. TAYLOR, AND R. M. GRAHAM. Similarity of the hemodynamic responses to static and dynamic exercise of small muscle groups. *Circ. Res.* 48: I87–I92, 1981.

BLOOR, C. M. Functional significance of the coronary collateral circulation. A. review. *Am. J. Pathol.* 76: 561–586, 1974.

BOERTH, R. C., J. W. COVELL, P. E. POOL, AND J. ROSS, JR. Increased myocardial oxygen consumption and contractile state associated with increased heart rate in dogs. *Circ. Res.* 24: 725–734, 1969.

BOHR, D. F. Vascular smooth muscle. In: *Peripheral Circulation*, edited by P. C. Johnson. New York: John Wiley & Sons, 1978.

BOWDITCH, H. P. Ueber die Eigenthumlichkeiten der Reizbarkeit, welchedie Muskelfasern des Herzens zeigen. Leipzig: *Berichte Math. Phys.* 23: 652, 1871.

BRADY, A. J. Electrophysiology of cardiac muscle. In: *The Mammalian Myocardium*, edited by G. A. Langer, and A. J. Brady. New York: John Wiley & Sons, 1974, p. 135–161.

BRADY, A. J. Mechanical properties of cardiac fibers. In: *Handbook of Physiology*, Sect. 2, edited by R. M. Berne, et al. Washington DC: American Physiological Society, 1979, vol. 1, p. 461–474.

BRAUNWALD, E., E. H. SONNENBLICK, J. ROSS, JR., G. GLICK, AND S. E. EPSTEIN. An analysis of the cardiac response to exercise. *Circ. Res.* 20–21 (Suppl. 1): 44–58, 1967.

BRAUNWALD, E., J. ROSS, JR., AND E. H. SONNENBLICK. *Mechanisms of Contraction of the Normal and Failing Heart*, 2nd ed. Boston: Little Brown, 1976.

BRAUNWALD, E. *Heart Disease. A Textbook of Cardiovascular Medicine.* Philadelphia: W. B. Saunders, 1980.

BRAUNWALD, E., AND J. ROSS, JR. Control of cardiac performance. In: *Handbook of Physiology*, Sect. 2, edited by R. M. Berne. Washington DC: American Physiological Society, 1979, vol. 1, p. 533–580.

BRECHER, G. *Venous Return.* New York: Grune & Stratton, 1956.

BRECHER, G. A., AND P. GALLETTI. Functional anatomy of cardiac pumping. In: *Handbook of Physiology*, Sect. 2: Circulation. Washington, D.C.: American Physiological Society, 1963, vol. 2, p. 759–798.

BROD, J. V. FENCL, Z. HEJL, AND J. JIRKA. Circulatory changes underlying blood pressure elevation during acute emotional stress in normotensive and hypertensive subjects. *Clin. Sci.* 18: 269–279, 1959.

BRONK, D. W., AND G. J. STELLA. Afferent impulses in carotid sinus nerve relation of discharge from single end organs to arterial blood pressure. *J. Cell. Comp. Physiol.* 1: 113–130, 1932.

BROOKS, C. McC., B. F. HOFFMAN, E. E. SUCKLING, AND O. ORIAS. *Excitability of the Heart.* New York: Grune & Stratton, 1955.

BROWN, A. M. Motor innervation of the coronary arteries of the cat. *J. Physiol.* (London) 198: 311–328, 1968.

BROWN, A. M., AND M. C. ANDERSEN. Plasticity of arterial baroreceptors in hypertension states. In: *Disturbances in Neurogenic Control of the Circulation*, edited by D. J. Reis. Bethesda: American Physiological Society, 1981.

BRUNS, D. L. A general theory of the causes of murmurs in the cardiovascular system. *Am. J. Med.* 27: 360–374, 1959.

BRUTSAERT, D. L., N. M. DE CLERCK, M. A. GOETHALS, AND P. R. HOUSMANS. Relaxation of ventricular cardiac muscle. *J. Physiol. (London)* 283: 469–480, 1978.

BUCKBERG, G. D., D. E. FIXLER, J. P. ARCHIE, AND J. I. T. E. HOFFMAN. Experimental subendocardial ischemia in dogs with normal coronary arteries. *Circ. Res.* 30: 67–81, 1972.

BUICH, G. E., AND T. WINSOR. *A Primer of Electrocardiography.* Philadelphia: Lea & Febiger,

BÜLBRING, E., A. F. BRADING, A. W. JONES, AND T. TOMITA (eds.). *Smooth Muscle.* London: Edward Arnold, 1970.

BULLER, A. J. The physiology of skeletal muscle. In: *Neurophysiology*, MTP International Review of Science. Baltimore: University Park Press, 1975, vol. 3, p. 279–302.

BURNS, J. W., J. W. COVELL, AND J. ROSS, JR. Mechanics of isotonic left ventricular contractions. *Am. J. Physiol.* 224: 725–732, 1973.

BURNS, J. W., AND J. W. COVELL. Myocardial oxygen consumption during isotonic and isovolumic contractions in the intact heart. *Am. J. Physiol.* 223: 1491–1497, 1972.

BURTON, A. C. Relation of structure to function of the tissues of the wall of blood vessels. *Physiol. Rev.* 34: 619–642, 1954.

BURTON, A. C. Importance of shape and size of heart. *Am. Heart J.* 54: 801–810, 1957.

BURTON, A. C. *Physiology and Biophysics of the Circulation*, ed. 2. Chicago: Year Book, 1972.

BURTON-OPITZ, R. Venous pressures. *Am. J. Physiol.* 9: 198–206, 1903.

BUTTERWORTH, J. S., M. R. CHASSIN, R. McGRATH, AND E. H. REPPERT. *Cardiac Auscultation—Including Audio-Visual Principles*, 2nd ed. New York: Grune & Stratton, 1960.

CARO, C. G., J. PEDLEY, R. C. SCHROTER, AND W. A. SEED. *The Mechanics of the Circulation.* New York: Oxford University Press, 1979.

CASE, D. B., J. M. WALLACE, H. J. KLEIN, M. A. WEBER, J. E. SEALEY, AND J. H. LARAGH. Possible role of renin in hypertension as suggested by renin sodium profiling and inhibition of converting enzyme. *N. Engl. J. Med.* 296: 641–655, 1977.

CASLEY-SMITH, J. R. Lymph and lymphatics. In: *Microcirculation*, edited by G. Kaley and B. M. Altura. Baltimore: University Park Press, 1977, vol. 1, p. 423–502.

CHAPMAN, J. B. Fluorometric studies of oxidative metabolism in isolated papillary muscle of the rabbit. *J. Gen. Physiol.* 59: 135–154, 1972.

CHATTERJEE, K., AND W. W. PARMLEY. The role of vasodilator therapy in heart failure. *Prog. Cardiovasc. Dis.* 19: 301–325, 1977.

CHIDSEY, C. A., E. BRAUNWALD, A. G. MORROW, AND D. T. MASON. Myocardial norepinephrine concentration in man: effects of reserpine and of congestive heart failure. *N. Engl. J. Med.* 269: 653–658, 1963.

CHIEN, S. Biophysical behavior of red cells in suspension. In: *The Red Blood Cells*, 2nd ed. New York: Academic Press, 1975, vol. 2, p. 1031–1133.

CHIEN, S. Role of the sympathetic nervous system in hemorrhage. *Physiol. Rev.* 47: 214–288, 1967.

CHIEN, S., S. VSAMI, H. M. TAYLOR, J. L. LUNDBERG, AND M. I. GREGERSEN. Effects of hematocrit and plasma proteins on human blood rheology at low shear rates. *J. Appl. Physiol.* 21: 81–87, 1966.

CIBA FOUNDATION SYMPOSIUM 24 (new series). *The Physiological Basis of Starling's Law of the Heart.* Amsterdam: Elsevier/North Holland Biomedical Press, 1974.

CLAUSEN, J. P. Effect of physical training on cardiovascular adjustments to exercise in man. *Physiol. Rev.* 57: 779–815, 1977.

CLAUSEN, J. P. The effects of physical conditioning. A hypothesis concerning circulatory adjustments to exercise. *Scand J Clin Invest.* 24: 305–313, 1969.

CLIFF, W. H., AND P. A. NICOLL. Structure and function of lymphatic vessels of the bat's wing. *Q. J. Exp. Physiol.* 55: 112–121, 1970.

COFFMAN, J. D., AND D. E. GREGG. Reactive hyperemia characteristics of the myocardium. *Am. J. Physiol.* 199: 1143–1149, 1960.

COKELET, G. R., H. J. MEISELMAN, AND D. E. BROOKS. *Erythrocyte Mechanics and Blood Flow.* New York: A. R. Liss, 1980.

COLEMAN, H. N., III. Role of acetylstrophanthidin in augmenting myocardial oxygen consumption: relation of increased O_2 consumption to changes in velocity of contraction. *Circ. Res.* 21: 487–495, 1967.

COLEMAN, T. G., A. W. COWLEY, AND A. C. GUYTON. Experimental hypertension and long-term control of arterial pressure. In: *Cardiovascular Physiology*, edited by A. C. Guyton and C. E. Jones. Baltimore: University Park Press, 1974, p. 259–297.

COLERIDGE, H. M., AND J. C. G. COLERIDGE. Afferent C-fibers and cardiorespiratory chemoreflexes. *Am. Rev. Respirat. Diseases* 115: 251–260, 1977.

COLERIDGE, J. G., J. C. G. COLERIDGE, AND D. G. BAKER. In: *Cardiac Receptors*, edited by R. J. Linden. London: Cambridge University Press, 1978, p. 117–137.

COLERIDGE, J. C. G., AND H. M. COLERIDGE. Chemoreflex regulation of the heart. In: *Handbook of Physiology: The Cardiovascular System*, edited by S. R. Geiger. Bethesda: American Physiological Society, 1979, p. 653–676.

COPENHAVER, W. M. (ed.). *Bailey Textbook of Histology.* Baltimore: Williams & Wilkins, 1967.

COURTICE, F. C. The lymphatic circulation. In: *Structure and Function of the Circulation*, edited by C. J. Schwartz, N. T. Werthessen, and S. Wolf. New York: Plenum Press, 1981, vol. 2, p. 487–602.

COVELL, J. W., J. S. FUHRER, R. C. BOERTH, AND J. ROSS, JR. Production of isotonic contractions in the intact canine left ventricle. *J. Appl. Physiol.* 27: 577–581, 1969.

COVELL, J. W., AND J. ROSS, JR. Nature and significance of alterations in myocardial compliance. *Am. J. Cardiol.* 32: 449–455, 1973.

COVELL, J. W., J. ROSS, JR., E. H. SONNENBLICK, AND E. BRAUNWALD. Comparison of force-velocity relation and ventricular function curve as measures of contractile state of intact heart. *Circ. Res.* 19: 364–372, 1966.

COWLEY, A. M., J. F. LAIRD, AND A. C. GUYTON. Role of baroreceptor reflex in the daily control of arterial pressure and other variables in the dog. *Circ. Res.* 32: 564–576, 1973.

CRAIGE, E. Heart sounds: Phonocardiography; Carotid, apex and jugular venous pulse tracings; and systolic time intervals. In: *Heart Disease, A Textbook of Cardiovascular Medicine*, edited by E. Braunwald. Philadelphia: W. B. Saunders, 1980.

CRANEFIELD, P. F. *The Conduction of the Cardiac Impulse.* New York: Futura Publishing Co., 1975.

CRANEFIELD, P. F., AND A. L. WIT. Cardiac arrhythmias. *Annu. Rev. Physiol.* 41: 459–472, 1979.

DALY, M., J. E. JAMES, AND G. ANGELL. Chemoreflexes in the circulation. Role of the arterial chemoreceptors in the control of the cardiovascular responses to breath-hold diving. In: *The Peripheral Arterial Chemoreceptors*, edited by M. J. Purves. London: Cambridge University Press, 1975, p. 387–405.

DAVIES, R. E. Role of ATP in contraction. In: *The Myocardial Cell: Structure, Function, and Modification by Cardiac Drugs*, edited by S. A. Briller and H. L. Conn. Philadelphia: University of Pennsylvania Press, 1966, p. 157–168.

DAWES, G. S. *Foetal and Neonatal Physiology: A Comparative Study of the Changes at Birth.* Chicago: Yearbook, 1968.

DAWES, G. S., AND J. H. COMROE, JR. Chemoreflexes from the heart and lungs. *Physiol. Rev.* 34: 167–201, 1954.

DeGEEST, H., M. N. LEVY, H. ZIESKE, AND R. I. LIPMAN. Depression of ventricular contractility by stimulation of the vagus nerves. *Circ. Res.* 17: 222–235, 1965.

DINTENFASS, L. *Rheology of Blood in Diagnostic and Preventive Medicine.* London: Butterworths, 1976.

DODGE, A. T., H. SANDLER, W. A. BAXLEY, AND R. R. HAWLEY. Usefulness and limitations of radiographic methods for determining left ventricular volume. *Am. J. Cardiol.* 18: 10–24, 1966.

DOMENECH, R. J., J. I. E. HOFFMAN, M. I. H. NOBLE, K. B. SAUNDERS, J. R. HENSON, AND S. SUBIJANTO. Total and regional coronary blood flow measured by radioactive microspheres in conscious and unanesthetized dogs. *Circ. Res.* 25: 581–596, 1969.

DONALD, D. E. Myocardial performance after excision of the extrinsic cardiac nerves in the dog. *Circ. Res.* 34: 417–424, 1974.

DONALD, D. E., AND J. T. SHEPHERD. Response to exercise in dogs with cardiac denervation. *Am. J. Physiol.* 205: 393–400, 1963.

DOW, P. Estimations of cardiac output and central blood volume by dye dilution. *Physiol. Rev.* 36: 77–102, 1956.

DOWNING, S. E., J. H. MITCHELL, AND A. G. WALLACE. Cardiovascular responses to ischemia, hypoxia and hypercapnia of the central nervous system. *Am. J. Physiol.* 204: 881–887, 1963.

DOWNING, S. E., J. P. REMENSNYDER, AND J. H. MITCHELL. Cardiovascular responses to hypoxic stimulation of the carotid bodies. *Circ. Res.* 10: 676–685, 1962.

DOWNING, S. E., AND E. H. SONNENBLICK. Cardiac muscle mechanics and ventricular performance: force and time parameters. *Am. J. Physiol.* 207: 705–715, 1964.

DOWNEY, J. M., AND E. S. KIRK. Distribution of the coronary blood flow across the canine heart wall during systole. *Circ. Res.* 34: 251–257, 1974.

DUDEL, J., AND W. TRAUTWEIN. Das Aktions potential under Mechanogramm des Herz muskels unter den Einfluss der Dehnung. *Cardiologia* 25: 344–362, 1954.

DULING, B. R., AND B. KLITZMAN. Local control of microvascular function: Role of tissue oxygen supply. *Ann. Rev. Physiol.* 42: 373–382, 1980.

DULING, B. R. Coordination of microcirculatory function with oxygen demand in skeletal muscle. In: *Advances Physiological Science.* vol. 7: *Cardiovascular Physiology. Microcirculation and Capillary Exchange*, edited by A. G. B. Kovach, J. Hamar, L. Szabo. Budapest: Pergamon Press, 1980, p. 1–16.

DUSTAN, H. P., R. C. TARAZI, AND E. L. BRAVO. Physiologic characteristics of hypertension. In: *Hypertension Manual*, edited by J. H. Laragh. New York: Yorke Medical Books, 1974, p. 227–256.

EBASHI, S. Excitation-contraction coupling. *Annu. Rev. Physiol.* 38: 293–313, 1976.

ECKENHOF, J. E., J. H. HAFKENSHIEL, C. M. LANDMESSER, AND M. HARMEL. Cardiac oxygen metabolism and control of the coronary circulation. *Am. J. Physiol.* 149: 634–649, 1947.

EINTHOVEN, W. Weiteres uber das Elektrokardiogramm. *Pflugers Arch.* 122: 517–584, 1908.

ENDO, M. Calcium release from the sarcoplasmic reticulum. *Physiol. Rev.* 57: 71–108, 1977.

ERDMAN, K. A. The active state and the force velocity relationship in cardiac muscle. In: *Heart Failure*, edited by H. Reindell, J. Kevl, and E. Doll. Stuttgart: Thieme Verlag, 1968, p. 133–138.

FABIATO, A., AND F. FABIATO. Calcium and cardiac excitation-contraction coupling. *Annu. Rev. Physiol.* 41: 473–484, 1979.

FAHRAEUS, R., AND T. LINDQVIST. The viscosity of blood in narrow capillary tubes. *Am. J. Physiol.* 96: 562, 1931.

FANBURG, B. L. Myocardial cell failure. In: *The Mammalian Myocardium*, edited by G. A. Langer and A. J. Brady. New York: John Wiley & Sons, 1974, p. 283–306.

FEIGENBAUM, H. *Echocardiography*, 2nd ed. Philadelphia: Lea & Febiger, 1976.

FEIGL, E. O. Parasympathetic control of coronary blood flow in dogs. *Circ. Res.* 25: 509–579, 1969.

FEIGL, E. O. Sympathetic control of coronary circulation. *Circ. Res.* 20: 262–271, 1967.

FENN, W. O. A quantitative comparison between the energy liberated and the work performed by the isolated sartorius muscle of the frog. *J. Physiol. (London)* 58: 175–203, 1923.

FISHMAN, A. P., AND D. W. RICHARDS (eds.). *Circulation of the Blood, Men and Ideas.* New York: Oxford Univ. Press, 1964.

FOLKOW, B., S. MELLANDER, AND O. OBERG. The range of effect of sympathetic vasodilator fibers with regard to consecutive sections of the muscle vessels. *Acta. Physiol. Scand.* 53: 7, 1961.

FOZZARD, H. A. Heart: Excitation-contraction coupling. *Annu. Rev. Physiol.* 39: 201–220, 1977.

FOZZARD, H. A. Conduction of the action potential. In: *Handbook of Physiology*, Sect. 2, edited by R. M. Berne, et al. Baltimore: Williams & Wilkins, 1979, vol. 1, p. 335.

FRANK, O. On the dynamics of cardiac muscle (translated by C. B. Chapman and E. Wasserman). *Am. Heart J.* 58: 282–317, 1959. (Original: Zur Dynamic des Herzmuskels. *Zeitschrift fur Biologie* 32: 370–447, 1895.)

FRANKLIN, J., AND A. DOELP. *Shocktrauma.* New York: St. Martin's Press, 1980.

FRONEK, A., M. COEL, AND E. F. BERNSTEIN. Quantitative ultrasonographic studies of lower extremity flow velocities in health and disease. *Circulation* 53: 957–960, 1976.

FRY, D. L. Acute vascular endothelial changes associated with increased blood velocity gradients. *Circ. Res.* 22: 165–197, 1968.

FUNG, Y. C. Rheology of blood in microvessels. In: *Microcirculation*, edited by G. Kaley and B. Altura. Baltimore: University Park Press, 1977, vol. 1, p. 279–298.

FUNG, Y. C. *Biomechanics. Mechanical Properties of Living Tissues.* New York: Springer-Verlag, 1981.

FUNG, Y. C. Introduction of biophysical aspects of microcirculation. In: *Microcirculation*, edited by G. Kaley and B. M. Altura. Baltimore: University Park Press, 1977, vol. 1, p. 253–254.

GABE, I. T., J. H. GAULT, J. ROSS, JR., D. T. MASON, C. J. MILLS, J. P. SCHILLINGFORD, AND E. BRAUNWALD. Measurement of instantaneous blood flow velocity and pressure in conscious man with a catheter-tip velocity probe. *Circulation* 40: 603–614, 1969.

GALLAGHER, K. P., G. OSAKADA, M. MATSUZAKI, W. S. KEMPER, AND J. ROSS, JR. Myocardial blood flow and function with critical coronary stenosis in exercising dogs. *Am. J. Physiol.* 12(5): H698–H707, 1982.

GANZ, W., K. TAMURA, H. S. MARCUS, R. DONOSO, S. YOSHIDA, AND H. J. C. SWAN. Measurement of coronary sinus blood flow by continuous thermodilution in man. *Circulation* 44: 181–195, 1971.

GAUER, O. H., J. P. HENRY, AND C. BEHN. The regulation of extracellular fluid volume. *Ann. Rev.* 32: 547–595, 1970.

GAULT, J. H., J. ROSS, JR., AND D. T. MASON. Patterns of brachial arterial blood flow in conscious human subjects with and without cardiac dysfunction. *Circulation* 34: 833–848, 1966.

GEDDES, L. A. *The Direct and Indirect Measurement of Blood Pressure.* Chicago: Year Book, 1970.

GEEST, H. DE., M. N. LEVY, AND H. ZIESKE. Reflex effects of cephalic hypoxia, hypercapnia, and ischemia upon ventricular contractility. *Circ. Res.* 17: 349–357, 1965.

GIBBS, C. L., AND J. B. CHAPMAN. Cardiac energetics. In: *Handbook of Physiology*, Sect. 2, edited by R. M. Berne et al. Washington DC: American Physiological Society, 1979, vol. 1, p. 775–804.

GIBBS, C. L., AND W. R. GIBSON. Effect of Quabain on the energy output of rabbit cardiac muscle. *Circ. Res.* 24: 951–967, 1969.

GIRLING, F. Vasomotor effects of electrical stimulation. *Am. J. Physiol.* 170: 131–135, 1952.

GLANTZ, S. A., AND W. W. PARMLEY. Factors which affect the diastolic pressure-volume curve. *Circ. Res.* 42: 171–180, 1978.

GOLDBERGER, E. *Unipolar Lead Electrocardiography.* Philadelphia: Lea & Febiger, 1947.

GOLDBERGER, A. L., AND E. GOLDBERGER. *Clinical Electrocardiography: A Simplified Approach*, 2nd ed. St. Louis: C. V. Mosby, 1981.

GOETZ, K. L., G. C. BOND, AND D. D. BLOXHAM. Atrial receptors and renal function. *Physiol. Rev.* 55: 157–205, 1975.

GOLDBLATT, H. Experimental renal hypertension mechanism of production and maintenance. *Circulation* 17: 642–647, 1958.

GOLDBLATT, H., J. LYNCH, R. F. HANZAL, AND W. W. SUMMERVILLE. Studies in experimental hypertension. I. The production of persistent elevation of systolic blood pressure by means of renal ischemia. *J. Exper. Med.* 59: 347–379, 1934.

GOODMAN, A. H., A. C. GUYTON, R. DRAKE, AND J. H. LOFLIN. A television method for measuring capillary red cell velocities. *J. Appl. Physiol.* 37: 126–130, 1974.

GRAHAM, T. P., JR., J. ROSS, JR., AND J. W. COVELL. Myocardial oxygen consumption in acute experimental cardiac depression. *Circ. Res.* 21: 123–138, 1967.

GRANDE, P. O., P. BORGSTROM, AND S. MELLANDER. On the nature of basal vascular tone in cat skeletal muscle and its dependence on transmural pressure stimuli. *Acta. Physiol. Scand.* 107: 365–376, 1979.

GRANT, R. T. Observations on local arterial reactions in rabbit's ear. *Heart* 15: 257–280, 1930.

GREENSPAN, K., AND R. E. EDMANDS. The inotropic effects

of digitalis. In: *Digitalis*, edited by C. Fisch and B. Surawicz. New York: Grune & Stratton, 1969.

GREGG, D. E. *Coronary Circulation in Health and Disease.* Philadelphia: Lea & Febiger, 1950.

GREGG, D. E. Some problems of the coronary circulation. *Verh. Deutsch Ges. Kreislaufforsch.* 21: 22–38, 1955.

GREGG, D. E., AND L. C. FISHER. Blood supply to the heart. In: *Handbook of Physiology.* Vol. II: Circulation, edited by W. F. Hamilton and P. Dow. Washington, D.C.: American Physiological Society, 1963, p. 1517–1584.

GREGG, D. E., E. M. KHOURI, AND C. R. RAYFORD. Systemic and coronary energetics in the resting unanesthetized dog. *Circ. Res.* 16: 102–113, 1965.

GROSSMAN, W., AND L. P. McLAURIN. Diastolic properties of the left ventricle. *Ann. Int. Med.* 84: 316–326, 1976.

GROSSMAN, W. (ed.). *Cardiac Catheterization and Angiography*, 2nd ed. Philadelphia: Lea & Febiger, 1980.

GUTTMAN, L., AND D. WHITTERIDGE. Effects of bladder distension on automatic mechanisms after spinal cord injuries. *Brain* 70: 361–404, 1947.

GUYTON, A. C. Determination of cardiac output by equating venous return curves with cardiac response curves. *Physiol. Rev.* 35: 123–129, 1955.

GUYTON, A. C. A concept of negative interstitial pressure based on pressures in implanted perforated capsules. *Circ. Res.* 12: 399–414, 1963.

GUYTON, A. C., B. ABERNATHY, J. B. LANGSTON, B. N. KAUFMAN, AND H. M. FAIRCHILD. Relative importance of venous and arterial resistance in controlling venous return and cardiac output. *Am. J. Physiol.* 196: 1008–1014, 1959.

GUYTON, A. C., T. G. COLEMAN, A. W. COWLEY, K. W. SCHEEL, R. D. MANNING, AND R. A. NORMAN. Arterial pressure regulation overriding dominance of the kidneys in long-term regulation and in hypertension. In: *Hypertension Manual*, edited by J. H. Laragh. New York: Yorke Medical Books, 1974, p. 111–134.

GUYTON, A. C., AND C. E. JONES. Central venous pressure: physiological significance and clinical implications. *Am. Heart J.* 86: 431–437, 1973.

GUYTON, A. S., C. E. JONES, AND T. G. COLEMAN. *Circulatory Physiology: Cardiac Output and its Regulation*, 2nd ed. Philadelphia: W. B. Saunders, 1973.

GUYTON, A. C., C. E. JONES, AND T. G. COLEMAN. Graphical analysis of cardiac output regulation. In: *Circulatory Physiology: Cardiac Output and its Regulation*, 2nd ed. Philadelphia: W. B. Saunders, 1973, ch. 14, p. 237–249.

HADDY, F. J., AND J. B. SCOTT. Active hyperemia, reactive hyperemia and autoregulation of blood flow. In: *Microcirculation*, edited by G. Kaley and B. M. Altura. Baltimore: University Park Press, 1977, vol. 2.

HALES, S. Statical essays. *Haemastatics*, vol. 2. London: Innys and Manby, 1733.

HANSON, J., AND J. LONG. Molecular basis of contractility in muscle. *Br. Med. Bull.* 21: 264–271, 1965.

HARLAN, J. C., E. E. SMITH, AND T. Q. RICHARDSON. Pressure-volume curves of systemic and pulmonary circuit. *Am. J. Physiol.* 213: 1499–1503, 1967.

HARRIS, C. C. *A Primer of Cardiac Arrhythmias: A Self-Instructional Program.* St. Louis: C. V. Mosby, 1979.

HARRIS, P., AND L. H. OPIE (eds.). *Calcium and the Heart.* New York: Academic Press, 1971.

HARVEY, W. *Exercitatio Anatomica de Motu Cordis et Sanguinis in Animalibus* (translated by C. D. Leake). Springfield, IL: Charles C Thomas, 1928.

HECHT, H. H. Electrical disorders of cardiac fibers. In: *Cardiovascular Disorders*, edited by A. N. Brest and S. J. Moyer. Philadelphia: F. A. Davis, 1968, p. 79–93.

HECHT, H. H., C. E. KOSSMANN, R. W. CHILDERS, R. LANGENDORF, M. LEO, K. M. ROSEN, R. D. PRUITT, R. C. TRUEX, H. N. UGLEY, AND T. B. WATT, JR. Atrioventricular and intraventricular conduction. *Am. J. Cardiol.* 31: 232–244, 1973.

HELVIG, E. B., AND F. K. MOSTOFI (eds.). *The Skin.* New York: R. E. Kreiger, 1980.

HENRY, P. D., G. G. AHUMADA, W. F. FRIEDMAN, AND B. E. SOBEL. Simultaneously measured isometric tension and ATP hydrolysis in glycerinated fibers from normal and hypertrophied rabbit heart. *Circ. Res.* 31: 740–749, 1972.

HERING, H. E. *Die Karotidssinus Reflexe auf Herz und Fegasse.* Leipzig: D. Steinkopff, 1927.

HESS, D. S., AND R. J. BACHE. Transmural distribution of myocardial blood flow during systole in the awake dog. *Circ. Res.* 38: 5–15, 1976.

HEYMANN, M. A., H. S. IWAMOTO, AND A. M. RUDOLPH. Factors affecting changes in the neonatal systemic circulation. *Annu. Rev. Physiol.* 43: 371–383, 1981.

HEYMANS, C., AND E. NEIL. *Reflexogenic Areas of the Cardiovascular System.* London: Churchill, 1958.

HEYMANS, C. J., J. J. BOUCKAERT, AND P. REGNIERS. *Le Sinus Carotidien.* Paris: Doin, 1973.

HEYMANS, C., J. J. BOUCKAERT, AND L. DAUTREBANDE. Au sujet du mechanisme de la bradycardie provoquee par la nicotine la lobeline le cyanure le sulfure de sodium, les nitrites et al morphine et de la bradycardie asphyxique. *Arch. Intern. Pharmcodyn.* 41: 261–289, 1931.

HIGGINS, C. B., S. E. VATNER, E. L. ECKBERG, AND E. BRAUNWALD. Alterations in the baroreceptor reflex in conscious dogs with heart failure. *J. Clin. Invest.* 51: 715–724, 1972.

HILL, A. V. Heat of shortening and dynamic constants of muscle. *Proc. R. Soc. London (Biol.), Ser. B.* 126: 136–195, 1939.

HILTON, S. M., AND K. M. SPYER. Central nervous regulation of vascular resistance. *Ann. Rev. Physiol.* 42: 399–411, 1980.

HINSHAW, L. B. Arterial and venous pressure-resistance relationships in perfused leg and intestine. *Am. J. Physiol.* 203: 271–274, 1962.

HODGKIN, A. L. The ionic basis of nervous conduction. *Science* 145: 1148–1154, 1964.

HOFFMAN, B. F. Effects of digitalis on electrical activity of cardiac fibers. In: *Digitalis: Mechanisms of the Inotropic Effect of Digitalis*, edited by C. Fisch and B. Surawicz. New York: Grune & Stratton, 1969, p. 93–109.

HOFFMAN, B. F. Physiology of atrioventricular transmission. *Circulation* 24: 506–517, 1961.

HOFFMAN, B. F., E. BINDLER, AND E. E. SUCKLING. Postextrasystolic potentiation of contraction in cardiac muscle. *Am. J. Physiol.* 185: 95–102, 1956.

HOFFMAN, B. F., AND P. F. CRANEFIELD. *Electrophysiology of the Heart.* New York: McGraw-Hill Book Company, 1960.

HOFFMAN, B. F., AND P. F. CRANEFIELD. The physiological basis of cardiac arrhythmias. *Am. J. Med.* 37: 670–684, 1964.

HOFFMAN, B. F., P. F. CRANEFIELD, AND A. G. WALLACE. Physiologic basis of cardiac arrhythmias. *Mod. Concepts Cardiovasc. Dis.* 35: 103–108, 1966.

HOFFMAN, J. T. E., AND G. D. BUCKBERG. The myocardial supply : demand ratio—a critical review. *Am. J. Cardiol.* 41: 327–332, 1978.

HOLT, J. P., E. A. RHODE, AND H. KINES. Ventricular volumes and body weight in mammals. *Am. J. Physiol.* 215: 704–715, 1968.

HONIG, C. R., AND J. BOURDEAU-MARTINI. Role of O_2 in control of the coronary capillary reserve. In: *Current Topics in Coronary Research*, edited by C. M. Bloor and R. A. Olsson. New York: Plenum Press, 1975, p. 55–71.

HUXLEY, A. F. Excitation and conduction in nerve: quantitative analysis. *Science* 145: 1154–1159, 1964.

HUXLEY, A. F. Muscular contraction. *J. Physiol. (London)* 243: 1–43, 1974.

HUXLEY, A. F. Local activation of muscle. *NY Ann. Acad. Sci.* 81: 446–452, 1959.

JAMES, T. Morphology of the human atrioventricular node with remarks pertinent to its electrophysiology. *Am. Heart J.* 62: 756–771, 1961.

JAMES, T. N. Anatomy of the conduction system of the heart. In: *The Heart*, 2nd ed., edited by J. W. HURST and R. B. LOGUE. New York: McGraw-Hill, 1970.

JAMES, T. N., AND L. SHERF. Specialized tissues and preferential conduction in the atria of the heart. *Am. J. Cardiol.* 28: 414–427, 1971.

JAMES, T. N., AND L. SHERF. Ultrastructure of myocardial cells. *Am. J. Cardiol.* 22: 289–416, 1968.

JAMES, T. N., L. SHERF, R. C. SCHLANT, AND M. E. SILVERMAN. Anatomy of the Heart. In: *The Heart*, edited by J. W. Hurst. New York: McGraw-Hill Book Co., 1982, p. 22–74.

JENSEN, C. R., AND G. FISHER. *Scientific Basis of Athletic Conditioning.* Philadelphia: Lea & Febiger, 1978.

JEWELL, B. R. A re-examination of the influence of muscle length on myocardial performance. *Circ. Res.* 40: 221–230, 1977.

JOHANSSON, B. Vascular smooth muscle reactivity. *Ann. Rev. Physiol.* 43: 355–370, 1981.

JOHNSON, E. A. Force-interval relationship of cardiac muscle. In: *Handbook of Physiology*, Sect. 2, edited by R. M. Berne, et al. Washington DC: American Physiological Society, 1979, vol. 1, p. 475–496.

JOHNSON, P. C. Review of previous studies and current theories of autoregulation. *Circ. Res.* 15(Suppl 2): 2–9, 1964.

JOHNSON, P. C. The role of intravascular pressure in regulation of the microcirculation. In: *Advances in Physiological Science.* Vol. 7: *Cardiovascular Physiology. Microcirculation and Capillary Exchange*, edited by A. G. B. Kovach, J. Hamar, L. Szabo. Budapest: Pergamon Press, 1980, p. 17–34.

JONES, D. R., AND W. J. PURVES. The carotid body in the duck and the consequences of its denervation upon the cardiac responses to immersion. *J. Physiol. (London)* 211: 279–294, 1970.

JOSEPHSON, M. E., AND S. F. SEIDES. *Clinical Cardiac Electrophysiology Techniques and Interpretations.* Philadelphia: Lea & Febiger, 1979.

JULIAN, F. J., AND M. R. SOLLING. Sarcomere length-tension relations in living rat papillary muscle. *Circ. Res.* 37: 299–308, 1975.

KAPLAN, N. M. Systemic hypertension: mechanisms and diagnosis. In: *Heart Disease. A Textbook of Cardiovascular Medicine*, edited by E. Braunwald. Philadelphia: W. B. Saunders Co., 1980, vol. 2.

KARNOVSKY, M. J. The ultrastructural basis of transcapillary exchanges. *J. Gen. Physiol.* 52: 64s–95s, 1968.

KATZ, A. M. Contractile proteins of the heart. *Physiol. Rev.* 50: 63–158, 1970.

KATZ, A. M. *Physiology of the Heart.* New York: Raven Press, 1977, ch. 6, 7, 9.

KATZ, A. M. Congestive heart failure: role of altered myocardial cellular control. *N. Engl. J. Med.* 293: 1184–1191, 1976.

KAUFMANN, R., T. BAYER, T. FURNISS, H. KRAUSE, AND H. TRITTHART. Calcium-movement controlling cardiac contractility. II. Analog computation of cardiac excitation-contraction coupling on the basis of calcium kinetics in a multicompartment model. *J. Mol. Cell. Cardiol.* 6: 543–559, 1974.

KETY, S. Theory of blood-tissue exchange and its application to measurement of blood flow. *Methods Med. Res.* 8: 223–227, 1960.

KEZDI, P. Resetting of carotid sinus in experimental renal hypertension. In: *Baroreceptors and Hypertension*, edited by P. Kezdi, New York: Pergamon Press, 1967, p. 301–308.

KHOURI, E. M., D. E. GREGG, AND C. R. RAYFORD. Effect of exercise on cardiac output, left coronary flow and myocardial metabolism in the unanesthetized dog. *Circ. Res.* 17: 427–437, 1965.

KIRK, E. S., AND C. R. HONIG. An experimental and theoretical analysis of myocardial tissue pressure. *Am. J. Physiol.* 125: 361–367, 1964.

KIRK, E. S., C. W. URSCHEL, AND E. H. SONNENBLICK. Problems in cardiac performance: regulation of coronary blood flow and the physiology of heart failure. In: MTP International Review of Science. Vol. I: *Physiology.* Baltimore: University Park Press, 1974, p. 299–334.

KLITZMAN, B., AND P. C. JOHNSON. Capillary network geometry and red cell distribution in hamster cremaster mus-

cle. *Am. J. Physiol.* 11(2): H211–H219, 1982.

KLOCKE, F. J., AND A. K. ELLIS. Control of coronary blood flow. *Annu. Rev. Med.* 31: 489–508, 1980.

KLOCKE, F. J., D. R. ROSING, AND D. E. PITTMAN. Inert gas measurements of coronary blood flow. *Am. J. Cardiol.* 23: 548–555, 1969.

KLOCKE, F. J., E. BRAUNWALD, AND J. ROSS, JR. Oxygen cost of electrical activation of the heart. *Circ. Res.* 18: 357–365, 1966.

KLOCKE, F. J., G. A. KAISER, J. ROSS, JR., AND E. BRAUNWALD. An intrinsic adrenergic vasodilator mechanism in the coronary vascular bed of the dog. *Circ. Res.* 16: 376–382, 1965.

KOCH-WESER, J., AND J. R. BLINKS. The influence of the interval between beats on myocardial contractility. *Pharmacol. Rev.* 15: 601–652, 1963.

KORNER, P. I. Control of blood flow to special vascular areas: brain, kidney, muscle, skin, liver, and intestine. In: *Cardiovascular Physiology*, edited by A. C. Guyton and C. E. Jones. Baltimore: University Park Press, 1974, p. 123–162.

KRAYER, O. The history of the Bezold-Jarisch effects. *Archiv. Exp. Pathol. Pharmakol.* 240: 361–368, 1961.

KROGH, A. *The Anatomy and Physiology of the Capillaries*, reprint edition. New York: Habner, 1959.

KROGH, A. The number and distribution of capillaries in muscles with calculations of the oxygen pressure head necessary for supplying the tissue. *J. Physiol. (London)* 52: 409–415, 1919.

KUNZE, D. L., AND A. M. BROWN. Sodium sensitivity of baroreceptors: reflex effects on blood pressure and fluid volumes in the cat. *Circ. Res.* 42: 714–720, 1978.

LANDIS, E. M. The capillary circulation. In: *Circulation of Blood: Men and Ideas*, edited by A. P. Fishman and D. W. Richards. New York: Oxford University Press, 1964.

LANDIS, E. M., AND J. H. GIBBON, JR. Effects of temperature and of tissue pressure on movement of fluid through human capillary wall. *J. Clin. Invest.* 12: 105–138, 1933.

LANDIS, E. M., AND J. R. PAPPENHEIMER. Exchange of substances through the capillary walls. In: *Handbook of Physiology.* Sect. 2: Circulation. Washington DC: American Physiological Society, 1963, vol. 2, p. 961–1034.

LANGER, G. A., AND A. J. BRADY. *The Mammalian Myocardium.* New York: John Wiley & Sons, 1974.

LANGER, G. A. The intrinsic control of myocardial contraction-ionic factors. *N. Engl J. Med.* 285: 1065–1071, 1971.

LARAGH, J. H. The biochemical regulation of blood pressure: pathogenetic and clinical implications. In: *Biochemical Regulation of Blood Pressure*, edited by R. L. Soffer. New York: John Wiley and Sons, 1981, p. 393–410.

LEAK, L. V., AND J. F. BURKE. Fine structure of the lymphatic capillary and the adjoining connective tissue area. *Am. J. Anat.* 118: 785–809, 1966.

LEAK, L. V., AND J. F. BURKE. Ultrastructural studies on the lymphatic anchoring filaments. *J. Cell Biol.* 36: 129–149, 1968.

LEAK, L. V. The structure of lymphatic capillaries in lymph formation. *Fed. Proc.* 35: 1863–1871, 1976.

LEFER, A. M., AND J. MARTIN. Relationship of plasma peptides to the myocardial depressant factor in hemorrhagic shock in cats. *Circ. Res.* 26: 59–69, 1970.

LEGATO, M. J. *The Myocardial Cell for the Clinical Cardiologist.* Mount Kisco: Futura Publishing Co., 1973.

LEGATO, M. J. The correlation of ultrastructure and function in the mammalian myocardial cell. *Prog. Cardiovasc. Dis.* 11: 391–409, 1969.

LEVY, M. N., N. G. MIL, AND H. ZIESKE. Functional distribution of the peripheral cardiac sympathetic pathways. *Circ. Res.* 19: 650–661, 1966.

LEVY, M. N. The cardiac and vascular factors that determine systemic blood flow. *Circ. Res.* 44: 739–746, 1979.

LEW, E. A. High blood pressure, other risk factors. The insurance viewpoint. In: *Hypertension Manual*, edited by J. H. Laragh. New York: Yorke Medical Books, 1974, p. 43–70.

LEWINTER, M. M., R. ENGLER, AND R. S. PAVELEC. Time-dependent shifts of the left ventricular diastolic filling relationship in conscious dogs. *Circ. Res.* 45: 641–653, 1979.

LEWIS, D. W. Phonocardiography. In: *Handbook of Physiology*, edited by W. F. Hamilton and P. Dow. Washington DC: American Physiological Society, 1962, sect. 2, vol. 1, p. 695–734.

LEWIS, T. *The Blood Vessels of the Human Skin and Their Responses*. London: Shaw, 1927.

LIPMAN, B. S., AND E. MASSIE. *Clinical Scalar Electrocardiography*, 4th ed. Chicago: Year Book, 1959.

LIPMAN, B. S., AND E. MASSIE. *Clinical Scalar Electrocardiography*, 6th ed. Chicago: Year Book Medical Publications, 1972.

LUNDGREN, O., AND M. JODAL. Regional blood flow. *Annu. Rev. Physiol.* 37: 395–414, 1975.

LUPU, A. N., M. H. MAXWELL, J. KAUFMAN, AND F. N. WHITE. Experimental unilateral renal artery constriction in the dog. *Circ. Res.* 30: 567–574, 1972.

MacGREGOR, D. C., J. W. COVELL, F. MAHLER, R. B. DILLY, AND ROSS, J., JR. Relations between afterload, stroke volume and descending limb of Starling's curve. *Am. J. Physiol.* 227: 884–890, 1974.

MacLEOD, R. D. M., AND M. J. SCOTT. The heart rate responses to carotid body chemoreceptor stimulation in the cat. *J. Physiol (London)* 175: 193–202, 1964.

McDONALD, D. A. *Blood Flow in Arteries*. Baltimore: Williams & Wilkins, 1974.

McKEEVER, W. P., D. E. GREGG, AND P. C. CANNEY. Oxygen uptake of the nonworking left ventricle. *Circ. Res.* 6: 612–623, 1958.

McNUTT, N. S., AND D. W. FAWCETT. Myocardial ultrastructure. In: *The Mammalian Myocardium*, edited by G. A. Langer and A. J. Brady. New York: John Wiley & Sons, 1974, p. 1–49.

McNUTT, N. S., AND R. S. WEINSTEIN. Membrane ultrastructure of mammalian intercellular functions. *Prog. Biophys. Mol. Biol.* 2: 45, 1973.

MAHLER, F., J. W. COVELL, AND J. ROSS, JR. Systolic pressure-diameter relations in the normal conscious dog. *Cardiovasc. Res.* 9: 447–455, 1975.

MARRIOTT, H. J. L., AND R. J. MYERBURG. Cardiac arrhythmias and conduction disturbances. In: *The Heart*, edited by J. W. Hurst. New York: McGraw-Hill, 1982, p. 519–556.

MARTINI, J., AND G. R. HONIG. Direct measurement of intercapillary distance in beating rat heart in situ under various conditions of O_2 supply. *Microvasc. Res.* 1: 244–256, 1969.

MASERI, A., G. A. KLASSEN, AND W. LESCH, M (eds.). *Primary and Secondary Angina Pectoris*. New York: Grune & Stratton, 1978.

MASON, D. T., GABE, I. T., GAULT, J. H., BRAUNWALD, E., ROSS, J. Jr., SHILLINGFORD, J. P. Applications of the catheter-tip electromagnetic velocity probe in the study of the central circulation in man. *Am. J. Med.* 49: 465–471, 1970.

MEERSON, F. Z. The myocardium in hyperfunction, hypertrophy and heart failure. *Circ. Res.* 25(suppl 2): 1–163, 1969.

MEHMEL, H. C., B. STOCKINS, K. RUFFMANN, K. VON OLSHAUSEN, G. SCHULER, AND W. KUBLER. The linearity of the end-systolic pressure-volume relationship in man and its sensitivity for assessment of left ventricular function. *Circulation* 63: 1216–1222, 1981.

MELLANDER, S., AND B. JOHANSSON. Control of resistance, exchanges, and capacitance functions in the peripheral circulation. *Pharmacol. Rev.* 20: 117–196, 1968.

MILLARD, R. W., C. B. HIGGINS, D. FRANKLIN, AND S. VATNER. Regulation of renal circulation during severe exercise in normal dogs and dogs with experimental heart failure. *Circ. Res.* 31: 881–888, 1972.

MILLS, L. J., AND J. H. MOYER. *Shock and Hypotension: Pathogenesis and Treatment*. New York: Grune & Stratton, 1965.

MILNOR, W. R. Pulmonary circulation. In: *Medical Physiology*, edited by V. B. Mountcastle. St. Louis: C. V. Mosby, 1968, vol. 1, p. 24–34, 209–220.

MILNOR, W. R. Arterial impedance as ventricular afterload. *Circ. Res.* 36: 565–570, 1975.

MIRSKY, I. Basic terminology and formulae for left ventricular wall stress. In: *Cardiac Mechanics: Physiological, Clinical and Mathematical Considerations*, edited by I. Mirsky, et al. New York: John Wiley & Sons, 1974, p. 3–10.

MIRSKY, I., D. N. GHISTA, AND H. SANDLER (eds.). In: *Cardiac Mechanics: Physiological, Clinical and Mathematical Considerations*. New York: John Wiley & Sons, 1974.

MITCHELL, J., AND W. SHAPIRO. Atrial function and the hemodynamic consequences of atrial fibrillation in man. *Am. J. Cardiol.* 23: 556–557, 1969.

MITCHELL, J. H., J. P. GILMORE, AND S. J. SARNOFF. The transport function of the atrium. Factors influencing the relation between mean left atrial pressure and left ventricular end-diastolic pressure. *Am. J. Cardiol.* 9: 237–247, 1962.

MITCHELL, J. H., AND K. WILDENTHAL. Static (isometric) exercise and the heart: physiological and clinical considerations. *Annu. Rev. Med.* 25: 369–381, 1974.

MONROE, R. B., AND G. N. FRENCH. Left ventricular pressure-volume relationships and myocardial oxygen consumption in the isolated heart. *Circ. Res.* 9: 362–374, 1961.

MONROE, R. G., W. J. GAMBLE, C. G. LAFARGE, A. E. KUMAR, AND F. J. MANASEK. Left ventricular performance at high end-diastolic pressures isolated, perfused dog hearts. *Circ. Res.* 26: 85–99, 1970.

MONROE, R. G., W. J. GAMBLE, C. G. LAFARGE, AND S. F. VATNER. Homeometric autoregulation. In: *The Physiological Basis of Starling's Law of the Heart*, Ciba Foundation Symposium No. 27. Amsterdam: Elsevier/North Holland Biomedical Press, 1974, p. 257–277.

MORGAN, H. E., AND J. R. NEELY. Metabolic regulation and myocardial function. In: *The Heart*, edited by J. W. Hurst. New York: McGraw-Hill Book Co., 1982, p. 128–142.

MOSHER, P., J. ROSS, JR., P. A. McFATE, AND R. F. SHAW. Control of coronary blood flow by an autoregulatory mechanism. *Circ. Res.* 14: 250–259, 1964.

MURPHY, Q. The influence of the accelerator nerves on the basal heart rate of the dog. *Am. J. Physiol.* 137: 727–730, 1942.

NADAS, A. S. Patent ductus revisited. *N. Engl. J. Med.* 295: 563–565, 1976.

NEELY, J. R., AND H. E. MORGAN. Myocardial utilization of carbohydrate and lipids. *Prog. Cardiovasc. Dis.* 15: 289–329, 1972.

NEIL, E., AND A. HOWE. Arterial chemoreceptors. In: *Handbook of Sensory Physiology*, edited by E. Neil. Vol. 3: *Enteroceptors*. Berlin and New York: Springer-Verlag, 1973, p. 47–80.

NETTER, F. H. Heart. In: *The Ciba Collection of Medical Illustrations*, edited by F. F. Yonkman. New York: Ciba Publications, 1969, vol. 5.

NICOLL, P. A., AND T. A. CORTESE, JR. The physiology of skin. *Annu. Rev. Physiol.* 34: 177–203, 1972.

NIH Report of Hypertension Task Force. *General Summary and Recommendations* 1: 79–1623, 1979.

NOBLE D. *The Initiation of the Heartbeat*. New York: Oxford University Press, 1979.

NOBLE, M. M., AND G. H. POLLACK. Controversies in cardiovascular research: molecular mechanisms of contraction. *Circ. Res.* 40: 333–342, 1977.

NOBLE, D. Application of Hodgkin-Huxley equations to excitable tissues. *Physiol. Rev.* 46: 1–50, 1966.

NOORDERGRAAF, A. *Circulatory System Dynamics*. New York: Academic Press, 1979.

OBERG, B. Overall cardiovascular regulation. *Ann. Rev. Physiol.* 38: 537–570, 1976.

OLSSON, R. A. Local factors regulating cardiac and skeletal muscle blood flow. *Ann. Rev. Physiol.* 43: 385–395, 1981.

O'ROURKE, M. F., AND M. G. TAYLOR. Vascular impedance of the femoral bed. *Circ. Res.* 18: 126–139, 1966.

PATTERSON, S. W., H. PIPER, AND E. H. STARLING. The regulation of the heartbeat. *J. Physiol. (London)* 48: 465–513, 1914.

PELLETIER, C. L. Circulatory responses to graded stimulation of carotid chemoreceptors in the dog. *Circ. Res.* 31: 431–443, 1972.

PEPINE, C. J., W. W. NICHOLS, AND C. R. CONTI. Aortic input impedance in heart failure. *Circulation* 58: 460–465, 1978.

PETERSON, D. F., AND A. M. BROWN. Pressor reflexes produced by stimulation of afferent fibers in the cardiac sympathetic nerves of the cat. *Circ. Res.* 28: 605–610, 1971.

POISEUILLE, J. L. M. Recherches experimentales sur le mouvement des liquides dans les tubes de tres petits diametres. *Paris: Mem. Acad. Sci.* 9: 433–544, 1846.

POOL, P. E., AND E. H. SONNENBLICK. The mechanochemistry of cardiac muscle. *J. Gen. Physiol.* 50: 951–965, 1967.

POULEUR, H., J. W. COVELL, AND J. ROSS, JR. Effects of nitroprusside on venous return and central blood volume in the absence and presence of acute heart failure. *Circulation* 61: 328–337, 1980.

POULEUR, H., J. W. COVELL, AND J. ROSS, JR. Effects of alterations in aortic input impendance on the force-velocity-length relationships in the intact canine heart. *Circ. Res.* 45: 126–136, 1979.

RACKLEY, C. E., V. S. BEHAR, R. E. WHALEN, AND H. D. McINTOSH. Biplane cineangiographic determinations of left ventricular function: pressure-volume relationships. *Am. Heart J.* 74: 766–799, 1967.

RAPELA, C. E., H. D. GREEN, AND A. B. DENNISON, JR. Baroreceptor reflexes and autoregulation of cerebral blood flow in the dog. *Circ. Res.* 21: 559–568, 1967.

REUTER, H. Exchange of calcium ions in the mammalian myocardium: mechanisms and physiological significance. *Circ. Res.* 34: 599–605, 1974.

REUTER, H. Properties of two inward membrane currents in the heart. *Annu. Rev. Physiol.* 41: 413–424, 1979.

ROBINSON, S. Physiology of muscular exercise. In: *Medical Physiology*, edited by V. Mountcastle. St. Louis: C. V. Mosby, 1974, vol. 2, p. 1273–1304.

ROOKE, G. A., AND E. O. FEIGL. Work as a correlate of canine left ventricular oxygen consumption and the problem of catecholamine oxygen wasting. *Circ. Res.* 50: 273–286, 1982.

ROSENBLUM, W. I. Viscosity: in vitro versus in vivo. In: *Microcirculation*, edited by G. Kaley and B. M. Altura. Baltimore: University Park Press, 1977, vol. 1, p. 325–334.

ROSS, J., JR., C. J. FRAHM, AND E. BRAUNWALD. The influence of the carotid baroreceptors and vasoactive drugs on systemic vascular volume and venous distensibility. *Circ. Res.* 9: 75–82, 1961.

ROSS, J., JR., C. J. FRAHM, AND E. BRAUNWALD. Influence of intracardiac baroreceptors on venous return, systemic vascular volume and peripheral resistance. *J. Clin. Invest.* 40: 563–572, 1961a.

ROSS, J., JR., AND E. BRAUNWALD. Studies on Starling's law of the heart. IX. The effects of impeding venous return on performances of the normal and failing left ventricle. *Circulation* 30: 719–727, 1964a.

ROSS, J., JR., AND E. BRAUNWALD. The study of left ventricular function in man by increasing resistance to ventricular ejection with angiotensin. *Circulation* 29: 739–749, 1964b.

ROSS, J., JR., J. W. LINHART, AND E. BRAUNWALD. Effects of changing heart rate by electrical stimulation of the right atrium in man: studies at rest, during muscular exercise, and with isoproterenol. *Circulation* 32: 549–558, 1965a.

ROSS, J., JR., E. H. SONNENBLICK, G. A. KAISER, P. L. FROMMER, AND E. BRAUNWALD. Electroaugmentation of ventricular performance and oxygen consumption by repetitive application of paired electrical stimuli. *Circ. Res.* 16: 332–342, 1965b.

ROSS, J., JR., J. W. COVELL, E. H. SONNENBLICK, AND E. BRAUNWALD. Contractile state of heart characterized by force-velocity relations in variably afterloaded and isovolumic beats. *Circ. Res.* 18: 149–163, 1966a.

ROSS, J., JR., J. H. GAULT, D. T. MASON, J. W. LINHART, AND E. BRAUNWALD. Left ventricular performance during muscular exercise in patients with and without cardiac dysfunction. *Circulation* 34: 597–608, 1966b.

ROSS, J., JR., E. H. SONNENBLICK, R. R. TAYLOR, AND J. W. COVELL. Diastolic geometry and sarcomere lengths in the chronically dilated canine left ventricle. *Circ. Res.* 28: 49–61, 1971.

ROSS, J., JR. Adaptations of the left ventricle to chronic volume overload. *Circ. Res.* 35(Suppl II): 64–70, 1974.

ROSS, J., JR. Electrocardiographic ST-segment analysis in the characterization of myocardial ischemia and infarction. *Circulation* 53(Suppl. I): 73–81, 1976a.

ROSS, J., JR. Afterload mismatch and preload reserve: a conceptual framework for the analysis of ventricular function. *Prog. Cardiovasc. Dis.* 18: 255–264, 1976b.

ROSS, J., JR., AND O'ROURKE, R. A. *Understanding the Heart and its Diseases.* New York: McGraw-Hill Book Co., 1976c.

ROSS, J., JR., AND K. L. PETERSON. Cardiac catheterization and angiography. In: Harrison's *Principles of Internal Medicine*, 9th ed. New York: McGraw-Hill Book Co., 1980.

ROSS, J., JR. Assessment of cardiac function and myocardial contractility. In: *The Heart*, edited by J. W. Hurst. New York: McGraw-Hill Book Co., 1982, p. 310–333.

ROVETTO, M. J., J. J. WHITMER, AND J. R. NEELY. Comparison of the effects of anoxia and whole heart ischemia on carbohydrate utilization in isolated working rat hearts. *Circ. Res.* 32: 699, 1973.

RUDOLPH, A. M., AND M. A. HEGMANN. Fetal and neonatal circulation and respiration. *Ann. Rev. Physiol.* 36: 187–207, 1974.

RUSHMER, R. F. *Cardiovascular Dynamics*, 2nd ed. Philadelphia: W. B. Saunders, 1961.

SABISTON, D. C., AND D. E. GREGG. Effect of cardiac contraction on coronary blood flow. *Circulation* 15: 14–20, 1957.

SAGAWA, K. Analysis of the ventricular pumping capacity as a function of input and output pressure loads. In: *Physical Bases of Circulatory Transport: Regulation and Exchange*, edited by E. B. Reeve and A. C. Guyton. Philadelphia: W. B. Saunders, 1967, p. 141–149.

SAGAWA, K. The ventricular pressure-volume diagram revisited. *Circ. Res.* 43: 677–687, 1978.

SAGAWA, K. The end-systolic pressure-volume relation of the ventricle: definition, modifications, and clinical use. *Circulation* 63: 1223–1227, 1981.

SARNOFF, S. J., R. B. CASE, G. H. WELCH, JR., D. W. BRAUNWA., AND W. N. STAINSBY. Performance characteristics and oxygen debt in a nonfailing metabolically supported, isolated heart preparation. *Am. J. Physiol.* 192: 141–147, 1958a.

SARNOFF, S. J., E. BRAUNWALD, G. H. WELCH, JR., R. B. CASE, W. N. STAINSBY, AND R. MACRUZ. Hemodynamic determinants of oxygen consumption of the heart with special reference to the tension-time index. *Am. J. Physiol.* 192: 148–156, 1958b.

SASAYAMA, S., J. ROSS, JR., D. FRANKLIN, C. BLOOR, S. BISHOP, AND R. B. DILLEY. Adaptations of the left ventricle to chronic pressure overload. *Circ. Res.* 38: 172–178, 1976.

SASAYAMA, S., D. FRANKLIN, AND J. ROSS, JR. Hyperfunction with normal inotropic state of the hypertrophied left ventricle. *Am. J. Physiol.* 232: H418–H425, 1977.

SCHALEKEMP, M. A., D. G. BEEVERS, J. D. BRIGGS, J. J. BROWN, D. L. DAVIES, R. FRASER, M. LEBEL, A. F. LEVER, A. MEDINA, J. J. MORTON, J. O. S. ROBERTSON, AND M. TREE. Hypertension in chronic renal failure: an abnormal relation between sodium and the renin-angiotensin system. In: *Hypertension Manual*, edited by J. H. Laragh. New York: Yorke Medical Books, 1974, p. 485–508,

SCHAPER, W. *The Collateral Circulation of the Heart.* New York: Elsevier, 1971, p. 176.

SCHER, A. M., AND M. S. SPACH. Cardiac depolarization and repolarization and the electrocardiogram. In: *Handbook of Physiology*, Sect. 2, edited by R. M. Berne. Washington DC: American Physiological Society, 1979, vol. 1, p. 357–392.

SCHEUER, J. Metabolism of the heart in cardiac failure. *Prog. Cardiovasc. Dis.* 13: 24–54, 1970.

SCHEUER, J., AND N. BRACHFELD. Coronary insufficiency: relations between hemodynamic, electrical and biochemical parameters. *Circ. Res.* 18: 178–189, 1966.

SCHMID-SCHONBEIN, H. Microrheology of erythrocytes, blood viscosity, and the distribution of blood flow in the microcirculation. *Int. Rev. Physiol.* 9: 1–62, 1976.

SCHLAGER, G. Spontaneous hypertension in laboratory animals: a review of the genetic implications. *J. Heredity* 63: 35–38, 1972.

SCHMID, P. G., D. L. LUND, AND R. ROSKOSKI, JR. Efferent autonomic dysfunction in heart failure. In: *Disturbances in Neurogenic Control of the Circulation*, edited by D. J. Reis. Bethesda: American Physiological Society, 1981.

SCHWARTZ, A. Is the cell membrane Na^+, K^+—ATPase enzyme system the pharmacological receptor for digitalis? *Circ. Res.* 39: 2–7, 1976.

SCHWARTZ, A. Active transport in mammalian myocardium. In: *The Mammalian Myocardium*, edited by G. A. Langer and A. J. Brady. New York: John Wiley & Sons, 1974, p. 81–104.

SCHWARTZ, C. J., N. T. WERTHESSEN, AND S. WOLF (eds.) *Structure and Function of the Circulation.* New York: Plenum Press, 1981, vol. 3.

SCHWARZ, F., J. SCHAPER, D. KITTSTEIN, W. FLAMENG, P. WALTER, AND W. SCHAPER. Reduced volume fraction of myofibrils in myocardium of patients with decompensated pressure overload. *Circulation* 63: 1299–1304, 1981.

SELKURT, E. E. Physiology of shock. In: *Physiology*, 4th ed., edited by E. E. Selkurt. Boston: Little, Brown, 1976, p. 419–434.

SHABETAI, R. *The Pericardium.* New York: Grune & Stratton, 1981.

SHABETAI, R., N. O. FOWLER, AND W. G. GUNTHEROTH. The hemodynamics of cardiac tamponade and constrictive pericarditis. *Am. J. Cardiol.* 26: 480–489, 1970.

SHEPHERD, J. T. *Physiology of the Circulation in Human Limbs in Health and Disease.* Philadelphia: W. B. Saunders, 1963.

SHEPHERD, J. T., AND P. M. VANHOUTTE. *Veins and Their Control.* London: W. B. Saunders, 1975.

SHEPHERD, J. T., AND P. M. VANHOUTTE. *The Human Cardiovascular System: Facts and Concepts.* New York: Raven Press, 1979.

SHIRATO, K., R. SHABETAI, V. BHARGAVA, D. FRANKLIN, AND J. ROSS, JR. Alteration of the left ventricular diastolic pressure-segment length relation produced by the pericardium: effects of cardiac distension and afterload reduction in conscious dogs. *Circulation* 57: 1191–1198, 1978.

SHOUKAS, A. A., AND K. SAGAWA. Control of total systemic vascular capacity by the carotid sinus baroreceptor reflex. *Circ. Res.* 33: 22–33, 1973.

SIEBENS, A. A., B. F. HOFFMAN, P. F. CRANEFIELD, AND C. M. BROOKS. Regulation of contractile force during ventricular arrhythmias. *Am. J. Physiol.* 197: 971–977, 1959.

SOMLYO, A. P., AND A. V. SOMLYO. Excitation and contraction in vascular smooth muscle. In: *Structure and Function of the Circulation*, edited by C. J. Schwartz and N. T. Werthessen. New York: Plenum Press, 1981, vol. 3, p. 239–286.

SOMMER, J. R., AND E. A. JOHNSON. Ultrastructure of cardiac muscle. In: *Handbook of Physiology*, Sect. 2, edited by R. M. Berne, et al. Washington DC: American Physiological Society, 1979, vol. 1, p. 113–186.

SONNENBLICK, E. H. Implications of muscle mechanics in the heart. *Fed. Proc.* 21: 975–990, 1962.

SONNENBLICK, E. H., AND S. E. DOWNING. Afterload as a primary determinant of ventricular performance. *Am. J. Physiol.* 204: 604–610, 1963.

SONNENBLICK, E. H., J. ROSS, JR., J. W. COVELL, G. A. KAISER, AND E. BRAUNWALD. Velocity of contraction as a determinant of myocardial oxygen consumption. *Am. J. Physiol.* 209: 919–927, 1965.

SONNENBLICK, E. H. The mechanics of myocardial contraction. In: *The Myocardial Cell*, edited by S. A. Briller and H. L. Conn, Jr. Philadelphia: Univ. of Pennsylvania Press, 1966, p. 173–250.

SONNENBLICK, E. H., W. H. PARMLEY, C. W. URSCHEL, AND D. L. BRUTSAERT. Ventricular function: evaluation of myocardial contractility in health and disease. *Prog. Cardiov-*

asc. Dis. 12: 449–456, 1970.

SPANN, J. F., JR., R. A. BUCCINO, E. H. SONNENBLICK, AND E. BRAUNWALD. Contractile state of cardiac muscle obtained from cats with experimentally produced ventricular hypertrophy and heart failure. *Circ. Res.* 21: 341–354, 1967.

SPERELAKIS, N. Origin of the cardiac resting potential. In: *Handbook of Physiology*, Sect. 2, edited by R. M. Berne, et al. Washington DC: American Physiological Society, 1979, vol. 1, p. 187–267.

SPIRO, D., AND E. H. SONNENBLICK. Comparison of contractile process in heart and skeletal muscle. *Circ. Res.* 15(Suppl. 2): 14–37, 1964.

SPOTNITZ, H. M., E. H. SONNENBLICK, AND D. SPIRO. Relation of ultrastructure to function in intact heart: sarcomere structure relative to pressure volume curves of intact left ventricles of dog and cat. *Circ. Res.* 18: 49–66, 1966.

STAINSBY, W. N., AND E. M. REKIN. Autoregulation of blood flow in resting skeletal muscle. *Am. J. Physiol.* 207: 117–122, 1961.

STAINSBY, W. N. Autoregulation of blood flow in skeletal muscle during increased metabolic activity. *Am. J. Physiol.* 202: 273–276, 1962.

STARLING, E. H. *The Linacre Lecture on the Law of the Heart* (Given at Cambridge, 1915.) London: Longmans, Green, 1918.

STEIN, P. D. *A Physical and Physiological Basis for the Interpretation of Cardiac Auscultation: Evaluations Based Primarily on the Second Sound and Ejection Murmurs.* Mount Kisco, NY: Futura Pub. Co., 1981.

STEGALL, H. F. Muscle pumping in the dependent leg. *Circ. Res.* 19: 180–190, 1966.

STONE, H. L., V. S. BISHOP, AND E. DONG, JR. Ventricular function in cardiac-denervated and cardiac-sympathectomized conscious dogs. *Circ. Res.* 20: 587–593, 1967.

STONE, H. O., H. K. THOMPSON, AND K. SCHMIDT-NIELSEN. Influence of erythrocytes on blood viscosity. *Am. J. Physiol.* 214: 913–918, 1968.

STREETER, D. D., JR. Gross morphology and fiber geometry of the heart. In: *Handbook of Physiology*, Sect. 2, edited by R. M. Berne, et al. Washington DC: American Physiological Society, 1979, vol. 1, p. 61–112.

STREETER, D. D., JR., H. M. SPOTNITZ, D. P. PATEL, J. ROSS, JR., AND E. H. SONNENBLICK. Fiber orientation in the canine left ventricle during diastole and systole. *Circ. Res.* 24: 339–347, 1969.

STULL, J. T., AND S. E. MAYER. Biochemical mechanisms of adrenergic and cholinergic regulations of myocardial contractility. In: *Handbook of Physiology*, Sect. 2, edited by R. M. Berne, et al. Washington DC: American Physiological Society, 1979, vol. 1, p. 741–774,

SUGA, H., K. SAGAWA, AND A. A. SHOUKAS. Load independence of the instantaneous pressure-volume ratio of the canine left ventricle and effects of epinephrine and heart rate on the ratio. *Circ. Res.* 32: 314–322, 1973.

SURAWICZ, B. Ventricular fibrillation. *Am. J. Cardiol.* 28: 268–287, 1971.

TAYLOR, M. G. An introduction to some recent developments in arterial haemodynamics. *Australas. Ann. Med.* 15: 71–86, 1966.

TAYLOR, R. R., J. W. COVELL, AND J. ROSS, JR. Volume-tension diagrams of ejecting and isovolumic contractions in left ventricle. *Am. J. Physiol.* 216: 1097–1102, 1969.

TAYLOR, S. R. Decreased activation in skeletal muscle at short muscle lengths. In: *The Physiological Basis of Starling's Law of the Heart*, Ciba Foundation Symposium 24. Amsterdam: Elsevier Scientific Pub. Co., 1974.

TENNANT, R., AND C. J. WIGGERS. The effect of coronary occlusion on myocardial contraction. *Am. J. Physiol.* 112: 351–361, 1935.

THEROUX, P., D. FRANKLIN, J. ROSS, JR., AND W. S. KEMPER. Regional myocardial function during acute coronary occlusion and its modification by pharmacologic agents. *Circ. Res.* 35: 896–908, 1974.

TOBIAN, L. A viewpoint concerning the enigma of hypertension. In: *Hypertension Manual*, edited by J. H. Laragh. New

York: Yorke Med. Books, 1974, p. 135–162.

TRAUTWEIN, W. Membrane currents in cardiac muscle fibers. *Physiol. Rev.* 53: 793–835, 1973.

TRUEX, R. C. Anatomy of the heart. In: *Encyclopedia Britannica.* Chicago: Encyclopedia Britannica, 1967, vol. 11, p. 220–222.

TSAKIRIS, A. G., D. E. DONALD, R. E. STURM, AND E. H. WOOD. Volume ejection fraction, and internal dimensions of left ventricle determined by biplane videometry. *Fed. Proc.* 28: 1358–1367, 1969.

VAN CITTERS, R. L., AND D. L. FRANKLIN. Cardiovascular responses in Alaska sled dogs during exercise. *Circ. Res.* 24: 33–42, 1969b.

VATNER, S. F., D. FRANKLIN, C. B. HIGGINS, T. PATRICK, AND E. BRAUNWALD. Left ventricular response to severe exertion in untethered dogs. *J. Clin. Invest.* 51: 3052–3060, 1972.

VATNER, S. F., D. FRANKLIN, R. L. VAN CITTERS, AND E. BRAUNWALD. Effects of carotid sinus nerve stimulation of blood flow distribution in conscious dogs at rest and during exercise. *Clin. Sci.* 15: 457–463, 1974.

VATNER, S. F., D. FRANKLIN, R. L. VAN CITTERS, AND E. BRAUNWALD. Effects of carotid sinus nerve stimulation on the coronary circulation of the conscious dog. *Circ. Res.* 27: 11–21, 1970.

VATNER, S. F., R. G. MONROE, AND R. J. McRITCHIE. Effects of anesthesia, tachycardia, and autonomic blockade on the Anrep effect in intact dogs. *Am. J. Physiol.* 226: 1450–1456, 1974.

VASSALLE, M. Cardiac automaticity and its control. *Am. J. Physiol.* 233: H625–H634, 1977.

VIVEROS, O. H., D. C. GARLICK, AND E. M. RENKIN. Sympathetic beta adrenergic vasodilatation in skeletal muscle of the dog. *Am. J. Physiol.* 215: 1218–1225, 1968.

WAGNER, M. L., R. LAZZARA, R. M. WEISS, AND B. F. HOFFMAN. Specialized conducting fibers in the interatrial band. *Circ. Res.* 18: 502–518, 1966.

WALLACE, A. G. Electrical activity of the heart. In: *The Heart,* 5th ed., edited by W. J. HURST. New York: McGraw-Hill Book Co., 1982, p. 115–127.

WADE, O. L., B. L. COMBES, A. W. CHILDS, A. COURNAND, AND S. BRADLEY, E. The effect of exercise on the splanchnic blood flow and splanchnic blood volume in normal man. *Clin. Sci.* 15: 457–463, 1956.

WEBER, K. T., J. S. JANICKI, R. C. REEVES, AND L. L. HEFNER. Factors influencing left ventricular shortening in isolated canine heart. *Am. J. Physiol.* 230: 419–426, 1976.

WEIDMANN, S. Heart: electrophysiology. *Ann. Rev. Physiol.* 36: 155–169, 1974.

WHELAN, R. F. *Control of peripheral circulation in man.* Springfield, IL: Charles C Thomas, 1967.

WHITMORE, R. I. *Rheology of the Circulation.* Oxford: Pergamon Press, 1968.

WIEDEMAN, M. P. Dimensions of blood vessels from distributing artery to collecting vein. *Circ. Res.* 12: 375–378, 1963.

WIEDERHIELM, C. A., AND B. V. WESTON. Microvascular, lymphatic, and tissue pressures in the unanesthetized mammal. *Am. J. Physiol.* 225: 992–996, 1973.

WIGGERS, C. J. *Physiology in Health and Disease.* Philadelphia: Lea & Febiger, 1954.

WILMORE, J. H. Acute and chronic physiological responses to exercise. In: *Exercise in Cardiovascular Health and Disease,* edited by E. A. Amsterdam, J. H. Wilmore, and A. N. DeMaria. New York: Yorke Medical Books, 1977, p. 53–69.

WINEGRAD, S. Electromechanical coupling in heart muscle. In: *Handbook of Physiology,* Sect. 2, edited by R. M. Berne, et al. Washington DC: American Physiological Society, 1979, p. 393–428.

WINSOR, T. *Primer of Vectorcardiography.* Philadelphia: Lea & Febiger, 1972.

WIT, A. L., AND P. F. CRANEFIELD. Effect of verapamil on the sinoatrial and atrioventricular nodes of the rabbit and the mechanism by which it arrests recurrent atrioventricular nodal tachycardia. *Circ. Res.* 35: 413–425, 1977.

WIT, A. L., M. R. ROSEN, AND B. F. HOFFMAN. Electrophysiology and pharmacology of cardiac arrhythmias. II. Relationship of normal and abnormal electrical activity of cardiac fibers to the genesis of arrhythmias. B. Re-entry, Section I. *Am. Heart J.* 88: 664–670, 1974.

WIT, A. L., M. B. WEISS, W. D. BERKOWITZ, K. M. ROSEN, C. STEINER, AND A. N. DAMATO. Patterns of atrioventricular conduction in the human heart. *Circ. Res.* 27: 345–359, 1970.

WOLFE, A. V. Demonstration concerning pressure-tension relations in various organs. *Science* 115: 243–244, 1952.

WOOD, E. *The Veins.* Boston: Little, Brown, 1965.

WOOD, E. H. Cardiovascular and pulmonary dynamics by quantitative imaging. *Circ. Res.* 38: 131–139, 1976.

WOOD, E. H., R. L. HEPPNER, AND S. WEIDMAN. I. Positive and negative effects of constant electric currents of current pulses applied during cardiac action potentials. II. Hypotheses: calcium movements, excitation-contraction coupling and inotropic effects. *Circ. Res.* 24: 436–445, 1969.

YOFFEY, J. M., AND F. C. COURTICE. *Lymphatics, Lymph and Lymphoid Tissue,* 2nd ed. Cambridge, MA: Harvard Univ. Press, 1956.

YORAN, C., L. HIGGINSON, M. A. ROMERO, J. W. COVELL, AND J. ROSS, JR. Reflex sympathetic augmentation of left ventricular inotropic state in the conscious dog. *Am. J. Physiol.* 10: H857–863, 1981.

YOUNG, M. The fetal and neonatal circulation. In: *Handbook of Physiology,* Sect. 2. Washington DC: American Physiological Society, vol. 2, 1964, p. 1619–1650.

ZAK, R., AND RABINOWITZ, M. Molecular aspects of cardiac hypertrophy. *Annu. Rev. Physiol.* 41: 539–552, 1979.

ZELIS, R., J. LONGHURST, R. J. CAPONE, AND G. LEE. Peripheral circulatory control mechanisms in congestive heart failure. *Am. J. Cardiol.* 32: 481–490, 1973.

ZINNER, S. H., P. S. LEVY, AND E. H. KASS. Familial aggregation of blood pressure in childhood. *N. Engl. J. Med.* 284: 401–404, 1971.

ZWEIFACH, B. W. Microcirculation. *Annu. Rev. Physiology.* 35: 117–150, 1973.

ZWEIFACH, B. W., AND A. FRONEK. The interplay of central and peripheral factors in irreversible hemorrhagic shock. *Prog. Cardiovasc. Dis.* 18: 147–180, 1975.

Blood and Lymph

CHAPTER 20

Blood and the Plasma Proteins: Functions and Composition of Blood

As blood moves through the capillaries of the organs and tissues, it performs its vital function of picking up and delivering a variety of materials whose transport through the circulation is required for the survival of complex multicellular organisms. First and foremost, it picks up O_2 from the lungs and delivers O_2 and glucose to all organs and tissues for the oxidative metabolic reactions essential for life. Of particular importance is an uninterrupted delivery of O_2 to the brain and heart. Blood also delivers amino acids, fatty acids, trace metals, and other substances to the cells for nutritive use or for incorporation into cellular components or secretory products. Blood brings to the cells the hormones and vitamins that modulate cellular metabolic processes. Through a constant exchange of molecules with the interstitial fluid, blood helps to maintain the pH and electrolyte concentrations of interstitial fluid within the ranges required for normal cell functions. Blood carries the waste products of metabolism to their organs of excretion: CO_2 to the lungs, bilirubin to the liver, non-protein nitrogenous products to the kidneys. In warm-blooded animals, blood transports heat generated in deep organs to the skin and lungs for dissipation. Blood also serves essential body protective functions. Its white blood cells combat invading microorganisms, mediate inflammation, and initiate immune responses to foreign materials. Its antibodies and complement components play vital roles in these defense responses. Its platelets and blood coagulation proteins maintain hemostasis.

Blood consists of cells suspended in a fluid called *plasma*. If blood is mixed with an anticoagulant powder to prevent clotting, the cells can be separated from the plasma by centrifugation. After vigorous centrifugation, a unit volume of normal human blood will consist of a volume of packed red blood cells of about 45%, on top of which is a volume of between 0.5 and 1% of white blood cells and platelets. The supernatant plasma will occupy the remaining volume of about 54%. If an anticoagulant is not added and blood is allowed to clot, the supernatant fluid that can be expressed from the clot is called *serum*. Serum differs from plasma in lacking fibrinogen (which has been converted to the fibrin of the clot), prothrombin, and other coagulation factor activities consumed during clotting and in containing minute but physiologically important amounts of materials released from platelets during clotting, *e.g.*, a platelet-derived growth factor for smooth muscle cells of the vascular wall.

Plasma consists of an aqueous solution of proteins, electrolytes, and small organic molecules. Plasma (100 ml, or 1 dl) will contain about 7 g of plasma proteins; about 900 mg of electrolytes, primarily sodium, chloride, and bicarbonate ions, but also physiologically critical amounts of potassium, calcium, magnesium, and phosphate ions; and a few hundred milligrams of small organic molecules, including about 100 mg of glucose, about 25 mg of nonprotein nitrogenous waste products, of which urea is the major constituent, and several hundred milligrams of lipids in the form of triglycerides, phospholipid, and cholesterol. Plasma from a fasting individual is translucent and is a pale yellow color due to small amounts of the pigment bilirubin; plasma from an individual who has eaten a fatty meal is opaque because of dietary fat, primarily triglycerides, suspended in the plasma as particles called chylomicrons.

THE PLASMA PROTEINS

Studies of the plasma proteins have yielded much information about physiologic processes and disease states. The accessibility of normal plasma and the development of powerful techniques for purifying proteins have made possible the determination of the structural and functional properties of a substantial fraction of the more than 100 individual plasma proteins recognized to date. The precise structure of the different classes of immunoglobulin molecules has been delineated (Chapter 22). The known components of the complex plasma proteolytic systems participating in inflammation and hemostasis, including many present in only trace amounts, have been purified, and their functions have been identified. A growing number of plasma protease inhibitors have also been purified, and their mechanisms and spectra of protease inhibition have been clarified. An increasing number of apolipoproteins participating in lipid transport and metabolism have also been delineated. Their study has advanced understanding of the most prevalent of serious diseases of Western civilization, atherosclerosis. Carrier proteins have been identified in plasma for many small molecules such as vitamins, hormones, trace metals, and drugs. A number of proteins have also been separated from plasma, whose physiologic functions remain unknown.

Studies of plasma proteins have also yielded important genetic information. Some plasma proteins exist in polymorphic forms differing in prevalence in different populations; they have proven a valuable tool in investigations of population genetics. At the molecular level, the determination of degrees of homology in amino acid sequences of purified proteins of related function, such as the vitamin K-dependent blood clotting factors, has broadened understanding of how multicomponent biologic systems develop in evolution. Similar studies of proteins of different species have clarified phylogenetic relationships.

Separation and Measurement of the Plasma Proteins

Although the approximately 7 g/dl of human plasma proteins consist of many proteins, one protein, albumin, makes up over 4 g of the total. Initially, albumin was measured by salting out techniques that separate albumin from all the other plasma proteins, the globulins. Later, electrophoretic techniques were developed that separate the plasma proteins into albumin and several globulin fractions that migrate at different rates in an electrical field (Ritchie, 1979). These globulin fractions were given Greek letter names: α_1, α_2, β, and γ. The vast majority of the γ globulins, which make up another 1.0 to 1.5 g of total plasma proteins, turned out to be immunoglobulins. A number of other proteins, whose concentrations range between about 50 and 300 mg/dl, have been identified within or between the other electrophoretically separated globulin fractions (Fig. 3.1). These proteins, which are of widely diverse functions, include: fibrinogen (which is not seen when serum is used for electrophoresis); two major protease inhibitors, α_1-pro-

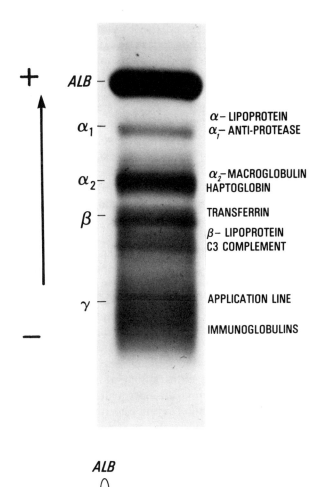

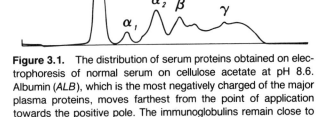

Figure 3.1. The distribution of serum proteins obtained on electrophoresis of normal serum on cellulose acetate at pH 8.6. Albumin (*ALB*), which is the most negatively charged of the major plasma proteins, moves farthest from the point of application towards the positive pole. The immunoglobulins remain close to the point of application. The intensity of the protein bands on the cellulose acetate strip can be measured with an instrument called a densitometer. The electrophoretic profile shown in the bottom half of the slide is the densitometer tracing obtained from the strip shown in the top half of the slide.

tease inhibitor (α_1-antitrypsin) and α_2-macroglobulin; the carrier proteins, haptoglobin and transferrin, two major lipoproteins, and the C_3 component of complement. An immunoelectrophoretic technique was also developed in which the plasma proteins are separated by electrophoresis and then precipitated by immunodiffusion against an antiserum to whole human serum. This sensitive technique resolves the plasma proteins into multiple precipitin arcs, each representing a different protein.

Serum albumin levels are frequently measured

(nowadays by automated dye binding techniques) as part of the screening laboratory evaluation of a new patient. Serum protein electrophoresis is also commonly performed, particularly when a disorder affecting immunoglobulins is suspected. Immunoelectrophoresis against whole serum antiserum has proven of more limited clinical application.

Numerous assays have been developed to measure small or even trace amounts of individual plasma proteins. These may be based upon the specific enzymatic activity of a protein or may involve measuring minute concentrations of antigen by radioimmunoassay or ELISA (enzyme-linked immunoabsorbant assay) methodology. Some of these trace plasma proteins, such as blood coagulation and complement components, have key extracellular functions. Others have no known extracellular function; they are intracellular enzymes that escape into the blood in small amounts as a result of normal cell function or the minor tissue insults of daily living. Plasma levels of such enzymes frequently are increased in disorders affecting their cells of origin, *e.g.,* elevation of amylase in pancreatic disease, alanine aminotransferase in hepatocellular disease, and creatine phosphokinase after injury to muscle. Measurement of individual trace plasma proteins has proven valuable in the diagnostic evaluation of many types of patients.

Synthesis of the Plasma Proteins

Albumin and the vast majority of the globulins are made in the liver. The major exception is the immunoglobulins, which are made by B-lymphocytes-plasma cells. In severe hepatocellular disease, the plasma level of albumin and many other plasma proteins falls. Cells other than hepatocytes or B-lymphocyte-plasma cells contribute significantly to the synthesis of a small number of plasma proteins. Thus, macrophages make substantial amounts of some components of the complement system; intestinal cells synthesize some apoproteins; and endothelial cells are the site of synthesis of von Willebrand factor, a plasma protein essential for normal platelet hemostatic function (Chapter 25).

Factors regulating the synthesis of albumin (Rothschild et al., 1977) are as yet poorly understood. The rise of plasma amino acids that follows ingestion of a meal containing protein stimulates the liver to synthesize albumin. A regulator of albumin synthesis sensitive to the colloid oncotic pressure of the plasma has been postulated. Inflammation stimulates the synthesis of a number of plasma proteins (see below) and in so doing appears to divert the protein synthetic apparatus of the

hepatocyte away from the synthesis of albumin. For this reason plasma albumin levels fall in all chronic diseases, not just in those disorders directly affecting hepatocytes.

Exposure to antigen stimulates synthesis of immunoglobulin. Experimental animals kept in a germ-free environment have virtually absent immunoglobulins. Conversely, immunoglobulin levels rise in chronic infectious disorders causing persisting antigenic stimulation of B-lymphocytes-plasma cells.

Dual mechanisms regulate the hepatic synthesis of a number of plasma proteins known as the acute phase proteins. Unknown factors, possibly different for each protein, maintain basal levels of production of the individual proteins in this group, whereas a common agent or agents probably stimulate the increased synthesis of these proteins after tissue injury or inflammation. Recently, interleukin 1, a material released by activated macrophages (Chapter 22), has been shown to stimulate hepatic synthesis of many acute phase proteins. Prostaglandins may also increase the synthesis of acute phase proteins, possibly through stimulating macrophage release of interleukin 1.

Distribution and Catabolism of the Plasma Proteins

When a radiolabeled plasma protein, such as radiolabeled albumin, is injected intravenously and the radioactivity of serial plasma samples is plotted against time on semilogarithmic paper, a plasma radioactivity decay curve is obtained whose shape corresponds to the sum of two decreasing exponentials. Two rate constants can be calculated from such curves: an initial larger rate constant for the equilibration of the protein between intravascular and extravascular compartments and a second smaller rate constant for the biologic decay of the protein. Analysis of such curves also provides an estimate of the distribution of the protein between the plasma and the interstitial fluid. From such data it is apparent that plasma proteins circulate not only intravascularly but also in a second slower pathway across capillary beds into the interstitial fluid and back into the plasma by way of lymphatic channels. About 5% of plasma albumin exchanges with albumin in interstitial fluid per hour. Moreover, such data indicate that about 60% of the total body mass of albumin is present within the extravascular fluid. Most other plasma proteins studied have similar high fractions of total mass within the extravascular fluid. Two plasma proteins are exceptions: IgM immunoglobulin and α_2-macroglobulin, which have difficulty in crossing capillary membranes because of their large size and are therefore largely confined to the intravascular compartment.

Plasma proteins appear to be broken down randomly, *i.e.*, a constant proportion of molecules are lost per unit time (first order kinetics), independent of how long individual molecules may have aged in the circulation. Intravascular biologic half-disappearance times vary widely for different proteins. At its normal plasma concentration, albumin has an intravascular half-disappearance time of 19 days, and IgG has an intravascular half-disappearance time of 21 days. These long half-times contrast with biologic intravascular half-disappearance times of only about 4 days for haptoglobin and fibrinogen and contrast with the remarkably short half-times for two of the blood coagulation proteins: 12 h for factor VIII and 5 h for factor VII.

The catabolism of albumin (Waldmann, 1977) may differ fundamentally from the catabolism of the other plasma proteins, since albumin is virtually the only plasma protein that is not a glycoprotein (see below). Tissue cells take up albumin by pinocytosis, which is thought to be the major mechanism for albumin catabolism. The albumin is broken down within the lysosomes of tissue cells to amino acids, which are then available for protein synthesis within the cells. A small fraction of albumin, less than 10%, is lost into the digestive tract, where it is broken down into amino acids that can be reabsorbed.

The vast majority of the other plasma proteins contain carbohydrate attached at asparagine residues of the polypeptide chains (Sharon and Lis, 1981). The carbohydrate residues form short branched chains that always terminate in sialic acid residues. When terminal sialic acid is removed *in vitro* by treatment of a plasma glycoprotein with an enzyme called neuramidase, the second residue from the end of the chain is exposed. It is always a galactose residue. In experiments first carried out with ceruloplasmin and then with other plasma glycoproteins, desialylated radiolabeled proteins were found to be rapidly removed from the circulation. Receptors on hepatocyte plasma membranes have been identified for galactose; these receptors bind the exposed galactose residues on desialylated plasma glycoproteins and remove the proteins from the circulation. Small amounts of desialylated proteins have recently been demonstrated in human serum, and it has been proposed that hepatic recognition of the exposed galactose on such molecules represents a major physiologic mechanism for plasma glycoprotein catabolism (Ashwell and Steer, 1981). However, desialylating enzymes have yet to be demonstrated in mammalian tissues and, for the present, the physiologic significance of this mechanism can only be tentatively accepted.

The function of a number of plasma proteins is to combine with other materials in plasma. The resultant complexes are usually removed more rapidly from the circulation than are the free proteins. Thus, when immunoglobulin molecules combine with antigen to form immune complexes, they may be removed rapidly from the circulation by the mononuclear phagocyte system. Hemoglobin-haptoglobin complexes are cleared from the circulation by specific receptors on hepatocyte membranes for the hemoglobin-haptoglobin complex. Antithrombin III-thrombin complexes and α_2-plasmin inhibitor-plasmin complexes have substantially shorter half-disappearance times than do free antithrombin III and free α_2-plasmin inhibitor. Thus, the catabolism of many plasma proteins varies with the extent to which they are called upon to carry out their physiologic functions.

Specific Plasma Proteins

ALBUMIN

As already mentioned, albumin makes up about 60% of the total plasma proteins, has a biologic intravascular half-time of 19 days, and is distributed two-fifths intravascularly and three-fifths extravascularly (Peters, 1975). In a normal 70-kg individual in the steady state, 14–17 g of albumin are made daily. Because of its plasma concentration and relatively low MW of 66,000, albumin is the principal protein responsible for the colloid osmotic pressure of plasma. Albumin also performs important carrier functions. Binding to albumin is the only physiologic mechanism for the transport of free fatty acids in plasma. Albumin transports bilirubin from mononuclear phagocytes, where it is formed from the heme of phagocytized red blood cells, to the liver for excretion. Albumin also serves as a secondary carrier for thyroxine, for cortisol, and for heme, when the capacities of their primary plasma transport proteins are exceeded. In addition, albumin binds many of the drugs given to patients. The tightness of binding and the competition for binding sites between drugs given simultaneously influences drug dosing.

Because of diminished synthesis, plasma albumin levels fall in any chronic illness. Very low levels are often found in patients with advanced hepatocellular disease and in patients with conditions in which diminished synthesis is accompanied by loss of large amounts of albumin into the urine (nephrosis) or into the gastrointestinal tract (protein-losing enteropathies).

PLASMA PROTEOLYTIC SYSTEMS

Four proteolytic systems are present in plasma: the complement system, the kinin system, the blood coagulation system, and the fibrinolytic system. The components of these systems are listed in Table 3.1. Their reactions are discussed in Chapters 23 and 25.

PLASMA PROTEASE INHIBITORS

The properties of a number of plasma protease inhibitors are listed in Table 3.2. α_1-protease inhibitor (α_1-antitrypsin) is the inhibitor normally present in plasma in the highest concentration. Although able to inhibit a wide spectrum of proteases, including the plasma proteases kallikrein, thrombin, and plasmin, α_1-protease inhibitor appears to function *in vivo* primarily as an inhibitor of pro-

Table 3.1

A summary of plasma proteolytic systems involved in inflammation, hemostasis, or both

A. *Complement system*
    Components of classical pathway: C1–C9
    Components of alternative pathway: Factors B, D, C3, C5–9
    Regulators
        Classical pathway: C1 inhibitor, factor I, C4 binding protein
        Alternative pathway: factor I, factor H, properdin
B. *Kinin system*
    Activators of kininogens
        Kallikreins released from blood cells
        Factors involved in forming plasma kallikrein
            Proenzymes: factor XII, prekallikrein
            Cofactor: HMW kininogen
    Kininogens: LMW kininogen, HMW kininogen
    Regulators: C1 inhibitor, α_2-macroglobulin
C. *Blood coagulation system*
    Factors involved in forming prothrombin activator
        Proenzymes
            Contact system proenzymes: factor XII, prekallikrein, factor XI
            Vitamin K-dependent proenzymes: factors VII, IX, X
        Cofactors
            Contact system: surface, HMW kininogen
            Tissue factor
            Factors V, VIII
            Platelet phospholipid
        Prothrombin
        Fibrinogen, factor XIII
        Regulators: Antithrombin III, Protein C
D. *Fibrinolytic system*
    Activators of plasminogen
        Released from endothelium: vascular activator, urokinase
        Proenzymes of weak plasma activators: factor XII, prekallikrein, factor XI
    Plasminogen
    Regulators: α_2-plasmin inhibitor, α_2-macroglobulin, histidine-rich glycoprotein

HMW, high molecular weight; LMW, low molecular weight.

teases in interstitial tissues released from white blood cells during inflammatory reactions. However, α_1-protease inhibitor does appear to function as the principal inhibitor of at least one plasma enzyme, the activated early blood coagulation factor, factor XI_a. Allelic structural genes for α_1-protease inhibitor exist (Pi genes) (Laurell and Jeppsson, 1975). They determine an individual's basal plasma level of α_1-protease inhibitor. The vast majority of individuals are homozygous for the Pi^m gene (MM phenotype); they have basal levels of about 250 mg/dl, which are capable of rising substantially in inflammation. Homozygotes for the uncommon Pi^z gene (ZZ phenotype) have very low plasma α_1-protease inhibitor levels. Many of these individuals develop pulmonary emphysema, a condition resulting from destruction of the alveolar structure of the lungs. These individuals apparently have insufficient α_1-protease inhibitor in the alveolar walls and spaces to prevent elastases, continually liberated from white blood cells in response to repeated minor pulmonary inflammatory events, from gradually destroying the alveoli (Gadek et al., 1982).

α_2-Macroglobulin is another major protease inhibitor of plasma that inhibits a wide spectrum of proteases (Harpel and Rosenberg, 1976). α_2-Macroglobulin is unique among protease inhibitors in its large size and in its ability to inhibit the proteolytic activity of proteases without inhibiting their enzymatic activity against small substrates such as synthetic esters. The physiologic functions of α_2-macroglobulin are not yet clear, although it has been shown to function as a back-up inhibitor for plasmin when the capacity of the primary inhibitor, α_2-plasmin inhibitor, is exceeded. α_2-Macroglobulin has been found to line the luminal surface of vascular endothelial cells (Becker and Harpel, 1976), and it has been postulated that α_2-macroglobulin could serve the general physiologic function of protecting cell surfaces from attack by serine protease enzymes.

CARRIER PROTEINS

A number of the carrier proteins of plasma and the substances that they bind are listed in Table 3.3. The carrier functions of albumin have already been described. The apolipoproteins are discussed in Chapter 48. Three carrier proteins, transferrin, haptoglobin, and hemopexin, are involved in processes affecting synthesis or catabolism of hemoglobin. The iron for new hemoglobin synthesis primarily comes from the iron of senescent red blood cells that are broken down within mononuclear phagocytes. *Transferrin* transports this iron from

Table 3.2
Properties of plasma protease inhibitors

Major Protease Inhibitors	Plasma, mg/dl	Concentration, μmol	Physiological inhibitor of	Manifestations of Deficiency
α_1-Protease inhibitor[a,b]	250	46	WBC elastase and collagenase, factor XI_a	Emphysema
α_1-Antichymotrypsin[a]	50	7.0	WBC cathepsin G	Unknown
α_2-Macroglobulin	250	3.5[c]	Plasmin, thrombin, kallikrein	Unknown
Inter-α-trypsin inhibitor[a]	50	3.0	Unknown	Unknown
Antithrombin III	15	2.5	Thrombin, factor X_a, IX_a	Thrombosis
C1 inhibitor[a]	18	1.5	Activated C1r, C1s, kallikrein	Hereditary angioneurotic edema
α_2-Plasmin inhibitor[a,d]	7	1.0	Plasmin	Fibrinolytic bleeding

Ranges are wide, and the concentrations listed are approximations.   [a] Concentration rises in tissue injury or inflammation.   [b] Formerly called α_1-antitrypsin.   [c] The low value results from a very high molecular weight of 725,000 daltons.   [d] Also called α_2-antiplasmin.

Table 3.3
Important carrier proteins of plasma

Protein	Materials Bound or Transported
Albumin	Primary carrier: fatty acids, bilirubin, many drugs
	Secondary carrier: heme, thyroxine, cortisol
Apolipoproteins	Triglycerides, phospholipids, cholesterol
Haptoglobin	Plasma hemoglobin from lysed erythrocytes
Hemopexin	Heme from plasma hemoglobin
Transferrin	Iron
Ceruloplasmin	Copper
Prealbumin	Thyroxine, vitamin A
Group specific (G) globulin	Vitamin D
Transcortin	Cortisol
Transcobalamins I and II	Cobalamin (vitamin B_{12})

the phagocytes through the plasma to developing erythroid precursors (normoblasts) in the bone marrow. Each molecule of transferrin can carry two atoms of iron, and plasma transferrin is normally about one-third saturated with iron. Normal plasma contains about 250 mg/dl of transferrin. In clinical laboratories its concentration is not measured directly, but is measured indirectly by determining the amount of iron needed to saturate the iron binding capacity of plasma. This value is called the total iron binding capacity (TIBC). Mean normal values are: for serum iron, 100 μg/dl; for TIBC, 300 μg/dl. Transferrin is distributed in a 60:40 ratio between the extravascular and intravascular compartments. Although transferrin moves into and out of the normoblast during iron delivery, this process apparently does not affect its catabolism. Transferrin has an intravascular biologic half-decay time of 10 days. Plasma transferrin concentration and, therefore, the TIBC, rises as a patient becomes iron deficient. Like albumin, transferrin synthesis is depressed in inflammation, and patients with chronic diseases usually have a low TIBC.

Haptoglobin forms complexes with dimers of hemoglobin. When red blood cells are lysed intravascularly, hemoglobin is released into the plasma and rapidly breaks up into α- and β-dimers. A single haptoglobin molecule can combine with two dimers, *i.e.*, with the equivalent of a single hemoglobin molecule. Hemoglobin-haptoglobin complexes are removed from plasma by receptors on hepatocytes, transported into hepatocyte lysosomes, and digested (Kino et al., 1980). Thus, haptoglobin functions to conserve body iron, conveying iron in plasma hemoglobin to the liver and preventing its loss in the urine through excretion of free dimers.

Normal plasma haptoglobin concentrations vary from about 40–180 mg/dl. In a 70-kg human, the daily breakdown of senescent erythrocytes results in a turnover of about 6 g of hemoglobin, of which about 0.6 g (representing 10% of erythrocyte breakdown) is released intravascularly. A haptoglobin concentration of 130 mg/dl in a 70-kg human can bind about 3 g of hemoglobin, or five times the amount liberated intravascularly each day. Plasma haptoglobin concentrations fall in patients with severe hepatocellular disease because the liver is the primary site of haptoglobin synthesis. Plasma haptoglobin levels also frequently fall in hemolytic disorders, because then most of the plasma haptoglobin is converted into hemoglobin-haptoglobin complexes and their rapid clearance does not stimulate compensatory increased haptoglobin production. However, tissue inflammation does stimulate increased haptoglobin production, and when hemolysis occurs in the setting of infection or other inflammatory conditions, plasma haptoglobin levels may not fall. In inflammatory states without hemolysis, high plasma haptoglobin levels are found.

Hemoglobin in plasma not only breaks up into dimers but into free oxidized heme (metheme) and globin chains. Whereas haptoglobin binds dimers, a second protein, *hemopexin*, binds free metheme.

The metheme-hemopexin complex is also rapidly removed from plasma by hepatocytes and catabolized. Hemopexin is normally present in plasma in a concentration of 50–100 mg/dl. When intravascular hemolysis results in the formation of amounts of metheme exceeding the binding capacity of hemopexin, metheme then binds to albumin, forming methemalbumin. Methemalbumin functions as a circulating heme storage protein. As new hemopexin is produced, methemalbumin gives up its heme to hemopexin, and the process of hepatic clearance of metheme-hemopexin complexes continues. In this way the body conserves iron in metheme that would otherwise be lost in the urine. Unlike its effect upon haptoglobin, inflammation does not stimulate an increased production of hemopexin.

The copper in plasma circulates bound to the protein *ceruloplasmin*. The normal plasma ceruloplasmin concentration is about 30 mg/dl. Estrogens react with steroid receptors on hepatocytes to stimulate ceruloplasmin synthesis, and plasma levels rise substantially in pregnancy or when women take oral contraceptive agents containing estrogens. Ceruloplasmin levels also rise in disorders producing inflammation or tissue injury. Low ceruloplasmin levels are found in patients with a hereditary disturbance of copper metabolism (Wilson's disease) in which biliary excretion of copper is impaired, and large amounts of copper accumulate in the body, damaging the liver and brain. Ceruloplasmin is an oxidase and may perform important functions in addition to carrying copper. Ceruloplasmin has been postulated to serve as the major extracellular scavenger of superoxide radicals generated by white blood cells during the inflammatory response (Goldstein et al., 1982). Ceruloplasmin may also function physiologically as a ferroxidase, catalyzing the conversion of ferrous iron in tissues to ferric iron, a reaction that must occur before transferrin can pick up and transport iron.

ACUTE PHASE PROTEINS

When tissue injury or infection results in local inflammation, interleukin 1 and probably other materials are formed that induce a systemic acute phase response (Kushner, 1982). This response usually includes fever, an increased release of certain hormones, and the increased synthesis of acute phase proteins. These include proteins whose increased synthesis has recognizable survival value after tissue injury or infection: the hemostatic factors, fibrinogen, and von Willebrand factor; the C_3 and factor B components of complement; haptoglobin; ceruloplasmin; and the protease inhibitors, α_1-protease inhibitor, α_1-antichymotrypsin, and α_2-antiplasmin. The synthesis of two proteins that are normally present in only trace amounts in human plasma increases many hundred fold during the acute phase response. One protein is C-reactive protein, so named because it was first identified as a material in plasma that precipitates a pneumococcal polysaccharide fraction known as the C fraction. The second protein is SAA (serum amyloid A protein), which was first recognized as the major component of secondary amyloid and was later identified, in a slightly larger molecular weight form, in plasma (Gorevic et al., 1982). C-reactive protein has recently been shown to bind to altered cell membranes. Once bound, like antibody, it can then activate the classical complement pathway. Macrophages have been shown to ingest particles coated with C-reactive protein and complement. Thus, C-reactive protein has been postulated to function in host defense in reactions that supplement but are distinct from the immunologic response (Gewurz, 1982). SAA has recently been shown to circulate in plasma attached to high density lipoprotein. A host defense function for SAA has not yet been identified.

Altered Plasma Proteins in Disease

Hereditary abnormalities of plasma proteins are an uncommon but well-recognized cause of disease. Thus, lifelong disorders of hemostasis, of which the hemophilias are the most important, result from hereditary abnormalities of the blood coagulation proteins. A hereditary deficiency of antithrombin III is associated with an increased risk for thromboembolic events. As already mentioned, a decreased plasma concentration of α_1-protease inhibitor in individuals homozygous for the Pi^z gene can result in the development of pulmonary emphysema. Failure to synthesize transferrin is a very rare cause of a refractory iron deficiency anemia.

Abnormalities of plasma proteins are common as manifestations of disease and, as discussed earlier, the plasma proteins are frequently examined in the diagnostic evaluation of patients.

CHAPTER 21

Hemopoiesis

The cells of certain tissues die or are consumed in their function. These tissues include skin, in which cells shed and are regenerated; gastrointestinal mucosa, in which cells slough into the lumen; and blood. The constant production of new cells for these tissues contrasts strikingly with the lack of production of new cells in other tissues, e.g., neural tissue, in which cells that die can not be replaced, or liver, in which hepatocytes proliferate to a limited degree only after injury.

Blood cells are made in early embryonic life primarily in the liver and, to a lesser extent, in the spleen. Hemopoiesis begins in the bone marrow at about the 20th embryonic week. As the fetus matures, hemopoiesis increases in the bone marrow and decreases in the liver and spleen. After birth, all blood cells except lymphocytes normally originate only from the bone marrow. In young children, active hemopoietic marrow is found throughout both the axial skeleton (cranium, ribs, sternum, vertebrae, and pelvis) and the bones of the extremities. In adults, hemopoietic marrow is confined to the axial skeleton and proximal ends of the femur and humerus. In chronic hemolytic disorders with continuous increased production of erythrocytes and in certain neoplastic blood cell disorders, hemopoiesis may again be found outside the bone marrow in the liver and spleen.

The normal bone marrow produces a dazzling array of blood cells; no other adult organ or tissue is called upon to produce so many different cells with specialized functions. These are divided into the erythrocyte (red blood cells (RBC), the white blood cells (WBC), of which there are five different types, and the platelet. The approximate concentrations of these cells in the blood are given in Table 3.4. *Erythrocytes* contain the body's oxygen transport protein, hemoglobin, and function, primarily to carry oxygen from the lungs to the tissues. The *WBC* are involved in the body's defense against microorganisms and other foreign materials. *Granulocytes* (polymorphonuclear neutrophils) are the principal mediators of bacterial defense and inflam-

mation. *Monocytes* and their tissue counterparts, *macrophages*, have varied functions, including: a primary role in resistance to certain microorganisms; the initiation and modulation of some types of lymphocyte responses; and a scavenger function in the removal of tissue debris and altered or foreign materials from blood. The functions of *eosinophils* and *basophils* are less well defined but include roles in resistance to parasitic infections and in atopic allergic reactions. *Lymphocytes* are the effectors of immune responses mediated both by antibodies and by cellular mechanisms. *Platelets* play key roles in the hemostatic reactions that prevent bleeding after minimal injury to blood vessels.

Animal studies described later and studies of chromosomal and enzyme markers in the blood cells of humans with chronic granulocytic leukemia suggest that blood cells originate from a single class of primitive cell, the *pluripotent stem cell*. This cell is thought to give rise to cells of both the myeloid and lymphocytic series. The term *myeloid*, used in this sense, refers to those cells series that arise exclusively within the bone marrow: erythrocytes, granulocytes, monocytes, eosinophils, basophils, and platelets. (Myeloid is an inexact word because it is also used to refer only to the granulocyte series in the bone marrow, as in the expression myeloid: erythoid ratio, which is the ratio of granulocytes and their precursors to nucleated RBC present in a bone marrow specimen). In contrast to myeloid cells, lymphocytes may originate in the bone marrow but are also produced in the thymus and, in response to immune stimuli, in the peripheral lymphoid tissues (lymph nodes, spleen, and the accumulations of lymphocytes in the wall of the gastrointestinal, urinary, and respiratory tracts).

Erythrocytes and platelets spend their entire life within the intravascular compartment; granulocytes, monocytes, and eosinophils spend only a few hours within the intravascular compartment before moving permanently into the tissues; lymphocytes can move back and forth between the blood and the peripheral lymphoid tissues. Cells remaining within

Table 3.4

Approximate concentrations and daily production rates of peripheral blood cells

Cell Type	Mean Concentration, per microliter	Daily Production Rate, per kg body weight
Erythrocytes		
Males	5.4×10^6	3.0×10^9
	$(4.7–6.1 \times 10^6)$	
Females	4.8×10^6	
	$(4.2–5.4 \times 10^6)$	
White blood cells		
Granulocytes	4500	1.6×10^9
	(2600–7000)	
Monocytes	300	1.7×10^{8a}
Eosinophils	150	Variable
Basophils	40	Unknown
Lymphocytes	2500	Unknown
	(1500–4000)	
Platelets	2.5×10^5	2.8×10^9
	$(1.5–4.0 \times 10^5)$	

Values in parenthesis are ranges.   [a] Based upon assuming production rate equals blood turnover rate and using an intravascular half-disappearance time of 8.4 h.

the intravascular space have a readily measurable life-span: 120 days for erythrocytes, about 10 days for platelets. Cells that spend most of their existence in the tissues have life-spans more difficult to determine, but granulocytes and eosinophils are thought to survive in the tissues for only a few days. Monocytes are transformed in the tissues into wandering and fixed macrophages, whose life-spans may vary from days to several months. As discussed in Chapter 22, lymphocytes consist of functional subclasses of cells with different life-spans influenced by the activity of the immune system. Some lymphocytes apparently survive for years.

Since erythrocytes, granulocytes, eosinophils, monocytes, and platelets have finite life-spans, fractions of these cells must be replaced daily. Approximate daily rates of production are listed in Table 3.4. In addition to meeting this steady demand for a variety of cells, the bone marrow must respond to intermittent increased demands for specific types of cells. For example, a bacterial infection creates a demand for increased numbers of granulocytes whereas acute blood loss creates a demand for increased numbers of erythrocytes. Responses must be appropriate both in cell type and in magnitude. Thus, blood cell production requires regulation of the proliferation and differentiation of cells capable of diverse genetic expression, i.e., stem cells, whose responses must be superbly ordered to maintain the integrity of the organism. Persisting failure of production in any major cell line (erythrocytes, granulocytes, monocytes, lymphocytes, or platelets) results eventually in death.

STEM CELLS AND PROGENITOR CELLS

Hemopoietic stem cells (Quesenberry and Levitt, 1979) are cells with an extensive proliferative capacity that possess two fundamental functional properties: first, an ability to give rise to new stem cells, i.e., self renewal and, second, an ability to differentiate into blood cells (Fig. 3.2). Stem cells cannot be distinguished morphologically from lymphocytes in the bone marrow. Their existence has been established in animals by experiments in which donor bone marrow was infused into lethally irradiated mice or into mice born with a hereditary stem cell defect (Boggs et al., 1982). Their existence has been inferred in man from observations in patients with hematologic malignancies and in patients receiving bone marrow transplants. The exact frequency of stem cells in human bone marrow is unknown but does not exceed 1–2/1000 nucleated cells. Small numbers of stem cells also circulate in the blood. When normal mice or humans with leukemia are given supralethal doses of radiation followed by an infusion of donor bone marrow cells, hemopoiesis is reestablished only in the marrow cavity and spleen of the mouse and in the marrow cavity of man. The specific supporting tissues in these organs that permit the growth and differentiation of stem cells are referred to as the *hemopoietic microenvironment*.

The experimental technique usually used to quantitate stem cells consists of irradiating mice to destroy their own stem cells and then injecting the

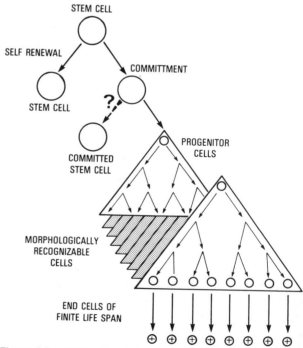

Figure 3.2.   Schematic of clonal hemopoiesis beginning with a stem cell.

mice with a limited number of bone marrow cells from donor animals of the same inbred strain. After several days, cellular aggregates (colonies) of differentiated blood cells are found in the spleen of irradiated animals. By the use of donor marrow containing chromosomal markers induced by giving the donor animal a small dose of radiation, it has been established that each spleen colony arises from a single stem cell. Therefore, counting the number of colonies visible in the spleen provides an estimate of the number of stem cells infused into an animal.

The property of self-renewal of stem cells can be demonstrated by harvesting cells from a spleen colony of an irradiated recipient mouse, injecting these cells into a second irradiated recipient mouse (serial passage), and observing the formation of new spleen colonies in the second mouse. Since all of the cells in a given spleen colony arise from a single stem cell, the spleen colonies in the second recipient mouse must have arisen from daughter stem cells present in the spleen colony from the first recipient mouse.

The process of self-renewal maintains the hemopoietic stem cell pool. Because daughter stem cells may have the same proliferative capacity as ancestor cells, the question arises: Is the stem cell immortal or is it programmed to cease self-renewal after a given number of divisions? This question has not been answered definitively, but in humans self-renewal prevents the marrow from running out of stem cells during a normal life-span. (With advanced age, the number or function of stem cells may be reduced, and normal hemopoiesis may then require expansion of other early marrow elements.)

When a multitide of cells arises from a single cell, as in the formation of a spleen colony from a single stem cell, that multitude of cells is called a clone. (Fig. 3.2 illustrates the clonal formation of a multitude of mature end cells from a single stem cell.) In normal hemopoiesis, mature blood cells originate from many stem cells. Normal hemopoiesis is thus *polyclonal*. In hematologic malignancies, the progeny of a single abnormal stem cell may overgrow and suppress the progeny of normal stem cells, and hemopoiesis may become *monoclonal*.

Bone marrow stem cells are usually in a quiescent state, i.e., they are not in the process of cell division. Mechanisms controlling self-renewal of stem cells are poorly defined. When bone marrow cells are infused into lethally irradiated mice, the infused stem cells appear initially to increase in number without giving rise to mature blood elements. Such experiments indicate that stem cells may proliferate in a mode in which self-renewal occurs without

differentiation. When an acute demand for increased blood cell production is produced in an animal as, for example, by the administration of cytotoxic drugs, stem cells are recruited into cell cycle, divide, and temporarily increase in number. When a chronic need for increased blood cell production exists, both pluripotent stem cells and cells committed to the production of one or a limited number of cell types (see below) are increased in number. The mechanisms that stimulate stem cells to proliferate are largely unknown, but *in vitro* studies suggest that both stimulatory and inhibitory humoral agents may be involved.

With beginning differentiation the stem cell loses its ability to give rise to all blood cells and becomes committed to the production of one or more cell lines (Fig. 3.3). Hypotheses of the mechanism regulating commitment differ (Till and McCulloch, 1980). One view is that extracellular inductive factors in the immediate microenvironment cause changes within a stem cell that result in commitment. Another view is that these changes occur randomly and that external factors influence only their probability. Such external factors are thought to include both humoral agents and interaction with cellular components of the hemopoietic microenvironment.

It is not known whether the process of commitment is gradual or sudden, i.e., whether a spectrum of stem cells committed to diminishing numbers of cell lines exists (Botnick et al., 1979) or whether a pluripotent stem cell is changed suddenly into a cell committed to a single cell line (as is implied in lineage diagrams such as those in Fig. 3.3). In one key experiment (Abramson et al., 1977) in which chromosome markers were induced in stem cells and the stem cells were serially passaged in lethally irradiated mice, some of the markers appeared in all blood cells, some only in myeloid cells, and a few only in thymic-dependent lymphocytes (T-lymphocytes). This observation suggests that there are at least three types of stem cells: a pluripotent stem cell giving rise to both the myeloid and lymphocytic series; a myeloid stem cell giving rise only to cells of the myeloid series (erythrocytes, granulocytes, monocytes, eosinophils, basophils, and platelets); and a T-lymphocyte stem cell. Other investigators believe that the genetic programs in stem cells for production of the various blood cells are lost in succession, e.g., lymphocyte programs first, granulocyte-monocyte programs second, erythrocyte or megakaryocyte programs third. Whether cells may become committed to a single myeloid cell line, e.g., to the production only of erythrocytes, and still retain the capacity for self renewal, i.e., remain

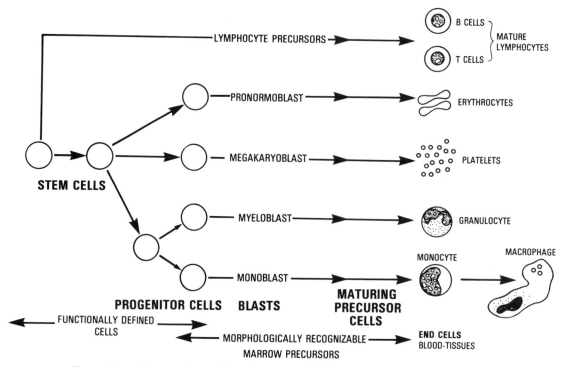

Figure 3.3.   A lineage diagram illustrating the development and maturation of blood cells.

stem cells, is not settled, for only minimal evidence for such stem cells has yet been obtained.

As the process of commitment continues, at some point it becomes irreversible and is accompanied by loss of the capacity for self-renewal. Such committed cells are no longer called stem cells but are called *progenitor cells.* Progenitor cells are thus defined by the following properties. First, when plated in *in vitro* cultures under proper conditions, a single progenitor cell can give rise to a colony of differentiated progeny, e.g., to a colony of erythroid precursors or to a colony of granulocyte-macrophages. Second, when such colonies are harvested and their cells are either replated *in vitro* or injected into irradiated mice, new colonies are generally not formed. Thus, additional progenitor cells are not usually demonstrable within the original colony, which means that the single progenitor cell giving rise to a colony does not undergo self renewal. Third, although committed to the production of a given myeloid cell line, a progenitor cell has not yet acquired distinctive features that permit its recognition by conventional morphological techniques.

Proliferation of bone marrow progenitors is at least in part under humoral control. *In vitro* studies have identified materials, listed in Table 3.5, which may function as specific growth factors regulating the proliferation and maturation of progenitor cells and also of morphologically recognizable, more differentiated precursor cells. These materials are discussed further below.

Table 3.5

Regulators of in vitro proliferation and maturation of the major myeloid cell lines

Cell Line	Material
Erythroid	
BFU-E	Burst-promoting activity
CFU-E	Erythropoietin
Granulocyte-monocyte	
CFU-GM	Colony-stimulating factors
Megakaryocytic	
CFU-M	Megakaryocyte-stimulating activity
Megakaryocytes	Thrombopoietin

BFU-E, burst-forming unit-erythroid; CFU-E, colony-forming unit-erythroid; CFU-M, colony-forming unit-megakaryocyte.

PRODUCTION OF MYELOID CELL LINES

Progenitor cells give rise to blast cells, which are the earliest morphologically recognizeable precursors of the different blood cell lines. Blasts from each of the major myeloid cell lines give rise to successively more differentiated cells within the marrow (Fig. 3.3). As differentiation proceeds, the proliferative capacity of the cell diminishes while the synthesis of cytoplasmic proteins specific for the differentiated cell increases. At some point within each differentiating cell line, DNA synthesis ceases, i.e., the cell enters a postmitotic compartment. In this compartment cytoplasmic maturation continues until a mature or almost mature cell is formed that is ready for release into the blood.

Subtle oscillations are detectable in the blood counts of normal persons, and more pronounced oscillations are sometimes noted in the blood counts of patients with certain hematologic disorders. These oscillations may reflect cyclic adjustments in rates of cell production in response to stimulatory and inhibitory influences within the bone marrow. Servomechanisms involving both positive and negative feedback are thought to be involved.

The above statements apply to all of the myeloid cell lines. Specific features of the production of individual myeloid cell lines are discussed in the following sections.

Erythropoiesis

STAGES OF ERYTHROPIESIS

Erythroid Progenitor Cells

Two types of erythroid progenitor cells (Testa, 1979), burst forming units-erythrocyte (BFU-E) and colony forming units-erythrocyte (CFU-E), have been recognized by the size of the colonies that they form on culture in vitro. BFU-E form very large colonies called bursts; a single BFU-E may give rise to thousands of nucleated erythroid precursors in culture. CFU-E form small colonies of up to 64 nucleated erythroid precursors. BFU-E are the most primitive known erythroid progenitor cell. In vivo, they are thought to give rise, through a continuum of intermediate forms (recognizable as cells producing intermediate-sized colonies in vitro) to CFU-E, which represent the most differentiated erythroid progenitor cell. BFU-E are present in the bone marrow, but also, like stem cells, circulate in small numbers in the blood. CFU-E are present only in the bone marrow. Neither BFU-E nor CFU-E have morphologic features permitting their distinction from each other or their recognition as members of the erythroid series.

BFU-E and CFU-E have different requirements for proliferation in culture. Formation of bursts from BFU-E in murine systems requires addition to cultures of media prepared by exposing mouse spleen cells to pokeweed mitogen. The active material in the media has been named burst-promoting activity (BPA). Whether its cellular source is the T-lymphocyte or the macrophage is not yet settled. A similar protein appears to be required to stimulate human BFU-E to form bursts. CFU-E, in contrast, do not require burst-promoting activity to form colonies in culture; however, they are dependent upon the presence of the glycoprotein erythropoietin. Erythropoietin is also required for the progeny of the nucleated erythroid precursors present in bursts to undergo cytoplasmic maturation in cultures, i.e., to begin hemoglobin synthesis.

When production of erythropoietin is suppressed in a rodent by elevating the animal's erythrocyte count with transfused erythrocytes, marrow CFU-E decrease in number whereas marrow BFU-E increase in number. This observation suggests that although BFU-E may proliferate in culture in the absence of added erythropoietin, they require erythropoietin to differentiate in vivo into CFU-E. From this observation and the observations cited above, one may construct a hypothesis of the regulation of erythroid differentiation. It is postulated that commitment of stem cells to the erythroid series and proliferation of primitive BFU-E reflect primarily the influence of materials from adjacent mononuclear cells in the marrow, e.g., burst-promoting activity. With continuing differentiation, BFU-E are transformed into a series of increasingly differentiated forms that are less responsive to locally produced burst-promoting activity and increasingly responsive to the hormonal influence of erythropoietin. Finally, CFU-E, the most differentiated erythroid progenitor, are formed, and they are responsive only to erythropoietin. As discussed below, the maturation of morphologically recognizable erythroid precursors is also modulated by erythropoietin.

Maturation of Normoblasts

The earliest morphologically recognizable member of the erythroid series is the *pronormoblast* (erythroblast). On conventional methylene blue-eosin stains of the bone marrow, e.g., Wright's stain, the pronormoblast appears as a cell of moderate to large size, with a round nucleus occupying most of the cell and containing fine punctate chromatin and small blue nucleoli, and with a rim of basophilic cytoplasm. This cell matures into a small orange-pink-staining, nonnucleated red blood cell. As maturation proceeds, the size of the cell progressively diminishes; the nucleus shrinks into a purple-black pyknotic mass that is finally extruded; and the cytoplasm changes color from blue, through intermediate shades, to orange-pink.

Three stages in this continuous morphologic sequence of maturation have been arbitrarily delineated. In the earliest, the *basophilic normoblast* stage, the nucleus has lost its visible nucleoli, the nuclear chromatin is beginning to condense, and the cytoplasm has become a deeper blue in color due to an increased RNA content. In the next, the *polychromatophilic normoblast* stage, the nucleus is smaller and contains chunky clumps of chromatin, and the cytoplasm has acquired a hue that varies in different cells from gray-purple to purple-pink, reflecting the combined effect of a decreasing RNA content and an increasing hemoglobin content. In the last, the *orthochromatic normoblast* stage, the nucleus is small, dense, and pyknotic, and the cytoplasm is more orange-pink than purple-gray. Normoblasts at different stages of maturation can be seen in Figure 3.5.

From three to five *cell divisions* take place during pronormoblast-normoblast maturation. Thus, one pronormoblast gives rise to from 8 to 32 red blood cells. Cell division stops at the late polychromatophilic normoblast stage, possibly because a rising cytoplasmic concentration of hemoglobin in some way signals the nucleus to cease DNA synthesis.

Hemoglobin synthesis, the major cytoplasmic event of erythroid maturation, has three requisites. First, adequate amounts of normal mRNAs for the polypeptide chains (α, β, and γ) of the major forms of fetal and adult globin must be transcribed and translated. (In some parts of the world, a high incidence of severe anemias, known collectively as the *thalassemias*, is seen because of genetic disorders resulting in failure of production of globin chains.) Second, the cell must synthesize normal amounts of the tetrapyrrole *protoporphyrin*, which involves a sequence of enzymatically catalyzed reactions that begin in the mitochondrion, move to the cytosol and, finally, move back to the mitochondrion. Third, *iron* must be available to the developing red blood cell. All developing erythroid forms, from the pronormoblast to the reticulocyte (see below), have surface membrane receptors for the plasma iron transport protein transferrin. Plasma transferrin binds to these surface receptors, is internalized by endocytosis, releases its iron and, in turn, is released by the cell to return to the plasma. The iron remains briefly within the cytosol as ferritin and then, within the mitochondrion, is inserted into protoporphyrin to give rise to *heme*. (Impaired erythropoiesis due to insufficient body iron for normal hemoglobin synthesis is the most common, worldwide cause for anemia.) The synthesis and structure of hemoglobin is discussed in more depth in Chapter 24.

Maturation of the Reticulocyte

When first formed by extrusion of the nucleus of an orthochromatic normoblast, a nonnucleated red blood cell still retains some cytoplasmic RNA, does not yet have its full complement of hemoglobin, and is slightly larger than a fully mature red blood cell. Therefore, a newly formed red blood cell can be recognized on a Wright's stained smear as a larger cell with a residual purplish tinge, i.e., as a

polychromatophilic macrocyte. Certain supravital dyes precipitate the RNA of a newly made red blood cell into beads and strands of dark blue material, forming a reticular network within the cytoplasm. Because of this staining phenomenon, the newly made red blood cell is called a *reticulocyte*. Normally, reticulocytes mature for 1–2 days in the marrow before entering the peripheral blood. During this time their size decreases while hemoglobin synthesis continues. Thus, the cell normally entering the blood after this further period of maturation in the marrow is no longer recognizable as a polychromatophilic macrocyte. However, it still retains enough cytoplasmic RNA during the first day of its circulating life-span to be recognizable on supravital staining as a reticulocyte. For this reason, counting reticulocytes on a blood smear provides a convenient clinical method for estimating the rate of red blood cell production.

The mature red blood cell is essentially a 33% (weight/volume) solution of hemoglobin enclosed within a flexible membrane. The cell has lost its ribosomes and mitochondria; therefore, it can not synthesize proteins and relies solely on anaerobic glycolysis for energy. However, the cell does contain the metabolic machinery necessary to maintain the integrity of its surface membrane, to keep its iron within heme in the ferrous form, and to prevent the oxidative degradation of its hemoglobin during the 120 days of its life-span. The processes whereby these are accomplished are discussed in Chapter 24.

REGULATION AND ASSESSMENT OF ERYTHROPOIESIS (Finch, 1982)

Although adjacent products of mononuclear cells such as burst-promoting activity probably influence the earliest stages of erythroid differentiation in vivo, the hormone *erythropoietin* functions as the major established physiologic regulator of erythropoiesis. Erythropoietin is made in the kidney, probably by cells in the glomerulus or juxtaglomerular body that sense *tissue hypoxia*. Tissue hypoxia may result from one of three general causes: a decreased blood hemoglobin content (anemia); a failure to oxygenate hemoglobin adequately in the lungs (lung diseases, certain congenital heart diseases, high altitude); or an impaired release of oxygen from hemoglobin at normal tissue oxygen tensions (an elevated blood carbon monoxide level, rare hereditary disorders of hemoglobin structure causing increased oxygen affinity).

Normal erythropoiesis requires a basal level of stimulation by erythropoietin of erythroid progenitor cells, particularly of CFU-E. When the basal production of erythropoietin is decreased in end stage kidney disease, patients become severely anemic. When the production of erythropoietin is augmented, as occurs, for example, after acute massive blood loss, the production of nucleated red blood cell precursors is increased. This can be recognized by an increase in the relative proportion of erythroid precursors to granulocytic forms in the bone marrow (a decreased myeloid:erythroid ratio), accompanied by an overall increase in marrow cellularity. Other effects of increased erythropoietin stimulation include: a reduced intermitotic time of developing normoblasts, an increased rate of hemoglobin synthesis within individual cells, skipped cell divisions during maturation, and an early release from the marrow of reticulocytes. These responses shorten the time for generation and delivery of new red blood cells. This is reflected in a shortened marrow radioiron transit time from its normal value of 3.5 days. (The marrow radioiron transit time is the half-time of appearance in the circulation of labeled red blood cells after an intravenous injection of radioactive iron.) Moreover, the early release of reticulocytes, before they can mature in the marrow, causes polychromatophilic macrocytes to be present on peripheral blood smears stained with Wright's stain. (This finding alerts one to the possibility of stimulated erythropoiesis in a patient.)

Total erythropoiesis can be evaluated from the *myeloid:erythroid ratio* of a bone marrow specimen and, since most of the plasma iron is incorporated into hemoglobin, from the *plasma iron turnover*. The latter is calculated from the serum iron level and the half-disappearance time of a tracer dose of radioactive iron injected intravenously. In healthy persons a small fraction of erythropoiesis is ineffective, *i.e.*, gives rise to defective cells that instead of circulating are phagocytosed by macrophages within the marrow. (In certain disorders of erythropoiesis, e.g., pernicious anemia, ineffective erythropoiesis may predominate.) *Effective erythropoiesis*, i.e., erythropoiesis producing red blood cells that circulate, may be estimated from the *absolute reticulocyte count* corrected for any early release of marrow reticulocytes (Finch, 1982) and from the curve of appearance of labeled red blood cells in the circulating blood following an intravenous injection of a tracer dose of radioactive iron. *Ineffective erythropoiesis* is estimated as the difference between total erythropoiesis and effective erythropoiesis; in practice it is usually evaluated by comparing the number of nucleated red blood cells seen in a bone marrow specimen with the number of reticulocytes counted in a blood sample.

When erythropoiesis is stimulated by erythro-

poietin and large amounts of iron for hemoglobin synthesis are readily available from the breakdown of red blood cells, as occurs in most hemolytic disorders, effective erythropoiesis may increase up to 6-fold. When erythropoiesis is stimulated by erythropoietin but only limited amounts of iron are readily available from normal body stores, as happens after acute blood loss, effective erythropoiesis usually increases only about 2 fold.

Production of Granulocytes and Monocytes-Macrophages

Progenitor cells are present in bone marrow and, in smaller numbers, in blood, which give rise on culture in vitro to colonies of granulocytes, macrophages, or both. Since granulocytes and macrophages are found in the same colonies and since all cells in a colony arise from a single progenitor cell, one may conclude that granulocytes and monocytes-macrophages arise from a common committed progenitor cell. This cell has been named the CFU-GM (colony-forming unit-granulocyte-macrophage). CFU-GM form colonies in culture only in the presence of specific glycoprotein molecules referred to, collectively, as CSF (colony-stimulating factor) or CSA (colony-stimulating activity (Burgess and Metcalf (1979) and Moore (1979)). In murine systems distinct CSFs have been delineated for macrophages, granulocytes, and eosinophils. In human systems CSFs with similar biologic activities are thought to exist but are less well characterized.

Monocytes, macrophages, and mitogen-stimulated T-lymphocytes are well-established cellular sources of CSF. Moreover, monocytes have been reported to secrete a soluble substance that stimulates T-lymphocytes to make CSF. Synthesis of CSF by endothelial cells has also been demonstrated and, indeed, many types of normal cells may have some capacity to produce CSF. Media conditioned by exposure to different human tissues, e.g., placenta, lung, or mitogen-stimulated leukocytes, have been used as experimental sources of human CSF. CSFs with different molecular weights and different biological activities have been partially purified from such conditioned media.

CSF promotes both the *proliferation* and *maturation* of normal progenitor cells in culture and is also required for survival in culture of colony-forming cells. CSF may promote maturation of granulocyte precursors in culture indirectly, by stimulating synthesis of a differentiation-inducing protein by other early cells in the culture. This sequence of events may couple proliferation and maturation in normal granulopoiesis. (CSF reportedly fails to induce synthesis of differentiation-inducing protein in some mouse myeloid leukemia cell lines (Lotem and Sachs, 1982).)

STAGES OF GRANULOCYTE MATURATION

Granulocyte-macrophage progenitor cells can not be recognized morphologically in the bone marrow. The earliest morphologically identifiable member of the granulocytic series is the *myeloblast*. On a bone marrow smear stained with Wright's stain, the myeloblast has a large round nucleus containing fine stippled chromatin and from 1 to 5 nucleoli. Its cytoplasm is blue and does not contain granules. Morphologic stages of the maturation of this cell into a segmented granulocyte have been arbitrarily defined. The first is the *promyelocyte* stage. The promyelocyte is a cell in which the nucleus still has visible nucleoli, the nuclear chromatin is just beginning to condense, and primary granules, which stain as azurophilic granules, are present in the cytoplasm. The promyelocyte matures into the *myelocyte*. At this stage, the cell still has a round nucleus, but its chromatin is further condensed, and nucleoli are no longer visible. The cytoplasm may vary in color in different portions of the cell as the blue color of the immature cell gives way to the yellow-pink color of the mature cell. Secondary granules, which are fine and pink-staining, can be seen in the cytoplasm. Azurophilic granules may still be visible but are reduced in number for two reasons. First, since primary granules are formed only in the promyelocyte and the cell has divided one or more times since that stage, individual cells contain fewer granules. Second, as the cell progressively matures, primary granules, although still present, no longer stain as visible azurophilic granules.

The myelocyte matures into the *metamyelocyte*. At this stage, the cell has an indented nucleus with heavily condensed chromatin. The cytoplasm is uniformly yellow-pink in color, and fine secondary granules are distributed throughout the cytoplasm. Azurophilic granules are no longer visible. As the metamyelocyte matures, its nucleus becomes increasingly indented, finally acquiring a crescent shape. At this stage the cell is called a *band*. Constrictions develop in the nucleus of the band. As these progress, the nucleus is finally segmented into lobes separated by fine filaments of chromatin; the cell is then considered a mature granulocyte.

PROPERTIES OF GRANULOCYTES

During maturation the granulocyte acquires the properties it needs for chemotaxis, phagocytosis, and killing of bacteria. This involves the packing of materials into granules, the acquisition of surface membrane receptors and enzymes, and the development of a contractile system. The *primary granules* of the granulocyte are lysosomal granules, which contain powerful enzymes that digest proteins and carbohydrates. The primary granules also contain the enzyme *myeloperoxidase*, which catalyzes killing of bacteria by reaction products of H_2O_2 and halide ions. When granulocytes phagocytose bacteria, the lysosomal granules discharge their enzymes into the phagocytic vacuole to attack the ingested bacteria. However, some of the enzymes escape from the cells, damaging the surrounding tissues and causing inflammation.

The *secondary granules* (also called specific granules or neutrophilic granules) contain materials that are readily released from the cell by lesser degrees of stimulation. These materials do not damage tissues. One material, an iron-chelating compound called *lactoferrin*, may inhibit bacterial growth by competing with bacteria for iron. Another material, alkaline phosphatase, has proven valuable clinically as a cytochemical marker to distinguish normal granulocytes from the granulocytes of chronic granulocytic leukemia, which are deficient in this enzyme. A third material, transcobalamin I, functions after its release from granulocytes as a plasma vitamin B_{12} binding protein. Its physiological function, if any, within the granulocyte is unknown. One enzyme, *lysozyme*, a mucopeptidase that attacks certain bacteria, is found in both primary and secondary granules.

As granulocytic precursors mature, they both gain and lose *surface membrane antigens*. Patterns of surface membrane antigens, determined by immunofluorescence and other immunologic techniques, can be used to characterize different stages of maturation. Maturing granulocytes acquire *surface membrane receptors*. These include a receptor for the Fc fragment of IgG and for an activated complement component, C3b; these receptors are required for efficient phagocytosis of bacteria coated with immunoglobulin and complement. The cell also develops membrane receptors for chemotactic materials that direct movement of granulocytes through vessel walls and into tissue sites. Enzymes are also incorporated into the surface membrane. One key surface membrane enzyme catalyzes a burst of oxygen consumption associated with phagocytosis that results in the formation of toxic oxygen metabolites needed for bacterial killing.

Granulocyte function involves mechanical events requiring directed cell contraction, e.g., as the granulocyte squeezes between endothelial cells to leave the blood; as portions of its membrane invaginate to form phagocytic vacuoles; and as granules move within the cell to fuse with phagocytic vacuoles. Cytoplasmic proteins that mediate the contractile process are synthesized during maturation. One protein, tubulin, is not directly involved in contraction but polymerizes to form microtubules that provide an internal skeleton to support contraction. Four other proteins are synthesized that, as in muscle, participate directly in the contractile process: actin, myosin, actin-binding protein, and gelsolin.

KINETICS OF GRANULOCYTE PRODUCTION (Cronkite, 1979)

When radiolabeled granulocytes are injected into a subject and a blood sample is drawn after equilibration, only 52–58% of the injected radioactivity can be accounted for in the circulating blood. Thus, only about one-half of the labeled granulocytes are circulating; the remainder are adherent to endothelial cells of postcapillary venules in a *marginal pool* that exchanges freely with the *circulating pool*. If the radioactive counts obtained in serial blood samples are plotted against time on semilogarithmic paper, a straight line of disappearance of radioactivity is obtained. This means that granulocytes are lost randomly from the intravascular compartment into the tissues. The half-time for the disappearance of cells from the intravascular compartment is about 6 h for DF^{32}P-labeled autologous cells and about 7.5 h for ^{3}H-thymidine-labeled isologous cells. A *blood granulocyte turnover rate* can be calculated from these values, and from the blood granulocyte count, the marginal pool size, and the blood volume. The mean turnover rate obtained with DF^{32}P cells is 1.6×10^{9} granulocytes/kg/day; the mean turnover rate obtained with ^{3}H-thymidine-labeled cells is 0.85×10^{9} granulocytes/kg/day. Thus, in a 70-kg man, about ten thousand-million granulocytes will pass each day through the blood into the tissues.

In the steady state, values for blood granulocyte turnover rate also represent values for the *rate of marrow granulocyte production*, if one neglects the small fraction of normal human granulopoiesis that is ineffective. The granulocytic series exists in the marrow in two pools: a *mitotic pool* of cells that divide and a *postmitotic* or *maturation pool* of cells that can no longer divide. The mitotic pool is made up of myeloblasts, promyelocytes, and myelocytes. For each myeloblast there are about 3 promyelocytes, and 13 myelocytes. From this and other observations it is thought that early granulocytic precursors undergo four or five divisions in the mitotic

compartment: one at the myeloblast stage, one or two at the promyelocyte stage, and two or three at the myelocyte stage. Data obtained from counting proportions of cells in mitosis and from pulse labeling cells in DNA synthesis with ^{3}H-thymidine have been used to calculate a compartment transit time, i.e., the time between a cell entering the compartment as a myeloblast and leaving the compartment as a metamyelocyte. Values of from 3 to 5 days have been obtained.

The postmitotic compartment is made up of metamyelocytes, bands, and granulocytes. The transit time through the postmitotic compartment can be determined by pulse labeling with ^{3}H-thymidine and by noting the elapsed time between the detection of labeled metamyelocytes in the marrow and the detection of labeled granulocytes in the blood. A value of 6.6 days has been obtained. Adding this value to the transit time for the mitotic compartment gives a total time of about 10 days for normal granulopoiesis, i.e., from the entrance of a cell into the mitotic compartment as a myeloblast to the appearance of its daughter granulocytes in the peripheral blood. This time shortens when the tissue demand for granulocytes increases. The postmitotic pool transit time may be reduced to 2 days in infectious and inflammatory states.

A *postmitotic pool size* can be calculated from the postmitotic pool transit time and the blood granulocyte turnover rate. Using values of 6.6 days for the transit time and 0.85×10^{9} granulocytes/kg/day for the turnover rate, one obtains a value for the postmitotic pool size of 5.8×10^{9} cells/kg, i.e., about 400 billion cells in a 70-kg man. The portion of this pool that is granulocytes, which is determined from differential cell counts of the marrow, is about three times the blood granulocyte turnover rate. Thus, the normal marrow contains a 3-day reserve of mature granulocytes available for rapid release into the blood.

The blood granulocyte count may rise because of redistribution of cells within the intravascular compartment, i.e., release of cells from the marginal pool into the circulating pool, as occurs after heavy exercise or an injection of epinephrine; because of increased release of cells from the postmitotic marrow pool into the intravascular compartment, as occurs at the onset of most acute inflammatory states; or because of both processes. When the count rises only because of release of cells from the marginal pool, the increased count is not associated with the appearance of bands in the blood. In contrast, when the count rises because of increased release of cells from the postmitotic marrow pool, bands will usually be released into the blood along with mature granulocytes. (An elevated blood granulocyte count with bands on the stained peripheral blood smear alerts one to look in a patient for a condition, e.g., an acute bacterial infection, stimulating an increased marrow release of granulocytes.)

REGULATION OF GRANULOPOIESIS

The extrapolation of *in vitro* data to the physiologic regulation of granulopoiesis is as yet largely speculative. Because progenitor cells are absolutely dependent upon CSF for growth and viability in culture, because radiolabeled CSF binds specifically to bone marrow cells, and because CSF is present in biological fluids, it is reasonable to conclude that CSF plays a role in the physiologic regulation of granulopoiesis. Nevertheless, direct evidence is lacking that administration of exogenous CSF to an animal stimulates granulopoiesis. Such experiments present formidable technical difficulties, for they require that CSF be purified from trace contaminants, e.g., endotoxin, capable of stimulating granulopoiesis; they require that an animal's endogenous CSF be suppressed; and they require that stimulation of granulocyte production be distinguished from accelerated granulocyte release from the postmitotic pool. Stimulators of granulopoiesis distinct from CSF have also been detected in vitro and could also have physiologic significance.

Figure 3.4, based largely upon in vitro data, illustrates how granulopoiesis could be modulated by positive and negative regulatory influences. The figure shows pathways whereby products of monocytes and granulocytes might feed back and inhibit the proliferative activity of CFU-GM. *Lactoferrin*, which as mentioned earlier is released from the secondary granules of mature granulocytes, has been shown to inhibit the formation of granulocyte-macrophage colonies in vitro. Initially, lactoferrin was thought to turn off production of CSF by monocytes but, more recently, it was reported that lactoferrin prevents monocytes from stimulating T-lymphocytes to produce CSF (Bagby et al., 1981). Lactoferrin is active in vitro only in its monomeric form, yet at the concentrations present in vivo, lactoferrin may exist primarily in a polymeric form. This possibility illustrates the hazards of extrapolating from in vitro observations to the in vivo regulation of granulopoiesis and the need for caution in accepting lactoferrin as a physiologic regulator of granulopoiesis.

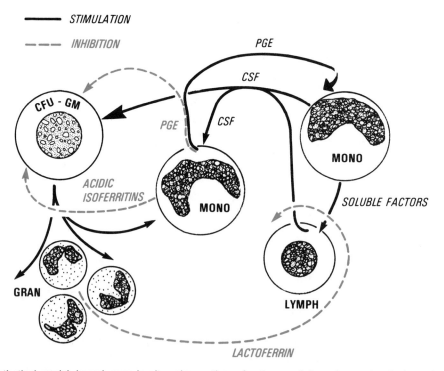

Figure 3.4. A hypothetical model, based upon in vitro observations, for the regulation of granulopoiesis and monopoiesis by the products of monocytes, granulocytes, and T-lymphocytes. In this model *CSF* is produced by monocytes and also by T-lymphocytes stimulated by soluble factors from monocytes. *CSF* acts on progenitor cells, stimulating proliferation and maturation, but *CSF* also stimulates monocytes to produce *PGE* (prostaglandin E). *PGE* decreases the sensitivity of progenitor cells to *CSF* but also stimulates monocytes to release *CSF*. Monocytes also produce acidic isoferritins which inhibit proliferation of progenitor cells. Granulocytes release lactoferrin, which acts as a feedback inhibitor of granulopoiesis by impairing monocyte-induced stimulation of *CSF* synthesis by T-lymphocytes.

Monocytes can synthesize and release prostaglandins of the E series (PGE). PGE decreases the responsiveness to CSF of progenitor cells in culture, possibly by elevating their cyclic AMP levels. CSF stimulates monocytes to release PGE and, conversely, PGE stimulates monocytes to release CSF. Thus, PGE may be viewed as both inhibiting granulopoiesis and monopoiesis, by decreasing the responsiveness of progenitor cells to CSF, and stimulating granulopoiesis and monopoiesis by increasing the release of CSF by monocytes. CSF, in turn, may be viewed as playing multiple roles in the regulation of granulopoiesis and monopoiesis: direct stimulation of proliferation of progenitor cells; indirect promotion of maturation through stimulation of early granulocytic precursors to produce differentiation-inducing protein; and indirect dampening of granulopoiesis and monopoiesis through stimulation of monocytes to release PGE.

Leukemic bone marrow was found to contain activity inhibiting the growth of normal granulocytic precursors but not inhibiting the growth of leukemic cells. This activity was later identified as a property of *acidic isoferritins* and was shown to be present in smaller amounts in normal bone marrow (Broxmeyer et al., 1982). It is not known yet whether acidic isoferritins function as a physiologic regulator of granulopoiesis.

Cell surfaces must be able to interact with modulating environmental factors for the regulation of granulopoiesis. Normal progenitor cells and myeloblasts have a surface Ia-like antigen, a gene product of the HLA system (see Chapter 22). The presence of surface Ia-like antigen appears linked to the ability of PGE and acidic isoferritins to inhibit granulopoiesis in vitro (Broxmeyer (1982) and Pelus (1982)). Thus, experimental conditions that cause loss of surface Ia-like antigen from progenitor cells in vitro are associated with loss of responsiveness of the cells to these inhibitory materials.

The progenitor cells of patients with chronic granulocytic leukemia lack surface Ia-like antigen and resist inhibition by PGE or acidic isoferritins in vitro. The growth of malignant cells in acute granulocytic leukemia is also resistant to inhibition by acidic isoferritins in vitro. Thus, the increased amounts of acidic isoferritins present in leukemic bone marrow plus the unresponsiveness of leukemic cells to inhibition by isoferritins could confer a growth advantage upon leukemic cells that would allow them eventually to replace the progeny of normal progenitor cells in the bone marrow.

MONOCYTES-MACROPHAGES

As already discussed, monocytes-macrophages and granulocytes arise from a common progenitor cell. Monocytes-macrophages mature in the following sequence: within the bone marrow, progenitor cell to monoblast to promonocyte to monocyte; within the tissues, monocyte to macrophage. The tissue macrophage pool, which far exceeds the combined size of the bone marrow and blood monocyte pools, includes the following: the von Kupffer cells lining the sinusoids of the liver; the macrophages within the substance and lining the sinusoids of the spleen and bone marrow; the macrophages lining the peripheral sinus of and distributed throughout lymph nodes; the alveolar macrophages of the lungs; and the macrophages free within the pleural and peritoneal cavities and scattered in variable numbers throughout most of the tissues. Tissue macrophages, particularly those in the liver, spleen, and bone marrow, were formerly referred to collectively as the reticuloendothelial system. However, the endothelial cells lining most of the microcirculation arise from endodermal cells whereas the macrophages lining the sinusoids of the liver, spleen, and bone marrow arise from mesodermal bone marrow elements. Therefore, the name reticuloendothelial system is inappropriate, and the macrophages of the tissues are now referred to as the *mononuclear phagocyte system*.

Monoblasts may be difficult to distinguish from myeloblasts in a normal bone marrow aspirate stained with Wright's stain. Promonocytes can be recognized as cells with a round to moderately indented nucleus, fine nuclear chromatin, and deeply basophilic cytoplasm containing a few granules. Mature monocytes are identified by their irregularly oval to deeply indented or "folded" nucleus, containing lacy chromatin, and by the gray-blue color and fine bluish-pink granules of their cytoplasm. The cytoplasm of promonocytes and monocytes contains sodium fluoride-inhibitable nonspecific esterases. Staining for these esterases has proven a reliable method for distinguishing promonocytes and monocytes from granulocytic precursors and from activated lymphocytes. In one study in which this staining technique was used, promonocytes were found to make up only about 0.05% and monocytes only about 1% of the nucleated cells in normal marrow. In another study, the monocytic series was reported to make up about 3% of the nucleated marrow cells.

When incubated in vitro with ^{3}H-thymidine, promonocytes are labeled but monocytes are not (van Furth et al., 1979). This indicates that promonocytes synthesize DNA, i.e., are actively dividing cells, whereas monocytes are not actively dividing cells. In vivo labeling studies with ^{3}H-thymidine in mice suggest that a monoblast divides to give rise to two promonocytes and that each promonocyte then divides only once to give rise to two monocytes. In mice and also in a single study in which hematologically normal humans were given ^{3}H-thymidine, labeled monocytes were found to leave the bone marrow compartment within 24 hours. No evidence was obtained for a bone marrow reserve of monocytes. Data on the intravascular residence time of monocytes in humans differ. In one study in which monocytes were labeled in vitro with ^{3}H-DFP and then reinfused, an intravascular half-time of 8.4 hours was found (Meuret and Hoffmann, 1973). The data also suggested the presence of a marginal pool 3½ times the size of the circulating pool. In contrast, in the study in which ^{3}H-thymidine was given to humans and the appearance and disappearance of labeled blood monocytes was followed, a much longer intravascular half-time of 71 hours was reported, and no evidence was obtained for a marginal pool (Whitelaw, 1972).

Whereas granulocytes arrive in the tissues fully differentiated to perform the specific tasks of phagocytosis and microbicidal killing, monocytes arrive in the tissues ready for further differentiation into cells capable of performing a remarkable variety of tasks. These include: the processing of antigen and modulation of lymphocyte activity in immune responses (see Chapter 22); the phagocytosis of certain types of microorganisms and of dead cells and denatured connective tissue matrix materials in inflammatory responses; the clearance from the circulating blood, by fixed macrophages lining the sinusoids of the liver, spleen, and bone marrow of denatured proteins, senescent red blood cells, and invading microorganisms; and important general functions related to homeostasis and body defense such as the secretion of CSF and interleukin 1 (see Chapter 22). Local environmental factors influence the differentiation of monocytes into cells capable of expressing these different functional activities.

In acute inflammation, first granulocytes and then macrophages accumulate in the inflamed tissue. More than 70% of the macrophages are newly derived from blood monocytes attracted to the inflamed site. Monocytes are retained at the site by processes that cause the cells to adhere to surfaces and differentiate. One such process involves a striking increase in the surface area of the cell, a phenomenon called spreading. Spreading results from reactions at the monocyte surface involving two components of the complement system, factors B and C5 (see Chapter 23) and the fibrinolytic enzyme plasmin. A second mechanism for adhesion and retention involves the binding of monocytes-macrophages to a protein of plasma, connective tissue matrix, and cell surfaces called fibronectin (Bevilacquia et al., 1981). Fibronectin binds avidly to denatured collagen and to fibrin at an inflammatory site (Mosher et al., 1981) and monocytes-macrophages, through specific surface receptors, bind to fibronectin. Binding to fibronectin also enhances the activity of monocyte-macrophage surface receptors involved in phagocytosis (Fc and C3b receptors) and induces secretion of proteases and plasminogen activator necessary for the scavenger function of macrophages.

Scanty data are available on the life-span of tissue macrophages. A turnover time of 21 days has been reported for von Kupffer cells in mice (Crofton et al., 1978). Studies of the replacement of host alveolar macrophages by donor alveolar macrophages after bone marrow transplantation in humans suggest a turnover time of alveolar macrophages of 80 days.

Production of Eosinophils, Basophils, and Mast Cells

EOSINOPHILS

Pure colonies of eosinophils can be obtained on culture of human bone marrow cells in vitro, which suggests that a separate progenitor cell exists for eosinophils. A CSF specific for eosinophils has been isolated in the mouse and probably also exists in man. Eosinophils mature in the bone marrow through stages similar to those described for neutrophilic granulocytes. Cells of the eosinophilic series make up about 3% of the nucleated blood cells in normal bone marrow. Most kinetic data on the production of eosinophils has been obtained in rodents; these have revealed a mean marrow transit time of 5.5 days and an intravascular half-time of 8–12 h. A marginal pool exists that is equal in size to the circulating pool. For each circulating eosinophil, 100 are reportedly present in the tissues, where they are found primarily in the skin and in the submucosa of the respiratory, gastrointestinal, and genitourinary tracts.

On preparations stained with Wright's stain, the mature eosinophil has a bilobed nucleus and cytoplasm that is a light blue color but is often difficult to see because of the many large orange-colored (eosinophilic) granules filling the cell. A second population of finer cytoplasmic granules is also present but is not visible by using usual techniques. Eosinophils contain different constituents than those present in the much more numerous neutrophilic granulocytes (Butterworth and David, 1981). These include a material called major basic protein (MBP), which makes up about 50% of the mass of the large granules. MBP is thought to play a key role in the eosinophil's ability to damage the larva-tissue stage of helminthic parasites. In addition, substantial amounts of another protein, called eosinophilic cationic protein (ECF), are present within the large granules. Its function and relation to MBP are not clear. The large granules also contain an eosinophil peroxidase with properties different from the myeloperoxidase of neutrophilic granulocytes. Within its finer granules the eosinophil contains some 10 to 20 times more aryl sulfatase B than is found in neutrophilic granulocytes. This enzyme, which hydrolyzes S-O ester linkages, can inactivate the sulfur-containing leukotrienes (slow reacting substance of anaphylaxis) that tissue mast cells liberate in immediate hypersensitivity reactions.

Factors influencing eosinophil production, release, and migration into tissues have been described (Mahmoud, 1980). Their relation to each other and their relative importance in regulating eosinophil kinetics under normal conditions and in conditions associated with eosinophilia are poorly understood. When mice were given a heterologous anti-eosinophil antiserum, their blood eosinophil levels fell; concomitantly, a low molecular weight peptide was demonstrable in their serum that stimulated eosinophil production when infused into other mice. This material was given the name eosinophilopoietin. Its cellular source is unknown.

The eosinophilia induced by parasitic infection in mice apparently requires the participation of a material or materials derived from T-lymphocytes. Thus, depletion of T lymphocytes by thymectomy and irradiation prevents eosinophilia following infection of mice with *Trichinella*. Moreover, when spleen lymphocytes obtained from Trichinella-infected mice are activated by exposure to Trichinella antigen, they release a soluble factor that stimulates the growth of eosinophils in murine marrow cultured in vitro. Mast cells release materials that are chemotactic for eosinophils and may also influence their marrow production and release; these materials include two tetrapeptides called the eosinophil chemotactic factor of anaphylaxis, higher molecular weight peptides, and histamine. Mast cell factors are thought to play a major role in the accumulation of eosinophils at sites of allergic tissue reactions.

Administration of adrenal glucocorticoids causes blood eosinophil levels to fall within about 2 hours, presumably because of margination or sequestration of circulating cells. Continued administration of adrenal glucocorticoids results in a continued eosinopenia associated with impaired release of eosinophils from the bone marrow. Eosinopenia is also found in acute bacterial infections, and this eosinopenia is also associated with a diminished release of marrow eosinophils. However, the eosinopenia of bacterial infections apparently does not result from increased secretion of adrenal glucocorticoids secondary to the stress of infection, since adrenalectomy in mice failed to prevent the eosinopenia of acute bacterial infection (Bass, 1975).

BASOPHILS AND TISSUE MAST CELLS

Basophils are multilobed cells that on Wright's stain contain large, purple-black metachromatic granules. They are the least numerous of human white blood cells (see Table 3.4), making up less than 0.1% of nucleated bone marrow cells. Basophils are produced in a maturation sequence similar to that for neutrophilic granulocytes and eosinophils. Basophils are thought to remain in the blood only for hours; their fate in the tissues is unknown.

The granules of the basophil contain histamine. The cell surface membrane contains receptors for the Fc fragment of IgE. These receptors are occupied to a varying degree with IgE molecules. When antigen reacts with IgE bound to basophils, reactions may be triggered that cause the basophils to release the contents of their granules. A related cell found in the tissues, the mast cell, also possesses metachromatic granules containing histamine and surface membrane receptors for IgE. Mast cells are found in abundant numbers in human lung tissue, skin, lymphoid tissue, and the submucosal layers of the digestive tract. Mast cells play a central role in triggering immediate hypersensitivity reactions (see Chapter 22).

Mast cells differ morphologically from basophils in possessing a round rather than a multilobed nucleus. Moreover, mast cell granules contain the proteoglycan heparin, whereas basophil granules contain chondroitin sulfate as their proteoglycan. Thus, unlike the blood monocyte to tissue macrophage sequence, blood basophils are not the precursors of tissue mast cells. However, mast cells, like basophils, may arise from a precursor cell in the bone marrow. When mice were irradiated and then transplanted with bone marrow cells from a strain having unusually large granules that can serve as a marker, the mast cells in the tissues of the irradiated animals were gradually replaced by mast cells containing the granule marker. Moreover, mice with a genetic stem cell defect causing reduced numbers of circulating blood cells also have markedly reduced numbers of mast cells in their tissues. When such mice were irradiated and then transplanted with normal stem cells, the number of mast cells in their tissues substantially increased (Kitamura et al., 1978).

Thrombopoiesis

STAGES OF THROMBOPOIESIS

Platelets are derived from cells in the bone marrow called megakaryocytes, which are readily recognized by their large-sized multilobulated nucleus and abundant cytoplasm (Fig. 3.5). Normal marrow has been estimated to contain about 1 megakaryocyte/500 nucleated red blood cells. With the development of techniques for culturing megakaryocytes in vitro, it became possible to identify a committed progenitor cell in bone marrow specific for megakaryocytes. This cell was named the CFU-M (colony-forming unit-megakaryocyte). One human CFU-M can give rise in culture to a pure colony of up to 50 megakaryocytes. CFU-M can not be identified by other properties.

CFU-M may divide repeatedly in culture and then give rise to small cells, about the size of a lymphocyte, called promegakaryoblasts. These cells also do not have a distinctive morphology but are identified in human systems through techniques utilizing either ultrastructural cytochemistry or immunofluorescence as cells with the beginnings of megakaryocytic differentiation. The former detects a marker, platelet peroxidase, in the endoplasmic reticulum; the latter detects a surface antigen reacting with an antibody against a platelet surface glycoprotein complex (GP II_a-III_b).

Promegakaryoblasts probably can no longer divide; instead, they undergo endoreduplication, i.e., replication of nuclear material without division of the cytoplasm. Megakaryoblasts, the first cells identifiable by conventional morphologic techniques as members of the megakaryocytic series,

are also capable of endoreduplication. Endoreduplication results in the formation of three major ploidy classes of megakaryocytes. (Ploidy refers to the number of sets of chromosomes in a cell; almost all human cells contain two sets of chromosomes, i.e., are diploid or 2n in class.) About ⅔ of the megakaryocytes are 16n with a chromosome complement 8 times that of a diploid cell; about ⅓ are 8n; and about ⅓ are 32n.

Endoreduplication ceases before significant maturation of megakaryocytes begins. As maturation proceeds, the megakaryoblast, which appears on bone marrow smears stained with Wright's stain as an intermediate-sized cell with an immature irregularly oval nucleus and sparse deeply basophilic cytoplasm, is changed into a much larger mature megakaryocyte with a multilobulated nucleus and plentiful cytoplasm filled with reddish-purple granules. The ploidy class of the megakaryocyte determines the size and degree of lobulation of its nucleus; each nuclear lobule corresponds approximately to 2n chromosomal material. The ploidy class also determines the amount of cytoplasm; megakaryocytes with a higher ploidy have more cytoplasmic mass and, therefore, give rise to increased numbers of platelets, larger platelets, or both. At the ultrastructural level, cytoplasmic maturation is characterized by the development of granules and mitochondria and by the appearance of an increasing amount of plasma membrane within the cytoplasm in the form of tubules and branching cysternae (demarcation membrane system) whose lumen communicates with the outside of the cell. The demarcation membrane system becomes the plasma surface membrane and open canalicular system (Fig. 3.30) of the platelet.

Mature megakaryocytes are ameboid cells that extend cytoplasmic pseudopods through the endothelial lining cells into the lumen of the marrow sinusoids, where the pseudopods then fragment. The bare nucleus of the megakaryocyte that has thus shed its cytoplasm is then engulfed and digested by macrophages in the bone marrow. The initial fragments of megakaryocyte cytoplasm may fragment further within the intravascular compartment and then acquire the typical discoid shape, marginal bundle of microtubules, and submarginal dense tubular system characteristic of platelets (see Fig. 3.30). Platelets remain within the intravascular space for their lifetime, but at any moment about ⅓ are pooled within an unique component of the intravascular compartment, the red pulp of the spleen. (In certain diseases with splenomegaly, up to 85% of the platelet mass may be pooled in a greatly expanded red pulp, with a resultant decrease in the peripheral blood platelet count.)

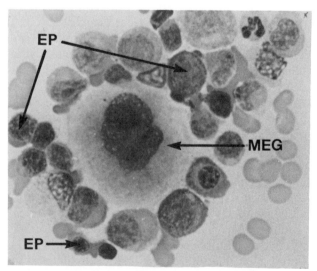

Figure 3.5. A microphotograph of a bone marrow smear showing a megakaryocyte with a polypoid nucleus (*MEG*), erythroid precursors at different stages of maturation (*EP*) and granulocyte precursors at various stages of maturation (unlabeled).

In rats, labeling studies of ^{3}H-thymidine suggest a transit time of a little less than 3 days for a cell to progress from the stage of a 2n morphologically unrecognizable precursor cell in cycle to the stage of a megakaryocyte with a bare nucleus. Such data are not attainable in humans because of the exposure to radiation that would be involved. However, it is known that about 5 days are required for the platelet count to return to normal in patients whose platelets have been depleted acutely, e.g., a patient with a massive gastrointestinal hemorrhage whose blood volume is replaced with stored blood that does not contain viable platelets.

REGULATION OF THROMBOPOIESIS

Thrombocytopenia, induced experimentally with heterologous antiplatelet antiserum, increases platelet production, whereas thrombocytosis, induced experimentally by platelet transfusions, suppresses platelet production. In acute thrombocytopenic states resulting from platelet destruction, platelet production may increase from 2- to 4-fold over several days; in chronic thrombocytopenic states resulting from platelet destruction, platelet production may be increased 6- to 10-fold (Harker, 1974).

When an experimental animal receives a severe thrombocytopenic stimulus, a small number of platelets, equivalent to only about 1 day's expected platelet production, are rapidly released into the blood. (This limited marrow reserve of platelets contrasts with the large marrow reserve of granulocytes.) Studies in which animals were given radioisotopes that are incorporated into platelet proteins, [35]sulfur (S) or [75]selenomethionine (Se), indicate that a reciprocal loss of mature megakaryocyte mass accompanies the rapid platelet release, as does an increase in the rate of megakaryocytic maturation.

Within 24 hours after induction of experimental thrombocytopenia in mice, both the number and size of the megakaryocytes in the marrow are increased. The increase in size reflects an extra mitosis during endoreduplication. The increase in numbers probably reflects an increase in divisions of CFU-M already in cycle and their development into morphologically recognizable early megakaryocytic forms. The total number of CFU-M, as measured in colony-forming assays, also increases but only after a delay of 2–3 days (Burstein et al., 1981).

Regulation of thrombopoiesis may be a 2-tiered process (Williams and Levine, 1982). Two humoral materials, one affecting proliferation and the other affecting maturation and release, have been identified by their functional activities. The first material to be identified, thrombopoietin, is assayed in vivo (Levin and Evatt, 1979). A test sample is infused into an animal whose endogenous platelet production has been suppressed by prior platelet transfusions, and the effect of the test sample upon the rate and extent of incorporation of ^{35}S or ^{75}Se into the animal's platelets is measured. An increased incorporation reflects stimulation of both platelet protein synthesis during maturation and platelet release. By use of this technique thrombopoietin has been demonstrated in the plasma of animals made acutely thrombocytopenic and in the plasma of some, but not all, patients with various types of thrombocytopenia.

The second humoral agent, discovered more recently, affects the early proliferative phase of megakaryopoiesis (Hoffman et al., 1981). It is assayed in vitro, as an activity in serum or a crude concentrate of urine that can increase the number and size of megakaryocyte colonies obtained on culture of bone marrow cells. This material has been named Meg-CSA (megakaryocytic-colony-stimulating activity). Since Meg-CSA has not been further purified, it has not been possible to determine whether Meg-CSA possesses thrombopoietin activity in vivo. However, experimental sources of thrombopoietin have not exhibited Meg-CSA activity. Meg-CSA is found in high titer in the serum of patients with severe thrombocytopenia and diminished numbers of marrow megakaryocytes. It has not been detected in the serum of patients with severe thrombocytopenia and increased numbers of marrow megakaryocytes, e.g., in the serum of patients with autoimmune thrombocytopenic purpura. The primary present source of Meg-CSA is the serum of patients with aplastic anemia.

The relation, if any, is unknown between Meg-CSA and an activity present in lectin-stimulated lymphocyte-conditioned medium that supports the growth in in vitro cultures of mixed myeloid cell lines that may include megakaryocytes. However, Meg-CSA has been shown to be distinct from erythropoietin, from burst-promoting activity, and from granulocyte-colony stimulating activity, none of which stimulate the growth of megakaryocytic colonies in culture.

The sensors triggering the synthesis and release of thrombopoietin and Meg-CSA, the sites and cellular sources of production of these materials, and their chemical compositions are unknown.

DISORDERS OF STEM CELLS

Aplastic Anemia

Since blood cells have finite life-spans, hematopoietic stem cells must continually supply new progenitor cells for the production of new blood cells.

Failure of this stem cell function results in aplastic anemia, a disorder in which the mature forms of all of the myeloid cell lines are markedly decreased in number in the peripheral blood (pancytopenia). Within 2 weeks of an event that destroys stem cells or severely impairs their function (e.g., exposure to a very high dose of radiation), an individual will become highly susceptible to bacterial infection because of severe granulocytopenia and to serious bleeding because of severe thrombocytopenia. Within about 2 months, the circulating red blood cell mass will fall to where red blood cells must be transfused to maintain tissue oxygenation. On bone marrow examination, fat will be found to have replaced normal marrow cellular elements (marrow aplasia).

Aplastic anemia may be caused by death of stem cells. Damage to the stem cell pool is usually so extensive that the remaining stem cells, if any, can not proliferate to replace those destroyed. The patient eventually dies of some complication of persisting pancytopenia, usually infection or bleeding. In young persons with a genetically matched sibling donor, bone marrow transplantation has proven an effective therapy for this otherwise highly fatal disease. Aplastic anemia may also result from suppression of stem cell growth by an abnormal immune response. This appears to be the mechanism for rare instances of aplastic anemia following acute viral hepatitis. Bone marrow function has been reported to improve in some of these patients following the administration of anti-thymocyte globulin, which is a heterologous antiserum made against human lymphocytes.

Myeloproliferative Disorders

Normal polyclonal hemopoiesis reflects the finely regulated activity of many stem cells. In the hematologic diseases that make up the myeloproliferative disorders, an event or events are triggered within a single stem cell that convey upon its descendents a selective growth advantage. As a result, hemopoiesis becomes monoclonal as the abnormal clone "takes over" blood cell production. Initially, normal stem cells may still be present, but their activity is suppressed; later, normal stem cells may be extinguished. The abnormal clone does not respond appropriately to regulatory influences, and blood cells are overproduced. Hemopoiesis may no longer be limited to the bone marrow; the spleen and sometimes the liver may enlarge as stem cells from the abnormal clone initiate hemopoiesis in these organs. Although increased proliferative activity of all blood cell lines is characteristic of the early stages of this group of disorders as a whole,

proliferation of one cell line usually predominates in the clinical manifestations of individual disorders, e.g., erythroid proliferation in *polycythemia vera*, granulocytic proliferation in *myeloid metaplasia*, and megakaryocytic proliferation in *essential thrombocythemia.* Qualitative abnormalities of the blood cells may also become apparent. Thus, in some patients the red blood cells may become increasingly misshapen with the formation of large tear drop-shaped cells, and the platelets may function defectively. In one of the more common disorders of this group, called *myelofibrosis with extramedullary myeloid metaplasia*, the bone marrow becomes increasingly occupied by fibrous tissue, probably because of abnormal release from megakaryocytes and platelets of a growth factor stimulating fibroblasts, while the spleen becomes massively enlarged because of extramedullary hemopoiesis.

Granulocytic Leukemias

Chronic granulocytic leukemia may be considered a myeloproliferative disorder in that normal hemopoietic cells are replaced by an abnormal clone arising from the transformation of a stem cell. It is not yet clear whether the transformation occurs in a pluripotent stem cell or in a stem cell common to all myeloid cell lines and to B-, but not to T-lymphocytes. The transformation apparently results from a somatic mutation; it produces a characteristic chromosomal marker, a shortened chromosome 22 (Philadelphia chromosome) demonstrable in all dividing cells of the abnormal clone. Chronic granulocytic leukemia differs from the myeloproliferative disorders mentioned above in the striking aggressiveness of the proliferation of the abnormal clone. Early in the disease megakaryocytes participate in the proliferative process, but as the disease progresses proliferative activity becomes increasingly confined to the granulocyte-monocyte cell line. These cells, particularly granulocytic forms, relentlessly increase in numbers, expanding and completely filling the marrow cavity. In an untreated patient the WBC count may rise into the hundreds of thousands per microliter range, and the spleen becomes of huge size. In some patients the disorder may abruptly transform into acute granulocytic leukemia or into acute lymphocytic leukemia.

The acute granulocytic leukemias are a group of closely related disorders in which the predominant cell in the bone marrow is an abnormal myeloblast. Increased numbers of promyelocytes or of aberrant forms with features of both early granulocytic and monocytic forms may also be found. Despite the large number of primitive cells in the marrow, only a small proportion of the cells are actively proliferating. The majority of the cells are quiescent and are apparently unable to differentiate into more mature granulocytic or monocytic cells. Normal hematopoietic cells are reduced or virtually absent, reflecting a suppressed normal stem cell activity. The etiology of acute granulocytic leukemia is unknown; occasionally, the disease follows an event known to damage stem cells, e.g., exposure to radiation or treatment with cytotoxic drugs. The pathogenetic mechanism may involve the malignant transformation of a stem cell, whose progeny then both lose programs for cell maturation and acquire resistance to most normal regulatory influences. In some patients, treatment with aggressive cytotoxic chemotherapy allows normal stem cells to function again and repopulate the marrow. The resulting remission may last for months or years before malignant cells grow back to replace the normal marrow elements.

CHAPTER 22

Lymphocytes and Immune Responses

PRODUCTION, DISTRIBUTION, AND FUNCTIONS OF LYMPHOCYTES

Lymphocytes are the effector cells for immune responses to antigens, i.e., to materials not recognized as self and therefore eliciting reactions designed to neutralize or destroy "nonself." Immune responses play vital roles in protecting the body against invasion by microorganisms. Immune responses also may provide surveillance against the development of malignant tumors by attacking cells whose surface antigens differ sufficiently from normal cells to be recognized as "foreign." Immune responses cause rejection of an organ transplant from an individual with different histocompatibility antigens from the host and also cause a converse reaction, graft vs. host disease, in which donor lymphocytes from a transplant attack host cells that they recognize as foreign.

Lymphocytes consist of two functional classes of cells, indistinguishable from each other by conventional light microscopy. One class, the T-lymphocyte, is involved in cellular immune processes and in the regulation of antibody synthesis. The other class, the B-lymphocyte, is involved exclusively in humoral immune processes; it is the precursor of the plasma cell, which is the principal antibody-forming cell of the body.

Organization of Lymphoid Tissue

Lymphocytes arise from pluripotent hematopoietic stem cells, which in the fetus may be in the liver, spleen, or bone marrow but which after birth are found in the bone marrow. Some evidence exists (see Chapter 21) for a separate, committed T-lymphocyte stem cell. Production of lymphocytes differs fundamentally from production of myeloid cell lines. As discussed in Chapter 21, the latter involves a steady, continuing commitment of stem cells to precursors of the myeloid cell lines, which mature within the bone marrow to form highly specialized

cells of finite life-span that can no longer divide after release into the blood. In lymphopoiesis, stem cells give rise in the bone marrow to primitive precursors of T-cells and B-cells that undergo further differentiation in sites termed the *central lymphoid tissues*. For T-cells the site is the thymus. For B-cells the site has been identified as a gut-associated organ called the bursa of Fabricius in the chicken. In mammals the site is as yet unknown but may be the bone marrow. Once processed in central lymphoid tissues, T- and B-cells are distributed as immunologically competent cells to the *peripheral lymphoid tissues*. In this way the peripheral lymphoid tissues acquire a large repertoire of T- and B-cells programmed to respond selectively to different antigens. Within the peripheral lymphoid tissues T- and B-cells then proliferate and differentiate further as a result of exposure to antigens.

The peripheral lymphoid tissues consist of the following: lymph nodes distributed throughout the body; the spleen; the ring of tonsillar tissues in the oropharynx (Waldeyer's ring); submucosal accumulations of lymphocytes in the respiratory tract, urinary tract and in the gut (particularly in the terminal ileum as Peyer's patches); and also a portion of the lymphocytes in the bone marrow. T- and B-cells occupy different sites within the peripheral lymphoid tissues. T-cells are found in the paracortical areas of the lymph nodes (Fig. 3.6), in the white pulp of the spleen as a dense sheath around the central arterioles, and in the interfollicular areas of Peyer's patches and other submucosal accumulations of lymphocytes in the gut, respiratory, and urinary tracts. T-cells also make up about 70% of the lymphocytes of peripheral blood. B-cells are found in the cortex of lymph nodes, in the white pulp of the spleen, and in the submucosa of the gut, and respiratory tract, where they form aggregates called follicles (Fig. 3.6). The majority of the small lymphocytes of the bone marrow are also B-cells.

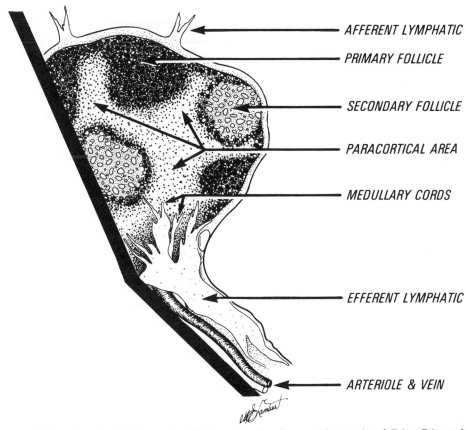

AFFERENT LYMPHATIC

PRIMARY FOLLICLE

SECONDARY FOLLICLE

PARACORTICAL AREA

MEDULLARY CORDS

EFFERENT LYMPHATIC

ARTERIOLE & VEIN

Figure 3.6.   Structure of a lymph node. B-cells are found in the cortex in primary and secondary follicles. Primary follicles are uniform aggregates of small, quiescent B-cells. Secondary follicles contain a central area of large, activated B-cells with a surrounding mantle of small B-cells. T-cells are found in the paracortical areas. Plasma cells accumulate in the medullary cords. Lymphocytes, primarily T-cells, leave the node via efferent lymphatic channels, eventually to enter the blood.

T-cells do not usually settle within a single fixed location in the peripheral lymphoid tissues. Many, but not all, T-cells cycle between the tissues and the blood, reentering the blood directly from the spleen and indirectly, via efferent lymphatics and the thoracic duct, from the other areas of the peripheral lymphoid tissues. T-cells move back into the peripheral lymphoid tissues primarily by passage between the endothelial cells of postcapillary venules. In rodents, two populations of circulating T-cells have been described, a short-lived population of T-cells not yet exposed to antigen, which homes to the spleen, and a long-lived population of T-cells that have been exposed to antigen, which homes to lymph nodes. The circulation of T-cells provides a mechanism for T-cell monitoring of antigens throughout the body. B-cells, which make up about 10 to 15% of peripheral blood lymphocytes, also recirculate between the peripheral lymphoid tissues and the blood but to a more limited extent.

Lymphocyte Proliferation in Peripheral Lymphoid Tissue

When quiescent, the T- and B-cells of the peripheral lymphoid tissues are small lymphocytes with a round nucleus, containing condensed chromatin and no visible nucleoli, and with a small rim of cytoplasm. However, when T- or B-cells are activated by exposure to specific antigen, the cells are transformed into large, metabolically active cells capable of proliferation, with a nucleus containing fine chromatin and nucleoli and with increased amounts of basophilic cytoplasm. Mitogenic stimuli arising during the immune response then stimulate the transformed cells to divide repeatedly with resultant clonal expansion of the activated cells. Normally, a base-line degree of continuing proliferative activity is present within the peripheral lymphoid tissues, reflecting a continuing exposure of the body to environmental antigens. The extent of continuing B-cell proliferative activity can be evaluated morphologically by looking for secondary follicles (Fig. 3.6), which contain germinal centers made up of enlarged, proliferating lymphocytes, in lymph nodes, the spleen, and the submucosa of the gut and respiratory tract.

When antigenic exposure is increased in a particular body area, proliferative activity in the peripheral lymphoid tissue adjacent to that area increases. For example, if one develops a bacterial infection of the arm, the lymph nodes of the corresponding axilla often become enlarged. Bacterial products are carried by the afferent lymphatics to these lymph nodes and stimulate the clonal expansion of T- and B-cells capable of responding to different antigenic determinants of the bacterial products. Numerous secondary follicles with large germinal centers are formed; the paracortical areas hypertrophy and contain large lymphocytes with blast-like features. Such clonal expansion is accompanied by differentiation into the effector cells of the T- and B-cell systems that regulate antibody formation, carry out cell-mediated immune responses, and synthesize antibody. Many plasma cells accumulate in the medullary area (Fig. 3.6) of the nodes. Activated T- and B-cells may also leave the nodes to seed more distant areas of the peripheral lymphoid tissue.

When stimulation by antigen ceases, proliferative activity regresses under the influence of cellular and humoral factors that suppress the immune response. Effector cells, which have finite life-spans, diminish in number. The nodes return towards their former state. However, a small number of T-cells, educated by their contact with the antigens involved, persist as immunologic memory cells. These memory T-cells probably circulate back and forth between the blood and lymph nodes for years as members of a known body pool of very long-lived T-lymphocytes. (This explains in part why a positive tuberculin skin test, which is a reaction triggered by T-cells previously exposed to antigens of the tubercule bacillus, persists for many years after

an initial exposure to tuberculosis.) Memory B-cells of unknown but lesser life-spans are also formed. These educated B-cells are programmed by their initial exposure to specific antigenic determinants to give rise, on subsequent exposure to the same determinants, to plasma cells that form high affinity antibodies of the IgG class. (Vaccination schedules take advantage of this phenomenon. They usually consist of at least two initial injections, separated by a short time interval, to obtain a higher titer of IgG antibodies, and occasional single booster injections, after long intervals, to maintain immunity, presumably by stimulating the formation of new memory B-cells.)

T-Cells

T-CELL DIFFERENTIATION AND FUNCTIONS

Stages in the differentiation of T-cells are illustrated in Figure 3.7. It shows T-cell precursors migrating from the bone marrow to the cortex of the thymus, wherein the cells undergo an extremely high rate of cell division. Most of the early thymocytes so formed die within the cortex. The reason for this high proliferative rate with its wastage of cells is unknown. It is possibly related to trial and error mechanisms that create a pool of cells, educated to distinguish between self and nonself histocompatability surface membrane determinants (discussed later), which can also, after distribution to the peripheral lymphoid tissues, respond selectively to specific immunogenic sequences of foreign antigenic determinants.

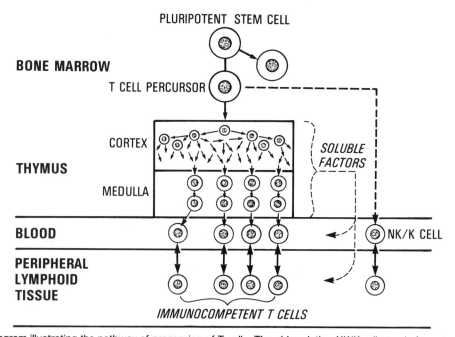

Figure 3.7.   A diagram illustrating the pathway of processing of T cells. The abbreviation *NK/K* cell stands for natural killer/killer cell.

A minority of lymphocytes from the dividing pool in the thymic cortex migrate to the medulla of the thymus, where they undergo further differentiation but no further cell division. Steps in further differentiation can be delineated by the appearance and disappearance of cell surface antigens detectable with monoclonal antibodies. When the process is completed, the medullary lymphocytes are then released as immunocompetent T-cells to the peripheral lymphoid tissues. Once released, T-cells, although able to circulate freely throughout the peripheral lymphoid tissues, never return to the thymus gland. However, the thymus elaborates soluble factors, which not only participate in the processing of resident thymocytes into postthymic T-cells but continue to exert a supportive "hormonal" effect upon T-cell function in the peripheral lymphoid tissues.

Newly formed T-cells not yet exposed to antigens in the peripheral tissues may be thought of as being composed of overlapping and functionally defined subpopulations (Fig. 3.8). One subpopulation can give rise to sensitized cytotoxic T-cells after exposure to foreign antigen on tissue cells in conjunction with class I mixed histocompatibility complex (MHC) coded products (defined later). The cytotoxic T-cells then react with the foreign antigenic determinants in conjunction with the MHC-coded products to lyse the tissue cells. A second functional subpopulation of T-cells proliferate after exposure to antigen on the surface of macrophages in conjunction with class II MHC-coded products (de-

fined later) and a soluble growth factor described in the next paragraph. They then may give rise to helper T-cells and to T-cells that mediate delayed hypersensitivity reactions. A third functional subpopulation of T-cells, which appear able to respond to soluble antigens in the absence of MHC-coded surface products, gives rise to suppressor T-cells. Helper and suppressor T-cells are key regulators of immune responsiveness, modulating both the production of antibody by B-cells and the function of cytotoxic T-cells.

Another functional subpopulation of T-cells depicted in Figure 3.8 may well overlap other subpopulations. These cells release a soluble factor called interleukin 2 (Il-2) after stimulation by a soluble factor released from macrophages called interleukin 1 (Il-1), which is described later. Interleukin 2 has also been called T-cell growth factor (TCGF) because of its effect upon T-cell growth in culture. When T-cells in culture are exposed to a plant lectin called phytohemagglutinin (PHA), they transform and begin to divide. Unlike an immunogenic antigenic determinant, which can activate only a T-cell programmed to respond to it, PHA acts as a nonspecific polyclonal T-cell activator. However, although an effective inducer of proliferation, PHA cannot sustain the growth of T-cells in culture. But if the PHA is then replaced by interleukin 2 in the culture medium, the activated T-cells will continue to grow in culture indefinitely. In vivo, exposure to antigen rather than PHA triggers T-cell proliferation by causing T-cells to trans-

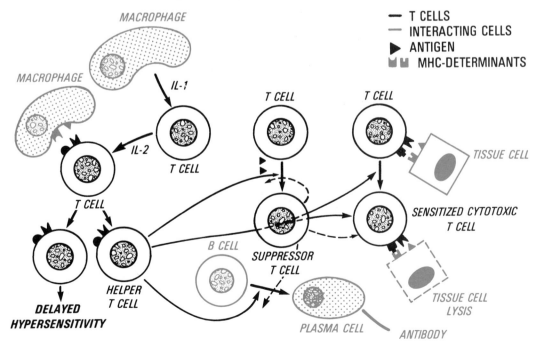

Figure 3.8.   The differentiation and interaction of subpopulations of T cells. Il-1 is interleukin-1; Il-2 is interleukin-2.

form and to become responsive to Il-2. The latter effect probably derives from the development of surface membrane receptors for Il-2. Il-2 can then modulate the extent of T-cell clonal expansions.

Another cell involved in cell-mediated immune processes, the NK/K cell (natural killer/killer cell) is also shown in Figure 3.7. The NK/K cell differs from a cytotoxic T-lymphocyte in that an NK/K cell may lyse a sensitive target cell on first contact, whereas a cytotoxic T-cell must first be sensitized by exposure to antigen on a target cell surface in conjunction with class I MHC-coded products. Furthermore, the NK/K cell has Fc receptors that can react with the Fc portion (defined later) of IgG antibody bound to a cell surface with resultant killing of the cell, a reaction termed antibody-dependent cell-mediated cytotoxicity (ADCC). (The two ways NK/K cells destroy target cells were once thought to be functions of separate NK cells and K-cells but are now thought to be functions of the same cell (Herberman, 1982); hence the combined name, NK/K cell.) NK/K cells do not require thymic conditioning for their formation, and they lack a surface antigen demonstrable on all differentiated T-cells with a particular monoclonal antibody. However, NK/K cells may have diverged at a very early stage from the T stem cell lineage, since continuous cultures of NK/K cells acquire surface markers found on mature T-cells. Although unequivocably shown to lyse certain types of target cells in vitro, particularly tumor cells and cells infected with virus, the in vivo significance of NK/K cells for cell-mediated immunity in humans is not yet clear.

CHARACTERIZATION OF T-CELLS BY SURFACE MARKERS

Cell surface markers may be used to quantitate numbers of T-cells and T-cell subpopulations in suspensions of lymphocytes prepared from peripheral blood, bone marrow, lymph nodes, and other tissues. Characterization of surface markers has proven valuable not only for experimental purposes but, increasingly, for the diagnosis and management of disease in patients. Mature T-cells have surface membrane receptors that react with sheep red cells. Therefore, T-cells will form rosettes, consisting of a central lymphocyte surrounded by several red blood cells, on incubation with sheep red cells. In early studies this was the principal technique used to identify T-cells. Now, monoclonal antibodies are available that will react with different antigenic determinants on the T-cell surface membrane. Patterns of reactivity to these antibodies can be used to identify thymocytes at different stages of T-cell processing and also to identify

subpopulations of differentiated T-cells. One particularly useful antibody reacts with all mature T-cells but not with immature T-cells. Another antibody reacts with surface antigen usually present on helper T-cells, whereas yet another antibody reacts with a surface antigen usually present on suppressor T-cells. T-cells have also been characterized by whether they possess a receptor for the Fc portion of IgM (T_μ-cells), a receptor for the Fc portion of IgG (T_γ-cells) or no Fc receptor (T null cells). As mentioned earlier, NK/K cells also have receptors for the Fc portion of IgG. NK/K cells may also be detected by their reaction with a specific monoclonal antibody.

EFFECTS OF AGE UPON T-LYMPHOCYTE PRODUCTION AND FUNCTION

The early development of a repertoire of T-lymphocytes capable of protecting the young of animal species against foreign antigens requires a vigorous processing of T-cells beginning in fetal life and extending through the newborn and growth period. This involves a steady stream of precursor cells moving from the bone marrow to the thymus and a steady stream of functional T-cells moving from the thymus to the peripheral lymphoid tissues. However, by 6 to 10 weeks of age in the rat, at which time the animal is still in the growth phase, the traffic of precursor cells from the bone marrow to the thymus ceases. T-cells continue to be supplied to the periphery apparently as a result of self-perpetuating divisions of cells within the thymic cortex.

Neonatal thymectomy in mice produces profound effects due to a deficiency of T-cells, including a "wasting disease" and atrophy of peripheral lymphoid tissues. Thymectomy in older mice produces less of an effect, although decreases are observed in T-cell-dependent antibody responses (described later), in the ability to reject foreign tissue transplants, and in the number of T-cells carrying surface markers identifying transitional cell populations destined to differentiate into helper and suppressor cells. The difference between the devastating effect of thymectomy in the newborn period and its lesser effect in the adult mouse undoubtedly reflects the building up in the interval of a population of peripheral lymphoid T-cells. This population appears able, at least partially, to sustain its functions in the absence of renewal of T-cells from the thymus and in the absence of the effects of thymic soluble factors.

The thymus atrophies with age. In humans the mass of the gland starts to decline after puberty and by age 50 is reduced to about 15% of its maximum. Aging in humans is associated with less effective immune responses requiring T-cell participation: delayed hypersensitivity responses, generation of cytotoxic T-cells, and stimulation of B-cells to produce high-affinity antibodies. Aging is also associated with diminished tolerance to autologous antigens, another phenomenon thought related to altered T-cell-mediated mechanisms. Diminished immune function increases the susceptibility to infection of older individuals and possibly also contributes to the increased incidence of malignancies found in older age groups. It is suspected but not proven that involution of the thymus, with consequent decreased renewal of T-cells and diminished function of existing T-cells, is causally related to senescence of the immune system in humans (Weksler, 1981; van de Griend et al., 1982).

B-Cells

SITE OF PROCESSING OF B-CELLS

Although the central lymphoid tissue of mammals that processes B-cells remains to be identified, the bone marrow is a leading candidate. It contains a fraction of lymphocytes undergoing both very rapid cell division and a high rate of cell death. This wastage of cells, which is similar to that found within the thymic cortex, could represent a consequence of the extensive germline DNA rearrangements (described later) that make possible the formation of a repertoire of B-cells capable of synthesizing immunoglobulin molecules with innumerably different antigenic specificities. The reassembly of germline DNA during differentiation probably results in the formation of many cells with aberrant DNA rearrangements or with DNA rearrangements coding for immunoglobulins reactive against self. Such cells presumably die in the processing tissue.

Further support for the bone marrow's role as a B-cell processing organ comes from the identification of pre-B cells in bone marrow. Pre-B cells are lymphocytes containing the heavy chain of IgM (μ chain) in their cytoplasm as evidence of their commitment to immunoglobulin synthesis. Cellular immunoglobulin markers are now known to appear and disappear in a sequence during the development of B-cells and their differentiation into plasma cells. Understanding this sequence and other features of B-cell development is helped by reviewing at this point the properties of immunoglobulins.

STRUCTURE OF IMMUNOGLOBULIN MOLECULES

Immunoglobulins are proteins possessing antibody activity. Their synthesis is unique; no other known body protein synthetic system requires dif-

ferent clones of cells to synthesize polypeptide chains that are structurally alike in one segment but structurally different in another segment. Yet the specificity of antibody molecules derives from the ability of any given clone of B-lymphocytes/plasma cells to synthesize an immunoglobulin molecule that differs in amino acid sequence at its antigen binding sites from the immunoglobulin molecules synthesized by all other clones of B-lymphocytes/plasma cells.

All immunoglobulin molecules are made up of a basic unit consisting of two light polypeptide chains and two heavy polypeptide chains linked by interchain disulfide bonds (Fig. 3.9). In any given immunoglobulin molecule the two light polypeptide chains are of identical amino acid composition and the two heavy polypeptide chains are of identical amino acid composition. Hybrid molecules containing different light chains or different heavy chains are never found. Most immunoglobulin molecules are monomers of this basic unit. However, some immunoglobulins are polymers of two units or of five units. A polymeric molecule is made up of identical basic units plus a single additional chain, called a J chain, that helps hold the polymer together. J chains are also synthesized by B-cells/plasma cells.

The amino acids of the light and heavy chains are not arranged linearly but form loops, which are called domains. The NH_2-terminal 110 amino acids of the chains, which contains a single domain, is known as the *variable region*. Within the variable region, segments have been identified called hypervariable regions; their overall amino acid sequences differ for each one of the many hundreds of thousands of different immunoglobulins that an individual synthesizes. As a result, the antigen binding site formed by the variable regions of an adjacent light and heavy chain of an immunoglobulin molecule (Fig. 3.9) binds preferentially to a single, specific antigenic determinant.

The carboxy-terminal portion of the light and heavy chains is called the *constant region*. Light chains have a constant region with a single domain; the larger heavy chains have a constant region of either three or four domains. The constant region of each type or subtype of light chain (see below) is almost invariant in its amino acid composition. Only minor differences, usually resulting from single amino acid substitutions, have been recognized between different individuals (allotypic differences).

The proteolytic enzyme papain cleaves immunoglobulin molecules at a region of the heavy chain called the hinge. Two antibody-combining fragments (Fab) are formed, each made up of a whole

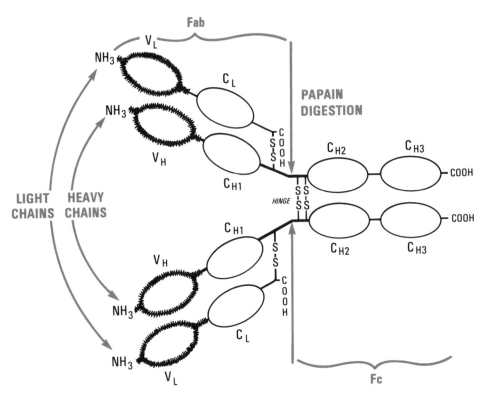

Figure 3.9.   A schematic model of an IgG molecule. The molecule consists of two identical light chains and two identical heavy chains held together by disulfide bonds. Each chain consists of an NH_3-terminal variable region, drawn with a wavy line, followed by a constant region, drawn with a straight line. The amino acids of the chains are folded by intrachain disulfide bonds (not shown) into globular regions called domains. The light chain has a single variable region domain, V_L, and a single constant region domain, C_L. The heavy chain has a single variable region domain, V_H, and three constant region domains, C_{H1}, C_{H2}, C_{H3}. Papain cleaves the heavy chains in an area known as the hinge, giving rise to two antibody-binding fragments, labeled Fab, and a single Fc fragment. Each Fab fragment is made up of a whole light chain and the NH_3-terminal portion of the heavy chain. The Fc fragment contains the carboxy-terminal portion of both heavy chains.

light chain and the variable and first constant region domain of a heavy chain (Fig. 3.9). A single fragment called the Fc fragment is also formed; it is made up of the remaining domains of the constant region of both heavy chains and determines the biologic properties of an immunoglobulin molecule. For example, the Fc fragment of some but not all immunoglobulin molecules contains a binding site for complement.

CLASSIFICATION AND BIOLOGIC ACTIVITIES OF IMMUNOGLOBULINS

A number of years ago, as a result of studies of the antigenic properties of monoclonal light chains excreted in the urine (Bence-Jones proteins) of patients with a plasma cell malignancy called multiple myeloma, two antigenically distinct groups of light chains were identified. They were given the names kappa (κ) and lambda (λ) light chains. Lambda light chains were later shown to be divisible further into several subtypes based upon single amino acid substitutions in the constant region of the light chain.

When similar studies were carried out of the antigenic properties of whole monoclonal immunoglobulins found in the plasma of patients with multiple myeloma and with a lymphoma called Waldenström's macroglobulinemia, five major kinds of heavy chains were identified. Immunoglobulins are divided into classes upon the basis of the kind of heavy chain they possess. These are as follows: IgM (μ chain), IgG (γ chain), IgA (α chain), IgD (δ chain), and IgE (ϵ chain). Immunoglobulin classes can be divided further into subclasses based upon serologic and physiochemical differences in their class-specific heavy chains. Thus, there are two subclasses of IgM, containing either μ_1 or μ_2 chains; two subclasses of IgA, containing either α_1 or α_2 chains; and four classes of IgG, containing γ_1, γ_2, γ_3, or γ_4 chains. Molecules with kappa or lambda light chains are distributed throughout each of the different classes and subclasses, with about two-thirds of human immunoglobulins being of the kappa light chain type and one-third being of the lambda light chain type.

The plasma of every normal member of the human species contains a spectrum of immunoglobulin *isotypes*, made up of each possible combination of one type or subtype of the light chains and one class or subclass of the heavy chains described above. Serologic differences have also been delineated between the immunoglobulins of different individuals. These result from the existence of alternative genes at a given structural locus for γ chains (Gm system), for α chains (Am system), and for kappa chains (Inv system). Thus, in populations, several *allotypes* may exist for a particular isotype. Finally, each immunoglobulin synthesized by a different B-lymphocyte/plasma cell clone differs antigenically because of the different amino acid composition of its hypervariable regions. Such antigenic differences are referred to as *idiotypic* differences. Each immunoglobulin synthesized is not only an antibody but also an antigen with its unique idiotype. The significance of this for regulation of antibody production is discussed later.

Characteristics of the different classes of immunoglobulins are listed in Table 3.6. IgM antibodies are the first antibodies formed after exposure to an antigen. IgM circulates as a pentamer with 10 antigen-binding sites on each molecule. Therefore, it functions as an efficient activator of the classical pathway of complement (Chapter 23). Because of its large size (900,000 daltons), plasma IgM is referred to as a macroglobulin. (In Waldenström's macroglobulinemia, such large amounts of IgM are secreted that the blood becomes very viscous and a patient may develop serious manifestations of hyperviscosity, such as coma, respiratory insufficiency, and bleeding.)

A minority of antigens will continue to elicit an IgM antibody response on continued exposure or on reexposure. However, for most antigens antibody production shifts to a different immunoglobulin class. Unusually, antibody production shifts to production of IgG, which is the major antibody of plasma and extracellular fluid and the only class of antibody that can pass the human placental barrier. (Maternal IgG that has crossed the placental bar-

rier provides passive immunity to a newborn infant until the infant begins making IgG antibody of its own at about 3 months of age.) The IgG_1 subclass, which constitutes about 70% of IgG, and the IgG_3 subclass bind complement efficiently after reacting with antigen, with resultant complement activation by way of the classical pathway (see Chapter 23). Neutrophilic granulocytes and monocytes/macrophages have receptors for the Fc portion of IgG_1 and IgG_3 and also for the C3b component of complement. Therefore, when antibodies of the IgG_1 or IgG_3 class bind to antigens on the surface of a microorganism or cell, and when this binding, in turn, brings down complement onto the surface, the microorganism or cell can be readily phagocytosed.

IgA is the major antibody found in secretions of the respiratory, gastrointestinal, and genitourinary tracts, where it undoubtedly functions to bind foreign antigens and impair their entry into the body. In secretions, IgA exists as a polymer of two immunoglobulin molecules plus a J chain plus yet another protein called secretory component. Secretory component is synthesized by mucosal epithelium and helps transport IgA molecules through epithelial cells into secretions. In plasma, IgA exists as either a monomer or a polymer containing a J chain but no secretory component.

IgD and IgE are found in trace amounts in plasma. The function of secreted IgD is unknown. As discussed below, IgD on the surface of B-cells serves as an antigen recognition site. IgE is secreted in increased amounts in patients with parasitic infections and in patients with atopic allergy. IgE binds by its Fc region to mast cells and can invoke mast cell-triggered immediate hypersensitivity reactions (discussed further later).

GENETIC MECHANISM OF IMMUNOGLOBULIN SYNTHESIS

To synthesize immunoglobulins B-cells must accomplish three genetically remarkable things. First, the cells must put together polypeptide chains in-

Table 3.6
Properties of immunoglobulins

	IgG	IgA	IgM	IgD	IgE
MW (daltons)	150,000	170,000	900,000	180,000	200,000
Plasma concentration (mg/dl)	700–1500	250	100	3	0.03
Intravascular t½ (days)	21	6	5	3	2
Intravascular distribution (%)	45	40	75	75	50
Major Ig of plasma and extravascular fluid	+	−	−	−	−
Major Ig in secretions	−	+	−	−	−
First antibody made to antigen	−	−	+	−	−
Binds to mast cells	−	−	−	−	+
Crosses placental barrier	+	−	−	−	−

finitely variable in the NH_2-terminal segment but of almost invariant composition in their carboxy-terminal segment. Second, since the immunoglobulin made by any given B-cell/plasma cell consists of two identical light chains and two identical heavy chains, the diploid cell must exclude one set of parental genes from the synthesis of each type and class of chain. Only one haploid set of genes are assembled and expressed in any given cell. Third, the B-cell/plasma cell, while continuing to make the same light chain and the same heavy chain variable region, must be able to switch the class of immunoglobulin that it makes, e.g., from IgM to IgG, by attaching a different heavy chain constant region to its heavy chain variable region (Davis et al., 1980).

The recent discovery of the ways in which immunoglobulin genes are assembled and transcribed represents a landmark advance in genetics. The older concept that all polypeptide chains are coded for by a single preformed germline gene has been replaced by a recombination model, which recognizes that several germline genes code for a single immunoglobulin polypeptide chain. A functional gene for the variable region of a heavy chain is now known to result from the assembly of three separate germline DNA fragments. One of many alternative V_H (variable) gene segments is joined to one of a family of DNA segments known as D_H (diversity) genes and then to one of six alternative J_H (junctional) genes. The mechanisms of joining genetic material coding for the variable region of heavy chains with genetic material coding for the constant region of heavy chains differs for the heavy chain classes. For μ and δ heavy chains, only a short intervening sequence separates the assembled variable region gene from DNA segments coding in succession for the constant region of the μ chain (C_μ gene) and the constant region of the δ chain (C_δ gene). Therefore, the assembled gene segments for the variable region, the intervening sequence, and the C_μ-C_δ segments can be transcribed as a unit. After transcription, the intervening sequence is deleted and the variable region RNA is spliced either to the C_μ RNA or the C_δ RNA. The mechanism for attaching constant gene genetic material for γ and α chains is discussed below.

The assembly and transcription of light chains is somewhat less complex. For kappa light chains one of several hundred different germline V_k segments is joined to one of five alternative J_k segments. After the variable region gene so formed is transcribed, its RNA is spliced to the RNA transcribed from a single constant region gene (C_k), thus forming the mRNA that is translated to produce a kappa

light chain. The variable region gene for a lambda light chain is also assembled by the joining of a germline V gene and a germline J gene. However, whereas the kappa gene system utilizes multiple alternative J_k genes with a single C_k gene, in lambda gene assembly a particular J_λ gene always seems to be associated with the same one of six possible duplicated lambda constant genes (C_λ genes).

The germline immunoglobulin genes are activated in a sequence. Early B-cell precursors can be identified that contain a $V_H/D_H/J_H$ recombination of heavy chain DNA segments but light chain genes still in germline configuration. If the $V_H/D_H/J_H$ combination is effective, then the cell begins to make cytoplasmic μ chains (and can be recognized as a pre-B cell by this more readily measureable marker of beginning B-cell differentiation). The cell also then begins the process of light chain gene recombination. It apparently always first attempts to assemble a kappa chain gene. If this succeeds, a B-cell is formed that produces a kappa type whole IgM molecule. If this fails on first attempt, a second attempt is made using the remaining haploid set of germline kappa chain DNA segments. Only if this too fails does the B-cell then attempt to recombine one of its two sets of lambda germline genes. If this succeeds, the cell makes a lambda type whole IgM molecule. If not, a second attempt is made. If both attempts at lambda gene recombination fail, then further maturation of the B-cell is aborted. The signals activating germline gene recombinations and the mechanism whereby one set of haploid genes is suppressed if the initial recombination is successful are as yet unknown.

The diversity of the antigen binding site of immunoglobulin molecules stems partly from the many combinations made possible by the assembly of different V, D, and J gene segments in forming heavy chain genes, by the assembly of different V and J genes in forming kappa or lambda light chain genes and by the different combinations of heavy and light chain genes used to make the whole immunoglobulin molecule. A further major source of diversity results from frame shifts that may occur at the sites of joining of the individual gene segments. Such frame shifts greatly expand the possibilities for further amino acid variability of the variable region of the polypeptide chains.

The B-cell/plasma cell uses two mechanisms for combining the same heavy chain variable region with different constant regions, i.e., to make more than one class of immunoglobulin molecule. During B-cell development, B-cells acquire both surface-bound IgM and surface-bound IgD (Fig. 3.10), which means that the cell must make both mole-

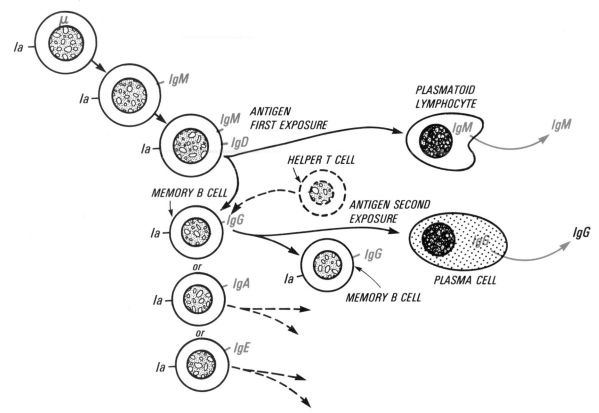

Figure 3.10. The development of B cells and their response to stimulation by antigen. Immunoglobulin B cell markers are printed in red. The symbol μ stands for the heavy chain of IgM. Ia is a synonym for class II mixed histocompatibility complex gene products.

cules. As already mentioned, the constant genes for IgM and IgD (C_μ and C_δ) are located next to each other on the heavy chain gene locus. Therefore, both are readily transcribed, and the initial RNA transcribed can be converted, by alternative sites of RNA splicing, into either a mRNA for IgM or a mRNA for IgD. However, the sites for the constant region genes for the IgG and IgA classes are located at a distance from the C_μ-C_δ sites on the germline heavy chain gene. Therefore, when a B-cell/plasma cell switches from making IgM to making IgG or IgA, a further DNA rearrangement is required. This involves DNA splicing between highly homologous switch sites that have been identified on the heavy chain gene preceding the C_μ-C_δ gene segment and preceding the gene segment for each one of the IgG and IgA subclasses. (For a more detailed discussion of the genetic mechanisms of immunoglobulin synthesis, the reader is referred to an immunology text, e.g., the recent chapter by Korsmeyer and Waldmann (1982).)

ANTIGEN RECOGNITION SITES AND OTHER B-CELL SURFACE MARKERS

During maturation each B-cell must acquire surface recognition sites for the specific haptenic determinant against which the cell, by its germline

DNA rearrangements, has become committed to make antibody. Moreover, the recognition sites must be acquired early enough for the cell to be silenced if its DNA rearrangements have programmed it to make antibody against self. The recognition sites are now known to be immunoglobulin, synthesized by the developing cell and embedded at its carboxy-terminus into the surface membrane of the cell (Hammerling, 1981). Thus, the surface-bound immunoglobulin extends to the external environment its variable region, ready to bind the specific antigenic determinants to which the cell is programmed to respond.

As already mentioned, the pre-B cell can be identified by μ chains in its cytoplasm. As the cell develops further, μ chains disappear from the cytoplasm and whole monomeric IgM molecules appear on the surface membrane (Fig. 3.10). This surface-bound IgM serves as the cell's first antigen recognition sites. These are particularly important for the early elimination of B-cells committed to make antibodies against self. With still further development, IgD appears on the B-cell surface, where it functions as a second class of antigen recognition sites. Both IgM and IgD recognition sites are apparently required for the B-cell to respond to the vast majority of antigens. When a B-cell containing both IgM and IgD on its surface is exposed to an appropriate antigen, the B-cell proliferates and differentiates into two types of cells. One is an IgM-secreting cell with properties intermediate between a lymphocyte and a plasma cell (plasmacytoid lymphocyte). The other is a B-cell whose surface IgM and IgD is replaced with surface IgG, IgA, or IgE (Fig. 3.10). With further exposure to antigen, these committed B-cells give rise to plasma cells that secrete IgG, IgA, or IgE and to an additional population of B memory cells containing IgG, IgA, or IgE on their surface.

Immunofluorescent staining for surface immunoglobulin markers is widely utilized both experimentally and clinically to identify B-cells in lymphocyte suspensions. Other surface antigens, including some defined by monoclonal antibodies, can also be used to characterize B-cells further. Thus, Ia antigen, a gene product of the HLA-D locus of the MHC (see below), is found on B-cells at all stages of development. (Ia is another name for class II MHC-coded surface molecules.) Mature B-cells also contain an Fc receptor reactive with aggregated immunoglobulins or antigen-antibody complexes, and some B-cells also possess complement receptors. When a B-cell differentiates into a plasma cell, it loses its surface Ia antigen and its surface immunoglobulin, although its cytoplasm then contains abundant amounts of the immunoglobulin that the cell is secreting.

MECHANISMS OF IMMUNE RESPONSES

Immune responses are divided into two categories: humoral immune responses, which are mediated by B-cells/plasma cells through production and systemic circulation of antibodies; and cell-mediated immune responses, which are mediated by T-cells through direct contact with target cells and through local liberation of soluble factors called lymphokines. Production of antibody is particularly important for protection against acute bacterial infections, e.g., pneumococcal pneumonia. Bacteria coated by antibody and, as a consequence, by complement are phagocytosed by granulocytes and by mononuclear phagocytes with greatly increased efficiency. Patients with diseases impairing antibody production have an increased risk of acute bacterial infections.

Cell-mediated immune responses are particularly important for combatting infections by microorganisms that reside and multiply within cells. These include viral infections and infections with bacteria that have an intracellular preference, e.g., the mycobacterium causing tuberculosis. Some microorganisms that usually do not cause human disease

will do so when cell-mediated immune responses are depressed. Infections with such organisms, e.g., cryptococcal meningitis, pneumocystis pneumonia, cytomegalovirus pneumonia, are referred to as opportunistic infections. Cell-mediated immune responses may also have a role in immune surveillance against body cells that have undergone premalignant changes.

Synthesis of Antibodies

The production of antibody usually requires interactions between macrophages, T-cells, and B-cells (Stobo, 1982). However, a few antigens, known as *thymus-independent antigens*, induce antibody formation by B-cells without cellular interaction. Such antigens are large molecules with repeating structural units, e.g., bacterial polysaccharide, in which the antigen apparently functions specifically through its haptenic determinants and nonspecifically, as a B-lymphocyte mitogen, through its repeating units. Thymus-independent antigens result in the synthesis of IgM antibody and induce little, if any, immunologic memory. The ABO blood group substances represent a clinically important example of thymus-independent antigens.

Production of antibody to the vast majority of antigens requires that both antigen and helper T-cells interact with B cells. Such antigens are therefore referred to as *thymus-dependent antigens.* An antigenic determinant of a thymus-dependent antigen consists of two parts: a haptenic component of precise specificity, which stimulates the B cell, and an immunogenic or "carrier" component of lesser specificity, which stimulates the T-cell (Benacerraf, 1981).

The development of helper T-cells for production of antibody against a thymus-dependent antigen requires that the T-cell first interact with a macrophage (Rosenthal, 1980). In this interaction the macrophage presents the immunogenic component of an antigenic determinant (immunogenic determinant) to the T-cell in association with or aligned with an Ia determinant (class II MHC-coded product). The T-cell must recognize both the immunogenic determinant and the Ia determinant. The structure of the receptor or receptors that the T-cell utilizes for this dual recognition is only partly understood. One receptor (or component of a receptor) is a surface-bound variable region of an immunoglobulin molecule, which, in a manner analogous to whole surface-bound immunoglobulin on B-cells, functions as the T-cell receptor for the immunogenic determinant. The structure of the second receptor (or component of a receptor), which recognizes the Ia determinant, is not yet known.

However, as a result of the T-cell "seeing" the immunogenic determinant in conjunction with the Ia determinant, the T-cell develops an additional surface receptor, namely, a receptor for the T-cell growth factor discussed earlier, interleukin 2 (Il-2).

Macrophages serve a second function in the production of antibody to a thymus-dependent antigen. They secrete a soluble factor, interleukin 1 (Il-1), which stimulates T-cells to secrete Il-2 (Farrar and Hilfiker, 1982). It is not yet clear whether the same T-cells can both secrete Il-2 and respond to its mitogenic effect. In any event, the dual phenomena of the development of receptors for Il-2 on the T-cell surface and the secretion of Il-2 by T-cells then results in the full activation of T helper cells.

As already mentioned, the B-cell must recognize both a specific haptenic determinant and helper T-cells in order to proliferate and differentiate into plasma cells producing antibody against a thymus-dependent antigen. The B-cell utilizes the idiotypic region of immunoglobulin molecules bound to its surface to recognize the haptenic determinant. Recognition of the helper T-cell involves direct contact between the B- and T-cells, which, in turn, involves the T-cell recognizing the Ia determinant on the B-cell. The helper T-cell then activates the B-cell through this direct contact. In addition, the helper T-cell may secrete a soluble factor or factors that stimulate the proliferating B-cells to differentiate into plasma cells.

A second population of T-cells, T suppressor cells, are also activated after exposure to thymus-dependent antigens. In some experimental situations, soluble antigen appears to activate T suppressor cells directly, requiring neither a preliminary reaction with macrophages nor presentation in conjunction with Ia. T suppressor cells dampen antibody production, both directly by inhibiting B-cell differentiation into plasma cells and indirectly by inhibiting T helper cells.

Feedback loops have been discovered that further illustrate the complexity of the cellular interactions involved in antibody production. Thus, whereas macrophages secrete Il-1, which activates T-cells, T-cells, in turn, secrete a soluble factor increasing the expression of Ia determinants on macrophages and thus enhancing their capability of presenting antigen to T-cells. Moreover, a subpopulation of T helper cells has been identified that not only augments antibody production but also augments the induction of T suppressor cells. Similarly, a subpopulation of T suppressor cells has been identified that not only suppresses antibody production but also suppresses further formation of T suppressor cells (a phenomenon referred to as contrasuppression).

Mechanisms other than interactions of B-cells with helper and suppressor T-cells also modulate antibody production. These include a changing availability of the challenging antigen. As the titer of antibodies rises, antibodies may increasingly bind antigen, forming immune complexes that are phagocytosed by the mononuclear phagocyte system. The resultant clearance of antigen may decrease the availability of antigen as a stimulus for further antibody production.

Antiidiotypic antibodies may also help to turn off antibody production (Jerne, 1975; Geha, 1981). When a host is exposed to a large protein antigen, a number of clones of B-cells are stimulated to make antibodies to different haptenic determinants of the antigen. Thus, a family of polyclonal antibodies are made, each member of which has its own different idiotypic region. As discussed earlier, the idioptypic region of an immunoglobulin molecule is itself an antigen. When sufficient immunoglobulin molecules of a given idiotype have accumulated, the host may then start to make antibodies against that idiotype. The antiidiotypic antibody will have an antigen-binding pocket with the same spatial configuration as that haptenic determinant of the antigen that stimulates formation of the original immunoglobulin molecule. Therefore, the antiidiotypic antibody can compete with the original haptenic determinant for binding to the B-cell immunoglobulin receptors of the clone making the original immunoglobulin molecule. This inhibits further proliferation of that B-cell clone. As increasing numbers of clones are inhibited in this manner, the overall antibody titer, which is the sum of the antibodies made to the individual haptenic determinants, should progressively fall.

The ability of B-cells/plasma cells to synthesize antiidiotypic antibodies means that for every B-cell programmed to make antibody to a specific foreign antigenic determinant, a counterpart B-cell must also be programmed to make antibody to the unique variable region of the first antibody. As one reflects upon the millions of foreign antigenic determinants to which B-cells can potentially make antibodies, and upon the millions of counterpart B-cells required to make antiidiotypic antibodies to the original antibodies, and then upon the millions of additional B-cells required for the possible next step in the sequence, namely, the production of anti-antiidiotypic antibodies—one is left awed by the enormous diversity of the communication network of the immune system.

Cell-Mediated Immune Responses

Cell-mediated immune responses include delayed

hypersensitivity reactions and cytotoxic reactions in which T-cells and NK/K cells kill target cells. A positive tuberculin skin test is a well studied example of a cutaneous delayed hypersensitivity reaction. The intradermal injection of tuberculin, an antigen extracted from the tubercle bacillus, results in the accumulation at the site of a small number of T-lymphocytes sensitized by previous exposure to the antigen. These sensitized lymphocytes interact with the antigen on the surface of a macrophage (Unanue, 1981) or of a related phagocytic epidermal cell, the Langerhans cell. Just as for initiation of antibody production to a thymus-dependent antigen, the interaction is genetically restricted, requiring that the T-cell also recognize the Ia determinants on the macrophage. This dual interaction then stimulates the T-lymphocyte to secrete lymphokines that bring about the further steps of the reaction.

In delayed hypersensitivity reactions lymphokines exert several important effects upon monocyte/macrophages. They cause monocytes from the blood to accumulate at the site and differentiate into macrophages (monocyte chemotactic factor or MCF); they inhibit macrophage migration from the site (macrophage inhibitory factor or MIF); and they activate the macrophage, enhancing its microbicidal properties and its ability to secrete proteolytic enzymes and generate toxic oxygen metabolites (macrophage-activating factor or MAF). Lymphokines also affect lymphocyte function, preparing nonsensitized lymphocytes to respond to the antigen (transfer factor) and causing both sensitized and nonsensitized lymphocytes to divide (mitogenic factor). Moreover, lymphokines called lymphotoxins directly damage tissue cells. As a result of all of these effects, by 24 to 48 hours after the injection of the antigen, an area of erythema and induration is found at the injection site. If it were biopsied, one would see an infiltrate of mononuclear cells, primarily macrophages, edema, and deposits of fibrin. The fibrin results from blood coagulation initiated by tissue factor activity (see Chapter 25) that develops on the surface of activated monocyte/macrophages.

Delayed hypersensitivity reactions, which provide a mechanism whereby large numbers of macrophages may accumulate and become activated at a site of antigen, particularly enhance the body's ability to fight infections by bacteria that can survive and multiply within unactivated monocyte/macrophages (e.g., the organisms that cause tuberculosis, brucellosis, and salmonella infections). Moreover, once activated, macrophages not only possess an enhanced microbicidal activity against the organism initiating the delayed hypersensitivity reaction but also an enhanced microbicidal activity against other organisms and an enhanced tumoricidal activity. (This is why producing delayed hypersensitivity reactions with multiple intradermal inoculations of an attenuated strain of tubercle bacillus, BCG, was tried as experimental treatment of certain human malignancies.)

A second cell-mediated immune response results in the lysis by cytotoxic T-cells of tissue cells infected with viruses. Unlike intracellular bacteria, which invade only monocyte/macrophages, viruses may invade a variety of tissue cells (e.g., hepatitis viruses invade hepatocytes and influenza viruses invade respiratory epithelial cells). When viruses infect tissue cells, viral antigens are translocated to the cell surface. Cytotoxic T-cells become sensitized to these viral antigens by "seeing" the viral antigen in conjunction not with Ia, which is absent from most tissue cells, but in conjunction with class I MHC-coded surface products (see below). Once sensitized, the cytotoxic T-cells can then lyse the infected cells in a reaction in which, once again, the cytotoxic T-cell must recognize both the viral antigen and the class I MHC-coded products on the cell surface. Such cell lysis, by preventing intracellular viral replication, may play an important role in resistance to viral infections. For example, cytomegalovirus pneumonia, which does not occur in patients with normal immune function, is a feared complication of acquired immunodeficiency states depressing cell-mediated immune reactions. However, the T-cell-mediated death of virus-infected cells is also responsible for major manifestations of disease induced by some viruses in experimental animals and, probably, also in man (Zinkernagel, 1978).

As discussed earlier, NK/K cells differ from cytotoxic T-cells in that NK/K cells can lyse target cells in vitro without the need for prior sensitization and independently of MHC-coded surface products. Most normal human cells are resistant to direct lysis by human NK/K cells in vitro, whereas some lymphoma cells and other tumor cells are sensitive to such lysis. Therefore, this cell-mediated reaction has been postulated to be important in vivo for immune surveillance against cells that may be changing to become tumor cells (Roder and Pross, 1982). Moreover, in studies in mice, the ability to reject a bone marrow transplant appears related to the state of activity of an animal's NK/K cells.

As also mentioned earlier, NK/K cells are the effector cells for yet another type of cell lysis, antibody-dependent cell-mediated cytotoxicity. In this reaction, IgG antibody binds by its variable region to a cell surface antigen; the NK/K cell is then brought into contact with the target cell by binding to the Fc portion of surface IgG molecules; and the proximity of the NK/K cell to the target cell triggers cell lysis. The importance of this reaction for normal immune function in humans is as yet unknown.

Activities of Lymphokines and Monokines

In the foregoing sections materials have been described that are secreted by lymphocytes and monocytes and that modulate activities of cells involved in immune responses (Rocklin et al., 1980). Called lymphokines and monokines, these materials are active in minute concentrations and are neither antigenically nor genetically restricted in their biologic effects. They have been identified primarily by investigating the biologic properties of supernatant fluids of in vitro cultures of mononuclear cells stimulated by antigen or nonspecific mitogen. The biochemical properties of the majority of these materials are as yet poorly characterized and, indeed, some of the activities described as separate factors could be different manifestations of the same material.

Interleukin 1 (Il-1) is a key monokine (Oppenheim et al., 1982) which, as already described, participates in immune responses by stimulating T-cells to produce interleukin 2 (Il-2). Il-1 also plays a role in inflammatory and body repair processes. Thus, it is now known that Il-1 is endogenous pyrogen, the material released from WBC after injection of endotoxin into experimental animals that alters hypothalamic function to induce fever. Moreover, Il-1 stimulates hepatocytes to produce acute phase proteins, e.g., fibrinogen and C-reactive protein, stimulates isolated rheumatoid synovial cells to produce collagenase and prostaglandins, and stimulates the growth of fibroblasts in culture.

The number of lymphokine activities involved in immune responses is large. These include the several activities involved in recruiting and activating macrophages in delayed hypersensitivity reactions; the cytotoxic material or materials, called lymphotoxin, that probably mediate the processes whereby cytotoxic T-cells and NK/K cells lyse target cells; Il-2, the T-cell growth factor; and a factor from T helper cells that induces B-cell differentiation into plasma cells. Other lymphokines, whose importance for normal immune responses is less clear, include transfer factor and interferon.

Some lymphocytes and monokines function as important modulators of hemopoiesis in in vitro culture systems. Thus, monocytes are a principal source of colony-stimulating factor. Monocytes also secrete a soluble factor that stimulates lymphocytes to produce colony-stimulating factor. Monocytes or T-lymphocytes produce burst-promoting activity (see Chapter 21) and yet another activity that stimulates mixed myeloid cell growth in in vitro cultures. Other biologically important lymphokines include a chemotactic factor for eosinophils and a factor stimulating osteoclast activity.

The Major Histocompatability Complex Genes and Their Products

The major histocompatability complex (MHC) is a single genetic region on human chromosome 6, which contains a tight cluster of genes coding for cell surface molecules that enable cells to recognize and communicate with each other in immune responses. It is called the MHC because antigenic determinants on these cell surface molecules determine whether organ or tissue transplants are recognized as self or foreign and, therefore, are accepted or rejected. In humans the MHC is usually refered to as the HLA (human leukocyte antigen) region, because it was first recognized by analysis of the patterns found on reacting peripheral blood lymphocytes with sera, obtained from multiply transfused patients or multiparous women, containing antibodies to leukocytes.

The HLA region is thought to consist of at least five loci for genes that code for cell surface molecules: HLA-A, HLA-B, HLA-C, HLA-D, and HLA-Dr. Genes at these loci code for two major classes of cell surface molecules with different structures and cellular distribution. Class I molecules, consist of a 44,000-dalton heavy chain in noncovalent association with a 12,000-dalton protein called microglobulin that has striking homology with the amino acid sequences found in some constant region domains of immunoglobulins. Class I molecules are present on virtually all cells. Class II molecules are made up of a 34,000-dalton chain and a 29,000-dalton chain. Class II molecules have so far only been identified on hemopoietic progenitor cells, macrophages, B-lymphocytes, and activated, but not quiescent, T-lymphocytes. Genes at the HLA-A, B, C, and probably Dr loci, which probably arose by tandem gene duplication, code for structurally similar class I molecules. Genes at the D locus code for class II molecules.

For reasons presumably related to survival advantage, the genes of the MHC system are highly polymorphic. Multiple allelic genes have been identified at each of the HLA loci. For example, a human may inherit from each parent one of 20 known alternative HLA-A genes and one of more than 40 known alternative HLA-B genes. Because of their proximity to each other on the chromosome, HLA genes are inherited as a unit or haplotype. An individual inherits a particular combination of alleles for all HLA loci (haplotype) from one parent and a different particular combination of alleles for all HLA loci (haplotype) from the other parent. HLA typing can be carried out serologically for the HLA-A, B, C, and Dr loci. In such typing, a suspension of peripheral blood lymphocytes is mixed with a panel of typing sera containing antibodies to different HLA antigens. If the cell possesses an antigen reacting with a particular serum, the cell is killed, which is recognized by its staining with the dye, trypan blue. The D antigens cannot be typed serologically. They are characterized in a test system, called the mixed leukocyte reaction (MLR), which is based upon the observation that when lymphocytes from individuals with different D antigens are cultured together, the lymphocytes activate each other and are transformed into large, blast-like cells that synthesize DNA and, therefore, take up ^{3}H-thymidine.

Although crucial in deciding the fate of organ transplants, the gene products of the MHC obviously did not arise in evolution for this purpose. They are involved in a broad spectrum of cellular recognition events in immune responses (McDevitt, 1980; Benacceraf, 1981). As already discussed, antibody production to thymus-dependent antigens requires that T-cells "see" immunogenic determinants of an antigen in conjunction with Ia determinants on the macrophage surface. Indeed, the particular immunogenic determinant that a T-cell "sees" may depend upon allelic differences in the Ia determinant. In mice, different strains exposed to a particular antigen recognize different antigenic determinants on that antigen. As a result, one particular strain may make high affinity IgG antibodies to the antigen, whereas another strain may make only minimal amounts of IgM antibody. Such differing responses have been found to be related to the particular allelic immune response genes that the mice possess. Similar differences in antibody production to antigens are thought to occur in humans and may reflect allelic differences at the HLA-D locus that govern expression of Ia determinants on macrophages.

Ia determinants are also involved in the interactions between helper T-cells and B-cells that occur during antibody production. However, once antibodies are formed, their effects—e.g., enhancement of phagocytosis of bacteria, lysis of cells due to antibody-complement-dependent reactions or due to antibody-dependent cell-mediated cytotoxicity—no longer require MHC coded products, i.e., are genetically unrestricted. Similarly, the initial reaction between sensitized T-cells and antigen on macrophages that sets off a delayed hypersensitivity reaction requires the participation of Ia determinants, whereas the subsequent actions of the lymphokines that are formed are genetically unrestricted. In contrast, class I MHC-coded products are involved both in the sensitization of cytotoxic T-cells to viral antigens and in the subsequent effector response in which the sensitized cytotoxic T-cells attack target cells possessing both the antigen and the class I molecules on their surface.

ABNORMALITIES OF IMMUNE RESPONSES IN DISEASE

Pathologic Manifestations of Humoral Immune Responses

Although humoral immune responses represent a key biological defense mechanism, under some circumstances antibodies cause manifestations of disease. Antibodies may damage or alter the function of cells or tissues in the following ways.

1. Immediate hypersensitivity reactions. When an antigen stimulates B-cells/plasma cells to make IgE, the antibodies will bind, nonspecifically by their Fc region, to the surface of tissue mast cells. On subsequent exposure, the antigen will bind to the variable regions of the antibodies on the surface of the mast cell. This causes the mast cell to undergo degranulation. Antigens from parasites stimulate production of IgE, a response thought helpful in combatting tissue parasitic invasion because one of the materials released from degranulated mast cells is chemotactic for eosinophils. Eosinophils drawn to the site then release materials that damage the parasite. However, in some individuals other antigens also stimulate production of IgE antibodies. An atopic person may develop IgE antibodies to inhaled antigens such as pollens or products of molds. On subsequent inhalation of the antigen, mast cells in the respiratory mucosa undergo degranulation, releasing mast cell histamine and leukotrienes (see Chapter 23) that may precipitate an attack of extrinsic bronchial asthma (Marquardt and Wasserman, 1982). Rare individuals may develop IgE or cytotropic IgG antibodies to a drug, e.g., penicillin, or to an antigen in bee venom. If the antigen then gains access to the blood stream (e.g., after swallowing a single penicillin tablet or after a bee sting), systemic mast cell degranulation may precipitate a potentially fatal anaphylactic reaction.

2. Interactions of autoantibodies with cellular or tissue antigens. Multiple mechanisms probably act to prevent an individual from making substantial amounts of antibodies to self: the "silencing" of differentiating B-cells that become programmed to make antibody to self; the education of developing T-cells to distinguish between MHC-coded products that are self and allogeneic MHC-coded products that are nonself and also to distinguish between self and nonself immunogenic determinants; and the continuous monitoring of B-cell and helper T-cell activities by suppressor T-cells. Nevertheless, these mechanisms sometimes fail. A variety of disorders are now recognized as resulting from the formation of antibodies reacting with normal body constituents (autoantibodies). Some autoantibodies cause cell or tissue damage, as, for example, antibodies to red cell membrane antigens in certain hemolytic anemias that activate complement on the red cell surface or antibodies to basement membrane that injure glomerular and pulmonary capillaries in the disorder known as Goodpasture's syndrome. Other autoantibodies cause disease not by destroying cells but by altering their function. In myasthenia gravis, severe voluntary muscle weakness and fatigue results from antibodies to acetylcholine receptors that reduce the number of functional receptors at the neuromuscular junctions of skeletal muscle. In an unusual type of diabetes mellitus, antibodies to insulin blockade the insulin receptors on cells, preventing insulin from stimulating these receptors to initiate intracellular processes. Conversely, in Graves' disease, antibodies to receptors on thyroid cells for thyroid-stimulating hormone (TSH) bind to the receptors and mimic the action of TSH. This causes hyperthyroidism due to excess production of thyroid hormone.

3. Precipitation of antigen-antibody complexes (immune complexes) in tissues. Depending upon their size, immune complexes that form in the blood are either phagocytosed by the mononuclear phagocytic system or continue to circulate, in which event they may become deposited in blood vessel walls. Precipitation of immune complexes in the glomerular capillary bed can cause inflammation of the glomeruli and eventual renal failure, as in the nephritis associated with the generalized autoimmune disease, systemic lupus erythematosus. Immune complexes deposited in the walls of small arteries cause a disorder, polyarteritis nodosa, in which severe inflammation of vessels results in areas of ischemic necrosis in many organs or tissues. Immune complexes may be made up of antibody and foreign antigen, as in the immune complexes that circulate in the prodromal phase of infection with hepatitis B virus, or may be made up of antibody and self-antigen, as in the complexes

of antibody to DNA and DNA that may be formed in systemic lupus erythematosus.

When a rabbit with circulating antibody is injected subcutaneously with the challenging antigen, the area of injection becomes inflamed as the result of formation of local antigen-antibody complexes (Arthus reaction). Acute pulmonary inflammatory reactions in individuals sensitive to the inhalation of organic dusts (hypersensitivity pneumonitis) are thought to represent a clinical counterpart of the experimental Arthus reaction.

Immune complexes formed or deposited in tissues produce inflammation by activating complement (see Chapter 23). An activated complement component, $C5_a$, is chemotactic for granulocytes. Granulocytes attracted to the site phagocytose the immune complexes. In the process, proteolytic enzymes escape from the granulocyte and injure the surrounding tissues. In addition, the activated terminal components of complement form a membrane attack complex that directly damages cell membranes.

Pathologic Manifestations of Cell-Mediated Immune Responses

Cell-mediated immune responses are by their nature to some extent autoaggressive. Lymphokines produced in a delayed hypersensitivity reaction may damage cells directly (lymphotoxins) and indirectly, through the release of proteolytic enzymes from activated macrophages. The severity of the resulting inflammatory response varies with the extent and duration of the reaction. In pulmonary tuberculosis, a continuing delayed hypersensitivity reaction may result in extensive necrosis of lung tissue. Environmental substances produce contact dermatitis in sensitized individuals through initiation of a delayed hypersensitivity reaction. This is the mechanism for the marked inflammatory response that exposure to the poison ivy plant provokes in many individuals.

When cytotoxic T-cells lyse target cells containing viral surface antigens, viral replication within that cell ceases but at the cost of the death of the cell. In at least one well studied experimental viral infection, lymphocytic choriomeningitis in the mouse, tissue damage stems not from the virus injuring the infected cells but from an attack of cytotoxic T-cells upon the infected cells (Zinkernagel, 1978). A similar mechanism may contribute to the progressive damage to hepatic parenchymal cells that occurs in chronic viral hepatitis. Cytotoxic cell-mediated processes may also contribute to autoimmune organ failure unrelated to infection, e.g., to the destruction of the mucosa of the stomach that may lead to the development of pernicious anemia.

Immunodeficiency States

A number of rare hereditary disorders have been recognized in which affected individuals develop repeated serious infections, unusual infections, or both as a consequence of defective immune function. The infections usually begin in infancy. In some disorders, combined antibody-mediated (B-cell) and cell-mediated (T-cell) immune failure may be found. In other disorders B-cell or T-cell function may be selectively impaired. Studies of mechanisms of immune failure in such patients have contributed substantially to our understanding of the physiology of human immune function. Of particular interest is the discovery that disturbed purine metabolism in lymphocytes due to a deficiency of the enzyme catalyzing conversion of adenosine to inosine (adenosine deaminase) is associated with a severe combined failure of both B-cell and T-cell function.

Acquired immunodeficiency states may arise in several clinical circumstances. Malignancies resulting from the unchecked proliferation of a clone of B-cells (e.g., poorly differentiated nodular lymphocytic lymphoma) or of plasma cells (multiple myeloma) impair polyclonal B-cell function, with resultant impaired synthesis of antibodies. Recipients of renal transplants must take drugs to suppress the immune response. Although necessary to prevent rejection of the renal allograft, this increases the patient's risk of infection. A new acquired immunodeficiency syndrome affecting cell-mediated immune responses and presumably transmitted by an infectious agent has been seen recently in promiscuous homosexual males, in intravenous drug abusers, in hemophilic patients who have received infusions of pooled plasma concentrates, and in a small number of immigrants to the U.S. from Haiti. Patients afflicted with this syndrome, which is often fatal, develop repeated opportunistic infections (cytomegalovirus pneumonia, pneumocystis infection, infection with unusual mycobacteria) and wasting. Some patients also develop a tumor of vascular endothelial cells called Kaposi's sarcoma.

CHAPTER 23

Inflammation and Phagocytosis

When microorganisms breach local defenses at skin and mucosal surfaces, systemic reactions are set off to destroy the invading microorganisms. These reactions include immune responses, described in Chapter 22, and responses described in the present chapter that lead to the ingestion and killing of pathogens by leukocytes. The ongoing synthesis of antibodies, which results in a normal range of plasma IgG of 0.7–1.5 g/dl; the opportunistic infections that threaten the individual whose cellular immune mechanisms are suppressed; and the virtual certainty of severe infection in the patient whose blood granulocyte count falls for any prolonged period below about 500/μl—all attest to the importance of a minimal continuing activity of these systemic reactions for the maintenance of normal natural resistance to infection.

The antibody response to invading microorganisms is linked to the phagocytic and microbicidal responses of granulocytes and macrophages through activation of a system of plasma proteins called the complement system (Müller-Eberhard and Schreiber (1980) and Reid and Porter (1981)). After its activation, terminal components of complement may form a surface membrane attack complex that can, in itself, lyse certain microorganisms. However, probably more important, the activation of complement generates mediators of vasodilatation and increased vascular permeability, chemotaxis, opsonization and phagocytosis, and the release of materials from granulocytes and macrophages that induce tissue inflammation (Kunkel et al., 1981). It is, therefore, not surprising that patients with very rare hereditary disorders producing a deficiency of the key complement component C3 have repeated life-threatening bacterial infections. The other proteolytic systems of plasma—the kallikrein-kinin system and the blood coagulation and fibrinolytic reactions—may also participate in the tissue responses to foreign materials but are of apparent lesser importance. Hereditary deficiencies of components of these proteolytic systems are not associated with an increased susceptibility to infection.

The introduction into tissues of microorganisms, antigen-antibody complexes, or precipitates of crystalline materials provokes local tissue injury and inflammation. Vasodilatation and hyperemia, edema from exudation of fluid from vessels with increased permeability, accumulation of granulocytes and also mononuclear phagocytes, deposition of fibrin, and damage to cells and connective tissue matrix may all occur. Although direct effects of invading microorganisms undoubtedly contribute to such findings, they usually stem primarily from the effects of mediators involved in the body's defense mechanisms. *The effectors of phagocytosis and microbicidal killing are also the effectors of inflammation.* Of major importance are preformed neutral proteases which may be released into the tissues from the lysosomal granules of phagocytes, and two classes of materials, toxic oxygen products and lipid mediators, which are generated when the surface membrane of the granulocyte is stimulated.

The linkage of phagocytosis and inflammation works both to the body's advantage and disadvantage. Release of materials from phagocytes into the surrounding tissues may facilitate the phagocytic attack upon microorganisms, particularly pathogens such as fungi and larval forms of parasites that are too large for single phagocytes to ingest. However, inflammatory tissue damage induced by products of phagocytes also represents an important pathogenetic mechanism of disease. For example, the release of proteases and toxic metabolites from granulocytes attempting to ingest the antigen-antibody complexes that form in the joint tissues of patients with rheumatoid arthritis or from granulocytes attempting to ingest crystals of uric acid that precipitate in the joint fluid of patients with gout probably represents a primary mechanism for the inflammation and tissue damage characteristic of these disorders (Weissmann et al., 1982).

THE COMPLEMENT SYSTEM

Nomenclature

The complement system consists of at least 20 plasma proteins that are involved in mediating biologic defense and inflammatory events on cell surfaces and in solution. Numbers, letters, and trivial names have all been used to identify different components of the system (Reid and Porter, 1981). A series of native molecules have been designated by the numbers C1 through C9, which are assigned almost but not quite in the order in which they have been found to participate in the activation reactions. (C4 reacts in the classical pathway before C2.) C1 was found to consist of a complex of three distinct proteins which were then given the names of C1q, C1r, and C1s. Some components have been given capital letter names. These include an enzyme, factor D, and a proenzyme, factor B, of the alternative pathway and four regulatory proteins: factor H, factor I, factor P, and the S protein. Five of the complement proteins are cleaved into fragments during activation. Fragments have been identified by adding lower case letters from the beginning of the alphabet to the symbol for the native molecule. For example, factor B is cleaved to a fragment, factor Bb; C3 is cleaved during activation to form C3a and C3b, and C3b may be further cleaved into fragments C3c, C3d, and C3e. Complexes of components are formed during activation; these are identified by listing their components in the order of their participation in the complex. Thus, the enzyme activity generated by the classical pathway that cleaves C5 (the classical pathway C5 convertase) is designated C4b,2a,3b. Finally, two regulatory proteins have trivial names but no number or letter symbols. These are $\overline{C1}$ inhibitor, which binds and inhibits $\overline{C1r}$ and $\overline{C1s}$, and the C4b binding protein. As the foregoing example illustrates, when, as in the activation of C1, activation does not result from the formation of a fragment, the active state is identified by placing a bar over the component.

Biosynthesis of the Complement Proteins

The liver is thought to be the primary site of synthesis of the majority of the plasma complement components. Monocytes-macrophages, fibroblasts, and epithelial cells also synthesize some of the complement components. Synthesis by these cells is not thought to be a significant source of plasma complement; it may be important, however, for local reactions involving complement on or in proximity to the surface of these cells.

C3, C4, and C5 are synthesized as single polypeptide chain precursor molecules but circulate in plasma as two chain (C3, C5) or three chain (C4) molecules in which the chains are held together by interchain disulfide bonds. C3, C4, and C5 have areas of amino acid homology and some common biological properties; they may have arisen from a common ancestral gene. C3 and C4, but not C5, contain a most unusual thio ester, which undergoes hydrolysis after activation of C3 and C4 to create a transient site for covalent bonding of these molecules to surface polysaccharides or proteins. C2 and factor B appear to be novel types of serine proteases. Each cleaves the same bonds in C3 and C5 as does the other. The structural genes for both C2 and factor B are located in proximity to each other in the HLA region of human chromosome 6. These molecules may also have arisen from a common ancestral gene.

C3, which may be viewed as the fulcrum protein of the complement system, is found in plasma in a concentration of about 120 mg/dl. Many of the other proteins are present in concentrations ranging from about 2.5 to 50 mg/dl. One protein, factor D, is a true trace protein, being present in plasma in a concentration of only 0.1 mg/dl.

A General Description of Complement Activation

Two activation pathways for complement have been delineated: the *classical pathway* and the *alternative pathway*. Binding of C1 to antigen-antibody complexes on membranes, in which the antibody is IgM or the γ_1 or γ_3 subclasses of IgG, triggers the classical pathway. It may also be initiated nonimmunologically by the binding of C1 to C-reactive protein on membranes or to DNA. The fibrinolytic enzyme plasmin may also activate C1 in solution.

The exposure of plasma to surface membranes with certain properties activates the alternative pathway. Repeating polysaccharide sequences are a structural feature of many such surfaces. (The lipopolysaccharide envelope of Gram-negative bacteria (endotoxin) and the cellophane coils of hemodialysis machines are clinically important examples of activating surfaces.) The alternative pathway may also be activated immunologically by IgA antibodies and by some IgG molecules.

Both activation pathways lead to the formation on the surface of cells, microorganisms, or other particles of complexes called C3 convertases. C3 convertases cleave C3 into C3a and a large fragment, C3b, which develops surface-binding and other important recognition sites. The C3 convertase of the classical pathway (C4b,2a) consists of an enzymatic subunit, C2a, and a nonenzymatic subunit, C4b. The C3 convertase of the alternative pathway (C3b,Bb) consists of an enzymatic subunit, Bb, and a nonenzymatic subunit, which is C3b itself. Because of this, activation of C3 by the alternative pathway has an amplification or positive feedback loop in which generation of new C3b may result in increased formation of alternative pathway C3 convertase which, in turn, cleaves more C3. This amplification loop is not present in the classical pathway; however, C3b formed in the classical pathway is also potentially available for incorporation into the alternative pathway C3 convertase.

With continuing C3 convertase activity, C3b molecules come down onto surfaces in increasing numbers. Some of these molecules bind in proximity to C3 convertases. The bound C3b has a recognition site for C5, and this site provides a mechanism for bringing C5 molecules into contact with the C3 convertases. Thus, as C5 molecules are attracted to the bound C3b, the C2a enzyme of the classical pathway convertase can begin to cleave C5. The C3 convertases are transformed into C5 convertases. Cleavage of C5 yields a small fragment, C5a, of major importance for the inflammatory process (see below) and a large fragment, C5b, which may bind to surfaces. The binding of C5b to a membrane initiates assembly on the membrane of the terminal complement components, which form a membrane attack complex, C5b,6,7,8,9, that can lyse biological membranes.

Multiple factors regulate complement activation. These include the properties of the membranes upon which C3b molecules are deposited; the lability of the covalent binding sites that develop on C3b and C4b, which limits their binding to membranes; the inherent instability of both C3 convertase complexes, which readily disassociate into their subunits; and the activity of several regulatory proteins that affect different steps in the activation process (Table 3.7).

A More Detailed Description of Activation of the Classical Pathway

Figure 3.11 is a diagram of the reactions of the classical pathway. The pathway is activated when C1q molecules in C1q,r,s, complexes bind to sites on the Fc portion of antibodies on membranes. C1q is an unusual protein with a central structure and six collagen-like triple helical filaments extending outwards from the central structure. Each filament ends in a globular region containing recognition sites for the Fc portion of IgM or the γ_1 or γ_3 subclasses of IgG, for C-reactive protein, and for DNA. When two or more globular regions on the same molecule bind to immunoglobulin on a surface

Table 3.7
Regulators of complement activity

Regulatory Agent	Function
$\overline{C1}$ inhibitor	Inactivates $\overline{C1r}$ and $\overline{C1s}$ stoichiometrically. In noncomplement systems, functions as major inactivator of kallikrein and activated Hageman factor.
C4 binding protein	Binds to C4b, rendering it susceptible to factor I.
Factor H	Binds to C3b, competing with B, displacing Bb, and rendering C3b susceptible to factor I.
Factor I (C3b inhibitor)	Cleaves and inactivates C3b in C3b-H complex and C4b in C4b-C4 binding protein complex.
Factor P (properdin)	Binds to and stabilizes the alternative pathway convertase, C3b,Bb.
S protein	Binds to C5b,6,7 modulating formation of membrane attack complex.
Serum carboxypeptidase N	Cleaves carboxy-terminal arginine residue of C3a and C5a, inactivating their anaphylotoxic activity.
Chemotactic factor inactivator	Plasma factor inactivating chemotactic activity of C5a.

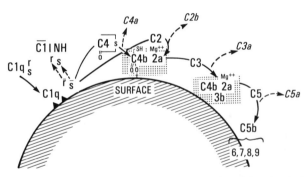

Figure 3.11. A diagram illustrating the activation of the classical pathway of complement. The reactions are initiated by the binding of C1q to the Fc region of two or more antibody molecules on a surface membrane. The first complex shown within a stippled area, the C4b,2a complex, is the classical pathway C3 convertase. The second complex shown within a stippled area, C4b,2a,3b, is the classical pathway C5 convertase. Further details of the reactions are given in the text. $\overline{C1}$ INH, $\overline{C1}$ inhibitor.

membrane, the conformation of the C1q,r,s complex is altered in a way that converts C1r into an active serine protease. $\overline{C1r}$ then activates C1s, the second serine protease of the C1 complex. $\overline{C1s}$, in reactions independent of the other members of the complex, cleaves both C4 and C2. $\overline{C1r}$ and $\overline{C1s}$ activities are regulated by an inhibitor, $\overline{C1}$ inhibitor which, by binding stoichiometrically to each of these molecules, inhibits their protease activity.

C4 is a large (206,000 daltons) triple-chain protein. $\overline{C1s}$ cleaves a small fragment, C4a from the chain of C4. This alters the large residual portion of the molecule, C4b, in a way that allows its thio ester bond to be broken. As a result, a labile acyl group is formed that can convalently bind C4b to antibody or other molecules on a membrane. However, this binding site is metastable. Although not shown in Figure 3.11 to reduce its complexity, the majority of C4b molecules do not bind to the surface but diffuse into the fluid phase. In the fluid phase, C4b molecules form complexes with a very large protein (about 700,000 daltons) called the C4b binding protein. Binding to C4b binding protein makes C4b susceptible to proteolysis by a regulatory enzyme, factor I, with resultant formation of inactive cleavage products. (Factor I is sometimes also referred to as the C3b inhibitor, because it was discovered to cleave C3b before it was discovered also to cleave C4b.)

C2 is a proenzyme (about 100,000 daltons) which is attracted to a second recognition site that develops when C4 is converted to C4b. C2 in proximity to C4b can then be cleaved by C$\overline{1}$s. This yields a 70,000-dalton fragment, C2a, that contains the active enzymatic site, and a 30,000-dalton fragment, C2b. C2b and most of the C2a molecules so formed diffuse away, but some of the C2a molecules remain associated with membrane-bound C4b, in a complex requiring the presence of Mg^{++} ions, to give rise to the classical pathway convertase C4b,2a.

As already mentioned, C4b,2a cleaves C3 into C3a, a small fragment derived from the amino terminal end of the α chain of C3, and the remainder of the molecule, which is now called C3b. C3a diffuses away into the fluid phase, as do most of the C3b molecules (also not shown in Figure 3.11 to reduce its complexity). Factor H binds to C3b in the fluid phase and, analogous to the binding of C4b to C4b binding protein, this binding renders C3b susceptible to proteolysis by factor I.

Some C3b molecules remain bound to the surface, in proximity to C4b,2a, through the metastable ester group formed by hydrolysis of the thio ester. C5 and factor H compete for recognition sites on the bound molecule. As also mentioned earlier, if C5 binds to C3b, then the C4b,2a convertase of C3 is transformed into a C4b,2a,3b convertase of C5, and the complement activation sequence continues. If factor H binds, then, as in the fluid phase, factor I cleaves C3b into inactive products, and further activation of complement stops.

C3b bound by its metastable binding site to a target membrane (e.g., the surface of a microorganism) or C3b that has escaped into the fluid phase may bind by a second recognition site to specific C3b receptors that are present on the membranes of granulocytes, monocyte-macrophages, B lymphocytes, and red blood cells (Fig. 3.12). Such binding to receptors on granulocytes and macrophages

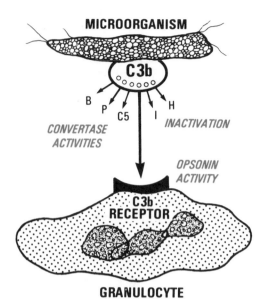

Figure 3.12. A diagram illustrating the multiple binding properties of C3b. C3b bound to a target cell membrane may bind by a second site to a C3b receptor on a granulocyte or macrophage; in this event the C3b acts as an opsonin facilitating phagocytosis of the target cell. C3b possesses binding sites for factor *B*, and factor *P*, and *C5*; binding to these materials allows expression of convertase activities; C3b also possesses binding sites for factors *H* and *I*; binding to these materials results in inactivation of C3b. Although not shown in the figure, C3b in the fluid phase may also bind to C3b receptors on granulocytes, macrophages, or red blood cell surfaces.

facilitates the phagocytosis of a target cell. However, binding of C3b to a C3b receptor, like binding of C3b to factor H, prevents C3b from participating further in the complement activation sequence. This has the teleologic advantage of protecting the membrane of the phagocyte from lysis by the membrane attack complex.

A More Detailed Description of the Activation of the Alternative Pathway

Figure 3.13 is a diagram of the reactions of the alternative pathway. This pathway is thought normally to undergo slow activation in the fluid phase. This stems from the spontaneous hydrolysis of the thio ester in a small number of C3 molecules, an event that then allows factor B to bind to the molecule. Once bound, factor B can be cleaved by the trace plasma protease factor D, which circulates in an active form. The resultant complex, sometimes referred to as the priming C3 convertase (Kazatchkine and Nydegger, 1982), can not itself bind to a surface membrane. Apparently, its metastable binding site has decayed. However, it can activate additional C3 molecules in the fluid phase, which can bind covalently to a surface through acyl groups newly formed after hydrolysis of the thio ester.

As shown in Figure 3.13, C3 in the priming C3 convertase is cleaved to C3b. Factor H can then bind to the complex, displacing Bb and rendering C3b susceptible to proteolysis by factor I. Factor I initially cleaves C3b at a single site, converting C3b into an altered molecule now called C3b$_i$. Further proteolysis then yields two fragments, C3c and C3d. C3c may undergo additional proteolysis, possibly induced by plasmin or some other plasma proteolytic enzymes, to form a fragment called C3e.

Whether or not the alternative pathway proceeds beyond an otherwise inconsequential cleavage of small numbers of C3 molecules in the fluid phase depends upon the properties of the surfaces onto which the resultant C3b molecules come down (Müller-Eberhard and Schreiber, 1980). In health, the surfaces will be nonactivating surfaces, which have the property of enhancing binding of factor H and diminishing binding of factor B to C3b. Binding to H then facilitates proteolysis of factor I with formation of the inactive products described above. Thus, despite its continuing slow fluid phase activation, the alternative pathway does not normally generate substantial amounts of C3b.

However, in pathologic conditions, e.g., the en-

trance of substantial amounts of endotoxin into the circulation during severe Gram-negative bacterial infection or the exposure of blood to artificial membranes during hemodialysis or open heart surgery, C3b molecules come down onto surfaces with activating properties. Such surfaces apparently prevent factor H from binding to C3b, facilitating instead the binding of factor B and its subsequent activation to Bb by factor D. The C3b,Bb complex that is thus formed is then stabilized by the binding of a third protein, factor P (properdin), to a recognition site on the C3b molecule. (It is possible that a similar stabilizing protein exists for the C4b,2a complex, but it is yet to be found.) The resultant C3b,Bb,P complex, sometimes referred to as the amplifying C3 convertase, then sets in motion the feedback loop described earlier that leads to a greatly increased activation of C3. As already described, when increasing numbers of C3b molecules come down onto the surface, the alternative C3 convertase, like its classical pathway counterpart, is transformed into a C5 convertase. Subsequent reactions, which are the same as for the classical pathway, lead to the formation of the C5b,6,7,8,9 membrane attack complex. A recently described

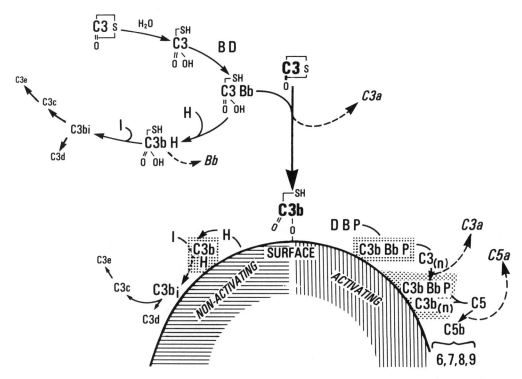

Figure 3.13. A diagram illustrating the activation of the alternative pathway of complement. The reactions in the *upper half* of the diagram take place in the fluid phase. The reactions that occur after C3b binds to a surface depend upon the properties of the surface. On a nonactivating surface the reactions shown on the left hand side of the lower portion of the diagram take place. These lead to further proteolysis and inactivation of C3b. On an activating surface, the reactions shown on the *righthand side* of the *lower half* of the diagram take place. These lead to the formation of a *C3 convertase (upper shaded area)* and then a *C5 convertase (lower shaded area).* Details of the reactions are given in the text.

protein called the S protein appears to regulate formation of the membrane attack complex by competing for membrane binding sites with the C5b,6,7 portion of the complex.

Biologic Activities of Activated Complement Components

Activation of complement results in a variety of biologic consequences (Jacob et al. (1981) and Kazatchkine and Nydegger (1982)). If activation proceeds to completion and the membrane attack complex forms on a biological membrane, then that membrane may undergo lysis. Such lysis seems to be important in combating certain bacterial infections, particularly meningococcal and gonococcal infections. Lysis of red blood cells by the membrane attack complex after certain types of antibodies react with red blood cell surface antigens represents an important mechanism of red blood cell destruction in immune hemolytic disorders.

As already mentioned, C3b bound by a labile binding site to target microorganisms or cells may then bind by a second site to a C3b receptor on the membrane of a granulocyte or monocyte-macrophage. Thus, C3b functions as a serum opsonin, i.e., a material that coats microorganisms and facilitates their phagocytosis by neutrophils or macrophages. The biologic significance of the binding of C3b to C3b receptors on B lymphocytes and on red blood cells is not yet clear.

Several fluid phase complement fragments possess biologic activity: C2 fragments, C3a, C5a, and C3e. C2 fragments, after further proteolysis in plasma, develop kinin activity that can directly increase vascular permeability and contract smooth muscle cells. (Excess C2-derived kinin in patients with a hereditary deficiency of C$\overline{1}$ inhibitor may result in life-threatening attacks of local tissue edema.) C3a and C5a are both anaphylatoxins, agents that also contract smooth muscle cells and enhance vascular permeability but which do so through degranulation and release of vasoactive substances from tissue mast cells. In addition, C5a, but not C3a, functions as a major chemoattractant of the inflammatory response, causing chemotaxis at low concentrations and aggregation and degranulation of granulocytes at higher concentrations (Jacob et al., 1980). The enzyme carboxypeptidase N cleaves the carboxy terminal arginine residue from C3a and C5a. This causes the molecules to lose their anaphylatoxic activity. C5a desarg, however, after interacting with an as yet poorly defined serum factor, partly retains its chemoattractant activity. C3e, a fragment formed from the further proteolysis of C3c, has been found to stimulate the release of granulocytes from the bone marrow.

Complement components have also been shown to participate in the reactions whereby macrophages are activated at inflammatory sites. Thus, one mechanism for inducing macrophage spreading and activation involves cleavage of factor B by plasmin and the subsequent interaction of Bb with C5 at the macrophage surface (Sundsmo and Götze, 1981). It is noteworthy that the macrophage can itself synthesize three of the necessary reactants for this process: plasminogen activator, factor B, and C5. C5a also appears important for the development of procoagulant activity (tissue factor) on monocyte surfaces after a variety of activating stimuli (Muhlfelder et al., 1979). The effect upon lymphocyte function of reactions of complement components on lymphocyte surfaces is an area of current investigative interest.

OTHER PLASMA PROTEOLYTIC SYSTEMS

Blood coagulation and fibrinolysis may be activated and kinins may be generated during inflammation, but the participation of these plasma proteolytic systems in the inflammatory response is probably of lesser importance than the participation of the complement system. Products of stimulated or damaged cells can activate each of these systems: the blood coagulation reactions by tissue factor on the surface of activated macrophages or damaged endothelial cells; fibrinolysis by tissue plasminogen activators released from endothelial cells, macrophages and, possibly, other cells; and the generation of bradykinin by kallikreins released from basophils, tissue mast cells, and neutrophils. In addition, each of these systems can be activated by a sequence of plasma enzymatic reactions called the Hageman factor-dependent pathway, because activation of a plasma protein called Hageman factor (factor XII) is the first step in the sequence (Cochrane (1982) and Kaplan et al. (1982)).

Figure 3.14 is a diagram of the Hageman factor-dependent reactions that take place when plasma is exposed to a negatively charged surface, such as a glass surface, in vitro. Hageman factor (HF) and high molecular weight kininogen (HMWK) come down onto the surface. Prekallikrein and coagulation factor XI circulate in plasma bound to HMWK and, therefore, are brought down onto the surface along with HMWK. It is not known how the initial activated Hageman factor (HF$_a$) molecules are formed on the surface; in one view, HF, after conformational changes resulting from its coming down onto a negatively charged surface, begins to

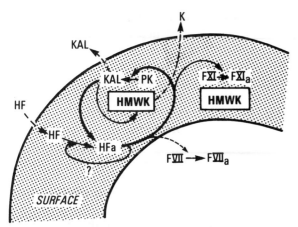

Figure 3.14. A diagram of reactions of the Hageman factor-dependent pathway that take place when plasma is exposed to a negatively charged surface such as that of glass. The *subscript a* denotes an activated factor. Factor *VII* and factor *XI* are two proteins of the blood coagulation system. Reactions within the *stippled area* take place on the surface. *Dashed lines* denote the movement of factors onto or off of the surface. Note the multiple enzymatic activities of HF_a, which catalyzes the activation of prekallikrein to kallikrein, factor XI to factor XI_a, F VII to F VII_a, and possibly the autoactivation of *HF* to *HF*$_a$. Note also the reciprocal activation loop in which *HF*$_a$ activates prekallikrein to kallikrein and kallikrein in turn activates *HF* to *HF*$_a$. Not shown in the diagram are additional reactions in which *HF*$_a$, factor *XI*$_a$, and kallikrein each activate plasminogen to plasmin. *HF*, Hageman factor; *HMWK*, high molecular weight kininogen; *PK*, prekallikrein; *Kal*; kallikrein; *K*, kinin.

activate itself by autodigestion (Kaplan et al., 1982). As HF_a molecules are formed they catalyze activation of prekallikrein to kallikrein. Kallikrein then splits bradykinin from HMWK and also from a low molecular weight kininogen of plasma that does not otherwise participate in the Hageman factor dependent pathway since it lacks the segment of the polypeptide chain of HMWK that possesses binding sites for surfaces and for prekallikrein and factor XI. Kallikrein also participates in an important positive feedback loop that substantially increases the rate of further formation of HF_a. Thus, the Hageman factor-dependent pathway contains a reciprocal activation step in which HF_a activates prekallikrein to kallikrein and kallikrein activates HF to HF_a.

Which, if any, tissue surfaces can function as activating surfaces comparable to a negatively charged surface in vitro is not clear. However, in vivo enzymatic mechanisms for activating HF in the fluid phase have been identified. Thus, endothelial cells have been shown to contain a proteolytic enzyme capable of activating HF, and elastase released from the primary granules of neutrophilic granulocytes has been shown to activate HF during inflammatory reactions in the lungs (Cochrane, 1982).

HF_a promotes blood coagulation in two ways: through the activation of factor XI in the contact-activation coagulation pathway and through the activation of factor VII in the extrinsic pathway (see Chapter 25). In reactions not shown in Figure 3.14, HF_a, kallikrein, and factor XI_a can each also function as weak intrinsic activators of plasminogen, with resultant activation of fibrinolysis. However, despite clear evidence for the role of the Hageman factor-dependent pathway in initiating blood coagulation, fibrinolysis, and kinin generation in vitro, individuals with a hereditary deficiency of HF, prekallikrein, or HMWK appear to be completely healthy. If blood coagulation, fibrinolysis, or kinin generation are impaired in such individuals in vivo, the degree of impairment is insufficient to produce recognized signs of disease. This observation raises questions about the importance of the Hageman factor-dependent pathway for homeostasis and for the inflammatory response.

Complement may interact with the other plasma proteolytic systems in the inflammatory response. As already alluded to, C5a may play a pivotal role in initiating blood coagulation at inflammatory sites through its participation in the generation of monocyte-macrophage tissue factor activity. Plasmin, thrombin, kallikrein, and HF_a can all activate factor B in vitro. Activation by plasmin may generate Bb in proximity to the macrophage surface, and formation of factor Bb appears to be a key step in one process that causes macrophage spreading and activation. As also mentioned earlier, plasmin interacts with complement in two additional ways: it may initiate the reactions of the classical complement pathway through the fluid phase activation of C1q, and it may cleave C2 fragments into material with kinin activity. The physiological significance of these reactions for the overall activity of complement in immunologic and inflammatory responses is unknown.

RESPONSES OF GRANULOCYTES

In phagocytic body defense reactions and their accompanying inflammatory tissue response, granulocytes adhere to vascular endothelium and emigrate from the blood vessels, accumulate at a tissue site, form aggregates, ingest microorganisms or other opsonized materials, activate biochemical reactions leading to microbial killing, and undergo degranulation (Drutz and Mills, 1982). During these processes mechanical events take place in which the granulocyte changes shape, rearranges its actin filaments, invaginates its plasma membrane to form and then close phagocytic vacuoles, fuses its primary and secondary granules with phagocytic vacuoles, and secretes the contents of its granules both intracellularly, into phagocytic vacuoles, and extracellularly (Fig. 3.15).

Binding of ligands, e.g., chemotactic peptides, complement components, or immune complexes, to receptors on the granulocyte surface initiates the biochemical events of granulocyte activation through altering the plasma membrane potential. This triggers signals that result in remodeling of membrane phospholipids and a rise in calcium ion concentration in the cytosol (Korchak et al., 1982). The increased calcium ion concentration is then thought to mediate many of the further biochemical responses. These include the activation of phospholipase A_2 with liberation of arachidonic acid and the oxidation of arachidonic acid, via cyclooxygenase and lipoxygenase pathways, to form prostaglandins and leukotrienes that act as inflammatory mediators (Figure 3.17). Another lipid mediator, acetyl glycerol ether phosphorylcholine, is also formed (Figure 3.18). At the same time, oxygen is consumed in a burst as a membrane-associated oxidase (Figure 3.19) is activated to generate superoxide anions (O_2^-), which in subsequent reactions are converted into H_2O_2, hydroxyl radicals, and other reduced oxygen species (Babior, 1978). These oxygen metabolites are utilized to kill microbes within phagocytic vacuoles by oxygen-dependent mechanisms, including one reaction in which the primary granule enzyme, myeloperoxidase, uses H_2O_2 to oxidize Cl^- to the microbicidal agent, HOCl. Microbes within phagocytic vacuoles are also killed by anerobic mechanisms involving the lowering of pH and the secretion into the vacuole of such primary (lysosomal) granule constituents as acid hydrolases and cationic proteins (Root and

Cohen, 1981). Neutral proteases from the lysosomal granules (elastase, cathepsin G, and collagenase), inflammatory lipid mediators, and reduced oxygen metabolites escape from the granulocyte. These materials injure the adjacent tissues, which become inflamed. Plasma protease inhibitors in extracellular fluid, primarily α_1-protease inhibitor, limit the tissue injury by inactivating neutral proteases. However, reduced oxygen metabolites may interfere with this protective response by oxidizing an essential methionyl thio ester side chain on α_1-protease inhibitor, with resultant loss of its inhibitory activity.

The above provides an overall description of the granulocyte responses of phagocytosis and inflammation. Specific events are discussed further in the following sections.

Chemotaxis

During chemotaxis, granulocytes migrate in the direction of an increasing concentration of materials called chemoattractants. A growing number of materials have been shown to function as chemoattractants in in vitro experiments in which granulocytes are placed in one side of a diffusion chamber, the test material is placed in the other side, and the ability of the test material to cause granulocytes to migrate from one side of the chamber to the other is measured. For in vivo relevance, a newly identified chemoattractant should induce chemotaxis in a concentration achievable physiologically and in the same low range as that of established physiological chemoattractants. Moreover, when injected into an experimental animal it should induce transient neutropenia due to adherence of granulocytes to vascular endothelium. Table 3.8 provides a list of chemoattractants considered potentially important for the inflammatory response. These include agents derived from diverse sources: from the breakdown of bacterial proteins and of mitochondrial proteins from damaged tissue cells (Carp, 1982); from the cleavage of the complement component, C5; from secreted constituents of the granules of platelets (Deuel et al., 1982); and from the release of lipid-derived mediators and other materials from stimulated WBC.

Synthetic oligopeptides, beginning with a formylated methionyl residue (e.g., *N*-formyl-Met-Leu-Phe), have proven particularly valuable as a source of purified structurally defined chemoattractants with which to delineate the characteristics of granulocyte chemoattractant receptors. It is of teleologic interest that mammalian granulocytes recognize such oligopeptides as chemoattractants. Protein synthesis directed by bacterial DNA and pro-

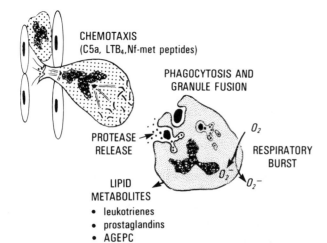

CHEMOTAXIS
(C5a, LTB₄, Nf-met peptides)

PHAGOCYTOSIS AND GRANULE FUSION

PROTEASE RELEASE

O_2

RESPIRATORY BURST

LIPID METABOLITES
- leukotrienes
- prostaglandins
- AGEPC

O_2^-

Figure 3.15. An illustration of the responses of granulocytes in inflammation. *LTB₄* is leukotriene *B₄*. *AGEPC* is acetyl glycerol ether phosphorylcholine.

Table 3.8
Chemoattractants of probable physiologic significance

Origin	Material
Bacterial protein breakdown	N-formyl-methionyl-oligopeptides
Mitochondrial protein breakdown	N-formyl-methionyl-oligopeptides
Complement cleavage product	C5a, C5a-des-Arg[a]
Platelet α granule secretion	Platelet factor 4
	Platelet-derived growth factor
Neutrophilic granulocyte activation	Metabolites of lipoxygenation of arachidonic acid (leukotriene B$_4$, 5-HETE, etc)
	AGEPC (acetyl glycerol ether phosphorylcholine)
	Crystal-induced chemotactic factor (CCF)[b]
Tissue mast cell degranulation	Eosinotactic tetrapeptides (eosinophil chemotactic factor of anaphylaxis)[c]
	HMW[d] neutrophil chemotactic factor of anaphylaxis
T-lymphocyte antigen or mitogen stimulation	Lymphokine chemoattractant for granulocytes

[a] C5a minus its carboxy terminal arginine. Requires an anionic helper factor present in serum. [b] A chemotactic factor produced when granulocytes phagocytose monosodium urate crystals. [c] Chemotactic to a lesser degree for neutrophilic granulocytes. [d] HMW is an abbreviation for high molecular weight.

tein synthesis directed by mitochondrial DNA (derived in evolution from bacteria) but not by nuclear DNA of eukaryocytes is initiated with an NH$_2$ terminal-formylated methionyl residue.

Different membrane receptors have been identified on granulocytes, including receptors for N-formylated methionyl peptides, for C5a, for a chemotactic leukotriene (LTB$_4$), and for a chemotactic factor released from granulocytes as they phagocytose sodium urate crystals (crystal-induced factor). One may suspect that additional receptors exist for other chemoattractants, such as the platelet-derived growth factor. Chemotactic factor receptors have been found to exist in a high affinity state, in which they transduce chemotactic signals, and in a low affinity state, in which they suppress chemotactic signals and instead transduce signals for superoxide production and lysosomal enzyme secretion (Snyderman and Goetzl, 1981). Guanine nucleotides have been reported to convert high affinity receptors to low affinity receptors. A chemoattractant, depending upon its concentration and upon the affinity state of its receptors, may induce either chemotaxis or granulocyte aggregation, degranulation, and superoxide anion production.

The morphology of the granulocyte changes strikingly during chemotaxis (Snyderman and Goetzl, 1981). The cell elongates and becomes polarized, developing a broad "head" called a lamellipodium which is oriented towards the concentration gradient of the chemotactic factor and a thin "tail" with terminal arborization called a uropod (Figure 3.15). On electron microscopy, increased numbers of actin microfilaments are seen beneath the plasma membrane in the lamellipodium and the uropod. Microtubules extend towards the poles of the cell from their origin in centrioles close to the lobes of the nucleus, providing a cytoskeletal support needed to maintain the orientation of the cell towards the chemotactic gradient. As the cell moves, the nucleus lags behind the center of the cell and is separated from its leading edge by the main mass of the cytoplasm. The cytoplasm of the leading edge of the lamellipodium appears free of granules, probably reflecting a degranulation of specific granules along the leading edge. Chemotactic receptors are concentrated in the surface membrane of the leading edge.

Calcium concentration rises rapidly in the cytosol of granulocytes exposed to a chemoattractant, and this event appears to couple chemotactic stimulation to cell migration. The increased calcium concentration reflects both a release of calcium from cell membranes and an increase in plasma membrane permeability to extracellular calcium. The increased cytoplasmic calcium concentration initiates linkage of actin to the protein gelsolin, with resultant shortening of actin filaments and the breakage of lattices between actin and actin-binding protein. This rearrangement of actin filaments plus the cell orientation provided by the rearrangement of microtubules permits directed locomotion.

Exposure to chemoattractants also triggers reactions affecting the phospholipids of granulocyte membranes. Methylation of membrane phospholipids (see below) is stimulated, as are the release and oxidation of arachidonic acid. How these biochemical events relate to granulocyte motility is not clear, but they apparently must take place. Thus, exposing the granulocyte to agents inhibiting methylation of membrane phospholipids or to agents inhibiting arachidonic acid oxidation inhibits chemotaxis.

As already mentioned, exposing granulocytes to a low concentration of a chemoattractant results in directed locomotion, whereas exposing granulocytes to a high concentration of the chemoattractant results in adherence, aggregation, and degranulation. Yet, in the inflammatory response adherence must precede migration, since granulocytes in the circulating blood must first adhere to vascular endothelium before they can migrate out of the vessels into the tissues. Adherence and aggregation apparently require the external secretion of lactoferrin from the specific granules. The lactoferrin then binds to receptors on the granulocyte surface, which alters the surface charge of the membrane and makes the granulocytes sticky (Boxer *et al.*, 1982). Thus, secretion from specific granules appears to be involved, both in the process whereby granulocytes migrate to an inflammatory site and in the process whereby granulocytes become immobilized and are retained at an inflammatory site.

Generation of Lipid Mediators

In many cell systems, stimulation of surface membrane receptors initiates methylation of membrane phospholipids (Hirata and Axelrod, 1980), activation of phospholipases, liberation of arachidonic acid, and subsequent oxidation of arachidonic acid by the lipoxygenase and cyclooxygenase pathways. In inflammation, occupancy of surface receptors by chemoattractants (Table 3.8), opsonins (IgG in antibody complexes, C3b) or other agents sets off these reactions in granulocytes, monocytes-macrophages, mast cells, platelets, and endothelial cells. It is likely that the methylation reactions are the same for all of these cells. However, the major end products of arachidonic acid oxidation—prostaglandins, thromboxanes, leukotrienes, and other derivatives of hydroperoxy-fatty acids—differ in the different cells (Goetzl, 1981).

Methylation of phospholipids takes place within seconds after stimulating surface membrane receptors. It begins on the cytoplasmic surface of the membrane, where an enzyme called methyltransferase I catalyzes the transfer of a methyl group from *S*-adenosyl-methionine to phosphatidylethanolamine. In the course of the reaction the substrate is translocated to the plasma surface of the membrane where, catalyzed by a second enzyme, methyltransferase II, two additional methyl groups are added to the phospholipid to yield phosphatidylcholine. As studied in rat mast cells, these methylation reactions are linked to but precede the influx of extracellular calcium into the cell and the subsequent activation of phospholipase A₂ (Hirata and Axelrod, 1980). The activation of phospholipase A₂ liberates arachidonic acid from phosphatidylcholine and also results in the formation of lysophosphatidylcholine (Fig. 3.16), a material with detergent-like properties that may be involved in the fusion of granule membranes and plasma membranes during degranulation.

A second phospholipase, phospholipase C, is also activated when leukocytes or platelets are stimulated. It differs from phospholipase A₂ in its inde-

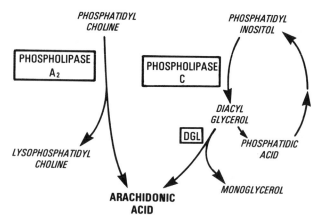

Figure 3.16. Mechanisms for the release of arachidonic acid from membrane phospholipids. Recycling of phosphatidic acid to phosphadityl inositol is also shown. The abbreviation *DGL* stands for the enzyme diglyceride lipase.

pendence from calcium and in its use as substrate of phosphatidyl inositol (PI), from which it splits the phosphate ester to form diacyl glycerol (Fig. 3.16). The latter is then converted either to monoglycerol and arachidonic acid or to phosphatidic acid (PA). The PA may be recycled to PI. In in vitro studies with liposomes, phosphatidic acid functions as a calcium ionophore, and it has been postulated that the PI-PA-PI cycle could function as a "gate" for regulating calcium flux across the surface membrane of stimulated cells. The relative contributions of phospholipase A₂ and phospholipase C to the release of arachidonic acid appears to vary in different cells. Neutrophilic granulocytes may primarily utilize phospholipase A₂, whereas platelets apparently utilize both phospholipases.

Oxygenation of arachidonic acid by way of the cyclooxygenase pathway results in formation of the endoperoxide PGG_2 which is then reduced to PGH_2. A reactive oxygen species is liberated in this reaction. As shown in Figure 3.17, PGH_2 is converted in different cells into different prostaglandin and thromboxane end products (As discussed in Chapter 25, production of thromboxane A_2 in the platelet and PGI_2 in the endothelial cell is particularly important for the regulation of hemostasis.) PGE_2 and PGD_2, the major prostaglandins formed by granulocytes, monocytes-macrophages, and mast cells, probably function as only minor vasodilators in the inflammatory response. Moreover, through raising cAMP levels, PGE_2 and PGD_2 decrease release of proteases from stimulated granulocytes and, so, are thought to function primarily as inhibitors of the inflammatory response. It seems paradoxical, therefore, that drugs impairing cyclooxygenase activity, such as aspirin or indomethacin, are effective clinically in diminishing symptoms of inflammation. This has led to speculation that the reactive oxygen species formed on reduction of PGG$_2$ to PGH$_2$, whose liberation would be prevented by agents inhibiting cyclooxygenase (Fig. 3.17), is a major inducer of tissue inflammation.

The leukotrienes, products of the lipoxygenase pathway of arachidonic acid oxidation, play dramatic roles in delayed hypersensitivity and inflammatory responses (Dahlen et al., 1981). In leukocytes (but not in platelets, in which arachidonic acid is oxidized to 12-hydroperoxy-fatty acid derivatives), lipoxygenase catalyzes the formation of 5-hydroperoxyeicosatetranoic acid (5-HPETE), which is a precursor of a material given the name leukotriene A$_4$ (the subscript 4 denotes its derivation from arachidonic acid, which has 4 double bonds). In mast cells and also in monocytes-macrophages, leukotriene A$_4$ (LTA$_4$) is conjugated with glutathione to give rise to leukotriene C$_4$ (LTC$_4$), which may then be converted to LTD$_4$, a molecule in which the glutamine residue has been cleaved from glutathione, and to LTE$_4$, a molecule in which the glycine residue is also lost from glutathione, leaving only its cysteinyl backbone. Mast cells have been known for many years to contain a potent agent called SRS-A (slow-reacting substance of anaphylaxis) that constricts smooth muscle and increases vascular permeability. SRS-A is thought to

mediate both the bronchospasm and the mucosal edema of bronchial asthma. The biologic activity of SRS-A is now known to stem from a mixture of the actions of LTC$_4$ and LTD$_4$ (Samuelsson, 1981). In neutrophilic granulocytes, LTA$_4$ is not conjugated with glutathione but is instead enzymatically converted into a molecule called LTB$_4$ (Figure 3.17). LTB$_4$ has been found to be among the most potent known inducers of leukocyte chemotaxis, aggregation, and degranulation. It represents a key mediator of the inflammatory response (Samuelsson, 1981).

Another potent lipid inflammatory mediator, acetyl glycerol ether phosphorylcholine, or AGEPC (Figure 3.18), has also recently been identified (Pinckard, 1982). It was originally called platelet-activating factor (PAF) because it was first recognized as a material causing platelet aggregation that was released from IgE-sensitized rabbit basophils stimulated by antigen. However, it was later found to be liberated after different types of stimulation of many cells, including neutrophils, monocytes-macrophages, and platelets. In addition to causing platelet aggregation, AGEPC induces chemotaxis, aggregation, degranulation, and superoxide production in granulocytes. When injected intravenously into baboons, it caused neutropenia, thrombocyto-

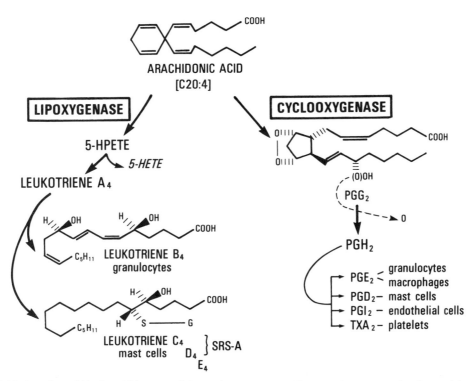

Figure 3.17. Oxidation of arachidonic acid by way of the cyclooxygenase pathway to form prostaglandins and thromboxane and by way of the lipoxygenase pathway to form leukotrienes. In the illustration of leukotriene C$_4$, the molecule attached at the 6 position that is represented by the symbol S——G is glutathione. In leukotriene D$_4$, the glutamine residue has been lost from glutathione. In leukotriene E$_4$, both the glutamine and glycine residues have been lost, leaving only the cysteinyl backbone of glutathione. SRS-A, slow-reacting substance of anaphylaxis; 5-HPETE, 5-hydroperoxyeicosatetraenoic acid; 5-HETE, 5-hydroxyeicosatetraenoic acid.

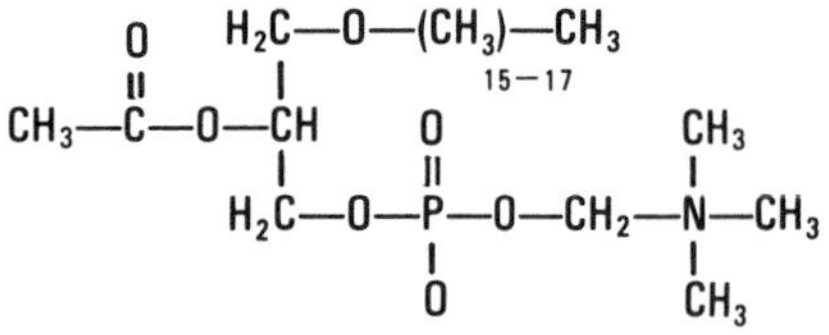

$$
\begin{array}{c}
\text{O} \qquad \text{H}_2\text{C}-\text{O}-(\text{CH}_3)-\text{CH}_3 \\
\quad\ \ \| \qquad\ \ | \qquad\quad {}_{15-17} \\
\text{CH}_3-\text{C}-\text{O}-\text{CH} \qquad \text{O} \qquad\ \ \text{CH}_3 \\
\qquad\qquad\quad\ | \qquad\quad\ \| \qquad\quad | \\
\qquad\quad\ \ \text{H}_2\text{C}-\text{O}-\text{P}-\text{O}-\text{CH}_2-\text{N}-\text{CH}_3 \\
\qquad\qquad\qquad\qquad | \qquad\qquad\quad | \\
\qquad\qquad\qquad\qquad \text{O} \qquad\qquad\quad \text{CH}_3
\end{array}
$$

Figure 3.18. The structure of acetyl glycerol ether phosphorylcholine.

penia, increased pulmonary artery pressure, and severe and prolonged systemic hypotension. AGEPC does not exist preformed in unstimulated neutrophilic granulocytes but forms within seconds after stimulation, possibly by acetylation of a precursor molecule, lysoglycerol 3-phosphorylcholine. The mechanism of release of AGEPC from cells is unknown. AGEPC is reported to induce platelet aggregation through stimulation of arachidonic acid oxidation via the cyclooxygenase pathway to form thromboxane A_2 (Chesney et al., 1982). Recently AGEPC has also been shown to stimulate arachidonic acid oxidation via the lipoxygenase pathway to form LTB_4 (Lin et al., 1982). To what extent production of LTB_4 can account for the effects of AGEPC as an inflammatory mediator is not yet known.

Reactive Oxygen Species

As already mentioned, granulocytes increase their consumption of oxygen during phagocytosis or after exposure to membrane-stimulating agents such as chemoattractants, a phenomenon referred to as the respiratory burst. Cyanide ion, which blocks mitochondrial energy-producing reactions, fails to prevent the respiratory burst. The increased oxygen uptake is associated with two biochemical events. The first is a 1 electron reduction of molecular oxygen to a form called superoxide anion (electron structure $\cdot\ddot{\text{O}}\!:\!\ddot{\text{O}}$), designated by the symbol O_2^-. The second is consumption of glucose in a pathway, the hexose monophosphate shunt, in which NADPH and reduced glutathione (GSH) are produced. It is now established that granulocytes must generate O_2^- as a step in the process whereby they kill certain bacteria in oxygen-dependent reactions (Babior, 1978; Klebanoff, 1980). In a hereditary disorder, chronic granulomatous disease, an inability to generate O_2^- leads in many of the affected children to recurrent, very troublesome bacterial infections.

Generation of O_2^- requires activation of a granulocyte oxidase. (An oxidase is an enzyme that transfers electrons to O_2.) The molecular mechanisms activating the oxidase are not yet clear. When responses of granulocytes activated by ligand-surface membrane receptor interactions were subjected to kinetic analysis, it was found that depolarization of the granulocyte membrane preceded O_2^- production by about 30 s, but that O_2^- began to form before the influx of extracellular calcium into the granulocyte (Weissmann et al., 1982). In other experiments, protease inhibitors were found to blunt O_2^- production after stimulation of granulocytes, which suggests that proteases may be involved in activation of the oxidase (Kitagawa et al., 1980).

Data have differed on the properties of the O_2^--forming oxidase, and it is possible that more than one oxidase is involved. However, most data support the view that the principle oxidase catalyzing the reaction is an enzyme which can be isolated in a particulate fraction obtained by differential centrifugation of homogenized granulocytes and which, in the intact cell, is located in the plasma membrane (Babior, 1978; Root and Cohen, 1981). This enzyme, which is thought to be a flavoprotein, can use either NADPH or NADP as an electron donor but prefers NADPH. The conversion of NADPH to NADP during O_2^- production and during a second reaction, discussed later, in which reduced glutathione (GSH) is generated for the reduction of H_2O_2, accounts for the activation of the hexose monophosphate shunt during the respiratory burst Fig. 3.19).

A severe deficiency of glucose-6-phosphate dehydrogenase, the enzyme catalyzing the first step in the hexose monophosphate shunt, is associated with repeated childhood bacterial infections similar to those seen in chronic granulomatous disease. This clinical observation suggests that NADH, whose production is independent of the hexose monophosphate shunt, can not substitute effectively for NADPH in the physiologic generation of O_2^-.

Generation of O_2^- results in the rapid formation of H_2O_2 in the granulocyte. This occurs through a reaction called a dismutation reaction because it involves the simultaneous reduction of one molecule of O_2^- and oxidation of a second molecule of O_2^-. Dismutation occurs spontaneously and rapidly at the acid pH of the phagocytic vacuole; however, at the neutral pH of the granulocyte cytoplasm, dismutation requires catalysis by an enzyme, superoxide dismutase (SOD). O_2^- and H_2O_2 may react together, in a reaction known as the Haber-Weiss reaction, to generate another reduced form of oxygen called hydroxyl radical (OH•), which is a potent oxidizing agent. A metal, such as iron, that can be reversibly oxidized and reduced promotes the Haber-Weiss reaction. It may be physiologically significant that lactoferrin, when it contains bound

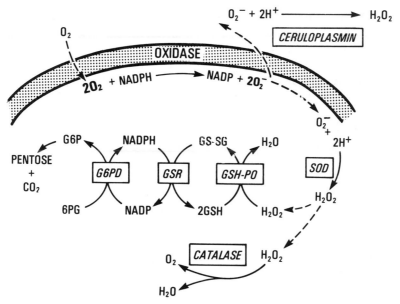

Figure 3.19.   Reactions associated with the respiratory burst in which superoxide anion is generated and converted to H_2O_2. The reactions of the hexose monophosphate shunt that provide NADPH as a source of the electron for superoxide anion and provide reduced glutathionine as a substrate for the detoxification of H_2O_2 by the enzyme glutathione peroxidase are also shown. In addition, the detoxification of H_2O_2 by the intracytoplasmic enzyme catalase and the conversion of extracellular superoxide anion to H_2O_2 by the plasma protein ceruloplasmin are illustrated. O_2^-, superoxide anion; *G6P*, glucose-6-phosphate; *6PG*, 6-phosphagluconate; *GSH*, reduced glutathione; *GS-SG*, the oxidized form of glutathione; *G6PD*, the enzyme glucose-6-phosphate dehydrogenase; *GSR*, the enzyme glutathione reductase; *GHS-PO*, the enzyme glutathione peroxidase; and *SOD*, the enzyme superoxide dismutase.

iron, can promote the production of OH• in the granulocyte.

How reactive oxygen species kill bacteria has been studied extensively but is still incompletely understood (Gabig and Babior, 1981; Root and Cohen, 1981; Fantone and Ward, 1982). O_2^-, itself, has little bactericidal activity. O_2^--dependent microbial killing apparently stems from the formation and action of its metabolites: H_2O_2, OH•, and probably other oxidizing reactive radicals. H_2O_2 is a comparatively weak microbicidal agent when acting alone. However, in the presence of the primary granule enzyme, myeloperoxidase (MPO), and a halide ion, H_2O_2 exhibits potent microbicidal activity, because MPO utilizes H_2O_2 to oxidize halide ions, particularly Cl^- but also I^-, to form potent microbicidal agents such as hypochlorous acid (HOCl) (Klebanoff, 1980). If, however, one accepts the MPO-H_2O_2-halide ion reaction as a primary physiological mechanism for oxygen-dependent bacterial killing, then one must assume that an additional potent "backup" mechanism also exists. This is necessary because 1 in 4000 individuals has a complete deficiency of myeloperoxidase (Parry et al., 1981), yet only a few individuals with myeloperoxidase deficiency develop severe infections. The majority have no clinical problems referable to the deficiency, which usually is discovered accidentally when a blood count is performed with an automated

counting technique that identifies granulocytes by a reaction involving myeloperoxidase. The evidence that OH•, a potent oxidizing agent, is involved in bacterial killing is as yet indirect. For example, addition of superoxide dismutase, which removes O_2^-, or addition of catalase, which removes H_2O_2, each independently inhibit bacterial killing in an experimental system. This has been interpreted to mean that each enzyme, by limiting the availability of one of the reactants of the Haber-Weiss reaction, prevents bacterial killing by preventing the generation of OH•.

In addition to inducing microbial killing, reactive oxygen metabolites that escape from the granulocyte can injure tissue cells and connective tissue matrix. Reduced oxygen species may damage cells and tissue both directly and indirectly, the latter through a chain reaction in which the peroxidation of lipids on cell membranes results in the generation of fatty acid radicals that can then react with other lipids, proteins, or free radicals in tissues (Fantone and Ward, 1982). Moreover, as mentioned earlier, reactive oxygen metabolites may enhance the tissue damage caused by neutral proteases from granulocytes by inactivating α_1-protease inhibitor, the major inhibitor of granulocyte elastase (Janoff and Carp, 1982).

Catalysis of the conversion of O_2^- to H_2O_2 by cytoplasmic superoxide dismutase and the subsequent reduction of H_2O_2 to water prevents reactive oxygen species from damaging the granulocyte itself. The granulocyte utilizes two mechanisms for reducing H_2O_2 to water (Fig. 3.19). In one, H_2O_2 is reduced by its utilization to oxidize GSH to GS-SG in a reaction catalyzed by glutathione peroxidase. This reaction is linked to the hexose monophosphate shunt as a source of NADPH to generate GSH. In the second, H_2O_2 is reduced by the enzyme catalase. In addition to these intracellular protective mechanisms, O_2^- released extracellularly from granulocytes may be reduced to H_2O_2 by the oxidase activity of the acute phase protein ceruloplasmin (Goldstein et al., 1982). The role of this reaction in limiting oxidant damage to tissues requires further evaluation, but it supplies a teleological reason for the rise in concentration of this acute phase protein in inflammatory states.

Phagocytosis and Degranulation

Granulocytes have on their surface recognition units, called Fc receptors, that may initiate phagocytosis when occupied by the Fc portion of IgG that has reacted with antigen on a particle or when occupied by the Fc portion of IgG in immune complexes that have become deposited upon a particle. As mentioned earlier, granulocytes also have a second surface recognition unit, the C3b receptor, which facilitates the recognition and adherence of particles coated with C3b to the granulocyte surface. Particles coated with both C3b and immunoglobulin are efficiently phagocytosed through mechanisms in which attachment via the C3b receptor brings the particle into contact with the granulocyte surface, and occupancy of the Fc receptor causes the engulfment of the particle. Particles coated only with IgG are also phagocytosed but less efficiently than when they are also coated with C3b. Particles coated only with C3b may or may not be phagocytosed. If they are, it is assumed that they also contain structural groups on their surface that can substitute for IgG and react with the Fc recognition site on the granulocyte membrane.

In the act of phagocytosis, the granulocyte surface membrane invaginates and extends pseudopods around the sides of the particle that fuse to enclose the particle within a vacuole, which is called a phagocytic vacuole. These mechanical events are associated with the assembly of actin into fine microfilaments and the assembly of tubulin into large microtubules. The latter are thought to form an internal skeleton against which the microfilaments can pull. Binding of actin filaments to actin binding protein along the inner surface of the plasma membrane is also required for closure of the phagocytic vacuole. Membrane hyperpolarization, an increase in calcium ion concentration in the cytosol, reactions involving membrane phospholipids causing release and oxidation of arachidonic acid, and the respiratory burst with generation of superoxide anion—all are associated with the act of phagocytosis. The rise in calcium concentration activates the cell contractile mechanism. The relation of the other biochemical events to the mechanical events of phagocytosis is less clear. As apparent from earlier discussion, stimuli that perturb the granulocyte surface membrane but do not cause phagocytosis also initiate these biochemical reactions.

The primary (lysosomal) and secondary (specific) granules of the granulocyte are attracted to the phagocytic vacuole. The membranes of the granules fuse with the membrane of the phagocytic vacuole and the contents of the granules are secreted into the phagocytic vacuole. At the same time the pH of the vacuole falls as a result of acidification mechanisms that are not yet understood. Thus, potent materials at an acid pH accumulate in proximity to the ingested microorganism but isolated from the rest of the cell. Conditions are achieved for an attack upon the microorganism by hydrolases such as glycosidases, arginase, and lysozyme; by cationic proteins that are bactericidal for Gram-negative organisms; and by lactoferrin, which by binding iron deprives microorganisms of this metal essential for their metabolic processes. In addition, superoxide anion and its metabolites diffuse into the phagocytic vacuole and initiate oxygen-dependent microbial killing.

Whereas, as just described, the specific granules secrete their contents intracellularly into phagocytic vacuoles during phagocytosis, they also secrete their contents extracellularly as part of the processes of chemotaxis and aggregation. The constituents of the specific granules do not injure tissues. Indeed, they may serve regulatory functions outside of the granulocyte, as, for example, the postulated role of lactoferrin in modulating granulopoiesis (Fig. 3.4). In contrast, the primary granules are thought to secrete their contents primarily intracellularly into phagocytic vacuoles during phagocytosis. Nevertheless, primary granule enzymes also escape into surrounding tissues because secretion begins before phagocytic vacuoles close (Weissmann et al., 1980). The extent of external secretion increases when the engulfment of particles is frustrated as, for example, by the large size of a particle. As mentioned earlier, the escape of primary granule contents brings neutral proteases, particularly the enzyme elastase, into close contact with substrates in the surrounding tissue that they can cleave and damage. Such release of granulocyte enzymes during phagocytosis represents an important mechanism for producing tissue damage in many acute clinical disorders and also in some serious chronic disorders. The latter include disorders affecting the joints (rheumatoid arthritis), the lung (pulmonary emphysema), and the kidney (some forms of glomerulonephritis).

CHAPTER 24

The Red Blood Cell

GENERAL DESCRIPTION, PRODUCTION, AND DESTRUCTION OF RED BLOOD CELLS

The red blood cell is the simplest cell of the human body. Formed as a nucleated cell in the bone marrow, it normally loses its nucleus before its release into the circulation. On entering the circulation, its still possesses residual ribosomes and mitochondria and a Golgi apparatus. These cytoplasmic organelles are lost after a day or so, and the red blood cell then assumes the shape of a flattened, indented sphere or biconcave disc (Fig. 3.20). Its dimensions are: diameter, 7.8 μm; thickness, 0.81 μm in the thin part and 2.6 μm in the thick part of the biconcave disc; suface area, 135 μm^2, and volume, 94 fl. The cell exhibits a remarkable plasticity, being able repeatedly to squeeze through capillaries only one-half of the diameter of the cell and then to return, undistorted, to its original biconcave shape. The high surface to volume ratio facilitates respiratory gas transfer, which is the red blood cell's sole known physiological function.

The mature red blood cell may be looked upon as a cell membrane enclosing proteins, electrolytes, and other components of energy systems. Ninety five percent of the protein of the cell is hemoglobin. The remaining proteins are largely enzymes of its energy systems, many of which, without means of replacement, must maintain at least minimal catalytic activity for the approximately 4-month lifespan of the red cell.

Production of Red Blood Cells

Erythropoiesis requires the presence, in a normal marrow microenvironment, of a normal population of hemopoietic stem cells; their differentiation and maturation under the influence of the erythropoietic growth factors, burst promoting factor, and erythropoietin; and the availability of three specific nutrients: the vitamins, folic acid and cobalamin (vitamin B$_{12}$), and iron. The characteristics of nucleated red blood cell precursors at increasing stages of maturation and of their end product, the nonnucleated marrow reticulocyte, have been described in Chapter 21. It was also pointed out that the newly made red blood cell can be distinguished from other red blood cells for about 24 hours after its release into the peripheral circulation because it still retains enough residual ribosomal RNA to be identified, by supravital staining, as a reticulocyte. Therefore, counting reticulocytes in a sample of peripheral blood provides a simple way of assessing red cell production clinically.

FOLIC ACID AND COBALAMIN IN ERYTHROPOIESIS

Erythropoiesis requires synthesis of DNA for cell division. In DNA, the pyrimidine base thymine, which is 5-methyl uracil, replaces its counterpart in RNA, uracil. Folate (a name used for folic acid or any of its derivatives) and cobalamin participate in coupled reactions that make available the methyl group needed for conversion of deoxyuridilate to deoxythymidilate in DNA synthesis.

Figure 3.21 is a diagram of the reactions thought to be involved. It shows folic acid being reduced by the addition of hydride ions to form tetrahydrofolate. Tetrahydrofolate is then converted to 5,10-methylenetetrahydrofolate, which, in turn, may either be converted into 5-methyltetrahydrofolate or be utilized to methylate deoxyuridilate in a reaction that oxidizes the 5,10-methylenetetrahydrofolate to dihydrofolate. Kinetic conditions favor the former reaction, and 5-methyltetrahydrofolate must be recycled to tetrahydrofolate to maintain an adequate supply of 5,10-methylenetetrahydrofolate for normal synthesis of deoxythymidilate. The recycling is coupled to a second reaction, the methylation of homocysteine to methionine, which is catalyzed by the enzyme methionine synthase. However, this second reaction also requires a coenzyme, cobalamin, which receives the methyl group from 5-methyltetrahydrofolate and transfers it to homocysteine. The reactions shown in Figure 3.21 may be impaired in different ways: by an insufficient supply of folate within the developing red cells; by drugs, such as the chemotherapeutic agent, methotrexate, that compete for the enzyme reducing dihydrofolate to tetrahydrofolate (dihydrofolate reductase); or by insufficient cobalamin to support the recycling of 5-methyltetrahydrofolate, which then traps folate in a form unusable for DNA synthesis (Herbert and Zalusky, 1962).

Folate molecules contain either one terminal glutamyl residue (monoglutamate) or several terminal glutamyl residues (polyglutamate). Plasma folate is in the monoglutamate form, whereas folate in tissues is in the polyglutamate form, an apparent requirement for its intracellular storage. In cobalamin deficiency, levels of cellular polyglutamate folate fall while the plasma level of monoglutamate

folate rises. The normal substrates for making polyglutamate tissue folate are formylfolates, which include the oxidized formylfolate, 5,10-methylene-tetrahydrofolate, and two folate derivatives not shown in Figure 3.21, 5-formyltetrahydrofolate (folinic acid) and 10-formyltetrahydrofolate. Neither 5-methyltetrahydrofolate nor tetrahydrofolate can apparently serve as substrates for making polyglutamate. The methyl group of methionine is thought to represent the major source of formate for formation of formylfolates. It has been suggested (Chanarin et al., 1981) that the impaired methionine synthesis of cobalamin deficiency interferes with DNA synthesis not primarily by trapping folate as 5-methyltetrahydrofolate but by limiting the supply of formylfolate available for maintaining intracellular folate stores. However, the recent demonstration of the accumulation of 5-methyltetrahydrofolate in cobalamin-deficient cell cultures provides direct evidence of trapping of 5-methyltetrahydrofolate in cobalamin deficiency (Fujii et al., 1982).

Folic acid is provided in green vegetables, many fruits, beans, nuts, and liver; intake varies widely between individuals depending upon eating habits and methods of preparing food. Folic acid in food is present as polyglutamates, which must be deconjugated by intestinal enzymes called conjugases to form monoglutamic folate, before folic acid can be absorbed. Normal tissue stores of folate are estimated as 5–10 mg. The daily folate requirement is

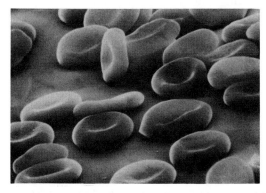

Figure 3.20. The biconcave shape of normal human red blood cells as seen by scanning electron microscopy [courtesy of Dr. Judith A. Berliner.]

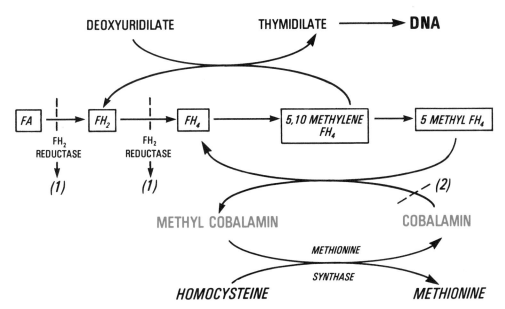

Figure 3.21. A diagram illustrating how the methylation of deoxyuridilate to thymdilate by 5,10-methylene FH$_4$ may be dependent upon the methylation of homocysteine to methionine. As shown by the *dashed arrow* labeled (2), cobalamin deficiency traps folate as 5-methyl FH$_4$ by impeding the methylation of homocysteine to methionine. As shown by the *dashed arrow* labeled (1), a drug competing for the enzyme, FH$_2$ reductase, would also impair DNA synthesis. *FA*, folic acid; *FH$_2$*, dihydrofolate; *FH$_4$*, tetrahydrofolate. See text for further details.

50–100 μg. Thus, failure of folate intake can lead to significant folate deficiency within about 3 months. Moreover, use of alcohol interferes with folate metabolism in as yet incompletely understood ways. Individuals drinking heavily and eating poorly for a prolonged period are particularly prone to develop serious folate deficiency.

Cobalamin is provided by animal protein in the diet; an average American diet provides 5–30 μg daily. A specific glycoprotein secreted by the parietal cells of the gastric mucosa, called intrinsic factor, must bind cobalamin before it can be absorbed from the gastrointestinal tract. Absorption takes place in the ileum as a result of the attachment of intrinsic factor to specific receptors in the brush border of mucosal cells. Cobalamin is transported to the tissues bound to a specific plasma transport protein called transcobalamin II. Cobalamin also binds to a second protein in plasma, transcobalamin I, which is synthesized within and secreted from granulocytes (see Chapter 21). The function of transcobalamin I is unknown; rare individuals with a hereditary absence of this protein apparently suffer no recognized ill effects (Carmel, 1982). Cobalamin is found in tissues in two forms, the coenzymes methylcobalamin and adenosylcobalamin. Body tissue content is between 2 and 3 mg, most of which is present in the liver. The daily requirement of cobalamin is 3 μg. Thus, several years of failure to absorb cobalamin are required to induce manifestations of cobalamin deficiency in a previously normal person. Atrophy of the gastric mucosa secondary to a suspected autoimmune reaction, with resultant lack of production of intrinsic factor, permanently impairs cobalamin absorption and results in the disease pernicious anemia.

Impaired DNA synthesis from either folate or cobalamin deficiency prevents the production of normal red blood cells and causes anemia. The red blood cells that are formed are larger than normal (macrocytic), with a mean volume exceeding 100 fl, and misshapen, with many oval and teardrop forms visible on a stained peripheral blood smear. The granulocytes are also larger than normal, and granulocytes containing an increased number of nuclear lobes (hypersegmentation) may be found. The nucleated red blood cell precursors in the bone marrow contain chromatin that appears abnormally fine and lacy with prominent parachromatin spaces (megaloblastic changes). A second abnormality, subacute combined degeneneration of the central nervous system, may also develop in patients with cobalamin deficiency but not in patients with folic acid deficiency. The metabolic cause of the nervous system lesion is not clear.

IRON METABOLISM

Iron is found within the body in heme, primarily in red blood cells and muscle cells, and as storage iron bound to ferritin or hemosiderin in mononuclear phagocytes and hepatic parenchymal cells. A normal 70-kg man has about 2.5 g of iron in hemoglobin, about 150 mg in muscle myoglobin, about 15 mg in trace heme tissue enzymes, and about 1 g of iron in stores. A very small amount of iron (about 3 mg) is also present in plasma bound to transferrin on its way to cells. Women have lower iron stores than men, ranging from none to about 500 mg. This reflects loss of iron from menstrual bleeding and pregnancy. (The iron cost of a pregnancy is about 500 mg, which is why a woman should be given therapeutic iron during pregnancy.)

Daily body iron loss from the gut and in urine and sweat is minimal, less than 1 mg. During lactation a woman loses an additional 0.5 to 1 mg daily. In contrast to these minimal losses from other sources, bleeding causes loss of substantial amounts of body iron. One gram of hemoglobin contains 3.4 mg of iron. In a normal individual with 15 g of hemoglobin per dl, 100 ml of blood contains approximately 50 mg of iron. Thus, removal of only 2 ml of blood results in the loss of 1 mg of iron, i.e., of an amount exceeding the normal daily iron loss from all other sources. A blood donation for transfusion (450 ml) depletes stores of 225 mg of iron. Normal menstrual bleeding has been estimated as 45 ml per month, but some normal women may lose up to 90 ml of blood per month.

The normal American diet contains an estimated 6 mg of iron per 1,000 calories, which varies in its availability for absorption. Iron in meat is more readily absorbed than iron in foods such as wheat or eggs. Normally, only enough iron is absorbed from the diet to balance iron loss (0.6 to about 2.0 mg/day) and to allow the accumulation over a lifetime of about 1 g of storage iron. However, the dietary intake of a substantial fraction of otherwise normal American women apparently does not contain sufficient iron to allow them to build up any iron stores before the menopause (Cook *et al.*, 1976).

Iron is absorbed in the upper small intestine. The amount absorbed depends not only upon the iron content of the diet but also upon a mucosal regulatory mechanism. Iron entering an intestinal mucosal cell either passes through the cell to enter the blood or is retained within the cell as ferritin. If the latter, the iron is eventually returned to the lumen of the gastrointestinal tract when the mucosal cell dies and is shed. What determines whether the cell lets iron through or holds it back is unknown.

Increased amounts are let through in iron deficiency anemia but also in certain anemias characterized by markedly increased but ineffective erythropoiesis and plentiful body iron stores. Mucosal cells fail to hold back iron normally in a genetic disorder called idiopathic hemochromatosis. When the body absorbs iron from a normal diet as avidly as possible, between 3 and 4 mg of iron are taken in daily. This is insufficient to restore to normal the body iron content of an individual with even a mild iron deficiency anemia (who may need replacement of say 750 mg of body iron), which is why such patients are treated with medicinal iron. Nevertheless, over many years the difference between taking in 3 or 4 mg of iron daily and losing 0.6 to 2.0 mg leads to the accumulation of grams of excess body iron, and the patient with idiopathic hemochromatosis usually develops evidence of organ failure from iron overload in mid-adult life.

Several tests can be used to evaluate body iron status. A bone marrow smear or biopsy can be examined for evidence of storage iron as granules of hemosiderin. The plasma transferrin concentration, which begins to rise as iron stores are depleted, can be measured (total iron-binding capacity or TIBC, see Chapter 20). Moreover, the trace amounts of serum ferritin in plasma can now be measured by radioimmunoassay. As understanding grows of the factors besides iron in stores that affect the serum ferritin level, particularly chronic infection and liver disease, this noninvasive and sensitive test will probably become the common clinical method for evaluating body iron status. A serum ferritin level of less than 12 µg/l is diagnostic of iron deficiency (Cook, 1982).

About 20 mg of iron are needed daily to make new hemoglobin to replace the hemoglobin catabolized in the breakdown of senescent red blood cells in mononuclear phagocytes. This iron is recycled from the catabolized hemoglobin, carried back by plasma transferrin from the mononuclear phagocytes to the normoblasts in the marrow. When red blood cell production is increased, additional iron must be brought to the developing red cells. If the increased red blood cell production is in response to an increased breakdown of red blood cells (hemolysis), then the additional iron is usually readily available within the mononuclear phagocytes for transport back to the bone marrow. However, if the stimulus for increased red cell production is acute blood loss, e.g., a major gastrointestinal hemorrhage, then the additional iron must be mobilized from less readily available iron in stores. This usually limits the increase in red blood cell production to about 2-fold. If the stores are empty, then, despite increased stimulation by erythropoietin, the marrow may be able to respond with only a very limited increase in red blood cell production.

When body iron deficiency progresses beyond storage iron depletion, insufficient iron is available to support a normal level of erythropoiesis. An individual then develops iron deficiency anemia. Since normal daily iron loss is small, iron deficiency anemia in an adult should not be attributed to insufficient dietary iron intake. A source for abnormal blood loss should always be sought. In contrast, in an infant who requires dietary iron for an expanding hemoglobin mass, insufficient dietary intake is the usual cause for iron deficiency anemia.

In iron deficiency the inability to synthesize normal amounts of hemoglobin causes red cells to be made that are smaller than normal, with a mean volume of less than 80 fl (microcytic). On a stained blood smear the cells appear to contain a diminished complement of hemoglobin and an expanded central area of pallor (hypochromic). With mild to moderate iron deficiency the mean concentration of hemoglobin within the red cells (mean corpuscular hemoglobin concentration or MCHC), which is calculated from measured values for mean red cell volume, red blood cell count and hemoglobin, may remain within the normal range of 32–34%. With severe iron deficiency anemia the value falls below 32%. In the uncomplicated patient the serum iron level will be below 30 µg/dl, the TIBC will be greater than 300 µg/dl, and the serum ferritin level will be less than 12 µg/l.

Destruction of Red Blood Cells

Normal senescence and death of red blood cells occur as a function of their age, but the molecular change that determines the death of the normal red cell is not known. It presumably alters the red cell membrane, since mononuclear phagocytes in the spleen, liver, and bone marrow recognize and remove the senescent cell. Hemoglobin, other proteins, and membrane lipids of the phagocytosed, senescent cell are catabolized within the mononuclear phagocyte. Heme is dissociated from the globin chains of hemoglobin, which are broken down into their amino acids. The heme is oxidized, in a reaction catalyzed by a microsomal enzyme, heme oxidase, with resultant opening up of the porphyrin ring structure and release of iron. As already mentioned, transferrin carries the iron back to the normoblast for incorporation into new hemoglobin. The scission of the porphyrin ring gives rise to a molecule of carbon monoxide, which is excreted in the lungs. (Breakdown of heme is the only reaction in the body in which carbon monoxide is formed.)

The remainder of the molecule, called biliverdin, is then reduced to bilirubin and transported through the plasma bound to albumin for excretion by the liver. In a 70-kg adult, about 200 mg of bilirubin, derived from the breakdown of 6 g of hemoglobin, are excreted by the liver daily.

About 3% of the lipids of red cell membranes consist of a glycolipid called globoside, which contains a short carbohydrate chain. During the degradation of globoside in mononuclear phagocytes, different enzymes cleave successive sugar molecules from the carbohydrate chain. A series of rare hereditary disorders called lipid storage diseases have been recognized as resulting from hereditary deficiencies of different cleaving enzymes. The most common disorder, Gaucher's disease, results from a deficiency of the enzyme β-glucosidase, which cleaves the last remaining sugar molecule of the chain. In Gaucher's disease macrophages clogged with partially digested globoside accumulate in ever increasing numbers in the spleen, liver, bone marrow, and other organs. As a consequence, affected individuals develop enormously enlarged spleens and deformities of the bones.

About 10% of daily red cell breakdown takes place not within mononuclear phagocytes but intravascularly, with resultant release, in a 70-kg person, of about 0.6 g of hemoglobin each day into the plasma. Hemoglobin in plasma breaks up into dimers; heme also disocciates from the globin chains. Two plasma proteins, described in Chapter 20, haptoglobin which binds dimers and hemopexin which binds heme, prevents the daily urinary loss of approximately 2 mg of iron that would otherwise occur from a continuous glomerular filtration of dimers and heme.

In pathologic states red blood cells may be destroyed at an accelerated rate. If production of new red blood cells can keep up with the accelerated rate of destruction, the red blood cell count will not fall (compensated hemolytic anemia). However, polychromatophilic macrocytes (see Chapter 21) will be seen on the peripheral blood smear, as evidence of an erythropoietin-stimulated premature release of marrow reticulocytes, and the reticulocyte count will be elevated. If the rate of destruction exceeds the capacity of the marrow to produce new red cells, then the red blood cell count will fall until a level is reached at which the number of cells being destroyed equals the number of cells being produced. In conditions with massive hemolysis, the red blood cell count may fall suddenly to very low levels. The body's compensatory mechanisms for delivering normal amounts of oxygen to the tissues in the face of a reduced hemoglobin level—an increase in cardiac output, redistribution of blood flow, and a shift to the right (see below) in the oxygen association curve of hemoglobin—may be inadequate to handle such a marked fall, and the patient may become very ill.

In many hemolytic processes altered red blood cells are removed primarily by mononuclear phagocytes (extravascular hemolysis). The unique circulation of the spleen allows this organ to play a particularly important role in extravascular hemolysis. As some of the small blood vessels enter the red pulp of the spleen they end in spaces called cords containing loosely meshed reticulin and mononuclear phagocytes. The red cells are emptied into the cords and must then percolate through the cords in intimate contact with mononuclear phagocytes and reenter venous sinuses in the red pulp by squeezing between endothelial cells. These conditions favor the removal of red blood cells with decreased deformability of the cell membrane or those containing rigid, cytoplasmic inclusion bodies such as Heinz bodies (see below). Because of an increased removal of red blood cells, the spleen usually enlarges in hemolytic disorders. In one disorder in which a hereditary red cell membrane defect causes cells to lose their biconcave shape and become microspheres with decreased deformability (hereditary spherocytosis), the spleen functions as the sole site of accelerated removal of the abnormal red cells. However, in most other extravascular hemolytic disorders, large numbers of damaged red blood cells are phagocytosed not only in the spleen but by mononuclear phagocytes in the liver and bone marrow.

In hemolytic states, the amount of bilirubin that the liver must excrete will usually increase several fold. The liver can clear this increased load only after the bilirubin concentration of the plasma has risen moderately. Thus, in many patients with hemolytic anemia the total bilirubin level is elevated from a normal value of about 1 mg/dl to a value in the 2–4 mg/dl range.

When large numbers of red blood cells break down within the circulating blood (intravascular hemolysis), enough hemoglobin may enter the plasma to give it a pink to purple color. The capacity of haptoglobin to bind hemoglobin dimers is exceeded, and hemoblobin dimers are excreted in the urine (hemoglobinuria). The capacity of hemopexin to bind heme is also exceeded, and the excess heme binds to albumin, giving rise to a new plasma pigment, methemalbumin. Moreover, in chronic states of intravascular hemolysis, renal tubule cells become filled with hemosiderin due to the absorption of dimers by the tubule cells and the breakdown of hemoglobin within the cells. Shedding into the urine of tubule cells filled with hemosiderin (hemosiderinuria) can be demonstrated by centrifuging the urine and staining the pellet for iron. This is sometimes used as a screening test for chronic intravascular hemolysis.

THE RED BLOOD CELL MEMBRANE

As a nonnucleated cell without organelles, the red blood cell has a plasma membrane but lacks the internal membranes enclosing the nucleus and organelles found in other cells. Red cell plasma membranes can be obtained in relatively pure state in large quantities by hypotonic lysis of red cells, followed by washing to release the cytosolic proteins, mostly hemoglobin. The membranes so obtained are composed of a lipid bilayer with proteins that may enter or traverse the bilayer or adhere to its cytoplasmic surface. Like the membranes of other cells, the red cell plasma membrane is a selective permeability barrier with specific pumps, channels, and gates. It exhibits a remarkable deformability, a necessary property of a cell whose function is to transport respiratory gases.

Membrane Lipids

The membrane lipids are of three types: phospholipids, cholesterol, and small amounts of glycolipids. The phospholipids have hydrophilic and hydrophobic moieties that form a bimolecular sheet, with the polar hydrophilic groups oriented towards the exterior and cytoplasmic surfaces, and the hydrophobic hydrocarbons oriented towards the interior of the bilayer. The choline-containing phospholipids, phosphatidyl choline (lecithin) and sphingomyelin, are primarily located in the outer layer of the bilayer, with their polar groups directed towards the exterior; the amino-containing phospholipids, phosphatidylethanolamine and phosphatidylserine, are primarily located in the inner layer, with their polar groups directed towards the cytoplasmic surface. The length and degree of saturation of the fatty acid residues of the phospholipids strongly influence the fluidity of the membrane. With increasing chain length and degree of saturation the fatty acids become less fluid, and the membrane becomes more viscous. Almost 55% of the membrane lipid is phospholipid and almost 45% is cholesterol, which is present in its free, nonesterified form. The cholesterol interacts with the phospholipid; if the ratio of cholesterol to phospholipid in the membrane rises, the membrane also becomes less fluid. The glycolipids of the membrane are made up of a lipid base called ceramide, consisting

of sphingosine and a long chain fatty acid, to which, as mentioned earlier, hexose molecules are attached.

The mature erythocyte cannot synthesize membrane lipids but can exchange its free cholesterol and outer layer phospholipids, lecithin and sphingomyelin, with plasma lipoproteins (Cooper, 1977). The activity of a plasma enzyme, lecithin-cholesterol acyltransferase (LCAT), influences the process. LCAT catalyzes transfer of a fatty acid from lecithin to esterify plasma cholesterol, thereby decreasing the availability for exchange with the red cell membrane of both free cholesterol and lecithin. In very advanced chronic liver disease (cirrhosis) plasma LCAT activity is decreased and red cell membranes may take up enough extra free cholesterol from plasma to increase substantially the cholesterol:phospholipid ratio of the membrane (Cooper, 1977). The cells become rigid, and they develop spur-like projections and undergo hemolysis (spur cell anemia) as they pass through the spleen.

Membrane Proteins

The membrane proteins are classified into two types, depending upon their ease of separation from the lipid bilayer: *peripheral proteins*, which are separable by mild means, and *integral proteins*, whose separation requires organic solvents or detergents. The integral proteins are embedded within or traverse the lipid bilayer, wherein they interact extensively with the hydrocarbon chains of the membrane lipids. The peripheral proteins, which do not penetrate the hydrophobic core of the lipid layer, are located primarily at the cytoplasmic face of the membrane (Fig. 3.22). They are bound by electrostatic or hydrogen bonds to the inner polar surface of the lipid bilayer and also, through organization of the membrane skeleton (see below) to the carboxy-terminal segment of an integral protein, band 3, that protrudes through the cytoplasmic face of the membrane.

When membrane proteins extracted from red cell ghosts are subjected to polyacrylamide gel electrophoresis in sodium dodecyl sulfate (SDS-PAGE) the proteins migrate different distances into the gel depending upon their size. Major and minor protein bands are visualized (Fig. 3.22) after staining for protein with Coomassie blue stain. These bands have been assigned arabic numbers, with decimals being used to identify bands thought to be subfractions of a major band or minor bands in close proximity to major bands. A number of bands now have more definitive names: band 5 has been identified as erythrocyte actin, band 6 as the enzyme

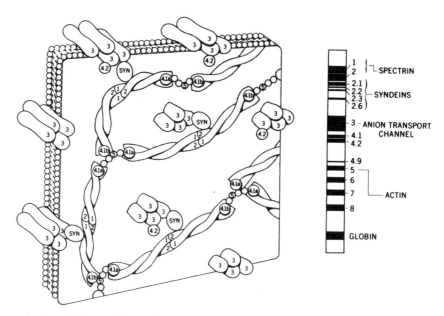

Figure 3.22. The organization of the membrane skeleton of the red blood cell. The integral protein, band 3, is shown protruding through the cytoplasmic face of the lipid bilayer and binding, through the syndeins, to the spectrin meshwork of the membrane skeleton. The heads of spectrin dimers associate to form spectrin tetramers at the sites marked *1* and *2* (for bands 1 and 2). The association of the 4.1 band proteins and of actin (labeled *5*) to the terminal portion of spectrin tetramers is also shown. The glycophorins, integral proteins that also penetrate the cytoplasmic surface of the lipid bilayer but which are not associated with the membrane skeleton, have not been illustrated. Also shown is an SDS-PAGE gel to illustrate the major membrane protein bands. [From Goodman, S. R., et al. Identification of the molecular defect in the erythrocyte membrane skeleton of some kindreds with hereditary spherocytosis. *Blood* 60: 772–784, 1982, by permission.]

glyceraldehyde-3-phosphate dehydrogenase (G-3-PD), bands 1 and 2 as the two chains of spectrin, and a series of bands, 2.1 and its proteolytic products 2.2, 2.3, and 2.6, as the syndeins (also called ankyrin). When such SDS gels are stained for carbohydrate with PAS stain, four additional bands are visualized which represent monomers or dimers of two glycoproteins called glycophorin A and glycophorin B.

The glycophorins and band 3 are the major integral proteins that pass from the exterior of the cell through the lipid bilayer and into the cell cytoplasm. Carbohydrate chains are attached to the external portion (NH$_2$-terminal segment) of the proteins. Glycophorin A, which makes up about 75% of red cell membrane glycoproteins, is particularly rich in carbohydrate, containing 16 oligosaccharide chains that represent 60% of the molecule. The complete amino acid sequence of glycophorin A has been determined. However, despite extensive knowledge of its structure, the function of glycophorin A remains unknown. Band 3 (Steck, 1978) contains 5–8% carbohydrate. Its amino acids are arranged in a tertiary structure that forms an anion channel, through which chloride ions enter and leave the red cell (chloride shift) as intracellular bicarbonate ion concentration changes with the CO$_2$ content of the blood (see below). The syndeins

(ankyrin), two glycolytic enzymes and, under some circumstances, hemoglobin bind to the cytoplasmic portion (carboxy-terminal segment) of band 3.

When red cell ghosts are extracted with nonionic detergents, several of the membrane proteins are not extracted but remain as an insoluble proteinaceous remnant called the red cell membrane skeleton. The membrane skeleton provides the structural integrity of the cell, determining at least partially the shape of the cell and its ability to change shape as it passes through the microcirculation. The principal protein of the membrane skeleton is spectrin; other proteins are erythrocyte actin and band 4.1 (a and b).

The spectrin molecule is made up of a 250,000-dalton polypeptide chain (band 1) and a 225,000-dalton polypeptide chain (band 2). The two chains, frequently referred to as the α and β chains, form a dimer consisting of two similar, flexible, fiber-like structures about 1000 Å in length, which lie parallel to each other and touch each other at multiple points. The β chain is phosphorylated on incubation in vitro with ATP; the functional significance of its phosphorylation in vivo is not clear. Recent evidence suggests that spectrin in the red cell membrane exists as a tetramer formed by the head to head association of two heterodimers. It has been suggested (Marchesi, 1983) that higher molecular weight oligomeric forms of spectrin are also present in the red cell membrane.

The spectrin tetramers (and possibly higher molecular weight oligomers) are linked to form the basic membrane skeleton network by erythrocyte actin filaments and the band 4.1 proteins, which are associated with the terminal ends of the spectrin tetramers. Through these bonds a cross-linked two-dimensional network of spectrin is formed just beneath the lipid bilayer (Fig. 3.22). Yet another protein, ankyrin (syndeins) associates with spectrin and also with the carboxy-terminal segment of band 3. This spectrin-ankyrin-band 3 association binds the spectrin meshwork to the inner surface of the membrane and also provides a potential link between events in the cytoplasm and the external surface of the red cell.

It is technically difficult to investigate abnormalities of the red cell membrane in the laboratory. Nevertheless, in certain of the hereditary hemolytic anemias, abnormalities of membrane proteins are beginning to be identified (Goodman *et al.*, 1982; Liu *et al.*, 1982). A defective spectrin has been described (Goodman *et al.*, 1982) in some kindreds with hereditary spherocytosis, a relatively common disorder in which the red cells are rigid microspherocytes and abnormally susceptible to lysis in hypotonic solutions. No clinical screening tests for red cell membrane defects are readily available; observations of abnormal red cell morphology and osmotic fragility, although useful, are far from specific.

A number of enzymes are present in red blood cell membrane preparations (Schrier, 1977). Some, like G3PD, the enzyme present in such large amounts that it forms band 6 on SDS gel electrophoresis, are found both in membrane preparations and in membrane-free cytosol. Others are enzymes whose activity is confined to the membrane. These include: an ATPase that is part of the Na^+-K^+ pump which maintains a high intracellular K^+ and low Na^+ relative to the extracellular fluid; an ATPase that is part of a Ca^{++} pump that extrudes Ca^{++} from the cell against a 50-fold concentration gradient; and protein kinases stimulated by cyclic AMP. The majority of the enzymes are found on the cytoplasmic surface of the membrane. However, some enzymes, including an acetylcholine esterase and several enzymes generally thought of as lysosomal enzymes (glycosidases, acid phosphatase), are externally oriented membrane enzymes. Their physiologic functions are not understood.

Membrane Antigens

Antigens are present on the external surface of the red blood cell membrane. The antigens of different individuals differ in their serologic reactivity as a result of minor structural differences between individuals in the carbohydrates, proteins, and lipids of the membrane. Antigens stemming from structural differences controlled by genes at the same locus (alleles) or closely linked loci are classified together as a blood group system. Over 20 different blood group systems containing about 400 antigens have been identified to date. Before a person can be given a blood transfusion, the antigenic composition of two key blood group systems, the ABO system and the Rh system, must be known. The red blood cells to be infused (donor cells) should be of the same type in these two systems as the red blood cells of the recipient.

The four phenotypes of the ABO system (A, B, AB, and O) result from small differences in the terminal sugars of carbohydrate chains of trace amounts of fucose-containing glycolipids in the red blood cell membrane. The ABO system is crucial in transfusion therapy because an individual has "natural antibodies" in the plasma to those ABO antigens that he or she does not possess. These antibodies arise in early infancy as a result of exposure to polysaccharides with ABO antigenic properties, which are present in microorganisms, seeds, and plants. The Rh system is typed primarily to determine whether an individual possesses (Rh positive) or does not possess (Rh negative) the D antigen of this system. The D antigen is highly immunogenic; antibodies to D may form following a single exposure of a D-negative individual to red blood cells containing the D antigen. It is not possible to carry out routine typing of other, less immunogenic antigens before a blood transfusion, since this would involve an enormous commitment of time and personnel. Instead, a sample of the recipient's serum is incubated with the red blood cells to be transfused (major side cross match) to assure that the serum does not contain antibodies that react with antigens on the donor red cells.

HEMOGLOBIN

Biosynthesis of Hemoglobin

The production of hemoglobin is a tightly regulated process requiring the coordinated synthesis of different polypeptide globin chains and of heme. Hemoglobin is synthesized primarily in the nucleated red cells of the marrow during the 6–8 days of erythroid cell maturation. Low levels of synthesis continue for a day or so in the reticulocyte; the mature erythrocyte cannot synthesize hemoglobin. The quantities of free globin subunits and of free heme in red cells are minute.

HEME SYNTHESIS

Heme, ferrous protoporphyrin IX, is synthesized in a series of steps that begin in the mitochondria with the condensation of glycine and succinyl CoA to form amino levulinic acid (ALA). This reaction, which is catalyzed by δ-amino levulinate synthetase and requires pyridoxal phosphate, is a committed step. Once ALA is formed reactions continue until heme is formed. Further steps take place in the cytoplasm where, catalyzed by a specific dehydrase, two moles of ALA condense to form porphobilinogen. Four porphobilinogens condense to form a linear tetrapyrole, which then cyclizes. Side chains of the tetrapyrole ring are successively reduced by additional enzymatic reactions to give rise to protoporphyrin IX. In the mitochondria, ferrous iron is inserted into the protoporphyrin ring, catalyzed by the enzyme ferochelatase. The product, heme (Fig. 3.23) represses the synthesis of the first enzyme of the process, δ-amino levulinate synthetase. Thus, heme regulates its own synthesis. Defects in enzymes of heme synthesis give rise to different clinical disorders known as the porphyrias.

Biosynthesis of Globin

Hemoglobin consists of two pairs of unlike chains: two α chains and two non-α chains, which may be β, γ, or δ chains. The major adult hemoglobin, hemoglobin A, is $\alpha_2\beta_2$; fetal hemoglobin is $\alpha_2\gamma_2$. The locus for the α chains is on chromosome 16 and is duplicated, i.e., a nomal individual has four α genes, two on each chromosome. The products of the duplicated α genes are identical. The non-α genes all lie closely linked on chromosome 11 (Fig. 3.24). The γ locus is also duplicated; the gene products are identical except for residue 136, which is a glycine at one locus (G_γ) and an alanine at the other (A_γ). The order of the non-α genes on chromosome 11 is G_γ, A_γ, δ, β. The amino acid sequence

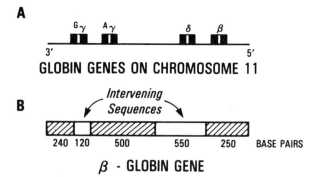

Figure 3.24. Globin genes. *A* illustrates the loci for the non-α globin genes on chromosome 11. *B* is a representation of the β-globin gene. The *hatched areas* indicate areas of base pairs that are translated. The *clear areas* are intervening sequences that are transcribed and eliminated in processing.

of the chains is determined by the triplet groups of bases in the DNA (codons). The sequences in DNA that code for the globin gene polypeptides are split (Fig. 3.24); each globin gene has two intervening sequences that are not translated into the amino acid sequence of the polypeptide chain. In β globin the small intervening sequence (intron) corresponds to an untranslated portion between the codons for amino acid residues 30 and 31 and the large intervening sequence between the codons for amino acid residues 104 and 105. The function of the intervening sequences is not known, but the total DNA sequence including the introns is transcribed into a large mRNA precursor which is then "processed" in the nucleus by addition of modified nucleotides at each end of the molecule, excision of the introns, and ligation of the remaining mRNA sequences. The mRNA thus created is transported to the cytoplasm where it associates with the translational machinery. Translation is a highly coordinated process requiring the interaction of mRNA and RNAs, proteins with specific enzymatic functions, and ribosomes. Translation may be considered to take place in three stages: initiation, elongation, and termination. Chain growth proceeds from the amino to the carboxyl end of the polypeptide chain; release factors that recognize a termination codon terminate the process.

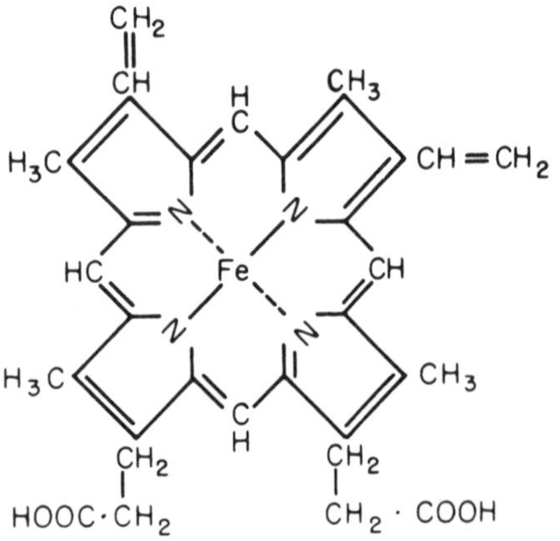

Hemoglobin synthesis is normally rather precisely coordinated by mechanisms that are not completely understood. Protein synthesis in reticulocytes is inhibited by heme deficiency. The inhibition appears to be based upon the activation of a protein kinase that phosphorylates an initiation factor. Heme normally regulates this kinase by inhibiting its activity.

There appears to be a minute free pool of α chains in developing erythroid cells. Both free α chains and free β chains are unstable, however, and may

Figure 3.23. The structure of heme.

precipitate in the cell and undergo proteolytic degradation.

During the last several years, techniques for establishing DNA sequences have provided powerful tools for the study of abnormalities of globin synthesis. Restriction endonucleases are enzymes that cleave DNA into specific fragments. By the use of several restriction enzymes of different specificities, a region of DNA can be sequenced. Enzymatic replication techniques and specific chemical cleavage by compounds that react with a specific base are also used to study DNA sequences.

Genetic abnormalities may give rise to structural hemoglobin variants, may impair polypeptide chain synthesis, or may cause both. Structural abnormalities most commonly are single amino acid substitutions that stem from substitution of a single base in a codon. However, in some structural variants, one or more amino acids have been deleted or a mutation in the normal termination codon has caused the elongation of a polypeptide chain.

Defects in the rates of synthesis of a specific polypeptide chain are expresssed clinically as a group of disorders called thalassemia. Alpha thalassemia (impaired synthesis of α chains) usually results from a deletion of one to four of the α genes. (Deletion of all four genes is fatal with death of the fetus in utero.) The non-α thalassemias may result either from deletions or from other gene abnormalities. In different β thalassemic disorders, part of the β gene, the entire β gene, or both the β and δ genes may be deleted. In some β thalassemias, β globin DNA is present but abnormal, giving rise to mRNAs that are defective, diminished, or absent. One of these abnormalities results from a mutation in a codon in an intron, another from a mutation converting the codon for amino acid residue 17 into a termination codon.

In yet other disorders a combination is found of decreased synthesis and structural abnormalities of a polypeptide chain. In hemoglobin E, a substitution in a codon near the small intron of the β gene results both in an amino acid substitution at residue 26 and a defect in RNA. In the Lepore hemoglobin syndromes, a crossing over between δ and β genes has produced a new chain of 46 amino acid residues in which the N terminal residues are of δ chain and the C terminal residues are of β chain origin.

The severest forms of thalassemia are characterized by anemia, hypochromic and microcytic red cells, and splenomegaly. The anemia, which often is severe enough to require transfusions, results both from inadequate red cell production and from damage to the red cells by the *normal* subunits which, lacking unlike subunits with which to form tetramers, precipitate in the red cells.

Functional Properties of Hemoglobin

Hemoglobin is found in an extraordinarily high concentration, 32 g/dl or a 5 mM solution, in the red blood cell. Within the cell, hemoglobin transports oxygen efficiently without exerting the large osmotic effect it would have as a plasma protein. Moreover, within the red blood cell a mechanism operates (discussed later) to reduce ferrihemoglobin (methemoglobin) and so keep the iron of hemoglobin in its oxygen-carrying ferrous form.

The important properties of hemoglobin as a transporter of oxygen—understood before much was known about the structure of hemoglobin—were elegantly summarized by Barcroft in 1928. These are as follows (see also Chapter 36).

1. The *oxygen affinity* of hemoglobin is so arranged that hemoglobin becomes fully saturated with oxygen in the lungs exposed to ambient air and delivers oxygen at the partial pressure of oxygen encountered in tissues. The oxygen affinities of different hemoglobins or different red cells can be compared by determining the partial pressure of oxygen at which half the hemoglobin is oxygenated and half is deoxygenated, i.e., the P_{50}.

2. The initial binding of oxygen to hemoglobin facilitates the further binding of oxygen to hemoglobin. This characteristic of hemoglobin binding is called *heme-heme interaction* because the oxygen binding of one heme affects the binding properties of other hemes. The changing oxygen affinity of hemoglobin with oxygenation results in a sigmoid curve (Fig. 3.25) when the degree of oxygenation or percent saturation of hemoglobin with oxygen is plotted against the partial pressure of oxygen (P_{O_2}). Such a plot is referred to as an oxygen dissociation curve of hemoglobin. The shape of its midportion is denoted by the value of n from the Hill equation: $\log(Y/(1 - Y)) = n \log(P_{O_2}/P_{50})$, where Y indicates fractional saturation with oxygen and n is an empiric constant without physical basis. Values of n for oxygen equilibria of normal mammalian hemoglobins are 2.8 to 3. For myoglobin the value of n is 1, which reflects the hyperbolic shape of the oxygen dissociation curve of myoglobin (Fig. 3.25). (The high oxygen affinity of myoglobin at normal tissue oxygen pressure allows myoglobin to function as an oxygen storage protein of muscle that releases its oxygen at the very low intracellular partial pressures of oxygen resulting from exercise.)

3. The oxygen affinity of hemoglobin changes with intracellular pH (Fig. 3.26). This prop-

erty of hemoglobin was first recognized by Christian Bohr in 1904, who noted the effect of P_{CO_2} upon the oxygen dissociation curve (Fig. 3.26). In the capillaries of metabolizing tissues, CO_2 enters the plasma and red cells. Red cells contain carbonic anhydrase that rapidly converts CO_2 to H_2CO_3, a weak acid that ionizes to H^+ and HCO_3^-, lowering intracellular pH. This increase in hydrogen ion concentration decreases the oxygen affinity of the hemoglobin (Bohr effect) and facilitates oxygen delivery to the tissues. As deoxyhemoglobin, a weaker acid than oxyhemoglobin and so

able to bind added protons, accumulates in the red cell, the deoxyhemoglobin binds the H^+ ions liberated from H_2CO_3. The increased HCO_3^- ions diffuse out of the red cell and are replaced by chloride ions in the "chloride shift." In the lungs the process is reversed; CO_2 is lost from the blood, pH rises, and the affinity of hemoglobin for oxygen increases. (The basis of relationships between O_2, CO_2, and H^+ binding have been analyzed in a classic article by Weyman 1964.)

Barcroft (1928) also suggested that an additional, then unidentified modulator of oxygen affinity of hemoglobin existed. Almost 40 years later this additional modulator was identified (Benesch and Benesch, 1967) as 2,3-diphosphoglycerate (2,3-DPG). Today, the *principal modulators* of oxygen affinity are recognized as hydrogen ion concentration (Bohr effect), temperature, and organic phosphates, particularly 2,3-DPG. ATP, the second most abundant organic phosphate in human red cells, is primarily bound to Mg^{++} and the Mg^{++} ATP complex has little effect on oxygen affinity (Bunn et al., 1971). The effect of temperature upon oxygen affinity appears physiologically appropriate: with increasing temperature, oxygen affinity decreases and with hypothermia oxygen affinity increases. The effect of temperature change is significant. A 10° increase in temperature nearly doubles the P_{50} of hemoglobin. Direct binding of carbon dioxide as in a carbamino complex also lowers oxygen affinity, but this effect is minor in human hemoglobin, where 10% or less of the CO_2 is transported in this form. The concentration of hemoglobin in the red cell also seems to have little effect upon oxygen affinity except for abnormal hemoglobins, notably sickle hemoglobin.

The mechanism of synthesis of 2,3-DPG in the red cell is discussed later. In cells other than the

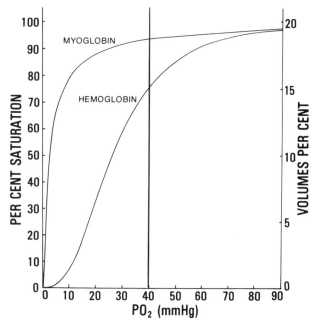

Figure 3.25. The difference between oxygen release from hemoglobin in whole blood and from myoglobin at 37°C and pH 7.4. The normal oxygen pressure of mixed venous blood is 40 mmHg. [From Bunn, H. F., et al. *Human Hemoglobins*. Philadelphia: Saunders, 1977, by permission.]

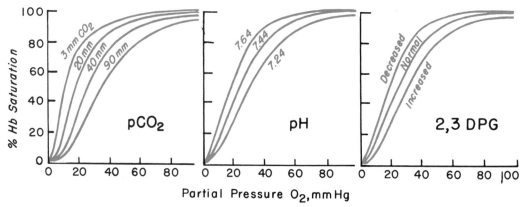

Figure 3.26. The influence of pCO_2, pH, and 2,3-DPG on the oxyhemoglobin dissociation curve.

erythrocytes, 2,3-DPG is present in minute concentrations; in the erythrocyte, the concentration of 2,3-DPG is about equimolar with hemoglobin, 5 mM. 2,3-DPG alters the oxygen affinity of hemoglobin by two mechanisms: by binding to deoxyhemoglobin as also discussed later and by its effect upon intracellular pH. Since 2,3-DPG is a highly charged impermeant anion with 5 titratable acid groups and 3.5 negative charges, it lowers intracellular pH relative to that of plasma. The further decrease of oxygen affinity of hemoglobin observed when the molar concentration of 2,3-DPG exceeds the molar concentration of hemoglobin reflects the contribution of 2,3-DPG to the Bohr effect.

Physiological factors that increase P_{50}, i.e., that shift the oxygen dissociation curve to the right, have an insignificant effect upon oxygen loading at the normal partial pressure of oxygen in arterial blood in the lungs (90 torr), but substantially increase oxygen release from hemoglobin at the mean end capillary partial pressure of oxygen in the tissues (40 torr). This is perhaps best illustrated when oxygen equilibria are plotted as O_2 content in volumes percent (rather than percent saturation) against P_{O_2}. As seen in Figure 3.27, for an oxygen dissociation curve with a normal P_{50} of 26.5 torr, 4.5 volumes of oxygen are delivered per 100 ml of blood. For a right shifted curve with a P_{50} of 36.5 torr, the O_2 content of arterial blood is reduced

slightly to 19 volumes percent, but each 100 ml of blood at an end capillary partial pressure of oxgyen of 40 torr has delivered 7.3 volumes of oxygen.

Structural Properties of Hemoglobin

The molecular weight of human hemoglobin is 64,400 daltons. The molecule is a tetramer of two pairs of unlike polypeptide chains, which undergo conformational isomerization with oxygenation. Hemoglobin, therefore, has two different stable structures, oxy and deoxy. Much of our present knowledge of hemoglobin is based upon the X-ray crystallographic studies of Perutz and his associates (Perutz, 1970, 1978, 1979) and upon the work of many investigators who delineated amino acid sequences.

As already mentioned, hemoglobin A, which accounts for more than 95% of normal adult hemoglobin, has two α and two β chains. Human fetal hemoglobin (hemoglobin F) has two α chains and two non-α chains, called γ chains, that differ from β chains by 39 amino acid residues out of a total of 146 residues for each chain (Table 3.9). A minor human hemoglobin, hemoglobin A_2, comprising only about 2–3% of total adult hemoglobin, is composed of two α and two δ chains; the δ chain differs in only 10 amino acid residues from the β chain.

The signal that determines the switch from fetal (γ chain) to adult (β-δ chain) hemoglobin synthesis is not known. It appears to be correlated with fetal maturity and occurs normally shortly before birth. The ability to turn off this switch mechanism would have major therapeutic implications in the common and serious hereditary anemias resulting from β chain defects, namely, the sickle cell disorders and the β thalassemias.

The individual chains of hemoglobin, α, β, and γ have oxygen-carrying properties similar to the single chain of myoglobin; a tetramer of β chains (hemoglobin H), which is found in a form of α thalassemia, has a hyperbolic oxygen dissociation

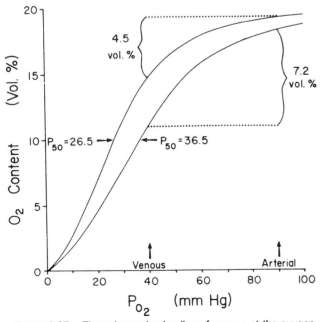

Figure 3.27. The enhanced unloading of oxygen at the oxygen pressure of mixed venous blood resulting from a right-shift in the oxyhemoglobin dissociation curve that increases the P_{50} from 26.5 mmHg to 36.5 mmHg. [From Klocke, R. A. Oxygen transport and 2,3-diphosphoglycerate (DPG). *Chest* 62: 79–85S, 1972, by permission.]

Table 3.9
Some characteristics of normal subunits of human hemoglobin

Polypeptide Chain	Number of Amino Acid Residues	N-Terminal Residues	Normal Hemoglobins in Which It Is Found	Site of Gene Responsible for Synthesis
α	141	Val-Leu-Ser	A F A_2 Gower II[a]	16
β	146	Val-His-Leu	A	11
γ	146	Gly-His-Phe	F	11
δ	146	Val-His-Leu	A_2	11

[a] Gower II is one of three embryonic hemoglobins synthesized in the yolk sac stage of development.

curve similar to that of myoglobin. Thus, the functional properties of hemoglobin that permit it to function as an effective oxygen transport protein require that the molecule be assembled as a tetramer of unlike chains, α with β, γ, or δ. The normal hemoglobins so formed, A, A$_2$, and F, differ little in their functional properties except for a lesser ability of hemoglobin F to bind organic phosphates.

Hemoglobin in its deoxy or T (tense) conformation has many salt bridges between and within subunits. As the hemoglobin molecule takes up successive O_2 molecules, these salt bridges are successively broken and the molecule reaches it stable fully oxygenated or R (relaxed) conformation. During oxygenation the position of $\alpha\beta$ dimers within the tetramer shift in a "ratchet" arrangement along the $\alpha_1\beta_2$ interface. The β chains move together by about 7 Å, narrowing the central cavity of the tetramer. The C terminal portion of the β chains move from an anchored positon between the F and H helices (Fig. 3.28) to a position in which their residues are available to the solvent. These reactions occur sequentially as the reaction of a heme with oxygen affects the structure around the remaining hemes in sequence. Thus, the heme-heme

interaction is an expression of the allosteric properties of hemoglobin.

2,3-DPG is bound in the central cavity of deoxyhemoglobin to specific chain residues (Arnone, 1972) where it, in effect, forms an additional salt bridge in deoxyhemoglobin. As the β chains move together in the oxy conformation, 2,3-DPG is expelled, and the opening becomes too narrow to permit binding at the specific deoxy site. The Bohr effect results from the high affinity of deoxyhemoglobin for protons. A change in the environment of three pairs of proton-binding groups mostly accounts for the effect. The largest contribution is from the C terminal histidines of the β chains.

Heme is embedded in a hydrophobic pocket between two helical segments (E and F) of each polypeptide chain of hemoglobin (Fig. 3.28). The heme is so oriented that its hydrophobic side chains are in the interior of the globin subunit and its polar side chains are on the surface. It is anchored in the chain by many interatomic contacts. The iron of the heme is covalently bound to hemoglobin by the proximal F8 histidine (Fig. 3.28). In the deoxy form, the heme iron is displaced 0.6 Å from the plane of the porphyrin ring, and the F8 heme-linked histidine is "tilted." On oxygenation, the iron moves into the plane of the ring, and the "tilt" changes. Oxygen is bound on the E7 (distal histidine) side of the heme in an area "guarded" by residue E11. The divalent heme iron—five coordinated in deoxyhemoglobin and six coordinated in oxyhemoglobin in which O_2 binds to the sixth position—binds oxygen without itself being oxidized.

Abnormalities of Hemoglobin as a Cause for Disease

GENETIC ABNORMALITIES OF HEMOGLOBIN

The thalassemias, a group of anemic disorders of varying severity, dependent upon the extent of impaired synthesis of a polypeptide chain of hemoglobin, have been discussed earlier. As already mentioned they arise from several genetic mechanisms, e.g., deletions, nonhomologous crossing over, or premature terminations.

Three common, clinically important structural hemoglobin variants result from single amino acid substitutions in the β chain: hemoglobin S (Pauling *et al.*, 1949) in which a valine replaces the normal glutamic acid at the sixth residue of the chain; hemoglobin C in which a lysine replaces glutamic acid at the same sixth residue site; and hemoglobin E in which a lysine replaces a normal glutamic acid at the 26th residue of the chain. As mentioned earlier, the mutation for hemoglobin E also appar-

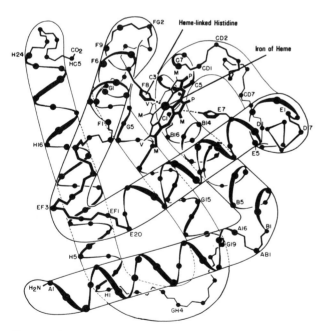

Figure 3.28. A three-dimensional representation of a hemoglobin subunit. The helical segments are assigned letters from A through H beginning at the NH$_2$ terminal segment of the molecule. Individual amino acids within segments are numbered consecutively. Interhelical segments bear the letters of the two adjacent helices (e.g., *EF*). The proximal heme-linked histidine and heme iron in a porphyrin ring are indicated. [From Dickerson: *The Proteins: Composition, Structure and Function*, 2nd ed., edited by H. Neurath, vol. 2. New York: Academic Press (1963), by permission.]

ently decreases β^E messenger RNA, and affected individuals have hypochromic, microcytic red cells as seen in thalassemia. Hemoglobin E is found most frequently in persons of southeast Asian ancestry.

Hemoglobin S (sickle hemoglobin) which consists of two normal α chains and two β^s chains, forms intracellular polymers on deoxygenation. With polymer formation the red cells containing hemoglobin S lose their normal deformability and assume a sickled shape, which reverts to normal on oxygenation. In the heterozygote state, called sickle cell trait, the red cells contain a mixture of hemoglobin A and hemoglobin S, and affected individuals, with rare exceptions, do not have overt clinical disease. In contrast, in the homozygous state, sickle cell anemia, the patient has a life-long anemia due to the short survival time of sickled cells and repeated painful episodes, called vasoocclusive crises, brought on by sickled cells occluding small blood vessels.

Hemoglobin C also has a reduced solubility but not to the same degree as hemoglobin S. Homozygotes for hemoglobin C have a compensated hemolytic state, unusual appearing red cells (target cells) on stained blood smears, and a minimally to moderately enlarged spleen. They do not have vasoocclusive crises. Both hemoglobin S and hemoglobin C are most often found in individuals of African or Mediterranean ancestry. The frequency of genes for thalassemia is also increased in these ethnic groups. Therefore, individuals who are heterozygotes for both hemoglobin S and hemoglobin C, and for hemoglobin S and thalassemia, are encountered. Such individuals have hemoglobin S/C disease or S-thal disease. Their red blood cells sickle in vivo, and they have vasoocclusive crises that may be as severe as those seen in sickle cell disease. However, their anemia is usually less severe than the anemia of sickle cell disease.

Hemoglobins S and C produce disease because they decrease hemoglobin solubility. Other hereditary structural abnormalities of globin affect other properties of hemoglobin: its stability, oxygen affinity, or ability to maintain iron in heme in the ferrous form. These variant hemoglobins are of little clinical importance because of their rarity, but their study has provided valuable information on the relation between globin chain structure and both the structural and functional integrity of hemoglobin.

Several *unstable hemoglobins*, arising either from single amino acid substitutions or deletions, have been characterized. Their globin chains are abnormally susceptible to denaturation, forming precipitates of hemoglobin within the cell that are de-

monstrable on supravital staining and are called Heinz bodies. Unlike aggregates of sickle hemoglobin which form when oxygen is released and become soluble again when oxygen is taken up, Heinz bodies are neither ligand-dependent for their formation nor reversible. Heinz bodies damage the red cell membrane and lead to accelerated red cell destruction, with a resultant anemia of variable severity.

Variant *hemoglobins with altered oxygen affinity* result from amino acid substitutions that stabilize or destabilize the oxy or deoxy conformation of hemoglobin. In very rare variants with decreased oxygen affinity, the oxygen dissociation curve is so far shifted to the right that hemoglobin cannot take up oxygen normally as blood passes through the lungs. The patient, therefore, appears cyanotic. However, despite incomplete oxygen loading in the lungs, oxygen is delivered to the tissues more effectively than normal because it is released at a higher partial pressure of oxygen, which facilitates its tissue diffusion. As a result the patient develops a moderate anemia (but not the pathophysiologic consequences of anemia) because the normal signal mechanism to produce erythropoietin is muted. Conversely, in the relatively more common variants with increased oxygen affinity, oxygen uptake in the lungs is normal but oxygen delivery to the tissues is impaired because of a left shift in the dissociation curve. Such patients develop an asymptomatic increase in their red blood cell count (erythrocytosis) because of an increased production of erythropoietin.

METHEMOGLOBINEMIA

The iron of heme normally is in the ferrous state; during reversible binding of oxygen the iron remains ferrous. When the heme iron is oxidized to the ferric state, the resulting methemoglobin is no longer capable of reacting with oxygen. In acid solutions, the oxygen binding site is occupied by water, in alkaline solutions by an OH^- group. Methemoglobin, which is brown in color, has a characteristic light absorption at 630 nm. Methemoglobinemia is encountered in three circumstances:

1. After exposure of normal red cells to toxic chemicals such as nitrites, aniline dyes, and certain oxidant drugs.
2. In heterozygotes (the homozygote state is lethal) for hemoglobin variants called M hemoglobins in which amino acid substitutions affecting the heme binding pocket (replacement of the proximal or distal heme linked histidine (F8 or E7) by tyrosine or replacement of E11 valine by glutamic acid) cause the iron of heme to be oxidized.

3. In homozygotes for deficiency of NADH-dependent methemoglobin reductase, which, as discussed later, normally reduces the small amounts of methemoglobin that form in red blood cells.

Levels of methemoglobin between 10 and 25% result in cyanosis but usually do not require treatment. At levels of 35%, patients may develop dyspnea and headache; levels of 70% are lethal. The toxicity of methemoglobin stems not only from its own inability to carry oxygen but from its effect upon the oxygen equilibria of hemoglobin tetramers. In hemoglobin molecules some of the hemes of a tetramer will be oxidized, while the remainder will be in the ferrous form. In this circumstance, the oxygen affinity of the functional ferrous hemes is increased, with a resultant "left shifted" oxygen dissociation curve and impaired delivery of oxygen to the tissues. As discussed later, toxic levels of methemoglobin may be treated with methylene blue.

CARBONMONOXY HEMOGLOBIN

Carbon monoxide (CO), like oxygen, binds to the sixth coordination position of the iron in heme, but the affinity of hemoglobin for CO is about 200 times that for oxygen. Carbon monoxide interferes with oxygen transport in two ways. In a portion of the hemoglobin molecules CO may occupy the iron of all of the hemes of the tetramer, and the molecule cannot bind oxygen. In other molecules, the hemoglobin is composed of mixed tetramers (e.g., α^{CO}, β^{CO}, α^O, β^O) with "left-shifted," high affinity, oxygen equilibria. Symptoms of carbonmonoxy hemoglobin toxicity appear at levels of 5–10% carbonmonoxy hemoglobin; at levels above 40% unconsciousness may proceed to death. At a given blood level, carbonmonoxy hemoglobin intoxication is more life-threatening than methemoglobinemia probably because CO affects heme enzymes in tissues as well as hemoglobin.

The high affinity of CO for hemoglobin results not from an increased rate of association but from a slow rate of dissociation of CO from heme. The half-disappearance time for CO hemoglobin in an individual with normal ventilatory function is about 4 hours; breathing pure oxygen will shorten the half-time to about 1 hour.

RED CELL METABOLISM

Although the anucleate red cell serving as a gas transporter has fewer metabolic needs than most other cells, it does require energy to maintain the shape and flexibility of the cell membrane, to maintain hemoglobin iron in its functional divalent form, and to preserve the high K^+, low Na^+, low Ca^{++} ionic milieu of the red cell against the gradient of the different ionic composition of plasma. An anaerobic pathway (Embden-Meyerhof pathway) in which ATP and NADH are generated, supplies this energy. The pathway has two shunts: an aerobic shunt, the hexose monophosphate shunt, in which NADPH is generated, and a second shunt, in which 2,3-DPG is produced. The pathway and its shunts are shown in Figure 3.29.

Glucose Metabolism

The main energy source of the red cell is glucose which is metabolized to pyruvate or lactate. Mature red cells do not have a citric acid cycle; only in reticulocytes which still possess mitochondria is pyruvate broken down to CO_2.

Glucose apparently enters the cell via a carrier protein; its transport is not insulin-dependent. Within the cell it is rapidly converted by hexokinase to glucose-6-phosphate. About 90% of the glucose-6-phosphate is processed via the anaerobic pathway in a series of enzymatic steps that require 2 moles of high energy phosphate (ATP) but generate up to 4 moles of ATP to provide a net potential yield of 2 moles of ATP per mole of glucose. Deficiencies, usually recessively inherited, of many of the enzymes of the anaerobic pathway (indicated by *asterisks* in Fig. 3.29) have been recognized in association with manifestations of hereditary nonspherocytic hemolytic anemia (Valentine, 1979). These are very rare except for deficiency of pyruvate kinase, which, although uncommon, ranks behind glucose-6-phosphate (see below) as the second most common inherited enzyme deficiency in human red blood cells. Deficiency of pyruvate kinase leads to hereditary anemia of variable severity that may be accompanied by jaundice and splenomegaly.

The Hexose Monophosphate Shunt

In the hexose monophosphate shunt, glucose-6-phosphate is oxidized with reduction of NADP to NADPH. High energy phosphate bonds are not generated in this shunt; its function includes the production of NADPH for use in reducing glutathione disulfide (GS-SG) to GSH (Beutler, 1983). As already mentioned, a deficiency of the enzyme glucose-6-phosphate dehydrogenase (G-6-PD), which catalyzes the oxidation of glucose-6-phosphate, is the most frequently encountered hereditary deficiency of a red cell enzyme.

Many different forms of G-6-PD have been recognized by electrophoretic and kinetic analyses; in only a few of these variants is the physiological enzymatic activity of the enzyme affected. G-6-PD

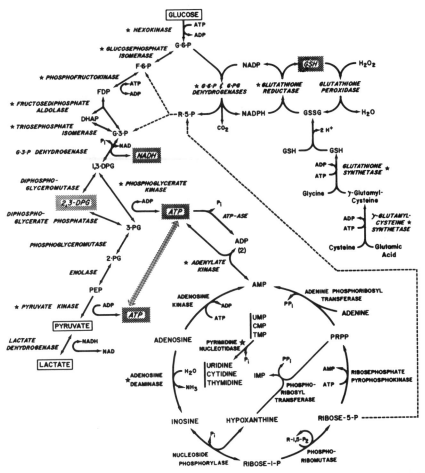

Figure 3.29. The reactions of human erythrocyte energy metabolism. The anaerobic glycolytic pathway (Embden-Meyerhof) is shown along the *left* side of the figure. Two shunts, the hexose monophosphate shunt, leading to the generation of GSH, and the 2,3-DPG shunt are depicted. Pathways of nucleotide salvage and degradation are shown in the *lower right* portion of the slide. *Asterisks* indicate enzymes found defective in association with hereditary hemolytic anemia [From Valentine, W. N. *et al.*: *The Metabolic Basis of Inherited Disease*, edited by Stanbury *et al.* New York: McGraw Hill (1983), by permission.]

activity normally declines as the red blood cell ages. In one variant, the A⁻ variant, enzymatic activity declines abnormally rapidly, and activity present in reticulocytes disappears in mature red cells. The A⁻ variant is inherited as a sex-linked trait with a high gene frequency in blacks; a deficiency of G-6-PD due to this variant is found in 11% of black American males. Affected individuals have no manifestations of the deficiency until exposed to certain stresses such as oxidant drugs or infections. Then, they may develop a Heinz body hemolytic anemia. Rare types of more severe G-6-PD deficiency have also been encountered in which hemolysis is not episodic but persistent.

In hemolytic episodes due to the common type of G-6-PD deficiency, young red cells, entering the circulation in increased numbers in response to hemolysis, possess enough G-6-PD to resist hemolysis. Therefore, an episode of hemolysis is usually self-limited. The mechanism of hemolysis after exposure to drugs is only partially understood. Exposure of red cells to some of the offending drugs yields low levels of hydrogen peroxide or free radicals. Hydrogen peroxide is detoxified in a reaction in which GSH is oxidized to GSSG (Fig. 3.19, Chapter 23). Glutathione reductase, in a reaction requiring NADPH, catalyzes reduction of the GSSG to replenish GSH. Red cells deficient in G-6-PD cannot generate NADPH at a normal rate to support this reaction. As a consequence, the products of oxidant drugs may damage both hemoglobin and the cell membrane. Precipitates of damaged hemoglobin that appear attached to the membrane (Heinz bodies) are formed in the cells. It is not known whether methemoglobin, which is often also demonstrated in episodes of G-6-PD-deficient hemolysis, is an initial compound in an oxidative denaturation pathway that leads to formation of Heinz bodies or is a second, independent effect of the drug upon hemoglobin.

The 2,3-Diphosphoglycerate Shunt

A second shunt, the 2,3-DPG shunt (Rapoport-Luebering shunt) is present in the main glycolytic pathway at the step of processing of 1,3-diphosphoglycerate (1,3-DPG). In the straight anaerobic glycolytic pathway, 1,3-DPG is converted to 3-phosphoglycerate (3-PG) with a gain of one ATP (two ATP for each starting glucose molecule, since one glucose molecule gives rise to two molecules of 1,3-DPG). In the shunt, an enzyme, diphosphoglycerate mutase, first catalyzes the formation of 2,3-DPG from 1,3-DPG and then catalyzes the loss of a phosphate from 2,3-DPG and its reentry into the glycolytic pathway as 3-PG (Fig. 3.29). ATP is not generated in these reactions. Therefore, when a molecule of glucose is processed to pyruvate by way of the 2,3-DPG shunt, two molecules of ATP are used up and two molecules of ADP are made. Glycolysis then generates NADH but no net gain in ATP.

2,3-DPG accounts for over half of the phosphorus in the erythrocyte and is about equimolar in quantity with hemoglobin. Several factors influence the amount of 2,3-DPG synthesized, including the ratio of cellular ADP and ATP levels and the overall rate of glycolysis. Alkalosis stimulates glycolysis because of the high pH optimum of the glycolytic enzyme, phosphofructokinase, whereas acidosis impairs glycolysis. Thus, the respiratory alkalosis induced by a rapid ascent to high altitudes, by stimulating glycolysis, increases red cell production of 2,3-DPG. Increased red cell levels of 2,3-DPG are also found in anemia, cardiac disease, and pulmonary disease. The stimulus for increased production of 2,3-DPG in these circumstances may arise from an increased binding of 2,3-DPG to deoxyhemoglobin, which shifts an increased proportion of 1,3-DPG into the shunt pathway.

The physiological usefulness of a glycolytic pathway shunt that prevents a gain in ATP was clarified when it was shown (Benesch and Benesch, 1967) that hemoglobin stripped of phosphate has a very high oxygen affinity, and that addition of 2,3-DPG to such a hemoglobin preparation lowers its oxygen affinity to that observed in a hemolysate prepared from normal red blood cells. It was suggested that 2,3-DPG and other red cell organic phosphates modulate oxygen affinity by binding to deoxyhemoglobin (Benesch and Benesch, 1969), a binding that has since been demonstrated directly by X-ray crystallography (Arnone, 1972). 2,3-DPG binds less well to the γ chains of fetal hemoglobin than to the β chains of adult hemoglobin, which benefits the fetus, since fetal red cells take up oxygen not at the partial pressure of oxygen in the lungs but at the partial pressure of oxygen in maternal blood flowing through the placenta.

The concentration of 2,3-DPG falls with aging of red cells, both in vivo and in vitro. The depletion of 2,3-DPG during storage of blood for transfusion in the acid citrate-dextrose anticoagulant formerly used for this purpose frequently led to administration of blood with a high oxygen affinity, which returned to normal in vivo only after 6–24 hours. Because of the potentially deleterious effects of this delay in many types of patients, most blood banks have changed to a citrate-phosphate-dextrose (CPD) anticoagulant of high pH, which better preserves red cell 2,3-DPG levels on storage.

Reduction of Methemoglobin

The products of red cell metabolism also interact with hemoglobin to reduce to ferrous hemoglobin the small amounts of methemoglobin that are constantly being produced in normal red cells. The red blood cell possesses both an NADH-dependent pathway and an NADPH-dependent pathway for reduction of methemoglobin. Only the NADH-dependent pathway functions physiologically. It utilizes the enzyme NADH methemoglobin reductase and cytochrome b5 as an electron acceptor to transfer an electron from NADH to heme. The NADPH pathway does not function physiologically despite the presence of the enzyme NADPH reductase because it lacks an electron acceptor. Methylene blue, an exogenous electron acceptor, can mobilize the NADPH pathway and is given in the treatment of patients with severe methemoglobinemia after exposure to an oxidant agent.

Nucleotide Metabolism

Nucleotides, e.g., adenosine monophosphate (AMP), are compounds consisting of a nitrogenous base, either a purine (adenine and guanine), or a pyrimidine (cytosine and uracil or thymine) joined to a sugar molecular (ribose or deoxyribose) and to one or more phosphate groups attached as an ester to the sugar molecule. *Nucleosides*, e.g., adenosine, are made up of a base plus a sugar molecule but lack an esterified phosphate group.

The developing red blood cell contains nucleotides in DNA and RNA. Nucleotides in DNA are lost from the cell when the nucleus is extruded. As already mentioned, ribosomal RNA is normally degraded within about 24 hours of the red blood cell entering the circulation. The cell can use the resulting purine ribonucleotides, but the resulting pyrimidine ribonucleotides are unwelcome, for they can compete with ATP and ADP in crucial reactions and so interfere with cell function (Valentine,

1979). An enzyme within the red cell, pyrimidine 5'-nucleotidase, catalyzes the dephosphorylation of the pyrimidine nucleotides; the nucleosides can then diffuse out of the cell. Rare patients with a hereditary deficiency of this enzyme have been identified. Their red cells contain large amounts of partially degraded ribosomes, which can be seen in many red blood cells of a Wright's stained blood smear as particles of basophilic material diffusely distributed throughout the cell (basophilic stippling). The patients have markedly elevated levels of pyrimidine nucleotides in their red blood cells and a moderate hemolytic anemia. The pathogenetic mechanism of the anemia is not clear but is presumably related to the metabolic abnormalities of the cell. Exposure of red blood cells to lead inactivates pyrimidine 5'-nucleotidase and may mimic the findings of the hereditary deficiency state. (The discovery of large numbers of red cells containing basophilic stippling on a patient's blood smear alerts one to the possibility of lead poisoning.)

The red blood cell must maintain an adequate pool of adenine nucleotides (ATP, ADP, AMP) to survive (Valentine and Paglia, 1980). The mature cell cannot synthesize adenine nucleotides de novo. They are conserved and possibly replenished by the metabolic reactions shown in the lower right hand portion of Figure 3.29, which are part of the "salvage pathways" of nucleotide metabolism.

When blood is preserved for transfusion in an anticoagulant solution to which adenine or adenosine has been added, the stored red blood cells maintain higher levels of ATP and have a better viability on transfusion than when blood is preserved in an anticoagulant solution to which adenine or adenosine has not been added. Adenine diffuses into the red blood cells and reacts with 5-phosphoribosyl-1-pyrophosphate (PRPP), in a reaction catalyzed by adenine phosphoribosyl transferase, to generate AMP. The PRPP for the reaction is formed from ribose 5-phosphate, which in turn may be generated from ribose 1-phosphate or from the catabolism of glucose by way of the hexose monophosphate shunt. (These reactions are shown, moving counterclockwise, from 6–12 o'clock in the circular representation of the salvage pathways provided in the *lower right hand portion* of Fig. 3.29.) Although demonstrated to occur when red cells are stored in an anticoagulant solution containing adenine in a transfusion bag, it is not known whether the trace amounts of adenine reportedly present in plasma diffuse into the red blood cell in vivo and contribute to the replenishment of adenine nucleotides. Patients with a rare hereditary deficiency of

the enzyme adenine phosphoribosyl transferase have no known hematologic abnormalities (Van Acker et al., 1977), which suggests that this pathway is not crucial for normal red blood cell function.

Adenosine diffusing into, or formed within, the red blood cell can take two pathways (Fig. 3.29). One pathway, which leads back to AMP through a phosphorylation reaction catalyzed by adenosine kinase, maintains or replenishes the adenine nucleotide pool. The second, competing pathway, which results in the deamination of adenosine to inosine by the enzyme adenosine deaminase, dead-ends at inosine monophosphate (IMP). It depletes the adenine nucleotide pool, since the red cell cannot convert IMP back into the adenine nucleotide pool (Fig. 3.29). An unusual hereditary hemolytic anemia has been described associated with a many-fold excess of the enzyme of the competing pathway, adenosine deaminase (Valentine et al., 1977). As a result, the deamination of adenosine to inosine is favored over the phosphorylation of adenosine to AMP. The reduced red cell life-span in these patients presumably results from an inability to maintain normal levels of the adenine nucleotide pool. One infers from this that diffusion of adenosine into the red blood cell in vivo and its conversion to AMP is a requirement for normal red blood cell viability (Valentine, 1979). In the rabbit, the liver may make adenosine that serves as a precursor of red blood cell adenine nucleotides (Lerner and Lowy, 1974).

ANEMIA AND ERYTHROCYTOSIS

A person is generally considered to be anemic when his or her hemoglobin level falls below 2 SDs of the mean for a normal population (below 14 g/dl for a man and 12 g/dl for a woman at sea level). When fully oxygenated, each gram of hemoglobin carries 1.39 ml of oxygen. Therefore, normal arterial blood with a hemoglobin concentration of 15 g/dl carries about 21 ml of oxygen, of which 4.5 ml are delivered to the tissues as the partial pressure of oxygen in blood flowing through a capillary bed falls to 40 torr (Fig. 3.27). In an anemic individual with a hemoglobin concentration of only half-normal, 7.5 g/dl, 100 ml of blood can carry only about 10 ml of oxygen. The body compensates for this loss of oxygen-carrying capacity by three mechanisms: by increasing the 2,3-DPG concentration within the red blood cell, which as already discussed, shifts the oxygen dissociation curve to the right and increases the amount of oxygen released from each gram of hemoglobin at normal capillary oxygen pressures; by an internal redistribution of blood flow that increases flow to tissues whose oxygen supply must be maintained; and by increasing cardiac output (Finch and Lefant, 1973).

Anemia may stem from an inability to make red blood cells normally, from loss of red blood cells due to bleeding or hemolysis, or from a combination of impaired production and increased loss. Causes of impaired production include: damage to or acquired proliferative abnormalities of stem cells or erythroid precursors; lack of erythropoietin due to destruction of kidney tissue; body iron deficiency; or deficiency of folic acid or cobalamin. Hemolysis may result from an intrinsic red blood cell defect, from an abnormality of the red cell environment, or both. Intrinsic defects include: red cell membrane defects (as in hereditary spherocytosis); defects of globin chain structure or synthesis that increase cell deformity or cause hemoglobin to precipitate within the cell (sickle cell anemia, thalassemia, unstable hemoglobins); and defects in enzymes of the red blood cell's energy systems (as in pyruvate kinase deficiency, which impairs production of ATP). Environmental causes of hemolysis include the formation of plasma antibodies that react with antigens on the red cell membrane and mechanical damage to red cells as blood flows through multiple small blood vessels in which, for one cause or another, strands of fibrin have been deposited (microangiopathic hemolytic anemia). In the relatively common cause for anemia in black males, glucose-6-phosphate dehydrogenase deficiency, anemia stems from a combination of the intrinsic defect, which impairs generation via the hexose monophosphate shunt of the hydride ion needed to form GSH, and environmental exposure to an oxidant drug that increases the need for GSH to protect hemoglobin from oxidant stress.

When someone becomes chronically ill, his or her hemoglobin level often falls 2 or 3 g/dl. This anemia of chronic disease has a complex pathogenesis. Apparently, mononuclear phagocytes are activated by the disease process and phagocytose red blood cells at a moderately increased rate. The marrow is unable to respond with an increase in red cell production and so the circulating mass of red cells falls until the number of cells being destroyed again equals the number being produced. The reasons why the marrow fails to increase production are not clear but include an impaired return of iron to normoblasts from the breakdown of hemoglobin in the mononuclear phagocytes and an inadequate erythropoietin response to the fall in hemoglobin concentration.

The terms *erythrocytosis* and *polycythemia* refer to a higher than normal red blood cell or hemoglobin concentration (greater than 18 g/dl in a man or than 16 g/dl in a woman at sea level). This may result from an increased circulating red blood cell mass, a contracted plasma volume, or both. An elevated hemoglobin concentration due solely to a contracted plasma volume is sometimes called *relative polycythemia*.

Erythrocytosis due to an increased circulating red blood cell mass usually stems from an increased production of erythropoietin. The increased production may be "appropriate," i.e., in response to some cause for impaired ability of hemoglobin to deliver oxygen to the tissues, or may be "inappropriate," arising from aberrant erythropoietin synthesis. Common causes of impaired arterial oxygen saturation include: residence at a very high altitude, chronic pulmonary disorders impeding uptake of oxygen as blood flows through the lungs, and congenital heart disorders that cause a substantial fraction of the circulating blood to be shunted from the right to the left side of the heart without passing through the lungs. Another common cause for an impaired ability of hemoglobin to deliver oxygen is an elevated carboxyhemoglobin concentration due to heavy smoking. As already mentioned, rare patients may have erythrocytosis because of inherited structural globin chain abnormalities that increase the affinity of hemoglobin for oxygen at normal end capillary oxygen pressure. *Inappropriate erythropoietin production* is not common and usually results from synthesis of erythropoietin by a malignant tumor. This is most often a renal tumor. Polycythemia not due to increased erythropoietin production occurs in a myeloproliferative disorder (see Chapter 21), called primary polycythemia or polycythemia vera, in which erythroid precursors arising from an aberrant clone no longer respond appropriately to normal regulatory influences.

CHAPTER 25

Hemostasis

When a blood vessel is damaged, a series of reactions are initiated to arrest bleeding, that is, to achieve and maintain hemostasis. The process involves at least four interrelated steps: contraction of the injured vessel; accumulation of platelets at the site of the lesion; activation of the blood coagulation reactions; and, as a secondary event, activation of fibrinolysis.

Vasoconstriction, which occurs immediately after injury and is usually transient, stems primarily from a direct effect of the injury upon vascular smooth muscle cells. Platelets accumulate as a result of reactions triggered when the endothelial cells lining the lumen of the vessel are disrupted and the platelets come into contact with underlying subendothelial tissues. Platelets adhere to the subendothelial tissue, undergo activation, and adhere to each other to form a growing increasingly compacted mass, the platelet hemostatic plug (Zucker, 1980). Simultaneously, exposure of trace plasma clotting proteins to the injured vessel wall initiates blood coagulation. This involves the sequential enzymatic activation of several serine protease proenzymes in reactions in which nonenzymatic plasma cofactors, phospholipid, and receptors on the surface membrane of activated platelets participate. The coagulant enzyme, thrombin, is generated. Thrombin splits small fibrinopeptides from fibrinogen and activates a cross-linking enzyme, factor XIII, with resultant formation of stabilized strands of fibrin that extend outwards from the platelet surface. Thus, a seal is formed that is made up of a fused mass of platelets reinforced by the meshwork of a fibrin clot.

Other processes are set in motion to limit growth of the platelet mass and fibrin clot. Endothelial cells bordering the injury site release materials that inhibit platelet aggregation (prostacyclin) and activate fibrinolysis (vascular plasminogen activator). The fibrinolytic enzyme plasmin dissolves fibrin strands, liberating soluble degradation products that may reenter the circulation. Over several days, fibrin continues to be both formed and dissolved at the injury site as a hemostatic seal is maintained and remolded while the proliferation of smooth muscle cells and fibroblasts, the deposition of new connective tissue matrix, and the ingrowth of a new luminal lining of endothelial cells repair the vessel wall.

Blood coagulation and platelet activation contribute to the mechanisms of tissue repair. Thrombin-activated factor XIII not only catalyzes cross-linking of fibrin molecules to each other but also catalyzes cross-linking of fibrin to an adhesive protein of plasma, connective tissue matrix, and cell surfaces called fibronectin (Mosher et al., 1981). Transitory binding of fibroblasts to successive molecules of fibronectin on fibrin strands facilitates the migration of fibroblasts into the area of injury. Platelets activated at the injury site secrete a polypeptide called platelet-derived growth factor, which stimulates the proliferation of smooth muscle cells and fibroblasts in the vessel wall.

Normal hemostatic function prevents excessive bleeding after the minor tissue injuries of daily living. Minimal stress to capillaries, e.g., the hydrostatic pressure exerted upon the capillaries of the lower legs while one stands upright, can cause gaps to form between the endothelial cells of the capillary wall. Platelets adhere to the exposed capillary basement membrane and plug these gaps. If an individual has a low platelet count (thrombocytopenia), red blood cells may leak from such gaps before they are plugged. Small bleeding spots about the size of freckles (petechiae) are formed. Persistent bleeding from erosions of the mucosal surfaces of the mouth and nose, gastrointestinal, and genitourinary tracts is another frequent manifestation of thrombocytopenic bleeding; platelet hemostatic plugs appear particularly important in controlling bleeding from the capillaries and small venules of such lesions. When an effective fibrin clot can not be formed, either because the blood coagulation reactions are impaired or because fibrinolysis is excessive, a trivial tissue injury may cause extensive bleeding. For example, a patient with severe hemophilia may loose a liter or more of blood into the soft tissues of an extremity from bleeding persisting for several days after a minor traumatic event that, in a normal person, might cause a bruise no larger than a 50-cent piece.

If an artery is torn in a severe crushing injury, the extensive trauma may induce persistent vasoconstriction that stems blood loss until other measures can be instituted. In contrast, if a large artery is severed by a lacerating injury, arterial spasm is brief, and blood flows out of the artery under such high pressure that the normal hemostatic reactions can not impede the flow. A tourniquet must be applied immediately to prevent the injured person from bleeding to death until the vessel can be surgically repaired.

During most surgical procedures small arteries are severed; each cut artery is occluded with a surgical instrument and then sutured. Very large numbers of arterioles, capillaries, and venules are also severed; bleeding from these vessels ceases spontaneously as the result of the hemostatic reactions. However, if hemostasis is impaired from any cause, bleeding from the myriads of small vessels that are not sutured may result in serious blood loss during surgery. Moreover, bleeding may reoccur during the first 7–10 days of the postoperative period.

Although essential for normal survival, the hemostatic reactions are harmful in pathologic circumstances in which they give rise to a blood clot within the lumen of a blood vessel. Such a blood clot, called a *thrombus*, may block arterial blood flow to a vital organ. This is the proximate cause of most heart attacks (myocardial infarctions) and strokes (cerebral infarctions). Thrombi may form in the veins of the legs of patients bedridden after surgery. Pieces of such thrombi may break off and be carried to the lungs where, lodging as emboli in branches of the pulmonary arteries, they seriously impair oxygenation of the blood. Patients with a high risk for thrombosis are sometimes given drugs prophylactically that impair hemostasis, e.g., drugs that interfere with platelet function or retard blood coagulation.

PLATELETS IN HEMOSTASIS

Platelets are nonnucleated cells about 2 to 4 μm in diameter, which are normally present in blood in a concentration of from 150,000 to 400,000/μl. Their production and distribution have been discussed in Chapter 21. One should recall: that platelets have a life-span of about 10 days; that, normally, about one-third of the circulating platelets are present at any given time within the red pulp of the spleen; that a substantial marrow reserve of platelets does not exist; and that if circulating platelets are suddenly depleted about 5 days are required for new platelet production to restore the platelet count to normal.

Properties of Platelets

PLATELET STRUCTURE

On light microscopy of a Wright's stained blood smear, platelets are seen as bits of light blue-staining cytoplasm containing purple granules. Electron microscopy reveals a complex ultrastructure (White and Gerrard, 1982). The unstimulated platelet circulates as a flattened disk (Fig. 3.30). Microtubules, which are tube-like structures made up of a polymerized protein called tubulin, can be seen encircling the platelet just below its surface membrane; they provide a structural support necessary for the unstimulated platelet to maintain its disk-like shape. Organelles are distributed throughout the platelet and include: α granules and lysosomal granules, which are morphologically indistinguishable, very dense granules called dense bodies, occasional mitochondria, and fine granules of glycogen. Two internal membranous systems can be recognized. One of these, the open canalicular system, is made up of sponge-like invaginations of the platelet surface membrane that provide multiple channels through which the platelet, when activated, can both take up external calcium ions and secrete the contents of its granules. The second system, the dense tubular system, is made up of channels of smooth endoplasmic reticulum that do not communicate with the exterior of the cell. The dense tubular system is thought to serve as an intracellular storage site for calcium ions.

The platelet surface membrane consists of an amorphous outer coat rich in carbohydrate (the glycocalix) and an underlying trilaminar unit membrane made up primarily of lipids and proteins. Three types of lipids are present: phospholipids, cholesterol, and small amounts of glycolipids. The membrane phospholipids are important for hemostasis both as a source of lipid activity for the blood coagulation reactions and as a source of arachidonic acid, the precursor of thromboxane A_2 (see below). Procoagulant phospholipid activity is oriented in the membrane in a way that prevents its reacting with plasma coagulation factors until the membrane undergoes reorganization after platelet stimulation.

Glycoproteins can be extracted from platelet membranes with detergents and separated by gel filtration according to size. Initially, three glycoprotein peaks of decreasing size were separated and

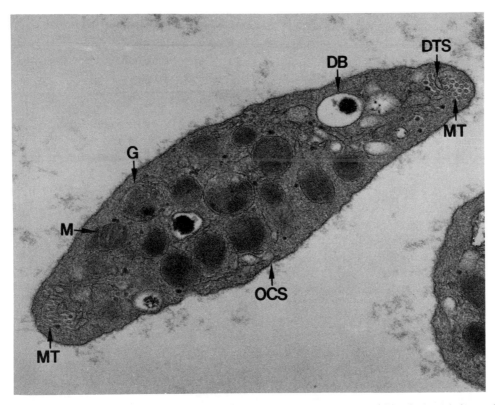

Figure 3.30. Electron micrograph of an unstimulated platelet. Note the disk-like shape, the bundles of microtubules encircling the cell beneath the surface membrane, and the absence of pseudopods. DB, dense body; DTS, dense tubular system; G, granule; M, mitochondrion; MT, microtubules; OCS, open canalicular system. [Courtesy of Dr. J. G. White.]

given the names GP (glycoprotein) I, II, and III. With improved techniques (Phillips and Agin, 1977), individual molecular species were identified within the three original glycoprotein peaks and given names derived by adding letters to the Roman numerals, e.g., GPIb, GPIIIa. Additional glycoproteins of smaller size were also recognized and given the Roman numeral names, GPIV and GPV. Receptor sites important for hemostasis have been identified on GPIb and in a complex formed by GP IIb-IIIa. GPIb also contains a receptor for the Fc fragment of IgG.

PLATELET METABOLIC PROCESSES

Glucose taken up by the cell or released from the breakdown of intracellular glycogen provides the main source of energy for the platelet. This is generated as ATP from both anaerobic glycolysis and the more efficient process of mitochondrial oxidative phosphorylation. Platelets also metabolize glucose by way of the hexose monophosphate shunt.

Platelets can synthesize glycogen and fatty acids. A metabolic need for the synthesis of the latter is not readily apparent. Platelets contain neither DNA nor RNA and so are unable to synthesize

proteins. Thus, when exposure of platelets to aspirin irreversibly inactivates the enzyme cyclooxygenase, the platelet can not synthesize new enzyme to replace the inactivated enzyme.

ATP and ADP are found in two pools within the platelets (Holmsen, 1982), a metabolic pool distributed throughout the cytoplasm and a storage pool confined to the dense bodies. The latter represents ADP and ATP packaged for secretion. In some circumstances stimulated platelets may secrete their dense body adenine nucleotides and continue to circulate as, for example, after blood passes through an extracorporeal oxygenator during open heart surgery. It is not clear as to whether such depleted platelets may replete their dense body ATP and ADP stores. When platelets are incubated in vitro with radioactively labeled inorganic phosphate, adenine nucleotides of the metabolic pool are rapidly labeled whereas adenine nucleotides of the storage pool are only very slowly labeled. This suggests that the adenine nucleotides of the metabolic pool are not readily available to replace ATP and ADP secreted from the storage pool.

The unstimulated platelet requires ATP to pump sodium from the cell, to maintain a number of platelet proteins in a phosphorylated state, and to support a continuous basal polymerization-depolymerization of platelet actin. Additional ATP is required after platelet stimulation to provide the energy for platelet shape change, contraction, and secretion. Increased breakdown of platelet glycogen and increased production of ATP by both the glycolytic and oxidative pathways accompanies platelet activation.

PLATELET CONTRACTILE SYSTEM

Platelets contain large amounts of actin, which constitutes about 10% of the total protein of the platelet, and small amounts of myosin with an actin:myosin molar ratio of about 100:1 (Adelstein and Pollard, 1978). Actin exists in two forms: as a small globular monomer called G actin and as a double-helical filament called F actin. In the unstimulated platelet, actin is found principally in its monomeric form. When platelets are activated, large amounts of actin are polymerized to F actin. This forms bundles of parallel microfilaments in the pseudopods that project from the activated platelet. In the body of the activated platelet, F actin forms a central filamentous meshwork, with strands extending to the platelet membrane, that persists after extraction of platelets with detergent as an insoluble cytoskeleton.

The globular heads of myosin bind to actin filaments; in this binding the myosin heads form repeating structures along the filament that look like arrowheads. This provides a way to orient the filament, since one can recognize an end towards which arrowheads point and an end towards which the arrowhead barbs face. Actin filaments grow primarily from their barbed end. Two proteins in the platelet stabilize actin filaments: actin binding protein, which cross-links actin filaments, and α-actinin, which may anchor filaments to the inner surface of the platelet membrane. Platelets also contain a protein called gelsolin that can destabilize actin filaments. When added to cross-linked F actin, gelsolin shortens the filaments, effectively cutting them between cross-links.

In addition to polymerization of actin, activation of platelets leads to increased association of myosin with actin filaments and to contraction of platelet actomyosin. A rise in free calcium ion concentration in the platelet cytoplasm causes calcium to bind to a calcium-binding protein called calmodulin (Cheung, 1980; Feinstein, 1982). The calcium-calmodulin complex then activates an enzyme, myosin light chain kinase, that phosphorylates the 20,000 dalton light chain of myosin. This phosphorylation permits an interaction of actin with myosin that activates myosin ATPase, with resultant genera-

tion of contractile force by platelet actomyosin threads.

Formation of Platelet Hemostatic Plugs

PLATELET ADHERENCE TO SUBENDOTHELIUM

When vascular endothelium is disrupted, platelets adhere to collagen in the exposed subendothelium (platelet adhesion), which is the initial step in the formation of platelet hemostatic plugs (Figure 3.31). A plasma protein, von Willebrand factor, is required for this reaction. (This protein was named after the individual who first described a hereditary bleeding disorder due to its deficiency. It is also referred to as factor VIII-WF because factor VIII coagulant protein circulates as a complex with von Willebrand factor protein, and at one time the two activities were thought to be properties of the same molecule.)

Von Willebrand factor binds both to collagen and to two receptors on the platelet surface membrane. Von Willebrand factor may be able to bind to one

FORMATION OF A STABLE HEMOSTATIC PLUG

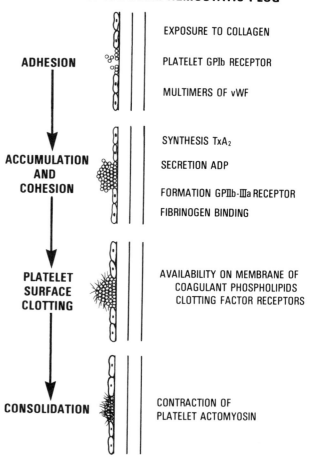

Figure 3.31.   Steps in the formation of a stable hemostatic plug. *TxA$_2$*, thromboxane A$_2$; *vWF*, von Willebrand factor.

receptor, located on GPIb, while the platelet is still in its unstimulated state. Binding of von Willebrand factor to this receptor is essential for normal platelet adhesion to collagen. Thus, the platelets of patients with a rare hereditary platelet disorder (Bernard-Soulier syndrome), in which platelets lack the GPIb glycoprotein, cannot adhere normally to collagen (Nurden and Caen, 1979). The second receptor, a complex of GPIIb-IIIa, does not form until the platelet is activated. The physiologic significance of binding of von Willebrand factor to the GPIIb-IIIa receptor is not clear.

The basic structural unit of von Willebrand factor is a 200,000-dalton molecule. Plasma von Willebrand factor consists of multimers built up from either dimers or tetramers of this 200,000 dalton unit; their molecular weights range from about 800,000 to 20,000,000 daltons. The nature of the bonds forming the doublets and holding the doublets together in multimers is unknown. Endothelial cells are the source of plasma von Willebrand factor (and also of deposits of von Willebrand factor demonstrable by immunofluorescent staining in vascular subendothelium). Normal plasma levels of von Willebrand factor are reported as 5–10 μg/ml. Plasma levels rise after stress, during pregnancy, and in acute inflammatory states.

The high molecular weight multimers of von Willebrand factor must be present in plasma for effective adhesion of platelets to collagen. Individuals in whom the total concentration of plasma von Willebrand factor is normal but the high molecular weight multimers are missing (a subtype of von Willebrand's disease known as type II$_a$ (Zimmerman and Ruggeri, 1982)) bleed abnormally because of impaired platelet adhesion. Megakaryocytes can also synthesize von Willebrand factor, which is a normal constituent of the platelet α granule and is secreted along with other granule constituents during platelet activation. A second subtype of von Willebrand's disease has recently been identified in which the high molecular weight multimers are missing from the plasma but present within the platelet. Since these patients also bleed abnormally, it appears that von Willebrand factor supplied by platelets can not substitute for plasma von Willebrand factor in supporting normal platelet adhesion to collagen.

PLATELET ACTIVATION

As platelets adhere to exposed subendothelium they become activated. Platelets arriving subsequently at the injury site also begin to activate, adhering to the platelets already present. Thus, a platelet mass starts to grow. Platelet activation consists of a series of progressive overlapping events: platelet shape change; cohesion of platelets; liberation and oxidation of arachidonic acid; secretion of α granule and dense granule contents; secretion of lysosomal granule contents; a reorganization of the platelet membrane that results in blood coagulation and formation of fibrin on the platelet surface; and an oriented centripetal contraction of actomyosin that consolidates the mass of aggregated platelets. As activation is carried through to completion, a stable platelet hemostatic plug is formed (Fig. 3.31).

Contact with collagen and the initial thrombin formed after tissue injury triggers platelet activation. An increasing concentration of thrombin at the injury site, supplemented by materials produced or liberated by the activated platelets themselves—ADP, thromboxane A$_2$, and possibly serotonin—maintain and amplify the activation. Thrombin is known to react with at least two platelet surface membrane glycoproteins: GPV, which it cleaves from the membrane, and GPIb, to which it binds in a reaction independent of the active enzymatic site of thrombin. However, the relation between the binding of thrombin to these membrane glycoproteins and thrombin-induced platelet activation is not understood. Specific receptors are also present on the platelet surface membrane for collagen, ADP, serotonin, and other agents, such as epinephrine and the prostaglandins PGI$_2$ and PGD$_2$, that can induce or modulate platelet activation. As yet, little is known about the properties of these receptors.

Coupling of Platelet Stimulation to Response

Stimuli that activate platelets cause calcium to flow from the exterior and from the vesicles of the dense tubular system into the cytosol (Fig. 3.32). As cytoplasmic calcium ion concentration rises, calcium-calmodulin complexes are formed that bind

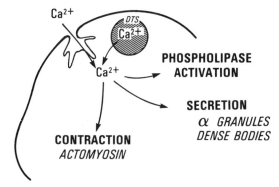

Figure 3.32. The flow of calcium into the cytoplasm of the platelet after stimulation and responses that this inititates. DTS, dense tubular system.

to and activate platelet enzymes (Cheung, 1980), and a calcium-dependent platelet protease is activated that may then activate other platelet enzymes by proteolysis (Feinstein, 1982). An increased cytoplasmic concentration of calcium ions may also directly affect intracellular processes (Lind and Stossel, 1982). Through such mechanisms the calcium flux following platelet stimulation initiates several platelet responses: *arachidonic acid release and oxidation*, by activation of phospholipase A$_2$ and other enzymes; *secretion of platelet granule contents*, the initial phase of which, a central movement of granules, may be facilitated by a calcium-regulated destabilization of actin filaments by gelsolin (Lind and Stossel, 1982); and *contraction of platelet actomyosin* which, as already discussed, results from reactions initiated by a calcium-calmodulin activation of myosin light chain kinase. However, at least part of the initial response to platelet stimulation, platelet shape change, does not appear to be calcium ion dependent, since platelets suspended in EDTA-containing buffer form pseudopods after exposure to activating stimuli. However, EDTA does block the step that follows, cohesion of platelets to form aggregates, which requires calcium ions for the assembly of a platelet receptor for fibrinogen (see below).

Platelet cyclic AMP levels influence the responsiveness of platelets to inducing agents, in part through regulation of a calcium pump that lowers cytoplasmic calcium ion levels. Prostacyclin (PGI$_2$), a prostaglandin secreted by endothelial cells, helps to keep circulating platelets in an unstimulated state through activation of platelet adenyl cyclase and a resultant rise in platelet cyclic AMP levels. Small amounts of PGD$_2$ generated within the platelets themselves may play a minor role in this process. Agents inhibiting phosphodiesterase, the enzyme that inactivates cyclic AMP, also dampen platelet responsiveness. (A drug called dipyridamole is sometimes given to patients with thrombotic disease for this purpose.)

Alterations of platelet membrane lipids affect platelet responsiveness. When plasma free cholesterol rises, free cholesterol in the platelet membrane also rises, with a resultant increase in platelet membrane fluidity and decrease in the threshold concentrations of stimuli inducing platelet activation (Shattil and Cooper, 1978). Consumption of a diet containing large quantities of fish oils decreases the arachidonic acid content of platelet membrane phospholipids and prolongs the bleeding time slightly, possibly as a reflection of impaired synthesis of thromboxane A$_2$ (Moncado, 1982). (Whether such effects contribute to the mechanisms whereby lowering plasma cholesterol levels and decreasing saturated fat intake reduce the risk of thrombosis is presently unknown.)

Platelet Shape Change

Within seconds of exposure to activating agents, platelets change shape from flattened disks to spheres with multiple projecting pseudopods (Fig. 3.33). This enables platelets to come into close contact with each other. As already mentioned, formation of pseudopods involves polymerization of actin. Actin microfilaments elongating from their barbed end, which faces the platelet membrane, push out the membrane ahead of them. The microfilaments line up in parallel through a process thought to involve cross-linking of microfilaments by phosphorylated actin binding protein. The microtubules encircling the cell just below the platelet membrane dissolve during shape change and reform centrally, surrounding platelet granules that have moved to the center of the cell prior to platelet secretion. Despite these profound morphologic alterations, platelet shape change is completely reversible. When inducing stimuli are weak, the platelet may revert to its unstimulated appearance.

Platelet Cohesion

The coming together of platelets to form aggregates (platelet cohesion) requires assembly on the

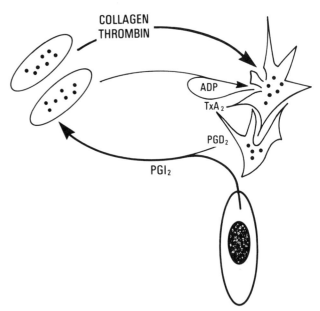

Figure 3.33. A schematic representation of the initiation of platelet activation by collagen and thrombin and of its amplification by *ADP* and thromboxane A$_2$(*TxA$_2$*). The drawing also illustrates that *PGI$_2$* generated in the endothelial cell and, to a lesser extent, *PGD$_2$* generated in the stimulated platelet inhibit platelet activation. As shown, the initial response of platelet activation, platelet shape change, is reversible.

platelet membrane of a receptor formed by a complex of GPIIb-IIIa. This complex functions as a receptor for fibrinogen, which must bind to the platelet surface before platelets can aggregate. Formation of the GPIIb-IIIa receptor requires the presence of calcium ions. However, it is apparently independent of platelet shape change, since the weak activating agent, epinephrine, causes platelets to aggregate without inducing an initial shape change.

Platelets secrete an α granule protein, thrombospondin, which also binds to the GPIIb-IIIa receptor. In addition, thrombospondin binds to fibrinogen (Leung and Nachman, 1982). In the gray platelet syndrome, a rare mild hereditary bleeding disorder in which platelets lack α granules, thrombin does not aggregate platelets normally, possibly because thrombospondin is missing. One can postulate a model for platelet cohesion in which thrombospondin binds to the GPIIb-IIIa receptor of one platelet and to a fibrinogen molecule attached to a GPIIb-IIIa receptor on another platelet. Alternatively, thrombospondin could form a bridge between platelets by binding to two fibrinogen molecules, each attached to a GPIIb-IIIa receptor on a different platelet. The GPIIb-IIIa complex also binds fibronectin and von Willebrand factor, but these proteins are not known to participate in platelet cohesion. When activating agents such as collagen, ADP, epinephrine, or thrombin are added to platelet-rich plasma deficient in von Willebrand factor, the platelets aggregate normally.

The GPIIb-IIIa receptor complex is essential for normal formation of platelet hemostatic plugs. Patients with a rare hereditary disorder called Glanzmann's thrombasthenia do not have a functional GPIIb-IIIa receptor on their platelets, either because these glycoproteins are missing or because they are functionally defective. These rare patients have profoundly impaired hemostasis. In the days before the availability of platelet concentrates for transfusion, patients with Glanzmann's thrombasthenia could bleed uncontrollably from a minor event such as a nose bleed.

Lipid Mediators of Platelet Responses

Platelet stimulation activates two phospholipases, phospholipase C and phospholipase A_2, that hydrolyze platelet membrane phospholipids (Broekman et al., 1980). The reactions by which these enzymes liberate arachidonic acid have been discussed earlier (see Chapter 23, Fig. 3.16), as have the lipoxygenase and cyclooxygenase pathways by which arachidonic acid is then oxidized (Fig. 3.17). In platelets the lipoxygenase pathway generates 12-hydroperoxy-eicosatetraenoic acid (HPETE), which is reduced to 12-hydroxy-eicosatetraenoic acid (HETE). These compounds have no known hemostatic functions. In WBC, the lipoxygenase pathway generates leukotrienes, powerful mediators of inflammatory and delayed hypersensitivity responses.

The cyclooxygenase pathway generates the endoperoxide PGG_2, which is rapidly reduced to PGH_2. A small amount of PGH_2 breaks down nonenzymatically in the platelet to form PGD_2, a prostaglandin that inhibits platelet aggregation. Although a specific platelet membrane receptor for PGD_2 has been identified, the significance of PGD_2 for hemostasis is questionable since it is formed in such small amounts. The major active product of PGH_2 metabolism in the platelet is thromboxane A_2, a potent but transient mediator of both platelet aggregation and vasoconstriction (Fig. 3.34). How thromboxane A_2 causes platelets to aggregate is not fully understood but apparently reflects the ability of thromboxane A_2 to mobilize intracellular calcium from the dense tubular system (Marcus, 1982). This results in two effects: first, the secretion of ADP, serotonin, thrombospondin, and other granule constituents; and, second, a fall in platelet cAMP with consequent increased responsiveness of the platelet to activating agents.

In endothelial cells, oxidation of arachidonic acid by the cyclooxygenase pathway generates PGI_2, a potent inhibitor of platelet aggregation (Fig. 3.34). A homeostatic balance has been postulated in which generation of PGI_2 counteracts generation of thromboxane A_2 (Moncado and Vane, 1979). In vitro, PGG_2 and PGH_2 released from platelets may be taken up and metabolized to PGI_2 by cultured endothelial cells (Marcus et al., 1980). A similar transfer of endoperoxides has been suggested as a mechanism for suppressing overgrowth of platelet hemostatic plugs as activated platelets at the borders of an area of injury come into contact with undamaged endothelium capable of synthesizing PGI_2.

Acetyl glycerol ether phosphorylcholine (AGEPC) has recently been identified as a potent lipid mediator of inflammation (Pinckard, 1982). As discussed in Chapter 23, many cells, including platelets, form AGEPC (Fig. 3.18) after stimulation. AGEPC activates platelets, inducing platelet aggregation and secretion. At present the mechanism of its platelet-activating effects is not clear, although some evidence has been obtained (Chesney et al., 1982) that AGEPC stimulates arachidonic acid oxidation via the cyclooxygenase pathway.

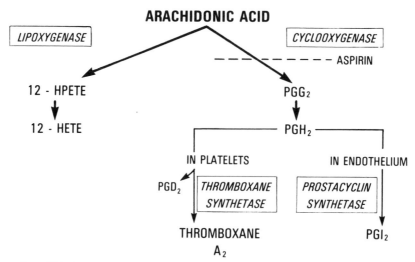

Figure 3.34. Pathways of arachidonic acid oxidation and products generated in the platelet. That the endoperoxide PGH_2 is a common substrate for the formation of thromboxane A_2 in the platelet, and PGI_2 (prostacyclin) in the endothelial cell is also shown.

A brief exposure to aspirin acetylates and inactivates cyclooxygenase and, since the platelet can not synthesize new enzyme, impairs production of thromboxane A_2 for the life-span of the platelet. Taking an aspirin tablet slightly prolongs the bleeding time of most normal individuals, e.g., from perhaps 4 minutes before aspirin to 5 or 6 minutes after aspirin. Aspirin inhibits the aggregation of platelets induced in vitro by a low concentration of collagen or thrombin but not by a higher concentration of collagen or thrombin. Rare patients have been identified with a hereditary defect of platelet cyclooxygenase function (Malmsten et al., 1975). They have a mild bleeding disorder in no way approaching the life-threatening bleeding of the patient with Glanzmann's thrombasthenia. From these observations it is clear that platelets possess an alternative mechanism or mechanisms for inducing secretion and aggregation that largely compensate for impaired synthesis of thromboxane A_2, when this occurs as an isolated defect.

However, if an individual has one abnormality affecting platelet function, then interfering with thromboxane A_2 synthesis by taking aspirin may strikingly retard the formation of hemostatic plugs. A patient with severe hemophilia can not generate thrombin normally, yet his bleeding time (which reflects the rate of formation of platelet hemostatic plugs independent of the rate of formation of a fibrin clot) is normal or only very slightly prolonged. Apparently, the reactions initiated by exposure to collagen suffice to permit the formation of hemostatic plugs strong enough to stop the bleeding from the very small vessels severed in the bleeding time test. If, however, a hemophilic patient takes a single aspirin tablet, his bleeding time may lengthen to over 30 minutes. Adding impaired col-

lagen-induced platelet activation to impaired generation of thrombin can apparently "paralyze" formation of platelet hemostatic plugs.

Platelet Secretion

During platelet activation, granules move to the center of the platelet, their membranes fuse with the membranes of the open canalicular system, and their contents are secreted through the open canalicular system into the surrounding tissues and plasma. As already mentioned, a rise in cytoplasmic calcium ion concentration, stemming in part from the action of thromboxane A_2, triggers these reactions.

Secretion from α granules slightly precedes secretion from dense granules. The contents of the α granules consist of a group of proteins present in much larger amounts in plasma—albumin, fibrinogen, fibronectin, von Willebrand factor, and factor V—and a group of proteins absent from plasma until secreted by platelets. The latter include platelet factor 4, β-thromboglobulin, platelet-derived growth factor, and thrombospondin. Secretion of proteins already present in plasma could represent no more than a vestigial evolutionary event. However, fibrinogen, von Willebrand factor, factor V, and fibronectin bind to the platelet surface during hemostasis and, conceivably, secretion from platelets makes these proteins available within platelet aggregates, where their availability might otherwise be limited.

The physiological functions of the proteins absent from plasma until secreted by platelets are as yet poorly understood. A possible role for thrombospondin in platelet aggregation has already been discussed. Secretion of platelet-derived growth factor at a site of vessel wall injury brings a stimulator of smooth muscle cell and fibroblast proliferation to where it is needed. Platelet factor 4, a cationic low molecular weight polypeptide, neutralizes the anticoagulant activity of heparin. Since heparin plays no known role in normal hemostasis, this can not represent the physiological function of platelet factor 4. Platelet factor 4 also binds to the glycosaminoglycan, heparan sulfate, which is present on the surface of endothelial cells and fibroblasts and in connective tissue matrix, but the significance of this binding is unknown. Both platelet-derived growth factor and platelet factor 4 are chemoattractants and so could play roles in the inflammatory response. β-thromboglobulin, another α granule cationic protein, has areas of homology with platelet factor 4 and possesses weak heparin-neutralizing activity. Its physiological function is also unknown. Plasma levels of platelet factor 4 and of β-thromboglobulin are sometimes measured as indicators of platelet activation in clinical thrombotic disorders.

The dense granules contain ADP, ATP, calcium ions, and serotonin. Serotonin is formed in the argentaffin cells of the gastrointestinal tract, is taken up from plasma by platelets and stored in the dense granules, and is then secreted from the platelets when they are activated. Although a vasoconstrictor and a weak platelet-activating agent, serotonin has no established physiological function in hemostasis. Large doses of the drug, reserpine, which deplete dense granules of serotonin, have no apparent effect upon hemostasis.

In contrast, ADP secreted from the dense granules is thought to play an important role in amplifying platelet activation and in recruiting new platelets into growing platelet aggregates. ADP functions as a primary aggregating agent and also acts synergistically with thrombin and collagen to enhance the activating effects of these agents. Patients with hereditary mild to moderate bleeding disorders have been described (Weiss, 1982) whose dense granules lack normal quantitites of ADP. When ADP is injected into the circulation of an animal, small platelet aggregates form that obstruct blood flow in the microcirculation. Similar ADP-induced platelet microthrombi have been postulated to form in certain clinical thrombotic disorders.

BLOOD COAGULATION

Nomenclature

Numbers, letters, and trivial names are used to identify different components of the blood coagulation reactions (Table 3.10). Some years ago an international committee assigned Roman numerals to the clotting factors. This nomenclature has taken hold for some but not all clotting factors; for example, fibrinogen is rarely referred to as factor I. Two proteins participating in both blood coagulation and kinin generation, prekallikrein and high molecular weight kininogen, have not been given Roman numeral names. Two recently identified vitamin K-dependent proteins were called protein C and protein S by their discoverers, and these

Table 3.10
Properties of agents involved in blood coagulation

Type	Name[a]	Synonym[a]	MW, daltons	Plasma Concentration μg/ml	Plasma Concentration nM	Biologic Half-time, days
Contact system proenzymes	F XII	Hageman f	80,000	29	360	2
	Prekallikrein	Fletcher f	88,000	45	510	
	F XI	PTA	160,000	4[b]	25[b]	2.5
Vitamin K-dependent coagulant proenzymes	F VII	Proconvertin	50,000	0.5	10	0.2
	F IX	Christmas f	57,000	4.0	70	1
	F X	Stuart f	57,000	8.0	140	1.5
	Prothrombin	Factor II	70,000	150	2.1×10^3	3
Cofactors	Tissue factor	Factor III		0	0	
	Platelet lipid			0	0	
	HMW kinogen		120,000	70	583	
	Factor V	Proaccelerin	330,000	7.0	21	1.5
	Factor VIII	AHF				0.5
Factors of fibrin deposition	Fibrinogen	Factor I	340,000	2,500	7.0×10^3	4.5
	Factor XIII	FSF	320,000	8	25	7
Inhibitors	Protein C		62,000	4	65	
	Antithrombin III		58,000	150	2.6×10^3	2.5
Unknown function	Protein S		69,000	25	348	

Approximate values derived from ranges reported by different investigators. [a] F or f, factor; PTA, plasma thromboplastin antecedent; AHF, antihemophilic factor; FSF, fibrin-stabilizing factor. [b] XI is a dimer with two active enzymatic sites per molecule. Therefore, the functional concentration is double the value given.

names will probably persist. When a plasma clotting factor with a Roman numeral name is activated by limited proteolysis, as occurs during blood coagulation, its activated form is identified by adding the lower case letter a, e.g., the activated form of factor IX is written factor IXa or, simply, IXa. The major protease inhibitor of blood coagulation, which neutralizes thrombin, factor Xa and factor IXa, is called antithrombin III, a name carried over from a time when the name antithrombin, followed by a Roman numeral, was used to distinguish between different thrombin-neutralizing activities in clotting mixtures. The international committee also assigned arabic numbers to platelet activities affecting the blood coagulation reactions. This nomenclature has been abandoned, but the name platelet factor 4 (PF4) perseveres for the cationic α granule protein neutralizing heparin activity. The name platelet factor 3 (PF3) is also sometimes used, although less frequently now, for platelet phospholipid procoagulant activity.

Production and Catabolism of Clotting Proteins

The plasma concentration of all clotting proteins except factor VIII falls in severe hepatocellular disease. Therefore, although activated monocytes can also synthesize small amounts of clotting factors (Østerud et al., 1980), the liver is considered the exclusive source of all of the clotting proteins in plasma except factor VIII. Factor VIII may also be made in the liver but, if so, other cells make substantial quantities of this protein as well, for plasma factor VIII levels are normal or even high in patients with advanced hepatocellular disease. Endothelial cells and megakaryocytes make von Willebrand factor. As mentioned earlier, factor VIII circulates as a complex with von Willebrand factor; however, attempts to demonstrate that endothelial cells also make factor VIII have failed.

Vitamin K is required for the synthesis of functional molecules of several serine protease proenzymes with established roles in the blood coagulation reactions: prothrombin, factor VII, factor IX, factor X, and protein C. An additional vitamin K-dependent protein, protein S, has also been isolated from plasma, but its function in blood coagulation, if any, is unknown. A number of residues of a unique amino acid, γ-carboxyglutamic acid (Gla) are present in the NH_2-terminal segment of the polypeptide chains of the vitamin K-dependent proteins. Vitamin K is needed for the postranslational reaction in which the Gla residues are formed by addition of a second carboxy group at the γ-carbon of selected glutamic acid residues. Gla residues acting together form strong calcium-binding sites

through which the vitamin K-dependent proteins, via calcium bridges, bind to phospholipid during blood coagulation.

The vitamin K-dependent proteins contain major areas of amino acid homology in the NH_2-terminal segment of their polypeptide chains, which suggests that they have evolved from a common ancestral gene. However, whereas one of the proteins, prothrombin, is present in substantial amounts in plasma (150 μg/ml), the other vitamin K-dependent proteins are trace plasma proteins with concentrations in the 0.5–8 μg/ml range. In vitamin K deficiency or when vitamin K is inhibited by an oral anticoagulant (e.g., warfarin), the vitamin K-dependent proteins can be demonstrated in plasma by immunologic techniques but, missing their Gla groups, are ineffective in the blood coagulation reactions.

The mechanisms regulating production of the plasma coagulation proteins are just beginning to be understood. Adding NH_2-terminal fragments of prothrombin to cultures of rat hepatoma cells stimulates synthesis of prothrombin and probably also of factor X. It has therefore been postulated (Graves et al., 1982) that NH_2-terminal fragments of prothrombin might serve as key regulatory elements inducing the synthesis of all of the vitamin K-dependent clotting proteins.

The mechanisms controlling basal production of fibrinogen are unknown. However, fibrinogen is an acute phase protein, and its increased production in inflammatory states has been shown to result from the release of interleukin 1 from activated monocytes. Von Willebrand factor and factor VIII are also acute phase proteins whose levels rise in inflammatory states. Moreover, vasoactive stimuli, such as hard exercise, injection of epinephrine, or injection of vasopressin, cause a sharp rise in plasma von Willebrand factor and factor VIII levels. This presumably reflects the release of preformed von Willebrand protein from vascular endothelial cells. Plasma concentrations of fibrinogen, von Willebrand factor, and factor VIII also rise in the third trimester of pregnancy, as do factor VII levels and, to a lesser extent, factor IX and X levels.

Analysis of plasma radioactivity decay curves after the intravenous injection of purified radioactively labeled fibrinogen reveals a biological intravascular half-time of 4–5 days for fibrinogen in normal subjects. Infusion of heparin to inhibit blood coagulation or tranexamic acid to inhibit fibrinolysis does not alter the half-time. About 75–80% of total body fibrinogen is in an intravascular pool. The small proportion of fibrinogen in an extravascular pool, 20–25%, as compared to 60% for albumin, may reflect the large size of the fibrinogen molecule (340,000 daltons), which could hinder its movement across capillary walls. Turnover of fibrinogen follows first order kinetics, i.e., a constant proportion of plasma fibrinogen disappears per unit time independent of plasma fibrinogen concentration.

Less extensive data are available for the biologic half-times of the other plasma clotting proteins. Several circulate wholly or partially in a complex with another protein: factor VIII with von Willebrand protein, factor XI with high molecular weight kininogen; prekallikrein, also with high molecular weight kininogen; factor XIII with fibrinogen; and protein S (which may or may not be a clotting protein) with the C4b binding protein of the complement system. Most half-time data have been obtained from following rates of loss of clotting factor activities after infusion of plasma products into patients with hereditary specific clotting factor deficiencies or from following rates of disappearance of the activity of the vitamin K-dependent clotting proteins after administration of a large dose of a vitamin K-inhibiting anticoagulant. Except for factor XIII, which appears to have an intravascular half-time of about 7 days, the other clotting factors have unusually short intravascular half-times, e.g., 3 days for prothrombin, 1.5 days for factor X, 12 hours for factor VIII, and only 5 hours for factor VII. The reasons for these rapid rates are unknown. They do not reflect inherent instability of the molecules themselves. Because of the short intravascular half-times of the clotting factors, a patient with a clotting factor deficiency who is bleeding usually requires repeated infusions of plasma products. For example, to prevent excessive bleeding after surgery in a patient with hemophilia A (factor VIII deficiency), one must administer an amount of factor VIII equivalent to one-half of the amount present in the circulation of a normal subject with a comparable plasma volume every 12 hours for from 7 to 10 days. Plasma concentrates must be used since use of whole plasma would quickly overexpand plasma volume.

A General Description of the Blood Coagulation Reactions

Blood coagulation may be viewed as occurring in three steps: an initial step in which a sequence of reactions results in the generation of an activator of prothrombin; a second step in which the prothrombin activator cleaves prothrombin to form thrombin; and a third step, in which the reactions of thrombin with fibrinogen and factor XIII lead to the deposition of cross-linked polymers of fibrin (Fig. 3.35). The first and second steps consist of successive reactions in which an inert or very minimally active precursor is converted into an active serine protease which then activates the next precursor in the sequence. Amplification occurs, so that formation of a few molecules of an initial serine protease culminates in formation of many molecules of thrombin (Jackson and Nemerson, 1980).

Cofactors are required for these reactions to occur fast enough to have physiologic significance. Cofactors appear to serve at least two functions. First, they confer specificity to the enzymatic reactions, targeting the serine proteases to specific substrates. Second, they provide a mechanism for localizing the reactions to the surfaces of damaged tissue cells or activated platelets, which helps to confine the reactions to an area of tissue injury.

Two pathways of initiating blood coagulation have been characterized. The first involves the factor XII (Hageman factor)-dependent reactions set off when blood is exposed to a negatively charged surface in vitro (Cochrane, 1982; Kaplan et al., 1982). These result in the generation of factor XIa. The second involves formation of a complex between factor VII and a lipoprotein activity called tissue factor present on the surface of altered or damaged cells (Nossel and Nemerson, 1982). As illustrated in Figure 3.35, both pathways of initiation lead to the activation of factor IX (Rapaport, 1981). In the next reaction, factor X is activated in two ways: by a complex of factor IXa, platelet phospholipid, and factor VIIIa and by the complex of tissue factor with factor VII or factor VIIa (discussed below). Factor Xa then forms a complex with platelet phospholipid and factor Va to form the physiological activator of prothrombin.

Positive feedback reactions have been recognized. These include: activation of minimally active native factor VII to fully active factor VIIa by factor Xa and, possibly, by factor IXa; and the autocatalytic actions of thrombin. Trace amounts of thrombin, by activating platelets and thus making available platelet procoagulant activity and also by activating the two cofactors, factor VIII and factor V, accelerate the subsequent formation of thrombin (Fig. 3.36).

Antithrombin III neutralizes the enzymatic activity of thrombin, factor Xa and factor IXa. Moreover, the cofactors factor VIIIa and factor Va rapidly lose their coagulant activity, both as a result of their inherent instability and as the result of their further proteolysis by activated protein C. These processes, plus the initiation of secondary fibrinolysis, prevent the blood coagulation reactions

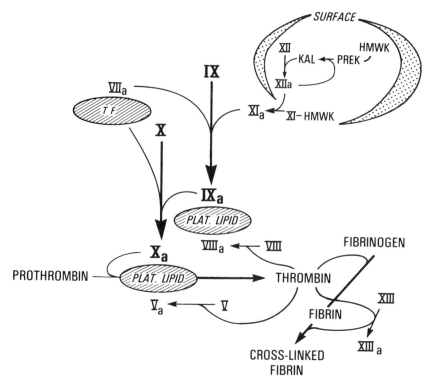

Figure 3.35.   An overall diagram of the blood coagulation reactions. The reactions initiated when blood is exposed to a negatively charged surface in vitro are summarized in the *upper righthand portion of the diagram*. These generate factor XI_a. Factor *IX* is shown to be activated in two ways: by factor XI_a and by a complex of tissue factor-factor VII_a. (To avoid further complexity, possible feedback reactions in which native factor *VII* is activated to factor VII_a are not shown but are illustrated in Fig. 3.37.) Factor *X* is also shown to be activated in two ways: by a complex of factor IX_a-platelet phospholipid-factor VIII_a and by a complex of tissue factor-factor VII_a. Prothrombin is shown to be activated on the platelet surface by a factor X_a-platelet phospholipid-factor V_a complex. Thrombin is shown to activate factor *VIII*, factor *V*, and factor *XIII* and to convert fibrinogen to fibrin. Also to avoid further complexity, the participation of calcium ions in the formation of all complexes, in the activation of factor IX_a by factor XI_a, and in the activation of factor XIII has not been shown. *HMWK*, high molecular weight kininogen; *KAL*, kallikrein; *PLAT. LIPID*, platelet procoagulant phospholipid; *PREK*, prekallikrein; *TF*, tissue factor.

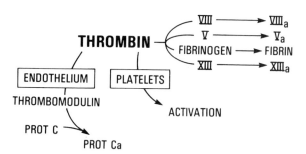

Figure 3.36   The multiple actions of thrombin in blood coagulation. *Prot C*, protein C; *Prot Ca*, activated protein C.

triggered at a site of injury from escaping from control and causing blood clots extending beyond a site of injury.

The above provides an overall description of blood coagulation. Individual reactions are described in more detail in the following sections.

Initiation of Blood Coagulation

The exposure of blood to a negatively charged surface in vitro initiates reactions involving four plasma proteins—factor XII, prekallikrein, factor XI, and high molecular weight kininogen—that have been described earlier in connection with the inflammatory response (see Chapter 23, Fig. 3.14). Reciprocal activation reactions take place in which factor XIIa activates prekallikrein to kallikrein and kallikrein activates factor XII. Two forms of activated factor XII molecules are generated. One, named α-XIIa, results from cleavage of a single peptide bond in factor XII; this form possesses the same molecular weight as the native molecule (80,000 daltons), remains surface bound, and activates factor XI efficiently. The second, referred to as β-XIIa, results from cleavage of two peptide bonds in factor XII; this form has a low molecular weight (28,000 daltons), is not retained on surfaces, and efficiently activates prekallikrein but is a poor activator of factor XI. As mentioned earlier, tissue surfaces capable of substituting in vivo for a negatively charged in vitro surface have not been definitively demonstrated, but enzymes have been identified in both endothelial cells and granulocytes that could activate factor XII in vivo (Cochrane, 1982).

Nevertheless, although important for initiating blood coagulation in a glass test tube, Factor XII, prekallikrein, and high molecular weight kininogen do not appear to play a role in normal hemostasis, since patients with isolated hereditary deficiencies of each of these factors do not bleed abnormally. Patients with hereditary factor XI deficiency are interesting. Although the majority have a mild bleeding disorder, occasional patients have not bled abnormally despite repeated hemostatic challenges. The mild bleeding tendency of most patients, plus the severe tissue bleeding experienced by a patient with factor XI deficiency who also developed an IgG inhibitor of factor XI (Stern and Nossel, 1982) establishes that factor XIa is required for physiologic hemostasis. Why some patients have not bled excessively despite plasma factor XI levels as low as those of patients who have bled (Rimon et al., 1976) remains a mystery. Also unknown is the presumed mechanism whereby factor XI is activated in vivo, independent of the contact activation reactions. Such an additional mechanism, yet to be discovered, has been postulated as the best explanation as to why factor XII, prekallikrein, and high molecular weight kininogen deficiency patients do not bleed abnormally whereas most factor XI deficiency patients do bleed abnormally.

Exposure of blood to tissue factor initiates a second pathway of blood coagulation. Tissue factor activity is a property of a specific membrane lipoprotein moiety present to a varying degree on all tissue cells after the cell membrane is in some way perturbed or damaged. Tissue factor activity is not normally present in blood, but circulating monocytes may acquire tissue factor activity after activation, for example, after exposure to endotoxin entering the blood in severe Gram-negative bacterial infections. Reagents with potent tissue factor activity that clot recalcified plasma within seconds are prepared for laboratory use by extracting brain tissue or lung tissue with physiologic saline. Fortunately, exposure of blood to tissue factor of this potency does not occur in normal hemostasis, although it can occur pathologically after trauma to brain tissue, with resultant intravascular coagulation so extensive that the circulating blood is depleted of fibrinogen and other clotting factors (Goodnight et al., 1974).

Endothelial cells possess low levels of a tissue factor activity that is not normally available to react with plasma coagulation factors (Maynard et al., 1977). Alterations of endothelial cell surface membranes making endothelial cell tissue factor available (Nossel and Nemerson, 1982) and disruption of endothelium so that blood is exposed to cells with tissue factor activity in the subendothelium are thought to occur at sites of vascular wall injury. The resultant activation of blood coagulation by low levels of tissue factor activity clearly represents a physiological mechanism for initiating blood coagulation at a site of tissue injury.

Tissue factor functions as a cofactor for factor VII, which must bind to tissue factor to function in blood coagulation (Fig. 3.37). The tissue factor-native factor VII complex, which possesses minimal coagulant activity (Zur and Nemerson, 1982), is thought to begin the activation of factor IX and factor X in hemostasis. However, several enzymes, including factor IXa and factor Xa (in the presence of phospholipid and calcium ions), factor XIIa, and thrombin, can cleave native one-chain factor VII to form a two-chain molecule, factor VIIa. A tissue factor-factor VIIa complex possesses at least a 25-fold greater enzymatic activity than a tissue factor-factor VII complex. The feedback activation of factor VII by the first molecules of factor IX and X converted to factor IXa and Xa by tissue factor-factor VII (Fig. 3.37) could well represent a significant amplification reaction in hemostasis, although proof of its operation in vivo is still lacking.

Patients with severe hereditary factor VII deficiency and essentially unmeasurable factor VII activity in their plasma have a very serious, frequently fatal, bleeding disorder. This observation testifies to the importance of the tissue factor pathway for initiating blood coagulation in normal hemostasis. However, patients with moderately severe hereditary factor VII deficiency and factor VII levels in

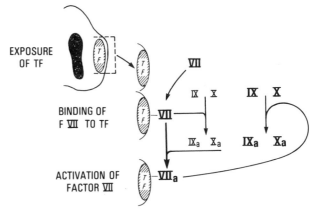

Figure 3.37. Steps in the tissue factor pathway of the initiation of blood coagulation. An alteration of the cell surface makes tissue factor activity available. Native factor *VII* binds to tissue factor on the cell surface, and the complex initiates activation of factors IX and X. The first molecules of factors IX$_a$ and X$_a$ formed feedback to activate factor VII to factor VII$_a$, which then greatly accelerates the further activation of factors IX and X.

the 3–10% range may experience little or no abnormal bleeding. Apparently, such reduced levels still provide enough factor VII molecules to saturate the limited amounts of tissue factor to which the blood is usually exposed in normal hemostasis.

Activation of Factor IX and Factor X

Both factor XIa and the tissue factor-factor VII complex convert native factor IX from its proenzyme single chain form into a two-chain activated molecule through two peptide bond cleavages that release an activation peptide from the native molecule (Fig. 3.38). Activation by factor XI can take place in the fluid phase and requires no cofactor except calcium ions which, by binding to factor IX and altering its configuration, markedly accelerate the reaction rate. In the tissue factor-factor VII activation of factor IX, which also requires calcium ions, factor IX binds to the complex on a cell surface, is activated on the surface and then is presumably released from the surface.

Since the tissue factor-factor VII complex also activates factor X in a reaction bypassing factor IX, the question arises as to whether the tissue factor-factor VII activation of factor IX has physiologic meaning. Observations in patients suggest that it does. The serious bleeding of patients with hemophilia, who lack either factor IX or factor VIII activity, the two proteins required to activate factor X by the pathway independent of tissue factor,

establishes that direct activation of factor X by tissue factor-factor VIIa does not suffice for normal hemostasis. The difference between the mild bleeding of the usual patient with factor XI deficiency and the severe bleeding of the usual patient with hemophilia B (factor IX deficiency) leads one to believe that a reaction for activating factor IX independent of factor XIa, such as the tissue factor-factor VII activation of factor IX, functions in normal hemostasis. Kinetic analyses of the tissue factor-factor VIIa activation of factors IX and X in purified systems support this view, but further studies are needed before the physiologic significance of the tissue factor-factor VII activation of factor IX can be fully assessed.

Factor IXa does not function as an effective activator of factor X in vitro unless factor VIIIa, phospholipid and calcium ions are also present in the clotting mixture. Native factor VIII will not support factor X activation in such mixtures but must first be activated by proteolysis induced by a trace of thrombin or possibly by factor Xa (Hultin, 1982). In in vitro clotting mixtures, suspensions of artificial phospholipid mixtures or platelets activated by a trace of thrombin can be used as the source of phospholipid (Hultin, 1982). In vivo the platelet is the only known source of procoagulant phospholipid for the reaction. One assumes, therefore, that activation of factor X by the factor IXa-factor VIIIa-phospholipid complex takes place on the platelet surface during normal hemostasis.

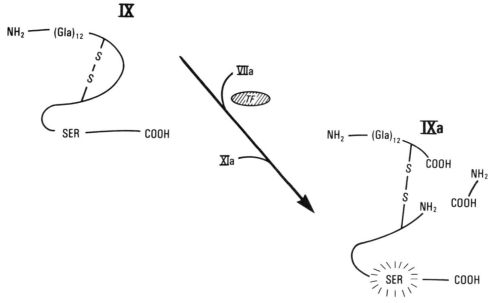

Figure 3.38. A schematic representation of the activation of factor IX. An activation peptide is cleaved from the native molecule to give rise to a two-chain activated molecule. As with all of the vitamin K-dependent clotting proteins, a number of γ-carboxyglutamic acid residues (Gla) are located in the amino terminal segment of the molecule, which becomes the light chain of the activated molecule. The active site serine is located in the carboxy terminal segment of the molecule, which becomes the heavy chain of the activated molecule. The two chains are held together by disulfide bonds. *TF*, tissue factor.

However, direct evidence that factor VIIIa or factor IXa binds to the platelet surface membrane has yet to be obtained.

Activation of Prothrombin

Analogous to the activation of factor X by factor IXa, the activation of prothrombin by factor Xa requires the participation as cofactors of factor V, phospholipid, and calcium ions. As with factor VIII in the activation of factor X, factor V must be activated by proteolysis before it can support the activation of prothrombin. A trace of thrombin is the only established activator of plasma factor V; an activity within platelets may also possibly activate platelet factor V (Østerud et al., 1977). The surface membrane of the activated platelet is the only known source of the phospholipid for the reaction. Factor Xa and factor Va have been shown to bind to the surface of activated platelets (Majerus and Milletich, 1978). Factor Va binds first and factor Xa then binds to the factor Va. The properties of the platelet surface receptor for factor Va are unknown, but it appears that it is distinct from the procoagulant phospholipid sites. Opinion differs as to whether the platelet must be activated before its factor Va receptor becomes available on the surface membrane.

Prothrombin binds to the platelet surface in two ways: by its Gla groups through calcium ion bridges to platelet phospholipid and by a factor Va binding site to factor Va on the platelet surface. During normal hemostasis, thrombin is formed principally, if not exclusively, upon the surface of activated platelets. Whether the prothrombin activator and prothrombin also come down together to generate

thrombin on other cell surfaces, such as monocyte cell surfaces, is not yet known.

Factor Xa induces two cleavages in prothrombin; one separates the NH2-terminal segment of the molecule containing the Gla groups and factor Va binding site (fragment 1-2) from the COOH-terminal segment containing the thrombin portion of the molecule (prethrombin). The second cleavage occurs in prethrombin and converts it to thrombin (see Fig. 3.39). Thrombin also cleaves prothrombin. Its cleavage site is located on the NH2-terminal segment between the Gla groups and the factor Va binding site. A fragment called fragment 1 is formed. Cleavage by thrombin may prevent prothrombin from subsequently forming thrombin by separating its phospholipid and factor Va binding sites before the molecule can be acted upon by factor Xa. Whether this negative feedback reaction occurs to any significant extent during hemostasis is unknown.

After prothrombin is converted to thrombin on the platelet surface, the thrombin portion of the molecule can move off of the surface, since it contains neither phospholipid nor factor Va binding sites. Thus, thrombin can diffuse into the surrounding plasma and extracellular fluid to catalyze conversion of fibrinogen to fibrin beyond the platelet surface.

Formation and Stabilization of Fibrin

Fibrinogen is a dimeric protein made up of two pairs of three nonidentical polypeptide chains, called the α, β, and γ chains. Electron microscopy reveals that the molecule possesses a trinodular structure. The central nodule is made up of the

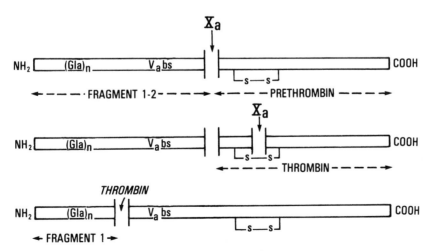

Figure 3.39. The limited proteolysis of prothrombin by factor X_a and by thrombin. Factor X_a cleaves prothrombin at two sites. Cleavage at one site gives rise to fragment 1-2 and prethrombin. Cleavage at a second site, on prethrombin, yields thrombin. Thrombin cleaves prothrombin to form fragment 1, but thrombin does not cleave prothrombin to form additional thrombin. *Gla*, γ-carboxyglutamic acid; $V_a bs$, factor V_a binding site.

NH₂-terminal ends of all six chains held together by disulfide bonds. Each distal nodule is made up of the carboxy segments of one pair of the three chains. In the intervening linear segments between the central nodule and each terminal nodule, the three chains are wound around each other to form a "coiled coil."

Thrombin cleaves small peptides from the NH₂-terminal of each α chain (fibrinopeptide A) and each β chain (fibrinopeptide B). This exposes new amino acid groups in the central nodule that function as contact sites for polymerization of the molecule. During polymerization the basic structure of the molecule does not change, but the newly exposed contact sites in the central nodule form non-covalent bonds with sites located in the distal nodules of other fibrinogen molecules. This takes place in a partially overlapping pattern that results in both lateral and end-to-end polymerization of fibrin molecules and the formation of insoluble fibrin strands (Doolittle, 1981).

Thrombin also activates factor XIII, an enzyme catalyzing the formation of covalent bonds between the polymerizing fibrin molecules (McDonagh, 1982). Plasma factor XIII is a tetrameric protein made up of two chains called *a* chains and two chains called *b* chains. The *a* chain contains the active enzymatic site; the *b* chain protects the *a*

chain in plasma in as yet unknown ways. Platelets contain factor XIII in their cytoplasm; in the platelet and in cells of the placenta, uterus, prostate, and liver that also contain factor XIII, the molecule is found as *a* chains without *b* chains.

Thrombin cleaves a peptide from the NH₂-terminal end of the *a* chain. Calcium ions must be present for the resultant thrombin-altered molecule to exhibit enzyme activity. Calcium binds to the *a* chain, causing it to separate from the *b* chain and also causing a conformational change that exposes a cysteine molecule at the active site (Fig. 3.40). The resultant activated molecule functions as a transglutaminase to catalyze formation of amide bonds between a γ-carbonyl group of glutamine and an ϵ-amino group of lysine.

The carboxy terminal segment of each γ chain of fibrinogen (fibrin) contains two cross-linking sites, one furnished by a glutamine residue and the other by a lysine residue. As fibrin molecules start to polymerize, two covalent bonds may be formed between γ chains on adjacent molecules oriented to each other in an antiparallel direction (Fig. 3.40). Additional cross-linking sites are present on the α chains. These α chain sites participate not only in the cross-linking of fibrin molecules to each other but in the cross-linking of fibrin to fibronectin and to α_2-plasmin inhibitor (discussed later). Factor

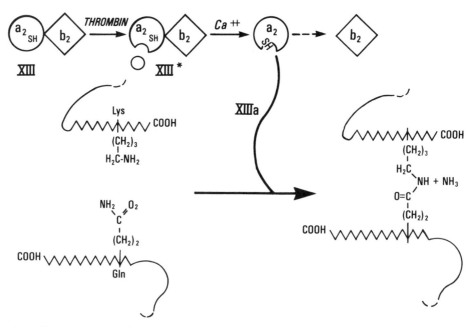

Figure 3.40.   A schematic representation of activation of plasma factor *XIII* by thrombin and calcium ions. Thrombin cleaves a peptide from the *a* chains of the molecule, which contain the active site. In the presence of calcium ions the *b* chains separate from the *a* chains. Calcium ions also induce a conformational change in the *a* chains that exposes an active site cysteine, which acti-

vates the enzyme. Also shown is the catalysis by factor *XIII*ₐ of the covalent cross-linking of the carboxy ends of two γ chains of adjacent fibrin molecules, oriented in an antiparallel direction, by the formation of an amide bond between the ϵ-amino group of lysine on one γ chain and the γ-carbonyl group of glutamine on the other chain.

XIIIa also cross-links fibronectin to collagen. Thus, through a cross-link between fibrin and fibronectin and a second cross-link between fibronectin and collagen, factor XIIIa may bind fibrin to collagen.

The function of platelet factor XIII is unknown. It is not secreted during platelet activation (McDonagh, 1982). Since actin and myosin are also substrates for factor XIIIa, it has been suggested that cross-linking of fibrin to platelet actin and myosin, possibly catalyzed by platelet factor XIIIa, could serve to strengthen the platelet hemostatic plug in the process of clot retraction.

The reactions catalyzed by plasma factor XIIIa stabilize fibrin and protect it from excessive fibrinolysis. They must take place for normal hemostasis, since patients with a hereditary deficiency of factor XIII have a moderate to severe bleeding disorder. A number of the patients also have impaired wound healing.

Regulation of Blood Coagulation

The circulating blood does not clot because it is not normally exposed to materials capable of initiating blood coagulation. Should activated clotting intermediates enter the circulation they are usually rapidly cleared by cellular mechanisms (Spaet et al., 1961). Thrombin infused into rabbits binds to sites on vascular endothelium, where it apparently reacts more rapidly with antithrombin III than when in solution. The thrombin-antithrombin III complexes then come off of the endothelium to enter the circulation and be removed by the liver (Lollar and Owen, 1980).

However, in some clinical situations, such as during major surgery, materials initiating blood coagulations may gain access to the blood and become trapped in areas of stasis, e.g., in the arcades of the deep veins of the calf in a patient lying motionless on his back. Because of stasis the blood coagulation reactions may proceed to completion, forming a small clot in the vessel despite the absence of prior vessel wall damage. Studies utilizing external counting of intravenously injected radioactive fibrinogen to detect such small thrombi reveal their presence in a surprisingly high number of older patients after major surgery. In most patients these small thrombi fail to enlarge to form clinically significant venous thrombi. This presumably reflects the local suppression of continuing clotting by plasma inhibitors, such as antithrombin III and activated protein C.

Antithrombin III is the primary inhibitor of thrombin, factor Xa, and factor IXa, the serine proteases of the intermediate and terminal steps of blood coagulation. Antithrombin III does not function as the primary inhibitor of the serine proteases initiating blood coagulation. α_1-protease inhibitor is the primary inhibitor of factor XIa. How the enzymatic activity of the tissue factor-factor VIIa complex is inhibited is unknown.

Antithrombin III forms a 1:1 stoichiometric complex with thrombin in a reaction involving the active site serine and an arginine group at the reactive site of the inhibitor. In solution in vitro antithrombin III inactivates thrombin over minutes. However, if the negatively charged polysaccharide anticoagulant heparin is added, antithrombin III inactivates thrombin virtually instantaneously. Heparin binds to positively charged ϵ-amino groups on lysyl residues of antithrombin, changing its conformation and making its reactive arginine more available to react with the active site serine of thrombin (Harpel and Rosenberg, 1976). Heparan sulfate, a substance with weak heparin-like anticoagulant activity, is present as a proteoglycan (a molecule containing a small protein core and long polysaccharide side chains) on the surface of endothelial cells and in connective tissue matrix. Binding of antithrombin III to heparan sulfate could possibly increase the rate of antithrombin III inactivation of serine proteases in vivo.

Radioactively labeled purified human antithrombin III injected intravenously into normal subjects has a biological intravascular half-life of between 2.5 and 3 days. When heparin is given, the intravascular half-life falls to a little more than 2 days (Collen et al., 1977). Antithrombin III is present in plasma in a modest molar excess over prothrombin (2.5 μM for antithrombin III, 2.0 μM for prothrombin). Families with hereditary antithrombin III deficiency have been reported in which multiple members with antithrombin III activity of about 40% of normal have experienced recurrent venous thrombosis. This observation confirms the importance of antithrombin III for the physiologic regulation of hemostasis and indicates that α_2-macroglobulin, which functions as a second line thrombin inhibitor, can not compensate completely for a reduced plasma antithrombin III level. Recently, a new heparin-dependent inhibitor of thrombin has been identified in plasma (Tollefsen and Blank, 1981); its physiologic function is not yet known.

Inactivation of the cofactors factor VIII and factor V represents another mechanism of control of blood coagulation. Although proteolysis of factor VIII and factor V by thrombin may be required for these molecules to participate in blood coagulation, the resultant activated forms are unstable and rapidly lose coagulant activity. Moreover, thrombin binds to a molecule on endothelium called thrombomodulin (Owen and Esmon, 1981) that changes thrombin's substrate specificity and enables it, in low concentrations, to activate protein C. Unlike the other activated vitamin K-dependent clotting proteins, activated protein C is an anticoagulant enzyme that rapidly cleaves and inactivates factors VIIIa and Va. A recent report of a family with thrombotic disease and reduced plasma antigenic levels of protein C (Griffin et al., 1981) focuses attention upon the possible importance of inactivation of factors VIIIa and Va by activated protein C in the physiologic regulation of blood coagulation.

FIBRINOLYSIS

General Description and Comparison With Blood Coagulation

Just as prothrombin is present in blood as the inert precursor of thrombin, so plasminogen is present in blood as the inert precursor of a second serine protease, plasmin, that cleaves fibrinogen (fibrin). The plasma concentration of the two proenzymes is similar (about 2 μM). However, whereas thrombin cleaves only small fibrinopeptides from fibrinogen, leaving the remainder of the molecule intact, plasmin catalyzes several cleavages in fibrin that break up the molecule into successively smaller soluble fragments called fibrin(ogen) degradation products.

Activation of prothrombin represents the last step in a sequence of reactions involving several serine proteases and their cofactors; activation of plasminogen is a simpler process (Fig. 3.41). Cells release plasminogen activators in an active state, ready to cleave plasminogen. At least two immunologically distinct activators from cells have been identified. The first, the *vascular plasminogen activator*, is the major activator released from vascular endothelium and from other tissues such as the heart and uterus. The second, called *urokinase* because it was first recognized in urine, is produced by epithelial cells, such as kidney epithelium, that line channels that must be kept free of fibrin to allow flow of secreted fluids (Hedner and Nilsson, 1981). Factor XIIa, kallikrein, and factor XIa, the serine proteases formed by the contact activation of blood, also function as plasminogen activators, but these "intrinsic" activators are weak and are probably of little physiologic significance (Collen, 1980).

The binding of plasma coagulation factors to tissue cell surfaces and platelet surfaces localizes the coagulation reactions. Two mechanisms localize fibrinolysis to the surface of fibrin. The first results from a change in a property of fibrinogen on conversion to fibrin that enhances the ability of plasminogen to bind to the molecule. The second results from a marked enhancement of the specific activity of vascular plasminogen activator (Collen, 1980) that follows its binding to fibrin. Although a weak activator in circulating blood, vascular plasminogen activator becomes a potent activator of plasminogen when both molecules are bound to a fibrin surface. Urokinase, in contrast, is a powerful activator in solution (which makes teleological sense since fibrinogenolysis in the circulating blood creates a dangerous bleeding tendency, whereas fibrinogenolysis in fluids excreted through epithelial channels prevents potential clogging of the channels).

Inhibition of intermediates of thrombin generation plays a major role in keeping the blood coagulation reactions under control. Inactivation of cellular plasminogen activators by plasma inhibitors plays at most a minor role in modulating fibrinolysis. Two mechanisms primarily regulate the activity of vascular plasminogen activator. One has already been mentioned, the need for its binding to fibrin to enhance its specific activity. The second is rapid clearance from the blood by the liver; the plasma half-life of vascular plasminogen activator in humans is 15 minutes.

Antithrombin III neutralizes thrombin slowly in

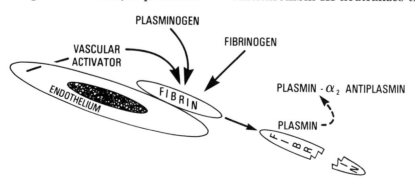

Figure 3.41. A simplified representation of the reactions of fibrinolysis induced during hemostasis by the release of vascular plasminogen activator from endothelial cells.

the absence of heparin. In contrast, α_2-antiplasmin inactivates plasmin very rapidly. Although plasmin bound to fibrin and involved in fibrin degradation is relatively protected, its estimated half-life is still only 10 s (Collen, 1980). Plasmin in solution is inactivated within 100 ms; the rate is among the fastest known for a protein-protein interaction. Inhibition by α_2-antiplasmin clearly represents a crucial regulatory mechanism preventing excessive fibrinolysis. Patients with a rare hereditary disorder in which only traces of α_2-antiplasmin are present in blood may bleed extensively into their tissues following minor trauma (Aoki et al., 1979).

Properties of the Components of the Fibrinolytic Reactions

PLASMINOGEN

Plasminogen, a single chain protein (93,000 daltons) is made in the liver. The factors regulating its synthesis are unknown. Native plasminogen has an NH_2-terminal glutamic acid (glu-plasminogen). Limited plasmic digestion cleaves a peptide from the NH_2-terminal segment of the molecule, giving rise to a smaller molecule with an NH_2-terminal lysine (lys-plasminogen). Glu-plasminogen has an intravascular half-time of 2–2.5 days; lys-plasminogen has an intravascular half-time of less than 1 day. The infusion of tranexamic acid does not lengthen the intravascular half-time of glu-plasminogen, which suggests that its normal catabolism is metabolic and not the result of continuing basal fibrinolysis. About 30% of total body plasminogen is within an extravascular pool.

Plasminogen has a high affinity lysine binding site through which it binds to fibrin. A plasma protein called histidine-rich glycoprotein can also bind to the high affinity lysine binding site of plasminogen. About 50% of plasma plasminogen is thought to circulate as a complex with histidine-rich glycoprotein (Lijnen et al., 1980). Occupancy of lysine binding sites by histidine-rich glycoprotein may be a mechanism for regulating fibrinolysis by limiting binding of plasminogen to fibrin.

α_2-Antiplasmin binds weakly to plasminogen, also through the lysine binding site. Histidine-rich glycoprotein reduces the extent of this reaction as well, thus increasing the free concentration of the inhibitor.

Through cleavage of a single arginine-valine bond, plasminogen activators convert inert single chain plasminogen into active two-chain plasmin (Robbins, 1982). The NH_2-terminal segment of plasminogen becomes the A chain of plasmin; it contains the lysine binding site. The carboxy terminal segment of the plasminogen molecule becomes the B chain of plasmin; it contains the active serine site. The first plasmin formed during plasminogen activation cleaves native glu-plasminogen into lys-plasminogen; in purified systems urokinase activates lys-plasminogen much faster than it activates glu-plasminogen. The importance of conversion of glu-plasminogen to lys-plasminogen in the activation of plasminogen in vivo is not clear.

PLASMINOGEN ACTIVATORS

Vascular plasminogen activator is a single chain protein (70,000 daltons) that may be cleaved to a two-chain molecule after release without apparent change in its specific activity in the presence of fibrin. Stress, exercise, and processes causing vasodilatation, such as venous occlusion, initiates the release of vascular plasminogen activator. Both a neurohumoral agent from the central nervous system and catecholamines are thought to stimulate release of activator from endothelial cells. Since the deposition of fibrin in vessels is usually followed by the initiation of fibrinolysis, the presence of fibrin on endothelium or the activity of intermediates of blood coagulation should also stimulate release of vascular plasminogen from endothelium. However, convincing evidence for this is still lacking.

As already mentioned, urokinase is released from kidney epithelium and is a normal constituent of urine. Vascular endothelium has also been shown to release small amounts of urokinase; the physiological significance of its release from endothelium is unknown. Urokinase prepared from cultured human kidney cells is available as a therapeutic agent.

Certain strains of hemolytic streptococci produce a nonenzyme protein that can indirectly activate plasminogen. This protein, called streptokinase, combines with plasminogen to form a complex that can then function to activate plasminogen. Despite its antigenic properties as a foreign protein, streptokinase is more often used for inducing therapeutic fibrinolysis than urokinase, because streptokinase is much less expensive.

PLASMIN INHIBITORS

α_2-Antiplasmin inactivates plasmin by combining both with its lysine binding site and with its active serine site. Since both of these sites are occupied when bound plasmin is degrading fibrin, α_2-antiplasmin inactivates free plasmin more rapidly than it inactivates plasmin on fibrin. Nevertheless, as mentioned earlier, it still inactivates the latter within 10 s, which suggests that active fibrinolysis requires constant production of new plasmin (Collen, 1980).

α_2-Antiplasmin inactivates plasmin in a 1:1 stoichiometric reaction. However, the plasma concentration of α_2-antiplasmin is 1 μM whereas that of plasminogen is 1.5–2 μM. Thus even a moderate fall in α_2-antiplasmin levels can impair the regulation of fibrinolysis. α_2-Macroglobulin serves as the backup inhibitor for plasmin; it is thought to function primarily after α_2-antiplasmin has been depleted. However, when plasma α_2-antiplasmin levels are low, as in patients with advanced liver disease or in patients with a rare hereditary deficiency of this inhibitor, initiation of fibrinolysis from whatever cause may precipitate serious bleeding due to an inability to keep the process under control.

Degradation of Fibrin(ogen) By Plasmin

Plasmin cleaves fibrinogen or fibrin in a sequence that gives rise to four degradation products, fragments X, Y, D, and E (Fig. 3.42). Fragment X is formed by cleavage of a long polar carboxy segment of the α chains and cleavage of a small peptide from the NH_2-terminal of the β chains. Fragment Y is formed from a cleavage in one linear segment, which separates the distal portion of that segment and its attached terminal nodule (fragment D) from the remainder of the molecule (fragment Y). A similar cleavage in the remaining linear segment of fragment Y then gives rise to a second fragment D and the residual central portion of the molecule, which is called fragment E. When fibrin has been cross-linked by factor XIIIa prior to its digestion by plasmin, degradation products are formed that contain combinations of the fragments, e.g., a degradation product containing two D fragments and one E fragment (D_2E).

When present in blood in large amounts, fibrin(ogen) degradation products interfere with hemostasis. The big fragments, X and Y, inhibit fibrin polymerization. The D domain of fibrinogen contains the site for binding of fibrinogen to the GP IIb-IIIa platelet receptor (Marguerie et al., 1982). Fragments X, Y, and D, by competing with intact fibrinogen for the receptor, inhibit platelet aggregation in vitro and, presumably, formation of platelet hemostatic plugs in vivo. Fibrin(ogen) degradation products can be readily measured immunologically as nonclottable fibrinogen-related antigen persisting in serum. Degradation products are frequently measured in clinical disorders in which one suspects that fibrin has been laid down in blood vessels.

Plasmin readily dissolves a freshly formed thrombus; however, after several days plasmin will no longer dissolve the thrombus. How fibrin, as it ages, acquires resistance to plasmin is not yet known. Factor XIIIa cross-links α_2-antiplasmin to fibrin; increased amounts of α_2-antiplasmin could become cross-linked to fibrin over time and therefore cause the resistance (Sakata and Aoki, 1980).

HEMOSTASIS IN DISEASE

Screening Tests of Hemostasis

Hemostatic function is examined in clinical medicine to evaluate the possibility of abnormal bleeding and to obtain information for the diagnosis and management of diseases whose manifestations include disturbed hemostasis. Frequently, two tests are carried out to screen for the adequacy of formation of hemostatic plugs—the platelet count and the bleeding time; and two tests are carried out to screen for adequacy of blood coagulation—the prothrombin time and the activated partial thromboplastin time (APTT).

If a patient's platelet count is very low, say 20,000 instead of 200,000/μl, one does not need to perform the bleeding time. One knows that insufficient platelets are available to form good hemostatic plugs. However, patients may have a normal count, yet their platelets may be unable to form hemostatic plugs normally. A prolonged bleeding time test will identify such a patient. In doing a bleeding time test, one inflates a blood pressure cuff on the patient's arm to 40 mmHg pressure and, using a template, makes an incision 1 cm long and 1 mm deep on the volar surface of the forearm. Bleeding from the incision stops when masses of platelets strong enough to hold against 40 mmHg back pres-

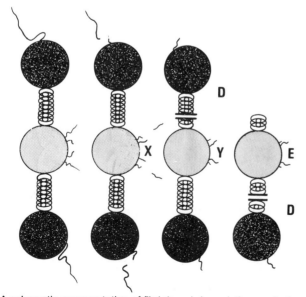

Figure 3.42. A schematic representation of fibrin(ogen) degradation products.

sure plug the ends of the tiny vessels that are severed.

In the prothrombin time test, recalcified plasma is clotted with a high concentration of tissue factor. (The test's name persists from an early era when the clotting time of plasma containing tissue factor was thought primarily to reflect the activity of prothrombin.) The tissue factor-factor VII complex activates both factor IX and factor X (Fig. 3.35). However, the high concentration of tissue factor used in the prothrombin time test leads to formation of an unphysiologically powerful tissue factor-factor VII complex, which directly activates factor X so rapidly that the test becomes independent of activation of factor IX. Therefore, the prothrombin time reflects the combined activity of the following clotting factors: factor VII, factor X, factor V, prothrombin, and fibrinogen.

In the APTT, plasma is recalcified in the presence of a reagent supplying a material like Celite powder that initiates contact activation and a suspension of phospholipids that substitute for platelet procoagulant phospholipids. (Such phospholipid preparations were once called partial thromboplastins, hence, the test's name.) The APTT reflects the combined activities of the following factors: the contact activation factors, factor XII, high molecular weight kininogen, prekallikrein, and factor XI; the factors missing in the hemophilias, factor VIII and factor IX; and factor X, factor V, prothrombin, and fibrinogen. Since tissue factor is not used in the test, the clotting time is independent of factor VII.

Usually, the platelet count, bleeding time, prothrombin time, and APTT will provide an adequate screen of hemostatic function. In certain circumstances (Rapaport, 1983) other tests are added, e.g., a second bleeding time after giving a patient aspirin, or tests of the stability of plasma clots incubated in saline and in a urea solution to look for evidence of excessive fibrinolysis and factor XIII deficiency. When screening tests are abnormal, the pattern of abnormality plus the patient's other clinical findings provide the information needed to proceed with specific tests of platelet function or specific coagulation factor assays.

Disorders Affecting Formation of Hemostatic Plugs

THROMBOCYTOPENIA

Hereditary thrombocytopenia is extremely rare; in contrast, acquired thrombocytopenia is the commonest cause of impaired hemostatic function. It may result from failure of platelet production or from accelerated platelet destruction. In addition, moderate thrombocytopenia, in the 60,000–120,000/μl range, may result from pooling of large numbers of platelets in an enlarged spleen.

Thrombocytopenia from decreased production usually reflects serious bone marrow disease, e.g., leukemia or marrow aplasia, or an effect upon marrow function of drugs used to suppress proliferation of malignant cells in cancer chemotherapy. Thrombocytopenia from accelerated peripheral destruction usually stems from an abnormal immune reaction. Patients may form autoantibodies to platelet antigens. Platelets coated with such antibodies and, frequently, complement are rapidly removed by mononuclear phagocytes, particularly in the spleen. This probably causes most instances of thrombocytopenia developing in adults without other evidence of illness (idiopathic thrombocytopenic purpura, or ITP). Antibodies may also bind to platelets not by combining with platelet antigens but as immune complexes that form in plasma and then bind to the Fc receptor of platelets. This may also lead to phagocytosis of platelets by mononuclear phagocytes and seems to be the mechanism of most instances of childhood ITP. Drugs may stimulate the formation of antibodies that form immune complexes in the presence of the drug, which then bind to platelets and cause thrombocytopenia. Quinidine is a commonly used drug producing this type of immune thrombocytopenia. Patients with severe Gram-negative bacterial infections frequently become thrombocytopenic. This apparently results from several processes: deposition of platelets on damaged vascular endothelium, activation by endotoxin of complement on the platelet surface, binding of immune complexes to platelets, and consumption of platelets in intravascular coagulation.

DISORDERS WITH NORMAL PLATELET COUNTS BUT IMPAIRED FORMATION OF HEMOSTATIC PLUGS

Hereditary disorders affecting the formation of hemostatic plugs include von Willebrand's disease, in which the abnormality is in the plasma, and a group of intrinsic platelet disorders. Von Willebrand's disease, an autosomal dominant disorder, is the most common hereditary hemostatic disorder. In most patients it stems from an inability to make normal amounts of von Willebrand factor; in occasional patients it stems from synthesis of abnormal von Willebrand molecules. Patients have a prolonged bleeding time, usually a reduced plasma level of von Willebrand antigen, and frequently a moderately reduced level of factor VIII activity.

The last reflects the circulation of factor VIII in plasma in a complex with von Willebrand protein. Most patients have about 30–50% of normal levels of von Willebrand factor and a mild bleeding disorder. Rare patients have very low levels of von Willebrand factor and serious bleeding.

Hereditary intrinsic platelet disorders can affect platelet adhesion to collagen or platelet activation and adherence to each other. As mentioned earlier, Bernard-Soulier disease is a disorder in which platelets lack GPIb on their plasma surface membrane and cannot adhere to collagen normally despite a normal plasma level of von Willebrand factor. A number of disorders have been identified which affect the stimulus-response arm of platelet activation. These include disorders of thromboxane A₂ synthesis, disorders in which platelets do not respond normally to exogenous thromboxane A₂, and disorders in which the constituents of platelet secretory granules, such as ADP, are decreased (Weiss, 1982). These dysfunctional states are associated with mild to moderate bleeding. In contrast, a disorder of the effector arm of platelet activation, Glanzmann's thrombasthenia, in which platelets fail to aggregate because they lack the GPIIb-IIIa receptor, may be associated with life-threatening bleeding. Patterns of platelet aggregation obtained on adding collagen, thrombin, epinephrine, sodium arachidonate, and ADP to a patient's platelet-rich plasma help distinguish between the different types of hereditary platelet functional disorders.

Platelet function may be impaired in a number of acquired conditions. In myeloproliferative disorders (Chapter 21), abnormal megakaryocytes may make defective platelets. Whereas autoantibodies to platelets usually cause thrombocytopenia, occasionally antibody-coated platelets are not phagocytosed but continue to circulate. Present in deceptively normal or only slightly depressed numbers, their function may, nevertheless, be seriously impaired. The monoclonal immunoglobulins that accumulate in plasma in malignant lymphoproliferative disorders (multiple myeloma, Waldenström's macroglobulinemia) may coat platelets and interfere with their function. Penicillin G and semisynthetic penicillin derivatives bind to the platelet surface and produce a dose-dependent inhibition of platelet function. The derivatives carbenicillin and ticarcillin, in particular, are sometimes given to patients in doses high enough to increase the risk of bleeding. Uremic patients frequently have a prolonged bleeding time despite a normal or moderately depressed platelet count. The cause for abnormal platelet function in uremia is not understood but is related to retention of products requiring intensive hemodialysis for removal from blood.

Disorders Affecting Blood Coagulation and Fibrinolysis

HEREDITARY DISORDERS

Hereditary deficiency states have been identified for each one of the known plasma coagulation factors. Indeed, the usual sequence was discovery of the patient first and the coagulation factor second, when the patient's blood coagulation findings could not be accounted for by the then known clotting factors. As mentioned earlier, factor XII deficiency, prekallikrein deficiency, and high molecular weight kininogen deficiency are not diseases. They prolong coagulation in glass tubes but not coagulation in vivo. The other disorders all are associated with abnormal bleeding. For unknown reasons, bleeding in hemophilia A and B is worse than in the other deficiency states. Sudden severe hemorrhage threatens the severe hemophilic patient throughout his life.

Except for the hemophilias and dysfibrinogenemias, the hereditary clotting factor disorders are autosomal recessive disease in which heterozygotes do not bleed abnormally. The dysfibrinogenemias are rare autosomal disorders in which heterozygotes do bleed abnormally. The structural genes for factor VIII and factor IX are located on the X chromosome. Therefore, the hemizygote male who inherits an abnormal gene from his mother is not protected by a normal X chromosome and develops the disease. For this reason, hemophilia is relatively common and is confined to males.

ACQUIRED DISORDERS

Acquired abnormalities of blood coagulation may develop in a variety of clinical circumstances that are summarized in Table 3.11. Many patients with advanced chronic liver disease die of uncontrollable gastrointestinal hemorrhage associated with severely impaired hemostasis.

Therapeutic Agents Affecting Hemostatic Function

Antiplatelet drugs have been evaluated in clinical trials of the prophylaxis of coronary and cerebral thrombotic disease. These include aspirin, the phosphodiesterase inhibitor dipyridamole (Persantin^R), and sulfinpyrazone, a uricosuric agent used in gout that also, by an unknown mechanism, lengthens the shortened platelet survival times found in some patients with arterial disease. The clinical trials have been inconclusive so far and, on

Table 3.11.
Disorders affecting blood coagulation or fibrinolysis

Condition	Defects
Hereditary	
Hemophilia A and B	Failure of synthesis or synthesis of defective factor VIII (Hemo A) or factor IX (Hemo B)
Deficiencies of other specific clotting factors (rare)	As above for factors XI, VII, V, X, XIII, prothrombin, and fibrinogen.
Hereditary excessive fibrinolysis (rare)	Failure of synthesis of α_2-antiplasmin
Acquired	
Disordered vitamin K metabolism	Defective synthesis of factors VII, IX, X, and prothrombin
No dietary intake plus antibiotics	
Impaired absorption due to small bowel disease or obstructive jaundice	
Oral anticoagulants	
Hepatocellular disease	Impaired synthesis of all clotting factors, except factor VIII. Impaired synthesis of α_2-antiplasmin
Disseminated intravascular coagulation	Depletion of plasma fibrinogen, factor VIII, factor V, and other factors, plus secondary fibrinolysis with circulating fibrin degradation products with antihemostatic properties
Complications of pregnancy	
Gram-negative infections	
Acute promyelocytic leukemia	
Metastic adenocarcinoma	
Traumatic brain tissue damage	
Acquired antibodies against clotting factors	Neutralization of coagulant activity of clotting factors (factor VIII antibodies, factor V antibodies, etc.).
Idiopathic (no known underlying disease)	
Complication of generalized autoimmune disorder, e.g., systemic lupus erythematosus	Formation of immune complexes retaining coagulant activity *in vitro* but removed *in vivo* by mononuclear phagocytes (prothrombin antibodies)
After transfusion of plasma products in patients with hereditary clotting factor deficiencies	
Primary fibrinogenolysis (rare)	Impaired synthesis of α_2-antiplasmin and impaired clearance of plasminogen activators in liver disease
Severe liver disease	
Metastatic carcinoma	Abnormal release of plasminogen activators in metastatic carcinoma

the whole, disappointing. Sulfinpyrazone or low dose aspirin may have a role, however, in keeping open vascular access shunts in uremic patients undergoing long-term hemodialysis.

Two drugs impeding blood coagulation, heparin and an oral anticoagulant (usually warfarin), are used to prevent or treat venous thrombosis and pulmonary embolism, mural thrombi within the chambers of the heart, clots on prosthetic heart valves, and arterial occlusion in some patients with progressive symptoms suggesting impending cerebral infarction. Treatment usually begins with heparin, since its anticoagulant effect is immediate. As discussed earlier, heparin acts by markedly enhancing the rate at which antithrombin III neutralizes thrombin, factor Xa, and factor IXa. Heparin must be given parenterally, usually intravenously in treating an established thrombus and subcutaneously for prophylaxis. Its intravascular half-life is about 90 min; therefore, it is often given by continuous intravenous drip. Its anticoagulant effect is commonly monitored by the activated partial thromboplastin time, which is kept at between 1.5 and 2 times normal. In unusual patients, heparin causes thrombocytopenia, so platelet counts are also checked periodically when heparin is given.

Drugs that impair platelet funtion, e.g., aspirin, must be avoided in patients receiving heparin.

Heparin is usually replaced with an oral anticoagulant when therapy is needed for more than a few days. As an oral anticoagulant acts by inhibiting synthesis of normal vitamin K-dependent clotting factors, its anticoagulant effect is delayed until the plasma levels of the normal factors fall. The prothrombin time test is used to monitor oral anticoagulant therapy. Factor VII, a vitamin K-dependent clotting factor influencing the prothrombin time test, has an intravascular half-life of only 5 hours whereas the other procoagulant vitamin K-dependent clotting factors have longer intravascular half-lives, e.g., 1.5 days for factor X. Therefore, the lengthening of the prothrombin time on beginning oral anticoagulant therapy primarily reflects the disappearance of normal plasma factor VII. Isolated deficiency of factor VII does not provide the protection against thrombosis afforded when all of the vitamin K factors are depressed to therapeutic levels. Regardless of the prothrombin time, therefore, the patient given an oral anticoagulant is not fully protected against thrombosis for about 5 days after starting treatment. Thus, heparin should be continued for several days after treatment with an oral anticoagulant is begun. During this period the activated partial thromboplastin time can be used to regulate heparin dosing, the prothrombin time to regulate oral anticoagulant dosing.

Two plasminogen activators, streptokinase and urokinase, are used clinically to dissolve thrombi. Fibrinolytic therapy has been advocated for patients with extensive venous thrombosis of the legs. However, such therapy is not without the risk of causing pieces of partially dissolved clot to break off and lodge as emboli in the lungs. Infusion of streptokinase, either intravenously or locally through a catheter inserted into the involved artery, is currently being evaluated as emergency therapy for acute coronary artery thrombosis. Acute thrombi can be dissolved, but the long-term benefit of this is yet to be evaluated. Systemic fibrinolytic therapy induces profoundly impaired hemostasis, and invasive procedures must be avoided in patients receiving such therapy.

BIBLIOGRAPHY

ABRAMSON, S., R. G. MILLER, AND R. A. PHILLIPS. The identification in adult bone marrow of pluripotent and restricted stem cells of the myeloid and lymphoid systems. *J. Exp. Med.* 145: 1567–1579, 1977.

ADELSTEIN, R. S., AND T. P. POLLARD. Platelet contractile proteins. In: *Progress in Hemostasis and Thrombosis*, edited by T. H. Spaet. New York: Grune & Stratton, 1978, vol. IV, p. 37–58.

AOKI, N., H. SAITO, T. KAMIYA, K. KOIE, Y. SAKATA, AND M. KOBAKURA. Congenital deficiency of α2-plasmin inhibitor associated with severe hemorrhagic tendency. *J. Clin. Invest.* 63: 877–884, 1979.

ARNONE, A. Xray diffraction study of binding of 2,3 DPG to human deoxyhemoglobin. *Nature* 237: 146–149, 1972.

ASHWELL, G., AND C. J. STEER. Hepatic recognition and catabolism of serum glycoproteins. *JAMA* 246: 2358–2364, 1981.

BABIOR, B. M. Oxygen-dependent microbial killing by phagocytes. *N. Engl. J. Med.* 298: 659–668, 721–725, 1978.

BAGBY, G. C., JR., V. D. RIGAS, R. M. BENNETT, A. A. VANDENBARK, AND S. G. HARINDER. Interaction of lactoferrin, monocytes, and T lymphocyte subsets in the regulation of steady-state granulopoiesis in vitro. *J. Clin. Invest.* 68: 56–63, 1981.

BARCROFT, J. *The Respiratory Function of the Blood. Part II. Haemoglobin.* London: Cambridge University Press, 1928.

BASS, D. A. Behavior of eosinophil leukocytes in acute inflammation. I. Lack of dependence on adrenal function. *J. Clin. Invest.* 55: 1229–1236, 1975.

BECKER, C. G., AND P. C. HARPEL. α_2-Macroglobulin on human vascular endothelium. *J. Exp. Med.* 144: 1–9, 1976.

BENACERRAF, B. Role of MHC gene products in immune regulation. *Science* 212: 1229–1238, 1981.

BENESCH, R., AND R. E. BENESCH. The effect of organic phosphates from the human erythrocyte on the allosteric properties of hemoglobin. *Biochem. Biophys. Res. Comm.* 26: 162–167, 1967.

BENESCH, R., AND R. E. BENESCH. Intracellular organic phosphates as regulators of oxygen release by hemoglobin. *Nature* 221: 618–622, 1969.

BEUTLER, E. Glucose-6-phosphate dehydrogenase deficiency. In: *The Metabolic Basis of Inherited Disease*, 5th ed., edited by J. B. Stanbury, J. B. Wyngaarden, D. S. Fredrickson, J. L. Goldstein and M. S. Brown. New York: McGraw Hill, 1983.

BEVILACQUIA, M.P., D. AMRANI, M. W. MOSESSON, AND C. BIANCO. Receptors for cold-insoluble globulin (plasma fibronectin) on human monocytes. *J. Exp. Med.* 153: 42–60, 1981.

BOGGS, D. R., S. S. BOGGS, D. F. SAXE, L. A. GRESS, AND D. R. CANFIELD. Hematopoietic stem cells with high proliferative potential. Assay of their concentration in marrow by the frequency and duration of cure of W/W^v mice. *J. Clin. Invest.* 70: 369–378, 1982.

BOTHWELL, T. H., R. W. CHARLTON, J. D. COOK, AND C. A. FINCH. *Iron Metabolism in Man.* Oxford:Blackwell, 1979.

BOTNICK, L. E., E. C. HANNON, AND S. HELLMAN. Nature of the hemopoietic stem cell compartment and its proliferative potential. *Blood Cells* 5: 195–210, 1979.

BOXER, L. A., R. A. HAAK, H. H. YANG, J. B. WOLACK, J. A. WHITCOMB, C. J. BUTTERICK, AND R. L. BAEHNER. Membrane-bound lactoferrin alters the surface properties of polymorphonuclear leukocytes. *J. Clin. Invest.* 70: 1049–1057, 1982.

BROEKMAN, M. J., J. W. WARD, AND A. J. MARCUS. Phospholipid metabolism in stimulated human platelets. Changes in phosphatidylinositol, phosphatidic acid and lyso-phospholipids. *J. Clin. Invest.* 66: 275–283, 1980.

BROXMEYER, H. A., J. BOGNACKI, P. RALPH, M. H. DORNER, L. LU, AND R. E. CASTRO-MALASPINA. Monocyte-macrophage-derived acidic isoferritins: normal feedback regulators of granulocyte-macrophage progenitor cells in vitro. *Blood* 60: 595–607, 1982.

BROXMEYER, H. E. Relationship of cell-cycle expression of Ia-like antigenic determinants on normal and leukemia human granulocyte-macrophage progenitor cells to regulation in vitro by acidic isoferritins. *J. Clin. Invest.* 69: 632–642, 1982.

BUNN, H. F., AND R. W. BRIEHL. The interaction of 2,3 diphosphoglycerate with various human hemoglobins. *J. Clin. Invest.* 49: 1088–1095, 1970.

BUNN, H. F., B. J. RANSIL, AND A. CHAO. The interaction between erythrocyte organic phosphate, magnesium and hemoglobin. *J. Biol. Chem.* 246: 5273–5279, 1971.

BUNN, H. F., B. G. FORGET, AND H. M. RANNEY. *Human Hemoglobins.* Philadelphia: Saunders, 1977.

BURGESS, A. W., AND D. METCALF. Review: the nature and action of granulocyte-macrophage colony stimulating factors. *Blood* 56: 947–958, 1980.

BURSTEIN, S. A., J. W. ADAMSON, S. K. ERG, AND L. A. HARKER. Megakaryocytopoiesis in the mouse: response to varying platelet demand. *J. Cell. Immunol.* 109: 333–341, 1981.

BUTTERWORTH, A. E., AND J. R. DAVID. Eosinophil function. *N. Engl. J. Med.* 304: 154–156, 1981.

CARMEL, R. A new case of deficiency of the R binder for cobalamin, with observations on minor cobalamin-binding proteins in serum and saliva. *Blood* 59: 152–156, 1982.

CARP, H. Mitochondrial N-formylmethionyl proteins as chemoattractants for neutrophils. *J. Exp. Med.* 155: 264–273, 1982.

CARROLL, R. C., AND J. M. GERRARD. Phosphorylation of platelet actin-binding protein during platelet activation. *Blood* 59: 466–471, 1982.

CHANARIN, I., R. DEACON, J. PERRY, AND M. LUMB. How vitamin B12 acts. *Brit. J. Haematol.* 46: 487–491, 1981.

CHESNEY, C. MCL., D. D. PIFER, L. W. BYERS, AND E. E. MUIRHEAD. Effect of platelet-activating factor (PAF) on human platelets. *Blood* 59: 582–585, 1982.

CHEUNG, W. Y. Calmodulin plays a pivotal role in cellular regulation. *Science* 207: 19–27, 1980.

COCHRANE, C. G. Plasma proteins and inflammatory disease. *Pharmacol. Rev.* 34: 39–42, 1982.

COLLEN, D. On the regulation and control of fibrinolysis. *Thromb. Haemostasis* 43: 77–89, 1980.

COLLEN, D., J. SCHETZ, F. DE COCK, E. HOLMER, AND M. VERSTRAETE. Metabolism of antithrombin III (heparin cofactor) in man: effects of venous thrombosis and of heparin administration. *Eur. J. Clin. Invest.* 7: 27–35, 1977.

COLLEN, D., AND B. WIMAN. Turnover of antiplasmin, the fast-acting plasmin inhibitor of plasma. *Blood* 53: 313–324, 1979.

COOK, J. D. Clinical evaluation of iron deficiency. *Semin. Hematol.* 19: 6–18, 1982.

COOK, J. D., C. A. FINCH, AND N. J. SMITH. Evaluation of the iron status of a population. *Blood* 48: 449–455, 1976.

COOPER, R. A. Abnormalities of cell-membrane fluidity in the pathogenesis of disease. *N. Engl. J. Med.* 297: 371–377, 1977.

CROFTON, R. W., M. M. C. DIESSELHOFF-DEN DULK, AND R. VAN FURTH. The origin, kinetics and characteristics of the Kupffer cells in the normal steady state. *J. Exp. Med.* 148: 1–17, 1978.

CRONKITE, E. P. Kinetics of granulocytopoiesis. *Clin. Haematol.* 8: 351–370, 1979.

DAHLEN, S. E., J. BJÖRK, P. HEDQVIST, K. E. ARFORS, S. HAMMARSTRÖM, J. A. LINDGREN, AND B. SAMUELSSON. Leukotrienes promote plasma leakage and leukocyte adhesion in postcapillary venules: *in vivo* effects with relevance to the acute inflammatory response. *Proc. Natl. Acad. Sci. USA* 78: 3887–3891, 1981.

DAVIS, M. M., S. K. KIN, AND L. HOOD. Immunoglobulin class switching: developmentally regulated DNA rearrangements during differentiation. *Cell* 22: 1–2, 1980.

DEUEL, T. F., R. M. SENIOR, J. S. HUANG, AND G. L. GRIFFIN. Chemotaxis of monocytes and neutrophils to platelet-derived growth factor. *J. Clin. Invest.* 69: 1046–1049, 1982.

DOOLITTLE, R. F. Fibrinogen and fibrin, *Sci. Am.* 245: 126–135, 1981.

DRUTZ, D. J., AND J. MILLS. Immunity and infection. In: *Basic and Clinical Immunology*, edited by D. P. Stites, J. D. Stobo, H. H. Fudenberg, AND J. V. Wells. Los Altos, CA Lange Medical Publications, 1982, p. 209–232.

FANTONE, J. C., AND P. A. WARD. Role of oxygen-derived free radicals and metabolites in leukocyte-dependent inflammatory reactions. *Am. J. Pathol.* 107: 397–418, 1982.

FARRAR, J. J., AND M. L. HILFIKER. Antigen-nonspecific helper factors in the antibody response. *Fed. Proc.* 41: 263–267, 1982.

FEINSTEIN, M. B. The role of calmodulin in hemostasis. In: *Progress in Hemostasis and Thrombosis*, edited by T. H. Spaet. New York: Grune & Stratton, 1982, vol. VI, p. 25–61.

FINCH, C. A., AND C. LENFANT. Oxygen transport in man. *N. Engl. J. Med.* 286: 407–415, 1973.

FINCH, C. A. Review: erythropoiesis, erythropoietin and iron. *Blood* 60: 1241–1246, 1982.

FUJII, K., T. NAGASAKI, AND F. M. HUENNEKENS. Accumulation of 50-methyltetrahydrofolate in cobalamin-deficient L 1210 mouse leukemia cells. *J. Biol. Chem.* 257: 2144–2146, 1982.

GABIG, T. G., AND B. M. BABIOR. The killing of pathogens by phagocytes. *Ann. Rev. Med.* 32: 313–326, 1981.

GADEK, J. E., G. A. FELLS, R. L. ZIMMERMAN, S. I. RENNARD, AND R. G. CRYSTAL. Antielastases of the human alveolar structures. Implications for the protease-antiprotease theory of emphysema. *J. Clin. Invest.* 68: 889–898, 1982.

GEHA, R. S. Current concepts in immunology. Regulation of the immune response by idiotypic-anti-idiotypic interactions. *N. Engl. J. Med.* 305: 25–28, 1981.

GEWURZ, H. Biology of C-reactive protein and the acute phase response. *Hosp. Prac.* 17: 67–81, 1982.

GOETZL, E. J. Oxygenation products of arachidonic acid as mediators of hypersensitivity and inflammation. *Med. Clin. North Am.* 65: 809–828, 1981.

GOLDSTEIN, I. M., H. B. KAPLAN, H. S. EDELSON, AND G. WEISSMANN. Ceruloplasmin: an acute phase reactant that scavenges oxygen-derived free radicals. *Ann. N.Y. Acad. Sci.* 389: 368–379, 1982.

GOODMAN, S. R., K. A. SHIFFER, L. A. CASORIA, AND M. E. EYSTER. Identification of the molecular defect in the erythrocyte membrane skeleton of some kindreds with hereditary spherocytosis. *Blood* 60: 772–784, 1982.

GOODNIGHT, S. H., G. KENOYER, S. I. RAPAPORT, M. J. PATCH, J. A. LEE, AND T. KURZE. Defibrination after brain-tissue destruction. *N. Engl. J. Med.* 290: 1043–1047, 1974.

GOREVIC, P. D., A. B. CLEVELAND, AND E. C. FRANKLIN. The biologic significance of amyloid. *Ann. N.Y. Acad. Sci.* 389: 380–394, 1982.

GRAVES, C. B., T. W. MUNNS, A. K. WILLINGHAM, AND A. W. STRAUSS. Rat factor X is synthesized as a single chain precursor inducible by prothrombin fragments. *J. Biol. Chem.* 257: 13108–13113, 1982.

GRIFFIN, J. A., B. EVATT, T. S. ZIMMERMAN, A. J. KLEISS, AND C. WIDEMAN. Deficiency of protein C in congenital thrombotic disease. *J. Clin. Invest.* 68: 1370–1373, 1981.

HAMMERLING, U. Differentiation of B lymphocytes in lineage development and in the immune response. Considerations derived from differentiation events induced in vitro. *Prog. Allergy* 28: 40–65, 1981.

HARKER, L. A. Control of platelet production. *Ann. Rev. Med.* 25: 383–400, 1974.

HARPEL, P. C., AND R. D. ROSENBERG. α_2-macroglobulin and antithrombin-heparin cofactor: modulators of hemostatic and inflammatory reactions. In: *Progress in Hemostasis and Thrombosis*, edited by T. H. Spaet. New York: Grune & Stratton, 1976, vol. III, p. 145–189.

HERBERMAN, R. B. Natural killer cells. *Hosp. Prac.* 17: 93–103, 1982.

HERBERT, V., AND R. ZALUSKY. Interrelation of vitamin B_{12} and folic acid metabolism: folic acid clearance studies. *J. Clin. Invest.* 41: 1263–1276, 1962.

HIRATA, F., AND J. AXELROD. Phospholipid methylation and biological signal transmission. *Science* 209: 1082–1090, 1980.

HOFFMAN, R., E. MAZUR, E. BRUNO, AND V. FLOYD. Assay of an activity in the serum of patients with disorders of thrombopoiesis that stimulates formation of megakaryocytic colonies. *N. Engl. J. Med.* 305: 533–538, 1981.

HOLMSEN, H. Biochemistry of the platelet: energy metabolism. In: *Hemostasis and Thrombosis: Basic Principles and Clinical Practice*, edited by R. W. Colman, J. Hirsh, V. J. Marder, E. W. Salzman. Philadelphia: J. B. Lippincott, 1982, p. 431–443.

HULTIN, M. B. Role of human factor VIII in factor X activation. *J. Clin. Invest.* 69: 950–958, 1982.

JACKSON, C. M., AND Y. NEMERSON. Blood coagulation. *Ann. Rev. Biochem.* 49: 765–811, 1980.

JACOB, H. S., P. R. CRADDOCK, D. E. HAMMERSCHMIDT, AND C. F. MOLDOW. Complement-induced granulocyte aggregation. An unsuspected mechanism of disease. *N. Engl. J. Med.* 302: 789–794, 1980.

JANOFF, A., AND H. CARP. Proteases, antiproteases, and oxidants: pathways of tissue injury during inflammation. In: *Current Topics in Inflammation and Infection*, edited by G. Majino, R. S. Cotran, and N. Kaufman. Baltimore: Williams & Wilkins, 1982, p. 62–82.

JERNE, N. The immune system: a web of V-domains. *Harvey Lecture Series* 70: 93–110, 1974–75.

KAPLAN, A. P., M. SILVERBERG, J. T. DUNN, AND B. GHEBREHIWET. Interaction of the clotting, kinin-forming, complement and fibrinolytic pathways in inflammation. *Ann. N.Y. Acad. Sci.* 389: 25–38, 1982.

KAZATCHKINE, M. D., AND U. E. NYDEGGER. The human alternative complement pathway: biology and immunopathology of activation and regulation. *Prog. Allergy* 30: 193–234, 1982.

KINO, K., H. TSUNOO, Y. HIGA, M. TAKAMI, H. HAMAGUCHI, AND H. NAKAJIMA. Hemoglobin-haptoglobin receptor in rat liver plasma membranes. *J. Biol. Chem.* 255: 9616–9620, 1980.

KITAGAWA, S., F. TAKAKU, AND S. SAKAMOTO. Evidence that proteases are involved in superoxide production by human polymorphonuclear leukocytes and monocytes. *J. Clin. Invest.* 65: 74–81, 1980.

KITAMURA, Y., S. GO, AND K. HATANAKA. Decrease of mast cells in W/W^v mice and their increase by bone marrow transplantation. *Blood* 52: 447–452, 1978.

KLEBANOFF, S. J. Oxygen metabolism and the toxic properties of phagocytes. *Ann. Int. Med.* 93: 480–489, 1980.

KLOCKE, R. A. Oxygen transport and 2,3-diphosphoglycerate (DPG). *Chest* 62: 79–85S, 1972.

KORCHAK, H. M., C. N. SERHAN, AND G. WEISSMANN. Activation of the human neutrophil: the roles of lipid remodeling and intracellular calcium. In: *Current Topics in Inflammation and Infection*, edited by G. Majin, R. S. Cotran, and N. Kaufman. Baltimore: Williams & Wilkins, 1982, p. 83–93.

KORSMEYER, S. J., AND T. A. WALDMANN. Immunoglobulins II: gene organization and assembly. In: *Basic and Clinical Immunology*, 4th ed., edited by D. P. Stites, J. D. Stobo, H. H. Fudenberg, and J. V. Wells. Los Altos, CA Lange Publications, 1982, p. 43–51.

KUNKEL, S. L., J. C. FANTONE, III, AND P. A. WARD. Complement mediated inflammatory reactions. *Pathobiol. Ann.* 11: 127–154, 1981.

KUSHNER, I. The phenomenon of the acute phase response. *Ann. N.Y. Acad. Sci.* 389: 39–48, 1982.

LAURELL, C. B., AND J. D. JEPPSSON. Protein inhibitors in plasma. In: *The Plasma Proteins. Structure, Function and Genetic Control*, 2nd ed., edited by F. W. Putnman. New York: Academic Press, 1975, vol. 1, p. 229–264.

LERNER, M. H., AND B. A. LOWY. The formation of adenosine in rabbit liver and its possible role as a direct precursor of erythrocyte adenine nucleotides. *J. Biol. Chem.* 249: 959–966, 1974.

LEVIN, J., AND B. L. EVATT. Humoral control of thrombopoiesis. *Blood Cells* 5: 105–121, 1979.

LIN, A. H., D. R. MORTON, AND R. R. GORMAN. Acetyl

glyceryl ether phosphorylcholine stimulates leukotriene B_4 synthesis in human polymorphonuclear leukocytes. *J. Clin. Invest.* 70: 1058–1065, 1982.

LIND, S. E., AND T. P. STOSSEL. The microfilament network of the platelet. In: *Progress in Hemostasis and Thrombosis*, edited by T. H. Spaet. New York: Grune & Stratton, 1982, vol. VI, p. 63–84.

LIJNEN, H. R., M. HOLYLAERTS, AND D. COLLEN. Isolation and characterization of a human plasma protein with affinity for the lysine binding sites in plasminogen. *J. Biol. Chem.* 255: 10214–10222, 1980.

LIU, S-G., J. PALEK, AND J. T. PRCHAL. Defective spectrin dimer-dimer association in hereditary eleptocytosis. *Proc. Natl. Acad. Sci. USA* 79: 2072–2076, 1982.

LOLLAR, P., AND W. G. OWEN. Clearance of thrombin from circulation in rabbits by high-affinity binding sites on endothelium. Possible role in the inactivation of thrombin by antithrombin III. *J. Clin. Invest.* 66: 1222–1230, 1980.

LOTEM, J., AND L. SACHS. Mechanisms that uncouple growth and differentiation in myeloid leukemia cells: restoration of requirement for normal growth-inducing protein without restoring induction of differentiation-inducing protein. *Proc. Natl. Acad. Sci. USA* 79: 4347–4351, 1982.

MAHMOUD, A. A. F. Eosinophilopoiesis. In: *The Eosinophil in Health and Disease*, edited by A. A. F. Mahmoud and K. F. Austen. New York: Grune & Stratton, 1980, p. 61–75.

MAJERUS, P. W., AND J. P. MILETICH. Relationships between platelets and coagulation factors in hemostasis. *Ann. Rev. Med.* 29: 41–49, 1978.

MALMSTEN, C., M. HAMBERG, J. SVENSSON, AND B. SAMUELSSON. Physiological role of an endoperoxide in human platelets. Hemostatic defect due to platelet cyclo-oxygenase deficiency. *Proc. Natl. Acad. Sci. USA* 72: 1446–1450, 1975.

MARCHESI, V. T. The red cell membrane skeleton: recent progress. *Blood* 61: 1–11, 1983.

MARCUS, A. J. Platelet lipids. In: *Hemostasis and Thrombosis: Basic Principles and Clinical Practice*, edited by R. W. Colman, J. Hirsh, V. J. Marder, and E. W. Salzman. Philadelphia: J. B. Lippincott, 1982, p. 472–485.

MARCUS, A. J., B. B. WEKSLER, E. A. JAFFE, AND M. J. BROEKMAN. Synthesis of prostacyclin from platelet-derived endoperoxides by cultured human endothelial cells. *J. Clin. Invest.* 66: 979–986, 1980.

MARGUERIE, G. A., N. ARDAILLOU, G. CHEREL, AND E. F. PLOW. The binding of fibrinogen to its platelet receptor. *J. Biol. Chem.* 257: 11872–11875, 1982.

MARGUERIE, G. A., T. S. EDGINGTON, AND E. F. PLOW. Interaction of fibrinogen with its platelet receptor as part of a multistep reaction in ADP-induced platelet aggregation. *J. Biol. Chem.* 255: 154–161, 1980.

MARQUARDT, D. I., AND S. I. WASSERMAN. Mast cells in allergic diseases and mastocytosis (Medical Progress). *West. J. Med.* 137: 195–212, 1982.

MEURET, G., AND G. HOFFMANN. Monocyte kinetic studies in normal and disease states. *Brit. J. Haematol.* 24: 275–285, 1973.

MONCADA, S., AND J. R. VANE. Arachidonic acid metabolites and the interactions between platelets and blood-vessel walls. *N. Engl. J. Med.* 300: 1152–1158, 1979.

MOORE, M. A. S. Humoral regulation of granulopoiesis. *Clin. Haematol.* 8: 287–309, 1979.

MOSHER, D. F., R. A. PROCTOR, AND J. E. GROSSMAN. Fibronectin: role in inflammation. *Adv. Inflamm. Res.* 2: 187–207, 1981.

MUHLFELDER, T. W., J. NIEMETZ, D. KREUTZER, D. BEEBE, P. A. WARD, AND S. J. ROSENFELD. C5 chemotactic fragment induces leukocyte production of tissue factor activity. A link between complement and coagulation. *J. Clin. Invest.* 63: 147–150, 1979.

MÜLLER-EBERHARD, H. J., AND R. D. SCHREIBER. Molecular biology and chemistry of the alternative pathway of complement. *Adv. Immunol.* 29: 1–53, 1980.

McDEVITT, H. O. Current concepts in immunology. Regulation of the immune response by the major histocompatibility system. *N. Engl. J. Med.* 303: 1514–1517, 1980.

McDONAGH, J. Structure and function of factor XIII. In: *Hemostasis and Thrombosis: Basic Principles and Clinical Practice*, edited by R. W. Colman, J. Hirsh, V. J. Marder, and E. W. Salzman. Philadelphia. J. B. Lippincott, 1982, p. 164–173.

NEMERSON, Y., AND H. L. NOSSEL. The biology of thrombosis. *Ann. Rev. Med.* 33: 479–488, 1982.

NIENHUIS, A. W., E. J. BENZ, JR., R. PROPPER, L. CORASH, W. F. ANDERSON, W. HENRY, AND J. BORER. Thalassemia major: molecular and clinical aspects. *Ann. Int. Med.* 91: 883–897, 1979.

NURDEN, A. T., AND J. P. CAEN. The different glycoprotein abnormalities in thrombasthenic and Bernard-Soulier patients. *Semin. Hematol.* 16: 234–250, 1979.

OPPENHEIM, J. J., B. M. STADLER, R. P. SIRAGANIAN, M. MAGE, AND B. MATHIESON. Lymphokines: their role in lymphocyte responses. Properties of interleukin 1. *Fed. Proc.* 41: 257–262, 1982.

ØSTERUD, B., U. LINDAHL, AND R. SELJELID. Macrophages produce blood coagulation factors. *FEBS Let.* 120: 41–43, 1980.

OWEN, W. G., AND C. T. ESMON. Functional properties of an endothelial cell cofactor for thrombin-catalyzed activation of protein C. *J. Biol. Chem.* 256: 5532–5535, 1981.

PARRY, M. F., R. K. ROOT, J. A. METCALF, K. K. DELANEY, L. S. KAPLOW, AND W. J. RICHAR. Myeloperoxidase deficiency. Prevalence and clinical significance. *Ann. Int. Med.* 95: 293–301, 1981.

PAULING, L., H. A. ITANO, S. J. SINGER, AND J. C. WELLS. Sickle cell anemia: a molecular disease. *Science* 110: 543–548, 1949.

PELUS, L. M. Association between colony forming units—granulocyte macrophage expression of Ia-like (HLA-DR) antigen and control of granulocyte and macrophage production. A new role for prostaglandin E. *J. Clin. Invest.* 70: 568–578, 1982.

PERUTZ, M. F. Stereochemistry of cooperative effects in hemoglobin. *Nature* 228: 726–739, 1970.

PERUTZ, M. F. Hemoglobin structure and respiratory transport. *Sci. Amer.* 239: 92–125, 1978.

PERUTZ, M. F. Regulation of oxygen affinity of hemoglobin: influence of structure of the globin on the heme iron. *Ann. Rev. Biochem.* 48: 327–386, 1979.

PETERS, T., JR. Serum albumin. In: *The Plasma Proteins. Structure, Function and Genetic Control*, 2nd ed., edited by F. W. Putnman. New York, Academic Press, 1975, vol. 1, p. 133–181.

PHILLIPS, D. R. Platelet membranes and receptor function. In: *Hemostasis and Thrombosis: Basic Principles and Clinical Practice*, edited by R. W. Colman, J. Hirsh, V. J. Marder, and E. W. Salzman. Philadelphia: J. B. Lippincott, 1982, p. 444–458.

PHILLIPS, D. R., AND P. P. AGIN. Platelet plasma membrane glycoproteins. Evidence for the presence of nonequivalent disulfide bonds using nonreduced-reduced two-dimensional gel electrophoresis. *J. Biol. Chem.* 252: 2121–2126, 1977.

PINCKARD, R. N. The "new" chemical mediators of inflammation. In: *Current Topics in Inflammation and Infection*, edited by G. Majino, R. S. Cotran, and N. Kaufman. Baltimore: Williams & Wilkins, 1982, p. 38–53.

QUESENBERRY, P., AND L. LEVITT. Hematopoietic stem cells. *N. Engl. J. Med.* 301: in 3 parts, 755–760, 819–823, 868–872, 1979.

RAPAPORT, S. I. The activation of factor IX by the tissue factor pathway. In: *Hemophilia and Hemostasis*, edited by D. Menache, D. Macn. Surgenor, and H. D. Anderson, New York: Alan R. Liss, 1981, p. 56–76.

RAPAPORT, S. I. Preoperative hemostatic evaluation: which tests if any. *Blood* 61: 229–231, 1983.

REID, K. B. M., AND R. R. PORTER. The proteolytic activation systems of complement. *Ann. Rev. Biochem.* 50: 433–464, 1981.

RIMON, A., S. SCHIFFMAN, D. I. FEINSTEIN, AND S. I. RAPAPORT. Factor XI activity and factor XI antigen in homozygous and heterozygous factor XI deficiency. *Blood* 48: 165–174, 1976.

RITCHIE, R. F. Specific proteins. In: *Todd-Sanford-Davidsohn. Clinical Diagnosis and Management by Laboratory Methods*, 16th ed., edited by J. B. Henry. Philadelphia: W. B. Saunders, 1979, vol. I, p. 228–258.

ROBBINS, K. C. The plasminogen-plasmin enzyme system. In: *Hemostasis and Thrombosis: Basic Principles and Clinical Practice*, edited by R. W. Colman, J. Hirsh, V. J. Marder, and E. W. Salzman. Philadelphia: J. B. Lippincott, 1982, p. 623–639.

ROCKLIN, R. E., K. BENDTZEN, AND D. GREINDER. Mediators of immunity: lymphokines and monokines. *Adv. Immunol.* 28: 55–136, 1980.

RODER, J. C., AND H. F. PROSS. The biology of the human natural killer cell. *J. Clin. Immunol.* 2: 249–262, 1982.

ROOT, R. K., AND M. S. COHEN. The microbicidal mechanisms of human neutrophils and eosinophils. *Rev. Infect. Dis.* 3: 565–598, 1981.

ROSENTHAL, A. S. Regulation of the immune response—role of the macrophage. *N. Engl. J. Med.* 303: 1153–1156, 1980.

ROTHSCHILD, M. A., M. ORATZ, AND S. S. SCHREIBER. Albumin synthesis. In: *Albumin Structure, Function and Uses*, edited by V. M. Rosenoer, M. Oratz, and M. A. Rothschild. New York: Permagon Press, 1977, p. 227–253.

SAMUELSSON, B. Leukotrienes: mediators of allergic reactions and inflammation. *Int. Arch. Allergy Appl. Immun.* 66(Suppl. 1): 98–106, 1981.

SCHRIER, S. L. Editorial review. Human erythrocyte membrane enzymes: current status and clinical correlates. *Blood* 50: 227–237, 1977.

SHARON, N., AND H. LIS. Glycoproteins: research booming on long-ignored, ubiquitous compounds. *Chem. Eng. News* 59: 21–44, 1981.

SHATTIL, S. J., AND R. A. COOPER. Role of membrane lipid composition, organization, and fluidity in human platelet function. In: *Progress in Hemostasis and Thrombosis*, edited by T. H. Spaet. New York: Grune & Stratton, 1978, vol. IV, p. 59–86.

SNYDERMAN, R., AND E. J. GOETZL. Molecular and cellular mechanisms of leukocyte chemotaxis. *Science* 213: 830–837, 1981.

SPAET, T. H., H. I. HOROWITZ, D. ZUCKER-FRANKLIN, J. CINTRON, AND J. J. BIEZENSKI. Reticuloendothelial clearance of blood thromboplastin by rats. *Blood* 17: 196–205, 1961.

STECK, T. L. The band 3 protein of the human red cell membrane: a review. *J. Supramolec. Struct.* 8: 311–324, 1978.

STERN, D. M., AND H. L. NOSSEL. Acquired antibody to factor XI in a patient with congenital factor XI deficiency. *J. Clin. Invest.* 69: 1270–1276, 1982.

STOBO, J. D. Cellular interactions in the expression and regulation of immunity. In: *Basic and Clinical Immunology*, 4th ed., edited by D. P. Stites, J. D. Stobo, H. H. Fudenberg, and J. V. Wells. Los Altos, CA Lange Publications, 1982, p. 89–96.

STYER, L. *Biochemistry*, 2nd ed. San Francisco: W. H. Freeman, 1982.

SUNDSMO, J. S., AND O. GÖTZE. Human monocyte spreading induced by factor Bb of the alternative pathway of complement activation. A possible role for C5 in monocyte spreading. *J. Exp. Med.* 154: 763–776, 1981.

TAUBER, A. I. Current views of neutrophil dysfunction. An integrated clinical perspective. *Am. J. Med.* 70: 1237–1246, 1981.

TESTA, N. D. Erythroid progenitor cells: their relevance for the study of haematological disease. *Clin. Haematol.* 8: 311–333, 1979.

TILL, J. E., AND E. A. McCULLOCH. Hematopoietic stem cell differentiation. *Biochem. Biophys. Acta* 605: 431–459, 1980.

TOLLEFSEN, D. M., AND M. K. BLANK. Detection of a new heparin-dependent inhibitor of thrombin in human plasma. *J. Clin. Invest.* 68: 589–596, 1981.

UNANUE, E. R. The regulatory role of macrophages in antigen stimulation. Part two: symbiotic relationships between lymphocytes and macrophages. *Adv. Immunol.* 31: 1–136, 1981.

VALENTINE, W. N. The Stratton lecture: hemolytic anemia and inborn errors of metabolism. *Blood* 54: 549–559, 1979.

VALENTINE, W. N., AND D. E. PAGLIA. The primary cause of hemolysis in enzymopathies of anaerobic glycolysis: a viewpoint. *Blood Cells* 6: 819–825, 1980.

VALENTINE, W. N., D. E. PAGLIA, A. P. TARTAGLIA, AND F. GILSANZ. Hereditary hemolytic anemia with increased red cell adenosine deaminase (45- to 70-fold) and decreased adenosine triphosphate. *Science* 195: 783–785, 1977.

VAN ACKER, K. J., H. A. SIMMONDS, C. POTTER, AND J. S. CAMERON. Complete deficiency of adenine phosphoribosyl transferase. Report of a family. *N. Engl. J. Med.* 297: 127–132, 1977.

VAN DE GRIEND, R. J., M. CARRENO, R. VAN DOORN, C. J. M. LEUPERS, A. VEN DEN ENDE, P. WIXERMANS, H. J. G. H. OOSTERHUIS, AND A. ASTALDI. Changes in human T lymphocytes after thymectomy and during senescence. *J. Clin. Immunol.* 2: 289–294, 1982.

VAN FURTH, R., J. A. RAEBURN, AND T. L. VAN ZWET. Characteristics of human mononuclear phagocytes. *Blood* 54: 485–500, 1979.

WALDMANN, T. A. Albumin catabolism. In: *Albumin Structure, Function and Uses*, edited by V. M. Rosenoer, M. Oratz, and M. A. Rothschild. New York: Permagon Press, 1977, p. 255–273.

WEISSMANN, G., C. SERHAN, H. M. KORCHAK, AND J. E. SMOLEN. Neutrophils: release of mediators of inflammation with special reference to rheumatoid arthritis. *Ann. N.Y. Acad. Sci.* 389: 11–24, 1982.

WEISSMANN, G., J. E. SMOLEN, AND H. M. KORCHAK. Release of inflammatory mediators from stimulated neutrophils. *N. Engl. J. Med.* 303: 27–34, 1980.

WEKSLER, M. E. The senescence of the immune system. *Hosp. Prac.* 16: 53–64, 1981.

WEYMAN, J., JR. Linked functions and reciprocal effects in hemoglobin: a second look. *Adv. Prot. Chem.* 19: 223–286, 1964.

WHITE, J. G., AND J. M. GERRARD. Anatomy and structural organization of the platelet. In: *Hemostasis and Thrombosis: Basic Principles and Clinical Practice*, edited by R. W. Colman, J. Hirsh, V. J. Marder, and E. W. Salzman. Philadelphia: J. B. Lippincott, 1982, p. 343–363.

WHITELAW, D. M. Observations on human monocyte kinetics after pulse labeling. *Cell Tissue Kinet.* 5: 311–317, 1972.

WILLIAMS, N., AND R. F. LEVINE. The origin, development and regulation of megakaryocytes. *Brit. J. Haematol.* 52: 173–180, 1982.

ZIMMERMAN, T. S., AND Z. M. RUGGERI. von Willebrand's disease. In: *Progress in Hemostasis and Thrombosis*, edited by T. H. Spaet. New York: Grune & Stratton, 1982, vol. VI, p. 203–236.

ZINKERNAGEL, R. M. Major transplantation antigens in host responses to infection. *Hosp. Prac.* 7: 83–92, 1978.

ZUCKER, M. B. The functioning of blood platelets. *Sci. Am.* 242: 86–103, 1980.

ZUR, M., R. D. RADCLIFFE, J. OBERDICK, AND Y. NEMERSON. The dual role of factor VII in blood coagulation. Initiation and inhibition of a proteolytic system by a zymogen. *J. Biol. Chem.* 257: 5623–5631, 1982.

Body Fluids and Renal Function

CHAPTER 26

Physiology of the Body Fluids

The primary function of the kidneys is the maintenance of the normal volume and composition of the body fluids. Thus, the kidneys are responsible for the excretion of excess water, ions, and waste products as well as for the conservation of solutes important to proper body function. In this chapter, an introduction to the physiology of the body fluids will be presented to provide a background for the study of renal function.

BODY WATER AND ITS SUBDIVISIONS

Water is by far the most abundant component of the body, constituting 45–75% of the body weight. This large variation in water content is primarily a function of variations in the amount of *adipose tissue*. Whereas skeletal muscle is over 75% water, skin is over 70% water, and organs such as the heart, lungs, and kidneys are approximately 80% water, adipose tissue contains less than 10% water (Table 4.1) (Skelton, 1927). Thus, the percentage of body weight that is water will vary inversely with the body's fat content. The total body water (TBW) is about 60% of body weight in normal young adult males and 50% of body weight in normal young adult females, who have a somewhat larger amount of subcutaneous fat. TBW may, however, be a much lower percentage of body weight in obese individuals or a somewhat greater percentage in extremely thin persons. In both sexes, the percentage of body weight that is water decreases with age (Table 4.2), a trend that can be attributed primarily to an increasing percentage of adipose tissue (Hays, 1980). During the *first year* of life, however, the marked decrease in the percentage of body weight that is water occurs principally because the cell mass (which contains at least 20% solids (see Table 4.1)) grows at a faster rate than the extracellular fluid volume (in which the percentage of solids is much less) (Friis-Hansen, 1957).

TBW is distributed into two major fluid compartments: *intracellular fluid* (ICF), which contains approximately 55% of the TBW, and *extracellular fluid* (ECF), which contains approximately 45% of the TBW (Table 4.3) (Edelman and Leibman, 1959). The ECF, in turn, is subdivided into several smaller compartments (Table 4.3). The most important ECF compartments are *plasma*, which contains about 7.5% of the TBW, and *interstitial fluid*, which contains about 20% of the TBW and includes the fluid between cells (i.e., in the interstitium) and in the lymphatics. Also classified as ECF compartments are the fluid crystallized in *bone* and the fluid in *dense connective tissues* such as cartilage, each containing about 7.5% of the TBW. Several minor extracellular fluid compartments, which together contain only about 2.5% of the TBW, include the fluid in the gastrointestinal, biliary, and urinary tracts, the intraocular and cerebrospinal fluids, and the fluid in the serosal spaces, such as the pleural, peritoneal, and pericardial fluids. The term *transcellular fluid* commonly is used to describe these minor ECF compartments because they are separated from the rest of the ECF by a layer of epithelial cells. While the transcellular fluid can be neglected in most experimental and clinical problems of fluid balance, its volume can be increased in certain disease states. For example, considerable fluid may accumulate in the *serosal spaces* in diseases involving the lung (pleural effusion), heart (pericardial effusion), or liver (ascites), or in the *gastrointestinal tract* in intestinal obstruction.

Although Table 4.3 accurately summarizes the distribution of body fluids among the various compartments, in clinical applications a somewhat simplified distribution often is used. For example, it is commonly stated that the ECF contains about one-third of the TBW (or 20% of the body weight of a normal adult male) and the ICF contains about two-thirds of the TBW (or 40% of the body weight of a normal adult male). This approximation is appropriate because the methods used to estimate ECF volume for clinical purposes exclude bone fluid and include variable amounts of dense connective tissue fluid and transcellular fluid (see below).

Table 4.1
Water content of body tissues

Tissue	% Water
Kidneys	83
Heart	79
Lungs	79
Skeletal muscle	76
Brain	75
Skin	72
Liver	68
Skeleton	22
Adipose tissue	10

From Skelton (1927).

Table 4.2
Approximate values for total body water in normal humans as percent of body weight

Age, yr	Male, %	Female, %
Newborn	80	75
1–5	65	65
10–16	60	60
17–39	60	50
40–59	55	47
60+	50	45

From Hays (1980).

Table 4.3
Body fluid compartments

Compartment	Percent of Total Body Water[a]	Percent of Total Body Weight — Normal Adult Male	Percent of Total Body Weight — Normal Adult Female
Intracellular fluid	55	33	27.5
Extracellular fluid	45	27	22.5
Interstitial	20	12	10
Plasma	7.5	4.5	3.75
Bone	7.5	4.5	3.75
Dense connective tissue	7.5	4.5	3.75
Transcellular	2.5	1.5	1.25
Total body water	100	60	50

[a] From Edelman and Leibman (1959).

Measurement of Body Fluid Compartments

The volumes of the various body fluid compartments cannot be measured directly, but estimates useful for experimental and clinical purposes can be obtained with the aid of *dilution methods*. These methods utilize marker substances that distribute in a specific body fluid compartment. If a known quantity of such a marker X is administered and given time to distribute throughout the compartment, then the volume of the compartment can be determined from the concentration of the marker in a sample of fluid from that compartment:

$$\text{Volume of compartment} = \frac{\text{Mass of X administered}}{\text{Concentration of X in compartment}} \quad (1)$$

Radioactive markers or markers whose concentrations can be assayed colorimetrically are generally used. A more precise application of the dilution method includes a correction for the amount of the marker lost (e.g., in urine) during the period of distribution:

$$\text{Volume of compartment} = \frac{\text{Mass of X administered} - \text{mass of X lost}}{\text{Concentration of X in compartment}} \quad (2)$$

Dilution methods can be used to estimate the volumes of TBW, ECF, and plasma (Table 4.4).

TOTAL BODY WATER

The volume of *TBW* is estimated using markers that distribute uniformly throughout *all* body fluids, such as *deuterated water* (D_2O) or *tritiated water* (HTO). The drug *antipyrine* also can be used, although it penetrates certain parts of the body water slowly and therefore tends to *underestimate* the TBW. It should be noted that since plasma is part of the TBW, the concentration of the marker in the TBW compartment (for *Eqs. 1* and *2*) can be obtained from a *plasma sample*.

EXTRACELLULAR FLUID

The estimation of *ECF* volume with dilution methods requires markers that can freely cross the capillary endothelium but that are predominantly excluded from cells. Unfortunately, no ideal marker substance for ECF is available (Edelman and Leibman, 1959). *Radioisotopes of ions* such as Na^+, Cl^-, Br^-, $SO_4^=$, and $S_2O_3^=$ (thiosulfate) can be used, but these enter cells to variable extents and therefore tend to *overestimate* the ECF volume. *Nonmetabolizable saccharides* such as inulin, mannitol, and raffinose also can be used, but these do not readily distribute throughout the entire extracellular compartment and therefore tend to *underestimate* the ECF volume. Furthermore, none of the ECF markers distributes into *bone fluid*, and many are

Table 4.4
Marker substances used to measure volumes of body fluid compartments

TBW	D_2O HTO Antipyrine
ECF	Radioisotopes of selected ions (Na^+, Cl^-, Br^-, $SO_4^=$, $S_2O_3^=$) Nonmetabolizable saccharides (inulin, mannitol, raffinose)
Plasma	^{131}I-albumin Evans blue dye (T-1824)

excluded from, or penetrate very slowly, *dense connective tissue fluid* and *transcellular fluid*. Dilution methods therefore give ECF volumes that range from approximately 27% of the TBW (for markers such as inulin that distribute primarily into plasma and interstitial fluid) to as much as 45% of the TBW (for markers such as Na^+ that, while not distributing into bone water, enter cells to some extent). Because such different volumes are obtained with different markers, the approximation that ECF contains about one-third of TBW is widely used for clinical purposes, as indicated above. Like the TBW, ECF includes plasma, so that the concentration of a marker in ECF can be obtained from a *plasma sample*.

PLASMA

The most commonly used markers for *plasma volume* take advantage of the fact that plasma proteins are distributed almost exclusively in the vascular compartment. Thus, the volume of plasma can be estimated using *radioisotopes of albumin* (e.g., ^{131}I-albumin) or *Evans blue dye* (also called T-1824), which binds tightly to albumin (the small amount of albumin that enters interstitial fluid generally can be neglected in dilution studies). A somewhat different method for estimating plasma volume involves the use of labeled erythrocytes (e.g., ^{51}Cr-erythrocytes) to determine the *blood volume* by dilution. The *plasma volume* can then be calculated as follows:

$$\text{Plasma volume} = \text{blood volume} \ (1 - \text{Hct}) \qquad (3)$$

where Hct is the *hematocrit*. It should be noted, however, that since the hematocrit measured from a peripheral vein slightly *overestimates* the actual hematocrit (primarily because the hematocrit in small capillaries is slightly less than that in larger vessels; see Chapter 7), the labeled erythrocyte method slightly *underestimates* the plasma volume.

ICF AND INTERSTITIAL FLUID

Dilution methods cannot be used to measure the volumes of ICF and interstitial fluid, since no markers that distribute exclusively in these compartments are available. However, once the volumes of the TBW and ECF have been estimated by dilution, the volume of *ICF* can be calculated as follows:

$$\text{ICF} = \text{TBW} - \text{ECF} \qquad (4)$$

Because different values for ECF volume are obtained with different markers, as indicated above, the ICF volumes calculated using *Eq. 4* also will vary. The approximation that ICF contains about

two-thirds of TBW is therefore widely used for clinical purposes (see above).

The volume of *interstitial fluid* can be calculated from the ECF and plasma volumes:

$$\text{Interstitial fluid} = \text{ECF} - \text{plasma} \qquad (5)$$

In general, the volume calculated using *Eq. 5* slightly *overestimates* the true interstitial fluid volume, since ECF as estimated by dilution methods includes not only interstitial fluid and plasma but also variable amounts of dense connective tissue fluid and transcellular fluid (see above).

COMPOSITION OF THE BODY FLUIDS

Units for Measuring Solute Concentrations

SI UNITS

While solute concentrations can be expessed in several different units, the *Système International (SI) Units* will be emphasized throughout this section. In SI units, concentrations are expressed in *moles per liter* (mol/l), where a mole of a solute is defined as the molecular weight (or atomic weight) of the solute in grams. For example, 1 mol of Na^+ (atomic weight 23) contains 23 g of Na^+, 1 mol of $CaCl_2$ (molecular weight 111) contains 111 g of $CaCl_2$, 1 mol of glucose (molecular weight 180) contains 180 g of glucose, and 1 mol of albumin (molecular weight 69,000) contains 69,000 g of albumin. Since the body fluids are relatively dilute, the units *millimoles per liter* (mmol/l) generally are more useful than mol/l, where each mmol represents 10^{-3} mol (1/1000 mol).

In spite of the preference given to SI units in this section, some concepts in renal physiology, particularly those presented in this chapter regarding the body fluids, require a consideration of the number of *electrical charges* per unit volume or the number of *discrete solute particles* per unit volume. For this reason, units that express solute concentrations in terms of *electrical equivalents* or *osmoles* also will be used in certain circumstances.

ELECTRICAL EQUIVALENTS

For electrolytes, concentrations often are expressed in *equivalents per liter* (eq/l). A mole of solute with valence v is equal to v equivalents of solute. For example, 1 mol of Na^+ (valence 1) is equal to 1 eq of Na^+, 1 mol of $CaCl_2$ (total valence 4: $Ca^{++} = 2$, $2Cl^- = 2$) is equal to 4 eq of $CaCl_2$, and 1 mol of albumin (valence 18 at pH 7.4) is equal to 18 eq of albumin. Since the body fluids are relatively dilute, the units meq/l are generally used, where each meq equals 10^{-3} eq. Note that by the above definition of an equivalent, solute concentrations expressed in mmol/l can be converted to meq/l by multiplying by the valence:

$$meq/l = mmol/l \times valence \qquad (6)$$

OSMOLES

As discussed in Chapter 2, the movement of water between different body fluid compartments is related to the concentration of discrete solute particles or *osmotically active particles*, regardless of their size or valence. Thus, for many problems involving the movement of water, solute concentrations are best expressed in *osmoles per liter* (osmol/l) or *osmoles per kilogram water* (osmol/kg H_2O). A mole of solute that dissociates into n discrete particles in solution is equal to n osmoles of solute. For example, 1 mol of Na^+ is equal to 1 osmol of Na^+, 1 mol of $CaCl_2$ is equal to 3 osmol of $CaCl_2$ (since $CaCl_2$ dissociates into three discrete solute particles), 1 mol of glucose is equal to 1 osmol of glucose, and 1 mol of albumin is equal to 1 osmol of albumin. The units of osmoles per liter and osmoles per kilogram water are respectively termed *osmolarity* and *osmolality*. In most physiological applications (and in this section), *osmolality* is the preferred unit since it is independent of the temperature of the solution and the volume occupied by the solutes in the solution. Again, since the body fluids are relatively dilute, the units mosmol/kg H_2O generally are used, where each mosmol equals 10^{-3} osmol. Note that by the above definition of an osmole, solute concentrations expressed in mmol/l can be converted to mosmol/kg H_2O as follows:

$$mosmol/kg\ H_2O = mmol/l \times n \qquad (7)$$

where n is the number of discrete particles into which the solute dissociates. It should be noted that n cannot always be determined by examining the chemical formula of the solute. For example, in the body fluids n for NaCl actually equals 1.75 instead of 2 because NaCl does *not* completely dissociate. It also should be noted that *Eq. 7* will be precise only if the units mmol/l refer to mmol/l of *water* instead of mmol/l of *solution*, since 1 liter of *water* does in fact equal 1 kg of water but 1 liter of *solution* may contain less than 1 kg of water if the solutes in the solution occupy a significant fraction of the solution volume (see below).

An alternative way of expressing the concentration of osmotically active solute particles in a solution is in terms of the *osmotic pressure* of the solution. At body temperature, the osmotic pressure and osmolality are related as follows:

Osmotic pressure (mmHg) = 19.3

$$\times osmolality\ (mosmol/kg\ H_2O) \qquad (8)$$

Table 4.5
Comparison of units for measuring solute concentrations

Solute	g/mol	eq/mol	osmol/mol[a]
Na^+	23	1	1
Cl^-	35.5	1	1
NaCl	58.5	2[b]	2
Ca^{++}	40	2	1
$CaCl_2$	111	4[c]	3
Glucose	180		1
Urea	60		1
Albumin	69,000	18[d]	1

[a] Values for osmoles per mole assume 100% dissociation and the properties of an ideal solution. [b] $Na^+ = 1$, $Cl^- = 1$. [c] $Ca^{++} = 2$, $2Cl^- = 2$. [d] At pH 7.4.

A comparison of the various units for measuring solute concentrations is presented in Table 4.5.

Although the body fluids contain a large variety of solutes, the *electrolytes* will be emphasized in the following discussion of body fluid composition. This is because electrolytes are the predominant solutes in the body fluids and also are primarily responsible for determining the distribution of water between the various compartments.

Extracellular Fluid

PLASMA AND INTERSTITIAL FLUID

These two major compartments of the extracellular fluid have very similar compositions (Fig. 4.1), with Na^+ as the predominant cation and Cl^- and HCO_3^- as the predominant anions. However, an important difference between plasma and interstitial fluid is the larger concentration of *proteins* in plasma. This difference exists because the capillary endothelium is freely permeable to water and to small solutes (the so-called *crystalloids*), such as inorganic ions, glucose, and urea, but has limited permeability to larger solutes (*colloidal particles*), such as large proteins and lipids (see Chapters 2 and 28).

Given the high permeability of the capillary endothelium to small solutes, one might anticipate that the concentrations of such solutes in plasma and interstitial fluid would be *identical*. However, plasma concentrations typically are measured in clinical laboratories as meq/l of *plasma volume* or mmol/l of *plasma volume*. In order to compare plasma and interstitial fluid concentrations, the plasma concentrations must be *corrected* to account for the significant fraction of the plasma volume occupied by proteins and lipids. The small solutes that can diffuse across the capillary endothelium are restricted to the *aqueous phase* of plasma, which normally occupies about 93% of the total plasma volume; the remaining 7% is occupied by plasma

Table 4.6
Approximate concentrations of solutes in body fluids

	Plasma, meq/l	Plasma Water, meq/l H₂O	Interstitial Fluid, meq/l H₂O	Intracellular Fluid (Skeletal Muscle), meq/l H₂O
Cations				
Na^+	142	153	145	10
K^+	4	4.3	4.1	159
Ca^{++}	2.5	2.7	2.4	<1
Mg^{++}	1	1.1	1	40
Total	149.5	161.1	152.5	209
Anions				
Cl^-	104	112	117	3
HCO_3^-	24	25.8	27.1	7
Proteins	14	15.1	<0.1	45
Other	7.5	8.2	8.4	154
Total	149.5	161.1	152.5	209
	mmol/l	mmol/l H₂O	mmol/l H₂O	mmol/l H₂O
Nonelectrolytes				
Glucose	4.7	5.0	5.0	
Urea	5.6	6.0	6.0	6.0

proteins and, to a lesser degree, lipids. Thus, plasma concentrations must be expressed as meq/l of *plasma water* or mmol/l of *plasma water* to be compared to interstitial fluid concentrations. Laboratory values expressed as meq/l or mmol/l of *plasma volume* can be converted to meq/l or mmol/l of *plasma water* by dividing by 0.93 (Table 4.6). Once such conversions are made, the plasma and interstitial fluid concentrations of small *nonelectrolytes* are in fact identical, but small *electrolytes* have slightly different concentrations in plasma and interstitial fluid. These differences can be attributed primarily to the *Gibbs-Donnan effect*.

Gibbs-Donnan Effect

Consider the hypothetical situation of equal sized plasma and interstitial fluid compartments separated by a capillary endothelium barrier (Fig. 4.2). Initially, five Na⁺ ions and five Cl⁻ ions are present in each compartment (Fig. 4.2A), and then a protein molecule that has five negative charges is added to the plasma compartment along with five additional Na⁺ ions to function as counterions to the negative charges on the protein (Fig. 4.2B) (it is assumed that the volume of the protein molecule is small and therefore does not significantly affect the volume of the aqueous phase of the plasma compartment). Since the capillary endothelium is permeable to Na⁺, Na⁺ diffuses from the plasma compartment to the interstitial fluid compartment along its concentration gradient, accompanied by Cl⁻ to maintain electrical neutrality. However, this migra-

tion will increase the concentration of Cl⁻ in the interstitial fluid compartment, thereby generating a concentration gradient that *opposes* the further diffusion of Cl⁻ from the plasma compartment to the interstitial fluid compartment. According to thermodynamic principles that are beyond the scope of this text, at equilibrium the distribution of Na⁺ and Cl⁻ between the plasma and interstitial fluid compartments is given by the *Gibbs-Donnan relationship*:

$$P_{Na} \cdot P_{Cl} = ISF_{Na} \cdot ISF_{Cl} \qquad (9)$$

where P and ISF designate the concentrations of the ions in the plasma and interstitial fluid compartments, respectively. If x Na⁺ ions and x Cl⁻ ions migrate from the plasma compartment to the interstitial fluid compartment before equilibrium is established, then at equilibrium $(10 - x)(5 - x) = (5 + x)(5 + x)$ so that x = 1, i.e., at equilibrium nine Na⁺ ions, four Cl⁻ ions, and the protein molecule with five negative charges will be present in the plasma compartment, while six Na⁺ ions and six Cl⁻ ions will be present in the interstitial fluid compartment (Fig. 4.2C). In other words, because of the anionic protein in the plasma compartment, the equilibrium state between the two compartments is characterized by three important features: (1) the small diffusible ions (Na⁺, Cl⁻) do *not* have equal concentrations in the two compartments, the *cation* (Na⁺) concentration being slightly higher in the protein-containing *plasma compartment* and the *anion* (Cl⁻) concentration being slightly higher in the interstitial fluid compartment; (2) the *total concentration* of equivalents of charge is greater in the protein-containing *plasma compartment*; and (3) in spite of the differences in ion concentrations and in the total concentration of equivalents of charge, *electrical neutrality* is maintained within each compartment, i.e., the total number of cationic charges equals the total number of anionic charges. Note that these same three features characterize the composition differences between plasma and interstitial fluid (Fig. 4.1, Table 4.6), although the situation is more complex than the example illustrated in Figure 4.2 due to the presence of additional small diffusible ions and the small amount of protein in interstitial fluid.

In spite of the differences in composition between plasma and interstitial fluid, these differences are sufficiently small that for most clinical purposes the electrolyte concentrations in plasma (which are readily measured) can be assumed to be representative of ECF in general, and a similar assumption will be made in subsequent chapters. It also should be noted that while the distinction between plasma concentrations expressed in terms of *plasma volume* and those expressed in terms of *plasma water* is important for a precise comparison of plasma and interstitial fluid compositions and for a consideration of the Gibbs-Donnan effect, in general the concentrations as measured by clinical laboratories (i.e., meq/l or mmol/l of *plasma volume*) are used in clinical medicine and will be used in subsequent chapters. However, in certain pathological conditions the failure to express solute concentrations in terms of plasma water can lead to the erroneous interpretation of laboratory data (Albrink et al., 1955). For example, consider the effects of severe hyperproteinemia (e.g., multiple myeloma) or hyperlipidemia (e.g., familial hyperlipidemia) on the plasma Na^+ concentration. In these conditions, water may represent considerably *less* than 93% of the plasma volume. As a result, even if the Na^+ concentration in *plasma water* is normal ($\simeq 153$ meq/l of plasma water), the Na^+ concentration as measured in the clinical laboratory (meq/l of plasma volume) may be reduced since each liter of plasma contains less water.

OTHER EXTRACELLULAR FLUID COMPARTMENTS

The electrolyte composition of the extracellular fluid in *bone* and *dense connective tissue* is believed to be similar to interstitial fluid. The various *transcellular fluids* have unique (and often varying) compositions that are considered in the sections dealing with the corresponding organ systems.

Intracellular Fluid

In contrast to plasma and interstitial fluid, ICF is not a continuous fluid phase, and the precise composition of ICF differs in different tissues. The data presented in Figure 4.1 and Table 4.6 are from mammalian skeletal muscle cells, but several characteristics of ICF in general can be identified (Manery and Hastings, 1939). In contrast to ECF, ICF contains relatively low concentrations of Na^+, Cl^-, and HCO_3^-. Instead, the predominant cation in ICF is K^+, while the predominant anions are *organic phosphates* (e.g., ATP) and *proteins* (Fig. 4.1). These striking composition differences between ICF and ECF can be attributed to several factors. First, the *Na^+-K^+ ATPase* in cell membranes actively transports Na^+ from and K^+ into cells, thereby accounting for the high Na^+ and low K^+ concentrations in ECF and the low Na^+ and high K^+ concentrations in ICF. Secondly, the cell membrane, which separates ICF from interstitial fluid, has very limited permeability to organic phosphates and proteins, resulting in the establishment of a *Gibbs-Donnan equilibrium* across the cell membrane. In analogy to the example presented in Figure 4.2, this Gibbs-Donnan equilibrium accounts for three important features that characterize the composition differences between ICF and intersti-

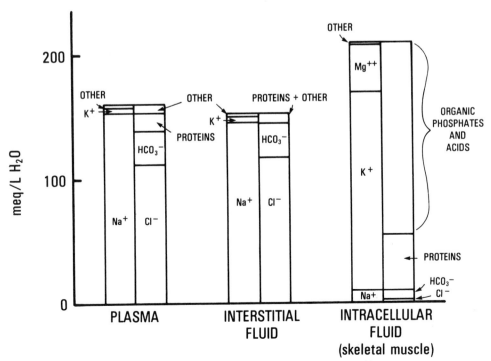

Figure 4.1. Electrolyte composition of the major body fluids.

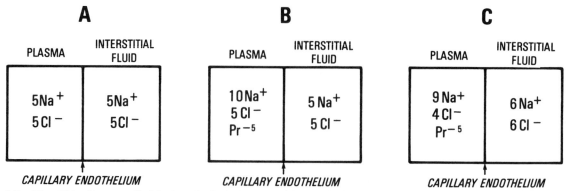

Figure 4.2. The Gibbs-Donnan effect in hypothetical plasma and interstitial fluid compartments. The capillary barrier is relatively impermeable to proteins and other large solutes. *A*, Initial condition of five Na^+ ions and five Cl^- ions in each compartment. *B*, A protein molecule with five negative charges is added to the plasma compartment, with five additional Na^+ to function as counterions. *C*, Equilibrium state is characterized by (1) a higher concentration of diffusible cations (Na^+) in the plasma compartment and a higher concentration of diffusible anions (Cl^-) in the interstitial fluid compartment; (2) a greater total concentration of equivalents of charge in the plasma compartment; and (3) electrical neutrality within each compartment.

tial fluid: (1) the concentration of small diffusible *cations* (Na^+, K^+) is greater in the protein-containing *ICF*, while the concentration of small diffusible *anions* (e.g., Cl^-) is greater in *interstitial fluid* (the distribution of cations is *not* governed solely by the Gibbs-Donnan equilibrium, however, due to the presence of the Na^+-K^+ ATPase); (2) the *total concentration* of equivalents of charge is greater in the protein-containing *ICF*; and (3) in spite of the differences in ion concentrations and in the total concentration of equivalents of charge, *electrical neutrality* is maintained in each compartment. Of course, electrical potential differences exist between the ICF and interstitial fluid (see Chapter 3), but these are created by excesses or deficits of such small numbers of ions (relative to the total number present) that a charge imbalance cannot be detected by measurements of electrolyte concentrations.

It should be noted that the ICF concentration values for a given tissue (such as those presented in Figure 4.1 and Table 4.6 for skeletal muscle cells) must be regarded as *mean* values for that tissue. This is because the various subcellular organelles (mitochondria, endoplasmic reticulum, nucleus, etc.) appear to differ somewhat in composition.

Measurement of Total Electrolyte Content of Body Fluids

Important information about the total content of a given ion in the body fluids can be obtained with dilution methods similar to those used to determine the volume of the body fluids, with a radioisotope of the ion employed as a marker substance. The total content of an ion as determined by such *isotope dilution* methods is termed the *exchangeable*

pool of the ion in question (Edelman and Leibman, 1959).

For Na^+, K^+, and Cl^-, the major ions in the body fluids, the exchangeable pool represents the majority of the total body contents of these ions as determined by analyses of cadavers. For example, *exchangeable Na^+* (Na_e^+) represents over 70% of the total body Na^+, while *exchangeable K^+* (K_e^+) represents over 90% of the total body K^+. Most of the *nonexchangeable* pools of Na^+ and K^+ are found in bone. In contrast, a relatively small percentage of the total body contents of Ca^{++}, Mg^{++}, and phosphate are exchangeable. For example, *exchangeable Ca^{++}* (Ca_e^{++}) represents only about 1% of the total body Ca^{++}, the remainder being part of the structure of bone (see Chapter 35).

OSMOLALITY OF THE BODY FLUIDS

Many clinical problems involving fluid and electrolyte balance require an understanding of the osmolality of the body fluids and how this affects the distribution of water between the extracellular and intracellular body fluid compartments. The contributions of the various solutes to the osmolalities of plasma, interstitial fluid, and ICF are illustrated in Figure 4.3. Note that in spite of the differences in composition, these fluids have essentially *identical* total osmolalities ($\simeq$290 mosmol/kg H_2O). This is because the capillary endothelium and almost all cell membranes are freely permeable to water, allowing the plasma, interstitial fluid, and ICF to be isoosmotic. Hence, the primary determinant of the distribution of water between the extracellular and intracellular fluid compartments is the number of *osmotically active solute particles* in each compartment. The isoosmolality of the body fluids

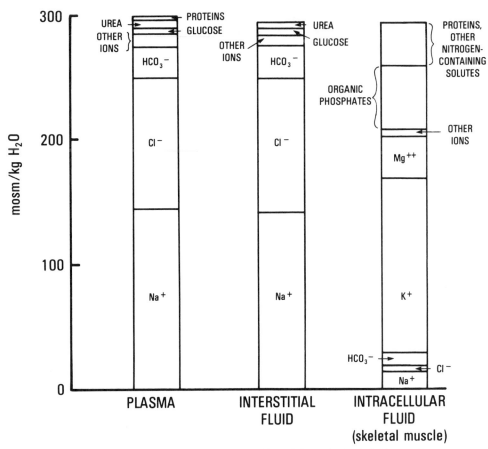

Figure 4.3. Osmotic composition of the major body fluids.

is *not* in conflict with the fact that plasma, interstitial fluid, and ICF contain different total concentrations of equivalents of charge (Fig. 4.1) because osmolality is a measure of the total concentration of discrete solute *particles* in solution. A divalent or polyvalent ion (e.g., Ca^{++}, Mg^{++}, $SO_4^{=}$, albumin) may contribute two or more equivalents of charge, even though it represents a *single* solute particle.

It should be noted that the isoosmolality principle applies primarily to the *major* body fluids. A few minor volumes of fluid have osmolalities that differ significantly from 290 mosmol/kg H_2O, including the peritubular interstitial fluid in the *renal medulla* (which can have an osmolality as high as 1200 mosmol/kg H_2O; see Chapter 31) and certain *transcellular fluids* (the most striking example being *urine*, whose osmolality can vary from 70 mosmol/kg H_2O to 1200 mosmol/kg H_2O; see Chapter 31). It also should be noted that although Figure 4.3 illustrates the osmolalities of the major body fluids to be identical, the osmolality of plasma actually exceeds that of interstitial fluid by approximately 1–2 mosmol/kg H_2O. This small osmolality difference can be attributed both to the osmotic contri-

bution of the plasma proteins and to the larger concentration of diffusible ions in plasma as a result of the Gibbs-Donnan effect (see Fig. 4.2 and Table 4.6). The deviation between the osmolality of plasma and that of interstitial fluid is so small that the major body fluids can be considered *isoosmotic* for most clinical problems involving the distribution of water between ECF and ICF. However, this deviation *does* have an important role in problems involving the distribution of water between plasma and interstitial fluid (Chapters 2 and 28). In such problems, the osmotic contribution of the plasma proteins and excess diffusible ions in plasma typically is expressed as an *osmotic pressure*, termed the *colloid osmotic pressure* or the *oncotic pressure* and designated by the symbol π. In normal plasma, the oncotic pressure is approximately 25 mmHg, of which about two-thirds is caused by the plasma proteins and one-third by the excess diffusible ions. Problems involving the distribution of water between plasma and interstitial fluid *also* require a consideration of the oncotic pressure resulting from the small amount of protein in *interstitial fluid*, normally about 3 mmHg. The relatively small magnitudes of these oncotic pressures is best illustrated by the fact that the *total* osmolality of plasma corresponds to an osmotic pressure of approximately 5600 mmHg (*Eq. 8*). An osmolality difference does *not* exist between interstitial fluid and ICF in spite of the osmotic contribution of the intracellular proteins and the Gibbs-Donnan effect between interstitial fluid and ICF (see above). This is because the Na^+-K^+ ATPase actively transports *three* Na^+ ions from ICF to interstitial fluid for every *two* K^+ ions transported from interstitial fluid to ICF, thereby preventing the osmolality of ICF from exceeding that of interstitial fluid.

Analysis of Plasma Osmolality and Relationship of Plasma Sodium Concentration to Osmolality

An important consequence of the isoosmolality principle is that the osmolality of the major body fluids can be studied by analyzing the osmolality of *plasma*. The plasma osmolality can be estimated as a sum of contributions from *electrolytes* plus contributions from *glucose* and *urea*:

$$P_{osm} \simeq \text{osmolality of electrolytes} + \text{osmolality of glucose}$$
$$+ \text{osmolality of urea} \quad (10)$$

ELECTROLYTE CONTRIBUTION

Since Na^+ and its attendant anions represent the major electrolytes in plasma, it can be assumed that

Osmolality of electrolytes

$$\simeq \text{Osmolality of } (Na^+ + \text{attendant anions}) \quad (11)$$

If it is further assumed that each Na^+ ion is paired with a *univalent* anion, then

Osmolality of electrolytes (mosmol/kg H_2O)

$$\simeq 2P_{Na} \text{ (mmol/l)} \quad (12)$$

The assumptions made deriving *Eq. 12* actually are less serious than might be expected, since they introduce opposing errors of essentially equal magnitude. First of all, the assumption that Na^+ and its attendant anions are the only electrolytes (*Eq. 11*) tends to *underestimate* the true osmotic contribution of the electrolytes, since it excludes other cations such as K^+, Ca^{++}, and Mg^{++}. But the assumption that each Na^+ ion is paired with a univalent anion (*Eq. 12*) tends to *overestimate* the true osmotic contribution of the electrolytes, since (1) Na^+ and its anions are *not* completely dissociated in the body fluids (even NaCl, for example, dissociates into only 1.75 particles, not 2 particles; see above); and (2) some Na^+ is paired with multivalent anions such as $SO_4^=$, $HPO_4^=$, and protein, resulting in a smaller number of solute particles than if all of the anions of Na^+ were univalent. Finally, *Eq. 12* tends to *underestimate* the true electrolyte con-

tribution, since P_{Na} is typically measured by clinical laboratories as meq/l or mmol/l of plasma *volume* instead of meq/l or mmol/l of plasma *water* as would be required to convert P_{Na} *precisely* to mosmoles per kilogram H_2O (*Eq. 7*). As previously noted, concentrations expressed as mmol/l of plasma *volume* are *less* than those expressed as mmol/l of plasma *water* (Table 4.6). Fortuitously, when these various errors are corrected, the osmolality of the electrolyte contribution is in fact closely approximated by $2P_{Na}$ (*Eq. 12*).

GLUCOSE AND UREA CONTRIBUTIONS

The contributions of glucose and urea to P_{osm} can be expressed as

$$\text{Osmolality of glucose (mosmol/kg } H_2O) \simeq P_G \text{ (mmol/l)} \quad (13)$$

and

$$\text{Osmolality of urea (mosmol/kg } H_2O) \simeq P_U \text{ (mmol/l)} \quad (14)$$

where the subscripts G and U refer to glucose and urea, respectively. If the plasma concentrations are expressed in terms of mmol/l of plasma *water* instead of mmol/l of plasma *volume*, Eqs. 13 and 14 become precise. Actually, many clinical laboratories measure plasma glucose and urea concentrations in the units milligrams of glucose per deciliter and milligrams urea nitrogen per deciliter, respectively. Since the molecular weight of glucose is 180, the molecular weight of urea nitrogen is 28 (i.e., 28 g of the 60 g in 1 mol of urea consists of nitrogen), and 1 l = 10 dl, the osmotic contributions of glucose and urea can be expressed as

$$\text{Osmolality of glucose (mosmol/kg } H_2O) \simeq P_G/18 \text{ (mg/dl)} \quad (15)$$

and

$$\text{Osmolality of urea (mosmol/kg } H_2O) \simeq P_{UN}/2.8 \text{ (mg/dl)} \quad (16)$$

where the subscript UN refers to urea nitrogen. When the electrolyte, glucose, and urea contributions are substituted into *Eq. 10*, the following equations for P_{osm} are obtained:

$$P_{osm} \text{ (mosmol/kg } H_2O) \simeq 2P_{Na} \text{ (mmol/l)}$$
$$+ P_G \text{ (mmol/l)} + P_U \text{ (mmol/l)} \quad (17)$$

and

$$P_{osm} \text{ (mosmol/kg } H_2O) \simeq 2P_{Na} \text{ (mmol/l)}$$
$$+ P_G/18 \text{ (mg/dl)} + P_{UN}/2.8 \text{ (mg/dl)} \quad (18)$$

Since the glucose and urea contributions normally account for only about 10 mosmol/kg H_2O, Eqs. 17 and 18 can be further simplified to

$$P_{osm} \text{ (mosmol/kg } H_2O) \simeq 2P_{Na} \text{ (mmol/l)} + 10 \quad (19)$$

i.e., in most cases a good estimate of P_{osm} can be

obtained from P_{Na}, an observation consistent with the fact that Na^+ and its attendant anions are the principal osmotically active solutes in plasma (Fig. 4.3).

It must be kept in mind, however, that while P_{Na} can be a good index of P_{osm}, P_{Na} is *not* a good index of either the *total* amount of osmotically active solute in the body or the TBW (Edelman et al., 1958). This is best illustrated by a set of examples involving changes in body solute or TBW. Figure 4.4*A* illustrates the volumes and osmolality of the ECF and ICF in a normal young adult 60-kg male. TBW is 36 l (60% of body weight) and, using the clinical approximation that ECF contains about one-third of TBW and ICF contains about two-thirds of TBW (see above), ECF and ICF are 12 l and 24 l, respectively. If P_{osm}, and hence the osmolality of the major body fluids, is 290 mosmol/kg H_2O, then the amount of osmotically active solute in the TBW, ECF, and ICF can be calculated as follows:

$$TBW_{osm} = \frac{\text{Total body osmotically active solute}}{\text{TBW}} \quad (20)$$

Total body osmotically active solute

$$= TBW_{osm} \times TBW$$

$$= 290 \text{ mosmol/kg } H_2O \times 36 \text{ l}$$

$$= 10{,}440 \text{ mosmol}$$

$$ECF_{osm} = \frac{\text{Extracellular osmotically active solute}}{\text{ECF}} \quad (21)$$

Extracellular osmotically active solute

$$= ECF_{osm} \times ECF$$

$$= 290 \text{ mosmol/kg } H_2O \times 12 \text{ l}$$

$$= 3480 \text{ mosmol}$$

$$ICF_{osm} = \frac{\text{Intracellular osmotically active solute}}{\text{ICF}} \quad (22)$$

Intracellular osmotically active solute

$$= ICF_{osm} \times ICF$$

$$= 290 \text{ mosmol/kg } H_2O \times 24 \text{ l}$$

$$= 6960 \text{ mosmol}$$

Figure 4.4*B* illustrates the changes in the volumes and osmolality of ECF and ICF that would result from the sudden loss of 360 mosmol of Na^+ and 360 mosmol of anions *without* changing the TBW. Since Na^+ is predominantly an *extracellular* ion, it can be assumed that this 720 mosmol of solute is lost exclusively from the ECF. The sudden loss of solute from the ECF without any change in TBW reduces the osmolality of the ECF. However, this decrease in ECF osmolality can be only *transient*,

since water will move from ECF to ICF until the osmolalities of all major body fluids are equal. After this redistribution of water is complete, the new osmolality of the body fluids can be calculated by substituting the *new* value for total body osmotically active solute into *Eq. 20*

$$TBW_{osm} = \frac{10440 \text{ mosmol} - 720 \text{ mosmol}}{36 \text{ l}}$$

$$= 270 \text{ mosmol/kg } H_2O$$

i.e., the net result is a decrease in the osmolality of *all* major body fluids, even though the solute is lost primarily from the ECF. The new volume of the ECF after the redistribution of water can be calculated by substituting the *new* values for extracellular osmotically active solute and osmolality into *Eq. 21*

$$ECF = \frac{3480 \text{ mosmol} - 720 \text{ mosmol}}{270 \text{ mosmol/kg } H_2O}$$

$$= 10.2 \text{ l}$$

i.e., the loss of 720 mosmol of Na^+ and attendant anions from the ECF results in the shift of approximately 1.8 l of the original 12 l of ECF to the ICF. While this example may seem unrealistic, a similar decrease in P_{osm} and shift of water from ECF to ICF can occur in patients who lose large quantities of Na^+ and water (e.g., as a result of copious diarrhea) and then replace the volume loss by drinking water.

Figure 4.4*C* shows the changes in the volumes and osmolality of the ECF and ICF that would result from the sudden loss of 360 mosmol of K^+ and 360 mosmol of anions, again without changing the TBW. Since K^+ is primarily an *intracellular* ion, it can be assumed that this 720 mosmol of solute is lost exclusively from the ICF. The sudden loss of solute from the ICF without any change in TBW reduces the osmolality of the ICF. However, this decrease in ICF osmolality can only be *transient*, since water will move from ICF to ECF until the osmolalities of all major body fluids are equal. After this redistribution of water is complete, the new osmolality of the body fluids can be calculated by substituting the new value for total body osmotically active solute into *Eq. 20*

$$TBW_{osm} = \frac{10440 \text{ mosmol} - 720 \text{ mosmol}}{36 \text{ l}}$$

$$= 270 \text{ mosmol/kg } H_2O$$

i.e., the net result is a decrease in the osmolality of *all* major body fluids, even though solute is lost primarily from the ICF. Note that the decrease in osmolality is the *same* as that resulting from the

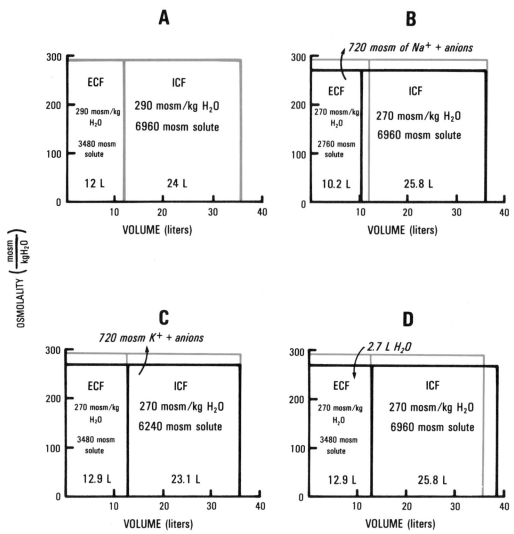

Figure 4.4. Effects of changes in body solute or TBW on osmolality of body fluids and on distribution of water between ECF and ICF in a 60-kg young adult male. *A*, Normal state, with a body fluid osmolality of 290 mosmol/kg H_2O and a TBW of 36 l (= 60% of total body weight), of which about one-third is in ECF (12 l) and two-thirds is in ICF (24 l). *B*, Following the sudden loss of 360 mosmol of Na^+ and 360 mosmol of anions from the ECF, with no change in TBW. *C*, Following the sudden loss of 360 mosmol of K^+ and 360 mosmol of anions from the ICF, with no change in TBW. *D*, Following the sudden addition of 2.7 l of distilled water to the ECF with no change in total body osmotically active solute. In *B*, *C*, and *D*, the original normal osmolality and distribution of water (from *A*) is outlined in *red*. Additional details are provided in the text.

loss of 720 mosmol of solute from the ECF (Fig. 4.4*B*), a result that could be anticipated from *Eq. 20*. The new volume of the ICF after the redistribution of water can be calculated by substituting the new values for intracellular osmotically active solute and osmolality into *Eq. 22*

$$ICF = \frac{6960 \text{ mosmol} - 720 \text{ mosmol}}{270 \text{ mosmol/kg } H_2O}$$

$$= 23.1 \text{ l}$$

i.e., the loss of 720 mosmol of K^+ and attendant anions from the ICF results in the shift of approximately 0.9 l of the original 24 l of ICF into the ECF.

Finally, Figure 4.4*D* illustrates the changes in the volumes and osmolality of the ECF and ICF that would result from the sudden addition of 2.7 l of distilled water to the ECF with no change in total body osmotically active solute. This addition of water reduces the osmolality of the ECF. However, this decrease in ECF osmolality can only be *transient*, since water will move from ECF to ICF until the osmolalities of all major body fluids are equal. After the redistribution of water is complete (and assuming for simplicity that none of the water is excreted), the osmolality of the body fluids can be calculated by substituting the new value for TBW into *Eq. 20*

$$TBW_{osm} = \frac{10440 \text{ mosmol}}{38.7 \text{ l}}$$

$$= 270 \text{ mosmol/kg } H_2O$$

The new volumes of the ECF and ICF can be calculated by substituting the new value for body fluid osmolality into *Eqs. 21* and *22* (note that the quantity of osmotically active solute in each compartment does *not* change)

$$ECF = \frac{3480 \text{ mosmol}}{270 \text{ mosmol/kg } H_2O}$$

$$= 12.9 \text{ l}$$

$$ICF = \frac{6960 \text{ mosmol}}{270 \text{ mosmol/kg } H_2O}$$

$$= 25.8 \text{ l}$$

i.e., approximately one-third of the added 2.7 l of water distributes into the ECF, while two-thirds distributes into the ICF. Note that the added water distributes between ECF and ICF in the same proportions as the original TBW, because the number of osmotically active solute particles in each compartment does not change.

Thus, in each of the three examples of changes in body solute and water illustrated in Figure 4.4, the osmolality of the major body fluids decreases from 290 mosmol/kg H_2O to 270 mosmol/kg H_2O. Furthermore, if it is assumed that the osmotic contribution of glucose and urea remains constant at approximately 10 mosmol/kg H_2O (see above), then P_{Na} decreases from 140 to 130 mmol/l in each case (*Eq. 19*). This result can also be predicted from the following considerations. Since the major body fluids are isoosmotic, *Eq. 20* can be used to calculate P_{osm} as well as TBW_{osm}:

$$P_{osm} = \frac{\text{Total body osmotically active solute}}{TBW} \quad (23)$$

The major osmotically active solutes in the body fluids are Na^+ and its attendant anions (primarily in ECF) and K^+ and its attendant anions (primarily in ICF). Furthermore, the osmotically active Na^+ and K^+ in the body fluids are closely approximated by the *exchangeable Na^+* (Na_e^+) and *exchangeable K^+* (K_e^+), respectively.* Thus

$$P_{osm} \simeq \frac{2Na_e^+ + 2K_e^+}{TBW} \quad (24)$$

where the multiplier 2 is included to account for the osmotic contributions of the attendant anions of Na^+ and K^+ (these anions are assumed for simplicity to be *univalent*; see above). But P_{osm} also is approximated by $2P_{Na}$ if the osmotic contributions of glucose and urea are neglected (*Eq. 19*). Substituting this approximation into *Eq. 24* gives

$$P_{Na} \simeq \frac{Na_e^+ + K_e^+}{TBW} \quad (25)$$

Thus, a decrease in body Na^+, a decrease in body K^+, *or* an increase in TBW could each result in a similar change in P_{Na}. *Eq. 25* and the examples illustrated in Figure 4.4 emphasize an important point about the use of P_{Na} as an index of P_{osm}, and hence body fluid osmolality: P_{Na} does *not*, *by itself*, give any information about either the total amount of osmotically active solute in the body or the TBW.

Figure 4.5 illustrates that the relationship expressed by *Eq. 25* is valid over a wide range of P_{Na} (Edelman et al., 1958). *Eq. 25* is particularly helpful in analyzing the changes in body solute or TBW that can occur in disease states, which can be considerably more complex than those illustrated in Figure 4.4. For example, sweating in a hot environment in the absence of water intake will decrease *both* Na_e^+ and TBW. But since the osmolality of sweat is *less* than that of plasma, the percentage decrease in TBW will be *greater* than the percentage decrease in Na_e^+. P_{Na} will therefore *increase* in spite of the decrease in Na_e^+ (*Eq. 25*).

While P_{Na} can represent a good index of P_{osm}, and hence the osmolality of the major body fluids,

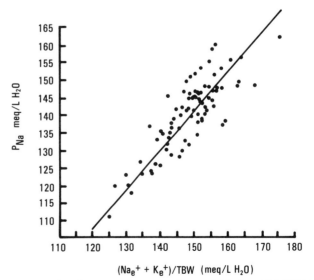

Figure 4.5. Correlation between P_{Na} and (Na_e^+ + K_e^+)/TBW, illustrating the validity of *Eq. 25*. [Reproduced from *The Journal of Clinical Investigation*, 1958, vol. 37, pp. 1236–1256, by copyright permission of The American Society for Clinical Investigation.]

---

* Actually, whereas virtually all of the K_e^+ is osmotically active, a small amount of Na_e^+ is absorbed to structures rich in mucopolysaccharides (e.g., tendons, cartilage) and is *not* osmotically active. However, the amount of this *residual Na^+* is so small that Na_e^+ represents a close approximation to the amount of osmotically active Na^+.

it must be kept in mind that P_{Na} may *not* provide an acceptable estimate of P_{osm} in certain special circumstances:

1. In the presence of elevated plasma levels of osmotically active solutes such as glucose or urea, P_{Na}, and hence the *estimated* P_{osm}, will be *normal* when P_{osm} *actually* is *high*. For example, the osmotic contribution of glucose or urea can become so large in uncontrolled diabetes mellitus or renal failure, respectively, that P_{osm} would have to be estimated with *Eq. 17* instead of *Eq. 19.*

2. In conditions such as severe hyperproteinemia or hyperlipidemia, P_{Na}, and hence the *estimated* P_{osm}, may be *low* when P_{osm} *actually* is *normal*. This is because the laboratory values of P_{Na} typically used in *Eq. 19* are expressed as mmol/l of plasma *volume* instead of mmol/l of plasma *water* (see above). While the use of typical laboratory values normally does not introduce a significant error (and in fact helps to *compensate* for other approximations made in the derivation of *Eq. 19*, as previously noted), in severe hyperproteinemia or hyper-

lipidemia P_{Na} expressed as mmol/l of plasma *volume* may be abnormally low, even though P_{Na} expressed as mmol/l of plasma *water*, and hence P_{osm}, is *normal*.

Isoosmolality vs. Isotonicity

Throughout this chapter, the major body fluids have been described as *isoosmotic* because they have virtually identical osmolalities. However, when discussing osmolalities of the body fluids or the osmolalities of solutions used clinically to replace body fluids, the terms *isotonic*, *hypotonic*, and *hypertonic* often are used. An *isotonic solution* is one in which normal body cells (e.g., red blood cells) can be suspended without a change in cell volume occurring. In contrast, cells suspended in a *hypotonic fluid* will swell (or even rupture) due to the entry of water, while cells suspended in a *hypertonic fluid* will shrink due to the exit of water. For example, a solution containing NaCl at an osmolality of 290 mosmol/kg H_2O is an isotonic solution that is used clinically as a replacement fluid. This solution commonly is referred to as *0.9% saline* (since it contains 0.9 g NaCl/dl) or "normal saline."

CHAPTER 27

Anatomy of the Kidneys

GROSS ANATOMY

The kidneys are bean-shaped organs located behind the peritoneal cavity, where they are protected from injury by a surrounding layer of fat. The two kidneys constitute about 0.5% of the total body weight, so that each kidney weighs approximately 150 g in a 60-kg man (Oliver, 1968). The concave surface of each kidney faces medially toward the vertebral column. Located in the middle of this surface is a longitudinal slit termed the *hilus*, through which a variety of structures, such as blood vessels, nerves, lymphatics, and the ureter, enter or exit the kidney (Fig. 4.6). The entire kidney is enclosed in a fibrous *capsule*.

Upon longitudinal section of the kidney (Asscher et al., 1982), two zones can be identified: an outer *cortex* and an inner *medulla* (Fig. 4.6). The medulla is composed of a number of pyramidal shaped structures, the *renal pyramids*, and can be divided into an *outer zone* (next to the cortex) and an *inner zone* (which forms the apexes of the pyramids, called *papillae*). The outer zone of the medulla, in turn, can be subdivided into an *outer stripe* and an *inner stripe* (Bulger, 1979). In some small animals, such as the rat or rabbit, each kidney has just one pyramid and is therefore termed *unipapillary*. In contrast, larger species such as the dog or man have *multipapillary* kidneys. Each human kidney, for example, contains 4 to 14 renal pyramids (Oliver, 1968). In multipapillary kidneys, the cortex forms not only a thick peripheral shell but also extends

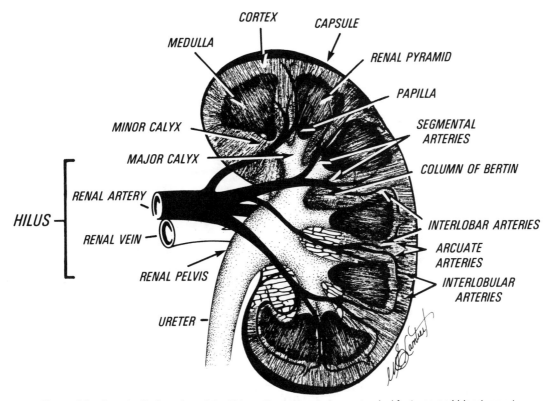

Figure 4.6. Longitudinal section of the kidney, illustrating major anatomical features and blood vessels.

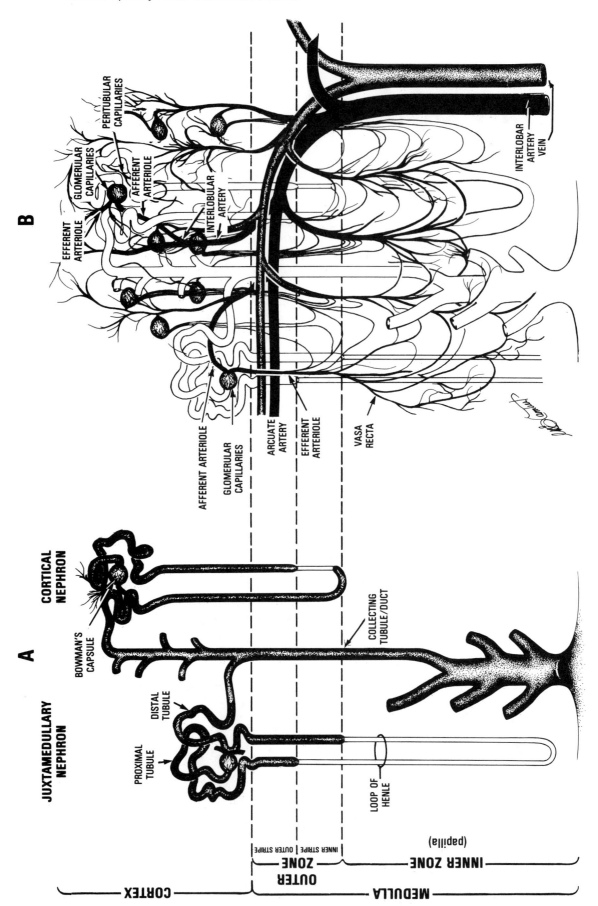

Figure 4.7. Cortical and juxtamedullary nephrons and their vasculature.

like a column toward the hilus between renal pyramids. These extensions of cortical tissue into the interior of the organ are the renal *columns of Bertin*.

Each papilla projects into a cup-shaped *minor calyx*. Several minor calyces join to form a *major calyx*; the major calyces unite into the funnel-shaped *renal pelvis* (Fig. 4.6). Urine continuously exits the tip of the papilla and is collected in the renal pelvis. From the renal pelvis, urine flows through the *ureter* to the *urinary bladder* for storage prior to intermittent voiding via the *urethra*. The structures from the minor calyces through the urethra commonly are termed the *urinary tract*.

The extensively regulated excretory functions of the kidneys are accomplished by individual functional units called *nephrons* (Fig. 4.7A). Approximately 1.0 to 1.5 million nephrons are packaged in the average human kidney (Oliver, 1968). Each nephon consists of two major structures, a *glomerular portion*, often termed *Bowman's capsule*, and a *tubule*. The tubule, in turn, can be divided into three major regions: the *proximal tubule*, the *loop of Henle*, and the *distal nephron*. Several adjacent distal nephrons share a common final segment, the *collecting tubule/duct*. In a longitudinal section of the kidney, the collecting tubules/ducts are evident as fanlike striations in the renal pyramids (Fig. 4.6).

Two basic types of nephrons can be distinguished. While all glomeruli are located in the cortex, nephrons whose glomeruli lie in the *outer* region of the cortex have short loops of Henle that remain exclusively in the cortex or that penetrate only the outer zone of the medulla. These nephrons are termed *cortical nephrons*. In contrast, nephrons whose glomeruli lie in the *inner* cortex near the corticomedullary junction have long loops of Henle that descend deep into the inner zone of the medulla, even as far as the tip of the papilla in some nephrons, before turning and ascending back to the cortex. These long-looped nephrons are termed *juxtamedullary nephrons*. In man, approximately 85% of the nephrons are cortical, with the remaining 15% being juxtamedullary (Valtin, 1977).

In general terms, the function of the *glomerular portion* of the nephron is the filtration of fluid and its dissolved constituents from plasma into the tubule, while the function of the *tubule* is to reduce the volume and modify the contents of the filtrate. The tubule both *reabsorbs* many compounds from the tubular fluid and *secretes* other substances into the tubular fluid, so that the final urine contains the constituents that must be excreted to preserve a normal body fluid volume and composition. The quantity of fluid and solute processed by the kidneys is *enormous*: the glomerular portions of the nephrons filter an average of 125 ml of plasma each minute or *180 l* of plasma each day, while the tubular portions reabsorb up to 99+% of the filtered water and essential solutes.

BLOOD SUPPLY

The renal vasculature is highly specialized to bring plasma to the glomerular portions of the nephrons for filtration and then to the tubular portions to take up reabsorbed water and solutes and to deliver substances to be secreted (Fig. 4.6, Fig. 4.7B). At the hilus, blood enters the kidney in the *renal artery* (in about 20% of human kidneys, more than one main renal artery is present). The main renal artery branches (Netter, 1973) a variable number of times into *segmental arteries*, which subdivide into *interlobar arteries*. Each interlobar artery penetrates the kidney through a column of Bertin, but before reaching the cortical surface, divides into arc-shaped *arcuate arteries*, which eventually run parallel to the surface. *Interlobular arteries* arise from the arcuate arteries and penetrate the cortex perpendicularly toward the surface. Branching from each interlobular artery are numerous small arterioles, the *afferent arterioles*. Each afferent arteriole leads blood toward a single nephron (the *a*fferent arteriole is *a*ttracted to the nephron).

The afferent arteriole intimately interacts with the glomerular portion of the nephron, where it breaks into a capillary network, the *glomerular capillary tuft*. About 20% of the water in plasma entering the afferent arteriole (along with ions and other small solutes dissolved in plasma water) filters out of the glomerular capillary into the cup-shaped collection area inside Bowman's capsule, often termed *Bowman's space*. The remaining 80% of plasma, along with the larger solutes (e.g., proteins and lipids) and cellular elements of blood, flows from the glomerular capillaries into an *efferent arteriole* (the *e*fferent arteriole *e*xits from the glomerular capillaries). The efferent arteriole disperses into a second capillary network that surrounds the tubular portions of the nephrons, delivering substances for tubules to secrete into the tubular fluid and taking up water and solutes reabsorbed by the tubules. Most of the capillaries in this second network surround the tubules in the cortex and are termed *peritubular capillaries* (Beeuwkes and Bonventre, 1975). However, some of the capillaries that are derived from the efferent arterioles of juxtamedullary nephrons descend to varying depths in the medulla (even as far as the tip of the papilla) before forming a capillary network. The capillaries reunite while turning around and ascend back to the cortex. These unique hair-pin-shaped capillaries are called *vasa recta* ("straight vessels") and have an important role in the mechanism for the concentration of urine (see Chapter 31).

From the peritubular capillaries and vasa recta, blood flows into *interlobular veins* and then leaves the kidney in veins flowing counter to, and named in accordance with, the adjacent arteries. Note the rather unusual feature of the renal vasculature, wherein blood traverses *two* arterioles and *two* capillary networks before being collected into the venous system.

THE NEPHRON

The anatomy of the nephron will now be considered in greater detail, with reference to its relationship to the specialized renal vasculature and its general function. The glomerular portion of the nephron, or *Bowman's capsule*, is intimately associated with the glomerular capillary tuft. In fact, nephrologists (specialists in diseases of the kidneys) and physiologists often use the term *glomerulus* to include *both* the nephron and capillary components, a convention that will be followed in this textbook as well. The anatomical relationship between nephron and capillaries in the glomerulus is rather difficult to visualize, and the following analogy may prove helpful. First, imagine putting your hand in

a rubber glove. Then, evert the portion of the glove covering the palm and back of your hand, so that it folds down over your fingers, while keeping your fingers in the fingers of the glove. In this analogy, your fingers represent the glomerular capillaries (with blood flowing into and out of your fingers), while the glove represents the epithelial cells of Bowman's capsule. During filtration, then, fluid and solutes move from your fingers (the glomerular capillaries) across the fingers of the glove (the epithelial cells of Bowman's capsule) into a collecting area formed by the everted hand of the glove (Bowman's space).

The actual filtration barrier is somewhat more complex than indicated by the above analogy (Bulger, 1979), consisting of three layers: capillary endothelium, basement membrane, and epithelium (Fig. 4.8). The *capillary endothelium* is similar to that lining capillaries elsewhere in the body, except that the glomerular endothelial cells have numerous holes or *fenestrae* in their cytoplasm. The *basement membrane*, composed of glycoproteins and mucopolysaccharides (but *no* cells), surrounds the glomerular capillaries. Overlying the basement membrane is the *epithelial cell layer* of Bowman's capsule. The epithelial cells that surround the capillary tuft (i.e., the fingers of the glove in the analogy used above) differ from those forming the rest of Bowman's capsule (the everted hand of the glove), primarily because they have numerous *foot proc-*

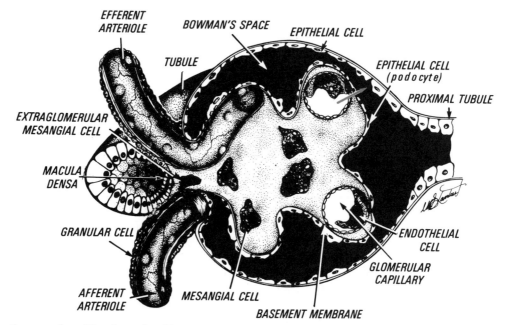

Figure 4.8.   Cross-section of the glomerulus. The pathway followed by water and small solutes as they are filtered from the glomerular capillaries into Bowman's space is indicated by the *red arrow*. The filtered molecules must cross the fenestrated *capillary endothelium*, the *basement membrane*, and the *epithelial cell layer* of Bowman's capsule. Also illustrated are the components of the specialized structure found where the tubule contacts its originating glomerulus, the so-called *juxtaglomerular apparatus* (JG apparatus).

esses, or *pedicels*; hence, these cells sometimes are called *podocytes*. The foot processes from adjacent podocytes interdigitate extensively with one another.

One final structural feature of the glomerulus remains to be described. Interspersed between adjacent capillaries, particularly in the center or *core* of the glomerulus, are clusters of *mesangial cells*. The function of these cells, which appear to provide support for the glomerular capillaries, is not known.

The *tubular portion* of the nephron consists of an *epithelial cell layer* that is continuous with the epithelial cell layer forming Bowman's capsule. A *basement membrane*, continuous with the glomerular basement membrane (Fig. 4.8), encases the entire tubule. As previously indicated, the tubule can be divided into three major regions: *proximal tubule*, *loop of Henle*, and *distal nephron*.

The epithelial cells in various regions of the tubule differ in many respects, such as size, shape, number of mitochondria, and number of microvilli on the luminal surface (Fig. 4.9). However, certain structural features are common to all tubular epithelial cells. For example, *cytoplasmic processes* extend from the lateral and basal surfaces of the cells to *interdigitate* with similar processes of adjacent cells, although the extent of this interdigitation varies somewhat in different regions. At the luminal surface, adjacent cells actually are *fused* together at a specialized junctional structure, the *zonula occludens* or *tight junction*. Below the tight junction, adjacent cells are separated by the so-called *lateral intercellular space*, which extends all the way to the bases of the cells. The lateral space is functionally continuous with the interstitial space surrounding the tubule and peritubular capillaries (the so-called

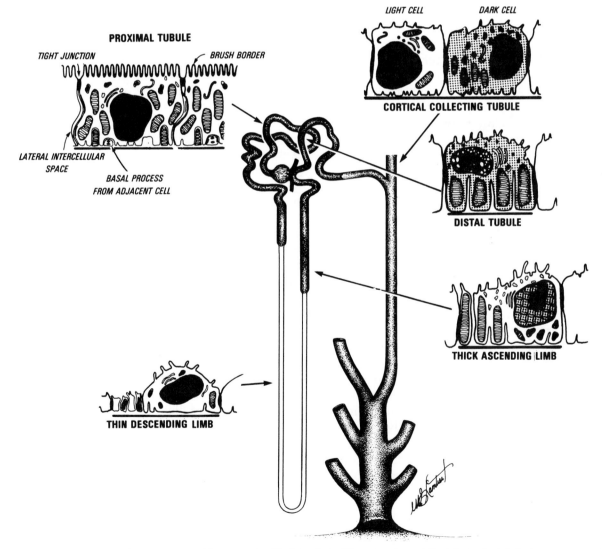

Figure 4.9. Comparison of major structural features of epithelial cells in different regions of the tubule. Certain features common to all regions (e.g., *tight junction, lateral intercellular space, basal process*) are identified on the proximal tubule illustration only.

peritubular space), as the basement membrane does not impede the diffusion of water or solutes. The tight junction therefore serves as a barrier to the movement of water and solutes between the lumen and peritubular space. However, the effectiveness of this barrier, or degree of "tightness," varies in different regions of the tubule (Fig. 4.9). Note that in discussions of the tubular epithelium, the luminal surface often is termed the *mucosal* or *apical* surface, while the basal surface often is termed the *serosal* or *peritubular* surface.

Proximal Tubule

The proximal tubule (Venkatachalam, 1980) has traditionally been divided into the *proximal convoluted tubule*, which contorts itself in an apparently random fashion in the cortex, and the *proximal straight tubule* (or *pars recta*), which dives toward the medulla. Recent studies suggest that the proximal convoluted tubule may be further subdivided on a functional basis into *early* and *late* convoluted segments, or part 1 and part 2, with the pars recta forming part 3 of the proximal tubule.

The epithelial cells of the proximal tubule have a *cuboidal* shape. A layer of closely packed *microvilli* covers the luminal surface, forming a *brush border*. The large *nuclei* are close to the basal surface. The cells have prominent *basal* and *lateral processes* and interdigitate extensively. *Mitochondria* are abundant, particularly in areas adjacent to the interdigitating processes.

The proximal tubule both *reabsorbs* substances from the tubular fluid and *secretes* substances into the tubular fluid. Thus, the proximal tubule is responsible for the initial processing of the glomerular filtrate. For example, approximately two-thirds of the filtered water and Na^+, and virtually all of the filtered glucose and amino acids, are *reabsorbed* in this region. Organic acids and bases, such as drugs and drug metabolites, are *secreted*. The epithelial cells of the proximal tubule are well adapted for the processes of reabsorption and secretion, with their large luminal surface area (created by the brush border), large basal surface area (created by the basal processes), and abundant mitochondria (which undoubtedly have a role in supplying energy for the various transport processes). The numerous peritubular capillaries that surround the proximal tubule take up the reabsorbed water and solutes and deliver the substances to be secreted.

Loop of Henle

The loop of Henle consists of a *descending limb* and an *ascending limb* (Jamison and Kriz, 1982). While some morphologists include part 3 of the proximal tubule, the pars recta, in the descending limb, most renal physiologists consider the loop of Henle to begin at the end of the pars recta. At this point, the cuboidal epithelial cells of the proximal tubule are replaced by flat, squamous epithelial cells with a small number of short microvilli and few mitochondria. This segment of the tubule is termed the *thin descending limb* because the epithelial cell layer is so flat, i.e., "thin" does *not* mean that the tubular lumen is diminished in diameter. In *cortical nephrons*, the hairpin turn that forms the tip of the loop of Henle occurs no deeper than the junction between the outer and inner zones of the medulla, whereas in *juxtamedullary nephrons* the hairpin turn may occur as deeply as the tip of the papilla. Thus, the thin descending limb of juxtamedullary nephrons is substantially longer than that of cortical nephrons (Fig. 4.7).

Differences between cortical and juxtamedullary nephrons also are present in the *ascending limb*. In *juxtamedullary nephrons*, the ascending limb begins with a segment of flat, squamous epithelial cells, termed the *thin ascending limb*. Although these cells resemble the cells of the thin descending limb when viewed by light microscopy, structural differences are discernible with the electron microscope. At the junction between the inner and outer zones of the medulla, the *thick ascending limb* begins. In this segment, the epithelial cells are cuboidal, extensively interdigitated, and contain numerous mitochondria like the epithelial cells of the proximal tubule. However, these cells lack a luminal brush border, although a small number of short microvilli are present. The thick ascending limb traverses the outer zone of the medulla (the *medullary portion* of the thick ascending limb) and then ascends through the cortex (the *cortical portion* of the thick ascending limb) to the level of its nephron's glomerulus. In *cortical nephrons*, the ascending limb consists *entirely* of cuboidal epithelial cells, i.e., cortical nephrons *lack* a thin ascending limb.

Like the proximal tubule, the loop of Henle both *reabsorbs* substances from the tubular fluid and *secretes* substances into the tubular fluid. For example, in juxtamedullary nephrons as much as 25% of the filtered water and Na^+ is reabsorbed in the loop of Henle, while urea, a major waste product, is added to the tubular fluid. The loop of Henle also has an important role in the mechanisms for both urinary concentration and urinary dilution.

Distal Nephron

The term *distal nephron* will be used to refer to all portions of the tubule from the end of the cortical thick ascending limb through the tip of the papilla. The junction between the cortical thick ascending limb and the distal nephron is marked by an astonishing fact and a unique structure. The cortex of each human kidney contains the glomeruli of 1.0 to 1.5 million nephrons. Remarkably, in each nephron the thick ascending limb returns to the glomerulus of its own origin (Fig. 4.7*A*)! The tubule contacts its originating glomerulus at the *vascular pole*, the region where the afferent and efferent arterioles, respectively, enter and leave the glomerulus. A specialized structure, the *juxtaglomerular apparatus* (JG apparatus), is found at this contact point (see below).

The distal nephron includes several anatomically and functionally distinct segments (Jacobson, 1981). Three major segments can be identified: the *distal tubule*, the *connecting tubule*, and the *collecting tubule/duct*. The collecting tubule/duct, in turn, can be subdivided into the *cortical collecting tubule*, the *medullary collecting duct*, and the *papillary collecting duct*. Actually, the term distal nephron is somewhat of a misnomer, because embryologically the distal tubule represents the last segment of the nephron proper. The collecting tubule/duct develops from the ureteral bud, and the connecting tubule is presumed to be the region where the embryonic nephron "connects" with the ureteral bud.

The *distal tubule* is a relatively short segment that extends outward from its contact point with the glomerulus toward the cortical surface. The *connecting tubule* has a somewhat convoluted shape and may join with adjacent connecting tubules before its junction with the cortical collecting tubule. The *cortical collecting tubule* begins as a short segment beyond the connecting tubule, continues as a slightly larger tubule formed by the union of two or more such segments, and finally becomes a straight duct penetrating toward the medulla.* The *medullary collecting duct* and *papillary collecting duct* represent the extensions of this straight duct into the outer medulla and papilla, respectively. Deep in the papilla, several papillary collecting ducts merge to form a common duct (often termed the *duct of Bellini*) that empties into a minor calyx.

The epithelial cells of the distal tubule, like those of the proximal tubule, are cuboidal, exhibit extensive basal and lateral interdigitations, and contain numerous mitochondria. However, these cells lack a luminal brush border, although a small number

of short microvilli are present as in the thick ascending limb. The epithelial cells of the connecting tubule resemble those of the distal tubule but are taller and have a more granular appearance.

In the collecting tubule/duct, the epithelial cells increase in height from a cuboidal shape in the cortical collecting tubule to a columnar shape in the papillary collecting duct. Two basic cell types can be distinguished (Jamison and Kriz, 1982). The predominant cell type, the *principal* or *light cell*, stains lightly because of a relative paucity of cellular organelles. In contrast, the *intercalated* or *dark cell* stains heavily because of abundant mitochondria. The epithelial cells of the collecting tubule/duct lack a brush border, although short microvilli may be seen as in the distal tubule.

The distal nephron is responsible for the final transformation of the tubular fluid into urine. Its functions include the reabsorption of Na^+ and Cl^-, the secretion of H^+ and K^+, and an important role in the concentration and dilution of urine.

Juxtaglomerular Apparatus

As indicated above, the *juxtaglomerular apparatus* (JG apparatus) is a specialized structure found in the region where the tubule contacts its originating glomerulus (Barajas, 1979). The JG apparatus consists of three components: macula densa, extraglomerular mesangial cells, and granular cells (Fig. 4.8). The *macula densa* is a row of tightly packed cuboidal epithelial cells lining the tubule at the site of contact. This row of cells defines the beginning of the distal nephron. The *extraglomerular mesangial cells* can be regarded as an extension of the mesangial cells of the glomerulus into the triangular region bounded by the afferent arteriole, efferent arteriole, and macula densa. The extraglomerular mesangial cells are sometimes called *agranular cells* to distinguish them from the *granular cells* of the JG apparatus. The *granular cells* (also called JG cells) are located both in the region of the extraglomerular mesangial cells and in the walls of the adjacent afferent and efferent arterioles. In fact, the granular cells are believed to be specialized types of smooth muscle cells derived from the walls of these arterioles.

The granular cells receive their name from the existence of secretory granules that contain *renin*, a proteolytic enzyme with a very important function. When secreted into the lumen of the arterioles, renin acts on a specific protein in plasma, *renin substrate* (also called angiotensinogen), to produce a decapeptide, *angiotensin I*. Angiotensin I, in turn, has two amino acids removed by *converting enzyme* to yield *angiotensin II*.

---

* The distal tubule, connecting tubule, and early cortical collecting tubule frequently are lumped together under the term "distal convoluted tubule." However, recent evidence from the best studied species in this regard (the rabbit) indicates that these three cortical segments have distinct anatomical and functional characteristics.

$$\text{Angiotensinogen} \xrightarrow{\text{renin}} \text{angiotensin I} \xrightarrow{\text{converting enzyme}} \text{angiotensin II}$$

Angiotensin II has at least two actions: it stimulates the adrenal cortex to secrete aldosterone, and it is a powerful constrictor of arteriolar smooth muscle. Because of the latter action, angiotensin II may be able to modulate both the flow of blood through and the hydrostatic pressure in the glomerular capillaries (see Chapter 28). In addition, the vasoconstrictor action of angiotensin II on *systemic* arterioles can increase the peripheral vascular resistance and thereby elevate the systemic blood pressure. Indeed, abnormal production of renin, and therefore angiotensin II, is a cause of the high blood pressure in a small percentage of patients with hypertension.

The association of vessels, glomerulus, and macula densa from one nephron in the JG apparatus in conjunction with the vasoactive angiotensin system represents an ideal anatomical and functional arrangement whereby the fluid reaching the distal nephron might "signal" the arterioles to alter the blood flow to and rate of filtration in the glomerulus (Chapter 28). The term *tubuloglomerular feedback* has been used to describe this postulated regulation of glomerular blood flow and filtration by the fluid delivered to the distal nephron. However, despite intensive investigation, the mechanism for such feedback remains to be established.

THE URINARY TRACT

As indicated above, the *urinary tract* consists of the minor and major calyces, renal pelvis, ureters, urinary bladder, and urethra (Netter, 1973). The entire urinary tract is lined by a transitional cell epithelium that does not further modify the composition of the urine leaving the distal nephron. Thus, the sole function of the urinary tract is the transmission and storage of urine.

In the calyces, renal pelvis, and ureters, the epithelial cell layer is surrounded by bundles of smooth muscle cells arranged in spiral configurations that contract when distended. Most importantly, distention of the renal pelvis and upper ureter initiates a *peristaltic contraction* that begins in the pelvis and spreads along the ureter to move urine toward the bladder. On the average, 1–5 such peristaltic contractions occur per minute, although the rate may increase to 20 or more per minute when distention is markedly increased due to high rates of urine production (Longrigg, 1982).

The ureters penetrate the bladder wall at an oblique angle, and although there are no anatomical sphincters at the junction between the ureter and the bladder (the *ureterovesicular junction*), this oblique penetration tends to keep the ureters closed except during ureteric peristalsis, thereby preventing reflux of urine from the bladder back into the ureters (Woodburne, 1964). A mechanism to prevent such reflux is important because the pressure in the bladder can exceed that in the ureter between periods of ureteric peristalsis and during voiding.

The urinary bladder, when distended, consists of a spherical portion (called the *body* or *fundus*) and an inferior cylindrical portion (called the *neck*). The epithelial cell layer lining the bladder is surrounded by the *detrusor muscle*, which consists of a randomly interwoven network of smooth muscle cell bundles. The smooth muscle in the bladder neck is mixed with a considerable amount of elastic tissue and functions as an *internal sphincter*. The urethra is surrounded by a band of skeletal (voluntary) muscle, the *external sphincter*.

During voiding (*micturition*), the detrusor muscle contracts, the neck of the bladder opens into a funnel shape, and the external sphincter is voluntarily relaxed. Once initiated, micturition normally continues until the bladder is entirely empty, although it can be interrupted by a powerful voluntary contraction of the external sphincter.

METHODS IN RENAL PHYSIOLOGY

An appreciation of major experimental methods available for the study of renal function can contribute significantly to an understanding of the material presented in subsequent chapters. These methods can be divided into four general categories: clearance measurements, micropuncture methods, perfusion of microdissected nephron segments, and model tissues.

Clearance Measurements

Clearance is a central concept in renal physiology, as it provides a way of evaluating the elimination of a substance by the kidneys (Smith, 1951). To understand the meaning of clearance, consider first the *rate of excretion* of a substance X, i.e., the mass of X excreted per unit time. This can be equated to the mass of X per unit volume of urine (i.e., the concentration of X in urine, U_X) multiplied by the volume of urine excreted per unit time (i.e., the urine flow rate, $\dot{V}$)

$$\frac{\text{Mass of X excreted}}{\text{Time}} = \frac{\text{Mass of X in urine}}{\text{Urine volume}} \times \frac{\text{Urine volume}}{\text{Time}}$$

or

$$\frac{\text{Mass of X excreted}}{\text{Time}} = U_X \dot{V}$$

Since an important function of the kidneys is the

removal of substances from plasma, it is useful to describe the elimination of X in a somewhat different way. Instead of talking about the rate of excretion of X in *urine*, consider the rate of removal of X from *plasma*. This can be equated to the concentration of X in plasma (P_X) multiplied by the volume of plasma from which X is completely removed, or "cleared," per unit time

$$\frac{\text{Mass of X removed from plasma}}{\text{Time}}$$
$$= \text{Plasma concentration of X}$$
$$\times \frac{\text{Volume of plasma "cleared" of X}}{\text{Time}}$$

The volume of plasma from which X is completely "cleared" per unit time is termed the *clearance of X* and is designated C_X. Therefore

$$\frac{\text{Mass of X removed from plasma}}{\text{Time}} = P_X C_X$$

It is important to note that the volume of plasma "cleared" of X is a *theoretical* volume rather than a volume that can be collected and directly measured. This is because no single milliliter of plasma has *all* of its X removed by the kidneys; instead, a certain *fraction* of the X in each milliliter of plasma is removed. But while C_X is only a theoretical volume, its value can be calculated from measurable quantities, since mass conservation requires that the rate of removal from plasma must equal the rate of excretion:

$$\frac{\text{Mass of X removed from plasma}}{\text{Time}}$$
$$= \frac{\text{Mass of X excreted in urine}}{\text{Time}}$$

Therefore

$$P_X C_X = U_X \dot{V}$$

and

$$C_X = \frac{U_X \dot{V}}{P_X} \qquad (26)$$

The great advantage of clearance measurements over other methods for studying renal function is that they require only urine and peripheral blood samples, i.e., clearance measurements are virtually *noninvasive*. With rare exceptions, then, the clearance technique is the only method available for the study of renal physiology in humans. But clearance measurements are useful only for evaluating the *overall* elimination of a substance by the kidneys, *not* for studying the function of *individual segments* of the nephron. Almost all statements made in this text about the function of individual nephron segments, therefore, are based on studies conducted in experimental animals using methods in the remaining three categories.

Micropuncture Methods

The technique of micropuncture, in which a tiny micropipette is inserted into a nephron segment or adjacent blood vessel, has resulted in major advances in the understanding of renal physiology. Five important examples of micropuncture methods will be considered briefly here (Deetjen et al., 1975; Jamison and Kriz, 1982):

1. Micropuncture can be used to aspirate fluid from accessible nephron segments for analysis of composition. For example, the composition of the fluid filtered into Bowman's space can be determined by direct analysis of fluid collected by micropuncture. In a variation of this method, fluid of known composition is injected into an accessible nephron segment and held stationary by prior injection of a droplet of oil to block forward flow. The fluid is later aspirated and analyzed to determine how the tubule has modified its composition.

2. Micropipette-sized pressure transducers can be used to measure the hydrostatic pressures in glomerular and peritubular capillaries, Bowman's space, and accessible nephron segments, allowing the forces involved in glomerular filtration to be studied.

3. Micropipette-sized glass electrodes can be used to measure the transepithelial potential difference and to study variations in this potential under different experimental conditions.

4. Micropuncture can be used to perfuse a single nephron distal to an oil block, a technique often termed *microperfusion*. For example, an oil droplet can be injected into part 1 of the proximal tubule and perfusion fluid injected in part 2. The perfusion fluid then flows through part 3 of the proximal tubule (the pars recta), the descending limb of the loop of Henle, and the ascending limb of the loop of Henle. The fluid is aspirated from the distal tubule and analyzed, allowing the actions of the perfused segments on the volume and composition of the tubular fluid to be determined. A variation of this microperfusion method involves the simultaneous injection of perfusion fluid into the tubular lumen and surrounding peritubular capillaries. In this manner, the composition of the fluid on *both* sides of the tubular epithelium can be controlled.

5. A small catheter can be advanced upstream from the calyceal area through a duct of Bellini and into the lumen of a papillary collecting duct. By sampling fluid at various levels of the

collecting duct, changes in the volume and composition of the tubular fluid as it passes through the duct can be determined. This technique, which does not actually entail puncture of the tubule, is termed *microcatheterization*.

The study of renal physiology by these micropuncture techniques is limited to regions of the nephron that are accessible from the surfaces of the kidney. Several regions of *cortical nephrons* can be micropunctured: the glomerular structures, parts 1 and 2 of the proximal tubule, and the distal tubule, connecting tubule, and cortical collecting tubule (however, much of the older micropuncture data obtained by distal puncture did not take into account the subdivision of the early distal nephron). In contrast, *none* of the above segments in juxtamedullary nephrons is accessible by micropuncture. On the other hand, from the papillary surface the thin descending and ascending limbs of the loop of Henle of *juxtamedullary nephrons*, but *not* cortical nephrons, can be punctured. Part 3 of the proximal tubule (the pars recta), the thick ascending limb of the loop of Henle, and the medullary collecting duct are not accessible to micropuncture in either cortical or juxtamedullary nephrons.

Perfusion of Microdissected Nephron Segments

The perfusion of microdissected nephron segments (Grantham and Burg, 1966) is a technique that can be used to study *all* nephron segments and has revealed significant functional specializations within the various segments of the nephron. Individual nephron segments are dissected by "teasing" and shaking small bits of tissue into smaller and smaller fragments while observing the process through a dissection microscope. The glomerulus and other landmarks such as the thin descending limb are used to identify the segments during the isolation procedure. The final isolated segment, which may be only 1–2 mm long and finer than a human hair, is placed in a bathing solution of known composition and then connected at both ends to micropipettes. Fluid is pumped through one pipette into the tubular lumen at a rate of a few nanoliters (10^{-9} l) per minute and is collected in the second pipette as it exits the nephron segment. Changes in volume and composition as well as in

the flux (movement) of radioactive solutes into or out of the tubular fluid can therefore be determined. In addition, a small electrode can be advanced through one of the micropipettes so that the transepithelial potential difference can be monitored. The rabbit kidney has been most extensively studied by these methods because of its relative ease of dissection. For unknown reasons, the rat kidney is virtually impossible to microdissect adequately. This is unfortunate, since most micropuncture data have been obtained in rats. Thus, a direct comparison between the findings obtained by micropuncture and those obtained in microdissected segments is hindered by potential species variations. A few segments of cortical collecting tubules have been dissected from human kidneys removed surgically as a consequence of renal disease (usually cancer of the kidneys), and the results obtained to date are similar to those obtained in cortical collecting tubules from rabbit kidneys.

Model Tissues

Much of what is known about transport and the hormonal regulation of transport across renal tubular epithelial cells has been learned from studies of epithelial cell layers from lower vertebrates, such as amphibians or reptiles. In these species, regulation of the ionic composition of the body fluids is not the exclusive domain of the kidneys, but also involves the skin (amphibians) and urinary bladder. The latter may be considered analogous to the collecting tubule/duct in mammalian kidneys, to which it is embryologically and functionally similar. The large size of the urinary bladder (over 10 cm^2 per animal) compared to the size of a tubule segment allows the transport of water and solutes such as Na^+, Cl^-, H^+, and urea to be monitored in a rapid and convenient fashion. The large size of the bladder also facilitates biochemical investigations, such as the study of cyclic AMP metabolism in relation to the action of antidiuretic hormone (see Chapter 31).

Another example of the use of model tissues in renal physiology has involved the study of fish. Some fish species, especially in the Antarctic ocean, are *aglomerular*, i.e., they lack glomeruli. Such a species was used to demonstrate unequivocally that tubular cells could secrete solutes into the tubular lumen (Shannon, 1938).

CHAPTER 28

Filtration and Blood Flow

Although the kidneys represent only 0.5% of the total body weight, they receive 20–25% of the total cardiac output. Thus, in an average adult with a cardiac output of 6 l/min, the *renal blood flow* (RBF) exceeds 1.2 l/min or 1700 l/day. This means that the total blood volume, which averages 6 l, passes through the renal vasculature nearly 300 times per day. The distribution of such a large share of the cardiac output to the kidneys is an important adaptation that enables the kidneys to regulate the normal quantity and composition of the body fluids.

Whereas the renal blood flow (RBF) represents the volume of *blood* flowing through the renal vasculature per unit time, the *renal plasma flow* (RPF) refers to the rate of *plasma* flow through the renal vasculature. If the hematocrit is 45%, RPF is 55% of RBF. The average RPF therefore exceeds 650 ml/min or 900 l/day.

As discussed in Chapter 27, approximately 125 ml of plasma are filtered each minute at the glomerulus, i.e., the *glomerular filtration rate* (GFR) is 125 ml/min or 180 l/day. Thus, approximately 20% of the RPF is filtered into Bowman's space. The ratio of GFR to RPF is termed the *filtration fraction* (FF).

PROPERTIES OF THE FILTRATION BARRIER

As discussed in Chapter 27, the filtration barrier consists of three layers: glomerular capillary endothelium, basement membrane, and epithelium. But in spite of its complex structure, the filtration properties of the barrier are qualitatively similar to those of other capillaries in the body. Thus, the barrier is freely permeable to water and to small solutes (*crystalloids*), such as ions, glucose, and urea. It has limited permeability to larger solutes (*colloidal particles*), such as large proteins and lipids, and is almost completely impermeable to the cellular elements of blood. Except for the absence of proteins and lipids, then, the glomerular filtrate is virtually identical to *plasma*. The term *ultrafiltrate* often is used when describing the glomerular filtrate, due to the exclusion of cellular elements *and* colloidal particles (a simple *filtrate* would only exclude the cellular elements of blood).

The size discrimination properties of the filtration barrier have been studied by comparing the concentration of a given substance in the filtrate to its concentration in plasma. Table 4.7 shows the ratio between the filtrate concentration and the plasma concentration for a variety of plasma constituents and infused substances. Note that the [filtrate]/[plasma] ratio is 1.0 for substances with molecular weights up to approximately 5000, whose effective radii are less than 15 Å, but falls sharply for larger molecules. For example, less than 1% of serum albumin ($\simeq$69,000 daltons; effective radius, $\simeq$36 Å) is filtered. Data such as these suggest that the filtration barrier behaves as if it had channels or *pores* up to approximately 60 Å in diameter (Navar, 1978a). The location of these pores in the filtration barrier has not been firmly established, however. The *holes* or *fenestrae* in the glomerular capillary endothelium (Chapter 27) are 500–1000 Å in diameter and are therefore much too large to prevent the filtration of large molecules. In the epithelial cell layer, the foot processes (pedicels) of adjacent podocytes appear to be separated by slits approximately 250 Å wide, also too large to account fully for the exclusion of large molecules. Present evidence suggests that the pores may be hydrated channels between the glycoprotein chains of the basement membrane. Such channels probably would be tortuous and might *not* be stable anatomical structures, which could explain why attempts to visualize pores by electron microscopy have not been successful.

While the [filtrate]/[plasma] ratios in Table 4.7 correlate fairly well with molecular weight and size, the extent to which a substance is filtered can depend on other factors as well. For example, the [filtrate]/[plasma] ratio for hemoglobin is three times larger than that for serum albumin, even though its molecular weight is just 3% less than that of albumin and its effective radius is just 8%

Table 4.7
Approximate [filtrate]/[plasma] ratios for selected substances

Substance	Molecular Weight, daltons	Effective Radius,[a] Å	[Filtrate]/ [Plasma]
Water	18	1.0	1.0
Urea	60	1.6	1.0
Glucose	180	3.6	1.0
Inulin	5000	14.8	0.98
Myoglobin	17,000	19.5	0.75
Egg albumin	43,500	28.5	0.22
Hemoglobin	68,000	32.5	0.03
Serum albumin	69,000	35.5	0.01

Adapted from Pitts (1974). [a] Estimated from diffusion coefficient.

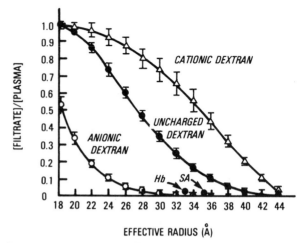

Figure 4.10. [Filtrate]/[plasma] ratios for uncharged dextran, anionic dextran, and cationic dextran as a function of effective radius. The [filtrate]/[plasma] ratios for hemoglobin (Hb) and serum albumin (SA) also are identified. [Adapted from Brenner et al., *American Journal of Physiology* 234: F455–F460, 1978.]

less than albumin's. One reason that hemoglobin is so much more readily filtered than serum albumin is *shape*: hemoglobin is a relatively compact, cylindrical molecule, whereas albumin is an asymmetrical ellipsoid. A second and more important reason involves *electrostatic factors*. Figure 4.10 illustrates how the [filtrate]/[plasma] ratio for *anionic* dextran is substantially less than that for *uncharged* dextran or *cationic* dextran of an equivalent size (Brenner et al., 1978). While both hemoglobin and albumin are anions and therefore have [filtrate]/ [plasma] ratios less than the corresponding uncharged or cationic dextran molecules (Fig. 4.10), albumin is *more* negatively charged than hemoglobin, which significantly hinders its filtration.

Substances with a [filtrate]/[plasma] ratio of 1.0 are said to be *freely filtered.* The amount of such a substance that is filtered per unit time can be readily calculated by multiplying its plasma concentration by the glomerular filtration rate. Thus, for a freely filtered substance X

$$\frac{\text{Mass of X filtered}}{\text{Time}} = \text{Plasma concentration of X}$$
$$\times \frac{\text{Volume of plasma filtered}}{\text{Time}}$$

or $$\frac{\text{Mass of X filtered}}{\text{Time}} = P_X \text{GFR} \qquad (27)$$

The term *filtered load* often is used when referring to the quantity $P_X \text{GFR}$.

While freely filtered substances must have a molecular weight of ≤ 5000 daltons (Table 4.7), a further requirement is the absence of significant binding to plasma proteins such as albumin. The filtration of certain ions and small solutes (e.g., Ca^{++}, Mg^{++}, bilirubin) is markedly reduced as a result of plasma protein binding. The filtered load of a substance that exhibits significant plasma protein binding is given by

$$\frac{\text{Mass of X filtered}}{\text{Time}} = P_X \text{GFR} \cdot F_X \qquad (28)$$

where F_X is the fraction of the substance in plasma that is *free* (i.e., unbound).

As indicated in Table 4.7, many of the small solutes in plasma have [filtrate]/[plasma] ratios of 1.0. However, if *very* precise concentration measurements are made, one finds that the filtrate and plasma concentrations of these solutes are *not* exactly the same. The reasons for such discrepancies have been introduced in Chapter 26. First, plasma concentration values typically are measured in clinical laboratories as millimoles per liter of *plasma volume.* In order to compare the plasma and filtrate concentrations of small solutes, the plasma concentrations must be *corrected* to account for the fact that these solutes are dissolved only in the *aqueous phase* of plasma, which normally represents about 93% of the plasma volume. Since only the aqueous phase is filtered, plasma concentrations must be expressed as millimoles per liter of *plasma water* in order to be compared to glomerular filtrate concentrations. Laboratory values expressed as millimoles per liter of *plasma volume* can be converted to millimoles per liter of *plasma water* by dividing by 0.93 (Chapter 26).

Once such conversions are made, the [filtrate]/ [plasma] ratios of small *nonelectrolytes* do in fact equal 1.0, but small *electrolytes* still have slightly different concentrations in plasma and filtrate. These remaining differences can be attributed primarily to the *Gibbs-Donnan effect* (Chapter 26). The presence of proteins in plasma, but not in the filtrate, results in the establishment of a Gibbs-

Donnan equilibrium between plasma and filtrate. Since the plasma proteins are negatively charged, *cations* such as Na^+ and K^+ have somewhat higher concentrations in *plasma water*, while *anions* such as Cl^- and HCO_3^- have somewhat higher concentrations in the *filtrate*. However, the concentration differences attributable to the Gibbs-Donnan effect are small and generally are neglected in renal physiology.

FORCES INVOLVED IN FILTRATION

The *Starling principle* provides a framework for analyzing fluid movement across capillaries. According to this principle, the rate and direction of fluid movement is determined by the balance of hydrostatic and oncotic pressures, as discussed in Chapter 6. The Starling principle can be expressed as follows:

$$\dot{q} = K_f[(P_c - P_i) - (\pi_c - \pi_i)] \qquad (29)$$

where* $\dot{q}$ = rate of fluid movement across the capillary, K_f = filtration coefficient, P_c = capillary hydrostatic pressure, P_i = interstitial hydrostatic pressure, π_c = plasma oncotic pressure, and π_i = interstitial oncotic pressure. The hydrostatic pressure gradient $(P_c - P_i)$ favors *filtration*, while the oncotic pressure gradient $(\pi_c - \pi_i)$ favors *reabsorption*. *Eq. 29* is written such that a *positive* value of $\dot{q}$ signifies *net filtration*, whereas a *negative* $\dot{q}$ signifies *net reabsorption*. Thus, the expression $[(P_c - P_i) - (\pi_c - \pi_i)]$ has been termed the *net filtration pressure*.

It is instructive to review how the net filtration pressure varies along the length of a typical systemic (*extrarenal*) capillary (Table 4.8*A*, Fig. 4.11*A*). Note that at the arterial end the balance of pressures favors *filtration* (net filtration pressure ≃ +16 mmHg). However, at the venous end *reabsorption* occurs (net filtration pressure ≃ −14 mmHg), since capillary hydrostatic pressure (P_c) declines markedly along the length of the capillary. The result is that filtration and reabsorption approximately balance. Actually, if the net filtration pressure is averaged along the capillary, a *mean* net filtration pressure of 8 mmHg is obtained. A slight excess of plasma therefore is filtered into the interstitium, although the net volume filtered in *all* extrarenal capillaries probably is less then 2 ml/min. This volume is returned to the systemic circulation via the lymphatics.

---

* Note that in *Eq. 29* and in the discussion pertaining to it, P is used to represent *hydrostatic pressures*, *not* plasma concentrations.

For *glomerular* capillaries, the Starling equation (*29*) can be rewritten as

$$GFR = K_f[(P_G - P_B) - (\pi_G - \pi_B)] \qquad (30)$$

where the subscripts G and B refer to the glomerular capillaries and Bowman's space, respectively. Since the amount of protein filtered into Bowman's space is negligible, π_B approaches zero and *Eq. 30* simplifies to

$$GFR = K_f[P_G - P_B - \pi_G] \qquad (31)$$

The variation in the net filtration pressure along the length of a glomerular capillary (Brenner and Humes, 1977) differs somewhat from that in extrarenal capillaries (Table 4.8*B*, Fig. 4.11*B*). Most importantly:

1. Capillary hydrostatic pressure remains relatively constant along the glomerular capillary, in contrast to its marked decline in extrarenal capillaries.
2. Because the volume of fluid filtered in glomerular capillaries is so large, the plasma oncotic pressure rises along the length of a glomerular capillary, whereas it remains relatively constant in extrarenal capillaries.

Thus, *filtration* occurs at the afferent end (net filtration pressure ≃ 10 mmHg), as in extrarenal capillaries. However, at the efferent end fluid movement *ceases* (net filtration pressure ≃ 0). *Filtration equilibrium* therefore is said to be present at the efferent end of glomerular capillaries.

The *mean* net filtration pressure in the glomerular capillaries cannot be determined precisely, since the exact point along the capillary at which filtration equilibrium is achieved is not known (Blantz, 1978). It has been estimated, however, that the mean net filtration pressure in glomerular capillaries is similar to that in extrarenal capillaries, ≃8 mmHg. The fact that the GFR averages 125 ml/min, whereas the total net filtration rate in *all* extrarenal capillaries is less than 2 ml/min, must therefore be attributed to differences in the *filtration coefficient* K_f. The larger filtration coefficient of glomerular capillaries probably is due primarily to the numerous fenestrae in the glomerular capillary endothelium (Chapter 27).

As indicated in Table 4.8*B*, the hydrostatic pressure in Bowman's space (P_B) is approximately 10 mmHg. It should be noted that this pressure not only is an important determinant of the net filtration pressure, but also serves to propel the glomerular filtrate through the proximal tubule and the remainder of the nephron.

Table 4.8

Net filtration pressure in extrarenal vs. glomerular capillaries

	A. Extrarenal Capillary			B. Glomerular Capillary	
	Arterial End	Venous End		Afferent End	Efferent End
P_c	40 mmHg	10 mmHg	P_G	45 mmHg	45 mmHg
P_i	2	2	P_B	10	10
π_c	25	25	π_G	25	35
π_i	3	3	π_B	0	0
Net filtration pressure = $(P_c - P_i) - (\pi_c - \pi_i)$	+16 mmHg	−14 mmHg	Net filtration pressure = $(P_G - P_B) - (\pi_G - \pi_B)$	+10 mmHg	≃0

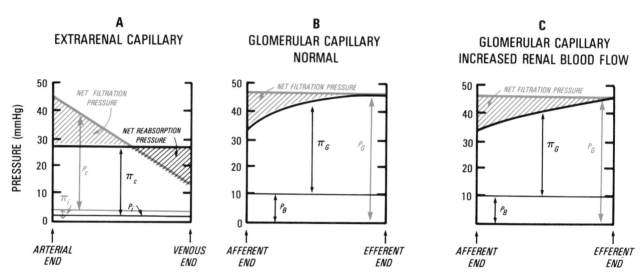

Figure 4.11. Variations in pressures affecting filtration along the length of extrarenal and glomerular capillaries. Pressures favoring *filtration* (the capillary hydrostatic pressure P_c and interstitial oncotic pressure π_i in *extrarenal* capillaries; the glomerular capillary hydrostatic pressure P_G in *glomerular* capillaries) are illustrated in *red*, while those favoring *reabsorption* (the interstitial hydrostatic pressure P_i and plasma oncotic pressure π_c in *extrarenal* capillaries; the hydrostatic pressure in Bowman's space P_B and glomerular capillary oncotic pressure π_G in *glomerular* capillaries) are illustrated in *black*. In *extrarenal* capillaries (A), *filtration* occurs at the arterial end and *reabsorption* occurs at the venous end. In *glomerular* capillaries (B), *filtration* occurs at the afferent end and *filtration equilibrium* is present at the efferent end. The point along the glomerular capillary at which filtration equilibrium is achieved is displaced toward the efferent end if RBF increases (C), since the glomerular capillary oncotic pressure π_G rises less rapidly.

REGULATION OF RBF AND GFR

Renal blood flow, like the blood flow to any organ, is controlled by the arteriovenous pressure difference across the vascular bed (the *perfusion pressure*) and the vascular resistance, as expressed by the general resistance equation (Chapter 5)

$$\text{RBF} = \frac{\Delta P}{R} \qquad (32)$$

The perfusion pressure for the renal vasculature is identified in Figure 4.12, which illustrates the normal hydrostatic pressure profile of the renal circulation (*curve A*). Note that the renal arterial pressure is essentially the same as that of other systemic arteries, with a mean normal value of 100 mmHg. Also note that the glomerular capillary hydrostatic pressure (P_G) is approximately 45 mmHg, as previously indicated (Table 4.8B). The large pressure drop that occurs in the afferent and efferent arterioles identifies these vessels as the major sites of

renal vascular resistance (Blantz, 1980). Changes in the diameters of these vessels represent the primary mechanism for adjusting the renal vascular resistance.

The factors that control *glomerular filtration rate* can be readily identified with the aid of the Starling equation (*Eq. 30*). In the normal kidney, changes in *glomerular capillary hydrostatic pressure* are an important cause of changes in GFR. Just as the hydrostatic pressure in a systemic capillary is controlled by the pressures and resistances in the adjacent arteriole and venule (Chapter 5), glomerular capillary hydrostatic pressure is controlled by the pressures and resistances in the adjacent afferent arteriole and efferent arteriole. Figure 4.12 illustrates examples of how changes in these pressures and resistances can affect P_G. Note that the effects of afferent arteriolar resistance changes on P_G, and hence GFR, are *opposite* to those of efferent arteriolar resistance changes. For example, an increase

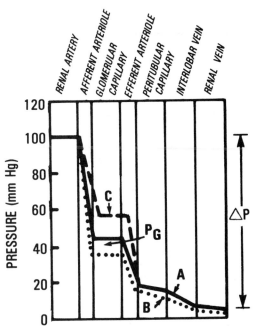

Figure 4.12. Hydrostatic pressure profile of the renal circulation. *A*, Normal profile. *B*, Following *afferent* arteriolar constriction and consequent *decrease* in glomerular capillary hydrostatic pressure (P_G). *C*, Following *efferent* arteriolar constriction and consequent *increase* in glomerular capillary hydrostatic pressure. ΔP represents the normal perfusion pressure of the renal circulation.

in *afferent* arteriolar resistance (e.g., due to afferent arteriolar constriction) *decreases* P_G (Fig. 4.12, *curve B*), whereas an increase in *efferent* arteriolar resistance *increases* P_G (Fig. 4.12, *curve C*). In contrast, the effects of afferent and efferent arteriolar resistance changes on RBF are the *same*. For example, an increase in either afferent *or* efferent arteriolar resistance will decrease RBF.

While changes in P_G are an important cause of changes in GFR, the other terms in the Starling equation also can affect the GFR, particularly in disease states (Tucker and Blantz, 1977). For example, the *filtration coefficient* (K_f) can be decreased by diseases that cause a thickening of the filtration barrier or decrease its surface area by destroying glomerular capillaries. The *hydrostatic pressure in Bowman's space* (P_B) increases in ureteral obstruction. Small changes in *glomerular capillary oncotic pressure* (π_G) can occur with dehydration (increased π_G) or hypoalbuminemia (decreased π_G). The *oncotic pressure in Bowman's space* (π_B) normally is negligible (*Eq. 31*), but it can be increased by diseases that increase the permeability of the filtration barrier, thereby allowing plasma proteins to leak into the filtrate. Furthermore, although not an explicit term in the Starling equation, *RBF* also can be an important determinant of GFR. This is because the rate at which plasma enters the glomerular capillaries will determine the

rate at which the plasma oncotic pressure (π_G) rises along the length of the glomerular capillary. For example, if filtration equilibrium normally occurs at a point two-thirds of the distance between the afferent and efferent ends of the glomerular capillary (Fig. 4.11*B*), an increase in RBF will increase GFR because π_G rises less rapidly, thereby displacing the point at which the net filtration pressure becomes zero further toward the efferent end of the capillary (Fig. 4.11*C*).

Autoregulation

From the above discussion, the changes in systemic arterial pressure that occur in different physiological situations would be expected to have significant effects on RBF and GFR. However, the maintenance of the normal quantity and composition of the body fluids by the kidneys is a continuous process that is performed most effectively if RBF and GFR remain relatively *constant* (Navar, 1978b). Figure 4.13 illustrates that RBF and GFR do in fact remain nearly constant even in the presence of large changes in mean systemic arterial pressure. This maintenance of a constant blood flow in spite of changes in systemic arterial pressure is termed *autoregulation*. While autoregulation also occurs in the heart, brain, and other organs (Chapter 6), the constancy of flow is quite striking in the kidneys. For example, in the dog RBF and GFR change by less than 10% when mean systemic arterial pressure is varied between 80 and 180 mmHg (Shipley and Study, 1951), and a similar autoregulatory range is believed to be present in man. Note that because of autoregulation, the *filtration fraction* (GFR/RPF) also remains relatively constant in this pressure range.

From *Eq. 32*, it is evident that in the autoregulatory range an increase in arterial pressure must be accompanied by an increase in renal vascular

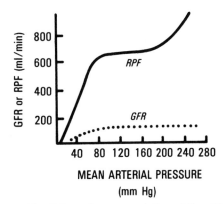

Figure 4.13. Schematic representation of the effect of mean arterial pressure on glomerular filtration rate (GFR) and renal plasma flow (RPF), illustrating the phenomenon of autoregulation.

resistance in order to maintain a constant RBF. But since GFR also is autoregulated, this increase in resistance must occur without significantly altering the glomerular capillary hydrostatic pressure. The *afferent arteriole* must therefore represent the major site of the autoregulatory resistance change, as illustrated in Figure 4.14.

In spite of intensive investigation, the mechanism for autoregulation is uncertain (Navar, 1978b). The mechanism is definitely *intrinsic* to the kidneys, since autoregulation can be demonstrated in transplanted, denervated kidneys as well as in isolated kidneys perfused in vitro. One theory, the *myogenic hypothesis*, attributes the increase in renal vascular resistance that accompanies an increase in systemic arterial pressure to contraction of the afferent arteriolar smooth muscle in response to stretch. The ability of arteriolar smooth muscle cells to contract when stretched has been demonstrated in vessels isolated from a variety of organs; in fact, the myogenic hypothesis originally was proposed as a mechanism for autoregulation in organs other than the kidneys (Chapter 6). The so-called *juxtaglomerular hypothesis* attributes autoregulation to changes in the rate of renin release from the granular cells, which in turn alters the local concentration of angiotensin II, a powerful vasoconstrictor (see below). One version of this

hypothesis suggests that it is the amount of Na^+ or other solutes delivered to the distal nephron, as "sensed" by the granular cells, that affects the rate of renin release.

As noted above, autoregulation represents an *intrinsic* mechanism for keeping RBF and GFR constant in the face of changing systemic arterial pressure. However, RBF and GFR also can be regulated by neural and hormonal influences. These *extrinsic factors* can *override* the autoregulatory mechanism, altering RBF and GFR even in the autoregulatory range of systemic arterial pressure.

Neural Regulation

Like most blood vessels, the vessels of the kidneys are innervated exclusively by vasoconstrictor fibers from the sympathetic division of the autonomic nervous system. At rest, sympathetic tone to the renal vasculature appears to be minimal, so that the vessels are almost maximally dilated. The renal vasoconstriction resulting from sympathetic stimulation causes a reduction in both RBF and GFR. However, with moderate sympathetic stimulation, GFR does not decrease as much as RBF; the vasoconstriction therefore must involve both the afferent *and* efferent arterioles. The filtration fraction, then, actually *increases* at moderate levels of sympathetic stimulation. With further sympathetic stimulation, the decreases in GFR parallel those in RBF; apparently, afferent arteriolar constriction predominates at these higher levels of stimulation.

It is important to note that the sympathetically mediated renal vasoconstriction represents part of the body's mechanism for controlling *systemic arterial pressure* rather than a mechanism for regulating RBF and GFR. Such vasoconstriction is initiated as a reflex response to a decrease in systemic arterial pressure, as sensed by the carotid sinus and aortic arch baroreceptors, e.g., during postural changes, hemorrhage, or syncope. A similar response can be initiated by decreases in pressure in the atria or pulmonary vessels, as sensed by baroreceptors located in their walls. In all of these situations, the renal vasoconstriction contributes to a rise in total peripheral vascular resistance, which in turn helps to restore the arterial pressure to its normal value (see Chapter 5). Sympathetic vasoconstriction also can be initiated in response to input from the CNS, for example during fright, pain, cold, exercise, and other stressful situations. The fact that GFR does not decrease as much as RBF at moderate levels of sympathetic stimulation can be regarded as an adaptation designed to maintain GFR as high as possible even if renal vascular

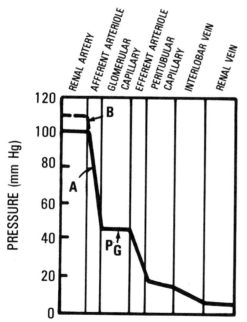

Figure 4.14. Changes in the hydrostatic pressure profile of the renal circulation during autoregulation. *A*, Normal profile. *B*, Following elevation of mean arterial pressure, afferent arteriolar constriction maintains constant glomerular capillary hydrostatic pressure (P_G).

resistance must increase to control systemic arterial pressure.

Hormonal Regulation

Many hormones and other endogenous substances can cause renal vasoconstriction or vasodilation. The substance most likely to have a physiological role in the regulation of RBF and GFR is *angiotensin II*, a potent vasoconstrictor that is synthesized in response to the release of renin from the granular cells (Chapter 27) (Wright and Briggs, 1979). Since renin is released in response to decreased renal perfusion and also in response to sympathetic stimulation of the granular cells (Chapter 32), at least part of the renal vasoconstriction occurring as a result of a decrease in systemic arterial pressure may be mediated by angiotensin II.

The *prostaglandins* also may have a physiological role in the regulation of RBF and GFR (Baer, 1981). The kidneys can synthesize several members of the prostaglandin family, including prostaglandin E_2, prostacyclins, leukotrienes, and thromboxanes. In terms of effects on RBF and GFR, the *vasodilator* prostaglandins (such as prostaglandin E_2) are believed to be most important. These prostaglandins may modulate the constrictor effects of sympathetic stimulation and angiotensin II; in fact, the synthesis of vasodilator prostaglandins appears to be *increased* by sympathetic stimulation or increased angiotensin II levels. A role for the vasodilator prostaglandins also is supported by the fact that inhibitors of prostaglandin synthesis (e.g., indomethacin) may cause a dramatic decrease in RBF and GFR in patients with diseases that impair perfusion of the kidneys (e.g., congestive heart failure, cirrhosis of the liver). Since RBF and GFR appear to be altered little by inhibitors of prostaglandin synthesis in normal individuals, it has been suggested that the vasodilator prostaglandins have their most significant effects on RBF and GFR when renal perfusion is compromised by abnormal conditions or disease. *Bradykinin*, a potent vasodilator, may have a similar role in modulating RBF and GFR.

Antidiuretic hormone (ADH), which is of major importance in the urinary concentrating mechanism (Chapter 31), is a vasoconstrictor and can markedly decrease RBF and GFR. *Serotonin* also causes renal vasoconstriction, reducing RBF and GFR. However, since the effects of ADH and serotonin have primarily been demonstrated with large, pharmacological doses, the physiological role of these substances in the regulation of RBF and GFR is uncertain.

INTRARENAL DIFFERENCES IN RBF AND GFR

Approximately 90% of the total RBF perfuses the *cortex*; only 10% perfuses the *medulla* (Barger and Herd, 1971). This marked discrepancy between cortical and medullary blood flow is not due to a difference in the number of blood vessels per unit of renal tissue, since the vascular volume is approximately 20% of the total tissue volume in both cortex and medulla. Instead, the reduced blood flow to the medulla apparently results from a relatively high vascular resistance in the vasa recta. Whether this high resistance is due to the length of these vessels, an increased viscosity of medullary blood, or other factors is unknown. It should be noted that the existence of a low medullary blood flow has considerable significance in the urinary concentrating mechanism (Chapter 31).

RENAL OXYGEN CONSUMPTION

As noted at the beginning of this chapter, the high blood flow to the kidneys is related to their role in regulating the quantity and composition of the body fluids. Thus, the kidneys differ from most other organs, in which blood flow is related to the oxygen requirements of the organ. Since the flow to the kidneys is so high relative to their oxygen needs, the renal arteriovenous oxygen difference is quite low, approximately 1.7 ml O_2/100 ml blood, compared to 4–5 ml O_2/100 ml blood for the body as a whole. The relationship between RBF, the renal arteriovenous oxygen difference, and renal oxygen consumption is expressed by the Fick equation (Chapter 6)

$$RBF = \frac{\dot{Q}_{O_2}}{RA_{O_2} - RV_{O_2}} \qquad (33)$$

where $\dot{Q}_{O_2}$ is the rate of renal oxygen consumption and $RA_{O_2} - RV_{O_2}$ is the renal arteriovenous oxygen difference.

In most organs, if the blood flow is varied, the *arteriovenous oxygen difference* changes in order to meet the oxygen needs of the organ (assuming, of course, that the metabolic activity of the organ, and hence its consumption of oxygen, remains constant). The kidneys are unique in that changes in blood flow are accompanied by parallel changes in *oxygen consumption*, with the arteriovenous oxygen difference remaining the same. The explanation for this unusual behavior is that a change in RBF generally is accompanied by a parallel change in GFR, and hence in the quantity of ions and other solutes that must be reabsorbed (Kiil et al., 1961). Solute reabsorption, in turn, requires oxygen for

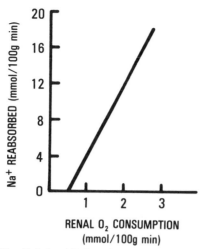

Figure 4.15. Relationship between renal oxygen consumption and amount of Na$^+$ reabsorbed by the kidneys. [Adapted from Deetjen and Kramer, *Pfluegers Archiv* 273: 636–650, 1961.]

energy, with as much as 75–85% of the renal oxygen consumption being used to support the active reabsorption of ions and other solutes, particularly Na$^+$. In fact, renal oxygen consumption is directly proportional to the amount of Na$^+$ reabsorbed (Fig. 4.15). Changes in *solute reabsorption*, then, are responsible for the changes in oxygen consumption that accompany changes in RBF.

Synthesis of Erythropoietin

Having considered renal *oxygen consumption*, it is appropriate to note that the kidneys also have a role in monitoring the adequacy of *oxygen delivery* to the tissues in general. In response to renal hypoxia, the kidneys synthesize *erythropoietin*, a glycoprotein hormone that regulates the production of red blood cells (erythrocytes) from precursor cells in the bone marrow (Fisher, 1980). For example, the elevated hematocrit (polycythemia) occurring upon exposure to high altitude or in chronic lung disease can be attributed to increased production of erythropoietin. Many diseases of the kidneys are associated with an erythropoietin deficiency, and hence anemia.

The exact site of erythropoietin synthesis in the kidney is not yet established, but has been postulated to be the glomerulus, renal medulla, and/or the JG apparatus. Actually, some investigators believe that the kidneys do not really synthesize erythropoietin, but instead produce an enzyme that acts on a plasma globulin to form erythropoietin.

MEASUREMENT OF GFR

The glomerular filtration rate can be measured with the aid of certain substances that are freely filtered at the glomerulus but are *not* reabsorbed or secreted by the tubule. *Inulin*, a fructose polymer with molecular weight $\simeq$5000 daltons, can be considered the "prototype" example of such a substance (Shannon and Smith, 1935). Since inulin is neither removed from the filtrate by reabsorption nor added to the filtrate by secretion, the entire amount filtered will be excreted in the urine. Hence, the rate of excretion must equal the filtered load

$$\frac{\text{Mass of inulin excreted}}{\text{Time}} = \frac{\text{Mass of inulin filtered}}{\text{Time}}$$

or
$$U_{In}\dot{V} = P_{In}GFR \qquad (34)$$

If *Eq. 34* is substituted into the expression for the *clearance* of inulin (*Eq. 26*), the following significant result is obtained

$$C_{In} = \frac{U_{In}\dot{V}}{P_{In}} = \frac{P_{In}GFR}{P_{In}}$$

or
$$C_{In} = GFR \qquad (35)$$

C_{In}, of course, can be calculated from measurements made on plasma and urine samples. Thus, *Eq. 35* defines a *noninvasive* method for measuring GFR.

That GFR can be equated to C_{In} is actually quite logical. If some of the inulin in the filtrate were reabsorbed, the volume of plasma "cleared" of inulin would be *smaller* than the volume filtered. On the other hand, if some of the inulin that escaped filtration were added to the filtrate by secretion, the volume of plasma "cleared" would be *greater* than the volume filtered. But inulin is neither reabsorbed nor secreted, so the volume of plasma "cleared" of inulin per unit time (C_{In}) is *equal* to the volume filtered per unit time (GFR).

Because C_{In} is equal to GFR, C_{In} and GFR often are used interchangeably in discussions of renal function. It should be emphasized that the key characteristic of inulin necessary for the equality of C_{In} and GFR is the special way that the kidneys handle inulin: inulin is freely filtered, but neither reabsorbed nor secreted. In addition, inulin is nontoxic, is not bound by plasma proteins, is neither metabolized nor synthesized by the kidneys, and has no effect on GFR.

Although inulin is considered the "prototype" substance for measuring GFR, other suitable substances have been identified, most notably *mannitol* and *iothalamate*, a contrast medium used in diagnostic radiology. However, all of these substances must be infused intravenously until a stable plasma concentration is achieved, an inconvenience for routine clinical determinations of GFR. A more common practice in the clinical setting is to use the endogenous substance *creatinine* to *estimate* GFR. Creatinine is an organic base formed during muscle protein metabolism as a degradation product of creatine phosphate. Like many other organic bases (see Chapters 29 and 30), creatinine is filtered at the glomerulus and secreted into the tubular lumen by the proximal tubular epithelial cells. However, at normal plasma levels in man the amount of creatinine secreted is only about 10–15% of the amount filtered. This means that

$$\frac{\text{Mass of creatinine excreted}}{\text{Time}} \simeq \frac{\text{Mass of creatinine filtered}}{\text{Time}}$$

or
$$U_{Cr}\dot{V} \simeq P_{Cr}\text{GFR} \qquad (36)$$

As an *approximation*, then, creatinine behaves like substances that are filtered only, such as *inulin* (*Eq. 34*).† The creatinine clearance can therefore be used clinically to obtain an approximate value for GFR

$$C_{Cr} = \frac{U_{Cr}\dot{V}}{P_{Cr}} \simeq \frac{P_{Cr}\text{GFR}}{P_{Cr}}$$

$$\simeq \text{GFR} \qquad (37)$$

Because a small amount of creatinine is secreted in man, C_{Cr} is of course slightly greater than C_{In}. The GFR value obtained from a C_{Cr} measurement is therefore slightly *greater* than the actual GFR, as determined from C_{In}. However, the size of the error is less than anticipated because the standard colorimetric methods for determining P_{Cr} measure other chromagens as well as true creatinine. The P_{Cr} values reported by clinical laboratories are therefore somewhat greater than the true P_{Cr} levels. This slight overestimate of P_{Cr} tends to reduce the C_{Cr} calculated with *Eq. 37* to a value closer to the actual GFR.

Clinically, when obtaining data for a creatinine clearance calculation, P_{Cr} typically is determined from a single plasma sample. However, a 24-hour urine collection generally is required for an accurate determination of $U_{Cr}\dot{V}$. Failure to collect a complete 24-hour urine sample is a common cause of erroneous C_{Cr} values. A simple method that can be used to check for a complete urine collection is to compare the C_{Cr} value determined from P_{Cr} and the 24-hour urine collection to the C_{Cr} value *estimated* from P_{Cr} *only*. Such an estimate is obtained from an empirical calculation that takes advantage of the dependence of creatinine production on muscle mass, which in turn is determined by age and body weight (Cockroft and Gault, 1976):

---

† In the dog, creatinine is handled *exactly* like inulin, i.e., it is freely filtered but neither reabsorbed nor secreted.

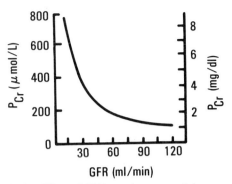

Figure 4.16. Effect of GFR on plasma creatinine concentration (P_{Cr}). Note that P_{Cr} is a sensitive indicator of GFR only at low GFRs.

$$C_{Cr} \simeq \frac{(140 - \text{age}) \times \text{body weight}}{72 \times P_{Cr}} \qquad (38)$$

When using this experimentally derived equation, age is expressed in *years*, body weight in *kg*, and P_{Cr} in *mg/dl*. The C_{Cr} value calculated from *Eq. 38* will be within 35% of the C_{Cr} value determined from P_{Cr} and the 24-hour urine collection in 95% of patients. *Eq. 38* is *not* accurate in conditions in which the ratio of muscle mass to body weight is distorted, such as pregnancy, extreme obesity, or diseases that cause muscular wasting.

While C_{Cr} represents a useful clinical approximation for GFR, important information about GFR actually can be obtained from a simple P_{Cr} measurement. This is because the rate of creatinine excretion must equal the rate of creatinine production, and therefore tends to be constant from day to day in the steady state, i.e.,

$$U_{Cr}\dot{V} = \text{constant}$$

Since $U_{Cr}\dot{V} = P_{Cr}C_{Cr}$, the quantity $P_{Cr}C_{Cr}$ also tends to be constant

$$P_{Cr}C_{Cr} = \text{constant}$$

or
$$P_{Cr}\text{GFR} \simeq \text{constant} \qquad (39)$$

P_{Cr} therefore is *inversely proportional* to the GFR, as illustrated in Figure 4.16.

An important feature of the GFR vs. P_{Cr} curve is its *shape*. Note that P_{Cr} changes little when GFR falls from its normal value of 125 ml/min to a value as low as 60 ml/min. However, P_{Cr} begins to increase markedly with further decreases in GFR. This means that P_{Cr} is a sensitive indicator of GFR only at fairly low GFRs. Thus, P_{Cr} is not useful for detecting small or even moderate decreases in GFR due to minor impairment of renal function, but is extremely useful for detecting the large decreases in GFR seen in severe renal dysfunction. The upper limit of normal for P_{Cr} is approximately 110 μmol/l (1.25 mg/dl) in men and 100 μmol/l (1.1 mg/dl) in women.

MEASUREMENT OF RPF

The Fick equation, introduced in the discussion of renal oxygen consumption (*Eq. 33*), provides the basis for measuring RPF. In a somewhat more general form, this equation can be written as

$$RBF = \frac{\dot{Q}_x}{RA_x - RV_x} \qquad (40)$$

where $\dot{Q}_x$ is the rate at which a substance X is consumed by the kidneys and RA_x and RV_x represent the concentrations of X in renal arterial and renal venous blood, respectively. When applied to renal *plasma* flow instead of renal *blood* flow, the equation becomes

$$RPF = \frac{\dot{Q}_x}{P_{RA_x} - P_{RV_x}} \qquad (41)$$

where P_{RA_x} and P_{RV_x} represent the concentrations of X in renal arterial and renal venous plasma, respectively. For substances that are not metabolized or synthesized by the kidneys, the rate at which the substance is "consumed" by the kidneys can be equated to its rate of *excretion*, $U_x\dot{V}$. Furthermore, if the substance is not "consumed" by tissues other than the kidneys, P_{RA_x} can be equated to P_X, the concentration of X measured in *systemic* arterial or venous plasma (a venous sample generally is used for convenience). Thus

$$RPF = \frac{U_X\dot{V}}{P_X - P_{RV_x}} \qquad (42)$$

A variety of substances could be used for the determination of RPF with *Eq. 42*. But whereas the rate of excretion $U_X\dot{V}$ and the systemic plasma concentration P_X are easily determined, renal venous plasma levels P_{RV_x} are *not* readily obtainable in routine clinical studies. The most ideal substance for a *noninvasive* determination of RPF with the Fick equation, then, would be one with a renal venous plasma concentration of *zero*, i.e., a substance that is *totally removed* from renal arterial plasma by the kidneys. For such a substance, the Fick equation (*Eq. 42*) reduces to the *clearance equation* (*Eq. 26*):

$$RPF = \frac{U_X\dot{V}}{P_X - 0}$$
$$= C_X \qquad (43)$$

This result is a logical one, since a renal venous plasma level of zero would indicate that the total RPF had been "cleared" of X.

What type of substance might have a renal venous plasma concentration of zero? Because only 20% of the RPF is filtered, the substance could *not* be completely "cleared" from renal plasma by filtration alone; it also would have to be added to the tubular fluid by secretion. Furthermore, this secretion would have to be efficient enough to remove virtually all of the substance that was not filtered. The substance that comes closest to meeting these criteria is *p-amino hippurate* (PAH). PAH is filtered at the glomerulus and, like other organic acids, is secreted by the proximal tubular epithelial cells, i.e., it is transported from the peritubular capillaries surrounding the proximal tubule into the tubular lumen. Furthermore, at low plasma levels of PAH this secretion is so efficient that virtually all of the PAH in the peritubular capillaries is secreted into the proximal tubular lumen. But even PAH is not *completely* "cleared" from renal plasma because 10–15% of the total RPF supplies portions of the kidneys that *cannot* remove PAH (or any other substance), e.g., the renal capsule, renal pelvis, perirenal fat, medulla, and papilla. Thus, the renal venous plasma concentration of PAH (or any other substance) is *not* zero. The so-called *extraction ratio* (E_{PAH}) is used to represent the fraction of PAH that is "cleared," or *extracted*, from renal plasma in a single passage through the kidneys:

$$E_{PAH} = \frac{P_{PAH} - P_{RV_{PAH}}}{P_{PAH}} \qquad (44)$$

Since 10–15% of the total RPF is not "cleared" of PAH, E_{PAH} has a value of 0.85–0.90.

But while C_{PAH} therefore only *approximates* the total RPF, it *does* accurately measure the rate of plasma flow to regions of the kidneys that *can* remove PAH, the so-called *effective renal plasma flow* (ERPF)

$$C_{PAH} = \frac{U_{PAH}\dot{V}}{P_{PAH}}$$
$$= ERPF \qquad (45)$$

By combining *Eqs. 42* and *44*, it is evident that

$$ERPF = RPF \cdot E_{PAH} \qquad (46)$$

For clinical purposes, the ERPF values obtained from C_{PAH} measurements provide a useful approximation for the total RPF. *Iodohippurate* has an extraction ratio similar to PAH and therefore is another substance whose clearance can be used to approximate the total RPF.

It should be noted that once *RPF* has been determined, *RBF* can be calculated from the following equation

$$RBF = \frac{RPF}{1 - Hct} \qquad (47)$$

where Hct is the *hematocrit*.

CALCULATION OF FILTRATION FRACTION

Table 4.9 summarizes the methods presented in this chapter for measuring the GFR and RPF. Once GFR and RPF have been determined, the *filtration fraction* can readily be calculated. The most accurate value of the filtration fraction would be obtained as follows:

$$FF = \frac{C_{In}}{RPF} \qquad (48)$$

Table 4.9
Methods for determining GFR and RPF

	Most Accurate Determination	Useful Approximation for Clinical Purposes
GFR	C_{In}	C_{Cr}
RPF	Fick equation	C_{PAH}

where RPF is determined with the Fick equation. However, an approximation useful for clinical purposes is

$$FF \simeq \frac{C_{Cr}}{C_{PAH}} \qquad (49)$$

CHAPTER 29

Introduction to Tubular Function

As indicated in Chapter 27, the tubular portion of the nephron modifies the contents of the glomerular filtrate so that the final urine contains only those constituents that must be excreted to preserve the normal volume and composition of the body fluids. This modification involves both *tubular reabsorption*, the process whereby water and other essential substances in the glomerular filtrate are *recovered* and returned to the body via the peritubular capillaries, and *tubular secretion*, the process whereby substances in the peritubular capillaries are transported across the tubular epithelium and *added* to the tubular fluid. In this chapter, a general introduction to these important functions of the tubule will be presented.

TUBULAR REABSORPTION

Since approximately 180 l of plasma are filtered each day in man, the tremendous importance of tubular reabsorption in conserving the body stores of water and essential solutes should be obvious. For a substance X, the amount reabsorbed per unit time can be calculated as the difference between the filtered load and the rate of excretion:

$$\frac{\text{Mass of X reabsorbed}}{\text{Time}}$$
$$= \frac{\text{Mass of X filtered}}{\text{Time}} - \frac{\text{Mass of X excreted}}{\text{Time}}$$

or
$$T_X = P_X GFR - U_X \dot{V} \qquad (50)$$

where T_X represents the rate of tubular reabsorption of X. Table 4.10 presents typical reabsorption rates for several major plasma constituents, emphasizing the magnitude of the reabsorption process. Also listed are typical *fractional reabsorption* values for these constituents, i.e., the fraction of the filtered load that is reabsorbed. The fractional reabsorption can be calculated as follows:

$$\text{Fractional reabsorption} = \frac{T_X}{P_X GFR} \qquad (51)$$

Note that the reabsorption of certain essential constituents, such as glucose and amino acids, is virtually *complete*, i.e., the *entire* filtered load is reabsorbed.

Eq. 50 can be used to derive an expression for the *clearance* of a substance that is filtered and then reabsorbed. Rearranging this equation gives

$$U_X \dot{V} = P_X GFR - T_X \qquad (52)$$

Eq. 52 can be substituted into the clearance equation $C_X = U_X \dot{V}/P_X$ to give

$$C_X = \frac{P_X GFR - T_X}{P_X}$$

or
$$C_X = GFR - \frac{T_X}{P_X} \qquad (53)$$

Therefore, for a substance that is filtered and reabsorbed

$$C_X < GFR$$

or, since GFR can be equated to the *clearance of inulin (Eq. 35)*

$$C_X < C_{In} \qquad (54)$$

This result is expected, since a smaller volume of plasma will be "cleared" of a substance that is filtered and then reabsorbed than will be "cleared" of a substance, such as inulin, that is filtered only. It should be noted that *Eq. 53* was derived primarily to illustrate the relationship between C_X and C_{In} (*Eq. 54*), not as a method for calculating C_X. C_X, of course, is most readily determined from the clearance equation (*Eq. 26*).

The transport mechanisms involved in the tubular reabsorption of substances such as those listed in Table 4.10 can be broadly classified as active or passive. In *passive* reabsorption, a substance diffuses from tubular lumen to peritubular capillary along an osmotic, electrical, and/or concentration gradient. Metabolic energy is not directly required, although energy may have been used to establish the initial gradient. In contrast, *active* reabsorption occurs against an electrical and/or concentration gradient, and requires the expenditure of metabolic energy. The reabsorption of various components of the glomerular filtrate will

Table 4.10
Typical rates of tubular reabsorption in normal man (GFR = 180 l/day)

	P_X, mmol/l	Filtered Load P_XGFR, mmol/day	Rate of Excretion $U_X\dot{V}$, mmol/day	Rate of Tubular Reabsorption $T_X = P_X$GFR $- U_X\dot{V}$, mmol/day	Fractional Reabsorption, T_X/P_XGFR $\times$ 100%
Na^+	140	25,200	100	25,100	>99%
Cl^-	105	18,900	100	18,800	>99%
K^+	4	720	50	670	93%
HCO_3^-	24	4,320	2	4,318	>99%
Glucose	5	900	0	900	100%
Alanine	0.35	63	0	63	100%
Urea	6	1080	432	648	60%
Water		180 l	1.5 l	178.5 l	>99%

be described in detail in later chapters. However, an important example of *active* reabsorption (T_m-limited reabsorption) and of *passive* reabsorption (urea) will be discussed here to illustrate certain general characteristics of active and passive reabsorption mechanisms.

Active Reabsorption: T_m-Limited Reabsorption

For many actively reabsorbed substances, the rate of reabsorption has a finite upper limit. The maximum rate that can be achieved is termed the *transport maximum* (T_m). Glucose, amino acids, phosphate, and sulfate are examples of substances that exhibit such T_m-*limited reabsorption*. Furthermore, the affinity of the transport system for many substances with T_m-limited reabsorption is so high that the *entire* filtered load is reabsorbed from the tubular fluid as long as the transport system is unsaturated. For example, the reabsorption of glucose and amino acids is virtually complete if the filtered load does not saturate the transport system.

It is instructive to consider the renal handling of a substance such as glucose from a quantitative standpoint. As noted above, when the filtered load of glucose does *not* saturate the transport system, the entire filtered load is reabsorbed, i.e., the rate of reabsorption equals the filered load

$$\frac{\text{Mass of glucose reabsorbed}}{\text{Time}} = \frac{\text{Mass of glucose filtered}}{\text{Time}}$$

or
$$T_G = P_G\text{GFR} \qquad (55)$$

Clearly, under such conditions no glucose is excreted. However, when the transport system is *saturated*, the excess load will be excreted. The transport system will be operating at its maximum rate under such conditions, i.e.,

$$\frac{\text{Mass of glucose reabsorbed}}{\text{Time}} = T_{m_G}$$

where T_{m_G} represents the *transport maximum* for

glucose. The rate of excretion therefore will be (cf. *Eq. 52*)

$$U_G\dot{V} = P_G\text{GFR} - T_{m_G} \qquad (56)$$

It is possible, then, to define a plasma concentration at which *glucosuria* first occurs. This critical plasma concentration is termed the *renal threshold* for glucose (P_{Th_G}). Theoretically, P_{Th_G} represents that plasma concentration at which the filtered load exactly saturates the transport system, i.e.,

$$P_{Th_G}\text{GFR} = T_{m_G}$$

or
$$P_{Th_G} = \frac{T_{m_G}}{\text{GFR}} \qquad (57)$$

GLUCOSE TITRATION CURVES

The above discussion can be summarized by examining how the renal handling of glucose varies with plasma glucose concentration. Since the filtered load (P_GGFR), rate of excretion ($U_G\dot{V}$), and rate of tubular reabsorption (T_G) all have the same units (e.g., mmol/min), plots of filtered load vs. P_G, rate of excretion vs. P_G, and rate of reabsorption vs. P_G can be drawn on a single graph (Fig. 4.17, *red curves*). The important features of these so-called *glucose titration curves* are as follows:

1. *Filtered load vs. P_G.* Since GFR is constant, the plot of filtered load (P_GGFR) vs. P_G is a straight line with slope GFR.
2. *Rate of Tubular Reabsorption vs. P_G.* At low P_G, the *entire* filtered load is reabsorbed. At higher P_G, the transport system becomes *saturated* and reabsorbs glucose at its maximum rate, T_{m_G}. In this example, T_{m_G} is approximately 2 mmol/min (375 mg/min), which is the average normal value in men (the average normal T_{m_G} in women is somewhat lower, about 1.7 mmol/min, or 303 mg/min).
3. *Rate of Excretion vs. P_G.* Since the entire filtered load is reabsorbed at low P_G, glucose is

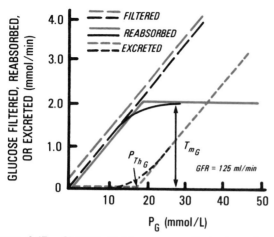

Figure 4.17. Glucose titration curves, illustrating effect of plasma glucose concentration (P_G) on filtered load, rate of excretion, and rate of tubular reabsorption. The maximum rate of tubular reabsorption (T_{m_G}) is about 2 mmol/min. *Red curves* are *idealized*, showing *abrupt* changes in the rate of tubular reabsorption and rate of excretion when the transport system is saturated, which is predicted to occur at a P_G of about 16 mmol/l (P_{Th_G}; see text). *Black curves* resemble *actual data* and exhibit gradual changes in the rate of tubular reabsorption and rate of excretion (*splay*).

virtually absent from the urine. Once the transport system becomes saturated, however, the excess filtered load is excreted (*Eq. 56*). In this example, in which GFR is 125 ml/min, glucosuria first occurs at a P_G of about 16 mmol/l (300 mg/dl), i.e., the *renal threshold* for glucose is 16 mmol/l (note that the same result also is obtained by substituting into *Eq. 57*). When P_G exceeds the renal threshold, the rate of excretion vs. P_G curve parallels the filtered load vs. P_G curve, i.e., it is a straight line with slope GFR, as predicted by *Eq. 56*.

It should be noted that the glucose titration curves drawn in *red* in Figure 4.17 are somewhat *idealized*, as the plots of rate of reabsorption vs. P_G and rate of excretion vs. P_G show an *abrupt* change when the rate of reabsorption reaches its maximum level. *Actual* glucose titration curves show a more *gradual* change in this region, as illustrated by the *black curves* in Figure 4.17 (Mudge, 1958). This feature of renal titration curves is called *splay*. Note that because of splay, the renal threshold for glucose is approximately 11 mmol/l (200 mg/dl) instead of 16 mmol/l (300 mg/dl), even though the transport system does not become saturated (i.e., the rate of tubular reabsorption does not reach its maximum level) until P_G is 16 mmol/l. The threshold value of 16 mmol/l shown by the *red curves* in Figure 4.17 (and calculated by *Eq. 57*) therefore can only be regarded as the *theoretical* renal threshold

for glucose. But although the measured renal threshold is closer to 11 mmol/l than to 16 mmol/l, it still is considerably larger than the normal P_G (about 5 mmol/l). Thus, glucose normally is absent from the urine.

At least two factors account for the splay phenomenon. First, although the affinity of the transport system for glucose is very high, some glucose will appear in the urine before the transport system is fully saturated. An analysis of the kinetics of glucose transport shows why this is so. Using the symbol R to represent the carrier responsible for the reabsorption of glucose, the reaction between glucose and the carrier can be written as GR $\rightleftharpoons$ G + R and can be described by the following mass action equation

$$K = \frac{[G][R]}{[GR]}$$

where K is the dissociation constant for the glucose-carrier reaction. Even though K is very small (i.e., the affinity of the carrier for glucose is very large), a finite concentration of glucose must remain in the tubular fluid in order to saturate the transport system. Some glucose therefore will be excreted prior to saturation.

A second factor contributing to the presence of splay is nephron heterogeneity. Considerable variability in the filtration capacity of glomeruli and in the reabsorptive capacity of renal tubules is believed to exist. For example, some nephrons may produce average volumes of filtrate but have subnormal reabsorptive capacities, and therefore are likely to excrete glucose at a lower P_G than the average nephron. In contrast, other nephrons may produce small volumes of filtrate but have high reabsorptive capacities, and therefore are likely to excrete glucose at a higher P_G than the average nephron.

The effects of P_G on filtered load, rate of excretion, and rate of tubular reabsorption have now been considered. It also is of interest to examine the effect of P_G on the *clearance* of glucose (Fig. 4.18). At low P_G, the entire filtered load is reabsorbed, so that no plasma is "cleared" of glucose

$$C_G = \frac{U_G \dot{V}}{P_G}$$
$$= 0 \qquad (58)$$

Once the renal threshold for glucose is exceeded and glucosuria begins to occur, the clearance of glucose gradually increases. A formal expression for C_G can be derived by substituting *Eq. 56* into the clearance equation $C_G = U_G \dot{V}/P_G$

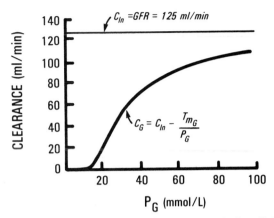

Figure 4.18. Effect of plasma glucose concentration (P_G) on clearance of glucose (C_G). C_G increases as P_G increases, asymptotically approaching the clearance of inulin (C_{In}).

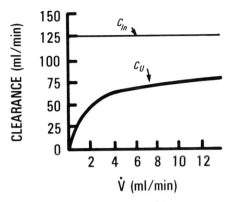

Figure 4.19. Effect of urine flow rate ($\dot{V}$) on clearances of urea (C_U) and inulin (C_{In}). As $\dot{V}$ exceeds 2 ml/min, C_U approaches two-thirds C_{In}. [Adapted from Chasis and Smith (1938).]

$$C_G = \frac{P_G GFR - T_{mG}}{P_G}$$

$$= GFR - \frac{T_{mG}}{P_G}$$

or
$$C_G = C_{In} - \frac{T_{mG}}{P_G} \qquad (59)$$

Thus, C_G asymptotically approaches C_{In} at high P_G, as illustrated in Figure 4.18. But C_G *always* is less than C_{In}, since some glucose always is reabsorbed (*Eq. 54*).

Passive Reabsorption: Urea

The renal handling of urea represents an important example of *passive reabsorption*. In many regions of the nephron, urea will be passively reabsorbed whenever its concentration in the tubular fluid exceeds that in the surrounding peritubular fluid. Such a urea concentration gradient is created in these regions by the reabsorption of *water* (Chapters 30 and 31).

Because urea reabsorption occurs as a consequence of the reabsorption of water, urea reabsorption varies markedly with the rate of urine flow (i.e., with the volume of water that is *not* reabsorbed). This dependence of reabsorption on urine flow rate is a unique characteristic of passively reabsorbed substances like urea. It is therefore instructive to compare the effects of urine flow rate on the renal handling of a substance such as *inulin* to the effects of flow on urea. Since inulin is neither reabsorbed nor secreted, the amount of inulin excreted always equals the amount filtered (*Eq. 34*)

$$U_{In}\dot{V} = P_{In}GFR$$

Thus, changes in $\dot{V}$ will result in reciprocal changes in U_{In}, so that the product $U_{In}\dot{V}$ always equals the filtered load $P_{In}GFR$. The *clearance* of inulin

($=U_{In}\dot{V}/P_{In}$) is therefore independent of urine flow rate and, as previously discussed, is always equal to the GFR (Chapter 28). In contrast, with *urea* the amount excreted increases with increasing urine flow rate (due to decreased reabsorption), so that the clearance of urea increases with increasing urine flow.

The effect of $\dot{V}$ on C_{In} and C_U is illustrated in Figure 4.19 (Smith, 1951). Note that C_U increases as $\dot{V}$ increases from less than 0.5 ml/min to approximately 1.5 ml/min, but then levels off as $\dot{V}$ exceeds 2 ml/min, approaching approximately two-thirds C_{In}.

TUBULAR SECRETION

Tubular secretion is the process whereby substances in the peritubular capillaries are transported across the tubular epithelium into the tubular lumen. Thus, whereas substances are *recovered* from the tubular fluid in tubular *reabsorption*, substances are *added* to the tubular fluid in tubular *secretion*. It is important to distinguish between secretion and *excretion*, the overall process by which the kidneys compose the final urine. For example, a substance could be filtered at the glomerulus, be partially reabsorbed in one or more regions of the nephron, never be secreted, and still be excreted.

For a substance X, the amount secreted per unit time can be calculated as the difference between the rate of excretion and the filtered load

$$\frac{\text{Mass of X secreted}}{\text{Time}}$$

$$= \frac{\text{Mass of X excreted}}{\text{Time}} - \frac{\text{Mass of X filtered}}{\text{Time}}$$

The rate of tubular secretion, like the rate of tubular reabsorption, commonly is designated by the symbol T_X. Hence

$$T_X = U_X \dot{V} - P_X GFR \qquad (60)$$

Eq. 60 can be used to derive an expression for the *clearance* of a substance that is filtered and then secreted. Rearranging this equation gives

$$U_X \dot{V} = P_X GFR + T_X \qquad (61)$$

Eq. 61 can be substituted into the clearance equation $C_X = U_X \dot{V}/P_X$ to give

$$C_X = \frac{P_X GFR + T_X}{P_X}$$

or

$$C_X = GFR + \frac{T_X}{P_X} \qquad (62)$$

Therefore, for a substance that is filtered and secreted

$$C_X > GFR$$

or

$$C_X > C_{In} \qquad (63)$$

This result is expected, since a larger volume of plasma will be "cleared" of a substance that is filtered and then secreted than will be "cleared" of a substance, such as inulin, that is filtered only. It should be noted that *Eq. 62* was derived primarily to illustrate the relationship between C_X and C_{In} (*Eq. 63*), not as a method for calculating C_X. C_X, of course, is most readily determined from the clearance equation (*Eq. 26*).

Like the mechanisms for tubular reabsorption, the mechanisms involved in tubular secretion can be broadly characterized as active or passive. The secretion of specific substances will be described in detail in later chapters. However, an important example of *active secretion* (T_m-limited secretion) will be discussed here as an illustration of certain general characteristics of active secretion mechanisms and because of its close correspondence to the active reabsorption example discussed above.

Active Secretion: T_m-Limited Secretion

In analogy to T_m-limited reabsorption, certain substances that are actively secreted exhibit T_m-limited secretion, i.e., the rate of secretion has a finite upper limit, termed the *transport maximum* (T_m). *p*-Amino hippurate (PAH), iodohippurate, and penicillin are among the substances whose secretion is T_m-limited. Some of these substances also exhibit a phenomenon that is somewhat analogous to the threshold phenomenon for reabsorption. Specifically, the affinity of the transport system for some secreted substances is so high that essentially *all* of the substance delivered to the

peritubular capillaries is secreted into the tubular fluid as long as the transport system is not saturated. For example, PAH and iodohippurate are almost completely removed from the peritubular capillaries when the transport system is unsaturated.

It is instructive to examine the renal handling of a substance such as PAH in a quantitative fashion. The rate of delivery of PAH to the peritubular capillaries is given by

$$\frac{\text{Mass of PAH delivered to}}{\text{peritubular capillaries}}{\text{Time}}$$

$$= P_{PAH} \times \frac{\begin{array}{c}\text{Volume of plasma delivered}\\\text{to peritubular capillaries}\end{array}}{\text{Time}} \qquad (64)$$

The rate of plasma delivery to the peritubular capillaries can be *approximated* by subtracting GFR from the total RPF, i.e.,

$$\frac{\begin{array}{c}\text{Volume of plasma delivered}\\\text{to peritubular capillaries}\end{array}}{\text{Time}} \simeq RPF - GFR \qquad (65)$$

However, as discussed in Chapter 28, approximately 10–15% of the total RPF supplies the renal capsule, renal pelvis, perirenal fat, medulla, and papilla, regions of the kidneys in which *no* filtration or secretion of PAH occurs. Thus, the rate of plasma delivery to the peritubular capillaries is more precisely given by

$$\frac{\begin{array}{c}\text{Volume of plasma delivered}\\\text{to peritubular capillaries}\end{array}}{\text{Time}} = ERPF - GFR \qquad (66)$$

where ERPF is the *effective renal plasma flow*, i.e., the rate of plasma flow to regions of the kidneys that *can* remove PAH (see *Eq. 45*). *Eq. 64* therefore can be expressed as

$$\frac{\begin{array}{c}\text{Mass of PAH delivered to}\\\text{peritubular capillaries}\end{array}}{\text{Time}} = P_{PAH}(ERPF - GFR) \qquad (67)$$

As noted above, when the secretory mechanism is not saturated, essentially *all* of the PAH delivered to the peritubular capillaries is secreted

$$\frac{\text{Mass of PAH secreted}}{\text{Time}}$$

$$= \frac{\begin{array}{c}\text{Mass of PAH delivered to}\\\text{peritubular capillaries}\end{array}}{\text{Time}}$$

i.e.,

$$T_{PAH} = P_{PAH}(ERPF - GFR) \qquad (68)$$

Under such conditions, then, PAH is almost completely removed from the plasma by the kidneys:

Approximately 20% of the PAH delivered to the kidneys enters the tubular fluid by glomerular filtration; the remaining 80% enters the peritubular capillaries and is secreted. This can be expressed formally by substituting *Eq. 68* into *Eq. 61*

$$U_{PAH}\dot{V} = P_{PAH}GFR + P_{PAH}(ERPF - GFR)$$
$$= P_{PAH}ERPF \qquad (69)$$

However, once the secretory mechanism is saturated, the rate of PAH secretion is fixed at its maximum value, $T_{m_{PAH}}$. The rate of PAH excretion then will be

$$U_{PAH}\dot{V} = P_{PAH}GFR + T_{m_{PAH}} \qquad (70)$$

PAH TITRATION CURVES

The renal handling of PAH, as described above, can be summarized using PAH titration curves (Fig. 4.20). The important features of these curves are as follows:

1. *Filtered Load vs. P_{PAH}*. Since GFR is constant, the plot of filtered load ($P_{PAH}GFR$) vs. P_{PAH} is a straight line with slope GFR.

2. *Rate of Tubular Secretion vs. P_{PAH}*. At low P_{PAH}, where the transport system is able to secrete essentially all of the PAH delivered to the peritubular capillaries, the rate of secretion is considerably greater than the filtered load. At higher P_{PAH}, the transport system becomes *saturated* and secretes PAH at its maximum rate, $T_{m_{PAH}}$. In this example, $T_{m_{PAH}}$ has a value of 0.4 mmol/min (80 mg/min), which is the average normal value in man.

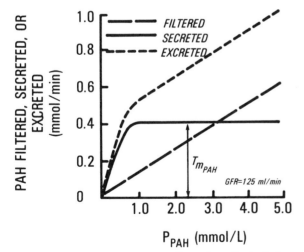

Figure 4.20. PAH titration curves, illustrating effect of plasma PAH concentration (P_{PAH}) on filtered load, rate of excretion, and rate of tubular secretion. The maximum rate of tubular secretion ($T_{m_{PAH}}$) is about 0.4 mmol/min. The rate of tubular secretion and rate of excretion change gradually as the transport system is saturated (*splay*).

3. *Rate of Excretion vs. P_{PAH}*. The curve rises steeply at low P_{PAH}. Once the transport system becomes saturated, however, the rate of excretion vs. P_{PAH} curve parallels the filtered load vs. P_{PAH} curve, i.e., it is a straight line with slope GFR, as predicted by *Eq. 70*.

Note that the PAH titration curves show a *gradual* change as the rate of tubular secretion reaches its maximum level. Thus, like the titration curves for glucose (Fig. 4.17, *black curves*), the PAH titration curves exhibit *splay*. The factors accounting for the splay in the PAH curves are analogous to those presented above for glucose.

The effect of P_{PAH} on the *clearance* of PAH now will be examined. At *low* P_{PAH}, essentially all of the PAH is removed from the ERPF by the kidneys, as noted above, i.e., the entire ERPF is "cleared" of PAH. This can be expressed formally by substituting *Eq. 69* into the clearance equation $C_{PAH} = U_{PAH}\dot{V}/P_{PAH}$

$$C_{PAH} = \frac{P_{PAH}ERPF}{P_{PAH}}$$

i.e.,
$$C_{PAH} = ERPF \qquad (71)$$

Eq. 71 was introduced previously, when the clinical use of PAH clearance measurements to approximate RPF was considered (*Eq. 45*). At *high* P_{PAH}, when the transport system is saturated, some of the PAH in the peritubular capillaries will *not* be secreted, i.e., a smaller volume of plasma is "cleared" of PAH. An expression for C_{PAH} under these conditions can be obtained by substituting *Eq. 70* into the clearance equation $C_{PAH} = U_{PAH}\dot{V}/P_{PAH}$

$$C_{PAH} = \frac{P_{PAH}GFR + T_{m_{PAH}}}{P_{PAH}}$$

$$= GFR + \frac{T_{m_{PAH}}}{P_{PAH}}$$

or
$$C_{PAH} = C_{In} + \frac{T_{m_{PAH}}}{P_{PAH}} \qquad (72)$$

Thus, C_{PAH} asymptotically approaches C_{In} at high P_{PAH}, as illustrated in Figure 4.21. But C_{PAH} *always* is greater than C_{In}, since some PAH always is secreted (*Eq. 63*).

BIDIRECTIONAL TRANSPORT

So far in this chapter, consideration has been given to substances that are either filtered and then *reabsorbed* or filtered and then *secreted*. Certain substances, however, are transported across the tubular epithelium in *both* directions, i.e., they are filtered and then both reabsorbed *and* secreted. For quantitative purposes, a substance that undergoes

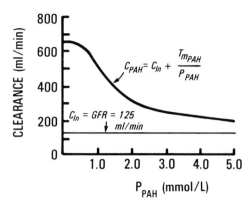

Figure 4.21. Effect of plasma PAH concentration (P_{PAH}) on the clearance of PAH (C_{PAH}). C_{PAH} decreases as P_{PAH} increases, asymptotically approaching the clearance of inulin (C_{In}).

such *bidirectional transport* is categorized according to whether there is *net* reabsorption or *net* secretion.

The renal handling of K^+ represents an important example of bidirectional transport. K^+ is reabsorbed in parts 1 and 2 of the proximal tubule, secreted in part 3 of the proximal tubule and/or the thin descending limb of the loop of Henle, reabsorbed in the thin and thick ascending limbs of the loop of Henle, and both reabsorbed and secreted in the distal nephron. In the nephron *as a whole*, K^+ generally undergoes *net reabsorption*. However, net secretion may occur under certain conditions, such as during the ingestion of a high K^+ diet (Chapter 34).

Another endogenous substance that exhibits bidirectional transport is *uric acid*, which is both reabsorbed *and* secreted in the proximal tubule, although *reabsorption* always predominates (Chapter 30).

Weak Organic Acids and Bases

The proximal tubular epithelium contains a special active transport system capable of secreting a variety of organic acids and another capable of secreting a variety of organic bases (Chapter 30). *p-Amino hippurate* (PAH), whose handling by the kidneys was discussed above, is an example of a substance that is secreted by the transport system for organic acids. It is important to emphasize that these transport systems only secrete organic molecules that are *ionized* in plasma. Whereas *strong* organic acids and bases are completely ionized in plasma (pH = 7.4), this is not necessarily true of *weak* organic acids and bases, which may exist partly or even primarily in the nonionized form at pH 7.4. Many weak organic acids and bases therefore undergo little active secretion. However, the

nonionized form of most weak organic acids and bases is sufficiently lipid-soluble to *passively* diffuse across the tubular epithelium. Since many drugs, drug metabolites, and other exogenous organic molecules are weak acids and bases, the effect of this passive diffusion on the renal handling of such substances is a topic of considerable significance.

Like other plasma constituents, a weak organic acid or base will be present in the filtrate at a concentration virtually identical with that in plasma. As water is reabsorbed, its concentration in the tubular fluid will exceed that in plasma, i.e., a concentration gradient favoring reabsorption develops. However, since only the *nonionized* form is sufficiently lipid-soluble to diffuse across the tubular epithelium, it is the concentration gradient of the *nonionized diffusible* form (*not* the *total* concentration gradient) that is significant. In any given region of the nephron, the concentration gradient of the nonionized form may favor either reabsorption *or* secretion, depending upon the relationship between the pH of the tubular fluid in that region and the pK_a of the acid or base. Weak organic acids and bases therefore can exhibit *bidirectional transport*.

As an example of such bidirectional transport, consider the renal handling of a *weak organic acid*, whose ionization reaction can be written as follows

$$HA \rightleftharpoons H^+ + A^-$$

If the tubular fluid is *acidic*, the reaction will shift to the left, favoring the formation of the *nonionized, diffusible* form HA (the exact relationship between the pH of the tubular fluid and the concentration of HA will depend on the pK_a of the acid). The concentration gradient for the reabsorption of the nonionized form created by the reabsorption of water (Fig. 4.22*A*) will be *augmented* by this shift of the equilibrium (Fig. 4.22*B*). Conversely, if the tubular fluid is *alkaline*, the formation of the *ionized, nondiffusible* form A^- is favored. The concentration gradient for the reabsorption of the nonionized form will therefore be small in spite of the reabsorption of water. In fact, depending on the relationship between the pH of the tubular fluid and the pK_a of the acid, the concentration of the nonionized form in the tubular fluid may become so small that a concentration gradient for the *secretion* of the nonionized form actually may develop (Fig. 4.22*C*). Hence, an *acidic* tubular fluid favors the *reabsorption* of weak organic acids, while an *alkaline* tubular fluid favors their *secretion*, and hence eventual *excretion*. The phenomenon whereby the *nonionized* form of a molecule (e.g.,

A   **B**   **C**

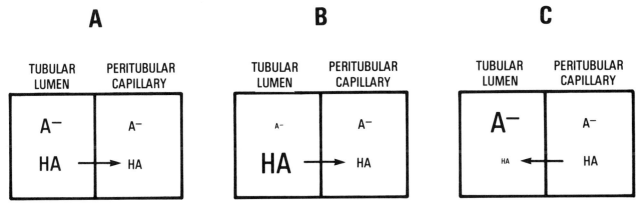

Figure 4.22.   Renal handling of weak organic acid. The relative concentrations of HA and A⁻ are indicated by the size of the letters. *A*, Reabsorption of water creates a concentration gradient for the reabsorption of the nonionized, diffusible form HA. *B*, Acidic tubular fluid *augments* the concentration gradient for the reabsorption of HA. *C*, *Alkaline* tubular fluid may produce a concentration gradient for the *secretion* of HA.

A   **B**   **C**

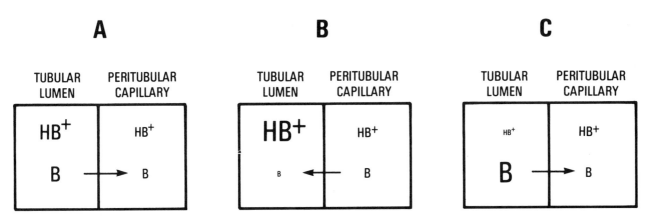

Figure 4.23.   Renal handling of weak organic base. The relative concentrations of HB⁺ and B are indicated by the size of the letters. *A*, Reabsorption of water creates a concentration gradient for the reabsorption of the nonionized, diffusible form B. *B*, Acidic tubular fluid may produce a concentration gradient for the *secretion* of B. *C*, Alkaline tubular fluid *augments* the concentration gradient for the reabsorption of B.

HA) can readily diffuse across the tubular epithelium while the *ionized* form (e.g., A⁻) cannot is termed *nonionic diffusion* or *diffusion trapping*.

The renal handling of a *weak organic base* is governed by similar principles

$$BH^+ \leftrightharpoons B + H^+$$

However, in this case an *alkaline* tubular fluid favors the formation of the *nonionized, diffusible* form (B), while an *acidic* tubular fluid favors the formation of the *ionized, nondiffusible* form (HB⁺). Thus, as illustrated in Figure 4.23 an *alkaline* tubular fluid favors the *reabsorption* of weak organic bases, while an *acidic* tubular fluid favors their *secretion*, and hence eventual *excretion*.

As previously noted, many drugs and drug metabolites are weak organic acids and bases. The fact that the tubular fluid pH can significantly affect the renal handling of weak acids and bases therefore has important implications for *drug excretion*. Urine normally is acidic (average pH <6.0; see

Chapter 33), which would favor the *reabsorption* of *acidic* drugs (Fig. 4.22*B*) and the *excretion* of *basic* drugs (Fig. 4.23*B*). Thus, in certain clinical situations the manipulation of urine pH to maximize or minimize drug excretion may be beneficial. For example, in cases of overdose involving certain *acidic* drugs (e.g., aspirin), *alkalinization* of the urine can promote drug excretion.

The *urine flow rate* also can have a significant effect on the excretion of weakly acidic and basic drugs. The reason is that the reabsorption of water has an important role in creating the concentration gradient for the passive reabsorption of the nonionized, diffusible form of the drug (Figs. 4.22*A* and 4.23*A*). With an increase in urine flow rate, this concentration gradient is diminished, promoting drug excretion. It should be noted that the effect of urine flow rate on the passive reabsorption of weak organic acids and bases is completely analogous to the effect of urine flow rate on the passive reabsorption of *urea* (see above).

FRACTIONAL EXCRETION AND FRACTIONAL REABSORPTION

Important information about the renal handling of a substance can be obtained from measurements of its concentration in plasma, in urine, and in tubular fluid samples obtained by micropuncture (Chapter 27). In the following discussion, such concentration data will be used to calculate the *fractional excretion*, *fractional reabsorption*, and other significant quantities.

The *fractional excretion* of a substance X (FE_X) can be defined as the fraction of the total amount of X in the glomerular filtrate that appears in the final urine. Thus

$$FE_X = \frac{\text{Mass of X excreted}}{\text{Mass of X filtered}}$$

Dividing both numerator and denominator by time gives

$$FE_X = \frac{\dfrac{\text{Mass of X excreted}}{\text{Time}}}{\dfrac{\text{Mass of X filtered}}{\text{Time}}}$$

or

$$FE_X = \frac{U_X \dot{V}}{P_X GFR} \qquad (73)$$

Several alternative expressions for FE_X can be derived from *Eq. 73*. For example, substituting C_X for $U_X\dot{V}/P_X$ and C_{In} for GFR gives

$$FE_X = \frac{C_X}{C_{In}} \qquad (74)$$

Alternatively, $U_{In}\dot{V}/P_{In}$ can be substituted for GFR in *Eq. 73* to give

$$FE_X = \frac{U_X \dot{V}/P_X}{U_{In}\dot{V}/P_{In}}$$

$$= \frac{U_X/P_X}{U_{In}/P_{In}} \qquad (75)$$

Eq. 75 is of particular interest because it allows FE_X to be determined *without* measuring the urine flow rate, $\dot{V}$.

The fractional excretion of X (FE_X) represents the fraction of the total amount of X in the filtrate that appears in *urine*. The fraction of the total amount of X in the filtrate that remains in the *tubular fluid* at some point in the nephron is termed the *fractional delivery* of X (FD_X), since it represents the fraction of X in the filtrate that is "delivered" to that point. By a derivation analogous to that for *Eq. 75*, the following expression for FD_X is obtained:

$$FD_X = \frac{TF_X/P_X}{TF_{In}/P_{In}} \qquad (76)$$

where TF refers to the concentration in the tubular fluid at the point in question. Such tubular fluid samples are obtained by micropuncture (Chapter 27).

Once the fractional excretion of a substance X is known (*Eq. 75*), the *fractional reabsorption* can be readily calculated, since

$$\text{Fractional reabsorption} = 1 - FE_X$$

i.e.,

$$\text{Fractional reabsorption} = 1 - \frac{U_X/P_X}{U_{In}/P_{In}} \qquad (77)$$

Similarly, once the fractional delivery of a substance X to a particular point in the nephron is known (*Eq. 76*), the fraction reabsorbed up to that point can be readily calculated

$$\begin{array}{l}\text{Fraction reabsorbed up to}\\ \text{point of sample}\end{array} = 1 - \frac{TF_X/P_X}{TF_{In}/P_{In}} \qquad (78)$$

The fractional reabsorption also can be calculated with *Eq. 51*, but the parameters required for the calculation are less readily obtained.

Table 4.11 includes typical U/P ratios and fractional excretion data for several plasma constituents from Table 4.10 as well as for inulin. To conform with the format for presenting such data in clinical medicine, the fractional excretion values have been converted to *percentages* by multiplying by 100%. Note that the fractional excretion of K^+ *exceeds* 100% when a vegetarian (high K^+) diet is ingested. This indicates that the amount of K^+ excreted is *greater* than the amount filtered, i.e., net *secretion* of K^+ has occurred. The fact that K^+ can undergo either net reabsorption or net secretion has been previously noted.

Fractional Excretion and Fractional Reabsorption of Water

The ratios U_{In}/P_{In} and TF_{In}/P_{In} are important because of their role in calculating the fractional excretion and fractional reabsorption of the various substances in the glomerular filtrate, as described above. However, it should be noted that these ratios *also* can be used to calculate the fractional excretion and fractional reabsorption of *water*.

Table 4.11

Typical fractional excretion values obtained from U/P ratios

	U_X/P_X	Fractional Excretion $U_X/P_X \div U_{In}/P_{In} \times 100\%$
Inulin	150	100%
Na^+	0.6	0.4%
Cl^-	0.75	0.5%
K^+		
Normal diet	10.5	7%
Vegetarian diet	180	120%
HCO_3^-	0.075	0.05%
Urea	60	40%

The fractional excretion of water (FE_{H_2O}) is defined by the expression

$$FE_{H_2O} = \frac{\text{Volume of water excreted}}{\text{Volume of water filtered}}$$

Dividing both numerator and denominator by time gives

$$FE_{H_2O} = \frac{\dfrac{\text{Volume of water excreted}}{\text{Time}}}{\dfrac{\text{Volume of water filtered}}{\text{Time}}}$$

or $\qquad FE_{H_2O} = \dfrac{\dot{V}}{GFR} \qquad (79)$

An expression for FE_{H_2O} that utilizes the U_{In}/P_{In} ratio is obtained by substituting $U_{In}\dot{V}/P_{In}$ for GFR

$$FE_{H_2O} = \frac{\dot{V}}{U_{In}\dot{V}/P_{In}}$$

i.e., $\qquad FE_{H_2O} = \dfrac{1}{U_{In}/P_{In}} \qquad (80)$

By an analogous derivation, it can be shown that the fraction of the total volume of water filtered that remains in the tubule at some point in the nephron, the *fractional delivery* of water (FD_{H_2O}), is given by

$$FD_{H_2O} = \frac{1}{TF_{In}/P_{In}} \qquad (81)$$

The *fractional reabsorption* of water can be calculated from the fractional excretion (*Eq. 80*) as follows:

$$\text{Fractional reabsorption} = 1 - FE_{H_2O}$$

or $\qquad$ Fractional reabsorption $= 1 - \dfrac{1}{U_{In}/P_{In}} \qquad (82)$

Similarly, the fraction of the filtered water reabsorbed up to some point in the nephron can be

Table 4.12
Summary of equations for calculating fractional excretion and fractional reabsorption

	Water	Other Substances
Fractional excretion (FE)	$\dfrac{1}{U_{In}/P_{In}}$	$\dfrac{U_X/P_X}{U_{In}/P_{In}}$
Fractional reabsorption	$1 - \dfrac{1}{U_{In}/P_{In}}$	$1 - \dfrac{U_X/P_X}{U_{In}/P_{In}}$
Fractional delivery to point of micropuncture sample (FD)	$\dfrac{1}{TF_{In}/P_{In}}$	$\dfrac{TF_X/P_X}{TF_{In}/P_{In}}$
Fraction reabsorbed up to point of micropuncture sample	$1 - \dfrac{1}{TF_{In}/P_{In}}$	$1 - \dfrac{TF_X/P_X}{TF_{In}/P_{In}}$

calculated from the fractional delivery (*Eq. 81*)

$$\text{Fraction reabsorbed up to point of sample} = 1 - FD_{H_2O}$$

or $\qquad$ Fraction reabsorbed up to point of sample $= 1 - \dfrac{1}{TF_{In}/P_{In}} \qquad (83)$

Table 4.12 summarizes the important equations introduced here that utilize U_{In}/P_{In} and TF_{In}/P_{In} ratios. It should be noted that U/P and TF/P ratios for *creatinine* can be substituted for the corresponding ratios for inulin if an approximate calculation is adequate, as in many clinical applications.

TF_{In}/P_{In} ratios will be used in later chapters, particularly when discussing the mechanisms for the concentration and dilution of urine (Chapter 31). Most importantly, TF_{In}/P_{In} will be used as an index of water reabsorption along the nephron. For example, at the end of the proximal tubule, TF_{In}/P_{In} is estimated to be approximately 3.0. Thus, approximately one-third of the filtered water is delivered to the end of the proximal tubule (*Eq. 81*), i.e., approximately two-thirds of the filtered water is reabsorbed in the proximal tubule (*Eq. 83*), as indicated in Chapter 27.

CHAPTER 30

The Proximal Tubule

As discussed in Chapter 27, the proximal tubule is responsible for the initial processing of the glomerular filtrate. Among its important functions are the reabsorption of approximately two-thirds of the filtered Na^+ and water, the reabsorption of virtually all of the filtered glucose and amino acids, and the secretion of organic acids and bases. These functions will be considered in detail in this chapter.

WATER PERMEABILITY

An important determinant of proximal tubular function is the high permeability of the proximal tubule to water, which prevents the establishment of an osmotic gradient across the proximal tubular epithelium. Even if an osmotic gradient is artificially imposed by placing a hypertonic or hypotonic solution in the tubular lumen via a micropipette, water moves in or out, respectively, so that within seconds no measurable osmotic gradient exists (Windhager et al., 1959). Physiologically, this means that the reabsorption of any solute by the proximal tubule will result in the reabsorption of water so that the tubular fluid osmolality remains equal to that of the peritubular fluid. The peritubular fluid surrounding the proximal tubule, i.e., the cortical peritubular fluid (Fig. 4.7), has an osmolality essentially identical to that of plasma (assumed here for simplicity to be approximately 300 mosmol/kg H_2O). Thus, the proximal tubular fluid has an osmolality identical to plasma as well, i.e., it is isotonic (Andreoli and Schafer, 1979).* Furthermore, since the original glomerular filtrate is isotonic, the reabsorbed fluid must also be isotonic.

SODIUM REABSORPTION

As previously noted, approximately two-thirds of the filtered Na^+ is reabsorbed by the proximal

_____

* Of course, the methods for measuring osmolality are not sufficiently precise to eliminate the possibility that the osmolality of the tubular fluid differs slightly from that of the peritubular fluid (e.g., by 1 mosmol/kg H_2O), but the convention is to state that the osmolality of the tubular fluid and peritubular fluid are the same.

tubule. Na^+ must be accompanied by an anion to maintain electrical neutrality; approximately 75% is accompanied by Cl^-, while the remaining 25% is accompanied by HCO_3^-. Although the reabsorption of any solute by the proximal tubule will result in the reabsorption of water, Na^+ and accompanying anions are the major solutes reabsorbed by the proximal tubule and therefore are primarily responsible for generating the osmotic driving force for water reabsorption. Thus, it is the reabsorption of two-thirds of the filtered Na^+ and accompanying anions by the proximal tubule that is responsible for the reabsorption of two-thirds of the filtered water.

It is important to keep in mind this general picture of the handling of Na^+ by the proximal tubule (reabsorption of about two-thirds of the filtered Na^+; 75% accompanied by Cl^-, 25% by HCO_3^-) throughout the following discussion of the mechanisms for proximal tubular Na^+ reabsorption (Windhager, 1979; Giebisch, 1978). Three major mechanisms have been identifed: unidirectional Na^+ transport (uniport), Na^+-H^+ exchange (antiport), and Cl^--driven Na^+ transport.

Unidirectional Na^+ Transport

Unidirectional Na^+ transport is the process whereby Na^+ passively enters the tubular epithelial cell across the luminal or brush border surface and then is actively extruded across the basal-lateral surfaces of the cell (Fig. 4.24A). The passive entry of Na^+ is carrier mediated and is driven by both a chemical gradient and an electrical gradient: the intracellular Na^+ concentration is quite low (probably less than 30 mmol/l) and the intracellular electrical potential is about 60 to 80 mV negative with respect to the lumen (Spring and Kimura, 1979). The active extrusion of Na^+ is accomplished by the Na^+-K^+ ATPase (Katz, 1982). It is this ATPase that generates both the low intracellular Na^+ concentration and the negative intracellular electrical potential. Thus, the energy for the unidirectional transport of Na^+, including the electro-

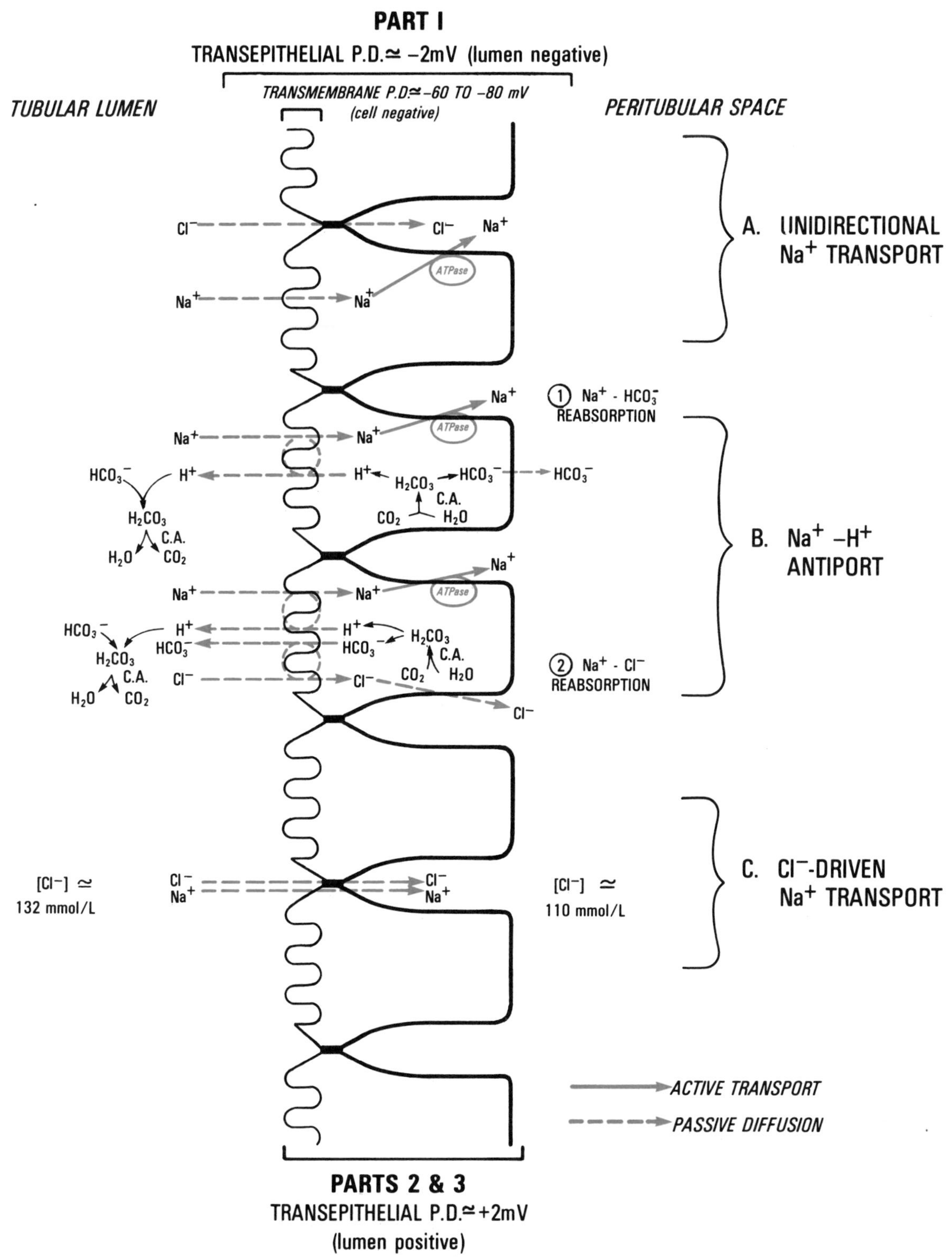

PART I

TRANSEPITHELIAL P.D.≃ −2mV (lumen negative)

TRANSMEMBRANE P.D.≃ −60 TO −80 mV
(cell negative)

TUBULAR LUMEN   *PERITUBULAR SPACE*

A. **UNIDIRECTIONAL Na⁺ TRANSPORT**

B. **Na⁺ −H⁺ ANTIPORT**

① Na⁺ - HCO₃⁻ REABSORPTION

② Na⁺ - Cl⁻ REABSORPTION

C. **Cl⁻-DRIVEN Na⁺ TRANSPORT**

[Cl⁻] ≃ 132 mmol/L   [Cl⁻] ≃ 110 mmol/L

→ *ACTIVE TRANSPORT*
⇢ *PASSIVE DIFFUSION*

PARTS 2 & 3

**TRANSEPITHELIAL P.D.≃ +2mV
(lumen positive)**

Figure 4.24. Mechanisms for Na⁺ reabsorption in the proximal tubule. *A*, Unidirectional Na⁺ transport (uniport). Na⁺ passively enters the cell along its electrochemical gradient and then is actively extruded across the basal-lateral surfaces by the Na⁺-K⁺ ATPase (illustrated here across the lateral surface). This reabsorption of Na⁺ generates a lumen-negative transepithelial potential difference (P.D.), but only of about −2 mV because Cl⁻ accompanies Na⁺ by diffusing across the "leaky" tight junctions. *B*, Na⁺-H⁺ exchange (antiport). The passive diffusion of Na⁺ into the cell is coupled (as indicated by the *dashed circle*) to the secretion of H⁺ into the tubular fluid. The Na⁺ is then actively extruded across the basal-lateral surfaces by the Na⁺-K⁺ ATPase, while the secreted H⁺ reacts with HCO₃⁻ in the tubular fluid to form carbonic acid, which then dissociates into CO₂ and water in a reaction catalyzed by carbonic anhydrase (C.A.) in the brush border. The HCO₃⁻ formed in the cell during the H⁺ secretion process can (1) diffuse across the *basal-lateral surfaces*, in which case the net result of the Na⁺-H⁺ antiport is *Na⁺-HCO₃⁻ reabsorption* (B①); or (2) diffuse across the *brush border* via a coupled exchange for Cl⁻ (as indicated by the *dashed circle*), in which case the net result of the Na⁺-H⁺ antiport is *Na⁺-Cl⁻ reabsorption* (B②). *C*, Cl⁻-driven Na⁺ transport. The Cl⁻ concentration in the tubular fluid increases to about 132 mmol/l in parts 2 and 3, resulting in a concentration gradient for the diffusion of Cl⁻ across the "leaky" tight junctions into the peritubular fluid, where the Cl⁻ concentration is about 110 mmol/l. This reabsorption of Cl⁻ generates a lumen-positive transepithelial P.D. in parts 2 and 3, but only of about +2 mV because Na⁺ accompanies Cl⁻ across the "leaky" tight junctions.

483

chemical gradient for the passive entry of Na^+, comes from ATP via the Na^+-K^+ ATPase. Such passive Na^+ entry and active Na^+ extrusion also occur in other cells, including erythrocytes, nerve, and muscle. However, the epithelial cells of the nephron are distinguished from these nonepithelial cells by their *polarity*: passive Na^+ entry occurs primarily at the *luminal* surface, while the Na^+-K^+ ATPase, and hence active Na^+ extrusion, occurs exclusively at the *basal-lateral* surfaces.

This unidirectional transport of Na^+ from tubular lumen to peritubular space will produce a transepithelial electrical potential difference (lumen negative with respect to peritubular space) whose magnitude will depend on the permeability of the tubular epithelium to one or more anions, which then could accompany Na^+. In part 1 of the proximal tubule, the high permeability of the proximal tubule to Cl^- effectively prevents the establishment of a large potential difference. Thus, the transepithelial potential difference in part 1 is only about -2 mV (lumen negative) at a time when a large quantity of Na^+ is being actively reabsorbed (Fig. 4.24). Because the high permeability of the proximal tubule to Cl^- prevents the establishment of a large potential difference, the proximal tubule is said to have a *low electrical resistance*. It is presumed that the low resistance pathway for the reabsorption of Cl^- includes the tight junctions between the epithelial cells (Fig. 4.24*A*). The proximal tubule, then, can reabsorb a large quantity of Na^+, Cl^-, and water *without* establishing *either* a large transepithelial electrical gradient *or* an osmotic gradient. The proximal tubule is therefore said to have "*leaky*" *tight junctions*.

Na$^+$-H$^+$ Exchange

The second mechanism for the reabsorption of Na^+ from the proximal tubular fluid involves the exchange of Na^+ in the tubular fluid for H^+ inside the epithelial cells (Fig. 4.24*B*) (Ullrich et al., 1977). This process can be demonstrated in *isolated brush border* preparations where the movement of Na^+ across the brush border surface (e.g., from side 1 to side 2) can be driven either by a favorable concentration gradient for Na^+ ($[Na^+]_1 > [Na^+]_2$) or for H^+ ($[H^+]_2 > [H^+]_1$) (Murer and Kinne, 1980). Thus, the Na^+-H^+ exchange mechanism appears to represent an *antiport* or *countertransport system* where the movement of a substance in one direction provides the energy for the coupled movement of a second substance in the opposite direction. In the *intact proximal tubule*, it is probable that the electrochemical gradient for the passive movement of Na^+ from tubular fluid into the epithelial cells (see above) provides the driving force for the entry of Na^+ and the coupled extrusion or *secretion* of H^+ into the tubular fluid. The Na^+ that enters the cells, in turn, is extruded into the peritubular fluid by the Na^+-K^+ ATPase, thereby maintaining the electrochemical gradient favoring further Na^+-H^+ exchange. Since this electrochemical Na^+ gradient is generated by the Na^+-K^+ ATPase, the energy for the Na^+-H^+ antiport is indirectly derived from ATP.

Because the Na^+-H^+ antiport involves the exchange of two cations, it may initially appear that this mechanism for Na^+ reabsorption would *not* produce a transfer of anions. However, a more detailed examination of the Na^+-H^+ antiport reveals that the Na^+ reabsorbed into the peritubular fluid actually is accompanied by HCO_3^- or Cl^-. To understand why, it is necessary to examine (1) the fate of the H^+ *ion* secreted into the tubular fluid; and (2) the fate of the HCO_3^- *ion* formed in the epithelial cells during the H^+ secretion process (Warnock and Rector, 1979).

1. *Fate of the H^+ Secreted into the Tubular Fluid.* As discussed in detail in Chapter 33, most of the H^+ that is secreted into the proximal tubular fluid will react with HCO_3^- in the tubular fluid to form carbonic acid (Fig. 4.24*B*):

$$H^+ + HCO_3^- \rightleftharpoons H_2CO_3 \underset{\text{C.A.}}{\rightleftharpoons} H_2O + CO_2$$

The carbonic acid, in turn, dissociates into CO_2 and water in a reaction catalyzed by *carbonic anhydrase* (C.A.), an enzyme found in large quantities in the brush border of the proximal tubule. As a result of H^+ secretion, then, the amount of HCO_3^- in the tubular fluid *decreases*.

2. *Fate of the HCO_3^- Formed During the H^+ Secretion Process.* The H^+ secretion process is described in Chapter 33, but its major features will be noted here (Fig. 4.24*B*). Carbonic anhydrase, which is found in the epithelial cells of the proximal tubule as well as in the brush border, catalyzes the formation of *carbonic acid* from CO_2 and water in the epithelial cells. The carbonic acid then dissociates into H^+, which is secreted into the tubular fluid in exchange for Na^+, and HCO_3^-, which can exit the cells. Most importantly, the HCO_3^- can diffuse across the *basal-lateral surfaces* into the peritubular fluid and peritubular capillaries. As illustrated in Figure 4.24*B*①, this means that the net result of the Na^+-H^+ antiport is the removal of Na^+ and HCO_3^- from the tubular fluid and the addition of Na^+ and

HCO_3^- to the peritubular fluid, a process that is referred to as *Na^+-HCO_3^- reabsorption*. It is important to note that the atoms comprising the HCO_3^- removed from the tubular fluid are *not* the same as those comprising the HCO_3^- returned to the peritubular fluid, but the *net* result is *equivalent* to HCO_3^- reabsorption.

HCO_3^- also may exit from the epithelial cells across the *brush border* into the tubular fluid, via a coupled exchange for Cl^- (Fig. 4.24B②) (Warnock and Yee, 1981). When this occurs, the net result of the Na^+-H^+ antiport is the removal of Na^+ and Cl^- from the tubular fluid and the addition of Na^+ and Cl^- to the peritubular fluid. Thus, the reabsorption of Na^+ and Cl^- can be accomplished by the *Na^+-H^+ antiport* mechanism as well as by the *unidirectional Na^+ transport mechanism*.

Cl^--Driven Na^+ Transport

In order to understand Cl^--driven Na^+ transport, the third mechanism for proximal tubular Na^+ reabsorption, it is necessary to consider the consequences of the first two mechanisms—unidirectional Na^+ reabsorption and the Na^+-H^+ antiport—upon the *composition* of the tubular fluid. The concentrations of HCO_3^- and Cl^- in the glomerular filtrate are approximately equal to their respective concentrations in plasma, 24 mmol/l and 110 mmol/l. If HCO_3^- and Cl^- were to accompany the Na^+ reabsorbed by the unidirectional Na^+ reabsorption and Na^+-H^+ antiport mechanisms in the ratio of 24 HCO_3^-:110 Cl^-, the concentrations of HCO_3^- and Cl^- in the tubular fluid would not change. However, the rate of H^+ secretion into the tubular fluid is such that the ratio of HCO_3^- reabsorbed to Cl^- reabsorbed is *greater* than 24:110. The concentration of HCO_3^- in the tubular fluid therefore decreases, to approximately 8 mmol/l, whereas the concentration of Cl^- increases, to approximately 132 mmol/l. The primary determinant of the rate of H^+ secretion, and hence the lowering of the tubular fluid HCO_3^- concentration, is the activity of the Na^+-H^+ antiport system, although some H^+ also may be secreted into the tubular fluid via an active proton pump in the brush border.

These reciprocal changes in tubular fluid HCO_3^- and Cl^- concentrations occur by the end of part 1 of the proximal tubule (Gottschalk et al., 1960). For the remainder of the proximal tubule, it appears that HCO_3^- and Cl^- are reabsorbed in the ratio of 8 HCO_3^-:132 Cl^-. The concentrations of HCO_3^- and Cl^- therefore do not change from their respective levels of 8 mmol/l and 132 mmol/l in parts 2 and 3 of the proximal tubule. Since the Cl^- concen-

tration in the peritubular fluid is approximately equal to that in plasma (110 mmol/l), this means that in parts 2 and 3 a concentration gradient for Cl^- exists between the tubular fluid and peritubular fluid. Because of the high permeability of the proximal tubule to Cl^- (see above), Cl^- will therefore diffuse passively from tubular lumen to peritubular space in parts 2 and 3, creating a transepithelial potential difference (lumen positive with respect to peritubular space). This transepithelial potential difference is only about +2 mV (lumen positive), however, because the proximal tubule is sufficiently permeable to Na^+ to allow Na^+ to accompany Cl^-. The Cl^- concentration gradient in parts 2 and 3 therefore results in a third mechanism for proximal tubular Na^+ reabsorption, *Cl^--driven Na^+ transport* (Fig. 4.24C). The low resistance pathway for Na^+ reabsorption, like that for Cl^- reabsorption, probably includes the "leaky" tight junctions between the tubular cells (Boulpaep and Sackin, 1977).

The relative proportions of Na^+ reabsorbed by the three major mechanisms for Na^+ transport in the proximal tubule have not yet been established. It is important to note, however, that the energy for all three mechanisms originates with the Na^+-K^+ ATPase, which establishes the electrochemical gradient for Na^+. This gradient, in turn, allows (1) the entry of Na^+ across the brush border in the *unidirectional Na^+ transport* mechanism; (2) the entry of Na^+ across the brush border in exchange for H^+ in the *Na^+-H^+ antiport* mechanism; and (3) the diffusion of Na^+ across the epithelium in the *Cl^--driven Na^+ transport* mechanism, with the Cl^- concentration gradient that causes this transport having been generated by the Na^+-H^+ antiport.

A general feature of proximal tubular Na^+ reabsorption that should be mentioned at this time is that approximately two-thirds of the filtered Na^+ is reabsorbed *regardless* of the filtered load of Na^+. For example, if the filtered load of Na^+ increases, an increased amount of Na^+ will be reabsorbed so that the *fraction* reabsorbed remains nearly constant (Burg and Orloff, 1968). This phenomenon, termed *load-dependent Na^+ reabsorption*, also is present in later regions of the nephron (Chapter 31) and helps to minimize any changes in Na^+ excretion caused by changes in GFR (Chapter 32). However, the fraction of the filtered Na^+ that is reabsorbed in the proximal tubule *can* be affected by the hydrostatic and oncotic pressures in the peritubular capillaries (part of the so-called *third factor* effect) and possibly by as yet unidentified factors (Chapter 32) (Green et al., 1981).

The concentrations of Na^+, Cl^-, HCO_3^-, and other solutes entering and leaving the proximal tubule in a normal person are shown in Table 4.13. TF_{In}/P_{In} ratios also are included. As indicated in Chapter 29, TF_{In}/P_{In} is particularly useful as an index of water reabsorption in the nephron, with the TF_{In}/P_{In} ratio of 3 at the end of the proximal tubule indicating that approximately one-third of the filtered water remains in the tubule at this point (*Eq. 81*).

WATER REABSORPTION

Thus far, the pathway for the movement of water across the proximal tubular epithelium has not been discussed. Indeed, there is uncertainty about the relative importance of two pathways: the *transcellular route* and the *paracellular route* (Fig. 4.25). As previously noted, the Na^+-K^+ ATPase is distributed along both the basal and lateral surfaces of the cell. However, since the lateral space between adjacent cells (the *lateral intercellular space*) is restricted in size (Welling and Welling, 1979), the osmolality of

Table 4.13
Approximate concentrations of substances entering and leaving the proximal tubule in normal man

	Entering the Proximal Tubule (via Glomerular Filtration)	Leaving the Proximal Tubule (to the Loop of Henle)
$[Na^+]$, mmol/l	140	140
$[Cl^-]$, mmol/l	110	132
$[HCO_3^-]$, mmol/l	24	8
[Urea], mmol/l	6	20
[Glucose, amino acids, other solutes], mmol/l	20	$\simeq 0$
Osmolality, mosmol/kg H_2O	300	300
TF_{In}/P_{In}	1	3

the fluid in the lateral space will be increased slightly by the extrusion of Na^+ from the cell into this space by the Na^+-K^+ ATPase. Any HCO_3^- or Cl^- that accompanies the extruded Na^+ also will contribute to this osmolality increase, as will any Na^+ or Cl^- that enters the lateral space by crossing the leaky tight junctions. The osmolality increase results in the movement of water from the tubular lumen into the lateral space. Some of the water goes from lumen to lateral space via the cell interior (the *transcellular* route), while the remainder moves from the lumen across the leaky tight junctions into the lateral space (the *paracellular* route) (Andreoli et al., 1979).

The increase in the volume of fluid in the confines of the lateral space produces an increase in the hydrostatic pressure in the lateral space which, in turn, drives water and dissolved solutes from the lateral space to the peritubular space, from which they are returned to the body via the peritubular capillaries. The rate of water and solute movement from peritubular space to peritubular capillaries can be markedly affected by the hydrostatic and oncotic pressures in the peritubular capillaries. This influence of peritubular capillary hydrostatic and oncotic pressure on proximal tubular reabsorption represents part of the third factor effect on Na^+ reabsorption (see above) and will be discussed in greater detail when the regulation of Na^+ excretion is considered (Chapter 32).

REABSORPTION OF GLUCOSE AND AMINO ACIDS

Over 99% of the filtered glucose normally is

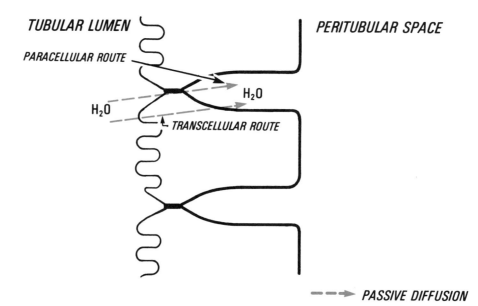

Figure 4.25. Pathways for water reabsorption in the proximal tubule.

reabsorbed by the proximal tubule. Thus, at a GFR of 180 l/day and a normal plasma glucose of 5 mmol/l (90 mg/dl), about 1 mol of glucose is reabsorbed daily from the proximal tubular fluid. What may seem surprising is that the reabsorption of glucose is coupled to the reabsorption of Na^+ (*cotransport*). Specifically, the carrier-mediated transport of glucose across the brush border is coupled to the passive entry of Na^+ across the brush border. The electrochemical gradient for Na^+ entry therefore provides the energy for the transport of glucose across the brush border. In fact, in isolated brush border preparations glucose can be transported *against* a concentration gradient by a Na^+ concentration gradient (Aronson and Sacktor, 1975).

The exact mechanism whereby the Na^+ electrochemical gradient facilitates the transport of glucose is not known. Nor has the glucose carrier been fully identified, although it can be selectively inhibited by *phlorizin* (Turner and Silverman, 1981). The administration of phlorizin to an animal will therefore decrease reabsorption of glucose by the proximal tubule and cause glucosuria, even though the plasma glucose concentration is *normal*. As discussed in Chapter 29, glucosuria may also occur when the plasma glucose concentration is *elevated*, thereby saturating the Na^+-glucose transport system.

The proximal tubule also reabsorbs over 99% of the filtered amino acids. The normal plasma concentration of free amino acids (i.e., amino acids *not* linked to one another to form peptides or proteins) is about 3 mmol/l. Thus, at a GFR of 180 l/day, nearly 540 mmol of amino acids are reabsorbed by the proximal tubule. Like glucose reabsorption, the reabsorption of amino acids by the proximal tubule is coupled to Na^+ reabsorption, with the carrier-mediated transport of amino acids across the brush border coupled to the passive entry of Na^+ across the brush border (Murer and Kinne, 1980). As with glucose, then, the electrochemical gradient for Na^+ entry provides the energy for amino acid transport across the brush border. The reabsorption of different classes of amino acids (neutral, basic, etc.) apparently occurs via different carriers, although the exact number of carriers is not known.

While the occurrence of glucose-Na^+ and amino acid-Na^+ cotransport at the *brush border* is fairly well established, there is less certainty about the mechanisms for the exit of glucose and amino acids across the *basal-lateral surfaces* of the proximal tubular cells. Presumably, some carriers are present in the basal-lateral surfaces that allow the glucose and amino acids to diffuse (probably down their concentration gradients) out of the cells into the peritubular fluid and peritubular capillaries (Murer and Kinne, 1980).

REABSORPTION OF PEPTIDES AND PROTEINS

As discussed in Chapter 28, peptides and proteins smaller than serum albumin are filtered to a variable extent (Table 4.7). Normally, these molecules are almost completely reabsorbed by the proximal tubule so that the urinary excretion of peptides and proteins is small. The reabsorption of peptides and proteins from the proximal tubular fluid occurs predominantly by *pinocytosis*. It is not certain as to whether these molecules bind to specific sites on the brush border surface prior to the pinocytosis. However, once inside the cell, they enter lysosomes, where they are degraded to their constituent amino acids. In other words, peptides and proteins are *not* returned intact to the peritubular capillaries (Boudreau and Carone, 1974). The degradation of proteins by the proximal tubular epithelial cells can be important in patients with diabetes mellitus. When an insulin-requiring diabetic develops renal disease, the dose of insulin necessary to control his plasma glucose may decrease since the diseased kidneys degrade less insulin.

SECRETION OF ORGANIC ACIDS AND BASES

The secretion of the organic acid *p*-amino hippurate (PAH) has already been introduced in Chapters 28 and 29. This secretion occurs exclusively in the *proximal tubule*. Specifically, PAH can be actively transported from the peritubular fluid across the basal-lateral surfaces of the tubular cells. Once inside the cell, the PAH can passively diffuse across the brush border into the proximal tubular fluid, probably by a carrier-mediated process (Tune et al., 1969).

The system that secretes PAH is *nonselective*, i.e., it can secrete a variety of different organic acids (Rennick, 1972). Table 4.14 lists some important examples of such compounds. Of particular interest

Table 4.14
Organic acids secreted by the proximal tubule

Endogenous Substances	Drugs and Other Exogenous Substances
Bile acids	Cephalothin
cAMP	Chlorothiazide
Hydroxyindoleacetic acid	Ethacrynic acid
Oxalic acid	Furosemide
Uric acid	Iodohippuric acid
	p-Amino hippuric acid (PAH)
	Penicillin
	Salicylic acid

is the large number of *antibiotics* and other drugs secreted by the organic acid system (Weiner and Mudge, 1964). A clinically useful *competitive inhibitor* of the organic acid secretory system, *probenecid*, can be administered in conjunction with rapidly secreted antibiotics such as penicillin to prolong their duration of action. In the presence of probenecid, penicillin is secreted (and therefore excreted) less rapidly.

A secretory system for organic bases also is present in proximal tubular cells (Rennick, 1981). Although less well studied, it presumably functions in a manner analogous to the organic acid system. Table 4.15 lists some of the important compounds secreted by the organic base system.

It is not known whether the same cells secrete both organic acids and organic bases or whether cellular specialization exists. In the case of the *organic acid system*, some *regional* specialization has been demonstrated in certain species.

REABSORPTION AND SECRETION OF URIC ACID

Uric acid is the end product of purine metabolism and is eliminated exclusively by the kidneys. The fractional excretion of uric acid in humans is approximately 10%, i.e., the amount excreted is about 10% of the amount filtered. In this regard man differs from most vertebrates, which excrete an amount of uric acid that is at least equal to the amount filtered (fractional excretion $\geq 100\%$) (Weiner, 1979). Furthermore, whereas in most animals the enzyme *uricase* metabolizes uric acid to allantoin, which is then excreted, man lacks uricase and therefore must excrete uric acid itself. Because of the unavailability of a suitable animal model for studying uric acid excretion (and other experimental problems, such as the difficulty in analyzing small quantities of uric acid), considerable uncertainty exists about the exact mechanisms involved in the renal handling of uric acid in humans. However, the proximal tubule is believed to both reabsorb and secrete uric acid, such that the filtered uric acid is gradually reabsorbed and the excreted uric acid is almost completely derived from secretion.

The reabsorption of uric acid is inhibited by *probenecid* (see below). Although probenecid also inhibits the secretion of organic acids, as previously discussed, the uric acid reabsorption system probably is not related to the organic acid secretory system. An experimental agent, *pyrazinoic acid*, inhibits the secretion of uric acid.

Although the fractional excretion of uric acid in

Table 4.15
Organic bases secreted by the proximal tubule

Endogenous Substances	Drugs and Other Exogenous Substances
Acetylcholine	Amiloride
Creatinine	Atropine
Dopamine	Cimetidine
Epinephrine	Hexamethonium
Histamine	Isoproterenol
Norepinephrine	Morphine
Serotonin	Neostigmine
Thiamine	Procaine
	Quinine
	Tetraethylammonium
	Trimethoprim

man averages about 10%, there is considerable individual variation. In persons with a lower fractional excretion, the plasma uric acid becomes elevated, thereby allowing the normal quantity of uric acid to be eliminated. This can be illustrated by considering an example of a person who must eliminate 3 mmol of uric acid daily ($U_{UA}\dot{V} = 3$ mmol/day). If the fractional excretion of uric acid (FE_{UA}) is 10% and the GFR is 180 l/day, the *clearance* of uric acid (C_{UA}) can be calculated from *Eq. 74*:

$$FE_{UA} = \frac{C_{UA}}{C_{In}}$$
$$= \frac{C_{UA}}{GFR}$$

Thus

$$C_{UA} = 18 \text{ l/day}$$

and

$$P_{UA} = \frac{U_{UA}\dot{V}}{C_{UA}} = 0.17 \text{ mmol/l}$$

If, however, FE_{UA} is only 3% but the GFR is still 180 l/day, by similar calculations $C_{UA} = 5.4$ l/day and $P_{UA} = 0.56$ mmol/l. A P_{UA} level of 0.56 mmol/l is sufficient to cause attacks of *gouty arthritis* in susceptible individuals. In such persons, *probenecid* can be used to decrease uric acid reabsorption, thereby increasing FE_{UA} and lowering P_{UA}.

Another potential problem with uric acid relates to its insolubility at acid pH (i.e., at the nomal urine pH in humans). In some individuals, uric acid *crystallizes* in the upper urinary tract to form *kidney stones*. Uric acid is involved in other diseases of the kidneys that are beyond the scope of this text.

UREA

Urea is the major end product of protein metabolism and, like uric acid, is eliminated exclusively by the kidneys. Neither the proximal tubule nor other regions of the nephron can actively reabsorb urea. However, as water is reabsorbed by the proximal tubule, the concentration of urea rises modestly, thereby establishing a concentration gradient for the passive diffusion of urea from tubular lumen to peritubular space. By the end of part 2 of the proximal tubule (the last part of the proximal tubule accessible to micropuncture), about 40% of the filtered urea has been reabsorbed by this mechanism.

Recent evidence suggests that urea may be actively *secreted* into the tubular fluid by part 3 of the proximal tubule (Kawamura and Kokko, 1976). It has been estimated that the urea concentration increases from about 6 mmol/l in plasma and glomerular filtrate to as high as 20 mmol/l by the end of the proximal tubule (Table 4.13).

OTHER FUNCTIONS OF THE PROXIMAL TUBULE

The proximal tubule has an important role in the handling of other ions in the glomerular filtrate. It reabsorbs K^+ (in parts 1 and 2), secretes K^+ (in part 3), and reabsorbs Ca^{++}, Mg^{++}, and phosphate. The renal handling of these ions is considered in greater detail in Chapters 34 and 35.

The proximal tubule also reabsorbs several miscellaneous organic molecules that are present in the glomerular filtrate, including lactate, citrate and other Krebs cycle intermediates, and water-soluble vitamins. In addition, proximal tubular cells synthesize ammonia, a compound that is important in H^+ excretion, as discussed in Chapter 33.

CHAPTER 31

The Loop of Henle and Distal Nephron

THE LOOP OF HENLE

In the loop of Henle, the Na^+Cl^--rich isotonic fluid that leaves the proximal tubule is reduced in volume and transformed into a hypotonic fluid in which a major osmotically active solute is urea. The loop of Henle accomplishes this by reabsorbing approximately 25% of the filtered Na^+ and 20% of the filtered water from the tubular fluid while adding substantial amounts of urea to the tubular fluid. In addition, as will be discussed at the end of this chapter, the loop of Henle has an important role in the mechanisms for both urinary concentration and urinary dilution (Burg and Stephenson, 1978).

The Medullary Gradient

The function of the loop of Henle must be considered in light of the somewhat unusual characteristics of the peritubular fluid in the medulla. First of all, whereas the osmolality of the peritubular fluid in the renal *cortex* is essentially identical with that of plasma (assumed here for simplicity to be approximately 300 mosmol/kg H_2O), the osmolality of the peritubular fluid in the *medulla* increases from about 300 mosmol/kg H_2O at the corticomedullary junction to approximately 1200 mosmol/kg H_2O at the tip of the papilla. Secondly, whereas Na^+ and Cl^- are the predominant osmotically active solutes in the *cortical* peritubular fluid, a substantial fraction of the osmotically active solutes in the *medullary* peritubular fluid is urea. In fact, the concentration of urea in the medullary peritubular fluid increases from the corticomedullary junction, where its concentration is similar to that in plasma, to the tip of the papilla, where it accounts for about half of the osmotically active solutes (Fig. 4.26) (Ullrich et al., 1961). The term *medullary gradient* has been used to describe the osmotic gradient that exists in the peritubular fluid between the corticomedullary junction and the tip of the papilla. The

mechanism for the establishment of this gradient will be considered at the end of this chapter.

Sodium, Chloride, and Water Reabsorption

The reabsorption of Na^+, Cl^-, and water in the loop of Henle is markedly dependent upon the *length of the loop* (Schmidt-Nielsen and O'Dell, 1961). Thus, solute and water reabsorption in *juxtamedullary nephron* loops, which penetrate to varying depths in the inner medulla, differs from that in *cortical nephron* loops, which penetrate to varying depths in the inner cortex and outer medulla. An additional reason that solute and water reabsorption in the loops of juxtamedullary nephrons differs from that in the loops of cortical nephrons involves differences in *ascending limb structure*, and hence function. As noted in Chapter 27, the ascending limb of juxtamedullary nephrons can be divided into two regions: the thin ascending limb, which begins at the hairpin turn, and the thick ascending limb, which begins at or near the junction between the inner and outer medulla. In contrast, cortical nephrons lack a thin ascending limb; the thick ascending limb therefore begins near the hairpin turn, i.e., at or above the junction between inner and outer medulla (see Fig. 4.7).

In this discussion, a "prototype" juxtamedullary nephron, whose loop reaches the tip of the papilla, and a "prototype" cortical nephron, whose loop reaches the junction between the outer and inner medulla, will be used to illustrate how the loop of Henle reabsorbs Na^+, Cl^-, and water (Fig. 4.26). "Prototype" solute concentrations and osmolalities in the tubular fluid and surrounding peritubular fluid as well as "prototype" TF_{In}/P_{In} ratios also will be given. However, it should be emphasized that these prototype concentrations, osmolalities, and TF_{In}/P_{In} ratios are *estimates* presented to illustrate the function of the loop rather than precise values

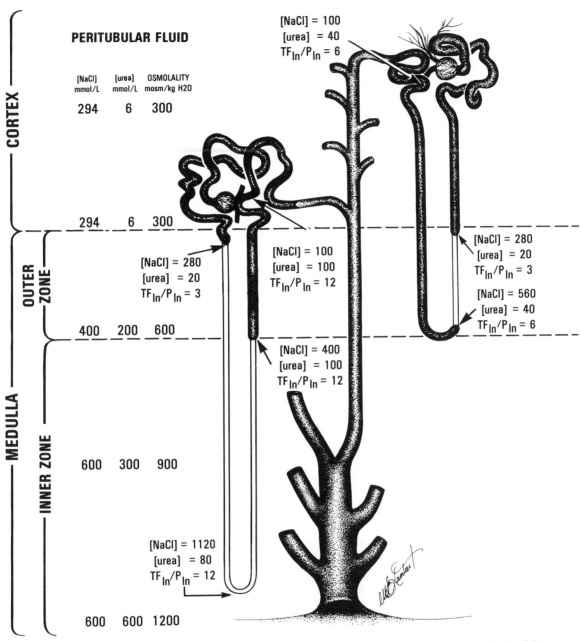

PERITUBULAR FLUID

[NaCl] mmol/L	[urea] mmol/L	OSMOLALITY mosm/kg H2O
294	6	300

[NaCl] = 100
[urea] = 40
TF_{In}/P_{In} = 6

294 6 300

[NaCl] = 280
[urea] = 20
TF_{In}/P_{In} = 3

[NaCl] = 280
[urea] = 20
TF_{In}/P_{In} = 3

[NaCl] = 100
[urea] = 100
TF_{In}/P_{In} = 12

[NaCl] = 560
[urea] = 40
TF_{In}/P_{In} = 6

400 200 600

[NaCl] = 400
[urea] = 100
TF_{In}/P_{In} = 12

600 300 900

[NaCl] = 1120
[urea] = 80
TF_{In}/P_{In} = 12

600 600 1200

Figure 4.26. Approximate solute concentrations, osmolalities, and TF_{In}/P_{In} ratios in loop of Henle tubular fluid and the surrounding peritubular fluid for a "prototype" cortical nephron (whose loop reaches the junction between the inner and outer medulla) and a "prototype" juxtamedullary nephron (whose loop reaches the tip of the papilla). The [NaCl] values given include small contributions from other electrolytes and nonurea solutes.

that should be mastered. Such estimation is necessary because micropuncture sampling of the tubular fluid and peritubular fluid is not possible in man. Furthermore, even in animal studies the only sites accessible to micropuncture are the cortical surface and near the tip of the papilla (Chapter 27).

THIN DESCENDING LIMB

The thin descending limb of the loop of Henle has a relatively low permeability to solutes and

lacks mechanisms for active solute transport but is highly permeable to water (Kokko, 1970). Thus, as the isotonic tubular fluid entering the thin descending limb flows downward through regions of increasingly hypertonic peritubular fluid (the medullary gradient), water is reabsorbed. In fact, the thin descending limb is so permeable to water that the osmolality of the tubular fluid becomes equal to that of the surrounding peritubular fluid, i.e., the thin descending limb resembles the proximal tubule insofar as an osmotic gradient cannot be maintained across its epithelium. The mechanism whereby the water reabsorbed from the thin descending limb is removed from the medulla without diluting or washing out the medullary gradient will be considered below.

Because of the reabsorption of water, the tubular fluid osmolality increases as fluid flows through the thin descending limb. In juxtamedullary nephrons whose loops reach the tip of the papilla (e.g., the "prototype" juxtamedullary nephron of Fig. 4.26), the tubular fluid osmolality increases from about 300 mosmol/kg H_2O at the start of the thin descending limb to as high as 1200 mosmol/kg H_2O at its termination at the hairpin turn. In cortical nephrons whose loops reach the junction between the outer and inner medulla (e.g., the "prototype" cortical nephron of Fig. 4.26), the tubular fluid osmolality is estimated to increase from about 300 mosmol/kg H_2O to about 600 mosmol/kg H_2O at the hairpin turn. The available micropuncture data indicate that these substantial increases in osmolality can in fact be accomplished primarily by the *reabsorption of water* from the tubular fluid rather than by the secretion of solutes into the tubular fluid. Specifically, TF_{In}/P_{In} is estimated to increase from about 3 at the end of the proximal tubule to approximately 12 at the hairpin turn in juxtamedullary nephrons whose loops reach the tip of the papilla, like the "prototype" juxtamedullary nephron considered here. Hence, approximately three-fourths of the water entering the thin descending limb of such juxtamedullary nephrons is reabsorbed there, i.e., the 4-fold increase in osmolality (from 300 mosmol/kg H_2O to 1200 mosmol/kg H_2O) can be attributed almost entirely to the reabsorption of water. Although the hairpin turn of cortical nephrons is not accessible to micropuncture, it is estimated that TF_{In}/P_{In} increases from about 3 at the end of the proximal tubule to approximately 6 at

the hairpin turn in the "prototype" cortical nephron considered here. According to this estimate, then, approximately one-half of the water entering the thin descending limb of this "prototype" cortical nephron is reabsorbed there, i.e., the 2-fold increase in osmolality (from 300 mosmol/kg H_2O to 600 mosmol/kg H_2O) can be attributed almost entirely to the reabsorption of water.

While the *osmolality* of the tubular fluid in the thin descending limb is equal to that of the surrounding peritubular fluid, there are important *composition* differences between the tubular fluid and the peritubular fluid, particularly in juxtamedullary nephrons. The isotonic tubular fluid that enters the thin descending limb contains approximately 280 mmol/l of electrolytes (predominantly Na^+ and Cl^-) and other solutes and approximately 20 mmol/l of urea (Tables 4.13 and 4.16). These solutes are *concentrated* in the thin descending limb as water is reabsorbed. In the juxtamedullary nephron example presented here, the solutes are concentrated approximately 4-fold in the thin descending limb, as noted above. The tubular fluid at the hairpin turn therefore contains about 1120 mmol/l of electrolytes (again, predominantly Na^+ and Cl^-) and 80 mmol/l of urea. The surrounding peritubular fluid, on the other hand, contains about 600 mmol/l of electrolytes and 600 mmol/l of urea (Fig. 4.26). In the cortical nephron example, the solutes are concentrated approximately 2-fold in the thin descending limb. The tubular fluid at the hairpin turn therefore is estimated to contain about 560 mmol/l of electrolytes and 40 mmol/l of urea, which compares with approximately 400 mmol/l of electrolytes and 200 mmol/l of urea in the surrounding peritubular fluid (Fig. 4.26). Thus, in both juxtamedullary nephrons and cortical nephrons the predominant solutes in the *tubular fluid* are Na^+ and Cl^-, but a substantial fraction of the solutes in the surrounding *peritubular fluid* is *urea*.

Table 4.16

Approximate concentrations of substances entering and leaving the loop of Henle in normal man

	Entering the Thin Descending Limb (from the Proximal Tubule[a])		Leaving the Thin Descending Limb	Leaving the Thin Ascending Limb	Leaving the Thick Ascending Limb (to the Distal Nephron)
[NaCl],[b] mmol/l	280	Cortical	560		100
		Juxtamedullary	1120	400	100
[Urea], mmol/l	20	Cortical	40		40
		Juxtamedullary	80	100	100
Osmolality, mosmol/kg H_2O	300	Cortical	600		140
		Juxtamedullary	1200	500	200
TF_{In}/P_{In}	3	Cortical	6		6
		Juxtamedullary	12	12	12

[a] See Table 4.13.   [b] Includes small contributions from other electrolytes and nonurea solutes.

THIN ASCENDING LIMB

After passing the hairpin turn, the tubular fluid enters the *ascending limb*. In *juxtamedullary nephrons*, the ascending limb begins with the *thin ascending limb*. In spite of its structural similarity to the thin descending limb, the thin ascending limb has completely different transport and permeability characteristics, being virtually impermeable to water but highly permeable to Na^+ and Cl^- and moderately permeable to urea (Imai and Kokko, 1974). Thus, as a result of the substantial concentration gradients for Na^+, Cl^-, and urea between the tubular fluid entering the thin ascending limb and the peritubular fluid (Fig. 4.26), Na^+ and Cl^- diffuse passively from lumen to peritubular space, while urea diffuses passively from peritubular space to lumen. Of course, the greater permeability of the thin ascending limb to Na^+ and Cl^- than to urea means that the number of moles of Na^+ plus Cl^- *leaving* the lumen *exceeds* the number of moles of urea *entering*. But in spite of this net solute exit, and in spite of the fact that the tubular fluid flows upward through a region of progressively decreasing peritubular fluid osmolality, the *volume* of the tubular fluid does not significantly change in the thin ascending limb due to the low water permeability. With net solute exit and no appreciable volume change, the osmolality of the tubular fluid in the thin ascending limb falls slightly below that of the surrounding peritubular fluid (Jamison et al., 1967). In the juxtamedullary nephron example presented here, the tubular fluid at the end of the thin ascending limb is estimated to contain about 400 mmol/l of electrolytes and 100 mmol/l of urea, for a total osmolality of approximately 500 mosmol/kg H_2O. In contrast, the surrounding peritubular fluid is estimated to contain approximately 400 mmol/l of electrolytes and 200 mmol/l of urea, for a total osmolality of 600 mosmol/kg H_2O (Fig. 4.26). Note that TF_{In}/P_{In} remains constant throughout this region, since the volume of tubular fluid does not change appreciably.

THICK ASCENDING LIMB

The tubular fluid enters the thick ascending limb from the thin ascending limb in juxtamedullary nephrons and from the thin descending limb in cortical nephrons (Fig. 4.26). Like the thin ascending limb, the water permeability of the thick ascending limb is negligible (Rocha and Kokko, 1973). However, the other functional characteristics of the thick ascending limb differ markedly from those of the thin ascending limb. First of all, the urea permeability of the thick ascending limb is quite low.

Second, and most important, the thick ascending limb *actively* transports Na^+ and Cl^- from lumen to peritubular space (Burg and Green, 1973). The combination of *low water permeability* and *active reabsorption of Na^+ and Cl^-* means that the thick ascending limb *lowers* both the osmolality of the tubular fluid and the concentration of Na^+ and Cl^- in the tubular fluid to levels below those in the surrounding peritubular fluid. The combination of *low water permeability* and *low urea permeability* means that the urea concentration is *unaltered* by the thick ascending limb. In the juxtamedullary nephron example presented here, the tubular fluid at the end of the thick ascending limb is estimated to contain about 100 mmol/l of electrolytes and 100 mmol/l of urea, for a total osmolality of approximately 200 mosmol/kg H_2O. In the cortical nephron example, the tubular fluid at the end of the thick ascending limb is estimated to contain approximately 100 mmol/l of electrolytes and 40 mmol/l of urea, for a total osmolality of 140 mosmol/kg H_2O (Fig. 4.26). Of course, since the thick ascending limb terminates in the *cortex* in both cortical and juxtamedullary nephrons, the solute concentrations and osmolality of the surrounding peritubular fluid are similar to those of plasma, i.e., approximately 294 mmol/l of electrolytes and other nonurea solutes and 6 mmol/l of urea, for a total osmolality of approximately 300 mosmol/kg H_2O. In *both* nephron types, then, the tubular fluid that leaves the loop of Henle is hypotonic and (particularly in juxtamedullary nephrons) has a urea concentration that is considerably greater than that of plasma and the surrounding peritubular fluid. It should be noted that TF_{In}/P_{In} remains constant throughout the thick ascending limb in both juxtamedullary and cortical nephrons, since the volume of the tubular fluid does not change significantly due to the low water permeability.

The mechanism for the active reabsorption of Na^+ and Cl^- in the thick ascending limb differs from that in other regions of the nephron. In contrast to other regions, the transport process is dependent upon the presence of Cl^- in the lumen and produces a transepithelial potential difference of approximately +6 to +10 mV (lumen positive). Thus, Cl^- is transported against both a concentration gradient (Fig. 4.26) and an electrical gradient, i.e., the actively transported ion appears to be Cl^- rather than Na^+. An apparent paradox exists, however, in that the active reabsorption of Na^+ and Cl^- in the thick ascending limb can be inhibited by *ouabain*, an agent believed to inhibit only the Na^+-K^+ *ATPase*. One model that presents a solution to this paradox is shown in Figure 4.27. The Na^+-K^+

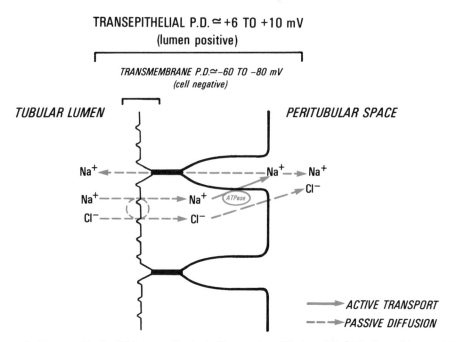

Figure 4.27.   Na⁺-Cl⁻ coupled transport in the thick ascending limb. The passive diffusion of Na⁺ into the cell is coupled to the entry of Cl⁻ (as indicated by the *dashed circle*). The Na⁺ is then actively extruded across the basal-lateral surfaces by the Na⁺-K⁺ ATPase, with Cl⁻ following passively to maintain electrical neutrality. Because the tight junction is slightly more permeable to Na⁺ than to Cl⁻, a small amount of Na⁺ leaks back into the lumen, generating a lumen-positive transepithelial potential difference (P.D.).

ATPase on the basal-lateral surfaces produces both a low intracellular concentration of Na⁺ and an intracellular electrical potential that is about 60–80 mV negative with respect to the lumen, i.e., an *electrochemical gradient* for passive Na⁺ entry similar to that in the proximal tubule is established. However, the entry of Na⁺ that occurs in the thick ascending limb differs from that in the proximal tubule insofar as it appears to be coupled to the entry of Cl⁻ (*cotransport*). Once inside the cell, the Na⁺ and Cl⁻ are believed to exit into the lateral intercellular space, the Na⁺ being actively extruded by the Na⁺-K⁺ ATPase and the Cl⁻ following passively to maintain electroneutrality. The model now postulates that the tight junction is *slightly* more permeable to Na⁺ than to Cl⁻. As a result, although most of the Na⁺ extruded into the lateral intercellular space is reabsorbed, a small amount leaks from the lateral space back across the tight junction into the tubular lumen (Fig. 4.27), thereby generating the lumen positive transepithelial potential difference. According to the model presented in Figure 4.27, then, the actively transported ion actually is Na⁺, *not* Cl⁻, even though Cl⁻ is reabsorbed against an electrochemical gradient.

An important characteristic of the Na⁺-Cl⁻ coupled transport system in the thick ascending limb is that an increased amount of Na⁺ will be reabsorbed by the thick ascending limb if an increased load of Na⁺ is delivered to this region, a phenomenon termed *load dependence*. For example, if the reabsorption of Na⁺ in the *proximal tubule* is decreased as a result of plasma volume expansion (part of the *third factor* effect; see Chapter 32), more Na⁺ will be delivered to the thick ascending limb and an *increased* amount of Na⁺ will be reabsorbed (Cortney et al., 1965). Such load-dependent Na⁺ reabsorption also is present in the *proximal tubule* (Chapter 30). The Na⁺-Cl⁻ coupled transport system in the thick ascending limb is *inhibited* by an important class of diuretic drugs, the so-called *loop diuretics*, whose members include *furosemide* and *ethacrynic acid*.

Table 4.16 summarizes the modifications in the composition of the tubular fluid that occur in the loop of Henle of the prototype cortical and juxtamedullary nephrons considered here. The permeability and transport characteristics of the various segments of the loop of Henle are summarized in Table 4.17.

Other Functions of the Loop of Henle

The loop of Henle secretes K⁺ (in the thin descending limb) and reabsorbs K⁺ (in the thin and thick ascending limbs). In addition, the thick ascending limb reabsorbs Ca⁺⁺. These functions are considered in greater detail in Chapters 34 and 35.

Table 4.17
Permeability and transport characteristics of loop of Henle and distal nephron[a]

| | Loop of Henle | | | Distal Nephron | | |
| | | | | | Collecting Tubule/Duct | |
	Thin Descending Limb	Thin Ascending Limb	Thick Ascending Limb	Distal Tubule and Connecting Tubule	Cortical Collecting Tubule and Medullary Collecting Duct	Papillary Collecting Duct
Na^+						
Permeability	0	+++	+	0	0	0
Active reabsorption	0	0	Yes	Yes	Yes	Yes
Urea permeability	0	+	0	0	0	+ADH +++ / −ADH 0
Water permeability	++++	0	0	0	+ADH ++++ / −ADH 0	+ADH ++++ / −ADH 0

[a] Relative permeabilities indicated by number of + symbols; 0 indicates low or negligible permeability.

THE DISTAL NEPHRON

The distal nephron produces the modifications in tubular fluid composition and volume necessary to complete the transformation of the tubular fluid into urine. These modifications include the reabsorption of as much as 8–10% of the filtered Na^+ (thereby making possible the recovery of virtually all of the filtered Na^+), the reabsorption of urea, the secretion of H^+ and K^+, and the reabsorption of up to 15% of the filtered water. Since the distal nephron includes several anatomically distinct segments (Chapter 27), it should not be surprising that the distal nephron is not a functionally homogeneous structure. Thus, the distal tubule, connecting tubule, and the various segments of the collecting tubule/duct (cortical collecting tubule, medullary and papillary collecting ducts) have specialized as well as common functions.

To appreciate the modifications in volume and composition that occur in the distal nephron, "prototype" solute concentrations and osmolalities in the tubular fluid and surrounding peritubular fluid as well as TF_{In}/P_{In} ratios will be given in the discussion that follows. Since the composition of the tubular fluid leaving the thick ascending limb is dependent upon the length of the loop of Henle (Fig. 4.26, Table 4.16), the composition of the tubular fluid entering the distal nephron will be different in different nephrons. However, because several adjacent nephrons share a common collecting tubule/duct, throughout much of the distal nephron the tubular fluid will be a *mixture* of contributions from different nephrons. It is therefore sufficient to consider *average* tubular fluid compositions in the distal nephron. The average tubular fluid entering the distal nephron is estimated to contain approximately 100 mmol/l of electrolytes and other nonurea solutes and 50 mmol/l of urea, for a total

osmolality of approximately 150 mosmol/kg H_2O, and to have a TF_{In}/P_{In} ratio of approximately 7 (Fig. 4.28). The average tubular fluid entering the distal nephron is therefore estimated to be quite similar to the tubular fluid leaving the thick ascending limb of the "prototype" cortical nephron discussed above (approximately 85% of the nephrons in man are cortical). Like the "prototype" concentrations and osmolalities for the loop of Henle, the distal nephron "prototype" concentrations and osmolalities must be regarded as *estimates* that illustrate the function of the kidneys rather than precise values that should be mastered.

It should be noted that whereas Na^+ and Cl^- are the primary nonurea solutes *entering* the distal nephron, other nonurea solutes (*nus*) become increasingly important *in* the distal nephron. This is because tubular reabsorption in the distal nephron can reduce the Na^+ and Cl^- concentrations to such low levels (see below) that the concentrations of solutes such as creatinine, uric acid, K^+, NH_4^+, Ca^{++}, Mg^{++}, and phosphate can no longer be neglected.

Sodium, Chloride, and Water Reabsorption

One transport function common to all segments of the distal nephron is the active reabsorption of Na^+. The mechanism for this Na^+ reabsorption is similar to the unidirectional Na^+ transport (uniport) mechanism found in the proximal tubule (Fig. 4.29; compare to Fig. 4.24). Na^+ passively enters the tubular epithelial cells across the luminal surface and then is actively extruded across the basal-lateral surfaces by the Na^+-K^+ ATPase (Schultz, 1981). As in the proximal tubule, the Na^+-K^+ ATPase provides the energy for the passive entry of Na^+ as well as for its active extrusion by generating the electrochemical gradient that favors Na^+ entry.

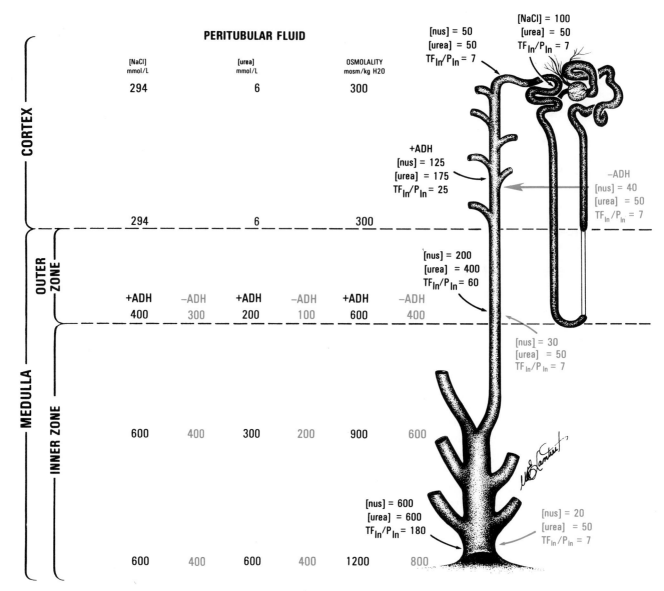

Figure 4.28.   Approximate solute concentrations, osmolalities, and TF_{In}/P_{In} ratios in distal nephron tubular fluid and the surrounding peritubular fluid in the presence of maximal ADH (*black*) and in the absence of ADH (*red*). Because the tubular fluid throughout much of the distal nephron is a mixture of contributions from different nephrons, only *average* tubular fluid compositions are shown. The [NaCl] values given include small contributions from other electrolytes and nonurea solutes; this term is replaced by [nus] (*nonurea solutes*) in the later parts of the distal nephron because the Na^+ and Cl^- concentrations can be reduced to such low levels that the concentrations of other electrolytes and nonurea solutes can no longer be neglected.

Such unidirectional Na^+ transport will produce a transepithelial electrical potential difference (lumen negative with respect to peritubular space) whose magnitude will depend on the permeability of the tubular epithelium to one or more anions, which then could accompany Na^+. In the distal nephron, the tubular epithelium is sufficiently permeable to Cl^- to allow much of the Na^+ to be accompanied by Cl^-. However, since the Cl^- permeability of the distal nephron is considerably lower than that of the proximal tubule, a substantial transepithelial potential difference can be gener-

ated in the distal nephron. In fact, the transepithelial potential difference can be as large as -70 mV (lumen negative) in the distal nephron (Fig. 4.29), depending upon the amount of Na^+ that is reabsorbed. Because the relatively low Cl^- permeability allows the establishment of such a large potential difference, the distal nephron is said to have a *high electrical resistance*, in contrast to the low electrical resistance of the proximal tubule. The high resistance properties of the distal nephron suggest that the tight junctions between epithelial cells are "tighter" than the "leaky" tight junctions of the

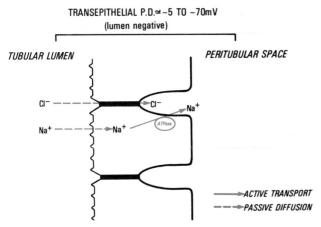

TRANSEPITHELIAL P.D.≅ −5 TO −70mV
(lumen negative)

TUBULAR LUMEN

PERITUBULAR SPACE

→ ACTIVE TRANSPORT
---→ PASSIVE DIFFUSION

Figure 4.29. Unidirectional Na⁺ transport in the distal nephron. Na⁺ passively enters the cell along its electrochemical gradient and then is actively extruded across the basal-lateral surfaces by the Na⁺-K⁺ ATPase. This reabsorption of Na⁺ generates a lumen-negative transepithelial potential difference (P.D.), which can be as large as −70 mV because the "tight" tight junctions of the distal nephron limit the ability of Cl⁻ to accompany Na⁺. It should be noted that some Cl⁻ may accompany Na⁺ via the transcellular route instead of across the "tight" tight junctions as shown; the exact pathway is not known.

proximal tubule. These "tight" tight junctions not only limit the ability of Cl⁻ to accompany Na⁺, but also prevent the Na⁺ pumped into the lateral intercellular space from leaking back into the tubular lumen to any significant extent.

An important characteristic of unidirectional Na⁺ transport in the distal nephron is that an increased amount of Na⁺ will be reabsorbed by the distal nephron if an increased load of Na⁺ is delivered to this region (Khuri et al., 1975). Like the proximal tubule and thick ascending limb, then, the distal nephron exhibits *load-dependent Na⁺ reabsorption*. It should be noted that because of this load dependence of distal nephron Na⁺ reabsorption, a larger transepithelial potential difference will be generated if delivery of Na⁺ to the distal nephron is increased. It also should be noted that in the collecting tubule/duct, Na⁺ reabsorption is stimulated by the hormone *aldosterone* (see Chapter 32). Thus, aldosterone represents an additional cause of a larger transepithelial potential difference.

DISTAL TUBULE AND CONNECTING TUBULE

The first two segments of the distal nephron, the *distal tubule* and *connecting tubule*, reabsorb Na⁺ by the unidirectional transport mechanism common to the entire distal nephron. However, certain characteristics of the distal tubule and connecting tubule closely resemble those of the thick ascending limb, including low permeability to water and urea

(Imai, 1979). The combination of *low water permeability* and *active Na⁺ reabsorption* (with Cl⁻ following passively) means that the distal tubule and connecting tubule *lower* both the osmolality of the tubular fluid and its electrolyte concentration. The combination of *low water permeability* and *low urea permeability* means that the urea concentration remains essentially *unchanged* in these segments. At the end of the connecting tubule, the tubular fluid is estimated to contain approximately 50 mmol/l of urea and 50 mmol/l of nonurea solutes (including electrolytes), for a total osmolality of 100 mosmol/kg H₂O (Fig. 4.28). Of course, the solute concentrations and osmolality in the surrounding cortical peritubular fluid are similar to those of plasma. Thus, the hypotonic tubular fluid entering the distal nephron has its osmolality further reduced in the distal tubule and connecting tubule, but the urea concentration remains considerably greater than that in plasma. TF_In/P_In remains constant in these regions with an approximate value of 7, since the volume of the tubular fluid does not change appreciably due to the low water permeability (Fig. 4.28).

COLLECTING TUBULE/DUCT

Like the distal tubule and connecting tubule, the various segments of the collecting tubule/duct (*cortical collecting tubule, medullary* and *papillary collecting ducts*) reabsorb Na⁺ by the unidirectional transport mechanism. However, the collecting tubule/duct differs from other segments of the nephron (as well as from all other tissues in the body) in that its water and urea permeability properties are regulated by a *hormone* (Imai, 1979). This hormone, *antidiuretic hormone* (abbreviated ADH; also called vasopressin), is a nonapeptide that is synthesized in the hypothalamus and stored in the posterior pituitary (Chapter 52). The factors that stimulate the release of ADH from the posterior pituitary will be discussed in Chapter 32, but the effects of ADH on the water and urea permeability properties of the collecting tubule/duct will be considered here (Schafer and Andreoli, 1978). In the *absence* of ADH, the collecting tubule/duct is relatively impermeable to water and urea. However, in the *presence* of ADH, the water permeability of the entire collecting tubule/duct and the urea permeability of the papillary collecting duct increase significantly. The changes in tubular fluid composition and osmolality that occur in the collecting tubule/duct will therefore be a function of the ADH level. In the discussion that follows, composition and osmolality changes will be examined first in the presence of *maximal* ADH and then in the *absence* of ADH (Kokko and Rector, 1972).

Maximal ADH

In the presence of maximal ADH, the high water permeability of the collecting tubule/duct prevents the establishment of an osmotic gradient across the tubular epithelium. Recall that the osmolality of the tubular fluid entering the collecting tubule/duct is only 100 mosmol/kg H_2O. Since the osmolality of the surrounding peritubular fluid ranges from approximately 300 mosmol/kg H_2O in the cortex to 1200 mosmol/kg H_2O at the tip of the papilla, this means that water is reabsorbed as the tubular fluid flows through the collecting tubule/duct. The continued reabsorption of Na^+ and Cl^- as the fluid flows through the collecting tubule/duct further promotes the reabsorption of water. In the *cortical collecting tubule*, over 70% of the water entering the collecting tubule/duct is reabsorbed, as evidenced by the fact that TF_{In}/P_{In} increases by a factor of approximately 3.5, from 7 to about 25. Since the high water permeability of the cortical collecting tubule is combined with low urea permeability, the urea concentration also increases by a factor of approximately 3.5, from about 50 mmol/l to about 175 mmol/l. The concentration of nonurea solutes increases less markedly, from about 50 mmol/l to about 125 mmol/l, because Na^+ and Cl^- are reabsorbed. At the end of the cortical collecting tubule, then, the osmolality of the tubular fluid is approximately 300 mosmol/kg H_2O, in osmotic equilibrium with the peritubular fluid at the corticomedullary junction (Fig. 4.28). The fact that the increase in osmolality (about 3-fold, from 100 to 300 mosmol/kg H_2O) is smaller than the increase in TF_{In}/P_{In} (3.5-fold) or urea concentration (3.5-fold) can be attributed primarily to the reabsorption of Na^+ and Cl^-.

In the *medullary collecting duct*, water continues to be reabsorbed as the tubular fluid flows downward through regions of increasingly hypertonic peritubular fluid (the medullary gradient) and as additional Na^+ and Cl^- are reabsorbed. It is estimated that over 50% of the water entering the medullary collecting duct is reabsorbed in this region, as evidenced by the fact that TF_{In}/P_{In} increases over 2-fold, from 25 to about 60. Since the high water permeability of the medullary collecting duct is combined with low urea permeability, as in the cortical collecting tubule, the urea concentration also increases over 2-fold, from about 175 to about 400 mmol/l. The concentration of nonurea solutes increases less markedly, from about 125 to

about 200 mmol/l, because Na^+ and Cl^- are reabsorbed. The osmolality of the tubular fluid at the end of the medullary collecting duct is therefore approximately 600 mosmol/kg H_2O, in osmotic equilibrium with the peritubular fluid at the junction between inner and outer medulla (Fig. 4.28). The fact that the osmolality increases less than TF_{In}/P_{In} or the urea concentration can be attributed primarily to the reabsorption of Na^+ and Cl^-.

Additional water is reabsorbed in the *papillary collecting duct* as the tubular fluid continues flowing downward through the medullary gradient. However, the papillary collecting duct differs from the cortical collecting tubule and the medullary collecting duct in that ADH increases its permeability to *urea* as well as to water. Since the concentration of urea in the tubular fluid entering the papillary collecting duct (about 400 mmol/l) exceeds that in the surrounding peritubular fluid (about 200 mmol/l at the junction between the inner and outer medulla), urea will be reabsorbed by the papillary collecting duct. This reabsorption of urea has an important role in the generation of the medullary gradient (see below) and also promotes the further reabsorption of water. It is estimated that approximately two-thirds of the water entering the papillary collecting duct is reabsorbed there, as TF_{In}/P_{In} increases about 3-fold, from approximately 60 to about 180. The concentration of nonurea solutes also increases approximately 3-fold, from about 200 to 600 mmol/l, because in spite of some Na^+ and Cl^- reabsorption the amount of nonurea solutes in the tubular fluid does not change significantly. Since urea is reabsorbed, its concentration increases less markedly, from approximately 400 to 600 mmol/l, i.e., at the end of the papillary collecting duct a transepithelial concentration gradient for urea no longer exists. The osmolality of the tubular fluid at the end of the papillary collecting duct is therefore approximately 1200 mosmol/kg H_2O, in osmotic equilibrium with the peritubular fluid at the tip of the papilla (Fig. 4.28). The fact that the osmolality increases less than TF_{In}/P_{In} or the concentration of nonurea solutes can be attributed primarily to the reabsorption of urea.

The tubular fluid from several papillary collecting ducts empties into a common duct of Bellini and from there flows through the minor calyx, major calyx, renal pelvis, and ureter into the urinary bladder (Chapter 27). No modifications in the composition or volume of the tubular fluid are believed to occur in any of these structures, i.e., the final urine is identical with the tubular fluid leaving the papillary collecting duct. In the presence of maximal ADH, then, the final urine has an osmolality of approximately 1200 mosmol/kg H$_2$O, of which about 600 mmol/l is urea and about 600 mmol/l represents nonurea solutes such as Na$^+$, Cl$^-$, creatinine, uric acid, K$^+$, NH$_4^+$, Ca^{++}, Mg^{++}, and phosphate. The U$_{In}$/P$_{In}$ ratio is approximately 180. Recall from Chapter 29 that U$_{In}$/P$_{In}$ can be used to calculate the fractional excretion of water (FE$_{H_2O}$) (*Eq. 80*):

$$FE_{H_2O} = \frac{1}{U_{In}/P_{In}}$$

In the presence of maximal ADH, then, only about 0.5% of the filtered water is excreted (1/180 ≃ 0.005); 99.5% is reabsorbed. If the GFR is 125 ml/min (180 l/day), this means that the rate of urine flow (V̇) will be only 0.6 ml/min (0.9 l/day) in the presence of maximal ADH.

The effect of ADH on the water permeability of the collecting tubule/duct involves the binding of ADH to receptors on the basal-lateral surfaces of the epithelial cells, the activation of adenylate cyclase, and the generation of cyclic AMP (Dousa and Valtin, 1976), which in turn leads to the insertion of protein-containing aggregates into the luminal membrane (Wade et al., 1977). These aggregates, which presumably function as channels for water movement, are believed to be components of tubular vesicles in the cells that fuse with the luminal membrane in the presence of ADH. The effect of ADH on the urea permeability of the papillary collecting duct appears to be mediated by a somewhat different mechanism, possibly involving a carrier for urea that is independent of the above-mentioned aggregates.

Absence of ADH

In the absence of ADH, the entire collecting tubule/duct is relatively *impermeable* to water. Since the thin ascending limb, thick ascending limb, distal tubule, and connecting tubule *always* are relatively impermeable to water, this means that in the absence of ADH the nephron is water-impermeable from the hairpin turn at the tip of the loop of Henle all the way to the end of the papillary collecting duct. As a result, the volume of the tubular fluid remains virtually unchanged not only in the ascending limb, distal tubule, and connecting tubule (see above), but in the collecting tubule/duct as well. In the absence of ADH, then, TF$_{In}$/P$_{In}$ remains approximately constant in the collecting tubule/duct, with a value of about 7. The urea concentration also remains approximately constant, with a value of about 50 mmol/l, since the

entire collecting tubule/duct has a low permeability to urea in the absence of ADH. However, the reabsorption of Na$^+$ and Cl$^-$ continues, thereby reducing the concentration of nonurea solutes from about 50 mmol/l at the beginning of the cortical collecting tubule to as low as 20 mmol/l at the end of the papillary collecting duct. The osmolality of the tubular fluid at the end of the papillary collecting duct is therefore as low as 70 mosmol/kg H$_2$O in the absence of ADH. Note that the transepithelial osmotic, Na$^+$Cl$^-$, and urea gradients become quite substantial in the collecting tubule/duct in the absence of ADH (Fig. 4.28), even though the medullary gradient is somewhat diminished (see below). The "tight" tight junctions of the distal nephron epithelium allow these gradients to be maintained.

Since the final urine is identical with the tubular fluid leaving the papillary collecting duct, this means that in the absence of ADH the final urine has an osmolality of approximately 70 mosmol/kg H$_2$O, of which about 50 mmol/l is urea and about 20 mmol/l represents nonurea solutes. The U$_{In}$/P$_{In}$ ratio is approximately 7, i.e., in the absence of ADH nearly 15% of the filtered water is excreted (1/7 ≃ 0.14); 85% is reabsorbed (*Eq. 80*). With a GFR of 125 ml/min (180 l/day), the rate of urine flow (V̇) will be greater than 15 ml/min (26 l/day). Because the volume of the tubular fluid remains virtually constant throughout the entire ascending limb and distal nephron in the absence of ADH, this urine flow rate also should represent the flow rate of tubular fluid at the tip of the loop of Henle. However, it should be noted that even in the absence of ADH the water impermeability of the collecting tubule/duct is *relative, not* absolute, so that a small fraction of the water flowing through the collecting tubule/duct will be reabsorbed. The actual urine flow rate may therefore be slightly less than the flow rate of tubular fluid at the tip of the loop.

Table 4.18 summarizes the modifications in the composition of the tubular fluid that occur in the distal nephron in the presence of maximal ADH and in absence of ADH. The permeability and transport characteristics of the various segments of the distal nephron are summarized in Table 4.17.

Other Functions of the Distal Nephron

The distal nephron has an essential role in the regulation of *K$^+$ excretion*. In fact, virtually *all* regulation of K$^+$ excretion occurs in the distal nephron, which can both reabsorb and secrete K$^+$. The distal nephron also has an essential role in regulating the body's *acid-base balance* via secretion of H$^+$. These important functions of the distal nephron

Table 4.18
Approximate concentrations of substances entering and leaving the distal nephron in normal man

	Entering the Distal Nephron (from the Loop of Henle)	Leaving the Connecting Tubule		Leaving the Cortical Collecting Tubule	Leaving the Medullary Collecting Duct	Leaving the Papillary Collecting Duct (Final Urine)
[nus],[a] mmol/l	100	50	+ADH	125	200	600
			−ADH	40	30	20
[Urea], mmol/l	50	50	+ADH	175	400	600
			−ADH	50	50	50
Osmolality, mosmol/kg H_2O	150	100	+ADH	300	600	1200
			−ADH	90	80	70
TF_{In}/P_{In}	7	7	+ADH	25	60	180
			−ADH	7	7	7

[a] nus, nonurea solutes.

are discussed in detail in Chapters 33 and 34. In addition, the distal nephron has a role in the regulation of Ca^{++} excretion, as discussed in Chapter 35.

URINARY CONCENTRATION AND DILUTION

The above discussion of Na^+, Cl^-, and water reabsorption in the distal nephron included a description of the events leading to the excretion of a concentrated urine and a dilute urine (Kokko and Rector, 1972). To summarize briefly, because the ascending limb reabsorbs Na^+ and Cl^- but is impermeable to water, the tubular fluid entering the distal nephron is *hypotonic* to plasma (average osmolality $\simeq$ 150 mosmol/kg H_2O). The osmolality of the tubular fluid is further reduced (to $\simeq$ 100 mosmol/kg H_2O) in the distal tubule and connecting tubule which, like the ascending limb, reabsorb Na^+ and Cl^- but are water-impermeable. Events in the collecting tubule/duct determine whether a *concentrated urine* or a *dilute urine* will be excreted. In the *presence of ADH*, the entire collecting tubule/duct is relatively permeable to water. Water is therefore reabsorbed, since the osmolality of the peritubular fluid surrounding the collecting tubule/duct ranges from approximately 300 mosmol/kg H_2O in the cortex to 1200 mosmol/kg H_2O at the tip of the papilla. With *maximal* ADH, the final urine has the same osmolality as the tubular fluid at the tip of the papilla (about 1200 mosmol/kg H_2O) and represents only about 0.5% of the filtered volume. In the *absence of ADH*, the entire collecting tubule/duct is relatively impermeable to water. The volume of the tubular fluid therefore does not change appreciably, but since solute is reabsorbed, the osmolality of the tubular fluid is further reduced (below 100 mosmol/kg H_2O). The final urine has an osmolality as low as 70 mosmol/kg H_2O and represents up to 15% of the filtered volume.

In all discussions of urinary concentration and dilution presented so far in this chapter, the presence of the *medullary gradient* was assumed. It is now appropriate to consider how this gradient is formed and maintained.

Formation of the Medullary Gradient: Countercurrent Multiplication

A key factor in the formation of the medullary gradient is the unusual anatomical configuration of the loop of Henle, which allows fluid to flow in opposite, or *countercurrent*, directions in its two limbs. Because of the specific transport and permeability properties of the ascending and descending limbs (Table 4.17), countercurrent flow allows a large osmotic gradient to be established in the peritubular fluid between the corticomedullary junction and the tip of the papilla. This is illustrated schematically in Figure 4.30, which demonstrates how an osmotic gradient could develop in the medullary peritubular fluid surrounding the loop of a *cortical nephron*. Recall that the loop of a cortical nephron consists of a *thin descending limb*, which is relatively impermeable to solutes but is highly permeable to water, and a *thick ascending limb*, which actively reabsorbs Na^+ and Cl^- but is relatively impermeable to water (cortical nephrons lack a thin ascending limb). Initially (i.e., at a hypothetical time zero), the osmolality of the fluid in both limbs of the loop and in the surrounding peritubular fluid would be similar to plasma; an approximate value of 300 mosmol/kg H_2O is used here for simplicity (Fig. 4.30*A*). Assume now that flow through the loop is momentarily stopped. Since the ascending limb actively reabsorbs Na^+ and Cl^- but is impermeable to water, the peritubular fluid osmolality increases and a small osmotic gradient is established between the ascending limb and the surrounding peritubular fluid at each horizontal level in the loop; the increased peritubular fluid osmolality, in turn, causes water to be reabsorbed from the water-permeable descending limb

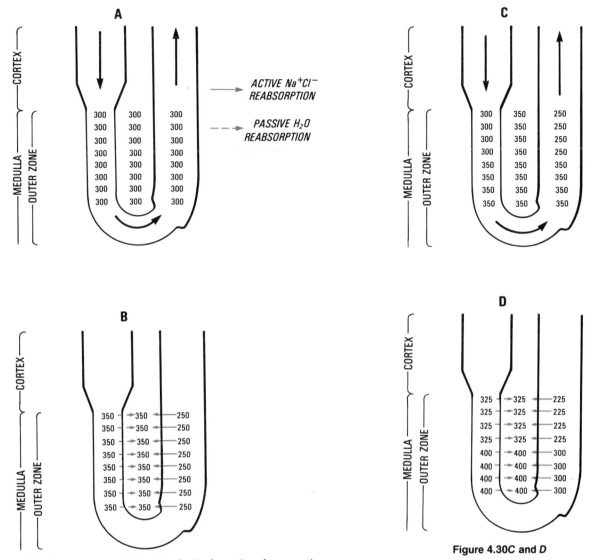

Figure 4.30. *A* and *B*, Mechanism for the formation of an osmotic gradient in the medullary peritubular fluid surrounding the loop of Henle of a cortical nephron, starting at a hypothetical time zero. The tubular fluid and peritubular fluid osmolalities are given in mosmol/kg H_2O. Details are provided in the text.

Figure 4.30C and D

(Fig. 4.30*B*). For illustrative purposes, it is arbitrarily assumed that a gradient of 100 mosmol/kg H_2O can be developed between the tubular fluid in the ascending limb and the fluid in the peritubular space and descending limb. In analogy to certain countercurrent flow applications in physical chemistry and engineering, this horizontal gradient often is referred to as the "single effect."

When flow continues (Fig. 4.30*C*), additional fluid with an osmolality of 300 mosmol/kg H_2O is introduced from the proximal tubule into the descending limb. Flow again is momentarily stopped, and a gradient of 100 mosmol/kg H_2O is again established between the fluid in the ascending limb

and the fluid in the peritubular space and descending limb (Fig. 4.30*D*). As the sequence repeats (Figs. 4.30*E* to 4.30*H*), note that because of countercurrent flow the *small* osmotic gradient between the ascending limb and the surrounding peritubular fluid at each *horizontal* level becomes "multiplied" considerably in the *vertical* direction, resulting in a *large* osmotic gradient between the peritubular fluid at the corticomedullary junction and the peritubular fluid surrounding the hairpin turn. This mechanism for the formation of a medullary osmotic gradient therefore is termed *countercurrent multiplication*. It should be evident that the basic requirements for the operation of the countercurrent multiplication mechanism are (1) countercurrent flow; (2) an active transport process allowing the establishment of the horizontal osmotic gradient; and (3) the different water permeability properties of

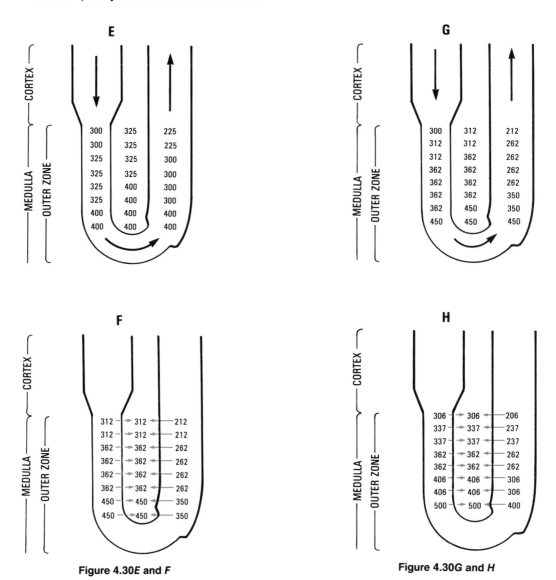

Figure 4.30*E* and *F*   **Figure 4.30*G* and *H***

the descending and ascending limbs. Note that although these basic requirements for countercurrent multiplication also are satisfied by the *cortical* part of the ascending limb and part 3 of the proximal tubule (the pars recta), the vertical osmotic gradient develops in the *medullary* peritubular fluid *only* (Fig. 4.30). This is because the relatively high blood flow to the renal cortex washes away the reabsorbed Na^+ and Cl^-, thereby preventing the establishment of an osmotic gradient in the cortical peritubular fluid.

In the cortical nephron example illustrated in Figure 4.30, the osmolality of the peritubular fluid surrounding the hairpin turn was increased to about 500 mosmol/kg H_2O by countercurrent multiplication. Recall that the hairpin turn in cortical nephrons occurs at or above the junction between the outer and inner medulla (Fig. 4.26). Since the per-

itubular fluid osmolality at this junction is estimated to be approximately 600 mosmol/kg H_2O (Fig. 4.26), Figure 4.30 represents a realistic example. It should be evident that the maximum peritubular fluid osmolality achieved by the countercurrent mechanism will depend upon the *length* of the loop. Thus, countercurrent multiplication in long-looped *juxtamedullary nephrons* is responsible for developing the very high osmolalities measured in the inner medulla, with values up to 1200 mosmol/kg H_2O at the tip of the papilla (Fig. 4.26). However, the countercurrent mechanism must operate somewhat differently in juxtamedullary nephrons. This is because a basic requirement of the mechanism is an active transport process along the entire length of the ascending limb to establish the horizontal osmotic gradient. In the thin ascending limb of juxtamedullary nephrons, a horizontal osmotic gradient can be generated as a result of *passive* reabsorption of Na^+ and Cl^-, as discussed previously, but this passive reabsorption process depends upon the *prior* existence of the medullary gradient to reabsorb water from the thin *descending* limb, thereby increasing the concentration of Na^+ and Cl^- in the tubular fluid delivered to the thin ascending limb. To understand how passive Na^+ and Cl^- reabsorption in the thin ascending limb might *first* be initiated, it is necessary to consider how *urea* becomes a component of the inner medullary gradient.

The steps leading to the addition of urea to the inner medulla are illustrated schematically in Figure 4.31. At a hypothetical time zero, the tubular fluid entering the thick ascending limb is similar to that leaving the proximal tubule, with a urea concentration of 20 mmol/l and a total osmolality of 300 mosmol/kg H_2O (Fig. 4.31, ①). As this fluid flows through the thick ascending limb, distal tubule, and connecting tubule (Fig. 4.31, ② and ③), the urea concentration remains *unchanged* due to the low water permeability and low urea permeability of these regions. However, Na^+ and Cl^- are actively reabsorbed, thereby reducing the osmolality of the tubular fluid. The cortical collecting tubule and medullary collecting duct also have a low urea permeability. Thus, if ADH is present, the concentration of urea will *increase* in these regions due to the reabsorption of water (Fig. 4.31, ④ and ⑤). Tubular fluid with a high urea concentration is

therefore delivered to the papillary collecting duct, which is relatively *permeable* to urea in the presence of ADH. As a result, urea is passively reabsorbed in the papillary collecting duct (Fig. 4.31, ⑥) and enters the peritubular fluid in the inner medulla. This addition of urea to the inner medulla is sufficient to initiate the passive reabsorption of Na^+ and Cl^- from the thin ascending limb. First, the urea causes water to be reabsorbed from the water-permeable thin descending limb (Fig. 4.31, ⑦), thereby increasing the concentration of Na^+ and Cl^- in the tubular fluid. Then, when this Na^+Cl^--rich fluid flows through the water-impermeable, Na^+Cl^--permeable thin ascending limb, Na^+ and Cl^- are *passively reabsorbed* (Fig. 4.31, ⑧). Although some of the urea added to the peritubular fluid in the inner medulla enters the thin ascending limb (which is somewhat permeable to urea) (Fig. 4.31, ⑨), the *net* result is solute *exit*. In other words, because of the urea added to the medullary peritubular fluid from the papillary collecting duct, a horizontal osmotic gradient can be created in the thin ascending limb *in spite* of the lack of an active transport process in this region. The countercurrent multiplication mechanism can therefore operate in *juxtamedullary nephrons* as well as in cortical nephrons.

Of course, even though the horizontal osmotic gradient in the inner medulla is created by a *passive* process, energy *is* expended in delivering a urea-rich tubular fluid to the papillary collecting duct

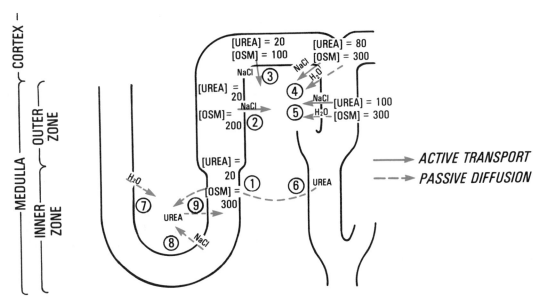

Figure 4.31. Mechanism for the addition of urea to the inner medullary gradient, starting at a hypothetical time zero. Tubular fluid urea concentrations and osmolalities are given in mmol/l and mosmol/kg H_2O, respectively. The events at the *circled numbers* are described in the text.

(Fig. 4.31), which in turn allows urea to enter the inner medulla. Most of this energy is expended by the *thick ascending limb*. Recall that the formation of a urea-rich tubular fluid depends on the reabsorption of water from the urea-impermeable cortical collecting tubule and medullary collecting duct in the presence of ADH. By actively reabsorbing Na^+ and Cl^-, the thick ascending limb allows water to be reabsorbed (1) from the *cortical collecting tubule* in the presence of ADH by delivering tubular fluid with a reduced osmolality to the distal nephron; and (2) from the *medullary collecting duct* in the presence of ADH by establishing an osmotic gradient in the peritubular fluid of the outer medulla. In fact, the *entire* medullary gradient will completely *vanish* within minutes after the administration of a *loop diuretic* such as furosemide, which inhibits the Na^+-Cl^- coupled transport system in the thick ascending limb (see above).

It also should be noted that the delivery of a urea-rich tubular fluid to the papillary collecting duct is dependent upon the presence of *ADH* as well as upon the active reabsorption of Na^+ and Cl^- by the thick ascending limb. This in part explains why the medullary gradient is *diminished* in the *absence* of ADH, as previously indicated (see Fig. 4.28); the failure of urea to enter the medullary interstitium in the absence of ADH also diminishes the gradient. Furthermore, it should be noted that once the medullary gradient is established, the urea concentration in the tubular fluid delivered to the papillary collecting duct will be even *higher* than that shown in Figure 4.31 (see Fig. 4.28). This is because (1) additional water will be reabsorbed from the medullary collecting duct when the medullary gradient is present; and (2) the urea that enters the thin ascending limb augments the urea concentration in the tubular fluid delivered to the thick ascending limb.

Maintenance of the Medullary Gradient: Countercurrent Exchange

Whereas countercurrent multiplication is the mechanism whereby the medullary gradient is *formed*, a second mechanism is required to *maintain* the medullary gradient. The importance of a mechanism for maintaining the medullary gradient is best illustrated by the fact that such a gradient could be diluted or washed out by *blood flow* to the medulla. For example, recall that even though the conditions for countercurrent multiplication are satisfied by the *cortical* part of the thick ascending limb and part 3 of the proximal tubule, an osmotic gradient does *not* form in the cortical peritubular fluid due to the relatively high blood flow to the cortex, which washes away the reabsorbed solute. *Medullary* blood flow is relatively low (only about 10% of the total RBF), but even this amount could significantly dilute the medullary gradient. That the medullary blood flow does *not* dilute the gradient can be attributed to the unusual anatomical configuration of the medullary blood supply.

Recall that the medullary blood supply is derived from the *vasa recta*, special hairpin-shaped capillaries that exit from the efferent arteriole of juxtamedullary nephrons (Fig. 4.7). Thus, *countercurrent flow* occurs not only in the loop of Henle, but in the capillaries supplying the medullary peritubular space as well. Figure 4.32 illustrates how this countercurrent blood flow helps to *maintain* (rather than dilute or wash out) the medullary gradient. It is assumed that the vasa recta, like other capillaries, are highly permeable to water and small solutes, and that the blood entering the vasa recta has an osmolality similar to systemic plasma; an approximate value of 300 mosmol/kg H_2O is used here for simplicity (Fig. 4.32, ①). Thus, as the blood flows downward through the increasingly hypertonic peritubular fluid (Fig. 4.32, ②), water passively diffuses out of the blood and solute diffuses in. The osmolality of the blood in the vasa recta therefore increases as the hairpin turn is approached. However, note that until the hairpin turn is reached, the osmolality of the blood always is slightly *below* that of the surrounding peritubular fluid. The reason is that the blood flow is sufficiently rapid to cause a small lag in equilibration between the blood and adjacent peritubular fluid. If the vasa recta did *not* have a hairpin configuration, the capillary would exit the kidney at the tip of the papilla (Fig. 4.32, ③) and would have *diluted* the gradient by adding water to and removing solute from the peritubular space. But because of the hairpin turn, the blood ascends through peritubular fluid of decreasing hypertonicity. Water and solute fluxes *opposite* to those in the descending limb therefore occur: water passively diffuses into the blood and solute diffuses out (Fig. 4.32, ④). Note that in the ascending limb, the osmolality of the blood is always slightly *greater* than that in the adjacent peritubular fluid due to a small lag in equilibration. Since the *volume* of blood leaving the medulla via the vasa recta is slightly greater than that entering (see below), some solute is removed from the peritubular space. However, the depletion of peritubular solutes is minimized because of countercurrent flow (*without* countercurrent flow, the blood would leave the medulla at ③). This mechanism for the preservation of the gradient is termed *countercurrent exchange*.

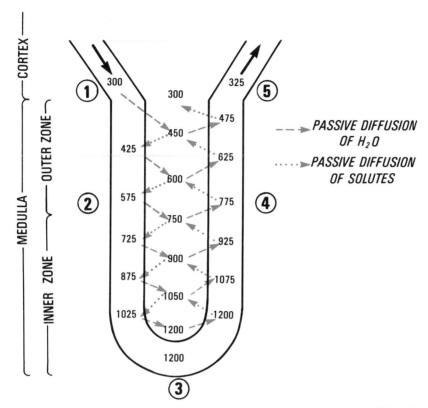

Figure 4.32. Maintenance of the medullary gradient by countercurrent exchange in the vasa recta. The tubular fluid and peritubular fluid osmolalities are given in mosmol/kg H_2O. The events at the *circled numbers* are described in the text.

As indicated above, the volume of blood leaving the medulla via the vasa recta is slightly greater than that entering. This increase in volume can be attributed primarily to the above-mentioned lag in equilibration between the vasa recta blood and the adjacent peritubular fluid. In the *descending* limb, the somewhat higher osmolality of the adjacent peritubular fluid tends to extract fluid from the capillary. However, oncotic forces exerted by plasma proteins *oppose* this fluid exit. In the *ascending* limb, the somewhat higher osmolality of the blood favors the uptake of fluid into the capillary. The oncotic forces exerted by plasma proteins now act *together* with the effect of the osmolality difference created by the lag in equilibration, to favor fluid uptake. As a result, fluid uptake by the ascending vasa recta exceeds fluid loss from the descending vasa recta. This serves to remove the water reabsorbed from the thin descending limb and medullary and papillary collecting ducts, which otherwise would dilute the gradient.

It should be emphasized that countercurrent exchange is a *passive* process, involving only the diffusion ("exchange") of solutes and water across the permeable walls of the vasa recta capillaries. The vasa recta have *no* role in creating the osmotic gradient, and in fact would rapidly dissipate the gradient if countercurrent multiplication were to

cease, e.g., following the administration of a loop diuretic (see above).

Factors Affecting the Concentrating and Diluting Mechanisms

In the normal kidney, the tubular fluid delivered to the collecting tubule/duct always is *hypotonic*. Then, in the collecting tubule/duct ADH regulates the final urine osmolality through its effects on water permeability. But the osmolality of the urine is not solely dependent upon the ADH level. Most importantly, both the concentrating and diluting mechanisms are markedly dependent on the proper function of the *thick ascending limb*. Even if plasma ADH is *high*, a concentrated (hypertonic) urine can be excreted *only* in the presence of the *medullary gradient* to provide the osmotic driving force for the reabsorption of water from the medullary and papillary collecting ducts. As discussed above, the active reabsorption of Na^+ and Cl^- by the thick ascending limb is the key factor in the formation of the medullary gradient. If ADH is *absent*, the osmolality of the tubular fluid is further reduced in the distal nephron and a dilute (hypotonic) urine is excreted. However, the urine will *not* be *maximally* dilute unless the tubular fluid entering the distal nephron is hypotonic, i.e., the production of a maximally dilute urine requires the active reabsorption of Na^+ and Cl^- from the water-impermeable thick ascending limb.

The concentrating and diluting mechanisms can be affected by other factors as well (Jamison et al., 1979). For example, although the proper function of the thick ascending limb is the most crucial factor in *establishing* the medullary gradient, the *maximum osmolality* of the gradient has several additional determinants (Burg and Stephenson, 1978):

1. *Length of the Loop of Henle and Percentage of Nephrons with Long Loops.* In man, 15% of the nephrons are long-looped juxtamedullary nephrons, and the maximum peritubular fluid osmolality is approximately 1200 mosmol/kg H_2O. But in the Saharan desert rat *Psammomys*, nearly 35% of the nephrons have very long loops, and the maximum peritubular fluid osmolality is as high as 5000 mosmol/kg H_2O (this means that the *urine* osmolality also can be as high as 5000 mosmol/kg H_2O, a very important water-conserving adaptation).

2. *Availability of Urea.* When a protein-deficient diet is ingested, the maximum peritubular fluid osmolality will be diminished (recall that urea is a protein breakdown product).

3. *Rate of Flow through the Collecting Duct.* A rapid flow tends to wash out the gradient, primarily by diluting the urea in the tubular fluid delivered to the papillary collecting duct and thereby reducing the concentration gradient for the reabsorption of urea into the inner medulla.

4. *Rate of Flow through the Vasa Recta.* With increased flow, the equilibration between peritubular fluid and vasa recta blood is impaired, and additional solute is removed from the medulla (see above). Thus, the maximum osmolality of the gradient is highest when flow through *both* the collecting duct and vasa recta is low.

5. *Presence of Prostaglandins.* Some prostaglandins may increase blood flow through the vasa recta, thereby reducing the maximum osmolality of the gradient (cf. no. 4 above).

Prostaglandins also impair the formation of a concentrated urine by inhibiting adenylate cyclase (Stokes, 1981), thereby inhibiting the effect of ADH on the water permeability of the collecting tubule/duct. Since ADH *stimulates prostaglandin synthesis* by the kidneys, this inhibition can be regarded as a form of *negative feedback.*

Quantitation of Urinary Concentration and Dilution

In order to assess quantitatively the operation of the concentrating and diluting mechanisms, an index of the ability of the kidneys to excrete a concentrated urine or a dilute urine is required. One simple index is the *urine osmolality*, U_{osm}. When measured at *high ADH* or in the *absence of ADH*, U_{osm} assesses the ability of the kidneys to excrete a *maximally concentrated* urine or a *maximally dilute* urine, respectively. As indicated above, in man U_{osm} approaches 1200 mosmol/kg H_2O at maximum ADH and can be as low as 70 mosmol/kg H_2O in the absence of ADH.

Another index for assessing the operation of the concentrating and diluting mechanisms is the so-called *free water clearance*. The free water clearance differs from other clearances discussed in this text since it is *not* defined by the standard clearance relationship, $C_X = U_X\dot{V}/P_X$ (*Eq. 26*). Instead, the free water clearance (C_{H_2O}) is defined as the difference between the rate of urine flow ($\dot{V}$) and the osmolar clearance (C_{osm}):

$$C_{H_2O} = \dot{V} - C_{osm} \qquad (84)$$

In order to understand the significance of free water clearance, it is helpful to examine the *osmolar clearance* more closely. The osmolar clearance is a standard clearance except that it refers to the *total* number of osmotically active solute particles instead of to a *single* substance. Thus, C_{osm} represents the volume of plasma completely cleared of all osmotically active solute particles per unit time and is calculated as $U_{osm}\dot{V}/P_{osm}$. But it is important to note that C_{osm} also represents the *hypothetical* rate of urine flow that *would* be measured if urine were *isotonic* to plasma, i.e., if U_{osm} were equal to P_{osm}. This can be shown by starting with the clearance equation $C_{osm} = U_{osm}\dot{V}/P_{osm}$. If $U_{osm} = P_{osm}$, then

$$C_{osm} = \dot{V} \qquad (85)$$

The free water clearance ($= \dot{V} - C_{osm}$) therefore represents the difference between the *actual* rate of urine flow ($\dot{V}$) and this *hypothetical* rate of *isotonic* urine flow (C_{osm}). When a *dilute* (hypotonic) urine is excreted, the *actual* urine flow rate is *greater* than the *hypothetical isotonic* urine flow rate. C_{H_2O} therefore is a *positive* number and represents the volume of distilled water, or *"free water"* (i.e., *solute*-free water), that would have to be added to the hypothetical isotonic urine to make the actual urine (Fig. 4.33A). Physiologically, this solute-free water is formed by reabsorbing solute in a water-impermeable region, such as the thick ascending limb,

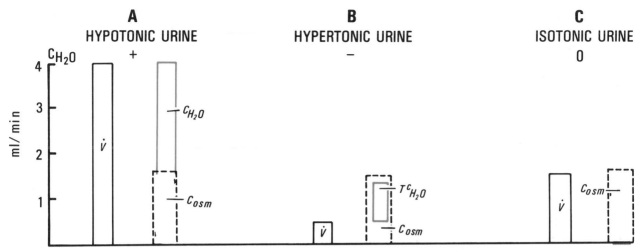

Figure 4.33. Use of *free water clearance* to quantitate urinary concentration and dilution. The free water clearance (C_{H_2O}, *solid red lines*) can be calculated by comparing the *actual* urine flow rate ($\dot{V}$, *solid black lines*) to the *hypothetical* flow rate that would be present if urine were isotonic (C_{osm}, *dashed black lines*). *A,* In *hypotonic* urine, C_{H_2O} is *positive* and represents the solute-free water that must be *added* to the hypothetical isotonic urine to make the actual urine. *B,* In *hypertonic* urine, C_{H_2O} is *negative* and represents the solute-free water that must be *removed* from the hypothetical isotonic urine to make the actual urine. *C,* In *isotonic* urine, C_{H_2O} is *zero* since the actual flow is the same as the hypothetical isotonic urine flow. Note that the negative free water clearance in *B* is designated $T^c_{H_2O}$ ($=(-C_{H_2O})$). C_{osm} is *independent* of urine tonicity with an average normal value of 1.5 ml/min.

distal tubule, connecting tubule, or (in the absence of ADH) collecting tubule/duct. When a *concentrated* (hypertonic) urine is excreted, the *actual* urine flow rate is *less* than the *hypothetical isotonic* urine flow rate. C_{H_2O} therefore is a *negative* number and represents the volume of "free water" (i.e., *solute*-free water) that would have to be removed from the hypothetical isotonic urine to make the actual urine (Fig. 4.33*B*). Physiologically, this solute-free water is removed by reabsorbing water in excess of solute, as in the cortical collecting tubule and medullary collecting duct when ADH is present. Of course, if the urine is *isotonic*, the actual urine flow rate is *equal* to the hypothetical isotonic urine flow rate (Fig. 4.33*C*), and the free water clearance is *zero*.

When a *concentrated* urine is excreted, reference often is made to the quantity $C_{osm} - \dot{V}$, a *positive* number, instead of to $\dot{V} - C_{osm}$, a *negative* number for a concentrated urine. This quantity is designated $T^c_{H_2O}$ (Fig. 4.33*B*) and is commonly termed the *negative free water clearance*, since $T^c_{H_2O}$ equals $(-C_{H_2O})$.

It should be noted in Figure 4.33 that C_{osm} is *independent* of the urine tonicity. This is because an average of 650 mosmol of electrolytes and other solutes must be excreted per day in normal man to eliminate waste products such as creatinine and urea and to maintain a normal electrolyte balance, i.e., the average $U_{osm}\dot{V}$ equals 650 mosmol/day or 0.45 mosmol/min (typical normal range: 400–900 mosmol/day or 0.3–0.6 mosmol/min). C_{osm} therefore has an average normal value of 1.5 ml/min (*Eq. 26*), as illustrated in Figure 4.33 (typical normal range: 1–2 ml/min).

CHAPTER 32

Regulation of Volume and Osmolality of the Body Fluids

Maintenance of the normal volume and osmolality of the body fluids requires that the *input* of solvent and solutes to the body equals the *output* of solvent and solutes from the body each day. This principle, input = output, is referred to as *balance* and applies to the normal, nongrowing, nonpregnant adult. In the normal person, of course, water is the only solvent of concern. In contrast, many different solutes contribute to the osmolality of the body fluids (Chapter 26). Since the major extracellular solute is Na^+ (along with its attendant anions), the regulation of the volume and osmolality of the *extracellular* fluid is dependent almost exclusively upon the regulation of the balances of water and Na^+. Moreover, since cell membranes are permeable to water, the volume and osmolality of the *intracellular* fluid also will be influenced by the balances of water and Na^+ in the extracellular fluid (Chapter 26). Therefore, this chapter on the regulation of volume and osmolality of the body fluids is directed at understanding the inputs and outputs of water and Na^+ and their regulation.

WATER

Input of Water

Water is added to the body fluids from three sources: the water content of food, water generated during the oxidation of food, and water consumed as a liquid (Table 4.19). The amount of water *in food* can vary widely, but the average diet of adult Americans contains about 800–1000 ml/day. About 300–400 ml of water are generated per day during the *oxidation of food*. For example, the oxidation of 1 mole of glucose (180 g) generates 6 moles of water (108 ml):

$$C_6H_{12}O_6 + 6O_2 \rightarrow 6CO_2 + 6H_2O$$

Neither the water obtained from food nor the water generated during the oxidation of food is an important variable in the regulation of water input. Thus, water input is regulated primarily by changes in the volume of water consumed *as a liquid*, which averages 1–2 l/day but can vary from less than 1 l/day to more than 20 l/day. For example, when large volumes of fluid are lost from the body, liquid consumption increases in response to the sensation of *thirst* (see below). However, it should be noted that when environmental conditions are normal and constant, liquid consumption is primarily determined by an individual's habits rather than by thirst. Man appears to be able to anticipate future requirements of water and to drink appropriate amounts of liquid (Fitzsimons, 1976).

Output of Water

Water is lost from the body via four routes (Table 4.19): insensible loss, sweat, feces, and urine. *Insensible loss* averages about 800–1000 ml/day, of which about half is lost as moisture in expired air and half is lost as water that evaporates through the skin (*water of transpiration*). The water of transpiration differs from *sweat*, which is produced by the sweat glands in response to thermal stress. The volume of sweat can vary from less than 200 ml/day in an individual at rest in a cool environment to as much as 8–10 l/day if the environmental temperature, humidity, and/or level of physical activity are increased. *Fecal loss* of water in normal man is only 100–200 ml/day, but it can exceed 5 l/day in diarrheal diseases such as cholera. Insensible loss, sweat, and fecal water do not represent important variables in the physiological regulation of water output (although sweat is of extreme importance in the regulation of *body temperature*; see Chapter 50). Thus, water output is regulated primarily by changes in the volume of *urine*, which averages 1–2 l/day but can vary from less than 1 l/day to more than 20 l/day. As discussed in Chapter 31, the urine volume is determined primarily by the plasma level of *ADH*.

Table 4.19
Major inputs and outputs of water[a]

Inputs	Average Volume per Day, ml/day	Outputs	Average Volume per Day, ml/day
Water content of food	800–1000	Insensible loss	800–1000
Water generated during oxidation of food	300–400	Sweat	200
Water consumed as a liquid	1000–2000	Feces	100–200
		Urine	1000–2000
Total	2100–3400	Total	2100–3400

Adapted from Weitzman and Kleeman (1980). [a] Inputs and outputs represent average basal values for an adult in a cool environment.

Because the body is unable to prevent insensible water loss, sweat, or fecal loss, these outputs of water commonly are termed *obligatory losses*. In addition, part of the *urine* output can be regarded as obligatory. As indicated in Chapter 31, an average of 650 mosmol of electrolytes and other solutes must be excreted per day to eliminate waste products such as creatinine and urea and to maintain a normal electrolyte balance. Thus, even if the urine is *maximally concentrated* ($U_{osm} \simeq 1200$ mosmol/kg H_2O), the *minimum* urine volume is 500–600 ml, i.e., approximately 500–600 ml of the daily urine volume represents an obligatory water loss.

Control of Water Balance

The input and output of water, then, are regulated primarily by changes in the volume of *liquid ingested* and the *urine volume*, which in turn are controlled by *thirst* and the plasma level of *ADH*, respectively. Both thirst and the release of ADH from the posterior pituitary are controlled by centers located in the hypothalamus that are stimulated primarily by two physiological conditions, increases in plasma osmolality and decreases in plasma volume:

1. Increases in *plasma osmolality* are sensed by cells within the hypothalamus termed *osmoreceptors* that are stimulated if water is removed from the cells, causing them to shrink. Hence, these cells respond to increases in plasma osmolality produced by (a) deficits of total body water; or (b) solutes that are predominantly excluded from the intracellular fluid, such as Na^+ and Cl^-, which cause water to shift from the intracellular compartment to the extracellular compartment. Solutes that enter cells readily, such as urea, do *not* change the volume of the osmoreceptors and therefore do not stimulate thirst or ADH release (Robertson et al., 1976).

2. Decreases in *plasma volume* are sensed by baroreceptors ("pressure receptors") located in both the low pressure regions (atria, pulmonary vessels) and the high pressure regions (carotid sinus, aortic arch) of the circulation (Schrier et al., 1979). A decrease in plasma volume inhibits the firing of these receptors, which in turn causes a reflex stimulation of thirst and ADH release. The baroreceptors can sense such changes in plasma volume because they actually respond to *stretch* rather than to pressure. In fact, they frequently are termed "volume receptors" rather than "pressure receptors."

Under normal conditions, thirst and ADH release are primarily under the control of the *osmoreceptors*. This is because the osmoreceptors are extremely sensitive to *small* changes in plasma osmolality. In fact, significant changes in ADH release occur when the plasma osmolality changes by less than 1%. In contrast, as much as a 10% change in plasma volume may be necessary before significant changes in ADH release are observed (Robertson et al., 1976). Thus, changes in plasma volume primarily affect thirst and ADH release in extreme circumstances such as severe dehydration, hemorrhage, or redistribution of body water away from the central circulation (e.g., edema of an extremity due to occlusion of venous outflow or loss of fluid into an infected area or body cavity).

Although the osmoreceptors are more *sensitive* than the volume receptors, the volume receptors can result in a *stronger* response. For example, a very low plasma volume will result in thirst and ADH release even in the presence of plasma *hypotonicity*. The fact that ADH release is *stimulated* at a *low* plasma osmolality if plasma volume is very low and, conversely, that ADH release is *suppressed* at a *high* plasma osmolality if plasma volume is very high is illustrated in Figure 4.34. The data presented in this figure are commonly summarized by the statement that *volume overrides tonicity*.

It should be noted that while thirst and ADH release are primarily determined by the plasma osmolality and volume, they can be affected by several additional conditions and pharmacological agents (Schrier et al., 1977), as summarized in Table 4.20. The question of whether angiotensin II has a physiological role in stimulating thirst and ADH release is not completely resolved.

To summarize the regulation of water input and output by thirst and ADH release, the pathway by which plasma osmolality and volume are returned

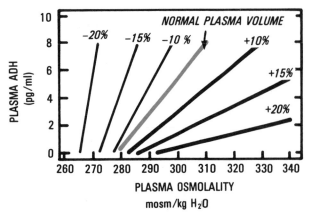

Figure 4.34. Relationship between plasma osmolality and plasma concentration of ADH at different plasma volumes. The *red line* represents the relationship at a normal plasma volume. *Light black lines* represent the relationship at plasma volumes that are *below* the normal plasma volume by the indicated percentages, illustrating that ADH release can be stimulated at *low* plasma osmolalities if the plasma volume is very low. *Dark black lines* represent the relationship at plasma volumes that *exceed* the normal plasma volume by the indicated percentages, illustrating that ADH release can be suppressed at *high* plasma osmolalities if the plasma volume is very high. [Adapted from *Kidney International* 10: 25–37, 1976, with permission.]

Table 4.20
Conditions that affect thirst and ADH release

Stimuli	Inhibitors
↑ Plasma osmolality	↓ Plasma osmolality
↓ Plasma volume	↑ Plasma volume
Angiotensin II	Ethanol[a]
Pain, emotion, stress[a]	α-Adrenergic agents
Cholinergic agents	Diphenylhydantoin[a]
β-Adrenergic agents	
Barbiturates	
Morphine[a]	
Nicotine[a]	

[a] Affects ADH release only.

to normal following an acute episode of dehydration is illustrated in Figure 4.35.

SODIUM

Input of Sodium

The input of Na^+ depends entirely upon the Na^+ content of food and water consumed (Table 4.21). While specially constructed diets may contain less than 10 mmol of Na^+ per day, individuals who use a salt shaker liberally may ingest more than 600 mmol/day. Thus, the daily input of Na^+ can vary markedly, depending upon an individual's diet and habits. Geographical factors also can influence Na^+ input by affecting the Na^+ content of water and agricultural products. The average diet of adult Americans contains 100–400 mmol Na^+/day. Al-

though an appetite for Na^+ can be demonstrated in animals, Na^+ input does not appear to be physiologically regulated in man.

Output of Sodium

Na^+ is lost from the body via three routes: sweat, feces, and urine (Table 4.21). The amount of Na^+ lost in *sweat* depends upon both the volume of sweat and the degree of adaptation to a hot environment. The concentration of Na^+ in sweat decreases as adaptation to heat occurs over a period of several days. Thus, Na^+ loss can vary from negligible in a person at rest in a cool environment to several hundred mmol/day in a nonheat-adapted person exercising in a hot environment. Under normal conditions, *fecal* Na^+ is negligible, although in diarrheal diseases such as cholera it may exceed 1000 mmol/day. Neither sweat nor fecal loss is an important physiological variable in the regulation of Na^+ output. Like water output, then, Na^+ output is regulated primarily by changes in the amount of Na^+ excreted in *urine*. Urinary Na^+ excretion averages 100–400 mmol/day in adult Americans, matching the average input, but can vary from almost negligible amounts to over 600 mmol/day.

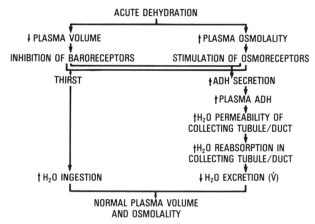

Figure 4.35. Pathway by which normal plasma volume and osmolality are restored following an acute episode of dehydration.

Table 4.21
Major inputs and outputs of sodium[a]

Inputs	Average Amount per Day, mmol/day	Outputs	Average Amount per Day, mmol/day
Food and water	100–400	Sweat	—[b]
		Feces	—[b]
		Urine	100–400
Total	100–400	Total	100–400

[a] Inputs and outputs represent average basal values for an adult ingesting a conventional American diet in a cool environment. [b] Negligible in basal state, but can become appreciable under certain conditions (see text).

Regulation of Sodium Balance

With no significant physiological regulation of Na^+ input in man, Na^+ balance must be achieved by changing Na^+ output to match Na^+ input. As indicated above, Na^+ output is regulated primarily by changes in the amount of Na^+ excreted in urine. In Chapters 30 and 31, the renal handling of Na^+ was discussed in detail. It is now appropriate to consider the mechanisms for regulating Na^+ excretion. Three major regulatory mechanisms have been identified: changes in GFR, aldosterone, and the so-called "third factor" effect (DeWardener, 1978).

CHANGES IN GFR

When Na^+ intake changes, compensatory changes in GFR occur that can affect Na^+ excretion. Recall from Chapter 28 that GFR is proportional to the *net filtration pressure* (*Eq. 31*):

$$GFR = K_f \, [P_G - P_B - \pi_G]$$

Changes in Na^+ intake can change the net filtration pressure and, hence, GFR. Consider, for example, the changes in net filtration pressure and GFR that could result from an *increase* in Na^+ intake (Fig. 4.36). Because Na^+ is primarily an *extracellular* solute, an increase in Na^+ intake causes an increase in plasma osmolality that stimulates the osmore-

ceptors (see above). The resulting stimulation of thirst and ADH release causes an expansion of the plasma volume, which in turn causes (1) an *increase* in *glomerular capillary hydrostatic pressure,* P_G (by the reflex pathway illustrated in Fig. 4.36); and (2) a *decrease* in *glomerular capillary oncotic pressure,* π_G (by diluting the plasma proteins). The increase in P_G and the decrease in π_G both tend to increase the net filtration pressure and GFR, thereby increasing the filtered load of Na^+ and promoting Na^+ excretion (Fig. 4.36).

However, it should be noted that such changes in GFR probably have a *minor* role in the regulation of Na^+ excretion, for at least two reasons. First of all, *autoregulation* and *tubuloglomerular feedback* (see Chapters 27 and 28) attenuate or even prevent any significant changes in GFR that might result from changes in Na^+ intake. Secondly, the amount of Na^+ reabsorbed by the proximal tubule, loop of Henle, and distal nephron varies directly with the load of Na^+ delivered to that region (see Chapters 30 and 31). The importance of this *load-dependent* reabsorption of Na^+ can be illustrated by the following example. If GFR is 180 l/day, P_{Na} is 140 mmol/l, and urinary Na^+ excretion ($U_{Na}\dot{V}$) is 126 mmol/day, then the rate of tubular reabsorption of Na^+ (T_{Na}) is (*Eq. 50*):

$$T_{Na} = P_{Na}GFR - U_{Na}\dot{V}$$
$$= 140 \text{ mmol/l} \times 180 \text{ l/day} - 126 \text{ mmol/day}$$
$$= 25{,}074 \text{ mmol/day}$$

i.e., 99.5% of the filtered load of Na^+ is reabsorbed. If GFR were now to increase by only 1%, to 181.8 l/day, *without* a change in T_{Na}, Na^+ excretion would increase 3-fold, to 378 mmol/day:

$$U_{Na}\dot{V} = P_{Na}GFR - T_{Na}$$
$$= 140 \text{ mmol/l} \times 181.8 \text{ l/day} - 25{,}074 \text{ mmol/day}$$
$$= 378 \text{ mmol/day}$$

However, such an increase in Na^+ excretion does *not* occur. Instead, because Na^+ reabsorption is load-dependent, 99.5% of the filtered load *continues* to be reabsorbed, so that T_{Na} increases to 25,325 mmol/l:

$$T_{Na} = 0.995 \times P_{Na}GFR$$
$$= 0.995 \times 140 \text{ mmol/l} \times 181.8 \text{ l/day}$$
$$= 25{,}325 \text{ mmol/day}$$

As a result, Na^+ excretion increases by less than 1%, to 127 mmol/day:

$$U_{Na}\dot{V} = P_{Na}GFR - T_{Na}$$
$$= 140 \text{ mmol/l} \times 181.8 \text{ l/day} - 25{,}325 \text{ mmol/day}$$
$$= 127 \text{ mmol/day}$$

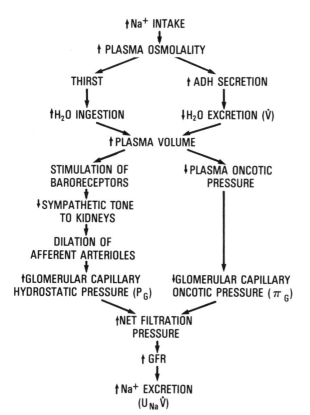

Figure 4.36. Pathway by which an increase in GFR could restore Na^+ balance following an increase in Na^+ intake.

This consequence of load-dependent reabsorption of Na^+, whereby a constant percentage of the filtered load of Na^+ continues to be reabsorbed as GFR varies in the physiological range, is termed *glomerulotubular balance*. The exact mechanism for load-dependent reabsorption and glomerulotubular balance is not completely understood but probably can be attributed in part to the very large capacity of the tubular reabsorptive mechanisms for Na^+.

Thus, changes in GFR are of minor importance in the regulation of Na^+ excretion. Such regulation must therefore be primarily accomplished by the remaining two mechanisms, aldosterone and the "third factor" effect.

ALDOSTERONE

A second factor regulating Na^+ excretion is *aldosterone*, a steroid hormone produced by the zona glomerulosa of the adrenal cortex. Increased plasma levels of aldosterone stimulate Na^+ reabsorption by stimulating the unidirectional Na^+ transport mechanism in the collecting tubule/duct, possibly via an increase in the entry of Na^+ into the cells from the tubular lumen (Chapter 31) (Petty et al., 1981). (Although this unidirectional Na^+ transport occurs *throughout* the distal nephron, aldosterone's effect appears to be limited to the *collecting tubule/duct*) (Stokes et al., 1981). Increased plasma aldosterone also stimulates H^+ and K^+ secretion in the collecting tubule/duct, as discussed in Chapters 33 and 34.

The secretion of aldosterone by the adrenal cortex is stimulated by four factors (Coghlan et al., 1979): (1) increased plasma angiotensin II, (2) increased P_K, (3) decreased P_{Na}, and (4) increased plasma adrenocorticotrophic hormone (ACTH). *ACTH* has a relatively minor effect on aldosterone secretion, although it does have an important role in regulating the secretion of adrenal glucocorticoids and androgens. *Increases in P_K* represent an important stimulus for aldosterone secretion (see Chapter 34), but do not represent a mechanism for changing aldosterone secretion when Na^+ intake changes. The increased aldosterone secretion seen with *decreased P_{Na}* represents an appropriate response for maintaining Na^+ balance:

$\downarrow P_{Na} \rightarrow \uparrow$ aldosterone

$\rightarrow \uparrow Na^+$ reabsorption by collecting tubule/duct

$\rightarrow \downarrow Na^+$ excretion

However, the effect of P_{Na} on aldosterone secretion is of minor importance in the regulation of Na^+ excretion for two reasons. First of all, decreases in P_{Na} have a relatively *weak* stimulatory effect on aldosterone secretion. Secondly, changes in Na^+ intake have minimal effects on P_{Na}. For example, while an increased Na^+ intake adds Na^+ to the extracellular fluid and produces a *transient* increase in P_{Na}, the *plasma osmolality* also increases, stimulating the osmoreceptors. The resulting stimulation of thirst and ADH release leads to expansion of the plasma volume (Fig. 4.36) and dilution of the ingested Na^+, so that the overall change in P_{Na} is small. Thus, the changes in aldosterone secretion that accompany changes in Na^+ intake must be primarily mediated by *angiotensin II*.

As indicated in Chapter 27, angiotensin II is an octapeptide that is formed from angiotensinogen in the following 2-step reaction:

$$\text{Angiotensinogen} \xrightarrow{\text{renin}} \text{angiotensin I}$$

$$\xrightarrow{\text{converting enzyme}} \text{angiotensin II}$$

Renin, the proteolytic enzyme that catalyzes the first step in this reaction, is secreted into plasma by the granular cells of the JG apparatus (Chapter 27). The secretion of renin appears to be stimulated by: (1) decreased renal perfusion pressure, as sensed by baroreceptors (stretch receptors) in the afferent arteriole; (2) changes in the volume or composition of the tubular fluid reaching the macula densa; (3) stimulation of the renal sympathetic nerves; and (4) a variety of other factors (Davis and Freeman, 1976). For example, since inhibitors of prostaglandin synthesis (e.g., indomethacin) impair renin release, it has been suggested that prostaglandins either directly stimulate renin secretion or perform a permissive function that facilitates the release of renin in response to other stimuli. To illustrate how changes in renin secretion, and hence angiotensin II production, aldosterone secretion, and Na^+ reabsorption, could accompany changes in Na^+ intake, consider the response of the renin-angiotensin-aldosterone system to an *increase* in Na^+ intake (Fig. 4.37). By mechanisms analogous to those illustrated in Figure 4.36, an increase in Na^+ intake results in an increase in plasma volume, which in turn causes a reflex decrease in sympathetic tone to the kidneys. This decrease in sympathetic tone decreases renin secretion both *indirectly* (by increasing renal perfusion pressure) and *directly* (by an effect on the granular cells).

Like the regulation of Na^+ excretion by changes in GFR, then, the regulation of Na^+ excretion by aldosterone occurs as a result of changes in the *plasma volume* caused by changes in Na^+ input. However, as previously indicated, aldosterone has

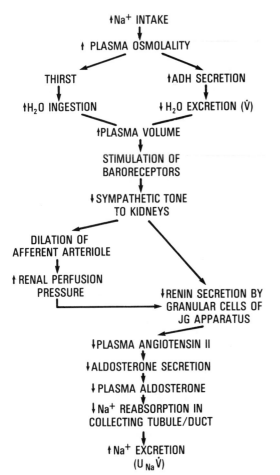

↑Na⁺ INTAKE

↑ PLASMA OSMOLALITY

THIRST   ↑ADH SECRETION

↑H₂O INGESTION   ↓ H₂O EXCRETION (V̇)

↑PLASMA VOLUME

STIMULATION OF BARORECEPTORS

↓SYMPATHETIC TONE TO KIDNEYS

DILATION OF AFFERENT ARTERIOLE

↑ RENAL PERFUSION PRESSURE

↓RENIN SECRETION BY GRANULAR CELLS OF JG APPARATUS

↓PLASMA ANGIOTENSIN II

↓ALDOSTERONE SECRETION

↓ PLASMA ALDOSTERONE

↓ Na⁺ REABSORPTION IN COLLECTING TUBULE/DUCT

↑ Na⁺ EXCRETION
(U _Na_ V̇)

Figure 4.37. Pathway by which a decrease in aldosterone secretion could restore Na⁺ balance following an increase in Na⁺ intake.

a much more important role in regulating Na⁺ excretion than changes in GFR.

THIRD FACTOR EFFECT

When the first two factors that regulate Na⁺ excretion, GFR and aldosterone, are controlled experimentally, an animal can *still* regulate Na⁺ excretion to match Na⁺ input. For example, in an experimental animal in whom a *constant GFR* is maintained by controlling blood flow to the kidneys and a high plasma concentration of *aldosterone* is maintained by administering large doses of the hormone, intravenous infusion of isotonic saline *still* will be followed by a decrease in Na⁺ reabsorption and, hence, an increase in Na⁺ excretion. The phenomenon whereby an increase in Na⁺ input can result in an increase in Na⁺ excretion *independent* of any significant increase in GFR or decrease in aldosterone level is termed the *third factor effect*. Conversely, a decrease in Na⁺ input can result in a decrease in Na⁺ excretion *independent* of any sig-

nificant change in GFR or aldosterone level, a phenomenon that can be referred to as the *absence* of third factor.

Despite intensive investigation, the mechanism for the third factor effect remains poorly understood (DeWardener, 1978). Actually, the third factor effect probably involves several different mechanisms, because changes in Na⁺ reabsorption in both the proximal tubule and distal nephron are observed. With *small* increases in Na⁺ intake, the third factor effect appears to be primarily due to a decrease in Na⁺ reabsorption in the *medullary collecting duct*. It has been postulated that this decrease in Na⁺ reabsorption is mediated by prostaglandins, bradykinin, or an as yet unidentified "natriuretic" hormone. When *large* quantities of Na⁺ are administered (e.g., by intravenous infusion), Na⁺ reabsorption in the *proximal tubule* is depressed as well. This decrease in proximal tubular Na⁺ reabsorption can be attributed at least in part to the dependence of proximal tubular water and solute reabsorption on the hydrostatic and oncotic pressures in the peritubular capillaries. According to the Starling principle, the rate of fluid movement from *capillaries* to *interstitial space* (i.e., the rate of *filtration*) is proportional to the difference between the hydrostatic and oncotic pressure gradients across the capillary wall, the so-called *net filtration pressure* (*Eq.* 29):

$$\text{Rate of filtration} \propto [(P_c - P_i) - (\pi_c - \pi_i)]$$

The rate of fluid movement from *interstitial space* to *capillaries* (i.e., the rate of *reabsorption*) is therefore proportional to

$$\text{Rate of reabsorption} \propto [(P_i - P_c) - (\pi_i - \pi_c)] \quad (86)$$

If *Eq.* 86 is applied to the reabsorption of fluid from the *peritubular* interstitial space into the *peritubular* capillaries, it is evident that an increase in peritubular capillary hydrostatic pressure (P_c) or a decrease in peritubular capillary oncotic pressure (π_c) will retard the reabsorption of fluid into the capillaries. The movement of fluid from the lateral intercellular space to the peritubular space will therefore be retarded, and the hydrostatic pressure in the lateral space will increase. This increased hydrostatic pressure, in turn, will impair the reabsorption of water and solutes by the proximal tubule, perhaps by allowing water and solutes that *already* have been transported into the lateral space to leak back (*"pump leak"*) into the tubular lumen. Figure 4.38 illustrates how the ingestion of a large quantity of Na⁺ could increase peritubular capillary hydrostatic pressure and decrease peritubular capillary oncotic pressure, thereby decreasing the reab-

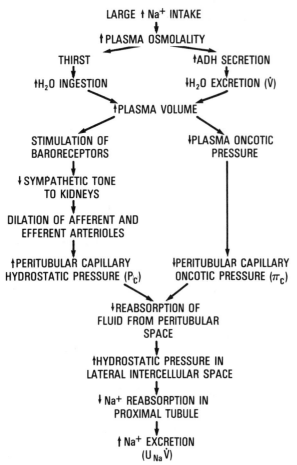

LARGE ↑ Na⁺ INTAKE

↑PLASMA OSMOLALITY

THIRST   ↑ADH SECRETION

↓H₂O INGESTION   ↓H₂O EXCRETION (V̇)

↑PLASMA VOLUME

STIMULATION OF   ↓PLASMA ONCOTIC
BARORECEPTORS   PRESSURE

↓SYMPATHETIC TONE
TO KIDNEYS

DILATION OF AFFERENT AND
EFFERENT ARTERIOLES

↑PERITUBULAR CAPILLARY   ↓PERITUBULAR CAPILLARY
HYDROSTATIC PRESSURE (P$_c$) ONCOTIC PRESSURE (π$_c$)

↓REABSORPTION OF
FLUID FROM PERITUBULAR
SPACE

↑HYDROSTATIC PRESSURE IN
LATERAL INTERCELLULAR SPACE

↓ Na⁺ REABSORPTION IN
PROXIMAL TUBULE

↑ Na⁺ EXCRETION
(U$_{Na}$V̇)

Figure 4.38.   Pathway by which an increase in peritubular capillary hydrostatic pressure (P$_c$) and a decrease in peritubular capillary oncotic pressure (π$_c$) could decrease proximal tubular Na⁺ reabsorption following a large increase in Na⁺ intake, thereby restoring Na⁺ balance. This pathway represents a possible mechanism for the *third factor effect* in the proximal tubule.

sorption of Na⁺ by the proximal tubule. Such changes in peritubular capillary hydrostatic and oncotic pressure also would decrease the reabsorption of water and other solutes by the proximal tubule, thus accounting for the observation that *all* proximal tubular reabsorption is decreased following the ingestion of a large quantity of Na⁺ or another cause of a plasma volume expansion. It should be noted that while the changes in hydrostatic and oncotic pressure illustrated in Figure 4.38 could account for the decrease in proximal tubular Na⁺ reabsorption following a large increase in Na⁺ intake, many investigators believe that a hormone also may be involved. It is not known if this "natriuretic" hormone is the same as the hormone postulated to decrease Na⁺ reabsorption in the medullary collecting duct in response to small increases in Na⁺ intake (see above).

Although the detailed mechanisms for the third factor effect are not completely understood, the

participation of both the proximal tubule and medullary collecting duct occurs in a logical manner. With *small* increases in Na⁺ intake, the third factor effect occurs in a region of the nephron that reabsorbs small quantities of Na⁺ and "fine tunes" the rate of Na⁺ excretion, i.e., the *medullary collecting duct*. With *large* increases in Na⁺ intake, the third factor effect also occurs in the region that reabsorbs the largest quantity of Na⁺, i.e., the *proximal tubule*.

Overall Response to Changes in Sodium Intake

It is important to emphasize that the three factors that regulate Na⁺ excretion in response to

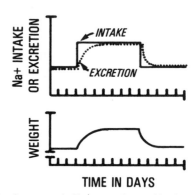

Figure 4.39.   Increases in Na⁺ excretion and body weight following an abrupt increase in daily Na⁺ intake. The increase in Na⁺ excretion parallels the increase in body weight, which in turn parallels an increase in plasma volume. In this example, the restoration of Na⁺ balance (Na⁺ input = Na⁺ output) requires approximately 3 days. The changes are reversed when daily Na⁺ intake returns to its original level. [Adapted from Seely and Levy (1981).]

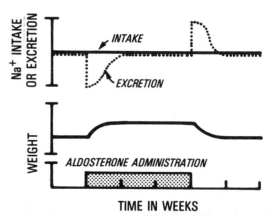

Figure 4.40.   The mineralocorticoid escape phenomenon. When aldosterone is administered to an experimental animal with a constant daily Na⁺ intake, Na⁺ excretion initially decreases, resulting in Na⁺ retention and an increase in body weight. However, after several days of Na⁺ retention, Na⁺ balance is again achieved (Na⁺ input = Na⁺ output) despite continued aldosterone administration. These changes are reversed when aldosterone is terminated. [Adapted from Seely and Levy (1981).]

changes in Na^+ input (changes in GFR, aldosterone, third factor effect) are activated *not* by a direct effect of the changes in Na^+ input, but instead by the resulting changes in *plasma volume* (Figs. 4.36–4.38). Even the "natriuretic" hormone(s) postulated to contribute to the third factor effect is (are) thought to be released in response to changes in plasma volume. Figure 4.39 illustrates that the increase in Na^+ excretion following an abrupt increase in daily Na^+ intake in fact occurs parallel to an increase in body weight, which in turn parallels an increase in plasma volume. Note that the complete restoration of Na^+ balance (Na^+ input = Na^+ output) requires several days (an average of 3–5 days in man). These changes are reversed when Na^+ intake returns to its original level.

A similar relationship between changes in Na^+ excretion and changes in plasma volume is seen in the phenomenon of *mineralocorticoid escape* (August et al., 1958). If a mineralocorticoid such as *aldosterone* is administered in an experimental setting, Na^+ excretion will, of course, initially decrease. However, after several days of Na^+ retention (and concomitant increase in plasma volume and weight), Na^+ excretion increases to its original level, matching Na^+ intake, i.e., Na^+ balance is again achieved *despite* the continued administration of the hormone (Fig. 4.40). Studies of experimental animals with mineralocorticoid escape indicate that the escape occurs because Na^+ reabsorption in the *medullary collecting duct* is decreased, even though mineralocorticoids (such as aldosterone) continue to *stimulate* Na^+ reabsorption *throughout* the collecting tubule/duct. Thus, mineralocorticoid escape appears to be related to the third factor effect following *small* increases in Na^+ intake (see above). The phenomenon of mineralocorticoid escape is clinically significant because it explains why patients with aldosterone-secreting tumors or hyperplasia of the adrenal cortex experience only moderate Na^+ retention and weight gain. However, such patients generally develop metabolic alkalosis and hypokalemia, since aldosterone continues to stimulate H^+ and K^+ secretion in the collecting tubule/duct (Chapters 33 and 34).

CHAPTER 33

Acid-Base Balance and Regulation of H$^+$ Excretion

The kidneys have a major role in regulating the H$^+$ ion concentration, or pH, of the body fluids. In a healthy individual, the pH of the ECF generally is maintained within a rather narrow range, with a mean normal value of 7.40 ± 0.02 in arterial plasma and 7.38 ± 0.02 in mixed venous plasma. The clinical evaluation of a patient's acid-base status generally is based on laboratory studies of *arterial* samples. Thus, an individual is considered to have an *acidosis* when the arterial pH falls below 7.38 and an *alkalosis* when the arterial pH rises above 7.42. Such precise control of pH is necessary because of the marked effects of pH changes on protein conformation, enzymatic reactions, and central nervous system function.

It should be noted that the use of pH rather than H$^+$ concentration is widespread in physiology and clinical medicine because clinical laboratories typically measure pH. However, since pH = −log [H$^+$], the H$^+$ concentration can be readily calculated if necessary, using the relation [H$^+$] = 10^{-pH}. Furthermore, in the pH range of 7.30–7.50 a *rapid* method for obtaining an *approximate* H$^+$ concentration is available. This method takes advantage of the fact that [H$^+$] is *40* nmol/l at pH *7.40* and that each *0.01 unit* change in pH from 7.40 corresponds to an *inverse* change of approximately *1 nmol/l* in [H$^+$] in the 7.30–7.50 pH range (Table 4.22). Thus, the normal arterial pH range of 7.38–7.42 corresponds to a [H$^+$] range of 42–38 nmol/l.

THREATS TO pH

Although small quantities of acid or base are present in certain foods and medications, the major threats to the pH of the body fluids are *acids* formed in *metabolic processes*. These metabolic acids can be conveniently divided into three categories:

1. *Volatile Acids: Carbon Dioxide.* CO$_2$, a major end product in the oxidation of carbohydrates, fats, and amino acids, can be regarded as an

acid by virtue of its ability to react with water to form *carbonic acid* (H$_2$CO$_3$), which in turn can dissociate to form H$^+$ and HCO$_3^-$:

$$CO_2 + H_2O \leftrightharpoons H_2CO_3 \leftrightharpoons H^+ + HCO_3^- \quad (87)$$

Because it is a gas (and can be eliminated by the lungs, as discussed below), CO$_2$ often is termed a *volatile acid.* Enough CO$_2$ is produced each day to add approximately 10,000 mmol of H$^+$ to the body fluids, and even greater amounts of CO$_2$ can be produced in exercise and hypermetabolic states such as thyrotoxicosis. Without compensating mechanisms, the addition of 10,000 mmol of H$^+$ would have a *catastrophic* effect on the pH of the body fluids. For example, in a 60-kg man with 36 l of total body water, without compensating mechanisms 10,000 mmol of H$^+$ would increase the H$^+$ concentration of the body fluids by nearly 300 mmol/l. This would represent more than a *7 million-fold* increase in the H$^+$ concentration, since the normal H$^+$ concentration at pH 7.4 is only 40 nmol/l or 0.00004 mmol/l.

2. *Fixed Acids: Sulfuric Acid and Phosphoric Acid.* Sulfuric acid is an end product of the oxidation of the sulfur-containing amino acids methionine and cysteine, while phosphoric acid is formed in the metabolism of phospholipids, nucleic acids, phosphoproteins, and phosphoglycerides. In contrast to CO$_2$, sulfuric acid and phosphoric acid are *nonvolatile* and

Table 4.22
Correlation of pH with estimated and actual levels of [H$^+$]

pH	Estimated [H$^+$], nmol/l	Actual [H$^+$], nmol/l
7.30	50	50.1
7.35	45	44.7
7.40	40	40
7.45	35	35.5
7.50	30	31.6

have therefore been termed *fixed acids*. The production of fixed acids varies with the diet, but typically results in the formation of 50–100 mmol of H⁺/day. While much less threatening to the body than the 10,000 mmol of H⁺ from CO_2, even the effect of 50–100 mmol of H⁺ would be disastrous without compensating mechanisms. For example, without compensating mechanisms the addition of just 50 mmol of H⁺/day to the body fluids of a 60-kg man would increase the H⁺ concentration of the body fluids by over 1 mmol/l, which would represent more than a *25,000-fold increase* in the H⁺ concentration.

3. *Organic Acids.* Organic acids such as lactic acid, acetoacetic acid, and β-OH butyric acid are formed during the metabolism of carbohydrates and fats. Normally, these acids are further oxidized to CO_2 and water and therefore do not directly affect the pH of the body fluids. However, in certain abnormal circumstances these organic acids may accumulate, causing an *acidosis*. For example, in hypovolemic and other forms of circulatory shock, *lactic acid* levels may increase markedly due to inadequate perfusion of tissues and the resulting increase in anaerobic glycolysis (*lactic acidosis*). In uncontrolled diabetes mellitus, *acetoacetic acid* and *β-OH butyric acid* may accumulate due to increased lipid catabolism and consequent overloading of the body's ability to metabolize acetyl CoA (*diabetic ketoacidosis*).

ACID-BASE BUFFER SYSTEMS

The kidneys and the lungs together share the responsibility for regulating the pH of the body fluids. However, it is the *buffer systems* of the body fluids that actually provide the most *immediate* defense against changes in pH. These buffers are *weak acids* that exist in a mixture of a protonated form and an unprotonated form in the physiological pH range. The dissociation reaction for such a weak acid HA is as follows:

$$HA \leftrightharpoons H^+ + A^- \qquad (88)$$

In order to understand how such an acid could attenuate, or "buffer," the pH changes caused by the addition or loss of H⁺ or OH⁻, it is necessary to examine reaction *88* more closely. The law of mass action for reaction *88* can be written as follows:

$$K_a = \frac{[H^+][A^-]}{[HA]} \qquad (89)$$

where K_a is the apparent dissociation constant for the acid. Taking the logarithm of *Eq. 89* results in the following expression:

$$\log K_a = \log \frac{[H^+][A^-]}{[HA]}$$

or

$$\log K_a = \log [H^+] + \log \frac{[A^-]}{[HA]} \qquad (90)$$

Eq. 90 can be rearranged to give

$$-\log [H^+] = -\log K_a + \log \frac{[A^-]}{[HA]} \qquad (91)$$

Substituting pH for $-\log [H^+]$ and defining $pK_a = -\log K_a$, the so-called *Henderson-Hasselbalch equation* is obtained:

$$pH = pK_a + \log \frac{[A^-]}{[HA]} \qquad (92)$$

This equation can be used to illustrate how a weak acid functions as a buffer. Consider a weak acid with a pK_a of 7.0. If in 1 l of solution the unprotonated form A⁻ and the protonated form HA are each present in a concentration of 10 mmol/l, then

$$pH = 7.0 + \log \frac{10}{10}$$

$$= 7.0 \qquad (93)$$

Assume now that 1 mmol of a *strong* acid HX is added to the solution. The strong acid will be completely dissociated:

$$HX \rightarrow H^+ + X^-$$

However, its H⁺ ion can combine with the unprotonated form of the weak acid A⁻, as follows:

$$H^+ + X^- + A^- \rightarrow HA + X^-$$

Since the weak acid HA is much less dissociated than the strong acid HX, it can be assumed that virtually all of the added H⁺ reacts with unprotonated A⁻. The concentration of A⁻ therefore will decrease to 9 mmol/l and the concentration of HA will increase to 11 mmol/l. According to the Henderson-Hasselbalch equation (*Eq. 92*), the pH of the solution then will be

$$pH = 7.0 + \log \frac{9}{11}$$

$$\simeq 6.9 \qquad (94)$$

i.e., the 1 mmol of strong acid has decreased the pH of the buffered solution by approximately *0.1 units*. This can be contrasted to a "*control*" in which the same amount of strong acid HX (1 mmol) is added to 1 l of an *unbuffered* solution such as *water*. Like the buffered solution considered above, water has a pH of 7.0. However, due to the absence of buffer the addition of 1 mmol of HX results in 1 mmol of

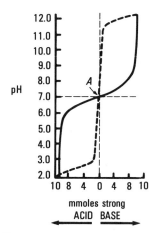

Figure 4.41. Changes in pH resulting from the addition of a strong acid or strong base to the weak acid buffer solution described in the text (*solid curve*) and to water (*dashed curve*). *Point A* represents the pH of the buffer solution and of water before the addition of acid or base (pH 7.0). The weak acid is most effective as a buffer in the vicinity of its pK_a (7.0).

H^+ being added to the solution, which originally contained only 0.0001 mmol of H^+ (pH 7.0). The final pH is therefore 3.0 (= $-\log 10^{-3}$ mol/l), i.e., the 1 mmol of strong acid has decreased the pH of the *unbuffered* solution by *4.0 units.*

A more general demonstration of the ability of the weak acid HA to function as a buffer is presented in Figure 4.41, which illustrates the changes in pH that would result from the addition of *varying* amounts of a strong acid *or* a strong base to the buffer solution discussed above. Point *A* represents the pH of the buffer solution *and* of water *before* the addition of acid or base (pH 7.0). The pH falls as increasing amounts of strong acid are added to the buffer solution (*solid curve to the left of A*), but not nearly as much as it would fall if the strong acid were added to water (*dashed curve to the left of A*). Similarly, the pH rises as increasing amounts of strong base are added to the buffer solution (*solid curve to the right of A*), but not nearly as much as it would rise if the strong base were added to water (*dashed curve to the right of A*). Note that the slope of the *solid curve* is flattest near point *A*, i.e., 1 mmol of strong acid or strong base causes the *smallest* change in pH when added to the buffer in the vicinity of pH 7.0. But recall that 7.0 also represents the pK_a of the weak acid used in these examples. This means that a weak acid is *most effective* as a buffer when the pH of the solution remains close to the pK_a of the acid.

A second factor that determines the buffering effectiveness of a weak acid is the total *concentration* of the weak acid in the solution. For example, if the HA buffer solution discussed above contained only 2 mmol/l of A^- and 2 mmol/l of HA (instead

of 10 mmol/l of A^- and 10 mmol/l of HA), its pH still would be 7.0 (compare to *Eq. 93*):

$$pH = 7.0 + \log \frac{2}{2}$$
$$= 7.0$$

However, the addition of 1 mmol of strong acid now would decrease the pH of the buffer solution by nearly *0.5 units* (instead of by only 0.1 units; see *Eq. 94*):

$$pH = 7.0 + \log \frac{1}{3}$$
$$\simeq 6.5$$

Buffer Systems of the Body Fluids

From the above discussion, it should be evident that the *optimal* buffer system for keeping the pH of the body fluids close to 7.40 would (1) have a pK_a close to 7.4; and (2) be present in a high concentration. Although there is no such optimal buffer in the body, the major body buffer systems, to be introduced here, satisfy these criteria to different extents.

HEMOGLOBIN AND OTHER PROTEINS

Proteins contain several ionizable groups that are weak acids, including C-terminal carboxyl groups, N-terminal amino groups, side chain carboxyl groups of glutamic and aspartic acid, side chain amino groups of lysine, and imidazole groups of histidine. While the pK_a of a given ionizable group will be different in different proteins, depending upon the local environment, in many proteins the *imidazole groups of histidine* and the *N-terminal amino groups* have pK_as sufficiently close to 7.4 to enable these proteins to function as effective buffers. Proteins are present in relatively high concentrations in cells and plasma, which further contributes to their effectiveness as buffers.

Hemoglobin is a particularly effective protein buffer and is a major buffer in blood. In fact, hemoglobin has approximately six times the buffering capacity of the plasma proteins in blood due to its high concentration (nearly four times greater than that of plasma proteins) and its 38 histidine residues (over twice as many as albumin, even though the molecular weights of albumin and hemoglobin are similar). Also contributing to the effectiveness of hemoglobin as a blood buffer is the fact that *deoxygenated* hemoglobin has imidazole groups with a somewhat higher pK_a (smaller K_a) than *oxygenated* hemoglobin (VanSlyke et al., 1922). As a result, once hemoglobin becomes deoxygenated in the capillaries, it is better able to bind the H^+ ions formed when CO_2 enters capillary blood from the tissues (Chapter 36). Thus, hemoglobin helps to minimize the fall in pH caused by the loading of CO_2 into capillary blood.

PHOSPHATE

The phosphate buffer system involves the weak acid $H_2PO_4^-$, which undergoes the following dissociation reaction

$$H_2PO_4^- \rightleftharpoons H^+ + HPO_4^= \qquad (95)$$

Since the pK_a for this reaction is 6.8, the phosphate buffer system is a suitable candidate for a body buffer. However, not all body fluids contain sufficient concentrations of phosphate for this buffer system to be effective. For example, the phosphate buffer system is much more effective in *ICF* than in *ECF*, since (1) the total concentration of phosphate is much greater in ICF than in ECF (Fig. 4.3); and (2) the intracellular pH generally is somewhat lower than the extracellular pH and therefore is closer to the pK_a of the phosphate buffer.

Phosphate also is effective in buffering the tubular fluid in the *kidneys*, as discussed below. Even though the tubular fluid is derived from plasma (i.e., from *ECF*), phosphate can be an effective buffer in the tubular fluid because (1) phosphate becomes greatly *concentrated* in the tubular fluid due to the reabsorption of water in excess of phosphate; and (2) the pH of the tubular fluid generally becomes more *acidic* than the pH of ECF and therefore is closer to the pK_a of the phosphate buffer.

BICARBONATE

The bicarbonate buffer system involves the weak acid H_2CO_3, which undergoes the following dissociation reaction

$$H_2CO_3 \rightleftharpoons H^+ + HCO_3^- \qquad (96)$$

The Henderson-Hasselbalch equation for this reaction is as follows:

$$pH = pK_a + \log \frac{[HCO_3^-]}{[H_2CO_3]} \qquad (97)$$

Since the pK_a for reaction *96* is 3.7, the bicarbonate buffer system appears to represent a very poor candidate for a body buffer. However, in the body fluids H_2CO_3 is in equilibrium with dissolved CO_2 and water (*Eq. 87*). Thus, the Henderson-Hasselbalch equation for the bicarbonate buffer system is most appropriately written as follows:

$$pH = pK_a' + \log \frac{[HCO_3^-]}{[CO_2]} \qquad (98)$$

where pK_a' includes the equilibrium constant for the formation of H_2CO_3 from CO_2 and water. But

the value of pK_a' is only 6.1, i.e., even when the equilibrium between H_2CO_3 and dissolved CO_2 is taken into account, the bicarbonate buffer system does not appear to represent an optimal body buffer. Nevertheless, the bicarbonate buffer system probably can be regarded as the most *important* buffer system in the body. This is because the concentrations of its components can be *independently regulated*, the CO_2 concentration by the lungs and the HCO_3^- concentration by the kidneys (see below). The significance of such regulation can best be appreciated by noting that according to the Henderson-Hasselbalch equation, the pH of a weak acid buffer solution is determined by the ratio of the concentrations of the two forms of the buffer. Thus, by regulating the CO_2 and HCO_3^- concentrations, the lungs and kidneys also regulate the pH of the body fluids (*Eq. 98*).

Although the Henderson-Hasselbalch equation applies to the other buffer systems of the body as well, in *none* of the other systems can the concentrations of the components be independently regulated. In other words, although the nonbicarbonate buffer systems can defend the body fluids against changes in pH, they cannot be used as a primary mechanism for adjusting the pH of the body fluids. Of course, any change in the $[HCO_3^-]/[CO_2]$ ratio will affect the concentration ratios for the *nonbicarbonate* buffers in the body fluids as well, since

$$pH = pK_{a_1} + \log \frac{[HCO_3^-]}{[CO_2]}$$

$$= pK_{a_2} + \log \frac{[HPO_4^=]}{[H_2PO_4^-]} = pK_{a_3} + \log \frac{[Hb^-]}{[HHb]}$$

$$= pK_{a_4} + \log \frac{[Prot^-]}{[HProt]} = \cdots \cdots \quad (99)$$

The interrelationship between the various buffer systems of the body, as expressed in *Eq. 99*, is termed the *isohydric principle*.

Before proceeding to describe the respiratory and renal regulation of the CO_2 and HCO_3^- concentrations, it should be noted that the Henderson-Hasselbalch equation for the bicarbonate buffer system most commonly is written in a form that allows the *concentration* of carbon dioxide, $[CO_2]$, to be replaced by the *partial pressure* of carbon dioxide, P_{CO_2}, a more conveniently measured quantity. According to Henry's law, the concentration of a gas dissolved in solution is directly proportional to its partial pressure. In plasma at 37°C, the relationship between the concentration of dissolved CO_2 (in mmol/l) and the partial pressure of CO_2 (in mmHg) is as follows (Chapter 36):

$$[CO_2] = 0.03 \times P_{CO_2} \qquad (100)$$

Substituting into *Eq. 98*, the Henderson-Hasselbalch equation for the bicarbonate buffer system becomes

$$pH = pK_a' + \log \frac{[HCO_3^-]}{0.03\ P_{CO_2}} \qquad (101)$$

Applying this equation to arterial plasma, the normal arterial pH of 7.40 corresponds to a $[HCO_3^-]$ of approximately 24 mmol/l and a P_{CO_2} of approximately 40 mmHg

$$pH = 6.1 + \log \frac{24}{0.03 \times 40}$$

$$= 7.40 \qquad (102)$$

In the remainder of this chapter, the plasma HCO_3^- concentration will be designated by $[HCO_3^-]_p$ (instead of by P_{HCO_3}, as in other chapters) to be consistent with the widespread use of square brackets to designate concentrations in the Henderson-Hasselbalch equation.

RESPIRATORY REGULATION OF pH

The lungs participate in the regulation of pH by regulating the partial pressure of CO_2 in arterial blood. As previously indicated, the CO_2 produced in metabolic processes can combine with water to form H_2CO_3, which in turn can dissociate to form H^+ and HCO_3^- (*Eq. 87*). However, this entire reaction is *reversed* in the lungs when CO_2 is eliminated, or "blown off," by ventilation, i.e., CO_2 is a *volatile acid*. In fact, in a person with normal lungs and a fixed rate of CO_2 production, the arterial P_{CO_2} is determined solely by, and is inversely proportional to, the alveolar ventilation, $\dot{V}_A$ (Chapter 36):

$$P_{CO_2} \propto \frac{1}{\dot{V}_A} \qquad (103)$$

Since enough CO_2 is produced each day to add approximately 10,000 mmol of H^+ to the body fluids, the elimination of CO_2 by the lungs has an *essential* role in regulating the pH of the body fluids. Of course, H^+ *is* added to the blood when CO_2 enters the capillaries for transport from the tissues to the lungs. However, virtually *all* of this H^+ reacts with blood buffers, primarily hemoglobin (see above), as evidenced by the fact that the venous pH is only slightly lower than the arterial pH (see above). The fact that the body's buffer systems provide the most *immediate* defense against changes in pH has been noted previously.

Because of the important relation between alveolar ventilation and arterial P_{CO_2} (*Eq. 103*), it should not be surprising that many of the mechanisms that allow the body to maintain a normal arterial pH are mediated by changes in alveolar ventilation. For example, chemoreceptors in the medulla oblongata,

aortic bodies, and carotid bodies respond to an increase in arterial P_{CO_2} by stimulating alveolar ventilation. Chemoreceptors in the carotid bodies also stimulate alveolar ventilation in response to decreases in arterial pH that occur *independently* of an increase in arterial P_{CO_2}. These regulatory mechanisms, which are discussed in greater detail in Chapter 39, are important not only in the day-to-day maintenance of a normal arterial pH but also in the body's response to certain acid-base disturbances (see below).

RENAL REGULATION OF pH

The kidneys participate in the regulation of pH by regulating the concentration of HCO_3^- in plasma, designated here by $[HCO_3^-]_p$ (see above). This regulation of $[HCO_3^-]_p$ involves two tasks. First, the kidneys regulate the amount of HCO_3^- *recovered* or *reabsorbed* from the glomerular filtrate. With a $[HCO_3^-]_p$ of 24 mmol/l and a GFR of 125 ml/min (180 l/day), approximately 3 mmol of HCO_3^- are filtered per minute (4320 mmol/day), i.e., without a mechanism for HCO_3^- recovery, the daily loss of HCO_3^- would be equivalent to that resulting from the addition of over 4000 mmol of a strong acid to the body fluids. Second, the kidneys *generate* HCO_3^- to replace the HCO_3^- lost in buffering the various strong acids formed in the body. Recall that as much as 100 mmol of H^+ is added to the body fluids each day from the *fixed* (i.e., *non-volatile*) *acids* produced during metabolism. The body's buffer systems, particularly the *bicarbonate buffer*, provide the most *immediate* defense against this H^+. For example, the following reaction illustrates how *sulfuric acid* produced during the metabolism of sulfur-containing amino acids is buffered by HCO_3^-:

$$2H^+ + SO_4^= + 2HCO_3^- \leftrightharpoons SO_4^= + 2H_2CO_3$$

$$\leftrightharpoons SO_4^= + 2CO_2 + 2H_2O \quad (104)$$

Since the CO_2 formed in this reaction is eliminated by the lungs, the net result of the reaction is the loss of two HCO_3^- ions from the body, which must be replaced by the kidneys. Even if the H^+ ions from the sulfuric acid were to initially react with a *different* body buffer, such as phosphate, by the isohydric principle (*Eq. 99*) a decrease in the concentration of the unprotonated form of *any* buffer will cause a decrease in the concentration of HCO_3^- as well. *Organic acids* such as lactic acid, acetoacetic acid, and β-OH butyric acid can similarly deplete the body's store of HCO_3^- whenever significant accumulation of these acids occurs (e.g., in lactic acidosis, diabetic ketoacidosis; see above).

A. PROXIMAL TUBULE

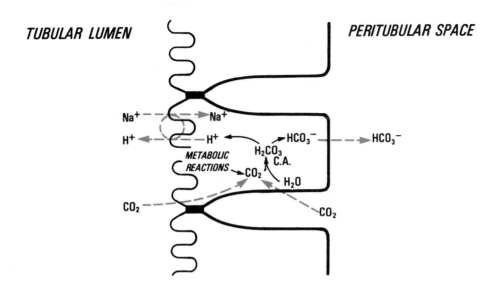

B. DISTAL NEPHRON

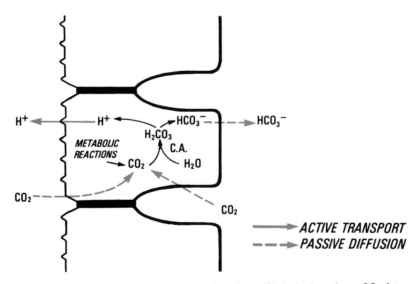

Figure 4.42. Mechanisms for H^+ secretion in the proximal tubule (*A*) and distal nephron (*B*). In both regions, CO_2 from metabolic processes, the peritubular capillaries, or the tubular lumen reacts with water to form H_2CO_3, a reaction catalyzed by carbonic anhydrase (C.A.) in the epithelial cells. The H_2CO_3 dissociates to form H^+ and HCO_3^-. The H^+ is secreted via the Na^+-H^+ exchange process (antiport) in the proximal tubule (indicated by the *dashed circle*) and via an active process in the distal nephron; the HCO_3^- diffuses across the basal-lateral surfaces into the peritubular space and peritubular capillaries, so that for every H^+ secreted, a HCO_3^- is returned to the systemic circulation.

It is important to note that both the reabsorption of filtered HCO_3^- and the generation of new HCO_3^- are accomplished by a *single* process, the *secretion of H⁺* by the tubular epithelial cells (Fig. 4.42) (Warnock and Rector, 1979). This H⁺ secretion occurs in the *proximal tubule* and *distal nephron.*

Although there are certain differences between these regions in the mechanism and regulation of H⁺ secretion, in both regions H⁺ secretion begins in the epithelial cells as CO_2 reacts with water to form H_2CO_3. The CO_2 for this reaction is either produced in the epithelial cells by metabolic proc-

PROXIMAL TUBULE

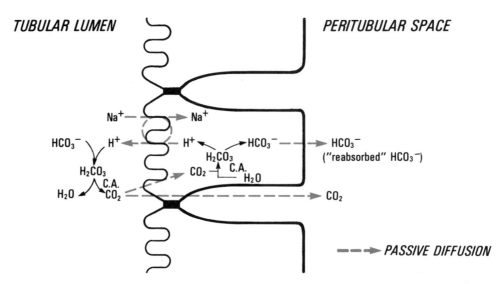

Figure 4.43. Reabsorption of filtered HCO_3^- via H^+ secretion. The secreted H^+ reacts with HCO_3^- in the tubular fluid to form H_2CO_3, which then dissociates into CO_2 and water, i.e., a HCO_3^- ion is *lost* from the tubular fluid. But since the H^+ secretion process *adds* a HCO_3^- ion to the peritubular fluid and peritubular capillaries, the *net result* is HCO_3^- *reabsorption*. The proximal tubule is illustrated here, since approximately 90% of the filtered HCO_3^- is reabsorbed in the proximal tubule. However, similar reactions occur in the distal nephron (except for the absence of carbonic anhydrase in the luminal surface) and are responsible for the reabsorption of most of the remaining HCO_3^- in the tubular fluid.

esses or diffuses into the cells from the peritubular capillaries or tubular lumen. While the formation of H_2CO_3 can proceed spontaneously, the epithelial cells of both the proximal tubule and the distal nephron contain the enzyme *carbonic anhydrase*, which catalyzes the reaction. The H_2CO_3 *dissociates* to form H^+ and HCO_3^-, and the H^+ is then secreted into the tubular lumen. In the *proximal tubule*, this secretion of H^+ is primarily coupled to the passive movement of Na^+ from the tubular lumen into the epithelial cells (Na^+-H^+ exchange, or *antiport*) (Fig. 4.42A) (Murer et al., 1976), although some active secretion may occur as well (Chapter 30). In contrast, H^+ secretion in the *distal nephron* is primarily an active process (Fig. 4.42B) (Al-Awqati, 1978). It should be noted that in spite of the reference to "distal nephron H^+ secretion," active H^+ secretion actually has been demonstrated in the *connecting tubule* and *collecting tubule/duct* only; conclusive evidence for active H^+ secretion in the distal tubule still is lacking.

An important feature of the H^+ secretory process illustrated in Figure 4.42 is that the HCO_3^- ion formed during the dissociation of H_2CO_3 can diffuse across the basal-lateral surfaces of the epithelial cells (Cohen et al., 1978) into the peritubular space and peritubular capillaries. Thus, *for every H^+ se-*

creted, a HCO_3^- is returned to the systemic circulation.*

Recovery of Filtered Bicarbonate

The kidneys must be able to reabsorb virtually *all* of the 4300+ mmol of HCO_3^- filtered each day. As indicated above, this reabsorption of filtered HCO_3^- is accomplished by *H^+ secretion*. To understand the relationship between H^+ secretion and HCO_3^- reabsorption, it is necessary to examine the fate of the secreted H^+ ions.

Most of the H^+ that is secreted into the tubular fluid will react with HCO_3^- in the tubular fluid to form H_2CO_3, which in turn dissociates into CO_2 and water (Fig. 4.43). Because all cellular membranes are freely permeable to CO_2, the CO_2 formed in this reaction can diffuse (1) into the *epithelial cells* to generate additional H^+ for secretion; or (2) through the epithelial cells into the *peritubular capillaries* for eventual elimination by the lungs. The water

---

* Actually, in the *proximal tubule* some of the HCO_3^- formed in the epithelial cells diffuses across the *brush border* and into the tubular fluid via a coupled exchange reaction for Cl^- (Fig. 4.24B). But since this exchange really represents a mechanism for the reabsorption of Na^+ and Cl^-, it will not be considered further in this chapter.

formed in the reaction mixes with the water in the tubular fluid, but its contribution to the total volume of the tubular fluid is negligible. It should be noted that in the *proximal* tubular fluid, the dissociation of H_2CO_3 into CO_2 and water is catalyzed by *carbonic anhydrase*, which is found in large quantities in the brush border of the proximal tubule (Lönnerholm, 1971). Thus, in the proximal tubule, carbonic anhydrase catalyzes not only the *formation* of H_2CO_3 in the epithelial cells, but also the *dissociation* of H_2CO_3 in the lumen.

As illustrated in Figure 4.43, each time a secreted H^+ reacts with a HCO_3^- in the tubular fluid, a HCO_3^- ion is *lost* from the tubular fluid. But the H^+ secretion process also results in the *addition* of a HCO_3^- ion to the peritubular fluid and peritubular capillaries. Thus, when H^+ reacts with HCO_3^- in the tubular fluid, the *net result* of H^+ secretion is the *reabsorption of HCO_3^-* (Fig. 4.43). Of course, the atoms comprising the HCO_3^- removed from the tubular fluid are *not* the same as those comprising the HCO_3^- returned to the peritubular fluid and capillaries, but the net result is *equivalent* to HCO_3^- reabsorption. The reaction of secreted H^+ with HCO_3^- in the tubular fluid, then, represents the mechanism whereby the kidneys reabsorb filtered HCO_3^-.

The fraction of the filtered HCO_3^- that is reabsorbed can be calculated using the methods described in Chapter 29. In normal man, approximately 90% of the filtered HCO_3^- is reabsorbed in the *proximal tubule*. Most of the remaining HCO_3^- is reabsorbed in the *distal nephron*. In fact, the *fractional excretion* of HCO_3^- typically is less than 0.1%, i.e., over 99.9% of the filtered HCO_3^- is normally reabsorbed. Since over 4300 mmol of HCO_3^- are filtered per day, this means that over 4300 mmol of H^+ must be secreted per day to reabsorb the filtered HCO_3^-.

Generation of New Bicarbonate

The kidneys must be able to generate 50–100 mmol of new HCO_3^-/day to replace the HCO_3^- lost in titrating the strong acids produced by the body (see above). Like the reabsorption of filtered HCO_3^-, the generation of new HCO_3^- is accomplished by the *H^+ secretion* process illustrated in Figure 4.42, in which a HCO_3^- ion is returned to the systemic circulation for each H^+ ion secreted. If the secreted H^+ reacts with HCO_3^- in the tubular fluid, the net result is *HCO_3^- reabsorption*, since the HCO_3^- returned to the systemic circulation simply replaces a HCO_3^- lost from the tubular fluid in the reaction with H^+ (Fig. 4.43). However, if

excess H^+ is secreted, the HCO_3^- returned to systemic circulation represents *new* HCO_3^-. The kidneys can therefore generate 50–100 mmol of new HCO_3^-/day simply by secreting 50–100 mmol of H^+ *in excess* of the 4300+ mmol of H^+ that must be secreted to reabsorb the filtered HCO_3^-.

The excess H^+ that is secreted in order to generate new HCO_3^- cannot, however, be *excreted* from the body as *free* H^+ ion. This is because the magnitude of the H^+ concentration gradient that can be maintained across the tubular epithelium is *limited*. The *proximal tubule*, with its "leaky" tight junctions, can maintain a pH gradient of only 0.5 pH units, i.e., the pH of the proximal tubular fluid cannot be reduced below $\simeq6.9$. The *distal nephron*, with its "tight" tight junctions, can maintain a pH gradient of almost 3 pH units, i.e., the pH of the tubular fluid in the distal nephron can be as low as 4.5. The minimum *urine* pH is therefore also 4.5. But even at this minimum urine pH, the concentration of free H^+ in urine is only 0.03 mmol/l. Thus, if the 50–100 mmol of excess H^+ secreted each day were to be excreted *solely* as free H^+ ion, over *1000 l* of urine would be required. The excess H^+ must therefore be excreted in combination with *buffers*.

The predominant buffer in the tubular fluid is, of course, HCO_3^-. However, since the reaction between H^+ and HCO_3^- results in the *reabsorption* of HCO_3^- (see above), *excess* H^+ cannot be excreted by combining with HCO_3^-. The excess H^+ must therefore combine with *nonbicarbonate* buffers in the tubular fluid, the most important of which are *ammonia* and *phosphate*.

AMMONIA

The dissociation reaction for the ammonia buffer system is as follows:

$$NH_4^+ \leftrightharpoons NH_3 + H^+ \qquad (105)$$

Since this reaction has a pK_a of 9.3, in *any* of the body fluids virtually *all* of the ammonia is present as the *protonated* ammonium ion (NH_4^+). For example, in plasma and the glomerular filtrate (pH = 7.4), the $[NH_3]/[NH_4^+]$ ratio is approximately 1:100. It might appear, then, that the amount of *unprotonated* NH_3 available to react with secreted H^+ would be insignificant. However, NH_3 is actually a *highly* effective buffer for secreted H^+ as a result of two important factors (Fig. 4.44*A*):

1. NH_3 is synthesized in the epithelial cells of the proximal tubule and distal nephron and secreted into the tubular fluid. In other words, the amount of NH_3 available to function as a

DISTAL NEPHRON

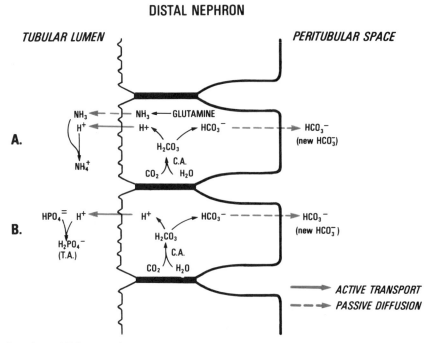

Figure 4.44.   Generation of new HCO_3^- via H^+ secretion. The HCO_3^- generated in the H^+ secretion process represents a *new* HCO_3^- if the secreted H^+ ion (*A*) reacts with NH_3 synthesized by the epithelial cells to form NH_4^+; or (*B*) reacts with HPO_4^- in the tubular fluid to form $H_2PO_4^-$, which represents the major component of the urinary titratable acids (T.A.). Although the distal nephron is illustrated here, similar reactions can occur in the proximal tubule (except that proximal tubular H^+ secretion is primarily coupled to Na^+ reabsorption; see Fig. 4.42*A*). However, most of the H^+ secreted in the proximal tubule reacts with HCO_3^- in the tubular fluid and therefore accomplishes the reabsorption of filtered HCO_3^- (Fig. 4.43) rather than the generation of new HCO_3^-.

buffer is *not* limited to the extremely small amount that enters the tubular fluid by glomerular filtration. It is estimated that approximately 60% of the NH_3 synthesized in the epithelial cells is derived from *glutamine* in the following reactions:

$$\text{Glutamine} \xrightarrow[\text{glutaminase}]{\quad \overset{NH_3}{\nearrow} \quad} \text{glutamic acid}$$

$$\xrightarrow[\substack{\text{glutamic} \\ \text{dehydrogenase}}]{\quad \overset{NH_3}{\nearrow} \quad} \alpha\text{-ketoglutarate}$$

(106)

The enzyme *glutaminase* is abundant in the mitochondria of tubular epithelial cells. The remaining 40% of the NH_3 synthesized in the epithelial cells is derived from other amino acids, particularly glycine and alanine (Pitts, 1971).

2. NH_3 and NH_4^+ have markedly different *solubility characteristics*. NH_3 is highly lipid-soluble and can passively diffuse across cellular membranes. In contrast, NH_4^+ is highly polar and crosses membranes poorly.

Thus, as NH_3 is synthesized and its concentration in the epithelial cells increases, NH_3 diffuses out of

the cells. Although NH_3 can diffuse into *either* the tubular lumen *or* the peritubular capillaries, NH_3 diffusion into the *lumen* (i.e., NH_3 secretion) is favored because the secreted NH_3 immediately reacts with previously secreted H^+ to form NH_4^+. The concentration of NH_3 in the tubular fluid therefore remains extremely low, maintaining a large concentration gradient for the further secretion of NH_3 into the lumen. Because of its polar character, the NH_4^+ stays in the tubular fluid and is excreted. The passive secretion of the *nonionized* NH_3 into the lumen with subsequent "trapping" of the *ionized* NH_4^+ in the tubular fluid represents an example of *nonionic diffusion* or *diffusion trapping*, a phenomenon that also is important in the renal handling of weak organic acids and bases (Chapter 29).

It is important to note that because the maintenance of a large concentration gradient for the secretion of NH_3 into the lumen is dependent upon the reaction between NH_3 and H^+, the rate of NH_3 secretion into the lumen is to some extent proportional to the amount of H^+ that is secreted. Sufficient NH_3 is therefore secreted to react with virtually *all* of the secreted H^+ that cannot be buffered by the low capacity phosphate buffer system (see below), allowing the 50–100 mmol of excess H^+ that must be *secreted* per day to then be *excreted* with a minimum fall in tubular fluid pH. Furthermore, it should be noted that the formation of NH_3 from glutamine in the epithelial cells is pH-dependent, increasing in acidosis and decreasing in alkalosis. For example, although only 30–70 mmol of H^+ normally is excreted as NH_4^+ per day, over 300 mmol of H^+ can be excreted as NH_4^+ in *chronic acidosis*. This *adaptation* of NH_3 production, which takes 3–5 days to develop fully, may be due to increased glutaminase activity (*Eq. 106*) and/or facilitation of glutamine transport into the mitochondria, where the conversion to glutamic acid occurs. The ability of the kidneys to augment NH_3 production and NH_4^+ excretion in acidotic conditions is of major importance in the body's response to acid-base disturbances (see below) (Simpson, 1971).

PHOSPHATE

The phosphate buffer system, described previously (*Eq. 95*), also represents an important buffer for secreted H^+ ions. Specifically, secreted H^+ can react with $HPO_4^=$ in the tubular fluid to form $H_2PO_4^-$, which is then excreted (Fig. 4.44*B*). Because the pK_a for the phosphate buffer system is 6.8, the ratio of $[HPO_4^=]/[H_2PO_4^-]$ is 4:1 in the glomerular filtrate (pH 7.4), i.e., most of the buffer is in the form that can react with H^+. However, in the *proximal tubule*, as in plasma (see above), the effectiveness of the phosphate buffer system is limited by its low concentration. In fact, in spite of the reabsorption of water in the proximal tubule, the phosphate concentration at the end of the proximal tubule may be even *lower* than in plasma and the glomerular filtrate, since 75–85% of the filtered phosphate normally is reabsorbed in the proximal tubule (see Chapter 35), compared to approximately two-thirds of the filtered water. But in the *distal nephron*, the phosphate concentration in the tubular fluid can be significantly increased due to the reabsorption of water in excess of phosphate. Thus, phosphate represents an important tubular fluid buffer primarily in the *distal nephron*. In a maximally acidic urine (pH 4.5), the $[HPO_4^=]/[H_2PO_4^-]$ ratio is less than 1:100, i.e., almost *all* of the $HPO_4^=$ has been converted to $H_2PO_4^-$. But in spite of such maximum utilization of the phosphate buffer system, in normal man the amount of phosphate delivered to the distal nephron allows only 12–40 mmol of H^+ to be excreted as $H_2PO_4^-$ each day. Furthermore, the amount of available phosphate remains relatively *constant*, even in *acidotic* conditions, when the kidneys must excrete additional H^+ ions

(see below). As previously indicated, the *ammonia* buffer system lacks these limitations of the phosphate system and therefore has a more important role in the excretion of excess H^+.

It should be noted that the excreted $H_2PO_4^-$ represents the major component of the so-called urinary *titratable acids* (TA), defined as those weak acids in urine that can be titrated when a strong base such as NaOH is used to bring the pH of an acidic urine back up to the pH of the glomerular filtrate (normally, pH 7.4). In addition to $H_2PO_4^-$, the titratable acid fraction of urine includes small amounts of several other weak acids that can function as urinary buffers, such as uric acid (pK_a = 5.75), creatinine (pK_a = 4.97), β-OH butyric acid (pK_a = 4.8), and acetoacetic acid (pK_a = 4.8). The titratable acid fraction does *not* include NH_4^+: since the pK_a of the ammonia buffer system is 9.3, insignificant amounts of NH_4^+ will be titrated by bringing the urine pH up to 7.4. It should be noted that β-OH butyric acid and acetoacetic acid are normally present in such small amounts that their role as urinary buffers is *negligible*. However, in *diabetic ketoacidosis* these acids are excreted in the urine in increased quantities and therefore can assume greater importance as urinary buffers (Schiess et al., 1948). Diabetic ketoacidosis, then, represents one example of an acidosis in which *both* the titratable acid fraction and the ammonia buffer system exhibit *adaptation* to allow increased H^+ excretion.

Quantitation of Acid Excretion

As previously indicated, whenever a H^+ ion is excreted as either NH_4^+ or titratable acid such as $H_2PO_4^-$, a new HCO_3^- ion is generated for the body. Thus

$$\begin{matrix} \text{New } HCO_3^- \\ \text{generated} \end{matrix} = U_{NH_4}\dot{V} + U_{TA}\dot{V} \qquad (107)$$

However, since the kidneys may fail to recover all of the filtered HCO_3^-, a more important quantity is the *net* amount of new HCO_3^- generated for the body. Since every HCO_3^- that is not recovered will be excreted in the urine, the net amount of new HCO_3^- generated can be calculated as follows:

$$\begin{matrix} \text{Net new } HCO_3^- \\ \text{generated} \end{matrix} = U_{NH_4}\dot{V} + U_{TA}\dot{V} - U_{HCO_3}\dot{V}$$

But because the *addition* of a new HCO_3^- ion to the body has an effect equivalent to the *elimination* of a H^+ *ion* from the body, the net amount of HCO_3^- generated is more commonly termed the *net acid excretion*, i.e.,

$$\begin{matrix} \text{Net acid} \\ \text{excretion} \end{matrix} = U_{NH_4}\dot{V} + U_{TA}\dot{V} - U_{HCO_3}\dot{V} \qquad (108)$$

In normal man, HCO_3^- excretion is negligible, so that the net acid excretion is approximately equal to $U_{NH_4}\dot{V} + U_{TA}\dot{V}$, i.e., the amount of H^+ secreted in generating new HCO_3^- for the body. Thus, net acid excretion in normal man is approximately 50–100 mmol/day.

Control of Hydrogen Secretion

Since both the recovery of filtered HCO_3^- and the generation of new HCO_3^- are accomplished by *H^+ secretion*, the kidneys can regulate the pH of the body fluids just by controlling this one process. The most important determinant of H^+ secretion by the epithelial cells of both the proximal tubule and distal nephron is the *intracellular pH*, H^+ secretion increasing as the intracellular pH falls and decreasing as the intracellular pH rises. The intracellular pH, in turn, is primarily determined by the *arterial pH* and the *plasma K^+ concentration*.

Arterial pH

Since changes in arterial pH produce corresponding changes in intracellular pH, H^+ secretion increases in acidosis and decreases in alkalosis. This relationship between H^+ secretion and arterial pH is of major importance in the renal regulation of the pH of the body fluids. For example, the primary cause of an *acidosis* could be either a decrease in $[HCO_3^-]_p$ or an increase in arterial P_{CO_2}. In both cases, the increased H^+ secretion caused by the acidosis generates additional HCO_3^- for the body, which in turn tends to return the arterial pH toward normal (see below). This *adaptation* of H^+ secretion to changes in arterial pH requires 4–5 days to develop fully. It should be noted that the intracellular pH, and hence H^+ secretion, is most *immediately* sensitive to arterial pH changes caused by changes in arterial P_{CO_2}. This is because CO_2 can cross cellular membranes much more readily than H^+ or HCO_3^-. In fact, it can be experimentally demonstrated that increases in arterial P_{CO_2} that occur while the arterial pH is kept *constant*, or even *increased*, by the simultaneous addition of HCO_3^- will still augment H^+ secretion.

Plasma K^+ Concentration

Intracellular pH is directly proportional to the plasma K^+ concentration. As a result, H^+ secretion increases in hypokalemia and decreases in hyperkalemia. The relationship between intracellular pH and plasma K^+ concentration derives from the fact that cells contain many large *anions*, such as proteins and organic phosphates (see Fig. 4.1), and therefore must contain a sufficient number of cat-

ions to maintain electrical neutrality. K^+, being the *major* intracellular cation, is primarily responsible for this maintenance of electrical neutrality (Fig. 4.1), but small changes in the role of the less abundant intracellular cations such as Na^+ or H^+ can be significant. Of particular relevance here is the fact that the roles of K^+ and H^+ are *reciprocally related*. For example, in *hypokalemia* K^+ leaves cells along its concentration gradient, and H^+ enters cells to maintain electrical neutrality. As a result, the intracellular pH falls, and H^+ secretion increases. Conversely, in *hyperkalemia* K^+ enters cells, and H^+ leaves to maintain electrical neutrality. The intracellular pH therefore rises, and H^+ secretion decreases.

It should be noted that because H^+ enters cells in *hypokalemia*, the *plasma* actually becomes slightly *alkalotic*, even though the *intracellular* pH *falls*. Conversely, in *hyperkalemia* the *plasma* becomes slightly *acidotic*, even though the *intracellular* pH *rises*. The relationship between plasma K^+ and arterial pH is discussed further in Chapter 34.

While intracellular pH is the most important determinant of H^+ secretion, another factor that can become significant in pathological states is the mass of *functioning renal tissue*. Patients with advanced renal disease frequently become acidotic due to a reduced capacity for H^+ secretion. H^+ secretion also can be significantly reduced by pharmacological agents that inhibit the enzyme *carbonic anhydrase*, which has an important role in the H^+ secretion process (Fig. 4.42). *Acetazolamide* represents the most important example of such an agent.

All of the preceding discussion of the control of H^+ secretion applies to *both* the proximal tubule and the distal nephron. However, certain mechanisms that affect H^+ secretion in just *one* of these regions also can be identified.

CONTROL OF H^+ SECRETION IN THE PROXIMAL TUBULE

Since H^+ secretion in the proximal tubule is primarily coupled to the passive movement of Na^+ from the tubular lumen into the epithelial cells, it should not be surprising that H^+ secretion in the proximal tubule can be affected by changes in proximal tubular *Na^+ reabsorption* (Malnic and Giebisch, 1979). Recall that Na^+ reabsorption in the proximal tubule can be influenced by changes in plasma volume, with Na^+ reabsorption decreasing in volume expansion (part of the so-called *third factor* effect) and increasing in volume contraction (Chapter 32). Given the coupling between Na^+ reabsorption and H^+ secretion, this means that H^+

secretion also decreases in plasma volume expansion and increases in plasma volume contraction. The increased H⁺ secretion in plasma volume contraction has an important effect on the renal response to certain acid-base disturbances (see below).

H⁺ secretion in the proximal tubule also can be affected by changes in the *filtered load of* HCO_3^- (Malnic and Giebisch, 1979). Because the "leaky" tight junctions of the proximal tubule can maintain a pH gradient of only 0.5 pH units (see above), only a small amount of the secreted H⁺ can remain in the tubular fluid as *free* H⁺ ion. Under normal conditions, this leakiness of the tight junctions does not affect proximal tubular H⁺ secretion, since virtually *all* of the secreted H⁺ reacts with buffers in the proximal tubular fluid, the most important of which is HCO_3^-. However, an abnormal decrease in the filtered load of HCO_3^- (e.g., due to a decrease in $[HCO_3^-]_p$) will decrease the amount of buffer available. Some of the H⁺ that is secreted into the tubular lumen will therefore leak back across the "leaky" tight junctions, thereby decreasing the *net* amount of H⁺ secreted.

H⁺ secretion in the proximal tubule is decreased somewhat by *parathormone* (PTH), but the physiological significance of this effect is unknown.

CONTROL OF H⁺ SECRETION IN THE DISTAL NEPHRON

Although H⁺ secretion in the distal nephron is an active process that does *not* appear to be coupled to Na⁺ reabsorption (see above), distal nephron H⁺ secretion, like proximal tubular H⁺ secretion, is markedly stimulated by increased Na⁺ reabsorption (Warnock and Rector, 1979). This is because a major factor affecting H⁺ secretion in the distal nephron is the magnitude of the *transepithelial potential difference*, with H⁺ secretion increasing as the lumen becomes more negative relative to the peritubular space (Ziegler et al., 1976). As indicated in Chapter 31, the transepithelial potential difference in the distal nephron is primarily determined by the amount of Na⁺ reabsorbed, with the lumen becoming more negative as Na⁺ reabsorption increases, e.g., due to aldosterone (Chapter 32) or increased Na⁺ delivery to the distal nephron (Chapter 31). Thus, distal nephron H⁺ secretion is markedly stimulated by aldosterone and increased Na⁺ delivery to the distal nephron. For example, the *aldosterone* secreted in response to a decrease in Na⁺ intake and plasma volume depletion will stimulate not only Na⁺ reabsorption but also H⁺ secretion (it should be noted, however, that aldosterone

may have a *direct* stimulatory effect on distal nephron H⁺ secretion (Ludens and Fanestil, 1972) in addition to this indirect stimulatory effect that occurs via stimulation of Na⁺ reabsorption). An important illustration of how *increased Na⁺ delivery* to the distal nephron can stimulate H⁺ secretion is provided by the *diuretic drugs*. Diuretics that inhibit Na⁺ reabsorption in the proximal tubule (*mannitol*) or in the thick ascending limb of the loop of Henle (*furosemide, ethacrynic acid*) will increase the delivery of Na⁺ to the distal nephron, resulting in increased Na⁺ reabsorption, an increased transepithelial potential difference, and increased H⁺ secretion. *Thiazide* diuretics, which inhibit Na⁺ reabsorption in the *very early* distal nephron, have a similar effect on H⁺ secretion, by increasing Na⁺ delivery to the collecting tubule/duct. Thus, all of these drugs increase H⁺ excretion, which contributes to the *metabolic alkalosis* seen with many diuretics (see below). In contrast, diuretics that impair Na⁺ reabsorption in the *distal nephron* (*spironolactone, triamterene, amiloride*) decrease the transepithelial potential difference in the distal nephron and therefore decrease H⁺ secretion.

Distal nephron H⁺ secretion also is increased if the tubular fluid delivered to the distal nephron contains a substantial concentration of anions such as sulfate and nitrate, to which the distal nephron epithelium has limited permeability. Since these so-called *impermeant anions* cannot accompany reabsorbed Na⁺ as readily as Cl⁻, the magnitude of the transepithelial potential difference generated by Na⁺ reabsorption increases, thereby promoting H⁺ secretion.

INTRODUCTION TO ACID-BASE DISTURBANCES

As previously indicated, the bicarbonate buffer system is the most important one in the body because the concentrations of its components can be independently regulated, P_{CO_2} by the lungs and $[HCO_3^-]_p$ by the kidneys. Thus, via the bicarbonate buffer system the lungs and kidneys not only have an important role in the day-to-day maintenance of a *normal* pH, but also can help to restore acid-base homeostasis when the pH becomes *abnormal*. The lungs and kidneys also can *cause* an abnormal pH by changing P_{CO_2} and $[HCO_3^-]_p$, respectively, as in certain pathological conditions (see below).

The *Henderson-Hasselbalch equation* for the bicarbonate buffer system provides a convenient starting point for the study of acid-base disturbances (*Eq. 101*). According to this equation, an *acidosis* can be produced by an increase in the arterial P_{CO_2} or a decrease in $[HCO_3^-]_p$, while an *alkalosis* can be produced by a decrease in the arterial P_{CO_2} or an increase in $[HCO_3^-]_p$. Since the arterial P_{CO_2} is regulated by the lungs, the acid-base disturbances resulting from a change in arterial P_{CO_2} are termed *respiratory disturbances*. In contrast, the acid-base disturbances resulting from a change in $[HCO_3^-]_p$ are termed *metabolic disturbances*, even though such changes in $[HCO_3^-]_p$ can be caused by abnormal renal or gastrointestinal function as well as by abnormal metabolic function (see below). The four so-called *primary acid-base disturbances*, then, are respiratory acidosis (increased arterial P_{CO_2}), respiratory alkalosis (decreased arterial P_{CO_2}), metabolic acidosis (decreased $[HCO_3^-]_p$), and metabolic alkalosis (increased $[HCO_3^-]_p$) (Table 4.23).

Several graphical methods have been developed to illustrate the changes in pH, P_{CO_2}, and $[HCO_3^-]_p$ that occur during the various acid-base disturbances. One of the most useful is the so-called *pH-bicarbonate diagram* (Fig. 4.45) (Davenport, 1974). Two types of curves are drawn on this graph:

1. *P_{CO_2} isobars*, which illustrate the relation between pH and $[HCO_3^-]_p$ at a *constant* P_{CO_2} (hence, the term *isobar*). Each isobar is obtained by selecting the P_{CO_2} and then using the Henderson-Hasselbalch equation to calculate how the pH changes when $[HCO_3^-]_p$ changes at that P_{CO_2}. Since the normal arterial P_{CO_2} is 40 mmHg and normal $[HCO_3^-]_p$ is 24 mmol/l, point *A* in Figure 4.45 represents the *normal arterial point*, with a pH of 7.4.

2. *Blood-buffer lines*, which illustrate the changes in pH and $[HCO_3^-]_p$ that occur when P_{CO_2} varies. Although pH and $[HCO_3^-]_p$ must change in such a way that the Henderson-Hasselbalch equation is satisfied, the Henderson-Hasselbalch equation cannot be used to calculate these changes since only one of the three variables in the equation (P_{CO_2}) is known. Thus, blood-buffer lines must be *experimentally determined*. The term *blood-buffer line* is used because the slope of the line can be regarded as an index of the buffering capacity of blood.† Consider, for example, two blood samples with blood-buffer lines X and Y, as shown in Figure 4.46. If P_{CO_2} increases from 40 mmHg to 80 mmHg, the pH falls 0.25

---

† It should be noted that the slope of the line measured on whole blood in vitro usually differs somewhat from that measured in vivo because of the buffering action of the interstitial fluid and other body tissues.

Table 4.23
Primary acid-base disturbances

Disturbance	Acute			Chronic		
	pH	P_{CO_2}	$[HCO_3^-]_p$	pH	P_{CO_2}	$[HCO_3^-]_p$
Respiratory acidosis	↓↓	↑↑[a]	↑	↓	↑↑	↑↑
Respiratory alkalosis	↑↑	↓↓[a]	↓	↑	↓↓	↓↓
Metabolic acidosis	↓↓	N[b]	↓↓[a]	↓	↓	↓↓↓
Metabolic alkalosis	↑↑	N[b]	↑↑[a]	↑	↑	↑↑↑

[a] Primary abnormality. [b] N, no significant change in hypothetical acute state before compensation has occurred (see text).

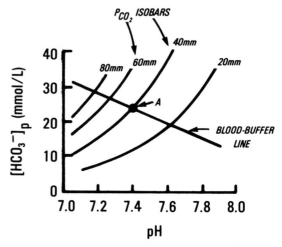

Figure 4.45. The pH-bicarbonate diagram. The P_{CO_2} isobars illustrate the relationship between pH and $[HCO_3^-]_p$ at the indicated P_{CO_2} values. The *blood-buffer line* illustrates the changes in pH and $[HCO_3^-]_p$ of normal blood that occur when P_{CO_2} is varied. *Point A* represents the *normal arterial point* (P_{CO_2} = 40 mmHg, $[HCO_3^-]_p$ = 24 mmol/l, pH = 7.4).

units in blood sample Y (to pH 7.15) compared with a fall of only 0.2 units in blood sample X (to pH 7.2). Thus, sample X is a *better buffer*. The primary determinant of the buffering capacity of the blood is the concentration of *hemoglobin*. In fact, line X represents the buffer line of blood with a *normal* hemoglobin concentration of 15 g/dl, while line Y represents the buffer line of blood with a hemoglobin concentration of only 5 g/dl. For comparison, line Z represents the buffer line of *plasma*. It should be noted that on all three buffer lines, $[HCO_3^-]_p$ increases slightly as P_{CO_2} increases because some additional H_2CO_3 is formed, which then dissociates. Of course, the pH falls in spite of this increase in $[HCO_3^-]_p$ because the *ratio* $[HCO_3^-]_p/P_{CO_2}$ falls.

In the following brief discussions of the four primary acid-base disturbances, the pH-bicarbonate diagram will be used to illustrate the changes in

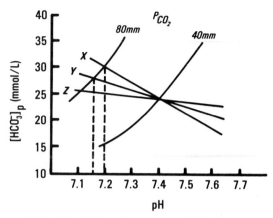

Figure 4.46. Blood-buffer lines as an index of the buffering capacity of blood. In blood with a normal hemoglobin concentration of 15 g/dl (X), the pH falls about 0.2 units (to 7.2) when P_{CO_2} increases from 40 to 80 mmHg. In contrast, blood with a hemoglobin concentration of 5 g/dl (Y) exhibits a 0.25 unit decrease in pH (to 7.15) when P_{CO_2} increases from 40 to 80 mmHg. The buffer line of plasma also is shown (Z). [Adapted from Davenport: *The ABC of Acid-Base Chemistry*, 6 ed. Chicago: University of Chicago Press, 1974. © 1947, 1949, 1950, 1958, 1969, 1974 by the University of Chicago.]

pH, P_{CO_2}, and $[HCO_3^-]_p$ that occur in each disturbance (Narins and Emmett, 1980; Masoro, 1982). It is necessary to consider what happens to pH, P_{CO_2}, and $[HCO_3^-]_p$ not only as a result of the primary disturbance itself, but also as a result of the secondary or *compensatory* responses that occur when the body attempts to restore the pH to normal. Thus, each acid-base disturbance is characterized by an *acute* or "uncompensated" phase followed by a *chronic* or "compensated" phase. It should be noted that in spite of the widespread use of the terms "uncompensated" and "compensated," the terms *acute* and *chronic* are preferable because the compensatory responses do *not* fully restore the pH to normal.

Respiratory Acidosis

The primary abnormality in respiratory acidosis is an increase in arterial P_{CO_2}. Given the important inverse relationship between arterial P_{CO_2} and alveolar ventilation (*Eq. 103*), a major cause of respiratory acidosis is *hypoventilation*. For example, barbiturates and other drugs that depress ventilation are a common cause of respiratory acidosis, particularly in the emergency setting. Respiratory acidosis also can be caused by lung diseases that impair gas exchange.

The changes in pH and $[HCO_3^-]_p$ that result from an *acute* increase in arterial P_{CO_2} can be shown readily on the pH-bicarbonate diagram, since the blood-buffer line illustrates how pH and $[HCO_3^-]_p$ change when the arterial P_{CO_2} changes (see above). For example, Figure 4.47 illustrates how pH and

$[HCO_3^-]_p$ change when the arterial P_{CO_2} increases *abruptly* from its normal value of 40 mmHg (*point A*) to 60 mmHg (*point B*). Or more generally, the *entire* blood-buffer line to the *left* of the normal arterial point depicts the pH and $[HCO_3^-]_p$ changes that occur in different cases of *acute respiratory acidosis*.

If the arterial P_{CO_2} *remains* elevated (i.e., if the cause of the respiratory acidosis *persists*), the body attempts to restore the pH to normal. This *compensatory response* is accomplished by the *kidneys*. Recall that the most important determinant of H⁺ secretion by the proximal tubule and distal nephron is the *intracellular pH*. With increased arterial P_{CO_2}, the intracellular pH falls and H⁺ secretion increases. The increased H⁺ secretion, in turn, generates additional HCO_3^- for the body, i.e., $[HCO_3^-]_p$ increases significantly. Thus, although the arterial P_{CO_2} remains elevated, the $[HCO_3^-]_p/P_{CO_2}$ ratio, and hence pH, increases toward normal. An example of such a compensatory response is depicted in Figure 4.47, which illustrates the renal compensation to the acute respiratory acidosis represented by *point B*. If the arterial P_{CO_2} remains elevated at 60 mmHg, the generation of new HCO_3^- by the kidneys causes $[HCO_3^-]_p$ to increase along the isobar for $P_{CO_2} = 60$ mmHg, and the pH increases toward normal (*point C*). *Point C*, then, represents a state of *chronic respiratory acidosis*. Note that the compensatory response is *not* complete, i.e., a mild acidosis remains.

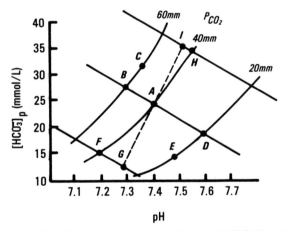

Figure 4.47. Changes in arterial pH, P_{CO_2}, and $[HCO_3^-]_p$ in the four primary acid-base disturbances and in the compensatory responses to these disturbances. *A*, Normal arterial point. *B*, Acute respiratory acidosis. *C*, Chronic respiratory acidosis. *D*, Acute respiratory alkalosis. *E*, Chronic respiratory alkalosis. *F*, Hypothetical acute metabolic acidosis. *G*, Chronic metabolic acidosis. *H*, Hypothetical acute metabolic alkalosis. *I*, Chronic metabolic alkalosis. Since the metabolic disturbances seldom develop acutely, the states of chronic metabolic acidosis and chronic metabolic alkalosis shown here are most likely to develop along the *dashed lines* connecting *A* to *G* and *A* to *I*, respectively. Details are provided in the text.

Respiratory Alkalosis

In respiratory alkalosis, the primary abnormality is a decrease in arterial P_{CO_2}. Virtually all cases of respiratory alkalosis result from *hyperventilation*. Thus, major causes of respiratory alkalosis include hypoxic conditions such as high altitude, certain CNS disorders, and psychological disturbances such as anxiety.

The changes in pH and $[HCO_3^-]_p$ that result from an *acute* decrease in arterial P_{CO_2} are depicted by the blood-buffer line to the *right* of the normal arterial point. For example, Figure 4.47 illustrates how pH and $[HCO_3^-]_p$ change when the arterial P_{CO_2} falls *abruptly* from its normal value of 40 mmHg (*point A*) to 20 mmHg (*point D*). If the arterial P_{CO_2} *remains* low, the body attempts to restore the pH to normal. As in respiratory acidosis, the compensatory response is accomplished by the *kidneys*. Specifically, the low arterial P_{CO_2} causes the intracellular pH to rise, thereby diminishing H^+ secretion. As a result, the kidneys not only fail to generate any new HCO_3^- to replace the HCO_3^- lost in titrating the strong acids produced by the body, but also fail to reabsorb all of the filtered HCO_3^-. $[HCO_3^-]_p$ can therefore fall significantly. Thus, although the arterial P_{CO_2} remains low, the $[HCO_3^-]_p/P_{CO_2}$ ratio, and hence pH, decreases toward normal. Figure 4.47 illustrates the renal compensation to the acute respiratory alkalosis represented by *point D*. If the arterial P_{CO_2} remains low at 20 mmHg, the failure of the kidneys to conserve HCO_3^- causes $[HCO_3^-]_p$ to fall along the isobar for $P_{CO_2} = 20$ mmHg, and the pH decreases toward normal (*point E*). *Point E*, then, represents a state of *chronic respiratory alkalosis*. As in respiratory acidosis, the compensatory response is *not* complete, i.e., a mild alkalosis remains.

Metabolic Acidosis

In metabolic acidosis, the primary abnormality is a decrease in $[HCO_3^-]_p$. Most cases of metabolic acidosis result from the abnormal accumulation of organic acids. Thus, the common causes of metabolic acidosis include hypovolemic and other forms of circulatory shock, in which lactic acid may accumulate (*lactic acidosis*), and uncontrolled diabetes mellitus, in which acetoacetic acid and β-OH butyric acid may accumulate (*diabetic ketoacidosis*) (see above). Metabolic acidosis also can occur in severe renal failure, since the kidneys are unable to secrete sufficient H^+ to generate new HCO_3^- to replace the HCO_3^- lost in titrating strong acids produced by the body or even to reabsorb all of the filtered HCO_3^-.

Although metabolic acidosis seldom develops *acutely*, it is instructive to consider a hypothetical case of an acute decrease in $[HCO_3^-]_p$. Figure 4.47 illustrates the fall in pH that would occur if $[HCO_3^-]_p$ dropped *abruptly* from its normal value of 24 mmol/l (*point A*) to 15 mmol/l (*point F*). Note that $[HCO_3^-]_p$ decreases along the isobar for $P_{CO_2} = 40$ mmHg, since respiratory function remains unchanged in this hypothetical acute state. The compensatory response to the decrease in pH is accomplished by the *lungs*. Specifically, the decreased arterial pH stimulates chemoreceptors in the carotid bodies, which in turn initiate a reflex stimulation of alveolar ventilation (Chapter 39). Because of the inverse relationship between alveolar ventilation and arterial P_{CO_2} (*Eq. 103*), this increase in alveolar ventilation results in a decrease in arterial P_{CO_2}. Thus, although $[HCO_3^-]_p$ remains low (and in fact becomes even *lower* as P_{CO_2} decreases), the $[HCO_3^-]_p/P_{CO_2}$ ratio, and hence arterial pH, increases toward normal. Figure 4.47 illustrates the respiratory compensation to the hypothetical acute metabolic acidosis represented by *point F*. The increased alveolar ventilation causes the arterial P_{CO_2} to fall, and the pH increases toward normal (*point G*). *Point G*, then, represents a state of *chronic metabolic acidosis*. Note that between *point F* and *point G* the arterial P_{CO_2} decreases along a blood-buffer line that *differs* from the buffer line of normal arterial blood, because the buffering capacity of arterial blood is *changed* in metabolic acidosis due to the decreased $[HCO_3^-]_p$. As in the respiratory acid-base disturbances, the compensatory response to a metabolic acidosis is *not* complete, i.e., a state of mild acidosis remains. The incompleteness of the compensatory response can be attributed to the decrease in arterial P_{CO_2} that occurs when alveolar ventilation is reflexly stimulated by the low arterial pH. This decrease in arterial P_{CO_2} is sensed by chemoreceptors in the medulla oblongata, carotid bodies, and aortic bodies, thereby attenuating the reflex stimulation of alveolar ventilation.

As indicated above, metabolic acidosis seldom develops acutely. Thus, respiratory compensation begins as soon as $[HCO_3^-]_p$ begins to fall. The state of chronic metabolic acidosis is therefore most likely to develop along the *dashed line* connecting *points A* and *G* in Figure 4.47, instead of from *point A* to *point F* to *point G*.

When metabolic acidosis is caused by the accumulation of organic acids (or other *nonrenal* factors), it should be noted that the *kidneys* as well as the lungs can have a role in the body's compensatory response. The mechanism for this renal correction of arterial pH is *not* increased H⁺ secretion secondary to the reduced arterial, and hence intracellular, pH. In fact, H⁺ secretion in metabolic acidosis generally does *not* increase significantly, or may actually *decrease*, because of the reduction in arterial P_{CO_2}. However, because of the low $[HCO_3^-]_p$ and consequent reduction in the filtered load of HCO_3^-, even a *diminished* rate of H⁺ secretion is sufficient not only to reabsorb all of the filtered HCO_3^- but also to generate additional HCO_3^- to replace the HCO_3^- lost in titrating the organic acids. The resulting increase in $[HCO_3^-]_p$ helps to increase the arterial pH toward normal.

Metabolic Alkalosis

In metabolic alkalosis, the primary abnormality is an increase in $[HCO_3^-]_p$. Most cases of metabolic alkalosis result from the loss of fluid from the body that contains little or no HCO_3^-: the body's store of HCO_3^- is therefore contained in a smaller volume, and $[HCO_3^-]_p$ increases. Thus, common causes of metabolic alkalosis include (1) vomiting and nasogastric suction, in which H⁺-rich gastric fluid is lost; and (2) diuretic drugs that result in the excretion of a large volume of acidic urine, such as the loop diuretics and thiazides (see above).

Like metabolic acidosis, metabolic alkalosis seldom develops *acutely*. However, it is still instructive to consider the hypothetical case of an acute increase in $[HCO_3^-]_p$. Figure 4.47 illustrates the increase in pH that would occur if $[HCO_3^-]_p$ increased *abruptly* from its normal value of 24 mmol/l (*point A*) to 35 mmol/l (*point H*); the increase occurs along the isobar for P_{CO_2} = 40 mmHg, since respiratory function remains unchanged in this hypothetical acute state. A compensatory response to the increased pH can be accomplished by the lungs, since the increased arterial pH is sensed by chemoreceptors in the carotid bodies, which in turn initiate a reflex decrease in alveolar ventilation. However, the magnitude of the respiratory compensation normally is quite *small*, since the response of the carotid body chemoreceptors to an *increase* in arterial pH is much less than their response to a *fall* in arterial pH, as occurs in metabolic acidosis. But even with a small increase in arterial P_{CO_2}, the $[HCO_3^-]_p/P_{CO_2}$ ratio, and hence pH, will decrease toward normal. Figure 4.47 illustrates a small respiratory compensation to the hypothetical acute metabolic alkalosis represented by *point H*. The decrease in alveolar ventilation causes the arterial P_{CO_2} to increase slightly and the pH to decrease toward normal (*point I*). *Point I*, then, represents a state of *chronic metabolic alkalosis*. As in metabolic acidosis, between *points H* and *I* the arterial

P_{CO_2} increases along a blood-buffer line that differs from the buffer line of normal arterial blood. Furthermore, as in metabolic acidosis any respiratory compensation would begin as soon as $[HCO_3^-]_p$ begins to change. The state of chronic metabolic alkalosis is therefore most likely to develop along the *dashed line* between *points A* and *I* in Figure 4.47, instead of from *point A* to *point H* to *point I*.

Given the limited *respiratory* response to metabolic alkalosis, the role of the *kidneys* in correcting the arterial pH must be explored. A possible mechanism for such a renal response is completely analogous to the mechanism for the renal correction of metabolic acidosis (see above). To summarize, H⁺ secretion may be normal or slightly elevated (due to the increased arterial P_{CO_2}), but because of the increased filtered load of HCO_3^- even a somewhat *increased* rate of H⁺ secretion is *insufficient* to recover all of the filtered HCO_3^-. HCO_3^- excretion therefore increases, $[HCO_3^-]_p$ falls, and the arterial pH decreases toward normal. However, it should be noted that this mechanism for the renal correction of metabolic alkalosis is of importance in a *limited* number of cases. This is because the most common causes of metabolic alkalosis involve a fluid loss (see above), i.e., most cases of metabolic alkalosis are accompanied by a *plasma volume contraction*. Recall that H⁺ secretion in both the proximal tubule and distal nephron is related to the amount of Na⁺ reabsorbed. To review, in the *proximal tubule* H⁺ secretion is directly coupled to Na⁺ reabsorption, while in the *distal nephron* H⁺ secretion is regulated in part by the transepithelial potential difference, which in turn is primarily determined by the amount of Na⁺ reabsorbed. Since plasma volume contraction is a potent stimulus to Na⁺ reabsorption in *both* the proximal tubule (via negation of the third factor effect) and distal nephron (via aldosterone), this means that plasma volume contraction causes a significant increase in H⁺ secretion as well. In fact, H⁺ secretion can increase so much that not only is all of the filtered HCO_3^- reabsorbed, but additional HCO_3^- is generated for the body, and $[HCO_3^-]_p$ actually may *increase*. Thus, when metabolic alkalosis is accompanied by plasma volume contraction, the renal response serves to *perpetuate* the alkalosis rather than to correct the arterial pH. In such cases of metabolic alkalosis, the kidneys will respond to the alkalosis by excreting HCO_3^- *only* after the plasma volume deficiency has been corrected by administering suitable fluids.

The compensatory responses that occur in each of the four primary acid-base disturbances are summarized in Table 4.23.

CHAPTER 34

Potassium Balance and the Regulation of Potassium Excretion

K^+ has an important role in the excitability of nerve and muscle, cell metabolism, and other physiological processes. Although K^+ is primarily an *intracellular* ion, as indicated in Chapter 26, the discussion of K^+ balance in this chapter will focus on the maintenance of a normal K^+ concentration in *extracellular* fluid (ECF), i.e., a normal *plasma* K^+ concentration (P_K). This is because the value of P_K is so low (average normal $P_K \simeq 4$ mmol/l) that even a small change in P_K can have significant adverse effects, particularly on the transmembrane potential of cardiac and skeletal muscle cells. For example, if an amount of K^+ equivalent to only 1% of the total body K^+ were added to the extracellular fluid, P_K would nearly double, and cardiac and skeletal muscle cells would be depolarized by 15 mV or more. Such depolarization is particularly serious in *cardiac* cells, leading to abnormal impulse conduction and, at a P_K above approximately 7 mmol/l, the possibility of life-threatening or even fatal arrhythmias (Chapter 9). Conversely, a decrease in P_K would hyperpolarize cells. Such hyperpolarization is particularly serious in *skeletal muscle*, leading to muscle weakness and, at a P_K below about 1 mmol/l, to paralysis.

The maintenance of a normal P_K requires, of course, that the input of K^+ equal the output of K^+. However, P_K also can be affected by the *distribution* of K^+ between the ECF and intracellular fluid (ICF). This chapter will therefore begin by examining the factors that affect the distribution of K^+ between ECF and ICF (Bia and DeFronzo, 1981). Then, the inputs and outputs of K^+ and their regulation will be considered.

DISTRIBUTION OF POTASSIUM

The high concentration of K^+ in ICF is generated by the Na^+-K^+ ATPase, which is present in the plasma membrane of all cells, including red and white blood cells, and platelets. Since resting cells are highly permeable to K^+, under normal conditions a steady state develops in which the quantity of K^+ *pumped into* cells by the Na^+-K^+ ATPase is equal to that passively *diffusing out* of cells. A change in P_K can therefore result from changes in either the active uptake of K^+ by cells or the passive diffusion of K^+ from cells, or both.

Changes in Active K^+ Uptake

Active K^+ uptake is accelerated by *insulin* (Andres et al., 1962), probably via stimulation of the Na^+-K^+ ATPase. In fact, insulin can be used clinically to lower P_K (glucose must be administered simultaneously, however, to prevent hypoglycemia). Uptake of K^+ by some cells is also increased by *β_2-adrenergic agonists* (Rosa et al., 1980). The mechanism for this effect is not certain, but like the insulin effect may involve stimulation of the Na^+-K^+ ATPase. The importance of the Na^+-K^+ ATPase in maintaining a normal P_K is further demonstrated by the finding that severe hyperkalemia may develop in patients who ingest large quantities of *digitalis* (Elkins and Watanabe, 1978), which inhibits the Na^+-K^+ ATPase.

Changes in Passive K^+ Diffusion

The passive diffusion of K^+ out of cells may be accelerated as a result of *cellular injury* or *death*. For example, tissues damaged as a result of severe burns or trauma can release sufficient K^+ to produce hyperkalemia. Rupture of red blood cells (*hemolysis*) also will increase P_K. If hemolysis occurs *after* a blood sample has been withdrawn from a patient with a *normal* P_K, the *measured* P_K will be factitiously elevated, i.e., P_K is normal but the in vitro hemolysis produces an abnormal laboratory value for P_K. The measured value for P_K also may be factitiously elevated due to in vitro rupture of platelets or white blood cells but only in patients with extremely high platelet counts (*thrombocytosis*) or extremely high white blood cell counts (*leukocytosis*, as in leukemia), respectively. Failure to recognize these potential factitious causes of hyperkalemia could lead to inappropriate treatment of a patient whose P_K actually is normal.

The passive diffusion of K^+ out of cells can also be influenced by the *arterial pH* (Leibman and Edelman, 1959), increasing as the arterial pH falls. In fact, in some types of metabolic acidosis every 0.1 unit decrease in arterial pH causes sufficient K^+ efflux to increase P_K by an average of 0.5–1.0 mmol/l, i.e., even a moderate acidosis can produce hyperkalemia. This relationship between arterial pH and P_K can be primarily attributed to the reciprocal relation between the roles of K^+ and H^+ in maintaining electrical neutrality inside cells (Adler and Fraley, 1977), introduced in Chapter 33. Recall that cells contain many large *anions* such as proteins and organic phosphates and therefore must contain a sufficient number of *cations* to maintain electrical neutrality. While K^+, being the *major* intracellular cation, is primarily responsible for this maintenance of electrical neutrality, the reciprocal relation between the roles of K^+ and H^+ can have important consequences. For example, in *acidosis* H^+ enters cells and protonates negatively charged groups on the large anions. K^+ therefore leaves cells to maintain electrical neutrality, augmenting P_K. Conversely, in *alkalosis* H^+ leaves cells, and K^+ enters to maintain electrical neutrality, lowering P_K. In fact, $NaHCO_3$ can be administered clinically to lower P_K in hyperkalemia (it should be noted, however, that an increase in P_{HCO_3} may *by itself* cause K^+ to enter cells, *independent* of any effect on arterial pH or protonation of intracellular anions) (Fraley and Adler, 1977).

INPUT OF POTASSIUM

The input of K^+, like the input of Na^+ (see Chapter 32), depends entirely upon the K^+ content of food and water consumed (Table 4.24). The average diet of adult Americans contains 50–100 mmol of K^+ per day, but the K^+ input can increase to over 500 mmol/day in persons whose diets contain large quantities of K^+-rich fruits and vegetables. At the other extreme, dietary K^+ input may be limited by starvation or disease (e.g., anorexia nervosa) to less than 10 mmol/day. Like Na^+ input, K^+ input does not appear to be physiologically regulated in man.

OUTPUT OF POTASSIUM

K^+, like Na^+ (see Chapter 32), can be lost from the body via three routes: sweat, feces, and urine

Table 4.24
Major inputs and outputs of potassium[a]

Inputs	Average Amount per Day, mmol/day	Outputs	Average Amount per Day, mmol/day
Food and water	50–100	Sweat	—[b]
		Feces	5–10
		Urine	45–90
Total	50–100	Total	50–100

[a] Inputs and outputs represent average basal values for an adult ingesting a conventional American diet in a cool environment. [b] Negligible in basal state, but can become appreciable under certain conditions (see text).

(Table 4.24). The average concentration of K^+ in *sweat* is similar to, or just slightly greater than, that in plasma. Thus, except in extreme conditions (e.g., an individual exercising in a hot environment), the amount of K^+ lost in sweat is negligible. *Fecal loss* of K^+ averages only 5–10 mmol/day but may exceed 100 mmol/day in diarrheal diseases such as cholera. Neither sweat nor fecal loss is an important variable in the physiological regulation of K^+ output. Like Na^+ output, then, K^+ output is regulated primarily by changes in the amount of K^+ excreted in *urine*. Urinary K^+ excretion averages 45–90 mmol/day in adult Americans, thereby maintaining K^+ balance (Table 4.24), but it can vary from less than 10 mmol/day to over 500 mmol/day.

REGULATION OF POTASSIUM BALANCE

With no physiological regulation of K^+ input, K^+ balance must be achieved by changing K^+ output to match K^+ input. As indicated above, K^+ output is regulated primarily by changing the amount of K^+ excreted in urine. The handling of K^+ in the various nephron segments has been noted briefly in previous chapters. To review, K^+ is reabsorbed in parts 1 and 2 of the proximal tubule, secreted in part 3 of the proximal tubule and/or the thin descending limb of the loop of Henle, reabsorbed in the thin and thick ascending limbs of the loop of Henle, and both reabsorbed and secreted in the distal nephron (Jamison et al., 1982). Thus, as indicated in Chapter 29, K^+ is one of the few substances that is reabsorbed *and* secreted.

The reabsorption of K^+ in parts 1 and 2 of the proximal tubule occurs in approximate proportion to the volume of water reabsorbed (LeGrimellec, 1975). Although the mechanism for this K^+ reabsorption is poorly understood, it is important to note that proximal tubular K^+ reabsorption does *not* vary with changes in K^+ balance, i.e., the quantity of K^+ reabsorbed in parts 1 and 2 of the proximal tubule is *independent* of K^+ input and output. The secretion of K^+ in part 3 of the proximal tubule and/or the thin descending limb of the loop of Henle is substantial (Jamison et al., 1976). In fact, micropuncture samples of the tubular fluid at the hairpin turn of juxtamedullary nephrons indicate that the quantity of K^+ reaching the hairpin turn can *exceed* the filtered load of K^+ (Battilana et al., 1978). K^+ secretion in part 3 and/or the thin descending limb increases when K^+ intake increases and therefore has a potential role in matching K^+ output to K^+ input. However, the reabsorption of K^+ in the thin and thick ascending limbs of the loop of Henle is so extensive that less than 10% of the filtered K^+ reaches the distal nephron *regardless* of K^+ input and output (Wright and Giebisch, 1978). It is likely that most of this reabsorption occurs in the *thick* ascending limb, where the lumen-positive transepithelial potential difference (generated by the Na^+-Cl^- coupled transport system; see Chapter 31) would favor K^+ reabsorption.

Since less than 10% of the filtered K^+ reaches the distal nephron *regardless* of K^+ input and output, the regulation of K^+ output to match K^+ input must occur almost exclusively in the *distal nephron,* which both reabsorbs and secretes K^+ (Malnic et al., 1964). Furthermore, K^+ *reabsorption* in the distal nephron appears to operate at a relatively *constant* rate, so that the regulation of K^+ output to match K^+ input occurs primarily by changing the rate of K^+ *secretion.* Under conditions of extremely low dietary K^+ intake, K^+ secretion is negligible

and *net K^+ reabsorption* occurs in the distal nephron. In contrast, with a high dietary K^+ intake, K^+ secretion increases markedly, and *net K^+ secretion* occurs. In adult Americans with an average dietary intake of 50–100 mmol/day, 45–90 mmol of K^+ must be excreted per day (Table 4.24). Assuming a filtered load of 720 mmol of K^+ per day ($P_K = 4$ mmol/l, GFR = 180 l/day), about 72 mmol of K^+ (= 10% of 720 mmol) reaches the distal nephron per day, i.e., an average dietary intake could result in *either* net K^+ reabsorption *or* net K^+ secretion in the distal nephron.

To understand how distal nephron K^+ secretion changes to regulate K^+ output, it is important to consider the secretory mechanism (Giebisch and Stanton, 1979). As illustrated in Figure 4.48, distal nephron K^+ secretion involves two steps: (1) active uptake of K^+ from the peritubular fluid into the cell by the Na^+-K^+ ATPase; and (2) diffusion of K^+ from the cell across the luminal surface into the tubular fluid (some K^+ may be *actively* transported across the luminal surface by a K^+ pump, as indicated in Fig. 4.48). The K^+ secretion process can therefore be controlled in a variety of ways. For example, K^+ secretion can be altered by changes in K^+ *uptake* by the tubular cells (e.g., due to changes in the activity of the Na^+-K^+ ATPase or in the number of Na^+-K^+ ATPase molecules), the *permeability* of the luminal membrane to K^+, the *intracellular K^+ concentration* (which affects the magnitude of the *concentration* gradient for diffusion),

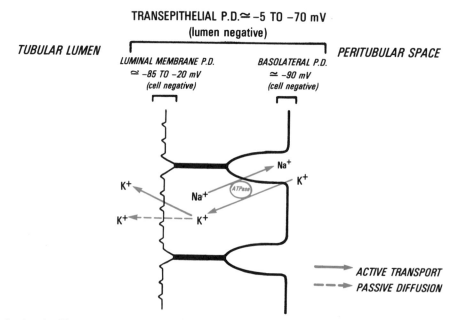

Figure 4.48.   Mechanism for K^+ secretion in the distal nephron. K^+ secretion is a two-step process, involving (1) active uptake of K^+ by the Na^+-K^+ ATPase; and (2) passive diffusion of K^+ across the luminal surface into the tubular fluid (some K^+ also may be *actively* transported across the luminal surface). Note that K^+ *reabsorption* in the distal nephron, which occurs at a relatively constant rate, is not illustrated here.

or the *transepithelial potential difference* (which affects the magnitude of the *electrical* gradient for diffusion). It should be noted that the electrical gradient across the luminal surface always *opposes* diffusion into the tubular fluid, but with a more negative transepithelial potential difference this opposing effect is minimized. For example, with a transepithelial potential difference of -5 mV (lumen negative), an electrical gradient of 85 mV opposes K^+ diffusion (since the intracellular electrical potential is 85 mV negative with respect to the lumen; see Fig. 4.48). In contrast, with a transepithelial potential difference of -70 mV (lumen negative), the electrical gradient opposing K^+ diffusion into the tubular fluid is only 20 mV.

Variations in dietary K^+ input cause several changes in the K^+ secretion process (Silva et al., 1977), some of which are mediated by *aldosterone*. Consider, for example, the changes in distal nephron K^+ secretion that result from an *increase* in dietary K^+ intake (Fig. 4.49). The increased K^+ input causes a slight increase in P_K, which in turn stimulates K^+ secretion by increasing K^+ *uptake* by the tubular cells. With chronic increases in K^+ input, this increased uptake is due, at least in part, to an increased number of Na^+-K^+ ATPase molecules. The increase in P_K also represents one of the major physiological stimuli for the secretion of aldosterone, which *further* stimulates K^+ secretion by (1) increasing the activity of the Na^+-K^+ ATPase, thereby increasing K^+ *uptake*; (2) stimulating Na^+ reabsorption in the distal nephron, thereby increasing the *transepithelial potential difference* (see below); and (3) increasing the *permeability* of the luminal membrane to K^+. Note that all of these changes in K^+ secretion can be attributed, either directly or indirectly, to the initial increase in P_K (Fig. 4.49).

Distal nephron K^+ secretion also can be influenced by factors that are *unrelated* to changes in K^+ input, including distal nephron Na^+ reabsorption, the presence of impermeant anions in the tubular fluid, the rate of tubular fluid flow, and acid-base status (Giebisch and Stanton, 1979).

Na$^+$ Reabsorption in the Distal Nephron

As indicated above, the rate of K^+ secretion can be altered by changes in the *transepithelial potential difference* in the distal nephron. Since the transepithelial potential difference is primarily a function of the amount of Na^+ reabsorbed (Chapter 31), this means that K^+ secretion, like H^+ secretion, is markedly stimulated by increased Na^+ reabsorption in the distal nephron, e.g., due to aldosterone or increased Na^+ delivery to the distal nephron (Chap-

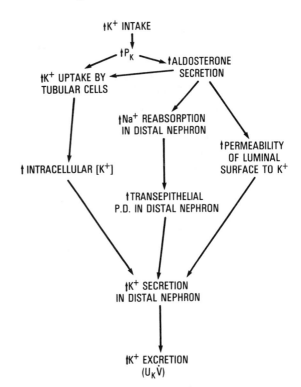

Figure 4.49. Pathway by which K^+ balance is restored following an increase in K^+ intake. P.D., potential difference.

ter 33). The fact that increased Na^+ reabsorption is responsible for part of the stimulatory effect of *aldosterone* on K^+ secretion has been noted above (Fig. 4.49). However, it should be emphasized that aldosterone can affect K^+ secretion not only in response to changes in K^+ intake (Fig. 4.49) but also in response to changes in *Na$^+$ intake* (Chapter 32). For example, the aldosterone secreted in response to a decrease in Na^+ intake and plasma volume depletion will stimulate K^+ secretion. An important illustration of how *increased Na$^+$ delivery* to the distal nephron can stimulate K^+ secretion is provided by the *diuretic drugs*. Diuretics that inhibit Na^+ reabsorption in the proximal tubule (*acetazolamide, mannitol*) or in the thick ascending limb of the loop of Henle (*furosemide, ethacrynic acid*) will increase the delivery of Na^+ to the distal nephron, resulting in increased Na^+ reabsorption, an increased transepithelial potential difference, and increased K^+ secretion. *Thiazide* diuretics, which inhibit Na^+ reabsorption in the *very early* distal nephron, have a similar effect on K^+ secretion, by increasing Na^+ delivery to the collecting tubule/duct. Thus, all of these drugs increase K^+ excretion and commonly cause *hypokalemia* as a side effect. In contrast, diuretics that impair Na^+ reabsorption in the *distal nephron* (*spironolactone, triamterene, amiloride*) decrease the transepithelial potential difference in the distal nephron and therefore decrease K$^+$ secretion. Consequently, such drugs decrease K$^+$ excretion and are referred to as *K$^+$-sparing diuretics*. Note that the effects of the diuretics on K$^+$ secretion are completely analogous to their effects on H$^+$ secretion (Chapter 33).

Impermeant Anions in the Tubular Fluid

K$^+$ secretion, like H$^+$ secretion, is increased if the tubular fluid delivered to the distal nephron contains a substantial concentration of anions such as sulfate and nitrate, which cannot accompany reabsorbed Na$^+$ as readily as Cl$^-$. In the presence of these *impermeant anions*, the magnitude of the transepithelial potential difference generated by Na$^+$ reabsorption is increased, thereby stimulating K$^+$ secretion (Chapter 33).

Rate of Tubular Fluid Flow

The concentration gradient for K$^+$ diffusion across the luminal surface is affected not only by the *intracellular K$^+$ concentration*, but also by the rate of *tubular fluid flow* through the distal nephron. For example, an *increased* rate of tubular fluid flow creates a more favorable concentration gradient by "washing out" the secreted K$^+$ (Good and Wright, 1979). In fact, the increased K$^+$ secretion seen with many diuretic drugs (see above) can be attributed in part to an increased rate of tubular fluid flow.

Acid-Base Status

Changes in acid-base balance can have significant effects on K$^+$ secretion (Gennari and Cohen, 1975). For example, an *acute alkalosis*, whether of respiratory or metabolic origin, stimulates K$^+$ secretion by causing K$^+$ to enter tubular cells, like other cells of the body (Fig. 4.50). As indicated above, this increased K$^+$ entry can be attributed to the reciprocal relationship between the roles of K$^+$ and H$^+$ in maintaining electrical neutrality inside cells. Acute alkalosis also stimulates K$^+$ secretion by increasing the *transepithelial potential difference* and the *permeability* of the luminal membrane to K$^+$, although somewhat indirectly (Fig. 4.50). By reducing H$^+$ secretion in the proximal tubule and distal nephron and thereby decreasing the reabsorption of HCO$_3^-$ from the tubular fluid (Chapter 33), acute alkalosis causes the tubular fluid in the distal nephron to have an elevated HCO$_3^-$ concentration and an elevated pH. The *excess HCO$_3^-$* functions as an *impermeant anion* and therefore stimulates K$^+$ secretion by increasing the transepithelial potential difference (see above). The *elevated pH* increases the K$^+$ permeability of the luminal membrane (Boudry et al., 1976), an effect similar to that of aldosterone. *Acute acidosis* causes a decrease in K$^+$ secretion by pathways opposite to those illustrated in Figure 4.50.

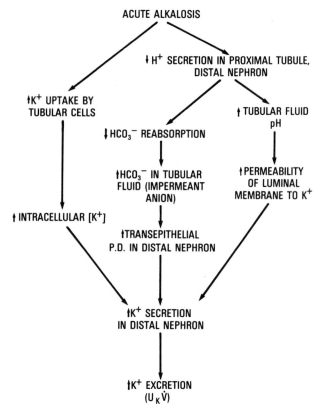

Figure 4.50. Pathway by which an acute alkalosis stimulates distal nephron K$^+$ secretion and, hence, K$^+$ excretion. P.D., potential difference.

Table 4.25
Major causes of increased K$^+$ secretion in distal nephron

↑ Dietary intake
Aldosterone
↑ Na$^+$ reabsorption in distal nephron
↑ Impermeant anions in tubular fluid delivered to distal nephron
↑ Tubular fluid flow
Acute alkalosis

The effects of *chronic* acid-base disturbances on K$^+$ secretion can be quite complex. For example, since alkalosis stimulates K$^+$ secretion and, hence, K$^+$ excretion, *chronic alkalosis* (days or weeks) may produce depletion of total body K$^+$ and, hence, a decrease in the concentration of K$^+$ in cells, including those of the distal nephron. Thus, chronic alkalosis may be accompanied by a *low* rate of K$^+$ secretion and K$^+$ excretion. Moreover, as K$^+$ is lost from cells, H$^+$ tends to enter to maintain electrical neutrality, which may result in the paradoxical situation of *extracellular alkalosis* and *intracellular acidosis*. Such an intracellular acidosis would stimulate H$^+$ secretion in the proximal tubule and distal nephron (Chapter 33), thereby *perpetuating* the extracellular alkalosis.

The major causes of increased distal nephron K$^+$ secretion are summarized in Table 4.25.

CHAPTER 35

Regulation of Calcium, Magnesium, and Phosphate Excretion

Ca^{++}, Mg^{++}, and phosphate are considered together in this chapter primarily because the same conditions and hormones tend to control the excretion of all three ions, a fact that is related to their role as the major constituents of bone mineral. The regulation of Ca^{++}, Mg^{++}, and phosphate balance differs from that of Na^+ and K^+ balance (Chapters 32 and 34) in several respects. First of all, the inputs of Na^+ and K^+ are equal to the amounts of these ions in food and water consumed and are not subject to physiological regulation. In contrast, the inputs of Ca^{++}, Mg^{++}, and phosphate do *not* equal the amounts of these ions in food and water consumed because (particularly in the case of Ca^{++}) gastrointestinal absorption is incomplete and is hormonally regulated. A second important difference between Na^+ and K^+ and the ions considered here is that Na^+ and K^+ balance are achieved primarily by adjusting urinary output to match dietary input. In contrast, Ca^{++}, Mg^{++}, and phosphate balance are determined not only by the inputs and outputs of these ions, but also by the factors that affect their deposition into *bone*. The importance of the bone deposits of Ca^{++}, Mg^{++}, and phosphate is emphasized by the large percentage of the total body stores of these ions that are found in bone (Table 4.26).

In contrast to Na^+ and K^+, then, the regulation of the body balance of Ca^{++}, Mg^{++}, and phosphate involves the *gastrointestinal tract* and *bone* as well as the kidneys. In this chapter, emphasis will be placed on the *renal component* of the regulation of the total body balance of these ions. The reader is referred to the gastrointestinal and endocrine sections of this text for further discussion of Ca^{++}, Mg^{++}, and phosphate balance.

CALCIUM

Although over 99% of the total body Ca^{++} is in bone (Table 4.26), the remaining Ca^{++}, which is primarily extracellular, has an important role in the excitation of nerve and muscle, muscle contraction, blood coagulation, and other physiological processes. The average normal concentration of Ca^{++} in plasma is 2.5 mmol/l (5 meq/l, 10 mg/dl). About 40–50% of the Ca^{++} in plasma is bound to plasma proteins; of the remaining 50–60% (1.25–1.5 mmol/l), about 90% is ionized and 10% is complexed with anions such as phosphate, sulfate, and citrate (Table 4.27) (Walser, 1973). It is the concentration of *ionized* Ca^{++} that affects physiological processes such as nerve and muscle function. However, at present clinical laboratories typically measure the *total* concentration of Ca^{++} in plasma rather than the concentration of *ionized* Ca^{++}. Accurate evaluation of P_{Ca} values reported by clinical laboratories therefore requires knowledge of the plasma protein concentration. For example, lowering of the plasma protein concentration by disease (e.g., cirrhosis of the liver) can lower the *total* concentration of Ca^{++} in plasma without necessarily altering the physiologically relevant concentration of *ionized* Ca^{++}.

Renal Regulation of Calcium Balance

The 40–50% of the plasma Ca^{++} that is protein bound cannot be filtered. Thus, at a GFR of 125 ml/min (180 l/day) and a total P_{Ca} of 2.5 mmol/l (i.e., a *filterable* P_{Ca} of 1.25–1.5 mmol/l), the filtered

Table 4.26
Distribution of calcium, magnesium, and phosphate

	Calcium, %	Magnesium, %	Phosphate,%
Bone	99	50–60	85
Nonbone			
Intracellular	<0.1	39–49	≃14.9
Extracellular	≃1	≃1	≃0.1

Adapted from Potts and Deftos (1974).

Table 4.27

Forms of calcium, magnesium, and phosphate in normal plasma

	Calcium	Magnesium	Phosphate	
Protein bound	40–50%	30%	Phospholipids Protein bound	} 67%
Free (filterable)	50–60%	70%	Acid-soluble (filterable)	33%
Ionized	≃90% of free	≃70% of free	Inorganic Ionized	≃90% of acid-soluble ≃45% of acid-soluble
Complexed	≃10% of free	≃30% of free	Complexed Organic	≃45% of acid-soluble ≃10% of acid-soluble
Concentration measured by clinical laboratories	*Total* plasma calcium	*Total* plasma magnesium	Acid-soluble phosphate (termed "plasma phosphate")	
Normal value	2.5 mmol/l (10 mg/dl)	0.75–1.25 mmol/l	1.0–1.6 mmol/l (3–5 mg phosphorus/dl)	

Includes data from Walser (1961a), Walser (1973), and Krane (1970).

load of Ca^{++} is approximately 225–270 mmol/day. In the normal adult, the maintenance of Ca^{++} balance involves the reabsorption of about 97–99% of the filtered Ca^{++}, i.e., about 2.5–7.5 mmol of Ca^{++} are excreted per day (Kesteloot and Geboers, 1982).

The renal handling of Ca^{++} is similar in many ways to the renal handling of Na^+ (Chapters 30–32). For example, about two-thirds of the filtered load of each ion is reabsorbed in the proximal tubule, 20–25% in the loop of Henle, and 8–10% in the distal nephron (Lassiter et al., 1963). The mechanisms for Ca^{++} reabsorption in the *proximal tubule* and *distal nephron* are not well understood (Suki, 1979). In the *loop of Henle*, the major portion of the Ca^{++} reabsorption occurs in the *thick ascending limb*, where the lumen-positive transepithelial potential difference generated by the Na^+-Cl^- coupled transport system (Chapter 31) favors Ca^{++} reabsorption (Boudreau and Burg, 1979). The similarities between the renal handling of Na^+ and Ca^{++} in the proximal tubule and loop of Henle are underscored by the finding that the two ions have virtually identical TF/P ratios in the early distal nephron, i.e., identical fractions of the filtered Na^+ and Ca^{++} are reabsorbed prior to the distal nephron (*Eq. 78*). Furthermore, in most cases a very good correlation exists between the *fractional excretions* of Na^+ and Ca^{++} (Walser, 1961b). For example, an increase in Na^+ intake will increase not only Na^+ excretion, but Ca^{++} excretion as well, an outcome that can be used clinically to treat hypercalcemia. The mechanism for this effect of Na^+ intake on Ca^{++} excretion probably includes the *third factor* effect in the proximal tubule, whereby *all* proximal tubular reabsorption is decreased following a large increase in Na^+ intake or another cause of plasma volume expansion (Chapter 32). Another important example of the correlation between Na^+ and Ca^{++} excretion is provided by the *loop diuretics*, which inhibit Na^+-Cl^- coupled transport in the thick as-

cending limb (Chapter 31). This inhibition increases not only Na^+ excretion but also Ca^{++} excretion by impairing the formation of the lumen-positive transepithelial potential difference and therefore inhibiting Ca^{++} reabsorption (see above). An *exception* to the general correlation between Na^+ and Ca^{++} excretion involves the *thiazide diuretics*, which increase the excretion of Na^+ but *decrease* Ca^{++} excretion, an effect used to treat some patients who form Ca^{++}-containing renal stones.

The regulation of Ca^{++} excretion to maintain Ca^{++} balance is accomplished primarily by *parathormone* (PTH), a polypeptide hormone secreted by the parathyroid glands in response to a decrease in the concentration of ionized Ca^{++} in plasma. PTH stimulates Ca^{++} reabsorption in the thick ascending limb (Suki et al., 1980) and distal nephron (Shareghi and Stoner, 1978), thereby decreasing Ca^{++} excretion and helping to restore plasma Ca^{++} to normal levels. Ca^{++} excretion also can be affected by factors not directly related to Ca^{++} balance (Massry and Coburn, 1973). For example, Ca^{++} excretion is increased by chronic adrenocorticosteroid therapy, chronic vitamin D excess, growth hormone, hypermagnesemia, and chronic metabolic acidosis as well as by increased Na^+ intake, plasma volume expansion, and the loop diuretics as noted above. Some of these factors may *not* have a *direct* effect on the renal handling of Ca^{++}. For example, an agent that stimulates Ca^{++} absorption from the gastrointestinal tract will elevate the concentration of ionized Ca^{++} in plasma and decrease the secretion of PTH, thereby increasing the renal excretion of Ca^{++}.

MAGNESIUM

About 50–60% of the total body Mg^{++} is in bone (Table 4.26). Most of the remaining Mg^{++} is found in ICF, where it represents an essential cofactor for enzymatic reactions and also has a role in nerve and muscle function (Wacker and Parisi, 1968). The normal concentration of Mg^{++} in plasma ranges from about 0.75–1.25 mmol/l (1.5–2.5 meq/l); why the normal plasma level of Mg^{++} varies by a greater percentage than the normal plasma levels of most other ions is unknown. About 30% of the Mg^{++} in plasma is bound to plasma proteins; of the remaining 70% (0.5–0.9 mmol/l), about 70% is ionized and 30% is complexed with the same anions (e.g., phosphate, sulfate, and citrate) that complex with Ca^{++} (Table 4.27) (Walser, 1973). As with Ca^{++}, clinical laboratories typically measure the *total* concentration of Mg^{++} in plasma, not just the concentration of ionized Mg^{++}.

Renal Regulation of Magnesium Balance

The 30% of the plasma Mg^{++} that is protein bound cannot be filtered. Thus, at a GFR of 125 ml/min (180 l/day) and a total P_{Mg} of 1 mmol/l (i.e., a *filterable* P_{Mg} of 0.7 mmol/l), the filtered load of Mg^{++} is approximately 125 mmol/day. In the normal adult, the maintenance of Mg^{++} balance involves the reabsorption of 90–99% of the filtered Mg^{++} (Quamme and Dirks, 1980), i.e., about 1–10 mmol of Mg^{++} are excreted per day.

The renal handling of Mg^{++} exhibits considerable variation from one species to another. While the details are not known (Quamme and Dirks, 1980), changes in Mg^{++} excretion generally parallel changes in Ca^{++} excretion (Massry and Coburn, 1973). For example, Mg^{++} excretion is decreased by *PTH*. However, the role of PTH in the regulation of Mg^{++} balance is uncertain. Even though PTH secretion is inhibited by an increase in the concentration of ionized Mg^{++} in plasma as well as by an increase in the concentration of ionized Ca^{++}, the parathyroid glands appear to be much less sensitive to changes in P_{Mg} than to changes in P_{Ca}.

It should be noted that although the renal excretion of Mg^{++} can be reduced to less than 1 mmol/day, it *cannot* be reduced to zero. As a result, individuals who ingest a low Mg^{++} diet (e.g., a chronic "diet" of ethanol) can become Mg^{++} depleted. In contrast, such a large quantity of Mg^{++} can be excreted by individuals with normal renal function that even the ingestion of excessive amounts of Mg^{++} (e.g., during chronic use of antacids) does not cause Mg^{++} overload.

PHOSPHATE

Approximately 85% of the total body phosphate is in bone (Table 4.26). Of the remainder, over 99% is intracellular and consists primarily of *organic phosphates* such as phospholipids, nucleic acids, nucleotides, phosphoproteins, and intermediates of carbohydrate metabolism. Emphasis in this chapter will be placed on the small amount of *inorganic phosphate* in the body fluids, which exists primarily as the anions $HPO_4^=$ and $H_2PO_4^-$. Inorganic phosphate is required for the synthesis of organic phosphates (e.g., ATP) and also is important as a body buffer (Chapter 33).

In plasma, about two-thirds of the total phosphate is in phospholipids or bound to plasma proteins. The remaining one-third is termed *acid-soluble phosphate* since it represents the phosphate that remains after plasma is treated with an acid such as trichloroacetic acid to precipitate plasma proteins and phospholipids. About 90% of the acid-soluble phosphate is inorganic phosphate, of which approximately half is ionized and half is complexed with ions such as Ca^{++} and Mg^{++} (see above); the remaining 10% represents organic phosphates such as ATP (Table 4.27). It is the *acid-soluble* phosphate, *not* the *total* phosphate, that is measured by clinical laboratories and that is referred to as *plasma phosphate*. Actually, clinical laboratories typically report plasma phosphate as *plasma phosphorus*. A plasma *phosphorus* concentration expressed in millimoles per liter is the same as a plasma *phosphate* concentration expressed in millimoles per liter, since one atom of phosphorus is found in each molecule of phosphate. However, plasma phosphorus concentrations often are expressed in milligrams per deciliter, in which case a plasma *phosphorus* concentration of 3.1 mg/dl is equal to a plasma *phosphate* concentration of 1 mmol/l, since each millimole of phosphate contains 31 mg of phosphorus. The normal plasma phosphate concentration (i.e., the *acid-soluble phosphate* concentration measured by clinical laboratories) ranges from about 1.0–1.6 mmol/l (3–5 mg phosphorus/dl). Like the plasma Mg^{++} concentration, then, the normal plasma phosphate varies by a greater percentage than the normal plasma levels of most other ions.

Renal Regulation of Phosphate Balance

Phospholipids and protein-bound phosphate cannot be filtered, i.e., only the *acid-soluble* phosphate is filtered. Thus, at a GFR of 125 ml/min (180 l/day) and a plasma phosphate of 1.3 mmol/l, the filtered load of phosphate is about 235 mmol/day (recall that what is termed *plasma phosphate* actually represents *acid-soluble phosphate*, i.e., filterable phosphate). In the normal adult, the maintenance of phosphate balance involves the reabsorption of 85–95% of the filtered phosphate, i.e., about 12–40 mmol of phosphate are excreted per day.

Normally, about 75–85% of the filtered phosphate is reabsorbed in the *proximal tubule* (Lang, 1980). Since the fractional excretion of phosphate is about 5–15%, it is generally assumed that about 10% of the filtered load is reabsorbed in the loop of Henle and/or distal nephron. However, it has also been proposed that a proximal tubular phosphate reabsorption of 75–85% is found only in *cortical nephrons* and that *juxtamedullary nephrons* reabsorb a much larger percentage of the filtered phosphate in their proximal tubules (Knox et al., 1977). By this theory, the combined effect of cortical nephrons and juxtamedullary nephrons could yield an overall fractional reabsorption of 85–95%.

Like glucose and amino acid reabsorption in the proximal tubule (Chapter 30), the reabsorption of phosphate in the proximal tubule is coupled to the reabsorption of Na^+ (*cotransport*) (Hoffmann et al., 1976). Furthermore, like glucose and amino acids, phosphate exhibits T_m-*limited reabsorption* (Chapter 29). Thus, variations in plasma phosphate can result in changes in phosphate excretion analogous to those illustrated in Figure 4.17 for glucose.

The regulation of phosphate excretion to maintain phosphate balance is accomplished primarily by *PTH*, which inhibits phosphate reabsorption in the proximal tubule such that 40% or more of the filtered phosphate can be excreted at high PTH levels and less than 5% in the absence of PTH (Knox et al., 1977). To understand how PTH can have a role in maintaining phosphate balance when the secretion of PTH is regulated primarily by the concentration of ionized Ca^{++} in plasma (see above), consider, for example, how phosphate balance is restored following an increase in plasma phosphate concentration (Fig. 4.51). Some of the excess phosphate will complex with ionized Ca^{++} in plasma, thereby lowering the plasma concentration of ionized Ca^{++}. The decrease in ionized Ca^{++}, in turn, stimulates the secretion of PTH, which then inhibits proximal tubular phosphate reabsorption, enabling the kidneys to excrete the excess phosphate. PTH also stimulates Ca^{++} reabsorption in the thick ascending limb and distal nephron, thereby enabling the plasma concentration of ionized Ca^{++} to return toward normal. Phosphate excretion also appears to be regulated by *dietary phosphate*, as subjects on an experimental low phosphate diet reabsorb nearly 100% of the filtered load of phosphate. The mediator(s) of this important homeostatic mechanism, which does *not* require PTH, is (are) unknown.

Phosphate excretion also can be affected by factors not directly related to phosphate balance (Lang, 1980). For example, phosphate excretion, like Ca^{++} excretion, is increased by *plasma volume expansion*, probably as a result of the *third factor* effect that depresses *all* proximal tubular reabsorp-

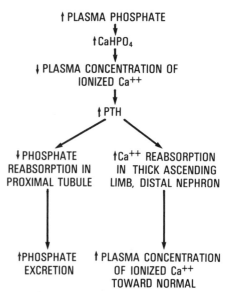

Figure 4.51.   Pathway by which phosphate balance is restored following an increase in plasma phosphate concentration.

tion in the presence of a plasma volume expansion. If the relationship between plasma phosphate and phosphate excretion is analyzed in a manner analogous to that illustrated in Figure 4.17, one can demonstrate that the *transport maximum* for phosphate ($T_{m_{PO_4}}$) is decreased by factors that inhibit phosphate reabsorption (e.g., PTH, plasma volume expansion) and increased by factors that increase phosphate reabsorption (e.g., ingestion of a low phosphate diet).

Actions of Parathormone

Since PTH can have effects on the renal handling of Ca^{++}, Mg^{++}, *and* phosphate, a brief description of its mechanism of action is appropriate. PTH acts via specific receptors on the basal-lateral surfaces of cells in the responsive segments of the nephron, including the proximal tubule, thick ascending limb, and distal nephron (Morel, 1981). The PTH-receptor complex activates adenylate cyclase, stimulating the production of cyclic AMP. Cyclic AMP, in turn, mediates the inhibition of phosphate reabsorption in the proximal tubule and the stimulation of Ca^{++} (and Mg^{++}) reabsorption in the thick ascending limb and distal nephron.

PTH also promotes an important metabolic reaction in the kidneys: the hydroxylation of 25-hydroxy vitamin D_3 to the more active 1,25-dihydroxy vitamin D_3 (*calcitriol*). Calcitriol stimulates the absorption of Ca^{++} and phosphate from the gastrointestinal tract and has an important role in regulating the total body balance of Ca^{++} and phosphate. In addition, calcitriol can decrease the renal excretion of Ca^{++} and phosphate, but the physiological significance of this effect is unknown.

BIBLIOGRAPHY

ADLER, S., AND D. S. FRALEY. Potassium and intracellular pH. *Kidney Int.* 11: 433–442, 1977.

AL-AWQATI, Q. H$^+$ transport in urinary epithelia. *Am. J. Physiol.* 235: F77–F88, 1978.

ALBRINK, M. J., P. M. HALD, E. B. MAN, AND J. P. PETERS. The displacement of serum water by the lipids of hyperlipemic serum. A new method for the rapid determination of serum water. *J. Clin. Invest.* 34: 1483–1488, 1955.

ANDREOLI, T. E., AND J. A. SCHAFER. Effective luminal hypotonicity: the driving force for isotonic proximal tubular fluid absorption. *Am. J. Physiol.* 236: F89–F96, 1979.

ANDREOLI, T. E., J. A. SCHAFER, S. L. TROUTMAN, AND M. L. WATKINS. Solvent drag component of Cl$^-$ flux in superficial proximal straight tubules: evidence for a paracellular component of isotonic fluid absorption. *Am. J. Physiol.* 237: F455–F462, 1979.

ANDRES, R., M. A. BALTZAN, G. CADER, AND K. L. ZIERLER. Effect of insulin on carbohydrate metabolism and on potassium in the forearm of man. *J. Clin. Invest.* 41: 108–114, 1962.

ARONSON, P. S., AND B. SACKTOR. The Na$^+$ gradient-dependent transport of D-glucose in renal brush border membranes. *J. Biol. Chem.* 250: 6032–6039, 1975.

ASSCHER, A. W., D. B. MOFFAT, AND E. SANDERS. *Nephrology Illustrated.* Philadelphia: W. B. Saunders, 1982.

AUGUST, J. T., D. H. NELSON, AND G. W. THORN. Response of normal subjects to large amounts of aldosterone. *J. Clin. Invest.* 37: 1549–1555, 1958.

BAER, P. G. The contribution of prostaglandins to renal blood flow maintenance is determined by the level of activity of the renin-angiotensin system. *Life Sci.* 28: 587–593, 1981.

BARAJAS, L. Anatomy of the juxtaglomerular apparatus. *Am. J. Physiol.* 237: F333–F343, 1979.

BARGER, A. C., AND J. H. HERD. The renal circulation. *N. Engl. J. Med.* 284: 482–490, 1971.

BATTILANA, C. A., D. C. DOBYAN, F. B. LACY, J. BHATTACHARYA, P. A. JOHNSTON, AND R. L. JAMISON. Effect of chronic potassium loading on potassium secretion by the pars recta or descending limb of the juxtamedullary nephron in the rat. *J. Clin. Invest.* 62: 1093–1103, 1978.

BEEUWKES, R., III, AND J. V. BONVENTRE. Tubular organization and vascular-tubular relations in the dog kidney. *Am. J. Physiol.* 229: 695–713, 1975.

BIA, M. J., AND R. A. DeFRONZO. Extrarenal potassium homeostasis. *Am. J. Physiol.* 240: F257–F268, 1981.

BLANTZ, R. C. Disorders of glomerular filtration. In: *Physiology of Membrane Disorders*, edited by T. E. Andreoli, J. F. Hoffman, and D. D. Fanestil. New York: Plenum Medical, 1978, p. 967–985.

BLANTZ, R. C. Segmental renal vascular resistance: single nephron. *Ann. Rev. Physiol.* 42: 573–588, 1980.

BOUDREAU, J. E., AND M. B. BURG. Voltage dependence of calcium transport in the thick ascending limb of Henle's loop. *Am. J. Physiol.* 236: F357–F364, 1979.

BOUDREAU, J. E., AND F. A. CARONE. Protein handling by the renal tubule. *Nephron* 13: 22–34, 1974.

BOUDRY, J., L. C. STONER, AND M. B. BURG. Effect of acid lumen pH on potassium transport in renal cortical collecting tubules. *Am. J. Physiol.* 230: 239–244, 1976.

BOULPAEP, E. L., AND H. SACKIN. Role of the paracellular pathway in isotonic fluid movement across the renal tubule. *Yale J. Biol. Med.* 50: 115–131, 1977.

BRENNER, B. M., T. H. HOSTETTER, AND H. D. HUMES. Glomerular permselectivity: barrier function based on discrimination of molecular size and charge. *Am. J. Physiol.* 234: F455–F460, 1978.

BRENNER, B. M., AND H. D. HUMES. Mechanics of glomerular ultrafiltration. *N. Engl. J. Med.* 297: 148–154, 1977.

BULGER, R. E. Kidney morphology. In: *Diseases of the Kidney*, edited by L. E. Earley and C. W. Gottschalk. Boston: Little, Brown and Company, 1979, p. 3–39.

BURG, M. B., AND N. GREEN. Function of the thick ascending limb of Henle's loop. *Am. J. Physiol.* 224: 659–668, 1973.

BURG, M. B., AND J. ORLOFF. Control of fluid absorption in the renal proximal tubule. *J. Clin. Invest.* 47: 2016–2024, 1968.

BURG, M. B., AND J. L. STEPHENSON. Transport characteristics of the loop of Henle. In: *Physiology of Membrane Disorders*, edited by T. E. Andreoli, J. F. Hoffman, and D. D. Fanestil. New York: Plenum Medical, 1978, p. 661–679.

CHASIS, H., AND H. W. SMITH. The excretion of urea in normal man and in subjects with glomerulonephritis. *J. Clin. Invest.* 17: 347–358, 1938.

COCKCROFT, D. W., AND M. H. GAULT. Prediction of creatinine clearance from serum creatinine. *Nephron* 16: 31–41, 1976.

COGHLAN, J. P., J. R. BLAIR-WEST, D. A. DENTON, D. T. FEI, R. T. FERNLEY, K. J. HARDY, J. G. MCDOUGALL, R. PUY, P. M. ROBINSON, B. A. SCOGINS, AND R. D. WRIGHT. Control of aldosterone secretion. *J. Endocrinol.* 81: 55P–67P, 1979.

COHEN, L. H., A. MUELLER, AND P. R. STEINMETZ. Inhibition of the bicarbonate exit step in urinary acidification by a disulfonic stilbene. *J. Clin. Invest.* 61: 981–986, 1978.

CORTNEY, M. A., M. MYLLE, W. E. LASSITER, AND C. W. GOTTSCHALK. Renal tubular transport of water, solute and PAH in rats loaded with isotonic saline. *Am. J. Physiol.* 209: 1199–1205, 1965.

DAVENPORT, H. W. *The ABC of Acid-Base Chemistry*, 6th ed. Chicago: University of Chicago Press, 1974.

DAVIS, J. O., AND R. H. FREEMAN. Mechanisms regulating renin release. *Physiol. Rev.* 56: 1–56, 1976.

DEETJEN, P., J. W. BOYLAN, AND K. KRAMER. *Physiology of the Kidney and of Water Balance.* New York: Springer, 1975.

DEETJEN, P., AND K. KRAMER. Die Abhängigkeit des O_2-Verbrauchs der Niere von der Na-Rückresorption. *Pflugers Arch.* 273: 636–650, 1961.

DeWARDENER, H. E. The control of sodium excretion. *Am. J. Physiol.* 235: F163–F173, 1978.

DOUSA, T. P., AND H. VALTIN. Cellular actions of vasopressin in the mammalian kidney. *Kidney Int.* 10: 46–63, 1976.

EDELMAN, I. S., AND J. LEIBMAN. Anatomy of body water and electrolytes. *Am. J. Med.* 27: 256–277, 1959.

EDELMAN, I. S., J. LEIBMAN, M. P. O'MEARA, AND L. W. BIRKENFELD. Interrelations between serum sodium concentration, serum osmolarity and total exchangeable sodium, total exchangeable potassium and total body water. *J. Clin. Invest.* 37: 1236–1256, 1958.

ELKINS, B. R., AND A. S. WATANABE. Acute digoxin poisonings: review of therapy. *Am. J. Hosp. Pharm.* 35: 268–277, 1978.

FISHER, J. W. Prostaglandins and kidney erythropoietin production. *Nephron* 25: 53–56, 1980.

FITZSIMONS, J. T. The physiological basis of thirst. *Kidney Int.* 10: 3–18, 1976.

FRALEY, D. S., AND S. ADLER. Correction of hyperkalemia by bicarbonate despite constant blood pH. *Kidney Int.* 12: 354–360, 1977.

FRIIS-HANSEN, B. Changes in body water compartments during growth. *Acta Paediatr.* 46(Suppl. 110): 1–68, 1957.

GENNARI, F. J., AND J. J. COHEN. Role of the kidney in potassium homeostasis: lessons from acid-base disturbances *Kidney Int.* 8: 1–5, 1975.

GIEBISCH, G. The proximal nephron. In: *Physiology of Membrane Disorders*, edited by T. E. Andreoli, J. F. Hoffman, and D. D. Fanestil. New York: Plenum Medical, 1978, p. 629–660.

GIEBISCH, G., AND B. STANTON. Potassium transport in the nephron. *Ann. Rev. Physiol.* 41: 241–256, 1979.

GOOD, D. W., AND F. S. WRIGHT. Luminal influences on potassium secretion: sodium concentration and fluid flow rate. *Am. J. Physiol.* 236: F192–F205, 1979.

GOTTSCHALK, C. W., W. LASSITER, AND M. MYLLE. Localization of urine acidification in the mammalian kidney. *Am. J. Physiol.* 198: 581–585, 1960.

GRANTHAM, J. J., AND M. B. BURG. Effect of vasopressin and cyclic AMP on permeability of isolated collecting tubules. *Am. J. Physiol.* 211: 255–259, 1966.

GREEN, R., R. J. MORIARTY, AND G. GIEBISCH. Ionic requirements of proximal tubular fluid reabsorption: flow dependence of fluid transport. *Kidney Int.* 20: 580–587, 1981.

HAYS, R. M. Dynamics of body water and electrolytes. In: *Clinical Disorders of Fluid and Electrolyte Metabolism*, edited by M. H. Maxwell and C. R. Kleeman. New York: McGraw-Hill, 1980, p. 1–36.

HOFFMANN, N., M. THEES, AND R. KINNE. Phosphate transport by isolated renal brush border vesicles. *Pflugers Arch.* 362: 147–156, 1976.

IMAI, M. The connecting tubule: a functional subdivision of the rabbit distal nephron segments. *Kidney Int.* 15: 346–356, 1979.

IMAI, M., AND J. KOKKO. Sodium-chloride, urea, and water transport in the thin ascending limb of Henle. Generation of osmotic gradients by passive diffusion of solutes. *J. Clin. Invest.* 53: 393–402, 1974.

JACOBSON, H. R. Functional segmentation of the mammalian nephron. *Am. J. Physiol.* 241: F203–F218, 1981.

JAMISON, R. L., C. M. BENNETT, AND R. W. BERLINER. Countercurrent multiplication by the thin loops of Henle. *Am. J. Physiol.* 212: 357–366, 1967.

JAMISON, R. L., AND W. KRIZ. *Urinary Concentrating Mechanism: Structure and Function.* Oxford: Oxford University Press, 1982.

JAMISON, R. L., F. B. LACY, J. P. PENNELL, AND V. M. SANJANA. Potassium secretion by the descending limb or pars recta of the juxtamedullary nephron *in vivo. Kidney Int.* 9: 323–332, 1976.

JAMISON, R. L., H. SONNENBERG, AND J. H. STEIN. Questions and replies: role of the collecting tubule in fluid, sodium, and potassium balance. *Am. J. Physiol.* 237: F247–F261, 1979.

JAMISON, R. L., J. WORK, AND J. A. SCHAFER. New pathways for potassium transport in the kidney. *Am. J. Physiol.* 242: F297–F312, 1982.

KATZ, A. I. Renal Na-K-ATPase: its role in tubular sodium and potassium transport. *Am. J. Physiol.* 242: F207–F219, 1982.

KAWAMURA, S., AND J. P. KOKKO. Urea secretion by the straight segment of the proximal tubule. *J. Clin. Invest.* 58: 604–612, 1976.

KESTELOOT, H., AND J. GEBOERS. Calcium and blood pressure. *Lancet* 1: 813–815, 1982.

KHURI, R. N., M. WIEDERHOLT, N. STRIEDER, AND G. GIEBISCH. Effects of graded solute diuresis on renal tubular sodium transport in the rat. *Am. J. Physiol.* 228: 1262–1268, 1975.

KIIL, F., K. AUKLAND, AND H. E. REFSUM. Renal sodium transport and oxygen consumption. *Am. J. Physiol.* 201: 511–516, 1961.

KNOX, F. G., H. OSSWALD, G. R. MARCHAND, W. S. SPIELMAN, J. A. HAAS, T. BERNDT, AND S. P. YOUNGBERG. Phosphate transport along the nephron. *Am. J. Physiol.* 233: F261–F268, 1977.

KOKKO, J. Sodium chloride and water transport in the descending limb of Henle. *J. Clin. Invest.* 49: 1838–1846, 1970.

KOKKO, J. P., AND F. C. RECTOR, JR. Countercurrent multiplication system without active transport in inner medulla. *Kidney Int.* 2: 214–223, 1972.

KRANE, S. M. Calcium, phosphate and magnesium. In: *International Encyclopedia of Pharmacology and Therapeutics*, ed-

ited by H. Rasmussen. London: Pergamon Press, 1970, p. 19–59.

LANG, F. Renal handling of calcium and phosphate. *Klin. Wochenschr.* 58: 985–1003, 1980.

LASSITER, W. C., C. W. GOTTSCHALK, AND M. MYLLE. Micropuncture study of renal tubular reabsorption of calcium in normal rodents. *Am. J. Physiol.* 204: 771–775, 1963.

LeGRIMELLEC, C. Micropuncture study along the proximal convoluted tubule. *Pflugers Arch.* 354: 133–150, 1975.

LEIBMAN, J., AND I. S. EDELMAN. Interrelations of plasma potassium concentration, plasma sodium concentration, arterial pH and total exchangeable potassium. *J. Clin. Invest.* 38: 2176–2188, 1959.

LONGRIGG, J. N. The upper urinary tract. *Br. J. Clin. Pharmacol.* 13: 461–468, 1982.

LÖNNERHOLM, G. Histochemical demonstration of carbonic anhydrase activity in rat kidney. *Acta Physiol. Scand.* 81: 433–439, 1971.

LUDENS, J. H., AND D. D. FANESTIL. Aldosterone stimulation of acidification of urine by isolated urinary bladder of the Colombian toad. *Am. J. Physiol.* 226: 1321–1326, 1972.

MALNIC, G., AND G. GIEBISCH. Cellular aspects of renal tubular acidification. In: *Membrane Transport in Biology. IVA. Transport Organs*, edited by G. Giebisch, D. C. Tosteson, and H. H. Ussing. Berlin: Springer, 1979, p. 299–355.

MALNIC, G., R. M. KLOSE, AND G. GIEBISCH. Micropuncture study of renal potassium excretion in the rat. *Am. J. Physiol.* 206: 674–686, 1964.

MANERY, J. F., AND A. B. HASTINGS. The distribution of electrolytes in mammalian tissues. *J. Biol. Chem.* 127: 657–676, 1939.

MASORO, E. J. An overview of hydrogen ion regulation. *Arch. Intern. Med.* 142: 1019–1023, 1982.

MASSRY, S. G., AND J. W. COBURN. The hormonal and non-hormonal control of renal excretion of calcium and magnesium. *Nephron* 10: 66–112, 1973.

MOREL, F. Sites of hormonal action in the mammalian nephron. *Am. J. Physiol.* 240: F159–F164, 1981.

MUDGE, G. H. Clinical patterns of tubular dysfunction. *Am. J. Med.* 24: 785–804, 1958.

MURER, H., U. HOPFER, AND R. KINNE. Sodium/proton antiport in brush-border-membrane vesicles isolated from rat small intestine and kidney. *Biochem. J.* 154: 597–604, 1976.

MURER, H., AND R. KINNE. The use of isolated membrane vesicles to study epithelial transport processes. *J. Memb. Biol.* 55: 81–95, 1980.

NARINS, R. G., AND M. EMMETT. Simple and mixed acid-base disorders: a practical approach. *Medicine* 59: 161–187, 1980.

NAVAR, L. G. Renal autoregulation: perspectives from whole kidney and single nephron studies. *Am. J. Physiol.* 234: F357–F370, 1978b.

NAVAR, L. G. The regulation of glomerular filtration rate in mammalian kidneys. In: *Physiology of Membrane Disorders*, edited by T. F. Andreoli, J. F. Hoffman, and D. D. Fanestil. New York: Plenum Medical, 1978a, p. 593–627.

NETTER, F. H. *Kidneys, Ureters, and Urinary Bladder. CIBA Collection of Medical Illustrations.* Summit, NJ: CIBA Pharmaceutical Company, 1973.

OLIVER, J. *Nephrons and Kidneys.* New York: Harper and Row, 1968.

PETTY, K. J., J. P. KOKKO, AND D. MARVER. Secondary effect of aldosterone on Na-K ATPase activity in the rabbit cortical collecting tubule. *J. Clin. Invest.* 68: 1514–1521, 1981.

PITTS, R. F. The role of ammonia production and excretion in regulation of acid-base balance. *N. Engl. J. Med.* 284: 32–38, 1971.

PITTS, R. F. *Physiology of the Kidney and Body Fluids*, 3rd ed. Chicago: Year Book Medical Publishers, 1974.

POTTS, J. T., AND L. J. DEFTOS. Parathyroid hormone, calcitonin, vitamin D, bone and bone mineral metabolism. In: *Diseases of Metabolism*, edited by P. K. Bondy and L. E. Rosenberg. Philadelphia: W. B. Saunders, 1974, p. 1225–1430.

QUAMME, G. A., AND J. H. DIRKS. Magnesium transport in the nephron. *Am. J. Physiol.* 239: F393–F401, 1980.

RENNICK, B. R. Renal tubular transport of organic cations. *Am. J. Physiol.* 240: F83–F89, 1981.

RENNICK, B. R. Renal excretion of drugs: tubular transport and metabolism. *Ann. Rev. Pharmacol.* 12: 141–156, 1972.

ROBERTSON, G. L., R. L. SHELTON, AND S. ATHAR. The osmoregulation of vasopressin. *Kidney Int.* 10: 25–37, 1976.

ROCHA, A. S., AND J. P. KOKKO. Sodium chloride and water transport in the medullary thick ascending limb of Henle. Evidence for active chloride transport. *J. Clin. Invest.* 52: 612–623, 1973.

ROSA, R. M., P. SILVA, J. B. YOUNG, L. LANDSBERG, R. S. BROWN, J. W. ROWE, AND F. H. EPSTEIN. Adrenergic modulation of extrarenal potassium disposal. *N. Engl. J. Med.* 302: 431–434, 1980.

SCHAFER, J. A., AND T. E. ANDREOLI. The collecting duct. In: *Physiology of Membrane Disorders*, edited by T. E. Andreoli, J. F. Hoffman, and D. D. Fanestil. New York: Plenum Medical, 1978, p. 707–737.

SCHIESS, W. A., J. L. AYER, W. D. LOTSPEICH, AND R. F. PITTS. The renal regulation of acid-base balance in man. II. Factors affecting the excretion of titratable acid by the normal human subject. *J. Clin. Invest.* 27: 57–64, 1948.

SCHMIDT-NIELSEN, B., AND R. O'DELL. Structure and concentrating mechanism in the mammalian kidney. *Am. J. Physiol.* 200: 1119–1124, 1961.

SCHRIER, R. W., T. BERL, AND R. J. ANDERSON. Osmotic and nonosmotic control of vasopressin release. *Am. J. Physiol.* 236: F321–F332, 1979.

SCHRIER, R. W., T. BERL, R. J. ANDERSON, AND K. M. MCDONALD. Nonosmolar control of renal water excretion. In: *Disturbances in Body Fluid Osmolality*, edited by T. E. Andreoli, J. J. Grantham, and F. C. Rector, Jr. Bethesda, MD: American Physiological Society, 1977, p. 149–178.

SCHULTZ, S. G. Homocellular regulatory mechanisms in sodium-transporting epithelia: avoidance of extinction by "flush-through." *Am. J. Physiol.* 241: F579–F590, 1981.

SEELY, J. F., AND M. LEVY. Control of extracellular fluid volume. In: *The Kidney*, edited by B. M. Brenner and F. C. Rector, Jr. Philadelphia: W. B. Saunders, 1981, p. 371–407.

SHANNON, J. A. The renal excretion of phenol red by the aglomerular fishes *Opsanus tau* and *Lophius piscatorius*. *J. Cell Comp. Physiol.* 11: 315–323, 1938.

SHANNON, J. A., AND H. W. SMITH. The excretion of inulin, xylose and urea by normal and phlorizinized man. *J. Clin. Invest.* 14: 393–401, 1935.

SHAREGHI, G. R., AND L. C. STONER. Calcium transport across segments of the rabbit distal nephron *in vitro*. *Am. J. Physiol.* 235: F367–F375, 1978.

SHIPLEY, R. E., AND R. S. STUDY. Changes in renal blood flow, extraction of inulin, GFR, tissue pressure, and urine flow with acute alterations of renal artery pressure. *Am. J. Physiol.* 167: 676–688, 1951.

SILVA, P., R. S. BROWN, AND F. H. EPSTEIN. Adaptation to potassium. *Kidney Int.* 11: 466–475, 1977.

SIMPSON, D. P. Control of hydrogen ion homeostasis and renal acidosis. *Medicine* 50: 503–541, 1971.

SKELTON, H. The storage of water by various tissues of the body. *Arch. Int. Med.* 40: 140–152, 1927.

SMITH, H. W. *The Kidney*. New York: Oxford University Press, 1951.

SPRING, K. R., AND G. KIMURA. Intracellular ion activities in *Necturus* proximal tubule. *Fed. Proc.* 38: 2729–2732, 1979.

STOKES, J. B. Integrated actions of renal medullary prostaglandins in the control of water excretion. *Am. J. Physiol.* 240: F471–F480, 1981.

STOKES, J. B., M. J. INGRAM, A. D. WILLIAMS, AND D. INGRAM. Heterogeneity of the rabbit collecting tubule: localization of mineralocorticoid hormone action to the cortical portion. *Kidney Int.* 20: 340–347, 1981.

SUKI, W. N. Calcium transport in the nephron. *Am. J. Physiol.* 237: F1–F6, 1979.

SUKI, W. N., D. ROUSE, R. C. K. NG, AND J. P. KOKKO. Calcium transport in the thick ascending limb of Henle. *J. Clin. Invest.* 66: 1004–1009, 1980.

TUCKER, B. J., AND R. C. BLANTZ. An analysis of the determinants of nephron filtration rate. *Am. J. Physiol.* 232: F477–F483, 1977.

TUNE, B. M., M. B. BURG, AND C. S. PATLAK. Characteristics of p-aminohippurate transport in proximal renal tubules. *Am. J. Physiol.* 217: 1057–1063, 1969.

TURNER, R. J., AND M. SILVERMAN. Interaction of phlorizin and sodium with the renal brush-border membrane D-glucose transporter: stoichiometry and order of binding. *J. Memb. Biol.* 58: 43–55, 1981.

ULLRICH, K. J., G. CAPASSO, F. RUMRICH, F. PAPVAS-SILIOUS, AND S. KLÖSS. Coupling between proximal tubular transport processes. *Pflugers Arch.* 368: 245–252, 1977.

ULLRICH, K. J., K. KRAMER, AND J. W. BOYLAN. Present knowledge of the countercurrent system in the mammalian kidney. *Prog. Cardiovasc. Dis.* 3: 395–431, 1961.

VALTIN, H. Structural and functional heterogeneity of mammalian nephrons. *Am. J. Physiol.* 233: F491–F501, 1977.

VAN SLYKE, D. D., A. B. HASTINGS, M. HEIDELBERGER, AND J. M. NEILL. Studies of gas and electrolyte equilibria in the blood. III. The alkali-binding and buffer values of oxyhemoglobin and reduced hemoglobin. *J. Biol. Chem.* 54: 481–506, 1922.

VENKATACHALAM, M. A. Anatomy and histology of the kidney. In: *Nephrology*, edited by J. H. Stein. New York: Grune & Stratton, 1980, p. 1–14.

WACKER, W. E. C., AND A. F. PARISI. Magnesium metabolism. *N. Engl. J. Med.* 278: 658–663, 1968.

WADE, J. B., W. A. KACHADORIAN, AND V. A. DISCALA. Freeze-fracture electron microscopy: relationship of membrane structural features to transport physiology. *Am. J. Physiol.* 232: F77–F83, 1977.

WALSER, M. Ion association. VI. Interactions between calcium, magnesium, inorganic phosphate, citrate, and protein in normal human plasma. *J. Clin. Invest.* 40: 723–730, 1961a.

WALSER, M. Divalent cations: physicochemical state in glomerular filtrate and urine and renal excretion. In: *Handbook of Physiology, Sect. 8: Renal Physiology*, edited by J. Orloff and R. W. Berliner. Bethesda, MD: American Physiological Society, 1973, p. 555–586.

WALSER, M. Calcium clearance as a function of sodium clearance in the dog. *Am. J. Physiol.* 200: 1099–1104, 1961b.

WARNOCK, D. G., AND F. C. RECTOR, JR. Proton secretion by the kidney. *Ann. Rev. Physiol.* 41: 197–210, 1979.

WARNOCK, D. G., AND V. J. YEE. Chloride uptake by brush border membrane vesicles isolated from rabbit renal cortex. *J. Clin. Invest.* 67: 103–115, 1981.

WEINER, I. M. Urate transport in the nephron. *Am. J. Physiol.* 237: F85–F92, 1979.

WEINER, I. M., AND G. H. MUDGE. Renal tubular mechanisms for excretion of organic acids and bases. *Am. J. Med.* 36: 743–762, 1964.

WEITZMAN, R., AND C. R. KLEEMAN. Water metabolism and the neurohypophyseal hormones. In: *Clinical Disorders of Fluid and Electrolyte Metabolism*, edited by M. H. Maxwell and C. R. Kleeman. New York: McGraw-Hill, 1980, p. 531–645.

WELLING, D. J., AND L. W. WELLING. Cell shape as an indicator of volume reabsorption in proximal nephron. *Fed. Proc.* 38: 121–127, 1979.

WINDHAGER, E. E. Sodium chloride transport. In: *Membrane Transport in Biology. IVA. Transport Organs*, edited by G. Giebisch, D. C. Tosteson, and H. H. Ussing. Berlin: Springer, 1979, p. 145–213.

WINDHAGER, E. E., G. WHITTEMBURY, D. E. OKEN, H. J. SCHATZMAN, AND A. K. SOLOMON. Single proximal tubules of the *Necturus* kidney. III. Dependence of H_2O movements on NaCl concentration. *Am. J. Physiol.* 197: 313–318, 1959.

WOODBURNE, R. T. Anatomy of the ureterovesical junction. *J. Urol.* 92: 431–435, 1964.

WRIGHT, F. S., AND J. S. BRIGGS. Feedback control of glomerular blood flow, pressure, and filtration rate. *Physiol. Rev.* 59: 958–1006, 1979.

WRIGHT, F. S., AND G. GIEBISCH. Renal potassium transport: contributions of individual nephron segments and populations. *Am. J. Physiol.* 235: F515–F527, 1978.

ZIEGLER, T. W., D. D. FANESTIL, AND J. H. LUDENS. Influence of transepithelial potential difference on acidification in the toad urinary bladder. *Kidney Int.* 10: 279–286, 1976.

Respiration

Uptake and Delivery of the Respiratory Gases

SCOPE OF RESPIRATORY PHYSIOLOGY

Man and other higher animals remove oxygen from the air and add carbon dioxide to it in the course of satisfying the metabolic demands of their tissues. The business of getting oxygen from the atmosphere into the cells and carbon dioxide out, that is, "gas exchange," is the essence of respiratory physiology.

We can identify several links in the chain of processes involved in gas exchange (Fig. 5.1):

1. Ventilation—the process of moving oxygen from the air into the alveoli in the depths of the lung (and carbon dioxide in the opposite direction).
2. Diffusion—movement of gases across the gas-blood barrier.
3. Matching of ventilation and blood flow—not easily shown on the diagram but critical for efficient gas exchange.
4. Pulmonary blood flow—to move the gases out of the lung.
5. Blood gas transport—carriage of oxygen and carbon dioxide in the blood.
6. Transfer of gases between the peripheral capillaries and the cells.
7. Utilization of oxygen (and production of carbon dioxide) in the cells.

Five other topics should also be included in the domain of respiratory physiology:

8. Structure-function relationships of the lung.
9. Lung mechanics—that is, the forces involved in supporting and moving the lung and chest wall.
10. Control of ventilation—the mechanism which regulates the gas exchange function of the lung.
11. Respiration in unusual environments—including high altitude, diving, space, and other situations where special problems are encountered.
12. Tests of lung function—these are important in assessing lung disease.

The organization of Section 5 is as follows:
Chapter 36 covers ventilation, pulmonary blood

flow, blood gas transport, and the delivery of oxygen to the cells (*1, 4, 5,* and *6* above). It might be more logical to discuss *items 5* and *6* later, but a discussion of the oxygen and CO_2 blood dissociation curves is an essential preliminary to an understanding of pulmonary gas exchange. In addition this chapter includes introductory sections on structure-function relationships of the lung (*8*), the physical gas laws, and the language of respiratory physiology.

Chapter 37 covers pulmonary gas exchange in-

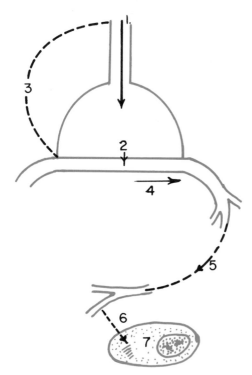

Figure 5.1. Diagram showing the various steps in the delivery of oxygen from air to tissues. *1*, ventilation; *2*, diffusion across the blood-gas barrier; *3*, matching of ventilation and blood flow; *4*, pulmonary blood flow; *5*, transport of gas in the blood; *6*, diffusion from capillary to cell; *7*, utilization of O_2 by mitochondria.

cluding diffusion and ventilation-perfusion relationships (*2* and *3* above).

Chapters 38, 39, and 40 discuss the mechanics of breathing (*9*), the control of ventilation (*10*), and respiration in unusual environments (*11*), respectively.

Intracellular oxidation (*7*) is mentioned in Chapters 36 and 1 but in the main is delegated to the biochemists. Tests of lung function are alluded to in various places, but for a full account the reader should refer to other texts (for example, Cotes, 1979; West, 1982).

STRUCTURE-FUNCTION RELATIONSHIPS OF THE LUNG

Throughout the body, the function of an organ is reflected in its structure; this is particularly true of the lung. Since the essential function of the lung is gas exchange let us start with the *blood-gas barrier* which separates the blood in the pulmonary capillary from the alveolar gas (Figs. 5.1 and 5.2). The barrier is composed of an alveolar epithelial cell, interstitial layer, and capillary endothelial cell. In addition, there is a layer of phospholipid (surfactant) on the epithelial cell not shown in this preparation (see Chapter 38). Note that the barrier is less than half a micron thick in places. Calculations indicate that the surface area is between 50 and 100 m^2 so that its structure is well suited to the business of gas exchange by diffusion through it.

Air is brought to one side of the barrier by ventilation, and blood to the other by the pulmonary circulation (Fig. 5.1). The *airways* or bronchi start at the trachea and form an intricate branching system (Fig. 5.3*A*) which leads gas into the depths of the lung. A few years ago, the number and size of the airways in a typical human lung were counted by the Swiss anatomist, Weibel, and his idealization is shown in Figure 5.3*B* (Weibel, 1963). Although it is an oversimplification, this model has been of great value in respiratory physiology.

Figure 5.3*B* shows that on the average, the airways branch approximately 16 times before any alveoli with their thin blood-gas barrier (Fig. 5.2) are seen. Since no gas exchange can occur in these generations, they comprise the *anatomic dead space*. Its volume is approximately 150 ml.* The smallest airways of this conducting zone are known as the terminal bronchioles. Beyond these lie airways with

---

* These figures for lung volumes, ventilation, etc. are only intended to give the reader a general idea. The values depend on age, sex, height, and weight and appropriate tables have been published (Cotes, 1979).

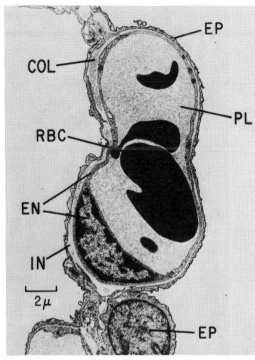

Figure 5.2. Electron micrograph showing a pulmonary capillary in the alveolar wall. Note the extremely thin blood-gas barrier of less than 0.5 μm thickness. *EP*, alveolar epithelium; *IN*, interstitial tissue; *EN*, capillary endothelium; *PL*, plasma; *RBC*, red blood cell; *COL*, collagen. [From Weibel (1973).]

a few alveoli in their walls, the respiratory bronchioles. These lead to alveolar ducts whose walls are composed entirely of alveoli and finally to the blindly ending alveolar sacs. All the gas exchange occurs in the alveolated portion of lung which is known as the respiratory zone. Its volume is about 3 l.

Blood is brought to the other side of the blood-gas barrier (Fig. 5.1) from the right heart by the pulmonary arteries which lead to the pulmonary capillaries (Fig. 5.4*A*). The capillaries lie in the walls of the alveoli and, seen from the vantage point of the alveolar space, form a dense network of interconnecting vessels (Fig. 5.4*B*). So dense is the network that the blood forms an almost continuous sheet in the alveolar wall. At normal capillary pressures, not all the bed is open, but recruitment of closed capillaries can occur if the pressure rises, for example on exercise. When all the capillaries are open, over 80% of the area of the alveolar wall is apparently available for gas exchange.

The lung has a second blood supply, the bronchial circulation via the bronchial arteries, which arise from the aorta. This flow is only about a hundredth of that in the pulmonary circulation, and its main purpose is to supply the walls of the large airways. There is also a small lymph flow from the lung.

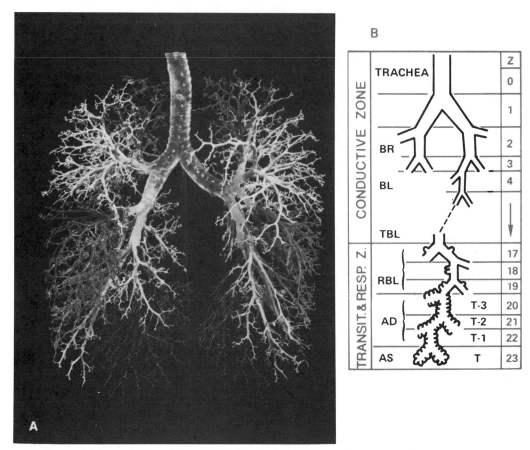

Figure 5.3. *A*, Cast of the airways of a human lung. *B*, Idealization of the human airways according to Weibel. *Z*, generation number; *BR*, bronchus; *BL*, bronchiole; *TBL*, terminal bronchiole; *RBL*, respiratory bronchiole; *AD*, alveolar duct; *AS*, alveolar sac. [From Weibel (1963).]

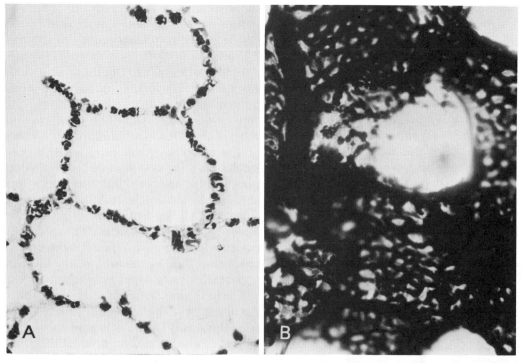

Figure 5.4. Microscopic appearances of capillaries in the alveolar walls of dog lung. *A* shows a thin cross-section; *B* shows the appearance of the network of capillaries in a thick section when the alveolar wall is seen face-on. [*A* from Glazier et al. (1969).]

The lymphatics run chiefly around the larger airways and blood vessels.

The alveolar tissue (parenchyma) contains fibers of elastin and collagen, and the fluid lining the alveoli has surface tension. As a result the lung is elastic and is held expanded by keeping the pressure around it (intrapleural pressure) lower than alveolar pressure. During inspiration the diaphragm contracts and moves downward, and the intercostal muscles increase the diameter of the rib cage. As a result the intrapleural pressure falls further and the lung expands. Expiration is normally passive and the lung returns to its resting position because of its elasticity. This topic is dealt with in more detail in Chapter 38.

BEHAVIOR OF GASES

Before we turn to the movement of gas into the lung, a brief review of the physical laws of gases may be useful.

Kinetic Theory of Gases

The molecules of a gas are in continuous random motion and are only deflected from their course by collision with other molecules or with the walls of a container. When they strike the walls and rebound, the resulting bombardment results in a pressure. The magnitude of the pressure depends on the number of molecules present and their speed.

BOYLE'S LAW

At constant temperature, the pressure (P) of a given mass of gas is inversely proportional to its volume (V), or

$$P_1V_1 = P_2V_2 \text{ (temperature constant)}$$

This can be explained by the fact that as the molecules are brought closer together (smaller volume), the rate of bombardment on a unit surface increases (greater pressure).

CHARLES'S LAW

At constant pressure, the volume is proportional to the absolute temperature (T), or

$$\frac{V_1}{V_2} = \frac{T_1}{T_2} \text{ (pressure constant)}$$

The explanation is that a rise in temperature increases the speed and momentum of the molecules, thus increasing the force of bombardment on the container. Another form of Charles's law states that at constant volume, the pressure is proportional to absolute temperature. (Note that absolute temperature is obtained by adding 273 to the centigrade temperature. Thus 37°C = 310° absolute.)

AVOGADRO'S LAW

Equal volumes of different gases at the same temperature and pressure contain the same number of molecules. Thus a gram molecule, for example, 32 g of oxygen, occupies 22.4 l at STPD (standard temperature 0°C, and pressure 760 mmHg, dry).

IDEAL GAS LAW

This combines the above three laws thus:

$$PV = nRT$$

where n is the number of gram molecules of the gas and R is the "gas constant." When the units employed are mmHg, liters, and degrees absolute, then R = 62.4. Real gases deviate from ideal gas behavior under certain conditions.

DALTON'S LAW

Each gas in a mixture exerts a pressure according to its own concentration, independently of the other gases present. That is, each component behaves as though it were present alone. The pressure of each gas is referred to as its *partial pressure* or tension. The total pressure is the sum of the partial pressures of all gases present. In symbols:

$$P_x = P \cdot F_x$$

where P_x is the partial pressure of gas x, P is the total pressure, and F_x is the fractional concentration of gas x (for example, if half the gas is oxygen, F_{O2} = 0.5). Conventionally the fractional concentration always refers to dry gas. Thus in gas which is saturated with vapor at 37°C where the water vapor pressure is 47 mmHg,

$$P_x = (P_B - 47) \cdot F_x$$

where P_B is the total (barometric) pressure and (P_B − 47) is the dry gas pressure.

Diffusion in the Gas Phase

Because of their random motion, gas molecules tend to distribute themselves uniformly throughout any available space until the partial pressure is everywhere the same. The process by which gas moves from a region of high to low partial pressure is known as diffusion. Light gases diffuse faster than heavy gases because the mean velocity of the molecules is higher. In fact the kinetic theory of gases states that the kinetic energy ($0.5 \, mv^2$) of all gases is the same (at a given temperature and pressure). From this it follows that the rate of diffusion of a gas is inversely proportional to the square root of its density (*Graham's Law*). Thus hydrogen (MW = 2) diffuses four times as rapidly as oxygen (MW = 32).

HENRY'S LAW

The volume of gas dissolved in a liquid is proportional to its partial pressure. Thus:

$$C_x = K \cdot P_x$$

where C is the concentration of gas in the liquid, for example, in milliliters of gas per 100 ml of blood and K is a solubility constant whose value depends on the units of C and P. Tables of solubility constants are available (Altman, 1961). Note that the partial pressure of a gas in solution is best defined as its partial pressure in a gas which is in equilibrium with that solution.

DIFFUSION OF DISSOLVED GASES

Gas moves from a region of high to low partial pressure in a liquid by diffusion, just as it does in the gas phase. Again the rate is inversely proportional to the square root of the molecular weight of the gas. In addition, the amount of gas that moves through a film of liquid (or a membrane) is proportional to the solubility of the gas.

LANGUAGE OF RESPIRATORY PHYSIOLOGY

Symbols and equations will be used sparingly in this section because many students are discouraged by them. Some kind of shorthand is essential, however, if the quantities are to be manipulated algebraically. Fortunately there is now general agreement on most of the symbols used (Pappenheimer, 1950), and these are shown in Table 5.1.

VENTILATION

Ventilation is the business of getting gas to and from the alveoli. Figure 5.5 is a highly simplified diagram of the lung to show the volumes and flows involved. The various bronchi which make up the conducting airways (Fig 5.3) are now represented by a single tube labeled anatomic dead space. This leads into the gas exchanging region of the lung which is bounded by the blood-gas interface and the pulmonary capillary blood. With each inspiration, about 500 ml of air enter the lung (tidal volume). Note how small a proportion of the total lung volume is represented by the anatomic dead space. Also note the very small volume of capillary blood compared with alveolar gas (compare Fig. 5.4). In this chapter we shall not consider the forces exerted by the lung and chest wall during ventilation; this subject is taken up in Chapter 38.

Lung Volumes

Before looking in detail at the movement of gas in the lung, it is useful to consider the static vol-

Table 5.1
Some symbols used in respiratory physiology

Primary symbols
  C Concentration of gas in blood
  F Fractional concentration in dry gas
  P Pressure or partial pressure
  Q Volume of blood
  Q̇ Volume of blood per unit time
  R Respiratory exchange ratio
  S Saturation of hemoglobin with O_2
  V Volume of gas
  V̇ Volume of gas per unit time
Secondary symbols for gas phase
  A Alveolar
  B Barometric
  D Dead space
  E Expired
  I Inspired
  L Lung
  T Tidal
Secondary symbols for blood phase
  a Arterial
  c Capillary
  c′ End-capillary
  i Ideal
  v Venous
  v̄ Mixed venous
Examples
  O_2 concentration in arterial blood Ca_{O_2}
  Fractional concentration of N_2 in expired gas $F_{E_{N_2}}$
  Partial pressure of O_2 in mixed venous blood $P\bar{v}_{O_2}$

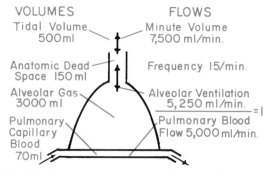

Figure 5.5. Diagram of a lung showing typical volumes and flows. There is considerable variation around these values. [From West (1977).]

umes of the lung. At one time a great deal of emphasis was placed on these measurements because they were thought to be valuable aids to the diagnosis of lung disease. Less attention is paid to them now, but they are still frequently determined.

SIMPLE SPIROMETRY

A spirometer is a light bell-shaped container which is immersed in a water tank which forms a seal (Fig. 5.6). Although it was invented as early as 1846 by Hutchinson to measure lung volumes in

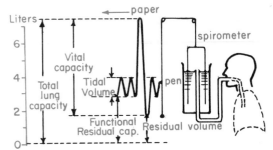

Figure 5.6. Lung volumes. Note that the functional residual capacity and residual volume cannot be measured with the spirometer.

various groups of people in London (including 121 sailors, 24 "pugilists and wrestlers," and 4 "giants and dwarfs"!) it is still used today. Modern modifications include dry spirometers made with a bellows, or a piston in a large cylinder. These often have an electrical output.

Figure 5.6 shows that as the bell moves up during exhalation, the pen moves down marking the chart. In the tracing shown, the subject first breathed normally and the excursion of the pen gave the *tidal volume*. Next he took a maximal inspiration followed by a maximal expiration. The exhaled volume is called the *vital capacity*. Normal breathing was then resumed.

Note that some gas remained in the lung after the maximal expiration; this is the *residual volume*. The volume of gas in the lung after a normal expiration is the *functional residual capacity* (FRC). Neither this nor the residual volume can be obtained by simple spirometry.

Additional terms are sometimes used. The maximum volume which can be inhaled from FRC is the *inspiratory capacity*. Subtracting the tidal volume from this gives the *inspiratory reserve volume*. The difference between the FRC and the residual volume is sometimes called the *expiratory reserve volume*.

MEASUREMENT OF FUNCTIONAL RESIDUAL CAPACITY

As stated above, the FRC and residual volume cannot be measured by simple spirometry because the lung cannot be emptied completely by a maximal exhalation. One method of determining the FRC is by helium dilution in a "closed circuit," that is, a system which does not allow gas to escape. The subject is connected to a spirometer containing a known concentration of helium which is almost insoluble in blood (Fig. 5.7A). After some breaths, the helium concentration in the spirometer and lung becomes the same (Fig. 5.7B). Since a neglible amount of helium has been lost, the amount of

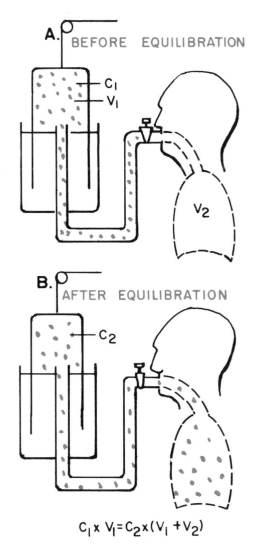

$$C_1 \times V_1 = C_2 \times (V_1 + V_2)$$

Figure 5.7. Measurement of the functional residual capacity by helium dilution.

helium present before equilibration (concentration × volume) is $C_1 \times V_1$ and equals the amount after equilibration, $C_2 \times (V_1 + V_2)$. Thus:

$$C_1 \times V_1 = C_2(V_1 + V_2), \text{ or:}$$

$$V_2 = \frac{V_1(C_1 - C_2)}{C_2}$$

In practice, the CO_2 produced by the subject is absorbed during equilibration, and oxygen is added to keep the spirometer volume constant.

The functional residual capacity can also be obtained by "open circuit" nitrogen washout. The subject breathes pure oxygen from a valve box which separates inspired and expired gas, and the expired gas is collected in a large spirometer. After about 7 min almost all the nitrogen has been washed out of a normal lung. The nitrogen concentration in the spirometer is then determined, and the volume of exhaled nitrogen is calculated. Knowing that the concentration in the lung before the washout was 80%, the lung volume can be computed. This method has the disadvantage that a very accurate measurement of nitrogen in the spirometer is required. In addition, patients with diseased lungs may not wash out all their nitrogen and an error is thus introduced.

A very different method of measuring FRC is with a body plethysmograph (Dubois et al., 1956). This is a large airtight box like a telephone booth in which the subject sits (Fig. 5.8). The pressure inside the box can be measured very accurately. The subject is asked to make respiratory efforts against a closed mouthpiece at a particular lung volume, for example, FRC. As he compresses the gas in his lung, the lung volume decreases slightly. As a result the volume of air in the box *increases* very slightly, and its pressure falls slightly. Boyle's law (see above) is then applied to the box gas. Knowing the change in pressure, the change in volume (ΔV) of the box gas, and hence the lung, can be determined.

Next, Boyle's law is applied to the gas in the lung. Suppose that the mouth pressures before and after the expiratory effort are P_1 and P_2, respectively. Then if V is the FRC of the lung,

$$P_1 V = P_2 (V - \Delta V), \text{ or}$$

$$V = \Delta V \cdot \frac{P_2}{P_2 - P_1}$$

Since P_1, P_2, and ΔV are known, V can be calculated.

It should be noted that the body plethysmograph and the gas dilution (or washout) method may measure different volumes. The body plethysmograph measures the total volume of gas in the lungs, including any which is trapped behind closed airways (for example, Fig. 5.71) and which therefore does not communicate with the mouth. By contrast, the helium dilution and nitrogen washout methods measure only communicating gas, or ventilated lung volume. In young normal subjects these volumes are virtually the same, but in patients with lung disease the ventilated volume may be considerably less than the total volume because of gas trapped behind obstructed airways.

Bulk Flow and Diffusion in the Airways

Figure 5.3 shows that the system of airways through which ventilation occurs branches successively at each generation. Using Weibel's data it is possible to calculate the cross-sectional area of each airway generation, and this is shown in Figure 5.9. Generation 0 is the trachea with a cross-sectional area of about 2.5 cm². Next comes generation 1 composed of the right and left main bronchi which have a total cross-section of 2.3 cm², and so on. Note that the area changes relatively slowly up to the region of the terminal bronchioles (the end of the conducting airways), but that it then suddenly increases very rapidly. In other words the combined airways have a shape something like a trumpet which is flared at the end.

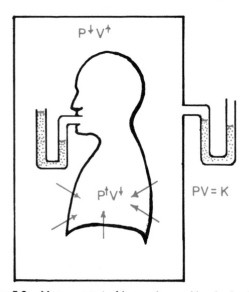

Figure 5.8. Measurement of lung volume with a body plethysmograph. When the subject makes an expiratory effort against a closed airway, he reduces the volume of his lung, airway pressure increases, and box pressure decreases. From Boyle's law, lung volume is obtained (see text).

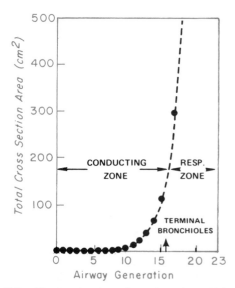

Figure 5.9. Diagram to show the extremely rapid increase in total cross-sectional area of the airways in the respiratory zone (compare Fig. 5.3). As a result, the forward velocity of the gas during inspiration becomes very small in the region of the respiratory bronchioles, and gaseous diffusion becomes the dominant mechanism of ventilation.

This geometry has an important influence on the mode of airflow (Cumming et al., 1966). In the airways down to the terminal bronchioles, gas moves predominantly by bulk flow, like water through a garden hose. However, since the same volume of gas traverses each generation, Figure 5.9 implies that the forward velocity of the inspired gas rapidly decreases as the gas enters the respiratory zone. In fact the velocity becomes so small that another mode of flow takes over as the dominant one. This is gaseous diffusion due to the random movement of the molecules. The rate of diffusion of the molecules is so rapid and the distance to be covered is so short (only a few mm) that differences in concentration along the terminal airways are virtually abolished within a second. Note however that inhaled dust particles diffuse very slowly because of their relatively large size and they therefore tend not to reach the alveoli, but to settle out in the region of the respiratory bronchioles.

Total and Alveolar Ventilation

Suppose the volume exhaled with each breath is 500 ml (Fig. 5.5) and there are 15 breaths/min. Then the total volume leaving the lung each minute is $500 \times 15 = 7,500$ ml/min. This is known as the *total ventilation* or *minute volume*. The volume of air entering the lung is slightly greater because more oxygen is taken in than carbon dioxide is given out.

However, not all the air that passes the lips reaches the alveolar gas compartment where gas exchange occurs. Of each 500 ml inhaled in Figure 5.5, 150 ml remains behind in the anatomical dead space. Thus the volume of fresh gas entering the respiratory zone each minute is $(500 - 150) \times 15$ or 5,250 ml/min. This is called the *alveolar ventilation* and is of key importance because it represents the amount of fresh inspired air available for gas exchange (Strictly, the alveolar ventilation is also measured on expiration, but the volume is almost the same.)

The total ventilation can easily be measured by having the subject breathe through a valve box and collecting all the expired gas in a large bag (sometimes called a Douglas bag). The alveolar ventilation, however, is more difficult to determine.

MEASUREMENT OF ALVEOLAR VENTILATION

One way to determine the alveolar ventilation is to measure the volume of the anatomic dead space (see below) and calculate the dead space ventilation (volume × respiratory frequency). This is then subtracted from the total ventilation.

This process can be summarized using the sym-

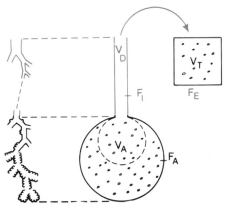

Figure 5.10. The tidal volume V_T is a mixture of gas from the anatomic dead space V_D and a contribution from the alveolar gas V_A. The concentrations of CO_2 are shown by the dots; F, fractional concentration; I, inspired; E, expired. Compare Figure 5.3B. [Adapted from Piiper.]

bols of Table 5.1. (also compare Fig. 5.10). If V denotes volume, and the subscripts T, D, and A refer to tidal, dead space and alveolar, respectively,† then:

$$V_T = V_D + V_A$$

Multiplying each term by n, the respiratory frequency, gives:

$$V_T \cdot n = V_D \cdot n + V_A \cdot n$$

whence

$$\dot{V}_E = \dot{V}_D + \dot{V}_A$$

where $\dot{V}$ means volume per unit time, $\dot{V}_E$ is total expired ventilation, and $\dot{V}_D$ and $\dot{V}_A$ are the dead space and alveolar ventilations, respectively. Thus

$$\dot{V}_A = \dot{V}_E - \dot{V}_D$$

A difficulty with this method is that the anatomical dead space is not easy to measure, although a value for it can be assumed with little error. A simple approximation for seated subjects is that the dead space in milliliters is equal to the subject's weight in pounds.

Another way of measuring alveolar ventilation in normal subjects is from the concentration of CO_2 in expired gas (Fig. 5.10). Since no gas exchange occurs in the anatomic dead space, there is no CO_2 there at the end of inspiration. (The extremely small amount of CO_2 in the atmosphere is neglected.) Thus, since all the expired CO_2 comes from the alveolar gas:

$$\dot{V}_{CO_2} = \dot{V}_A \cdot F_{A_{CO_2}}$$

---

† Note that V_A here means the volume of alveolar gas in the tidal volume, not the total volume of alveolar gas in the lung.

where $\dot{V}_{CO_2}$ means the volume of CO_2 exhaled per unit time, and $F_{A_{CO_2}}$ is the fractional concentration of CO_2 in alveolar gas. (In other words if the concentration of CO_2 is 5%, the fractional concentration is 0.05). Therefore:

$$\dot{V}_A = \frac{\dot{V}_{CO_2}}{F_{A_{CO_2}}}$$

Thus the alveolar ventilation can be obtained by dividing the CO_2 output by the alveolar concentration of this gas. The CO_2 output is derived by collecting expired gas in a bag and analyzing it for CO_2. The alveolar concentration can be determined by means of a rapid CO_2 analyzer which samples gas in a mouthpiece through which the subject breathes. The last portion of a single expiration is

representative of alveolar gas (see discussion of Fig. 5.11).

Note that the partial pressure of CO_2 (denoted P_{CO_2}) is proportional to the concentration of the gas in the alveoli, or:

$$P_{A_{CO_2}} = F_{A_{CO_2}} \cdot K$$

where K is a constant. Therefore:

$$\dot{V}_A = \frac{\dot{V}_{CO_2}}{P_{A_{CO_2}}} \cdot K$$

Since the P_{CO_2} of alveolar gas and arterial blood are almost identical in normal subjects:

$$\dot{V}_A = \frac{\dot{V}_{CO_2}}{P_{a_{CO_2}}} \cdot K$$

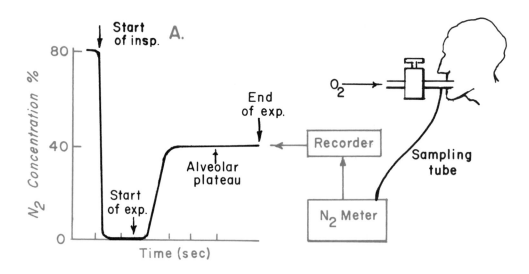

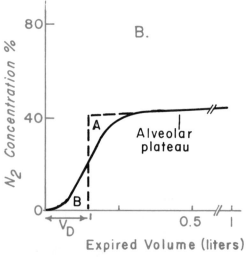

Figure 5.11. Fowler's method of measuring the anatomic dead space with a rapid N_2 analyzer. *A* shows that following a test inspiration of 100% oxygen, the nitrogen concentration rises during expiration to an almost level "plateau" representing pure alveolar gas. *B* shows a plot of nitrogen concentration against expired volume. The dead space is the expired volume up to the vertical broken line which makes the areas *A* and *B* equal.

and thus the arterial P_{CO_2} can be used to determine alveolar ventilation. Note that if the alveolar ventilation is halved (and the CO_2 production remains unchanged as it will do under steady state conditions), the alveolar and arterial P_{CO_2} will double.

Anatomical Dead Space

We have seen that the anatomic dead space is the volume of the conducting airways (Figs. 5.3 and 5.5). The normal value is in the region of 150 ml, and it increases with large inspirations because of the traction exerted on the bronchi by the surrounding lung parenchyma. The dead space also depends on the size and posture of the subject.

The volume of the dead space can be measured by *Fowler's method* (Fowler, 1948). The subject breathes through a valve box and the sampling tube of a rapid nitrogen analyzer continuously samples gas at the lips (Fig. 5.11). During air breathing, the concentration of nitrogen changes little between inspiration and expiration. To perform the test, the subject takes a single inspiration of 100% oxygen and then exhales fully. The nitrogen concentration changes that occur are shown in Figure 11A. During inspiration, the nitrogen concentration is zero. On expiration, first pure dead space (100% oxygen) is sampled, and then the nitrogen concentration rises as the dead space is increasingly washed out by alveolar gas. Finally an almost uniform gas concentration is seen representing pure alveolar gas. This phase is often called the "alveolar plateau," although in normal subjects it is not quite flat and in patients with lung disease it may rise steeply.

Expired volume is recorded at the same time, and then the nitrogen concentration is plotted against this. (Alternatively nitrogen concentration and volume can be recorded together on an XY recorder). The dead space if found by drawing a vertical line such that area *A* is equal to area *B* on Figure 5.11*B*. The dead space is then the volume exhaled up to the vertical line. In effect this method measures the volume of the conducting airways down to the midpoint of the transition from dead space to alveolar gas. The volume of the anatomical dead space is not frequently measured, partly because the dead space changes little in disease, and partly because it requires a very rapid N_2 analyzer.

Physiological Dead Space

Another method of measuring dead space was described by Bohr (1891). Figure 5.10 shows that all the expired CO_2 comes from the alveolar gas and none from the dead space. Therefore:

$$V_T \cdot F_{E_{CO_2}} = V_A \cdot F_{A_{CO_2}}$$

where V_A indicates the component of the tidal volume which comes from the dead space.

But as stated earlier:

$$V_T = V_A + V_D$$

Therefore

$$V_A = V_T - V_D$$

and by substitution

$$V_T \cdot F_{E_{CO_2}} = (V_T - V_D) \cdot F_{A_{CO_2}}$$

whence

$$\frac{V_D}{V_T} = \frac{F_{A_{CO_2}} - F_{E_{CO_2}}}{F_{A_{CO_2}}}.$$

Because the partial pressure of a gas is proportional to its concentration,

$$\frac{V_D}{V_T} = \frac{P_{A_{CO_2}} - P_{E_{CO_2}}}{P_{A_{CO_2}}} \text{ (Bohr equation)}$$

In normal subjects the P_{CO_2} in alveolar gas and arterial blood are almost identical so that the equation is often written:

$$\frac{V_D}{V_T} = \frac{Pa_{CO_2} - P_{E_{CO_2}}}{Pa_{CO_2}}$$

The normal ratio of dead space to tidal volume is in the range 0.2–0.35 during resting breathing. The ratio decreases on exercise but increases with age.

It should be emphasized that Fowler's and Bohr's methods measure somewhat different things. Fowler's method measures the volume of the conducting airways down to the level where the rapid dilution of inspired gas occurs with gas already in the lung. This volume is determined by the geometry of the rapidly expanding airways (Figs. 5.3 and 5.9), and because it reflects the morphology of the lung, it is called the anatomical dead space. Bohr's method measures the volume of the lung which does not eliminate CO_2. Because this is a functional measurement, the volume is called the physiological dead space. In normal subjects the volumes are very nearly the same. In patients with lung disease, however, the physiological dead space may be considerably larger because of inequality of blood flow and ventilation within the lung (Chapter 37).

Analysis of Respiratory Gases

CHEMICAL ABSORPTION METHODS

The original methods of measuring the composition of respiratory gases depended on absorption of the CO_2 and oxygen and the corresponding changes in volume of the gas (at constant pressure). In *Haldane's method* (Peters and Van Slyke, 1932),

10 ml of gas is drawn into a gas buret, and the CO_2 is absorbed by adding potassium hydroxide and shaking the mixture. After recording the change in volume, potassium pyrogallate or some other oxygen absorber is added, and the change in volume is again determined. The remainder of the gas is then nitrogen (plus argon which is included with nitrogen in respiratory studies). In *Scholander's micromethod* (1947), only 0.5 ml of gas is required, and the volume changes are measured in a buret fitted with a micrometer gauge.

PHYSICAL METHODS

The above methods are still used for calibration but have been replaced by physical methods for most purposes. The *respiratory mass spectrometer* (Fowler, 1969) separates the components of a gas mixture according to their mass by accelerating the ions in electric and magnetic fields. The results of the analysis are immediately available on a pen recorder, and the time of analysis is typically only 0.1 s. These analyzers are expensive but are being increasingly used.

Infrared analyzers depend on the absorption of infrared radiation from the test gas. They are extensively employed for CO_2 and CO. *Thermal conductivity analyzers* measure the rate at which heat is lost from a hot wire immersed in the test gas. This principle is particularly suitable for helium. *Radiation emission* can be used to measure nitrogen when the gas is electrically excited (Fig. 5.11). Oxygen is sometimes measured by a *paramagnetic analyzer* or a *polaragraphic electrode* (see below). Finally *gas chromatography* is being increasingly used for the analysis of individual gas samples.

PULMONARY CIRCULATION

Just as ventilation brings oxygen to the blood-gas barrier where gas exchange occurs, so the pulmonary circulation picks up the oxygen in the alveoli and delivers it to the left heart from where it is distributed to the rest of the body.

The pulmonary circulation begins at the main pulmonary artery which receives the mixed venous blood pumped by the right ventricle. This artery then branches successively like the system of airways (Fig. 5.3), and indeed the pulmonary arteries accompany the bronchi down the centers of the secondary lobules as far as the terminal bronchioles. Beyond that they break up to supply the capillary bed which lies in the walls of the alveoli (Fig. 5.4). The pulmonary capillaries form a dense network in the alveolar wall which makes an exceedingly efficient arrangement for gas exchange (Figs. 5.2 and 5.4). So rich is the mesh that some

physiologists feel that it is misleading to talk of a network of individual capillary segments and prefer to regard the capillary bed as a sheet of flowing blood interrupted in places by posts, rather like an underground parking garage. The oxygenated blood is then collected from the capillary bed by the small pulmonary veins which run between the lobules and eventually unite to form the four large veins (in man) which drain into the left atrium.

At first site, this circulation appears to be simply a small version of the systemic circulation which begins at the aorta and ends in the right atrium. Indeed, the pulmonary circulation is often called the "lesser circulation." However, there are important differences between the two circulations, and confusion frequently results from attempts to emphasize similarities between them.

Pressures within the Pulmonary Blood Vessels

The pressures in the pulmonary circulation are remarkably low. The mean pressure in the main pulmonary artery is only about 15 mmHg; the systolic and diastolic pressures are about 25 and 8 mmHg, respectively (Fig. 5.12). The pressure is therefore very pulsatile. By contrast the mean pressure in the aorta is about 100 mmHg—about six times more than in the pulmonary artery. The pressures in the right and left atriums are not very dissimilar—about 2 and 5 mmHg, respectively. Thus the pressure differences from inlet to outlet of the pulmonary and systemic systems are about $(15 - 5) = 10$ and $(100 - 2) = 98$ mmHg, respectively—a factor of 10.

In keeping with these low pressures, the walls of the pulmonary artery and its branches are remarkably thin and contain relatively little smooth muscle (they are easily mistaken for veins). This is in striking contrast to the systemic circulation where the arteries generally have thick walls, and the

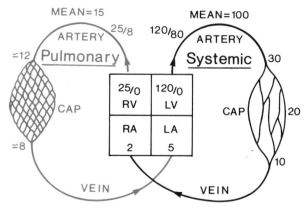

Figure 5.12. Comparison of pressures (mmHg) in the pulmonary and systemic circulations. Hydrostatic differences modify these.

arterioles in particular have abundant smooth muscle.

The reasons for these differences become clear when the functions of the two circulations are compared. The systemic circulation regulates the supply of blood to various organs, including those which may be far above the level of the heart (the upstretched arm, for example). By contrast, the lung is required to accept the whole of the cardiac output at all times. It is rarely concerned with directing blood from one region to another (an exception is localized alveolar hypoxia, see below), and its arterial pressure is therefore as low as is consistent with lifting blood to the top of the lung. This keeps the work of the right heart as small as is feasible for efficient gas exchange to occur in the lung.

The pressure within the pulmonary capillaries is uncertain. Several pieces of evidence suggest that it lies about halfway between pulmonary arterial and venous pressure, and some work indicates that much of the pressure drop occurs within the capillary bed itself. Certainly the distribution of pressures along the pulmonary circulation is far more symmetrical than in its systemic counterpart, where most of the pressure drop is just upstream of the capillaries (Fig. 5.12). In addition, the pressure within the pulmonary capillaries varies considerably throughout the lung because of hydrostatic effects (see below).

Pressures around the Pulmonary Blood Vessels

The pulmonary capillaries are unique in that they are virtually surrounded by gas (Figs. 5.2 and 5.4). It is true that there is a very thin layer of epithelial cells lining the alveoli, but the capillaries receive little support from this and, consequently, they are liable to collapse or distend, depending on the pressures within and around them. (The latter is very close to atmospheric pressure; indeed, during breath holding with the glottis open, the two pressures are identical.) Under some special conditions, the effective pressure around the capillaries is reduced by the surface tension of the fluid lining the alveoli. But usually, the effective pressure is alveolar pressure, and when this rises above the pressure inside the capillaries, they collapse. The pressure difference between the inside and outside of the vessels is often called the *transmural pressure*.

What is the pressure around the pulmonary arteries and veins? This can be considerably less than alveolar pressure. As the lung expands, these larger blood vessels are pulled open by the radial traction of the elastic lung parenchyma which surrounds them (Figs. 5.13 and 5.14). Consequently, the effec-

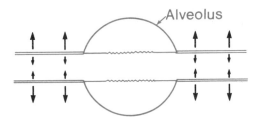

Figure 5.13. "Alveolar" and "extra-alveolar" vessels. The first are mainly the capillaries and are exposed to alveolar pressure. The second are pulled open by the radial traction of the surrounding lung parenchyma, and the effective pressure around them is therefore lower than alveolar pressure. [From Hughes et al. (1968).]

tive pressure around them is low; in fact, there is some evidence that this pressure is even less than the pressure around the whole lung (intrapleural pressure). This paradox can be explained by the mechanical advantage which develops when a relatively rigid structure like a blood vessel or bronchus is surrounded by a rapidly expanding elastic material such as lung parenchyma. At any event, both the arteries and veins increase their caliber as the lung expands.

The behavior of the capillaries and the larger blood vessels is so different that they are often referred to as alveolar and extra-alveolar vessels (Fig. 5.13). Alveolar vessels are exposed to alveolar pressure and include the capillaries and the slightly larger vessels in the corners of the alveolar walls. Their caliber is determined by the relationship between alveolar pressure and the pressure within them. Extra-alveolar vessels include all the arteries and veins which run through the lung parenchyma. Their caliber is greatly affected by the lung volume, since this determines the expanding pull of the parenchyma on their walls. The very large vessels near the hilum are outside the lung substance and are exposed to intrapleural pressure.

Pulmonary Vascular Resistance

It is useful to describe the resistance of a system of blood vessels as: vascular resistance = (input pressure − output pressure)/blood flow (see Chapter 5). This number is certainly not a complete description of the pressure-flow properties of the system. For example, the number usually depends on the magnitude of the blood flow. Nevertheless, it often allows a helpful comparison of different circulations, or the same circulation under different conditions.

We have seen that the total pressure drop from pulmonary artery to left atrium in the pulmonary circulation is only some 10 mmHg against about 100 mmHg for the systemic circulation. Since the

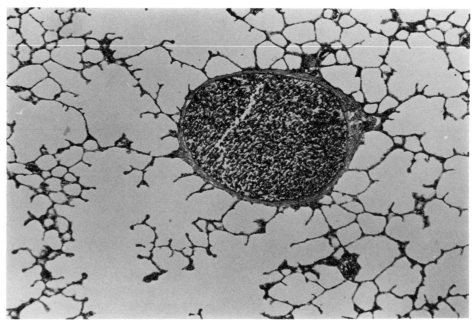

Figure 5.14. Section of lung showing many alveoli and an extra-alveolar vessel (in this case a small vein) with its perivascular sheath.

blood flows through the two circulations are virtually identical it follows that the pulmonary vascular resistance is only one-tenth of that of the systemic circulation. The pulmonary blood flow is about 6 l/min so that, in numbers, the pulmonary vascular resistance = (15 − 5)/6 or about 1.7 mmHg/l/min. The high resistance of the systemic circulation is largely caused by the muscular arterioles which allow the regulation of blood flow to various organs of the body. The pulmonary circulation has no such vessels and appears to have as low a resistance as is compatible with distributing the blood in a thin film over a vast area in the alveolar walls.

Although the normal pulmonary vascular resistance is extraordinarily small, it has a remarkable facility for becoming even smaller as the pressure within it rises. Figure 5.15 shows that an increase in either pulmonary arterial or venous pressures causes pulmonary vascular resistance to fall. Two mechanisms are responsible for this. Under normal conditions, some capillaries are either closed, or open but with no blood flow. As the pressure rises, these vessels begin to conduct blood, thus lowering the overall resistance. This is termed *recruitment* (Fig. 5.16) and is apparently the chief mechanism for the fall in pulmonary vascular resistance which occurs as the pulmonary artery pressure is raised from low levels. The reason why some vessels are unperfused at low perfusing pressures is not fully understood, but perhaps is caused by random differences in the geometry of the complex network (Fig. 5.4*B*) which result in preferential channels for flow.

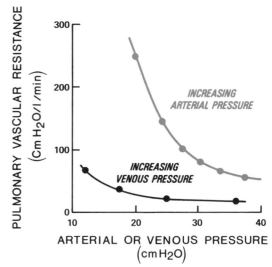

Figure 5.15. Fall in pulmonary vascular resistance as the pulmonary arterial or venous pressure is raised. When the arterial pressure was changed, the venous pressure was held constant at 12 cm H_2O, and when the venous pressure was changed, the arterial pressure was held at 37 cm H_2O. (Data from an excised dog lung preparation.)

At higher vascular pressures widening of individual capillary segments occurs. This increase in caliber or *distension* is hardly surprising in view of the very thin membrane which separates the capillary from the alveolar space (Fig. 5.2). Distension is apparently the predominant mechanism for the fall in pulmonary vascular resistance at relatively high vascular pressures.

Another important determinant of pulmonary vascular resistance is *lung volume*. The caliber of the extra-alveolar vessels (Fig. 5.13) is determined

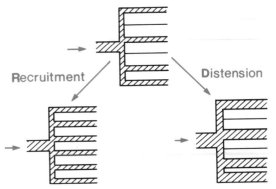

Figure 5.16. Recruitment (opening up of previously closed vessels) and distention (increase in caliber of vessels). These are the two mechanisms for the decrease in pulmonary vascular resistance which occurs as vascular pressures are raised.

by a balance between various forces. As we have seen, they are pulled open as the lung expands. As a result their vascular resistance is low at large lung volumes. On the other hand, their walls contain smooth muscle and elastic tissue which resist distention and tend to reduce their caliber. Consequently they have a high resistance when lung volume is low (Fig. 5.17). Indeed, if the lung is completely collapsed, the smooth muscle tone of these vessels is so effective that the pulmonary artery pressure has to be raised several centimeters of water above downstream pressure before any flow at all occurs. This is called a *critical opening pressure* (Chapter 6).

Is the vascular resistance of the capillaries influenced by lung volume? This depends on whether alveolar pressure changes with respect to the pressure inside the capillaries, that is, whether their transmural pressure alters. If alveolar pressure rises with respect to capillary pressure, the vessels tend to be squashed and their resistance rises. This usually occurs when a normal subject takes a deep inspiration because the pulmonary vascular pressures fall. (The heart is surrounded by intrapleural pressure.) However, the pressures in the pulmonary circulation do not remain steady after such a maneuver. An additional factor is that the caliber of the capillaries is reduced at large lung volumes because of stretching of the alveolar walls. Thus even if the transmural pressure of the capillaries is not changed with large lung inflations their vascular resistance increases (Fig. 5.17).

Because of the role of smooth muscle in determining the caliber of the extraalveolar vessels, drugs which cause contraction of the muscle increase pulmonary vascular resistance. These include serotonin, histamine, and norepinephrine. These drugs are particularly effective vasoconstrictors when the lung volume is low and the expanding forces on the vessels are weak. Drugs which can

relax smooth muscle in the pulmonary circulation include acetylcholine and isoproterenol.

Measurement of Pulmonary Blood Flow

Since the whole of the cardiac output normally goes through the lungs, pulmonary blood flow can be obtained from the Fick principle or indicator dilution methods as described in Chapter 13. The Fick and dye methods give the average flow over a number of heart cycles. It is also possible to measure *instantaneous pulmonary blood flow* using the body plethysmograph (Fig. 5.18). For this application the subject inhales a gas mixture containing 79% nitrous oxide and 21% O_2 from a rubber bag inside the box. Nitrous oxide is a very soluble gas, and as it is taken up by the blood, the box pressure

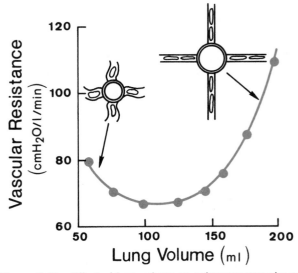

Figure 5.17. Effect of lung volume on pulmonary vascular resistance when the transmural pressure of the capillaries is held constant. At low lung volumes, resistance is high because the extra-alveolar vessels become narrow. At high volumes, the capillaries are stretched and their caliber is reduced. (Data from a dog lobe preparation.)

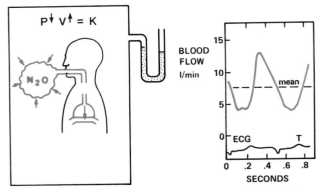

Figure 5.18. Measurement of instantaneous capillary blood flow by recording nitrous oxide uptake in the body plethysmograph. Calculated blood flow is shown on the *right* with the electrocardiogram.

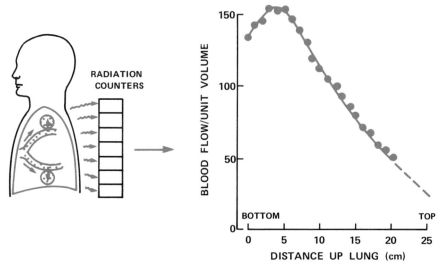

Figure 5.19.   Measurement of the distribution of blood flow in the upright human lung using radioactive xenon. The dissolved xenon is evolved into alveolar gas from the pulmonary capillaries. The units of blood flow are such that if flow were uniform, all values would be 100. Note the small flow at the apex. [Redrawn from Hughes et al. (1968).]

falls in a series of small steps which are synchronous with the heart beat. Since the uptake of nitrous oxide is flow-limited (Fig. 5.36), instantaneous blood flow can be calculated. In normal subjects, there is considerable pulsatility of pulmonary capillary blood flow, and this is altered by disease.

Distribution of Blood Flow

So far we have been assuming that all parts of the pulmonary circulation behave identically. However, considerable inequality of blood flow exists within the human lung. This can be shown by using the technique shown in Figure 5.19. The subject is seated with a bank of counters or a radiation camera mounted behind the chest. Radioactive xenon gas is dissolved in saline and injected into a peripheral vein. When the xenon reaches the pulmonary capillaries it is evolved into alveolar gas because of its low solubility, and the distribution of radioactivity can be measured by the counters during breath holding.

In the upright human lung, blood flow decreases almost linearly from bottom to top reaching very low values at the apex (Fig. 5.19). This distribution is affected by change of posture and exercise. When the subject lies supine, the apical zone blood flow increases, but the basal zone flow remains virtually unchanged, with the result that the distribution from apex to base becomes almost uniform. However, in this posture, blood flow in the posterior (dependent) regions of the lung exceeds flow in the anterior parts. Measurements on men suspended upside-down shows that apical blood flow may exceed basal flow in this position. On mild exercise,

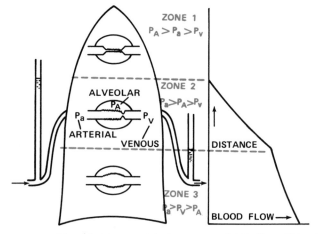

Figure 5.20.   Model to explain the uneven distribution of blood flow in the lung based on the pressures affecting the capillaries. [From West et al. (1964).]

both upper and lower zone blood flows increase, and the regional differences become less.

The uneven distribution of blood flow can be explained by the hydrostatic pressure differences within the blood vessels. If we consider the pulmonary arterial system as a continuous column of blood, the difference in pressure between the top and bottom of a lung 30 cm high will be about 30 cm water, or 23 mmHg. This is a large pressure difference for such a low pressure system as the pulmonary circulation (Fig. 5.12) and its effects on regional blood flow are shown in Figure 5.20.

There may be a region at the top of the lung (*zone 1*) where pulmonary arterial pressure falls below alveolar pressure (normally close to atmospheric pressure). If this occurs, the capillaries are squashed flat and no flow is possible. This zone 1 does *not* occur under normal conditions, because the pulmonary arterial pressure is just sufficient to raise blood to the top of the lung, but may be present if the arterial pressure is reduced (following severe hemorrhage, for example) or if alveolar pressure is raised (during positive pressure ventilation). This ventilated but unperfused lung is useless for gas exchange and is called *alveolar dead space.*

Further down the lung (*zone 2*), pulmonary arterial pressure increases because of the hydrostatic effect and now exceeds alveolar pressure. However, venous pressure is still very low and is less than alveolar pressure, and this leads to remarkable pressure-flow characteristics. Under these conditions blood flow is determined by the difference between arterial and alveolar pressures (not the usual arterial-venous pressure difference). Indeed venous pressure has no influence on flow unless it exceeds alveolar pressure. This situation can be modeled with a flexible rubber tube inside a glass chamber (Fig. 5.21). When chamber pressure is greater than downstream pressure, the rubber tube collapses at its downstream end, and the pressure in the tube at this point limits flow. The pulmonary capillary bed is clearly very different from a rubber tube, but nevertheless the overall behavior is similar and is often called the Starling resistor, sluice or waterfall effect. Since arterial pressure is increasing down the zone but alveolar pressure is the same throughout the lung, the pressure difference responsible for flow increases. In addition, increasing recruitment of capillaries occurs down this zone.

In *zone 3*, venous pressure now exceeds alveolar

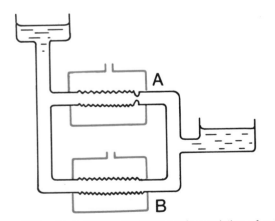

Figure 5.21. Two Starling resistors each consisting of a thin rubber tube inside a container. When chamber pressure exceeds downstream pressure as in *A*, flow is independent of downstream pressure. However, when downstream pressure exceeds chamber pressure as in *B*, flow is determined by the upstream-downstream difference.

pressure, and flow is determined in the usual way by the arterial-venous pressure difference. The increase in blood flow down this region of the lung is apparently caused chiefly by distention of the capillaries. The pressure within them (lying between arterial and venous) increases down the zone, while the pressure outside (alveolar) remains constant. Thus their transmural pressure rises, and indeed measurements show that their mean width increases. Recruitment of previously closed vessels may also play some part in the increase in blood flow down this zone.

The scheme shown in Figure 5.20 summarizes the role played by the capillaries in determining the distribution of blood flow. At low lung volumes, the resistance of the extraalveolar vessels becomes important, and a reduction of regional blood flow is seen starting first at the base of the lung where the parenchyma is least expanded (see Fig. 5.71). This reduced blood flow can be explained by the narrowing of the extraalveolar vessels which occurs when the lung around them is poorly inflated (Fig. 5.17).

Hypoxic Vasoconstriction

We have seen that passive factors dominate the vascular resistance and the distribution of flow in the pulmonary circulation under normal conditions. However, a remarkable active response occurs when the P_{O_2} of alveolar gas is reduced. This consists of contraction of smooth muscle in the walls of the small arterioles in the hypoxic region. The precise mechanism of this response is obscure, but it occurs in excised isolated lung and so does not depend on central nervous connections. Excised segments of pulmonary artery can be shown to constrict if their environment is made hypoxic, so that it may be a local action of the hypoxia on the artery itself. One hypothesis is that cells in the perivascular tissue release some vasoconstrictor substance in response to hypoxia, but an intensive search for the mediator has not been successful. Interestingly, it is the P_{O_2} of the alveolar gas, not the pulmonary arterial blood, which chiefly determines the response. This can be proved by perfusing a lung with blood of a high P_{O_2} while keeping the alveolar P_{O_2} low. Under these conditions the response occurs.

The vessel wall presumably becomes hypoxic through diffusion of oxygen over the very short distance from the wall to the surrounding alveoli. Recall that a small pulmonary artery is very closely surrounded by alveoli (compare the proximity of alveoli to the small pulmonary vein in Fig. 5.14). The stimulus-response curve of this constriction is very nonlinear. When the alveolar P_{O_2} is altered in the region above 100 mmHg, little change in vascular resistance is seen. However when the alveolar P_{O_2} is reduced below approximately 70 mmHg, marked vasoconstriction may occur, and at a very low P_{O_2} the local blood flow may be almost abolished.

The vasoconstriction has the effect of directing blood flow away from hypoxic regions of lung. These regions may result from bronchial obstruction, and by diverting blood flow the deleterious effects on gas exchange are reduced. At high altitude, generalized pulmonary vasoconstriction may occur, leading to a large rise in pulmonary arterial pressure and a substantial increase in work for the right heart. But probably the most important situation where this mechanism operates is at birth. During fetal life, the pulmonary vascular resistance is very high, partly because of hypoxic vasoconstriction, and only some 15% of the cardiac output goes through the lungs (see Chapter 19). When the first breath oxygenates the alveoli, the vascular resistance falls dramatically because of relaxation of vascular smooth muscle, and the pulmonary blood flow enormously increases.

Other active responses of the pulmonary circulation have been described. A low blood pH causes vasoconstriction, especially when alveolar hypoxia is present. There is also evidence that the autonomic nervous system exerts a weak control, an increase in sympathetic outflow causing stiffening of the walls of the pulmonary arteries and vasoconstriction.

Water Balance in the Lung

Since only $\frac{1}{2}$ μm of tissue separates the capillary blood from the air in the lung (Fig. 5.2), the problem of keeping the alveoli free of fluid is critical. Fluid exchange across the capillary wall is believed to obey Starling's law (Chapter 6). The force tending to push fluid *out* of the capillary is the capillary hydrostatic pressure minus the hydrostatic pressure in the interstitial fluid, or $P_c - P_i$. The force tending to pull fluid *in* is the colloid osmotic pressure of the proteins of the blood minus that of the proteins of the interstitial fluid, or $\pi_c - \pi_i$. This force depends on the reflection coefficient σ which indicates the effectiveness of the capillary wall in preventing the passage of proteins across it. Thus

$$\text{net fluid out} = K[(P_c - P_i) - \sigma (\pi_c - \pi_i)]$$

where K is a constant called the filtration coefficient.

Unfortunately, the practical use of this equation is limited because of our ignorance of many of the values. The colloid osmotic pressure within the

capillary is about 28 mmHg. The capillary hydrostatic pressure is probably about halfway between arterial and venous pressure but is much higher at the bottom of the lung than at the top. The colloid osmotic pressure of the interstitial fluid is not known but is about 20 mmHg in lung lymph. However, this value may be higher than that in the interstitial fluid around the capillaries. The interstitial hydrostatic pressure is unknown but is thought by some physiologists to be substantially below atmospheric pressure. It is probably that the net pressure of the Starling equation is outward, causing a small lymph flow of perhaps 20 ml/hr in man under normal conditions.

Where does fluid go when it leaves the capillaries? Figure 5.22 shows that fluid which leaks out into the interstitium of the alveolar wall tracks through the interstitial space to the perivascular and peribronchial spaces within the lung. Numerous lymphatics run in the peribronchial and perivascular spaces, and these help to transport the fluid to the hilar lymph nodes. In addition, the pressure in these perivascular spaces is low, thus forming a natural sump for the drainage of fluid (compare Fig. 5.13). The earliest form of pulmonary edema is characterized by engorgement of these peribronchial and perivascular spaces and is known as interstitial edema.

In a later stage of pulmonary edema, fluid crosses the alveolar epithelium into the alveolar spaces (Fig. 5.22). When this occurs the alveoli fill with fluid one by one, and since they are then unventilated, no oxygenation of the blood passing through them is possible. What causes fluid to start moving across the alveolar spaces is not known, but it may be that this occurs when the maximal drainage rate

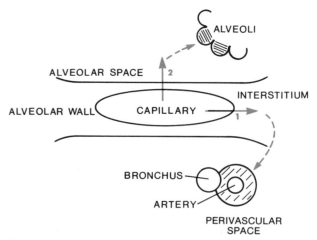

Figure 5.22.   Two routes for loss of fluid from pulmonary capillaries. Fluid which enters the interstitium initially finds its way into the perivascular and peribronchial spaces. Later fluid may cross the alveolar wall, filling alveolar spaces.

through the interstitial space is exceeded and the pressure there rises too high. The normal rate of lymph flow from the lung is only a few milliliters per hour, but it can be shown to increase greatly if the capillary pressure is raised over a long period. Alveolar edema is much more serious than interstitial edema because of the interference with pulmonary gas exchange.

Other Functions of the Pulmonary Circulation

The chief function of the pulmonary circulation is to move blood to and from the blood-gas barrier so that gas exhange can occur. However, it has other important functions too. One is to act as a reservoir for blood. We saw that the lung has a remarkable ability for reducing its pulmonary vascular resistance as its vascular pressures are raised through the mechanisms of recruitment and distension (Fig. 5.16). The same mechanisms allow the lung to increase its blood volume with relatively small rises in pulmonary arterial or venous pressures. This occurs, for example, when a subject lies down after standing. Blood then drains from the legs into the lung.

Another function of the lung is to filter blood. Small blood thrombi are removed from the circulation before they can reach the brain or other vital organs. There is also evidence that many white blood cells are trapped by the lung, although the value of this is not clear.

Metabolic Functions of the Lung

The lung has important metabolic functions in addition to gas exchange, and this is a convenient place to discuss them briefly. One of the most important of these is the synthesis of phospholipids such as dipalmitoyl lecithin, which is a component of pulmonary surfactant. This is considered further in Chapter 38. Protein synthesis is also clearly important, since collagen and elastin form the structural framework of the lung. Under abnormal conditions, proteases are apparently liberated from leukocytes or macrophages in the lung, causing breakdown of proteins and thus emphysema. Another significant area is carbohydrate metabolism, especially the elaboration of mucopolysaccharides of bronchial mucus.

A number of vasoactive substances are metabolized by the lung as shown in Figure 5.23. Since the lung is the only organ which receives the whole circulation, it is uniquely suited to modifying blood-borne substances. The only known example of biologic activation by passage through the pulmonary circulation is the conversion of the relatively inac-

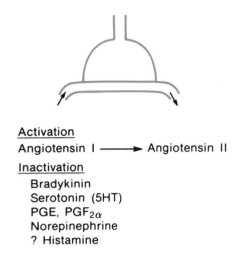

Activation
Angiotensin I ——→ Angiotensin II
Inactivation
    Bradykinin
    Serotonin (5HT)
    PGE, $PGF_{2\alpha}$
    Norepinephrine
    ? Histamine

Figure 5.23. Pulmonary metabolism of vasoactive substances. The inactivation of bradykinin, serotonin, PGE, and $PGF_{2\alpha}$ is highly effective but only partial for norepinephrine and histamine.

tive polypeptide, angiotensin I, to the potent vasoconstrictor, angiotensin II. The latter, which is up to 50 times more active than its precursor, is unaffected by passage through the lung. The conversion of angiotensin I is catalyzed by an enzyme (angiotensin-converting enzyme) which is located on the walls of the capillary endothelial cells.

Many vasoactive substances are completely or partially inactivated during passage through the lung. Among those that are almost completely (more than 80%) inactivated or removed are bradykinin, serotonin (5-hydroxytryptamine), and the prostaglandins E_1, E_2, and $F_{2\alpha}$. Norepinephrine and possibly histamine are taken up to a lesser degree. In case of some substances, such as serotonin and norepinephrine, the loss of activity is mainly due to uptake and storage. For other materials, inactivation is by the action of an enzyme, for example the peptidase which breaks down bradykinin. It is of interest that the same enzyme which activates angiotensin I also inactivates bradykinin.

Some vasoactive materials pass through the lung without significant gain or loss of activity. These include epinephrine, prostaglandins A_1 and A_2, angiotensin II, and vasopressin (ADH).

Several vasoactive substances are normally synthesized or stored within the lung but may be released into the circulation in pathological conditions. For example, in anaphylaxis, or during an asthma attack, histamine, bradykinin, prostaglandins, and "slow-reacting substances" are discharged into the circulation. Other conditions in which the lung may release potent chemicals include pulmonary embolism and alveolar hypoxia.

There is also evidence that the lung plays a role in the clotting mechanism of blood under normal and abnormal conditions. For example, there are large numbers of mast cells containing heparin in the interstitium. Finally, the lung is able to secrete special immunoglobulins, particularly IgA, in the bronchial mucus which contribute to its defenses against infection. The significance of some of these metabolic functions is still obscure, but it is clear that the lung has important functions in addition to its main role of gas exchange.

CARRIAGE OF OXYGEN AND CARBON DIOXIDE BY THE BLOOD

In some ways it would be more logical to consider gas exchange within the lung before dealing with the transport of oxygen and CO_2 by the blood to the periphery of the body. However, knowledge of the oxygen and CO_2 blood dissociation curves is essential for an understanding of gas exchange, so we shall deal with this topic now.

Oxygen

Oxygen is carried in the blood in two forms—dissolved and in combination with hemoglobin.

DISSOLVED OXYGEN

This obeys Henry's law, that is, the amount dissolved is proportional to the partial pressure (see *bottom broken line* in Fig. 5.24). For each mmHg of P_{O_2} there is 0.003 ml of O_2/100 ml of blood (sometimes written 0.003 vol%). Thus normal arterial blood with a P_{O_2} of 100 mmHg contains 0.3 ml of oxygen/100 ml.

It is easy to see that this way of transporting oxygen must be inadequate for man. Suppose that on exercise the maximal cardiac output is 25 l/min (25,000 ml/min). Then the total amount of oxygen

which could be delivered to the muscles would be only $25,000 \times 0.3/100$ or 75 ml of oxygen/min, assuming that all the oxygen could be unloaded. But the maximal oxygen consumption on exercise is of the order of 3,000 ml/min, so clearly an additional method of transporting oxygen is required.

OXYGEN DISSOCIATION CURVE

At this point the reader may wish to review those aspects of the biochemistry of hemoglobin which were covered in Chapter 24. Oxygen forms an easily reversible combination with hemoglobin (Hb) to give oxyhemoglobin:

$$O_2 + Hb \rightleftharpoons HbO_2$$

Suppose we take a number of glass containers (tonometers), each containing a small volume of blood, and add gas with various concentrations of oxygen. After allowing time for the gas and blood to reach equilibrium, we measure the P_{O_2} of the gas and the oxygen content of the blood. Knowing that 0.003 ml of oxygen will be dissolved in each 100 ml of blood/mmHg of P_{O_2}, we can calculate the oxygen combined with Hb (Fig. 5.24). Note that the amount of oxygen carried by the Hb increases rapidly up to a P_{O_2} of about 50 mmHg, but at a higher P_{O_2} the curve becomes much flatter. The maximum amount of oxygen that can be contained with Hb is called the *oxygen capacity*. It can be measured by exposing the blood to a very high P_{O_2} (say 600 mmHg) and subtracting the dissolved oxygen. One gram of pure Hb can combine with 1.39 ml of oxygen‡ and since normal blood has about 15 g of Hb/100 ml the oxygen capacity is about 20.8 ml of O_2/100 ml of blood.

The *oxygen saturation* of Hb is given by:

$$\frac{O_2 \text{ combined with Hb}}{O_2 \text{ capacity}} \times 100$$

The oxygen saturation of arterial blood with a P_{O_2} of 100 mmHg is about 97.5%, while that of mixed venous blood with a P_{O_2} of 40 mmHg is about 75%. It is important to grasp the relationships between P_{O_2}, O_2 saturation, and oxygen content (or concentration). For example, suppose a severely anemic patient with a Hb concentration of only 7.5 g/100 ml of blood has normal lungs and an arterial P_{O_2} of 100 mmHg. His oxygen saturation will be 97.5% (at normal pH, P_{O_2}, and temperature), but the oxygen combined with Hb will be only 10.4 ml/100

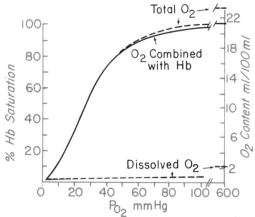

Figure 5.24. Oxygen dissociation curve (*solid line*) for pH 7.4, P_{CO_2} 40 mmHg and temperature 37°C. The total blood oxygen content is also shown for a hemoglobin concentration of 15 g/100 ml of blood.

‡ Some authorities give 1.34 or 1.36 ml/100 ml. The reason is that under the normal conditions of the body, some of the hemoglobin is in forms such as methemoglobin which cannot combine with oxygen.

ml. Dissolved oxygen will contribute 0.3 ml giving a total oxygen content of 10.7 ml/100 ml of blood.

The curved shape of the oxygen dissociation curve has several physiological advantages. The nearly flat upper portion assists in the diffusion of oxygen across the blood-gas barrier in the lung and thus the loading of oxygen by the blood; this will become clearer when we consider pulmonary gas exchange in Chapter 37. An additional advantage is that small falls in the P_{O_2} of alveolar gas will hardly affect the oxygen content of arterial blood and thus the amount of oxygen available to the tissues. The steep lower part of the dissociation curve means that the peripheral tissues can withdraw large amounts of oxygen for only a small drop in capillary P_{O_2}. This maintenance of blood P_{O_2} assists the diffusion of oxygen into the tissue cells.

Oxygenated Hb is bright red but reduced Hb is purple, so that a low arterial oxygen saturation causes *cyanosis*. This is not a reliable clinical sign of mild desaturation, however, because its recognition depends on so many variables, such as lighting conditions and skin pigmentation. Since it is the amount of reduced Hb that is important, cyanosis is often marked when polycythemia is present, but it may be difficult to detect in anemic patients.

Several factors may shift the position of the oxygen dissociation curve, including changes in the pH, P_{CO_2}, and temperature of the blood, and the concentration of organic phosphates within the red blood cell. Figure 5.25 shows that a fall in pH, rise in P_{CO_2}, and rise in temperature all shift the curve

to the right. Opposite changes shift it to the left. Most of the effect of P_{CO_2}, which is known as the *Bohr effect* (1904), can be attributed to its action on pH. As discussed in Chapter 24, an increase in H^+ concentration slightly alters the configuration of the Hb molecule and thus reduces the accessibility of oxygen to the heme groups. The consequent rightward shift enhances the unloading of oxygen for a given P_{O_2} in a tissue capillary. A simple way of remembering these shifts is that an exercising muscle is acid, hot, and has a high P_{CO_2}. Consequently, it benefits from the increased unloading of oxygen in its capillaries.

An increase in organic phosphates, particularly 2,3-diphosphoglycerate (2,3-DPG), within the red blood cell also shifts the oxygen dissociation to the right and thus assists unloading of oxygen (Benesch and Benesch, 1969; Chanutin and Curnish, 1967). Some biochemical aspects of this subject are discussed in Chapter 24. An increase in red cell 2,3-DPG occurs in chronic hypoxia, for example, after 2 days of ascent to an altitude of 4,500 m (about 15,000 feet) (Lenfant et al., 1970). Calculations suggest that the resultant shift of the dissociation curve increases the amount of oxygen released from the blood by about 10%. This may be a useful feature of acclimatization to moderately high altitudes, although it is less important than other factors such as hyperventilation (see Chapter 40). At much higher altitudes the advantages of the increase in 2,3-DPG disappear because the loading of oxygen in the pulmonary capillaries is adversely affected by the rightward shift. Other hypoxic states in which a useful increase in 2,3-DPG are seen include chronic lung disease, cyanotic heart disease, and severe anemias. The resultant enhanced unloading of oxygen in the peripheral tissues can be regarded as an intrinsic adaptive mechanism in response to hypoxemia.

In blood stored for transfusions, a slow decrease in 2,3-DPG brings about increased oxygen affinity, with the result that after transfusion of large quantities the release of oxygen to the tissues is subnormal. The change in the blood is influenced by the preservative used and can be retarded by the addition of inosine (Bunn et al., 1969).

A useful measure of the position of the oxygen dissociation curve is the P_{O_2} for 50% oxygen saturation. This is known as the P_{50}. The normal value for human blood is about 26 mmHg at P_{CO_2} 40 mmHg, pH 7.4, and temperature 37°C. The P_{50} differs between species of animals, often being appreciably lower in animals with large body weight (for example the small rhesus monkey and the gorilla have P_{50} values of approximately 32 and 25

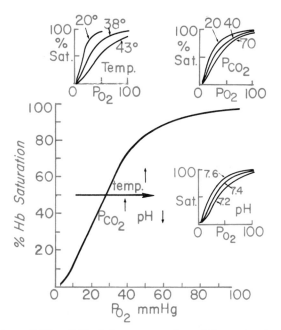

Figure 5.25. Shift of the oxygen dissociation curve by pH, P_{CO_2}, and temperature (*temp.*); *Sat.*, saturation.

mmHg, respectively). The changes in P_{50} can presumably be explained by variations in composition of the globin portion of the molecule. The term *oxygen affinity* is also used to refer to the position of the oxygen dissociation curve; the affinity increases as the P_{50} decreases.

Carbon Monoxide

Carbon monoxide interferes with the oxygen transport function of blood by combining with Hb to form carboxyhemoglobin (COHb). The dissociation curves of oxy- and carboxyhemoglobin have basically similar shapes. However, CO has about 250 times the affinity of oxygen for Hb; this means that CO will combine with the same amount of Hb as oxygen when the CO partial pressure is 250 times lower. For example, at a P_{CO} of 0.16 mmHg, 75% of the Hb is combined with CO as COHb. Thus when plotted on the same scale (Fig. 5.26) the CO dissociation curve has almost a right-angled bend near a P_{CO} of zero.

The higher affinity of CO for Hb means that people inadvertently exposed to small concentrations of CO in the air (for example, during a fire in a building) may have a large proportion of their Hb tied up as COHb, and thus unavailable for oxygen carriage. If this happens, the Hb concentration and P_{O_2} of the blood may be normal, but its oxygen content is grossly reduced. Small amounts of COHb also shift the oxygen dissociation curve to the left, thus making it difficult for the blood to unload the oxygen that it does carry. Figure 5.26 shows also the oxygen dissociation curve of blood containing 60% COHb as contrasted to the curve for anemic blood.

A patient with CO poisoning is treated by giving him pure oxygen or 95% O_2 with 5% CO_2 to breathe. The addition of 5% CO_2 to the inspired gas increases the rate of washout of CO from the blood. In specialized medical centers, hyperbaric oxygen therapy is often given because by raising the inspired P_{O_2} to 2,000 mmHg, all the oxygen required by the body tissues can be carried by the blood in the dissolved form (see Chapter 40).

Carbon Dioxide

CARBON DIOXIDE CARRIAGE

Carbon dioxide is carried in the blood in three forms—dissolved, as bicarbonate, and in combination with proteins as carbamino compounds (Fig. 5.27).

Dissolved CO_2

Like oxygen, dissolved CO_2 obeys Henry's law, but because CO_2 is about 20 times more soluble than O_2, the dissolved form plays a more significant role in the normal carriage in the blood. In fact about 10% of the gas which is evolved into the lung from the mixed venous blood is in the dissolved form (Fig. 5.28).

Bicarbonate

This is formed in the blood by the following sequence:

$$CO_2 + H_2O \overset{C.A.}{\rightleftharpoons} H_2CO_3 \rightleftharpoons H^+ + HCO_3^-$$

The first reaction is very slow in plasma but fast within the red blood cell because of the presence there of the enzyme *carbonic anhydrase* (CA). This is a zinc-containing protein which is present in a high concentration in the red cell but not in the

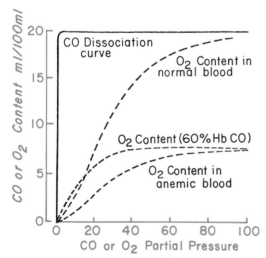

Figure 5.26. Comparison of dissociation curves for CO and oxygen. Note that the hemoglobin (Hb) is almost completely saturated with CO at a very low partial pressure. Note also that the CO dissociation curve for 60% HbCO is displaced to the *left* compared with the oxygen dissociation curve for anemic blood.

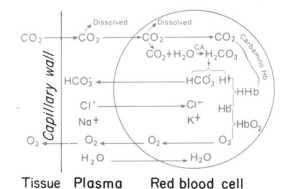

Figure 5.27. Scheme of the uptake of O_2 and liberation of carbon dioxide in systemic capillaries. Exactly opposite events occur in the pulmonary capillaries.

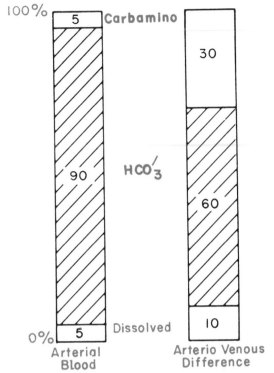

Figure 5.28. The column on the *left* shows the proportions of the total CO_2 content in arterial blood. The column on the *right* shows the proportions which make up the arterial-venous difference.

plasma. It is also found in other tissues, including gastric and intestinal mucosa, renal cortex, and muscle. It can be inhibited by the administration of acetazolamide (Diamox) which causes diuresis in man through its action on the kidney (Chapter 32). In experimental animals, very large doses of acetazolamide interfere with CO_2 transport by slowing the hydration reaction (Roughton, 1964).

Ionization of the carbonic acid within the red cell occurs rapidly and does not need an enzyme. When the concentration of hydrogen and bicarbonate ions in the cell rises, HCO_3^- diffuses out, but H^+ cannot move out easily because the cell membrane is relatively impermeable to cations. Thus in order to maintain electrical neutrality, Cl^- ions diffuse into the cell from the plasma, the so-called *chloride shift* (Fig. 5.27). The movement of chloride is in accordance with the Gibbs-Donnan equilibrium (Chapter 2).

Some of H^+ ions which are liberated are bound to hemoglobin:

$$H^+ + HbO_2 \rightleftharpoons H\cdot Hb + O_2$$

This reaction occurs because reduced Hb is a better proton acceptor (less "acid") than the oxygenated form. Thus the presence of reduced Hb in the peripheral blood helps with the loading of CO_2,

while the oxygenation which occurs in the pulmonary capillary assists in the unloading. The fact that the deoxygenation of the blood increases its ability to carry CO_2 is often known as the *Haldane effect* (Christiansen et al., 1914).

These events associated with the uptake of CO_2 by the blood increase the osmolar content of the red cell, and consequently water enters the cell, thus increasing its volume. When the cells pass through the lung, they shrink a little.

Carbamino Compounds

These are formed by the combination of CO_2 with terminal amine groups of blood proteins. The most important protein is the globin of hemoglobin, and the reaction can be represented:

$$CO_2 + Hb\cdot NH_2 \rightleftharpoons Hb\cdot NH\cdot COOH \rightleftharpoons Hb\cdot NH\cdot COO + H^+$$

This occurs very rapidly without an enzyme, and most of the carbamic acid is in the ionized form. Reduced Hb can bind much more CO_2 than HbO_2. Thus, again unloading of oxygen in peripheral capillaries facilitates the loading of CO_2 while oxygenation enhances the unloading of CO_2 in the lung (Haldane effect).

The relative contributions of the various forms of CO_2 in blood to the total CO_2 content are shown in Figure 5.28. Note that the great bulk of the CO_2 is in the form of bicarbonate. The amount in the dissolved form is small as is that as carbaminohemoglobin. However, these proportions do not reflect the changes that take place when CO_2 is loaded or unloaded by the blood. Of the total venous-arterial difference, about 60% is attributable to HCO_3^-, 30% to carbamino compounds, and 10% to dissolved CO_2.

CO_2 DISSOCIATION CURVE

The relationship between the PCO_2 and the total CO_2 content of blood is shown in Figure 5.29. Note that the CO_2 dissociation curve is much more linear than the oxygen dissociation curve (Fig. 5.24). Note also that the lower the saturation of Hb with oxygen the larger the CO_2 content for a given PCO_2. As mentioned earlier, this Haldane effect can be explained by the better ability of reduced Hb to mop up the H^+ ions produced when carbonic acid dissociates and the greater facility of reduced Hb to form carbaminohemoglobin.

Figure 5.30A shows that the CO_2 dissociation curve is considerably steeper than that for oxygen. For example, in the range 40–50 mmHg, the CO_2 content changes by about 4.7 ml/100 ml compared with the oxygen content of only about 1.7 ml/100 ml.

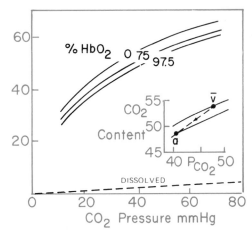

Figure 5.29. CO_2 dissociation curves for blood of different oxygen saturations. Note that oxygenated blood carries less CO_2 for the same P_{CO_2}. The "physiological" curve between arterial and mixed venous blood is also shown in the inset.

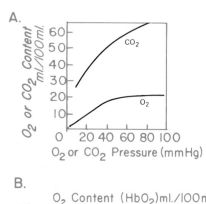

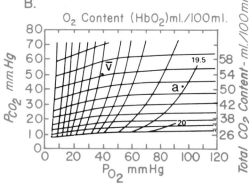

Figure 5.30. *A.*, Typical oxygen and CO_2 dissociation curves plotted with the same scales. Note that the CO_2 curve is much steeper. *B.*, O_2-CO_2 diagram showing lines of equal oxygen and CO_2 content. The lines are not parallel to the X and Y axes because of the Bohr and Haldane effects (see text).

A useful way of displaying the interactions between the oxygen and CO_2 dissociation curves is by means of the O_2-CO_2 diagram (Rahn and Fenn, 1955) (compare Fig. 5.45). In this diagram (Fig. 5.30*B*) the X and Y axes show the partial pressures of oxygen and CO_2, respectively, in samples of blood. Note that the lines of oxygen and CO_2 content are not straight and parallel to the axes as they

would be if the contents were simply proportional to the partial pressures. Choose a P_{O_2} on the X axis (say 50 mmHg). Follow this P_{O_2} vertically (increasing P_{CO_2}) and lines of decreasing oxygen content will be encountered (Bohr effect). The same procedure following a given P_{CO_2} to the right (increasing P_{O_2}) will give decreasing CO_2 contents (Haldane effect). This diagram will be used in the next chapter for analysis of pulmonary gas exchange.

It should be emphasized that the events involved in the transport of CO_2 by the blood have a profound effect on the acid-base status of the body. This important subject is discussed in detail in Chapter 33.

ANALYSIS OF BLOOD GASES

Removal of Blood Samples

Arterial or venous blood can be drawn with a syringe and needle. Arterial blood is commonly obtained by puncture of the brachial or radial artery after infiltrating the skin with local anesthetic. The dead space of the syringe is filled with dilute heparin solution to prevent clotting of the blood. At the end of the procedure, pressure is applied to the site to prevent extravasation of blood. Venous blood can be obtained from an arm vein for determination of the hematocrit and electrolyte concentrations. Such a sample is of little value for blood gases, however, because it refelects only the local conditions.

To obtain mixed venous blood from the pulmonary artery, a long catheter must be passed through the right heart. Fine plastic catheters, sometimes with a small balloon on the end (Swan-Ganz), can be floated in from a peripheral vein along the blood stream without much difficulty. However, very occasionally an abnormal heart rhythm develops, and the procedure is not as innocuous as an arterial puncture.

Blood P_{O_2}, P_{CO_2} and pH

The determination of P_{O_2}, P_{CO_2} and pH in blood has been revolutionized by the introduction of blood gas electrodes which enable these measurements to be made in a few minutes.

Oxygen partial pressure is measured with a polarographic oxygen electrode (Fig. 5.31). The principle is that if a small voltage (0.6 V) is applied to a platinum electrode immersed in a buffer solution, the current which flows is proportional to the P_{O_2}. In practice the buffer is separated from the blood by a semipermeable membrane through which oxygen diffuses. Oxygen is consumed by the electrode; consequently, the P_{O_2} falls with time. The electrode

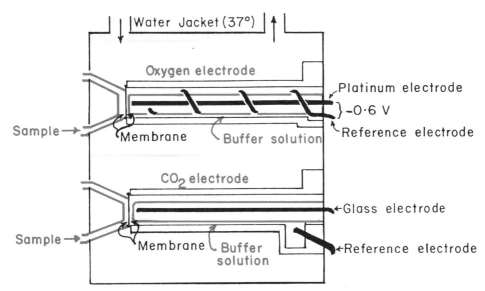

Figure 5.31. Oxygen and CO_2 electrodes for measuring the partial pressures of these gases in blood. For the measurement of oxygen, the gas diffuses through the semipermeable membrane and causes a flow of current which is proportional to P_{O_2}. For CO_2, the gas diffuses through the semipermeable membrane and changes the pH of the buffer solution. This is measured by the glass electrode.

is calibrated with gas or with a solution of known P_{O_2}.

Carbon dioxide partial pressure is measured with a P_{CO_2} electrode (Fig. 5.31). This is essentially a glass pH electrode surrounded by a bicarbonate buffer which is separated from blood by a thin membrane. CO_2 diffuses through the membrane and alters the pH of the buffer. This is measured by the electrode which reads out P_{CO_2} directly. Calibration is by means of CO_2 gas mixtures.

Blood pH can be measured with a glass pH electrode immersed in the blood. The reading gives the plasma pH, the value inside the cells being 0.1–0.2 pH units lower. In practice the P_{O_2}, P_{CO_2}, and pH electrodes are part of the same piece of equipment, and all three measurements can be made on 1 ml of blood.

Oxygen Saturation

Blood oxygen saturation can be determined by spectrophotometry because of the different absorption of light of certain wavelengths by oxygenated and reduced hemoglobin. The sample of blood is placed in a cuvette through which light is passed, and the oxygen saturation is read out directly on a meter. Other devices measure the light which is reflected from the surface of a blood sample. The amount of carboxyhemoglobin can sometimes be measured in the same instrument because of its special absorption characteristics.

The *ear oximeter* is a specialized small spectrophotometer which measures the oxygen saturation of "arterialized" blood flowing through the heated ear lobe. Because of the consequent vasodilation, the ear lobe blood flow increases so much that the oxygen saturation of venous blood approaches that of arterial blood. This is not an accurate measurement, especially when the oxygen saturation is very low, but it is often useful in clinical investigation.

Oxygen and CO_2 Contents

These can be determined with the Van Slyke manometric apparatus (Peters and Van Slyke, 1932). The gases are liberated from the blood by adding lactic acid, potassium ferricyanide, and saponin, and vigorously shaking the mixture in a partial vacuum. The evolved gas is then analyzed by absorption with sodium hydroxide and an anthraquinone mixture. Oxygen content can also be derived from devices employing polarographic electrodes or fuel cells to consume the oxygen.

GAS TRANSPORT TO THE TISSUES

Diffusion from Capillary to Cell

Oxygen and CO_2 move between the systemic capillary blood and the tissue cells by diffusion from regions of high to low partial pressures. The principles governing diffusion (Fick's law) will be discussed in detail in the next chapter. Here it should be pointed out that the distance to be covered by diffusion in the peripheral tissues is considerably greater than in the lung. For example, the distance between open capillaries in resting muscle is of the order of 50 μm, whereas the thickness of the blood-gas barrier in the lung is only 1/100 of this. On the other hand, during exercise when the oxygen consumption of the muscle increases, additional capillaries open up, thus reducing the diffusion distance and increasing the cross-sectional area available for diffusion. Because CO_2 diffuses about 20 times faster than oxygen through tissue (see Fig. 5.35) elimination of CO_2 poses less of a problem than oxygen delivery.

Recent measurements suggest that the movement of oxygen through some tissues in vitro is too fast to be accounted for by simple passive diffusion. It is possible that *facilitated diffusion* occurs under some circumstances (see Chapter 2), possible carriers being cytochrome P-450 or, within muscle cells, myoglobin. Another possibility is that convective processes ("stirring") occur on a small scale, perhaps within cells.

Tissue Partial Pressures

Various models have been suggested to account for the way in which the P_{O_2} falls in tissue between adjacent open capillaries. Figure 5.32A shows a hypothetical cylinder of tissue between capillaries in which blood comes in with a P_{O_2} of 100 mmHg. As the oxygen diffuses from the peripheral capillary into the center of the cylinder, the oxygen is consumed and the P_{O_2} falls. In *1.* the balance between oxygen consumption and delivery (determined by the capillary P_{O_2} and the intercapillary distance)

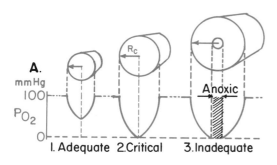

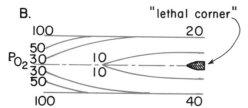

Figure 5.32. Scheme showing the fall in P_{O_2} between adjacent open capillaries. In *A.* three hypothetical cylinders of tissue are shown. In *2.* the cylinder has a critical radius R_c while in *3.* the radius of the cylinder is so large that there is an anoxic zone in the middle of the cylinder. *B.* shows a section along the hypothetical cylinder of tissue. The P_{O_2} in the blood adjacent to the tissue is assumed to fall from 100 to 20 mmHg along the capillary. Lines of equal P_{O_2} are shown. Note the possibility of a "lethal corner" in the middle of the cylinder at the venous end.

results in an adequate P_{O_2} in all the tissue. In *2.*, the intercapillary distance or the oxygen consumption has been increased until the P_{O_2} at one point in the tissue falls to zero. This is referred to as a *critical* situation. In *3.* there is an anoxic region where aerobic (that is, oxygen-utilizing) metabolism is impossible. Under these conditions, the tissue turns to anaerobic glycolysis with the formation of lactic acid.

The situation *along* the cylinder is shown in Figure 5.32B. Here it is assumed that the P_{O_2} in the capillaries at the periphery of the tissue cylinder falls from 100 to 20 mmHg. As a consequence the P_{O_2} in the center of the cylinder falls toward the venous end of the capillary. It is clear that, on the basis of this model, the most vulnerable tissue is that furthest from the capillary at its downstream end. This was referred to as the "lethal corner" by Krogh. Direct measurements of the P_{O_2} in tissues of experimental animals (for example, the surface of the cerebral cortex), have been made by means of oxygen microelectrodes. These have confirmed the steep descent of P_{O_2} between open blood vessels.

How low can the P_{O_2} fall before oxygen utilization ceases? Probably the value depends on the type of tissue involved. For example, there is evidence that the cells of the central nervous system cease functioning at a higher P_{O_2} than liver cells. In measurements on suspensions of liver mitochondria in vitro, oxygen consumption continues at the same rate until the P_{O_2} falls to the region of 1–3 mmHg. In general it appears that the much higher P_{O_2} in capillary blood is to ensure an adequate pressure for diffusion of oxygen to the mitochondria and that at the sites of oxygen utilization the P_{O_2} may be very low.

Oxygen Delivery and Utilization

The amount of oxygen theoretically available to a tissue per minute is given by:

$$\dot{Q} \times Ca_{O_2}$$

where $\dot{Q}$ is the blood flow to the tissue and Ca_{O_2} is the arterial oxygen content (see Table 5.1).

The amount of oxygen consumed is given by:

$$\dot{Q} \times (Ca_{O_2} - Cv_{O_2})$$

where Cv_{O_2} is the oxygen content of the blood draining from the tissue. The ratio of the amount consumed to the amount available is sometimes called the oxygen utilization. This is given by:

$$\frac{\dot{Q}\,(Ca_{O_2} - Cv_{O_2})}{\dot{Q} \cdot Ca_{O_2}} = \frac{Ca_{O_2} - Cv_{O_2}}{Ca_{O_2}}$$

The utilization varies widely from organ to organ,

being as small as 10% in the kidney, 60% in the coronary circulation, and over 90% in exercising muscle. For the body as a whole at rest, the value is about 25%, rising to 75% on severe exercise.

In most types of tissue hypoxia, the oxygen utilization is high. This is true of (1) "hypoxic hypoxia" when the P_{O_2} of the arterial blood is low, for example in pulmonary disease; (2) "anemic hypoxia" when the oxygen-carrying capacity of the blood is reduced as in anemia or CO poisoning; and (3) "circulatory hypoxia" when tissue blood flow is reduced, as in shock or local obstruction. In (4) "histotoxic hypoxia," however, such as cyanide poisoning, the arterial-venous oxygen difference and the oxygen utilization are very low. This is because the cyanide prevents the use of O_2 by cytochrome oxidase within the tissues.

A different situation is seen with poisoning by dinitrophenol. Here oxygen is consumed giving energy as heat, but adenosine triphosphate production is prevented because of uncoupling of phosphorylation. As a consequence, useful work by the muscles is not possible.

Tissue Survival

The body has only small stores of oxygen which can be called upon during complete anoxia or asphyxia. The total store is only about 1500 ml of oxygen—enough to maintain life for only 6 min, if it were appropriately distributed. Tissues vary considerably in their ability to withstand oxygen deprivation, depending on how easily they can utilize anaerobic glycolysis. The cerbral cortex and the myocardium are particularly vulnerable to anoxia; in man, cessation of blood flow to the cerebral cortex results in loss of function within 4–6 s, loss of consciousness in 10–20 s, and irreversible changes in 3 to 5 min. The newborn of many species are less vulnerable to complete ischemia than the adults. Also there is great variation from one species to another.

Pulmonary Gas Exchange

OXYGEN TRANSPORT FROM AIR TO TISSUES

Figure 5.33 shows a simple scheme of the partial pressures of oxygen as it moves from the air in which we live to the mitochondria of the tissue cells where it is utilized. The scheme is shown for a hypothetical perfect lung. This does not exist but we shall use it as a model to show how the actual lung falls short of this perfect scheme, both in health and disease. In practice there are four causes of hypoxemia, that is, an abnormally low P_{O_2} in arterial blood, and each will be considered in turn.

First look at the P_{O_2} in the air that is warmed and moistened as it is inhaled into the upper airways. If the body temperature is 37°C, the partial pressure of water vapor will be 47 mmHg. Thus, assuming a barometric pressure of 760 mmHg, the total dry gas pressure is $760 - 47 = 713$ mmHg. Since the fractional concentration of oxygen in dry air is 0.2093, the P_{O_2} of moist inspired air is 0.2093×713 or 149 mmHg (say, 150 mmHg).

The scheme of Figure 5.33 is similar to the problem of designing a hydroelectric power station in a town which is 150 m below the level of a water reservoir. There is nothing we can do about the figure of 150, and we are therefore concerned about losses of pressure in the pipes leading from the reservoir to the turbines where the power is generated. In the same way a physiologist (or physician) is concerned about losses in the partial pressure of oxygen from the atmosphere to the mitochondria.

The first surprising feature of Figure 5.33 is that by the time the air has reached the alveolar gas, it has already lost about one-third of the available oxygen pressure. What is the reason for this apparent extravagance? The level of P_{O_2} in the alveolar gas is set by a balance between two processes. First, there is the addition of oxygen by alveolar ventilation which was examined in the last chapter. This occurs with each breath and is therefore a discontinuous process. However, it can be looked upon as continuous with little error because the fluctuations in alveolar P_{O_2} during the normal breathing cycle

are only 2 or 3 mmHg. This is because the tidal volume is only about ½ l, which is added to some 3 l of gas already in the lung. Moreover, only about one-third of the oxygen which is added with each breath is taken up. The second process is the removal of oxygen from the lung by the pulmonary capillary blood. This again can be regarded as a continuous process with little error, though in fact it is pulsatile because of fluctuations in the capillary flow (Fig. 5.18). It is the balance between these two opposing processes which sets the normal level of P_{O_2} at 100 mmHg.

In this hypothetical perfect lung, the blood draining from the lung will have the same partial pressure as the alveolar gas. When the blood reaches the systemic capillaries, however, oxygen diffuses out to the tissues where the P_{O_2} is much lower. Actually it is an oversimplification to talk of a single value for tissue P_{O_2} because it probably varies considerably within a particular tissue (Fig. 5.32) and between tissues in different parts of the body. In addition, there is evidence that the P_{O_2} at the mitochondria is extremely low, perhaps about 1 mmHg. Nevertheless, it is useful to include this last link in the chain because it reminds us that, other things being equal, any losses of partial pressure along the chain must inevitably depress the P_{O_2} available in the tissues where it is used.

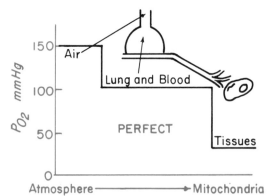

Figure 5.33. Scheme of the oxygen partial pressures from air to tissues. This depicts a hypothetical perfect situation which does not exist in practice.

Hypoventilation

The first cause of hypoxemia to be discussed is hypoventilation (Fig. 5.34). Recall that the level of P_{O_2} in alveolar gas is determined by a balance between two processes. On the one hand, oxygen is added by alveolar ventilation, and on the other, oxygen is removed by the blood flow. Generally the oxygen uptake by the body at rest is substantially constant. It is true that under exceptional conditions, such as hypothermia during cardiac surgery, for example, it is possible to reduce resting oxygen consumption, but by and large there is little variation in the level of oxygen uptake at rest. Therefore, if for any reason the level of alveolar ventilation falls, so does the P_{O_2} in the alveolar gas and therefore in the arterial blood (Fig. 5.34). Essentially, the same argument shows that the P_{CO_2} must rise, since this is determined by a balance between the rate at which CO_2 is added to alveolar gas by the blood flow, and the rate at which it is washed out by alveolar ventilation. Thus hypoventilation always causes a fall in alveolar and arterial P_{O_2} and a rise in P_{CO_2}.

How far does the P_{CO_2} rise? In the last chapter we saw that there is a very simple relationship between the alveolar ventilation $\dot{V}_A$ and the alveolar P_{CO_2} which is known as the *alveolar ventilation equation:*

$$\dot{V}_A = \frac{\dot{V}_{CO_2}}{P_{A_{CO_2}}} \cdot K$$

where K is a constant. This expression can be rearranged:

$$P_{A_{CO_2}} = \frac{\dot{V}_{CO_2}}{\dot{V}_A} \cdot K$$

and since, in normal lungs, the alveolar and arterial P_{CO_2} are almost identical we can write:

$$Pa_{CO_2} = \frac{\dot{V}_{CO_2}}{\dot{V}_A} \cdot K$$

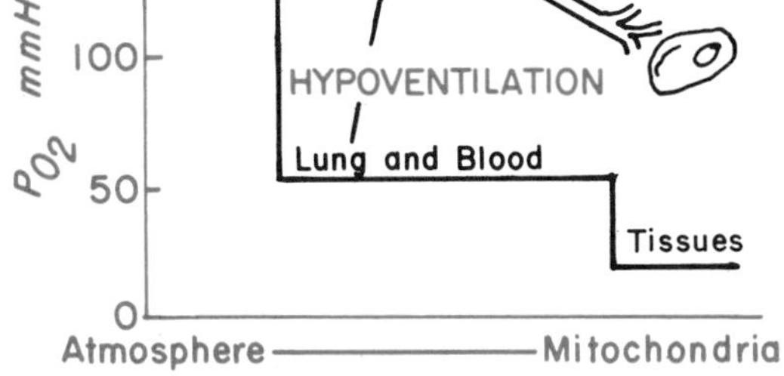

Figure 5.34. Scheme showing the result of hypoventilation (compare Fig. 5.33). Note the depression of P_{O_2} in alveolar gas and, therefore, in the arterial blood and tissues.

This very important equation indicates that the level of P_{CO_2} in alveolar gas or arterial blood is inversely related to the alveolar ventilation. For example, if the alveolar ventilation is halved, the P_{CO_2} doubles. This is true only after a steady-state is re-established so that the CO_2 production rate is the same as before. In practice the P_{CO_2} will rise over 10–20 min, rapidly at first and then more slowly, as the body stores of CO_2 are gradually filled.

How far will the P_{O_2} fall during hypoventilation? One way to look at this is by means of an equation which is analogous to the alveolar ventilation equation for CO_2 given above,

$$(\dot{V}_{A_I} \cdot P_{I_{O_2}} - \dot{V}_A \cdot P_{A_{O_2}}) = \dot{V}_{O_2} \cdot K$$

where $\dot{V}_{A_I}$ and $\dot{V}_A$ are the inspired and expired alveolar ventilation, respectively. If the inspired and expired alveolar ventilations are assumed to be the same, this equation reduces to:

$$P_{I_{O_2}} - P_{A_{O_2}}{}^* = \frac{\dot{V}_{O_2}}{\dot{V}_A} \cdot K$$

(Compare the alveolar ventilation equation for CO_2). This means that the difference between the inspired and alveolar P_{O_2} is inversely related to the alveolar ventilation. However, this equation is not strictly correct because the inspired and expired alveolar ventilations are generally slightly different because less CO_2 is eliminated than oxygen is taken up (respiratory exchange ratio is less than 1). The asterisk in the equation means that it is only an approximation.

Another way of calculating the alveolar P_{O_2} during hypoventilation is from the *alveolar gas equation:*

$$P_{A_{O_2}} = P_{I_{O_2}} - \frac{P_{A_{CO_2}}}{R} + F$$

where R is the respiratory exchange ratio ($\dot{V}_{CO_2}/\dot{V}_{O_2}$) and F is a correction factor which is generally small during air breathing (1–3 mmHg) and can be ignored. Suppose a man with normal lungs takes an overdose of a barbiturate drug which depresses his alveolar ventilation. Typically his alveolar P_{CO_2} might rise from 40 to 60 mmHg (the actual value will be determined by the alveolar ventilation equation). Assuming that his respiratory exchange ratio is one, his alveolar P_{O_2} will be given by:

$$P_{A_{O_2}} = P_{I_{O_2}} - \frac{P_{A_{CO_2}}}{R} + F$$

$$= 149 - \frac{60}{1}$$

$$= 89 \text{ mmHg}$$

Note that his alveolar P_{O_2} falls by 20 mmHg which

is the same amount that his P_{CO_2} rises. If $R = 0.8$, which is a more typical resting value:

$$P_{A_{O_2}} = 149 - \frac{60}{0.8} + F$$

$$= 74 \text{ mmHg}$$

In this case, the P_{O_2} falls more than the P_{CO_2} rises. Note, however, that in practical terms the hypoxemia is generally of minor importance compared with the hypercapnia (CO_2 retention) and consequent respiratory acidosis.

A patient with alveolar hypoventilation (and normal CO_2 production) always has an increased arterial P_{CO_2}. However, the arterial P_{O_2} may be normal or high if he is receiving supplementary oxygen. Suppose the man with barbiturate intoxication referred to above is given 30% oxygen to breathe. Assuming that his ventilation remains unchanged, his alveolar P_{O_2} will rise from 74 to about 139 mmHg (try the sum yourself). Thus a small increase in inspired P_{O_2} is very effective in eliminating the hypoxemia of hypoventilation.

Causes of hypoventilation include drugs such as morphine and barbiturates which depress the respiratory center, paralysis of the respiratory muscles, trauma to the chest wall, and any situation which greatly increases the resistance to breathing such as might occur when a diver breathes very dense gas at a great depth. If there is any doubt as to whether a patient is hypoventilating, measure the arterial P_{CO_2}. If it is not raised he is not hypoventilating.

Diffusion

Figure 5.33 and 5.34 imply that the partial pressures of oxygen in the alveolar gas and in the blood that drains from the lung are identical. However, in practice the P_{O_2} in the blood is always slightly lower, and the disparity may become large in disease. To understand how this occurs we must consider the movement of gases across the blood-gas barrier.

It is now generally believed that oxygen and CO_2 cross the blood-gas barrier by simple diffusion. In the early part of this century, however, there were physiologists such as Bohr and Haldane who believed that the alveolar epithelium could actively secrete oxygen and CO_2 against a partial pressure gradient (Bohr, 1909; Haldane, 1922). Indeed, in the 1935 edition of Haldane and Priestley's book *Respiration* they devoted a whole chapter to the evidence for oxygen secretion. Such a mechanism was thought to be responsible for the very high P_{O_2} that occurs in the swim bladder of some fish, although it is now thought that this is probably the

result of countercurrent exchange. However, early in this century a number of experiments were carried out at high altitudes in an attempt to prove the theory of oxygen secretion by the lung.

LAW OF DIFFUSION

The evidence now available suggests that the diffusion of oxygen and CO_2 through the blood-gas barrier obeys *Fick's law* as shown in Figure 5.35. This states that the volume of gas per unit time moving across a tissue sheet is directly proportional to the area of the sheet and the difference in partial pressures between the two sides but inversely proportional to the tissue thickness:

$$\dot{V}_{gas} = \frac{A}{T} \cdot D \cdot (P_1 - P_2)$$

This law was originally formulated in terms of concentrations but partial pressures are more convenient in this context; the two are directly related by Henry's law.

It is clear that the properties of a tissue sheet which would enhance diffusion are a large surface area and a small thickness. We saw in the previous chapter that the surface area of the blood-gas barrier of the human lung is between 50 and 100 m^2, with a thickness of less than 0.5 μm in many places. Its geometry is therefore ideally suited to rapid diffusion. The Fick equation also contains a diffusion constant (D) often called the diffusivity. This depends upon the structure of the sheet and the species of gas. For a given tissue, the diffusivity is proportional to the solubility (S) of the gas in the sheet and inversely proportional to the square root of the molecular weight (MW) of the gas:

$$D \propto \frac{S}{\sqrt{MW}}$$

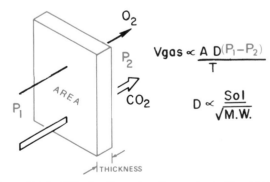

Figure 5.35. Diffusion through a tissue sheet. The amount of gas transferred is proportional to the area (A), a diffusion constant (D), and the difference in partial pressures ($P_1 - P_2$), and is inversely proportional to the thickness (T). The constant is proportional to the gas solubility (Sol), but inversely proportional to the square root of its molecular weight (MW).

If we compare the diffusion of oxygen and CO_2, we find that the latter has a great advantage. The solubility of CO_2 in normal saline at 37°C is approximately 24 times greater than that of O_2. Their MWs are not far apart, at 44 for CO_2 and 32 for oxygen, so that when the square root of the quotient is taken, the result is 1.2. Thus, CO_2 diffuses about 20% slower by virtue of its high MW but some 24 times faster by virtue of its high solubility. The net result is that the diffusion rate of CO_2 through a tissue sheet is about 20 times that of oxygen.

DIFFUSION AND PERFUSION LIMITATIONS

Consider what happens when a given quantity of blood enters the pulmonary capillary of an alveolus that contains a foreign gas such as CO or nitrous oxide. How rapidly will the partial pressure in the blood rise? Fig. 5.36 shows the time courses as the blood moves through the capillary, a process which takes about 0.75 s. Look first at CO. When the red cells enters the capillary, CO moves rapidly across the extremely thin blood-gas barrier from the alveolar gas into the red cell. As a result, the partial pressure of CO in the cell rises. However, as we saw in the last chapter, blood can combine with a large amount of CO for a very small rise in partial pressure (Fig. 5.26). This is because of the tight bond which CO forms with hemoglobin. Thus as the blood moves along the capillary, the P_{CO} in the blood hardly changes; consequently, no appreciable back pressure develops, and the gas continues to

move rapidly across the alveolar wall. It is clear, therefore, that the amount of CO which gets into the blood is limited by the diffusion properties of the blood-gas barrier, not by the amount of blood available.* The rate of transfer of carbon monoxide is therefore said to be *diffusion limited*.

Contrast the time course of nitrous oxide. When this gas moves across the alveolar wall into the blood, no combination with hemoglobin takes place. As a result the blood has nothing like the avidity for nitrous oxide that it had for CO, and the partial pressure rises rapidly. Indeed Figure 5.36 shows that the partial pressure of nitrous oxide in the blood virtually reaches that of the alveolar gas by the time the red cell is only one-fourth of the way along the capillary. After this point almost no nitrous oxide is transferred. Thus, the amount of this gas taken up by the blood depends entirely on the amount of available blood flow and not at all on the diffusion properties of the blood-gas barrier. The transfer of nitrous oxide is therefore *perfusion limited*.

Now consider oxygen. Its time course lies between that of CO and nitrous oxide. Oxygen combines with hemoglobin (unlike nitrous oxide) but with nothing like the avidity of CO. In other words, the rise in partial pressure when oxygen enters a red blood cell is much greater than is the case for the same volume of CO. Figure 5.36 shows that the P_{O_2} of the red cell as it enters the capillary is already about four-tenths of the alveolar value because of the oxygen in mixed venous blood. Under resting conditions the capillary P_{O_2} virtually reaches that of alveolar gas when the red cell is about one-third of the way along the capillary. Therefore, under these conditions oxygen transfer is perfusion limited like nitrous oxide. In some abnormal circumstances, however, when the diffusion properties of the lung are impaired (for example, because of thickening of the alveolar wall by disease), the blood P_{O_2} does not reach the alveolar value by the end of the capillary, and now there is some diffusion limitation as well.

OXYGEN UPTAKE ALONG THE PULMONARY CAPILLARY

Figure 5.37 shows time courses for P_{O_2} in the pulmonary capillary of the normal lung and in a lung in which the blood-gas barrier is thickened by

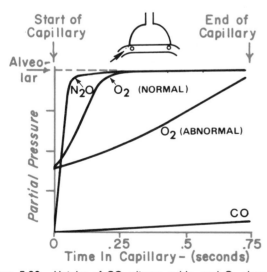

Figure 5.36. Uptake of CO, nitrous oxide, and O_2 along the pulmonary capillary. Note that the blood partial pressure of nitrous oxide virtually reaches that of alveolar gas very early in the capillary so that the transfer of this gas is perfusion limited. By contrast, the partial pressure of CO in the blood is almost unchanged so that its transfer is diffusion limited. O_2 can be perfusion or partly diffusion limited, depending on the conditions.

*This introductory description of CO transfer is not completely accurate because of the limit imposed by the rate of reaction of CO with hemoglobin (see later). Also, the example assumes that the partial pressure of CO in the alveolar gas is very small, typically about 1 mmHg.

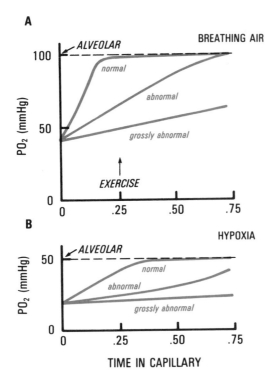

A

BREATHING AIR

ALVEOLAR

normal

abnormal

grossly abnormal

PO₂ (mmHg)

100

50

0

EXERCISE

0   .25   .50   .75

B

HYPOXIA

ALVEOLAR

normal

abnormal

grossly abnormal

PO₂ (mmHg)

50

0

0   .25   .50   .75

TIME IN CAPILLARY

Figure 5.37. Oxygen time courses along the pulmonary capillary. *A* shows the situation when the alveolar P$_{O_2}$ is 100 mmHg and the diffusion properties of the blood-gas barrier are normal, abnormal, or grossly abnormal because of thickening of the alveolar wall. Note that exercise reduces the time available for oxygenation. *B* shows the situation when the alveolar P$_{O_2}$ is abnormally low. Note the slower time courses. A combination of alveolar hypoxia and exercise might result in hypoxemia.

disease. Note that the P$_{O_2}$ in the red cell as it enters the capillary is normally about 40 mmHg. Across the blood-gas barrier, less than 0.5 μm away, is the alveolar P$_{O_2}$ of 100 mmHg (compare Figs. 5.2 and 5.33). Oxygen floods down this large pressure gradient, and the P$_{O_2}$ in the red cell rises rapidly; indeed, we have seen that it very nearly reaches the P$_{O_2}$ of the alveolar gas by the time the red cell is only one-third of its way along the capillary. Thus, under normal circumstances the difference in P$_{O_2}$ between alveolar gas and end-capillary blood is immeasurably small—a mere fraction of a mmHg (see also Fig. 5.40). In other words, the diffusion reserves of the normal lung are enormous.

During severe exercise the pulmonary blood flow is greatly increased, and the time normally spent by the red cell in the capillary, about 0.75 s, may be reduced to as little as one-third of this. This reduced time for oxygenation greatly stresses the diffusion reserves of the lung, but in normal subjects breathing air, there is generally still no measurable reduction in end-capillary P$_{O_2}$. If the blood-gas barrier is markedly thickened, however, so that oxygen diffusion is impeded, the rate of rise of P$_{O_2}$

in the red blood cell is correspondingly slower, and the P$_{O_2}$ may not reach that of alveolar gas before the time available for oxygenation in the capillary has run out. In this case a measurable P$_{O_2}$ difference between alveolar gas and end-capillary blood may occur.

Another way of stressing the diffusion properties of the lung is to lower the alveolar P$_{O_2}$ (Fig. 5.37*B*). Suppose that the P$_{O_2}$ has been reduced to 50 mmHg, either by the subject going to high altitude, or by giving the subject a low oxygen mixture to breathe. Now the rise in P$_{O_2}$ along the capillary is relatively slow, and failure to reach the alveolar P$_{O_2}$ is more likely. The slow rate of rise of P$_{O_2}$ is associated with the fact that we are operating on a much steeper part of the O₂ dissociation curve (Fig. 5.24) and the hemoglobin is therefore more avid for the O₂ (compare the CO curve in Fig. 5.36). It can be shown that the rate of rise of partial pressure is determined by the ratio of the solubility of the gas in the blood-gas barrier, to the effective "solubility" of the gas in blood, that is, the slope of the dissociation curve (Wagner, 1977). Thus when the P$_{O_2}$ is low and the O₂ dissociation curve is steep, there is a slower rate of rise of P$_{O_2}$ in the capillary. Heavy exercise at very high altitude is one of the few situations where diffusion impairment of oxygen in normal subjects can be convincingly demonstrated. By the same token, a patient with a thickened blood-gas barrier will be most likely to show evidence of diffusion impairment if he breathes a low oxygen mixture, especially if he exercises as well.

MEASUREMENT OF DIFFUSING CAPACITY

In order to measure the diffusion properties of the lung, we must clearly choose a gas whose uptake is diffusion limited. Although this is true of oxygen under some very special conditions, and although at one time this gas was therefore used to measure the diffusing capacity of the lung, CO is the obvious choice.

Fick's law (Fig. 5.35) states that:

$$\dot{V}_{gas} = \frac{A}{T} D (P_1 - P_2)$$

However, for a complex structure like the blood-gas barrier of the lung, it is not possible to measure the area and thickness during life. Instead, the equation is rewritten:

$$\dot{V}_{gas} = D_L \cdot (P_1 - P_2)$$

where D$_L$ is called the diffusing capacity of the lung, a term which includes the area, thickness, and diffusion properties of the tissue sheet and the gas concerned. Thus the diffusing capacity for carbon monoxide is given by:

$$D_L = \frac{\dot{V}_{CO}}{P_1 - P_2}$$

where P_1 and P_2 are the partial pressures of alveolar gas and capillary blood, respectively. But as we saw in Figure 5.36, the partial pressure of CO in capillary blood is so small that it usually can be neglected, although exceptions to this sometimes occur in smokers who have significant levels of CO in their blood. In general, however:

$$D_L = \frac{\dot{V}_{CO}}{P_{A_{CO}}}$$

Or, in other words, the diffusing capacity of the lung for CO is the volume of CO transferred in milliliters per minute per mmHg alveolar partial pressure.

Several techniques for making this measurement are available. In the *single breath method*, a single inspiration of a dilute mixture of CO is made, and the rate of disappearance of CO from the alveolar gas during a breath holding period of 10 s is calculated. This is usually done by measuring the inspired and expired concentrations of CO with an infrared analyzer. The alveolar concentration of CO is not constant during the breath holding period but allowance can be made for that. Helium is also added to the inspired gas to give a measurement of lung volume by dilution. This method is extensively used in clinical lung function laboratories.

In the *steady state method*, the subject breathes a low concentration of CO (about 0.1%) for ½ minute or so until a steady state has been reached. The constant rate of disappearance of CO from alveolar gas is then measured for a further short period along with the alveolar concentration. The latter can be obtained by sampling gas at the end of each expiration, or by making allowance for the dead space using the Bohr method (see Chapter 36).

The normal value for the diffusing capacity for CO at rest is about 25 ml/min/mmHg, and it increases to two or three times this value on exercise. It is reduced by diseases which thicken the blood-gas barrier, such as interstitial lung diseases, including sarcoidosis and asbestosis.

REACTION RATES WITH HEMOGLOBIN

So far we have assumed that all the resistance to the movement of O_2 and CO resides in the barrier between blood and gas. However, Figure 5.2 shows that the path length from the alveolar wall to the center of a red blood cell exceeds that in the wall

itself so that some of the diffusion resistance is located within the capillary. In addition, there is another type of resistance to gas transfer which is most conveniently discussed at this point, that is, the resistance caused by the finite rate of reaction of oxygen or CO with hemoglobin inside the red blood cell.

When oxygen is added to blood, its combination with hemoglobin is quite fast, being well on the way to completion in 0.2 s. However, oxygenation occurs so rapidly in the pulmonary capillary (Fig. 5.37) that even this rapid reaction significantly delays the loading of oxygen by the red cell. Thus, the uptake of oxygen can be regarded as occurring in two stages: (a) diffusion of oxygen through the blood-gas barrier (including the plasma and red cell interior); and (b) reaction of the oxygen with hemoglobin (Fig. 5.38). Each stage can be regarded as contributing a resistance to the movement of oxygen, and it is possible to sum the two resultant resistances to produce an overall "diffusion" resistance (Roughton and Forster, 1957).

The diffusing capacity of the lung can be written as:

$$D_L = \frac{\dot{V}_{gas}}{P_A - P_C} \quad \text{or} \quad P_A - P_C = \frac{\dot{V}_{gas}}{D_L}$$

where P_C is the partial pressure inside the red cell. By analogy, the diffusing capacity of the membrane can be written:

$$D_M = \frac{\dot{V}_{gas}}{P_A - P_{pl}} \quad \text{or} \quad P_A - P_{pl} = \frac{\dot{V}_{gas}}{D_M}$$

where P_{pl} means the plasma partial pressure. Furthermore, we can write a similar equation to represent the rate of reaction of oxygen (or CO) with hemoglobin. This rate can be described by a variable θ whose units are milliliters of oxygen per

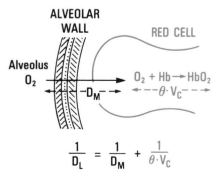

Figure 5.38. The diffusing capacity of the lung D_L is made up of two components, that due to the diffusion process itself and that attributable to the time taken for oxygen (or CO) to react with hemoglobin (Hb). θ, rate of reaction of oxygen (or CO) with hemoglobin; V_c, volume of capillary blood; D_M, diffusing capacity of the membrane.

minute per mmHg per milliliter of blood. This is analogous to the "diffusing capacity" of 1 ml blood, and when multiplied by the volume of capillary blood (V_c) gives the effective "diffusing capacity" of the rate of reaction of oxygen with hemoglobin:

$$\theta \cdot V_c = \frac{\dot{V}_{gas}}{P_{pl} - P_c} \quad \text{or} \quad P_{pl} - P_c = \frac{\dot{V}_{gas}}{\theta \cdot V_c}$$

now

$$P_A - P_c = (P_A - P_{pl}) + (P_{pl} - P_c)$$

substituting

$$\frac{\dot{V}_{gas}}{D_L} = \frac{\dot{V}_{gas}}{D_M} + \frac{\dot{V}_{gas}}{\theta \cdot V_c}$$

and therefore

$$\frac{1}{D_L} = \frac{1}{D_M} + \frac{1}{\theta \cdot V_c}$$

This holds for both O_2 and CO.

When the diffusing capacity for CO uptake is measured, the value of θ depends on the prevailing P_{O_2} in the alveolar gas because of the competition of oxygen and CO for hemoglobin. Therefore, by performing the diffusing capacity measurement at different levels of alveolar P_{O_2}, two equations for D_L are obtained with two unknowns, D_M and V_c. In this way D_M and V_c can be separately evaluated.

In practice, the resistances offered by the membrane and blood components to the uptake of CO are approximately equal. Thus, diseases which cause the capillary blood volume to fall can reduce the diffusing capacity of the lung. This discussion also applies to oxygen uptake along the pulmonary capillary, although the values of θ for this gas are different.

CARBON DIOXIDE TRANSFER ALONG THE PULMONARY CAPILLARY

We have seen that the rate of diffusion of CO_2 through tissue is about 20 times faster than oxygen because of the much higher solubility of CO_2 (Fig. 5.35). At first sight, therefore, it seems unlikely that CO_2 elimination could be adversely affected by diffusion difficulties, and, indeed, for many years that was held to be so. The reaction of CO_2 with blood, however, is complex, involving the hydration of CO_2, the chloride shift across the red cell membrane, and the formation of carbamino compounds (see Fig. 5.27). Some of these reactions are relatively slow. Moreover, the CO_2 dissociation curve is quite steep (Fig. 5.30A) and, as we have seen, this tends to slow the rate of fall of partial pressure of a gas in the pulmonary capillary (Fig. 5.37B).

Recent work suggests that a partial pressure difference for CO_2 can develop between end-capillary

blood and alveolar gas if the blood-gas barrier is diseased. Fig. 5.39 shows the calculated normal time course for CO_2 and how this might be affected by thickening of the blood-gas barrier. Note that the P_{CO_2} of the blood as it enters the capillary is about 45–47 mmHg and that the normal P_{CO_2} of alveolar gas is about 40 mmHg. The diagram shows that the time taken for the blood to reach virtually the same partial pressure as alveolar gas is similar to that for oxygen under normal conditions (Fig. 5.37A) so that a considerable time reserve is available. However, when the diffusing capacity of the membrane is reduced to one-fourth of its normal value, a small difference between end-capillary blood and alveolar gas develops. To what extent the elimination of CO_2 can be limited by such a mechanism is unknown at the present time.

Shunt

This is the third cause of hypoxemia. A shunt is defined as any mechanism by which blood which has not been through ventilated areas of lung is added to the systemic arteries. This is shown schematically in Figure 5.40. Even in the normal lung there is some blood in this category. For example, some of the bronchial arterial blood which has supplied the walls of the large airways finds its way into the pulmonary veins downstream of the pulmonary capillaries. Since this blood has been partly depleted of oxygen, its addition to the pulmonary venous blood reduces the P_{O_2} there. Again, there is a small amount of blood which drains from the myocardium of the left ventricle directly into the cavity of the ventricle via Thebesian veins. This blood contributes to a lowering of systemic arterial P_{O_2}.

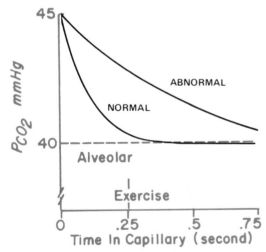

Figure 5.39. Change in P_{CO_2} along the pulmonary capillary when the diffusion properties are normal and abnormal (compare the change in P_{O_2} in Fig. 5.37). [From Wagner and West (1972).]

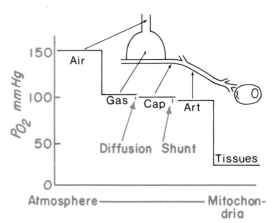

Figure 5.40. Scheme of oxygen transfer from air to tissues showing the depression of arterial P_{O_2} caused by diffusion and shunt. [From West (1977).]

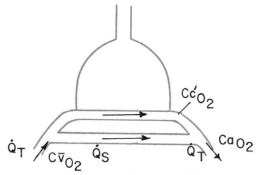

Figure 5.41. Measurement of shunt flow. The oxygen carried in the arterial blood equals the sum of the oxygen carried in the capillary blood and that in the shunted blood (see text). $\dot{Q}_T$, total blood flow; Ca_{O_2}, oxygen concentration in arterial blood; $\dot{Q}_S$, shunted blood flow; Cc'_{O_2} and $C\bar{v}_{O_2}$ are oxygen concentrations in end-capillary and mixed venous blood, respectively.

In normal subjects the reduction of arterial P_{O_2} from these sources amounts to only 5 mmHg or so. In some patients, however, especially those with congenital heart disease who have blood moving from the right to the left sides of the heart through septal defects, the depression of arterial P_{O_2} by shunt may be very severe indeed.

When the shunt is caused by the addition of mixed venous blood (pulmonary arterial) to blood draining from the capillaries (pulmonary venous), it is possible to calculate the amount of shunt flow, as shown in Figure 5.41. The total amount of oxygen leaving the system is the total blood flow ($\dot{Q}_T$) multiplied by the oxygen concentration in the arterial blood (Ca_{O_2}):

$$\dot{Q}_T \times Ca_{O_2}$$

This must equal the sum of the amounts of oxygen in the shunted blood:

$$\dot{Q}_S \times C\bar{v}_{O_2}$$

and end-capillary blood:

$$(\dot{Q}_T - \dot{Q}_S) \times Cc'_{O_2}$$

Thus,

$$\dot{Q}_T \times Ca_{O_2} = (\dot{Q}_S \times C\bar{v}_{O_2}) + (\dot{Q}_T - \dot{Q}_S) \times Cc'_{O_2}$$

Rearranging, we have:

$$\frac{\dot{Q}_S}{\dot{Q}_T} = \frac{Cc'_{O_2} - Ca_{O_2}}{Cc'_{O_2} - C\bar{v}_{O_2}}$$

This is known as the *shunt equation*.

The oxygen concentration in arterial blood can be obtained by arterial puncture. End-capillary blood cannot be sampled directly, but it can be calculated if the P_{O_2} of alveolar gas is known, assuming no alveolar to end-capillary difference caused by diffusion impairment. Mixed venous

blood can be obtained only by means of a catheter in the pulmonary artery. In practice the denominator in the equation is sometimes assumed to be 5 ml oxygen/100 ml blood, but this assumption can lead to substantial errors if the cardiac output is abnormally high or low.

Notice that this equation is only strictly true if the shunted blood has the same composition as mixed venous blood. This is not always the case, for example, when the shunted blood comes from the bronchial veins. In such a case it may be useful to calculate an "as if" shunt, that is, what the shunt *would* be if all the observed depression of arterial oxygen content were caused by the addition of true mixed venous blood.

An important feature of a shunt is that if the subject breathes 100% oxygen, the arterial P_{O_2} does not rise to the level it does in a normal subject. This is because the shunted blood which bypasses ventilated alveoli is never exposed to the high alveolar P_{O_2} resulting from oxygen breathing, so that this blood continues to depress the arterial P_{O_2}. Some elevation of the arterial P_{O_2} occurs, however, because of the oxygen added to the capillary blood. Most of the added oxygen is in the dissolved form rather than attached to hemoglobin because the blood which is perfusing ventilated alveoli is nearly fully saturated (see Fig. 5.24).

Figure 5.42 shows why a relatively small shunt results in a large depression of arterial P_{O_2} during 100% oxygen breathing. A small shunt depresses the oxygen *content* of arterial blood (see the shunt equation given above), and when the alveolar P_{O_2} is very high, the curve relating oxygen content and partial pressure is nearly flat. In fact, the slope reflects only the addition of dissolved oxygen (compare Fig. 5.24). Therefore, a given reduction in oxygen content causes a much larger fall in arterial

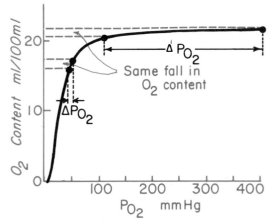

Figure 5.42. Reason why a shunt causes a much larger fall in arterial partial pressure of oxygen (P_{O_2}) when the subject is breathing a high oxygen mixture, as compared with breathing a low O_2 mixture. For the same fall in oxygen content in the arterial blood, the fall in P_{O_2} is much greater when the alveolar P_{O_2} is high because of the change in shape of the oxygen dissociation curve.

P_{O_2} when the alveolar P_{O_2} is high than when the alveolar P_{O_2} is normal or low.

Does a shunt result in an elevated arterial P_{CO_2}? This might be expected since the mixed venous blood is relatively rich in CO_2. In practice, however, the arterial P_{CO_2} is usually not raised because any tendency for it to increase stimulates ventilation (Chapter 39). The result is that the P_{CO_2} in end-capillary blood is reduced until the arterial P_{CO_2} is normal. In fact, some patients with a large shunt have an abnormally low arterial P_{CO_2} because of the excessive stimulation of ventilation by the hypoxemia.

Ventilation-Perfusion Inequality

The fourth cause of hypoxemia is both the commonest in practice and also the most difficult to understand. It is the mismatching of ventilation and blood flow within the lung; the key to understanding how this affects gas exchange is the ventilation:perfusion ratio.

VENTILATION:PERFUSION RATIO

Consider a model of a lung unit (Fig. 5.5) in which the uptake of oxygen is being mimicked using dye and water (Fig. 5.43). Powdered dye is continuously poured into the unit to represent the addition of oxygen by alveolar ventilation. (In practice, ventilation is discontinuous but it makes little difference to the result.) Water is pumped continuously through the unit to represent the blood flow which removes the oxygen. A stirrer mixes the alveolar contents, a process normally accomplished by gaseous diffusion. The key question is: What

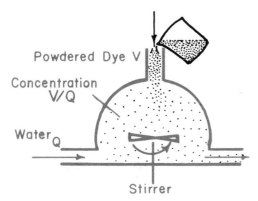

Figure 5.43. Model to illustrate how the ventilation:perfusion ratio determines the P_{O_2} in a lung unit. Powdered dye is added by ventilation of the rate V and removed by blood flow Q to represent the factors controlling alveolar P_{O_2}. The concentration of dye is given by V/Q. [From West (1977).]

determines the concentration of dye (or oxygen) in the alveolar compartment and therefore in the effluent water (or blood)?

It is clear that both the rate at which the dye is added (ventilation) and the rate at which water is pumped (blood flow) will affect the concentration of dye in the model. What may not be intuitively clear is that the concentration of dye is determined by the ratio of these rates. In other words, if dye is added at the rate of V g/min and water is pumped through at Q l/min, the concentration of dye in the alveolar compartment and effluent water is V/Q g/l.

In exactly the same way, the concentration of oxygen (or, better, P_{O_2}) in any lung unit is determined by the ratio of ventilation to blood flow, and this is true not only for oxygen but also for CO_2, nitrogen, or any other gas which is present under steady-state conditions. This is the reason why the ventilation:perfusion ratio plays such a key role in pulmonary gas exchange.

EFFECT OF ALTERING THE VENTILATION:PERFUSION RATIO OF A LUNG UNIT

Consider first a lung unit with a normal ventilation:perfusion ratio of about 1 (compare Fig. 5.5). This is shown in Figure 5.44A. The inspired air has a P_{O_2} of 150 mmHg (Fig. 5.33) and a P_{CO_2} of zero. The mixed venous blood entering the unit has a P_{O_2} of 40 mmHg and P_{CO_2} of 45 mmHg. The alveolar P_{O_2} of 100 mmHg is determined by a balance between the addition of oxygen by ventilation and its removal by blood flow. The normal alveolar P_{CO_2} of 40 mmHg is set similarly.

Now suppose that the ventilation:perfusion ratio of the unit is gradually reduced by obstructing its

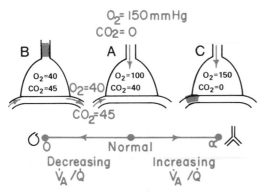

Figure 5.44. Effect of alterations of the ventilation:perfusion ratio on the P_{O_2} and P_{CO_2} in a lung unit. $\dot{V}_A/\dot{Q}$, ventilation:perfusion ratio. [From West (1977).]

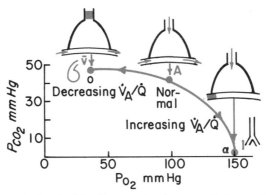

Figure 5.45. O_2-CO_2 diagram showing a ventilation:perfusion ratio line. The P_{O_2} and P_{CO_2} of a lung unit move along this line from the mixed venous point $\bar{v}$ to the inspired gas point I as its ventilation:perfusion ratio is increased (compare Figure 5.44). [From West (1977).]

ventilation, leaving its blood flow unchanged (Fig. 5.44*B*). It is clear that the oxygen in the unit will fall and the CO_2 will rise, although the relative changes of these two are not immediately obvious.† It is easy to predict, however, what will eventually happen when the ventilation is completely abolished (ventilation:perfusion ratio of zero). Now the oxygen and CO_2 of alveolar gas and of end-capillary blood must be the same as those of mixed venous blood. (In practice, completely obstructed units eventually collapse, but such long-term effects can be neglected at the moment.) Note that in this model we are assuming that what happens in one lung unit out of a very large number does not affect the composition of the mixed venous blood.

Suppose instead that the ventilation:perfusion ratio is increased by gradually obstructing blood flow (Fig. 5.44*C*). Now the alveolar oxygen rises and the CO_2 falls, eventually reaching the composition of inspired gas when blood flow is abolished (ventilation:perfusion ratio of infinity). Thus, as the ventilation:perfusion ratio of the unit is altered, its gas composition approaches either that of mixed venous blood or that of inspired gas.

A convenient way of depicting these changes is to use the O_2-CO_2 diagram (Rahn and Fenn, 1955; Fig. 5.45). Here the P_{O_2} is plotted on the X axis and P_{CO_2} on the Y axis. First, locate the normal alveolar gas composition, point *A* ($P_{O_2} = 100$, $P_{CO_2} = 40$). If blood has in fact equilibrated with alveolar gas along the capillary (Fig. 5.37 and 5.39), this point can equally well represent the end-capillary blood.

Next, find the mixed venous point $\bar{v}$ ($P_{O_2} = 40$, $P_{CO_2} = 45$). The bar above v means "mixed" or "mean." Finally, find the inspired point I ($P_{O_2} = 150$, $P_{CO_2} = 0$). Also note the similarities between Figs. 5.44 and 5.45.

The line joining $\bar{v}$ to I passing through A shows the changes in alveolar gas (and end-capillary blood) composition that can occur when the ventilation:perfusion ratio is either decreased below normal (A → $\bar{v}$) or increased above normal (A → I). Indeed, this line indicates *all* the possible alveolar gas compositions in a lung which is supplied with gas of composition I and blood of composition $\bar{v}$. For example, such a lung could not contain an alveolus with a P_{O_2} of 70 and P_{CO_2} of 30 mmHg, since this point does not lie on the ventilation:perfusion ratio line. This alveolar composition could exist, however, if the mixed venous blood or inspired gas were changed so that the line then passed through this point.

REGIONAL GAS EXCHANGE IN THE LUNG

In many diseases, such as the very common chronic obstructive lung diseases (including chronic bronchitis and emphysema), there is evidence of great mismatching of ventilation and blood flow throughout the lung. As a consequence lung units will be dispersed most of the way along a ventilation:perfusion ratio line as shown in Figure 5.45. Unfortunately, however, little is known as yet about the distribution of ventilation:perfusion ratios in these diseased lungs. For this reason it is helpful to look at the pattern of regional gas exchange in the normal upright lung. Here the distribution of ventilation:perfusion ratios follows a simple topographical pattern, and the effects on regional gas exchange are easily appreciated. In addition there is

---

† The alveolar gas equation is not applicable here because the respiratory exchange ratio is not constant. The appropriate equation is

$$\frac{\dot{V}_A}{\dot{Q}} = \frac{8.63 \cdot R \cdot (Ca_{O_2} - C\bar{v}_{O_2})}{P_{A_{CO_2}}}$$

This is called the ventilation:perfusion ratio equation.

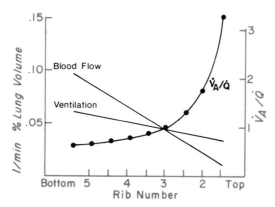

Figure 5.46. Distribution of ventilation and blood flow down the upright lung (compare Fig. 5.69). Note that the ventilation:perfusion ratio ($\dot{V}_A/\dot{Q}$) decreases down the lung. [From West (1977).]

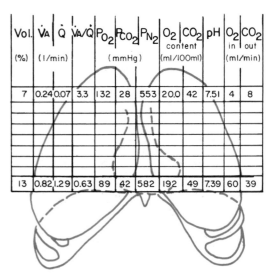

Vol. (%)	$\dot{V}_A$ (l/min)	$\dot{Q}$	$\dot{V}_A/\dot{Q}$	P_{O_2}	P_{CO_2}	P_{N_2} (mmHg)	O_2 content (ml/100ml)	CO_2	pH	O_2 in (ml/min)	CO_2 out
7	0.24	0.07	3.3	132	28	553	20.0	42	7.51	4	8
13	0.82	1.29	0.63	89	42	582	19.2	49	7.39	60	39

Figure 5.47. Regional differences in gas exchange down the normal lung. Only the apical and basal values are shown for clarity. $\dot{V}_A$, alveolar ventilation; $\dot{Q}$, blood flow; P_{O_2}, partial pressure of oxygen; P_{CO_2}, partial pressure of CO_2; P_{N_2}, partial pressure of nitrogen.

some evidence that these regional differences affect the localization of some types of disease, for example, adult tuberculosis.

It was pointed out earlier that blood flow is unevenly distributed in the normal upright human lung (Fig. 5.19). At normal lung volumes, blood flow increases markedly from apex to base. Ventilation also increases down the lung (see Chapter 38), but the regional differences are less marked than those for blood flow. The approximate distributions of blood flow and ventilation are depicted in Figure 5.46, where both are taken to change linearly when plotted against rib number. Note that the ventilation:perfusion ratio increases from a low value near the base of the lung to a very high value near the apex.

Since the ventilation:perfusion ratio determines the gas exchange in any lung unit, the resulting regional differences in gas exchange can be calculated from the principles shown in Figures 5.45 and 5.46. The details of such calculations need not be given here; suffice it to say that they are greatly complicated by the O_2 and CO_2 dissociation curves which are nonlinear and interdependent (Figs. 5.24 and 5.29). Originally the solutions to the ventilation:perfusion ratio equation were obtained graphically (Rahn and Fenn, 1955), but they are now more easily found by numerical analysis with a computer (West and Wagner, 1977).

Figure 5.47 shows typical values calculated for a lung model divided into nine imaginary horizontal slices. (Naturally there will be variations between subjects; the chief aim of this approach is to describe the principles underlying gas exchange). Note first that the volume of the lung in the slices is less near the apex than the base. Ventilation is less at the top than the bottom, but the differences of blood flow are much more marked. Consequently,

the ventilation:perfusion ratio decreases down the lung, and all the differences in gas exchange follow from this. Note that the P_{O_2} changes by over 40 mmHg while the difference in P_{CO_2} between apex and base is much less. (Incidentally, the high P_{O_2} at the apex probably accounts for the predilection of adult tuberculosis for this region since it provides a more favorable environment for this organism.) The variation in P_{N_2} is, in effect, by default since the total pressure in the alveolar gas is the same throughout the lung.

The regional differences of P_{O_2} and P_{CO_2} imply differences in the end-capillary contents of these gases which can be obtained from the appropriate dissociation curves (Figs. 5.24 and 5.29). Note the surprisingly large difference in pH down the lung which reflects the considerable variation in P_{CO_2} of the blood in the presence of a constant base excess. The minimal contribution to overall oxygen uptake made by the apex can be mainly attributed to the very low blood flow there. The differences in CO_2 output are much less since these can be shown to be more closely related to ventilation. As a result, the respiratory exchange ratio (CO_2 output:oxygen uptake) is higher at the apex than the base. During exercise, when the distribution of blood flow becomes more uniform, the apex assumes a more appropriate share of oxygen uptake.

VENTILATION:PERFUSION INEQUALITY AS A CAUSE OF HYPOXEMIA

While the regional differences in gas exchange discussed above are of interest, more important to

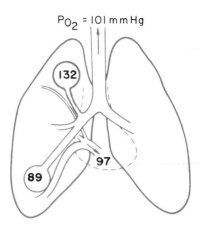

$P_{O_2} = 101\,mmHg$

132

97

89

Figure 5.48. Depression of the arterial P_{O_2} by ventilation:perfusion inequality. In this diagram of the upright lung, only two groups of alveoli at the apex and base are shown. The relative sizes of the airways and blood vessels indicate their relative ventilations and blood flows. Because most of the blood comes from the poorly oxygenated base, depression of the blood P_{O_2} is inevitable. [From West (1963).]

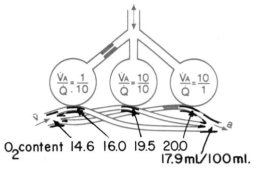

$\dfrac{\dot{V}_A}{\dot{Q}} = \dfrac{1}{10}$ $\dfrac{\dot{V}_A}{\dot{Q}} = \dfrac{10}{10}$ $\dfrac{\dot{V}_A}{\dot{Q}} = \dfrac{10}{1}$

O_2content 14.6 16.0 19.5 20.0
17.9 ml/100ml.

Figure 5.49. Additional reason for the depression of arterial P_{O_2} by mismatching of ventilation and blood flow. The lung units with a high ventilation:perfusion ratio add relatively little oxygen to the blood compared with the decrement caused by alveoli with a low ventilation:perfusion ratio. $\dot{V}_A$, alveolar ventilation; $\dot{Q}$, blood flow. [From West (1977).]

the body as a whole is whether uneven ventilation and blood flow affect the overall gas exchange of the lung, that is, its ability to take up O_2 and put out CO_2. It transpires that a lung with ventilation:perfusion inequality is not able to transfer as much oxygen and CO_2 as a lung which is uniformly ventilated and perfused, other things being equal. Or, if the same amounts of gas are being transferred (because these are set by the metabolic demands of the body), the lung with ventilation:perfusion inequality cannot maintain as high an arterial P_{O_2} or as low an arterial P_{CO_2} as a homogeneous lung, again other things being equal.

The reason why a lung with uneven ventilation and blood flow has difficulty oxygenating arterial blood can be illustrated by looking at the differences down the upright lung (Fig. 5.48). Note that the P_{O_2} at the apex is some 40 mmHg higher than at the base of the lung. However, the major share of the blood leaving the lung comes from the lower zones where the P_{O_2} is low. This has the result of depressing the arterial P_{O_2}. By contrast, the expired alveolar gas comes more uniformly from apex and base because the differences of ventilation are much less than those for blood flow (Fig. 5.46). By the same reasoning, the arterial P_{CO_2} will be elevated since this is higher at the base of the lung than at the apex (Fig. 5.47).

An additional reason why uneven ventilation and blood flow depress the arterial P_{O_2} is shown in Figure 5.49. This depicts three groups of alveoli with low, normal, and high ventilation:perfusion ratios. The oxygen contents of the effluent blood are 16, 19.5, and 20 ml/100 ml, respectively, and thus the units with the high ventilation:perfusion

ratio add relatively little oxygen to the blood compared with the decrement caused by the alveoli with the low ventilation:perfusion ratio. Thus, the mixed capillary blood has a lower oxygen content than blood from units with a normal ventilation:perfusion ratio. This can be explained by the nonlinear shape of the oxygen dissociation curve, which means that although units with a high ventilation:perfusion ratio have a relatively high P_{O_2}, this does not increase the oxygen content of their blood very much. This additional reason for the depression of P_{O_2} does not apply to the elevation of the P_{CO_2}, since the CO_2 dissociation curve is almost linear in the working range.

The net result of these mechanisms is a depression of the arterial P_{O_2} below that of the mixed alveolar P_{O_2}—the so-called alveolar-arterial oxygen difference. In the normal upright lung this is of trivial magnitude, being only about 4 mmHg as a result of ventilation:perfusion inequality. Its development is described here only to illustrate how uneven ventilation and blood flow must result in depression of the arterial P_{O_2}. In lung disease, the lowering of arterial P_{O_2} by this mechanism can be extreme.

Oxygen breathing is very effective in raising the arterial P_{O_2} of a patient whose hypoxemia is caused by ventilation:perfusion inequality. The reason is that lung units which are ventilated (if only poorly) will eventually have all their nitrogen washed out; their alveolar P_{O_2} is then given by the expression:

$$P_{O_2} = P_B - P_{H_2O} - P_{CO_2}$$

The basis for this equation is that the sum of all the partial pressures must equal the barometric pressure. Since the P_{CO_2} is limited by the value in mixed venous blood (Fig. 5.45), the alveolar P_{O_2} of all units generally will now exceed 600 mmHg. In practice, however, the arterial P_{O_2} may take many minutes to complete its rise because the nitrogen is washed out of poorly ventilated regions so slowly.

VENTILATION:PERFUSION INEQUALITY AS A CAUSE OF CARBON DIOXIDE RETENTION

Imagine a lung which is uniformly ventilated and perfused and which is transferring normal amounts of oxygen and CO_2. Suppose that in some magical way the matching of ventilation and blood flow is suddenly disturbed while everything else remains unchanged. What happens to gas exchange? It transpires that the effect of this "pure" ventilation:perfusion inequality (that is, everything else held constant) is to reduce *both* the oxygen uptake and CO_2 output of the lung. In other words, the lung becomes less efficient as a gas exchanger for both gases. Thus, mismatching ventilation and blood flow must cause both hypoxemia and hypercapnia (CO_2 retention), other things being equal.

In practice, however, patients with undoubted ventilation:perfusion inequality often have a normal arterial P_{CO_2}. The reason for this is that whenever the chemoreceptors sense a rising P_{CO_2}, there is an increase in ventilatory drive (Chapter 39), with a consequent increase in ventilation to the alveoli. This usually effectively returns the arterial P_{CO_2} to normal. Such patients, however, can only maintain a normal P_{CO_2} at the expense of this increased ventilation to their alveoli; the ventilation in excess of what they would normally require is sometimes referred to as *wasted ventilation* and is necessary because the lung units with abnormally high ventilation:perfusion ratios are inefficient at eliminating CO_2. Such units are said to constitute an *alveolar dead space*.

While the increase in ventilation to a lung with ventilation:perfusion inequality is usually effective at reducing the arterial P_{CO_2}, it is much less effective at increasing the arterial P_{O_2}. The reason for the different behavior of the two gases lies in the shapes of the CO_2 and oxygen dissociation curves (Figs. 5.24 and 5.29). The CO_2 dissociation curve is almost straight in the physiological range, with the result that an increase in ventilation will raise the CO_2 output of lung units with both high and low ventilation:perfusion ratios. By contrast, the almost flat top of the oxygen dissociation curve means that only units with moderately low ventilation:perfusion ratios will benefit appreciably from the increased ventilation. Those units which are very high on the dissociation curve (high ventilation:perfusion ratio) increase the oxygen content of their effluent blood very little because the hemoglobin is almost fully saturated with oxygen anyway (Fig. 5.49). Those units which have a very low

ventilation:perfusion ratio continue to put out blood which has an oxygen content close to that of mixed venous blood. The net result is that the mixed arterial P_{O_2} rises only modestly, and some hypoxemia always remains.

MEASUREMENT OF VENTILATION:PERFUSION INEQUALITY

Relatively little information is presently available about the distribution of ventilation:perfusion ratios in diseased lungs. It is possible to use radioactive gases, for example, xenon-133, to define topographical differences in ventilation and blood flow (Ball et al., 1962), especially in the normal lung. The distribution of ventilation is obtained by having the subject inhale a single breath of the gas and then recording the counting rates over different regions of the chest (Fig. 5.68). The distribution of blood flow can be determined in a similar way if a solution of xenon-133 is injected into a vein (Fig. 5.19). The patterns of ventilation and blood flow shown in Figure 5.47 are based on this type of measurement. In most diseased lungs, however, so much inequality of ventilation and blood flow exists between closely adjacent lung units that methods based on external counting of radioactivity have inadequate resolution.

A recently described technique depends on the pattern of elimination by the lung of a series of foreign gases infused into a vein in solution (Wagner et al., 1974). This method reveals narrow distributions of ventilation:perfusion ratios in normal subjects, while patients with lung disease have broader dispersions and sometimes bimodal patterns. Much information about normal and diseased lungs has been obtained with this new method but it is too complicated to use in the routine clinical laboratory.

In practice the method most commonly used for assessing the amount of ventilation:perfusion inequality is based on measurements of the P_{O_2} and P_{CO_2} of arterial blood and expired gas (Riley and Cournand, 1949). First, the *alveolar-arterial oxygen difference* is measured. This is obtained by subtracting the arterial P_{O_2} from the so-called "ideal" alveolar P_{O_2}. The latter is the P_{O_2} which the lung *would* have if there were no ventilation:perfusion inequality and gases were exchanging at the same respiratory exchange ratio as in the real lung. It is derived from the alveolar gas equation in its full form:

$$P_{A_{O_2}} = P_{I_{O_2}} - \frac{P_{A_{CO_2}}}{R} + P_{A_{CO_2}} \cdot F_{I_{O_2}} \cdot \frac{1-R}{R}$$

The arterial P_{CO_2} is used for the alveolar value.

An increased alveolar-arterial oxygen difference

is caused both by abnormally low and abnormally high ventilation:perfusion ratios within the lung, though chiefly by the former. It is possible separately to assess the approximate contribution of these two groups to the impairment of gas exchange. The effect of the units with low ventilation:perfusion ratios is determined by calculating the *physiologic shunt*. Here we pretend that all the hypoxemia is caused by blood passing through unventilated alveoli (although this is a gross oversimplication). The shunt equation is then used in the form:

$$\frac{\dot{Q}_{PS}}{\dot{Q}_T} = \frac{C_{i_{O_2}} - C_{a_{O_2}}}{C_{i_{O_2}} - C_{\bar{v}_{O_2}}}$$

where $\dot{Q}_{PS}$ refers to physiologic shunt and $C_{i_{O_2}}$ de-notes the O_2 content of blood draining from "ideal" alveoli. The latter is found from the ideal alveolar P_{O_2} and the oxygen dissociation curve.

The effect of lung units with abnormally high ventilation:perfusion ratios is assessed by calculating the *physiologic dead space*. Here we pretend that all the lowering of P_{CO_2} in expired gas is caused by unperfused alveoli together with the anatomic dead space. The Bohr dead space equation is used in the form:

$$\frac{V_{D_{phys}}}{V_T} = \frac{P_{a_{CO_2}} - P_{E_{CO_2}}}{P_{a_{CO_2}}}$$

where $V_{D_{phys}}$ refers to physiologic dead space. Most patients with chronic obstructive lung disease, for example, have increases in both the physiologic shunt and dead space.

CHAPTER 38

Mechanics of Breathing

The subject of the mechanics of breathing includes the forces that support and move the lung and chest wall, together with the resistances they overcome and the resulting flows. The subject is sometimes considered along with ventilation (Chapter 36), which is the process by which gas gets to and from the alveoli. The field of mechanics is so large and important, however, that it is convenient to discuss it in a separate chapter. For reviews of areas of mechanics, the reader is referred to Mead (1973), Hoppin and Hildebrandt (1977), and Pedley et al. (1977).

MUSCLES OF RESPIRATION

Inspiration

In normal quiet breathing inspiration is active, but expiration passive. In other words, the inspiratory muscles distort the lung and chest wall from their equilibrium position, and the elastic properties of the system return them to their resting position during expiration. The most important muscle of inspiration is the *diaphragm*. This is a thin sheet of muscle in the shape of a dome which is attached to the lower ribs, sternum, and vertebral column. When the diaphragm contracts two things happen (Fig. 5.50). First, the abdominal contents are forced downward, thus enlarging the vertical dimension of the chest. Second, the rib margins are moved upwards and outwards. At first sight it seems paradoxical that shortening of the diaphragm should lead to an increase in the transverse diameter of the rib cage. This comes about because when the rib cage is lifted by the leverage action of the diaphragm on the abdominal contents, the ribs also move out because of the way they are hinged to the spine.

The diaphragm is supplied by two phrenic nerves, one to each lateral half. The fibers come from the spinal cord at cervical levels 3 and 4. During normal tidal breathing the dome moves about a centimeter or so, but on forced inspiration and expiration, a total excursion of up to 10 cm may be seen. If one-half of the diaphragm is paralyzed because the phrenic nerve supplying it is damaged, this portion moves up rather than down with inspiration when the intrathoracic pressure falls. This is called paradoxical movement and can be demonstrated at fluoroscopy by asking the patient to sniff.

The effectiveness of the diaphragm for changing the dimensions of the chest is related to the strength of its contraction and to its shape when relaxed. Normal descent of the diaphragm is impeded by advanced pregnancy, extreme obesity, or tight abdominal garments. In addition, patients who have had upper abdominal surgery often have limited diaphragmatic movement because of pain. Patients with advanced pulmonary emphysema often have very large lung volumes with almost flat diaphragms. Under these circumstances the diaphragm works very inefficiently and the work of breathing is increased. Recent work suggests that fatigue of the diaphragm may also occur in some patients with severe lung disease.

The next most important group of muscles of inspiration are the *external intercostals*. Figure 5.51 shows that these run downward and forward between adjacent ribs. When they contract the shortening of the muscle causes the ribs to rise. As a result the anteroposterior diameter of the chest is increased because of the downward slanting angle of the ribs. In addition, the ribs can be thought of as forming a bucket-handle shape hinged both at the spine and on the sternum anteriorly. For this reason the lateral diameter of the chest also increases as the ribs are raised.

The extent to which a given rib moves when the external intercostal muscles contract depends on the relative stability or fixation of adjacent ribs. The uppermost ribs are supported by the shoulder girdle, and as a result, contraction of the muscles tends to raise the whole of the rib cage.

The intercostal muscles are supplied by intercostal nerves which come off the spinal cord at the same level. Paralysis of the intercostal muscles alone by transection of the spinal cord in the lower cervical region below the origin of the fibers sup-

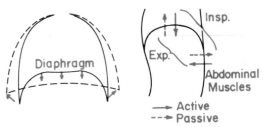

Figure 5.50. On inspiration (*Insp.*), the dome-shaped diaphragm contracts, the abdominal contents are forced down and forward, and the rib cage is lifted. Both increase the volume of the thorax. On forced expiration (*Exp.*), the abdominal muscles contract and push the diaphragm up.

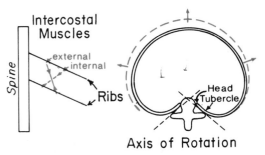

Figure 5.51. When the external intercostal muscles contract, the ribs are pulled upward and forward, and they rotate on an axis joining the tubercle and head of rib. As a result both the lateral and anteroposterior diameters of the thorax increase. The internal intercostals have the opposite action.

plying the phrenic nerves does not seriously affect breathing because the diaphragm is so effective.

The last group of inspiratory muscles is called the *accessory muscles* of inspiration. They have this name because they make little if any contribution during normal quiet breathing, but on exercise or forced respiratory maneuvers they may contract vigorously. They include the scalene muscles in the neck which elevate the first two ribs, and the sternocleidomastoids which insert into the top of the sternum. Patients with breathlessness at rest because of severe lung disease can be seen to be using these muscles. Other muscles which sometimes come into play include the alae nasi which cause flaring of the nostrils (particularly striking in the horse), and the small muscles of the head and neck.

Expiration

As indicated above, this is normally passive during quiet breathing. On exercise and during forced respiratory maneuvers, however, the expiratory muscles contract. The most important are those of the *abdominal wall* (Fig. 5.50). When they contract the intraabdominal pressure is raised and the diaphragm is pushed upward into the chest, thus reducing its volume. The abdominal muscles include

the rectus abdominus, internal and external oblique muscles, and transversus abdominus. These muscles also contract forcefully during coughing, vomiting, and defecation.

The other expiratory muscles are the *internal intercostals*. Their action is opposite to that of the external intercostal muscles (Fig. 5.51), so that as they shorten the ribs are pulled downward, backward, and inward, thus decreasing both the anteroposterior and lateral dimensions of the thoracic cage. Electromyographic studies show that these muscles contract also during straining, thus stiffening the intercostal spaces and preventing them from bulging outward.

ELASTIC PROPERTIES OF THE LUNG

Pressure-Volume Curve

Figure 5.52 shows how the pressure-volume curve of an excised dog lung can be measured. The bronchus is cannulated, and the lung is placed inside a jar in which the pressure is gradually reduced in

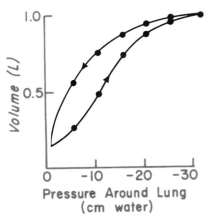

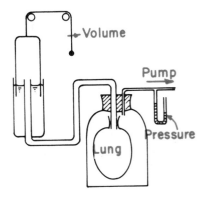

Figure 5.52. Measurement of the pressure-volume curve of excised lung. The lung is held at each pressure for a few seconds while its volume is measured. The curve is nonlinear and becomes flat at high expanding pressures. Note that the inflation and deflation curves are not the same; this is called hysteresis.

steps of say 5 cm of water by means of a pump. Notice first that when there is no pressure distending the lung there is a small volume of gas in it. Then as the pressure is gradually reduced in steps, the volume increases. Initially this occurs rapidly, but after the expanding pressure exceeds about 20 cm of water the volume changes are much less. In other words, the lung is much stiffer when it is well expanded. During the reverse procedure, when the pressure in the jar is allowed to return to atmospheric pressure in similar steps the volume decreases. Notice, however, that at any given pressure, say −10 cm of water, the volume on the descending limb of the pressure-volume curve exceeds that obtained while the lung was being expanded. This failure of the lung to follow the same course during deflation as it did during inflation is called *hysteresis.*

It is important to be clear about what is meant by these negative values of pressure. The pressures here are recorded with respect to atmospheric pressure. In other words, zero is the pressure recorded when both limbs of the water manometer are at the same level. In absolute units zero pressure is about 760 mmHg so that what is called −10 cm of water here is about 753 mmHg (10 cm of water is equivalent to 10/1.36 or 7.4 mmHg). This use of atmospheric pressure as a zero reference is a convenient and universal practice. Imagine the confusion that would be caused if a patient's blood pressure were reported as 880 mmHg instead of 120!

Suppose the lung of Figure 5.52 is expanded, with a pressure of say −20 cm of water, and then the line leading from the lung to the spirometer is clamped. Then the pump is turned off so that the pressure in the jar rises to atmospheric pressure, and the pressure inside the lung is measured by attaching the saline manometer to the airway. What pressure would be recorded? The answer is 20 cm of water above atmospheric pressure. This is because the pressure that matters is the difference between the inside and outside of the lung. In fact an identical pressure-volume curve can be obtained by inflating the lung with positive pressure while the pressure outside the lung is atmospheric. The pressure difference across the lung is often called the *transpulmonary pressure* (compare the term "transmural pressure" which is often used to indicate the pressure difference across a blood vessel wall).

In Figure 5.52 the pressure outside the lung was reduced by exhausting the gas in the jar with a pump. However, the same result could have been obtained if the jar itself had been expanded like a concertina. This is what happens in the intact animal. As the volume of the thoracic cage is increased through the action of the diaphragm and the intercostal muscles, the pressure within it is reduced and the lung is expanded. The fact that the space between the lung and chest wall is almost nonexistent does not matter. The normal intrapleural space contains no gas and only a few milliliters of liquid. Nevertheless, it is possible to measure a pressure in this space by inserting a small balloon catheter into the space in experimental animals. If this is done a pressure-volume curve can be drawn just like the one shown in Figure 5.52 for an excised dog lung. In practice a measurement of pressure inside the esophagus is a useful approximation to the intrapleural pressure and can fairly easily be obtained by having a subject swallow a small balloon on the end of a catheter. This is how pressure-volume curves are measured in the clinical pulmonary function laboratory.

Compliance

Figure 5.52 shows that the pressure-volume curve of the lung is nonlinear. It is useful in practice, however, to refer to the slope of the pressure-volume curve over a particular range of interest. For example, we might measure the change in pressure over the liter above functional residual capacity on the descending limb of the pressure-volume curve. The volume change per unit pressure change is known as the *compliance* of the lung. It is clearly not a complete description of the pressure-volume properties of the lung, but nevertheless it is useful in practice as a measure of the comparative stiffness of lungs. The stiffer the lung the less the compliance. The compliance is reduced by diseases which cause an accumulation of fibrous tissue in the lung, or by edema in the alveolar spaces which prevents the inflation of some lung units. The compliance of the lung is increased in pulmonary emphysema, and also with age, probably because of alterations in the elastic tissue in both instances. The normal value of the compliance in man is about 200 ml/cm of water. In other words, to inspire a normal tidal volume of about 500 ml, intrapleural pressure has to fall by 2–3 cm of water.

The compliance of a lung clearly depends on its size. Thus a mouse lung will have a much smaller volume change per unit change in transpulmonary pressure than an elephant lung. In the same way, a patient who has had one lung removed surgically will have a reduced compliance. Thus if compliance is used to get information about the elastic properties of the lung one must take account of size. Sometimes this is done by calculating the specific compliance, which is the compliance per unit volume. Another index of elastic properties that is sometimes useful in the pulmonary function laboratory is the transpulmonary pressure at total lung capacity. This is abnormally low in emphysema in spite of the fact that the lung volume is usually abnormally high.

What is responsible for the elastic behavior of the lung, that is, its tendency to return to its original position after distortion? There are two factors. One is the elastic components of the tissue of the lung which are visible in microscopic sections (Fig. 5.53). This figure shows a thick section of normal adult lung stained to show the fibers of elastin in the pulmonary parenchyma. In addition, not shown in this section, there are numerous collagen fibers distributed throughout the lung tissue. Elastin fibers are easily stretched while collagen fibers are not. However, the elastic behavior of the lung probably does not depend on simple elongation of elastic fibers but more on their geometric arrangement. This has been called "nylon stocking" elasticity. The analogy is apt because in a stocking the nylon threads are very difficult to stretch, and the elastic behavior of the stocking is the result of distortion of the geometrical arrangement of the knitted fibers. Again, the changes in the pressure-volume properties of lungs that occur with age may have to do more with changes in the geometrical arrangement of the fibers because the actual

amounts of collagen and elastin apparently do not change.

Surface Tension

The second factor responsible for the elastic behavior of the lung is the surface tension of the liquid film lining the alveoli. Figure 5.54A indicates that the surface tension is defined as the force (in dynes for example) acting across an imaginary line 1 cm long in a liquid surface. This tension develops because the cohesive forces between adjacent liquid molecules are greater than the forces between the molecules of liquid and gas outside the surface. One way to think of surface tension is to imagine that the surface consists of a thin rubber membrane under stretch. If an incision 1 cm long were made in this membrane it would gape, and if sutures were put in to bring the two cut sides together the total force of the sutures would be equal to the surface tension.

An important property of surface tension is that it generates a pressure in a bubble (Fig. 5.54B). Again imagine the surface as a thin rubber membrane under stretch like a balloon. The tension in the wall develops a pressure just as a hoop encircling a barrel tends to crush it inward. The relationship between the surface tension in the wall and the pressure developed in a soap bubble is given by Laplace's law (Fig. 5.54C). This states that for each surface of the bubble, the pressure is equal to twice the tension divided by the radius, or for both surfaces P = 4T/r. Notice that in two bubbles with the same surface tension the pressure developed by the smaller bubble will exceed that of the larger bubble; if they are connected the smaller bubble will therefore blow up the larger bubble.

This raises the question of why the lung is not unstable. At first sight it can be regarded as a series of 300 million bubbles all connected by airways and all containing a gas-liquid interface with surface

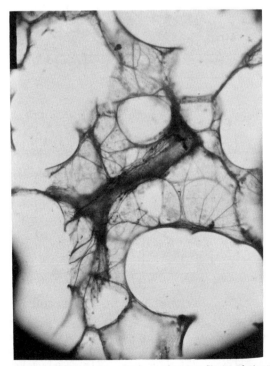

Figure 5.53. Section of human lung showing fibers of elastin in the alveolar walls and around blood vessels. These contribute to the elastic behavior of the lung [Courtesy of Dr. R. R. Wright.]

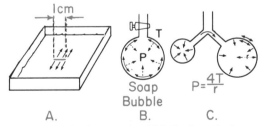

Figure 5.54. *A.,* Surface tension (T) is the force in dynes acting across an imaginary line 1 cm long in a liquid surface. *B.,* Surface forces in a soap bubble tend to reduce the area of the surface and generate a pressure (P) within the bubble. *C.,* Because the smaller bubble generates a larger pressure, it blows up the larger bubble. r, radius.

tension. It is true that each "bubble" or alveolus is surrounded by many other alveoli which tend to confer some stability (see later). Nevertheless why is such a mass of "bubbles" not unstable like a soap foam?

The answer to this question leads to a fascinating story in the development of pulmonary mechanics. As long ago as 1929, von Neergaard showed that lungs inflated with saline are easier to distend than air-filled lungs. Figure 5.55 shows the results of a typical experiment using excised cat lungs; it can be seen that at a given distending volume, the pressure developed by the air-inflated lungs is considerably greater than that for saline-filled lungs. Since presumably the only difference between the two lungs is the abolition of the gas-liquid interface by saline filling, this experiment demonstrates that a large part of the retractive force comes from the surface tension. Some years later Radford (1954) made some calculations of the pressure-volume properties that a lung would have if the surface tension of the fluid lining the alveoli was that of plasma, namely about 70 dynes/cm. He found that the transpulmonary pressures that would be needed to inflate the lung were far too high, and concluded that the assumptions he had made about the geometry of the air spaces were incorrect.

At about this time Pattle (1955) was working in a government defense laboratory on the effects of chemical gases on the lung. In studying the edema foam that comes from the lungs of animals exposed to noxious gases he noticed that the bubbles of this foam were extremely stable. He recognized that they therefore had an extremely low surface tension, since small bubbles with a normal surface tension shrink rapidly (this is because the large pressure developed inside them is sufficient to force gas out of them by diffusion). Pattle's observation led to the discovery of pulmonary *surfactant*, which is a phospholipid secreted by alveolar cells and which has the property of reducing the surface tension of the liquid lining layer.

Clements (1962) showed how the property of lung extracts to a lower surface tension could be demonstrated in a surface balance as shown in Figure 5.56A. In this device the material to be investigated is added to a surface that is alternately compressed and expanded by a barrier across the trough. The tension of the surface is measured by dipping a platinum strip into it and attaching this to a force transducer. Figure 5.56B shows that if the trough contains pure water, the surface tension obtained is independent of the area of the surface and has a value of about 70 dynes/cm. If detergent is added to the water, the surface tension falls considerably, but it is still independent of the area of the surface. However, when a lung extract is obtained (by rinsing out the airways with saline) and this is added

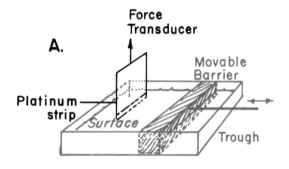

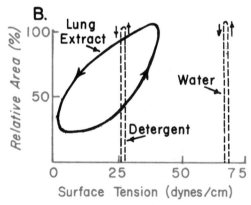

Figure 5.56. *A.*, Surface balance. The area of the surface is altered and the surface tension is measured from the force exerted on a platinum strip dipped into the surface. *B.*, Plots of surface tension and area obtained with a surface balance. Note that lung washings show a change in surface tension with area and that the minimum tension is very small.

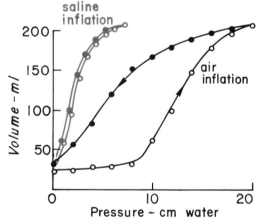

Figure 5.55. Comparison of pressure-volume curves of air-filled and saline-filled lungs (cat). *Open circles*, inflation; *closed circles*, deflation. Note that the saline-filled lung has a higher compliance and also much less hysteresis than the air-filled lung. [From Radford, (1957).]

to the trough, a curve is drawn as the area of the surface is altered.

Two features of this curve should be noted. One is that, in general, the smaller the area the lower the value of the surface tension. Thus if we extrapolate this behavior to the lung and assume that all alveoli secrete the same amount of surfactant, the smaller alveoli will presumably have a lower surface tension than the larger alveoli, and therefore the tendency for the smaller alveoli to inflate the larger alveoli (Fig. 5.54C) will be lessened. Note also that the curve shows the same kind of hysteresis as the pressure-volume curve of the lung; that is, for a given surface area the surface tension during expansion is much greater than during compression. It seems likely that most of the hysteresis of the intact lung (Fig. 5.52) can be explained by the behavior of the surface tension. Additional evidence for this statement is that the pressure-volume curve for saline-filled lungs shows little hysteresis (Fig. 5.55).

The composition of pulmonary surfactant is not completely known, but it contains the phospholipid, dipalmitoyl lecithin. This material shows the same kind of surface tension behavior on the surface tension balance. The surfactant is secreted by the type 2 alveolar cells. Osmiophilic lamellated bodies can be seen in electron micrographs of type 2 cells in the alveolar walls (Fig. 5.57). These lamellated bodies are extruded to form surfactant. The type 2 cells synthesize the phospholipid from fatty acids that are either extracted from the blood or are themselves synthesized in the lung. Synthesis is fast, and there is a rapid turnover of surfactant. Three chemical pathways are available for the incorporation of fatty acids into lecithin, and there apparently are differences between the pathways used by the adult and by the fetal lung. Because surfactant is formed relatively late in fetal life, babies born without adequate amounts develop respiratory distress and may die.

What are the physiological advantages of pulmonary surfactant? First, a low surface tension in the alveoli increases the compliance of the lung and

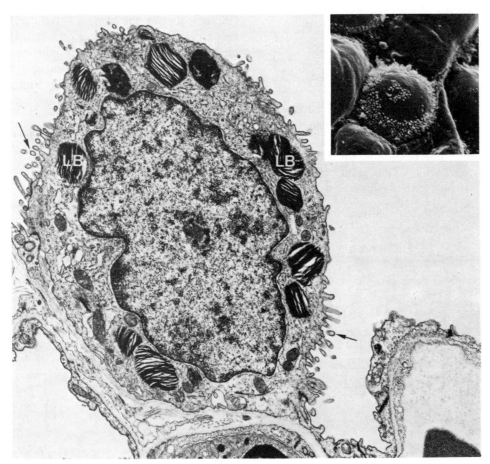

Figure 5.57.   Electron micrograph of type 2 epithelial cell (×10,000). Note the osmiophilic lamellated bodies (*LB*), large nucleus, and microvilli (indicated by *arrows*). The *inset* at top right is a scanning electron micrograph showing the surface view of a type 2 cell with its characteristic distribution of microvilli (×3400). [From Weibel and Gil (1977).]

reduces the work of expanding it with each breath. Second, stability of the alveoli is promoted. Third, surfactant helps to keep the alveoli dry. This action is more difficult to grasp but the inward-acting forces that tend to collapse the alveoli also tend to pull water into the alveolar spaces from the pulmonary capillaries in the alveolar wall. This tendency is reduced by pulmonary surfactant. Indeed, some physiologists believe that this may be the major role of pulmonary surfactant.

In some pathologic conditions there is an absence of pulmonary surfactant. A good example is the respiratory distress syndrome of the newborn, which is particularly likely to develop in premature infants whose surfactant system is poorly developed. As a result the lungs are stiff and difficult to inflate, they contain areas of alveolar atelectasis, and also areas of edema.

Interdependence

Another mechanism which apparently contributes to the stability of the alveoli in the lung has been described (Mead et al., 1970). Figure 5.58 illustrates how all the alveoli (except those immediately adjacent to the pleural surface) are surrounded by other alveoli and are therefore supported by each other. Furthermore it has been shown that in a structure such as this with many connecting links, any tendency for one group of units to reduce or increase its volume relative to the rest of the structure is opposed. For example, if a group of alveoli has a tendency to collapse, large expanding forces will be developed on them by the surrounding parenchyma which then tends to be overexpanded. In fact, calculations and measurements show that surprisingly large expanding

forces can be developed on a portion of collapsed lung as the lung around it is expanded.

This support offered to lung units by those surrounding them is termed *interdependence*. The same mechanism is responsible for the development of low pressures around large blood vessels and airways as the lung expands; in this case the fact that the relatively stiff blood vessels do not expand to the same extent as the parenchyma around them is responsible for the development of low perivascular pressures. Interdependence also may well play an important role in the prevention of atelectasis and in the opening up of areas that have collapsed for some reason. In fact some physiologists believe that the role of this mechanism may be more important than that of surfactant in maintaining the stability of the small air spaces.

ELASTIC PROPERTIES OF THE CHEST WALL

Just as the lung is elastic, so is the thoracic cage. Although it is not so immediately obvious, the chest wall is pulled in by the elastic retractive forces of the lung just as the lung is expanded by the elastic force developed by the chest wall. This becomes clearer if we consider what happens when air is introduced into the intrapleural space and a pneumothorax is formed as shown in Figure 5.59. When the lung collapses and it no longer pulls the chest wall in, the wall springs out.

In clinical practice it is often possible to detect differences in expansion of the chest wall by careful

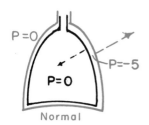
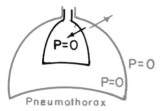

Figure 5.59. The tendency of the lung to recoil to its deflated volume is balanced by the tendency of the chest cage to bow out. As a result, the intrapleural pressure is subatmospheric. Pneumothorax allows the lung to collapse and the thorax to spring out. P, pressure.

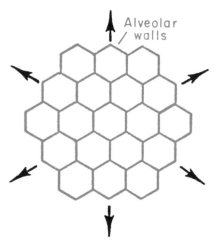

Figure 5.58. Diagram of an array of alveoli emphasizing how each unit is supported by surrounding units. This is referred to as interdependence and is important in maintaining alveolar stability.

examination of the chest or by looking at a chest radiograph and noticing whether the ribs are crowded or not. Typically a patient with a reduced compliance caused by fibrosis of one lung will have the chest wall pulled in to some extent on that side. By contrast a patient with a unilateral pneumothorax often has an overexpanded chest wall on the affected side.

The interaction between the elasticity of the lung and the chest wall can be summarized in what is known as a relaxation pressure-volume diagram (Rahn et al., 1946). This is shown in Figure 5.60, which contains a large amount of information that may be rather confusing at first sight. The relaxation pressure is the airway pressure obtained when the subject is completely relaxed, that is, when he is not attempting either to inflate or deflate his lung and chest wall by muscle activity. In can be measured as shown on the left half of the figure by having the subject inhale or exhale to a particular volume as measured by the spirometer, then closing the tap, and then measuring the airway pressure with the subject completely relaxed. Incidentally, this is difficult for untrained subjects to do.

Look first at the continuous line for the lung and chest wall. Notice that at functional residual capacity (FRC) the relaxation pressure is zero. That is to say, this is the equilibrium position of the lung and chest wall when airway pressure is the same as atmospheric pressure. As the volume is increased, the relaxation pressure becomes positive, since the lung and chest wall tend to return to the equilibrium FRC position. At total lung capacity, the relaxation pressure is about 30 cm of water. Below FRC, the residual volume is gradually approached with a very

low relaxation pressure. Under these conditions the lung and chest wall tend to spring out when the expiratory muscles are relaxed, thus generating the negative relaxation pressure.

Now look at the broken line for the lung alone. This is the same curve on the human lung as can be obtained on a dog lung when it is inflated by positive pressure (compare Fig. 5.52). Notice that at FRC, a positive pressure of some 5 cm of water is developed as the lung tries to collapse. At total lung capacity this pressure has increased to about 25 cm of water. Below FRC the relaxation pressure falls to zero at what is known as the minimal volume of the lung, that is, the volume that the lung takes up when there is no expanding pressure on it. However, as we saw previously, some gas remains in the lung under these conditions.

Finally, turn to the broken line for the chest wall alone. We can imagine this being measured on a man with a normal chest wall and no lung! Notice that at FRC the chest wall develops a negative relaxation pressure. In other words it is tending to spring out at FRC just as the lung is tending to collapse inward. In fact the negative relaxation pressure of the chest wall and the positive relaxation pressure of the lung are identical, thus making this the equilibrium position for the lung and chest wall together. Indeed at any volume, the curve for the combined lung and chest wall can be explained by the addition of the individual lung and chest wall curves. Notice that it is not until the volume of the chest wall is increased to about 70% of vital capacity that it no longer tends to spring out. In fact, this is the resting position of the chest wall, and at volumes above this the tendency of the chest

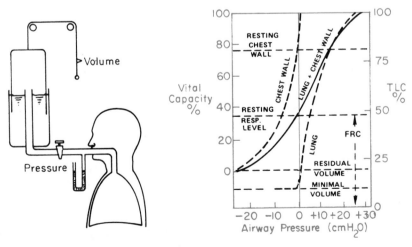

Figure 5.60. Relaxation pressure-volume curve of the lung and chest wall. The subject inspires (or expires) to a certain volume from the spirometer, the tap is closed, and he then relaxes his chest. The curve for lung + chest wall can be explained by the addition of the individual lung and chest wall curves. See text for details. FRC, functional residual capacity; TLC, total lung capacity. [From Rahn et al. (1946).]

wall to collapse back results in positive relaxation pressures.

AIRWAY RESISTANCE

Airflow through Tubes

If air flows through a tube (Fig. 5.61), a difference of pressure exists between the ends. The pressure difference depends on the rate and pattern of flow. At low flow rates, the stream lines may everywhere be parallel to the sides of the tube (*A*). This is known as laminar flow. As the flow rate is increased, unsteadiness develops especially at branches. Here separation of the stream lines from the wall may occur with the formation of local eddies (*B*). At higher flow rates still, complete disorganization of the stream lines is seen; this is turbulence (*C*).

The pressure-flow characteristics for *laminar flow* were first described by the French physician Poiseuille. In smooth straight circular tubes, the volume flow rate $\dot{V}$ is given by:

$$\dot{V} = \frac{\Delta P \pi r^4}{8nl}$$

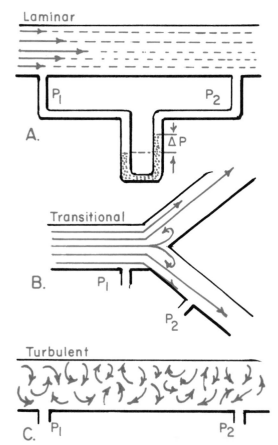

where ΔP is the driving pressure, r radius, n viscosity, and l length. It can be seen that driving pressure is proportional to flow rate, or $\Delta P = KV$. Since flow resistance R is driving pressure divided by flow we have

$$R = \frac{8nl}{\pi r^4}.$$

Note the critical importance of tube radius; if the radius is halved, the resistance increases 16-fold! However, doubling the length only doubles resistance. Note also that the viscosity of the gas but not its density affects the pressure-flow relationship.

Another feature of laminar flow when this is fully developed is that the gas in the center of the tube moves twice as fast as the average velocity. Thus, a spike of rapidly moving gas travels down the axis of the tube (Fig. 5.61*A*). This changing velocity across the diameter of the tube is known as the velocity profile, and its shape is a parabola. The axial velocity, that is, the velocity in the center of the stream, is twice the mean flow velocity, but the gas immediately adjacent to the wall of the tube is not moving at all. This latter is known as the boundary layer. This parabolic velocity profile is only seen if the tube is long enough for it to develop, and the pattern is then called "fully developed laminar flow." If gas enters the tube from a larger chamber, for example, the length of the tube has to be many times its diameter before this pattern becomes completely established.

Turbulent flow has different properties. Here pressure is not proportional to flow rate but, approximately, to its square: $P = K\dot{V}^2$. In addition, the viscosity of the gas becomes less important, but an increase in gas density increases the pressure drop for a given flow. Turbulent flow does not have the very high axial flow velocity that is characteristic of laminar flow.

Whether flow will be laminar or turbulent depends to a large extent on Reynold's number, Re. This is given by

$$Re = \frac{2rvd}{n}$$

where d is density, v average velocity, r radius, and n viscosity. In straight smooth tubes, turbulence is probable when Reynold's number exceeds 2,000. The expression shows that turbulence is most likely to occur when the velocity of flow is high and the tube diameter large (for a given velocity). Note also that a low density gas like helium tends to produce less tubulence.

In such a complicated system of tubes as the bronchial tree with its many branches, changes in

Figure 5.61. Patterns of airflow in tubes. In *A*, the flow is laminar, in *B*, transitional with eddy formation at branches, and in *C* turbulent. Resistance is equal to $(P_1 - P_2)$/flow.

caliber during ventilation, and irregular wall surfaces, application of the above principles is difficult. In practice, for laminar flow to occur the entrance conditions of the tube are critical. If eddy formation occurs upstream at a branch point, this disturbance is carried downstream some distance before it disappears. Thus, in a rapidly branching system like the lung, fully developed laminar flow (Fig. 5.61A) probably only occurs in the very small airways where the Reynold's numbers are very low (approximately 1 in terminal bronchioles). In most of the bronchial tree, flow is transitional (B), although true turbulence may occur in the trachea, especially on exercise when flow velocities are high. In general, driving pressure is determined by both the flow rate and its square:

$$\Delta P = K_1 \dot{V} + K_2 \dot{V}^2.$$

For a review of this area the reader is referred to Pedley et al. (1977).

Pressures during the Breathing Cycle

Figure 5.62 shows the intrapleural and alveolar pressures during a normal breathing cycle. Intrapleural pressure can be measured approximately by means of a balloon catheter placed in the esophagus. Alveolar pressure is more difficult to obtain but can be deduced from measurements made in a

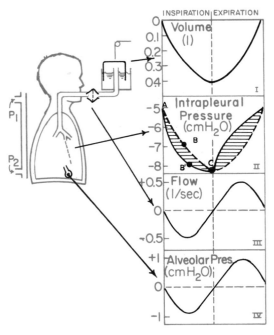

Figure 5.62. Pressures during the breathing cycle. If there were no airway resistance, alveolar pressure would remain at zero and intrapleural pressure would follow the broken line, *ABC*, which is determined by the elastic recoil of the lung. Airway (and tissue) resistance contributes the cross-hatched portion of intrapleural pressure (see text).

body plethysmograph. Panel I on the right shows that spirometer volume decreases on inspiration and then increases during expiration. Panel III shows that inspiratory flow rate (downward deflection) increases and then decreases during inspiration; expiratory flow rate follows. Alveolar pressure (panel IV) is measured with respect to atmospheric pressure.

Before inspiration begins there is no flow and therefore alveolar pressure is zero. For inspiratory flow to occur, alveolar pressure falls and then returns to zero at the end of inspiration. On expiration, alveolar pressure rises above atmospheric pressure in order to overcome the resistance of the airways during expiration. If airway resistance remained constant during the breathing cycle, the shapes of the curves for flow and alveolar pressure (panels III and IV) would be identical.

Now look at intrapleural pressure. Before inspiration, this is −5 cm of water with respect to atmospheric pressure because of the elastic recoil forces of the lung. Intrapleural pressure falls during inspiration for two reasons. The first is that lung volume increases and the elastic recoil forces become greater. If these were the only forces operating, intrapleural pressure would fall along the broken line *ABC*, and if the compliance of the lung were constant, the shape of this line would be identical to that of the volume change (panel I). An additional fall of pressure is required, however, to move gas along the airways; the result is that intrapleural pressure falls along the line *AB'C*. Thus the cross-hatched portion of the intrapleural pressure curves represents the pressure necessary to overcome airway resistance. In addition, there is a contribution made by the viscous resistance of the lung tissues, that is, the forces required to move one layer of tissue over another; this will be discussed a little later.

The total pressure drop between the mouth and intrapleural space is thus made up of two components: (1) that caused by the pressure drop along the airways due to their resistance to air flow, and (2) the pressure drop across the lung caused chiefly by the static recoil forces. In addition, there is a small component of tissue viscous resistance. As an equation of pressures:

(mouth − intrapleural) = (mouth − alveolar)
+ (alveolar − intrapleural)

On expiration opposite changes occur such that intrapleural pressure is now less negative than it would be in the absence of flow because alveolar pressure rises to overcome airway resistance. Indeed, with a forced expiration intrapleural pressure may easily go positive, that is, above atmospheric pressure.

Measurement of Airway Resistance

For the complete system of airways, resistance (R_{aw}) is defined as:

$$R_{aw} = \frac{P_{mouth} - P_{alv}}{\dot{V}}$$

where P_{mouth} and P_{alv} are the pressure in the mouth and alveoli, respectively. Mouth pressure is easy to measure, but alveolar pressure can be obtained only indirectly. One method of doing this is with a body plethysmograph (Fig. 5.63).

The subject sits in the air-tight box and breathes rapidly and shallowly. Before inspiration (*A*), the box pressure is atmospheric. At the onset of inspiration, the pressure in the alveoli falls, and the alveolar gas expands by a volume ΔV. This compresses the gas in the box, and from its change in pressure ΔV can be calculated (compare Fig. 5.8). If lung volume is known, ΔV can be converted into alveolar pressure using Boyle's law. Flow is measured simultaneously and thus airway resistance is obtained. A measurement can be made during expiration in the same way. Lung volume is determined with the plethysmograph as described in Figure 5.8.

Airway resistance can also be measured during normal breathing from an intrapleural pressure record as obtained with an esophageal balloon (Fig. 5.62). However, in this case tissue viscous resistance is included as well, and the result is often referred to as pulmonary resistance. As we saw above, intrapleural pressure reflects two sets of forces, those opposing the elastic recoil of the lung, and those overcoming the resistance to air and tissue flow. It is possible to subtract the pressure caused by the recoil forces during quiet breathing because this is

proportional to lung volume (if compliance is constant). The subtraction is done with an electrical circuit so as to leave a plot of pressure against flow which gives (airway + tissue) resistance. This method is not satisfactory in lungs with severe airway disease because the uneven time constants prevent all regions from moving together (Fig. 5.73).

Chief Site of Airway Resistance

As the airways penetrate toward the periphery of the lung, they become more numerous but much narrower (Figs. 5.3 and 5.9). Based on Poiseuille's equation with its (radius)[4] term, it would be natural to think that the major part of the resistance lies in the very narrow airways. Indeed, this was thought to be the case for many years. However, it has been shown by direct measurements of the pressure drop along the bronchial tree that the major site of resistance is the medium-sized bronchi, and that the very small bronchioles contribute relatively little resistance (Macklem and Mead, 1967). Figure 5.64 shows that most of the pressure drop occurs in the airways up to approximately the seventh generation. Less than 20% can be attributed to airways less than 2 mm in diameter. The reason for this apparent paradox is the prodigious number of small airways. In other words, the resistance of each airway is relatively large, but there are so many airways arranged in parallel that the combined resistance is small.

The fact that the peripheral airways contribute

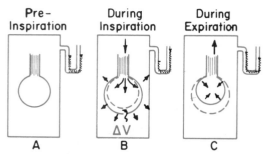

Figure 5.63. Measurement of airway resistance with a body plethysmograph. During inspiration, the alveolar gas expands and box pressure therefore rises. From this, alveolar pressure can be calculated, which, when divided by the flow gives airway resistance (see text). ΔV, change in volume. [Adapted from Comroe et al. (1962).]

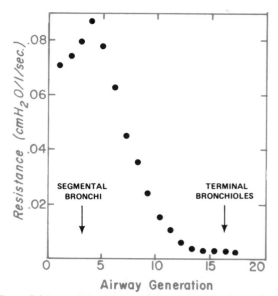

Figure 5.64. Location of the chief site of airway resistance. Note that the intermediate sized bronchi contribute most of the resistance and that relatively little is located in the very small airways. [From Pedley et al. (1970).]

so little resistance is important in the detection of early airway disease. Because they constitute a "silent zone" it is probable that a considerable degree of small airway disease can be present before the usual measurements of airway resistance can pick up an abnormality. As a consequence, special methods aimed at detecting the resistance of small airways have been explored. It has been suggested that the measurement of "closing volume" (Fig. 5.72) might be a sensitive test of small airways disease.

Factors Determining Airway Resistance

Lung volume has an important effect on airway resistance. Because the bronchi which run within the lung parenchyma are supported by the guy-wire action of the surrounding lung tissue, their caliber is increased as the lung expands. This can be appreciated at bronchoscopy, a form of examination of the bronchial tree in which a lighted tube is passed down into the large bronchi via the mouth or nose. With each inspiration, the bronchi can be seen to enlarge. Figure 5.65 shows that as lung volume is reduced, airway resistance rises rapidly. This relationship is very nonlinear. However, if the reciprocal of resistance, known as *conductance*, is calculated, this increases approximately linearly with lung volume.

It is clearly important therefore to standardize lung volume if a measurement of airway resistance is being reported in a patient. Many patients who have disease of the airways and therefore a high airway resistance (for example, patients with bronchial asthma) often breathe at a high lung volume, and this helps to reduce their resistance in the

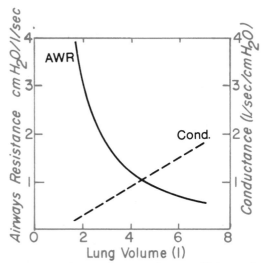

Figure 5.65. Variation of airway resistance with lung volume. If the reciprocal of airway resistance (AWR), that is, conductance (Cond.), is plotted, the graph is a straight line. [From Briscoe and Dubois (1958).]

direction of normal. At very low lung volumes, the small airways in the region of the terminal and respiratory bronchioles may actually close completely, thus trapping gas in the alveoli. As will be explained shortly, this is especially likely to occur at the bottom of the lung where the lung is least well expanded.

Another important determinant of the caliber of the airways is the *tone of the bronchial smooth muscle*. This is under autonomic control. Sympathetic stimulation causes dilatation, whereas parasympathetic activity causes bronchial constriction. Drugs such as isoproterenol and epinephrine which stimulate the β-adrenergic receptors in the airways cause bronchial dilatation and are used to treat the bronchial constriction which occurs in asthma. A fall in alveolar P_{CO_2} as a result of hyperventilation or a local reduction of pulmonary blood flow also causes bronchoconstriction, apparently by a direct action on the smooth muscle of the airways. Bronchial constriction also occurs reflexly through stimulation of receptors in the trachea and large bronchi by irritants such as cigarette smoke, inhaled dusts, and noxious gases (Chapter 39). Motor innervation is by the vagus nerve. The injection of histamine into a pulmonary artery causes constriction of smooth muscle located in the mouths of the alveolar ducts. A similar response occurs when some types of microemboli are injected into the pulmonary circulation. Swelling of the bronchial mucosa caused by inflammation will also increase airway resistance.

The *density and viscosity* of the gas which is breathed affect the pressure drop along the airways because flow is transitional in most regions of the bronchial tree (Fig. 5.61*B*). In diving at great depths under the sea where the density of the gas is enormously increased, large pressures are required to move the gas, but if a helium-oxygen mixture is substituted, the pressures are considerably smaller. In the past, patients with severe airway disease were occasionally treated with helium-oxygen mixtures in order to reduce their work of breathing. The fact that changes in density rather than viscosity have such an influence on resistance is evidence that flow is not purely laminar in the medium-sized airways where the main site of resistance lies (Fig. 5.64).

Dynamic Compression of Airways

Suppose we ask a subject to inspire to his total lung capacity and then breath out as hard and as fast as he can down to residual volume. We can then plot expiratory flow rate against lung volume as shown in Figure 5.66, the so-called *flow-volume*

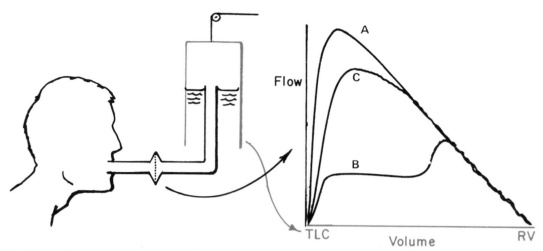

Figure 5.66. Flow-volume curves. In *A*, a maximal inspiration was followed by a forced expiration. In *B*, expiration was initially slow and then forced. In *C*, expiratory effort was submaximal. In all three, the descending portions of the curves are almost superimposed. TLC, total lung capacity; RV, residual volume.

curve. The result will be a curve like *A* showing that flow rises very rapidly to a high value and then declines over most of expiration.

A curious feature of this flow-volume envelope is that it is virtually impossible to penetrate it. For example, suppose that the subject starts breathing out slowly from total lung capacity as shown in curve *B*, and then half way through expiration accelerates as hard as he can. We would find that the latter part of the flow-volume curve is almost superimposed on the first one that was inscribed. Finally, suppose he exhales from total lung capacity but this time makes much less effort overall. We would obtain a curve *C* where the maximum flow rate is somewhat reduced, but over much of expiration the flow-volume curve will be almost superimposed on that obtained with the forced expiration.

Two conclusions follow from these suprising observations. The first is that something powerful is limiting the flow rate that can be achieved at a given lung volume, since it is impossible to penetrate the flow-volume envelope no matter what tricks are tried. Second, that over most of expiration the flow rate is virtually independent of effort, in that whether we try very hard or very much less the flow rates are the same.

More information can be obtained about this curious state of affairs by plotting the data in another way as shown in Figure 5.67. To obtain this figure the subject inhales to total lung capacity and then exhales fully, with varying degrees of effort. We then plot flow rate and intrapleural pressure at the *same* lung volume for each different inspiration and expiration and thus obtain these so-called isovolume pressure flow curves. The figure

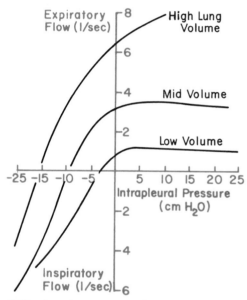

Figure 5.67. Isovolume pressure-flow curves drawn for three lung volumes. Each of these was obtained from a series of forced expirations or inspirations (see text). Note that at the high lung volume, a rise in intrapleural pressure (obtained by increasing expiratory effort) resulted in a greater expiratory flow. However, at mid- and low volumes, flow becomes independent of effort after a certain intrapleural pressure has been exceeded. [From Fry and Hyatt (1960).]

shows that at high lung volumes the flow rate continues to increase as the intrapleural pressure is raised, as we might expect. At midvolume, however, there is surprisingly no further increase in flow rate once an intrapleural pressure of some 5–10 cm of water has been exceeded. This phenomenon is even more marked at low lung volumes. Thus at these low and intermediate lung volumes, flow is effort-independent.

The reason for these remarkable findings is that

compression of the larger airways effectively limits the flow rate. Figure 5.68 shows the mechanism. In *A* before inspiration, intrapleural pressure is −5 cm of water, and since there is no air flow the pressure throughout the airways is atmospheric. If we assume that the pressure outside the large airways is intrapleural, the transmural pressure across these airways is 5 cm of water tending to hold them open. Now in *B* at the start of inspiration, intrapleural pressure falls to −7 cm of water, and alveolar pressure falls to −2 cm of water. We are assuming at this point there has been a negligible change in lung volume, so that the pressure difference between intrapleural and alveolar spaces remains at 5 cm of water. There will be a pressure drop along the airways because of resistance to air flow; suppose that at a particular point the pressure is −1 cm of water. At this place we see that the transmural pressure holding the airway open has now increased to 6 cm of water. *C*, shows the situation at the end of inspiration. Intrapleural pressure is −8 cm of water, and since the air has stopped moving the pressure inside the airways at every point is atmospheric. There is now a pressure of 8 cm of water holding the airway open.

A striking change occurs at the beginning of forced expiration as shown in *D*. Intrapleural pressure rises abruptly to 30 cm of water. The pressure difference between intrapleural and alveolar spaces is still 8 because lung volume has changed by a negligible amount so early in the expiration. Thus alveolar pressure is now 38 cm of water. Again there will be a pressure drop along the airways because

of their flow resistance, and at the point we are interested in the pressure is 19 cm of water. It can be seen, however, that now there is a pressure of 11 cm of water across the airway tending to *close* it. The airway therefore collapses.

We now have a situation in which flow occurs through a collapsible tube when the outside pressure exceeds downstream pressure. This was discussed earlier as the "vascular waterfall" which determines blood flow in zone 2 of the lung (Fig. 5.21). Under such conditions, flow is independent of downstream pressure, being determined only by the difference between upstream pressure and the pressure outside the collapsible tube. In the case of the lung, this becomes alveolar pressure minus intrapleural pressure.

Two important conclusions follow from this. One is that no matter how forceful the expiration, the flow rate cannot be increased because as intrapleural pressure goes up it takes alveolar pressure with it. Thus the driving pressure (alveolar-intrapleural pressure) remains constant. This explains why flow rate becomes effort-independent. Second, maximum flow rate will be determined partly by the elastic recoil forces of the lung because these are what generate the difference between alveolar and intrapleural pressure. Notice that these elastic recoil forces will decrease as lung volume becomes smaller and that this is one reason why the maximum flow rate decreases as lung volume falls (Fig. 5.66). Another reason is that the resistance of the peripheral airways increases as lung volume reduces (Fig. 5.65).

Several factors will exaggerate this flow-limiting mechanism. One is any increase in the resistance of the airways peripheral to the collapse point, since this will magnify the pressure drop along them and thus decrease the intrabronchial pressure during expiration. Notice that once collapse occurs, the resistance of the airways proximal (mouthwards) of the point of collapse becomes irrelevant. Flow rate is determined by the lung elastic recoil pressure divided by the resistance of the peripheral airways up to the collapse point. Another factor which exaggerates this dynamic compression is a reduction in radial traction on the airways offered by surrounding lung tissue. Such a reduction occurs in emphysema, which is characterized by destruction of alveolar walls. Also in the emphysematous lung the elastic recoil forces are reduced (increased compliance). As a result the driving pressure under conditions of dynamic compression (alveolar minus intrapleural pressure) is reduced. Indeed, while this type of flow limitation is seen only during forced expiration in normal subjects, it may well occur

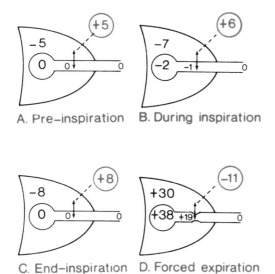

A. Pre-inspiration   B. During inspiration

C. End-inspiration   D. Forced expiration

Figure 5.68. Scheme showing why airways are compressed during a forced expiration. Note that the pressure across the airways is holding it open except during the forced expiration. See text for details.

during normal expiration in patients with severe emphysema. By contrast in restrictive lung disease, e.g., caused by fibrosis, the maximal flow rate for a given lung volume may be somewhat higher than normal because of the increased retractive forces.

INEQUALITY OF VENTILATION

Topographical Differences within the Lung

Suppose a seated normal subject inhales a breath of radioactive xenon gas (Fig. 5.69). When the xenon-133 enters the counting field, its radiation penetrates the chest wall and can be recorded by a bank of counters or a radiation camera. In this way the volume of the inhaled xenon going to various regions can be determined (Ball et al., 1962).

Figure 5.69 shows results obtained in a series of normal volunteers using this method. It can be seen that ventilation per unit volume is greater near the bottom of the lung and becomes progressively smaller toward the top. Other measurements made when the subject is in the supine position show that the difference disappears with the result that apical and basal ventilations become the same. In that posture however, the ventilation of the lowermost (posterior) part of the lung exceeds that of the uppermost (anterior) part. Again in the lateral position (subject on his side) the dependent lung is the better ventilated.

The cause of these topographical differences in ventilation lies in the distortion that occurs in the lung as a consequence of its weight (Milic-Emili et al., 1966). Figure 5.70 shows that the intrapleural pressure is less negative at the bottom than at the top of the upright lung. These differences are apparently caused by the weight of the lung. Anything which is supported requires a larger pressure below it than above it to balance the downward acting weight forces, and the lung which is supported by the chest wall and diaphagm is no exception. Thus the pressure near the base is higher (less negative) than at the apex.

Figure 5.70 shows the way in which the volume of a portion of lung (for example, a lobe) expands as the pressure around it is decreased (compare Fig. 5.52). The pressure inside the lung is the same as atmospheric pressure. Note that the lung is easier to inflate at low volumes than at high volumes where it becomes stiffer. Because the expanding pressure at the base of the lung is small, this region has a small resting volume. However, because it is situated on a steep part of the pressure-volume curve, it expands well on inspiration. By contrast, the apex of the lung has a large expanding pressure, a big resting volume, and small change in volume on inspiration.

Now when we talk of regional differences in ventilation, we mean the change in volume per unit resting volume. It is clear from Figure 5.70 that the base of the lung has both a larger *change* in volume and smaller *resting* volume than the apex. Thus, its ventilation is greater. Note the paradox that although the base of the lung is relatively poorly expanded compared with the apex, it is better ventilated. The same explanation can be given for the large ventilation of dependent lung in both the supine and lateral positions.

A remarkable change in the distribution of ventilation occurs at low lung volumes. Figure 5.71 is similar to Figure 5.70 except that it represents the situation at residual volume (that is, after a full expiration). Now the intrapleural pressures are less negative because the lung is not so expanded and the elastic recoil forces are smaller. Nevertheless, pressure differences between apex and base are still present because of the weight of the lung. Note that

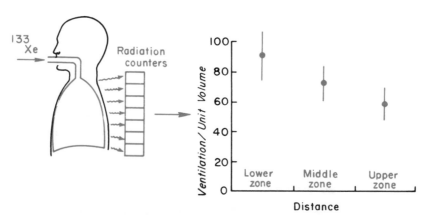

Figure 5.69.   Measurement of regional differences in ventilation with radioactive xenon. When the gas is inhaled, its radiation can be detected by counters outside the chest. Note that the ventilation decreases from the lower to upper regions of the upright lung.

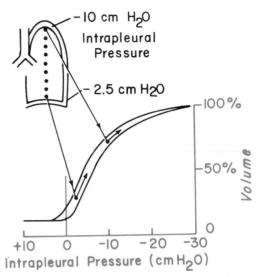

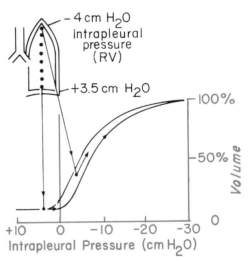

Figure 5.70. Cause of the regional differences of ventilation down the lung. Because of the weight of the lung, the intrapleural pressure is less negative at the base than at the apex. As a consequence, the basal lung is relatively compressed in its resting state but expands more on inspiration than the apex. [From West (1977).]

Figure 5.71. Distribution of ventilation at very low lung volumes. Now intrapleural pressures are generally less negative, and the pressure at the base actually exceeds airway (atmospheric) pressure. As a consequence airway closure occurs in this region, and no gas enters with small inspirations. RV, residual volume. [From West (1977).]

the intrapleural pressure at the base now actually exceeds airway (atmospheric) pressure. Under these conditions the lung at the base is not being expanded but compressed, and ventilation is impossible until the local intrapleural pressure falls below atmospheric pressure. By contrast the apex of the lung is on a favorable part of the pressure-volume curve and ventilates well. Thus the normal distribution of ventilation is inverted, the upper regions ventilating better than the lower zones.

Airway Closure

Figure 5.71 shows that the compressed region of lung at the base does not have all its gas squeezed out. In practice, small airways, probably in the region of respiratory bronchioles (Fig. 5.3), close first, thus trapping gas in the distal alveoli (compare Fig. 5.52). This airway closure occurs only at very low lung volumes in young normal subjects. In elderly apparently normal men, however, airway closure in the lowermost regions of the lung occurs at somewhat higher volumes and may be present at functional residual capacity. The reason for this is that the aging lung loses some of its elastic recoil and the intrapleural pressures therefore become less negative, thus approaching the situation shown in Figure 5.71. In these circumstances, dependent regions of the lung may be only intermittently ventilated, and this leads to defective gas exchange. A similar situation frequently develops in patients with chronic lung disease.

Closing Volume

We have seen that during a full expiration a volume is reached at which the airways at the base of the lung begin to close. This is known as the *closing volume*. It can be measured as follows. First the subject breathes out to residual volume, and then he takes a vital capacity breath of pure oxygen. During the subsequent slow full expiration, the nitrogen concentration at the lips is measured with a rapid nitrogen meter. Figure 5.72 shows the pattern obtained.

Four phases can be recognized (McCarthy et al.,

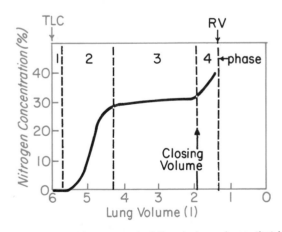

Figure 5.72. Measurement of the closing volume, that is, the lung volume at which the airways at the bottom of the lung close. The subject makes an inspiration of 100% oxygen and then exhales to residual volume. Four phases can be recognized (see text). The last is caused by preferential emptying of the apex of the lung after the lower zone airways have closed. The apex has a higher nitrogen concentration because it receives less oxygen (compare Fig. 5.11).

1972). First pure dead space gas is exhaled (phase *1*), followed by a mixture of dead space and alveolar gas (phase *2*), and then pure alveolar gas (phase *3*). The same sequence is shown in Fig. 5.11. Toward the end of expiration, an abrupt increase in nitrogen concentration is seen (phase *4*). This signals closure of airways at the base of the lung. The reason why the nitrogen concentration rises at this point is that when the breath of oxygen was taken, the alveoli at lung apex were initially larger and therefore increased their volume less during the inspiration. (Fig. 5.71). Therefore the apex received less oxygen. As a consequence, the nitrogen in the apex was less diluted with oxygen, and the preferential emptying of the apex in phase 4 caused a rise in nitrogen concentration to be seen.

In young normal subjects, the closing volume is about 10% of the vital capacity. It increases steadily with age and is equal to about 40% of the vital capacity, that is, the functional residual capacity, at about the age of 65 years. There is some evidence that small amounts of disease of the small airways increase the closing volume, although the mechanism of this is not fully understood. Cigarette smokers may have increased closing volumes when other tests of lung function are still normal. Unfortunately the repeatability of the test is often poor. There is also some question about whether the airways of dependent lung really do close at the onset of phase 4. Nevertheless the test is sometimes used to detect early disease of the small airways.

Time Constants and Dynamic Compliance

In addition to the topographical inequality referred to above, the diseased lung often shows uneven ventilation as a result of regional differences in airways resistance and compliance. These factors also probably operate to a small extent in the normal lung.

Suppose we regard a lung unit (Fig. 5.5) as an elastic chamber connected to the atmosphere by a tube. Then the amount of ventilation will depend on the compliance of the chamber and the resistance of the tube. In Figure 5.73 unit *A* has a normal distensibility and airway resistance. Its volume change on inspiration is large and rapid so that it is complete before expiration for the whole lung begins (*vertical broken line*). By contrast unit *B* has a low compliance, and its change in volume is rapid but small. Finally, unit *C* has a large airway resistance so that inspiration is slow and not complete before the lung begins to exhale. Note that the shorter the time available for inspiration (fast breathing rate), the smaller the inspired volume.

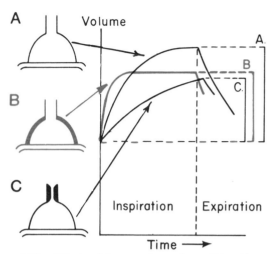

Figure 5.73.   Effects of decreased compliance (*B*) and increased airways resistance (*C*) on ventilation of lung units compared with normal (*A*). In both instances the inspired volume is abnormally low. [From West (1977).]

Such a unit is said to have a long *time constant*, the value of which is given by the product of the compliance and resistance. Thus, inequality of ventilation can result either from alterations in local distensibility or in airway resistance.

Since the tidal volume of a lung with uneven time constants decreases as the time available for inspiration is reduced, that is, as the breathing frequency increases, it is sometimes said that its *dynamic compliance* is reduced. This is a somewhat loose term because the reduced tidal volume is caused in part by the resistance of the airways rather than the elastic properties of the lung. This frequency dependence of dynamic compliance can be used, however, as a sensitive test of increased airway resistance (Woodcock et al., 1969).

We saw earlier (Fig. 5.64) that very little of the resistance of the normal bronchial tree is located in the very peripheral airways. This region, therefore, constitutes a "silent zone" in which considerable disease probably can occur without being detectable by measurements of total airway resistance. Nevertheless, changes of resistance in peripheral airways do cause uneven time constants and therefore a fall in dynamic compliance with increased breathing frequency. In practice, the dynamic compliance is measured at a series of frequencies from about 10 up to 120 breaths/min.

Stratified Inhomogeneity

Another possible mechanism of uneven ventilation is incomplete diffusion within the airways of the respiratory zone. We saw in Chapter 36 that the dominant mechanism of ventilation of the lung beyond the terminal bronchioles is diffusion. Normally this occurs so rapidly that differences in gas concentration within the secondary lobule are virtually abolished within a fraction of a second (Cumming et al., 1966). However, if there is dilatation of the airways in the region of the respiratory bronchioles as in some disease such as centrilobular emphysema, the distance to be covered by diffusion may be enormously increased. In these circumstances, inspired gas is not distributed uniformly within the respiratory zone because of uneven ventilation *along* the lung units. This is known as stratified inhomogeneity.

Measurement of Ventilatory Inequality

The single breath nitrogen test (Figs. 5.11 and 5.72) can be used to measure the degree of uneven ventilation in the lung. In normal subjects, the alveolar plateau (phase *3*) is almost flat because the nitrogen in the lung is almost uniformly diluted by the breath of oxygen. In patients with uneven ventilation, however, the nitrogen concentration rises in phase *3*. This can be explained by the fact that poorly ventilated regions (which receive relatively little oxygen and therefore have a relatively high nitrogen concentration) empty last. This late emptying is caused, in part, by the increased time constants (Fig. 5.73C) of the poorly ventilated regions.

Uneven ventilation can also be detected by a multibreath nitrogen washout. The subject breathes pure oxygen via a one-way valve, and the expired nitrogen concentration is measured at the lips. When the nitrogen concentration is plotted against the number of breaths on semilogarithmic paper, an almost straight line is found in normal subjects. This is because the successive dilution of the nitrogen in the lung by each breath of oxygen causes an exponential decay in nitrogen concentration. In the presence of uneven ventilation, however, the line becomes successively flatter, because different regions of the lung lose their nitrogen at different rates until finally only the nitrogen in the worst-ventilated spaces is being washed out. Various indices of uneven ventilation can be derived from the washout pattern.

TISSUE RESISTANCE

When the lung and chest wall are moved, some pressure is required to overcome the viscous forces within the tissues as they slide over each other. Thus, part of the cross-hatched portion Figure 5.62 should be attributed to these tissue forces. This tissue resistance is only about 20% of the total (tissue + airway) resistance in young normal subjects, although it may increase in some diseases. This total resistance is sometimes called *pulmonary resistance* to distinguish it from airway resistance.

WORK OF BREATHING

Work is required to move the lung and chest wall. In this context, it is most convenient to measure work as pressure × volume.

Work Done on the Lung

This can be illustrated on a pressure-volume curve (Fig. 5.74). During inspiration the intrapleural pressure follows the curve *ABC* and the work done on the lung is given by the area *OABCD*. Of this, the trapezoid *OAECD* represents the work required to overcome the elastic forces, and the cross-hatched area *ABCEA* represents the work overcoming viscous (airway and tissue) resistance (compare Fig. 5.62). The higher the airway resistance or the inspiratory flow rate, the more negative (rightward) would be the intrapleural pressure excursion between *A* and *C* and the larger the area.

On expiration, the area *AECFA* is the work required to overcome airway (+ tissue) resistance. Normally this falls within the trapezoid *OAECD*, and thus this work can be accomplished by the energy stored in the expanded elastic structures and released during a passive expiration. The difference between the areas *AECFA* and *OAECD* represents the work dissipated as heat.

The higher the breathing rate, the faster the flow rates and the larger the viscous work area *ABCEA*. On the other hand, the larger the tidal volume, the larger the elastic work area *OAECD*. It is of interest that patients who have a reduced compliance (stiff lungs) tend to take small rapid breaths, while patients with severe airway obstruction often breathe

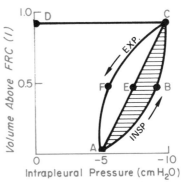

Figure 5.74. Pressure-volume curve of the lung showing the inspiratory work done overcoming elastic forces (area OAECDO) and viscous forces (hatched area ABCEA). INSP, inspiration; EXP, expiration.

slowly. These patterns tend to reduce the work done on the lungs.

Total Work of Breathing

The total work done moving the lung and chest wall is difficult to measure, although estimates have been obtained by artificially ventilating paralyzed patients (or "completely relaxed" volunteers) in an iron lung type of respirator. Alternatively, the total work can be calculated by measuring the oxygen cost of breathing and assuming a figure for the *efficiency* as given by:

$$\begin{array}{c} \text{Efficiency} \\ \text{in } \% \end{array} = \frac{\text{useful work}}{\begin{array}{c}\text{total energy expended} \\ \text{(or } O_2 \text{ cost)}\end{array}} \times 100$$

The efficiency is believed to be about 5–10%.

The oxygen cost of quiet breathing is extremely small, being less than 5% of the total resting oxygen consumption. With voluntary hyperventilation it is possible to increase this to 30%. In patients with obstructive lung disease, the oxygen cost of breathing may limit their exercise ability.

CHAPTER 39

Control of Ventilation

We have seen that the chief function of the lung is to exchange oxygen and carbon dioxide between blood and gas and thus maintain normal levels of P_{O_2} and P_{CO_2} in arterial blood. In this chapter we shall see that, despite widely differing demands for oxygen uptake and carbon dioxide output made by the body, the arterial P_{O_2} and P_{CO_2} are normally kept within close limits. This remarkable regulation of gas exchange is possible because the level of ventilation is so carefully controlled.

The three basic elements of the respiratory control system (Fig. 5.75) are:

1. *Sensors* which gather information and feed it to the
2. *Central controller* in the brain which coordinates the information and, in turn, sends impulses to the
3. *Effectors* (respiratory muscles) which cause ventilation.

We shall see that increased activity of the effectors generally ultimately decreases the sensory input to the brain, for example, by decreasing the arterial P_{CO_2}. This is an example of negative feedback.

CENTRAL CONTROLLER

The normal automatic process of breathing originates in impulses which come from the brainstem. The cortex can override these centers if voluntary control is desired. Additional input from other parts of the brain occurs under certain conditions.

Brainstem

The periodic nature of inspiration and expiration is controlled by neurons located in the pons and medulla. These have been designated the *respiratory centers*. However, these should not be thought of as comprising a discrete nucleus but rather as a somewhat poorly defined collection of neurons with various components.

Three main groups of neurons are recognized.

1. Medullary respiratory center in the reticular formation of the medulla. This is comprised of two identifiable areas. One group of cells in the dorsal region of the medulla (dorsal respiratory group) is associated chiefly with inspiration; the other in the ventral area (ventral respiratory group) is mainly for expiration. One popular (though not universally accepted) view is that the cells of the *inspiratory area* have the property of intrinsic periodic firing and they are responsible for the basic rhythm of ventilation. When all known afferent stimuli have been abolished, these inspiratory cells generate repetitive bursts of action potentials which result in nervous impulses going to the diaphragm and other inspiratory muscles.

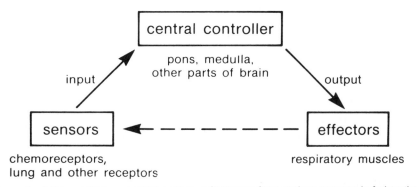

Figure 5.75. Basic elements of the respiratory control system. Information from various sensors is fed to the central controller, the output of which goes to the respiratory muscles. By changing ventilation, the respiratory muscles reduce perturbations of the sensors (negative feedback).

The intrinsic rhythm pattern of the inspiratory area starts with a latent period of several seconds during which there is no activity. Action potentials then begin to appear, increasing in a crescendo over the next few seconds. During this time inspiratory muscle activity becomes stronger in a "ramp"-type pattern. Finally, the inspiratory action potentials cease, and inspiratory muscle tone falls to its preinspiratory level.

The inspiratory ramp can be "turned off" prematurely by inhibiting impulses from the *pneumotaxic center* (see below). In this way inspiration is shortened and, as a consequence, the breathing rate increases. The output of the inspiratory cells is further modulated by impulses from the vagal and glossopharyngeal nerves. Indeed, these terminate in the tractus solitarius, which is situated close to the inspiratory area.

The *expiratory area* is quiescent during normal quiet breathing because ventilation is then achieved by active contraction of the inspiratory muscles (chiefly the diaphragm), followed by passive relaxation of the chest wall to its equilibrium position (Chapter 38). However, in more forceful breathing, for example, on exercise, expiration becomes active as a result of the activity of the expiratory cells. It should be noted that there is still not universal agreement on how the intrinsic rhythmicity of respiration is brought out by the medullary centers.

2. Apneustic center in the lower pons. This area is so named because if the brain of an experimental animal is sectioned just above this site, prolonged inspiratory gasps (apneuses) interrupted by transient expiratory efforts are seen. Apparently, the impulses from the center have an excitatory effect on the inspiratory area of the medulla, tending to prolong the ramp action potentials. Whether this apneustic center plays a role in normal human respiration is not known, although in some types of severe brain injury, this type of abnormal breathing is seen.

3. Pneumotaxic center in the upper pons. As indicated above, this area appears to "switch off" or inhibit inspiration and thus regulate inspiratory volume and, secondarily, respiratory rate. This has been demonstrated experimentally in animals by direct electrical stimulation of the pneumotaxic center. Some investigators believe that the role of this center is "fine tuning" of respiratory rhythm because a normal rhythm can exist in the absence of this center.

Cortex

Breathing is under voluntary control to a considerable extent, and the cortex can override the function of the brainstem within limits. It is not difficult to halve the arterial P_{CO_2} by hyperventilation, although the consequent alkalosis may cause tetany with contraction of the muscles of the hand and foot (carpopedal spasm). Halving the P_{CO_2} increases the pH by about 0.2 units (Chapter 33).

Voluntary hypoventilation is more difficult. The duration of breath-holding is limited by several factors, including the arterial P_{CO_2} and P_{O_2}. A preliminary period of hyperventilation increases breath holding time, especially if oxygen is breathed. However, factors other than chemical are involved. This is shown by the observation that if at the breaking point of breath holding, a gas mixture is inhaled which *raises* the arterial P_{CO_2} and *lowers* the P_{O_2}, a further period of breath holding is possible.

Other Parts of the Brain

Other parts of the brain, such as the limbic system and hypothalamus, can affect the pattern of breathing, for example, in affective states such as rage and fear.

EFFECTORS

The muscles of respiration include the diaphragm, intercostal muscles, abdominal muscles, and accessory muscles such as the sternomastoids. The actions of these muscles were described at the beginning of Chapter 38. In the context of the control of ventilation, it is crucially important that these various muscle groups work in a coordinated manner, and this is the responsibility of the central controller. There is evidence that some newborn children, particularly those that are premature, have uncoordinated respiratory muscle activity, especially during sleep. For example, the thoracic muscles may try to inspire while the abdominal muscles expire. This may be a factor in the "sudden infant death syndrome."

SENSORS

Central Chemoreceptors

A chemoreceptor is a receptor which responds to a change in the chemical composition of the blood or some other fluid around it. The most important receptors involved in the minute-by-minute control of ventilation are those situated near the ventral surface of the medulla in the vicinity of the exit of the 9th and 10th nerves. In animals, local application of H^+ or dissolved carbon dioxide to this area stimulates breathing within a few seconds. For some time it was thought that the medullary respiratory center itself was the site of action of CO_2, but most physiologists now believe that the chemoreceptors are anatomically separate. Some evidence suggests that they lie about 200–400 μm below the ventral surface of the medulla (Fig. 5.76).

The central chemoreceptors are surrounded by brain extracellular fluid and respond to changes in its H^+ concentration. An increase in H^+ concentration stimulates ventilation while a decrease inhibits it. The composition of the extracellular fluid around the receptors is governed by the cerebrospinal fluid (CSF), local blood flow, and local metabolism.

Of these, the CSF is apparently the most important. It is separated from the blood by the blood-brain barrier which is relatively impermeable to H^+ and HCO_3^- ions, although molecular CO_2 diffuses across it easily. When the blood P_{CO_2} rises, CO_2 diffuses into the CSF from the cerebral blood vessels, liberating H^+ ions which stimulate the chemoreceptors. Thus, the CO_2 level in blood regulates ventilation chiefly by its effect on the pH of the CSF. The resulting hyperventilation reduces the P_{CO_2} in the blood and, therefore, in the CSF. The cerebral vasodilation which accompanies an increased arterial P_{CO_2} enhances diffusion of CO_2 into the CSF and the brain extracellular fluid.

The normal pH of the CSF is 7.32, and since it contains much less protein than blood, it has a much lower buffering capacity. As a result the change in CSF pH for a given change in P_{CO_2} is greater than that in blood. If the CSF pH is displaced over a prolonged period, a compensatory change in HCO_3^- occurs, as a result of transport across the blood-brain barrier, thus tending to restore the pH over the course of 36–48 hours. The pH of the CSF is therefore returned toward normal more promptly than the pH of arterial blood by renal compensation, a process which takes several days. The resetting of the CSF pH in relation to blood pH results in its predominating influence on ventilation and arterial P_{CO_2}.

One example is a patient with chronic lung disease and CO_2 retention of long standing who may have a nearly normal CSF pH and, therefore, an abnormally low ventilation for his arterial P_{CO_2}. A similar situation is seen in normal subjects who are exposed to an atmosphere containing 3% CO_2 for some days.

Peripheral Chemoreceptors

These are located in the carotid bodies at the bifurcation of the common carotid arteries, and in the aortic bodies above and below the aortic arch. The carotid bodies are the most important in man. They contain glomus cells of two or more types

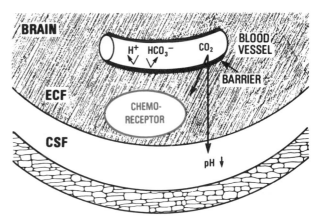

Figure 5.76. Scheme of the environment of the central chemoreceptors. They are bathed in brain extracellular fluid (ECF) through which CO_2 easily diffuses from blood vessels to CSF. H^+ and HCO_3^- ions cannot easily cross the blood-brain barrier.

which show an intense fluorescence staining because of their large content of dopamine. The mechanism of chemoreception is not yet understood. A popular view has been that glomus cells themselves are chemoreceptors. Another more recent hypothesis is that they are inhibitory interneurons and that the impulses are generated in the afferent terminals of the carotid sinus nerve. This theory proposes that impulses generated at the nerve fiber terminal release an excitatory transmitter that causes the glomus cell to release dopamine. In turn, dopamine acts on the nerve fiber terminal to inhibit impulse generation. The nerve in which impulses travel from the carotid body is known as Hering's nerve; it is a branch of the glossopharyngeal nerve.

The peripheral chemoreceptors respond to decreases in arterial P_{O_2} and pH, and increases in arterial P_{CO_2}. They are unique among tissues of the body in that their sensitivity to changes in arterial P_{O_2} begins around 500 mmHg. Figure 5.77 shows that the relationship between firing rate and arterial P_{O_2} is very nonlinear; relatively little response occurs until the arterial P_{O_2} is reduced below 100 mmHg but then the rate rapidly increases. The carotid bodies have a very high blood flow for their size (20 ml/min/g tissue) and consequently a very small arterial-venous O_2 difference despite a very high metabolic rate. As a consequence, they respond to arterial rather than venous P_{O_2}. The response of these receptors can be very fast; indeed, their discharge rate can alter during the respiratory cycle as a result of the small cyclic changes in blood gases. The peripheral chemoreceptors are responsible for all the increase of ventilation which occurs in man in response to arterial hypoxemia. Indeed, in the absence of these receptors, severe hypoxemia depresses respiration, presumably through a direct

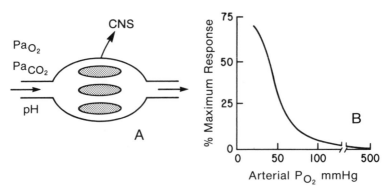

Figure 5.77. *A*, a carotid body which responds to changes of P_{O_2}, P_{CO_2}, and pH in arterial blood. Impulses travel to the central nervous system (*CNS*) through Hering's nerve. *B*, the nonlinear response to arterial P_{O_2}. Note that the maximum response occurs below a P_{O_2} of 50 mmHg.

effect on the respiratory centers. Complete loss of hypoxic ventilatory drive has been shown in patients with bilateral carotid body resection.

The response of the peripheral chemoreceptors to arterial P_{CO_2} is much less important than that of the central chemoreceptors. For example, when a normal subject is given a mixture of CO_2 in O_2 to breathe, less than 20% of the ventilatory response can be attributed to the peripheral chemoreceptors. However, their response is more rapid, and they may be useful in matching ventilation to abrupt changes in P_{CO_2}.

In man, the carotid but not the aortic bodies respond to a fall in arterial pH. This occurs regardless of whether the cause is respiratory or metabolic. Interaction of the various stimuli occurs. Thus, increases in chemoreceptor activity in response to decreases in arterial P_{O_2} are potentiated by increases in P_{CO_2} and, in the carotid bodies, decreases in pH.

Lung and Other Receptors

Lung receptors are of three main types.

PULMONARY STRETCH RECEPTORS

These are believed to lie within airway smooth muscle. They discharge in response to distention of the lung, and their activity is sustained with lung inflation, thus showing little adaptation. The impulses travel in the vagus nerve via large myelinated fibers.

The main reflex effect of stimulating these receptors is a slowing of respiratory frequency due to an increase in expiratory time. This is known as the Hering-Breuer inflation reflex. It can be well demonstrated in a rabbit preparation where the diaphragm has a slip of muscle from which recordings can be made without interfering with the other respiratory muscles. Classical experiments showed

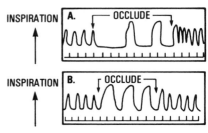

Figure 5.78. Hering-Breuer reflexes. The records show contractions of an isolated strip of rabbit diaphragm. In *A*, the trachea was occluded at the end of inspiration. Note the inhibition of further inspiratory contraction (inflation reflex). In *B*, the trachea was occluded at the end of expiration; this occlusion is followed by prolonged inspiratory contractions (deflation reflex).

that inflation of the lungs tended to inhibit further inspiratory muscle activity (Fig. 5.78). The opposite response is also seen, that is, deflation of the lungs tends to initiate inspiratory activity (deflation reflex). Thus these reflexes can provide a self-regulatory mechanism or negative feedback.

The Hering-Breuer reflexes were once thought to play a major role in ventilation by determining the rate and depth of breathing. This could be done by using the information from these stretch receptors to modulate the "switching off" mechanism in the medulla. Thus, bilateral vagotomy in most animals causes slow, deep breathing. However, more recent work indicates that the reflexes are largely inactive in adult man unless the tidal volume exceeds one l, as in exercise. It has been shown that transient bilateral blockade of the vagi by local anesthesia in awake man does not change either breathing rate or volume. There is some evidence that these reflexes may be more important in newborn babies.

IRRITANT RECEPTORS

These are thought to lie between airway epithelial cells, and they are stimulated by noxious gases, cigarette smoke, inhaled dusts, and cold air. The impulses travel up the vagus in myelinated fibers, and the reflex effects include bronchoconstriction and hyperpnea. Some physiologists prefer to call these receptors "rapidly adapting receptors" because they are apparently involved in additional mechanoreceptor functions as well as responding to noxious stimulants on the airway walls. It is possible that irritant receptors play a role in the bronchoconstriction of asthma attacks as a result of their response to released histamine.

J RECEPTORS

The term "juxta-capillary," or J, is used because the receptors are believed to be in the alveolar walls close to the capillaries (Paintal, 1970). The evidence for this location is that they respond very quickly to chemicals injected into the pulmonary circulation. Identification of the receptors is uncertain, but sparse nonmyelinated fibers have been demonstrated in alveolar walls in histologic sections. The impulses pass up the vagus nerve in slowly conducting nonmyelinated fibers and can result in rapid shallow breathing, although intense stimulation causes apnea. These receptors do not show a firing pattern that can be associated with normal respiration. There is evidence that engorgement of pulmonary capillaries and increases in the interstitial fluid volume of the alveolar wall activate these receptors. They may play a role in the dyspnea (see below) associated with left heart failure, interstitial lung disease, pneumonia, and microembolism. It has been suggested that stimulation of the J receptors may contribute to the increase in ventilation which occurs on exercise, but that suggestion is largely speculative.

A fourth group of receptors has recently been described. These are lodged in the airways and are innervated by nonmyelinated fibers, but their role is unclear.

Other receptors outside the lung include the following.

NOSE AND UPPER AIRWAY RECEPTORS

The nose, nasopharynx, larynx, and trachea contain receptors that respond to mechanical and chemical stimulation. Various reflex responses have been described, including sneezing, coughing, and bronchoconstriction. Laryngeal spasm may occur if the larynx is irritated mechanically, for example, during insertion of an endotracheal tube with insufficient local anesthesia.

JOINT AND MUSCLE RECEPTORS

Impulses from moving limbs are believed to be part of the stimulus to ventilation during exercise, especially in the early stages.

γ SYSTEM

Many muscles, including the intercostal muscles and diaphragm, contain muscle spindles which sense elongation of the muscle. This information is used to reflexly control the strength of contraction. These receptors may be involved in the sensation of dyspnea which occurs when unusually large respiratory efforts are required to move the lung and chest wall, for example, because of airway obstruction.

ARTERIAL BARORECEPTORS

Increase in arterial blood pressure can cause reflex hypoventilation or apnea through stimulation of the aortic and carotid sinus baroreceptors (Chapter 16). Conversely, a decrease in blood pressure may result in hyperventilation. A possible advantage of this reflex is enhancement of venous return following severe hemorrhage. However, the duration of these reflexes is typically short lived.

PAIN AND TEMPERATURE

Stimulation of many afferent nerves can bring about a change in respiration. Pain often causes a period of apnea followed by hyperventilation. Heating the skin may result in hyperventilation. The increased ventilation seen in fever is thought to be due in part to stimulation of hypothalamic thermoreceptors.

INTEGRATED RESPONSES

Now that we have looked at the various units which make up the respiratory control system (Fig. 5.75), it is useful to consider the overall responses of the system to changes in the arterial CO_2, O_2, and pH, and to exercise.

Response to Carbon Dioxide

The most important factor in the control of ventilation under normal conditions is the P_{CO_2} of the arterial blood. The sensitivity of this control is remarkable. In the course of daily activity with periods of rest and exercise, the arterial P_{CO_2} is probably held to within 3 mmHg. During sleep it may rise a little more.

The ventilatory response to CO_2 is normally measured by having the subject inhale CO_2 mixtures or rebreathe from a bag so that the inspired P_{CO_2} gradually rises. In one technique which can easily be used in the pulmonary function laboratory, the subject rebreathes from a bag which is prefilled with 7% CO_2 and 93% O_2. As he rebreathes, he adds metabolic CO_2 to the bag, but the O_2 concentration remains relatively high. In such a procedure the P_{CO_2} of the bag gas increases at the rate of about 4 mmHg/minute.

Figure 5.79 shows the results of experiments carried out in which the inspired mixture was adjusted to yield a constant alveolar P_{O_2}. (In this type of experiment on normal subjects, alveolar end-tidal P_{O_2} and P_{CO_2} are generally taken to reflect the arterial levels.) It can be seen that with a normal P_{O_2} the ventilation increases by about 2–3 l/min for each 1 mm Hg rise in P_{CO_2}. Lowering the P_{O_2} produces two effects: there is a higher ventilation for a given P_{CO_2} and, also, the slope of the line becomes steeper. It should be emphasized that there is considerable variation between subjects.

Another way of measuring respiratory drive is to record the inspiratory pressure during a brief period of airway occlusion (Whitelaw et al., 1975). The subject breathes through a mouthpiece attached to a valve box, and the inspiratory port is provided with a shutter. This is closed during an expiration (the subject being unaware), so that the first part of his next inspiration is against an occluded airway. The shutter is opened after about 0.5 sec. The pressure generated during the first 0.1 s of attempted inspiration (known as $P_{0.1}$) is taken as a measure of respiratory center output. This is largely unaffected by the mechanical properties of the respiratory system, although it may be influenced by lung volume. This method can be used to study the respiratory sensitivity to CO_2, hypoxia, and other variables as well.

A reduction in arterial P_{CO_2} is very effective in reducing the stimulus to ventilation. For example, if the reader hyperventilates voluntarily for a few seconds, he will find that he has no urge to breathe for a short period. An anesthetized patient will frequently stop breathing for a minute or so if he is first overventilated by the anesthesiologist.

The ventilatory response to CO_2 is reduced by sleep, increasing age, and genetic, racial, and personality factors. Trained athletes and divers tend to have a low CO_2 sensitivity. Various drugs depress the respiratory center, including morphine and barbiturates. Patients who have taken an overdose of one of these drugs often have marked hypoventilation. The ventilatory response to CO_2 is also reduced if the work of breathing is increased. This can be demonstrated by having normal subjects breathe through a narrow tube. The neural output of the respiratory center is not reduced, but it is not so effective in producing ventilation. The abnormally small ventilatory response to CO_2 and the CO_2 retention in some patients with lung disease can be partly explained by the same mechanism. In such patients, reducing the airway resistance with bronchodilators often increases their ventilatory response. There is also some evidence that the sensitivity of the respiratory center is reduced in these patients.

As we have seen, the main stimulus to increase ventilation when the arterial P_{CO_2} rises comes from the central chemoreceptors which respond to the increased H^+ concentration of the brain extracellular fluid near the receptors. An additional stimulus comes from the peripheral chemoreceptors, both because of the rise in arterial P_{CO_2} and the fall in pH.

Response to Oxygen

The way in which a reduction of P_{O_2} in arterial blood stimulates ventilation can be studied by having a subject breathe hypoxic gas mixtures. The end-tidal P_{O_2} and P_{CO_2} are used as a measure of the arterial values. Figure 5.80 shows that when the alveolar P_{CO_2} is kept at about 36 mmHg the alveolar P_{O_2} can be reduced to the vicinity of 50 mmHg before any appreciable increase in ventilation occurs. Raising the P_{CO_2} (by altering the inspired mixture) increases the ventilation at any P_{O_2} (compare Fig. 5.79). Note that when the P_{CO_2} is increased, a reduction in P_{O_2} below 100 mmHg causes

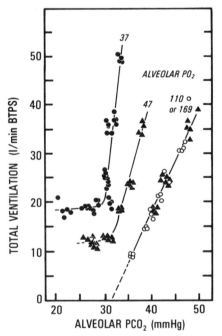

Figure 5.79. Ventilatory response to CO_2. Each curve of total ventilation against alveolar P_{CO_2} is for a different P_{O_2}. In this study, no difference was found between an alveolar P_{O_2} of 110 and one of 169 mmHg, although some investigators have found that the slope of the line is slightly less at the higher P_{O_2}. BTPS, body temperature, pressure, and saturation. [From Nielsen and Smith (1951).]

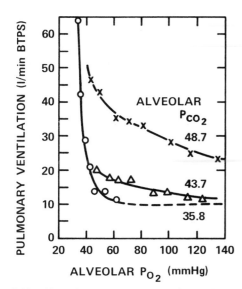

Figure 5.80. Hypoxic response curves. Note that when the P_{CO_2} is 36 mmHg, almost no increase in ventilation occurs until the P_{O_2} is reduced to about 50 mmHg. BTPS, body temperature, pressure, and saturation. [Adapted from Loeschke and Gertz (1958).]

some stimulation of ventilation, unlike the situation when the P_{CO_2} is normal. Thus, the combined effects of both stimuli exceed the sum of each stimulus given separately; this is referred to as interaction between the high CO_2 and low O_2 stimuli.

Various indices of hypoxic sensitivity are in use. One is the increment in ventilation when the arterial (or alveolar) P_{O_2} is reduced from 150 to 40 mmHg (so-called $\dot{V}_{40}$). The mean value in normal subjects is about 35 l/min, but the normal range is very large. Another index that is sometimes used in the pulmonary function laboratory is based on the finding that there is a nearly linear relationship between arterial O_2 saturation and ventilation as the inspired P_{O_2} is reduced. Since the O_2 saturation can conveniently be measured with an ear oximeter, the hypoxic sensitivity can be expressed as the change in ventilation per 1% fall in O_2 saturation.

Large differences in hypoxic ventilatory response occur between individuals, and the sensitivity is, in part, genetically determined. Interestingly, the sensitivity is considerably reduced or "blunted" in persons who have been hypoxemic since birth, such as those born at high altitude, or patients with cyanotic congenital heart disease ("blue babies"). When someone who was born and has lived for years at high altitude descends to sea level, the blunted response persists for many years.

We have seen (Fig. 5.80) that the P_{O_2} can normally be reduced a long way without evoking any increase in ventilation. Therefore, the role of this hypoxic drive in the day by day control of ventilation is small. However, on ascent to high altitude, a large and persistent increase in ventilation occurs in response to the hypoxia. This increase is further considered in Chapter 40.

In some patients with severe lung disease, the hypoxic drive to ventilation becomes very important. These patients have chronic CO_2 retention, and the pH of their brain extracellular fluid has returned to near normal despite a raised P_{CO_2}. Thus, they have lost most of their increase in the CO_2 stimulus to ventilation. In addition, the initial depression of blood pH has been nearly abolished by renal compensation so that there is little pH stimulation of the peripheral chemoreceptors (see below). Thus, the arterial hypoxemia becomes the chief stimulus to ventilation. If such a patient is given a high O_2 mixture to breathe to relieve his hypoxemia, his ventilation may become grossly depressed. His ventilatory state is best monitored by measuring his arterial P_{CO_2}.

As we have seen, hypoxemia reflexly stimulates ventilation by its action on the carotid and aortic body chemoreceptors. It has no action on the central chemoreceptors; indeed, in the absence of peripheral chemoreceptors, hypoxemia depresses respiration. However, prolonged hypoxemia can cause mild cerebral acidosis, which in turn can stimulate ventilation.

Response to pH

A reduction in arterial blood pH stimulates ventilation. In practice, it is often difficult to separate the ventilatory response resulting from a fall in pH from that caused by an accompanying rise in P_{CO_2}. However, in experimental animals in whom it is possible to reduce the pH at a constant P_{CO_2}, the stimulus to ventilation can be convincingly demonstrated. Patients with a partly compensated metabolic acidosis (such as in diabetes mellitus) who have a low pH and low P_{CO_2} show an increased ventilation. Indeed, this is responsible for the reduced P_{CO_2}.

As we have seen, the chief site of action of a reduced arterial pH is probably the peripheral chemoreceptors. It is also possible that the central chemoreceptors or the respiratory center itself are affected by a change in blood pH if it is large enough. In this case, the blood-brain barrier becomes partly permeable to H^+ ions.

Response to Exercise

On exercise, ventilation increases promptly and during strenuous exertion may reach very high levels. A fit young man who attains a maximum O_2 consumption of 4 l/min may have a total ventilation of 120 l/min, that is, about 15 times his resting level. This increase in ventilation closely matches the increase in O_2 uptake and CO_2 output. It is remarkable that the cause of the increased ventilation on exercise remains largely unknown.

The arterial P_{CO_2} does not increase during most forms of exercise; indeed, during severe exercise it typically falls slightly. The arterial P_{O_2} usually increases slightly, although it may fall at very high work levels. The arterial pH remains nearly constant for moderate exercise, although during heavy exercise it falls because of the liberation of lactic acid through anaerobic metabolism. If this occurs, ventilation is further stimulated by the increased H^+ concentration. However, it is clear that none of the mechanisms we have discussed so far can account for the large increase in ventilation observed during light to moderate exercise.

Other stimuli have been suggested. *Passive movement of the limbs* stimulates ventilation in both anesthetized animals and awake man. This is a reflex with receptors presumably located in joints or muscles. It is probably responsible for the abrupt increase in ventilation which occurs during the first few seconds of exercise. A recent hypothesis is that *oscillations in arterial P_{O_2} and P_{CO_2}* may stimulate the peripheral chemoreceptors even though the mean level remains unaltered. These fluctuations are caused by the periodic nature of ventilation and increase when the tidal volume rises, as on exercise. Another theory is that the central chemoreceptors increase ventilation to hold the *arterial P_{CO_2} constant* by some kind of servomechanism, just as a thermostat can control a furnace with little change in temperature. The objection that the arterial P_{CO_2} often *falls* on exercise is countered by the assertion that the preferred level of P_{CO_2} is reset in some way. Proponents of this theory believe that the ventilatory response to inhaled CO_2 may not be a reliable guide to what happens on exercise.

Yet another hypothesis is that ventilation is linked in some way to the additional *CO_2 load* presented to the lungs in the mixed venous blood during exercise. In animal experiments, an increase in this load produced either by infusing CO_2 into the venous blood or by increasing venous return has been shown to correlate well with ventilation. However, a problem with this hypothesis is that no suitable venous receptor has been found.

Additional factors which have been suggested include the *increase in body temperature* during exercise which stimulates ventilation. Finally, it has been proposed that "irradiation" of the respiratory centers by *impulses from the motor cortex* or *hypothalamus* is responsible for linking ventilation to muscle activity during exercise. However, none

of the theories proposed so far is completely satisfactory.

ABNORMAL PATTERNS OF BREATHING

Subjects with severe hypoxemia often exhibit a striking pattern of periodic breathing known as *Cheyne-Stokes respiration*. This pattern is characterized by periods of apnea of 5–20 s separated by approximately equal periods of hyperventilation when the tidal volume gradually waxes, then wanes. This pattern is frequently seen at high altitude, especially at night during sleep. It may be associated with marked cyclic changes in the arterial oxygen saturation caused by the variations in the alveolar P_{O_2} as the ventilation decreases and increases again (Fig. 5.81). The pattern is also found in some patients with severe heart disease or brain damage.

The pattern can be reproduced in experimental animals by lengthening the distance through which blood travels on its way to the brain from the lung. Under these conditions, there is a long delay before the central chemoreceptors sense the alteration in P_{CO_2} caused by a change in ventilation. As a result, the respiratory center hunts for the equilibrium condition always overshooting it. However, not all instances of Cheyne-Stokes respiration can be explained on this basis.

DYSPNEA

A feeling of shortness of breath is one of the most important symptoms of lung disease. Dyspnea refers to the sensation of difficulty with breathing and should be distinguished from simple tachypnea (rapid breathing) or hyperpnea (increased ventila-

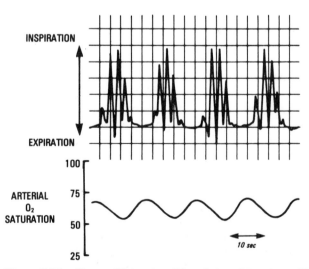

Figure 5.81. Cheyne-Stokes breathing during sleep at an altitude of 6300 m. Note the waxing and waning of breathing and the consequent fluctuation of arterial O_2 saturation.

tion). Because dyspnea is a subjective phenomenon, it is difficult to measure, and the factors responsible for it are poorly understood. Broadly speaking, dyspnea occurs when the *demand for ventilation* is out of proportion to the patient's *ability to respond* to that demand. As a result, breathing becomes difficult, uncomfortable, or labored.

An *increased demand for ventilation* is often caused by changes in the blood gases and pH. High ventilations on exercise are commonly seen in patients with inefficient pulmonary gas exchange, especially those with large physiological dead spaces who tend to develop CO_2 retention and acidosis unless they achieve high ventilations. Another important factor is stimulation of intrapulmonary stretch receptors or J receptors. These factors may explain, in part, the increased ventilation of patients with pulmonary fibrosis and congestive heart failure, respectively.

A *reduced ability to respond* to the ventilatory needs is generally caused by abnormal mechanics of the lung or chest wall. Frequently, increased airway resistance is the culprit, as in asthma, but other causes include a stiff chest wall, as in the patient with a hunchback (kyphoscoliosis).

The assessment of dyspnea is difficult. Usually, exercise tolerance is determined from a standard questionnaire which grades breathlessness according to how far the patient can walk on the level or upstairs without pausing for breath. Occasionally, ventilation is measured at a standard level of exercise and then related to the patient's maximum voluntary ventilation in an attempt to obtain an index of dyspnea. However, it should be remembered that dyspnea is something that only the patient feels, and as such it cannot be accurately measured.

CHAPTER 40

Respiration in Unusual Environments

The surface area of the lung is approximately 50–100 m^2, compared with only some 2 m^2 for the skin. Therefore, in one sense the lung serves as our principal physiological link with the environment. Respiratory physiologists have always been intrigued by the challenges of unusual environments, as illustrated, for example, by a long and colorful history of physiological expeditions to high altitudes. Very recently the first measurements of respiratory function were made at the highest point on earth, the summit of Mount Everest where the inspired P_{O_2} is about 43 mmHg. At the other end of the pressure spectrum, there is intense research activity at the present time in the area of hyperbaric physiology. Divers connected with the exploration and recovery of oil in the North Sea area, for example, are exposed to great hazards, and physiological accidents are disturbingly common.

Nearer home, everyone is exposed to atmospheric pollution which continues to increase in most parts of the world. Moreover, the composition of the pollutants changes as man attempts to clean the environment, and new problems therefore arise all the time. Even those who live at normal pressures breathing clean air have, at one stage, coped with the most momentous challenge of all—that of changing from placental to lung respiration in a few seconds at the time of birth!

HIGH ALTITUDE

Hypoxia of High Altitude

The barometric pressure decreases with distance above the earth's surface in an approximately exponential manner (Fig. 5.82). The pressure at 18,000 ft is only one-half the normal 760 mmHg, so the P_{O_2} of moist inspired gas is $(380 - 47) \times 0.2093 = 70$ mmHg (Table 5.2). (Recall that 47 mmHg is the partial pressure of water vapor at body temperature—see Chapter 36). At the summit of Mount Everest (altitude 8,848 m; 29,028 ft) where the barometric pressure is approximately 253 mmHg the inspired P_{O_2} is $(250 - 47) \times 0.2093 = 43$ mmHg. For a person who has the misfortune to bail out of an aircraft flying at 63,000 ft at a barometric pressure of 47 mmHg, the P_{O_2} of inspired gas in the trachea will be zero! Indeed the tissue fluids vaporize ("boil") near this pressure.

It might be noted that even a modest ascent for example in a modern skiing resort, can cause appreciable hypoxemia. Thus at a cable car station at 3,400 m (11,000 ft) the inspired P_{O_2} is about 95 mmHg. If one assumes an alveolar P_{CO_2} of 35 mmHg, respiratory exchange ratio of 0.8 and alveolar-arterial oxygen difference of 3 mmHg, the alveolar gas equation (Chapter 36) gives an arterial P_{O_2} of only 53 mmHg. No wonder it is difficult to adjust one's skis! In terms of oxygen transport, however, it should be remembered that because of the shape of the oxygen dissociation curve, the arterial oxygen saturation even then will be approximately 89%, assuming a base excess of zero.

In spite of the hypoxia associated with high altitude, some 15 million people live at elevations over 10,000 ft, and permanent residents live higher than 16,000 ft in the Peruvian Andes. A remarkable degree of acclimatization occurs when man ascends to these altitudes; indeed, climbers have lived for several days at altitudes that would cause unconsciousness within a minute or two in the absence of acclimatization. For example, mountaineers have spent over a week above 7,600 m (25,000 ft) at an oxygen pressure which usually produces unconsciousness within about 2 min in persons abruptly exposed to this low pressure.

In all discussions of the effects of hypoxia caused

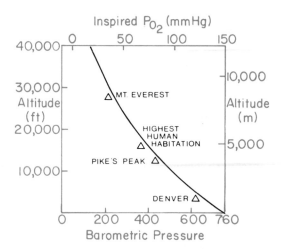

Figure 5.82. Relationship between altitude and barometric pressure. Note that at 1500 m (5,000 ft) (Denver), the P_{O_2} of moist inspired gas is about 130 mmHg, but it is only 43 mmHg on the summit of Everest.

Table 5.2
Barometric pressure (P_B) at various altitudes. The inspired P_{O_2} is for moist inspired gas and is calculated from $0.2093 \times (P_B - 47)$.

Altitude			Barometric Pressure, mmHg	Inspired P_{O_2}, mmHg
m	ft	Location		
0	0	Sea level	760	149
2,240	7,350	Mexico City (Olympic games 1968)	580	112
3,400	11,000	Skiing resort	500	95
5,300	17,500	Highest permanent habitation	390	72
8,848	29,028	Summit of Mt. Everest	253	43
19,000	63,000	High performance aircraft	47	0

by low barometric pressure a clear distinction should be made between (1) sudden exposure to low pressure such as occurs when an aircraft loses cabin pressure, (2) exposure over several weeks when a lowlander mountaineer acclimatizes during a high climb, and (3) the lifelong exposure of a permanent resident at high altitude. In general, the tolerance to high altitude, as reflected for example in the amount of physical activity that can be accomplished, increases with the length of stay as a result of the process of acclimatization.

ACUTE MOUNTAIN SICKNESS

Newcomers to high altitude frequently complain of headache, fatigue, dizziness, palpitations, nausea, loss of appetite, and insomnia. This is known as acute mountain sickness. The cause is not known, although a combination of the hypoxemia and respiratory alkalosis has been suggested. Some investigators believe that the condition is due to mild cerebral edema. There is evidence that cerebral blood flow may be considerably increased as a result of the severe hypoxemia, although the increased arterial pH tends to inhibit this. Measurements of retinal blood flow show that this can double in lowlanders moved to an altitude of 5,300 m (17,500 ft) (Frayser et al., 1974). Retinal hemorrhages apparently frequently occur, although they usually heal leaving no visual defects. Periodic or Cheyne-Stokes breathing is common at high altitude especially during sleep (Chapter 39).

Long-term residents at high altitude sometimes develop an ill-defined syndrome characterized by fatigue, reduced exercise tolerance, severe hypoxemia and excessive polycythemia. This is called *chronic mountain sickness* or Monge's disease.

Acclimatization

HYPERVENTILATION

This is one of the most beneficial physiological responses to exposure to a low oxygen partial pressure. Its value can be seen by considering the alveolar gas equation (see Chapter 37) for a climber breathing air on the summit of Mount Everest. If his ventilation were normal giving an alveolar P_{CO_2} of 40 mmHg, his alveolar P_{O_2} would be:

$$PA_{O_2} = PI_{O_2} - \frac{PA_{CO_2}}{R} + F$$
$$= 43 - \frac{40}{0.8} + 2$$
$$= -5 \text{ mmHg!}$$

In this calculation the respiratory exchange ratio is assumed to be 0.8; under these conditions the small correction factor is 2 mmHg.

Clearly something must be done about this! In practice the climber increases his alveolar ventilation approximately 5-fold and thus reduces his alveolar P_{CO_2} to 7.5 mmHg. The equation then becomes:

$$PA_{O_2} = 43 - \frac{7.5}{0.8} + 0.4$$
$$= 34 \text{ mmHg}$$

The arterial P_{O_2} will be several mmHg below this. In fact, calculations show that there is severe diffusion limitation of oxygen transfer under these conditions of extreme hypoxia (compare Fig. 5.37B), and the arterial P_{O_2} is believed to be about 28 mmHg. Moreover, because of the diffusion limitation, it will fall further during exercise. Nevertheless, climbers have now reached the summit of Mt. Everest without supplementary oxygen, though their accounts (e.g., Messner, 1979) show that they are extremely close to the limit of human tolerance.

Less striking degrees of hyperventilation occur

at intermediate altitudes. Permanent residents who live at 4,600 m (15,000 ft) in the Peruvian Andes have an arterial P_{CO_2} of about 33 mmHg (Hurtado, 1964). A group of physiologists who wintered for some 4 mo in the Himalayas at 5,800 m (19,000 ft) had an average alveolar P_{CO_2} of about 23 mmHg (West et al., 1962). Figure 5.83 shows the changes in alveolar P_{O_2} and P_{CO_2} that occur in acclimatized subjects and also in subjects acutely exposed to a low barometric pressure. Note that the latter typically do not reduce their P_{CO_2} (that is, hyperventilate) until the pressure is reduced below that corresponding to an altitude of about 3,000 m (about 10,000 ft).

The cause of the hyperventilation is hypoxic stimulation of the peripheral chemoreceptors (see Chapter 39). The resulting low arterial P_{CO_2} and alkalosis of the cerebrospinal fluid (CSF) and blood tend to inhibit this increase in ventilation initially. After a day or so, however, the pH of the CSF is reduced to some extent by outward movement of bicarbonate, and after 2 or 3 days the pH of the arterial blood is returned to near normal by the renal excretion of bicarbonate. These brakes on ventilation then diminish and ventilation tends to increase further. However, the cause of the sus-

tained hyperventilation at high altitude is still not fully understood.

Interestingly, people born at high altitude have a diminished ventilatory response to hypoxia which is apparently not corrected by subsequent residence at sea level. Conversely, those born at sea level who move to high altitudes for a prolonged period retain their hypoxic response intact. Thus, this sensitivity is determined very early in life (Lahiri et al., 1969). Patients with congenital heart disease and large right-to-left shunts causing severe hypoxemia also have a blunted ventilatory response to hypoxia.

POLYCYTHEMIA

Another feature of acclimatization to high altitude is an increase in the red blood cell concentration of the blood. The resulting rise in hemoglobin concentration, and therefore oxygen-carrying capacity, means that although the arterial P_{O_2} and oxygen saturation are diminished, the oxygen *content* of the arterial blood may be normal or even above normal. For example, in permanent residents at 15,000 ft in the Peruvian Andes, the arterial P_{O_2} is only 45 mmHg and the corresponding arterial oxygen saturation only 81% (Hurtado, 1964). Ordinarily this would considerably decrease the arte-

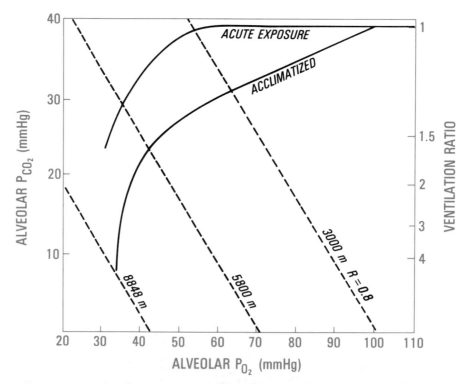

Figure 5.83. O_2-CO_2 diagram showing the effects of acutely exposing subjects to low pressures, and the results of acclimatization to high altitude. Note that acutely exposed subjects do not increase their ventilation until the pressure is reduced below the equivalent of about 3000 m. The P_{CO_2} on the summit of Everest (8848 m) is about 7.5 mmHg. This corresponds to an approximately 5-fold increase in ventilation (*right-hand scale*). The diagonal lines indicate the possible values for P_{CO_2} and P_{CO_2} at a given altitude if the subject maintains a respiratory exchange ratio (R) of 0.8. [From Rahn and Otis (1949) with additional data from West et al. (1983).]

rial oxygen content, but because of the polycythemia, the hemoglobin concentration is increased from 15 to 19.8 g/100 ml, giving an arterial oxygen content of 22.4 ml/100 ml which is above the normal sea level value.

The polycythemia also tends to maintain the P_{O_2} of mixed venous blood; typically in Andean permanent residents at 15,000 ft this P_{O_2} is only 7 mmHg below normal as shown in Figure 5.84. Note how little the P_{O_2} falls from the arterial to the mixed venous blood. This is partly a reflection of the polycythemia, which means that the blood P_{O_2} falls less for a given reduction in oxygen content in the peripheral capillaries, and partly due to the steepness of the oxygen dissociation curve in this region.

The stimulus for the increased production of red blood cells is hypoxia, which releases erythrpoietin from the kidney, which in turn stimulates the bone marrow. Polycythemia is also seen in many patients with chronic hypoxemia caused by lung or heart disease.

Recent work suggests that the polycythemia of high altitude may not be wholly advantageous. The resulting increase in viscosity of the blood (Chapter 7) increases the work of the heart and may cause uneven blood flow in systemic capillaries. Indeed some physicians have advocated hemodilution, that is removal of some red cells with replacement by cell-free fluid, in people at high altitude. Certainly this procedure does not appear to reduce exercise tolerance. It may be that high altitude polycythemia

is an inappropriate response to tissue hypoxia. However, the issue is not yet settled.

OTHER FEATURES OF ACCLIMATIZATION

These include a *shift to the right of the oxygen dissociation curve* (at pH of 7.4) which results in a better unloading of oxygen in venous blood at a given P_{O_2}. The cause of the shift is an increase of the concentration of 2,3-diphosphoglycerate within the red cells, which develops because of the hypoxia and alkalosis. This shift has been observed to occur after 2 days of residence at 4600 m (15,000 ft) and is calculated to increase the amount of oxygen released from the blood by about 10% (Lenfant et al., 1970). At higher altitudes the benefit is less because of interference with the loading of oxygen in the pulmonary capillaries. Indeed, during exercise at extreme altitudes where oxygen transfer across the pulmonary capillary is diffusion-limited, a left-shifted curve is advantageous. In practice, a left-shifted curve is often seen under in vivo conditions because of the respiratory alkalosis. This is especially true at extreme altitudes but has also been described in permanent residents at 4600 m altitude.

There is also evidence that the *number of capillaries* in peripheral tissues increases, and also that changes occur in the *oxidative enzymes* inside the cells. These conclusions are chiefly based on animal studies. The *maximum breathing capacity* increases because the air is less dense (see Chapter 38) and makes possible the very high ventilations (up to 200 l/min) which occur on exercise. The maximum oxygen uptake, however, declines rapidly above 4500 m. This is improved to some extent if 100% oxygen is breathed but does not return to sea level values for reasons which are not understood.

Pulmonary vasoconstriction frequently occurs at high altitudes as a result of the alveolar hypoxia (see Chapter 36). As a consequence of the hypoxic vasoconstriction the pulmonary arterial pressure rises, as does the work done by the right heart. This is exaggerated by the polycythemia which increases the viscosity of the blood. Hypertrophy of the right heart is seen with characteristic changes in the electrocardiogram. There seems to be no physiological advantage to this response except that the topographical distribution of blood flow becomes more uniform (see Fig. 5.20) and the degree of ventilation-perfusion inequality is therefore slightly reduced.

The pulmonary hypertension is occasionally associated with *high altitude pulmonary edema*. Typically a climber or skier who has ascended to high altitude, perhaps without an adequate period of

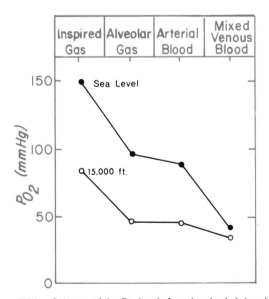

Figure 5.84. Scheme of the P_{O_2} levels from inspired air to mixed venous blood at sea level and in residents at an altitude of 4,600 m (15,000 ft). Note that in spite of the much lower inspired P_{O_2} at altitude, the P_{O_2} of the mixed venous blood is only 7 mmHg lower. [From Hurtado (1964).]

acclimatization, develops shortness of breath and may cough up pink frothy sputum. The attack often comes on at night. The cause of this condition is not understood, but the fact that measurements of pulmonary artery wedge pressure are normal implies that the pulmonary venous pressure is not raised. One hypothesis is that the hypoxic vasoconstriction is uneven, and leakage of fluid therefore occurs in those regions of the capillary bed which are not protected from the high pulmonary arterial pressure. The treatment is to move the patient to a lower altitude and also to give oxygen if this is available.

EXERCISE TOLERANCE AT HIGH ALTITUDE

Permanent residents at an altitude of 4,600 m (15,000 ft) in the Peruvian Andes have a remarkably high work capacity in spite of their hypoxemia. Barcroft (1925) in a colorful early study of these people expressed astonishment at their athletic ability, for example, during a game of football. More extensive studies by Hurtado (1964) have confirmed the remarkable exercise tolerance of these permanent residents. It is of interest that the South American Incas maintained two armies—one that was kept at high altitude and therefore remained acclimatized, and another for fighting on the plains.

Lowlanders who acclimatize to an altitude of 4,600 m (15,000 ft) can achieve only about 75% of the maximal oxygen consumption that they had at sea level. Above this altitude their exercise tolerance falls rapidly, being about 60% of the sea level at 5,800 m (19,000 ft) and less than 40% at 7,400 m (24,400 ft). When acclimatized subjects breathe a mixture with the same inspired P_{O_2} as exists on the summit of Mt. Everest, their maximal oxygen consumption is about 1 l/min, that is less than one quarter of their sea level value. Under these extremely hypoxic conditions, the maximal oxygen uptake is exquisitely sensitive to changes in inspired P_{O_2}. It has even been suggested that whether a climber can scale Mount Everest breathing ambient air may depend critically on the barometric pressure for that day! Certainly it is a remarkable coincidence that the P_{O_2} at the highest point on earth is just sufficient for man to survive.

SPACE FLIGHT

Compared to the high altitude climber exposed to low oxygen, cold, and wind, the astronaut is indulged because he controls his own atmosphere and climatic conditions. In the Apollo flights to the moon, the environment was 100% oxygen with a cabin pressure of one-third of an atmosphere; this gave an inspired P_{O_2} of about 260 mmHg. The atmosphere of the Space Shuttle is air at 760 mmHg.

A number of physiological problems have been encountered by the astronauts. Motion sickness because of the disorientation can be very disabling. Diuresis with loss of circulating blood volume is seen early in space flight. This is caused by movement of blood from the lower body into the thorax in the weightless state and consequent stimulation of the atrial volume receptors (Chapter 16). Cardiovascular "deconditioning" is liable to occur; as a result of this, postural hypotension may be seen when the astronaut returns to earth. Decalcification of bone occurs through disuse. There may also be a small reduction in red cell mass apparently related to the reduction in bodily activity. Presumably the distribution of ventilation and blood flow in the lung will become more uniform (Figs. 5.19 and 5.69) in the gravity-free environment, and as a result there will be small improvement in pulmonary gas exchange, but no measurements have yet been made.

Breathing pure oxygen is associated with several hazards; this is a convenient place to discuss them.

Oxygen Toxicity

The usual problem is getting enough oxygen into the body, but it is possible to have too much. When high concentrations of oxygen are breathed for many hours, damage to the lung may occur (Clark and Lambertsen, 1971). If guinea pigs are placed in 100% oxygen at atmospheric pressure for 48 h they develop pulmonary edema. Some of the first pathologic changes are seen in the endothelial cells of the pulmonary capillaries (Fig. 5.2). It is not easy to administer very high concentrations of oxygen to patients. However, this is sometimes done in intensive care units where the patients are mechanically ventilated with a respirator via an endotracheal tube or tracheostomy. Under these conditions evidence of impaired gas exchange has been demonstrated after 30 h of inhalation of 100% oxygen (Fig. 5.85). Normal volunteers who breathe 100% oxygen at atmospheric pressure for 24 h complain of substernal distress which is aggravated by deep breathing, and they develop a diminution of vital capacity of 500–800 ml. This is probably caused by absorption atelectasis (see below).

Another hazard of breathing 100% oxygen is seen in premature infants who develop blindness because of retrolental fibroplasia, that is fibrous tissue formation behind the lens. Here the mechanism is local vasoconstriction caused by the high P_{O_2} in the incubator. It can be avoided if the oxygen concentration is less than 40%, or if the arterial P_{O_2} is kept below 140 mmHg.

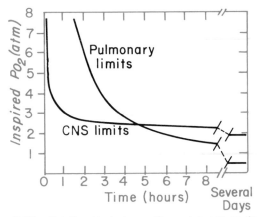

Figure 5.85. Relationship between P_{O_2} and exposure time responsible for oxygen toxicity. CNS, central nervous system. [From Lambertsen (1974).]

Absorption Atelectasis

This is another danger of breathing 100% oxygen. Suppose that an airway is completely obstructed by mucus (Fig. 5.86*A*). The total pressure in the trapped gas is close to 760 mmHg (it may be a few mmHg less as it is absorbed because of elastic forces in the lung). But the sum of the partial pressures in the mixed venous blood is far less than 760 mmHg. This is because the P_{O_2} of the mixed venous blood remains relatively low even when oxygen is breathed. In fact the rise in oxygen *content* of arterial and mixed venous blood when oxygen is breathed will be the same if the cardiac output and oxygen consumption remain unchanged. This follows from the Fick principle:

$$\dot{V}_{O_2} = \dot{Q}(Ca_{O_2} - C\bar{v}_{O_2})$$

as discussed in Chapter 36. The arterial oxygen content will rise by only some 1.5 ml/100 ml due to the increased dissolved oxygen since the hemoglobin is normally nearly fully saturated (Fig. 5.24). As a result, the increase in venous P_{O_2} is only about 10–15 mmHg because of the shape of the oxygen dissociation curve, in spite of the fact that the arterial P_{O_2} increases several hundred mmHg (compare Fig. 5.42). Thus, since the sum of the partial pressures in the alveolar gas greatly exceeds that in the venous blood, gas diffuses into the blood, and rapid collapse of the alveoli occurs. Reopening such an atelectatic area may be difficult because of surface tension forces in such small units (Fig. 5.54).

Absorption collapse also occurs in a blocked region, even when air is breathed, although here the process is slower. Figure 5.86*B* shows that again the sum of the partial pressures in venous blood is less than 760 mmHg because the fall in P_{O_2} from arterial to venous blood is much greater than the rise in P_{CO_2} (this is a reflection of the steeper slope

of the CO_2 dissociation curve compared with that of oxygen—see Fig. 5.30*A*). Since the total gas pressure in the alveoli is nearly 760 mmHg, absorption is inevitable. Actually the changes in the alveolar partial pressures during absorption are somewhat complicated, but it can be shown that the rate of collapse is limited by the rate of absorption of nitrogen. Since this gas has a low solubility, its presence acts as a "splint" which, as it were, supports the alveoli and delays collapse. Even relatively small concentrations of nitrogen or other poorly soluble inert gases in alveolar gas have a useful splinting effect. Nevertheless, postoperative atelectasis is a common problem in patients who are treated with high oxygen mixtures. Collapse is particularly likely to occur at the bottom of the lung where the parenchyma is least well expanded (Fig. 5.70) or the airways are actually closed (Fig. 5.71). The same basic mechanism of absorption is responsible for the gradual disappearance of a pneumothorax, or a gas pocket introduced under the skin.

INCREASED PRESSURE

During diving the pressure increases by 1 atmosphere for every 33 ft of descent. Pressure by itself is relatively innocuous down to 1,000–1,500 ft of depth so long as it is balanced. However, if a gas cavity such as the lung, middle ear, or intracranial sinus fails to communicate with the outside, the pressure difference may cause compression on descent or overexpansion on ascent. For example, it is very important for a scuba diver to exhale as he ascends in order to prevent overinflation and possible rupture of alveolar walls of his lung. The increased density of the gas at depth increases the work of breathing (see Chapter 39). This may result in CO_2 retention, especially on exercise.

Decompression Sickness (Dysbarism)

During diving or working in caissons at high pressure the high partial pressures of nitrogen force this poorly soluble gas into solution in body tissues. This occurs particularly in fat, which has a relatively high nitrogen solubility. The blood supply of adipose tissue is meager, however, and the blood can carry little nitrogen. In addition, the gas diffuses slowly because of its low solubility (Fig. 5.35). As a result, equilibration of nitrogen between the tissues and the environment takes hours. Nevertheless, while the total volume of nitrogen in the body after equilibration is only about one l at sea level, this increases to about 10 l (measured at sea level pressure) in a diver at a depth of 300 ft.

During ascent, nitrogen is slowly removed from the tissues since the partial pressure there is higher

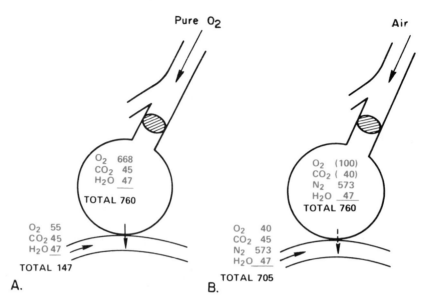

Figure 5.86. Reasons for atelectasis of alveoli beyond blocked airways (*A*) when oxygen and (*B*) when air is breathed. Note that in both cases, the sum of the gas partial pressures in the mixed venous blood is less than in the alveoli. In *B*, the P_{O_2} and P_{CO_2} are shown in parentheses because these values change with time. However, the total alveolar pressure remains within a few mmHg of 760.

than in the alveolar gas and arterial blood. If decompression is unduly rapid, bubbles of gaseous nitrogen are released, just as CO_2 comes out of solution when a bottle of champagne is opened. The first observation of this phenomenon was made by Robert Boyle in 1670 when he decompressed a viper and noticed that a bubble formed in the eye. Some small bubbles can apparently occur without physiological disturbances, but large numbers of bubbles cause pain, especially in the region of joints ("bends"). Bubbles of gas in the pulmonary circulation may cause the "chokes"—a feeling of shortness of breath often accompanied by coughing. In severe cases there may be neurological disturbances such as deafness, impaired vision, vestibular disturbances, and even paralysis caused by bubbles which obstruct blood flow in vessels in the central nervous system (CNS). In overwhelming decompression sickness, the patient may collapse and die. Professional divers or caisson workers sometimes develop avascular necrosis of the head of the femur or other long bones late in life (that is, the bone dies because of obstruction to its blood supply). This is possibly due to intermittent obstruction by nitrogen bubbles over many years.

Decompression sickness can be avoided if the diver ascends slowly enough to prevent bubbles from becoming unduly large. In practice he uses tables which indicate the depths at which he should pause for periods of time during ascent, a process known as "staging." Table 5.3 shows an example of a decompression table as used by the U.S. Navy when the diver breathes compressed air. Note that

for a 40 min dive at a depth of 200 ft, the total decompression time is almost 2 h. If a dive at 200 ft is extended to 90 min, the decompression time is over 5 h! By contrast, a 4 h dive at 40 ft requires only 12 min of decompression time.

Tables of the type shown in Table 5.3 were originally developed by Haldane (Boycott et al., 1908) and were based partly on observations of decompression sickness in experimental animals, and partly on calculations of how rapidly nitrogen is washed out of body compartments. Figure 5.87 shows the rate of washout of three typical tissues. There is considerable variation, and these numbers should be considered illustrative only. Note that the half-time of blood is about 2 min (that is, half the nitrogen is removed in this time), that of muscle is about 20 min, while fat has a half-time of about 1 h. The rates of washin of nitrogen when the diver is at depth are similar. As a result the events which take place in the various tissues when a diver descends to 300 ft for 20 min and then starts his ascent (Table 5.3) are complicated and depend on the relative washin and washout rates of the various tissues. It should be pointed out that while Figure 5.87 indicates the behavior of typical tissues, the body actually consists of a spectrum of half-times determined by the nitrogen solubility and blood supply of the various components.

For many years it was thought that the value of slow decompression was in preventing the formation of nitrogen bubbles. Recent work, however, for example with ultrasonic bubble detectors, suggests that small bubbles usually occur in the blood during

Table 5.3
Typical decompression table showing the depth and time of the stops required to prevent decompression sickness after dives to various depths. Ascent rate assumed to be 60/min.

Depth of Dive, ft	Time on Bottom, min	Stops					Total Time, min[a]
		50 ft	40 ft	30 ft	20 ft	10 ft	
40	250					11	12
50	200					35	36
60	160					48	49
70	120				4	47	52
80	120				17	56	74
90	100				21	54	77
100	90			3	23	57	85
110	80			7	23	57	89
120	70			9	23	55	89
130	60			9	23	52	86
140	60			16	23	56	97
150	50			12	23	51	89
160	50		2	16	23	55	99
170	40		1	10	23	45	82
180	40		3	14	23	50	93
200	40	2	8	17	23	59	112

This table is for illustrative purposes only and should not be used for recreational diving.   [a] Approximate total decompression time.

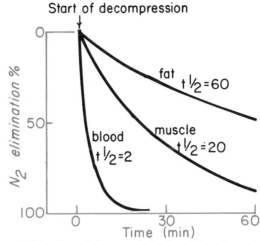

Figure 5.87. Rate of nitrogen elimination during decompression following a dive for three typical tissues. The half times (*t½*) shown depend on the blood flow which may vary. These are therefore illustrative curves only.

normal decompression even in the absence of symptoms. Thus the modern objective is to prevent these bubbles which are inevitably present from growing large enough to cause decompression sickness (Hills, 1969). This change of emphasis has affected the design of decompression tables so that divers now tend to spend more time decompressing at greater depths but actually reach the surface faster.

The treatment of decompression sickness is by recompression in a pressure chamber. This reduces the volume of the bubbles, forces them back into solution, and often results in a dramatic reduction of symptoms. Industrial operations involving deep dives, for example, deep-sea drilling platforms, have compression chambers available. The amateur scuba diver who develops bends is not so fortunate, however, and there may be long delay in transporting him to an appropriate facility. Oxygen inhalation during decompression can be used to accelerate the washout of nitrogen.

The risk of decompression sickness following very deep dives can be reduced if a helium-oxygen mixture is breathed during the dive. Helium is about one-half as soluble as nitrogen so that less is dissolved in tissues. In addition, it has one-seventh of the molecular weight of nitrogen and therefore diffuses more rapidly through tissue (Fig. 5.35). Both of these factors reduce the risk of bends after long dives. Another advantage of a helium-oxygen mixture for divers is its low density which reduces the work of breathing. Pure oxygen or enriched oxygen mixtures cannot be used at depth because of the dangers of oxygen toxicity (see below).

Aviators also may develop decompression sickness due to nitrogen bubble formation if they are suddenly decompressed at very high altitudes as a result of cabin failure. This can be prevented if they breathe 100% oxygen prior to the flight and thus wash out the nitrogen.

SATURATION DIVING

To avoid the long periods of decompression associated with deep dives, some divers have lived in underwater habitats for periods of 2 wk or so. Under these conditions, the tissues equilibrate with nitrogen at that depth, often about 50 ft. It is then possible to make excursion dives to, say, 150 ft and return to the habitat with little or no decompression time (compare Table 5.3). Of course when the divers eventually return to the surface, they must decompress for a long period. In present day commercial diving in the North Sea, divers sometimes live for several days in a pressure chamber on the tender at a simulated depth of 300 ft. This allows them to make dives to depths of 600 ft.

Inert Gas Narcosis

Although nitrogen is usually regarded as a physiologically inert gas, at high partial pressures it affects the CNS. At a depth under water of about 150 ft there is a feeling of euphoria (not unlike that following a martini or two), and divers have been known to offer their mouthpiece to a fish! At higher partial pressures, loss of coordination and eventually coma may develop. The mechanism of action is not understood but may be related to the high fat/water solubility ratio of nitrogen which is a general property of anesthetic agents. Other gases such as helium and hydrogen can be used at much greater depths without narcotic effects.

Oxygen Toxicity

Earlier in this chapter it was noted that inhalation of 100% oxygen at 1 atmosphere can damage the lung. Another form of oxygen toxicity affects the CNS leading to convulsions when the inspired P_{O_2} considerably exceeds 760 mmHg. The convulsions may be preceded by premonitory symptoms such as nausea, ringing in the ears, and twitching of the face.

The likelihood of convulsions depends on the inspired P_{O_2} and the duration of exposure (Fig. 5.85), and it is increased if the subject is exercising. At a P_{O_2} of 4 atmospheres, convulsions frequently occur within 30 min. For increasingly deep dives, the oxygen concentration must be progressively reduced to avoid toxic effects, and may eventually be less than 1% for a normal inspired P_{O_2}! The amateur scuba diver should *never* fill his tanks with oxygen because of the danger of a convulsion underwater. However, pure oxygen is sometimes used by the military for shallow dives because a closed breathing circuit with a CO_2 absorber leaves no tell-tale bubbles. The biochemical basis for the deleterious effects of a high P_{O_2} on the CNS is not fully understood, but is probably the inactivation of certain enzymes, especially dehydrogenases containing sulfhydryl groups.

Scuba Diving

Self-contained underwater breathing apparatus (scuba) is now used extensively by sport divers as well as professionals. The hazards of decompression sickness and inert gas narcosis have been discussed above. Additional dangers include lung rupture and consequent *air embolism* which may occur if a diver ascends without exhaling. Under these conditions the volume of gas in the lungs expands so much that the alveolar walls tear, air escapes into the small pulmonary veins, and some of it travels to the cerebral and coronary circulations. Unconsciousness and collapse may rapidly supervene. Treatment is by recompression; oxygen breathing hastens the absorption of bubbles in the blood. Scuba divers should always be accompanied (buddy system) and be familiar with the physiological principles of diving (see, for example, Council for National Cooperation in Aquatics, 1974).

Shallow Water Blackout

There have been a number of accidents in relatively shallow water, for example a swimming pool, where the victim loses consciousness during ascent from a breath-hold dive. One reason for this is that a diver may hyperventilate before a dive to reduce the ventilatory drive caused by CO_2 accumulation. He can then stay submerged for a long period until the arterial P_{O_2} falls to a low level and stimulates breathing. As he ascends, the ambient pressure falls and the alveolar, and therefore arterial, P_{O_2} become so low that he loses consciousness. Hyperventilation prior to a long breath-hold dive should therefore be discouraged, and another adult should ideally be present.

Drowning

Drowning is suffocation by submersion, usually in water. Studies of drowned or nearly drowned people, and of experimental animals, show that the most important blood gas changes are severe hypoxemia combined with hypercapnia and respiratory acidosis (Modell, 1971). After recovery, a metabolic acidosis may persist. While the initial cause of the hypoxemia is bloodflow through unventilated lung, aspiration of fresh water apparently interferes with the function of pulmonary surfactant and leads to areas of alveolar atelectasis in people who recover from near drowning. This atelectasis may cause persistent hypoxemia. Aspiration of sea water moves additional fluid into the lung from the blood through osmotic forces. This fluid increases the proportion of unventilated lung.

A few victims of drowning do not inhale water into their lungs because of reflex contraction of their laryngeal muscles; they die from acute asphyxia. Transient changes in blood volume commonly occur in drowning. The blood volume increases in fresh water drowning due to the rapid transfer of hypotonic fluid into the circulation. Opposite changes occur in sea water drowning. At one time it was thought that changes in blood electrolytes were of great importance in drowning, but more recent work shows that in most cases the electrolytes are within the normal range.

Inert Gas Counterdiffusion

An interesting complication of exposure to high pressures has been described by Lambertsen and his colleagues (Graves et al., 1973). Men in compression chambers who breathe one inert gas, for example, nitrogen, while surrounded by another, for example, helium, sometimes develop itching and swollen patches in their skin. This is apparently caused by the formation of bubbles which occur without any change in ambient pressure. The cause of the bubbles is localized supersaturation; that is, the sum of the partial pressures of the gases exceeds the pressure in the surrounding tissues and consequently the gases come out of solution. This occurs at lipid-water interfaces because one gas diffuses particularly rapidly through the lipid, while the other gas diffuses fast through the water. The result is that the sum of the partial pressures at the interface becomes very high. In experimental animals, the bubbles can grow so large that gas embolism can occur resulting in death. Although this phenomenon was first noticed during pressure breathing, it can also occur at normal pressure.

Hyperbaric Oxygen Therapy

Increasing the arterial P_{O_2} to a very high level is useful in some clinical situations. One is severe CO poisoning where most of the hemoglobin is bound to CO and therefore unavailable to carry oxygen. By raising the inspired P_{O_2} to 3 atmospheres in special chambers, the amount of dissolved oxygen in arterial blood can be increased to about 6 ml/100 ml (Fig. 5.24), and thus the needs of the tissues can be met without functioning hemoglobin. Occasionally, an anemic crisis is managed in this way. Hyperbaric oxygen is also useful for treating gas gangrene because the organism cannot live in a high P_{O_2} environment. A hyperbaric chamber is also useful for treating decompression sickness.

Fire and explosions are serious hazards of a 100% oxygen atmosphere, especially at increased pressure. For this reason, oxygen in a pressure chamber is given by mask, and the chamber itself is filled with air.

Artificial Gills

Various ingenious devices (Paganelli et al., 1967) have been proposed to enable a diver to remove dissolved oxygen from water, a process which is normally accomplished by the fish gill. In Figure 5.88 the diver is connected to a bag with a large gas-exchanging surface made of a silicon polymer which is highly permeable to oxygen and CO_2. Oxygen from the water diffuses into the bag and CO_2 diffuses out. At first sight the system appears to offer an unlimited period of submersion, but it eventually fails for an interesting reason. Because the volume of CO_2 exhaled is generally less than the volume of O_2 consumed, the nitrogen in the bag is gradually concentrated and its partial pressure rises. As a result, nitrogen gradually diffuses into the water and the bag progressively empties.

Oddly enough nature uses the same principle to allow the water boatman *Corixa* to remain submerged for long periods. This beetle dives with a tiny bubble of air adhering to its ventral surface and connected with the respiratory air tubes (spiracles) which lead to the tissues. Oxygen diffuses into the bubble from the water and CO_2 diffuses out, but eventually the bubble collapses through loss of nitrogen.

A way around this difficulty was suggested by Rahn when he designed a chamber for fish watching in the Niagara River (Paganelli *et al.*, 1967). This has a frame which supports the permeable membrane and thus allows the chamber pressure to fall by a small amount (Fig. 5.89). The result is that the nitrogen partial pressure is in equilibrium with that of the water, and the system is in a true steady state of gas exchange.

Again nature has priority with a similar solution. Some insects exchange gas under water by means of a "plastron," that is, a bubble supported on stiff hairs which protrude from the insect's body (Crisp, 1964; Rahn and Paganelli, 1968). By means of surface tension, the framework of hairs prevents the bubble from collapsing even when the pressure inside it falls. Essentially the same steady state of gas exchange as that shown in Figure 5.89 is thus developed.

Liquid Breathing

As explained earlier, in very deep diving the carrier gas (usually nitrogen) is responsible for most of the problems, including decompression sickness and inert gas narcosis. An exotic solution therefore is to do away with this vehicle gas entirely and breathe the oxygen dissolved in liquid!

Kylstra and his colleagues (1962) were the first to show that it is possible for mammals to survive for some hours breathing liquid instead of air. They immersed mice in saline in which the oxygen concentration was increased by exposure to 100% oxygen at 8 atmospheres pressure. Subsequently, mice, rats, and dogs have survived a period of breathing fluorocarbon exposed to pure oxygen at 1 atmosphere. This liquid has a high solubility for both oxygen and CO_2. The animals successfully returned to air breathing.

Because liquids have a much higher density and viscosity than air, the work of breathing is enormously increased. Adequate oxygenation of the arterial blood can be obtained, however, if the inspired concentration is raised sufficiently. Interestingly, a serious problem is eliminating CO_2. In Chapter 36 we saw that diffusion within the airways is chiefly responsible for the gas exchange which occurs between the alveoli and the terminal or respiratory bronchioles where bulk flow takes over. Because the diffusion rates of gases in liquid are several orders of magnitude slower than in the gas phase, this means that a large partial pressure difference for CO_2 between the alveoli and terminal bronchioles must be maintained if CO_2 is to be eliminated adequately. Animals breathing liquid,

Figure 5.88. Proposed artificial gill consisting of a bag made of a material highly permeable for oxygen and CO_2. Exchange occurs with the gases in the surrounding water; however, eventually the bag empties because nitrogen diffuses out (see text). Hz, Hertz.

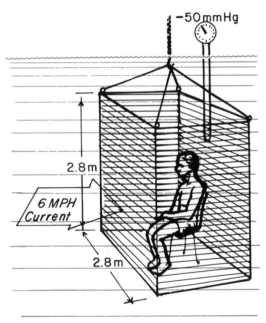

Figure 5.89. Artificial gill designed by Rahn for fish watching in the Niagara River. The permeable membrane is supported on a frame, and under steady-state conditions the pressure in the box is less than that in the surrounding water, so that nitrogen exchange is in a steady state. [From Paganelli et al. (1967).]

therefore, commonly develop CO_2 retention and acidosis. Note that the diffusion pressure for oxygen can always be raised by increasing the inspired P_{O_2}, but the converse option is not available to assist in the elimination of CO_2.

Apart from its possible use in deep diving, liquid breathing has other potential applications. In Chapter 38 we saw that premature babies some-

times develop a respiratory distress syndrome characterized by patchy atelectasis, pulmonary edema, and stiff lungs. This is caused, in large part, by the absence of pulmonary surfactant. A suggested method of treatment is to ventilate the lungs of these infants with fluorocarbon until the surfactant system matures. This would abolish the liquid-air interface and thus the high surface tension forces which are chiefly responsible for the abnormalities. Preliminary experiments in premature experimental animals have given promising results (Schaffer et al., 1976).

Another possible application of liquid breathing is in future space flight where very large accelerations may be necessary. The lung is vulnerable to high accelerations because of the different densities of air and blood within it. Filling the lung with saline confers a great degree of protection against injury by acceleration.

Finally, it is of interest that washing out one lung with normal saline to remove accumulated secretions is now an accepted form of therapy in some types of disease, for example, alveolar proteinosis. The saline is introduced by means of a special catheter (Carlens) which separates the two lungs, and the patient uses the dry lung for gas exchange while the other is being rinsed.

OTHER UNUSUAL ENVIRONMENTS

Perinatal Respiration

All mammalian lungs undergo the transition from a liquid-filled to an air-filled condition at birth. During fetal life, gas exchange takes place through the placenta. Maternal blood enters from the uterine arteries and surges into small spaces called the intervillous sinusoids. Fetal blood is supplied through the umbilical arteries to capillary loops which protrude into the intervillous spaces. Gas exchange occurs across the blood-blood barrier approximately 3.5 μm thick.

This arrangement is much less efficient for gas exchange than is the adult lung. Maternal blood apparently swirls around the sinusoids somewhat haphazardly, and there are probably large differences in P_{O_2} within these blood spaces. Contrast this situation with the air-filled alveoli where rapid gaseous diffusion maintains an almost uniform composition in each lung unit. The result is that the blood leaving the placenta in the uterine vein has a P_{O_2} of only about 30 mmHg (Fig. 5.90). Note the very low P_{O_2} of about 22 mmHg in the descending aorta. However, the blood going to the head has a somewhat higher P_{O_2} of 25 mmHg. These differences of P_{O_2} are the result of streaming within the fetal heart (see Chapter 19). For example, much of

the relatively well oxygenated blood from the inferior vena cava crosses the patent foramen ovale into the left atrium and is available for distribution to the fetal head.

THE FIRST BREATH

The emergence of the baby into the outside world is perhaps the most cataclysmic event of his life. He is suddenly bombarded with a variety of external stimuli. In addition, the process of birth interferes with placental gas exchange with resulting hypoxemia and hypercapnia which stimulate the chemoreceptors. As a combined result of these stimuli, he makes his first gasp.

The fetal lung is not collapsed but is inflated with liquid to about 40% of total lung capacity. This fluid is apparently continuously secreted by alveolar cells during fetal life and has a low pH. Some of it is squeezed out as the infant moves through the birth canal, but the remainder has an important role in the subsequent inflation of the lung. As air enters the lung, large surface tension forces have to be overcome. Since the larger the radius of curvature, the lower the forces (Fig. 5.54), this preinflation reduces the pressures required. Nevertheless, the intrapleural pressure during the first breath may fall to −40 cm of water before any air enters the lung (Fig. 5.91), and peak pressures as low as −100 cm of water during the first few breaths have been recorded. These very large transient pressures are partly caused by the high viscosity of the lung liquid compared with air. The inspiratory gasps may be augmented reflexly, the afferent impulses coming from stretch receptors in the stiff lung. Recent work shows that the fetus makes small rapid breathing movements in the uterus over a considerable period before birth.

Expansion of the lung is very uneven at first. However, pulmonary surfactant which is formed relatively late in fetal life is available to stabilize open alveoli, and the lung liquid is removed by the lymphatics and capillaries. Within a few moments the functional residual capacity has almost reached its normal value, and an adequate gas-exchanging surface has been established. It is several days, however, before uniform ventilation is achieved.

The circulatory changes which occur at birth were discussed in Chapter 19.

Polluted Atmospheres

A polluted atmosphere is, unfortunately, hardly an unusual environment in modern urban areas. One of the chief sources of pollutants is the automobile, which accounts for some 40% of all pollutants by weight in the United States. It produces major portions not only of CO but also of hydro-

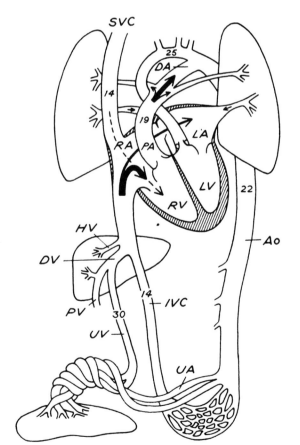

Figure 5.90. P_{O_2} values in the fetal circulation. *UV*, umbilical vein; *PV*, portal vein; *DV*, ductus venosus; *HV*, hepatic vein; *RA*, right atrium; *SVC*, superior vena cava; *RV*, right ventricle; *PA*, pulmonary artery; *DA*, ductus arteriosus; *LA*, left atrium; *LV*, left ventricle; *Ao*, aorta; *UA*, uterine artery. [From Comroe (1974).]

carbons and nitrogen oxides. Electrical utilities and industrial plants are major sources of sulfur oxides because the coal and oil fuels contain sulfur as an impurity. Table 5.4 lists the sources of major air pollutants in this country in 1974.

TYPES OF POLLUTANTS

Carbon monoxide is the largest pollutant by weight; it is produced by the incomplete combustion of carbon in fuels, principally in automobile engines. The propensity of CO to tie up hemoglobin was discussed in Chapter 36. A commuter using a Los Angeles freeway may have up to 10% of his hemoglobin bound to CO, especially if he is a cigarette smoker, and there is evidence that this can impair his mental skills (Coburn et al., 1976). The emission of CO and other pollutants by automobile engines can be reduced by installing a catalytic converter which processes the exhaust gases.

Sulfur oxides are corrosive, poisonous gases produced when the sulfur-containing fuels are burnt, chiefly by power stations. Sulfur oxides irritate the upper respiratory tract and the eyes, and long-term exposure causes chronic bronchitis. The best way to reduce these emissions is to use low sulfur fuels, but these are more expensive.

Hydrocarbons, like carbon monoxide, represent

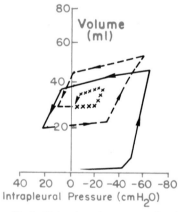

Figure 5.91. Typical intrapleural pressures developed in a newborn human infant during the first three breaths after birth. [From Avery (1968).]

unburned wasted fuel. They are not toxic, however, at concentrations normally found in the atmosphere. Their hazardous nature stems from their role in forming photochemical oxidants under the influence of sunlight.

Particulate matter includes particles with a wide range of sizes up to visible smoke and soot. Major sources are power stations and industrial plants. Often their emission can be reduced by processing the waste air stream by filtering or scrubbing, although removing the smallest particles is often expensive.

Nitrogen oxides are produced when fossil fuel is burned at very high temperatures in power stations and automobiles. These gases irritate the eyes and upper respiratory tract and are responsible for the yellow haze of smog.

Photochemical oxidants include ozone and other substances such as peroxyacyl nitrates, aldehydes, and acrolein. They are not primary emissions but are produced by the action of sunlight on hydrocarbons and nitrogen oxides. They cause eye and lung irritation, damage to vegetation, offensive odors, and they contribute to the thick haze of smog.

The concentration of atmospheric pollutants is often greatly increased by a temperature inversion, especially in a low-lying area surrounded by hills such as the Los Angeles basin. The inversion prevents the normal escape of warm surface air to the upper atmosphere.

BEHAVIOR OF AEROSOLS

Many pollutants exist as *aerosols*, that is, very small particles which remain suspended in the air. When an aerosol is inhaled, its fate depends on the size of the particles (Fig. 5.92). Large particles are removed by *impaction* in the nose and pharynx. This means that the particles are unable to turn corners rapidly because of their inertia, and as they impinge on the wet mucosa they are trapped. Many medium-sized particles deposit in small airways because of their weight. This is called *sedimentation* and tends to occur particularly where the flow velocity is suddenly reduced because of the enor-

Table 5.4

Estimated emissions of major air pollutants (in millions of tons per year) in U.S.A., 1974, as reported by the Environmental Protection Agency

Source	CO	Sulfur Oxides	Hydro-carbons	Par-ticulates	Nitrogen Oxides	Total
Transportation sources (especially automobiles)	73.5	0.8	12.8	1.3	10.7	99.1
Stationary sources (especially power stations)	0.9	24.3	1.7	5.9	11.0	43.8
Industrial processes	12.7	6.2	3.1	11.0	0.6	33.6
Other, including solid waste disposal, forest fires	7.5	0.1	12.8	1.3	0.2	21.9
Total	94.6	31.4	30.4	19.5	22.5	198.4

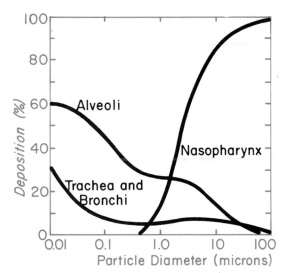

Figure 5.92. Deposition of aerosol particles of various sizes in different regions of the respiratory tract. The largest particles impact in the nasopharynx, while the smallest particles can reach the alveoli.

mous increase in combined airway cross-section (Fig. 5.9). For this reason deposition is heavy in the terminal and respiratory bronchioles, and this region of a coal miner's lung shows a heavy dust concentration. The smallest particles (less than 0.1 μm in diameter) reach the alveoli where some deposition occurs through *diffusion* to the walls. Many small particles are not deposited at all but are exhaled with the next breath.

CLEARANCE OF DEPOSITED PARTICLES

Fortunately the normal lung is very efficient at removing particles which are deposited within it. Particles which fall on the bronchial walls are swept up by a moving staircase of mucus propelled by cilia. The mucus is secreted by specialized cells of the bronchi, including goblet cells which make up part of the bronchial mucosa, and cells of mucous glands which lie deeper in the bronchial wall and secrete the mucus through ducts. There is evidence that the normal mucous film has two layers, a superficial gel layer which is relatively viscous and is efficient at trapping deposited particles, and a deeper layer which is less viscous and in which the cilia can beat easily. It may be that the retention of secretions that occurs in disease is caused in part by changes in the composition of the mucus so that it cannot be propelled by the cilia.

The cilia are 5 to 7 μm in length and they beat in a synchronized fashion at between 1,000 and 1,500 times a minute. This moves the mucus carpet upward at a speed of 1 to 3 cm/min in the large airways. Eventually the particles reach the level of the pharynx where they are swallowed. In very

dusty environments, mucus secretion may be increased so much that cough and expectoration assist in the clearance. The normal operation of the mucociliary system can be adversely affected by atmospheric pollution. For example, tobacco smoke, sulfur oxides, and nitrogen oxides may paralyze the cilia and also change the character of the mucus.

Particles which are deposited in the alveoli are not cleared by the mucociliary system since it does not extend down to that level. Instead the particles are engulfed by phagocytic macrophages which roam around the alveoli by ameboid action. Laden macrophages leave the lung via the lymphatics or blood, while some migrate to the mucociliary escalator. Normal macrophage activity can be impaired by cigarette smoke, oxidant gases, radiation, and the ingestion of alcohol.

Oxygen Therapy

In the treatment of patients with severe lung disease, the environment is often changed by the physician who increases the concentration of inspired oxygen, or controls ventilation by means of a mechanical ventilator. Oxygen therapy is usually very effective in relieving hypoxemia, except when this is caused by a shunt (blood flow through unventilated alveoli). In this case the added oxygen does not have access to the shunted blood. The oxygen content of the nonshunted blood will nevertheless rise somewhat, chiefly because of the addition of dissolved oxygen, and the arterial P_{O_2} may therefore rise by a few mmHg. By contrast, the hypoxemia of hypoventilation, diffusion impairment, or ventilation-perfusion inequality is readily corrected by oxygen therapy, although in the last instance, the arterial P_{O_2} may not reach a steady level for many minutes because poorly ventilated regions of lung wash out their nitrogen so slowly.

Oxygen is now normally administered by small cannulas inserted into the nostrils, or by means of a plastic oronasal mask. This may be of the Venturi type in which the oxygen flow entrains air and thus allows a preset concentration of oxygen to be inspired. Oxygen tents are rarely used except for children. Newborn babies can be nursed in incubators in which the P_{O_2} can be controlled.

There are dangers in giving too much oxygen. Reference has already been made to the pulmonary edema which occurs in lungs exposed to 100% oxygen for more than 2 or 3 days. Other hazards include a propensity for atelectasis, and retrolental fibroplasia in neonates.

Some patients with respiratory failure develop serious CO_2 retention when given high concentrations of oxygen to breathe. These are patients with compensated respiratory acidosis whose ventilatory drive comes chiefly from the effect of the severe hypoxemia on their peripheral chemoreceptors (see Chapter 39). When this hypoxemia is relieved, the patient's ventilation falls, and he may develop dangerous CO_2 retention and acidosis. The proper management is to give a judicious concentration of oxygen, usually 24–28% initially, and to monitor the arterial P_{O_2}, P_{CO_2}, and pH frequently.

Mechanically Assisted Ventilation

While hypoxemia can usually be relieved by administering oxygen, CO_2 retention can be corrected only by increasing the ventilation. Mechanical ventilators are now frequently used in the management of patients in respiratory failure. These are of two types: constant pressure and constant volume respirators. The former generate a preset pressure for each inspiration and are relatively simple to use. A gas tank is employed as the source of the pressure. If the lungs or chest wall are unusually stiff (low compliance), however, constant volume ventilators are more reliable because they deliver a preset ventilation even if there are changes in the mechanical properties of the lung.

Before a ventilator can be used, an airtight connection to the lungs must be made. This can be done by means of a plastic tube inserted through the mouth or nose into the trachea (endotracheal tube). An inflated cuff on the end gives an airtight seal. Alternatively a tracheostomy tube may be inserted through the neck and anterior wall of the trachea. The inspired gas should be humidified to prevent drying of the upper airways. Sometimes a small pressure of about 5 cm of water or more is applied to the airways at the end of expiration (positive end expiratory pressure or PEEP). This tends to prevent atelectasis from occurring and often increases the arterial P_{O_2} in patients with respiratory failure. Undue pressure, however, may interfere with venous return to the chest.

There are numerous problems in ventilator therapy. These include the accumulation of secretions in the lung, damage to the trachea by the endotracheal cuff, and malfunction of the equipment. Effective therapy requires a skilled team of personnel in a specialized intensive care unit.

Resuscitation

It seems reasonable to expect a respiratory physiologist to be as familiar with emergency resuscitation as, say, a Boy Scout. The procedure may be life-saving for victims of near drowning, electric shock, or a heart attack (myocardial infarct).

The aim of cardiopulmonary resuscitation is to establish a circulation of oxygenated blood to the brain and heart (*Journal of the American Medical Association*, 1980). The A, B, Cs of basic life support are therefore *Airway, Breathing, and Circulation*. First the rescuer should determine that the victim is truly unresponsive and not just asleep. Then in rapid sequence, he must establish an airway, determine if there is adequate spontaneous respiration (and if not, provide artificial ventilation), and determine whether there is adequate circulation (and if not, give external cardiac compression). Time should not be wasted in moving the victim because seconds, not minutes, count.

AIRWAY

The victim should be rolled on to his back on a hard, flat surface. In this position the tongue of the unconscious patient tends to drop back and obstruct the upper airway, especially when loss of muscle tone allows forward flexion of the head. To establish an airway, lift the victim's neck with one hand and tilt the forehead back with the other (Fig. 5.93A). Then displace the mandible (jaw bone) forward by pulling strongly at the angle of the jaw. Clear the mouth and pharynx of mucus, blood, vomitus, or foreign material. Determine if the patient is breathing by inspecting the chest for movements and listening in front of the mouth.

BREATHING

If the victim is not breathing, or if breathing is very shallow or irregular, artificial ventilation should be started. The rescuer should grasp the patient's nose between his thumb and index finger completely to seal the nostrils. He then takes a deep breath, makes a tight seal around the patient's mouth with his mouth, and blows. At the end of the breath, he removes his mouth and allows the victim to exhale passively. If it is necessary to use one hand to pull the mandible forward to obtain an adequate airway, the rescuer must seal the victim's nose with his cheek.

If three to five attempts to inflate the lung in this way fail, the patient should be rolled on to this side and a sharp blow delivered between the shoulder blades in an attempt to clear the airway. If dentures are causing obstruction, they should be removed. Continued obstruction may necessitate an emergency tracheostomy if trained personnel are available.

Now feel the carotid or femoral artery for pulsations. The carotid artery is found by placing the index and middle finger on the larynx and then sliding the fingers laterally into the groove between

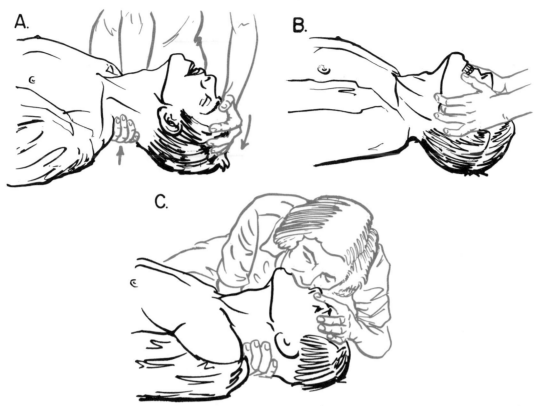

Figure 5.93. Resuscitation by means of mouth-to-mouth breathing. *A.* shows how the head is tilted and *B.* shows how the jaw is pulled forward to ensure an open airway. *C.* shows the technique of mouth-to-mouth breathing.

the trachea and neck muscles. If pulsations are present, indicating that the heart is beating, continue mouth-to-mouth breathing at the rate of 12–15 breaths/min. Allow 2 s for inspiration and 3 s for expiration. Artificial ventilation should be continued until spontaneous respiration returns, or as long as the pulses remain palpable, or previously dilated pupils remain constricted.

CIRCULATION

If carotid or femoral pulsations are absent at the onset, or if they disappear, external cardiac compression should be begun immediately. Place the heel of one hand on the lower third of the sternum just above the level of the xiphoid. With the heel of the other hand on top of it, apply firm vertical pressure sufficient to force the sternum about 1½ to 2 inches downward (less in children) about once a second (60–80/min).

If only one rescuer is present, two quick, deep lung inflations should be given after every 15 sternal compressions. When two rescuers are available,

one mouth-to-mouth inflation should be given after every fifth compression. The breath should be delivered between sternal compressions. The carotid pulse should be checked after 1 min and every 5 min subsequently. Check the pupils periodically. Pupils that remain widely dilated are an indication of cerebral hypoxia and brain damage. Resuscitation must be continuous during transportation to the hospital.

If a person is seen to collapse and develop impalpable pulses in the presence of the rescuer, this is known as a *witnessed cardiac arrest*. In this instance it can be assumed that the primary cause of the heart stoppage is ventricular fibrillation, not generalized myocardial hypoxia. A *precordial thump* should be given within 1 min of the arrest. To do this, the rescuer holds his closed fist 8 to 12 inches above the midpoint of the victim's sternum with the fleshy (ulnar) side of the fist toward the chest. He then delivers one quick, strong thump to the sternum. This may restore normal ventricular contractions. If there is no immediate response, cardiopulmonary resuscitation should be begun at once.

BIBLIOGRAPHY

ALTMAN, P. L. *Blood and Other Body Fluids.* Washington, D.C.: Federation of American Societies for Experimental Biology, 1961, p. 12.

AVERY, M. E. *The Lung and Its Disorders in the Newborn Infant.* Philadelphia: Saunders, 1968.

BALL, W. C., JR., P. B. STEWART, L. G. S. NEWSHAM, AND D. V. BATES. Regional pulmonary function studied with xenon[133]. *J. Clin. Invest.* 41: 519–531, 1962.

BARCROFT, J. *The Respiratory Function of the Blood. Part 1. Lessons From High Altitudes.* London: Cambridge Univ. Press, 1925.

BARER, G. R., P. HOWARD, AND J. W. SHAW. Stimulus-response curves for the pulmonary vascular bed to hypoxia and hypercapnia. *J. Physiol. (Lond.)* 211: 139–155, 1970.

BATES, D. V., P. T. MACKLEM, AND R. V. CHRISTIE. *Respiratory Function in Disease,* ed. 2. Philadelphia: Saunders, 1971.

BENESCH, R., AND R. E. BENESCH. Intracellular organic phosphates as regulators of oxygen release by hemoglobin. *Nature (Lond.)* 221: 618–622, 1969.

BOHR, C. Über die Lungenathmung. *Skand. Arch. Physiol.* 2: 236–268, 1891.*

BOHR, C. Über die spezifische Tatigkeit der Lungen bei der respiratorischen Gasaufnahme und ihr Verhalten zu der durch die Alveolarwand stattfindenden Gasdiffusion. *Skand. Arch. Physiol.* 22: 221–280, 1909.*

BOHR, C., K. A. HASSELBALCH, AND A. KROGH. Ueber einen in biologischer Beziehung wichtigen Einfluss, den die Kohlensaurespannung des Blutes auf dessen. Sauerstoffbinding übt. *Skand. Arch. Physiol.* 16: 402–412, 1904.*

BORN, G. V., G. S. DAWES, J. C. MOTT, AND J. G. WIDDICOMBE. Changes in the heart and lung at birth. *Cold Spring Harbor Symp. Quant. Biol.* 19: 102–108, 1954.

BOYCOTT, A. E., G. C. C. DAMANT, AND J. S. HALDANE. The prevention of compressed air illness. *J. Hyg. (Camb.)* 8: 342–443, 1908.

BOYLE, R. Continuation of the experiments concerning respiration. *Philos. Trans. R. Soc. Lond.* 5: 2035–2056, 1670.

BRISCOE, W. A., AND A. B. DUBOIS. The relationship between airway resistance, airway conductance, and lung volume in subjects of different age and body size. *J. Clin. Invest.* 37: 1279–1285, 1958.

BUNN, H. F., M. H. MAY, W. F. KOCHOLATY, AND C. E. SHIELDS. Hemoglobin function in stored blood. *J. Clin. Invest.* 48: 311–321, 1969.

CAMPBELL, E. J. M., E. AGOSTONI, AND J. N. DAVIS (eds.). *The Respiratory Muscles: Mechanics and Neural Control,* 2nd ed. Philadelphia: Saunders, 1970.

CHANUTIN, A., AND R. R. CURNISH. Effect of organic and inorganic phosphates on the oxygen equilibrium of human erythrocytes. *Arch. Biochem.* 121: 96–102, 1967.

CHERNIACK, N. S., AND G. S. LONGOBARDO. Cheyne-Stokes breathing: an instability in physiologic control. *N. Engl. J. Med.* 288: 952–957, 1973.

CHERNIACK, R. M., L. CHERNIACK, AND A. NAIMARK. *Respiration in Health and Disease.* Philadelphia: Saunders, 1972.

CHRISTIANSEN, J., C. G. DOUGLAS, AND J. S. HALDANE. The absorption and dissociation of carbon dioxide by human blood. *J. Physiol. (Lond.)* 48: 244–271, 1914.

CLARK, J. M., AND C. J. LAMBERTSEN. Pulmonary oxygen toxicity: a review. *Pharmacol. Rev.* 23: 37–133, 1971.

CLEMENTS, J. A. Surface phenomenon in relation to pulmonary function. *Physiologist* 5: 11–28, 1962.

COBURN, R. E., E. R. ALLEN, A. AYERS, D. BARTLETT, JR., S. M. HORWATH, L. H. KULLER, V. G. LATIES, L. D. LONGO, AND E. P. RADFORD, JR. *The Biologic Effects of Carbon Monoxide.* Washington, D.C.: National Academy of Sciences, National Research Council, 1976.

COMROE, J. H. *Physiology of Respiration.* 2nd ed. Chicago, Year Book, 1974, p. 234–271.

COMROE, J. H., R. E. FORSTER, A. B. DUBOIS, W. A. BRISCOE, AND E. CARLSEN. *The Lung—Clinical Physiology and Pulmonary Function Tests,* 2nd ed. Chicago: Year Book, 1962.

COTES, J. E. *Lung Function,* 3rd ed. Oxford: Blackwell, 1979.

COUNCIL FOR NATIONAL COOPERATION IN AQUATICS. *The New Science of Skin and Scuba Diving,* 4th ed. New York: Association Press, 1974.

CRISP, D. J. Plastron respiration. In: *Recent Progress in Surface Science,* edited by J. F. Danielli, K. G. A. Pankhurst, and A. C. Riddiford. New York: Academic Press, 1964, vol. 2, p. 377–425.

CUMMING, G., J. CRANK, K. HORSFIELD, AND I. PARKER. Gaseous diffusion in the airways of the human lung. *Respir. Physiol.* 1: 58–74, 1966.

DARLING, R. C., A. COURNAND, AND D. W. RICHARDS. An open circuit for measuring residual air. *J. Clin. Invest.* 19: 609–618, 1940.

DUBOIS, A. B., S. Y. BOTELHO, G. N. BEDELL, R. MARSHALL, AND J. H. COMROE, JR. A rapid plethysmographic method for measuring thoracic gas volume. *J. Clin. Invest.* 35: 322–326, 1956.

FOWLER, K. T. The respiratory mass spectrometer. *Phys. Med. Biol.* 2: 185–199, 1969.

FOWLER, W. S. Lung function studies. II. The respiratory dead space. *Am. J. Physiol.* 154: 405–416, 1948.

FRAYSER, R., G. W. GRAY, AND C. S. HOUSTON. Control of the retinal circulation at altitude. *J. Appl. Physiol.* 37: 302–304, 1974.

FRY, D. L., AND R. E. HYATT. Pulmonary mechanics: a unified analysis of the relationship between pressure, volume and gas flow in the lungs of normal and diseased human subjects. *Am. J. Med.* 29: 672–689, 1960.

GILL, M. B., J. S. MILLEDGE, L. G. C. E. PUGH, AND J. B. WEST. Alveolar gas composition at 21,000 to 25,700 feet. *J. Physiol. (Lond.)* 163: 373–377, 1962.

GLAZIER, J. B., J. M. B. HUGHES, J. E. MALONEY, AND J. B. WEST. Measurements of capillary dimensions and blood volume in rapidly frozen lungs. *J. Appl. Physiol.* 26: 65–76, 1969.

GRAVES, D. J., J. IDICULA, C. J. LAMBERTSEN, AND J. A. QUINN. Bubble formation resulting from counterdiffusion supersaturation: a possible explanation for inert gas "urticaria" and vertigo. *Phys. Med. Biol.* 18: 256, 1973.

GURTNER, G., AND R. BURNS. Physiological evidence consistent with the presence of a specific O_2 carrier in the placenta. *J. Appl. Physiol.* 39: 728–730, 1975.

HALDANE, J. S. *Respiration.* New Haven: Yale Univ. Press, 1922.

HALDANE, J. S., AND J. G. PRIESTLEY. *Respiration.* New Haven: Yale Univ. Press, 1935.

HILLS, B. A. Thermodynamic decompression. In: *The Physiology and Medicine of Diving and Compressed Air Work,* edited by P. B. Bennett and D. H. Elliot. London: Baillière, Tindall and Cassell, 1969.

HOPPIN, F. G., JR., AND J. HILDENBRANDT. Mechanical properties of the lung. In: *Bioengineering Aspects of the Lung,* edited by J. B. West. New York: Marcel Dekker, 1977.

HUGHES, J. M. B., J. B. GLAZIER, J. E. MALONEY, AND J. B. WEST. Effect of lung volume on the distribution of pulmonary blood flow in man. *Resp. Physiol.* 4: 58–72, 1968.

HURTADO, A. Animals in high altitudes: resident man. In: *Handbook of Physiology, Adaptation to the Environment,* Sect. 4, edited by D. B. Dill. Washington, D.C.: American Physiological Society, 1964, p. 843–860.

HUTCHINSON, J. On the capacity of the lungs, and on the respiratory functions, with a view of establishing a precise and

---

* English translations of these papers can be found in: WEST, J. B. (ed.). *Translations in Respiratory Physiology.* Stroudsburg: Dowden, Hutchinson and Ross, 1975.

easy method of detecting disease by the spirometer. *Med. Chir. Trans. (Lond.)* 29: 137, 1846.

JOURNAL OF THE AMERICAN MEDICAL ASSOCIATION. Standards and guidelines for Cardiopulmonary Resuscitation (CPR) and Emergency Cardiac Care (ECC). 244: 453–509, 1980.

KYLSTRA, J. A., M. O. TISSING, AND A. VAN DER MAEN. Of mice as fish. *Trans. Am. Soc. Artif. Intern. Organs* 8: 378–383, 1962.

LAHIRI, S., F. F. KAO, T. VELASQUEZ, C. MARTINEZ, AND W. PEZZIA. Irreversible blunted respiratory sensitivty to hypoxia in high altitude natives. *Respir. Physiol.* 6: 360–374, 1969.

LAMBERTSEN, C. Respiration. In: V. Mountcastle, *Medical Physiology.* St. Louis: C. V. Mosby, 1974, p. 1361–1597.

LAMBERTSEN, C. J. Therapeutic gases: Oxygen, carbon dioxide, and helium. In: *Drill's Pharmacology in Medicine*, 3rd ed., edited by J. R. DiPalma. New York: McGraw Hill, 1965.

LENFANT, C., J. D. TORRANCE, R. D. WOODSON, P. JACOBS, AND C. A. FINCH. Role of organic phosphates in the adaptation of man to hypoxia. *Fed. Proc.* 29: 1115–1117, 1970.

LOESCHCKE, H. H., AND K. H. GERTZ. Einfluss des O_2-Druckes in der Einatmungsluft auf die Atemtätigkeit des Menschen, geprüft unter Konstanthaltung des alveolaren CO_2-Druckes. *Pflugers Arch. Ges. Physiol.* 267: 460–477, 1958.

MACKLEM, P. T., AND J. MEAD. Resistance of central and peripheral airways measured by a retrograde catheter. *J. Appl. Physiol.* 22: 395–401, 1967.

McCARTHY, D. S., R. SPENCER, R. GREEN, AND J. MILIC-EMILL. Measurement of "closing volume" as a simple and sensitive test for early detection of small airway disease. *Am. J. Med.* 52: 747–753, 1972.

McDONALD, D. M., AND R. A. MITCHELL. The innervation of glomus cells, ganglion cells and blood vessels in the rat carotid body: a quantitative ultra-structural analysis. *J. Neurocytol.* 4: 177–230, 1975.

McMICHAEL, J. A rapid method for determining lung capacity. *Clin. Sci.* 4: 167–173, 1939.

MEAD, J. Respiration: pulmonary mechanics. *Ann. Rev. Physiol.* 35: 169–192, 1973.

MEAD, J., T. TAKISHIMA, AND D. LEITH. Stress distribution in lungs: a model of pulmonary elasticity. *J. Appl. Physiol.* 28: 596–608, 1970.

MESSNER, R. The Mountain. In: *Everest: Expedition to the Ultimate.* London: Kay and Ward, 1979.

MILIC-EMILI, J., J. A. M. HENDERSON, M. B. DOLOVICH, D. TROP, AND K. KANEKO. Regional distribution of inspired gas in the lung. *J. Appl. Physiol.* 21: 749–759, 1966.

MODELL, J. H. *The Pathophysiology and Treatment of Drowning and Near-Drowning.* Springfield, Ill.: Charles C Thomas, 1971.

NIELSEN, M., AND H. SMITH. Studies on the regulation of respiration in acute hypoxia. (With an appendix on respiratory control during prolonged hypoxia.) *Acta Physiol. Scand.* 24: 293–313, 1952.

PAGANELLI, C. V., N. BATEMAN, AND H. RAHN. Artificial gills for gas exchange in water. In: *Proceedings of the Third Symposium on Underwater Physiology*, edited by C. J. Lambertsen. Baltimore: Williams and Wilkins, 1967, ch. 38.

PAINTAL, A. S. The mechanism of excitation of type J receptors, and the J reflex. In: *Breathing: Hering-Breuer Centenary Symposium*, edited by R. Porter. London: Churchill, 1970, p. 59–71.

PAPPENHEIMER, J. R. Standardization of definitions and symbols in respiratory physiology. *Fed. Proc.* 9: 602–605, 1950.

PATTLE, R. E. Properties, function and origin of the alveolar lining layer. *Nature* 175: 1125–1126, 1955.

PEDLEY, T. J., R. C. SCHROTER, AND M. F. SUDLOW. The prediction of pressure drop and variation of resistance within the human bronchial airways. *Respir. Physiol.* 9: 387–405, 1970.

PEDLEY, T. J., R. C. SCHROTER, AND M. F. SUDLOW. Gas flow and mixing in the airways. In: *Bioengineering Aspects of the Lung*, edited by J. B. West. New York: Marcel Dekker, 1977.

PETERS, J. P., AND D. D. VAN SLYKE. *Quantitative Clinical Chemistry 2, Methods.* Baltimore: Williams and Wilkins, 1932.

PUGH, L. G. C. E., M. B. GILL, S. LAHIRI, J. S. MILLEDGE, M. P. WARD, AND J. B. WEST. Muscular exercise at great altitudes. *J. Appl. Physiol.* 19: 431–440, 1964.

RADFORD, E. P., JR. Method for estimating respiratory surface area of mammalian lungs from their physical characteristics. *Proc. Soc. Exp. Biol. Med.* 87: 58–61, 1954.

RADFORD, E. P., JR. Recent studies of mechanical properties of mammalian lungs. In: *Tissue Elasticity*, edited by J. W. Remington, Washington, D.C.: American Physiological Society, 1957, p. 177–190.

RAHN, H., AND W. O. FENN. *A Graphical Analysis of the Respiratory Gas Exchange.* Washington, D.C.: American Physiological Society, 1955.

RAHN, H., AND A. B. OTIS. Man's respiratory response during and after acclimatization to high altitude. *Am. J. Physiol.* 157: 445–462, 1949.

RAHN, H., A. B. OTIS, L. E. CHADWICK, AND W. O. FENN. The pressure-volume diagram of the thorax and lung. *Am. J. Physiol.* 146: 161–178, 1946.

RAHN, H., AND V. PAGANELLI. Gas exchange in gas gills of diving insects. *Respir. Physiol.* 5: 145–164, 1968.

READ, D. J. C. A clinical method for assessing the ventilatory response to carbon dioxide. *Australas. Ann. Med.* 16: 20–32, 1967.

RILEY, R. L., AND A. COURNAND. "Ideal" alveolar air and the analysis of ventilation-perfusion relationships in the lungs. *J. Appl. Physiol.* 1: 825–847, 1949.

ROUGHTON, F. J. W. Transport of oxygen and carbon dioxide. In: *Handbook of Physiology*, sect. 3, edited by W. O. Fenn and H. Rahn. Washington, D.C.: American Physiological Society, 1964, vol. 1, p. 767.

ROUGHTON, F. J. W., AND R. E. FORSTER. Relative importance of diffusion and chemical reaction rates in determining rate of exchange of gases in the human lung, with special reference to true diffusing capacity of pulmonary membrane and volume of blood in the lung capillaries. *J. Appl. Physiol.* 11: 290–302, 1957.

SCHAFFER, T. H., D. RUBENSTEIN, G. D. MOSKOWITZ, AND M. DELIVORIA-PAPADOPOULOS. Gaseous exchange and acid balance in premature lambs during liquid ventilation since birth. *Pediatr. Res.* 10: 227–231, 1976.

SCHOLANDER, P. F. Analyzer for accurate estimation of respiratory gases in one half cubic centimeter samples. *J. Biol. Chem.* 167: 235–250, 1947.

TASK GROUP ON LUNG DYNAMICS. Deposition and retention models for internal dosimetry of the human respiratory tract. *Health Physics* 12: 173–207, 1966.

VON NEERGAARD, K. Neue Auffassungen über einen Grundbegriff der Atemmechanik: Die Retraktionskraft der Lunge, abhängig von der Oberflächenspannung in den Alveolen. *Zeit. ges. exper. Med.* 66: 373–394, 1929.*

WAGNER, P. D. Diffusion and chemical reaction in pulmonary gas exchange. *Physiol. Rev.* 57: 257–312, 1977.

WAGNER, P. D., R. B. LARAVUSO, R. R. UHL, AND J. B. WEST. Continuous distributions of ventilation-perfusion ratios in normal subjects breathing air and 100% O_2. *J. Clin. Invest.* 54: 54–68, 1974.

WAGNER, P. D., AND J. B. WEST. Effects of diffusion impairment on O_2 and CO_2 time course in pulmonary capillaries. *J. Appl. Physiol.* 33: 62–71, 1972.

WEIBEL, E. R. *Morphometry of the Human Lung.* New York: Academic Press, 1963.

WEIBEL, E. R. Morphological basis of alveolar-capillary gas exchange. *Physiol. Rev.* 53: 419–495, 1973.

WEIBEL, E. R., AND J. GIL. Structure-function relationships at the alveolar level. In: *Bioengineering Aspects of the Lung*, edited by J. B. West. New York: Marcel Dekker, 1977.

WEST, J. B. Blood-flow, ventilation, and gas exchange in the lung. *Lancet* 2: 1055–1058, 1963.

WEST, J. B. *Pulmonary Pathophysiology: The Essentials*, 1st ed. Baltimore: Williams & Wilkins, 1977a.

WEST, J. B. *Pulmonary Pathophysiology: The Essentials*, 2nd ed. Baltimore: Williams and Wilkins, 1982.

WEST, J. B. *Ventilation/Bloodflow and Gas Exchange*, 3rd ed. Oxford: Blackwell, 1977b.

WEST, J. B., S. LAHIRI, M. B. GILL, J. S. MILLEDGE, L. G. C. E. PUGH, AND M. P. WARD. Arterial oxygen saturation during exercise at high altitude. *J. Appl. Physiol.* 17: 617–621, 1962.

WEST, J. B., AND P. D. WAGNER. Pulmonary gas exchange. In: *Bioengineering Aspects of the Lung*, edited by J. B. West. New York: Marcel Dekker, 1977.

WEST, J. B., P. H. HACKETT, K. H. MARET, J. S. MIL-LEDGE, R. M. PETERS, JR., C. H. PIZZO, AND R. M. WINSLOW. Pulmonary gas exchange on the summit of Mt. Everest. *J. Appl. Physiol.: Respirat. Environ. Exercise Physiol.* 55:678–687, 1983.

WHITELAW, W. A., J-P DERENNE, AND J. MILIC-EMILI. Occlusion pressure as a measure of respiratory center output in conscious man. *Resp. Physiol.* 23: 181–199, 1975.

WOOLCOCK, A. J., N. J. VINCENT, AND P. T. MACKLEM. Frequency dependence of compliance as a test for obstruction in the small airways. *J. Clin. Invest.* 1097–1106, 1969.

Gastrointestinal System

CHAPTER 41

Overview of Gastrointestinal Function

OVERVIEW OF GASTROINTESTINAL FUNCTION

The gastrointestinal system consists of the alimentary canal from the mouth to the anus and those organs which empty their contents into the canal; the salivary glands, biliary tract, and pancreas. The liver parenchymal cells produce the primary secretion of bile, which may be modified by absorption or secretion in the bile ducts or gallbladder. Other functions of the liver, such as protein synthesis, storage or release of sugars and fatty acids, and synthesis of urea, are traditionally considered under intermediary metabolism. Similarly, the islets of Langerhans in the pancreas produce insulin, but the endocrine (as opposed to exocrine) function of pancreas is usually discussed in the physiology of the endocrine system.

The gastrointestinal system, while capable of performing its functions in vitro, is subject to control by both extrinsic nerves and endocrine glands. Therefore, some attention must be given to nerves connecting the system to the brain (the vagi and sympathetic nerves) and the organization of those parts of the brain exerting control of digestive function (the visceral nervous system, or limbic system, including the phylogenetically older portions of the cerebral cortex).

The endocrine glands also influence digestion. The anterior pituitary exerts trophic effects on the gut, adrenal corticosteroids influence the transport of water and electrolyte across the intestinal mucosa, and thyroid hormones modify transit of intestinal content through the gut.

The gastrointestinal tract is also an important part of the body's immune system. Both humoral antibodies and the cellular immune system involving lymphocytes in the Peyer's patches of the intestine play an important role in protecting the body against microorganisms in the lumen of the gut and foreign proteins capable of acting as antigens.

The contents of the digestive tract during interdigestive periods serve as a kind of body fluid compartment, not truly within the body, yet to some extent in equilibrium with other body fluids and available for the exchange of water and electrolytes.

Lastly, the gastrointestinal system, like all those in the body, is dependent upon the circulatory system for gas exchange and nutrients. Changes in activity of cellular processes in the alimentary canal are usually accompanied by appropriate changes in blood flow.

The major physiologic processes which occur in the gastrointestinal system are *secretion*, *digestion*, and *absorption*. They are dependent upon the contents of the gut reaching the appropriate site for digestion. The unabsorbed residue is expelled from the anus. These latter functions are accomplished by the contractile and relaxing activity of the smooth muscle of the alimentary canal and its sphincters in the esophagus (upper and lower esophageal sphincters), stomach (pylorus), ileocecal junction, and anus. In addition, the sphincter of Oddi exerts an important control over the secretion of bile.

Secretion is the transport of fluid, electrolyte, peptides, or proteins from cells or tissues into blood, body fluids, or body cavities such as the lumen of the gut. Originally, the term was used to describe movement against concentration gradients, but that is no longer a necessary consideration. Transport between cells (paracellular transport) may be considered as secretion. Excretion refers to a similar process whereby the substances transported are waste products, e.g., bilirubin. The purpose of secretions into the intestine is to modify the contents of the digestive tract by reducing the size of nutrient particles and to solubilize constituents in water. Enzymes in salivary and pancreatic secretions begin the digestion of carbohydrates and proteins in the lumen of the gut while enzymatic digestion is completed by enzymes in the brush border of the small intestinal epithelium. Pancreatic lipase and colipase from the pancreas and bile acids from the liver digest triglycerides and form water-soluble micelles in the gut lumen.

There is about 2500 ml of secretions added daily to the lumen of the alimentary canal. About 36 g of protein is contained in the secretion and about 2 g of sodium. With the exception of saliva, secretions are usually isotonic with respect to plasma.

Digestion, or the breakdown of food constituents into smaller structures in preparation for absorption, can occur in the lumen of the gut or on the brush border of intestinal epithelial cells. It occurs for the most part in the proximal portion of the small intestine. Some of the earliest stages, such as the emulsification of fat, probably occur as a result of repetitive contractions in the distal end of the stomach.

Absorption is the entry of constituents of the content of the gut into the mucosa. It is often the net result of both lumen-to-blood and blood-to-lumen movements. Most of the absorption of sugars, amino acids, and bile acids is sodium dependent and involves a process of cotransport in which sodium-potassium ATPases serve as the energy-producing sources for active transport. The process of lipid absorption appears to involve movement of fatty acids out of micelles and across an unstirred water layer to the absorptive cell surface, where they enter by passive diffusion. Long-chain fatty acids are trapped by fatty-acid-binding proteins and are subsequently reincorporated into triglycerides. The latter are incorporated into chylomicrons along with lipoproteins.

In the course of the normal day, about 6–11 l of fluid enter the alimentary canal. Only about 1–2 l enter the colon, and less than 200 g of feces leave the rectum. These feces normally contain less than 5 g of fatty acids and 1.5 g of nitrogen per day on a diet containing 100 g of protein, 300 g of carbohydrate, and 120 g of fat.

Movement of the intestinal stream depends upon whether the alimentary canal contains recently eaten food or is in an interdigestive period. In the latter case, bursts of depolarization (migrating motor complex (MMC)) move from the esophagus to the ileocecal valve at frequencies of one every 1½–2 h. The depolarizations are coincident with a wave of contractions. The result is a propulsive movement which empties the upper gastrointestinal tract to the cecum. Eating abolishes the migrating motor complex. It is replaced by irregular contractions which have the effect of mixing intestinal content and advancing the intestinal stream toward the colon over short segments. There are two check points: the pylorus and the ileocecal valve, where aboral movement is restricted. Both sphincters normally prevent movement from the distal to the proximal side. The esophageal sphincters have a similar function.

Movement through the colon occurs at a much slower rate. The normal transit time through the alimentary canal is 65 ± 8 h.

These processes permit the gastrointestinal system to accomplish its function of absorbing water and nutrients and at the same time conserving losses in the feces. In addition to these specific functions, the gastrointestinal system has other important roles in the function of the body. Food and water intake is regulated in such a way that the body weight remains within fixed normal limits in middle adult life. Some of the signals which lead to hunger, thirst, or satiety originate in the alimentary canal. The epithelial lining of the stomach serves as a protective barrier against injury from acids, digestive enzymes, micro-organisms, and injurious substances in food and drink. Acid secretion in the stomach reduces the numbers of bacteria entering from the oral cavity and the ileocecal valve keeps colonic bacteria out of the small intestine. The migrating motor complex may be a contributing factor to maintaining relative sterility in the small intestine.

Individuals can now survive for indefinite periods of time on intravenous feedings, provided that the necessary nutrients, including vitamins and trace elements, are supplied. Therefore, the absorptive function of the intestine is not essential for life. Removal of the pancreas can be tolerated if insulin is given and substitution treatment with pancreatic enzymes is supplied. Removal of the liver, however, is fatal, unless a new liver is transplanted, due to its role in metabolic functions.

CHAPTER 42

Control of the Gastrointestinal System

Isolated cells can now be prepared from secretory, absorptive, and motor (muscular) tissues of the gastrointestinal system. It is clear that each of them can function in the absence of nerves or hormones. However, the rate of secretion and absorption can be modified, and muscle cells can be induced to shorten in response to hormones or neurotransmitters. In general, clusters or clumps of secretory cells seem to be more responsive to stimulation by hormones than dispersed individual cells.

Hormones or neurotransmitters bind to receptor sites on the surface of individual cells. This interaction leads to release of a second messenger which initiates intracellular events. In secretory cells the second messenger may involve the activation of cyclic AMP by adenylcyclase, as in the case of histamine-2 agonists and parietal cells, or secretin and pancreatic acinar cells. Another system involves the intracellular release of calcium, as in the case of cholecystokinin, or acetylcholine and pancreatic acinar cells, and gastrin and acetylcoholine and gastric parietal cells. Activation of protein kinases by phosphorylation then leads to the final stages of the secretory process.

Bitar and his associates (1982) showed that receptors for acetylcholine, gastrin, enkephalins, and vasoactive intestinal polypeptide (VIP) exist on isolated smooth muscle cells from the human gastric antrum. Shortening was induced by the first three. Activation of cyclic AMP occurred in the contracting cells.

Transport in isolated intestinal mucosa can be studied by mounting tissue in Ussing chambers so that ionic movements and potential changes can be followed simultaneously. In general α-adrenergic agonists lead to increases in the absorption of sodium chloride, while acetylcholine induced secretion. Vasoactive intestinal peptide also stimulated secretion.

The role of intrinsic nerves can be studied by using segments of intestine or isolated organs such as the stomach, esophagus, or common bile duct in vitro. Electrical field stimulation at the appropriate parameters will act only via nerves. Such studies with the use of blocking agents can identify neurotransmitters and differentiate between contractions of longitudinal and circular smooth muscle.

In the esophagus, field stimulation leads to contractions in the body of the esophagus and relaxation of the lower esophageal sphincter.

Figure 6.1 shows a diagrammatic representation of the enteric nervous system. In the stomach the submucosal plexus is only sparsely represented.

The isolated intestine is capable of coordinated movements as shown by its ability to expel a pellet in response to distention (the peristaltic reflex). Isolated perfused intestinal segments or everted gut sacs show transport of glucose or L-amino acids from mucosa to serosa against a concentration gradient.

Cholecystokinin-gastrin peptides (CCK), vasoactive intestinal polypeptide, somatostatin, substance P, and enkephalins have all been demonstrated in nerves in the gastrointestinal tract. They are therefore neuropeptides. Most are distributed throughout the intestine. Vasoactive intestinal polypeptide is present in greater density in sphincters (Alumets et al., 1979). At present, it is very difficult to assess the functional significance of intestinal neuropeptides. Substance P is present in nerves connecting the intestine to prevertebral ganglia and may serve a sensory function in other enteric neurons.

The gastrointestinal tract contains many endocrine cells scattered among other epithelial cells on the intestinal mucosa. Under electron microscopy secretory granules of various sizes and density can be seen between the nucleus and the basal pole of the cell. Immunocytochemistry shows that the granules may contain immunoreactive gastrin,

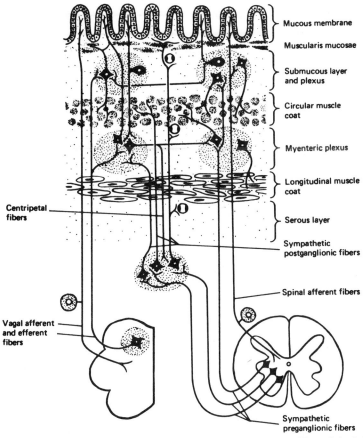

Figure 6.1. Extrinsic and intrinsic nerves in the wall of the digestive tract. [From Schofield (1968).]

CCK, secretin, serotonin, motilin, and neurotensin. Gastrin is contained in cells in the gastric antrum and proximal duodenum. Secretin and CCK-like immunoreactivity is found in cells of the duodenum and jejunum. Motilin-containing cells are found mostly in the upper small intestine.

Levels of immunoreactive gastrin, CCK, secretin, and neurotensin rise after a meal. Exogenous hormones injected in amounts sufficient to produce similar blood levels act to increase acid secretion, pancreatic enzyme secretion, and pancreatic bicarbonate secretion, respectively, and to inhibit gastric secretion. Motilin concentrations rise with the onset of the gastric migrating motor complex in the fasting state.

Pearse proposed that gut endocrine cells originated from the neural crest. He also considered them to be APUD cells (amine precursor uptake and decarboxylation). Roth has reported that chemical messengers are present in unicellular organisms and suggests that gut peptides are just another example of cell-to-cell messengers present from the early stages of development (Pearse, 1982; Roth et al., 1982a; Roth et al., 1982b).

The significance of enterochromaffin cells in con-

trol of the gastrointestinal system is still uncertain. Gastrin-containing cells may also contain an ACTH-like peptide.

Bombesin is a peptide from frog skin which releases gastrin into the blood stream. A similar peptide known as a gastrin-releasing peptide has been found in mammals. It is present in the stomach and may be released from neurons in response to vagal stimulation. There is a cholinergically mediated release of gastrin from the antrum, but vagally stimulated gastrin release is even greater after the anticholinergic substance atropine. A bombesin-like peptide may be the excitatory neurotransmitter.

In general gut peptides are present in blood and tissue in forms of varying molecular weights. The larger molecular forms are thought to be precursors and are usually less potent (Irvine and Murphy, 1981). There are two major groups of gut hormones: gastrin and CCK, as well as secretin, VIP, glucagon, and gastric inhibitory peptide (GIP). The first share a common C-terminal tetrapeptide; the latter have similar amino acids in various portions of the molecules. Table 6.1 shows the structure of some of the peptides.

Table 6.1
Amino acid structure of GI peptides

	CCK	CCK-33	CCK-OP	Gastrin G-34	Gastrin G-17	Gastrin G-14	Secretin	VIP	Motilin	
39	TYR									
38	ILE									
37	GLN									
36	GLN									
35	ALA									
34	ARG			GLP						
33	LYS	LYS		LEU						
32	ALA	ALA		GLY						
31	PRO	PRO		PRO						
30	SER	SER		GLN						
29	GLY	GLY		GLN						
28	ARC	ARC		HIS						
27	VAL	VAL		PRO			HIS	HIS	PHE	1
26	SER	SER		SER			SER	SER	VAL	2
25	MET	MET		LEU			ASP	ASP	PRO	3
24	ILE	ILE		VAL			GLY	ALA	ILE	4
23	LYS	LYS		ALA			THR	VAL	PHE	5
22	ASN	ASN		ASP			PHE	PHE	THR	6
21	LEU	LEU		PRO			THR	THR	TYR	7
20	GLN	GLN		SER			SER	ASP	GLY	8
19	SER	SER		LYS			GLU	ASN	GLU	9
18	LEU	LEU		LYS			LEU	TYR	LEU	10
17	ASP	ASP		GLN	GLP		SER	THR	GLN	11
16	PRO	PRO		GLY	GLY		ARG	ARG	ARG	12
15	SER	SER		PRO	PRO		LEU	LEU	MET	13
14	HIS	HIS		TRP	TRP	TRP	ARG	ARG	GLN	14
13	ARG	ARG		LEU	LEU	LEU	ASP	LYS	GLU	15
12	ILE	ILE		GLU	GLU	GLU	SER	GLN	LYS	16
11	SER	SER		GLU	GLU	GLU	ALA	MET	GLU	17
10	ASP	ASP	ASP	GLU	GLU	GLU	ARG	ALA	ARG	18
9	ARG	ARG	ARG	GLU	GLU	GLU	LEU	VAL	ASN	19
8	ASP	ASP	ASP	GLU	GLU	GLU	GLN	LYS	LYS	20
7	TYR-SOH	TYR-SOH	TYR-SOH	ALA	ALA	ALA	ARG	LYS	GLY	21
6	MET	MET	MET	TYR-R	TYR-R	TYR-R	LEU	TYR	GLN	22
5	GLY	GLY	GLY	GLY	GLY	GLY	LEU	LEU		23
4	TRP	TRP	TRP	TRP	TRP	TRP	GLN	ASN		24
3	MET	MET	MET	MET	MET	MET	GLY	SER		25
2	ASP	ASP	ASP	ASP	ASP	ASP	LEU	ILE		26
1	PHE-NH$_2$	PHE-NH$_2$	PHE-NH$_2$	PHE-NH$_2$	PHE-NH$_2$	PHE-NH$_2$	VAL	LEU		27
							NH$_2$	ASN		28
								NH$_2$		
Residues	39	33	8	34	17	14	27	28	22	
Mol Wt	3918	1143	1352	3919	2173	1913	3056	3381	2700	

A number of the gut neuropeptides have also been found in the brain. They include CCK, VIP, substance P, and secretin. While this may be coincidental, the fact that in the brain they are present in highest concentrations in the limbic system suggests that functional relationships may be present.

Gastrin is the most potent stimulant of gastric acid secretion known. It also exerts a trophic effect on the gastric mucosa. It is released by vagal stimulation, by distention of the body of the stomach, by the topical application of acetylcholine to antral mucosa, by bombesin, intravenously, by peptides in the stomach, and by a meal, especially one rich in proteins. Its release is inhibited by acidification of the antrum and by somatostatin. High blood values are found in patients unable to secrete acid.

Cholecystokinin-pancreozymin stimulates the secretion of pancreatic enzymes, contracts the gallbladder, and relaxes the sphincter of Oddi. It exerts a trophic effect on the pancreas. It is probably released by amino acids, especially tyrosine and phenylalanine in the duodenum. Sensitive specific radioimmunoassays for CCK 33, CCK 39, and CCK-OP have yet to be developed, so that the response to a meal or other secretagogues is controversial. The gastrin-CCK family of peptides are the oldest phylogenetically (Rehfeld, 1981).

Secretin stimulates secretion of bicarbonate by

the ducts of the pancreas and by the bile ducts. It is released by acid in the duodenum. After a meal it rises intermittently as the duodenal pH falls below 3.5.

Serotonin stimulates intestinal propulsive activity and the secretion of water and electrolytes. Its physiologic role is still uncertain.

Somatostatin is an inhibitor of gastrin release, of gastric acid and pepsin secretion, of pancreatic secretion, of fat absorption, and of gastric contractions and emptying. It has been identified by immunocytochemistry in cells adjacent to gastrin-containing cells where the cells appear to make contact through elongated processes of the somatostatin cells. This makes somatostatin a candidate for acting through a paracrine action as well as a classic hormone.

Motilin is a candidate to initiate migrating motor complexes in the stomach. MMCs in the duodenum may appear in the absence of elevation of plasma motilin. Its release ceases with feeding.

Neurotensin inhibits gastric secretion and contractions. It is a candidate to mediate the inhibitory effects of fat in the duodenum.

The extrinsic nerves supplying the gastrointestinal system are shown in Figure 6.2. Sensory fibers constitute the great majority of fibers in the abdominal vagus. They are mostly unmyelinated. Sensory fibers in the sympathetics arise from receptors near blood vessels in the mesentery. Mechanoreceptors, chemoreceptors, and osmoreceptors have all been demonstrated in vagal sensory endings. Sympathetic nerves carry mechanoreceptors and probably nerves responding to painful stimuli. There are fibers connecting the enteric nervous system to prevertebral ganglia containing substance P.

The vagus innervates the enteric nervous system from the esophagus to the midtransverse colon. Electrical stimulation of the distal crushed, or cut, ends of the vagus leads to major changes in exocrine and endocrine secretions, intestinal transport, and movements of the alimentary canal. Since none of the efferent vagal fibers reach the effector cells, the wide variety of actions depends upon the neurons in the enteric nervous system with which the vagal fibers form synaptic connections. Acetylcholine is the presumed neurotransmitter released from vagal efferents. However, neurons in the enteric plexuses

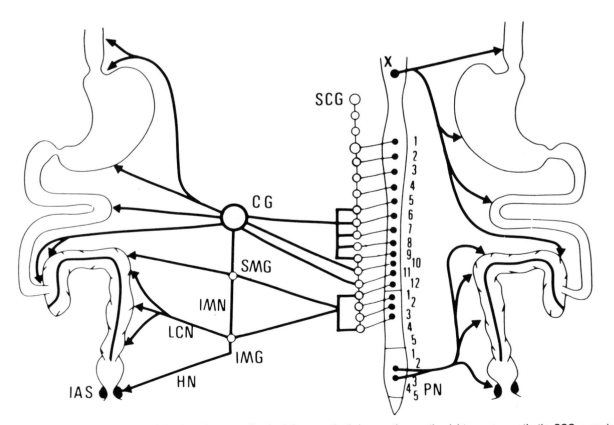

Figure 6.2. Extrinsic innervation of the digestive tract. On the *left* sympathetic innervation, on the right, parasympathetic. *SCG*, superior cervical ganglion; *CG*, celiac ganglion; *SMG*, superior mesenteric ganglion; *IMG*, inferior mesenteric ganglion; *IMN*, intermesenteric nerve; *LCN*, lumar colonic nerves; *HN*, hypogastric nerves; *X*, dorsal motor nucleus of the vagus; *PN*, pelvic nerves; *IAS*, internal anal sphincter. [From Roman and Gonella (1981).]

may release acetylcholine, adenine nucleotides (purinergic nerves), VIP (peptidergic nerves), and probably others (Burnstock, 1972; Daniel, 1978).

The depolarization of enteric neurons induced by vagal release of acetylcholine can be reduced by norepinephrine released from sympathetic neurons in synaptic contact with the same enteric neurons (adrenergic inhibitory neurons).

Acetylcholine effects can be blocked at two levels: at the junction between the vagi and the enteric ganglia (the nicotinic action of acetylcholine) or at the junction between the enteric neuron and the effector cell (the muscarinic action of acetylcho-

line). Recent evidence suggests that both muscarinic and nicotinic receptors may be present on the same postganglionic neurons (Goyal and Rattan, 1975). Adrenergic blockers may act either against α- or β-receptor sites. Typical examples are phentolamine and propanolol, respectively. Agonists for the respective adrenergic receptors are norepinephrine and isoprenaline.

The vagus nerve trunks contain immunoreactive CCK octapeptide, VIP, substance P and enkephalins (Lundberg et al., 1979 and Dockray et al., 1981). The first three are present in sensory fibers, the latter in motor (efferent fibers). The peptides have

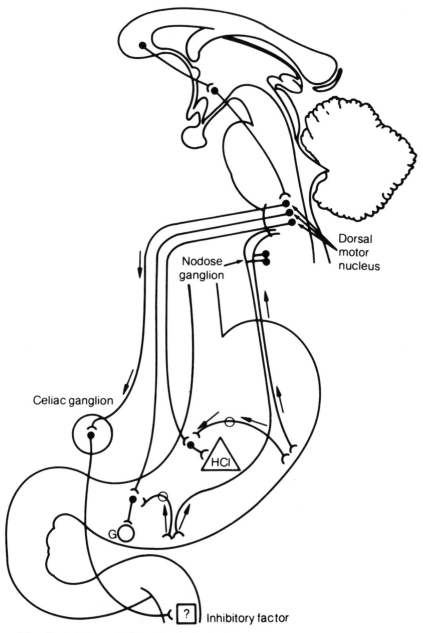

Figure 6.3. Control of vagal efferent by the brain. G, gastrin. [From Brooks and Rosato (1974).]

been shown to migrate along the nerve fibers, but their significance is unknown.

Vagovagal reflexes involve sensory neurons with cell bodies in the nodose ganglion, internuncial neurons in the brain stem, and nerve cells in the dorsal motor neurons of the vagus. Extracellular recordings from internuncial neurons show discharges in synchrony with cyclic inflation and deflation of a balloon in the stomach of anesthetized animals (Burks and Barber, 1982). Electrical stimulation of the dorsal motor nucleus evokes gastric acid secretion and phasic antral contractions similar to those seen in response to stimulation of peripheral vagal trunks, either sensory or motor (Lombardi et al., 1982).

Vagal sensory fibers also are capable of responding to glucose in the intestine or in portal blood. These responses can be relayed to cells in the hypothalamus and globus pallidus. Efferent responses can then excite the vagal nuclei. Cells in the hypothalamus are also sensitive to low levels of glucose in the blood and glucose utilization within the cells. Such connections can account for gastric acid secretion and phasic antral contractions in response to low blood sugars following insulin or 2-deoxy-D-glucose-induced glucocytopenia (Kadekaro et al., 1977). Variations in basal acid output from the stomach during a 12-h cycle correlate inversely with levels of the blood sugar in man (Moore, 1980).

Reflex stimulation via the V, VII, and X, nerves elicits vagus induced acid secretion. Electrical stimulation of the ventromedial regions of the hypothalamus in rats reduced the acid secretory response to insulin hypoglycemia, presumably through an inhibitory action of the lateral hypothalamus (Misher and Brooks, 1967). Ventromedial electrolytic lesions inceased basal acid-pepsin secretion (Ridley and Brooks, 1965).

Cutting the optic or auditory nerves in dogs reduced the acid secretory response to insulin hypoglycemia in dogs (Schapiro et al., 1970, a and b; 1971, 1973). Acute cerebral decortication, but not decerebration, reduced acid secretion and phasic antral contractions to electrical efferent vagal stimulation in cats (Feng et al., 1983).

These observations suggest a series of levels of organization within the brain, with the vagal nuclei as the final common pathway and with a second integration point in the hypothalamus.

Figure 6.3 illustrates a proposed integration of central control mechanisms for the nervous control of gastric function.

Much less is known about the control of the gastrointestinal system by the sympathetic nervous system. Vagosympathetic pathways to the stomach have been described and stimulation of the splanchnic nerves under selected circumstances can inhibit secretion and motor activity (Reed et al., 1971a). Figure 6.2 shows a diagram of the sympathetic innervation of the gut. Most importantly from a clinical point of view is that sensory neurons traveling with sympathetic nerves mediate visceral pain.

Detailed information on the role of hormones and nerves in the control of gastrointestinal system will be found under the specific processes affected.

CHAPTER 43

Movements and Muscular Activity in the Gastrointestinal System

It is matter of personal choice as to which of the major processes in the gastrointestinal system should be dealt with first. At present the role of nerves is better understood in the control of movements than of secretion or absorption. This is partly because movements can be recorded on-line simultaneously with nerve impulses. Hormones, on the other hand, have a much less developed role in the control of motor activity. In any case much of what is applicable to one process will be relevant to the others.

MASTICATION

The first step in the digestive process is the mastication or chewing of food (Hightower and Janowitz, 1979). In addition to breaking down food particles into smaller ones, it mixes food with saliva and allows salivary amylase to penetrate into the moistened mass of food, now known as a food bolus.

The mandible moves in all planes and also rotates. The force of a bite is legendary. In man molars can exert a force greater than 122 kg. Most chewing is in response to reflex stimulation. The impairment of chewing in edentulous patients is probably more important in producing symptoms of "indigestion" than many physicians appreciate.

PHYSIOLOGY OF THE ESOPHAGUS

Swallowing: Phases

Swallowing is traditionally divided into three phases: the buccal, pharyngeal and esophageal stages. The buccal phase is accomplished by elevating the front of the tongue against the surface of hard palate. The tongue is retracted and depressed, displacing the food bolus into the pharynx. Mastication and respiration influence pressures in the oral cavity. Negative pressures as low as -300 cm of water can be recorded near the incisor teeth while positive pressures as high as 100 cm of water can be found in the posterior part of the oral cavity.

The initiation of swallowing is usually subject to voluntary control, but it is essentially a reflex action. Saliva is necessary for normal swallowing (Brooks, 1970).

The second or pharyngeal phase occurs when the food bolus comes in contact with the pillars of the fauces, tonsils, soft palate, base of the tongue, and posterior wall of the pharynx. This initiates an involuntary reflex accomplished in less than 1 second. The pharyngeal muscles initiate a propulsive contraction over the pharynx. As a result, the food bolus is propelled toward the esophagus. At the same time, the soft palate presses against the posterior pharyngeal wall to seal off the nasopharynx, and the larynx is closed off by approximation of the vocal cords. The epiglottis is deflected horizontally and the larynx is carried forward and upward under the base of the tongue. These movements prevent food from entering the airway. Figure 6.4 shows the movements in the first and second phases of swallowing, as seen in selected frames of a lateral cine radiographic study (Davenport, 1977 adapted from Rushmer and Hendron, 1951).

The third or esophageal phase involves the relaxation of the upper esophageal sphincter and the initiation of a primary peristaltic wave which traverses the entire esophagus. The act of swallowing, once the primary esophageal peristaltic wave is initiated, is independent of extrinsic nerves. The complexity of the nervous control of the first and second phases of swallowing is illustrated in Figure 6.5 (Doty, 1968).

Properties of Esophageal Muscle: Responses to Myogenic and Nervous Stimuli

The human esophagus is a hollow muscular tube lined with stratified squamous epithelium. The muscular layers consist of an inner circular and an outer longitudinal layer. The muscle of the proximal third of the esophagus is striated and under the control of somatic nerves. The distal two thirds consist of smooth muscle. The upper esophageal sphincter coincides with the cricopharyngeal muscle. The lower esophageal sphincter has no anatomic counterpart (Liebermann-Meffert et al., 1979). Figure 6.6 shows these relationships.

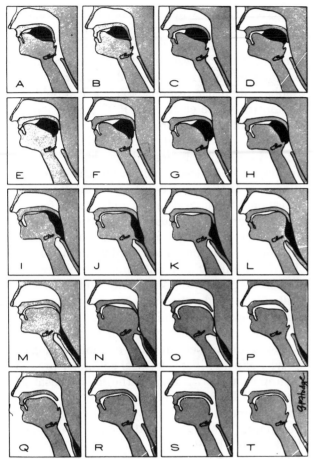

Figure 6.4. Cineradiographic frames swallowing in man. [From Davenport (1977).] Adapted from Rushmer and Hendron, 1951.

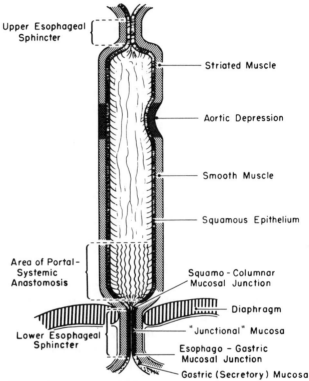

Figure 6.6. Anatomy of the esophagus and gastroesophageal junction. [From Cohen and Harris (1976).]

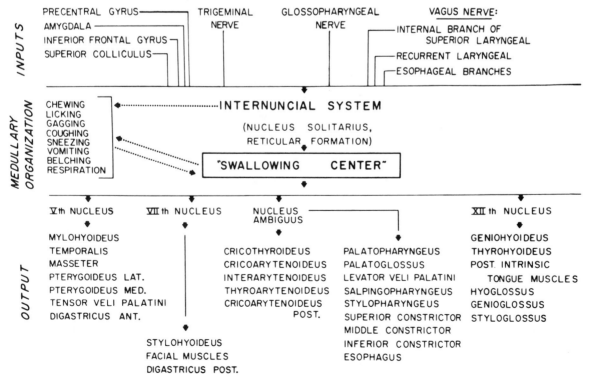

Figure 6.5. Nervous control of muscular events in swallowing. [From Doty (1968).]

Circular and longitudinal esophageal smooth muscle have differing electrophysiologic properties. The lower esophageal sphincter muscle has unique properties compared to muscle immediately above the sphincter or in the adjacent stomach (McKirdy et al., 1981). Strips of esophageal muscle mounted in vitro can be stimulated electrically (field stimulation) (Christensen and Lund, 1969; Schulze-Delrieu and Lepsien, 1982). The results are shown in Figure 6.7. The circular muscle of the lower esophageal sphincter relaxed. Longitudinal muscle strips from the body of the esophagus contracted as long as the stimulation was applied (duration response). Strips of circular muscle from the body of the esophagus contracted only when the stimulus was turned off (off response). Extrapolating these observations to the intact esophagus, it is a reasonable hypothesis to suggest that the lower esophageal sphincter has a high level of resting contractility (tone). In response to the nervous initiation of swallowing, contraction is transiently inhibited. The longitudinal muscle of the body of the esophagus contracts in response to reflex excitation. The circular muscle contracts as the excitatory stimulation ends. If stimulation begins in the proximal esophagus and spreads distally, the sequential spread of the off response could account for the primary peristaltic wave.

The response to field stimulation can be blocked

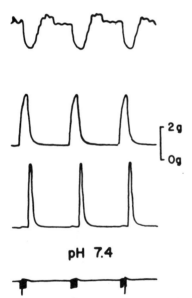

pH 7.4

Figure 6.7. Responses of isolated muscle strips from opossum esophagus to electrical field stimulation. *Top record,* Relaxation of muscle from the lower esophageal sphincter. *Middle record,* contraction during stimulation of longitudinal muscle from the body of the esophagus ("duration" response). *Lower record,* contraction of circular muscle of the body of esophagus ("off" response). *Bottom line,* stimulus signal. [From Schulze-Delrieu and Lepsien (1982).]

by tetrodotoxin, which blocks sodium channels. However, basal lower esophageal sphincter tone is unaffected (Goyal and Rattan, 1976). On the other hand, contraction results in the opening of calcium channels in smooth muscle. Calcium is required for esophageal muscle response to electrical field stimulation (de Carle et al., 1977). The contraction of the body of the human esophagus, including the action of longitudinal and circular muscle, is blocked by anticholinergic drugs. Hence, the neurotransmitter is acetylcholine. On the other hand relaxation of the lower esophageal sphincter (LES) in response to nerve stimulation is little affected by anticholinergics. The neurotransmitter is unknown but is not norepinephrine. Vasoactive intestinal polypeptide is a candidate, based upon the presence of a dense innervation with immunoreactive VIP, as seen by immunocytochemical studies (Uddman et al., 1978), and prevention of relaxation by anti-VIP antiserum (Biancani et al., 1982). A component of cholinergic excitatory control exists, as shown by the ability of bethanechol, a stable analogue of acetylcholine, to increase basal sphincter pressures (Roling et al., 1972) and by the fall in basal LES pressure (LESP) induced by anticholinergic drugs. Enkephalinergic nerves may inhibit contraction by presynaptic blockade of adrenergic transmission (Uddman et al., 1980a).

The role of hormones is largely confined to their part in maintaining resting lower esophageal sphincter pressure. A great variety of gastrointestinal peptides either increase or decrease LESP. None have been shown to be a dominant factor under physiological conditions. Perhaps of more importance is that protein-containing meals increase resting LESP while meals high in fat reduce it (Castell, 1975). Perhaps it is the resultant of many hormones released synchronously which will prove to be of central importance (Snape and Cohen, 1976). Female sex hormone changes may account for the decrease in lower esophageal sphincter pressure in the last trimester of pregnancy (Fisher et al., 1978).

Methods of Study of Esophageal Motility

The ability to measure pressures within the esophagus by perfused catheters resulted in an explosion of information about esophageal motility. Figure 6.8 shows a typical assembly. Three catheter tips are mounted together 5 cm apart. Figure 6.9 shows a primary peristaltic wave after swallowing, recorded at six sites along the esophagus, and the relaxation in sequence of the upper and lower esophageal sphincters. More commonly, the lower esophageal sphincter pressure is determined by

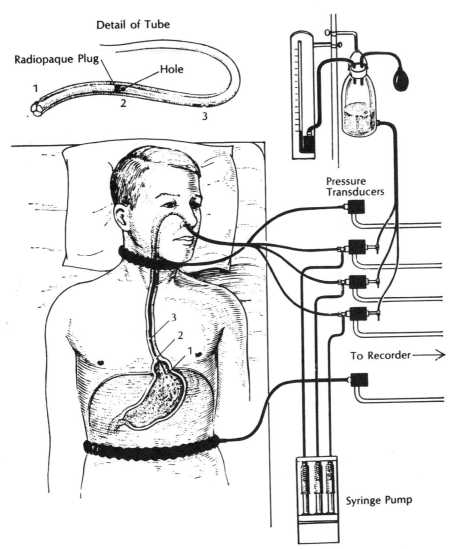

Figure 6.8. Diagram of the method of intraluminal esophageal manometry in man. Details of manometry tube shown at *upper left*. Small plastic tubes are continuously perfused with saline at a low rate. Bellows around neck registers swallowing; another around waist registers breathing. [From Harris and Jackson (1968).]

slowly pulling a perfused catheter tip from the stomach into the esophagus. Figure 6.10 shows a high pressure zone extending over several centimeters in length. This is the lower esophageal sphincter. Note the changes in pressure with respiration and the reversal of direction with inspiration and expiration as the catheter tip enters the thoracic cavity.

Esophageal Manometry

Several characteristics of the lower esophageal sphincter deserve emphasis. When a subject swallows, the LES relaxes completely so that there is no pressure gradient between the esophagus and the stomach as shown in Figure 6.11. Secondly, the

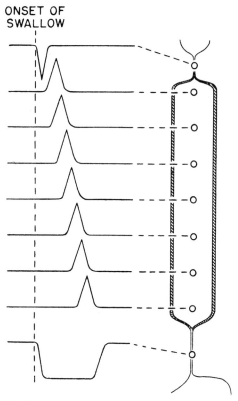

Figure 6.9.   Sequence in time of pressure changes in esophageal sphincters and body during swallowing. [From Cohen and Harris (1976).]

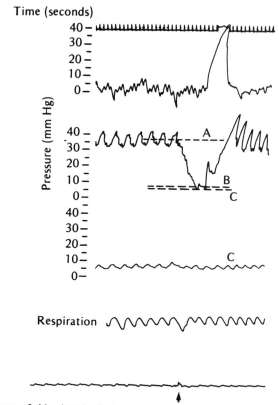

Figure 6.11.   Intraluminal pressures recorded in the human esophagus 5 cm above the lower esophageal sphincter, (*top record*). Pressure in the lower esophageal sphincter (*second record*). *A*, Resting level; *B*, Level in response to swallowing; *C*, pressure in stomach. Intragastric pressure (third record). Respiration (*fourth record* and *bottom record*). [From Cohen (1978).]

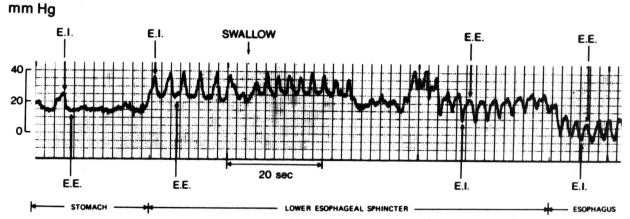

Figure 6.10.   Recording of intraluminal pressures during pull through of perfused tube from stomach to esophagus in a human subject. Note direction of end inspiratory (*EI*) and end expiratory pressure (*EE*), changes, and their reversal at the diaphragm. [From Hurwitz et al. (1979).]

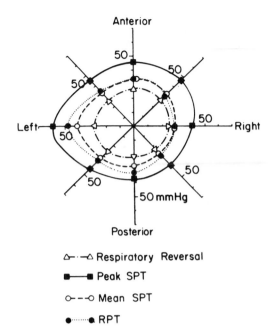

Figure 6.12. Radial determination of intraluminal pressure in normal human subjects in the lower esophageal sphincter. *RPT*, Rapid pull through; *SPT*, Slow pull through. Highest peak pressure, mean pressure, and pressure point of respiratory reversal shown. Note asymmetry of sphincter. [From Welch and Drake (1980).]

LES is not a perfect sphere. Figure 6.12 shows recordings made with radically oriented sensors at the same level. The pressure is relatively higher anteriorly and to the left (Welch and Drake, 1980). The upper esophageal sphincter is also asymmetrical (Winans, 1972; Gerhardt et al., 1978).

Recent studies indicate that the LES periodically relaxes inappropriately in healthy subjects particularly after meals (Dent et al., 1980). Studies also indicate that the migrating myoelectric complex (MMC) begins in the lower esophageal sphincter on most occasions (Itoh et al., 1978b). Recently, electrical spiking activity has been recorded from the LES which correlated with cholinergically stimulated contractions (Reynolds et al., 1982).

The primary peristaltic wave is initiated by swallowing. However, a similar wave can be elicited by inflating a balloon in the esophagus without swallowing (secondary peristalsis). The central nervous pathways of the primary peristaltic wave involve the medullary swallowing center. If the secondary peristaltic wave traverses a sufficient length of the esophagus, it will continue over the remainder of the esophagus in the absence of extrinsic nerves. Local nervous or muscular factors appear to be sufficient to sustain it (Diamant and El-Sharkawy, 1977; Sarna et al., 1977).

Pathophysiology of the Esophagus

The most common disorder of the esophagus is reflux esophagitis. In this condition the lower esophageal sphincter fails to maintain a barrier between the stomach and the esophagus. Acid gastric juice flows into the esophagus, particularly when the patient is lying down or when intragastric pressure is elevated by a large meal, heavy lifting, pregnancy, or constricting clothing. Acid reflux sets up an inflammatory reaction in the lower esophageal mucosa and leads to the symptom of heartburn. In severe cases, ulceration, bleeding and a fibrous stricture can result. Most patients respond to simple medical measures such as elevating the head of the bed and suppressing gastric acidity by neutralizing agents or antisecretory drugs. A few patients will require reconstructive surgery by wrapping the stomach around the lower end of the esophagus.

Other esophageal diseases include achalasia, cancer of the esophagus, diffuse esophageal spasm, and scleroderma. In achalasia the lower esophageal sphincter fails to relax with swallowing, and primary peristalsis in the body of the esophagus is replaced by low amplitude nonpropulsive contractions. The result is a functional obstruction of the lower end of the esophagus with dilatation of the esophagus, difficulty in swallowing (dysphagia) and, in some cases, aspiration into the lungs and pneumonia. Pathologically there is a disappearance of ganglion cells in the myenteric plexus. Treatment involves stretching the LES with an air-filled balloon or incising the muscular wall down to the mucosa at surgery (Heller operation).

Esophageal cancers also obstruct the esophagus and produce dysphagia, particularly for solids. Diffuse esophageal spasm is characterized by pain beneath the sternum on swallowing (odynophagia). It is characterized by large amplitude nonpropulsive contractions in the body of the esophagus. Some patients get relief with anticholinergic drugs. Calcium channel-blocking drugs are under intensive study. Scleroderma begins as a nerve degeneration in the esophagus and progresses to a replacement of smooth muscle with fibrous tissue. Patients may complain of heartburn and/or dysphagia. The LES pressure is low, and primary peristalsis is absent. There is no satisfactory treatment for the long-term course of the disease.

Diseases of the upper portion of the esophagus supplied by striated muscle include regurgitation of food through the nose and mouth and aspiration into the lungs in patients with cerebrovascular disease. Paralysis of the muscles involved in the oropharyngeal stages of swallowing results from destruction of the cell bodies of nerves supplying the muscles. Less commonly, the upper esophageal sphincter may fail to relax due to unknown reasons. Surgical cutting of the cricopharyngeus muscle may provide relief.

GASTRIC MOTILITY

Receptive Relaxation

The stomach functions as a reservoir and as a pump. The fundus and corpus serves as a reservoir and the antrum as a pump. When swallowing occurs, the fundus and corpus resting pressure falls (Abrahamsson and Jansson, 1969). A similar effect can be obtained by inflating a balloon in the esophagus. This may occur coincident with a rise in the amplitude of phasic contractions in the esophagus. Both vago-vagal and vagosympathetic reflex pathways are involved in this phenomenon, which is known as receptive relaxation (Abrahamsson, 1973). In the vago-vagal pathway the inhibitory neurotransmitter to the smooth muscle is unknown: VIP, adenine nucleotide, and dopamine are all under consideration (Fahrenkrug et al., 1978; Valenzuela, 1976a). For the vagosympathetic pathway, the neurotransmitter is norepinephrine acting to inhibit excitation of the enteric ganglia by acetylcholine from vagus nerves.

Another consequence of this system is that cutting the vagal nerves to the stomach shifts the pressure-volume relationship of the stomach to the right, making it more resistant to distention by intragastric liquids (Stadaas, 1975). Therefore, a small increase in volume can lead to a large rise in intragastric pressure.

Response to a Meal and Interdigestive Migrating Motor Complex

The activity of the stomach changes dramatically when a meal enters the stomach. During the interdigestive period the stomach passes through cycles of peristaltic waves. The individual waves occur at a rate of 3/min, and the cycle lasts about 5 min in man (Hinder and Kelly, 1977).

The burst of contractions then moves on down the digestive tract to the ileocecal valve. This requires approximately 1.5–2.0 h. When the cycle ends at the ileocecal valve, a new burst of contractions occurs in the stomach. The cycle can be broken down into four phases: phase I is characterized by the absence of contractions, phase II by irregular contractions, phase III by a burst of regular large amplitude contractions, and phase IV by a return to quiescence (Code and Marlett, 1975). The effect of this activity is to move the contents

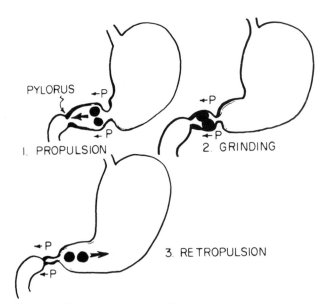

Figure 6.13. Antral pumping. The consequences of proximal and distal phasic antral contractions as seen by cineradiography in dogs with radio-opaque spheres in the stomach. ←P, Antral peristalsis. [From Kelly (1981).]

of the stomach to the ileocecal valve and into the colon.

As soon as food enters the stomach this activity ceases and is replaced by absence of activity in the body of the stomach and low amplitude phasic contractions in the antrum. Contractions in the antrum can either propel material (especially solids) into the duodenum, grind it within the antrum, or propel it retrograde into the body of the stomach (Fig. 6.13). The effect of the grinding activity is to reduce the size of solid food particles and to promote the emulsification of fat.

Cutting the vagal trunks abolishes phasic antral contractions, although after 6 months to a year it returns to normal. Anticholinergic drugs abolish phasic contractions in both the antrum and corpus of the stomach so that acetylcholine is presumed to be the excitatory neurotransmitter.

Properties of the Pyloric Sphincter

The pylorus, in contrast to the LES, is a readily seen and felt muscular mass at the outlet to the stomach. Attempts to demonstrate a high pressure zone by pull-through manometric techniques have yielded conflicting results, but since negative results are often inconclusive, many investigators believe it exists (Winans, 1976). Whether the pylorus behaves similarly to the antrum or has properties of its own is also controversial. Vagal stimulation has diphasic effects on the pylorus in cats with the inhibitory neurotransmitter still not identified (Edin et al., 1979a and b, 1980a). Enkephalins are plentiful in nerves at the pylorus, as determined by immunocytochemistry. Naloxone, an opioid receptor blocking agent, blocks the contractions of the pylorus to vagal stimulation and to enkephalins (Edin et al., 1980b). Substance P is another projected neurotransmitter in exciting contraction of the pylorus (Lidberg et al., 1982). VIP, on the other hand, increases transpyloric flow and may relax the sphincter. In man, insulin hypoglycemia, a powerful vagal stimulant, increases pyloric pressure. After using various adrenergic agonists and blocking drugs, β-adrenergic excitation appears to be the most likely mediator, but some α-agonist activity is also required (Phaosawasdi et al., 1981). Release of secretin or glucagon cannot account for the effects of insulin (Phaosawasdi and Fisher, 1982).

Based upon physical properties of the cat's pylorus, at low levels of stretch, the pylorus acts like a sphincter and prevents duodenogastric reflux, but when stretched to a diameter of 9–14 mm it acts as a fixed structure (Biancani et al., 1980).

Electrophysiological Properties of the Stomach

The foregoing motor activities of the stomach depend upon the electrical activity of smooth muscle. The fundic portion of the stomach is electrically silent, but the remainder of the stomach shows two types of depolarizations: a slow partial depolarization at the rate of 3/min and a series of rapid greater depolarizations superimposed on the others. There is still no consensus for terminology, but here we shall refer to the first as slow waves (pacesetter potentials, electrical control potentials) and the second as spikes. The first is an inherent property of smooth muscle related to changes in permeability of the cell membrane to sodium. It occurs in the absence of contractions. The second is associated with contractions and represents action potentials. Since spikes do not occur in the absence of slow waves, the latter determine the maximal rate of

contractions. However, the percent of slow waves accompanied by spikes is subject to wide variation.

The slow waves originate in smooth muscle cells located high on the greater curvature of the stomach. These cells constitute the pacemaker. From this site, the slow waves sweep down the stomach to the pylorus. Figure 6.14 shows the location of the pacemaker. Figure 6.15 shows an example of the extracellular recording of triphasic gastric slow waves in man. Figure 6.16 shows a recording when spikes were also present and contractions occurred in a dog's stomach. Figure 6.17 shows the sequential distal progression of slow waves in a human stomach. Figure 6.18 shows a similar progression of mechanical contractions in a dog's stomach. During the interdigestive period bursts of migrating myoelectric activity begin in the esophagus and stomach. Figure 6.19 shows the first three phases of such a cycle in the dog.

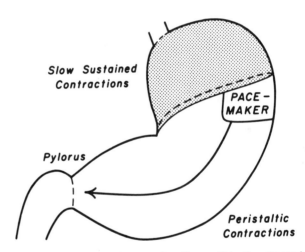

Figure 6.14. Regional motor functions within the stomach. Pacemaker is area of origin of gastric electrical "slow" waves. [From Kelly (1976).]

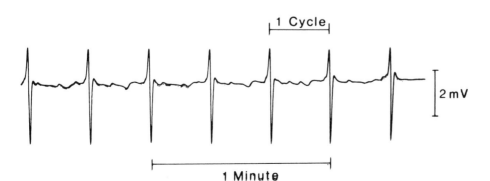

Figure 6.15. Gastric "slow" waves recorded from the human stomach. [From Hinder and Kelly (1977).]

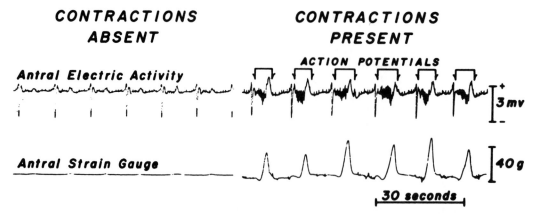

CONTRACTIONS ABSENT

CONTRACTIONS PRESENT

ACTION POTENTIALS

Antral Electric Activity

3 mv

Antral Strain Gauge

40 g

30 seconds

Figure 6.16. Gastric antral myoelectrical activity and phasic contractions in a dog. Note that slow waves are present at all times but that spikes appear only during mechanical contractions. *Spikes*, Antral action potentials. [From Kelly (1976).]

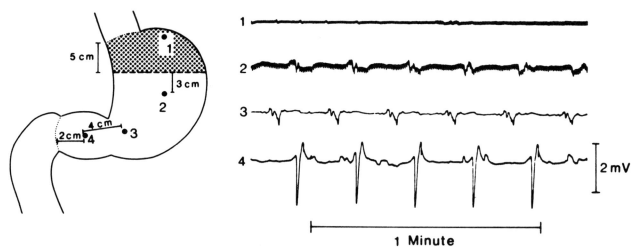

Figure 6.17. Electrical activity from the fundus, body, and antrum of the human stomach; only slow waves are displayed. [From Hinder and Kelly (1977).]

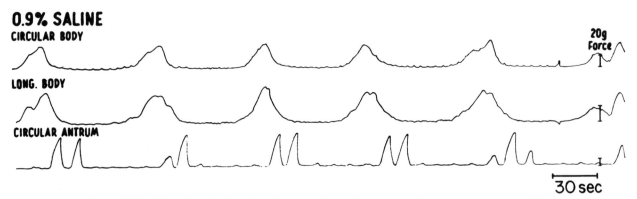

0.9% SALINE

CIRCULAR BODY

20g Force

LONG. BODY

CIRCULAR ANTRUM

30 sec

Figure 6.18. Spread of phasic contractions over the dog's stomach. [From Anderson et al. (1968).]

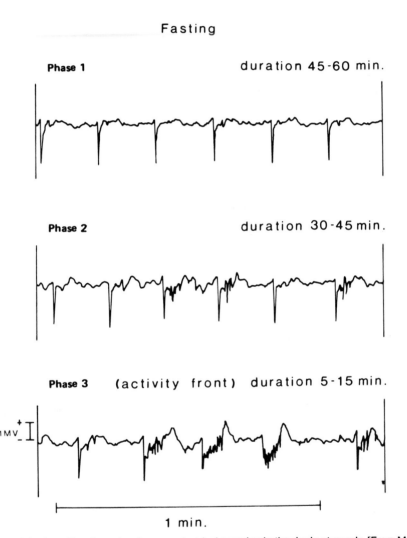

Figure 6.19. Three phases of the interdigestive migrating myoelectrical complex in the dog's stomach. [From Mroz and Kelly (1977).]

Cellular Basis Of Gastric Motility

With microelectrodes it is possible to record intracellular potentials from gastric smooth muscle cells in isolated muscle strips. The form of resting and spontaneous spikes varies, depending upon the location in the stomach. Figure 6.20 shows such recordings in the dog stomach. Figure 6.21 shows similar recordings from the antrum of a human stomach. The relationship of spikes to contractions is shown for the canine corpus in Figure 6.22 and for the antrum in Figure 6.23.

When neurotransmitters and hormones are added to the medium, the shape of the spikes and

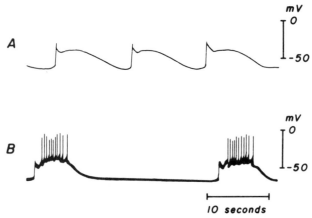

Figure 6.21. Intracellular membrane potentials and action potentials from human antral circular muscle. [From Szurszewski (1981).]

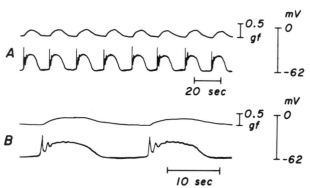

Figure 6.22. Simultaneously recorded spontaneous mechanical and intracellular electrical activity of circular muscle from the middle of the body (corpus) of dog stomach. Note pacemaker potential (slow fall from −62 mV) during relaxation. *Top trace,* Force of contraction. *Lower trace,* intracellular potentials. Each action potential is preceded by pacemaker potential. [From Szurszewski (1981).]

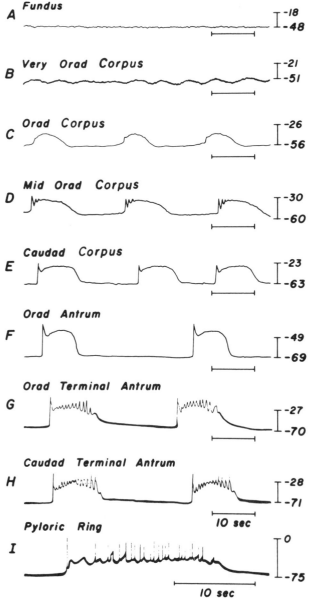

Figure 6.20. Resting intracellular membrane potentials and action potentials from different parts of the dog stomach. [From Szurszeweski (1981).]

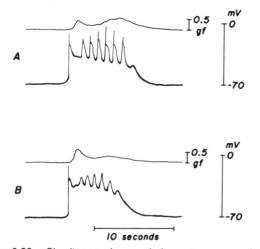

Figure 6.23. Simultaneously recorded spontaneous mechanical and intracellular electrical activity of circular muscle of canine terminal antrum. *Top trace,* Contractions; *bottom trace,* Intracellular potentials. A, Oscillations on plateau potential leading to spikes and phasic contractions. B, Oscillations in plateau potential but no spikes and a sustained contraction associated with plateau potential. [From Szurszewski (1981).]

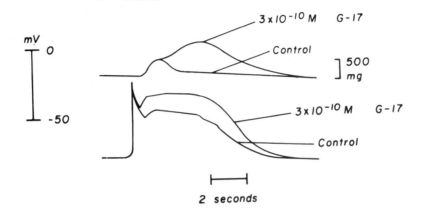

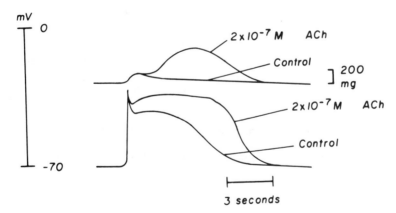

Figure 6.24. Effects of G-17 gastrin and acetylcholine (*ACh*) on electromechanical coupling of dog circular muscle of proximal antrum. *Top traces*, contraction. *Bottom trace*, intracellular action potential. [From Szurszewski (1981).]

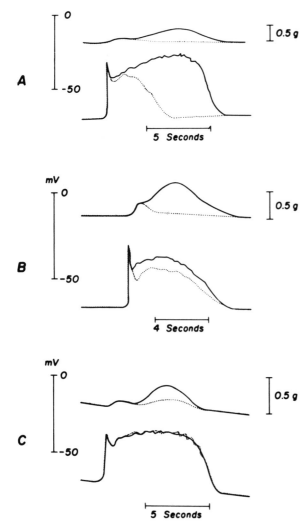

Figure 6.25. Comparison of the effects of norepinephrine (*A*), neurotensin (*B*), and vasoactive intestinal polypeptide (*C*) on electromechanical coupling of circular muscle of dog antrum. [From Szurszewski (1981).]

the amplitude of the contractions can be modified. In Figure 6.24, acetylcholine increases the amplitude and duration of the spike and the amplitude of contraction. Gastrin has a similar effect. Conversely, in Figure 6.25 norepinephrine and neurotensin decreases the amplitude of the plateau of the spike and reduces the force of contraction. However, not all inhibitors of gastric contractions reduce spikes. In Figure 6.25 VIP reduced the force of contraction but had no effect on the spikes (Szurszewski, 1981). While the role of the vagus nerves in gastric motor activity is well established, the physiologic significance of hormones is uncertain, with the exception of motilin's role in initiating the gastric MMC.

Gastric Emptying

The end result of gastric motor activity is emptying of the stomach. In recent years scintigraphic methods have become available to monitor the emptying of liquid and solid meals in normal human subjects. It is also possible to distinguish between lipid and nonlipid phases.

Liquids leave the human stomach at an exponential rate (Fig. 6.26). Four factors have been found to delay the emptying of liquids from the stomach by stimulating receptors in the duodenum. The first is osmolality (Barker et al., 1974). Hyperosmolar solutions delay gastric emptying in proportion to their hyperosmolality. There are also acid receptors which appear to titrate hydrogen ion to a pH of 5.0 (Hunt and Knox, 1969). Fatty acids delay gastric emptying independently of their hydrogen ion concentrations, but coincidentally the effect is still proportional to their molar concentration in a manner similar to that of acid inhibition (Hunt and

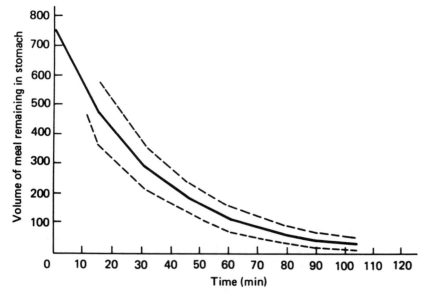

Figure 6.26. Volume of a liquid meal remaining in the human stomach plotted against time. Mean and standard deviation in 19 subjects. [From Hunt (1951).]

Knox, 1968). Tryptophane has an inhibitory effect on gastric emptying of liquids which is not found with other amino acids (Fisher and Hunt, 1977).

The site of receptors varies. Table 6.2 shows the sites in the upper small intestine. The action of receptors is influenced by vagal innervation since after truncal vagotomy the delaying effects of hyperosmolar solutions and fat are greatly reduced.

Another factor is the caloric density of the meal. In conscious monkeys McHugh and Moran (1979) found that glucose solutions left the stomach at a rate inversely related to the caloric density of the solution, as shown in Figure 6.27. The mechanism

of this effect is unknown, but it is of great interest to students of obesity.

In a series of meals in man comparing simultaneously the emptying of solids and liquids, solids emptied more slowly. However, lipids emptied more slowly than nonlipid aqueous and solid phases (Fig. 6.28). Other factors include degree of homogenization of solid meals and particle size. Figure 6.29 shows that a homogenized meal emptied more rapidly than a solid-liquid meal, especially in the first hour. Liver particles <0.25 mm in size emptied more rapidly than those 10 mm in size (Wiener *et al.*, 1981). Inhibition of digestion of liver by the administration of the H_2 blocker cimetidine to inhibit acid-pepsin secretion delayed gastric emptying (Ohashi and Meyer, 1980). The role of antral contractions in the emptying of solids is shown by the minimal antral motor activity during emptying of liquids alone and the pronounced activity after a solid-liquid meal (Figs. 6.30 and 6.31).

Gastric emptying involves the secretions added by the stomach as well as the meal itself. Figure 6.32 shows the relative rates of emptying of meal water and gastric secretions. The absolute rate of gastric emptying reflects the rate of duodenal-gas-

Table 6.2
Localization of receptors for inhibition of gastric emptying in the upper small intestine

	Duodenum		
Receptor	First 5 cm	2nd–4th parts	Jejunum
Osmoreceptor	−	−	+
Acid	+	−	+
Fat	−	−	+
Tryptophan	+	+	+

From Cook (1977). The + or − signs indicate the presence (+) or absence (−) of the receptor.

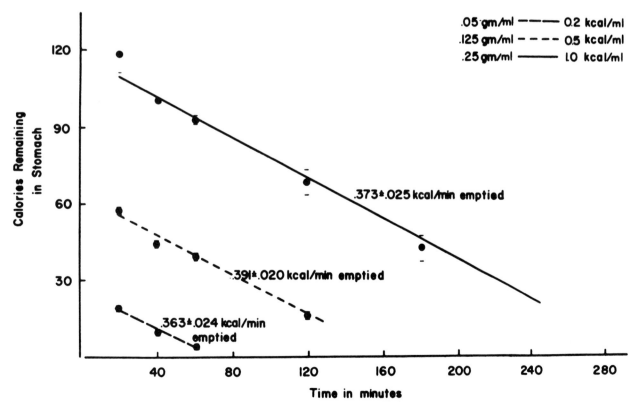

Figure 6.27. Gastric emptying of 150 ml liquid meals containing glucose in varying concentrations in conscious monkeys. Note slope of emptying in terms of calories remaining in the stomach was the same for meals with three different caloric densities. [From McHugh and Moran (1979).]

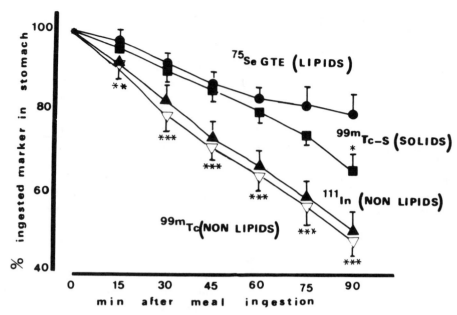

Figure 6.28. Simultaneous determinations of three different phases of a mixed meal by using labeled markers for solids (^{99m}Tc-S), lipids (^{99m}Tc-S), nonlipid solids (^{99m}Tc) and liquids (^{111}In). Lipids emptied most slowly, followed by solids. Nonlipids, both solids and liquids, emptied faster than all solids or lipids, as determined by external scintigraphy in man. [From Jian et al. (1982).]

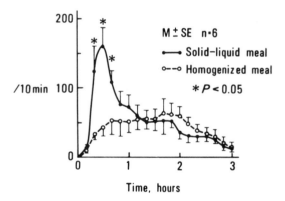

Figure 6.29. Comparison of the gastric emptying of fluids in man after nonlipids, both solids and liquids, emptied faster than all solids or lipids as determined by external scintigraphy in man. [From Jian et al. (1982).] Mixed solid-liquid meals and homogenized meals. [From Rees et al. (1979).]

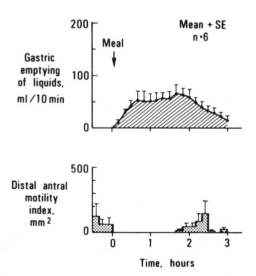

Figure 6.30. Relationship between gastric emptying of liquids (*upper graph*) and distal antral motor activity (*lower graph*) after homogenizing a mixed meal. [From Rees et al. (1979).]

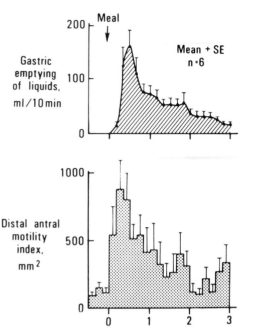

Figure 6.31. Relationship between gastric emptying of liquids (*upper graph*) and distal antral motor activity (*lower graph*) after a mixed solid-liquid meal in normal human subjects. [From Rees et al. (1979).]

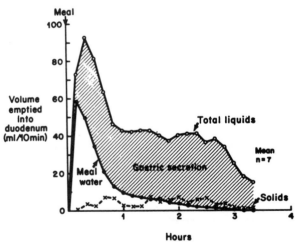

Figure 6.32. Gastric emptying of fluids including gastric secretions after solid-liquid meals in normal human subjects. [From Malagelada et al. (1977).]

tric reflux across the pylorus as well as secretions and a meal (Sonnenberg et al., 1982).

Gastric emptying is rate limiting for the absorption of glucose and probably for amino acids. On the other hand, diets containing over 120 gm of fat per day result in increased amounts of fat lost in the stools due to exceeding the capacity of the intestine for absorption rather than to an excessive rate of gastric emptying.

Pathophysiology of Gastric Motility

Disorders of gastric emptying include the increased rate of emptying of liquids and the decreased rate of emptying of solids after truncal vagotomy. Some patients with diabetes mellitus empty slowly, perhaps because of degeneration of the vagus nerves. The most interesting disorder of gastric emptying is the group of patients with non-ulcer dyspepsia or functional indigestion. Some of these patients have marked delays in the emptying of liquids and solids. In a few, tachygastria with marked increases in the frequency of spike activity has been demonstrated. Metoclopramide improves gastric emptying and relieves symptoms in some patients, probably by releasing acetylcholine from nerve endings.

Some patients with duodenal ulcer have excessively rapid emptying of solid meals. As a result the buffering capacity of gastric content due to proteins in the meals falls rapidly. Antacids must be given more frequently to neutralize gastric acid secretion. In some gastric ulcer patients, gastric emptying may be delayed, even though the ulcer is not near the pylorus. The pylorus fails to prevent reflux from the duodenum, and bile acids refluxed into the stomach may contribute to injury of the mucosa.

SMALL INTESTINAL MOTILITY

The function of small intestinal movements is different during the digestive and interdigestive periods. Within the interdigestive phase the small intestinal motor activity functions to move secretions and shed cells and bacteria into the colon, in cycles of propulsive contractions sweeping from the stomach to the ileocecal valve. Some cycles begin in the small intestine rather the stomach.

After a meal it is the function of small intestinal motility to mix intestinal content, expose it to a large absorptive surface, and gradually move the intestinal stream toward the cecum. Our knowledge of the nature of these functions in man has been obtained primarily by the use of cineradiography, intraluminal pressure recordings, and recordings of changes in electrical potentials in the intestinal wall.

Methods of Assessing Small Intestinal Motor Function

Cineradiography employs essentially a motion picture of fluoroscopic images as a barium meal passes through the intestine. Figure 6.33 is a drawing of single frames. Note that contractions may be either concentric or eccentric. In general, the small intestine divides into a series of barium-containing

Dog's jejunum segmenting

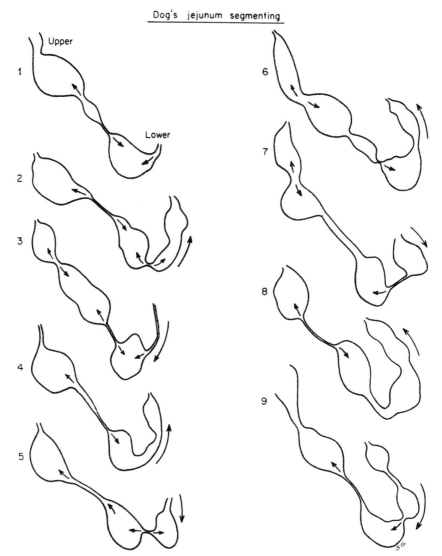

Figure 6.33. Frames from a continuous cinefluorographic recording after a meal mixed with barium had entered a dog's upper jejunum. There are nine frames at 1-s intervals. *Arrows* indicate direction in which jejunal contents were moving. Note that aboral movement occurs during segmenting contractions in the absence of peristalsis. [From Davenport (1975).]

segments, most of which are no more than a foot in length. Propulsive waves travel over these short segments and then cease, usually lasting 2–12 s (Hightower, 1968). The net effect is one of mixing and exposing a greater absorptive surface. Rhythmic contractions occur only 25% of the time (Friedman et al., 1964). In the duodenum, changes in tone in addition to phasic contractions occurred predominantly in the distal duodenum (Friedman et al., 1965).

Recordings of changes in electrical potentials in the wall of the small intestine can be made with suction electrodes or in a few postoperative states with implanted electrodes. Spike potentials indicate contractile activity. The amplitude of pressure changes correlated inversely with the interval between spikes. The duration of contractile activity on cineradiography correlated with the duration of spikes. About 10–50% spike potentials occurred in the absence of visible pressure waves (Oigaard and Dorph, 1974).

Migrating Interdigestive Myoelectric and Motor Complexes

Migrating myoelectrical complexes were recorded from fasted human subjects and showed the four phases already described. The length of intestine simultaneously activated measured about 40 cm. The duration of the complex was over 100 min as it migrated from duodenum to cecum (Fig. 6.34). Once the complex reached the cecum, a new complex appeared in the duodenum. Phase III lasted for 3 min in the duodenum, increasing to 5 min in the jejunum. Phase I lasted for 20–90 min and phase

Electrode
no.

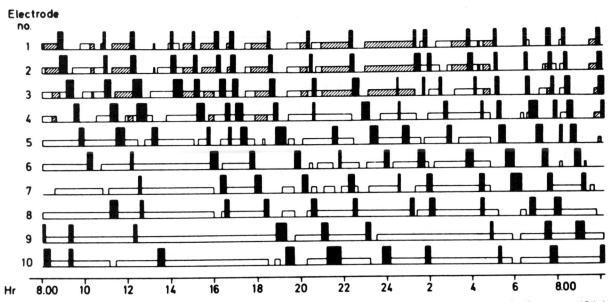

Figure 6.34. Diagramatic representation of recordings of spike potentials at 10 sites from the duodenum to the ileum over 12 h in a fasted human subject. *Base line*, Phase I; *open column*, Phase II; *closed column*, Phase III; *Hatched column*, minute rhythm of the migrating myoelectrical complex. Note how many MMCs appear to migrate aborally. "Minute rhythm," bursts of action potentials at regular intervals of about 1/100 s during periods of more than 5 min. [From Fleckenstein and Oigaard (1978a).]

II for 35–135 min. The mean conductive velocity of the phase and the length of intestine activated simultaneously decreased distally (Fleckenstein, 1978; Fleckenstein and Oigaard, 1978).

Studies with pressure-sensing devices are more common and presumably reflect electrical spiking activity. The migrating motor complex is associated with effects on secretion as well as motility. There was an increase in gastric acid and pancreatic and biliary secretion in relation to phase III (Keane et al., 1980; Vantrappen et al., 1979a; and DiMagno et al., 1982). Perfusion data showed phase durations of 7, 11, and 12 min for phase III in the jejunum, ileum, and terminal ileum, respectively. The velocity of MMC was 5, 1, and 1 cm/min, respectively. The period between MMCs was 84–100 min. About 14% of the MMCs originated distal to the ligament of Treitz (Kerlin and Phillips, 1982). A liquid test meal disrupted MMCs at all levels. The duration of the fed pattern was 3–10 h. Marked inter- and intraindividual variations occurred.

A similar study with dilution markers showed that nutrient liquid test meals increased flow promptly and then fluctuated until the onset of the next MMC. By that time over 90% of the dye dilution indicator had been aspirated. In the fasting state 85% of the periods of high flow were associated with an activity front (MMC). About half of the flow could be attributed to phase 3 and the MMC, although this occupied less than a third of the recording time (Kerlin et al., 1982). These stud-

ies indicate that the MMC is an important factor in propulsive activity during the interdigestive phase.

Sham-feeding delays the onset of the next MMC for over 2 h (Defilippi and Valenzuela, 1981). Lipid meals inhibited the MMC for longer periods than carbohydrates or proteins (Rees et al., 1982).

Observations over 24 h with pressure-sensing capsules indicate considerable variation in the length of phases of the MMC within and between subjects (Thompson et al., 1980). The duration of phase 2 was significantly reduced during sleep (Thompson et al., 1980; Thompson et al., 1981). Migrating motor complexes originating in the duodenum were more common in sleep (Finch et al., 1982).

Figure 6.35 shows a migrating motor complex in man. The most likely candidate for the initiation of MMCs in man is a rise in the plasma concentration of immunoreactive motilin (Vantrappen et al., 1979b). Peak levels occurred during phase 3 (You et al., 1980). There are two molecular forms of motilin in human blood: the smaller resembles porcine motilin (Christofides et al., 1981).

For further insight into the mechanism and control of MMCs we must turn to experimental animals. Many studies have been carried out in conscious dogs with chronically implanted electrodes or force transducers.

Using force transducers, Itoh et al. (1978a) found that migrating motor complexes began simultane-

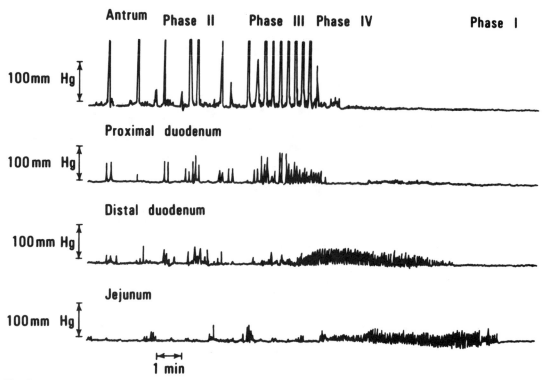

Figure 6.35. Recording of intraluminal pressures by perfused catheters in the stomach, duodenum, and jejunum in fasting normal human subjects. Phases refer to phases of the migrating motor complex. [From Rees et al. (1982).]

ously in the stomach and duodenum, and in all but 18 of 150 cycles reached the ileum. The frequency of contractions was slowest in the stomach (1/min) but lasted the longest (22 min). The contractile force was greatest in the antrum (Fig. 6.36). Figure 6.37 shows comparable data for myoelectrical activity.

Exogenous motilin induced phase 3 activity in dogs in the stomach and duodenum but did not correlate with its continued migration (Lee et al., 1978). Somatostatin prevented the MMC in response to motilin (Ormsbee et al., 1978). Feeding abolished the MMC in an autotransplanted gastric proximal pouch as well as the duodenum in situ, suggesting a humoral mediator. Irrigation of an antral pouch with acetylcholine in a similar preparation suppressed the MMC and elevated the serum gastrin even if acid secretion was inhibited by cimetidine (Thomas et al., 1980). Gastrin and CCK-OP converted interdigestive electrical activity to a fed pattern but had no effect on plasma motilin levels (Lee et al., 1980a). Plasma motilin levels do not correlate with the initiation of MMCs in the jejunum (Poitras et al., 1980). Lipids or medium chain length triglycerides inhibited the MMC for longer periods of time than glucose or casein and also reduced the percentage of slow waves accom-

panied by spikes (Schang et al., 1978; Schang et al., 1981).

A variety of isolated jejunal loops have been used to test the relative roles of hormones and nerves in control of the MMC. In general, fed patterns could be induced in loops but not in the main intestine. On the other hand, feeding failed to interrupt the MMC in the transplanted loops (Eeckhout et al., 1980; Sarr and Kelly, 1981).

The relation of the MMC to propulsive activity and absorption was also studied in similar preparations. During fasting, the volume of liquids passing through a jejunal segment per unit time was greatest during phase 3 and least in phase 1. After feeding, the movement of liquids was similar to that during phase 2. Spheres followed a similar pattern, but at a slower rate (Sarr and Kelly, 1980). Studies on absorption showed that the absorption of water was greatest in phase I and least in phase II. Absorption was greater during feeding than fasting, but transit times were slower.

Experiments in which loops were denervated of both vagal and sympathetic supplies showed slower conduction and shorter intervals between MMCs in the loop. Nevertheless, the qualitative response of MMCs was similar. These findings suggest that hormonal factors are primary in the initiation of

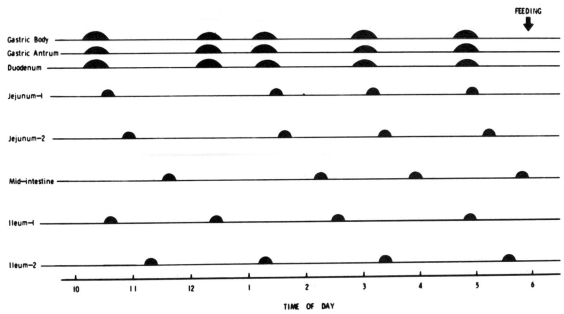

Figure 6.36.   Diagrammatic representation of forces recorded by eight extraluminal force transducers along the stomach and small intestine over a 8-h period in conscious dogs. Note second MMC did not progress beyond the duodenum. The fasting pattern was interrupted by feeding at the end of the period of study. [From Itoh et al. (1978).]

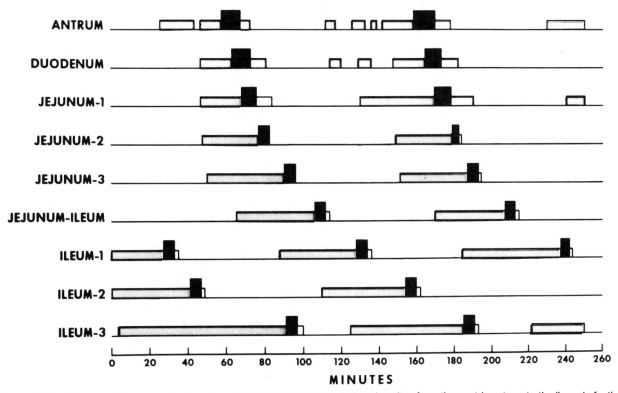

Figure 6.37.   Diagrammatic representation of myoelectrical activity recorded at nine sites from the gastric antrum to the ileum in fasting conscious dogs over 6 h. *Stippled lines* indicate beginning appearance of action potentials on slow waves; *black lines* indicate intense spiking activity (phase III). *Nondelineated* cycles indicate periods when spike activity was dying out. [From Davenport (1977); adapted from Code and Marlett (1975).]

the MMC (Itoh et al., 1981b). Autografts of jejunal loops exhibit MMCs independently of the remaining small intestine (Aeberhard et al., 1980). A role for the vagus is indicated by the cessation of MMCs in the stomach, but not in the duodenum, during cooling of vagal trunks in the neck (Hall et al., 1982). Close intraarterial injections of atropine blocked further propagation of the MMC (Sarna et al., 1981b). Both intrinsic and extrinsic nerves are needed for temporal regularity of the MMC (Ormsbee et al., 1981).

Experiments involving excisions of parts of the proximal duodenum showed that the pacemaker for duodenal slow wave is situated in the first 5–6 mm of duodenum and not at the ampulla of Vater (Hermon-Taylor and Code, 1971). Excision of the celiac and superior mesenteric ganglia lead to larger bursts of spikes simultaneously at several sites which did not migrate. All dogs had diarrhea (Marlett and Code, 1979).

These studies in dogs indicate that while hormones initiate and inhibit migrating motor complexes, intrinsic and extrinsic nerves are required for normal temporal relationships of the cycle. They also show that phase 3 is associated with propulsion of intestinal content and secretion of acid, bile, and pancreatic juice. Absorption is relatively reduced. Studies in rats suggest that disruption of the MMC by morphine increases growth of enteric organisms (Scott and Cahall, 1982).

Properties of Small Intestinal Muscle

The wall of the small intestine contains three muscle coats: the muscularis mucosae, the circular layer of the muscularis propria, and the longitudinal layer of the muscularis propria located from the lumen outward. Smooth muscle cells are small spindle-shaped cells with a single nucleus. They have a high surface:volume ratio which makes available about 1 m^2 of cell surface per gram of tissue for exchange with other cells. Intestinal cells function as a unit. This is made possible by special junctions between cells. One is formed by the attachments of myofilaments to the cell membrane on adjacent cells, intermediate junctions. They seem to form direct mechanical links between cells. They transmit force from bundles of myofilaments. The other junction is the gap junction or nexus, where cell membranes fuse and probably contain channels for the exchange of ions and molecules. They establish electrical contact between cells (Gabella, 1981).

Unlike striated muscle there is no obvious arrangement of thick and thin sliding filaments. However, myosin can be identified in dense bodies and actin in thin filaments. The ratio of thin:thick filaments is 12:1, compared to 2:1 in skeletal muscle. This reflects the low concentration of myosin in smooth muscle. Nevertheless, visceral smooth muscle can generate a tension per unit of transverse sectional area that is as large as that of skeletal muscle.

Another difference lies in the 3–4 times greater amount of collagen in a visceral smooth muscle, compared to skeletal muscle. It may function as a kind of intramuscular tendon.

Some of the most striking features of intestinal smooth muscle are its ability to maintain tension at an extremely low cost of energy and its ability to shorten to 0.3–0.4 of its resting length. The tension cost at equivalent temperature is more than 100 times greater in terms of ATP for the mouse soleus compared to the guinea pig taenia coli (Paul, 1981). Many smooth muscles maintain a continuous low level of activity (tone) and may also exhibit spontaneous rhythmic contractions.

The biochemical basis for contraction of intestinal smooth muscle is that isometric force is determined by the level of actinomyosin activity, which in turn is determined by the intracellular calcium concentration.

Electrical Properties of Small Intestinal Smooth Muscle

Electrical slow waves are generated in the longitudinal muscle layer. In vitro longitudinal cat intestinal muscle shows a resting membrane potential of −68 mV. It generates rhythmic spontaneous slow waves and spike potentials. The mechanism of the generation of slow waves is still controversial. It may reflect changes in sodium or chloride conductivity or the activity of an oscillating electrogenic pump (El-Sharkawy and Daniel, 1975).

The ionic basis for the action potential or spike is initially a depolarization mediated by a calcium-dependent inward sodium current. This is followed by a plateau potential in which both sodium and calcium may be involved as current carriers. Repolarization is attributable to an outward potassium current. The latter current begins early in the phase of depolarization and may account for the failure of an overshoot of the action potential to occur (Szurszewski, 1981).

The slow waves in longitudinal muscle spread to circular muscle, where they may be amplified. Circular muscle is important for supplying the excitatory current necessary for normal propagation (Conner et al., 1979).

There is a gradient in the frequency of slow waves in the small intestine. Figure 6.38 shows the frequency of slow waves in the normal human duodenum at increasing ages obtained with suction electrodes. Figure 6.39 shows similar data from anesthetized cats (Davenport, 1977). There may be differences in species, but the principle of control by more proximal segments is widely applicable. It has been compared to a series of oscillators, each having its characteristic frequency. In dogs, it is possible to drive the slow waves by stimulating electrodes from various sites in the small intestine. The mean maximum driven frequency is highest near the pylorus. In the proximal small intestine, antiperistalsis can occur (Sarna and Daniel, 1975).

Innervation of the Small Intestine

The intrinsic nerve supply of the small intestine consists of the myenteric plexus (Auerbach) between the longitudinal and circular muscle layers and the submucosal plexus (Meissner) beneath the circular muscle layer (Fig. 6.40) (Furness and Costa, 1980). In recent years, a number of peptides have been identified in the plexuses which may function as neurotransmitters. Table 6.3 shows the percent of ganglion cells in the two nerve plexuses which contain the neuropeptides substance P, vasoactive intestinal polypeptide, enkephalin, and somatosta-

tin. Acetylcholine is a common excitatory neurotransmitter at the neuromuscular junction, but at the ganglia it may activate inhibitory neurons. Norepinephrine acts as a neurotransmitter almost entirely at the level of the ganglion cells in the myenteric plexus, where it inhibits the action of acetylcholine released from vagal or pelvic nerve endings. The major exception is where it innervates vascular smooth muscle and directly excites vasoconstriction (Gillespie and Maxwell, 1971).

The ganglia of the enteric plexuses more closely resemble neural tissue of the central system than other autonomic ganglia. It is possible that the major role of the enteric plexuses in the control of circular muscle is to inhibit the force of contraction, while for longitudinal muscle it may have a predominantly excitatory role (Wood, 1975). Recordings from the myenteric plexus with intra and extracellular electrodes disclose three general types of neurons: Burst-type, mechanosensitive neurons, and single-spike neurons (Wood, 1981). Burst type neurons appear to possess cholinergic receptors, but their discharge is driven by an unknown neurotransmitter. Mechanosensitive neurons are probably interneurons rather than primary afferents.

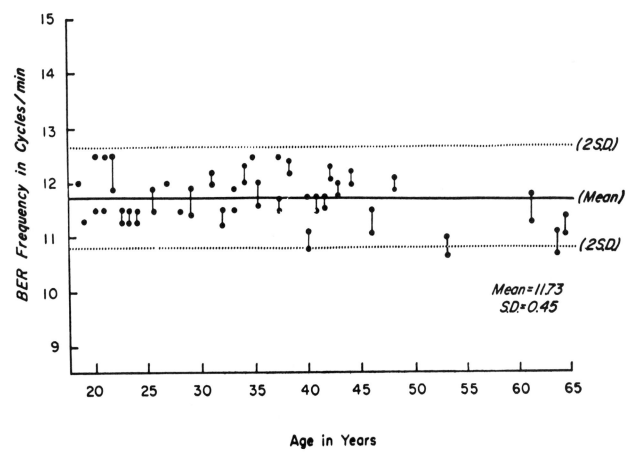

Figure 6.38. Frequency of slow waves in the human duodenum as related to the age of the subjects. [From Christensen et al. (1964).]

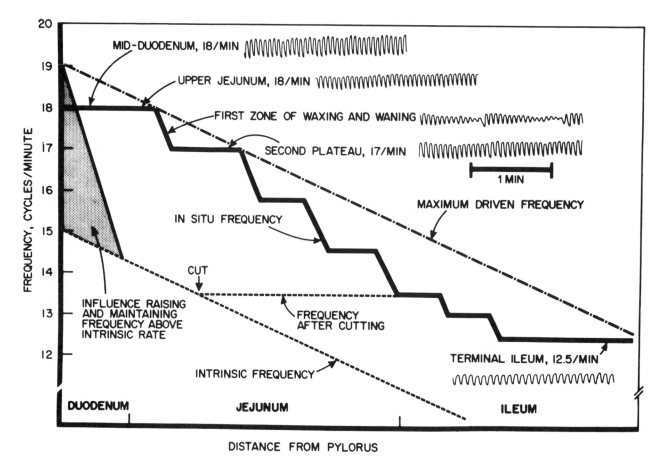

Figure 6.39. Frequency of slow waves in the small intestine of anesthetized cats. *Stepwise heavy line* shows the observed frequency. Each plateau is separated by a zone of waxing and waning slow waves. The *lower slanting dashed line* shows the intrinsic frequency of slow waves in short isolated segments of the intestine. The *upper slanting dotted line* indicates the maximal driven frequency of slow waves for the various points along the intestine. If the intestine is transected at the point labeled *CUT*, the frequency distal to the cut falls to that of the intrinsic frequency at the point of transsection. [From Davenport; adapted from Diament and Bortoff (1969).]

Single-spike neurons behave like postganglionic parasympathetic neurons. Some are driven by serotonin-releasing neurons.

The role of the enteric nervous system in segmenting or mixing movements appears to involve reciprocal contraction and relaxation of narrow bands of circular muscle, with nerves providing inhibition and myogenic slow waves triggering contraction in the no longer inhibited muscle. For propulsive or peristaltic activity, more complicated circuitry is involved. Figure 6.41 shows a tentative model. The basic action is to activate inhibitory neurons ahead of the bolus and to inactivate inhibitory neurons behind it. The latter effect is shown to be mediated by purinergic neurons. The bolus is shown as activating mechanosensitive neurons by distention. AH/type 2 motor interneurones are characterized by high resting membrane potentials and low input resistance and may be tonic type units activated by serotonin. In response to distention, the sensory neuron releases serotonin, which acts on the motor interneuron. The interneuron, through processes projecting proximally and distally in the intestine, acts by a direct cholinergic mechanism to contract circular muscle proximally and longitudinal muscle distally. At the same time, through nonadrenergic (? purinergic) motor neurones responding to acetylcholine, other interneurons inhibit longitudinal muscle proximally and circular muscle distally.

The peristaltic reflex in response to distention can be demonstrated in an isolated segment of ileum (Kosterlitz, 1968). Contraction of the longitudinal muscle precedes that of circular muscle.

EXTRINSIC INNERVATION

The small intestine is innervated by motor (efferent fibers) of the vagus nerves and by the splanchnic nerves. There is little information on the effects of stimulating these nerves in man. In experimental animals, electrical efferent vagal stimulation induced large amplitude phasic con-

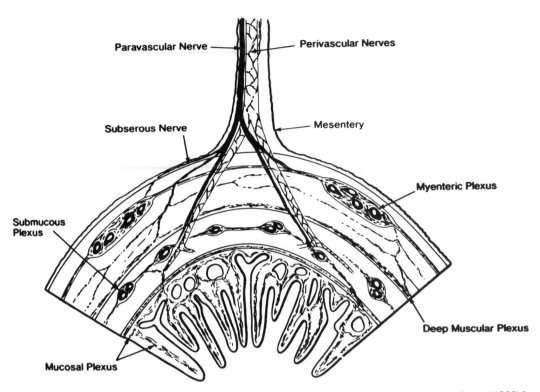

Figure 6.40. Diagrammatic representation of the enteric nervous system. [From Furness and Costa (1980).]

Table 6.3
Percent of ganglion cells in the enteric nervous system with immunoreactivity for specific peptides

	Substance P		VIP		Enkephalin		Somatostatin	
	SP[a]	MP	SP	MP	SP	MP	SP	MP
Corpus of Stomach	0	11.7	20	6.6	0	12.0	0	1.4
Ileum	5.1	2.8	27.3	7.8	0	15.2	19.7	3.8
Colon	4.3	12.9	27.0	2.3	0	12.9	20.4	2.6

Data from Schutzberg (1980). [a]SP, submucosal plexus; MP, myenteric plexus.

tractions in the duodenum of anesthetized dogs (Mir et al., 1978) and ferrets (Collman et al., 1982). The effect is due to cholinergic excitation. Under other circumstances, cholinergically mediated vagal stimulation can cause inhibition of phasic contractions, especially in the duodenum (Gidda and Goyal, 1980). Electrical field stimulation of isolated muscle strips from circular or longitudinal muscle produced relaxation in circular muscle and contraction of longitudinal muscle. Relaxation was mediated by nonadrenergic noncholinergic nerves.

The effects of vagotomy on the small intestine in man are controversial but probably of little significance in most patients. There is the occasional patient who develops explosive episodes of diarrhea.

Afferent fibers from the small intestine travel in the vagi and in the splanchnic and sympathetic nerves. There are also afferents which transmit impulses to the celiac and superior mesenteric ganglia. Mechanoreceptors are certainly present with fibers in the vagi and in fibers to the celiac ganglion. Chemoreceptors are probably also present.

REFLEXES IN THE SMALL INTESTINE

The instillation of hydrochloric acid, soap, sodium deoxycholate, and bile into the small intestine induces a propulsive movement characterized by relaxation below the point of stimulation and phasic contractions proximal to it (Thomas and Baldwin, 1971). This is a mucosal reflex which can be abolished by topical anesthetics. There is also a muscular reflex which can be abolished by hypoxia (Hukuhara et al., 1961).

Inhibitory influences on other parts of the intestine in response to distention of a loop of small intestine can be demonstrated. These are the intestino-intestinal inhibitory reflexes (Johansson et al., 1965) which can be abolished by splanchnicectomy (Friedman et al., 1975).

The Ileocecal Valve

A high pressure zone exists at the ileocecal junction zone in man (Cohen et al., 1968). Inflating a balloon in the colon increases pressure in the ileocecal sphincter, but inflation in the ileum de-

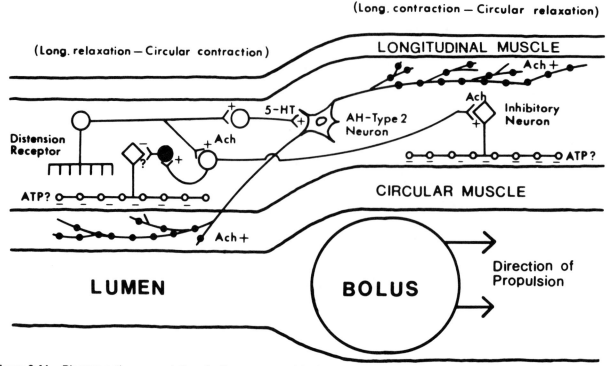

Figure 6.41. Diagrammatic representation of reflex nervous activity in the enteric nervous system during propulsion of an intraluminal bolus: +, an excitatory synapse; −, an inhibitory synapse. Neurotransmitters are indicated when known. *ACh*, Acetylcholine, *5-HT*, 5-Hydroxytryptamine; *ATP*, Purinergic transmitter. [From Wood (1981).]

creases sphincter pressure. Such an organization of the sphincter would function primarily to prevent retrograde flow from cecum to ileum. Studies of transit time in patients after resection of the ileocecal junctional zone show a shortened transit time.

In dogs, resection of both the distal ileum and the junctional zone shortened transit time much more than either alone (Singleton et al., 1964). Muscle strips from the ileum in cats and oppossums possess unique properties such as relaxation in response to electrical field stimulation. Those from the colon showed little response (Conklin and Christensen, 1975).

In anesthetized cats, electrical efferent stimulation of both the vagi and splanchnics increases pressure in the ileocecal sphincter (Rubin et al., 1981). Norepinephrine appears to be the neurotransmitter mediating the splanchnic excitatory response. Acetylcholine is the neurotransmitter at both the ganglionic and postganglionic pathway excited by the vagi, but enkephalins released by acetylcholine may act on receptors on the smooth muscle to produce contraction (Ouyang et al., 1982).

Phasic but not tonic contractions induced by efferent vagal stimulation were accompanied by electrical spiking activity. The frequency of slow waves in the sphincter was similar to that in the ileum but faster than that in the adjacent colon (Ouyang et al., 1981).

The human ileocecal sphincter contracts in response to pentagastrin, but the physiologic significance of gastrin-peptides in controlling the sphincter is unknown (Castell et al., 1970). At present, nervous control mechanisms appear to be more important.

The major roles of the ileocecal sphincter in man appear to be in delaying transit into the colon and preventing bacterial overgrowth in the ileum.

Hormones and Small Intestinal Motility

The best documented evidence for a role of hormones in the control of small intestinal motility is the temporal correlation between a rise in the plasma immunoreactive motilin level and the appearance of phase 3 of the MMC in dog and man. Recent studies suggest a possible role for pancreatic polypeptide in the inhibition of the MMC in the dog (Bueno et al., 1982). Neurotensin can convert a fasting type pattern of activity in the duodenum and jejunum to a fed pattern in the dog and a few human subjects (Thor et al., 1982).

Intestinal extracts can cause vigorous contractions and alternating relaxations of the finger-like villi of the small intestine. Villikinin is the name given to a proposed hormone contained in the extracts (Kokas, 1974). Such a substance could be of importance in increasing absorptive surface and increasing lymph flow.

Propulsion

The mechanism by which the small intestine propels its contents toward the colon remains uncertain. There may be significant differences between the more proximal small intestine and the ileum. Fluid injections into vascularly perfused intestinal segments of ileum showed an initial aboral and adoral ejection from the segment, followed by a strong aboral ejection. Duodenal and midjejunal segments in general showed no net movement. These experiments emphasize the importance of changes in resistance to flow in adjacent segments in determining net movement (Weems and Seygal, 1981; Weems, 1982).

The Circulation and Small Intestinal Motility

The effects of small intestinal motility on the circulation to the intestine is not nearly so marked as those of cardiac muscle on coronary blood flow. Small increases in intraluminal pressure have little effect on vascular resistance (Walus and Jacobson, 1981).

Pathophysiology of Small Intestinal Motility

Primary motor disorders of the small intestine are mainly limited to diseases affecting smooth muscle or the enteric nervous system. Scleroderma is an example of a disease which in its earlier stages affects primarily the enteric nervous system with loss of reflex responses but later causes atrophy and weakness of smooth muscle. Intestinal pseudoobstruction can be a molecular disease of smooth muscle with an inherited defect. Resection of the ileocecal sphincter and ileum shortens intestinal transit. Vagotomy, particularly complete abdominal vagotomy, occasionally is followed by abrupt severe watery diarrhea. The mechanism is still uncertain. Intestinal obstruction or stasis leads to bacterial overgrowth. The migrating motor complex during the interdigestive stage is absent in some patients. It is difficult to prove that diarrhea is a purely motor disturbance, but toxins such as cholera toxin, and the enterotoxin produced by some *Escherichia coli* strains produce propulsive movements as well as stimulating intestinal secretion.

COLONIC MOTILITY

The anatomy of the human colon differs from that of the small intestine. The longitudinal muscle is largely confined to three bands or taenias on the serosal surface of circular muscle from the cecum to the junction of the sigmoid colon and rectum (Pace and Williams, 1969). The colon is also divided into saccular segments or haustra. Although some haustra disappear when the taenias are cut, others persist until an incision is made in the muscularis. The internal diameter of the colon is largest in the cecum and diminishes distally.

The motor functions of the colon are to increase the exposure of intestinal content to the absorptive surface by mixing or segmenting movements, to provide storage for intestinal content, and to periodically propel intestinal content toward the rectum. The contents of the cecum are normally semiliquid. The stool takes on its normal consistency only in the descending and sigmoid colon.

Most of the total transit time through the gastrointestinal tract is taken up in the colon. Approximately 2–3 days may pass before the residue of a meal entering the cecum is evacuated from the rectum. During most of this time there is little propulsive activity. Usually, once or twice in a 24-hr period, intestinal content moves from the right to the left colon—the "mass movement" (Hertz and Newton, 1913).

Colonic Movements

In contrast to the small intestine, the contractile activity of the human colon is irregular in time and amplitude, especially in the sigmoid, which is the portion most extensively studied due to its easy accessibility. Much of our knowledge of colonic movements comes from X-ray studies, sometimes combined with measurements of pressures. Considering the colon as a whole, some kind of contractile activity is usually present, although individual segments are often quiescent (Ritchie, 1968a). Shifting of contents between individual haustra was the only form of movement seen about 40% of the time. During the rest of the period of observation, contents moved through two or more haustra. Retropulsion through haustra occurred about two-thirds as often as aboral movement. The rate of propulsion was about 8 cm/h in the aboral direction and 3 cm/h in the adoral direction, leaving a net distal movement of 5 or 6 cm/h (Ritchie, 1968b; Ritchie, 1970). Activity was greater in the transverse colon, with a faster rate of propulsion.

Figure 6.42 shows a model of the action of segmental contractions in the colon.

On a few occasions, a mass movement has been seen fluoroscopically in which the contents of the colon moved from the right to the left side (Hertz and Newton, 1913). The subjects were unaware of the event. Figure 6.43 is a drawing of such an

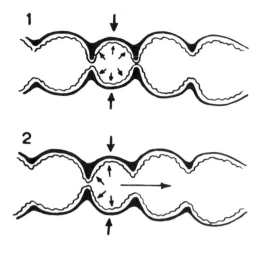

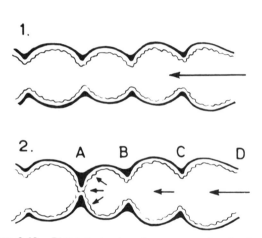

Figure 6.42. Diagrams to show how pressure localized to one haustrum may initiate movement of colonic content (*upper diagram*) or halt movement through haustra. [From Painter *et al.* (1965).]

observation. A similar movement documented by radiotelemetering pressure-sensing capsules and opaque markers can be seen with certain laxatives (Holdstock et al., 1970). Characteristically, the propulsive movement ends 45 cm above the anus.

Contractile activity can be measured with small balloons or perfused catheter tips. Figure 6.44 shows such a recording system with these sensors in the distal colon. Quantitative analysis of the records is based upon frequency, amplitude, and duration, or a combination of the last two, which is called the motility index (Fig. 6.45). Attempts to correlate recordings with function have been disappointing in that movements occur in the absence of changes in pressure. The changes in pressure

accompanying mass movements can only be inferred from the response to laxatives. They are not identifiable apart from nonpropulsive changes. Pressure records do indicate increased contractile activity in response to meals, physical activity, and cholinergic agonists and a decrease during rest and sleep (Connell, 1961; Deller and Wangel, 1965; Rosenblum and Cummins, 1954; and Holdstock et al., 1970).

Electrophysiology of the Colon

As a result of the limitations of pressure-sensing methodology, many investigators turned to recording changes in electrical potentials by means of electrodes clipped to the mucosa or, in some instances, by electrodes implanted on the serosa at the time of elective laparotomies. Unfortunately, the electrical properties of the colon are quite different from those of the small intestine, and controversy over recording techniques and physiological significance abounds. Figure 6.46 shows recordings at two sites in the rectosigmoid with clip electrodes. Two types of activity are present—slow waves and spikes. The slow waves vary in amplitude and frequency. At 12 cm from the anus they occur at a frequency of 5/min. Contractions coincident with spikes are seen only at 14 cm from the anus. These observers reported a predominant slow wave frequency of 6 cycles/min, comprising 87.5–92.5% of the total slow activity in the rectosigmoid. The remaining slow wave activity was at 3 cycles/min. Similar observations were reported by others (Taylor et al., 1975). Methods of analysis of recordings are now improved with filters and tapes. Fast Fourier transform analysis can be used. Two cyclic rhythms could still be seen using subserosal electrodes implanted at the time of laparotomy. The incidence of two rhythms was twice as great in the transverse colon as in the ascending colon (Stoddard et al., 1979). Other workers found variable frequencies of slow waves which were not phase-locked to those in adjacent segments. Bursts of spikes at the rate of 25–40 cycles/min were seen in association with longlasting contractions and slow waves. In addition, bursts of spikes at 25–40 cycles per minute were seen which lasted about 20 s and were propagated to adjacent electrodes. They were independent of slow waves and could propagate in either direction. They were not always associated with detectable pressure changes (Sarna et al., 1982). Figure 6.47 shows an example. The same investigators implanted subserosal electrodes at four sites in the taenia at laparotomy. They found variable frequency and amplitude of slow waves, especially in the ascending and sigmoid colon. In

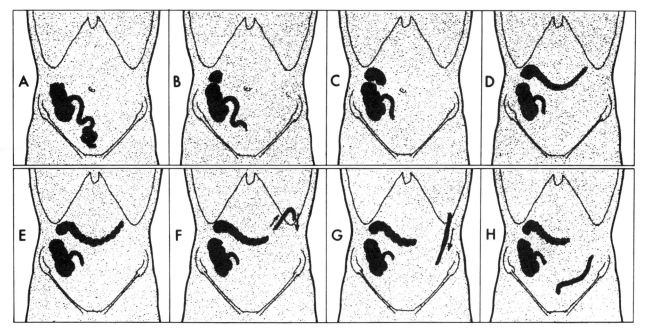

Figure 6.43. Original fluoroscopic observations of a mass movement in the human colon. *A*, 5 h after a barium meal; *B*, The end of a meal; *C*, immediately after end of meal; *D*, Barium advanced through transverse colon; *E*, Haustra appear; *F*, Barium passes splenic flexure 5 min later, *G* and *H*, Barium moves to sigmoid. [From Hertz and Newton (1913), adapted by Davenport (1977).]

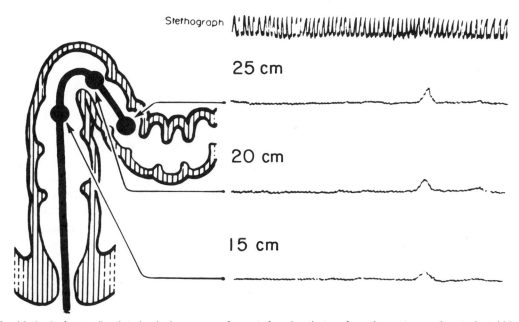

Figure 6.44. Method of recording intraluminal pressures from perfused catheters from the rectum and rectosigmoid in man. [From Parks (1965).]

the proximal colon the dominant frequency was low, and the slow waves were not phase locked in either the circular or longitudinal direction (Fig. 6.48). The middle portion of the colon showed a dominant high frequency of slow waves, and they were phase-locked most of the time. The distal colon exhibited a dominant slow frequency, and the slow waves were not phase-locked (Sarna et al.,

1980). In general then there is an upward gradient of frequency of slow waves in the right colon and a downward gradient in the left colon. They also found the migrating contractile complex described earlier in the proximal, middle, and distal colon. Its duration decreased distally (Sarna et al., 1981a).

Studies on human colonic smooth muscle strips or segments of colon in vitro have not resolved

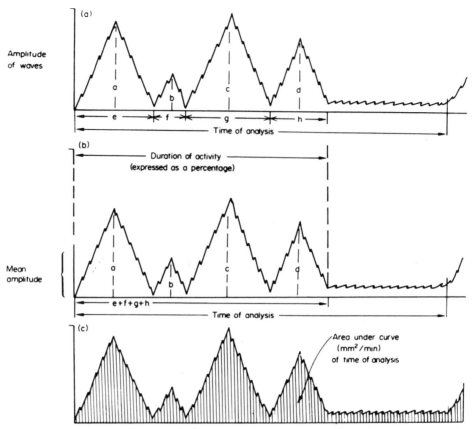

Amplitude
of waves

Mean
amplitude

Figure 6.45. Methods of quantitative analysis of colonic motor activity. Total activity may be expressed as (*a*) sum of products of amplitude and duration of each wave or (*b*) the produce of the mean amplitude of waves and duration of activity expressed as a percentage of the time of analysis. [From Parks (1965).]

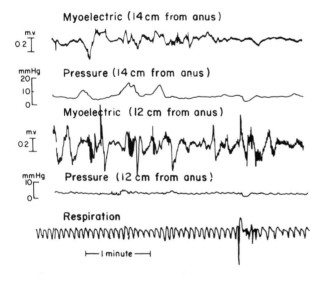

Figure 6.46. Recording of myoelectrical activity and intraluminal pressures in the rectum and rectosigmoid in a normal human subject. Slow waves are present intermittently at a frequency of 6/min. Spike activity is best seen in the recording 12 cm from the anus. Contractions are seen only at 14 cm from the anus. [From Snape et al., 1977.]

these problems. Opened segments of colon muscle in vitro show varying frequencies of slow waves which can be interpreted as the result of poorly coupled oscillators or harmonics of a single fundamental frequency (Chambers et al., 1981).

The differences in structure between the human colon and the colon in animals make it hazardous to transfer observations, but many important techniques are suitable only for experimental animals. For example, extraluminal force transducers can be applied to the serosa of the colon to record changes in force directly in the muscle under the transducer. In conscious monkeys, contractions of circular muscle in the left colon occur about 20–30% of the time at a rate of about 20 per 30 min. Electrical spiking activity occurs coincident with contractions, but there may be no changes in intraluminal pressure (Brodribb et al., 1979). In the case of the right colon, contractions could be recorded at the rate of $3.3 \pm 1.5/6$ min (Sillin et al., 1978).

The most extensive studies of colonic electrical and motor activity have been made in anesthetized cats. In addition to the slow waves and spikes, the

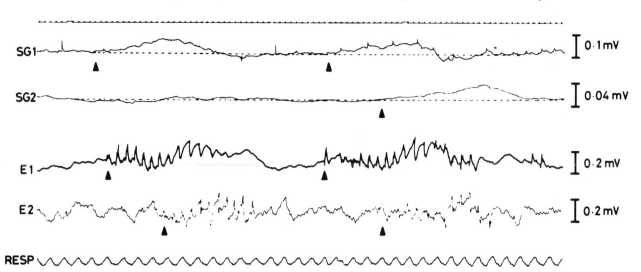

Figure 6.47. Recordings of mechanical force and myoelectrical activity during two distally propagated spike bursts at electrodes E1 and E2 in healthy human subjects. Both bursts are associated with tonic lumen-occluding contractions. The onsets of spike bursts and contractions are indicated by *arrowheads*. Each tick in the top trace represents 1 s. [From Sarna et al. (1982).]

PHASE LAG VARIATION
OF COLONIC ECA

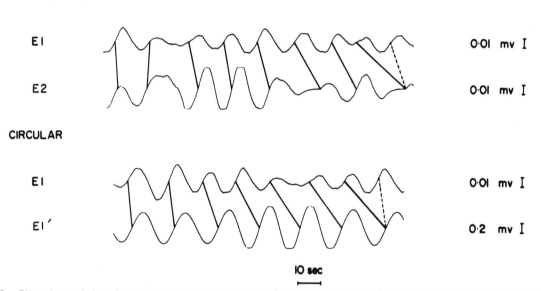

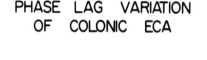

Figure 6.48. Phase lag variation of colonic electrical control activity. Slow wave recordings from human longitudinal and circular smooth muscle of the colon in vivo. The electrodes E1 and E2 were located 5 cm apart on the same taenia. E1 was at the same level but on the adjacent taenia. The slow waves were not phase-locked in either the circular or longitudinal direction. [From Sarna et al. (1980).]

cat exhibits two forms of contractile activity. The first are spike-associated slow waves which show an increasing frequency gradient from the ascending to the mid transverse colon. The second is migrating spike bursts independent of slow waves which move from the hepatic flexure to the sigmoid. Figure 6.49 shows a model relating these phenomena to mixing and propulsive activity. The relation to man of these observations in the cat is uncertain.

Properties of Colonic Smooth Muscle

The myoelectrical activity in the colon differs from that in the stomach and small intestine. The slow waves are irregular in timing and variable in amplitude, shape, and polarity, especially in the distal colon. Slow waves propagate only short distances (<5 cm). They propagate more rapidly in a transverse than a longitudinal axis. In man slow

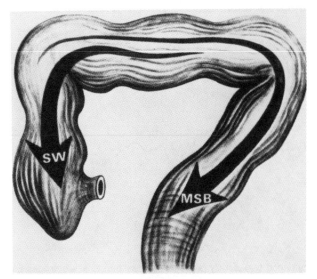

Figure 6.49. Diagram relating flow of colonic contents to slow waves and migrating spike bursts. Slow waves spread toward the cecum away from a pacemaker located in the midportion of the transverse colon. Since the slow waves pace rhythmic contractions, the contractions should produce a flow with a polarity in the same direction (*arrow Sw*). The polarity of slow wave spread is probably not fixed. The migrating spike bursts begin at a variable position in the middle or proximal colon and migrate toward the rectum. Since contractions accompany migrating spike bursts, such contractions should produce flow in the same direction (*arrow MSB*). The migrating spike burst also has the capacity for reversal of direction. [From Christensen et al. (1974).]

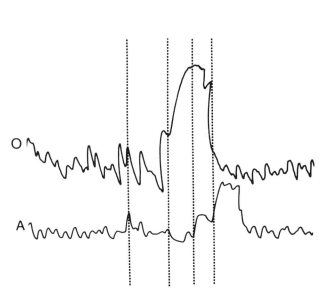

Figure 6.50. The peristaltic reflex in the colon. Recordings from strain gauges on longitudinal and circular muscle *above O* and, *below A* a balloon inflated in an isolated cat colon. [From Costa and Furness (1976).]

waves originate in longitudinal muscle (Szurszewski, 1981). Both calcium and cyclic AMP appear to be involved in the generation of slow waves. Cyclic AMP reduces the frequency and amplitude of slow waves in cat colon longitudinal muscle strips in vitro. The effect could be reduced by increasing calcium levels (Anuras, 1982).

The migrating spike bursts are independent of slow waves. Some investigators regard both slow waves and spikes as action potentials leading to contractions (Szurszewski, 1981).

Slow waves are properties of circular muscle in the dog, while spikes appear in longitudinal muscle. When the two muscle layers were in contact, the recordings resembled those in longitudinal muscle. In the dog circular colonic muscle contraction may be independent of spikes (Hara and Szurszewski, 1981).

Innervation of the Colon

The principle difference between the enteric plexus in the colon and that in the small intestine is that the myenteric plexus is concentrated beneath the taenia. There is little information available about activity within the neurons of this plexus. There is pharmacologic evidence from in vitro circular and longitudinal muscle strips from the human sigmoid colon that α-agonists contract both muscles (Gagnon et al., 1972). Electrical field stimulation of longitudinal and circular muscle strips from the descending colon in the presence of atropine always relaxed the muscle, implicating a nonadrenergic inhibitory neurotransmitter (Crema et al., 1968).

The colon in vitro responds to distention or to a pellet by a coordinated peristaltic reflex. The initial event is a contraction of the longitudinal muscle proximally followed by a contraction of circular muscle. Relaxation occurs distally. Figure 6.50 shows such a response in the cat colon (Costa and Furness, 1976). A model of the intrinsic nerve circuits involved is shown in Figure 6.51.

The extrinsic innervation of the colon is shown on Figure 6.2. There is a dual innervation of both the parasympathetic supply (vagi and pelvic nerves) and the sympathetic (splanchnics and lumbar colonic nerves). The physiologic significance of the vagus supply is uncertain. The pelvic nerves contain both cholinergic excitatory and inhibitory neurones. The latter presumably act by way of nonadrenergic inhibitory (? purinergic) nerves in the enteric nervous system (Roman and Gonella, 1981).

The splanchnic nerves are inhibitory to the proximal colon while the lumbar colonic nerves are inhibitory to the distal colon. There are also inhibitory inputs from both the spinal cord and the inferior mesenteric ganglion as shown in Figure 6.52.

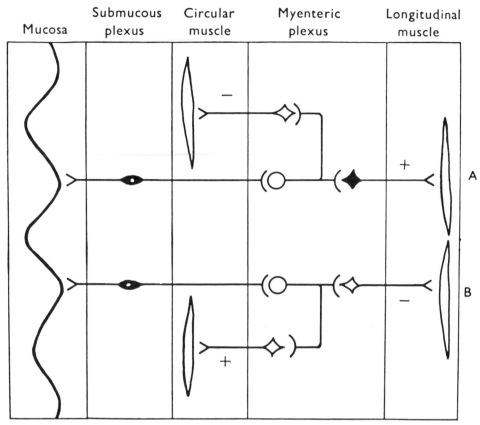

Figure 6.51. Diagram of a possible arrangement of sensory, motor (+), and inhibitory (−) nerves in the intrinsic nerve plexus which may cause reciprocal activation of the two muscle layers. [From Kottegoda (1969).]

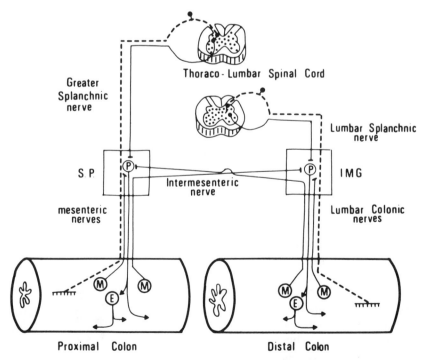

Figure 6.52. Diagrammatic representation of possible neural connections of proximal and distal segments of guinea pig colon with prevertebral and spinal cord. *SP*, Solar plexus; *IMG*, Inferior mesenteric ganglion; *P*, Postganglionic sympathetic neurons; *E*, Cholinergic excitatory intramural neurons; *M*, Mechanoreceptors. The *dashed lines* represent the direct afferent connection from gut wall to spinal cord. [From Roman and Gabella (1981); adapted from Szursewski (1977).]

An intestinocolic inhibitory reflex can be demonstrated in cats mediated through the prevertebral ganglia. Candidate neurotransmitters include VIP, enkephalins, substance P, and somatostatin. Destruction of the lumbosacral cord in man abolishes all reflex activity, including the intestinointestinal inhibitory reflex (Connell et al., 1963).

The right side of the colon in man responds differently in vivo to intravenous cholinergic agonists. There are contraction of the right colon and abolition of phasic contractions on the left side (Kern and Almy, 1952). Resection of the pelvic nerves in man slowed transit through the colon. A bypass from the right colon to the rectum returned transit to normal (Devroede and Lemarche, 1974). There may be dopaminergic excitatory nerves to the colon in man (Lanfranchi et al., 1978).

The Colonic Response to Feeding

There is an increase in spiking activity and phasic contractions in the rectosigmoid of man after feeding a 1000-calorie meal (Snape et al., 1978). The most important ingredient in the meal for exciting the response is fat (Wright et al., 1980). Both cholinergic and hormonal influences are involved, but the hormone(s) have not been identified (Snape et al., 1979; Renny et al., 1982). Recently, enkephalins have been implicated since the response can be blocked by naloxone (Sun et al., 1982). An afferent limb of the pathway can be blocked by a topical anesthetic in the stomach, but the response persists after total gastrectomy (Duthie, 1978). The physiological role of hormones in the control of colonic motility in general is uncertain although some, such as cholecystokinin and glucagon, do have effects when given exogenously.

Defecation

The act of defecation is usually subject to conscious nervous control. It is likely that only the contents of the lower 40 cm of rectum and colon are evacuated. The stimulus is thought to come from mechanoreceptors in the rectum responding to distention. Central nervous sensory inhibitory mechanisms are able to suppress the local sensation unless it is very pronounced or reinforces behaviorally conditioned responses. Sensations must also arise from muscles in the pelvic floor. Arrival of stool in the anal canal signals relaxation of the external sphincter (Duthie, 1982). Distention of the rectum elicits a reflex relaxation of the internal sphincter (Fig. 6.53). Increased intra-abdominal pressure forces the walls of the upper portion of the anal canal apart. The descent of the muscles of the pelvic floor pulls the walls of the canal apart lat-

erally. Anal continence depends upon the pelvic floor as well as the anal sphincters (Fig. 6.54).

Pathophysiology of Colonic Motility

The best example of a colonic motility disorder which can be satisfactorily explained by its pathophysiology is Hirschsprung's disease, in which an aganglionic segment of colon results in a functional obstruction. Diverticular disease, at least in some subjects, seems to be related to abnormally high pressures within haustra due to forceful mixing contractions. This forces the mucosa out in areas of the wall which are relatively weak, usually where the vascular supply enters. The irritable bowel has been associated with differences in the percentage of slow and fast slow wave frequencies, compared to that of normals. This is a controversial obser-

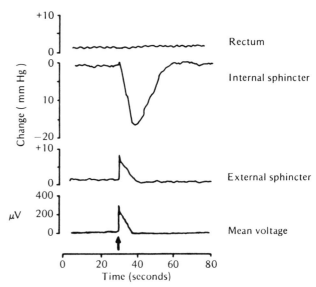

Figure 6.53. Recordings from balloons in the rectum, internal and external anal sphincters, and mean voltage in external anal sphincter in a normal human subject in response to distention of the rectal balloon. [From Tobon et al. (1968).]

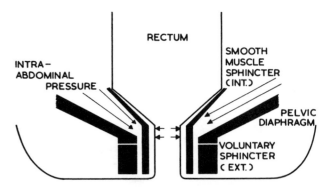

Figure 6.54. Role of the pelvic floor in maintaining anal continence. *Arrows* indicate direction of forces generated by internal anal sphincter and intra-abdominal pressure. [From Phillips and Edwards (1965).]

vation at present. Finally, constipation is associated with delayed transit through the colon which can be corrected by bulk or fiber in the diet in some patients. Patients with the irritable bowel syndrome who have diarrhea as a principal symptom have more fast rectosigmoid contractions than do normal subjects or patients with constipation.

THE MOTOR FUNCTION OF THE BILIARY TRACT

Movements and Pressures in the Biliary Tract and Sphincter of Oddi

The motor activity of the gallbladder, bile ducts, and sphincter of Oddi functions to deliver concentrated bile into the duodenum in response to a meal. The bile then promotes the digestion of fat. Between meals, the sphincter of Oddi prevents reflux of duodenal content into the bile and pancreatic ducts. Recent evidence suggests that it retains most of the bile acid pool within the biliary tract during the interdigestive period and interrupts the enterohepatic circulation with resulting effects on the composition of hepatic bile.

There are microscopic observations showing that the bile canaliculi contract at about 6-min intervals and expel the canalicular contents from the lumen (Phillips et al., 1982). Studies of the bile ducts themselves show no contractile activity, but the sphincter of Oddi exhibits both tonic and phasic contractions. Most anatomists find little smooth muscle in the wall of the extrahepatic bile ducts.

Pressures within the biliary and pancreatic ducts can be obtained with perfused catheters. In man, the pressure in the common bile duct in eight subjects was 11.4 ± 1.3 mmHg, compared to 32.5 ± 3.3 mmHg in the pancreatic duct of five subjects.

The peak pressure in the sphincter of Oddi was 110 ± 10.6 mmHg (Csendes et al., 1979).

Pressure in the sphincter of Oddi can be measured by a pull-through technique similar to that used in the lower esophageal sphincter. Figure 6.55 is an example of such a recording (Geenen et al., 1980). The mean pressure in the common bile duct was 8 ± 0.6 mmHg above duodenal pressure while the pressure in the sphincter was 4–5 mm higher than in the bile duct. Note the phasic contractions in the sphincter.

Recent studies in man during the interdigestive state show a correlation between the output of bile acids into the duodenum and the migrating myoelectrical complex (Peeters et al., 1980). The output of bile acids increased 4-fold in normal human subjects before the onset of phase III in the proximal small intestine.

Long-term studies (13 h) in conscious dogs with force transducers sewn to the serosa of the antrum, gallbladder, and small intestine show cyclic contractions in relation to the MMC during the interdigestive period. Figure 6.56 shows diagramatically the tonic and phasic contractions in the gallbladder and gastric antral walls in the fasting and fed states (Itoh et al., 1982a). The rate of the rhythmic phasic contractions was the same (4/min) (Takahashi et al., 1982a). There is a tonic rise in gallbladder wall force at the onset of phase II of the MMC in the duodenum (Itoh et al., 1982b). During phase III, gallbladder contraction returned to normal (Itoh et al., 1981). The role of motilin in mediating the interdigestive contractions of the gallbladder is uncertain (Takahashi et al., 1982b).

There is considerable controversy over the extent of bile flow into the duodenum during fasting in intact human subjects or after removal of the gall-

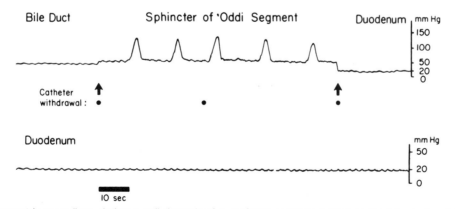

Figure 6.55. Manometric recordings during a pull through of a perfused catheter system in the bile duct, sphincter of Oddi, and duodenum in man. Each *dot* represents a 2- to 3-mm withdrawal of the catheter. The margins of the sphincter of Oddi are shown by *arrows*. Within the sphincter of Oddi, a basal pressure is recorded that is only a few millimeters of mercury greater than the pressure in the common bile duct. Superimposed on the basal sphincter pressure are phasic pressure waves about 6.5 mm Hg above basal. [From Geenan et al. (1980).]

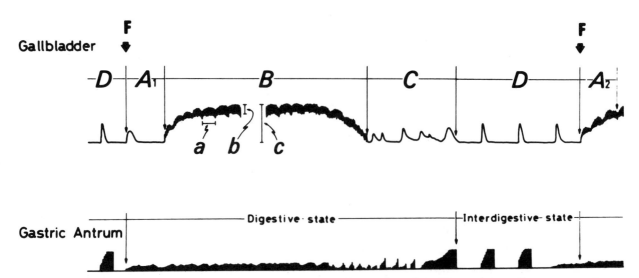

Figure 6.56.   Schematic illustration of the force of contraction of the gallbladder and of the gastric antrum for 24 hours in the conscious dog using extraluminal force transducers. A_1 and A_2 indicate immediate postprandial periods. *B*, Period of rhythm contractions; *C*, Period of irregular contractions; *D*, interdigestive contraction period; *a*, group of contractions; *b*, amplitude of rhythmic contractions; and *c*, maximal tonic contractions. [From Itoh (1982).]

bladder. Recent studies comparing bile flow into the duodenum in postcholecystectomy patients with and without a T-tube projecting from the bile duct into the duodenum shows that bile acid output was reduced >67% when the sphincter was intact (Askin et al., 1978). This indicates that sphincter is an important factor in interrupting the enterohepatic circulation.

The development of scintiscanning techniques with ^{99m}-technetium labeled contrast media makes it possible to quantitate emptying of the human gallbladder in response to stimuli, such as food and hormones. Healthy subjects evacuate about 60% of the contents of the gallbladder after a fatty meal (Krishamurthy et al., 1981). The response is best mimicked by the sequential administration of small (5 ng/kg) doses of CCK octapeptide rather than a single large (20 ng/kg) dose intravenously (Krishamurthy et al., 1982). The emptying of the gallbladder after a liquid fatty meal is a rapid and smooth response achieving 75% emptying at 30 min, as shown in Figure 6.57. Note that the gallbladder empties more rapidly than the stomach (Fisher et al., 1982). A similar response can be obtained with an intravenous infusion of crude CCK as seen in Figure 6.58 (Spellman et al., 1979).

Using ultrasound or oral cholecystography to evaluate emptying of the gallbladder, the intramuscular injection of CCK octapeptide produced greater contraction of the gallbladder than did a single intravenous bolus (Lalyre et al., 1981). The maximally effective dose was 100 ng/kg. Pure CCK-33 given by intravenous infusion produced nearly 90% contraction after 10 min. After 40 min, the

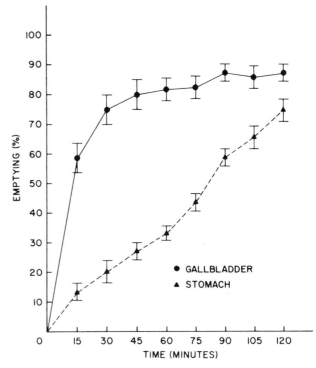

Figure 6.57.   Emptying of the human gallbladder and stomach in response to a meal. [From Fisher et al. (1982).]

gallbladder had refilled (Lilja et al., 1982). The determination of blood levels of immunoreactive CCK-OP and CCK-33 by radioimmunoassay during such studies will help to clarify the physiological role of CCK.

The explanation for the greater emptying of the gallbladder after intravenous infusions or intramuscular injections of CCK, as compared to that of

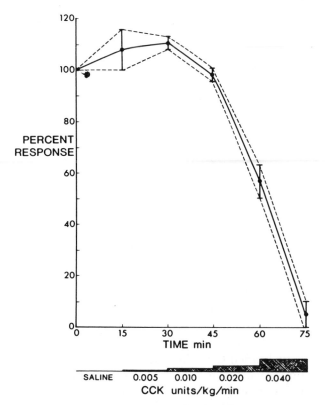

Figure 6.58. Contraction of the human gallbladder in vivo in response to increasing doses of crude cholecystokinin. [From Spellman et al. (1979).]

an intravenous bolus, may be the contraction of the infundibular portion of the gallbladder and cystic duct seen by cinematography after a large bolus of CCK (Torsoli, 1981).

Some Properties of the Sphincter of Oddi

As noted above, the sphincter of Oddi exhibits both tonic and phasic contractions. The phasic contractions progress toward the duodenum and appear to be propulsive, as shown in Figure 6.59 (Toouli et al., 1982). The phasic contractions can be abolished by a 20 ng/kg intravenous bolus of CCK octapeptide. At the same time, the pressure in the sphincter of Oddi falls by nearly 50%, as shown in Figure 6.60 (Hogan et al., 1982).

Studies in the opossum in vivo show a correlation between electrical spike bursts recorded from the sphincter of Oddi and the migrating myoelectrical complex in the upper small intestine in the fasted state. Four phases of spike bursts can be seen: (a) low frequency bursts over a period of 20 min; (b) a progressive increase in frequency over the next 40–45 min; (c) a peak frequency of bursts for 5 min; (d) a decline to base line over 15 min. The entire cycle lasts 87 ± 11 min SD. It coincides with the MMC. Feeding abolishes the cycle, replacing it with 5–6 bursts/min for 3 h. In the opossum, the spikes

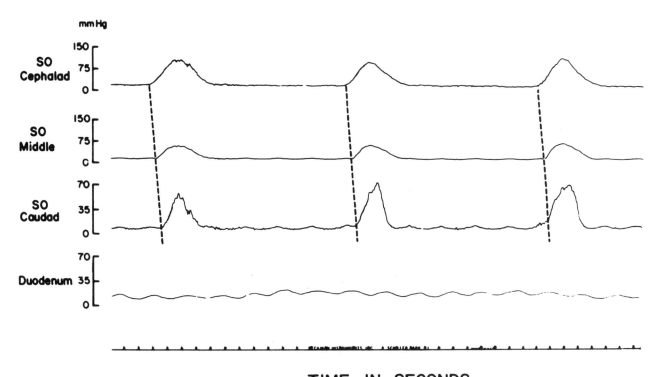

TIME IN SECONDS

Figure 6.59. Recording of propulsive antegrade contractions in the human sphincter of Oddi from a triple lumen-perfused catheter with recording sites 2 mm apart. All three contractions propagate toward the duodenum and are independent of pressure changes in the duodenum. [From Toouli et al. (1982).]

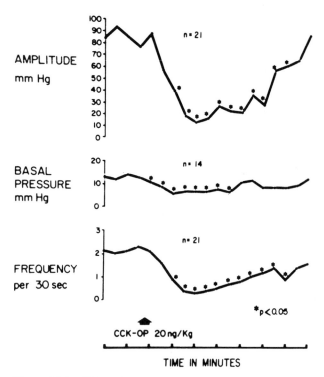

Figure 6.60. Effect of an intravenous bolus of cholecystokinin octapeptide on the amplitude and frequency of phasic contractions in the sphincter of Oddi. Values are shown at 30-s intervals. [From Hogan et al. (1982).]

correlate closely with mechanical contractions. Slow waves on the other hand vary in frequency and are sometimes not seen (Toouli et al., 1980; Honda et al., 1982).

Hormonal Control

The importance of CCK in the control of biliary tract motility has been emphasized. Recent evidence in man indicates that the exocrine pancreas and the gallbladder are equally sensitive to the action of CCK, if the latter is given in addition to a background infusion of secretin (MacGregor et al., 1976). This is likely to occur in the course of a response to a meal. In vitro studies with gallbladder muscle strips from cats show potentiation of contraction in response to CCK by subthreshold doses of secretin (Chowdhury et al., 1975). In vivo in the cat, CCK-33 and CCK octapeptide contract the gallbladder and relax the sphincter of Oddi. As in humans, CCK abolishes phasic contractions in the sphincter of Oddi and decreases tonic pressures (Behar and Biancani, 1980).

Using ultrasound to outline the gallbladder, the fasting volumes can be found to be significantly increased during pregnancy and correlated with the level of serum progesterone. The emptying of the

gallbladder is slowed. Both factors return to normal postpartum (Everson et al., 1982). Similar delays in emptying have been noted during the normal menstrual cycle (Nilsson and Stattin, 1967).

The mechanism of action of cholecystokinin on the gallbladder appears to be a direct one on smooth muscle in both the cat and guinea pig (Chowdhury et al., 1975; Yau et al., 1973). Its action on the sphincter of Oddi in dogs is also myogenic (Lin, 1975). In the cat, CCK increases intracellular levels of cAMP in the wall of the gallbladder, but increases phosphodiesterase activity in cells of the sphincter of Oddi (Persson, 1972).

Role of the Nervous System

The most generally accepted evidence for a role of the nervous system in the control of motor activity of the human biliary tract comes from the results of truncal and selective vagotomies. The volume of the gallbladder at rest and after CCK is greater after vagotomy, probably due to cutting the hepatic branch (Bouchier, 1970). Another interesting observation in man is the demonstration of a rich innervation of the gallbladder wall with immunoreactive VIP containing nerve fibers (Sundler et al., 1977).

For more details, we must turn to experimental animals, and of these the cat has been most extensively studied. Electrical stimulation of the vagus potentiated the rise in gallbladder pressure in response to CCK (Pallin and Skoglund, 1964) and contracted the gallbladder (Behar and Biancani, 1980). Splanchnic nerve stimulation on the other hand relaxed the gallbladder which had contracted in response to CCK (Persson, 1973). VIP relaxes the gallbladder (Jansson, 1979).

In the dog, a pylorocholecystic reflex can be demonstrated with contraction of the gallbladder in response to distention of the antrum. It is blocked by atropine, hexamethonium (a ganglionic blocker), and vagotomy (Debas and Yamagishi, 1979).

It would appear that nervous control of motility in the biliary tract is less important than hormonal control, but there may be interactions which are not yet clearly defined.

Pathophysiology of the Biliary Tract

Gallstones usually form in the gallbladder, yet hepatic bile composition in patients with cholesterol gallstones favors precipitation of cholesterol. Recent observations demonstrate delayed or incomplete emptying of the gallbladder in response to CCK. In the prairie dog model of cholesterol gallstones, induced by diets high in cholesterol, sphincterotomy reduced the volume of the gallbladder, and the incidence of stones was reduced significantly. These observations suggest that delayed emptying of the biliary tract may be a factor in the pathophysiology of cholesterol gallstones. The association of pregnancy and oral contraceptives with gallstones may be due in part to stasis in the gallbladder induced by progesterone.

For many years, a syndrome of right upper quadrant pain resembling biliary colic has been attributed to a dysfunction of the nervous or humoral control of evacuation of the gallbladder and bile ducts (biliary dyskinesia). Our ability to identify patients with increased pressures in the sphincter of Oddi may represent an objective criterion for the diagnosis. Fibrosis of the sphincter of Oddi has been considered a potential cause of pancreatitis. Again, measurements of the sphincter pressure will establish the presence of significant obstruction.

CHAPTER 44

Secretion

GENERAL CONSIDERATIONS

Secretion in the gastrointestinal tract refers to the net movement of water and electrolyte, lipids, and proteins into the lumen of the alimentary canal or the ducts of the digestive glands (salivary glands, pancreas, biliary system). It includes endocrine secretion into the blood by endocrine cells in the mucosa of the digestive tract, such as gastrin. There is also evidence for secretion of peptides into the intercellular space and into the lumen of the digestive tract—paracrine secretion. Figure 6.61 shows diagrammatic examples of the different types of secretion (Larsson, 1979). Often there are bidirectional fluxes of water and electrolyte from blood to lumen and from lumen to blood. The presence of a quantitative excess of the former leads to net secretion. A great deal of information is available about transport membranes in vitro. If movement occurs against an electrochemical gradient, then "active transport" occurs. Special cases include the sodium-dependent secretion of bicarbonate by the ductal cells of the pancreas, the surface epithelial cells of the stomach, and the biliary canaliculi of the hepatocytes. The metabolic energy is at least in part supplied by the ubiquitous sodium-potassium ATP-ase in the contraluminal cell membrane (Flemstrom, 1981; Schulz, 1981; Erlinger, 1982).

Unfortunately, in the case of fluid and electrolyte transport by the exocrine pancreas, salivary glands, and liver, it is not possible to measure the electrical potential at the secretory surface. However, many of the modern concepts and methodologies of epithelial transport in vitro can be applied to secretion by the mucosa of the stomach and intestine. At the same time, secretion of water and electrolyte can be studied in isolated organ systems, such as the perfused intestine, pancreas, or liver. On a more physiological scale, the control of secretion by hormones and nerves can be examined in anesthetized or conscious animals and in man.

Secretion of Proteins

The secretion of proteins is another important general biological phenomenon which is necessary in the gastrointestinal system for providing digestive enzymes, protective substances such as glycoproteins, and binding proteins such as intrinsic

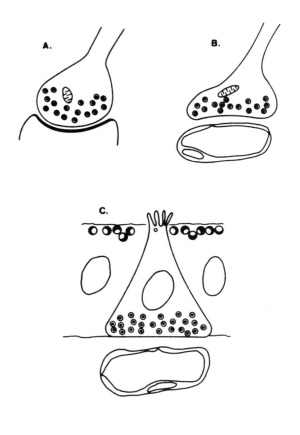

Figure 6.61. Diagram illustrating four different mechanisms by which peptides released from their cell of origin can influence other cells. *A.*, Classical neurotransmission. *B.*, Neuroendocrine secretion with release of neurotransmitter into blood strain. *C.*, Classical endocrine secretion. *D.*, Paracrine secretion [From Larsson (1979).]

factor (Hopkins, 1979). Exportable proteins are synthesized by the ribosomes of the rough endoplasmic reticulum (RER) and segregated into the cisternal space of the RER. Transport continues through the cisternae to the Golgi apparatus. The mechanism of entry into the Golgi is controversial: whether by the shunting of vesicles or by way of tubular connections. Pores have been demonstrated in the membrane on the external side of the Golgi membrane, while condensing vacuoles appeared on the internal side, possibly representing tubular connections. The condensing vacuoles are then transformed into mature zymogen granules with a glycoprotein-containing membrane (Kern et al., 1979). Finally, the zymogen granule fuses with the cell membrane, and the contents are extruded by a process of exocytosis (Borgese et al., 1979) (Fig. 6.62). The lost membrane may be recovered by invagination and enclosure within condensing vacuoles or lysosomes as in Figure 6.63 (Herzog and Miller, 1979).

Recent studies have focused on the molecular events as the protein precursors leave the microsomes and enter the cisternae of the rough endo-

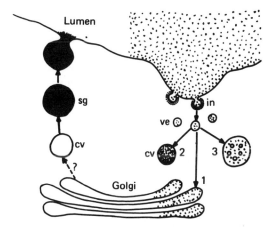

Figure 6.63. Diagrammatic representation of interactions between cellular components and the cell surface in salivary secretory cells. On the *left* the route followed by secretory products from condensing vacuoles (*cv*) to secretory granules (*sg*) is shown. The granules fuse with the apical cell surface. On the *right* is shown the route taken by membrane retrieved from the cell surface after exocytosis as traced by dextrans. Patches of surface membrane with a coat on their cytoplasmic face invaginate (*in*) and pinch off, losing their coats in the process. The apical vesicles (*ve*) can therefore move either to the rims of the Golgi cisternae to condensing vacuoles or to liposomes. [From Herzog and Farquhar (1977).]

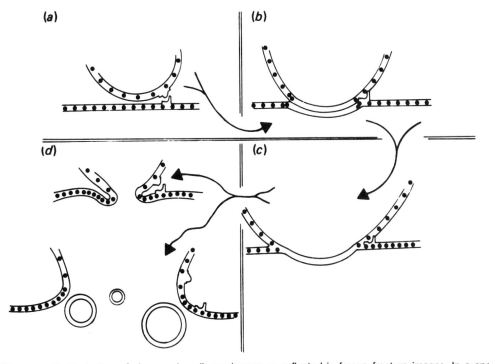

Figure 6.62. Diagrammatic illustration of changes in cell membranes as reflected in freeze fracture images. In *a* specific sites are recognized in both the apical membrane and the secretory granule. In *b* the two membranes engage at areas which are free of intramembrane particles (IMP). The stage of membrane opposition is probably reversible. In *c* the process continues with formation of a hybrid. IMP-free membrane formed by elimination of the cytoplasmic leaflets of the granule and the plasma membranes in their region of interaction. By *d* the hybrid membrane collapses, and continuity is established between the interior of the granule and the extracellular space. If the exocytotic opening is small as in the upper portion of *d* the molecules involved in membrane fusion-fission can be largely reaccommodated within the fused membrane, and no shedding of membrane material is noted. However, when the opening is large, the relocation of membrane components is insufficient, and these components are released into the extracellular space where they give rise to IMP-free vesicles. The latter may be an artifact of the methodology. [From Borgese et al. (1980).]

plasmic reticulum. Figure 6.64 shows a proposed model. The newly forming precursor secretory molecule contains a sequence of amino acids known as a signal peptide. The signal peptide enters a channel in the large (60s) unit of the microsome (Fig. 6.64). The signal peptide and the ribosome bind to receptors on the external surface of the cisternal membrane of the RER. As a result, a pore develops in the membrane. On the inner surface of the membrane is a signal peptidase which cleaves the signal portion from the newly formed precursor secretory protein. The signal peptide is left within the membrane, and the new protein enters the lumen of the cisternum. The ribosome now dissociates from the membrane, and the pore is eliminated by lateral diffusion of its components in the membrane (Blobel et al., 1979).

This view of secretion of proteins is not unchallenged. Zymogen granules, except for those in proximity to the islets of Langerhans, are thought to contain a uniform composition of precursor enzymes. Yet in vivo striking changes in the ratio of individual enzymes have been observed. An alternative mechanism for enzyme secretion proposes a series of intracellular compartments within the cytoplasm across which equilibration of proteins can occur. They could serve as a mixing chamber for enzymes from other pools and as a direct precursor pool for secretion independent of exocytosis (Rothman, 1975).

The control of secretion of proteins by extracellular mechanisms involves the binding of neurotransmitters and/or hormones to receptors on the secretory cells. Much has been learned from isolated secretory cells or secretory units such as pancreatic acini. In general, neurotransmitters have their own specific receptors. Hormones have different receptors, but other hormones may compete for the same receptor and reduce or completely block the excitatory effect of stimulatory hormones. Furthermore, the occupation of the receptor may initiate a series of events leading to activation of intracellular events resulting in secretion. The best studied examples are activation of cyclases and subsequently the intracellular formation of cyclic nucleotides. This is followed by the activation of protein kinases and finally the actual secretion of electrolyte or protein.

Role of Calcium

The second major control system involves release of intracellular calcium which leads to secretion independently of cyclic nucleotides. Figure 6.65 shows a pancreatic acinar cell with receptors and secretory mechanisms. Note that acetylcholine and CCK activate a calcium-mediated enzyme secretion, while secretin and VIP act via a cAMP pathway. These mechanisms were developed using guinea pig acinar cells. In man, secretin and VIP have little effect on enzyme secretion. It is not known if human acinar cells have receptors for secretin and VIP on their surfaces.

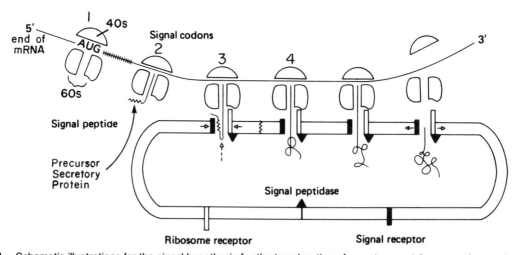

Figure 6.64. Schematic illustrations for the signal hypothesis for the translocation of secretory protein across the rough endoplasmic reticulum membrane. Various stages of the translation of a messenger RNA for a secretory protein on a membrane-bound polysome are indicated. The polysome contains six ribosomes. The first two ribosomes near the 5' end of the mRNA are not yet bound to the membrane. The nascent chain is indicated to grow in a tunnel in the large ribosomal subunit. The signal peptide portion is indicated as a zig-zagged line at the amino-terminus of the nascent chain (see second and third ribosome) or cleaved from the nascent chain and located in the membrane (see membrane between ribosome 3 and 4). The signal-peptidase site of the nascent chain is indicated by an upward pointing *arrowhead* connected to a *dashed line.* Ribosome-receptor and signal-receptor activity are represented arbitrarily by two different integral membrane proteins. Alternatively these two activities could be represented as two separate domains on one integral membrane protein. [From Blobel et al. (1979).]

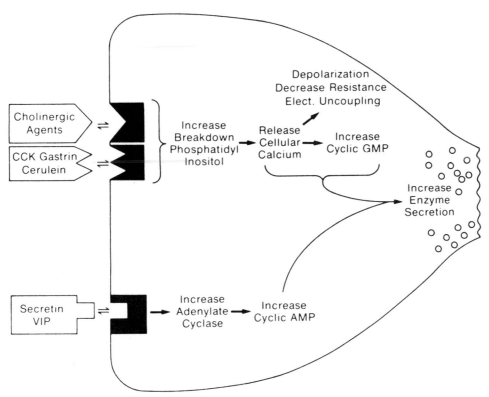

Figure 6.65. Diagram of a pancreatic acinar cell showing receptors for hormones and neurotransmitters and the two major pathways by which occupation of receptors initiates secretion of enzymes. [From Gardner (1979).]

Actions of Hormones and Nerves

The action of hormones and nerves in intact organ systems or intact subjects is more difficult to interpret, but closer to the physiological condition. Perfused isolated organ experiments indicate that electrical nerve stimulation or the addition of hormones can evoke or inhibit secretion. Vascular perfusion of oxygenated blood or chemically defined media can be used and the hormone added to the perfusate. Specific blocking agents for neurotransmitters or sodium channel blockers (tetrodotoxin) can be used to study nervous control. Antibodies raised against specific hormones can be given to determine the role of hormones in basal or stimulated secretion.

The blood levels by radioimmunoassay of hormones in intact animals or human subjects can be determined in response to normal stimuli such as eating. The blood levels can be compared with those obtained following exogenous hormone administration required to produce specific effects on secretion. If the blood levels are similar, physiologic action is possible. It must be remembered, however, that release of neurotransmitters or paracrine secretions will not be reflected in blood levels. Also, physiologically significant events usually alter the release of many hormones.

Removal of specific endocrine glands in the digestive tract is often impossible because of the heterogenous distribution of endocrine cells in the mucosa of the intestine, but nerves can be cut and the resulting deficit in secretion defined. However, denervation may affect a variety of both nervous and endocrine control systems.

Very little has been learned about the myenteric nervous system in the control of secretion. Most of what we know comes from the electrical field stimulation of isolated segments of the digestive tract or the response of isolated mucosae to neurotransmitters.

SALIVARY GLAND SECRETION

The particular interest of secretion of the salivary glands for physiologists is their nearly pure secretory function and the fact that it is under nervous control almost exclusively. Salivary glands in man secrete water and electrolyte, a starch splitting enzyme, amylase, and glycoproteins. The function of the salivary glands is to lubricate food for swallowing, to solubilize material so it can be tasted, to cleanse the mouth and teeth, and to begin the digestion of starch, and to serve as an excretory route for certain organic and inorganic substances.

Composition of Salivary Secretion

The total amount of saliva secreted in man is about 1 to 1.5 l/day (1 ml/min), much of which is secreted at meal time. The resting level of salivary secretion is less than 0.025 ml/min. At rest the parotids secrete about 25%, the mandibular glands 70%, and the sublingual glands 5% of the total salivary secretion. Saliva is distinctive among digestive gland secretions because it is somewhat hypotonic. The composition of saliva depends upon the rate of secretion and the specific stimulus (Hightower and Janowitz, 1979). Representative examples of the composition of stimulated (reflex) and unstimulated parotid and submaxillary secretion in man are shown in Tables 6.4 and 6.5. Figure 6.66 shows the relationship of electrolyte concentrations to flow in mammalian salivary glands. In man, unstimulated secretion from the parotid shows a negative correlation between flow and total solids, viscosity, and the concentrations of sodium and magnesium. Sodium concentration was correlated

Table 6.4
Composition of normal unstimulated human parotid saliva

Potassium	36.7 mEq/l
Sodium	2.6
Calcium	3.0
Magnesium	0.3
Chloride	24.8
Bicarbonate	1.0
Phosphate	15.3
Urea	19.6 mg/100 ml
Uric acid	9.5
Glucose	1.0

From Mandel and Wotman (1976).

Table 6.5
Composition of stimulated human parotid and mandibular saliva

	Parotid	Mandibular
Potassium	20mEq/l	17mEq/l
Sodium	23	21
Calcium	2	3.6
Magnesium	0.2	0.3
Chloride	23.0	20.0
Bicarbonate	20.0	18.0
Phosphate	6.0	4.5
Urea	15.0 mg/100 ml	7.0 mg/100 ml
Uric acid	3	2.0
Glucose	<1	<1
Total lipids	2.8	2.0
Proteins	250	150.0

From Mandel and Wotman (1976).

positively (Shannon, 1969). In stimulated parotid saliva, calcium and sodium concentrations and amylase concentration correlated positively with flow rate, while inorganic phosphate concentration showed a negative correlation. Potassium concentration was unrelated to flow (Blomfield *et al.*, 1976). These results show the importance of reflex (oral citrate) stimulation on primary secretion in the parotid. There was a substantial secretion of calcium independent of amylase secretion.

Structure-Function Relationships

The salivary glands are tubuloalveolar structures as shown in Figure 6.67. The cells lining the alveoli are known as end piece cells (Young and van Lannep, 1978). They consist of three cell types: serous cells, mucous cells, and seromucinous cells (Fig. 6.68). The first type have small granules containing amylase and variable amounts of mucopolysaccharides. The second contain droplets of mucus and acid mucopolysaccharides. Seromucinous cells contain both acidic and neutral mucopolysaccharides. The parotid is largely a serous cell gland (Fig. 6.69), the sublingual largely mucous, and the mandibular a mixed gland. The serous cells in these glands tend

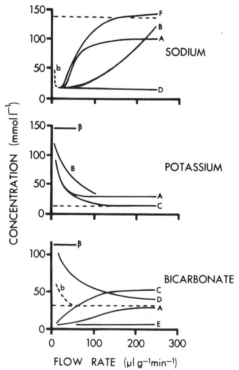

Figure 6.66. Relationship of concentration of salivary electrolytes and flow rate of saliva. *Horizontal broken lines* indicate what the concentrations of each ion might be in primary salivary secretion. *A, B, C, D, E* and *F* = shape of curves in different salivary glands (parotid, mandibular, and lingual glands) and in different species. Curve *B* in the *upper panel*, curve *A* in the *middle panel*, and curve *C* in the *lower panel* represent the relationships seen in human parotid and mandibular glands. *b* in the *upper panel* indicates the pattern at low flow rates from the human parotid and *b* in the *lower panel* the pattern in the rat mandibular gland. [From Young (1979).]

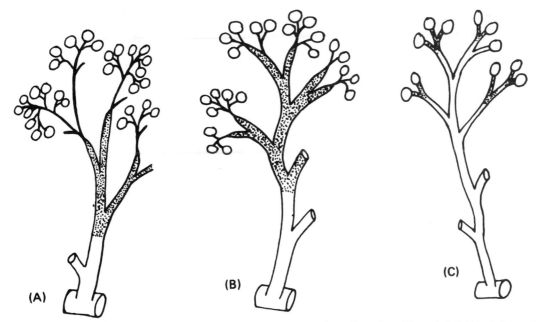

Figure 6.67. Ramifications of human salivary duct systems. *A*, parotid; *B*, submaxillary; *C*, sublingual. *Solid black*, intercalated ducts; *stippled*, striated ducts; *unshaded*, excretory and interlobular ducts. [From Leeson (1967).]

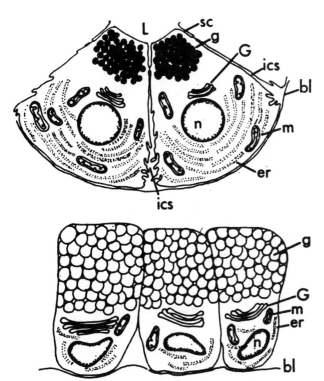

Figure 6.68. Diagram illustrating the relationship between secretion and shape of secretory cells and end pieces. *bl*, basal lamina; *er*, rough endoplasmic reticulum; *G*, Golgi complex; *g*, secretory granule; *L*, lumen; *ICS*, lateral intercellular space; *m*, mitochondria; *n*, nucleus; *sc*, secretory canaliculus. *Top*, Serous cells with proteins in secretory granules. Basal region is expanded to accommodate the rough endoplasmic reticulum (*bottom*). Epithelial cells containing large mucus-containing secretory granules and relatively little er. These cells have a columnar shape. [From Young and van Lannep (1978).]

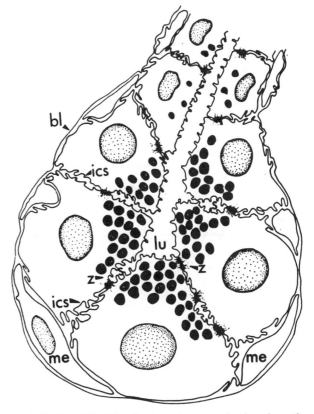

Figure 6.69. End piece of a typical serous gland such as the parotid, showing secretory canaliculi opening into lumen. The canaliculi abut junctional complexes which divide the canaliculi and the lateral intercellular spaces (*ics*). Myoepithelial cells (*me*) and the basement membrane (*bl*) are also shown. [From Young and van Lennep (1978).]

to be at the opposite pole of the acinus from the duct, forming a crescent or demilune. They usually communicate with a lumen via a system of secretory capillaries which pass between the mucous cells.

The ducts of the salivary glands are known as intercalated, striated, and excretory ducts. The intercalated ducts are lined with a low epithelium in which the nuclei occupy much of the cell. The striated ducts take their name from the appearance of basal striations which represent extensive infolding of the basal cell membrane around the mitochondria, arranged in a series of parallel columns. They resemble certain renal tubular cells. The excretory ductal cells line the more distal ducts emptying into the major duct.

The structural arrangement is compatible with a primary secretion of the end piece modified by absorption and/or secretion in the ducts. Micropuncture has been possible only at the level of the intercalated ducts so that the composition of fluid in the end piece is unknown. The structure of the salivary glands is notoriously variable between species.

One interesting cellular structure in the salivary glands is the myoepithelial cell shown in Figure 6.70 which surrounds the ducts and end pieces and can compress them, forcing saliva toward the main duct. A similar cell is found in the ducts of the breasts. They must be considered in interpreting increases in the flow of saliva during sympathetic stimulation.

The blood supply to the salivary glands is characterized by a rich plexus around the ducts. This is compatible with an important role for the ducts in modifying secretion from the alveoli. Figure 6.71 shows a model of the formation of saliva, emphasizing the importance of a dual capillary bed.

The innervation of the salivary glands is supplied by parasympathetic cholinergic nerve fibers and sympathetic adrenergic fibers. Recent studies with immunocytochemical techniques have demonstrated the presence of immunoreactive VIP in nerve fibers in the salivary glands of rat, cat, and man. They can be seen around the acini, blood vessels, and ducts (Uddman et al., 1980b). Substance P is also present in nerves of the salivary glands (Young, 1979).

Figure 6.72 shows the parasympathetic innervation of the salivary glands. The sympathetic supply is derived from the lateral horn in the spinal cord between the first and second thoracic segments. Axons from the lateral horn leave the cord via the ventral roots and enter the paravertebral sympathetic trunk, continuing cephalad to the superior cervical ganglion where they synapse with ganglion cells. From these neurons, axons arise and follow the arterial blood supply to the salivary glands.

Nerve fibers to the end pieces and ducts are classified as epilemmal if they do not penetrate the epithelial basement membrane and hypolemmal if they do so. Epilemmal nerve fibers are about 100 nm from the cell surface; hypolemmal are 20 nm or less. Most hypolemmal fibers are cholinergic (Garrett, 1975). Myoepithelial cells are supplied mostly by sympathetic nerve fibers (Darke and Smaje, 1971). In glands where the nerves were close to the

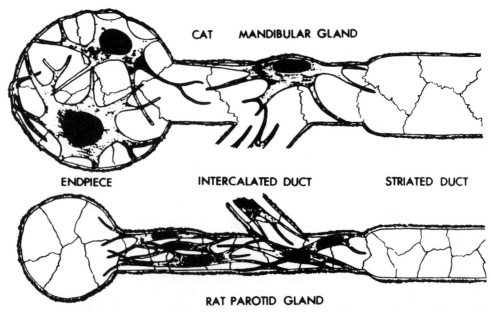

CAT   MANDIBULAR GLAND

ENDPIECE   INTERCALATED DUCT   STRIATED DUCT

RAT PAROTID GLAND

Figure 6.70.   Diagrammatic representation of the arrangement of myoepithelial cells seen in the cat mandibular gland above, and the rat parotid below. [From Young and Lannep (1978), adapted from Garrett (1976).]

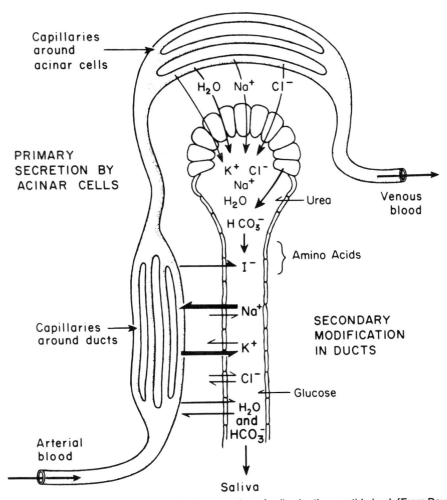

Figure 6.71.   A scheme of electrolyte exchanges during secretion of saliva by the parotid gland. [From Davenport (1977).]

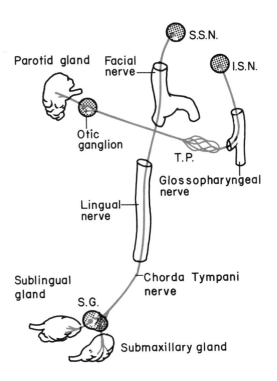

Figure 6.72. Diagram of the parasympathetic nerve supply of the salivary glands. *S.S.N.*, superior salivary nucleus; *I.S.N.*, inferior salivary nucleus; *S.G.*, submaxillary ganglion; *T.P.*, tympanic plexus. [From Hightower (1966).]

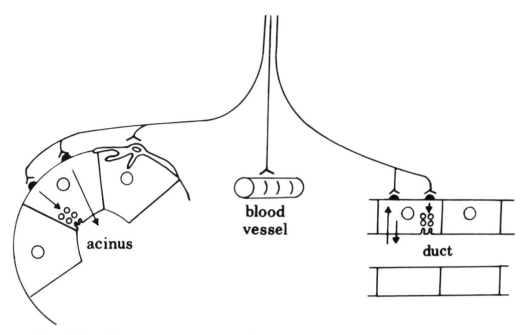

Figure 6.73.   Diagram showing distribution of nerves to a salivary gland. [From Emmelin (1981).]

secretory cells, single electrical impulses initiated a response, but where the distance was greater, multiple impulses were required (Garrett, 1974) (Fig. 6.73).

Micropuncture Studies

The relation of structure to function has been clarified in part by micropuncture studies. Fluid from the intercalated ducts in experimental animals indicate that at this level the fluid consists of a plasma-like isotonic fluid. This represents the primary secretion (Martinez, 1964). In the ducts, absorption of sodium, but not water, leads to hypotonicity. Potassium is secreted in the ducts (Young et al., 1967).

Cellular Aspects of Salivary Secretion

The mechanism of these shifts in electrolytes is not clear, but presumably there is an electrogenic sodium-potassium pump in the ducts. The primary secretion is not due to filtration because it continues at ductal pressures equal to the systemic blood pressure.

In man there is no significant secretion from the major salivary glands during sleep, so that spontaneous secretion in the true sense does not exist. Secretion occurs in response to stimulation by way of the parasympathetic and sympathetic nerves. The neurotransmitters at the neurosecretory junction are acetylcholine and norepinephrine. End piece cells are supplied by both (Fig. 6.73). The flow of saliva depends presumably upon the number of

end piece cells secreting. When norepinephrine interacts with α-adrenergic receptors adenylcyclase is activated, and the production of intracellular cyclic AMP is increased. This second messenger leads to the production of primary secretion of water and electrolytes. At present, we cannot distinguish between secretion from the end piece cells and those in the intercalated ducts. Acetylcholine acts to stimulate primary secretion in most salivary glands. Its mechanism of action in stimulating the secretion of water and electrolyte is unknown.

Studies with in vitro vascularly perfused rabbit mandibular glands show that neither chloride nor bicarbonate in the perfusate are essential for primary secretion. At present, it is not possible to distinguish between electrically neutral cotransport of sodium chloride and a double counter transport (sodium for hydrogen and chloride for bicarbonate) as a mechanism for primary secretion (Case et al., 1981). Cholinergic stimulation evoked a brisk flow of saliva and isoprenaline stimulated secretion rich in potassium, bicarbonate, and protein. Sympathetic nerve stimulation and phenylephrine (an α-agonist) had little effect (Case et al., 1980). Using acetylcholine in the perfusate, an initial release of potassium into the perfusate was followed by a period of increased uptake. Therefore, changes occur in both the permeability of the luminal and contraluminal membranes of the cells. In this preparation one cannot distinguish between end piece and ductal cells (Peterson, 1970). The end piece epithelium is quite "leaky," and most of the salt and water probably enters by way of paracellular channels (Fig. 6.71).

While the rate of salivary secretion is determined primarily by end piece secretion, its composition reflects events in the ducts. The principal role of the ducts is in absorption, but secretion also occurs. The ductal epithelium is very impermeable to water. The absorption of sodium and chloride proceeds at a faster rate than the secretion of potassium and bicarbonate. As a result, saliva becomes hypotonic. There is evidence that chloride transport out of the ducts occurs passively, while there may be active transport of bicarbonate into the ducts and of sodium out of the ducts (Young, 1979). The relatively high intracellular concentration of chloride in cells of salivary glands is not yet explained (Peterson, 1981). The secretion of potassium and bicarbonate into the ducts occurs in response to both sympathetic and parasympathetic nerve stimulation (Schneyer et al., 1972). As wherever bicarbonate secretion is found, the presence of carbonic anhydrase in ductal and serous cells may be significant (Leder and Tritschler, 1966).

Table 6.6 lists the proteins in parotid and mandibular saliva. Secretion of proteins occurs in response to both parasympathetic nerve stimulation. Amylase is the protein which has been studied most extensively. It consists of at least two families of isoenzymes which can be differentiated by electrophoresis from pancreatic amylase (Keller et al., 1971). It can serve as a marker for end piece secretion much as inulin marks glomerular filtration in the kidney. The mechanism of amylase secretion is quite similar to that in the exocrine pancreas (see section on Secretion), except that the rate of secretion is slower and storage is more prolonged (Castle et al., 1972). However, here too, there is evidence that amylase may pass directly from the endoplasmic reticulum to the lumen without passing into secretory granules (Garrett et al., 1977). Glycoproteins in saliva include neutral sulfated and sialic acid-containing mucins (Tabak et al., 1982). Figure 6.74 shows an example of a portion of the peptide chain of a glycoprotein and its sugar side chains. The protein portion of mucin is quite constant, but the sugars vary. Blood group substances represent one group of salivary glycoproteins. They are produced in the mucus cells (Fig. 6.68). Secretory piece is made either in the ductal or seromucinous end piece cells. It may bind to the IgA dimer at the basolateral membrane of the cell and then move to the apical portion for secretion (Young and van Lennep, 1979). Kallikrein, a serine protease, is found within intracellular membrane-bound granules of ductal cells of the striated ducts of the

Table 6.6
Proteins in human parotid and mandibular saliva

	Parotid	Mandibular
Acinar cells	Amylase[c]	Amylase[a]
	Cationic glycoproteins[c]	Cationic glycoproteins[a]
	Anionic glycoproteins[a]	Anionic glycoproteins[c]
	Secretory piece	
	Kallikrein	
	Lactoperoxidase	
	Lactoferrin	
Non acinar cells or unknown	Lysozyme[a]	Lysozyme[b]
	Ribonucleases[b]	Ribonucleases[a]
	Secretory IgA	
	Phosphatases	
	Esterases	
	β-glucuronidase	
	Lactic dehydrogenase	

From Mandel and Wotman (1976).   [a] Relatively low concentrations. [b] Moderate concentrations.   [c] High concentrations.

mandibular glands of cats. It is secreted into the ducts in response to reflex stimuli (Bhoola and Dorey, 1971; Maranda et al., 1978). The process of secretion of kallikrein probably involves exocytosis (Emmelin, 1981).

The synthesis of amylase in salivary end piece cells can be dissociated from its secretion during stimulation with epinephrine. Cycloheximide inhibits protein synthesis, but not secretion under these circumstances. Both epinephrine and dibutyryl cyclic AMP can stimulate protein synthesis directly and independently of effects on secretion (Grand and Gross, 1969).

Studies with dispersed rat parotid acinar cells show that isoproterenol, a β-adrenergic agonist, and epinephrine both stimulate amylase secretion into the medium and increase the formation of intracellular cyclic AMP. The effect of epinephrine can be blocked by propanolol, a β-blocker, but not by phentolamine, an α-blocker. Propanolol also blocks the action of isoproterenol, a β-agonist. These observations indicate that β-adrenergic receptors control amylase secretion in end piece cells. Dibutyryl cyclic AMP stimulates amylase secretion even in the presence of β-blockers, and, therefore, must act beyond the site of adrenoreceptors (Mangos et al., 1975a).

Cholinergic substances in the same preparation cause an efflux of potassium. The effect can be blocked by atropine, but not β-blockers (Mangos et al., 1975b).

Calcium uptake in isolated rat parotid exocrine cells is stimulated by α- or β-adrenergic agonists. The release of amylase by α-agonists is dependent upon external calcium in the medium, but that by

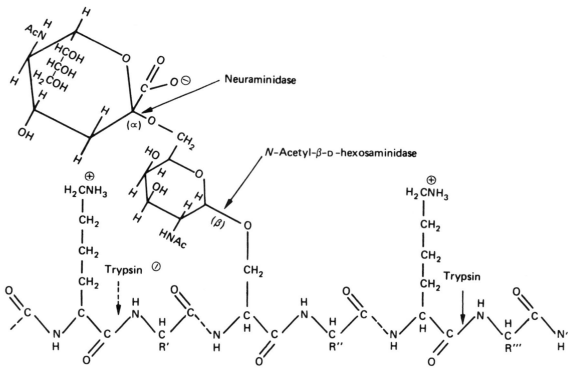

Figure 6.74. Diagrammatic segment of an ovine salivary mucus carbohydrate side chain and the site of cleavage by various enzymes. [From Gottschalk (1966).]

β-agonists is not (Randle et al., 1979). There is a very brief period of calcium efflux followed by an increased uptake. The location of the intracellular pool of calcium is probably in or close to the plasma membrane of the cell (Putney et al., 1981). Two peptides from frog skin, physalaemin and eloidosin, and the neuropeptide substance P increase amylase secretion from end piece cells (Thulin, 1976; Liang and Cascieri, 1979).

Human parotid and mandibular saliva contains both calcium and phosphorus. The concentration of calcium in mandibular saliva is about twice that of parotid saliva. About two-thirds of the calcium is dialyzable, the rest is protein-bound (Mandel et al., 1964). In rat parotid saliva, electrical stimulation of both the parasympathetic and sympathetic nerve supplies caused secretion of saliva with a high calcium concentration, but the secretion of amylase was low in the former and high in the latter, so that sympathetic stimulation leads to the secretion of amylase and calcium in parallel (Schneyer et al., 1977). The volume of saliva and output of calcium were much greater with parasympathetic saliva, but the output of amylase was much greater with sympathetic stimulation. The calcium and amylase content of the parotid was reduced twice as much by sympathetic stimulation. With stimulation of the parasympathetic nerves, calcium was transferred

from the plasma to saliva without packaging in granules, while with sympathetic stimulation all of the calcium was packaged along with amylase (Schneyer et al., 1978). Parotid saliva was secreted at increasing rates as the frequency of electrical stimulation of the nerve increased. Calcium concentration correlated with the rate of flow of saliva, but amylase concentration was independent of flow (Schneyer, 1979). From these results we can conclude that calcium in saliva is found free or associated with amylase and that the ratio of the two will depend upon the nature of the stimulus to secretion.

Studies with isolated fragments of rat parotid or mandibular glands show that adrenergic agonists regulate protein secretion by a partially calcium-independent route involving β-receptors and cyclic AMP. Cholinergic agonists increase cyclic GMP rather than AMP and require calcium (Rossignol et al., 1977). Figure 6.75 shows a model for the action of norepinephrine and acetylcholine from autonomic nerves on the secretion of amylase and water and electrolyte from end piece cells. Acetylcholine, norepinephrine, and substance P receptors can all activate primary secretion by way of calcium-mediated events. The β-adrenergic receptor, however, acts primarily to activate secretion of amylase by way of a cyclic AMP mechanism independent of calcium (Peterson, 1981).

MAMMALIAN SALIVARY GLAND

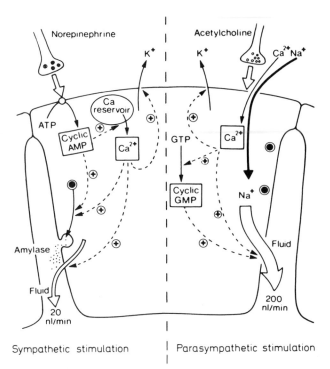

Figure 6.75. Model to account for action of norepinephrine and acetylcholine on end piece cells of rat parotid gland. Norepinephrine, released from sympathetic nerve terminals, interacts with α-adrenergic receptors in basal plasma membrane and stimulates synthesis of cyclic AMP, which then initiates release of amylase at luminal plasma membrane. One action of cyclic AMP may be to release calcium ions from intracellular reservoirs which, in turn, may play a role in the release of amylase, but also initiate an increase in potassium conductance at the basal membrane and provoke secretion of some fluid at the luminal membrane. Acetylcholine, released from parasympathetic nerve terminals interacts with basally located muscarinic receptors and induces an increase in membrane sodium and calcium conductance, which leads to an influx of these ions into the cell. There is a concomitant or subsequent increase in potassium conductance of the basal membrane and efflux of potassium from the cell to the interstitium. These events are associated in some way with increased intracellular cyclic GMP, which, with calcium ions, seems to be responsible for causing vigorous secretion of fluid across the luminal membrane. For simplicity, interaction of norepinephrine with an α-receptor, which mimics the effect of acetylcholine on a muscarinic receptor, is not shown. [From Berridge (1975).]

Electrophysiology of Salivary Secretion

The electrophysiology of the salivary glands is an important example of the relation of electrical properties of cells to their secretion. Salivary end piece cells are electrically unexcitable with respect to secretion. Stimulation of parasympathetic or sympathetic nerves or electrical field stimulation of the salivary glands will produce a biphasic response with hyperpolarization followed by depolarization in end piece cells (Lundberg, 1958; Gallacher, 1979; Peterson, 1981). Figure 6.76 shows an example of

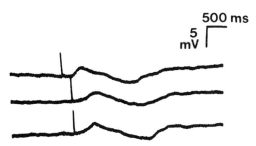

Figure 6.76. Oscilloscope tracing of membrane potential responses to three single shock stimuli from the same parotid acinar cell. The resting membrane potential is −65 mV in each case. *Vertical bars* indicate stimulus artifacts. [From Gallacher and Peterson (1980).]

intracellular recordings in response to field stimulation (Gallacher and Peterson, 1980). Under other circumstances, only depolarization is seen. Field stimulation acts by release of acetylcholine since it is blocked by atropine. These results can be incorporated into a model of primary secretion and the secretion of amylase by end piece cells (Fig. 6.77).

Hyperpolarization appears to be the result of increased permeability of the contraluminal cell membrane to potassium and a reduction in cell membrane resistance. No action potentials result (Peterson, 1981). The depolarization results from an influx of sodium into the cell. The resting membrane potential of cat parotid end piece cells depends upon the activity of a sodium-potassium activated ATPase (Fritz, 1972). Substance P, acetylcholine, and epinephrine all reduce input resistance with an initial depolarization in isolated segments of rat parotid gland (Gallacher and Peterson, 1980).

The nature of the electrical message in nerves stimulating the salivary glands under normal conditions is an important problem. Stimulating the parasympathetic nerves at a frequency of 4–8 Hz seemed to produce a flow similar to that during a meal (Emmelin, 1967). On the other hand, the decrease in vascular resistance on cat mandibular glands and the release of immunoreactive VIP was greater with short bursts of electrical stimulation at high frequencies, e.g., 20 Hz (Anderson et al., 1982).

Nervous Control of Salivary Secretion in Vivo

Direct electrical stimulation of the parasympathetic or sympathetic nerves can be carried out only in experimental animals. Parasympathetic nerve stimulation elicits higher flows than sympathetic stimulation. End piece cells may be more sensitive to acetylcholine than to norepinephrine. It is known that several parasympathetic axons innervate the same end piece cell. However, if sympathetic stim-

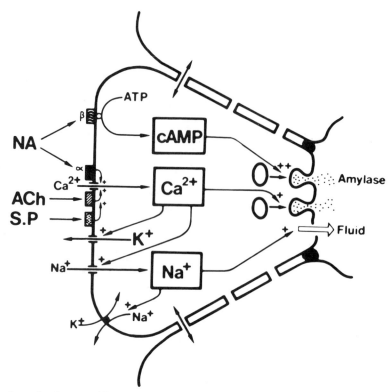

Figure 6.77.   Shows a schematic diagram of intracellular events after excitation of cholinergic, α- and β-adrenergic, and peptidergic receptors in parotid acinar cells. *NA*, norepinephrine; *SP*, substance P. [From Peterson (1981).]

ulation is superimposed on a background of submaximal parasympathetic stimulation an increase in flow results (Emmelin, 1981).

In experimental animals the salivary glands can be stimulated by intraarterial or intravenous doses of acetylcholine and catecholamines, particularly β-adrenergic agonists. In the cat, pilocarpine, a cholinergic agonist, elicits profuse salivary secretion. The rat parotid gland responds to epinephrine, norepinephrine, and phenylephrine (an α-agonist) by secreting the same pattern of proteins determined electrophoretically. However, the dose of each agent shifted the pattern, depending upon whether high or low doses were given (Abe et al., 1980). In man norepinephrine and epinephrine probably do not function physiologically as hormones.

Stimulation of the two different divisions of the autonomic nervous system can lead to the secretion of different proteins in mandibular saliva in the cat. Peroxidase was secreted in response to both parasympathetic and sympathetic electrical stimulation, but acid phosphatase was secreted only in response to parasympathetic stimulation (Garrett and Kidd, 1977).

Based upon information obtained by micropuncture of the intercalated ducts in rat parotid, the amylase and potassium concentration in primary

secretion was much greater after stimulation with acetylcholine than with isoproterenol (Mangos et al., 1973).

Ductal Secretion

The secretory function of the ducts can be studied using an in vitro perfused duct system. Cholinergic agonists reduce net ductal absorption of sodium and chloride (Martin et al., 1973). β-Adrenergic agonists increase sodium absorption, but decrease potassium secretion without changing net water movement (Schneyer and Thavornthon, 1973). These results indicate that nervous control can exert an important control on the composition of saliva through an action on the ducts.

Other control mechanisms in the ducts include the role of calcium in the perfusate. Calcium binders in the perfusate cause a marked increase in net calcium absorption from the ducts and a smaller net secretion of potassium (Schneyer, 1974a). The cellular cytoskeleton is also involved, since inhibitors of microfilaments in the apical region of ductal cells reduce the secretion of potassium (Schneyer, 1974b).

The function of the ducts can be explored by injecting inhibitors of transport in a retrograde fashion into the ducts. Inhibitors abolish the usual relationship of sodium and chloride concentrations to flow by inhibiting the reabsorption of sodium in the ducts (Yoshimura, 1967; Mangos et al., 1973).

Reflex Control of Salivary Secretion

Under physiological conditions most nervous excitation of salivary secretion is reflex induced in response to either unconditioned or conditioned reflexes. Conditioned salivary reflexes were popularized by Pavlov, who conditioned dogs to salivate at the sound of a bell. Mechanoreceptors and chemoreceptors in the oral cavity respond to chewing and taste excitants with an unconditioned reflex salivary response appropriate to the stimulation, e.g., dry solids stimulate a profuse watery flow of saliva. The pathway of the reflexes includes an integrating center in the medulla. The area concerned with the parotid lies caudal to those for the mandibular glands in the cat. These integrating centers are subject to control from the cerebral cortex, amygdala, and hypothalamus in relating salivation to feeding patterns. The area of the medulla where the salivary nuclei are located is close to those nuclei regulating respiration and vomiting as shown in Figure 6.78.

The phenomenon of "paralytic" secretion in the salivary glands following denervation is an important example of sensitization of secretory cells to neurotransmitters following denervation. The secretion following parasympathetic denervation is best seen in response to acetylcholine (Ekstrom, 1980; Emmelin, 1981).

Hormones play little role in the short-term control of salivary secretion. However, aldosterone can influence the ratios of sodium to potassium concentrations in saliva. Angiotensin may also influence the composition of saliva (Young and Van Lennep, 1979). Hypophysectomy results in atrophy of the salivary glands.

Response to Food

The most important physiological stimulus to salivation is food. Rabbits secrete much larger amounts of saliva in response to pellets containing cellulose than to carrots. The response was mediated by cholinergic nerves. Frequencies of nerve stimulation required to duplicate the salivary response were 1–5 Hz for the parasympathetic nerves and 1 Hz for the sympathetic nerves (Gjorstrup, 1980). Amylase was secreted at approximately the same rate after eating carrots or pellets.

Cell Turnover, Growth, and Adaptation in Salivary Glands

Changing from a liquid to a solid diet resulted in an increase in DNA, RNA, and weight of rat parotid

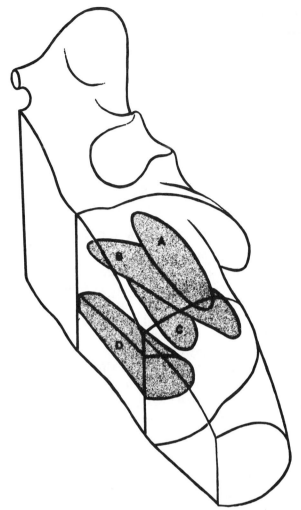

Figure 6.78. Perspective illustration of anatomical relationships within the medulla of responsive regions for *A*, spasmodic respiratory movement; *B*, salivation; *C*, vomiting; and *D*, forced inspiration. [From Borison and Wang (1949).]

glands. Elimination of either division of the autonomic nervous division reduced the response, but cutting the parasympathetic supply abolished it (Schneyer and Hall, 1976). Rat parotid glands in animals on a liquid diet weigh less, but are more sensitive to stimulation with cholinergic agonists than those of rats on a solid diet (Ekstrom and Templeton, 1977). Amputation or extraction of the inferior incisors in rats leads to hypertrophy of the mandibular glands in rats (Perec et al., 1964). Isoproterenol, a β-adrenergic agonist, given subcutaneously to rats for 3–11 days will produce hypertrophy of both the parotid and mandibular gland (Pohto and Paasonen, 1964). The increase in DNA synthesis in mouse parotid glands during the administration of isoproterenol is preceded by a sharp increase in the incorporation of uridine into cytoplasmic RNA (Novi and Baserga, 1972). These observations demonstrate the importance of neurotransmitters in mediating trophic influences on the salivary glands and in participating in adaptive mechanisms.

Blood Flow and Salivary Secretion

The relation of blood flow to secretion in the salivary glands is an example of the increase in blood flow which occurs in response to the local increased metabolic demands of the secretory process. In the salivary glands this is mediated by noncholinergic vasodilator nerve fibers. It now appears that these nerves release vasoactive intestinal peptide (VIP) (Bloom and Edwards, 1980).

Pathophysiology of Salivary Secretion

Distribution of the salivary glands in the course of irradiation treatment of malignant tumors of the head and neck or loss of secretory function of the glands in Sjögren's disease can be expected to exacerbate dental caries and impair swallowing. Cystic fibrosis or mucoviscidosis is associated with increased viscosity of exocrine secretions. The parotid glands of patients with this disease secrete salivas with higher than normal sodium concentrations, suggesting a deficit in sodium absorption in the ducts (Marmar et al., 1966; Mandel and Wotman, 1976).

GASTRIC SECRETION

The function of gastric secretion is to maintain relative sterility of the stomach, to digest food, to promote the absorption of vitamin B_{12} and some forms of iron, and to protect the gastric mucosa from injury.

Normally, the contents of the stomach contain only a few ($<10^3$ colonies/ml) organisms of the type normally present in the oral cavity. The small intestinal content contains similarly few organisms, while distal to the ileocecal sphincter counts rise to 10^{11} to 10^{12} colony-forming units/ml. Loss of the ability to secrete hydrochloric acid by the stomach is followed by bacterial overgrowth in both the stomach and upper small intestine. After partial surgical removal of the acid-secreting mucosa of the stomach, bacterial infections with certain enteric organisms become more common and the number of ingested organisms required to produce infection is reduced.

Hydrochloric acid and pepsinogens begin the digestion of proteins, but there are few data on their quantitative importance. Nevertheless, the digestion of animal tissues which can occur in the stomach is very impressive. A lingual lipase may contribute importantly to the digestion of fat in the newborn (Hamosh, 1979). The normal absorption of vitamin B_{12} (cyanocobalamine) is dependent upon secretion of acid and the glycoprotein intrinsic factor. In the absence of secretion of intrinsic factor by the oxyntic cells in man, a profound and eventually fatal anemia due to arrested naturation of red blood cells will develop (pernicious anemia).

Hydrochloric acid secretion is necessary for the normal absorption of nonheme dietary iron and for the ferric form of iron (Powell and Halliday, 1981). An old problem is how does the stomach avoid digesting its own mucosa? One of the protective factors is the secretion of bicarbonate by surface epithelial cells in the stomach (Flemstrom and Garner, 1982). Another is the secretion of mucus. A layer of mucus delays the diffusion of hydrogen ion to the cell surface, especially at levels of hydrogen ion concentration found in the fasting stomach (Bahari et al., 1982).

Structural Basis for Gastric Secretion

The gastric mucosa, like the musculature, is divided into two portions: the proximal two-thirds which contains the fundic acid and pepsin-secreting glands, and the distal third which contain bicarbonate-secreting pyloric glands. The endocrine cells which secrete gastrin are found in the pyloric glands. Figure 6.79 shows a fundic gland and the three types of cells present: acid-secreting oxyntic or parietal cells, chief cells which secrete pepsinogens, and mucus neck cells, which are the precursor cells for those lining the surface of the stomach. The pyloric glands are branched glands with the gastrin-containing cells. An individual gastrin-containing cell is shown in Figure 6.80. There are at least eight other endocrine type cells in the gastric mucosa, including those containing somatostatin, glucagon, and bombesin-like material (Ito, 1981).

The morphology of oxyntic cell changes dramatically from the resting to the stimulated state. Figure 6.81 shows the ultrastructure of cells under both conditions. At rest the oxyntic cells contain large amounts of tubulovesicular structures and a few collapsed canaliculi. During stimulation the tubulovesicular structures disappear, and the canaliculi dilate (Frexinos et al., 1971). The simplest explanation is that the tubulovesicles fuse with the canaliculi, and the membrane then everts (Forte et al., 1981). The morphologic changes are an essential feature of actively secreting cells, but may occur in the absence of secretion, as for example, in response to thiocyanate.

The structure of the chief cell resembles that of pancreatic acinar cells. Figure 6.82 shows a diagrammatic representation of a chief cell. The hypothesis is suggested that accumulation of zymogen

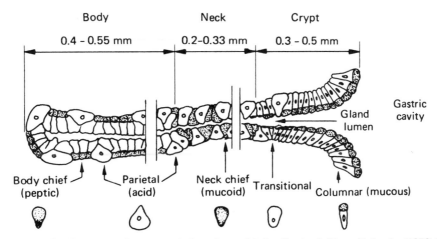

Figure 6.79.   Diagrammatic representation of a gastric fundic gland. [From Hollander (1952).]

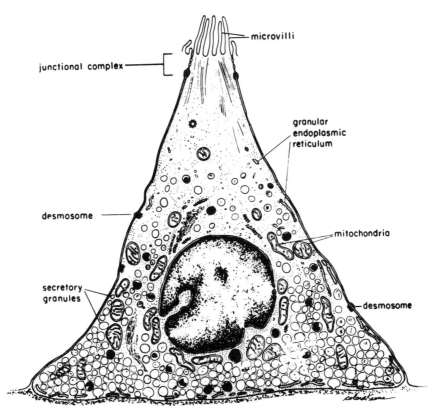

Figure 6.80.   Diagrammatic representation of a gastrin G cell from a pyloric gland. Note that the apical surface reaches the gastric lumen. [From Ito (1981).]

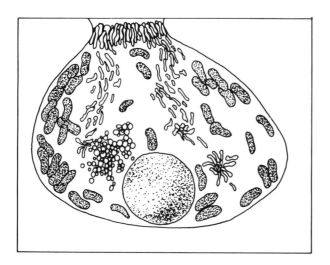

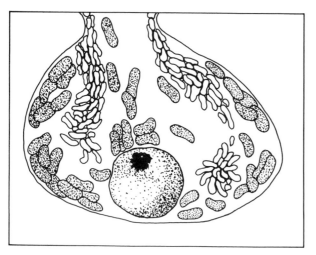

Figure 6.81. Diagrammatic representations of electron micrographs of human gastric parietal (oxyntic) cells. *Upper micrograph* of a resting cell, note tubulovesicles. *Lower micrograph* of a stimulated cell, tubulovesicles have largely disappeared, but canaliculi have become more prominent. [Drawn from Frexinos et al. (1978).]

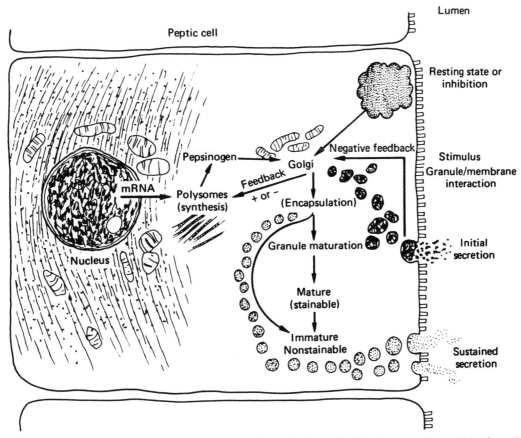

Figure 6.82. A schematic representation of the cellular mechanisms for the synthesis, storage, and secretion of pepsinogen. [From Hirschowitz (1967).]

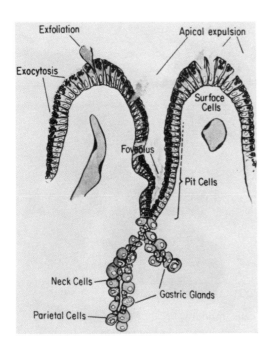

Figure 6.83. Diagrammatic representation of the canine gastric epithelium. Mucus pit cells line the foveolus and during migration are gradually transformed into surface mucus cells lining the interfoveolar surface. Apical expulsion and cell exfoliation most often occur in interfoveolar areas, while exocytosis occurs throughout the gastric epithelium. Sulfated glycoprotein is represented by darker granules within the mucus neck cells of the foveolus. The foveolus opens into the gastric gland whose neck cells, unlike mucus pit cells, stain only for neutral glycoproteins. [From Zalewsky and Moody (1979).]

granules in the resting state inhibits synthesis of the enzymes and that a nonstainable immature form of pepsinogens exists in the cytoplasm.

Mucus-secreting cells are formed as mucus neck cells in the fundic glands and move up on to the surface of the pits or foveoli and onto the surface of the gastric mucosa. Mucous granules are extruded by exocytosis (Fig. 6.83). There is also exofoliation of mucous cells and finally apical expulsion by fusion of granules and pinching off the apical membrane (Zalewsky and Moody, 1979). Intrinsic factor is associated with tubulovesicles of human oxyntic cells at rest and migrates to the periphery of the secretory canaliculi with the onset of secretion (Levine et al., 1981).

Table 6.7
Quantitative analysis of cells in the body mucosa of five human stomachs

Total Weight of Stomach	Weight of Body of Stomach as % of Stomach Weight	Weight of Mucosa of Body of Stomach as % of Stomach Weight
143 ± 18 g	83 ± 2%	38 ± 2%

Cellular Composition as % of the Mucosa of the Body of the Stomach

Oxyntic Cells	Chief Cells	Surface Mucus Cells	Mucus Neck Cells
32 (24–38)	26 (20–30)	17 (12–20)	6 (4–9 range)

From Hogben et al. (1974).

Quantitative assessment of the human stomach gave the data shown in Table 6.7 for the various cell types (Hogben et al., 1974). There is a linear or slightly curvilinear correlation between the number of oxyntic cells and the acid output after a maximally stimulatory dose of histamine as shown in Fig. 6.84 (Card and Marks, 1960). Mucosal cell replacement in the gastric mucosa occurs within 4–6 days. Renewal of parietal and chief cells is much slower (Lipkin, 1981). Growth of the gastric mucosa is subject to humoral control. Elevated levels of gastrin in the peripheral blood are associated with increased number of oxyntic cells in patients with gastrinomas. Gastrin is a powerful trophic agent in rodents increasing the number of oxyntic cells. However, in normal man, large pharmacologic doses of pentagastrin were needed to induce proliferation of the fundic mucosa. There was no change in the antral mucosa (Hansen et al., 1976).

The Composition of Gastric Secretion

Gastric secretion includes hydrochloric acid, sodium and potassium, bicarbonate, pepsinogens, intrinsic factor, and glycoproteins. When collecting gastric content by stomach tube, it should be recalled that there will be swallowed saliva and duodenal content as well as endogenous gastric secretion. The composition of gastric secretion varies significantly during a 24-h cycle. Figure 6.85 shows

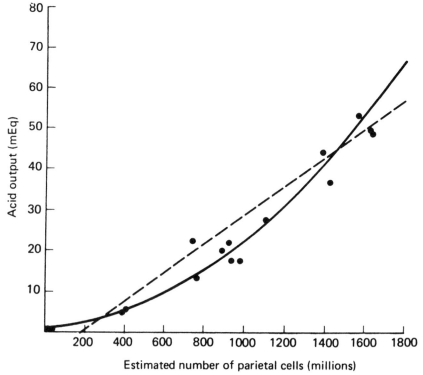

Figure 6.84.   Relationship of acid output to number of gastric parietal cells (oxyntic cells). [From Card and Marks (1960).]

the changes in acid output in man over 24 h. The explanation for the peak between 8 P.M. and midnight is unknown, but is clearly not related to changes in immunoreactive gastrin in the blood. There are probably short lasting increases in acid output which occur coincident with the interdigestive migrating motor complex (MMC). Basal acid output averages 0.5–2.0 mM/h with a concentration of 10–40 mM/l. Figure 6.86 shows representative data (Feldman and Richardson, 1981).

By far the most important stimulus to acid secretion is food. Two techniques have been developed to measure acid output in response to a meal: intragastric titration and double perfusion studies of the stomach and duodenum. In general, both show that the response approaches the maximal acid-secreting capacity for the stomach: 25–35 mM/h. Figure 6.87 shows the response to a meal in healthy subjects using intragastric titration with alkali (Fordtran and Walsh, 1973). The meal is in liquid form, and the pH is held constant at pH 5.5, so that the serum gastrin rises to higher than normal levels. The acid output to a mixed solid and liquid meal in six healthy subjects is shown in Figure 6.88 using the marker perfusion technique. Note that the concentration of hydrogen (titratable acidity) ion remains elevated for over 4 h.

There is an inverse relationship between hydrogen ion and sodium ion concentrations in the stomach. In general, as the rate of gastric secretion increases, the concentration of hydrogen ion rises until it approaches 160 mM/l. The chloride ion concentration always exceeds that of hydrogen ("neutral chloride"). The difference becomes small at maximal rates of acid secretion. As noted earlier,

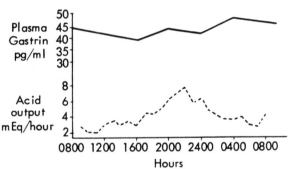

Figure 6.85. Diurnal variation in fasting human basal gastric acid secretion and serum immunoreactive gastrin. [From Moore et al. (1973).]

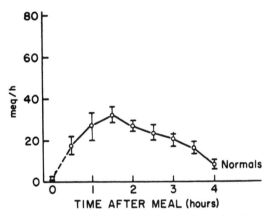

Figure 6.87. Acid output in normal man, determined by intragastric titration, after a steak meal. [From Fordtran and Walsh (1973).]

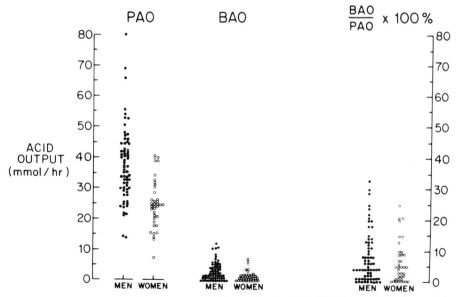

Figure 6.86. Peak acid output (*PAO*), basal acid output (*BAO*), and ratio of BAO:PAO in 105 healthy volunteers. Ages averaged 29 years for men and 31 years for women. [From Feldman and Richardson (1981).]

acid-secreting capacity depends upon the number of oxyntic cells, and it is assumed that oxyntic cells respond in an all-or-none fashion. The nonacid secretion of the stomach has a far more limited capacity. Therefore, maximal acid secretory responses are diluted to only a small extent by nonacid secretions.

Potassium is secreted by the stomach in concentrations higher than in the plasma. Pepsinogens are large protein molecules. Pepsinogen I is secreted only by chief cells in the fundic mucosa. Pepsinogen II is secreted by chief cells in both fundic and pyloric gland-containing mucosa (Samloff, 1982). In the lumen of the stomach, in the presence of acid, 42 amino acids at the amino end of pepsinogen

are removed, and the molecule becomes an active protease.

Recently, the structure of gastric mucus has been defined in its native state. Figure 6.89 shows the structure of native mucus and the action of pepsins in hydrolysing it in the gastric lumen (Allen and Garner, 1980). The sialic acid portion of the side chains appears to determine the viscous properties of mucus.

Properties of the Gastric Mucosa

The physical properties of the acid-secreting mucosa and the pyloric mucosa differ, with the latter being more permeable or "leaky." This probably reflects intercellular junctions which selectively permit passage of cations (Powell, 1981). Figure 6.90 shows the structures participating in the mucosal barrier. The metabolic integrity of the mucosal cells also permits them to dispose of potentially injurious ions. The mucosal surface maintains a negative electrical potential difference (PD) with respect to the serosa. The PD is -5 mV in the duodenum, -25 at the pylorus, -35 in the body of the stomach, and -12 in the esophagus (Geall *et al.*, 1970). This probably reflects active transport of chloride into the lumen by surface epithelial cells. The presence of a mucus layer on the surface of human gastric mucosa in vitro reduced the hydrogen ion concentration from 5.6 mM in the lumen to 10^{-4} at the junction between the mucus and the surface epithelium (Bahari et al., 1982).

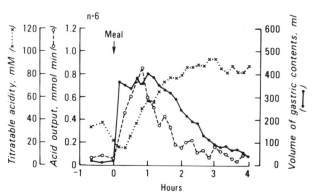

Figure 6.88. Acid output, volume of secretion, and acid concentration in healthy human subjects after ingestion of a solid-liquid meal. [From Malagelada (1981).]

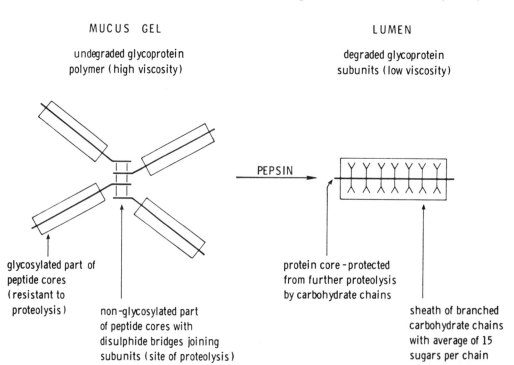

MUCUS GEL

undegraded glycoprotein polymer (high viscosity)

LUMEN

degraded glycoprotein subunits (low viscosity)

PEPSIN

glycosylated part of peptide cores (resistant to proteolysis)

non-glycosylated part of peptide cores with disulphide bridges joining subunits (site of proteolysis)

protein core - protected from further proteolysis by carbohydrate chains

sheath of branched carbohydrate chains with average of 15 sugars per chain

Figure 6.89. Peptic hydrolysis of gastric mucus glycoprotein. [From Allen (1980); adapted from Allen (1978).]

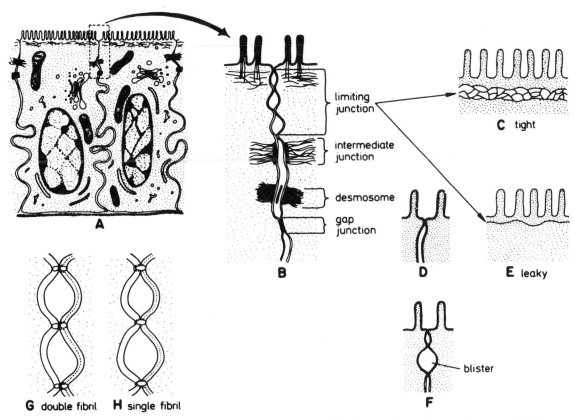

Figure 6.90. Intercellular junctions of epithelia. Limiting or tight junctions differ between tight (*C*) and leaky (*E*) epithelia. Tight epithelia have junctions composed of several strands of integral membrane protein, whereas leaky epithelia have only one or two strands. Hypertonic mucosal solutions can cause blisters or blebs in tight junctions (*F*). Junctional fibrils may have single or double fibrils. [From Powell (1981).]

Cellular Physiology of Gastric Secretion

ACID AND ELECTROLYTE SECRETION

The cellular physiology of acid secretion has been worked out using four experimental preparations: isolated gastric mucosa, isolated oxyntic cells, isolated vesicles from oxyntic cells, and isolated fundic glands. Studies on isolated gastric mucosa established histamine as a potent stimulant to acid secretion. Recent results indicate that the initial acid secretory response to histamine in piglet mucosa is preceded by a rise in tissue cAMP and a decrease in electrical resistance; however, once acid secretion is established, the levels of cAMP fall despite continued acid secretion (Machen et al., 1982). Therefore, cAMP can be regarded as a trigger turning on the acid secretory response, but other messengers must be involved to continue it. The major limitation of this methodology is the uncertainty over which type of cell is responding, given the heterogenous composition of the mucosa.

Isolated oxyntic cells have permitted a vigorous analysis of receptors, even though hydrogen ion secretion cannot be measured directly because of the concurrent release of hydroxyl ions into the medium. Oxygen consumption, changes in the appearance of oxyntic cells, and accumulation of weak bases such as aminopyrine within the canaliculi can be taken as evidence of acid secretion. Three types of specific receptors can be demonstrated: those for histamine, acetylcholine, and gastrin. Potentiation between all three agonists can be shown. Assuming that endogenous histamine and acetylcholine are always available to oxyntic cells, the effects of giving a single exogenous secretogogue in vivo should be easily seen. Both H-2-blocking agents such as cimetidine and anticholinergics such as pirenzepine reduce the response of acid secretion to a variety of stimulants in vivo as would be expected from the in vitro results. Figure 6.91 shows a model of receptors on the oxyntic cell and how they might mediate the secretory process. Cyclic AMP is the initiating messenger for histamine-stimulated acid secretion. Cyclic AMP is not involved in the action of gastrin or acetylcholine. The action of the latter is dependent upon calcium entry into the cell and can be blocked by lanthanum which reduces calcium flux across membranes and displaces surface-bound calcium. Anticholinergics such as atropine are specific blocking agents for

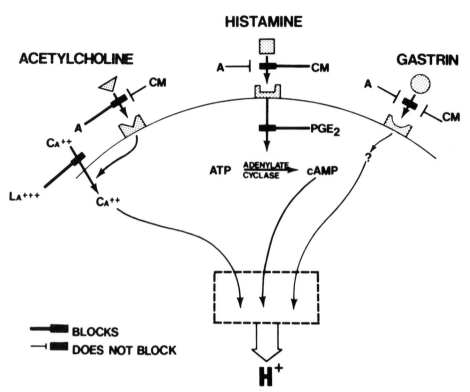

Figure 6.91. Shows a model for the actions, interactions, and effectors for secretagogue stimulation of the function of isolated parietal cells. Atropine (*A*) and cimetidine (*CM*) act as inhibitors at the receptor sites. Acetylcholine increases calcium (*C*A) influx while lanthanum (*L*A) blocks it. The role of calcium in the action of gastrin is uncertain. Histamine stimulates adenylate cyclase to form cAMP. Prostaglandin E_2 inhibits stimulation. [From Soll (1981).]

acetylcholine. Calcium probably is required for the action of gastrin as well. Cimetidine blocks the action of histamine (Soll, 1981). Mast cells from canine fundic mucosa can synthesize histamine and may be the source of histamine in canine mucosa (Beaven et al., 1982). Histamine-stimulated increased production of cAMP as well as acid secretion can be blocked by prostaglandins of the E_2 series. However, the effect of dibutyryl cAMP is not prevented, indicating that prostaglandins act prior to cyclic AMP. It is possible that prostaglandins have a local role in the modulation of acid secretion. Evidence for a role for microfilaments in acid formation is provided by inhibition of acid secretion in isolated piglet oxyntic cells by cytochalasin B (Black et al., 1982). Despite these clear results in animals, it should be noted the human oxyntic cells have not yet been tested in vitro for their responses to secretogogues or inhibitors.

Vesicles prepared from isolated parietal cell microsomes are oriented in such a way that the outer surface represents the normal cytosol-facing surface. In the presence of ATP and of potassium inside the vesicle, vesicles take up hydrogen ion. This preparation lead to the isolation of an enzyme, H^+-K^+-ATPase, which provides the major energy

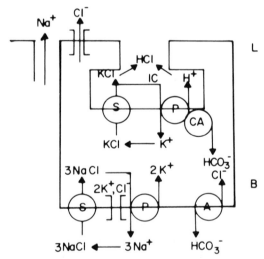

Figure 6.92. Model of secreting parietal cell showing the various pathways present on apical, canalicular, and basolateral surfaces. *IC*, intracellular canaliculus; *L*, lumen; *B*, blood; *S*, symport; *A*, antiport; *P*, pump;][, conductance. [From Sachs and Berglindh (1981).]

source for the secretion of HCl. Subsequent studies have demonstrated that this enzyme is specific for the parietal cell and located on the surface of the canalicular membrane. Figure 6.92 shows a model of the system for hydrogen ion secretion. On the basolateral surface of the oxyntic cell, there are two exchange mechanisms: a bicarbonate-chloride and a sodium-potassium exchange as well as a portal for the entry of sodium-chloride. On the canalicular membrane, there is a port for the exit of potassium chloride and the H^+-K^+-ATPase pump for the secretion of hydrogen ion. Carbonic anhydrase is located near the ATPase, which provides a mechanism for converting the remaining hydroxyl ion to bicarbonate (Sachs and Berglundh, 1981).

The last preparation to be considered is the isolated fundic glands. The secretion of acid by the glands can be assessed by the same criteria used for isolated oxyntic cells. They respond readily to histamine and acetylcholine. Addition of a phosphodiesterase inhibitor is necessary to bring out the response to gastrin. Thiocyanate inhibits the accumulation of aminopyrine in response to histamine, but not the increase in tissue cAMP or structural changes in the oxyntic cells. It probably increases a proton leak (Hersey et al., 1981). The Na^+-K^+-ATPase inhibitor ouabain inhibits accumulation of aminopyrine in this preparation. The effect can be blocked by increasing the extracellular concentrations of potassium or by decreasing the concentration of sodium. The potassium content is probably regulated by a Na^+-K^+-ATPase (Koelz et al., 1981a). Prostaglandin E_2 reduced aminopyrine accumulation (acid secretion) in response to histamine in rabbit fundic glands (Levine et al., 1982). Gastrin stimulates accumulation of aminopyrine and potentiates the effects of histamine (Chew and Hersey, 1982). Perhaps of greater interest is the observation that a substituted benzimidazole compound inhibits accumulation of aminopyrine and hence acid secretion in human isolated fundic glands by inhibiting the H^+-K^+-ATPase (Olbe et al., 1982).

Chloride secretion by frog gastric mucosa occurs as a result of active transport, a passive leak, or exchange diffusion. The active transport of chloride is linked to hydrogen ion transport (Forte, 1969). Exchange of chloride for bicarbonate across the basolateral membrane is essential for acid secretion (Durbin, 1977). Both hydrogen and chloride secretion are dependent on sodium in the bathing medium (Machen and McLennan, 1980). The cellular sources of potassium and chloride outside the oxyntic cell have not been demonstrated in vitro. However, an extensive series of studies in vivo have shown that the surface epithelial cells and the cells of the pyloric glands can secrete a fluid resembling an ultrafiltrate of plasma.

These observations lead to the two component hypothesis which proposes that gastric secretion represents a mixture of parietal (oxyntic) and nonparietal components. If we assume a fixed composition of the parietal component then the amount and composition of the nonparietal component can be calculated. There are differing opinions of the extent to which the nonparietal component can vary in its rate of secretion (Makhlouf, 1981).

The best evidence for attributing secretion of nonparietal electrolytes to specific cells relates to the secretion of bicarbonate by surface epithelial cells in both the body and antrum of the stomach. Both in vitro and in vivo secretion of bicarbonate can be stimulated by acetylcholine, calcium, and prostaglandin E_2. In vitro, it is inhibited by anoxia, cyanide, and carbonic anhydrase inhibitors. Secretion of bicarbonate requires chloride on the luminal side, suggesting that a bicarbonate-chloride exchange may take place. Since there is little change in transmucosal potential difference, it is likely that secretion is an electrically neutral exchange at the luminal cell membrane. Quantitative estimates of the amount of bicarbonate secretion during the basal state show that it is equal to about 5–10% of the maximal hydrogen ion-secreting capacity for a given species. Some ultrafiltration of bicarbonate across the mucosa probably also occurs (Flemstrom, 1981).

PEPSINOGEN AND INTRINSIC FACTOR SECRETION

The cellular physiology of the secretion of pepsinogens has not been explored to the extent of that of other enzymes. From organ culture studies in the rabbit, it is known that the mucosa can incorporate ^{14}C-leucine into protein and that about 65–90% of the protein is pepsinogen. Acetylcholine stimulated protein secretion and in small amounts potentiated secretion in response to pentagastrin and CCK octapeptide (Sutton and Donaldson, 1975). In these experiments, the medium was buffered. When the medium was acidified, pepsinogen secretion doubled (Kapadia and Donaldson, 1978). In contrast to pepsinogen, the control of intrinsic factor (IF) secretion in the same organ system was sensitive to histamine and cyclic AMP. A phosphodiesterase inhibitor (IBMX) increased the secretion of IF 4-fold, while increasing tissue levels of cAMP. Dibutyryl cyclic AMP increased IF secretion to the same extent. Similarly, IBMX potentiated the secretion of IF in response to histamine. This action could be blocked by cimetidine (Kapadia et al., 1979). Acidification of the medium blocked the secretion of IF. In isolated rabbit fundic glands, both carbachol and isoproterenol (a β-agonist) stimulated pepsinogen release through a cyclic AMP-dependent, calcium-independent mechanism (Koelz et al., 1981a). These results emphasize the role of cholinergic mechanisms and acid in the secretion of pepsinogens. Intrinsic factor secretion is linked to histamine and cAMP.

GLYCOPROTEIN SECRETION

The synthesis of glycoproteins in surface epithelial cells proceeds like that of other proteins until the proteins reach the rough endoplasmic reticulum and Golgi apparatus, where the sugar side chain is added. The sugars are added stepwise one after another. The reactions are catalyzed by enzymes known as glycosyltransferases. There is a separate enzyme for every sugar linkage. As many as 16 different enzymes can be involved in the biosynthesis of a carbohydrate side chain of human gastric mucus glycoprotein. Some of the glycoproteins have blood group activity. Only the enzymes specific for the subject's blood group will be present in the mucosa to add the appropriate terminal sugars (Allen, 1981).

In the Golgi apparatus, the glycoproteins are incorporated into zymogen granules. It takes about 2 h for the labeled amino acids and sugars to reach the granules. Contractile protein filaments and microtubules may play an important role in the release of secretory granules from surface epithelial cells (Ito, 1981) (Fig. 6.93). Little is known of nervous or hormonal control mechanisms.

Gastrointestinal Hormones: Gastrin and Somatostatin

The cellular physiology of the secretion of gastrointestinal hormones is difficult to study, because the cells of origin are scattered among other cell types in the mucosa. We shall consider only two in the stomach: gastrin and somatostatin. Most of these experiments have been done in rats. Immunoreactive gastrin can be demonstrated within granules 150–200 mμ in diameter in the space between the nucleus and the basal surface of cells in the antral mucosa of man (Greider et al., 1972). Somatostatin-like material is found in D type (pancreas) cells (Chiba et al., 1981). In normal man, antral biopsies showed a ratio of 4.4 ± 0.4 gastrin to somatostatin cells (Arnold et al., 1982). Gastrin-containing cells from rat antra can be cultured in vitro. Release of gastrin can be accelerated by di-

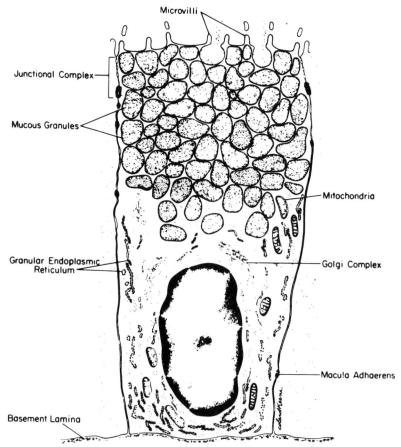

Figure 6.93. Surface mucous cell of small intestinal mucosa. [From Ito (1981).]

butyryl cAMP and phosphodiesterase inhibitors (DeSchryver-Kecskemeti et al., 1977). Involvement of microfilaments and microtubules is suggested by the effects of colchicine and deuterium oxide (DeSchryver-Kecskemeti et al., 1980).

Rat antral tissue in organ tissue has been shown to incorporate ^{3}H tryptophane into immunoreactive gastrin. Cyclic AMP increased both synthesis and secretion of gastrin (Harty et al., 1977). Secretion of gastrin into the medium was increased by the cholinergic agonist carbachol. Synthesis of gastrin also increased. An anticholinergic agent had no effect on basal secretion of gastrin, but inhibited carbachol-induced secretion in a dose-related manner (Harty and McGuigan, 1980). The release of gastrin by dibutyryl cyclic AMP is calcium-dependent, and calcium ionophores increase gastrin release (Harty et al., 1981a). Bovine serum albumin and 10% peptone also release gastrin (Lichtenberger et al., 1978).

Somatostatin inhibits gastrin release from rat antral cells in tissue culture secreted in response to arginine (DeSchryver-Kecskemeti and Greider, 1976). In the rat organ culture preparation, bovine serum albumin and peptides increase the release of somatostatin from the antrum (Lichtenberger et al., 1981). Exogenous somatostatin inhibited the basal release of gastrin without affecting protein synthesis. It also blocked carbachol-stimulated gastrin release (Harty et al., 1981b). Antisomatostatin antibodies caused a disappearance of free somatostatin from the medium while increasing gastrin release (Chiba et al., 1981). Somatostatin added to the medium decreased gastrin release. These results indicate a paracrine action of somatostatin in the inhibition of gastrin release as suggested by Larsson (1980) based upon anatomical considerations (Fig. 6.94). Note that somatostatin can inhibit acid secretion by either a direct action on oxyntic cells or indirectly through inhibition of gastrin release according to this model.

The isolated perfused rat stomach can be used to study the secretion of gastrin and somatostatin. Several groups of investigators have shown that addition of exogenous somatostatin to the perfusate inhibits gastrin release (Saffouri et al., 1980). Addition of antibodies to somatostatin to the perfusate resulted in disappearance of free somatostatin from the venous effluent, while not changing or increasing the level of gastrin (Saffouri et al., 1981; Chiba et al., 1981). Secretin and vasoactive intestinal polypeptide (VIP) increased somatostatin release. There is a reciprocal relationship between gastrin and somatostatin release: methacholine increases gastrin release while inhibiting the release of so-

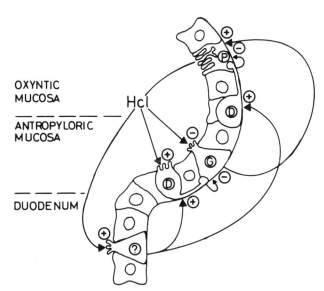

Figure 6.94. Diagram of oxyntic cells (*P*), somatostatin-containing cells (*D*), and gastrin-containing cells (*G*) in gastric and duodenal mucosa. [From Larsson (1980).]

matostatin (Schubert and Makhlouf, 1982). Prostaglandins have the opposite effects. About 80% of the gastrin was secreted into the blood and 20% into the lumen of the stomach (Saffouri et al., 1980). These results have been confirmed (Martindale et al., 1982). Bombesin, on the other hand, stimulated the release of both (DuVal et al., 1981). Electrical stimulation of vagal secretory fibers reduced the level of somatostatin in the portal effluent to control levels during an intravenous infusion of gastric inhibitory peptide (GIP) (McIntosh et al., 1979). This could be prevented by atropine.

The neural control of the release of gastrin and somatostatin is controversial. There is agreement that the release of gastrin is stimulated by both bombesin and acetylcholine, while somatostatin inhibits gastrin release directly. Somatostatin release was inhibited by carbachol. The disagreement is over the role of ganglionic (nicotinic) cholinergic receptors. One group finds no involvement of nicotinic receptors in the action of carbachol (Martindale et al., 1982). The other group finds that nicotinic receptors can be stimulated to release somatostatin by a noncholinergic pathway, while releasing gastrin by both cholinergic and noncholinergic pathways (Schubert and Makhlouf, 1982).

The ex vivo perfused dog stomach releases gastrin in response to electrical stimulation of vagal secretory fibers (Lanciault et al., 1975; Lefebvre et al., 1978). Exogenous somatostatin prevented the release of gastrin by carbamylcholine (carbachol) and also reduced the release of glucagon (Lefebvre et al., 1981). Experiments in cats also showed increased release of gastrin and decreased release of somatostatin during electrical vagal stimulation (Uvnas-Wallensten et al., 1980). The release of gastrin by both carbachol and bombesin from antral mucosa in vitro has been noted in dog antra (Weiss et al., 1977).

The control of gastrin and somatostatin release is reciprocally related with cholinergic transmitters favoring gastrin release and inhibition of somatostatin release. Both bombesin and acetylcholine released from nerve terminals may act directly on the gastrin-releasing cell (Fig. 6.95).

There is increasing evidence of a role for adrenergic nerves in the control of gastrin release. Distention is a well recognized stimulus to gastrin release as measured by increases in serum gastrin levels in the intact human stomach (Soares et al., 1975). Release can be blocked by the β-blocker propanolol and increased by atropine and phentolamine, an α blocker. These effects can be obtained in the absence of an effect on gastric acid secretion. The results suggest that there are secretory β-adrenergic and inhibitory α-adrenergic pathways to the gastrin cell in the human stomach (Peters et al., 1982).

Other mechanisms stimulating gastrin release include the topical application of acetylcholine to the antral mucosa and the presence of small peptides or selected amino acids in gastric content (Taylor et al., 1982a). The latter probably accounts for the elevation of serum immunoreactive gastrin in man after a protein mean (Richardson et al., 1976). Fasting for 4–10 days depresses the level of plasma gastrin (Uvnas-Wallensten and Palmblad, 1980). Presumably, there are chemoreceptors in the gastric mucosa which either act through reflexes in the wall of the stomach or by way of vago-vagal reflexes to release gastrin. The most important luminal factor in the control of gastrin release is the hydrogen ion concentration. An intragastric pH of less than 2.5 prevents the rise in plasma gastrin in response to amino acids (Walsh et al., 1975). Raising the pH to 5.5 results in the removal of inhibition of gastrin release. The signal to reduce gastrin secretion presumably reaches the gastrin cell by way of the microvillous apical surface, which is exposed to contents in the lumen. Proximal gastric vagotomy in patients with duodenal ulcer results in a basal fasting plasma gastrin level about a third higher than before the vagotomy. Gastrin levels in response to a meal approximately double before vagotomy and nearly triple after vagotomy, even if the gastric pH is maintained at the same level throughout (Feldman and Richardson, 1981).

Stimulation of the efferent fibers of the vagus

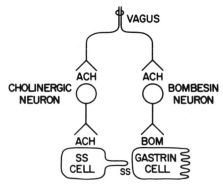

Figure 6.95.   Model for control of the gastrin release from the G cell. [From DuVal et al. (1981).]

nerve results in release of gastrin into the blood. In man, this is demonstrated by a modified technique of sham feeding in which the patient chews food, but expectorates it. The mean increase in serum gastrin is about 15 pg/ml. Other means of vagal stimulation in man, such as insulin hypoglycemia or 2-deoxy-D-glucose, also elevate serum gastrin levels. In anesthetized experimental animals, the motor vagal fibers can be stimulated electrically. The serum gastrin levels approximately double. In the cat, a rise in the small molecular species of gastrin accounts for most of the elevated gastrin in arterial blood (Bauer et al., 1977). The increase in gastrin can be blocked by acid perfusion of the stomach or by antrectomy in dogs (Lanciault et al., 1973).

In man, gastrin is found in cells of the antral glands and in the duodenum. The principal molecular species of gastrin in the antrum is the 17-amino acid peptide G-17. There is also a large amount of the N-terminal half of the G-34 molecule produced by hydrolysis of G-34 with trypsin (NTG-34). This fragment is secreted into the plasma along with G-17 and G-34 and presumably packaged in the same zymogen granules (Pauwels and Dockray, 1982). These observations support the concept that gastrin is synthesized as G-34 and subsequently subjected to limited proteolysis in a manner similar to that of insulin and proinsulin (Dockray, 1981).

In the duodenum, the predominant form of gastrin is G-34 (Lamers et al., 1982). This is also the form preferentially released into the blood from the human duodenum. The metabolism of gastrin is an important determinant of the concentration of gastrin in the blood. Most of the important studies have been done in the dog, but with human synthetic gastrins. They show that small carboxyl terminal fragments, such as pentagastrin, are selectively removed from the circulation in the liver and that about half of the labeled, but biologically inactive material, appears in the bile (Stagg et al., 1971). Most of the larger molecular forms of gastrin are removed in the mesenteric circulation (Temperley et al., 1971; Strunz et al., 1978). In dogs, the half-life of human G-34 was 16 min compared to 3 min for G-17. On the other hand, G-17 was 5 times more potent on a molar basis (Walsh et al., 1974). Similar studies showed that human minigastrin (G-14) and little gastrin (G-17) had similar half times in the circulation and were equivalent in potency (Carter et al., 1979). More recent studies of blood from the antral or portal veins in dogs showed that less than 10% of the total immunoreactivity could be attributed to G-17, and about 10% to G-34. The remainder was largely due to fragments of G-17. It may be that endogenous G-17 in the dog is cleaved into fragments while passing from the G-cell into the antral veins by enzymes in the endothelial cells. These fragments may be the most important physiologic secretogogues in the dog (Dockray et al., 1982).

Somatostatin is present in higher concentrations in the alimentary tract in the stomach. There are two molecular forms of somatostatin, SS-14 and SS-28. The larger form is probably a prohormone. There is a rise in immunoreactive plasma somatostatin after a meal, but little is known about the control of its release under these circumstances. The half-life in the circulation is estimated at about 2 min (Walsh, 1981).

Nervous Control of Gastric Secretion

INTRINSIC NERVES

Little has been written about the relationship between nerve endings and oxyntic cells. Recently, acetylcholine esterase-positive nerves were demonstrated in direct contact with oxyntic cells having varying amounts of acetycholine-esterase in their cytoplasm in rabbits. This suggests a direct cholinergic innervation of the oxyntic cells (Hanker et al., 1977). The submucosal enteric plexus is poorly developed in the stomach. Attempts have been made to selectively destroy the submucosal nerve supply by injecting alcohol into the submucosa of vagally denervated pouches of dogs. The gastric acid secretory response to both histamine and a meal were markedly reduced (Schapiro et al., 1977). As yet, no study of the electrical activity in the myenteric plexus of the stomach is available. Cholinergic neurons excite secretion of acid, contractions of smooth muscle, release of gastrin, and increase of mucosal blood flow. There is also evidence for cholinergic inhibition of gastrin release and for relaxation of gastric tonic contractions in the body and fundus of the stomach.

Norepinephrine is released from adrenergic nerves in the stomach. There is little direct evidence for their role in the control of acid secretion under normal circumstances. In experimental animals, norepinephrine usually inhibits acid secretion, presumably by inhibiting the action of acetylcholine at the level of ganglion cells of the enteric nervous system.

There is immunocytologic evidence for VIP, enkephalins, and substance P in intrinsic nerves of the stomach, but their role in gastric secretion is uncertain. The ability of the H_2 blocker cimetidine to inhibit all types of stimulation of gastric acid secretion implies that histamine participates in the action of nerves on oxyntic cells. One possibility is that cholinergic nerves release histamine from mast-like cells in the mucosa.

Cholinergic nerves excite the secretion of pepsinogen and intrinsic factor from chief cells and oxyntic cells, respectively. As noted earlier, the intracellular pathways for stimulation of pepsinogens and intrinsic factor are different in the rabbit. Cholinergic nerves also stimulate an alkaline secretion, resembling an ultrafiltrate of plasma and mucus in dogs (Altamirano, 1963). In amphibia cholinergic stimulation increases the permeability of surface epithelial cells to chloride (Shoemaker et al., 1970). Little is known of the physiologic actions of other neurotransmitters on the secretion of macromolecules or nonparietal secretion.

EXTRINSIC NERVES

The major extrinsic nerves for the control of gastric acid secretion are the vagi and the splanchnics. The most direct approaches to the role of extrinsic nerves is to determine the effect of electrical stimulation on secretion and the deficits following cutting the nerves (Bonfils et al., 1979). Electrical vagal stimulation can elicit near maximal achievable acid outputs in the anesthetized cat (Bauer, 1977; Sjodin, 1977). The same result is obtained whether or not the sympathetic component is included in the stimulated nerve trunk. The acid secretory response can be blocked by either cimetidine or atropine. In the ex vivo perfused dog stomach, both vagal stimulation and a permissive level of gastrin are required for an acid secretory response (Adair et al., 1974). In order to study vagal stimulation in intact unanesthetized man, it is necessary to resort to sham feeding, insulin hypoglycemia, or glucocytopenia induced by 2-deoxy-D-glucose (Emas, 1973). In man, sham feeding stimulates gastric secretion. The magnitude of the response is affected by the subject's perception of the desirability of the food (Janowitz et al., 1950). Resection of the antrum and duodenal bulb reduced the acid output to less than half of preoperative levels. Infusions of gastrin did not restore output to control levels, suggesting that in man, direct vagal innervation of the stomach is responsible for a significant portion of the acid secretory response (Knutson and Olbe, 1974). Truncal vagotomy in patients with duodenal ulcer prevented the acid response to sham feeding, but the integrated gastrin output (calculated by the area under the curve of gastrin concentration over time minus the basal acid output) still increased (Feldman et al., 1980; Stenquist et al., 1979). Sham feeding accounted for about one-third of the total acid output in response to a meal (Richardson et al., 1977). Proximal gastric vagotomy, which selectively denervates the acid-secreting mucosa, blocks the acid secretory response to sham feeding (Olbe and Knutson, 1976), suggesting that the direct vagally mediated action on oxyntic cells is the important factor in stimulating acid secretion. On the other hand, others have concluded that the rise in plasma gastrin with sham feeding can contribute to the acid secretory response since acidifying the antrum (inhibiting gastrin release) reduces the acid secretory response to sham feeding by 40–50% (Feldman and Walsh, 1980).

Insulin hypoglycemia stimulates acid secretion, presumably by exciting glucose-sensitive neurons in the hypothalamus, which in turn, stimulate cells in the dorsal motor nucleus of the vagus. Horseradish peroxidase studies trace vagal efferents from the wall of the stomach to the dorsal motor nucleus and the nucleus of the solitary tract in cats (Yamamoto et al., 1977). The motor nuclei to the digestive tract, cardiac, pulmonary, and tracheobronchial branches, appear to be intermixed (Kalia and Mesulam, 1980). Substance P immunoreactivity is present in the periphery of the dorsal motor nucleus (Maley and Elde, 1981). Electrical stimulation of the dorsal motor nucleus increases acid secretion (Wyrwicka and Garcia, 1979). Stimulation is prevented or reduced by vagotomy, atropine, or cimetidine (Carter et al., 1976). The changes in plasma gastrin are affected little, possibly because adrenergic stimulation still occurs. The peak response to an intravenous infusion of insulin is equivalent to that obtainable by pentagastrin or histamine (Farooq and Isenberg, 1975).

Similar results can be obtained with 2-deoxy-D-glucose which produces a block in glucose metabolism within the cell, although the blood glucose may rise. These results indicate that the acid secretory response to sham feeding, insulin hypoglycemia, and 2-deoxy-D-glucose can be used to detect incomplete surgical vagotomies. In rats, the frequency of spike potentials in the central cut end of the vagus

(efferents) approximately doubled after either insulin or 2-deoxyglucose (Hirano and Niijima, 1980). The activity could be halted by giving glucose into the portal vein, suggesting hepatic vagal afferents may carry excitatory signals to the brain in response to hypoglycemia (Sakaguchi and Yamaguchi, 1979).

In dogs, it has been shown that there is both a critical threshold of hypoglycemia and a dose-related effect of insulin in determining acid output in response to a bolus of intravenous insulin (Spencer and Grossman, 1971). Supramaximal doses of insulin inhibit acid secretion.

The effect of vagotomy on acid secretion in man is to reduce basal and peak acid outputs to histamine and pentagastrin, and to abolish an increase in acid output in response to insulin hypoglycemia. In the case of histamine, the effect is to increase the threshold-stimulating dose about 8 times and to shift the dose response to the right (Elder et al., 1972; Rosato and Rosato, 1971). Because of the side effects of large doses of histamine in man, the effect can be seen better in the dog (Hirschowitz, 1977). The decreased response to pentagastrin after vagotomy can be restored to normal by bethanechol, an analogue of acetylcholine (Hirschowitz and Helman, 1982). The acid secretory response to a meal is also reduced by vagotomy. In the dog, the decrease is due primarily to the loss of the early centrally mediated vagal response (Guldvog and Gedda-Dahl, 1980). There is, however, evidence for a loss of a tonic vagally mediated inhibitory influence. Following truncal vagotomy, a vagally denervated fundic pouch increased its acid output in response to a meat meal (Sugawara et al., 1972). In dogs with vagal denervation of the antrum, a subsequent proximal vagotomy increased acid secretion in response to a meal, suggesting that nerve pathways to the fundus mediate the tonic vagal inhibition (Debas et al., 1981). Proximal gastric vagotomy increased gastrin release in response to bombesin 2-fold, indicating that there is also a tonic vagally mediated inhibition of gastrin release (Hirschowitz and Gibson, 1978). There is also an increased release of gastrin after proximal vagotomy in response to 2-deoxy-D-glucose in dogs (Hirschowitz and Gibson, 1979).

The role of the more cephalad portions of the central nervous system in the efferent nervous control of acid secretion is still poorly understood. Removal of the cerebral cortex in cats reduces the acid secretory response to electrical vagal stimulation, while decerebration had no effect (Feng et al., 1983). A single decorticate human subject failed to respond to insulin hypoglycemia (Doig et al., 1953). The intravenous injection of labeled 2-deoxy-D-glucose in the rat can be coupled to scanning techniques which showed that the rate of glucose utilization was increased in the lateral nucleus of the solitary tract, the dorsal motor nucleus, the medial forebrain bundle, the superior olivary nucleus, and the interstitial nucleus of the stria terminalis (Kadekaro et al., 1980). Similarly, labeled insulin injected into the left lateral ventricle of rats showed intense reactions over the medial eminence and the ventral arcuate nucleus (van Hoyten and Posner, 1980). Presently, a vigorous research effort is underway using intraventricular injections and injections into discrete areas of the brain of neuropeptides to determine the physiologic significance of these putative neurotransmitters.

In addition to the release of gastrin by electrical stimulation of efferent vagal fibers, there is evidence that other immunoreactive peptides are released, including insulin, 5-hydroxytryptamine, bombesin, and vasoactive intestinal polypeptide (Uvnas-Wallensten, 1978; Ahlman et al., 1978; Uvnas-Wallensten and Walsh, 1980; Edwards et al., 1978). There is release of gastrin, serotonin, and somatostatin into the lumen of the stomach, as well as into the blood (Hengels et al., 1980; Andersson and Nilsson, 1974; Zinner et al., 1982).

Enkephalin-like immunoreactivity is present in the efferent vagus nerve fibers to the stomach. Evidence that enkephalinergic nerves participate in the control of acid secretion is based upon the observation that the opioid antagonist naloxone reduces basal acid secretion and decreases the acid response to sham feeding by about 50%. It also reduces the response to pentagastrin (Feldman and Cowley, 1982).

In the dog, a pylorooxyntic reflex has been described based upon distention of a vagally innervated antral pouch. Graded increases in serum gastrin concentrations and acid output from a gastric fistula resulted. The increase in gastrin could be prevented by acidification of the antral pouch, but acid output is decreased only modestly. On the other hand, vagal denervation of the antral pouch prevents the acid response of the main stomach to distention with an acidified solution. Therefore, the response to antral distention has two components: acid secretion in response to gastrin release mediated by local reflexes in the antrum, and acid secretion in response to pylorooxyntic reflexes in the wall of the stomach or by way of vago-vagal reflexes (Debas et al., 1975). In man, antral distention with a 150-ml balloon inhibits acid secretion stimulated by pentagastrin. Distention alone inhibits basal acid secretion (Schoon et al., 1978). Somatostatin levels in the blood were unchanged. The evidence that these are nervous reflexes is entirely indirect,

and their quantitative significance in acid secretion in response to a meal is unknown.

The evidence for vagal pathways for the stimulation of secretion of nonacid components in gastric secretion is sparse. Insulin-hypoglycemia does stimulate secretion of a nonparietal secretion in man, which differs from that secreted in response to histamine (Fisher and Hunt, 1950). Insulin hypoglycemia in dogs with total gastric pouches increased both the secretion of aminopolysaccharides and glycoproteins. The effect was abolished by vagotomy (Wise and Ballenger, 1971). Vagotomy in man reduces the increase in secretion of intrinsic factor normally seen with pentagastrin (Vatn et al., 1975). In man, vagal stimulation along with histamine and pentagastrin stimulate the secretion of pepsinogens in parallel with that of hydrochloric acid.

Phases of Gastric Acid Secretion

In order to emphasize the origin of excitatory influences in the sequential stimulation of acid secretion during the course of a meal, gastric acid secretion can be classified as cephalic, gastric, and intestinal. The cephalic phase is mediated through the vagus nerves and the release of gastrin. The stimuli are the sight, smell, and taste of food. The gastric phase is mediated by local gastric and vago-vagal reflexes and the release of gastrin. The stimuli are distention and probably the chemical action of peptides and selected amino acids (Byrne et al., 1975). In dogs the chemical stimuli act directly on the fundic mucosa (Debas and Grossman, 1975; Konturek et al., 1976a). The intestinal phase is mediated through the release of an unknown peptide from the small bowel mucosa. A peptide designated as enterooxyntin, with the appropriate stimulatory properties, has been extracted from porcine intestinal mucosa (Vagne and Mutt, 1980). The stimuli are peptones and amino acids, such as glycine, di, tri, and tetraglycines (Isenberg et al., 1977; Wald and Adibi, 1982).

Inhibition of Gastric Secretion

The stimulatory control of gastric acid secretion is balanced by inhibitory control mechanisms, especially those originating in the duodenum. There is little direct evidence to implicate nerves in this process. The very first portion of the duodenum in dogs appears to be the source of a mucosal factor which can inhibit acid secretion. It is not secretin and has been named "bulbogastrone" (Uvnas, 1971). Inhibition of pentagastrin-stimulated secretion was lost when the duodenal bulb was denervated in dogs, suggesting an enterogastric reflex mechanism (Konturek and Johnson, 1971). The normal stimulus would presumably be hydrogen ions. In dogs antibodies to secretin reduced the plasma immunoreactive secretin level to nearly zero, while the acid secretory response to a mixed meal rose by about 25% (Chey et al., 1981). This suggests that in the dog, secretin is a physiologic inhibitor of acid secretion. In man it may be that secretin is more important as an inhibitor in the fasting state (Schaffalitzky de Muckadell et al., 1981).

One of the first examples of intestinal inhibition of gastric acid secretion was the action of fat in the duodenum. In dogs oleic acid in the duodenum really delays the gastric secretory response of a vagally innervated gastric pouch to a meat meal rather than abolishing it (Long and Brooks, 1965). Saturated and unsaturated fats were equally effective (Brooks et al., 1962). In man, fat in the jejunum inhibited acid secretion in response to a steak meal without altering plasma gastrin levels (Christiansen et al., 1976). Inhibition by fat is reduced by vagotomy. These results suggest a direct inhibitory effect, partially mediated by nerves.

However, in the classical experiments fat inhibited acid secretion from a transplanted pouch in dogs. The blood-borne factor was called enterogastrone. More recently the term has been applied to any humoral agent from the small intestine which inhibits acid secretion. The identification of the hormone(s) has proven to be difficult. CCK was considered a likely candidate, but lack of a suitable radioimmunoassay has made detection in the blood difficult. Assays for enteroglucagon, gastric inhibitory peptide and vasoactive intestinal polypeptide (VIP) failed to clearly identify a specific inhibitor (Christiansen et al., 1979). More recently neurotensin has been suggested as the "enterogastrone." It is released into the blood in response to a meal (Mashford et al., 1978). Intralipid more than doubled blood levels of neurotensin-like immunoreactivity, while amino acids and glucose had little effect (Rosell et al., 1979). During intraduodenal infusions of oleic acid, the blood levels of neurotensin correlated with inhibition of pentagastrin-induced acid secretion (Kihl et al., 1981). The chyme from a mixed meal reinfused into the jejunum inhibited acid secretion in response to a meal. Oleic acid and maltose each inhibited acid secretion, but not casein hydrolysate (Miller et al., 1981).

There is some evidence that the colon can release inhibitors of gastric acid secretion. Perfusions of the colon with glucose and mannitol inhibited acid secretion in response to pentagastrin in man (Jian et al., 1981). Hypertonic glucose and mannitol inhibited pepsinogen secretion.

As a result of the inhibitory mechanisms initiated in the small intestine and the decline in stimulatory factors, the gastric acid concentration declines and eventually reaches its base-line level ending the digestive phase of acid secretion.

Distribution of Blood Flow

Gastric mucosal blood flow at rest accounts for about three-quarters of the total gastric blood flow. In the dog the mucosa of the body of the stomach receives about 63% of the total blood flow to the stomach and the antrum receives about 11% (Shoor et al., 1979). The antrum, therefore, differs from the body of the stomach with respect to mucosal blood flow as it does in motor activity and secretion.

Methods for determining gastric mucosal flow are still not widely used. The aminopyrine clearance method is based upon the principle that a weak base passes through lipid barrier readily in its unionized form, but not when ionized. Aminopyrine diffuses readily from the blood into the oxyntic cells and into the canaliculi. When the canaliculi are secreting acid it becomes trapped and is "cleared" into gastric juice. It can be seen that clearance into gastric juice depends upon maintaining an acid milieu in the stomach.

An important corollary is that the concentrations in plasma and gastric juice should be related according to the following equation based upon the Henderson-Hasselbalch equation.

$$R = \frac{1 + 10^{(\text{pKa}-\text{pH gastric})}}{1 + 10^{(\text{pKa}-\text{pH plasma})}}$$

The pKa of aminopyrine is 5. Assuming a gastric pH of 1 and a plasma pH of 7, the value for R becomes 10^4. Experimentally, the value for R turns out to be in the range of 30–40. Therefore, it is likely that the clearance of aminopyrine is limited by blood flow. If the mucosa clears the drug from the circulation completely in a single passage through the tissue, it can be used to measure gastric mucosal blood flow (Guth, 1981). The R value becomes important when blood flow and acid secretion are both changing. If the R value remains constant when acid secretion is increasing, then it is likely that the increase in blood flow is secondary to the increase in acid secretion. When acid secretion is decreased secondary to a reduction in blood flow, R will fall.

Although there are concerns about toxicity in man, the aminopyrine method has been used by labeling the aminopyrine with [14]carbon. Aminopyrine clearance in five normal subjects was 45 ml/

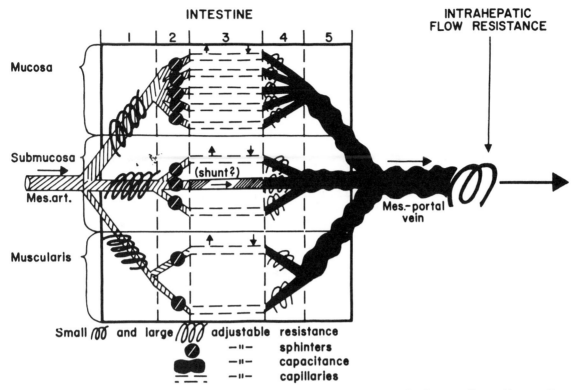

Figure 6.96. Schematic illustration of gastric circulation: *1*, Precapillary resistance vessels; *2*, precapillary sphincters; *3*, exchange vessels; *4*, postcapillary resistance vessels; and *5*, capacitance vessels. [From Folkow (1967).]

min at rest and 150 ml/min after stimulation with pentagastrin. The R value was 55 during stimulation (Guth et al., 1978). The clearance of neutral red by the stomach can also be used as an indication of gastric mucosal blood flow in man. There is a 4-fold increase during stimulation with pentagastrin (Knight and McIsaac, 1977).

In experimental animals gastric mucosal blood flow can be measured by injecting isotopically labeled microspheres into the left ventricle, then killing the animal and determining the radioactivity in the tissues. This method correlated well with the aminopyrine method in anesthetized dogs when histamine was used to increase blood flow, but not with isoproterenol, which increases blood flow but not acid secretion (Archibald et al., 1975).

A new method for determining changes in gastric mucosal blood flow in man is based upon using a coaxial fiber optic bundle introduced through the biopsy channel of a fiber optic gastroscope to detect the spectrum of hemoglobin. This method measures mucosal blood volume. That of the corpus is greater than that of the antrum (Kamada et al., 1982).

Other methods can be used in animals such as the application of electromagnetic flow meters to arteries supplying the stomach and drop recorders to register venous effluent (Guth and Ballard, 1981).

GASTRIC MICROCIRCULATION

The gastric microcirculation can now be visualized in vivo in animals by transillumination. Fluorescent dyes can be conjugated to serum albumin and injected intravenously. Figure 6.96 shows a diagram of the microcirculation. Note the resistance, exchange, and capacitance vessels and the precapillary sphincters. Despite the interest in submucosal shunts in the control of mucosal blood flow, they appear to be of little importance physiologically (Guth and Ballard, 1981).

CONTROL OF THE GASTRIC CIRCULATION

The nervous system is the primary means for the direct control of blood flow to the stomach. In experimental animals, electrical stimulation of the splanchnic nerves reduces gastric blood flow with a relatively greater effect on the mucosal vessels (Svanik and Lundgren, 1977). The systemic blood pressure had to be maintained constantly in order to see the reduction in mucosal flow during splanchnic stimulation (Reed and Sanders, 1971). Adrenalectomy had no effect, so that the action of splanchnic stimulation is a direct nervous effect on the blood vessels. In rats interruption of sympathetic nerves by injecting a local anesthetic into the celiac ganglion or cutting the splanchnics increased the amount of India ink contained in capillaries of the gastric mucosa (Arabehety et al., 1959).

During stimulation of acid secretion with pentagastrin, splanchnic stimulation reduces gastric mucosal blood flow and acid secretion in parallel so that the R value did not change (Blair et al., 1975).

Electrical vagal stimulation of so-called high threshold fibers (see vagus in gastric motility) increases gastric blood flow, due largely to an increase in mucosal flow (Martinson, 1965). There is an increase in capillary permeability and the capillary filtration coefficient (Jansson et al., 1970). Truncal and proximal vagotomy, on the other hand, decreases gastric mucosal blood flow in both man and animals without a change in the R value (Knight et al., 1978; Bell and Battersby, 1968; Hunter et al., 1979). In dogs the reduction persisted for at least 3 months (Nakamura et al., 1974).

In vivo microscopy of the gastric mucosa in rats showed vasodilation in response to vagal stimulation and vasoconstriction during sympathetic stimulation (Guth and Smith, 1977). The neurotransmitter in the sympathetics is norepinephrine. In the case of the vagus, the neurotransmitter is unknown. Recent studies on dog gastric arteries in vitro show that electrical field stimulation caused contraction which was blocked by tetrodotoxin and an α-adrenergic blocker. Acetylcholine alone was without effect, but caused relaxation during field stimulation. It did not relax contraction in response to norepinephrine. It is possible, therefore, that cholinergic nerves decrease release of norepinephrine by sympathetic nerves. Atropine, during electrical stimulation, increases tension in vascular smooth muscle (Van Hee and Vanhoutte, 1978). After vagotomy the dose of histamine required to produce half maximal dilatation of mucosal vessels was reduced by two orders of magnitude (Koo, 1982).

The central nervous system can also influence gastric blood flow. In dogs with electromagnetic flow meters on the left gastric artery, stimulation of the caudal hypothalamus reduced blood flow. The effect was blocked by excision of the celiac ganglion. Stimulation in the rostral hypothalamus increased blood flow (Leonard et al., 1964). In the rat, stimulation of the lateral hypothalamus increased gastric mucosal blood flow. The effect could be blocked by injecting norepinephrine into the lateral ventricle (Osumi et al., 1977).

Histamine greatly increases gastric mucosal blood flow. In dogs it has been shown that this is an effect on the body of the stomach, but not on the antrum (Menguy, 1962). Cimetidine, a potent H_2 blocking agent, has no effect on gastric mucosal blood flow independent of the reduction in gastric secretion in man and animals (Knight et al., 1977; Smy, 1981). A histamine-1 agonist increased gastric mucosal blood flow without an effect on acid secreting, suggesting a role for H_1 receptors in mediating the increase in blood flow (Guth and Ballard, 1981). Norepinephrine in dogs decreased gastric blood flow during a histamine infusion (Cummings et al., 1963).

Gastrointestinal hormones can influence gastric blood flow, but mainly secondarily to their effect on gastric acid secretion. Gastrin and histamine increase blood flow in cats along with acid secretion (Harper et al., 1968). Pitressin and norepinephrine reduce both (Cowley and Code, 1970). Vasoactive intestinal peptide reduced blood flow in dogs with a secondary inhibitory effect on acid secretion (Konturek et al., 1965b). In conscious cats somatostatin decreased blood flow and acid-secretion (Albinus et al., 1977). There was no change in the R value in similar experiments in dogs (Konturek et al., 1976c). It appears that hormones have little direct action on gastric blood flow.

Prostaglandins have mixed effects on gastric blood flow. The PGE series applied topically to the mucosa increases blood flow in the resting stomach of the dog while reducing it along with acid secretion during histamine infusion (Cheung, 1980). Given intraarterially, PGE_1 can increase mucosal blood flow while decreasing acid secretion. Prostacyclins in rats increased gastric mucosal flow while inhibiting acid secretion during stimulation with pentagastrin or a peptone meal. Under similar circumstances PGE_2 reduced gastric mucosal blood flow (Konturek et al., 1979). However, in conscious dogs prostacyclin reduced acid secretion and gastric mucosal blood flow in parallel (Kauffman et al., 1979). These results indicate that prostaglandins acting locally can modify gastric mucosal blood flow.

Autoregulation is a readjustment of vascular resistance to maintain perfusion constant. Gastric blood flow shows this phenomenon in dogs with hypoxia, i.e., increased flows, decreased resistance, and increased oxygen uptake during histamine stimulation (Perry et al., 1982).

Pulsatile perfusion is not required to maintain normal blood flow in anesthetized dogs (Shoor et al., 1979).

As a factor in mediating the gastric response to a meal in conscious dogs, gastric mucosal blood flow nearly tripled in response to a meal (Bond et al., 1979).

Gastric blood flow is controlled by nervous mechanisms, but usually respond secondarily to an increase in acid secretion. Only with marked reduction in mucosal blood flow does it become rate-limiting for gastric secretion.

Pathophysiology of Gastric Secretion

Gastric acid-pepsin secretion plays a permissive role in chronic gastric ulcers and probably is a major contributing aggressive factor in duodenal ulcer. In some patients with duodenal ulcer, there is an increased number of oxyntic cells, an increased sensitivity to many stimuli, and a failure to inhibit gastrin release at the usual level of pH. Acute gastroduodenal erosions are more likely to be related to a failure of local defense mechanisms, such as mucus or sodium secretion or decreased blood flow. The ability of H_2 blocking agents to heal peptic ulcers while blocking all stimuli to acid secretion emphasizes the central role of histamine in acid secretion. Preliminary reports suggest an equally important role for H^+-K^+-ATPase inhibitors. Accelerated gastric emptying removes buffers from the stomach after a meal and results in a more rapid fall in gastric pH after a meal. Hypergastrinemia occurs in two situations at opposite poles of the secretory spectrum: the achlorhydria of pernicious anemia with fundic gland atrophy and the hypersecretion of a gastrinoma.

Pancreatic Secretion

The secretions of the pancreas are involved in the hydrolysis of triglycerides, starch, and proteins. They are largely responsible together with bile for luminal digestion. The secretion of bicarbonate by the pancreas is an important factor in neutralizing gastric hydrochloric acid in the duodenum and in maintaining a pH appropriate for enzymatic digestion. Recent investigations show that blood from the islets of Langerhans flows first by acinar cells. Hence, acinar cells near the islets are perfused with relatively high concentrations of islet cell hormones. This is probably important in the control of acinar cell function. The pancreas is a major site of protein synthesis, and the relationship of synthesis to secretion can be studied in the pancreas to good advantage. Much of our knowledge of the secretion of proteins has come from studies of the exocrine pancreas.

The Composition of Exocrine Pancreatic Secretion

The human pancreas secretes approximately 0.2–0.3 ml/min of a clear colorless liquid during the interdigestive phase and can increase to 3.15 ml/min in response to secretin and presumably in response to a meal (Osnes et al., 1979; Rinderknecht

et al., 1978). Pancreatic juice obtained directly from the pancreatic duct by endoscopic retrograde cannulation (ERCP) shows a basal bicarbonate concentration of 70–80 mM rising to 122 ± 6 mM after stimulation by duodenal acidification. The bicarbonate output rose from 15–23 μmol/min in the basal state to 1.4 ± 0.4 mM/min with duodenal acidification (Hanssen et al., 1978). The maximal capacity of the pancreas to secrete bicarbonate in response to secretin was about 1 mM/min (Farooq et al., 1974). In most normal subjects, the bicarbonate secretory response to exogenous secretin exceeds the acid secretory response to histamine (Banks et al., 1967). In dogs there is a linear correlation between the weight of the pancreas and the maximal bicarbonate output to secretin as seen in Figure 6.97 (Hansky et al., 1963). The normal human pancreas weighs 85–90 g, so that the maximal bicarbonate output per gram of pancreas in man is higher than in the dog (Gorelick and Jamieson, 1981).

Pancreatic secretion contains protein in a concentration of about 7 mg/ml during stimulation by cholecystokinin and secretin. Most of the protein is made up of enzymes. Using two dimensional isoelectric focusing, 19 different proteins were identified in human pure pancreatic juice. They included two α-amylases, a lipase, four procarboxypeptidases, three trypsinogens, a chymotrypsinogen, two proelastases, two colipases, a prophospholipase A_z, and four glycoproteins of unknow functional significance (Scheele et al., 1981). One of the procarboxypeptidases was found in only 4 of 10 subjects. Amylase and lipase are active in pure pancreatic juice, but all of the proteolytic enzymes and colipase are secreted as inactive proenzymes. They are activated by hydrolysis of a portion of the molecule in the lumen of the intestine (see section on digestion). Trypsin and chymotrypsin are endopeptidases which hydrolyze internal peptide bonds, while carboxypeptidases are exopeptidases hydrolyzing only terminal peptide bonds.

Other proteins present in small amounts in pancreatic juice include immunoglobulins (Soto et al., 1977; Shah et al., 1982), kallikrein (Mann et al., 1980), RNase, DNase, trypsin inhibitor (Keller and Allan, 1967), lysosomal enzymes (Rinderknecht et al., 1979), and alkaline phosphatase (Dyck et al., 1978). Small amounts of serum albumin can also be detected.

The concentrations of sodium and potassium in pancreatic secretion are close to those in plasma and independent of the flow rate. Bicarbonate and chloride concentrations vary directly and inversely, respectively, with the rate of flow. The sum remains

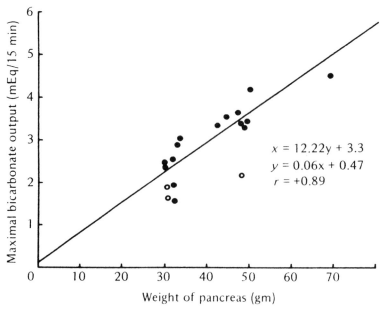

Figure 6.97. Relationship between weight of the pancreas and the maximal bicarbonate output with secretin in dogs. [From Hansky et al. (1963).]

constant (Fig. 6.98). Pancreatic secretion is isomotic with plasma. After prolonged exogenous stimulation at high flow rates, this relationship between flow and bicarbonate and chloride concentrations may no longer hold.

Protein outputs in pancreatic secretions vary with the nature of the stimuli. Cholecystokinin is a potent stimulant to protein secretion, while secretin is a weak one. However, the two secretogogues potentiate each other in the secretion of both bicarbonate and protein at submaximal doses.

A problem of both practical and theoretical importance is the question of nonparallel secretion of proteins in pancreatic secretion. Changes in the relative amounts of lipolytic, proteolytic, and amylotic activities in pancreatic secretion in response to changes in the amounts of carbohydrate, protein, and fat in the diet over a period of days are well recognized, but whether or not short-term alterations occur is controversial. Recent evidence suggests that changes in the ratios of amylase to chymotrypsin and to lipase can occur in man in response to crude preparations of secretin and CCK (Dagorn et al., 1977).

Structural Basis for Pancreatic Exocrine Secretion

The pancreas, like the salivary glands, is a tubuloalveolar gland draining into two major ducts: one joining the common bile duct and the other, the lesser or accessory duct, entering the duodenum independently. The functional unit consists of an

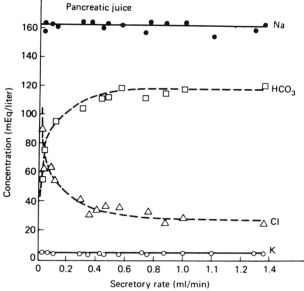

Figure 6.98. Relationships between electrolyte concentrations in pancreatic juice and the volume secreted. [From Bro-Rasmussen et al. (1956).]

acinus and its draining duct. The acinus is lined by cells with zymogen granules, which contain the proenzymes. While many early immunocytochemical studies indicated that all zymogen granules contained a full complement of enzymes, recent studies in rats show that those acinar cells near the islets of Langerhans contain different ratios of proenzymes in their zymogen granules than other acinar cells (Malaisse-Logue et al., 1975).

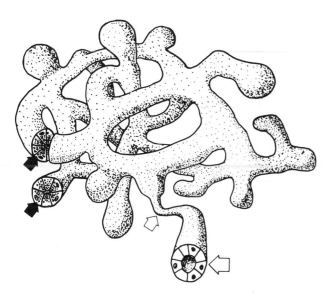

Figure 6.99. Diagrammatic representation of the arrangement of ducts of the exocrine pancreas in the dog. [From Bockman (1978).]

By analogy with the salivary glands one might expect that the acini would secrete a primary solution which would then be modified in the ducts. This is supported by the results of micropuncture of the ducts in rabbits, but acinar fluid has still not been examined (Caflisch et al., 1980). Perfusion studies of the main duct in vitro also indicate that there is a loss of bicarbonate and a gain of chloride, probably in response to exchange diffusion (Scratcherd, 1980).

Recent reconstructions of the pancreatic ducts indicate a high degree of anastomosing between the small ducts in the pancreas of dogs as seen in Figure 6.99 (Bockman et al., 1978). The human pancreas shows a similar arrangement.

The acinar cell is the fundamental unit for enzyme secretion. Figure 6.100 shows a diagram of an electron photomicrograph of an acinar cell. The cells near the islets contain a larger volume of zymogen granules and a larger cytoplasm and nu-

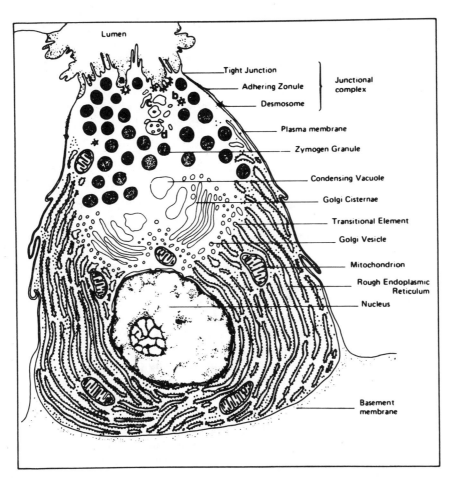

Figure 6.100. Diagram of a pancreatic acinar cell. The sequences of events (*a–d*) illustrated in the apical zone of the cell represents the features of secretion of enzymatic protein *a*, caveolus; *b*, coated vesicle; *c*, endocytotic vesicle; *d*, multivesicle body. [From Case (1978).]

cleus. The centroacinar cells are seen at the junction of acini and ducts. They are smaller than acinar cells and contain large and abundant mitochondria, but no zymogen granules. Both acinar and centroacinar cells have microvilli on their apical surfaces. The enzyme carbonic anhydrase, often related to bicarbonate formation, has been localized by histochemical techniques to centroacinar cells and epithelial cells of the ducts (Kumpulainen and Jalovaara, 1981).

The structural basis for the secretion of proteins by the exocrine pancreas has already been considered. Much of the evidence rests on the time sequence of radiolabeled amino acids as they are incorporated into proteins in the microsomes and then move into the cisternae of the rough endoplasmic reticulum to the Golgi apparatus and into zymogen granules. Figure 6.101 shows the distribution of radiolabeled proteins over time in intracellular organelles after pulse-labeling followed by unlabeled amino acids. Condensing vacuoles are probably newly forming zymogen granules. It follows from this concept that pancreatic proenzymes and enzymes should be secreted in the same ratios as they exist in the zymogen granules. Exceptions to the "parallel" secretion of enzymes exist, and this has been a stimulus to alternative mechanisms of protein secretion. Rothman's equilibrium hypothesis provides for the diffusion of proenzymes from the granules to other compartments and hence exit from the cell by means other than exocytosis (Rothman, 1980).

The processes of protein synthesis, zymogen formation, and exocytosis can be dissociated by pancreatic necrosis-producing agents, such as a choline-deficient diet with ethionine in mice. Under these conditions before necrosis can be detected, exocytosis ceases, zymogen granules accumulate, and yet protein synthesis continues (Koike et al., 1982).

The innervation of the pancreas is by way of the vagi and splanchnic nerves. Most adrenergic nerve terminals end in relation to blood vessels (Legg, 1968). Cholinergic nerve endings can be seen in relation to acinar cells (Watari, 1968; Lenninger, 1974). Recently, VIP-immunoreactive nerve terminals have been seen close to the acini and blood vessels (Sundler et al., 1978). Nerve endings on cell bodies of the small ganglion cells in the pancreas of the cat were found to contain substance P, VIP, enkephalin, and CCK-like immunoreactivity (Larsson and Rehfeld, 1979). These findings suggest that cholinergic fibers are the predominant type of nerves directly innervating acinar cells. Peptidergic nerves may modulate the release of acetylcholine

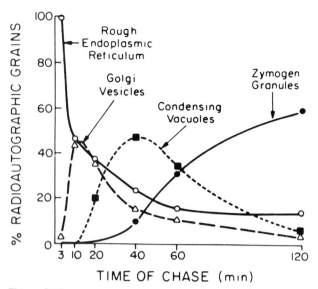

Figure 6.101. Vectorial movement of secretory proteins in the pancreatic acinar cell derived from pulse-chase experiments using radiolabeled amino acids. [From Gorelick and Jamieson (1981).]

from these nerves. Cholinergic nerves also innervate the smooth muscle of the pancreatic duct in the cat (Garrett et al., 1973). There is morphologic evidence for a sphincter of the main pancreatic duct in the dog (Keane et al., 1979).

While morphology has contributed much to our knowledge of the secretion of proteins, it has not been very helpful in the case of water and electrolyte secretion.

Electrophysiological Properties of Pancreatic Acinar Cells

The application of cholinergic agonists or electrical field stimulation in vitro or electrical vagal stimulation in vivo will produce a fall in the membrane potential of pancreatic acinar cells (Davison and Ueda, 1977; Davison and Pearson, 1979; Peterson, 1981). There is a concomitant fall in membrane resistance and an influx of sodium ions. Cholecystokinin has a similar effect. Release of intracellular calcium is thought to open channels in the cell membrane. Under some circumstances, these agents may produce biphasic effects on membrane potentials with hyperpolarization (Kanno, 1972). Both sodium and calcium must be present in the medium for this phenomenon to occur in vitro. Individual acinar cells within an acinus are electrically coupled, and it is estimated that the current spreads to about 500 cells in a acinus unit. There is an increased efflux of chloride ions as well as an influx of sodium due to an opening of weakly positively charged channels. Depolarization in response to acetylcholine saturates at a lower dose than the inward current response. The inward current is largely carried by sodium as determined by voltage clamp experiments (McCandless et al., 1981). There is a linear relationship between membrane potential and membrane current. The action of acetylcholine and CCK and related peptides is on the surface membrane, since intracellular injection was without effect. The depolarization of acinar cells can be dissociated from secretion of enzymes. Atropine blocks acetylcholine-stimulated enzyme secretion, but not the depolarization. The resting acinar cell membrane is electrically inexcitable. Direct determination of intracellular calcium and sodium by calcium and sodium selective microelectrodes shows an increase in cytosolic calcium and sodium in response to acetylcholine-induced depolarization in mouse pancreatic acinar cells (O'Doherty and Stark, 1982). The significance of these changes for electrolyte secretion by acinar cells will be considered later.

Mechanism of Protein Synthesis and Secretion by the Exocrine Pancreas

The morphologic basis for exocrine secretion of proteins has already been discussed (Case, 1978a; Palade, 1975; Rothman, 1975; Singh and Webster, 1976). The initial step is the uptake of amino acids into the acinar cell. This is probably by a process of facilitated diffusion or cotransport with sodium (Case, 1978b). The synthesis of pancreatic secretory proteins occurs at a rate of about 5% of the total stored amount per hour in the rat (Solomon, 1981). Using rat pancreatic acini in vitro, both cholecystokinin (CCK) and carbachol (a cholinergic agonist) increased the incorporation of ^{3}H-phenylalanine into acinar proteins (Korc, 1982). The possibility exists that pancreatic enzyme secretion by exocytosis stimulates protein synthesis and that the presence of enzymes in the intestine can exert a negative feedback by way of CCK release on both pancreatic secretion and protein synthesis.

Once proteins are synthesized they are transported through the cell, packaged for the most part in membrane-bound granules and, therefore, effectively outside the cellular milieu. Exocytosis is a calcium-dependent process. The bulk of evidence favors an uptake and reutilization of cell membrane left over after exocytosis (Meldolesi, 1974; Gorelick and Jamieson, 1981).

Studies in man on pure pancreatic juice after the intravenous injection of ^{75}Se-methionine show a minimum transit time for appearance of labeled proteins in the juice of 36 ± 8 min. Cerulein (a CCK-like peptide from frog skin) greatly decreased

the specific activity of secreted proteins, suggesting that previously synthesized proteins were diluting the newly produced enzymes. This would indicate the presence of at least two intracellular separable pools for enzyme secretion (Robberecht et al., 1977). Zinc is included in the structure of carboxypeptidases. In the pig it requires about 3 h for ^{65}Zn to pass through the acinar cells and emerge in pancreatic secretion (Pekas, 1971).

Pancreatic acinar cells contain prominent microfilaments (actin) in relation to their microvilli and microtubules near the Golgi apparatus. Agents which disrupt microfilaments, such as cytochalasin B and those which reduce the number of microtubules, such as vinblastin and colchicine, also reduce the secretion of pancreatic proteins in experimental animals (Bauduin et al., 1975; Williams and Lee, 1976). This suggests a role for the cytoskeleton in protein secretion.

Recently, the activity of the enzyme protein carboxyl methylase has been correlated with the amylase secretory response to CCK and carbachol (Povilaitis et al., 1981). This enzyme transfers methyl groups to free carboxyl groups or protein substrates, neutralizing negative charges. It could facilitate the likelihood of collisons between zymogen granules and the plasma membrane.

It seems unlikely that acinar cells secrete proteins in the absence of a water phase, but the composition of the fluid is unknown. It has been suggested recently that the transmembrane current flow in acinar cells in response to various stimuli could account for an uptake of sodium chloride sufficient to supply the salt and consequently the water for acinar secretion. This neutral fluid secretion is calcium-dependent (Peterson et al., 1981). Older in vivo studies in cats showed that the output of amylase occurred in a constant volume of 0.5 ml regardless of the rate of flow of water of electrolyte (Greenwell and Scratcherd, 1974).

Using the in vitro rabbit pancreas when flow rates were reduced by both osmotic gradients and by ouabain (a Na^+-K^+-ATPase inhibitor) or a low sodium content in the bath, flow and protein secretion continued in parallel, suggesting a fluid secretion by the acinar cells (Ho and Rothman, 1979, 1982).

Amylase Secretion by Isolated Pancreatic Acinar Cells or Acini

The availability of methods for isolating pancreatic acinar cells has stimulated much research on the cellular control of amylase secretion. Later, groups of cells still retaining their organization as an acinus proved to be more sensitive to stimulation of secretion (Gardner and Jensen, 1981). It should be borne in mind that there are significant differences in the response to secretagogues by acinar cells of different species. Human acinar cells have not been studied.

A great variety of receptors for hormones and neuropeptides have been identified. They can be classified into those which act to release amylase by way of intracellular calcium release and those which activate adenyl-cyclase and cyclic AMP. In the first category are acetylcholine and CCK and in the second are secretin, VIP, substance P, and a new peptide, histidine N-terminal, isoleucine C-terminal (PHI) (Jensen et al., 1981). In the case of VIP and secretin there are two types of receptors: high and low affinity receptors. Only occupation of the high affinity receptor results in amylase secretion; yet both activate cyclic AMP. It also appears that the cyclic AMP available to the high affinity receptor is in a different cellular compartment that that accessible to the low affinity receptor (Gardner et al., 1982). See Figure 6.65 for a diagram of the intracellular mechanisms for amylase release.

There are three phenomena of interest demonstrated with in vitro acini. The first is potentiation of amylase release by combinations of secretagogues. When two secretagogues act by either cyclic AMP or calcium release, but not when both share the same mechanism, potentiation occurs. By potentiation I mean that amylase output after exposure to the two secretagogues is greater than the sum of the maximal response to each acting alone. A reduced response to stimulation following the initial exposure to a secretagogue such as CCK is known as desensitization or tachyphylaxis. This may be seen after supramaximal stimulating doses (restricted stimulation) and may partly explain the biphasic dose-response curve of amylase output to increasing doses of CCK in vivo. Peak desensitization is not seen until 60 minutes after CCK (Barlas et al., 1982a). A third observation is the persisting stimulation (60–90 min) of amylase secretion by CCK even after vigorous washing of the cells. This is specific for CCK among secretagogues and results from persistent occupation of CCK receptors. It is referred to as "persistent stimulation" (Collins et al., 1981).

It was also discovered by serendipity that dibutyryl cyclic GMP is a competitive antagonist for the occupation of CCK receptors. The ability of cyclic nucleotide analogues to stimulate amylase release does not correlate with their ability to compete for receptors of CCK (Barlas et al., 1982b).

The secretion of amylase by isolated acini offers a good opportunity to study the relation of structure

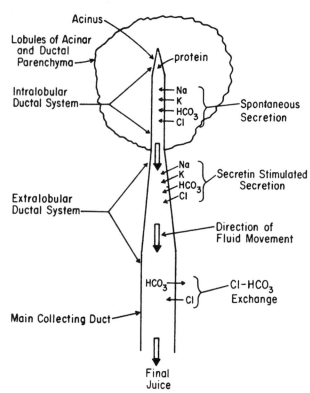

Figure 6.102. Diagram of pancreatic acinus and duct system. [From Swanson and Solomon (1973).]

of stimulating peptides to their potency as secretagogues. In general, the potency of CCK-related peptides resides in the C-terminal portion of the molecule while for the secretin-VIP peptides the N-terminal portion is more important. It follows that C-terminal fragments of CCK are more potent than the natural CCK-33 or CCK-39, while most of the molecule is required for the action of secretion and VIP.

Recent evidence also shows that insulin binds to receptors on isolated pancreatic acinar cells and increases the uptake of glucose (Williams et al., 1981). In animals made diabetic, insulin increased protein synthesis and amylase secretion (Korc et al., 1981; Sankaran et al., 1981; Goldfine et al., 1981).

The Role of Calcium in Pancreatic Acinar Cell Function

Calcium acts at several points in stimulus-secretion coupling. It may be responsible for the increase in membrane permeability which leads to the sodium influx and depolarization of acinar cells in response to acetylcholine and CCK. Calcium release coincides with the breakdown of phosphatidyl inositol. The changes in membrane composition could release calcium from calmodulin (Williams, 1980). There is still controversy over the physiologic importance of the early efflux of calcium from the cell in response to acetylcholine and CCK and the later influx of calcium (Dormer et al., 1981; Wakasugi et al., 1981). Nevertheless, the results are compatible with the ability of the perfused pancreas to secrete enzymes for a short time in the absence of calcium in the perfusate. The source of the intracellular calcium released by secretagogues has been attributed to microsomes, mitochondria, or the cell membrane (Dormer et al., 1981; Otsuki and Williams, 1982; Schulz et al., 1981). The protein synthesis-inhibiting agent cycloheximide increases the sensitivity of acinar cells for amylase secretion in response to CCK and acetylcholine. This suggests that once synthesis of amylase has occurred, the packaging of enzymes into zymogen granules and release by exocytosis are independent of continued synthesis. Cycloheximide had no effect on calcium efflux. This can be interpreted as evidence for tonic inhibition of synthesis of a regulatory protein such as calmodulin (Otsuki and Williams, 1982). The uptake of calcium by isolated microsomes from pancreatic acini is an energy-dependent ATP-requiring process stimulated by CCK (Ponnappa et al., 1981). Sodium is required in the bathing medium for amylase release in response to cholinergic and CCK-like secretagogues. A calcium ionophore was ineffective in the absence of sodium, suggesting a role for sodium in intracellular calcium release (Williams, 1975). The calcium ionophore is also effective in stimulating amylase release for only about 10 min in the absence of calcium in the medium (Gardner et al., 1980).

In vivo, acute hypercalcemia in man increases pancreatic enzyme output on a background of secretin stimulation and enhances the effect of CCK (Goebell et al., 1972a; Malagelada et al., 1976). In experimental animals, chronic hypercalcemia increased the output of both calcium and enzymes in response to CCK. The calcium output independent of protein was unchanged (Goebell et al., 1972b). In cats the fluid and bicarbonate response to secretin was reduced. The calcium output in pancreatic juice in response to secretin doubled, but in these experiments there was no change in enzyme outputs, and calcium in pancreatic juice reflected ionized calcium in the plasma (Layer et al., 1982). It appears that acute and chronic hypercalcemia do alter pancreatic secretory responses to secretagogues.

Cyclic Nucleotides and Pancreatic Secretion

The role for cyclic nucleotides in the control of pancreatic secretion of amylase in isolated cells and acini has already been discussed. The species differences in amylase release in response to secretin and cyclic AMP in vitro make it rather hazardous to extrapolate to man. Studies in the perfused cat pancreas implicate cyclic AMP in the action of secretin on pancreatic secretion (Case and Scratcherd, 1972). In the guinea pig, electrical field stimulation increased tissue levels of cyclic AMP and GMP. The effect could be mimicked by vasoactive intestinal peptide and hence suggests that VIP may have served as a neurotransmitter (Pearson et al., 1981).

Cholecystokinin and Pancreatic Secretion

Cholecystokinin was the term given to an extract of intestinal mucosa which when given parenterally contracted the gallbladder. Later another extract which increased pancreatic enzyme secretion was prepared and designated pancreozymin. With the final purification of these extracts it became clear that a single molecule possessed both properties. As originally isolated from porcine intestine, CCK was a 33 amino acid peptide with the same C-terminal tetrapeptide as gastrin (For the structure see Table 6.1). The sulfate group on the 7th C-terminal amino acid (tyrosine) is critical to its biologic properties. All of these properties reside in the C-terminal heptapeptide. The octapeptide is more potent than CCK-33. Another variant, CCK-39, has been isolated from hog mucosa.

Present concepts suggest that CCK-33 is the precursor hormone present within endocrine (APUD) cells in the proximal intestinal mucosa, mostly in the crypts (Buffa et al., 1976; Polak et al., 1975). In the process of entering the bloodstream some of the CCK-33 may be reduced to CCK-8. In the blood CCK-8 is present by radioimmunoassay and probably is the active form of the hormone, although the presence and significance of CCK-33 awaits a more satisfactory immunoassay (Calam et al., 1982; Walsh et al., 1982).

The fate of CCK in the bloodstream is uncertain, but in patients with renal failure its action is prolonged, and its decline in the blood as determined by radioimmunoassay is prolonged (Owyang et al., 1982a).

The release of CCK into the bloodstream is uncertain because of a lack of a suitable radioimmunoassay (RIA). It does increase after a meal by both RIA and bioassay. Using pancreatic secretion of enzymes as an assay, the intestinal perfusion of essential amino acids, and particularly L-phenylalanine and tryptophan, releases CCK in man and dog (Go et al., 1970; Murthy et al., 1983; Singer et al.,

1980a). Micellar fat solutions also increase enzyme output. Comparison of the enzyme response of the pancreas in situ and a transplanted pancreas in dogs to intestinal perfusions or meals and to exogenous CCK shows that while the denervated pancreas does indeed respond, its sensitivity to endogenous CCK is much reduced in the case of the transplant. The responsiveness of the pancreas in situ can be similarly reduced by atropine, a cholinergic antagonist (Solomon and Grossman, 1979; Singer et al., 1980b). Similarly, the latency of the response to intraduodenal stimulants is shorter than that to injection of CCK into the portal vein (Singer et al., 1980a). The question of the role of local reflexes in the release of CCK in man is still controversial and awaiting a suitable RIA for CCK (Ertan et al., 1975). The conclusion of these indirect experiments is that while CCK is a likely mediator of pancreatic protein secretion, local reflexes play an important role as well.

IN VITRO ACTION OF CCK ON THE PANCREAS

Cholecystokinin octapeptide (CCK-OP), CCK-33, and CCK-39 all increase amylase secretion and calcium efflux from dispersed guinea pig acinar cells but CCK-8 is 10 to 30 times more potent (Sjodin and Gardner, 1977). The structural requirements of CCK for amylase secretion and for desensitization are the same (Villanueva et al., 1982a). There is no relation to the ability to cause residual stimulation. These results are compatible with two binding sites for CCK: one prefers CCK-33 and dissociates from the peptide rapidly. The other when occupied dissociates slowly. Site 1 must be occupied for site 2 to be occupied. When site 2 is occupied the secretion of amylase is submaximal. The rate of dissociation of CCK from its receptors can be increased by replacing the tyrosine sulfate residue by hydroxynorleucine (Villanueva et al., 1982b). Residual stimulation by CCK-8 can be reversed by dibutyryl cyclic AMP (Collins et al., 1981a).

Cholecystokinin octapeptide (CCK-8) has a biphasic effect on protein synthesis in mouse pancreatic acini: it stimulates at low doses and inhibits at higher ones. It is effective in acini from fed, but not fasted rats, and it does not increase the uptake of amino acids. Cholinergic agents have similar effects. The calcium ionophore inhibits protein synthesis, suggesting that the stimulatory action of CCK occurs independently of calcium (Korc et al., 1982).

The potentiation of amylase secretion stimulated

by CCK by VIP can be augmented by an inhibitor of phosphodiesterase illustrating postreceptor modulation of the action of VIP by calcium-releasing secretagogues in rat pancreatic acini (Collen et al., 1982).

The preparation of a ^{125}I-cholecystokinin makes it possible to study binding of CCK to isolated mouse pancreatic acini directly. There are two binding sites: one of high and one of low affinity. The high affinity receptor mediates amylase release; the low one controls uptake of sugars. The calcium pool released by the high affinity receptor may stimulate enzyme synthesis, while a different pool released by the low affinity receptor may inhibit it. This accounts for the biphasic action on protein synthesis of CCK (Sankaran et al., 1982). Electron autoradiomicrographs show the specific binding of CCK to mouse acinar cells. With CCK-33 it is half maximal at 2 min and maximal after 10 min. After 2 min 90% of the total cellular radioactivity remains as intact CCK, but after 30 min it declines to 65%. The fraction of bound hormone that fixed to the acini is 84% at 2 min, declining to 78% after 30 min. The radioactivity concentrates over the basolateral membranes. At 2 min 13% of the grains were in the cell interior increasing to 42% at 30 min. Most of the grains localized over the rough endoplasmic reticulum. This could exert an effect on protein synthesis (Williams et al., 1982).

IN VIVO ACTION OF CHOLECYSTOKININ

Pure cholecystokinin (CCK-33), CCK-39, CCK-8, and cerulein all produce the same maximal protein secretion and low bicarbonate output from the dog pancreas. The output of protein is potentiated by acid in the duodenum. Sodium oleate increases the bicarbonate output above that achievable with CCK alone (Debas and Grossman, 1973). No data are available on the actions of pure CCK-33 in man, but CCK-8 quadruples enzyme output when given on a secretin background (Regan et al., 1979).

In animals CCK has a trophic action on the pancreas, increasing both the size and number of acinar cells. For the action of CCK on gastric emptying, gastric secretion, the motility of the biliary tract, and colonic motility, see the appropriate sections.

Pancreatic Secretion of Water and Electrolytes

The role of the acinar cell in the secretion of salt and water by the exocrine pancreas has already been discussed. As is the case in the salivary glands, the ducts of the pancreas can modify profoundly the water and electrolyte composition of acinus secretion. Micropuncture studies show that in the unstimulated gland chloride concentrations of pancreatic fluid increased from 60 to 90 mM between the intralobular and main duct. During secretin-induced flow there was a slight decrease from 65 to 50 mM in the rabbit pancreas (Schulz, 1981). In the cat during secretin stimulation the chloride concentration fell from 112 ± 2 mM in the intralobular ducts to 46 ± 8 in the main duct (Lightwood and Reber, 1977). In one of the few micropuncture studies where bicarbonate concentrations were measured directly, it varied directly with flow from 70 to 110 mM from the intralobular to the extralobular ducts in the rabbit pancreas (Swanson and Solomon, 1973). Figure 6.102 shows a model based upon micropuncture studies. The main pancreatic collecting duct has the capacity for a passive chloride-bicarbonate exchange. Sodium and potassium concentrations do not change significantly along the ducts. These studies implicate the small ducts as the principal site of bicarbonate secretion in response to secretin.

Unfortunately, the preparation of isolated pancreatic ductal cells has not yet achieved the success of acinar cells. Preliminary results in rats after destruction of acinar cells during a copper-deficient diet supplemented with penicillamine show an increase in bicarbonate concentration in vivo from 36 to 64 mM and a fall of chloride concentration from 127 to 98 mM (Folsch and Creutzfeldt, 1977). Preparations of isolated ductal cells from such glands are awaited.

Epithelial cells lining the ducts of the cat pancreas were heavily labeled with 3H-oubain-binding sites on the basolateral membranes. Since ouabain markedly reduces the flow of pancreatic juice, this suggests that the ductal cells have a major role in the secretion of water and electrolyte involving a Na^+K^+-activated ATPase (Bundgaard et al., 1981).

The major unresolved question is the site of the active transport mechanism. Current hypotheses favor an active transport of hydrogen ion, since weak acids can substitute for bicarbonate in the perfused cat pancreas. In this model carbon dioxide diffuses into the ductular cells and is converted to carbonic acid in the presence of carbonic anhydrase. Carbonic acid dissociates into bicarbonate and hydrogen ions at the luminal membrane. Bicarbonate ions diffuse into the lumen, and hydrogen ions are actively transported back into the cell by means of a proton pump ($Mg^{2\pm}$-ATPase). It is then transported across the basolateral membrane by a H^+-Na^+ exchange carrier. Energy for this comes from the sodium gradient between the cell and the extracellular fluid. This gradient is maintained by the Na^+-K^+-ATPase also located in the basolateral membrane. These mechanisms are depicted in Figure 6.103 (Schulz, 1981; Scratcherd et al., 1981).

Secretin and Pancreatic Secretion

Secretin is a 27 amino acid peptide originally isolated from the mucosa of hog proximal small intestine. As noted earlier its biologic activity depends primarily on the N-terminal portion of the molecule (Bodansky et al., 1977). Immunocytochemical studies indicate that secretin is present in granules of cells in the mucosa of proximal small intestinal crypts. Their apical surface makes contact with the lumen, while the granules are located mainly at the contraluminal side of the cell (Polak et al., 1971).

Both endogenous and exogenous secretin had a half-life of about 2.5–3.0 min in the dog (Curtis et al., 1976). In human experiments a half-life of 3.8 min has been found after exogenous secretin (Schaffalitzky de Muckadell et al., 1978). As in the case of CCK, patients with renal failure have a delayed disappearance of secretin from the blood after exogenous administration (Rhodes et al., 1975).

After a period of uncertainty, radioimmunoassays for secretin seem to have settled upon a low level of fasting immunoreactive secretin, which increases intermittently after duodenal acidification and after a meal (Chang and Chey, 1980; Schaffalitzky de Muckadell et al., 1978; 1979a, 1979b).

The release of secretin occurs in response to duodenal acidification. Judging by the pancreatic response rather than by direct measurement of plasma secretin levels, secretin release during a duodenal pH of <1–3 in dogs depends on the load of titratable acidity delivered to the duodenum (Meyer et al., 1970a and b). The absolute pH level for secretin release appears to be between 4.5 and 5.0. Whether or not bile acids are physiological releasers of secretin is uncertain.

In the dog, truncal vagotomy or atropine decreased pancreatic secretion in response to endogenous secretin release or physiological doses of exogenous secretin (Chey et al., 1979a). In man immunoreactive plasma secretin levels after jejunal acidification were the same with or without atropine, but bicarbonate outputs were less with atropine (You et al., 1982). Atropine also shifted pancreatic secretory dose responses to secretin to the right in dogs (Singer et al., 1981). These results implicate local cholinergic mechanisms in the control of secretin release.

Sodium oleate but not triolein released secretin in dogs. There was a greater increase in pancreatic

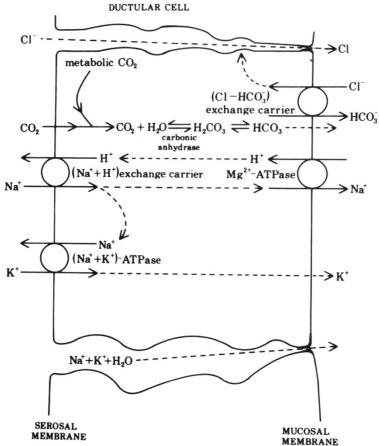

Figure 6.103. A schematic representation of the proposed cellular mechanisms for pancreatic secretion by ductular cells. [From Scratcherd (1981).]

bicarbonate outputs than could be accounted for by the increase in plasma secretin. This suggests that CCK was also released and potentiated the secretin response (Fiachney et al., 1981).

EFFECTS OF SECRETIN ON PANCREATIC CELLS IN VITRO

The problems of interpretation of amylase secretion by acinar cells in response to secretin have already been discussed. It is of interest that the ability of secretin to activate cyclic AMP in acinar cells resides in the N-terminal portion of the molecule (Gardner et al., 1977). Using isolated duct fragments, secretin also increased tissue levels of cyclic AMP (Folsch et al., 1980).

THE ACTION OF SECRETIN IN VIVO

Secretin is the most potent known stimulus of water and bicarbonate secretion. During collections of pure pancreatic juice in man, a maximal concentration of 135 ± 9 mM of bicarbonate was reached and an output of 383 ± 79 μM/kg/h at a dose of 129 ng/kg/h of synthetic secretin. There was a

parallel increase in cyclic AMP in the pancreatic juice (Domschke et al., 1976). Similar results were obtained by collecting duodenal content after 1.25 clinical units/kg/h. The relationship between secretin dose and bicarbonate output was linear (Vagne et al., 1969). When such a dose was given as a bolus, amylase output was less than 0.5 units/min, the plasma immunoreactive secretin rose to 700 pmol/l, and the half-life of secretin was 2.5 min (Beglinger et al., 1982).

Somewhat higher concentrations of bicarbonate can be obtained in dogs: about 155 mM with a chloride concentration of 25 mM (Baron et al., 1963). The infusion of acid into the duodenum can elicit a pancreatic secretory response equal to the maximal response to secretin in dogs (Konturek et al., 1970). Vasoactive intestinal peptide (VIP) stimulated a bicarbonate output only 17% of that achievable with secretin in the dog (Konturek et al., 1975). It is likely that VIP's physiological role is as a neurotransmitter rather than a hormone.

In the cat there was no significant difference in the bicarbonate output from the pancreas between intravenous and intraportal administration of secretin, indicating a lack of inactivation of secretin in the liver (Konturek et al., 1977).

Secretin has other pharmacologic actions, but their physiologic significance is uncertain. Antisecretin antibodies prevent the normal rise in immunoreactive secretin after a meat meal in dogs and reduce pancreatic bicarbonate output by 80%. There was no change in the protein concentration in pancreatic juice (Chey et al., 1979b). Serum gastrin levels were also higher after the antisecretin serum (Lee et al., 1980b). Secretin increases intestinal blood flow, but at doses higher than those required to stimulate pancreatic secretion (Beijer et al., 1979). It also inhibits gastric acid secretion while stimulating pepsin secretion. In the dog its action as an inhibitor of gastrin-induced acid secretion appears to be in the physiologic range (Johnson and Grossman, 1969).

Secretin inhibits gastrin antral contractions and delays gastric emptying in the dog. It is a potent stimulant of hepatic bile flow. It has a trophic effect on the rat pancreas. In dogs secretin reduces the pressure in the pancreatic duct and in the duodenum, with a resulting decrease in resistance to pancreatic flow (DiMagno et al., 1981).

Secretin clearly has a role in the control of pancreatic secretion of bicarbonate both as a direct stimulant and as a potentiator with cholecystokinin.

Other Regulatory Peptides and Pancreatic Secretion

Probably the most interesting regulatory peptide with a classical hormonal action for pancreatic secretion is pancreatic polypeptide (PP). It was originally recognized as a contaminant of insulin. It contains 36 amino acids and can be readily assayed in the blood by a radioimmunoassay. Over 90% of PP is found in the pancreas. The cell of origin is a pancreatic D cell present in both the parenchyma and the islets (Heitz et al., 1976). PP levels in the blood rise 3–10 fold after a meal (Villanueva et al., 1978). The mean ½ disappearance time from plasma is about 7 min in man and 5 min in the dog (Adrian et al., 1978; Taylor et al., 1979). The levels of PP in plasma are increased in patients with renal disease (Hallgren et al., 1977). PP release is subject to control by cholinergic mechanisms and the vagi. Atropine reduced PP release in response to vagal stimulation, while acetylcholine stimulated PP release in the isolated perfused porcine pancreas (Schwartz et al., 1978). Atropine also blocked PP release in response to a meal in man (Taylor et al., 1978). Truncal vagotomy was followed immediately by a reduction in PP release

in response to a meal in man. Vagotomy and antrectomy totally abolished the response (Schwartz et al., 1976). The effect of antrectomy may be due to an interruption of local reflexes to the pancreas (Taylor et al., 1981). β-Adrenergic stimulation releases PP in man and may account for the residual rise in response to a meal after vagotomy (Sive et al., 1980). If the vagotomy was performed months or years before, a rise in PP is seen after a meal (Adrian et al., 1977). In dogs the rise in PP was only 2% of prevagotomy levels after a meal at 1 mo, but rose to 21% at 6 mo (Taylor et al., 1982b). There is also an intestinal phase of PP release as shown by a rise during perfusion of an isolated intestinal loop (Thiry-Vella Loop) with liver extract in dogs (Modlin et al., 1981).

The unresolved problem is the physiologic action of PP. It does reduce the output of pancreatic enzymes in response to duodenal perfusion of amino acids, oleic acid, glucose, CCK-8, secretin, and pentagastrin in man (Owyang et al., 1982). It could be a physiologic antagonist of CCK. The blood levels required for inhibition are within the range of those seen after a meal (Konturek et al., 1982).

Human synthetic gastrin may have a physiologic role as a stimulant of pancreatic protein secretion (Valenzuela et al., 1976b). Calcitonin, glucagon, and thyroid-releasing hormone all have inhibitory effects on pancreatic secretion of uncertain physiologic significance. Pancreatic secretion in response to CCK and secretin is reduced in patients with adrenocortical insufficiency and returns to almost normal with treatment (Gullo et al., 1982). It may contribute to the excessive loss of fat in the stools sometimes seen. Substance P produces slight stimulation of basal pancreatic secretion, but inhibits stimulated secretion in dogs (Konturek et al., 1981). Prostaglandins have no role in stimulus-secretion coupling in the pancreas (Stenson and Lobos, 1982).

Somatostatin inhibits the volume of secretin-stimulated pancreatic secretion in man (Domschke et al., 1977). Bombesin stimulates the volume and enzyme outputs of the human pancreas (Basso et al., 1975). In experimental animals, extracts of intestinal mucosa have been reported to have specific enzyme-releasing properties (chymodenin) or anticholecystokinin properties (pancreotone).

Pancreatic Secretion and Intestinal Motility

Studies in dogs indicate that pancreatic secretion increases in volume and the output of enzymes during phase III of the migrating motor complex in the duodenum (DiMagno et al., 1979; Keane et al.,

1981; Itoh et al., 1981a). The mechanism is not known, but it is not due to the release of motilin, since this peptide has no effect on pancreatic secretion.

Nervous Control of Pancreatic Secretion

The role of local intestinal reflexes in mediating pancreatic secretion in response to a meal has already been noted. As in the case of gastric secretion, there is a cephalic phase of pancreatic secretion mediated by the vagus. This may be in part secondary to the release of gastrin, but a direct component also exists (Sarles et al., 1968). The modified sham feeding discussed under gastric secretion had no effect (Read et al., 1978).

A gastric phase of pancreatic secretion has been demonstrated in dogs. Antral distention stimulated both the pancreatic outputs of bicarbonate and protein during acidification of an antral pouch, which presumably blocked the release of gastrin. The response was blocked by atropine and vagotomy, implicating a vago-vagal reflex (Debas and Yamagishi, 1978).

Vagal stimulation by means of insulin hypoglycemia or 2-deoxy-D-glucose, or electrical stimulation of vagal efferents in experimental animals stimulates primarily pancreatic protein secretion. The pig is an exception in that the volume of juice secreted is large. It has been suggested that this is mediated by VIPergic nerves. In dogs gastrin, VIP, and somatostatin were released into the portal circulation during electrical vagal stimulation (Guzman et al., 1979). There was no release of secretin or CCK. Afferent vagal stimulation can elicit pancreatic amylase secretion by a vago-vagal reflex (Harper et al., 1959). Proximal gastric vagotomy probably has little effect on pancreatic secretion in man (Ramus et al., 1982). The effect of truncal vagotomy is controversial. There is no effect on secretin release (Ward and Bloom, 1975). Most investigators studying dogs have found a decreased pancreatic secretory response to intraduodenal stimulants or a meal. The mechanism is unknown. Some remarkable trophic effects have been reported on the pancreas in the rat following truncal vagotomy. The weight of the gland increased almost 100% due almost entirely to acinar cell hyperplasia (Tiscornia and Perec, 1981). Others have reported pancreatic hyperplasia and hypertrophy in rat pancreas with cholinergic stimulation (Carling and Templeton, 1982).

Stimulation of the splanchnic nerves reduced pancreatic secretion in cats, probably due to reduced blood flow (Greenwell et al., 1967). Catecholamines have similar effects. In the (in vitro) rabbit pancreas, norepinephrine and isoproterenol appeared to have a direct inhibitory action (Hubel, 1970).

In general, cholinergic agonists are stimulatory to pancreatic protein secretion, and atropine reduces the response to a variety of stimuli, including exogenous secretin (Thomas, 1964). However, acetylcholine in the duodenum did not stimulate pancreatic secretion in dogs (Sum et al., 1969). Nicotine inhibits pancreatic secretion, probably by inducing vasoconstriction through the release of catecholamines (Solomon et al., 1974).

Adaptation

A large number of experiments in rats have shown that the distribution of enzymes in pancreatic secretion can be altered over a period of days by changing the composition of the diet. Low carbohydrate-high protein diets reduced amylase secretion and increased the secretion of proteases. Insulin has been implicated as a signal in the adaptive changes in amylase secretion. Similar kinds of changes occurred in response to high fat diets (Saraux et al., 1982). More significant is the decrease in the ability of pancreatic acini in vitro to metabolize glucose if the rats had been fed a high fat diet. The effect was due entirely to a reduction in glucose transport into the cells. There may have been a reduction in the number of glucose carrier proteins in the cell plasma membrane (Bazin and Lavan, 1982). A few reports suggest that adaptation can occur in dogs and man. More rapid selective enzyme secretion changes are controversial.

Growth

Increases in size of pancreatic acinar cells (hypertrophy) and in number (hyperplasia) occur under a number of circumstances. Most of these have been demonstrated in rats and mice. Thus CCK and gastrin given over days or months can induce pancreatic hyperplasia and hypertrophy. In some cases increased secretory capacity has also been demonstrated (Peterson et al., 1978). Secretin potentiated the effect of a CCK analogue (Solomon et al., 1978). Lactation in rats also induces pancreatic hyperplasia. Extensive small bowel resections in rats result in pancreatic hyperplasia (Haegel et al., 1981). Chronic perfusion of the duodenum with hydrochloric acid, phenylalanine, and tryptophan, or combinations of both produced hyperplasia of the pancreas (Johnson et al., 1980).

Development

The duodenal fluid of newborn infants contains no amylase activity, negligible lipase activity, and low levels of protease activity. Administration of CCK did not stimulate secretion of enzymes. By 2 yr of age, a full response to CCK was present (Lebenthal and Lee, 1980). In rats pancreatic responsiveness to a cholinergic agonist did not appear until 3–15 days of life. Term fetal rat pancreas did not respond to cholinergic agonists or CCK, but did to a calcium ionophore, suggesting a deficit of receptors on the cell membrane (Werlin and Grand, 1979). Before weaning in rats hydrocortisone produces pancreatic hyperplasia. Cholecystokinin-like agents, on the other hand, did not produce hyperplasia until after weaning (Morisset and Jolicoeur, 1980; Morisset et al., 1981).

Control of Pancreatic Enzyme Secretion by Negative Feedback

In rodents the secretion of proteases is increased by diverting pancreatic secretion and decreased by introducing proteases into the intestine. Similar increases have been seen with feeding protease inhibitors with the production of pancreatic hyperplasia. Such a mechanism has not been found in normal man (Krawisz et al., 1980).

Response of the Exocrine Pancreas to a Meal

The most important stimulus for pancreatic secretion is a meal. In the dog it equalled 20–40% of the maximum secretory capacity. Recently, techniques have been developed to monitor pancreatic secretion in conscious dogs over long periods of time. While the initial response, rich in enzymes, occurred within 2–3 h of feeding. A larger volume response with a high bicarbonate, but low enzyme concentration, occurred 11 h after feeding. Secretion returned to base-line after 16 h. The authors attributed the late peak to the slow rate of emptying of solids from the stomach because it was not seen with liquid meals (Itoh et al., 1980). In man the pancreatic enzyme response to a solid-liquid meal exceeds the maximal response to CCK. The factors controlling the pancreatic response include both the chemical composition and physical properties of the meal. The strongest stimulants to pancreatic enzyme secretion are fatty acids and monoglycerides. By themselves they can stimulate maximal enzyme output. Proteins are next, while carbohydrates have little stimulatory action. These nutrients exert their action in the duodenum. Diverting the chyme at the ligament of Treitz did not reduce enzyme output nor did reintroduction increase it. The most important factor appears to be the area of contact of nutrient with the mucosa. Therefore, it is the load of nutrient in the case of fat or protein delivered to the duodenum, rather than the concen-

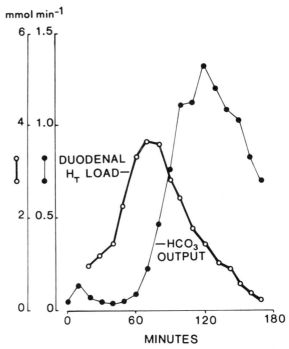

Figure 6.104. Changes in the delivery of titratable acid and bicarbonate to the duodenum after a meal in man. Note that the bicarbonate response follows the delivery of the acid load. [From Moore et al. (1979).]

tration, which determines the magnitude of stimulation. Since food buffers acid, the duodenal release of secretin usually comes later than the release of CCK. The delay in bicarbonate secretion is shown in Figure 6.104. In fact, the enzyme output to a meal is not significantly reduced in subjects receiving cimetidine in doses sufficient to reduce acid output by more than 50% (Malagelada, 1981). Homogenized meals stimulate pancreatic secretion for shorter periods of time, since they leave the stomach more rapidly (Malagelada et al., 1979).

Pancreatic Blood Flow

Most of the pancreatic blood flow passes through the islets before reaching the acini, which provides for a high concentration of islet cell regulatory peptides reaching the acini (Lifson et al., 1980). The pancreatic circulation in dogs is subject to postocclusive reactive hyperemia and active autoregulation, mainly by a myogenic action. When arterial pressure falls, blood flow decreases despite a fall in arterial resistance. The arteriovenous oxygen difference increases, and oxygen uptake remains unchanged. The filtration coefficient also increases, probably as a result of metabolic regulation (Kvietys et al., 1982). Pancreatic secretagogues such as secretin also increase pancreatic blood flow (Goodhead et al., 1970).

Enteroinsular Axis

The possibility that pancreatic polypeptide, glucagon, and insulin exert important influences on the exocrine pancreas remain to be settled. The main action of gastric inhibitory peptide (GIP) is to release insulin.

Pathophysiology of Exocrine Pancreatic Secretion

Diseases of the pancreas are poorly understood. The one common thread in acute pancreatitis is autodigestion. Any factors which could lead to intrapancreatic conversion of proenzymes to active enzymes could produce the clinical syndrome. It is interesting that hypercalcemia during parenteral nutrition has been associated with acute pancreatitis. The response of the pancreas to secretin (bicarbonate output) is the most sensitive test of normal exocrine pancreatic function. There is an endocrine as well as an exocrine secretion of pancreatic enzymes, and a rise in plasma levels is the most important laboratory test in acute pancreatitis.

HEPATIC BILE SECRETION

The function of bile is to participate in the digestion of fat, to solubilize monoglycerides in water prior to absorption, and probably to dispose of a portion of the body's pool of cholesterol. Under normal conditions most of the bile secreted by liver cells is stored in the gallbladder and emptied into the duodenum in response to a meal. Hepatic bile is concentrated in the gallbladder, but apart from acidifying the bile there is little change in the amount of electrolytes or solutes between hepatic and gallbladder bile. Many individuals live normal lives after their gallbladder has been removed surgically.

Composition of Bile in Man

Table 6.8 shows the composition of hepatic bile obtained from T tubes in the common bile duct or by aspiration at the time of surgical exploration of the abdomen. It will be noted that most of the cations are present in concentrations similar to that in plasma but that the bicarbonate concentration may be substantially higher. Proteins are present in low concentrations in bile and probably represent plasma protein for the most part. The yellow color of bile is due to the presence of bilirubin, a product of the metabolism of hemoglobin, which is present in bile in much higher concentrations than in plasma (<1.0 mg/dl). The most characteristic components of bile are the bile acids, phospholipids (mostly lecithins), and cholesterol. Collectively they are referred to as the bile lipids. The bile acids are partly responsible for the digestion and solubilization of fats. Since they are derived from cholesterol, they serve as excretory metabolites of cholesterol. Both bile acids and phospholipids solubilize cholesterol in micelles. Bile is isomotic with respect to plasma. This is in spite of concentrations of ions which may exceed plasma osmolality. The formation of micellar aggregates with bile acids including the accompanying cations reduce osmotic activity and thereby maintains isosmolality.

It is estimated that the human liver secretes about 600 ml of hepatic bile a day at a rate of about 0.4 ml/min (Erlinger, 1982b). There is an increase in hepatic secretion of bile after meals due to the return of more bile acids to the liver by way of the portal vein (Wagner et al., 1974).

The most important factor in determining the

Table 6.8

Composition of human hepatic bile

	Erlinger (1982)	Thureborn (1962): T Tube Drainage from 7 Patients with Gallstones	Van der Linden and Norman (1967): Aspiration from 101 Patients with Gallstones
Sodium	132–165 mEq/l	146–165	
Potassium	4.2–5.6	2.7–4.9	
Calcium	1.2–4.8	2.5–4.8	
Magnesium	1.4–3.0	1.4–3.0	
Chloride	96–126	88–115	
Bicarbonate	17–56	27–55	
Bilirubin		0.25–1.2 mM/l	1.5 ± 0.7 SE
Protein		25–140 mg/100 ml	

	Shaffer et al. (1972): T Tube Drainage from 7 Patients with Cholesterol Gallstones	Shaffer et al. (1972): T Tube Drainage from 5 Patients without Gallstones
Total bile acids	73.4 ± 0.9 SE, moles%	76.2 ± 1.2 SE, moles%
Phospholipids	19.0 ± 0.8	17.5 ± 6.3
Cholesterol	7.7 ± 0.3	6.3 ± 0.8

composition and rate of hepatic bile secretion is the enterohepatic circulation of bile acids. If more than 20% of bile is diverted to outside the body in monkeys after removal of the gallbladder, there is a decrease in bile lipid concentrations, with bile acid concentrations falling relatively more than those of phospholipids and cholesterol (Dowling et al., 1971). With greater diversions, bile flow also decreases markedly.

Bile acids are steroids with cholesterol-like structures modified by a side chain (carbon atoms 20–24 in Fig. 6.105). Hydroxyl groups are added at carbon atoms 3, 7, and 12. The hydroxyl groups result in a polar or water-soluble aspect of the molecule. The tertiary structure shows folding of the molecule at the site of each hydroxyl group. Figure 6.106 shows the composition of solutes in both hepatic and gallbladder bile. It can be seen that cholates and chenodeoxycholates account for about a third of bile acids, while lithocholate accounts for about 5% (Carey, 1982). Cholic acid, a trihydroxy bile acid, and chenodeoxycholic acid, a dihydroxy bile acid, are both synthesized in the liver and are known as primary bile acids. In the intestine, bacteria convert cholic acid to deoxycholic acid and chenodeoxycholic acid to lithocholic acid by removing the 7α-hydroxyl group. These are known as secondary bile acids (Fig. 6.107). Each of the bile acids can be conjugated with either glycine or taurine (Fig. 6.108). Lithocholic acid is relatively insoluble in water and is sulfated in one pass through the portal circulation in man. Sulfated lithocholate is not reabsorbed from the intestine. Lithocholate can inhibit the secretion of bile in experimental animals.

The primary bile acids and deoxycholic acid form micelles above a critical micellar concentration. Micelles are aggregates of bile acids and phospholipids on the outside with cholesterol or monoglycerides in the center (Fig. 6.109). Depending on the lecithin:bile acid ratio, micelles may be small and simple or large and mixed (Mazer et al., 1980). The hydroxyl groups of the bile acids face the surface, and the phospholipid molecules are arranged so that their polar groups protrude from the surface. Cholesterol molecules or other lipids are held within. Micellar solutions are clear. One theory proposes that the micelles are formed as the bile acids pass through the membrane of the bile canaliculi. This would account for the parallel secretion of bile acids, phospholipids, and cholesterol.

The relation of bile acids and phospholipids to cholesterol in bile is of critical importance in maintaining cholesterol in solution. By plotting the percent of total molar lipid concentrations on three

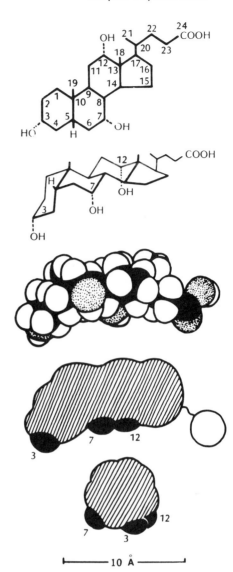

Figure 6.105. Structure of cholic acid including tertiary structure and molecular models in longitudinal and cross-section. [From Carey and Small (1972).]

coordinate axes, the zones of cholesterol saturation and supersaturation can be defined. The latter can be further separated into a metastable zone where precipitation of cholesterol crystals occur in the presence of a nidus and a zone where cholesterol crystals precipitate spontaneously. Figure 6.110 defines these zones.

This discussion of bile acids is an oversimplification. Some bile acids such as lithocholic acid can be formed by alternate pathways. Hydroxylation may begin on the side chain. Some bile acids in meconium contain 20–22 carbon atoms, and cholic acid can be formed from cholesterol by microorganisms. There are multiple branch points in the synthesis of cholic and chenodeoxycholic acid. The 7α-hydroxylation of cholesterol considered rate-

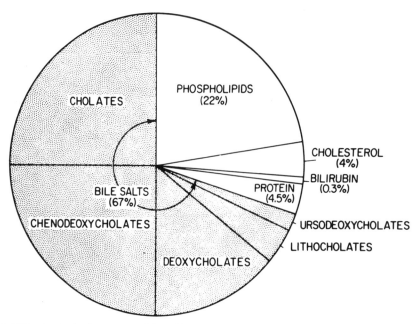

Figure 6.106. Typical solute composition of hepatic and gallbladder bile in man. [From Carey (1982).]

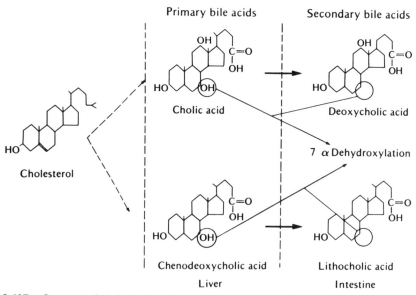

Figure 6.107. Structure of cholesterol and primary and secondary bile acids. [From Williams (1974).]

limiting in the synthesis of primary bile acids can be bypassed in the synthesis of other bile acids. Appreciable quantities of sulfated and glucuronidated bile acids can be found in the urine. Not all bile acids form micelles (Lester et al., 1983). Nevertheless, these experimental bile acids are of unknown physiological significance, so that the generalizations noted earlier still are useful.

About half of all the protein in human bile was made up of serum proteins. The total daily secretion was 160–900 mg/day (Hardwicke et al., 1964). Several enzymes can be detected in rat bile: amylase (Donaldson et al., 1979), alkaline phosphatase (Crofton and Smith, 1981; Hatoff and Hardison, 1982), and lysosomal enzymes (LaRusso and Fowler, 1979). The importance of the latter lies in the possible significance of the exocytosis of lysosomes and their contents as a pathway of secretion. It is interesting that protein secretion and the secretion of lysosomal enzymes were unaffected by interrupting the enterohepatic circulation in rats. Detergents increase the secretion of both lysosomal enzymes

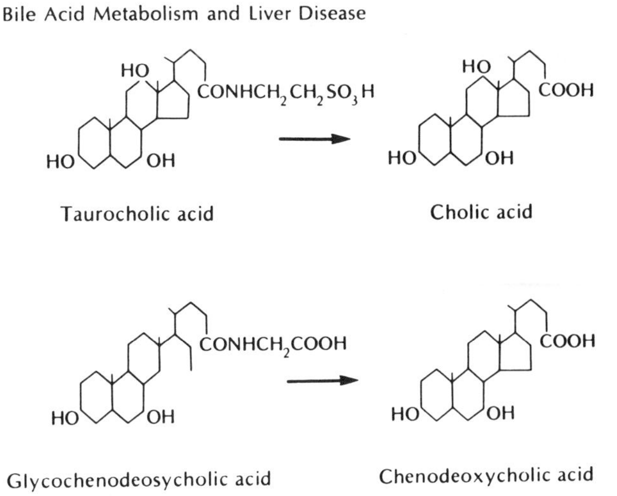

Bile Acid Metabolism and Liver Disease

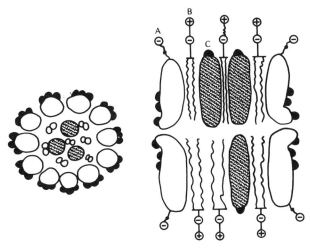

Figure 6.108.   Conjugated bile acids and their deconjugation by bacteria. [From Williams (1974).]

and biliary lipids which may indicate that endocytosis and exocytosis occur in parallel (LaRusso et al., 1982).

Alkaline phosphatase secretion into rat bile is increased by intravenous bile acids. The effectiveness of individual bile acids paralleled their ability to elute alkaline phosphatase from the surface membranes of liver cells (Hatoff and Hardison, 1982).

The appearance of immunoglobulins in bile leads to morphologic studies in rats showing vascular transport of IgA across the hepatocyte after binding to secretory piece on the sinusoidal cell membrane to the canaliculi (Renston et al., 1980; Wilson et al., 1980). This does not appear to be a significant pathway in man where secretory piece is found only in the cell membranes of ductal cells (Dooley et al., 1982).

Glycoproteins can be transported from the sinusoids to the canaliculus with a transit time proportional to their molecular weight and independent of receptors on the cell membrane. They presumably enter by a paracellular pathway (Thomas et al., 1982). Glycoproteins may be constituents of aggregates (micelles) of bile lipids and bilirubin (Bouchier and Cooperband, 1967).

Bile is an important route of excretion of copper in man (van Berge Henegouwen et al., 1977). Its secretion is unrelated to that of bile acids.

Other factors influencing the composition of hepatic bile include a diurnal cycle in the secretion of biliary lipids (Metzger et al., 1973) and diet. Increasing the caloric intake or the protein intake in patients with T tubes in the common bile duct increased bile cholesterol concentrations (Sarles et al., 1970).

Figure 6.109.   Model of a mixed micelle. [From Soloway (1978).]

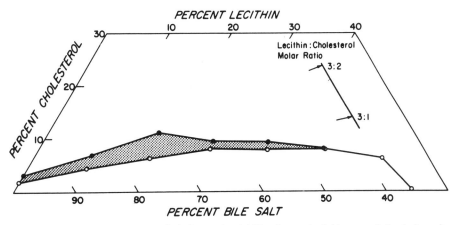

Figure 6.110.   Tricoordinate graph showing area of cholesterol solubility, the metastable zone (*stippled area*), and the zone where precipitation of cholesterol crystals occurs. Coordinates are in molar percent of bile lipids. [From Carey and Small (1978).]

Structure-Function Relationships

Bile acids are both synthesized on microsomes within hepatocytes from cholesterol and transported from the blood in sinusoids to the canaliculus. Figure 6.111 shows a diagram of a hepatocyte, including microfilaments and microtubules (French et al., 1982). Agents which disrupt microfilaments and microtubules decrease bile flow in response to bile acids in rats (Dubin et al., 1980; Gregory et al., 1978). The possibility of a paracellular pathway between the sinusoids and the canaliculi is suggested by the demonstration that lanthanum can pass from the sinusoids to the canaliculi (Layden et al., 1978). Bile acid infusions in rats produced blisters near the canaliculi, suggesting osmotic gradients (Boyer et al., 1979a). Figure 6.112 shows the relationships of the cellular and paracellular pathways (Erlinger, 1982b).

The functional unit of the liver is the acinus with hepatocytes arranged around blood and bile passing from the portal triad to the central vein. The hepatocytes in zone 1 transport most of the incoming bile acids and make the greatest contribution to bile flow (Fig. 6.113) (Gumucio and Miller, 1981). The major portion of the bile acids in portal blood is removed by the first six to nine periportal liver cells (Jones et al., 1980).

A role for the bile ducts in hepatic bile formation is suggested by the microvilli on the epithelial cells.

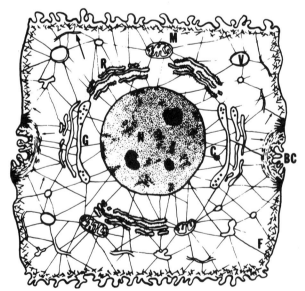

Figure 6.111. Diagrammatic representation of a hepatocyte showing proposed cytoskeleton. The bile canaliculi (*BC*) are flanked by desmosomes and intermediate filaments (*small arrows*). Nuclei (*N*), golgi (*G*), centriole (*C*), rough endoplasmic reticulum (*R*), mitochondria (*M*), vesicles (*V*), microtubules (*large arrows*), and microfilaments (*F*) are shown. [From French et al. (1982).]

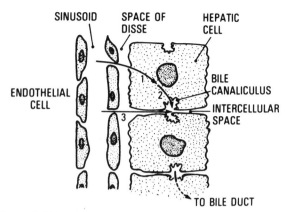

Figure 6.112. Anatomic pathways for entry of solutes from blood to bile. Solutes can enter bile through transcellular pathway after uptake and transport by the hepatocyte (*1*) and secretion by the canalicular membrane (*2*) or through the paracellular pathway via the intercellular junction (*3*). [From Erlinger (1982).]

The smallest divisions end in blind channels. They gradually increase in diameter as they approach the extrahepatic ducts. It is not known whether each canaliculus has its own biliary ductule (canal of Hering) (Steiner and Carruthers, 1961).

Mechanisms of Hepatic Bile Formation: Canalicular Bile Flow

Bile formation begins in the canaliculi. Certain molecules such as erythritol enter the biliary tract by way of the canaliculi and can serve as markers of canalicular secretion. In man this accounts for about 40% of bile flow (Prandi et al., 1973). Bile acids given systemically increase bile flow and the secretion of bile acids in parallel (Lindblad and Schersten, 1976). This represents bile acid-dependent bile flow.

Bile Acid-Independent Bile Flow

Correlations between the secretion of bile and the output of bile acids in bile suggested that in most species there was a linear increase in bile flow as the output of bile acids increased. If the relationship was extrapolated to zero output of bile acids, bile flow still occurred. This became known as the bile acid-independent fraction of bile. Figure 6.114 shows the relationship graphically. Unfortunately, there are little data on bile flow at very low bile acid outputs so that it is possible that the line falls rapidly and never cuts the Y-axis. Results in rats and monkeys suggest this is the case (Baker et al., 1979). It is possible that bile acids alter the permeability of the canalicular membrane.

Bile in the ducts is modified by the secretion of bicarbonate and the absorption of water and electrolyte. In the dog bicarbonate secretion in the

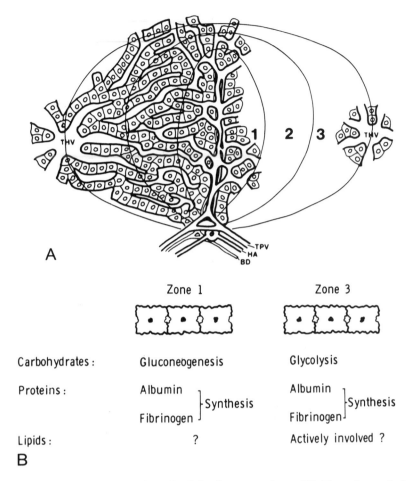

Zone 1   Zone 3

Carbohydrates: Gluconeogenesis   Glycolysis

Proteins: Albumin   Albumin
                         ⎤Synthesis   ⎤Synthesis
                    Fibrinogen⎦   Fibrinogen⎦

Lipids: ?   Actively involved ?

B

Figure 6.113. The hepatic acinus, the microvascular unit of the liver parenchyma (*A*). The acinar axis is formed by the terminal branches of the portal venule (*TPV*), the hepatic arteriole (*HA*), and the bile ductule (*BD*). Blood enters the acinus sinusoids in zone *1* and flows sequentially through zone *2* and into acinar zone *3*, where it exits via the terminal hepatic venules (*THV*). The sinusoids of zone 1 are highly anastomotic, while those of zone 3 are straight and empty into the THV in a radial arrangement. *Bottom diagram* (*B*) shows zonal contribution to bile formation hepatocytes of zone 1 transport most of the load of incoming bile salts and presumably therefore make the greatest contribution to the secretion of the BSDF. Hepatocytes of zone 3 may contribute significantly to the secretion of the bile salt nondependent bile flow BSNDF. [From Gumucio and Miller (1981).]

common bile duct occurs against an electrochemical gradient. In man it is estimated that about 0.15 ml of bile per minute is contributed by canalicular bile acid-dependent flow and an equal amount by bile acid-independent flow. The bile ducts add another 0.11 ml/min (Erlinger, 1982a). Bile from the common duct of conscious dogs after the gallbladder has been removed approaches that from the gallbladder after an overnight fast so that considerable absorption of water and electrolyte can occur. Glucose absorption also occurs in the ducts (Guzelian and Boyer, 1974).

Cellular Aspects of Hepatic Bile Secretion

Bile acids and other organic anions such as bilirubin are taken up by hepatocytes from plasma in the sinusoids. Evidence from experiments with isolated hepatocytes, isolated perfused livers, and intact animals indicate that the uptake of bile acids is a specific saturable process (Scharschmidt and Stephens, 1981; Reichen and Paumgartner, 1976; Glasinovic et al., 1975). A receptor, probably a glycoprotein, for bile acids on the surface membrane of rat liver cells has been demonstrated (Accantino and Simon, 1976). It shows a saturable binding. An inhibitor of protein synthesis reduced bile acid binding sites (Gonzalez et al., 1979). Another protein-binding bilirubin has been partially characterized from rat liver cell plasma membrane (Wolkoff et al., 1979; Wolkoff and Chung, 1980).

Once inside the hepatocyte both bile acids and bilirubin can bind to cytosolic proteins. The best characterized example is ligandin or Y-protein which binds bilirubin and reduces back diffusion into the sinusoidal blood (Scharschmidt et al., 1975). Others have purified Y-binding proteins which bind bile acids (Kaplowitz et al., 1982). Both microfilaments and microtubules may be involved

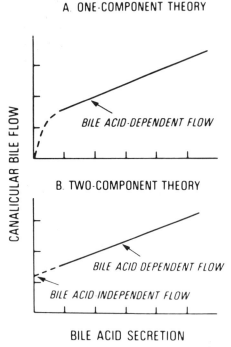

A. ONE-COMPONENT THEORY

CANALICULAR BILE FLOW

BILE ACID-DEPENDENT FLOW

B. TWO-COMPONENT THEORY

BILE ACID DEPENDENT FLOW

BILE ACID-INDEPENDENT FLOW

BILE ACID SECRETION

Figure 6.114. Canalicular bile flow. *Upper graph* shows a one-component theory where bile acids account for all canalicular flow. *Lower graph* illustrates a two-component theory where a fraction of the bile is secreted in the absence of bile acids. [From Erlinger (1982).]

in the uptake of bile acids by liver cells, since agents which disrupt them interfere with the uptake of bile acids by isolated rat liver cells (Reichen et al., 1981).

Little is known of the path of bile acids within the liver cell and the mechanism by which they leave the canaliculus. There are two main hypotheses: vesicular transport in a manner similar to that for proenzymes in the pancreas or cotransport with sodium. The energy is provided by a sodium-potassium ATPase in the basolateral membranes. Bile acids are probably formed in the microsomes. The hydroxylation of cholesterol is performed by enzymes in the microsomes and mitochondria (Okishio and Nair, 1966). Electron micrographs show changes in the Golgi zone and the presence of pericanalicular vacuoles in rat livers actively secreting bile (Boyer et al., 1979b; Jones et al., 1979). In view of the differences between rats and man in the vesicular transport of immunoglobulins, it would seem appropriate to reserve judgment until further evidence becomes available.

The demonstration that a sodium-potassium ATPase was present on basolateral membranes, rather than on canalicular membrane as initially proposed, required new concepts of its role (Blitzer

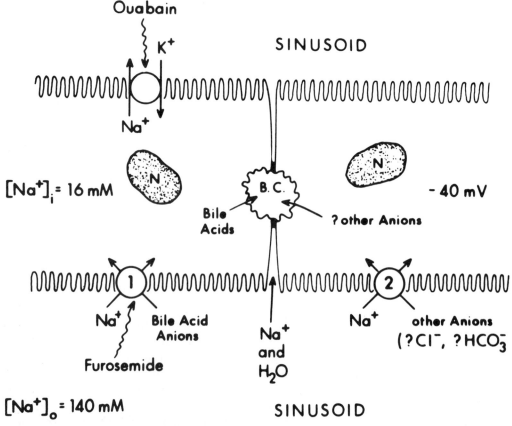

Ouabain

K⁺

SINUSOID

Na⁺

$[Na^+]_i = 16\ mM$

N

B.C.

Bile Acids

?other Anions

N

- 40 mV

Na⁺   Bile Acid Anions

Furosemide

Na⁺ and H₂O

Na⁺   other Anions (?Cl⁻, ?HCO₃⁻)

$[Na^+]_o = 140\ mM$

SINUSOID

Figure 6.115. Proposed model for hepatocellular bile formation. *B.C.*, bile canaliculus; *N*, nucleus; $[Na^+]_i$, intracellular sodium concentration; $[Na^+]_o$, extracellular or blood sodium concentration. [From Blitzer and Boyer (1982).]

and Boyer, 1978; Poupon and Evans, 1979). However, by analogy with other secretory systems, the possibility of cotransport with sodium became an attractive hypothesis. Since it is not possible to measure the electrochemical gradient across the bile canaliculus directly, active electrogenic transport of bile acids cannot be excluded. Figure 6.115 shows a model of the cotransport hypothesis. Sodium and the bile acids share a common carrier to enter the liver cell across the sinusoidal membrane. Sodium is then pumped back across the membrane by the Na$^+$-K$^+$-ATPase. The bile acids then diffuse across the canalicular membrane in response to the electrochemical gradients. Once in the canalicular lumen, the bile acids exert an osmotic force to draw water into the canaliculus (Sperber, 1965). Solvent drag and diffusion will also lead to movement of electrolytes into the canaliculi. In the isolated perfused rat liver it can be shown that more cations enter bile than would be required to preserve electrical neutrality. The bile acids differ in their effects, depending upon their osmotic properties and their effects on the permeability of membranes (Anwer and Hegner, 1982). Micellar aggregates secreted by the canaliculus can act as "sinks" for further secretion of anions (Scharschmidt and Schmid, 1978). Bile acids can be shown to increase membrane fluidity (a change in the ability of molecules, especially proteins, to move within the cell membrane) as determined by fluorescence polarization and electron spin resonance (Scharschmidt et al., 1981).

As noted in Figure 6.115, the sodium and water entering the canaliculus in response to bile acid transport may traverse a paracellular pathway. Certain bile acids such as ursodeoxycholic stimulate bicarbonate secretion (Dumont et al., 1980).

Recent results with isolated vesicles prepared from plasma membranes indicate that an inward directed sodium chloride gradient leads to electronegative transport of taurocholate into the vesicles, confirming the presence of cotransport (Duffy et al., 1981; Inoue et al., 1982).

The bile acid-independent bile flow can also act through the neutral transport of bicarbonate or chloride across the sinusoidal membrane, with the sodium-potassium ATPase providing the energy (Fig. 6.116).

The evidence for a bile acid-independent bile flow has already been discussed. Substitution of ions in the isolated perfused liver suggest that active transport of bicarbonate contributes to bile acid-independent bile flow (BAIF) (Van Dyke et al., 1982). Removal of bicarconate from the perfusate reduced BAIF by 50% (Hardison and Wood, 1978). Evi-

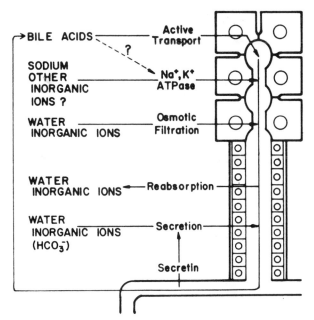

Figure 6.116. Schematic view of bile secretion. Canalicular bile flow includes bile formed in response to active bile acid secretion (bile acid-dependent flow) and in response, probably, to active secretion or inorganic electrolytes, mostly Na$^+$, K$^+$, ATPase-mediated ion transport (bile acid-independent flow). Bile acids may modulate the activity of the Na$^+$, K$^+$, ATPase (*dotted arrows*) and thereby influence bile acid-independent flow. The bulk of water and inorganic ions enter the canalicular bile as modified in the ductules and/or ducts by net absorption and secretion of water and inorganic ions. [From Erlinger (1982).]

dence supporting a role for Na$^+$-K$^+$-ATPase in the BAIF is supplied by the observation that in obstruction of the bile ducts the BAIF and the ATPase activity increase in parallel (Wannagat et al., 1978).

The major secretory mechanism localized to the bile ducts is a secretin-induced secretion of bicarbonate. There is a fall in chloride secretion but little change in bile acid output (Waitman et al., 1969). The response has not been correlated with plasma levels of immunoreactive secretin. Figure 6.116 shows a diagram of bile secretion.

Secretion of Bilirubin in Hepatic Bile

As noted earlier, bilirubin is taken up from the sinusoids by a saturable and specific mechanism. Once inside the cytoplasm it is conjugated with glucuronide and secreted at the canaliculus largely as the diglucuronide. It is the transport into the canaliculus which is rate-limiting. There is controversy over the steps resulting in the formation of the diglucuronide. One group of investigators attributes diglucuronide formation to glucuronyl transferase (Gollan and Schmid, 1982) as shown in Figure 6.117, while another believes that a dismu-

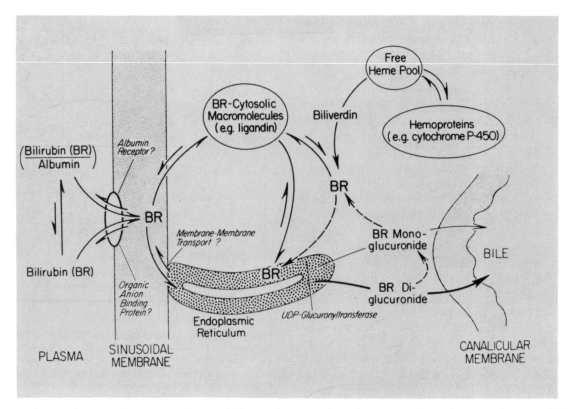

Figure 6.117.   Scheme of bilirubin uptake, metabolism, and transport in the hepatocyte. [From Gollan and Schmid (1982).]

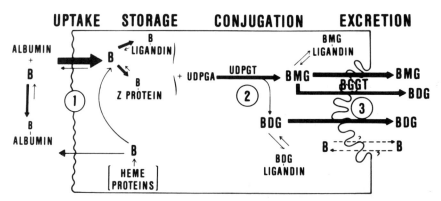

Figure 6.118.   Scheme of hepatic transport of bilirubin (*B*). In the circulation bilirubin is tightly bound to albumin. This complex disassociates, and bilirubin enters the liver cell by facilitated diffusion ①. A fraction of the bilirubin within the cell is derived from the breakdown of heme. Bilirubin inside the cell is bound to the cytosolic proteins ligandin and Z protein which prevents its efflux from the cell. Bilirubin is conjugated with glucuronic acid ② in the presence of bilirubin UDP- glucuronyl transferase (*UDPGT*) and UDP glucuronic acid (*UDPGA*) to form bilirubin monoglucuronide (*BMG*) and bilirubin diglucuronide (*BDG*), both of which may bind to ligandin. The hepatocyte plasma membranes are enriched in an enzyme (*BGGT*) which dismutes BMT to BDG and unconjugated bilirubin. [From Chowdhury et al. (1982).]

tase converts bilirubin monoglucuronide to the diglucuronide (Chowdhury et al., 1982) (Fig. 6.118).

Based upon studies in patients following cholecystectomy for gallstones, virtually all of the bilirubin in hepatic bile is conjugated with glucuronide. A T tube equipped with an occludable balloon and a port for reinfusion of bile into the common bile duct ensured complete collections yet maintained the enterohepatic circulation. Bilirubin secretion was linearly and positively correlated with the secretion of bile acids and bile flow. About one-third of the output of bilirubin was associated with the bile acid-independent fraction, while bile acid-dependent flow accounted for the rest. Presumably micelle formation is not necessary for the secretion of bilirubin glucuronide in the bile acid-independent fraction (Shull et al., 1977).

Control of Hepatic Bile Secretion: Enterohepatic Circulation of Bile Acids

As noted earlier, the most important factor in the control of hepatic bile secretion is the enterohepatic circulation of bile acids. Table 6.9 summarizes the quantitative aspects of bile acid metabolism in man. The location of the bile acid pool is predominantly in the gallbladder in the interdigestive state and shifts into the small intestine after a meal. The rate of bile acid synthesis in the liver is determined by the level of bile acids in portal blood. It is suppressed by high levels such as occur after a meal and accelerated by low levels present during the fasting state. Normally bile acid synthesis balances the amount of bile acids lost in the stool. With interruption of the enterohepatic circulation, however, bile acid synthesis can only increase 2- to 3-fold. Therefore, the bile acid pool is reduced, and the rate of bile secretion falls (Fig. 6.119).

Bile acids complete an enterohepatic cycle about three times with each meal or 6–15 times a day. As the size of the bile acid pool falls, the rate of

Table 6.9
Kinetics of bile acid metabolism in human subjects

Bile Salt	Pool Size, mg	Turnover Rate, days	Daily Rate of Synthesis, mg	Daily Secretion Rate, g	Cycling Frequency, days	Absorption, %
Cholate	500–1500	0.2–0.5	180–360			
Deoxycholate	200–1000	0.2–0.3				
Chenodeoxycolate	500–1400	0.2–0.3	100–250			
Lithocholate	50–100	1.0				
Total bile salts	1250–4000			11–40	11–15	93–99

From Carey (1982).

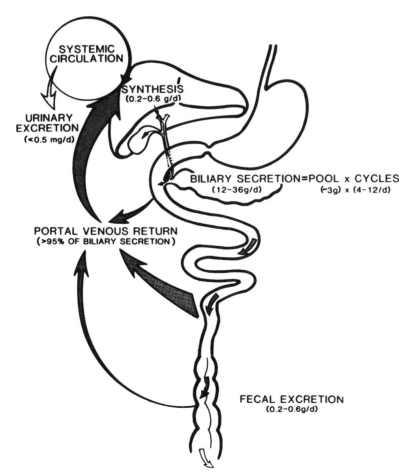

Figure 6.119. Enterohepatic circulation of bile acids showing typical values for healthy human subjects. [From Carey (1982).]

recycling increases. The evidence suggests that the rate of recycling is the primary event, and the bile acid pool changes in response to the recycling rate. Recycling determines the bile acid return to the liver and its uptake by liver cells. The frequency of gallbladder contractions and the transit rate of bile acids through the intestine to the major site of absorption in the ileum will, therefore, control bile acid return in portal venous blood. Bile acid synthesis will then vary inversely with bile acid uptake in the liver (Carey, 1982). Multiple small meals or a high carbohydrate diet, which has minimal effects on gallbladder contraction, will result in a decreased bile acid pool from decreased bile acid synthesis, while anticholinergic drugs will increase primary bile acid pools (Duane and Hanson, 1978; Duane, 1978; Duane and Bond, 1980).

Hormonal Control of Bile Secretion

A number of hormones are known to influence the composition and rate of flow of hepatic bile, but only secretin is likely to play a role under physiological conditions. It leads to the secretion of a dilute bile with a high concentration of bicarbonate, presumably by activating a bicarbonate pump in the bile ducts. Glucagon increases bile flow in experimental animals and in man (Kaminski et al., 1980; Dyck and Janowitz, 1971). It also increases the output of bilirubin in bile (Jarrett et al., 1979). Estradiol increased the cholesterol saturation index in young men and produced stasis in the bile ducts in rats (Anderson et al., 1980; Kreek et al., 1969). In pregnancy, the bile acid-independent flow was reduced nearly two-thirds in the hamster (Reyes and Kern, 1979). Somatostatin also reduced bile flow in the rat (Ricci and Fevery, 1981). The effect on bile acid-independent flow was greater than that on bile acid-dependent flow.

Nervous Control of Bile Secretion

There is little evidence that electrical stimulation of the vagus nerves in animals has any significant effect on hepatic bile secretion. The effects of vagotomy are controversial. Insulin hypoglycemia stimulates hepatic bile secretion, but its mechanism of action is uncertain (Smith et al., 1981).

Response of Hepatic Bile Secretion to a Meal

In the dog, hepatic bile flow increases after a meal even when the enterohepatic circulation is interrupted. This could be the result of secretin-induced bicarbonate secretion (Brooks, 1976). As noted earlier with an intact enterohepatic circulation, feeding increases the enterohepatic circulation of bile acids. During a meal in man, some of the portal venous bile acids escape uptake by the liver and appear in the peripheral blood.

Hepatic Bile Secretion and Hepatic Blood Flow

In the isolated perfused rat liver, decreased perfusion rates selectively decreased bile acid-independent flow without changing bile acid-dependent flow (Tavaloni et al., 1978). If the portal venules were blocked with microspheres, bile flow was reduced by a third (Avner et al., 1981).

Pathophysiology of Hepatic Bile Secretion

One of the characteristic features of hepatic bile in patients with cholesterol gallstones is that the bile is supersaturated with respect to cholesterol. Correcting this by feeding chenodeoxycholic acid or ursodeoxycholic acid will dissolve the stones in some patients. Increased secretion of cholesterol in bile is found in obese patients as well as some patients with gallstones. Oral contraceptives containing estrogens predispose to cholesterol gallstones. Interrupting the enterohepatic circulation of bile by surgical removal of the ileum results in supersaturation of hepatic bile with cholesterol and increases the risk of cholesterol gallstones.

The rise in systemic venous blood levels of bile acids after a meal can be used as a test of liver cell function. Rises to greater than normal levels indicate impaired uptake. Radionuclides are available that are selectively taken up by the liver and excreted into the bile. They can be detected by scintigraphy to outline the bile ducts.

Disorders of bilirubin metabolism leading to jaundice (hyperbilirubinemia) occur at each stage of the secretion process: uptake (Gilbert's syndrome), conjugation (Crigler-Najjar syndrome) and secretion across the canaliculus (Dubin-Johnson syndrome).

INTESTINAL SECRETION

The secretions of the small and large intestine have been the subject of much controversy. In the 1930s the "succus entericus" was thought to play an important role in digestion. An extract of intestinal mucosa, found to stimulate small intestinal secretion, was designated "enterocrinin" (Nasset, 1938). By the 1950s the succus entericus had virtually vanished and only the secretion of Brunner's glands in the duodenum remained. Then came the recognition that the voluminous stools of cholera were the result of a functional hypersecretion of the small intestinal mucosa. The purification of cholera and other bacterial toxins has provided a valuable tool to stimulate intestinal secretion and to study the mechanism of secretion.

The physiological significance of intestinal secretion remains uncertain. It is difficult to characterize in part because it may occur in individual segments of the gut and because it is then absorbed distally. When intraluminal contents contain high molecular weight solutes, secretion may be necessary to maintain fluidity. With low molecular weight solutes, osmotic forces may be sufficient. Local reflexes responding to distention can stimulate secretion. Secretions may be useful to dilute and wash away potentially injurious substances. Regulatory peptides may exert some of their actions by acting on the surface of the mucosa (Field, 1981).

Composition of Intestinal Secretion

The composition of intestinal secretion in man under normal circumstances is difficult to determine. Perfusion of jejunal segments in normal subjects show net absorption of water, sodium, potassium, and chloride. Water follows absorption of electrolyte. There is probably an active secretion of hydrogen ion which in the lumen reacts with bicarbonate to form CO_2. The CO_2 is then absorbed (Turnberg et al., 1970a). In the ileum there is also net absorption, but there is secretion of bicarbonate, probably in exchange for chloride (Turnberg et al., 1970b).

It has been shown clearly by double isotope studies that the net movement of water and electrolyte in the intestine is the resultant of blood-to-lumen and lumen-to-blood fluxes. The term secretion as applied to the intestine usually refers to net blood-to-lumen transport without regard for demonstration of movement against an electrochemical gradient. Figure 6.120 shows the consequences of altering one or both of the unidirectional fluxes.

In cholera the large volume of stool is due almost entirely to an increase in the blood-to-lumen flux of water in the small intestine. Lumen-to-blood transport of glucose remains normal and can be used therapeutically to increase the absorption of water and electrolyte. If the hypersecretion of the small intestine in cholera can be taken as an exaggeration of a normal mechanism of intestinal transport, then the composition of the cholera stool should approach that of intestinal secretion.

In conscious dogs with denervated intestinal loops open at both ends onto the abdominal wall (Thiry-Vella loops) intestinal content can be collected by blowing air through the loop either under basal conditions or during stimulation with an agent such as glucagon. The composition of the loop contents differs in jejunal loops from ileal loops with respect to chloride and bicarbonate (Barbezat and Grossman, 1971). Table 6.10 shows a comparison of the composition of cholera stools with secretion from jejunal and ileal loops in dogs.

The quantitation of the secretion of proteins in

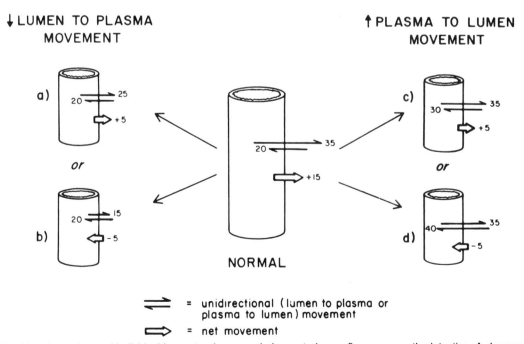

Figure 6.120. Net absorption and individual lumen-to-plasma and plasma-to-lumen fluxes across the intestine. A decrease in lumen-to-plasma movement or an increase in the plasma-to-lumen movement will result either in decrease in net absorption (*a* and *c*) or net secretion (*b* and *d*). The magnitude of both the existing absorptive process and the altered transport process will determine whether there is net secretion or net absorption. [From Binder (1980).]

Table 6.10
Electrolyte concentration of intestinal secretions

	Na	K	HCO$_3$	Cl
Stools from cholera patients	126 ± 9	19 ± 9	47 ± 10	94 ± 9
Secretion from jejunal loops in dogs[a]				
Basal	145 ± 5	11.7 ± 1.7	14 ± 3	146 ± 1
Stimulated by glucagon	144 ± 4	7.5 ± 0.5	14 ± 3	151 ± 1
Secretion from ileal loops in dogs[a]				
Basal	147 ± 3	8.1 ± 0.6	84 ± 3	67 ± 6
Stimulated by glucagon	150 ± 2	5.8 ± 0.5	86 ± 3	80 ± 2

From Field (1981). [a] From Barbezat and Grossman (1971).

intestinal secretion is still open to investigation. In a study of secretion from canine intestinal loops, there was no increase in protein output during spontaneous secretion (Reed et al., 1979). In rats, cholera toxin produces a 4- to 5-fold increase in immunoreactive mucin secretion from the intestine. There is also an increase in the rate of synthesis of glycoprotein (Forstner et al., 1981).

Bicarbonate Secretion in the Duodenum

The study of bicarbonate secretion in the duodenum of laboratory mammals is complicated by the presence of Brunner's glands. In man when the duodenum is perfused with solutions of varying pH, the contents are neutralized in the first portion of the duodenum if the perfusate has a pH of 2 or more. Below that, some hydrogen ion enters the more distal duodenum. A decrease in osmolality occurred with perfusate with pHs of 1 to 2, due to the formation of CO_2. Acid is a stimulant of duodenal bicarbonate secretion when bile and pancreatic juice are excluded (Winship and Robinson, 1974).

Bullfrogs have no Brunner's glands in the proximal duodenum, but the first 2 cm of the mucosa can secrete bicarbonate in vitro (Flemstrom et al., 1982a). An electrogenic mechanism accounts for about half, passive diffusion for 40%, and secretion of endogenous bicarbonate for less than 10% (Simson et al., 1981).

In cats, rats, and dogs, prostaglandins, glucagon, and GIP cause the secretion of bicarbonate into the duodenum. The secretion is inhibited by indomethacin (an inhibitor of prostaglandin synthesis) and by acetazolamide (an inhibitor of carbonic anhydrase). Secretin had no effect (Flemstrom and Garner, 1982). Figure 6.121 shows paracellular and transcellular pathways. In vivo perfusion studies in guinea pigs and cats show that acidification of the duodenum stimulates bicarbonate secretion at least in part mediated by prostaglandins and cAMP by means of an electrogenic mechanism. Another neutral bicarbonate secretory pathway involving a Cl⁻-

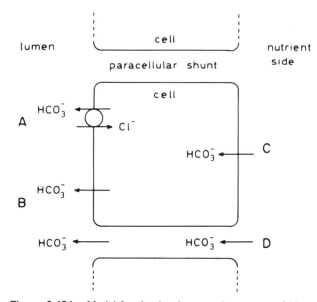

Figure 6.121. Model for duodenal mucosal transport of bicarbonate. In *A*, transport involves Cl⁻-HCO$_3$ exchange. In *B*, transport of bicarbonate is independent of luminal chloride. This process is stimulated by prostaglandins and dibutyryl cAMP. In *C* the anion carrier also displays affinity to chloride. In *D* there is passive migration of bicarbonate through shunt pathways which are sensitive to variations in transmucosal hydrostatic pressure. [From Flemstrom and Gardner (1982).]

HCO$_3$⁻ exchange is stimulated by glucagon and GIP (Flemstrom et al., 1982b). Calculations suggest that secretion of bicarbonate by the mucosa can account for most of the neutralization of gastric acid rather than the secretion of Brunner's glands.

Water and Electrolyte Secretion in the Jejunum

Secretion of chloride against an electrochemical gradient can be seen in perfusion studies of the human jejunum, particularly if bicarbonate is omitted from the perfusate (Davis et al., 1981a). Secretion of sodium and water in the perfused human jejunum can be induced by L-arginine in isotonic saline. The effect was not seen in adjacent segments and is, therefore, probably not hormonally mediated (Hegarty et al., 1981). It is inhibited by parenteral chlorpromazine, which has been shown to inhibit the action of secretagogues acting by way of cAMP in animals (Holmgren et al., 1978).

Cyclic AMP and Intestinal Secretion

Studies in small intestinal mucosa from animals in vitro implicate cAMP in the secretory process. In addition to stimulating secretion, it also inhibits the absorption of sodium chloride in the ileum. Sodium and chloride enter the cell by a process of cotransport. The sodium enters by carrier-facilitated diffusion and is pumped out by the ubiquitous Na^+-K^+ ATPase in the basolateral membrane. Chloride shares the same carrier and diffuses out of the cell in response to a concentration gradient. The secretory action of cyclic AMP is thought to be due to the entry of chloride across the basolateral membrane coupled to sodium. The chloride permeability of the luminal membrane is increased, and chloride passively enters the lumen. The sodium which entered the cell with chloride is recycled across the basolateral membrane and then enters the lumen by way of a paracellular route. These ion movements are shown in Figure 6.122. It is assumed that the absorptive process is occurring in villous cells and the secretory process in crypt cells. This is supported by the action of inhibitors of protein synthesis which selectively inhibit secretion and reduce mitoses in crypts (Serebro et al., 1969) and by agents which damage the villi while sparing the crypts. Intestinal secretion in response to cholera toxin is unimpaired (Roggin et al., 1972).

The action of agents such as prostaglandin PGE and vasoactive intestinal polypeptide activating the cyclic AMP system is shown diagrammatically in Figure 6.123. All agents which increase the concentration of cAMP in the intestinal mucosa increase secretion. It is interesting that caffeine converts net absorption to net secretion in the human jejunum. Caffeine is an inhibitor of phosphodiesterase (Wald et al., 1976). The role of cyclic GMP in intestinal secretion is uncertain.

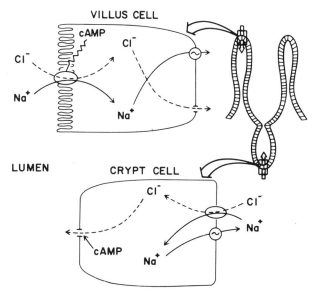

Figure 6.122. The postulated separate actions of cAMP on ion transport in intestinal villus and crypt cells. [From Field (1981).]

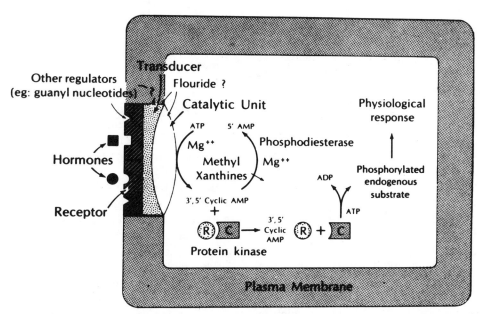

Figure 6.123. A schematic model of the adenylate cyclase system. *R* represents the regulatory components and *C* the catalytic components. [From Kimberg (1974).]

Intestinal Secretion in the Ileum

Many of the studies of small intestinal secretion have been performed in isolated rabbit ileal mucosa. This tissue contains a neutral Na^+Cl^--absorptive mechanism which has not been shown in other parts of the intestine. Bicarbonate secretion can be demonstrated in the rabbit ileum in vitro, but not in response to dibutyryl cAMP. It probably results from a $Cl^--HCO_3^-$ exchange (Dietz and Field, 1973; Hubel, 1969). There is also an exchange of sodium for hydrogen ion (Podesta and Mettrick, 1977). Histamine stimulates secretion in rabbit ileal mucosa by inhibiting net chloride absorption and also reduced $Cl^--HCO_3^-$ exchange (Linaker et al., 1981).

Calcium and Intestinal Secretion

The demonstration that a calcium ionophore, which permits the entrance of calcium into cells of the ileal mucosa, stimulated chloride secretion launched a new approach to the cellular control of secretion. The effect was similar to that of cAMP, but there was no increase in the tissue level of cAMP (Bolton and Field, 1977). It is possible that calcium may mediate the response to cAMP. Figure 6.124 shows a proposed schema for the control of secretion. Two different pathways are depicted. Cyclic AMP-activating stimuli would release calcium stores within the cell. Other secretagogues could increase calcium influx. The final result would be an increase in cytosolic calcium concentration. This results in the "activation" of the calcium-binding protein calmodulin, which increases the permeability of the apical membrane to chloride. VIP and prostaglandins act by way of cAMP, while acetylcholine and serotonin increase calcium influx. Somatostatin inhibits secretion by preventing the activation of calmodulin or by blocking its action. Enkephalins may block the entry of calcium into the cell (Dobbins and Binder, 1981).

Colonic Secretion of Water and Electrolyte

The human colon is normally an absorptive organ. Evidence for its secretory activity comes largely from in vitro studies on animal tissues. There is no coupled absorption of NaCl, so that active chloride secretion in response to cAMP is the main means by which the nucleotide stimulates secretion. Figure 6.125 is a diagram of the ionic shifts (Binder, 1981).

As in the case of the small intestine, calcium ionophore stimulates active electrogenic chloride secretion in the rabbit colon. It is due to an increase in the serosa to mucosa flux of chloride. If calcium is removed from the bath, the ionophore is without

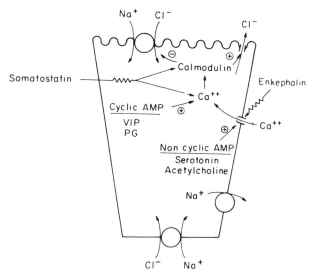

Figure 6.124. Hypothetical role for calcium in control of intestinal secretion. Serotonin and acetylcholine act independently of cAMP and increase calcium influx, while VIP and prostaglandin E, acting through cAMP, mobilize intracellular stores of calcium and increase cytosolic calcium concentration, which activates calmodulin. Activated calmodulin inhibits neutral sodium and chloride influx and increases the permeability of the brush border membrane to chloride. Somatostatin may work by blocking the calcium activation of calmodulin or blocking the effect of activated calmodulin. Enkephalin may work by blocking calcium entry. [From Dobbins and Binder (1981).]

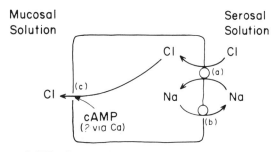

Figure 6.125. Proposed model of cAMP stimulation of active chloride secretion. [From Binder (1981).]

effect on secretion (Frizzell, 1977). Cyclic AMP increases intracellular concentration of calcium and stimulates Cl^- secretion. It may act by the release of calcium from intracellular stores (Frizzell and Heintze, 1979).

In vivo perfusion studies on rat colon show a secretion of potassium in the proximal colon (Kliger et al., 1981). A high potassium diet increased potassium secretion by the rat colon both in vivo and in vitro (Fisher et al., 1976).

The rabbit colon can secrete bicarbonate while absorbing chloride into the lumen, according to the diagram shown in Figure 6.126 (Schultz, 1979). In perfusion studies in the intact human colon, the secretion of bicarbonate only accounted for 45% of

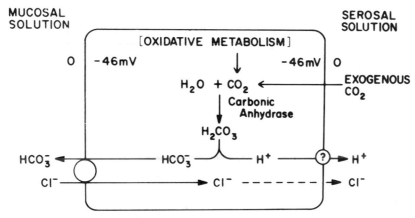

Figure 6.126. Model for absorption of chloride and secretion of bicarbonate by the mammalian colon. [From Schultz (1981).]

the absorption of chloride. The colon was virtually impermeable to potassium (Davis et al., 1983).

Structural Basis of Intestinal Secretion

As is the case with other exocrine secretions, morphologic studies are more helpful in the understanding of the secretion of proteins than of the secretion of water and electrolyte. Figure 6.127 shows examples of merocrine and apocrine secretion of granular material from undifferentiated crypt cells from human small intestine. In merocrine secretion the granules exit by exocytosis, while in apocrine secretion a portion of the cytoplasm is pinched off as well (Trier, 1964).

Goblet cells are mucus-containing cells in the epithelial surface of the small and large intestine. Figure 6.128*A* shows a resting goblet cell from the rat (Neutra and Leblond, 1966). Figure 6.128*B* in the *lower panel* shows a similar cell from a rabbit after stimulation by acetylcholine (Neutra et al., 1982). Note that the limiting membranes of the granules remain intact.

Paneth cells are found in the base of the crypts of the Lieberkuhn. They are characterized by eosinophilic granules in their cytoplasm. The granules are secreted by exocytosis in response to cholinergic agents (Trier and Madara, 1981). The secretion contains immunoglobulins and lysozyme (Sandow and Whitehead, 1979). These secretions may be bactericidal.

Enterochromaffin cells in the epithelium contain serotonin which is probably secreted into the lumen. Those in the basal lamina probably are secreted into the bloodstream (Pentilla, 1968). Endocrine cells in the small intestinal mucosa contain secretin, CCK, gastrin, motilin, neurotensin, somatostatin, and enteroglucagon as determined by immunocytochemistry. These peptides circulate in the blood and are released in response to a meal.

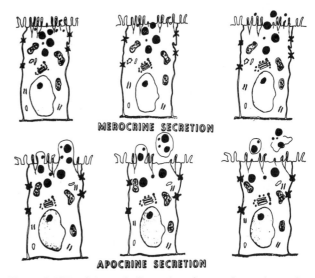

Figure 6.127. Schematic illustration of merocrine and apocrine secretion by undifferentiated intestinal crypt cells. [From Trier (1964).]

Protein Secretion

In the rabbit small intestine about 20% of all newly synthesized protein was IgA (Trier, 1976). In man, plasma cells in the lamina propria contained mostly IgA, while secretory piece was present in crypt epithelium (Brown et al., 1976). Secretory piece probably serves as a receptor of IgA (Brown et al., 1977). Cholinergic agonists in rats released IgA into the perfusate of ileal loops, while anticholinergics blocked their action (Wilson et al., 1982).

Alkaline phosphatase is released into the lumen of small intestinal loops in response to secretin and CCK (Dyck et al., 1973). Most of the alkaline phosphatase as detected by staining in cells in the human small intestine is in the villous tips (Lev and Griffiths, 1982). There are soluble forms which are released in response to feeding fat (Shields et al., 1982). There is also a rise in serum alkaline phosphatase (Young et al., 1981).

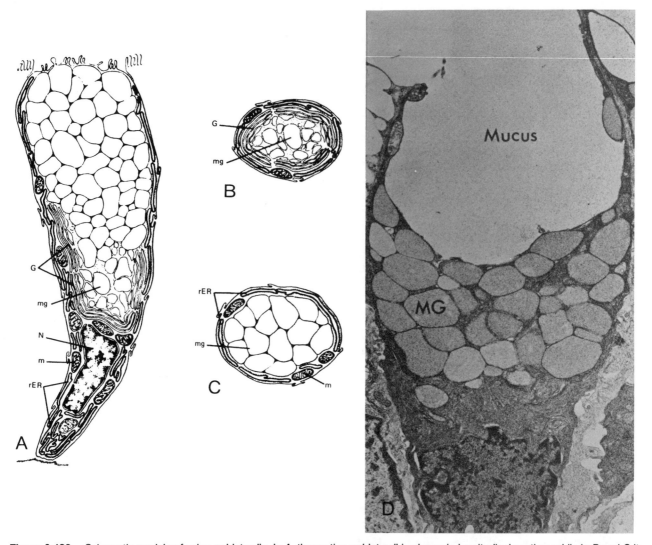

Figure 6.128. Schematic models of colon goblet cells. In *A*, the resting goblet cell is shown in longitudinal section, while in *B* and *C* it is shown in cross-section at the supranuclear and supra-Golgi levels *G*, golgi complex; *N*, nucleus; *MG*, mucinogen granules; *m*, mitochondria; *rER*, rough endoplasmic reticulum. [From Neutra and LeBlond (1966).] In *D* secretion has been stimulated by acetylcholine. The deep cavitation of the apical cell surface is due to exocytosis involving incorporation of mucous granule membranes into luminal plasma membrane. *MG*, basal mucous granule not yet secreted. [From Neutra et al. (1982).]

Secretion of Glycoproteins

Glycoproteins appear in the Golgi region and reach the apical surface of the cell in 3 or 4 h (Forstner, 1970). Incorporation of ^{14}C-glucosamine into mucus appears first in the microsomes, then in mitochondria and the brush border, and finally it appears in luminal contents after 2.5 h. Incorporation of ^{14}C-glucosamine into glycoproteins is increased by β-adrenergic agonists (Forstner et al., 1973). In man, the mucins of the goblet cells of the small intestine cell are mostly neutral, while those of the colon are acidic (Subbaswamy et al., 1971).

Intestinal secretion, but not synthesis of glycoproteins, is increased by acetylcholine in the human rectum (MacDermott et al., 1974). Dibutyryl cAMP

increased synthesis of glycoproteins in rabbit colon in organ culture (LaMont and Ventola, 1977). Exocytosis occurred in crypt goblet cells in response to acetylcholine given to rats in vivo (Specian and Neutra, 1982).

Intestinal glycoproteins in the pig have fewer subunits than those in the stomach (Mantle et al., 1981).

Control of Intestinal Secretion: Hormones

It is not possible to describe the physiologic mechanisms for the control of intestinal secretion beyond the qualitative effects of what may be pharmacologic doses of neuropeptide, hormones, and neurotransmitters. However, the stimulation of se-

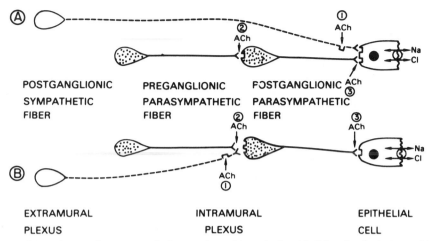

POSTGANGLIONIC   PREGANGLIONIC   POSTGANGLIONIC
SYMPATHETIC   PARASYMPATHETIC   PARASYMPATHETIC
FIBER   FIBER   FIBER

EXTRAMURAL   INTRAMURAL   EPITHELIAL
PLEXUS   PLEXUS   CELL

Figure 6.129. Two possible schemes for autonomic innervation of intestinal epithelial cells. In *A* the cell is innervated by both postganglionic sympathetic and parasympathetic nerves with acetylcholine serving as the neurotransmitter at both neurosecretory junctions. In *B*, the sympathetic neuron ends on the postganglionic neuron of the parasympathetic neuron and modulates release of acetylcholine from the preganglionic parasympathetic neuron. Acetylcholine may act at three sites: ①, either the nicotinic stimulatory or muscarinic inhibitory receptor on the postganglionic sympathetic fiber; ②, cholinergic synapse between pre- and postganglionic parasympathetic neurons; and ③, cholinergic receptor on the epithelial cell. [From Powell and Tapper (1979).]

cretion by VIP, secretin, glucagon, GIP, gastrin, and thyrocalcitonin in man is worth noting (Schwartz et al., 1974; Warnes et al., 1974; Hicks and Turnberg, 1974; Helman and Barbezat, 1977; Modigliani et al., 1976; Gray et al., 1976). VIP and secretin increase the formation of cAMP (Binder, 1980). High, but not low, doses of VIP stimulate intestinal secretion in man (Davis et al., 1981b). Somatostatin inhibits intestinal secretion stimulated by glucagon (Barbezat and Reasbeck, 1981).

Calcium channel blockers prevented secretion induced by serotonin in rabbit ileal mucosa in vitro (Donowitz et al., 1980). Prostaglandin PGE$_1$ stimulates secretion from the jejunum of man during perfusion studies (Matuchansky et al., 1976). In rats, agents which disrupt microtubules blocked the effects of prostaglandins in stimulating intestinal secretion (Notis et al., 1981).

Nervous Control of Intestinal Secretion

There is little direct evidence that nerves control intestinal secretion under physiological conditions, but the difference between cholinergic agonists and adrenergic agonists is striking. Acetylcholine leads to secretion and norepinephrine to absorption of water and electrolyte in vivo and in vitro (Isaacs et al., 1976; Field and McColl, 1973). Electrical field stimulation of rabbit ileum in vitro produced intestinal secretion of chloride by increasing calcium entry into epithelial cell (Hubel and Callanan, 1980). Similar results were obtained with human ileal mucosa (Hubel and Shirazi, 1982). Muscarinic (cholinergic) receptors have been demonstrated on

isolated rat colonic epithelial cells (Zimmerman and Binder, 1982). Electrical vagal stimulation released serotonin into the lumen of cat jejunal loops (Zinner et al., 1982). Figure 6.129 indicates two possible models for the nervous control of intestinal secretion (Powell and Tapper, 1979a). A paralytic secretion from the colon occurred after preganglionic sympathectomy which could be blocked with an anticholinergic drug. Stimulation of the pelvic nerves stimulated colonic secretion (Powell and Tapper, 1979b).

Role of Intraluminal and Venous Pressures in Intestinal Secretion

Acute blood volume expansion in rats resulted in net blood-to-lumen movement of sodium, probably by a paracellular route (Humphreys and Early, 1971). Similar changes were seen in dogs (Higgins and Blair, 1971). At mesenteric venous pressures above 30–35 cmH$_2$O, secretion of water and electrolyte appeared (Yablonsky and Lifson, 1976). Negative lumen pressures of −20 cmH$_2$O or more also lead to secretion (Hakim et al., 1977). Increasing venous pressure so that capillary filtration pressure increased by 12–15 mmHg resulted in secretion in cats (Granger et al., 1977). Reduction in plasma proteins to <3.0 g/dl resulted in secretion in isolated dog gut segments (Duffy et al., 1978).

Pathophysiology of Intestinal Secretion

Large volume watery diarrhea is frequently due to small intestinal hypersecretion such as seen in cholera and *Escherichia coli* infections. Peroral glucose solutions can be absorbed and are useful in treatment. Prostaglandin-induced bicarbonate secretion in the duodenum may contribute to the cytoprotective effects of prostaglandins in acute stress ulcers. Cyclic AMP is implicated in the hypersecretory actions of some GIP peptides. The rare syndrome of congenital chloridorrhea is due to a lack of ileal bicarbonate-chloride exchange. Islet-cell tumors producing VIP, thyrocalcitonin from medullary thyroid cancers, and prostaglandins may result in massive watery diarrhea and hypokalemia. Villous adenomas of the colon can secrete sufficient potassium to result in hypokalemia. Mucus can be seen in the diarrheal discharges of cholera or in the formed stools of "mucus colitis," a form of the irritable bowel syndrome. Perhaps the failure to develop intestinal secretion during portal venous hypertension is due to the slow development of increased venous pressure in diseases, such as cirrhosis of the liver, which allows adaptive changes to occur.

CHAPTER 45

Digestion

Digestion, which gives its name to the alimentary canal, refers to the process by which dietary constitutents are reduced in size and solubilized for presentation to the small intestinal mucosa for absorption (Brooks, 1980). It is also likely that biologically active peptides secreted into the lumen of the digestive tract are inactivated by digestion. Normal diets in modern industrialized societies consist of carbohydrates, proteins, and fats in the ratio of 2:1:1 in grams. Carbohydrates are present in the diet as starches and sugars such as sucrose and lactose (milk sugar). Both of the latter are disaccharides.

Proteins are large molecules of several thousands in molecular weight, consisting of amino acids linked by peptide bonds. Dietary fat is composed almost entirely of triglyceride containing three long chain fatty acids linked by ester bonds to glycerol. The fatty acids contain 16–18 carbon atoms and may be saturated (palmitic or stearic acid) or unsaturated (oleic or linoleic acid). Medium chain length fatty acids containing 6–12 carbon atoms compose about 10% of dietary fat.

Digestion occurs in the lumen of the digestive tract and at the surface of the intestinal mucosa. We shall consider luminal and membrane digestion separately.

Luminal digestion starts in the stomach when food is exposed to salivary amylase, pharyngeal lipase, and gastric pepsins. The role of mastication and antral contractions in digestion has already been discussed under motility. The magnitude of gastric digestion is difficult to quantitate, but its effect in accelerating gastric emptying indicates that it is of physiological importance.

LUMINAL DIGESTION IN THE SMALL INTESTINE

Chyme enters the duodenum from the stomach. It elicits a flow of bile and pancreatic juice and also the secretion of Brunner's glands and the duodenal mucosa. The secretion of Brunner's glands is not thought to have an important role in digestion (Cooke, 1967). It is difficult to dissociate the secretion of Brunner's glands from that of the duodenal mucosa in mammals, but the mucosa is thought to be a more important source of bicarbonate (see intestinal secretion). Regardless of the pH of ingested food, by the time the chyme enters the jejunum the pH approaches neutrality. Both hypotonic meals (steak) and hypertonic meals (coffee and doughnuts) are brought to isotonicity in the duodenum. This is important in establishing an optimal milieu for digestion and absorption (Fordtran and Locklear, 1966).

Digestion of Starches and Sugars in the Lumen of the Small Intestine

Dietary carbohydrate consists of about 60% starch, 30% sucrose (table sugar), and 10% lactose. Dietary starch is a mixture of about 20% amylose and 80% amylopectin. Amylose is made up of long linear chains of glucose molecules linked together through an oxygen atom between the first and sixth carbon atoms by an oxygen bridge (Fig. 6.130).

Alpha amylase, predominantly from the exocrine pancreas, attacks the carbon 1-4 linkages between glucose molecules in the interior of starch molecules. The result, in the case of amylose, is the formation of maltose, maltotriose, maltotetrose, and maltopentose (Fig. 6.131). The latter two sugars are then hydrolyzed to maltose and maltotriose. In the case of amylopectin, α-limit dextrins containing one or more of the branching points are also formed during digestion by α-amylase (Fig. 6.132). The number of glucose units varies with an average of 8. The pH optimum is about 6–7 and the presence of chloride ion is required (Gray, 1983).

Digestion of Proteins and Peptides in the Lumen of the Small Intestine

Proteins and peptides enter the duodenum and are exposed to the action of pancreatic proteases. These enzymes are secreted as inactive precursors. On entering the duodenum the terminal portions of the precursors are cleaved by a brush border enzyme, enterokinase. Enterokinase is localized to the brush border and glycocalyx of the human duo-

745

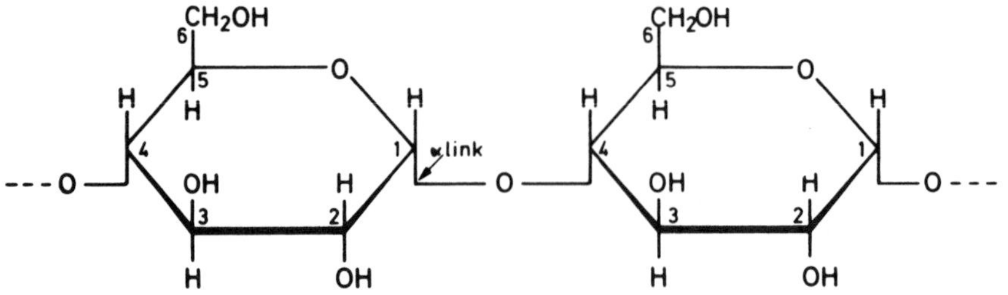

Figure 6.130.   Structure of 1, 4 glucose links in amylase and amylopectin. Numbers refer to carbon atoms. [From Gray (1970).]

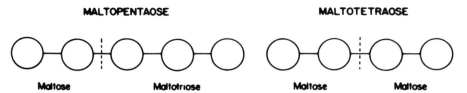

Figure 6.131.   Tetra- and pentasaccharide products of amylase digestion. Each circle represents a glucose molecule. The bridging lines indicate 1,4 links. *Dotted lines* are located at sites of amylase action on the interior of the molecule, yielding tri- and disaccharides as the final hydrolytic products. [From Gray (1970).]

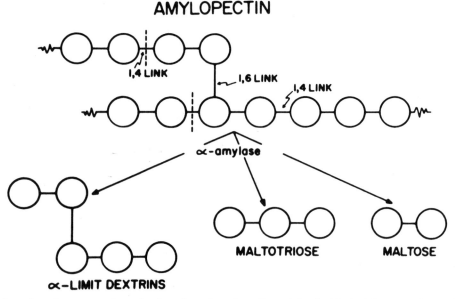

Figure 6.132.   Action of amylase on amylopectin. A portion of amylopectin and the final hydrolytic products are shown with glucose molecules (*open circles*) joined by 1,6 (*vertical connecting lines*) links. [From Gray (1975).]

denum and proximal jejunum (Hermon-Taylor et al., 1977). There is evidence that it is released into the lumen in response to secretin and CCK (Moss et al., 1979) and to bile acids (Rinderknecht et al., 1978). Once trypsinogen has been converted to trypsin, the latter acts autocatalytically to form more trypsin. Trypsin also converts chymotrypsinogen to chymotrypsin, proelastase to elastase, and procarboxypeptidase to carboxypeptidase in a manner similar to the cascade of blood coagulation (Fig. 6.133). The sites of cleavage are shown in Figure 6.134.

The active proteases can be divided into endo and exopeptidases. Their action on peptides is shown in Figure 6.135. The endopeptidases produce peptides which can be attacked by the exopeptidases (Gray and Cooper, 1971). The final products are neutral and basic amino acids and small peptides of two to six amino acid units.

Digestion of Lipids in the Lumen of the Small Intestine

Dietary fat is present as triglycerides and is insoluble in water. By digesting the triglycerides to fatty acids and monoglycerides and by forming micelles with bile acids, the fat is converted into

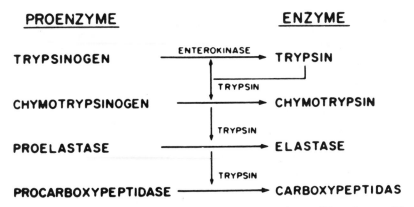

Figure 6.133. Activation of pancreatic proteases in the lumen of the duodenum. [From Gray and Cooper (1971).]

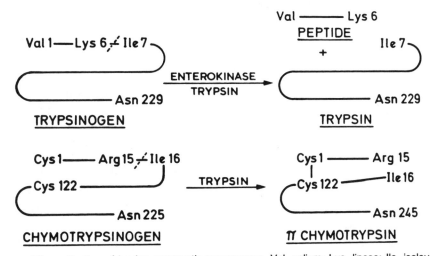

Figure 6.134. Diagram of the activation of bovine pancreatic proenzymes. Val, valium; Lys, lipase; Ile, isoleucine; Asn, asparagine; Cys, cysteine. The numbering refers to the positions on the protein chain beginning at the N-terminal amino acid. [From Gray and Cooper (1971).]

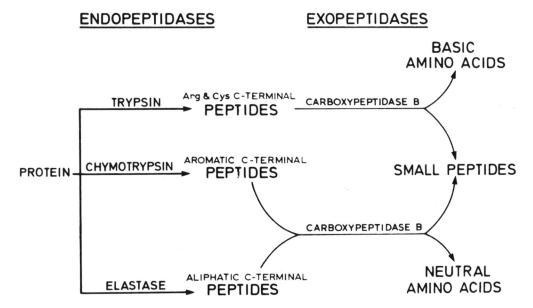

Figure 6.135. Intraduodenal sequential action of pancreatic endopeptidases and exopeptidases on dietary protein. The final products on the *right* are the substrates which must be handled by the intestinal cell. Arg, arginine; Cys, cysteine. [From Gray and Cooper (1971).]

water-soluble material. The structure of a lipid micelle has already been shown (see Fig. 6.109). Pancreatic lipase cannot hydrolyze triglyceride in the presence of bile acids unless a pancreatic protein colipase is also present (Patton, 1981). Presently a sequential course of events proposes that colipase binds first to triglyceride at the oil-water interphase in the presence of bile acids. Then lipase binds to the colipase at the substrate interphase as shown in Figure 6.136 (Desnuelle, 1976) and hydrolysis begins. Lipase acts to hydrolyze the ester bonds at the 1 and 3 positions, leaving a single fatty acid as a β-monoglyceride. Colipase is secreted as an inactive precursor by the pancreas in parallel with lipase. It is cleaved by trypsin and then binds to triglyceride. Recent studies with human duodenal content in vitro indicates that not enough endogenous colipase is secreted to provide maximum release of monoglyceride in the presence of bile acids in a concentration above the critical micellar concentration. Exogenous colipase increased hydrolysis of triglyceride. Therefore, colipase may provide the fine control for lipolysis (Gaskin et al., 1982).

Once free fatty acids and/or monoglycerides are formed and released they can be incorporated into mixed micelles with bile acids and phospholipids. The size of these mixed micelles is estimated at a Stokes radius of 2–4 mm. For each mole of bile acid there are about 1.4 mol of fatty acids, 0.15 mol of lecithin, and 0.06 mol of cholesterol (Mansbach et al., 1975).

Duodenal content after a fatty meal separates into several phases upon standing. The oil phase consists of spheres of triglyceride. During hydrolysis calcium soaps can be seen forming as brittle

shells on the surface of fat droplets. A second phase consists of long strings and blebs of lipid bilayers—the viscous isotropic phase. The clear watery phase consists of mixed micelles. The action of lipase can be viewed as "unzippering" the oil phase by opening water channels as shown in Figure 6.137.

Pancreatic prophospholipase A_z is activated by trypsin in the presence of bile acids and calcium. It acts on mixed bile acid-phospholipid micelles to produce one molecule of lysophospholipid and one molecule of free fatty acid. There is also a nonspecific pancreatic lipase.

MEMBRANE DIGESTION IN THE SMALL INTESTINE

At the end of the luminal phase of digestion the small intestine contains disaccharides and oligosaccharides, small peptides and amino acids, and micelles containing fatty acids and monoglycerides. An important study was carried out using meals of solutions of albumin, starch, and triglycerides (Borgstrom et al., 1957). The conclusion was reached that the luminal phase of digestion was completed in the proximal jejunum. With larger amounts of albumin, perfusion studies in man showed that peptide digestion was continuing in the ileum (Adibi and Mercer, 1973).

Membrane Digestion of Oligosaccharides

The final stage of digestion of carbohydrates is completed by enzymes of the surface membrane of the intestinal mucosal cells covering the villi. These consist of two groups: β-galactosidases and α-glucosidases. The former is represented by lactase which hydrolyses lactose to glucose and galactose.

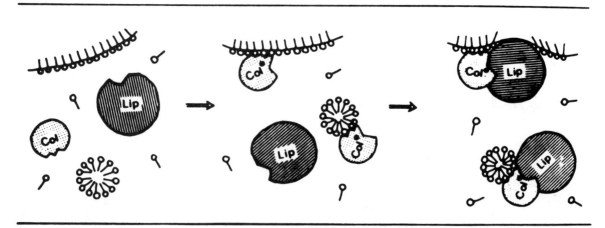

Figure 6.136. Role of colipase in the fixation of lipase to a hydrophilic interface (micelle or triglyceride interface coated with an amphipath). Col* designates the cofactor form in which a binding site for lipase has been created by previous adsorption to micelle or interface. [From Desnuelle (1976).]

LIPOSOMAL OR MICELLAR PHASE   VISCOUS ISOTROPIC PHASE   OIL PHASE

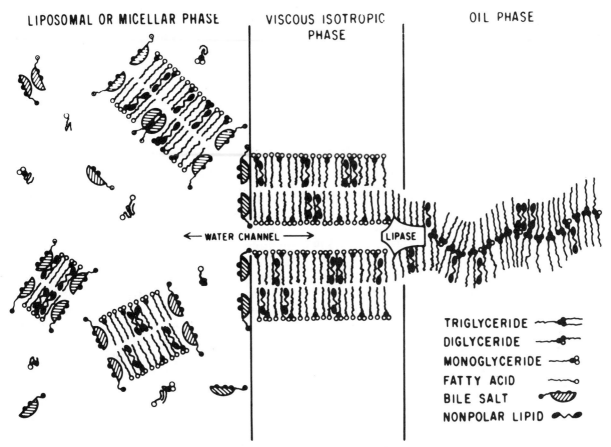

TRIGLYCERIDE
DIGLYCERIDE
MONOGLYCERIDE
FATTY ACID
BILE SALT
NONPOLAR LIPID

WATER CHANNEL

LIPASE

Figure 6.137. Schematic "zipper model" of fat digestion showing the preservation of the hydrocarbon domain during lipase hydrolysis of oil (tryglyceride) and the subsequent solubilization of the VI phase by bile salts. [From Patton (1981).]

Except for persons of northern European background most individuals experience a decrease in lactase activity after weaning, which results in a relative intolerance to lactose manifested by bloating, flatulence, and loose stools after drinking milk or eating dairy products. There are two α-glucosidases: glucoamylase, which hydrolyzes maltose, maltotriose, and longer oligosaccharidases with carbon 1 to 4 linkages up to 9 glucose molecules, and sucrase. Sucrase acts at the nonreducing α (1–4) end links of glucosyl oligosaccharides produced by the digestion of amylose and amylopectin by amylase. It is also required for removal of the nonreducing α (1–6)-linked glucose in α-limit dextrins. The term sucrase indicates its role in the hydrolysis of sucrose to glucose and fructose, while the term sucrase-isomaltase indicates its broader hydrolytic activities. The presence of two different active sites on a single polypeptide molecule is unique (Sjostrom et al., 1980). Sucrase-isomaltase is responsible for about 75% of the total maltase activity. Glucoamylase is responsible for the rest. Disaccharidases are synthesized on the brush border membrane as the cells move from the crypts to the villus tips.

In contrast to the other disaccharides where there is more than sufficient enzyme for digestion, the rate of hydrolysis of lactose is slower than the transport capacity of the absorptive cells so that surface hydrolysis is rate-limiting for the assimilation of lactose.

Membrane Digestion of Peptides

The digestion of peptides by the mucosa of the small intestine is somewhat more complicated than the digestion of oligosaccharides, because there are peptidases both on the brush border and within the cytoplasm of the absorptive cells. Nearly 90% of the epithelial cell peptidases reside within the cytosol of the cell (Adibi and Kim, 1981).

There are two important oligopeptidases in the brush border. One is an amino oligopeptidase which removes N-terminal amino acid from peptides having 2–6 amino acid residues. It has maximum specificity for tri- and tetrapeptides, particularly if the amino acid residues contain bulky side chains such as leucine and methionine. Its action is blocked by a proline residue in the penultimate position. The hydrolysis of tetrapeptides is almost totally dependent upon surface hydrolysis by amino oligopeptidase. The enzyme also at least partially hydrolyzes many di- and tripeptides containing residues with bulky side chains such as leucine and methionine. A second class of peptidases consists of at least two dipeptidases. One which has been well characterized is a dipeptidyl peptidase with very high specificity for N-terminal dipeptide units of oligopeptides, especially when proline is located at the next to the last position. Brush border peptidases account for about 12% of the digestion of dipeptides but as much as 60% of tripeptide activity (Sleisenger and Kim, 1979).

Recent investigations indicate that precursors of brush border peptidases are synthesized on the rough endoplasmic reticulum of the mucosal epithelial cells and pass through the Golgi apparatus before insertion into the plasma membrane. At least 60% of these glycoproteins are first incorporated into the basolateral membranes before reaching the brush border (Feracci et al., 1982). The cytosol precursor peptidases exist as monomers, dimers, and tetramers with a common subunit (Maze and Gray, 1980).

The hydrolysis of peptides and the transport of di- and tripeptides is rapid, but the transport into the cell of amino acids is rate-limiting for the fraction of dietary protein digested all the way to amino acids.

Control of Membrane Digestion

There is a large body of evidence in rodents that the activity of intestinal disaccharidases increases in response to the feeding cycle and also in response to increases in the proportion of carbohydrate in the diet. The relevance of these observations to man is uncertain.

PATHOPHYSIOLOGY OF DIGESTION

Disorders which reduce the concentration of bile acids in the duodenum below the critical micelle-forming level will impair the digestion of fats such as interruption of the enterohepatic circulation or bacterial overgrowth in the small intestine with organisms which can deconjugate and dehydroxylate bile acids. Pancreatic inflammation or blockage of the main pancreatic duct will result in decreased hydrolysis of triglycerides due to a lack of lipase and colipase. Excretion of fat in the stool will increase (steatorrhea). Excessive acid secretion such as occurs with gastrin-secreting tumors will reduce the duodenal pH and may inactivate pancreatic enzymes. A relative lack of the enzyme lactase is the most common cause of milk intolerance. Inflammatory disease of the wall of the small intestine such as Crohn's disease can reduce the amount of brush border digestive enzymes and lead to reduced membrane digestion. Rarely congenital lack of specific enzymes or the presence of abnormal inactive enzyme can produce specific maldigestion; examples include deficiencies of sucrase-isomaltase, enterokinase, colipase, and lipase. Reduction of the surface area of the small intestinal mucosa as in celiac sprue leads to decreased membrane digestion and also to reduced pancreatic secretion in response to meals as a result of decreased mucosal stores of secretin and CCK.

CHAPTER 46

Absorption

Absorption is the most important function of the digestive tract. Until the development of methods for long-term intravenous feeding, absorption from the small intestine was essential for life. Most absorption occurs in the small intestine. However, normal absorption of water and electrolyte requires the presence of the colon. Table 6.11 shows the difference in amounts and concentrations of electrolyte and water between fluid entering the colon and leaving in the stool (Turnberg, 1980). Therefore, about 80% of the water and 90% of the sodium and chloride ions which enter the colon each day are absorbed (Schultz, 1981a). In patients with ileostomies after removal of the colon, the daily loss of electrolyte is at least 50 mEq, while the normal rate is 5–15 mEq (Phillips, 1969). The patient with an ileostomy must maintain an intake of salt and water sufficient to compensate for these excessive losses. The normal stool weight in the USA is <200 g/day.

There are specialized absorptive functions for certain parts of the digestive tract. Active transport mechanisms for the absorption of iron and calcium are present in the duodenum and for bile acids and vitamin B_{12} in the ileum. The ileum is less "leaky" than the jejunum and therefore less permeable to fluids and solutes. Chloride ion is absorbed in exchange for bicarbonate ion in the ileum. Nutrients are absorbed poorly from the colon.

Table 6.11
Daily colonic salt and water turnover in normal human subjects (average values)

	Intestinal Content: Entering Colon		Colonic Content: Leaving in Stool		
	Concentration (mmol/l)	Amount (ml or mmol)	Concentration (mmol/l)	Amount (ml or mmol)	Amount Absorbed (ml or mmol)
H_2O		1500		100	1400
Na	140	210	40	4	206
K	6	9	90	9	nil
Cl	70	105	16	2	103
HCO_3	50	75	30+		

From Turnberg (1980).

We shall consider water and electrolyte absorption followed by the absorption of nutrients from the small and large intestine. Then we shall consider water and electrolyte absorption from the gallbladder, the role of immunity in the absorption of macromolecules, and finally the behavior of gases in the digestive tract.

WATER AND ELECTROLYTE ABSORPTION FROM THE SMALL INTESTINE

About 4500 ml of fluid enter the small intestine per day. Of this, 2000 ml is taken in as food and drink, 1000 ml is added as saliva, and 1500 as gastric secretion. Another 2500 ml is added in the duodenum by the secretion of pancreatic juice and the delivery of bile for a total of 7000 ml/day (Turnberg, 1980). It has been estimated that the maximal capacity of the human intestine to absorb water is about 20 l/day (Love et al., 1968). Absorption of water occurred at a rate of 0.035 ml/min/cm in the small intestine. Sodium absorption was 4.72 μEq and potassium absorption 0.183 μEq/min/cm and was constant along the small intestine (Palma et al., 1981). Stools were passed only when the terminal ileum emptied more than 6 ml/min into the colon. At that time net absorption of water by the colon was occurring at the rate of 2.7 ± 0.3 ml/min.

In man during perfusion studies of the jejunum, sodium is absorbed from a solution of isotonic saline by diffusion. In the presence of bicarbonate, absorption of sodium occurs against a concentration gradient. There is probably an exchange of hydrogen ion (secretion) for sodium (absorption) (Turnberg et al., 1970a). Potassium absorption occurs in response to concentration gradients and to solvent drag (Turnberg, 1971).

In the human ileum, sodium absorption occurs against very steep electrochemical gradients (Fordtran et al., 1968). Chloride concentrations in the lumen of the ileum are lower and those of bicarbonate higher than in the jejunum or plasma. This may be the result of an exchange of chloride for bicarbonate. There is also evidence for an exchange of sodium for hydrogen (Turnberg et al., 1970b). Channels for fluid exchange in the ileum have a smaller diameter, which reduces solvent drag and any possible back flux of ions after active absorption (Turnberg, 1980).

In Vitro Studies of Water and Electrolyte Absorption from the Small Intestine

Intestinal mucosa can be mounted in Ussing chambers and the movement of water and electrolyte examined under strictly controlled physicochemical conditions. The mucosa is mounted between two chambers so that the concentrations of ions in the bathing media can be varied on either side. Recording electrodes determine the electrical potential difference between the mucosal and serosal surfaces. By passing a current across the mucosa sufficient to reduce the electrical potential to zero, the short circuit current can be determined. By determining these physiochemical properties, the small intestinal epithelium can be shown to be relatively leaky. About 95% of the electrical conductance of the mucosa is due to passive paracellular pathways (Schultz, 1981b). Therefore, the "tight" junctions between cells must be relatively permeable. The paracellular channels are cation-selective, but with little difference between alkali metals. The channels have an estimated diameter of 4–8Å.

With the use of this preparation, the bulk of transcellular absorption in the upper small intestine has been shown to occur by way of an electrically neutral process, with the overall effect of a one for one transport of sodium and chloride from the mucosal side into the serosal solution as shown in Figure 6.138 in rabbit intestinal mucosa. Replacement of sodium in the mucosal solution with choline abolishes chloride transport, and replacement of chloride with sulfate inhibits sodium absorption. Increases in intracellular levels of cAMP inhibit the absorption of sodium and chloride equally. None of these manipulations affects the transepithelial electrical potential difference with the exception of the removal of sodium. Sodium is extruded from the cell by the sodium-potassium pump on the basolateral cell membrane. The exit of chloride follows its electrochemical gradient, which accounts for part of the efflux, but the precise mechanism is unclear (Schultz, 1981b). Recent studies showed rat brush border membrane vesicles favor a double exchange of Na^+ for H^+ and Cl^- for HCO_3- rather than cotransport (Liedtke and Hopfer, 1982a,b).

Another important mechanism for the absorption of sodium in the small intestine is that in associa-

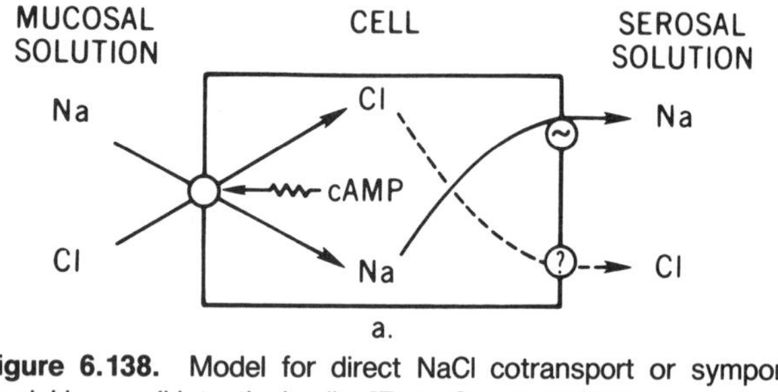

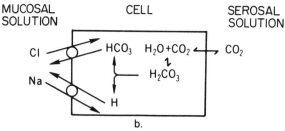

Figure 6.138. Model for direct NaCl cotransport or symport model by small intestinal cells. [From Schultz (1981).]

Figure 6.140. Model for dual Na-H and Cl-HCO₃ countertransport or antiport model. [From Schultz (1981).]

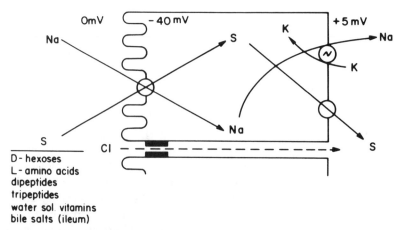

Figure 6.139. Model for sodium-coupled absorption of organic solutes (S) by the small intestine. [From Schultz (1981).]

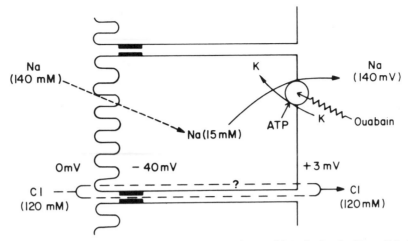

Figure 6.141. Model for uncoupled sodium absorption by the small intestinal cells. [From Schultz (1981).]

tion with the absorption of sugars and amino acids. The downhill movement of sodium into the cell provides energy for the uphill entry of the organic solutes. Important evidence for this has been provided by the demonstration of sodium-dependent glucose transport into vesicles prepared from absorptive cell membranes (Kimmich, 1981).

Figure 6.139 shows a model of sodium transport coincident with the absorption of hexoses, amino acids, short peptides, and water-soluble vitamins. Sodium is pumped out of the cell by the basolateral cellular Na^+-K^+-ATPase, and potassium enters. This maintains the electrochemical gradient across the mucosal membrane. The absorption of sodium makes the serosal solution electrically positive with respect to the mucosal solution and provides a driving force for the diffusion of chloride across the mucosa by way of paracellular channels.

The situation in the rabbit ileum is somewhat different, as was noted in the human ileum in vivo. Figure 6.140 shows a model for two neutral countertransport processes: a Na^+-H^+ and a Cl^--HCO_3^- exchange with endogenous or exogenous CO_2 being converted to H_2CO_3 by the action of the enzyme carbonic anhydrase which provides H^+ and HCO_3^- for exchange with mucosal Na^+ and Cl^-. The H^+ and HCO_3^- which enter the lumen vanish as H_2O and CO_2. The net effect is a one for one absorption of sodium and chloride. The double exchange could be controlled by the intracellular pH restricting the amount of HCO_3^+ or H^+ available for exchange.

Another mechanism for sodium chloride absorption in rabbit ileum is by uncoupled sodium absorption as shown in Figure 6.141. Sodium enters the cell in response to the electrical gradient of $-40mV$ and an intracellular sodium activity, which is one-tenth that in the mucosal solution. It exits from the basolateral membrane as a result of energy provided

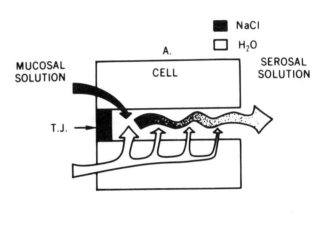

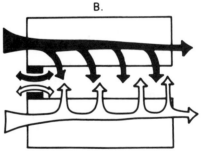

Figure 6.142. Model for isotonic fluid absorption considering extreme leakiness of the junctional complexes and the high hydraulic conductivities of the limiting membrane. *A,* Earlier model. *B,* Present model with NaCl entering all along intercellular space. [From Schultz (1981).]

by the hydrolysis of ATP by the enzyme Na^+-K^+-ATPase. This enzyme and the absorption of sodium is inhibited by ouabain applied only on the serosal side. Since the pump extrudes more sodium than it takes in potassium, the negative electrical gradient within the cell provides a driving force for the passive flow of chloride from mucosal to serosal surface. The principal route of chloride absorption is through paracellular channels, but some small transcellular contribution cannot be excluded.

Human small intestinal mucosa obtained at surgery also demonstrates active transport of sodium into the serosal fluid in specimens from both jejunum and ileum (Corbett et al., 1977). Glucose increased sodium absorption in both. Active transport of chloride into the serosal fluid was seen only in the ileum. Bicarbonate secretion in the ileum was independent of glucose. Similar results were seen with human jejunum where the effect of glucose was seen in bicarbonate and chloride-free solutions (Binder, 1974).

Absorption of fluids by leaky epithelia is always isotonic with fluid in the mucosal compartment. Figure 6.142 shows a model which accounts for isotonic transport. Solute enters the intercellular space either through tight junctions between cells or across basolateral membranes. This sets up an osmotic gradient which is followed by water through the same routes. The relative proportion entering by transcellular and paracellular routes is unknown.

Control of Small Intestinal Absorption of Water and Electrolyte

Catecholamines can increase the absorption of sodium and chloride and inhibit the secretion of bicarbonate in rabbit ileum in vitro. The effect is reduced by α but not β blockers. Norepinephrine stimulates fluid absorption of rat small intestine in vivo. Calcium may be involved (Field, 1981). Somatostatin has similar effects. Enkephalins have similar effects on guinea pig ileum. Calcitonin and substance P also inhibit sodium chloride absorption in rat ileum in vitro (Walling et al., 1977). With human jejunal mucosa in vitro, indomethacin, an inhibitor of prostaglandin E_2, increased mucosa to serosa flux of sodium. PGE_2 was a potent inhibitor of sodium chloride absorption (Bukhave and Rask-Madsen, 1980). Unfortunately, the physiologic role of these agents in the control of sodium and water absorption is unknown.

WATER AND ELECTROLYTE ABSORPTION FROM THE COLON

The epithelial surface of the colon is tighter than that of the small intestine. As a result, it is easier to maintain concentration differences across the wall of the colon. The maximal capacity of the colon in man to absorb water is estimated at 2500 ml/day. Perfusion of the cecum with 2 or 4 l of fluid lead to mild diarrhea only in the latter case. The net maximal absorption in the colon was 5700 ml of water, 816 mEq of sodium, and 44 mEQ of potassium in four healthy subjects (Debongnie and

Phillips, 1978). In contrast to the contents of the small intestine, fecal water contents are hypotonic with respect to plasma in the case of sodium and chloride, while the concentrations of potassium and bicarbonate are higher in stool water. The potential difference across the colon is -10–50 mV, mucosa negative to serosa. These findings support an active transport absorptive mechanism for sodium chloride and bicarbonate (Devroede and Phillips, 1969). The mechanism for potassium transport in vivo is controversial, with some evidence for and against potassium secretion by active transport (Hayslett et al., 1982). Some of the controversy may be the result of failure to distinguish between parts of the colon.

Absorption of Water and Electrolyte from the Colon in Vitro

Using primarily colonic mucosa in Ussing chambers, three mechanisms have been demonstrated for the absorption of sodium and chloride: an electrogenic sodium absorption best seen in rabbit colon, neutral absorption of sodium chloride in rat colon, and electrically neutral chloride absorption coupled to bicarbonate secretion best seen in rabbit distal colon. Figure 6.143 shows a model of double membrane of active electrogenic (uncoupled) sodium absorption (Schultz, 1981a). Sodium enters down an electrochemical gradient by simple diffusion. However, to account for the relationship between the rate of absorption of sodium and the concentration of sodium in the bathing media, it is necessary to postulate that the sodium concentration within the cell exerts a negative feedback effect on the permeability of the apical cell membrane to sodium. Sodium is pumped out of the cell at the basolateral membrane by the ouabain-sensitive Na-K^+-ATPase. Amiloride abolishes sodium entry while amphotericin, which disrupts the apical membrane, increases the entry of sodium and results in a loss of the relationship between the concentration of sodium and flux of sodium. Aldosterone enhances sodium entry (Fig. 6.143).

In the isolated rat colon, a major fraction of the absorption of sodium chloride occurs by an electrically neutral mechanism similar to that shown in the small intestine (Fig. 6.139). However, the rat colon shows marked differences in absorptive properties in different parts of the colon. The rates of sodium and chloride absorption are much greater (2-fold) in the upper colon than in the rectum. It is possible that the neutral cotransport mechanism for sodium chloride absorption in the rat is a property of the proximal colon, while the uncoupled electrogenic sodium transport is seen in the distal

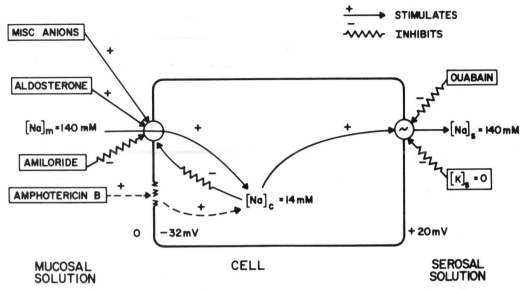

Figure 6.143.   Model for sodium absorption by mammalian colon cells. [From Schultz (1981).]

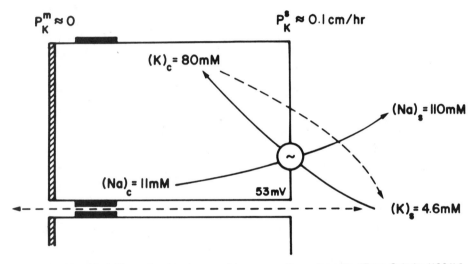

Figure 6.144.   Model for potassium transport by mammalian colon cells. [From Schultz (1981).]

colon and rectum, similar to that in the descending colon of the rabbit (Yau and Makhlouf, 1975; Schultz, 1981a). In the rabbit colon in vitro there is evidence that the absorption of chloride is balanced by the secretion of bicarbonate in an obligatory anion exchange mechanism (see Fig. 6.127). Bicarbonate is derived from CO_2 produced by oxidative metabolism.

Potassium movements in rabbit colon in vitro in part reflect passive diffusion through the paracellular route and in part diffusion across the apical cell membrane after potassium has been pumped into the cell by the ATPase at the basolateral cell membrane. Figure 6.144 shows a model (Schultz, 1981a).

Similar studies with human colonic tissue also

support a bicarbonate-chloride exchange (Hawker et al., 1978).

Control of Water and Electrolyte Absorption from the Colon

Serotonin and vasoactive intestinal peptide reduce the absorption of sodium chloride by decreasing the mucosal to serosal flux. VIP increased the tissue levels of cAMP (Racusen and Binder, 1977; Dharmsathaphorn et al., 1980). Somatostatin increased net absorption of chloride without changing levels of cAMP. There are noradrenergic neurons in rat colonic mucosa, but their functional significance is unknown (Wu and Garginella, 1981). Aldosterone enhances sodium absorption from the colon. None of these control mechanisms have been shown to significantly alter absorption under normal circumstances in intact subjects or animals (Turnberg, 1980).

ABSORPTION OF IRON

Iron is absorbed by an active transport mechanism in the duodenum. There are three steps: uptake by the mucosa, transport into the epithelial cells, and release into the plasma. Within the cell, a variable portion of the iron may be stored as ferritin and lost into the lumen when the cell exfoliates. The mean uptake of iron in normal subjects was 12%, of which only 3.6% was transferred to plasma. Intracellular iron-binding proteins have been isolated, but their role in iron absorption is unclear (Powell and Holliday, 1981). Iron-deficient subjects show an increased absorption of iron: those with iron overload show less absorption of iron than normal. Acute stimulation of erythropoiesis increases iron absorption; depression of erythropoiesis decreases iron absorption. A genetic strain of mice exist with a specific deficiency of iron uptake by proximal intestinal mucosal cells, although transfer to the blood is normal. In another strain iron uptake is normal, but transfer is faulty.

When the absorption of iron is increased, little or no ferritin is formed, and the iron is delivered rapidly to the plasma. When iron absorption is depressed, more iron is trapped in ferritin and retained in the cell. The release of iron from mucosal cells into plasma is poorly understood, but it does not appear to be rate-limiting or dependent on specific iron receptors in the plasma. Probably the initial influx across the brush border is rate-limiting. It appears that there are specific receptors for iron on the brush border and also a transport protein within the cell, closely resembling the iron-binding protein, transferrin, in the blood, which transports iron across the cell to the plasma membrane.

Healthy human subjects ingest about 15 mg of iron daily. Of this about 7 mg is solubilized and about 4 mg taken up by the intestinal mucosa. Finally, about 1 mg is absorbed. About 50% of dietary iron is present in heme (mainly from myoglobin and hemoglobin in meat). The extra iron absorbed from heme-fortified meals was 3 times greater than that from ferrous sulfate. It has also been noted that three times as much iron is absorbed from meals containing meat as those without meat (Hallberg et al., 1979).

During pregnancy in mice, iron uptake and transfer were both increased. In pregnant rats, the placenta takes up a significant amount of iron (Batey and Gallagher, 1977a). Other attempts to demonstrate humoral control of iron metabolism have led to conflicting results (Batey and Gallagher, 1977b).

CALCIUM ABSORPTION

Calcium absorption also involves an active transport mechanism in the proximal small intestine and a control system which provides increased absorption in the presence of calcium deficiency. In perfusion studies in man, glucose increased calcium absorption 3-fold, but this apparently resulted from a higher calcium concentration secondary to an increase in the absorption of water (Norman et al., 1980). There is a small tendency for luminal calcium disappearance to plateau at higher concentrations, suggesting a carrier-mediated transport (Ireland and Fordtran, 1973). This was not seen with intestinal biopsy material (Duncombe et al., 1980). On balance it is likely that the bulk of calcium absorption of man occurs by passive calcium uptake in adults (Duncombe and Reeve, 1981).

A calcium-binding protein has been identified by immunofluorescence in small intestinal biopsies from vitamin D-deficient and uremic subjects. There was strong immunofluorescence for it in the basement membrane and the apical portion of mucosal cells. Immunofluorescence was absent in rachitic babies but returned after treatment with 25-hydroxycholecalciferol ($250HD_3$) (Helmke et al., 1974). The principal site of action of $1,25 (OH)_2D_3$ is the small intestine, where it promotes calcium and phosphate absorption. It enters the cell membrane of the intestinal absorptive cell and binds to a high affinity receptor in the cell cytoplasm. The hormone-receptor complex then migrates to the cell nucleus where activation of selected genes results in messenger RNA transcription and subsequent protein synthesis. Proteins known to be induced by $1,25 (OH)_2D_3$ include calcium-binding protein (CaBP), calcium-activated ATPase, and alkaline phosphatase. In addition, D_3 can alter the permeability of the brush border membrane independently of protein synthesis (Rasmussen et al., 1979). It seems to alter membrane phospholipids. The increase in permeability may be due to a change in membrane lipid structure (Max et al., 1978). It is likely that the calcium-binding protein binds intracellular protein in order to prevent excessive concentrations if ionized calcium which might be toxic. Calcium-activated ATPase is located on the basolateral cell membrane and the brush border membrane. Its physiological significance is uncertain.

In summary, vitamin D_3 ($1,25(OH)_2 D_3$) increases the permeability to calcium and thereby increases the flux of calcium into the intestinal absorption cell. Calcium-binding protein is synthesized to bind intracellular calcium and to prevent concentrations of calcium from rising to toxic levels. Calcium is then transported out of the cell across the basolateral membrane, possibly by an active transport mechanism involving calcium-activated ATPase (Duncombe and Reeve, 1981).

In rats, the transport of calcium varies in the different portions of the small intestine: calcium efflux is higher in the ileum, but influx is higher in the duodenum (Younoszai and Schedl, 1972). Vitamin D increases the net absorption of calcium in the duodenum but not in the ileum. It appears to act at both the brush border and the basolateral cell membrane (Younoszai et al., 1973). Calcium absorption also occurs in the rat cecum, but secretion occurs distally (Petith and Schedl, 1976; Petith et al., 1978). Rachitic rats secreted calcium in the colon which was converted to absorption by vitamin D. The vitamin also increases the calcium-binding activity in the colon (Petith et al., 1979; Favus et al., 1981). Newborn rats adapt to a low calcium diet to a greater degree than adult rats (Ormbrecht et al., 1979).

Isolated intestinal microvillous membranes from rats contain three vitamin D-dependent activities: calcium binding, calcium-dependent ATPase, and *p*-nitrophenyl phosphatase (Kowarski and Schachter, 1980). Basolateral cell membrane vesicles from rat intestine accumulate calcium in response to ATP and release Ca in response to calcium ionophore. The ATP-dependent calcium uptake is stimulated by the calcium regulatory protein, calmodulin, in a dose-dependent manner. Calmodulin increases the maximal transport rate and the calcium affinity of the transport mechanism. Calmodulin may modulate transepithelial calcium absorption.

Studies with membrane vesicles indicate that, both at the brush border and basolateral membranes, the transcellular calcium and inorganic phosphate fluxes are coupled to sodium. In the brush border a cotransport of phosphate and Na leads to cellular accumulation of phosphate. At the basolateral membrane cotransport of calcium and sodium leads to extrusion of calcium. A $Ca^- - ATPase$ may also be involved (Murer and Hildmann, 1981). In rats, dietary restriction of phosphate reduced calcium absorption in vitamin D-deficient animals (Brautbar et al., 1981).

ABSORPTION OF CARBOHYDRATE

As a result of the intraluminal and brush border digestive processes discussed in Chapter 45, the monosaccharides glucose, galactose, and fructose are available for uptake into the cells of the mucosa.

Fructose enters in response to its concentration gradient by facilitated diffusion. Glucose and galactose share a common specific transport mechanism. Probably an integral membrane protein in the brush border functions as a carrier. In some respects integral membrane proteins are more like "gates" than carriers (Freel and Goldner, 1981). It has a high binding affinity for glucose and galactose and also binds sodium at a ratio of two sodium ions for each molecule of glucose. Once the carrier approaches the inner surface of the membrane its affinity for glucose and sodium falls, and they are released into the cytosol. The energy for the uphill movement of glucose is provided by the sodium pump in the basolateral membrane which extrudes sodium into the intercellular space in exchange for potassium. The energy is derived from ATP by an ATPase as shown in Figure 6.145. This is a form of cotransport. An important part of the evidence for this mechanism depends upon the availability of specific inhibitors such as phlorizin, which blocks the uptake of glucose at the brush border (Gray, 1981).

In the early stages after a meal, the concentration of monosaccharides in the intestinal lumen is much greater than that in the cell or in the extracellular space. Under these conditions only a mechanism of entry and exit is necessary. Later, however, the glucose must be absorbed against a lumen-to-blood concentration difference. Perfusion studies in man suggest that about 50% of glucose absorption can occur independently of cotransport with luminal sodium (Bieberdorf et al., 1975).

Glucose leaves the absorption cell by three mechanisms. About 15% reenters the lumen by reverse transport on the protein carrier or transporter. Another 25% leaves by passive diffusion across the relatively permeable basolateral membrane. The bulk of glucose exit occurs by way of a sodium-independent carrier in the basolateral membrane (see Fig. 6.145) (Freel and Goldner, 1981).

Much of the evidence for this mechanism of transport has been obtained from studies with isolated chicken intestinal epithelial cells. The maximal transport capacity ($\dot{V}_{max}$) of the basolateral membrane system for glucose is greater than that of the brush border carrier in the steady state when sugar has accumulated within the cell. It is sodium-independent and is selectively blocked by phloretin and cytochalasin B (Kimmich, 1981).

After a mixed meal in man, nearly twice as much glucose was absorbed from loops of jejunum as from loops of ileum of equal length (Fordtran and Inglefinger, 1968). The absorption of glucose was essentially completed in the first 50–100 cm of jejunum.

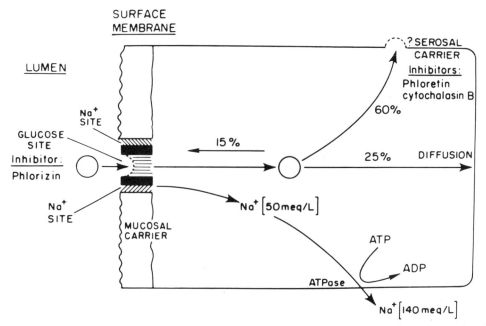

Figure 6.145.   Model for active transport of glucose or galactose. The membrane carrier has high affinity at the luminal surface for both monosaccharide and sodium. Based upon findings of 2:1 stoichometry of transport for sodium and glucose, there are probably two sodium sites on the carrier. The polar glucose may pass through a hydrophilic core of the carrier to be discharged into the cell interior by virtue of lowered affinity for both sodium and glucose at the inner side of the surface membrane. The energy for monosaccharide transport is provided by the Na^+-K^+-ATPase pump at the basolateral membrane, and exit of glucose appears to be facilitated by a serosal carrier. Although the carrier proteins have not been isolated their functional role has been studied in intact cells and isolated membrane vesicles with the aid of the potent inhibitors for the brush border and serosal carriers. [From Gray (1981).]

ABSORPTION OF AMINO ACIDS AND PEPTIDES

In contrast to sugars, many small peptides are not digested to amino acids until they enter the intestinal epithelial cells. A few dipeptides enter the portal circulation intact. Nearly 90% of epithelial cell digestive peptidases reside in the cytosol fraction of the cell (Gray, 1983). Dipeptide and tripeptide digestive products, especially if they contain glycine or proline residues, enter the cell by means of a carrier or transporter in the brush border membrane by a mechanism that is probably partially sodium-dependent (Fig. 6.146). Of the peptides that are assimilated intact into the interior of the absorption cell, at least 90% are hydrolyzed to free amino acids before entering the portal venous system. Diglycine can be detected in peripheral venous blood after intestinal perfusion studies in man, and hydroxyproline peptides have been found in plasma after a gelatin meal (Adibi and Kim, 1981).

Amino acids are absorbed by cotransport with sodium in a fashion similar to that of glucose at the brush border. There appear to be at least four specific transport mechanisms for neutral amino acids (aromatic amino acids), dibasic amino acids (diamino-acids, e.g., lysine or arginine), dicarbox-

ylic amino acids (glutamic or aspartic acid), and immunoacids and glycine (proline and hydroxyproline glycine). Transport across the basolateral membrane has been analyzed for only alanine, lysine, and proline, and appears to be rate-limiting (Munck, 1981). Other studies are controversial, and much remains uncertain.

Compared to the absorption of di- and tripeptides, the absorption of amino acids by the small intestine appears to be relatively slow. It is probably rate-limiting for the assimilation of the smaller fraction of dietary protein that is cleaved to amino acids within the lumen of the small intestine. Patients with certain specific defects in the absorption of amino acids are able to absorb dipeptides, and no significant defect in nutrition results. The results of perfusion tests in man indicate a more complete absorption of amino acids from oligopeptides and amino acids than from amino acids alone. More recent studies indicate that the advantage is limited to certain amino acids and for relatively high concentrations (Hegarty et al., 1982a and b). The peptide carrier transport system for di- and triglycine solutions seems to be more efficient than that of amino acid carrier systems (Fig. 6.146).

A vascularly perfused small intestine of amphibia has been utilized to demonstrate routes of uptake of amino acids from dipeptides. One is shared with

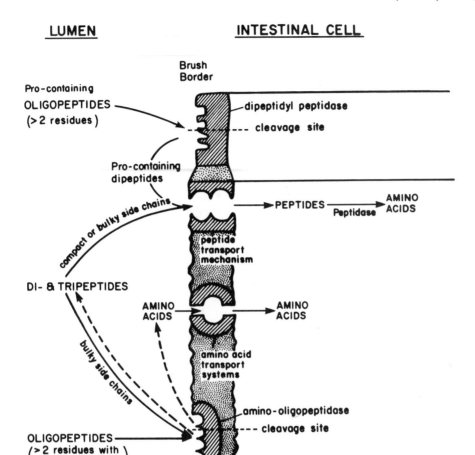

Figure 6.146. Entry of peptides and amino acids into absorption cells. Role of cytosolic peptidases. Both surface hydrolysis and intact transport occur, depending upon the number of amino acids in the peptide and the length of the side chain in the residues. [From Gray (1983).]

free amino acids and is sodium-dependent; another is sodium independent (Cheeseman, 1980).

There is greater dipeptidase activity in ileal than jejunal aspirates in man and a slower rate of absorption of dipeptides and amino acids in the ileum. This results in a greater accumulation of free amino acids in the lumen of the ileum. The amount of hydrolysis and absorption of peptides in the ileum probably depends upon the amount delivered to the ileum.

The absorption of folates by the small intestine is dependent upon the digestion of pteroylpolyglutamates in the diet to pteroylmonoglutamate by α-glutamylhydrolase, or folate conjugase, which is present in the mucosa. There is evidence for folate conjugase activity in both the brush border-membrane and within the cell, probably in lysosomes. The uptake into the cell is saturable and pH-sensitive. The rate-limiting step is probably the transport of pteroylmonoglutamates into the cell. The entire small intestine contains enzymes for deconjugating polyglutamyl folate, but absorption occurs

preferentially in the proximal jejunum (Rosenberg, 1981).

The absorption of cobalamine (vitamin B_{12}) by the mucosa of the terminal ileum is an example of a complex mechanism to permit transport of a large molecule. The role of intrinsic factor has already been mentioned (see Chapter 44). Once the complex of cobalamine and intrinsic factor has formed in the stomach, the cobalamine is protected from digestion in the small intestine. When the complex reaches the terminal ileum, the absorptive cells contain receptors on the brush border, to which the complex binds. By poorly understood mechanisms, cobalamine enters the cell while intrinsic factor remains in the lumen (Donaldson, 1981).

ABSORPTION OF FAT

As a result of intraluminal hydrolysis of fat, the intestinal lumen contains bile acids, fatty acids, phospholipids, cholesterol, and monoglycerides as mixed micelles. Only the system containing monoglyceride and fatty acid has been partially charac-

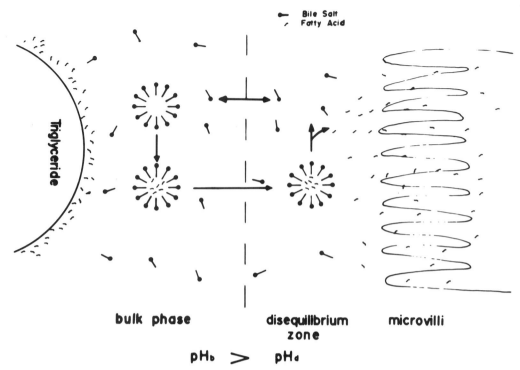

Figure 6.147. Schematic diagram showing postulated role of a low pH disequilibrium zone in the uptake of lipolytic products from mixed micelles. Acidic disequilibrium zone facilitates micellar diffusion and dissociation. Fatty acid diffuses into epithelial cells: bile acid monomers return to the bulk phase to reform mixed micelles. Disassociation of micelles may explain why lipolytic products are absorbed proximally and bile acid distally. [From Shiau (1981).]

terized (Thomson and Dietschy, 1981). The spontaneous dissociation of fatty acids from bile acids allows the fatty acids to cross the unstirred water layer on the surface of the mucosa. This further promotes the release of fatty acids from the mixed micelles since micellar fatty acid is in equilibrium with the free fatty acids in the unstirred water layer. Another factor is the acidic microclimate adjacent to the epithelial cells attributed to intestinal secretion of hydrogen ion. A low pH facilitates micellar diffusion as well as micellar dissociation. Converting a nonprotonated fatty acid to its protonated form in a low pH microclimate further facilitates diffusion across the lipid membrane of the cell (Shiau, 1981) (Fig. 6.147). Crossing the unstirred water layer is the rate-limiting step in the uptake of medium and long chain fatty acids. The mucous coat overlying the brush border membrane is another barrier which must be penetrated (Gray, 1983). Uptake into the cell is a passive process dependent upon lipid solubility, although there is suggestive evidence of binding or carrier-mediated transport (Thomson and Dietschy, 1981).

The absorption of cholesterol also involves movement through an obligatory aqueous phase after dissociation from micelles. Endogenous cholesterol contributes about 50% of the total cholesterol delivered to the lumen of the digestive tract. Esterified cholesterol is hydrolyzed by pancreatic cholesterol esterase. Cholesterol is absorbed as free cholesterol rather than as cholesterol esters. Bile acids are required for cholesterol absorption by increasing its solubility, providing a reservoir for uptake, and also possibly by affecting the permeability of the brush border membrane.

Fat-soluble vitamins are relatively nonpolar molecules and, like cholesterol, depend upon micellar solubilization for absorption across the unstirred water layer. They are largely absorbed from the proximal small intestine in a dose-dependent fashion.

Lipids are absorbed principally from the proximal small intestine. Biliary cholesterol is absorbed similarly, but dietary cholesterol is absorbed all along the small intestine, preferentially in the second and third quarters of rat intestine.

Once fatty acids enter the absorptive cells, long chain saturated and unsaturated fatty acids are utilized at different rates, despite the fact that the rate of uptake is similar. A soluble intracellular fatty acid-binding protein has been demonstrated, which has different affinities for saturated and unsaturated fatty acids. It has a molecular weight of 12,000 and possesses a higher affinity for unsaturated fatty acids. It has virtually no affinity for medium or short chain fatty acids. This may explain why medium chain fatty acids are not reesterified during absorption. Fatty acid-binding protein also serves a protective function because unbound fatty acids are potentially cytotoxic. It facilitates the removal of long chain fatty acids from the cell membrane and the transfer of fatty acids from the brush border membrane to the smooth endoplasmic reticulum. It may be important in facilitating the esterification of long chain fatty acids.

Most of the dietary triglyceride enters the absorptive cells as monoglyceride and unesterfied fatty acid. These components are converted back to triglycerides. There are two pathways for triglyceride biosynthesis: one involves the conversion of fatty acids to triglycerides by way of phosphatidic acid derived from α- glycerophosphate, the α-glycerophosphate pathway. The other and major pathway is the direct acylation of monoglyceride to diglyceride and then to triglyceride in the endoplasmic reticulum (the monoacyl glycerol pathway) (Fig. 6.148). Both pathways require activated fatty acid as the acyl donor. The enzyme responsible for this reaction is microsomal acyl-CoA synthetase. The acyl receptor for the phosphatidic acid pathway is α-glycerophosphate derived from glucose metabolism. The acyl receptor for the monoglyceride pathway is monoglyceride derived from intestinal absorption. Figure 6.148 shows the various biochemical pathways in absorption, reesterification,

and exit from the cell as chylomicrons for long chain fatty acids. Recent electron microscopic studies on human small bowel biopsies indicate that linolenate at low luminal concentration may be absorbed and transported into the portal vein. Such low concentrations should be present after a mixed meal (Surawicz et al., 1981).

The monoacyl glycerol pathway normally accounts for more than 70% of the total intestinal triglyceride synthesis during the absorptive state. The contribution of the α-glycerol phosphate pathway to total esterification in the ileum decreases in adult life in rats.

Chylomicra are droplets of apolar lipids surrounded by a monolayer of polar lipids and protein. The nascent chylomicra migrate, contained in Golgi-derived secretory vesicles, toward the lateral cell membrane where they fuse and are extruded by exocytosis. At the basolateral membrane, coated pits appear in the areas where the chylomicra exit, probably derived from the fusion of secretory vesicles with the plasma membrane.

From the lateral intercellular space, the chylomicra pass through gaps in the basement membrane, cross the lamina propria, and enter the lymphatics through openings between endothelial cells. The entire process of fat absorption can be seen within 30 min after feeding fat (Thomson and Dietschy, 1981).

The final assembly and exocytosis of chylomicrons are the rate-limiting steps in the sequence of the delivery of fat to the circulation. The synthesis

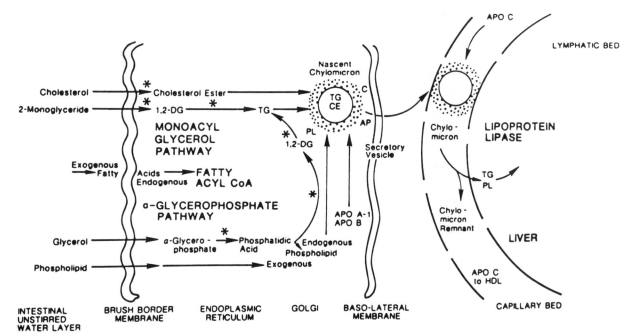

Figure 6.148. Intestinal intracellular events in the metabolism and exit of lipids from absorptive cells. [From Thomson and Dietsky (1981).]

of apoproteins is critical for chlyomicron formation, since inhibitors of protein synthesis prevent their formation. Apoproteins appear in the apical portion of intestinal absorptive cells and increase in density during fat absorption. Phospholipids may also be required. Rat chylomicrons contain about 7% lecithin. Microtubules may be involved in the pathway of vesicles in the cell, since colchicine leads to the accumulation of lipoprotein particles in small intestinal epithelial cells.

The major portion of fat absorption occurs in the jejunum, but when large amounts of fat are given, fat is absorbed in increasing amounts in the ileum. The capacity of the intestine to absorb fat is more limited than for carbohydrate and proteins. It is possible to exceed the capacity of ileal loops in dogs to absorb fat. In rats the maximal absorption capacity for long chain triglyceride was one-fifth that of medium chain triglyceride. The latter is absorbed directly into the portal venous system (Brooks, 1970).

ABSORPTION OF BILE ACIDS

Unlike the absorption of monoglycerides and fatty acids, the principal site of absorption of bile acids is in the distal ileum. Bile acid micelles and monomers diffuse across the unstirred water layer. Monomers in the aqueous phase are taken up at the surface of the absorptive cell in proportion to the mass of lipid in the aqueous phase. The function of the micelle is to act as a solubilizing vehicle, which conveys the water-insoluble lipid molecules across the unstirred water layer up to the cell membrane (Wilson, 1981).

In the jejunum bile acids are absorbed by passive diffusion. The permeability of the apical membranes of intestinal absorptive cells to bile acids is decreased by the addition of a hydroxyl group, glycine or taurine, to the steroid nucleus. On the other hand, removal of the negative charge increased permeability. There is controversy over the quantitative aspects of jejunal absorption of bile acids in rats.

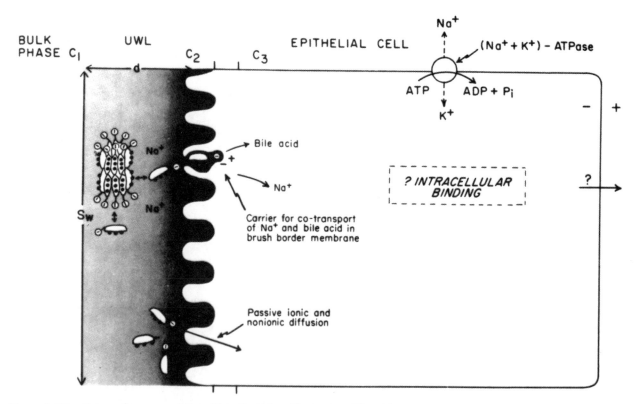

Figure 6.149. Schematic representation of intestinal bile acid transport. Bile acid monomers and micelles diffuse from bulk phase (C_1) across unstirred water layer (UWL) of thickness (d), and surface area (S_w) to the aqueous-membrane interface at C_2. Bile acid monomers are transported across brush border membrane either by passive diffusion throughout the intestine or by active transport in the ileum. The ileal recognition site is depicted to contain a steroid recognition component, a positive charge to react with negatively charged bile acid side chains, and an anionic site to interact with sodium. Energy for active transport arises from the sodium gradient across the brush border membrane. Low intracellular sodium concentration necessary for continued operation of this system is maintained through Na^+-K^+ exchange by an ATPase located on the basolateral membrane. Intracellular binding components and mechanisms for the cellular exit of bile acids are unknown. [From Wilson (1981).]

In the ileum, active transport of bile acids occurs as indicated by absorption against a concentration gradient (Fig. 6.149). It is sodium-dependent and represents another example of cotransport. The first step is the association between bile acid and a membrane receptor. The uptake of bile acid by isolated ileal absorptive cells is inhibited by ouabain or the absence of sodium in the medium. The inhibition is uptake by oubain is paralleled by a decrease in Na^+-K^+-ATPase activity. Uptake in vesicles prepared from ileal, but not jejunal, absorptive cells was sodium-specific. The results indicated that the action of sodium was due to flux coupling by way of a cotransport system rather than electrical coupling to preserve overall electrical neutrality.

ABSORPTION FROM THE GALLBLADDER

Gallbladder bile is concentrated about 10-fold due to the selective removal of water and inorganic ions. Table 6.12 shows a comparison of hepatic and gallbladder bile based on human, dog, and cat bile (Rose, 1981). Note that acidification also occurs. The concentration of bile in the gallbladder occurs through the net absorption of sodium chloride and sodium bicarbonate with water followed by osmotic coupling.

Most of the studies on the concentrating function of the gallbladder have focused on the gallbladder in vitro. There are some rather striking differences between species. Normal human gallbladders have not been obtained for these studies. Therefore, some caution should be taken in the application of data obtained from other species to man.

The gallbladder sustains a gradient of about 10,000 to 1 between bile and plasma with respect

Table 6.12
Composition of hepatic and gallbladder bile[a]

	Hepatic	Gallbladder
Na (mEq/l)	160	270
K	5	10
Cl	90	15
HCO_3	45	10
Ca	4	25
Mg	2	
Bilirubin (mM)	1.5	15
Protein (mg/100 ml)	150	
Bile acids (mM)	50	150
Phospholipids (mM)	8	40
Cholesterol (mM)	4	18
Total solids (mg/l)		125
pH	8.0	6.5

From Rose (1981).   [a] Values presented are averages for humans, dog, and cat.

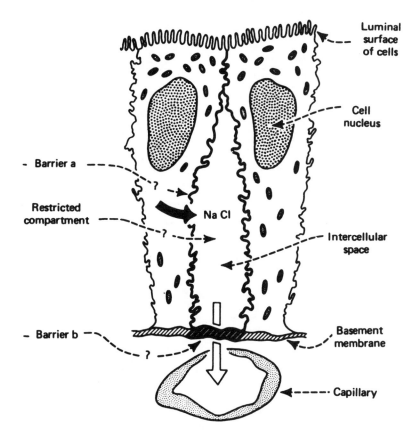

Figure 6.150. Dilated intercellular spaces in gallbladder epithelium during fluid absorption. Water follows sodium pumped into the intercellular space across the basolateral membrane. [From Dietchy (1966).]

to organic constituents, yet it is classified as a "leaky" epithelium comparable to the small intestine in the case of water and inorganic ions. Therefore, it has highly selective permeability properties. The primary barrier is the epithelium rather than the more superficial layers of the gallbladder wall. The epithelium is more permeable to potassium than to sodium. Bromide and chloride both cross the tissue by exchange diffusion. As much as 95% of the transepithelial ionic diffusion occurs between cells rather than transcellularly. The electrical transepithelial resistance is one-tenth that of the luminal or basolateral cell membrane resistance. Positive mucosal streaming potentials (the flow of solution through a membrane with fixed electrical charges carries with it an excess of ions of opposite charge to that on the membrane and results in a potential difference) in response to hyperosmotic solutions on the mucosal side indicate that the water channels are negatively charged. This tends to reduce the permeability to chloride ions.

Contrary to the older concept of a neutral sodium chloride pump, the gallbladder is more permeable to sodium than chloride. There is a low selectivity

for alkali cations, indicating a very hydrated pathway across the epithelium. Both cations and water share the pathway. Ouabain, an inhibitor of Na^+-K^+-ATPase, when added to the serosal surface, inhibits net water flow. Substitution of K^+ for Na^+ has a similar effect. Therefore, solute transport is the primary event, and water follows in response to an osmotic gradient. The absorbed solution is isotonic to the bathing solution over a wide range of osmolarities.

When the gallbladder is actively absorbing fluids, the lateral intercellular channels become dilated as shown in Figure 6.150. Conversely when no transport is occurring they are collapsed. The sodium-potassium ATPase is distributed all along the lateral intercellular space. These findings are consistent with a primary cotransport of sodium chloride. In the human gallbladder in vitro the mucosa is 3–8 mV negative with respect to the serosa. The rate of sodium transport across the gallbladder accounts for the short circuit current. In the absence of a transmural potential difference, there is no net chloride absorption.

The coupling of sodium and chloride fluxes oc-

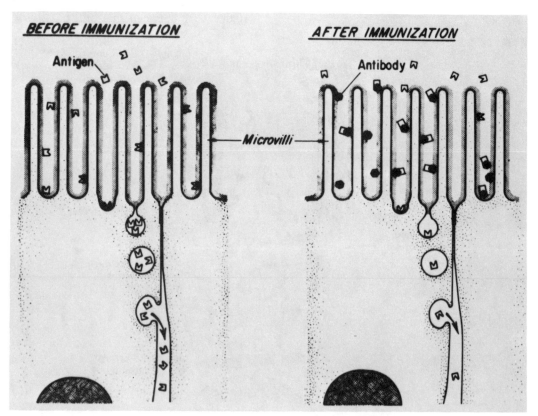

Figure 6.151. Diagrammatic representation of antigen absorption in the small intestine before and after immunization. Before immunization antigens are continuously absorbed by epithelial cells and transported into the intercellular space. After immunization a local immune response results in specific antibodies appearing in the glycocolyx and within the intestinal lumen. Antigen absorption is inhibited by the formation of antigen-antibody complexes at the cell surface. Antigens cannot attach to the cell membrane. Therefore pinocytosis cannot occur, and the antigens are broken down to small fragments in the lumen. [From Walker (1981).]

curs at the mucosal membrane. The driving force is the extrusion of sodium by the Na^+-K^+-ATPase at the basolateral membranes. The sodium chloride cotransport mechanism is highly specific for sodium, but absorbs bicarbonate and other anions in place of chloride. The entry of sodium is now thought to be electrogenic as in the case of the ileum (see Fig. 6.141). Chloride follows from the mucosa to the serosa in response to the electrically positive serosal solution.

Presumably water follows sodium into the intercellular space in response to an osmotic gradient, but this has not been determined directly. Potassium enters the absorptive cell at the basolateral membrane through the action of a Na^+-K^+-ATPase. Bicarbonate is necessary for maximal fluid absorption by the in vitro gallbladder. It is likely that there is a Cl^--HCO_3^- exchange in the basolateral membrane.

Secretion in the cat gallbladder in vivo can be induced by intravenous VIP. The gallbladder can absorb certain amino acids and sugars by a sodium-dependent active transport system similar to that in the small intestine. Unconjugated but not conjugated bilirubin is absorbed in the gallbladder by diffusion.

ABSORPTION OF MACROMOLECULES

Large molecules such as bovine serum albimin can be absorbed from the small intestine of certain mammals, including human infants, early in the neonatal period. Antigens to food proteins can also cross the small intestine in the first 3 mo of life. Figure 6.151 illustrates diagrammatically how immunization can reduce the absorption of antigens by preventing the adherence of antigens to the epithelial cell membrane, thereby interfering with pinocytosis and entry into the cell.

THE INTESTINAL CIRCULATION AND ABSORPTION

In the small intestine, the arrangement of the arterial and venous drainage of the villus is such to permit the action of a countercurrent multiplier system similar to that in the kidney (Fig. 6.152). So far the only important mechanism related to the countercurrent mechanism is the increased susceptibility of the villous tips to ischemic necrosis. The shunting of oxygen could further reduce oxygenation of cells at the tip of the villus when the flow of blood was already reduced due to hypovolemic or hemorrhagic shock (Tepperman and Jacobson,

COUNTERCURRENT   EXCHANGE

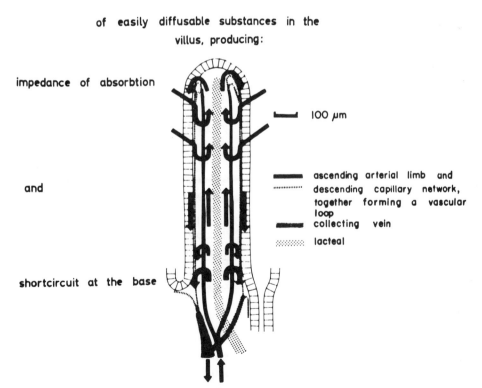

of easily diffusable substances in the villus, producing:

impedance of absorbtion

and

shortcircuit at the base

—— 100 μm

—— ascending arterial limb and
········· descending capillary network,
together forming a vascular loop
▬▬ collecting vein
░░░░ lacteal

Figure 6.152.   A model of the countercurrent mechanism in the small intestinal villus. As in the kidney, a gradient of osmolality develops within the villus here in response to sodium drawing water from the central artery. Entry of lipid-soluble substances at the villous tip is retarded by the countercurrent exchange. [From Folkow (1967).]

Table 6.13
Composition of flatus: %

No. of Subjects	Diet	N_2	O_2	CO_2	CH_4	H_2
20	Regular	59.0[a]	3.9	9.0	7.2	20.9
		(39.7–88.2)[b]	(0–15.7)	(1.2–15.0)	(0–30.3)	(3.1–34.0)
6	Regular	82.7	1.2	9.8	0.9	5.5
		(77.1–86.4)	(0–4.3)	(7.3–12.7)	(0.1–2.0)	(2.4–9.9)
5	Regular	61.2	3.6	8.1	7.3	19.8
5	Baked beans	19.4	1.1	49.1	22.4	8.0

From Levitt et al. (1981).   [a] Mean.   [b] Range.

1981). Blood flow to the splanchnic circulation increases after a meal (Brandt et al., 1955). Fat is the most potent component of the diet in producing the hyperemia in jejunal loops in dogs (Siregar and Chou, 1982).

INTESTINAL GAS

Contrary to popular belief, the normal intestine contains relatively small amounts of gas. Using an argon washout method, the normal human intestine contained only 30–199 ml of gas (Levitt et al., 1981). Gas in the digestive tract can be derived from (1) swallowed atmospheric air, (2) the liberation of CO_2 when hydrogen and bicarbonate ions interact, (3) by the production of gas by intestinal bacteria and (4) by diffusion from the blood. Individuals vary greatly in their ability to swallow air; amounts up to 1300 ml/min have been reported. Significant amounts of CO_2 are produced in the duodenum by the reaction of pancreatic bicarbonate with gastric hydrochloric acid. Most of it is absorbed in the small intestine. Any CO_2 appearing in flatus is probably due to bacterial fermentation. Bacterial flora can produce hydrogen, methane, and CO_2. Studies in germ-free animals and newborn infants indicate that hydrogen is produced solely from bacterial metabolism and almost entirely in the colon in normal subjects. Beans are notorious for causing high hydrogen excretion in flatus due to the amount of undigestible oligosaccharides, which reach the colon and are subject to fermentation by bacteria. Breath hydrogen can be used to measure transit time to the cecum. Methane in flatus varies enormously in human subjects. About one-third produce up to 0.4 ml/min. This phenomenon is constant over months or years. It occurs in families. If one parent is a methane producer, the children have a 50% likelihood of producing methane. If both parents produce methane the likelihood approaches 100%. However, there is no relation of methane production between spouses. The tendency is established early in life rather than a hereditary phenomenon. Methane is trapped in stool and hence the stools tend to float. The only gas which might

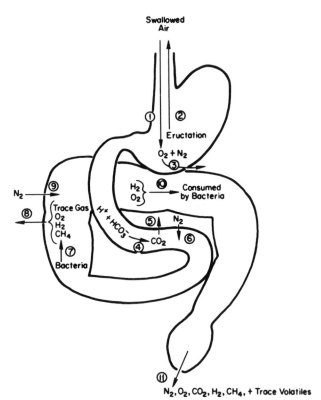

Figure 6.153. Mechanisms influencing rate of accumulation of gas in the gastrointestinal tract. Air is swallowed ① and a sizeable fraction is then belched back ②. The oxygen of gastric air diffuses into the blood draining the stomach ③. The reaction of hydrogen and bicarbonate yields carbon dioxide ④ which rapidly diffuses into the blood ⑤ while nitrogen ⑥ diffuses into the lumen down a gradient established by the carbon dioxide production. In the colon bacteria produce CO_2, H_2, and CH_4 ⑦ which diffuse into the blood perfusing the colon ⑧. Bacteria also consume O_2 and H_2 ⑨. N_2 diffuses into the colon ⑩ down a gradient established by the bacterial production of CO_2, H_2, and CH_4. The net result of all these processes determines the composition and rate of passage of gas passed by rectum ⑪. [From Levitt et al. (1981).]

contribute significantly to intestinal gas by diffusion from the circulation is nitrogen. It is not possible to distinguish between nitrogen in flatus derived from diffusion and that from swallowed air.

The composition of flatus is shown in Table 6.13. Note the effect of feeding beans. Normal flatus production varies from 200–300 ml/day. If lactase-deficient subjects eat lactose there is a dramatic increase in flatus due to fermentation in the colon by unabsorbed lactose. These principles are illustrated in Figure 6.153.

PATHOPHYSIOLOGY OF ABSORPTION

Malabsorption can be due to failure to digestion or failure of water, electrolytes, and nutrients to cross the epithelial cells and enter portal venous blood or (in the case of fat) the lymph. A decrease in the absorptive surface available due to disease such as celiac disease or surgical removal leads to malabsorption. Removal of the distal ileum leads to malabsorption of cobalamine and bile acids. The ileum is able to compensate by hypertrophy and hyperplasia after removal of jejunum, but the jejunum is unable to compensate for extensive removal of the ileum. Rare congenital diseases such

as lack of the active transport system for glucose-galactose can result in malabsorption. Malabsorption of fat can be due to inability to form chylomicrons as in abeta-lipoproteinemia or obstruction in the lymphatics due to lymphangiectasia or cancer. Malabsorption of fat can be due to a decrease in the number of absorptive cells or to diseased cells. The latter may be unable to resynthesize triglycerides.

Malabsorption of water and electrolyte may occur in the small intestine or colon. Massive volumes of diarrheal stool indicate small intestinal malabsorption or hypersecretion. Malabsorption in the colon can lead to stool volumes of up to a liter per day. Osmotic diarrhea is due to bacterial fermentation and the production of more osmotically active particles in the colon. Bacterial overgrowth in the small intestine can have similar effects. No true example of excessive absorption causing constipation has been found.

BIBLIOGRAPHY

ABE, K., K. YONEDA, R. FUJITA, V. YOKOTA, AND C. DAWES. The effect of epinephrine, norepinephrine and phenylephrine on the types of proteins secreted by rat salivary glands. *J. Dent. Res.* 59: 1627–1634, 1980.

ABRAHAMSSON, H. Studies on the inhibitory nervous control of gastric motility. *Acta Scand. Physiol. (Suppl.)*: 390: 1–38, 1973.

ABRAHAMSSON, H., AND G. JANSSON. Elicitation of reflex vagal relaxation of the stomach from pharynx and esophagus in the cat. *Acta Physiol. Scand.* 77: 172–178, 1969.

ACCANTINO, L., AND F. R. SIMON. Identification and characterization of a bile acid receptor in isolated liver surface membranes. *J. Clin. Invest.* 57: 496–508, 1976.

ADAIR, L. S., J. E. SHAW, J. URQUHART, AND F. P. BROOKS. Permissive role of gastric in vagally mediated acid secretion. *Physiologist* 17: 169, 1974.

ADIBI, S. A., AND D. W. MERCER. Protein digestion in human intestine as reflected in luminal, mucosal, and plasma amino acid concentration after meals. *J. Clin. Invest.* 52: 1586–1594, 1973.

ADIBI, S. A., AND Y. S. KIM. Peptide absorption and hydrolysis. In: *Physiology of the Gastrointestinal Tract*, edited by L. R. Johnson. New York: Raven Press, 1981, p. 1073–1095.

ADRIAN, T. E., H. S. BESTERMAN, T. J. C. COOKE, S. R. BLOOM, A. S. BARNES, R. C. G. RUSSELL, AND R. G. FABER. Mechanism of pancreatic polypeptide release in man. *Lancet* 1: 161–163, 1977.

ADRIAN, T. E., G. R. GREENBERG, H. S. BESTERMAN, AND S. R. BLOOM. Pharmacokinetics of pancreatic polypeptide in man. *Gut* 19:907–909, 1978.

AEBERHARD, P. F., L. D. MAGNENOT, AND W. A. ZIMMERMAN. Nervous control of migratory myoelectric complex of the small bowel. *Am. J. Physiol.* 238: (Gastrointest. Liver Physiol. 1): G102–G108, 1980.

AHLMAN, H., H. N. BHARGAVA, P. E. DONAHUE, B. NEWSON, T. K. DAS, AND L. M. NYHUS. The vagal release of 5-hydroxytryptamine from enterochromaffin cells in the cat. *Acta Physiol. Scand.* 104: 262–270, 1978.

ALBINUS, M., E. L. BLAIR, R. M. CASE, D. H. COY, A. GOMEZ-PAN, B. H. HIRST, J. D. REED, A. V. SCHALLY, B. SHAW, P. A. SMITH, AND J. R. SMY. Comparison of the effect of somatostatin on gastrointestinal function in the conscious and anesthetized cat and on the isolated cat pancreas. *J. Physiol. (London)* 269: 77–91, 1977.

ALLEN, A. Structure and function of gastrointestinal mucus. In: *Physiology of the Gastrointestinal Tract*, edited by L. R. Johnson. New York: Raven Press, 1981, p. 617–639.

ALLEN, A., AND A. GARNER. Mucus and bicarbonate secretion in the stomach and their possible role in mucosal protection. *Gut* 21: 249–262, 1980.

ALTAMIRANO, M. Alkaline secretion produced by intra-arterial acetylcholine. *J. Physiol. (London)* 168: 787–803, 1963.

ALUMETS, J., O. SCHAFFALITZKY DE MUCKADELL, J. FAHRENKRUG, F. SUNDLER, R. HAKANSON, AND R. UDDAMN. A rich VIP nerve supply is characteristic of sphincters. *Nature* 280: 155–156, 1979.

ANDERSON, A., O. F. W. JAMES, H. S. MACDONALD, S. SNOWBALL, AND W. TAYLOR. The effect of ethynyl oestradiol on biliary lipid composition in young men. *Eur. J. Clin. Invest.* 10: 77–80, 1980.

ANDERSON, J. J., R. J. BOLT, B. M. ULLMAN, AND P. BASS. Effect of bile and fat on gastric motility under the influence of various stimulants. *Am. J. Digest. Dis.* 13: 157–167, 1968.

ANDERSSON, P. O., S. R. BLOOM, A. V. EDWARDS, AND J. JARHULT. Effects of stimulation of the chorda tympani in bursts on submaxillary responses in the cat. *J. Physiol. (London)* 322: 469–483, 1982.

ANDERSSON, S., AND G. NILSSON. Appearance of gastrin in perfusates from the isolated gastric antrum of dogs. *Scand. J. Gastroent.* 9: 619–621, 1974.

ANURAS, S. cAMP and calcium in generation of slow waves in cat colon. *Am. J. Physiol.* 242 (Gastrointest. Liver Physiol. 5): G124–G127, 1982.

ANWER, M. S., AND D. HEGNER. Importance of solvent drag and diffusion in bile acid-dependent bile formation: ion substitution studies in isolated perfused rat liver. *Hepatology* 2: 580–586, 1982.

ARABEHETY, J. T., H. A. DOLCINI, AND S. J. GRAY. Sympathetic influences on circulation of the gastric mucosa of the rat. *Am. J. Physiol.* 197: 915–922, 1959.

ARCHIBALD, L. H., F. G. MOODY, AND M. A. SIMONS. Comparison of gastric mucosal blood flow as determined by aminopyrine clearance and γ-labeled microspheres. *Gastroenterology* 69: 630–635, 1975.

ARMBRECHT, H. J., T. V. ZENSER, M. E. BRUNS, AND B. B. DAVIS. Effect on age on intestinal calcium absorption and adaptation to dietery calcium. *Am. J. Physiol.* 236 (Endocrinol. Metab. Gastrointest. Physiol. 5): E769–E774, 1979.

ARNOLD, R., M. V. HULST, CH. NEUHOF, H. SCHWARTING, H. D. BECKER, AND W. CREUTZFELDT. Antral gastrin-producing G-cells and somatostatin-producing D-cells in different states of gastric acid secretion. *Gut* 23: 285–291, 1982.

ASKIN, J. R., D. T. LYON, D. T. SHULL, C. I. WAGNER, AND R. D. SOLOWAY. Factors affecting delivery of bile into the duodenum in man. *Gastroenterology* 74: 560–565, 1978.

AVNER, D. L., M. M. BERENSON, P. E. CHRISTIAN, F. L. DATZ, AND R. G. LEE. A model to study altered intrahepatic perfusion: validation, quantitation and effect on bile flow. *Hepatology* 1: 493, 1981.

BAHARI, H. M. M., I. N. ROSS, AND L. A. TURNBERG. Demonstration of a pH gradient across the mucus layer on the surface of human gastric mucosa in vitro. *Gut* 23: 513–516, 1982.

BAKER, A. L., R. A. B. WOOD, A. R. MOOSA, AND J. L. BOYER. Sodium taurocholate modifies the bile acid-independent fraction of canalicular bile flow in the rhesus monkey. *J. Clin. Invest.* 64: 312–320, 1979.

BANKS, P. A., W. P. DYCK, D. A. DREILING, AND H. D. JANOWITZ. Secretory capacity of stomach and pancreas in man. *Gastroenterology* 53: 575–578, 1967.

BARBEZAT, G. O., AND M. I. GROSSMAN. Intestinal secretion stimulation by peptides. *Science* 174: 422–423, 1971.

BARBEZAT, G. O., AND P. G. REASBECK. Somatostatin inhibition of glucagon-stimulated jejunal secretion in the dog. *Gastroenterology* 81: 471–474, 1981.

BARKER, G. R., G. MCL. COCHRANE, G. A. CORBETT, J. N. HUNT, AND S. K. ROBERTS. Actions of glucose and potassium chloride on osmoreceptors slowing gastric emptying. *J. Physiol. (Lond.)* 237:183–186, 1974.

BARLAS, N., R. T. JENSEN, AND J. D. GARDNER. Cholecystokinin-induced restricted stimulation of pancreatic enzyme secretion. *Am. J. Physiol.* 242(*Gastrointest. Liver Physiol* 5): G464–G469, 1982a.

BARLAS, N., R. T. JENSEN, M. C. BEINFELD, AND J. D. GARDNER. Cyclic nucleotide antagonists of cholecystokinin: structural requirements for interaction with the cholecystokinin receptor. *Am. J. Physiol.* 242 (*Gastrointest. Liver Physiol.* 5): G161–G167, 1982b.

BARON, J. H., C. V. PERRIER, H. D. JANIWITZ, AND D. A. DREILING. Maximum alkaline (bicarbonate) output of the dog pancreas. *Amer. J. Physiol.* 204: 251–256, 1963.

BASSO, N., S. GIRI, G. IMPROTA, E. LEZOCHE, P. MELCHIORRI, M. PERCOCO, AND V. SPERANZA. External pancreatic secretion after bombesin infusion in man. *Gut* 16: 994–998, 1975.

BATEY, R. G., AND N. D. GALLAGHER. Role of the placenta in intestinal absorption of iron in pregnant rats. *Gastroenterology* 72: 255–259, 1977a.

BATEY, R. G., AND N. D. GALLAGHER. Study of the subcellular localization of ^{59}Fe and iron-binding proteins in the duodenal mucosa of pregnant and nonpregnant rats. *Gastroenterology* 73: 267–272, 1977b.

BAUDUIN, H., C. STOCK, D. VINCENT, AND J. F. GRENIER. Microfilamentous system and secretion of enzyme in the exocrine pancreas. Effect of cytochalasin B. *J. Cell Biol.* 66: 165–182, 1975.

BAUER, R. F. Electrically-induced vagal stimulation of gastric acid secretion. In: *Nerves and the Gut*, edited by F. P. Brooks and P. W. Evers. Thorofare, N.J.: Charles B. Slack, 1977, p. 86–95.

BAUER, R. F., J. E. MCGUIGAN, AND F. P. BROOKS. Gastrin release by electrical vagal stimulation in cats. *Gastroenterology* 72: 1028, 1977.

BAZIN, R., AND M. LAVAN. Effects of high-rat diet on glucose metabolism in isolated pancreatic acini of rats. *Am. J. Physiol.* 243 (*Gastrointest. Liver Physiol.* 6): G448–454, 1982.

BEAVEN, M. A., A. H. SOLL, AND K. J. LEWIN. Histamine synthesis by intact mast cells from canine fundic mucosa and liver. *Gastroenterology* 82: 254–262, 1982.

BEGLINGER, C., K. GYR, B. WERTH, U. KELLER, AND J. GIRARD. Comparative effects of synthetic and natural secretin on pancreatic secretion and on secretin, insulin, and glucagon levels in man. *Digest. Dis. Sci.* 27: 231–233, 1982.

BEHAR, J., AND P. BIANCANI. Effect of cholecystokinin and the octapeptide of cholecystokinin on the feline sphincter of Oddi and gallbladder. *J. Clin. Invest.* 66: 1230–1239, 1980.

BEIJER, H. J. M., F. A. S. BROUWER, AND G. A. CHARBON. Time course and sensitivity of secretin-stimulated pancreatic secretion and blood flow in the antesthetized dog. *Scand. J. Gastroenterol.* 14: 295–300, 1979.

BELL, P. R. F., AND C. BATTERSBY. Effect of vagotomy on gastric mucosal blood flow. *Gastroenterology* 54: 1032–1037, 1968.

BERRIDGE, N. J. The interaction of cyclic nucleotides and calcium in the control of cellular activity. In: *Advances in Cyclic Nucleotides Research*, edited by P. Greengard and G. A. Robinson. New York: Raven Press, 1975, vol. 6, p. 1–98.

BHOOLA, K. D., AND G. DOREY. Kallikrein, trypsin-like proteases and amylase in mammalian submaxillary glands. *Brit. J. Pharmacol.* 43: 784–793, 1971.

BIANCANI, P., J. H. WALSH, AND J. BEHAR. VIP antiserum inhibits neurally mediated relaxation of the in vitro cat lower esophageal sphincter (LES). *Gastroenterology* 82: 1017, 1982.

BIANCANI, P., M. P. ZABINSKI, M. D. KERSTEIN, AND J. BEHAR. Mechanical characteristics of the cat pylorus. *Gastroenterology* 78: 301–309, 1980.

BIEBERDORF, F. A., S. MOROWSKI, AND J. S. FORDTRAN. Effect of sodium, mannitol and magnesium on glucose, galactose, 3-O-methyl glucose and fructose absorption in the human ileum. *Gastroenterology* 68: 50–57, 1975.

BINDER, H. J. Sodium transport across isolated human jejunum. *Gastroenterology* 67: 231–236, 1974.

BINDER, H. J. Net fluid and electrolyte secretion: the pathophysiological basis of diarrhea. *Viewpoints on Digestive Diseases* 12, 1980.

BINDER, H. J. Colonic secretion. In: *Physiology of the Gastrointestinal Tract*, edited by L. R. Johnson. New York: Raven Press, vol. 2, p. 1003–1020.

BITAR, K. N., B. SAFFOURI, AND G. M. MAKHLOUF. Cholinergic and peptidergic receptors on isolated human antral smooth muscle cells. *Gastroenterology* 82: 832–837, 1982.

BLACK, A., T. M. FORTE, AND J. G. FORTE. The effects of microfilament disrupting agents on HCl secretion and ultrastructure of piglet gastric oxyntic cells *Gastroenterology* 83: 595–604, 1982.

BLAIR, E. L., E. R. GRAND, J. D. REED, D. J. SANDERS, G. SANGER, AND B. SHAW. The effect of sympathetic nerve stimulation on serum gastrin, gastric acid secretion and mucosal blood flow responses to meat extract stimulation in anesthetized cats. *J. Physiol. (London)* 253: 493–504, 1975.

BLITZER, B. L., AND J. L. BOYER. Cytochemical localization of Na^+, K^+-ATPase in the rat hepatocyte. *J. Clin. Invest.* 62: 1104–1108, 1978.

BLITZER, B. L., AND J. L. BOYER. Cellular mechanisms of bile formation. *Gastroenterology* 81: 341–357, 1982.

BLOBEL, G., P. WALTER, C. N. CHANG, B. M. GOLDMAN, A. H. ERICKSON, AND V. R. LINGAPPA. Translocation of proteins across membranes: the signal hypothesis and beyond. In: *Secretory Mechanisms*, edited by C. R. Hopkins and C. J. Duncan. Cambridge:Cambridge University Press, 1979, p. 9–36.

BLOMFIELD, J., A. R. RUSH, AND H. M. ALLARS. Interrelations between flow rate, amylase, calcium, sodium, potassium and inorganic phosphate in stimulated human parotid saliva. *Arch. Oral Biol.* 21: 645–650, 1976.

BLOOM, S. R., AND A. V. EDWARDS. Vasoactive intestinal peptide in relation to atropine-resistant vasodilation in the submaxillary gland of the cat. *J. Physiol. (London)* 300: 41–53, 1980.

BOCKMAN, D. E. Anastomosing tubular arrangement of dog exocrine pancreas. *Cell Tiss. Res.* 189: 497–500, 1978.

BODANSKY, M., M. L. FINK, AND G. BODEN. The role of some polar amino acid residues in secretin. *Gastroenterology* 72: 801–802, 1977.

BOLTON, J. E., AND M. FIELD. Ca ionophore-stimulated ion secretion in rabbit ileal mucosa: relation to actions of cyclic 3′, 5′-AMP and carbamylcholine. *J. Membrane Biol.* 35: 159–173, 1977.

BOND, J. H., R. A. PRENTISS, AND M. D. LEVITT. The effects of feeding on blood flow to the stomach, small bowel and colon of conscious dog. *J. Lab. Clin. Med.* 93: 594–599, 1979.

BONFILS, S., M. MIGNON, AND C. ROZE. Vagal control of gastric secretion. *International Rev. Physiol.* 19: 59–106, 1979.

BORGESE, N., P. DE CAMILLI, Y. TANAKA, AND J. MELDOLESI. Membrane interaction in secretory cell systems. In: *Secretory Mechanisms*, edited by C. R. Hopkins and C. J. Duncan. Cambridge: Cambridge University Press, 1979, p. 117–144.

BORGSTROM, B., A. DAHLQUIST, G. LUNDH, AND J. SJOVALL. Studies of intestinal digestion and absorption in the human. *J. Clin. Invest.* 36: 1521–1536, 1957.

BORISON, H. L., AND S. C. WANG. Functional localization of central coordinating mechanisms. *J. Neurophysiology* 12: 305–313, 1949.

BOUCHIER, I. A. D. The vagus, the bile and gallstones. *Gut* 11: 799–803, 1970.

BOUCHIER, I. A. D., AND S. R. COOPERBAND. Isolation and characterization of macromolecular aggregate associated with bilirubin. *Clin. Chem. Acta.* 15: 291–302, 1967.

BOYER, J. L., E. ELIAS, AND T. J. LAYDEN. The paracellular pathway and bile formation. *Yale J. Biol. Med.* 52: 49–60, 1979a.

BOYER, J. L., M. ITABASHI, AND Z. HRUBAN. Formation of pericanalicular vacuoles during sodium dehydrocholate choleresis—a mechanism for bile acid transport. In: *The Liver: Quantitative Aspects of Structure and Function*, edited by R. Preisig and J. Bircher. Berne, Switzerland: Editio Cantor Arlendorf, 1979b, p. 163–178.

BRANDT, I. L., L. CASTLEMAN, H. D. RUSKIN, I. J. GREENWALD, AND J. KELLEY. The effect of oral protein and glucose feeding on splanchnic blood flow and oxygen utilization in normal and cirrhotic subjects. *J. Clin. Invest.* 34: 1017–1025, 1955.

BRAUTHBAR, N., B. S. LEVINE, M. W. WALLING, AND J. W. COBURN. Intestinal absorption of calcium: role of dietary phosphate and vitamin D. *Am. J. Physiol.* 241 (*Gastrointest. Liver Physiol.* 4): G49–G53, 1981.

BRODRIBB, A. J. M., R. E. CONDON, V. COWLES, AND J. J. DeCOSSE. Effect of dietary fiber on intraluminal pressure and myoelectrical activity of left colon in monkeys. *Gastroenterology* 77: 70–74, 1979.

BROOKS, F. P. *Control of Gastrointestinal Function.* New York: Macmillan Co., 1970, p. 16–31.

BROOKS, F. P. Anatomy and physiology of the gallbladder and bile ducts. In: *Gastroenterology*, 3rd ed. Vol III: *Liver*, edited by H. L. Bockus. Philadelphia: W.B. Saunders, 1976, p. 611–650.

BROOKS, F. P. Mechanisms of digestion and absorption. In: *Scientific Foundations of Gastroenterology*, edited by W. Sircus and A. N. Smith. London: William Heinemann Medical Books, 1980, p. 389–397.

BROOKS, F. P., AND E. F. ROSATO. The vagus nerves: their significance for gastrointestinal function and peptic ulcer disease. *Viewpoints on Digestive Diseases.* Vol. 6, No. 4, September, 1974.

BROOKS, F. P., D. J. SANDWEISS, AND J. F. LONG. The relationship betwen peptic ulcer and coronary occlusion. *Trans. Am. Clin. Climat. Assoc.* 74: 51–59, 1962.

BRO-RASMUSSEN, F., S. A. KELLMAN, AND J. H. THAYSEN. The composition of pancreatic juice as compared to sweat, parotid saliva and tears. *Acta Physiol. Scand.* 37: 97–113, 1956.

BROWN, W. R., Y. ISOBE, AND P. K. NAKANE. Studies of translocation of immunoglobulins across intestinal epithelium. II. Immunoelectricon microscopic localization of immunoglobulins, and secretory component in human intestinal mucosa. *Gastroenterology* 71: 985–995, 1976.

BROWN, W. R., K. ISOBE, AND P. K. NAKANE. Studies on translocation of immunoglobulins across intestinal epithelium. IV. Evidence for binding of IgA and IgM to secretory component in intestinal epithelium. *Gastroenterology* 73: 1333–1339, 1977.

BUENO, L., J. FIORAMONTI, V. RAYNER, AND Y. RUCK-EBUSCH. Effects of motility, somatostatin and pancreatic polypeptide on the migrating myoelectric complex in pig and dog. *Gastroenterology* 82: 1395–1402, 1982.

BUFFA, R., E. SOLCIA, AND V. L. W. GO. Immunohistochemical identification of the cholecystokinin cell in the intestinal mucosa. *Gastroenterology* 70: 528–532, 1976.

BUKHAVE, K., AND J. RASK-MADSEN. Saturation kinetics applied to in vitro effects of low prostaglandin E_2 and F_2 concentrations on ion transport across human jejunal mucosa. *Gastroenterology* 78: 32–42, 1980.

BUNDGAARD, M., M. MOLLER, AND J. H. POULSEN. Localization of sodium pump sites in cat pancreas. *J. Physiol. (London)* 313: 405–414, 1981.

BURKS, T. F., AND W. D. BARBER. The brain-gut connection: central projections of gastric afferent neurons. *Gastroenterology* 82: 1027, 1982.

BURNSTOCK, G. Purinergic nerves. *Pharmacol Rev* 24: 509–581, 1972.

BYRNE, W. J., D. L. CHRISTIE, M. E. AMENT, AND J. H. WALSH. Acid secretory response in man to 18 individual amino acids. *Clin. Res.* 25: 108A, 1977.

CAFLISCH, C. R., S. SOLOMON, AND W. R. GALEY. In situ micropuncture study of pancreatic duct pH. *Am. J. Physiol.* 238 (*Gastrointest. Liver Physiol.* 1): G263–G268, 1980.

CALAM, J., A. ELLIS, AND G. J. DOCKRAY. Identification and measurement of molecular variants of cholecystokinin in duodenal mucosa and plasma: diminished concentration in patients with celiac disease. *J. Clin. Invest.* 69: 218–225, 1982.

CARD, W. I., AND I. N. MARKS. The relationship between the acid output of the stomach following "maximal" histamine stimulation and the parietal cell mass. *Clin. Sci.* 19: 147–163, 1960.

CAREY, M. C. The enterohepatic circulation in the liver. In: *The Liver: Biology and Pathobiology*, edited by I. M. Arias, H. Popper, D. Schachter, and D. Schafritz. New York: Raven Press, 1982, p. 429–466.

CAREY, M. C., AND D. M. SMALL. Micelle formation by bile salts. *Arch. Intern. Med.* 130: 509–527, 1972.

CAREY, M. C., AND D. M. SMALL. The physical chemistry of cholesterol solubility in bile. *J. Clin. Invest.* 61: 998–1026, 1978.

CARLING, R. C. J., AND D. TEMPLETON. Cholinergic control of rat pancreatic acinar cell division. *J. Physiol. (London)*, in press, 1983.

CARTER, D. C., J. A. FORREST, R. A. LOGAN, I. ANSELL, G. LIDGARD, R. C. HEADING, AND D. J. C. SHEARMAN. Effect of the histamine H^2-receptor antagonist cimetidine on gastric secretion and serum gastrin during insulin infusion in man. *Scand. J. Gastroenterol.* 11: 565–570, 1976.

CARTER, D. C., I. L. TAYLOR, J. ELASHOFF, AND M. I. GROSSMAN. Reappraisal of the secretion potency and disappearance rate of pure human minigastrin. *Gut* 20: 705–708, 1979.

CASE, R. M. Synthesis, intracellular transport and discharge of exportable proteins in the pancreatic acinar cell and other cells. *Proc. Aust. Physiol. Pharmacol. Soc.* 9: 29–42, 1978a.

CASE, R. M. Synthesis, intracellular transport and discharge of exportable proteins in the pancreatic acinar cell and other cells. *Biol. Rev.* 53: 211–354, 1978b.

CASE, R. M., A. D. CONIGRAVE, M. HUNTER, I. NOVAK, C. H. THOMPSON, AND J. A. YOUNG. Secretion of saliva by the rabbit manidublar gland in vitro: the role of anions. *Philos. Trans. Roy. Soc. (Lond.)* B 196: 179–192, 1981.

CASE, R. M., A. D. CONIGRAVE, I. NOVAK, AND J. A. YOUNG. Electrolyte and protein secretion by the perfused rabbit mandibular gland stimulated with acetylcholine or catecholamines. *J. Physiol. (London)* 300: 467–487, 1980.

CASE, R. M., AND T. SCRATCHERD. The actions of dibutyryl cyclic adenosine $3'5'$-monophosphate and methyl xanthines on pancreatic exocrine secretion. *J. Physiol. (London)* 223: 649–668, 1972.

CASTELL, D. O. Diet and the lower esophageal sphincter. *Am. J. Clin. Nutr.* 28: 1296–1298, 1975.

CASTELL, D. O., S. COHEN, AND L. D. HARRIS. Response of human ileocecal sphincter to gastrin. *Am. J. Physiol.* 219: 712–715, 1970.

CASTLE, J. D., J. D. JAMIESON, AND G. E. PALADE. Radioautographic analysis of the secretion process in the parotid acinar cell of the rabbit. *J. Cell Biol.* 53: 290–311, 1972.

CHAMBERS, M. M., K. L. BOWES, Y. D. KINGMA, C. BANNISTER, AND K. R. COTE. In vitro electrical activity in human colon. *Gastroenterology* 81: 502–508, 1981.

CHANG, T-M., AND W. Y. CHEY. Radioimmunoassay of secretin: a critical review and current status. *Digest. Dis. Sci.* 25: 529–552, 1980.

CHEESEMAN, C. I. The transport of dipeptides by the small intestine. *Can. J. Physiol. Pharmacol.* 58(11): 1326–1333, 1981.

CHEUNG, L. Y. Topical effects of 16, 16-dimethyl prostaglandin E_2 on gastric blood flow in dogs. *Am. J. Physiol.* 238 (*Gastrointest. Liver Physiol.* 1): G514–G519, 1980.

CHEUNG, L. Y., AND S. F. LOWRY. Effects of intra-arterial infusion of prostaglandin E_1 on gastric secretion and blood flow. *Surgery* 83: 699–704, 1978.

CHEW, C. S., AND S. J. HERSEY. Gastrin stimulation of isolated gastric glands. *Am. J. Physiol.* 242 (*Gastrointest. Liver Physiol.* 5): G504–G512, 1982.

CHEY, W. Y., M. S. KIM, AND K. Y. LEE. Influence of the vagus nerve on release and action of secretin in dog. *J. Physiol. (London)* 293: 453–446, 1979a.

CHEY, W. Y., M. S. KIM, K. Y. LEE, AND T-M. CHANG. Effect of rabbit antisecretin serum on post-prandial pancreatic secretion in dogs. *Gastroenterology* 77: 1268–1275, 1979b.

CHEY, W. Y., M. S. KIM, K. Y. LEE, AND T-M. CHANG. Secretin is an enterogastrone in the dog. *Am. J. Physiol.* 240 (*Gastrointest. Liver Physiol.* 3): G239–244, 1981.

CHIBA, T., S. KADOWAKI, T. TAMINATO, K. CHIHARA, Y. SEINO, S. MATSUKURA, AND T. FUGITA. Effect of antisomatostatin γ globulin on gastrin release in rats. *Gastroenterology* 81: 321–326, 1981.

CHOWDHURY, J. R., J. M. BERKOWITZ, M. L. PRAISSMAN, AND J. W. FARA. Interaction between octapeptide-cholecystokinin, gastrin, and secretin on cat gallbladder in vitro. *Am. J. Physiol.* 229: 1311–1315, 1975.

CHOWDHURY, J. R., A. W. WOLKOFF, AND I. M. ARIAS. Heme and bile pigment metabolism. In: *The Liver: Biology and Pathobiology*, edited by I. Arias, H. Popper, D. Schachter, and D. A. Shafritz. New York: Raven Press, 1982, p. 309–332.

CHRISTENSEN, J., S. ANURAS, AND R. L. HAUSER. Migrating spike bursts and electrical slow waves in the cat colon: effects of sectioning. *Gastroenterology* 66: 240–247, 1974.

CHRISTENSEN, J., AND G. F. LUND. Esophageal responses to distention and electrical stimulation. *J. Clin. Invest.* 48: 408–418, 1969.

CHRISTENSEN, J., H. P. SCHEDL, AND J. A. CLIFTON. The basic electrical rhythm of the duodenum in normal human subjects and in patients with thyroid disease. *J. Clin. Invest.* 43: 1659–1667, 1964.

CHRISTIANSEN, J., J. F. REHFELD, AND F. STADIL. Effect of intrajejunal fat on meal-stimulated acid and gastric secretion in man. *Scand. J. Gastroenterol.* 11: 673–676, 1976.

CHRISTIANSEN, J., A. BECH, J. FAHRENKRUG, J. J. HOLST, K. LAURITSEN, A. J. MOODY, AND O. SCHAFALITZKY DE MUCKADELL. Fat-induced jejunal inhibition of gastric acid secretion and release of pancreatic glucagon, enteroglucagon, gastric inhibitory polypeptide and vasoactive intestinal polypeptide in man. *Scand. J. Gastroenterol.* 14: 161–166, 1979.

CHRISTOFIDES, N. D., M. G. BRYANT, M. A. GHATEI, S. KISHIMOTO, A. M. J. BUCHAN, J. M. POLAK, AND S. R. BLOOM. Molecular forms of motilin in the mammalian and human gut and human plasma. *Gastroenterology* 80: 292–300, 1981.

CODE, C. F., AND J. A. MARLETT. The interdigestive myoelectric complex of the stomach and small bowel of dogs. *J. Physiol. (London)* 246: 289–309, 1975.

COHEN, S. Diseases of the esophagus. In: *Gastrointestinal Pathophysiology*, edited by F. P. Brooks. New York: Oxford University Press, 1978, p. 68–99.

COHEN, S., AND L. D. HARRIS. Anatomy and normal functional physiology of the esophagus and pharynx. In: *Disorders of the Gastrointestinal Tract, Disorders of the Liver and Nutritional Disorders*, edited by J. M. Dietschy. New York: Grune & Stratton, 1976, p. 1–3.

COHEN, S., L. D. HARRIS, AND R. LEVITAN. Manometric characteristics of the human ileocecal junctional zone. *Gastroenterology* 54: 72–75, 1968.

COLLEN, M. J., V. E. SUTLIFF, G. Z. PAN, AND J. D. GARDNER. Post receptor modulation of action of VIP and secretin on pancreatic enzyme secretion by secretagogues that mobilize cellular calcium. *Am. J. Physiol.* 242 (*Gastrointest. Liver Physiol.* 5): G423–G428, 1982.

COLLINS, S. M., S. ABDELMOUMENE, R. T. JENSEN, AND J. D. GARDNER. Reversal of cholecystokinin-induced persistent stimulation of pancreatic enzyme secretion by dibutyryl cyclic GMP. *Am. J. Physiol.* 240 (*Gastrointest. Liver Physiol.* 3): G466–G471, 1981a.

COLLINS, S. M., S. ABDELMOUMENE, R. T. JENSEN, AND J. T. GARDNER. Cholecystokinin-induced persistent stimulation of enzyme secretion from pancreatic acini. *Am. J. Physiol.* 240 (*Gastrointest. Liver Physiol.* 3): G459–G465, 1981b.

COLLMAN, P. I., D. GRUNDY, AND T. SCRATCHERD. Vagal influences on small intestinal motility in the anesthetized ferret. *J. Physiol. (London)* 328: 51P, 1982.

CONKLIN, J. L., AND J. CHRISTENSEN. Local specialization at ileocecal junction of the cat and opossum. *Am. J. Physiol.* 228: 1075–1081, 1975.

CONNELL, A. M. The motility of the pelvic colon. *Gut* 2: 175–186, 1961.

CONNELL, A. M., H. FRANKEL, AND L. GUTTMAN. The motility of the pelvic colon following complete lesions of the spinal cord. *Paraplegia* 1: 98–115, 1963.

CONNOR, J. A., A. W. MANGEL, AND B. NELSON. Propagation and entrainment of slow waves in cat small intestine. *Am. J. Physiol.* 237 (*Cell Physiol.* 6): C237–C246, 1979.

COOKE, A. R. The glands of Brunner. In: *Handbook of Physiology.*, sect. 6: *Alimentary Canal*, vol 2: Secretion, edited by C. F. Code. Washington, DC: Am. Physiol. Soc., 1967, p. 1087–1095.

COOKE, A. R. Localization of receptors inhibiting gastric emptying in the gut. *Gastroenterology* 72: 875–880, 1977.

CORBETT, G. L., P. E. T. ISAACS, A. K. RILEY, AND L. A. TURNBERG. Human intestinal ion transport. *Gut* 18: 136–140, 1977.

COSTA, M., AND J. B. FURNESS. The peristaltic reflex: an analysis of the nerve pathways and their pharmacology. *Nauny-Schmiedeberg's Archiv. Pharmacol.* 294: 47–60, 1976.

COWLEY, D. J., AND C. F. CODE. Effects of secretory inhibi-

tors on mucosal blood flow in the nonsecreting stomach of conscious dogs. *Am. J. Physiol.* 218: 270–274, 1970.

CREMA, A., M. DEL TACCA, G. M. FRIGO, AND S. LECCHINI. Presence of a non-adrenergic inhibitory system in the human colon. *Gut* 9: 633–637, 1968.

CROTTON, P., AND A. SMITH. High molecular-mass alkaline phosphatase in serum and bile: physical properties and relationship with other high-molecular-mass enzymes. *Clin. Chem.* 27: 860–866, 1981.

CSENDES, A., A. KRUSE, P. FUNCH-JENSEN, M. J. OSTER, J. ARNSHOLT, AND E. AMDRUP. Pressure measurements in the biliary and pancreatic duct systems in controls and in patients with gallstones, previous cholecystectomy or common bile duct stones. *Gastroenterology* 77: 1203–1210, 1979.

CUMMINGS, J. D., A. L. HAIGH, E. H. L. HARRIES, AND E. M. NUTT. A study of gastric secretion and blood flow in the anesthetized dog. *J. Physiol. (London)* 168: 219–233, 1963.

CURTIS, P. J., H. R. FENDER, P. L. RAYFORD, AND J. C. THOMPSON Disappearance half-time of endogenous and exogenous secretin in dogs. *Gut* 17: 595–599, 1976.

DAGORN, J. C., J. SAHEL, AND H. SARLES. Non-parallel secretion of enzymes in human duodenal juice and pure pancreatic juice collected by endoscopic retrograde catheterization of the papilla. *Gastroenterology* 73: 42–45, 1977.

DANIEL, E. E. Peptidergic nerves in the gut. *Gastroenterology* 75: 142–144, 1978.

DARKE, A. C., AND L. H. SMAJE. Myoepithelial cell activation in the submaxillary salivary gland. *J. Physiol. (London)* 219: 89–102, 1971.

DAVENPORT, H. W. A digest of digestion. Chicago: Year Book Medical Publishers, 1975.

DAVENPORT, H. W. *Physiology of the Digestive Tract*, 4th ed. Chicago: Year Book Medical Publishers, 1977.

DAVIS, G. R., C. A. SANTA ANA, S. MORAWSKI, AND J. S. FORDTRAN. Active choloride secretion in the normal human jejunum. *J. Clin. Invest.* 66: 1326–1333, 1981a.

DAVIS, G. R., C. A. SANTA ANA, S. MORAWSKI, AND J. S. FORDTRAN. Effect of vasoactive intestinal polypeptide on active and passive transport in the human jejunum. *J. Clin. Invest.* 67: 1687–1694, 1981b.

DAVIS, G. R., S. G. MORAWSKI, C. A. SANTA ANA, AND J. S. FORDTRAN. Evaluation of chloride/bicarbonate exchange in the human colon in vivo. *J. Clin. Invest.* 71: 201–207, 1983.

DAVISON, J. S., AND G. T. PEARSON. Electrophysiological effects of field stimulation on mouse pancreatic acinar cells. *J. Physiol. (London)* 295: 45P, 1979.

DAVISON, J. S., AND N. UEDA. Depolarization of rat pancreatic acinar cells by cervical vagal stimulation. *J. Physiol. (London)* 266: 36P, 1977.

DEBAS, H. T., AND M. I. GROSSMAN. Pure cholecystokinin: pancreatic protein and bicarbonate response. *Digestion* 9: 469–481, 1973.

DEBAS, H. T., AND M. I. GROSSMAN. Chemicals bathing the oxyntic gland area stimulate acid secretion in dog. *Gastroenterology* 69: 654–659, 1975.

DEBAS, H. T., J. H. WALSH, AND M. I. GROSSMAN. Evidence for oxyntopyloric reflex for release of antral gastrin. *Gastroenterology* 68: 687–690, 1975.

DEBAS, H. T., A. M. SEAL, C. CORK, AND J. H. WALSH. Vagal distribution for stimulation and inhibition of gastrin release. *Gastroenterology* 80: 1133, 1981.

DEBAS, H. T., AND T. YAMAGISHI. Evidence for pyloropancreatic reflex for pancreatic exocrine secretion. *Am. J. Physiol.* 234 (*Endocrinol. Metab. Gastrointest. Physiol.* 3): E468–E471, 1978.

DEBAS, H. T., AND T. YAMAGISHI. Evidence for a pylorocholecystic reflex for gallbladder contraction. *Ann. Surg.* 190: 170–175, 1979.

DEBONGNIE, J. C., AND S. F. PHILLIPS. Capacity of the human colon to absorb fluid. *Gastroenterology* 74: 698–703, 1978.

DE CARLE, D. J., J. CHRISTENSEN, A. C. SZABO, D. C.

TEMPLEMAN, AND D. R. McKINELEY. Calcium dependence of neuromuscular events in esophageal smooth muscle of the opossum. *Am. J. Physiol.* 232 (*Endocrinol. Metab. Gastrointest. Physiol.* 1): E547–E552, 1977.

DeFILIPPI, C., AND J. VALENZUELA. Sham feeding disrupts the interdigestive motility complex in man. *Scand. J. Gastroenterol.* 16: 977–979, 1981.

DELLER, D. J., AND A. G. WANGEL. Intestinal motility in man. I. A study combining the use of intraluminal pressure recording and cineradiography. *Gastroenterology* 48: 45–57, 1965.

DENT, J., W. J. DODDS, R. H. FRIEDMAN, T. SEKIGUCHI, W. J. HOGAN, R. C. ARNDORFER, AND D. J. PETRIE. Mechanisms of gastroesophageal reflux in recumbent asymptomatic human studies. *J. Clin. Invest.* 65: 256–267, 1980.

DE SCHRYVER-KECSKEMETI, K., AND M. H. GREIDER. Effect of somatostatin on arginine-induced gastric release in tissue culture. *Clin. Res.* 24: 534A, 1976.

DE SCHRYVER-KECSKEMETI, K., M. H. GREIDER, E. REEDERS, AND J. E. McGUIGAN. Studies on gastrin secretion in vitro from cultures of rat pyloric antrum: effects of agents modifying the microtubular-microfilament system. *Gastroenterology* 78: 339–345, 1980.

DE SCHRYVER-KECSKEMETI, K., M. H. GREIDER, M. K. SAKS, E. REEDERS, AND J. E. McGUIGAN. The gastrin producing cells in tissue cultures of the rat pyloric antrum. *Lab. Invest.* 37: 406–410, 1977.

DESNUELLE, P. The lipase-colipase system. In: *Lipid Absorption: Biochemical and Clinical Aspects*, edited by K. Rommel and H. Goebel. Lancaster, UK: MTP Press, 1976, p. 23–26.

DEVROEDE, G., AND J. LAMARCHE. Functional importance of extrinsic parasympathetic innervation to the distal colon and rectum in man. *Gastroenterology* 66: 273–280, 1974.

DEVROEDE, G. J., AND S. F. PHILLIPS. Conservation of sodium, chloride and water by the human colon. *Gastroenterology* 56: 101–109, 1969.

DHARMSATHAPHORN, K., L. RACUSEN, AND J. W. DOBBINS. Effect of somatostatin on ion transport in the rat colon. *J. Clin. Invest.* 66: 813–820, 1980.

DIAMANT, N. E., AND T. Y. EL-SHARKAWY. Neural control of esophageal peristalsis: a conceptual analysis. *Gastroenterology* 72: 546–556, 1977.

DIETSCHY, J. M. Recent developments in solute and water transport across the gallbladder epithelium. *Gastroenterology* 50: 692–707, 1966.

DIETZ, J., AND M. FIELD. Ion transport in rabbit ileal mucosa IV bicarbonate secretion. *Am. J. Physiol.* 225: 858–861, 1973.

DI MAGNO, E. P., J. C. HENDRICKS, V. L. W. GO, AND R. R. DOZOIS. Relationships among canine fasting and pancreatic secretions, pancreatic duct pressure, and duodenal phase III motor activity—Boldyreff revisited. *Digest. Dis. Sci.* 24: 689–693, 1979.

DI MAGNO, E. P., J. C. HENDRICKS, R. R. DOZOIS, AND V. L. W. GO. The effect of secretin on pancreatic duct pressure, resistance to pancreatic flow and duodenal motor activity in the dog. *Digest. Dis. Sci.* 26: 1–6, 1981.

DOBBINS, J. W., AND H. J. BINDER. Pathophysiology of diarrhea: alterations in fluid and electrolyte transport. *Clin. Gastroenterol.* 10: 605–626, 1981.

DOCKRAY, G. J. Multiple molecular forms of hormones: significance and methods of study. In: *Gut Hormones*, 2nd ed. edited by S. R. Bloom and J. M. Polak. Edinburgh: Churchill Livingstone, 1981, p. 43–48.

DOCKRAY, G. J., R. A. GREGORY, H. J. TRACY, AND W-Y. ZHU. Transport of cholecystokinin-octapeptide-like immunoreactivity toward the gut in afferent vagal fibers in cat and dog. *J. Physiol.* 314: 501–511, 1981.

DOCKRAY, G. J., R. A. GREGORY, J. H. TRACY, AND W-Y. ZHU. Post secretory processing of heptadecapeptide gastrin: conversion to C-terminal immunoreactive fragments in the circulation of the dog. *Gastroenterology* 83: 224–232, 1982.

DOIG, R. K., S. WOLF, AND H. G. WOLFF. Study of gastric function in a "decorticate:" man with a gastric fistula. *Gastroenterology* 23: 40–44, 1953.

DOMSCHKE, S., W. DOMSCHKE, W. ROSCH, S. J. KONTUREK, E. WUNSCH, AND L. DEMLING. Bicarbonate and cyclic AMP content and pure human pancreatic juice in response to graded doses of synthetic secretin. *Gastroenterology* 70: 533–536, 1976.

DOMSCHKE, S., W. DOMSCHKE, W. ROSCH, S. J. KONTUREK, W. SPRUGEL, P. MITZNEGG, W. WUNSCH, AND L. DEMLING. Inhibition by somatostatin of secretin-stimulated pancreatic secretion in man: a study with pancreatic juice. *Scand. J. Gastroenterol.* 12: 59–63, 1977.

DONALDSON, L. A., S. N. JOFFE, W. Mc INTOSH, AND M. J. BRODIE. Amylase activity in human bile. *Gut* 20: 216–218, 1979.

DONALDSON, R. M., JR. Intrinsic factor and the transport of cabalamin. In: *Physiology of the Gastrointestinal Tract*, edited by L. R. Johnson. New York: Raven Press, 1981, vol. 2, p. 642–658.

DONOWITZ, M., N. ASARKOF, AND G. PIKE. Calcium dependence of serotonin-induced changes in rabbit ileal electrolyte transport. *J. Clin. Invest.* 66: 341–352, 1980.

DONOWITZ, M., AND J. L. MADARA. Effect of extracellular calcium depletion on epithelial structure and function in rabbit ileum: a model for selective crypt or villus epithelial cell damage and suggestion of secretion by villus epithelial cells. *Gastroenterology* 83: 1231–1243, 1982.

DOOLEY, J. S., B. J. POTTER, H. C. THOMAS, AND S. SHERLOCK. A comparative study of the biliary secretion of human dimeric and monomeric IgA in the rat and in man. *Hepatology* 2: 323–327, 1982.

DORMER, R. L., J. H. POULSEN, V. LICKO, AND J. A. WILLIAMS. Calcium fluxes in isolated pancreatic acini: effects of secretagogues. *Am. J. Physiol.* 240 (*Gastrointest. Liver Physiol.* 3): G38–G49, 1981.

DORMER, R. L., AND J. A. WILLIAMS. Secretogogue-induced changes in subcellular Ca^{++} distribution isolated pancreatic acini. *Am. J. Physiol.* 240 (*Gastrointest. Liver Physiol* 3): G130–G140, 1981.

DOTY, R. W. Neural organization of deglutition. In: *Handbook of Physiology*, sect. 6: *Alimentary Canal*, vol. 4: *Motility*, edited by C. F. Code. Washington, DC: Am. Physiol. Soc., 1968, p. 1861–1902.

DOWLING, R. A., E. MACK, AND D. M. SMALL. Biliary lipid secretion and bile composition after acute and chronic interruption of the enterohepatic circulation in the rhesus monkey. IV. Primate biliary physiology. *J. Clin. Invest.* 50: 1917–1926, 1971.

DUANE, W. C. Simulation of the defect of bile acid metabolism associated with cholesterol cholelithiasis by sorbital ingestion in man. *J. Lab. Clin. Med.* 91: 969–978, 1978.

DUANE, W. C., AND J. H. BOND. JR. Prolongation of intestinal transit and expansion of bile acid pools by propantheline bromide. *Gastroenterology* 78: 226–230, 1980.

DUANE, W. C., AND K. C. HANSON. Role of gallbladder emptying and small bowel transit in regulation of bile acid pool size in man. *J. Lab. Clin. Med.* 92: 859–872, 1978.

DUBIN, M., M. MAURICE, G. FELDMANN, AND S. ERLINGER. Influence of colchicine and phalloidin on bile secretion and hepatic ultrastructure in the rat. Possible interaction between microtubules and microfilaments. *Gastroenterology* 79: 646–654, 1980.

DUFFY, M. C., B. L. BLITZER, AND J. L. BOYER. Direct determination of the driving forces for taurocholate (TC) uptake into liver plasma membrane (LPM) vesicles. *Hepatology* 1: 507, 1981.

DUFFY, P. A., D. N. GRANGER, AND A. E. TAYLOR. Intestinal secretion induced by volume expansion in the dog. *Gastroenterology* 75: 413–418, 1978.

DUMONT, M., S. ERLINGER, AND S. UCHMAN. Hypercholeresis induced by ursodeoxycholic acid and 7-ketolithocholic acid in the rat: possible role of bicarbonate transport. *Gastroenterology* 79: 82–89, 1980.

DUNCOMBE, V. M., AND J. REEVE. Calcium homeostasis in digestive disorders. *Clin. Gastroenterol.* 10A: 653–670, 1981.

DUNCOMBE, V. M., R. W. E. WATTS, AND T. J. PETERS. In vitro calcium uptake by jejunal specimens from patients with idiopathic hypercalciuria. *Lancet* II: 1334–1336, 1980.

DURBIN, R. P. Chloride transport and acid secretion in stomach. *Gastroenterology* 73: 937–930, 1977.

DUTHIE, H. L. Colonic response to eating. *Gastroenterology* 75: 527–528, 1978.

DUTHIE, H. L. Defaecation and the anal sphincters. *Clin. Gastroenterol.* 11: 621–631, 1982.

DU VAL, J. W., B. SAFFOURI, G. C. WEIR, J. H. WALSH, A. ARIMURA, AND G. M. MAHKLOUF. Stimulation of gastrin and somatostatin secretion from the isolated rat stomach by bombesin. *Am. J. Physiol.* 241 (*Gastrointest. Liver Physiol.* 4): G242–G247, 1981.

DYCK, W. P., AND H. D. JANOWITZ. Effect of glucagon on hepatic bile secretion in man. *Gastroenterology* 60: 400–404, 1971.

DYCK, W. P., F. F. HALL, AND C. R. RATLIFF. Hormonal control of intestinal alkaline phosphatase secretion in the dog. *Gastroenterology* 65: 445–450, 1973.

DYCK, W. P., A. M. SPIEKERMAN, AND C. R. RATLIFF. Pancreatic secretory isoenzyme of alkaline phosphatase. *Proc. Soc. Exper. Biol. Med.* 159: 192–194, 1978.

EDIN, R., H. AHLMAN, AND J. KEWENTER. The vagal control of the feline pyloric sphincter. *Acta Physiol. Scand.* 107: 169–174, 1979a.

EDIN, R., J. M. LUNDBERG, H. AHLMAN, A. DAHLSTROM, J. FAHRENKRUG, T. HOKFELT, AND J. KEWENTER. On the VIP ergic innervation of the feline pylorus. *Acta Physiol. Scand.* 107: 185–187, 1979b.

EDIN, R., J. LUNDBERG, L. TERENIUS, A. DAHLSTROM, T. HOKFELT, J. KEWENTER, AND H. AHLMAN. Evidence for vagal enkephalinergic neural control of the feline pylorus and stomach. *Gastroenterology* 78: 492–497, 1980a.

EDIN, R., A. AHLMAN, A. DAHLSTROM, AND J. KEWENTER. The transmission mechanism of the vagal control of the feline pylorus. *J. Neural Transm.* 48: 177–188, 1980b.

EDWARDS, A. V., P. M. M. BIRCHANI, S. J. MITCHELL, AND S. R. BLOOM. Changes in the concentration of vasoactive intestinal peptide in intestinal lymph in response to vagal stimulation in the calf. *Experientia* 34: 1186–1187, 1978.

EECKHOUT, C., I. DE WEVER, G. VANTRAPPEN, AND J. JANSSENS. Local disorganization of interdigestive migrating complex by perfusion of a thirty-vella loop. *Am. J. Physiol.* 238 (*Gastrointest. Liver Physiol.* 1): G509–G513, 1980.

EKSTROM, J. Sensitization of the rat parotid gland to secretagogues following either parasympathetic denervation or sympathetic denervation or decentralization. *Acta Physiol. Scand.* 108: 253–261, 1980.

EKSTROM, J., AND D. TEMPLETON. Difference in sensitivity of parotid glands brought about by disuse and overuse. *Acta Physiol. Scand.* 101: 329–335, 1977.

ELDER, J. B., G. GILLESPIE, E. H. G. CAMPBELL, J. E. GILLESPIE, G. P. CREAN, AND A. W. KAY. The effect of vagotomy on the lower part of the acid dose-response curve to pentagastrin in man. *Clin. Sci.* 43: 193–200, 1972.

EL-SHARKAWY, T. Y., AND E. E. DANIEL. Ionic mechanisms of intestinal electrical control activity. *Am. J. Physiol.* 229: 1287–1288, 1975.

EMAS, S. Vagal influences on gastric acid secretion. *Scand. J. Gastroenterol.* 8: 1–4, 1973.

EMMELIN, N. Nervous control of salivary glands. In: *Handbook of Physiology*, sect. 6: *Alimentary Canal*, vol. 2: *Secretion*, edited by C. F. Code. Washington, DC: Am. Physiol. Soc. 1967, p. 595–632.

EMMELIN, N. Secretory mechanisms. In: *Scientific Foundations of Gastroenterology*, edited by W. Sircus and A. N. Smith. London: Wm. Heinemann Medical Books, 1980, p. 219–225.

EMMELIN, N. Nervous control of mammalian salivary glands. *Phil. Trans. R. Soc. Lond. B* 296: 27–35, 1981.

ERLINGER, S. Hepatocyte bile secretion: current views and controversies. *Hepatology* 1: 352–359, 1981.

ERLINGER, S. Secretion of bile. In: *Diseases of the Liver*, 5th ed., edited by L. Schiff and E. R. Schiff. Philadelphia: J. B. Lippincott, 1982a, p. 93–118.

ERLINGER, S. Bile flow in the liver. In: *Biology and Pathobiology*, edited by I. M. Arias, H. Popper, D. Schachter, and D. Shafritz. New York: Raven Press, 1982b, p. 407–427.

ERTAN, A., F. P. BROOKS, D. ARVAN, AND C. N. WILLIAMS. Mechanism of release of endogenous cholecystokinin by jejunal amino acid perfusion in man. *Am. J. Digest. Dis.* 20: 813–823, 1975.

EVERSON, G. T., C. McKINLEY, M. LAWSON, M. JOHNSON, AND F. KERN, JR. Gallbladder function in the human female: effect of the ovalatory cycle, pregnancy and contraceptive steroids. *Gastroenterology* 82: 711–719, 1982.

FAHRENKRUG, J., A. HAGLUND, M. JODAL, O. LUNDGREN, L. OLBE, AND O. B. SCHAFFALITZKY DE MUCKADELL. Nervous release of vasoactive intestinal polypeptide in the gastrointestinal tract of cats: possible functional implications. *J. Physiol. (London)* 284: 291–305, 1978.

FAICHNEY, A., W. Y. CHEY, Y. C. KIM, K. Y. LEE, M. S. KIM, AND T. M. CHANG. Effect of sodium oleate on plasma secretin concentration and pancreatic secretion in dog. *Gastroenterology* 81: 458–462, 1981.

FAROOQ, O., R. A. L. STURDEVANT, AND J. I. ISENBERG. Comparison of synthetic and natural porcine secretins on human pancreatic secretion. *Gastroenterology* 66: 204–209, 1974.

FAROOQ, O., AND J. I. ISENBERG. Effect of continuous intravenous infusion of insulin versus rapid intravenous injection of insulin on gastric acid secretion in man. *Gastroenterology* 68: 683–686, 1975.

FAVUS, M. J., S. C. KATHPALIA, AND F. L. COE. Kinetic characteristics of calcium absorption and secretion by rat colon. *Am. J. Physiol.* 240 (*Gastrointest. Liver Physiol.* 3): G350–G354, 1981.

FELDMAN, M., AND Y. M. COWLEY. Effect of an opiate antagonist (Nalaxone) on the gastric acid secretory response to sham feeding, pentagastrin and histamine in man. *Digest. Dis. Sci.* 27: 308–310, 1982.

FELDMAN, M., AND C. T. RICHARDSON. Gastric acid secretion in humans. In: *Physiology of the Gastrointestinal Tract*, edited by L. R. Johnson. New York: Raven Press, 1981, p. 693–707.

FELDMAN, M., AND J. H. WALSH. Acid inhibition of sham feeding-stimulated gastrin release and gastric acid secretion: effect of atropine. *Gastroenterology* 78: 772–776, 1980.

FENG, Y-S., C. A. ARONCHICK, AND F. P. BROOKS. Decortication and decerebration effects compared on efferent electrical vagal stimulation of the stomach in cats. *Clin. Res.* 31: 530A, 1983.

FERACCI, H., A. BERNASOC, J. P. GORVEL, AND S. MAROUX. Localization by immunofluorescence and histochemical labeling of aminopeptidase N in relation to its biosynthesis in rabbit and pig enterocytes. *Gastroenterology* 82: 317–324, 1982.

FIELD, M. Secretion of electrolytes and water by mammalian small intestine. In: *Physiology of the Gastroinetesinal Tract*, edited by L. R. Johnson. New York: Raven Press, 1981, p. 963–982.

FIELD, M., AND I. McCOLL. Ion transport in rabbit in rabbit ileal mucosa. III. Effect of catecholamines. *Am. J. Physiol.* 225: 852–857, 1973.

FINCH, P. M., D. M. INGRAM, J. D. HENSTRIDGE, AND B. N. CATCHPOLE. Relationship of fasting gastroduodenal motility to the sleep cycle. *Gastroenterology* 83: 605–612, 1982.

FISHER, K. A., H. J. BINDER, AND J. P. HAYSLETT. Potassium secretion by colonic mucosal cells after potassium adaptation. *Am. J. Physiol.* 231: 987–994, 1976.

FISHER, M., AND J. N. HUNT. Effects of hydrochlorides of amino acids in test meals on gastric emptying. *Digestion* 16: 18–22, 1977.

FISHER, R. B., AND J. N. HUNT. The inorganic components of gastric secretion. *J. Physiol. (London)* 111: 138–149, 1950.

FISHER, R. S., G. S. ROBERTS, C. J. GRABOWSKI, AND S. COHEN. Inhibition of lower esophageal sphincter circular muscle by female sex hormones. *Am. J. Physiol.* 234 (*Endocrinol. Metab. Gastrointest. Physiol.* 3): E243–E247, 1978.

FISHER, R. S., F. STETZER, E. ROCK, AND L. S. MALMUD. Abnormal gallbladder emptying in patients with gallstones. *Digest. Dis. Sci.* 27: 1019–1024, 1982.

FLECKENSTEIN, P. Migrating electrical spike activity in the fasting human small intestine. *Digest. Dis. Sci.* 23: 769–775, 1978.

FLECKENSTEIN, P., AND A. OIGAARD. Electrical spike activity in the human small intestine: a multiple electrode study of fasting and diurnal variations. *Am. J. Digest. Dis.* 23: 776–780, 1978.

FLEMSTROM, G. Gastric secretion of bicarbonate. *Physiology of the Gastrointestinal Tract*, edited by L. R. Johnson. New York: Raven Press, 1981, p. 603–616.

FLEMSTROM, G., AND A. GARNER. Gastroduodenal HCO_3^- transport: characteristics and proposed role in activity regulation and mucosal protection. *Am. J. Physiol.* 242 (*Gastrointest. Liver Physiol.* 5): G182–G193, 1982.

FLEMSTROM, G., J. R. HEYLINGS, AND A. GARNER. Gastric and duodenal HCO_3^- transport in vitro: effects of hormones and local transmitters. *Am. J. Physiol.* 242 (*Gastrointest. Liver Physiol.* 5): G100–G110, 1982a.

FLEMSTROM, G., A. GARNER, O. NYLANDER, B. C. HURST, AND J. R. HEYLINGS, JR. Surface epithelial HCO_3^- transport by mammalian duodenum in vitro. *Am. J. Physiol.* 243 (*Gastrointest. Liver Physiol.* 6): G348–G358, 1982b.

FOLKOW, B. Regional adjustments of intestinal blood flow. *Gastroenterology* 52: 423–432, 1967.

FOLSCH, U. R., AND W. CREUTZFELDT. Pancreatic duct cells in rats: secretory studies in response to secretin cholecystokinin-pancreozymin and gastrin in vivo. *Gastroenterology* 73: 1053–1059, 1977.

FOLSCH, U. R., H. FISHER, H. D. SOLING, AND W. CREUTZFELDT. Effects of gastrointestinal hormones and carbamylcholine on cAMP accumulation in isolated pancreatic duct fragments from the rat. *Digestion* 20: 277–292, 1980.

FORDTRAN, J. S., AND F. J. INGLEFINGER. Absorption of water, electrolytes, and sugars from the human gut. In: *Handbook of Physiology*, sect. 6: *Alimentary Canal*, vol. 3: *Absorption*, edited by C. F. Code. Washington, DC: Am. Physiol. Soc., 1968, p. 1457–1490.

FORDTRAN, J. S., AND T. W. LOCKLEAR. Ionic constituents and osmolarity of gastric and small intestinal fluids after eating. *Am. J. Digest. Dis.* 11: 503–521, 1966.

FORDTRAN, J. S., AND J. H. WALSH. Gastric acid secretion rate and buffer content of the stomach after eating. *J. Clin. Invest.* 52: 645–657, 1973.

FORDTRAN, J. S., F. C. RECTOR, J. R., AND N. W. CARTER. The mechanisms of sodium absorption in human small intestine. *J. Clin. Invest.* 47: 884–900, 1968.

FORSTNER, G. (^{1-14}C) glucosamine incorporation by subcellular fractions of intestinal mucosa. *J. Biol. Chem.* 245: 3584–3592, 1970.

FORSTNER, G., M. SHIH, AND B. LUKIE. Cyclic AMP and intestinal glycoprotein synthesis: the effect of β adrenergic agents, theophylline and dibutyryl cyclic AMP. *Can. J. Physiol. Pharmacol.* 51: 122–129, 1973.

FORSTNER, J. F. Intestinal mucins in health and disease. *Digestion* 17: 234–263, 1978.

FORSTNER, J. F., N. W. ROOMI, R. E. F. FAHIM, AND G. G. FORSTNER. Cholera toxin stimulates secretion of immunoreactive intestinal mucin. *Am. J. Physiol.* 240 (*Gastrointest. Liver Physiol.* 3): G10–G16, 1981.

FORTE, J. G. Three components of Cl^- flux across isolated bullfrog gastric mucosa. *Am. J. Physiol.* 216: 167–714, 1969.

FORTE, J. G., J. A. BLACK, T. M. FORTE, T. E. MACHEN, AND J. M. WOLOSIN. Ultrastructural changes related to functional activity in gastric oxyntic cells. *Am. J. Physiol.* 241 (*Gastrointest. Liver Physiol.* 4): G349–358, 1981.

FREEL, R. W., AND A. M. GOLDNER. Sodium-coupled nonelectrolyte transport across epithelia: emerging concepts and directions. *Am. J. Physiol.* 241 (*Gastrointest. Liver Physiol.* 4): G451–460, 1981.

FRENCH, S. W., I. KONDO, T. IRIE, T. J. IHRIG, N. BENSON, AND R. MUNN. Morphologic study of intermediary filaments in rat hepatocytes. *Hepatology* 2: 29–38, 1982.

FREXINOS, J., M. CARBALLIDO, A. LOUIS, AND A. RIBET. Effects of pentagastrin-stimulation on human parietal cells. An electron microscopic study with quantitative evaluation of cytoplasmic structures. *Am. J. Digest. Dis.* 16: 1065–1074, 1971.

FRIEDMAN, G., J. D. WAYE, L. A. WEINGARTEN, AND H. D. JANOWITZ. The pattern of simultaneous intraluminal pressure changes in the human proximal small intestine. *Gastroenterology* 47: 258–268, 1964.

FRIEDMAN, G., B. S. WOLF, J. D. WAYE, AND H. D. JANOWITZ. Correlation of cine radiographic and intraluminal pressure changes in the human duodenum: an analysis of the functional significance of monophasic waves. *Gastroenterology* 49: 37–49, 1965.

FRIEDMAN, M. H. F. The entero-enteric reflexes. In: *Functions of the Stomach and Intestine*, edited by M. H. F. Friedman, Baltimore: University Park Press, 1975, p. 57–74.

FRITZ, M. E. Cationic dependence of resting membrane potentials of parotid acinar cells. *Am. J. Physiol.* 223: 644–647, 1972.

FRIZZELL, R. A. Active chloride secretion by rabbit colon: calcium dependent stimulation by ionophore A23187. *J. Membrane Biol.* 35: 175–187, 1977.

FRIZZELL, R. A., AND K. HEINTZE. Electrogenic chloride secretion by mammalian colon. In: *Mechanisms of Intestinal Secretion*, edited by H. J. Binder. New York: Alan R. Liss, 1979, p. 101–110.

FURNESS, J., AND M. COSTA. Types of nerves in the enteric nervous system. *Neuroscience* 5: 1–20, 1980.

GABELLA, G. Structure of muscles and nerves in the gastrointestinal tract. In: *Physiology of the Gastrointestinal Tract*, edited by L. R. Johnson. New York: Raven Press, 1981, p. 197–241.

GAGNON, D. J., G. DEVROEDE, AND S. BELISLE. Excitatory effects of adrenaline upon isolated preparations of human colon. *Gut* 13: 645–657, 1972.

GALLACHER, D. V. Parotid acinar cells: membrane potential changes induced by electrical field stimulation. *J. Physiol.* 295: 46P, 1979.

GALLACHER, D. V., AND O. H. PETERSEN. Electrophysiology of parotid acini: effects of electrical field stimulation and ionophoresis of neurotransmitters. *J. Physiol. (London)* 305: 43–57, 1980a.

GALLACHER, D. V., AND O. H. PETERSEN. Substance P increases membrane conductance in parotid acinar cells. *Nature* 283: 393–395, 1980b.

GARDNER, J. D., T. P. CONLON, H. C. BEYERMAN, AND A. VAN ZON. Interaction of synthetic 10-Tyrosil analogues of secretin with hormone receptors on pancreatic acinar cells. *Gastroenterology* 73: 52–56, 1977.

GARNER, J. D., AND R. T. JENSEN. Regulation of pancreatic enzyme secretion in vitro. In: *Gastrointestinal Physiology*, edited by L. R. Johnson. New York: Raven Press, 1981, p. 831–872.

GARDNER, J. D., M. D. WALKER, AND A. S. ROTTMAN. Effect of A23187 on amylase release from dispersed acini prepared from guinea pig pancreas. *Am. J. Physiol.* 238: (*Gastrointest. Liver Physiol.* 1): G458–G466, 1980.

GARDNER, J. D., L. Y. KORMAN, M. D. WALKER, AND V. E. SUTLIFF. Effects of inhibitors of cyclic nucleotide phosphodiesterase on the actions of vasoactive intestinal peptide and secretin on pancreatic acini. *Am. J. Physiol.* 242 (*Gastrointestinal. Liver Physiol* 5): G547–G551, 1982.

GARRETT, J. R. Innervation of salivary glands, morphological considerations. In: *Secretory Mechanisms of Exocrine Glands, Alfred Benzon Symposium VII*, edited by N. A. Thorn and O. H. Petersen. Copenhagen: Munksgaard, 1974, p. 17–27.

GARRETT, J. R. Recent advances in physiology of salivary glands. *Brit. Med. Bull.* 31: 152–155, 1975.

GARRETT, J. R., AND A. KIDD. Effects of secretory nerve stimulation on acid phosphatase and peroxidase in submandibular saliva and acini in cats. *Histochem. J.* 9: 435–451, 1977.

GARRETT, J. R., P. ALM, AND S. LENNINGER. Smooth muscle of the pancreatic duct of the cat and its innervation. *Experientia* 29: 842–844, 1973.

GARRETT, J. R., T. J. HARROP, A. KIDD, AND A. THULIN. Nerve-induced secretory changes in salivary glands. In: *Nerves and the Gut*, edited by F. P. Brooks and P. W. Evers. Thorofare, NJ: Charles B. Slack, 1977, p. 14–40.

GASKIN, K. J., P. R. DURIE, R. E. HILL, L. M. LEE, AND G. G. FORSTNER. Colipase and maximally activated pancreatic lipase in normal subjects and patients with steatorrhea. *J. Clin. Invest.* 69: 427–434, 1982.

GEALL, M. G., S. F. PHILLIPS, AND W. H. J. SUMMERSKILL. Profile of gastric potential difference in man: effects of aspirin, alcohol, bile and endogenous acid. *Gastroenterology* 58: 437–443, 1970.

GEENAN, J. E., W. J. HOGAN, W. J. DODDS, E. T. STEWART, AND R. C. ARNDORFER. Intraluminal pressure recording from the human sphincter of Oddi. *Gastroenterology* 78: 317–324, 1980.

GERHARDT, D. C., T. J. SHUCK, R. A. BORDEAUX, AND D. H. WINSHIP. Human upper esophageal sphincter: response to volume, osmotic and acid stimuli. *Gastroenterology* 75: 268–274, 1978.

GIDDA, J. S., AND R. K. GOYAL. Influence of vagus nerves on electrical activity of opossum small intestine. *Am. J. Physiol.* 239 (*Gastrointest. Liver Physiol.* 2): G406–410, 1980.

GILLESPIE, J. S., AND J. D. MAXWELL. Adrenergic innervation of sphincteric and non-sphincteric smooth muscle in the rat intestine. *J. Histochem. Cytochem.* 19: 676–681, 1971.

GJORSTRUP, P. Parotid secretion of fluid and amylase in rabbits during feeding. *J. Physiol.* (*London*) 309: 101–116, 1980.

GLASINOVIC, J. C., M. DUMONT, M. DUVAL, AND S. ERLINGER. Hepatocellular uptake of bile acids in the dog: evidence for a common carrier-mediated transport system. *Gastroenterology* 69: 973–981, 1975.

GO, V. L. W., A. F. HOFMANN, AND W. H. J. SUMMERSKILL. Pancreozymin bioassay in man based on pancreatic enzyme secretion: potency of specific amino acids and other digestive products. *J. Clin. Invest.* 49: 1558–1564, 1970.

GOEBELL, H., CH. STEFFEN, G. BALTZER, K. A. SCHLOTT, AND CH. BODE. Stimulation of pancreatic enzyme secretion by acute hypercalcemia. *Dtsch. Med. Wochenschr.* 97: 300–301, 1972a.

GOEBELL, H., CH. STEFFEN, AND CH. BODE. Stimulatory effect of pancreozymin-cholecystokinin on calcium secretion in pancreatic juice of dogs. *Gut* 13: 477–482, 1972b.

GOLDFINE, I. D., B. M. KRIZ, K. Y. WONG, G. HIADEK, A. L. JONES, AND J. A. WILLIAMS. Insulin action in pancreatic acini from steptozotocin-treated rats. IV. Electron microscope autoradiography of ^{125}I-insulin. *Am J. Physiol.* 240 (*Gastrointest. Liver Physiol.* 3): G69–G75, 1981.

GOLLAN, J. L., AND R. SCHMID. Bilirubin update: formation, transport and metabolism. In: *Progress in Liver Diseases*, edited by H. Popper and F. Schaffner. New York: Grune & Stratton, 1982, vol. 7, p. 261–284.

GONZALEZ, M. D., E. SUTHERLAND, AND F. R. SIMON. Regulation of hepatic transport of bile salts. *J. Clin. Invest.* 63: 684–694, 1979.

GOODHEAD, B., H. S. HIMAL, AND J. ZANBILOWICZ. Relationship between pancreatic secretion and pancreatic blood flow. *Gut* 11: 62–68, 1970.

GORELICK, F. S., AND J. D. JAMIESON. Structure-function relationships of the pancreas. In: *Physiology of the Gastrointestinal Tract*, edited by L. R. Johnson. New York: Raven Press, 1981, p. 773–794.

GOTTSCHALK, A. Glycoproteins: their composition, structure and function. Amsterdam: Elsevier, 1966.

GOYAL, R. K., AND S. RATTAN. Nature of the vagal inhibitory innervation to the lower esophageal sphincter. *J. Clin. Invest.* 55: 1119–1126, 1975.

GOYAL, R. K., AND S. RATTAN. Genesis of basal sphincter pressure: effect of tetrodotoxin on lower esophageal sphincter pressure in opposum in vivo *Gastroenterology* 71: 62–67, 1976.

GRAND, R. J., AND P. R. GROSS. Independent stimulation of secretion and protein synthesis in rat parotid gland. *J. Biol. Chem.* 244: 5608–5615, 1969.

GRANGER, D. N., N. A. MORTILLARO, AND A. E. TAYLOR. Interactions of intestinal lymph flow and secretion. *Am. J. Physiol.* 232: E13–E18, 1977.

GRAY, G. M. Carbohydrate digestion and absorption. *Gastroenterology* 58: 96–107, 1970.

GRAY, G. M. Carbohydrate digestion and absorption: role of a the small intestine. *N. Engl. J. Med.* 292: 1225–1230, 1975.

GRAY, G. M. Carbohydrate absorption and malabsorption. In: *Physiology of the Gastrointestinal Tract*, edited by L. R. Johnson. New York: Raven Press, 1981, p. 1063–1072.

GRAY, G. M. Mechanisms of digestion and absorption of food. In: *Gastroenterology*, 3rd ed., edited by M. Slessinger and J. S. Fordtran. Philadelphia: W. B. Saunders, 1983.

GRAY, G. M., AND H. L. COOPER. Protein digestion and absorption. *Gastroenterology* 61: 535, 1971.

GRAY, T. K., P. BRANNAN, D. JUAN, S. G. MORAWSKI, AND J. S. FORDTRAN. Ion transport changes during calcitonin-induced intestinal secretion in man. *Gastroenterology* 71: 392–398, 1976.

GREENWELL, J. R., AND T. SCRATCHERD. The kinetics of pancreatic amylase secretion and its relationship to volume flow and electrical conductance in the anesthetized cat. *J. Physiol.* (*London*) 239: 443–457, 1974.

GREENWELL, R., A. A. HARPER, AND T. SCRATCHERD. Effects of splanchnic nerve stimulation on pancreatic secretion, blood flow and electrical conductance. *Gut* 8: 635, 1967.

GREGORY, D. H., Z. R. VLAHCEVIC, M. F. PRUGH, AND L. SWELL. Mechanism of secretion of biliary lipids: role of a microtubular system in hepatocellular transport of biliary lipids in the rat. *Gastroenterology* 74: 93–10, 1978.

GREIDER, M. H., V. STEINBERG, AND J. E. McGUIGAN. Electron microscopic identification of the gastrin cell of the human antral mucosa by means of immunochemistry. *Gastroenterology* 63: 572–583, 1972.

GULDVOG, I., AND D. GEDDE-DAHL. Comparison of physiological and pharmacological stimulation of acid secretion in vagally innervated and denervated gastric pouches in the same dog. *Scand. J. Gastroenterology.* 15: 929–937, 1980.

GULLO, L., P. PRIORI, AND G. LABO. Influence of adrenal cortex on exocrine pancreatic function. *Gastroenterology* 83: 92–86, 1982.

GUMUCIO, J. J., AND D. L. MILLER. Functional implications of liver cell heterogeneity. *Gastroenterology* 80: 393–403, 1981.

GUTH, P. H., AND W. BALLARD. Physiology of the gastric circulation. In: *Physiology of the Gastrointestinal Tract*, edited by L. R. Johnson. New York: Raven Press, 1981, p. 709–731.

GUTH, P. H., AND E. SMITH. Nervous regulation of the gastric microcirculation. In: *Nerves and the Gut*, edited by F. P. Brooks and P. W. Evers. Thorofare, NJ: Charles B. Slack, 1977, p. 365–376.

GUTH, P. H., H. BAUMANN, M. I. GROSSMAN, D. AURES, AND J. ELASHOFF. Measurements of gastric mucosal blood flow in man. *Gastroenterology* 74: 831–834, 1978.

GUZELIAN, P., AND J. L. BOYER. Glucose reabsorption from bile. *J. Clin. Invest.* 53: 526–535, 1974.

GUZMAN, S., J. A. CHAYVIALLE, W. A. BANKS, P. L. RAYFORD, AND J. C. THOMPSON. Effect of vagal stimulation on pancreatic secretion and on blood levels of gastrin, cholecystokinin secretin, vasoactive intestinal peptide and somatostatin. *Surgery* 86: 329–336, 1979.

HAEGEL, P., C. STOCK, J. MARESCAUX, B. PETIT, AND J. F. GRENIER. Hyperplasia of the exocrine pancreas after small bowel resection in the rat. *Gut* 22: 207–212, 1981.

HAKIM, H. A., C. B. PAPLEAX, J. B. LANE, N. LIFSON, AND M. E. YABLONSKI. Mechanism of production of intestinal secretion by negative luminal pressure. *Am. J. Physiol.* 233 (*Endocrinol. Metab. Gastrointest. Physiol.* 2): E416–E421, 1979.

HALL, K. E., T. Y. EL-SHARKAWY, AND N. E. DIAMANT. Vagal control of migrating motor complex in the dog. *Am. J. Physiol.* 243 (*Gastrointest. Liver Physiol.* 6): G276–284, 1982.

HALLBERG, L., E. BJORN-RASMUSSEN, L. HOWARD, AND L. ROSSANDER. Dietary heme iron absorption—a discussion of possible mechanisms for the absorption-promoting effect of meat and for the regulation of iron absorption. *Scand. J. Gastroenterol.* 14: 769–780, 1979.

HALLGREN, R., G. LUNDQUIST, AND R. E. CHANCE. Serum levels of human pancreatic polypeptide in renal disease. *Scand. J. Gastroenterol.* 12: 923–927, 1977.

HAMOSH, M. A review: fat digestion in the newborn: role of lingual lipase and preduodenal digestion. *Pediat. Res.* 13: 615–622, 1979.

HANKER, J. S., E. J. TAPPER, AND W. W. AMBROSE. Enzyme cytochemical correlates of the nervous control of the gastric parietal cell. In: *Nerves and the Gut*, edited by F. P. Brooks and P. W. Evers. Thorofare, NJ: Charles B. Slack, 1977, p. 65–68.

HANSEN, O. H., T. PEDERSEN, J. K. LARSEN, AND J. F. REHFELD. Effect of gastrin on gastric mucosal cell proliferation in man. *Gut* 77: 536–541, 1976.

HANSKY, J., O. M. TISCORNIA, D. A. DREILING, AND H. D. JANOWITZ. Relationship between maximal secretory output and weight of the pancreas in the dog. *Proc. Soc. Exp. Biol. Med.* 114: 654–656, 1963.

HARA, Y., AND J. H. SZURSZEWSKI. Mechanical and intracellular electrical activity of smooth muscle of the canine colon. *Gastroenterology* 80: 1169, 1981.

HARDISON, W. G. M., AND C. A. WOOD. Importance of bicarbonate in bile salt independent fraction of bile flow. *Am. J. Physiol.* 235 (*Endocrinol. Metab. Gastrointest. Physiol.* 4): E158–E164, 1978.

HARDWICKE, J., J. G. RANKIN, K. J. BAKER, AND R. PREISIG. The loss of protein in human and canine hepatic bile. *Clin. Sci.* 26: 509–517, 1964.

HARPER, A. A., C. KIDD, AND T. SCRATCHERD. Vago-vagal reflex effects on gastric and pancreatic secretion and gastrointestinal motility. *J. Physiol. (London)* 148: 417–436, 1959.

HARPER, A. A., J. D. REED, AND J. R. SMY. Gastric blood flow in anesthetized cats. *J. Physiol. (London)* 194: 795–807, 1968.

HARRIS, L. D., AND B. T. JACKSON. When to hospitalize for hiatal hernia. *Hosp. Pract.* p. 47–53, April 1968.

HARTY, R. F., AND J. E. McGUIGAN. Effect of carbachol and atropine on gastrin secretion and synthesis in rat antral organ culture. *Gastroenterology* 78: 925–930, 1980.

HARTY, R. F., J. C. van der VYVER, AND J. E. McGUIGAN. Stimulation of gastrin secretion and synthesis in antral organ culture. *J. Clin. Invest.* 60: 51–60, 1977.

HARTY, R. F., D. G., MAICO, AND J. E. McGUIGAN. Role of calcium in antral gastrin release. *Gastroenterology* 80: 491–497, 1981a.

HARTY, R. F., D. G. MAICO, AND J. E. McGUIGAN. Somatostatin inhibition of basal and carbachol-stimulated gastrin release in rat antral organ culture. *Gastroenterology* 81: 707–712, 1981b.

HASSEN, L. E., M. OSNES, AND J. MYREN. Pancreatic secretion obtained by endoscopic cannulation of the main pancreatic duct and secretion release after duodenal acidification in man. *Scand. J. Gastroenterol.* 13: 325–330, 1978.

HATOFF, D. E., AND W. G. M. HARDISON. Bile-acid-dependent secretion of alkaline phosphatase in rat bile. *Hepatology* 2: 433–439, 1982.

HAWKER, P. C., K. E. MASHITER, AND L. A. TURNBERG. Mechanisms of transport of Na, Cl, and K in the human colon. *Gastroenterology* 74: 1241–1247, 1978.

HAYSLETT, J. P., J. HALEVY, P. E. PACE, AND H. J. BINDER. Demonstration of net potassium absorption in mammalian colon. *Am. J. Physiol.* 242 (3) (*Gastrointest. Liver Physiol.* 5): G209–G214, 1982.

HEGARTY, J. E., P. D. FAIRCLOUGH, M. L. CLARK, AND A. M. DAWSON. Jejunal water and electrolyte secretion induced by L-arginine in man. *Gut* 22: 108–113, 1981.

HEGARTY, J. E., P. D. FAIRCLOUGH, K. J. MORIARTY, M. J. KELLY, AND M. L. CLARK. Effects of concentration on in vivo absorption of a peptide containing protein hydrolysate. *Gut* 23: 304–309, 1982a.

HEGARTY, J. E., P. D. FAIRCLOUGH, K. J. MORIARTY, M. J. KELLY, AND A. M. DAWSON. Comparison of plasma and intraluminal amino acid and profiles in man after means containing a protein hydrolysate and equivalent amino acid mixture. *Gut* 23: 670–674, 1982b.

HEITZ, P. H., J. M. POLAK, S. R. BLOOM, AND A. G. E. PEARSE. Identification of the D-cell as the source of human pancreatic polypeptide (HPP). *Gut* 17: 755–758, 1976.

HELMAN, C. A., AND G. O. BARBEZAT. The effect of gastric inhibitory polypeptide on human jejunal water and electrolyte transport. *Gastroenterology* 72: 376–379, 1977.

HELMKE, K., K. FEDERLIN, P. PIAZOLO, J. STRODER, R. JESCHKE, AND H. E. FRANZ. Localization of calcium-binding protein in intestinal tissue by immunofluorescence in normal, vitamin D deficient and uraemic subjects. *Gut* 15: 875–879, 1974.

HENGELS, H. K., J. E. MULLER, T. SCHOLTEN, AND W. P. FRITSCH. Evidence for the secretion of gastrin into human gastric juice. *Gut* 21: 760–765, 1980.

HERMON-TAYLOR, J., AND C. F. CODE. Localization of the duodenal pacemaker and its role in the organization of duodenal myoelectric activity. *Gut* 12: 40–47, 1971.

HERMON-TAYLOR, J., J. PERRIN, D. A. W. GRANT, A. APPLEYARD, M. BUBEL, AND A. I. MAGEE. Immunofluorescent localization of enterokinase in human small intestine. *Gut* 18: 259–265, 1977.

HERSEY, S. J., C. S. CHEW, L. CAMPBELL, AND E. HOPKINS. Mechanism of action of SCN in isolated gastric glands. *Am. J. Physiol* 240 (*Gastrointest. Liver Physiol.* 3): G232–G238, 1981.

HERTZ, A. F., AND A. NEWTON. The normal movement of the colon in man. *J. Physiol. (London)* 57: 57–65, 1913.

HERZOG, V., AND M. G. FARQUHAR. Luminal membrane retrieved after exocytosis reaches most Golgi cisternae in secretory cells. *Proc. Natl. Acad. Sci. USA* 74: 5073–5077, 1977.

HERZOG, V., AND F. MILLER. Membrane retrieval in secretory cells. In: *Secretory Mechanisms*, edited by C. R. Hopkins and C. J. Duncan. Cambridge: Cambridge University Press, 1979, p. 101–116.

HICKS, T., AND L. A. TURNBERG. Influence of glucagon on the human jejunum. *Gastroenterology* 67: 1114–1118, 1974.

HIGGINS, J. R., JR., AND N. P. BLAIR. Intestinal transport of water and electrolytes during extracellular volume expansion in dogs. *J. Clin. Invest.* 50: 2569–2579, 1971.

HIGHTOWER, N. C. Salivary secretion. In: *Physiological Basis of Medical Practice*, 8th ed., edited by C. H. Best and N. B. Taylor, Baltimore: Williams & Wilkins, 1966.

HIGHTOWER, N. C., JR. Motor action of the small bowel. In: *Handbook of Physiology*, sect. 6: *Alimentary Canal*, vol. 4: *Motility*, edited by C. F. Code. Washington, DC: Am. Physiol. Soc., 1968, p. 2001–2024.

HIGHTOWER, N. C., AND H. D. JANOWITZ. Salivary secretion. In: *Best & Taylor's Physiological Basis of Medical Practice*, 10th ed., edited by J. R. Brobeck. Baltimore: Williams & Wilkins, 1979, p. 2–24–23–37.

HINDER, R. A., AND K. A. KELLY. Human gastric pacesetter potential: site of origin, spread and response to gastric transection and proximal gastric vagotomy. *Am. J. Surg.* 133: 29–33, 1977.

HIRANO, T., AND A. NIIJIMA. Effects of 2-deoxy-D-glucose, glucose and insulin on efferent activity in gastric vagus nerve. *Experientia* 36: 1197–1198, 1980.

HIRSCHOWITZ, B. I. Secretion of pepsinogen. In: *Handbook of Physiology*, sect. 6: *Alimentary Canal*, vol. 2: *Secretion*, edited by C. F. Code. Washington, DC: Am. Physiol. Soc., 1967, p. 889–918.

HIRSCHOWITZ, B. I. The vagus and gastric secretion. In: *Nerves and the Gut*, edited by F. P. Brooks and P. W. Evers, Thorofare, NJ: Charles B. Slack, 1977, p. 96–118.

HIRSCHOWITZ, B. I. AND R. G. GIBSON. Stimulation of gastrin release and gastric secretion: effect of bombesin and a nonapeptide in fistula dogs with and without fundic vagotomy. *Digestion* 18: 227–239, 1978.

HIRSCHOWITZ, B. I., AND R. G. GIBSON. Augmented vagal release of antral gastrin by 2-deoxy-glucose after fundic vagotomy in dogs. *Am. J. Physiol.* 236 (*Endocrinol. Metab. Physiol.* 5): E173–E179, 1979.

HIRSCHOWITZ, B. I., AND C. A. HELMAN. Effects of fundic vagotomy and cholinergic replacement on pentagastrin dose responsive gastric acid and pepsin secretion in man. *Gut* 23: 675–682, 1982.

HO, J. J. L., AND S. S. ROTHMAN. Intracellular osmotic events accompanying protein secretion by the exocrine pancreas. *Am. J. Physiol.* 238: (*Gastrointest. Liver Physiol.* 1): G289–G297, 1979.

HO, J. J. L., AND S. S. ROTHMAN. Nature of flow dependence of protein secretion by the exocrine pancreas. *Am. J. Physiol.* 242 (*Gastrointest. Liver Physiol.* 5): G32–G39, 1982.

HOGAN, W. J., W. J. DODDS, J. TOOULI, J. E. GEENAN, J. HELM, AND R. C. ARNDORFER. Motility of the choledochoduodenal junction. In: *Motility of the Digestive Tract*, edited by M. Wienbeck. New York: Raven Press, 1982, p. 387–396.

HOGBEN, C. A. M., T. H. KENT, P. A. WOODWARD, AND A. J. SILL. Quantitative histology of the gastric mucosa: man, dog, cat, guinea pig and frog. *Gastroenterology* 67: 1143–1154, 1974.

HOLDSTOCK, D. J., J. J. MISIEWICZ, T. SMITH, AND E. N. ROWLANDS. Propulsion (mass movements) in the human colon and its relationship to meals and somatic activity. *Gut* 11: 91–99, 1970.

HOLLANDER, F. Current views on the physiology of the gastric secretions. *Amer. J. Med.* 13: 52–63, 1952.

HOLMGREN, J., S. LANGE, AND I. LONROTH. Reversal of cyclic AMP-mediated intestinal secretion in mice by chlorpromazine. *Gastroenterology* 75: 1103–1108, 1978.

HONDA, R., J. TOOULI, W. J. DODDS, S. SARNA, W. J. HOGAN, AND Z. ITOH. Relationship of sphincter of Oddi spike bursts to gastrointestinal myoelectric activity in conscious opossums. *J. Clin. Invest.* 69: 770–778, 1982.

HOPKINS, C. R. The secretory process. In: *Outline in Secretory Mechanisms*, edited by C. R. Hopkins and C. J. Duncan. Cambridge: Cambridge University Press, 1979, p. 1–8.

HUBEL, K. A. Effect of luminal chloride concentration on bicarbonate secretion in rabbit ileum. *Am. J. Physiol.* 217: 40–45, 1969.

HUBEL, K. A. Response of rabbit pancreas in vitro to adrenergic agonist and antagonists. *Am. J. Physiol.* 219: 1590–1594, 1970.

HUBEL, K. A., AND D. CALLANAN. Effects of Ca^{2+} on ileal transport and electrically induced secretion. *Am. J. Physiol* 239 (*Gastrointest. Liver Physiol.* 2): G18–G22, 1980.

HUBEL, K. A., AND S. SHIRAZI. Human ileal ion transport in vitro: changes with electrical field stimulation and tetrodotoxin. *Gastroenterology* 83: 63–68, 1982.

HUKUHARA, T., S. SUMI, AND S. KOTANI. Comparative studies on the intestinal reflexes in rabbits, guinea pigs, and dogs. *Jap. J. Physiol.* 11: 205–211, 1961.

HUMPHREYS, M. H., AND L. E. EARLEY. The mechanism of decreased intestinal sodium and water absorption after acute volume expansion in the rat. *J. Clin. Invest.* 50: 2355–2367, 1971.

HUNT, J. N., AND M. T. KNOX. A relation between the chain length of fatty acids and the slowing of gastric emptying. *J. Physiol.* (*London*) 194: 327–336, 1968.

HUNT, J. N., AND M. T. KNOX. The slowing of gastric emptying by nine acids. *J. Physiol.* (*London*) 201: 161–179, 1969.

HUNT, J. N., AND W. R. SPURRELL. The pattern of emptying of the human stomach. *J. Physiol.* (*London*) 113: 157–168, 1951.

HUNTER, G. C., J. GOLDSTONE, R. VELLA, AND L. W. WAY. Effect of vagotomy upon intragastric redistribution of microvascular flow. *J. Surg. Res.* 26: 314–319, 1979.

HURWITZ, A. L., A. DURANCEAU, AND J. K. HADDAD. *Disorders of Esophageal Motility*. Philadelphia: W.B. Saunders, 1979, p. 1–179.

INOUE, M., R. KINNE, T. TRAN, AND I. M. ARIAS. Taurocholate transport by cat liver sinusoidal membrane vesicles: evidence of sodium cotransport. *Hepatology* 2: 572–579, 1982.

IRELAND, P., AND J. S. FORDTRAN. Effect of dietary calcium and age on jejunal calcium absorption in humans studied by intestinal perfusion. *J. Clin. Invest.* 52: 2672–2681, 1973.

IRVINE, G. B., AND R. F. MURPHY. Multiple forms of gastroenteropancreatic hormones. *Gut* 22: 1048–1069, 1981.

ISACCS, P. E. T., C. L. CORBETT, A. K. RILEY, P. C. HAWKER, AND L. A. TURNBERG. In vitro behavior of human intestinal mucosa: the influence of acetylchloride on ion transport. *J. Clin. Invest.* 58: 535–542, 1976.

ISENBERG, J. L., IPPOLITI, A. F., AND V. L. MAXWELL. Perfusion of the proximal small intestine with peptone stimulates gastric acid secretion in man. *Gastroenterology* 73: 746–752, 1977.

ITO, S. Functional gastric morphology. In: *Physiology of the Gastrointestinal Tract*, edited by L. R. Johnson. New York: Raven Press, 1981, p. 517–550.

ITOH, Z., AND I. TAKAHASHI. Periodic contractors of the canine gallbladder during the interdigestive state. *Am. J. Physiol.* 240: (*Gastrointest. Liver Physiol.* 3): G183–G189, 1981b.

ITOH, Z., S. TAKEUCHI, I. AIZAWA, AND R. TAKAYANAGI. Characteristic motor activity of the gastrointestinal tract in fasted conscious dogs measured by implanted force transducers. *Digest. Dis. Sci.* 23: 229–238, 1978a.

ITOH, Z., I. AIZAWA, R. HONDA, K. HIWATASHI, AND E. F. COUCH. Control of lower-esophageal sphincter contactile activity by motilin in conscious dogs. *Dig. Dis. Sci.* 23: 341–345, 1978b.

ITOH, Z., R. HONDA, AND K. KIWATASHI. Biphasic secretory response of exocrine pancreas to feeding. *Am. J. Physiol.* 238: (*Gastrointest. Liver Physiol.* 1): G332–G337, 1980.

ITOH, Z., I. TAKAHASHI, M. NAKAYA, AND T. SUZUKI. Variation in canine exocrine pancreatic secretory activity during the interdigestive state. *Am. J. Physiol.* 241 (*Gastrointest. Liver Physiol.* 4): G98–G103, 1981a.

ITOH, Z., I. AIZAWA, AND S. TAKEUCHI. Neural regulation of interdigestive motor activity in canine jejunum. *Am. J. Physiol.* 240: (*Gastrointest. Liver Physiol.* 3): G324–330, 1981b.

ITOH, Z., I. TAKAHASHI, AND T. SUZUKI. Contractile patterns of the gallbladder between meals in the dog. In: *Motility of the Digestive Tract*, edited by M. Wienbeck, New York: Raven Press, 1982a, p. 405–413.

ITOH, Z., I. TAKAHASHI, M. NAKAYA, T. SUZUKI, H. ARAI, AND K. WAKABAYASHI. Interdigestive gallbladder bile concentration in relation to periodic contraction of gallbladder in the dog. *Gastroenterology* 83: 645–651, 1982b.

JANOWITZ, H. D., F. HOLLANDER, D. ORRINGER, M. H. LEVY, A. WINKELSTEIN, M. R. KAUFMAN, AND S. G. MARGOLIN. A quantitative study of the gastric secretory response to sham feeding in a human subject. *Gastroenterology* 16: 104–116, 1950.

JANSSON, R. Effects of gastrointestinal hormones on concentrating function and motility in the gallbladder. *Acta Physiol. Scand. Suppl.* 456: 1–38, 1979.

JANSSON, G., O. LUNDGREN, AND J. MARTINSON. Neurohormonal control of gastric blood flow. *Gastroenterology* 58: 425–429, 1970.

JARRETT, L. N., R. JACKMAN, AND G. D. BELL. Intravenous glucagon increases biliary lipid output and promotes bilirubin excretion in man. *Gut* 20: A935, 1979.

JENSEN, R. T., K. TATEMOTO, V. MUTT, C. F. LEMP, AND J. D. GARDNER. Actions of a newly isolated intestinal peptide PHI on pancreatic acini. *Am. J. Physiol.* 241 (*Gastrointest. Liver Physiol.* 4): G498–G502, 1981.

JIAN, R., H. S. BESTERMAN, D. L. SARSON, C. AYMES, J. HOSTEIN, S. R. BLOOM, AND J. C. RAMBAUD. Colonic inhibition of gastric secretion in man. *Digest. Dis. Sci.* 26: 195–201, 1981.

JIAN, R., N. VIGNERON, Y. NAJEAN, AND J. J. BERNIER. Gastric emptying and intragastric distribution of lipids in man. A new scintigraphic method of study. *Digest. Dis. Sci.* 27: 705–711, 1982.

JOHANSSON, B., O. JONSSON, AND B. LJUNG. Supraspinal control of the intestino-intestinal inhibitory reflex. *Acta Physiol. Scand.* 63: 442–449, 1965.

JOHNSON, L. R., AND M. I. GROSSMAN. Characteristics of inhibition of gastric secretion by secretin. *Am. J. Physiol.* 217: 1401–1404, 1969.

JOHNSON, L. R., S. J. DUDRICK, AND P. D. GUTHRIE. Stimulation of pancreatic growth by intraduodenal amino acids and HCl. *Am. J. Physiol.* 239 (*Gastrointest. Liver Physiol.* 2): G400–G405, 1980.

JONES, A. L., D. L. SMUCKER, J. S. MOONEY, R. K. OCKNER, AND R. K. ADLER. Alterations in hepatic pericanalicular cytoplasm during enhanced bile secretory activity. *Lab. Invest.* 40: 512–517, 1979.

JONES, A. L., G. T. HDRADEK, AND R. H. RENSTON. Autoradiographic evidence for hepatic lobular concentration gradient of bile acid derivative. *Am. J. Physiol.* 238 (*Gastrointest. Liver Physiol.* 1): G233–G237, 1980.

KADEKARO, M., C. TIMO-IARIA, AND M. DELL M., VINCENTI. Control of gastric secretion by the central nervous system. In: *Nerves and the Gut*, edited by F. P. Brooks and P. W. Evers. Thorofare, N J: Charles B. Slack, 1977, p. 380–429.

KADEKARO, M., H. SAVAKI, AND L. SOKOLOFF. Metabolic mapping of neural pathways involved in gastrosecretory response to insulin hypoglycemia in the rat. *J. Physiol. (London)* 300: 393–407, 1980.

KALIA, M., AND M. M. MESULAM. Brain stem projections of sensory and motor components of the vagus complex in the cat. II. Laryngeal, tracheobronchial, pulmonary, cardiac and gastrointestinal branches. *J. Comp. Neurol.* 193: 467–508, 1980.

KAMADA, T., N. SOTO, S. KAWANO, H. FUSAMOTO, AND H. ABE. Gastric mucosal hemodynamics after thermal or head injury. A clinical application of reflectance spectrophotometry. *Gastroenterology* 83: 535–540, 1982.

KAMINSKI, D. L., W. H. BROWN, AND Y. G. DESHPANDE. Effect of glucagon on bile cAMP secretion. *Am. J. Physiol.* 238 (*Gastrointest. Liver Physiol.* 1): G119–G123, 1980.

KANNO, T. Calcium-dependent amylase release and electrophysiological measurements in cells of the pancreas. *J. Physiol. (London)* 226: 353–371, 1972.

KAPADIA, C. R., AND R. M. DONALDSON, JR. Macromolecular secretion by isolated gastric mucosa. *Gastroenterology* 74: 535–539, 1978.

KAPADIA, C. R., D. E. SCHAFER, R. M. DONALDSON, JR., AND E. R. EBERSOLE. Evidence for involvement of cyclic nucleotides in intrinsic factor secretion by isolated rabbit gastric mucosa. *J. Clin. Invest.* 64: 1044–1049, 1979.

KAPLOWITZ, N., T. YAMADA, AND Y. SUGIYAMA. Comparison of bile acid binding to purified rat liver Y and Y' proteins. *Gastroenterology* 82: 1232, 1982.

KAUFFMAN, G. L., B. J. R. WHITTLE, D. AURES, J. R. VANE, AND M. I. GROSSMAN. Effects of prostacyclin and a stable analogue, 6β-PGI$_1$, on gastric acid secretion, mucosal

blood flow, and blood pressure in conscious dogs. *Gastroenterology* 77: 1301–1306, 1979.

KEANE, F. B., E. P. DI MAGNO, D. G. KELLY, R. R. DOZOIS, AND V. L. W. GO. Physiologic and morphologic evidence for a canine pancreatic sphincter. Am. Pancr. Ass., Chicago, Nov. 1–2, 1979.

KEANE, F. B., E. P. DI MAGNO, R. R. DOZOIS, AND V. L. W. GO. Relationships among canine, interdigestive exocrine, pancreatic and biliary flow, duodenal motor activity, plasma, pancreatic polypeptide, and motilin. *Gastroenterology* 78: 310–316, 1980.

KEANE, F. B., R. R. DOZOIS, V. L. W. GO, AND E. P. DI MAGNO. Interdigestive canine pancreatic juice composition and pancreatic reflux, and pancreatic sphincter anatomy. *Digest. Dis. Sci.* 26: 577–584, 1981.

KELLER, P. J., AND B. J. ALLAN. The protein composition of human pancreatic juice. *J. Biol. Chem.* 242: 281–287, 1967.

KELLER, P. J., D. L. KAUFMAN, B. J. ALLAN, AND B. L. WILLIAMS. Further studies on the structural differences between the isoenzymes of human parotid α amylase. *Biochemistry* 10: 4867–4874, 1971.

KELLY, K. A. Gastric motility after gastric operations. *Surg. Ann.* 6: 103–123, 1974.

KELLY, K. A. Gastric motility in health and after gastric surgery. *Viewpoints on Digestive Diseases* 8: No. 2, 1976.

KELLY, K. A. Motility of the stomach and gastroduodenal junction. In: *Physiology of the Gastrointestinal Tract*, edited by L. R. Johnson. New York: Raven Press, 1982, p. 393–410.

KERLIN, P., AND S. PHILLIPS. Variablity of motility of the ileum and jejunum in healthy humans. *Gastroenterology* 82: 694–700, 1982.

KERLIN, P., A. ZINSMEISTER, AND S. PHILLIPS. Relationship of motility to flow of contents in the human small intestine. *Gastroenterology* 82: 707–710, 1982.

KERN, F., JR, AND T. P. ALMY. The effects of acetylcholine and methacholine upon the human colon. *J. Clin. Invest.* 31: 555–560, 1952.

KERN, H. F., W. BIEGER, A. VOLKL, G. ROHR, AND G. ADLER. Regulation of intracellular transport of exportable proteins in the rat exocrine pancreas. In: *The Secretory Mechanisms*, edited by C. R. Hopkins and C. J. Duncan. Cambridge: Cambridge University Press, 1979, p. 79–100.

KIHL, B., A. ROKAEUS, S. ROSSELL, AND L. OLBE. Fat inhibition of gastric acid secretion in man and plasma concentrations of neurotensin-like immunoreactivity. *Scand. J. Gastroenterol.* 16: 513–526, 1981.

KIM, Y. S., Y. W. KIM, H. D. GAINES, AND M. H. SLEISENGER. Zymogram studies of human intestinal brush border and cytoplasmic peptidases. *Gut* 20: 981–991, 1979.

KIMBERG, D. V. Cyclic nucleotides and their role in gastrointestinal secretion. *Gastroenterology* 67: 1023–1064, 1974.

KIMMICH, G. A. Intestinal absorption of sugar. In: *Physiology of the Gastrointestinal Tract*, edited by L. R. Johnson. New York: Raven Press, 1981, p. 1035–1061.

KLIGER, A. S., H. J. BINDER, C. BASTL, AND J. P. HAYSLETT. Demonstration of active potassium transport in the mammalian colon. *J. Clin. Invest.* 67: 1189–1196, 1981.

KNIGHT, S. E., AND R. L. Mc ISAAC. Neutral red clearance as an estimate of gastric mucosal blood flow in man. *J. Physiol. (London)* 272: 62–63P, 1977.

KNIGHT, S. E., R. L. Mc ISAAC, AND L. P. FIELDING. The effect of the histamine-H$_2$-antagonist cimetidine on gastric mucosal blood flow. *Gut* 18: A948, 1977.

KNIGHT, S. E., R. L. Mc ISAAC, AND L. P. FIELDING. The effect of highly selective vagotomy on the relationship between gastric mucosal blood flow and acid secretion in man. *Brit. J. Surg.* 65: 721–723, 1978.

KNUTSON, U., AND L. OLBE. The effect of exogenous gastrin on the acid sham-feeding response in antrum-bulb-resected duodenal ulcer patients. *Scand. J. Gastroenterol.* 9: 231–238, 1974.

KOELZ, H. R., C. S. CHEW, G. SACHS, AND S. J. HERSEY.

Cholinergic and β adrenergic pepsinogen release by isolated rabbit glands. *Gastroenterology* 80: 1194, 1981a.

KOELZ, H. R., G. SACHS, AND T. BERGLINDH. Cation effects on acid secretion in rabbit gastric glands. *Am. J. Physiol.* 241 (*Gastrointest. Liver Physiol* 4): G431–G442, 1981b.

KOIKE, H., M. L. STEER, AND J. MELDOLESI. Pancreatic effects of ethionine: blockade of exocytosis and appearance of crinophagy and autophagy precede cellular necrosis. *Am. J. Physiol.* 242 (*Gastrointest. Liver Physiol.* 5): G297–G307, 1982.

KOKAS, E. Villikinin. *Gastroenterology* 67: 750–752, 1974.

KONTUREK, S. J., AND L. R. JOHNSON. Evidence for an enterogastric reflex for the inhibition of acid secretion. *Gastroenterology* 61: 667–674, 1971.

KONTUREK, S. J., J. TASLER, J., AND W. OBTULOWICZ. Pancreatic dose response to synthetic secretion and intraduodenal acid in dogs. *Am. J. Digest. Dis.* 15: 987–992, 1970.

KONTUREK, S. J., P. THOR, A. DEMBINSKI, AND R. KROL. Comparison of secretion and vasoactive intestinal peptide on pancreatic secretion in dogs. *Gastroenterology* 68: 1527–1535, 1975.

KONTUREK, S. J., J. TASLER, W. OBTULOWICZ, AND M. CIESZKOWSKI. Comparison of amino acids bathing the oxyntic gland area with stimulation of gastric secretion. *Gastroenterology* 70: 66–69, 1976a.

KONTUREK, S. J., A. DEMBINSKI, P. THOR, AND R. KROL. Comparison of vasoactive intestinal peptide (VIP) and secretin in gastric secretion and mucosal blood flow. *Pfluger's Archiv.* 361: 175–181, 1976b.

KONTUREK, S. J., J. TASLER, M. CIESZKOWSKI, D. H. COY, AND A. V. SCHALLY. Effect of growth hormone release-inhibiting hormone on gastric secretion, mucosal blood flow, and serum gastrin. *Gastroenterology* 70: 737–741, 1976c.

KONTUREK, S. J., S. DOMSCHKE, W. DOMSCHKE, E. WUNSCH, AND L. DEMLING. Comparison of pancreatic responses to portal and systemic secretin and VIP in cats. *Am. J. Physiol.* 232 (*Endocrinol. Metab. Gastrointest. Physiol.* 1): E156–E158, 1977.

KONTUREK, S. J., C. LANCASTER, A. J. HANCHAR, J. E. NEZAMIS, AND A. ROBERT. The influence of prostacyclin on gastric mucosal blood flow in resting and stimulated canine stomach. *Gastroenterology* 76: 1173, 1979.

KONTUREK, S. J., J. JAWOREK, J. TASLER, M. CIESZKOWSKI, AND W. PAWLIK. Effect of substance P and its C-terminal hexapeptide on gastric and pancreatic secretion in the dog. *Am. J. Physiol.* 241 (*Gastrointest. Liver Physiol.* 4): G74–G81, 1981.

KONTUREK, S. J., C. A. MEYERS, N. KWIECIEN, W. OBTULOWICZ, J. TASLER, J. OLESKY, B. KOPP, D. H., COY, AND A. V. SCHALLY. Effect of human pancreatic polypeptide and its C-terminal hexapeptide on pancreatic secretion in man and in the dog. *Scand. J. Gastroenterol.* 17: 395–399, 1982.

KOO, A. Vagotomy attenuates histamine-induced vasodilatation in rat gastric microcirculation. *J. Physiol.* (*London*) 325: 23–24P, 1982.

KORC, M. Regulation of pancreatic protein synthesis by cholecystokinin and calcium. *Am. J. Physiol.* 243 (*Gastrointest. Liver Physiol.* 6): G69–G75, 1982.

KORC, M., A. C. BAILEY, AND J. A. WILLIAMS. Regulation of protein synthesis in normal and diabetic rat pancreas by cholecystokinin. *Am. J. Physiol.* 241 (*Gastrointest. Liver Physiol.* 4): G116–G121, 1981a.

KORC, M., Y. IWAMOTO, H. SANKARAN, J. A. WILLIAMS, AND I. D. GOLDFINE. Insulin action in pancreatic acini from streptozotocin-treated rats. I. Stimulation of protein synthesis. *Am. J. Physiol.* 240 (*Gastrointest. Liver Physiol.* 3): G56–G63, 1981b.

KOSTERLITZ, H. W. Intrinsic and extrinsic nervous control of motility of the stomach and the intestines. In: *Handbook of Physiology*, sect. 6: *Alimentary Canal*, vol 4: *Motility*, edited by C. F. Code. Washington, DC: Am. Physiol. Soc., 1968, p. 2147–2172.

KOTTEGODA, S. R. An analysis of possible nervous mechanisms involved in the peristaltic reflex. *J. Physiol.* (*London*) 200: 687–712, 1969.

KOWARSKI, S., AND D. SCHACHTER. Intestinal membrane calcium-binding protein: vitamin-D dependent membrane component of the intestinal calcium transport mechanism, *J. Biol. Chem.* 255(22): 10834–10840, 1980.

KRAWISZ, A. R., L. J. MILLER, E. P. DI MAGNO, AND V. L. W. GO. In the absence of nutrients, pancreatic-biliary sections do not exert feedback control of human pancreatic or gastric function. *J. Lab. Clin. Med.* 95: 13–18, 1980.

KREEK, M. J., R. E. PETERSON, M. H. SLEISENGER, AND G. H. JEFRIES. Effects of ethinylestradiol-induced cholestasis on bile flow and biliary excretion of estradiol and estradiol glucuronide by the rat. *Proc. Soc. Exper. Biol. Med.* 131: 646–650, 1969.

KRISHAMURTHY, G. T., V. R. BOBBA, AND E. KINGSTON. Radionuclide ejection fraction: a technique for quantitative analysis of motor function of the human gallbladder. *Gastroenterology* 80: 482–490, 1981.

KRISHNAMURTHY, G. T., V. R. BOBBA, E. KINGSTON, AND F. TURNER. Measurement of gallbladder emptying sequentially using a single dose of ^{99m}TC-labeled hepatobiliary agent. *Gastroenterology* 83: 773–776, 1982.

KUMPULAINEN, T., AND P. JALOVAARA. Immunohistochemical localization of carbonic anhydrase isoenzymes in the human pancreas. *Gastroenterology* 80: 796–799, 1981.

KVIETYS, P. R., J. M. Mc LENDON, G. B. BULKLEY, M. A. PERRY, AND D. N. GRANGER. Pancreatic circulation: intrinsic regulation. *Am. J. Physiol.* 242 (*Gastrointest. Liver Physiol.* 5): G596–G602, 1982.

LALYRE, Y., D. E. WILSON, J. KIDAO, Ch. HALL, AND V. CAPEK. Comparison of intravenous and intramuscular sincalide (C-terminal octapeptide of cholecystokinin) on gallbladder contraction in man. *Digest. Dis. Sci.* 26: 214–217, 1981.

LAMERS, C. B., J. H. WALSH, J. B. JANSEN, A. R. HARRISON, A. F. IPPOLITI, AND J. H. van TONGEREN. Evidence that gastrin 34 is preferentially released from the human duodenum. *Gastroenterology* 83: 233–239, 1982.

LAMONT, J. T., AND A. VENTOLA. Stimulation of colonic glycoprotein synthesis by dibutyryl cyclic AMP and theophylline. *Gastroenterology* 72: 82–86, 1977.

LAMONT, J. T., AND A. S. VENTOLA. Purification and composition of colonic epithelial mucin. *Biochim. Biophys. Acta* 626: 234–243, 1980.

LANCIAULT, G., C. BONOMA, AND F. P. BROOKS. Vagal stimulation, gastrin release and acid secretion in anesthetized dogs. *Am. J. Physiol.* 225: 546–552, 1973.

LANCIAULT, G., J. E. SHAW, J. URQUHART, L. ADAIR, AND F. P. BROOKS. Response of the isolated perfused stomach of the dog to electrical vagal stimulation. *Gastroenterology* 68: 294–300, 1975.

LANFRANCHI, G. A., L. MARZIO, C. CORTINI, AND E. M. ASSET. Motor effect of dopamine on human sigmoid colon: evidence for specific receptors. *Am. J. Digest. Dig.* 23: 257–263, 1978.

LARSSON, L.-I. Structure and function of putative paracrine cells. In: *Gut Peptides*, edited by A. Miyoshi. Amsterdam: Elsevier, 1979, p. 112–117.

LARSSON, L. I. Peptide secretory pathways in GI tract: cytochemical contributions to regulatory physiology of the gut. *Am. J. Physiol.* 239 (*Gastrointest. Liver Physiol.* 2): G237–G246, 1980.

LARSSON, L. T., AND J. F. REHFELD. Peptidergic and adrenergic innervation of pancreatic ganglia. *Scand. J. Gastroenterol.* 14: 433–437, 1979.

LA RUSSO, N. F. AND S. FOWLER. Coordinate secretion of acid hydrolases in rat bile: hepatocyte exocytosis of liposomal protein. *J. Clin. Invest.* 64: 948–954, 1979.

LA RUSSO, N. F., L. J. KOST, J. A. CARTER, AND S. S. BARHAM. Triton WR 1339, a liposomotrophic compound, is excreted into bile and alters the biliary excretion of liposomal enzymes and lipids. *Hepatology* 2: 209–215, 1982.

LAYDEN, T. J., E. ELIAS, AND J. L. BOYER. Bile formation in the rat. The role of the paracellular shunt pathway. *J. Clin. Invest.* 62: 1375–1386, 1978.

LAYER, P., J. HOTZ, H. P. SCHMITZ-MOORMANN, AND H. GOEBELL. Effects of experimental chronic hypercalcemia on feline exocrine pancreatic secretion. *Gastroenterology* 82: 309–316, 1982.

LEBENTHAL, E., AND P. C. LEE. Development of functional response in human exocrine pancreas. *Pediatrics* 66: 556–560, 1980.

LEDER, O., AND P. TRITSCHLER. Nachweis hochster carbohydrase-aktivitation in epithel der speicheldrusengange. *Pflugers Arch.* 292: 229–238, 1966.

LEE, K. Y., W. Y. CHEY, H. H. TAI, AND H. YAJIMA. Radioimmunoassay of motilin: validation and studies on the relationship between plasma motilin and interdigestive myoelectric activity of the duodenum of dog. *Am. J. Digest. Dis.* 23: 789–795, 1978.

LEE, K. Y., M. S. KIM, AND W. Y. CHEY. Effects of a meal and gut hormones on plasma motilin and duodenal motility in dog. *Am. J. Physiol.* 238 (*Gastrointest. Liver Physiol.* 1): G280–G283, 1980a.

LEE, K. Y., M. S. KIM, Y. C. KIM, J. ROMINGER, AND W. Y. CHEY. Secretin is an enterogastrone in dog. *Regulatory Peptides* 1: S66, 1980b.

LEESON, C. R. Structure of salivary glands. In: *Handbook of Physiology*, sect. 6: *Alimentary Canal*, vol.2: *Secretion*, edited by C. F. Code. Washington, DC: Am. Physiol. Soc., 1967, p. 463–495.

LEFEBVRE, P. J., LUYCKX, A. S., AND A. H. BRASSINNE. Vagal stimulation and its role in eliciting gastrin but not glucagon release from the isolated perfused dog stomach. *Gut* 19: 185–188, 1978.

LEFEBVRE, P. J., A. S. LUYCKX, AND A. H. BRASSINNE. Inhibition by somatostatin of carboxylcholine induced gastrin and glucagon release from the isolated perfused canine stomach. *Gut* 22: 793–797, 1981.

LEGG, P. G. Fluorescence studies on neural structures and endocrine cells in the pancreas of the cat. *Z. Zellforsch* 88: 487–495, 1968.

LENNINGER, S. The autonomic innervations of the exocrine pancreas. *Med. Clin. North Am.* 58: 1311–1318, 1974.

LEONARD, A. S., D. LONG, L. A. FRENCH, E. T. PETER, AND O. H. WANGENSTEEN. Pendular pattern in gastric secretion and blood flow following hypothalamic stimulation origin of the stress ulcer. *Surgery* 56: 109–120, 1964.

LESTER, R., J. ST. PYREK, J. M. LITTLE, AND E. W. ADROCK. What is meant by the term "bile acid"? *Am. J. Physiol.* 244 (*Gastrointest. Liver Physiol.* 7): G107–110, 1983.

LEV, R., AND W. C. GRIFFITHS. Colonic and small intestinal alkaline phosphatase. *Gastroenterology* 82: 1427–1435, 1982.

LEVINE, J. S., P. K. NAKANE, AND R. H. ALLEN. Human intrinsic factor secretion: immunocytochemical demonstration of membrane-associated vesicular transport in parietal cells. *J. Cell. Biol.* 90: 644–655, 1981.

LEVINE, R. A., K. R. KOHEN, E. H. SCHWARTZEL, JR., AND C. E. RAMSAY. Prostaglandin E_2-histamine interactions on cAMP, cGMP, and acid production in isolated fundic glands. *Am. J. Physiol.* 242 (*Gastrointest. Liver Physiol.* 5): G21–G26, 1982.

LEVITT, M. D., J. H. BOND, AND D. G. LEVITT. Gastrointestinal gas. In: *Physiology of the Gastrointestinal Tract*, edited by L. R. Johnson, New York: Raven Press, 1981, p. 1301–1316.

LIANG, T., AND M. A. CASCIERI. Substance P stimulation of amylase release by isolated parotid cells and inhibition of substance P induction of salivation by substance P. *Molec. Cell. Endocrinol.* 15: 151–162, 1979.

LICHTENBERGER, L. M., J. M. SHOREY, AND J. S. TRIER. Organ culture studies of rat antrum: evidence for an antral inhibitor of gastrin release. *Am. J. Physiol.* 235 (*Endocrinol. Metab. Gastrointest. Physiol.* 4): E410–E415, 1978.

LICHTENBERGER, L. M., L. S. SHAW, AND R. B. BAILEY. Release of somatostatin from rodent antral mucosae maintained in organ culture. *Am. J. Physiol.* 24 (*Gastrointest. Liver Physiol.* 4): G59–G66, 1981.

LIDBERG, P., R. EDIN, J. M. LUNDBERG, A. DAHLSTROM, S. ROSELL, K. FALKERS, AND H. AHLMAN. The involvement of substance P in the vagal control of the feline pylorus. *Acta Physiol. Scand.* 114: 307–309, 1982.

LIEBERMANN-MEFFERT, D., M. ALLGOWER, P. SCHMID, AND A. L. BLUM. Muscular equivalent of the lower esophageal sphincter. *Gastroenterology* 76: 31–38, 1979.

LIEDTKE, C. M., AND U. HOPFER. Mechanism of Cl^- translocation across small intestinal brush-border membrane I. Absence of Na^+-Cl^- cotransport. *Am. J. Physiol.* 242 (*Gastrointest. Liver Physiol.* 5): G263–G271, 1982a.

LIEDTKE, C. M., AND U. HOPFER. Mechanism of Cl^- translocation across small intestinal brush-border membrane. II. Demonstration of Cl^--OH^- exchange and Cl^- conductance. *Am. J. Physiol.* 242 (*Gastrointest. Liver Physiol.* 5): G272–G280, 1982b.

LIFSON, N., R. G. KRAMLINGER, R. R. MAYRAND, AND E. J. LENDER. Blood flow to the rabbit pancreas with special reference to the islets of Langerhans. *Gastroenterology* 79: 466–473, 1980.

LIGHTWOOD, R., AND H. A. REBER. Micropuncture study of pancreatic secretion in the cat. *Gastroenterology* 72: 61–66, 1977.

LILJA, P., C. J. FAGAN, I. WIENER, K. INOUE, L. C. WATSON, P. L. RAYFORD, AND J. C. THOMPSON. Infusion of pure cholecystokinin in humans. *Gastroenterology* 83: 256–261, 1982.

LIN, T. M. Actions of gastrointestinal hormones and related peptides on the motor function of the biliary tract. *Gastroenterology* 69: 1006–1022, 1975.

LINAKER, B. D., J. S. Mc KAY, N. B. HIGGS, AND L. A. TURNBERG. Mechanisms of histamine stimulated secretion in rabbit ileal mucosa. *Gut* 22: 964–970, 1981.

LINDBLAD, L., AND T. SCHERSTEN. Influence of cholic and chenodeoxycholic acid on canalicular bile flow in man. *Gastroenterology* 70: 1121–1124, 1976.

LIPKIN, M. Proliferation and differentiation of gastrointestinal cells in normal disease states. In: *Physiology of the Gastrointestinal Tract*, edited by L. R. Johnson. New York: Raven Press, 1981, p. 145–161.

LOMBARDI, D. M., F. S. FENG, AND F. P. BROOKS. Disassociation of secretory and motor responses to stimulation of the dorsal motor nucleus of the vagus in anesthetized cats. *Gastroenterology* 82: 1120, 1982.

LONG, J. F., AND F. P. BROOKS. Relation between inhibition of gastric secretion and absorption of fatty acids. *Am. J. Physiol.* 209: 447–451, 1965.

LOVE, A. H. G., T. G. MITCHELL, AND R. A. PHILLIPS. Water and sodium absorption in the human intestine. *J. Physiol.* (*London*) 195: 133–140, 1968.

LUNDBERG, A. Electrophysiology of the salivary glands. *Physiol. Revs.* 38: 21–40, 1958.

LUNDBERG, J. M., T. HOKFELT, J. KEWENTER, G. PETTERSON, H. AHLMAN, R. EDIN, A. DAHLSTROM, G. NILSSON, L. TERENIUS, K. UVNAS-WALLENSTEN, AND S. SAID. Substance P, VIP and enkephalin-like immunoreactivity in the human vagus nerve. *Gastroenterology* 77: 468–471, 1979.

MacDERMOTT, R. P., R. M. DONALDSON, JR., AND J. S. TRIER. Glycoprotein synthesis and secretion by mucosal biopsies of rabbit colon and human rectum. *J. Clin. Invest.* 54: 545–554, 1974.

MacGREGOR, I. L., J. R. KNILL, AND J. H. MEYER. Similar sensitivities of pancreatic and biliary secretion to cholecystokinin plus secretin infusion. *Am. J. Digest. Dis.* 21: 641–644, 1976.

MACHEN, T. E., AND W. L. Mc LENNAN. Na$^+$-dependent H$^+$ and Cl$^-$ transport in in vitro frog gastric mucosa. *Am. J. Physiol.* 238 (*Gastrointest. Liver Physiol.* 1): G403–G413, 1980.

MACHEN, T. E., M. J. RUTTEN AND E. P. M. EKBLAD. Histamine, cAMP, and activation of piglet gastric mucosa. *Am. J. Physiol.* 248 (*Gastrointest. Liver Physiol.* 5): G79–G84, 1982.

MAKHLOUF, G. M. Electrolyte composition of gastric secretion. In: *Physiology of the Gastrointestinal Tract*, edited by L. R. Johnson. New York: Raven Press, 1981, p. 551–556.

MALAGELADA, J-R. Quantification of gastric solid-liquid discrimination during digestion of ordinary meals. *Gastroenterology* 72: 1264–1267, 1977.

MALAGELADA, J-R. Gastric pancreatic and biliary responses to a meal. In: *Physiology of the Gastrointestinal Tract*, edited by L. R. Johnson. New York: Raven Press, 1981, p. 893–924.

MALAGELADA, J-R., K. H. HALTERMULLER, G. W. SIZEMORE, AND V. L. W. GO. The influence of hypercalcemia on basal and cholecystokinin-stimulated pancreatic, gallbladder, and gastric functions in man. *Gastroenterology* 71: 405–408, 1976.

MALAGELADA, J-R., V. L. W. GO, AND W. H. J. SUMMERSKILL. Different gastric pancreatic and biliary responses to solid-liquid or homogenized meals. *Digest. Dis.* 24: 101–110, 1979.

MALAISSE-LAGAE, F., M. RAVAZZOLA, P. ROBBERECHT, A. VANDERMEERS, W. J. MALAISSE, AND L. ORCI. Exocrine pancreas: evidence for topographic partition of secretory function. *Science* 190: 795–797, 1975.

MALEY, B., AND E. ELDE. Localization of substance P-like immunoreactivity in cell bodies of the feline dorsal vagal nucleus. *Neurosci. Lett.* 27: 187–191, 1981.

MANDEL, I. D., AND S. WOTMAN. The salivary secretions in health and disease. *Oral Sciences Rev.* 8: 25–47, 1976.

MANGOS, J. A., N. R. McSHERRY, AND S. N. ARVANITAKIS. Autonomic regulation of secretion and transductal fluxes of ions in the rat parotid. *Am. J. Physiol.* 225: 683–688, 1973.

MANGOS, J. A., N. R. McSHERRY, T. BARBER, S. N. ARVANITAKIS, AND V. WAGNER. Dispersed rat parotid acinar cells. II. Characterization of adrenergic receptors. *Am. J. Physiol.* 229: 560–565, 1975a.

MANGOS, J. A., N. R. McSHERRY, AND T. BARBER. Dispersed rat parotid acinar cells. III. Characterization of cholinergic receptors. *Am. J. Physiol.* 229: 566–569, 1975b.

MANN, K., B. LIPP, J. GRUNST, R. GEIGER, AND H-J. KARL. Determination of kallikrein by radioimmunoassay in human body fluids. *Agents Actions* 10: 329–333, 1980.

MANSBACH, C. M., R. S. COHEN, AND P. B. LEFF. Isolation and properties of the mixed lipid micelles present in intestinal content during fat digestion in man. *J. Clin. Invest.* 56: 781–791, 1975.

MANTLE, M., D. MANTLE, AND A. ALLEN. Polymeric structure of pig small intestinal mucus glycoprotein. *Biochem. J.* 195: 277–285, 1981.

MARANDA, B., J. A. A. RODRIGUES, M. SCHACHTER, T. K. SHNITKA, AND J. WEINBERG. Studies on kallikrein in the duct systems of the salivery glands of the cat. *J. Physiol. (London)* 276: 321–328, 1978.

MARLETT, J. A., AND C. F. CODE. Effects of celiac and superior mesenteric ganglionectomy on interdigestive myoelectric complex in dogs. *Am. J. Physiol.* 237 (*Endocrinol. Metab. Gastrointest. Physiol.* 6): E432–E436, 1979.

MARMAR, J., G. J. BARBERO, AND M. S. SABINGA. The pattern of parotid gland secretion in cystic fibrosis of the pancreas. *Gastroenterology* 50: 551–556, 1966.

MARTIN, C. J., C. FROMTER, B. GEBLER, H. KNAUF, AND J. A. YOUNG. The effects of carbachol on water and electrolyte fluxes and transepithelial electrical potential differences of the rabbit submaxillary main duct perfused in vitro. *Pflugers Arch.* 341: 131–142, 1973.

MARTINDALE, R., G. L. KAUFFMAN, S. LEVIN, J. H. WALSH, AND T. YAMADA. Differential regulation of gastrin and somatostatin secretion from isolated perfused rat stomachs. *Gastroenterology* 83: 240–244, 1982.

MARTINEZ, J. R. Mikropunktionsuntersuchungen an der glandula submaxillaris du ratte. *Pfluger Archiv. ges Physiol.* 279: R17, 1964.

MARTINSON, J. The effect of graded vagal stimulation on gastric motility, secretion and blood flow in the cat. *Acta Physiol. Scand.* 65: 300–309, 1965.

MASHFORD, M. L., G. NILSSON, A. ROKEAUS, AND S. ROSELL. The effect of food ingestion on circulating neurotensin-like immunoreactivity (NTLI) in the human. *Acta Physiol. Scand.* 104: 244–246, 1978.

MATUCHANSKY, C., J-Y. MARY, AND J-J. BERNIER. Further studies on prostaglandin E$_1$-induced jejunal secretion of water and electrolytes in man, with special reference to the influence of ethacrynic acid, furosamide, and aspirin. *Gastroenterology* 71: 274–281, 1976.

MAX, E. E., D. B. P. GOODMAN, AND H. RASMUSSEN. Purification and characterization of chick intestine brush border membrane. Effects of 1 α (OH) vitamin D$_3$ treatment. *Biochim. Biophys. Acta* 511: 224–239, 1978.

MAZE, M., AND G. M. GRAY. Intestinal brush-border amino-oligopeptidases: cytosol precursors of the membrane enzyme. *Biochem.* 19: 2351–2358, 1980.

MAZER, N. A., G. B. BENEDEK, AND M. C. CAREY. Quasielastic light-scattering studies of aqueous biliary lipid systems. Mixed micelle formation in bile-salt-lecithin solutions. *Biochem.* 19: 601–605, 1980.

McCANDLES, M., A. NISHIYAMA, O. H. PETERSEN, AND H. G. PHILPHOTT. Mouse pancreatic acinar cells: voltage-clamp study of acetylcholine-evoked membrane current. *J. Physiol. (London)* 318: 57–72, 1981.

McHUGH, P. R., AND T. H. MORAN. Calories and gastric emptying: a regulatory capacity with implications for feeding. *Am. J. Physiol.* 236 (*Regulatory Integrative Comp. Physiol.* 5): R254–R260, 1979.

McINTOSH, C. H. S., R. A. PEDERSON, H. KOOP, AND J. C. BROWN. Inhibition of GIP stimulated somatostatin-like immunoreactivity (SLI) by acetylcholine and vagal stimulation. In: *Gut Peptides*, edited by A. Miyoski. Amsterdam: Elsevier, 1979, p. 100–104.

McKAY, J. S., B. D. LINAKER, N. B. HIGGS, AND L. A. TURNBERG. Studies of the antisecretory activity of morphine in rabbit ileum in vitro. *Gastroenterology* 82: 243–247, 1982.

McKIRDY, H., R. W. MARSHALL, AND H. L. DUTHIE. In vitro identification of human oesophagogastric junction. *Ztschr Gastroenterologie* XIX: 446–447, 1981.

MELDOLESI, J. Dynamics of cytoplasmic membrane in guinea pig pancreatic acinar cells. I. Synthesis and tunover of membrane proteins. *J. Cell Biol.* 61: 1–13, 1974.

MENGUY, R. Effects of histamine on gastric blood flow. *Am. J. Digest. Dis.* 7: 359–383, 1962.

METZGER, A. L., R. ADLER, S. HEYMSFIELD, AND S. M. GRUNDY. Diurnal variation in biliary lipid composition. *N. Engl. J. Med.* 288: 333–336, 1973.

MEYER, J. H., L. W. WAY, AND M. I. GROSSMAN. Pancreatic response to acidification of various lengths of proximal intestine in the dog. *Am. J. Physiol.* 219: 971–977, 1970a.

MEYER, J. H., L. W. WAY, AND M. I. GROSSMAN. Pancreatic bicarbonate response to various acids in the duodenum of the dog. *Am. J. Physiol.* 219: 964–970, 1970b.

MILLER, L. J., J-R. MALAGELADA, W. F. TAYLOR, AND V. L. W. GO. Intestinal control of human post-prandial gastric function: the role of components of jejuno-ileal chyme in regulating gastric secretion and gastric emptying. *Gastroenterology* 80: 763–769, 1981.

MIR, S. S., G. R. MASON, AND H. S. ORMSBEE, III. Vagal influence on duodenal motor activity. *Am. J. Surg.* 135: 97–101, 1978.

MISHER, A., AND F. P. BROOKS. Electrical stimulation of hypothalamus and gastric secretion in the albino rat. *Am. J. Physiol.* 211: 403–406, 1967.

MODIGLIANI, R., J-Y, MARY, AND J. J. BERNIER. Effects of synthetic human gastrin I on movements of water, electrolytes and glucose across the human small intestine. *Gastroenterology* 71: 978–984, 1976.

MODLIN, I. M., D. ALBERT, R. CROCHETT, A. SANK, AND B. M. JAFFE. Evidence for an intestinal mechanism of pancreatic polypeptide release. *Digest. Dis. Sci.* 26: 587–590, 1981.

MOORE, E. W., H. J. VERINE, AND M. I. GROSSMAN. Pancreatic bicarbonate response to a meal. *Acta Hepatogastroenterol.* 26: 30–36, 1979.

MOORE, J. G. The relationship of gastric acid secretion to plasma glucose in five men. *Scand. J. Gastroenterol.* 15: 625–632, 1980.

MOORE, J. G., AND M. WOLFE. The relation of plasma gastrin to the circadian rhythm of gastric acid secretion in man. *Digestion* 9: 97–105, 1973.

MORAN, T. H., AND P. R. McHUGH. Distinctions among three sugars in their effects on gastric emptying and satiety. *Am. J. Physiol.* 241 (*Regulatory Integrative Comp. Physiol* 10): R25–R30, 1981.

MORRISSET, J., AND L. JOLICOEUR. Effect of hydrocortisone on pancreatic growth in rats. *Am. J. Physiol.* 239 (*Gastrointest. Liver Physiol.* 2): G95–G98, 1980.

MORRISSET, J., L. JOLICOEUR, P. GENIK, AND A. LORD. Interaction of hydrocortisone and caerulein on pancreatic size and composition in the rat. *Am. J. Physiol.* 241 (*Gastrointest. Liver Physiol.* 4): G37–G42, 1981.

MOSS, S., G. R. BIRCH, R. W. LOBLEY, AND R. HOLMES. The release of enterokinase following secretin and cholecystokinin-pancreozymin in man. *Scand. J. Gastroenterol.* 14: 1001–1007, 1979.

MROZ, C. T., AND K. A. KELLY. The role of the extrinsic antral nerves in the regulation of gastric emptying. *Surg. Gynec. & Obstet.* 145: 369–377, 1977.

MUNCK, B. G. Intestinal absorption of amino acids. In: *Physiology of the Gastrointestinal Tract*, edited by L. R. Johnson. New York: Raven Pres, 1981, p. 1097–1122.

MURER, H., AND B. HILDMANN. Transcellular transport of calcium and inorganic phosphate in the small intestinal epithelium. *Am. J. Physiol.* 240 (*Gastrointest. Liver Physiol.* 3): G409–G416, 1981.

MURTHY, S. N. S., V. P. DINOSO, AND H. R. CLEARFIELD. Bile acid and pancreatic trypsin outputs are parallel during intraduodenal infusion of essential amino acids. *Dig. Dis. Sci.* 28: 27–32, 1983.

NAKAMURA, K., K. ISHI, M. KUSANO, AND S. HAYASHI. Acute and long-term effects of vagotomy on gastric muscosal blood flow. In: *Vagotomy. Latest Advances with Special Reference to Gastric and Duodenal Ulcer*, edited by F. Halle and S. Andersson. New York: Springer-Verlag, 1974, p. 109–111.

NASSET, E. S. Enterocrinin, a hormone which excites the glands of the small intestine. *Am. J. Physiol.* 121: 481–487, 1938.

NELLANS, H. N., AND D. V. KIMBERG. Anomalous calcium secretion in rat ileum: role of paracellular pathway. *Am. J. Physiol.* 236 (*Endocrinol. Metab. Gastrointest. Physiol.* 5): E473–481, 1979.

NELLANS, H. N., AND J. E. POPVITCH. Calmodulin-regulated, ATP-driven calcium transport by basolateral membranes of rat small intestine. *J. Biol. Chem.* 256: 9932–9936, 1981.

NEUTRA, M., AND C. P. LEBLOND. Synthesis of the carbohydrate of mucus in the Golgi complex as shown by electron microscope radioautography of goblet cells from rats injected with glucose-H^3. *J. Cell Biol.* 30: 119–136, 1966.

NEUTRA, M. R., L. J. O'MALLEY, AND R. D. SPECIAN. Regulation of intestinal goblet cell secretion. II. A survey of potential secretagogues. *Am. J. Physiol.* 242 (*Gastrointest. Liver Physiol.* 5): G380–G387, 1982.

NILSSON, S., AND S. STATTIN. Gallbladder emptying during the normal menstrual cycle. *Acta Chir. Scand.* 133: 648–652, 1967.

NORMAN, D. A., S. G. MORAWSKI, AND J. S. FORDTRAN. Influence of glucose, fructose, and water movement on calcium absorption in the jejunum. *Gastroenterology* 78: 22–25, 1980.

NOTIS, W., S. A. ORELLANA AND M. FIELD. Inhibition of intestinal secretion in rats by colchicine and vinblastine. *Gastroenterology* 81: 766–772, 1981.

NOVI, A. M., AND R. BASERGA. Correlation between synthesis of ribosomal RNA and stimulation of DNA synthesis in mouse salivary glands. *Lab. Invest.* 26: 540–547, 1972.

O'DOHERTY, J., AND R. J. STARK. Stimulation of pancreatic acinar secretion: increases in cytosolic calcium and sodium. *Am. J. Physiol.* 242 (*Gastrointest. Liver Physiol.* 5): G513–G521, 1982.

OHASHI, H., AND J. H. MEYER. Effect of peptic digestion on emptying of cooked liver in dogs. *Gastroenterology* 79: 305–310, 1980.

OIGAARD, A.A, AND S. DORPH. The relative significance of electrical spike potentials and intraluminal pressure waves as quantitative indicators of motility. *Am. J. Digest. Dis.* 19: 797–803, 1974.

OKISHIO, T., AND P. P. NAIR. Studies on bile acids: some observations on the intracellular localization of major bile acids in rat liver. *Biochem.* 5: 3662–3668, 1966.

OLBE, L., AND U. KNUTSON. Gastric acid response to sham-feeding in the dog and the duodenal ulcer patient. *Acta Hepatogastroent.* 23: 455–458, 1976.

OLBE, L., U. HAGLUND, R. LETH, T. LIND, C. CEDERBERG, G. EKENVED, B. ELANDER, E. FELLENIUS, P. LUNDBERG, AND B. WALLMARK. Effects of substituted benzimidazole (H-149/94) on gastric acid secretion in humans. *Gastroenterology* 83: 193–198, 1982.

ORMSBEE, H. S., III, S. L. KOEHLER, JR., AND G. L. TELFORD. Somatostatin inhibits motilin induced interdigestive contractile activity in the dog. *Am. J. Digest. Dis.* 23: 789–795, 1978.

ORMSBEE, H. S., G. L. TELFORD, C. M. SUTER, P. D. WILSON, AND G. R. MASON. Mechanisim of propagation of canine migrating motor complex—a reappraisal. *Am. J. Physiol.* 240 (*Gastrointest. Liver Physiol.* 3): G141–G146, 1981.

OSNES, M., L. E. HANSSEN, AND S. LARSEN. The unstimulated pancreatic secretion obtained by endoscopic cannulation and the plasma secretin levels in man. *Scand. J. Gastroenterol.* 14: 503–509, 1979.

OSUMI, Y., S. AIBARA, K. SAKAE, AND M. FUJIWARA. Central noradrenergic inhibition of gastric mucosal blood flow and acid secretion in rats. *Life Sci.* 20: 1407–1416, 1977.

OTSUKI, M., AND J. A. WILLIAMS. Protein synthesis inhibitors enhance secretagogue sensitivity of in vitro rat pancreatic acini. *Am. J. Physiol.* 243 (*Gastrointest. Liver Physiol.* 6): G285–G290, 1982.

OUYANG, A., W. J. SNAPE, JR., AND S. COHEN. Myoelectric properties of the cat ileocecal sphincter. *Am. J. Physiol.* 240 (*Gastrointest. Liver Physiol.* 3): G450–G458, 1981.

OUYANG, A., C. J. CLAIN, W. J. SNAPE, JR., AND S. COHEN. Characterization of opiate-mediated responses of the feline ileum and ileocecal sphincter. *J. Clin. Invest.* 69: 507–515, 1982.

OWYANG, C., L. J. MILLER, E. P. DI MAGNO, J. C. MITCHELL, III, AND V. L. W. GO. Pancreatic exocrine function in severe human chronic renal failure. *Gut* 23: 357–361, 1982a.

OWYANG, C., J. H. SCARPELLO, AND A. I. VINIK. Correlation between pancreatic enzyme secretion and plasma concentration of human pancreatic polypeptide in health and in chronic pancreatitis. *Gastroenterology* 83: 55–62, 1982b.

PACE, J. L., AND I. WILLIAMS. Organization of the muscular wall of the human colon. *Gut* 10: 352–359, 1969.

PAINTER, N. S., S. C. TRUELOVE, G. M. ARDRAN, AND M. TUCKEY. Segmentation and the localization of intraluminal pressures in the human colon, with special reference to the pathogenesis of colonic diverticula. *Gastroenterology* 49: 169–177, 1965.

PALADE, G. Intracellular aspects of the process of protein synthesis. *Science* 189: 347–358, 1975.

PALLIN, B., AND S. SKOGLUND. Neural and humoral control

of the gallbladder-emptying mechanism in the cat. *Acta Physiol. Scand.* 60: 358–362, 1964.

PALMA, R., N. VEDON, AND J. J. BERNIER. Maximal capacity for fluid absorption in human bowel. *Digest. Dis. Sci.* 26: 929–934, 1981.

PARKS, T. G. Colonic motility in man. *Postgrad. Med. J.* 49: 90–99, 1973.

PATTON, J. S. Gastrointestinal lipid digestion. In: *Pathology of the Gastrointestinal Tract*, edited by L. R. Johnson. New York: Raven Press, 1981, p. 1123–1146.

PAUL, R. J. Smooth muscle: mechanochemical energy conversion relations between metabolism and contractile. In: *Physiology of the Gastrointestinal Tract*, edited by L. R. Johnson. New York: Raven Press, 1981, p. 269–288.

PAUWELS, S., AND G. J. DOCKRAY. Identification of NH_2-terminal fragments of big gastrin in plasma. *Gastroenterology* 82: 56–61, 1982.

PEARSE, A. G. E. Letter to the editor. *N. Engl. J. Med.* 307: 629–630, 1982.

PEARSON, G. T., J. SINGH, M. S. DAOUD, J. S. DAVISON, AND O. H. PETERSEN. Control of pancreatic cyclic nucleotide levels and amylase secretion by noncholinergic, nonadrenergic nerves: a study employing electrical field stimulation of guinea pig segments. *J. Biol. Chem.* 256: 11025–11031, 1981.

PEETERS, T. L., G. VANTRAPPEN, AND J. JANSSEN. Bile acid output and the interdigestive migrating motor complex in normals and in cholecystectomy patients. *Gastroenterology* 79: 678–681, 1980.

PEKAS, J. C. Pancreatic incorporation of ^{65}Zn and histidine-^{14}C into secreted proteins of the pig. *Am. J. Physiol.* 220: 799–803, 1971.

PENTTILA, A. Distribution and intensity of monomine oxidase activity in the mammalian duodenum. *Acta Physiol. Scand.* 73: 121–127, 1968.

PEREC, C. J., A. B. HOUSSAY, A. V. PERONACE, AND M. N. FABIANO. Submaxillary and retrolingual hypertrophy by inferior incisor extractions in the rat. *Acta Physiol. Latino Americana* 14: 211–214, 1964.

PERRY, M. A., D. MURPHREE, AND D. N. GRANGER. Oxygen uptake as a determinant of gastric blood flow autoregulation. *Digest. Dis. Sci.* 27: 675–679, 1982.

PERSSON, C. G. A. Adrenergic, cholecystokinetic and morphine induced effects on extrahepatic biliary motility. *Acta Physiol. Scand. Suppl.* 383: 1–32, 1972.

PERSSON, C. G. Dual effects on the sphincter of Oddi and gallbladder induced by stimulation of the right great splanchnic nerve. *Acta Physiol. Scand.* 87: 334–343, 1973.

PETERS, M. N., J. H. WALSH, J. FERRARI, AND M. FELDMAN. Adrenergic regulation of distention-induced gastrin release in humans. *Gastroenterology* 82: 659–663, 1982.

PETERSEN, H., T. SOLOMON, AND M. I. GROSSMAN. Effect of chronic pentagastrin, chlecystokinin and secretin on pancreas of rats. *Am. J. Physiol.* 234 (*Endocrinol. Metab. Gastrointest. Physiol.* 3): E286–E293, 1978.

PETERSEN, O. H. Some factors influencing stimulation-induced release of potassium from the cat submandibular gland to fluid perfused through the gland. *J. Physiol. (London)* 208: 431–447, 1970.

PETERSEN, O. H. Electrophysiology of exocrine gland cells. In: *Physiology of the Gastrointestinal Tract*, edited by L. R. Johnson. New York: Raven Press, 1981, p. 749–772.

PETERSEN, O. H., Y. MARUYAMA, J. GRAF, R. LAUGIER, A. NISHIYAMA, AND G. T. PEARSON. Ionic currents across pancreatic acinar cell membranes and their role in fluid secretion. *Philos. Trans. R. Soc. London B, Biol. Sci.* 296: 151–166, 1981.

PETITH, M. M., AND H. P. SCHEDL. Intestinal adaptation to dietary calcium restriction: in vivo cecal and colonic calcium transport in the rat. *Gastroenterology* 71: 1039–1042, 1976.

PETITH, M. M., J. R. WENGER, AND H. P. SCHEDL. Vitamin D dependence and aboral gradient of in vivo intestinal calcium transport in the rat. *Am. J. Digest. Dis.* 23: 943–947, 1978.

PETITH, M. M., H. D. WILSON, AND H. P. SCHEDL. Vitamin

D dependence of in vivo calcium transport and mucosal calcium binding protein in rat large intestine. *Gastroenterology* 76: 99–104, 1979.

PHAOSAWASDI, K., AND R. S. FISHER. Hormonal effects on the pylorus. *Am. J. Physiol.* 243 (*Gastrointest. Liver Physiol.* 6): G330–G335, 1982.

PHAOSAWASDI, K., R. GOPPOLD, AND R. S. FISHER. Pyloric pressure response to insulin-induced hypoglycemia in humans. *Am. J. Physiol.* 241 (*Gastrointest. Liver Physiol.* 4): G321–G327, 1981.

PHILLIPS, S. F. Absorption and secretion by the colon. *Gastroenterology* 56: 966–971, 1969.

PHILLIPS, S. F., AND D. A. W. EDWARDS. Some aspects of anal continence and defaecation. *Gut* 6: 396–406, 1965.

PHILLIPS, M. J., C. OSHIO, M. MIYAIRI, H. KATZ, AND C. R. SMITH. A study of bile canalicular contractions in isolated hepatocytes. *Hepatology* 2: 763–768, 1982.

PODDAR, S., AND S. JACOB. Mucosubstance histochemistry of Brunner's glands, pyloric glands and duodenal goblet cells in the ferret. *Histochemistry* 65: 67–81, 1979.

PODESTA, R. B., AND D. F. METTRICK. HCO_3 and H^+ secretion in rat ileum in vivo. *Am. J. Physiol.* 232 (*Endocrinol. Metab. Gastrointest. Physiol.* 1): E574–E579, 1977.

POHTO, P., AND M. K. PAASONEN. Studies on the salivary gland hypertrophy induced in rats by isoprenaline. *Acta Pharmacol. Toxicol.* 21: 45–50, 1964.

POITRAS, P., J. H. STEINBACH, G. VAN DEVENTER, C. F. CODE, AND J. H. WALSH. Motilin-independent ectopic fronts of the interdigestive myoelectric complex in dogs. *Am. J. Physiol.* 239 (*Gastrointest. Liver Physiol.* 2): G215–G200, 1980.

POLAK, J. M., S. BLOOM, I. COULLING, AND A. G. E. PEARSE. Immunofluorescent localization of secretin in the canine duodenum. *Gut* 12: 605–610, 1971.

POLAK, J. M., S. R. BLOOM, P. L. RAYFORD, A. G. E. PEARSE, A. M. BUCHAN, AND J. C. THOMSPON. Identification of cholecystokinin-secretory cells. *Lancet* 2: 1016–1018, 1975.

PONNAPPA, B. C., R. L. DORMER, AND J. A. WILLIAMS. Characterization of an ATP-dependent Ca^2 uptake system in mouse pancreatic microsomes. *Am. J. Physiol.* 240 (*Gastrointest. Liver Physiol.* 3): G122–G129, 1981.

POUPON, R. E., AND W. H. EVANS. Biochemical evidence that Na^+K^+-ATPase is located at the lateral region of the hepatocyte surface membrane. *FEBS Lett.* 8: 374–378, 1979.

POVILAITIS, V., C. GAGNON, AND S. HEISLER. Stimulus-secretion coupling in exocrine pancreas: role of protein carboxyl methylation. *Am. J. Physiol.* 240 (*Gastrointest. Liver Physiol.* 3): G199–G205, 1981.

POWELL, D. W. Barrier function of epithelia. *Am. J. Physiol.* 241 (*Gastrointest. Liver Physiol.* 4): G275–G288, 1981.

POWELL, D. W., AND E. J. TAPPER. Autonomic control of intestinal electrolyte transport. In: *Frontiers of Knowledge in the Diarrheal Diseases*, edited by H. D. Janowitz and D. B. Sachar. Upper Montclair, NJ: Projects in Health Inc., 1979a, p. 37–52.

POWELL, D. W., AND E. J. TAPPER. Intestinal transport: cholinergic-adrenergic interactions. In: *Mechanisms of Intestinal Secretion*, edited by H. J. Binder. New York: Alan R. Less, 1979b, p. 175–192.

POWELL, L. W., AND J. W. HALLIDAY. Iron absorption and iron overload. *Clin. Gastroenterol.* 10: 707–736, 1981.

PRANDI, D., S. ERLINGER, J. C. GLASINOVIC, AND M. DUMONT. Canalicular bile production in man. *Digestion* 8: 437, 1973.

PUTNEY, J. W., J. POGGIOLI, AND S. J. WEISS. Receptor regulation of calcium release and calcium permeability in parotid gland cells. *Phil. Trans. R. Soc. Lond. B* 296: 37–45, 1981.

RACUSEN, L. C., AND H. J. BINDER. Alteration of large intestinal electrolyte transport by vasoactive intestinal polypeptide in the rat. *Gastroenterology* 73: 790–796, 1977.

RACUSEN, L. C., AND H. J. BINDER. Effect of prostaglandin

on ion transport across isolated colonic mucosa. *Digest. Dis. Sci.* 25: 900–904, 1980.

RAMUS, N. I., R. C. N. WILLIAMSON, J. M. OLIVER, AND D. JOHNSTON. Effect of highly selective vagotomy on pancreatic exocrine function on cholecystokinin and gastric release. *Gut* 23: 553–557, 1982.

RANDLE, P. J., S. FODEN, AND P. KANAGASUNTHERAM. Studies of cell and mitochondrial calcium pools in rat hepatocytes and parotid exocrine cells. In: *Secretory Mechanisms*, edited by C. R. Hopkins and C. J. Duncan. Cambridge: Cambridge Univ. Press, 1979, p. 199–223.

RASMUSSEN, H., O. FONTAINE, O., E. E. MAX, AND D. B. P. GOODMAN. The effect of 1 α hydroxyvitamin D_3 administration in calcium transport in chick intestine brush border membrane vehicles. *J. Biol. Chem.* 254: 2993–2999, 1979.

READ, N. W., K. COOPER, AND J. S. FORDTRAN. Effect of modified sham-feeding on jejunal transport and pancreatic and biliary secretion in man. *Am. J. Physiol.* 234 (*Endocrinol. Metab. Gastrointest. Physiol.* 4): E417–E420, 1978.

READ, N. W., G. J. KREJS, R. M. BARKELY, S. G. MORAWSKI, AND J. S. FORDTRAN. Spontaneous secretion from the dog small intestine in vivo. *J. Clin. Med.* 93: 381–389, 1979.

REDINGER, R. N. The effect of loss of gallbladder function on biliary lipid composition in subjects with cholesterol gallstones. *Gastroenterology* 71: 470, 474, 1976.

REED, J. D., AND D. J. SANDERS. Pepsin secretion, gastric motility, and mucosal blood flow in anesthetized cat. *J. Physiol. (London)* 216: 159–170, 1971.

REED, J. D., D. J. SANDERS, AND V. THORPE. The effect of splanchnic nerve stimulation on gastric acid secretion and mucosal blood flow in the anaesthetized cat. *J. Physiol. (London)* 214: 1–13, 1971.

REES, W. D. W., V. L. W. GO, AND J-R MALAGELADA. Antroduodenal motor response to solid-liquid and homogenized meals. *Gastroenterology* 76: 1438–1442, 1979.

REES, W. D. W., J-R. MALAGELADA, L. J. MILLER, AND V. L. W. GO. Human interdigestive and postprandial gastrointestinal motor and gastrointestinal hormone patterns. *Digest. Dis. Sci.* 27: 321–329, 1982.

REGAN, P. T., V. L. W. GO, AND E. P. DI MAGNO. Exocrine and endocrine pancreatic function in man in response to cholecystokinin (CCK) and octapeptide of CCK (CCK-OP). *Gastroenterology* 76: 1224, 1979.

REHFELD, J. Four basic characteristics of the gastrin-cholecystokinin system. *Am. J. Physiol.* 240 (*Gastrointest. Liver Physiol.* 3): G255–G266, 1981.

REICHEN, J., AND G. PAUMGARTNER. Uptake of bile acids by perfused rat liver. *Am. J. Physiol.* 231: 734–742, 1976.

REICHEN, J., M. D. BERMAN, AND P. D. BERK. Role of microfilaments and of microtubules in taurocholate uptake by isolated rat liver cells. *Biochim. Biophys. Acta* 643: 126–133, 1981.

RENNY, A., W. J. SNAPE, S. COHEN, E. A. SUN, AND R. LONDON. The interaction of multiple receptors in the gastrocolonic response. *Clin. Res.* 30: 288A, 1982.

RENSTON, R. H., D. G. MALONEY, A. L. JONES, G. T. HRADEK, K. Y. WONG, AND I. D. GOLDFINE. Bile secretory apparatus evidence for a vesicular, transport mechanism for proteins in the rat, using horseradish peroxidase and (^{125}I) insulin. *Gastroenterology* 78: 1373–1388, 1980.

REYES, H., AND F. KERN, JR. Effect of pregnancy on bile flow and biliary lipids in the hamster. *Gastroenterology* 76: 144–150, 1979.

REYNOLDS, J. C., A. OUYANG, AND S. COHEN. Electrically coupled intrinsic responses of feline lower esophageal sphincter. *Am. J. Physiol.* 243 (*Gastrointest. Liver Physiol.* 6): G415–G423, 1982.

RHODES, R. A., W. Y. CHEY, H. H. TAI, AND H. TABECHIAN. Hypersecretinemia associated with renal failure. *Clin. Res.* 23: 255A, 1975.

RICCI, G. L., AND J. FEVERY. Cholestatic action of somatosta-

tin in the rat: effect on the different fractions of bile secretion. *Gastroenterology* 81: 552–562, 1981.

RICHARDSON, C. T., J. H. WALSH, M. I. HICKS, AND J. S. FORDTRAN. Studies on the mechanisms of food-stimulated gastric acid secretion in normal human subjects. *J. Clin. Invest.* 623–631, 1976.

RICHARDSON, C. T., J. H. WALSH, K. A. COOPER, M. FELDMAN, AND J. S. FORDTRAN. Studies on the role of cephalic vagal stimulation in the acid secretory response to eating in normal human subjects. *J. Clin. Invest.* 60: 435–441, 1977.

RIDLEY, P. T., AND F. P. BROOKS. Alterations in gastric secretion following hypothalamic lesions producing hyperphagia. *Am. J. Physiol.* 209: 319–323, 1965.

RINDERKNECHT, H., M. R. NAGARAJA, AND N. F. ADHAM. Effect of bile acids and pH on the release of enteropeptidase in man. *Am. J. Digest. Dis.* 23: 332–336, 1978a.

RINDERKNECHT, H., I. G. RENNER, A. P. DOUBLAS, AND N. F. ADHAM. Profiles of pure pancreatic secretions obtained by direct pancreatic duct cannulation in normal healthy human subjects. *Gastroenterology* 75: 1083–1089, 1978b.

RINDERNECHT, H., I. G. RENNER, AND H. H. KOYAMA. Lysosomal enzymes in pure pancreatic juice from normal healthy volunteers and chronic alcoholics. *Digest. Dis. Sci.* 24: 180–186, 1979.

RITCHIE, J. A. Colonic motor activity and bowel function. Part I. Normal movement of contents. *Gut* 9: 442–456, 1968a.

RITCHIE, J. A. Colonic motor activity and bowel function. *Gut* 9: 502–511, 1968b.

RITCHIE, J. A. The transport of colonic contents in the irritable colon syndrome. *Gut* 11: 668–672, 1970.

ROBBERECHT, P., M. CREMER, AND J. CHRISTOPHE. Discharge of newly synthesized proteins in pure juice collected from the human pancreas: indication of more than one pool of intracellular digestive enzymes. *Gastroenterology* 72: 417–420, 1977.

ROGGIN, G. M., J. G. BANWELL, J. H. YARDLEY, AND T. R. HENDRIX. Unimpaired response of rabbit jejunum to cholera toxin after selective damage to villus epithelium. *Gastroenterology* 63: 981–989, 1972.

ROLLING, G. T., R. L. FARRELL, AND D. O. CASTELL. Cholinergic response of the lower esophageal sphincter. *Am. J. Physiol.* 222: 967–972, 1972.

ROMAN, C., AND J. GONELLA. Extrinsic control of digestive tract motility. In: *Physiology of the Gastrointestinal Tract*, edited by L. R. Johnson. New York: Raven Press, 1981, p. 319–326.

ROSATO, E. F., AND F. E. ROSATO. Effect of truncal vagotomy on acid and pepsin responses to histamine in duodenal ulcer subjects. *Ann. Surg.* 173: 63–66, 1971.

ROSE, R. C. Absorptive functions of the gallbladder. In: *Physiology of the Gastrointestinal Tract*, edited by L. R. Johnson. New York: Raven Press, 1981, p. 1021–1033.

ROSELL, S., AND A. ROKAEUS. The effect of ingestion of amino acids, glucose and fat on circulating neurotensin-like immunoreactivity (NTLI) in man. *Acta Physiol. Scand.* 104: 263–267, 1979.

ROSENBERG, I. H. Intestinal absorption of folate. In: *Physiology of the Gastrointestinal Tract*, edited by L. R. Johnson. New York: Raven Press, 1981, p. 1221–1230.

ROSENBLUM, M. J., AND A. J. CUMMINS. The effect of sleep and of amytal on the motor activity of the human sigmoid colon. *Gastroenterology* 27: 445–450, 1954.

ROSSIGNOL, B., G. KERYER, G. HERMAN, A. M. CHAMBAUT-GUERIN, AND C. COHOREAU. Function of cholinergic and adrenergic receptors in the control of metabolism and protein secretion in rat salivary glands. In: *First International Symposium on Hormonal Receptors in Digestive Tract Physiology*, edited by S. Bonfils, et al. Amsterdam: Elsevier, 1977, p. 311–323.

ROTH, J., D. LE ROITH, J. SHILOACH, J. C. ROSENZWEIG, M. A. LESNIAK, AND J. HAVRANKOVA. The evolutionary origins of hormones, neurotransmitters, and other extracellular chemical messengers. *N. Engl. J. Med.* 306: 525–527, 1982a.

ROTH, J., D. LE ROITH, J. SHILOACH, J. L. ROSENZWEIG, M. LESNIAK, AND J. HAVRANKOVA. Letter to editor. *N. Engl. J. Med.* 307: 630, 1982b.

ROTHMAN, S. S. Protein transport by the pancreas. *Science* 190: 747–753, 1975.

ROTHMAN, S. S. Passage of proteins through membranes—old assumptions and new perspectives. *Am. J. Physiol.* 238 (*Gastrointest. Liver Physiol.* 1): G391–G402, 1980.

RUBIN, M. R., J. FOURNET, W. J. SNAPE, JR., AND S. COHEN. Adrenergic regulation of ileocecal sphincter function in the cat. *Gastroenterology* 78: 15–21, 1980.

RUBIN, M. R., B. A. CARDWELL, A. OUYANG, W. J. SNAPE, JR., AND S. COHEN. Effect of bethanechol or vagal nerve stimulation on ileocecal sphincter pressure in the cat. *Gastroenterology* 80: 974–979, 1981.

SACHS, G., AND T. BERGLINDH. Physiology of the parietal cell. In: *Physiology of the Gastrointestinal Tract*, edited by L. R. Johnson. New York: Raven Press, 1981, p. 567–602.

SAFFOURI, B., C. C. WEIR, K. N. BITAR, AND G. M. MAKH-LOUF. Gastrin and somatostatin secretion by perfused rat stomach: functional linkage of antral peptides. *Am. J. Physiol.* 238 (*Gastrointest. Liver Physiol.* 1): G495–G501, 1980.

SAFFOURI, B., J. W. Du VAL, A. ARIMURA, AND G. M. MAKHLOUF. Effects of VIP and secretin on gastrin and somatostatin secretion by the isolated rat stomach. *Gastroenterology* 80: 1267, 1981.

SAKAGUCHI, T., AND K. YAMAGUCHI. Changes in efferent activities of the gastric vagus nerve by administration of glucose in the portal vein. *Experientia* 35: 875–876, 1979.

SAMLOFF, I. M. Pepsinogens I and II: purification from gastric mucosa and radioimmunoassay in serum. *Gastroenterology* 82: 26–32, 1982.

SANDOW, M. J., AND R. WHITEHEAD. The paneth cell. *Gut* 20: 420–431, 1979.

SANKARAN, H., Y. IWAMOTO, M. KORC, J. A. WILLIAMS, AND I. D. GOLDFINE. Insulin action in pancreatic acini from streptozotocin-treated rats. II. Binding of ^{125}I-insulin to receptors. *Am. J. Physiol.* 240 (*Gastrointest. Liver Physiol.* 3): G63–G68, 1981.

SANKARAN, H., I. D. GOLDFINE, A. BAILEY, V. LICKO, AND J. A. WILLIAMS. Relationship of cholecystokinin receptor binding to regulation of biological functions in pancreatic acini. *Am. J. Physiol.* 242 (*Gastrointest. Liver Physiol.* 5): G250–G257, 1982.

SARAUX, B., A. GIRARD-GLOBA, M. OUAGUED, AND D. VACHER. Response of the exocrine pancreas to quantitative and qualitative variations in dietary lipids. *Am. J. Physiol.* 243 (*Gastrointest. Liver Physiol.* 6): G10–G15, 1982.

SARLES, H., R. DANI, G. PREZELIN, C. SOUVILLE, AND C. FIGARELLA. Cephalic phase of pancreatic secretion in man. *Gut* 9: 214–221, 1968.

SARLES, H., J. HAUTON, N. E. PLANCHE, H. LAFONT, AND H. GEROLAMI. Diet cholesterol gallstones and composition of the bile. *Am. J. Digest. Dis.* 15: 251–260, 1970.

SARNA, S. K., AND E. E. DANIEL. Electrical stimulation of small intestinal electrical control activity. *Gastroenterology* 69: 660–667, 1975.

SARNA, S. K., E. E. DANIEL, AND W. E. WATERFALL. Myogenic and neural control systems for esophageal motility. *Gastroenterology* 73: 1345–1352, 1977.

SARNA, S. K., B. L. BARDAKJIAN, W. E. WATERFALL, AND J. F. LIND. Human colonic electrical control activity (ECA). *Gastroenterology* 78: 1526–1536, 1980.

SARNA, S. K., W. E. WATERFALL, B. L. BORDAKJIAN, AND J. F. LIND. Types of human colonic electrical activities recorded post-operatively. *Gastroenterology* 81: 61–70, 1981a.

SARNA, S., C. STODDARD, L. BELBECK, AND D. MCWADE. Intrinsic nervous control of migrating myoelectric complexes. *Am. J. Physiol.* 241 (*Gastrointest. Liver Physiol.* 4): G16–G23, 1981b.

SARNA, S., P. LATIMER, D. CAMPBELL, AND W. E. WA-TERFALL. Electrical and contractile activities of the human rectosigmoid. *Gut* 23: 698–705, 1982.

SARR, M. G., AND K. A. KELLY. Patterns of movement of liquids and solids through canine jejunum. *Am. J. Physiol.* 239 (*Gastrointest. Liver Physiol.* 2): G497–G503, 1980.

SARR, M. G., AND K. A. KELLY. Myoelectrical activity of the autotransplanted canine jejunoileum. *Gastroenterology* 81: 303–310, 1981.

SARR, M. G., K. A. KELLY, AND S. F. PHILLIPS. Canine jejunal absorption and transit during interdigestive and digestive motor states. *Am. J. Physiol.* 239 (*Gastrointest. Liver Physiol.* 2): G167–G172, 1980.

SCHAFFALITZKY DE MUCKADELL, O. B., J. FAHRENK-RUG, S. WATT-BOOLSEN, AND H. WORNING. Pancreatic response and plasma secretin concentration during infusion of low dose secretion in man. *Scand. J. Gastroenterol.* 13: 305–312, 1978.

SCHAFFALITZKY DE MUCKADELL, O. B., J. FAHRENK-RUG, AND S. J. RUNE. Physiological significance of secretin in the pancreatic bicarbonate secretion. I. Responsiveness of the secretin-releasing system in the upper duodenum. *Scand. J. Gastroenterol.* 14: 79–83, 1979a.

SCHAFFALITZKY DE MUCKADELL, O. B., J. FAHRENK-RUG, P. MATZEN, S. J. RUNE, AND H. WORNING. Physiological significance of secretin in the pancreatic bicarbonate secretion. II. Pancreatic bicarbonate response to a physiological increase in plasma secretin concentration. *Scand. J. Gastroenterol.* 14: 85–90, 1979b.

SCHAFFALITZKY DE MUCKADELL, O. B., J. FAHRENK-RUG, J. NIELSEN, I. WESTPHALL, AND H. WORNING. Meal-stimulated secretin release in man: effect of acid and bile. *Scand. J. Gastroenterol.* 16: 981–988, 1981.

SCHANG, J., J. DAUCHEL, P. SAVA, F. ANGEL., P. BOUCHET, A. LAMBERT, AND J. F. GRENIER. Specific effects of different food components on intestinal motility. *Eur. Surg. Res.* 10: 425–432, 1978.

SCHANG, J-C., K. A. KELLY, AND V. L. W. GO. Post-prandial hormonal inhibition of canine interdigestive gastric motility. *Am. J. Physiol.* 240 (*Gastrointest. Liver Physiol.* 3): G221–G224, 1981.

SCHAPIRO, H., L. D. WRUBLE, L. G. BRITT, AND T. A. BELL. Sensory deprivation on visceral activity. I. The effect of visual deprivation on canine gastric secretion. *Psychosom. Med.* 32: 379–396, 1970a.

SCHAPIRO, H., C. W. GROSS, T. NAKAMURA, L. D. WRU-BLE, AND L. G. BRITT. Sensory deprivation on visceral activity. II. The effect of auditory and vestibular deprivation on canine gastric secretion. *Psychosom. Med.* 32: 515–521, 1970b.

SCHAPIRO, H., L. G. BRITT, C. W. GROSS, AND K. J. GAINES. Sensory deprivation on visceral activity. III. The effect of olfactory deprivation on canine gastric secretion. *Psychosom. Med.* 33: 429–435, 1971.

SCHAPIRO, H., L. G. BRITT, AND R. H. DOHRN. Sensory deprivation on visceral activity. IV. The effect of temporary visual deprivation on canine gastric secretion. *Digest Dis.* 18: 573–575, 1973.

SCHAPIRO, H., H. D. McDOUGAL, F. E. ROSATO, G. CAR-WELL, AND N. J. JACKSON. Canine gastric secretion regulation by the submucosal ganglia and nerve plexi. In: *Nerves and the Gut*, edited by F. P. Brooks and P. W. Evers, Thorofare, NJ: Charles B. Slack, 1977, p. 430–438.

SCHARSCHMIDT, B. F., AND R. SCHMID. The micellar sink. A quantitative assessment of the association of organic anions with mixed micelles and other macromolecular aggregates in rat bile. *J. Clin. Invest.* 62: 1122–1131, 1978.

SCHARSCHMIDT, B. F., AND J. E. STEPHENS. Transport of sodium, chloride, and taurocholate by cultured rat hepatocytes. *Proc. Natl. Acad. Sci. USA* 78: 986–990, 1981.

SCHARSCHMIDT, B. F., J. G. WAGGONER AND P. D. BERK. Hepatic organic anion-uptake in the rat. *J. Clin. Invest.* 56: 1280–1292, 1975.

SCHARSCHMIDT, B., E. B. KEEFFE, D. A. VESEY, N. M.

BLANKENSHIP, AND R. K. OCKNER. In vitro effects of bile salts on rat liver plasma membrane, lipid fluidity and ATPase activity. *Hepatology* 1: 137–145, 1981.

SCHEELE, G., D. BARTELT, AND W. BIEGER. Characterization of human exocrine pancreatic proteins by two-dimensional isoelectric focusing/sodium dodecyl sulfate gel electrophoresis. *Gastroenterology* 80: 461–473, 1981.

SCHNEYER, L. H. Effects of calcium on Na, K transport by perfused main duct of rat submaxillary gland. *Am. J. Physiol.* 226: 821–826, 1974a.

SCHNEYER, L. H. Differential effects of cytochalasin β on Na and K transport in a perfused salivary duct. *Am. J. Physiol.* 227: 606–612, 1974b.

SCHNEYER, L. H., AND T. THAVORNTHON. Isoproterenol induced stimulation of sodium absorption in perfused salivary duct. *Am. J. Physiol.* 224: 136–139, 1973.

SCHNEYER, L. H., J. A. YOUNG, AND C. A. SCHNEYER. Salivary secretion of electrolytes. *Physiol. Rev.* 52: 720–777, 1972.

SCHNEYER, C. A. Calcium, amylase, and flow rate of rat parotid saliva with diverse frequencies of parasympathetic nerve stimulation. *Proc. Soc. Exper. Biol. Med.* 162: 405–409, 1979.

SCHNEYER, C. A., AND H. D. HALL. Neurally mediated increase in mitosis and DNA of rat parotid with increase in bulk of diet. *Am. J. Physiol.* 230: 911–915, 1976.

SCHNEYER, C. A., C. SUCANTHAPREE, AND L. SCHNEYER. Neural regulation of calcium and amylase of rat parotid saliva. *Proc. Soc. Exper. Biol. Med.* 156: 132–135, 1977.

SCHNEYER, C. A., C. SACANTHAPREE, L. H. SCHNEYER, AND D. JIRAKLILSOMCHOK. Total calcium and amylase output of rat parotid with electrical stimulation of autonomic innervation. *Proc. Soc. Exper. Biol. Med.* 159: 478–483, 1978.

SCHOFIELD, G. C. The enteric plexus of mammals. In: *International Review of General and Experimental Zoology*, edited by W. J. L. Felts and R. J. Harrison, New York: Academic Press, 1968, p. 53–116.

SCHOON, I-M, S. BERGEGARDH, U. GROTZINGER, AND L. OLBE. Evidence for a defective inhibition of pentagastrin-stimulated gastric acid secretion by antral distention in the duodenal ulcer patient. *Gastroenterol.* 75: 363–367, 1978.

SCHUBERT, M. L., AND G. M. MAKHLOUF. Regulation of gastrin and somatostatin secretion by intramural neurons: effect of nicotinic receptor stimulation with dimethyl-phenylpiperaginium. *Gastroenterol.* 83: 626–632, 1982.

SCHULTZ, S. G. Mechanisms of intestinal secretion. In: *Frontiers of Knowledge in the Diarrheal Diseases*, edited by H. D. Janowitz and D. B. Sachar. Upper Montclair, NJ: Projects in Health, Inc., 1979, p. 3–13.

SCHULTZ, S. G. Ion transport by mammalian large intestine. In: *Physiology of the Gastrointestinal Tract*, edited by L. R. Johnson. New York: Raven Press, 1981a, p. 991–1002.

SCHULTZ, S. G. Salt and water absorption by mammalian small intestine. In: *Physiology of the Gastrointestinal Tract*, edited by L. R. Johnson, New York: Raven Press, 1981b, p. 983–989.

SCHULZ, I. Electrolyte and fluid secretion in the exocrine pancreas. In: *Physiology of the Gastrointestinal Tract*, edited by L. R. Johnson. New York: Raven Press, 1981, p. 795–819.

SCHULZ, I., T. KIMURA, H. WAKASUGI, W. HAASE, AND A. KRIBBEN. Analysis of calcium fluxes and calcium pools in pancreatic acini. *Philos. Trans. R. Soc. Lond. B. Biol. Sci.* 296: 105–114, 1981.

SCHULZE-DELRIEU, K., AND G. LEPSIEN. Depression of mechanical and electrical activity in muscle strips of opossum stomach and esophagus by acidosis. *Gastroenterology* 82: 720–724, 1982.

SCHUTZBERG, M., T. HOKFELT, G. NILSSON, L. TERENIUS, J. F. REHFELD, M. BROWN, R. ELDE, M. GOLDSTEIN, AND S. SAID. Distribution of peptide- and catecholamine-containing neurons in the gastrointestinal tract of rat and guinea pig: immunohistochemical studies with anti-sera to substance P, vasoactive intestinal polypeptide, enkephalins, somatostatin, gastrin/cholecystokinin, neurotensin and dopamine β hydroxylase. *Neuroscience* 5: 689–744, 1980.

SCHWARTZ, C. J., D. V. KIMBERG, H. E. SHEERIN, M. FIELD, AND S. I. SAID. Vasoactive intestinal peptide stimulation of adenylate cyclase and active electrolyte secretion in intestinal mucosa. *J. Clin. Invest.* 54: 536–544, 1974.

SCHWARTZ, T. W., F. STADIL, R. E. CHANCE, J. F. REHFELD, L. I. LARSSON, AND N. MOON. Pancreatic polypeptide response to food in duodenal-ulcer patients before and after vagotomy. *Lancet* 1: 1102–1105, 1976.

SCHWARTZ, T. W., J. HOLST, J. FAHRENKRUG, J. LINDKAER, S. JENSEN, O. V. NIELSEN, J. F. REHFELD, O. B. SCHAFFALITZKY DE MUCKADELL, AND F. STADIL. Vagal cholinergic regulation of pancreatic polypeptide secretion. *J. Clin. Invest.* 61: 781–789, 1978.

SCOTT, L. D., AND D. L. CAHALL. Influence of the interdigestive myoelectrical complex on enteric flora in the rat. *Gastroenterology* 82: 737–745, 1982.

SCRATCHERD, T. Pancreatic secretory mechanisms. In: *Scientific Foundations of Gastroenterology*, edited by W. Sircus and A. N. Smith. London: Wm. Heinemann Med. Books, 1980, p. 579–591.

SCRATCHERD, T., D. HUTSON, AND R. M. CASE. Ionic transport mechanisms underlying fluid secretion by the pancreas. *Philos. Trans. R. Soc. Lond. B. Biol. Sci.* 296: 167–178, 1981.

SEREBRO, H. A., F. L. IBER, J. H. YARDLEY, AND T. R. HENDRIX. Inhibition of cholera toxin action in rabbit by cycloheximide. *Gastroenterology* 56: 506–511, 1969.

SHAFFER, E. A., J. W. BRAASH, AND D. M. SMALL. Bile composition at and after surgery in normal persons and patients with gallstones. *N. Engl. J. Med.* 287: 1317–1322, 1972.

SHAH, P. C., S. FREIER, B. H. PARK, P-C. LEE, AND E. LEBENTHAL. Pancreozymin and secretin enhance duodenal fluid antibody levels to cow's milk proteins. *Gastroenterology* 83: 916–921, 1982.

SHANNON, I. L. Effect of rate of gland function on pH, viscosity, total solids, calcium and magnesium in unstimulated parotid fluid. *Proc. Soc. Exper. Biol. Med.* 130: 874–878, 1969.

SHEERIN, H. E., AND M. FIELD. Ileal HCO$_3$ secretion: relationship to Na and Cl transport and effect of theophylline. *Am. J. Physiol.* 228: 1065–1074, 1975.

SHIAU, Y. F. Mechanism of intestinal fat absorption. *Am. J. Physiol.* 240 (*Gastrointest. Liver Physiol.* 3): G1–G9, 1981.

SHIELDS, H. M., F. A. BAIR, M. L. BATES, S. T. YEDLIN, AND D. H. ALPERS. Localization of immunoreactive alkaline phosphatase in the rat small intestine at the light microscopic level by immunocytochemistry. *Gastroenterology* 82: 39–45, 1982.

SHOEMAKER, R. L., G. M. MAKHLOUF, AND G. SACHS. Action of cholinergic drugs on nicturas gastric mucosa. *Am. J. Physiol.* 219: 1056–1060, 1970.

SHOOR, P. M., L. D. GRIFFITH, R. B. DILLEY, AND E. F. BERNSTEIN. Effect of pulseless perfusion on gastrointestinal blood flow and its distribution. *Am. J. Physiol.* 236 (*Endocrinol Metab. Gastrointest. Physiol.* 5): E28–E32, 1979.

SHULL, S. D., C. I. WAGNER, B. W. TROTMAN, AND R. D. SOLOWAY. Factors affecting bilirubin excretion in patients with cholesterol or pigment gallstones. *Gastroenterology* 72: 625–629, 1977.

SILK, D. B. A., J. A. NICHOLSON, AND Y. S. KIM. Relationships between mucosal hydrolysis and transport of two phenylalanine dipeptides. *Gut* 17: 870–876, 1976a.

SILK, D. B. A., J. A. NICHOLSON, AND Y. S. KIM. Hydrolysis of peptides within lumen of small intestine. *Am. J. Physiol.* 231: 1322–1329, 1976b.

SILLIN, L. F., W. J. SCHULTE, R. E. CONDON, J. H. WOODS, P. BASS, AND V. L. W. GO. Gastrocolic response: is it GIP? *Proc. Sixth Inter. Symp. GI Motility.* Edinburgh, 1978, p. 361–362.

SIMSON, J. N. L., A. MERHAV, AND W. SILEN. Alkaline secretion by amphibian duodenum. I. General characteristics. *Am. J. Physiol.* 240 (*Gastrointest. Liver Physiol.* 3): G401–G408, 1981.

SINGER, M. V., T. E. SOLOMON, J. WOOD, AND M. I. GROSSMAN. Latency of pancreatic enzyme response to intraduodenal stimulants. *Am. J. Physiol.* 238 (*Gastrointest. Liver Physiol.* 1): G23–G29, 1980a.

SINGER, M. V., T. E. SOLOMON, AND M. I. GROSSMAN. Effect of atropine on secretion from intact and transplanted pancreas in dog. *Am. J. Physiol.* 238 (*Gastrointest. Liver Physiol.* 1): G18–G22, 1980b.

SINGER, M. V., T. E. SOLOMON, H. RAMMERT, F. CASPARY, W. NIEBEL, H. GOEBELL, AND M. I. GROSSMAN. Effect of atropine on pancreatic response to HCl and secretin. *Am. J. Physiol.* 240 (*Gastrointest. Liver Physiol.* 3): G376–G380, 1981.

SINGH, M., AND P. D. WEBSTER. A review of macromolecular transport and secretion at the cellular level. *Am. J. Digest. Dis.* 21: 346–355, 1976.

SINGLETON, A. O., D. C. REDMOND, AND J. E. McMURRAY. Ileoceal resection and small bowel transit and absorption. *Ann. Surg.* 169: 690–694, 1964.

SIREGAR, H., AND C. C. CHOU. Relative contribution of fat, protein, carbohydrate and ethanol to intestinal hyperemia. *Am. J. Physiol.* 242 (*Gastrointest. Liver Physiol.* 5): G27–G31, 1982.

SIVE, A. A., A. I. VINIK, AND N. LEVITT. Adrenergic modulation of human pancreatic polypeptide (HPP) release. *Gastroenterology* 79: 665–672, 1980.

SJODIN, L. Electrical stimulation of nerves and secretion from digestive glands. In: *Nerves and the Gut*, edited by F. P. Brooks and P. W. Evers. Thorofare, NJ: Charles B. Slack, 1977, p. 1–13.

SJODIN, L., AND J. D. GARDNER. Effect of cholecystokinin variant (CCK$_{39}$) on dispersed acinar cells from guinea pig pancreas. *Gastroenterology* 73: 1015–1018, 1977.

SJOSTROM, H., O. NOREN, L. CHRISTIANSEN, H. WACKER, AND G. SEMENZA. A fully active, 2-active-site, single-chain sucrase (EC 3.2.1.48)-isomaltase (EC 3.2.1.10) from pig small intestine: implications for the biosynthesis of a mammalian integral stalked membrane protein. *J. Biol. Chem.* 255: 11332–11338, 1980.

SLEISENGER, M. H., AND Y. S. KIM. Protein digestion and absorption. *N. Engl. J. Med.* 300: 659–663, 1979.

SMITH, R. B., J. P. EDWARDS, AND D. JOHNSTON. Effect of vagotomy on exocrine pancreatic and biliary secretion in man. *Am. J. Surg.* 141: 40–47, 1981.

SMY, J. R. The effect of locally induced hyperaemia on cimetidine reduction of gastric acid secretory responses to gastrin pentapeptide in anaesthetized cats. *J. Physiol. (London)* 317: 69–70P, 1981.

SNAPE, W. J., JR., AND S. COHEN. Hormonal control of esophageal function. *Arch. Intern. Med.* 136: 538–542, 1976.

SNAPE, W. J., G. M. CARLSON, AND S. COHEN. Human colonic myoelectric activity in response to prostigmine and the gastrointestinal hormones. *Am. J. Digest. Dis.* 22: 881–887, 1977.

SNAPE, W. J., S. A. MATARAZZO, AND S. COHEN. Effect of eating and gastrointestinal hormones on human colonic myoelectrical and motor activity. *Gastroenterology* 75: 373–378, 1978.

SNAPE, W. J., JR, S. H. WRIGHT, W. M. BATTLE, AND S. COHEN. The gastrocolic response: evidence for a neural mechanism. *Gastroenterology* 77: 1235–1240, 1979.

SOARES, E. C., S. ZATERKA, AND J. H. WALSH. Acid secretion and serum gastrin at graded intragastric pressures in man. *Gastroenterology* 72: 676–679, 1975.

SOLL, A. H. Physiology of isolated canine parietal cells: receptors and effectors regulating function. In: *Physiology of the Gastrointestinal Tract*, edited by L. R. Johnson. New York: Raven Press, 1981, p. 673–691.

SOLOMON, T. E. Regulation of exocrine pancreatic cell prolif-

eration and enzyme synthesis. In: *Physiology of the Gastrointestinal Tract*, edited by L. R. Johnson. New York: Raven Press, 1981, p. 873–892.

SOLOMON, T. E., AND M. I. GROSSMAN. Effect of atropine and vagotomy on response of transplanted pancreas. *Am. J. Physiol.* 236 (*Endocrinol. Metab. Gastrointest. Physiol.* 5): E186–E190, 1979.

SOLOMON, T. E., N. SOLOMON, L. L. SHANBOUR, AND E. D. JACOBSON. Direct and indirect effects of nicotine on rabbit pancreatic secretion. *Gastroenterology* 67: 276–283, 1974.

SOLOMON, T. E., H. PETERSEN, J. ELASHOFF, AND M. I. GROSSMAN. Interaction of caerulein and secretin on pancreatic size and composition in the rat. *Am. J. Physiol.* 235 (*Endocrinol. Metab. Gastrointest. Physiol.* 4): E714–E719, 1978.

SOLOWAY, R. D. Gallstone disease. In: *Gastrointestinal Pathophysiology*, edited by F. P. Brooks, 2nd ed. New York: Oxford University Press, 1978, p. 235–254.

SONNENBERG, A., S. A. MULLER-LISSNER, H. F. WEISER, W. MULLER-DUYSUNG, F. HEINZEL, AND A. L. BLUM. Effect of liquid meals on duodenogastric reflux in humans. *Am. J. Physiol.* 243 (*Gastrointest. Liver Physiol.* 6): G42–G47, 1982.

SOTO, J., J. BRAMIS, A. H. AUFSES, AND D. A. DREILING. The pancreas and immunoglobulins. I. Immunoglobulin levels in pancreatic secretion of patients with normal function: preliminary studies. *Am. J. Gastroenterol.* 67: 345–347, 1977.

SPECIAN, R. S., AND M. R. NEUTRA. Regulation of intestinal goblet cell secretion. I. Role of parasympathetic stimulation. *Am. J. Physiol.* 242 (*Gastrointest. Liver Physiol.* 5): G370–G379, 1982.

SPELLMAN, S. J., E. A. SHAFFER, AND L. ROSENTHAL. Gallbladder emptying in response to cholecystokinin. *Gastroenterology* 77: 115–120, 1979.

SPENCER, J., AND M. I. GROSSMAN. The gastric secretory response to insulin: an "all-or-none" phenomenon? *Gut* 12: 891–896, 1971.

SPERBER, I. Biliary secretion of organic anions and its influence on bile flow. In: *The Biliary System*, edited by W. Taylor, Oxford, UK: Blackwell, 1965, p. 457–467.

STADAAS, J. O. Intragastric pressure/volume relationship before and after proximal gastric vagotomy. *Scand. J. Gastroenterol.* 10: 129–134, 1975.

STAGG, B. H., J. M. TEMPERLEY, AND J. H. WYLLIE. The fate of pentagastrin. *Gut* 12: 825–829, 1971.

STEINER, J. W., AND J. S. CARRUTHERS. Studies on fine structure of terminal branches of the biliary tree. *Am. J. Pathol.* 38: 639–661, 1961.

STENQUIST, B., J. F. REHFELD, AND L. OLBE. Effect of proximal gastric vagotomy and anticholinergics on acid and gastrin responses to sham-feeding in duodenal ulcer patients. *Gut* 20: 1020–1027, 1979.

STENSON, W. F., AND E. LOBOS. Metabolism of arachidonic acid by pancreatic acini: relation to amylase secretion. *Am. J. Physiol.* 242 (*Gastrointest. Liver Physiol.* 5): G493–G497, 1982.

STODDARD, C. J., H. L. DUTHIE, R. H. SMALLWOOD, AND D. A. LINKENS. Colonic myoelectric activity in man: comparison of recording techniques and methods of analysis. *Gut* 20: 476–483, 1979.

STRUNZ, U. T., J. H. WALSH, AND M. I. GROSSMAN. Removal of gastrin by various organs in dogs. *Gastroenterology* 74: 32–33, 1978.

SUBBASWAMY, S. G. Patterns of mucin secretion in human intestinal mucosa. *J. Anatomy* 108: 291–294, 1971.

SUGAWARA, K., R. CHOWLA, AND M. M. EISENBERG. Temporal relationship of gastric acid secretion to a meat meal before and after vagotomy. *Surg. Gynecol. Obstet.* 134: 307–310, 1072.

SUM, F. T., H. L. SCHIPPER, AND R. M. PRESHAW. Canine gastric and pancreatic secretion during intestinal distention and intestinal perfusion with choline derivatives. *Canad. J. Physiol. Pharmacol.* 47: 115–118, 1969.

SUN, E. A., W. J. SNAPE, JR., S. COHEN, AND A. RENNY. The role of opiate receptors and cholinergic neurons in the gastrocolic response. *Gastroenterology* 82: 689–693, 1982.

SUNDLER, F., J. ALUMETS, R. HAKANSON, S. INGEMANSSON, J. FAHRENKRUG, AND O. B. SCHAFFALITZKY DE MUCKADELL. VIP innervation of the gallbladder. *Gastroenterology* 72: 1375–1377, 1977.

SUNDLER, F., J. ALUMETS, R. HAKANSON, J. FAHRENKRUG, AND O. B. SCHAFFALITZKY DE MUCKADELL. Peptidergic (VIP) nerves in the pancreas. *Histochem.* 55: 173–176, 1978.

SURAWICZ, C. M., D. R. SAUNDERS, J. SILLERY, AND C. E. RUBIN. Linoleate transport by human jejunum: presumptive evidence for portal transport at low absorption rates. *Am. J. Physiol.* 240 (*Gastrointest. Liver Physiol.* 3): G157–G162, 1981.

SUTTON, D. R., AND R. M. DONALDSON, JR. Synthesis and secretion of protein and pepsinogen by rabbit gastric mucosa in organ culture. *Gastroenterology* 69: 166–174, 1975.

SVANIK, J., AND O. LUNDGREN. Gastrointestinal circulation. In: *Gastrointestinal Physiology II*, edited by R. K. Crane. Baltimore: Univ. Park Press, 1977, vol. 12, p. 8–10.

SWANSON, C. H., AND A. K. SOLOMON. A micropuncture investigation of the whole tissue mechanism of electrolyte secretion by the in vitro rabbit pancreas. *J. Gen. Physiol.* 62: 407–429, 1973.

SZURSZEWSKI, J. H. Electrical bases for gastrointestinal motility. In: *Physiology of the Gastrointestinal Tract*, edited by L. R. Johnson, New York: Raven Press, 1981, p. 1435–1466.

TABAK, L. A., M. J. LEVINE, I. D. MANDEL, AND S. A. ELLISON. Role of salivary mucins in the protection of the oral cavity. *J. Oral Pathol.* 11: 1–17, 1982.

TAKAHASHI, I., M. NAKAYA, T. SUZUKI, T., AND Z. ITOH. Post prandial changes in contractile activity and bile concentration in gallbladder of the dog. *Am. J. Physiol.* 243 (*Gastrointest. Liver Physiol.* 6): G365–G371, 1982a.

TAKAHASHI, I., T. SUZUKI, T. I. AIZAWA, AND Z. ITOH. Comparison of gallbladder contractions induced by motilin and cholecystokinin in dogs. *Gastroenterology* 82: 419–424, 1982b.

TAVALONI, N., J. S. REED, AND J. L. BOYER. Hemodynamic effects on determinants of bile secretion in isolated rat liver. *Am. J. Physiol.* 234 (*Endocrinol. Metab. Gastrointest. Physiol.* 3): E584–E592, 1978.

TAYLOR, I., H. L. DUTHA, R. SMALLWOOD, AND O. LINKENS. Large bowel myoelectrical activity in man. *Gut* 16: 808–814, 1975.

TAYLOR, I. L., M. FELDMAN, C. T. RICHARDSON, AND J. H. WALSH. Gastric and cephalic stimulation of human pancreatic polypeptide release. *Gastroenterology* 75: 432–437, 1978.

TAYLOR, I. L., T. E. SOLOMON, J. H. WALSH, AND M. I. GROSSMAN. Pancreatic polypeptide, metabolism and effect on pancreatic secretion in dogs. *Gastroenterology* 76: 524–528, 1979.

TAYLOR, I. L., G. L. KAUFFMAN, JR., J. H. WALSH, H. TROUT, P. CHEW, AND J. W. HARMON. Role of the small intestine and gastric antrum in pancreatic polypeptide release. *Am. J. Physiol.* 240 (*Gastrointest. Liver Physiol.* 3): G387–G391, 1981.

TAYLOR, I. L., W. J. BYRNE, D. L. CHRISTIE, M. E. AMENT, AND J. H. WALSH. Effect of individual L-amino acids on gastric acid secretion and serum gastrin and pancreatic polypeptide release in humans. *Gastroenterology* 83: 279–284, 1982a.

TAYLOR, I. L., M. SINGER, AND G. L. KAUFFMAN. Time-dependent effects of vagotomy on pancreatic polypeptide release. *Digest. Dis. Sci.* 27: 491–494, 1982b.

TEMPERLEY, J. M., B. H. STAGG, AND J. H. WYLLIE. Disappearance of gastrin and pentagastrin in the portal circulation. *Gut* 12: 372–376, 1971.

TEPPERMAN, B. L., AND E. D. JACOBSON. Mesenteric circulation. In: *Physiology of the Gastrointestinal Tract*, edited by L. R. Johnson. New York: Raven Press, 1981, p. 1317–1336.

THOMAS, J. E. Mechanism of action of pancreatic stimuli studied by means of atropine-like drugs. *Am. J. Physiol.* 206: 124–128, 1964.

THOMAS, J. E., AND M. V. BALDWIN. The intestinal mucosal reflex in the unanesthetized dog. *Am. J. Digest. Dis.* 16: 642–647, 1971.

THOMAS, P. A., AND K. A. KELLY. Hormonal control of interdigestive motor cycles of canine proximal stomach. *Am. J. Physiol.* 237 (*Endocrinol. Metab. Gastrointest. Physiol.* 6): E192–E197, 1979.

THOMAS, P. A., J-C. SCHANG, K. A. KELLY, AND V. L. W. GO. Can endogenous gastrin inhibit canine interdigestive gastric motility? *Gastroenterology* 78: 716–721, 1980.

THOMAS, P., C. A. TOTH, AND N. ZAMCHECK. The mechanism of biliary excretion of α_1 acid glycoprotein in the rat: evidence for a molecular weight-dependent nonreceptor-mediated pathway. *Hepatology* 2: 800–803, 1982.

THOMPSON, G. D., D. L. WINGATE, L. ARCHER, M. J. BENSON, W. J. GREEN, AND R. J. HARDY. Normal patterns of human upper small bowel motor activity recorded by prolonged radiotelemetry. *Gut* 21: 500–506, 1980.

THOMPSON, D. G., L. ARCHER, W. J. GREEN, AND D. L. WINGATE. Fasting motor activity occurs during a day of normal meals in healthy subjects. *Gut* 22: 489–492, 1981.

THOMSON, A. B. R., AND J. M. DIETSCHY. Intestinal lipid absorption: major extracellular and intracellular events. In: *Physiology of the Gastrointestinal Tract*, edited by L. R. Johnson. New York: Raven Press, 1981, p. 1147–1220.

THOR, K., S. ROSELL, A. ROKAEUS, AND L. KAGER. Neurotensin changes the motility pattern of the duodenum and proximal jejunum from a fasting-type to a fed type. *Gastroenterology* 83: 569–574, 1982.

THULIN, A. Secretory and motor effects in the submaxillary gland of the rat on intra-arterial administration of some polypeptides and autonomic drugs. *Acta Physiol. Scand.* 97: 343–348, 1976.

THUREBORN, E. Human hepatic bile: composition changes due to altered enterohepatic circulation. *Acta Chir. Scand. Suppl.* 303: 1–63, 1962.

TISCORNIA, O. M., C. J. PEREC, AND D. A. DREILING. Chronic truncal vagotomy: effect on weight and function of the rat pancreas. *Gastroenterol. Clin. Biol.* 5: 947, 1981.

TOBON, F., N. C. R. W. REID, J. L. TALBERT, AND M. M. SCHUSTER. Nonsurgical test for the diagnosis of Hirschsprung's disease. *N. Engl. J. Med.* 278: 188–194, 1968.

TOOULI, J., R. HONDA, W. J. DODDS, W. J. HOGAN, J. M. ORLOWSKI, S. SARNA, AND R. C. ARNDORFER. Manometric and electromyographic features of the opossum sphincter of Oddi. *Gastroenterology* 78: 1279, 1980.

TOOULI, J., J. E. GEENEN, W. J. HOGAN, W. J. DODDS, AND R. C. ARNDORFER. Sphincter of Oddi motor activity: a comparison between patients with common bile duct stones and controls. *Gastroenterology* 82: 111–117, 1982.

TORSOLI, A. The function of biliary "sphincters." *J. Roy. Coll. Surg.* (*Edinburgh*) 16: 270–273, 1971.

TRIER, J. S. Studies of small intestinal crypt epithelium. II. Evidence for and mechanisms of secretory activity by undifferentiated crypt cells of the human small intestine. *Gastroenterol.* 47: 480–495, 1964.

TRIER, J. S. Organ-culture methods in the study of gastrointestinal-mucosal function and development. *N. Engl. J. Med.* 295: 150–155, 1976.

TRIER, J. S., AND J. L. MADARA. Functional morphology of the mucosa of the small intestine. In: *Physiology of the Gastrointestinal Tract*, edited by L. R. Johnson, New York: Raven Press, 1981, p. 925–961.

TURNBERG, L. A. Potassium transport in the human small bowel. *Gut* 12: 811–818, 1971.

TURNBERG, L. A. Water and electrolyte metabolism. In: *Scientific Foundations of Gastroenterology*, edited by W. Sircus and A. N. Smith. London: William Heinemann Medical Books, 1980, p. 397–407.

TURNBERG, L. A., J. S. FORDTRAN, N. W. CARTER, AND F. C. RECTOR, Jr. Mechanism of bicarbonate absorption and its relationship to sodium transport in the human jejunum. *J. Clin. Invest.* 49: 548–556, 1970a.

TURNBERG, L. A., F. A. BIEBERDORF, S. G. MORAWSKI, AND J. S. FORDTRAN. Interrelationships of chloride, bicarbonate, sodium and hydrogen transport in human ileum. *J. Clin. Invest.* 49: 557–567, 1970b.

UDDMAN, R., J. ALUMETS, L. EDVINSSON, R. HAKANSON, AND F. SUNDLER. Peptidergic (VIP) innervation of esophagus. *Gastroenterology* 75: 5–8, 1978.

UDDMAN, R., J. ALUMETS, R. HAKANSON, F. SUNDLER, AND B. WALLES. Peptidergic (enkephalin) innervation of the mammalian esophagus. *Gastroenterology* 78: 732–737, 1980a.

UDDMAN, R., J. FAHRENKRUG, L. MALM, J. ALUMETS, R. HAKANSON, AND F. SUNDLER. Neuronal VIP in salivary glands: distribution and release. *Acta Physiol. Scand.* 110: 31–38, 1980b.

UVNAS, B. Role of duodenum in inhibition of gastric acid secretion. *Scand. J. Gastroenterol.* 6: 113–125, 1971.

UVNAS-WALLENSTEN, K., AND G. NILSSON. Quantitative study of insulin release induced by vagal stimulation in anesthetized cats. *Acta Physiol. Scand.* 102: 137–142, 1978.

UVNAS-WALLENSTEN, K., AND J. PALMBLAD. Effect of food deprivation (fasting) on plasma gastrin levels in man. *Scand. J. Gastroenterol.* 15: 187–192, 1980.

UVNAS-WALLENSTEN, K., AND J. WALSH. Release of bombesin-like immunoreactivity (BLI) into the portal vein of cats induced by electrical stimulation of the vagal nerves. *Regulatory Peptides Suppl.* 1: S117, 1980.

UVNAS-WALLENSTEN, K., S. EFENDIC, A. ROOVETE, AND C. JOHANSSON. Decreased release of somatostatin into the portal vein following electrical vagal stimulation in the cat. *Acta Physiol. Scand.* 109: 393–398, 1980.

VAGNE, M., L. DESCOS, AND P. MARTIN. Maximum bicarbonate secretion of the human pancreas. *Compt. Rendu. Soc. Biol. (Paris)* 163: 1403–1406, 1969.

VAGNE, M., AND V. MUTT. Entero-oxyntin: a stimulant of gastric acid secretion extracted from porcine intestine. *Scand. J. Gastroenterol.* 15: 17–22, 1980.

VALENZUELA, J. E. Dopamine as a possible neurotransmitter in gastric relaxation. *Gastroenterology* 71: 1019–1022, 1976.

VALENZUELA, J. E., J. H. WALSH, AND J. I. ISENBERG. Effect of gastrin on pancreatic enzyme secretion and gallbladder emptying in man. *Gastroenterology* 71: 409–412, 1976.

VAN BERGE HENEGOUWEN, G. P., T. N. TANGEDAHL, A. F. HOFFMANN, T. C. NORTHFIELD, N. F. LA RUSSO, AND J. T. McCALL. Biliary secretion of copper in healthy man: quantitation by an intestinal perfusion technique. *Gastroenterology* 72: 1228–1231, 1977.

VAN DER LINDEN, W., AND A. NORMAN. Composition of human hepatic bile. *Acta Chir. Scand.* 133: 307–313, 1967.

VAN DYKE, R. W., J. E. STEPHENS, AND B. F. SCHARSCHMIDT. Effects of ion substitution on bile acid-dependent and independent bile formation by rat liver. *J. Clin. Invest.* 70: 505–517, 1982.

VAN HEE, R. H., AND P. M. VANHOUTTE. Cholinergic inhibition of adrenergic neurotransmission in the canine gastric artery. *Gastroenterology* 74: 1266–1270, 1978.

VAN HOUTEN, AND B. I. POSNER. Insulin binding sites localized to nerve terminals in rat median eminence and arcuate nucleus. *Science* 207: 1081–1083, 1980.

VANTRAPPEN, G. R., T. L. PEETERS, AND J. JANSSENS. The secretory component of the interdigestive migrating motor complex in man. *Scand. J. Gastroenterol.* 14: 663–667, 1979a.

VANTRAPPEN, G. R., J. JANSSENS, T. L. PEETERS, S. R. BLOOM, N. P. CHRISTOFIDES, AND J. HELLEMANS. Motilin and the interdigestive migrating motor complex in man. *Digest. Dis. Sci.* 24: 497–500, 1979b.

VATN, M. H., L. S. SEMB, AND E. SCHRUMPF. The effect of atropine and vagotomy on the secretion of gastric intrinsic factor (IF) in man. *Scand. J. Gastroenterol.* 10: 59–64, 1975.

VILLANUEVA, M. L., J. A. HEDO, AND J. MARCO. Fluctuations of human pancreatic polypeptide in plasma: effect of normal food ingestion and fasting. *Proc. Soc. Exper. Biol. Med.* 159: 245–248, 1978.

VILLANUEVA, M. L., S. M. COLLINS, R. T. JENSEN, AND J. P. GARDNER. Structural requirements for action of cholecystokinin on enzyme secretion from pancreatic acini. *Am. J. Physiol.* 242 (*Gastrointest. Liver Physiol.* 5): G416–G422, 1982a.

VILLANUEVA, M. L., J. MARTINEZ, M. BODANSKY, S. M. COLLINS, R. T. JENSEN, AND J. D. GARDNER. Selective modification of the ability of cholecystokinin to cause residual stimulation of pancreatic enzyme secretion. *Am. J. Physiol.* 243 (*Gastrointest. Liver Physiol.* 6): G214–G217, 1982b.

WAGNER, C. I., R. S. SOLOWAY, B. W. TROTMAN, AND E. F. ROSATO. Effects of meals and fasting on biliary lipid output and frequency of enterohepatic circulation of bile salts in cholecystectomized man. *Gastroenterology* 66: 792, 1974.

WAITMAN, A. M., W. P. DYCK, AND H. D. JANOWITZ. Effect of secretin and acetazolamide on the volume and electrolyte composition of hepatic bile in man. *Gastroenterology* 56: 286–294, 1969.

WAKASUGI, H., H. STOLZE, W. HAASE, AND I. SCHULZ. Effect of La^{3-} on secretagogue-induced Ca^{2+} fluxes in rat isolated pancreatic acinar cells. *Am. J. Pathol* 240 (*Gastrointest. Liver Physiol.* 3): G281–G289, 1981.

WALD, A., AND S. A. ADIBI. Stimulation of gastric acid secretion by glycine and related oligopeptides in humans. *Am. J. Physiol.* 242 (*Gastrointest. Liver Physiol.* 5): G85–G88, 1982.

WALD, A., C. BARK, AND T. M. BAYLESS. Effect of caffeine on the human small intestine. *Gastroenterology* 71: 738–742, 1976.

WALKER, W. A. Intestinal transport of macromolecules. In: *Physiology of the Gastrointestinal Tract*, edited by L. R. Johnson. New York: Raven Press, 1981, p. 1271–1289.

WALLING, M. W., T. A. BRASITUS, AND D. V. KIMBERG. Effects of calcitonin and substance P on the transport of Ca, Na, and Cl across rat ileum in vitro. *Gastroenterology* 73: 89–94, 1977.

WALSH, J. H. Gastrointestinal hormones and peptides. In: *Physiology of the Gastrointestinal Tract*, edited by L. R. Johnson. New York: Raven Press, 1981, p. 59–144.

WALSH, J. H., H. T. DEBAS, AND M. I. GROSSMAN. Pure human gastrin: immunochemical properties, disappearance half time, and acid stimulation action in dogs. *J. Clin. Invest.* 54: 477–485, 1974.

WALSH, J. H., C. T. RICHARDSON, AND J. S. FORDTRAN. pH dependence of acid secretion and gastric release in normal and ulcer subjects. *J. Clin. Invest.* 55: 462–468, 1975.

WALSH, J. H., C. B. LAMERS, AND J. E. VALENZUELA. Cholecystokinin-octapeptide-like immunoreactivity in human plasma. *Gastroenterology* 82: 438–444, 1982.

WALUS, K. M., AND E. D. JACOBSON. Relation between small intestinal motility and circulation. *Am. J. Physiol.* 241 (*Gastrointest. Liver Physiol.* 4): G1–G15, 1981.

WANNAGAT, F-J., R. D. ADLER, AND R. K. OCKNER. Bile acid-induced increase in bile acid-independent flow and plasma membrane Na K-ATPase activity in rat liver. *J. Clin. Invest.* 61: 297–307, 1978.

WARD, A. S., AND S. R. BLOOM. Effect of vagotomy on secretion release. *Gut* 16: 951–956, 1975.

WARNES, T. W., P. HINE, AND G. KAY. The action of secretin and pancreozymin on small-intestinal alkaline phosphatase. *Gut* 15: 39–47, 1974.

WATARI, N. Fine structure of nervous elements in the pancreas of some vertebrates. *Z. Zellforsch.* 85: 291–314, 1968.

WEEMS, W. A. Intestinal wall motion, propulsion, and fluid movement: trends toward a unified theory. *Am. J. Physiol.* (*Gastrointest. Liver Physiol.* 6): G177–G188, 1982.

WEEMS, W. A., AND G. E. SEYGAL. Fluid propulsion by cat intestinal segments under conditions requiring hydrostatic work. *Am. J. Physiol.* 240 (*Gastrointest. Liver Physiol.* 3): G147–G156, 1981.

WEINER, K., L. S. GRAHAM, T. REEDY, J. ELASHOFF, AND J. H. MEYER. Simultaneous gastric emptying of two solid foods. *Gastroenterology* 81: 257–266, 1981.

WEISS, B. A., U. T. STRUNZ, AND J. H. WALSH. Gastrin release by bombesin from canine antral mucosa in vitro. *Clin. Res.* 25: 113A, 1977.

WELCH, R. W., AND S. T. DRAKE. Normal lower esophageal sphincter pressure: a comparison of rapid vs. slow pull-through techniques. *Gastroenterology* 78: 1446–1451, 1980.

WERLIN, S. L., AND R. J. GRAND. Development of secretory mechanisms in rat pancreas. *Am. J. Physiol.* 236 (*Gastrointest. Liver Physiol.* 5): E446–E450, 1979.

WILLIAMS, C. N. Bile and metabolism and liver disease. In: *Gastroenterology Pathophysiology*, edited by F. P. Brooks, New York: Oxford University Press, 1974, p. 161.

WILLIAMS, J. A. Na⁺ dependent of in vitro pancreatic amylase release. *Am. J. Physiol.* 229: 1023–1026, 1975.

WILLIAMS, J. A. Regulation of pancreatic acinar cell function by intracellular calcium. *Am. J. Physiol.* 238 (*Gastrointest. Liver Physiol.* 1): G269–G279, 1980.

WILLIAMS, J. A., AND M. LEE. Microtubules and pancreatic amylase release by mouse pancreas in vitro. *J. Cell Biol.* 71: 795–806, 1976.

WILLIAMS, J. A., H. SANKARAN, M. KORC, AND I. D. GOLDFINE. Receptors for cholecystokinin and insulin in isolated pancreatic acini: hormonal control for secretion and metabolism. *Fed. Proc.* 40: 2497–2502, 1981.

WILLIAMS, J. A., H. SANKARAN, E. ROACH, AND I. D. GOLDFINE. Quantitative electron microscope autoradiograph of ¹²⁵I-cholecystokinin in pancreatic acini. *Am. J. Physiol.* 243 (*Gastrointest. Liver Physiol.* 6): G291–G296, 1982.

WILSON, F. A., V. L. SALLEE, AND J. M. DIETSCHY. Unstirred water layers in the intestine: rate determinant of fatty acid absorption from a micellar solutions. *Science* 174: 1031–1033, 1971.

WILSON, F. A. Intestinal transport of bile acids. *Am. J. Physiol.* 241 (*Gastrointest. Liver Physiol.* 4): G38–G92, 1981.

WILSON, I. D., M. WONG, AND S. L. ERLANDSEN. Immunohistochemical localization of IgA and secretory component in rat liver. *Gastroenterology* 79: 924–930, 1980.

WILSON, I. D., R. D. SOLTIS, R. E. OLSON, AND S. L. ERLANDSEN. Cholinergic stimulation of immunoglobulin A secretion in rat intestine. *Gastroenterology* 83: 881–888, 1982.

WINANS, C. S. The pharyngoesophageal closure mechanism: a manometric study. *Gastroenterology* 63: 768–777, 1972.

WINANS, C. S. The fickle pylorus. *Gastroenterology* 70: 622–623, 1976.

WINSHIP, D. H., AND J. E. ROBINSON. Acid loss in human duodenum. *Gastroenterology* 66: 181–188, 1974.

WISE, L., AND W. F. BALLINGER. Effect of vagal stimulation and inhibition on gastric mucus and sulfated aminopolysaccharide secretion. *Ann Surg.* 174: 976–982, 1971.

WOLKOFF, A. W., AND C. T. CHUNG. Identification, purification and partial characterization of an organic anion binding protein from rat liver cell plasma membrane. *J. Clin. Invest.* 65: 1152–1161, 1980.

WOLKOFF, A. W., C. A. GORESKY, J. SELLIN, Z. GATMAITAN, AND I. M. ARIAS. Role of ligandin in transfer of bilirubin from plasma into liver. *Am. J. Physiol.* 236 (*Gastrointest. Liver Physiol.* 5): E638–E648, 1979.

WOOD, J. D. Neurophysiology of Auerbach's plexus and control of intestinal motility. *Physiol Revs.* 55: 307–324, 1975.

WOOD, J. D. Physiology of the enteric nervous system. In: *Physiology of the Gastrointestinal Tract*, edited by L. R. Johnson, New York: Raven Press, 1981, p. 1–37.

WRIGHT, S. H., W. J. SNAPE, W. BATTLE, S. COHEN, AND R. L. LONDON. Effect of dietary components on gastroco-

Ionic response. *Am. J. Physiol.* 238 (*Gastrointest. Liver Physiol.* 1): G228–G232, 1980.

WU, ZE-AI C., AND T. S. GARGINELLA. Functional properties of noradrenergic nervous system in rat colonic mucosa: uptake of tritium-labeled norepinephrine. *Am. J. Physiol.* 214 (*Gastrointest. Liver Physiol.* 4): G137–G142, 1981.

WYRWICKA, W., AND R. GARCIA. Effect of electrical stimulation of the dorsal nucleus of the vagus nerve on gastric acid secretion in cats. *Exper. Neurol.* 65: 315–325, 1979.

YABLONSKI, M. E., AND N. LIFSON. Mechanism of production of intestinal secretion by elevated venous pressure. *J. Clin. Invest.* 57: 904–915, 1976.

YAMAMOTO, T., H. SATOMI, H. ISE, AND K. TAKAHASHI. Evidence of the dual innervation of the cat stomach by the vagal dorsal motor and medial solitary nuclei as demonstrated by the horseradish peroxidase method. *Brain Res.* 122: 125–131, 1977.

YAU, W. M., G. M. MAKHLOUF, L. E. EDWARDS, AND J. T. FARRAR. Mode of action of cholecystokinin and related peptides on gallbladder muscle. *Gastroenterology* 65: 451–456, 1973.

YAU, W. M., AND G. M. MAKHLOUF. Comparison of transport mechanisms in isolated ascending and descending rat colon. *Am. J. Physiol.* 228: 191–195, 1975.

YOSHIMURA, H. Secretory mechanism of saliva and nervous control of its ionic composition. In: *Secretory Mechanisms of Salivary Glands*, edited by L. H. Schneyer and C. A. Schneyer. New York: Academic Press, 1967, p. 56–74.

YOU, C. H., W. Y. CHEY, AND K. Y. LEE. Studies on plasma motilin concentration and interdigestive motility of the duodenum in humans. *Gastroenterology* 79: 62–66, 1980.

YOU, C. H., J. M. ROMINGER, AND W. Y. CHEY. Effects of atropine on the action and release of secretin in humans. *Am. J. Physiol.* 242 (*Gastrointest. Liver Physiol.* 5): G608–G611, 1982.

YOUNG, G. P., S. FRIEDMAN, S. T. YEDLIN, AND D. H. ALPERS. Effect of fat feeding on intestinal alkaline phosphatase activity in tissue and serum. *Am. J. Physiol.* 241 (*Gastrointest. Liver Physiol.* 4): G461–G468, 1981.

YOUNG, J. A. Salivary secretion of inorganic electrolytes. In: *International Rev. Physiol.: Gastrointestinal Physiol. 3*, edited by R. K. Crane. Baltimore: University Park Press, 1979, vol. 19, p. 1–58.

YOUNG, J. A., AND E. W. VAN LENNEP. The morphology of salivary glands. London: Academic Press, 1978.

YOUNG, J. A., AND E. W. VAN LENNEP. Transport in salivary and salt glands in membrane transport. In: *Membrane Transport in Biology. Transport Organs*, edited by G. Giebisch. Berlin: Springer, 1979, vol. 4B, p. 563–664.

YOUNG, J. A., E. FROMTER, E. SCHOGEL, AND K. F. HAMANN. Micropuncture and perfusion studies of fluid and electrolyte transport in the rat submaxillary gland. In: *Secretory Mechanisms of Salivary Glands*, edited by L. H. Schneyer and C. A. Schneyer. New York: Academic Press, 1967, p. 11–31.

YOUNOSZAI, M. K., AND H. P. SCHEDL. Intestinal calcium transport: comparison of duodenum and ileum in vivo in the rat. *Gastroenterology* 62: 565–571, 1972.

YOUNOSZAI, M. K., E. URBAN, AND H. P. SCHEDL. Vitamin D and intestinal calcium fluxes in vivo in the rat. *Am. J. Physiol.* 225: 287–292, 1973.

ZALEWSKY, C. A., AND F. G. MOODY. Mechanisms of mucus release in exposed canine gastric mucosa. *Gastroenterology* 77: 719–729, 1979.

ZIMMERMAN, T. W., AND H. J. BINDER. Muscarinic receptors on rat isolated colonic epithelial cells. *Gastroenterology* 83: 1244–1251, 1982.

ZINNER, M. J., B. M. JAFFE, L. DE MAGISTRIS, A. DAHLSTROM, AND H. AHLMAN. Effect of cervical and thoracic vagal stimulation on luminal serotonin release and regional blood flow in cats. *Gastroenterology* 83: 1403–1408, 1982.

Metabolism

CHAPTER 47

Regulation of Carbohydrate Metabolism

VITAL IMPORTANCE OF PLASMA GLUCOSE HOMEOSTASIS

Maintenance of plasma glucose levels within fairly narrow limits is of vital importance in the mammalian organism. If the plasma glucose rapidly falls to low levels (below 40–50 mg/dl) and stays low, even for 5 or 10 min, the consequences can be dramatic and drastic. This is because under ordinary circumstances the central nervous system depends absolutely upon a continuing minute-to-minute supply of glucose. Whereas most tissues can readily utilize free fatty acids or other blood-transported substrates when glucose becomes unavailable, nerve tissue depends absolutely on glucose, the only energy substrate it can utilize at a significant rate. Consequently, sustained hypoglycemia (e.g., after an overdose of insulin) can lead to coma and, if uncorrected, death. Prompt intervention to correct the hypoglycemia will save the patient's life, but if the hypoglycemia has been profound and prolonged, there will be irreversible brain damage that can be extensive and totally disabling.

Abnormal elevation of plasma glucose levels (*hyperglycemia*) does not pose an analogous acute threat, yet prolonged hyperglycemia is also ultimately life-threatening. If levels above 300 or 400 mg/dl are sustained for days, the patient will lose large amounts of glucose in the urine (*glucosuria*). This will entail an obligatory loss of water and electrolytes, leading to progressive dehydration, decrease in blood volume (*hypovolemia*), hypotension, shock, and coma. Prolonged hyperglycemia also, then, can ultimately lead to death.

These dramatic consequences of extreme acute departures from the norm illustrate the importance of glucose homeostasis. It should be stressed, however, that even relatively small departures from the norm may be deleterious if sustained chronically. For example, a growing body of evidence indicates that even modest hyperglycemia over a period of years may account for the dysfunction of the nervous system, blood vessels, kidneys, and other tissues associated with diabetes mellitus (the so-called late complications of diabetes mellitus). Repeated episodes of hypoglycemia, no one of which approaches the life-threatening severity discussed above, may nevertheless have a cumulative effect and lead to nervous system damage.

Because glucose homeostasis is so obviously vital, this chapter will deal primarily with the mechanisms that operate to maintain normal plasma glucose levels. Of course, there are many additional facets of carbohydrate metabolism of importance in the body economy. For example, synthesis of ribose and deoxyribose is sine qua non for DNA synthesis; glycolipids play essential roles in membranes generally; complex carbohydrates are fundamental components of connective tissue matrices. These, however, are more the province of biochemistry and pathophysiology. Here we confine ourselves to a consideration of the "physiological chemistry" of glucose homeostasis. This is best done by an input-output analysis in which we consider in turn the various *sources* of the pool of plasma glucose and the various *sinks*, i.e., the exit routes for glucose from the plasma compartment.

It is useful to deal with the problem at two levels. First, we simply ask where the glucose comes from and where it goes without regard to the intracellular enzymatic machinery that operates to govern transport and utilization—a "black box" approach. Second, we look into the black boxes (organs and cells) to see what can be said about enzymes and their regulation. One reason for keeping approaches at these two levels separate is that the levels of our certainty about them differ. At the higher level of organization—at the level of physiological chemistry—matters are more nearly settled, and it is this level that we deal with in clinical situations. For example, there is absolutely no doubt that insulin favors deposition of liver glycogen; on the other hand, there is still uncertainty about the precise molecular mechanisms by which insulin increases glycogen synthase activity. Again, there is no doubt that insulin favors the transport of glucose across plasma membranes and into the bulk of body tissues, namely, muscle and adipose tissue; the molecular mechanisms involved are still incompletely defined. Consequently we shall discuss glucose homeostasis primarily first at the physiological level and secondarily at the level of cell biology and intracellular mechanisms.

SOURCES OF THE PLASMA GLUCOSE POOL

These are shown in Figure 7.1. There are just three sources of the plasma glucose. The two major sources are: (A) intestinal absorption of dietary glucose and its precursors and (B) release of glucose from the liver. Under ordinary circumstances, the kidney is a relatively minor third source; however, in prolonged starvation it becomes a significant source (Owen et al., 1969).

Dietary Sources

These are shown in Table 7.1. Very few foods contain significant amounts of free glucose. Significant quantities of glucose are presented in the form of disaccharides, especially sucrose, and in the form of polysaccharides (mainly, starch in plant foods and some glycogen in animal foods). Disaccharides are rapidly hydrolyzed; the fructose and galactose moieties are rapidly absorbed and converted to glucose. Consequently, monosaccharides and disaccharides cause a prompt increase in plasma glucose concentration whereas glucose presented in the form of polysaccharides enters the blood stream more slowly and causes less of a spike

in plasma glucose concentration. This difference becomes important in the dietary management of patients with diabetes mellitus, in whom spikes in plasma glucose concentration are to be avoided.

Amino acids derived from dietary protein (with the exception of leucine) can all contribute to de novo glucose formation. Some amino acids can contribute all of their carbon atoms to the formation of glucose. For example, alanine, after transamination to yield pyruvate, can be converted quantitatively to glucose under the right conditions. This generation of glucose from protein (or any other nonglucose) precursors is designated *gluconeogenesis*. Other amino acids can contribute some proportion of their carbon atoms for gluconeogenesis. For example, during the degradation of tyrosine, four carbon atoms become available as fumaric acid which can enter the Krebs cycle and be converted to oxaloacetic acid. The latter in turn can be converted to phosphoenolpyruvate on the pathway toward glucose-6-phosphate formation. Leucine, on the other hand, cannot contribute carbons for glu-

Table 7.1
Dietary sources of plasma glucose

1. Glucose per se (minor source)
2. Glucose-containing disaccharides
   Sucrose (fructosyl-glucose)
   Lactose (galactosyl-glucose)
   Maltose (glucosyl-glucose)
3. Glucose-containing polysaccharides (major source)
   Starch
   Glycogen
4. Sugars readily converted to glucose
   Fructose
   Galactose
5. Gluconeogenic amino acids
6. Glycerol moiety of triglycerides

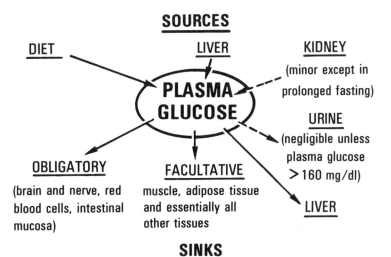

Figure 7.1. Sources and sinks for plasma glucose.

coneogenesis; its degradation yields CO_2, acetyl CoA, and acetoacetic acid. In the mammalian organism, acetate cannot make any *net* contribution to glucose formation. Acetate can enter the Krebs cycle, but the Krebs cycle can be thought of as simply a complicated way of converting acetate carbons to 2 moles of CO_2. After one turn of the cycle, no carbon atoms are left from which to generate precursors for net glucose formation. Thus fatty acids, like leucine, cannot make any *net* contribution to gluconeogenesis via the Krebs cycle (although radioactive carbons from isotopically labeled fatty acids or labeled leucine will eventually find their way into glucose and glycogen). In prolonged fasting or in diabetic ketoacidosis, acetone, generated by decarboxylation of acetoacetate, has the potential to contribute carbons to glucose formation, but this is probably a minor pathway. Thus, the triglycerides in the diet are a rather insignificant source of carbon atoms for gluconeogenesis, since only the carbon atoms in the glycerol moiety can contribute. The glycerol released from the triglycerides can be readily converted to glucose in the liver. Thus about 10% of the carbon atoms in a triglyceride constitute a source for glucose formation. While fatty acids do not contribute carbon atoms for gluconeogenesis, they nevertheless strongly stimulate gluconeogenesis from *other* precursors (see Chapter 49 and Fig. 7.19).

In summary, then, input of glucose or glucose precursors from the diet is one way to maintain plasma glucose levels. This happens only during a small portion of the day, i.e., for the few hours after each meal. How do we introduce glucose into the plasma pool during the rest of the day?

Hepatic Input

If we make measurements of the concentration of plasma glucose in various vascular beds during the fasting state, we find that the arteriovenous difference is positive (i.e., arterial concentration exceeds venous concentration), except across the liver and the kidney. From the magnitude of the A-V difference and simultaneous measurements of blood flow, we can calculate the rate of delivery of glucose into the plasma and show that the contribution of the liver far exceeds that of the kidney. The latter only makes a quantitatively important contribution during prolonged fasting. The greatest bulk of glucose input during the periods between meals and during an overnight fast must come from the liver. This is dramatically demonstrated when a total hepatectomy is performed. Very quickly, the plasma glucose level begins to drop, and the animal goes into hypoglycemic convulsions unless glucose

is infused intravenously (Mann and Magath, 1924). The kidney can and does make a small contribution under most circumstances, but its contribution is normally too small to prevent fatal hypoglycemia when hepatic output is abruptly cut off. During an extended period of fasting the kidney can account for as much as one-third of the total glucose production, but total production is much reduced. If we look into the "black boxes," we find that the liver and the kidney are the only two organs that contain significant levels of glucose-6-phosphatase activity, the enzyme needed to convert glycogen degradation products or glycolytic intermediates to free glucose so that they can be channeled out to the plasma.

The liver can contribute glucose to the plasma by two general mechanisms (Fig. 7.2): (1) by breakdown of stored glycogen (glycogenolysis) and (2) by the formation of glucose from nonglucose precursors (gluconeogenesis). When glucose is readily available from dietary sources, the liver increases its store of glycogen, but this store can only reach a maximum of about 75 g. Between meals and during the early hours of fasting, this stored glycogen is broken down to provide glucose to the plasma, but the supply is very limited. The absolute minimum total body requirement for glucose during a fast is approximately 125–150 g/day in order to supply the brain, for which glucose is an obligatory substrate. Smaller additional amounts (30–40 g) are needed for peripheral nerves, red and white blood cells and the renal medulla. Thus, the glucose available in stored hepatic glycogen is only sufficient to provide glucose requirements for less than 12 h. If fasting continues, other sources must be invoked, and the mechanisms of hepatic gluconeogenesis become all important (Fig. 7.3).

The most important source of carbons for gluconeogenesis during fasting is the amino acids re-

HEPATIC CONTRIBUTION TO PLASMA GLUCOSE

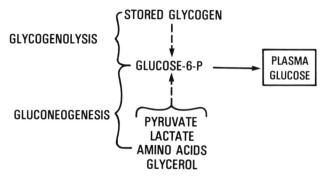

Figure 7.2. Hepatic contributions to plasma glucose via glycogenolysis and gluconeogenesis.

CARBON SOURCES FOR GLUCONEOGENESIS

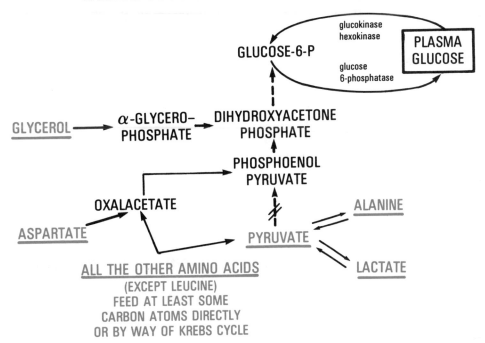

Figure 7.3. The major sources of carbon atoms contributing to gluconeogenesis.

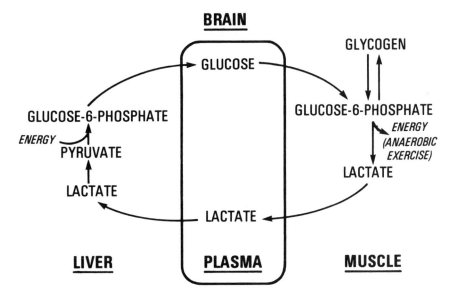

Figure 7.4. The Cori cycle, by which lactate produced in muscle is converted to glucose by the liver and then returned to muscle.

leased from tissue protein, predominantly muscle proteins. During starvation, gluconeogenesis from amino acids contributes about 75 g of glucose daily. Most of this derives from breakdown of muscle protein, but there is some protein degradation in virtually every organ. All of the amino acids are available to some extent to help provide the needed glucose. Plasma concentrations of amino acids increase acutely during starvation, but the relative magnitude of the increase varies, and the efficiency of extraction of these amino acids by the liver

varies. From studies of A-V differences, it is clear that alanine and glutamine are quantitatively the most important amino acids in the gluconeogenesis of starvation. The contribution of alanine is out of proportion to the amount of alanine in muscle proteins. A significant amount of the alanine brought to the liver during starvation probably represents alanine formed from pyruvate by transamination using amino groups from other amino acids.

A second source of carbons for gluconeogenesis

is the lactate released from muscle as a result of anaerobic glycolysis. As shown in Figures 7.4, this lactate can be carried to the liver and there converted to glucose. If this glucose returns to the muscle and undergoes glycolysis there, there has clearly been no *net* contribution to the glucose stores of the body. However, the muscle has been provided with a further source of energy for anaerobic glycolysis; that energy was provided by catabolism of other substrates in the liver. This cycling of carbons from lactate in the muscle to glucose in the liver and back to muscle is designated the *Cori cycle* (Fig. 7.4). Alternatively, some of the glucose generated in the liver from muscle lactate can be channeled up to the brain and thus can help provide the much needed glucose to maintain central nervous system function during fasting.

Finally, glycerol released from adipose tissue during the mobilization of stored triglycerides can be almost quantiatively converted to glucose in the liver. Even during total starvation, however, this triglyceride glycerol can only account for about 20 g of new glucose generated daily. As discussed above, the free fatty acids released from stored triglycerides cannot make any net contribution toward gluconeogenesis. Nevertheless, the free fatty acids delivered to the liver play a crucial role in gluconeogenesis because they act as a trigger to stimulate gluconeogenesis from *other carbon sources.*

The mammalian organism simply does not have mechanisms for storing important quantities of glucose. As indicated above, the amount stored in liver glycogen can only maintain plasma glucose levels for 12 h at most. Muscle and adipose tissue contain glycogen, but that glycogen cannot be released directly to the plasma as glucose (these tissues lack glucose-6-phosphatase), and only a small fraction of it can contribute via the Cori cycle. During fasting, then, we depend primarily on gluconeogenesis from the amino acids released from muscle protein. The potential for such conversion of amino acids to glucose has been demonstrated by elegant isotopic studies of the pathways of degradation of the various amino acids. However, the earliest demonstration that a given amino acid can give rise to a net increase in available glucose came from physiologic studies, particularly studies utilizing phlorizin as a tool. This toxic glycoside blocks the reabsorption of glucose in the renal tubule so that the renal threshold for glucose falls to almost zero. Consequently, glucose formed in the body is rapidly and almost quantitatively lost into the urine. Administration of a single amino acid to a phlorizin-treated animal and measurement of the consequent

net increase in urinary glucose provide a rather good indication of the effectiveness of that amino acid as a gluconeogenic precursor. Knowing that the available stores of glucose are almost completely exhausted after less than 12 h of starvation, we can easily infer the tremendous potential for gluconeogenesis from observation of patients with uncontrolled diabetes mellitus. These patients will maintain plasma glucose levels of over 400 mg/dl for many days, even though they have nausea and vomiting, and may take in no food at all. Furthermore, they excrete as much as 200–300 g of glucose daily for many days on end. These findings vividly demonstrate that gluconeogenesis *must* be occurring at the rate of several hundred grams daily in these patients.

SINKS FOR PLASMA GLUCOSE

These are shown in Figure 7.1. All body tissues can and do utilize plasma glucose. Some are *obligatory* users, i.e., they cannot use alternative substrates when glucose is unavailable. Nerve tissue cannot use free fatty acids (FFA), the major alternative circulating fuel, at a significant rate. Hence the serious consequences of acute hypoglycemia. The nervous system requires about 125–150 g of glucose daily under most conditions. However, in prolonged starvation (5–6 wk) the brain undergoes an interesting metabolic switch that allows it to utilize ketone bodies (β-hydroxybutyrate and acetoacetate) in place of over 50% of its usual glucose requirement (Owen et al., 1967). During such starvation, fatty acids are continually mobilized from the huge stores of adipose tissue triglycerides, and a portion of them is continually converted to ketone bodies in the liver. This adaptation in metabolism of the brain ensures that it can survive with less drastic depletion of muscle protein to provide substrate for gluconeogenesis.

A few other tissues utilize glucose almost exclusively. These include red blood cells, the intestinal mucosa, and the renal medulla. Their use of glucose is largely or exclusively via anaerobic glycolysis. Most of the body tissues, however, are *facultative* users of glucose. When FFA levels are high and glucose and insulin levels are low, as in the fasting state, these tissues can and do switch to use FFA as their primary metabolic fuel.

The liver is both a source and a sink for plasma glucose. In fact, both uptake and release are going on at all times. The *net balance* is under multiple controls that determine whether at any given time it represents a source (net input into plasma) or a sink (net uptake from plasma). In the absence of supervening hormonal or neural signals, the liver shows a net output of glucose when the plasma level is low (below 120–150 mg/dl) and a net uptake when it is high (above 120–150 mg/dl). This "autoregulation" can be overridden by changes in hormonal balance.

Under normal conditions no glucose is lost by renal excretion. Glucose is rapidly filtered into the glomerular fluid, but it is normally reabsorbed very efficiently by the renal tubules. The T_m is about 340 mg/min. If the glomerular filtration rate is normal (120 ml/min), spillage will only occur when the plasma glucose reaches 160–190 mg/dl (renal threshold for glucose). Generally the amount of glucose excreted in the urine will be a reasonably good indicator of the degree of hyperglycemia. However, there are significant and instructive exceptions. For example, during pregnancy, when the mother's glomerular filtration rate (GFR) is normally elevated, glucosuria may occur even at normal or only slightly elevated plasma glucose levels. Conversely, with renal damage, such as can occur as a late complication of diabetes mellitus, the GFR may be abnormally low. Under these circumstances, the urine may remain free of glucose even though plasma glucose levels are high. Under these conditions, quantification of the glucose in the urine will underestimate the severity of the hyperglycemia. Finally, mention should be made of certain rare inherited disorders in which renal tubular transport mechanisms are defective, including tubular glucose reabsorption (e.g., Fanconi's syndrome). Because of the low renal threshold, glucosuria occurs without hyperglycemia.

NORMAL PLASMA GLUCOSE LEVELS AND GLUCOSE TOLERANCE

Normal plasma glucose levels after an overnight fast range from 70 to 110 mg/dl. Values in the older literature, representing whole blood glucose concentrations, are about 15% lower than plasma concentrations. In evaluating the older literature, one must also be aware that some values are falsely high because they were based on nonspecific methods that included reducing substances other than glucose. Today, almost without exception, *plasma* glucose is measured and measured specifically, e.g., using enzymatic methods based on glucose oxidase.

Plasma glucose increases with the ingestion of each meal, rising to a peak at 30–60 min and returning to basal values at about 120 min. The magnitude of the response and its duration depends on the size and composition of the meal. Between meals the glucose level is remarkably constant with a slight downward drift during the night. Precise measurements show minor phasic changes of a few

milligrams per deciliter, reflecting pulsatile release of regulatory hormones, but these changes are too small to be of practical significance.

The changes in plasma glucose levels in response to ingestion of a glucose load have long been used clinically and in research to evaluate the effectiveness of glucose homeostatic mechanisms. This is the oral glucose tolerance test (OGTT). As currently employed, a 75–100 g dose of glucose is given after an overnight fast. Plasma glucose levels are determined on venous blood samples taken just before the glucose load, at 1 h and at 2 h. In normal subjects the fasting level is below 115, the 1-h value is below 200, and the 2-h value is below 140 mg/dl. If both the 1- and 2-h values are more than 200, the test is definitely abnormal, indicating primary or secondary diabetes mellitus. Intermediate test results indicate "impaired glucose tolerance," which may or may not have clinical implications (National Diabetes Data Group, 1979).

The sequence of events that determine the shape of the OGTT curve illustrates some of the basic elements involved in glucose homeostasis (Fig. 7.5). Plasma glucose begins to rise within a few minutes of glucose ingestion, reflecting the rapid efficient absorption of glucose from the upper duodenum. Administration of glucose by mouth is thus very effective in quickly correcting hypoglycemia. Glucose is absorbed by active transport at a rate that is a function of the concentration at the mucosal surface. When the concentration is high, the transport mechanism becomes saturated, and the rate of absorption does not increase further with increasing concentration in the intestine. The large dose of glucose used in the OGTT is enough to maximize the initial rate of absorption in the proximal duodenum. The rate at which the plasma glucose level

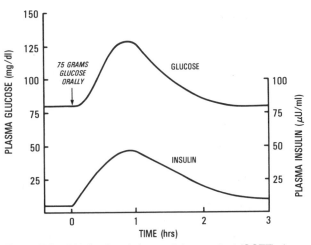

Figure 7.5. Idealized oral glucose tolerance test (OGTT) showing a typical normal response in plasma glucose level and plasma insulin level after a 75-g oral dose of glucose.

rises and the peak value reached will be determined in part by this input rate. For example, patients with hyperthyroidism absorb glucose more rapidly, and this plays a role in the abnormal OGTT seen in a significant number of them (although other factors are also involved). If the full glucose load is delivered into the intestine too rapidly, as in patients with partial gastrectomy or gastrojejunostomy, the OGTT curve rises rapidly to an abnormally high level at 30–60 min but generally returns to normal by 2 h or even overshoots to hypoglycemic values.

Returning to the normal OGTT curve, we see that a peak value is reached at 30–60 min, yet intestinal glucose absorption is still going on very actively at that time. The levels begin to fall despite continuing absorption because the rate of glucose removal from plasma has progressively increased and now exceeds the rate of absorption. This is mainly due to enhanced secretion of insulin, which begins almost immediately on ingestion of the glucose (Fig. 7.5). Even before plasma glucose levels rise appreciably, the output of insulin into the portal vein has already started to increase. This initial response is triggered by hormones released from the gastrointestinal mucosa in response to food ingestion. Several gastrointestinal hormones have been shown to have the *potential* of stimulating insulin secretion (including secretin, pancreozymin, gastrin, glucagon, and gastric inhibitory polypeptide). Most of these act only at very high concentrations; gastric inhibitory peptide (GIP), however, is effective at concentrations actually reached *in vivo* and is probably importantly involved (Dupre et al., 1969, 1973). The importance of the intestinal hormones in augmenting the insulin response was nicely demonstrated by showing that the integrated response to intravenous glucose is less than that to the equivalent dose given orally, even though arterial glucose levels are lower in the latter case (Perley and Kipnis, 1967). Once the glucose level begins to rise, by all odds the major stimulus for the continuing increase in insulin output from the pancreas is the direct action of glucose on the B-cells. However, there are many additional factors that can modulate the magnitude of the insulin response and the shape of the OGTT curve.

Most of the load of glucose is initially taken up by the liver, although all body tissues participate to some extent. The dominant role of the liver is understandable since it is bathed in the portal vein blood and is therefore exposed to both the newly absorbed glucose and the newly secreted insulin at concentrations higher than those found anywhere else in the body. The glucose taken up and stored in the liver is either converted directly to glycogen or converted to fatty acids and then stored as triglycerides. In the periphery, some of the extra glucose is stored in adipose tissue, some as glycogen but most as triglycerides (see Chapter 48); another portion is stored as muscle glycogen. Finally, some of the extra glucose is immediately oxidized, substituting for free fatty acids in those tissues that are facultative for glucose utilization, especially cardiac and skeletal muscle.

A wide variety of metabolic and hormonal pertubations can in principle cause an abnormal OGTT by interfering with the efficient disposition of the glucose load. An absolute or relative defect in insulin response underlies the abnormal test in primary diabetes mellitus. However, there are many bases for an abnormal test. For example, liver disease severe enough to limit the rate of hepatic glycogenesis can cause a "diabetic" type of glucose tolerance curve. Anything that affects the sensitivity of tissues to insulin can potentially yield an abnormal OGTT. For example, adrenal glucocorticoids decrease insulin sensitivity; thus in many acute stress situations associated with excess glucocorticoid output (febrile illnesses, trauma, surgery) the OGTT may be abnormal. Increased sympathetic nervous system activity and adrenal medullary secretion of catecholamines also contribute by inhibiting insulin output.

An alternative method for evaluating glucose homeostasis, primarily for research purposes, is the intravenous glucose tolerance test (IGTT). A 25-g dose is given rapidly by vein, plasma glucose levels are measured at 10-min intervals for 1 h, and the results plotted on semilogarithmic graph paper. The slope of the disappearance curve gives a rate constant, k, proportional to the efficiency of glucose disposition. In normal individuals the k value is 1.2 or greater. The IGTT is obviously independent of one set of the variables affecting the OGTT, namely, those influencing glucose absorption. Thus it may be useful in evaluating patients with gastrointestinal disorders.

This brief discussion of the events determining the response to a glucose load will provide a context for the following discussion of mechanisms regulating glucose uptake and utilization. One last point should be made in regard to the remarkable efficiency of the process. If 75 g of glucose, the dose used most commonly in the standard oral test, were suddenly introduced into the extracellular fluid volume of a 70-kg man (14 l), the glucose level would rise by over 500 mg/dl. In normal individuals, the observed rise is often less than 50 mg/dl. Partly this is due to the fact that the absorption is stretched over an extended time period. However, in diabetics, most of whom have normal rates of glucose absorption, glucose levels after an OGTT do rise by as much as 300 mg/dl because of inefficient uptake and utilization. (Diagnostic OGTTs are not necessary in frank diabetes mellitus with fasting hyperglycemia and are not advisable because of the marked hyperglycemia induced.)

REGULATION OF GLUCOSE UPTAKE FROM PLASMA

Glucose Transport across Cell Membranes

Different mechanisms for glucose transport are found in different tissues. *Active transport*, i.e., an energy-dependent process that can transport against a concentration gradient, occurs in the intestinal epithelium and in the renal tubular epithelium. Neither intestinal absorption of glucose nor tubular reabsorption of glucose by this active transport mechanism is insulin dependent.

In muscle, adipose tissue and other insulin-dependent tissues, glucose uptake occurs by a *carrier-mediated transport* mechanism. Glucose cannot be transported against a concentration gradient by this mechanism. It is a form of passive transport but more is involved than simple diffusion. This is evident from four characteristic properties of such a *facilitated diffusion* process:

(a) *Saturation kinetics.* Transport increases with external glucose concentration but only to a certain maximum rate.

(b) *Stereospecificity.* The unnatural isomer (L-glucose) is not transported.

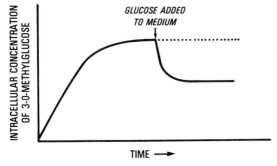

Figure 7.6 Idealized representation of an experiment demonstrating the phenomenon of countertransport (Morgan et al., 1964). The glucose analog, 3-O-methylglucose, enters the cell via the same carrier-mediated transport system involved in glucose uptake. In the absence of added glucose, the intracellular concentration of the analog rises to reach a plateau value which is maintained. The analog is not a substrate for phosphorylation and is not metabolized. At the time indicated by the *arrow*, glucose is added to the medium, and the intracellular concentration of the analog falls to reach a new lower plateau level (see text for further discussion).

(c) *Competition.* Sugars of similar structure competitively inhibit glucose transport.

(d) *Countertransport.* By taking advantage of the fact that a nonmetabolizable glucose analogue (3-O-methylglucose) is transported by the same system it can be shown (using an isolated rat diaphragm as an example of muscle tissue) that the carrier system is operating bidirectionally (Morgan et al., 1964). This principle is illustrated in Figure 7.6. If only 3-O-methylglucose is present in the medium, its intracellular concentration builds up and stabilizes when the concentration in the cytoplasm equals that outside. At this point, the rate of movement of the analogue into the cell equals its rate of movement out. If now glucose itself is added to the medium, the intracellular concentration of the analogue suddenly begins to fall. This appears to be paradoxical since the analogue is moving out against a concentration gradient. The explanation lies in the fact that on the outside glucose is now competing with the analogue for a limited number of carrier molecules in the membrane and, thus, the inward movement of the analogue is slowed down. On the inside of the membrane there is no comparable competition because the glucose molecules are phosphorylated almost as soon as they enter the cell, and glucose-6-phosphate cannot bind to the carrier. Consequently, the *outward* movement of the analogue continues for a while at about the same rate that prevailed when the glucose was added. As the concentration of the analogue on the inner side of the membrane falls, its rate of transport out falls until a new steady state is established (Fig. 7.6).

All of these properties can be accounted for if one postulates a set of carriers within the plasma membrane that shuttle from outer to inner surfaces and back, able to bind glucose reversibly at either surface. If the number of carriers is finite, there will be a maximum rate of transport when all are occupied (*saturation kinetics*). If the configuration of the carrier is sharply defined, it will combine only with molecules of certain configurations (*stereospecificity*). If the configuration is not absolutely specific, some molecules of very closely related structure will also bind (*competition*). Even though the system operates symmetrically, differential metabolism of one of two competing ligands can lead to a paradoxical transport against a gradient (*countertransport*).

In muscle and adipose tissue, the rate-limiting step in the uptake of glucose is its transport across the cell membrane by the carrier-mediated mechanism. The rate of glucose entry increases with the concentration of glucose presented to the cell. However, even at very high levels the rate at which entering glucose is phosphorylated to form glucose-6-phosphate is so rapid that there is an almost unmeasurably low concentration of free glucose inside the cell (Fig. 7.7). Insulin, the most important regulator of glucose transport in these tissues, can accelerate the inflow to the point that intracellular free glucose can be demonstrated (which supports the proposition that free glucose and not a derivative of it is delivered at the inner surface of the cell membrane).

In contrast to muscle and adipose tissue membranes, liver cell membranes appear to be permeable to glucose, and the rate-limiting step for uptake is the rate at which the free glucose delivered into the cytoplasm can be phosphorylated (Fig. 7.7). In the liver, glucose is phosphorylated by two major types of kinases: hexokinase, a relatively nonspecific enzyme saturated at very low glucose concentrations (K_m about 10^{-5} M); glucokinase, a specific enzyme with a K_m value near the physiologic range of glucose concentrations (K_m about 10^{-2} M). The latter is only partially saturated at normal glucose levels so that flux can increase when glucose levels increase and fall when glucose levels fall. Glucokinase is not inhibited by glucose-6-phosphate; thus the latter can build up and drive glycogen synthesis without inhibiting the continuing hepatic uptake of glucose when plasma levels are high. The activity of glucokinase decreases in starvation and in diabetes mellitus (Weinhouse, 1976).

Insulin does not *directly* enhance membrane transport in liver, but it definitely does so *indirectly*. This effect depends on induced glucokinase and the fact that glucose-6-phosphate *is* a feedback inhibitor of *hexokinase*. Reducing glucose-6-phosphate concentration accelerates glucose phosphorylation and, thus, glucose transport. Insulin reduces glucose-6-phosphate levels by stimulating its conversion to glycogen (by favoring activation of glycogen synthase) and by stimulating glycolysis. Insulin definitely stimulates hepatic glucose uptake and reduces hepatic glucose output, even though the hepatic plasma *membrane* is not insulin responsive.

Factors Influencing Rates of Glucose Uptake

The systems regulating glucose uptake (and also glucose production) are highly interactive. For purposes of discussion, we dissect them but recognize that rarely does a *single* factor act independently. Indeed, the successful operation of the homeostatic system depends on a balance of simultaneously interacting stimuli and counterstimuli.

PLASMA GLUCOSE LEVEL: AUTOREGULATION

In both insulin-sensitive and insulin-insensitive tissues, the glucose concentration available is a key factor determining rate of uptake. As discussed above, the transport mechanism is saturable but only at extremely high glucose concentrations. The maximum rate can be increased under the influence of insulin, and insulin will enhance uptake at any glucose concentration. However, even in the complete absence of insulin, glucose uptake continues in all tissues. In the diabetic deprived of all insulin, the plasma glucose level rises to very high values. At those high levels, and driven by them, the *absolute* rate of glucose uptake and oxidation (milligrams of glucose per hour) can be as high as it is in a normal subject; however, the *fractional* rate of glucose uptake (percentage of plasma pool per hour) is much lower because of the lesser efficiency of the transport mechanism.

Autoregulation in the liver has been demonstrated by perfusing the isolated organ with media containing different concentrations of glucose. Below about 150 mg/dl the liver shows a net release of glucose; above this level it shows a net uptake. *In vivo*, however, changes in glucose levels will also induce changes in key regulatory hormones, especially insulin and glucagon, and these superimposed hormonal stimuli are of overriding importance. For example, in the insulin-deprived diabetic the liver continues to show net release of glucose even in the face of plasma glucose concentrations above 500 mg/dl.

FREE FATTY ACIDS

In the presence of high plasma levels of free fatty acids (FFA), the rate of glucose uptake by skeletal and cardiac muscle at any given level of insulin is reduced (*Randle effect*). In the fasting state, FFA levels are elevated because of rapid mobilization of adipose tissue triglycerides. Thus, the utilization of glucose by peripheral tissues is inhibited, helping

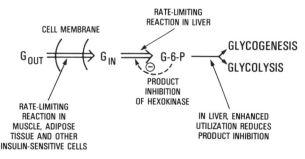

Figure 7.7. Schematic representation of glucose transport and phosphorylation.

to conserve the limited supply of glucose, which is needed to supply obligatory substrate for the brain. In the fed state, arteriovenous difference studies across the heart show that glucose oxidation accounts for the bulk of myocardial oxygen consumption; in the fasting state, glucose utilization drops, and FFA oxidation accounts for most of the oxygen consumed. Again, hormone effects are superimposed: insulin levels are lower in the fasted state, reducing glucose utilization in muscle but not affecting it in the brain.

MUSCULAR WORK

At any given glucose concentration, the rate of glucose uptake into muscle (skeletal and cardiac) is enhanced by muscular contraction. This effect does not depend on an increase in insulin concentration. Electrical stimulation of isolated skeletal muscle strips *in vitro* enhances glucose uptake; increasing the work of the isolated perfused heart by increasing the filling pressure does the same thing. The effects of low levels of insulin and low levels of exercise are additive. This phenomenon is of practical importance in the management of diabetic patients. Their insulin requirement is less on days when they engage in vigorous muscular exercise; failure to adjust the dose downward may result in a hypoglycemic reaction.

Prolonged anaerobiosis will also increase glucose uptake at a given level of glucose and insulin. This effect can be dissociated from the effect of exercise *per se* and is of lesser magnitude.

HORMONAL EFFECTS

Insulin is without question the key hormone controlling rates of glucose removal from plasma. In the bulk of the body tissues—muscle and adipose tissue—it directly enhances the rate of glucose transport into the cells. In the liver, it both enhances glucose uptake and inhibits glucose release. The dominant role of insulin is clearly shown by the profound hypoglycemia produced by injection of a large dose of this hormone, despite the recruitment of a battery of counterregulatory responses. The cellular mechanisms involved in the action of insulin are discussed in Chapter 49.

Glucocorticoids decrease glucose uptake in peripheral tissues, both basal and insulin-stimulated. Part of the effect is associated with a decrease in the number of insulin receptors, an effect that appears, however, to be transient. In addition, cortisol plays a permissive role in hormone-stimulated release of free fatty acids from adipose tissue which, as pointed out above, would also tend to reduce glucose

transport (Randle effect). However, the *major* basis for the hyperglycemic effect of glucorticoids is enhancement of gluconeogenesis.

Catecholamines tend to reduce glucose uptake. A direct effect on glucose uptake into diaphragm in vitro has been demonstrated, which may reflect increased glycogenolysis and inhibition of hexokinase by glucose-6-phosphate. In addition, catecholamines have an indirect effect by increasing FFA levels (Randle effect). However, the major effect, as with cortisol, is not on glucose uptake but on hepatic glucose production (see below).

Growth hormone has biphasic effects on glucose uptake. Acutely it has an insulin-like effect, increasing glucose uptake, but this lasts only an hour or two. Long-term elevation of growth hormone levels, as in acromegaly, decreases glucose uptake by muscle and adipose tissue and induces a form of secondary diabetes mellitus. Sensitivity to insulin is reduced and is only partially compensated for by increased production of insulin.

The decreased sensitivity to insulin may relate to competition between insulin and the somatomedins (insulin-like growth factors I and II) that can bind with low affinity to the insulin receptor but may not have as marked an effect on glucose transport (see Chapter 52). An additional inhibition of glucose uptake is referable to increased FFA levels secondary to more rapid lipolysis in adipose tissue.

Factors Influencing Rates of Hepatic Release of Glucose

There are four critical points at which metabolic and hormonal control of hepatic glucose production can be exercised. These control points, as in almost every regulated metabolic system, are points in the pathway at which the forward and reverse reactions are catalyzed by *different*, independently controlled enzymes. The four sites of control are indicated in the simplified scheme shown in Figure 7.8. Two general categories of control are involved: (1) regulation of glycogen synthesis and degradation; (2) regulation of glycolysis and gluconeogenesis.

GLYCOGEN SYNTHESIS AND DEGRADATION

The rate-limiting step for synthesis is the incorporation of glucose from uridine diphosphoglucose (UDPG) into glycogen, catalyzed by glycogen synthase; degradation is catalyzed by glycogen phosphorylase. This is the classic example of a "push-pull" system in which there is coordinate regulation of the two enzymes involved. For example, glucagon simultaneously activates phosphorylase and deactivates glycogen synthase, as briefly reviewed in Figure 7.9. The same cyclic AMP-dependent protein kinase leads to the phosphorylation of both glycogen synthase, which it acts on directly, and phosphorylase, through an intermediate step (i.e., phosphorylation and activation of phosphorylase kinase). The latter in turn phosphorylates phosphorylase *b*, converting it to its *active* form, phosphorylase *a*. The phosphorylation of glycogen synthase, on the other hand, converts it to a *less* active

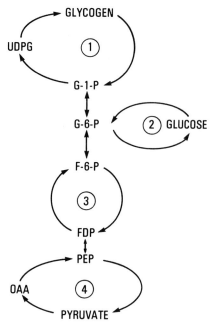

Figure 7.8. Skeleton outline of the pathways for glucose metabolism indicating the four major control points at which the forward and reverse reactions are catalyzed by separate enzymes.

form or forms. Thus, the hormone effect is maximized by accelerating breakdown while simultaneously inhibiting synthesis of glycogen.

Glucagon is the most important physiologic stimulus for glycogen degradation and release of glucose from the liver. It acts through its specific membrane receptor to increase adenylate cyclase activity and, through cyclic AMP, protein kinase activity (see Fig. 7.9).

Catecholamines, like glucagon, enhance glycogen degradation but the mechanism is now recognized to be different (Exton, 1979). The *major* glycogenolytic action of epinephrine is due to its α-adrenergic effects. They appear to be mediated not by changes in cAMP but by control of calcium ion concentration which regulates the activity of phosphorylase kinase. The β-adrenergic effects of catecholamines are exercised, like those of glucagon, through the adenylate cyclase system. Concurrently, the catecholamines inhibit insulin release, reducing the action of the major antagonist of catecholamine effects on glycogen metabolism.

Insulin favors glycogen deposition and inhibits glycogen degradation. Thus insulin counteracts the effects of glucagon and the catecholamines, favoring glucose uptake into and inhibiting glucose release from the liver. The mechanism is not yet firmly established. Under some circumstances, insulin can counter the effects of glucagon and catecholamines on cAMP formation, but under other circumstances this does not adequately account for the effect, either in the liver or in adipose tissue. Recent studies suggest that insulin interaction with its receptor may generate a second messenger, not

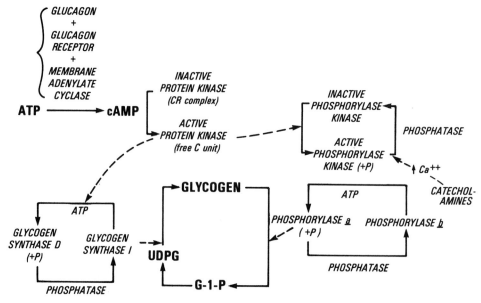

Figure 7.9. Schematic representation of glycogen metabolism in the liver and the enzymes involved in its regulation.

yet definitively characterized, and that this second messenger may be responsible for the antagonism. One attractive hypothesis is that the insulin second messenger may stimulate protein phosphatase activity (see Figure 7.9).

GLUCONEOGENESIS AND GLYCOLYSIS

Gluconeogenesis can be stimulated either directly or indirectly. Direct effects are brought about by affecting the activities of the enzymes at control points 2, 3, or 4 shown in Figure 7.8. Indirect effects are brought about by affecting the delivery of gluconeogenic substrates to the liver. Under ordinary circumstances, the rate of delivery of gluconeogenic substrates to the liver (amino acids, lactate, glycerol) does not saturate the capacity of the liver to convert them to glucose. Consequently, changes in the extrahepatic tissues that increase the concentrations of these substrates reaching the liver can increase gluconeogenesis even without changes in hepatic enzymes (e.g., increased protein degradation, increased anaerobic glycolysis in muscle, and increased lipolysis in adipose tissue with release of glycerol). The effect of enhanced uptake of FFA to increase gluconeogenesis is in a special category. The FFA are not themselves substrates for gluconeogenesis. As detailed below, increased uptake of FFA shifts the metabolic stances of the liver and channels *other* substrates toward glucose-6-phosphate (glycerol, lactate, amino acids).

Glucagon has a major direct effect through its regulation at control point 4 (Fig. 7.8). Pyruvate kinase, catalyzing the essentially irreversible conversion of phosphoenolpyruvate (PEP) to pyruvate, is converted to an inactive form by a cAMP-dependent reaction that is enhanced by glucagon. This event inhibits glycolysis and of itself tends to drive the reverse (gluconeogenic) pathway. At the same time glucagon stimulates conversion of pyruvate to PEP, probably by increasing the activity of PEP carboxykinase (conversion of oxaloacetate to PEP). In addition, at least in the diabetic state, glucagon may play a significant role in FFA mobilization and indirectly may enhance gluconeogenesis.

Epinephrine also enhances gluconeogenesis and, like glucagon, stimulates pyruvate conversion to PEP. However, the effects on gluconeogenesis are not tightly linked to changes in tissue cAMP levels, so the mechanism remains uncertain. In the intact animal, the powerful effect of epinephrine on FFA mobilization provides an indirect stimulus to gluconeogenesis and also provides more substrate in the form of glycerol.

Insulin suppresses gluconeogenesis, and several mechanisms are operative. Studies in perfused livers show that there is an acute direct effect that may relate to decreases in cyclic AMP levels. Long-term insulin deprivation induces adaptive changes in the levels of gluconeogenic enzymes that favor glucose production. In addition to these direct effects, insulin reduces gluconeogenesis by suppressing the flow of substrates to the liver. It inhibits protein degradation in muscle and lipolysis in adipose tissue, reducing the flow of FFA to the liver, thus removing a "trigger" for gluconeogenesis. A deficiency of insulin, reciprocally, favors gluconeogenesis.

Cortisol increases gluconeogenesis by increasing protein degradation and thus the flow of amino acids to the liver (indirect effect). It also induces synthesis of gluconeogenic enzymes.

Free Fatty Acids

While not themselves a substrate for net new glucose formation, high FFA levels channel gluconeogenic substrates to glucose-6-phosphate. Several effects are involved:

1. Generation of high intrahepatic concentrations of acetyl CoA, which inhibits oxidation of pyruvate to acetyl CoA via pyruvate dehydrogenase and in this way channels pyruvate to oxalacetate via pyruvate carboxylase (see Fig. 7.19).

2. Generation of high intrahepatic concentrations of fatty acyl CoA, which inhibits pyruvate kinase (PEP to pyruvate) and, at control point 3 (Fig. 7.8), inhibition of phosphofructokinase due to high levels of citrate and ATP generated during FFA oxidation (Garland et al., 1963). The rapid oxidation of FFA also provides reduced pyridine nucleotide needed for gluconeogenesis.

SUMMARY

Plasma glucose levels are affected by a wide variety of factors—metabolic, neural, and hormonal. The multiple influences and their mechanisms of action are summarized in Table 7.2. How these hormones are secreted and how that secretion is regulated is discussed in Chapters 52, 54, and 49.

The interplay involved can be illustrated taking the response to insulin-induced hypoglycemia as an example. If the plasma glucose begins to drop rapidly, the patient feels cold and anxious. He is pale and shows a tremor. These responses signal a sympathetic discharge with release of catecholamines, a major counterregulatory response. The catecholamines tend to correct the hypoglycemia by the several mechanisms listed in Table 7.2. The hypoglycemia *per se* suppresses further endogenous in-

Table 7.2
Summary of hormonal effects on plasma glucose levels

Hormones Tending to Raise Levels	Major Mechanisms	Minor Mechanisms
Glucagon	Increases glycogenolysis (via cAMP)	
	Increases gluconeogenesis (cAMP deactivation of pyruvate kinase).	Increases FFA levels (which inhibits glucose uptake via the Randle effect and stimulates gluconeogenesis).
Catecholamines	Increase glycogenolysis (α-adrenergic effect via Ca^{++}; β-adrenergic effect via cAMP	Increase FFA levels
	Inhibit insulin release (α-adrenergic effect)	Some direct effect on glucose uptake into muscle
Glucocorticoids	Increase gluconeogenesis (by increasing flow of substrate amino acids from muscle protein degradation; by inducing hepatic synthesis of gluconeogenic enzymes (e.g., transaminases, pyruvate carboxylase, glucose-6-phosphatase)	Increase FFA levels (by permissive effect favoring lipolytic action of catecholamines).
	Decrease glucose uptake by decreasing insulin sensitivity (downregulation of receptors)	
Growth hormone	Decreases glucose uptake by decreasing insulin sensitivity (downregulation of receptors?; competition between somatomedins and insulin?) (N.B.; for an hour or so after administration, *lowers* glucose levels but dominant, long-term effect is to raise them)	Increases FFA levels
Insulin	Increases glucose uptake in all insulin-sensitive tissues (direct effect on transport; increase of pyruvate dehydrogenase activity) and also in liver (secondary to effects on glycogen metabolism and glycolysis)	Decreases FFA levels (counteracts lipolytic effects of catecholamines and glucagon).
	Decreases glucose release by liver (decreases activity of phosphorylase and glucose-6-phosphatase)	
	Decreases gluconeogenesis (by reducing flow of amino acid substrate from muscle; reducing FFA levels; decreasing activity of glucose-6-phosphatase and other gluconeogenic enzymes)	
Somatostatin	Inhibits glucagon release (and also insulin release)	
	Inhibits intestinal absorption of glucose.	

sulin secretion and favors glucagon secretion. Insulin secretion is also suppressed by the direct α-adrenergic effect of catecholamines on the pancreas. Output of growth hormone and of glucocorticoids is enhanced, both of which tend to increase plasma glucose levels (Table 7.2). Finally, the patient feels hungry, ingests food, and thus corrects the hypoglycemia—a complex behavioral mecha-

nism, still poorly understood at the molecular level, but very effective!

The reader is referred to several excellent reviews for further elaboration and documentation (Cahill and Owen, 1968; Felig, 1980; Hers, 1976; Larner et al., 1968; Philips and Vassilopoulo-Sellin, 1980; Pilkis et al., 1978).

CHAPTER 48

Regulation of Lipid and Lipoprotein Metabolism

INTRODUCTION

In the preceding chapter, it was pointed out that the total available stores of glucose are strictly limited; except during the several hours following each meal, most body tissues depend upon oxidation of fatty acids, the other major caloric substrate, to provide energy. Those fatty acids are mobilized from the huge stores of triglyceride in adipose tissue depots and transported through the blood stream. The first part of this chapter deals with the regulation of fatty acid mobilization from adipose tissue, the transport mechanisms involved, and the metabolic fate of the mobilized *free fatty acids* (FFA). In addition, there is an elaborate system for the transport of triglycerides, phospholipids, cholesterol, and some less common lipids in several classes of *plasma lipoproteins*. The second part of this chapter deals with the nature of this complex lipoprotein system, the ways in which it is regulated, and the relationship between it and atherosclerosis.

THE ADIPOSE TISSUE "ORGAN"

Adipose tissue accounts for about 20% of the total body weight of a normal young adult—about 15 kg in the average person. There is (unfortunately for some) almost no limit to the extent to which this store of adipose tissue can be increased. Almost all of the increase in body mass of the extremely obese represents enlargement of the stores of adipose tissue. Thus, in an obese individual weighing 140 kg, more than 50% of body weight is represented by adipose tissue. About 90% of the mass of the adipose tissue represents stored triglycerides (over 13 kg in normal adult). Thus there is enough triglyceride available to provide fuel for 2 mo or more at average rates of calorie consumption. This is in striking contrast with the limited amounts of stored glucose available, only enough to provide energy for less than a day. The extensive store of adipose tissue triglycerides is distributed widely throughout the mammalian organism, the pattern differing from species to species. Migrating birds deposit huge quantities of fat in the abdomen (omental fat) in the days just before their arduous southward flight, which may involve as much as 36 h continuously in the air. Their body weight may increase by over 50% while "stuffing" for the trip, and every bit of the extra fat is burned up en route (Odum, 1965).

In man, adipose tissue is found subcutaneously over the entire body, with some extra deposits in the areas of the buttocks and breasts. There are large deposits of adipose tissue in the mesentery, some around the kidneys, and some around the heart. While careful studies have revealed some quantitative differences in metabolism of adipose tissue from various anatomic sites, these differences are subtle, and it is almost correct to say that the adipose tissue can be regarded as a homogeneous "organ" that happens to be anatomically widely distributed (like the skin) (Wasserman, 1965). There are some regional differences as shown, for example, by the selective deposition of fat in hips, buttocks, and breasts under the influence of estrogenic hormone. In addition to its physiologic role as a store of calories, adipose tissue plays a structural role, for example, in the cushioning of viscera (such as the kidney) and as an insulating layer reducing the rates of loss of body heat.

The mature adipose tissue cell (*adipocyte*) consists of a large structureless droplet of lipid surrounded by a very thin rim of cytoplasm. The area containing the nucleus is somewhat thickened, giving the cell the appearance of a signet ring in microscopic section. In the embryo and early fetus, the adipocytes contain very little stored triglyceride, and they do not have this characteristic appearance. Nevertheless, careful histologic examination can define nests of cells that are in fact differentiated and destined to become mature triglyceride-filled adipocytes in the adult. The adipocyte is a unique cell in its ability to continue to take up and store large quantities of fatty acids in the form of triglycerides. Other cells have at most a very limited ability to increase their triglyceride stores and that increased storage is never in the form of a unilocular central fat droplet.

The bulk of the adipose tissue in man is *white fat* as just described; in addition, particularly in the infant, there is a small amount of so-called *brown fat*. This form of adipose tissue is literally brown because of its high concentration of mitochondria which contain cytochromes and other pigments. Brown adipose tissue can also store large quantities of triglycerides, but the storage occurs in multiple droplets in the cytoplasm rather than in the form of a single, central fat droplet. Brown adipose tissue is particularly prominent in hibernating mammals and is believed to play a role in the generation of heat when the animal is roused from its hibernating state (Joel, 1965). Some recent data suggest the possibility that metabolism in brown fat may be relevant to maintenance of energy balance in man (Jung et al., 1979; Himms-Hagen, 1979).

Enlargement of the mass of the adipose tissue organ can occur: (1) because the total number of fat cells of a given size is increased, or (2) because the size of the fat cells is increased, or (3) a combination of these two. In most obese adults, an increase in cell size without an increase in cell number is found. Especially in individuals becoming obese during childhood, there tends to be both an increase in cell size and an increase in cell number. The total number of potential fat cells is affected by early nutrition, with overnutrition tending to increase and undernutrition tending to decrease the potential number of adipocytes in the mature individual (Hirsch and Batchelor, 1976; Salans, 1980).

METABOLISM OF ADIPOSE TISSUE

For many years adipose tissue was considered to be a relatively inert tissue. This was based in part on the very low rate of oxygen consumption of adipose tissue when expressed per gram of total wet weight. What was overlooked was the fact that 90% of the weight of the adipose tissue represents inert stored triglycerides. When adipose tissue oxygen consumption is expressed per milligram of cell protein, the value turns out to be surprisingly high—comparable to that of the liver and other metabolically highly active organs. Another long-standing misconception about adipose tissue was that the

triglycerides stored there were metabolically inert. The classical studies of Schoenheimer (1942) using stable isotopes dispelled that notion over 40 years ago by demonstrating the continuing turnover of the fatty acids in adipose tissue. We now know that the deposition and mobilization of fatty acids goes on all the time, even when the mass of adipose tissue is not changing.

The importance of the adipose tissue in sustaining the body during starvation is evident even to the casual observer. Yet it was quite mysterious for many decades as to just how the adipose tissue triglycerides were mobilized for utilization in the various body tissues. Most of the lipids in the plasma are present in the form of *lipoproteins*, and about 90% of the triglycerides are found in the *very low density lipoproteins* (VLDL). Consequently, efforts were made to demonstrate the release of lipoproteins and/or triglycerides from adipose tissue, but all such efforts gave negative results. In the mid-1950s, the presence of a small but, as it turned out, highly significant concentration of *free fatty acids* (FFA) was demonstrated in plasma, i.e., fatty acids not covalently bonded to any other compound (Gordon and Cherkes, 1956; Dole, 1956). The rate at which these free fatty acids turned over was enormously fast, with the half-life being only 2 or 3 min. Thus, even though the concentration was very low—less than 10% of the concentration of very low density lipoprotein triglycerides—the turnover of this fraction in the fasting state could account for transport of enough fatty acids to equal the daily energy utilization. Subsequent studies of adipose tissue *in vitro* confirmed that indeed stored triglycerides were mobilized as free fatty acids and free glycerol, i.e., prior hydrolysis was required. Release of free fatty acids represents the only significant output from the adipose tissue triglyceride fatty acid stores (Steinberg and Vaughan, 1965).

A highly simplified input-output scheme of adipocyte metabolism is shown in Figure 7.10. Hydrolysis of the stored triglycerides is under the regulation of a unique lipase, *hormone-sensitive lipase*. This enzyme can degrade triglycerides all the way to glycerol and FFA. Some of the FFA can be reincorporated into depot triglycerides, after being activated to fatty acyl coenzyme A, by reaction with glycerol-3-phosphate. Part of the FFA released is always thus reincorporated; the remainder is released to the plasma. The glycerol released from hydrolyzed triglycerides is lost from the cell almost quantitatively because the adipocyte contains little or no glycerophosphokinase. In the liver, which is rich in glycerophosphokinase, free glycerol can be very rapidly incorporated into glucose or

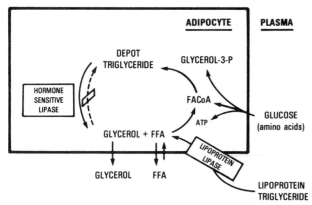

Figure 7.10. Schematic representation of the major inputs to the triglyceride stores of the adipocyte, the continuing lipolysis and reesterification of fatty acids, and the output exclusively as free fatty acids and glycerol.

into triglycerides after first being converted to glycerophosphate. Because adipose tissue cannot reutilize the liberated glycerol, measurements of the rate of glycerol release constitute an index of the true rate of triglyceride hydrolysis; the rate of release of FFA will always underestimate it to some extent because of concurrent reesterification (Steinberg and Vaughan, 1965).

The sources of adipose tissue triglycerides are several (Fig. 7.10). Glucose contributes carbons both to the formation of glycerol-3-phosphate and, after glycolysis, to the synthesis of fatty acids. Glucose in addition provides energy in the form of ATP generated from its own degradation, which can be used to synthesize triglycerides and add them to the depot stores. Amino acids can also be taken up by the adipose tissue and metabolized to yield both fatty acids (after generation of acetyl coenzyme A) and, in the case of certain amino acids, glycerophosphate (glyceroneogenesis).

Another important input is from chylomicron triglycerides and from the triglyceride-rich very low density lipoproteins. The chylomicrons represent exogenous fat, newly absorbed from the intestine; the very low density lipoproteins represent endogenous fat, released from the liver. These triglyceride-rich lipoproteins are acted on by a second lipase in which the adipose tissue is rich—*lipoprotein lipase*. This enzyme is located predominantly on the capillary endothelium of the adipose tissue (and of other tissues), but it is synthesized by the adipocyte and then transported to its location on the endothelial cells. There it can act on the triglycerides contained in triglyceride-rich lipoproteins, initially generating free fatty acids and 2-monoglycerides because of the positional specificity of the enzyme. Spontaneous isomerization, however, can make the remaining fatty acid available. A large fraction of

the triglycerides in chylomicrons is taken up into adipose tissue by this mechanism while the remainder of the chylomicron, including a residual small fraction of the triglyceride but almost all of the cholesterol, is taken up predominantly in the liver (see Havel et al., 1980, for further discussion).

The characteristics of plasma free fatty acids (FFA) are summarized in Table 7.3. The term "free" indicates that these fatty acids are not covalently bonded (in contrast to the fatty acids esterified to cholesterol or to glycerol, in cholesterol esters and triglycerides, respectively). While they are free in the strict chemical sense, they are actually very tightly bound, albeit *noncovalently*, to serum albumin and transported as albumin-FFA complexes. The affinity of FFA for binding sites on albumin is so great that less than 0.5% of the total FFA in the plasma is present in unbound form. Each albumin molecule can bind two molecules of FFA very tightly, with an association constant of approximately 10^7; an additional four molecules can be bound with a somewhat lower affinity (Goodman, 1958). Thus in a normal individual with an albumin concentration of about 7 g/dl, corresponding to approximately 1 μeq of albumin per ml, 2 μeq of FFA can be tightly bound per milliliter of plasma. As the total concentration of FFA rises, the percentage present in unbound form (at any given albumin level) increases. Even at the very highest concentrations reached under any ordinary circumstances, however, less than 1% of the total FFA is present in unbound form. Yet, the physical form of FFA that actually leaves the plasma must be the unbound FFA; the rate at which albumin leaves the plasma compartment is orders of magnitude slower than the rate at which FFA leaves. Despite the relatively modest concentration of FFA normally present, its turnover is so rapid that an enormous amount of substrate is potentially channeled through this fraction in the fasting individual. Indeed, as indicated in Table 7.3, the daily transport of FFA in the fasting state is more than sufficient to account for total daily caloric expenditure. Some

Table 7.3
Characteristics of plasma free fatty acids

Chemical Form: Not *covalently* bonded; hence, designated free fatty acids (FFA), unesterified fatty acids (UFA), or nonesterified fatty acids (NEFA).

Physical Form: Tightly bound, but noncovalently, to serum albumin; association constants 10^6–10^7; <1% unbound, i.e., in free solution.

Concentration: Normally about 0.5 μEq/ml (12.5 mg/dl)

Turnover: $t_{1/2}$ 2–3 min; plasma pool size 1750 μEq; total turnover in fasting state, 550 μEq/min or over 200 g/day fraction of transported FFA is redeposited in lipid esters (in the liver, for example) and not oxidized. In the fed state, when glucose provides a major portion of caloric substrate, FFA levels are low, and turnover is correspondingly reduced. As discussed below, the major if not exclusive control point for regulating FFA turnover is through regulation of mobilization. The half-life of FFA appears to be relatively independent of its concentration, and the fractional uptake into various tissues appears to be relatively constant.

The metabolic fate of mobilized FFA is depicted schematically in Figure 7.11 (Steinberg, 1966). The liver is the organ most active in removal of FFA from the plasma. About one-third of the total FFA in the plasma passing through the liver is extracted during a single passage. The FFA taken up is in part reesterified and stored temporarily in the form of triglycerides, phospholipids, or cholesterol esters. Another portion is oxidized completely to CO_2 and water. Another fraction is converted to ketone bodies (incomplete oxidation), which are released from the liver and then oxidized completely to CO_2 and water by the extrahepatic tissues. (The liver lacks the enzymes necessary for oxidation of ketone bodies.) Finally, either immediately or subsequent to the initial deposition, fatty acids that have been incorporated into lipid esters can be secreted from the liver again in the form of lipoprotein-borne lipids. As discussed above, the lipoprotein-borne lipids, especially the triglycerides in very low density lipoproteins, can be redeposited into adipose tissue through the action of lipoprotein lipase. This completes a "fatty acid transport" cycle. When

mobilization of FFA from adipose tissue occurs in excess of that needed immediately for fuel, the liver will increase its stores of lipid esters. Those stored esters can at a later time be oxidized, or they can be transported back to the adipose tissue in lipoproteins, restoring the balance. Finally, some FFA is taken up by essentially every tissue (with the notable exception of brain and nerve tissue), providing them with substrate for immediate combustion or, under some circumstances, for storage in ester form until needed at a later time. The amount of this temporary storage is, however, very limited.

From the above discussion, one can predict certain consequences of overly rapid and excessive mobilization of FFA from adipose tissue. These include the following: (1) deposition of triglycerides in the liver and a tendency to develop a *fatty liver*; (2) increased rates of production of ketone bodies and a tendency to develop *ketoacidosis*; (3) an increased output of lipoproteins from the liver, especially triglyceride-rich lipoproteins, and a tendency to develop *hyperlipidemia*; (4) inhibition of the rate of glucose uptake in peripheral tissues (Randle effect) and a tendency toward *hyperglycemia*. Two other consequences should be mentioned, although their importance is less firmly established. First, there may be some increase in total body oxygen consumption, since excessively high FFA levels may uncouple oxidative phosphorylation in mitochondria; second, there may be an increased tendency for intravascular thrombosis because of the effects of free fatty acids and/or the consequent hyperlipoproteinemia on platelet function and other aspects of the clotting mechanism.

REGULATION OF FFA MOBILIZATION

The triglycerides in adipose tissue are always in a dynamic state, as mentioned above, even when the total mass of triglyceride is not changing. The balance can be shifted by altering the rate of lipolysis, i.e., by changing the activity of hormone-sensitive lipase. It can also be shifted by altering the rate of re-esterification of released FFA back into triglycerides. The latter can be increased by anything that favors activation of FFA (e.g., providing additional ATP) or increases the availability of glycerol-3-phosphate as acceptor. Glucose provides both ATP and glycerophosphate and therefore strongly stimulates the re-esterification arm of the cyclic process. Insulin, by favoring glucose uptake at any given glucose concentration, will act to inhibit FFA release in part by this mechanism. However, the more important effect of insulin is directly on the state of activation of hormone-

PLASMA COMPARTMENT

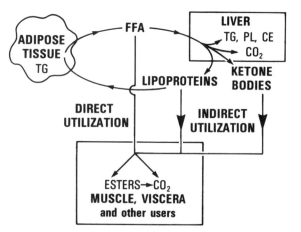

Figure 7.11. The fatty acid transport cycle, illustrating the fate of free fatty acids released from adipose tissue, their metabolism in liver and extrahepatic tissues and their reentry into the adipose tissue stores.

sensitive lipase and thus on the degradation of stored triglycerides (see below).

There are four major modes of regulation of fat mobilization (see Steinberg, 1966, for review).

Metabolic Control. As already indicated, glucose will suppress mobilization by favoring re-esterification. Amino acids can do the same, although less potently.

Neural Control. Adipose tissue is richly innervated, and stimulation of its nerve supply causes release of norepinephrine at sympathetic endings. This in turn activates hormone-sensitive lipase and promotes FFA mobilization.

Direct Hormonal Control. A surprisingly large number of different hormones can act on adipose tissue in vitro to enhance FFA release. Two of these are clearly established to play a role in regulation in vivo: norepinephrine and epinephrine. Intravenous infusion of these catecholamines elicits a prompt increase in plasma FFA levels. Glucagon is a potent stimulus to FFA release *in vitro*. Infusion of glucagon *in vivo*, however, does not normally increase FFA levels. This is probably because of the accompanying glycogenolytic action of glucagon, leading to an increase in plasma glucose levels and a consequent increase in plasma insulin levels. Glucagon also has a direct effect in stimulating insulin release. These effects in normal subjects mask any tendency to increase fatty acid release. On the other hand, in Type I diabetic patients who cannot respond with an increase in insulin levels, or in experimental subjects in whom insulin response has been suppressed by infusion of somatostatin, a glucagon effect on FFA mobilization has been demonstrated. Thus, glucagon regulation of fat mobilization may only be significant in the fasting state and in diabetic ketoacidosis.

The other hormones shown to stimulate FFA release *in vitro* include ACTH, vasopressin, melanocyte-stimulating hormone, and a number of others. However, there is no conclusive evidence as yet that changes in these hormones within the physiologic range have a significant impact on FFA release in vivo.

Indirect Hormonal Control. Cortisol and other glucocorticoids have little or no direct effect in stimulating FFA release from adipose tissue in vitro. Yet, an intact adrenal cortex is necessary in order to obtain an optimal response to lipolytic hormones such as catecholamines in vivo. This is another example of the so-called "permissive effect" of the glucocorticoids. Presumably, the continuing action of the glucocorticoid is necessary to maintain appropriate levels of one or more of the enzymes or other elements involved in the acute lipolytic re-

sponse to catecholamines and other lipid-mobilizing hormones.

Thyroid hormone, again, has little direct activity in mobilizing free fatty acids but is essential in order to maintain responsiveness to rapidly acting fat-mobilizing hormones. Again the action can be described as predominantly a permissive one. Thyroid hormone modulates a number of other cyclic AMP-dependent hormonal responses in a similar way (see Chapter 53).

HORMONE-SENSITIVE LIPASE

This enzyme has been purified from adipose tissue of the rat and chicken and fully characterized (Huttunen et al., 1970; Berglund et al., 1980; Belfrage et al., 1980; reviewed in detail by Steinberg, 1976). It is distinct from lipoprotein lipase, as discussed above, and from the lipase activities found in lysosomes. The latter are optimally active at acid pH whereas hormone-sensitive lipase is optimally active at neutral pH. The enzyme is found predominantly in the final supernatant fraction in the form of a very high molecular weight lipid-rich complex. It can be dissociated from these lipids by appropriate detergent treatment and purified. A single enzyme protein appears to be responsible for hydrolysis of triglycerides, diglycerides, and cholesterol esters. It may also be active against monoglycerides, but adipose tissue also contains a separate monoglyceride hydrolase as well.

The mechanism of activation is shown schematically in Figure 7.12 (Butcher et al., 1965; Huttunen et al., 1970; Corbin et al., 1970). It follows the now classic pattern for activation by phosphorylation through the action of a cyclic AMP-dependent protein kinase. Hormone-receptor interaction increases the activity of membrane-bound adenylate

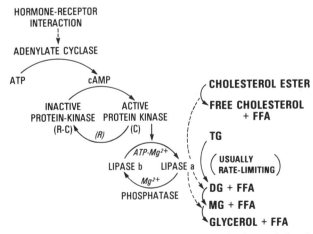

Figure 7.12. Mechanisms involved in the activation and deactivation of hormone-sensitive lipase in adipose tissue.

cyclase. The increased concentrations of intracellular cyclic AMP dissociate protein kinase from its inactive form (R-C complex) to its active form (free C unit) and the active protein kinase in turn catalyzes the phosphorylation of lipase from the less active form (lipase *b*) to the more active form (lipase *a*). Deactivation has been shown to be catalyzed by protein phosphatases (Severson et al., 1977). Thus, changes in the activity of a "lipase phosphatase" could equally regulate fatty acid mobilization, but hormonal regulation of that activity has yet to be demonstrated.

During the mobilization of triglycerides, there is very little accumulation of diglycerides or monoglycerides in adipose tissue, indicating that the rate-limiting reaction is the hydrolysis of the first ester bond in the triglyceride. Hormone-sensitive lipase can catalyze the complete hydrolysis of triglycerides to glycerol and FFA. There are some differences in the rates at which different fatty acyl ester bonds are hydrolyzed, but these are generally small differences. Therefore, the composition of the liberated FFA is rather close to that of the bulk of adipose tissue stores. In addition to hydrolyzing triglycerides, the same hormone-sensitive lipase also hydrolyzes cholesterol esters. The significance of this is not certain, but it could prevent progressive buildup in cholesterol ester content during the mobilization of adipose tissue triglycerides.

Adipose tissue contains glycogen and the full complement of glycogen-metabolizing enzymes, analogous to those found in the liver. In animals that are fasted and refed, the glycogen content of adipose tissue can rise to very high levels. Presumably this provides a mechanism to help dispose of large quantities of ingested glucose which can later be converted to fat. The glycogen synthase and phosphorylase of adipose tissue are under hormonal regulation analogous to that in the liver. Thus, catecholamines activate phosphorylase and convert glycogen synthase to its D or relatively inactive form; conversely, insulin increases the activity of glycogen synthase and decreases the activity of glycogen phosphorylase. Thus, when glucose and insulin are available in abundance, the adipose tissue will tend to take up glucose and deposit it as glycogen for later utilization.

Insulin is the only major hormone that tends to *suppress* the mobilization of FFA. Injection of antibody against insulin, creating a sudden insulin deficiency, is associated with an almost immediate rise in plasma FFA levels. In chronic insulin deficiency, such as occurs as in young diabetic subjects, FFA levels reach values higher than those seen under any other circumstances and are associated with many adverse metabolic consequences, discussed later. The mechanism by which insulin blocks FFA mobilization includes its effects to enhance fatty acid reesterification through promoting glucose uptake. Much more important, however, is its direct effect on the state of activation of the enzyme: insulin prevents the epinephrine-induced conversion of lipase *b* to lipase *a* (Steinberg, et al., 1975). Under some conditions, this effect can be explained by insulin regulation of cyclic AMP levels (Butcher et al., 1966; Loten and Sneyd, 1970; Manganiello and Vaughan, 1973). However, under other circumstances it is clear that the insulin must have an additional effect since it can reduce the activity of hormone-sensitive lipase without much effect on the tissue concentrations of cyclic AMP (Khoo et al., 1973). It has been suggested that this effect might be mediated through control of lipase phosphatase, but direct evidence is still lacking (Severson et al., 1977).

The effects of cortisol, as mentioned above, seem to be permissive. In the isolated adipose tissue, cortisol can enhance FFA release when presented in combination with growth hormone. The effects are blocked by puromycin or by actinomycin D, indicating that the effects relate to induction of new enzyme synthesis (Fain et al., 1965). The actions of the rapidly acting hormones (catecholamines and glucagon) are not blocked by inhibitors of protein synthesis or inhibitors of RNA synthesis.

LIPOPROTEIN LIPASE

The uptake of triglycerides from chylomicrons and very low density lipoproteins is catalyzed by lipoprotein lipase, and the activity of this enzyme is evidently rate-limiting. Its activity is increased by insulin and glucose. Thus its activity is high in the fed and low in the fasted state—the reciprocal of the changes seen in hormone-sensitive lipase. Again, catecholamines, which *increase* hormone-sensitive lipase activity, *decrease* lipoprotein lipase activity. This reciprocal control led to speculation that the two major lipases in adipose tissue might represent mirror images of the same enzyme protein, having different substrate specificities and different functions at different times. However, both enzymes have now been isolated and fully characterized, and they are clearly independent. Furthermore, patients with inherited complete deficiency of lipoprotein lipase (familial hyperchylomicronemia) do not have any difficulty in regulating transport of free fatty acids into and out of adipose tissue.

Lipoprotein lipase is not confined to adipose tissue. In fact, lipoprotein lipase has been demonstrated in almost every tissue. In all tissues studied, the enzyme appears to act at the capillary surface, as in the adipose tissue. Thus, triglyceride-rich lipoproteins entering the microcirculation bind to the capillary-bound enzyme, for which they have a high-affinity, and there they undergo partial hydrolysis. The lipolytic products can then enter the parenchymal cells of the organ involved. For example, the uptake of chylomicron and VLDL triglycerides by cardiac muscle is mediated by the action of lipoprotein lipase. Whereas fasting causes a *decrease* in the lipoprotein lipase level in adipose tissue, it causes an *increase* in the levels of the enzyme in the heart and in many other tissues. Thus, in the fasting state there is a shunting of newly ingested triglyceride to the tissues that require substrate and away from storage in the adipose tissue, an appropriate response. In the fed state, the reciprocal changes occur, and the triglycerides tend to be deposited in the adipose tissue (reviewed by Havel et al., 1980).

THE PLASMA LIPOPROTEINS

Lipids are found in the plasma at concentrations far above their solubility in aqueous systems. For example, normal plasma cholesterol concentrations in middle age are over 200 mg/dl whereas the solubility of free cholesterol in water is less than 0.1 mg/dl and that of cholesterol esters even less. Triglycerides and phospholipids, the other major lipid components in plasma, are also present at concentrations far above their solubility. This is only possible because of the way in which lipids are associated with specialized apoproteins to form large spherical lipoprotein complexes. The highly nonpolar lipids (cholesterol esters and triglycerides) are concentrated almost exclusively in the central core while the more polar lipids, mainly phospholipids, and the protein moieties make up a polar shell. The physical properties of the package are those of proteins, not lipids, and the lipoproteins can be resolved by precipitation, electrophoresis, or differential centrifugation using methods like those used with ordinary proteins. The interaction of lipoproteins with cell membrane receptors, and thus their metabolic fate, is also dictated largely by the nature of their protein moieties.

Lipoprotein Classes

Five major classes of plasma lipoproteins are currently recognized (Table 7.4). However, these classes do not constitute stoichiometrically precise complexes. Each can be subdivided into subclasses or even regarded as a spectrum of molecules. Nevertheless, the crude classification is sufficient for

Table 7.4
The major classes of plasma lipoproteins

Class of Lipo-protein	Density (g/ml)	Flotation Constant (S_f)	Electro-phoretic Mobility (Paper or Agarose)
Chylomicrons	<0.96	>400	Remain at origin
Very low density (VDL)	<1.006	20–400	pre-β (α)
Intermediate density (IDL)	1.006–1.019	12–20	β to pre-β
Low density (LDL)	1.019–1.063	0–12	β
High density (HDL)	1.063–1.21	(Analyzed separately)	α

many purposes, including broad definition of metabolic functions and relationship to disease states (Havel et al., 1980).

DENSITY CLASSES

Lipids have densities much lower than those of proteins. For example, triglycerides have densities less than 1.00 g/ml, and cholesterol has a density of 1.05; proteins have densities of 1.25–1.35. The densities of the lipoproteins thus vary according to the proportions of lipid and protein. As a general class they all have densities less than that of all the other plasma proteins and can be separated from them on that basis. If enough salt is added to plasma to raise its background density (density of the non-protein fluid phase) to 1.21 g/ml and if the sample is centrifuged overnight in the ultracentrifuge at >100,000 × g, all the lipoproteins rise to the surface, well separated from all the other plasma proteins, which sediment toward the bottom of the tube. If plasma is centrifuged without changing its background density (which is 1.006), any chylomicrons present will float after just 30 min at 100,000 × g. A subsequent overnight centrifugation brings the VLDL class to the top. Successive centrifugation steps at salt densities adjusted up to 1.019, then to 1.063, and finally to 1.21 float up the intermediate density (IDL), low density (LDL), and high density lipoprotein (HDL) classes, respectively. This separation by preparative ultracentrifugation is widely used in research since the individual fractions can be obtained in quantity for further characterization (Havel et al., 1955).

FLOTATION CLASSES

In the analytic ultracentrifuge one normally measures the rate at which protein boundaries sediment under specified conditions and designates a Svedberg sedimentation coefficient (S). Using a higher medium density, lipoproteins are made to float (i.e., move centripetally), and the rate of flotation is specified in analogous fashion as a Svedberg flotation coefficient (S_f). The rate of flotation increases according to the difference between the density of the particle and that of the medium; the large, lipid-rich particles float fastest. The S_f values shown in Table 7.4 apply when the medium has a density of 1.063.

ELECTROPHORETIC MOBILITY CLASSES

The several classes can also be resolved by electrophoresis. By staining specifically for lipid, the lipoproteins can be located readily, even in the presence of all the other serum proteins. The mobilities of the several classes of lipoproteins, shown in Table 7.4, are determined primarily by the apoproteins associated with them. Size also plays a role, especially in the case of the chylomicrons, which are too large to penetrate most gels and do not leave the origin. Electrophoretic separation, being the simplest, is the most widely used method for clinical purposes.

SIZE AND CHEMICAL COMPOSITION

Chylomicrons consist predominantly of triglycerides (85–95% by weight) surrounded by a thin shell of protein (1–2% by weight) and phospholipid. They also contain newly absorbed cholesterol, mainly in esterified form. The particles are greater than 80 nm in diameter, visible in a dark field in the light microscope.

VLDL, except after a fat-containing meal, account for most of the triglycerides in plasma. They vary tremendously in size (30–80 nm), mainly because of different triglyceride content (20–80% by weight).

LDL (about 20 nm) carry most (about two-thirds) of the cholesterol in normal plasma. The LDL is about 50% by weight cholesterol and 20% by weight protein.

HDL molecules (about 8 nm) are about 50% protein by mass and only 50% lipid—hence, their high density.

The clinical laboratory usually analyzes only for total cholesterol content and total triglyceride content of the plasma. Only for relatively sophisticated studies is a further breakdown into relative concentrations of different lipoprotein classes necessary. However, knowing the general composition of the lipoproteins, one can draw useful inferences about lipoprotein patterns knowing only total cholesterol

and triglyceride values. For example, a marked elevation of cholesterol level with a normal triglyceride level cannot be due to an increase in VLDL alone; since VLDL is rich in triglycerides, an increased VLDL sufficient to raise cholesterol level would necessarily increase triglyceride levels even more. One can conclude if cholesterol is elevated and triglyceride normal that the patient has an increase in LDL concentration. (Elevation of HDL sufficient to cause a marked increase in cholesterol level can occur but is seldom seen).

APOPROTEINS

The protein moieties associated with the lipoproteins play not only a structural role, stabilizing the lipid particles and conferring protein-like properties on them, but also a variety of functional roles. The nomenclature is not particularly helpful because letter designations were applied chronologically as the apoproteins were identified. In several instances, heterogeneity has been discovered subsequently, and several different apoproteins with no apparent genetic or functional relationship are grouped into a given "family." For example, the apoprotein associated with HDL was originally designated "apoprotein A." Subsequent studies established that there are two major apoproteins, now designated apo A-I and apo A-II, that differ considerably in structure and function. Both have been completely sequenced. Apo A-I, with a molecular weight of about 27,000, is a cofactor for the enzyme lecithin-cholesterol acyltransferase; apo A-II, with a molecular weight of about 16,000, consists of two identical polypeptides joined by a disulfide bridge and has no established functional role. Apoprotein E, found in both VLDL and HDL, has turned out to include at least three different isoforms separable by isoelectric focusing. This is more nearly what one would designate a "family" since they derive from allelic genes at a given locus. The apo C group, on the other hand, consists of proteins of quite different structure and different function. For example, apoprotein C-II is an obligatory cofactor for the action of lipoprotein lipase; the other C apoproteins, apo C-I and apo C-III, cannot substitute for it. Thus, several kindreds have been discovered in which apo C-II is specifically deleted, and these families suffer from familial hyperchylomicronemia because of the inability to clear chylomicrons from the blood at an adequate rate.

In addition to the enzyme and co-factor functions mentioned above, the most important role of the apoproteins is to target lipoproteins to the appropriate cells and tissues by virtue of receptor-apoprotein interactions. The best established of these is the receptor recognizing LDL, which binds apoprotein B with high affinity (Goldstein and Brown, 1977). Deletion of this receptor results in extremely high plasma concentrations of LDL and therefore of cholesterol (familial hypercholesterolemia). This same receptor also recognizes and binds apoprotein E with high affinity (Innerarity and Mahley, 1978). A separate receptor for apoprotein E has been described in the liver but not thus far in other tissues (Sherrill et al., 1980). Apoprotein-receptor interactions are being actively investigated, and a clearer picture should emerge over the next few years.

Lipoprotein Metabolism

At the outset it should be recognized that there are important species differences in the composition and pattern of plasma lipoproteins and in the ways in which they are metabolized. For example, most animals have extremely low plasma concentrations of LDL; man has the highest levels found in any animal species. In the following discussion, whenever possible, we will present patterns of lipoprotein metabolism as they occur in man. Several detailed reviews are available (Goldstein and Brown, 1977; Steinberg, 1979; Havel et al., 1980; Stanbury et al., 1983).

VLDL, IDL, AND LDL

The liver is the major source of VLDL, with the intestine making a relatively small contribution (about 15% or less). The large, triglyceride-rich VLDL particles emerging from the liver are acted upon by lipoprotein lipase as they pass through the capillary beds. This enzyme breaks down VLDL triglycerides, and the products are taken up into the various extrahepatic tissues, as shown in Figure 7.13. At the same time, some of the apoproteins of the VLDL are lost. Over a relatively short time (the half-life of VLDL in the plasma is only about 60 min), a large fraction of the triglyceride has been removed, and the size of the particle has decreased to that characteristic of an intermediate density lipoprotein (IDL). Some of the IDL molecules are removed intact from the circulation, but in man most are further degraded either by the action of hepatic lipase and/or further action of lipoprotein lipase to generate LDL. This final IDL-to-LDL transformation is associated with the loss of any remaining apoproteins other than apoprotein B and with the loss of some additional residual triglyceride from IDL. During this elaborate transformation, the amount of apoprotein B per lipoprotein particle remains constant. In other words, the apoprotein B

of the VLDL particle is the linchpin—the constant that defines the VLDL-IDL-LDL pathway. During this interconversion, most of the VLDL cholesterol is also retained.

LDL is taken up and degraded by tissues throughout the body. The liver is the major site of removal, accounting for about one-half of all LDL degradation; the remainder is distributed among a variety of tissues, no one of which approaches the liver in individual importance. Weight for weight, however, the adrenal cortex and the gonads are extremely active in LDL degradation, which provides cholesterol precursor for the biosynthesis of the steroid hormones they produce. These tissues also synthesize cholesterol endogenously and can produce their steroid hormones in the absence of LDL, but maximal steroid hormone production may depend upon LDL being provided. About two-thirds of LDL degradation *in vivo* is mediated by the LDL receptor. That receptor recognizes both apo B and apo E and is also designated the B/E receptor. In the rat, it appears that HDL plays an important role analogous to this role of LDL, i.e., HDL, recognized by its apoprotein E moiety, delivers a significant amount of cholesterol to adrenal and gonads.

The uptake of IDL into the liver appears to depend on recognition of its apoprotein E moiety rather than the apoprotein B. This conclusion is based on the inherited abnormality of apoprotein E in families with dysbetalipoproteinemia (type III hyperlipoproteinemia), in which there is an accumulation of lipoproteins with characteristics similar to those of IDL. These patients are characterized by homozygosity for one of the isoforms of apoprotein E, apo E_2, an isoform that is evidently not well recognized by receptors. These patients, like those with familial hypercholesterolemia, suffer premature atherosclerosis leading to myocardial infarctions and strokes.

CHYLOMICRONS

Chylomicrons generated during the digestion and absorption of fat have a very short lifetime in the plasma (half-life about 10 min). Their metabolism, like that of VLDL, depends upon the action of lipoprotein lipase, and the pattern is rather similar (see Fig. 7.13). Apoprotein C-II is an obligatory cofactor. Chylomicrons as they enter the blood stream from the thoracic duct have a very different apoprotein composition from that of circulating chylomicrons. Apoproteins are transferred by exchange from the VLDL and HDL fractions to the chylomicron fraction, and this is essential to optimize the action of lipoprotein lipase on the chylomicron particles. Chylomicrons contain a form of apoprotein B that is synthesized only by the intestine in man (but synthesized also by the liver in the rat). This form of apoprotein B has a lower molecular weight than that synthesized in the liver, but its physical properties and immunochemical properties are closely related. The intestinal form has been designated B48 and the hepatic form B100, designations roughly indicating their relative mo-

lecular weights. Since no B48 is found in the LDL fraction in man, it is evident that in man essentially all of the chylomicron apoprotein B is removed from the plasma compartment before degradation to the molecular size of LDL can take place. In other words, the scheme for catabolism of chylomicrons would be analogous to that shown in Figure 7.13, except that we would substitute "chylomicron remnant" for "IDL" and there would be no final conversion to LDL or its equivalent.

HDL

HDL is synthesized in the liver and in the intestine. It emerges into plasma in the form of flattened discs, essentially bilayers of phospholipid containing some cholesterol, mainly free, and some apoproteins. In the plasma compartment, the nascent HDL is acted upon by lecithin-cholesterol acyltransferase (LCAT), which converts free cholesterol to ester cholesterol by transfer of a fatty acid from lecithin (Fig. 7.14). Most of the lipoproteins secreted into the plasma contain only limited amounts of cholesterol ester; most of the cholesterol ester found in the plasma at steady state is the result of the action of this enzyme. Cholesterol esters can be transferred among lipoprotein classes, a process mediated by an exchange carrier protein

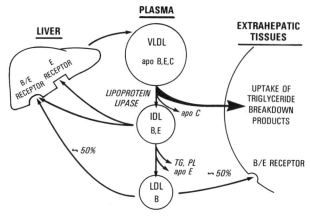

Figure 7.13. Schematic representation of the metabolic interconversions of very low density lipoprotein (*VLDL*) to intermediate density lipoprotein (*IDL*) and on to low density lipoprotein (*LDL*) (see text for elaboration).

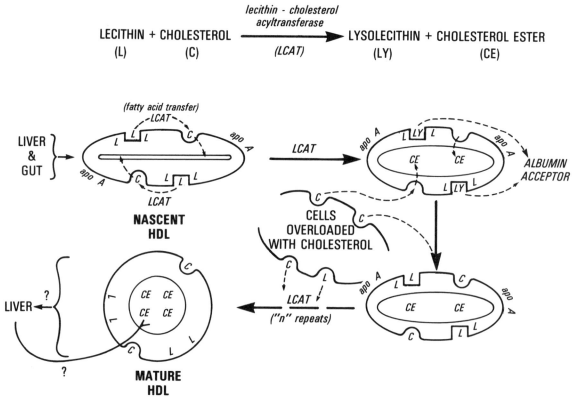

Figure 7.14. Schematic representation of nascent HDL from the liver, its conversion to mature HDL by the action of lecithin-cholesterol acyltransferase (LCAT), and the postulated role of HDL in reverse cholesterol transport (see text).

present in normal human plasma and in the plasma of most animals (absent in the rat). The result is that the lipoproteins undergo a great deal of "remodeling" after their secretion.

As a result of the action of LCAT, HDL gradually assumes a spherical configuration with cholesterol esters in the central core of the molecule. In the process, it becomes capable of accepting free cholesterol from tissues, as indicated in Figure 7.14, which may be of central importance in preventing overload of cholesterol in various peripheral tissues. All tissues participate to some extent in the uptake and degradation of LDL as discussed above, but none of them except for the gonads and adrenal cortex can further catabolize cholesterol. Furthermore, all tissues synthesize cholesterol de novo at some rate. Therefore, there must be some mechanism for returning cholesterol to the liver for excretion in order to avoid progressive buildup of tissue cholesterol levels. HDL probably plays a key role in this "reverse cholesterol transport."

During the action of lipoprotein lipase on chylomicrons or on VLDL, sections of their phospholipid shell become redundant as the core of triglycerides gets smaller. These surface components make an additional contribution to HDL formation. In man the HDL fraction includes at least two subfractions, HDL_2 and HDL_3, the latter being smaller and more dense. The addition of lipids from chylomicrons or VLDL can convert HDL_3 to HDL_2. Epidemiologic studies leave no doubt that subjects with low concentations of HDL_2 are much more likely to develop premature atherosclerosis than subjects with high concentrations of HDL_2. Whether this "protective effect" is accounted for by the mechanism of reverse cholesterol transport we have just discussed is not fully established.

Lipoprotein Uptake at the Cellular Level

Lipoprotein uptake into the cell can occur by one of three classes of mechanism: (1) receptor-mediated endocytosis; (2) adsorptive endocytosis not mediated by a specific receptor; and (3) fluid endocytosis (not involving adsorption to the cell membrane).

The best established receptor-mediated mechanism is that involved in the uptake of LDL. The work of Brown and Goldstein establishing the nature and function of this receptor was a milestone in the development of our understanding of lipoprotein metabolism. The receptor is widely distributed among all body tissues, including the liver, and about two-thirds of total body LDL degradation is mediated by this receptor. Its function is schematically indicated in Figure 7.15. LDL binds with high affinity to the receptor, which tends to cluster in specialized areas of the plasma membrane known as "coated pits." The LDL concentrated in the coated pit is taken into the cell by invagination of the membrane (endocytosis), and a vesicle containing LDL pinches off from the plasma membrane. That vesicle ultimately fuses with a lysosome. There the lysosomal hydrolyases degrade the various components of the LDL, including the protein and cholesterol esters. The products of lysosomal hydrolysis are released, and the free cholesterol formed regulates cholesterol metabolism of the cell. It inhibits cholesterol biosynthesis by suppressing the production of HMG CoA reductase; it stimulates the rate of cholesterol esterification so that any excess cholesterol is temporarily stored in the form of cholesterol esters, primarily cholesterol oleate. Finally, it in some way inhibits the further production of LDL receptors. The latter effect provides a feedback control so that the cell can limit the amount of cholesterol delivered to it in the form of LDL.

Adsorptive endocytosis not involving the LDL receptor occurs at all times. In patients who totally lack the LDL receptor (homozygous familial hypercholesterolemia), there is no receptor-mediated uptake, yet the total uptake and degradation of LDL per day actually exceeds that in a normal individual. This must all obviously occur by pathways independent of the LDL receptor, but these pathways are incompletely characterized.

Finally, some incorporation and degradation takes place by the random imbibition of LDL molecules when the cell takes up the surrounding medium by fluid or bulk endocytosis. The amounts taken up by this mechanism are limited and are strictly in proportion to the concentration of LDL in the surrounding medium.

Regulation of the LDL receptor is an important mechanism for controlling plasma LDL concentrations. For example, one of the most widely used drugs for treating hypercholesterolemia is cholestyramine, which sequesters bile acids in the intestine. It has been shown that the drug is effective in part because it leads to an induction of hepatic LDL receptors and thus an increase in the rate of hepatic degradation of LDL. Other drugs have also been shown to affect LDL receptor number.

The LDL receptor recognizes not only apo B, the only significant apoprotein of LDL, but also apo E. The latter apoprotein is found in some subfractions of HDL, and those subfractions are avidly adsorbed and taken up by the so-called LDL receptor. It might better be referred to as the apo B/E receptor since it recognizes both.

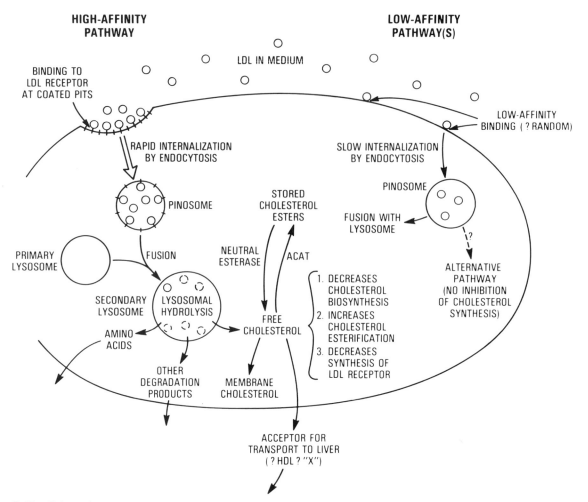

Figure 7.15. Schematic representation of the mechanism for uptake of *LDL* by endocytosis, including both the specific receptor-mediated mechanism and low-affinity nonspecific mechanisms (see text).

The liver, in addition to the apo B/E receptor, expresses a receptor involved in the binding and uptake of chylomicron remnants and VLDL remnants. Strong evidence suggests that the binding to this receptor depends upon apoprotein E. In fact, it appears to bind only two of the three major isoforms of apoprotein E—apo E_3 and apo E_4—with high affinity. Apo E_2 binds very poorly, and patients who are homozygous for apo E_2 may have accumulation of remnants (type III hyperlipoproteinemia). However, hyperlipidemia is *not* always associated with this pattern of apo E isoforms, although all of the patients who do have familial dysbetalipoproteinemia are homozygous for apo E_2. Additional genetic or environmental factors are needed to evoke the hyperlipidemia.

HDL appears to be taken up and degraded in part by saturable mechanisms, but the receptors have not been fully characterized. Again, there is suggestive evidence that the low molecular weight form of apo B (B48) may be recognized by a receptor different from the classical apo B/E receptor for LDL, but ultimate proof is lacking.

Physiologic and Pathophysiologic Significance of Lipoprotein Transport

The role of the lipoproteins in triglyceride transport and cholesterol transport is readily apparent. Chylomicrons are responsible for transporting whatever quantity of fat one ingests daily. In the average American diet, this amounts to about 100 g daily. Triglyceride transport in VLDL is more modest by far, amounting to about 10–15 g daily. However, this retransport of fatty acids from liver to extrahepatic tissues in the form of VLDL triglycerides is essential to avoid excessive accumulation of triglycerides and development of a fatty liver (see Fig. 7.11).

The importance of delivery of cholesterol in LDL to tissues producing steroid hormones was discussed above. In addition, delivery of cholesterol may be essential in other tissues under certain circumstances. For example, during growth or tissue repair the requirements for cholesterol may exceed the capacity of the cells involved to synthesize cholesterol de novo. Thus, during growth LDL may play a more important role as a vehicle for transport of cholesterol than it does in the adult at steady state.

From a pathophysiologic point of view, the lipoproteins are extremely important. There is no question about the strong positive correlation between LDL levels and susceptibility to atherosclerosis. The myocardial infarction rate and stroke rate in patients with very high LDL and high total plasma cholesterol levels is much greater than that in other patients. The best evidence that elevated LDL levels are in themselves atherogenic is provided by the families with inherited deficiency of the LDL receptor (Goldstein and Brown, 1977). This disorder is monogenic. Consequently, if one accepts the one gene-one protein hypothesis, all of the phenotypic expressions must be related ultimately to the single underlying genetic error. In this case, the genetic error is the deletion of the LDL receptor, and its immediate consequence is an elevation of the plasma concentrations of LDL. The development of atherosclerosis in these patients, then, must be linked to this elevation of LDL, although the precise mechanisms of that linkage remain to be established. The causative LDL-atherosclerosis connection is further strengthened by the recently described rabbit model of LDL receptor deficiency. These animals have plasma cholesterol levels over 400 mg/dl on their usual diet and develop atherosclerotic changes as early as age 3 mo (Watanabe, 1980; Tanzawa et al., 1980).

Other lipoprotein fractions have also been implicated as possibly contributing to atherosclerosis, although the epidemiologic correlations are less striking. Patients with hypertriglyceridemia only (i.e., having elevation of chylomicrons and/or VLDL) have been found in some studies to show a higher than average incidence of atherosclerosis and its complications, but other epidemiologic studies find no such increased risk. If one assesses the risk associated with elevated cholesterol levels and "corrects" for it, then there is little or no additional risk associated with being hypertriglyceridemic. On the other hand, from a mechanistic point of view, the presence of lipoproteins enriched in triglycerides may not be without consequences and implications for atherogenesis.

Finally, as mentioned previously, elevated levels of HDL are associated with a *decreased* atherosclerotic tendency. This negative correlation is just as strong as the positive correlation between atherogenesis and *increased* LDL levels. One hypothesis has been mentioned above, namely, that HDL may play a role in reverse cholesterol transport. By accepting cholesterol from cholesterol-laden cells in the periphery and carrying it back to the liver for excretion, HDL may defer the development of atherosclerosis. Another hypothesis relates to the competition between certain subclasses of HDL and LDL for uptake and degradation. Apoprotein E-containing HDL competes effectively with LDL for binding to the apo B/E receptor. Thus, a high concentration of HDL might reduce the rate of uptake of LDL. There may be competition also for nonsaturable binding and uptake. Finally, there is evidence that HDL may protect the endothelial lining against damage induced by LDL.

Space does not allow a complete discussion of the complexities of the atherogenic process. Several comprehensive reviews are available (Goldstein and Brown, 1977; Benditt, 1978; Wissler, 1978; Ross, 1981; Steinberg, 1981). Abnormal lipoprotein patterns are by no means the only cause, of course. Many individuals develop serious atherosclerosis in the presence of normal lipoprotein patterns. Other factors that are known to be associated with increased risk include: (1) hypertension; (2) hyperglycemia and diabetes mellitus; (3) obesity; (4) cigarette smoking; (5) being male rather than female; and (6) aging. Just how these so-called risk factors relate to the disease process at the cellular level remains to be established. Damage to the endothelium, such as might occur in association with hypertension, is believed to be an important factor. The tendency toward platelet aggregation with consequent release of factors stimulating cell growth and favoring local thrombosis is believed to be a factor. Elevated lipoprotein levels and some of these other mechanisms may relate closely to one another. For example, high concentrations of LDL can themselves be damaging to endothelial cells and can initiate the consequences of endothelial denudation. Again, high concentrations of plasma lipoproteins tend to favor platelet aggregation and the consequences of that atherogenic process.

In summary, the involvement of LDL in atherogenesis is clearly established by experimental studies, epidemiologic studies and, most importantly, genetic studies in patients and animals lacking the LDL receptor. The apo B/apo E-containing remnants of incomplete VLDL degradation are also rather clearly implicated. Other lipoproteins are probably involved as well, but the data are less clear. Effective intervention to decrease the toll of myocardial infarction and stroke requires attention to a series of risk factors, most notable of which are hyperlipoproteinemia, hypertension, and cigarette smoking (Levy, 1981).

CHAPTER 49

The Endocrine Pancreas

INTRODUCTION

The normal human pancreas, weighing about 80 g, is a flattened, elongate organ lying against the posterior wall of the upper abdomen (Fig. 7.16). The vast bulk of the gland consists of acinar cells that synthesize and secrete digestive enzymes that enter the duodenum via the pancreatic duct system. This *exocrine* function of the pancreas is discussed in Chapter 44.

Scattered more or less randomly through the pancreas and accounting for only 1–2% of its weight are several hundred thousand microscopic nests of cells, the islets of Langerhans. These cells do not connect with the system of pancreatic ducts. The islets collectively constitute the *endocrine pancreas*, secreting several critically important hormones directly into the bloodstream. This arrangement of an endocrine organ buried within an exocrine organ has no established functional significance. In fact, in some lower species the islet tissue occurs anatomically separated from the exocrine pancreas. For example, the islet tissue of the angler fish (and other bony fish) is concentrated in one or two discrete tiny organs ("principal islets") that can be as large as a pea. Many early studies of islet physiology were done using these fish so that the endocrine cells could be investigated separately. Later, techniques were developed for isolating islets from mammalian pancreas. The independence of the islets and the exocrine pancreas is nicely illustrated by the fact that ligation of the pancreatic duct causes atrophy of all the acinar tissue (because of back pressure), yet the islets survive and continue to function. Banting and Best (1922) exploited this differential atrophy in order to isolate insulin for the first time. Previous attempts had failed because the enormous amounts of digestive enzymes in the homogenates of the pancreas rapidly degraded the small amounts of insulin; in homogenates of the atrophied pancreas this was much less of a problem. Also Banting and Best used acidic ethanol to extract the insulin, and this inactivated any residual proteolytic enzymes.

Each islet in the mammalin pancreas contains several cell types that differ in morphology, staining properties, and functions (Orci, 1976; Munger, 1981; Orci, 1982). The use of specific antibodies and immunofluorescent labeling techniques has identified four major cell types and the hormones each secretes:

1. A cells, secreting glucagon (also designated α or α_2 cells)
2. B cells, secreting insulin (β cells)
3. D cells, secreting somatostatin (α_1 cells)
4. F cells, secreting pancreatic polypeptide (PP cells)

The "missing" C and E cells have been tentatively identified on morphologic grounds in some animal species but the limited data available do not establish them as unique cell types.

The four major cell types are not randomly distributed within the islet. The B cells, the predominant cell type, make up the large central mass of the islet, as diagrammatically illustrated in Figure 7.16. The A cells, the second most common cell type (about 20% of the total), form a rim around this central mass in most of the islets. However, in the inferior portion of the head of the pancreas islets contain very few A cells and many more F cells, i.e., the islets are morphologically and possibly functionally different in this portion of the pancreas. D cells are also found almost exclusively near the outer rim, in close proximity to both A cells and the inner core of B cells. As we shall see, the hormones secreted by these different cell types can affect the rate of hormone secretion by their neighbors. Thus *local* release of somatostatin from D cells, for example, could inhibit secretion of glucagon and insulin by A and B cells, respectively. The physiologic importance of this *paracrine* regulation is under current investigation. In addition, the different islet cell types have been shown to communicate directly through "gap junctions," providing the potential for a second, more direct form of mutual control.

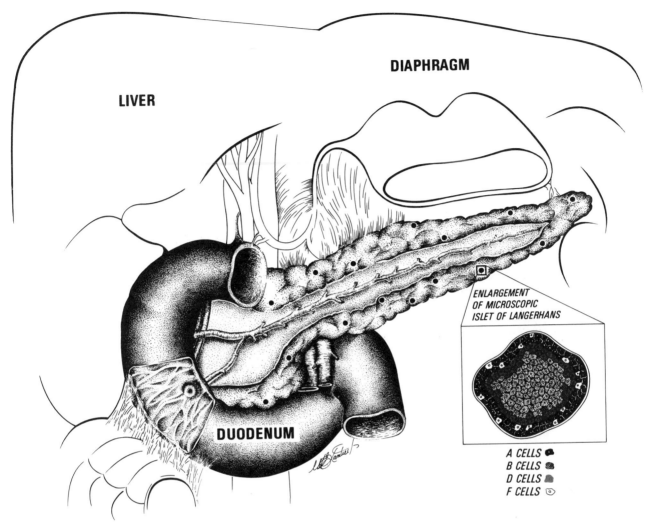

Figure 7.16. Schematic representation of the anatomy of the pancreas. *Inset,* Arrangement of the various cell types in a typical islet of Langerhans.

INSULIN

Synthesis and Secretion

The biosynthesis and secretion of insulin follows the general pattern that applies to polypeptide hormones generally (see Chapter 51). The mature messenger RNA moves from the nucleus to take up its position on ribosomes dotting the external or cytoplasmic surfaces of the rough endoplasmic reticulum. There it is translated to yield a single, long polypeptide chain, designated *preproinsulin.* The N-terminal sequence of preproinsulin (signal peptide) is cleaved as the nascent protein is secreted into the cisternae of the endoplasmic reticulum and the cleaved product, proinsulin, moves along the cisternae down to the Golgi apparatus. There it is packaged into secretory vesicles that bud from the Golgi membrane. The secretory vesicles of the B cells contain mainly insulin but also a small percentage of proinsulin. The vesicle contains a high concentration of zinc, which is known to form tight complexes with insulin, but the physiologic significance of zinc is not yet established. The vesicles are moved to the inner aspect of the plasma membrane by a process that involves the microtubular system, and they then fuse with the membrane, discharging their content into the extracellular space (exocytosis or emiocytosis).

The proinsulin molecule (Fig. 7.17) is coiled and cross-linked internally by two specific disulfide bonds. In the Golgi apparatus, and continuing after packaging into secretory vesicles, proinsulin undergoes proteolytic cleavage by enzymes with a high degree of specificity. They attack arginine and lysine residues that are paired at either end of a 31-residue "C-peptide" in human proinsulin; two basic amino acids at either end are released in free amino acid form. Thus, a total of 35 residues, constituting the "connecting peptide," are excised from the middle of the proinsulin molecule (see Fig. 7.17). The disulfide bridges are not affected. This generates the mature hormone, which consists of two polypeptide chains linked only by the disulfide bridges. There is, in addition, an intrachain disulfide bridge connecting residues six and eleven in the A chain. The C-peptide accumulates with insulin in the secretory granules and is released into the plasma with the insulin in approximately equimolar amounts (Horwitz et al., 1976). Normally the conversion of proinsulin to insulin is almost completed prior to emiocytosis, but proinsulin may account for as much as 15% of the secreted product.

The biosynthesis of insulin as a single polypeptide chain is essential in order that the mature hormone have the appropriately placed disulfide bridges. If the disulfide bridges in insulin are chemically reduced and then reoxidized *in vitro*, random coupling of half-cystine residues takes place, and only a small percentage of the biological activity is restored. Reduction and reoxidation of proinsulin, on the other hand, restores almost all of the biological activity. The specific configuration of the proinsulin molecule, dictated by its primary amino acid sequence, presumably brings into close apposition the correct pairs of half-cystine residues that are to be linked through disulfide bridges in the mature insulin molecule.

Human insulin has been produced in bacteria using recombinant DNA techniques (Crea et al., 1978; Goeddel et al., 1979). From the known amino acid sequences and the genetic "code book," DNA strands were chemically synthesized to code separately for the A and B chains. These synthetic "genes" were cloned in *Escherichia coli* (linked to the β-galactosidase gene), and large quantities of galactosidase-A chain and galactosidase-B chain were produced. These were then chemically cleaved and the separate A and B chains joined. The final product did not differ in chemical or biological properties from pure pancreatic human insulin (Chance et al., 1981).

The genomic DNA sequences coding for preproinsulin in man and in the rat have been completely determined (Bell et al., 1979; Cordell et al., 1979). Complementary DNA has been prepared from messenger RNA (from a human insulinoma and from isolated rat islets), cloned in bacteria, and the base sequences of the cloned cDNA determined. The amino acid sequences predicted for proinsulin agreed precisely with those previously established for the purified prohormone. The 72-base sequence preceding the sequence coding for proinsulin established for the first time the full amino acid sequence of the "signal" peptide or *pre* region of preproinsulin. Finally, the actual genomic DNA for rat preproinsulin has been isolated in amounts sufficient for sequencing. Two slightly different forms of rat insulin (differing at just two positions on the B chain) have been identified, and two nonallelic genes have been identified. (Lomedico et al., 1979). The genes for preproinsulin I and preproinsulin II both contain introns, i.e., sequences transcribed into primary messenger RNA but then spliced out before mRNA translation and therefore not represented in the protein product. Both genes contain an intron of 119 residues upstream from the *pre* region; only the preproinsulin II gene contains an additional long intron (499 residues) within the connecting peptide region. The base sequences of the two genes upstream from the 5'-capping site, where mRNA transcription starts, show striking homology. These areas presumably participate in regulation of gene expression, and thus synthesis of the two forms of rat insulin may be under equivalent control.

Two genetic variants of human insulin have been described, and more undoubtedly will be discovered. In one case the linkage between the C-peptide and

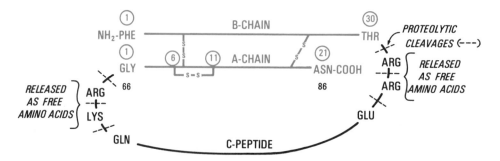

Figure 7.17.   General structure of proinsulin and of the mature insulin derived from it (*black*). The B-chain of insulin represents the first 30 residues of the proinsulin molecule and the A-chain the last 21. The intervening 35 amino acid-residue peptide, the connecting peptide, is split out posttranslationally. Additional peptide cleavages take place with the loss of four free amino acids (three arginines and one lysine) leaving the so-called C-peptide, 31 residues long. (See text for further discussion).

the B chain resists cleavage, leading to hyperproinsulinemia but with normal carbohydrate metabolism, presumably because of the biological activity contributed by high proinsulin levels (Gabbay et al., 1976). In the other example, the patient produced both normal insulin and insulin in which leucine was substituted for phenylalanine at residue 24 or 25 in the B chain (Olefsky et al., 1980). The biological activity of the mutant insulin was very low and, in addition, it competed with the normal insulin for binding to receptors. Thus the patient was diabetic and "hyperinsulinemic" (since the abnormal insulin was recognized by standard radioimmunoassay). For reviews of insulin synthesis and secretion see Steiner, 1977, Permutt, 1981, and Tager et al., 1981.

Insulin Assay

Hormone concentrations are most accurately and most sensitively determined by radioimmunoassay (RIA) as first developed for insulin assay by Berson and Yalow (1959). An unknown sample of unlabeled insulin is allowed to compete with a known amount of radioactive insulin for binding to an antibody against insulin. The degree of displacement of labeled insulin from the antibody by the unlabeled unknown is measured by separating bound from free insulin and assaying the radioactivity in one or both fractions. Several variant methods are in use, including enzyme-linked assays, but the principle is the same.

Since the chemical structure of insulin is fully established, concentrations can be expressed in nanomolar terms or in nanograms per milliliter. By convention, however, concentrations are still most commonly stated in terms of *in vivo* biological activity. This was defined many years ago for the U.S. Pharmacopoeia in terms of the potency in lowering the blood glucose level of rabbits! The U.S.P. standard has a potency of about 22 U/mg, but it is not 100% pure. Biological activity has also been assayed *in vitro* by measuring stimulation of glucose uptake by isolated rat diaphragm or rat adipose tissue (epididymal fat pad). The fat pad bioassay for serum insulin yields a considerably higher value than RIA because of the presence in plasma of insulin-like hormones. Antibodies against insulin itself suppress the biological activity of serum insulin but not that of the insulin-like hormones, which were thus designated *nonsuppresible insulin-like activity* (NSILA). It is now known that some or all of NSILA can be attributed to somatomedins (insulin-like growth factors; IGF) which are produced under growth hormone regula-

tion and are recognized by insulin receptors, although with low affinity (see Chapter 52).

The RIA for insulin is accurate and specific. However, because the structure of insulin in mammals is highly conserved, antibodies prepared against one mammalian insulin cross-react with insulin from other species. Proinsulin also cross-reacts—understandably since its structure embraces both the chains of the mature insulin molecule—but with lower affinity than insulin itself. In standard assays, about 25–50% of proinsulin will be detected. C-peptide, on the other hand, does not cross-react with antiinsulin antibodies. C-peptide levels can be assayed by RIA using specific anti-C-peptide antibodies. This can be helpful clinically under some circumstances. For example, it is useful in assessing insulin secretion in patients whose serum contains high levels of antibodies against insulin, which precludes assay by conventional RIA. Since C-peptide and insulin are secreted in approximately equimolar amounts, C-peptide levels reflect rates of insulin secretion (although C-peptide levels are somewhat higher because it is more slowly removed from the plasma) (Horwitz et al., 1976). RIA for C-peptide can also distinguish hyperinsulinemia due to endogenous secretion (from an insulin-secreting tumor, for example) from that due to exogenous insulin administration (from surreptitious dosage). In the former, C-peptide levels will be elevated; in the latter they will be low.

Factors Regulating Insulin Release

These are shown in Table 7.5. The routing of ingested nutrients and of metabolic intermediates derived from them is generally stated to be under the control of the endocrine system. For example, insulin facilitates the disposition of ingested glucose by stimulating its uptake into liver, muscle, and adipose tissue. However, it would be equally true to say, reciprocally, that the endocrine system is under the control of foodstuffs and metabolic intermediates. Thus, ingestion of glucose stimulates the output of insulin and suppresses the output of glucagon. This interlocking system works well because during evolution there has been selection for an optimal hormone response to metabolites and an optimal reciprocal response of metabolites to hormones. Insulin is intimately involved in the regulation not only of glucose metabolism but also of protein and fat metabolism. One might predict, then, that any and all of the major foodstuffs might play some role in regulating insulin release, and that is indeed the case.

After each meal the rate of insulin secretion

Table 7.5
Factors influencing insulin release

Stimulation	Inhibition
Physiologic	
Glucose	
Amino acids	
Gastrointestinal peptide hormones (esp. GIP)	
Ketone bodies (esp. in starvation)	
Glucagon	Somatostatin
Parasympathetic stimulation	Sympathetic stimulation (splanchnic nerve)
β-Adrenergic stimulation	α-Adrenergic stimulation
Pharmacologic and experimental	
Cyclic AMP	α-Deoxyglucose
Theophylline	Mannoheptulose
Sulfonylureas	Diazoxide
Salicylates	Prostaglandins
	Diphenylhydantoin
	β cell poisons: Alloxan, streptozotocin

increases, and the plasma insulin levels rise. During the overnight fasting period insulin levels tend to drift downward and are usually below 10–20 μU/ml by morning; after a meal they may reach peak values as high as 100 μU/ml. These levels, measured in peripheral plasma, are much lower than those in the portal circulation for two reasons. First, the newly secreted insulin enters the portal circulation initially, and only later is it diluted into the larger volume of the systemic circulation. Second, the liver is highly efficient in the uptake of insulin, taking up in one circulation about 50% of what is delivered to it. Thus the liver is normally exposed to insulin concentrations in portal vein blood 3- to 10-fold higher than those to which other tissues are exposed. When the diabetic patient is treated with subcutaneous injections of insulin, however, there is no such differential exposure. Consequently, the response to exogenous insulin may be different from that to endogenously secreted insulin.

The normal basal rate of insulin secretion in man is estimated to be 0.5–1.0 U/h. Because of the postprandial bursts of secretion, the total daily secretion may be as much as 40 U/day. During fasting, less is secreted, and plasma levels drift down progressively. For a review of factors regulating insulin release see Gerich et al, 1976.

GLUCOSE

Without question the plasma level of glucose is the most important determinant of the rate of insulin release. Both synthesis and secretion are stimulated when plasma glucose levels rise and inhibited when they fall. Glucose has a direct effect that is readily demonstrated using the perfused pancreas or isolated islet tissue. In addition, glucose exerts indirect stimulation when given orally, by stimulating the release of peptide hormones from the gastrointestinal tract (see below).

Two major mechanisms have been proposed to explain the direct stimulatory effect of glucose on insulin secretion. The *glucose receptor theory* proposes that the β cells of the islet express specific receptors that recognize glucose and respond to it by increasing insulin synthesis and secretion, perhaps by generating a "second messenger." The second hypothesis holds that the metabolism of glucose in the β cell is necessary and that one or more of the metabolites of glucose represent the direct stimulus to an increased synthesis and secretion of insulin. Evidence for both hypotheses is available, and it may be that both are operative.

The response of the islet cell to glucose stimulation occurs in two phases. With very little lag time there is an immediate response, and the rate of insulin release reaches a peak within a minute or two. This first phase is transient, and the rate of release drops sharply back toward normal levels over the next 5 min or so. Then there follows a second phase, the rate of release increasing again over the following hour or so. This second phase can be blocked by inhibitors of protein synthesis and presumably reflects a late stimulation of new insulin formation. The first phase represents primarily the release of preformed insulin present in the secretory granules.

AMINO ACIDS

Plasma insulin levels rise after a meal consisting exclusively of protein. This response is partly attributable to a direct effect of higher plasma amino acid levels on the β cell. Insulin release is increased by intravenously administered amino acids, and isolated islet tissue responds directly. In addition, a protein meal stimulates insulin release indirectly by causing secretion of hormones derived from the gastrointestinal tract, as does glucose. The potency of the different amino acids in stimulating insulin secretion varies. The most potent are arginine, leucine and lysine. Intravenous doses of these amino acids have been used to evaluate the ability of the pancreas to secrete insulin. Other amino acids, such as valine and histidine, for example, are much less potent.

GASTROINTESTINAL PEPTIDE HORMONES

The insulin response to orally administered glucose is considerably greater than the response to the same amount of glucose given intravenously at a rate to match the plasma glucose levels reached after ingestion of glucose. The same disparity holds for amino acids given orally vs. those given intravenously. The greater effectiveness of the oral dose is due to the release from the gastrointestinal tract of hormones that reinforce the stimulus of the metabolite and boost the insulin response of the pancreas. Many different gastrointestinal hormones have been shown to have the *potential* of stimulating insulin release, but it is not yet clear which of these are important physiologically. Secretin, gastrin, and pancreozymin are all effective, but the amounts released in response to meals do not appear to be sufficient to account for the incremental insulin response to oral glucose. Gastric inhibitory polypeptide (GIP), a 43-amino acid residue peptide, is a very potent stimulus, and it is effective at plasma concentrations comparable to those reached under physiologic conditions (Dupre et al., 1973). Its release is stimulated not only by glucose but also by fats and amino acids. Thus, the ingestion of a meal sends an "anticipatory" signal to the pancreas to increase insulin release even before substrate levels have risen very much, and this amplifies the response.

KETONE BODIES AND FATTY ACIDS

Ketone bodies (acetoacetate and β-hydroxybutyrate) and free fatty acids administered intravenously enhance insulin release in experimental animals, but evidence for their effectiveness in man is limited. In starvation, when free fatty acid and ketone body levels are elevated, insulin levels are low, indicating that these stimuli are probably of secondary importance.

GLUCAGON AND SOMATOSTATIN

Glucagon stimulates and somatostatin inhibits insulin release. These effects are readily demonstrated by administering the hormones intravenously. Thus systemic levels of these hormones may be relevant to regulation of insulin release. However, regulation within the islet itself is probably more significant (*paracrine* control). As summarized in Figure 7.18, the three major hormones produced by islet cells affect each other's synthesis and secretion in a complex, interactive manner. Local concentrations of the hormones within the islet are probably very high and could therefore exercise control locally more effectively than they do after release into the general circulation with the consequent dilution. In addition, some islet cells have been shown to have *gap junctions*, i.e., open connections that allow transfer of materials from

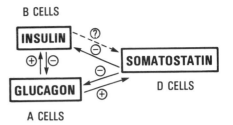

Figure 7.18. Diagrammatic representation of currently known regulatory interactions that may be important in paracrine regulation of islet hormone production and release.

cell to cell without escape into the extracellular milieu. How much of the interaction that occurs relates to such cell-cell connections is not known. Somatostatin inhibits the release of both insulin and glucagon. Advantage has been taken of this somatostatin suppression to study independently the metabolic effects of insulin and of glucagon. Constant intravenous infusion of somatostatin reduces release of both insulin and glucagon to very low levels. Then the infusion of insulin alone or of glucagon alone allows the investigator to study the effects of one of these hormones without the complicating, counteracting effects of the other. The extent to which somatostatin operates under physiologic conditions as a regulator remains uncertain.

Insulin suppresses glucagon release. Consequently, an increase in insulin release will tend to be accompanied by a decrease in glucagon release. The increase in insulin release in response to glucose may be an important factor in decreasing glucagon levels. Conversely, when insulin levels fall, as during fasting, glucagon levels will tend to increase.

Glucagon stimulates the release both of somatostatin and of insulin. The former will tend to *decrease* insulin release, and it becomes apparent that there is no simple way to predict quantitatively the responses to a given stimulus. The interactions are obviously complex, and much remains to be learned about the detailed kinetic interactions and their relative importance in determining patterns of hormonal and metabolic responses. For reviews of the roles of glucagon and somatostatin see Unger and Orci, 1981, and Arimura, 1981.

AUTONOMIC NERVOUS SYSTEM REGULATION

Parasympathetic stimulation via the vagus nerve increases, whereas sympathetic stimulation via the splanchnic nerve inhibits, insulin release. Administration of epinephrine or norepinephrine systemically inhibits insulin release. This effect is exercised via interaction with α-receptors. Agonists that affect primarily the β-adrenergic receptors actually stimulate release of insulin, but the α effect of the naturally occurring catecholamines dominates under ordinary circumstances. The decrease in insulin release associated with stress (such as severe infections, exercise, or hypothermia) may be attributable in part to an increase in sympathetic discharge. Administration of phentolamine, an α-adrenergic blocking agent, increases insulin release, suggesting that a tonic suppressive effect of the sympathetic nervous system is constantly operative.

The central nervous system also plays a role in regulating insulin secretion, and the outflow probably occurs via the hypothalamus and the autonomic nervous system (Louis-Sylvestre, 1976; Powley, 1977; Berthaud et al., 1981). Stimulation of the ventromedial nuclei of the hypothalamus suppresses insulin release, whereas destruction of these nuclei causes hyperinsulinism. While difficult to document and quantify, it is likely that the emotional state of the patient may importantly alter insulin release by way of the hypothalamus and its autonomic outflow tract.

The mechanism of action of the catecholamines on β cell secretion is not fully established. Cyclic AMP will stimulate insulin release, and the effects of β-adrenergic stimuli may be exercised in part by increases in cyclic AMP. There is also evidence that changes in cytosolic concentrations of calcium ion may play a role and that calmodulin, a specific calcium-binding protein, may be involved. The exact relationships among cyclic AMP, calcium ion, and calmodulin remain to be worked out.

PHARMACOLOGIC AND EXPERIMENTAL FACTORS

Theophylline stimulates insulin release, presumably through its ability to inhibit phosphodiesterase and increase cyclic AMP concentrations. The sulfonylureas, orally effective hypoglycemic drugs, are effective primarily because of their ability to directly stimulate insulin secretion. Finally, salicylates are effective hypoglycemic agents. These drugs inhibit the conversion of arachidonic acid to prostaglandins. The prostaglandins have been demonstrated to inhibit insulin release.

Insulin release is inhibited by mannoheptulose and 2-deoxyglucose, substances that interfere with the utilization of glucose by the islet cells. Diazoxide is a potent inhibitor of insulin release, sufficiently potent to be sometimes helpful in managing patients with insulin-secreting tumors prior to resection or in patients with metastatic insulin-secreting tumors. Diphenylhydantoin, a drug in widespread use for control of epileptic seizures, inhibits insulin

release and significantly increases the insulin requirement of diabetic patients. Finally, certain cytotoxic agents with a high degree of specificity for the β cell can be used to produce experimental diabetes in animals (alloxan and streptozotocin). The latter compound has been used to control intractable hypoglycemia in patients with insulin-secreting tumors.

Biological Effects of Insulin

Because of its dramatic effects on glucose metabolism and because these were historically the first to be described, there has been a tendency to categorize insulin as a hormone regulating carbohydrate metabolism. Actually insulin exercises important controls over the metabolism of all the major foodstuffs—carbohydrates, fats, and proteins. These effects are reasonably well summarized by saying that insulin favors anabolism and storage. It favors the synthesis and deposition of glycogen in the liver, the synthesis of fatty acids in liver and adipose tissue and their deposition and retention in adipose tissue as stored triglycerides, and the uptake of amino acids and their incorporation into proteins in muscle and other tissues. When insulin levels are high, as in the hours following the ingestion of a meal, nutrients are thus directed to appropriate storage sites; when insulin levels are low, as in the fasting state, all of these processes are reversed and metabolic substrates are mobilized in the form of glucose, free fatty acids and amino acids. In addition insulin stimulates synthesis of RNA and DNA. These effects are somehow selective although the exact mechanisms that are operative remain unknown. Thus some of the same enzymes that are acutely regulated (e.g., by changes in the state of phosphorylation of the enzyme protein) are also regulated in a sustained, long-term fashion by increasing (or decreasing) the *amount* of enzyme protein available.

In Table 7.6 and in the following sections, the metabolic effects of insulin are dealt with individually. While useful as a first approach, this is quite artificial. *In vivo* all of these effects are highly interactive—with one another and with the simultaneous influence of many other metabolic, hormonal, and neural factors. For example, a number of so-called counterregulatory hormones (including the catecholamines, glucagon, glucocorticoids, and growth hormone) modulate the effects of insulin. Furthermore, the net impact of insulin will depend on the number of receptors available, on the levels of substrate available, and on a number of substrate-determined autoregulatory controls.

Table 7.6
Biological effects of insulin

A. On carbohydrate metabolism
  1. Reduces rate of release of glucose from liver
     a. by inhibiting glycogenolysis.
     b. by stimulating glycogen synthesis.
     c. by stimulating glucose uptake.
     d. by stimulating glycolysis.
     e. by indirectly inhibiting gluconeogenesis via inhibition of fatty acid mobilization from adipose tissue.
  2. Increases rate of uptake of glucose into all insulin-sensitive tissues, notably muscle and adipose tissue
     a. directly, by stimulating glucose transport across the plasma membrane.
     b. indirectly, by reducing plasma-free fatty acid levels.
B. On lipid metabolism
  1. Reduces rate of release of free fatty acids from adipose tissue.
  2. Stimulates de novo fatty acid synthesis and also conversion of fatty acids to triglycerides in liver.
C. On protein metabolism
  1. Stimulates transport of free amino acids across the plasma membrane in liver and muscle.
  2. Stimulates protein biosynthesis and reduces release of amino acid from muscle.
D. On ion transport
E. On growth and development

INSULIN-RECEPTOR INTERACTION

Almost all of the metabolic effects of insulin are initiated by the interaction of the insulin molecule with a highly specific receptor on the plasma membrane (Stadie et al., 1953; Roth et al., 1975; Kahn et al., 1981). This binding is saturable, of very high affinity, and highly specific. Most important, there is excellent correlation between the binding of various chemically modified forms of insulin to the receptor, on the one hand, and their biological effects, on the other. Most if not all of the biological effects are probably initiated by insulin-receptor interaction independent of subsequent internalization of the receptor-bound insulin, although it remains possible that hormone action continues (or may even be expressed differently) after the hormone-receptor complex is internalized. Evidence that internalization of the insulin molecule is not necessary comes from studies in which insulin was covalently bonded to large, inert beads that could not themselves be internalized. Insulin attached to such beads nevertheless was active on cultured cells (Cuatrecasas, 1969). Another piece of evidence favoring this view is the finding that antibodies against the insulin receptor mimic the biological effects of insulin itself. Monovalent F_{ab} fragments are ineffective, suggesting that clustering of the receptors is necessary to elicit biological activity.

By use of affinity chromatography and by pho-toaffinity labeling the insulin receptor has been partially purified and characterized (reviewed by Czech, 1981). It is a membrane protein of high molecular weight (about 350,000) made up of four disulfide-linked subunits: two identical α chains of about 125,000 daltons and two β chains of 45,000–90,000 daltons. It has been pointed out that the structure of the receptor is reminiscent of that of the immunoglobulins. The receptor appears to be able to bind 2 moles of insulin per mole, which may relate to the unusual "negative cooperativity" observed for insulin effects as a function of concentration, i.e., the binding of the first mole of insulin to the receptor makes it more difficult to bind the next.

Of great interest is the recent finding that the binding of insulin induces autophosphorylation of the receptor protein itself and that it is a tyrosine residue that becomes phosphorylated (Kasuga et al., 1982; Roth and Cassell, 1983). Regulation of enzyme function by phosphorylation of the enzyme protein is common, but it is generally a serine residue that is phosphorylated. The first example of tyrosine phosphorylation was discovered in relation to the oncogenic viruses that insert the *src* gene into the host's genome (see review by Houslay, 1981). The gene product is a membrane protein that phosphorylates tyrosine residues. Furthermore, epidermal growth factor (EGF) also causes phosphorylation of tyrosine residues and itself undergoes tyrosine autophosphorylation. The exact mechanisms linking tyrosine phosphorylation to growth control remain to be worked out, but these findings strongly suggest that it must be of general importance. Effects of insulin on growth are reviewed in Chapter 56 and by Straus (1981).

Precisely how insulin-receptor interaction leads to the wide spectrum of biological effects produced by insulin is not known. The stimulation of glucose transport, amino acid transport, and ion transport are easily visualized as more or less immediate consequences of changes in membrane configuration induced by interactions with the membrane receptor. If the effects on intracellular enzymes are to be accounted for also on the basis of interaction with the membrane-bound receptor, however, it becomes necessary to postulate the generation of a second messenger that carries information into the cell. For example, an immediate effect of insulin is the activation of pyruvate dehydrogenase, a mitochondrial enzyme. Pyruvate dehydrogenase is an interconvertible enzyme, i.e., it can be covalently modified to affect its activity. Phosphorylation of the enzyme protein deactivates, and removal of the phosphate activates. Recent studies have shown that insulin can interact with a plasma membrane fraction to generate a low molecular weight, heat-stable compound or compounds that can in turn act on isolated mitochondria to activate this enzyme, probably through activation of pyruvate dehydrogenase phosphatase (Seals and Jarrett, 1980). The chemical nature of the messenger has not been established. The fact that trypsin can mimic some of the effects of insulin on intact cells suggests that a proteolytic fragment generated at the membrane surface (possibly from the receptor itself) may represent the insulin second messenger. Again there is preliminary evidence that insulin-receptor interaction may generate an oligopeptide that can mimic insulin effects on glycogen synthase activity and inhibit protein kinase activation, but attempts to characterize it are still in progress (Larner et al., 1979).

The functional importance of the insulin receptor in vivo has been established beyond doubt in a number of ways. Recently a number of diabetic patients with marked resistance to the action of insulin have been shown to be deficient in the number of insulin receptors; others have been shown to have a normal complement of receptors but fail to respond because of circulating autoantibodies against their own insulin receptors (Kahn et al., 1976). Like many other receptors, the insulin receptor is subject to downregulation. If for any reason insulin levels in the plasma remain elevated for an extended period of time (hours), the number of insulin receptors expressed at the cell membrane decreases. A given dose of insulin in the down-regulated state will have a smaller effect than it has in the control or "upregulated" state (Roth et al., 1975). This concept that response depends not only on the dose of hormone administered but also on the number of receptors currently expressed is, of course, of general applicability in endocrinology. Changes in receptor number on circulating blood cells (lymphocytes or red blood cells) appear to reflect changes in receptor number occurring generally.

EFFECTS ON CARBOHYDRATE METABOLISM

Insulin reduces plasma glucose levels both by stimulating uptake of glucose into tissues and by inhibiting the production and release of glucose from the liver. Glucose uptake by muscle and adipose tissue is stimulated directly, by enhancing the carrier-mediated transport (Morgan et al., 1964; Park et al., 1968), and indirectly, by inhibiting FFA release and reducing plasma FFA levels. Plasma FFA tends to inhibit glucose uptake (Randle effect),

and thus a decrease in FFA levels favors glucose uptake.

Production and release of glucose by the liver is reduced in a concerted manner involving several points of attack. Insulin stimulates glycogen synthesis by enhancing the activity of glycogen synthase and simultaneously inhibiting glycogen breakdown by decreasing the activity of glycogen phosphorylase (see Fig. 7.9 and accompanying discussion in Chapter 47). Insulin simultaneously promotes glycolysis and inhibits gluconeogenesis. Directly or indirectly, insulin influences each of the four key control points at which forward and reverse reactions are catalyzed by different enzymes (Fig. 7.8). Conversion of glucose to glucose-6-P, fructose-6-P to fructose diphosphate, and phosphoenolpyruvate to pyruvate are all favored. In addition, insulin strongly activates pyruvate dehydrogenase, shunting pyruvate on to acetyl CoA which can be converted to fatty acids and stored or enter the Krebs cycle and be oxidized.

A centrally important indirect mechanism by which insulin influences hepatic gluconeogenesis is through its effects on FFA mobilization. When insulin levels are low and FFA are being taken up avidly by the liver, gluconeogenesis is stimulated by the acetyl CoA derived from fatty acid oxidation. As shown in Figure 7.19, acetyl CoA tends to shunt pyruvate back up the gluconeogenic pathway by simultaneously exerting a feedback inhibition of pyruvate dehydrogenase and powerfully stimulating pyruvate carboxylase. At the same time utilization of the acetyl CoA for resynthesis into fatty acyl CoA is inhibited because of feedback inhibition. Additionally, flow of substrate is directed up to glucose because FDPase is enhanced and PFK is inhibited. Thus, pyruvate and any substrates that contribute to the pyruvate pool are directed up to glucose-6-P. Under these conditions (starvation or diabetic ketoacidosis), alanine and other gluconeogenic amino acids released from muscle become key sources of glucose.

EFFECTS ON LIPID METABOLISM

Insulin is remarkably potent in suppressing the release of FFA from adipose tissue. In fact, this effect occurs at insulin concentrations below those needed to stimulate glucose uptake in most tissues. Furthermore, the effect on FFA mobilization is immediate—even faster than the effect on plasma glucose levels. Under normal conditions insulin is constantly exerting a "braking" effect on FFA mobilization. Thus, when antiinsulin antiserum is injected intravenously into an animal the plasma FFA

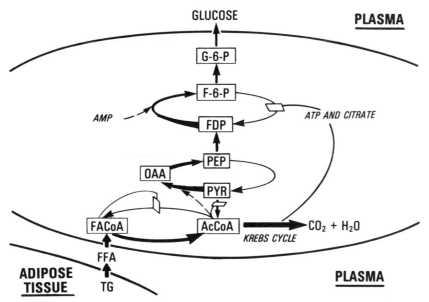

Figure 7.19. Key control points at which rapid metabolism of free fatty acids in the liver leads to stimulation of gluconeogenesis (see text for discussion).

levels begin to rise immediately. Since there do not appear to be any important controls over the rate of utilization of FFA (other than FFA levels per se), this insulin effect is centrally important. Insulin is the key hormone monitoring the availability of glucose—suppressing FFA release when glucose is available and permitting rapid mobilization when glucose levels fall and lipids are needed to provide metabolic fuel. Insulin acts at the level of the fatty acid-mobilizing enzyme, hormone-sensitive lipase. In vivo this enzyme is under tonic control by catecholamines such that it is always at least partly in the active (phosphorylated) state. Insulin counteracts these effects, inhibiting conversion of lipase *b* to lipase *a*. Under some conditions this insulin effect is exercised at the level of cyclic AMP, but other mechanisms must also be involved, perhaps through control of lipase phosphatase activity.

Insulin stimulates fatty acid biosynthesis and also the incorporation of fatty acids into triglycerides in the liver and in adipose tissue. In addition, by increasing lipoprotein lipase activity it favors the uptake and deposition of triglycerides from very low density lipoproteins and chylomicrons into adipose tissue. Thus insulin channels substrate into lipid stores in the fed state; in the fasted state, when insulin levels are low, fat mobilization is favored.

EFFECTS ON PROTEIN METABOLISM

Insulin has a direct stimulatory effect on the transport of amino acids across the plasma membrane, an effect that is primary, not secondary to its effect on glucose transport. In addition, it directly stimulates protein biosynthesis by other mechanisms. In part this effect is at the translational level, involving initiation or the state of aggregation of polysomes. In addition, insulin stimulates RNA and DNA synthesis, effects presumably related to its growth-promoting properties. The interrelationships between insulin, growth hormone, and the insulin-like growth factors (somatomedins) are discussed in Chapter 52).

EFFECTS ON ION TRANSPORT

One of the immediate effects of insulin is to cause hyperpolarization of the plasma membrane, an effect that must reflect changes in ion fluxes. *In vivo* it alters the distribution of sodium and potassium between extracellular and intracellular spaces. It favors movement of potassium into cells, and this is of great practical significance when treating patients with insulin deficiency. Vigorous treatment with insulin, without appropriate replacement of potassium, can cause serious or even fatal hypokalemia. There is some evidence that insulin acts directly on membrane-bound Na^+-K^+ ATP-ase (Hougen et al., 1978).

Insulin increases cytosolic calcium ion levels by effecting release of calcium from bound form within the cell and/or favoring entrance from the extracellular fluid. Since calcium ion, via calmodulin, exercises control over a number of the enzyme systems regulated by insulin, this mechanism may have more general significance.

GLUCAGON

Synthesis, Secretion, and Assay

Circulating glucagon is a simple polypeptide of 29 amino acid residues (Mr 3,485) (Bromer et al., 1957). It has no cross-linkages and exists in solution as a random coil. It is synthesized in the A cells as a prohormone of much higher molecular weight (about 9,000) which is then cleaved prior to secretion.

The general pattern for biosynthesis and secretion is like that for insulin. Preformed glucagon is stored in secretory granules that can disgorge their content into the surrounding medium rapidly under appropriate stimulation.

The earlier literature on glucagon is confusing because it was not at first recognized that the intestine secretes polypeptides that cross-react with many or most antisera against pancreatic glucagon but do not replace glucagon functionally. Highly specific antisera are now available that react only with true "pancreatic glucagon." In some species there are cells in the gastrointestinal tract that actually do secrete the same polypeptide secreted by the α cells, but in man the pancreas appears to be the sole source of true glucagon. None is found in the plasma after total pancreatectomy; glucagon-like immunoreactivity, however, persists, as it does in most mammalian species. The physiologic significance of the latter is not clear.

Like insulin, glucagon enters the portal circulation and is first presented to the liver. However, the efficiency of hepatic extraction of glucagon is considerably less than that for insulin, and so the portal:systemic ratio is much less (about 1.5:1 instead of almost 10:1). Normal peripheral concentrations in the basal state are about 100–150 pg/ml; basal rates of secretion are about 100–150 μg/day.

Factors Regulating Glucagon Release

The effects of glucose on glucagon secretion are reciprocal to those on insulin secretion—hypoglycemia stimulates and hyperglycemia suppresses release. It has been suggested that in addition to direct glucose effects on the A cells, glucose may suppress glucagon release indirectly through its stimulation of insulin release (Table 7.7). Insulin is an inhibitor of glucagon release and could act within the islet (paracrine control; cf. Fig. 7.18).

Glucagon secretion, like that of insulin, is stimulated by amino acids. The rise in glucagon level together with the rise in insulin stimulated by

Table 7.7

Factors influencing glucagon release

Stimulation	Inhibition
Amino acids	Glucose
Gastrointestinal polypeptide hormones	Insulin
Catecholamines (exercise)	Free fatty acids
Growth hormone	
Glucocorticoids	

amino acids may act to prevent hypoglycemia when food intake is primarily protein. Again like insulin, glucagon is secreted in response to a number of gastrointestinal polypeptides, including pancreozymin.

A number of factors that *inhibit* release of insulin and/or antagonize its biological activities *stimulate* glucagon release. These include catecholamines, growth hormone, and glucocorticoids. Glucagon levels tend to rise during exercise, whereas insulin levels fall; both may be due to increased catecholamine release secondary to increased sympathetic nervous system activity during exercise.

The half-life of plasma glucagon is about 10 min—2 to 3 times as long as that of insulin. Still the duration of action is quite short and acute responses can be aborted quickly.

Biological Effects of Glucagon

In most respects the biological effects of glucagon are opposed to those of insulin. Thus, glucagon tends to raise rather than lower plasma glucose levels, mostly by its potent stimulation of hepatic glycogenolysis. This effect is mediated via stimulation of adenylate cyclase and activation of cAMP-dependent protein kinase. The latter simultaneously activates phosphorylase (via phosphorylase kinase) and inactivates glycogen synthase.

Glucagon increases hepatic gluconeogenesis, again opposing the effect of insulin. The glucagon effect is exerted in part at the level of pyruvate kinase, which it deactivates. In addition, it stimulates the conversion of pyruvate to phosphoenolpyruvate. Finally, it may increase the uptake of amino acids into the liver to provide substrate for gluconeogenesis.

Glucagon has the potential of enhancing lipolysis in adipose tissue (whereas insulin suppresses it). While the glucagon effects are readily demonstrated with isolated adipose tissue, the response of plasma FFA levels to glucagon administered *in vivo* is usually minimal or even, paradoxically, a fall rather than a rise. This is readily explained on the basis of the sharp increase in glucose levels that accompanies administration of glucagon and the consequent rise in insulin levels. These override any tendency of glucagon to mobilize FFA. However, in diabetic subjects or in subjects that have fasted for some time, the glucagon effects on FFA may become important.

Other effects of glucagon have been described (including effects on the gastrointestinal tract, on calcium metabolism, on ion transport, and on myocardial function), but their physiologic significance remains to be fully established.

An assessment of the relative importance of glucagon and insulin in respect to control of carbohydrate metabolism and, in particular, of their relative importance in diabetes mellitus is difficult to make. To some extent they may be regarded as paired hormones, and some argue that the ratio of insulin to glucagon is more relevant in evaluating the diabetic tendency than the concentration of either one alone. However, we know that pancreatectomy, which removes both, leads to gross deterioration of glucose homeostasis. This would seem to argue for the predominant importance of insulin. Patients with glucagon-secreting tumors do have glucose intolerance, but it tends to be moderate compared to that of patients with insulin deficiency. Furthermore, the diurnal fluctuations in glucagon level are small, and the glucagon response to a glucose meal is relatively small; small increases in plasma glucose may increase insulin levels without much effect on glucagon levels. On the other hand, as discussed above, when there is insulin deficiency (absolute or relative) the severity of the disturbance in glucose homeostasis may be importantly determined by glucagon levels.

SOMATOSTATIN

Somatostatin is a tetradecapeptide hormone first discovered and characterized as a potent hypothalamic inhibitor of the release of growth hormone from the pituitary (Brazeau et al., 1973). Studies of its biological effects when administered intravenously revealed, unexpectedly, that it also potently inhibited secretion of both insulin and glucagon (Koerker et al., 1974; Yen et al., 1974). It is now clear that somatostatin has a much broader biological significance (Arimura, 1981). It inhibits release of both growth hormone and TSH from the pituitary and under some circumstances also inhibits release of prolactin and ACTH. It has been shown to occur throughout the gastrointestinal tract and to act as an inhibitor of many gastrointestinal functions: the secretion of gastric acid and pepsin; the secretion of pancreatic digestive enzymes; intestinal motility and intestinal absorption (including the absorption of glucose); secretion of many

gastrointestinal hormones (secretin, pancreozymin, vasoactive intestinal peptide, gastric inhibitory polypeptide). Somatostatin is widely distributed throughout the central and peripheral nervous system and probably plays a role in neurotransmission. In the present context we shall limit consideration to the role of somatostatin in the regulation of the function of the pancreatic islets where it is synthesized and secreted by the D cells.

Structure and Synthesis

Somatostatin is a 14-amino acid peptide with a carboxy-terminal cysteine residue in disulfide linkage to another cysteine residue at position 3. More recently a 28-amino acid residue peptide has been purified from hypothalamus and intestine in which the carboxy-terminal 14 residues are identical to somatostatin. This larger peptide is much more potent than somatostatin (as much as 10 times more potent in inhibiting growth hormone and insulin release) and may represent the form active within cells. Other, still larger forms have been reported. Recently the complete sequence of the preprohormone has been deduced by cloning the cDNA for human somatostatin (using mRNA from a human somatostatin-secreting tumor) (Shen et al., 1982). The coding region predicts a 116 residue precursor protein with the sequence of somatostatin-28 at the carboxy-terminus. A highly nonpolar region at the amino-terminus presumably represents a signal peptide; whether or not the prohormone has biological functions different from those of somatostatin remains to be determined. The cDNA for somatostatin has also been cloned from the angler fish (Hobart et al., 1980). These fish have discrete islets separate from the pancreas, and some of these are relatively enriched in D cells. Two cDNA forms have been identified. In one, the carboxy-terminal 14-amino acid somatostatin has the same sequence as the mammalian peptide, but the other differs at two residues and there is some evidence that they may differ functionally. It is possible that local effects of "somatostatin" may relate to portions of the prohormone structure or other peptides than the 14-amino acid or even the 28-amino acid forms studied intensively in the past. It is also possible that more than one gene is involved.

Biological Effects (in Relation to Islet Function)

Somatostatin probably has no intrinsic direct effects on glucose metabolism analogous to the effects of insulin and glucagon. It does, however, inhibit the intestinal absorption of glucose, and this can result in apparent effects on glucose tolerance.

Its major effects are probably mediated through its inhibition of insulin and glucagon secretion. Because local concentrations within the islet are so much higher than systemic concentrations it is likely that local, paracrine regulation is most important.

Intravenous infusion of somatostatin into man causes a fall in plasma levels of both insulin and glucagon. Glucose levels also fall, and this, in the face of lower insulin levels, provides evidence that glucagon does normally play a role maintaining plasma glucose levels. Continuous somatostatin infusion has been useful as a research tool for evaluating the roles of glucagon and insulin separately, maintaining the levels of one or the other of them by infusion.

That somatostatin exerts tonic control over insulin and glucagon secretion within the islet has been demonstrated using antisomatostatin antibodies. Added to isolated islets such antibodies can cause an increase in the rate of secretion of insulin and glucagon. It remains to be determined whether and how somatostatin can exert *selective* control. The anatomical relationships and/or intercellular connections via gap junctions could play a role. Since it acts to inhibit secretion of two antagonistic hormones, its impact on metabolism cannot be predicted until more is known about selective mechanisms.

Somatostatin secretion *in vitro* is regulated by many of the same factors that regulate insulin and glucagon secretion, but the patterns are difficult to interpret, as just mentioned. Glucose, amino acids, and gastrointestinal peptides stimulate somatostatin secretion. Glucagon also stimulates, but no effect of insulin has been established. Acetylcholine stimulates, and epinephrine inhibits. The latter effect is via α-adrenergic mechanisms. Pure β-adrenergic agonists stimulate release via cyclic AMP. The physiologic relevance of these various potential regulatory factors remains to be established (Gerich, 1981).

PANCREATIC POLYPEPTIDE

Specialized cells in the islets, tentatively designated F cells, secrete a basic polypeptide of 36 amino acid residues—pancreatic polypeptide (PP). This peptide, first discovered as a minor contaminant of avian insulin, has now been found in many avian and mammalian species, including man. Complete sequences are available for human, bovine, ovine, and porcine PP, and they are very similar. Plasma levels of PP in man (60–100 pg/ml) are comparable to those of glucagon, and they respond to a number of metabolic stimuli, especially protein-containing meals. Responses to intravenous amino acids or glucose are much smaller and variable, suggesting the mediation of gastrointestinal polypeptide hormones. Despite a large and growing literature on the chemistry, secretion, and metabolic effects of PP, it is still uncertain just what its physiologic or pathophysiologic significance may be (Hazelwood, 1981) (see p. 723).

DIABETES MELLITUS

Perhaps the best way to bring together and to interrelate the multiple, complex effects of insulin and glucagon on carbohydrate, lipid, and protein metabolism is to consider the disturbances in the animal or patient with diabetes mellitus. Patients with type I diabetes mellitus (juvenile-onset diabetes mellitus) have an absolute insulin deficiency. Their basal insulin levels are very low, and they respond poorly or not at all to stimuli that normally increase plasma insulin levels. Animal models for the disease can be produced by selective destruction of the β cells using alloxan or streptozotocin, which are toxic to the β cells (but at higher dosages have some effect on other tissues as well). The metabolic derangements in diabetes mellitus are mutliple, and many are secondary rather than immediate consequences of insulin deficiency (or glucagon excess). Yet almost without exception these disturbances can be traced back to the primary inability to secrete insulin. A similar but not identical array of metabolic disturbances can be enountered in another category of diabetes mellitus, type II diabetes mellitus (adult-onset diabetes mellitus). Here insulin levels are normal or, commonly, elevated, but the responses of the tissues to insulin is for some reason below normal, i.e., there is a *relative* rather than an *absolute* insulin deficiency. There are undoubtedly many different mechanisms underlying an apparent resistance or insensitivity to insulin, some involving receptor-insulin interactions and some involving subsequent steps (postreceptor defects) (Olefsky and Kolterman, 1981). The following discussion of mechanisms is limited to the patient with type I diabetes mellitus.

Major Signs and Symptoms in Patients with Absolute Insulin Deficiency

A typical history for a type I patient with diabetes mellitus often begins with *polyuria*. The child urinates frequently and copiously, often having to get up several times during the night. He drinks large quantities of water (*polydipsia*) and also increases his food intake strikingly (*polyphagia*). Despite this increase in food intake he loses weight, a paradoxical situation seen in diabetes and in hyperthyroidism but rarely in other clinical situations.

If untreated the child becomes listless and drowsy. He may complain of nausea. Appetite now dwindles, and weight loss becomes more marked. Later there may be vomiting and severe abdominal pain. Eventually he slips into a coma, unresponsive to ordinary stimuli. Breathing is deep and regular (Kussmaul breathing), a characteristic finding in metabolic acidosis.

In the emergency room it is seen that the child is severely dehydrated. His blood pressure is dangerously low. The urine is strongly positive for glucose (*glucosuria*) and acetoacetic acid (*ketonuria*); plasma levels are also markedly elevated (*hyperglycemia* and *ketonemia*). The plasma is milky in appearance, and the triglyceride level is enor-

mously elevated due to accumulation of very low density lipoproteins (VLDL) and chylomicrons (*hyperlipemia*). Plasma pH is below 7.0, and plasma bicarbonate is also very low (*metabolic acidosis*). Plasma FFA levels are as much as 5-fold elevated. Plasma sodium and potassium are low (*hyponatremia* and *hypokalemia*). Insulin levels, finally, are very low or unmeasurable.

In the following paragraphs, with the help of the flow diagrams in Figures 7.20A–7.20C, we show how all of these disturbances can be accounted for as ultimate consequences of insulin deficiency. It should be pointed out, however, that when the deficiency has persisted for any length of time it may not be enough to treat with insulin alone. For example, it may be absolutely essential to restore as rapidly as possible the large volumes of fluid and

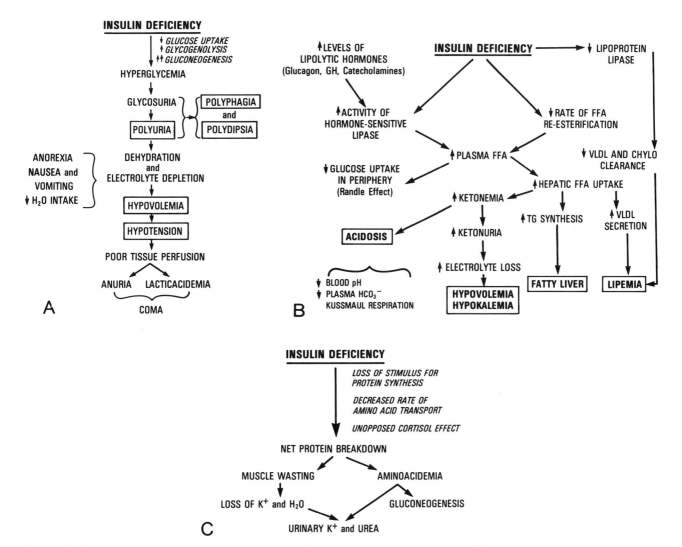

Figure 7.20. *A*, Flow diagram indicating the sequence of events by which impaired carbohydrate metabolism in diabetes mellitus is expressed pathophysiologically. *B*, Flow diagram indicating the sequence of events by which disturbances in lipid metabolism express themselves pathophysiologically in diabetes mellitus. *C*, Flow diagram indicating how disturbances in protein metabolism in diabetes mellitus express themselves pathophysiologically.

the large amounts of electrolytes that have been lost over a period of days.

Consequences of Disturbed Carbohydrate Metabolism

These are shown in Figure 7.20*A*. Polyuria, polydipsia, and polyphagia (the three Ps of diabetes) are the consequences of sustained hyperglycemia with the accompanying loss of large quantities of glucose in the urine (because the renal threshold is exceeded). Since not all of the glucose can be reabsorbed there is an accompanying obligatory loss of water into the collecting tubules (osmotic diuresis). This loss of water also entails some accompanying loss of electrolytes as well. The quantities of glucose lost can be enormous—hundreds of grams per day. To maintain energy balance the patient takes in larger quantities of food. When nausea and vomiting supervene there is negative caloric balance and weight loss. Not only adipose tissue stores but also muscle protein become depleted. In the absence of insulin, hormone-sensitive lipase is fully active and FFA release is maximal. Protein breakdown predominates over protein synthesis. The amino acids mobilized to the liver provide most of the substrate for gluconeogenesis. As indicated in Figure 7.20*A*, with the onset of nausea and vomiting dehydration is accentuated and additional electrolyte losses are incurred. Total blood volume falls, and eventually compensatory mechanisms fail to sustain blood pressure. In the extreme, tissue perfusion falls below critical values; there is tissue anoxia and production of excess lactic acid. Finally, renal function is shut down partially or completely.

Consequences of Disturbed Lipid Metabolism

These are shown in Figure 7.20*B*. The acidosis stems from an overproduction of ketone bodies (β-hydroxybutyric acid, acetoacetic acid, and acetone). These have their origin in the liver, to which FFA are being delivered at an enormous rate. As shown in Figure 7.21, acetyl coenzyme A derived from fatty acid oxidation is generated at a rate that exceeds the ability of the liver to oxidize it completely. This leads to synthesis of acetoacetyl coenzyme A and from it free acetoacetic acid. Glucagon enhances ketogenesis by increasing hepatic carnitine levels, making it possible for fatty acids to be transported more rapidly across the mitochondrial membrane barrier (as fatty acylcarnitine). Thus, the maximal rate of formation of acetoacetate is increased (McGarry and Foster, 1977). The latter cannot be oxidized by liver, because the enzymes needed for its activation are not present. It leaves the liver as acetoacetate and also in the reduced form, β-hy-

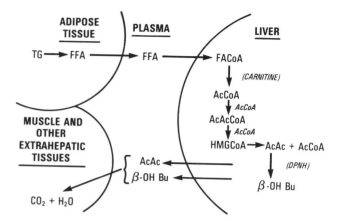

Figure 7.21. Diagrammatic representation of the pathways involved in generation of ketone bodies in the liver and their catabolism by extrahepatic tissues. *TG*, triglyceride; *FFA*, free fatty acid; *FACoA*, fatty acyl CoA; *AcCoA*, acetyl CoA; *AcAcCoA*, acetoacetyl CoA; *HMGCoA*, β-hydroxy-β-methylglutaryl CoA; *β-OH Bu*, β-hydroxybutyric acid; *AcAc*, acetoacetic acid.

droxybutyrate. Muscle and other extrahepatic tissues *can* activate acetoacetate to its coenzyme A derivative, convert it to acetyl CoA by thiolytic cleavage, and oxidize it via the Krebs cycle. Some acetoacetate is nonenzymatically oxidized to acetone and CO_2; the acetone is excreted in part via the lungs and can sometimes be smelled on the patient's breath. It is primarily the accumulation of acetoacetic acid and β-hydroxybutyric acid that causes the blood pH to fall. With the onset of acidosis there is stimulation of the respiratory center, and the increased loss of CO_2 tends to prevent the pH from falling, but this compensatory mechanism eventually fails to keep up. Meanwhile the ketone bodies are being shed in the urine. In fact the renal threshold for them is quite low (and so ketonuria of mild degree does not necessarily signal severe acidosis). The renal loss of these acids is a major basis for the accompanying loss of sodium and potassium.

The unbridled mobilization of FFA has other unfortunate consequences. Through the mechanisms illustrated in Figure 7.19, it stimulates gluconeogenesis and in this way indirectly contributes to the consequences of hyperglycemia. In addition, it is the basis for the hyperlipemia. When FFA are delivered to the liver at a high rate, fatty acid oxidation and triglyceride synthesis are stimulated. Secretion of the newly synthesized triglycerides in the form of very low density lipoproteins (VLDL) is one mechanism called into play to prevent development of a fatty liver. Thus an increased rate of VLDL secretion is one basis for lipemia. The clearance of VLDL, and also of chylomicrons, depends on lipoprotein lipase. The synthesis and secretion of this enzyme is stimulated by glucose and insulin. In the diabetic there is a relative lipoprotein deficiency, and so the rate of removal of triglyceride-rich lipoproteins is abnormally slow, accentuating the lipemia.

Consequences of Disturbed Protein Metabolism

These are shown in Figure 7.20C. The net breakdown of muscle protein causes blood amino acid levels to rise and, as discussed above, hepatic uptake of these feeds gluconeogenesis. The loss of muscle mass also entails losses of cellular potassium; urinary nitrogen and potassium are increased.

Consequences of Chronic Insulin Deficiency

With modern methods of management, diabetic patients do not frequently slip into severe diabetic ketoacidosis like that discussed above. If they do, they can almost always be successfully treated. Disability and death in diabetes much more often relate to atherosclerosis, which occurs prematurely, or to the so-called late complications of diabetes. These include damage to the kidneys, eyes, nervous system, and skin. The precise cause of these devastating chronic changes, which may only appear after decades, is not understood. However, there is a good deal of experimental evidence and some clinical evidence to suggest that they may be the result of hyperglycemia *per se* sustained over many years. It is known that hemoglobin can combine nonenzymatically with glucose, and the hemoglobin conjugate, hemoglobin A_{1c}, is present at elevated levels in diabetics (Rahbar, 1968; Rahbar et al., 1969). It increases in concentration roughly in proportion to the degree of sustained hyperglycemia and is used clinically as a guide to the degree of control of the patient's hyperglycemia (Koenig et al., 1976; Bunn, 1981). Other plasma proteins and a number of tissue proteins are also more heavily glucosylated in diabetics. Whether such glucosyla-

tion of proteins is involved in causing complications is still not known. In any case the weight of evidence linking complications to hyperglycemia is such that current best management calls for maintenance of normoglycemia at all times as long as this does not entail risks to the patient (e.g., from intermittent hypoglycemia).

Consequences of Insulin Excess

Overdosage with insulin can produce symptomatic hypoglycemia which, if severe and persistent, can go on to coma and death. If the plasma glucose level falls rapidly, as it does after a dose of regular insulin (rapidly absorbed from the injection site), the symptoms include anxiety, hunger, tremulousness, and cold sweat. These symptoms may be recognized as those accompanying an injection of epinephrine, and indeed they reflect a sympathetic discharge, a key response to hypoglycemia. This will tend to prevent the developing hypoglycemia by stimulating release of glucose from the liver (activation of phosphorylase) and reducing utilization in the periphery (conserving glucose for the CNS). Glucagon levels rise, again in an attempt to correct glucose levels. Other so-called counter-regulatory hormones include glucocorticords and growth hormone, but their effects are more important in long-term control. If the acute fall in plasma glucose continues, the patient may become comatose and suffer irreversible brain damage. The same chain of events can occur in patients with insulin-secreting tumors (insulinoma).

If the plasma glucose falls only gradually, the symptoms referable to sympathetic discharge may not occur. Instead the patient becomes listless, mentally confused, slurred of speech, and slips gradually into coma. These symptoms all reflect the progressive deterioration of CNS functions, since glucose is the only substrate the brain can utilize at a significant rate (except during prolonged starvation).

CHAPTER 50

Energy Balance

THERMODYNAMIC CONSIDERATIONS

The advertisements for miracle weight loss programs ("Eat All You Want and Lose Weight") raise false hopes that the first law of thermodynamics has been repealed or at least amended. Unfortunately, the human body is subject to the same fundamental laws of thermodynamics that govern any other physicochemical system. The first law states that energy can neither be created nor destroyed. The various forms of energy (thermal, chemical, electrical, mechanical) are interconvertible, but the *total* energy of any closed system is immutable. Thus, the total energy of the body will only decrease if it does work on the external environment or transfers heat or waste materials to the external environment. This basic notion of energy balance can be illustrated by a theoretical experiment. We place a man in a box that isolates him completely from his environment. To simplify matters we will not permit any external work to be done. Now we need to concern ourselves only with changes in the physicochemical state of the man, the box, and heat exchanges with the environment. Let us define the physicochemical state at the beginning (state A) in terms of the temperature, pressure, chemical composition, phases, etc. of the total system, as customary in physical chemistry. Then we introduce into the box an amount of glucose which, when completely combusted, will yield exactly 2000 kcalories, and we have our man in the box ingest this amount over the ensuing 24 h. During this time the only energy loss from the system is heat, with body temperature being above that of the environment. At the end of the 24 h we find that the physicochemical state of the system (state B) is identical to the initial state (state A), i.e., the man has neither gained nor lost weight, his body temperature is the same, and the composition of his tissues is unchanged. From precise measurements of the heat transfers taking place we find that the net heat loss was exactly 2000 kcal! Had our subject been able to do work on the external environment, such as lifting bricks and setting them on a table outside his box, we would then have had to subtract a work term in our calculations, and our subject would weigh less. Had we introduced more than 2000 cal and had the subject produced the same amount of heat and done no work, then we would have found that the physicochemical state B was *not* identical to state A. The difference would lie in the deposition of some uncombusted glucose carbons, mostly in the form of fat. Note that energy used to do work *within* the body over our 24-h experiment will have been reconverted to heat energy if state A equals state B. For example, the heart does do mechanical work in pumping blood, but that reverts to heat as the kinetic energy of the blood, through frictional forces, is expended. Similarly, energy is used to maintain ionic gradients—ATP is consumed—but if the gradients do not change, all is expressed ultimately as heat produced.

In the case of a substrate like glucose, which is completely absorbed and completely converted to carbon dioxide and water, matters are relatively simple. Had we done our experiment with foodstuffs not completely absorbed or not completely combusted, we would have had to measure the residual caloric equivalents appearing in feces and urine. Actually, the absorption of most foodstuffs is highly efficient, and the residual caloric equivalents in feces is generally small (less than 5% of calorie intake). Combustion of proteins, however, is incomplete. The nitrogen is excreted in the form of urea, and some of the complex proteins leave other uncatabolized residues which must be taken into account when striking an energy balance.

The metabolic pathways by which glucose and other substrates are catabolized can be very complex. Glucose, for example, can be oxidized by way of the Embden-Meyerhof glycolytic pathway or by way of the pentose shunt; it can be oxidized directly or can first enter glycogen and only later enter the glycolytic pathway. However, as long as the final products are the same, the amount of energy released will be the same *independent of the pathway followed*. This fundamental law, the law of Hess, greatly simplifies how we deal with energy balance problems.

The applicability of the laws of thermodynamics to the living system does not have to be taken on faith. It has been experimentally demonstrated. The first quantitative demonstration was that of Lavoisier and Laplace in 1780 (see Mendelsohn, 1964). They measured the amount of carbon dioxide produced by a guinea pig over a 10-h period by trapping the CO_2 in lime water (calcium hydroxide). Carbon dioxide had only been recently been discovered by Joseph Black and was called "fixed air." They also measured the amount of heat produced by the guinea pig over that same 10-h period. The latter measurement was made using a crude ice calorimeter, a device in which heat produced causes ice in a surrounding jacket to melt. The volume of ice melted is proportional to the heat produced. They knew that "fixed air" was also produced by the ordinary burning of carbon. What they proceeded to show was that the amount of heat produced divided by the amount of "fixed air" produced was the same whether the combustion was by burning charcoal or by allowing the guinea pig to burn its own body fuels. In other words, they showed that the combustion of foodstuffs by an animal to generate carbon dioxide yielded an amount of heat comparable to that obtained by ordinary combustion. This same basic experiment has been done many, many times in the last 200 years with increasing refinement in the accuracy and precision of the measurements. All of the results confirm the applicability of the first law. For example, Rubner (1894) directly measured heat production of dogs in a specially designed large calorimeter and calculated the amounts of substrates burned from measurements of gaseous exchange (see section on Indirect Calorimetry). The results agreed with an error of less than 3%.

Man is not exempt. The applicability of the first law to man was first demonstrated at the turn of the century by Atwater and Benedict (1905). These investigators constructed a large metabolic chamber in which a man could live for days and in which total heat production could be measured accurately along with rates of oxygen consumption and carbon dioxide production. They carried out meticulous balance studies in which the caloric values of foods were determined directly (or calculated from gaseous exchange measurements) as the input; the heat produced was measured precisely, and the caloric values of urine and feces were measured. Heat production was measured by monitoring the in-

crease in temperature of water circulating in pipes that jacketed the chamber and arranging matters so that no other heat losses occurred. A very important element in assessing total heat production is the measurement of the heat used for vaporization of water lost through the skin and lungs. This process of "insensible water loss" (i.e., conversion of water from liquid phase to gas phase in the absence of sweating) is a major mechanism for loss of heat from the body. The heat capacity of water is very large (0.58 kcal/g), and approximately 700 ml of water are converted to vapor by the abovementioned process every day. As much as 25% of basal calorie production daily occurs by this mechanism. When all of these various inputs and outputs were precisely measured by Atwater and coworkers they found that they could achieve a total energy balance to within less than 2%, a remarkable engineering feat for the time and a direct confirmation that the human system obeys the laws of thermodynamics (Atwater, 1904; Benedict, 1905).

CALORIC VALUES OF FOODSTUFFS

The energy released during the catabolism of carbohydrates and fats in the body is identical to that released by their ordinary combustion because the catabolism is complete, i.e., the products are carbon dioxide and water. The caloric values for various dietary foodstuffs are given in Table 7.8. Values are expressed in kilocalories (kcal or Cal = 1000 cal = the amount of heat necessary to raise the temperature of 1000 g of water by 1°, from 14.5 to 15.5°C). While the different forms of carbohydrate have somewhat different caloric values, the mixture of carbohydrates in most diets is such that an approximation using 4.1 kcal/g is satisfactory for most purposes. The high concentration of calories in ethanol is sometimes overlooked in assessing energy balance. Like refined sugar, ethanol in the diet adds calories without essential nutrients (amino acids, vitamins, and minerals). These "empty calories" can be a basis for poor nutrition.

Table 7.8
Calorie values of foodstuffs (Cal/g)

Carbohydrates (avg 4.1)	
glucose	3.7
starch	4.2
ethanol	7.1
Fats (avg 9.3)	
stearic acid	9.5
glycerol	4.3
acetic acid	3.5
Proteins (avg 4.1)	
glycine	2.1
leucine	5.9

The only quantitatively important lipid in the diet comes in the form of triglycerides, although there are small quantities of sterols, phospholipids, and other lipids in some foodstuffs. Fatty acids account for about 90% of the mass of the triglycerides. Their caloric equivalent is about 9.5 kcal/g. The other 10% of the triglyceride molecule is the glycerol backbone. Glycerol is rapidly converted to glucose in the body, and its metabolism is thus that of a carbohydrate. The caloric equivalent of glycerol is 4.32 kcal/g, similar to that of other carbohydrates. The ingestion of "fats" in the diet therefore contributes a small amount of carbohydrate precursor as well. As discussed in Chapter 47, when triglycerides are mobilized from adipose tissue during starvation the glycerol released along with the fatty acids makes a small but real contribution to gluconeogenesis.

For nutritional purposes a value of 9.3 kcal/g can be used to estimate the caloric contribution of dietary fats. This much higher value per gram for fats, as compared to that for carbohydrates, relates to the lower state of oxidation of the carbon atoms in fatty acids contrasted with the partially oxidized state of the carbon atoms in carbohydrates. By the same token, however, the amount of oxygen needed to combust a gram of fat is considerably larger than that needed to completely combust a gram of carbohydrate. Because of these relationships, the amount of energy released *per liter of oxygen utilized* is very nearly the same, whether the substrate is fat or carbohydrate:

$$C_6H_{12}O_6 + 6O_2 \rightarrow 6CO_2 + 6H_2O$$
(glucose; 180 g)

Energy yield: 670 kcal; 112 kcal/mole O_2 or 5 kcal/l

$$C_{16}H_{32}O_2 + 23O_2 \rightarrow 16CO_2 + 16H_2O$$
(palmitic acid, 256 g)

(1)

Energy yield: 2430 kcal; 105 kcal/mole O_2 or 4.7 kcal/l

Proteins are not completely catabolized in vivo, and therefore their caloric value as nutrients is less than the energy released by ordinary combustion. For example, the complete combustion of glycine yields 3.12 kcal/g whereas the caloric value of glycine catabolized in the body is only 2.12 kcal/g. The difference represents the caloric value of the urea formed during metabolism and excreted in the urine. Proteins also contain other substituents that are incompletely combusted. Furthermore, caloric values of different amino acids range widely—from 2.12 kcal/g for glycine to 5.9 kcal/g for leucine. However, the average amino acid composition of a diet that includes a wide variety of protein sources tends toward an average that can be used for making reasonable estimates of caloric intake. The nutritional average for protein is exactly like that for carbohydrate, 4.1 kcal/g.

RESPIRATORY QUOTIENT

The respiratory quotient (RQ) is defined as the rate of production of carbon dioxide divided by the rate of utilization of oxygen:

$$RQ = \frac{\dot{V}_{CO_2}}{\dot{V}_{O_2}}$$

This value will be 1.0 when carbohydrate is the only substrate being oxidized; it will be 0.7 when only fatty acids are being oxidized. A valid measurement requires that these gases appear in the expired air at the same rate at which they are being produced by metabolic processes. If the subject overbreathes, then the rate of CO_2 output will be accelerated, and a false high value obtained; if the patient breathes too shallowly, there will be some CO_2 retention and a false low value will be obtained.

If only carbohydrates and fats are being catabolized, an estimate of the relative contribution of these two metabolic fuels can be made from a measurement of the respiratory quotient. This will cover a range of values from 0.7 to 1.0, depending upon whether fat or carbohydrate is the exclusive fuel and with a spectrum of values between for different ratios of fat to carbohydrate. If protein is also being metabolized, this simple approach cannot be applied. However, it is possible to make a correction for the contribution of protein in the diet by measuring nitrogen excretion in the urine and calculating from it the amount of protein being catabolized. The amounts of oxygen and of carbon dioxide attributable to catabolism of that amount of protein can be subtracted from the total measured oxygen consumption and carbon dioxide production. This reduces the data so that a "nonprotein respiratory quotient" can be calculated and from that the contribution of fat and carbohydrate can be determined (see Richardson, 1929, for review).

CALORIMETRY

Direct Calorimetry

Total body heat production in man and animals has been measured using a variety of techniques. The Atwater-Benedict metabolic chamber has already been described above. Less cumbersome devices using various heat sensor systems have been utilized, but these are largely confined to sophisticated research studies. For most purposes the indirect method is adequate.

Indirect Calorimetry

The heat produced from the combustion of 1 gram of carbohydrate is considerably less than that produced from the combustion of 1 g of fat. However, the amounts of oxygen needed for combustion of these two substrates are in an inverse relationship, as shown by *Eq.* 7.1. Consequently, the amount of heat produced when 1 l of oxygen is utilized is similar whether that oxygen is used to combust carbohydrate (5.0 kcal/l) or to combust fat (4.7 kcal/l). The estimated value when protein is the exclusive substrate is not too different (4.3 kcal/l). If precise values are needed for research purposes, it is necessary to take account of these differences. That can be done by measuring urinary nitrogen and calculating the amount of protein undergoing catabolism. Knowing that, one can correct respiratory gas data to derive the nonprotein respiratory quotient, as discussed above. In this way, it is possible to determine just what mixture of metabolic foodstuffs is being combusted and calculate the heat production attributable to each class. However, for many purposes the values do not need to be calculated so exactly. In the postabsorptive state, after an overnight fast, the mixture of metabolic fuels is rather constant from individual to individual. The postabsorptive respiratory quotient is about 0.82, indicating that fat is the predominant fuel being oxidized.

The respiratory quotient for the complete combustion of various substrates is readily determined from the chemical equation for their oxidation. Thus it can be readily shown that the respiratory quotient for complete combustion of lactate is 1.0 and for the complete combustion of glycerol 0.86. The respiratory quotient for protein oxidation is not so readily determined because of its incomplete combustion, but an empirical average respiratory quotient of about 0.80 will generally apply. Some anomalies arise when oxidation is incomplete. For example, if fatty acids are partially oxidized, only to the level of acetoacetic acid, the respiratory quotient will be less than 0.7. If carbohydrate is converted to fat and stored, the respiratory quotient will be greater than 1.0. Ordinarily, it is assumed that one is dealing with a steady state, i.e., without any storage or accumulation of intermediates, and then the values discussed above apply. For detailed discussion of physiologic methods of measurement and calculation of gaseous exchange, see Consolazio et al. (1963).

BASAL METABOLIC RATE (BMR)

The basal metabolic rate is defined as the rate of calorie consumption in the postabsorptive state (after an overnight fast), in the absence of any muscular activity, at a comfortable environmental temperature, and with the patient resting comfortably (but not sleeping). The measurement is a useful research tool in physiology. It was at one time widely used in clinical medicine for the assessment of thyroid function, but it has been displaced by more direct measurements based on immunoassay of hormone levels in the plasma.

The BMR is determined using indirect calorimetry, measuring the rate of oxygen consumption. This can be measured in a closed system, such as the Benedict-Roth apparatus, in which the patient breathes from an oxygen reservoir and expired CO_2 is trapped. The decrease in volume of the reservoir reflects utilization of oxygen. Alternatively, the measurement can be made in an open system using the Pauling oxygen analyzer, which directly measures the partial pressure of oxygen in a stream of air supplied to the patient's mask (or into a chamber) and also measures the concentration in the expired air. Coupled with volume-flow measurements this allows a calculation of rates of oxygen utilization. In all cases the measurements are corrected to standard temperature and pressure (0°C and 760 mm of pressure). In the basal state it is assumed that the respiratory quotient is 0.82 and that the consumption of 1 l of oxygen corresponds to the expenditure of 4.82 kcal.

Obviously, calorie consumption will be greater per unit time in a large person than in a small person, and one way of normalizing is to express the results in terms of kcal/kg/h. Even better normalization is obtained if the results are expressed in relationship to the estimated surface area of the body (kcal/m^2/h). Surface areas can be estimated from height and weight using the empirical formula developed by DuBois and DuBois (1916): (area in cm^2 equals $71.8 \times$ weight$^{0.425} \times$ height$^{0.725}$), or by using a nomogram or appropriate tables. As an example, an 80-kg man 180 cm tall has a body surface area of exactly 2 m^2. The difference between using weight and surface area as the reference denominator is striking when one compares animals of very different sizes. For example, the BMR of the mouse is 654 kcal/kg/day and that of man about 35 kcal/kg/day. When related to surface area, however, the values are much closer, i.e., 1888 vs 1042 kcal/m^2/day (see Kleiber (1947) for review).

A convenient rough value to remember for BMR in normal man is about 1 kcal/kg/h, or 35–40 kcal/m^2/h. Some feeling for amounts of heat involved comes from comparison with a familiar source. The rate of heat production of an 80-watt light bulb is about the same as that of a 70-kg man in the basal state.

A number of factors influence basal metabolic rate. It decreases with age, showing approximately a 10% decrease between the ages of 20 and 60. The BMR is slightly higher in males than in females (about 10%). If the body temperature rises (in a very hot climate or as a result of a fever), the BMR can increase considerably. On exposure to cold, rats and other animals increase their metabolic rate even without muscle tensing or frank shivering. Man, however, does not appear to be capable of this *nonshivering thermogenesis*, and only increases metabolic rate significantly by virtue of muscle tensing or shivering. The level of thyroid hormone activity is a key determinant of metabolic rate, rising in hyperthyroidism and falling in hypothyroidism (see Chapter 53). Values as much as 50% above or below the norm can be observed. In part this thyroid effect is mediated via increases in ion flux associated with increased activity of membrane-bound Na-K-ATPase (Smith and Edelman, 1979). The catecholamines increase basal metabolic rate, hence the importance of avoiding stress when attempting to measure BMR. Caffeine and theophylline also will increase BMR.

BMR is always measured after an overnight fast because there is an increment in energy production in response to ingestion of foods (diet-induced thermogenesis or "specific dynamic action"). This added heat production, which can amount to as much as 20% of the nominal caloric value of the foods ingested, is not related exclusively to gastrointestinal activity because it can be elicited by intravenous injection of nutrients. The possibility that there may be variations in the magnitude of food-induced thermogenesis from individual to individual is currently being closely examined to see if it can explain differences in tendencies toward obesity.

METABOLIC "COSTS" OF PHYSICAL ACTIVITY

The total daily caloric expenditure can be considered as the sum of basal caloric expenditure plus that attributable to physical activity. Actually, the rate of caloric expenditure during sleep is somewhat less than that observed during a formal measurement of BMR (by about 10%), but during most of the day it is well above BMR. During sleep, caloric expenditure may be about 1 kcal/min; running hard, one can expend as much as 30–40 kcal/min. Obviously, the physical costs of activity range very widely (Table 7.9). Precise measurements of these energy costs are difficult. Most of the data have

Table 7.9
Approximate energy costs of various physical activities

Activity	Rate of Energy Expenditure (Cal/min)
Sleeping	ca 1.0
Dressing, washing, and shaving	3.8
Walking at 2 mph (70-kg man)	3.2
Walking at 4 mph (70-kg man)	5.8
Dancing the foxtrot	5.2
Dancing the rumba	7.0
Skiing	10–18
Woodchopping	11–13

Data from Passmore and Durnin (1955).

been obtained by having the subject breathe from a reservoir of oxygen while engaging in various activities (measuring the actual oxygen consumption) and then calculating energy expenditure by indirect calorimetry. Eight hours of hard work by a miner may cost up to 3000 kcal whereas a clerk may only expend an additional 300 kcal in connection with his 8 h of work. While we can measure BMR rather exactly and while we can measure caloric intake exactly by careful accounting for food intake, it is very difficult to assess exactly the total caloric expenditure of an individual over any given 24-h period, let alone his expenditure over a more extended period of time. The importance of this fact in connection with studies of energy balance and studies of obesity becomes apparent when we look at the narrow margin of error that is allowable.

Consider an individual who is in caloric balance, i.e., his weight is stable over an extended period of time with his usual food intake and his usual energy expenditure. Suppose that individual increases his caloric intake by only 200 kcal/day, i.e., an increase of only about 10%. Over a 1-yr period he will take in 73,000 extra kcal. Now, a pound of adipose tissue (450 g), which is 90% triglyceride by mass, has a caloric equivalent of about 3500 kcal. By the first law of thermodynamics the extra calories taken in over 1 yr will result in the deposition of about 20 lbs, and if continued for 5 yr, 100 lbs! Conversely, seemingly trivial differences in physical activity would also result in large changes in body weight if not accompanied by appropriate changes in calorie intake. For example, the difference in calorie expenditure when simply standing quietly and when standing and engaging in animated conversation is about 0.3 kcal/min. If an individual became talkative for 4 h/day instead of being taciturn, he might be expending 70 extra kcal daily. Over the period of a year, again, he might *lose* as much as 17 lbs if he did not increase calorie intake. Normal individuals manage to maintain their body weight within rather narrow limits, even though they change their physical activity very drastically, far beyond the narrow limits we have outlined above. For example, a college student may work intensively as a lumberjack during the summer and return to the relatively sessile existence of the college campus during the rest of the year. However, he adjusts his food intake appropriately and usually does not show huge changes in weight. The basis for this fine tuning that adjusts calorie intake to calorie expenditure is poorly understood. Regulation at the level of hypothalamus is certainly involved, as discussed in Chapter 69, but the precise mechanisms remain to be established. It is essential to keep in mind that small discrepancies between caloric intake and caloric expenditure, discrepancies that are at the very limits of our ability to quantify them, can account for the development of obesity. It is not necessary to break the first law. For a review of the metabolic costs of physical activity, see Passmore and Durnin, 1955.

OBESITY

Obesity represents a major medical problem in the United States and in most of the developed countries. The obese as a group are more likely to develop a number of diseases (e.g., hypertension, gallstones, and diabetes mellitus), and their life span is significantly shortened. Obesity is defined in terms of an increase in the amount of adipose tissue, which normally accounts for about 15% of total body weight. A football player, however, can be *overweight* (i.e., above the tabulated norm for his height) yet not be *obese* because all of the excess weight represents muscle, not fat. Generally, however, increases in weight reflect increases in adipose tissue, i.e., body fat. The exact relationship between degree of obesity and threat to health is still not fully established, but there is no doubt about the life-threatening implications of moderately severe obesity.

Anything that leads to a continuing imbalance between food intake and energy consumption (positive energy balance) will lead to obesity. However, to state that obesity is due to excess intake of food is no more profound or enlightening than to say that alcoholism is due to excess intake of alcohol! *Why* do some individuals manage to maintain a nice balance throughout a lifetime, despite wide swings from time to time in their energy expenditure, while others become seriously obese? Food-seeking behavior is a complex, highly integrated process. Consider the multiple steps involved in deciding to rise from the desk, walk to the kitchen, take down the peanut butter jar, make a sandwich,

chew, swallow, etc., etc. What determines the frequency with which such a behavioral pattern recurs? The amounts eaten each time? Somehow, it must normally be linked to energy expenditure, but the precise linkage mechanisms are still incompletely understood. Clearly, central nervous system control is involved and, as discussed in Chapter 69, the hypothalamic centers are known to play a key role.

Bilateral destructive lesions in the ventrolateral nuclei of the hypothalamus can totally arrest food-seeking behavior, causing the animal to actually starve to death with food right there in the cage with it! Bilateral destructive lesions in the ventromedial nuclei, on the other hand, induce voracious eating and gross obesity. Direct hypothalamic damage secondary to tumors or trauma has on occasion caused marked obesity in patients, but this is extremely rare. Certainly most obese patients have no evidence of central nervous system dysfunction. Whether more subtle dysfunction occurs in hypothalamic sensitivity to various input signals or in effectiveness in sending appropriate output control signals is not known.

The hypothalamic centers are obviously subject to modulating influences originating in the cortex. The mental image of a peanut butter jar on the shelf or the smell of hamburgers grilling has behavioral consequences. Many elegant psychological studies reveal potentially significant behavior differences between the obese and the lean with respect to perception of and the response to appetite.

A number of metabolic and endocrine abnormalities have been observed in obese subjects. For example, they have abnormally high plasma insulin levels, and their insulin response to the ingestion of glucose is greater than normal. However, normal individuals who deliberately gain weight for experimental purposes by forcing themselves to overeat also develop hyperinsulinemia; when they revert to their normal eating habits and return to their normal weight, their hyperinsulinemia disappears. Thus, hyperinsulinemia and most of the "metabolic abnormalities" described in the obese represent *results* of obesity rather than causes (Sims et al., 1973).

Obesity is almost certainly not a single disorder but a common manifestation of a wide variety of disorders that affect energy balance in some way. Whatever the cause or causes, behavior modification to decrease food intake and moderate exercise to increase caloric expenditure represent effective management. A useful overview of many basic and clinical aspects of obesity can be found in the NIH conference *Obesity in Perspective* (Bray, 1973).

BIBLIOGRAPHY

ARIMURA, A. Recent progress in somatostatin research. *Biomed. Res.* 2: 233–257, 1981

ATWATER, W. O. Neue Versuche uber Stoff- und Kraftwechsel im menschlichen Korper. *Ergebnisse der Physiol.* 3: 497, 1904.

ATWATER, W. O. AND F. G. BENEDICT. A respiration calorimeter. Publ. No. 42, Carnegie Institution of Washington, 1905.

BANTING, F. G., AND BEST, C. H. Internal secretion of pancreas. *J. Lab. Clin. Med.* 7: 251, 1922.

BELL, G. I., W. F. SWAIN, R. PICTET, B. CORDELL, H. M. GOODMAN, AND W. J. RUTTER. Nucleotide sequence of a cDNA clone encoding for human preproinsulin. *Nature* 282: 525–527, 1979.

BENDITT, E. P. The monoclonal theory of atherogenesis. *Atheroscler. Rev.* 3: 77–86, 1978.

BERGLUND, L., J. C. KHOO, D. JENSEN, AND D. STEINBERG. Resolution of hormone-sensitive triglyceride/diglyceride lipase from monoglyceride lipase of chicken adipose tissue. *J. Biol. Chem.* 255: 5420–5428, 1980.

BERSON, S. A., AND R. S. YALOW. Quantitative aspects of reaction between insulin and insulin-binding antibody. *J. Clin. Invest.* 38: 1996–2016, 1959.

BERTHAUD, H. R., D. A. BEREITER, E. R. TRIMBLE, E. C. SIEGEL, AND B. JEANRENAUD. Cephalic phase reflex insulin secretion: neuroanatomical and physiological characterization. *Diabetologia* 20 (Suppl.): 393–401, 1981.

BRAY, G. A. (ed.). Proceedings of a Conference Sponsored by the John E. Fogarty International Center for Advanced Study in the Health Science. DHEW Publication No. (NIH) 75-708, 1973.

BRAZEAU, P., W. VALE, R. BURGUS, N. LING, M. BUTCHER, J. RIVIER, AND R. GUILLEMIN. Hypothalamic polypeptide that inhibits secretion of immunoreactive pituitary growth hormone. *Science* 179: 77–79, 1973.

BROMER, W. W., L. G. SINN, A. STAUB, AND O. K. BEHRENS. The amino acid sequence of glucagon. *Diabetes* 6: 234, 1957.

BUNN, H. F. Nonenzymatic glycosylation of protein: relevance to diabetes. In: *Diabetes Mellitus*, edited by J. S. Skyler and G. F. Cahill, Jr. New York: Yorke Medical Books, 1981, p. 173–178.

BUTCHER, R. W., R. J. HO, H. C. MENG, AND E. W. SUTHERLAND. Adenosine 3′,5′-monophosphate in biological materials. *J. Biol. Chem.* 240: 4515–4523, 1965.

BUTCHER, R. W., J. G. T. SNEYD, C. R. PARK, AND E. W. SUTHERLAND, JR. Effect of insulin on adenosine 3′, 5′-monophosphate in the rat epididymal fat pad. *J. Biol. Chem.* 241: 1651–1653, 1966.

CAHILL, G. F., JR., AND O. E. OWEN. Some observations on carbohydrate metabolism in man. In: *Carbohydrate Metabolism*, edited by F. Dickens, P. S. Randle, and W. J. Whelan. London: Academic Press, 1968, vol. I, p. 457–522.

CHANCE, R. E., E. P. KROEFF, J. A. HOFFMAN, AND B. H. FRANK. Chemical, physical and biological properties of biosynthetic human insulin. *Diabetes Care* 4: 147–154, 1981.

CORDELL, B., G. BELL, E. TISCHER, F. M. DeNOTO, A. ULLRICH, R. PICTET, W. J. RUTTER, AND H. M. GOODMAN. Isolation and characterization of a cloned rat insulin gene. *Cell* 18: 533–543, 1979.

CONSOLAZIO, C. F., R. E. JOHNSON, AND L. J. PECORA. Physiologic measurements of metabolic function in man. New York: McGraw-Hill Book Co., 1963.

CORBIN, J. D., E. M. REIMANN, D. A. WALSH, AND E. G. KREBS. Activation of adipose tissue lipase by skeletal muscle cyclic adenosine 3′, 5′-monophosphate-stimulated protein kinase. *J. Biol. Chem.* 245: 4849–4857, 1970.

CREA, R., A. KRASZEWSKI, H. TADASKI, AND K. ITAKURA. Chemical synthesis of genes of human insulin. *Proc. Natl. Acad. Sci. (USA)* 75: 5765–5769, 1978.

CUATRECASAS, P. Interaction of insulin with the cell membrane: the primary action of insulin. *Proc. Natl. Acad. Sci. (USA)* 63: 450, 1969.

CZECH, M. P. Insulin action. In: *Diabetes Mellitus*, edited by J. S. Skyler and G. F. Cahill, Jr. New York: Yorke Medical Books, 1981, p. 64–72.

DENTON, R. M., R. W. BROWNSEY, AND G. J. BELSHAM. A partial view of the mechanism of insulin action. *Diabetologia* 21: 347–362, 1981.

DOLE, V. P. A relation between non-esterified fatty acids in plasma and the metabolism of glucose. *J. Clin. Invest.* 35: 150, 1956.

DuBOIS, D., AND E. F. DuBOIS. Clinical calorimetry. X. A formula to estimate the approximate surface area if height and weight be known. *Arch. Intern. Med.* 17: 863, 1916.

DUPRE, J., J. D. CURTIS, R. H. UNGER, R. W. WADDELL, AND J. C. BECK. Effects of secretin, pancreozymin, or gastrin on the response of the endocrine pancreas to administration of glucose or arginine in man. *J. Clin. Invest.* 48: 745–757, 1969.

DUPRE, J., S. A. ROSS, D. WATSON, AND J. D. BROWN. Stimulation of insulin secretion by a gastric inhibitory polypeptide in man. *J. Clin. Endocrinol. Metab.* 37: 826–828, 1973.

EXTON, J. H. Mechanisms involved in effects of catecholamines on liver carbohydrate metabolism. *Biochem. Pharmacol.* 28: 2237–2240, 1979.

FAIN, J. N., V. P. KOVACEV, AND R. O. SCOW. Effects of growth hormone and dexamethasone on lipolysis and metabolism in isolated fat cells of the rat. *J. Biol. Chem.* 240: 3522–3529, 1965.

FELIG, P. Disorders of carbohydrate metabolism. In: *Metabolic Control and Disease*, edited by P. K. Bondy and L. E. Rosenberg. Philadelphia: W. B. Saunders, 1980, pp. 276–392.

FOA, P. P., J. S. BAJAJ, AND N. L. FOA (eds.). *Glucagon: Its Role in Physiology and Clinical Medicine.* New York: Springer-Verlag, 1977.

FREDRIKSON, G., P. STRALFORS, N.-O. NILSSON, AND P. BELFRAGE. Hormone-sensitive lipase of rat adipose tissue. *J. Biol. Chem.* 256: 6311–6320, 1981.

GABBAY, K. H., K. DeLUCA, J. N. FISHER, JR., M. E. MAKO, AND A. H. RUBENSTEIN. Familial hyperproinsulinemia. *N. Engl. J. Med.* 294: 911–915, 1976.

GARLAND, P. B., P. J. RANDLE, AND E. A. NEWSHOLME. Citrate as an intermediary in the inhibition of phosphofructokinase in rat heart muscle by fatty acids, ketone bodies, pyruvate, diabetes and starvation. *Nature (London)* 200: 167–170, 1963.

GERICH, J. E., M. A. CHARLES, AND G. GRODSKY. Regulation of pancreatic insulin and glucagon secretion. *Annu. Rev. Physiol.* 38: 353, 1976.

GERICH, J. E. Somatostatin and diabetes. In: *Diabetes Mellitus*, edited by J. S. Skyler and G. F. Cahill, Jr. New York: Yorke Medical Books, 1981, p. 48–55.

GOEDDEL, D. V., D. G. KLEID, F. BOLIVAR, H. L. HEYNEKER, D. G. YANSURA, R. CREA, T. HIROSE, A. KRASZEWSKI, K. ITAKURA, AND A. RIGGS. Expression in *Escherichia coli* of chemically synthesized genes for human insulin. *Proc. Natl. Acad. Sci. (USA)* 76: 106–110, 1979.

GOLDSTEIN, J. L., AND M. S. BROWN. The low-density lipoprotein pathway and its relation to atherosclerosis. *Annu. Rev. Biochem.* 46: 897–930, 1977.

GOODMAN, D. S. The interaction of human serum albumin with long-chain fatty acid anions. *J. Am. Chem. Sci.* 80: 3892, 1958.

GORDON, R. S., JR., AND A. CHERKES. Unesterified fatty acid in human plasma. I. *J. Clin. Invest.* 35: 206, 1956.

HAVEL, R. J., J. L. GOLDSTEIN, AND M. S. BROWN. Lipoproteins and lipid transport. In: *Metabolic Control and Disease*, edited by P. K. Bondy and L. E. Rosenberg. Philadelphia: W. B. Saunders, 1980, p. 393–494.

HAVEL, R. J., H. A. EDER, AND J. H. BRAGDON. The distribution and chemical composition of ultracentrifugally separated lipoproteins in human serum. *J. Clin. Invest.* 34: 1345–1353, 1955.

HAZELWOOD, R. L. Synthesis, storage, secretion, and significance of pancreatic polypeptide in vertebrates. In: *The Islets of Langerhans*, edited by S. J. Cooperstein and D. Watkins. New York: Academic Press, 1981.

HERS, H. G. The control of glycogen metabolism in the liver. *Annu. Rev. Biochem.* 45: 167–189, 1976.

HIMMS-HAGEN, J. Obesity may be due to a malfunctioning of brown fat. *J. Canad. Med. Assoc.* 121: 1361–1364, 1979.

HIRSCH, J., AND B. BATCHELOR. Adipose tissue cellularity in human obesity. *Clin. Endocrinol. Metab.* 5: 299–311, 1976.

HOBART, P., R. CRAWFORD, L.-P. SHEN, R. PICTET, AND W. J. RUTTER. Cloning and sequence analysis of cDNAs encoding two distinct somatostatin precursors found in the endocrine pancreas of anglerfish. *Nature* 288: 137, 1980.

HORWITZ, D. L., H. KUZUYA, AND A. H. RUBENSTEIN. Circulating serum C-peptide. *N. Engl. J. Med.* 295: 207, 1976.

HOUGEN, T. J., B. E. HOPKINS, AND T. W. SMITH. Insulin effects on monovalent cation transport and Na-K-ATPase activity. *Am. J. Physiol.* 234: C59, 1978.

HOUSLAY, M. D. Membrane phosphorylation: a crucial role in the action of insulin, EGF, and pp.60^{src}? *Bioscience Reports* 1: 19–34, 1981.

HUTTUNEN, J. K., D. STEINBERG, AND S. E. MAYER. Protein kinase activation and phosphorylation of purified hormone-sensitive lipase. *Biochem. Biophys. Res. Commun.* 41: 1350, 1970.

INNERARITY, T. L., AND R. W. MAHLEY. Enhanced binding by cultured human fibroblasts of apo E-containing lipoproteins as compared with low density lipoproteins. *Biochem.* 17: 1440–1447, 1978.

JOEL, C. D. The physiological role of brown adipose tissue. In: *Adipose Tissue. Handbook of Physiology*, edited by A. E. Renold and G. F. Cahill, Jr. Washington, DC: American Physiological Society, 1965, sect. 5, p. 87–100.

JUNG, R. T., P. S. SHETTY, W. P. T. JAMES, M. A. BARRAND, AND B. A. CALLINGHAM. Reduced thermogenesis in obesity. *Nature* 279: 322–323, 1979.

KAHN, C. R., J. S. FLIER, R. S. BAR, J. A. ARCHER, P. GORDEN, M. A. MARTIN, AND J. ROTH. The syndromes of insulin resistance and acanthosis nigricans: insulin-receptor disorders in man. *N. Engl. J. Med.* 294: 739, 1976.

KAHN, C. R., K. L. BAIRD, J. S. FLIER, C. GRUMFELD, J. T. HARMON, L. C. HARRISON, F. A. KARLSSON, M. KASUGA, G. L. KING, U. C. LANG, J. M. PODSKALNY, AND E. VAN OBBERGEHEN. Insulin receptors, receptor antibodies, and the mechanism of insulin action. *Recent Prog. Horm. Res.* 37: 477–538, 1981.

KASUGA, M., Y. ZICK, W. L. BLITHE, M. CRETTAZ, AND C. R. KAHN. Insulin stimulates tyrosine phosphorylation of the insulin receptor in a cell-free system. *Nature (London)* 298: 667, 1982.

KHOO, J. C., D. STEINBERG, B. THOMPSON, AND S. E. MAYER. Hormonal regulation of adipocyte enzymes: the effects of epinephrine and insulin on the control of lipase, phosphorylase kinase, phosphorylase, and glycogen synthase. *J. Biol. Chem.* 248: 3823–3830, 1973.

KLEIBER, M. Body size and metabolic rate. *Physiol. Rev.* 27: 511, 1947.

KOENIG, R. J., C. M. PETERSON, R. L. JONES, C. SAB-

DEK, M. LEHRMAN, AND A. CERAMI. Correlation of glucose regulation and hemoglobin A$_{1c}$ in diabetes mellitus. *N. Engl. J. Med.* 295: 417, 1976.

KOERKER, D. J., W. RUCH, E. CHIDECKEL, J. PALMER, C. J. GOODNER, J. ENSINCK, AND C. C. GALE. Somatostatin: hypothalamic inhibitor of the endocrine pancreas. *Science* 184: 482–484, 1974.

LARNER, J., C. VILLAR-PALASI, N. D. GOLDBERG, S. J. BISHOP, F. HUIJING, J. I. WENGER, H. SASKO, AND N. P. BROWN. Hormonal and non-hormonal control of glycogen synthesis—control of transferase phosphatase and transferase I kinase. *Adv. Enzyme Regul.* 6: 409–423, 1968.

LARNER, J., G. GALASKO, K. CHENG, A. A. DePALOI-ROACH, L. HUANG, P. DAGGY, AND J. KELLOGG. Generation by insulin of a chemical mediator that controls protein phosphorylation and dephosphorylation. *Science* 206: 1408, 1979.

LEVY, R. I. Declining mortality in coronary heart disease. *Arteriosclerosis* I: 312–325, 1981.

LOMEDICO, P., N. ROSENTHAL, A. EFSTRATIADIS, W. GILBERT, R. KOLODNER, AND R. TIZARD. The structure and evolution of the two nonallelic rat preproinsulin genes. *Cell* 18: 545–558, 1979.

LOTEN, E. G., AND J. G. SNEYD. An effect of insulin on adipose tissue adenosine 3′, 5′-monophosphate diesterase. *Biochem. J.* 120: 187, 1970.

LOUIS-SYLVESTRE, J. Preabsorptive insulin release and hypoglycemia in rats. *Am. J. Physiol.* 230: 56–60, 1976.

MANGANNIELLO, V., AND M. VAUGHAN. An effect of insulin on cyclic adenosine 3′,5′-monophosphate phosphodiesterase activity in fat cells. *J. Biol. Chem.* 248: 7164–7170, 1973.

MANN, F. C., AND T. B. MAGATH. Die Werkung der totalen Leberextirpation. *Ergeb. Physiol.* 23: 212–262, 1924.

McGARRY, J. D., AND D. W. FOSTER. Hormonal control of ketogenesis. *Arch. Intern. Med.* 137: 495, 1977.

MENDELSOHN, E. *Heat and Life.* Cambridge, MA: Harvard Univ. Press, 1964, p. 148–153.

MORGAN, H. E., D. M. REGEN, AND C. R. PARK. Identification of a mobile carrier-mediated sugar transport system in muscle. *J. Biol. Chem.* 239: 369, 1964.

MUNGER, B. L. Morphological characterization of islet cell diversity. In: *The Islets of Langerhans*, edited by S. J. Cooperstein and D. Watkins. New York: Academic Press, 1981, p. 1–49.

NATIONAL DIABETES DATA GROUP. Classification and diagnosis of diabetes mellitus and other categories of glucose intolerance. *Diabetes* 28: 1039, 1979.

ODUM, E. P. Adipose tissue in migrating birds. In: *Adipose Tissue. Handbook of Physiology*, edited by A. E. Renold and G. F. Cahill, Jr. Washington, DC: American Physiological Society, 1965, sect. 5, p. 37–43.

OLEFSKY, J. M., AND O. G. KOLTERMAN. Mechanisms of insulin resistance in obesity and noninsulin-dependent (type II) diabetes. In: *Diabetes Mellitus*, edited by J. S. Skyler and G. F. Cahill, Jr. New York: Yorke Medical Books, 1981, p. 73–90.

OLEFSKY, J. M., M. SAEKOW, H. TAGER, AND A. H. RUBENSTEIN. Characterization of a mutant human insulin species. *J. Biol. Chem.* 255: 6098–6105, 1980.

ORCI, L. The microanatomy of the islets of Langerhans. *Metabolism* 25 (Suppl. 1): 1303, 1976.

ORCI, L. Macro- and micro-domains in the endocrine pancreas. *Diabetes* 31: 538–565, 1982.

OWEN, O. E., A. P. MORGAN, H. G. KEMP, J. M. SULLIVAN, M. G. HERRERA, AND G. F. CAHILLL, JR. Brain metabolism during fasting. *J. Clin. Invest.* 46: 1589–1595, 1967.

OWEN, O. E., P. FELIG, A. P. MORGAN, J. WAHREN, AND G. F. CAHILL, JR. Liver and kidney metabolism during prolonged starvation. *J. Clin. Invest.* 48: 574–583, 1969.

PARK, C. R., O. B. CROFFORD, AND T. KONO. Mediated (nonactive) transport of glucose in mammalian cells and its regulation. *J. Gen. Physiol.* 52: 296, 1968.

PASSMORE, R., AND J. V. G. A. DURNIN. Human energy expenditure. *Physiol. Rev.* 35: 801–840, 1955.

PERLEY, M. J., AND D. M. KIPNIS. Plasma insulin responses to oral and intravenous glucose: studies in normal and diabetic subjects. *J. Clin. Invest.* 26: 1954–1962, 1967.

PERMUTT, M. A. Biosynthesis of insulin. In: *The Islets of Langerhans*, edited by S. J. Cooperstein and D. Watkins. New York: Academic Press, 1981, p. 75–95.

PHILLIPS, L. S., AND R. VASSILOPOULOU-SELLIN. Somatomedins. *N. Engl. J. Med.* 302: 371–438, 1980.

PILKIS, S. J., C. R. PARK, AND T. H. CLAUS. Hormonal control of hepatic gluconeogenesis. *Vitam. Horm.* 36: 383, 1978.

POWLEY, T. The ventromedial hypothalamic syndrome, satiety and a cephalic phase hypothesis. *Psychol. Rev.* 84: 89–126, 1977.

RAHBAR, S. An abnormal haemoglobin in red cells of diabetes. *Clin. Chem. Acta* 22: 296, 1968.

RAHBAR, S., D. BLUMENFELD, AND H. M. RANNEY. Studies of the unusual hemoglobin in patients with diabetes mellitus. *Biophys. Res. Commun.* 36: 838, 1969.

RICHARDSON, H. B. The respiratory gradient. *Physiol. Rev.* 9: 61, 1929.

ROSS, R. Atherosclerosis: a problem of the biology of arterial wall cells and their interaction with blood components. *Arteriosclerosis* 1: 293–311, 1981.

ROTH, R. A., AND D. J. CASSELL. Insulin receptor: evidence that it is a protein kinase. *Science* 219: 299–301, 1983.

ROTH, J., C. R. KAHN, M. A. LESNIAK, P. GORDEN, P. DeMEYTS, K. KEGGESI, D. M. NEVILLE, JR., J. R. GARVIN III, A. H. SOLL, P. FREYCHET, I. D. GOLDFINE, R. S. BAR, AND J. A. ARCHER. Receptors for insulin, NSILA-s and growth hormone: application to disease states in man. *Recent Prog. Horm. Res.* 31: 95, 1975.

RUBNER, M. Die Quelle der thierischen Warme [the source of animal heat]. *Ztschn. Biol.* 30: 73, 1894.

SALANS, L. B. Obesity and the adipose cell. In: *Metabolic Control and Disease*, edited by P. K. Bondy and L. E. Rosenberg. Philadelphia: W. B. Saunders, 1980, p. 495–521.

SCHOENHEIMER, R. The dynamic state of body constituents. Cambridge, MA: Harvard University Press, 1942.

SEALS, J. R., AND L. JARETT. Activation of pyruvate dehydrogenase by direct addition of insulin to an isolated plasma membrane/mitochondria mixture: evidence for generation of insulin's second messenger in a subcellular system. *Proc. Natl. Acad. Sci. (USA)* 77: 77, 1980.

SHEN, L-P, R. L. PICTET, AND W. J. RUTTER. Human somatostatin. I. Sequence of the cDNA. *Proc. Natl. Acad. Sci. (USA)* 79: 4575, 1982.

SEVERSON, D. L., J. C. KHOO, AND D. STEINBERG. The role of phosphoprotein phosphatases in the reversible deactivation of chicken adipose tissues hormone-sensitive lipase. *J. Biol. Chem.* 252: 1484–1489, 1977.

SHERRILL, B. C., T. L. INNERARITY, AND R. W. MAHLEY. Rapid hepatic clearance of the canine lipoproteins containing only the E apoprotein by a high affinity receptor. *J. Biol. Chem.* 255: 1804–1807, 1980.

SIMS, E. A. H., E. DANFORTH, JR., E. S. HORTON, G. A. BRAY, J. GLENNON, AND L. A. SALANS. Effects of experimental obesity in man. *Recent Prog. Horm. Res.* 29: 457–496, 1973.

SKYLER, J. S., AND G. F. CAHILL, JR. (eds.). *Diabetes Mellitus.* New York: Yorke Medical Books, 1981.

SMITH, T. J., AND I. S. EDELMAN. The role of sodium transport in thyroid thermogenesis. *Fed. Proc.* 38: 2150–2153, 1979.

STADIE, W. C., N. HAUGAARD, AND M. VAUGHAN. The quantitative relation between insulin and its biological activity. *J. Biol. Chem.* 200: 745–781, 1953.

STANBURY, J. B., J. B. WYNGAARDEN, D. S. FREDRICKSON, J. L. GOLDSTEIN, AND M. S. BROWN. (eds). *The Metabolic Basis of Inherited Disease. Part 4. Disorders of Lipoprotein and Lipid Metabolism.* New York: McGraw-Hill Book Co., 1983, p. 589–710.

STEINBERG, D. Catecholamine stimulation of fat mobilization and its metabolic consequences. *Pharmacol. Rev.* 18: 217–235, 1966.

STEINBERG, D. Interconvertible enzymes in adipose tissue regulated by cyclic AMP-dependent protein kinase. *Adv. Cyclic Nucleotides Res.* 7: 157–198, 1976.

STEINBERG, D. Origin, turnover and fate of plasma low density lipoprotein. *Prog. Biochem. Pharmacol.* 15: 166–199, 1979.

STEINBERG, D. Metabolism of lipoproteins at the cellular level in relation to atherogenesis. In: *Lipids, Lipoproteins and Atherosclerosis*, edited by N. E. Miller and B. Lewis, Amsterdam: Elsevier, 1981, p. 31–48.

STEINBERG, D., AND M. VAUGHAN. Release of free fatty acids from adipose tissues *in vitro* in relation to rates of triglyceride synthesis and degradation. In: *Adipose Tissue. Handbook of Physiology*, edited by A. E. Renold and G. F. Cahill. Washington, DC: American Physiological Society, 1965, sect. 5, p. 335.

STEINBERG, D., S. E. MAYER, J. C. KHOO, E. A. MILLER, R. E. MILLER, B. FREDHOLM, AND R. EICHNER. Hormonal regulation of lipase, phosphorylase, and glycogen synthase in adipose tissue. *Adv. Cyclic Nucleotide Res.* 5: 549–568, 1975.

STEINER, D. F. Insulin today (Banting lecture). *Diabetes* 26: 322, 1977.

STEINER, D. F., AND N. FREINKEL (eds.). *Endocrine Pancreas Handbook of Physiology.* Washington, DC: American Physiological Society, 1972, sect. 7, vol. 1.

STRAUS, D. S. Minireview: effects of insulin on cellular growth and proliferation. *Life Sci.* 29: 2131, 1981.

TAGER, H. S., D. F. STEINER, AND C. PLATZELT. Biosynthesis of insulin and glucagon. In: *Methods in Cell Biology.* New York: Academic Press, 1981, p. 73.

TANZAWA, K., Y. SHIMADA, M. KURODA, Y. TSUJITA, M. ARAI, AND Y. WATANABE. WHHL-rabbit: a low density lipoprotein receptor-deficient animal model for familial hypercholesterolemia. *FEBS Lett.* 118: 81–84, 1980.

UNGER, R. H., AND L. ORCI. Glucagon and the A cell. *N. Engl. J. Med.* 304: 1518, 1575, 1981.

WASSERMAN, F. The development of adipose tissue. In: *Adipose Tissue. Handbook of Physiology*, edited by A. E. Renold and G. F. Cahill, Jr. Washington, DC: American Physiological Society, 1965, sect. 5, p. 87–100.

WATANABE, Y. Serial inbreeding of rabbits with hereditary hyperlipidemia (WHHL-rabbits). *Atherosclerosis* 36: 261–268, 1980.

WEINHOUSE, S. Regulation of glucokinase in liver. *Curr. Top. Cell. Regul.* 11: 1–50, 1976.

WISSLER, R. W. Current status of regression studies. *Atherosclerosis Rev.* 3: 213–230, 1978.

YEN, S. S. C., T. M. SILER, AND G. W. DeVANE. Effects of somatostatin in patients with acromegaly: suppression of growth hormone, prolactin, insulin and glucose levels. *N. Engl. J. Med.* 290: 935–938, 1974.

Endocrine System

CHAPTER 51

Principles of Hormone Action and Endocrine Control

INTRODUCTION

Even the most primitive forms of life regulate the expression of their genetic material in a sophisticated manner. With increasing complexity and differentiation of structure and function, regulatory mechanisms have assumed increasing importance in the survival of the organism, in its adaptation to the environment, and in its reproduction. The two primary communication systems, the nervous system and the endocrine system, serve as a biological communication network for integration of the organism's response to a changing environment. Many nervous system functions are mediated by hormones, and the endocrine system is regulated to a large degree by the nervous system. Together the nervous and endocrine systems institute alterations in metabolism, behavior, and development to conform to internal and external environmental demands.

The endocrine glands synthesize and secrete hormones, discrete chemical substances which transfer information from one set of cells to another. The term "hormone" was first used by E. H. Starling in 1905 to describe the gastrointestinal peptide secretin. The derivation is from the Greek word meaning "set in motion" or "excite." Hormones have either of two major chemical structures: peptide or steroid. The molecular mechanisms of action of biogenic amines such as epinephrine, which is derived from the amino acid tyrosine, resemble the peptide group of hormones, whereas thyroid hormone, which is also derived from tyrosine, more closely resembles the steroid group of hormones.

In contrast to the nervous system, where neurotransmitters are released from axons in direct proximity to their target, the endocrine glands secrete hormones into the blood stream, which carries them to distal targets. The distance between hormone producer cell and hormone responder cell may be large, as from pituitary to gonads; moderate, as

from hypothalamus to pituitary; or small, as from δ to β cell in pancreatic islets. The latter is designated as a paracrine system to distinguish it from the first two, which are endocrine systems. At their site of action, hormones serve as regulators and integrators of the target cell response with the demands of the organism as a whole.

HORMONES AS ALLOSTERIC EFFECTORS

Monod et al. (1963) defined regulatory control of protein structure and consequent function by small molecule effectors. The small molecules are not metabolized by interaction with specific proteins, but by binding to one site on the protein, they alter the tertiary configuration and subsequent function of a distal site. Proteins can thus exist in various structural conformations and states of activity; their activity can be modified by regulatory molecules which are not direct substrates of the protein. Proteins subject to this form of regulation are termed "allosteric proteins," and the binding molecules are termed "allosteric effectors." All hormones are allosteric effectors. The allosteric proteins to which hormones bind are termed hormone receptors. Although the kinetics of interaction between hormone and receptor may be complex, the interaction can be represented overall as a bimolecular reaction:

$$\text{Hormone} + \text{Receptor} \underset{k_2}{\overset{k_1}{\rightleftharpoons}} \text{Hormone} \cdot \text{Receptor} \quad (8.1)$$

$$[H] \qquad [R] \qquad\qquad [HR]$$

The binding of hormone to receptor protein involves hydrophobic interactions, hydrogen bonds, Van der Waals forces, and salt bridges but not covalent linkages. Hence allosteric regulation can be quickly terminated once the allosteric effector is reduced in concentration. In the mass action equation shown, formation of the active [HR] complex depends on the concentration of both the hormone and the receptor, as well as on the intrinsic affinity of the receptor for the hormone. Classical endocrinology dealt with variations in the concentration of hormone as the primary factor determining formation of the active [HR] complex and, thus, biological activity. When an endocrine gland is ablated and hormone concentration decreases, [HR] decreases, and an endocrine deficiency state ensues; conversely, when excess hormone is present, increased [HR] is formed, and an endocrine excess syndrome results. Extensive control systems regulate the synthesis, secretion, and transport of hormones. These systems control the concentration of hormone in body fluids and, thus, hormone-dependent biological responses. The receptor is equally important in formation of the active [HR] complex. Like hormones, receptors are subject to extensive regulation. Both genetic and acquired disease states involving hormone receptors have been described. Patients with a decrease in a specific hormone receptor manifest a deficiency state despite normal or elevated concentrations of hormones. When receptor concentrations increase, an endocrine excess state may occur despite normal circulating concentrations of hormone.

Receptor proteins contain at least two principal domains: a hormone binding site and an activity site. The hormone binding site demonstrates high specificity and high affinity for the proper hormone ligand. Specificity is sufficient to distinguish between two hormone molecules with only minor structural modifications such as the presence or absence of a hydroxyl group on a steroid molecule or an amino acid change in a peptide. Affinity is high with equilibrium dissociation constants for hormone-receptor interactions of 10^{-8} to 10^{-12} M. Specificity and affinity of receptors are sufficient for recognition of the nanogram to picogram quantities of hormones present per ml of blood (1 part in a million to 1 part in a billion of total protein present). The activity site on receptor proteins couples them directly or indirectly to enzyme systems and biological responses. The activity site serves to distinguish receptors from other proteins, such as the hormone transport proteins of plasma, which have a binding site for recognition but no known biological activity. Hormone receptors are therefore identified by their ability to bind natural and synthetic hormones with high specificity and affinity in the same order and concentration at which these hormones activate target cell biological responses.

The distribution of hormone receptors determines the ability of cells to respond to a particular hormone. Receptors for certain hormones are lo-calized to few cell types, whereas receptors for other hormones are widely distributed. For example, ACTH receptors are localized to the adrenal cortex and, to a lesser extent, to fat cells; only these target cells respond to ACTH. Estrogen receptors are high in uterus, ovary, and breast; these are principal sites of estrogen action. Receptors for cortisol and thyroid hormone are widely distributed, and most cell types respond to these hormones.

Reaction with a specific receptor is the initial event in hormone action. The induced conformational change and altered activity of the receptor then causes a series of biochemical changes which produces the physiological response. At a biochemical level these changes include regulation of enzyme activity, regulation of the biosynthesis and degradation of nucleic acids and proteins, and regulation of the ability of target cells to grow and to replicate. Hormones thus control the differentiated function and growth potential of their target tissues. In general, hormones regulate the degree of expression of the genetic material of a differentiated target cell but do not change the type of genetic information expressed. For example, hormones such as epinephrine, insulin, and cortisol profoundly alter the metabolic activity of liver cells; liver cell-specific functions are, however, always maintained.

There are two principal classes of hormone receptors. Receptors for peptide hormones are located in the cell membrane, with the hormone binding site expressed on the exterior cell surface (Fig. 8.1).

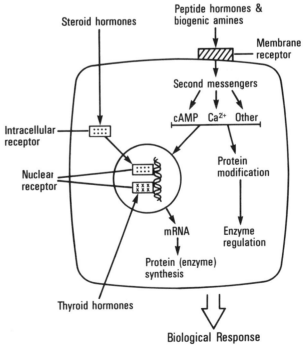

Figure 8.1. Pathways through which hormones function.

Binding of circulating peptide hormones to these cell surface receptors alters the activity of membrane-bound enzymes and transport processes. These in turn change the cellular concentration of regulatory molecules such as cAMP and Ca^{2+}, which then function as the second messenger of hormone action. Enzyme activity is then altered either directly by modifying existing enzyme molecules or by altering the synthesis and concentration of enzyme molecules. Receptors for steroid hormones are located within the cell. Hydrophobic, lipid-soluble steroid hormone molecules diffuse readily into cells from the circulation to bind to specific intracellular receptors. The steroid hormone-receptor complex acts at a nuclear level to initiate the biochemical events specifying the biological effect. Thyroid hormone resembles steroid hormones in its molecular mechanism of action, except that unoccupied thyroid hormone receptors are located in the nucleus, in contrast to unoccupied steroid hormone receptors, which are located in the cytoplasm. Action at a nuclear level alters the biosynthesis and concentration of enzyme molecules. Alterations in enzyme activity via changes in concentration or by modification underlie the biological response of each cell, which is integrated with that of other cell types in the organism, into the physiological response.

CELL SURFACE RECEPTORS FOR PEPTIDE HORMONES

A large number of peptide hormones have been identified, including classical pituitary, pancreatic, and parathyroid hormones, neurohormones, gastrointestinal hormones, cell growth factors, and biogenic amines. Each peptide hormone is recognized by a specific and distinct cell surface receptor which translates hormone recognition into a biological response. Although each hormone receptor is a unique molecular species, there are certain features which are similar among peptide hormone receptors. Peptide hormone receptor proteins are asymmetric with a hormone binding region exposed on the external cell surface, a hydrophobic region which anchors them within the lipid bilayer of the cell membrane, and an activity site which interacts either with other membrane proteins or with the interior of the cell. Peptide hormone receptors are glycosylated, often contain more than one subunit, and require phospholipid or a membrane environment for biological activity. These receptors move within the plane of the cell membrane. Mobility was initially demonstrated for catecholamine receptors by fusing a cell containing β-receptors but no adenylate cyclase (turkey erythrocyte or rat glioma)

with a cell containing adenylate cyclase but no β-receptors (Friend leukemia or mouse Y-1 adrenal cells) (Schramm et al., 1977). In membranes from the resulting hybrid cell, isoproterenol readily stimulated formation of cAMP, indicating that β-receptors and catalytic cyclase units from the two different parents had moved together into a functional complex. Mobility of receptors was directly demonstrated using hormones conjugated to fluorescent indicators (Schlessinger et al., 1978). Cytoskeletal elements which are anchored at the cell membrane may contribute to receptor mobility.

There are approximately 10^4–10^5 receptors on a target cell. These receptors are constantly turning over. The insulin receptor, for example, has a half-life of ~7 h (Krupp and Lane, 1981). After synthesis, receptors are diffusely inserted into the cell membrane. When hormones are bound, they cluster and localize in membrane invaginations (Fig. 8.2). Receptors with bound hormone are then internalized by adsorptive endocytosis. Endocytic vesicles fuse with lysosomes, and the peptide hormone and receptor are largely degraded by lysosomal hydrolases. Some receptors escape degradation and are recycled to insert into the cell membrane along with newly synthesized receptors. Most available evidence indicates that the hormonal signal is transmitted at the cell membrane, although hormone-induced clustering of receptors may in some cases be required for signal generation (Kahn et al., 1978). Internalization is clearly involved in degradation of peptide hormones and peptide hormone receptors. Although it is possible that hormone signals are transmitted after internalization, evidence for this is indirect.

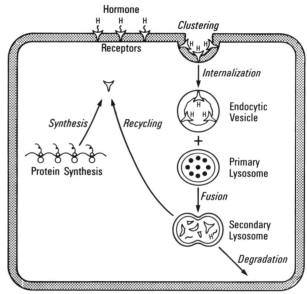

Figure 8.2. Metabolism of peptide hormone receptors.

Peptide hormone receptors are dynamic molecules which are extensively regulated. Regulation of receptors by their homologous hormone results most commonly in downregulation, a decrease in the number of receptors on each cell. Chronic exposure to hormone causes a decrease in receptor concentration which results in decreased sensitivity to circulating hormone (desensitization, tachyphylaxis). Downregulation involves conformational changes in receptors and internalization. Insulin resistance in type II diabetes mellitus and desensitization to repeated administration of epinephrine are examples where this occurs. Conversely, receptor concentrations rise when homologous hormone levels are low, resulting in increased cell sensitivity. Regulation of receptors by heterologous hormones is a major mechanism for integration of responses by various cell types. Examples include FSH induction of LH receptors in ovarian granulosa cells so that LH now stimulates steroid hormone production, thyroid hormone elevation of β-adrenergic receptors so that increased sympathetic activity occurs in thyrotoxicosis without alterations in catecholamine concentrations, and estrogen induction of progesterone receptors in uterus and breast tissues so that enhanced responses occur during pregnancy. Both homologous and heterologous hormones primarily control the concentration of receptors present in target tissues. Less commonly, the affinity of the receptor for the hormone may be altered. Insulin receptors exhibit negative cooperativity, which decreases receptor affinity as occupancy by hormone increases (DeMeyts et al., 1976); membrane active agents such as phorbol esters affect the affinity of epidermal growth factor receptors for their ligand (King and Cuatrecasas, 1982).

The concentration of hormone required to induce the biological response half-maximally is frequently less than the concentration required for half-maximal receptor occupancy (Fig. 8.3*A*). Because full biological responses occur with occupancy of only a fraction of available receptors, the additional receptors are termed "spare." This dissociation between biological response and receptor occupancy is the predicted one when a second mediator is generated which itself binds reversibly to an intracellular receptor (effector) protein (Strickland and Loeb, 1981). Receptors are spare, not for generation of the second messenger (cAMP), but for the biological response elicited by this second allosteric effector. Second messengers thus provide amplification of the initial hormone signal. Such a system with spare receptors provides high sensitivity to changes in circulating hormone concentrations. In systems with spare receptors, a reduction in receptor concentration results in a requirement for more hormone to elicit the same response, i.e., decreased sensitivity but no decrease in the maximum biological effect possible (Fig. 8.3*B*). In systems where receptors are directly coupled to the biological response, a decrease in receptors results in a decrease in the maximal biological response attainable.

INTRACELLULAR MEDIATORS OF HORMONE ACTION

Biological information contained in the active peptide hormone-receptor complex is transmitted inside the cell via second messengers. The concept of second messengers was first proposed by E. W. Sutherland following his landmark discovery of cAMP in 1957 (Sutherland and Rall, 1958). cAMP is the intracellular mediator for a number of hormones, including adrenocorticotropin (ACTH), antidiuretic hormone (ADH), β-adrenergic catecholamines, calcitonin, follicle-stimulating hormone (FSH), glucagon, human chorionic gonadotropin (HCG), luteinizing hormone (LH), melanocyte-

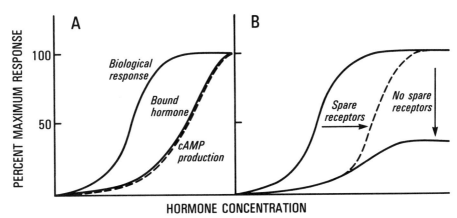

Figure 8.3. *A,* Relationship between receptor occupancy, second messenger production, and biological response. *B,* Effect of decreasing receptor concentration in "spare" and "nonspare" conditions.

stimulating hormone (MSH), parathyroid hormone (PTH), and thyroid-stimulating hormone (TSH). An equally large number of peptide hormones do not use cAMP as an intracellular mediator; these include angiotensin II, α-adrenergic catecholamines, gonadotropin releasing-hormone (GnRH), growth hormone, fibroblast, epidermal and nerve growth factors, insulin and insulin-like hormones, oxytocin, placental lactogen, prolactin, and somatostatin. Whereas the biology of cAMP is well understood, mechanisms through which the latter group of hormones function are incompletely defined.

cAMP

cAMP is formed from ATP by the membrane-bound enzyme adenylate cyclase. Hormone-sensitive adenylate cyclase is made up of three components: the peptide hormone receptor, a coupling unit (G), and catalytic cyclase (C) (Ross and Gilman, 1980) (Fig. 8.4). The hormone receptor does not interact directly with the catalytic cyclase but interacts through the coupling unit (Rodbell, 1980). When hormone binds to receptor, there is rapid interaction of $H \cdot R$ with G to form $H \cdot R \cdot G$ (*Eq. 8.2*).

$$H \cdot R + G \rightleftharpoons H \cdot R \cdot G$$
$$H \cdot R \cdot G + GTP \rightleftharpoons H \cdot R \cdot G \cdot GTP \qquad (8.2)$$
$$H \cdot R \cdot G \cdot GTP + C \rightleftharpoons H \cdot R \cdot G \cdot C$$

GTP
Active Complex

In this form, G becomes activated and preferentially binds GTP. G is active only when GTP is bound; it is inactive when GDP is bound. Formation of $H \cdot R \cdot G$ increases removal of inhibitory GDP and facilitates GTP binding (Cassel and Selinger, 1978). The active $H \cdot R \cdot G \cdot GTP$ complex then binds to the catalytic cyclase to activate enzymatic formation of cAMP.

Reversal of activation of adenylate cyclase results from GTP hydrolysis and reversal of the reactions shown in *Eq. 8.1* and *8.2*. GTP is hydrolyzed to GDP, which induces a confirmation of G unable to bind to C. Cholera toxin, a ubiquitous stimulator of cAMP formation, acts via ADP ribosylation of G to inhibit GTP hydrolysis. By preventing GTP hydrolysis, cholera toxin causes prolonged activation of adenylate cyclase. Once GTP is bound to G, the affinity of the receptor for the hormone is reduced to also favor reversal of the active state.

Hormone responses thus depend not only on the hormone and its receptor but also on the coupling protein G and the cellular concentrations of GTP and GDP. G protein and catalytic cyclase are the same in all hormone-sensitive adenylate cyclase

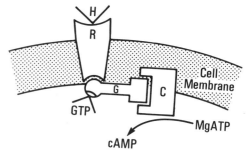

Figure 8.4. Hormone-sensitive adenylate cyclase. [Adopted from Rodbell (1980).]

systems. Molecular differences reside in receptor proteins, but these too may contain a common activity region which binds to G. In pseudohypoparathyroidism, the genetic defect in G protein results in inadequate responses to a number of peptide hormones, including PTH, glucagon, and ADH (Farfel et al., 1980).

By coupling hormone receptors to an enzyme, the initial hormonal signal is greatly amplified. Each hormone molecule stimulates formation of many molecules of cAMP. cAMP, in turn, interacts with a specific intracellular allosteric receptor protein which is coupled to an enzyme to give further amplification. The receptor for cAMP is the regulatory subunit of cAMP-dependent protein kinase (Gill and Garren, 1971) (Fig. 8.5). Binding of cAMP induces a conformation change in the regulatory subunit so that restraints on the catalytic kinase subunit are removed. The active catalytic subunit then catalyzes the transfer of the γ phosphate of ATP to substrate proteins. Because phosphorylation of proteins alters their activity, this covalent modification is a major mechanism through which enzyme activity is modified. Phosphorylation of proteins is catalyzed by a number of kinase enzymes in addition to cAMP-dependent protein kinase; many of these are also regulated by hormones.

There are two isoenzyme forms of cAMP-dependent protein kinase which are similar in all cells. Specific biological effects therefore depend on the genetic expression of a particular cell type, i.e., the potential substrates available for phosphorylation. The first discriminant is the expression of cell surface peptide receptors. Only target cells with specific receptors are activated. The second discriminant is the cell type. Elevations in cAMP in hepatic cells result in glycogenolysis and gluconeogenesis, whereas elevations in cAMP in adrenal cells result in increased formation of steroid hormones. In cells such as fat cells, where multiple hormones such as β-adrenergic agonists, glucagon, and ACTH all increase cAMP formation, stimulation of lipolysis is the common resultant. The long

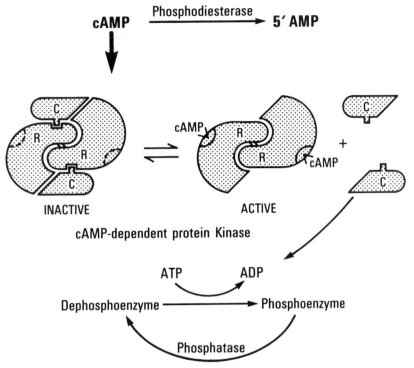

Figure 8.5. Mechanism of action of cAMP.

series of biochemical steps between the hormonal first messenger and the ultimate biological effect provide for quantitative and graded responses, amplification, diversification, and interaction with other regulatory signals.

The hormone-induced increase in cAMP is terminated by degradation of cAMP to 5′-AMP. Although cAMP concentrations are determined primarily by the rate of synthesis, degradative phosphodiesterase enzymes are also regulated. Drugs such as methyxanthines, which are inhibitors of phosphodiesterase, are frequently used clinically to augment cAMP responses. Phosphate groups are removed from proteins by phosphatase enzymes which are themselves subject to regulation.

Calcium and Other Mediators

Mechanisms of action are less well defined for peptide hormones which do not stimulate adenylate cyclase. Three principal mechanisms have been identified: calcium-dependent processes, activation of membrane-bound enzymes, and cGMP. Other important pathways will undoubtedly be discovered because none of these three adequately explains the action of all hormones.

Calcium plays a regulatory role in muscle contraction, neuromuscular transmission, secretion, and hormone action. Hormone-receptor interactions alter intracellular calcium concentrations by facilitating calcium uptake and redistribution from

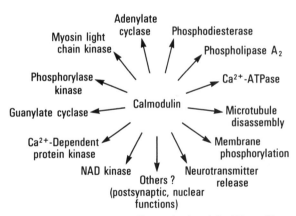

Figure 8.6. The biological effects of calmodulin. [From Cheung (1980).]

cellular storage sites. Intracellular calcium concentrations are low so that large changes in cellular concentration occur when calcium is transported from either extracellular or sequestered intracellular pools. One intracellular receptor for calcium is the protein calmodulin, which is closely related to troponin C, the calcium binding protein involved in muscle contraction (Means and Dedman, 1980). Occupancy of calcium binding sites results in a conformational change in calmodulin so that this regulatory protein interacts with a number of cellular enzymes to alter their activity (Cheung, 1980) (Fig. 8.6). Calcium-calmodulin activates myosin light chain kinase to stimulate contraction in smooth muscle and other cells. Calcium-calmodulin stimulates phosphorylase kinase to increase glycogen breakdown. The actions of angiotensin II, α-adrenergic catecholamines, and vasopressin to stimulate hepatic glycogenolysis may occur via this mechanism. Calcium-calmodulin activates cyclic nucleotide phosphodiesterase and guanylate cyclase; it also activates adenylate cyclase in some cell types. Calcium-calmodulin activates membrane-bound enzymes, such as phospholipase A_2, and is an important component of the mitotic apparatus, where it increases microtubule disassembly. Calcium-calmodulin affects additional cellular processes, suggesting that a number of enzymes have evolved with calmodulin recognition sites. As with cAMP-dependent protein kinase, which catalyzes phosphorylation of a number of substrates, calmodulin interacts with a number of targets to induce diverse cellular effects dependent on the genetic expression of each cell type. Calcium·calmodulin interactions with cyclic nucleotide systems provides a mechanism for intracellular communication.

Several membrane-bound enzymes are activated by hormone receptors. Epidermal growth factor stimulates a membrane-bound protein phosphokinase which catalyzes phosphorylation of proteins at tyrosine residues (Ushiro and Cohen, 1980). Because the transforming proteins of several RNA tumor viruses are also tyrosine-specific protein kinases and because several tumor-produced growth factors also stimulate tyrosine-specific kinase activity, this class of enzymes may mediate cell growth (Bishop, 1980; Todaro et al., 1980). Insulin also stimulates tyrosine protein kinase activity associated with the insulin receptor (Kasuga, 1982). A membrane-associated protein phosphokinase which is dependent on calcium, phosphatidylserine, and unsaturated diacylglycerol derived from phosphatidylinositol turnover also appears activated by peptide hormone binding (Nishizuka et al., 1981). β-Adrenergic catecholamine-receptor interaction increases methylation of membrane phosphatidylethanolamine; this increases membrane fluidity and may initiate changes in activities of other membrane enzymes (Hirata and Axelrod, 1980). Insulin stimulates a membrane bound protease to generate small peptides which have biological activity within the cell (Popp et al., 1980). Hormone·receptor interactions likely directly affect additional membrane-associated enzymes, as well as exerting effects on cellular transport processes.

cGMP rises within target cells stimulated by muscarinic cholinergic agents, oxytocin, and several other hormones. A metabolic system similar to that for cAMP exists, consisting of guanylate cyclase,

cGMP-dependent protein kinase, and phosphodiesterase. The exact biological role of cGMP, however, remains uncertain because few specific biochemical processes subject to regulation by this system are known (Gill and McCune, 1979).

INTRACELLULAR RECEPTORS FOR STEROID AND THYROID HORMONES

Steroid Hormones

Steroid hormones are made in adrenal and gonadal tissues from cholesterol. The adrenal cortex produces cortisol, aldosterone, and androgen precursors; the ovary produces estrogen and progesterone; and the testis produces testosterone. Vitamin D is also synthesized from precursor cholesterol and may be considered a steroid hormone which regulates mineral metabolism. Although each steroid hormone has a basic cholesterol nucleus to which different modifications and substitutions have been made, the receptor proteins for each possess sufficient specificity and affinity to provide recognition for a single class of steroid hormones.

In contrast to peptide hormone receptors which are located in cell membranes, receptors for steroid hormones are found within the cell (Fig. 8.7). Hydrophobic steroid hormones are soluble in the lipid environment of the cell membrane and likely enter cells by passive diffusion. They are retained within cells by binding to specific receptor proteins; cells

Steroid Hormone Responsive Cell

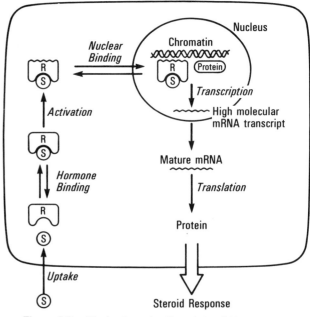

Figure 8.7.   Mechanism of action of steroid hormones.

which contain these receptors are the target for that particular steroid hormone, changing their function in response to the hormone. Steroid receptor proteins are asymmetric molecules present at 3000–100,000 copies/target cell. Biological responses are proportional to the number of receptors with bound steroid hormone, suggesting that there are no intermediate second messengers, as with polypeptide hormones. Some of the steroid hormone receptors, such as those for progesterone, are composed of subunits.

After binding the steroid hormone with high specificity and affinity, the receptor protein is activated. Activation, which requires heat in addition to substances present in cytoplasm, results in a physical alteration in the receptor so that it binds tightly in the nucleus to chromatin, which is the site of action of steroid hormones. After activation, steroid hormone-receptor complexes have different charges, sizes, and affinities for nuclei. The binding of activated steroid hormone-receptor complexes to chromatin results in changes in the production of specific messenger RNAs which in turn direct synthesis of enzyme proteins which catalyze the reactions which underlie the physiological effects.

The activated steroid hormone-receptor complex specifically stimulates the rate of transcription of messenger RNAs (Ringold et al., 1977). The change from the basal to the stimulated rate of transcription of specific messenger RNAs may occur as rapidly as within 5–15 minutes (Swaneck et al., 1979) and involves increased synthesis of high molecular weight nuclear messenger RNA precursors containing both exons present in mature messenger RNA and introns which are removed during nuclear processing to the mature messenger RNA. Steroid hormone effects on messenger RNA processing and stability may also occur but are quantitatively less important. Because less than 1% of the messenger RNAs of the cell are altered, the initial effects of steroid hormones on gene expression are specific. This specificity may arise from specific chromosomal proteins which bind the activated steroid hormone-receptor complex and from specific DNA sequences which are recognized by the activated receptor. Secondary effects may then arise from the increases in specific enzyme proteins specified by these initial messenger RNAs.

Specific regions of DNA have been identified as the sites for steroid hormone receptor action. In well-studied cases, these are located 5′ to the hormone-inducible gene (Groner et al., 1983). These DNA sequences can be linked to other genes with the result that these genes then come under steroid hormone regulation.

The most common effect of steroid hormones is to increase the synthesis of messenger RNA. This has been well documented in many cases, for example, in cortisol induction of tyrosine aminotransferase, tryptophane oxygenase, and growth hormone; estrogen induction of ovalbumin, vitellogenin, and prolactin; and progesterone induction of avidin, α_2-globulin, and aldolase. Even when the biological effect is an inhibitory one, as in cortisol-induced inhibition of the inflammatory response, specific proteins are induced (Blackwell et al., 1980). Steroid hormones may decrease mRNA synthesis as well. The inhibitory effect of cortisol on production of the messenger RNA for the ACTH precursor is a well-studied example (Nakanishi et al., 1977).

Although receptors exhibit a high degree of specificity for their particular steroid, crossover may occur. For example, aldosterone receptors may be occupied by cortisol so that higher concentrations of the latter hormone cause sodium retention similar to that induced by aldosterone. Steroid hormone receptors may also be occupied by antagonists as well as by agonists. Antagonist occupancy of binding sites induces an inactive receptor configuration. For example, spironolactone, a mineralocorticoid antagonist, forms a spironolactone-mineralocorticoid receptor complex which does not bind to nuclear components (Marver et al., 1974). Tamoxifen, an estrogen antagonist, allows nuclear binding, but the receptor is not recycled and replenished as it is when estrogen is bound to the receptor (Clark et al., 1978).

Thyroid Hormone

Like steroid hormones, thyroid hormone acts by binding to specific intracellular receptors. The intracellular receptors for thyroid hormone are located in the nucleus bound to chromatin (Fig. 8.1). Unlike steroid hormone receptors, thyroid hormone is not required for receptors to bind to chromatin (Baxter and Funder, 1979). Nuclear thyroid hormone receptors bind T_3, the metabolically active form of thyroid hormone, with an affinity greater than that for T_4, the precursor of T_3. Thyroid hormone occupancy of nuclear receptor sites increases expression of genetic information by increasing the synthesis of specific messenger RNAs which code for enzyme proteins specifying the biological response. The pattern of thyroid hormone-induced proteins depends not only on the particular target cell but also on the nutritional status of the animal (Oppenheimer and Dillmann, 1979). This provides integrated appropriate responses among various hormonal signals.

BIOSYNTHESIS AND
METABOLISM OF HORMONES

Although regulation of hormone synthesis and metabolism will be considered in chapters dealing with individual endocrine glands, general features of these processes, which are common to all hormones of the peptide and steroid classes, will be considered here. Thyroid hormones and biogenic amines will be specifically covered in Chapters 53 and 54.

Peptide Hormones

Because peptide hormones are secretory proteins, their biosynthesis and secretion follow pathways of protein synthesis and packaging common to secretory proteins. The genetic information in DNA which codes for peptide hormones is discontinuous with intervening sequences between regions which code for protein (Fig. 8.8). The gene is initially transcribed into a large molecular weight messenger RNA precursor. Intervening sequences (introns) are progressively removed, and the coding portions (exons) are spliced together into the mature messenger RNA, which is capped and tailed. This genetic arrangement is thought to have facilitated evolution through recombination. Differential splicing also provides for generation of more than one

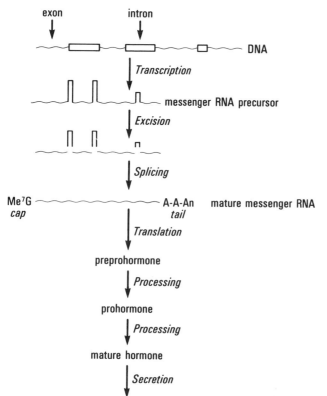

Figure 8.8. Biosynthesis of peptide hormones.

hormone from a single gene region. The calcitonin gene product is differentially spliced in the C cells to yield calcitonin and in the hypothalamus to yield a different peptide product (Amara et al., 1982). The major regulation of protein hormone biosynthesis by steroid hormones, polypeptide hormones, and other environmental signals is through increases or decreases in the rate of transcription of the gene into messenger RNA.

Messenger RNA is transcribed into protein which is in some cases considerably larger than the final mature circulating hormone. There is a "pre" sequence, coded from the 5′ end of the messenger RNA, which contains many hydrophobic amino acids. This leader or signal sequence is characteristic of secretory proteins and specifies vectoral synthesis of the protein into the inside of the membrane of the endoplasmic reticulum (Lingappa and Blobel, 1980). The signal sequence is removed within the membrane space. The remaining prohormone is transferred within the cell inside the membranes of the endoplasmic reticulum to the Golgi complex, which packages the protein into secretory granules. The prohormone is converted by further proteolytic activity to the mature hormone within the Golgi and secretory granules. The extent and location of the pro-protein removed varies among different peptide hormones. In proinsulin the connecting, or "C", peptide which joins the A and B chains and facilitates proper disulfide bonding is removed from the middle of the prohormone, and mature insulin containing two peptide chains is secreted (Chan et al., 1976). ACTH is processed from a large pro-form which contains information for lipotropin, melanocyte-stimulating hormone, and endorphins on the carboxy-terminal and information which may affect steroid synthesis on the amino-terminal side of ACTH (Mains et al., 1977). The pro-form of smaller peptide hormones may thus contain sequences for more than one biologically active peptide hormone which may be generated during proteolytic processing of the prohormone. Certain larger polypeptide hormones such as prolactin and growth hormone have little or no pro-sequences.

Secretory granules fuse with the plasma membrane to discharge the mature peptide hormone into the circulation. This process is under hormonal regulation and requires calcium. When a producer cell is highly stimulated, secretory granules which store hormone are largely depleted; conversely, when the cell is quiescent, secretory granules accumulate.

Peptide hormones are short lived within the circulation. Small neurohormones are so rapidly degraded that significant concentrations are reached only in the hypothalamic-pituitary portal circulation; insulin has a half-life of about 7 min and ACTH has a half-life of 3 min; the glycoprotein hormones such as TSH, LH, and FSH have the longest half-lives, ranging up to 180 min for FSH. Peptide hormones are destroyed by circulating proteases, via lysosomal proteases after receptor binding, and for the glycoproteins after binding to the hepatic asialoglycoprotein receptor. It is evident that continuous synthesis and secretion must occur to maintain circulating concentrations of these short-lived species. It is also evident that a form of continuous parenteral administration is required when peptide hormones are given to replace glandular secretion.

Steroid Hormones

All steroid hormones are derived from cholesterol. Many of the modifications of the cholesterol nucleus are common between classes of steroid hormones, although sufficient differences exist for specific recognition by receptors and consequent specific biological action. Cholesterol is obtained either from the circulation via low density lipoprotein uptake (high density lipoprotein in some species) or via biosynthesis from acetate. Steroidogenic tissues require large amounts of cholesterol and have increased numbers of LDL receptors and possess many lipid droplets containing cholesterol esterified to fatty acids. The principal modification of cholesterol precursor involves hydroxylations which require NADPH and molecular oxygen and which are catalyzed by specific cytochrome P-450 enzymes. Dehydrogenase, isomerase, and aromatase enzymes also modify the cholesterol nucleus.

In all steroidogenic tissues (adrenal cortex, ovary, testis), the rate-limiting step in steroid hormone biosynthesis which is under trophic hormone (ACTH, LH, FSH) control lies in the conversion of cholesterol to pregnenolone (Fig. 8.9). This step involves the mitochondrial cytochrome P-450 enzyme which hydroxylates and cleaves the side chain of cholesterol. Once this slow step is speeded up, the flow of substrate through the biosynthetic enzyme pathway is rapid. Steroidogenic tissues contain large numbers of specialized mitochondria and an extensive network of endoplasmic reticulum. Substrates flow from mitochondria to endoplasmic reticulum and back to mitochondria as the molecule is progressively modified. Modifications depend on the particular biosynthetic enzymes which are expressed in each gland. For example, 11β-hydroxylase is largely localized to the adrenal cortex while aromatase is largely localized to the ovary. In contrast to peptide hormones, steroid hormones are not stored but are secreted as synthesized. Increased secretion therefore directly represents increased synthesis.

After secretion into the circulation, steroid hormones are bound to transport proteins. The transport glycoproteins are synthesized in the liver and bind most circulating steroid hormones. Corticosteroid binding globulin (CBG or transcortin) binds more than 90% of circulating cortisol, and sex steroid binding globulin binds more than 90% of circulating testosterone and lesser amounts of estradiol. Aldosterone is weakly bound to albumin rather than to a specific transport protein. The free unbound fraction of circulating steroid is the active species which enters cells to bind to specific receptor proteins. Plasma binding provides a reservoir of hormone protected from metabolism and renal

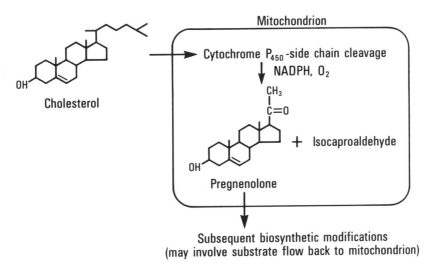

Figure 8.9. The rate-limiting step of steroid hormone biosynthesis.

clearance which can be released to cells. The circulating half-life of steroid hormones is thus significantly longer than that of peptide hormones. Transport proteins which exhibit specific binding are not, however, required for the action of steroid hormones. Genetic defects in transport proteins are compatible with normal endocrine function, and a number of active synthetic analogs bind poorly to transport proteins.

Metabolism of steroid hormones occurs principally in the liver and results in molecules which are biologically inert. These are conjugated with sulfate or glucuronide to render them water soluble for excretion. The principal exception to the general rule that metabolism leads to inactivation is testosterone. In target tissues testosterone is converted to dihydrotestosterone by the enzyme 5α-reductase. This conversion is required in many target tissues in which dihydrotestosterone is the active androgen species preferentially bound by receptors.

ORGANIZATION OF ENDOCRINE CONTROL SYSTEMS

One principal function of the endocrine system is to maintain internal homeostasis despite changes in the external environment. The second principal function is reproduction. In both of these processes, multiple hormones cooperate to bring about appropriate biochemical and physiological responses. For example, the primitive signal of glucose (substrate) lack has expanded to the broader signal of stress or fright and evokes a coordinated neural and endocrine response (Tomkins, 1975). Increases in glucagon and catecholamines stimulate glycogenolysis to release glucose from the liver into the circulation for immediate use by critical organs such as the brain. Cortisol is produced in greater amounts by the adrenal cortex and induces gluconeogenesis so that hepatic glucose production is maintained from muscle and other body proteins. Lipolysis is stimulated by a number of hormones, including epinephrine, glucagon, ACTH, and growth hormone, and free fatty acids are released into the circulation to be used as an alternate metabolic fuel. Insulin, which signals glucose utilization, is suppressed. This metabolic adaptation is coordinated with a number of physiological and behavioral changes, such as increased sympathetic nervous system activity causing peripheral vasoconstriction and increased blood pressure.

Reproduction requires coordinated endocrine changes. During the menstrual cycle the ovarian follicle develops in response to the pituitary hormones LH and FSH so that ovulation can occur during the midcycle surge of LH. The uterus undergoes coordinated changes in response to ovarian estrogen and after ovulation to ovarian progesterone produced by the corpus luteum. This allows preparation of a uterine implantation site for a fertilized egg. Sexual behavior is strongly affected by changes in endocrine function.

Mechanisms which underlie these coordinated responses are many and include neural input to the endocrine system as well as hormonal modulation of nervous function, regulation of hormone biosynthesis and secretion, alterations in hormone receptors, and changes in metabolism of hormones.

During evolution many endocrine glands became distant from the nervous system. A hierarchy of neuroendocrine control developed, with the hypothalamus serving as the proximal neural signaling station to produce neurohormones which reach the pituitary gland via a short portal circulation (Fig. 8.10). The anterior pituitary produces a number of metabolic and reproductive peptide hormones which are trophic, i.e., which control the production of other hormones by distal endocrine glands. This arrangement provides for integration with the nervous system and for an expansion and modulation of the effects of the initial signal.

Once a hormonal response has been initiated and an appropriate metabolic or reproductive response has occurred, the signal must be terminated. Feedback control is one of the principal mechanisms through which this occurs. As shown in Figure 8.10,

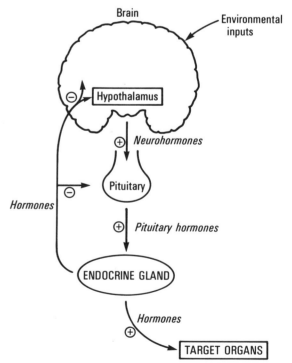

Figure 8.10.   Feedback control of hypothalamic-pituitary-peripheral endocrine gland function.

the hormonal product of the peripheral endocrine gland, such as the thyroid, adrenal cortex, testis, or ovary, exerts negative feedback control over the production and secretion of the stimulatory pituitary hormone. Feedback occurs at the level of the pituitary producer cell and likely at the level of the central nervous system and hypothalamus as well. Conversely, when concentrations of the peripheral endocrine gland hormone are low, feedback inhibition is lessened so that increased production of pituitary hormone occurs to stimulate production of the lowered hormone—and thus reestablish homeostasis. Feedback control occurs as well in endocrine glands not under pituitary control. For example, an elevation in serum Ca^{2+} feedback inhibits production of parathyroid hormone. A de-

crease in serum Ca^{2+} concentration results in elevated PTH which acts on bone, kidney, and gastrointestinal tract to reestablish a normal serum Ca^{2+} concentration.

The sophisticated endocrine control system which will be considered in subsequent chapters evolved from more primitive systems whose function was similar—regulation. cAMP is found in bacteria and serves as a symbol of glucose lack. Polypeptide hormones such as insulin and ACTH are made by paramecia. These primitive signals have evolved into the more complex systems necessitated by organ specialization. This communication system coordinates behavior with metabolism and reproduction in a manner most conducive to survival of the individual and species.

CHAPTER 52

The Hypothalamic-Pituitary Control System

INTRODUCTION

The pituitary is the proximal endocrine gland link to the central nervous system for neuroendocrine control of both metabolism and reproduction. Neurohormones synthesized in the hypothalamus reach the anterior lobe of the pituitary via a specialized portal vascular system. These neurohormones control synthesis and secretion of the six major peptide hormones of the anterior pituitary; these anterior pituitary hormones regulate peripheral endocrine glands (thyroid, adrenal, gonads), as well as growth and lactation. The posterior lobe of the pituitary is composed of axons derived from neuronal cells of the hypothalamus and serves as a storage site for two peptide hormones which are synthesized in the hypothalamus but which act in the periphery to regulate water balance, uterine contraction, and milk ejection.

Isolation, characterization, and synthesis of neurohormones has been a formidable task because these peptides are synthesized in only a small number of cells which are distributed in several areas of the hypothalamus. Only small amounts of specific neurohormones are produced by these cells, and significant concentrations occur only in local areas such as the hypophyseal portal vascular system. Collection of over 1 million animal hypothalami and a Herculean labor of protein isolation were required to yield the presently known neurohormones (Table 8.1) (Burgus et al., 1969; Schally et al., 1971; Brazeau et al., 1973; Vale et al., 1981; Guillemin et al., 1982; Rivier et al., 1982). Because these are small peptides, once the structure was known, chemical synthesis yielded large amounts of both natural and modified neurohormones, and by using immunological and pharmacological techniques, it was found that these hypothalamic neurohormones are produced in the periphery as well as in the hypothalamus and function in local paracrine systems, especially in the gastrointestinal tract; they also function as specialized neurotransmitters in the central nervous system.

Identified neurohormones and their target cells in the anterior pituitary are shown in Figure 8.11. Two patterns of action are evident: first, control of more than one target cell by a single neurohormone and, second, both positive and negative regulation of particular anterior pituitary cell types. For example, thyrotropin releasing hormone (TRH) stimulates production and secretion of both thyroid-stimulating hormone (TSH) and prolactin. Under pathological conditions TRH may also stimulate production and secretion of growth hormone. Somatostatin (SS) also affects more than one cell type, exerting negative control over growth hormone and TSH synthesis and secretion. Growth hormone-producing cells provide an example of dual control: these cells are stimulated by growth hormone-releasing hormone (GRH) and are inhibited by SS. The rate of production of growth hormone depends on the relative strength of these two stimuli.

Neurohormones are peptides, except for the biogenic amine dopamine, which is the principal prolactin-inhibiting hormone. Regulation of most anterior pituitary function depends primarily on positive stimulatory signals from the hypothalamus. When the anterior pituitary is removed from its hypothalamic connections and transplanted in a distal site such as the kidney capsule, production of all anterior pituitary hormones except prolactin ceases. Increased prolactin production indicates that dopamine is the principal regulator of lactotrophs because when the negative influence of dopamine is removed, prolactin production increases.

ANATOMY

Neurohormones are synthesized in specialized neurosecretory cells which are concentrated within the hypothalamus. Numerous nerve fibers termi-

Table 8.1
Hypothalamic-releasing hormones

Neurohormone	Structure*	Pituitary Site of Action	Effect	Extrahypothalamic Synthesis	Extrapituitary effects
Thyrotropin releasing hormone (TRH)	PyroEHPNH₂	Thyrotroph Prolactotroph	Stimulate Stimulate	Brain and GI tract	Yes
Growth hormone-releasing hormone (GRH)	YADAIFTKSYRKVLGQLSARKLLQDI-MSROOGESNQERGARARL	Somatotroph	Stimulate	Pancreas	
Somatostatin (SS)	AGCKNFFWKTFTSC	Somatotroph Thyrotroph	Inhibit Inhibit	Brain, pancreas, and gastrointestinal tract	Yes
Gonadotropin-releasing hormone (GnRH)	PyroEHYSYGLRPGNH₂	Gonadotroph	Stimulate	Gonads	Yes
Corticotropin-releasing hormone (CRH)	SQEPPISLDLTFHLLREVLEMTKAD-QLAQQAHSNRKLLDIANH₂	Corticotroph	Stimulate	Pancreas	Yes
Dopamine	$HO\!-\!\langle\rangle\!-\!CH_2CH_2NH_2$ HO	Prolactroph	Inhibit	Adrenal medulla and nervous system	Yes

* Single letter symbols for amino acids are used.

$CH_2CH_2NH_2$

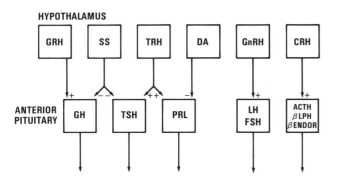

Figure 8.11. Regulation of anterior pituitary cell function by hypothalamic neurohormones. DA, dopamine; GH, growth hormone.

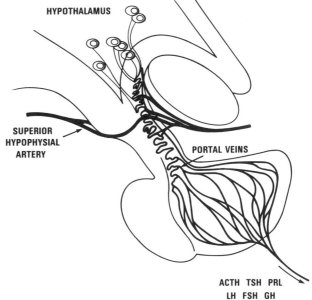

Figure 8.12. Functional connection of hypothalamus and anterior pituitary via the hypophyseal portal vascular system.

nate around these cells to bring information from other parts of the nervous system important for regulating production of neurohormones. The hypothalamus, which is located below the thalamus, lies behind the optic chiasm, between the optic tracts and anterior to the mammillary bodies. Anatomic localization of neurosecretory cells which produce a specific neurohormone is not precise, except for the large cells of the supraoptic and paraventricular areas, which synthesize antidiuretic hormone (ADH) and oxytocin, which are stored in the posterior lobe of the pituitary gland.

Neurohormones reach the anterior pituitary gland via a specialized portal vascular system. Nerve fibers originating in neurosecretory cells terminate in the median eminence of the hypothalamus located below the third ventricle, and the highest concentrations of hypothalamic neurohormones are found there. Blood supply to the anterior pitui-

tary is through branches of the internal carotid artery, principally the superior hypophyseal artery (Fig. 8.12). Branches of the superior hypophyseal artery form a capillary plexus in the median eminence and upper portion of the pituitary stalk. The region of the median eminence has a poorly developed blood-brain barrier so that blood which reaches the anterior pituitary via the portal vessels has first been in contact with the nervous tissue of the median eminence. The portal vessels, which arise from this capillary network and carry neurohormones, terminate in sinusoidal capillaries in the anterior lobe of the pituitary. This specialized circulation with two capillary networks delivers physiologically significant concentrations of neurohormones to their target cells in the anterior pituitary, but the concentrations of neurohormones which reach the peripheral circulation are insignificant. Retrograde flow may also occur via the hypophyseal portal system so that anterior pituitary hormones can also reach the hypothalamus.

In contrast to the anterior lobe of the pituitary, which is connected to the hypothalamus via the hypophyseal portal vascular system, the posterior lobe of the pituitary is an elongated extension of the ventral hypothalamus. Neurons from neurosecretory cells in the supraoptic and paraventricular nuclei of the hypothalamus terminate partially on the median eminence and primarily in the posterior lobe of the pituitary, where secretory granules containing ADH or oxytocin are in proximity to adjacent capillaries. Secretion of posterior pituitary hormones occurs in response to nerve impulses originating in hypothalamic cell bodies.

The pituitary gland which weighs approximately 500 mg is centrally located in the sella turcica, a bony cavity within the sphenoid (Fig. 8.13). The anterior lobe, which comprises about 80% of glandular weight, is derived from oral ectoderm, whereas the posterior lobe is derived from an outpouching of neural tissue from the region of the third ventricle. The pituitary stalk extends to the hypothalamus through a dural reflection. An intermediate lobe located between anterior and posterior lobes is present in certain species and during fetal development, but it is vestigial in adult man. Because surrounding structures are vital ones, expansion of the pituitary due to tumor formation can result in superior extension with compression of the optic chiasm and visual loss or lateral extension with compression of cavernous sinuses containing cranial nerves.

THE ANTERIOR PITUITARY

Adrenocorticotropin (ACTH) and Related Peptides

BIOSYNTHESIS AND STRUCTURE

ACTH, the principal regulator of the adrenal cortex is a 39-amino acid single chain polypeptide hormone which is derived by proteolytic processing from a larger precursor protein made in corticotrophic cells of the anterior pituitary (Eipper and Mains, 1980). The structure of the precursor protein deduced from the nucleotide sequence of cloned DNA complementary to messenger RNA of the

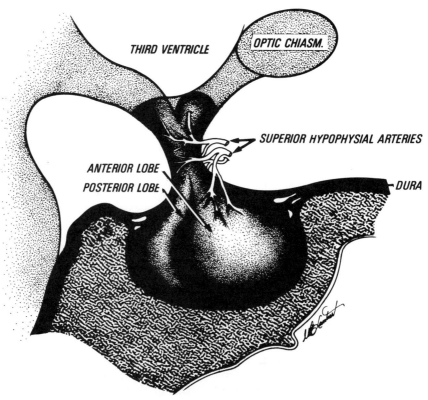

Figure 8.13. Anatomy of the sella turcica and surrounding structures (sagittal view).

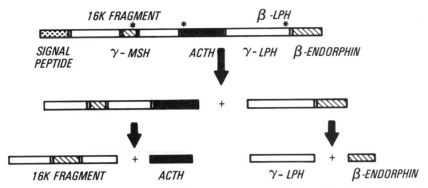

Figure 8.14. Structure of the ACTH precursor molecule. MSH core sequences are indicated by *. γ-MSH is located within the 16,000 M_r amino terminal fragment; α-MSH is located at the amino terminus of ACTH; β-MSH is located at the carboxy terminus of γ-LPH. Dibasic amino acid residues which serve as the recognition site for proteolytic processing enzymes are indicated by *open spaces* in the bars. [Adapted from Nakanishi et al. (1979) and Eipper and Mains (1980).]

bovine species is shown in Figure 8.14 (Nakanishi et al., 1979). A 26-amino acid signal sequence containing a middle region rich in hydrophobic residues directs synthesis of the precursor protein into membrane structures for ultimate secretion. The core sequence for melanocyte-stimulating hormone (MSH) is reiterated within the structure of the precursor, suggesting that it arose during evolution by gene duplication. A 16,000 M_r fragment is located on the amino-terminal side, and β-lipotropin is located on the carboxy-terminal side of the ACTH sequence. Dibasic amino acid residues serve as recognition signals for proteolytic enzymes which cleave the precursor protein into the final hormonal products. The first step is separation of the β-LPH containing carboxy-terminus. The amino-terminus then undergoes further processing to yield mature ACTH, and the 16,000 M_r fragment, whereas β-LPH is processed into γ-LPH and β-endorphin. Although β-endorphin contains the sequence for met-enkephalin, β-endorphin is the final product which accounts for opiate agonist activity derived from this pathway.

Biological activity resides in the first 20 amino acids of the ACTH sequence, which are identical in all mammalian species studied (Fig. 8.15) (Riniker et al., 1972). Species variations occur in the carboxy-terminal 15 amino acids which are not required for biological activity. The amino-terminal 13 amino acids are essential for biological activity, and the sequence of basic amino acids lys[15]-lys[16]-arg[17]-arg[18] is important for binding to adrenocortical receptors (Gill, 1979). Residues 1–13 of ACTH correspond to the sequence of α-MSH and account for the intrinsic skin darkening effects of the ACTH molecule.

Synthesis and secretion of ACTH are regulated by corticotropin-releasing hormone (CRH), a 41-amino acid peptide produced in hypothalamic neu-

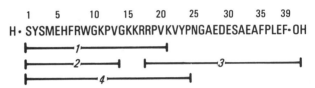

Figure 8.15. The amino acid sequence of human ACTH. Numbered segments indicate: *1*, the first 20 amino acids which possess full biological activity; *2*, sequence of α-MSH; *3*, sequence of CLIP; *4*, the major synthetic ACTH available for clinical use. Single letter symbols for amino acids are used.

rosecretory cells located near the supraoptic and paraventricular nuclei (see Fig. 8.32, Chapter 54) (Vale et al., 1981). CRH has significant sequence homology with Sauvagene, a peptide isolated from frog skin which was previously found to have CRH activity (Erspamer and Melchiorri, 1980). ACTH secretion is also stimulated by ADH, and under certain conditions CRH and ADH synergize to give maximal ACTH production and secretion (Yasuda et al., 1982). CRH binds to receptors on ACTH-producing cells to stimulate formation of cAMP (Labrie et al., 1982) and, as expected, CRH stimulates coordinate production and secretion of both ACTH and β-endorphin. Feedback inhibition of ACTH and β-endorphin production by cortisol occurs at the level of the corticotroph cell of the anterior pituitary where CRH and cortisol have opposing effects (Allen et al., 1978). Glucocorticoids decrease CRH-stimulated cAMP formation (Bilezikjian and Vale, 1982) and production of messenger RNA for the ACTH precursor protein (Nakanishi et al., 1977). Inhibition of hypothalamic CRH production by cortisol has also been implied from indirect studies (Dallman, 1979).

The intermediate lobe of the pituitary is composed almost entirely of cells which synthesize the ACTH precursor protein. In the intermediate lobe, ACTH is processed to α-MSH in which the amino-terminal serine is acetylated and the amino-terminal valine is amidated from the adjacent glycine at residue 14 and to corticotropin-like intermediate peptide (CLIP, $ACTH_{18-39}$). Processing of the precursor protein thus varies between anterior and intermediate lobes, yielding ACTH in the anterior pituitary and α-MSH and CLIP in the intermediate lobe. Regulation also differs. The intermediate lobe is controlled by neural inputs, including dopamine as an inhibitory neurohormone, but responds poorly to CRH and glucocorticoids. Although the intermediate lobe is vestigial in man after fetal life, some cases of Cushing's disease with overproduction of ACTH are resistant to suppression with glucocorticoids but suppress with dopamine, a pattern similar to that which occurs in intermediate lobe cells (Lamberts et al., 1982).

THE CRH·ACTH·CORTISOL AXIS

The response to stress which involves multiple hormones, including ACTH, catecholamines, ADH, angiotensin, glucagon, and growth hormone, is integrated by the nervous system. The CRH·ACTH·cortisol axis is central to these integrated responses to stress. The most primitive signal was substrate lack, and the gluconeogenic actions of cortisol are essential for adaptation to fasting. The signal has become more complex, and this axis has become an important part of a general neuroendocrine response to a variety of noxious environmental stimuli such as pain, trauma, and hypovolemia (Gann et al., 1978). CRH neurosecretory cells are regulated by a number of different afferent nervous impulses which arise in various regions of the nervous system and go through several synaptic intermediates prior to connection with CRH neurons. Several neurotransmitters, including norepinephrine, serotonin, acetylcholine, dopamine, and GABA, act on intermediates in this complex pathway. When CRH is released, it activates the sympathetic nervous system, in addition to stimulating ACTH biosynthesis and secretion (Brown et al., 1982). Centrally administered CRH increases blood pressure, heart rate, and behavioral responses characteristic of stress (Fisher et al., 1982). Activation of the sympathetic nervous system includes the adrenal medulla and its catecholamines. In response to CRH, both ACTH and β-endorphin are released in equimolar quantities (β-endorphin is present in secreted β-LPH and as mature β-endorphin). ACTH acts principally on the adrenal cortex to stimulate cortisol production, although ACTH also stimulates lipolysis in fat cells. Cortisol, through its actions on target organs, is the major regulator of adaptive responses to stress (Chapter 54). The coordinate actions of cortisol and catecholamines mobilize substrate to maintain blood glucose and energy metabolism during the period of stress. Cardiovascular and behaviorial responses are integrated with these metabolic adaptations in response to the CRH signal. Fluid volume and circulation are maintained in concert. ADH potentiates CRH actions on ACTH secretion and causes renal water conservation. ACTH and angiotensin increase adrenocortical aldosterone synthesis to enhance salt and water retention. These responses mediated through the hypothalamic-pituitary-adrenal-sympathetic nervous system allow survival under numerous adverse conditions.

The physiological role of the regions of the precursor protein, other than that of ACTH, are unclear. Skin pigmentation which accompanies states of ACTH excess is due to the intrinsic MSH activity of the ACTH molecule ($ACTH_{1-13}$) and to β-LPH and γ-LPH because β-MSH does not appear as a product of the ACTH precursor molecule in man. β-Endorphin does not exert analgesia in the periphery but may exert central effects, and ACTH-β-endorphin-containing cells are located within the nervous system (Krieger et al., 1980). Products of the ACTH precursor protein may also affect nervous system responses. Both the amino- and carboxy-terminal regions of the ACTH precursor affect adrenal steroidogenesis, but whether this occurs under physiological conditions is uncertain (Chapter 54).

Glycoprotein Hormones

The pituitary glycoprotein hormones, thyroid-stimulating hormone, luteinizing hormone (LH), follicle-stimulating hormone (FSH), and the placental hormone, chorionic gonadotropin (CG), are each composed of two subunits (α and β). The α subunits of all four glycoprotein hormones have the same amino acid sequence, whereas the sequences of the β subunits differ. Isolated subunits are inactive but can be combined to give fully active molecules which have the biological specificity of the β subunit (Pierce et al., 1971). Branched carbohydrate side chains are attached to α and β subunits (Pierce and Parsons, 1981). The α subunits each contain two oligosaccharides; β subunits contain one or two oligosaccharides. HCG of placental origin resembles LH with a 30 amino acid extension at the carboxy-terminus where additional carbohydrate is attached. The carbohydrate substitutions contribute to both the folding of α and β subunits into mature hormone (Weintraub et al., 1980) and to the relatively long circulating half-lives of glycoprotein hormones, compared to those of peptide hormones. Half-lives range from 30 min for LH to many hours for HCG.

Synthesis of TSH occurs in thyrotrophs which contain small basophilic storage granules and which constitute about 5% of anterior pituitary cells. The α and β subunits of TSH are synthesized on separate messenger RNAs and glycosylated prior to folding into the active dimer in secretory granules (Kourides et al., 1979). More α than β subunits are made, and an excess of free α subunits is found in plasma under certain conditions (Kourides et al., 1974). Mature TSH has a molecular weight of 28,000 and contains about 16% carbohydrate.

The two pituitary gonadotropins, LH and FSH are synthesized in the same cells. α and β subunits are synthesized on separate messenger RNAs, glycosylated in endoplasmic reticulum, and folded into the mature $\alpha\beta$ fully glycosylated structure prior to secretion. The interaction between subunits, while noncovalent, is very strong. As indicated for TSH, the intact $\alpha\beta$ dimer is required for biological activity.

THE TRH-TSH-THYROID HORMONE AXIS

TSH regulates the structure and function of the throid gland. TSH synthesis and secretion are controlled by neuroendocrine signals from the hypothalamus and by circulating thyroid hormones from the periphery (Table 8.2; see also Fig. 8.26 in Chapter 54). Thyrotropin-releasing hormone is the major hypothalamic hormone which stimulates synthesis and secretion of TSH (Morley, 1981). TRH, a tripeptide with the structure pyroglutamyl-histidyl-prolineamide, binds to receptors not only on thyrotrophs to stimulate TSH but also on lactotrophs to stimulate synthesis and secretion of prolactin (Fig. 8.11). Although cAMP reproduces some of the effects of TRH, it does not appear to be the intracellular mediator of TRH action, and there is better evidence for an important role of calcium in TRH action (Geras et al., 1982). From studies of the effects of TRH on prolactin synthesis, it is clear that the tripeptide rapidly induces major changes in the rate of gene transcription (Potter et al., 1981). TRH thus has major regulatory effects not

only on secretion ("releasing hormone") but on biosynthesis. Synthesis and secretion of TSH are inhibited by somatostatin, dopamine, and cholecystokinin, which reach the anterior pituitary via the hypophyseal portal system to act on the anterior pituitary; these also exert regulatory effects within the hypothalamus.

The feedback effects of thyroid hormone occur at the level of the thyrotroph cells of the anterior pituitary. Increased concentrations of thyroid hormone decrease biosynthesis and secretion of TSH in part by decreasing the number of TRH receptors (Hinkel et al., 1981). When thyroid hormone concentrations are high, TRH is ineffective in stimulating TSH secretion. When thyroid hormone concentrations decrease, this feedback inhibition is removed, and TSH synthesis and secretion increase in response to unopposed actions of TRH. Glucocorticoids and androgens have minor inhibitory, whereas estrogen has minor stimulatory, effects on TSH secretion.

Although under most conditions changes in circulating thyroid hormone concentrations are the most important determinant of plasma TSH concentrations (see Chapter 53), changes in TRH concentrations may also be important. In newborn humans and in many animals, acute exposure to cold increases plasma TSH; subsequent increases in thyroid hormone are important for thermogenesis. Because anti-TRH antibodies block cold-induced increases in TSH, increased TRH probably mediates this response (Szabo and Frohman, 1977). Only minor TSH responses to cold occur in adult humans. There are examples of hypothalamic hypothyroidism in which low TSH levels are reversed by administration of TRH. Plasma TSH concentrations exhibit a circadian rhythm with peak concentrations around midnight preceding the onset of deep sleep; this circadian rhythm in TSH concentrations may be due to changes in TRH.

THE HYPOTHALAMIC-PITUITARY-GONADAL AXIS

This axis, which is essential to the reproductive process in both males and females, is discussed fully in Chapters 57 and 58. Synthesis and secretion of both LH and FSH are stimulated by a single hypothalamic neurohormone, GnRH, a decapeptide (Fig. 8.11, Table 8.1). In females, elevations in GnRH initiate the midcycle LH surge which causes ovulation. Feedback regulation by ovarian estrogens and testicular androgens is complex. During the follicular phase of the menstrual cycle, estrogen exerts positive feedback effects, augmenting gonadotroph responses to GnRH (Yen et al., 1975). Es-

Table 8.2
Control of TSH synthesis and secretion

Classification	Stimulated by	Inhibited by
Major	TRH	T_4, T_3
Minor	Estrogen	Somatostatin
		Cholecystokinin
		Glucocorticoids
		Androgens

trogen also exerts negative feedback control over gonadotropin synthesis. This effect, which is also essential to normal progression of the menstrual cycle, is utilized in oral and injectable steroidal contraceptives. Androgens also exert negative feedback effects essential to normal spermatogenesis.

In males, LH acts on Leydig cells of the testis to stimulate testosterone biosynthesis. FSH acts on Sertoli cells and is essential for seminephrous tubule function and spermatogenesis. In females, FSH causes follicular growth in a process which depends on both LH and estrogens. Ovarian granulosa cells are the principal target for FSH. The LH surge at midcycle initiates ovulation, and LH regulates corpus luteum formation and function following ovulation.

Growth Hormone
STRUCTURE AND BIOSYNTHESIS

Human growth hormone is a 191 amino acid single chain polypeptide containing two intrachain disulfide bridges. It has major sequence homology with the placental hormone chorionic somatomammotropin (CS) (83%) and lesser homology with prolactin (16%) (Niall et al., 1973). These three peptides arose from a common ancestral gene and, although distinct, each causes some of the biologic effects of the other. Prolactin evolved with growth hormone and somatomammotropin from a common ancestor by gene duplication about 400 million years ago (Cooke et al., 1980). The growth hormone and somatomammotropin genes are located on chromosome 17, whereas the prolactin gene is located on chromosome 6 (Owerbach et al., 1980; Owerbach et al., 1981). Human growth hormone has a high degree of sequence homology with growth hormones from other species, but only human or other primate growth hormone is active in humans. Therefore, over the past 25 years, growth hormone for human use has been obtained by extraction from human pituitaries removed at autopsy. Using molecular cloning technology, the human gene has been now expressed in bacteria (Goeddl et al., 1979), and if safe and effective in human trials, this source of human growth hormone may replace human pituitaries.

Like other mammalian genes, that for growth hormone consists of exons interrupted by introns. Several forms of growth hormone, in addition to the 191 amino acid native peptide, are present in material extracted from human pituitaries (Lewis et al., 1980). One form which contains a deletion of 15 amino acids between residues 32 and 46 corresponds to removal of one exon; this form stimulates growth but lacks other actions of growth hormone. In somatotrophs of the anterior pituitary, growth hormone is made as a prohormone and after removal of the signal peptide is stored in large secretory granules. Somatotrophs are abundant in the anterior pituitary, and growth hormone constitutes up to 10% of the dry weight of the gland. After secretion, growth hormone has a circulating half-life of about 20 min.

REGULATION OF SYNTHESIS AND SECRETION

Growth hormone-releasing hormone (GRH) is the major stimulator, and somatostatin is the major inhibitor of the synthesis and secretion of growth hormone. GRH is a 40 to 44 amino acid peptide which was originally isolated from a human pancreatic tumor (Table 8.1) (Guillemin et al., 1982; Rivier et al., 1982). Somatostatin is a 14 amino acid peptide which inhibits the synthesis and secretion of both growth hormone and TSH (Vale et al., 1975). Somatostatin is found not only in the hypothalamus but also in other areas of the brain, as well as outside the central nervous system. It is synthesized in δ cells of the pancreas and inhibits secretion of both insulin and glucagon. Somatostatin also inhibits secretion of gastrin, secretin, and renin, and it decreases gastrointestinal motility. It functions therefore not only as a hypothalamic neurohormone but also in paracrine systems in the periphery and as a neuromodulator within the central nervous system.

Growth hormone synthesis and secretion are regulated by a number of factors (Table 8.3). Metabolic substrates, neuropharmacologic agents, and peripheral hormones all affect growth hormone secretion; many act via hypothalamic regulation of GRH and

Table 8.3

Control of growth hormone synthesis and secretion

Stimulated by	Inhibited by
Neurohormone	
GRH	Somatostatin
Central effects	
Hypoglycemia	Hyperglycemia
Amino acids (arginine, lysine, leucine)	Free fatty acids
Onset of deep sleep	
α-Adrenergic agonists	β-Adrenergic agonists
Dopamine	
Serotonin	
Enkephalins	
Pituitary effects	
Thyroid hormone	Progesterone
Cortisol (physiologic concentrations)	Cortisol (supraphysiologic concentrations)
Estrogen	

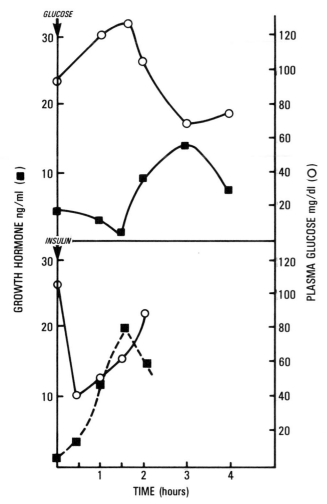

Figure 8.16. Response of plasma growth hormone to a meal and to insulin-induced hypoglycemia.

somatostatin, but some act directly on anterior pituitary somatotrophs. Several of the stimuli are used clinically to assess growth hormone secretory capacity in children with short stature or adults with suspected hypopituitarism, and several of the inhibitors are used to determine suppressibility of growth hormone secretion in suspected acromegaly or gigantism.

Plasma levels of growth hormone are high at the end of fetal life and during the first weeks after birth but then fall to normal resting values of 2–5 ng/ml. Under physiological conditions, growth hormone secretion is suppressed by elevated plasma glucose concentrations and is therefore low early after food intake (Fig. 8.16). When plasma glucose falls in response to postprandial increases in insulin, growth hormone rises and functions to prevent excessively low plasma glucose concentrations. Insulin-induced hypoglycemia is a more major stimulus to growth hormone secretion, but this procedure may cause excessive hypoglycemia when hy-

popituitarism is present. Growth hormone secretion is also stimulated by several amino acids, and infusion of L-arginine is a safe procedure for evaluating growth hormone secretory capacity. Fatty acids inhibit growth hormone secretion.

The metabolic factors act via the hypothalamus, as do most neuropharmacologic agents. During a 24-h period, elevations in growth hormone occur after each meal, but the major peak of growth hormone, accounting for up to 70% of total daily secretion, occurs during sleep. Growth hormone secretion occurs primarily during deep sleep (stages III and IV) and is triggered by the same neural mechanisms which initiate deep sleep (Weitzman et al., 1975). Nocturnal secretion of growth hormone is particularly high in children at puberty, when nocturnal rises in LH, FSH, and TSH are also especially marked. α-Adrenergic agonists, L-dopa, serotonin, and enkephalins stimulate growth hormone secretion, whereas β-adrenergic agonists inhibit secretion.

Thyroid hormone is required for growth hormone synthesis. It acts directly on pituitary cells to enhance the rate of production of growth hormone messenger RNA (Martial et al., 1977). In hypothyroidism the release of growth hormone in response to hypoglycemic and arginine stimuli is diminished; treatment with thyroid hormone restores normal responsiveness. Thyroid hormone is also necessary for stimulation of growth hormone synthesis by glucocorticoids (Samuels et al., 1977). Effects of glucocorticoids are biphasic. At physiological concentrations, cortisol stimulates growth hormone synthesis and secretion, but at high concentrations cortisol inhibits growth hormone production. Estrogens facilitate, whereas progesterone inhibits, growth hormone secretion.

BIOLOGICAL EFFECTS

Growth hormone has two major actions: stimulation of somatic growth and regulation of metabolism. Excesses or deficits in growth hormone in man lead to striking changes in somatic growth. Growth hormone deficiency in childhood results in lack of growth and proportional dwarfism (ateliotic dwarfism). Treatment with growth hormone results in marked acceleration of growth so that height, bone age, and body size become more correct for chronological age. Conversely, excesses of growth hormone as occur with certain tumors result in marked acceleration of growth. If excess growth hormone is present before puberty, gigantism will result. Often the growth hormone-secreting pituitary tumor will compress and destroy surrounding pituitary tissue or interfere with the hypophyseal portal circulation and delivery of releasing hormones so that other pituitary hormones will be deficient. Without adequate gonadotropins, sex steroids will be deficient, epiphyses will not close, and linear growth will continue (Chapter 56). If excessive growth hormone is present in adult life after epiphyseal closure, height will not increase but acral bone width, soft tissue, and organ growth will occur, producing the clinical syndrome of acromegaly.

Growth is mediated in large part by somatomedins, insulin-like growth factors whose synthesis is controlled by growth hormone (Philips and Vassilopoulou-Sellin, 1980). An astute observation led to the discovery of somatomedins. It was found that the addition of growth hormone to plasma obtained from hypophysectomized animals failed to stimulate cartilage explants, but if growth hormone were administered to hypophysectomized animals, their plasma would stimulate cartilege explants (Salmon and Daughaday, 1957). This implied that a substance was generated in response to growth hormone and that this substance, rather than growth hormone itself, directly stimulated growth. Two somatomedins (A and C) have been characterized as two insulin-like growth factors (IGF II and IGF I) (Svoboda et al., 1980). The primary structure of the IGF is similar to that of insulin and includes a portion equivalent to the connecting peptide of proinsulin (Fig. 8.17) (Blundell et al., 1978). There are distinct receptors for insulin and for insulin-like growth factors although, like the peptides, the receptors have similar structures. Somatomedins bind to insulin receptors with a lower affinity than insulin and exert insulin-like effects. Conversely, insulin binds to somatomedin receptors with lower affinity than somatomedin and exerts growth effects.

Somatomedin concentrations depend on growth hormone (Fig. 8.18). In acromegaly, where growth hormone levels are high, somatomedin levels are high, and in hypopituitarism, where growth hor-

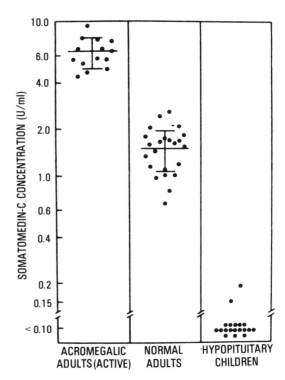

Figure 8.18. Somatomedin C concentrations in growth hormone excess and deficiency states. [From Van Wyk and Underwood (1978).]

mone levels are low, somatomedin levels are low. In one form of dwarfism, the Laron syndrome, growth hormone concentrations are normal to elevated, but somatomedin concentrations are low, suggesting that dwarfism results from sometomedin deficiency (Daughaday et al., 1969). Somatomedin production is also dependent on insulin and on nutritional status. In poorly controlled diabetics and in protein malnutrition, growth hormone concentrations are elevated, but somatomedin concentrations are low. Growth is poor under both conditions. Correction of the defects by either insulin administration to diabetics or by protein addition to deficient diets results in rises in somatomedin, even though growth hormone concentrations decrease. As discussed in Chapter 56, growth is an integrated process requiring multiple hormonal signals in addition to adequate nutrition; one level of integration is at the level of somatomedin generation. Somatomedins are produced in the liver and in fibroblasts. They circulate bound to carrier proteins whose synthesis is also under growth hormone control.

Metabolic effects of growth hormone overlap with growth effects but may result directly from growth hormone interactions with its receptor in addition to effects which are due to somatomedins. Metabolic effects of growth hormone are biphasic.

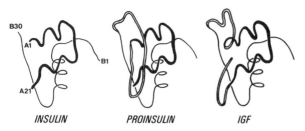

Figure 8.17. Schematic representation of 3-dimensional structures based on x-ray analysis of insulin and on model building of proinsulin and IGF-I. The A chain of insulin is shown by a thickened line, the B chain by a *solid line* and the connecting peptide by a *dashed line.* [From Blundell et al. (1978).]

Acutely, growth hormone exerts insulin-like effects. It increases glucose uptake in muscle and adipose tissue, stimulates amino acid uptake and protein synthesis in liver and muscle, and inhibits lipolysis in adipose tissue. Several hours after administration of growth hormone, these initial effects disappear, and the more profound metabolic effects of growth hormone occur. These effects which persist with prolonged elevations in plasma growth hormone are antiinsulin-like. Glucose uptake and utilization are inhibited so that blood glucose rises, and lipolysis is increased so that plasma-free fatty acids rise. Growth hormone, which rises during fasting, is thus important in the body's adaptation to lack of food. Along with cortisol, epinephrine, and glucagon, it maintains blood glucose for cental nervous system use and mobilizes fat depots to provide alternate metabolic fuel.

Effects of growth hormone are appropriate to the needs of the whole body. When adequate food is ingested and insulin is active, growth is appropriate, and the major effects of growth hormone are anabolic and growth promoting. Under these conditions, growth hormone stimulates macromolecular synthesis and growth in a number of organs. Insulin and adequate nutrition facilitate growth hormone induction of somatomedins. During fasting, when insulin is low, growth hormone increases blood glucose and free fatty acids as part of the integrated hormonal response to substrate lack. Generation of somatomedins is impaired when insulin is low and

nutrient intake is poor, so that growth is not enhanced. In insulin deficiency these growth hormone responses may worsen the diabetic state.

Prolactin

STRUCTURE AND BIOSYNTHESIS

Prolactin is a single chain polypeptide hormone of 198 amino acids with three intrachain disulfide bridges (Shome and Parlow, 1977). Lactotrophs constitute approximately 30% of the cells of the anterior pituitary, and an increase in lactotrophs accounts for most of the enlargement of the pituitary which occurs during pregnancy. Prolactin turns over more rapidly than growth hormone so that pituitary stores are significantly less. The circulating half-life of prolactin is 30–50 min.

REGULATION OF SYNTHESIS AND SECRETION

In humans the major function of prolactin is regulation of milk production. During pregnancy, prolactin rises progressively to concentrations 10-fold higher than those in the nonpregnant state (Fig. 8.19) (Riggs and Yen, 1977). Increased estrogen concentrations of pregnancy are the major stimulator of this increase in prolactin production and lactotroph growth. After parturition, prolactin levels fall, but periodic rises in prolactin concentrations occur with each period of suckling. These increases in prolactin maintain milk production for subsequent feeding periods. Prolactin concentra-

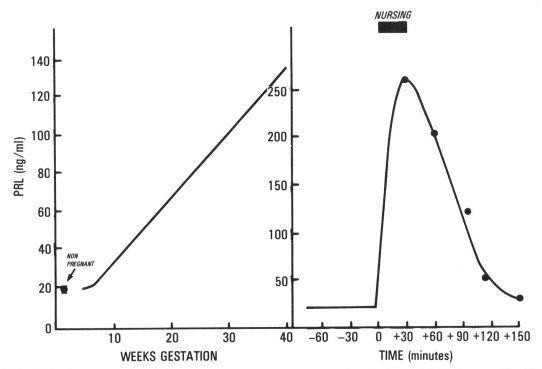

Figure 8.19. Prolactin concentrations during pregnancy and nursing. [Adapted from Noel et al. (1974); Riggs and Yen (1977).]

Table 8.4
Control of prolactin synthesis and secretion

Inhibitors	Stimulators
Dopamine	Dopamine antagonists
Dopamine agonists	Phenothiazines
Bromoergocryptine and other	Butyrophenones
ergot alkaloids	Metoclopramide
Glucocorticoids	Catecholamine depletors
Thyroid hormone	Reserpine
	α-Methyldopa
	TRH
	Estrogen
	Epidermal growth factor
	Sleep
	Pregnancy
	Suckling
	Sexual intercourse (females only)
	Exercise
	Hypoglycemia
	Arginine
	Serotonin
	Histamine H_2 antagonists
	Opiate agonists

tions are high in fetuses during the last 10 wk of intrauterine life, high in amniotic fluid, and high in umbilical vein blood (Aubert et al., 1975). After parturition, infant prolactin levels fall to adult values over a 1–2-mo period.

Prolactin synthesis is controlled principally by increasing or decreasing the rate of transcription of the prolactin gene. The major regulator is dopamine, which inhibits prolactin synthesis (Fig. 8.11, Table 8.4). When the pituitary stalk is severed, as may occur with acceleration-deceleration head injuries, or when the anterior pituitary is transplanted away from its hypothalamic connections, prolactin synthesis increases while synthesis of all other anterior pituitary hormones decreases. Lactotrophs are thus unique among anterior pituitary cells in being under major hypothalamic restraint. Dopamine agonists such as bromoergocryptine are widely used to inhibit prolactin synthesis. Bromoergocryptine also frequently inhibits growth of prolactin-secreting pituitary tumors, suggesting that dopamine provides a physiological brake on lactotroph growth as well as prolactin synthesis (Thorner et al., 1980). Because a number of drugs either interfere with the synthesis of dopamine or act as dopamine antagonists, increases in prolactin commonly occur, leading to alterations in the menstrual cycle and to abnormal lactation. The antihypertensives reserpine and α-methyldopa deplete catecholamines, including dopamine, to increase prolactin. Phenothiazines and butyrophenones, potent antipsychotic drugs, are dopamine antagonists and increase prolactin synthesis by blocking the re-

straining influence of dopamine. Metoclopramide is another dopamine antagonist which increases prolactin synthesis and secretion. Glucocorticoids and thyroxine in higher concentrations inhibit prolactin synthesis by direct action on the lactotroph.

Although prolactin synthesis is under the major restraining influence of dopamine, hormone production responds to a number of stimulatory factors. TRH binds to lactotroph membrane receptors to stimulate prolactin gene transcription. Estrogen is the major stimulator of prolactin synthesis during pregnancy, acting directly on lactotrophs. Epidermal growth factor also increases prolactin synthesis although its role in normal physiology is uncertain (Schonbrunn et al., 1980).

Prolactin secretion is maximal during sleep, with peak levels occurring 5–8 hr after the onset of sleep. Even during pregnancy when prolactin concentrations are increased, diurnal peaks of secretion of prolactin continue. Nipple stimulation, in addition to suckling, causes increased prolactin production; increases in prolactin also occur with some chest wall lesions involving the nipple areas. Prolactin secretion also increases with exercise, with sexual intercourse (in females only), and with hypoglycemia and arginine administration. Serotonin, histamine H_2 antagonists, and opiate agonists are additional stimulators of prolactin secretion.

Prolactin is the most frequent hormone which is produced in excess by pituitary tumors. Because prolactin synthesis and secretion are influenced by many agents, evaluation of patients with hyperprolactinemia requires a careful history and physical examination to differentiate between factors included in Table 8.4 and a pituitary tumor. Clinical evaluation is especially important because stimulation or suppression tests have been of little value in the differential diagnosis of hyperprolactinemia. Extremely high levels of prolactin (>200 ng/ml) indicate a pituitary tumor, but with levels this high the tumor is usually evident from radiographic procedures.

BIOLOGICAL EFFECTS

The breast is the principal site of action of prolactin in humans. Development of the breast requires estrogen, progesterone, glucocorticoids, and insulin in addition to prolactin. Chorionic somatomammotropin, which is synthesized in high concentrations by the placenta, also stimulates mammary gland development. During pregnancy there are both ductal and alveolar-lobular growth. Even though prolactin concentrations increase greatly, lactation does not occur because of inhibitory effects of progesterone and estrogen. After parturition progesterone and estrogen levels fall, and pro-lactin-dependent lactogenesis and milk secretion occur. Lactation is maintained by periodic suckling which, through a neural arc, stimulates prolactin, which acts on mammary epithelium to increase milk production. In the prepared differentiated mammary epithelial cell, prolactin exerts major regulatory effects so that over 50% of the total protein synthesized by the cell is milk protein.

Throughout human evolution lactation has provided the principal nourishment during the first 1–2 yr of life. In addition to stimulating lactation, prolactin inhibits synthesis and secretion of LH and FSH. This results in amenorrhea which may persist throughout the period of nursing. Although not a reliable method of contraception, historically nursing has played a major role in limiting additional pregnancy so that adequate nutrition and resources are available to the living child and mother. Non-nursing elevations in prolactin, such as occur with pituitary tumors and in response to drugs listed in Table 8.4, also inhibit LH and FSH, resulting in the clinical syndromes of amenorrhea and galactorrhea. In males, high concentrations of prolactin result in impotence via this same process of gonadotropin inhibition. Galactorrhea in males with hyperprolactinemia is uncommon because estrogen-dependent breast growth and differentiation have not occurred.

In species other than man, prolactin has major effects on organs other than the breast. In pigeons, prolactin stimulates the crop sac which produces glandular secretion for feeding the young. In rats and mice, prolactin has direct actions on the ovary in conjunction with LH and FSH to prolong life of the functioning corpus luteum. In fish, especially those which have a life cycle involving both fresh water and salt water, prolactin is a major osmoregulatory hormone.

Specific high affinity prolactin receptors are located in breast, liver, kidney, adrenals, and gonads (Posner et al., 1974). However, in man it is uncertain as to whether prolactin significantly effects organs other than the breast. In mammary epithelial cells, prolactin stimulates milk protein synthesis not only by increasing the rate of gene transcription but also by significantly prolonging the half-life of the milk protein messenger RNAs (Rosen et al., 1980). The latter effect may be due to increased polyadenylation which stabilizes messenger RNA.

THE POSTERIOR PITUITARY

Structure and Biosynthesis of ADH and Oxytocin

The posterior pituitary secretes antidiuretic hormone (ADH, vasopressin) and oxytocin, two pep-

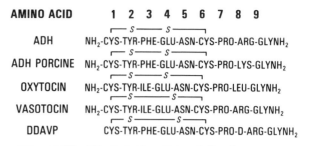

Figure 8.20. Structure of posterior pituitary hormones.

tide hormones which are synthesized in the supraoptic and paraventricular nuclei of the hypothalamus. Each hormone is made up of nine amino acids with a six-amino acid ring formed by a disulfide bond between cysteine residues at positions 1 and 6 and a three-amino acid tail containing a carboxy-terminal glycineamide (Fig. 8.20). ADH in all mammals except pigs contains arginine at residue 8; pig ADH contains lysine at this position. Oxytocin differs from ADH by having isoleucine at position 3 and leucine at position 8. Nonmammalian vertebrates synthesize vasotocin which has an isoleucine at position 3. Because vasotocin is found in the nervous system of early aquatic species, it is thought to be the ancestral hormone from which both ADH and oxytocin evolved (Sawyer and Pang, 1977).

The disulfide bridge is required for biological activity. A basic residue in position 8 is necessary for antidiuretic activity while an isoleucine at position 8 promotes oxytocic activity. Among the many analogs synthesized, 1-desamino-8-D-arginine vasopressin (DDAVP) is used clinically to treat ADH deficiency states. Removal of the amino group from cysteine at position 1 and substitution of D-arginine at position 8 favor antidiuretic activity over pressor activity and result in a molecule which is less susceptible to proteolytic degradation.

Although ADH and oxytocin are synthesized in discrete cells, producer cells for each hormone are located in both the supraoptic and paraventricular nuclei areas. Each peptide is synthesized as part of a larger precursor protein and after processing remains noncovalently bound to a portion of the precursor termed neurophysin. The 166 amino acid precursor contains a 19 amino acid signal sequence at the amino-terminus followed by the sequence for ADH, which is connected to the sequence of neurophysin by gly-lys-arg (Land et al., 1982). This glycine provides for amidation of glycine 9 of ADH, and the dibasic region provides a site for cleavage of ADH from neurophysin. A 39 amino acid glycopeptide present in the precursor is located on the carboxy-terminal side of neurophysin. After synthesis in the cell body, ADH and oxytocin bound to

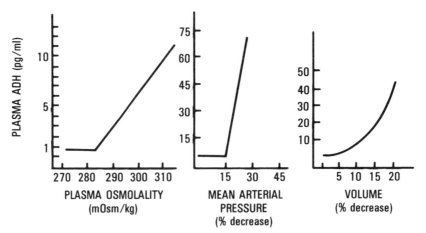

Figure 8.21. Control of plasma ADH concentration by osmolality (*left panel*), mean arterial pressure (*center panel*), and circulatory volume (*right panel*). Note changes in scale on the ordinate. [Adapted from Robertson et al. (1973) and Dunn et al. (1973).]

their specific neurophysins are transported down the axons to be stored in secretory granules in nerve terminals in the posterior pituitary. In response to nerve impulses there is concordant secretion of both ADH and its neurophysin or oxytocin and its neurophysin. The hormones rapidly dissociate from their neurophysin after secretion; subsequent functions of the neurophysins are unknown. Like other small peptide hormones, ADH and oxytocin have short half-lives in the circulation of approximately 10 min. ADH is degraded by a postproline cleaving enzyme (position 7), tryptic activity (position 8), aminopeptidase (position 1), and reduction of the disulfide bond (Walter and Simmons, 1977). Metabolism occurs principally in the kidney and liver.

Regulation of Synthesis and Secretion of ADH

The major action of ADH is to promote water conservation by the collecting duct of the kidney. In higher concentrations, it also causes vasoconstriction.* ADH, like aldosterone, importantly maintains normal fluid homeostasis and vascular and cellular hydration. The osmotic pressure of body water is the principal variable controlling secretion of ADH. Osmotic pressure is sensed by osmoreceptor cells in the hypothalamus which transmit signals to ADH-synthesizing cells in the supraoptic and paraventricular areas. Small increases in plasma osmolality of approximately 1–2% stimulate increased secretion of ADH and subsequent water conservation to reestablish normal plasma osmolality which in man is approximately

287 mOsmol/kg (Fig. 8.21). Volume depletion is the second major stimulus to ADH secretion. As shown in Figure 8.21, more major changes in plasma volume and mean arterial pressure are required to stimulate ADH secretion, but the magnitude of the rise in plasma ADH concentration is larger. For example, loss of 500 ml of blood, or approximately 7% of total blood volume, does not alter plasma ADH in recumbent subjects, but if these subjects stand and reduce their central blood volume by an additional 10%, ADH rises (Robertson, 1977). Changes in circulatory volume and mean arterial pressure are sensed by baroreceptors in the left atrium, pulmonary veins, carotid sinus, and aortic arch and are transmitted to the central nervous system through vagal and glossopharyngeal nerves. Changes in circulatory volume will increase plasma ADH regardless of the plasma osmolality. However, ADH secretion responds to osmolality over a wide range of circulatory volumes, with the rise in ADH steeper when circulatory volume is low.

ADH is a first line of defense of water balance. When water intake is decreased, osmolality rises, and ADH is secreted to promote renal water conservation and reestablishment of normal osmolality. With major losses of intravascular volume, as in hemorrhage, ADH, which is secreted in large amounts, promotes water conservation to reestablish intravascular volume and causes vasoconstriction to maintain blood pressure. When water intake is increased, osmolality is decreased, and ADH is suppressed so that the kidney excretes free water to reestablish normal osmolality.

A number of factors in addition to osmolality and circulatory volume affect ADH secretion. Nausea is a potent stimulus, as are pain, stress, hypoglycemia, and exercise. ADH secretion is stimulated by cholinergic agonists, β-adrenergic agonists, angioten-

---

* Water retention equals antidiuresis, and a more appropriate nomenclature than antidiuretic hormone might be water conserving hormone. Vasopressin is also an unfortunate choice of a name because water conservation is the major function of the hormone rather than vasoconstriction.

sin, prostaglandins, and several other pharmacologic agents. ADH secretion is inhibited by alcohol, α-adrenergic agonists, glucocorticoids, and several other pharmacologic agents.

Thirst is importantly entrained to ADH control of water balance (Fitzsimons, 1972). Hypothalamic osmoreceptors are connected to hypothalamic thirst centers as well as to ADH synthesizing neurons. When water intake decreases, a rise in osmolality of about 1% results in an increase in ADH to stimulate renal water conservation. A 2–3% rise in osmolality results in significant thirst and water-seeking behavior. The osmotic threshold for thirst of about 295 mOsmol/kg is higher than that for ADH secretion so that thirst and water intake provide a second line of defense after renal water conservation. A decrease in circulatory volume and blood pressure strongly stimulates thirst, but this stimulus operates only under severe conditions. Angiotensin is a potent stimulus to thirst, but its role under normal physiological conditions is incompletely defined.

Biological Effects of ADH

In the normal kidney 85–90% of the 200 l/day of glomerular filtrate are reabsorbed isosmotically in the proximal tubule. There is selective reabsorption of Na^+ in the ascending limb of Henle's loop so that the approximately 10% (20 l) of glomerular filtrate which reaches the distal nephron is hypotonic. The amount of hypotonic fluid reaching the distal nephron varies, depending on the glomerular filtration rate and sodium intake. When sodium intake and GFR are high, the volume reaching the distal nephron is higher than 20 l, whereas when sodium intake and GFR are low, the volume reaching the distal nephron is lower. ADH regulates the amount of water which is further reabsorbed from the hypotonic fluid which reaches the collecting duct. Without ADH the collecting duct is impermeable to water so that large volumes of urine (~16 ml/min) as dilute as 50 mOsmol/kg are excreted (positive free water clearance). With maximum concentrations of ADH the collecting duct is freely permeable so that water moves along an osmotic gradient from the hypotonic lumen to the hypertonic medullary interstitium. Small volumes of urine (0.5 ml/min) maximally concentrated to about 1200 mOsmol/kg are excreted (negative free water clearance). Variations in urine osmolality over the full range of 50–1200 mOsmol/kg occur in response to changes in plasma ADH from 1–10 pg/ml.

High affinity receptors for ADH which are located in epithelial cells of the distal nephron stimulate cAMP production. cAMP-mediated changes in membrane proteins and in cytoskeletal elements anchored to the membrane control water permeability channels (Dousa and Valtin, 1976). ADH also increases prostaglandin production, and prostaglandins exert a local negative feedback effect on ADH action (Dunn and Hood, 1977). Inhibitors of prostaglandin synthesis thus augment ADH action.

ADH also stimulates smooth muscle contraction and causes generalized vasoconstriction. Concentrations of ADH higher than those causing maximum renal water conservation are required so that vasopressor actions are important primarily when volume depletion stimulates secretion of high concentrations of ADH (Fig. 8.11). ADH has been used as an adjunct to control bleeding esophageal varices by causing splanchnic vasoconstriction. Because ADH may cause coronary arteries to constrict, high doses must be used with caution. ADH also increases ACTH secretion and has been used as a test of pituitary reserve.

Deficits and Excesses of ADH

Diabetes insipidus results from either lack of ADH or from inability of the kidney to respond to normal amounts of ADH (nephrogenic diabetes insipidus). Hypophysectomy does not usually result in permanent diabetes insipidus because, although significant retrograde degeneration of hypothalamic producer cells occurs, a number of neurons terminate on the median eminence, and those cells continue to function. With ADH deficiency, the thirst mechanism assumes a primary role in maintenance of water balance, and most patients with diabetes insipidus maintain their serum osmolality at the thirst threshold of about 295 mOsmol/kg by ingestion of large amounts of water each day. When free access to water is interrupted, these patients develop severe dehydration because of continued urinary water losses. DDAVP and vasopressin tannate in oil are effective replacement therapy for ADH deficiency. Nephrogenic diabetes insipidus is more difficult to control, although thiazide diuretics have been useful in reducing urine volume by decreasing the volume of fluid reaching the distal nephron.

The response of 24 normal adults to a water load of 20 ml/kg given over 30 min is shown in Figure 8.22. The dilution of body solutes reduces osmolality, decreases ADH, decreases urine osmolality, and results in excretion of more than 90% of the water load within 4 hr. An excess of ADH, defined as an inappropriately high concentration relative to plasma osmolality, would prevent the fall in urine osmolality and excretion of free water so that pro-

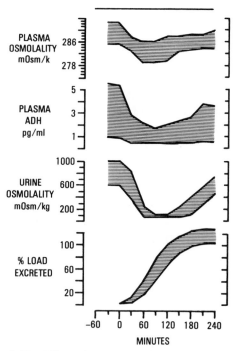

Figure 8.22. Effect of water ingestion on ADH secretion and water balance. A water load of 20 ml/kg was ingested between 0 and 30 min. [Adapted from Robertson (1981).]

gressive hypo-osmolality would develop. Because water is freely permeable across most cell membranes, plasma hypo-osmolality reflects cellular overhydration. When osmolality falls below 250 mOsmol/kg, brain cell swelling and disturbances of mental function occur. When the expansion of body water reaches 10%, the kidney rejects sodium in the proximal tubule so that urinary sodium excretion increases. Suppression of the renin-angiotensin-aldosterone system by volume expansion also contributes to urinary sodium excretion. The paradoxical excretion of sodium in the presence of a low serum sodium provides an important clue that the hyponatremia is caused by dilution of normal amounts of sodium rather than a body deficit of sodium. The syndrome of inappropriate antidiuretic hormone secretion (SIADH) occurs in a variety of clinical circumstances and is recognized by the triad of low plasma osmolality, inappropriately high urine osmolality, and significant amounts of sodium in the urine. When this occurs, it is necessary to restrict water intake. Declomycin, which causes a reversible form of nephrogenic diabetes insipidus, may also be useful.

Biological Effects of Oxytocin

The two principal targets for oxytocin are myoepithelial cells of the breast and smooth musle cells of the uterus. Myoepithelial cells are specialized smooth muscle cells surrounding the alveoli of the mammary gland. In response to oxytocin, these cells contract and move milk from the alveoli to the large sinuses in the process of milk ejection. Oxytocin release occurs in response to sensory impulses arising from the nipple during suckling but may also occur when women play with their infants without suckling. ADH does not change, but prolactin also rises in response to suckling to maintain milk production.

Oxytocin stimulates contraction of uterine smooth muscle cells. This contractile response to oxytocin is dependent on estrogen and is antagonized by progesterone. The role of oxytocin in initiation of labor is uncertain because plasma oxytocin concentrations do not clearly increase with parturition. During the first two trimesters of pregnancy, the uterus is relatively resistant to oxytocin, but during the latter part of the third trimester, uterine sensitivity to oxytocin increases significantly so that changes in target tissue responsiveness or in local oxytocin concentrations, rather than changes in circulating concentrations, may be important. During pregnancy a proteolytic oxytocinase which cleaves the amino terminal cysteine appears in uterus and placenta, and its concentration rises in plasma. This may prevent oxytocin from stimulating uterine contractions until parturition. Oxytocin is used clinically to increase uterine contraction in certain carefully defined obstetrical conditions.

NEUROHORMONES OUTSIDE THE HYPOTHALAMUS

Neurohormones initially identified in the hypothalamus by their effects on anterior pituitary function are also synthesized in several locations throughout the body, especially within the gastrointestinal tract. Conversely, peptide hormones initially identified in the gastrointestinal tract are also synthesized in the hypothalamus and other central nervous system areas. These peptide hormones therefore have many more actions than those initially identified. Generally, these small peptide hormones act near their site of synthesis, rather than being released into the general circulation. It has been more difficult to define local paracrine actions because classical endocrine gland ablation followed by hormone replacement cannot be carried out. The hormones have been localized by immunochemical methods and by the ability of isolated cells to make the hormone *in vitro*. A functional role has been implied by either local application or by neutralizing endogenous hormones with injected antisera. Because local actions often modulate effects of other hormones and neurotransmitters, experimental approaches used have only partially defined the effects of these peptide hormones.

Somatostatin, initially identified as an inhibitor of growth hormone release, is made in δ cells of the pancreatic islets and in the gut. The δ cells are in close proximity to β cells which synthesize insulin and α cells which synthesize glucagon. Somatostatin inhibits secretion of both insulin and glucagon and reduces a variety of digestive functions to slow the rate of nutrient entry. Somatostatin may thus serve as a paracrine signal to coordinate the rate of absorption of nutrients from the gut with pancreatic islet function (Unger et al., 1978).

GnRH, initially identified as the stimulator of LH and FSH synthesis and secretion, has direct effects on the ovary and testis. Acute effects are stimulatory, whereas chronic effects are inhibitory (Hseuh and Jones, 1981). Material with GnRH activity (gonadocrinin) has been isolated from gonadal tissue and is thought to play a local modulatory role within the gonads (Ying et al., 1981).

TRH is widely distributed throughout the gastrointestinal tract, and the gastrointestinal active hormones cholecystokinin, secretin, vasoactive intestinal peptide, bombesin, and substance P are found in the hypothalamus and central nervous system. Somatostatin and GnRH are also found in extrahypothalamic areas of the nervous system. These peptides appear to function as specialized neurotransmitters acting at a limited number of sites. A large number of behavioral effects have been observed with central nervous system administration. Changes in body temperature, sexual behavior, pain perception, mood, learning, and memory all occur (Moss, 1979). It is likely that these central nervous system effects are important in coordinating nervous system function and behavior with metabolic and reproductive effects of the endocrine system.

CHAPTER 53

The Thyroid Gland

INTRODUCTION

Thyroid hormone regulates metabolic processes in most organs and is essential for normal nervous system development. In contrast to many hormones whose concentrations fluctuate rapidly in response to environmental signals, thyroid hormone exhibits remarkable stability. It is synthesized in the largest endocrine gland, stored in large quantities within that gland, secreted as a less active prohormone which is tightly bound to plasma proteins, and is converted into its most active form primarily in peripheral tissues.

STRUCTURE

The normal adult thyroid gland, which weighs approximately 20 g, is composed of two lobes joined by a midline isthmus (Fig. 8.23). The gland is located in the lower part of the neck, anterior to the trachea, between the sternocleidomastoid muscles. It receives an extensive arterial blood supply from superior and inferior thyroid arteries on each side. The four parathyroid glands are located posterior to the thyroid gland, as are the recurrent laryngeal nerves. Microscopically, the thyroid gland is composed of closely packed follicles. The wall of each follicle is composed of thyroid cells which become taller as their metabolic activity increases. The interior of the follicle is filled with colloid, a proteinaceous material containing principally thyroglobulin. In addition to supporting cells, the thyroid also contains C cells of neuroectodermal origin which synthesize calcitonin (Chapter 55).

BIOSYNTHESIS OF THYROID HORMONES

Thyroid hormones are iodinated derivatives of the amino acid tyrosine. As shown in Fig. 8.24, thyroxine (T_4), the major secretory product of the thyroid gland, consists of two phenyl rings linked via an ether bridge with an alanine side chain on the inner ring. It contains four iodine atoms attached at carbons 3 and 5 of the inner ring and 3' and 5' of the outer ring. These substitutions impose a three-dimensional structure on the molecule in which the planes of the aromatic rings are perpendicular to each other (Cody, 1978). 3,5,3'-Triiodothyronine, or T_3, the active form of thyroid hormone, is principally derived by peripheral removal of one iodine from the outer ring. Removal of iodine from carbon 5 of the inner ring yields reverse T_3, an inactive metabolite.

Hormone synthesis occurs in thyroid follicular cells which exhibit marked functional polarity from their basal to apical sides. The biosynthetic process shown in Figure 8.25 has been divided into a number of steps which have been demonstrated either from genetic defects or by the use of inhibitors. The first step is active transport of iodide into the thyroid cell. This process occurs at the basal surface and requires ATP and Na^+/K^+ ATPase-mediated sodium transport (DeGroot and Niepomniszcze, 1977). By this active transport process, the thyroid gland efficiently extracts iodide, even when plasma concentrations are low; through action of this transport system, ratios of thyroid to plasma iodide may exceed 100. Other anions, such as perchlorate and pertechnetate, are also transported. Perchlorate has been used as a functional test to displace inorganic iodide from the cell, and pertechnetate as $^{99m}TcO_4^-$ is used to image the thyroid. Iodide is also actively concentrated by salivary glands, gastric mucosa, small intestine, skin, breast, and placenta but to a lesser extent; it is not significantly incorporated into protein within those tissues.

Within the thyroid cell, iodide is rapidly oxidized and incorporated into tyrosine residues in thyroglobulin molecules. This results in monoiodotyrosine (MIT) and diiodotyrosine (DIT) residues within the protein. There is subsequent coupling of these iodinated tyrosines in the thyroglobulin molecule to yield T_4 and T_3. Both organification and coupling reactions are catalyzed by thyroid peroxidase and require H_2O_2 (Taurog, 1970). These reactions, which occur at the apical surface of the thyroid cell, take place on the thyroglobulin molecule. Thyroglobulin, a large glycoprotein of 660,000

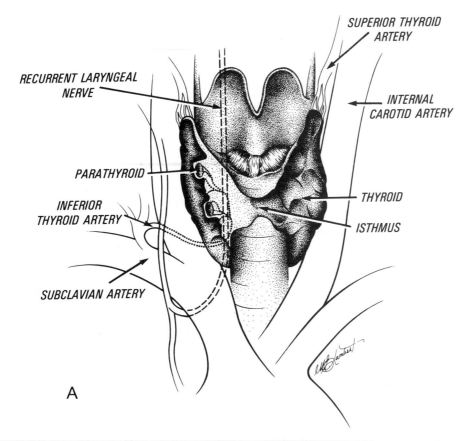

SUPERIOR THYROID ARTERY

RECURRENT LARYNGEAL NERVE

INTERNAL CAROTID ARTERY

PARATHYROID

THYROID

INFERIOR THYROID ARTERY

ISTHMUS

SUBCLAVIAN ARTERY

A

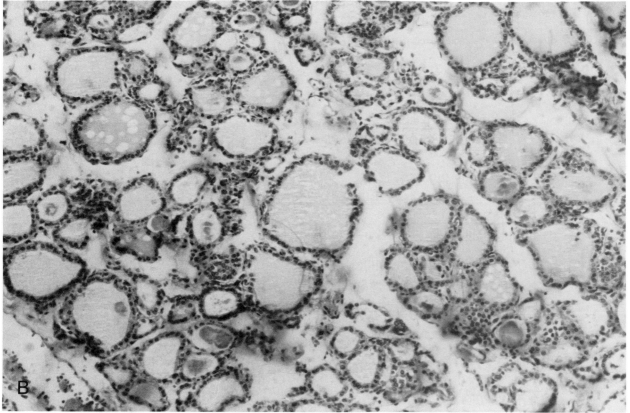

B

Figure 8.23.   Gross and microscopic anatomy of the thyroid gland.

M_r, constitutes about 50% of the total protein synthesized by the thyroid gland. It is not unusually rich in tyrosine residues nor is it rich in thyroid hormone. Normally iodinated thyroglobulin contains about 26 atoms of iodine with 3 to 4 T_4 and 0.2 T_3 residues per molecule (Izumi and Larsen, 1977). The unique feature of thyroglobulin appears to be its tertiary structure which exposes tyrosine residues for iodination and favors the coupling reaction.

Mature thyroglobulin is stored as colloid in the lumen of the thyroid follicles. Formation of active thyroid hormones requires reabsorption of colloid by endocytosis, fusion of endocytic vesicles with lysosomes, and degradation of thyroglobulin via

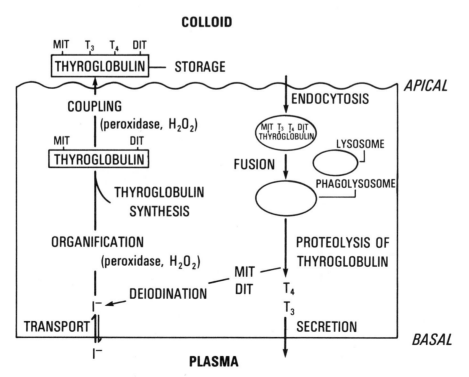

HO⟨⟩—O—⟨⟩—CH₂-CH-COOH   THYROXINE (T₄)

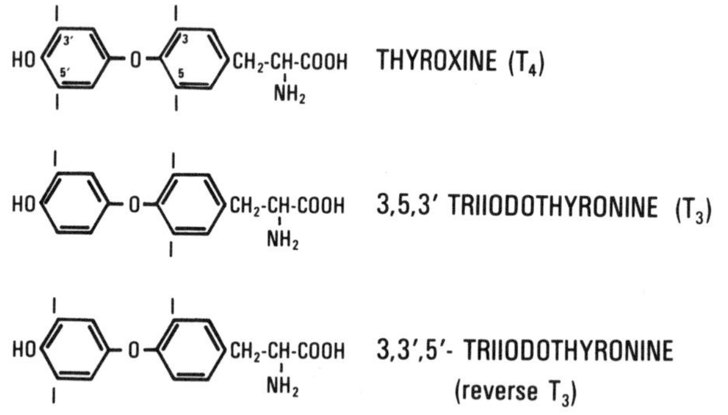

Figure 8.24. Structures of the major circulating thyroid hormones.

lysosomal proteases. T_4 and T_3 are secreted into the circulation in a ratio of about 10:1. This is somewhat less than the ratio in thyroglobulin and may reflect some conversion of T_4 to T_3 within the thyroid gland (Laurberg, 1978). The iodine in MIT and DIT is reclaimed for hormone biosynthesis by a thyroid deiodinase enzyme. This deiodination pathway reclaims approximately 50% of the iodine of thyroglobulin and is quantitatively important in thyroid gland economy, as demonstrated by the fact that congenital defects in deiodinase result in goitrous hypothyroidism.

REGULATION OF THE THYROID GLAND

Hypothalamic-Pituitary-Thyroid Axis

Thyroid stimulating hormone (TSH) is the major regulator of thyroid gland function (Fig. 8.26). Central nervous system control is exerted via the hypothalamic neurohormone thyrotropin releasing hormone (TRH), which binds to specific cell surface receptors on pituitary thyrotrophs to stimulate TSH synthesis and secretion. Somatostatin reduces TSH secretion and may transmit inhibitory signals from the hypothalamus. The glycoprotein TSH consists of α and β subunits, with biological specificity residing in the β subunit (Pierce and Parsons, 1981). TSH increases thyroid gland production of thyroid hormones which exert negative feedback control at the level of the pituitary thyrotroph (Vale

Figure 8.25. Pathway of biosynthesis of thyroid hormones.

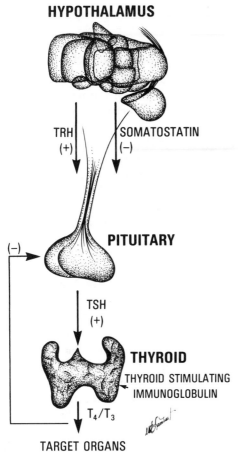

HYPOTHALAMUS

TRH
(+)

SOMATOSTATIN
(−)

(−)

PITUITARY

TSH
(+)

THYROID

THYROID STIMULATING
IMMUNOGLOBULIN

T₄/T₃

TARGET ORGANS

Figure 8.26.   The hypothalamic-pituitary-thyroid axis.

et al., 1968). Variations in circulating concentrations of thyroid hormone are the principal regulator of TSH through this feedback loop. The pituitary gland contains an unusually active 5′-deiodinase system which converts T_4 to T_3 so that circulating T_4, in addition to T_3, is an important feedback inhibitor of TSH secretion (Larsen and Silva, 1981). For example, during fasting when peripheral conversion of T_4 to T_3 is reduced, TSH and T_4 remain normal because of maintenance of intrapituitary conversion of T_4 to T_3 (Larson et al., 1981). One inhibitory effect of thyroid hormone is reduction of TRH receptors, which reduces the biological effectiveness of a given concentration of TRH (Perrone and Hinkle, 1978).

Precise quantitation of TSH, T_4, and T_3 in plasma provides the basis for accurate diagnosis of many thyroid diseases. In hypothyroidism due to thyroid gland disease, TSH concentrations are elevated, whereas in hypothyroidism due to hypothalamic or pituitary disease, TSH concentrations are low (Fig. 8.27). Because elevations in TSH rarely cause hyperthyroidism, elevated concentrations of T_4 and T_3 due to either thyroid-stimulating immunoglobulin or glandular autonomy result in low TSH concentrations. Reestablishment of normal TSH has also been useful in defining physiological replacement doses of thyroid hormone for treatment of hypothyroidism. Because thyroid hormone

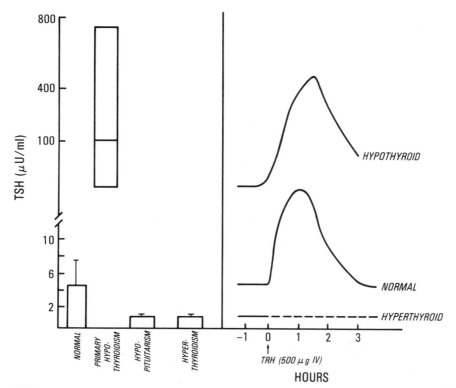

Figure 8.27.   Plasma TSH concentrations in various forms of thyroid disease. In the *right panel*, TSH responses to 500 µg of intravenous RH are shown.

feedback is at the level of the pituitary, administration of TRH causes little or no rise in TSH in hyperthyroid states, whereas it causes an exaggerated rise in TSH in hypothyroidism due to diseases of the thyroid gland (Fig. 8.27). Measurement of the response to TRH may thus be helpful in identifying mild abnormalities of thyroid gland function.

An intact hypothalamic-pituitary-thyroid axis is well established by 12–16 weeks of gestation. Fetal thyroid function depends on this axis and is independent of maternal thyroid function because neither thyroid hormones nor TSH cross the placenta in significant amounts (Fisher et al., 1977). Attempts to supply the fetus via maternal ingestion of large amounts of thyroid hormone have been unsuccessful. Both iodide and antithyroid drugs do cross the placenta and maternal ingestion of these may affect fetal thyroid function. T_3 concentrations in the fetus and cord blood at term are low, whereas reverse T_3 is high (Chopra et al., 1975), suggesting that 5′-deiodinase activity is low in utero. At birth there is a sharp rise in TSH, probably due to exposure to the cold outside world, followed by an increase in T_4 and T_3, which are maximum at 24 h. Normal metabolism of T_4 and T_3 is established soon after birth.

TSH binds to specific high affinity cell surface receptors on thyroid cells to activate adenylate cyclase and increase cellular concentrations of cAMP (Dumont et al., 1978). TSH increases the entire pathway of hormone biosynthesis, including iodide transport, organification, thyroglobulin synthesis, coupling, endocytosis, and thyroglobulin proteolysis. In Graves' disease thyroid-stimulating immunoglobulin recognize the TSH receptor and increase glandular function via cAMP and the same metabolic pathways utilized by TSH (Volpe, 1981). In vivo elevated concentrations of TSH are associated with progressive growth of the thyroid gland unless destruction of thyroid cells has occurred. Intermittent increases in TSH in nontoxic nodular goiters and continuous increases in TSH in iodide deficiency and in enzymatic defects in the biosynthetic pathway result in large glands which may reach 5–10 times normal size and obstruct the trachea, neck veins, and esophagus. Following hypophysectomy, atrophy of the thyroid gland occurs. It is not certain as to whether TSH directly stimulates thyroid gland growth or whether a concomitant increase in delivery of growth factors is required. Studies in cell culture have confirmed the trophic effect of TSH on differentiated function, but data on a direct growth stimulatory role are

conflicting (Westermark et al., 1979; Ambesi-Impiombato, et al., 1980).

Autoregulation

Autoregulation of the thyroid gland is essential because dietary intake of iodine may vary greatly. Iodine is obtained principally from seafood in the diet. In inland areas of the world, especially mountainous ones where, over millions of years, iodine has been washed from the soil, dietary iodine deficiency occurs. Although the thyroid iodide trapping mechanism becomes remarkably efficient, in areas where iodine intake is less than 60 μg/day, goiter is common. Growth of the thyroid gland, which is dependent on TSH, augments the ability of the gland to trap iodide and synthesize sufficient hormone to maintain a euthyroid state. In areas where iodine intake is less than 20 μg/day, very large glands are common; compensation may be incomplete, and endemic cretinism results. In iodine deficiency, glandular production of hormone also changes so that T_3 becomes the major secretory product (Greer et al., 1968). This adaptation provides a more active hormone containing less iodine.

Normal iodine intake in Western cultures is approximately 250 μg/day but may be greater because of use of iodized salt and use of iodates in making bread. Approximately 50 μg is used for thyroid hormone biosynthesis, and 200 μg is excreted by the kidneys (Fig. 8.28). Additional iodide for hormone biosynthesis is derived from metabolism of thyroid hormones. Autoregulatory mechanisms which enhance TSH-mediated adaptations to reduced iodine intake also protect against excessive iodine ingestion so that plasma thyroid hormone concentrations remain constant. In response to large quantities of iodine (2 mg or more) there is a sharp decrease in organification (Wolff and Chiakoff, 1948). This decrease in organification depends

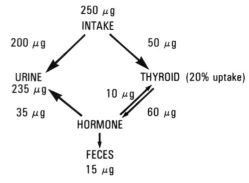

Figure 8.28. Daily iodine balance in the USA. [From Robbins (1980).]

on a high concentration of iodide within the gland and results in decreased formation of thyroid hormone. Decreased organification is an acute response to high cellular iodide concentrations. When excessive iodide ingestion continues, the thyroid gland adapts by decreasing active transport of iodide into the thyroid cell (Braverman and Ingbar, 1963). Increased cellular concentrations of iodide are reduced and the block to organification is removed. In normal people, ingestion of large amounts of iodine thus does not adversely affect thyroid function. Because large iodide pools dilute any radioactive iodine ingested, iodine supplementation has been recommended as protection against radioactive iodine. In abnormal thyroids, autoregulatory adaptations may not occur. If inhibition of organification does not occur, iodine ingestion will result in hyperthyroidism (Jod-Basedow effect). If organification is inhibited, but transport does not adaptively decrease, high cellular concentrations of iodide will persist and result in hypothyroidism (iodide goiter and hypothyroidism) (Wolff, 1969).

Very large doses of iodine also inhibit release of thyroglobulin-bound thyroid hormones from the gland. This distinct inhibitory effect was used in the past to treat thyrotoxicosis and is still useful in preparing patients for thyroid surgery.

TRANSPORT AND METABOLISM

In a normal adult the thyroid gland secretes approximately 80 μg of T_4 per day. Approximately 6 μg of T_3 are secreted by the thyroid gland each day, with 80% of total daily production of T_3 arising from peripheral deiodination of T_4 (Schimmel and Utiger, 1977; Chopra et al., 1978). Conversion of T_4 to T_3, which occurs primarily in liver and kidney, is catalyzed by 5'-deiodinase, a microsomal enzyme which requires reduced sulfhydryl groups from glutathionine and other compounds for activity (Fig. 8.29). In addition to conversion of T_4 to the active hormone T_3, T_4 also undergoes deiodination in the inner ring to yield the inactive metabolite reverse T_3. Inactivation of the hormone occurs by further deiodinations, deamination-decarboxylation of the alanine side chain, and conjugation with glucuronide.

Circulating concentrations of T_3 are determined not only by glandular production of T_4 and T_3 but also by the activity of 5'-deiodinase. As shown in Figure 8.29, a number of conditions are associated with decreased 5'-deiodinase activity which result not only in decreased production of T_3 but also in increased concentrations of reverse T_3 due in part to the inability of reverse T_3 to be further metabo-

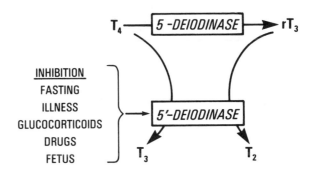

Figure 8.29. Peripheral metabolism of thyroid hormones. [From Schimmel and Utiger, 1977.]

lized. The best studied paradigm is fasting; changes occurring during illness may relate to similar metabolic factors. During fasting there is a 20% decrease in plasma T_3 by the end of the first day and a 50% decrease at the end of 3 days with a reciprocal increase in reverse T_3 (Wartofsky and Burman, 1982). With prolonged fasting of more than 3 wk, reverse T_3 tends to normalize, but T_3 remains low. Refeeding rapidly reverses these changes. The decrease in T_3 production with fasting may be an important adaptive response contributing to the observed decreased metabolic rate and conservation of body tissues because, clearly, maintenance of normal or elevated rates of metabolism would be inappropriate when substrate is not available. Decreased production of T_3 is accompanied by adaptive changes in the pattern of enzymes induced in the liver so that T_3 induction of α-glycerol phosphate dehydrogenase is maintained but T_3 induction of malic enzyme is not (Oppenheimer et al., 1979). Hepatic T_3 nuclear receptors are reduced (Burman et al., 1977). Even though T_3 concentrations are reduced during fasting, T_4, free T_4, and TSH concentrations remain normal. This serves to maintain normal glandular production and prevents the thyroid from inappropriately compensating for the physiologically important reduction in T_3.

Upon secretion into the circulation, T_4 and T_3 are tightly bound to serum proteins so that only a small fraction of the total amount circulates in the free form to enter cells and to exert metabolic control (Table 8.5). Binding proteins serve as a protected reservoir to prevent renal clearance and provide free hormone as needed. T_4 is bound to three proteins: thyroid-binding globulin (TBG), thyroid-binding prealbumin (TBPA), and human serum albumin (HSA), all of which are synthesized in the liver (Robbins et al., 1978). TBG, a glycoprotein of 63,000 M_r, has the highest affinity for thyroid hormones; it binds approximately 75% of T_4

Table 8.5.
Quantitative aspects of thyroid hormone metabolism

	Production, μg/day		Mean Concentration in Plasma, ng/ml		Half-life, days
	Gland	Periphery	Total	Free	
T_4	78		80	0.03	6.5
T_3	6	24	1.2	0.004	1.3
rT_3	3	27	0.4	0.00098	0.1

[Data kindly provided by Dr. Inder Chopra.]

and 50% of T_3 in the circulation. Affinity constants are 2×10^{10} M^{-1} for T_4 and 2×10^8 M^{-1} for T_3. TBPA has a lower affinity for T_4, being 100-fold less than that of TBG but has a much greater capacity to bind primarily T_4. HSA has an even weaker affinity but is present in high concentrations. Protein binding, principally to TBG, results in the long half-life of circulating thyroid hormones (Table 8.5). T_4 has a half-life of about 1 wk while T_3, which is bound less tightly to proteins, has a half-life of 1.3 days.

The concentration of the free active fraction is determined by the mass action equation shown in Chapter 51, equation 8.1. Present methodology is sufficient to accurately quantitate total T_4 and T_3 concentrations in plasma, but measurement of the free fraction is tedious and difficult. Usual clinical measurements of total hormone concentrations must, therefore, be interpreted with knowledge of factors likely to affect hepatic synthesis of TBG and other binding proteins. During pregnancy, TBG concentrations increase because of estrogen stimulation of TBG production. An increase is evident by 1 month and is maximal by the 3rd month of pregnancy. When TBG increases, the initial effect will be to shift hormone from the free to the bound state. The total concentration will increase, first, because of decreased metabolic clearance and, second, because of increased glandular production due to the hypothalamic-pituitary-thyroid axis, which responds to the initial decreased concentration of free hormones. The free concentration will be restored to normal, with equilibrium established at higher concentrations of bound hormone. The pregnant woman thus has normal free hormone levels and normal thyroid hormone function but an elevated total plasma concentration of hormones because of an increase in the bound fraction. Conversely, androgens and certain illnesses such as hepatic cirrhosis may decrease TBG and TBPA production. Again, patients will maintain normal free T_4 concentrations, although the total will be decreased. Measurement of TBG or available TBG binding sites may occasionally be necessary to ac-

curately interpret measured total thyroid hormone concentrations.

BIOLOGICAL EFFECTS OF THYROID HORMONE

The multiple effects of thyroid hormone result from occupancy of nuclear receptors with subsequent effects on gene expression. Receptors appear the same in all tissues, with differing biochemical effects resulting from the pattern of gene expression in each cell type. Receptors specifically bind T_3. Under physiological conditions 90% of receptor bound hormone is T_3; the affinity of nuclear receptors for T_3 of 2.9×10^{-11} M exceeds that for T_4 10-fold (Samuels and Tsai, 1973); and 5- to 7-fold lower concentrations of T_3 than T_4 are required to induce biological effects. T_3 receptor complexes increase messenger RNA and protein production which mediate observed biological responses (Baxter et al., 1979).

Growth and Development

Thyroid hormone is required for normal nervous system development and linear growth. In the absence of thyroid hormone, axonal and dendritic development and myelination of the nervous system are defective. In humans, thyroid hormone appears especially critical during the immediate postnatal period. With early hormone replacement, hypothyroid newborns attain normal intellectual development; if therapy is delayed by 6–12 months, mental retardation is permanent. Routine laboratory screening of newborn infants is now carried out to allow treatment of this preventable form of mental retardation, which occurs in about 1 in 4400 live births. Once nervous system development is complete, hypothyroidism causes mental slowing, but this is completely reversible with thyroid hormone therapy.

When thyroid hormone is deficient, linear growth is slowed, maturation of growing epiphyseal end plates is retarded, and tooth eruption is delayed. In cell culture, T_3 stimulates proliferation of several cell types, reflecting the direct growth-promoting effects of T_3 on many organs. In addition, T_3 is required for production of both growth hormone and insulin-like growth factors. It is also required for action of insulin-like growth factors on epiphyseal cartilage (Froesch et al., 1976).

Energy Metabolism

Thyroid hormone stimulates the basal rate of metabolism, oxygen consumption, and heat production. Biochemical mechanisms underlying this fundamental property of thyroid hormone, which was first recognized in 1895 by Magnus-Levy, is incompletely understood. At least a part of the T_3-induced increase in oxygen consumption in tissues such as liver and diaphragm muscle results from increased activity of Na^+/K^+ ATPase (Ismail-Beigi and Edelman, 1970). T_3 induces increased synthesis of Na^+/K^+ ATPase, which consumes ATP during its activity to transport Na^+ and K^+ across the cell membrane (Smith and Edelman, 1979). Generation of ATP to support the increased activity of this enzyme depends on oxidative metabolism. The stimulatory effects of thyroid hormone on general protein synthesis and on several metabolic pathways also contribute to oxygen consumption.

Thyroid hormone is important for thermogenesis in homeotherms. In humans, thyroid hormone deficiency results in a reduction in basal metabolic rate and in core temperature. When excessive thyroid hormone is present, the basal metabolic rate is increased, and core temperature is elevated. Sympathetic nervous system activity contributes importantly to temperature control as discussed below.

Organ Systems

Thyroid hormone importantly regulates sympathetic nervous system activity. Although T_3 does not appear to alter production or concentration of catecholamines, it does induce synthesis of β-adrenergic receptors (Lefkowitz, 1979). As indicated by the mass action equation shown in Chapter 51, Eq. 8.1, an increase in receptors will increase formation of active hormone-receptor complexes and biological activity. Although many of the effects of T_3 on adrenergic receptors have been identified in animal studies, it is likely that this is also an important action of thyroid hormone in man. Many of the signs and symptoms of hyperthyroidism reflect increased β-adrenergic activity. Patients exhibit tachycardia, increased cardiac output, wide pulse pressure, peripheral vasodilatation, frequent movement, diffuse anxiety, and lid retraction. There is heat dissipation with warm skin and increased sweating. Many of these signs and symptoms of excessive thyroid hormone can be abolished by β-adrenergic blockade, although increased doses are often required to saturate the increased number of receptors. Conversely, in hypothyroidism, where β-adrenergic receptor synthesis is impaired, α-adrenergic activity may predominate, resulting in peripheral vasoconstriction and increased blood pressure.

T_3 affects most organs of the body. Although the effects of thyroid hormone are most easily appreciated by analysis of deficiency and excess states, it is important to remember that physiologically T_3 acts as a metabolic regulator to maintain homeostasis between these two extremes. In addition to effects on the heart via adrenergic receptors, T_3 maintains normal myocardial contractility, in part through stimulating the expression of the most active isoenzyme form of myosin ATPase (Hjalmarsen et al., 1970). In hypothyroidism less active isoenzyme predominates; administration of T_3 restores the normal isoenzyme (Hoy et al., 1977). T_3 also restores normal myocardial contractility. Normal skeletal muscle function also requires thyroid hormone. Muscle weakness occurs both in hypothyroidism and in hyperthyroidism; in the latter it may result from excessive catabolism of muscle proteins.

In hypothyroidism, weight gain is common because of decreased metabolic rate. Hypomotility of the colon and constipation are common. Conversely, in hyperthyroidism there is increased motility and hyperdefecation. The increased metabolic rate results in weight loss despite a significant increase in caloric intake. In hypothyroidism, hypoventilation occurs because of muscle weakness, decreased respiratory effort and accumulation of pleural fluid. In hypothyroidism the skin is not only cool but also coarse and dry, and subcutaneous accumulation of mucopolysaccharides may result in characteristic myxedema. In hyperthyroidism the skin is warm, smooth, and moist. Excesses or deficiencies of thyroid hormone also lead to alteration in function of other endocrine systems. This results in infertility in both hypo- and hyperthyroid states.

The altered metabolic states resulting from either an excess or a deficiency of thyroid hormone affect the metabolism of a number of other hormones, vitamins, and drugs. T_3 increases the metabolic clearance of cortisol; it increases sex steroid-binding globulin to decrease the metabolic clearance of estrogen and testosterone; it increases both the utilization and clearance of vitamins; and it increases the clearance of a number of drugs, including digitalis. Consequently, patients with hypothyroidism are exquisitely sensitive to a number of drugs and may exhibit toxic effects at usual therapeutic doses. Patients with hyperthyroidism require a larger vitamin intake and often require increased doses of a drug to achieve a therapeutic response.

PHARMACOLOGY

Because most T_3 is derived under normal physiological conditions from deiodination of T_4 in peripheral tissues, T_4 is the preferred hormone for treatment of hypothyroidism, regardless of etiology. The long circulatory half-life of T_4 provides remarkable stability once a therapeutic dose is established. Approximately 150 μg/day (100–200 μg) provides full replacement with restoration of normal physiology, including normalization of TSH concentrations. The shorter half-life of T_3 makes it less appropriate for long-term use; it has an additional disadvantage of bypassing regulatory mechanisms controlling T_4 to T_3 conversion. Regardless of the degree of hypothyroidism—mild, moderate, or severe—the goal is ultimate provision of a full replacement dose of T_4. In thyroid disease, residual production of T_4 may decrease and may require excessive TSH and thyroid enlargement to sustain so that maintenance of a euthyroid state with suboptimal replacement dosages is uncertain, and full replacement is recommended once a decision to use thyroid hormone is made. It is clear from consideration of the normal hypothalamic-pituitary-thyroid axis that a "little" thyroid cannot be given to increase metabolism, cause weight loss, or regularize menses. With normal thyroid function, replacement of 50% of T_4 production by exogenous hormone will lead to a 50% reduction of endogenous production so that free T_4 and T_3 concentrations will remain normal. Full replacement will fully shut off endogenous thyroid synthesis. Thyroid function can only be increased long-term by administering supraphysiological doses which create a hyperthyroid and pathological state. T_4 is optimally used in full physiological replacement doses only for hypothyroid states or to suppress TSH-dependent growth of thyroid tissue.

Antithyroid drugs presently used to treat hyperthyroidism are thioamides. Propylthiouracil and methimazole are used in the United States, and carbimazole is available in England. These drugs inhibit the biosynthesis of thyroid hormones by blocking the organification reaction (Marchant et al., 1978). Full return to a euthyroid state requires 4–6 wk because of the long circulating half-life of T_4 and the large glandular stores of thyroglobulin in colloid which must be depleted. Propylthiouracil has an additional peripheral action to block the deiodination of T_4 to T_3 and may thus have a somewhat more rapid onset of action. Antithyroid drugs are effective in restoring a euthyroid state to patients with both Graves' disease and toxic nodular goiters. In Graves' disease, approximately 50% of patients who are treated for 1 yr will remain euthyroid after discontinuing the drug. The other 50% will relapse and thus require ultimate treatment with radioactive iodine or surgery. Toxic nodular goiters do not usually remit, so antithyroid drugs are used only to control hyperthyroidism and prepare a patient for ultimate therapy with radioactive iodine or surgery.

Radioactive isotopes of iodine are metabolized identically to stable [127]I. They are concentrated by the thyroid iodide transport system and organified in thyroid cells. Radioactive iodine thus specifically delivers radioactive energy primarily to one cell type and is useful for imaging and for radiation therapy. [131]I, which has a half-life of 8 days and emits both x-rays and β particles, is the principal isotope used to treat hyperthyroidism. [131]I is administered orally and, in doses sufficient to correct hyperthyroidism, delivers little radiation to any other cells. The major disadvantage of radioactive iodine therapy is hypothyroidism, which occurs in about 20% of treated patients in the first year with an incidence of about 1% per year thereafter (Becker et al. 1971). By comparison the incidence of hypothyroidism in surgically treated patients is 20–30%. Because many papillary and follicular thyroid cancers also concentrate iodide, radioactive iodine can be used to treat metastatic disease with minimal scatter of radiation to surrounding tissues (Beierwaltes, 1978). [125]I with a half-life of 13 h and emission of x-rays is a good isotope for imaging the thyroid gland because little radiation is delivered.

CHAPTER 54

The Adrenal Gland

THE ADRENAL CORTEX

Introduction

The adrenal glands are composed of two organs of separate embryological derivation: an outer cortex of mesodermal origin and an inner medulla of neuroectodermal origin. In a broad sense, both the steroid hormones synthesized by the cortex and the catecholamines synthesized by the medulla mediate adaptive responses of the organism to a changing environment. The model of "stress" which evokes secretion of both steroid hormones and catecholamines may have evolved as a response to the primitive signal of substrate lack (Baxter and Rosseau, 1979). When food and water intake are not possible, cortisol stimulates gluconeogenesis to maintain blood glucose for nervous system use; aldosterone stimulates sodium retention to maintain intravascular volume and cellular hydration; and epinephrine stimulates both mobilization of energy-providing substances and cardiovascular and neuromuscular adaptation. Because the environment is constantly changing for free living animals, including man, such adaptive hormones have evolved as major regulators of homeostasis and control a number of metabolic processes in a wide variety of cell types.

Structure

The adrenal glands are paired, pyramid-shaped organs each weighing approximately 5 g in man, located above the upper pole of each kidney (Fig. 8.30). The adult adrenal cortex, which makes up approximately 90% of the gland, is composed of three zones. The outer subcapsular glomerulosa zone synthesizes aldosterone while the middle fasciculata and inner reticularis zones synthesize cortisol and androgen precursors. Blood supply is from several arteries which enter the outer portion of the cortex, and blood flows centrally through fenestrated capillaries to the inner medulla. Venous drainage is through a single vein into the vena cava on the right and into the renal vein on the left.

Integrated functioning of the adrenal gland depends on this anatomical arrangement. The adrenal cortex regenerates with normal zonation when only a thin rim of glomerulosa calls remain attached to the capsule or when cells are transplanted to other sites (Ingle and Higgins, 1938; Srougi et al., 1980). Zonation appears to depend on centripetal blood flow, which generates a gradient of steroid hormone concentration which increases from subcapsular cells to those more centrally located (Hornsby and Crivello, 1983). These steroids inhibit selected steroidogenic enzymes so that biosynthesis of aldosterone is limited to outer glomerulosa cells, and precursor flow into androgens is facilitated in inner reticularis cells. The medullary enzyme phenylethanolamine N-methyl transferase, which catalyzes conversion of norepinephrine to epinephrine, is regulated by cortisol (Wurtman and Axelrod, 1966), which is delivered in high concentrations in centrally flowing sinusoidal blood.

Biosynthesis and Metabolism of Adrenocortical Hormones

Cholesterol which is utilized for steroid hormone synthesis is derived from cholesterol esters which are stored in abundant cytoplasmic lipid droplets in adrenocortical cells and from circulating cholesterol carried in low density lipoprotein particles. In response to ACTH, increased precursor cholesterol is made available by activation of cholesterol esterase which releases free cholesterol from ester storage depots (Beckett and Boyd, 1977) and by facilitated uptake of LDL cholesterol via LDL receptors which are especially abundant in adrenocortical tissue (Brown et al., 1979). Free cholesterol is transported to mitochondria, where it is metabolized to pregnenolone and isocaproaldehyde by the cytochrome P-450 side chain cleavage enzyme located in the inner mitochondrial membrane (see Fig. 8.9, Chapter 51) (Burstein and Gut, 1971). The rate-limiting step in steroid hormone biosynthesis is the interaction of free cholesterol with cytochrome P-450 side chain cleavage enzyme (Simpson et al.,

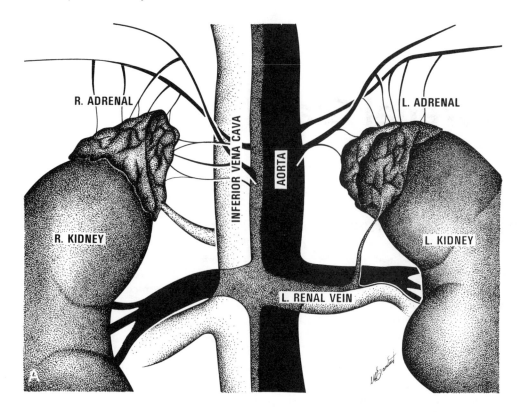

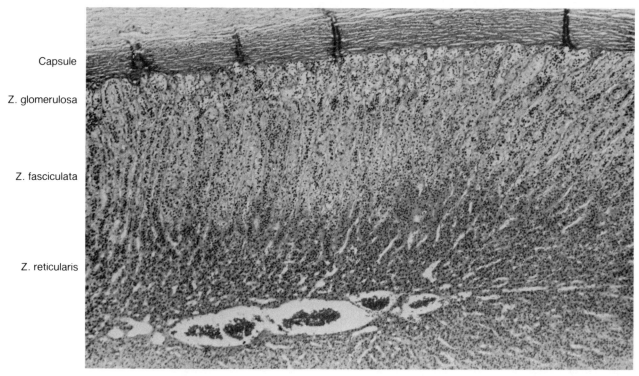

Capsule

Z. glomerulosa

Z. fasciculata

Z. reticularis

Figure 8.30. Gross and microscopic anatomy of the adrenal gland.

1978). This step requires continuing protein synthesis and may involve formation of polyphosphoinositol to facilitate this interaction. The limiting step in steroid hormone biosynthesis is thus availability of cholesterol substrate to the mitochondrial enzyme; ACTH and all steroidogenic stimuli increase this interaction of substrate with enzyme to acutely increase steroid hormone synthesis.

The cholesterol transformation (which involves hydroxylations at carbons 20 and 22) and subse-

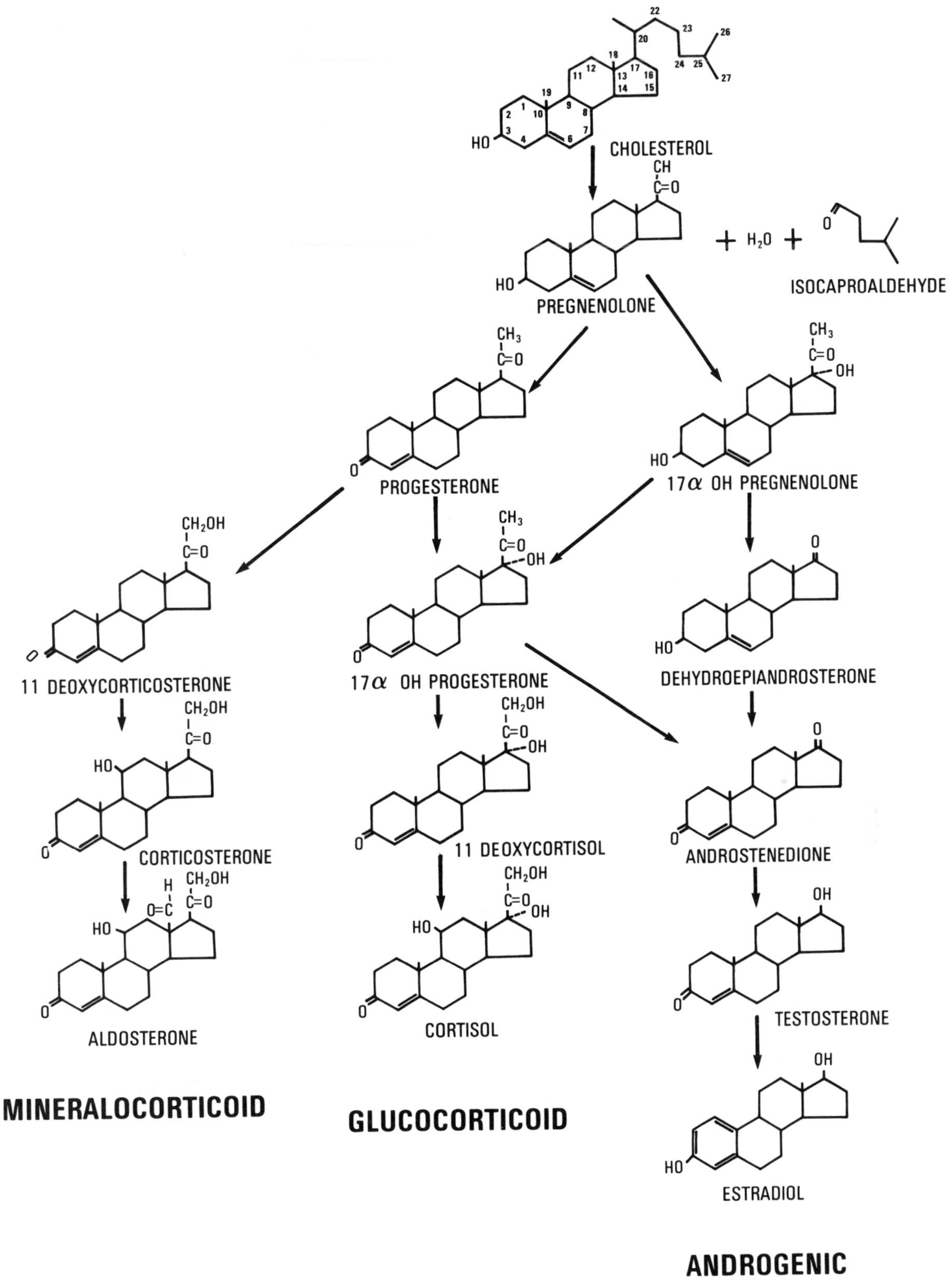

Figuree 8.31.   Pathway for biosynthesis of adrenocortical hormones.

quent steroid hydroxylations require NADPH, molecular oxygen, and specific cytochrome P-450 enzymes. Pregnenolone is rapidly removed from mitochondria and sequentially modified to yield the

three major classes of adrenocortical steroids (Fig. 8.31). In addition to the major transformations shown in Figure 8.31, other intermediates, including sulfated ones, may exist (Wolfson and Lieberman, 1979). In endoplasmic reticulum, pregnenolone is hydroxylated at the 17 position to yield 17α-OH pregnenolone or is converted to progesterone by a 3β-hydroxysteroid dehydrogenase, $\Delta^{4,5}$-isomerase enzyme complex, which also converts 17α-OH pregnenolone to 17α-OH progesterone and dehydroepiandrosterone to androstenedione. 17α-OH progesterone is then hydroxylated at carbon 21 to yield 11-deoxycortisol. The final 11β-hydroxylation of cortisol is catalyzed by mitochondrial 11β-hydroxylase. Reactants flow through this highly organized biosynthetic pathway from cytosol to mitochondria to endoplasmic reticulum to mitochondria.

Formation of aldosterone in the zona glomerulosa proceeds from progesterone by hydroxylations at carbons 21, 11, and 18. 17α-Hydroxylation is not required, and the activity of this enzyme in glomerulosa cells is very low. The 18-hydroxyl group is oxidized to an aldehyde in mitochondria to yield aldosterone.

Dehydroepiandrosterone (DHEA), which is secreted primarily as the sulfated derivative, and androstenedione are the principal androgenic steroids produced by the adrenal cortex. 17α-Hydroxylation of either pregnenolone or progesterone permits 17,20-lyase activity to remove the side chain from carbon 17. The adrenal androgens have little intrinsic biological activity and are primarily active after conversion in peripheral tissues to testosterone and estrogen.

Daily production rates and plasma concentrations of the principal secretory products of the adrenal cortex are shown in Table 8.6. Cortisol is the feedback regulator of the hypothalamic:pituitary:adrenocortical system and determines the rate of production of all steroids except aldosterone via this feedback axis. In plasma, more than 90% of cortisol is bound, principally to corticosteroid-binding globulin (CBG), a glycoprotein synthesized in the liver (Burton and Westphal, 1972). CBG is not required for biological effects of cortisol but serves as a reservoir to protect cortisol from renal clearance and degradation. CBG, which contains one high affinity binding site per molecule, is normally present in concentrations sufficient to bind about 250 ng of cortisol per milliliter. When cortisol production is elevated and the above capacity is exceeded, cortisol is bound weakly to serum albumin, and the urinary concentration of free cortisol rises sharply. CBG production is increased by estrogens so that by the third trimester of pregnancy CBG concentrations are twice those of the nonpregnant state. When CBG is increased, the biologically active free fraction will be maintained at normal levels via the hypothalamic:pituitary:adrenocortical feedback axis. Total plasma cortisol as measured in clinical situations will be elevated due to the increase in the bound fraction; free cortisol concentrations will, however, remain normal. CBG is decreased in certain liver diseases, hypothyroidism, and nephrosis, but again free cortisol concentrations are normally maintained. Interpretation of measured total plasma cortisol concentrations therefore requires knowledge of the metabolic state of the patient as well as of the time of sampling in relation to the circadian rhythm.

Aldosterone is only weakly bound to plasma proteins so that most aldosterone is metabolically inactivated during a single passage through the liver. DHEA and androstenedione are only weakly bound to plasma proteins, but testosterone and estrogen are tightly bound to a specific sex steroid binding globulin.

Cortisol has a circulating half-life of about 90 min, whereas aldosterone, which is weakly protein bound, has a half-life of about 15 min. The principal site of metabolism is the liver, where inactivation of steroid hormones occurs and conjugation with glucuronic acid and sulfate render metabolites water soluble for excretion by the kidney. The principal alteration of cortisol is reduction of the double bond between carbons 4 and 5 and reduction of the ketone at carbon 3. The resulting inactive tetrahy-

Table 8.6.
Production rates and plasma concentrations of major adrenocortical hormones

Class	Steroid	Production Rate, mg/day	Plasma concentration, ng/ml
Glucocorticoid	Cortisol	8–25	40–180
	Corticosterone	1–4	2–4
Mineralocorticoid	Aldosterone	0.05–0.15	0.15
	Deoxycorticosterone	0.6	0.15
Androgenic	DHEA	7–15	5
	DHEA-S		1200
	Androstenedione	2–3	1.8

Concentrations vary with time of day, sex, and stage of menstrual cycle for particular steroids.

drocortisol is conjugated with glucuronic acid at carbon 3 for renal excretion. Reduction of the ketone at carbon 20 yields the cortol series of metabolites. Aldosterone is metabolized to tetrahydroaldosterone and conjugated with glucuronic acid at the 3 position or is conjugated with glucuronic acid at carbon 18. DHEA and androstenedione can be converted to the more active androgen testosterone in the periphery. However, most of these steroids are metabolized by reduction of the double bond between carbons 4 and 5 and by hydroxylations at carbons 7 and 16.

Clinical assessment of adrenocortical function is principally done by quantitation of circulating steroids using competitive protein binding methods such as radioimmunoassay. Traditionally, production has been inferred by quantitation of urinary metabolites by chemical methods. The Porter-Silber measurement of the 17,21-dihydroxy-20-ketone group (17-hydroxysteroids) quantitates about one-third of cortisol metabolites and is an accurate reflection of cortisol production in both excess and deficiency states. The Zimmermann reaction, which measures 17-ketosteroids, provides an assessment of adrenal androgen production. Testosterone, the principal active androgen is quantitated by competitive protein binding methods.

Regulation

THE HYPOTHALAMIC:PITUITARY: ADRENOCORTICAL AXIS

Production of cortisol and androgen precursors is controlled by ACTH, whereas aldosterone production is additionally regulated by angiotensin and potassium. Corticotropin-releasing hormone (CRH), a neuropeptide of 41 amino acids (Vale et al., 1981), is synthesized in the hypothalamus and reaches ACTH-producing cells of the anterior pituitary via the hypophyseal portal system (Fig. 8.32). In response to CRH, corticotroph cells of the pituitary synthesize and secrete ACTH, which circulates and specifically binds to high affinity receptors on the surface of adrenocortical cells to stimulate synthesis and secretion of cortisol (Buckley and Ramachandran, 1981). ACTH also stimulates synthesis of aldosterone and of adrenal androgens. Cortisol, but not other adrenal steroids, exerts negative feedback control on ACTH synthesis by suppressing transcription of the ACTH gene in the pituitary and by suppressing formation of CRH in the hypothalamus. Like other endocrine systems, the hypothalamic:pituitary:adrenocortical axis tends to maintain its own homeostatic set point

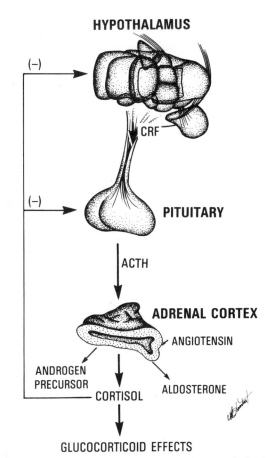

Figure 8.32. Hypothalamic:pituitary:adrenocortical axis.

unless strongly driven by continuous environmental signals received by the nervous system. Net synthesis of ACTH is a result of the relative strength of stimulatory (CRH) and inhibitory (cortisol) signals. When cortisol or a synthetic derivative such as dexamethasone is given in pharmacologic amounts, ACTH synthesis is decreased. However, when patients are given suppressive amounts of dexamethasone prior to major surgery, the stress of surgery is sufficient to override the suppression and to increase ACTH synthesis.

Secretion of ACTH and of cortisol exhibits a circadian rhythm, with the highest rate occurring around the time of morning awakening (Fig. 8.33). Both ACTH and cortisol are secreted in episodic bursts, with the integrated sum of these bursts giving the characteristic circadian rhythm where mean morning plasma ACTH and cortisol concentrations exceed evening plasma concentrations approximately 2-fold. This circadian rhythm appears importantly tied to sleep:wake cycles and to food intake. In rats, which are nocturnal feeders, peak concentrations of ACTH and corticosterone occur prior to initiation of the feeding period (Miyabo et al., 1980). The circadian rhythm thus favors higher

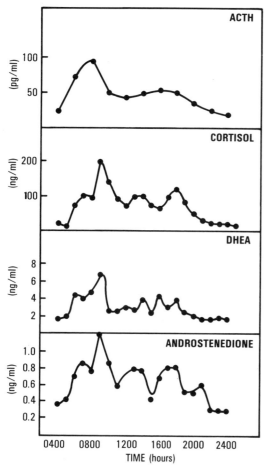

Figure 8.33. Diurnal variation in ACTH and adrenocortical steroid hormones. Data derived by hourly sampling in a 24-yr-old woman. DHEA-S levels (not shown) do not vary because of slow clearance.

cortisol concentrations during the major period of fasting each 24 h, whether that occurs during the day or night.

THE RENIN:ANGIOTENSIN:ALDOSTERONE AXIS

Aldosterone, the principal sodium-retaining steroid hormone in man, maintains normal fluid balance and circulatory volume. Aldosterone production is regulated by the renin:angiotensin system, as well as by ACTH and potassium. The enzyme renin is produced principally in juxtaglomerular cells located in the media of afferent arterioles at their entry into the renal glomerulus (Fig. 8.34). Renin, a proteolytic enzyme, cleaves the amino terminal decapeptide from angiotensinogen or renin substrate, a glycoprotein which is synthesized in the liver. The decapeptide angiotensin I is further processed by converting enzyme which removes the two carboxy terminal amino acids to give the active

octapeptide angiotensin II. Angiotensin converting enzyme is widely distributed in the endothelium of vascular beds, especially in the lung. Although angiotensin II is the principal active peptide hormone, angiotensin III (des Asp^1 angiotensin II) is also active on zona glomerulosa adrenocortical cells.

The concentration of renin in blood is the principal regulator of this system. Juxtaglomerular cells synthesize and secrete renin in response to several signals: baroreceptor, β-adrenergic, prostaglandin, and the fluid composition of the distal nephron at the macula densa. The juxtaglomerular cells, located in the wall of the afferent glomerular arteriole, respond as baroreceptors and secrete renin in response to changes in renal perfusion pressure: decreased perfusion pressure stimulates renin secretion, whereas increased perfusion pressure inhibits renin secretion. Renal sympathetic nerves, which terminate at the juxtaglomerular and afferent arteriolar smooth muscle cells, increase renin secretion via β-adrenergic receptors. Prostaglandins also increase renin secretion. The macula densa, which is a specialized area of the distal nephron in proximity to juxtaglomerular cells, monitors tubular fluid composition and mediates renin release. There is controversy as to which signal is sensed (Na^+, Cl^-, Ca^{2+}) and how it is transmitted to the juxtaglomerulosa cell (Davis and Freeman, 1976).

When intravascular volume is decreased, as occurs with dehydration or hemorrhage, renal perfusion pressure is decreased, the sympathetic nervous system is activated, and renin synthesis and secretion are increased (Fig. 8.34). Under most conditions adequate concentrations of angiotensinogen are present in blood, and increased renin produces increased angiotensin I, which is rapidly converted to angiotensin II by the large amount of converting enzyme available. Angiotensin II is an extremely potent vasoconstrictor through its action on arterial smooth muscle cells. It acts not only systemically to raise blood pressure but also locally to decrease filtration at the glomerulus. Angiotensin II and III bind to specific high affinity receptors in the adrenal cortex to stimulate synthesis and secretion of aldosterone. Aldosterone acts on its targets, primarily the distal nephron, to cause sodium reabsorption and consequent expansion of intravascular and extracellular fluid volumes. When intravascular volume and renal perfusion pressure are increased, renin secretion is suppressed.

Potassium is also a potent regulator of aldosterone synthesis with small changes in plasma [K$^+$] within the physiological range altering aldosterone

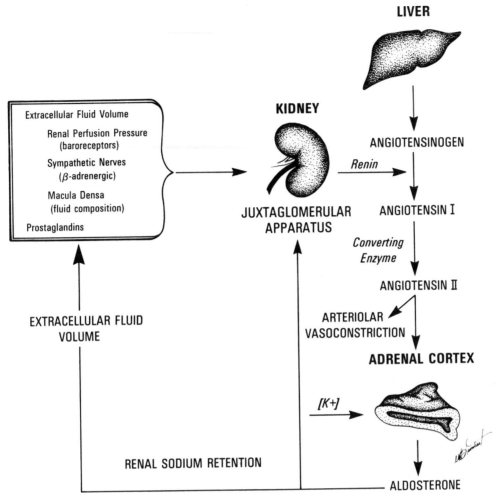

Figure 8.34. Renin:angiotensin:aldosterone axis.

synthesis. Increases in plasma [K$^+$] elevate aldosterone synthesis, whereas decreases in plasma [K$^+$] lower aldosterone synthesis. Because aldosterone promotes Na$^+$ readsorption by facilitating K$^+$ and H$^+$ secretion, this regulatory system is a homeostatic one to maintain normal plasma [K$^+$].

Several additional regulators of zona glomerulosa and zone fasciculata steroid hormone synthesis have been identified, but the physiological role of these is questionable. β-Lipotropin and β-melanotropin (a fragment of β-lipotropin), which are derived from the ACTH precursor, preferentially stimulate aldosterone synthesis in glomerulosa cells (Matsuoka et al., 1981). The concentration required exceeds that usually found in plasma. The 16,000 M$_r$ fragment and the γ-MSH portion of that derived from the ACTH precursor potentiate low concentrations of ACTH to stimulate corticosterone synthesis (Pedersen and Brownie, 1980). Although the concentration required is within the physiological range, the total effect is small.

Effects of ACTH and Angiotensin on the Adrenal Cortex

ACTH works via cAMP. The binding curve for ACTH is superimposable on the concentration curve for cAMP formation; however, only a small fraction of receptors need to be occupied to give a maximal steroidogenic response (Hornsby and Gill, 1978) (see Chapter 51). Proof of the hypothesis that cAMP mediates the effects of ACTH was provided by showing that mutant adrenocortical cells with altered cAMP-dependent protein kinase have responses to ACTH that are altered in parallel (Rae et al., 1979). Acutely, ACTH stimulates steroid hormone biosynthesis by facilitating interaction of free cholesterol with cytochrome P-450 side chain cleavage enzyme. All steroidogenic stimuli acutely increase pregnenolone formation; the conversion of pregnenolone to final steroid products depends on the enzyme complement of the cell. Chronically, ACTH induces the enzymes of the steroidogenic pathway to increase biosynthetic capacity. ACTH increases the quantity of 17α-, 11β-, and 21-hydroxylase and 3β-hydroxysteroid dehydrogenase $\Delta^{4,5}$-isomerase enzymes so that cortisol is the principal product of the biosynthetic pathway. Following hypophysectomy there is atrophy of the adrenal cortex, with a corresponding decrease in the enzymes of the steroidogenic pathway; administration of ACTH restores adrenocortical structure and steroidogenic enzymes to normal (Liddle et al., 1962; Purvis et al., 1973). When ACTH concentrations are elevated, there is enlargement of the cortex and increased steroidogenic responsiveness. Induction of steroid biosynthetic enzymes by ACTH requires concomitant cellular hypertrophy which depends on the availability of growth factors and nutrients (Gill et al., 1982). ACTH and cAMP inhibit cell replication (Masui and Garren, 1971) so that in conjunction with growth factors the adrenocortical cell is hypertrophied with increased functional capacity (Gill et al., 1979). Inhibition of cell division is of advantage for a system which is continuously changing its function, because hypertrophy is a readily reversible process whereas cell replication is reversible only by cell death exceeding cell division. Cell replication occurs only with prolonged ACTH treatment and significant receptor desensitization (Hornsby and Gill, 1977). The link coordinating adrenal delivery of growth factors with ACTH is not known. Following unilateral adrenalectomy, growth of the remaining gland is mediated via neural pathways, suggesting that neural signals are important growth promoters for the adrenal cortex (Engeland and Dallman, 1975).

When cortisol or synthetic derivatives are used clinically, ACTH production is diminished, and atrophy of the adrenal cortex occurs. When high-dose glucocorticoids are administered for about 4 wk or longer, recovery of the normal feedback axis is delayed. A similar pattern is observed following removal of cortisol-secreting adrenal and ACTH-secreting pituitary tumors. As shown in Figure 8.35, recovery of ACTH is delayed. When ACTH production resumes, it exhibits a diurnal secretory pattern, even though adrenal production of cortisol is low. Recovery of the adrenal cortex follows pituitary recovery so that the entire process may require 6–9 mo (Graber et al., 1965). Because the zona glomerulosa is also regulated by angiotensin, exogenous or endogenous elevations of cortisol do not suppress aldosterone production. Patients receiving pharmacologic amounts of glucocorticoids, therefore, maintain normal fluid and electrolyte balance via the renin:angiotensin:aldosterone axis.

Angiotensin principally affects the zona glomerulosa to increase aldosterone synthesis. Angioten-

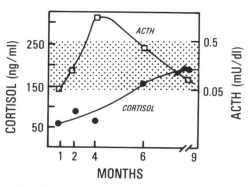

Figure 8.35. Recovery of the hypothalamic:pituitary:adrenocortical axis after exposure to long-term elevated concentrations of glucocorticoids. All measurements were made at 0600 h and *closed circles* represent median values from 14 patients. The *dotted area* represents the range for normal subjects. [From Tyrrell and Baxter (1981), based on original data of Graber et al. (1975).]

sin does not appear to work via cAMP but may use calcium as a second messenger (Garrison et al., 1979). Both the early step of pregnenolone formation from cholesterol and the later step of aldosterone formation from corticosterone are increased (Kramer et al., 1980). Angiotensin exerts a trophic effect as well as an acute stimulatory effect on the zona glomerulosa. The glomerulosa zone widens during sodium depletion, and angiotensin II stimulates growth of adrenal cells in culture (Gill et al., 1977). The boundary between the zona glomerulosa and zona fasciculata moves inward or outward, depending on the long-term need for mineralocorticoid or glucocorticoid. Inward movement with enlargement of the zona glomerulosa occurs with sodium depletion and angiotensin elevation via cell growth and inhibition of conversion of glomerulosa to fasciculata cells. Stimulation with ACTH, which acutely increases aldosterone synthesis, moves the boundary outward by facilitating conversion of glomerulosa to fasciculata cells. As noted, zonation depends on the blood flow and may involve gradients of steroid hormones and free radical effects on enzymes.

Life History of the Adrenal Cortex

In man and other primates the fetal adrenal gland is very large relative to body size; it approaches the size of the fetal kidney at term. The gland is composed of an outer subcapsular definitive or adult zone, which is the precursor of the three zones of the postnatal cortex, and a large inner fetal zone, which constitutes about 80% of the fetal adrenal cortex (Lanman, 1953). The cells of the fetal zone lack 3β-hydroxysteroid dehydrogenase,$\Delta^{4,5}$-isomerase (Fig. 8.31) so that precursors are converted to dehydroepiandrosterone, which is sulfated and secreted as the principal steroid product (Diezfalusy,

1964). The fetal zone synthesizes large amounts of steroids (100–200 mg/day at term) which serve as precursors for synthesis of estrogens by the placenta (Sitteri and McDonald, 1966). This fetoplacental unit is terminated at parturition, and after birth there is rapid involution of the fetal zone and establishment of the three zones of the definitive cortex from subcapsular cells. Fetal ACTH is required for maintenance of both fetal and definitive zones (Bernischke, 1956). When fetal zone cells are placed into cell culture, ACTH induces the full steroid biosynthetic pathway of the definitive cortex (Simonian and Gill, 1981), suggesting that, in utero, local conditions specifically suppress 3β-hydroxysteroid dehydrogenase.

Cortisol production remains constant throughout life, as measured under both basal and ACTH-stimulated conditions (Pintor et al., 1980). During early childhood to age 5, adrenal androgen secretion is low. Production of adrenal androgens, which contribute to the development of axillary and pubic hair (adrenarche), then increases progressively after age 5 to reach a maximum around age 13. This rise coincides with progressive development of the inner reticularis zone (Dhom, 1973), which produces more androgens relative to cortisol than does the fasciculata zone (O'Hare et al., 1980). Although a specific adrenal androgen-stimulating hormone has been proposed, it has not been identified. Alternatively, autoregulatory mechanisms within the adrenal cortex may cause increased adrenal androgen production. The enlarging reticularis zone is exposed via centripetal blood flow to increased concentrations of steroid hormones produced by outer zones; these steroids may preferentially decrease steroidogenic enzymes such as cytochrome $P\text{-}450_{11\beta}$ and 21 to divert precursors into androgen end products (Hornsby, 1980). After puberty there is a progressive decline in adrenal androgen secretion, with DHEA declining about 5-fold from ages 30 to 80.

Biological Effects of Glucocorticoids

Cortisol, the principal glucocorticoid in man, has widespread effects on most organs of the body to regulate metabolism of protein, nucleic acid, and fat, as well as carbohydrate. Physiologically, cortisol mediates adaptive responses to stress and fasting; pharmacologically, cortisol is used to suppress the inflammatory response. In fasting, which is the best-studied paradigm, cortisol maintains blood glucose by stimulating gluconeogenesis. Glucose is the required substrate for nervous system metabolism until chronic adaptation to fasting allows ketone bodies to be utilized. In the absence of food intake, the approximately 75 gm of glycogen stored in the liver are sufficient to maintain blood glucose

for only 12–24 h. The large amounts of muscle and other body proteins, as well as fat, provide a much greater potential substrate reserve to maintain blood glucose and nervous system function. As shown in Figure 8.36, the liver is the principal site of gluconeogenesis, although in prolonged fasting the kidneys also contribute. Cortisol has anabolic effects on the liver, inducing the synthesis of a number of enzymes involved in transamination of amino acids and in gluconeogenesis (Baxter, 1979). Cortisol also increases glycogen synthesis and accumulation in the liver. The effects of cortisol on most other organs are catabolic and provide substrate for hepatic glucose production. In muscle, which is quantitatively the most important source, cortisol inhibits protein and nucleic acid synthesis and enhances protein breakdown to provide amino acids, principally alanine, for use by the liver (Felig, 1975). Cortisol blocks both glucose and amino acid uptake in the periphery to further enhance gluconeogenesis and blood glucose levels. Cortisol increases lipolysis and enhances the effects of other lipolytic stimuli such as catecholamines. Free fatty acids are an alternate fuel source via conversion to ketone bodies and provide reducing equivalents to sustain gluconeogenesis.

The effects of cortisol on individual organ systems are based on this pattern of reactions. Clinically, the role of cortisol is most evident when it is present in excessive amounts. Its role in reestablishing homeostasis after transient disturbances must often be inferred from analysis of deficiency or excess states. Because adrenal insufficiency involves loss of additional hormones, the effects of cortisol are most easily seen when cortisol is produced in excessive amounts or when it is administered in pharmacological doses.

The principal clinical use of cortisol and synthetic glucocorticoids is to suppress inflammatory responses. Cortisol inhibits migration of polymorphonuclear leukocytes, monocyte-macrophages, and lymphocytes at the site of inflammation; it inhibits release of vasoactive, proteolytic, and other peptides; it inhibits fibroblast growth and wound healing. Although the mechanisms through which cortisol exerts its anti-inflammatory effects are not well understood, it may work in part through increased synthesis of "macrocortin" (Blackwell et al., 1980; Hirata et al., 1980). This polypeptide inhibits phospholipase A_2, thereby decreasing membrane release of arachidonic acid, the precursor of both prostaglandins and leukotrienes, two mediators of inflammation. Cortisol suppresses cell-mediated immunity with greater effects on T and B lymphocytes. A lympholytic effect is used in treatment of leukemia and lymphoma but is probably

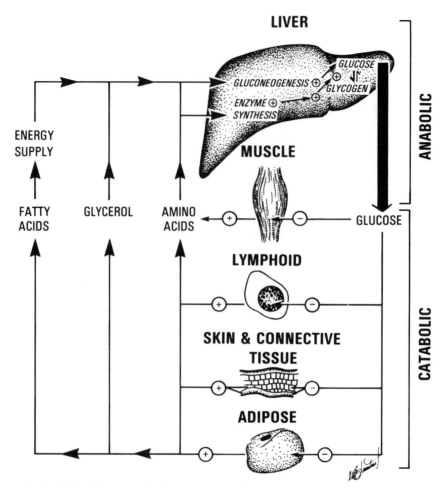

Figure 8.36. Glucocorticoid effects. The *arrows* indicate the general flow of substrate in response to glucocorticoids. The + and − signs indicate stimulation or inhibition, respectively. [From Baxter (1979).]

not a prominent effect under normal conditions. Cortisol decreases circulating lymphocytes, monocytes, eosinophils, and basophils. It increases circulating polymorphonuclear leukocytes, primarily by release from storage in vessel walls. Antibody production is not usually affected by glucocorticoids.

The catabolic effects of high concentrations of cortisol are evident on muscle where there is loss of muscle mass and development of weakness, on skin where there is thinning and separation of subcutaneous tissues to result in striae, and on bone where there is osteoporosis and a proclivity to bone fractures. Glucocorticoids inhibit bone cell function and deposition of the collagen matrix; they also inhibit gastrointestinal calcium absorption so that calcification of the matrix is impaired. Glucocorticoids may interfere with gastric synthesis of prostaglandins, which are required for maintenance of the normal protective barrier against gastric acid and pepsin. Capillary fragility is increased, and bruising with minor trauma is common. High concentrations of glucocorticoids inhibit linear growth

and skeletal maturation in children. Growth hormone is suppressed, but the major factor in reduced growth is inhibition of protein synthesis in many cell types.

In fed humans the effects of high concentrations of glucocorticoids may result in hyperglycemia. Cortisol also stimulates appetite so that weight gain is common. Insulin rises in response to hyperglycemia and may accentuate a peculiar central deposition of fat in the face, neck, and abdomen.

In response to severe stress, adrenal production of cortisol may rise as much as 10-fold. This is an adaptive response which favors survival. The mineralocorticoid, deoxycorticosterone, was introduced into clinical medicine before cortisol. This advance improved survival of patients with adrenal insufficiency 3- to 5-fold. The introduction of cortisol had a much greater impact, increasing survival more than 10-fold (Dunlop, 1963). Knowledge of basal production rates and of the capacity of the gland to increase production in response to stress allows ready calculation of replacement doses of cortisol for patients with adrenal insufficiency.

In addition to the effects of cortisol mentioned above, it is required ("permissive") for the action of a number of hormones (e.g., catecholamines) and for synthesis of others (e.g., growth hormones). Cortisol is also important during development. Cortisol increases surfactant synthesis in fetal lung, glutamine synthetase in neural retina, and hepatic enzymes in fetal liver, and is required for breast development. It is required for normal growth. When cortisol concentrations are elevated over prolonged periods as in Cushing's disease or with pharmacological use, catabolic effects are sustained. Therapy may yield desired anti-inflammatory effects, but a price is paid by all organ systems.

Biological Effects of Mineralocorticoids

Aldosterone, the principal mineralocorticoid, stimulates reabsorption of Na^+ in the collecting tubules of the kidney. When Na^+ is reabsorbed from tubular fluid, electrical neutrality must be maintained by secretion of K^+ or H^+ or by concomitant reabsorption of an anion such as Cl^-. The exchange of K^+ and H^+ for Na^+ will be greater when the quantity of Na^+ reaching the collecting tubule is increased. When extracellular fluid volume is decreased as, for example, by gastrointestinal losses or by hemorrhage, activation of the renin:angiotensin system increases aldosterone to promote Na^+ conservation. Because Na^+ is the principal extracellular substance active as an osmole, water is also retained to expand the depleted extracellular fluid volume (Fig. 8.34). With adrenal insufficiency, Na^+ conservation in the distal tubule is defective so that excessive Na^+ is lost in the urine, with resultant hyponatremia and contraction of the extracellular fluid volume. Although most Na^+ is reabsorbed in the proximal tubule, with aldosterone deficiency up to 240 mEq of Na^+ may be lost per day because of failure of Na^+ conservation by the collecting tubules. Because cation exchange is not occurring, hyperkalemia and acidosis result. The diminished extracellular fluid volume results in decreased blood pressure and reduced cardiac output. Cortisol also exerts important effects on the circulation and is required both for normal cardiac output and for vasoconstrictor effects of catecholamines. If adrenal insufficiency is uncorrected, circulatory collapse and death occur. Mineralocorticoid deficiency is a strong contributor to the muscle weakness characteristic of adrenal insufficiency and also contributes to poor growth in children. In adrenal insufficiency ADH concentrations are increased to facilitate water conservation and cellular hydration so that hyponatremia represents both Na^+ deficiency and dilution of extracellular $[Na^+]$. Patients with adrenal insufficiency are characteristically unable to excrete an administered load of free water because of elevated concentrations of ADH and lack of effects of cortisol on the collecting duct. Aldosterone also regulates ion transport in sweat glands, salivary glands, and the gastrointestinal tract. When exposed to increased ambient temperatures, patients with adrenal insufficiency continue to produce sweat with a high Na^+ content and are unable to adapt to climate changes.

When aldosterone is produced or administered in excessive amounts, there is increased Na^+ reabsorption and facilitated K^+ and H^+ exchange, resulting in hypokalemia and alkalosis. Hypokalemia and alkalosis are more pronounced when increased Na^+ reaches the collecting tubule for aldosterone driven exchange. Initially, isotonic volume expansion of 2–4 l will occur over 2–3 days. After this, intrarenal compensatory mechanisms are activated to result in reduced Na^+ reabsorption in the proximal tubule. The body thus "escapes" from the effects of aldosterone, and Na^+ excretion then equals Na^+ intake. Facilitated cation exchange in the collecting tubule continues, and hypokalemic alkalosis is sustained. The modest volume expansion also continues. Although edema does not occur, hypertension results when excess aldosterone is present for a long time. Magnesium and calcium excretion are increased when aldosterone is elevated because of decreased proximal tubular reabsorption in parallel with decreased proximal tubular reabsorption of Na^+ in the compensated state. When aldosterone is elevated because of autonomous production by adenomas or hyperplasia of the glomerulosa zone, the renin:angiotensin system will be suppressed because of volume expansion. As occurs more commonly with processes such as congestive cardiac failure and liver cirrhosis, the renin:angiotensin system is activated, and secondary hyperaldosteronism results. In these conditions, edema results from Na^+ retention and the primary disease process.

As for all steroid hormones, aldosterone acts via a specific receptor protein to regulate specific gene expression. In renal collecting tubules, Na^+ enters at the luminal membrane and is extruded at the serosal surface into the interstitial fluid by action of the Na^+/K^+ ATPase enzyme, which moves a Na^+ ion out of the cell and a K^+ ion into the cell. Aldosterone induces synthesis of at least four mitochondrial enzymes involved in energy generation to drive the Na^+/K^+ ATPase (Edelman and Marver, 1980). It may also increase Na^+ channels, with the ion gradient being a major driving force for activity of Na^+/K^+ ATPase (Ludens and Fanestil, 1979). Aldosterone also appears to increase the H^+ antiporter and may increase K^+ secretion into tubular fluid by either increasing K$^+$ channels or by means of a potassium pump.

Pharmacology

In adrenal insufficiency, replacement of glucocorticoid and mineralocorticoid hormones sustains normal life. However, the principal clinical use of adrenal steroids is to suppress undesirable inflammatory reactions, and more than 7 million people in the United States receive them each year. Because glucocorticoid receptors appear identical in all cell types, it has been impossible to synthesize a steroid molecule which has anti-inflammatory but not other glucocorticoid effects. Because mineralocorticoid receptors are distinct from glucocorticoid receptors, it has been possible to synthesize steroid molecules with relatively pure glucocorticoid or mineralocorticoid effects. As shown in Figure 8.37, a number of substitutions can be made on the cortisol molecule. Introduction of a double bond between carbons 1 and 2 increases glucocorticoid potency (prednisone and prednisolone). A halogen substitution at carbon 9 increases both glucocorticoid and mineralocorticoid activities. Because this substitution increases mineralocorticoid activity especially, it results in an orally active mineralocorticoid (9α-fluorocortisol) which is used clinically because the degradation of aldosterone in the liver precludes its use as an oral agent. Methylation at carbon 16 greatly reduces mineralocorticoid activity. 9α-Fluoro-16α-methylprednisolone (dexamethasone) is thus a potent glucocorticoid with essentially no mineralocorticoid activity. Methylation at carbon 6 increases water solubility. A variety of combinations of substitutions at these indicated positions results in the many available steroid products.

Glucocorticoid antagonists are not available, but the mineralocorticoid antagonist spironolactone, which binds to the mineralocorticoid receptor to induce an inactive conformation, is used in conditions of aldosterone excess.

It is evident that long-term high dose glucocorticoid therapy results in side effects and prolonged suppression of the hypothalamic:pituitary:adrenocortical axis. A number of strategies have been useful in minimizing these two unwanted effects. Strategies include use of minimal doses which are effective in treating the disease process, timing of administration to morning or every other day schedules to allow function of the normal circadian rhythm, and local application. Because the renin:angiotensin system is not affected by administered glucocorticoids, mineralocorticoid function remains normal.

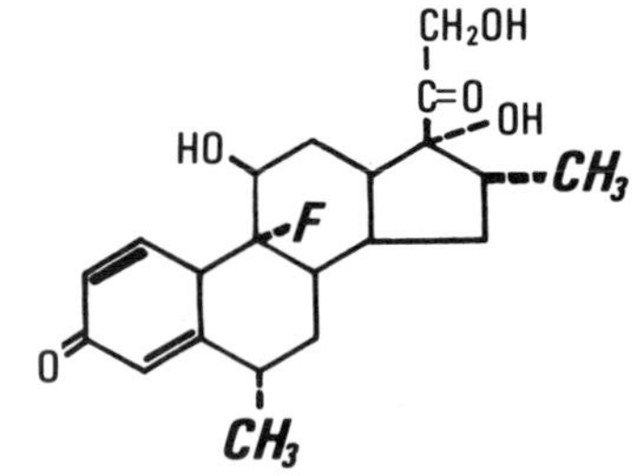

Figure 8.37.   Pharmacologic modifications of the cortisol molecule.

THE ADRENAL MEDULLA

The adrenal medulla is both a part of the autonomic nervous system and the endocrine system. The sympathetic chromaffin granule-containing cells of the adrenal medulla are equivalent to postganglionic neurons. The adrenal medulla is innervated by cholinergic preganglionic fibers from the greater splanchnic nerve. Acetylcholine released from these preganglionic neurons activates cells of the adrenal medulla to synthesize and secrete catecholamines. In contrast to other postganglionic sympathetic neurons, which release norepinephrine from axon termini at their site of action, adrenal medullary cells release epinephrine into the circulation. Adrenal medulla cells are thus neuronal cells which function as an endocrine gland.

Because the adrenal medulla is properly considered as an integral part of the sympathetic nervous system, full discussion of catecholamine synthesis, metabolism, and action is included in Chapter 69. The adrenal medulla is unique in containing phenylethanolamine-N-methyltransferase, an enzyme which transfers a methyl group to the amino terminus of norepinephrine to form epinephrine. Epinephrine is the principal hormonal product of the adrenal medulla, although smaller amounts of norepinephrine are also released. In addition to catecholamines, the adrenal medulla also produces opioid peptides, including met-enkephalin, leu-enkephalin, and related heptapeptide and octapeptide sequences (Noda et al., 1982).

Epinephrine functions as an integral part of the sympathetic nervous system, stress, and metabolic responses. Secretion is coordinated with cortisol production via the CRH axis described in Chapter 52. Despite the important physiologic role of the adrenal medulla, it is not essential for survival. Following adrenalectomy, glucocorticoid and mineralocorticoid replacement are essential, but epinephrine is not required because the sympathetic nervous system and norepinephrine are sufficient.

CHAPTER 55

Hormonal Regulation of Mineral Metabolism

INTRODUCTION

Three hormones, vitamin D, parathyroid hormone (PTH), and calcitonin are the principal regulators of mineral metabolism. The endocrine system composed of these three hormones serves to maintain a remarkably constant concentration of ionized calcium in extracellular fluids. This constancy, which is necessary for proper bone mineralization, neuromuscular excitability, membrane integrity, cellular biochemical reactions, stimulus-secretion coupling, and blood coagulation is maintained in spite of wide variations in dietary intake.

The role of this endocrine control system can be appreciated by considering normal calcium balance. A normal adult body contains about 1 kg of calcium. Approximately 99% is present in the skeleton as hydroxyapatite $[Ca_{10}(PO_4)_6(OH)_2]$ and 1% in soft tissues and extracellular fluids. About 1% of skeletal calcium is exchangeable with extracellular fluid, but this 10 gm is large relative to the 900 mg present in extracellular fluid. A normal diet provides from 200 to 2000 mg with an average of 1000 mg of calcium per day (Fig. 8.38). Of this, approximately 300 mg is absorbed, the majority in the ileum because of its large absorptive surface. Calcium is also secreted into the intestinal tract in bile, pancreatic juice, and intestinal secretions so that net absorption of calcium equals about 175 mg/day. The total extracellular fluid space contains about 900 mg of calcium which is in dynamic equilibrium with other compartments. Approximately 500 mg of calcium is deposited into and reabsorbed from bone in the ongoing process of bone remodeling. Sixty percent of serum calcium is ultrafilterable so that approximately 10,000 mg is filtered at the renal glomerulus daily. Renal reabsorption of calcium is extremely efficient, so that only a small amount of calcium equal to that absorbed in the gut is excreted in the urine daily. As shown in Fig. 8.38, calcium balance is zero, with excretion in feces and urine exactly equaling dietary intake.

In spite of these large movements of calcium between body compartments, serum calcium is maintained constant at about 10 mg/dl (2.5 mM). The essential fraction is the ionized one which equals 50% of the total serum calcium (1.3 mM). Forty percent of total serum calcium is protein-bound, principally to albumin, and 10% is complexed with diffusible anions. Because total serum calcium is usually measured, knowledge of the concentration of serum albumin and of pH may be required to estimate the ionized fraction. A decrease in serum albumin of 1 g/dl (normal = 4 g/dl) results in a decrease in total serum calcium of 0.8 mg/dl without affecting the ionized fraction significantly. Because calcium is bound to carboxyl groups on albumin, pH changes affect ionized calcium. Acidosis decreases binding and increased ionized calcium, whereas alkalosis increases binding and decreases ionized calcium. Acidosis thus protects against hypocalcemia by shifting albumin-bound calcium to the ionized form, and alkalosis such as that which occurs with acute hyperventilation decreases ionized serum calcium by increasing binding to serum albumin.

The constancy of extracellular fluid concentrations of calcium is maintained by hormonal regulation of the absorption of calcium from the gastrointestinal tract, the mobilization of skeletal calcium, and the reabsorption of filtered calcium in kidney tubules. Vitamin D, PTH, and calcitonin are the regulators of these processes.

VITAMIN D

Structure and Biosynthesis

Although initially discovered as a fat-soluble vitamin which would prevent the bone disease rickets, vitamin D is a classical steroid hormone having both dietary and endogenous precursors. Historically, rickets was a major health problem accentuated by poor nutrition and by limited exposure to

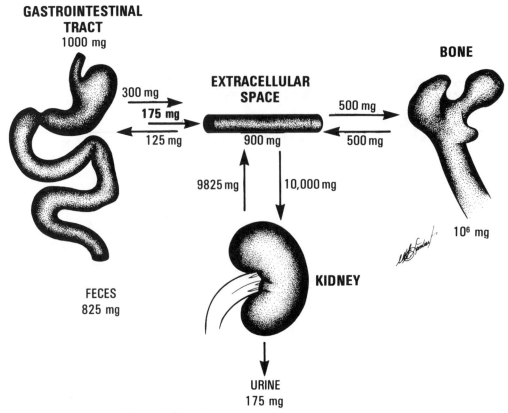

Figure 8.38. Calcium balance on an average diet.

sunlight; it increased with the urban migration and the environment created by the industrial revolution. Poor nutrition and limited exposure to sunlight which were worsened during World War I led to a corresponding increase in clinically evident rickets—and increased study of this disease. It was found that rickets could be treated either by exposure to sunlight or by use of the folk remedy cod liver oil. Vitamin D was isolated in 1930 and is now widely used as a dietary supplement, so that simple vitamin D-deficient rickets is now distinctly unusual.

Vitamin D_3, cholecalciferol, is formed by ultraviolet irradiation of precursor 7-dehydrocholesterol present in skin (Fig. 8.39). This reaction can provide adequate vitamin D so a dietary source is not essential. Adequate exposure to UV light is necessary, with more exposure being required for darker skinned races. Dark-skinned children in northern urban environments where there is less sunlight exposure therefore have an especial requirement for dietary sources of vitamin D. Ergocalciferol, vitamin D_2, is the compound formed in plants from precursor ergosterol and differs from vitamin D_3 by the presence of a double bond between carbon 22 and 23 and a methyl group at carbon 24. Vitamin D_2 and D_3 are not abundant in foods (with the exception of fish liver), and irradiation to convert

precursors to vitamin D_2 and D_3 are required. Both vitamin D_2 and D_3 are prohormones and require subsequent modification to yield active hormone. Because hormones derived from either vitamin D_2 or D_3 are equally active in man, the generic term vitamin D will be used.

Formation of vitamin D from 7-dehydrocholesterol precedes via a 6,7-cis isomer intermediate, previtamin D (Holick and Clark, 1978). Vitamin D is transported in plasma bound to a specific vitamin D-binding protein which binds vitamin D with a greater affinity than previtamin D. Previtamin D thus preferentially remains in skin which serves as a storage depot after irradiation, while vitamin D is removed by the binding protein. Vitamin D-binding protein, an α-globulin synthesized in the liver, functions as do other steroid-binding proteins to provide a reservoir of hormone protected from renal clearance (Haddad and Walgate, 1976). It binds vitamin D, 25-(OH)D, 24,25-(OH)$_2$D, and 1,25-(OH)$_2$D. Vitamin D is metabolized in liver by hydroxylation at carbon 25 to yield 25-(OH)D. This is the major circulating form of vitamin D with a serum concentration of about 25 ng/ml and a half-life of 15 days.

The final metabolic conversion of vitamin D occurs in the kidney where hydroxylation at carbon 1 yields the active hormone 1,25-(OH)$_2$D (Deluca and

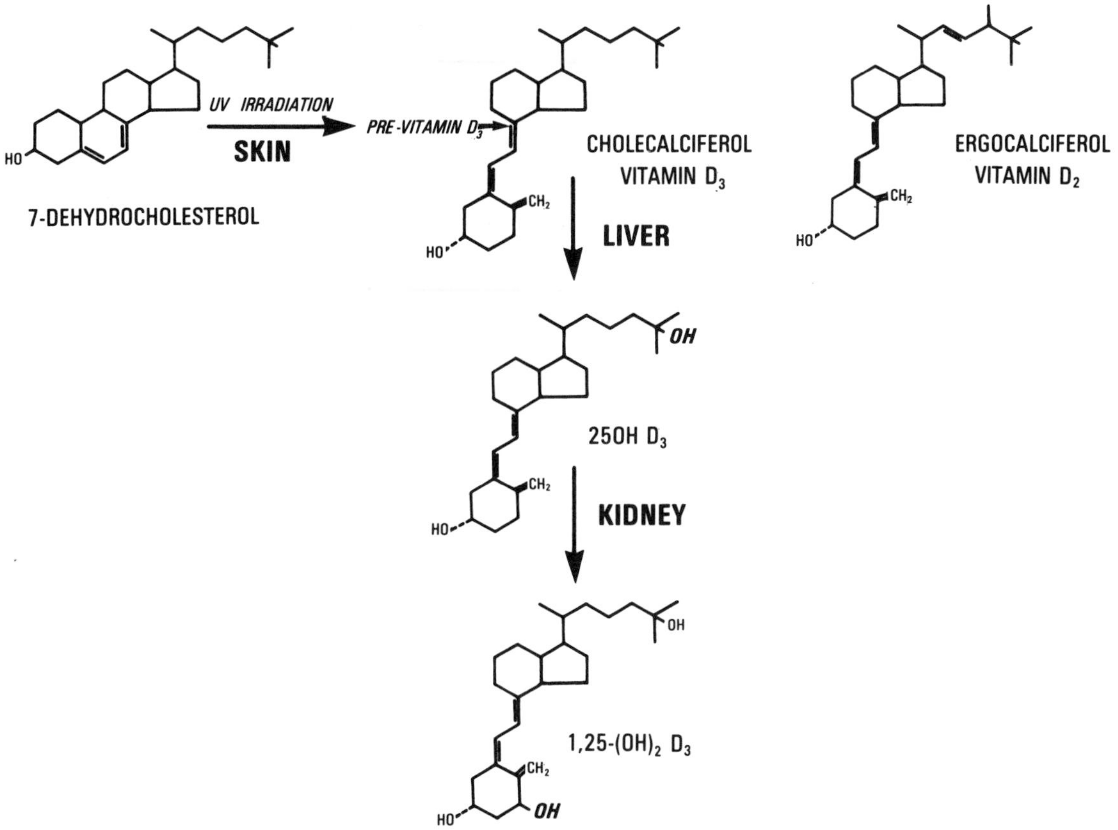

Figure 8.39.   Biosynthesis of metabolically active 1,25-(OH)₂D.

Schnoes, 1976; Fraser, 1980). The 1α-hydroxylation is catalyzed by a mitochondrial mixed function oxidase similar to adrenocortical steroid hydroxylases. This last enzymatic step in the formation of biologically active 1,25-(OH)₂D is the principal site of regulation. 1,25-(OH)₂D has a circulating half-life of about 15 h and a plasma concentration of 20–50 pg/ml.

Several additional modifications of the vitamin D steroid nucleus occur. Hydroxylation at carbon 24 is favored when hydroxylation at carbon 1 is low, and conversely hydroxylation at carbon 24 is low when hydroxylation at carbon 1 is high, suggesting that formation of 24,25-(OH)₂D represents an inactivation pathway. It is possible, however, that 24,25-(OH)₂D has some biological effects on bone (Bordier et al., 1978). Hydroxylation at carbon 26 appears to result in an inactive compound.

Regulation of 1,25-(OH)₂D Formation

To maintain normal calcium homeostasis, the portion of dietary calcium which is absorbed varies. When calcium intake is low the absorbed fraction may be as high as 90%, whereas, when calcium intake is high, much lower fractional absorption occurs. Intestinal absorption of calcium and proper adaptation to dietary intake are mediated by vi-

tamin D. Regulation of 1,25-(OH)₂D is the principal control point for this homeostatic system (Haussler and McCain, 1977).

When oral intake of calcium, phosphate, and vitamin D are decreased, serum levels of calcium and phosphate decrease. Parathyroid hormone secretion is stimulated in response to lowering of ionized serum calcium concentrations and acts to reestablish normal concentrations of serum-ionized calcium by stimulating bone resorption and renal tubular calcium reabsorption. PTH also increases renal phosphate excretion and lowers serum phosphate concentrations. These two signals, elevated PTH and low serum phosphate, act on renal tubular cells to stimulate 1α-hydroxylase activity which increases formation of 1,25-(OH)₂D (Fig. 8.40). This active form of the vitamin increases intestinal absorption of calcium and phosphate so that serum calcium concentrations return to normal, the hypocalcemic stimulus for PTH secretion is removed, and homeostasis is maintained. 1,25-(OH)₂D may also directly decrease PTH formation via a classical feedback loop.

Dietary intake and UV exposure determine the availability of precursor vitamin D₂ and D₃. Intestinal absorption occurs primarily in the ileum and because vitamin D is fat-soluble requires bile salts.

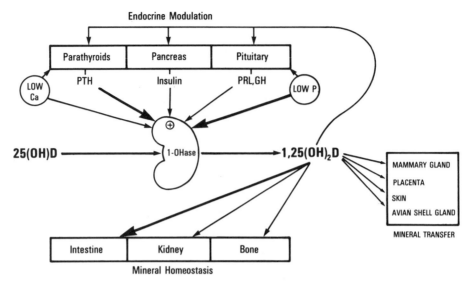

Figure 8.40.   Function and regulation of 1,25-(OH)₂D. [From Haussler and McCain (1977).]

Daily requirements are estimated to be 400 U/day (10 μg). Vitamin D deficiency may occur with chronic biliary obstruction and steatorrhea and lead to osteomalacia. Formation of 25-(OH)D depends largely on the concentration of precursor vitamin D_2 and D_3 and is not a site of major regulation of plasma concentrations of 1,25-(OH)₂D. The activity of renal 1α-hydroxylase controls the concentration of 1,25-(OH)₂D. Because the kidney is the only significant source of 1α-hydroxylase, inadequate formation of 1,25-(OH)₂D occurs in renal failure (Stanbury, 1978). Not only is the mass of kidney tissue and therefore enzyme decreased, but also with renal failure, phosphate excretion is reduced and serum phosphate rises. Increased phosphate inhibits 1α-hydroxylation so that little 1,25-(OH)₂D is formed. Acidosis, a frequent result of renal failure, also impairs 1α-hydroxylase activity (Lee et al., 1977). Deficiency of the active form of vitamin D causes osteomalacia, a prominent feature of renal osteodystrophy. Therapy is directed toward use of 1,25-(OH)₂D, reduction of serum phosphate, and correction of acidosis, so that residual 1α-hydroxlyase can be expressed.

During growth, pregnancy, and lactation the body's requirement for calcium increases. Circulating concentrations of 1,25-(OH)₂D increase to facilitate increased calcium absorption. During pregnancy and lactation, the increase in plasma concentrations is due in part to estrogen-mediated increases in the vitamin D-binding protein and to effects of prolactin on 1α-hydroxylase activity (6). Insulin may also facilitate 1,25-(OH)₂D formation.

Biological Effects

The intestine is the principal target for 1,25-(OH)₂D, although the hormone also affects bone and to a smaller extent kidney. Like other steroid hormones, 1,25-(OH)₂D binds with high specificity and affinity to a receptor protein which functions to increase gene expression. The net effect of 1,25-(OH)₂D is to increase intestinal absorption of calcium.

At a normal calcium intake of 1000 mg about 15% of calcium absorption occurs by passive diffusion (Fig. 8.38). The amount of calcium absorbed via this mechanism is not sufficient to sustain calcium balance, even when calcium intake is significantly increased. The major component of calcium absorption occurs by an active transport process which is regulated by 1,25-(OH)₂D. The rate-limiting step is uptake of calcium at the mucosal surface; 1,25-(OH)₂D increases this process. In intestinal as in other cells the cytoplasmic concentration of calcium is low (μM) relative to that in the extracellular fluid (mM). Calcium which enters the cell must therefore be sequestered as it moves from the mucosal surface to the luminal surface where it is transported out of the cell into the extracellular compartment. Calcium is sequestered inside the cell in mitochondria and by a 15,000 M_r calcium-binding protein whose concentration is regulated by 1,25-(OH)₂D (Wasserman et al., 1976). Vitamin D also induces alkaline phosphatase and calcium-dependent ATPase. Although all of these identified mechanisms contribute to 1,25-(OH)₂D-induced intestinal absorption of calcium, none are sufficient to explain the increased mucosal transport and additional biochemical events are undoubtedly induced by 1,25-(OH)₂.

1,25-(OH)₂D also increases intestinal phosphorus absorption by increasing the active transport of phosphorus. Phosphorus is abundant in the diet, and its absorption is less tightly regulated than that of calcium. A normal diet provides about 1400 mg of phosphorus per day with a net phosphorus absorption of about 900 mg/day. Five hundred milligrams is excreted in feces and 900 mg in urine to maintain phosphorus balance. Phosphorus balance is positive during growth.

Bone is the second major target for vitamin D. Mineralization of bone is a complex process involving both simple and complex molecules in addition to calcium and phosphate (Boskey, 1981). The collagen matrix at the growing end of long bones is synthesized and secreted by chondrocytes. Influx of calcium and phosphate as well as other substances to the extracellular matrix is regulated via these cells. Mineralization is initiated on the extracellular matrix, and crystal growth then fills the matrix with mineral. Blood vessels invade the calcified cartilage matrix, and bone remodels to achieve optimal strength. Bone remodeling is a process which continues throughout life, long after epiphyseal fusion and cessation of linear growth of bone. Bone remodeling, which consists of bone formation and bone resorption, occurs continually, and a significant fraction of bone is replaced each year. Osteoblasts are the primary cells concerned with synthesis of new bone. These cells, which cover bone-forming surfaces, produce bone osteoid which will subsequently undergo calcification. Osteocytes are mature bone cells which are less active than osteoblasts. Osteoclasts are multinucleated cells derived from macrophages which function to resorb bone.

Vitamin D deficiency is characterized by defective mineralization of bone osteoid. By increasing intestinal absorption of calcium and phosphate, vitamin D provides sufficient calcium and phosphate to initiate the crystallization process at bone surfaces. 1,25-$(OH)_2D$ also directly affects bone cells to facilitate mineralization even without measurable changes in the plasma concentrations of calcium and phosphate (Boris et al., 1978). 1,25-$(OH)_2D$ increases alkaline phosphatase enzyme activity in osteoblast-like cells (Manolagas et al., 1981). In vitro, 1,25-$(OH)_2D$ also acts on bone to increase resorption, an effect potentiated by parathyroid hormone. Although this resorptive effect of vitamin D appears antithetical to its major action to increase bone mineralization, it may be important in the dynamic process of bone remodeling.

Vitamin D facilitates calcium reabsorption in the distal nephron. There are several other tissues which translocate calcium which have documented 1,25-$(OH)_2D$ receptors, including skin, mammary gland, placenta, and the avian shell gland. Parathyroid, pituitary, and pancreas also contain 1,25-$(OH)_2D$ receptors. Although the precise role of 1,25-$(OH)_2D$ in these tissues is uncertain, effects on calcium and phosphate transport and feedback effects on overall calcium homeostasis are likely.

PARATHYROID HORMONE

Structure and Biosynthesis

Parathyroid hormone (PTH) is an 84 amino acid peptide synthesized within chief cells of parathyroid glands. There are four parathyroid glands in man, each weighing about 40 mg, located behind the thyroid gland (Fig. 8.23, Chapter 53). Occasionally, inferior parathyroid glands may migrate into the mediastinum during development, and occasionally accessory glands are present. In parathyroid chief cells, PTH is synthesized as part of a larger precursor protein containing a 25 amino acid hydrophobic residue-rich signal sequence and a 6 amino acid prosequence preceding the amino terminus of PTH (Habener and Potts, 1978). These portions of the precursor are sequentially removed within the cell, and mature 84 amino acid PTH is secreted from storage granules. A large parathyroid secretory protein which is cosecreted with PTH may function as a PTH carrier during maturation of the PTH precursor in chief cells (Majzoub et al., 1979). Biosynthesis is closely coupled with secretion: the amount of stored hormone is small and would be exhausted under stimulated conditions within 1–2 h without a concomitant increased rate of hormone synthesis. After secretion, PTH is cleaved between residues 33 and 34. The amino terminal fragment possesses the full biological activity of the intact molecule, whereas the carboxy terminal fragment is biologically inactive. Because clearance of the carboxy terminal fragment from the circulation is relatively slow, it provides a convenient marker for PTH secretion. Radioimmunoassay quantitation of the carboxy terminal fragment correlates well with PTH overproduction states provided renal function is normal. In renal failure the fragment accumulates because of diminished clearance as well as overproduction.

Regulation of Synthesis and Secretion

Synthesis and secretion of PTH are regulated by the concentration of ionized calcium in serum. When ionized serum calcium decreases below the physiological set point of 1.3 mM, PTH synthesis and secretion increase to reestablish homeostasis. When hypocalcemia is prolonged, significant parathyroid growth occurs to augment biosynthetic capacity. When ionized serum calcium increases above 1.3 mM, PTH synthesis is suppressed and serum calcium diminishes. A low rate of residual PTH synthesis and secretion persist even with very high levels of serum calcium.

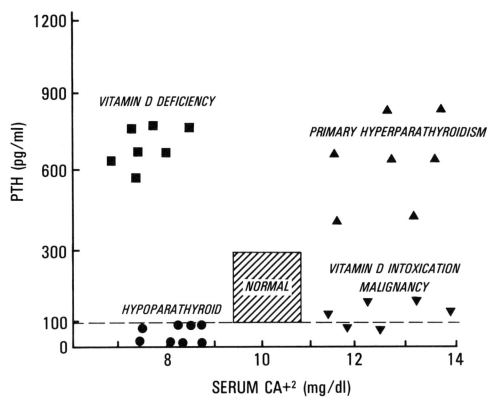

Figure 8.41. Simultaneous PTH and calcium concentrations in patients with various forms of hypo- and hypercalcemia with normal renal function.

By simultaneously measuring serum calcium and PTH, the feedback axis can be accurately evaluated. When hypocalcemia results from PTH deficiency, PTH concentrations are low, whereas when hypocalcemia results from vitamin D deficiency, PTH concentrations are elevated as a compensatory response to hypocalcemia (Fig. 8.41). In hypercalcemia due to malignancy and vitamin D intoxication, PTH concentrations are low. In contrast in primary hyperparathyroidism where PTH secretion from adenomatous or hyperplastic parathyroid glands is autonomous, PTH concentrations are elevated even though hypercalcemia is present.

PTH synthesis is also stimulated by β-adrenergic agonists, but this is a minor effect physiologically compared to that of ionized calcium (Brown et al., 1977). Phosphate does not affect PTH synthesis. Magnesium is required for PTH synthesis, and in severe hypomagnesemia PTH synthesis is impaired.

Biological Effects

Bone and kidney are the two principal target organs affected by PTH. PTH cell surface receptors are coupled to adenyl cyclase, and cAMP is the intracellular mediator of PTH action (Chase and Aurbach, 1967). In one form of pseudohypoparathy-roidism characterized by tissue resistance to the biological effects of PTH, a genetic defect exists in the GTP-binding protein which couples hormone receptors to catalytic cyclase subunits (Farfel et al., 1980). Ineffective generation of cAMP results in apparent resistance to the biological effects of PTH.

The major effect of PTH is to maintain normal ionized serum calcium concentrations. PTH stimulates bone resorption, releasing calcium into extracellular fluid. The principal target of PTH is the osteoblast. PTH inhibits osteoblast synthesis of new bone, but increases osteoblast-initiated recruitment of osteocytes and osteoclasts which participate in bone resorption. The overall effect is increased bone resorption and decreased new bone formation. When PTH concentrations are persistently elevated, significant bone loss occurs with subperiosteal resorption, formation of bone cysts, and spontaneous fractures.

PTH has two major effects on the kidney: to increase calcium reabsorption in the distal nephron and to decrease proximal tubular reabsorption of phosphate. Both responses function to maintain a normal ionized serum calcium. Of the approximately 10,000 mg of calcium filtered at the glomerulus each day, more than 90% is reabsorbed in the proximal tubule and loop of Henle in a process strongly linked to sodium. Reabsorption of the approximately 1000 mg which reaches the distal tubule is regulated by PTH. When PTH is increased, more calcium is reabsorbed in the distal nephron, whereas, when PTH is decreased, less calcium is reabsorbed, and urinary calcium excretion rises. Calcium excretion thus depends both on the amount of calcium which reaches the distal tubule and on the concentration of circulating PTH. The amount of calcium which reaches the distal nephron in turn depends on the filtered load of calcium and on sodium intake. In hypercalcemia the filtered load of calcium increases so that more calcium is delivered to the distal tubule. In hyperparathyroidism PTH increases fractional reabsorption of calcium, but because of the large calcium load delivered, hypercalcuria results. Calcium reabsorption in the proximal tubule is linked to sodium reabsorption. When sodium intake is high, more sodium escapes reabsorption in the proximal tubule to reach the distal nephron. Under these conditions more calcium also reaches the distal tubule and is

excreted. Increased sodium intake is therefore used in many forms of hypercalcemia to increase calcium excretion and lower serum calcium.

The reabsorption of phosphate which occurs in the proximal tubule is controlled by PTH. PTH decreases proximal tubular reabsorption of phosphate, so that increased urinary excretion of phosphate occurs. Sustained elevations in PTH thus result in hypophosphatemia in addition to hypercalcemia. Both PTH and hypophosphatemia stimulate 1α-hydroxylase activity to increase formation of 1,25-$(OH)_2D$ (Fig. 8.40). PTH effects on gastrointestinal absorption of calcium and phosphate are mediated via 1,25-$(OH)_2D$.

CALCITONIN

Structure and Biosynthesis

Calcitonin is a 32 amino acid polypeptide with an amino terminal disulfide bridge linking positions 1 and 7 and an amidated carboxy terminus (Fig. 8.42). Its two major biological effects are to lower

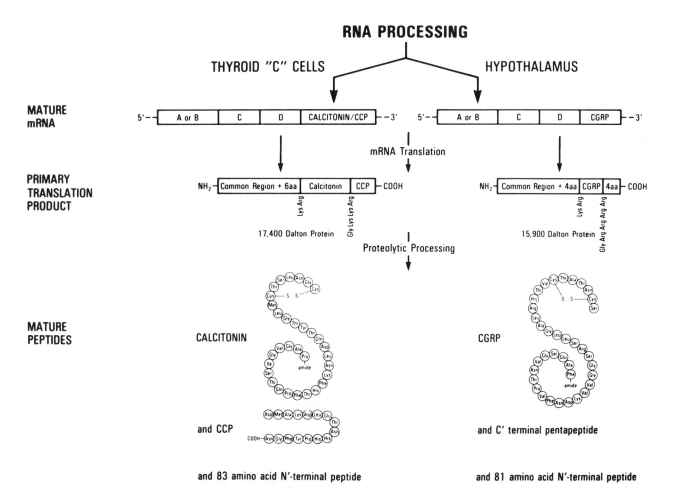

Figure 8.42. Differential processing of the calcitonin gene into calcitonin in thyroid C cells and into calcitonin gene-related product in hypothalamus. *CCP*, calcitonin carboxy-terminal peptide; *CGRP*, calcitonin gene-related peptide. [From Amara et al. (1982).]

serum calcium and phosphate. Calcitonin is synthesized in C cells of neuroendocrine origin which are located primarily within the thyroid gland but to a lesser extent in thymus. In lower vertebrates calcitonin is synthesized in a separate gland, the ultimobranchial body. Like other secretory proteins, calcitonin is synthesized as a portion of a larger precursor protein (Amara et al., 1980). The function of the large 76 amino acid amino terminal fragment and of the smaller 16 amino acid carboxy terminal fragments processed from the calcitonin precursor is not known, but recent evidence indicates that the carboxy terminal fragment (CCP) acts in concert with calcitonin to lower serum calcium (MacIntyre et al., 1982). In other tissues such as hypothalamus an alternate exon is retained in the precursor, giving rise to a calcitonin-gene-related peptide (CGRP) of undefined function (Amara et al., 1982). Circulating concentrations of immunoreactive calcitonin are less than 100 pg/ml; it has a circulatory half-life of about 10 min (Austin and Heath, 1981).

The sequence of calcitonin varies significantly between humans, pigs, sheep, rats, and salmon. All have a 7-membered amino terminal disulfide-bonded ring and a carboxy terminal prolineamide, but differ greatly in the central three-fourths of the molecule. Salmon calcitonin, which has increased biological potency in humans, is the form available for human use.

Regulation of Synthesis and Secretion

Synthesis and secretion of calcitonin are controlled by the concentration of ionized serum calcium. When serum calcium rises above about 9 mg/dl, calcitonin secretion increases in a linear fashion. The stimulus for calcitonin synthesis is thus opposite that for PTH secretion. When ionized serum calcium decreases, PTH secretion rises to restore serum calcium, whereas calcitonin falls to remove a hypocalcemic stimulus. When ionized serum calcium rises, PTH synthesis is suppressed, and calcitonin synthesis is increased to provide a hypocalcemic signal. As noted below the physiological effects of calcitonin on serum calcium in man are minor, and the change in PTH synthesis in response to variations in ionized serum calcium is the major homeostatic adjustment.

Calcitonin synthesis and secretion are stimulated by gastrin, cholecystokinin, glucagon, and β-adrenergic agonists (Cooper et al., 1978). The effects of these hormones on calcitonin synthesis may represent a mechanism for integration of disposition of dietary calcium or may be a reflection of the neuroectodermal origin of C cells.

Biological Effects

Calcitonin reduces bone resorption by inhibiting osteoclast function. This results in a decrease in the concentration of both calcium and phosphate in serum. cAMP serves as the second messenger for calcitonin action. In spite of the reproducible effects of calcitonin on bone and serum mineral content, the hormone does not appear to be a major physiological regulator. After removal of the thyroid-containing C cells and documented calcitonin deficiency or after development of medullary thyroid carcinoma, a neoplasm of C cell origin which results in markedly elevated calcitonin production, serum calcium, and phosphate concentrations remain normal. There is decreased osteoclast function in medullary thyroid carcinoma, indicating that calcitonin is biologically active, but it does not affect serum calcium concentrations. Effects of calcitonin are most marked when the rates of bone turnover and osteoclast function are highest, as in the young or in diseases such as Paget's disease. The major recognized use of calcitonin is thus in treatment of Paget's disease, where the hormone reduces the accelerated rate of bone turnover.

Calcitonin increases the urinary excretion of calcium, phosphate, sodium, potassium, and magnesium. This renal effect of calcitonin contributes to the hypocalcemic effects of calcitonin but persists only as long as calcitonin concentrations are elevated.

HYPO- AND HYPERCALCEMIC STATES

The integrated axis involving vitamin D, PTH, and, to a lesser extent, calcitonin maintains a constant ionized serum calcium concentration in response to either hypocalcemic or hypercalcemic challenge. When these challenges are severe and persistent or when the endocrine control system is deranged, clinically evident disorders of mineral homeostasis result.

When dietary calcium intake is decreased or when fecal or urinary calcium excretion is increased, the tendency to systemic hypocalcemia is counteracted by increased PTH and 1,25-$(OH)_2$D. In response to transient hypocalcemia, PTH synthesis and secretion increase with resultant mobilization of calcium and phosphate from bone and increased renal tubular reabsorption of calcium. PTH and lowered phosphate increase synthesis of 1,25-$(OH)_2$D which increases calcium and phosphate absorption from the gastrointestinal tract. Initial deficits in calcium are thus ultimately replaced from dietary sources, and overall bone mineral content is protected. When the endocrine defense system is impaired, as with either vitamin D or PTH deficiency, hypocalcemia results. As noted, vitamin D deficiency, whether due to inadequate intake, inadequate absorption, lack of UV irradiation, or failure to convert precursor to $1,25\text{-}(OH)_2D$, results in hypocalcemia and inadequate mineralization of bone osteoid. Sustained elevations of PTH occur in response to hypocalcemia. PTH partially corrects extracellular fluid hypocalcemia by mobilization of bone calcium and increased renal tubular reabsorption of calcium. Maintenance of serum calcium near normal values occurs only at the expense of bone, so that demineralization is severe due both to lack of $1,25\text{-}(OH)_2D$ and to compensatory hyperparathyroidism. PTH deficiency results from either removal of or damage to parathyroid glands (hypoparathyroidism) or from resistance to the biological effects of PTH (pseudohypoparathyroidism). With PTH deficiency, hypocalcemia results from lack of calcium mobilization from bone and failure of renal calcium conservation. In addition, both the PTH and hypophosphatemic stimuli for formation of $1,25\text{-}(OH)_2D$ are missing, and secondary vitamin D deficiency results (Sinha et al., 1977). Like most other peptide hormones, PTH is ineffective when given orally because of degradation in the gastrointestinal tract and only transiently effective when given parentally because of the short circulating half-life. To prevent signs and symptoms of hypocalcemia, including neuromuscular irritability, muscle cramps, paresthesias, tetany, laryngospasm, and seizures, vitamin D therapy is utilized. Because formation of $1,25\text{-}(OH)_2D$ is impaired, massive doses of vitamin D_3 must be given (up to 1000 times normal daily intake) or active $1,25\text{-}(OH)_2D$ must be used. Calcium supplementation is useful but ineffective unless adequate active $1,25\text{-}(OH)_2D$ is provided.

Hypercalcemia is avoided under physiological conditions when calcium intake rises by a fall in serum PTH and a resulting decrease in formation of $1,25\text{-}(OH)_2D$. This reduces both gastrointestinal absorption of calcium and bone resorption, and increases urinary calcium excretion. Hypercalcemia will result from excessive PTH secretion or from excessive intake of vitamin D. It may result also from malignancy when bone dissolution delivers so much calcium to the extracellular space that renal excretory capacity is exceeded. A number of factors mediate bone resorption in malignancy, including an osteoclast-activating factor (Mundy et al., 1974), prostaglandins (Seyberth et al., 1976), and others (Stewart et al., 1980).

In hyperparathyroidism, where autonomous excessive synthesis and secretion of PTH occur, hypercalcemia and hypophosphatemia result. Bone resorption is increased leading to decreased mineral and osteoid, and the increased filtered load of calcium predisposes to renal stones and renal calcification. Hypercalcemia may result in muscle weakness, lethargy, coma, constipation, nausea, and resistance to ADH, with inability to concentrate the urine. The latter may lead to dehydration and acceleration of the hypercalcemic state. Therapy is directed toward surgical removal of abnormal parathyroid tissue. Short-term control of hypercalcemia can be achieved with saline diuresis and oral phosphate which tends to favor calcium-phosphate deposition in bone. Calcitonin has occasionally been used to reverse hypercalcemia.

Excessive vitamin D results from ingestion or from increased sensitivity in sarcoidosis. Hypercalcemia results from increased absorption from the gastrointestinal tract and from bone resorption. In addition to cessation of vitamin D, glucocorticoids counteract hypercalcemia by opposing the effects of vitamin D.

In hypercalcemia resulting from malignancy involving bone, both PTH and $1,25\text{-}(OH)_2D$ are appropriately decreased; therapy must therefore be directed toward increasing renal calcium excretion and toward treatment of the malignant process. Oral phosphates and mitramycin, a cytotoxic antibiotic, are useful in addition.

CHAPTER 56

Hormonal Regulation of Growth and Development

INTRODUCTION

Ordered, controlled growth is essential for development and maintenance of normal humans. The mechanisms which control growth are selective for different types of cells, for example, wound healing, growth of the contralateral kidney or adrenal cortex after unilateral nephrectomy or adrenalectomy, and hepatic regeneration following partial hepatic resection. Growth is orderly, and there are strict controls on the extent of proliferation of cells in any organ. Loss of strict controls on cell proliferation are manifest in cancer.

Growth of individual cells, termed hypertrophy, is a process resulting in reversible amplification of the cells' differentiated function, whereas cell division, termed hyperplasia, results in an increased number of functional cells. Both hypertrophy and hyperplasia are controlled by hormones and are dependent on availability of adequate nutrients and ions. To study growth, simpler tissue cultures of varying cell types have been frequently used, and results are extrapolated to more complex in vivo situations. Using cell culture systems, a frequent experimental approach has been to arrest cell growth by removing essential compounds from culture media and by measuring the effects of added compounds on growth of resting cells. Under most conditions of deprivation, mammalian cells reversibly cease growing with a diploid complement of DNA (G_1 growth arrest). Addition of stimulatory compounds results in cellular hypertrophy followed by DNA replication and cell division. Using this approach, a number of new hormones termed growth factors have been identified. Because normal cells exhibit orderly growth in culture and cease growing at saturation (density-dependent inhibition of cell proliferation), endogenous oncogenes have been identified by their ability to impose abnormalities on cell growth control and cause piling up of cells beyond contact inhibition (Cooper et al., 1980; Shih, 1981).

REQUIREMENTS FOR CELLULAR GROWTH

Nutrients and Ions

Adequate nutrients and ions are required for growth of any population of cells whether bacterial or mammalian. By creating deficiency states in vitro, a requirement for a number of small molecular weight nutrients can be demonstrated. These include ions such as calcium, phosphate and magnesium, amino acids, glucose, fatty acids, vitamins, transferrin and bound iron, ceruloplasmin and bound copper, and nucleic acids. Because of extensive metabolic transformations, normal man requires only 20 essential organic compounds in addition to a source of calories and water. Eight amino acids, 11 vitamins, and linoleic acid are essential. Deficiencies of any single element are unusual. The most generalized deficiencies in humans are caloric malnutrition (marasmus) and protein malnutrition (kwashiorkor). In caloric malnutrition, protein may be utilized as metabolic fuel, resulting in combined protein-caloric malnutrition; similarly, in protein malnutrition, interference with the ingestion of the usually bulky carbohydrate diet often results in a combined deficiency state. These deficiencies are widespread in underdeveloped countries and are more common than usually recognized among disadvantaged, elderly, chronically ill, and alcoholic persons in developed countries. In children with protein-caloric malnutrition, growth is markedly retarded and sexual maturity is delayed. In people of all ages, processes involving cell growth are impaired, including wound healing and function of the immune system, so that death from infection is common.

In nutrient deficiency a number of hormonal adaptations occur. Adaptations to fasting have been discussed in other chapters and include a reduction in insulin secretion, decreased peripheral conversion of T_4 to T_3, elevations in cortisol, epinephrine, and glucagon, and increased growth hormone but impaired somatomedin production. With nutrient lack, growth is obviously inappropriate, and hormonal changes mediate mobilization of stored energy and suppress cellular growth. Reproduction is also inappropriate, and low gonadotropins and suppression of ovarian and testicular function occur.

Adequate nutrient intake is required for growth, but excess intake does not cause growth of tissues other than fat. Utilization of nutrients for growth of cells and organs requires hormones, and these regulate the process of growth, provided adequate nutrients are available. Growth factors may facilitate cellular uptake of nutrients required for hypertrophy and division of one cell into two (Holley, 1975).

Growth Factors

Animal serum has been used for many years to support growth of cells in culture. The essential macromolecular components in serum are hormones, and, by using an appropriate mixture of hormones, cell growth can be optimized without serum (Barnes and Sato, 1980). Even though cell growth in vivo is specific, i.e., after partial hepatectomy only liver grows, no organ-specific growth factors have been identified. Rather a combination of hormones regulate optimal growth, and delivery of this combination to tissues along with other factors such as nutrient availability, the extracellular matrix on which cells grow, the concentration of hormone receptors, and the concentration of antigrowth factors control organ growth. In addition to well characterized hormones, a number of new polypeptide hormone growth factors have been identified which stimulate cell growth and division (Table 8.7).

Epidermal growth factor (EGF), a 53 amino acid polypeptide, was initially identified by its ability to accelerate eye opening in newborn mice (Cohen and Taylor, 1974). It was independently isolated later from human urine, based on its ability to inhibit gastric acid secretion (Gregory, 1975). Production of EGF in submaxillary glands is stimulated by androgens and adrenergic agonists. EGF stimulates proliferation of a large number of cells, including epithelial, fibroblastic, and mesodermally derived cells (Carpenter and Cohen, 1979). The EGF receptor possesses tyrosine residue-specific protein kinase activity, and EGF increases phosphorylation of cellular proteins at tyrosine residues (Cohen et al., 1982).

Some tumor cells produce polypeptide growth factors which not only stimulate proliferation of the tumor producer but also other cells (Todaro et al., 1979). These transforming growth factors

Table 8.7
Growth factors

I. Epidermal growth factor (EGF)
  A. Synthesized in submaxillary gland; synthesis regulated by androgens and by adrenergic agonists
  B. Single chain peptide of 6045 M_r (53 AA); identical to urogastrone, initially identified by its ability to inhibit gastric acid secretion
  C. Stimulates growth of epithelial, fibroblast, and granulosa cells
  D. 170,000 M_r receptor has tyrosine protein kinase activity
II. Transforming growth factor (TGF)
  A. Produced by a variety of tumor cells, both virally transformed and spontaneously occurring
  B. Polypeptide with a primary sequence different from EGF
  C. Stimulates growth and confers the transformed phenotype on untransformed fibroblasts
  D. Some, but not all, TGFs bind to EGF receptors
III. Platelet derived growth factor (PDGF)
  A. Found in α-granules in platelets
  B. 30,000 M_r cationic peptide with two polypeptide chains linked by disulfide bonds
  C. Released during platelet release reaction and stimulates proliferation of several cell types, including glial, smooth muscle, and fibroblast
  D. Distinct receptor which contains tyrosine residue-specific protein kinase activity
IV. Tumor angiogenesis factors (TAF)
  A. Produced by both normal and tumor cells
  B. Several different substances
  C. Stimulates capillary ingrowth and neovascularization
  D. Action facilitated by heparin and inhibited by protamine
V. Fibroblast growth factor (FGF)
  A. Purified from pituitary and brain tissue; brain FGF is a fragment of myelin basic protein
  B. Single chain polypeptide 13,400 M_r
  C. Stimulates growth of a variety of mesodermally derived cell types: fibroblasts, adrenocortical cells, granulosa and luteal cells, myoblasts, smooth muscle cells, vascular endothelial cells, and chondrocytes; not active on ectodermally derived cells
  D. Synergistic with glucocorticoids
VI. Nerve growth factor (NGF)
  A. Synthesized in submaxillary gland; also present in snake venom
  B. Dimeric protein of 26,500 M_r; sequence homology with insulin
  C. Stimulates replication in embryonic nervous tissue; in mature tissue, stimulates hypertrophy, axon growth, and enzyme induction, but does not stimulate DNA synthesis; target is ganglion cells
  D. Receptors on plasma and nuclear membrane
VII. Somatomedin; insulin-like growth factor (IGF)
  A. Made in liver, but may be synthesized in fibroblasts and other cells
  B. Single chain peptide of 7,000 M_r; several forms identified with strong sequence homology with insulin
  C. Stimulates DNA synthesis in chondrocytes, fibroblasts, and other cells
  D. Synthesis is dependent on growth hormone

(TGF) reversibly confer the transformed phenotype on untransformed fibroblasts (Todaro et al., 1980). Local production of TGF may provide the tumor cells with a high concentration of growth factor and a selective advantage over their regulated counterparts. Although TGF has a different structure from EGF, it binds to the EGF receptor. A second TGF acts independently of the EGF receptor (Roberts et al., 1982).

A 30,000 M_r cationic polypeptide, platelet-derived growth factor (PDGF), is the major mitogen found in serum (Ross et al., 1979). It is stored in α-granules in platelets and released during the platelet release reaction. PDGF stimulates proliferation of several cell types, including glial, smooth muscle, and fibroblast. Because PDGF is transported and released from platelets, it appears to have a role in wound healing and may also contribute to growth of arterial plaques in arteriosclerosis. Binding of PDGF to specific high affinity cell surface receptors stimulates tyrosine residue-specific protein kinase activity in a reaction analogous to that induced by EGF and TGF binding to their receptors (Ek et al., 1982).

Growth of new capillary blood vessels is important in normal processes such as development of the embryo, formation of the corpus luteum, and wound healing, as well as in pathologic processes such as chronic inflammation and tumor growth. Because nutrient and growth factors diffuse out of blood vessels to adjacent tissue cells over only a limited distance, organ or tumor growth occurs only when blood supply is maintained by capillary ingrowth. Soluble, diffusible factors which stimulate capillary proliferation were initially identified in extracts from tumor cells (Folkman et al., 1971). These tumor angiogenesis factors (TAF) appear to consist of several distinct growth factors. Production of such factors is not unique to tumors, but is a property of normal cells and is required for organ regeneration and wound healing. For example, cyclic changes in the ovary involve vascular changes mediated in part by angiogenic factors produced by the corpus luteum (Gospodarowicz and Thaknal, 1978). Heparin facilitates TAF-stimulated capillary growth and protamine inhibits it (Taylor and Folkman, 1982).

Fibroblast growth factor (FGF), a basic polypeptide with a 13,400 M_r, has been purified from pituitary and brain (Gospodarowicz, 1975; Gospodarowicz et al., 1978). FGFs from the two sources are not identical; FGF from brain represents a fragment of myelin basic protein. FGF stimulates growth of a variety of cells derived from embryonic mesoderm (Gospodarowicz et al., 1978). FGF stimulates cell migration, increases cloning efficiency, extends the life-span of normal cells in culture, and increases production of the extracellular matrix protein fibronectin as well as increases cell growth (Gill et al., 1979).

Nerve growth factor (NGF) is an insulin-like protein which induces morphologic and metabolic differentiation of sympathetic and sensory neurons (Levi-Montalcini and Angeletti, 1968; Bradshaw et al., 1974). Sensory neurons are stimulated by NGF only during a short period of embryonic development, whereas sympathetic cells retain sensitivity even in adult animals. NGF stimulates neurite outgrowth with accompanying cellular hypertrophy and neurotubule polymerization. It also stimulates noradrenergic neurotransmitter synthesis and regeneration of adrenergic fibers in brain. Because mature nerve cells do not divide, the principal effect of NGF is stimulation of cellular hypertrophy and differentiated function. NGF like EGF is synthesized in large amounts in salivary glands, and NGF possesses serine protease activity (Orenstein et al., 1978). Receptors for NGF have been demonstrated on both the cell surface and nuclear membrane and appear to interact with tubulin (Levi-Montalcini et al., 1974; Andres et al., 1977).

The somatomedin:insulin-like growth factors (IGF) are under the control of growth hormone (Chapter 52). A number of additional growth factors have been described, including erythropoietin, T and B cell growth factors, and others which are less completely defined. Although the complex process of growth can be stimulated by growth factors, classical hormones are also required. Insulin, thyroid hormone, and cortisol have growth-promoting effects and potentiate the action of the growth factors shown in Table 8.7.

Growth of cells depends on a number of factors in addition to growth factors and nutrients. The presence of neighboring cells, the extracellular matrix, blood supply, and antigrowth factors are all important. Antigrowth factors, termed chalones, may be produced locally to limit tissue growth (Lozzio et al., 1975). For example, cartilage-produced factors inhibit growth of both cartilage and capillaries, and epithelial cells produce a factor which inhibits epithelial cell growth (Holley et al., 1978). Hormones such as ACTH control both structure and function of the adrenal cortex. ACTH inhibits DNA replication in adrenocortical cells, but, in the presence of growth factors, increases the differentiated function of steroid hormone production in hypertrophied cells. Because the requirement for increased steroid hormone production is intermittent, this mechanism allows amplification by reversible cellular hypertrophy. Hyperplasia, which is reversible only when cell death exceeds cell generation, occurs only when concentrations of

ACTH are elevated over long periods and receptors decrease (Hornsby and Gill, 1977).

HUMAN GROWTH AND DEVELOPMENT

In intrauterine life, hormonal signals are important for growth, development, and maturation of organs (for example, see Chapter 59). The orderly process of postnatal growth is also critically regulated by hormones as well as genetic makeup and nutrition. During the first year of postnatal life, the rate of growth is rapid, though less than in utero. By 1 yr, birth length doubles and birth weight triples. Thyroid hormone is essential for myelinization and brain growth during this time. By 6–8 mo of postnatal life, cell multiplication in the brain is completed, and, if thyroid hormone is deficient during this time, irreversible mental retardation occurs. Deficiency of thyroid hormone beginning later in life results in growth retardation and mental slowing, but these processes are reversible with thyroid hormone.

As shown in Figure 8.43, the rate of growth then slows during childhood years until puberty. During childhood, height increases 5–7.5 cm/yr, and weight increases 2–2.5 kg/yr. Charting of measured increases in height and weight on graphs which include 95% confidence limits for a large number of normal children is a valuable method for continued assessment of a child's progress. Failure to grow at a normal rate provides an important clue to underlying abnormalities. Impaired nutrition, including diseases such as chronic diarrheal states, psychological disturbances, and hormonal derangements must be considered. The most common endocrine

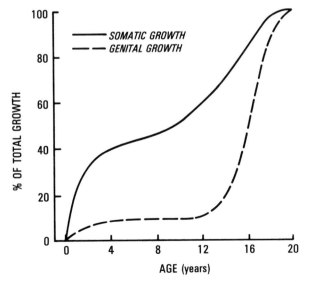

Figure 8.43. Relative rate of somatic and genital growth during normal human development.

deficiencies are of growth hormone, thyroid hormone, and insulin.

Puberty and maturation of gonadal systems results in an increase in growth rate. In females growth rate increases to 9 cm/yr and in males to 10 cm/yr. The increase in growth rate is due to the gonadal steroid hormones, estrogen, and testosterone, which accelerate the rate of growth but also cause ultimate fusion of the epiphyseal end plate growth center to the end of bones so that growth is terminated. Radiographic assessment of epiphyseal end plates provides an accurate indication of bone age for comparison with chronological age and allows prediction of the potential for further height increases (Greulich and Pyle, 1959).

In females puberty begins at approximately age 11 with breast budding followed by menarche at age 13. Secondary sexual characteristics develop, and psychological and behavioral changes ensue. Regular ovulatory menstrual cycles are established 2–3 yr after menarche, and linear growth ceases at age 16–17. In males testicular size begins to increase at age 12 followed 1 year later by penis enlargement and increased rates of linear growth. Secondary sexual characteristics occur at approximately age 14–15, and growth ceases at approximately age 18.

Abnormalities in gonadotropin or gonadal steroid production will alter this orderly program of growth and sexual development. When sex steroid deficiencies occur, growth of long bones may continue, so that a typical eunuchoid stature consisting of an arm span greater than height and of a lower body segment from symphysis pubic to heel greater than the upper body length occurs. Adequate genital growth and development of secondary sexual characteristics are lacking. With precocious puberty or with excessive production of adrenal androgen precursors resulting from enzyme defects in cortisol synthesis, early growth will be accelerated, but premature fusion of epiphyseal end plates will result in ultimate short stature.

In adulthood growth is limited to repair and renewal processes. Change in a number of hormones, i.e., excess cortisol, deficient thyroid hormone, will impair processes such as wound healing.

ABNORMALITIES OF GROWTH

Detailed analysis of inadequate rates of growth and abnormal patterns of sexual maturity requires careful assessment of nutritional intake, endocrine gland function, psychosocial background, the presence of chronic disease, and genetic makeup. Any or all of these may be responsible for abnormalities in postnatal growth and development. Because appropriate hormonal therapy is available, identification of endocrine abnormalities is especially valuable. Similarly, identified endocrine abnormalities contributing to impaired adult growth processes such as wound healing are correctable.

In cancer, restraints on normal growth are relaxed, so that overgrowth and metastasis of tumor cells occurs. Tumor arising in some organs continue to respond to hormones which stimulate growth of the normal cells of the tissue of origin. For example, many breast cancers grow in response to estrogen as does the normal breast; reduction of estrogen concentration and use of antiestrogens slow growth and cause regression of metastases in many breast cancers which contain estrogen receptors (Heuson and Coure, 1981). Prostate cancer, like normal prostate tissue, may grow in response to androgens. Reduction in androgens and estrogen therapy may retard the progress of this disease.

The mechanisms through which growth factors and hormones stimulate cell proliferation are being actively investigated because these may provide essential information about the mechanisms through which malignant processes subvert normal growth control. For example, the transforming gene product of several RNA tumor viruses is a protein kinase with specificity for tyrosine residues in proteins. Transformation to the malignant state results from excessive production of this tyrosine residue-specific protein kinase which, at lower concentrations, is a normal cell constituent. EGF and PDGF both directly stimulate tyrosine residue-specific protein kinase activity in their proliferative action on cells, and TGF, which also activates this kinase, imposes a reversible transformed phenotype on cells (Buss and Gill, 1983). Identification of cellular oncogenes provides the first step toward identification of their protein products and the role of these proteins in cell growth control.

CHAPTER 57

Hormonal Regulation of Testicular Function

The testes, like the ovaries, produce both gametes and steroid hormones. These two activities of the testes are segregated anatomically with spermatogenesis in the seminiferous tubules and androgen biosynthesis occurring in the Leydig cells. The anterior pituitary participates in control of both of these functions through its secretion of the gonadotropins, follicle-stimulating hormone (FSH) and luteinizing hormone (LH). Although androgen biosynthesis and spermatogenesis are intimately related, for convenience, they will be discussed independently as the pituitary-Leydig cell axis and the pituitary-seminiferous tubular axis (Figs. 8.44 and 8.45). The discussion which follows is directed primarily at primate physiology and pathophysiology; however, when appropriate, studies from a variety of species are discussed for illustrative purposes.

PITUITARY-LEYDIG CELL AXIS

The essential features of the pituitary-Leydig cell axis are shown schematically in Figure 8.44. The components of the axis will be discussed separately followed by a discussion of its integrated control.

Hypothalamo-Pituitary Unit

The intimate association of the pituitary gland with the hypothalamus offered by the hypophyseal portal system provides a pathway for substances from the brain to contol anterior pituitary function. Even though the structure and function of several hypothalamic peptides are known, the factors which control their secretion are not entirely understood. A recent examination of the angioarchitecture of the hypothalamopituitary unit has indicated that it is much more complex than once appreciated. An extensive study of several species has suggested that the vasculature between the brain and pituitary is arranged in such a manner that each of these organs can control the secretion of the other. In Figure 8.46 several options for blood flow within the pituitary and adjacent structures are shown. Blood can move from the hypothalamus to the pituitary as previously suggested by many investigators but, in addition, flow in the portal vessels may be reversible. Furthermore, it is possible for the pituitary to secrete directly into the cavernous sinus, carotid artery, and to the cerebrospinal fluid. Secretion of pituitary hormones to the brain may account for the impotence associated with high prolactin levels in men (see below).

LUTEINIZING HORMONE-RELEASE HORMONE (LHRH)

LHRH is a decapeptide which is synthesized in the hypothalamus. This peptide is secreted into the hypophyseal portal system and is transported down the pituitary stalk to the pituitary (Fig. 8.46) where its stimulates secretion of two separate hormones (LH and FSH). In the pituitary, LHRH binds to specific receptors on the surface of gonadotrophs which, in turn, activate calcium influx and subsequent LH and FSH secretion. It should be emphasized that LHRH is found in other parts of the brain in addition to the hypothalamus. These observations, along with the fact that behavior is modified following LHRH administration to animals, suggest that this peptide may have physiological effects on the central nervous system beyond those implied by its name.

LUTEINIZING HORMONE

LH derives its name from its action in the female where it facilitates ovulation and converts the ovulated follicle into a functioning corpus luteum. When this hormone was first discovered in the male, it was called interstitial cell-stimulating hormone (ICSH) because of its effect on testicular Leydig cells. It is now recognized that LH and ICSH are identical so the term LH is used in both sexes. This pituitary hormone is one of two glycoprotein

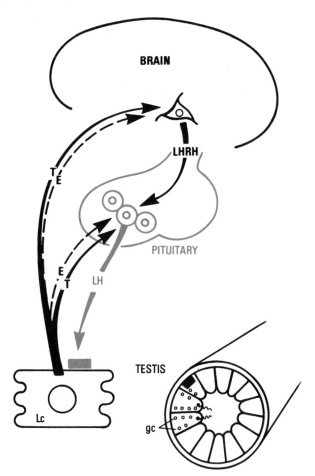

Figure 8.44. The pituitary-gonadal axis is composed of the hypothalamus and its neural connections with the rest of the brain, pituitary, and the testes. The testis function is divided into Leydig cells (*Lc*) and the seminiferous tubules which are made up of Sertoli cells and their accompanying germ cells. For the purpose of discussion, the pituitary-gonadal axis is divided into a pituitary-Leydig cell axis shown above and pituitary seminiferous tubular axis shown in Figure 8.45. *LHRH*, LH releasing hormone; *LH*, luteinizing hormone; *T*, testosterone; *E*, estradiol. [From Bardin (1978).]

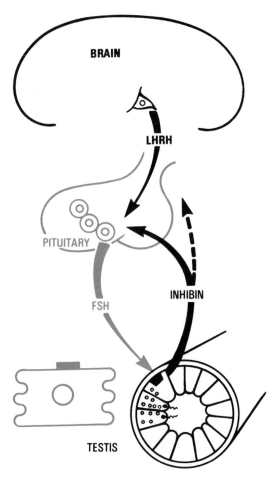

Figure 8.45. The pituitary seminiferous tubular axis is composed of the hypothalamus, its neural connections with the rest of the brain, the pituitary, and the seminiferous tubules. *LHRH*, luteinizing hormone-releasing hormone; *FSH*, follicle-stimulating hormone. [From Bardin (1978).]

hormones secreted by the anterior pituitary in response to the hypothalamic LH-releasing hormone (LHRH). LH like FSH, thyrotropin (TSH), and human chorionic gonadotropin (hCG), contains two peptide chains designated α and β. The α subunit of all hormones in a given species appears to be identical. By contrast, the β subunit determines both the structural and biological differences between each of these glycoprotein hormones. These conclusions are derived from studies in which a hybrid formed by combining the α chain of TSH and the β chain of LH possess the activity of luteinizing hormone. Similarly, the combination of α-LH and β-TSH results in a molecule that stimulates the thyroid. The biological activity of LH is the same as hCG, the major gonadotropin secreted by the placenta. Furthermore, the structure of LH

is almost identical to hCG except for an additional 30 amino acids on the carboxy terminal of the β chain of the latter hormone. Although it was once supposed that the major functional difference between LH and hCG was their sites of synthesis in the pituitary and placenta, respectively, recent studies have demonstrated that very small amounts of hCG may also be made in the testis, pituitary, and other tissues. The functional significance of hCG in these organs remains to be established (Pierce et al., 1971; Ross, 1977).

The secretion of LHRH into the hypophyseal portal system is episodic, rather than constant, and this results in turn in the pattern of LH release illustrated in Figure 8.47. The hormone is secreted in a series of secretory pulses which have amplitudes ranging from 35–270% from nadir to peak and which occur with a frequency of 2–4/6 h. A variety of studies indicate that sex steroids (estrogen, progesterone, and testosterone) alter both the frequency and the amplitude of LH secretory pulses

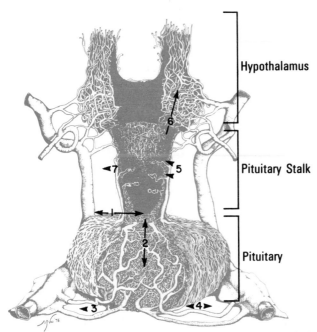

Figure 8.46. The vascular anatomy of the monkey hypothalamus and pituitary. Several options and the direction of blood flow within the pituitary and adjacent structures are shown by the *numbered arrows* as follows: *1*, Blood flow in short portal vessels must be reversible. *2*, Blood flow within the neurohypophysis must be reversible. *3*, Confluent pituitary veins may carry secretions directly to the cavernous sinus, *but*, at other times the connected upper limbs may carry secretions from the adenohypophysis to the neurohypophysis. *4*, In some arteries linking the pituitary to the circle of Willis, flow may be reversible. *5*, The internal plexus provides the tanycyte ependyma blood which has already passed through the external plexus which may allow for tanycyte transport of hormones into the ventricle. *6*, Capillaries may carry secretions from the infundibulum to the hypothalamus. *7*, Fenestrations within long portal vessels may allow leakage of pituitary secretions into the subarachnoid space surrounding the circle of Willis. Thus, of the potential efferent routes from the neurohypophysis, one is directed to the systemic circulation, one is directed to the adenohypophysis, and four are directed to the brain. [From Bergland and Page (1977).]

in men. A major problem posed by the variable secretion of LH is that single samples cannot be used to evaluate gonadotropin physiology or for diagnosis and management of patients with endocrine disorders. As a consequence, LH levels are measured in multiple plasma samples obtained at 20-min intervals for 3–6 h or in timed urine samples. Both of these approaches estimate "mean" or "integrated" LH secretion per unit of time (Santen and Bardin, 1973).

Following administration of exogenous LH, its disappearance from blood is described by two linear exponentials, with a mean initial phase half-time of 40 min and a second phase half-time of 120 min. Prolonged retention of LH in the bloodstream almost certainly relates to its sialic acid content. Several studies have demonstrated that asialo gly-

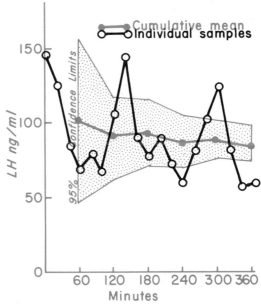

Figure 8.47. The *solid line* and *dotted area* are the cumulative mean LH levels and the 95% confidence limits of that mean at hourly intervals for 6 h in a single subject. For comparison, the *open circles* represent the actual estimates of serum LH in samples obtained at 20-min intervals. [From Santen and Bardin (1973).]

coproteins (including LH) are rapidly cleared by the liver. Detailed studies of LH metabolism indicate that this hormone has a metabolic clearance rate of 25 ml of plasma/min which is unrelated to gonadal function. Most of the LH secreted into the blood is degraded, since only a small fraction of the daily production rate appears in the urine.

Leydig Cells

The interstitial cells of Leydig are located in the connective tissue stroma of the testis between the seminiferous tubules. These cells differentiate and secrete androgens during the 7th week of fetal life in the human. The activation of Leydig cells in the male fetus correlates with the onset of androgen-dependent differentiation. Since Leydig cells appear to regress after birth, it has been assumed that their initial development results from gonadotropin stimulation. If such a tropin is required, then hCG is probably the hormone responsible for fetal differentiation since functional Leydig cells develop in anencephalic infants born without pituitary glands. During childhood, Leydig cells revert to an undifferentiated state and are activated again at the time of puberty as plasma gonadotropins increase. During differentiation of Leydig cells in the fetus, in the pubertal boy, and following stimulation with either exogenous LH or hCG, androgen secretion correlates with development of smooth endoplasmic reticulum which contains the enzymes necessary for steroid synthesis.

Luteinizing hormone stimulation of testosterone synthesis and secretion is initiated by hormone binding to specific LH receptors on Leydig cell membranes. This hormone-receptor interaction stimulates membrane-bound adenylate cyclase which catalyzes the synthesis of cAMP. In addition to LH, fluoride ion, cholera toxin, and guanyl nucleotides (GTP and analogues) stimulate cAMP synthesis. Studies of these stimulators have led to the conclusion that the adenylate cyclase system is composed of three distinct membrane-bound units. These include the hormone receptor, which contains the specific site for recognition of LH; the catalytic unit of adenylate cyclase, which converts ATP to cAMP and PP; and the guanyl nucleotide regulatory subunit, which binds GTP and couples the hormone receptor to adenylate cyclase. It is of note that the adenylate cyclase systems for other hormones (ACTH, FSH, etc.) differ from the one on Leydig cell only in the specificity of their receptors. Once cAMP is released into the cytoplasm of the Leydig cell it binds to the regulatory subunit of protein kinase which then dissociates from and, in turn, activates the catalytic subunit of this enzyme. The subsequent phosphorylation of Leydig cell proteins is believed to facilitate cholesterol conversion to pregnenolone and to increase androgen synthesis. Although initial studies failed to demonstrate a correlation between intracellular cyclic AMP levels and testosterone secretion, recent experiments have indicated that steroid synthesis can be directly correlated with the occupancy of intracellular binding sites for cyclic AMP. Although the immediate effects of LH on steroid synthesis are believed to be mediated by the cyclic AMP protein kinase system, it is not known whether growth-promoting and differentiating effects of LH are operative through this mechanism (Christensen, 1975).

TESTICULAR ANDROGEN BIOSYNTHESIS

An androgen is a substance which stimulates growth of the male reproductive tract. It is important to recognize that this is a biological and not a chemical definition. Most potent androgens, however, are steroids, and several are shown in Figure 8.48. Extensive studies have demonstrated that testosterone is the major androgen produced in testes and secreted into the systemic circulation. The pathway of androgen biosynthesis is summarized in Figure 8.49. Small amounts of other potent steroids such as dihydrotestosterone (Fig. 8.48) are secreted, but they contribute very little to the overall androgen content of the blood in men. This is in marked contrast to women where dihydrotestosterone is a significant fraction of total circulating androgens. Androstenedione, a testosterone precursor (Fig. 8.49), is also secreted by the testes, but its only importance in man is as a plasma precursor

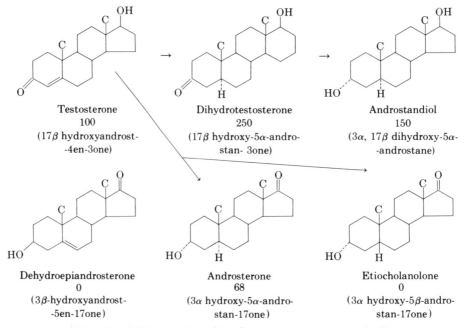

Figure 8.48. Androgens and ketosteroids. In the *top* portion of the figure are the most important naturally occurring androgens. The relative biological potency of each is shown, with testosterone having a potency of 100. Testosterone is metabolized intracellularly to dihydrotestosterone and androstanediol. On the *bottom* half are the relatively nonandrogenic 17-ketosteroids which are secreted by the adrenal. Testosterone is also metabolized to ketosteroids.

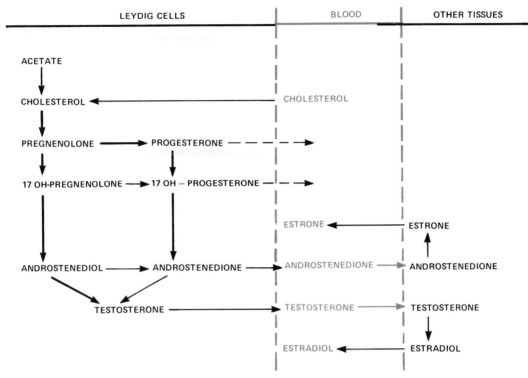

Figure 8.49. Steroid biosynthesis in the testis. Testosterone is the major secretory product of the testis but other precursors such as androstenedione are secreted as well. Testosterone and androstenedione are converted in other tissues to estradiol and estrone, which enter blood to become the major estrogens in men.

for estrogen. Other testosterone precursors such as 17-hydroxyprogesterone and progesterone are also secreted by the testis, but their biological functions, if any, are not known. The mean blood levels, urinary levels, and production rates of testosterone and other steroids in man are shown in Table 8.8.

Leydig cells are the primary site of testosterone biosynthesis, and the testes secrete 95% of the testosterone in the blood of normal men. In the Leydig cells, cholesterol is either synthesized from acetate or accumulated from the circulating cholesterol in low density lipoprotein. Leydig cell mitochondria, like those in adrenal, ovary, and placenta, are unique in their ability to cleave the cholesterol side chain to produce pregnenolone. The latter C-21 steroid is the precursor for steroid synthesis in all these organs. In the testes, pregnenolone is converted to testosterone by several microsomal enzymes as shown in Figure 8.49 (Christensen, 1975).

TESTOSTERONE TRANSPORT IN PLASMA

Since testosterone is relatively insoluble in aqueous solutions, it is transported in the blood bound to plasma proteins. One of the proteins which binds testosterone with high affinity is testosterone-estradiol-binding globulin (TeBG) (Fig. 8.50). This is a β-globulin which is distinct from

Table 8.8
Androgen levels and production rates

	Men	Women
Testosterone		
Plasma (ng/dl)	700	40
Urine (μg/day)	70	4
Production rate (μg/day)	7,000	300
Androstenedione		
Plasma (ng/dl)	120	170
Production rate (μg/day)	2,400	3,400
Dihydrotestosterone		
Plasma (ng/dl)	45	20
Production rate (μg/day)	300	60

cortisol-binding globulin, the major transport protein for cortisol, corticosterone, and progesterone.

The physical properties of TeBG have now been elucidated following the isolation of this protein in fully active form from plasma. TeBG is a 94,000-dalton dimer that has 30% carbohydrate and one binding site per molecule. The concentration of TeBG in plasma is increased 5- to 10-fold by estrogens and is decreased 2-fold by testosterone. Interestingly, excess thyroid hormone also increases TeBG, and decreased levels of this protein are found in hypothyroid individuals. Finally, TeBG is also elevated in the plasma of patients with abnormal liver function (Rosen, 1976; Bardin et al., 1981).

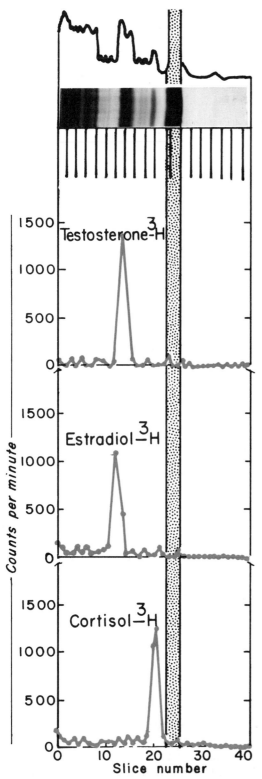

Figure 8.50. Electrophoretic demonstration of testosterone-estradiol-binding globulin and cortisol-binding globulin (*bottom chart*). Radioactive testosterone, estradiol, and cortisol were incubated with three samples of human plasma which was electrophoresed on polyacrylamide gels. The gels were sliced and counted. The peaks indicate the mobilities of the specific steroid-binding globulins. The *dotted bar* indicates the mobility of albumin. Under the conditions of the study, steroid binding to albumin was not demonstrated.

In blood of men, only 3% of testosterone is free, while 42% is bound to TeBG, 39% to albumin, and 16% to other plasma proteins. The rate of testosterone metabolism correlates directly with the quantity of free plus albumin-bound testosterone. These observations suggested that TeBG-bound androgen is not readily available for in vivo metabolism. Since target tissue metabolism of testosterone is required for expression of its biological activity in reproductive tissues, it was suggested that TeBG-bound testosterone is inactive. However, the recent observations that TeBG can enter androgen-responsive cells of the male reproductive tracts indicate that this binding protein may influence the action on metabolism of testosterone in these tissues. This postulate is supported by recent in vitro studies on human prostate indicating that TeBG determines the metabolic fate of testosterone, independent of its effect on plasma binding. Regardless of these considerations, the fact that TeBG is not present in the plasma of all species suggests that this binding protein does not play an essential role in the mechanism of action of androgens.

TESTOSTERONE METABOLISM

Steroid metabolism may be defined as the irreversible removal of hormone from blood. When testosterone enters a cell, it may be metabolized either to a more active steroid which exerts its action on that cell or on some other tissue or to a less active steroid which is conjugated or excreted in the bile or urine. It is conceivable that both types of metabolism could occur in a sequential manner in the same cell. In any event, irreversible removal from blood is accomplished in both instances by a change in steroid structure. Examples of testosterone metabolism are shown in Figures 8.48 and 8.49. In the first instance, the biological activity of testosterone is amplified by 5-α-reductase, an enzyme which reduces the major blood androgen to dihydrotestosterone (Fig. 8.48). This latter hormone is more potent than testosterone, and its production occurs in tissue of the male reproductive tract and skin where androgens stimulate cell division. Secondly, androgens are metabolized to relatively inactive steroid sulfates and glucuronides which are excreted. This latter type of metabolism occurs primarily in liver (Fig. 8.48). Finally, the biopotency of testosterone is modified by aromatization to a potent estrogen, estradiol. This reaction occurs in brain, breast, and other tissue, providing significant local concentrations of estrogens in individual tissues as well as contributing significantly to blood estrogens (Fig. 8.49) (Lipsett et al., 1966).

THE ORIGIN OF BLOOD ESTROGENS IN MEN

Estrone and estradiol are the important blood estrogens in men (Fig. 8.49). Following orchiectomy, the levels of both of these steroids decrease. These observations suggested that the testes secrete estrogenic hormones. Attempts, however, to identify estrone and estradiol in spermatic venous blood indicated that only 10–20% of these steroids are directly secreted by the testes. Similarly, measurements of estrogen in adrenal venous blood suggested that this endocrine gland could account for only a small fraction of the total amount present in blood. The origin of the major portion of blood estrogens in men was first suggested by the observation that patients treated with large doses of testosterone for hypogonadism developed gynecomastia. It was subsequently demonstrated that intravenously administered testosterone is converted to estradiol which reenters the blood. Many nonendocrine tissues are known to have the cytochrome P-450-dependent aromatase required for conversion of androgens to estrogens, and whole body kinetic studies have demonstrated conclusively that the major portions of blood estradiol and estrone in normal men are derived from blood testosterone and androstenedione, respectively (Fig. 8.49). The quantitative aspects of this extraglandular estrogen production from androgens is summarized in Figure 8.51 (Siiteri and MacDonald, 1973).

In men with endocrinopathies that suggest estrogen dominance (such as gynecomastia), the origin of plasma estrogens is indeed complex. In some individuals, total estrogen production from all sources is normal, and the only demonstrable hormonal abnormality is decreased testosterone secretion. In these instances, feminization relates to the increased amount of estrogen relative to androgen rather than the absolute level of each class of hormone. By contrast, some feminized individuals have increased plasma estrogen levels which are clearly related to increased secretion of estrogens or estrogen precursors, such as androstenedione. The feminization associated with Klinefelter's syndrome is a representative example of this latter condition and is believed to represent either increased testicular aromatase or increased substrate formation (Fig. 8.52). Finally, a generalized increase of aromatase in peripheral tissues has been described. In a man with this disorder, feminization occurs because a large fraction of blood androgens is converted to estrogens in peripheral tissues (MacDonald et al., 1971).

THE BIOLOGICAL ACTIONS OF ANDROGENS

Testosterone and other androgens have some biological activity on virtually every tissue in the body. Their more important functions are to (1) produce differentiation of Wolffian ducts, external

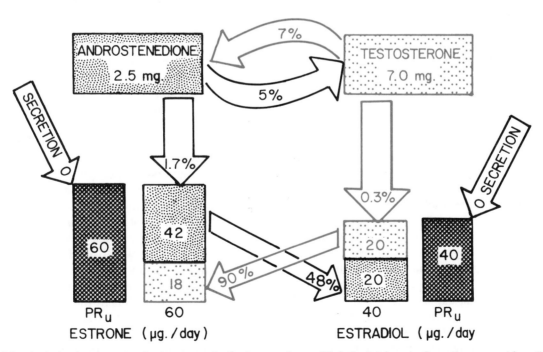

Figure 8.51. Analysis of androgen and estrogen production in normal men. PR_u is the total production rate measured from the specific activity of urinary metabolites. The *upper boxes* indicate the production rate of androstenedione and testosterone. The *arrows* indicate the fraction of each steroid which is converted to the indicated product. In the case of normal men, virtually all of the estrone and estradiol are derived from androstenedione and testosterone, and direct secretion is very low or zero. [From MacDonald et al. (1971).]

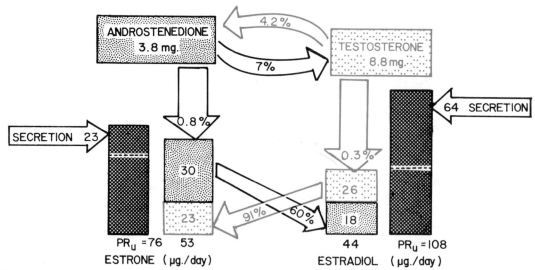

Figure 8.52. Analysis of androgen and estrogen production in an 18-year-old male with gynecomastia and mosaic Klinefelter's syndrome. Androgen production was normal, but excessive estradiol production arose by testicular secretion. The symbols are the same as in Figure 8.51. [From Siiteri and MacDonald (1973).]

genitalia, and brain in the male fetus (see Chapter 60); (2) stimulate linear body growth, nitrogen retention, and muscular development in the adolescent and mature male; (3) stimulate adult maturation of the external genitalia and accessory sexual organs including the penis, scrotum, prostate, and seminal vesicles; (4) induce enlargement of the larynx and thickening of the vocal cords which results in the low-pitched voice; (5) stimulate beard and axillary and pubic hair growth as well as stimulate temporal hair recession and balding; (6) facilitate libido and sexual potentia; (7) stimulate or suppress organ-specific proteins in tissues such as the liver, kidney, and salivary glands; and (8) produce aggressive warlike behavior. All of these effects are considered physiological responses to androgens with the possible exception of the effects on behavior. This is considered a toxic effect of testosterone by some investigators.

The biological actions of testosterone and its metabolites have been classified according to their site of action. All effects which relate to growth of the male reproductive tract or development of secondary sexual characteristics are attributed to the "androgenic" action of these steroids, while the effects on somatic tissues are termed "anabolic." Earlier studies suggested that these might be two independent biological actions of the same class of steroids. Subsequent experiments, however, indicated that these are organ-specific responses and that the molecular mechanisms which initiate androgenic responses are the same as those which stimulate anabolic actions. At present, it is customary, therefore, to classify the action of steroids according to the molecular mechanisms of action.

THE MECHANISMS OF TESTOSTERONE ACTION

As noted above, testosterone stimulates responses in many tissues. The molecular basis for these reactions is summarized schematically in Figure 8.53. Most of the actions of testosterone are mediated by way of androgen receptors (Fig. 8.53*A* and *B*); however, in some tissues, estrogen receptors (Fig. 8.53*B*) and other intracellular effectors (Fig. 8.53*D*) are important (Mainwaring, 1977; Bardin and Catterall, 1981).

Since testosterone enters the cell by passive diffusion, the amount in a given tissue is determined by plasma concentration and by intracellular enzymes and binding proteins. Following entry, testosterone per se is bound by the androgen receptor and is the dominant intracellular androgen in organs such as brain, pituitary, and kidney (Fig. 8.53*A*). In other tissues, testosterone is metabolized by the enzyme, 5-α-reductase. The product of this metabolic conversion is dihydrotestosterone, a steroid which is more potent than testosterone in some bioassay systems. There is sufficient 5-α-reductase activity in skin, prostate, seminal vesicle, and epididymis to metabolize most of the testosterone to dihydrotestosterone which is the dominant intracellular androgen in these tissues. As noted in Chapter 60, the 5-α-reductase in the adult is strikingly different from that in the fetus.

One of the first steps in androgen action is the binding of testosterone or dihydrotestosterone to a specific intracellular protein termed the androgen receptor. The presence of this protein, in part, determines whether a tissue will respond to androgen. Once the androgen-receptor complex is

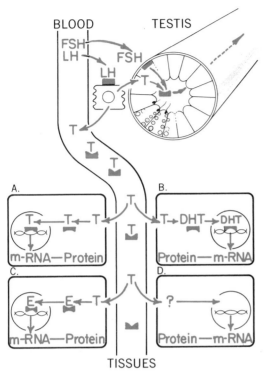

Figure 8.53. Secretion and mechanism of action of testosterone. Testosterone is synthesized in Leydig cells and secreted into the blood where it is transported bound to testosterone-estradiol-binding globulin (*T*) and other proteins. Testosterone enters tissues by passive diffusion where it exerts its action either directly as testosterone, via the androgen receptor before (panel *A*) or after conversion to dihydrotestosterone (*DHT* in panel *B*). In other tissues, such as the brain, testosterone is aromatized to estradiol (*E* in panel *C*) which is active by way of estrogen receptors. In other tissues (panel *D*), testosterone mediates its effect independent of androgen or estrogen receptors. *FSH*, follicle-stimulating hormone; *LH*, luteinizing hormone; *m-RNA*, messenger ribonucleic acid. [From Bardin (1978).]

formed, it is transferred to the nucleus where it binds to specific sites on chromatin. The binding site for the steroid-receptor complex has been termed the nuclear acceptor site. Interaction of the steroid-receptor complex with its acceptor results in a striking increase in nuclear metabolism. This includes an increase in chromatin activity, an increase in the activity of ribonucleic acid (RNA) polymerases, an increase in the number of initiation sites on chromatin, and an increase in synthesis of all classes of RNA. These events lead to increased transfer of RNA to cytoplasm, which results in protein synthesis, cellular growth, and differentiated function (Fig. 8.53*A* and *B*).

Although most of the effects of testosterone are mediated by way of the androgen receptor, in some tissues testosterone is aromatized to estradiol, and the action of this metabolite is mediated via the estrogen receptor (Fig. 8.53*C*). Both testosterone and dihydrotestosterone (a nonaromatizable androgen) stimulate uterine growth. The endometrium

in the testosterone-treated animals becomes hyperplastic, whereas that in the dihydrotestosterone-treated animals remains atrophic. These observations suggest that the endometrial hypertrophy in the testosterone-treated animals resulted from estrogenic metabolites as shown in Figure 8.49. A similar mechanism may, in part, explain the masculinizing effect of testosterone on the brain. Androgen administration to the neonatal rodent produces a maturing effect on the central nervous system which allows for constant rather than cyclic ovarian function in the adult. Evidence indicating that this effect on the brain is produced by estrogenic products of testosterone is suggested by the following: (1) nonaromatizable androgens do not masculinize; (2) animals with testicular feminization which have estrogen receptors but no androgen receptors in the brain partially masculinize in response to testosterone; (3) the normal female rat does not masculinize in response to endogenous estrogens because of high levels of the estrogen binder, α-fetoprotein, in the brain; (4) large doses of exogenous estradiol produce effects similar to testosterone. It should be emphasized that these observations have been made only in rodents, and it is not entirely clear whether they are duplicated in other species (see Chapter 60).

Some effects of testosterone are not mediated by either an androgen or estrogen receptor. That several actions of testosterone on prostate are mimicked by dibutyryl cyclic AMP suggests that the adenylate cyclase-protein kinase system may be an important mediator of some phases of this class of steroids. Recent studies have also demonstrated that androgens can stimulate microsomal protein synthesis by a receptor-independent mechanism in the liver. The molecular basis of this latter effect is not known (Fig. 8.53*D*) (Mainwaring, 1977; Bardin and Catterall, 1981).

Hormonal Control of the Pituitary-Leydig Cell Axis

The rate of testosterone synthesis and secretion by Leydig cells is primarily dependent upon the LH secreted into the blood. Increased LH secretion results in Leydig cell hypertrophy and increased testosterone secretion, whereas lowered LH levels such as occur following hypophysectomy is associated with reduced activity of testicular Leydig cells and decreased testosterone secretion. Similarly, LH secretion by the anterior pituitary is reciprocally controlled by the direct action of gonadal steroids on the central nervous system and the anterior pituitary. A fall in testosterone concentrations results in increased release of LH into the general circulation. In response to the rise in blood LH, the

Leydig cells secrete testosterone which, in turn, inhibits LH secretion (Fig. 8.44). In this manner, the hypothalamus, the pituitary, and the Leydig cells form a functional unit which serves to maintain mean testosterone blood levels relatively constant from day to day and from week to week. In men there is, however, a slight diurnal variation in blood testosterone with the levels at 8 A.M. being 20 to 25% higher than those at 6 P.M. (Baker et al., 1975). This variation is more marked in monkeys.

As noted above, some of the actions of testosterone on differentiation of the brain are mediated via its metabolite, estradiol. Since both testosterone and estradiol inhibit LH secretion, it is possible that androgen effects on gonadotropin release are induced by estrogenic metabolites. It should be noted, however, that even though both of these steroids suppress mean or integrated LH secretion, each produces a different effect on the pattern of LH release from the pituitary. An acute infusion of estradiol reduces the mean LH level by decreasing the amplitude of each LH discharge. This is a direct effect of estrogen achieved in part by decreasing the sensitivity of the pituitary to LHRH. By contrast, an acute testosterone infusion decreases the mean plasma LH level by reducing the frequency of each LH discharge, and this effect is not associated with change in the pituitary responsivity to LHRH. These latter observations imply that the suppressive effects of testosterone are mediated by way of the hypothalamus. The divergent effects of acute testosterone and estradiol administration on the pattern of LH secretion, along with the observation that dihydrotestosterone can also suppress LH, indicated that aromatization of androgens is not a prerequisite for their action on the central nervous system. In addition, these studies suggest that sex steroids act on the central nervous system to regulate the quantity of LHRH secreted into the hypophyseal system and, in addition, to regulate the sensitivity of the pituitary to this releasing factor.

THE PITUITARY-SEMINIFEROUS TUBULAR AXIS

Follicle-Stimulating Hormone (FSH)

FSH is secreted from the PAS-positive basophils in the anterior pituitary. Although it is released in response to LHRH, the potency of this peptide on inducing FSH secretion in the adult human is only one-fifth that of its LH-releasing activity. As a consequence, the minute to minute coupling of FSH secretion to each pulse of LHRH is not as evident as it is for LH. Although some studies suggest that there may be a separate FSHRH, the existence of

such a peptide must await isolation and structure determination. Once FSH is released into the blood its only known function is to stimulate growth and development of the testes (Fig. 8.45).

The Seminiferous Tubules

GERM CELLS

The major mass of the testis is comprised of tightly coiled seminiferous tubules which may reach 70 cm in length in man. Both ends of these tubules empty into the rete testis. The tubules are lined by spermatogonia which undergo mitotic division to give rise to primary spermatocytes (Fig. 8.54). These cells mature through several stages and subsequently undergo meiotic division so that the resulting secondary spermatocytes contain 23 rather than 46 chromosomes. These latter cells rapidly mature into spermatids which are found near the lumen of the tubule and are transformed without further division into spermatozoa (Fig. 8.54). Each of these cell types within the seminiferous epithelium are arranged in well defined cellular associations, which succeed one another in a cyclic fashion along the length of the seminiferous tubule. Each complete sequence of changes in cellular associations is called the spermatogenic cycle. In man, the cycle consists of six stages as compared to 12 and 14 in the monkey and rat, respectively. The ability to recognize individual germ cells and their cellular associations in various stages of the spermatogenic cycle is the best way of classifying germinal cell disorders at present (Steinberger and Steinberger, 1975)

In man, approximately 74 days are required for conversion of the relatively undifferentiated spermatogonia into the highly specialized sperm cell, which is independently motile and possesses the enzymes required for penetration of the ovum (Fig. 8.54). A seminiferous tubule is functionally the first portion of the male reproductive tract in which sperm development and maturation occurs. This entire tract is comprised of the seminiferous tubules, rete testis, efferent ductules, epididymis, vas deferens, urethra, and penis, along with their associated structures, including the prostate and seminal vesicles (Steinberger and Nelson, 1955).

Both LH and FSH are required for spermatogenesis. It is currently held that all of the effects of LH are mediated by way of testosterone from the Leydig cells. As a consequence, testosterone and FSH are the hormones which act directly on the seminiferous tubular epithelium. An understanding of the hormonal requirements for sperm development was complicated for many years by the fact that the initiation and maintenance of the process

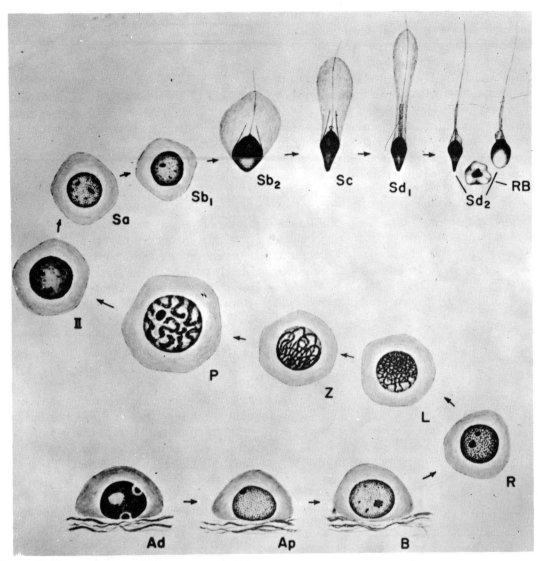

Figure 8.54. Drawings illustrating the main steps of spermatogenesis in man. Spermatogenesis starts at the *bottom left* with the dark, type A, spermatogonium (*Ad*) considered as the stem cell and terminates with the step d (*Sd₂*) spermatid or spermatozoon at the *top right*. In sequence the other elements are: *Ap*, pole type A spermatogonium; *B*, type B spermatogonium; *R*, resting (or preleptotene) spermatocytes; *L*, leptotene spermatocyte; *Z*, zygotene spermatocyte; *P*, pachytene spermatocyte; *II*, secondary spermatocyte; *Sa*, *Sb₁*, *Sb₂*, *Sc*, *Sd₁*, spermatids at various steps of spermiogenesis. [From Heller and Clermont (1964).]

appear to be under independent control. In the immature animal, initiation of spermatogenesis requires both testosterone (or LH) and FSH. Once the normal germinal epithelium is established, testosterone alone can maintain sperm production in the hypophysectomized animal, provided treatment is begun immediately after removal of the pituitary. If, however, the seminiferous tubular epithelium is allowed to regress, then both testosterone and FSH are again required to reinitiate spermatogenesis (Steinberger, 1971).

SERTOLI CELLS

The "backbone" of the seminiferous tubule is formed by Sertoli cells which rest on the basal lamina and are attached to one another by special-ized junctional complexes near the basal portion of each cell. The arrangements of these cells are shown diagrammatically in Figure 8.55. The Sertoli cell junctional complexes limit the transport of fluid and molecules from the interstitial space into the seminiferous tubular lumen and vice versa. The functional significance of the anatomical arrangement of these cells was first suggested by studies in which dyes administered intravenously failed to enter the seminiferous tubular lumen. These and similar observations suggested that there is a "blood-testis barrier" similar to that which exists between the vascular space and the brain (Fawcett, 1975).

Sertoli cells and spermatogonia are the only cellular elements of the seminiferous tubular epithelium which are in contact with the basal lamina. As a consequence, most germinal development occurs behind the blood testis barrier with the germ cells in intimate contact with Sertoli cell plasma membrane (Fig. 8.55). The appreciation of this anatomical arrangement led to the conclusion that Sertoli cells were required to nourish germ cells as they moved away from the basal lamina toward the lumen of the tubule. In addition to their role in germ cell maturation, Sertoli cells also have a number of other important functions including (1) maintenance of the blood-testis barrier; (2) phagocytosis of damaged germ cells; (3) production of proteins which are secreted into the seminiferous tubular lumen; and (4) estrogen production before puberty. These functions are illustrated diagrammatically in Figure 8.55. In addition, the Sertoli cells are a major site of hormone action in testes.

Relatively little was known about Sertoli cell

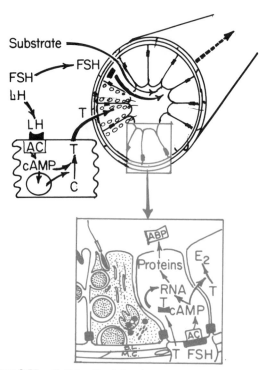

Figure 8.55. Follicle-stimulating hormone (*FSH*) from the blood and testosterone (*T*) from Leydig cells stimulate the seminiferous tubules of the testis (*upper panel*). The backbone of the seminiferous tubules is the pyramidal-shaped Sertoli cells (*lower panel*). The Sertoli cells rest on a basal lamina (*BL*) which is, in turn, surrounded by myoid cells (*MC*). The intracellular spaces between Sertoli cells are closed by junctional complexes (designated by the red and black squares) which prevent various substrates from entering the seminiferous tubular lumen other than via Sertoli cell cytoplasm. In the inset (*lower panel*) germinal cells develop surrounded by Sertoli cell cytoplasm (*left*), and the actions of testosterone and FSH on Sertoli cell cyclic AMP metabolism and protein secretion (*right*) are shown. A major secretory protein of the Sertoli cells is androgen-binding protein (*ABP*). FSH also stimulates testosterone conversion to estradiol (*E₂*). For other abbreviations, see Figure 8.53.

secretion until recent studies demonstrated that "androgen binding protein" (ABP) is synthesized by those cells and secreted into the seminiferous tubular fluid. This protein is transported into the epididymis where it is concentrated in the proximal portion of the caput and is apparently degraded. Highly purifed ABP prepared from rabbit and human epididymides is structurally and immunologically similar to TeBG in plasma of these species. In the rat which does not synthesize TeBG, ABP is secreted both into the seminiferous tubular lumen and into the blood. The function of ABP, like that of TeBG, is not certain but is believed to transport androgens into reponsive cells in the reproductive tract (Bardin et al., 1981).

Even though the function of ABP is not completely understood at present, this protein nonetheless serves as a marker for Sertoli cell function. Both testosterone and FSH stimulate ABP synthesis by different mechanisms (Fig. 8.55). In addition, FSH is believed to be required for ABP secretion into the seminiferous tubular lumen. Besides serving as an index of hormone response, ABP secretion is also affected by primary testicular disorders such as cryptorchidism and hereditary seminiferous tubular failure. The use of specific proteins as markers of testicular function is an important tool for understanding pathophysiology. Since most recent studies indicate that ABP is secreted into the blood, it will now be possible to study seminiferous tubular secretion without testicular biopsy.

ABP is believed to be one of many testis-specific proteins. For many years it was known that the fluid in the seminiferous tubular lumen also contained serum proteins, but it was not known how they could get into this space, since it was presumed that they had to cross the blood-testis barrier. It is now known that the Sertoli cells can synthesize both testis-specific and serum proteins. The synthesis and secretion of these products is not only controlled by hormones but by myoid and germinal cells which are in contact with Sertoli cells. Thus, the seminiferous tubule is a functional unit whose cellular elements work in concert under hormonal control to produce nonmotile spermatozoa. These germ cells are washed out of the tubule into the rete testis by the tubular fluid. This latter structure is a collection chamber from whence the sperm are transferred to the epididymis via the efferent ductules (Fig. 8.56).

THE EPIDIDYMIS

The epididymis is a structure that is formed by a single duct which is coiled upon itself. The major functions of this organ are to assist in the extrates-

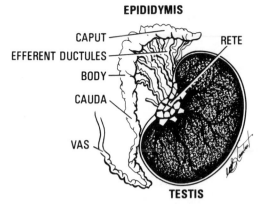

EPIDIDYMIS

CAPUT

EFFERENT DUCTULES

BODY

CAUDA

VAS

RETE

TESTIS

Figure 8.56. The testis and epididymis.

ticular maturation of spermatozoa and to provide a storage site for these germ cells between ejaculations. The growth and function of the epididymis are dependent upon testosterone and upon factors that are transported from the testis in tubular fluid.

When the spermatozoa arrive in the caput epididymis (Fig. 8.56) they are nonmotile and therefore not capable of fertilization. As they migrate through the caput and corpus, an androgen-dependent maturation process occurs so that by the time sperm arrive in the cauda epididymis they are able to move. This maturation process is facilitated by a product(s) which is secreted by the epididymal epithelium which binds to sperm. Upon ejaculation, sperm from the cauda epididymis are mixed with secretory products of the prostate and seminal vesicles to form semen. It is of note that it is the products of these latter organs rather than the testes and epididymides that form the bulk of seminal plasma. Once sperm are deposited inside the female tract, further maturation occurs which is called capacitation. This is described in Chapter 59.

Hormonal Control of the Pituitary-Seminiferous-Tubular Axis

The fact that FSH secretion from the pituitary rises following orchiectomy indicates that a factor from the testis controls the secretion of this gonadotropin much in the same way that androgens and estrogens control LH secretion. The facts, however, that testosterone administration has little effect on FSH secretion and supraphysiological doses of estrogen are required to suppress FSH suggest that another testicular secretory product in addition to or instead of sex steroids is required for control of FSH. This assumption was supported by early observations indicating that testicular extract devoid of steroids could suppress FSH levels. The principle in these extracts was termed "inhibin" and the rise of FSH in individuals with damaged seminiferous tubular epithelium was attributed to defective in-

hibin secretion (Fig. 8.45). Although initial observations suggested that inhibin was produced by germ cells, subsequent in vivo and in vitro studies indicate that the Sertoli cells are the most likely origin of this material. Further studies on the importance of this agent for seminiferous tubular function depend upon the development of an assay for inhibin which is sensitive enough to measure this hormone in blood (Franchimont et al., 1977).

Prolactin and Testicular Function
PROLACTIN SECRETION

Although the existence of prolactin in rodents and sheep was recognized for many years it was only much later that this pituitary hormone was isolated from primates. In females, this molecule stimulates milk production and is required for establishing a functional corpus luteum in some species. It was only with the advent of specific radioimmunoassays that it was realized that prolactin circulated in the plasma of males and that levels were generally comparable or somewhat lower in those observed in females. Furthermore, plasma levels of prolactin in the male (like the female) increase with puberty, decrease following castration, and can be restored by treatment with gonadal steroids. The factors which alter the synthesis and secretion of prolactin in men are similar to those in women. Some of the factors which regulate prolactin secretion are summarized in Figure 8.57.

One of the most convincing arguments that prolactin exerts an effect on the male reproductive tract was derived from studies in hereditary dwarf mice. Two recessive mutations in the mouse, sw and df, result in an absence of acidophil cells in the pituitary gland. Animals homozygous (dw/dw; df/df) for either of these genes produce little or no prolactin and reproduce poorly. Treatment of male dwarf mice with purified prolactin increases sperm production—and improves fertility. Studies in hypophysectomized animals indicate that prolactin alone exerts very little effect on the male reproductive tract but significantly potentiates the effect of LH on the Leydig cell (Fig. 8.57). That the action of prolactin is directly on Leydig cell was suggested by studies indicating that the receptor sites for this molecule are exclusively in this interstitial area of the testis (Bartke, 1977).

In addition to its action on the testis, prolactin stimulates the male reproductive tract in some species. When prolactin is administered to castrate hypophysectomized animals, it is essentially inactive. However, when administered with testosterone, there is a significant increase in the effect on the prostate and seminal vesicle above that expected with the androgen alone. The male repro-

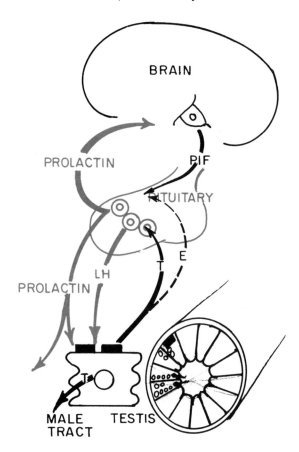

Figure 8.57.   The mechanism of action and control of prolactin secretion. Prolactin-inhibiting factor (*PIF*) secreted by the hypothalamus partially inhibits the release of prolactin from the pituitary. Prolactin binds to specific receptors on Leydig cell membranes and facilitates the secretion of testosterone (*T*) by luteinizing hormone (*LH*). Testosterone and prolactin then synergize on the male tract. Testosterone and estradiol (*E*) inhibit LH secretion but facilitate prolactin secretion. As noted in the text, there is some evidence that prolactin may also act directly on the brain. [From Bardin (1978).]

ductive tract and other androgen-responsive tissues have membrane receptors for prolactin. Since prolactin has not been shown to act through a cyclic AMP-protein kinase system, as is the case of LH and FSH, it was of interest to determine how this peptide facilitated androgen-induced differentiation. Studies by several investigators indicate that prolactin increases the amount of androgen receptor complex which is transferred to the nucleus of androgen-responsive tissues (Baker et al., 1977).

HYPERPROLACTINEMIA

As noted above, a considerable number of studies suggest that prolactin synergizes with LH and testosterone to increase reproductive function in the male. It was, therefore, of considerable interest to find in male rats that prolactin-producing pituitary tumors result in testicular involution and decreased testosterone levels. In these studies, decreased reproductive function was associated with prolactin-induced lowering of testosterone by an unknown mechanism. By contrast, almost all *men* with small prolactin-producing pituitary tumors are impotent regardless of their plasma testosterone levels. In addition, sexual function is restored by lowering prolactin levels with bromocryptine or by surgical removal of the pituitary tumor. These latter observations suggest that prolactin can produce impotence independent of its action on testosterone secretion. Although the mechanism by which prolactin influences sexual drive is not known, it is intriguing to speculate that this pituitary hormone can act directly on the brain to influence behavior. The vascular channels which could transport prolactin to the brain are shown in Fig. 8.46 (Thorner et al., 1977).

CHAPTER 58

Hormonal Regulation of the Ovary

Reproductive activity in the human female and in subhuman primates depends on regular cyclic activity of the hypothalamus, pituitary, ovaries, and uterus. The last decade has seen great progress toward our understanding of the complex interrelation of these diverse organs (diZerega and Hodgen, 1981; Knobil, 1980; Van Look and Baird, 1980; Greep, 1976). The regular monthly vaginal bleeding observed in most primates is directly dependent upon the cyclic release of the steroid hormones, estrogen and progesterone, from the ovaries. Ovarian steroid synthesis is, in turn, coordinated in a cyclic manner by the central nervous system through the secretion of pituitary follicle-stimulating hormone (FSH) and luteinizing hormone (LH). The release of LH and FSH depends on the hypothalamic peptide luteinizing hormone-releasing hormone (LHRH).

During an ovarian cycle, a cohort of follicles develops in the ovaries. Growth of germ cells and estrogen synthesis occurs in the follicles during the first half of the cycle. A single follicle gains dominance and matures to release an ovum, whereas all other follicles belonging to the same cohort undergo atresia. At the site of ovulation a corpus luteum is formed from the ruptured follicle and is responsible for progesterone and estrogen secretion during the latter part of the cycle. The corpus luteum is maintained for a fixed period of time, and as its function decreases, a new set of follicles begins to develop, and a new ovarian cycle begins.

During the ovarian cycle the endometrium proliferates under the influence of ovarian estrogen and progestins. With regression of the corpus luteum, the endometrium is sloughed, and a menstrual period occurs. The period of time between the beginning of one menstrual period and the beginning of the next is termed the menstrual cycle.

Although in the strict sense the ovarian cycle and the menstrual cycle refer to cyclic changes in the ovary and endometrium, respectively, these terms are often used synonymously with "reproductive cycle," which implies the cyclic, morphologic, and hormonal changes which occur in the central nervous system, anterior pituitary, ovary, and uterus. There is a great variation in the reproductive cycles of different species. As a consequence, in the discussion which follows, the hormonal control of the primate ovary is presented when possible.

MORPHOLOGIC CHANGES DURING THE REPRODUCTIVE CYCLE

The Ovary

The ovary is anatomically divided into three separate compartments: the follicles, the corpus luteum, and the interstitium.

OOCYTE AND FOLLICULAR MATURATION

In the human ovary the process of oogenesis is completed before birth, and the population of germ cells, or oogonia, reaches a maximum of 6.8 million 5 mo after conception. By the time of birth the number is reduced to approximately 2 million (Baker, 1963). If one considers that reproductive capacity in a woman lasts approximately 40 yr, no more than 480 ova will be successfully ovulated. The remainder of the original oocytes undergo progressive atresia throughout the reproductive years.

Between the 8th and 13th wk of fetal life, some of the oogonia cease to undergo mitosis, and initiation of nuclear changes characteristic of prophase of the first meiotic divison marks the conversion of these cells into primary oocytes. In these cells, the nucleus and the chromosomes persist in prophase of the first meiotic division until the time of ovulation when meiosis is resumed, and the first polar body is formed. Thus, some of the oocytes may remain in this stage of division for up to 40 years. There is evidence that inhibition of meiosis throughout most of follicular maturation depends upon the follicular milieu. Thus, in vitro, meiosis is resumed in oocytes aspirated from mature follicles and incubated in balanced salt solutions. Moreover, in vivo, extrusion of the polar body has been observed in follicles in which granulosa cells are undergoing degeneration, suggesting that granulosa cells participate in intrafollicular inhibition of meiosis. Consistent with this supposition, a substance which inhibits resumption of meiosis in isolated oocytes in vitro has been identified in antral fluid and in extracts of granulosa cells from porcine ovarian follicles (Tsafriri et al., 1980).

Although the chromosomes remain condensed in the nucleus throughout the life of primary oocytes, these cells, which are 9 to 25 μm in diameter when meiosis begins, undergo remarkable growth and accumulation of cytoplasm. The diameter of the follicular structure around the oocyte enlarges in parallel. However, once the oocyte reaches a mean maximal diameter of 80 μm, follicular growth continues so that diameters of preovulatory follicles range from 10 to 20 mm (Igram, 1962, and Edwards, 1965). At ovulation, the oocyte and surrounding cells in the *cumulus oophorus* are extruded. Completion of the first meiotic metaphase and formation of the first polar body converts the primary oocyte into the secondary oocyte, which enters the Fallopian tube. After another nuclear division, the second polar body is eliminated. As a result of two nuclear divisons with only one replication of the chromosomes, the ovum nucleus contains a haploid set of chromosomes at the time of sperm penetration at fertilization (Mossman and Duke, 1973).

FOLLICULAR MORPHOLOGY

In the fetus, the primary oocytes become closely associated with a single layer of spindle-shaped cells which are the precursors of granulosa cells. The oocyte and associated granulosa cells are separated from the surrounding stroma by a membrane, the "basal lamina." The oocyte and spindle-shaped cells enclosed by the basal lamina are called a "primor-

dial follicle." During the 5th and 6th mo of gestation, follicular maturation is initiated when granulosa cell precursors in primordial follicles differentiate into a single layer of cuboidal cells that begin to divide. This granulosa cell proliferation gives rise to multiple layers of cells which contribute to enlargement of the follicle. Granulosa cells secrete mucopolysaccharides, which give rise to a translucent halo called the "zona pellucida" that surrounds the oocytes (Fig. 8.58). The cytoplasmic processes of granulosa cells traverse the zona pellucida and are thus electrically coupled with the plasma membrane of the oocyte. Shortly after granulosa cells begin to proliferate in primary follicles, cortical stromal cells outside the basal lamina differentiate the theca interna and externa cells, which are important for steroid synthesis. Granulosa cell proliferation, differentiation and hypertrophy of theca cells, and growth of the oocyte all combine to increase the diameter of maturing follicles. When the sagittal diameters reach 100–200 μm, localized accumulation of fluid appears among the granulosa cells in some primary follicles. These loculi increase in size and become confluent and give rise to a central fluid-filled cavity called the "antrum." Antrum formation transforms the primary follicle into the secondary or "Graafian follicle" (Fig. 8.59), in which the oocyte occupies an eccentric position, surrounded by 2–3 layers of granulosa cells, collectively called the "cumulus oophorus" (Harrison, 1962; Lintern-Moore et al., 1974).

ANTRAL FLUID

Antral or follicular fluid contains sex steroid hormones, plasma proteins, mucopolysaccharides, and electrolytes. While plasma proteins, gonadotropins, prolactin, and proteins unique to granulosa cells reach the antral fluid by diffusion from vas-

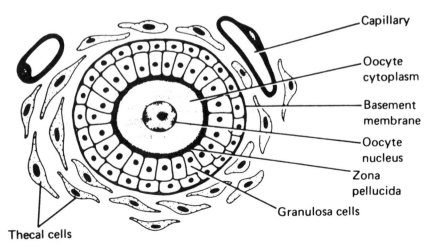

Figure 8.58.   Primary follicle. [Adapted from Begley et al. (1980).]

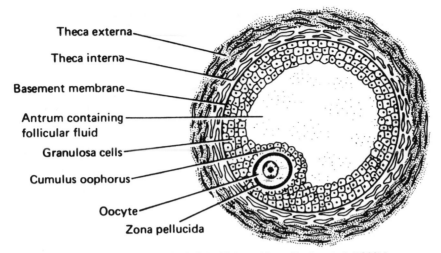

Theca externa
Theca interna
Basement membrane
Antrum containing follicular fluid
Granulosa cells
Cumulus oophorus
Oocyte
Zona pellucida

Figure 8.59.   Secondary follicle. [Adapted from Begley et al. (1980).]

cular spaces outside the basal lamina, and muco-polysaccharides are secretory products of granulosa cells, the origin of antral fluid steroid hormones is less clear. Certainly, some of the sex steroid hormones are secreted by theca and interstitial cells and diffuse across the basal lamina into antral fluid. It also seems probable that some of the antral fluid steroid hormones are secreted by granulosa cells, and variations in antral fluid hormone concentration among follicles suggest that regulation is more complicated than can be ascribed to diffusion alone (Edwards, 1974).

ATRESIA

Prior to the appearance of follicles during fetal ovarian development, many germ cells are eliminated during oogonial divison, but once follicles appear in the fetal ovary, a follicular degenerative process called atresia begins and continues until the last oocyte is removed from the ovary at the time of menopause. The number of primordial follicles declines from 2 to 4 million at birth to an estimated level of around 400,000 at menarche. Since there is no ovulation prior to menarche, reduction of oocytes must result exclusively from atresia. However, even after ovulation begins, more than 99% of oocytes in the human ovary are removed by atresia. The fact that estrogens are secreted by ovaries of premenarcheal girls indicates that follicles destined to undergo atresia may secrete hormones which are active in stimulating development of secondary sexual characteristics (Baker, 1963; Winter and Faiman, 1973).

After the menarche, atresia persists and out of a cohort of developing follicles, usually only one follicle ovulates during each menstrual cycle. The interval between beginning maturation and com-

pleting ovulation of a follicle in a given cycle is unknown. However, a minimum figure for the time required for a follicle to mature to the point of ovulation can be inferred from the observation that ovulation induction requires 10–12 days of preovulatory stimulation with exogenous LH and FSH in women whose ovaries contain only primary follicles.

OVULATION

Mechanically, ovulation consists of rapid enlargement, followed by protrusion from the surface of the ovarian cortex and rupture of the follicle, with extrusion of an oocyte and adhering cumulus oophorus. In the human ovary, this sequence occurs over a 5- to 6-day period prior to the preovulatory LH surge (see below) that marks the end of the follicular phase of the cycle and precedes actual rupture by as much as 16 h. A conical stigma on the surface of the protruding follicle precedes rupture. Rupture of this stigma is accompanied by gentle rather than explosive expulsion of the oocyte and antral fluid, suggesting that the latter is not under high pressure. Currently, stigma formation and rupture are thought to result from effects of hydrolytic enzymes acting locally on protein substrates in the basal lamina. This conclusion is based on the observations that instillation of protease inhibitors into the antral fluid inhibits ovulation and suggests that these or similar enzymes participate in this process (Doyle, 1951; Espey, 1974).

THE CORPUS LUTEUM

After rupture of the follicle and release of the ovum, granulosa cells increase in size and accumulate a yellow pigment, lutein, which lends its name to this new anatomical unit, the corpus luteum.

During the first 3 days after ovulation, the granulosa cells proliferate, and the corpus luteum reaches a diameter of approxiamtely 1 cm. Capillaries penetrate into the granulosa layer, reach the cavity, and fill it with blood. From days 4 to 9 after ovulation, vascularization continues and reaches a peak at about day 9. Soon thereafter the corpus luteum begins its regression, which is first marked by a decrease in the amount of blood present in the capillaries.

An ovarian cycle is begun when a new set of follicles begins to grow as the corpus luteum begins to regress. As will be seen in the next section, this is also the time during which plasma levels of FSH increase. It is apparent, therefore, that a new cycle begins during the luteal phase of the preceding cycle.

Endometrium

After the menstrual period, under the influence of estrogen, the epithelium and stroma of the endometrium undergo mitotic proliferation (proliferative period, Fig. 8.60*A*). The endometrium becomes considerably thicker. The spiral arteries elongate as the endometrial glands increase in length and size. This is primarily a period of rapid growth, and only a modest increase in glycogen content of the endometrial cells occurs.

Following ovulation, under the influence of progesterone and estrogen produced by the corpus luteum, functional differentiation of the endometrium occurs, but rapid proliferation of the endometrium ceases. The endometrial glands become tortuous and irregular, and their lumens are filled with secretion. Glycogen contents of epithelial cells and secretory products are markedly increased (secretory period). The stroma accumulates less water and is less dense, and alkaline phosphatase increases around the blood vessels. These morphologic and functional changes are required for the endometrium to support implantation of the fertilized ovum.

Following estrogen and progesterone withdrawal during the decline of corpus luteum function, endometrial desquamation and necrosis is preceded by intense spasmodic constrictions of the spiral arteries. The precise mechanisms for the initiation of the vasospasm are unknown. However, the results of several recent studies indicate that local production of prostaglandins may have a vasoconstricting action. These studies indicate that prostaglandins are found in high concentrations in the endometrium prior to menstruation. Furthermore, the release of prostaglandins may also explain the increased uterine contraction which occurs at the time of menstrual flow (Markee, 1940; Pickles, 1967).

Other studies indicate that local fibrinolytic activity in the blood vessels of the endometrium reaches maximal levels at the time of menstruation. The noncoagulability of the menstrual blood has been attributed to the presence of this fibrinolytic system in the shed endometrium (Todd, 1964).

REPRODUCTIVE HORMONES

The major hormones involved in the control of the male and female reproductive system are the decapeptide LHRH, the protein hormones LH and FSH, and several classes of steroids (estrogens, progestins, and androgens). The structure, metabolism, and excretion of LH and FSH have been reviewed in Chapters 52 and 57. The function and mechanism of action of these hormones in the female are discussed below.

Luteinizing Hormone-Releasing Hormone

The structure of LHRH was determined independently in the laboratories of Schally (Matsuo *et al.*, 1971) and Guillemin (Burgus *et al.*, 1971). LHRH is a decapeptide (pyro-Glu_1-His_2-Trp_3-Ser_4-Tyr_5-Gly_6-Leu_7-Arg_8-Pro_9-Gly_{10}-NH_2) and is found in the hypothalamus of all mammalian species studied so far. The major biological function of LHRH is to control the release of pituitary LH and FSH.

The mechanism of action by which LHRH stimulates the secretion of pituitary gonadotropins is believed to consist of three steps (Conn *et al.*, 1981). The first step is the interaction of LHRH with specific receptors on the cell membrane of the pituitary. The second step involves the mobilization of ionic calcium (Ca^{++}), which is believed to fulfill the requirements of an intracellular mediator for LHRH action. The opening of Ca^{++} channels in the plasma membrane may lead to a net influx of Ca^{++} ions and an increase in Ca^{++} levels in the cytosol which subsequently promotes the third step in the mechanism, the migration of hormone-containing secretory granules from the cytoplasm to the cell membrane. There, in a process that also requires Ca^{++} ions, the membrane of the granule fuses with the cell membrane and releases its content by exocytosis into the intercellular space. LHRH action is very rapid. It can promote the release of gonadotropins from the "release pool" in less than a minute. There is evidence for a second kind of LH pool in the pituitary, the "reserve pool," which requires sustained LHRH stimulation before its release is initiated. The exact role of LHRH in the biosynthesis of these hormones has not been elucidated.

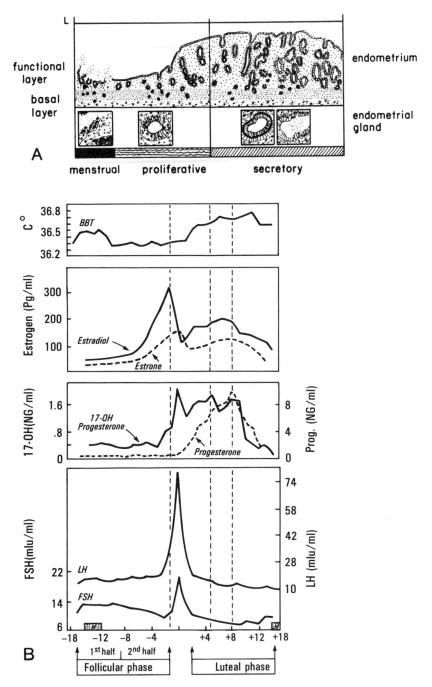

Figure 8.60. The hormonal events of the reproductive cycle are centered on the midcycle LH peak (day 0) and are shown in relation to endometrial changes (*A*) of nonpregnant women throughout the menstrual cycle. Basal body temperatures (*BBT*) are shown at the top of *B*. Plasma levels of luteinizing hormone (*LH*) and follicle-stimulating hormone (*FSH*) are shown on the bottom two lines of *B*. Gonadal steroid levels (estradiol, estrone, 17-hydroxyprogesterone, and progesterone) are shown in the center two panels of *B*. M, days of menstrual bleeding. [Adapted from Ross et al. (1970), Korenman et al. (1970), and Hafez (1978).]

Substitution or omission of certain amino acids of the LHRH molecule can result in analogs with agonistic or antagonistic action. Superactive analogs have been used for ovulation induction, for ovulation inhibition, and as contraceptives. LHRH is useful in the treatment of delayed puberty and various infertility problems in men and women.

LHRH challenge tests have clinical application for the evaluation of hypothalamic-pituitary function.

Luteinizing Hormone

The postulated mechanism of action of LH is shown in Figure 8.61. At the target cell surface, LH interacts with a specific membrane receptor which

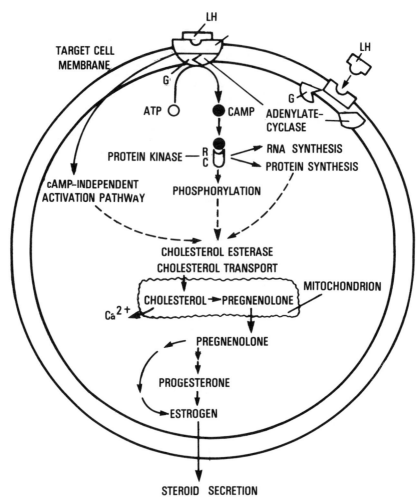

Figure 8.61. Diagram of postulated and demonstrated mechanisms involved in LH action on the ovary. For details, see text. *ATP*, Adenosine triphosphate; *cAMP*, Cyclic adenosine monophosphate; *G*, Guanyl nucleotide regulatory subunits; RNA, Ribonucleic acid; R, Regulatory and *C*, Catalytic subunit of protein kinase.

is coupled to adenylate cyclase by the guanyl nucleotide regulatory subunit (G). Adenylate cyclase catalyzes the conversion of ATP to cyclic AMP, which activates a class of enzymes—protein kinases—by binding to their regulatory subunit (R) and thereby freeing the catalytic subunit (C) to catalyze the phosphorylation of specific proteins. At the same time, an increase in the Ca^{++} permeability of the cell membrane and in the efflux of Ca^{++} ions from the mitochondria is achieved which leads to a rise in cytosolic Ca^{++} concentration which has various positive and negative feedback effects on the products of cyclic AMP-dependent protein kinases and on the permeability of the cell membrane to monovalent cations.

Some of the phosphorylated proteins are believed to facilitate the hydrolysis of cholesterol esters and the transport of free cholesterol into the mitochondrium, where it is metabolized to pregnenolone, the precursor of progesterone synthesis in the corpus luteum and of estrogen synthesis in the thecal cells.

The possibility of a cyclic AMP-independent activation pathway in addition to the one described has not been ruled out.

There is no simple relationship between receptor occupancy and hormone response. Occupancy of only a small percentage of receptors can lead to maximal response, and a small dose of hormone can potentiate a secondary response, while the administration of a large dose is followed by a lack of response to a second hormone administration (desensitization) and later by a loss of unoccupied receptors (downregulation) (Dorrington, 1979).

Follicle-Stimulating Hormone

Compared to other glycoprotein hormones, there is little known about the mechanism of FSH action. FSH binds to a specific receptor on its target cell, activates adenylate cyclase through coupling with the guanyl nucleotide regulatory subunit, and increases cAMP concentrations. cAMP then activates protein kinase, which results in phosphorylation of proteins which may act at the transcriptional and/or translational level to influence protein synthesis similar to the steps described for LH action (see Fig. 8.61).

In the female, FSH binds only to the granulosa cells of the ovarian follicles. It stimulates the mitotic activity of granulosa cells and converts the surrounding stroma into a layer of theca cells. FSH induces an aromatizing enzyme that converts androgens to estradiol, which acts synergistically with FSH to further stimulate follicular development and to increase the number of FSH receptors in the granulosa cells. Granulosa cells in small follicles contain only FSH receptors. As the follicle grows from antral to preovulatory stage, FSH induces the appearance of LH receptors so that Graafian follicles bind both gonadotropins. The plasma levels of FSH throughout the reproductive cycle are shown in Figure 8.60.

Estrogens

In the nonpregnant, sexually mature female, estrogens are secreted almost exclusively from the ovarian follicle. Contrary to the placenta, the human ovary is capable of synthesizing estrogens de novo from acetate. However, as is the case for other steroid synthesizing glands, the major precursor of ovarian steroids is plasma LDL cholesterol, from which pregnenolone is synthesized. In the ovary, the conversion of cholesterol to pregnenolone is accelerated by LH. Figure 8.62 shows the two major pathways of steroidogenesis. The synthetic scheme that uses progesterone and 17-hydroxyprogesterone is called the Δ^4 pathway and the one that uses pregnenolone and 17-hydroxypregnenolone, the Δ^5 pathway (Fig. 8.62). It should be noted that the same synthetic schemes for steroid synthesis exist in the testis (Chapter 57).

In the ovary the Δ^5 pathway is the major route for estrogen synthesis in the follicles, whereas progesterone and estrogen synthesis in the corpus luteum follow predominantly the Δ^4 pathway. All intermediates in the biosynthesis of estrogens are secreted in varying amounts. In the early follicular phase, testosterone is secreted at a higher rate (260 μg/day) than estrogens (60 μg/day). By the late follicular phase, estrogen secretion has surpassed that of testosterone, reaching a secretion rate of 400–900 μg/day. During the luteal phase, approximately 300 μg/day of estrogen are secreted. At that time, progesterone is secreted in milligram amounts (25 mg/day), surpassing all other ovarian steroids. The most important estrogen secreted by the ovary is estradiol. Estrone is also secreted; its biological activity is less than that of estradiol. There are also

many nonsteroidal compounds, such as diethylstilbestrol, which have estrogenic activity. The important biological activities of estrogen include: (a) stimulation of growth of both the myometrium and endometrium; (b) maintenance of a thick vaginal mucosa and indirectly the acidic vaginal pH; (c) stimulation of cervical glands to secrete copious quantities of viscous mucous; (d) stimulation of breast growth and development; (e) deposition of subcutaneous fat which results in a characteristic feminine habitus; (f) regulation of gonadotropin secretion including both "negative" and "positive" feedback (see Hormonal control of the reproductive cycle); (g) sensitization of the ovaries to gonadotropins; and (h) retardation of linear body growth in association with facilitation and epiphyseal closure.

Estradiol and estrone are synthesized in the ovary from steroids with 19 carbon atoms. The biosynthetic pathway up to the point of androstenedione and testosterone is essentially the same as that in the testis (Fig. 8.49). In the final series of reactions the steroid A ring is aromatized with the loss of one methyl group to form the C-18 steroid estrogens, estradiol, and estrone (Fig. 8.62). The day-to-day fluctuations of plasma estradiol and estrone during the reproductive cycle are shown in Figure 8.60.

Progestins

Progestin was the name originally given to the crude extract of the corpora lutea which could prepare and maintain a secretory endometrium during the latter half of the reproductive cycle and during pregnancy. Progesterone is the most active principle in these extracts. Hormones which have progestational activity are referred to as progestins or gestagens. Progesterone is the secretory end product of steroid metabolism in corpora lutea (Fig. 8.62) and in the placenta (see Chapter 59). Plasma progesterone levels throughout the menstrual cycle are shown in Figure 8.60. During the reproductive cycle and the first trimester of pregnancy, progesterone is secreted almost exclusively by the corpus luteum. During the latter two-thirds of pregnancy the placenta assumes this function. The adrenal gland also secretes a small quantity of progesterone. In all of these organs, progesterone is synthesized from cholesterol and pregnenolone.

The biological actions of progesterone and other gestagens include: (a) antagonism of the growth-promoting effect of estrogen on the endometrium and conversion of this rapidly proliferating organ into a secretory structure capable of maintaining an implanted blastocyst; (b) conversion of the cervical mucus from a very viscous to a nonviscous

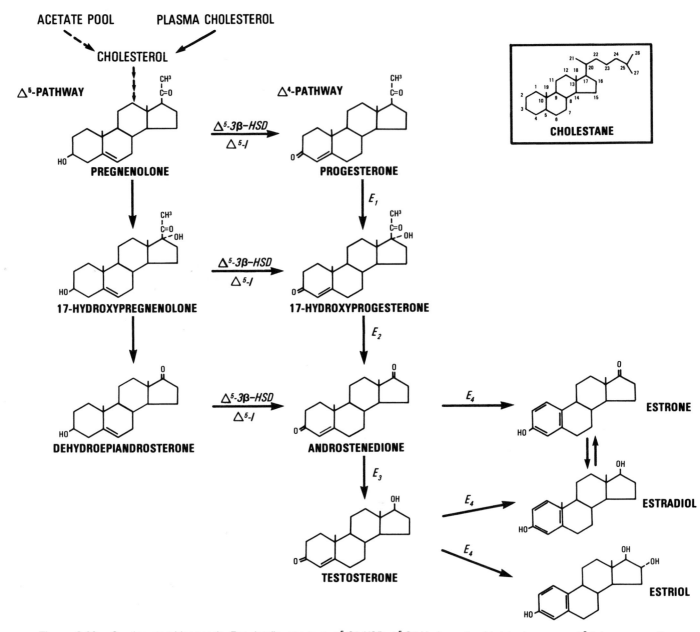

Figure 8.62. Ovarian steroidogenesis. For details, see text. Δ^5-3β-*HSD*, Δ^5-3β-Hydroxysteroid dehydrogenase; Δ^5-*I*, Isomerase; E_1: C_{21} steroid 17α-hydroxylase; E_2, 17-20-Desmolase; E_3, 17β-hydroxysteroid dehydrogenase; E_4, Aromatizing enzymes. The top formula, cholestane, shows the proper number for each carbon atom.

fluid; (c) stimulation of mammary gland growth and development; (d) inhibition of uterine motility; and (e) regulation of gonadotropin secretion.

Mechanism of Steroid Hormone Action

Estradiol, like other steroids, is thought to exert its action directly on the nucleus of the cell (Figure 8.7, Chapter 51). As a consequence, an estrogen-response tissue must have estrogen receptor and nuclear acceptor sites to which activated receptor can bind. Upon entry into the cytosol of the cell (by diffusion), estradiol is bound to a specific receptor (ERc). In the cytoplasm, the estrogen receptor complex is activated (ERn) and translocated to the nucleus. This complex binds to acceptor sites in chromatin and enhances processes associated with differentiated functions of the responsive tissue, which include the production and utilization of messenger and other classes of RNA needed for the synthesis of constituent enzymic and secretory proteins, as well as the receptor itself (Jensen, 1979). In some cells replication of DNA is also stimulated, followed by cell division. The concentration of estrogen receptors in most tissues is constitutive, but in some instances it is increased by estradiol.

The concentration of progesterone receptors in

the uterus and other progesterone responsive tissues is markedly increased by estrogen. In fact, one of the recognized actions of estradiol is to stimulate synthesis of progesterone receptors. The induction of progesterone receptors with estrogen can explain the synergistic action of these two hormones on the uterus. The general scheme for progesterone action via its receptor is similar to that described for estrogens above. However, the mechanism of action of this steroid has been most thoroughly studied in the chick oviduct (O'Malley and Schrader, 1976; Chan and O'Malley, 1976). The progesterone receptor from this tissue has been purified and studied in great detail. The native receptor is a dimer composed of A and B subunits, each of which has a steroid binding site. Once the progesterone-charged receptor is transferred to the nucleus, it binds to chromatin through a binding site on the B subunit. The subunits dissociate, and the A subunit, still bearing its hormone, attaches to the DNA, causes the helix to open, and allows a molecule of RNA polymerase to occupy an initiation site. This results in synthesis of new mRNA.

HORMONAL CHANGES DURING THE REPRODUCTIVE CYCLE

The changes in the plasma levels of estrogens, progestins, and gonadotropins during a reproductive cycle are shown in Figure 8.60. For descriptive purposes the reproductive cycle can be divided into a follicular phase (which can be subdivided into a first and second half), an ovulatory phase, and a luteal phase (Ross et al., 1970; Vande Wiele et al., 1970).

Follicular Phase (First Half)

This period begins in the late luteal phase of the preceding cycle with a rise in the plasma levels of FSH and a concomitant initiation of follicular growth. LH levels also rise during this phase, but the increase starts several days later than the increase in FSH. It is important to note that during the first half of the follicular phase, there is no change in the blood levels of estrogens or progestins (Fig. 8.60).

Studies in rhesus monkeys have shown that at midfollicular phase (about 6 days after the onset of menses) the follicle destined to ovulate (dominant follicle) has been selected (diZerega and Hodgen, 1981). The mechanism by which the dominant follicle achieves its preeminence and inhibits the maturation of other follicles is not yet fully understood. One explanation is that the dominant follicle secretes a substance that inhibits the maturation of other follicles on both ovaries. If this is the case,

then the dominant follicle must be able to thrive in a milieu not conducive to the final maturation of any other follicles (diZerega and Hodgen, 1981).

Follicular Phase (Second Half)

This portion of the follicular phase begins approximately 7–8 days before the preovulatory LH surge and is characterized by an increase in plasma estrogen levels (estradiol and estrone) and a progressive increase in the sensitivity of the pituitary to LHRH. Estradiol increases slowly at first and then rapidly reaches a maximum on the day before the LH peak. The initial rise in estrone is not as great, but estrogen levels reach a peak concomitant with the peak of LH. While plasma estrogen levels are increasing, plasma FSH decreases, and plasma LH slowly and steadily increases. Plasma LHRH levels in cycling women are highest from 4 days before until 2 days after the LH surge (Miyake, 1980). Several days before ovulation there is a rise of 17-hydroxyprogesterone, which reaches a maximum on the day of the LH surge. It is significant that plasma progesterone levels do not increase during this period.

Ovulatory Phase

During this period there is a rapid rise in plasma LH levels which leads to the final maturation of the Graafian follicle and follicular rupture some 16–24 h after the LH peak. Although there is a simultaneous increase in plasma FSH levels, the role of this pituitary hormone on the ovulatory process is not understood. Soon after the beginning of the LH surge and prior to ovulation, plasma estradiol levels drop precipitously, and plasma progesterone increases slightly (Fig. 8.60).

The preovulatory peaks of estradiol, and 17-hydroxyprogesterone correlate with rapid thecal development and follicular maturation. During the final rapid growth of the follicle, the theca interna enlarges and assumes a functional appearance. Following ovulation the theca interna cells persist in the corpus luteum. Both estrogen and 17-hydroxyprogesterone secretion decrease rapidly in the ovulatory period, but both increase again in the luteal phase as the corpus luteum, with both its theca interna cells and luteinized follicular cells, becomes vascularized. While these observations do not prove that estrogens and 17-hydroxyprogesterone are secreted by the theca interna cells, they are nonetheless consistent with that hypothesis (Strott et al., 1969).

To distinguish steroids secreted by the ovary from those produced by metabolism in peripheral tissues from precursors secreted by the adrenal gland, hormone levels have been compared in ovarian and peripheral bloods. These studies indicate that the ovaries secrete many steroids including pregnenolone, progesterone, 17α-hydroxyprogesterone, dehydroepiandrosterone, androstenedione, testosterone, estrone, and 17β-estradiol. Early in the follicular phase of the cycle, no significant differences have been found in steroid hormone concentration in blood from the two ovaries, but in the late follicular phase or in the luteal phase concentrations of all steroids measured are higher in blood collected from the ovary containing the dominant follicle or the corpus luteum. These, and other observations, indicate that the dominant preovulatory follicular complex and the corpus luteum secrete 17β-estradiol, androstenedione, 17α-hydroxyprogesterone, and progesterone, whereas the stroma appears to secrete only androstenedione (Mikhail, 1970; Lloyd et al., 1971; Baird et al., 1974).

Luteal Phase

As noted above, luteinization is initiated by LH. Recent studies also have indicated that ova exert an inhibitory influence on follicle luteinization. When the ovum is removed from a mature follicle, granulosa cells immediately luteinize and form a corpus luteum. By contrast, when the ovum is removed from an immature follicle, the granulosa cells degenerate, and the follicle becomes atretic (Nalbandov, 1970). From these studies it was concluded that luteinization occurs spontaneously following expulsion of the ovum from the mature follicle since only the ovectomized follicles luteinized and the neighboring follicles remained unchanged. Further studies suggest that follicular fluid contains a humoral agent which prevents luteinization. These observations are consistent with those of Channing (1969), who observed that follicular cells placed in tissue culture luteinized spontaneously in the absence of all hormonal stimuli. However, the life-span and function of follicular cells which are luteinized spontaneously either in vivo or in vitro are reduced. LH is required for normal corpus luteum survival and function. Furthermore, LH administered in high doses can luteinize follicular cells even without ovulation and can produce a luteinized follicle with a trapped ovum (Ludwig and Horowitz, 1969). These observations indicate that either LH or ovum removal can luteinize the ovarian follicle; however, during a normal reproductive cycle both mechanisms are probably operative.

The sine qua non of the luteal phase is the marked increase in progesterone secretion which reaches a maximum about 8 days after the midcycle LH peak. 17-hydroxyprogesterone has a double peak during the luteal phase at 5 and 8 days, respectively, and as noted previously, estrone and estradiol plasma levels increase in parallel with those of the progestins. Concomitant with the increase of gonadal steroids, plasma LH and FSH decline throughout most of the luteal phase until 3 or 4 days before menses. When the corpus luteum begins to regress, and concomitantly with the decline in progesterone levels, plasma FSH levels rise, initiating the development of a new cohort of follicles.

The factors which control the life-span of the corpus luteum in the normal reproductive cycle are ill defined. In many species, termination of luteal function is under control of a luteolytic factor produced by the uterus. In the human and monkey there is no evidence for the existence of such a factor (Rowson, 1970).

THE CYCLIC CONTROL OF OVARIAN FUNCTION

As mentioned above, the cyclic ovarian function depends on the interactions of the hypothalamic hormone LHRH, the pituitary gonadotropins LH and FSH, and the secretion of ovarian steroids. The hypothalamus is the site of the integrated control for the cyclic LHRH secretion. This part of the brain contains the necessary receptors and transducers for transmitting signals from the pituitary and the ovary, as well as from other parts of the central nervous system into effective LHRH release. Signals from the arcuate nucleus trigger the release of a bolus of LHRH into the pituitary portal blood approximately once per hour. In response, the pituitary gonadotrophs release pulses of LH and FSH. In contrast to the hypothalamus, the pituitary is an isolated unit, has little if any independent function, and is ultimately controlled by LHRH secreted by the hypothalamus. One possible mechanism by which the synthesis and release of LHRH may, in turn, be regulated by the "short loop" feedback of pituitary LH to the hypothalamus is by way of reverse blood flow through the pituitary-hypophyseal portal system (Bergland and Page, 1978).

It is now generally accepted that ovarian steroids (most importantly estradiol and progesterone) can modulate the release of pituitary gonadotropins by negative and positive feedback effects exerted at the level of the anterior pituitary and the hypothalamus (Fig. 8.63). Estradiol is the most potent gonadotropin inhibitor. This has been demonstrated by the fact that LH and FSH rise following

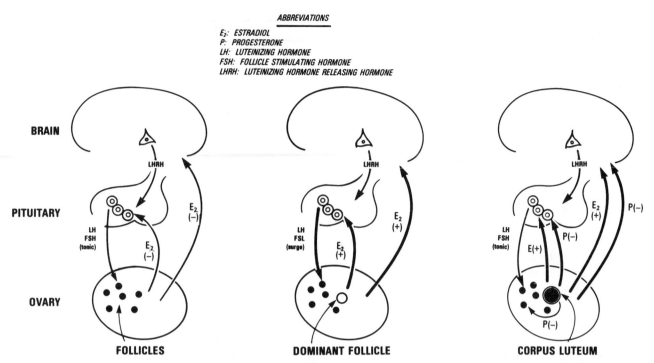

ABBREVIATIONS

E_2: ESTRADIOL
P: PROGESTERONE
LH: LUTEINIZING HORMONE
FSH: FOLLICLE STIMULATING HORMONE
LHRH: LUTEINIZING HORMONE RELEASING HORMONE

Figure 8.63. (*Left*) Early follicular phase. Immature follicles secrete small amounts of E_2 which inhibit LH and FSH secretion to tonic levels. The negative feedback mechanism of E_2 is exerted at the level of the pituitary and the hypothalamus. (*Middle*) Time of LH surge. The growing follicles and the dominant follicle secrete larger amounts of E_2. When levels of 150 pg/ml of plasma are exceeded for more than 36 h, the negative feedback loop is interrupted, and positive feedback takes effect, inducing the preovulatory LH and FSH surges. (*Right*) Luteal phase. High levels of P inhibit the positive feedback action of E_2. P also inhibits follicular development. When luteolysis occurs and the corpus luteum regresses, the inhibitory action of P is removed and a new cohort of follicles begins to grow, and LH and FSH concentrations increase.

ovariectomy and are suppressed following estradiol infusion. Low circulating levels of estradiol, such as are present during the early follicular phase of the cycle, cause maximal suppression of LH (Fig. 8.63).

In addition to the negative feedback of estradiol, this steroid also is responsible for the development of the "positive" or stimulatory feedback effect of ovarian steroids on the secretion of gonadotropins. The positive feedback action is minimal when estradiol levels are low. It becomes greater as plasma levels of estradiol increase toward preovulatory values (Fig. 8.63). A number of studies of positive feedback have indicated that the strength and duration of the estrogenic effect are of prime importance (Keye and Jaffe, 1975; Van Look and Baird, 1980). As a consequence, the factors responsible for the sequential incremental rise in estradiol during the follicular phase have been studied in detail. This cascade of events occurs in the ovary and has led some investigators to propose that the "clock" which determines the time of ovulation is of ovarian origin in primates rather than in the central nervous system, as is the case for rodents. The sequence of events in the ovary may proceed as follows. Initially FSH stimulates a progressive in-

crease in FSH receptor. This action is augmented by gradually increasing levels of endogenous estradiol. This is followed by a delayed but pronounced increase in LH receptors. Thus, estradiol initially acts to promote the ability of FSH to increase its own receptors and, finally, promotes the ability of FSH to stimulate and maintain LH receptors. The appearance of LH receptors in the granulosa cells may be responsible for progesterone production. The preovulatory secretion of progesterone, even though very small, may exert a feedback influence on the estrogen-primed pituitary and hypothalamic neurons. Although both estradiol and progesterone may promote a midcycle LH surge in primates, the interrelationship of these two steroids in stimulating this effect in intact women is not known (Richards et al., 1976). There is good evidence, however, to suggest that progesterone may be essential for inducing the preovulatory surge of FSH (Van Look and Baird, 1980).

The availability of synthetic LHRH demonstrated conclusively that estradiol exerts an effect directly on the pituitary as well as on the neuronal elements of the hypothalamus. By mimicking the pulsatile LHRH release from the hypothalamus by repeated injections of this hypothalamic peptide, several investigators have indicated that the pituitary can detect small increments or decreases in circulating LHRH. In addition, varying the amount of prior exposure to estradiol influences the pituitary response to a given dose of LHRH. Thus, the amount of LHRH delivered to the pituitary and the estradiol level combine to determine the pituitary-gonadotropic function (Krey and Everett, 1973; Yen et al., 1972; Jaffe and Keys, 1974).

The progressive nonlinear increase in circulating estradiol during the follicular phase acts at the level of the pituitary as well as the hypothalamus. Although estradiol may not trigger the LH surge directly, it does modify the cellular activity of both the pituitary and the hypothalamus so that the stage is set for the midcycle release of LH and FSH. At the level of pituitary, estradiol augments the responsiveness of the gonadotrophs to LHRH.

Following ovulation, the corpus luteum produces progesterone throughout the duration of the luteal phase. Injection of progesterone into postmenopausal women does not decrease gonadotropin levels, suggesting that progesterone has no direct negative feedback action. Progesterone does reduce the frequency of LH secretory pulses. In addition it blocks the positive feedback effect of estradiol (Fig. 8.63). This is important since plasma estradiol levels are relatively high during the luteal phase and, in the absence of progesterone, would cause stimulation of LH release (positive feedback). Since the positive feedback mechanism of estradiol is not operative during the luteal phase, LH and FSH secretion are controlled only by the negative feedback of estradiol and are lowest during the midluteal phase, when corpus luteum secretion of estradiol and progesterone are maximal. Near the end of the luteal phase, when luteolysis occurs and the inhibitory feedback effect of estradiol action is removed, LH and FSH levels rise, and the growth of a new set of follicles is initiated.

TRANSPORT AND METABOLISM OF ESTRADIOL AND PROGESTERONE

The transport of estradiol and progesterone in plasma differs somewhat from that of testosterone and cortisol in that the former steroids are transported primarily by albumin. The affinity of testosterone-estradiol-binding globulin (TeBG) for estradiol is much less than for testosterone. Since there is much more testosterone in plasma, it will occupy most of the binding sites on TeBG. Similarly, progesterone and cortisol bind with approximately equal affinities to cortisol-binding globulin (CBG). Since cortisol always circulates in at least a 10-fold greater concentration in plasma than progesterone,

relatively more of the binding sites on CBG will be occupied by cortisol than by progesterone. Therefore, even though progesterone can bind to a specific plasma transport protein, the majority of progesterone is transported by albumin since the specific binding sites are occupied by another steroid.

Estradiol and its metabolic products are conjugated with sulfate and/or glucuronic acid for excretion. The keto groups of progesterone are reduced to form pregnanediol, which is then excreted as a conjugate. Unlike testosterone, estradiol and progesterone are not metabolized to more active steroid products, as both of these hormones are biologically active on target tissues in the form in which they are secreted into the blood.

Origin of Testosterone in Women

The origin of testosterone in women, illustrated in Figure 8.64, is analogous to the production of estrogens in men. Testosterone is secreted into the blood stream from both the ovary and the adrenal cortex. Both of these endocrine glands, in addition, produce relatively nonandrogenic steroid precursors or prehormones such as androstenedione and dehydroepiandrosterone. Steroid prehormones are then converted to testosterone in liver and peripheral tissue. In normal women androstenedione and dehydroepiandrosterone serve as precursors for 50 and 15%, respectively, of plasma testosterone. Since the origin of blood testosterone in normal women is diverse, one can anticipate similar but variable diversity of testosterone production in women with virilizing disorders (Bardin and Mahoudeau, 1970).

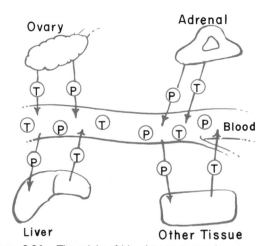

Figure 8.64. The origin of blood testosterone in women. Testosterone (*T*) is secreted into the blood by the ovary and adrenal gland. These endocrine glands also secrete precursor steroids or prehormones (*P*) into the blood which are transported to the liver and other tissues, where they are converted to testosterone which re-enters the blood.

HORMONAL CONTROL OF FERTILITY

The most effective method of contraception, the birth control pill, is based on oral administration of steroids. Estrogens and progestins are used either combined or, as with the "minipill," progestins are used alone. In addition, various combinations of steroids can also be administered as long-acting injectable preparations or via intrauterine systems.

Combined Pill

Over 20 different brands of combination estrogen-progestin oral contraceptives are available in the USA. For the pills to be effective via the oral route, estradiol and progesterone cannot be used since they are metabolized in the gastrointestinal tract and liver. As a consequence, synthetic estrogens such as mestranol (50–100 μg/day) or ethinylestradiol (20–50 μg/day) are used in combination with various synthetic progestins, such as norethindrone, norethindrone acetate, norgestrel, ethinodiol diacetate or norethynodrel (0.3–100 mg/day). The hormones are given in a cyclic fashion for 21 days, beginning on day 5 of the menstrual cycle, followed by 7 days of placebo treatment or no pills.

The elevated estrogen and progestin levels inhibit the midcycle LH surge and ovulation by exerting negative feedback effects on the hypothalamus. Irregular LH peaks are sometimes observed, while FSH levels are usually suppressed. Ovarian progesterone production is diminished, but estrogens continue to be secreted. The effects on the endometrium are variable and depend on the type and dosage of the contraceptive. Rapid progression from proliferation to early secretory changes can be observed within a few days from the start of daily intake, followed by regressive changes. Secretory activity is either minimal or absent. The pregnancy rate for combined pills is approximately 2%.

Progestin-Only Pill

The progestin-only method, or "minipill," uses either 350 μg of norethindrone or 75 μg of norgestrel per day taken continuously without interruptions. Failure rates are higher with this method as compared to that of the combined pills, especially if pills are missed. The minipill is preferable for women who are breast feeding since it has fewer adverse effects on lactation than combined estrogen-progestin pills. Progesterone-only pills inhibit ovulation in only 60% of cycles. An alternative mechanism of antifertility action is their effect on the endometrium where the induced changes prevent egg-sperm interaction.

CHAPTER 59

Fertilization, Pregnancy, and Lactation

The survival of a species ultimately depends upon fertilization, maintenance of pregnancy, and a feeding strategy which will permit survival of the newborn. This chapter will discuss the mechanics of ovum penetration by the sperm which in turn leads to cell division and the development of the blastocyst. This will be followed by a presentation of factors involved in maintenance of pregnancy, the onset of labor, and the control of lactation.

FERTILIZATION

When spermatozoa are released from the germinal epithelium into the lumen of the seminiferous tubule they are fully formed but do not have the capacity for progressive motility which is required for fertilization. The sperm cells are washed out of the tubule into the rete testis and the epididymis by the fluid drive generated by Sertoli cells. Motility is acquired by an androgen-dependent step during passage of the spermatozoa through the head and body of the epididymis. Although the exact role played by the epididymal cells in this maturational process is not understood, recent studies suggest that there are a variety of changes which take place on the surface of spermatozoa during epididymal transit. It is not clear how these changes are related to the acquisition of motility. Spermatozoa are stored in the tail of the epididymis until ejaculation, when they are deposited in the vagina.

The spermatozoa present in the ejaculate are not capable of fertilization despite their rapid motility. A further change occurs in the female reproductive tract where the ability to fertilize is acquired through a process called capacitation. Capacitation appears to be a universal phenomenon among mammals and can only be defined as the functional changes that spermatozoa undergo in the female tract which allows them to fertilize an egg. A large number of studies have attempted to demonstrate the morphological, chemical, and motility corre-

lates of capacitation. Recent evidence suggests that a critical aspect of capacitation may involve membrane changes which facilitate the influx of calcium. Other correlates of capacitation are activated motility and the acrosome reaction.

Activated motility was first described in capacitated hamster spermatozoa. When activation occurs, the smooth undulation of the tail of the spermatozoa changes to a whip-like motion. While activated motility is correlated with the ability to penetrate ova, it is not clear whether this type of motion is a general phenomenon and can be observed in all species. While this is present in many mammals, in other species such as the human, activated motility is not obvious. There are several reasons which could explain these differences. It is possible that activation occurs in all species, but the differences between activated and nonactivated motility are subtle and cannot be observed readily without quantitative measurements such as can be made by a high speed video camera. Even though activated motility appears to be associated with capacitated spermatozoa, its role in fertilization or even capacitation is not fully understood.

Another correlate of capacitation is the acrosome reaction. This was initially defined as the fusion between the plasmalemma of the spermatozoon and the outer acrosomal membrane. Once this fusion occurs, fenestrations in the fused membranes take place which allow acrosomal enzymes to be released into the surrounding media. Subsequently, the entire acrosome is eventually lost, and the inner acrosomal membrane with some associated enzymes, is exposed to the environment over a large portion of the sperm head. The plasmalemma, however, remains intact over the posterior aspect of the sperm head. In this regard, the acrosome reaction is believed to be analogous to the release of the contents of a large secretory granule.

Classically, the acrosome reaction and loss is

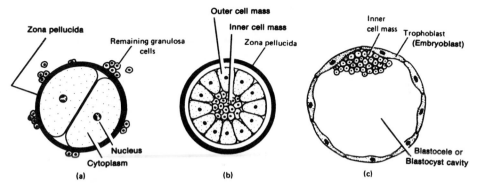

Figure 8.65. Preimplantation stages of the conceptus. *a*, Two cell stage with some residual granulosa (cumulus) cells over the outer surface of the zona pellucida. *b*, Morula. *c*, Blastocyst showing inner cell mass. [Adapted from Begley et al. (1980).]

believed to occur in the oviduct near the cumulus oophorus, which is the covering of granulosa cells that remain on the ovulated egg (Fig. 8.65). One argument for this concept comes from observations of spermatozoa which are flushed from the oviduct. Those in oviductal fluid have intact acrosomes, while those in the cumulus oophorus have altered or missing acrosomes. Another line of evidence which is consistent with the idea that the acrosome reaction and loss occurs in the cumulus comes from the observation that cells of the cumulus mass are held together by hyaluronidase-sensitive material. Hyaluronidase is a major acrosomal enzyme, and antibodies directed against hyaluronidase block the cumulus dispersal prior to fertilization. The common interpretation of these observations is that the acrosome is lost from the fertilizing spermatozoa during its passage through the cumulus before arrival at the zona pellucida. Recent studies, however, have shown that sperm binding to zona pellucida requires an intact acrosome. It is therefore possible that some spermatozoa undergo the acrosome reaction while passing through the cumulus so as to release acrosomal enzymes which are responsible for release of cumulus cells from the outside of the ovulated egg. Other spermatozoa pass through the dispersed cell mass and bind to the zona pellucida. The enzymes released from these spermatozoa through the fenestrations in the fused membranes provide a means of zona penetration. By the time the fertilizing spermatozoa reach the surface of the egg the outer acrosomal membrane has been lost. The sperm cell then makes contact with the microvilli of the egg over the posterior part of the sperm head. Soon after fusion of the egg and the sperm plasma membranes microvilli of the egg come up over the broad side of the sperm head. The head then flexes as it is taken into the cytoplasm of the egg. Over the next several hours the entire spermatozoon including the tail, is taken into the egg,

and the process of fertilization is complete. Division begins to form the early embryo (Fig. 8.65).

PREGNANCY

Implantation

In the human female fertilization occurs in the fallopian tube. Early cleavage of the fertilized egg takes place in the oviduct. As the conceptus or zygote makes its way into the uterine cavity, it develops from a 2-cell stage into a morula consisting of approximately 60 cells (Fig. 8.65). In the uterus, the morula is transformed into a blastocyst, which is a large fluid-filled cavity (primary yolk sac) with a surface layer of outer cells which develop into the trophoblast, and an inner cell mass which is the early stage of the embryo. During its early development the conceptus is still surrounded by the zona pellucida. At approximately day 21 of the menstrual cycle (7 days after ovulation), the zona pellucida is dissolved by enzymatic digestion and the blastocyst is ready for implantation.

Pregnancy is definitely established with implantation which is the process by which the blastocyst becomes embedded in the inner lining of the uterus (endometrium) and establishes connections with the maternal circulation. For convenience, pregnancy is sometimes dated from ovulation or from the date of the last menstrual period. Implantation and growth of the zygote requires an appropriate hormonal milieu (see below), which is achieved by the precise synchronization of ovarian, embryonic, and endometrial functions during the early days of pregnancy. The three phases of implantation which lead to the establishment of the embryo in the amniotic cavity, where it is supplied with nutritional, metabolic, and circulatory necessities through the placenta from the mother, are shown diagrammatically in Figure 8.66.

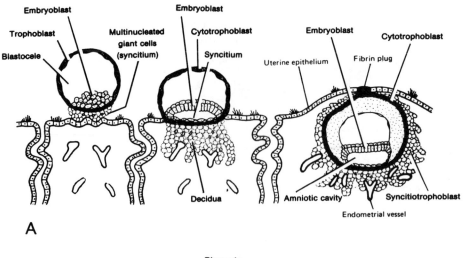

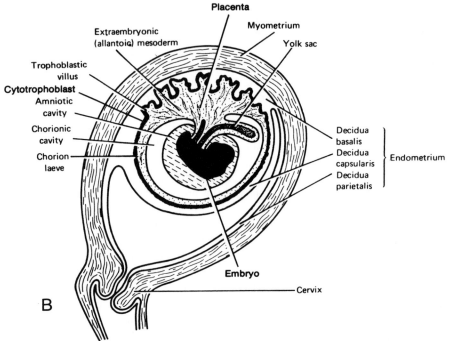

Figure 8.66. Three phases of implantation (*A*) and implanted embryo (*B*). Note the expansion of the syncytiotrophoblast from the embryonic pole to cover the entire surface, eroding maternal vessels as it grows. [Adapted from Begley et al. (1980).]

Development of the Fetoplacental Unit

By day 12 after ovulation the human blastocyst is completely embedded in the endometrium (Fig. 8.66). The cells of the endometrium adjacent to the zygote enlarge and form a compact cell mass called the decidua. The wall of the blastocyst facing the decidua thickens and forms the cytotrophoblast and the syncitium which will later develop into the placenta. As the syncitiotrophoblast expands by pushing between the epithelial cells of the endometrium, the entire blastocyst sinks into the endometrial epithelium, eroding maternal vessels as it grows. Simultaneously, the surface area of the conceptus is greatly increased, the spaces within the syncitiotrophoblast, the lacunae, fill with maternal blood and thus provide nutrients and oxygen for the embryo. By day 17 after ovulation placental blood vessels and interaction between maternal and fetal circulation are established. As the placenta matures, it serves endocrine, immunological, and nutritive functions. Its definitive structure is attained by 12 weeks of pregnancy (Villee, 1969). At this point the fetus and the placenta function as a unit (fetoplacental unit) to carry out certain functions that each alone are incapable of completing. The interdependence of the fetus and the placenta can best be illustrated by fetoplacental steroidogenesis. In this instance both the fetus and placenta are required to synthesize the steroid hormones of pregnancy. (For details see Estrogens under Hormonal Maintenance of Pregnancy.)

Hormonal Maintenance of Pregnancy

From the moment of fertilization, all mechanisms involved in the establishment, maintenance, and termination of pregnancy are controlled by an interplay of hormones produced in the mother and fetus. Three or 4 days after fertilization, the dividing zygote is capable of synthesizing a variety of hormones (Klopper and Diczfalusy, 1969; Davies and Ryan, 1972; Fuchs and Klopper, 1977; Sauer, 1979). In women and other primates, the early conceptus exerts a direct luteotropic effect on the ovaries by secreting human chorionic gonadotropin (hCG) which stimulates the corpus luteum to increased progesterone production and prolongs its life-span until (at approximately 5 weeks of gestation) placental progesterone and estrogen synthesis is sufficiently developed to maintain pregnancy even in the absence of the ovaries. Thus, the fetus and placenta are also responsible for the production of a number of steroid hormones which exert a variety of metabolic effects both on the fetus and the mother. The most important hormones are discussed below.

HUMAN CHORIONIC GONADOTROPIN

hCG was first discovered in 1927 in the urine of pregnant women, and its detection has been used as a clinical test for pregnancy ever since. This hormone was measured by bioassay in toads, mice, or rabbits until simpler immunological assays most commonly using complement fixation became available. Using a highly sensitive radioimmunoassay, hCG can be detected in blood or urine of pregnant women as early as 7 days after fertilization (Lau, 1975). This glycoprotein hormone is produced by the trophoblasts of the placenta. Plasma levels rise throughout the early weeks of pregnancy and peak at 9 weeks of gestation when the number of cytotrophoblasts is maximal. Subsequently, hCG levels decline, but remain measurable throughout the remainder of gestation (Fig. 8.67).

The carbohydrate portion of hCG makes up approximately 30% of the molecule. Each molecule contains 20 sialic acid residues, linked to six carbohydrate chains. Biological activity in vivo is reduced when sialic acid is removed by neuraminidase due to increased clearance of the desialated hormone from the blood. The fact that desialated hCG does retain activity in vitro emphasizes that sialic acid influences recognition of hCG by receptors responsible for homone metabolism but not by those that determine biological activity. Like other glycopro-

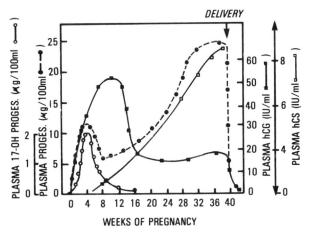

Figure 8.67. Maternal plasma levels of 17-hydroxyprogesterone, progesterone, human chorionic gonadotropin (*hCG*) and human chorionic somatomammotropin (*hCS*) during pregnancy.

tein hormones (LH, FSH, and TSH) hCG consists of two noncovalently bonded subunits, α and β, with molecular weights of approximately 18,000 and 28,000, respectively. The α-subunit of hCG is not distinguishable from that of LH, FSH, and TSH, while the β-subunit (hCGβ) is unique to hCG, even though many regions of its amino acid sequences are homologous with LHβ. hCGβ is particularly distinctive in having 32 additional amino acids on its carboxyterminus as compared to LHβ. Neither of the subunits have biological activity either in vivo or in vitro.

The measurement of hCG is used not only for the diagnosis of pregnancy, but also for management of women who are believed to be at risk of spontaneous abortion. Patients with an unexpected decline or with consistently low urinary hCG levels are more likely to abort. However, missed abortion cannot always be reliably diagnosed by hCG assay, since trophoblastic tissue may survive for long periods of time following fetal death and maintain hCG levels similar to those of pregnancy. hCG determinations are therefore, useful in detection of retained placental tissue, and in the diagnosis of malignant and nonmalignant trophoblastic neoplasms and other tumors which synthesize hCG. Once it has been established that a cancer secretes hCG, then this hormone may be used to determine the success of therapy.

PROGESTERONE

The steroid hormone, progesterone, is a secretory product of the corpus luteum (see Chapter 58, Fig. 8.62) and the placenta. Together with estrogen it induces the secretory phase of the uterine glands in preparation for ovum implantation. From the time of fertilization until just after implantation, the progesterone required for endometrial maintenance is derived from the corpus luteum which is under the control of pituitary LH. This period corresponds to the second half of the reproductive cycle. Pituitary control of the corpus luteum and endometrium lasts 10–13 days. Unless placental hCG is produced in sufficient quantities to maintain the corpus luteum beyond this period, the pregnancy will terminate several days after implantation as a consequence of luteolysis, followed by a fall in progesterone concentration and menstruation. In normal pregnancy, however, corpus luteum maintenance is transferred from pituitary to placental control, and progesterone secretion continues. Both progesterone and 17-hydroxyprogesterone productions increase until peak concentrations are attained at 3–4 wk of gestation (Fig. 8.67). Subsequently plasma concentrations of both steroids decline at the same rate, the plasma progesterone level reaching a nadir at 6–8 wk after which it increases again as a result of placental secretion. By contrast, 17-hydroxyprogesterone levels continue to decline. Since the placenta does not synthesize significant quantities of 17-hydroxyprogesterone, this steroid is a good index of corpus luteum function during pregnancy.

As mentioned above, pregnancy maintenance is initially dependent on ovarian hormones in all mammals. But in many animals with long periods of gestation, such as human, the placenta becomes capable of producing sufficient progesterone to maintain gestation even in the absence of ovaries. Studies in rhesus monkeys have also shown that the corpus luteum of early pregnancy regresses and is inactive during midpregnancy. However, there is hormonal and morphological evidence suggesting a reactivation of luteal function and luteal progesterone synthesis during the third trimester (Koering et al., 1973). This mechanism is most likely also present in other primates, including the human female.

Progesterone is necessary for the maintenance of a quiescent, noncontractile uterus. The mechanism by which it prevents premature uterine contractions and abortion is not clear. Both membrane and cytoplasmic receptor-mediated events have been proposed. The observation that a competitive inhibitor of progesterone binding to receptor will terminate pregnancy suggests that the latter of these possibilities is correct. It has been postulated that progesterone inhibits the spread of contractive waves within the myometrium, thereby preventing coordinated and forceful uterine contractions. This action has been termed the progesterone block (Csapo, 1969), and it can be easily demonstrated in

lower animals such as the rabbit. In this latter species a decline in progesterone concentration is associated with the onset of uterine contraction and the end of pregnancy. There are serious theoretical objections to the progesterone block theory (Kao, 1967), since it cannot be demonstrated in humans and in certain other animals with long gestation periods. Unlike the situation in the rabbit, uterine contractions at term in the human are not associated with falling blood progesterone levels, nor does progesterone administration, even at high doses, inhibit labor. However, proponents of the progesterone block theory point out that blood levels in the human do not accurately reflect the local level of progesterone, which is synthesized in the placenta and transmitted to the adjacent myometrium. Parenterally administered progesterone, even at high doses, may not reach the myometrium to afford levels comparable with those supplied locally by the placenta (see Parturition below).

In the human, the placenta is the major site of progesterone production during pregnancy. At term, the placenta secretes approximately 250 mg of progesterone per day as estimated from direct measurements of progesterone in umbilical and uterine veins. Progesterone concentrations in maternal peripheral plasma are elevated during pregnancy and rise as gestation progresses (Fig. 8.67), reaching values ranging from 11–32 μg/100 ml of plasma (LeMaire et al., 1970), as compared with nonpregnancy levels varying from 0.1–2 μg/100 ml of plasma (Johansson, 1969) at different times of the menstrual cycle. The increase in plasma progesterone levels during pregnancy is reflected in increasing maternal urinary excretion of pregnanediol, one of its major metabolites. From amounts averaging about 10 mg/24 h at the end of the first trimester, pregnanediol excretion rises to an average of about 45 mg/24 h at the 36th wk, after which it remains relatively constant (Diczfalusy and Troen, 1961). Wide variations occur, both among individuals and from day to day in the same individual. The myometrial progesterone concentration is highest subjacent to the placenta and falls in the more peripheral areas (Zander et al., 1969). Absolute concentrations vary greatly among individuals, and no significant changes have been found in association with the onset of labor.

In the human trophoblast, cholesterol from the maternal compartment is the main substrate for progesterone synthesis (Fig. 8.68). Placental synthesis of progesterone from acetate is quantitatively insignificant. In addition, fetal precursors are not important, as suggested by the observation that placental progesterone production persists in the

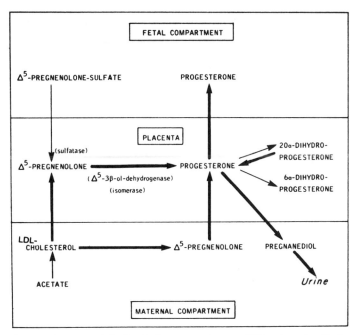

Figure 8.68. Main pathways of progesterone metabolism in human pregnancy (*LDL* = low-density lipoprotein). [Redrawn from Thau and Lanman (1975).]

human for up to several weeks following intrauterine fetal death or in the presence of an anencephalic fetus with hypoplastic adrenals or following removal of the fetus with the placenta left in situ, as occurs in cases of ectopic pregnancy.

Placental progesterone passes both to the mother and the fetus. In the mother, pregnanediol glucuronide is formed in the liver and excreted in the maternal urine. Although this is the principal maternal metabolite, its excretion accounts for less than 4% of the total progesterone produced. Of the remainder, a large, but not precisely measured, fraction passes to the fetus, where it is metabolized in the fetal adrenal, liver, and other tissues. In the adrenal, hydroxylation in the 11, 17, and 21 positions and sulphation lead to the production of a variety of corticosteroid sulfates; 6α and 16-OH-progesterone sulfates are also formed. In the fetal liver, the principal metabolic product is 20α-dihydroprogesterone which occurs also predominantly as sulfate. The significance of these metabolites, if any, in fetal physiology is not known. In the mother progesterone is also metabolized by many tissues.

Measurements of maternal blood progesterone levels have been suggested as a means of assessing the health of the fetoplacental unit. These measurements primarily reflect placental progesterone production. This is evidenced by the fact that maternal ovarian and placental progesterone production can continue and maintain blood levels near normal even after fetal death. Thus estimates of progesterone secretion provide very little information about fetal health.

Preterm birth has, in some studies, been statistically associated with lower maternal progesterone levels than in term births. In individual cases, however, maternal progesterone levels have no prognostic value for determining whether a pregnancy will terminate prematurely.

ESTROGENS

As noted in Chapter 58, in the nonpregnant, sexually mature female estrogens are secreted almost exclusively by the ovarian follicles. They have numerous physiological effects; most importantly, they act to stimulate growth of fallopian tubes, uterus, cervix, vagina, and breasts. Under the combined influence of estrogens and progesterone, the endometrium proliferates and the endometrial glands grow, providing a suitable environment for the maintenance of the fertilized egg before nidation and an appropriate surface for blastocyst implantation. Estrogen-induced hyperplasia and hypertrophy of the uterus provide, at least initially, for the accommodation of the growing fetus.

Estrogens also produce a number of metabolic effects on the liver which may increase markedly during pregnancy. These include increased levels of blood cholesterol and other lipids, increased activity of the renin-angiotensin system, and increased coagulability of blood. These effects are usually minor but may account for some of the morbidity of pregnancy.

In most, if not all, mammals the placenta serves as a source for the increase in estrogen synthesis and secretion during pregnancy. In humans, the placental contribution is particularly large, and is reflected in maternal urinary estrogen levels which rise progressively during pregnancy from below 0.1 mg/day in nonpregnant women to a daily output of up to 33 mg/day near term. At this time, maternal peripheral plasma levels are 150 ng/ml, as compared to 0.06 ng/ml during the midluteal phase of the menstrual cycle.

As noted in Chapter 58, the mature mammalian ovary, under appropriate gonadotropic stimulus, is capable of synthesizing estrogens de novo from acetate and from HDL cholesterol from the circulation. Thus, in the ovary all steps of the biosynthetic process after cholesterol synthesis occur in the follicle. In the human, ovarian steroid production declines by the ninth week of pregnancy, but total estrogen production continues to increase, and the placenta becomes the main source for this hormone. Placental estrogen biosynthesis is distinguished from that in the ovary in that it depends on precursors supplied by the fetus. Both the fetus and placenta have incomplete steroidogenic systems so that each complements the other in the process of estrogen production. The major precursor for estrogen biosynthesis in the placenta is

dehydroepiandrosterone sulfate (DHEAS) which is produced primarily by the fetal adrenal (Fig. 8.69). Before reaching the placenta, most of the DHEAS is hydroxylated in the 16α position by the fetal liver. It then passes to the placenta, where it is hydrolyzed by the enzyme sulphatase, converted to a Δ^4-3-ketone, and aromatized to estriol. A portion of fetal DHEAS bypasses the liver, is not 16α hydroxylated, and is converted to androstenedione and testosterone which are, in turn, aromatized to estrone and estradiol (see also Chapter 58, Fig. 8.62). DHEAS is not converted to estrogens in the fetus due to limited aromatase activity. This precursor is not produced in the placenta because of its inability to convert progesterone to 17α-OH progesterone. As mentioned above, interdependence of fetus and placenta in completing the pathway for estrogen biosynthesis has led to the term "fetoplacental unit." Although the placenta also receives DHEAS which originates in the maternal adrenal, this accounts for only 10% of the estrogen excreted in late pregnancy.

Estrogens formed in the placenta are secreted into the maternal bloodstream in the free form. Before urinary excretion, they are conjugated to glucuronides and, to a lesser extent, to sulphates by the maternal liver. Conjugation diminishes the biological activity of the estrogens but increases their solubility in aqueous medium, permitting their excretion in urine. Most animals excrete estrogens as

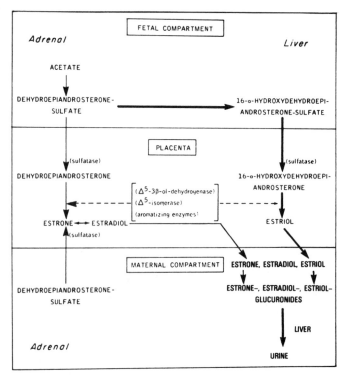

Figure 8.69.   Main pathways of estrogen metabolism in human pregnancy. [Adapted from Thau and Lanman (1972).]

conjugates of estrone or estradiol, but in the human, the major estrogen of pregnancy is estriol, accounting for 80–95% of the total synthesis in late pregnancy. No other known animal synthesizes this much estriol, and some primates, such as the macaque, make little or none of this estrogen. Since estriol has relatively modest estrogenic activity and its conjugates still less, the excretion of large amounts of estrogens during human pregnancy has less biological significance than the total numerical value would imply. Like the other estrogens, estriol is conjugated in the maternal liver before excretion.

The extraordinary rise in total estrogen production during human pregnancy reflects increasing production of DHEAS by the fetal adrenal. In the human (though not in most animals) the fetal adrenal is hypertrophied owing to the striking development of an internal zone (fetal zone) which regresses after birth. Its enlargement appears to result from stimulation by ACTH from the fetal pituitary. Increased activity of the fetal pituitary is probably occasioned by feedback mechanisms activated by loss of fetal adrenal glucocorticoids. Loss occurs by transfer across the placenta to the mother and by metabolism in the fetal liver; transcortin levels rise progressively in the maternal circulation during pregnancy, creating a transplacental gradient across which fetal corticosteroids are lost. Further maternal-fetal disparity is created by the competition of progesterone for cortisol binding sites with relatively higher levels of progesterone in the fetal circulation. As in other situations with increased ACTH stimulation of the adrenal, the production of adrenal C-19 compounds, in this case predominantly DHEAS, is also increased, providing the precursor for the large amounts of estriol synthesized by the placenta.

Because estriol, estradiol, and estrone have their origin predominantly in the fetoplacental unit, it is easy to understand why maternal estrogen assays provide an index of the health of the fetus and placenta. Estriol assays are used to follow mothers whose fetuses are believed to be at increased risk for fetal growth retardation, toxemia, and prolonged pregnancies. Unfortunately, wide variations of estriol values occur normally among pregnant women, so that serial determinations are necessary to evaluate a given situation. The decline or the failure of estriol levels to rise signals an abnormality of the fetoplacental unit. For example, fetal death is accompanied by a sharp drop in urinary estriol. In addition, women with anencephalic fetuses have maternal estrogen levels which are one-tenth of those of normal women. These fetuses have hypoplastic adrenals secondary to defective pitui-taries. The low maternal estrogen values reflect the relative lack of estrogen precursors from the unstimulated fetal adrenals (Fig. 8.69).

HUMAN CHORIONIC SOMATOMAMMOTROPIN

Prolactin-like activity was found in extracts from human placenta in the early 1960s and was called human placental lactogen (hPL), because it was lactogenic in the pigeon crop sac assay and promoted milk production in pseudopregnant rabbits (Josimovich and MacLaren, 1962). Growth hormone-like activity was found in the same kind of preparation (Grumbach et al., 1968). A polypeptide containing both activities was subsequently characterized, and the name human chorionic somatomammotropin (hCS) was adopted. The hormone has a single polypeptide chain which shows multiple homologies with human growth hormone (Li et al., 1971). hCS, like prolactin, is believed to be part of a growth hormone multigene family.

A substance similar to hCS was extracted from the placentas of two monkeys, macaques, and baboons. Macaque CS has partial immunological cross-reactivity with hCS and, to a greater extent, with human growth hormone (hGH). hCS, however, is relatively low in growth-promoting activity, having less than $\frac{1}{100}$ of the biological activity of hGH.

hCS has two opposing effects on carbohydrate metabolism (Samann et al., 1968); it is diabetogenic, probably producing insulin antagonism as does hGH. It also stimulates insulin secretion, simulating an effect observed in hGH-deficient children given hGH. Plasma-free fatty acids are also increased by hCS.

Many of these effects observed upon administration of hCS are changes which occur during human pregnancy. It is difficult to attribute any one of them specifically to hCS independently of concurrently acting hormones, such as hGH, estrogens, and corticosteroids. The observed effects which are consistent with the action of hCS, have been interpreted as favoring the growth of the fetus largely through decreasing maternal utilization of carbohydrate and making increased nutritional supplies available in the fetus. This would serve to provide an energy source for fetal protein anabolism.

In spite of the numerous effects of hCS on a variety of systems, it is difficult to assess its role in pregnancy. As with the chorionic gonadotropins, the restriction of chorionic somatomammotropins to primates indicates that other mammalian species do not have a general requirement for this protein hormone, in contrast to the generally essential nature of estrogen and progesterone. This conclusion is supported by studies of an individual with a complete deficiency of hCS prior to birth. Analysis of the DNA from this person indicated deletion of the hCS gene on both chromosomes. The association between gene deletion and a normal growth pattern in this individual indicates that hCS is not required for fetal or extrauterine growth (Werzel et al., 1982).

hCS is synthesized in the syncitial trophoblast, and, like hCG, is secreted predominantly into the maternal circulation. It becomes detectable in maternal plasma after the first month of pregnancy and subsequently rises to values of about 8 μg/ml at term (Fig. 8.67), when its production rate has been estimated at about 1 g/day (Kaplan et al., 1968). These values for both concentration and production rate are far higher than those for any other human polypeptide hormone. Unlike hCG, very little hCS appears in maternal urine. The half-life of hCS in blood is 29 min.

The concentration of hCS in maternal blood affords a way of detecting placental dysfunction (Genazzani et al., 1972). Pregnancies ending in abortion are associated with low or decreasing hCS levels. The concentrations of this hormone are also lower than normal in preeclampsia, fetoplacental dystrophy, multiple placental infarcts, prolonged pregnancy with chronic fetal distress, and preceding intrauterine fetal death in diabetic pregnancies. High levels are found in rhesus-isoimmunization with hydrops fetalis, in twin pregnancies, and in nondiabetic pregnancies with excessively large babies. There is a significant correlation between fetal weight and plasma hCS values up to a few days before parturition.

HUMAN CHORIONIC THYROTROPIN

The normal placenta secretes a thyroid-stimulating glycoprotein, termed human chorionic thyrotropin (hCT), that cross-reacts immunologically with bovine but not with human, pituitary thyroid-stimulating hormone (TSH). Its molecular weight is approximately 30,000, and its chromatographic and electrophoretic properties are similar to those of TSH.

The physiological role of hCT in pregnancy is uncertain. hCT could play a role in the incidence of thyroid enlargement during pregnancy. However, the fact that thyroid-stimulating activity in the serum of pregnant women is highest during early pregnancy and decreases in subsequent months, while hCT levels are higher at term (Hennen et al., 1969) argues against this possibility. Patients with hydatidiform moles and other trophoblastic tumors have been shown to have increased thyroid function. While this could be due to hCT, convincing

evidence indicates that most of the thyrotropic activity in the serum of patients with this disease is due to the inherent TSH activity of hCG.

PRO-OPIOMELANOCORTIN (POMC)

POMC is the precursor peptide for ACTH and β-lipotropin in the anterior pituitary. In the intermediate lobe these latter peptides are further processed to α-MSH plus CLIP and γ-lipotropin plus β-endorphin. The placenta also has the ability to synthesize POMC. In this organ POMC is processed more like in the intermediate lobe than in the anterior pituitary. The role of the peptides derived from this precursor in the placenta remains to be established.

HYPOTHALAMIC PEPTIDES

Several peptides which until recently were thought of as pituitary-releasing hormones have also been found in the placenta. LHRH-like material and higher molecular forms of this peptide have been demonstrated in extracts from placental tissues and in the medium of placental tissue maintained in culture (Siler-Khodr and Khodr, 1978; Gautron et al., 1981). Thyrotropin-releasing hormone has also been identified in placental tissues. The function of these peptides in the placenta or the uterus remains to be established. The ability of the placenta to synthesize these as well as other hormones demonstrates the enormous versatility of this organ.

Nonendocrine Functions of the Placenta

In addition to being a site for hormone production, the placenta serves several other functions that are essential for the well-being and normal development of the fetus: transfer (gas exchange, fetal nutrition, and excretion) and immunological functions.

TRANSFER AND BARRIER FUNCTIONS

The placenta controls the interchange between fetal and maternal compartments by providing selected permeability for different materials and thus facilitating or restricting transfer of various compounds to and from the fetus. The transfer of blood cells is severely restricted, whereas the transfer of many essential nutrients is accelerated by a variety of transport mechanisms. The placenta serves as a fetal lung, and gases are transferred between fetal and maternal compartments by simple diffusion. The amount of oxygen delivered to the fetus depends on the concentration in the maternal blood and placental blood flow. Oxygen uptake by fetal blood is favored by a higher hemoglobin concentration (12% in the fetus vs. 10% in the mother). The transfer of carbon dioxide from the fetal to maternal circulation is rapid because of its high diffusion constant (20-fold that of oxygen). Carbon dioxide is carried in blood as bicarbonate (62%), bound to fetal hemoglobin (30%), and in solution (8%). The release of carbon dioxide from the fetal hemoglobin in the placenta is facilitated by the simultaneous uptake of oxygen.

The exchange rate of a number of electrolytes has been studied. The efficiency of sodium exchange across the placenta increases as pregnancy progresses, reaching a maximum at approximately 35 wk of gestation, followed by a decrease toward term. Many actions are controlled by active transport mechanisms, that is, transport against the gradient.

Glucose, a major energy source for the fetus, is transferred to the fetus by diffusion, facilitated by a transport mechanism. Facilitated diffusion involves a carrier molecule that shuttles back and forth to speed the rate of transport across the membrane without requiring energy supply. Thus it differs from active transport which requires the expenditure of energy and is capable of transport against a concentration gradient.

Amino acids are delivered by active transport mechanisms to the fetus, where they are essential for protein synthesis. As with glucose, placental transfer of most amino acids is also stereospecific.

Some maternal plasma proteins are transferred to the fetus by pinocytosis, a transport process which becomes more efficient toward term. Maternal immunoglobulins are transferred by receptor-mediated endocytosis. This selective transfer mechanism leads to a more rapid transfer rate for immunoglobulins than for other plasma proteins.

The transfer to the fetus of maternal lipids is generally slow. Phospholipids and cholesterol which are transported in plasma as lipoproteins (such as LDL and HDL) are released from the plasma proteins. Phospholipids are hydrolyzed by the placenta. The lipids are resynthesized into phospholipids by the fetal liver.

Unconjugated steroids are transferred rapidly in both directions, while the transfer of sulphates and glucosiduronates is restricted. Sulfates are hydrolyzed in the placenta to the unconjugated form, but there are no hydrolases for the glucosiduronates in the placenta.

The health of the placenta is very important for normal fetal growth and development. Defects in placental transfer of one or a number of the above products can be caused by changes in diffusion and transport mechanisms through thickening of the basement membrane, deposition of fibrinoid or edema between the placenta and uterus, separation of the placenta, or placental infarcts. Any one of these can retard fetal growth and maturation.

IMMUNOLOGICAL FUNCTIONS

Two basic immunological functions are attributed to the normal human placenta: it protects the fetus which is histoincompatible with the mother against maternal immunological attack and it provides the fetus transiently with immunological defenses by the transfer of selected maternal immunoglobulins which protect the newborn infant until its own immunological defense mechanisms have been developed. The newborn's antibody complement is also supplemented by the transfer of maternal immunoglobulins during breast feeding (see lactation).

The implanted placenta has been likened to a graft, but despite the intimacy of fetal and maternal tissues in the placenta they do not share a common circulation. This separation of vasculature is maintained by a continuous layer of tissue composed of fetal trophoblasts and a sheet of fibrinoid at the boundary between fetal and maternal tissue. This physical separation of mother and fetus is probably essential for the survival of the fetus, which is histoincompatible with the mother because of the paternal component of its genetic constitution. How the fetal trophoblast which is in direct contact with the maternal circulation avoids immunological rejection is not clear.

Parturition

The timing of birth is the culmination of a gestational period that is well characterized for every type of eutherian mammal. As with other mammals, pregnancy length in the human is remarkably constant (40 ± 2 wk from the last menstrual period). The factors controlling timing of intrauterine development and its termination by parturition at the appropriate time for each species have been the object of extensive investigations and speculations. Malpas (1933) was the first to suggest that the fetus plays an active role in the timing of its own birth. It is now known that the onset of parturition is regulated by the interaction of fetal and maternal endocrine, neural, and mechanical factors. The striking variations observed between species have precluded a universal explanation. The sheep is the best studied animal model but differs in many respects from the situation in the human (Challis, 1980; Liggins et al., 1977).

The discussion which follows focuses on the major factors involved in the induction of labor, including the role of the fetus and of the hormones progesterone, estrogen, oxytocin, and prostaglandins. When possible special emphasis is given to human parturition.

THE ROLE OF THE FETUS

As mentioned above, pregnancy length in normal women is remarkably constant. The mean gestation time for anencephalic fetuses (provided patients with polyhydramnios are excluded) is similar (39.7 wk) but shows much wider variations about the mean, suggesting that the endocrine status of the human fetus plays a role in the "fine tuning" of the initiation of parturition.

In the sheep, contrary to the human, the role of the fetal pituitary adrenal axis in the initiation of labor has been clearly established. In this animal, cortisol and ACTH concentrations in fetal plasma rise during the final 2–3 days of gestation (Rees et al., 1975). In this species infusion of cortisol or ACTH into the intact fetus leads to premature parturition (Kendall et al., 1977). It is therefore believed that the large physiological rise in ACTH and cortisol which occurs in the fetal sheep during the final 2–3 days of gestation triggers labor.

PROGESTERONE

Progesterone is essential throughout pregnancy to maintain a quiescent, noncontractile uterus. In mammals in which gestation is longer than the lifespan of the corpus luteum during the luteal phase of the menstrual cycle, various means are utilized to assure the supply of progesterone during pregnancy: the functional life-span of the corpus luteum of pregnancy may be prolonged (usually by the action of a placental gonadotropin), and/or the progesterone-producing capacity may be transferred from the ovary to the placenta. The human female belongs to the latter group, and placental progesterone production is sufficient to maintain pregnancy even in the absence of the ovaries after the 6th to 9th wk of pregnancy.

Contrary to ovarian dominated animals, like the sheep, the rabbit, and the rat, no conclusive evidence for a fall in plasma progesterone levels prior to parturition or the onset of spontaneous abortion has been found in women. It has been suggested that effective progesterone concentrations within the pregnant uterus may be controlled at the local level, especially in the fetal membranes, where specific progesterone receptors regulate the effective progesterone concentrations (Schwarz et al., 1976). A decrease in specific progesterone receptors could produce a local progesterone withdrawal (without a change in the blood level) and trigger an increase of prostaglandin production. The latter is another factor in the multicomponent mechanism involved in the induction of labor (see below).

ESTROGENS

In several animals there is a significant rise in maternal blood estrogen levels 1 or 2 days before the onset of labor. The rise is striking in sheep, but in women no increase has been found beyond the continuation of the rise which extends throughout pregnancy. Estradiol levels in the amniotic fluid rise at an increased rate during the final 20 days of human pregnancy. The concentration of estrone sulfate, the major estrogen in the amniotic fluid of rhesus monkeys and humans, increases by more than 100% during the last 15–20 days of gestation. At present, there is only suggestive evidence that estrogens play a role in human parturition. More information on the site of action of estrogen has to be obtained before a conclusion can be drawn.

OXYTOCIN

In women, the oxytocin release increases during labor. Fetal production contributes to this increase and it has been suggested that the protracted labor seen in anencephalic fetuses may be a consequence of the lack of the fetal stimulus. In spite of the frequent use of oxytocin to initiate labor, the role of this peptide in the onset of normal parturition is not well understood; one reason for this is the fact that an enzyme that degrades oxytocin (oxytocinase) is highest at term. A possible mechanism of action of oxytocin is to regulate prostaglandin synthesis and to act in concert with prostaglandins to facilitate myometrial contractions.

PROSTAGLANDINS (PG)

PGs and PG metabolites appear to be implicated in human parturition as well as in that of lower animals. Possible mechanisms of action of this class of compounds and of other hormones are shown schematically in Figure 8.70. Estrogens and progesterone, respectively, act through receptor-mediated processes to stimulate and suppress the synthesis of mRNA essential for the production of PG-synthetase in the endometrium. The local decrease in the effectiveness of progesterone associated with continued action of estrogen serves to increase PG levels in the uterine fluid. In addition, PG production is increased by oxytocin which also acts via receptor binding to the endometrium. PGs and oxytocin then act directly on the uterine musculature to stimulate contractions. Activation of the

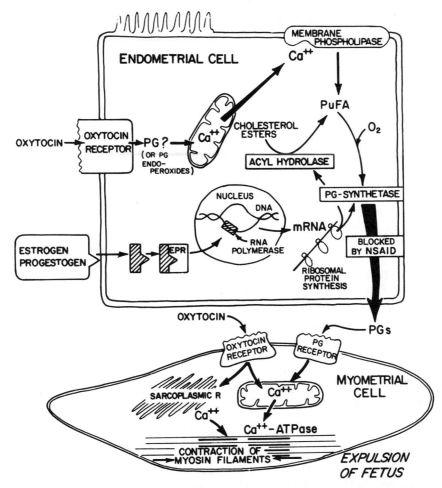

Figure 8.70. *Role of Prostaglandins in the Uterus.* The endometrium appears to be a major source of prostaglandins, and prostaglandin production is increased by both estrogen and oxytocin in the uterus. A direct action of prostaglandins on the uterine musculature is to stimulate contraction. This action of prostaglandins may be receptor-mediated and appears to involve an increase in available calcium concentration which activates the actin-myosin contractile process. [From Caldwell and Behrman (1981).]

contractile processes appears to require an increase in available calcium concentrations, which is achieved by decreasing ATP-dependent binding of calcium to subcellular membranes (Carsten, 1973) and by influx of calcium into the cell.

Urinary concentrations of PG metabolites increase gradually throughout pregnancy (Hamberg, 1974). A terminal rise of PGF levels was observed in amniotic fluid and of 13,14-dihydro-15-oxo $PGF_{2\alpha}$ concentrations in peripheral plasma. The role of prostaglandins during normal parturition is supported by the observation that indomethacin, a PG synthetase inhibitor, prolongs gestation in monkeys and results in protracted labor (Novy et al., 1974). Similar observations were made in women who had taken large amounts of aspirin during their pregnancy.

In conclusion, there are no major alterations in plasma steroid concentrations prior to the initiation of human parturition. However, changes in target organ responsiveness to these steroids may lead to altered tissue responsiveness which in turn may lead to the initial rise in PG production in the fetal membranes which is followed by activation of the actin-myosin contractile processes initiated by the combined action of PGs and oxytocin.

LACTATION

Lactation is a key factor in the successful reproduction of mammals. Until the relatively recent and massive experimentation with alternatives to human milk, breastfeeding was essential to human reproduction. There is increasing interest in the immunological, nutritional, and psychological consequences of human lactation and concern about possible long-term deleterious effects of alternatives to human milk. For instance, cross-species comparisons indicate substantial differences in composition of colostrum and of milk, both with regard to major and minor nutrients, immunoglobulins, cells, and nonspecific immune factors. It appears that the milk of each species is specifically tailored to the nutritional and immunological requirements of its young. Additionally, much cross-cultural evidence (Jelliffe and Jelliffe, 1978) and recent experimental evidence indicate that the human usually remains in an anovulatory state during the period of full lactation (carefully defined by Short, 1983), which has obvious consequences to birth spacing in many parts of the world.

Structure of the Mammary Gland

The fully developed mammary gland is a compound tubuloalveolar gland which is divided into 15 to 25 lobes, each of which is subdivided into lobules of various orders. The duct system of each lobe ultimately empties, via a lactiferous duct, at the tip of the nipple. There is an enlargement of the lactiferous duct, called the sinus lactiferous, immediately underlying the areola (the pigmented area surrounding the nipple). Important sensory nerve endings are found in the skin of the nipple which form the afferent limb of the neuroendocrine reflex arcs and through which the stimulation caused by suckling is transmitted to the brain, causing the onset and maintenance of copious milk production postpartum via oxytocin and prolactin secretion.

The secretory parenchyma of the mammary gland is composed of a simple cuboidal or low columnar epithelium lining individual alveoli and alveolar ducts. Lying between the basal lamina and the alveolar epithelium are myoepithelial cells which form a basket-like network around the alveoli, and which contract in response to circulating oxytocin, ejecting milk from the alveoli and alveolar ducts, and increasing intramammary pressure. The myoepithelial cells are thus the effector cells for the efferent limb of the neuroendocrine milk-ejection reflex.

The lobes of the mammary gland are surrounded by adipose tissue and dense connective tissue. The intralobular connective tissue contains no fat. The connective tissue stroma of the gland becomes increasingly infiltrated with lymphocytes, plasma cells, and eosinophils during the second half of pregnancy. In addition to cells, the connective tissue stroma of the mammary gland contains lymphatics, blood vessels, and their accompanying sensory and sympathetic nerves. While there is adrenergic motor innervation to smooth muscle of the blood vessels and nipples, there is no motor innervation to the myoepithelial cells. Parasympathetic innervation is absent from the gland, and it is generally believed that there is no direct secretomotor innervation in the parenchyma (Hebb and Linzell, 1970), the entire control of secretory events being hormonally mediated (see below).

Mammary Gland Growth and Development

The mammary gland is a complex organ which for optimal development and ultimate function requires the interaction of many hormones, including prolactin, estrogen, progesterone, adrenal steroids, insulin, growth hormone, and thyroid hormone. The process by which the mammary gland is prepared for lactation can be divided into four stages: prior to birth, from birth to puberty, puberty, and pregnancy.

During the embryonic and fetal period some aspects of mammary gland development appear to be independent of the hormonal environment. For instance, interactions between the mesenchyme and the epithelium have a role in mammary development, as has been shown in experiments with explants from rabbits and mice, in which isolated epithelium requires the presence of mesenchyme for induction of mammary buds.

The mammary gland exhibits sexual dimorphism as early as the fetal period. Depending on the species, this dimorphism may be expressed in the rate of growth of the male vs. the female gland, relative to the rate of increase in body size, and it may be expressed by differences in nipple development, or even by destruction of the mammary rudiments in males, as is seen in some strains of mice. Prior to puberty, breast growth is isometric to that of body growth. At puberty, both the ductal system and the supporting adipose tissue grow much faster than the body in general due to increased secretion of estrogens. Women without ovaries, therefore, develop breasts of normal size and consistency only when treated with estrogens. Growth of the duct system requires estrogen, which acts synergistically with growth hormone, prolactin, and adrenal corticosteroids. The development of the lobuloalveolar system is dependent on the presence of estrogen, progesterone, and prolactin.

The multihormonal interaction involved in the growth of the mammary gland and in the initiation of lactation is depicted in Figure 8.71. In contrast to the mammary glands of other species, the breasts of nonpregnant adult women are differentiated enough to begin synthesizing milk and lactating within as little as 2 wks of treatment with estrogens, followed by stimulation of prolactin secretion (Tyson et al., 1976).

Mammary Gland Development during Pregnancy

The final stage of breast development occurs during pregnancy. This phase is initiated by extension and branching of the duct system in the first trimester, followed early in the second trimester by the formation of new lobules and alveoli which

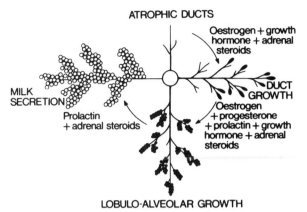

ATROPHIC DUCTS

Oestrogen + growth hormone + adrenal steroids

DUCT GROWTH

MILK SECRETION

Prolactin + adrenal steroids

Oestrogen + progesterone + prolactin + growth hormone + adrenal steroids

LOBULO-ALVEOLAR GROWTH

Figure 8.71. The multihormonal interaction on the growth of the mammary gland and in the initiation of lactogenesis and lactation, delineated in the hypophysectomized-ovariectomized-adrenalectomized rat. [From Lyons: *Proc R Soc (B)* 149:303, 1958; Yen and Jaffe (1978).]

depend on the secretion and/or presence of estrogen and progesterone. The breast shows some secretory activity by midpregnancy and continues its development through the third trimester.

The hormonal requirements for growth of mouse mammary gland and milk synthesis have been well-defined in tissue culture and indicate the need for multiple hormones. Growth of ducts and alveoli at midpregnancy can be achieved by administration of estrogen and progesterone to the rodent provided the pituitary and adrenals are intact. These organs provide growth hormone and glucocorticoids, respectively, which are required for mammary gland growth. If the mammary gland of an animal in midpregnancy is placed in tissue culture, insulin, glucocorticoids, and prolactin are required to produce mammary gland differentiation and milk synthesis (Topper, 1970). While there are species differences in the hormonal requirement for mammary growth and development, gonadal, adrenal, and pituitary hormones have been necessary in all animals studied to date.

In the human, the development of the secretory apparatus of the breast is stimulated by prolactin, placental lactogen, estrogens, and progesterone, but during pregnancy milk production is minimal and lactation does not occur. This is probably due to the presence of high plasma levels of estrogen and progesterone, which act to stimulate mammary development, but inhibit the actual formation of milk, possibly at the level of the glandular alveolar and terminal ductule epithelial cells.

Onset of Lactation and Milk Letdown

Although milk synthesis and secretion in women begin about the 5th month of pregnancy, the initiation of copious lactation is a consequence of delivery. Full lactation appears to require a fall in the level of estrogen and progesterone, which in the human occurs only after the delivery of the placenta.

It is difficult to relate specific hormonal changes to the onset of copious milk secretion because of the experimental problems involved in the simultaneous determination of changing hormone levels in frequent plasma samples and the correlation of such changes with changing rates of milk synthesis or with the uptake of milk precursors. However, it is clear that two major hormonal mechanisms are involved in the onset of lactation. First, the fact that the mammary gland is freed at parturition from the inhibitory effects of estrogen and progesterone and second, that at parturition there is a rise in the concentrations of prolactin and adrenal corticosteroids, both lactation-promoting hormones.

Maintenance of Lactation

Copious milk secretion and the maintenance of full lactation depend on suckling. Prolactin is released from the pituitary in response to nursing or mechanical stimulation of the nipple. Sustained milk secretion is usually achieved by 2 weeks after delivery. Prolactin alone can maintain lactation in women, and ovariectomy does not prevent lactation. Frequent suckling or nipple stimulation can induce lactation even in parous postmenopausal women and in men.

One of the actions of prolactin is to control both estrogen and prolactin receptors. In addition, estrogen in large amounts can diminish prolactin binding, which may be the reason why lactation normally does not occur before parturition. After delivery, when estrogen and progesterone levels have declined, prolactin action is unopposed, and its anabolic effects on the mammary gland are manifest. At this time, basal prolactin levels are not elevated, but periodic surges of prolactin-release induce lactogenesis (Fig. 8.72).

Milk release or letdown from the breast at the time of nursing is accomplished by the action of oxytocin on the myoepithelial cells of the aveoli. Oxytocin is produced in the hypothalamus and is released from the neurohypophysis. The milk-ejection reflex makes milk stored within the breast available for the baby. Suckling evokes milk ejection and is involved in the maintenance and augmentation of lactation because it promotes the release of both oxytocin and prolactin.

Milk Composition

The secretion obtained from the mammary gland in the first few days after parturition has been

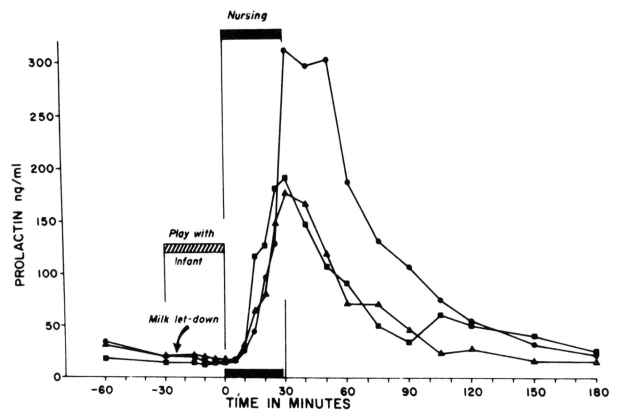

Figure 8.72. Plasma prolactin concentrations during anticipation of nursing and nursing in three women who were between 22 and 26 days postpartum. The mothers played with their infants for 30 min before suckling began. Milk let-down occurred in each case. [From Noel et al. (1974); Yen and Jaffe (1978).]

termed "colostrum." The composition of colostrum differs from mature milk in containing more protein, sodium, and chloride and less lactose and potassium. With successive sucklings the composition changes rapidly to that of mature milk (Cowie et al., 1980).

Milk composition differs significantly between species. In addition to gross differences in proteins, fat, and lactose content, there are also subtle variations in all constituents of milk, such as albumins, immunoglobulins, cholesterol, water, and salts. Variations also occur between individuals and within an individual according to the stage of lactation.

Lactation and Health

Lactation is associated with several beneficial phenomena. It provides both a correctly balanced diet and immunological protection for the infant and a variable period of infertility for the mother, thus postponing pregnancy and increasing birth spacing to the benefit of both mother and child.

Mothers' milk is the best way to nurture the growing infant. Although babies may be raised on cows' milk without apparent deleterious side effects, some delayed, previously unsuspected problems such as allergies may develop (Buisseret, 1978). The growing trend in many countries to substitute commercially available products for breastfeeding deprives the child of specific protection provided by mothers' milk and colostrum. Colostrum and milk are important vehicles for the transfer of immunoglobulins from mother to child. There is increasing evidence that secretory IgA in the colostrum and milk may act within the lumen of the neonate's gut to protect against infection. This represents an important defense mechanism for the newborn, whose own immune system is not yet fully developed.

CHAPTER 60

Sexual Differentiation

Members of our society are taught from early years to distinguish men from women; however, sex identification is a comprehensive concept implying more than gender recognition. Sexual differentiation begins at the time of fertilization and continues into early adult life. This is a complex process which is influenced independently by genetic, hormonal, and behavioral determinants. When we speak of the sex of an individual we mean the composite of the chromosomal sex, gonadal sex, genital sex, and psychological sex. In this chapter the development and interrelationship of each of these parameters are discussed. At the outset it is important to recognize that each individual has the potential for either male or female differentiation at each step of the developmental progression which ultimately determines gender. An understanding of the factors which regulate this process has been derived in large part from studies of animals and patients with genetically determined disorders of sexual differentiation.

GENETIC, CHROMOSOMAL, OR NUCLEAR SEX

For reviews of this topic see Ohno, 1967 and Mittwoch, 1973. Genetic sex is determined at the time of fertilization when the spermatozoon and ovum unite and depends upon which sex chromosome is present in the sperm cell. All extragonadal cells in the human contain 23 pairs of chromosomes (22 pairs of autosomes and one pair of sex chromosomes). In the male the sex chromosomes are X and Y, whereas in the female there are two X chromosomes. During the process of ova and sperm development, germ cells divide by a process called meosis so that pairs of chromosomes are split between two daughter cells. As a result each gamete contains 22 autosomes and 1 sex chromosome. In the testis, myosis produces sperm which contain X and others containing Y chromosomes. Ova, however, contain only X chromosomes. At the time of fertilization, the sex of the resulting zygote will be determined by the sex chromosome carried by the fertilizing sperm. If an ovum is fertilized by a sperm bearing an X chromosome, then the combination XX results, and a female develops. On the other hand, if the fertilizing sperm bears a Y chromosome, then the XY combination produces a male.

Techniques have been developed for examining human chromosomes. Cells in metaphase are spread, and the chromosomes are stained for counting or photography and mounting. A systematized array of chromosomes from a single cell is called a karyotype (Figs. 8.73 and 8.74). The meaning of this term is usually extended to imply that the chromosomal pattern of the cells examined is typical of all the cells of the individual. To determine accurately the karyotype, however, one must characterize the chromosomes from several tissues, and at least 100 metaphase cells from each tissue should be counted. The karotype of a normal woman is designated 46, XX (Fig. 8.73); that of a normal male 46, XY (Fig. 8.74). The karyotype of an individual with an extra X chromosome is 47, XXY.

Biological function of the sex chromosomes has been inferred from the examination of patients with an abnormal number of sex chromosomes. In individuals with an XXY (Ford et al., 1959), or an XXXXY (Barr et al., 1963) complement of sex chromosomes, testes develop indicating that the Y chromosome carries strong male-determining genes which can induce fetal testicular development even in the presence of two or more X chromosomes. Without a Y chromosome, testicular development does not usually occur. The pericentric region of the Y chromosome has been implicated as the source of genes directing testicular differentiation whereas regions on the long arm (Yq) have been associated with spermatogenesis. With the exception of genes affecting stature and spermatogenesis, there is little evidence for the presence of other genes on the Y chromosome of man (McKusick and Ruddle, 1977). The X chromosome is also important to spermatogenesis in that it must be inactivated prior to meiosis (see section on sex chromatin below).

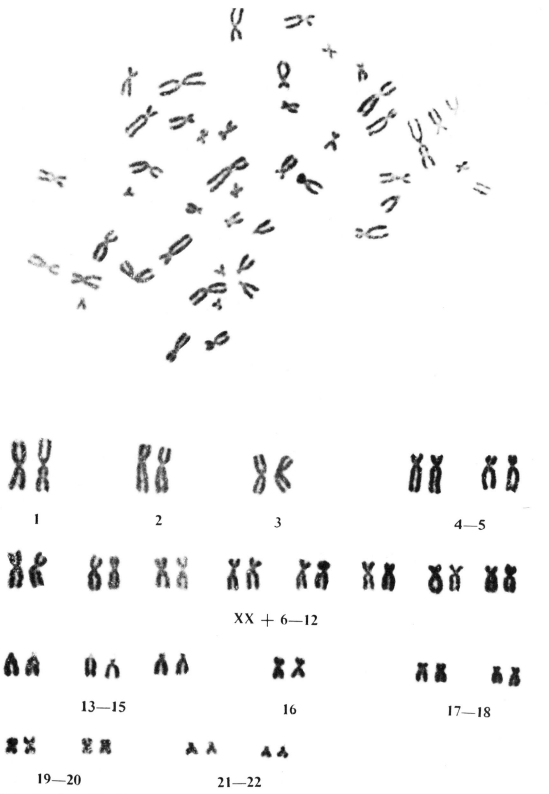

Figure 8.73.   Karyotype of blood lymphocytes from a woman. The pair of X chromosomes is placed at the *left* of the second row.

The H-Y antigen has been studied extensively for its possible role as the primary determinant of testicular development in the fetus. This is a protein which is on the surface of all male tissues and originally was described as a histocompatibility antigen since it was responsible for rejection. The relationship between the H-Y locus and a testis-determining gene on the Y chromosome is not

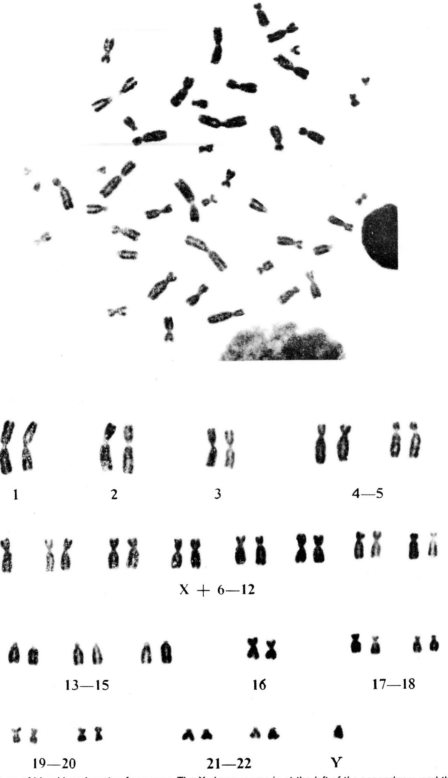

Figure 8.74. Karyotype of blood lymphocytes for a man. The X chromosome is at the *left of the second row*, and the Y chromosome is on the *bottom right*. Chromosomes can be stained or banded with a trypsin-Giemsa technique. Homologous chromosomes can be identified and paired according to size and binding pattern. Chromosomes are positioned so that their shorter p arms point upward and their longer q arms point downward.

entirely clear, but they are believed to be closely linked, if they are not identical. The finding that males with two Y chromosomes have more H-Y antigen than normal males supports the localization of the H-Y gene on the Y chromosome. Evidence that the H-Y antigen itself plays a testis-determining role is derived from the observations that this protein is detected in most but not all instances where a testis is present, including sex-reversed males. Some of these latter individuals are well-masculinized males that have a testis and 46, XX karyotypes. The discrepancy between gonadal and chromosomal sex is possibly due to translocation of the H-Y genetic material to one of the X chromosomes or to an autosome. The finding of some XY females without testes has posed a problem concerning the assignment of the gene for the H-Y antigen to the Y chromosome. Examples of situations in which testes did not develop in the presence of H-Y antigen have been explained by postulating a defect in the H-Y receptor protein which rendered the target cells unable to respond.

The H-Y antigen appears to have been conserved during evolution at least to the extent that a similar cross-reactive antigen has been found in humans, mice, rats, guinea pigs, rabbits, birds, and amphibia. In birds, females are heterogametic ZW. In amphibia, the male is the heterogametic sex in *Rana pipiens*, while it is the female in the *Xenopus laevis.* In each case, the presence of a cell surface component, cross-reacting with antiserum to mouse H-Y antigen, is present in the heterogametic sex. From results obtained in the nonmammalian species, it has become apparent that the H-Y antigen should more correctly be described as the director of differentiation of the heterogametic gonad (Ohno, 1976; Wachtel et al., 1975).

The biological functions of the X chromosomes are more diverse. The requirement for a pair of X chromosomes for ovarian differentiation is suggested by the fact that ovaries do not develop in women with an unpaired X chromosome (gonadal dysgenesis: 45, XO). This observation does not hold for all species since functional ovaries develop in XO mice. Deletions of portions of the X chromo

some also lead to gonadal dysgenesis. The short arm (Xp) contains genes that escape inactivation, and deletions within this region result in dysgenetic ovaries. As will be noted below, the germ cells in the ovary require two active X chromosomes for normal development. The loss of germ cells (germ cell atresia) may, therefore, result from deletion of Xq due to lack of the genes required for germ cell development. This results in gonadal dysgenesis and infertility. In contrast to the Y chromosome, which is mainly concerned with sex differentiation, the X chromosome contains genes which influence the differentiation of a wide variety of extragonadal characteristics. The X chromosome tends to be highly conserved in evolution, whereas the Y chromosome is not. The function assigned to the X chromosome in one species tends to be X-linked in others while functions related to the Y chromosome may be distributed about the genomes of mammals. The XO male vole, *Ellobius lutescens*, is the extreme example of the widespread distribution of male-determining genes since this animal has no identifiable Y chromosome at all (Gordon and Ruddle, 1981).

Sex Chromatin (Nuclear Sex)

Since the chromosomal composition of the male differs from that of the female by only a single chromosome, it was surprising to observe a morphologic difference in nuclei from male and female animals. The difference between the male and the female lies in a special mass of chromatin (sex chromatin or Barr body) in the female. The sex chromatin is typically a planoconvex mass about 1 μm in diameter which stains positively for deoxyribonucleic acid and which lies close to or in contact with the nuclear membrane (Fig. 8.75). It is now generally accepted that the sex chromatin is comprised of one of the X chromosomes which remains

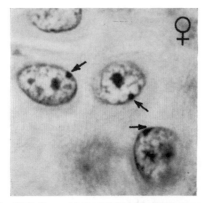 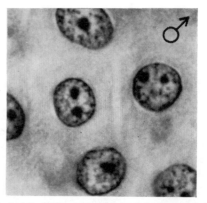

Figure 8.75.   Nuclei of cells in the epidermal spinus layer of a genetic female (*left*) and a genetic male (*right*). The sex chromatin bodies are indicated by *arrows* in the female. [From Grumbach and Barr (1958).]

genetically inactive. Once inactivation has occurred, the same X chromosome is inactivated in all descendants of that cell. Chromatin-positive cells which contain two X chromosomes (46, XX and 47, XXY) have a single Barr body, whereas those with more than two X chromosomes may have two chromatin bodies. It follows, therefore, that the number of chromatin bodies in any diploid nucleus is one less than the total number of X chromosomes (Lyon, 1972).

GONADAL SEX

Until the 5th or 6th week of embryonic life, the gonad in both sexes develops in an identical fashion, with the potential to develop into either a testis or an ovary. At the 6th week the embryonic gonads appear as ridges on either side of the dorsal mesentery immediately adjacent to the developing adrenal gland. If the indifferent gonad is to become a testis due to the presence of Y chromosomes and its associated genes, the somatic cells are organized into tubular structures at the 7th week of fetal life. Primary germ cells migrate into the fetal gonad and are incorporated into the developing seminiferous tubules. At about the 8th week Leydig cells appear in the medullary mesenchyme, and their function ensures subsequent male differentiation of the genital ducts and external genitalia. As testicular differentiation proceeds, the medulla of the testis progressively enlarges, and the cortex regresses. In the mature testis the only remnant of the embryonic cortex is the tunica vaginalis.

Differentiation of the indifferent gonad into an ovary occurs at about 9 wk of fetal life. It is of interest to note that primary germ cells of a fetus with a 46, XX karyotype have a sex chromatin body (inactive X chromosome) prior to their migration into a genital ridge. Once inside the embryonic gonad, however, the sex chromatin body is lost, suggesting that both X chromosomes in the primordial germ cell are active (Witschi, 1957). As noted above, *two* active X chromosomes are believed to be necessary for differentiation of an ovary. In the developing ovary, the embryonic cortex proliferates to form the major portion of this gonad. The medulla regresses and forms the hilum of the mature ovary.

From these observations it is apparent that the gonadal sex is to a large extent determined by the chromosomal sex and, in most instances, both are the same. By contrast, genital sex secondary sexual characteristics and most of the other dimorphic features of men and women are hormonally determined and can easily be made inconsistent with the genetic sex by treatment with sex steroids (Haseltine and Ohno, 1981).

GENITAL SEX

Differentiation of the Genital Ducts

At the 7th week of embryonic life the fetus contains primordia of the genital duct systems for both the male (Wolffian ducts) and the female (Müllerian ducts). If a male develops, the Wolffian duct system will give rise to the epididymis, vas deferens, seminal vesicles, and the ejaculatory duct; whereas, in the female the Müllerian ducts serve as the anlage of the Fallopian tubes and uterus. During the 3rd month of fetal life either the Müllerian or the Wolffian ducts complete their differentiation while the others regress.

The mechanism whereby one set of ducts dominates over the other to form the internal genitalia of the developing embryo was demonstrated in a series of experiments by Jost (1953). In these studies embryonic rabbits were either gonadectomized or treated with steroids in utero and the subsequent effect upon Müllerian and Wolffian development observed. The results of these studies are schematically summarized in Figure 8.76. In the presence of a testis, Wolffian structures develop, and Müllerian structures regress (Fig. 8.76, *A* and *B*). The development of normal Müllerian structures is not dependent upon the ovary since this ductal system differentiates in both the castrate female and male, even if the castration is unilateral (Fig. 8.76*C–E*). Transplantation of a testis into the female adjacent to the ovary results in unilateral Wolffian differ-

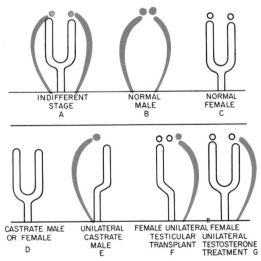

Figure 8.76. Schematic summary of Wolffian and Müllerian development in rabbit embryos. Ovaries and Müllerian structures are indicated by *open figures*, the Wolffian structures by *shaded* areas (see text for explanation).

entiation associated with Müllerian regression on the side of the transplant (Fig. 8.76*F*). By contrast, unilateral testosterone treatment of the female with a crystal of steroid implanted adjacent to one ovary resulted in ipsilateral Wolffian development without Müllerian regression (Fig. 8.76*G*). From these observations, the hormonal control of gonaduct differentiation was formulated as follows:

Androgens stimulate Wolffian development, and without the influence of this class of steroids, this ductal system regresses. As will be noted below, however, there are two androgens (testosterone and dihydrotestosterone) which have different effects on androgen-dependent differentiation.

The Müllerian duct system develops normally in the absence of a hormonal stimulus from either gonad, but regression of this ductal system is due to a glycoprotein hormone from the seminiferous tubule which has been termed Müllerian duct inhibiting factor. Müllerian duct regression commences shortly after differentiation of the seminiferous tubules, and consequently the formation of this inhibiting substance constitutes the first endocrine function of the embryonic testis (Wilson et al., 1981). This hormone is believed to stimulate growth of the mesenchymal cells which surround the Müllerian ducts, causing their active regression (Jost, 1953; Jost, 1971; Josso et al., 1977).

Differentiation of the External Genitalia

The external genitalia of both sexes are identical up until the end of the 8th fetal week and have the potential to differentiate into the organs of either sex. At this stage the major components of the external genitalia are the genital tubercle, urethral folds, urogenital slit, and labioscrotal swelling (Fig. 8.77). Under the influence of androgens from the fetal testes (or from exogenous administration) male differentiation of the external genitalia occurs. The genital tubercle forms the penis, the urogenital folds, the penile urethra, the labioscrotal swellings and the scrotum. In the absence of androgens female differentiation of the external genitalia proceeds with the development of a clitoris, labia minora, and labia majora. The structures which develop from the indifferent genital primordia in the presence and in the absence of androgens are summarized in Table 8.9. It is significant that the androgen-dependent differentiation of the external genitalia can only occur during fetal life. Even though the female structures are not hormone dependent, once the vagina and labia have developed they cannot be differentiated into analogous male structures even if exposed to very large quantities of androgens (Grumbach and Van Wyk, 1974).

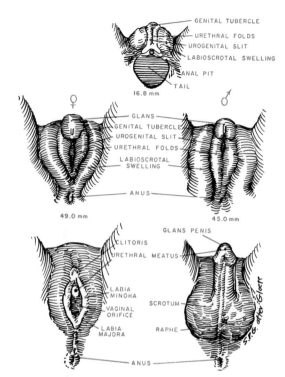

Figure 8.77. Differentiation of male and female external genitalia from indifferent primordia. Male differentiation occurs only in the presence of androgenic stimulation during the first 12 weeks of fetal life. [From Grumbach and Van Wyck (1974).]

Table 8.9
Differentiation of the external genitalia from indifferent primordia

Primordial Structure	Structures Which Develop *with* Androgen Exposure	Structures Which Develop *without* Androgen Exposure
Genital tubercle	Penis	Clitoris
Urethral folds	Corpus spongiosum (penile urethra)	Labia minora
Labioscrotal swellings	Scrotum	Labia majora
Urogenital sinus	Prostatic utricle	Lower two-thirds of vagina
	Bulbourethral glands	Skene's glands Bartholin's glands

THE IMPORTANCE OF TESTOSTERONE AND DIHYDROTESTOSTERONE IN MALE DIFFERENTIATION

From the above discussion, it is clear that female development occurs in the absence of gonadal hormones. By contrast, male differentiation requires Müllerian duct inhibiting factor and androgens. Recent studies on the mechanism of action of this class of steroids have emphasized that both testosterone and its metabolite, dihydrotestosterone, are important in this process. The mechanism of action of androgens in different tissues is illustrated diagrammatically in Figure 8.53 in Chapter 57. In some tissues, testosterone enters the cell and is bound by a recognition molecule, termed the androgen receptor. Testosterone receptor complex is then transferred to the nucleus of the cell, where it is bound to chromatin. The receptor chromatin interaction facilitates ribonucleic acid (RNA) synthesis which, in turn, facilitates protein synthesis and subsequent differentiation. In other tissues, testosterone is metabolized to dihydrotestosterone by the enzyme 5-α-reductase. Dihydrotestosterone then interacts with the androgen receptor, which is transferred to the nucleus of the cell, where it ultimately results in androgen-dependent differentiation (Mainwaring, 1977).

The necessity of the androgen receptor for androgen action is demonstrated in animals and humans with the sex-linked recessive defect, testicular feminization. Individuals with this disorder lack androgen-dependent differentiation of both Wolffian duct derivatives and of the external genitalia. As a consequence, genetic males appear as phenotypic females with abdominal testes. Müllerian ducts regress due to normal Müllerian duct inhibiting factor. Early studies demonstrated that mice and rats with this disorder could not transfer testosterone or dihydrotestosterone to the nuclei of androgen responsive tissue. These observations suggested an abnormality of the androgen receptor. This hypothesis was confirmed with specific receptor assays (Fig. 8.78). Most animals and humans with this disorder have a small amount of "residual receptor," although this is not detectable in most routine receptor assays. Subsequent studies demonstrated that testicular feminization is a heterogeneous condition, with some individuals having a reduced amount of normal receptor, others a receptor with reduced affinity, and others a receptor that will bind to chromatin but not activate nuclear RNA metabolism (Bardin, et al., 1973).

The possibility that there may be differential requirements for testosterone and dihydrotestosterone in various tissues was suggested by studies indicating that the 5-α-reductase activity in the adult was strikingly different from that in the fetus. The studies, summarized in Figure 8.79, indicate that in the developing embryo, there is very little 5-α-reductase in the primordial epididymides and seminal vesicles until after androgen-dependent differentiation of the male reproductive tract has occurred. By contrast, prostate and the derivatives of the genital tubercle have 5-α-reductase activity well before the onset of androgen biosynthesis in the fetal testis. It was concluded that some fetal

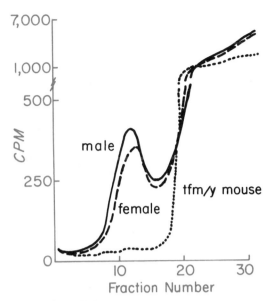

Figure 8.78. ³H-androgen binding in tissues from normal male, female, and androgen-insensitive testicular feminized (tfm/y) rodents. ³H-testosterone was incubated with mouse kidney cytosol, and the cytosol was sedimented through sucrose gradients. In this assay, the smaller peak, sedimenting in about fraction 12, represents the androgen receptor. The absence of specific androgen binding in tissues from tfm/y animals is in marked contrast to their presence in normal males and females. [From Bullock and Bardin (1972).]

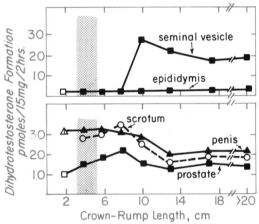

Figure 8.79. 5-α-reductase activity in urogenital tracts of male human embryos as a function of age. Enzyme activity was evaluated by the formation of 5-α-dihydrotestosterone from testosterone. Tissues in the *upper panel* are derived from the Wolffian ducts, and in the *lower panel* from the urogenital sinus (prostate), urogenital swelling (scrotum), and urogenital tubercle (penis). The *stippled bar* represents the period of masculine genital differentiation. [Adapted from Siiteri and Wilson (1974).]

tissue may require testosterone and some dihydrotesterone for differentiation.

The importance of dihydrotestosterone formation by fetal tissues is illustrated by men with an inherited 5-α-reductase deficiency. In an isolated community in the Dominican Republic, where 5-α-

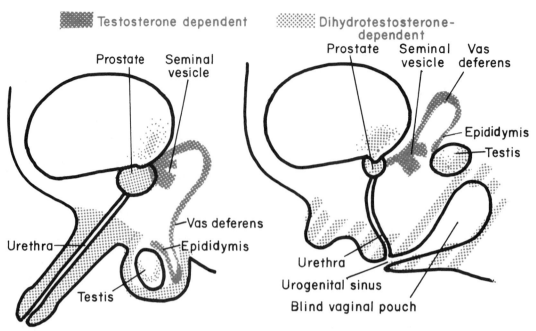

Figure 8.80. Evidence for the role of testosterone and 5-α-dihydrotestosterone in sexual differentiation. Genitalia from normal (*left*) and 5-α-reductase-deficient (*right*) males are depicted. In affected males, differentiation of testosterone-dependent, but not dihydroxy-testosterone-dependent, tissue occurs. [From Imperato et al. (1976).]

reductase deficiency is observed in many families, the local inhabitants use the term "quevedoces" (penis at 12) for this disorder. This relates to the fact that affected males have ambiguity of the external genitalia at birth but undergo masculine development at the time of puberty. These men have normal testosterone levels but cannot form dihydrotestosterone in tissues. At birth they have testes, epididymides, vas deferens, and seminal vesicles but lack a prostate, penis, and well-formed scrotum. These observations are consistent with the hypothesis that dihydrotestosterone is required for differentiation of some tissues in the fetus while testosterone will suffice for others. The observations that 5-α-reductase activity is low in adult testes and that normal sperm develop in men with 5-α-reductase deficiency indicate that dihydrotestosterone is not required for spermatogenesis. The observations are summarized in Figure 8.80.

BRAIN DIFFERENTIATION

Even though the central nervous system is anatomically remote from the external genitalia, the differentiation of both of these structures is similar since both require fetal androgen exposure if a masculine pattern of development is to occur. In the adult female, pituitary gonadotropins are released in a cycle fashion. In the mature male, gonadotropins are secreted in relatively constant rates, even when an ovary is transplanted into an orchiectomized host. Androgen treatment of female

rat fetuses, either in utero or in the early neonatal period, results in a noncyclic pattern of gonadotropin secretion after puberty, even in the presence of a functioning ovary. From these observations it is apparent that in the absence of fetal androgens, the hypothalamus develops a cyclic center for gonadotropin release, whereas fetal androgen exposure produces a noncyclic center. A discussion of the androgens involved in brain differentiation is presented in Chapter 58. Another way that the development of the hypothalamus is similar to that of external genitalia is that only a brief androgen exposure is required to effect an irreversible change.

Recent studies have extended these observations by demonstrating neuroanatomical, neurophysiological, and behavioral dimorphism. There is a growing list of examples of sex differences in central nervous system morphology. These differences fall into three different categories: (a) ultrastructural differences in cellular or synaptic organelles; (b) differences in synaptic or dentritic organization; and (c) differences in the gross volume of defined cell groups (MacLusky and Naftolin, 1981). Some examples of these dimorphic traits include the fact that males have fewer synapses in some areas of the brain than do females. Castration of male animals within 12 h of birth is associated with an increase in the female number of synapses, while testosterone treatment of female animals on day 4 of life decreases the number of such synapses in the male range. Furthermore, electrophysiological studies show that more of the cells projecting to the medial basal hypothalamus receive synaptic connections from the amygdala in males than in normal female animals or neonatally castrated males, while female animals treated with testosterone have an intermediate number. Additionally, other neurons fire twice as fast in normal female and neonatally castrated males as do these same cell types in normal males and females treated neonatally with testosterone. Neonatal female rats have much higher titers of plasma luteinizing hormone (LH) and follicle-stimulating hormone (FSH) than do males of equivalent age and show increased staining of LH-releasing hormone-producing neurons in various parts of the brain than do male animals.

Most of the above studies have been done in rodents, and it is not clear that similar considerations apply to primates. Female primates exposed to perinatal androgen or human female patients with congenital and adrenal hyperplasia (who have been exposed to high androgen concentrations in utero) display cyclic gonadotropin release and female sexual behavior. In the human, however, a sexual dimorphism in neural organization underlying cognition has been demonstrated in tests of spatial processing: boys perform in a manner consistent with right hemisphere specialization as early as 6 years of age, whereas girls show evidence of bilateral representation until age 13. These observations may reflect differences in the rate of sexual maturity, since regardless of sex, early maturing adolescents perform better in tests of verbal than spatial abilities, whereas late maturing ones show the opposite pattern. In monkeys, lesions in the orbital prefrontal region of the brain performed in infancy were associated with impairment in behavioral tests in male animals at 2½ mo of age, whereas similar defects were not detected in females with comparable lesions until 15–18 mo of age. These studies, therefore, suggest that even in the primate species, a sex-dependent difference in brain organization is present, although obviously no studies on the effects of neonatal castration or hormone replacement have been done (Krieger, 1978).

PSYCHOLOGICAL SEX OR GENDER ROLE

Gender is selected at birth in accord with the apparent genital sex. This becomes the legal sex and also sets the behavioral patterns which influence psychosexual identification of the individual with others in society. It is not clear that gender role is an inherent characteristic of the fetus or that it is established by a direct hormone action on the central nervous system similar to that responsible for differentiation of the gonadotropin release

centers. In spite of the sexual differences in the brain described above, a child can identify as a male or female prior to the age of 2 years. Thereafter, psychosexual differentiation or gender role is established by sex orientation through dress, hair style, toilet habits, and constant comparison with other children and adults. Through this exposure a child is continually reminded of his gender and comes to identify with one of the sexes. From psychological studies of girls who were mistakenly raised as boys, and vice versa, it is now clear that gender role is established early in life and maintained, even in the face of abnormal external genitalia and subsequent development of secondary sexual characteristics which are opposite to those of the identified sex.

From these and similar studies, it was suggested that it would be very difficult, and perhaps unwise, to change the gender of an individual after 2 years of age. Observations on patients with 5-α-reductase deficiency question this generally held assumption. These individuals are born with ambiguous genitalia and are raised initially as girls. At puberty, when the phallus enlarges, they switch gender and behave as men thereafter (Imperato-McGinley and Peterson, 1976).

Sexually Dimorphic Behavior

From an examination of children, adolescents, and adults who have been exposed prenatally to high levels of sex hormones, investigators have sought to determine whether androgens, estrogens, and progestins have subsequent effects on behavioral development of humans, as they do in animals. If so, androgenic steroids should produce some male type behavior; progesterone might be expected to antagonize the effects of androgens, and estrogens may have either masculinizing effects or feminizing effects, depending on the compound and dose. The behavioral categories of sexually dimorphic behavior that have been the focus of these are as follows (Ehrhardt and Meyer-Bahlbury, 1981).

ENERGY EXPENDITURE

In human beings this has been measured as active outdoor play and athletic skills. In nonhuman primates the comparable behavioral aspect is rough-and-tumble play. Intense physical energy expenditure of this type is believed to be an essential aspect of psychosocial development, and it appears to be influenced by sex steroid variations before birth.

SOCIAL AGGRESSION

In nonhuman primates, such behavior categories as pursuit, threat behavior in childhood, and fighting are included. For human beings, aggressive behavior includes physical and verbal fighting in childhood and adolescence.

PARENTING REHEARSAL

In humans, children engage in doll play (rather than in play with cars and trucks), playing "house," playing "mother" and "father," participating in infant care, and fantasizing about having children of their own. Observations of lower mammalian species suggest that the threshold and intensity of parenting is dimorphically related to the type and level of prenatal hormones.

PEER CONTACT, FRIENDS, AND PATTERNS OF GROUP INTERACTION

In nonhuman primates, dominance behavior and sex segregation of play groups have received special attention. In human behavior the categories would be preference of playmates by sex.

GENDER ROLE LABELING

A given child is labeled by his or her social environment on the basis of play behavior patterns and gender role preference, for example, as "tomboy" or "sissy."

GROOMING BEHAVIOR

Humans adorn themselves, as indicated by clothes preference, jewelry, makeup, hairdo, and the like. The analogous behavior aspect in nonhuman primates may be grooming. It is doubtful that prenatal hormonal levels have any direct effect on the intensity of this type of grooming behavior in human beings.

When the above types of behavior have been compared in normals and in individuals exposed to hormones prenatally, the evidence suggests that human psychosexual differentiation is influenced by prenatal hormones, albeit to a limited degree. Where there are effects, they are similar to animal experiments on the role of prenatal androgens, progestogens, and estrogens in the expression of sex-dimorphic behavior. Best established are the effects of prenatal androgens on physical energy expenditure and play rehearsal of parenting behavior; effects on peer preferences and grooming behavior may be related to the former. The evidence for antiandrogenic action of prenatal progesterone and related compounds seems insufficient as yet; observations from various independent studies are in agreement concerning a slight demasculinizing effect of progestogens. Any conclusions drawn on the basis of human research studies have to remain tentative, but it is already obvious that prenatal

hormones have to be considered, along with other factors that interact in exerting their influence on the expression of sex-dimorphic behavior.

The evidence for the role of prenatal hormones in the development of sexual orientation is inconclusive. The data available on intersex patients indicate that prenatal hormones do not rigidly determine sexual orientation. More subtle contributions cannot be ruled out, however, and this possibility is strengthened by recent findings of differences between heterosexual and homosexual individuals in neuroendocrine regulatory characteristics (Ehrhardt and Meyer-Bahlbury, 1981).

Puberty

The secondary sexual characteristics are the last features to develop in both sexes and occur as a result of the hormonal changes at the time of puberty. The process of puberty is typified by an increase in growth rate and the appearance of striking somatic differences between boys and girls. The onset of these changes actually antedates the appearance of secondary sex characteristics by a few years. Thus, if bodily changes are considered, sexual maturation may be a longer process than ascribed to the period of visible changes in the external genitalia.

The first readily detectable genital change during puberty in boys is an increase in rate of testicular growth that is first discernible at a mean age of 11.6 years. This change occurs approximately 6 months later than the first changes in girls, which is usually breast development. Thus, the time of onset of the pubertal process may be similar for boys and girls, even though the development of secondary sexual characteristics may progress at different rates. For example, pubic hair appears approximately 1½ yr later in boys than in girls. The pattern of pubertal changes is summarized in Table 8.10.

Although androgen secretion by the testes and estrogen secretion by the ovaries are the primary

Table 8.10
Pattern of pubertal changes in boys and girls

Pubertal Event	Mean Age at Onset	
	Boys	Girls
Breast development		11.2
Testicular enlargement	11.6	
Pubic hair development	13.4	11.7
Peak height velocity	14.1	12.1
Menarche		13.5
Adult pubic hair configuration	15.2	14.4
Adult type breast		15.3

[From Santen and Kulin (1976).]

modulators of the somatic changes which appear during puberty, other hormones do play a role. In particular, growth hormone appears to be necessary in order to realize the full growth-promoting effects of the gonadal steroids. For example, growth hormone-deficient boys will exhibit a growth spurt when exposed to testosterone, but to a somewhat lesser degree than in the presence of growth hormone. Furthermore, the pubertal process usually remains incomplete in the absence of thyroid hormone.

The precise physiological changes leading to the onset of puberty are a complex process involving maturation of the brain and gonad. Several phenomena, however, regulating gonadotropin control do change during adolescence, presumably as a result of altered central nervous system function. One of the most striking of these is the negative feedback of gonadotropin secretion. Studies that provide evidence for the existence of negative feed-back prior to puberty include: (a) elevated LH and FSH levels in castrated prepubertal individuals, and (b) prepubertal gonadal hypertrophy following removal of one gonad. This latter observation suggests that a decrease in gonadal secretory products in such individuals causes a secondary rise in gonadotropins which further stimulates the remaining gonad. It is of interest that when estrogens are administered to prepubertal children, considerably lower doses are required for depression of gonadotropins than in adults. Thus, the negative feedback present before puberty is not only operative but highly sensitive. A change in the sensitivity of the negative feedback system occurs during sexual maturation in man. The adult type interaction of gonadal and pituitary hormones is attained in mid- to late-puberty, as levels of gonadotropins and gonadal steroids progressively increase. The site of the negative feedback control is probably at both the hypothalamic and pituitary levels.

BIBLIOGRAPHY

ALLEN, R. G., R. HERBERT, M. HINMAN, H. SHIBUYA, AND C. B. PERT. Coordinate control of corticotropin, β-lipotropin and β-endorphin release in mouse pituitary cell cultures. *Proc. Natl. Acad. Sci. U.S.A.* 75: 4972–4976, 1978.

AMARA, S. G., V. JONAS, M. G. ROSENFELD, E. S. ONG, AND R. M. EVANS. Alternate RNA processing in calcitonin gene expression generates mRNAs encoding different polypeptide gene products. *Nature* 298: 240–244, 1982.

AMARA, S. G., M. G. ROSENFELD, R. S. BIRNBAUM, AND B. A. ROOS. Identification of the putative cell-free translation product of rat calcitonin mRNA. *J. Biol. Chem.* 255: 2645–2648, 1980.

AMBESI-IMPIOMBATO, F. S., L. A. M. PARKS, AND H. G. COON. Culture of hormone-dependent functional epithelial cells from rat thyroids. *Proc. Natl. Acad. Sci. U.S.A.* 77:3455–3459, 1980.

ANDRES, R. Y., I. JENG, AND R. A. BRADSHAW. Nerve growth factor receptors: identification of distinct classes in plasma membranes and nuclei of embryonic dorsal root neurons. *Proc. Natl. Acad. Sci. U.S.A.* 74: 2785–2789, 1977.

AUBERT, M. L., M. M. GRUMBACH, AND S. L. KAPLAN. The ontogenesis of human fetal hormones. III. Prolactin. *J. Clin. Invest.* 56: 155–164, 1975.

AUSTIN, L. A., AND H. HEATH, III. Calcitonin. Physiology and pathophysiology. *N. Engl. J. Med.* 304: 269–278, 1981.

BAIRD, D. T., P. E. BURGER, G. D. HEAVON-JONES, AND R. J. SCARAMUZZI. The site of secretion of androstenedione in nonpregnant women. *J. Endocrinol.* 63: 201–212, 1974.

BAKER, H. W. G., R. J. SANTEN, H. G. BURGER, D. M. DE KRETSER, B. HUDSON, R. J. PEPPERELL, AND C. W. BARDIN. Rhythms in the secretion of gonadotropins and gonadal steroids. *J. Steroid Biochem.* 6: 793–801, 1975.

BAKER, H. W. G., T. J. WORGUL, R. J. SANTEN, L. S. JEFFERSON, AND C. W. BARDIN. Effect of prolactin on nuclear androgens in perfused male accessory sex organs. In: *The Testis in Normal and Infertile Men*, edited by P. Troen and H. Nankin. New York: Raven Press, 1977, p. 379.

BAKER, T. G. A quantitative and cytological study of germ cells in human ovaries. *Proc. R. Soc. (Lond.) [Biol.] Ser. B.* 158: 417–433, 1963.

BARDIN, C. W. Pituitary testicular axis. In: *Reproductive Endocrinology*, edited by S. S. C. Yen and R. B. Jaffe. Philadelphia: W. B. Saunders, 1978, p. 110–125.

BARDIN, C. W., L. P. BULLOCK, R. J. SHERINS, I. MOWSZOWICZ, AND W. R. BLACKBURN. Part II. Androgen metabolism and mechanism of action in male pseudohermaphroditism: a study of testicular feminization. *Recent Prog. Horm. Res.* 29: 65–109, 1973.

BARDIN, C. W., AND J. F. CATTERALL. Testosterone: a major determinant of extragenital sexual dimorphism. *Science* 211: 1285–1294, 1981.

BARDIN, C. W., AND J. A. MAHOUDEAU. Dynamics of androgen metabolism in women with hirsutism. *Ann. Clin. Res.* 2: 251–262, 1970.

BARDIN, C. W., N. MUSTO, G. GUNSALUS, N. KOTITE, S.-L. CHENG, F. LARREA, AND R. BECKER. Extracellular androgen binding proteins. *Annu. Rev. Physiol.* 43: 189–198, 1981.

BARNES, D., AND G. SATO. Serum-free cell culture: a unifying approach. *Cell* 22: 649–655, 1980.

BARR, M. L., D. H. CARR, J. POZSONYI, R. A. WILSON, H. G. DUNN, T. S. JACOBSON, J. R. MILLER, M. LEWIS, AND B. CHOWN. The XXXXY sex chromosome abnormality. *Can. Med. Ass. J.* 87: 891–901, 1962.

BARTKE, A. Prolactin and the physiological regulation of the mammalian testis. In: *The Testis in Normal and Infertile Men*, edited by P. Troen and H. Nankin. New York: Raven Press, 1977, p. 367.

BAXTER, J. D. Glucocorticoid hormone action. In: *Pharmacology of Adrenal Cortical Hormones*, edited by G. N. Gill. Oxford: Pergamon Press, 1979, p. 67–121.

BAXTER, J. D., N. L. EBERHARDT, J. W. APRILETTI, L. K. JOHNSON, R. D. IVARIE, B. S. SCHACHTER, J. A.

MORRIS, P. H. SEEBURG, H. M. GOODMAN, K. R. LATHAM, J. R. POLANSKY, AND J. A. MARTIAL. Thyroid hormone receptors and responses. *Recent Prog. Horm. Res.* 35: 97–153, 1979.

BAXTER, J. D., AND J. W. FUNDER. Hormone receptors. *N. Engl. J. Med.* 301: 1149–1161, 1979.

BAXTER, J. D., AND C. G. ROSSEAU. In *Glucocorticoids and the Metabolic Code in Glucocorticoid Hormone Action*, edited by J. D. Baxter and C. G. Rosseau. Heidelberg: Springer-Verlag, 1979, p. 613–627.

BECKER, D. V., W. M. MCCONAHEY, AND B. M. DOBYNS. In: *Further Advances in Thyroid Research*, edited by K. Fellinger and R. Hofer. Vienna: Verlag der Miener Medizinischen academie, 1971, vol. 1, p. 603–609.

BECKETT, G. J., AND G. S. BOYD. Purification and control of bovine adrenal cortical cholesterol ester hydrolase and evidence for the activation of the enzyme by a phosphorylation. *Eur. J. Biochem.* 72: 223–233, 1977.

BEGLEY, D. J., J. A. FIRTH, AND J. R. S. HOULT. *Human Reproduction and Developmental Biology*. New York: MacMillan Press Ltd., 1980.

BEIERWALTES, W. H. The treatment of thyroid cancer with radioactive iodine. Semin. *Nucl. Med.* 8: 79–94, 1978.

BENIRSCHKE, K. Adrenals in anenchephaly and hydrocephaly. *Obstet. Gynecol.* 8: 412–425, 1956.

BERGLAND, R. M., AND R. B. PAGE. Can the pituitary secrete directly to the brain? (affirmative anatomical evidence). *Endocrinology* 102: 1325–1338, 1978.

BERGLAND, R. M., S. L. DAVIS, AND R. B. PAGE. Pituitary secretes to brain. *Lancet* 2: 276–277, 1977.

BILEZIKJIAN, L. M., AND W. W. VALE. Stimulation of adenosine 3′,5′-monophosphate production by growth hormone-releasing factor and its inhibition by somatostatin in anterior pituitary cells, in vitro. *Endocrinology* 113: 657–662, 1983.

BISHOP, J. M. The molecular biology of RNA tumor viruses: a physician's guide. *N. Engl. J. Med.* 303: 675–682, 1980.

BLACKWELL, G. J., R. CARNUCCIO, M. DI ROSA, R. J. FLOWER, L. PARENTE, AND P. PERSICO. Macrocortin: a polypeptide causing the anti-phospholipase effect of glucocorticoids. *Nature* 287: 147–149, 1980.

BLUNDELL, T. L., S. BEDARKAR, E. RINDERKNECHT, AND R. E. HUMBEL. Insulin-like growth factor: a model for tertiary structure accounting for immunoreactivity and receptor binding. *Proc. Natl. Acad. Sci. U.S.A.* 75: 180–184, 1978.

BORDIER, P., H. RASMUSSEN, P. MARIE, L. MIRAVET, J. GUERIS, AND A. RYCKWAERT. Vitamin D metabolites and bone mineralization in man. *J. Clin. Endocrinol. Metab.* 46: 284–294, 1978.

BORIS, A., J. F. HURLEY, T. TRMAL, J. P. MALLON, AND D. S. MATUSZEWSKI. Evidence for the promotion of bone mineralization by 1α,25-dihydroxycholecalciferol in the rat unrelated to the correction of deficiencies in serum calcium and phosphorus. *J. Nutr.* 108: 1899–1906, 1978.

BOSKEY, A. L. Current concepts of the physiology and biochemistry of calcification. *Clin. Orthop. Relat. Res.* 157: 225–257, 1981.

BRADSHAW, R. A., R. A. HOGUE-ANGELETTI, AND W. A. FRAZIER. Nerve growth factor and insulin: evidence of similarities in structure, function, and mechanism of action. *Recent Prog. Horm. Res.* 30: 575–596, 1974.

BRAVERMAN, L. E., AND S. H. INGBAR. Changes in thyroidal function during adaptation to large doses of iodide. *J. Clin. Invest.* 42: 1216–1231, 1963.

BRAZEAU, P., W. VALE, R. BURGUS, N. LING, M. BUTCHER, J. RIVIER, AND R. GUILLEMIN. Hypothalamic polypeptide that inhibits the secretion of immunoreactive pituitary growth hormone. *Science* 179: 77–79, 1973.

BROWN, E. M., S. HURWITZ, C. J. WOODARD, AND G. D. AURBACH. Direct identification of beta-adrenergic receptors on isolated bovine parathyroid cells. *Endocrinology* 100: 1703–1709, 1977.

BROWN, M. R., L. A. FISHER, J. RIVIER, J. SPIESS, C. RIVIER, AND W. VALE. Corticotropin-releasing factor: effects on the sympathetic nervous system and oxygen consumption. *Life Sci.* 30: 207–210, 1982.

BROWN, M. S., P. T. KOVANEN, AND J. L. GOLDSTEIN.

Receptor-mediated uptake of lipoprotein-cholesterol and its utilization for steroid synthesis in the adrenal cortex. *Recent Prog. Horm. Res.* 35: 215–257, 1979.

BUCKLEY, D. I., AND J. RAMACHANDRAN. Characterization of corticotropin receptors on adrenocortical cells. *Proc. Natl. Acad. Sci. U.S.A.* 78: 7431–7435, 1981.

BUISSERET, P. D. Common manifestations of cow's milk allergy in children. *Lancet* I: 304–305, 1978.

BULLOCK, L. P., AND C. W. BARDIN. Androgen receptors in testicular feminization. *J. Clin. Endocrinol. Metab.* 35: 935–937, 1972.

BURGUS, R., M. BUTCHER, AND N. LING. Structure moleculaire du facteur hypothalamique (LRF) d'origine ovine controlant la secretion de l'hormone gonadotrope hypophysaire de luteinisation (LH). *Compt. Rend. Acad. Sci. Paris* 273: 1611–1633, 1971.

BURGUS, R., T. F. DUNN, D. DESIDERIO, W. VALE, AND R. GUILLEMIN. Derives polypeptidiques de synthaese doues d'activite hypophysiotrope TRH: nouvelles observation. *C. R. Acad. Sci. [D] Paris* 269: 154, 1969.

BURMAN, K. D., Y. LUKES, F. D. WRIGHT, AND L. WARTOFSKY. Reduction in hepatic triiodothyronine binding capacity induced by fasting. *Endocrinology* 101: 1331–1334, 1977.

BURSTEIN, S, AND M. GUT. Biosynthesis of pregnenolone. *Recent Prog. Horm. Res.* 27: 303–349, 1971.

BURTON, R. M., AND U. WESTPHAL. Steroid hormone-binding proteins in blood plasma. *Metabolism* 21: 253–276, 1972.

BUSS, J. E., AND G. N. GILL. The role of phosphorylation of tyrosine residues in proteins in cell growth control. In: *Growth Factors*, edited by R. Bradshaw, R. Ross, and G. J. Todaro. New York, Academic Press, 1984.

CALDWELL, B. V., AND H. R. BEHRMAN. Prostoglandins in reproductive processes. *Med. Clin. North Am.* 65: 927–936, 1981. Physiol. III. *Intern. Rev. Physiol.* 22: 277–324, 1980.

CARPENTER, G., AND S. COHEN. Epidermal growth factor. *Annu. Rev. Biochem.* 48: 193–216, 1979.

CARSTEN, M. E. Prostaglandins and cellular calcium transport in the pregnant uterus. *Am. J. Obstet. Gynecol.* 117: 824–832, 1973.

CASSEL, D., AND Z. SELINGER. Mechanism of adenylate cyclase activation through the β-adrenergic receptor: catecholamine-induced displacement of bound GDP by GTP. *Proc. Natl. Acad. Sci. U.S.A.* 75: 4155–4159, 1978.

CHALLIS, J. R. G. Endocrinology of late pregnancy and parturition. In: *Reproductive Physiology III*, edited by R. O. Greep. Baltimore: University Park Press, 1980, vol. 22, p. 277–324.

CHAN, L., AND B. W. O'MALLEY. Mechanism of action of the sex steroid hormones. (Three parts). *N. Engl. J. Med.* 294: 1322–1328, 1372–1381, 1430–1437, 1976.

CHAN, S. J., P. KEIM, AND D. F. STEINER. Cell-free synthesis of rat preproinsulins: characterization and partial amino acid sequence determination. *Proc. Natl. Acad. Sci. U.S.A.* 73: 1964–1968, 1976.

CHANNING, C. P. Studies on tissue culture of equine ovarian cell types: pathways of steroidogenesis. *J. Endocrinol.* 43: 403–414, 1969.

CHASE, L. R., AND G. D. AURBACH. Parathyroid function and renal excretion of 3'5'-adenylic acid. *Proc. Natl. Acad. Sci. U.S.A.* 58: 518–525, 1967.

CHEUNG, W. Y. Calmodulin plays a pivotal role in cellular regulation. *Science* 207: 19–27, 1980.

CHOPRA, I. J., J. SACK, AND D. A. FISHER. Circulating 3,3',5'-triiodothyronine (reverse T_3) in the human newborn. *J. Clin. Invest.* 55: 1137–1141, 1975.

CHOPRA, I. J., D. H. SOLOMON, U. CHOPRA, S. Y. WU, D. A. FISHER, AND Y. NAKAMURA. Pathways of metabolism of thyroid hormones. *Recent Prog. Horm. Res.* 34: 521–567, 1978.

CHRISTENSEN, A. K. Leydig cells. In: *Handbook of Physiology*, edited by R. O. Greep and E. B. Astwood. Washington, D.C.: American Physiological Society, 1975, sect. 7, vol. 5, p. 57.

CLARK, J. H., J. W. HARDIN, S. A. McCORMACK, AND H.

A. PADYKULA. Mechanism of action of estrogen antagonist: relationship to estrogen receptor binding and hyperestrogenization. *Prog. Cancer Res. Ther.* 10: 107, 1978.

CODY, V. Thyroid hormones: crystal structure, molecular conformation, binding, and structure-function relationship. *Recent Prog. Horm. Res.* 34: 437–475, 1978.

COHEN, S., H. USHIRO, C. STOSCHECK, AND M. CHINKERS. A native 170,000 epidermal growth factor receptor-kinase complex from shed plasma membrane vesicles. *J. Biol. Chem.* 257: 1523–1531, 1982.

COHEN, S., AND J. M. TAYLOR (Part I), S. COHEN, AND C. R. SAVAGE, JR. (Part II). Part I. Epidermal growth factor: chemical and biological characterization. Part II. Recent studies on the chemistry and biology of epidermal growth factor. *Recent Prog. Horm. Res.* 30: 533–574, 1974.

CONN, P. M., J. MARIAN, M. McMILLIAN, J. STERN, D. ROGERS, M. HAMBY, A. PENNA, AND E. GRANT. Gonadotropin-releasing hormone action in the pituitary: a three step mechanism. *Endocrinol. Rev.* 2: 174–185, 1981.

COOKE, N. E., D. COIT, R. I. WEINER, J. D. BAXTER, AND J. A. MARTIAL. Structure of cloned DNA complementary to rat prolactin messenger RNA. *J. Biol. Chem.* 255: 6502–6510, 1980.

COOPER, C. W., R. M. BOLMAN, III, W. M. LINEHAN, AND S. A. WELLS, JR. Interrelationships between calcium, calcemic hormones and gastrointestinal hormones. *Recent Prog. Horm. Res.* 34: 259–283, 1978.

COOPER, G. M., S. OKENQUIST, AND L. SILVERMAN. Transforming activity of DNA of chemically transformed and normal cells. *Nature* 284: 418–421, 1980.

COWIE, A. T., I. A. FORSYTH, AND I. C. HART. Hormonal control of lactation. New York: Springer-Verlag, 1980.

CSAPO, A. In: *Progesterone: Its Regulatory Effects on the Myometrium*. New York: Ciba Foundation Study Group No. 34, 1969.

DALLMAN, M. F. Adrenal feedback on stress-induced corticoliberin (CRF) and corticotropin (ACTH) secretion. In: *Interactions Within the Brain-Pituitary-Adrenocortical System*, edited by M. T. Jones, B. Gillham, M. F. Dallman, and S. Chattopadahyay. New York: Academic Press, 1979, p. 149–162.

DAUGHADAY, W. H., Z. LARON, A. PERTZELAN, AND J. N. HEINS. Defective sulfation factor generation: a possible etiological link in dwarfism. *Trans. Assoc. Am. Physician* 82: 129–140, 1969.

DAVIES, I. J., AND K. J. RYAN. Comparative endocrinology of gestation. *Vitam. Horm.* 30: 223–279, 1972.

DAVIS, J. O., AND R. H. FREEMAN. Mechanisms regulating renin release. *Physiol. Rev.* 56: 1–56, 1976.

DEGROOT, L. J., AND H. NIEPOMNISZCZE. Biosynthesis of thyroid hormone: basic and clinical aspects. *Metabolism* 26: 665–718, 1977.

DELUCA, H. F., AND H. K. SCHNOES. Metabolism and mechanism of action of vitamin D. *Annu. Rev. Biochem.* 45: 631–666, 1976.

DE MEYTS, P., A. R. BIANCO, AND J. ROTH. Site-site interactions among insulin receptors. *J. Biol. Chem.* 251: 1877–1888, 1976.

DHOM, G. The prepuberal and puberal growth of the adrenal (adrenarche). *Beitr. Pathol.* 150: 357–377, 1973.

DICZFALUSY, E., AND P. TROEN. Endocrine functions of the human placenta. *Vitam. Horm.* 19: 229–311, 1961.

DIEZFALUSY, E. The feto-placental unit. *Excerpta Med. Int. Congr. Ser.* 183: 65–109, 1964.

DiZEREGA, G. S., AND G. D. HODGEN. Folliculogenesis in the primate ovarian cycle. *Endocr. Rev.* 2: 27–49, 1981.

DORRINGTON, J. H. Pituitary and placental hormones. In: *Reproduction and Human Welfare: A Challenge to Research*, edited by C. R. Austin and R. V. Short. Cambridge: Cambridge University Press, 1979.

DOUSA, T. P., AND H. VALTIN. Cellular actions of vasopressin in the mammalian kidney. *Kidney Int.* 10: 46–63, 1976.

DOYLE, J. B. Exploratory culdotomy for observation of tubo-ovarian physiology at ovulation time. *Fertil. Steril.* 2: 475–486, 1951.

DUMONT, J. E., J. M. BOEYNAEMS, C. DECOSTER, C. ERNEAU, F. LAMY, R. LECOCQ, J. MOCKEL, J. UNGER, AND J. VAN SANDE. Biochemical mechanisms in the control of thyroid function and growth. *Adv. Cyclic Nucleotide Res.* 9: 723–734, 1978.

DUNLOP, D. Eighty-six cases of Addison's disease. *Br. Med. J.* 2: 887–891, 1963.

DUNN, F. L., T. J. BRENNAN, A. E. NELSON, AND G. L. ROBERTSON. The role of blood osmolality and volume in regulating vasopressin secretion in the rat. *J. Clin. Invest.* 52: 3212–3219, 1973.

DUNN, M. J., AND V. L. HOOD. Prostaglandins and the kidney. *Am. J. Physiol.* 233: F169–F184, 1977.

EDELMAN, I. S., AND D. MARVER. Mediating events in the action of aldosterone. *J. Steroid Biochem.* 12: 219–224, 1980.

EDWARDS, R. G. Maturation *in vitro* of mouse, sheep, cow, pig, Rhesus monkey, and human ovarian oocytes. *Nature* 208: 349–351, 1965.

EDWARDS, R. G. Follicular fluid. *J. Reprod. Fertil.* 37: 189–219, 1974.

EHRHARDT, A. A., AND H. F. L. MEYER-BAHLBURG. Effects of prenatal sex hormones on gender-related behavior. *Science* 211: 1312–1318 1981.

EIPPER, B. A., AND R. E. MAINS. Structure and biosynthesis of proadrenocorticotropin/endorphin and related peptides. *Endocr. Rev.* 1: 1–27, 1980.

EK, B., B. WESTERMARK, A. WASTESON, AND C- H. HELDIN. Stimulation of tyrosine-specific phosphorylation by platelet-derived growth factor. *Nature* 295: 419–420, 1982.

ENGELAND, W. C., AND M. F. DALLMAN. Compensatory adrenal growth is neurally mediated. *Neuroendocrinology* 19: 352–362, 1975.

ERSPAMER, V., AND P. MELCHIORRI. Active polypeptides: from amphibian skin to gastrointestinal tract and brain of mammals. *Trends Pharmacol. Sci.* 1: 391–395, 1980.

FARFEL, Z., A. S. BRICKMAN, H. R. KASLOW, V. M. BROTHERS, AND H. R. BOURNE. Defect of receptor-cyclase coupling protein in pseudohypoparathyroidism. *N. Engl. J. Med.* 303: 237–242, 1980.

FAWCETT, D. W. Ultrastructure and function of the Sertoli cell. In: *Handbook of Physiology*, edited by R. O. Greep and E. B. Astwood. Washington, D.C.: American Physiological Society, 1975, sect. 7, vol. 5, p. 21.

FELIG, P. Amino acid metabolism in man. *Annu. Rev. Biochem.* 44: 933–955, 1975.

FISHER, D. A., J. H. DUSSAULT, J. SACK, AND I. J. CHOPRA. Ontogenesis of hypothalamic-pituitary-thyroid function and metabolism in man, sheep, and rat. *Recent Prog. Horm. Res.* 33: 59–116, 1977.

FISHER, L. A., J. RIVIER, C. RIVIER, J. SPIESS, J. VALE, AND M. R. BROWN. Corticotropin-releasing factor (CRF): central effects on mean arterial pressure and heart rate in rats. *Endocrinology* 110: 2222–2224, 1982.

FITZSIMONS, J. T. Thirst. *Physiol. Rev.* 52: 468, 1972.

FOLKMAN, J., E. MERLER, C. ABERNATHY, AND G. WILLIAMS. Isolation of a tumor factor responsible for angiotenesis. *J. Exp. Med.* 133: 275–288, 1971.

FORD, C. E., K. W. JONES, O. J. MILLER, U. MITTWOCH, L. S. PENROSE, M. RIDLER, AND A. SHAPIRO. The chromosomes in a patient showing both mongolism and the Klinefelter syndrome. *Lancet* 1: 709–710, 1959.

FRANCHIMONT, P., S. CHARI, M. T. HAZEE-HAGELSTEIN, M. L. DEBRUCHE, AND S. DURAISWAMI. Evidence for the existence of inhibin. In: *The Testis in Normal and Infertile Men*, edited by P. Troen and H. Nankin. New York: Raven Press, 1977, p. 253–270.

FRASER, D. R. Regulation of the metabolism of vitamin D. *Physiol. Rev.* 60: 551–613, 1980.

FROESCH, E. R., J. ZAPF, T. K. AUDHYA, E. BEN-PORATH, B. J. SEGEN, AND K. D. GIBSON. Nonsuppressible insulin-like activity and thyroid hormones: Major pituitary-dependent sulfation factors for chick embryo cartilage. *Proc. Natl. Acad. Sci. U.S.A.* 73: 2904–2908, 1976.

FUCHS, F., AND A. KLOPPER. *Endocrinology of Pregnancy.* New York: Harper & Row, 1977.

GANN, D. S., D. G. WARD, AND D. E. CARLSON. Neural control of ACTH: a homeostatic reflex. *Recent Prog. Horm. Res.* 34: 357–400, 1978.

GARRISON, J. C., M. K. BORLAND, V. A. FLORIO, AND D. A. TWIBLE. The role of calcium ion as a mediator of the effects of angiotensin II, catecholamines, and vasopressin on the phosphorylation and activity of enzymes in isolated hepatocytes. *J. Biol. Chem.* 254: 7147–7156, 1979.

GAUTRON, J. P., E. PATTOU, AND C. KORDON. Occurrence of higher molecular forms of LHRH in fractionated extracts from rat hypothalamus, cortex and placenta. *Mol. Cell. Endocr.* 24: 1–15, 1981.

GENAZZANI, A. R., F. POCOLA, P. NERI, AND P. FIORETTI. Human chorionic somatomammotropin (HCS): plasma levels in normal and pathological pregnancies and their correlation with placental function. *Acta Endocrinol. Suppl.* 167: 1–39, 1972.

GERAS, E., M. J. REBECCHI, AND M. C. GERSHENGORN. Evidence that stimulation of thyrotropin and prolactin secretion by thyrotropin-releasing hormone occur via different calcium-mediated mechanisms: studies with verapamil. *Endocrinology* 110: 901–906, 1982.

GILL, G. N. ACTH regulation of the adrenal cortex. In *Pharmacology of Adrenal Cortical Hormones*, edited by G. N. Gill. Oxford: Pergamon Press, 1979.

GILL, G. N., J. F. CRIVELLO, AND P. J. HORNSBY. Growth, function, and development of the adrenal cortex: insights from cell culture. *Cold Spring Harbor Conf. Cell Proliferation* 9: 461–482, 1982.

GILL, G. N., AND L. D. GARREN. Role of the receptor in the mechanism of action of adenosine 3′:5′-cyclic monophosphate. *Proc. Natl. Acad. Sci. U.S.A.* 68: 786–790, 1971.

GILL, G. N., P. J. HORNSBY, AND M. H. SIMONIAN. Regulation of growth and differentiated function of cultured bovine adrenocortical cells. *Cold Spring Harbor Conf. Cell Proliferations* 6: 701–715, 1979.

GILL, G. N., C. R. ILL, AND M. H. SIMONIAN. Angiotensin stimulation of bovine adrenocortical cell growth. *Proc. Natl. Acad. Sci. U.S.A.* 74: 5569–5573, 1977.

GILL, G. N., AND R. W. MCCUNE. Guanosine 3′,5′-monophosphate-dependent protein kinase. *Curr. Top. Cell. Regul.* 15: 1–45, 1979.

GOEDDEL, D. V., H. L. HEYNEKER, T. HOZUMI, R. ARENTZEN, K. ITAKURA, D. G. YANSURA, M. J. ROSS, G. MIOZZARI, R. CREA, AND P. H. SEEBURG. Direct expression in *Escherichia coli* of a DNA sequence coding for human growth hormone. *Nature* 281: 544–548, 1979.

GORDON, J. W., AND F. H. RUDDLE. Mammalian gonadal determination and gametogenesis. *Science* 211: 1265–1271, 1981.

GOSPODAROWICZ, D. Purification of a fibroblast growth factor from bovine pituitary. *J. Biol. Chem.* 250: 2515–2520, 1975.

GOSPODAROWICZ, D., H. BIALECKI, AND G. GREENBURG. Purification of the fibroblast growth factor activity from bovine brain. *J. Biol. Chem.* 253: 3736–3743, 1978.

GOSPODAROWICZ, D., G. GREENBURG, H. BIALECKI, AND B. R. ZETTER. Factors involved in the modulation of cell proliferation in vivo and in vitro: the role of fibroblast and epidermal growth factors in the proliferative response of mammalian cells. *In Vitro* 14: 85–118, 1978.

GOSPODAROWICZ, D., AND K. K. THAKRAL. Production of a corpus luteum angiogenic factor responsible for proliferation of capillaries and neovascularization of the corpus luteum. *Proc. Natl. Acad. Sci. U.S.A.* 75: 847–851, 1978.

GRABER, A. L., R. L. NEY, W. E. NICHOLSON, D. P. ISLAND, AND G. W. LIDDLE. Natural history of pituitary-adrenal recovery following long-term suppression with corticosteroids. *J. Clin. Endocrinol. Metab.* 25: 11–16, 1965.

GREEP, R. O. The female reproductive system. In: *Reproduction and Human Welfare: A Challenge to Research*, edited by R. O. Greep, M. A. Koblinsky, and F. S. Jaffe. London: MIT Press, 1976, p. 81–164.

GREER, M. A., Y. GRIMM, AND H. STUDER. Qualitative changes in the secretion of thyroid hormones induced by iodine deficiency. *Endocrinology* 83; 1193–1198, 1968.

GREGORY, H. Isolation and structure of urogastrone and its relationship to epidermal growth factor. *Nature* 257: 325–327, 1975.

GREULICH, W. W., AND S. E. PYLE. *Radiographic Atlas of Skeletal Development of the Hand and Wrist.* Stanford, CA: Stanford University Press, 1959.

GRONER B., H. PONTA, M. BEATO, AND N. E. HYNES. The proviral DNA of mouse mammary tumor virus—its use in the study of the molecular details of steroid hormone action. *Mol. Cell. Endocr.*, 32: 101–116, 1983.

GRUMBACH, M. M., AND M. L. BARR. Cytologic tests of chromosomal sex in relation to sexual anomalies in man. *Recent Prog. Horm. Res.* 14: 255–334, 1958.

GRUMBACH, M. M., S. L. KAPLAN, J. J. SCIARRA, AND I. M. BURR. Chorionic growth hormone-prolactin (CGP): secretion, disposition, biologic activity in man, and postulated function as the "growth hormone" of the second half of pregnancy. *Ann. N.Y. Acad Sci.* 148: 501–531, 1968.

GRUMBACH, M. M., AND J. J. VANWYK. Disorders of sex differentiation. In: *Textbook of Endocrinology*, edited by R. H. Williams. Philadelphia: W. B. Saunders, 1974, ch. 8.

GUILLEMIN, R., P. BRAZEAU, P. BOHLEN, F. ESCH, N. LING, AND W. B. WEHRENBERG. Growth hormone-releasing factor from a human pancreatic tumor that caused acromegaly. *Science* 218: 585–587, 1982.

HABENER, J. F., AND J. T. POTTS, JR. Biosynthesis of parathyroid hormone. *N. Engl. J. Med.* 299: 580–585, 635–644, 1978.

HADDAD, J. G., JR., AND J. WALGATE. 25-Hydroxyvitamin D transport in human plasma. Isolation and partial characterization of calcifidiol-binding protein. *J. Biol. Chem.* 251: 4803–4809, 1976.

HAFEZ, E. S. E. Human reproductive physiology. *Ann Arbor Science* 94, 1978.

HAMBERG, M. Quantitative studies on PG synthesis in man. 3. Excretion of the major urinary metabolite of $PGF_{1\alpha}$ and $F_{2\alpha}$ during pregnancy. *Life Sci.* 14: 247–252, 1974.

HARRISON, R. J. The structure of the ovary. C. Mammalian. In: *The Ovary*, edited by S. Zuckerman, A. M. Mandl, and P. Eckstein. London: Academic Press, 1961, p. 143–182.

HASELTINE, F. P., AND S. OHNO. Mechanisms of gonadal differentiation. *Science* 211: 1272–1278, 1981.

HAUSSLER, M. R., AND T. A. McCAIN. Basic and clinical concepts related to vitamin D metabolism and action. (Two parts) *N. Engl. J. Med.* 297: 974–983, 1041–1050, 1977.

HAUSSLER, M. R., J. W. PIKE, J. S. CHANDLER, S. C. MANOLAGAS, AND L. J. DEFTOS. Molecular action of 1,25-dihydroxyvitamin D_3: new cultured cell models. *Ann. N.Y. Acad. Sci.* 372: 501–517, 1981.

HEBB, C., AND J. L. LINZELL. Innervation of the mammary gland. A histochemical study in the rabbit. *Histochem. J.* 2: 491–505, 1970.

HELLER, C. G., AND Y. CLERMONT. Kinetics of the germinal epithelium in man. *Recent Prog. Horm. Res.* 20: 545–575, 1964.

HENNEN, G., J. G. PIERCE, AND P. FREYCHET. Human chorionic thyrotropin: further characterization and study of its secretion during pregnancy. *J. Clin. Endocrol. Metab.* 29: 518–594, 1969.

HEUSON, J. C., AND A. COURE. Hormone-responsive tumors. In: *Endocrinology and Metabolism*, edited by P. Felig, J. D. Baxter, A. E. Broadus, AND L. A. Frohman. New York: McGraw-Hill, 1981, p. 1275–1303.

HINKLE, P. M., M. H. PERRONE, AND A. SCHONBRUNN. Mechanism of thyroid hormone inhibition of thyrotropin-releasing hormone action. *Endocrinology* 108: 199–205, 1981.

HIRATA, F., AND J. AXELROD. Phospholipid methylation and biological signal transmission. *Science* 209: 1082–1090, 1980.

HIRATA, F., E. SCHIFFMANN, K. VENKATASUBRAMAN-IAN, D. SALOMON, AND J. AXELROD. A phospholipase A_2 inhibitory protein in rabbit neutrophils induced by glucocorticoids. *Proc. Natl. Acad. Sci. U.S.A.* 77: 2533–2536, 1980.

HJALMARSON, A. C., C. F. WHITFIELD, AND H. E. MORGAN. Hormonal control of heart function and myosin ATPase activity. *Biochem. Biophys. Res. Commun.* 41: 1584–1589, 1970.

HOLICK, M. F., AND M. B. CLARK. The photobiogenesis and metabolism of vitamin D. *Fed. Proc.* 37: 2567–2574, 1978.

HOLLEY, R. W. Control of growth of mammalian cells in cell culture. *Nature* 258: 487–490, 1975.

HOLLEY, R. W., R. ARMOUR, AND J. H. BALDWIN. Density-dependent regulation of growth of BSC-1 cells in cell culture: growth inhibitors formed by the cells. *Proc. Natl. Acad. Sci. U.S.A.* 75: 1864–1866, 1978.

HORNSBY, P. J. Regulation of cytochrome P-450-supported 11β-hydroxylation of deoxycortisol by steroids, oxygen, and antioxidants in adrenocortical cell cultures. *J. Biol. Chem.* 255: 4020–4027, 1980.

HORNSBY, P. J., AND J. F. CRIVELLO. Role of lipid peroxidation and biological antioxidants in the function of the adrenal cortex. *Mol. Cell. Endocr.* 30: 1–20, 30: 123–147, 1983.

HORNSBY, P. J., AND G. N. GILL. Hormonal control of adrenocortical cell proliferation. Desensitization to ACTH and interaction between ACTH and fibroblast growth factor in bovine adrenocortical cell cultures. *J. Clin. Invest.* 60: 342–352, 1977.

HORNSBY, P. J., AND G. N. GILL. Characterization of adult bovine adrenocortical cells throughout their life span in tissue culture. *Endocrinology* 102: 926–936, 1978.

HOY, J. F. Y., P. A. McGRATH, AND P. T. HALE. Electrophoretic analysis of multiple forms of rat cardiac myosin: effects of hypophysectomy and thyroxine replacement. *J. Mol. Cardiol.* 10: 1053–1076, 1977.

HSUEH, A. J. W., AND P. B. C. JONES. Extrapituitary actions of gonadotropin-releasing hormone. *Endocr. Rev.* 2: 437–461, 1981.

IMPERATO-McGINLEY, J., AND R. E. PETERSON. Male pseudohermaphroditism: the complexities of male phenotypic development. *Am. J. Med.* 61: 251, 1976.

INGLE, D. K., AND HIGGINS, G. M. Regeneration of adrenal gland following enucleation. *Am. J. Med. Sci.* 196: 232–240, 1938.

INGRAM, D. L. Atresia. In: *The Ovary*, edited by S. Zuckerman, A. M. Mandl, and P. Eckstein. London: Academic Press, 1962, p. 247–273.

ISMAIL-BEIGI, F., AND I. S. EDELMAN. The mechanism of thyroid calorigenesis: role of active sodium transport. *Proc. Natl. Acad. Sci. U.S.A.* 67: 1071–1078, 1970.

IZUMI, M., AND P. R. LARSEN. Triiodothyronine, thyrozine, and iodine in purified thyroglobulin from patients with Graves' disease. *J. Clin. Invest.* 59: 1105–1112, 1977.

JAFFE, R. B., AND W. R. KEYE, JR. Estradiol augmentation of pituitary responsiveness to gonadotropin-releasing hormone in women. *J. Clin. Endocrinol. Metab.* 39: 850–855, 1974.

JELLIFFE, D. B., AND E. F. P. JELLIFFE. *Human Milk in the Modern World.* Oxford, UK: Oxford University Press, 1978.

JENSEN, E. V. The oestrogens. In: *Reproduction in Mammals. 7. Mechanism of Hormone Action*, edited by C. R. Austin and R. V. Short. Cambridge, UK: Cambridge University Press, 1979.

JOHANSSON, E. D. B. Progesterone levels in peripheral plasma during the luteal phase of the normal human menstrual cycle measured by a rapid competitive protein binding technique. *Acta Endocrinol.* 61: 592–606, 1969.

JOSIMOVICH, J. B., AND J. H. MacLAREN. Presence in the human placenta and term serum of a highly lactogenic substance immunologically related to pituitary growth hormone. *Endocrinology* 71: 209–220, 1962.

JOSSO, N., J-Y. PICARD, AND D. TRAN. The antimuellerian hormone. *Recent Prog. Horm. Res.* 33: 117–167, 1977.

JOST, A. Problems of fetal endocrinology: the gonadal and

hypophyseal hormones. *Recent Prog. Horm. Res.* 8: 379–418, 1953.

JOST, A. Embryonic sexual differentiation. In: *Hermaphroditism, Genital Anomalies and Related Endocrine Disorders*, edited by H. W. Jones and W. W. Scott. Baltimore: Williams & Wilkins, 1971.

KAHN, C. R., K. L. BAIRD, D. B. JARRETT, AND J. S. FLIER. Direct demonstration that receptor crosslinking or aggregation is important in insulin action. *Proc. Natl. Acad. Sci. U.S.A.* 75: 4209–4213, 1978.

KAO, C. Y. In: *Cellular Biology of the Uterus*, edited by R. M. Wynn. New York: Appleton-Century-Croft, 1967.

KAPLAN, S. L., E. GURPIDE, J. J. SCIARRA, AND M. M. GRUMBACH. Metabolic clearance rate and production rate of chorionic growth hormone-prolactin in late pregnancy. *J. Clin. Endocrinol. Metab.* 28: 1450–1460, 1968.

KASUGA, M., Y. ZICK, D. L. BLITH, F. A. KARLSSON, H. U. HARING, AND C. R. KAHN. Insulin stimulation of phosphorylation of the β subunit of the insulin receptor. *J. Biol. Chem.* 257: 9891–9894, 1982.

KENDALL, J. Z., J. R. G. CHALLIS, I. C. HART, C. T. JONES, M. D. MITCHELL, J. W. K. RITCHIE, J. S. ROBINSON, AND G. D. THORBURN. Steroid and prostaglandin concentrations in the plasma of pregnant ewes during infusion of adrenocorticotrophin or dexamethasone to intact or hypophysectomized foetuses. *J. Endocrinol.* 75: 59–71, 1977.

KEYE, W. R., JR, AND R. B. JAFFE. Strength-duration characteristics of estrogen effects on gonadotropin response to gonadotropin-releasing hormone in women. I. Effects of varying duration of estradiol administration. *J. Clin. Endocrinol. Metab.* 41: 1003–1008, 1975.

KING, A. C., AND P. CUATRECASAS. Resolution of high and low affinity epidermal growth factor receptors. *J. Biol. Chem.* 3053–3060, 1982.

KLOPPER, A., AND F. DICZFALUSY. *Foetus and Placenta.* Oxford, UK: Blackwell, 1969.

KNOBIL, E. The neuroendocrine control of the menstrual cycle. *Recent Prog. Horm. Res.* 36: 53–88, 1980.

KOERING, M. J., R. WOLF, AND R. K. MEYER. Morphological and functional evidence for corpus luteum activity during late pregnancy in the Rhesus monkey. *Endocrinology* 93: 686–693, 1973.

KORENMAN, S. G., D. TULCHINSKY, AND L. W. EATON, JR. Radio-ligand procedures for estrogen assay in normal and pregnancy plasma. Acta Endocrinol. Suppl. 147: 291–304, 1970.

KREY, L. C., AND J. W. EVERETT. Multiple ovarian responses to single estrogen injections early in rat estrous cycles: impaired growth, luteotropic stimulation and advanced ovulation. *Endocrinology* 93: 377–384, 1973.

LAU, H. L. Testing for pregnancy. In: *Practice of Medicine.* New York: Harper & Row, 1975, vol. 3, ch. 29, p. 1–13.

LeMAIRE, W. C., P. W. CONLY, A. MOFFETT, AND W. W. CLEVELAND. Plasma progesterone secretion by the corpus luteum of term pregnancy. *Am. J. Obstet. Gynecol.* 108: 132–134, 197.

LEVI-MONTALCINI, R., AND U. ANGELETTI. Nerve growth factor. *Physiol. Rev.* 48: 53–569, 1968.

LEVI-MONTALCINI, R., R. REVOLTELLA, AND P. CALISSANO. Microtubule proteins in the nerve growth factor mediated response. *Recent Prog. Horm. Res.* 30: 635–669, 1974.

LEWIS, U. J., R. N. P. SINGH, G. F. TUTWILER, M. B. SIGEL, E. F. VANDERLAAN, AND W. P. VANDERLAAN. Human growth hormone: a complex of proteins. *Recent Prog. Horm. Res.* 36: 477–508, 1980.

LI, C. H., J. S. DIXON, AND D. CHUNG. Primary structure of the human chorionic somatomammotropin (HCS) molecule. *Science* 173: 56–58, 1971.

LIDDLE, G. W., D. ISLAND, AND C. K. MEADOR. Normal and abnormal regulation of corticotropin secretion in man. *Recent Prog. Horm. Res.* 18: 125–166, 1962.

LIGGINS, G. C., R. J. FAIRCLOUGH, AND S. A. GRIEVES. Parturition in the sheep. In: *The Fetus and Birth*, edited by

M. O'Connor and J. Knight. Amsterdam and New York: Elsevier/Excerpta Medica, 1977, p. 5–30.

LINGAPPA, V. R., AND G. BLOBEL. Early events in the biosynthesis of secretory and membrane proteins: the signal hypothesis. *Recent Prog. Horm. Res.* 36: 451–475, 1980.

LINTERN-MOORE, S., H. PETERS, G. P. M. MOORE, AND M. FABER. Follicular development in the infant human ovary. *J. Reprod. Fertil.* 39: 53–64, 1974.

LIPSETT, M. B., H. WILSON, M. A. KIRSCHNER, S. G. KORENMAN, L. M. FISHMAN, G. A. SARFATY, AND C. W. BARDIN. Studies on Leydig cell physiology and pathology: secretion and metabolism of testosterone. *Recent Prog. Horm. Res.* 22: 245–281, 1966.

LLOYD, C. W., J. LOBOTSKY, D. T. BAIRD, J. A. McCRACKEN, AND J. WEISZ. Concentration of unconjugated estrogens, androgens, and gestagens in ovarian and peripheral venous plasma of women: the normal menstrual cycle. *J. Clin. Endocrinol. Metab.* 32: 155–166, 1971.

LOZZIO, B. B., C. B. LOZZIO, E. G. BAMBERGER, AND S. V. LAIR. Regulators of cell division: endogenous mitotic inhibitors of mammalian cells. *Int. Rev. Cytol.* 42: 1–47, 1975.

LUDENS, J. H., AND D. D. FANESTIL. The mechanism of aldosterone function. In: *Pharmacology of Adrenal Cortical Hormones*, edited by G. N. Gill. Oxford: Pergamon Press, 1979, p. 143–185.

LUDWIG, K. S., AND M. HOROWITZ. Anovular corpus luteum in a woman taking an oral contraceptive. *Obstet. Gynecol.* 33: 696–702, 1969.

LYON, M. F. X-chromosome inactivation and development patterns in mammals. *Biol. Rev.* 47: 1, 1972.

MacDONALD, P. C., J. M. GRODIN, AND P. K. SIITERI. Dynamics of androgen and oestrogen secretion. In: *Control of Gonadal Steroid Secretion*, edited by D. T. Baird and J. A. Strong. Edinburgh: Edinburgh Univ. Press, 1971, p. 158.

MacINTYRE, I., C. J. HILLYARD, P. K. MURPHY, J. J. REYNOLDS, R. E. G. DAS, AND R. K. CRAIG. A second plasma calcium-lowering peptide from the human calcitonin precursor. *Nature* 300: 460–462, 1982.

MacLUSKY, N. J., AND F. NAFTOLIN. Sexual differentiation of the central nervous system. *Science* 211: 1294–1303, 1981.

MAINWARING, W. I. P. The mechanism of action of androgens. In: *Monographs in Endocrinology*. New York: Springer Verlag, 1977.

MAINS, R. E., B. A. EIPPER, AND N. LING. Common precursor to corticotropins and endorphins. *Proc. Natl. Acad. Sci. U.S.A.* 74: 3014–3018, 1977.

MAJZOUB, J. A., H. M. KRONENBERG, J. T. POTTS, JR., A. RICH, AND J. F. HABENER. Identification and cell-free translation of mRNA coding for a precursor of parathyroid secretory protein. *J. Biol. Chem.* 254: 7449–7455, 1979.

MALPAS, P. Postmaturity and malformation of the foetus. *J. Obstet. Gynecol. Br. Empire* 40: 1046–1053, 1933.

MANOLAGAS, S. C., D. W. BURTON, AND L. J. DEFTOS. 1,25-Dihydroxyvitamin D_3 stimulates the alkaline phosphatase activity of osteoblast-like cells. *J. Biol. Chem.* 256: 7115–7117, 1981.

MARCHANT, B., J. F. H. LEES, AND W. D. ALEXANDER. Antithyroid drugs. *Pharmacol. Ther. [B]* 3: 305–348, 1978.

MARKEE, J. E. Menstruation in intraocular endometrial transplants in the Rhesus monkey. *Contrib. Embrol.* 24: 223, 1940.

MARTIAL, J. A., J. D. BAXTER, H. M. GOODMAN, AND P. H. SEEBURG. Regulation of growth hormone messenger RNA by thyroid and glucocorticoid hormones. *Proc. Natl. Acad. Sci. U.S.A.* 74: 1816–1820, 1977.

MARVER, D., J. STEWART, J. W. FUNDER, D. FELDMAN, AND I. S. EDELMAN. Renal aldosterone receptors: studies with [^{3}H]aldosterone and the antimineralocorticoid [^{3}H]spirolactone (SC-26304). *Proc. Natl. Acad. Sci. U.S.A.* 71: 1431–1435, 1974.

MASUI, H., AND L. D. GARREN. Inhibition of replication in functional mouse adrenal tumor cells by adrenocorticotropic hormone mediated by adenosine 3':5'-cyclic monophosphate. *Proc. Natl. Acad. Sci. U.S.A.* 68: 3208–3210, 1971.

MATSUO, H., Y. BABA, R. M. G. NAIR, A. ARIMURA, AND A. V. SCHALLY. Structure of the porcine LH- and FSH-releasing hormone. I. The proposed amino acid sequence. *Biochem. Biophys. Res. Commun.* 43: 1334–1339.

MATSUOKA, H., P. J. MULROW, R. FRANCO-SAENZ, AND C. H. LI. Effects of β-lipotropin and β-lipotropin-derived peptides on aldosterone production in the rat adrenal gland. *J. Clin. Invest.* 68: 752–759, 1981.

McKUSICK, V. A., AND F. H. RUDDLE. The status of the gene map of the human chromosomes. *Science* 196: 390–405, 1977.

McNUTT, N. S., AND A. L. JONES. Observations on the ultrastructure of cytodifferentiation in the human fetal adrenal cortex. *Lab. Invest.* 22: 513–517, 1970.

MEANS, A. R., AND J. R. DEDMAN. Calmodulin—an intacellular calcium receptor. *Nature* 285: 73–77, 1980.

MIKHAIL, G. Hormone secretion by the human ovary. *Gynecol. Invest.* 1: 5–20, 1970.

MITTWOCH, U. *Genetics of Sex Differentiation.* New York: Academic Press, 1973, p. 1–253.

MIYABO, S., K.-I. YANAGISAWA, E. OOYA, T. HISADA, AND S. KISHIDA. Ontogeny of circadian corticosterone rhythm in female rats: effects of periodic maternal deprivation and food restriction. *Endocrinology* 106: 636–642, 1980.

MIYAKE, A., Y. KAWAMURA, T. AONO, AND K. KURACHI. Changes in plasma LRH during the normal menstrual cycle in women. *Acta Endocrinol.* 93: 257–263, 1980.

MONOD, J., J. P. CHANGEUX, AND F. JACOB. Allosteric proteins and cellular control systems. *J. Mol. Biol.* 6: 306–329, 1963.

MORLEY, J. E. Neuroendocrine control of thyrotropin secretion. *Endocr. Rev.* 2: 396–436, 1981.

MOSS, R. L. Actions of hypothalamic-hypophysiotropic hormones on the brain. *Annu. Rev. Physiol.* 41: 617, 1979.

MOSSMAN, H. W., AND K. L. DUKE. *Comparative Morphology of the Mammalian Ovary.* Madison: University of Wisconsin Press, 1973.

MUNDY, G. R., L. G. RAISZ, R. A. COOPER, G. P. SCHECHTER, AND S. E. SALMON. Evidence for the secretion of an osteoclast stimulating factor in myeloma. *N. Engl. J. Med.* 291: 1041–1046, 1974.

NAKANISHI, S., A. INOUE, T. KITA, M. NAKAMURA, A. C. Y. CHANG, S. N. COHEN, AND S. NUMA. Nucleotide sequence of cloned cDNA for the bovine corticotropin-β-lipotropin precursor. *Nature* 278: 423–427, 1979.

NAKANISHI, S., T. KITA, S. TAII, H. IMURA, AND S. NUMA. Glucocorticoid effect on the level of corticotropin messenger RNA activity in rat pituitary. *Proc. Natl. Acad. Sci. U.S.A.* 74: 3283–3286, 1977.

NALBANDOV, A. V. Comparative aspects of corpus luteum function. *Biol. Reprod.* 2: 7–13, 1970.

NIALL, H. D., M. L. HOGAN, G. W. TREGEAR, G. V. SEGRE, P. HWANG, AND H. FRIESEN. The chemistry of growth hormone and the lactogenic hormones. *Recent Prog. Horm. Res.* 29: 387–416, 1973.

NISHIZUKA, Y., AND Y. TAKAI. Calcium and phospholipid turnover in a new receptor function for protein phosphorylation. *Cold Spring Harbor Conf. Cell Proliferation* 8: 237, 1981.

NODA, M., Y. FURUTANI, H. TAKAHASHI, M. TOYOSATO, T. HIROSE, S. INAYAMA, S. NAKANISHI, AND S. NUMA. Cloning and sequence analysis of cDNA for bovine adrenal preproenkephalin. *Nature* 295: 202–206, 1982.

NOEL, G. L., H. K. SUH, AND A. G. FRANTZ. Prolactin release during nursing and breast stimulation in postpartum and nonpostpartum subjects. *J. Clin. Endocrinol. Metab.* 38: 413–423, 1974.

NOVY, M. J., J. M. COOK, AND L. MANAUGH. Indomethacin block of normal onset of parturition in primates. *Am. J. Obstet. Gynecol.* 118: 412–416, 1974.

O'HARE, M. J., E. C. NICE, AND A. M. NEVILLE. Regulation of androgen secretion and sulfoconjugation in the adult human adrenal cortex: studies with primary monolayer cell cultures. In: *Adrenal Androgens,* edited by A. R. Genazzani, J. H. H. Thijssen, and P. K. Siiteri. New York: Raven Press, 1980, p. 7–25.

OHNO, S. *Sex Chromosomes and Sex Linked Genes,* New York: Springer-Verlag, 1967.

OHNO, S. A hormone-like action of H-Y antigen and gonadal development of XY-XX mosaic males and hermaphrodites. *Hum. Genet.* 35: 21–25, 1976.

O'MALLEY, B. W., AND W. T. SCHRADER. The receptors of steroid hormones. *Sci. Am.* Feb: 32–43, 1976.

OPPENHEIMER, J. H., AND W. H. DILLMANN. Nuclear receptors for thyroid hormone action. *Fed. Proc.* 38: 2154–2161.

OPPENHEIMER, J. H., W. H. DILLMANN, H. L. SCHWARTZ, AND H. C. TOWLE. Nuclear receptors and thyroid hormone action: a progress report. *Fed. Proc.* 38: 2154–2160, 1979.

ORENSTEIN, N. S., H. F. DVORAK, M. H. BLANCHARD, AND M. YOUNG. Nerve growth factor: a protease that can activate plasminogen. *Proc. Natl. Acad. Sci. U.S.A.* 75: 5497–5500, 1978.

OWERBACH, D., W. J. RUTTER, N. E. COOKE, J. A. MARTIAL, AND T. B. SHOWS. The prolactin gene is located on chromosome 6 in humans. *Science* 212: 815–816, 1981.

OWERBACH, D., W. J. RUTTER, J. A. MARTIAL, J. D. BAXTER, AND T. B. SHOWS. Genes for growth hormones, chorionic somatomammotropin, and growth hormone-like gene on chromosome 17 in human. *Science* 209: 289–292, 1980.

PEDERSEN, R. C., AND A. C. BROWNIE. Adrenocortical response to corticotropin is potentiated by part of the amino-terminal region of procorticotropin/endorphin. *Proc. Natl. Acad. Sci. U.S.A.* 77: 2239–2243, 1980.

PERRONE, M. H., AND P. M. HINKLE. Regulation of pituitary receptors for thyrotropin-releasing hormone by thyroid hormones. *J. Biol. Chem.* 253: 5168–5173, 1978.

PHILLIPS, L. S., AND R. VASSILOPOULOU-SELLIN. Somatomedins. (Two parts). *N. Engl. J. Med.* 302: 371–380, 438–446, 1980.

PICKLES, V. R. Prostaglandins in human endometrium. *Int. J. Fertil.* 12: 335, 1967.

PIERCE, J. G., T-H. LIAO, S. M. HOWARD, B. SHOME, AND J. S. CORNELL. Studies on the structure of thyrotropin: its relationship to luteinizing hormone. *Recent Prog. Horm. Res.* 27: 165–212, 1971.

PIERCE, J. G., AND T. F. PARSONS. Glycoprotein hormones: structure and function. *Annu. Rev. Biochem.* 50: 465–495, 1981.

PINTOR, C., A. R. GENAZZANI, G. CARBONI, T. FANNI, S. ORANI, F. FACCLINETTI, AND R. CORDA. Adrenal androgens and pubertal development in physiological and pathological conditions. In: *Adrenal Androgens,* edited by A. R. Genazzani, J. H. H. Thijssen, and P. K. Sitteri. New York: Raven Press, 1980, p. 173–182.

POPP, D. A., F. L. KIECHLE, N. KOTAGAL, AND L. JARETT. Insulin stimulation of pyruvate dehydrogenase in an isolated plasma membrane-mitochondrial mixture occurs by activation of pyruvate dehydrogenase phosphatase. *J. Biol. Chem.* 255: 7540–7543, 1980.

POSNER, B. I., P. A. KELLY, R. P. C. SHIU, AND H. G. FRIESEN. Studies of insulin, growth hormone and prolactin binding: tissue distribution, species variation and characterization. *Endocrinology* 95: 521–531, 1974.

POTTER, E., A. K. NICOLAISEN, E. S. ONG, R. M. EVANS, AND M. G. ROSENFELD. Thyrotropin-releasing hormone exerts rapid nuclear effects to increase production of the primary prolactin mRNA transcript. *Proc. Natl. Acad. Sci. U.S.A.* 78: 6662–6666, 1981.

PURVIS, J. L., J. A. CANICK, J. I. MASON, R. W. ESTABROOK, AND J. L. McCARTHY. Lifetime of adrenal cytochrome P-450 as influenced by ACTH. *Ann. N.Y. Acad. Sci.* 212: 319–343, 1973.

RAE, P. A., N. S. GUTMANN, J. TSAO, AND B. P. SCHIMMER. Mutations in cyclic AMP-dependent protein kinase and corticotropin (ACTH)-sensitive adenylate cyclase affect adrenal steroidogenesis. *Proc. Natl. Acad. Sci. U.S.A.* 76:

1896–1900, 1979.

REES, L. H., P. M. B. JACK, A. L. THOMAS, AND P. W. NATHANIELSZ. Role of foetal adrenocorticotrophin during parturition in sheep. *Nature* 253: 274–275, 1975.

RICHARDS, J. S., J. J. IRELAND, M. C. RAO, G. A. BERNATH, A. R. MIDGLEY, JR., AND L. E. REICHERT, JR. Ovarian follicular development in the rat: hormone receptor regulation by estradiol, follicle stimulating hormone and luteinizing hormone. *Endocrinology* 99: 1562–1570, 1976.

RIGGS, L. A., AND S. S. C. YEN. The pattern of increase in circulating prolactin levels during human gestation. *Am. J. Obstet. Gynecol.* 129: 454–456, 1977.

RINGOLD, G. M., K. R. YAMAMOTO, J. M. BISHOP, AND H. E. VARMUS. Glucocorticoid-stimulated accumulation of mouse mammary tumor virus RNA: increased rate of synthesis of viral RNA. *Proc. Natl. Acad. Sci. U.S.A.* 74: 2879–2883, 1977.

RINIKER, B., P. SIEBER, W. RITTEL, AND H. ZUBER. Revised amino-acid sequences for porcine and human adrenocorticotrophic hormone. *Nature New Biol.* 235: 114–115, 1972.

RIVIER, J., J. SPIESS, M. THORNER, AND W. VALE. Characterization of a growth hormone-releasing factor from a human pancreatic islet tumour. *Nature* 300: 276–278, 1982.

ROBBINS, J. Iodine deficiency, iodine excess and the use of iodine for protection against radioactive iodine. *Thyroid Today* 3, 8: 1–5, 1980.

ROBBINS, J., S-Y. CHENG, M. C. GERSHENGORN, D. GLINOER, H. J. CAHNMANN, AND H. EDELNOCK. Thyroxine transport proteins of plasma: molecular properties and biosynthesis. *Recent Prog. Horm. Res.* 34: 477–519, 1978.

ROBERTS, A. B., M. A. ANZANO, L. C. LAMB, J. M. SMITH, C. A. FROLIK, H. MARQUARDT, G. J. TODARO, AND M. B. SPORN. Isolation from murine sarcoma cells of novel transforming growth factors potentiated by EGF. *Nature* 295: 417–419, 1982.

ROBERTSON, G. L. The regulation of vasopressin function in health and disease. *Recent Prog. Horm. Res.* 33: 333–385, 1977.

ROBERTSON, G. L. Disease of the posterior pituitary. In: *Endocrinology and Metabolism*, edited by P. Felig, J. D. Baxter, A. E. Broadus, and L. A. Frohman. New York: McGraw-Hill, 1981.

ROBERTSON, G. L., E. A. MAHR, S. ATHAR, AND T. SINHA. Development and clinical application of a new method for the radioimmunoassay of arginine vasopressin in human plasma. *J. Clin. Invest.* 52: 2340–2352, 1973.

RODBELL, M. The role of hormone receptors and GTP-regulatory proteins in membrane transduction. *Nature* 284: 17–22, 1980.

ROSEN, J. M., R. J. MATUSIK, D. A. RICHARDS, P. GUPTA, AND J. R. RODGERS. Multihormonal regulation of casein gene expression at the transcriptional and posttranscriptional levels in the mammary gland. *Recent Prog. Horm. Res.* 36: 157–193, 1980.

ROSNER, W. The binding of steroid hormones in human serum. In: *Trace Complements of Plasma, Isolation and Clinical Significance*, edited by G. A. Jamieson and T. J. Greenwalt. New York: A. R. Liss, 1976, vol. 5, p. 377.

ROSS, E. M., AND A. G. GILMAN. Biochemical properties of hormone-sesitive adenylate cyclase. *Annu. Rev. Biochem.* 49: 533–564, 1980.

ROSS, G. T. Clinical relevance of research on the structure of human chorionic gonadotropin. *Am. J. Obstet. Gynecol.* 129: 793–808, 1977.

ROSS, G. T., C. M. CARGILLE, M. B. LIPSETT, P. L. RAYFORD, J. R. MARSHALL, C. A. STROTT, AND D. RODBARD. Pituitary and gonadal hormones in women during spontaneous and induced ovulatory cycles. *Recent Prog. Horm. Res.* 26: 1–62, 1970.

ROSS, R., A. VOGEL, P. DAVIES, E. RAINES, B. KARIYA, M. J. RIVEST, C. GUSTAFSON, AND J. GLOMSET. The platelet-derived growth factor and plasma control cell proliferation. *Cold Spring Harbor Conf. Cell Proliferation* 6: 3–16,

1979.

ROWSON, L. E. A. The evidence for luteolysin. *Br. Med. Bull.* 26: 14–16, 1970.

SALMON, W. D., JR., AND W. H. DAUGHADAY. A hormonally controlled serum factor which stimulates sulfate incorporation by cartilege in vitro. *J. Lab. Clin. Med.* 49: 825, 1957.

SAMAAN, N., S. C. C. YEN, D. GONZALEZ, AND O. H. PEARSON. Metabolic effects of placental lactogen (HPL) in man. *J. Clin. Endocrinol. Metab.* 28: 485–491, 1968.

SAMUELS, H. H., Z. D. HORWITZ, F. STANLEY, J. CASANOVA, AND L. E.. SHAPIRO. Thyroid hormone controls glucocorticoid action in cultured GH_1 cells. *Nature* 268: 254–257, 1977.

SAMUELS, H. H., AND J. S. TSAI. Thyroid hormone action in cell culture: demonstration of nuclear receptors in intact cells and isolated nuclei. *Proc. Natl. Acad. Sci. U.S.A.* 70: 3488–3492, 1973.

SANTEN, R. J., AND C. W. BARDIN. Episodic luteinizing hormone secretion in man: pulse analysis, clinical interpretation, physiologic mechanisms. *J. Clin. Invest.* 52: 2617–2628, 1973.

SANTEN, R. J., AND H. E. KULIN. The male reproductive system. In: *Practice of Pediatrics.* Hagerstown, MD: Harper & Row, 1976, ch. 53, p. 1–44.

SAUER, M. J. Review: hormone involvement in the establishment of pregnancy. *J. Reprod. Fert.* 56: 725–743, 1979.

SAWYER, W. H., AND P. K. T. PANG. Evolution of neurohypophyseal hormones and their function. In: *Neurohypophysis: International Conference on the Neurohypophysis*, edited by A. M. Moses and L. Share. Basel: Karger, 1977.

SCHALLY, A. V., A. ARIMURA, J. KASTIN, H. MATSUO, Y. BABA, T. W. REDDING, R. M. G. NAIR, L. DEBELJUK, AND W. F. WHITE. Gonadotropin-releasing hormone: one polypeptide regulates secretion of luteinizing and follicle-stimulating hormones. *Science* 173: 1036–1038, 1971.

SCHIMMEL, M., AND R. D. UTIGER. Thyroid and peripheral production of thyroid hormones. *Ann. Intern. Med.* 87: 760–768, 1977.

SCHLESSINGER, J., Y. SHECHTER, M. C. WILLINGHAM, AND I. PASTAN. Direct visualization of binding, aggregation, and internalization of insulin and epidermal growth factor on living fibroblastic cells. *Proc. Natl. Acad. Sci. U.S.A.* 75: 2659–2663, 1978.

SCHONBRUNN, A., M. KRASNOFF, J. M. WESTENDORF, AND A. H. TASHJIAN, JR. Epidermal growth factor and thyrotropin-releasing hormone act similarly on a clonal pituitary cell strain. *J. Cell. Biol.* 85: 786, 1980.

SCHRAMM, M., J. ORLY, S. EIMERL., AND M. KORNER. Ccoupling of hormone receptors to adenylate cyclase of different cells by cell fusion. *Nature* 268: 310–313, 1977.

SCHWARZ, B. E., L. MILEWICH, J. M. JOHNSTON, J. C. PORTER, AND P. C. MacDONALD. Initiation of human parturition. V. Progesterone binding substance in fetal membranes. *Obstet. Gynecol.* 48: 685–689, 1976.

SEYBERTH, H. W., G. V. SEGRE, P. HAMET, B. J. SWEETMAN, J. T. POTTS, AND J. A. OATES. Characterization of the group of patients with the hypercalcemia of malignancy who respond to treatment with prostaglandin synthesis inhibitors. *Trans. Assoc. Am. Physicians* 89: 92, 1976.

SHIH, C., L. C. PADHY, M. MURRAY, AND R. A. WEINBERG. Transforming genes of carcinomas and neuroblastomas introduced in mouse fibroblasts. *Nature* 290: 261–264, 1981.

SHOME, B., AND A. F. PARLOW. Human pituitary prolactin (hPRL): the entire linear amino acid sequence. *J. Clin. Endocrinol. Metab.* 45: 1112–1115, 1977.

SHORT, R. V. The biological basis for the contraceptive effects of breast feeding. Workshop on Breast Feeding and Fertility Regulation. Geneva, in press, Feb. 1982.

SIITERI, P. K., AND P. C. MacDONALD. Placental estrogen biosynthesis during human pregnancy. *J. Clin. Endocrinol. Metab.* 26: 751–761, 1966.

SIITERI, P. K., AND P. C. MacDONALD. Role of extraglandular estrogen in human endocrinology. In: *Handbook of Physiology*, edited by R. O. Greep and E. B. Astwood. Baltimore: Williams & Wilkins, 1973, sect. 7, vol. 2, part 1, p. 615.

SIITERI, P. K., AND J. D. WILSON. Testosterone formation and metabolism during male sexual differentiation in the human embryo. *J. Clin. Endocrinol. Metab.* 38: 113–125, 1974.

SILER-KHODR, T. M., AND G. KHODR. Luteinizing hormone releasing factor content of the human placenta. *Am. J. Obstet. Gynecol.* 130: 216–219, 1978.

SIMONIAN, M. H., AND G. N. GILL. Regulation of the fetal human adrenal cortex: effects of adrenocorticotropin on growth and function of monolayer cultures of fetal and definitive zone cells. *Endocrinology* 108: 1769–1779, 1981.

SIMPSON, E. R., J. L. McCARTHY, AND J. A. PETERSON. Evidence that the cycloheximide-sensitive site of adrenocorticotropic hormone action is in the mitochondrion. *J. Biol. Chem.* 253: 3135–3139, 1978.

SINHA, T. K., H. F. DELUCA, AND N. H. BELL. Evidence for a defect in the formation of 1α,25-dihydroxy vitamin D in pseudohypoparathyroidism. *Metabolism* 26: 731–738, 1977.

SMITH, T. J., AND I. S. EDELMAN. The role of sodium transport in thyroid thermogenesis. *Fed. Proc.* 38: 2150–2153, 1979.

SROUGI, M., R. F. GITTES, AND R. H. UNDERWOOD. Influence of exogenous glucocorticoids and ACTH on experimental adrenal autografts. *Invest. Urol.* 17: 265–268, 1980.

STANBURY, S. W. Vitamin D and the syndromes of azotaemic osteodystrophy. *Contrib. Nephrol.* 13: 132–146, 1978.

STEINBERGER, E. Hormonal control of mammalian spermatogenesis. *Physiol. Rev.* 51: 1, 1971.

STEINBERGER, E., AND W. O. NELSON. The effect of hypophysectomy, cryptorchidism, estrogen and androgen upon the level of hyaluronidase in the rat testis. *Endocrinology* 56: 429–444, 1955.

STEINBERGER, E., AND A. STEINBERGER. Spermatogenic function of the testis. In: *Handbook of Physiology*, edited by R. O. Greep and E. B. Astwood. Washington, DC: American Physiological Society, 1975, sect. 7, vol. 5, p. 1.

STEWART, A. F., R. HORST, L. J. DEFTOS, E. C. CADMAN, R. LANG, AND A. E. BROADUS. Biochemical evaluation of patients with cancer-associated hypercalcemia. *N. Engl. J. Med.* 303: 1377–1383, 1980.

STRICKLAND, S., AND J. N. LOEB. Obligatory separation of hormone binding and biological response curves in systems dependent upon secondary mediators of hormone action. *Proc. Natl. Acad. Sci. U.S.A.* 78: 1366–1370, 1981.

STROTT, C. A., T. YOSHIMI, G. T. ROSS, AND M. B. LIPSETT. Ovarian physiology: relationship between plasma LH and steroidogenesis by the follicle and corpus luteum, effect of HCG. *J. Clin. Endocrinol.* 29: 1157–1167, 1969.

SUTHERLAND, E. W., AND T. W. RALL. Fractionation and characterization of a cyclic adenine ribonucleotide formed by tissue particles. *J. Biol. Chem.* 232: 1077–1091, 1958.

SVOBODA, M. E., J. J. VAN WYK, D. G. KNAPPER, R. E. FELLOWS, F. E. GRISSOM, AND R. J. SCHLUETER. Purification of somatomedin C from human plasma: chemical and biological properties, partial sequence analysis and relationship to other somatomedins. *Biochemistry* 19: 790, 1980.

SWANECK, G. E., J. L. NORDSTROM, F. KREUZALER, M-J. TSAI, AND B. W. O'MALLEY. Effect of estrogen on gene expression in chicken oviduct: evidence for transcriptional control of ovalbumin gene. *Proc. Natl. Acad. Sci. U.S.A.* 76: 1049–1053, 1979.

SZABO, M., AND L. A. FROHMAN. Suppression of cold-stimulated thyrotropin secretion by antiserum to thyrotropin-releasing hormone. *Endocrinology* 101: 1023–1033, 1977.

TAUROG, A. Thyroid peroxidase and thyroxine biosynthesis. *Recent Prog. Horm. Res.* 26: 189–247, 1970.

TAYLOR, S., AND J. FOLKMAN. Protamine is an inhibitor of angiogenesis. *Nature* 297: 307–312, 1982.

THAU, R. B., AND J. T. LANMAN. Endocrinology aspects of placental function. In: *The Placenta and Its Maternal Supply Line*, edited by P. Greenwald. Lancaster, PA: Medical and

Technical Publishing Co., 1975.

THORNER, M. O., W. H. MARTIN, A. D. ROGOL, J. L. MORRIS, R. L. PERRYMAN, B. P. CONWAY, S. S. HOWARDS, M. G. WOLFMAN, AND R. M. MacLEOD. Rapid regression of pituitary prolactinomas during bromocriptine treatment. *J. Clin. Endocrinol. Metab.* 51: 438–445, 1980.

THORNER, M. O., C. R. W. EDWARDS, J. P. HANKER, G. ABRAHAM, AND G. M. BESSER. Prolactin and gonadotropin interaction in the male. In: *The Testis in Normal and Infertile Men*, edited by P. Troen and H. Nankin. New York: Raven Press, 1977, p. 351.

TODARO, G. J., J. E. DELARCO, H. MARQUARDT, M. L. BRYANT, S. A. SHERWIN, AND A. H. SLISKI. Polypeptide growth factors produced by tumor cells and virus-transformed cells: a possible growth advantage for the producer cell. *Cold Spring Harbor Conf. Cell Proliferation* 6: 113–127, 1979.

TODARO, G. J., C. FRYLING, AND J. E. De LARCO. Transforming growth factors produced by certain human tumor cells: polypeptides that interact with epidermal growth factor receptors. *Proc. Natl. Acad. Sci. U.S.A.* 77: 5258–5262, 1980.

TODD, A. S. Localization of fibrinolytic activity in tissues. *Br. Med. Bull.* 20: 210, 1964.

TOMKINS, G. M. The metabolic code. *Science* 189: 760–763, 1975.

TOPPER, Y. J. Multiple hormone interactions in the development of mammary gland *in vitro*. *Recent Prog. Horm. Res.* 26: 287–308, 1970.

TSAFRIRI, A., S. H. POMERANTZ, AND C. P. CHANNING. Follicular control of oocyte maturation. In: *Ovulation in the Human*, edited by P. G. Crosignami and D. R. Mishell. London: Academic Press, 1976, p. 31.

TYRREL, J. B., AND J. C. BAXTER. Glucocorticoid therapy. In: *Endocrinology and Metabolism*, edited by P. Felig, J. D. Baxter, A. E. Broadus, and L. A. Frohman. New York: McGraw-Hill, 1981, p. 620.

TYSON, J. E., M. KHOJANDI, J. HUTH, AND B. ANDREASSEN. The influence of prolactin secretion on human lactation. *J. Clin. Endocrinol. Metab.* 40: 764–773, 1975.

UNGER, R. H., R. E. DOBBS, AND L. ORCI. Insulin, glucagon, and somatostatin secretion in the regulation of metabolism. *Annu. Rev. Physiol.* 40: 307, 1978.

USHIRO, H., AND S. COHEN. Identification of phosphotyrosine as a product of epidermal growth factor-activated protein kinase in A-431 cell membranes. *J. Biol. Chem.* 255: 8363–8365, 1980.

VALE, W., P. BRAZEAU, C. RIVIER, M. BROWN, B. BOSS, J. RIVIER, N. BURGUS, N. LING, AND R. GUILLEMIN. Somatostatin. *Recent Prog. Horm. Res.* 31: 365–397, 1975.

VALE, W., R. BURGUS, AND R. GUILLEMIN. On the mechanism of action of TRH: effects of cycloheximide and actinomycin on the release of TSH stimulated in vitro by TRH and its inhibition by thyrosin. *Neuroendocrinology* 3: 34–46, 1968.

VALE, W., J. SPIESS, C. RIVIER, AND J. RIVIER. Characterization of a 41-residue ovine hypothalamic peptide that stimulates secretion of corticotropin and β-endorphin. *Science* 213: 1394–1397, 1981.

VAN LOOK, P. F. A., AND D. T. BAIRD. Review article: regulatory mechanisms during the menstrual cycle. *Eur. J. Obstet. Gynecol. Reprod. Biol.* 11: 121–144, 1980.

VAN WYK, J. J., AND L. E. UNDERWOOD. The somatomedins and their action. In: *Biochemical Actions of Hormones*, edited by G. Litwack. New York: Academic Press, 1978, vol. V.

VANDE WIELE, R. L., J. BOGUMIL, I. DYRENFURTH, M. FERIN, R. JEWELEWICZ, M. WARREN, T. RIZKALLAH, AND G. MIKHAIL. Mechanisms regulating the menstrual cycle in women. *Recent Prog. Horm. Res.* 26: 63–103, 1970.

VILLEE, D. B. Development of endocrine function in the human placenta and fetus. (two parts). *N. Engl. J. Med.* 281: 473–484, 533–541, 1969.

VOLPE, R. Autoimmunity in the endocrine system. In: *Monographs in Endocrinology*, No. 2. Heidelberg: Springer-Verlag, 1981.

WACHTEL, S. S., S. OHNO, G. C. KOO, AND E. A. BOYSE.

Possible role for H-Y antigen in the primary determination of sex. *Nature* 257: 235–236, 1975.

WALTER, R., AND W. H. SIMMONS. Metabolism of neurohypophyseal hormones: considerations from a molecular viewpoint. In: *Neurohypophysis: International Conference on the Neurohypophysis*, edited by A. M. Moses and L. Share. Basel: Karger, 1977.

WARTOFSKY, L., AND K. D. BURMAN. Alterations in thyroid function in patients with systemic illness: the "euthyroid sick syndrome." *Endocr. Rev.* 3: 164–217, 1982.

WASSERMAN, R. H., R. A. CORRADINO, AND C. S. FULLMER. Some aspects of vitamin D action, calcium absorption and vitamin D-dependent calcium-binding protein. *Vitam. Horm.* 32: 299–324, 1976.

WEINTRAUB, B. D., B. S. STANNARD, D. LINNEKIN, AND M. MARSHALL. Relationship of glycosylation to *de novo* thyroid-stimulating hormone biosynthesis and secretion by mouse pituitary tumor cells. *J. Biol. Chem.* 255: 5715–5723, 1980.

WEITZMAN, E. D., R. M. BOYAR, S. KAPEN, AND L. HELLMANN. The relationship of sleep and sleep stages to neuroendocrine secretion and biological rhythms in man. *Recent Prog. Horm. Res.* 31: 399–446, 1975.

WESTERMARK, B., F. A. KARLSSON, AND O. WALINDER. Thyrotropin is not a growth factor for human thyroid cells in culture. *Proc. Natl. Acad. Sci. U.S.A.* 76: 2022–2026, 1979.

WILSON, J. D., F. W. GEORGE, AND J. E. GRIFFIN. The hormonal control of sexual development. *Science* 211: 1278–1284, 1981.

WINTER, J. S., AND C. FAIMAN. Pituitary-gonadal relations in female children and adolescents. *Pediatr. Res.* 7: 948–953, 1973.

WITSCHI, E. Sex chromatin and sex differentiation in human embryos. *Science* 126: 1288–1290, 1957.

WOLFF, J. Iodide goiter and the pharmacologic effects of excess iodide. *Am. J. Med.* 47: 101–124, 1969.

WOLFF, J., AND I. L. CHAIKOFF. The inhibitory action of iodide upon organic binding of iodine by the normal thyroid gland. *J. Biol. Chem.* 172: 855–856, 1948.

WURTMAN, R. J., AND J. AXELROD. Control of enzymatic synthesis of adrenaline in the adrenal medulla by adrenal cortical steroids. *J. Biol. Chem.* 241: 2301–2305, 1966.

WURZEL, J. M., J. S. PARKS, J. E. HERD, AND P. V. NIELSEN. A gene deletion is responsible for absence of human chorionic somatomammotropin. *DNA* 1: 251–257, 1982.

YASUDA, N., M. A. GREER, AND T. AIZAWA. Corticotropin-releasing factor. *Endocr. Rev.* 3: 123–140, 1982.

YEN, S. S. C., AND R. B. JAFFE. (eds.). *Reproductive Endocrinology, Physiology, Pathophysiology and Clinical Management.* Philadelphia: W. B. Saunders, 1978.

YEN, S. S. C., B. L. LASLEY, C. F. WANG, H. LEBLANC, AND T. M. SILER. The operating characteristics of the hypothalamic-pituitary system during the menstrual cycle and observations of biological action of somatostatin. *Recent Prog. Horm. Res.* 31: 321–363, 1975.

YEN, S. S. C., C. C. TSAI, G. VANDENBERG, AND R. REBAR. Gonadotropin dynamics in patients with gonadal dysgenesis: a model for the study of gonadotropin regulation. *J. Clin. Endocrinol. Metab.* 35: 897–904, 1972.

YING, S-Y., N. LING, P. BOHLEN, AND R. GUILLEMIN. Gonadocrinins: peptides in ovarian follicular fluid stimulating the secretion of pituitary gonadotropins. *Endocrinology* 108: 1206–1215, 1981.

ZANDER, J., K. HOLZMANN, A. M. VON MUSTERMANN, B. RUNNENBAUM, AND W. SILVER. In: *The Foeto-placental Unit*, edited by A. Pecile and C. Finzi. Amsterdam: Excerpta Medica Foundation, 1969.

Neurophysiology

CHAPTER 61

Sensory Processes

SIZING UP THE HUMAN BRAIN

Chapter 3 (pp. 28–57) revealed neurophysiology in terms of unitary cellular mechanisms. It showed how nerve cells are structurally and functionally distinguished from other cells and how they perform a specialized function for delivering signals rapidly and reliably without decrement, even over long distances. The following chapters relate how nerve cells, glia, cerebral blood vessels, and ensheathing membranes compose this physiologically distinctive and powerful organ system which literally *is us*. The brain is where we live, where we look after ourselves and cope with the world.

Approximate Numbers of Neurons

By rough approximation, the human nervous system operates some 10 million afferent (input) neurons, 20 billion central neurons, and ½ million efferent (output) neurons (Blinkof and Glezer, 1968). This reveals an overall convergence between sensory intake and motor outflow. More important is the fact that *for every peripheral neuron—sensory and motor—several thousand central neurons function in an exclusively interneuronal universe.* They have only indirect access to either receptors or effectors. Sensory signals enter, motor messages emerge, while beyond this input and output, which looms so large in our consciousness, there exists momentous freedom of choice residing within the central nervous system. It is exactly this freedom of choice that has been expanding in the evolutionary emergence of larger-brained creatures like ourselves. It provides priorities for rapid turnaround responses and also allows for time for reflection, using imagination and memory.

Importance of Evolution and Development

Everything concerning the nervous system can usefully be considered in light of requirements for evolution and development. Three agents, each high in information content, are interdependent: linear molecules (DNA, RNA, and protein) which deliver the information necessary for the organization of an adult; the processes by which that delivery is accomplished; and the ensuing behavior of that individual. The behavior, of course, is dependent on the neurophysiological substrates for sensory processing, perception, judgment, and motor performance.

These three information-laden systems vault over one another in cartwheel fashion throughout evolution. Every revolution is hammered out on the hard anvil of life's experiences. It is *behavior* that is crucial as to whether the DNA will be passed along, as to whether the DNA prescribes the substrate for behavior, as to whether the behavior, in turn, is dependent on adequate development, and so on. And, we are fitted for life not only by the adequacy of our genetic inheritance but also by the adequacy of caregiving by the preceding generation during the process of our early development, including whatever other life experiences we have encountered.

Nervous System Variations through Normal Inheritance

In order to comprehend better the human brain, we need to consider its variability according to the dimensions of normal gene sorting and other factors contributing to individuality. Varieties of offspring of a single human pair are reckoned as follows: random sorting of gametes results in more than 8.39 million varieties of sex cells potentially available in each mate (2^{23}). Fertilization multiplies these two potentialities to yield more than 70.37 trillion different possibilities for each child of the mated pair ($2^{23} \times 2^{23}$). This number of potentially genetically different individuals out of a single human mating is probably greater than the total number of individuals in the hominid line who reached reproductive age throughout the past 5 million years!

This is not the whole of genetic individuality, for there are many individualizing processes that occur during development and throughout life: genetically normal equal crossovers, random, at a high rate; unequal crossovers, massive mutations, the equivalent of antibody diversity; somatic mutations, 10^{-7} per cell division (with a vast number of cell divisions in the brain, there will be many such effects); and alterations by virus particles, including "jumping genes," which abound, operating thousands of times in a lifetime and involving thousands of different viruses; insertion sequences and gene conversion; plus all other mutations induced by thermal agitation, chemical mutagens, radiation, infectious diseases, etc.

Other Factors Increasing Individuality

Furthermore, other factors, such as nutrition, especially during the period of embryogenesis and throughout the whole of gestation and the first 4 years of life, when the brain is growing enormously rapidly, can interfere with an individual living up to his or her genetic potential. For example, about 8% of all monozygotic twins have significantly different body and brain weights due to a process of placental steal, which can occur with monocotyledonous placentas. It is not that one twin gets smarter; the other one does not live up to his or her genetic potential by reason of a relatively inadequate supply of oxygen and nutrients. If zinc and pyridoxal phosphate, among other essential nutrients, are not available on a continuing basis, cells cannot divide, proliferate processes, and build synapses. Some windows of opportunity for growth and development of the brain are only open during certain intervals, if the opportunity is lost, for any reason, it may not be possible to make up for that loss. Each of the effects of deleterious influences and want of critical building materials *multiplies or at least adds to the already astronomical 70.37 trillion individual varieties.* Since roughly half of the human genetic dowry involves instructions for organizing the nervous system, these observations guarantee an extraordinary individuality that must contribute to structurally and functionally distinctive brains. If this were not enough, it is obvious that everything we experience and learn throughout our lives induces further brain individuation.

It is well known to students of anatomy that even gross neuroanatomy is various among presumably normal human brains. The anatomy laboratory and neurosurgical demonstrations illustrate this fact; it is often difficult to find the central sulcus separating primary sensory and motor cortex; a neurosurgeon wants to turn a big flap and probably wants also to stimulate the cortex of a waking patient if he or she considers that it may be necessary to resect brain tissue in some especially critical area.

Stereotactic atlases for estimating the location of subcortical and brain stem structures are usually based on only a very few brains but nevertheless include precautions respecting normal variability. We have produced cinemorphology films on nearly 70 normal human brains and have found no two alike, even on gross morphological criteria. The variability includes structures in all regions of the brain.

Characterization of Global Nervous System Functions

What structural-functional characteristics distinguish the nervous system as an organ system? First, the nervous system has certain structure-function priorities. It is the initial organ to differentiate during embryogenesis; it remains the fastest growing—hence, it is the largest organ system throughout gestation. It begins to govern the rest of the body already *in utero.* In fact, the hypothalamus signals to the pituitary when the baby is ready to be born, and an endocrine signal from the baby is what ordinarily initiates labor.

Throughout life, the nervous system must gather information from and coordinate activities for all other organ systems. The nervous system is designed to serve the *functional integrity* of one individual and to provide a variety of combined visceral and somatic competences for survival and reproductive functions, including the capacity to contribute to the survival of mate and offspring (MacLean, 1982).

Second, the nervous system is composed of by far the largest variety of structurally and functionally differentiated elements, nerve cells which are remarkably specialized in form, chemistry, physiology, pharmacology, immunology, and other properties, including unique responses to injury. It is conservatively estimated that *the human nervous system contains more than 50 million different kinds of nerve cells,* meaning clearly distinguishable cell types (Bullock, 1981).

Third, the principal function of the nervous system is *information processing:* this is a continuous space-time transaction governed by central nervous system activity in order to evaluate sensory and motor events, mainly to regulate both internal and external contributions to subjective existence. What conditions in the outside world present threat or opportunity to this individual at this time? What are the ongoing general and detailed conditions of the body? What physiological events are contributing to states of satisfaction and dissatisfaction? How can visceral and somatic events, and events in the environment be organized so as to optimize internal satisfactions? What priorities can be applied to competing appetites and behavioral options in order that satisfactions may be optimized for the longer-range future?

Fourth, the nervous system is engaged in dynamic mapping strategies concerned with experiences, plans, and behavior regarding parts of the body and parts of the world (Mountcastle, 1975). These strategies are partly based on genetic information, partly on the outcome of structural and developmental processes, and partly on information which the nervous system itself generates, gathers from other sources, evaluates, and stores for the better governance of organs of the body and strategies for the better pursuit of internal satisfactions. The mapping strategies involve sensory and motor fields, with numerous maps for each sensory modality and each motoric strategy. There are maps relating back and forth between cerebrum and cerebellum, between the brain stem and thalamus, between the thalamus and cortex, and throughout the limbic system which, relatively, is rather remote from sensory input and motor outflow.

To accomplish all these balancing-evaluating-governing tasks, the nervous system must monitor signals generated by the nervous system itself, by receptors embedded in various tissues of the body, and by distance receptors lodged near the surface of the body (also embedded in body tissues) which are peculiarly sensitive to certain environmental events. The prime purpose of this massive information gathering, evaluating, and deciding is to secure biological and biologically derived satisfactions for a particular individual at a particular time and to secure advantages toward future satisfactions. As the philosopher Santayana contended, all actions, even the most exalted and altruistic, aim at internal satisfactions.

Information Processing by the Nervous System

What is the nature of nervous system information processing? The most important consideration about information relates to what it is not: information is not matter, and it is not energy, nor is it a combination of the two. Information is entirely different from matter and energy, and it is only secondarily dependent on them for its generation, transmission, storage, and expression.

The role of information in the nervous system is different from that in genetic material and that in computers. Genetic information is represented by a one-dimensional map which specifies the three-dimensional organization of an adult. DNA bears this information in a compact and stable form and is not obliged to manipulate that information; indeed, it is strictly constrained against risking loss of that information. Information in DNA is acted out in serial progressions once in a generation. Information can be added to DNA only adventitiously, and DNA information is normally radically altered only by sexual union. DNA is an example of a "read-only" information system. Computer information, at least until now, lacks a genetic-reproductive or mapping capability independent of outside operators and therefore depends for its design and operation on external information. The nervous system, by contrast, is truly a self-organizing, self-regulating information system which continues to differentiate and to operate with original structure, modified by experience, and with gains in memory stores and programs for action, in favor of its own systems of internal reward and punishment, throughout a lifetime.

Thus, individuality in nervous systems is expressed by genetic uniqueness, structural and functional self-organizing characteristics, and cumulative stores of developmental and life experience. The brain, therefore, becomes progressively, increasingly differentiated during one's life-span.

The shaping of sensory pathways by individuating experience means that the brain develops individualistic (and culture-bound) sensory "lenses" that are invisible to us (because perceptual processes appear to lie some synaptic relays behind them). This implies that there may be as many perceptual worlds as there are individuals and that interindividual differences are more or less exaggerated insofar as individual (and cultural) experiences are more or less different. This creates problems, not only for the sciences that deal with perception but also for those concerned with interindividual and cross-cultural communication and understanding. A great deal of what may be considered skullduggery in the world may be at bottom introduced by differentially conditioned perceptual mismatches. This concept identifies an important "catch 22" for human existence that has only recently become identifiable.

If we understood the biological, psychological, and cultural mechanisms affecting perception, judgment, and behavior, it would make it easier for us to be more tolerant, more willing to be tentative rather than hasty in our responses, and better able to sort out ambiguous situations carefully. We would be better able to comprehend features of the world that do not correspond to our present limited world view. We would realize more vividly than we do already that processes essential to education and maturation, as individuals and as societies, depend on our being tentative about our views of the biological, physical, and social world around us. This introductory section, Sizing Up the Human Brain, therefore is pertinent both for introducing concepts of how we conceive the brain functions and for introducing some of the potentialities of neurosciences, broadly conceived.

The Residence of Consciousness

There is no longer doubt that consciousness is mortal and depends on a living brain; the challenge is to discover how the brain generates and controls consciousness and how consciousness operates in the brain in both receptive and projective modes (Hilgard, 1977). Converging disciplines in the neurosciences continue to contrive more powerful techniques for investigating higher nervous system structure-function relations—for examples, from synaptic plasticity and remodeling of neuronal circuitry to functional organization of primary sensory and motor cortex, integrative processes in multimodality association cortex, brain waves associated with subjective and cognitive functions, and differences in information processing in the left and right cerebral hemispheres. These and related contributions improve professional and public understanding of what constitutes "human nature" and establish foundations for a scientific study of human values (Sperry, 1972).

Neuroscience has traditionally built solid groundwork for convergence between the physical and life sciences, in recent decades between these disciplines and psychology and, increasingly, between all of these disciplines and anthropology and sociology. Artists and thoughtful writers in all cultures of all times, philosophers, historians, anthropologists, sociologists, and psychologists are allies with neuroscientists in exploring the ranges of perception, consciousness, and behavior that reflect both the good and bad news devolving from these observations. It's a bum joke that all we can do in the world is work muscles and glands. It is correct that we can do naught but contract, relax and secrete but, it's precisely when and which muscles we contract and relax and when and where we point our secretions that truly matters. To discriminate in this way we possess a greater degree of freedom and, with freedom, a greater degree of responsibility for ourselves and others than is possessed by any other species, or than is exercised by any culture.

We are above all *human beings*, and it is the functioning of our brain that presents opportunities to us to live up to the claim we make by calling ourselves *Homo sapiens*.

Human behavior controls human evolution. Mankind has long been empirically intervening in the evolution of his own and many other plant and animal species. Through scientific advancement he has gained increasingly self-conscious control over evolutionary processes. It remains to be seen as to whether human wisdom will forefend potential man-made, man-controlled physical devastation that could be globally catastrophic for all higher forms of life. There is no mandate. Our brains put no ceiling on cooperation, mutual faith, trust, empathy, and altruism. It may well be that ignorance of the degrees of freedom available to us holds us in thrall.

Neurosciences provides a moving scientific disclosure of nervous system functions, particularly of higher neural processes once thought beyond reach, beginning to bypass that old conundrum, "How can the brain/mind explore itself?" These endeavors are helping to develop a deeper understanding of the risks and potentialities of human existence. Adequate knowledge of brain processes relating to social perceptions, judgments, and behavior can come none too soon for, as neuroscientist Paul MacLean remarked, "It is too bad the brain wasn't cracked before the atom!"

SENSORY PROCESSING

The adaptation of man to his environment and to the society of his fellow men requires the constant processing of information received from the environment and from individuals within the environment. Any form of activity constitutes an interaction process wherein any kind of movement or action produces change. The detection of such a change via the sensory systems permits the regulation and control of subsequent actions. Much of consciousness, even in the absence of significant sensory inputs, seems to involve sensory imagery. We think in terms of visual and auditory imagery and imagery involving the other senses. Although in the analytic study of sensory processes there is often a tendency to consider them as independent dimensions, our perception of the environment and our relations to it depend in fact upon many complex interactions of the individual sensory modalities (Gibson, 1966). It is our purpose in the material which follows to consider the various important sensory processes individually for the purpose of understanding their nature, acknowledging the fact that they do not function independently but provide a collective basis for consciousness and the control of our activities.

Receptor Organs

Different sensory modalities are characterized by the location and nature of specialized receptors which mediate responses to stimuli appropriate for the transmission of information in each sensory domain. These cells set in motion a chain of events that involves sensory processing and may ultimately give rise to conscious perception. Most cells are affected by certain wavelengths of light, specific ions, protons, CO_2 and various forms of mechanical perturbation. Receptors have specialized through evolutionary selection to be especially sensitive to particular aspects of the stimulating world. They still retain the ability to respond to more prosaic influences, but they are specialized, and their function of converting some physical event into membrane phenomena that can set in train sensory signals is called *transduction* (Changeux and Dennis, 1982; Hammes et al., 1973). Subsequent alterations of the signal, although perhaps less dramatic, are nonetheless capable of being quite subtle and interesting.

Sensory transducers may be spontaneously active, as are photoreceptors, even in the dark, and this makes possible their differential activation by very subtle stimuli. Photoreceptors in the dark-adapted eye are credited with being capable of responding to a single photon. This is all the more remarkable because a photon is one of the lightest of all particles and carries no charge. In the case of hair cell receptors in the cochlea, in acoustically favorable circumstances, the threshold for best tone detection is obtained with vibration of the tympanic membrane less than half the diameter of the hydrogen atom. Many sense receptors are so sensible that they verge on responding to thermal noise. Many characteristics of receptor organs and their innervation are arranged to improve signal-to-noise ratios.

Sensory receptors are sometimes divided among each of three classifications, the *exteroceptors*, the *interoceptors*, and the *proprioceptors*. The *exteroceptors* respond to stimuli that arise outside of the body, such as light which stimulates the eye, sound pressure which stimulates the ear, various forms of mechanical pressure or contact with objects of a temperature different from that of the skin, which stimulate the receptors of touch, and minute quantities of chemical substances which stimulate the olfactory sense organs and the receptor organs of taste. The *interoceptors* lie within the mucous linings of the respiratory or digestive tract and respond to materials ingested or inhaled and to changes in pressure. The *proprioceptors* are sensory receptor endings that respond to stimuli generated by muscular movement of the body or changes in muscular tension. They reside in skeletal muscle, in tendons, in joints, in the heart, in the blood vessels, and in the gastrointestinal wall. The vestibular receptors of the nonauditory labyrinth respond to linear and angular accelerations that may be imposed upon the passive organism or may be generated by motor acts of the organism. These receptors work in close harmony with the proprioceptors of muscle and are sometimes included with this category. They are also extremely important in the control of eye movements and, hence, are important for visual perception.

The Stimulus

Classically, the physical form of stimulation that is most natural or appropriate for a given sensory receptor has been called the *adequate*, or *appropriate*, stimulus. The physical dimensions of the stimulus and the range of energies to which the receptor responds characterize the transduction process.

CONSCIOUS CONCOMITANTS OF SENSATION

The prominence of a sensory dimension in our consciousness varies specifically from one sensory modality to another. Prominence can not be assigned any fixed order either; it may vary with circumstances.

For someone listening with concentration to a musical work or engrossed in a conversation, the auditory dimension may be more prominent than the visual. For someone engaged in any form of athletic competition which involves complex movements of the body, the vestibular sense will be playing an extremely critical role but may have no conscious concomitant.

Olfactory and gustatory senses may be prominent under certain circumstances, but their prominence tends to be relatively infrequent and is associated with such actions as eating and exposure to unusual sources of stimulation. Other classes of receptors that respond to chemical stimuli, those in the carotid sinus and the aortic bodies, afford an important basis for control of physiological processes but never intrude into consciousness. Receptors for the sensation of pain, whether they reside in the skin or deeper within the bodily structure, can provide a compelling conscious component of sensation but, fortunately, under normal circumstances they are relatively inactive. Their role is usually a protective one in providing information to assist the organism in removing itself from sources of potential injury.

Although it is convenient to think of the senses categorically, it is most important to recognize that our perceptions of the physical world in which we live are dependent upon multisensory inputs and interactions among the senses. Thus, the interpretation of visual stimuli may frequently be influenced by vestibular stimuli, proprioceptive stimuli and, possibly, auditory stimuli. The developing organism learns to expect certain correlations among different stimulus dimensions. As we shall see, certain unique situations which may result in a noncorrespondence of different stimuli with expected relationships may be extremely disconcerting or even debilitating. Such noncorrespondences may be encountered between visual and vestibular inputs during exposure to the motions of a ship on a heavy sea or the complex flight path of a maneuvering aircraft. Noncorrespondence between visual and tactual or muscle senses may arise during attempts to move about in a "distorted room."

Methods of Studying Sensory Systems

PSYCHOPHYSICAL STUDIES

It is possible to explore the appropriate stimulus in terms of its physical dimensions along various continua. In the case of vision, the subject's task may simply be to indicate when he is able to detect the presentation of a stimulus. The investigator may then explore such variables as the wavelength of the stimulus, its physical intensity, the region of the retina on which the stimulus is projected, the size of the stimulus, and the temporal duration of its exposure. Obviously, these various dimensions may be expected to interact, and their interactions can also be studied.

Similar investigations of the physical dimensions of the stimulus have been carried out for hearing, touch, and the acceleration forces which stimulate the vestibular mechanism. It is not possible to isolate physical continua associated with olfactory and gustatory stimulation and, so, investigators have attempted to catalog and look for similarities among chemical stimuli which provide similar qualitative experiences in these sensory dimensions. In addition to the minimum required physical energy, or concentration of a chemical substance, necessary for the threshold detection of stimulation, the investigator may wish to explore the effect of increasing stimulation along an intensive or concentration dimension and to note the effect on the subject. Limits of stimulation may be defined at which pain is experienced or at which further increase in the appropriate dimension of the stimulus is accompanied by no further increase in the subjective intensity of the sensation. Such studies may be carried out by having subjects adjust a matching stimulus to the experimental stimulus or by requiring them to make a subjective rating of stimulus intensity.

ELECTROPHYSIOLOGICAL STUDIES

It has been possible in man to record gross electrical activity arising in the retina as a result of photic stimulation. Such activity, recorded as an electroretinogram, permits the examination of some of the peripheral processes in the visual system in man.

Unfortunately, similar indices of peripheral activity are not readily available for the other sensory systems.

Sensory processes may be studied in animals, with measurement of response at any desired point within the nervous system.

In the visual system, in at least one organism, intracellular recordings have been obtained from each of the known cell types of the retina (Dowling, 1970). Electrodes need not be inserted into a cell in order to record the activity of a single cell. Relatively small electrodes, with tip diameter of from 1 to 5 or more μm. and located in the vicinity of a single cell or a single fiber, may pick up signals which reflect the activity of that cell or fiber alone.

SENSORY CODING

Studies of the activity of single cells and single fibers throughout the various sensory systems provide important insights into the nature of sensory processing. As sensory information is transmitted from the primary receptor along various neural elements which link the receptor to cortical areas for the analysis of sensory signals, there is selective divergence and convergence as well as lateral control exerted at nearly every central sensory relay. By means of parallel path signaling and distribution of signals to multiple different subcortical and cortical areas for distinctive analytical processing, several distinctive features of the stimulating event may be segregated out.

For example, there are cells in the cortex of higher animals which respond to a moving stimulus with a certain orientation in the visual world. In hearing, specific cells may respond to certain patterns of change in the frequency or intensity of an auditory stimulus. Others may respond maximally when a stimulus is presented to one ear shortly before it is presented to the other, as is the case for a sound source located on one side of the sensing organism. Such cells selective to the sequence of lateral stimulation have even been reported for the olfactory system. It thus appears that the nervous system is abstracting out certain dimensions of the stimulus world. In some way the brain succeeds in putting these dimensions back together in a fashion that permits us to achieve veridical perceptions of the world in which we live.

ANIMAL RESEARCH

Receptor cell specialization occurred very early in evolution, and it is a commonplace to use receptors from invertebrates as well as vertebrates to investigate problems of transduction and initiation of sensory nerve impulses. Mechanisms of integration are similarly advantageously exploited through the use of lower animal forms. However, some sense organs and sensory pathways are found only in higher animals, and there is always the requirement to prove the pertinence of neurophysiological analysis to our own kind. By and large, what has been found neurophysiologically in invertebrates has been echoed in some analogous form in higher mammals.

Presumably, the closer to man phylogenetically the experimental animal under study is, then the more valid may be extrapolations from electrophysiological studies of that animal to the sensory process of man. Students of sensory systems, however, have not been content with this kind of reasoning. Another approach is to study the whole animal just as man himself is studied in sensory experiments. The animal cannot verbalize his responses, but he can be trained to respond affirmatively or negatively if the experimenter provides him with an adequate reason to do so. The reason may be a food reward, the avoidance of a shock, a few drops of water, or some other appropriate *reinforcement* for a correct response. Studies then can be conducted in which animals will signal appropriately if they detect a stimulus. If, in behavioral studies of sensory processes with animals, the results are similar to, or identical with, those found in man, it is more reasonable to assume that the underlying physiological processes studied by means of electrical recording techniques may be the same as those which exist in man.

In behavioral studies, it is usual to train an animal to perform some special response when it makes a desired sensory discrimination. Some dimension of the stimulus, such as the wavelength of light or the frequency of sound, may then be changed, and the threshold for a discrimination may be redetermined under these new conditions.

In order to compare electrophysiological studies with behavioral studies of this sort, a fixed criterion is also adopted. In the case of electrophysiological studies, the criterion may be a certain rate of discharge of impulses by a single cell somewhere in the nervous system. The intensity of an auditory stimulus may be adjusted in order to achieve this rate of discharge for various frequencies of stimulation, or the intensity of the visual stimulus may be adjusted for each of a number of wavelengths in order to achieve this rate of discharge if the cell is one in the visual system. In general, results appear to be more nearly comparable with behavioral stud-

ies when such procedures are used in electrophysiological studies. An alternative approach is to hold the energy of the stimulus constant while such dimensions as frequency or wavelength are varied and to record the neural discharge frequency as the dependent variable.

ELECTICAL STIMULATION

In man as well as animals, direct stimulation of sensory systems has sometimes been employed as a method of bypassing the primary receptor process. This may provide one useful way in which limitations of the sensory responses imposed by the peripheral receptor can be revealed e.g., in the study of the temporal resolution capacity of the visual system, results with rapidly repeated electrical stimuli could be perceived as varying in time at higher rates than the maximum rate perceived as "flickering" when photic stimulation was employed. In animal preparations, in addition to its value as a method of bypassing the primary receptor, electrical stimulation has also been used in an effort to determine the significance of certain efferent pathways associated with various sensory systems.

SURGICAL INTERVENTION

One test of the significance of certain components in the visual system which can readily be applied with animals depends upon the removal or destruction of the component. Various regions of the brain may be ablated or destroyed. Nerve tracts may be sectioned. Animals trained to respond to stimulation in a given way prior to such procedures may then be tested to see whether they have retained their prior ability. If they have not, the conclusion that the component in question or the pathway in question is essential to the kind of discrimination under study may be justified. It is always important, however, to carry out retraining procedures. Sometimes, surgical intervention may disrupt behavior only temporarily. The demonstration of relearning ability of a given sensory discrimination may reflect such temporary disruption, or it may reveal the presence of alternative mechanisms which can be called into play following damage or destruction to the primary mechanisms which subserve the response in a normal animal.

One surgical procedure that provides a measure of control within a single organism is the split-brain technique pioneered by Roger Sperry (1961). The optic chiasm, the corpus callosum, and other cerebral commissures are sectioned medially. Communication between the right half of the brain and the left half of the brain is thereby presumably eliminated. When a visual sensory discrimination is then studied by stimulation of the eye on one side, this discrimination will be mediated by the visual nervous system on that same side. The effects of surgical intervention at certain locations within the visual system on one side may be observed and compared with performance of the animal in the same kind of discrimination task when the other side is employed. A lack of communication between the two halves of the system is supported by studies which show no transfer of discriminations learned on one side in subsequent testing of the other side. The situation however, is not as clear-cut as one might hope. Studies that have failed to demonstrate any transfer have been based on the use of food as a reward for correct performance. Other studies have proven that when punishment is used to induce performance, there may be transfer, even when positive food rewards have failed to elicit any evidence of transfer (Sechzer, 1964).

Although it is not possible to create a split-brain experimentally in man, it sometimes occurs that there is a congenital absence of the corpus callosum. Such a case has been studied by Sperry (1974). In other instances, the corpus callosum has been deliberately cut for specific neurosurgical purposes, e.g., tumor removal. A few cases of severe unilateral epilepsy with spread to the contralateral side were treated with callosal section (Bogen, 1968) and were studied by Sperry and Gazzaniga (1967). Most of what has been learned about lateralization of function in the two hemispheres has been gained by application of knowledge from animal and neurosurgical experiences and by investigation of a relatively few fortuitous opportunites. It speaks well for the investigators and their subjects that remarkably fine patient cooperation has been maintained, and a great deal has been learned in the process.

COMPUTER ANALYSIS

Simple computer methods for administering iterative stimuli and averaging many responses (to improve signal to noise ratios) have been utilized for decades. With the advent of laboratory scale computers it became possible to provide randomized stimuli of various kinds and sort out and average the individual responses to each kind, in corresponding bins. More recently, easily programmable, high-speed, powerful laboratory computers have become commonplace and have become incorporated into the research protocol in high sophisticated ways, revolutionizing data gathering, analysis, and display. Rapid change of choice of stimuli that may be contingent on the response of an animal or human subject, even a cascade of contin-

gent choices, has become feasible. It is now possible to plan a whole "tree" program which branches in accordance with the neurophysiological response and to run complicated contingent choice pathways in real time, sorting the data correspondingly.

Neural Response to Stimulation

GRADED POTENTIAL

The nature of the initial response of a receptor system to stimulation varies with the sensory modality. In most instances, the initially detectable electrical response is in the form of a graded potential. That is, the amplitude of the response varies, or is graded, dependent upon the amplitude of stimulation. Such a response may be relatively sustained during the time course of stimulation, although frequently it will show an inital maximum level which may reduce somewhat with continuation of a sustained stimulus. The origin of such graded potentials is not clearly understood, although they may arise in primary receptors. They differ from the spike discharge of neurons in that their amplitude may vary and that they are sustained in duration, rather than occurring repetitively with sustained stimulation. Polarity of the graded potential may vary, depending on the nature of the stimulus (Fig. 9.1). Records obtained in the region of the retinal receptors show a difference in the polarity of the response, depending on the wavelength of

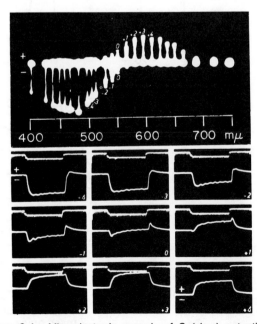

Figure 9.1. Microelectrode records of S (slow) potentials in retina of fish. *Upper portion* shows relative polarity of response as a function of wavelength of stimulus. *Lower portion* records show time course, amplitude, and polarity of responses to 300-ms stimuli. [From MacNichol and Svaetichin (1958).]

the stimulus (MacNichol and Svaetichin, 1958). Only in the case of the olfactory system does the primary receptor serve also as the primary neuron, discharging with impulses, the frequency of which varies with the concentration of the stimulating substance.

THE GENERATOR POTENTIAL

Before a cell will fire, there must be a depolarization of the cell membrane to a certain "threshold" level. When this level is reached, the spike discharge follows. In the wake of a spike discharge, the sustained stimulation of the cell by contacting elements results in a repetition of the process. If the depolarization does not reach the required level, the cell will not discharge (Fig. 9.2). It is evident that as the level of stimulation is increased, the generator potential reaches the critical amplitude more rapidly, permitting a more rapid frequency of discharge of the cell. The spike discharge occurs on top of the measurable generator potential. The generator potential is monophasic, of relatively short latency and, as suggested above, may be graded in amplitude. It thus is distinguished from the propagated action potential, which is of roughly constant amplitude for a given nerve fiber.

NERVE FIBER DISCHARGE

The pattern of a nerve fiber's discharge frequency may vary considerably, depending on the nature of stimulation, as well as the specific modality under consideration. With the initiation of a stimulus, spike discharge frequency may be very rapid after a short latency, followed by a sustained discharge frequency at a higher rate than the resting level, until the stimulus is terminated. Upon termination of the stimulus, there may be a brief period of inhibition of any response prior to resumption of some resting level of response. Some fibers may be quiescent and may show no detectable resting level of discharge. The relation of discharge frequency to intensity of stimulation is illustrated in Figure 9.3 for a single optic nerve fiber of the horseshoe crab. It is evident that as the stimulus intensity is reduced, latency of response grows longer, the initial frequency of discharge is lower, and the repetitive discharge of the unit, while the stimulus remains on, becomes slower and less regular.

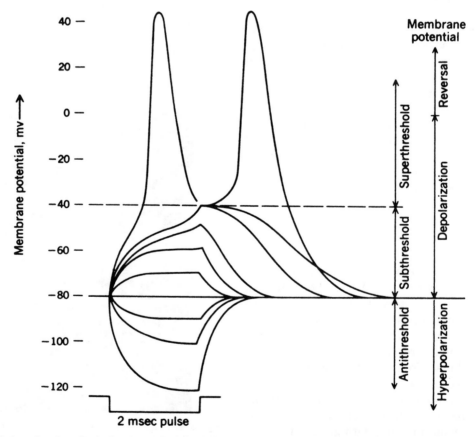

Figure 9.2. Initiation of an impulse by local membrane polarizations. Current pulses of fixed duration and various size and polarity induce variations in membrane potentials shown by family of curves. At shock strength of about 35 mv, the membrane depolarization required for spike discharge is reached, and the nerve fiber fires. [From Katz (1966).]

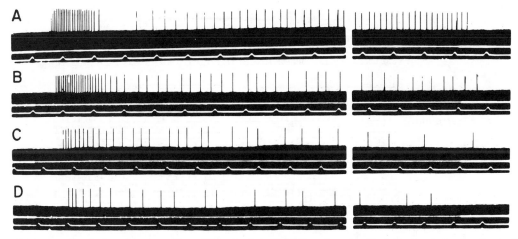

Figure 9.3. Discharge of a single optic nerve fiber of *Limulus* for stimulation at each of four levels of stimulus illumination: *A*, 6.3 × 10⁴; *B*, 6.3 × 10³; *C*, 6.3 × 10²; and *D*, 6.3 × 10 meter candles. [From Hartline and Graham, (1933).]

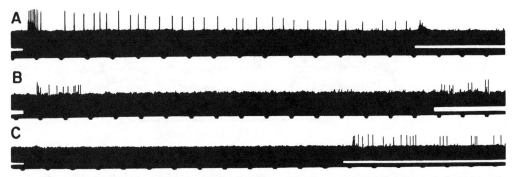

Figure 9.4. Three different types of retinal fiber discharge in the frog. Stimulus onset and duration is indicated by disappearance of white line above the 200-msec time marker. [From Hartline (1938).]

By reason of interactive processes within sensory systems, not all cells respond with heightened discharge activity upon stimulation. Some may be inhibited by a stimulus (Fig. 9.4). The *top trace* shows a unit which responds initially with a rapid burst of activity and then a sustained level of activity while the stimulus remains on. The *middle trace* illustrates a unit which responds with a burst of activity upon initiation of the stimulus, a relative quiescence while the stimulus is maintained, and then additional activity upon termination of the stimulus. The *bottom trace* shows a unit which does not fire but is inhibited when a stimulus is presented. Upon termination of the stimulus, a burst of activity occurs and is sustained for some time prior to return of the unit to its resting level.

As noted earlier, some units in sensory systems are spontaneously active. This provides a means of detecting subtle differences, the onset of a weaker stimulus, or a shift in stimulus intensity or quality, more than could be detected if the receptor had to be excited to some threshold value before it could discharge. Figure 9.5 shows neuron responses at the thalamic relay (lateral geniculate nucleus) along

the visual pathway. The geniculate unit is receiving signals via the third synaptic relay, along the most direct throughput channel of information. It is a fourth order unit: receptor, bipolar cell, ganglion cell, and lateral geniculate neuron.

Both excitatory and inhibitory impulses converge on the lateral geniculate neuron along a path that is spontaneously active. This unit is closely linked in its response pattern to the preceding ganglion cells serving that part of the visual field. The upper part of the figure shows analog responses in order along the time axis for 10 successive trials of a spot of light directed at the center of that unit's receptive field. Each time the spot of light is flashed for 50 ms, the unit shows a strong burst of impulses. When the same spot is flashed in the same location but is coupled with an annular surrounding light (see *inset*, Fig. 9.5, for diagrammatic representation) and the annular stimulus is continued for 500 ms, the unit responds with less vigor and does not resume its usual background activity until the annular stimulus is turned off, when there is a moderate "off" response. When the annular stimulus is shown alone, the unit is dramatically inhibited and re-

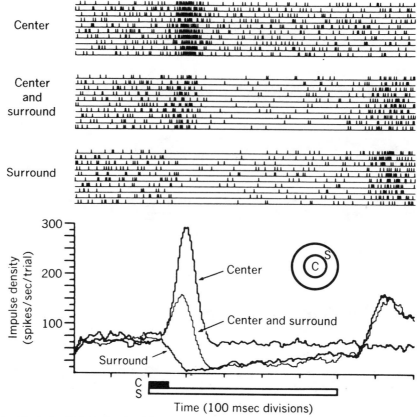

Figure 9.5. An important feature of integration involves convergence of excitatory and inhibitory influences on neurons. A unit in the lateral geniculate nucleus relay between the retina and cortex displays background activity prior to stimulus onset. When a small spot of light strikes the center of its receptive field, excitation is followed by return to background activity. When light is applied to center and surround together, excitation is less marked. It is followed by inhibition and a rebound "off response." When surround light is applied alone, pure inhibition is followed by an "off response." Because the unit is spontaneously active it can be raised or reduced in rate of activity and thereby can register more subtle stimulus differences. *C*, center spot, 0.6° in diameter with flash duration of 50 ms; *S*, surround 1–2.7° in inner-outer diameter with flash duration of 500 ms. [From Poggio et al (1969).]

mains inhibited, gradually escaping, until the annular light is turned off when there is a more marked "off" response. One single unit, then, has monitored (and relayed) considerable information: *where* on the retina, *when* there has been a change in lighting, and *whether* the light was a local spot, a surrounding ring of light, or a combination.

Obviously, for any cell to show a decrement of response to some form of stimulation, as compared with the absence of any stimulation, there must be a resting level of activity which can be decreased.

NEURAL CODING OF STIMULUS INFORMATION

It is evident that a particular physical event in the visual world can trigger changes in photoreceptors and a train of successive neurons to whatever end station may be identified in the central nervous system. Once the physical event has been transduced to membrane-transmittable signals, there is

no longer anything left of the light; everything is now transformed into the common currency of nerve-graded responses and action potentials. This is called neural coding. Perkel and Bullock (1968) have written an authoritative analysis of this subject.

We can see that part of the code is represented by *which neurons* are activated: the "labeled line." The *timing* of the physical event is signaled (with a certain latency) by onset of a *change in background, signaling*: a "doorbell" code. Additional information is coded by *whether the unit is made more or less active* and *how it responds to the intensity and duration* of the stimulating events. Furthermore, the unit conveys information relating to *differential physical effects within that unit's field.* Not shown is the fact that when any other region of the retina is illuminated, outside the receptive field of this lateral geniculate unit, its "spontaneous" activity is not altered. In other words, it conveys information that *there has been no (local) change.*

COMPOUND ACTION POTENTIAL

When a bundle of nerve fibers is transmitting information in response to sensory stimulation, a compound action potential may be recorded from the bundle with a gross electrode. This shows changes in voltage with time which reflect the summed electrical activity of all of the components within the bundle. When a large number of fibers is firing simultaneously or at nearly the same time, the amplitude of the action potential will be high. Compound action potentials typically show several peaks, associated with classes of neurons having roughly comparable conduction velocities. Thus, a wave of activity for a given class of fibers reaches the electrode at very nearly the same time for all of the fibers within the class. These fibers are then represented by a specific component of the compound response (Fig. 9.6).

Some Problems of Central Integration

In thinking about nervous system functions, consider the implications of this "compound action potential" seen in peripheral nerves (Fig. 9.6.). Given that nerve conduction varies with axon diameter, degree of myelination, temperature, and other factors; given also that conduction velocity slows at points of axonal branching (where collaterals may be given off), and given that nerve impulses slow down in regions of axonal termination, it is obvious that conduction times for any message, from origin to destination, will show a similar dispersal of action potentials, spreading out farther with increasing lengths of conduction pathways.

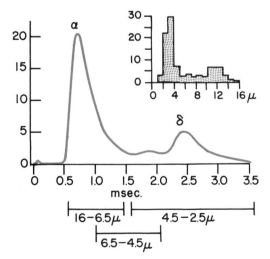

Figure 9.6. Compound action potential in a human sensory nerve. Counts of fiber diameters encountered in that nerve are shown. Voltage (arbitrary units) as a function of time illustrates how impulses spread out in time in accordance with fiber diameter [From Gasser (1960).]

At each synaptic junction there is a modest (0.1 to 0.5 ms) and somewhat variable *synaptic delay*. Most circuits have a spectrum of temporal variables, in addition to sometimes exhibiting a range of numbers of synaptic relays along the course of what is ostensibly a single pathway, that is, having first, second, or nth order neurons in the same bundle of fibers, as in the case of the spinal cord dorsal columns, for example. Second order neurons often contribute additional complications, by changing the phase lag or otherwise altering the "code" in responding to messages delivered by first order neurons. Receptors introduce finite delays, and the efferent command involves further delays between the nerve impulse and the effector response, in addition to mechanical delays in muscle and tendon, limb inertia, etc.

You recognize immediately (and with startling clarity, if you did not know already) that there must be a time-distributed dispersal of nerve impulses occurring along every neuronal pathway. Then, how does the nervous system interpret *simultaneity*? How can we tap finger and foot simultaneously, a linked gesture that is easily achievable to within about ½ ms, roughly the interval of a single synaptic delay?

In the outside world, events impinge on different sense receptors and on different regions of the body, yet we can estimate, to a good approximation, to within 1 ms or so, whether such events are simultaneous. *Perceptual processes must be able to "read" these time-dispersed central events very precisely to be able to recognize that they represent environmental events that occurred simultaneously.* Also, how is central command organized in order to achieve such a high degree of accuracy in the ordering of simultaneous gestures and events in different parts of the body, in the jaw and foot, for example?

Therefore, both in perception and command, and throughout all central transactions, *the nervous system must take into account the variables of conduction velocity occurring along every peripheral and central pathway.* The brain must take into account in sensory perception, central processing, and motor command all of the time variables occurring in these various pathways during, for example, spinal cord conduction between upper and lower limb segments. Messages for foot tapping, for example, must be dispatched well beforehand and dispersed over a range of some 8–12 ms. This would approximately account for spinal interlimb conduction differences, so as to dispatch correctly commands for "simultaneous" movement of foot and hand. For such commands, the central nervous system must dispatch a barrage of impulses that are precisely, calculatedly dispersed in space and time. As for sensory recognition of simultaneity, our perceptual apparatus must make allowances for an analogous centripetal dispersal of sensory pathway conduction intervals.

The truth is that we know much less than we would like to know about this question of solving simultaneity. What is known will be supplied in the chapters on individual sensory modalities because the problems are somewhat different in each one, as well as in the chapters on various kinds of neuronal control. An easy answer is that evolution would not have given advantage to a nervous system that could not solve such problems. It is a problem that had to be solved by primitive nervous systems, and we are their legacy. In thinking about the brain, we need to recognize that any and all functions are distributed in space and time; there is no focal point where perception or command are put completely together. It is a moment of disappointment to most people when they learn that there is no central station where all attributes involved in perception and command, for seeing, listening, reading, or speaking are concentrated, or where all the attributes of a given perceived object ("Mom," for example) are located. In short, the nervous system deals with ionic and molecular events that are remote from the outside world and that have spatial, temporal, and content representations radically different from what we think we understand of the physical universe.

SOME ANATOMICAL CORRELATES OF SENSORY INTEGRATION

Obviously, structure is important for all sensory systems, for example, the nature of the sensory field, a retina, a sheet of olfactory receptors, or a muscle tendon organ. These systems dictate how the physical or chemical stimulus can get at the receptor. Transduction may occur in a specialized cell such as a photoreceptor, or in a naked nerve ending, or in some specialized organ associated with the nerve terminal (e.g., Merkel's disk). Directly, or indirectly, the process of transduction is converted into nerve impulses which are variously coded: instant of firing, phase locking to the stimulus, frequency of discharge (instantaneous frequency, increment above background, rate of change, weighted average), temporal pattern of impulses, number of impulses or duration of a burst of impulses, and so forth (Perkel and Bullock, 1968).

At the initial junction between receptor and nerve terminal, or at the first synaptic relay in a neuron that has been directly activated, a host of

integrative effects may be applied. There can be immediate feedback from the second order unit to the first, reciprocal "push-me-pull-you" synaptic relations, or straight throughput of signals, with or without effects from neighboring or laterally integrating interneurons. In any event, integration takes place at each such station.

Figure 9.7 illustrates how afferent neuron preterminal arborizations distribute the spent results of action potentials in the form of graded responses. This involves slowing of the rate of spread of membrane responses at each branching point and eventually results in multiple synaptic events affecting a variety of cells in accordance with the time-space distribution of the waning signals. Figure 9.7*A*

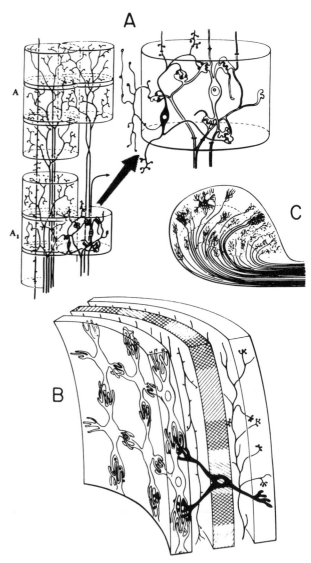

Figure 9.7. A comparison of the preterminal arborizations of specific sensory afferents that gives some idea of the possibilities for convergence and divergence. *A*, The lateral geniculate nucleus; *B*, the medial geniculate nucleus; *C*, the ventroposterolateral nucleus. [From Szentagothai (1982).]

shows the long penetration and multiple arborizations of ganglion cell axons entering the lateral geniculate nucleus. The inset drawing shows an enlarged version of one of the planes of section. Close observation of the connections reveals that there are both *divergence* and *convergence* of signaling at this relay. In Figure 9.7*B* are shown layers of cells in the medial geniculate nucleus, the major thalamic relay for auditory signals en route to auditory cortex. Incoming axons enter alternating layers and distribute signals to many cells in these layers with effects that are relayed to other layers once removed, by interneurons, one of which is shown in black. Figure 9.7*C* shows the arrangement of afferent fibers in the somatic sensory system entering the ventroposterolateral nucleus which relays that modality to primary sensory cortex.

The most important and obvious point is that these three thalamic sensory relays are organized very differently in terms of access of incoming signals to intrinsic neurons, both interneurons and neurons that will relay messages to cortex. Obviously, different geometries are conducive to different patterns of integration.

Analogous extremely simplified diagrams are presented in Figure 9.8 in order to allow rough comparisons to be made between several sensory modalities and their counterpart structural organization. These diagrams are deliberately scaled differently to make easier comparisons between first order, second order, and nth order neurons along each vertical assembly. At the bottom are the three major thalamic relay nuclei we have been considering, lateral geniculate nucleus (LGB), which provides visual relay; medial geniculate nucleus (MGB), which provides auditory relay; and ventroposterolateral relay (VPL), which provides somatic sensory relay.

Initial relays in olfactory and visual systems are remarkably similar, with strong lateral integrative controls. Olfaction, however, does not relay in the thalamus before going to cortex. This system— olfactory apparatus, pathways, and olfactory cortex—is phylogenetically older than the thalamus. Secondary routings involve the thalamus at a later stage of olfactory discrimination, but this is like the secondary association relays through thalamus in other sensory modalities.

The LGB relay includes descending projections (interrupted axons entering thalamus from the side opposite entrance of ganglion cell axons from the retina). The auditory system has complex integration going on close to the sense organ but conspicuously less lateral integration at higher levels. The auditory system also includes strong descending controls, only the lower levels of which are depicted

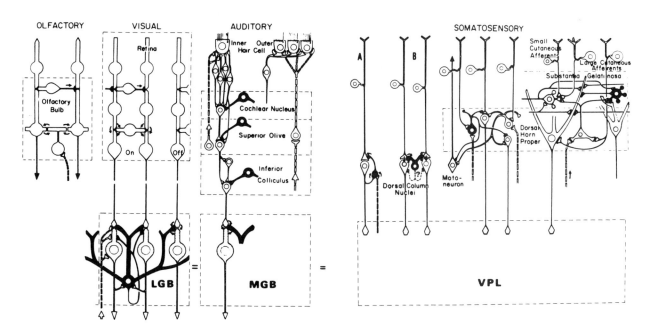

Figure 9.8. Simplified diagram to compare elementary neuron couplings in the four main sensory pathways. Inhibitory interneurons are indicated by *thick solid lines*. Only "main line" couplings can be indicated, and convergences and divergences are largely neglected. Descending control is indicated by connections drawn in *dashed* neurons. In the primary afferent-dorsal column pathway, diagram of the somatosensory system *(middle right)*, two possible interneuron couplings are indicated. *A* demonstrates the original concept of an elementary recurrent feedback coupling, while *B* shows an arrangement using dendritic synapses with triads. *LGB*, lateral geniculate body; *MGB*, medial geniculate body; *VPL*, ventroposterolateral nucleus (Szentagothai and Shepherd, 1970.). [From Szentagothai and Arbib (1982).]

here. For the auditory pathway, there are more relays along the ascending trajectory between cochlea and auditory cortex. There are also more connections crossing from one side of the brain stem to the other, meaning more bilaterality of representation than with other modalities.

The somatic sensory diagram reveals three different patterns. From left to right, there is the phylogenetically most recent—dorsal column system, with a long first order neuron (the longest neurons in the body), with tightly concentrated integration at the first central relay in lower brain stem. In the center is a phylogenetically intermediately old somatic sensory system with large-scale integration occurring in the sensory nuclei at the segmental level in the dorsal horn of the spinal cord. The diagram on the right shows the phylogenetically oldest somatic sensory group, which conveys a wide range of somatic sensations in addition to pain and temperature. Integration in this oldest system is closer to the periphery than in either of the other somatic sensory systems. This illustrates a general principle, that phylogenetically newer sensory systems reach further toward the brain to begin integrative processing; they include larger diameter fibers and thus indicate priority for getting their messages reliably and quickly to a high level. In the same strategic context, the phylogenetically newer motor pathways have the longest axons, extending from motor cortex directly, monosynaptically, to spinal motoneurons.

Table 9.1
A comparison between some formerly held and presently held views of neurophysiology

Formerly Held Views	Presently Held Views
All-or-nothing action potentials (spikes) are the only means of signaling	Spikes are necessary for long distance decrementless conduction; the main business of the nervous system, transactions among nerve cells, takes place where nerve membranes have graded responses.
Transmission between neurons is affected solely by electrical transmission or (some held equally vehemently) solely by chemical transmission (1940s–1950s).	Transmission between neurons may involve either electrical (bidirectional or unidirectional) or chemical transmission, or both.
Synaptic transmission involves only a limited number of chemical neurotransmitters, two or four (early 1950s) ditto, but perhaps a dozen (early 1960s).	Synaptic transmission implicates scores of neurotransmitters, including amino acids, small peptides, medium-sized peptides, neurohormones, neurohumors, neuromodulators
Transactions between neurons is directional at a given nuclear relay station and is directional along a sequence of nuclear relays.	Microcircuits allow for feedback as well as feedforward at given local junctions, and there is often (perhaps usually) feedback from station to station along a given sequence of relays.
Communication between nerve cells took place only at synaptic contacts.	The whole microenvironment of neurons and glia is sensitive to specific variable ionic fluxes and electrotonic field forces, both generated by the nervous system and imposed from outside.
Endocrine and nervous system mechanisms are largely autonomous (until late 1930s and 1940s); the nervous system (hypothalamus) controls the posterior pituitary (1950s); the hypothalamus dispatches factors by way of the portal circulation in the pituitary stalk that controlled the anterior pituitary.	The nervous system is almost as bona fide an endocrine system as are the "ductless glands"; both traffic many of the same trophic factors, hormones, and neuroendocrine signaling factors; nervous and endocrine systems are engaged in intimate mutual transactions at various locations throughout the body.
A neuron has only one neurotransmitter (Dale's law).	A neuron may have more than one to several neurotransmitters.
A neuron gets committed early in maturation to producing one and only one neurotransmitter.	During development, some neurons may shift from commitment to one neurotransmitter to another, depending on environmental conditions.
All processes involved in communication happen at the nerve membrane, and the effects are prompt (in milliseconds).	Nerve membranes have receptors, the excitation of which may signal a "second messenger" which can direct changes in the interior of the cell that can affect excitability. The effects may be long lasting, from milliseconds to the lifetime of the cell. Cells have sugars, proteins, and mucopolysaccharide structures on the surface of the cell membrane which reach out (dynamically) long distances and may be sensitive to specific ions, electrotonic field forces, and the like. Some structures in cell membranes connect through the membrane into the cytoplasm and can affect "signals" that may have meaningful effects apart from ordinary neurotransmission.
Neurons and glia are not only separate derivatives, but the glia only contribute structural and perhaps nutritional and "trophic" influences, phagocytic functions, and the like.	Glia are affected by changes in ion flux and in turn influence neuronal excitability; some may act as an ionic "sink" and "source" to conserve ions for neuronal use locally. Glia support neurons, provide channels for growth, exert control over local blood supply, ensheath neurons in myelin in dynamically interactive ways, and may participate in shifting fine structures around and constituting barriers or buffers between neurons.
Neuron sends messages from cell body to axon to dendrites and cell body in relays.	*Bona fide* directional synaptic junctions are found to signal from axon to axon, both presynaptically and *en passant*, from axon to dendrite, dendrite to axon, dendrite to dendrite, axon to soma, soma to axon or dendrite, or soma to soma; there are "push-me-pull-you" synapses sending and receiving neurotransmitters in close juxtaposition, electrical (unidirectional and bidirectional) junctions, and specialized junctions for exchange of a variety of molecular agents; any combination is possible, but the organization is not random.
Neurons are excited to some threshold and are converted from being silent to being active, reverting to silence again when not specifically being excited.	Neurons and some receptors (e.g., photoreceptors) are often spontaneously active; this allows a much finer discrimination than threshold controls; also, many neurons constitute pacemakers, initiators, and governors, which are controllers of large-scale effects by virtue of their intrinsic firing capabilities, and these may be controlled or triggered by light, hormones, or other state-controlling agents.
Neuron signaling are graded by shifts in frequency of firing.	Frequency of firing is not the only or even dominant method of signal "coding"—instances of the following have been found to provide functional codes: "Labeled lines" from or to a specified element; time of occurrence, "doorbell code" (instant of firing, phase locking to stimulus); frequency (weighted average, increment or decrement from background, rate of change, firing or omitting firing at fixed intervals); temporal pattern of impulses or durations of burst; irregularly vs. regularly spaced intervals.

Table 9.1—*continued*

A comparison between some formerly held and presently held views of neurophysiology

Formerly Held Views	Presently Held Views
At least in higher mammals, neurons that will be formed are present at birth.	Some neurons, even in humans, keep on dividing, at least through the first year of life; some neuron replacement may occur later; neurons appear to grow throughout life, with continuing plastic modeling of circuitry as affected by activity, at least at local and regional levels, continuing indefinitely.
Neuronal relays, at least in main "through conduction" pathways (sensory paths to cortex, for example), pass along signals with high fidelity, close to one-to-one.	Neuronal integration characteristically occurs at all synaptic relays, and even within a given neuron before generating an action potential and before delivering its messages to the next neuron; there are usually feedback and feedforward, as well as lateral integrative operations at most relays, with open loop up-and-down controls being exerted from relay to relay.
In general, Cartesian principles of cause and effect and direction of time effects are operative.	Cooperativity at the quantum mechanical level and at all levels of greater complexity appear to be operating and to contribute importantly to self-organization and self-regulation of systems; many of these operate to improve signal to noise ratios (e.g., shearing forces generated by tectorial membrane in cochlea have horizontal thrust on inner hair cells and radial thrust on outer hair cells, which improves signal-to-noise ratios).
Nutrition requires sufficient calories, vitamins, and minerals (with rising knowledge of the existence, quantitative requirements, and sources of these).	Nutrition and especially zinc (not stored), vitamin B6 (needed for protein synthesis), and trace minerals are especially important during embryogenesis and throughout gestation and early childhood when the brain is rapidly growing; also, the above are important whenever there is nerve or CNS damage, because sprouting, new synapse formation, and modeling begin within hours and remain active well into prolonged rehabilitation; it is implied that nutrition is equally conspicuously needed during any transformation of circuitry, as in learning and the exercise of memory.

CHAPTER 62

Hearing

THE SIGNIFICANCE OF HEARING

Hearing rivals vision as a means for extracting highly complex and detailed information from the environment. Hearing depends on sounds that can be analyzed in physical terms of component sine waves of certain amplitudes and frequencies. Our aim is to describe how sounds reach the acoustic epithelium and are analyzed by neurophysiological circuits. How *hearing* takes place is only partly understood, but the information gained is already rewarding.

The most interesting and important sounds are not simple vibrations. Speech and music, for example, are exceedingly complex. It's a rich combination of physical events and neurophysiological consequences that renders speech and music recognizable, intelligible, and enjoyable. Further, when we hear people speaking a language we do not comprehend we are struck by the important insight that the *information* conveyed by sound depends largely on an individual's past experience. In fact, the information conveyed to us by *any* sound is not uniquely dependent on the physical parameters of that sound. The limits of human auditory apparatus and associated neural pathways, human capacities for speech and language, a given individual's pertinent history, are all important to medical practice, for helping people with problems of acoustic reception, processing, perception, and behavior relating to hearing.

Hearing and speech, an interdependent combination, provide the principal means for organizing social transactions. A loss of hearing, or congenital absence of hearing, is a social and communications disaster that must be compensated as well as possible early in life, or as soon as hearing is lost, by alternative means for communication. Otherwise, that individual will suffer devastating social isolation. It will affect a child's participation in all aspects of life, cognitive as well as social. Increasing social isolation also disturbs elderly persons with encroaching deafness.

Hearing is even more important than vision with respect to *language and its development*. Hearing is our fundamental access to language. Much of our thinking is auditory. We may even engage in talking to ourselves, subvocally perhaps, but nonetheless it is part of the process of communicative thinking. We say we are "thinking *aloud*." From the beginning of our lives to the end, we are immersed in social and cognitive functions that involve speech and hearing, or, in the case of the deaf, some equivalent cognitive process, such as sign language.

Within a few days of birth, as soon as a newborn can be tested, he or she manifests preference for hearing a tape recording of his or her own mother's voice reading a passage as compared with another mother's voice reading the same passage (DeCasper and Fifer, 1980). By the age of 6 months the normal infant is beginning to babble sounds characteristic of any given language. One of the great transformations of life takes place when an infant discovers there is special social meaning in some sounds, and begins to exercise control over his or her environment by rendering imitative sounds that become increasingly meaningful.

Deaf children born of hearing parents have grave problems. Neither they nor their parents know how to communicate with one another. Unless the parents learn sign language quickly and adequately, the two generations will find no equivantly rich and dynamic medium through which to express their desires and ideas. The child will have difficulties in cognition and learning as well as in communication, and the consequent frustration is likely to be manifested by antisocial behavior.

A deaf child with deaf parents, however, develops normally. The parents present sign language, and this becomes an adequate substitute for speech and hearing. The child learns to comprehend signs and to progress along much the same course by which a hearing child develops language (Klima and Bellugi, 1980).

THE PHYSICAL CHARACTERISTICS OF SOUND

Hearing is the subjective experience of exposure to sound. Sound consists of waves of alternating condensation and rarefaction in an elastic medium such as air, water, metal, etc. By arbitrary definition, the term sound refers to the limited range of normal human hearing. Ultrasound is beyond the nominal human upper frequency limit of hearing but can be heard by many vertebrates, including many mammals, even primates, and many invertebrates, especially insects.

Normal human hearing covers a range from about 20 cycles/s (cps, Herz, or Hz), to somewhat above 20,000 Hz, or 20 kHz. Some youths can hear up to 25 kHz and perhaps higher, but hearing is gradually lost from higher to lower registers, during normal aging. By midlife, hearing is usually limited to below about 4 or 5 kHz. The most valuable range for purposes of speech and hearing is from about 200 Hz to 4.5 kHz. The physical dimensions of sound frequency can be more easily grasped in terms of spatial dimensions: 20 Hz has a wavelength in air of about 15 m, whereas an ultrasonic 50 kHz has a wavelength of about 6 mm. By convention, vibra-

tions above 16 kHz are called *ultrasound*, while those below 20 Hz are *infrasound*. Sound reverberates very differently according to wavelength; hence, difficulties are encountered in trying to design an auditorium that should be acoustically appropriate for a wide range of sound frequencies.

Both vestibular and auditory structures, and the lateral line organ in fish and amphibia, derive from the same anlagen and share similar receptor cells, called hair cells. These are cuboidal epithelial cells which receive their name from the fact that stereocilia protrude from their outer surface. Both auditory and vestibular hair cells are innervated by first order afferent nerve fibers from the VIIIth cranial nerve. Figure 9.9 presents the gross anatomy of the companion organs, cochlea, and vestibular apparatus, and their innervation.

The vestibular system can respond to vibrations, but because of the arrangement of its sensory epithelium within vestibular saccules and canals—all encased in a hard, bony box—its best response is

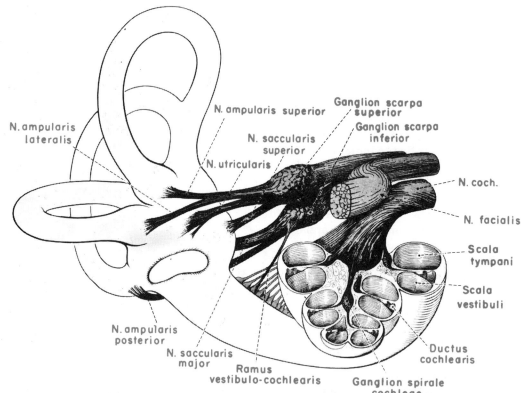

Figure 9.9. General schematic depiction of the inner ear showing three-dimensional relations of vestibular apparatus and cochlea with their respective branches of the VIIIth cranial nerve. Each ampulla of the three vestibular semicircular canals has its own nerve bundle, and large bundles also serve the utricles and the saccule. All vestibular afferent neurons have their cell bodies in the two Scarpa's ganglia, and they convey a sense of balance and motion of the head in three-dimensional space. The cochlea has been sectioned through the modiolus, its central bony axis, to show three turns of the cochlea, composed of three spiraling canals, scala vestibuli, which connects with the vestibular apparatus, scala tympani, and between, the cochlear partition—the scala media or cochlear duct. Afferent innervation of the cochlea comes from the spiral ganglion which winds through the spiraling core of the modiolus. The oval window to which middle ear ossicles deliver vibrations interpreted by the nervous system as hearing can be seen as an indented oval outline on the base of the labyrinth. The oval window communicates with the scala vestibuli. Note the close proximity of the important facial nerve which may be implicated by inflammatory or neoplastic diseases affecting the inner ear, or by cold exposure to wind against the ear, leading to paralysis of facial muscles on one side, Bell's palsy. [From Melloni (1957).]

to linear and angular accelerations. It is motion of the encasing bone and inertial drag of fluid in the membranous channels that gives rise to vestibular receptor activity. Fluid in the labyrinth, without having an exit, cannot very effectively be put into vibratory motion.

Access to vibrations is by way of the oval window, which has direct physical linkage to the outside world via the middle ear ossicles and the tympanic membrane. The acoustic epithelium in the cochlea is similarly embedded in bone-enclosed chambers. But it can be put into motion because there is a flexible relief diaphragm, the round window, which covers a hole in the bony wall of the scala tympani of the cochlea. Vibratory stimuli delivered to the oval window directly affect the scala vestibuli, which conveys vibrations to the round window by moving the intervening membranes, the delicate vestibular membrane, and the basilar membrane, which enclose the cochlear duct (Figure 9.16). It is on this basilar membrane that the sensory epithelium with its hair cell receptors and nerve terminals rides the sound waves. The active oval window and the passive round window enable the system to deliver vibrations to this partition.

Figure 9.10 shows the tympanic membrane, en face, *A*, and in a cutaway side view, *B*, together with the middle ear ossicles. The ossicular chain has a slight mechanical advantage, approximately 1.3. The difference between the dimensions of the most responsive area in the tympanic membrane (about 40 mm^2), and the area of the oval window in the cochlea by which the acoustic epithelium has access to vibrations (about 3 mm^2), gives a substantial

mechanical advantage, approximately 13:1. This provides a remarkably good *impedance match* between air and the fluid chambers of the cochlea; hence, it allows efficient transmission from vibrations occurring in air outside the head to the acoustic epithelium. The impedance match varies between 50 and 75% efficient throughout the range of human speech.

Specification of Sounds

Referring to Figure 9.11 we can appreciate that the dynamic range of sound *intensities* to which the ear can respond normally is extremely broad. At the frequency of maximum sensitivity, between 1,000 and 3,000 Hz, the ratio of maximum intensity that can be heard without damage to the ear, to the minimum intensity that can be detected, is approximately 10^{12}. The minimum power to which the ear can respond is of the order of 10^{-16} watt/cm^2. This is a standard reference value; in fact, many individuals can detect lesser amounts of power. This power level is comparable to a sound pressure of approximatley 0.0002 dyne/cm^2 at 20°C.

The tremendous range of physical energy to which the ear responds as well as the nature of the response of the auditory system renders the use of a logarithmic scale desirable for the specification of sound intensity. Such a scale represents increases in ratio steps. The unit used in audition is the decibel. A unit of 1 Bel (named after Alexander Graham Bell) represents a ratio increase by a factor of 10, 2 Bels a factor of 100, etc. There are thus only 12 Bels over the range from approximate threshold to the level at which sound becomes

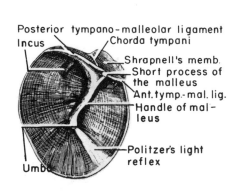

A.

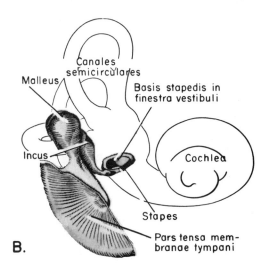

B.

Figure 9.10. *A,* View of outside of tympanic membrane as seen with an otoscope (Portman, 1951). Position of handle of malleus is visible. *B,* Cutaway section of tympanic membrane to show attachment of malleus for maximum excursion of membrane. Malleus articulates with incus, so the combination of two leverages applies to piston-like motion of stapes against flexing membrane which covers oval window of labyrinth. [From Melloni (1957).]

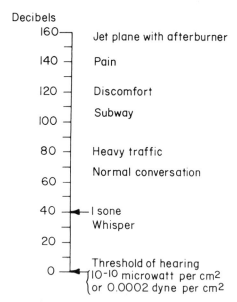

Figure 9.11. A decibel (dB) scale showing approximate acoustic power encountered in the presence of various sources. The sone is a unit of perceived loudness based on ratio estimates for which the loudness of a 1,000-Hz tone at 40 dB provides reference. [From Stevens (1958).]

painful in the frequency range of maximum sensitivity. The decibel (dB) was devised as a unit which would provide a finer index of sound level. As the name implies, there are 10 dB to the Bel.

The decibel notation, which covers an extended power range, was first found useful for comparing power, voltage, and current in electrical systems. It always needs to be compared to some standard reference level. The usual reference standards for measuring sound meters, hearing impairment, and so forth, is to a sound pressure level (spl) of 0.0002 dynes/cm², or 10^{-16} W/cm². A newer standard is designated in *newtons*, and may refer to an spl of 2×10^{-5} N/m².

One or another of these standard references are encountered wherever measurements are made using sound meters, as in the measurement of hearing abnormalities. These standards represent a level of sound near threshold for the frequency of maximum sensitivity in the normal ear. A reference level of 1 dyne/cm² is also used by some investigators. In addition, decibel notation may be employed to represent the amount of attenuation of a signal below the maximum level that can be obtained from a given amplifier. The levels in decibels of some typical sources of sound are shown in Figure 9.11 relative to the standard reference. The *sone* is a unit of loudness based on a subjective ratio scaling technique.

ANATOMY AND PHYSIOLOGY OF THE EAR

The organ of hearing consists of outer, middle,

and inner ear. A thin tympanic membrane forms the partition between outer and middle ear. It has a reflective sheen that provides the physician a useful diagnostic index; a red or bulging tympanic membrane is indicative of middle ear disorder. The middle ear cavity, normally air-filled, is connected to the throat by a narrow passage called the Eustachian tube. This passage expels air from the middle ear cavity more easily than it opens for air; hence, rapid rise in altitude (reduction in ambient pressure) allows air to escape from the middle ear via the Eustachian tube, whereas rapid descent may lead to an unpleasant inward pressure of air against the drum, associated with diminished hearing. This can usually be cleared by swallowing and by thrusting the larynx and trachea downward and forward. Nasal and throat congestion may make this maneuver more difficult.

As mentioned earlier, the ear responds to frequencies between approximately 20 and 20,000 Hz. It is not equally responsive to all of these frequencies, however, as illustrated in Figure 9.12. Optimum response occurs at approximately 2,000 Hz. The nature of this frequency response is determined in large measure by the physical characteristics of the receiving organ: the nature of the external auditory canal, the tympanic membrane, the chain of small bones which links the tympanic membrane through the middle ear to the fluids of the inner ear.

In the middle ear are three ossicles: the *malleus* (hammer), *incus* (anvil), and *stapes* (stirrup), which are articulated with one another in that order, from the tympanic membrane to the oval window (to which the footplate of the stapes is attached by a

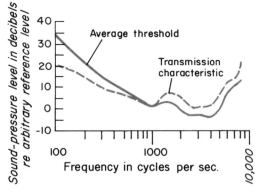

Figure 9.12. Transmission characteristics of the ear based on the summed impedances of the mechanical chain: tympanic membrane, malleus, incus, and stapes; the middle ear cavities; cochlea and round window membrane as calculated by Zwislocki, compared with an average subjective acoustic threshold curve. There is a close parallel which indicates the importance of the physical characteristics of the ear as these respond to different frequencies and thereby influence hearing. [From Zwislocki (1965).]

flexing membrane). Referring again to Figures 9.9 and 9.10, note how motion of the broad tympanic membrane utilizes leverage of the malleus favorably on leverage of the incus to impart a piston-like thrust to the stapes. When the eardrum is vibrated, the ossicles are set in motion so the stapes, with benefit of this large mechanical advantage, transmits its vibrations to the fluid-filled *scala vestibuli*.

Attenuation of Sound by Contraction of Middle Ear Muscles

There are two middle ear muscles (MEM): the *tensor tympani*, which attaches to the handle of the malleus, and is innervated by the trigeminal (Vth cranial nerve), and the *stapedius*, which attaches to the stapes and is innervated by the facial (VIIth cranial nerve). The effect of their contraction is to alter and to attenuate sound prior to its reaching the cochlea. The muscles act consensually to sound delivered to either ear, just as the pupils react to light entering either eye; that is, effective sound delivered in one ear induces contralateral as well as ipsilateral contraction of MEMs, and binaural stimulation is more effective than monaural. They can contract to enhance sound transmission in the vicinity of 1–2 kHz while attenuating responses to frequencies below and above that level (see Starr, 1969). This enhances frequencies in the central range of voice communication and reduces auditory activation by other frequencies.

MEMs have long been assumed to provide *reflex protection* of the cochlea from damaging loud sounds. This is not all they do, and it is perhaps not the most important of their functions. Since latency for acoustic reflex contraction of MEMs is about 15 ms, they offer no protection against damage from abrupt sounds such as a pistol shot, but they may serve the purpose of protecting against the continuing sounds of jet engines, for example. They respond to acoustic stimuli at 70–80 dB; psychophysical experiments indicate that perception of frequencies around 2 kHz are enhanced by MEM contractions (Reger, 1960; Smith, 1943). MEMs are active during REM sleep. MEMs participate in other activities, contracting prior to and during vocalization, thus *attenuating responses to self-initiated sounds*. MEM contractions can be elicited by sensory stimulation of the external auditory canal, pinna, face, and neck, and by yawning, chewing, swallowing, and other movements, including movements of the trunk and limbs. Every step in walking, for example, is associated with MEM contractions. They may well protect the ossicular chain during such movements. The MEMs can be contracted voluntarily, resulting in a subjective sense of faint roar or flutter. These facts suggest why it

may be an advantage when listening intently, to stand "stock still."

Note that MEM contractions associated with vocalization are not reflexes because they begin before the voice; MEM contractions associated with other movements do likewise. Nonacoustic, nonreflexive, MEM contractions are equally active in deafened animals. The organization of MEM control is complex, as implied by the different patterns of MEM responses following lesions introduced at different levels along the auditory pathway (Baust and Berlucchi, 1963). Their activity is also modified by classical conditioning (Simmons et al., 1959), by recent prior acoustic experience, a sound's significance, and purposeful activity directed to other sensory modalities (Carmel and Starr, 1962).

MEMs are active in response to prolonged sound stimulation, gradually relaxing during the course of continuing sound exposure except while contracting during bodily movements (Figure 9.13). Note the conspicuous rise in integrated activity recorded at the round window due to relaxation of the MEMs. This muscle relaxation is not compelled by fatigue, because with every gross bodily movement, the MEMs contract more strongly, simultaneously cutting down on cochlear activity as indicated by the accompanying sharp reductions in round window response.

MEM contractions in cats result in up to a 20-dB reduction in cochlear microphonics (Galambos and Rupert, 1959), and in man, with both acoustic and nonacoustic activation, MEM contractions result in up to 15-dB elevation of hearing thresholds (Salomon, 1963). Altogether, it is evident that the nervous system employs the MEMs actively to control its own acoustic input in accordance with idiosyncratic past experiences, expectations, and purposes. To this degree, subjective auditory experiences must be idiosyncratic. It is only when one establishes laboratory conditions, with controlled acoustics, similar subjects who are relaxedly attentive and given similar instructions, that there is a fair comparability of subjective acoustic experience.

Patients lacking MEMs experience a deterioration in speech intelligibility during increasing sound intensities at a level that is about 20 dB below what is experienced by normal subjects (Borg and Jackrisson, 1973). This indicates that MEMs improve discriminative hearing in the presence of loud sounds. If the middle ear ossicles are lost, of course, sound waves can travel to the oval and round windows via air in the middle ear; but this will impact both windows nearly simultaneously, and the system thereby loses the advantage of ossicular impedance matching. There is a consequent reduction in sensitivity of hearing of about 30 dB,

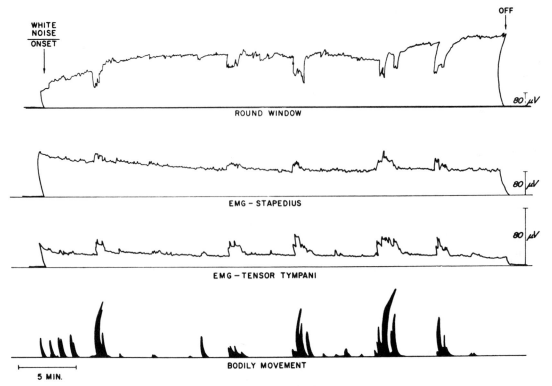

Figure 9.13. Middle ear muscle responses (cat) associated with movement, during long continued loud sound. Integrated records of electrical activity recorded at the round window, within the stapedius and tensor tympani muscles, and from a piezo-electric element recording bodily movements. Continuous white noise of same intensity throughout. The round window response, mainly summed cochlear microphonics, shows an abrupt rise at onset of the noise, and a gradual further increase throughout the duration of approximately 45 min of sound stimulation. This indicates that more sound was being received, over time, by the cochlea. This is due to the fact that the middle ear muscles are strongly but not maximally contracted throughout the entire period and are both gradually relaxing during the whole time span, especially the stapedius. Neither muscle is relaxing due to fatigue, because each major bodily movement (shown below) is associated with increase activity of the middle ear muscles, and, accompanying these contractions, the round window response is sharply attenuated. Not shown here is evidence that the relaxation of middle ear muscles during long continued loud sound is at least partially compensated by central neural attenuation that is reflected by integrated responses recorded from successively higher centers along the auditory pathway. [From Carmel and Starr (1963).]

equivalent to reducing the sound of a loud voice to a barely audible whisper.

If the oval window becomes ossified, as it does in a disorder known as *otosclerosis*, the acoustic epithelium has to depend solely on vibrations transmitted through bone. In that event, hearing loss is even more severe, because there is a very poor impedance match between air and bone. This was apparently Beethoven's problem (Goodhill, 1983). By the time he was composing his Ninth Symphony, Beethoven could barely hear cannon fire, and had therefore to compose entirely from auditory memory and musical discipline. His hearing was limited to bone conduction, so he used to hold a stick between his teeth against the sound board of his piano to be able to hear some semblance of the sounds. Modern hearing aids, which apply electromagnetically driven vibrations directly over bone that is near the skin surface, usually over the mastoid process, and which are designed to contrib-

ute a close impedance match to bone, can restore considerable hearing to persons similarly afflicted.

The Inner Ear

As noted previously, the functional interface between middle ear and inner ear is provided by the two small flexible openings: one in the vestibule between vestibular apparatus and cochlea, the *oval window*, is put into vibration by the footplate of the stapes, and the other, the *round window*, compensates for displacement of fluid in the three canals.

These canals are called *scala vestibuli, scala media* or cochlear duct, and *scala tympani* (Figure 9.16). The scala vestibuli and scala tympani contain *perilymph*, whereas the scala media contains *endolymph* which bathes the acoustic epithelium and has a different chemical composition, being much higher in potassium, and lower in sodium, chloride, and protein (see Table 9.2). This difference in electrolyte and protein composition is contributed by

Table 9.2
Composition of fluids of inner ear

	Spinal Fluid	Perilymph	Endolymph
Potassium (meq/l)	4.2	4.8	144.4
Sodium (meq/l)	152.0	150.3	15.8
Chloride (meq/l)	122.4	121.5	107.1
Protein (meq/100 ml)	21.0	50.0	15.0

From Smith, Lowry, and Wu (1954).

the *stria vascularis* which lies on the outer wall of the same endolymphatic chamber.

This introduces a steady potential difference between endolymph and perilymph. The potential is dependent on active metabolic processes and is quickly lost if circulation to the stria is compromised. The net effect of this activity is to boost the electrical potential across the basilar membrane so that it is about double that across ordinary cell membranes (Tasaki, 1954). Of course, the cochlea receives important autonomic innervation, in this context, noradrenergic fibers from the sympathetics, which control the vascular supply (Densert, 1974).

The two chambers containing perilymph, scala vestibuli and scale tympani, connect freely with one another at the top of the cochlear spiral, via the *helicotrema*. Reissner's membrane separates scala media from scala vestibuli, so it must be the first to be displaced by vibrations introduced into the oval window at the base of the cochlea, at the opposite extremity from the helicotrema. Since these chambers are all fluid-filled and confined within the hard bony box of the cochlear shell, there must be means for compensation when compression and rarefaction are introduced into scala vestibuli. This compensation is available through the *round window*, a membranous partition between the base of the scala tympani and the middle ear cavity. When pressure is applied against the scala vestibuli by inward motion of the footplate of the stapes, the round window bulges outward.

The actual sensory apparatus, the *organ of Corti*, rests on the *basilar membrane* between scala media and scala tympani. The basilar membrane is a sheet of radially oriented collagen fibers stretched between a bony partition that extends part way across the cochlea from its origin on the bony modiolus, the central axis of the helical spiral of the cochlea, to the outer wall of the cochlea. This membrane with its sensory epithelium capped by the tectorial membrane constitutes the *cochlear partition*.

Sound is coupled into the basilar membrane almost solely at its base. The physical characteristics of the basilar membrane are such that it is relatively stiff at the base and becomes less so along its length

toward the apex. The change in stiffness is by a factor of 100 along the entire length. At the same time, the membrane becomes broader toward its apex. This is true even though the cochlea as a whole is becoming narrower.

Vibrations are transmitted out along the membrane but are attenuated as they approach its tip. The amount of attenuation increases with the frequency of the sound. As the membrane becomes less stiff, higher frequencies of sound are damped to a greater extent.

There are no fixed, individual points along the membrane which, at a given frequency of stimulation, represent minimum amounts of excursion, i.e., nodal points. Rather, a wave is propagated along the membrane, with the maximum amplitude of excursion at any point corresponding to some value defined by an envelope, the shape of which is characteristic of the specific driving frequency (Figure 9.14). The lower the frequency of the driving sound pressure wave, the greater the proportion of the membrane which is set in motion. For frequencies of up to about 20–30 Hz, the entire membrane is set in motion and it vibrates in phase. That is, all along its length it is moving either down toward the vestibular canal or up toward the tympanic canal at the same time. For frequencies of up to about 200 Hz, the entire membrane is set in motion, but there will be at least one phase change along its length. Part of it will be moving in one direction and another part in the opposite direction. As suggested earlier, the wave will move along the membrane as the motion occurs. For frequencies above the 200-Hz level, less than the entire membrane is set in motion, and the higher the frequency, the less will be the length of the membrane which moves. The envelope of the traveling wave which moves along the membrane is relatively constant

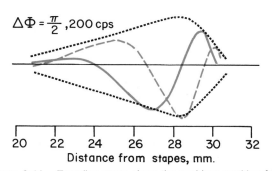

Figure 9.14. Traveling wave along the cochlear partition for a 200-Hz tone. The *solid line* indicates the deformation pattern at a given instant. The line with the *short dashes* shows the same traveling wave one-fourth of a cycle later. The *envelope* shows the maximum displacement at each point. [From von Békésy (1947).]

for a given condition of stimulation. In Figure 9.14 the envelope of the traveling wave is shown, and the nature of the excursion of the basilar membrane at any given point is illustrated by the lines drawn within the *dashed lines* which represent the limiting envelope. The *short-dashed line* and the *solid line* represent two different points in time and two different degrees of progression of the traveling wave along the membrane. It is thus clear that the nature of the motion pattern of the basilar membrane is determined by the frequency characteristics of the sound stimulation.

In summary, virtually all of the energy transmitted to the inner ear is absorbed by the basilar membrane at its stiff basal end, and motions of the basilar membrane are propagated along its length toward the broader, less stiff apex most efficiently for lower frequencies that are less damped by this organ. The nature of the motion pattern is a characteristic of the stimulating frequency, hence, the pattern of excitation of receptors distributed along the length of the membrane is also determined by the nature of the stimulus. The nature of the motion will be complex for a complex stimulus. The propagation of traveling waves from base to apex of the cochlea requires nearly 3 ms, so successive points along the organ of Corti are stimulated with increasing delay. Motion of the cochlear partition is entirely damped within that time.

For a given frequency of stimulation, the nature of the motion pattern will be altered by the amplitude of stimulation. The greater the amplitude, the greater will be the amplitude of excursion of the basilar membrane, and the greater will be the length of the membrane involved in the motion for those frequencies higher than that for which the entire length of the membrane is ordinarily set in motion. As illustrated in Figure 9.14, the envelope of the traveling wave shows how damping results in the falling off of amplitude of motion to zero beyond a certain region of the membrane at a given frequency. The nature of the motion of the basilar membrane follows from its physical characteristics. Actually, observations by von Békésy (1960) of the motion provided important clues to the physical characteristics of this organ. Figure 9.15 shows two methods by which von Békésy demonstrated movements of the basilar membrane. Amplitude of vibration is shown above, and resonance curves determined by stroboscopic light below (Békésy, 1960).

Khanna and Leonard (1982) placed an optically flat, gold crystal to serve as a mirror at known locations along the membrane. By using a laser-mirror amplifying system, they succeeded in making more refined measurements of mechanical responses of the basilar membrane. They determined the sound pressure level requirements to obtain a vibration amplitude of 10^{-8} cm in the basal turn and plotted these as a function of frequency. This revealed a shallow minimum in the region of 250 Hz to 3 kHz, a second plateau between 5 and 14 kHz, and a narrow sensitivity trough in the region of 20 kHz followed by a nearly vertical rise at higher frequencies. The more the authors were able to prevent damage to the basilar membrane, the more the curves they obtained resembled tuning curves for first order auditory afferent neurons. This suggests that the main filter in transducing acoustic stimuli may be the basilar membrane itself.

Movement of the basilar membrane is translated into neuronal firing by action of the receptors—hair cells—in the organ of Corti on the basilar membrane. There are about *3,500 inner hair cells* and somewhat more than *20,000 outer hair cells.* The inner hair cells form a single file (Figure 9.16), and the outer hair cells form a bank of three or more rows on the outer margin of the tunnel of Corti. Differences in innervation patterns between inner and outer hair cells are depicted in Figures 9.17 and 9.19. Inner hair cells have only about 50 stereocilia, whereas outer hair cells have about 100. Cochlear hair cells of all adult mammals lack kinocilia.

Recordings have recently been made from single inner hair cells (Russell and Sellick, 1981) and from single outer hair cells (Dallos et al., 1982). Hair cells in the cochlea and in the vestibular portion of the inner ear are quite similar, except for the persistence of the kinocilium on vestibular hair cells. You may refer to Chapter 63, Vestibular Functions, for a more complete account of how hair cells perform as mechanoelectric transducers. The viscous coupling of hair cells with the tectorial membrane in the cochlea plays a significant role by statically patterning stereocilia displacement. The stereocilia of inner hair cells seems barely to touch the tectorial membrane, whereas the stereocilia of outer hair cells are partly embedded in that structure. Since the tectorial membrane is hinged independently from the basilar membrane and has a free border, its action can be presumed to be different in relation to the two receptor cell populations.

The general innervation pattern of the organ of Corti is shown in Figures 9.17 and 9.19. There is a single afferent nerve ending for each inner hair cell. Other afferents serve a few outer hair cells, all of which are located toward the basal turn of the cochlea from where the afferents enter the organ of Corti. Efferent neurons, which number about 1200, are distributed more diffusely. They are shown in Figure 9.17 with the stippled synaptic enlargements

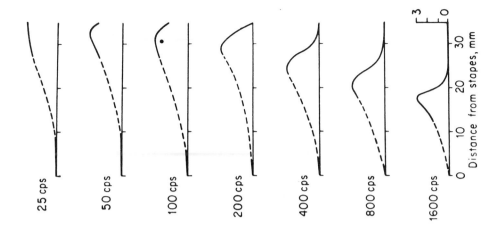

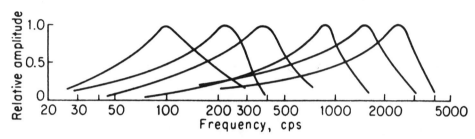

Figure 9.15. Patterns of vibration of cochlear partition of a cadaver specimen for various frequencies and distances from stapes. *Above,* Amplitudes of partition vibration to different frequencies were measured millimeter by millimeter along the cochlea and graphed. Note that the upper part of the cochlear partition does not vibrate at frequencies above 800 cps. The *dashed lines* represent similar amplitude measurements made closer to the stapes. These observations required cutting away more and more from the apex. *Below,* Resonance curves for six positions along the cochlear partition made using stroboscopic illumination. [From Bekesy (1960).]

applied to afferent axons near the point of afferent departure from the inner hair cells. Other efferents have multiple large diameter terminations which synapse with large numbers of outer hair cells.

Sound causes the cochlear duct to swing alternately to and from the scala tympani and the scala vestibuli, which motion causes displacement of the basilar membrane in relation to the tectorial membrane. This introduces shearing forces which bend the stereocilia and provide an adequate stimulus to these receptors. Because of the unusually high standing potential, introduced by metabolic activity of the stria medularis, between endolymph and the interior of the receptor cell, amounting to about 180 mV, changes in membrane conductance induced by the shearing forces are accompanied by a more than usually rapid influx or efflux of ions. These in turn generate the receptor potential that initiates the release of neurotransmitter at the synaptic junctions the receptor cells share with afferent nerve endings.

Cochlear Potentials

Apart from the standing potential across the organ of Corti, there are three principal *dynamic* electrical phenomena that accompany sound stimuli delivered to the cochlea: *cochlear microphonics* (CM), positive and negative *summation potentials* (+SP and −SP), and *nerve action potentials* (AP) of the auditory nerve, the cochlear action potential.

Cochlear microphonics (CMs) relate to the instantaneous pattern of vibration of the cochlear partition and represent two summating potentials relating to up and down displacements of the cochlear partition as it follows the envelope of excursion of acoustic stimuli. CMs are well-named, because their electrical responses are faithful to the stimulus in the manner of a physical microphone. The summation potentials relate separately to bending of the stereocilia (+SP) and generation of nerve activity (−SP). The AP depends on nearly synchronous discharge of first order afferents in response to brief transient acoustic stimuli. The AP represents the first nervous system access to the acoustic world.

Recording from the round window, or in humans with an electrode that penetrates the tympanic membrane, one sees CMs followed by large ampli-

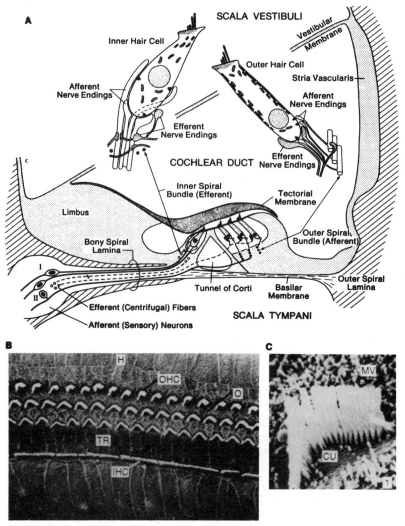

Figure 9.16. *A,* Cellular organization of organ of Corti in the guinea pig. Enlarged diagrams, located by arrows, show fine structure of inner and outer hair cells. *B,* Scanning electron micrograph looking down on hair cells, showing differences in arrangement of inner (IHC) and outer (OHC) hair cells and their stereocilia. *H,* Henson's cell; *TR,* tunnel rod. *C,* Scanning electron micrograph of stereocilia of an outer hair cell. *MV,* microvilli (stereocilia); *CU,* cuticular plate. [*A,* redrawn from Smith in Brodal (1981); Smith, in Eagles (1975); Shephard (1983).]

tude negative-positive-negative sequences of the AP (Figure 9.18). The negative AP waves are referred to as N_1, N_2, and N_3. If the stimulus polarity is reversed from condensation-rarefaction to rarefaction-condensation, the CMs reverse polarity, as would a microphone, but the AP does not. If a continuing, varying sound or noise is introduced, as is likely in the natural environment, the cochlear microphonics tend to override the neural responses so that the latter cannot be recorded. That problem can be evaded, however, by using a computer both to generate the desired varying sounds used in the stimulus *and* to subtract those same wave forms from the microphonic responses, with an appropriately scaled and delayed pattern that corresponds to the sound stimulus (Clark and Livingston, 1962).

COCHLEAR MICROPHONICS (CMs)

There is uncertainty as to origin of the CMs, but they probably represent a weighted sum of activities in many hair cells. The phase shift of CM as measured along the basilar membrane (Tasaki et al., 1952) probably coincides with motion of the basilar membrane and reveals that the phase (wave) velocity is exponential and independent of frequency of the sound (Honrubia, 1970). The CM cannot account for the sharp tuning curves exhibited by first order neurons. As indicated above, the basilar membrane is probably largely responsible for the characteristic sharp critical frequency (CF) sensitivity. The electric current of the CMs must flow through the body of the hair cells and may contribute to

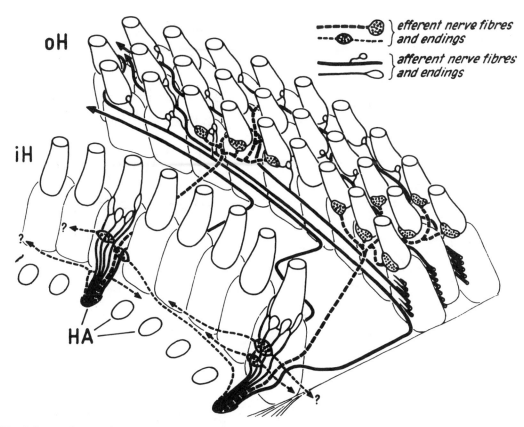

Figure 9.17. Schema of general innervation pattern of the organ of Corti. Afferent innervation of outer hair cells goes spirally a long distance in one direction only, basalward. Efferent synaptic endings are abundant, inhibiting afferents shortly after they leave their attachment to the inner hair cells and directly ending on outer hair cells. Efferents come from both ipsilateral and contralateral superior olives. iH, inner hair cell; oH, outer hair cell; HA, habenula. [From Spoendlin (1970).]

receptor mechanisms. This represents an electrical effect generated by cellular activity the sum of which (in this case, the CMs) has a feedback effect upon the very same cells that contributed to the sum in the first place, another example of integration.

CMs cannot be more than *indirectly* responsible for nerve firing, for the following reasons: CMs have virtually no latency, auditory nerve impulses do; CMs have no refractory period, auditory nerves do; CMs do not fatigue, neurons do; CMs follow stimulus rates in excess of those which can be followed by auditory neurons; there is a linearity in CMs following shifts of sound intensity which is not equivalently matched by neural responses. CMs, therefore, are preliminary to nerve responses, and perhaps only remotely related to their initiation.

SUMMATING POTENTIALS (+SP AND −SP)

Additional complex cochlear potentials are known as positive and negative *summating potentials* (+SP and −SP). These can be recorded best from electrodes in the scala tympani. The two sum-

mating potentials can be separated in time and space and are believed to have independent origins. The +SP are thought to be a consequence of bending hairs of the hair cells (Davis, 1958). Measurements of latency of the −SP and of its responses during hypoxia leads to the notion that −SP may have a neural origin (Kupperman, 1966, 1970). Moreover, −SP can be masked, as neural response can, whereas neither +SP nor CMs can be masked.

NEURONAL ACTION POTENTIALS (AP)

There are about 30,000 primary afferent neurons with cell bodies in the spiral ganglion which lies within the bony modiolus along the axis of the cochlea. This gives a ratio of about 10:1 primary afferents to the number of inner hair cells. The afferent neurons are bipolar cells with axonal branches that extend peripherally to innervate the cochlea and centrally to innervate the cochlear nuclei. Peripheral branches acquire myelin as they leave the organ of Corti. Since the unmyelinated region is therefore quite long, there is an opportunity for integration among graded responses to

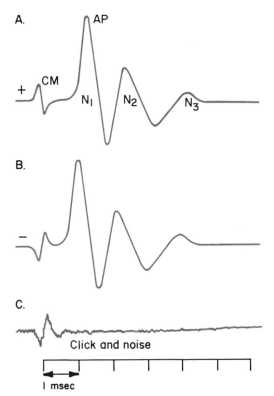

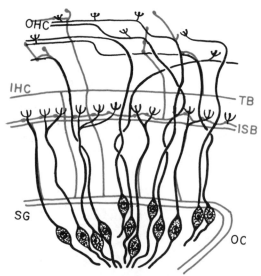

Figure 9.19. Schematic illustration of the innervation of cochlear hair cells. Afferent fibers are in *solid black lines;* efferent fibers are represented by red lines. *OHC,* outer hair cells; *IHC,* inner hair cells; *SG,* spiral ganglion; *OC,* olivocochlear bundle; *ISB,* internal spiral bundle; *TB,* tunnel bundle. [From Smith and Rasmussen (1963).]

Figure 9.18. The electrical response to clicks as recorded from the round window of the cochlea of the cat. *CM,* cochlear microphonic potential; *AP,* action potential of the auditory nerve. The three components of the action potential are represented by N_1, N_2, and N_3. A reversal of polarity of the stimulus between *A,* at the top of the figure, and *B* results in an inversion of the cochlear microphonic but no change in the action potential except for a slight shift in latency. At *C,* the action potential is completely masked by white noise, while the cochlear microphonic remains unchanged.

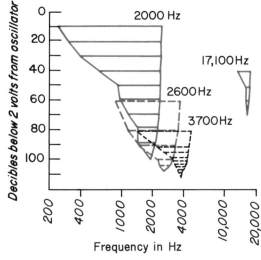

Figure 9.20. Response areas of four neurons in the cochlear nucleus of the cat. Each neuron has a specific frequency at which a minimal acoustic energy is required to excite it. Each responds also to progressively lower frequencies at higher intensities. There is a slight extension of the range of response into higher frequencies, but the cut-off is much sharper on the high frequency side. Some distortion of the shape of these cut-off functions as a result of the logarithmic frequency plot must be taken into account. [From Galambos and Davis (1943).]

occur within a single axon and perhaps interneuronal effects among bare axons traveling together.

In the cat 95% of the afferent neurons have a single terminal which ends on a single inner hair cell (IHC Figures 9.16 and 9.17). Each inner hair cell may be innervated by 8–10 different first order neurons. Most of the remaining afferents cross over the bottom of the tunnel of Corti to spiral toward the base of the cochlea a distance of about 1 mm, using their last 0.2 mm to innervate several outer hair cells. A fractional subpopulation of spiral ganglion cells (0.5%) consists of large myelinated axons which do not cross the tunnel but branch and course toward the apex as well as toward the base of the cochlea and innervate up to 10 inner hair cells each.

Most if not all primary auditory fibers of mammals are spontaneously active, exhibiting from a few to 100 or more spikes/sec. (Liberman, 1978). Each auditory afferent has a critical frequency (CF) which subtends a narrow frequency band of greatly reduced threshold (Figure 9.20). This lowest CF

threshold is a way of characterizing that neuron, even though it may respond to a broad range of frequencies at substantially higher stimulus intensities. The CF for a given neuron has functional importance like that of the sensory *field* for a first order tactile unit or the visual *receptive area* for a unit in the retina. Like those designations, CF is useful for identifying units which occupy higher stations throughout central representations of the auditory system. Like tactile and visual fields, CF is only one parameter of potential auditory analysis, but one of prime importance.

The overall frequency response curve of a given neuron may be narrow or broad, and symmetrical or asymmetrical. The frequency response curve for a typical first order auditory neuron is abruptly steep on the high-frequency side with a generally broader shoulder on the low-frequency side (Figure 9.21). Even though the overall frequency response curves may be broad, the CF identification is important because the aggregate of all such CF profiles constitutes the neural equivalent of the audiogram, the threshold for hearing over the entire auditory range. The sharp tuning of these units is not only dependent on basilar membrane responses but also on an adequate supply of oxygen (Evans, 1973). A modest reduction of oxygen tension distinctly elevates the CF threshold.

When two tones are used as the stimulus rather than one, there appear discrete patterns of interference, presumably lateral inhibition exerted between afferent neurons. The term inhibition in this case is used by inference rather than in confidence. True inhibition between first order auditory afferents has not been certified. The difficulty is that there is no way, with the auditory system being attached to a moving partition to eliminate the hypothesis that what has been interpreted as lateral

inhibition is due solely and entirely to mechanical phenomena. If so, the auditory system is the only one that does not manifest lateral inhibition among its first order neurons. We may assume for the present that there are undoubtedly mechanical effects that resemble lateral inhibition, and that there probably also exists lateral inhibition among first order neurons in the auditory system. In any case, the phenomenal consequence is that individual unit frequency response curves become even more individualistic and narrower (Figure 9.22). All primary auditory neurons show such two-tone suppression. In addition to likely interneuronal inhibitory-type processes, there is also apparent recruitment of one fiber by another, influencing the one to fire when it wouldn't otherwise have done so, or to increase its rate of spontaneous firing.

It has been estimated that when the ear is stimulated by a 40-dB tone at 1,000 Hz, approximately 3,000 hair cells may be activated. With a change in frequency by a just discriminable amount, it is probable that less than 2% of these hair cells are no longer activated and have been supplanted by others which were formerly inactive. A great deal of evidence suggests that it is the overall pattern of stimulation of many receptors rather than the activity of just a few which is involved in even the relatively simple discriminations of pure frequencies from one another.

At onset of noise or tonal stimulation that includes appropriate frequencies, the spontaneous rate of firing of an individual afferent neuron will

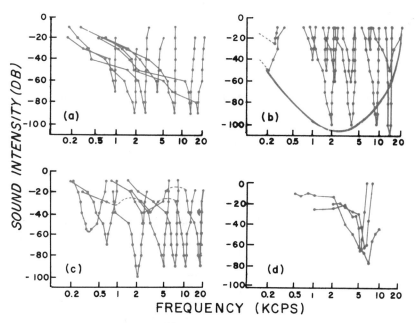

Figure 9.21. Response areas of single neurons obtained from *a*, cochlear nerve; *b*, inferior colliculus; *c*, trapezoid body; and *d*, medial geniculate body of the cat. [From Katsuki, 1961).]

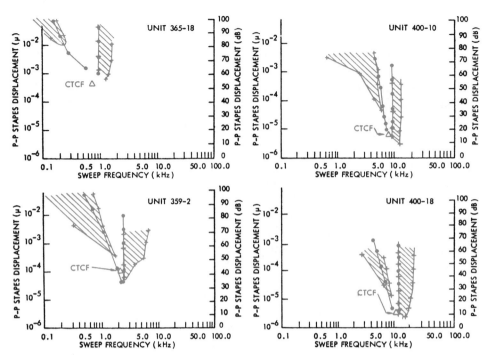

Figure 9.22. Response and inhibitory areas for four fibers in the cat auditory nerve. The characteristic frequency (*CTCF*) is marked by a *triangle* and represents that frequency to which the unit is most sensitive. The boundary of the response area, marked by *solid circles*, is defined by the amplitude of a sweep frequency which results in an increase in rate by 20% over the level of spontaneous activity. Ordinate axes represent peak to peak stapes displacement in micra and in decibels. The *hatched areas* represent frequencies and amplitudes in which the sweep frequency reduces the response rate by 20% or more below the rate obtained with the characteristic frequency alone. These inhibitory areas are defined by presenting the characteristic frequency continuously along with sweep frequency, the latter varying exponentially in time between two frequency values which differ by one or two decades. [From Sachs and Kiang (1968).]

increase. If the tone is continued, the unit will shortly adapt to a lower level of activity which resembles its spontaneous activity but at a higher average rate. If a pure tone is presented, there is *phase-locking*, that is, units will tend to fire at a certain phase of the stimulus but not necessarily regularly with every cycle of the tone. Whether a unit fires or not remains probabilistic, as it was during spontaneous activity; but when the unit does fire, it is more likely to fire in time (phase locking) with a particular part of the ongoing stimulus cycle. Phase locking of responses in primary afferents in cats and monkeys can follow sinusoidal stimuli up to about 5,000 Hz. The actual unitary firing rate is considerably lower, but phase locking in nonetheless obvious (Kiang et al., 1965).

ACCOUNTING FOR THE DYNAMIC HEARING RANGE

We are able to distinguish speech and other sounds in the middle range of audible frequencies over a dynamic range of intensities from the faintest whisper to the loudest shout, a range of about 100 dB. This is most remarkable—a range of 10^{12} (Fig. 9.11). How does the biological system manage

this at the input stages? Most individual first order neurons have a dynamic range that seldom exceeds 30–50 dB. Evans found that not all cochlear fibers have restricted dynamic ranges; many have dynamic ranges in excess of 60 or 70 dB and some are not completely saturated (that is, do not reach their highest firing rate) at even greater dynamic ranges. Because of lateral inhibition, and mechanical interferences affecting the organ of Corti, not all cochlear fibers ever become fully saturated. Moreover, dynamic range could be coded by some combined index of activity, i.e., rate and degree of phase locking.

Evans found also that units activated by frequencies remote from their CF are more indifferent about phase locking. There is a shift of a unit's CF toward lower frequencies at higher intensities if the CF is above 1 kHz, and a shift of the CF toward higher frequencies at higher intensities if the CF is below 1 kHz. This means that at higher intensities individual units shift their CF, and the net effect is to move their CFs toward an acoustic center around 1,000 Hz. These intensity-imposed shifts are considerable, up to ½ octave the lower frequency shoulder along with shifting of the CF (Evans, 1981).

Temporary Threshold Shift Following Loud Sound

The natural environment is relatively noisy. Following prolonged exposure to loud noise at a given frequency, there is an elevation of auditory threshold for some minutes. Oddly enough, the effects are greatest for frequencies *higher than the frequency of loud sound exposure*, by from ½ to 1 octave. Long duration background noise biases units to shift their CFs toward higher frequencies. That this is not a question of fatigue is shown by the fact that similar noise interspersed with tone bursts does not show this effect. The CF shift in the presence of long duration background noise therefore is probably an adaptive mechanism, some sort of automatic gain control in the graded response area of the terminal axon prior to the probabilistic spike generator.

Summing up Cochlear Plasticity

Leaving aside disease processes which may affect the external, middle, and inner ear, and such interferences as water or wax damping the tympanic membrane, what systematic alterations of characteristics of the physical stimulus occur before first order neurons can deliver messages to the central nervous system?

The human ear responses in frequency and threshold conform to physical limitations of the peripheral auditory mechanism. The ossicular chain is remarkably but not completely efficient as an impedance matching device.

The middle ear muscles are dynamically active in anticipatory as well as in reflex operations; they act differently on the basis of recent acoustic experience and are conditionable as well as under some degree of voluntary control. They may be dynamically active during steady-state conditions, quiet or noisy, and they seem to operate on behalf of interests of the host that introduce individualistic distortions into incoming signals.

The basilar membrane has imperfect response characteristics which may be especially distorting in the context of abruptly shifting, complex sounds. Basilar membrane contributes to the sharp best frequency profile of first order neurons. Mechanical events in the cochlear partition manifest interference phenomena which strongly resemble neuronal lateral inhibition and may cross-react with such. First order lateral inhibition which would have the effect of sharpening best frequency profiles probably exists but is difficult to demonstrate because of physical dynamics of the cochlea. Both the sensory epithelium and its surroundings in the cochlear duct are continuously dependent on an adequate supply of oxygen.

There are important dynamic as well as standing potentials, and these partly derive from and partly influence cochlear receptors and nerve endings, as well as interact with one another.

First order neurons are spontaneously active, and this influences their responsivity (as in phase locking), as well as serving the main purpose of enabling detection of subtle changes in acoustic signals. First order afferent neurons exhibit a variety of frequency response curves and phase locking propensities under idealized pure tone conditions. These are drastically affected by two-tone stimulation, and even more radically by complex sounds. The critical frequency as well as more general aspects of a unit's frequency range are altered by hypoxia, tonal interactions, differences in sound intensity, background noise, and recent past acoustic history. Thresholds, critical frequency, and phase locking activities are all shifted by recent environmental sounds. The shifts are substantial for perceptual processing, in the range half to whole octaves.

Somewhat more spectacularly, *efferent* neurons to the cochlea may add a further plasticity to receptor and first order neuron activity and responses, a matter of central nervous system control over sensory receptors and sensory neurons, which topic will be taken up in the next section.

What this admittedly incomplete inventory of factors which contribute to cochlear plasticity shows is that first order auditory neurons are not only marvelously sensitive and especially acutely sensitive to a critical frequency, but that their modifiability as units and as an ensemble cannot be adequately defined by what is learned in the course of presentation of pure tones at near-threshold levels. The ear as a hearing organ is dynamically modified in its performance by commonplace background and louder noises that assault our ears, but is, nevertheless, highly resilient and, under controlled conditions, extremely reliable.

Efferent Control of the Cochlea

A discrete bundle of *efferent* fibers leaves the brain stem in company with the vestibular branch of the VIIIth cranial nerve. This divides in two to innervate both vestibular and cochlear sensory epithelia. The neurophysiology of vestibular efferents is described in the next chapter. Figures 9.16, 9.17, 9.19, and 9.27 illustrate how cochlear efferents terminate. The efferent system is presumed to exert a continuing, dynamic central nervous system control over auditory input. Some efferent fibers separate before the bundle leaves the brain stem, and these innervate the cochlear nuclei. Altogether, this outflow system exerts powerful influences on cochlear receptors, afferent nerve endings, and first-to-second-order auditory relay. The effects are notable all along auditory pathways, from cochlea to auditory cortex, and downward along auditory feedback channels. Efferent control of the cochlea has been the subject of general reviews (Klinke and Galley, 1974; Iurato, 1974).

Axons that comprise the efferent bundle originate from both ipsilateral and contralateral neurons in a cluster of brain stem nuclei known as the superior olivary complex (Rasmussen, 1960). Although the *olivocochlear bundle* (OCB) is made up of only about 1200 axons, these branch extensively and provide multiple dense synaptic endings on all 20,000 or so outer hair cells. Their activation attenuates responses of outer air cells and causes reduction or failure of afferent nerve impulses arising from this part of the cochlea. This amounts to *presynaptic inhibition.*

The pattern of efferent innervation relating to the inner hair cells is quite different. These efferent fibers rarely synapse directly on the hair cells but instead synapse *en passant* with first order afferent neurons close to their sensory attachment to the inner hair cells. In this location, efferent signals attenuate or block afferent nerve impulse generation. This amounts to *postsynaptic inhibition.*

All contacts of OCB axons, *en passant,* and at terminal endings, contain distinctive granular synaptic vesicles which are about 30 nm in diameter. These large, grainy looking synaptic regions characterize *efferent* synapses and aid in interpretation of fine cochlear circuitry. On the basis of histological and pharmacological evidence, they involve cholinergic transmission (Guth et al., 1972). Choline acetyltransferase is present in OCB terminals, indicating that acetylcholine is manufactured locally. Moreover, stimulation releases acetylcholine into the perilymph (Fex, 1968). When the system is physiologically active, there is a DC shift in both endocochlear and summating potentials.

Crossing olivocochlear fibers travel superficially beneath the floor of the IVth ventricle and can be stimulated or surgically interrupted in that location. Galambos stimulated the crossing fibers and obtained transient suppression of cochlear N_I and N_{II} first order afferent responses. Cutting the bundle distal to the point of stimulation eliminated this effect (Galambos, 1956).

Although the olivocochlear bundle can be activated reflexly, it is influenced to a greater extent by central descending controls. Descending im-

pulses converge on the OCB system fairly directly from auditory stations in the brain stem, and are relayed via those nuclei from diencephalic, and cortical stations—the medial geniculate nucleus, and temporoinsular cortex. Neurons that give rise to the crossed OCB receive impulses bilaterally from the contralateral cochlear nuclei, inferior colliculus, and nucleus of the lateral lemniscus, that is, from each of the major contralateral auditory nuclei in the brain stem. The uncrossed bundle receives only from the cochlear nuclei, bilaterally.

Sound can elicit activation or inhibition of this system. These effects can be elicited by sound in either ear; the latency varies but is usually greater than 5–10 ms, and the effects build up during 100–200 ms and are longer lasting. In steady-state conditions, OCB effects may include reduction or elimination of spontaneous activity among auditory neurons. With sound-evoked activity, the auditory neurons are strongly inhibited, with reductions in response up to 25 dB. Inhibition is greatest in the area of an afferent neurons's critical frequency (CF). Sharp profiles may be blunted more than 20%. The characteristic effect is elevation of cochlear microphonics (CM) by a few decibels, and simultaneous reduction in amplitude and slight increase in latency of afferent responses. Sounds most effective in eliciting OCB responses are in the middle frequencies of the auditory range, the greatest effects on CM being at 1 kHz, and on afferent nerve attenuation at 5–10 kHz. Simultaneously with inhibition of afferents, there is a marked reduction of responses throughout the ascending auditory pathways (Desmedt, 1962).

If the cochlear action potential is being masked by another sound, activation of the OCB improves signal to noise ratios in the cochlea (Dewson, 1967). Dewson (1968) trained monkeys to discriminate between human speech sounds presented in the presence of noise of different intensities; there was significant impairment in discrimination following sectioning of the crossed OCB fibers. Sectioning interfered with refined "perceptual" signal to noise discrimination, as it had with "physical" signal processing at the level of the cochlea.

In summary, by means of the olivocochlear bundle, the central nervous system seems to be able to exert control over peripheral acoustic receptor mechanisms and central sensory relays for the benefit of cognitive processes at higher levels. The net effect seems to be a relative exclusion of some sounds and an improvement of signal extraction from others. Further, the olivocochlear bundle system appears to function in some ways analogous to the middle ear muscles, one step closer to the central nervous system. There is a need to discover how closely their central control patterns may be linked. The overall physiological design of both of these centrifugal systems for the analysis of sounds allows the central nervous system to be more selective about what it hears, and in the process, to hear better what it wants to hear.

CENTRAL AUDITORY PATHWAYS

Central auditory pathways are arbitrarily divided into *ascending* and *descending* components and are described in that order. It is necessary to bear in mind, however, that these pathways actually constitute shorter and longer *feedforward and feedback loops* which control both "ascending" and "descending" messages as these are being relayed in either direction. More is involved, particularly at brain stem levels, because the auditory system is also heavily engaged in *brain stem and spinal acoustic reflexes, brain stem mechanisms relating to audiovisual coordination, and auditory participation in signals being exchanged between the cerebral and cerebellar hemispheres.*

Imagine a hierarchy of potentialities in which conscious auditory experience and conscious acoustic-related performance are at the top. At the bottom, an acoustic startle reaction can trigger involuntary brain stem and spinal reflexes prior to any conscious participation. Auditory input has direct access to mechanisms of *arousal, a variety of acoustic reflexes, and rapid acoustic-motor* operations achievable through highly practiced auditory skills, as well as to increasingly conscious progressions from lowest (most automatic) to highest (most cognitive) levels of utilization of acoustic information.

Decisive switching between operations that use short neuronal loops at lowest levels to operations that use longer-loop circuits is accomplished in the brain stem, mostly automatically and semiautomatically, with little deliberately conscious intervention or modification of decision switching possible. The ascending pathways provide longer loop circuits and, ultimately, vast telencephalic regions for subjective participation. No doubt we are fortunate that we aren't required to operate lower level loop processes under conscious control.

EXAMPLE OF PLASTICITY OF RESPONSES TO ACOUSTIC STIMULI

It is possible to implant permanent indwelling bipolar electrodes in the brains of experimental animals in many different regions, auditory and otherwise. Bipolar electrodes ensure that responses recorded will be due primarily to local events, whereas monopolar electrodes pick up any activity

between their location and the "indifferent" or ground electrode. Oddly enough, with bipolar electrodes, in fully awake animals, it is possible to pick up large amplitude responses to acoustic stimuli in areas of the brain not ordinarily considered to be auditory. It depends on whether the experimenter makes the acoustic stimuli "significant."

Figure 9.23 illustrates evoked potentials recorded with bipolar electrodes implanted in the caudate nucleus (part of the extrapyramidal motor system—the basal ganglia), intralaminar nuclei in the thalamus (part of the arousal mechanism of the brain), midbrain (nucleus of Bechterev, the superior vestibular nucleus), and a classical auditory relay nucleus, the medial geniculate body (Galambos, 1961). The illustration demonstrates the effects of simple Pavlovian conditioning on activity at each of these several areas.

In this study, clicks were presented once per 5 s until "habituated." When clicks are first presented, the click-evoked responses tend to be widespread, and the amplitude of click-evoked potentials is large, at least throughout the classical auditory pathway. As the same acoustic stimulus is repeated many times, the response fluctuates but gradually declines to some steady-state lower level of amplitude. It is then said to be "habituated." An "unconditioned stimulus" is then applied, in this case a mild puff of air directed at the monkey's face. This is by no means painful, but it is a bit annoying. The puff of air is timed with the clicks. The immediate effect is the generation of widespread large amplitude responses to the puff-click combination. Even when the puff of air is discontinued, the click-evoked responses remain elevated for some time. The middle part of Figure 9.23 shows "extinction trials," that is, trials with clicks alone after the puff of air to the face is discontinued. The bottom traces show that after 60–80 extinction trials it is difficult to see any evoked potential to the clicks, even in the medial geniculate body recording. However, if only one puff of air to the face is introduced, the click-evoked responses are widespread again and have large amplitudes. The effects are now more difficult to habituate. This suggests that the brain of the monkey is manifesting interest in whatever may be associated or related by conditioning to the annoying stimulus. Similar kinds of experiments have been done including rewarding rather than annoying stimuli, and the effects are analagous (Galambos et al., 1956; Hernández-Peón et al., 1956; Hernández-Peón, 1960). An animal's brain manifests interest in this way to either rewarding or punishing events and *specific interest in events linked to biologically meaningful stimuli.* Also, the

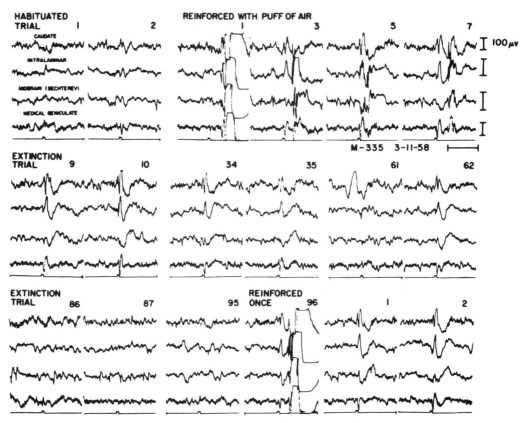

Figure 9.23. Evoked potentials recorded with bipolar electrodes implanted in monkey in basal ganglia (caudate), thalamus (intralaminar), midbrain (bechterev), and auditory pathway (medial geniculate) during simple pavlovian conditioning: Clicks presented about once per 5 s until habituated; reinforced with accompanying puff of air in the face and becoming habituated again when no longer reinforced. The large amplitude evoked responses are readily restored by a single reinforcement. [Galambos, 1961.]

brain seems able to bring into activity widespread regions that do not have "wired-in" connections with the pathway along which the conditional stimulus (the click) normally travels.

This line of evidence suggests that the circuits of the brain are more modifiable than we ordinarily presume and that routing of signals depends at least in part on what signals seem to be important to a particular animal at a particular time in a particular context. For instance, a similar monkey, similarly implanted, given a first puff of air associated with click would probably show large, widespread responses, but habituation would be different from this animal which has been reinforced after habituation. So the past history of events affecting a sensory pathway—even indirectly—are important.

The auditory system is a good example of *transactional mechanisms*. The word "transaction" has special significance in neurophysiology. It indicates much more than interaction, as between two interdependent components. Everyday language tends to limit our imagination to interactions, to subject-predicate progressions, step at a time, even when addressing complex processes. *Transaction means*

multiple, mutually interdependent systems in simultaneous action. Transactional mechanisms constitute irreversible thermodynamic processes that are indispensable for all living systems (Prigogine, 1977).

Ascending Auditory Pathways: Cochlea to Cortex

Ascending auditory pathways (Figure 9.24) comprise a succession of neuronal relays for transmission of information that climaxes in hearing. The process of cochlear transduction of sound into neuronal events gives origin to detailed, dynamic spatiotemporal representations in the nervous system of complex motions of the cochlear partition. Frequency, duration, time interval, and other time-dependent distributions are delivered to the central nervous system by neuronal signals. From this information, individuals with normal hearing discriminate pitch, loudness, localization of sound in space, and the time, intensity, and frequency modulations that are essential to the interpretation of complex sounds, as in speech.

Even a low intensity pure tone activates a great

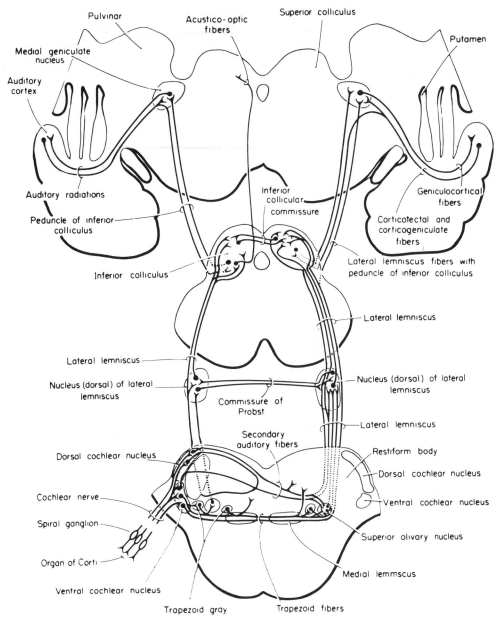

Figure 9.24 Major central pathways of the auditory system. [From Crosby et al. (1962).]

length of the cochlear partition; hence, there cannot ever be reasonably local neuronal activation as there can be in vision and touch. Two *just discriminable tones* excite two extensive and nearly completely overlapping neuronal populations. Such subtle differences must be distinguished against background "spontaneous activity" occurring among almost all auditory neurons. The problem is how does the nervous system differentiate patterns of activity that differ by so little? Similarly, how can we distinguish two *just discriminable intensities* that have the same tonal pattern? How can we locate a sound in space that must be distinguished by *just discriminable differences in intensity*

and latency (intensity, because of the sound shadow of the head, and latency, because of differences in time of transit of sound to the near and far ears, with sound traveling 330 m/s in air)? Further, how do perceptual processes project the significance of these representations beyond the the ear into the surrounding three-dimensional space?

COCHLEAR NUCLEI (CN)

First order axons from cell bodies in the spiral ganglion of the cochlea bifurcate on their way to the brain stem and deliver impulses to both *dorsal and ventral CN* (Figure 9.24). First order and second order neurons maintain a strict cochleotopic (tonotopic) pattern of representation, creating dynamic three-dimensional nuclear "maps" of sounds. A comprehensive review of auditory mechanisms in the CN and superior olivary complex is provided by Brugge and Geisler, 1978.

The CN are organized somewhat like cerebellar cortex, with intrinsic granule cells which receive input from cochlear afferents, from cells intrinsic to the CN, and from higher auditory stations. The granule cells send parallel fibers to synapse with rows of apical dendrites of pyramidal cells, which in turn project to higher centers as far as the inferior colliculus.

In the CN, the first central auditory relay, many cells have "primary neuron-like" responses, and others exhibit more complicated firing patterns because they reflect greater influence by intrinsic neurons and descending terminals. Both types of CN cells project separately to the superior olivary complex. They also project in parallel to distinctive destinations in higher centers, mainly via the contralateral inferior colliculus. This initiates a pathway that preserves an approximation of the characteristics of primary afferent firing all the way to cortex. This system finds representation in the primary auditory cortex, among several other kinds of inputs, and contributes to what is considered *"core" auditory representation*, which is confined to that primary cortical area. More complicated and "cosmopolitan" ascending messages contribute at the cortical level to *"belt" auditory representation*. This identification derives from the fact that these cortical areas form a circumferential auditory cortical zone around the "core" primary auditory cortex (Figure 9.25).

Fibers from the dorsal cochlear nucleus pass over the inferior cerebellar peduncle, cross the midline, and reach all the way to the inferior colliculus. Some fibers, along the way, give off collaterals or terminate in the contralateral reticular formation, superior olivary complex, and nuclei of the lateral lemniscus. The dorsal cochlear nucleus thus provides a predominantly crossed auditory pathway. Fibers from the ventral cochlear nucleus cross beneath the inferior cerebellar peduncle, give off collaterals to the reticular formation bilaterally, and terminate bilaterally in the superior olivary complex. The inferior cochlear nucleus thus contributes more locally and bilaterally. Superior olivary axons contribute massively to the main ascending auditory pathway in the brain stem, the lateral lemniscus (lemniscus = ribbon), and pass either directly to the inferior colliculus or relay upwards via the nucleus of the lateral lemniscus. *Thus, the "main ascending auditory line" arises from cochlear afferents which divide to serve two cochlear nuclei. The*

dorsal CN contributes a "throughway" that projects directly and indirectly to the inferior colliculus. This "throughway" system contributes heavily, but not exclusively, to "core" representation in auditory cortex. The ventral CN contributes a "byway" that goes mainly bilaterally to trapezoid body, superior olivary complex, and reticular formation. This becomes a polysynaptically relayed ascending auditory pathway that carries along with it much else besides acoustic information. This "byway" auditory system contributes mainly to "belt" representation in auditory cortex.

SUPERIOR OLIVARY COMPLEX (SOC)

This complex provides a major ascending auditory relay, which is at least third order, and it also gives rise to the efferent projections of the olivocochlear bundle (OCB), described in a previous section, and integrates binaural stimuli.

As a relay station, the SOC transmits acoustic messages bilaterally, preponderantly contralaterally, to nuclei of the lateral lemniscus and to the inferior colliculus (see Figure 9.24). It receives second order inputs from the ventral cochlear nuclei on each side of the brain stem, and from the dorsal cochlear nucleus on the contralateral side. It also receives information important for comparative timing of events in the two ears from axons of neurons in the trapezoid body and from collaterals of the cochlear axons that serve trapezoid neurons. Since the OCB strongly influences primary and second order auditory neurons, the SOC participates in a kind of braided peripheral and central feedback system. This weaves in trapezoidal contributions at the brain stem level, while at the same time, it is involved with both ascending and descending auditory pathways which contribute impulses that converge back again upon the SOC. The SOC feedback system thus has both horizontal control (cf. spinal segmental reflex) braid, and vertical control (cf. spinal longitudinal reflex) components.

The olivary complex is organized tonotopically, somewhat differently in different regions of the complex, as is appropriate for locally different kinds of information analysis. Most olivary units resemble first and second order neurons in their response characteristics. Some, however, introduce a new type of signal, *"off"* or afterresponses, that is, they become active or more active when a sound is *discontinued*. In this, they resemble "off" response units in the visual system.

OLIVARY ANALYSIS OF SOUND LOCATION IN SPACE

Everyday experience indicates that the *direction*

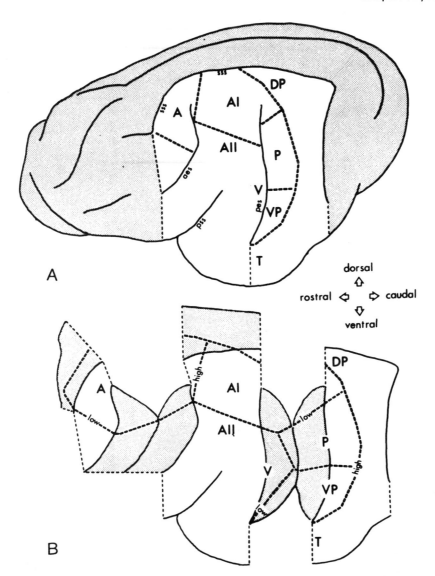

dorsal
⇧
rostral ⇦   ⇨ caudal
⇩
ventral

Figure 9.25. The core auditory field of the cat divided into functional subdivisions, *AI, A, P, VP*, receives inputs from medial geniculate nucleus of thalamus by which information from two ears reach cortex, delivering best frequency and binaural maps of acoustic world. This field receives elaboration of this thalamocortical input from same and opposite hemispheres, with alternating dense bands of contralateral summation and dominant suppression, all tonotopically organized. Ipsilateral projections from field *A* and *P* into *AI* relate to the binaural analytic functions of *AI*. A surrounding belt of cortex, *AII, V, DP, T*, including cortical areas deep within the sulci, shown below opened out to the surface, contribute additional discriminative functions [From Imig and Reale (1981).]

of a sound source can be fairly accurately estimated. The difference in path length δ l is $d \cdot$ sine α, where d is the distance between the two ears and α is the angle between the sound source and a coronal plane through the ears. The difference in arrival times δ $t = \delta$ l/c, where c is the speed of sound.

When $\alpha = 30°$ and the distance $d = 0.17$ m between the ears, the time delay is δ $t = 0.085/340 = 2.5 \cdot 10^{-4}$ s. The human auditory system can detect much smaller binaural differences, delays as short as 10^{-5} s under optimal conditions, corresponding to an angle of about 3° between sound source and the midsagittal plane of the head (Schmidt, 1978).

It is not entirely known how sound is localized in space, but the SOC provides a neurophysiological first step in the process of making such interpretations. Binaural interactions can be demonstrated by evoking "binaural beat" potentials when tones of different frequencies are applied independently to each ear. Binaural beat potentials can be recorded from the SOC, but its characteristics there do not entirely correspond to perceptual experience; hence it can be said that we probably perceive binaural beats at some higher level (Wernick and Starr, 1968).

The SOC is the first station along the ascending auditory pathways that receives bilateral input. The medial superior olive (part of the olivary complex) has an important organizational strategy for location of sound in space (Morest, 1973). Its pattern of dendritic arborizations and the axonal endings contacting medial SOC neurons correspond to *an auditory frequency map* along the dorsoventral axis of the nucleus and to *mechanisms for analysis of time.*

Cells showing bilateral responses have thresholds and best frequencies that are almost identical for the two ears. This helps to match differential sound latency and intensity in the two ears with potential sound sources.

Specialized cells in the lateral part of the SOC respond to latency differences between impulses initiated in the two ears. This provides a first step in subjective location of sound in space and a basis for correct brain stem reflex movements of eyes and head toward sources of unexpected sound. These cells are inhibited by neuronal signals from the contralateral ear: glycine is the likely inhibitory neurotransmitter (Moore and Caspary, 1983). Delay of the contralateral input is minimized by being relayed through cells in the CN which lie closest to cochlear afferent input. These cells have large axons that cross the midline and relay via cells in the trapezoid body, which are among the largest in the brain. The CN-trapezoid synaptic junction is also impressive: each CN axon forms a cup-like, calyx ending that encapsulates the somata of principal neurons in the trapezoid body. The total conduction time from the contralateral ear is nearly identical with that from the ipsilateral ear, despite the longer distance and the additional synapse.

The neurophysiology of brain stem integration of binaural stimuli was initiated by Galambos et al., 1959, and extended by Goldberg and Brown, 1969. Among binaurally responsive cells, some are excited by stimulation of each ear (excitatory-excitatory units). The discharge rate of these neurons is dictated primarily by the *average binaural intensity*. Other SOC neurons are excited by sound to one ear but inhibited by sound to the opposite ear (excitatory-inhibitory units). Changes in interaural intensity elicit large response changes because each side augments shifts in the other. These units appear to be driven by *interaural intensity differences*. This is important in *locating the source of sounds at high frequency*.

Goldberg and Brown (1969) found that many SOC neurons are responsive to differences in time of arrival of tones at the two ears. With low tones, when there is likely to be phase locking among primary afferents, these units discharge maximally when binaural coincidence is closest, and at pro-

gressively lower response rates when binaural latencies are increasingly separated, as when a sound source moves away from a listener's midline. For stimulation at best frequency, the interaural delay time is not much affected by shifts in either interaural or binaural intensities. Many SOC neurons have a characteristic delay which remains fixed for a variety of frequencies within a broad band low-frequency response zone. These *characteristic delay units* are capable of encoding information about the *location of signal sources that may embody several different low-frequency sounds.*

All this is evidence for preservation in the central nervous system of a *topologically valid "map" of the sensory epithelium* with which *high-fidelity transmission timing* combines to establish *new useful information.*

NUCLEUS OF THE TRAPEZOID BODY

This projection system is also tonotopically organized. Each principal cell receives one axonal terminal which provides but one calyx. The calyx embraces the trapezoid neuron with axonal expansions like tulip petals, covering half of the surface of the recipient cell with synapses. The trapezoid cells also receive some additional synapses from collaterals of the CN neurons that innervate the calyx, as though to make assurance doubly sure. The calyx junction is chemical, and it is electrophysiologically secure. Trapezoid cells project to olivary relay cells and to the cells from which the olivocochlear bundle (OCB) originates. Neurons of the OCB, in the periolivary group, receive bilaterally from CN, inferior colliculus (and indirectly from lateral geniculate nuclei and inferotemporal cortex via inferior colliculus), and trapezoid nuclei, as well as from other groups in the superior olivary complex. Their important projections and functional role have been depicted previously under Efferent Control of the Cochlea.

NUCLEI OF THE LATERAL LEMNISCUS

These constitute groups of scattered nuclei which cluster along the path of the lateral lemniscus. Cells in these nuclei ascend to the inferior colliculus or cross the midline to ascend to the inferior colliculus on the opposite side, or terminate in the counterpart contralateral nucleus of the lateral lemniscus. The crossing fibers constitute a minor commissure of the auditory system (Figure 9.24).

Inferior Colliculus

The inferior colliculus (IC) is the most important brain stem junction for the integration of ascending and descending auditory signals (Figure 9.24). It is an obligatory station for relay of auditory impulses traveling up and down, except that auditory cortex bypasses IC for its projections to midbrain and pons. The IC is the functional link between brain stem and telencephalic organization of auditory information (Figure 9.26). The superior and inferior colliculi interchange visual and auditory information; both project to midbrain mechanisms for arousal; both contribute to cerebral-cerebellar control of bodily performance in an environment that has dynamic sights and sounds. There is a major commissure at this level, crossing between the two ICs and enabling each IC to contribute to contralateral auditory cortex.

The IC divides conveniently into two main functional regions: a "core," laminated *central nucleus* and a "belt" region that includes *pericentral and external nuclei* (Syka et al., 1981). The central nuclear lamellae are tonotopically "tuned" from low to high frequencies ongoing from superficial to deep layers. The central nucleus sends *core information* to the laminated part of the medial geniculate body (MGB) which relays on *to primary auditory cortex*, bilaterally, with a greater number of fibers ascending on the same side. Since the IC already consists of largely crossed projections, this means that although there is bilateral representation, auditory functions are predominantly crossed.

Correlations begin to be made with other sensory modalities early in the centrifugal trajectory of all sensory systems. In the case of hearing, the pericentral and external nuclei of IC receive somesthetic information from collaterals of the spinothalamic tracts and medial lemniscus. Similarly, they receive visual signals from the superior colliculus. In its turn, IC projects auditory information to deep layers of the superior colliculus. The information derived from such intermodality correlations is kept separate from core information. The belt projection system combines visual and somesthetic with auditory information and projects to nonlaminated parts of the medial geniculate body, which in turn transmit to the belt area of auditory cortex that surrounds the core area of primary auditory cortex (Figure 9.25). This provides *separate parallel ascending "auditory" systems for integration of intramodality and intermodality sensory processing.*

IC neurons respond to monaural and binaural stimulation, and cells in the central IC nucleus of the cat are spatially segregated according to categories of binaural response characteristics, i.e., excited by both ears, excited by one ear and inhibited by the other, excited by one ear and not by the other, etc. (Semple and Aitkin, 1979). This means

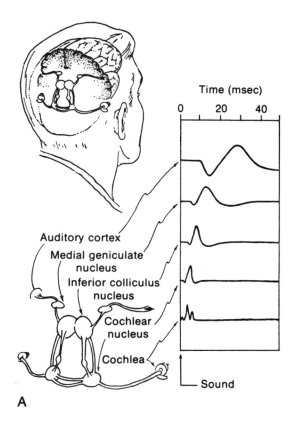

A

Time (msec)

Auditory cortex
Medial geniculate nucleus
Inferior colliculus nucleus
Cochlear nucleus
Cochlea

Sound

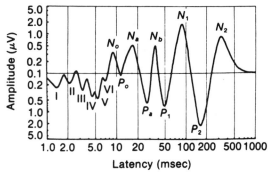

B

Figure 9.26. Schematic representation of central ascending auditory pathways and their event-related potentials (ERPs) from cochlea to cortex in humans. Approximated three-dimensional representation of main auditory pathway in a cutaway diagram of the brain and head. This schema is enlarged beneath, where *arrows* indicate the approximate form of evoked potentials at successively higher anatomical levels. Time scale on *right* indicates roughly the latencies for the early waves reflecting mass activity at the cochlea, cochlear nucleus, inferior colliculus, medial geniculate nucleus, and auditory cortex. Below is relative amplitude of summed brain stem waves designated I–VI, followed by a series of wavelets designated by *N* (negative peaks) and *P* (positive deflections). These auditory ERPs have clinical importance because the earlier ones are affected by brain stem lesions and the later ones by brain lesions, anesthesia, and even ongoing psychological processes (Picton et al., 1974). [From Bullock (1977).]

that the neural mechanisms for location of sound in space are further analyzed at the IC level.

Projections to the IC come from auditory cortex, bilaterally, from the medial geniculate body, from contralateral IC, ipsilaterally from some and bilaterally from other nuclei of the lateral lemniscus, from the ipsilateral SOC, and from CN bilaterally. IC sends fibers to join output from the superior colliculus going to midbrain reticular formation and periaqueductal gray matter, there contributing to *acoustic and visual arousal*. These pathways are known as *tectoreticular* and *tectopontine* pathways, so named because they arise from the tectum of the midbrain. The tectopontine pathway, which relays in the dorsolateral pontine nucleus and projects to the midline (vermis) region of cerebellar cortex, provides for coordination of audiovisual information within the cerebellum. These circuits are probably involved in auditorily and visually guided motor behavior from the startle response at the simplest level, to speech at a level of great complexity.

The IC projects to all caudal auditory nuclei, with the exception of cells involved in the calyx junctions in the trapezoid body, to pontine and midbrain reticular formation, periaqueductal gray, contralateral IC, to the medial geniculate body, and to ipsilateral auditory cortex. Every part of IC projects to the medial geniculate body. The core projects to the laminated part of the medial geniculate body, and the belt pericentral and external nuclei project to nonlaminated parts of the medial geniculate.

The inferior colliculus is not only rich in distributive and integrative responsibilities, it is exceptionally busy. The inferior colliculus has the highest metabolic rate per unit weight of any region in the nervous system (Sokoloff, 1961).

HUMAN AUDITORY BRAIN STEM-EVOKED POTENTIALS

At this point of transition between brain stem and forebrain, it is desirable to examine human brain stem auditory evoked potentials. This will help make events relating to the ascending auditory pathway to the level of the IC more explicit. Refer to Figure 9.26 which simulates diagrammatically the auditory pathway from cochlea to auditory cortex, locates it within the three-dimensional space of the human head, and illustrates the nature of the auditory evoked potentials as these have been assigned anatomical correspondence along that pathway. According to Buchwald (1983), the last of the brain stem responses evoked by repeated acoustic stimulation and averaged for a thousand or more repetitions is associated with activity induced in the IC (wave V–VI in the lower part of Figure 9.26).

A method of stimulation and recording, called "far field" recording, discovered by Jewett and Williston (1971), permits remote electrode pickup of very weak potentials which, by averaging more than thousands of transactions, become identifiable signals. The waves of interest to our consideration of the brain stem take place within the first 10 ms of the click stimuli. Note that the latency scale is a log scale. Wave I originates from the auditory nerve response to the click. Waves of opposite polarity which can be recorded at each mastoid between 2 and 4 ms following the click are thought to be generated in the CN and SOC (Galambos et al., 1959; Erulkar, 1972). Both lateral lemniscus (Webster, 1971) and inferior colliculus (Kitahata et al., 1969) show relatively stable responses at high rates of stimulation, like wave V.

The tracing at the bottom of Figure 9.26 from Picton et al. (1974) represents mean data from eight subjects and is somewhat diagrammatic. Clicks were 60 dBSL presented monaurally at 1/s, averaging 1024 responses from vertex and mastoid electrodes (Picton et al., 1974). We shall refer again to Figure 9.26 when we consider the far field click-evoked responses which are generated by acoustic activation of the forebrain.

Medial Geniculate Body

As noted above, core projections are relayed via the laminated portion of this auditory relay nucleus of the thalamus (MGB). Here there is a detailed tonotopical cochlear representation (Aitkin et al., 1981). Connections between IC and MGB units in this core projection system are essentially point to point between the same best frequencies at each relay. In belt regions of MGB there is convergence of somatic, vestibular, and auditory input on single neurons (Blum et al., 1979). The dorsal division of MGB which relays belt projections responds to tones in a labile fashion, often firing to a broad range of frequencies with long latencies (Aitkin et al., 1981). A medial division of the MGB provides axons to all areas of auditory cortex, including core and belt areas, and in addition, to the second sensory (somesthetic) area of cortex.

The Auditory Cortex

CAT AUDITORY CORTEX

A central auditory koniocortex can be identified in all mammalian species. We present data relating to cat, which has been intensively studied, monkey, and human. Koniocortex receives from the laminated portion of MGB a complete and orderly representation of audible frequencies (Figure 9.27)

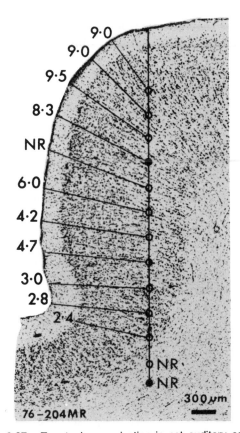

Figure 9.27. Tonotopic organization in cat auditory cortex is investigated by insertion of microelectrode and determining best frequency responses, locus by locus, along the electrode tract, as depicted here. Best frequency values in Hz are projected to surface along lines parallel to radial cell columns intersecting these sites. *Solid circles* mark the center of lesions placed for localization purposes at two recording locations along the penetration. [From Reale and Imig (1980).]

which provides monaural and binaural maps of best frequency. These maps are made up of isofrequency strips. The isofrequency strips of the primary auditory cortex in the cat appear to constitute successions of *isofrequency columns* which encode the same frequency for all units encountered during penetration through the entire depth of cortex (Merzenich et al., 1975). *Binaural interaction columns* have also been described, exhibiting either binaural summation or inhibition (Imig and Adrian, 1977). The summation and inhibition columns occupy alternating bands oriented across the isofrequency strips. Within the binaural summation columns, there is a mosaic of minor *ipsi- or contralateral dominance columns.*

Distributed throughout this area of core representation are corticocortical projections, ipsilateral and contralateral, which affect responses elicited in areas identified in Figure 9.25 as *AI*, *A* (anterior), and *P* (posterior), and *VP* (ventral posterior). These represent distinctive regions within primary

auditory cortex (Imig and Reale, 1981). These regions are surrounded by a peripheral auditory belt which includes *AII*, *V* (ventral), and *DP* (dorsposterior). Patterns of ipsilateral and contralateral projection among these regions are organized into alternating bands of dense and sparse innervation that relate to binaural summation or contralateral-dominant suppression. Dense projections from the opposite hemisphere, via the corpus callosum, are associated with summation columns which augment activity resulting from frequency and timing coincidences generated in the two ears. Areas of sparse callosal innervation are asssociated with dense ipsilateral innervation and correspond to contralateral-dominant suppression.

The anatomical patterns and neurophysiological responses obtained with anesthetized animals provide only a bare skeleton of activity observed in waking animals (see note by Goldstein and Abeles, 1975). Neuronal coding of tonal frequency in cortical neurons is a very complex process, influenced not only by thalamocortical core projections and by other afferent inputs and corticocortical connections, but also by cells intrinsic to the cortical columns in which detailed analysis takes place.

Whereas direct throughput pathways are relatively resistant to anesthesia, the corticocortical and intrinsic cortical columnar neurons appear to be especially susceptible to the effects of anesthetics. In waking animals these components may be affected by states of alertness and direction of attention (Evans, 1974). On another occasion, Evans cautioned against comparing waking humans with anesthetized animals because the effects of descending auditory control mechanisms are probably also eliminated by anesthesia.

MONKEY AUDITORY CORTEX

In the monkey, auditory cortex consists of a central koniocortex surrounded by at least four different belt areas, with a more elaborate system of cortical association connections than are found in the cat. A progression of connections is seen from the primary auditory cortex; each step involves reciprocal connections. The first step projects from primary auditory cortex to the cortical auditory belt areas and to area 8a in the frontal lobe which contributes to voluntary direction of gaze. Step two projects from the belt area to adjacent area 22 in the superior temporal gyrus and from 8a to the more frontally located area 9. Step three involves further progression from the supratemporal gyrus to the supratemporal sulcus and from area 9 to areas 10 and 12 near the frontal pole. This march of progressive auditory associations and signal elaborations contributes additional intersensory integration while migrating spatially through the cortex to the frontal lobes which are concerned with forward planning of behavior. With multiple sensory and motor patterns combined, at that point, the images may be more readily converted into commands for performance.

HUMAN AUDITORY CORTEX

According to the comprehensive cytoarchitectonic study of von Economo and Horn (1930), the auditory field of the human cerebral cortex occupies the transverse temporal gyrus (Heschl's gyrus) and a variable amount of the caudally situated planum temporale—that is, a field corresponding approximately to Brodmann's areas 41 and 42. Auditory sensations can be produced by electrical stimulation of Heschl's gyrus in patients (Penfield and Jasper, 1954). The sensations are usually described as ringing, humming, clicking, buzzing, and so forth, and are most often referred to the opposite ear. Celesia found that in fully anesthetized patients, no auditory evoked potentials could be recorded outside the primary auditory field of von Economo and Horn. In alert patients, with local anesthesia, however, potentials with longer latencies could be recorded from the superior temporal gyrus and from both frontal and parietal opercula (Celesia, 1976). The opercular responses did not relay through the primary auditory cortex because the evoked potentials persisted after temporal lobectomy (required for compelling clinical reasons).

HUMAN HEMISPHERIC DIFFERENCES

von Economo and Horn were among the first to point out the pronounced individual variation and the considerable right-left asymmetries in the size and convolution of the supratemporal plane. Heschl's gyrus, they found, is usually solitary and longer on the left side. In general, the planum temporale, caudal to the Heschl's gyrus, is larger on the left than on the right side, but in some brains the situation may be reversed (Witelson and Pallie, 1973). These asymmetries may be visible even in computerized axial tomography (Galaburda et al., 1978). All this speaks in favor of a functional differentiation between the auditory cortex in the right and the left hemispheres. A functional difference between the auditory functions of the two hemispheres is apparent also from *dichotic listening experiments*. Scores for the two ears indicate the relative domination of the contralateral hemisphere (test delivered randomly to either ear through earphones coupled to a dual channel tape recorder).

Usually the right ear (left hemisphere) reveals a better score for verbal tests, while the left ear (right hemisphere) is better for recognition of music (Kimura, 1961, 1964). Broadbent finds that the two hemispheres can be seen as performing different parts of an overall integrated task, sensory or motor, rather than completely separate and parallel operations (Broadbent, 1974).

Descending Auditory Pathways

Paralleling the ascending auditory pathway there is a stepwise, descending pathway with auditory cortex at top and ending with the cochlear hair cells (Figure 9.28). Relays along this projection are influenced by ascending as well as lateral communications. This establishes feedforward and feedback loops of various lengths and complexities, which are not yet functionally understood.

Auditory cortex sends three descending tracts: directly to thalamus (MGB) and midbrain (IC), and, by a single relay, to pons (to reproject to cerebellar vermis). *The corticogeniculate projection* is reciprocal on a point-to-point basis to the geniculocortical projection (Oliver and Hall, 1978b). Each auditory cortical area projects to the sources of its particular afferent fibers in the medial geniculate nucleus (Figure 9.28).

The core auditory cortex sends *corticocollicular projections* directly to the inferior colliculus, bilaterally. Belt auditory cortical areas send much less robust descending projections to the IC. The corticocollicular projection goes mainly to the pericentral region and not to the central nucleus (Massopust and Ordy, 1962; Rasmussen, 1964; Oliver and Hall, 1978a, b).

Auditory cortex also sends fibers to the *superior colliculus*, mainly from the auditory belt area. This part of the tectum also receives a different set of fibers directly from primary auditory cortex destined for relay there as part of corticopontine projections that ultimately affect audiovisual representation in the vermis of the cerebellum.

The next major link in the descending auditory pathway arises from the IC (Figure 9.28). The IC neurons that project downward are in the same location as IC neurons that relay along the ascending pathway. The descending fibers project to the superior olivary complex and to the cochlear nuclei, bilaterally (Rasmussen, 1964; Casseday et al., 1976; Oliver and Hall, 1978b).

The *colliculoolivary projection* terminates in the periolivary nuclei, preferentially ipsilaterally, and in the regions giving rise *to the crossed olivocochlear bundle*. The cells of origin of the olivocochlear

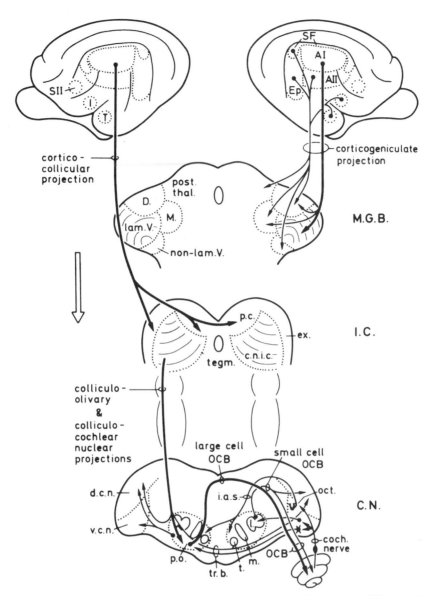

Figure 9.28. Diagram of the descending auditory pathway in the cat. Primary auditory cortex (*AI*), secondary auditory area (*AII*), posterior ectosylvian area (*Ep*), and suprasylvian fringe (*SF*), medial geniculate body (*MGB*), inferior colliculus (*IC*), cochlear nuclear complex (*CN*), olivocochlear bundle (*OCB*), periolivary body (*p.o.*), trapezoid body (*tr.b.*). [From Brodal (1981).]

bundle can thus be influenced by both ascending and descending auditory fibers. The cochlear nuclei only receive and do not send fibers to the cochlea. The cochlea receives its efferent nerve supply entirely from the olivocochlear bundle.

Cerebral Auditory Event-Related Potentials (ERP)

Referring again to Figure 9.26, we can see that following V–VI which was reasonable to assign to activity ascending the lateral lemniscus or taking place within the IC, there is a succession of waves labeled N_o, P_o, N_a, P_a, N_b, P_1, N_1, P_2, and N_2. Picton et al. (1974) pointed out that the P_a component is similar to a potential recorded in primates and

humans from widespread areas of association cortex in parietal, frontal, and temporal regions. They concluded that middle latency components (N_o, P_o, N_a, P_a, and N_b), 8–80 or 100 ms, "largely represent potentials from thalamus and cortex."

The later ERP components are widespread over the frontocentral scalp areas and are thought to be generated by cortex rather than by thalamus or cephalic brain stem. The late components (P_1, N_1, P_2, N_2) are of larger amplitude over frontocentral regions and may be composed of "parallel late waves" generated by both specific (including auditory) and nonspecific cortex. This conclusion is supported by intracranial recordings from human and primate brains. Recording from primate frontal cortex exhibits responses from auditory stimuli that seem exactly homologous to human recordings and have been shown to be of cortical origin (Hardin and Castellucci, 1970). It is presumed that the primary receiving (auditory) cortex exerts a considerable measure of control over association cortex response through corticocortical and corticothalamocortical connections. The relationship to auditory stimulation is confirmed by the fact that changing parameters of stimulus spatial origin and pitch alters the form of nN_1–P_2 components (Butler, 1972).

This means that the auditory far field-evoked potentials allow access to neural mechanisms involved in transmission all along the entire auditory pathway, from cochlea to cortex. This new technique provides a noninvasive method for measuring the integrity of the auditory path and for utilizing an objective measure of the substrates for hearing in neurological diagnosis. Furthermore, since the late waves are greatly affected by processes of human attention and perception, it opens a neurophysiological window to the mind.

Event-Related Potentials Associated with Expectancy

Figure 9.29 shows the effects on auditory evoked potentials of subjective expectancy. Human subjects are given a train of clicks or tone pips at regular intervals and asked to count the number of occasions (randomized) when one of the sound stimuli is omitted. While counting events that did *not* occur, the subject manifests a large amplitude positive wave at about 300 ms. This is called *the P300 wave*. Such expectancy potentials may be, as in this instance, considerably larger than responses to the acoustic stimulus. The potential certainly represents a brain event, but what kind of brain event is it? It may be recognition of novelty, a mismatch of expectation while looking for a nonevent (which thereby becomes a brain event), or perhaps a brain event associated with the resolution of prior uncertainty. The P300 appears to relate to the *task* and can be recorded in response to mental decision or resolution associated with detection and problem-solving experiences in relation to any kind of sensory stimulation—regardless of modality, auditory, visual, or tactile (Galambos, 1974).

The P300 in Schizophrenics

Schizophrenia has long been characterized as a disintegration of psychic processes that renders schizophrenics "incapable of holding the train of thought in the proper channel" (Bleuler, 1924, translated 1951). Long latency auditory-evoked po-

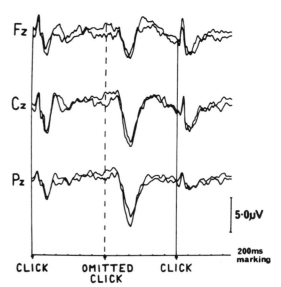

Figure 9.29. Cerebral electrical responses to an expected but nonoccurring event. Event-related (in this case nonevent) potentials recorded from vertex and mastoid and averaged over 64 responses in human subjects presented a train of clicks or tones at regular intervals (1.1 s). An occasional member of the train is omitted. Subject anticipates a stimulus at each appropriate interval and is asked to count the number of times the expected stimulus does *not* occur. A large positive wave at 300 ms (P300) occurs regularly in the place of the expected signals. This is obviously a brain event, perhaps a recognition of novelty, or associated with matching expectation while looking for a nonevent, perhaps the resolution of prior uncertainty. Similar electrical events at this 300-ms interval have been recorded in relation to various detection and problem-solving auditory, visual, and tactile experiments. Hence the P300 appears to be related to central activity aroused by the task, regardless of the particular sensory pathway. [From Picton et al. (1974).]

tentials now suggest that with tasks of detecting occasional targets among standard tones in one ear and ignoring all tones presented to the other ear, and especially with tasks of "divided" attention requiring detection of all occasional targets in both ears, schizophrenics (matched for age, sex, and educational background with controls) were slower and less accurate. The N_1 and P300 potentials were both significantly larger for control subjects. The schizophrenic P300 abnormality is apparently not due to a lack of stimulus-set, arousal, attention, or motivation, but to a more general abnormality of cerebral processing related to the amount of task-relevant information that is being processed (Baribeau-Braun et al., 1983).

Potentials Associated with the Unexpected

The effects of the unexpected reveals unusual late potentials. This has been demonstrated for instance by Neville (1982) and is exemplified in Figure 9.30. Her purpose was 2-fold, to conduct parallel experiments on cognitive processing with

A. Oddball Sequence

B. Oddball plus Novel Sequence

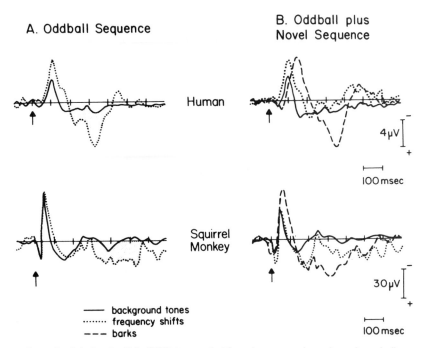

Human

4 μV

100 msec

Squirrel Monkey

30 μV

100 msec

—— background tones
········· frequency shifts
– – – barks

Figure 9.30. *A,* Averaged event-related potentials (ERPs) recorded from human scalp and monkey skull screw electrodes to frequent background tones (*solid line*) and infrequent shifts in tone frequency (*dotted line*). All tones were 190 ms in duration. *B,* ERPs from same electrodes to subsequent presentations of frequent tones (*solid line*), infrequent shifts in tone frequency (*dotted line*), and substituted dog barks (*dashed line*). Note that the unusual (oddball sequence) is associated with augmented and longer lasting, more elaborate responses, and the unexpected (novel sequence) elicits even more elaborate responses (Neville, 1982). [From Galambos and Hillyard (1981).]

human subjects and experimental animals. The cognitive processes can be investigated in the humans and questions can be extrapolated to primates which then can be explored experimentally. The second purpose is to have some physiological token of the cognitive process for confirmation during experimental interventions. In this experiment, tones are repeated so that many events can be averaged. Infrequently, randomly, there is a computer-controlled shift in tone frequency. These events are accumulated by the computer and averaged to compare with responses to the regular tone sequences. The "oddball" events give rise to larger amplitude, more elaborate far field auditory evoked late cerebral responses. Now, occasionally and randomly the sound of a dog bark is introduced in time with the rhythm of the tone stimuli. This is apparently perceived by human subjects and squirrel monkeys as being unexpected and out of context. These novel sequences are accumulated and compared with routine and "oddball" sequences. The dashed line for dog bark responses are of still greater amplitude and physiological consequence.

Linguistic Unexpectancy

Kutas and Hillyard (1980) discovered that event-related potentials can be modified by the subject having an incongruous cognitive experience. Figure 9.31 shows an example of such incongruity in the form of a "linguistic unexpectancy." The events consist of individual words flashed once per second, so this is a visual rather than an auditory example, but the analogy is pertinent here to the discussion of late evoked potentials relating to cognitive functions. The words are composed into seven-word sentences, some examples of which are presented in the figure. By averaging event-related potentials over 40 such sentences a reasonably consistent tracing is obtained. Occasionally and randomly a sentence is given a semantically incongruous twist, as in the last example, "I take coffee with cream and dog." Such semantic mismatch is regularly associated with a novel late wave at about 400 ms which is designated N400. Such late potentials, evoked by cognitively incongruent stimuli, are again related to the subjective experience independent of the sense modality employed as events to elicit event-related potentials. With hearing or vision, the experience of incongruity is associated with N400 potentials over the vertex.

Basic Auditory Functions and Hearing Impairment

A variety of properties of the auditory system have been investigated by utilizing psychophysical procedures. These include studies of the way in which the characteristics of the stimulus must be

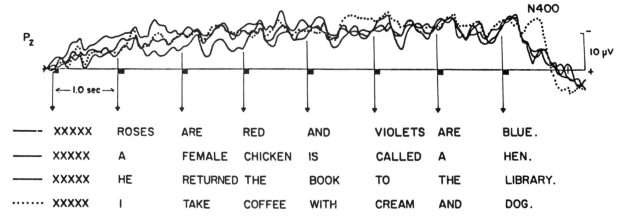

Figure 9.31. Event-related potentials (ERPs) recorded throughout the presentation of 7-word sentences, with words flashed individually at 1/s. Each tracing is the averaged ERP across 40 sentences in which the last word was either appropriate to the context (*solid lines*) or semantically inappropriate (*dotted lines*). Samples of the 160 different sentences that were presented to this subject are shown below. The "semantic mismatches" occurred in 25% of the sentences and elicited a prominent N400 component (Kutas and Hillyard, 1980). [From Galambos and Hillyard (1981).]

varied for a listener to be able to discriminate between two frequencies (roughly speaking, pitch discrimination) or intensities (roughly speaking, loudness) or the loudness of two tones of different frequencies. These approaches test the system as a whole and depend greatly on having standard and favorable conditions (background sound level, etc.) and standard instructions.

The auditory system changes with development as we have seen, and it changes with aging as we shall see. Long sustained motivation and training may make a substantial difference. For example, there is evidence that *blind people may have superior hearing* (Niemyer and Starlinger, 1981). They compared 18 chronically blind subjects and 18 normally sighted subjects with normal hearing. The blind had better speech discrimination, especially with sentence comprehension tests, both with and without competing stimulating noise. Especially interesting is the fact that the blind subjects had a significantly shorter latency for the N_1, 140 ms instead of 160 ms ($P < 0.0002$). It is possible that keen and practiced attention can improve the performance of the auditory pathway, and that learning processes may be shaping descending as well as input pathways to favor the goal of improved hearing.

ABSOLUTE THRESHOLD

The smallest amount of sound power that the ear can detect in any given frequency is necessarily a statistical concept. Attempts to measure such an amount produce variable results which are influenced by differences in the subjects of the test, differences from time to time in a given subject, and a variety of procedural factors. Nonetheless,

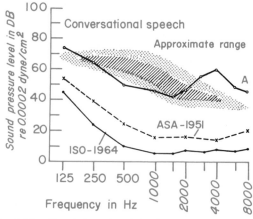

Figure 9.32. The International Audiometric Zero (*ISO-1964*) curve compared with the American Standards Association (*ASA*) 1951 curve. The ISO curve represents a better approximation of the thresholds of normal young adults than does the ASA 1951 curve. The *stippled area* indicates the pressure-frequency region in which most conversational speech sounds occur. Curve *A* represents thresholds typical of a case of moderate hearing impairment. [From Davis (1965b).]

minimum sound power levels must be selected which can serve as a standard representative of a large population of normal listeners. Two functions which are representative of the minimum threshold in sound pressure are illustrated in Figure 9.32, including the international audiometric zone curve (International Standards Organization, 1964) which serves as a current standard.

DIFFERENTIAL THRESHOLDS

Differential thresholds, at least for the discrimination of intensity differences, may be considered as an extension of the absolute threshold determination when the background is changed from one of quiet to one with a controlled continuous sound level. In many sensory modalities it has been demonstrated that there is a rough relation between the level of intensity increment necessary for discrimination and the intensity level of the background against which the discrimination is to be made. This relation is known as the Weber-Fechner law, but is really only an approximate description over a limited range of the variables which may be studied. In audition, the intensity increment required for discrimination at background levels near thresholds is relatively large but declines in relative amplitude up to a background level of approximately 50 dB above threshold.

When the change in frequency of a sound necessary for discrimination of the fact that frequency has been changed is studied as a function of frequency, there is an approximately constant ratio between the frequency increment and the reference frequency above 1,000 Hz. The absolute increment in frequency is roughly constant at 2 or 3 Hz from 60 to 1000 Hz for sound pressure levels above 30 dB.

The nature of the frequency response of the human ear can be examined from the level of threshold for detection of sound at a given frequency up to the maximum comfortable level of sound stimulation by obtaining data on the matching of loudness at a variety of levels over the entire frequency spectrum. There is considerable flattening of the frequency response with increase in intensity level and hence, loudness. This is particularly true at the low frequency end of the spectrum. This is true to a lesser extent at the highest frequencies. It is more difficult to obtain data at these high-frequency levels for a large number of subjects. It is clear that no simple metric for the addition of loudnesses of different frequencies is practical.

MASKING

From our discussion of differential thresholds for sound intensity, it is clear that with elevation of a background sound the increment which can just be discriminated must also be increased. The background sound thus serves, to a degree, to mask the increment. The phenomenon becomes more complex when one examines the masking effects of various frequencies on a given frequency or the effects of a wide band of frequencies or noise on a given frequency. In general, masking is more effective the closer the frequency of the masking tone to that of the tone being masked, although the precise nature of the relation here may be complicated by the beating phenomenon. Low-frequency tones mask high-frequency tones more effectively

than the converse. The latter effect might be expected from our knowledge of the motions of the basilar membrane. Interestingly enough, and of some practical significance in the measurement of hearing disability, there is relatively little masking effect when a masking tone is presented to one ear and a test tone to the other. This phenomenon is useful in suppressing the better ear when one wishes to test an ear in which unilateral hearing impairment is suspected. A fairly broad spectrum noise serves as a useful mask for a wide range of frequencies in these circumstances.

Types of Hearing Impairment

The causes of deafness are divided into five classes (Davis, 1951). In the first of these, there is some interference with the conduction of sound to the neural mechanism of the inner ear. Hearing difficulties resulting from these causes are labeled *conduction deafness*. The second category is that resulting from damage to the cochlear mechanism or the auditory nerve, and this is labeled *nerve deafness*. The other three classes include *central deafness*, *diplacusis* or a false sense of pitch which may be caused by edema of the labyrinth, and *tinnitus* associated with ringing in the ears which is usually caused by hypersensitivity of hair cells or their nerves.

CONDUCTION DEAFNESS

Conduction deafness may be caused by such a simple problem as the accumulation of wax in the external ear canal. It is more significant, however, when it occurs by reason of such factors as the pathological hardening of the tympanic membrane, perforation of that membrane, or a loss of mobility or destruction of part of the ossicular chain. A small perforation of the drum is often temporary and may heal, but dysfunction of the ossicular chain requires some kind of intervention for its correction. In *otosclerosis*, there is a growth of bone which may tend to anchor the footplate of the stapes such that it is unable to transmit motion to the oval window of the scala vestibuli. Initial effects may be manifested by a progressive loss of hearing for the low frequencies, with possible subsequent losses in high frequency which accompany pathological changes in the cochlea itself.

A variety of acute effects may result in hearing impairment. Several of these cause conduction deafness. If the eustachian tube is blocked for any reason during a rapid change in ambient pressure, the balance of pressure between the external meatus and the middle ear will be upset. Rupture of the drum will not occur unless the pressure differential reaches the range of 100–500 mm of mercury. The problem is perhaps more frequently encountered by the diver during descent into deep water. A pressure differential which can cause rupture of the drum can easily be reached, but it will generally be preceded by warning pain which will result in the diver's ascent before rupture occurs. In spite of the warning which pain provides, transient injury to the drum is a frequent problem for overzealous skin divers.

If the eustachian tube remains closed for any length of time, the air trapped in the middle ear is gradually absorbed. The lowered partial pressure of oxygen in the cavity may be accompanied by the entrance of fluid into the middle ear when gas pressure falls below hydrostatic pressure in the capillary bed. The accumulated fluid increases friction and decreases the transmission of sound.

NERVE DEAFNESS

The most common cause of sensorineural impairment is probably exposure to high level sound intensities (Graham, 1966). As mentioned above, there is a temporary shift in threshold after exposure to a loud sound or noise (Fig. 9.33). Exposure to loud noises yields temporary and may lead to lasting hearing loss. Loud noises may also affect neural coding processes. This may be one basis for perceptual disabilities following even modest intervals of exposure to loud sounds, even when sounds can be readily heard (Fig. 9.33). Loud sound induces alterations among excitatory and inhibitory patterns so that units which were spontaneously active

may be inactivated, and others which responded only to "onset" of a stimulus may be continuously active (Willott and Lu, 1982).

If the sound is loud enough and the period of exposure is long enough or repeated frequently enough, the elevation of threshold may be permanent. In general, sensorineural deafness affects hearing for higher frequencies to a much greater extent than low frequencies, and there appears to be a particular vulnerability in the region of 4,000 Hz.

When caused by exposure to high level sound, such deafness is bilateral. When the shift in threshold is of the order of 40–60 dB, the loss may be accompanied by *tinnitus*, a sustained ringing sound which is attributed to spontaneous discharge of injured hair cells.

Hearing loss associated with injury to the hair cells may be accompanied by the phenomenon of *recruitment*. That is, at the frequencies of greatest impairment there is a more rapid growth of loudness than normal (compare the rapid growth of loudness in the normal ear at low frequencies) such that the balance of sounds over a broad range of frequencies, including the range of impairment, may be normal at higher sound intensities.

If sensorineural impairment is the result of infection, it may be unilateral. Menière's syndrome is accompanied by hearing loss in the low frequencies, and by sometimes violent attacks of vertigo. This probably is the result of an involvement of the other sensory elements in the labyrinth as well as the cochlea.

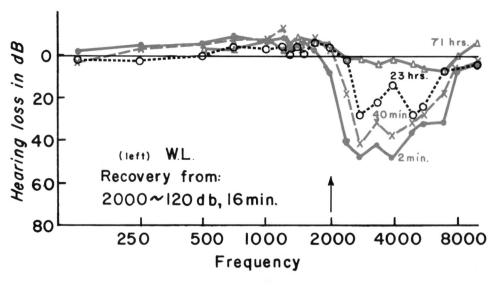

Figure 9.33. Temporary threshold shift (*TTS*) after exposure to a tone of 2,000 Hz at 120 decibels (dB) sound pressure level for 16 min. Recovery is shown over the course of time up to 71 h after exposure. Note that the shift of threshold is greatest for frequencies above that of the exposure tone. [From Davis et al. (1950).]

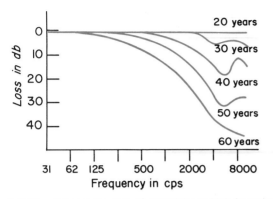

Figure 9.34. Progressive loss of sensitivity at high frequencies with increasing age. The normal audiogram at 20 years of age serves as a reference for the losses represented at greater ages. [From Licklider (1951).]

Finally, just as vision appears to show progressive degradation with increased age, so does hearing (Fig. 9.34), particularly for higher frequencies. Fortunately, the discrimination of speech depends primarily on frequencies between about 500 and 3,000 to 5,000 Hz. Therefore, the characteristic high-frequency loss with increasing age causes only a minimum of difficulty in most circumstances (French and Steinberg, 1947).

Measurement and Diagnosis of Hearing Impairment

AUDIOMETRY

The standard clinical tool for the measurement of hearing, the audiometer, consists of an oscillator for the production of selected frequencies usually ranging from 125 to 12,000 Hz, an amplifier, an attenuator or volume control, and headphones which can be driven individually or with various balances of excitation. The output of the oscillator is adjusted at each of the frequencies such that when the attenuation control is set at zero, the output of the headphones represents a signal corresponding to the sound pressure level of a standard reference curve (see Fig. 9.32). The attenuation control is usually calibrated in 5-dB steps. An increase in the reading of the attenuation control corresponds to an increase in the sound power level at the headphones by an equivalent amount. Thus, if an appropriate attenuator setting can be determined at which a subject just hears the tone at a given frequency, then the reading of the attenuator control represents the sound power level over a normal standard which is required for that patient at that frequency. An audiogram can then be plotted which represents hearing loss at each of the selected frequencies in terms of the increased sound

power required in decibels as compared with the normal level.

Several methods have been developed for the testing of hearing without the requirement for any instruction or special performance on the part of the patient.

Clinical Use of Auditory Evoked Responses

One of the most advantageous methods of clinical diagnosis is that of auditory evoked potentials (see Callaway et al., 1978; Moore, 1983). Brain stem-evoked potentials can help distinguish and localize peripheral and brain stem interruptions of the auditory pathway. These can aid in the distinction of various kinds of coma, drug-induced, traumatic, and neurological disease processes (Starr, 1976). They can also be used in assessment of the neurological status of premature infants (Schulman-Galambos and Galambos, 1975).

Middle latency components are more difficult to use as measures for hearing evaluation because of large intersubject variability in comparing evoked potential thresholds with behavioral measures (Polich and Starr, 1983). Nevertheless, Galambos and colleagues find that potentials obtained with 40 stimulus presentations/s gives auditorily sensitive and reliable measures of the middle latency components (Galambos et al., 1981). Robinson and Rudge (1977, 1978) have studied patients with multiple sclerosis, a disorder that interferes intermittently and in scattered regions with myelinated nerve conduction. They found that 12% of their MS patients may have normal brain stem evoked potentials but show abnormal middle latency auditory evoked potentials. Serial recordings in these patients could distinguish between active and quiescent phases of the disease.

New methods for scanning of head and spine by computed X-ray, positron emission, and nuclear magnetic resonance tomography, coupled with new methods for eliciting and measuring (averaging and quantifying) biological potentials greatly improve opportunities for accurate neurological diagnosis. For nearly a century biological opacity has been penetrated by X-rays, but now this can be accomplished in multiplanar sections rather than in collapsed conical shadows of three-dimensional neuroanatomy. These methods promise a better premortem correlation between structure and function of the central nervous system than was ever previously possible.

BONE CONDUCTION TESTS

A variety of procedures is available which assist in the diagnosis of hearing impairment. In addition to the headphones used with an audiometer, a small vibrator may also be used. This is placed either on the forehead or on the mastoid prominence behind the ear. Vibrations of the skull are transmitted to the cochlea and oval window directly. For frequencies up to about 1,500 Hz, the nature of suspension of the ossicles in the middle ear is such that they do not tend to move with skull vibration. Thus, there is a relative motion between the footplate of the stapes and the oval window. Sound is heard just as it would be if the stapes itself were vibrating. At higher frequencies, the skull does not vibrate as a rigid body but rather in sections. As vibrations are transmitted to the bony walls of the cochlea, the contained fluids are set in motion, and hair cells are stimulated. With either mechanism of bone conduction, the pathways of the external meatus and the middle ear are bypassed. Thus, a loss which may be registered when measurements are made with air conduction by reason of conduction deafness will not appear.

The audiograms illustrated in Figure 9.35 provide an example of the diagnostic value of employing both air conduction and bone conduction measure-ments. In the curve on the *left*, a near normal audiogram is obtained by bone conduction, while there is a better than 60-dB attenuation with air conduction. Clearly, the nature of the impairment here involves conduction to the inner ear. On the other hand, in the figure on the *right*, the roughly equal impairments demonstrated by air conduction and bone conduction signify some form of sensori-neural involvement. Problems may arise in testing unilateral deafness with bone conduction tech-niques. Sound may be conducted around the head to the good ear when the bone conduction vibrator is located on the mastoid bone behind the impaired ear. The patient can then provide a correct report as to the presence or absence of the tone near normal levels in spite of his impairment. This prob-lem is overcome by presenting a masking sound to the good ear. As indicated above, the contralateral effect of masking is relatively minimal. The appro-priate masking level may be adjusted on the basis of comparative measurements of air conduction thresholds in each of the ears. When the good ear is masked, bone conduction thresholds can confi-dently be made on the impaired ear.

SPEECH AUDIOMETRY

For practical purposes, particularly the extent to which understanding of speech may be affected by a hearing loss, the absolute threshold for the entire range of frequencies may be of little concern, and even the absolute threshold in that range of fre-quencies which is most important for the discrimi-nation of speech may be of little importance. For this reason, a variety of tests of hearing based on the recognition of spoken words has been devel-oped. Selected test words may include spondees, words of two syllables which are equally stressed, or monosyllabic words. List of words are presented to the patient from phonographic or tape record-ings, such that the loudness can be standardized and controlled. Performance is based on the num-ber of words that can be repeated correctly at

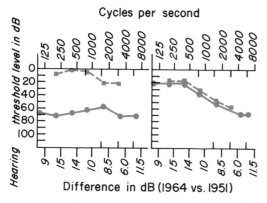

Figure 9.35. Audiograms illustrating conductive impairment of hearing on the *left* and sensorineural impairment of hearing on the *right*. The *solid curve* is for air conduction, the *dashed curve* for bone conduction. Note that bone conduction and air conduction curves are quite similar in sensorineural impairment.

Table 9.3
Classes of hearing handicap ISO-1964

dB	Class	Degree of Handicap	Average Hearing Threshold Level for 500 and 2,000 Hz, in the Better Ear		Ability to Understand Speech
			More Than, dB	No More Than, dB	
25	A	Not significant		25 (ISO)	No significant difficulty with faint speech
40	B	Slight handicap	25 (ISO)	40	Difficulty only with faint speech
55	C	Mild handicap	40	55	Frequent difficulty with normal speech
70	D	Marked handicap	55	70	Frequent difficulty with loud speech
90	E	Severe handicap	70	90	Can understand only shouted or amplified speech
	F	Extreme handicap	90		Usually cannot understand even amplified speech

From Davis (1965b).

various intensity levels. Lists of words are selected to include appropriate representation of the various speech sounds in conversational English. The percentage of the words which the patient is able to repeat correctly is his discrimination score or articulation score. An index of the adequacy of hearing for normal social interaction may be derived from such scores (Davis, 1965b).

In the absence of elaborate tests of hearing which involve the spoken words, the elevation of threshold at frequencies of 500, 1,000, and 2,000 Hz provides a useful indication of the significance of the impairment (see Table 9.3).

Hearing Aids

Hearing aids improve the conduction of sound to the inner ear. They may thus be useful in aiding someone with a conduction deafness or even with sensorineural impairment when the resulting deafness is not complete. No effort is made to match the response characteristics of the hearing aid to the frequency response characteristics of the loss of the patient. Such "prescription" correction has not been found to be of any benefit. With severe sensorineural defects, there may be little benefit derivable from the use of any type of hearing aid.

CHAPTER 63

Vestibular Function

Vestibular function is provided by a generally unobtrusive system of sensorimotor controls which operates reflexly to stabilize vision and coordinate movements and balance of head, body, and limbs. The vestibular system operates effectively in a dy-

ice skates or "hot dogging" with skis). It is also useful for animals which make free use of three-dimensional space: birds, fishes, amphibia, marine mammals (hang-gliders, SCUBA divers) etc., in order for them to maintain the flying and swimming

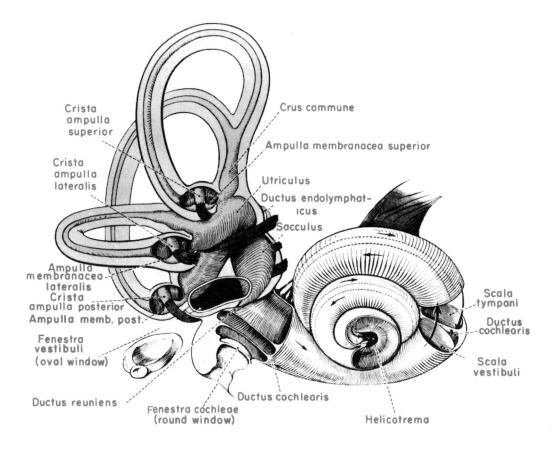

Figure 9.36. A schematic illustration of relations among the semicircular canals, utricle, saccule, and the cochlear duct on the right side. The membranous semicircular canals are shown within their surrounding bony structure. Nerve connections into the ampullae, utricle, and saccule are shown. [From Melloni (1957).]

namic three-dimensional environment that has a consistent gravitational force of 1 g. The vestibular system provides short-term inertial guidance that helps land animals resist gravity and avoid falling down during risky maneuvers (e.g., jumping with

trajectories they intend and to keep track of their whereabouts over the short term. The vestibular system works effectively in controlling the coordination of astronauts in a zero gravity environment.

The vestibular apparatus (Fig. 9.36) is a very old

mechanism in vertebrate history, and there are analogous sensory counterparts in the bodies of most invertebrates, testifying to the extraordinary evolutionary utility of having continuing information about relative motion of parts and the body as a whole in space.

We tend to think of the world as we perceive it, at least when we are studying sensory processes, as a world made up of dimensions that correspond to our various sensory systems. For man, the vestibular sense hardly has an independent identity by reason of its lack of any conscious concomitant most of the time. In fact, our perceptions of the world depend in large degree on interactions among the senses, and the stability of the visual world in particular is dependent upon the coordination of motor responses involving postural mechanisms, the extraocular muscles, and sensory information provided by the vestibular mechanism.

ANATOMY AND PHYSIOLOGY OF THE VESTIBULOAUDITORY APPARATUS

Peripheral Components of the Vestibular System

THE SEMICIRCULAR CANALS

The vestibular apparatus includes three toroidal, membranous, *semicircular canals*, each with an individual enlargement, the *ampulla*, which contains a small patch of sensory epithelium. The ampullae of the horizontal or lateral canal, and the superior or anterior canal are located close to one another anteriorly where they enter the vestibule (Fig. 9.36). The ampulla of the posterior or inferior canal enters the vestibule at its posterior end. The two vertical canals, superior and posterior, share a common crus at the point of their departure from the vestibule. There is a mirror image trio of canals on the contralateral side of the head (Fig. 9.37). The canals on each side are functionally linked in a mutually reinforcing way via the central nervous system.

When the head is held erect the horizontal canals are actually inclined downward toward the rear by approximately 30°. If the head is tilted forward such that the opening of the external meatus and the outer canthus of the eyelids are on the same horizontal line, then the horizontal canals are very nearly in a horizontal plane (Fig. 9.37). Since the anterior canal on one side is in a plane parallel to that of the posterior canal on the other side, the contralateral anterior and posterior canals are maximally stimulated by rotations about a common axis. A knowledge of these orientations is helpful for isolation of specific canals during testing. The semicircular canals evolved and developed ade-

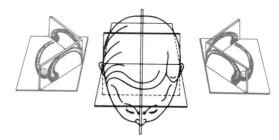

Figure 9.37. Relations of the semicircular canals to the median sagittal, transverse frontal, and horizontal planes of the head.

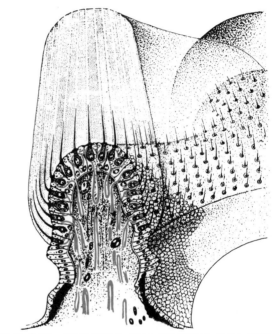

Figure 9.38. Illustration in three dimensions of the crista ampullaris. [From Wersäll (1956).]

quately *to detect and measure angular accelerations in any direction*.

THE UTRICLE AND SACCULE

The vestibule contains two additional regions which support sensory epithelia, the *utricle* (utriculus) and, on a stalk from the utricle, the *saccule* (sacculus). Each has a patch of neuroepithelium, and within each patch are slabs of hair cells that are oriented systematically to be sensitive to displacements in any of several determined directions (see Figs. 9.42 and 9.44). The utricular and saccular organs evolved and developed so that the four organs in the two ears can *monitor gravity* as well as detect and *measure linear accelerations and to some extent vibrations* in practically any plane. When the head is tilted backwards there is greater ambiguity as to the position of the head in space relative to when it is held in other positions.

THE SPIRAL COCHLEA

The cochlea—the hearing portion of the inner ear—is also attached to the vestibule. In between the vestibular and auditory components of the inner ear are two flexible membranous areas in the wall of the vestibule which are indispensable for the acoustic reception system, the oval and the round windows. They are separated from direct communication with one another by the cochlear partition. Initiation of movement at the oval window imparts motion to the cochlear partition which is compensated by flexible deformation at the round window. (see Chapter 62, Hearing). The cochlear partition bears a sixth region of sensory epithelium in the inner ear. It evolved and developed to be effective over a limited frequency range to provide information on which *frequency* (roughly pitch) and *intensity* (roughly loudness) and *more complex acoustic analyses* are based. Each of these six patches of neuroepithelia depends on mechanoelectric transduction by *hair cells*, and each sends and receives *afferent* and *efferent* communications with the central nervous system by way of the VIIIth cranial nerve.

PHYSIOLOGY OF THE SEMICIRCULAR CANALS

Each semicircular canal constitutes a continuous fluid channel in one plane, and the three planes lie nearly exactly at right angles to one another. Each ampulla contains a *crista* with its important *sensory epithelium*, surmounted by a gelatinous *cupula* into which the protruding stereocilia of the hair cells penetrate (Fig. 9.38). The crista and cupula reach from side to side and from floor to ceiling of the ampulla, presenting a deformable barricade that allows inertial push rather than frank flow; it checks the extent of movement of fluid in the semicircular canal. When dye is introduced into the canal on one side of the cupula, although the cupula bends in response to acceleration, the dye does not escape to the other side (Steinhausen, 1933). When the stereocilia are displaced by as little as 100 pm (trillionths of a meter, equivalent to atomic diameters) in the direction of the kinocilium, the cell responds by partial depolarization and subsequent release of neurotransmitter.

Figure 9.39 shows that physical displacement of the stereocilia over a small range (ca. 2 μm) gives rise to large fluctuations in receptor cell membrane potential (ca. 15 mV). It also shows that the kinocilium is not essential to this receptor response and that the stereocilia are sufficient (Hudspeth and Jacobs, 1979).

The canals contain *endolymph* which is rich in potassium and low in sodium. The endolymph is confined within a tortuous, tubular, membrane-ensheathed chamber—hence *membranous labyrinth*. The enclosing membrane separates endolymph from *perilymph*. Perilymph lies outside the membranous labyrinth and is itself bounded by the periosteum and connective tissue that covers the dense temporal bone. As with cerebrospinal fluid, perilymph contributes extracellular fluid for all tissues of the inner ear except the stria vascularis in the cochlea which is bathed in endolymph. The perilymph is in communication with cerebrospinal fluid by a thin duct and has a nearly identical ionic composition to CSF (Table 9.2, page 993). As will be indicated shortly, the distinctive ionic composition of these two fluid compartments contributes importantly to the sensitivity of the hair cells.

Each of the hair cells has from 60 to 100 stereocilia or individual hairs which emerge from its distal surface. A single *kinocilium* is located on one side of the cluster of stereocilia. The kinocilium is the longest of the cilia, and the stereocilia are graded in length with the longest being closest to the kinocilium. The two types of hair cells are distributed over the entire surface of the crista, but the bottle-shaped cell (Fig. 9.40) is concentrated on the vertex of the crista. In the cristae of the horizontal canals, the kinocilia of the hair cells are always located on the side toward the utricle. On the other hand, in the case of both the anterior and the posterior vertical canals, the kinocilia are always located on the side of the hair cells away from the utricle. It has been demonstrated that bending of the cilia of the hair cells in the direction of the kinocilium results in *depolarization* of the cells and heightened neural activity. Bending in the opposite direction results in *hyperpolarization* and reduction of the base level of activity (Fig. 9.41). Thus, when patterns of motion are such that the endolymph tends to move with respect to the membranous wall of the canals, there will be excitation in the member of a pair of canals on one side of the head and inhibition on the other side. Rotation of the body to the right about a vertical axis with head erect will cause bending of the cupula in the right horizontal canal toward the utricle or vestibule. This will result in excitation. The horizontal canal on the left side will be inhibited as its cupula is bent away from the opening into the utricular chamber.

Depending upon the orientation of the axis of angular acceleration, the two anterior vertical canals may both be inhibited, both excited, one inhibited and one excited, or neither stimulated. It must be remembered that they do not operate as a

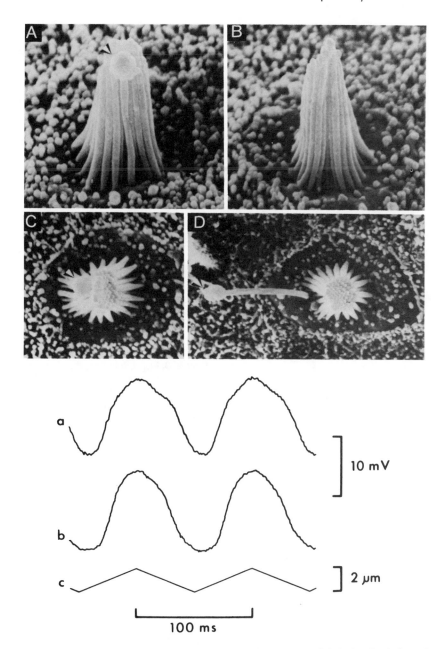

Figure 9.39. Experimental demonstration that stereocilia generate the receptor potentials in hair cells. *A,* Scanning electron micrograph showing stereocilia on a single hair cell after removal of the single kinocilium. *D,* View from above showing how the kinocilium (to the *left*) was deflected and removed, *C. a* and *b,* similar intracellular receptor potentials obtained from a normal cell (*a*) and from a hair cell with kinocilium deflected (*b*). Stimulation by stereocilia bundle deflection by a vibrating probe (*c*). [From Hudspeth and Jacobs (1979).]

pair, however. The anterior canal on one side and the posterior canal on the other work as a pair, and they are so arranged that whenever one is excited, the other will be inhibited. In the case of both the anterior and posterior vertical canals, rotations which result in bending of the cupula away from the opening into the utricle cause excitation. The point of major importance is that the pattern of excitation and inhibition of the various canals is unique for any specific axis of rotation.

The backward tilted position of all three of the canals when the head is erect results in excitation of two vertical canals and one horizontal canal for any rotation about a vertical axis.

Any change in the resting level of activity of the neural elements in the system, an increase due to excitation or a decrease due to inhibition, may convey information. In the normal mechanism, excitation in one canal will be matched by inhibition in the paired canal. If for some reason the function of one canal is lost, the paired canal may furnish sufficient information to offset the loss, when both its inhibition and excitation are taken into account.

Excitation and inhibition of the appropriate canals commence with the beginning of angular *acceleration*. These responses are reversed with angular *deceleration*. If a constant angular velocity is maintained, responses from the canals may continue for 20 s or more after the end of angular acceleration, and then cease. Sustained angular velocity is very unusual outside of laboratory studies or such specialized activities as figure skating, however.

The semicircular canal portion of the vestibular system is organized and oriented in 3-space within the skull during development so as to be admirably suited for the *detection and measurement of angular accelerations around any axis in three-dimensional space.*

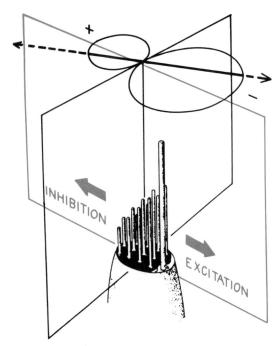

Figure 9.41. Excitation, the depolarization response which results when hair cells are bent in the direction of the kinocilium or inhibition, the polarization response for bending in the opposite direction. For a given direction of bending, the amplitude of effect is approximated by the length of a line drawn (*upper arrows*) from the point of tangency of the two circles to the intersection of either the circle representing excitation or the circle representing inhibition. [From Flock (1965).]

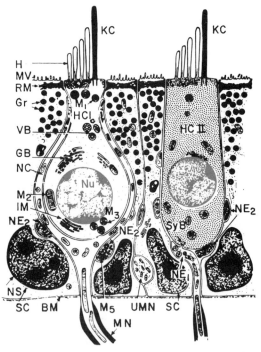

Figure 9.40. Two types of vestibular sensory cells. The flask-shaped type I cell (*HCI*) is surrounded by a nerve calyx (*NC*). Nerve endings presumed to be efferent (*NE 2*) make contact with the nerve calyx and directly with the type II cylindrical sense cell (*HC II*). Type II cells are supplied by both NE 1 and NE 2 types of nerve endings. Several types of mitochondria (*M₁* to *M₅*) are found in the sensory cells and neural elements. Additional elements identified in the figure include kinocilia (*KC*), stereocilia or hairs (*H*), microvilli (*MV*), granules (*Gr*), cell nuclei (*Nu*), supporting cells (*SC*), unmyelinated nerve fibers (*UMN*), myelinated fibers (*MN*), basement membrane (*BM*), Golgi complex (*GB*), synaptic bar (*SyB*), recticular membranes (*RM*), intracellular membrane (*IM*), vesiculated body (*VB*), and nucleus of supporting cell (*NS*). [From Engström et al. (1965b).]

THE OTOLITH ORGANS

The function of otolith organs is fundamentally different from the function of semicircular canals. Whereas the latter are excited by reason of the inertial properties of endolymph within the membranous canals, in the otolith organs, stimulation depends upon the difference in density of the otoconia or statoconia, small calcium carbonate crystals, with respect to the gelatinous substrate which supports them.

The specific gravity of the otoconia is approximately 2.9. This is considerably higher than the specific gravity of the supporting structure. Thus, the action of a component of linear acceleration on the otoconia results in the exertion of a force for deformation of the otolithic membrane and consequently, possible bending of the hair cells. The direction of bending will depend upon the orientation of the resultant acceleration force.

For a given position of the head, otoconia may exert a force which will maintain hair cells in a bent position, even though the head is maintained stationary. In the case of the semicircular canals, a force acts upon the cupula only during angular acceleration. At constant velocity of rotation, the cupula returns to its resting position, and excitation of the hair cells ceases. At least some cells of the otolith organs respond continuously for a given head position.

THE UTRICLES

The macula of the utricle is similar to the crista, with both types of hair cells and supporting cells. The cupula is replaced by the otolithic membrane, however, into which extend the shorter hairs of the utricular hair cells. The densely packed utricular stones or calcite crystals are embedded in the upper surface of the otolithic membrane. The macular surface of the utricle is located on the lower and forward walls of the utricular sac, with its posterior two-thirds in a plane approximately parallel to that of the horizontal canal. The anterior third is turned up slightly. The right and left utricular maculas are in the same orientation and are symmetrical in shape and organization.

Unlike the hair cells of the cristae, the hair cells of the utricular maculas do not all have the same orientation with respect to location of their kinocilia. Virtually every orientation is represented in each of the macular surfaces. The *arrows* in Figure 9.42 represent orientations of the hair cells over the surface of the macula. It is evident that tilting of the head from a vertical position in any direction will result in the bending of hairs of at least some of the hair cells in the direction of their kinocilia.

Individual functional units give maximal responses in accordance with the kind of orientation pattern depicted in Figure 9.42. In addition to the response of the system to the orientation of the linear acceleration vector of the earth's gravitational field, there is a displacement of the otoconia whenever the body is subjected to linear accelerations of any sort under dynamic conditions.

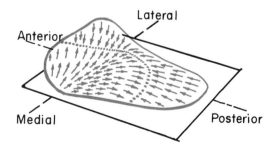

Figure 9.42. The pattern of polarization of sensory cells in the utricular macula of the guinea pig. *Arrows* indicate the direction of bending of stereocilia for excitation. [From Spoendlin (1965).]

THE SACCULES

The saccule is connected to the cochlear duct by way of the ductus reuniens. The macula of the saccule is located on the medial wall of this cavity in an approximately vertical orientation in an anteroposterior plane. The otoliths (Fig. 9.43) are lateral, with the macular substrate medial. There is a slight concavity of the macular surface as viewed from its lateral aspect on a given side. The surface and its shape along with the direction of orientation of the kinocilia of the hair cells are illustrated in Figure 9.44. Unlike the macular surface of the utricle, the hair cells on either side of a line representing a reversal in orientation have kinocilia oriented away from each other rather than toward each other. The saccules are otherwise similar to the utricles anatomically and are probably similar in function.

HAIR CELL RECEPTOR PHYSIOLOGY

Furukawa and Ishii (1967a, 1967b) have succeeded in recording from single afferent nerve fibers which innervate the macula of the saccule. The primary sensory fibers showed spontaneous activity. The otolith membrane can be trimmed to overlie a small region of the macula and can be moved by means of piezoelectric probe. The hair cells in the saccule project their stereocilia into the gelatinous undersurface of the membrane that bears the otoliths. When this is displaced in one direction, the nerve fiber discharges, whereas displacement in the opposite direction is without effect (Fig. 9.45).

The globular cells are nearly completely surrounded by a calyx-type afferent nerve ending, presumably for better security of receptor-neuron transmission. There may be electrical as well as chemical transmission between the hair cell and the calyx-type afferent ending (Flock, 1971). The *apical surfaces* of both kinds of receptor cells contain stereocilia which face into the membranous labyrinth, and thus are bathed by endolymph. It may be noted that plasma membrane covers each of the stereocilia, like closely fitting gloves, providing a very greatly expanded membrane for potential ion exchange. The *basal surfaces* of the hair cells are surrounded by supporting cells which are all apparently freely bathed by perilymph in the same way that central nervous system cells are bathed by cerebrospinal fluid. Tight junctions form a collar around the apical end of hair cells and prevent the two fluids from intermingling.

Because of differences in composition of the two fluids (see Table 9.2), there is a marked difference

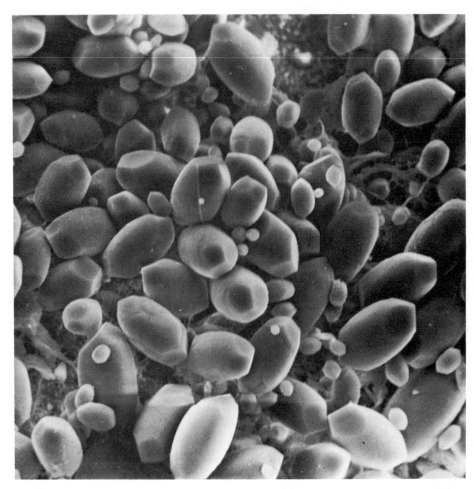

Figure 9.43. Otoconia of the macula of man as revealed by a scanning electron microscope. Hexagonal prisms may be discerned clearly in unsectioned material. Photomicrograph provided through the courtesy of Hans Engström. See Engström, 1968.

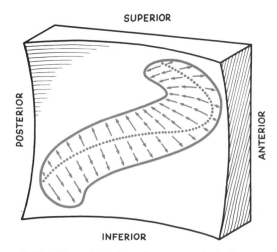

Figure 9.44. The polarization of sensory cells in the saccular macula of the guinea pig. Note that although the direction for depolarization is reversed in the region of the middle of the structure as was true for the utricle, the direction of bending for excitation is reversed as contrasted with that shown in Figure 9.42. [From Spoendlin (1965).]

in the *concentration of different species of ions* on the outside of the plasma membrane *in different regions of the same hair cell* (Hudspeth, 1983). This means that the membrane will be easily *depolarized* (e.g., from -60 mV to -40 mV), by admittance of potassium ions from endolymphatic fluid (rich in potassium ions, poor in sodium ions) at the apical end of the cell when the stereocilia are pushed in the direction of receptor activation. Adequate movement is presumed to result in opening of ion channels in the apical plasma membrane. The high concentration of potassium ions in endolymph will then force a high rate of migration of those ions across the membrane, perhaps across plasma membrane covering the stereocilia. By observation, the prompt potential changes seem to be of highest amplitude in the upper parts along the lengths of the stereocilia (Hudspeth, 1983).

Depolarization alters the receptor so that it promptly admits calcium ions in the basal region

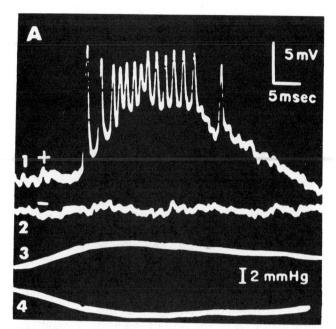

Figure 9.45. Recording from an afferent nerve fiber in the saccula. Displacement of the otolith in one direction (*3*) caused depolarization of the nerve fiber and discharge of nerve impulses (*1*), whereas a shift in the opposite direction (*4*) had no effect (*2*). [From Furukawa and Ishii (1967).]

of the cell, just as a terminal axon admits calcium ions on arrival of an action potential. The effect of calcium ions in the presynaptic region of the cell is to mobilize synaptic vesicles near the presynaptic membrane, induce fusion of vesicles to the plasma membrane, and release neurotransmitter molecules into the synaptic cleft.

A reverse of the potential shift may occur when the hair bundle is pushed in the opposite direction, away from the kinocilium. Already opened ion channels are presumably closed, and, quickly thereafter, due to the continuing action of the cell membrane to pump potassium ions out of the cell, the membrane potential becomes *hyperpolarized* (e.g., goes from −60 mV to −65 mV). Hyperpolarization reduces calcium ion influx into the base of the cell and reduces the amount of neurotransmitter being released. This will have the effect of reducing spontaneous activity among primary afferent neurons.

Several aspects of this receptor function deserve emphasis: The amount of *bending* of the stereocilia can be very small, about 1 μm distance, bending the stereocilia about 3°. A very small shift in membrane potential has large scale *amplifying* consequences. Activation of the receptor is *probabilistic* rather than a threshold process. This allows detection of extremely small displacements. The process is extremely *rapid*; cell membrane potentials begin to

respond to movement of stereocilia within a few tens of microseconds (millionths of a second). Furthermore, the stereocilia *adapt quickly*. If they are displaced by one micrometer, within a few tenths of a second the membrane potential has been largely restored to status prior to stimulation. This allows the same cell to be nearly maximally excitable when another displacement occurs, even though the previous displacement has endured.

In most cases, when the receptor cell stereocilia are displaced in the opposite direction, presumably some of the spontaneously open ionic channels are caused to close. (If the channels are already closed, there will be no effect with reverse direction displacement of stereocilia. This is presumably the case as illustrated in Figure 9.45.) This shifts the internal potential toward increased hyperpolarization, typically from −60 mV to −65 mV. If the stereocilia are displaced at right angles to the optimal direction there is also no effect.

FLUID DYNAMICS IN ANGULAR ACCELERATION

Endolymph contributes its inertial mass during angular acceleration of the head. Because each canal lies in a single plane, this inertial mass moves relative to the walls of a given canal only when motion takes place around an axis that is approximately at right angles to the plane of that canal.

Rotate water in a glass: note how the water lags behind the motion of the glass. When you maintain steady rotation, the water begins to "follow" the glass and will "catch up" with motion of the glass. Stop rotation of the glass, and the water moves again, in an opposite direction relative to the glass. Slide the glass across a smooth surface and observe that the same inertial mass of water "rides along" with the glass. If the glass is filled, covered, and sealed, there is no relative motion between glass and water except with rotation.

As the head is rotated actively or passively around an appropriate axis, the inertial mass of the endolymph causes it to lag behind the motion of the surrounding membranous canal. If turning continues at the same rate the endolymphatic inertial mass catches up with the canal and comes to hold the same position in relation to its boundary; there is then no further relative motion—until the turning halts, when the relative movements reverse themselves.

When angular acceleration of the head occurs in respect to an appropriate axis of rotation, the inertial mass of endolymph pushes against the crista and cupula barricade, from one side or the other according to the direction of angular acceleration, and the cupula is displaced, thus exerting mechanical force against the stereocilia which are embedded in the base of the cupula. In one direction, when displacement of the hairs is toward the kinocilium, the cells will be excited; in the other direction, they will be inhibited. From these observations we may anticipate some of the subjective and reflex consequences of angular acceleration, for example, that there will be consequences during start-up but not during continuing acceleration, *and* there will be physiological consequences during and following deceleration as well.

AFFERENT INNERVATION OF THE VESTIBULAR APPARATUS

Figure 9.40 illustrates schematically the two types of hair cells found throughout the sensory epithelium of the vestibular apparatus. Individual receptor cells provide a wide range of sensitivities in both cristae and maculae. Dynamic forces affecting the cristae are confined to each of three canals which lie in orthogonal planes (Figs 9.36 and 9.37). The two choice direction of movement of endolymphatic fluid in each canal, in respect to the orientation of the receptors in each crista (Fig. 9.38), dictates whether a given angular acceleration will be excitatory or inhibitory.

The systematic distribution of differently oriented receptors in the macular sensory epithelium (see Figs. 9.42 and 9.44) provides the basis for mapping the direction of effective mechanical forces, static, linear, and rotary, which act across the two curved surfaces, each of which approximates a single plane (roughly horizontal in the utricle and roughly sagittal in the saccule).

Figure 9.40 shows afferent (NE$_1$) and efferent (NE$_2$) axon terminals innervating both flask-shaped and cuboidal receptors. Both afferent and efferent fibers belong to the VIIIth cranial nerve. The axons provide central nervous system input from and output to all five sensory epithelia of the vestibular portion of the inner ear.

Terminal *afferent* axons form either a calyx-type ending (NC) which embraces the flask-shaped receptor cell bodies (to the *left* in Fig. 9.40) or relatively bulbous bouton endings which are applied to the basal regions of the cuboidal receptors (NE$_1$). On the basis of morphological evidence, the calyx-type endings receive from very numerous chemical and electrical synaptic junctions. It is safe to assume that this is a very secure one-to-one transmission relationship. Afferent endings arise from bipolar cells that are nestled in the vestibular ganglion between the internal auditory meatus and the vestibular apparatus. Their peripherally directed axonal branch innervates hair cells; their centrally directed axonal branch innervates second order cells in the vestibular nuclei in the brain stem.

Primary vestibular afferents from each of the cristae of the semicircular canals deliver information that is comparable to that of a "bidirectional angular accelerometer." Afferents from the utricle and saccule contribute information comparable to that of "multidirectional linear accelerators," in each of two major planes. Since both of the macular surfaces are curved, the motion detection is all the more comprehensive. This combination of information from angular and linear accelerators amounts to the equivalent of an "inertial guidance system." Since the vestibular apparatus in the two ears are systematically linked together functionally in the central nervous system, the reliability of information from the system as a whole is excellent.

EFFERENT INNERVATION OF THE VESTIBULAR APPARATUS

Efferent components of the vestibular nerve (Fig. 9.40 (*NE 2*) and Figs. 9.46 and 9.47) arise from three small cell groups: one originates from cells in the lateral vestibular nucleus, and the other two arise *bilaterally* from cells in the reticular formation lateral to the abducent (VIth cranial nerve) nucleus, ventromedial to the lateral vestibular nucleus (Gacek and Lyon, 1974; Warr, 1975). Vestibular efferents innervate all five sensory epithelia of the vestibular apparatus, and presumably affect afferent input from all of the hair cells in those epithelia (Smith and Rasmussen, 1967).

Efferent axons terminate in boutons that contain abundant synaptic vesicles (*NE 2* in Fig. 9.40; vesiculated endings in Fig. 9.46). These have the morphological characteristics of presynaptic terminals, and degenerate when the vestibular nerve is cut. (Primary afferents do not degenerate because their cell bodies lie distally to the zone of nerve section which therefore does not interrupt the peripheral portion of their bipolar axon.) Some efferents terminate on the outside of the afferent calices surrounding the flask-shaped receptors. There they are in a position to *modulate graded responses in primary afferents.* Since the primary afferents are spontaneously active, efferent influence could be very subtle. Other efferent axons attach directly to the base and sides of the cuboidal receptors, thus presumably *inhibiting the receptors directly.* The receptors show appropriate postsynaptic morphology at the site of these efferent junctions. All ves-

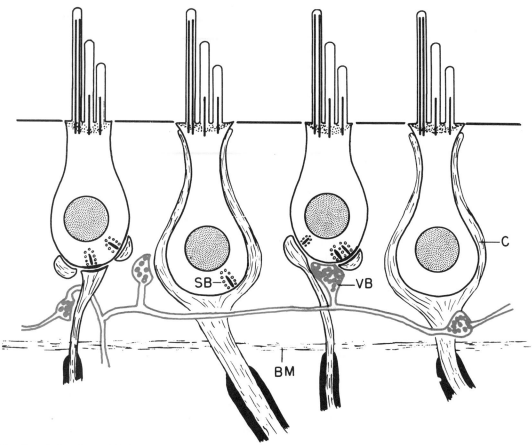

Figure 9.46. Four hair cells and their nerve endings shown in relation to vesiculated boutons (*VB*), chalice terminals (*C*), and other boutons and nerve fibers in the chinchilla macula. *BM*, basement membrane; *SB*, synpatic bar. The horizontal plexus is believed formed by efferent nerves. [From Smith and Rasmussen (1967).]

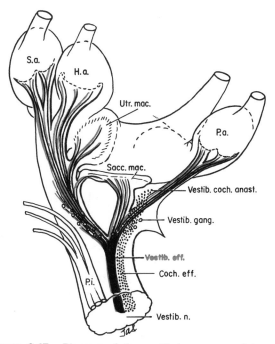

Figure 9.47. Diagram of the vestibular nerve and branches illustrating the course and distribution of the efferent vestibular fibers to the superior ampulla (*S.a.*), the horizontal ampulla (*H.a.*), and the posterior ampulla (*P.a.*). [From Gacek (1967).]

tibular efferents apparently deliver acetylcholine as their neutrotransmitter, and its effects are evidently entirely inhibitory.

Neurophysiology of the Vestibular Apparatus

Rasmussen (1940) found an average of 18,500 nerve fibers in human vestibular nerves, a total that includes efferent as well as afferent axons. The number of efferents in humans is unknown, but judging from the ratios of afferent to efferent fibers in other species (where there is quite a variation), a reasonable guess would be that between 1–2% of the total, say 200, may be efferents. The efferents may have more effect than their numbers might imply. From the auditory system, where the efferent innervation is better understood, we know that even a small number of efferents can branch sufficiently to establish boutons directly on all receptor cells or their primary afferent terminals. The few numbers of fibers and many branches implies, nonetheless, that the effects of vestibular efferents are likely to be fairly diffuse and nondiscriminating (Fig. 9.47).

The resting activity of primary afferent neurons averages around 90 impulses/s in the monkey. This provides that there can be very subtle time and motion discriminations by the central nervous system. This is another instance where a sense organ modulates ongoing activity rather than makes use of threshold phenomena.

Afferents from the utricular macula show tonic, phasic-tonic, and phasic responses to changes in head position (Precht, 1979). Different units convey magnitude and rate of change of gravity and linear accelerations of head position and motion in various directions. The tonic neurons are nonadapting. The phasic-tonic units respond to gravity and show significant adaptation. Phasic units respond only when the head is moving; they have no position sensitivity over a wide range. It has been shown in a variety of species that between 0.1 and 1 Hz of oscillatory stimulation, crista afferents carry a signal relating to head velocity (Precht, 1979).

The central processes of afferent cells in the vestibular ganglion divide into ascending and descending branches when they enter the brain stem. They innervate the four main vestibular nuclei, the floccular lobe of the cerebellum, and the reticular formation in the neighborhood of the abducent (VIth cranial nerve) nucleus. The four main vestibular nuclei are: *superior*, *lateral* (or nucleus of Deiters), *medial*, and *descending* (or inferior). The incoming fibers that branch anteriorly pass to the superior nucleus and the cerebellum. The descending branches pass to the medial and descending vestibular nuclei. The semicircular canals are represented primarily in the superior nucleus and rostral part of the medial nucleus, with some fibers going to the descending nucleus. Afferents from the utricular macula appear to go mainly to the lateral vestibular nucleus, with some fibers to the medial and descending but none to the superior (Brodal, 1981).

Projections of the Vestibular Nuclei

In general, the vestibular nuclei project to the *spinal cord, cerebellum, reticular formation, and superior colliculus* (see Figs. 9.48, 9.49, and 9.50). Roughly half the neurons in the vestibular nuclei generate reflex control of eye movements in relation to motion of the head in 3-space in the context of a directionally uniform gravitational field. The other half generate reflex control of movements of the head and limbs on a dynamic center of gravity of the body moving in three-dimensional space as affected by a steady gravitational force. The result is a vestibulocentric balance and orientation in respect to the individual as a whole.

The more *caudal parts* of the vestibular nuclei

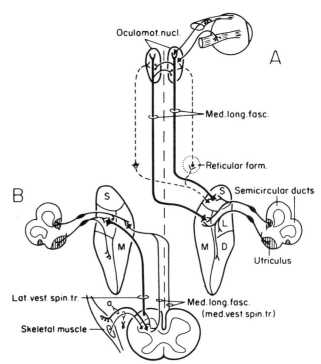

Figure 9.48. Simplified diagrams of main features in the organization of primary vestibular circuits to the vestibular nuclei and from there the ascending (*A*) and descending (*B*) projections to eye motor and spinal motor controls. *D, L, M,* and *S*: Descending, lateral, medial, and superior vestibular nuclei. [From Brodal (1981).]

project to the spinal cord, as shown on the *left* (*B*) in Brodal's diagram (Fig. 9.48). Lateral (L), descending (D), and medial (M) vestibular nuclei contribute descending axons via the lateral vestibulospinal tract, medial vestibulospinal tract, and medial longitudinal fasciculus (MLF). The MLF extends from the level of nuclei of the IIIrd cranial nerve to upper thoracic spinal levels. The majority of fibers in the medial longitudinal fasciculus come from the vestibular nuclei. The MLF interconnects three pairs of motor nuclei, cranial nerves III, IV, and VI, which innervate the extraocular muscles. It interconnects the vestibular-ocular system, with motor nuclei serving cervical and upper thoracic motor nuclei, thereby controlling turning of the head in coordination with directing the gaze. Head and eye coordination of this sort is necessary in order to achieve stability of vision in any self-originated movements, movements of the environment, and for visual grasping of moving or nonmoving targets of interest. The combination of ascending and descending vestibular projections *control head and eye turning in relation to* a vestibulocentric balance, and eye and body control system. The MLF contributes importantly to vestibu-

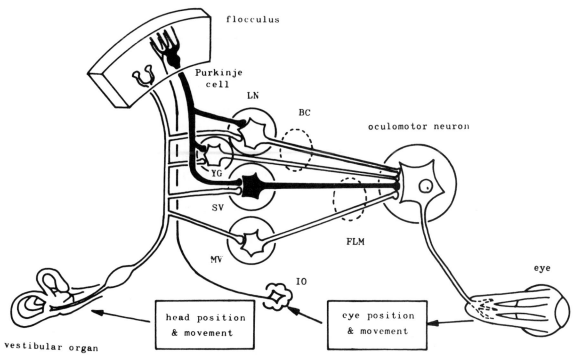

Figure 9.49. Neuronal connections between the vestibuloocular reflex and cerebellar cortex. Inhibitory neurons are filled in *black*; excitatory neurons are indicated by hollow structures. The vestibular branch of the VIIIth nerve has direct projections via mossy fibers to vestibulocerebellum, the flocculus, as well as to vestibular nuclei. Vestibuloocular three-neuron reflex (VOR) begins with signals from the semicircular canals and otolith organs, which are relayed by second order neurons in the vestibular nuclei, and project to the eye motor nuclei. Inhibitory action is largely relayed by the superior vestibular nucleus (*SV*) and excitatory action by the medial vestibular nucleus (*MV*). Both of these nuclear groups are under inhibitory influences of the cerebellum. Y group of vestibular nuclear complex (*YG*) and lateral vestibular nucleus (*LN*) receive from both VIIIth nerve excitation and cerebellar inhibition. The cerebellum appears to improve a feedforward control of eye movements in response to vestibular stimulation. This establishes a two-way relationship between the VOR and the flocculus. *IO,* inferior olive; *BC,* brachium conjunctivum; *FLM,* medial longitudinal fasciculus. [From Ito (1974).]

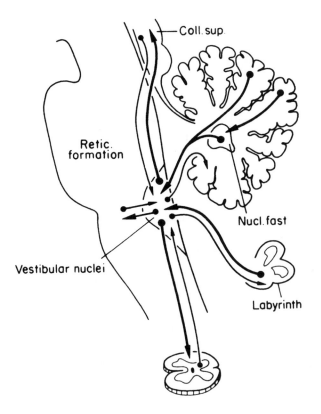

Figure 9.50. Note that the vestibular nuclei have reciprocal connections with the spinal cord, the labyrinth, the cerebellum, the reticular formation, and the superior colliculus. The connections indicated by *heavy lines* are more massive than those shown as *thin lines*. It can be seen that the major afferent input to the vestibular nuclei comes from the labyrinth and cerebellum, while their main output goes to the spinal cord and mesencephalon (mainly the eye motor nuclei). [From Brodal (1981).]

lar directed head and eye motor coordination and to the individual's perception of self in three-dimensional space. Additional ascending and descending vestibular contributions outside the MLF serve *skeletal muscle* responses along the entire longitudinal column of *brain stem and spinal motor nuclei*, adjusting the body to gravity and other static and acceleratory forces (Gernandt et al., 1957, 1959).

The vestibular nuclei also send axons and axon collaterals to widespread regions in the brain stem reticular formation, and the reticular formation in turn sends axons to each of the vestibular nuclei. There is a relatively private system of projections from the lateral vestibular nucleus (Deiter's nucleus) to the lateral reticular nucleus which in turn projects to the cerebellum.

Ascending axons from medial and superior vestibular nuclei (Fig. 9.48*A*) pass upward within the MLF to oculomotor nuclei in the midbrain. The medial nucleus sends such axons predominantly contralaterally, while the superior nucleus sends such axons predominantly ipsilaterally. The superior vestibular nucleus also contributes to eye motor control, bilaterally, indirectly, by way of the reticular formation (Fig. 9.48*A*).

Vestibular Reflexes

VESTIBULOOCULAR REFLEXES

Our eyes ride in a moving head. If the eyes were fixed when the head is moved actively or passively, there would be sweeping movements of the visual scene across each retina. The visual system has no way readily to detect the source of such shifts in retinal activity. Head movements are thus a potential hazard to stable vision and are compensated by reflex eye movements in an opposite direction—by an equal amount—at an equivalent rate. This obviously desirable control is brought about by *vestibuloocular reflexes* (*VOR*).

Input to the VOR, induced by head movement, is generated in the sensory epithelium of the vestibular apparatus. As noted earlier, primary vestibular impulses provide information that defines motions of the head. The vestibular nuclei, with an assist from cortex of the cerebellum, generate signals to the oculomotor apparatus which activate the extraocular muscles and provide a compensatory eye movement (Fig. 9.49). VOR operates, as do head and eye movements, mainly in the horizontal plane. A stable retinal image is accounted for almost exclusively by the VOR; other mechanisms such as the optokinetic reflex and pursuit mechanisms con-

tribute only minimal compensations (Miles and Lisberger, 1981).

The shortest VOR path involves three synapses: first order to second order neurons in the vestibular nuclei, vestibular to eye motor neurons in the IIIrd, IVth and VIth cranial nerve nuclei, and neuromuscular junctions in the extraocular muscles. The VOR is considered an *open loop control system* because the output of the reflex, retinal images and eye motor events, do not influence the vestibular receptors which initiated the action. The optokinetic reflex system is comparatively slow, 80 ms between shift of images on the retina and compensatory eye movements. Because execution of the VOR does not depend on visual feedback, it can operate in contexts of low illumination, and quite as effectively in the dark. If a light *were* then turned on, vision would already be stabilized.

The *pursuit* system allows tracking of small objects across a textured background, which requires overriding the optokinetic reflex. When there are no visual targets, as and in sky, open water, and in darkness, optokinetic and pursuit systems would be of no assistance. The VOR is considered a *feedforward system* that works to ensure that the retina will be stable in any given context for the perception of any target of opportunity. Even when both optokinetic and pursuit systems operate in concert, the latency for their responses is of the order of 80 ms; hence, dependence on visual feedback would disable effective visual control during rapid head movements (for a review of this subject, see Miles and Lisberger, 1981).

When the upright head is rotated to the *left*, the VOR compensates by promptly rotating the eyes to the *right* at a corresponding rate. This compensatory eye movement stabilizes the retinal image. The VOR system thus prevents head movements from disturbing retinal images. This compensation allows the gaze to fix on the nonmoving surround, even though the head is moving, a compensation that will succeed for about 60° of head rotation. If head rotation continues, the eye motor control system introduces a saccadic eye movement so that the eyes skip rapidly forward to stabilize at a new point in the visual scene. This gives rise to a rhythmic sequence of slow and rapid eye movements which are known as *nystagmus*.

The VOR is high-speed, accurate, and consistent, almost machine-like in its performance. There is only a 12-ms delay from head movement to eye motor response. Over a frequency range of 0.1–1.0 Hz, the gain of the monkey's VOR, peak eye velocity divided by peak head velocity is between 0.9 and 1.0.

DYNAMIC ADAPTATIONS OF VESTIBULOOCULAR REFLEXES

If the VOR regularly fails to stabilize retinal images during head movements, the reflex undergoes lasting adaptive changes that restore visual stability. Information that defines the operational requirements for this reflex adaptation must originate from *vision*. The VOR system can be induced to adjust to radically different visual demands (image magnification and minification—which create *motion-velocity problems* for visual stability; and left-right, up-down, and combined image reversals which create *motion-direction problems* for the VOR). How is this radical but remarkably useful adaptation accomplished along a three-neuron reflex arc?

The VOR has been experimentally altered in this way by means of reversing prism spectacles, in animals as well as in humans. Sophisticated experiments have been carried out with humans, monkeys, rabbits, rats, birds, and fishes (reviewed by Miles and Lisberger, 1981). It is evident that changes in the VOR following the introduction of displacing or magnifying lenses are due to modifications of the basic three-neuron reflex pathway and *not* due to some learning strategy that is improvised to adjust the reflex. Miles and Lisberger summarize the evidence:

1. Neither rate of adaptation to lenses nor rate of reversal when the lenses are removed suggest a learning strategy. Changes in gain are almost machine-like in acquisition and recovery, showing a similar time course in different individuals and in the same individual with repeated exposures.
2. After the VOR has adapted to a given experimentally imposed requirement, it shows long persisting alterations in caloric tests of vestibuloocular functions that closely match the visually induced adaptations.
3. In experiments where the head is immobilized after the VOR has become adapted, the adaptation is retained at least several days without further vestibular reinforcement.
4. Abrupt, unexpected head movements reflect the adapted state. Moreover, following adaptation to magnifying spectacles, trials in which zero gain would be preferable indicate that the VOR, in the adapted state, is fixed and immutable at least for considerable time.

MECHANISMS OF ADAPTATION OF VESTIBULOOCULAR REFLEXES

Arguments have been advanced that the floccular lobe of the cerebellum is the site of physiological adaptation of the VOR. But experiments reveal that although the flocculus probably contributes to VOR gain in rabbits (Ito, 1974), in monkeys the flocculus appears to govern adaptive gain control of eye movements induced by the visual pursuit system, adapting to a gaze velocity signal rather than to a head velocity signal. In monkeys, changes of activity in floccular Purkinje neurons were a consequence rather than a cause of changes in the VOR gain. *Evidence indicates that the same neurons may be used for different physiological strategies in different species.*

A pertinent additional principle of adaptation was discovered in the flatfish. During metamorphosis, this fish changes from a normally oriented dorsal-ventral–up-down fish to being tilted 90° to one side when they become bottom-adapted adult flatfish. In this process, the vestibular apparatus is rotated 90° with respect to the direction of the gravitational field. The former horizontal semicircular canals become vertical canals. Graf and Baker (1983) have shown that an adaptive change in the VOR occurs in second order vestibular neurons, which receive input from the horizontal semicircular canal and would ordinarily project to horizontal extraocular motor units. Instead, these neurons develop extensive axonal arborizations which terminate on extraocular motor neurons which innervate *vertical* extraocular muscles. Second order vestibular neurons relating to the horizontal canal had not previously been shown to terminate on vertical extraoculomotor neurons.

The important lesson here is that *the nervous system has a remarkable capacity to adapt to radical environmental changes. This capacity is to be found in many* circuits, but it is by no means ubiquitous. Each specific instance seems to be "designed" (by evolution and development) to solve a unique problem, which in this example is retinal stability in a moving eye, in a moving head, on a moving body, with the stability depending on information of many different kinds from many different sources all converging on a three-neuron arc!

Not many examples have been studied, and none is fully explained, but the VOR illustrates some important neurophysiological principles: The adaptive circuits are all goal-seeking, and the goal includes requirements for precision, speed, gain control, coordination, sometimes subordination, in relation to other goal-seeking control systems. My surmise is that building mechanisms for adaptation in systems like the VOR which must perform swiftly, accurately, and dependably, for example, in order to achieve visual stability, requires three things: (1) convergence of many diverse but strictly pertinent systems controls onto the pathway that is subject to adaptation, (2) arrangement for rapid access of information required for all aspects of the adaptation, and precise amplification of signals along only the reflex circuit (in this case, affecting two synaptic junctions), and (3) some still mysterious mechanisms for altering circuit performance as a whole against goal-stipulated criteria such as visual stability, despite ordinary, everyday perturbations. Persistent discrepancies call for this reorganizational step to be reapplied. Perhaps everyday experiences involving minor perturbations are continually shaping the adaptive mechanism toward more perfect achievement. It implies that skill acquisition, in athletics, music, and other technical capabilities may involve adaptations at such "low level" circuits, including primary input and final output components. We can appreciate from consideration of the VOR how demanding are the consequences, how elegant the achievements, and what a considerable time and biological engineering effort must be involved in the construction of the necessary adaptive controls, and in their readaptation to an earlier operative control status.

Adaptive gain control of the VOR is lost following ablation of the floccular lobe of the cerebellum. Removal of the flocculus disrupts this reflex even though (in primates, at least) the flocculus makes only an ancillary contribution to the VOR. This emphasizes a basic principle of neurophysiology, namely, that *removal* (or cooling, or obtunding with drugs) of one part of the nervous system does not reveal the function of *that* part: quite a lot less specifically helpfully, it tells us only *what the nervous system does in the absence of that part.* Think about that one because it is antiintuitive but yet perfectly reasonable, and it is of fundamental importance to all neurological interpretations.

LOCATION OF THE MECHANISMS OF ADAPTATION OF VESTIBULOOCULAR REFLEXES

By studying optokinetic and pursuit systems interactively along with the VOR, it has been possible to examine which parts of the three-neuron VOR are implicated in adaptive processes involving each of these functionally important mechanisms. This can be appreciated in a rudimentary way by examining Figure 9.49 which reveals the minimum connections of the VOR and illustrates its relations to the flocculus in the cerebellum. Neurons which are filled in with black are inhibitory. Primary vestibular afferents project to the four main vestibular nuclei *and* to the floccular cortex in the cerebellum. The medial and superior vestibular nuclei project to eye motor nuclei by way of the FLM, while the

lateral nucleus (LN) and associated nuclear groups (YG) contribute via reticular pathways to the same motor nuclei. The "oculomotor nucleus" represents all three cranial nerve nuclei (III, IV, and VI) which serve the six extraocular muscles. Feedback from eye position and movement to the inferior olive (IO) returns via climbing fibers to floccular Purkinje neurons. By means of these connections, the cerebellum receives simultaneous representation of vestibular and oculomotor activity from which it influences the three-neuron VOR pathway.

Visual messages which will be discussed more fully in the next chapter contribute an "error" message for the VOR. This includes *retinal slip* and *eye velocity*, to which may be added a proprioceptive contribution also relating to eye velocity from extraocular muscle afferents (eye position and movement in Fig. 9.49). This integrated visual information converges on the floccular cerebellar cortex. The combination of information involving vestibular, eye proprioception, retinal slip, eye velocity, and cerebral eye motor commands further integrated by the cerebellar Purkinje cells, relays an "error" message to neurons in the VOR pathway. This message manifests any inappropriate gain in the VOR circuit. Thereafter begins the slow, corrective adaptation process. Correction obviously involves processes that can converge onto and control synaptic performance at each of the central synaptic junctions and perhaps also, in a not very precise way, modify vestibular primary afferents through influences of vestibular efferents. Nothing is known about how these converging elements actually revise performance of the primary circuitry, only about how long it takes and what may be the limits of adaptation.

Optokinetic signals (from visual cortex) influence the first synaptic relay of the VOR system, and visual pursuit signals influence the pathway at the second synaptic relay, in the eye motor nuclei (Miles and Lisberger, 1981).

OTHER VESTIBULAR SYSTEMS OF CONTROL

Figure 9.50 illustrates the main projections and functional domains of the vestibular system. The peripheral vestibular apparatus projects mainly to the vestibular nuclei, and these interact reciprocally with the brain stem reticular formation. There is a particular focus of vestibular projection to the reticular formation in the vicinity of the VIth cranial nerve nucleus. This region is sometimes called a "center for lateral gaze," because the VIth cranial nerve innervates the lateral rectus muscle of the eye which turns the eye horizontally to the ipsilateral side. Excitation of the reticular formation in this vicinity elicits not just a single muscle pull but a coordinated, conjugate deviation of both eyes to the ipsilateral side, with reciprocal innervation of the medial rectus on the same side and double reciprocal innervation of the pair of horizontal tractors on the contralateral side of the brain stem. This elaborate conjugate deviation of the eyes is under bilateral control of the vestibular nuclei on each side—a *dual bilateral vestibular conjugate eye motor control system.*

Cerebellar contributions to eye movements which act partly through the vestibular nuclei are confined to ipsilateral vestibular oculomotor influences. For conjugate deviation of the eyes to the ipsilateral side, floccular signals disinhibit vestibular neurons which project to the ipsilateral lateral rectus (by inhibiting an inhibitory neuron), and directly inhibit the ipsilateral medial rectus (Ito, 1973). The contralateral flocculus has its own ipsilateral influence. Thus, the cerebellum does *not* exert a dual eye motor control on top of the vestibular dual eye motor control system.

The vestibular nuclei also project strongly to the lateral part of the reticular formation which in turn projects to the cerebellar cortex. As we have already

observed, the vestibular nuclei send direct fibers to the flocculus (indicated by the *small, hooked arrow.* Note in Figure 9.50 that the vestibular nuclei are themselves under powerful inhibitory influence from the cerebellar cortex, indicated by the *two heavy arrows* descending from the cerebellar hemisphere. One arrow goes directly from cerebellar cortex to the vestibular nuclei. The other arrow projects to the vestibular nuclei via the fastigial nucleus, one of the "roof nuclei" of the cerebellum (roof: meaning overhead in relation to the fourth ventricle).

The vestibular nuclei have powerful projections to the midbrain, particularly to the superior colliculus and to each eye motor nucleus (IIIrd, IVth, and VIth cranial nerve nuclei). We have already considered the vestibuloocular reflex system. Now we shall consider how all this integrates with vestibulospinal reflexes.

VESTIBULOSPINAL REFLEXES

First order vestibular afferents, as we have seen, project to vestibular nuclei, brain stem reticular formation, and the floccular lobe in the cerebellum. Figure 9.51 illustrates the vestibulospinal reflex (VSR) schematically. The primary vestibular neu-

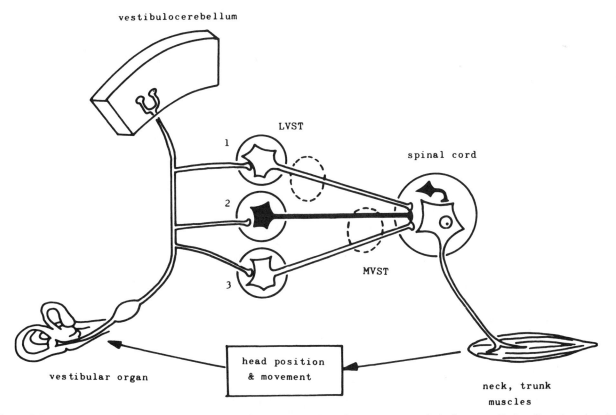

Figure 9.51. Neuronal organization of the three-neuron vestibulospinal reflex arc composed of primary vestibular afferents, secondary vestibular neurons, and spinal motor neurons. *LVST*, lateral vestibulospinal tract; *MVST*, medial vestibulospinal tract. The whole column of motor nuclei the length of the brain stem and spinal cord is under powerful control by vestibulospinal reflexes. [From Ito (1974).]

ron is shown projecting mossy fiber terminals (knob endings) in the cerebellum, with collaterals (or other axons) represented as ending on three types of second order sensory cells. These have been categorized by Ito (1974): (1) fast velocity lateral vestibulospinal tract neurons (LVST) which excite motor neurons, (2) slow velocity medial vestibulospinal tract neurons (MVST) which inhibit spinal motor neurons, and (3) fast velocity excitatory medial vestibulospinal tract neurons.

The primary afferent vestibular neuron, as depicted, is intended to represent all varieties of vestibular input from all five vestibular sensory epithelia. This overall contribution is like an inertial guidance system, and it is distributed immediately to second order neurons which are engaged in patterns of neuromuscular control relating to all skeletal musculature. The second order sensory neurons, partly in the vestibular nuclei and partly in the reticular formation, make different integrative contributions depending on what they are receiving from many other sources. Second order vestibular neurons also receive from the flocculus and the vermis of the cerebellum, from midbrain collicular neurons relating to head and eye coordination, from the VOR system, and from somatic sensory fields. Thus, *second order vestibular neurons constitute a point of convergence from many different sources of information that relate to the control of head, body, and limbs in space.* Vestibular input, contributing to this busily integrating pool of neurons, delivers its important messages relating to the position and motion of the head. *The output of these second order neurons, in the vestibular nuclei and reticular formation, goes to all motor neurons for skeletal muscles, along the entire length of the neuraxis, from eye muscles to the external anal sphincter.*

Why all the convergence onto second order vestibular neurons? Because information from many sources may be needed for adequate response under command of the vestibular apparatus. The vestibulospinal reflex (VSR), like the VOR, is a three-neuron system which constitutes a rapid response system, with fast latency fibers in both lateral and medial vestibulospinal tracts (Fig. 9.51). The urgency of this command system is that with either active or passive movement of the body in a gravitational field there is always the risk of falling down. So the vestibular apparatus which controls balance and equilibrium, in order to be able to avert disaster, must have swift access to all skeletal muscle systems to command what is necessary to maintain and recover balance and equilibrium under almost any contingencies. This is as important,

perhaps even more important, than stability of visual images under the control of the VOR. To meet this need some of the vestibulospinal neurons and some of the corresponding reticulospinal neurons are among the largest diameter and most rapidly conducting of any vertebrate neurons.

The goal of both VOR and VSR systems is to achieve motor performances that match requirements introduced by head and body movements. In the case of VOR, maintenance of visual stability despite head movements is the goal. In the case of VSR, maintenance of balance and equilibrium despite body movements is the goal. Since the magnitude of errors and the compensatory adaptations vary with different optical requirements for the VOR, and different imposed inertial loads (e.g., carrying a backpack while skiing) for the VSR, it is necessary to have mechanism by which experience can modify these circuits. Separately adjustable temporal frequency channels might provide greater flexibility for adaptation to the overall reflex requirements. There is evidence (Lisberger et al., 1983) that there may be multiple channels in the VOR, and by implication, perhaps also in the VSR, which segregate different rate response components into subsystems (channels) that have different dynamic response characteristics for control of the gain between vestibular input and error signals from visual and somatic sensory sources.

SPINOVESTIBULOSPINAL REFLEXES

We now need to inquire into how somatic sensory information from the body affects the vestibular reflex apparatus in relation to the maintenance of balance and equilibrium. This system is known as the spinovestibulospinal reflex (SVS) system. Figure 9.52 diagrammatically illustrates these mechanisms which operate through a minimum of a four-neuron reflex arc. *The somatic input serves as both feedback and feedforward contributions to the VSR circuits.* Thus, somatic information can provide corrective information that relates to what body and limb activities actually occurred (to compare with the vestibular command for their control). This is an example of somatic feedback to VSR. It also feeds forward information about body movements which occur (actively or passively) prior to motion of the head, hence in advance of vestibular activation. This sets the stage for a more appropriate VSR. *Somatic sensory information can also instruct VSR sequences,* for instance, in respect to tonic neck reflexes (see Chapter 70, Segmental Controls). Tonic neck reflexes relate to the sequence of getting up from a reclining position. First, the head is

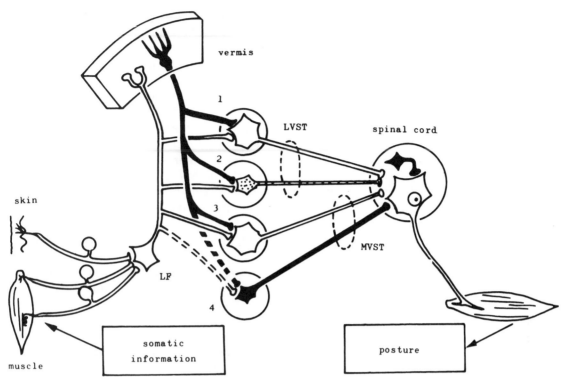

Figure 9.52. Relationship between four-neuron spinovestibulospinal reflex arc and the cerebellar vermis. Excitatory spinal input to Deiters nucleus, source of major vestibulospinal projections, also project to cerebellar vermis which in turn exerts an inhibitory influence on the spinovestibulospinal reflex arc. This establishes a two-way relationship between the spinovestibulospinal reflex arc and the cerebellar vermis. *LF*, lateral funiculus of cervical cord, other abbreviations as in Figure 9.52. [From Ito (1974).]

raised. This activates appropriate patterns among vestibular afferents. Cervical somatic sensory input, activated simultaneously, especially afferent input from vertebral joint receptors, further informs VSR circuits concerning the action of elevation of head and neck. Following this initiation, somatic messages to VSR elicit a cervical to caudal ordering of successive skeletal muscle contractions appropriate for fulfillment of the act of arising. One normally gets up head first, and somatosensory contributions to VSR combine to provide the organization of axial and limb muscle contractions in proper sequence.

Figure 9.52 shows how the vermis of the cerebellum, which has somatotopic representation, makes its contribution to this VSR system. In effect, the cerebellum does not initiate but does contribute as a "servomechanism" to boost the VSR in its execution of smoothly coordinated motor control of balance and equilibrium.

On the left are shown cutaneous receptors and receptors from tendon organs and muscle spindles (of which more will be found in Chapter 70, Segmental Controls). These first order somatic afferents convey information as to the position of axial and limb musculature, the set point of skeletal

muscle readiness, the degree of tension being exerted, and the cutaneous components that may be responding to body and limb position and strain. These somatic afferent signals are relayed by second order neurons in the dorsal horn of the spinal cord whose axons ascend in the lateral funiculus (LF) of the spinal cord. They project to the vermis of the cerebellum and give off collaterals, and other axons in the tract contribute terminals, to the three types of vestibulospinal projections described above and depicted in Figure 9.51, plus a fourth type of neuron (*4*, shown in Fig. 9.52) which projects by way of relatively slowly conducting inhibitory neurons via the medial vestibulospinal tract (MVST) to motor neurons serving skeletal muscles all along the neuraxis. Again, there is a grand convergence of pertinent sources of information to the third order vestibular and reticular units concerned in vestibulospinal outflow.

Vestibular Threshold Determinations

There is a minimum angular rate of acceleration which can be detected by a human subject. Under optimum conditions, that rate is probably of the order of $0.10°/s^2$ (Clark, 1967; Clark and Stewart,

1968). In order to achieve such low threshold values, it is necessary to employ a reference light point which rotates with the subject. Apparently, vestibular inputs to the extraocular muscles in response to the low acceleration result in some eye movement and hence a relative movement of the image of the light spot on the retina which can be detected. If subjects are rotated in complete darkness, thresholds of the order of $0.05°/s^2$ or slightly lower are reported (DeVries and Schierbeck, 1953; Meiry, 1966).

Threshold accelerations for the stimulation of nystagmus may be observed directly or recorded in terms of the electrical signal between two electrodes located on the temples. Such thresholds do not depend upon subjective report and this may be considered an advantage. Threshold stimulus for nystagmus is probably of the same order as threshold stimulus for detection of rotation under optimum conditions (Timm, 1953; Benson, 1974).

The otolith organs, particularly the utricles upon which detection of head tilt is believed to depend, may be studied by tilting the body very slowly from an erect posture in a chair device to some position which deviates from the vertical. If the rate of tilt is sufficiently slow to prevent stimulation of the semicircular canals, then the question may be asked how much change in orientation of the body with respect to the vertical can be detected by the subject. A common procedure for studies of this sort is the presentation of an illuminated line of light, the orientation of which is controlled by the subject. If the subject has been tilted to the right or left and is asked to position the illuminated line so that it is in a vertical orientation, then any deviation of the line from the true vertical provides a measure of the extent to which the observer is deceived with respect to the orientation of his own body. Probably by reason of the orientation of the utricles within the head, the human subject is most sensitive to slight amounts of tilt away from the vertical.

Analogous with the nystagmus response to stimulation of the semicircular canals is the counterrolling of the eyes which occurs when the head is tilted away from the vertical. The phenomenon does not lend itself readily to precise threshold determinations, however. Counterrolling in the earth's gravitational field is approximately 10° for 60° of tilt away from the vertical. No electrical index of counterrolling is readily recordable. The dependence of counterrolling on otolith organs, which in turn are dependent upon a gravitational vector, has been demonstrated by studies of counterrolling in response to tilt in normal and labyrinthine defective subjects both on the earth's surface at $1 \times g$

and under zero gravity conditions during flight through a ballistic trajectory in an aircraft. Labyrinthine defective subjects show little or no counterrolling under any circumstances. Normal subjects show no counterrolling of the eyes with tilt of the body when they are in a zero gravity trajectory (Miller et al., 1966). At zero gravity, the relatively high specific gravity of the otoconia is of no consequence for stimulation of the hair cells embedded in the macular substrate. Change in the orientation of the body is not accompanied by a change in a linear acceleration vector.

It is appropriate to assume that the threshold angle of the otolith organs for tilt will vary with the magnitude of the prevailing gravitational acceleration force. A functional relation between counterrolling in minutes of arc and degrees of tilt at each of three levels of acceleration is illustrated in Figure 9.53 for both normal subjects and labyrinthine defective subjects (Miller and Graybiel, 1965). With increase in the gravitational force above the normal gravitational force at the earth's surface, there is an increase in counterrolling of the eyes directly proportional to the increased acceleration force.

Temporal Duration of Effects of Vestibular Stimulation

When a subject is exposed to constant angular acceleration about a given axis, a sensation of rotation is induced along with nystagmus movements of the eyes. There appears to be an increase in the perceived angular velocity for approximately 30 s followed by an impression of gradual decrease in

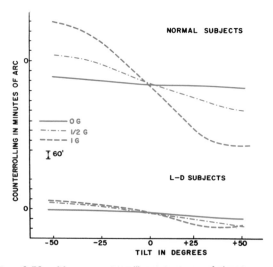

Figure 9.53. Mean counterrolling response of the eyes measured in minutes of arc as a result of tilting in angular degrees with respect to a resultant G force. Results are presented for both normal and labyrinthine-defective (*L-D*) subjects. [From Miller and Graybiel (1965).]

angular velocity. Continuation of a nystagmus response with continuation of angular acceleration after reduction of the oculogyral illusion before termination of angular acceleration suggests that central factors may override the output of the peripheral sensing device, as this influences conscious perception.

For the usual exposures to angular acceleration for relatively short durations followed by the maintenance of a constant angular velocity for varying amounts of time, the perception of rotation and nystagmus both continue for 20 or 30 s after angular acceleration has stopped in a normal subject. This observation fits with the notion of the cupula as a damped torsion pendulum which gradually resumes its normal undeflected position when angular velocity is constant. On the other hand, if rotation occurs about a horizontal axis, the perception of rotation continues for the entire duration of rotation even at constant angular velocity (Benson and Bodin, 1966). The explanation of this difference might depend upon the fact that with rotation about a horizontal axis there is a continuously changing orientation of the linear acceleration acting on the otolith organs. This provides a continuing sensory input which corresponds to the continuing rotation, even though the velocity is constant.

A curious aspect of this situation is that with rotation about a horizontal axis there is an immediate cessation of the impression of rotation when physical rotation is stopped. In other words, although the semicircular canal stimulation resulting from deceleration should produce a reversing signal for some seconds after rotation is stopped, the perception of rotation stops immediately.

AVIATION

As Adrian (1943) has pointed out, the vestibular mechanism is not ideally suited to situations in which one is subjected to protracted angular motions. Prior to the advent of high speed vehicles, particularly aircraft which can maneuver relatively freely in three-dimensional space, man was not ordinarily subjected to prolonged angular rotation. The vestibular system may be ideally suited to coordination of the movements of the body and fixation of the eyes for short-term relatively low amplitude rotations, but this is clearly not the case for prolonged rotation. After prolonged rotation at constant angular velocity, any deceleration results in sensory signals from the canals which are appropriate to rotations in the opposite direction and quite inappropriate to the actual motions involved. This gives rise to difficulty of interpretation of the nature of the motion path of an aircraft in flight and has been blamed for innumerable aircraft accidents.

SPACE FLIGHT

The zero gravity environment of space is inconvenient from the standpoint of manipulating liquid foods and other aspects of daily living, and also presents certain physiological hazards with respect to venous return of blood to heart and calcium metabolism. For these reasons, it has been suggested that an artificial gravity might be useful for space stations and for space travel. Such an artificial gravity could be achieved only by the maintenance of a constant angular velocity of the space station with occupants living at an appropriate radius so that the centrifugal force would provide an artificial gravity. Problems arise by reason of the fact that the radius must necessarily be relatively short. In moving about in such an environment, individuals will tilt their heads about various axes which may be orthogonal to the axis of rotation. Such rotations may give rise to Coriolis accelerations, the effects of which can be most disturbing with respect to orientation and at worst may contribute to motion sickness.

Investigations have been conducted in which volunteer subjects have lived in rotating environments on the earth's surface with the rotation about a vertical axis. Free movement within such an environment is most disturbing and induces motion sickness in many subjects.

ADAPTATION

Adaptation to an environment such that the effects of movement are no longer disturbing can occur if the rate of rotation is less than 6 or 7 rpm (Newsom et al., 1966). For rates of rotation above 10 rpm long-term adaptation is impossible (Graybiel et al., 1960). Such a situation may actually be more disturbing than the situation which would exist in a rotating space station. On the surface of the earth the major component of linear accelerations is the earth's gravitational component, and a linear acceleration resulting from rotation is relatively small and constantly changing with movement of the individual along the radius of rotation.

There appears to be little adaptation effect for the otolith organs when these are tilted to new positions in the earth's gravitational field. Whereas the semicircular canals are stimulated only by angular acceleration and hence do not respond to constant angular velocity, the otolith organs are stimulated by shearing forces resulting from the pattern of depression of the macular substrate by the otoconia.

If an observer is seated at a given radius on a centrifuge, rotating about a vertical axis, the otolith organs will respond to the resultant force of the centrifugal acceleration and the earth's gravitational field. The result of this kind of stimulation is an impression of tilting of the visual world. This is kown as the oculogravic illusion (Clark, 1967).

Tumblers and divers maintain a sense of their position in space while performing elaborate rotational movements which would be completely disorienting to anyone lacking in experience, if he could possibly perform the motions in the first place. Figure skaters are capable of high speed rotation, as high as 280 rpm for some seconds, following which they can immediately go into another complicated skating routine or stand immobile. They are able to suppress nystagmus motions of the eyes almost completely if they have an opportunity to look at their visual surroundings (Collins, 1966). Nystagmus movements have been detected by telemetry techniques in skaters, however, while they are in motion and after rotation if their eyes are closed. It has also been demonstrated fairly conclusively that habituation is highly selective (Guedry, 1965). That is, a figure skater may be able to maintain complete control for rotation in a given direction with the head in an accustomed orientation. On the other hand, if direction of rotation is reversed or if the head is tilted into an unusual posture, the result may be severe disorientation of the skater following the rotation.

Nystagmus

Nystagmus movements of the eyes usually associated with high angular acceleration consist of relatively slow components in the direction opposite to that of the acceleration and fast components in the direction of acceleration.

It is clear that the nystagmus effect continues for some 20 s or more after the angular acceleration has stopped. Cessation of the nystagmus in the initial direction is followed by a secondary nystagmus with the slow phase in the opposite direction.

For clinical tests using a rotating chair, a usual procedure is to rotate the subject at a rate of once every 2 s for a total of 10 revolutions. The chair is then stopped abruptly. Nystagmus motions are observed which have been induced by the abrupt deceleration of the chair. The slow phase is thus in the same direction as the original direction of chair rotation. If during rotation the head is nearly erect but tilted forward approximately 30°, then the nystagmus movement of the eyes will be from side to side or horizontal. If the head is tilted down nearly 90° onto one shoulder with rotation about a vertical axis, then the direction of nystagmus will be up and down with respect to the head. If the head is tilted back 60° or down 120° from the erect position, then the nystagmus will be rotary, that is, rather than a linear motion there will be a torsional movement of the eyes. This is slightly more difficult to observe but can readily be detected by referring to some unique characteristic of the iris or a vessel within the sclera. A wide variety of nystagmus motions is possibly depending on the position of the head with respect to the axis of rotation. It is evident that the three pairs of semicircular canals work in concert to drive the appropriate extraocular muscles for compensation of motion of the head about any axis of rotation.

Nystagmus may also be induced by the movement of objects in a given direction before the eyes. Thus, if the eyes are open during rotation, the *optokinetic* nystagmus will be in the same direction as the vestibular nystagmus, and they will serve to enhance each other. Postrotational nystagmus is reduced if the eyes have been open during rotation. Also, with the eyes open during deceleration, optokinetic nystagmus will tend to cancel out the vestibular nystagmus. In observing nystagmus effects with the rotating Bárány chair, subjects are instructed to keep their eyes closed until the rotation of the chair has been stopped, at which point the eyes are opened for observation.

ORIENTING RESPONSES

The vestibular stimulation which results in nystagmus movements of the eyes influences other musculature as well. In a number of animals, nystagmus movements of the whole head may be observed. Whole body orientation in the direction opposite to that of rotation may also occur. There is a tendency for increased tone in the extensors on the side of the body in the direction of rotation and increased tone in the flexors on the side away from the direction of rotation. The direction of these effects is, of course, reversed following deceleration. Thus, if a subject is instructed to stand up and walk as soon as the rotation of the Bárány chair is stopped, he will tend to stagger or fall in the direction of the rotation.

LABYRINTHINE DYSFUNCTION

Any abrupt interference with the normal function of the labyrinth organs, particularly the vestibular mechanism, may be accompanied by the disturbing subjective symptoms of vertigo and dizziness along with ataxia and gross impairments of coordination. A variety of conditions may be classified as labyrinthitis, accompanied by vertigo and other symptoms. Inflammation of the vestibular branch of the 8th nerve, local infection within the vestibular mechanism, and abnormal ionic balance within the system may be responsible for disturbances. Treatment of such conditions is usually symptomatic. Drugs which are effective in the prevention of motion sickness may be useful (see above).

Gross vestibular impairment accompanied by paroxysmal attacks of vertigo, subjective noise (tinnitus aurium), and a progressive impairment of hearing of the sensorineural type characterize a syndrome which is known as Menière's disease. Involvement of both vestibular function and the hearing mechanism suggests a localization of the disease process within the labyrinth. It is generally attributed to disturbed ionic balance within the inner ear, and distention of the cochlear duct has been observed upon postmortem examination of patients who had suffered from the disease. Symptoms may be relieved by reducing the sodium intake. Intractable cases have been treated by surgical destruction of the labyrinth, intracranial section of the vestibular division of the 8th nerve, or ultrasonic irradiation of the vestibular end organs for their destruction.

Labyrinthotomy performed under these circumstances has grossly different effects than those observed in animals. The reason is probably that gross disturbance of the mechanism already exists before such intervention is carried out in man. In a normal animal, unilateral destruction of the labyrinthine mechanism has a result similar to excitation of the contralateral vestibular system. Slow nystagmus movements are observed in the direction of the damaged side, and there is a tendency for increased tonus of extensors on the undamaged side and of flexors on the damaged side. Immediately after destruction of the labyrinth, the animal is grossly disoriented. This disorientation is followed by a gradual recovery. If the remaining labyrinth is destroyed, the animal again becomes disoriented and then recovers gradually.

Vestibular function may be impaired by the administration of specific toxic substances such as streptomycin. In doses of 2 g or more daily over a period of weeks, the occurrence of vestibular symptoms is fairly frequent. Animal studies suggest a direct action on the peripheral sense receptors, although it has been suggested that in man the effect may be on the ventral cochlear nuclei. The effects of streptomycin ingestion are usually bilat-

eral. If medication is stopped sufficiently early the effects may be reversible. Vestibular function is impaired before there is gross evidence of hearing loss. Hearing impairments are greater for frequencies of 4,000 Hz and above so that the deficit may not be noticed, as these frequencies are of relatively little importance for the perception of speech.

Diagnosis of vestibular impairment is accomplished by artificial stimulation of the vestibular mechanism. This may be accomplished by rotation in a chair or by caloric stimulation. Following rotation of the head about various axes, nystagmus movements of the eyes are timed, and their duration is compared with normal standards. Impaired function is frequently associated with a greatly reduced duration of nystagmus. Similar observations are made following caloric stimulation. Caloric stimulation has the advantage of permitting selective unilateral stimulation. The selection of canals to be stimulated may be controlled to a limited extent by positioning of the head during stimulation. The effect of stimulation was believed to be based on convection currents, but recent experiments in the space shuttle indicate that this is not so.

ROTATORY VESTIBULAR TESTING OF THE LABYRINTH

The need for improved diagnosis of disorders of the peripheral vestibular apparatus has led to the development of improved tests. Figure 9.54 shows the difference between a normal subject and a patient with unilateral labyrinthine nerve section done to relieve symptoms of spontaneous dizziness. Both subjects underwent a series of sinusoidal rotations in the dark (to avoid visual tracking). The peak velocity was 60°/s and at slow oscillatory frequencies below 0.2 Hz. Nystagmus was recorded to analyze eye velocity during the slow counterrotation phase. Note that with one labyrinth removed from controlling the VOR nystagmus, responses are more variable, smaller in amplitude, asymmetrical, and nonlinear. Phase lag was 5-fold normal, and responses to clockwise and to counterclockwise rotations showed very different zero crossings. These changes are characteristic of patients with unilateral labyrinthine disorder (Honrubia et al., 1982).

A further analysis of vestibular function is made possible by analysis of the vestibuloocular reflex pathway. Figure 9.55 shows responses relating to each component of the VOR pathway. The upper trace compares head acceleration at three different frequencies of slow sinusoidal oscillation. The second trace compares responses (changes in firing

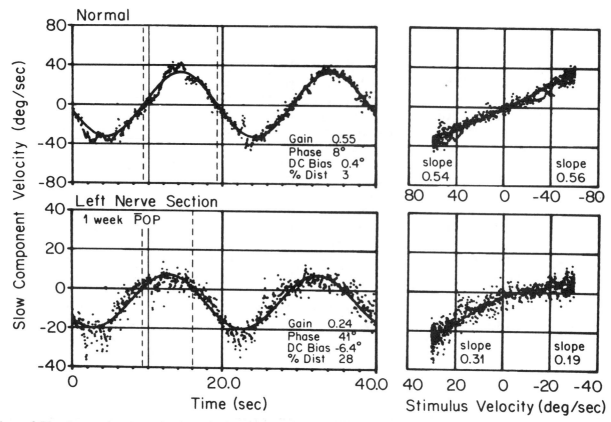

Figure 9.54. Averaged nystagmus responses of a normal subject (*top*) and a patient (*bottom*) 1 week after a unilateral left labyrinthine nerve section to relieve spontaneous dizziness. Both subjects underwent a series of sinusoidal rotations in the dark. The normal subject shows a wide amplitude shift in slow component velocity which is quite uniform from trial to trial, nearly symmetrical and linear; the patient's responses are quite variable, smaller in amplitude, asymmetrical, and nonlinear. The responses to clockwise rotation and to counterclockwise rotation have very different zero crossings (*vertical dashed lines*). [From Honrubia et al. (1982).]

rate) simultaneously with acceleration of the head which are occurring in a primary afferent neuron in the eighth nerve. The third trace compares responses in second order neurons in the vestibular nucleus. The bottom trace compares eye velocity consequences of the vestibuloocular reflex.

An ideal machine would have zero phase lag and zero gain between head movement and counterrotating direction of gaze. The afferent neuron shows a progressive increase in phase lag with an increase in frequency. At the vestibular nucleus, there is another, somewhat smaller, shift. But the delay is greater than one would expect would be due to synaptic transmission across one synapse. The eye velocity measurement shows a further increase in phase lag. Note that *because of cross-connections between the right vestibular apparatus and the left abducent motor neurons, the eye velocity trajectory is reversed.* The significance of this connection is that at ordinary velocities, the eye velocity closely reverses that of the head and *thus the VOR stabilizes the retina in accordance with the direction and velocity of motion of the head* (Honrubia et al., 1982).

SOME RELATIONS BETWEEN VESTIBULAR STIMULATION AND SENSATION

Threshold for horizontal motion, in a parallel swing, for normal subjects is about 2–6 cm/s². The threshold is not appreciably affected when subjects are immersed in water. It is considerably higher in patients in whom both labyrinths are deficient, but normal if one labyrinth remains functioning. Thresholds for horizontal motion sensation are also high (up to 20 cm/s²) in deaf children with non-functioning vestibular labyrinths. Figure 9.56 illustrates the approximate motion of the parallel swing and shows curves describing velocity and acceleration. Below is illustrated the subjective sensation associated with motion in the parallel swing. Sensations are further disturbed when the subject's head is tilted backwards into the position for the "blind spot of the otoliths" (Jongkees, 1974).

In order to reduce the effects of limb inertia on sensations of motion, that is, in order more closely to approximate pure vestibular sensations, subjects can be immersed in water, and the container can

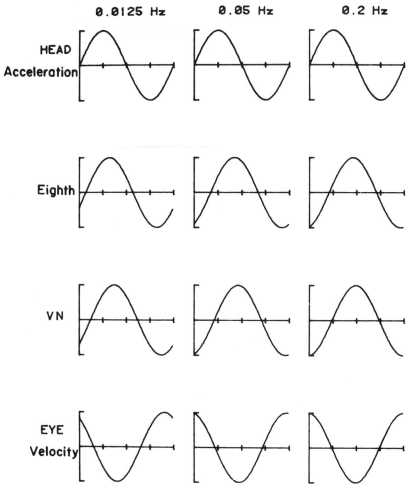

Figure 9.55. Computer-generated graph to compare the temporal course during head acceleration of changes of firing rate in a single vestibular afferent neuron (eighth), a second order neuron in vestibular nucleus, and velocity of horizontal rotation of the eye in response to vestibuloocular (*VOR*) reflex. With increase in oscillation frequency, primary neuron shows increase in phase lag which is mainly due to canal dynamics. In vestibular nucleus, the further shift in phase lag is greater than could be expected from synaptic decay. Eye motor response shows an additional phase lag and, because of crossed innervation between vestibular nuclei and abducent motor neurons, eye motion is reversed. Thus, the vestibuloocular reflex stabilizes vision with respect to the environment during head movement. [From Honrubia et al. (1982).]

be put into rotation. If the rotation is maintained at a constant rate, the effects from the semicircular canals becomes steady, leaving "pure" otolith sensory signals. Figure 9.57 depicts such submersion maneuvers. Rotation occurs around an axis that passes between the subject's ears. During rotations at less than 20 rpm, the subjective experience is one of rotating as if sitting in a ferris wheel. During rotations between 20 and 55 rpm, the subjective experience is one of riding an eliptical ferris wheel. Above that speed of rotation, the sensation is one of strictly vertical rise and fall (Mayne, 1974).

Vestibular Sensations Including Motion Sickness

Responses of the human subject to vestibular stimulation must include consideration of the per-

ceptual processes. What we perceive is not determined solely by the events in the physical environment, the physical mode of delivery to sensory cells, or the response of those cells to stimulation. Perception is compounded of memory and conditioning and present mind set, along with the harvest of incoming sensory signals. Moreover, as we have seen, past experiences, expectations, and purposes can modify incoming sensory signals at the level of the receptors and at relay stations along the ascending sensory trajectory.

Angular acceleration of the vestibular apparatus may not cause nystagmus, depending on such factors as whether the room is dark or lighted, whether the individual is drowsy or alert, whether attentive to sensory input or interested in something else,

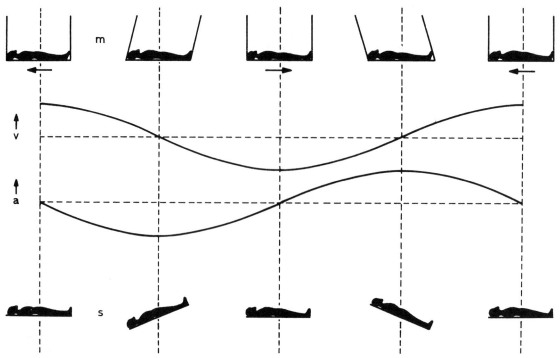

Figure 9.56. Schema of placement, motion, acceleration, and subjective position of subject in a parallel swing. *m,* movement; *v,* velocity; *a,* acceleration; *s,* sensation. [From Jongkees (1974).]

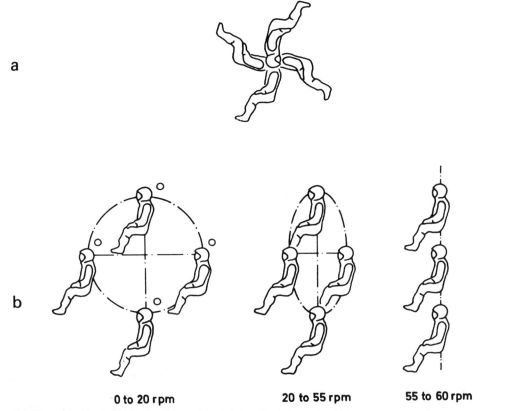

0 to 20 rpm   20 to 55 rpm   55 to 60 rpm

Figure 9.57. Rotation of subjects submerged in water, moving through successive positions depicted in *a*. Axis of rotation is approximately through the subject's ears. Submersion in water practically eliminates limb inertial cues. Conditions of rotation make subjects dependent on otolith sensory data to determine either their position or motion in space. Illusory movements sensed by subjects at different rates of rotation are depicted in *b*. [From Mayne (1974).]

whether the context involves threat or security, etc. We are concerned here with neurophysiological processes affected by age, past experiences, expectations, and purposes. Furthermore, processes such as reflex adaptations, discussed previously, affect the earliest (lowest level) of neuronal processing of sensory signals and therefore have consequences throughout ascending sensory pathways, directly affecting the data with which perception is created, and motor and sensory mechanisms are affected as a consequence of adaptation of fundamental reflexes, such as the VOR. Since reflex adaptation and restoration take considerable time, there will be discrepancies in perception and sensorimotor control during periods of reflex adaptation and return to normal status.

Normal perception can be better understood by investigating malfunctions and maladaptations in which there may be illusions in conflict with reality as interpreted by other sensory modalities or by external objectifying sources of information. Vestibular neurophysiology provides a rich *resource for the study of sensory organization*: how different components of the inner ear, and other sensory sources, combine to create an internal perception of reality, true or false.

MOTION SICKNESS

Motion sickness is not well understood physiologically. Prevailing notions emphasize overstimulation or hypersensitivity of sensory epithelium, or in the case of motion sickness in the weightless state, insufficient stimulation of the sensory epithelium. Conflict may be generated among different sensory modalities. It may be that motion sickness derives from neurophysiological conflict among messages that cannot be integrated into a meaningful whole (Mayne, 1974).

Conflicts are commonplace in relation to the vestibular system inasmuch as so many different sensory inputs converge on vestibular reflexes and on higher sensory processing. Intermodality discrepancies imply conflict among central integrative processes, and such conflicts may be urgent and perhaps overwhelming when there is danger of falling or being damaged by things moving strangely in the environment.

Not all perceptual conflicts are disturbing; some may be interesting or amusing. The sensation of spinning of the visual world following rotation has been termed the oculogyral illusion. This may not be an unpleasant experience and has been extremely useful as an index of the course of effects of vestibular stimulation. The sensation of a turn-

ing environment is also associated with vertigo. This term is usually applied under circumstances where the sensed rotations are disturbing. Vertigo may also include blurring of vision, faintness, and dizziness. The general feeling of *dizziness* may or may not include vertigo and is associated with unsteadiness and disequilibrium. Vertigo may be a serious problem for the pilot of an aircraft if it results in confusion on his part with respect to his position or movement in space. It is also serious as a concomitant of motion sickness. Unusual motion patterns in which the head is rotated about two axes simultaneously can produce motion sickness in anyone. Some individuals are more resistant than others, but none is immune. Motion sickness may also occur when visual and vestibular stimuli are not in accord. Such a situation occurs for someone on board ship who is unable to see the stable horizon. His visual environment moves with him and does not behave as would a stable visual world for the patterns of motion to which he is exposed by the pitching and rolling of the ship. Motion sickness may also be induced when the body is not subjected to motion, but the visual world, as seen in a motion picture film for example, changes as it might during airplane acrobatics or a roller coaster ride. Head motions may accompany such stimulation. An important element in the induction of motion sickness is apparently lack of accord between visual and vestibular inputs. The symptoms of motion sickness reflect the reaction of the autonomic nervous system.

Motion sickness may be eliminated by removal of the vestibular apparatus, by sectioning the VIIIth cranial nerve (Johnson et al., 1951), and by removal of the flocculonodular lobe of the cerebellum (Tyler and Bard, 1949).

Angular accelerations and/or linear accelerations can induce motion sickness. The most effective stimulus for motion sickness is simultaneous multiplanar angular accelerations. Visual disturbances in the absence of motion can induce nausea and some of the other symptoms of motion sickness. Weightlessness during space flight may be associated with dizziness and nausea which is aggravated by quick head movements. It may be that the labyrinth becomes sensitive due to the absence of adequate gravitational stimulation of the otoliths (Berry, 1970). The objective signs of motion sickness include pallor, cold perspiration, yawning, salivating, vomiting, vertiginous disorientation, and a lurking fear of, or wish for, death.

The symptoms may vary from feeling mildly out of sorts to being seriously incapacitated, with imbalance, incoordination, and disorientation, accompanied by vomiting, progressive dehydration, electrolyte imbalance, and related disabilities. It may be life-threatening because of dehydration, reduced circulating blood volume, reduced blood pressure, cardiac arrhythmias from electrolyte imbalance, and other effects.

It cannot be surprising that so many acute and widespread symptoms should appear if there is confusion or conflict among sensory signals and compelling reflexes. Which sensory signals in conflict are to be trusted? Which reflexes? Which perceptual experiences? Even within one system, e.g., semicircular canals and otolith organs, there may be contradictory, conflicting sense data. Perhaps the resulting effect is one of rejection of the conflicting sources (vestibular and visual, for example) and retreat to as nearly motionless and visually stable an environment as can be found.

CHAPTER 64

Vision

LIGHT AND THE OPTICS OF THE EYE

Introduction: Light

Within the entire spectrum of electromagnetic radiation, a narrow range of frequencies activates retinal photoreceptors. The subjective effects accompanying the chain of neurological events that result from this phenomenon are what we call "light." Within the visible light range, special photoreceptors are differentially affected by particular photic energies (frequencies, wavelengths), which we know as "colors." Without photoreceptors and nervous systems there would be neither light nor color in the universe.

Light is on the same continuum with gamma rays, X-rays, and broadcast bands of radiation. The distribution of frequencies and wavelengths of electromagnetic radiation is illustrated in Figure 9.58. Because all objects in the universe which are above absolute zero temperature are radiating, absorbing, and partly reflecting electromagnetic energies in all directions at once, vision requires the selection and ordering of these reverberating radiant phenomena to make possible the formation of biologically meaningful images.

The first step is provided by the light-gathering powers of transparent tissues of the eye. These, curiously enough, are easily damaged by electromagnetic energies that are only marginally above and below the range of visible light. Image formation is provided by an orderly arrangement of photoreceptors which lie on the far side of the path of light as it enters the retina. Correspondences ("mappings") between the retina and various central nervous system representations involve simultaneous parallel sensory processing. According to these biological events, distributed over space and time, we infer the detailed geometry and movement as well as the brightness, color, and surface texture of objects in the subjectively visible universe. We confidently "project" these internal inferences outside ourselves, as belonging to specifiable events in the space-time world we live in and explore.

Vision involves information transformations which begin with the transduction of photoreceptors by light. It involves the extraction of information from patterned light, and the internal representation and analysis of that information in ways that are designed to contribute to goal-seeking behavior. We are not passive observers. Vision is a cognitive process which is composed to suit motor performance. In the process, our past experiences, anticipations, and intentions are continually involved. Our visual experiences are "taylor-made." In this respect each individual's visual world is unlike anyone else's. It is only in laboratory situations, with simple visual targets and standard instructions, that strict visual comparability is observed.

The study of vision is important because of what it tells us concerning the limits and opportunities of our perceptual world. It is also important because of the insight it provides into brain structure-function relations. It contributes precise and useful knowledge because pathways for vision and eye motor control traverse many different regions of the brain, and therefore a careful analysis of visual and oculomotor disturbances reveals a great deal about where neurological pathology exists and of what it likely consists.

REFRACTION OF LIGHT

By application of the wave theory of transmission of electromagnetic energy, it can be demonstrated that light which passes from one medium into another will have its direction of propagation altered if it enters the new medium along a path which is other than perpendicular to the interface between the two media. The ratio of the sine of the angle formed between the path of the light in the less dense medium and a perpendicular to the surface between the media to the sine of the angle formed by the path in the more dense medium with respect to the same perpendicular is equal to the index of refraction for the two media (Snell's law) (Fig. 9.59).

The principle of refraction is extremely important in a variety of ways. If the surface between two

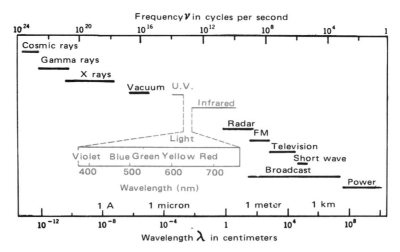

Figure 9.58. The electromagnetic spectrum. The region visible to the human eye is expanded. [From McKinley (1947).]

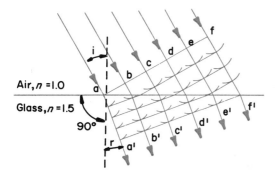

Figure 9.59. Refraction of light at surface between glass and air. Angle *i* is the angle of incidence and angle *r* is the angle of refraction. Index of refraction, *n*, is given for each medium. The rotation of wave front, *abcdef*, as it enters a medium of higher index of refraction at an angle other than normal to the interface is shown by wave front *a'b'c'd'e'f'*.

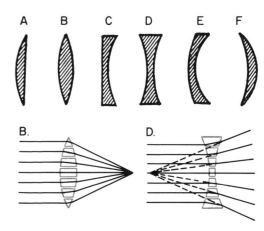

Figure 9.60. Cross-sections of lenses. *A*, planoconvex; *B*, Biconvex; *C*, Planoconcave; *D*, Binconcave; *E*, Convexoconcave; *F*, Concavoconvex. The way in which refraction of a beam at curved surfaces may result in convergence or divergence is illustrated for *B*. and *D*.

media is not a plane but is curved, then rays of light which reach this surface at different points in a beam of large cross-section will be refracted by different amounts and will be propagated through the new medium in different directions. It is on the basis of this principle that lenses may be formed from glass. These can cause parallel beams of light to converge or diverge, depending upon the design of the lens (Fig. 9.60). Light of shorter wavelength is refracted a greater amount than light of longer wavelength. Thus, a beam of white light which consists of a continuous distribution of energy for all visible frequencies from the longest to the shortest will be spread out if it enters glass at an angle other than perpendicular, and the components of different frequencies will travel along different pathways. This principle is turned to practical advantage by the use of the prism which permits the separation of the spectral components of light (Fig. 9.61).

Although light is often specified in terms of its

wavelength, it is more sensible to specify it in terms of frequency. The reason for this is that the frequency of light is not altered when it passes from one medium into another of different refractive index. Velocity does change, and thus wavelength must also change a proportional amount.

Light-Gathering Properties of the Eye

Prior to reaching the photosensitive layer of the retina, light which stimulates the eye must pass through the cornea, lens, anterior chamber, the vitreous humor, and a number of nonsensitive retinal layers (Fig. 9.62). The distance from the corneal surface to the retina in the region of the foveal center of the eye is approximately 24 mm. The optical elements of the eye must therefore provide a fairly high optical power in order to focus parallel rays which enter the eye at the retina. The power of the eye is nearly 60 diopters.

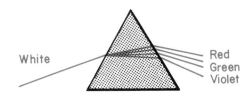

Figure 9.61. Dispersion of light by a prism.

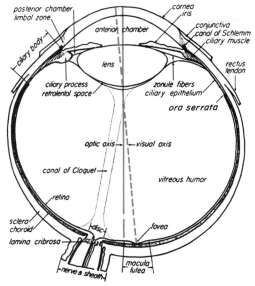

Figure 9.62. Horizontal section of the right human eye. [From Walls (1942).]

Most of the 60-diopter refractive power of the eye is provided by the curvature of the cornea. This avascular structure is about 1 mm thick. It consists of a stroma of collagen fibrils of mesodermal origin. Instead of criss-crossed fibrils as in the rest of the sclera—the capsule of the eye—corneal fibrils lie in parallel, a contribution to transparency. The cornea transmits light from about 300 nm, slightly into ultraviolet, to about 2,000 nm, slightly into infrared. Most of the entering infrared radiation is absorbed by the aqueous and vitreous humors.

Thorough dehydration of the cornea is essential for corneal transparency. If the sclera is similarly dehydrated, it too becomes transparent. Some corneal dehydration takes place across Bowman's membrane, a transparent "glassy membrane" on the outer surface of the cornea, which permits evaporation through the overlying epithelium from the corneal surface into air. The corneal epithelium heals readily if it is damaged. This healing process does not interfere with the optics, but if Bowman's membrane is damaged, the cornea develops a scar that can distort the passage of light. The inner surface of the cornea is supported by an elastic membrane (Descemet's) which is covered with a single layer of endothelial cells. This endothelium

performs the major task of dehydrating the corneal stroma. It creates osmotic dehydration *and* operates a metabolic pump which actively bails water out of the cornea. As long as this endothelial contribution is successful, the cornea remains transparent; when it fails, the cornea becomes chalky white like the sclera.

AQUEOUS HUMOR

The transparent fluid which fills the anterior and posterior chambers on either side of the iris diaphragm is formed from blood plasma. The blood supply is that to the ciliary body (Davson, 1950). Production of aqueous humor occurs continuously, and there must be a route for its drainage from the eye. That route is the canal of Schlemm (see Fig. 9.63). A complete turnover of aqueous humor occurs approximately once every hour. If for any reason the outflow is blocked, or aqueous production is increased, there may be increased intraocular pressure. Normally, the intraocular pressure is approximately 15–18 mmHg higher than the intracranial pressure (Adler, 1959). This pressure differential helps to maintain the shape of the eye and, consequently, the appropriate spacing of its optical components.

Aqueous humor closely approximates a dialysate of blood plasma and provides the principal source of metabolites for the lens and cornea. Carbonic anhydrase inhibitors reduce the rate of aqueous production and may be of assistance in controlling glaucoma (increased intraocular pressure).

Additional refractive power, up to 9 or 10 diopters, is provided by the crystalline lens which is made up of similarly transparent collagen fibrils. The shape of the lens can be dynamically altered by contraction of radial fibers which pull the ciliary body forward, reducing tension on the zonal fibers (Fig. 9.63). This relaxes the lens capsule and allows

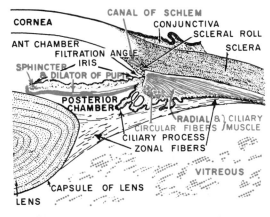

Figure 9.63. Detail of the anterior segment of the human eye. [Redrawn from Weymouth (1955).]

the lens surface to become more convex, a process that can accommodate sharp image resolution over a range from a remote viewpoint (for practical purposes more than about 6 m) to a near point of about 10 cm, depending on age.

Almost any optical system suffers from certain imperfections. Variation in the refractive index with the wavelength of light results in a different lens power for different wavelengths, and therefore chromatic aberration distorts images for illumination which represents a broad band of wavelengths. In addition, the eye suffers from spherical aberration. That is, rays of light which enter the lens near its periphery may be brought to a focus at a slightly different depth than other rays entering near the center of the lens. These two types of aberration are illustrated in Figure 9.64.

THE PUPIL

The optical path can be shuttered by closure of the eyelids which, varying somewhat according to pigmentation, admit only about 1% of the incident light. With eyes open, the amount of light entering is limited by the iris diaphragm or pupil which varies in diameter, depending on background illumination from about 2 to 8 mm. This 1:4 change in diameter corresponds to a 1:16 change in pupillary cross-sectional area, thereby controlling that range

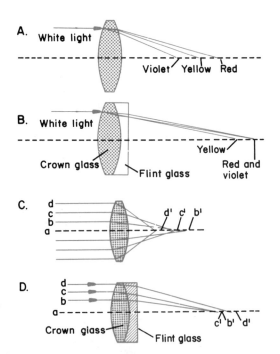

Figure 9.64. Chromatic and spherical aberration. *A.*, Chromatic aberration in simple lens; *B.*, Achromatic lens partially correcting chromatic aberration; *C.*, Spherical aberration in a simple lens; *D.*, Achromatic lens partially correcting spherical aberration. [From Riggs (1965a).]

of retinal illumination. Parasympathetic innervation of a sphincter muscle provides active contraction of the pupil, and sympathetic innervation of radial dilator muscles provides active dilation of the pupil. An abrupt increase in external illumination yields a brisk reflex contraction of both pupils, even though only one retina may be illuminated (reflex consensual contraction). The diameter of the pupil is promptly stopped down to a smaller size, and, although the illumination may remain constant, the pupil slowly dilates slightly to a steady-state aperture. A decrease in external illumination yields a slower dilation of the pupil.

When the pupil constricts, spherical and chromatic aberrations of the optical system are reduced, and the depth of focus is increased. During observation of a near object, the lens accommodates by increasing its curvature (the accommodation reflex), while the pupil simultaneously constricts. Both effects contribute to a sharper retinal image. However, there is a limit to the advantage gained by pupillary constriction. When light passes through a very small aperture, diffraction effects occur, wave rarefaction and interference, and these being to degrade the image. An optically optimum pupil diameter in bright illumination is probably between 2 and 3 mm.

CHANGES IN THE LENS WITH AGING

In childhood the lens is capable of altering the resolving power of the eye over a range of 9 or 10 diopters. The lens grows by adding layers, like an onion, reaching adult size by age 13–14, when it begins to lose water content and become increasingly rigid. The range of change of its curvature is thereby gradually reduced. Beginning in the fifth or sixth decade, many people require additional refractive power to compensate for the lessened maximal curvature of the lens. This aging effect is practically solved by adding a convex lens. There is also gradual loss of lens transparency which may be aggravated by trauma, radiation, or metabolic problems such as uncontrolled diabetes. An eye with considerable opacity in the lens is said to have a cataract. X- and gamma-rays induce lens opacities, and neutrons are even more damaging. An increasingly opaque lens may need to be removed to preserve sight, and a corresponding power in a convex lens substituted.

VITREOUS HUMOR

The vitreous humor is a gel-like substance containing a network of thin fibers of a highly hygroscopic vitrein protein similar to gelatin (Duke-Elder, 1937). With the aid of an electron microscope, it is possible to distinguish the fibrils and interfibrillar substances of the vitreous body of the human eye. Fibrils are approximately 6.7 nm and deviate remarkably little from this size. Spherical particles of 10 nm in diameter have also been found (Schwarz, 1961).

OPTICAL DEFECTS

In a normal eye with relaxed accomodation, objects at an optical distance of infinity should be brought to focus on the retina. Such an eye is said to be emmetropic. If the point of focus is located out in front of the retina, it is necessary to move the object toward the eye, displacing the image further posteriorly until it falls on the retina, in order to see it sharply. Such an eye is said to be myopic. If the image of an object at optical infinity falls behind the retina, then increased accommodation of the lens is necessary in order to bring it forward onto the retinal surface. Such an eye is said to be hypermetropic.

Myopia is usually attributed to an elongation of the eyeball, although it may well involve abnormalities of curvature of the optical elements. Conversely, hypermetropia is attributed to a foreshortening of the axial length of the eye ball. It, too, may involve deviations from normal of the optical elements, however. Myopia is correctible by the use of a negative or diverging lens in front of the eye in order to permit distance vision. Frequently a myope without too severe a deficit may be able to read perfectly well without correction. The hypermetrope will require a positive lens for reading, or, at least, will have to start wearing glasses at an earlier than normal age as his accommodative reserve is lost (Fig. 9.65).

DISTRIBUTION OF RETINAL RECEPTORS DEFINES SPATIAL RESOLUTION OF IMAGES

Unlike a photographic film, the resolving power of the photosensitive layer of the eye, the retina, is not uniform over its entire surface. A condition of optimum clarity is achieved for images which fall on or in the neighborhood of the visual center of the eye, the fovea (see Fig. 9.62). The distribution of receptors over the retina is nonuniform, with the greatest concentration of cones in the foveal center. The concentration of rods increases from a point just outside the rod-free fovea to a maximum at a distance of some 20° from the foveal center (Fig. 9.66). The distribution of receptor density is discussed in more detail in the following section. In addition to the nonuniform distribution of receptors, there is a region in the normal eye which is devoid of any receptors. That is the so-called optic

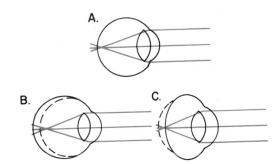

Figure 9.65. Comparison of shapes of the eyeball which may be associated with *A.*, Emmetropia; *B.*, Myopia; *C.*, Hypermetropia.

disk (Fig. 9.62), where nerve fibers leave the eye to form the optic nerve and where the retinal circulation enters and leaves the eye. This is a region of 3° or more in diameter centered at a point 15–17° from the foveal center and with its center just slightly below the nasal horizontal meridian. The presence of the blind spot is usually not noticed, even for monocular viewing. The reason for this is the fact that it is not associated with a dark region in the visual field but is seen as identical with the surrounding region. It is as if the surrounding region "fills in" the blind area in conscious perception. The blind spot may readily be demonstrated as shown in Figure 9.67. If the left eye is covered and the small cross is fixated, the black dot will disappear as the page is brought to within approximately 14 in. from the eye. It will remain invisible until the cross is to within approximately 10 in. of the eye. The vertical extent of the blind spot may be explored by scanning the page vertically along an imaginary line through the fixation cross with the page held at a distance of approximately 12 in.

ADAPTATIONS FOR BEST VISUAL ACUITY

As noted, blood vessels which serve the inner half of the neural network of the retina enter and leave the globe at the optic disk. Vessels which serve the pigment layers, the photosensitive epithelium, and the outer half of the neural network of the retina penetrate the sclera nearer the equator of the globe.

The arteries and veins which branch out from the disk to cover the inner surface of the retina lie directly in the light path from cornea to retina. They therefore contribute optical distortions. Two regions are spared such interference by blood vessels: the central fovea and the horizontal axis of the eye, which corresponds to the horizon of the world when the head and eyes are in an upright, level-headed position of gaze. (Fig. 9.68).

The fovea is not only spared a vascular overlay, but all parts of cells and fibers that are not engaged

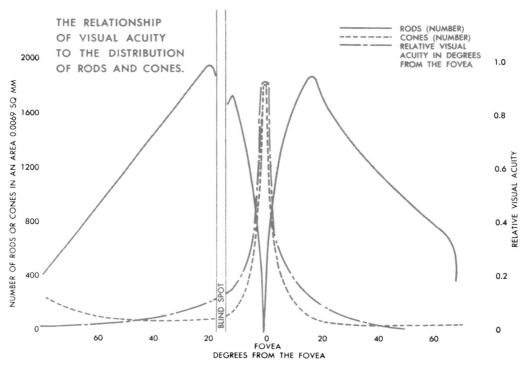

Figure 9.66. Distribution of rods and cones along a horizontal meridian. *Parallel verticle lines* show the limits of the blind spot. Visual acuity for a high luminance as a function of retinal location is included for comparison. [From Woodson (1954); data from Osterberg (1935) and Wertheim (1894).]

Figure 9.67. Demonstration of the blind spot. With the left eye closed, fixate the small cross. Vary the distance of the page until the black spot disappears. The horizontal extent of the blind spot is indicated indirectly by the range of depth in which the black spot is invisible. Some indication of the vertical extent of the blind spot may be obtained by scanning the page along an imaginary vertical line through the fixation cross.

in photoreception are shifted to regions of the retina outside the fovea (Fig. 9.69). Ganglion cell axons that are in transit to the optic nerve detour around the fovea. The bulk of retinal cells and fibers as well as blood vessels are thus kept out of the optical path at least in the region of the fovea. Furthermore, there are no rods in this vicinity so that cones can transmit data relating to bright illumination and color vision. The narrowest cones, which are in the center of the fovea, have a diameter of 1–1.5 μm in humans, and the outer segment, containing photopigment, is approximately doubled in length, thus maximizing visual sensitivity as well as acuity in the center of the fovea (Fig. 9.70).

There are about 35,000 cones in this rod-free region at the center of the fovea, subtending an angle of about 1°. Apart from possible modulating effects of lateral connections affecting foveal transmission, there is relatively direct "private line"

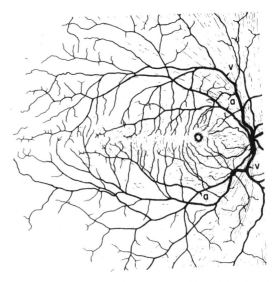

Figure 9.68. Distribution of optic nerve fibers (*thin lines*) and blood vessels (arteries, *a*; veins, *v*) in the retina of an adult Rhesus macaque. The disc of the optic nerve is at the right margin of the figure. The central fovea, free of vessels, appears to the left of the disc. [From Polyak (1941).]

access, consisting of only two synaptic junctions between single cones in central fovea and single ganglion cells. All of the above features contribute to particularly excellent foveal visual data extraction.

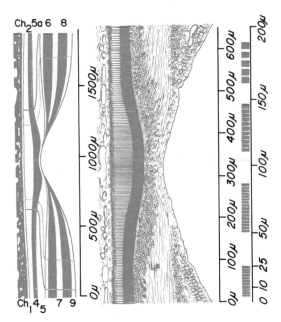

Figure 9.69. Central fovea of the human retina. The upper part of the figure is a diagrammatic illustration of the arrangement of the layers in the central region. *Ch,* choroid coat. Numbered layers correspond with the retinal layers as described in the text. [From Polyak (1957).]

ANATOMY OF THE VISUAL SYSTEM

The Extraocular Muscles

The roundish eyeball has a specific gravity which approximates that of the soft orbital tissues. The globe is loosely, elastically attached to the base of the eyelids by the conjunctivae anteriorly. It is reined by six extraocular muscular tendons which insert in the anterior half of the globe, and tethered posteriorly by the long, flexible optic nerve. Tenon's capsule attaches to the margins of the orbits and forms a hammock-like partition which keeps orbital fat where it belongs, to the sides and behind the globe. This capsule attaches in a ring anterior to the insertion of the tendons of the extraocular muscles and provides separate capsular sheaths around each of these muscles.

The globe is thus loosely wrapped and supported by its almost neutral buoyancy within the bony orbits. It thus presents negligible inertial drag during rapid head and eye movements: the eyeballs do not ballotte about in the orbit during rapid head movements, and the optical axes can be swiftly and efficiently swept about according to reflex and pursuit commands.

Fixation can be maintained on a moving object only by relatively precise disparate movements of the eyes. With changes in the depth of an object viewed, the eyes must converge or diverge appropriately, depending on whether the object is ap-

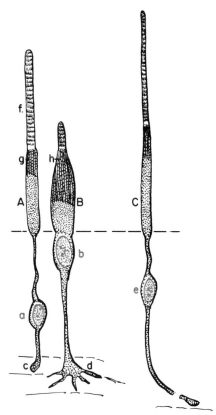

Figure 9.70. Drawings of human rods and cones as seen under light microscopy. *A,* Rod; *a,* Nucleus; *c,* Central connections; *g,* Ellipsoid; *f,* Outer segment. *B,* Cone; *b,* Nucleus; *d,* Central pedicle; *h,* Ellipsoid; *C,* Cone in the fovea centralis, *e,* Nucleus [From Cajal (1904).]

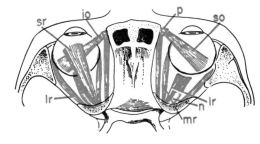

Figure 9.71. Oculomotor muscle of man, as seen from above in a dissected head. On the *left,* a portion of the superior oblique has been cut away to reveal the inferior oblique; on the *right,* the superior rectus has been removed to permit a view of the inferior rectus; *io,* inferior óblique; *ir,* inferior rectus; *lr,* lateral rectus; *mr,* medial rectus; *n,* optic nerve; *p,* pulley for tendon of superior oblique; *so,* superior oblique; *sr,* superior rectus. [From Walls (1942).]

proaching or receding. The immediate control of movement of each eye is effected by six extraocular muscles (Fig. 9.71). Four rectus muscles control movements about two approximately perpendicular axes which are in turn nearly perpendicular to the line of sight with eyes directed straight ahead. The lateral or external rectus is located on the temporal side of the eye, the medial or internal rectus is on the nasal side of the eye, with superior rectus at the top of the eye and the inferior rectus at the bottom of the eye. Two additional muscles, the superior oblique and the inferior oblique, cause torsional rotation of the eye about the visual axis with some depression or elevation (Fig. 9.72).

The Eye

The eye is enclosed by three membranes, the outer of which consists of the opaque sclera over most of the globe which is continuous with the transparent cornea of the anterior surface of the eye. The major blood supply of the eye lies in the heavily pigmented choroid coat directly below the sclera. The pigmentation of this layer serves to absorb light within the eye, preventing back scatter, and to reduce entry into the eye except through the cornea and lens. At the anterior pole of the eye the choroid coat is continuous with the ciliary body and iris diaphragm. The third, innermost layer of the eye is the retina itself. Visual receptors, neural connections of these receptors, and additional vascular tissue lie in the retina.

THE CORNEA

The cornea is a highly transparent tissue, devoid of blood vessels and covered by a single layer of epithelium which is supplied with unmyelinated, undifferentiated nerve endings from the ophthalmic branch of the trigeminal nerve (Fig. 9.73). Stimulation of the corneal surface normally triggers a blink reflex. Even though it lacks specialized nerve endings, the cornea is responsive to light touch and is capable of differentiating lightly applied shape differences. Although protectively sensitive, the cornea can be exposed without discom-

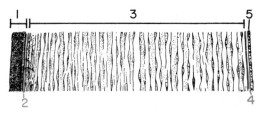

Figure 9.73. Vertical section of human cornea (lying horizontally); *1*, corneal epithelium; *2*, anterior elastic lamina; *3*, substantia propria; *4*, posterior elastic lamina; *5*, endothelium of the anterior chamber.

fort for seeing underwater, and properly fitted contact lenses are tolerated. The ability of free nerve endings to serve a variety of sensations in other parts of the body is well established.

Transparency of the cornea depends on proper osmotic and hydrostatic pressures in the aqueous humor and contiguous tissues. Corneal transparency is threatened by physical abrasion, too great an elevation of intraocular pressure, osmotic imbalance, severe temperatures, and by ultraviolet and infrared radiation. Edema of the corneal epithelium leads to increased light scattering by that layer which results in the appearance of halos around bright lights. Fortunately, the corneal epithelium and trigeminal innervation are repaired within a few days following mild ultraviolet light injury (sunburn, snow blindness) and superficial trauma.

Penetrating wounds leave scars that may interfere with refraction and transmission of light, but corneal transplantation is relatively more successful than with other tissues because of its lack of blood supply. Although it lacks blood supply, the cornea consumes oxygen at a high rate as a result of its endothelial activity. As noted earlier, this activity is indispensable for maintaining corneal transparency by continuously bailing water out of the collagen stroma. Oxygen is supplied by diffusion from air and from surrounding tissues, and CO_2 leaves via the same paths. Because contact lenses diminish respiration of the cornea, it is advisable that they be removed for several hours each day. The supply of glucose and elimination of lactic acid involves diffusion through the aqueous humor and adjacent tissues.

THE PUPIL

The relation of the iris diaphragm and pupil to the lens is portrayed in Figure 9.63. The iris diaphragm, which is supported by an annular attachment to the ciliary body, is a highly pigmented disk which controls a dynamic pupillary aperture. *Pupillary constriction to light* is effected by contraction of a sphincter muscle which encircles the inner

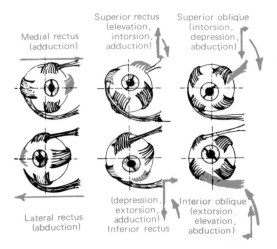

Figure 9.72. Action of the three pairs of antagonistic eye muscles. [From Adler (1965).]

margin of the iris. Retinal afferents, excited by incident light, leave the optic tract to enter the cephalic brain stem, whence they travel rostrally to a pretectal pupilloconstrictor zone. This region, near the junction of the aqueduct and third ventricle, is especially vulnerable to damage from elevated intracranial pressure. Therefore, sluggishness or disappearance of the pupilloconstrictor response to light is one of the most important vital signs to monitor when there is any possibility of increasing intracranial pressure.

From this pupilloconstrictor zone, impulses are relayed to the Edinger-Westphal nuclei *bilaterally*. Each pupil is served by preganglionic parasympathetic fibers from the ipsilateral Edinger-Westphal nucleus. Its axons course with the third cranial nerve to the ciliary ganglion, whence postganglionic parasympathetic fibers pass to smooth muscle fibers in the iris sphincter (Fig. 9.63). Inhibition of tonic activity in the Edinger-Westphal nucleus provides a major contribution to pupillary dilation.

Pupilloconstrictor reflexes have been examined from an engineering viewpoint by Stark (1968). He and his colleagues stimulate the retina by means of a spot of white light which enters the center of the pupil where it will not be intercepted by movement of the iris diaphragm, and measure changes in pupillary responses by means of infrared light to which the retina is unresponsive. This analysis can be pursued without discomfort in alert human subjects. By such means, the pupilloconstrictor reflex has been identified as a stable-zero servomechanism, and it has also been used for analysis of the effects of drugs on the participating neuronal circuits.

The pupil of the human eye undergoes continuous small fluctuations in diameter, "hippus," even under steady illumination, and this can be observed on close inspection. Since there is nearly 100% correlation between oscillations in the two pupils at rest, in response to increasing incident light on the retina in one eye, and in response to visual accommodation, the oscillations must involve part of the circuit that is shared by both eyes, probably the Edinger-Westphal nucleus. The stability and oscillatory behavior of the pupil, however, cannot be attributed to any individual component of the pupilloconstrictor system but must be understood as deriving from the *system as a whole*. This is a prime example of neuronal integration.

If gaze is shifted from a distant object to a near one, *accommodation-convergence* accompanied by *pupillary constriction* occurs. This interesting and useful reflex involves both smooth and skeletal muscle responses. Impulses originating in the retina

are relayed to visual cortex, whence additional relays pass downward to an "accommodation-convergence region" in the vicinity of the oculomotor nuclei. The oculomotor response thus initiated brings the two optical axes to bear on the near object. The loop from retina through visual cortex to the oculomotor accommodation region continues thereafter to "search" for binocular correspondences among visual details of that object. The neighboring Edinger-Westphal nuclei are simultaneously activated. Parasympathetic outflow from each Edinger-Westphal nucleus follows the pupilloconstrictor efferent path to the iris sphincter.

Thus, there are two distinctive initiatory pathways for pupilloconstrictor reflexes, responding to light and accompanying accommodation, which are widely separated in their retinodiencephalic and retinocortical relays and may be separately affected by disease. Central nervous system syphilis, for example, can result in a narrow pupil that does not constrict to light but responds to accommodation and convergence (Argyll-Robertson pupil).

Radially oriented pupillodilator muscles in the iris are served by an upper thoracic spinal center, with relay through the superior cervical sympathetic ganglion. Active pupillary dilation in addition to inhibition of tonic activity in the Edinger-Westphal nuclei, occurs reflexly on shading the eye and also accompanies severe pain and strong emotional states. Pupillary dilation is also associated with emotional interest and has been used as an involuntary indicator of affectional regard. This provides, in contrast to a "lie detector"—a "love detector." The efferent pathway proceeds from frontal cortex to posterior hypothalamus to neurons in the reticular formation of the lower brain stem which in turn project to the intermediolateral (visceral) column of motor neurons in upper thoracic spinal levels. Impulses from this latter region reach radial dilator muscles in the iris by relay through the superior cervical sympathetic ganglion.

THE LENS

The crystalline lens is a biconvex, avascular body which is composed of collagen fibers enclosed within an elastic lens capsule (Fig. 9.63). Energy for lens metabolism is derived from glucose oxidation. Oxygen, ascorbic acid, and glutathione are supplied from the aqueous humor (Davson, 1949). New lens stroma is laid down more slowly after age 13–14, and water is correspondingly slowly withdrawn so that the size of the lens remains relatively constant thereafter. But the dehydration process leads to a gradual increase in lens rigidity. This accounts for the loss of ability for visual accommodation with increasing age. The lens capsule is held in a state of tension and flattened lens curvature by the pull of elastic suspensory ligaments which are anchored to the ciliary body. This in turn attaches to the sclera, which is prevented from collapsing by virtue of sustained intraocular pressure and by the turgidly hydrated vitreous humor which fills the posterior chamber and supports the lens from behind.

Without active contraction of the ciliary muscle, the normal eye is focused at infinity. For near vision, accommodation is achieved by contracting the ciliary muscle which relaxes the suspensory ligaments and permits the lens to become more convex, thus increasing its refractive power. The degree of contraction of the ciliary muscle is dictated by information relating to the degree of noncorrespondence obtaining between binocular foveal representations of objects in striate cortex. Accommodation thus depends on integrity of the primary visual cortex (area 17 of Brodmann). The refractive power of the lens is continually shifting even when the subject attends a fixed target, a "searching" process that approximates a linear servomechanism.

Disparate images from the two eyes, compared in primary visual cortex, also drive the oculomotor system so that the axes of the two globes will cross one another to obtain the best correspondence of images of objects within the range of visual interest. Blurring of images from the two retinas thus serves simultaneously to bring the two globes to bear on the same point in three-dimensional space, and to exercise the accommodation reflex whereby the lenses are driven to seek the sharpest focus.

The reduced refractive power of the lens as compared with the cornea partly corrects for chromatic and spherical aberrations of the optical system. Each surface through which light passes reduces light transmission by about 4%, which means that by the time light has traversed the lens it has been reduced nearly 16% from that incident on the cornea. Spectacles, of course, reduce light by another 8%. Some light is absorbed by the lens, particularly in the shorter wavelengths. Absorbed infrared and ultraviolet light may cause cumulative damage to the stroma of the lens, and a lens that has become opaque in this way or by trauma is called a cataract.

THE RETINA

During embryogenesis, eyes derive from the emergence on each side of the forebrain of an optic vesicle. The anterior and posterior walls of the protruding vesicle are coapted (like a pushed-in hollow rubber ball) so that the tip becomes cup-shaped. As the optic cup approaches the surface epithelium of the face, it induces formation by the epithelium of a translucent cornea and lens. The inner wall of the cup becomes retina which sends its fibers back along the optic stalk, while the outer wall becomes the pigmented layer of the retina. The retina is thus a true sensory outreach of the brain, deriving its cellular organization directly from an outpouching of the neural tube. The retina, like the olfactory bulb, is, strictly speaking, a "ganglion of the brain." This explains why the optic nerve and olfactory tract have myelin sheaths that are wrapped by central nervous system oligodendrocytes rather than by peripheral nervous system Schwann cells.

An orderly arrangement of retinal cells represents a hierarchy of information processing which proceeds from *photoreceptors* to *bipolar cells* to *ganglion cells* (Fig. 9.74). Each human eye contains about 120 million receptors, with a ratio of about 20:1 of rods to cones. The count of bipolar cells is much smaller, and ganglion cells number somewhat over 1 million. Within the retina, therefore, there is an obligatory functional convergence of about 100:1. Each ganglion cell sends its axon centralward by way of the optic nerve.

The main "throughput" of neurons (plus their feedback via local microcircuits)—from photoreceptors to bipolar cells to ganglion cells—is further modulated by two layers of distinctive cell types which make lateral connections within the retina. These are (a) *horizontal cells*, which lie within an

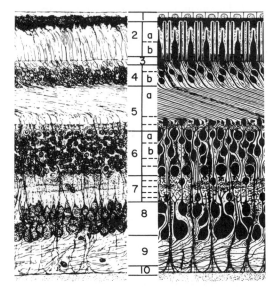

Figure 9.74. Vertical sections of the human retina from the central area midway between the areal periphery and the central fovea. *Right* and *left* halves of the figure represent different methods of staining. The various layers are described in the text. [From Polyak (1957).]

outer nuclear layer, at the level of the junction between photoreceptors and bipolar cells, and (b) *amacrine cells*, which lie within an inner nuclear layer at the junction between bipolar cells and ganglion cells (Fig. 9.75).

All horizontal cells have synaptic junctions with many photoreceptors and in turn modulate both the receptors (by feedback) and the bipolar cells (by feedforward). Horizontal cells have receptive fields that extend well beyond their anatomical boundaries: This is made possible by gap junctions which interconnect horizontal cells serially with even more remote horizontal cells. One class of horizontal cells, which lack axons, has connections preponderantly with rods and has to do mainly with dim light vision. Another class of horizontal cells, which possesses axons, has connections preponderantly with cones and has to do mainly with bright illumination and color vision. Amacrine cells receive from bipolar cells and feed back modulating

impulses to them as well as feed forward signals to ganglion cells. The two classes of cells involved in lateral integration within the retinal thus contribute to sensory processing by taking into account both local and remote events taking place throughout the retina. When signals leave the retina by way of axons in the optic nerve, they are already complexly representative of time- as well as space-related image organization.

We have considered how the pupil regulates the amount of light entering the eye. Additional controls of light flux operate within the retina. These preserve the full range of retinal responses in accordance with local retinal illumination and local luminance contrasts, for any given level of overall incident light intensity, from noon at the beach to dusk under the trees. First, photochemical processes in individual receptors adjust the efficacy of photochemical transduction, over seconds to minutes, so that they operate discriminatively within a wide range of intensities, depending on any given time-averaged light intensity at that cell location. Further, horizontal and amacrine cells rapidly modulate retinal throughput in accordance with the average of light falling on the retina as a whole, as well as occurring locally. This is accomplished in relation to a wide range of regional luminance differences and local space-time alterations in luminance (Fig. 9.76).

Obviously it is not feasible for each bipolar and each ganglion cell to respond to all nuances of all photoreceptors with which they are linked. The net effect of horizontal and amacrine cell contributions is to preserve a broad range of response capabilities for local ganglion cells in relation to whatever may be the amount of general and local ambient illumination. This enables the retina to form high-contrast, sharply defined neural images throughout an extremely wide range of incident light. Converging graded responses from different widespread and local retinal origins combine into a remarkably differentiated message that is relayed to the central nervous system by ganglion cell axons. This message takes into account a multitude of factors occurring across the retinal nerve network, while at the same time conveying quite specific and detailed information relating to each particular locale. Altogether, this constitutes *retinal integration* of visual information.

While recording electrical responses of individual nerve cells in the retina, Kuffler opened a new level of understanding when he discovered the most effective ways to elicit responses from individual retinal cells. He found that a small spot of light, about 2 mm in diameter on the retina, was effective in

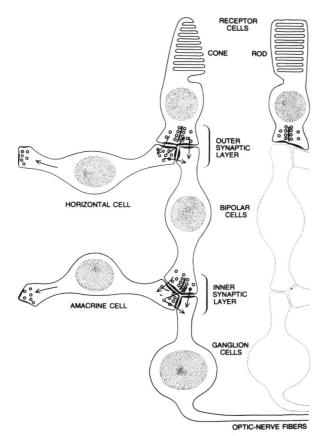

Figure 9.75. Circuitry of the retina was deduced from electron micrographs by Dowling, depicted schematically to show the direction of neurotransmitter release and hence the direction of excitatory or inhibitory influence from receptor to ganglion cells. Bipolar and ganglion cells pass along signals based on interaction of vertical and lateral cells; lateral interneurons transmit back to cells that drive them, across to one another and on to the succeeding main-line transmission cells. [From Werblin (1973).]

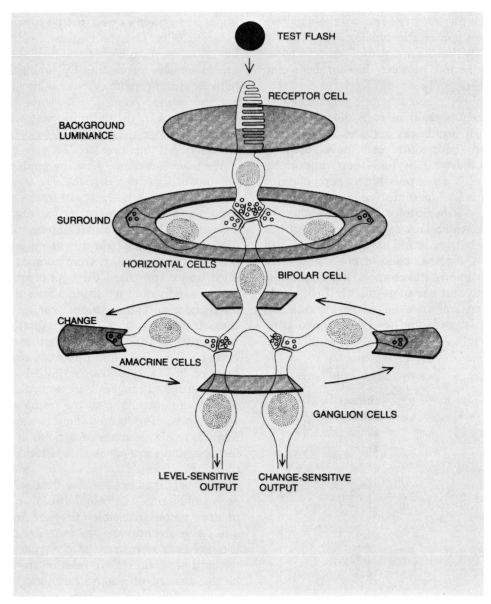

Figure 9.76. Retinal sensitivity is determined by three distinct properties of the visual scene, each activating a specific mechanism. Average background luminance affects photochemical processes in receptor cells, partly by feedback from horizontal cells which directly and indirectly survey a wide territory. Spatiotemporal change in surrounding regions affects interactions mediated by amacrine cells on the bipolar cell terminals. At any local and widespread background luminance, ganglion cells carry signals that are correspondingly modified. [From Werblin (1973).]

defining a given cell's receptive field. The spot of light increases or decreases the corresponding neuron's rate of firing, depending upon its location within that cell's receptive field.

Kuffler defined two basic types of receptive field: on center, and off center. An *on center receptive field* is characterized by vigorous responses when its center is fully illuminated by a spot of light, and is most completely inhibited when its ring-shaped perimeter is fully illuminated and its center is left dark. An *off center receptive field* yields maximal inhibition when its center is fully illuminated, and

it generates maximum activity when an annular surround is illuminated and the center of its field is left dark. When either an on-center or an off-center neuron is inhibited and then abruptly relieved of inhibition by discontinuation of the inhibiting illumination, that neuron responds with a short burst of impulses. For both on-center and off-center fields, activity elicited by appropriate spot and annular illuminations, are internally antagonistic, i.e., with each receptive field type, if both center and surround are illuminated simultaneously, the effects tend to cancel one another. This explains why general illumination of the retina is relatively ineffectual in terms of generating retinal responses.

Thus, ganglion cells have two basic kinds of receptive fields and are themselves characterized as being on-center or off-center neurons. Closely neighboring ganglion cells tend to process information in a similar way, and this characteristic holds generally throughout the central projections of the visual system. Nonetheless, because receptive fields extensively overlap, even a small (0.1 mm) spot of light will strike many receptive fields simultaneously and will therefore generate two such antagonistic patterns of response in many different ganglion cells. Receptive fields near the central region of the retina are generally smaller than those in the periphery, the range being from 0.5 to 8° diameter for the center zone of central as compared with peripheral regions of the retina.

Two types of on-center neurons and off-center neurons are encountered: one exhibits sustained discharges, and their axons conduct relatively slowly (X cells); the other type manifests transient responses, and its axons are more rapidly conducting (Y cells). X cells provide information about relatively steady patterns of illumination, while Y cells can better register transient events such as movements. Y cells may also provide a smoothing inhibitory effect in advance of the arrival of X cell messages. Another class of cells, W, appears to respond to both light and dark spot stimulation in the center of their receptive fields. Their (W) axons have still slower conduction velocities.

Kuffler's studies also provided a key to understanding the significance of luminous signal patterns represented centrally. A receptive field for any neuron throughout the visual system is defined as that area of the retina or that corresponding part of the visual field from which the discharges of that neuron can be most effectively influenced. By definition, light projected outside that receptive field has no effect. Within the receptive field, illumination of some areas produces increased firing, and illumination of others produces inhibition of that cell's activity. In lightly anesthetized cats and monkeys, particular parts of the retina can be stimulated by shining spots or patterns of light onto a screen the animal is facing, at a distance for which the eye is appropriately refracted. By this means, not only the retina, but the entire visual system can be investigated for individual neuronal responses to patterns and colors of illumination. Hubel and Wiesel shared the Nobel Prize for the elegance and comprehensiveness of their investigative studies based on this systematic approach.

Characteristically, receptor cells and other units in the retina, including most ganglion cells, and most cells throughout the visual system, are spontaneously active, exhibiting graded responses and spike discharges, even in the absence of illumination. Such background activity enables the nervous system to respond to more subtle shifts in visual stimulus characteristics, and subtler changes in the organization of visual information, than would be the case if some fixed threshold of stimulus intensity were required to induce activity on the part of each unit at each successive level of visual circuitry. Modulation of activity, therefore, involves simply increasing or decreasing already ongoing neuronal activities. This permits a more discriminating, graded response to changes, whether these are induced by stimuli from the environment or by activities taking place among synaptic junctions elsewhere throughout visual and related systems.

RETINAL RECEPTORS

Retinal receptors are of two kinds: *rods*, for grey level discrimination and dark-adapted (scotopic) vision, and *cones*, for light and color vision. Both kinds of receptors are segmented, the outer segment being a modified cilium which contains many photoreceptive discs. A typical rod (Fig. 9.70*A*) has a long, slender outer segment; a typical cone (*B*) has a short, squat outer segment. But, in the fovea, cones (*C*) have long, slender outer segments with thousands of discs, an advantage for light-gathering purposes. Foveal cones are only 1 or 2 μm in diameter, an advantage for spatial discrimination in central vision.

The outer segment of both rods and cones is filled with double-membrane discs which incorporate photopigment. Figure 9.77 depicts the manner of development of the outer segment of cones and rods. Depending on their lengths, the outer segments in fully developed (adult) receptors contain hundreds to thousands of discs, with a total disc surface of hundreds of thousands of square microns. The discs are stacked so that their broad surfaces present normal to incoming light. They undergo rapid turnover, being replaced at a rate of about 30 discs per day from specialized plasma membrane. Discs are not only formed from plasma membrane, but remain structurally and functionally connected with the plasma membrane. Figure 9.78 is a three-dimensional representation that shows how the discs relate to the plasma membrane. They are connected to the cytoplasmic surface by narrow tube-like processes.

Rhodopsin, a globular molecule is embodied in the disc membrane and undergoes a *cis* to *trans*

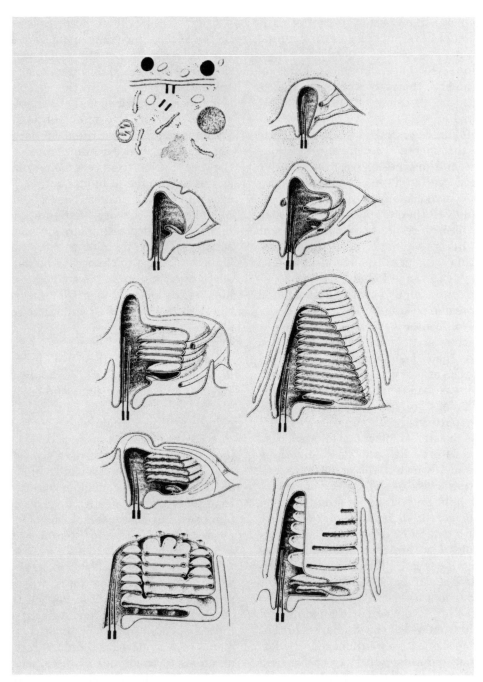

Figure 9.77.   Schematic drawing illustrating the development of the cone outer segment of the tadpole (upper six sketches). Similar sketch illustrating development of the rod outer segment of the tadpole (lower three sketches). [From Nilsson (1964); from L. M. Beidler and W. E. Reichardt (1970).]

isomerization when exposed to light (Figs. 9.79 and 9.80). Isolated rod outer segments contain the biochemical machinery for both transduction and light adaptation (Hubbell and Bownds, 1979). The main action of light appears to be a 10-fold activation of a cyclic GMP phosphodiesterase (DPE). Bleaching of one rhodopsin molecule can lead to the disappearance of 5×10^4 molecules of cyclic GMP, a large amplification of the light signal. The dark

level of GMP concentration is restored in less than 1 min after light is extinguished.

Exactly how the receptor plasma membrane is activated by disc events is not known, but the effect, surprisingly, is *hyperpolarization* (an increase in internal negativity). It is presumed that sodium permeability of the plasma membrane of the outer segment is maintained in the dark by GMP-mediated protein phosphorylation. When light strikes a

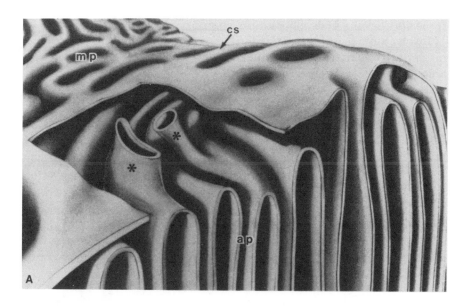

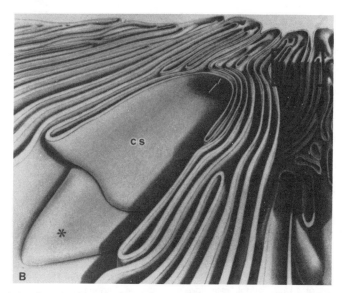

Figure 9.78. Three-dimensional drawings to show the form of cone outer segment sheath membrane. *A,* Looks down on cytoplasmic surface (*cs*) which forms microplicae (*mp*) that appear as interconnected ridges and valleys. The microplicae are the origins of tube-like connections that grow out to form apical projections (*ap*) which appear as longitudinal pleated folds in the lower part of the drawing. Part of apical membrane has been cut away to show the origins of two apical projections (*asterisks*). *B,* Shows a cone sheath (*cs*) in longitudinal section. Some of the apical projections are seen tangentially sectioned close to apical surface (*brackets*). *Arrow* shows cap at tip of cone outer segment which overlies a leaf-like process (*asterisks*). [From Fisher and Steinberg (1982).]

rhodopsin molecule in a disc membrane, there is transient suppression of sodium permeability of the rod plasma membrane, and the rod cell then hyperpolarizes. Addition of calcium ions suppresses receptor permeability in the dark, thus mimicking the effect of light. The calcium ion effect may be mediated by the cyclic GMP system.

The direction of shift in polarity across the receptor cell membrane is opposite to that seen in other sensory receptors (Kaneko, 1979). This means that the photoreceptor cell membrane must be actively *depolarized in the dark*. It implies that

there is a continuous (quantal) emission of neurotransmitters. Although this extra ion-pumping and transmitter release requires additional energy, it also provides extraordinary sensitivity to weak light stimulation. The inner segment is believed to maintain the ionic composition by an active ionic pump, the energy for which is supplied by mitochondria which are densely packed in this part of the photoreceptor. The inner segment evidently releases neurotransmitter at a near maximal rate in the dark. During bright illumination, transmitter release is almost stopped. Transmitter release from photoreceptors in the dark results in depolarization of horizontal cells and hyperpolarization of bipolar cells. The transmitter itself has not been identified.

Early receptor potentials have been dissociated into their constituent component contribution in relation to the steps of rhodopsin conversion during bleaching with light. This is depicted in Figure 9.80 which schematizes the physical-chemical events. Figure 9.81 shows the corresponding shifts in electrical potential associated with the successive steps in the photochemical conversion process.

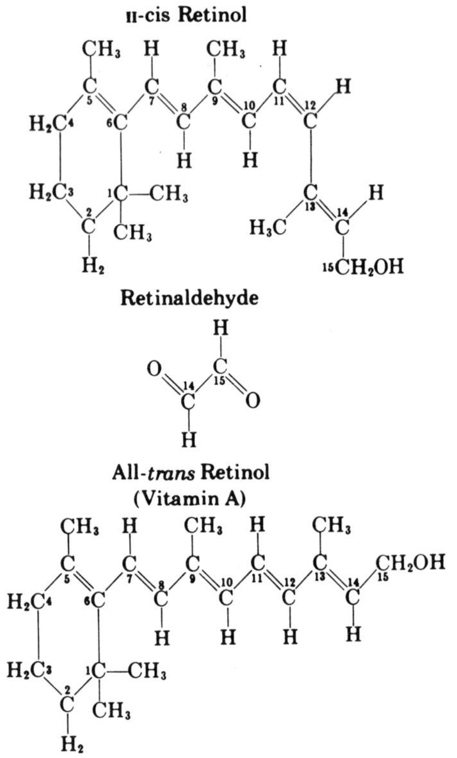

ELECTRORETINOGRAM

The electroretinogram (ERG) is generated by electrical activity of retinal cells responding in synchrony to large-scale illumination (Fig. 9.82). The ERG is used to study retinal physiology and to diagnose retinal disorders. An initial, brief *a-wave*, negative on the vitreal side of the retina, reflects *receptor cell responses to light*. A slightly later vitreal-positive *b-wave* presumably arises from *glial cells* in deep layers of the retina, and it also corresponds closely to *ganglion cell activity*. A large, slow *c-wave* (not shown in Fig. 9.82), is triggered by receptors and sustained by potassium leakage which prolongedly hyperpolarizes the pigmented epithe-

Figure 9.79. Molecular structures of 11-*cis* retinol, retinaldehyde, and all-*trans* retinol (vitamin A).

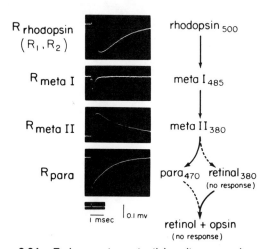

Figure 9.81. Early receptor potential as it appears in a dark-adapted rat eye and at three successive intervals of time following a bleaching exposure. The outline depicts in order which spectral intermediates appear during the rhodopsin cycle. The wavelength of maximum absorption of each spectral intermediate is designated by subscripts. Responses were recorded from the excised eye at 37°C, except $R_{meta I}$ which was recorded at 5°C. Corneal positive responses upward. [From Cone and Cobbs (1969); from Biedler and Reichardt (1970).]

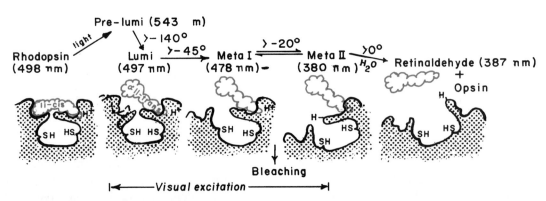

Figure 9.80. Stages in the bleaching of rhodopsin. The chromophore of rhodopsin, 11-*cis* retinal, fits closely a section of the opsin structure. Light isomerizes retinal from 11-*cis* to the *trans* (prelumirhodopsin) configuration. The opsin opens progressively to metarhodopsin I and II by which time visual excitation has occurred. The product decays to retinal and opsin. [From Wald and Brown (1965).]

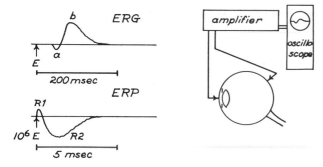

Figure 9.82. Diagram shows method of recording an electroretinogram (*ERG*) and early receptor potential (*ERP*) in a dark-adapted eye. Each potential fluctuation is biphasic, cornea-positive (upward), and cornea-negative (downward). The ERP has no measurable latency but requires much more intense illumination. [From Wald (1972).]

lium. The origin of a *dc component* of the ERG has not been established.

When the ERG is recorded with an active electrode placed closed to the fovea, there is a reduction in both the b-wave and dc component and an enhancement of the a-wave, as expected from the fact that in this region the inner layers of the retina are displaced laterally, so that the foveal ERG reflects mainly cone receptor responses. When retinal circulation is obstructed, the dc component and the b-wave are promptly abolished, but the a-wave and c-wave persist because receptors and pigmented epithelium are sustained by the choroidal circulation.

Rod and cone receptor potentials have different time parameters; cone receptor potentials rise and decay more rapidly than rod receptor potentials. This has practical effects on how well two closely spaced light flashes can be distinguished from a single flash. The lowest frequency of subjective fusion, when repetitive flashes appear to fuse together into a single stimulus, is called the critical fusion frequency (CFF). At favorable levels of illumination, CFF for the fovea is about 50 Hz, reflecting central vision, principally the response recovery characteristics of cones. This suggests that we are near the threshold for distinguishing the intermittence of gas discharge tubes (fluorescent lights) at 60 cycles, more definitely with European 50-cycle electric currents. The CFF for flashes 20° away from the center of the fovea, with a larger share of rod receptors in the field, is about 20 Hz. At low levels of flash intensity, the CFF falls off sharply to levels of 10 Hz and below.

Motion picture perception depends upon CFF, and at customary projection rates, images are satisfactorily fused, out to about 10° outside the center of vision, at normal projection rates of 24 frames

per second. Rod and cone rates of recovery at levels of high intensity illumination are more sluggish, the basis for positive afterimages, as when the image of a camera flash bulb is seen to persist for several scores of seconds following the flash.

Gap junctions link neighboring cones of the same color-receptor type. This provides an electrical coupling which enables spatial summation within a radius of about 40 μm for such functionally similar cells. Spatial summation among rods covers a still wider area, with a radius of some hundreds of micrometers. Illumination of even wider zones diminishes local receptor potentials, indicating that there is a very large-scale retinal center-surround antagonism. This latter effect is attributable to horizontal cells which act back upon receptors through reciprocal synapses. Antagonistic center-surround organization is further reinforced by subsequent data processing at the level of bipolar, amacrine, and ganglion cell transactions, and constitutes a fundamental principle of retinal and central visual integration.

Tomita and co-workers (1968) have achieved microelectrode penetration of single cones. When they illuminated the retina and scanned through the visual spectrum during many such penetrations, 16% of the units peaked in the blue, 10% in the green, and 74% in the red. This is consistent with the trichromatic theory of color vision.

Furthermore, when Tomita and his colleagues plotted response amplitudes against relative light intensity, responses in the weak intensity range were linear, and the line extrapolated down to a light intensity of 1 quantum of absorption, at which level the response amplitude was still on the order of a few microvolts. This reinforces earlier interpretations based on human subjective responses under ideal conditions to faint light stimuli in which it was presumed by Hecht and co-workers that a single photon could provide sufficient stimulation to be perceived as a visual event (Hecht et al., 1942). Under ideal conditions, therefore, the eye appears to be capable of responding to the ultimately minimal photic stimulus, a single photon.

Neural Connections of the Visual System

RETINA

Retinal throughput involves receptor, bipolar, and ganglion cells in that order (Fig. 9.75). These cells are arranged in layers, as is characteristic of sensory, and more generally, of neuronal circuits organized for signal analysis. Bipolar cell bodies contribute a middle nuclear layer, and as their name implies, one protoplasmic process reaches outward to make direct synaptic connections with receptors and horizontal cells in an outer nuclear layer, while the other process reaches inward to make direct synaptic connections with amacrine and ganglion cells in an inner nuclear layer. All components in the retina, except ganglion cells, react with graded responses rather than with fully regenerative action potentials. Ganglion cells are the only cells in the retina that have typical axons, and these convey all visual information from retina to brain.

Horizontal cells make synaptic contacts with a large number of receptor cells and receive indirect influences via gap junctions with other horizontal cells, from many more remote receptors. Because horizontal cells have reciprocal synapses with receptors, they can provide dynamic, large-scale spatial interactions among receptors. A ring of light falling on the retina a few degrees distant from given receptors reduces the amplitude of their graded response by feedback inhibition through horizontal cells. Moreover, horizontal cell feedback to cones is color-specific. *Receptors and horizontal cells together thus contribute the first steps in antagonistic center-surround integration, and color segregation, which characterize retinal, thalamic, and early cortical processing of visual information.* It is within the outer plexiform layer that the *orientation of* light-dark and color *boundaries* is first established.

Bipolar cells contact a number of receptor cells directly and are influenced by more remote receptors via horizontal cells. *Large bipolars receive from both rods and cones; small bipolar receive exclusively from cones.* In much of the fovea, midget bipolar cells have multiple dendritic synapses with only one cone, and activate corresponding individual midget ganglion cells. This provides a direct, "private line" to the brain from single cones in central visual areas.

Responses of both large and small bipolar cells are of two kinds: *On-center bipolars* are *depolarized* by a spot of light at the center of their receptive area (and *hyperpolarized* by illumination of an annular surround). *Off-center bipolars* are *hyperpolarized* by center spot illumination (and *depolarized* by annular surround illumination). These two types of bipolar cells connect directly and through amacrine cells to two classes of ganglion cells which are each specifically *activated* either by on-center spot lighting or by off-center (dark center) events. The corresponding ganglion cells are named "on-center neurons" and "off-center neurons," as are subsequent units in the thalamus and cortex which respond to these distinctive light contrast signals.

Categories of Ganglion Cells

Ganglion cells are classified according to distinctive anatomical and functional properties: Large X cells show a brisk response onset and alter their discharge according to changes of light intensity during continuing stimulation. They are most densely represented in the foveal region and have small receptive fields. X cells are thought to contribute primarily to high acuity vision.

Y cells show a brisk onset but respond only transiently to continuing stimulation. They seem to be most affected when the stimulus goes on or off. Their density is greatest in an annular distribution just outside the fovea, tapering off in abundance both toward the center and toward the periphery. X and Y cells combine to contribute temporal discrimination of retinal events, and Y cells also respond to the direction of boundary movements. Both X and Y categories include both on-center and off-center neurons.

A smaller category of ganglion cells, W, are slowest conducting. W cells, like X, are most abundant in the fovea. They may have center-surround organization and give sustained *or* transient responses, and some of them respond both when a center spot is turned on *and* when it is turned off. These are known as on-off-center phasic W cells, capable of detecting local edges. Still other W cells have directional sensitivity. W and Y cells have similarly large receptive fields.

A small population of ganglion cells shows a miscellany of attributes: Less than 1% show direction selectivity, responding to a certain direction of contrast movement, but not to movement in the opposite direction. Less than 1% maintain continuous activity that is suppressed by the presence of any form of contrast within their receptive field. Less than 5% behave oppositely to this latter group, being excited by any contrast within their receptive field.

All types of ganglion cells send projections, mostly by way of collaterals, to the pretectal area where they participate in pupilloconstrictor reflexes. W and Y cells also project by way of collaterals to the superior colliculus and there contribute to command and control of eye movements. X cells do not contribute to the SC. All three main cell types, W, X, and Y, have parallel and separate pathways which project to different laminae in the lateral geniculate body (LGN) from which the same categories are independently projected to visual cortex.

In frogs, birds, and rabbits quite elaborate integration takes place in the retina: detection of dimming, convex edges, and motion, which in frogs contributes to retinal "bug detection" (Maturana et al., 1960). In birds an *efferent* control exerted on the retina (fibers originating in the thalamus with axon terminals on amacrine cells) improves visual discrimination in dim light (Miles, 1972). It is still moot, in humans, whether there is a centrifugal control of the retina, arising from the brain, as there is in many fishes, reptiles, amphibia, birds, and lower mammals. Note that in humans, at postmortem, years after enucleation of an eye, the optic nerve has been found to contain about 10% of the normal count of axons (Wolter, 1965, 1968). This implies, but does not prove, that such a centrifugal system exist in humans. It has been found recently that visual attention alters electroretinographic (ERG) responses to visual stimuli in humans (Eason, 1981, 1983). This suggests strongly that such centrifugal influences do exist and that, moreover, subjective visual attention can affect retinal activity. There are analogous central to peripheral efferents which influence other sensory receptors and ascending sensory relays. It is by such centrifugal means that the nervous system modulates its own earliest stages of sensory processing (Livingston, 1978).

Central Projections of the Retina

The convex surfaces of cornea and lens cause light rays to converge at a point between the lens and the retina, and rediverge in reversed order so that topographic relations between the visual field and its image on the retina, like a camera's image on photographic film, are reversed left to right and up to down. A vertical axis of the retina which passes through the center of the fovea divides retinal output into nasal and temporal halves (Fig. 9.83). Ganglion cells in each of the right and left retinal halves of each eye send their axons to the contralateral lateral geniculate nucleus (LGN). Thus, the receptive fields of cells in the LGN are all projected to the opposite side of the visual world.

A horizontal line divides these halves into upper and lower quadrants. The inverted image in each

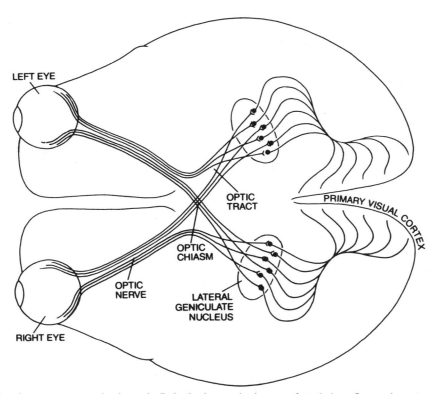

Figure 9.83. Visual pathway represented schematically in the human brain, seen from below. Output from the retina is conveyed by ganglion cell axons via the optic nerve and, after partial crossing, the optic tract to the lateral geniculate nuclei. Each half of the visual scene is projected to the geniculate of the opposite hemisphere. Neurons in the geniculates send their axons to the primary visual cortex. Enucleation of one eye or section of the optic nerve obviously yields an ipsilateral blind eye. Section of fibers crossing in the optic chiasm leads to "tunnel vision" by blinding both nasal retinal halves. Section of the optic tract or destruction of the geniculate or visual cortex causes loss of vision for the contralateral half visual field. The pathways are so orderly that quadratic and even smaller deficits can localize lesions to specific parts of the optic radiations and cortex. [From Hubel and Wiesel (1979).]

retina is relayed in a straightforward way through the LGN to visual cortex so that *the upper portion of visual cortex in each cerebral hemisphere represents the lower quadrant of the contralateral visual field.*

In general, topological organization of ganglion cells in the retina are preserved throughout the several different parallel visual pathways. In this way, retinal patterns are mapped and remapped for a variety of analytic purposes in several different regions of the brain.

Ganglion cell axons sweep across the retina and converge at the optic papilla, where they exit the globe as a bundle of fibers, the *optic nerve* (Fig. 9.83). As they continue along their course centralward they are identified as the *optic chiasm*, where fibers from the two retinal halves part company, and later as the *optic tract* which enters the lateral geniculate nucleus (LGN).

Altogether, the optic chiasm and the optic tract send off collaterals to at least six brain regions: (1) the hypothalamus (*suprachiasmatic nucleus (SN)*); (2) the *accessory optic system (AOS)*; (3) the pretectum; (4) the optic tectum (*superior colliculus (SC)*); (5) the ventral thalamus (*ventral lateral geniculate nucleus (LGNv)*); and (6) the lateral thalamus (*dorsal lateral geniculate nucleus (LGNd)*). It is in this latter station, in primates, where axons thought of as the classical visual pathway terminate. However, each of these several destinations of ganglion cell axons, and *their* onward projections, plays an important role in various sensorimotor mechanisms relating to vision. For example, Sprague et al. (1981) show that in cats, removal of striate and peristriate cortex (areas 17 and 18) in animals trained and tested before and after cortical ablation, there is excellent retention of pattern and form discrimination and only minimal deficiencies in visual attention, orientation, depth discrimination, and visuomotor activity, postoperatively. The authors interpret their studies to mean that the suprasylvian cortex which receives predominantly from pulvinar relays in the thalamus must be providing exceedingly important information in visual perception. We are forewarned that all perceptual and other cognitive functions are distributed, and that some traditional assumptions about what is primary and what is secondary in perceptual processing may soon be subjected to radical revision.

How is Distributed Visual Representation Integrated? Central representation of visual information involves up-down and left-right reversals, also elastic distortions of areal representation and rerepresentation, and it delivers axon collaterals from the same and different ganglion cells to regions which analyze visual information differently. An obvious question rooted in these anatomical and physiological features constitutes a provocative challenge to human thinking: *How, really, do we perceive the world around us?* In particular, how does this combination of precise yet segregated, spatially and temporally displaced circuitry and related systems yield coherent visual perception? How is that which we perceive as a coherent scene assembled or reassembled from this distributed array? Even putting the question this way is already oldfashioned: there is evidently nowhere (and there is no need for) a place where there exists any "scene assembled."

A commonplace illusion which we have got to abandon is that there is some sort of an homunculus, a little "me" that is a "witness" at some station in the brain: that there is some sort of composite of the individual persona which confronts a composite representation of his/her self within a similarly composite representation of his/her environment (see Crick, 1979). Rather, *perception provides an active outward projection which relates primarily to options for motor performance.* This is derived from cognitive processing of visual information during the course of its parcellation and distribution. *Perception is obliged to operate in this coherent way perforce of requirements for motor performance.* We are obliged to *behave*, by means of a given body in a given world. *Organisms are built for action.* My interpretation is that it is only secondary in terms of evolution that a conscious role in behavior gained advantageous momentum and out of that emerged perception. From the beginning perception has been conditioning behavior.

Vision, like other active sensory processes, must inevitably conform to certain physical facts: what visual information is distributed centrally is inevitably, intrinsically, functionally linked to sensory surfaces and to motor effectors. More than that, the way sensory surfaces relate inwardly to the body wall and outwardly to the environment, and the articulation of body parts together with all the options for motor performance, all contribute to the foundations of visual sensory processing and to the formation of visual percepts. Active outward projection onto "ourselves" and onto an "outside world" results from cognitive operations in the same way that motor command and control funnel dynamic, three-dimensional, parcellated activities taking place in the central nervous system into performance of complex coherent movements.

These dynamic, three-dimensional, space-time processes constitute a puzzle of perception which provides one of the prime challenges to neurosciences, indeed, to any branch of science or philosophy. The key to this understanding is that perception is organized for behavior. Perception apparently *evolves* in the species and *develops* in the individual as necessary to take actions. Perception is cognitively composed into *active projection* representing events *outside* ourselves, primarily in relation to provision of *options for behavior* (Livingston, 1978).

Suprachiasmatic Circadian Control

The suprachiasmatic nuclei (SN) lie dorsal to the optic chiasm and immediately lateral to the third ventricle. Each nucleus receives bilateral projections from retinal fibers, mostly collateral branches of axons that are following the optic tract. The SN exerts neural control over a number of circadian rhythms: eating and drinking, locomotor activity, sleep/wakefulness, and levels of pineal secretion of serotonin and adrenal corticosterone (Moore, 1974). SN synchronizes these rhythms and, with some hysteresis, regulates their schedule according to the intensity and duration of environmental light. The highest glucose consumption of the SN is in daytime, even without light cues. Bilateral ablation of the SN disrupts these cycles and also interferes with reproductive behavior.

PRETECTAL PUPILLOCONSTRICTOR AND OTHER OCULAR REFLEXES

The pretectal region, immediately anterior to the superior colliculus, receives fibers from many sources: retina, superior colliculus, the ventral part of the lateral geniculate nucleus, striate and extrastriate visual cortex, and frontal eye fields. These inputs are preponderantly retinotopically organized. The retinal input is bilateral and from all known ganglion cell types. On center tonic X cells contribute importantly to the pupilloconstrictor reflex previously described, and these and other inputs contribute to *ocular fixation reflexes* and to *optokinetic nystagmus.* The latter, the gaze-following reflex, depends on integrity of visual cortex.

Accessory Optic System for Integration of Visual-Vestibular Reflexes

Terminal nuclei in the pretectal region receive axon collaterals from the optic tract which respond selectively to slow vertical movements of targets and visual surround. The neighboring nucleus of the optic tract receives collaterals responsive to slow movements in the horizontal plane. These nuclei relay their distinctive visual messages to an *accessory optic system (AOS) which integrates visual, vestibular, and gaze control information for per-* *ception and analysis of movements in the environment in relation to the position and motion of head and eyes in three-dimensional space.*

Optic fibers which contribute to the AOS originate from large, directionally selective retinal ganglion cells which possess very large receptive fields, up to 90°. Cells in the terminal and optic nuclei respond optimally to slow movements of broad, textured patterns. They send axons across the midline to enter the cerebellar cortex in the flocculonodular lobe as mossy fibers. Other axons go uncrossed to the ipsilateral inferior olive which relays impulses across the midline to enter the same cortex via climbing fibers. Floccular cortex receives ipsilateral vestibular information by way of mossy fibers, directly from the vestibular apparatus and indirectly by relay through the vestibular nuclei. It provides oculomotor centers with velocity command signals in support of pursuit movements.

Retinal impulses influence cerebellar cortex with respect to movements that sweep across large areas of the visual field, especially movements along the horizontal and vertical axes of the retina. Separate Purkinje cells respond to horizontal and vertical visual movements, and both kinds of Purkinje cells are affected by head movements, especially by rotations in the planes of the three semicircular canals. *The cerebellum thus integrates horizontal and vertical motions of visual targets and surround, with head and eye movements indexed to three-dimensional space.*

AOS activation, by surround and target motion, contributes to stabilizing gaze at frequencies too low for the vestibular system to be effective. Motion of the entire visual surround influences perception of motion, and nystagmus, and affects the firing rates of neurons in the vestibular nuclei. It makes more veridical vestibular contributions relating to slow head movements as compared with what occurs with vestibular input alone (without AOS input), and the resulting, integrated signal, representing head velocity in three-space, is transmitted to thalamus and cortex.

By similar means the AOS contributes to *vestibuloocular reflexes (VOR) which function to stabilize the gaze during active and passive movements of the head.* Intention to fix the gaze modifies neuronal activity in the vestibular nuclei, and this suggests the way by which motion sickness may be partially ameliorated by employing deliberate visual activity. The readiest known access of visual signals to vestibular nuclei is through the AOS via the flocculus. These constitute good examples of intersensory functional integration and illustrate how the central nervous system contributes to integration of its own input at the earliest stages of sensory processing.

The AOS coordinates relative head and eye movements in the following way: If the head is actively or passively rotated horizontally, the eyes are turned reflexly in the opposite direction at a rate that enables eye fixation to a nonmoving environment. This reflex (the vestibuloocular reflex, VOR) is overridden when the eyes are tracking a target that is moving through the visual field. The cerebellar cortex adjusts an appropriate gain for this system, by matching the relative velocities of head and eyes to the visual field, and any moving target within the visual field.

Plasticity in the AOS system is manifested when a human or animal subject wears reversing lenses for a few weeks. The eye reflexes to head rotations reverse so as to enable fixation to a now contrarily moving visual field. This conditional reflex operates in the newly entrained way for both active and passive head rotations, *even in the dark*. When the reversing lenses are removed, it takes considerable time for readjustment before the original reflex is restored. Surgical removal of the flocculonodular lobe eliminates this remarkable integrative plasticity (Melvill Jones, 1977), and it also reduces or eliminates motion sickness (Bard, 1948).

Superior Colliculus and General Visual-Somatic Sensorimotor Integration

The cephalic half of the tectum of the midbrain, the superior colliculus (SC), is organized for vision and general visual-somatic sensorimotor integration. The SC resembles cortex by reason of its richly layered structure. Collateral fibers from the optic tract, including axons from a variety of retinal ganglion cells (except X cells), distribute terminals within the surface layers of the SC. The colliculus receives additional topologically organized projections from both striate and extrastriate visual cortex. These project to layers of the SC immediately beneath the topologically corresponding retinal projections (Ingle and Sprague, 1975).

Both retinal and cortical maps are segregated into bands or slabs, similar to those seen in visual cortex. Deeper lying collicular layers receive a variety of other visual and related contributions, including (a) projections from the frontal eye fields (area 8 of Brodmann) from which conjugate eye movements can be elicited, (b) topologically organized projections from sensorimotor and auditory cortex, and (c) projections from cells in the substantia nigra whose axons are likewise organized into bands. These contributions imply a wealth of opportunities for the *integration of visual explora-*

tions, body imagery, the acoustic surround, and somatic motor controls.

The SC receives additional inputs from the hypothalamus (presumably conveying information about motivation and neuroendocrine status), the lateral geniculate nucleus (conveying information about the visual world), diffusely projecting thalamic nuclei (conveying effects relating to arousal and directed attention), from the inferior colliculus (conveying spatially organized acoustic information), from the periaqueductal gray, from the cerebellum, and from spinal and trigeminal sources both of which preserve a somatotopic representation of the body image.

The circuitry of the SC is obviously organized for integration of visual with nonvisual information and for generation of appropriate head and eye movements, and other bodily responses toward objects of potential visual interest. Visual cortex is needed to search for visual targets when there are no localizing clues, and it probably makes use of the SC in the process. Once localizing clues, acoustic, somesthetic, or visual, are available, the colliculus can direct the gaze binocularly toward such sources, using centrifugal eye, head, and body command and control.

Of the two top-layered SC maps, one originates from the retina and the other from striate and extrastriate visual cortex. These contribute to binocularly driven, visually directionally selective cells which lie in deeper layers of the colliculus and which manifest properties that reflect a combination of retinal and cortical visual response characteristics. These deeper lying SC cells direct body, head, and conjugate eye movements toward that part of the visual field which corresponds to the retinotopic map at the particular collicular site stimulated. This provides a *visual orienting response* which can be initiated by visual, acoustic, or somesthetic excitation or by voluntarily directed attention. SC activation is aroused by stimulation of the reticular activating system. With these informational contributions the SC generates a *visual grasp reflex*, an undoubtedly important mechanism for evolutionary success.

The superior colliculus gives rise to a *crossed tectospinal tract* which terminates in gray matter of the dorsal horn in the cervical spinal cord, *by means of which the SC contributes to head turning and head-eye coordination.* Both crossed and uncrossed tectal fibers project to brain stem interneurons which relate to the eye motor nuclei (cranial nerves III, IV, and VI). These tectooculomotor projections contribute importantly to *binocular fixation.* Similar projections reach the cerebellum by relay through both pontine nuclei and the inferior olives.

The SC also contributes to two major thalamo-cortical visual systems: One SC contribution consists of projections to the *classical visual pathway via both ventral and dorsal components of the lateral geniculate nucleus.* The second contribution (which also preserves its retinotopic organization) projects to the *lateral-posterior part of the pulvinar from which next-order projections are relayed, preserving topological correspondence, onto striate and extra-striate visual cortex.* This latter retinocolliculopul-vinocortical pathway constitutes a circuit that parallels the classical retinogeniculocortical pathway, with but one additional synaptic relay.

Local Cerebral Blood Flow Affected by Visual Activity

Local cerebral blood flow (LCBF) values were measured using xenon inhalation and computed X-ray tomogram scans in normal human subject with eyes closed in a darkened room compared with eyes open in a brightly lighted room (Meyer et al., 1981). The visual system showed an increase of +16.8% ($P < 0.01$), whereas frontotemporal cortex and basal ganglia showed no significant change. Compared to values measured with eyes closed and EEG showing α (resting) activity, eyes open LCBF values were higher in SC by +11.5%, LGN by +27.4%, striate cortex by +11.3% and occipital cortex by +4.2%.

LATERAL GENICULATE NUCLEUS

By the time visual information leaves the retina, significant analyses have already occurred. Signals have passed through two to four direct-path synaptic relays, influenced by selective and analytic lateral microcircuits which contribute to early visual processing. Impulses from distinctive retinal events have combined at each junction into new patterns which take collateral influences into account. The results of this succession of integrative operations are then transmitted to visual cortex by way of a major thalamic relay, the lateral geniculate nucleus (LGN). This omits, for the moment, pathways that go to the brain stem prior to transmission to LGN, including the alternate pathway to both striate and extrastriate cortex, which passes by way of the superior colliculus with its projections to thalamic relays in both LGN and pulvinar nuclei.

Partial crossing of the optic nerves in the optic chiasm means that both LGN and brain stem traffic of visual information is related to retinal events in the corresponding half-retinas, split vertically, i.e., the temporal retina projects ipsilaterally and the nasal retina projects contralaterally. Each of these four half-retinas responds to events in half the visual field, and the corresponding information is projected as closely corresponding data to the two hemispheres by way of both brain stem and thalamic circuits. Left LGN and left occipital cortex thus receive impulses which originate from two left half-retinas, each of which is processing information from the right half of the visual field; while the converse holds for the right half-retinas which project to right LGN and cortex and have to do with events in the left half of the visual world (Fig. 9.83).

Whereas each retina involves convergence from about 100 photoreceptors to one ganglion cell, the thalamic relay is approximately one-to-one, i.e., there are about a million cells in each optic tract and a million cells in each LGN. Individual optic tract fibers enter the LGN and divide into several branches, each of which may terminate on a different geniculate neuron. LGN organization provides for divergence, convergence, and overlap at this relay.

Organization of the Lateral Geniculate Nucleus

Cells of the LGN are organized into six layers: the first, fourth, and sixth receive fibers from the contralateral eye; the second, third, and fifth receive fibers from the ipsilateral eye (Fig. 9.84). The two reduced images below show the effects of removal of the right eye. The corresponding laminae in left and right lateral geniculate bodies show severe atrophy. Precise retinotopic mapping in the LGN is indicated by the fact that *small retinal lesions result in transneuronal degeneration of discrete clusters of geniculate neurons, confined to the appropriate laminae.* Moreover, a precise correspondence is maintained throughout the LGN between the two corresponding retinal projections as they engage their assigned laminae. Corresponding points from the two retinas line up over one another across the stack of laminae, i.e., following the *solid curve* through the enlarged image at the *top* in Figure 9.84. Even though signals from the two eyes are affected by the same visual events and project in registration to adjacent LGN laminae, there is no evidence in the LGN of comparison being made between signals coming from the two retinal projections. This occurs only after the information reaches striate cortex.

Responses of cells in the LGN to light patterns cast on the retina are remarkably similar to those of retinal ganglion cells. Almost all LGN units respond best to small light spots with "on-center" or "off-center" illumination, roughly circular, and of a particular size and location on the retina. Opposite effects are elicited by changes of illumination in the annular surround. A typical on-center geniculate neuron responds to a spot of light applied

RIGHT EYE

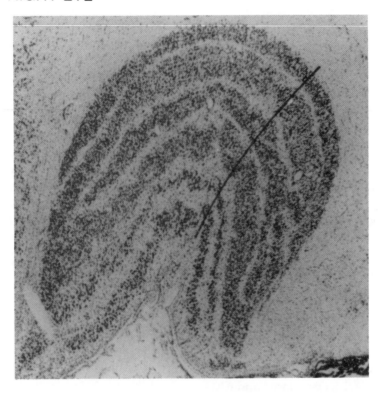

LEFT EYE   RIGHT EYE

Figure 9.84.   Lateral geniculate nucleus of a normal monkey *(top)* is a layered structure in which cells in layers 1, 4, and 6 (numbered from *bottom* to *top*) receive their input from the contralateral eye, and those in layers 2, 3, and 5 receive information from the eye on the same side. The retinogeniculate maps are in register, so that neurons along any radius *(black line)* receive signals from the same part of the visual scene. The two lateral geniculates in the slightly reduced micrographs to the *below* are from an animal with vision limited to the left eye only. Note that in each of these geniculates the three layers with inputs from the right eye have atrophied. [From Hubel and Wiesel (1979).]

to the center of its receptive field with a rapid increase of impulse above the level of spontaneous activity. When light is extended across both center and surround, the activity is increased only mod-

erately. When light is applied only to the annular surround, the unit is strongly inhibited, which inhibition is followed by "rebound." When light lasts more than about ½ s, there is adaptation which

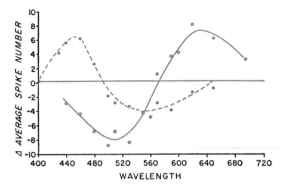

Figure 9.85. Difference between average number of spike discharges at stimulus onset and average number of spike discharges at stimulus offset as a function of wavelength (nm) of the stimulus for each of two single neural units in the lateral geniculate body of the monkey. The results illustrated with *solid circles* and a *dashed line* show a maximum "on" response in the blue region and a maximum "off" response in the yellow region. Results presented with *open circles* and a *solid line* indicate a maximum "off" response in the green region and a maximum "on" response in the red region of the spectrum. [From DeValois (1960).]

varies according to which part of the receptive field is illuminated.

Such a single unit can convey considerable information: the exact stimulus location, its approximate shape and size, its intensity, various temporal features including the rate of onset and offset of the light as well as its duration, measured not only by the interval between onset and offset but also by the process of adaptation during steady illumination. Some LGN cells can signal whether the stimulus is steady in intensity or not, whether moving or not, and perhaps the direction of motion. Most LGN cells also include information about the wavelength of the light.

Such cells respond well to light or dark line segments in any orientation which project onto the entire center and intrude along only a narrow path across the surround. They reflect integration of local light intensity levels in comparison with the average illumination in the neighboring region, and they are generally not affected by changes in illumination across the entire retina.

Many individual LGN cells respond by increasing their activity on exposure to all wavelengths of light. Other cells, called color-opponent cells, show increased activity during stimulation in one wavelength region and suppression of their resting level of activity during presentation of light in another wavelength region. Some cells may be excited by relatively long wavelengths of light and inhibited by middle wavelengths. These are characterized as "red-on, green-off." Other cells may be characterized as "green-on, red-off." Still other cells may be characterized as "yellow-on, blue-off" and "blue-on,

yellow-off" (Fig. 9.85). The precise nature of the response of these cells depends upon the conditions of adaptation of the eye.

The Importance of Visual Experiences Early in Life

An interesting consequence of experience during development of the visual system is manifested in the LGN. If one eye is simply occluded for some months early in life, rather than being removed, the corresponding laminae of the contralateral (1,4,6) and ipsilateral (2,3,5) LGN do not disappear but are markedly retarded in their development. They look pale in stained sections, and there is a pronounced reduction in the number of physiologically identifiable Y cells. If the nondeprived eye is now removed and the deprived eye remains occluded, there is no restoration of the Y-cell population. If, however, the nondeprived eye is removed and the deprived eye is then given 3 months of visual experience, the percentage of Y-cell encountered recovers to a normal range.

Such restoration of functional competence can be induced in kittens if the occlusion is effected at about 5 wk and the removal of the nondeprived eye and the beginning of visual experience for the deprived eye take place at about 4 mo of age. If a similar procedure is undertaken between 12 and 16 mo, well into young adult cat life, some reversal can still be achieved. Nonetheless, provision of visual experience for the deprived eye and simply occluding (rather than removing) the nondeprived eye does *not* yield any recovery (Geisert et al, 1982). It is as though deprivation by occlusion interferes with functional development, but removal of the eye eliminates some kind of functional competition that affects geniculate neurons. Comparable changes result from visual deprivation in monkeys (Hubel and Wiesel, 1979).

Altogether, this means that there is a limited window of time within which considerable functional restoration following deprivation is possible, *provided that appropriate visual experiences take place.*

Clinically, it is a commonplace that interference anywhere along the light path in human infants needs to be corrected in childhood in order to obtain adequate vision. It is also true that useful binocular vision depends upon *early binocular visual experiences.* If a child has defective conjugate eye motor control that is not corrected early in life, even though each eye may have normal vision, binocular vision may be permanently impaired unless the optical axes can be brought to track together on visual targets. *The visual system for competent monocular and binocular vision is already developed in the newborn, but its enduring functional integrity depends on appropriate visual experiences early in life.*

Lateral Geniculate Projections to Visual Cortex

Axons of relay neurons in the LGN contribute optic radiations which project exclusively to striate cortex. This pathway maintains strict retinotopic representation so that quadratic and even smaller defects can be detected clinically by testing visual fields when there is partial destruction of geniculostriate projections. Moreover, small lesions in striate cortex result in retrograde degeneration of narrow sectors extending through all six layers of the LGN, indicating that the geniculostriate projections maintain precise retinotopic relations with both corresponding retinal halves as the projections invade cortex. The close juxtaposition of the two retinal representations facilitates comparison of visual information coming from the two eyes, a process which occurs first in striate cortex. This image matching is essential for fine eye motor control of binocular foveal fixation and for appropriate accommodation of the lens.

STRIATE CORTEX

Striate cortex in humans (area 17 of Brodmann) is located largely in the posterior half of the medial surface of the occipital lobe. Central vision projects into the most posterior limits of that region. In most mammals, visual cortex lies on the convexity of the hemisphere, and in monkeys the area of central vision is still exposed laterally. Aggregate receptive fields grow larger as visual targets are projected farther into the retinal periphery; the same millimeter of progress along striate cortex represents an increasingly larger segment of visual field; this corresponds roughly to the differences in density of ganglion cells between paracentral and peripheral regions of retina. A few thousand geniculate fibers enter each millimeter of cortex. Based on an approximately one-to-one relay in the LGN, this means an equivalent of a few thousand retinal ganglion cells. With over a million ganglion cells, this allows a few thousand millimeters of cortical display.

X, Y, and W cells have separate channels to striate cortex (Van Essen, 1979). Retinal maps of various kinds in most cortical and subcortical circuits preserve topological order. Neighboring cells have neighboring or overlapping receptive fields. As emphasized, central retinal areas are magnified compared to representations of retinal periphery. Within this topology, however, there are finer grained distinctions. For example, as striate cortical cells are sampled in electrode progression (sloping tangentially) across the cortex, about every millimeter of lateral progress is associated with an abrupt displacement of the aggregate receptive fields.

Powell and associates have shown that the number of neurons along a vertical line perpendicular to the surface of the cortex, that is, across its thickness, in a 30-μm diameter circle, is remarkably constant in most areas of cortex and in most mammals, at about 110 (Rockel, Hiorns, and Powell, 1974). The same count for striate cortex in primates, however, is approximately double (260), and the thickness of striate cortex is also increased. Thickness depends on development of dendritic arborizations and on fine circuitral organization, terminals, synapses, spines, etc. The characteristics of *columnar organization* in striate cortex will be described shortly.

Striate cortex is defined by sharply demarcated boundaries within which cortical layering is particularly distinctly visible, hence the adjective "striate." Striations consist of alternating fiber and cell layers as in other regions of cerebral cortex, but in primary visual cortex one layer, IV, is particularly thick. The middle part of this layer, IVb, consists mainly of fibers which are so heavily myelinated that this layer can be seen macroscopically.

Afferent Input to Striate Cortex

The main afferent input to striate cortex comes from the lateral geniculate nucleus which distributes abundant terminals to layer IV. Geniculate axon terminals establish synaptic endings on spines of dendrites and dendritic surfaces of pyramidal cells and to a lesser extent on stellate cells in layer IV. Afferents from the pulvinar nucleus end in layer I. Since this is the most superficial layer of cortex, offering synaptic terminals on the farthest branches of apical dendrites, the influence of pulvinar input to cortex is less powerful, probably providing modulating influences rather than compelling control over firing of pyramidal cells.

The only interhemispheric projections terminate in a narrow zone at the posteriormost margin of area 17, along its border with area 18. These corpus callosum projections contribute to matching the midline between the two visual half-fields which are separated into two hemispheric maps by the partial crossing of fibers in the optic chiasm.

Efferent Output from Striate Cortex

Cells with different response characteristics and different degrees of complexity are located in different cell layers. This is important because the efferent projections from different layers of cortex are destined for different cortical and subcortical projections. After an unknown number of local intracortical synaptic transits, the same region of cortex that receives geniculostriate radiations sends retinotopically organized messages back to influence the originating geniculate relay cells. This is another example of sensory feedback cortex shaping its own input. Striate cortex also sends efferents to several other cortical areas for additional visual discrimination and analysis, and to several subcortical stations concerned with further visual processing and with visually related motor control. For example, layer VI of striate cortex projects, probably exclusively, to LGN, providing the direct sensory feedback circuit. Layer V projects, also precisely retinotopically, to the superior colliculus, contributing to eye motor control, particularly to keep both eyes fixed on the target of interest, and to keep both of them in precise focus. Layers II and III provide corticocortical projections to a variety of other cortical regions.

Functional Characteristics of Intrinsic Striate Neurons

Hubel and Wiesel divided visual cortical cells into categories of *simple, complex, and hypercomplex*. They discovered that cells with similar characteristics of complexity: *receptive field location, line segment orientation, and ocular dominance,* tend to be grouped together. Many "on-center" and "off-center" cells respond like geniculate cells, with circularly symmetrical receptive fields. These are found in layer IV, in the vicinity of the geniculostriate terminals, and are presumably directly influenced by that input.

Another large population of neurons in the striate cortex respond preferentially not to spots of light but to *line segments having a specific orientation.*

Oriented Line Segments

Line segment discrimination naturally specifies a particular orientation. By identifying the orientation to which a neuron responds best, one can establish that they discriminate differences between orientations of about 10–20°, i.e., a clockwise or counterclockwise shift of orientation of only about 5–10° to either side of best response orientation will reduce or abolish the response. When you think that the angular rotation of the hour hand of a clock is 30° for 1 h, you can appreciate that the discrimination of cortical cells for line segment orientation in the visual field is rather precise.

Figure 9.86 localizes neurons that respond to a pattern of vertical stripes and shows how they are distributed over an area of striate cortex in the monkey. Such stimuli are segregated (except for layer IV) in curved bands of uniform width. Similar patterns, slightly displaced, would be found for all other arbitrary orientations of line segments. Within a local region of cortex, therefore, all orientations are differentially represented so that each region of the visual field can generate localized cortical activation according to given specifications of line segments and hence to any outlines of objects in view.

Figure 9.87 illustrates schematically the way in which it may be assumed that retinal center-surround ganglion cells can inform *simple cells* in striate cortex via the lateral geniculate nucleus. Those neurons which respond to an oriented line in a restricted location are called *simple cells.* They are found in abundance in the vicinity of layer IV. A series of retinal ganglion cells with, for example, on-center receptive fields, relay information in parallel to cortical cells. If a straight file of such ganglion cells were connected to a single neuron in the striate cortex, it would behave as if it were concerned with a straight line segment having a particular orientation at that particular location on the retina. Obviously tracing and proving this exact circuitry is presently too formidable a task, but inasmuch as many striate neurons behave as if they were connected in this way, the conjecture is justified.

Complex cells are found in all layers of striate cortex except layers I and IV. Their responses are less strictly bound by retinal location. They respond best to lines of a particular orientation when these lines are moving across their receptive field. Somewhat more complex units respond only to specifically oriented lines when they are moved in one direction only, and are inhibited by lines of similar orientation moved in the opposite direction. Some neurons show a strong preference for a limited length of line segment and thus contribute to corner detection as well as to edge detection.

Hypercomplex cells respond best to two or more borders presented anywhere within a large retinal area, regardless of the size of the image (Fig. 9.88). Cells having these latter characteristics provide a basis for complex boundary detection and detection of movement of contrasting figures without necessarily being restricted to particular regions of the visual field.

Figure 9.88 shows how a mixture of ganglion cells in a localized region of the retina, connected with corresponding cellular elements in the LGN, would yield in the cortex not only a series of local line segment detectors which have the same orientation, but how those cells could combine information nec-

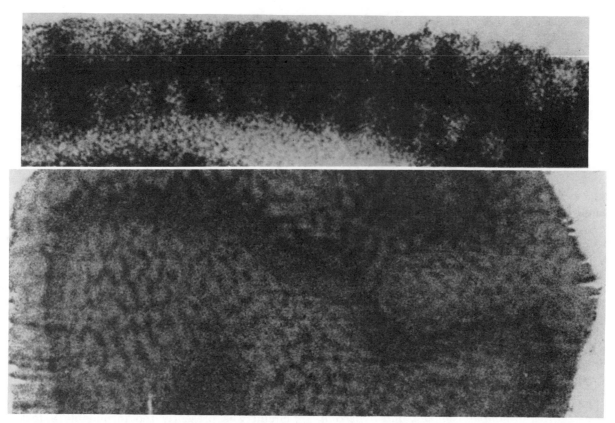

Figure 9.86. Orientation columns in the primary visual cortex of the monkey, visualized by deoxyglucose autoradiography, seen in cross-section (*top*) and tangential section (*bottom*). Immediately after injection of the radioactively labeled deoxyglucose the animal was stimulated by a pattern of vertical stripes so that the cells responding to vertical lines were most active and accumulated greater stores of the incompletely metabolized glucose. Active cell regions constitute narrow bands about 0.5 mm apart. Layer IV, with no orientation preference, is uniformly radioactive. Seen in the tangential section (*bottom*) the large oval darker region represents layer IV. In the other layers the vertical orientation columns constitute intricately curved bands, like the walls of a maze seen from above, and the distance from one band to next is uniform. [From Hubel and Wiesel (1979).]

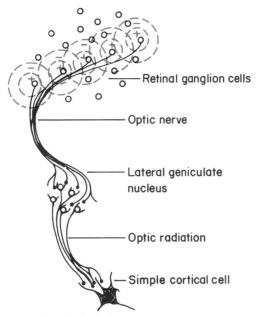

Figure 9.87. Connections from retinal ganglion cells through cells of the lateral geniculate body of the simple cells of the cortex which respond to line element stimulation of the retina.

essary for *complex cells* to respond to directional motion of line segments. By combining information from two such complex cells, data would be available by which to account for the performance characteristics of *hypercomplex cells*. These can respond to two or more borders in motion.

Table 9.4 summarizes the incremental gain of differentiated capabilities of neurons at successive levels of the visual system, from receptors to striate cortex. All cells in striate cortex are presumed to derive their differentiated visual response characteristics from similar kinds of specific connections received from initial input signals that resemble those of retinal ganglion cells, modulated by LGN as influenced by corticogeniculate projections, and further modulated by superior colliculus contributions relayed through the pulvinar. This latter contribution contains information concerning direction of gaze, together with integrated intermodality sensory information. Such examples indicate how still higher order cells might be connected for the detection of increasingly specific and complex

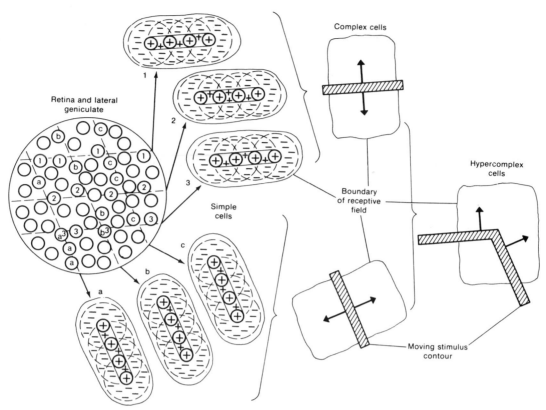

Figure 9.88. Response patterns and extraction of signal patterns in visual pathway. Many units are shown at the level of lateral geniculate body (at which responses are essentially like those of the retina). Each cell has a small concentric field with an on or off spot center and opposing surround; here all are assumed to be on-center units. If all cells marked 1 converge on a single *simple cell* in the primary visual cortex, this cell will have a larger, linear receptive field, specifically responsive to a thin bar of light or a border at orientation shown, with light falling on the centers of the field of first order units and darkness covering only part of the surround. Stimuli at other orientations are relatively ineffective or inhibitory. Other simple cells are illustrated, 2, 3, a, b, and c. The same lateral geniculate cells could be contributing at the same time, in different orientations. If several simple cells, each responsive to stimuli at the same orientation, converge on a *complex cell*, the resultant receptive field will be larger and specific for stimulus orientation and, often, movement or direction of movement, but the stimulus would not have to fill any particular part of the receptive field to be excitatory. Two or more complex cells may provide information for a *hypercomplex cell* in which stimulation is most effective when there are two or more borders whose orientation is determined by inputs. With continuing convergence of this sort, still higher order cells can be programmed to respond only to highly specific and complex shapes falling anywhere within a large retinal area, irrespective of the absolute size of the image. [From Bullock (1977).]

shapes. Furthermore, similar principles hold for detection of shapes that may not be localized to any particular region of the visual field, and for detection of shapes regardless of their size.

Of course, each step to higher order complexity is attended by a slightly longer processing time. Progress is being made to determine the succession of stages of sensory processing (see Marr, 1982; Julesz, 1965; Julesz and Schumer, 1981) as well as to establish the regions of the brain in which such processing takes place.

Ocular Dominance Patterns

When geniculate axons terminate in visual cortex, inputs from the right and left eyes are still segregated. Axons stemming from LGN layers 1, 4, and 6 (contralateral eye) and from layers 2, 3, and 5 (ipsilateral eye) terminate in alternating files about 0.4 mm wide. Figure 9.89 shows such *ocular dominance columns* in two views at right angles to one another, perpendicular to the cortical surface, and sliced tangentially through the cortex. Adjacent parts of the ocular dominance slabs relating to the two eyes are closely matched to one another so that corresponding regions of the two retinas lie side by side. Thus a general map of one half of the visual field, suitably magnified, spreads across occipital cortex. Ocular dominance from the two eyes exerts itself prominently in layer IV.

All simple cells in the striate cortex appear to respond only monocularly. About half of all complex cortical cells respond to stimulation of corresponding points in each half-retina. *Cells which respond binocularly have identical positions in re-*

Table 9.4
Characteristics of receptive fields at successive levels of the visual system

Type of Cell	Shape of Field	What is Best Stimulus?	How good Is Diffuse Light as a Stimulus?	Is Orientation of Stimulus Important?	Is Position of Stimulus Important?	Are There Distinct "On" and "Off" Areas within Receptor Fields?	Are Cells Driven by Both Eyes?	Can Cells Respond Selectively to Movement in One Direction?
Receptor	⊗	Light	Good	No	Yes	No	No	No
Ganglion		Small spot or narrow bar over center	Moderate	No	Yes	Yes	No	No
Geniculate		Small spot or narrow bar over center	Poor	No	Yes	Yes	No	No
Simple		Narrow bar or edge	Ineffective	Yes	Yes	Yes	Yes (except in monkey layer IV)	Some can
Complex		Bar or edge	Ineffective	Yes	No	No	Yes	Some can
Hypercomplex		Line or edge that stops; corner or angle	Ineffective	Yes	Yes	Yes	Yes	Some can

From Kuffler and Nicholls (1976). Information is collected from retinal areas of varying size, emphasizing contrast between center and surround, and at higher levels, defining edges, corners, and directional motion of contrast boundaries.

spect to the *visual field, respond to similar complexity of target illumination, and have the same orientation and preference for direction of movement.* Although units may respond more actively to one eye than to the other, they appear to have equivalent access to similar arrays of simpler units which are themselves responding in a similar way, but solely to monocular activation. Figure 9.90 provides a map of ocular dominance. It is a photomontage reconstruction of a part of monkey visual cortex, with bright zones representing the contralateral and dark zones representing the ipsilateral eye.

Orientation-specific neurons are disposed throughout the cortex in an orderly progression of orientation preferences. These are arranged in strips that course more or less orthogonally to the ocular dominance zones, so that each member of a local ocular dominance pair (representing corresponding loci in each retina) contains a complete 180° sequence of line segment orientations. For given neurons in such a pair, the left-right dominance is determined by afferent input, while the degree of that dominance and specific line segment orientation are determined intracortically. The

combination of local systems analysis for binocularity and orientation occupies an area of about 800 μm × 800 μm and constitutes a *macrocolumn* which is characteristic for the entire cortical representation of the visual field.

Because of the paucity of corticocortical connections within striate cortex, the lack of callosal communications except for a narrow strip along the vertical meridian, and the fact that this is the least interconnected of all cortical regions, it is probable that visual perception takes place elsewhere as a more distributed function. An animal with an intact striate area but without other cortical visual and visual associative areas would probably be perceptually blind (Edelman and Mountcastle, 1978). This intricacy of circuitry is innate: it is found in the newborn, but evidently it requires *binocular visual experience* during development in order to become "hard-wired" in the adult.

Poggio and Fisher (1977) (and Fisher and Poggio, 1979) have identified neurons in area 17 of the waking, behaving monkey that respond to *receptive field disparities.* They found excitatory and inhibitory neurons which were responsive to depth cues

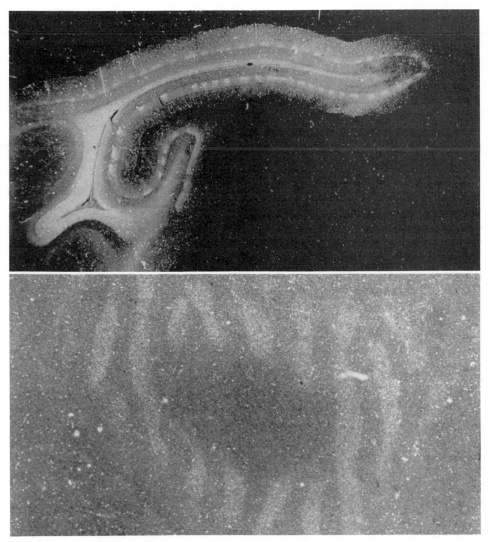

Figure 9.89. Ocular dominance columns in the primary visual cortex of the monkey seen in cross-section (*top*) and tangential section (*bottom*). Bright bands are columns dominated by cells conveying right eye information. A radioactively labeled amino acid was injected into the right eye, picked up and transported by ganglion cells of the retina to the lateral geniculate nucleus whence it was picked up and transported to ocular dominance cells in the visual cortex where the radioactive particles accumulated and are seen as slab-like ocular dominance columns (about 0.4 mm diameter) from side *(top)* and facing *(bottom)* views. The dark intervals separating the bright regions represent ocular dominance columns relating to the left (uninjected) eye. [From Hubel and Wiesel (1970).]

in the visual field. This would facilitate stereoscopic vision and contribute to the maintenance of visual fixation. These neurons are presumably involved in the continuing "search" for best binocular matching of images coming from the two eyes. Such depth-tuned neurons were found widely distributed throughout the thickness of striate cortex, but most densely in layers V and VI which project subcortically to superior colliculus and LGN. Neurons sensitive to larger degrees of binocular disparity, in front of the plane of fixation ("near neurons") or behind it ("far neurons") were more numerous in layers II and III, known to project to nearby cortical fields (areas 18 and 19) where *stereoscopic percep-* *tion* (as contrasted with sensory processing) appears to take place.

COLUMNAR ORGANIZATION OF STRIATE CORTEX

Mountcastle, who first characterized the physiological significance of columnar organization of cortex in his classical analysis of somatosensory cortical functions, wrote later that "the basic unit of operation in the neocortex is a vertically arrayed group of cells heavily interconnected along that vertical axis, sparsely so horizontally" (Edelman and Mountcastle, 1978, p. 16). The human cortex

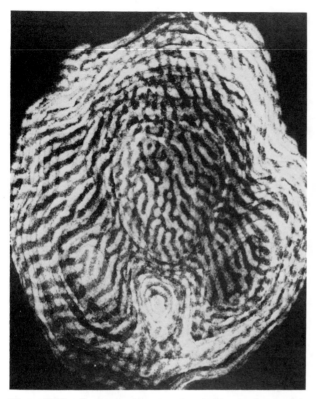

Figure 9.90. Ocular dominance pattern in about a fourth of the visual cortex of a macque monkey, a montage produced by LeVay from tissue slices. One eye was injected with tritium-labeled proline. The radioactive amino acid was transported from eye to lateral geniculate nucleus and thence to visual cortex. Bright stripes due to exposure of photographic emulsion reveal cortical zones that represent the contralateral eye. The dark stripes represent the uninjected eye. Each stripe is about 350 μm in width. [From Constantine-Paton and Law (1982).]

is composed of an estimated 50 billion neurons, with a surface area of about 4,000 cm². The number of columns is estimated at 600 million, each with a diameter of about 30 μm. These columns belong to larger processing units, estimated at 600,000 in number, which vary from 500 to 1,000 μm in diameter in different cortical areas. The cortical column is an input-output device which processes and distributes information. It is modular in dimensions, although particularly complicated analysis apparently requires expanded modular organization. Not only are the horizontal connections of a column sparse, but an individual column is ordinarily dynamically isolated by lateral, pericolumnar, inhibition.

Columnar arrangement enables two-dimensional mapping of several variables simultaneously within an extended region of cortex. Within a column, divergent connections to different output cells belonging to that column enable selective information processing, "feature extraction," among certain of that column's input signals and route the extracted

information to particular output destinations. Specific connections on the part of distinctive subsets of columns within a general sensory or motor region contribute to other sorts of explicit information processing in areas of subsequent cortical and subcortical projection.

Columnar organization for feature extraction has been established for *distance* (Hubel and Wiesel, 1970), *direction* (Blakemore and Pettigrew, 1970), *movement* (Zeki, 1974), and *color* (Zeki, 1973, 1977). Figure 9.91 illustrates columnar organization of color-specific neurons in superficial layers of striate cortex, excited by specific wavelengths at the center of their receptive fields and inhibited by opponent colors in the surround (Livingstone and Hubel, 1983).

OTHER VISUAL PROJECTIONS TO STRIATE CORTEX

We have already noted that there is a retinotopic longer circuiting visual pathway that goes by way of the *superior colliculus* to the LGN and thereby rejoins the classical visual pathway. This roundabout pathway contributes to the classical visual pathway by having gained information from the superior colliculus which is a main focal point for gathering information from other sensory and motor sources. Another projection via the superior colliculus is by way of the lateral posterior pulvinar and thence to striate and extrastriate visual cortex. The pulvinar also receives reciprocal feedback projections from both striate and extrastriate visual cortex.

There is a less abundant projection from retina to the *pretectal area*, which relays on the lateral pulvinar. This relates, reciprocally, only to extrastriate visual cortex.

Visual Mechanisms Relating to Arousal

A third system of projections important to forebrain visual processing concerns the *midbrain reticular formation*. The reticular formation is readily activated by changes in visual input, as by a flash of light or sudden darkness, or by any other significant change in visual pattern. Visual arousal is predominantly controlled by retinal projections to the superior colliculus and projections from there to the midbrain reticular formation. *The reticular formation, in turn, activates diffusely projecting nuclei in the thalamus:* the centrum medianum, intralaminar nuclei, and reticularis nucleus, which relay cephalically to arouse the entire forebrain, including, of course, striate and extrastriate visual cortex. Effective forebrain arousal can also be achieved by marshalling visual attention which is based on cog-

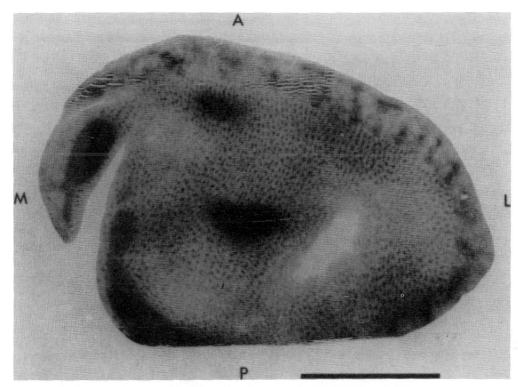

Figure 9.91.   Color contrast cluster of cells in layers 2 and 3 of monkey visual cortex. The stippled appearance of this tangential slice of occipital cortex derives from cytochrome oxidase stain. These "blobs" are found to be excited by certain wavelengths of light in the receptive field center and turned off by opponent color light in the receptive field surround. They are not orientation-specific but project to regions which integrate colors and object detection. [From Livingstone and Hubel (1982).]

nitive processing of visual signals (Galambos and Hillyard, 1981). It is understood that cognitive arousal depends on *extrastriate* visual processing and activation of extrastriate corticifugal projections to the midbrain reticular formation.

EXTRASTRIATE CORTICAL AREAS CONTRIBUTING TO VISION

A partial map of striate and extrastriate cortical areas concerned with vision is presented in Figure 9.92. The owl monkey has relatively fewer convolutions and hence is easier to expose and to map; yet the primate features are probably fairly comparable to those we might expect in human cortex. Certainly, judging by other comparative neuroscientific evidence, we can expect greater rather than less diversity and complexity of visual projection systems in the human brain.

The figure is based upon anatomical and electrophysiological evidence for visual projection systems, based on studies by Allman and Kaas (1972). Twelve separate areas are depicted, each of which has a distinctive magnification, orientation, and specialization of connections and functions. Nine of these areas constitute maps in which the topological order is obviously preserved. Yet it is easy to see that, e.g., the "horizon" in the visual field, seen to the *lower left* and designated by *solid squares*, appears in the brain maps to be repeated many times in a variety of startling different projections—as a horizontal curve, as a vertical line, as two opposing curves, etc. The same can be said for the representation of other contours. In other words, although topology is preserved in these nine fields, they are projected in a variety of configurations and orientations, sometimes as mirror-images abutting one another.

The organization rules presumably derive from very practical constraints: the need to obtain a variety of analytic representations that can extract visual information which has proven—during the course of evolution—useful to extract; to conserve space; to take the shortest path for projection from subcortical to cortical fields and the shortest path for projection from one cortical field to another; to take advantage of successive stages of information extractions by making additional combinatorial and integrative steps for further advantageous extraction; to contribute to and compare information from a variety of subcortical and cortical representations, including some—but, in fact, not all—counterpart representations in the contralateral hemisphere.

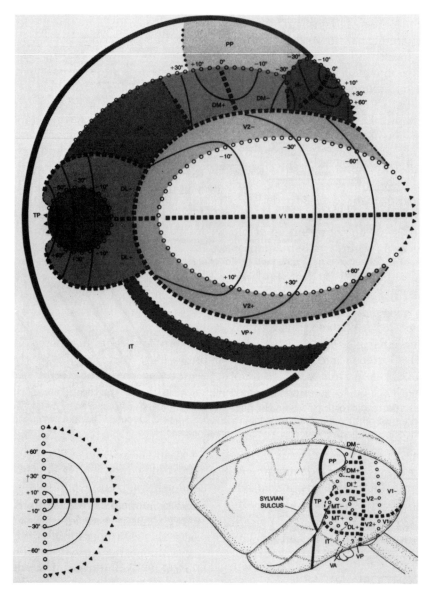

Figure 9.92.   Visual cortex of the owl monkey exemplifies the tendency of the cerebral cortex to be "mapped into areas topologically related to their function. Nine areas are depicted, based on studies of Allman and Kaas, which are orderly maps of the monkey's visual field (and those areas that respond to stimuli from the visual field but do not seem to represent it in an orderly way). The top map depicts schematically the posterior third of left hemisphere cortex unfolded and oriented so that it can be viewed from above. The nine orderly areas are: first visual (*V1*), second visual (*V2*), dorsolateral crescent (*DL*), the middle temporal (*MT*), dorsointermediate (*DI*), dorsomedial (*DM*), medial (*M*), ventral posterior (*VP*), and ventral anterior (*VA*). The three apparently unorganized visual areas are the posterior parietal (*PP*), temporoparietal (*TP*), and inferotemporal (*IT*). The chart at the *bottom left* shows the right half of the visual field. *Solid squares* mark the horizontal meridian, *open circles* the vertical meridian, and *solid triangles* the outer boundaries of the visual field. These symbols are superimposed on corresponding parts of the sketch of the left hemisphere and on the schematic cortical map. Plus signs indicate the upper part of the visual field, minus signs the lower part [From Crick (1979).]

For example, as one area is reprojected onto another, the point-to-point relations are efficiently preserved (and perhaps developmental strategies also simplified) by making them anatomical mirror images. Examination of the physiological evidence indicates that each map has a somewhat different "trick" for analytic purposes which is characterized by its distinctive input-output connections.

Figure 9.93 shows two such projection areas re-

lating to vision, in the monkey. This investigative approach seeks to establish structure-function relations by testing for behavioral capacities, before and after bilateral surgical removal of a particular region of cortex. Two extrastriate cortical regions relating to vision are depicted in this figure (Weiskrantz, 1974). On each side of the figure are indicated the approximate dimensions of maximal and shared areas of cortical lesions as reconstructed

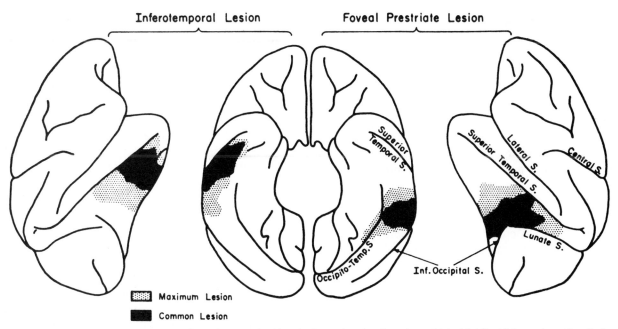

Figure 9.93. Lateral and ventral view of monkey cerebral hemisphere showing foveal prestriate (*right*) which receives directly from striate visual areas and relays to inferotemporal (*left*). The prestriate area contributes to retinal disparity detection and interhemispheric integration. Lesions here interfere with small-scale brightness discrimination, with object and pattern discrimination, learning and relearning, and with primitive visual spatial ability. The inferotemporal cortex is directly influenced by interhemisphere cortical connections via corpus callosum. Deficits following lesions here relate to object discrimination learning, color discrimination, and memory for learned visual discrimination as well as faulty selection and maintenance of attention to visual cues. This region also apparently has to do with categorization of visual memories and their management. [From Weiskrantz (1974).]

histologically from several animals. On the right side are shown lesions relating to an area in which the fovea is directly remapped from striate cortex, the *foveal prestriate area*. This region also receives strong interhemispheric connections from visual areas. On the left side are shown lesions in the *inferotemporal region*, an area that is less directly connected with primary visual projections, and is also strongly influenced by interhemispheric connections.

Animals with lesions in the foveal prestriate area showed postoperative loss of their ability to discriminate patterns and objects, to learn and relearn visual tasks, to discriminate brightness of small objects, and to carry out simple visual spatial tasks. This region is known to contribute to retinal disparity detection which may account for some of the behavioral deficits. Monkeys with lesions in the inferotemporal region showed faulty object and color discrimination learning, faulty memory for previously learned visual discriminations, and faulty selection and maintenance of attention to visual cues.

Fuster and Jervey (1982), recording from single cells in the inferotemporal region in monkeys during performance of tasks requiring attention to color and memory for color cues, found neurons involved in both attentive and mnemonic visual

processes. This and other evidence suggests that inferotemporal cortex contributes to the categorization of visual memories and their management.

Visual Input to Visuomotor Mechanisms in Monkey Parietal Cortex

Neurons in the parietal cortex in waking monkeys are activated by visual stimuli, perhaps via retinocollicular projections. This afferent input is thought to provide the visual cues for activating the visuomotor mechanisms of the parietal lobe for the direction of visual attention (Yin and Mountcastle, 1977, 1978).

The responses of light-sensitive neurons in the parietal cortex are strongly influenced by the position of the eyes (Andersen and Mountcastle, 1983). In waking trained monkeys, there is a controlling effect of the angle of gaze upon visually evoked responses of parietal neurons that is dependently related to the state of *directed visual attention*. More than 60% of parietal neurons show this property when the animals are in the behavioral state of attentive fixation. The effect is powerful: the activity of average light-sensitive neurons influenced by the angle of gaze was increased more than three times by a shift of gaze of 20% in the direction optimal for this effect.

FRONTAL EYE FIELDS

In addition to more than a score of cortical areas which relate to sensory processing for visual perception, there is a region in the frontal lobes, corresponding to area 8 of Brodmann, from which pupillary dilatation and conjugate eye movements can be elicited. Ocular movements, mainly contralateral conjugate deviations of the eyes, have been elicited by electrical stimulation in this area in monkeys, apes, and humans by numerous investigators. Eye movements may be accompanied by turning of the head in the same direction.

There are no direct cortical projections to the eye motor nuclei, so cortical influences on eye movements are always indirect. The frontal eye fields are concerned with voluntary eye movements. Single neurons in the frontal eye fields of unanesthetized monkeys have been found to fire during voluntary eye movements (Bizzi and Schiller, 1970; Bizzi, 1974). It is likely that voluntary eye movements are initiated elsewhere and that the motivating and initiating impulses, from wherever they may be generated, fulfill their program by activating eye motor control neurons in area 8. Moreover, like other voluntary actions, the direction of gaze may be interfered with by reflex mechanisms, in this case relating to head and neck reflexes, the labyrinth, and cerebellum. Figure 9.94 illustrates some of the circuits relating to reflex and voluntary control of eye movements.

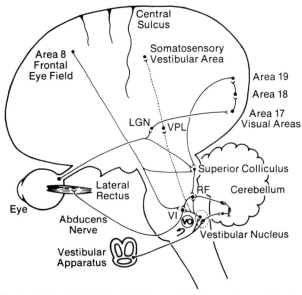

Figure 9.94. Circuits involved in control of eye movements. For simplicity, only control of the lateral rectus muscle is illustrated. Similar connections are involved in control of other extraocular muscles. Abducens nucleus (*VI*), reticular formation (*RF*), lateral geniculate nucleus (*LGN*), thalamic ventroposterior lateral nucleus (*VPL*), vestibuloocular reflex (*VO*). [From Shepherd (1983), adapted from many sources.]

The frontal eye fields, among other sensory and motor fields relating to vision, sends retinotopically organized fibers to the superior colliculus and pretectal area (Kuenzle and Akert, 1977). Several nonvisual cortical areas that also project to the superior colliculus include primary sensorimotor cortex, temporal and parietal cortex, and parts of auditory cortex. The corticifugal projections from area 8 end chiefly in layer IV of the superior colliculus, beneath axonal terminals from retina and striate cortex.

A destructive lesion in area 8 is followed by paralysis of conjugate gaze toward the opposite side. With respect to lateral movements, each hemisphere appears to tonically facilitate gaze toward the contralateral side. Therefore, when one frontal eye field is destroyed the normal balance is upset, and the eyes at rest may be conjugately deviated toward the side of the lesion. Paralytic deviation of the eyes occurs frequently with vascular lesions of the internal capsule, because of interference with projections between the frontal eye fields and the brain stem. When asked to look to the side opposite such a lesion, the patient is unable to do so. If, however, the patient is asked to look at an object that is slowly moved across the visual field from the "good" to the paralyzed side, the eyes can pursue a normal range of movements. If the patient is asked to fixate on a point directly ahead, and the head is then turned slowly toward the side of the lesion, the gaze can be maintained, even though this involves directing the eyes toward the paralytic side. Therefore, *it is voluntary direction rather than reflex following that is lost.*

Such paralysis is not ordinarily permanent but recovers after a few weeks. If the cause is gradual (as with slow growth of a tumor), paralysis may not occur. There is evidence for compensation by the frontal eye field in the contralateral hemisphere.

Eye movements can also be elicited by stimulation of striate cortex, and the resulting movements direct and hold binocular gaze to the region of the visual field which is represented in that part of the visual cortical "map." Occipitally controlled eye movements exhibit the characteristics of a *cortical reflex mechanism* rather than resembling voluntary direction of gaze. Whereas *searching for visual targets depends on the frontal eye fields, following visual targets depends on integrity of striate cortex.*

Eye movements elicited by stimulation of striate cortex resemble smooth pursuit movements, conjugate gaze directed toward the location in the visual field that corresponds to the locus of cortical stimulation. Since the hemisphere stimulated relates to the contralateral visual field, the eyes are directed contralaterally but not so abruptly and not

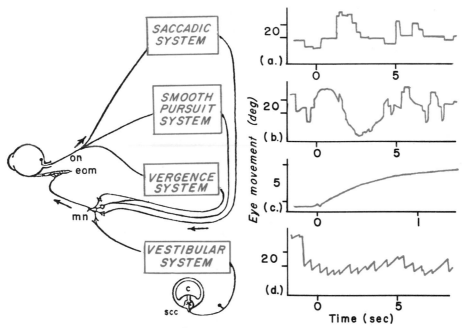

Figure 9.95. The primate eye movement control system. The types of eye movements controlled by the saccadic system are illustrated at *a*.; the smooth pursuit system, at *b*.; the vergence system, at *c*.; and the vestibular system, at *d*. Each system converges on the motor nuclei (*mn*) which innervate the extraocular muscles (*eom*) and move the eyeball. *on*, optic nerve; *c*, cupula;, *scc*, semicircular canal. [From Robinson (1968).]

so predominantly in the horizontal plane as with stimulation of the frontal eye fields.

Figure 9.95 schematizes four different types of eye movements that are of particular clinical importance. The saccadic system involves abrupt, conjugate, darting of the eyes in a movement that resembles directed search for a visual target. Saccadic eye movements can be observed by watching someone reading. It is an advantage to give them something with a full page line so that the eyes will come to rest twice or three times in the course of reading from left to right. Then they will dart back to the start of the next line, and may need to make small adjusting saccades before locking into the reading, making short saccades along the next line of text. The speed of motion is swift, and visual perception is momentarily occluded during the transit.

Smooth pursuit movements occur in the course of following a moving target of interest through the visual field. The conjugate movements are slower but may speed up if the target advances swiftly. Motor control of the axes of vision is capable of such rapid movements that no object more than a few meters distant from the observer can possibly go faster than the axes can follow, hence pursuit is relatively slow compared with saccades.

The vergence system is very much slower compared with saccades or pursuit movements, the eyes conjugately converge or diverge as the depth of fixation is shifted from far to near and near to far

targets. The vestibular system (considered in Chapter 63) affects mainly movements that have a slow phase followed by a rapid recovery, *nystagmus*.

Saccades are probably largely governed by the frontal eye fields. Pursuit and vergence movements are controlled by the occipital cortex, probably largely by neurons in area 17. These three systems exert their control mainly through the superior colliculus. The vestibular system has its own direct and indirect access to the eye motor nuclei and also utilizes longer circuiting through the superior colliculus. All four systems ultimately converge on motor neurons in the IIIrd, IVth, and VIth cranial nerve nuclei.

TOPOLOGICAL DISTRIBUTIONS AND PERCEPTUAL COMPLETENESS

The preservation of functional integrity in neurophysiological "maps" of sensory and motor functions does not require that representations be equal in area, but does require that certain topological relations be preserved. This permits differential magnification of central retinal representations while maintaining point to point relations. Thus, the foveal region of the retina has a primary cortical representation which is 35 times expanded as compared with that of the extreme periphery of the retina. Topology permits a succession of retinal, geniculate, and cortical representations to undergo elastic distortions without gaps, shearing displacements, or intrusions.

But wait! The optic chiasm has placed a tremendous cleft through an otherwise continuous map of the visual world. The partial crossing of optic nerve fibers has imposed a division of information between two halves of the visual field in both brain stem and cortical projections. This provides a prime example of a neurological puzzle: How is it that *perception can present the visual field as a whole, whereas only half of the visual field is represented in each hemisphere*? There is no perceptual interruption in the midline. The two representational halves are stitched together without a seam so completely and neatly that perception across the midline is spatially unambiguous, and behavior across the visual field is pursued with sublime confidence.

Processes of *projection* encapsulate this mystery. This interhemispheric visual division probably provides the best example with which to explore the very important issue of *how do we perceive our bodies and the surrounding environment as entirely unfragmented entities, even though the central nervous system representations of these entities are obviously distributed*?

Although the scale of physical separation is larger between the two hemispheres, the problem is exactly the same as that which attends the question of how a multitude of different sensory projections and different sorts of analytic operations in the course of a single hemisphere become integrated in perception. First, by the time sensory signalling acquires even the humblest hint of awareness, it is already "projected" onto a world outside our nervous system, onto our body, and onto a changeable surrounding environment. *Nothing in the way of sensory processing is appropriate for behavior that is not already projected outward.* Even monosynaptic reflexes manifest outward projection. It is clear that *sensory projection is implicit in every aspect of behavior, from the humblest reflex to the most deliberately elaborate performance.*

The general principle is well illustrated by the present example. How the hemispheric split in the visual pathway is composed into a perceptually intact scene is understandable in an evolutionary context. Projection of an intact world is forced upon organisms by requirements for behavior. Perceptual guidance of behavior is obliged to maintain intact all varieties of distributed sensory representations of body image and images of the surrounding world, however displaced through cortical and subcortical areas, no matter how widely separated and seemingly untidily scattered they may seem to us to be.

Distributed sensory and motor representations in higher nervous systems evolved only after integrated projection systems were already staunchly, comprehensively built into functioning systems. The evolutionary pressure to foster distributed functions, to obtain more powerful analysis of sensory and motor functions, could be fulfilled only on foundations of fully integrated systems. The necessity for maintaining sensory processing sufficient for integrity was persistently hammered out on the hard anvil of evolutionary experience. The requirement was demanding *integration*, always obviously necessary for coherent behavior, always operatively compelling for perception.

Distributed Representation of Central Sensory and Motor Mechanisms

An obvious riposte is that "there *must* be some location where the dispersed information is all brought together." But that is not the case. Self-organizing systems can operate as a whole and at the same time have widely distributed rather than localized mechanisms which subserve the processes of overall control (Prigogine, 1977; Jantsch, 1980).

Nevertheless, there are some liaison structures that probably contribute to operation of the system as a whole: There is a narrow band of overlapping duplication of representation on both hemispheres at the median vertical line which separates the two halves of the visual field. Retinal ganglion cells along that line distribute axons to both LGNs which relay the duplication to both hemispheres. There is cross-correlation between these margins of duplicate representation between the striate cortex in each hemisphere via a band of axons in the posterior part of the corpus callosum. Additionally, there are potentially contributory junctions between the two halves of the visual field in the superior colliculus, and also through multiple projections to the brain stem reticular formation.

It is useful to consider how important a role the reticular formation plays in integrative convergence of information from multiple sources controlling motor patterns. The reticular formation, in brain stem and spinal cord, operates to resolve this motor convergence problem as it impinges on efferent neurons. It may be that the same system plays a role during the first stages of sensory processing, when the reticular formation has initial access to data from receptor surfaces, and at later stages of integrative operations on sensory information, during perceptual implementation, when the reticular formation plays a central role in mechanisms of attention.

The important point to understand unambiguously is that there does not need to be, and apparently there is not, any comprehensive, whole-scale representation of any single sensory modality or any single comprehensive motor act, or any single multisensorimotor domain where representation of the body as an entirety and the environment as an entirety can be brought into juxtaposition with one another. That is an inevitably *distributed process.* The difficulty of distributed command and control can be appreciated when the problem of governance of head-eye-hand coordination is addressed, and when, on reflection, it is realized how multifaceted and distributed representation of the environment as a whole must be. In Chapter 61, Sensory Processes, we dealt briefly with the problem of space-time coordination within sensory and motor pathways in the nervous system; the same interpretations are directly pertinent and helpful here.

Nervous system fulfillment of behavioral requirements has always taken place in an evolutionary context, the same context that shaped the sensory and motor mechanisms that have become increasingly widely distributed through central nervous system representation and rerepresentation. Behavior cannot occur even at the lowest level of organizational simplicity without involving such coherent projection. Sensory projection and behavior emerged simultaneously during evolution and are causally linked to one another. That is, projection is essential for successful behavior, and behavior continually tests and maintains the adequacy of projection. Plasticity of the nervous system ensures that corrections and adjustments are made according to changes in status of the body as well as changes in status of the environment. *It does not matter what the geometries of central representation are, both sensation and behavior maintain an obligatory outward projection of a dynamic body as a whole into a dynamic environment as a whole.*

What is especially intriguing about the large number of extrastriate visual cortical areas that have recently begun to be defined anatomically and physiologically is that they appear to belong to "family clusters" which are more intimately intracortically connected among one another and with their special thalamic connections than with striate visual cortex and the classical visual "association" areas (Crick, 1979; Graybiel and Berson, 1981). These extrastriate areas also appear to have more to do with conscious visual experience than does striate cortex. In short, instead of there being a dominating principal visual pathway, from retina via LGN to striate cortex, from which cortex corticocortical projections distribute to "visual association cortex," there are three or more relatively separate but parallel projection systems transmitting between retina and cortex: retinothalamic (LGN relay), retinotectothalamic (superior colliculus-LGN relay), and retinopretectothalamic (pulvinar relay) pathways to cortex. Added to this is the still more roundabout transmission of visual signals relating to arousal that pass by way of the midbrain reticular formation.

Some of the corresponding cortical fields seem to be relatively remote from influence by striate cortex. They also seem to participate in visual processing more by way of their subcortical input-outputs than by what has been classically interpreted as visual corticocortical association mechanisms. From a clinical point of view this might explain how some visual capabilities may be preserved even though significant parts of the classical visual pathway have been damaged. It also may explain why early experiments concerning visual memory (Lashley, 1950) were interpreted as indicating that behavioral deficits were not so much due to the removal of localized areas of cortex as they were due to the overall amount of cortex removed. This earlier interpretation gave rise to notions of mass action of cortex, whereas it now is increasingly evident that there are multiple, distributed, but nonetheless explicitly connected, reentry pathways that can better account for such evidence (Edelman and Mountcastle, 1978).

Visual Representation in the Claustrum

A thin shell of subcortical gray matter lying beneath the insular cortex of the hemisphere is the *claustrum.* The posterior part of the claustrum has neurons that are primarily concerned with vision (LeVay and Sherk, 1981), neurons which receive from and send to striate cortex. Retinotopy is preserved in each direction. The claustrum appears to have no subcortical projections and only a small contralateral projection. Claustrocortical axons terminate in all cortical layers, but most heavily in layers IV and VI, whereby the claustrum can influence geniculocortical input (IV) and corticogeniculate output (VI) for all the striate cortex. With these important connections, the claustrum can filter or impose a gate at these two important, functionally related, junctures.

EFFECTS OF EARLY VISUAL PATTERN DEPRIVATION AND RESTRICTION

Postnatal development of mammalian visual pathways is greatly affected by early visual experiences. Deprivation of patterned vision during early life, even with provisions for adequate diffused light to reach the retina, leads to alterations in neural connections and to severe and enduring reductions in visual capabilities. There is a window of time (until approximately 3 mo of age) during which such effects are easy to induce in kittens and monkeys, and beyond which the effects are difficult, if not impossible, to reverse. Such deprivations, in the adult animal, do not have appreciable deleterious effects. *The effects of deprivation and restriction early in life on the structure and function of visual pathways is a general phenomenon which has been demonstrated in a variety of species, including humans* (for reviews, see Movshon and Van Sluyters, 1981; Sherman and Spear, 1982).

Monocular deprivation induces anatomical and physiological changes in the lateral geniculate nucleus, as described earlier. Retinal cells seem not to be affected, but LGN X and Y cells are affected differently and presumably have different requirements for postnatal visual stimulation. Mainly, there is strong competition between the two eyes for control of striate neurons.

Within the geniculocortical pathway there is little or no evidence for functional effects of deprivation on W cell activity. Geniculate X cells in deprived laminae show some loss of spatial resolution, but mainly these cells seem unable to drive many striate neurons, so X cell deficits are more obvious at the cortical level. Y cells are significantly affected, and this is obvious already at the geniculate level. During monocular deprivation, apparently, binocular competition and some noncompetitive effects can be observed. Competition dominates development of the geniculate Y cells, but a noncompetitive mechanism interferes with X cell development.

Wiesel and Hubel (1974) demonstrated that in very young, visually inexperienced monkeys, the map of the visual field displays the same receptive field types, selectivity for stimulus orientation, and orderly arrangement of orientation columns that are encountered in the striate cortex of adult animals. *But ocular dominance columns in the neonate are only partially* developed, a process that is essentially completed by 3 to 6 wk of postnatal age (LeVay et al., 1981).

If the eyelids in one eye are sutured closed at birth, the ocular dominance columns in the corresponding regions of striate cortex are of unequal width: Those for the eye that has had visual experience are about three times broader than those for the eye that has been closed. Later monocular eye closure, and reverse suturing of the eyelids, demonstrate that afferents that have already partly segregated in the ocular dominance columns can reexpand. The authors believe that the effects of monocular deprivation should be viewed as a distortion of normal growth processes and not as evidence for an instructional role of visual experience in cortical development.

When one eye is pattern-deprived at birth, segregation of geniculate terminals in layer IVC proceeds, but afferents from the deprived eye undergo process retraction, leaving the seeing eye's afferents in possession of abnormally wide IVC layer bands. The process seems to involve a redistribution of axon terminals rather than bulk retrenchment of entire axons.

Reversed suturing of eyelids, if performed early in life, i.e., after the effects of initial suturing have taken their toll, yet before early plasticity has begun to wane, permits the initially sutured eye to reexpand its occupancy of layer IVC and, indeed, to go on to become the physiologically dominant eye. Ocular dominance patterns in a normally developing monkey can be influenced by eye closure at 1 year of age, but not in the adult animal.

In cats it has been shown that functional deficits in lid-sutured and dark-reared kittens do not apparently affect the retina but do affect the lateral geniculate and striate cortical levels (reviewed by Sherman and Spear, 1982). The usually narrow time window can be moved forward by *prolonged rearing in the dark*, for a year or longer, after which monocular deprivation can still be differentiated after one eye is opened for as short a time as 3 days (Cynader and Mitchell, 1980).

In the course of binocular suture experiments, noncompetitive mechanisms appear to be quite severe, and more deleterious than are the consequences of monocular suturing. There is evidence that visual deprivation can induce some synaptic circuitry *to degenerate*, some *to slow or arrest its development*, and some *to develop but to become extraordinarily vulnerable to competition.*

There is evidence that beside the effects of competition and withdrawal, there may be important compensatory improvements in complementary circuits. It has been known for many years that removal of cortex in young animals is less damaging to adult behavior than is damage to similar cortical fields in the adult. This is taken to be evidence for conspicuous plasticity in early life, and evidence that this plasticity is functionally designed to compensate functionally for the deficits.

For example, if newborn kittens receive damage to striate and adjacent visual cortex (removal of the equivalent of areas 17, 18, and 19), the resultant behavioral deficits are far less severe than when similar lesions are made in adult cats. If, however, the neonatal lesions include the suprasylvian gyri in addition to striate and peristriate cortex, the cats are much more impaired in performing form and pattern discriminations. The evidence indicates that after neonatal removal of the primary visual areas, striate and peristriate, some functional compensation occurs, involving both the visual pathways and the lateral suprasylvian visual area (Spear et al., 1980).

EFFECTS OF RESTRICTED VISUAL EXPERIENCES

Depending on whether or not patterned vision is part of an animal's experience in early life, the specificity of response patterns in cortical neurons varies greatly. With normally patterned experience, specificity develops normally. As we have seen, when animals are deprived of patterned vision (reared in the dark or with sutured eyelids), the specificity that is present at birth begins to deteriorate. Even a relatively brief exposure (6 h) to patterned information after 6 wk of dark rearing largely reverses this loss of specificity.

Still more important is the fact that *experience with distorted visual patterns early in life leads to distortion of the specificities of visual neuronal responses.* Held (1967) found that young monkeys whose limb movements were restricted from their own view made inappropriate visually guided gestures when opportunity was given for such sensorimotor performance. He also found that kittens

reared in the dark required visual experiences with gestures of their own paw toward objects in view in order to be competent to execute such visually guided gestures correctly at a later time.

According to Blakemore (1974), there is a definite time window in early life when adaptation to visual experience is critical with respect to orientation preferences of striate cortical neurons (Fig. 9.96). He raised kittens in the dark, and exposed them either to horizontal lines or to vertical lines for a few hours each day. Such kittens developed striate neurons that responded preferentially to the orientation of the line segments to which they had been exposed. The proportion of such cellular commitment depended in part on the neonatal period of exposure and in part on the duration of exposure (Fig. 9.96). He interpreted this to mean that the *probability* of exposure to specific line segments determines the orientation preference of a given striate neuron.

When Blakemore exposed individual kittens to alternating experiences with horizontal *and* vertical line segments, their cortical cells displayed preference for either horizontal or vertical stimuli and were strikingly unresponsive to stimuli having any alternative orientation (Fig. 9.97). Evidently, each neuron develops preference for a stimulus feature to which it has most often been exposed. Such

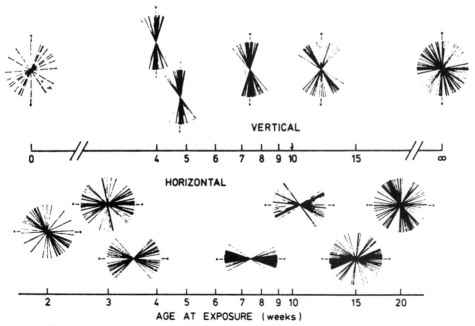

Figure 9.96. Formation of feature extracting neurons in kittens. The critical period of environmental modification of the visual cortex is illustrated. Kittens were kept in the dark and exposed to vertical (*upper half*) or horizontal (*lower half*) for a few hours each day, at various ages. On the upper abscissa the results that appear above zero age refer to a binocularly deprived kitten where the orientation preferences (*parallel vertical lines*) were very vague, and those at infinity on the abscissa are for a normal adult animal kept in vertical stripes over a 4-mo period. The scale is logarithmic for display convenience and has no theoretical implications. [From Blakesmore (1974).]

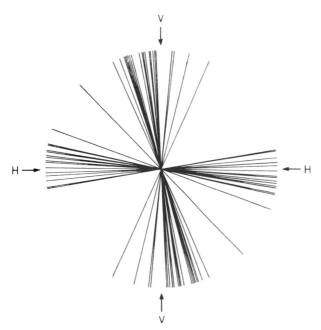

Figure 9.97. Entrainment of cortical neurons by early life experience. Illustration of orientation preferences for 37 cortical cells from a kitten exposed alternatively for 2 h to horizontal, 2 h to vertical, until it had experienced more than 50 h of each. [From Blakemore (1974).]

plasticity has real adaptive value: it ensures that an animal develops a visual system that is optimally matched to the visual world in which it finds itself.

Altogether, this evidence indicates that early life experiences within any given environment shapes the brain in accordance with that particular environment. *Depending on the degree of singularity of individual experiences, brains must be uniquely distinguished from an early age.* Thereafter, behavior continues to test the reliability of visual representations against ongoing experiences. Feedback from such experiences continues to contribute to further plastic modeling of visual representations. Visual circuits gradually become more committed and "hard-wired" in this way, as long as the environment remains relatively stable.

Plastic adaptations are powerful and swift during early life, especially during a time period of special biological adaptability such as we have been discussing. As an animal grows older, the capacity and the rate of such adaptations may slow, but considerable plasticity endures lifelong and is rapidly reinvoked whenever there is brain injury or some occasion for radically different environmental experiences (e.g., when a person starts wearing distorting lenses or prisms).

BASIC VISUAL FUNCTIONS

Vision is an information-processing task that is organized to provide *behaviorally valuable represen-*

tations of the visible universe. Behavior utilizes visual representations in the process of seeking biological satisfactions. The consequences of practical success and failure of behavior forces its weighted influence throughout evolution and during individual development.

It is the consequences of behavior guided by the *image* that are vitally important to survival of the individual, the society, and the species. We bet the future on subjective images, whether we are steering a car, selecting a mate, or voting. The public-held images of man and society and the sociology of knowledge lie at the fountainhead of history, economics, and politics, and the pursuit of politics by any and all means (Boulding, 1956).

But what is the *image* made of, strictly speaking; in this case the visual image? There are certain fundamental capabilities that we share with a long ancestral line: the ability to discriminate differences in *brightness*; to distinguish light based on its *wavelength*; to measure *movement*; and to resolve *spatial depth*.

BRIGHTNESS DISCRIMINATION

Discrimination of brightness involves fine spatial and temporal interpretations. In foveal acuity, we can correctly guage vernier lines down to a gap that is less than the distance between centers of the narrowest cones. We can integrate information over short line segments of two limbs of a vernier, even when the lines are curved rather than straight, even though they are moving or the subject's eyes are moving, even though three dots are substituted for the vernier, and the dots are not presented simultaneously (up to 20 ms temporal dispersion), and even though the duration of exposure is limited to 1.5 ms (Crick et al., 1981).

We are somewhat less adept at measuring temporal differences in brightness. Because of local inhibitory surround, we can interpret rapid changes in brightness at any given point in the retina and throughout the overall pattern of stimulation of the retina, without difficulty. The interval of integration is matched nicely with the rates of eye movement and the average duration of visual fixation on a given point.

The eye is remarkable in its ability to adapt appropriately so as to obtain good vision over an astonishing range of brightness intensities, over a billion-fold. The adaptation is rather more rapid for changes in flux from low to high intensities than for changes from high to low intensities. Two or three minutes are adequate for adaptation to increasing brightness, whereas a half hour is needed for dark adaptation.

COLOR VISION

Perception of distinctive colors is a function that is concentrated increasingly toward the center of the retina and involves activation of cones in relatively bright light. Color coding is not emphasized in striate cortex, although color contrast cells have been found to cluster in layers 2 and 3 throughout that region (Fig. 9.91). Zeki (1973) has found an additional area, in the anterior bank of the lunate sulcus dorsally and continuing onto the posterior bank of the inferior occipital sulcus in the monkey in which color is emphasized. All cells responding to retinal stimulation were color-coded, responded best to one wavelength, and weakly or not at all to other wavelengths or to white light. No cells were encountered in this area which responded equally well to all wavelengths or to white light of different intensities. There was a range of requirements for shape and position, some preferring particular shapes, whereas others were activated by color regardless of shape. What was common was the tendency for cells that preferred a particular wavelength to be grouped together. It was implied that such color-coded neurons are organized into columns in this area of cortex, as they are in striate cortex (Fig. 9.91).

MOTION DETECTION

Motion detection is important for species evolution and individual development because animal movement is critical to survival, and motion pervades the visual world (Allman, 1977). Evolution of vision has required conjoint analysis of movement from the point of view of sense organs in the body wall, including the vestibular apparatus as well as the retina. The latter is affected both by movement of the body and the eyes in relation to the body. Information about movement contributes to establishing the shape and structure of an object, and that information helps in predicting its future trajectory. In analysis of motion, time is of the essence, more so than in any other aspect of vision. Old images of the state of a moving body (the self or another) are rapidly outdated.

Center-surround systems are ideal for assembling mechanisms for measurement of motion (Marr, 1982). Analysis of motion and analysis of contours are probably combined. W cells probably provide an important pathway and Y cells contribute information about isolated edges, bars, and slits moving at moderate velocities. More complicated discriminations relating to object motion can be built up from these. Zeki has found cells that respond to changes in image size and disparity in the posterior bank of the superior temporal sulcus in the monkey and which signal information about the important phenomena of motions towards and away from the animal.

DEPTH DISCRIMINATION

The ability to make spatial resolution in depth involves a number of clues of which the most important is the disparity of timing and location of events coming from the two eyes as optically and neuronally projected within the visual system. Although other areas may be involved, primary visual cortex is certainly engaged in this discrimination, and both binocular fixation and lens accommodation for optimal tuning to viewing distances are dependent on striate cortex. Additional clues by which objects are perceived to exist at different depths in space probably involve analysis in other regions as well.

Functions Relating to Brightness

BRIGHTNESS DISCRIMINATION

The data in Figure 9.98 represent the test intensity required for the detection of a light against a background adapting intensity for a human observer and in the retina of the monkey. Data from the retina of the monkey are based on a selected criterion response level of the late receptor potential (Brown, 1968); this is recorded with a tiny electrode inserted into the retina at an appropriate depth to provide an index of the activities of the receptors themselves. It should be recognized, however, that the late receptor potential as recorded represents the activity of many receptors. It is undoubtedly also influenced by interactions which occcur as a result of connections of the horizontal cells.

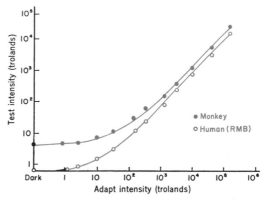

Figure 9.98. Log of threshold retinal illumination (trolands) as a function of the adapting level. The *solid circles* represent a 10 μv criterion response level of the late receptor potential of a monkey. The *open circles* are based on human foveal thresholds measured with the same apparatus. [From Boynton and Whitten (1970).]

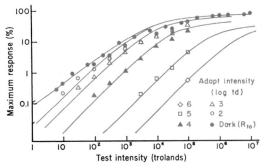

Figure 9.99. Late receptor potential amplitude (percent of maximum) as a function of test flash intensity for each of five adaptation intensities and for dark adaptation. Average values of six monkeys. [From Boynton and Whitten (1970).]

For any given level of adaptation the nature of the response at the retina is remarkably similar to that at other adapting luminances. In Figure 9.99, the response in terms of percentage of the maximum late receptor potential is plotted as a function of the test intensity for each of six conditions of adaptation. The functions, with the exception of that for dark adaptation, are remarkably similar and are simply shifted to higher ranges of test intensity with increase in the level of the adapting intensity (Boynton and Whitten, 1970). The range of response of the system thus remains approximately the same for any condition of adaptation, but as the adapting conditions are elevated, there is an expansion of the range of luminances over which gradations in response may be obtained. The "gain" of the system is thus reduced as higher levels of background luminance are encountered, and this occurs in the early processes which take place within the retina itself. This kind of process is undoubtedly involved in the ability of the eye to function over an intensity range of greater than 10^9:1.

DARK ADAPTATION

There are important changes which occur within the visual system when illumination is reduced from daylight levels to very low levels before it is possible for even the crude low luminance or "scotopic" vision to become optimum. The process of "dark adaptation" requires a half an hour or longer after an abrupt transition from a high illumination surrounding to one of relative darkness (Bartlett, 1965b). In Figure 9.100 the minimum amount of light which can just be detected is represented on the ordinate, and time in the dark is represented on the abscissa. The ordinate axis is plotted on a logarithmic scale, thus greatly amplifying the lower portion of this relationship. For a period of from 8–12 min after a high luminance adapting field has been removed, the threshold value at first reduces

rapidly, and then the rate of reduction in threshold is slowed, reaching a minimum at about 10 min. A little later there is a subsequent additional reduction in threshold which again occurs at a decreasing rate and approaches a minimum at some time after 30 min in the dark.

For various reasons, the higher threshold portion of the curve in Figure 9.100 has been associated with the cones, and the lower threshold portion has been associated with rods. In the first place, when a small test stimulus is employed which is restricted to the fovea, two branches are not obtained. In the second place, if a red test light is used for which the rods are no more sensitive than the cones, the second branch of the curve does not appear, even though the test light may be presented to a region of the retina which is rich in rods. If a subject in a dark adaptation experiment is asked to identify the color of a test flash, he is able to do so during most of the time for the first 8 or 10 min of dark adaptation, corresponding to the initial segment of the curve in Figure 9.100, but is unable to do so unless the test light is restricted to extremely long wavelengths at later times during dark adaptation. Thus, color discrimination is approximately associated with the "cone" portion of the curve, and the inability to discriminate color is associated with the "rod" portion of the curve.

It was believed for many years that the time course of dark adaptation could be understood in terms of the time required for regeneration of the photopigments of the visual receptors which are bleached by any high level adapting light. The time course of dark adaptation certainly is associated with the regeneration of photopigments, but the level of threshold cannot be explained directly in terms of the concentration of photopigment present

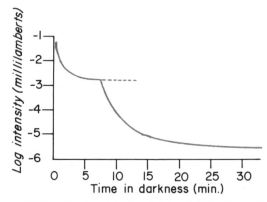

Figure 9.100. The time course of reduction in a light detection threshold during dark adaptation. The initial branch of the curve is associated with photopic vision and is attributed to cone function. The later branch is often called the scotopic branch and is associated with rod function.

or the quantum catching capability of the retinal pigments.

Because red light does not so radically exhaust rod responses as ωo other wave lengths, dark adaptation may be preserved by wearing red goggles so that only red light can enter the eye. Hence, radiologists who must go in and out of a darkened fluoroscopy room can preserve much of their dark adaptation and yet be able to work at intervals in a lighted environment. For the same reason, lights on instruments in the cabin of an aircraft and on the bridge of a ship at night are illuminated with red light.

The bleached photopigments do regenerate, but as Rushton has shown, the logarithm of change in threshold is more nearly proportional to the change in concentration of photosensitive material than is the change in threshold itself. The tremendous range of change during adaptation is most readily explained in terms of changes in retinal organization which permit increased summation of energy over area and time as well as to some mechanism which increases the responsiveness or "gain" of the system at very low levels of illumination.

It should be evident that the precise mechanisms which underlie the adaptability of the visual system to such a broad range of conditions of illumination are as yet unknown. It has been clearly demonstrated, however, that with increasing dark adaptation there is increased spatial and temporal summation at the retinal level. These changes permit more efficient utilization of small quantities of energy.

The increased temporal and spatial summation capabilities permit detection and utilization of low light levels. They do not preclude fine spatial and temporal resolution if sufficient light is presented to the eye when it is in a dark-adapted state. That is, the dark-adapted eye is capable of just as high a level of spatial or temporal acuity as is the light-adapted eye, given sufficient luminance to work with. Of course, if the necessary luminance remains available for any length of time, then the state of adaptation is changed from one of dark adaptation to some level of light adaptation. Nonetheless, fine spatial detail can be discriminated quite well in a very brief flash of sufficient luminance presented to the dark-adapted eye (Brown et al., 1953).

LIGHT ADAPTATION

Just as the eye adjusts when the visual environment is darkened, so does it also adjust when the environment is made brighter. The nature of the response is similar, as illustrated in Figure 9.99, when it is considered on a log plot, but, as mentioned earlier, this implies a reduction in the gain

of the system. A given amount of change in response requires a proportionately greater change in the level of stimulation when the adapting level is raised. We have seen that it may require more than 30 min for the eye to achieve maximum sensitivity in darkness after it has been exposed to a bright adapting light. The time course of adaptation to a lighter environment is usually much more rapid. Under most circumstances, light adaptation is complete in from 3 to 5 min, and the eye adjusts to a relatively steady-state condition of responsiveness within that time.

The time course of light adaptation has been measured in a variety of ways. Early measurements were performed by exposing the eye to an adapting light after it had been completely dark-adapted and then determining its sensitivity by turning off the adapting light and quickly presenting a test light (Wald and Clark, 1937).

The results of such studies indicated that with increased duration of exposure to an adapting light, there is an increase in the threshold for detection of subsequent test light as adapting exposures increase up to 3 to 5 min and sometimes much longer.

The initial presentation of an adapting light obviously has a significant, albeit transient, effect on sensitivity to a test light. This effect is attributed to neural events in the retina. Abrupt elevation of retinal illumination probably results in a level of neural activity considerably in excess of that which corresponds to the state of affairs after the adapting light has been on for some time. Thus, the system cannot handle increment information as efficiently immediately after exposure to an adapting light as it can after the light has been on for a time. The slower rise of threshold for the test light which follows the initial rapid elevation and subsequent decline is attributed to the more gradual change in photochemical balance within the retina.

One aspect of light adaptation is that activation of cones has an inhibitory effect on rods. This suppression of rod activity acts to reduce possible rod interference, with activity taking place along the cone path. In cases of total color blindness, in the absence of cones and cone activity, the inhibitory effect on rods is absent. This may account for the photosensitivity and the experience of dazzle by lights, and the need for people with total color blindness to use dark glasses even in situations of moderate illumination.

SPATIAL RESOLUTION OF THE VISUAL SYSTEM

The light-gathering powers of the normal eye cast a reduced image of an enormous visual field onto the tiny retina where some 120 million receptor cells are accordingly activated. The retinal and central nervous events associated with sensory processing are projected outward so that perceptual experience manifests to us large-scale real objects: our own bodies and a visual world beyond our grasp.

One of the puzzling aspects is the fact that the quality of the visual image for the normal observer is better than one should expect. If all the aberrations of the optical system, the density of receptors, the size of receptive fields, etc. are considered, the clarity of vision is surprisingly good. Perception is smoothed over, as in the case of our obliviousness of the blind spot (see Fig. 9.67), except when such sensory deficits are deliberately looked for.

More importantly, spatial discrimination is sharpened by interactive processes already considered. One effect is edge enhancement, whereby the boundary between a bright field and a dark field is emphasized, and yet neuronal activity on either side, in the bright field and in the dark field, is only slightly biased toward light-dark differences. Activity defining the brightness contrast is confined almost entirely to the boundary. The bright and dark fields are only slightly shifted from the level of expected spontaneous activity.

Edge enhancement thus presents an enormous economy for central nervous representation. In effect, neurons do not have to "paint all surfaces," but only borders and an indication of the direction of displacement of brightness on either side of borders. Thus, signals relating to edges are relatively enhanced, as are signals relating to orientation, spatial frequency, size, color, and movement, all economically achieved in the course of providing retinotopic representation through several successive serial and parallel visual projections (Cowey, 1981). The enhancement effects are all accomplished by lateral inhibition which occurs in the retina and at many successive stations.

Visual Acuity

In the case of vision, the term *acuity* is reserved to refer to spatial resolution capability. Visual acuity is formally defined as the reciprocal of the minimum visual angle in minutes subtended at the eye by an object in space which can be resolved by the visual system. Visual acuity is not an invariant quantity for a given eye or even under a given level of illumination; it depends importantly upon the way in which it is measured.

The simplest form of visual acuity is the so-called *minimum detectable* visual acuity. This is the discrimination of a small dark spot against a lighter background, or a dark line against a contrasting background. Under ideal conditions, the finest line

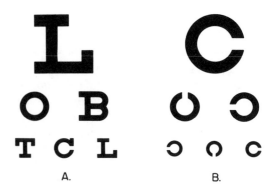

Figure 9.101. *A.*, Representative letters as used in a Snellen chart for the testing of visual acuity. The lines and serifs of the letters are one-fifth the height or width of the entire letter. *B.*, Landolt ring test symbols. The thickness of the line and the width of the gap are both one-fifth of the outer diameter of the ring. [From Riggs (1965b).]

which can be resolved subtends a visual angle of less than 1 s of arc. This represents a visual acuity by the definition given above of greater than 60. The diameter of a dark spot against a lighter background must be somewhat greater than the width of a dark line in order to be discriminated.

Visual acuity measurements made clinically may depend upon a test of the minimum separable threshold. The Landolt ring is a commonly used target on charts for the measurement of acuity (Fig. 9.101*B*). The gap in the ring is positioned at random from one ring to the next on a given chart, and the patient must identify the location of the gap. The smallest ring size for which the gap can be correctly located determines the visual acuity with this test. For the normal observer, the width of the gap subtends a visual angle of approximately 1 min of arc at the standard measuring distance of 20 ft. Normal visual acuity for such a target is therefore *one*. It is more often expressed in relation to the distance at which it can be correctly detected for a given observer in relation to a normal populations. Thus, an acuity of 20/200 implies that a given individual can resolve a test figure at a distance of 20 ft which the normal observer could resolve at 200 ft. This represents a severe defect. A visual acuity of 20/10 indicates that an observer can resolve a test pattern at 20 ft which the normal observer must view at a distance of 10 ft in order to resolve it.

When visual acuity is tested with Snellen letter charts (Fig. 9.101*A*), the procedure is very much the same. The size of those elements which permit discrimination of the letters determine their visibility, and visual acuity is calculated from the sizes of these elements for those letters which can just be resolved at the normal distance of 20 ft. Again, a visual acuity of one, corresponding to a visual angle subtended by elements of detail of approximately 1 min or arc, is the norm. Acuity is expressed as the distance at which the patient can identify the letters in a given line relative to the distance at which a "normal" observer can discriminate the same letters.

TEMPORAL DISCRIMINATION CAPABILITY OF THE VISUAL SYSTEM

Flicker Fusion Frequency

One of the oldest ways of studying temporal resolution capability of the eye has been the technique of presenting a series of flashes to the eye and determining the rate of flash presentation at which the stimulation appears to fuse and is seen as a steady light. This is known as the *flicker-fusion frequency or critical frequency of fusion* (Brown, 1965b). The results of such a study are presented in Figure 9.102. This shows the critical frequency as a function of the logarithm of retinal illuminance (Hecht and Verrijp, 1933). Except at very high luminances, there is a tendency for the frequency of fusion to increase as retinal illuminance increases. That is, the eye is better able to resolve temporal discontinuities with brighter stimuli. Although it might seem more "reasonable" for a brighter light to be more persistent, this is obviously not the case.

Perceptual Duration

At fairly high levels of light adaptation, a black outlined figure on a white background has a perceptual duration of approximately 250 ms. This was ascertained by adjusting the rate at which the black outlined target had to be presented repetitively in order to appear to be on continuously. It was ac-

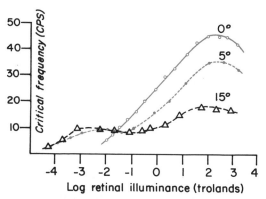

Figure 9.102. The relation between critical fusion frequency and log retinal illuminance for white light for each of three different retinal locations; at the fovea and 5 and 15° above the fovea. [From Hecht and Verrijp (1933).]

tually on for a duration of approximately 20 ms, but observers reported that it was seen continuously so long as it was presented within 250 ms of the preceding presentation on each successive presentation (Haber and Standing, 1969). When the observer is dark-adapted and there is no steady background illumination, the "persistence" of a presentation of the same short duration and at the same luminance is increased to over 400 ms.

In general, results of experiments of this sort indicate that a light flash of shorter than 250 ms presented to the light-adapted eye will appear to be of nearly 250 ms duration regardless of its duration. For the dark-adapted eye, the apparent duration will be somewhat longer. For longer light flashes, the apparent duration will conform fairly closely to the actual physical duration. It is interesting to note that an interval of approximately 250 ms corresponds to the duration of individual fixations of the eye on a printed page. Evidence indicates that an interval of from 200 to 250 ms is required for the processing of information derived from each visual fixation (Haber and Nathanson, 1969).

The temporal resolution characteristics of the visual system are thus quite complex and include a brief limitation at the level of the retina for integration of energy and a more severe limitation which almost certainly involves higher centers in the brain for sequential discrimination of intelligent information.

Discrimination Based on Wavelength

COLOR VISION

The Trichromacy of Color Vision

The eye is not nearly as good a discriminator of the wavelength or frequency of light as is the ear of the frequency of sound stimulation. Within certain limits, any color sensation may be matched by the mixture of just three primaries if these are appropriately selected spectral lights. The only stringent requirement in selecting primaries is that no two of them may be combined in any way to match the third. For practical reasons in the matching of colors associated with extreme wavelengths and having relatively high luminance, primaries are usually selected to include a red, a green, and a blue or violet element. In fact, precise matches of monochromatic lights cannot be achieved by simple addition of all three of the primaries in the opposite side of a bipartite field. It is sometimes necessary to add one of the primaries to the wavelength to be matched in order to achieve a perfect match between the two halves of the field. The appearance of the wavelength to be matched is thus altered slightly when a small amount of one of the primaries is added to it. It is accepted as a convention that when a given amount of a primary is added to the field to be matched, it is treated as a negative value in the overall addition process.

Actually, quite good ranges of colors from violet to deep red may be achieved by simple additive mixture of primaries. It is not possible to match all spectral colors without the subtractive technique described above, however.

Opponents Processes

There appear to be inhibitory interactions between certain of the primary color discrimination mechanisms at stages central to the primary receptors. When a red light is mixed with a bluish green light, the combination tends to be achromatic if the components are appropriately selected. Similarly, addition of a yellow light and a blue light may yield an achromatic product. Individual wavelengths, which when mixed together tend to neutralize each other or produce a white stimulus, are said to be mixture complements. It is as if they are stimulating elements within the color vision-detecting system which work in opposition and hence tend to cancel out the effects of complementary stimuli when these are presented simultaneously.

If the eye is exposed for a period of 5 or 10 s to a bright red light and this is followed by a white field, the latter will appear greenish in hue. If the original red stimulation was smaller in size than the subsequent white field, a green region will be seen against the white field corresponding in size to that of the original red stimulus. Such successive color contrast effects can also be seen dramatically following exposure to a green light which induces a red hue in a white field, or a yellow light which induces blue, or a blue light which induces yellow. The precise afterimage hue depends upon the wavelength distribution of the initial "inducing" stimulus. As reported earlier, electrophysiological processes have been found at the level of the lateral geniculate body which indicate that on certain cells the effect of turning off a red stimulus is the same as the effect of turning on a green stimulus. For other cells the reverse may be true. Similar effects in individual cells are found for yellow and blue stimulus lights (DeValois, 1973).

If the region of the retina which permits identification of various hues associated with given wavelengths of stimulation is studied carefully, an interesting result is found. Red and green may be identified out to a certain distance from the fovea which is very nearly the same for each of these colors. On the other hand, wavelengths which typically elicit

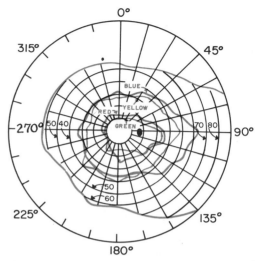

Figure 9.103. Retinal regions in which red, green, blue, yellow, and achromatic lights can be detected. Note that the zones in which color may be discriminated are smallest for red and green, somewhat larger for yellow and blue, and largest for the simple detection of light independent of color. [From Optical Society of America (1953).]

the response "yellow" or "blue" can be presented at a greater distance from the fovea and will evoke the characteristic color-naming response. In a word, the color-discrimination ability of the retina for yellow and blue extends out further into the periphery than is the case for red and green. These relations are illustrated in Figure 9.103. Finally, if a small patch of color is surrounded by a neutral field and the reflectance of the neutral field is adjusted appropriately, the neutral field will appear tinged with a hue corresponding to the complement of the color patch. This is known as simultaneous color contrast (Graham and Brown, 1965).

CHROMATIC PROPERTIES

Rhodopsin has been extracted from human rod outer segments by maceration of the retina and differential centrifugation. The absorption of light at different wavelengths in a suspension of rod particles is plotted in Figure 9.104. In the same graph is displayed the measure of dark-adapted spectral sensitivity, corrected for the light absorbed by the ocular media so that it resembles sensitivity to light at the retinal surface. A third curve was obtained in a subject who had the lens, the principal absorbing structure in the optical path, removed for cataract. The nearly exact superposition of these three curves indicates that rhodopsin provides the molecular basis for light sensitivity on the part of rods.

Cone pigments, which are implicated in conditions of higher illumination, absorb light at three

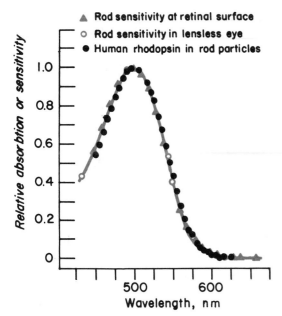

Figure 9.104. Absorption spectrum of human rhodopsin, measured in a suspension of rod outer segments, compared with the spectral sensitivity of human rod vision. The latter data have been corrected for ocular transmission or are based on measurements of the spectral sensitivity in aphakic (lensless) eyes. [From Wald and Brown (1958).]

distinctive peak wavelengths, indicating that different pigments are associated with different cones. The curves in Figure 9.105 show average light absorbance from single cones from the excised eyes of humans and monkeys. The *solid* and *open circles* measure reflectivity of pigments as established in normal human subjects. The overlap is convincing that the light absorbance *in vitro* and light reflectivity *in vivo* stem from the same molecular process. By retinal reflectance of different wavelengths of lights, blue-green and red, two distinctive curves are obtained, indicating responses from two different populations of cones.

The Young-Helmholtz theory is generally accepted as a suitable basis for observations that any color sensation can be experienced by normal color vision subjects by mixing three pure primary colors from blue, green, and red components of the visual spectrum. The four color sensations, blue, green, yellow, and red can be elicited by appropriate mixtures of the primary colors. Yellow can be experienced by mixing pure red and green colors, and purple by mixing red and blue. These sensations are achieved by mixing colors *additively*. Surfaces reflect color because they absorb all light except that which is reflected. Therefore, the selection of appropriate colors in paints is achieved by *subtractive* color mixing. Thus, in practice different rules are followed, but the principle remains the same

from the point of view of spectral densities reaching the retina.

COLOR BLINDNESS

The evidence favoring opponents processes that involve red and green as apparently having opposite effects, and yellow and blue as apparently having opposite effects, is lent further support by the nature of color blindness. Although some color blindness is complete, most is not, in that most individuals who suffer from deficiency in color discrimination are able to discriminate some colors although not all. Where three primaries are required for the normal observer, many of these color-blind individuals require only two. They are therefore called *dichromats* as opposed to *trichromats*, the term employed for the normal observer, or *monochromats*, the term applied to the completely color blind.

Monochromats

There are individuals whose vision is entirely monochromatic; that is, they can match any spectral distribution with a single monochromatic light, provided they are permitted to adjust its luminance. Any monochromatic wavelength will serve perfectly well. Monochromats, sometimes called achromats, are for the most part found to lack normal cones. The vision of these rod monochromats is scotopic, and their luminosity function corresponds to that of the dark-adapted peripheral retina of the normal (Sloan, 1954). Visual acuity is low, high luminances are disturbing, and there is chronic nystagmus. A

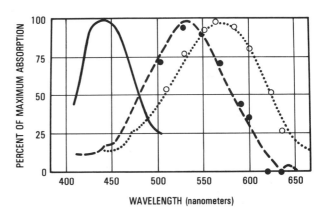

Figure 9.105. The cone pigments of normal color vision absorb lights of different wavelengths. The *curves* show average spectral absorbance from single cones in excised eyes of humans or monkeys scaled to equivalent maxima. *Open circles* represent the reflectivity of pigments in red-sensitive cones, *closed circles* the green-sensitive cones in the living human eye. Coincidence of the measurements demonstrates that single cones contain single pigments. Curves, from left to right indicate blue-, blue-green-, and red-sensitive receptor populations. [From Rushton (1975).]

small percentage of monochromats apparently have one type of cone in addition to rods. It has been suggested that these individuals may see in a single color under photopic conditions and in shades of gray under scotopic conditions. If this is true, they probably should not be called achromats, although the unitary character of their photopic vision can hardly provide any useful chromatic information.

Dichromats

Both protanopes and deuteranopes tend to confuse reds and greens. The colors which they perceive are all in shades of yellow and blue as these are seen by normals. This conclusion is based on evidence from cases of monocular color blindness (Judd, 1948; Graham et al., 1961). In such individuals the appearance of various stimuli presented to the dichromatic eye may be matched in the normal eye and thus provide an indication of what the color-blind eye sees relative to normal color vision. The color confusions of the tritanope involve yellows and blues.

Extensive studies have been performed on the spectral response functions of color-blind subjects. In general, they show relatively little loss in luminosity, and are capable of responding to levels of

energy which differ only slightly from those required to stimulate the eye of the normal observer. In addition, their visual acuity is almost normal. These two facts lead to the conclusion that their deficiency is not the absence of an entire class of cones. Rather, it suggests that they may be lacking in one or another of the photopigments found in normal cones.

The three cone pigments and the rod pigment (which contributes to achromatic sensation) constitute the spectral light-absorbing systems of normal human photoreceptors. If the rod or cone pigments are bleached in the human eye by illumination with an appropriate wavelength of light, there is naturally a change in reflectivity from the retina. Bleaching with red light yields reflectivity that demonstrates a difference of absorption, hence reflectivity, as influence by the red-absorbing pigment. This imitates in the normal eye the *erythrolabe* spectrum difference observed in deuteranopes (curve on the *right* in Fig. 9.106). Bleaching with blue-green light reveals the spectrum difference of *chlorolabe*, imitating the condition encountered in protanopes. The green-absorbing pigment, chlorolabe, can be determined by measuring and comparing light reflectance after partial bleaching with red

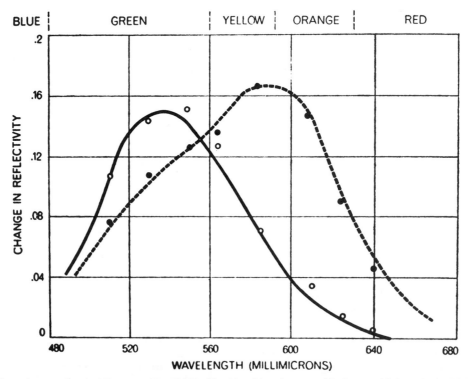

Figure 9.106. Two pigments in normal cones demonstrated by bleaching the eye with deep red lights, then with blue-green lights, and recording the change in reflectivity of the fovea. Bleaching with red lights yields results shown by *solid circles* and coincides with the erythrolabe spectrum difference (*dashed curve*) as measured in the deuteranope. When the bleaching light is blue-green, the foveal reflectivity is shown by *open circles*, conforming to the difference spectrum of chlorolabe (*solid curve*) as measured in the protonope. [From Rushton (1962).]

light *and* blue-green light. Since the protanope's fovea responds in the same way following both bleaches, exhibiting the "chlorolabe difference spectrum," the protanope's eye evidently contains only one pigment. Bleaching with white light, which shows the total pigment present, shifts foveal reflectivity upward at each higher wavelength and measures the protanope's total range of sensitivity to white lights. Other measures of bleaching also suggest that protanope cones contain only one pigment (Rushton, 1962).

Color Anomaly

There is another class of color deficiency which is termed anomalous trichromacy. People with this condition require three primaries to match colors and yet the mixtures which match for them differ in proportions from those made by normal observers. There is an anomalous condition for each of the dichromacies; they are therefore categorized as protanomalous, deuteranomalous or tritanomalous. In these categories individuals may be found ranging all the way between normal and the dichromat.

Hereditary Aspects

Color vision deficiency is known to be genetically linked. More men are color blind than women, and there are different incidences in different races. Among Caucasians approximately 8% of all men are dichromats or anomalous trichromats; only about 0.64% of women are so affected (Hsia and Graham, 1965). Approximately 1% of men and 0.02% of women are protanopes. A slightly higher percentage of men, 1–2%, are deuteranopes; only 0.01% of women are deuteranopes. The occurrence of tritanopes is really too small to be assigned any percentage value.

Figure 9.107 presents George Wald's 1972 proposal for the arrangement of genes on the X-chromosome which would determine normal green and red pigment production for normal human color vision. Various deficits in those genes would result in reduction or elimination of production of the requisite visual pigments. Since males do not have an extra X-chromosome, any deficit will be reflected in some form of color blindness. Blue-blindness occurs in about equal frequency in the two sexes and is evidently autosomal.

Dicromacy, as we have seen, appears in two main types: *protanopes* (*red-missing*) *and deuteranopes* (*green-missing*). Psychophysical evidence suggests that some classically diagnosed dichromats may actually have three cone types rather than two. The anomalous cones, previously thought to be absent, are less sensitive than normal cones to both spectral

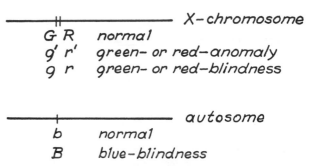

Figure 9.107. Diagram suggesting hypothetical arrangement of genes that specify the three opsins which, with 11-*cis* retinal, form the blue-, green-, and red-sensitive pigments (*B*, *G*, *R*) of normal human color vision. Two such genes on the X-chromosome are assumed to determine *G* and *R*. Mutations in these genes that result in the production of less effective pigments account for color anomaly; other mutations that result in the failure to form functional visual pigments result in color blindness. The normal condition is dominant to red or green anomaly which in turn is dominant to red or green blindness. Blue blindness is almost equally distributed between males and females and seems to be autosomal. [From Wald (1972).]

and temporal variations, and their spectral sensitivities resemble those of the abnormal cones in anomalous trichromats (Frome et al., 1982).

Perception of Depth

There are a variety of cues which afford information as to differences in depth, several of which are operative in monocular vision and two of which require binocular vision.

RELATIVE SIZE

The further an object is from the eye, the smaller is its image on the retina. Retinal image size thus could afford a cue to depth. Nevertheless, retinal image size alone is not a particularly effective cue in the estimation of absolute distance (Gogel et al., 1957), even for familiar objects.

LINEAR PERSPECTIVES

The relative image size of individual objects as a cue to depth may be extended to the image of the entire visual field. The texture of objects which recede from the observer in depth grows finer the greater the distance of the particular point under observation. Images of parallel lines comprising the edges of three-dimensional rectangular objects tend to converge as they recede from the observer in depth. Linear perspective refers to the geometry of projection of a three-dimensional world on a two-dimensional plane and is reasonably considered to be an extension of the relative size factor to the continuum of visual space.

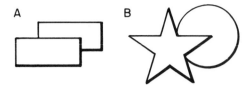

Figure 9.108. Two line drawings which represent the phenomenon of interposition as a cue to the discrimination of relative distance. At *A*, one rectangle appears to be cutting off a portion of a second rectangle located at a greater distance. The contours of the first interrupt the contours of the second rectangle. At *B*, a star appears to be located in front of a circle. In this part of the figure there are discontinuities in the outline boundaries of both star and circle at their intersection. The discontinuities in the outline of the star, however, are characteristic of its shape. [Adapted from Graham (1965b).]

INTERPOSITION

When two objects are located at different distances in space, the nearer may intrude between the observer and the more distant object. The result will be the partial occlusion of the more distant object and an interruption of its familiar form. Ratoosh (1949) expressed the notion formally, stating that a discontinuity in the first derivative of an outline contour affords a cue for discrimination of relative depth of the object forming the contour. An illustration of this concept is presented in Figure 9.108. In part *A* of the figure, one rectangle appears to be located in front of the second in a manner such that it is cutting off the lower left portion of the more distant rectangle. The boundaries of the rectangle that appears to be more distant are interrupted by the boundaries of the rectangle that appears closer. The contour lines which define what is presumed to be the rear rectangle are thus discontinuous at their point of intersection with the contour lines which define the rectangle which appears to be closer.

It is unsatisfactory, however, to attempt to explain interposition entirely in terms of such notions as contour discontinuity. A rectangle has four discontinuities, one at each corner, but once the outline is perceived as a rectangle, the discontinuities at the corner have no relevance to depth; they merely confirm rectangularity. It is more appropriate to define interposition in terms of the interruption of contours which are expected on the basis of a reasonable assessment of the nature of the figure. This is illustrated in part *B* of Fig. 9.108. This appears to show a star located in front of a circle. Actually, there are discontinuities in the outline contour of the star as well as that of the circle at the point of intersection of the contours of the two figures. The discontinuities in the outline of the star conform to expectation on the basis of its shape

as a star. The discontinuities in the outline of the circle, however, are incompatible with its being a circle, unless the star is located in front of it. Otherwise, the circle is not a circle but a circular form with a pie-shaped cut-out which just conforms to one of the points of the star. For most observers the latter possibility is much less likely than that of a full circle with a star located in front of it (Chapanis and McCleary, 1953). It is thus evident that the cue of interposition is very much dependent upon the observer's knowledge of outline contours of the objects involved.

MONOCULAR MOVEMENT PARALLAX

Whenever there is motion of an observer or of objects which are located at different distances with respect to an observer, there is usually an accompanying change in perspective of the objects from the point of view of the observer. This is illustrated in Fig. 9.109 for a situation where an observer is moving along a line indicated by the arrow. Two objects, *a* and *b*, are located in space at the positions indicated. The locations of the images of these objects on the retina of the observer are indicated by a' and b'. It is evident that the relative locations of these images is reversed with motion of the observer from O_1 to O_3. To the passenger who looks out the side window of a moving car or train, objects nearby, such as telephone poles, appear to be moving rapidly in the direction opposite to that of the

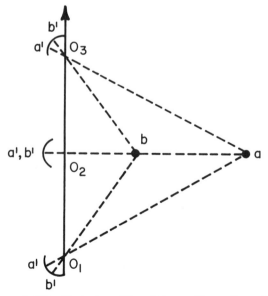

Figure 9.109. Illustration of the relative positions of two objects located at different distances from an observer as he moves along the path represented by the *arrow*. Relative changes in the locations of retinal images of the two objects are shown. Such changes in retinal image positions provide the monocular movement parallax cue for distance discrimination.

motion of the vehicle. The more distant the object, the less rapid is its apparent motion in the opposite direction. Objects at a middle distance fixated by the observer may appear to be stationary, and more distant objects will then appear to be moving in the same directions as the observer. Even the slightest motion of the head introduces mononocular movement parallax effects which provide important cues to the relative depth of objects in the immediate surroundings. Smaller movements are necessary for the discrimination of monocular movement parallax in the horizontal direction than in the vertical (Tschermak-Seysenegg, 1939; Graham et al., 1948).

ACCOMMODATION

The accommodative response of the eye which alters the physical characteristic of the lens in order to achieve a maximally sharp image on the retina could, logically, serve as a cue to depth. If appropriate feedback information is available as to the nature of the motor response necessary to achieve optimun. focus of the retinal image, this might provide a cue to the depth of the object viewed. When all cues are eliminated other than that of accommodation, however, it is found that this alone does not provide a very accurate basis for the determination of depth (Graham, 1965b).

CONVERGENCE

In binocular vision, vergence movements of the eyes result in stimulation of the two foveas by light reflected from the point of regard in the visual world. As this point recedes in space, the eyes must diverge. As it approaches the observer, the eyes must converge. Vergence movements are under the control of the extraocular muscle system, and the relative tension in the musculature involved is a possible basis for information as to relative depth of the point of fixation. Convergence and accommodation change together in binocular vision. With increased convergence, the eye must accommodate for nearer objects in order to maintain a sharp retinal image.

BINOCULAR RETINAL DISPARITY

The two eyes are displaced in space by approximately 65 mm. Thus they provide two different views of the visual world. Objects located at different depths are seen in different relative perspectives by the two eyes. This can be demonostrated fairly simply as illustrated in Figure 9.110 by fixating the forefinger of the right hand while the forefinger of the left hand is held behind it. The more distant finger is seen as double. It is clear from the figure that for the left eye, the more distant finger is

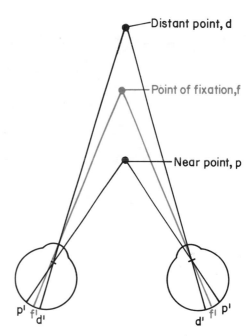

Figure 9.110. Illustration of the doubling of images of objects seen binocularly at depths which differ from the point of fixation. For additional information, see text.

located to the left of the point of fixation, while for the right eye it is located to the right of the point of fixation. The point of fixation is imaged on the foveas of the two eyes. The more distant fingertip is imaged on noncorresponding points in the two eyes, to the right of the fovea in the left eye and to the left of the fovea in the right eye. (Corresponding points in the two eyes are defined as those points which would be in contact if the two retinas were superposed with the two foveas in contact. Thus the temporal retina of the right eye corresponds with the nasal retina of the left eye, etc.) If fixation is maintained on the right forefinger and the left forefinger is then brought in closer to the observer, it will again be seen as double, but this time the left eye image will appear to the right of the point of fixation and the right eye image will appear to the left of the point of fixation. Again, noncorresponding points of the two retinas are stimulated by the light from the nearer object.

In the example given, the effect is a fairly dramatic one in that double images are seen. Normally, they are not seen, although with the exception of the object of fixation, parts of the images of most objects in the visual field stimulate noncorresponding points on the two retinas. Such retinal image disparity appears to provide a cue as to relative depth. When disparity is not great, or when the disparate imates are sufficiently removed from the region of the fovea, objects are not seen as double but are seen to be located closer or further away

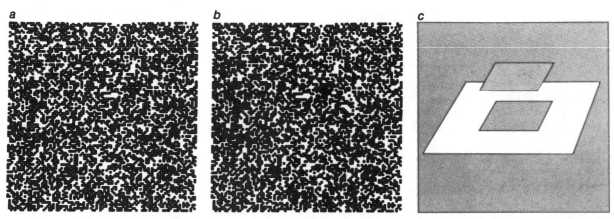

Figure 9.111. Stereoscopic image pair (*left*) consist of random dot patterns generated by computer. When these two images are viewed at a reading distance with gaze at infinity through the images, or with a stereoscope, a center panel is suddenly perceived to float an inch or so above the surface of the page. A tilted perspective view of this perception is illustrated at right. [From Julesz (1965).]

than the point of fixation, depending upon the nature of the disparity. Geometrically, there are a number of points in visual space other than the point of fixation which will provide retinal stimulation of corresponding points in the two retinas. The locus of these points in space is known as the *geometric horopter*. The derivation of the term implies the limiting locus for observation of single images. In fact, most points which do not stimulate precisely corresponding retinal locations in the two eyes are not seen as double but are seen as at different depths. The geometric horopter was worked out in detail by Johannes Müller (1826). This conception of the horopter is oversimplified; some of the complications which must be considered were worked out by Helmholtz (1924) and by Hering (1861, 1864). A good review of the subject is presented by Ogle (1950).

The recent studies reported by Barlow et al. (1967) and by Hubel and Wiesel (1970), which demonstrate the existence of cortical cells that respond differentially to certain amounts of disparity, particularly in the horizontal dimension, provide a neurophysiological basis for the perception of depth as a result of binocular retinal disparity.

Forms can be discriminated in depth solely on the basis of retinal disparity in the absence of any monocular cues as to their identity. Julesz (1964) has demonstrated this with the aid of stereo pairs which are comprised of random dot patterns (Fig. 9.111). One member of a pair consists of a randomly distributed pattern of dots. The other member is identical except that all of the dots within an imaginary outline of some form are transposed laterally

with respect to the rest of the dots. Dots intruded upon by the transposition are erased. The vacated portion of the pattern is refilled with random dots. When the two eyes fuse either the dots which have been transposed or the background dots, the resulting retinal disparity of dots not fused, is accompanied by the appearance of the form. It may appear either in front of or behind the plane of the background, depending on the direction of the disparity.

Julesz and Spivack (1967) interpreted this kind of stereo fusion to mean that perceptual experience is based on an analysis of vernier cues which are then applied to the process of stereopsis. However, Nishihara and Poggio (1982) find that center-surround ganglion cells, each integrating over angular extents of several arc minutes, can detect such coarse monocular structures in the Julesz-type stereograms and other kinds of images as well. These and other authors (see an investigative treatment of human visual perception in Marr, 1982) suggest a simpler solution: In early visual sensory processing, images are obligatorily filtered through several channels, each with a different center-surround receptive field size, and stereomatching of all types may be based on detection of boundaries between regions of positive and negative response (zero-crossings). Certainly nonvernier, monocular cues are present very early in visual sensory processing and can be used along with vernier cues for successful stereomatching. Present evidence suggests that human perception of binocular vision may be simplified by such a general strategem early in visual sensory processing.

CHAPTER 65

Olfaction and Taste

CHEMICAL SENSES IN GENERAL

Olfaction and taste have to do with the most vital, intimate, and emotion-inducing sensory experiences in human existence. These two chemical senses are directly engaged in indispensable guidance of behavior on behalf of survival and reproduction. Just because chemical appreciation and selection are not easily put into words does not relegate these functions to any secondary position in terms of biological, physiological, and psychological importance.

One of the prime features of all living systems is that they are *open systems*. They depend on resources in the environment and they initiate a throughput of chemical substances upon which life itself depends. In order to accomplish this favorable metabolic transaction, living systems must select and incorporate from the environment those particular substances which are required in order to satisfy individual, continuous but fluctuating, metabolic needs. This demands discriminating chemical sensibility and selectivity, which all living systems must possess in order to survive and reproduce.

Every cell participates continuously, of necessity, in chemoacceptance and chemoelimination. Even cells in multicellular organisms retain this selective, indispensable, self-sufficient capability. Individual cells have to possess chemoreceptive and chemoselective competence; the metabolic machinery of each cell depends on the capacity of that cell to recognize and to incorporate what it requires from the local environment.

For all sizeable organisms there must be additional chemoreceptor systems to regulate the distribution and elimination of chemicals in the process of bodily chemical throughput. Such *interoceptive controls* operate to satisfy tissue, organ, and whole body conditions, as necessary to maintain an appropriate cellular access to nutrients and elimination of wastes. Specialized control systems are organized for governance of digestion, circulation, respiration, temperature regulation, excretion, etc., and hypothalamic mechanisms which guide overall behavior as affected by internal states involving appetite and satiety. A remarkable feature of chemical sensibility is that the ensemble contributes information that has some effects back on chemical receptor mechanisms with respect to which chemicals may at the time be most needed.

Large multicellular organisms have local and regional chemosensitive mechanisms (interoceptors and associated reflexes) which facilitate tissue and organ access to oxygen and nutrients and the disposal of carbon dioxide and wastes. Examples of clinical importance are the carotid body—sensitive to arterial blood levels of O_2, and associated reflex mechanisms for regulating oxygen transport—and medullary receptors—sensitive to plasma levels of CO_2, and reflex mechanisms governing elimination of that product. Central integrative interoceptive chemosensitive systems in the hypothalamus measure appetite and satiety mechanisms which serve to guide the whole organism toward suitable water and food intake, and reproductive activities.

In all vertebrates, in addition to such *interoceptors* with their associated reflex mechanisms and behavioral guidance, there are two *exteroceptive* chemosensory systems, *olfaction and taste*. Each has its own neuroepithelium, separate central pathways, and fields of cortical representation. Together, they contribute exteroceptive chemoselectivity, *olfaction for distance-to-neighboring chemoselection, and taste for contact chemoselection*, which guide behavior relating to smelling (tracking), eating, drinking, and reproduction.

Figure 9.112 shows the relationship between the olfactory neuroepithelium located in the roof of the nasal cavity and the overlying cribriform plate through which olfactory nerves reach the olfactory bulb. Figure 9.113A illustrates the relationship between the tongue, palate, and glottis fields of taste receptors and their respective innervations.

When olfactory and gustatory systems were initially investigated physiologically, it was assumed

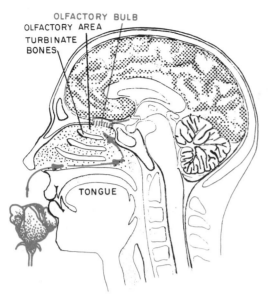

Figure 9.112. Parasagittal cross-section of the head showing the three turbinate bones, the olfactory area, and the termination of the endings of nerves from the olfactory bulb in the olfactory epithelium. The passage of air-bearing odorous substances is shown by *arrows*. [From Amoore et al. (1964).]

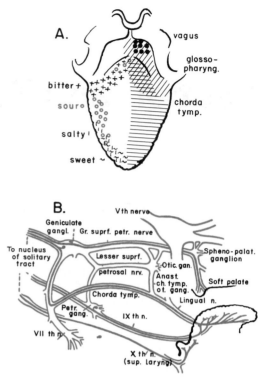

Figure 9.113. *A,* Sensitivity of the tongue in different regions for four primary taste qualities. *Diagonal cross-hatching* represents the region innervated by the glossopharyngeal nerve. *Solid dots* show the region innervated by the vagus. *Horizontal lines* represents the region innervated by the chorda tympani (von Békésy, 1966). *B,* The pathways of the nerve supplies are shown in greater detail and a sagittal cross-sectional view of the tongue. [Adapted from Brodal (1959).]

that there would be a relatively small number of specialized classes of receptors and separate pathways for the representation of categorizable odors and tastes. This view was reinforced by clinical evidence that four fundamental taste categories, sweet, salt, sour, and bitter, have separate regions of special receptivity on the human tongue. However, when olfaction and taste were investigated using single cell recordings, it became obvious that these systems are far more subtle and versatile than this would suppose.

Single olfactory receptor neurons were found to respond differentially to many different odorants and mixtures of odorants. Single taste receptors were similarly found to respond differentially to various complex taste stimuli. Both systems are strongly affected by recent past receptor history, and they show adaptation to persisting stimulus states. A valuable insight into molecular mechanisms that can account for such complex chemoreception became available through the study of *bacterial chemotropism*. Although at first glance this seems remote from sensory receptors, it provides valuable insight into the kinds of molecular machinery that initiate sensory processing in both olfaction and taste. It also provides understanding of the way interoceptors and individual cellular chemoselectivity likely operate. Mechanisms of chemoreception also have implications for how neurons, with their molecular organization for neurotransmitter reception, may combine that function with intracellular signalling. Bacterial chemotaxis

thus turns out to have heuristic value for understanding interdependent behavior among cells in multicellular organisms and molecular means for chemical exteroception and some of the integrative processes characteristic of nervous systems.

Chemoreception and Molecular Signalling

Koshland and others (see review 1980; Russo and Koshland, 1983) have shown that single bacteria have about 30 receptor molecules, each of which responds relatively specifically to a limited number of chemical compounds important to metabolism and survival. Each receptor molecule presents its chemoselective surface outside the cell, while the bulk of the molecule extends through the cell membrane whereby it *signals information relating to the state of the receptor to the cytoplasm.* Prompt covalent modification of the molecule in the cytoplasm resets the chemoselective capacity of the receptor to zero, thus improving bacterial ability continuously to monitor chemical opportunities in the environment.

The ensemble of receptor molecules in a single

cell integrates responses from two or more simultaneous stimuli. Responses to attractants show additive effects, and unfavorable stimuli (increases in concentration of repellents or decreases in attractants) have subtractive effects. Most receptor molecules are genetically determined, but a certain range of receptor competences can be induced according to the growth conditions to which the organism is exposed. By this adaptation, the bacterium improves the match between its chemoselective competence and given environmental conditions.

In addition to immediate receptor response mechanisms, the same molecular system adapts differentially to a steady-state and to a changing environment. Further, the same molecular system signals bacterial motility—swimming in a consistent direction—or random tumbling to obtain another direction—leading again to swimming in a consistent direction. Direct molecular linkage between selective chemoreception and selective motor performance provides both approach and avoidance behaviors. The system continually guides the organism into environments that are chemoselectively more favorable for survival. In this way, chemoselection provides "sensible" guidance for bacterial behavior; analogous "sensible" chemoselective guidance for human behavior is provided by olfactory and gustatory systems.

It is important to note that *there is no requirement to metabolize the molecules that activate the chemoreceptors*; nutrients are taken into the cell by other selective channels, as required. In these respects the bacterial chemoreceptor signalling system is functionally quite analogous to the separate and specialized vertebrate systems of olfaction and taste.

The bacterial chemoreceptive system also provides an important analogy to nerve cells. All neurons manifest chemoselection, adaptation, and response—the latter in terms of electrical and chemical discharges. *Chemoreceptive molecular mechanisms in neurons serve analogous integrative functions; they respond to a specific chemical message and signal the interior of the cell.* The most spectacular difference arises from the fact that most neurons are rather stretched out for purposes of decrementless signal conduction, to link up at least two and perhaps a multitude of primary exchanges of signals.

OLFACTION

Nasal Passages and Olfactory Innervation

Five separate sources of innervation of epithelia in the interior of the mammalian nose are all af-

fected by odorants; at least four of these are bona fide chemoselective sensory systems: (1) general nasal mucous membranes innervated by the trigeminal nerve; (2) epithelium innervated by the terminal nerve; (3) the septal organ (of Masera); (4) the vomeronasal organ (of Jacobson); and (5) the classical olfactory neuroepithelium.

TRIGEMINAL INNERVATION

The trigeminal nerve innervates all respiratory mucous membranes along nasal and oropharyngeal pathways, including the olfactory neuroepithelium itself. Trigeminal innervation is confounding because it responds to touch, temperature, and pain stimuli, and it reacts strongly to a wide variety of noxious odorants. It makes indirect and partial contributions to the perceptions we identify as smell and taste.

TERMINAL NERVE INNERVATION

The terminal nerve (TN) spreads delicate fibers and bare nerve endings along surface epithelia, including the olfactory epithelium, near the roof of the nasal passages. Bipolar neuronal cell bodies of the TN lie along the medial aspect of the olfactory nerve and bulb. These cells contain luteinizing hormone-releasing factor (LH-RH) (Springer, 1982). Their central projections terminate bilaterally in ventral gray matter of the frontal lobes overlying the anterior commissure, the supracommissural nucleus. Oddly enough, they also project collaterals to the inner plexiform layer of the retina and there may "prejudice" the eye of the beholder with olfactory information!

Evidence implies that the TN may play a role in mediating sensory and behavioral responses to sexual pheromones. Electrical stimulation of the TN in fishes leads to behavioral responses similar to normal courtship (Døving and Selset, 1980); transection of the TN drastically reduces male behavioral responses to pheromones of reproductively active females; stimulation of the TN in reproductively active males leads to sperm release. Antidromic stimulation of the TN collateral to the retina elicits consistent sperm release, whereas stimulation of other orbital structures is without such effects (Demski and Northcutt, 1983).

VOMERONASAL ORGAN INNERVATION

The vomeronasal organ is confined to a tubular cul de sac which has a narrow entrance filled with mucus emptying into the nasal cavity from the medial wall of the septum. The walls of the vomeronasal organ are lined with neuroepithelium similar to that of the main olfactory epithelium; vomeronasal receptor neurons, however, have microvilli but are without cilia. The neural receptors of the vomeronasal organ *respond to similar odorants as does the main olfactory neuroepithelium,* although at higher thresholds. This means that microvilli must be adequate for chemoreception and suggests that cilia may provide advantages by having much greater surface area for exposure of receptor molecules.

The vomeronasal chemoreceptor neurons project to the accessory olfactory bulb, which in turn has further central projections to the amygdala and olfactory cortex, all of which remain independent of the central projections of the main olfactory epithelium and olfactory bulb. These vomeronasal projection sites, in turn, send discrete projections to the ventromedial nucleus of the hypothalamus. Thus, the vomeronasal organ functions as a separate chemoreceptive system which is especially concerned with pheromone detection and which has its own separate central destinations relating to sex-related neuroendocrine functions and behaviors.

The functional role of the vomeronasal system appears to be similar and conservative across many classes of vertebrates. Evidence for pheromone detection on the part of the vomeronasal organ and the role of its projections in sexual responses have been well established in mammals (Meredith et al., 1980; Power et al., 1979) and have been more extensively studied in garter snakes by Halpern and colleagues (1983). In snakes, prey identification, and prey and conspecific trail following, courtship, and social aggregation depend on integrity of the vomeronasal system. Associated with these snake behaviors is tongue-flicking which physically delivers odorants to the mouth of the vomeronasal organ. In mammals, pheromonally mediated, sex-related social behaviors depend on the vomeronasal system and not on the main olfactory system. They involve species-specific responses to chemical substances emitted or secreted by animals; these responses appear to be comparatively stereotypical and resistant to conditioning. It may be that the main olfactory system, responding to air-borne odorants, provides the necessary chemosensorily motivated behavior in order for the animal to close in on appropriate odorant sources which then, by means of the vomeronasal system, become identified in respect to sexual and other important forms of behavior.

THE SEPTAL ORGAN

The septal organ, discovered by Masera in 1943 also resembles olfactory epithelium. Its neurons are bipolar cells, the dendritic ends of which act as chemoreceptors, the axons of which send impulses to the olfactory bulb (Masera, 1943). Its functional role is not known.

The Principal Olfactory Neuroepithelium

Olfactory neuroepithelium appears as a yellowish patch visible in the upper posterior nasal passages on the lateral wall, partly covering the superior turbinate, and extending medially onto the nasal septum. In humans, the sensory surface occupies an area of approximately 250 mm^2 in each nostril. Its general location beneath the cribriform plate and overlying olfactory bulb is illustrated in Figure 9.112. Diffusion of odorants into the nasal passages during normal breathing involve eddy currents that sweep air past the olfactory epithelium from posterior to anterior. More effective exposure is obtained by more forceful, short staccato respiratory movements—of inspiration and expiration—as in sniffing. Stimulation of the olfactory mucosa also occurs during normal expiration and may contribute importantly to what we perceive as the "taste" of food. If olfaction is blocked, eating an apple cannot be distinguished from eating a potato or even an onion.

Olfactory epithelium consists of *primary olfactory neurons, stem cells, and supporting cells,* with the cell bodies all being located in the middle of a pseudostratified epithelium. The surface of this neuroepithelium consists of the apical borders of slender olfactory neurons surrounded by supporting cells with which they share tight junctions (Fig. 9.114). Tight junctions are identified in this figure as "terminal bars." Tight junctions seal olfactory neuroepithelium against penetration of substances from the surface and create a cross-epithelial resistance of a few thousand ohms/cm^2. As described below, primary olfactory cells are normally recycled. During replacement, new olfactory neurons, daughters of stem cells, push up between the supporting cells. New tight junctions are formed in the process, maintaining structural integrity of the olfactory epithelium. The supporting cells are linked to one another by gap junctions which stabilize their electrical potentials against local perturbations.

The nasal surface of the supporting cells bear microvilli which form a thick carpet that cushions the olfactory neuronal cilia. Each olfactory neuron extends 10–15 long whip-like cilia, presumably motile, into the overlying mucous as seen in Figure 9.114. The olfactory nerve membrane encloses each of these cilia, like the fingers of a glove, and this provides an enormous extent of exposed neuronal surface. The cilia bear numerous *intramembranous*

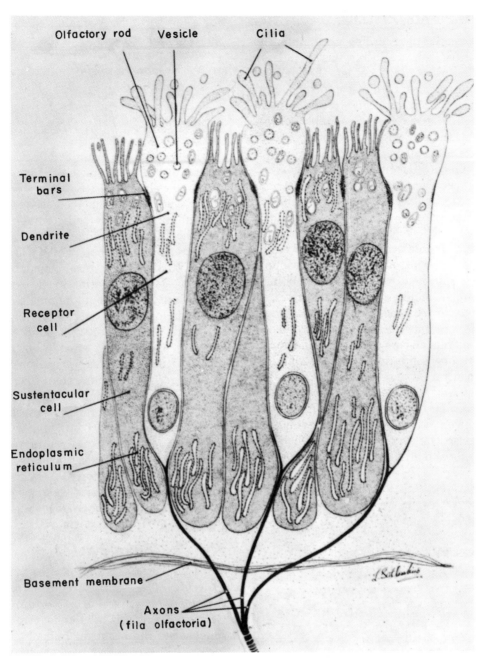

Figure 9.114. Olfactory mucosa showing the relationship of various cell types based on electron microscopic observations. [From de Lorenzo (1963).]

particles which constitute the *molecular receptive apparatus*. The density of these particles is on the order of 10^4 μm^2, severalfold more dense on olfactory cilia than on cilia of any other cells along the respiratory tract. *Contact with this particle-bearing ciliary surface by odorous molecules is the adequate stimulus for smell.*

Olfactory Neuron Responses

Odorant activation of an olfactory neuron results in a relatively slow, long lasting potential (Ottoson, 1956), which is both a *receptor potential* and a *generator potential* inasmuch as *the olfactory neuron is both a receptor and an afferent neuron.* There is ordinarily considerable spontaneous activity in these units, and odorant activation generates additional spike potentials. Following activation, there may be postexcitatory inhibition. From the basal surface of the olfactory neuron a fine (about 0.2 μm in diameter) unmyelinated axon emerges which passes upward through the cribriform plate to convey action potentials to the olfactory bulb.

THE ELECTROOLFACTOGRAM

Ottoson pioneered in recording gross potentials across the mucosal surface in response to odorant stimulation. This response represents the combined activity of many thousands of receptors. Such averaged response patterns are highly complex (Ottoson, 1971). On introduction of a suitable odorant, near threshold, there is widespread prolonged excitation. Figure 9.115 shows a combination of electroolfactogram recorded from the nasal sensory mucosa and extracellular unitary responses recorded from the olfactory bulb. Note that a very effective odorant, butanol, gives a relatively slow mucosal response takeoff, followed by a gradual reduction in amplitude until there occurs a break and a more rapid falloff on discontinuation of the stimulus. This pattern is reflected by roughly corresponding activity in bulbar unit responses (lower traces). Note unit spontaneous activity, an initial burst followed by a silent pause during the early phase of the stimulus at higher concentrations of odorant, and postexcitatory inhibition. With a 5-fold increase in concentration of odorant there is less of a peak at onset and more continuity of response during continuation of the stimulus. With another doubling of concentration, there is a high amplitude mucosal response, while the individual bulbar units discharge briskly and briefly and then are inhibited

for a few seconds. At higher concentrations of the odorant, the initial discharge may be briefer, followed by powerful suppression of spontaneous activity. Responses to some odorants consist mainly of suppression.

SINGLE ODORANT MOLECULES MAY TRIGGER OLFACTORY RESPONSES

The best known example of olfactory sensitivity is that of the male silk moth *Bombyx mori* which responds behaviorally to a conspecific female odorant concentration that corresponds to about one molecule per sensory cell (Schneider, 1969). It is probable that mammalian olfaction approaches this theoretical limit, although obstacles to making certain of this possibility are formidable.

There is great variation in the extent of olfactory epithelium in different mammals, and this is reflected by differences in their ability to smell. Allison and Warwick (1949) found that the rabbit olfactory epithelium, about 4.5 cm^2 in area, contains about 50 million olfactory neurons. Mozell (1976) estimated that with more than 10 cilia per olfactory neuron, a surface area per cilium of 30 μm^2 and a density of intramembranous particles per unit ciliary membrane surface of $10^4/\mu$m^2, there are roughly 10^{14} chemoreceptive sites in total for the rabbit.

The German Shepherd dog has nearly 40 times the total area of olfactory epithelium as compared with the rabbit. The dog's *olfactory ciliary surface* is estimated to be nearly 8 m^2, i.e., several times the total area of the dog's entire body surface. If only one-tenth of the available receptor molecular sites were accessible to the odorant, there would still be about 10^{11} receptor sites available for each molecule of odorant at threshold. The absolute threshold for detection of certain fatty acids by German Shepherds is at concentrations about 100 times lower than that for humans, but this advantage may lie largely in the comparatively much more extensive olfactory neuroepithelium in the dog (72 times). The dog therefore probably has no great advantage over humans at the single receptor level. A fair estimation at present is that under ideal conditions, one molecule of a suitable odorant is probably sufficient to excite a single human olfactory neuron. This agrees with previous independent estimates for the dog and for humans (Moulton, 1976).

There is as yet no compelling evidence for territorial distribution of odorant preferences along the surface of the neuroepithelium, although the nose must be separating vapors that have different physical parameters in ways similar to the function of a

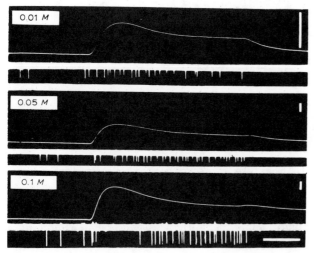

Figure 9.115. Effect on olfactory bulb neurons of increasing strength of odor stimulation. *Upper trace* shows slow potential recorded from sensory layer in the mucosa; *lower trace*, discharge of single bulbar unit. Odor: butanol. Note stimulatory effect of dilute butanol, with very slight pause during rise of peak mucosal response at 0.05 M concentration, and longer inhibitory phase with 0.1 M concentration, followed by a brisk discharge up until the end of butanol exposure. Time bar, 5 s; vertical bar in each record, 1 mV, in reference to receptor potential. [From Døving; from Ottoson and Shepherd (1967).]

gas chromatograph. If olfactory discrimination involves spatiotemporal patterns of activation, and if these patterns are generated by active inhalation and sniffing behavior, then an individual may be able to obtain better quality olfactory information through coupling olfactory searching with respiratory motor control (Macrides, 1976). Moreover, information from differences in concentration of odorants in the two nostrils must be exceedingly important in olfactory tracking. Spatial factors have such obvious importance in other sensory systems that well defined spatial organization is likely to be found for olfaction. As shown below, there is no question that among projections into the bulb and throughout the remainder of the olfactory system there is definite differential topographic organization.

RESPONSES BY SINGLE PRIMARY OLFACTORY NEURONS

More recently, it has been found possible to record intracellularly from primary olfactory neurons, as shown in Figure 9.116. Note the spontaneous activity; near absolute threshold, there is a barely discernible pause, activity, and slowing. At successively higher concentrations, there is a more definitive and prompt response followed by more prolonged postexcitatory inhibition.

According to Mathews (1972) individual olfactory neurons are selectively sensitive in response to a specifiable list of odorants and not to others. No two olfactory neurons seem to respond to the same group of odorants, and there does not seem to be any obvious chemical signature to the way odorants cluster in their ability to activate individual neurons. Odorants with very different properties may activate the same cell, and chemicals with similar properties may not do so.

The olfactory system seems to be organized in such a way as to encode an extremely large number of different odorants and combinations of odorants, as indicated by the enormous variety of different activity profiles which are observed when recording from single primary afferents (Gesteland et al., 1965). There are evidently no pure odorant single olfactory neurons. It appears that evolution did not have to cope with such utter simplicity. After all, it is only in the laboratory setting that olfactory systems are presented with "odorless" air into which "single pure odorants" may be introduced. In natural circumstances, a great many odorants are carried along with every inhalation. The wide range of chemorecognition potentialities in the intramembranous particles, their enormous number and den-

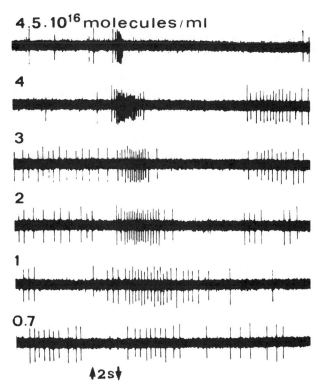

$4.5 \cdot 10^{16}$ molecules/ml

Figure 9.116. Responses of primary olfactory neurons to various concentrations of anisole. [From Holley and Døving (1977).]

sity, and the vast extent of ciliary surfaces, may account for this complexity at the single receptor level.

Odors for any particular species need to be operationally defined in accordance with their conspecific and prey-preditor specific sources and their potential roles in the regulation of behavior and endocrine functions. Neuronal coding probably depends on species-specific receptor molecules in respect to the physicochemical and structural properties of odorants. Subsequent synaptic processing must involve the extraction of information which represents increasingly higher order aspects of both behavior and endocrine physiology. It is likely that among different species the mechanisms for neuronal extraction of information may be comparable, although what particular information is utilized for what behavioral and endocrine purposes is undoubtedly species-specific. During normal olfactory exploration, the neurological consequences will depend not only on concentration of the odorant (and competing odorants in the same atmosphere) but on physical interactions of odorants with the neuroepithelium, solubility of the odorant, location of the olfactory receptor neurons in the epithelium, the sniffing pattern, etc. (Macrides, 1976).

DIFFERENTIAL RESPONSES AMONG SINGLE PRIMARY OLFACTORY NEURONS

It is important to grasp the fact that the nervous system, during reception of action potentials from sense organs has a great many options for information analysis. Figure 9.117 illustrates the fact that two single primary olfactory neurons can generate different patterns of response in relation to the same odorant, amyl acetate, delivered at the same concentration, with the same respiratory force and rhythm. A hamster is connected to double tracheotomy tubes for controlled artificial sniffing. The left and right sides of this figure show responses on the part of each of two single primary olfactory neurons. The top tracings show individual single action potentials in response to the odorant. It is evident that the second unit has a larger amplitude and somewhat more prolonged action potential.

The middle tracings show continuous unitary responses to repeated successive respiratory cycles. Odorant was present in the inhaled air during the 30 s indicated by a broad white slanted bar to the left of the composite tracings. The rate of air movement during forced inspiration and passive expiration is shown in the tracing at the bottom of these composites. It is obvious here that the unit with the smaller action potential nonetheless presents a more concerted, longer lasting burst at the onset of exposure to the odorant. Its smoother spontaneous activity is obviously inhibited in the postexcitatory phase, as is the more rugged and irregular spontaneous activity of the other unit.

At the bottom are two frequency response histograms which indicate that the first unit has a low, regular rate of spontaneous discharge. It shows greatly increased activity as a result of rhythmic bursts during exposure to the odorant and shows conspicuously increased and less regular spontaneous activity following discontinuation of the odorant. The unit depicted on the right exhibits a great deal of irregular spontaneous activity. During the period of exposure to the odorant, the average of its activity is actually reduced, and there is a slow recovery of the previous spontaneous activity. This evidence from Macrides (1976) indicates some of the subtleties of temporal order which add to spatial and other factors involved in chemical analysis of the environment.

What the neurophysiologist chooses to extract, average, and analyze on any given occasion, and considers to be significant, undoubtedly represents only a small fraction of information that could be utilized or ignored by central neural circuits for sensory processing and perception. Integrative

RESPONSE SURFACES

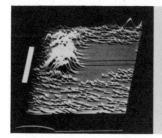
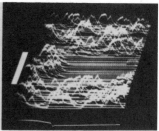

FREQUENCY RESPONSE HISTOGRAMS

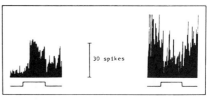

30 spikes

Figure 9.117. Frequency response histograms for hamster single unit olfactory neuron (*top*) in response to amyl nitrate puffs during artificial sniffing. Tracing below middle histograms indicates nasal air flow. Height of *slanted bar* indicates duration of introduction of amyl acetate. Variations in bursting pattern within inhalation cycles are batched into representative "overall" responses as recorded from response surfaces. Two units depicted below show differential responses to amyl acetate. [From Macrides (1976).]

mechanisms at higher levels of nervous organization must be increasingly engaged with the problem of what sensory neural activities represent biologically important features of the environment and more particularly, how they might be exploited for guidance of behavioral and endocrine responses toward immediate and longer range internal satisfactions.

THE PRIMITIVE NATURE OF OLFACTORY EPITHELIA

An important distinction from all other sense organs is that *olfactory neuron cell bodies lie at the surface of the outside world,* shielded only by a thin layer of mucus. They provide an immediately available sensory receptor membrane; their cell bodies are directly engaged with the stimulating world. In contrast, other sensory neurons have cell bodies that lie in sheltered ganglia and send out more or less long dendrites to the periphery. Many of the dentritic endings don't even then get involved directly with stimuli, but make synaptic junctions with specialized sensory receptor cells. This peripheral distribution of sensory cell bodies in olfactory neuroepithelium is characteristic of the organization of invertebrate nervous systems and *suggests that vertebrate olfactory mechanisms are primitive holdovers.*

A further distinction is that in three olfactory epithelia, classical olfactory, vomeronasal, and septal, *all olfactory neurons undergo a natural turnover:* they develop, mature, and die in a matter of weeks and are replaced by primitive neuroblast-type stem cells which also replace olfactory neurons when neuroepithelium is injured. This means that there is a continuous death and regeneration process going on, including the projection of replacement axons and synaptic junctions in first order olfactory relays in the olfactory bulb and in the accessory olfactory bulb (Graziadei, 1976). This is again a characteristic that is more often encountered in primitive nervous systems. It is generally held that no other nerve cells are regenerated in the adult nervous system. The olfactory system therefore provides an interesting model for study of the principles of central synaptic regeneration.

Electron microscopic as well as light microscopic features of olfactory neurons also indicate that they belong to a class of very primitive neurons, reinforcing the notion that olfaction may be the most primitive of all our senses.

There is good evidence for topological order in the projection from olfactory neuroepithelium to glomeruli in the olfactory bulb (Land et al., 1971; Land, 1973). Other evidence for spatial gradients of activity in the olfactory epithelium and in the olfactory bulb has been reviewed by Moulton, 1976. Exposure of rats to strong odor stimulation for periods of a week or more leads to transneuronal degeneration of tufted and mitral cells in restricted regions of the olfactory bulb, and these regions vary with different odors (Døving and Pinching, 1973). Despite the transneuronal degeneration in the bulb, there were no pathological changes found, using electron microscopy, in the olfactory nerve fibers. Sharp et al., 1975, analyzed 2-deoxyglucose accumulation in rats exposed to strong odorant stimulation for 45 min and found restricted regions of activity in the olfactory bulb. Surprisingly, in animals breathing room air, they found occasional, small dense loci of activity limited to the glomerular layer.

THE OLFACTORY BULB

Axons of primary olfactory neurons engage directly synaptically with *mitral and tufted cells* which are output neurons for the olfactory bulb. There is an enormous convergence in this process. As noted earlier, the rabbit olfactory neuroepithelium has

about 50 million receptor neurons on each side. Correspondingly, there are about 1,900 glomeruli, and each glomerulus receives approximately 26,000 axons from the periphery. All the olfactory afferents make synapses on mitral and tufted cell *dendrites*, implying that this afferent control is not quite compelling. There are about 25 mitral cells for each glomerulus, so there is an additional 1000-fold convergence within the bulb.

Sensory processing in the olfactory bulb involves two types of cells intrinsic to the bulb, *periglomerular and granule cells.* Shepherd (1970) recognized similarities in functional organization of the bulb to that in the retina, as shown in Figure 9.118. The olfactory short-axon periglomerular cell is analogous to the retinal short-axon horizontal cell, and the olfactory axonless granule cell is analogous to the retinal axonless amacrine cell.

The direct olfactory pathway consists of primary afferents to mitral and tufted cells. *The modulating circuit, as in the retina, is organized in two tiers:* the first (periglomerular) influences *bulbar input;* the second (granular) influences *bulbar output.* Integration is obtained by reciprocal synaptic (microcircuit) connections, just as it is in the retina. The neurotransmitter for both intrinsic cell types is thought to be GABA (Price, 1977).

Apart from these local circuits, and of course the primary olfactory afferent terminals, *additional endings in the bulb come from each of the same areas of higher olfactory representation to which the mitral and tufted cells project.* These centrifugal projections provide higher centers with a channel to control their own input. Their function is analogous to other central sensory control systems which influence primary receptors and successive ascending sensory relays. *These centrifugal neurons influence mitral and tufted cells indirectly through the control they exert on periglomerular and granule cells.*

There are two additional bulbar projections (aminergic in nature), from the brain stem, *serotonin (5-HT) fibers from raphé nuclei,* and *norepinephrine (NE) fibers from the locus coeruleus.* The aminergic projections from the brain stem probably function by altering the general state of the olfactory bulb as during arousal and attention, and possibly by modifying synaptic efficacy on the basis of cumulative experience.

FUNCTIONAL DEVELOPMENT OF THE OLFACTORY BULB

Olfaction is important to mammals in relation to suckling behavior and intergenerational bonding. Restricted parts of the olfactory bulb mature earlier than does the remainder; in contrast to relatively

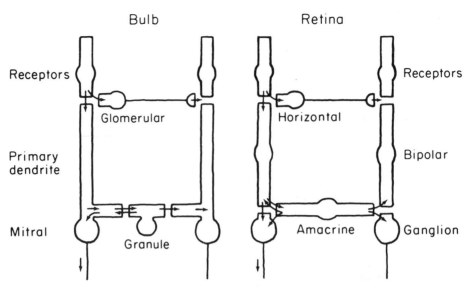

Figure 9.118. Basic circuit of bulb and retina compared. On the *left* is a schematic diagram of synaptic connections within the olfactory bulb for comparison with a similar diagram on the *right*, of the retina. Receptor-neurons project directly to the olfactory bulb where they excite primary dendrites of mitral (and tufted) cells. They also affect short axon periglomerular (glomerular) cells that provide lateral inhibition which modulates mitral and tufted cell inputs. Output of mitral and tufted cells is modulated by reciprocal synaptic microcircuitry involving axonless granule cells. In the retina, receptors activate both bipolar and horizontal cells. Horizontal cells provide lateral inhibition that modulates retinal output. Bipolar output and that of ganglion cells are further modulated by the axonless amacrine cells. The organization of longitudinal and horizontal connections in a two-tiered system of sensory control is closely similar. Note that in each case throughput of sensory signals is being affected at the first synaptic relay and that refinement and analysis of sensory information is already taking place within the bulb and retina. [From Shepherd (1970).]

indistinct glomeruli in the main bulb, these early maturing glomeruli are already histologically well defined at birth. According to Teicher et al., 1980, this *modified glomerular region* lies close to the border between the main bulb and the accessory olfactory bulb, and it is apparently activated by pheromones associated with suckling behavior.

Using 2-deoxyglucose uptake, Greer et al. (1982) showed that both receptor cell terminals and their postsynaptic targets in this restricted part of the olfactory bulb are active within a few hours of birth, at a time when glomerular inhibition may not yet be functioning, and when both 2DG evidence and behavioral evidence indicate that general sensitivity to odorants is low. Thus, a newborn mammal evidently has special developmental precocity in a sensory modality that contributes instrumentally to an essential newborn behavior.

ELECTRICAL ACTIVITY OF THE OLFACTORY BULB

Grossly recorded electrical activity of the olfactory bulb reveals high-frequency, somewhat irregular spontaneous activity which is suppressed by anesthesia but persists after destruction of the olfactory neuroepithelium (Adrian, 1950). When odorants are presented, lower frequency waves with a regular rhythm (ranging from 10–60 Hz) are

induced. Unit discharges occur in synchrony with these waves. There is rapid adaptation to repeated olfactory stimuli, and this *adaptation is virtually abolished when the efferent paths to the bulb are sectioned* (Moulton, 1963).

MECHANISMS OF PATTERN DISCRIMINATION IN OLFACTION

Differential receptor capabilities entail patterns of great complexity, probably even more complex than those in vision. The organization of the bulb, with its intrinsic inhibition affecting bulbar reception and relay, may subserve analogous functions to lateral inhibition in the retina. In the visual system, this type of organization contributes to contrast enhancement, directional selectivity, detection of motion, adaptation, etc.

Like the retina, olfactory neuroepithelium is a sheet of receptors. There probably is differential regional sensitivity of receptor cells to odorants within the sheet, and this may be enhanced by controlled, reiterated movements of odor-laden air across the sheet (as in gas chromatography). Sniffing relates to olfactory perception in the way that actions of the iris, lens, and extraocular muscles regulate exposure of the retinal receptor sheet to stimulus patterns.

Bulbar inhibitory patterns, organized in the glomeruli, provide additional means for sensory analysis. Bulbar excitatory and inhibitory circuits are implicated in rhythmic activity of the bulb which is largely under centrifugal control. At times, the bulb undergoes seizure-like activity which takes on added significance in light of the vulnerability of olfactory cortex to seizure activity (see below).

Central Connections of the Olfactory Bulb

Organization of the olfactory system has been skillfully conceptualized, simplified and arranged in anatomical order by Shepherd (1983) as illustrated in Fig. 9.119. The olfactory bulb is properly represented as a part of the brain. Incoming primary afferents enter bulbar glomerular layer where they contact mitral and tufted neurons (designated m). The glomerular relay is modulated by periglomerular and granule cell actions which in turn are under the influence of centrifugal neurons (*dashed lines*) projecting outward from areas of higher olfactory representation and also from the brain stem. The major projection pathways, with destinations of mitral and tufted cell terminals, are depicted in due order.

Mitral and tufted cells send axons along the *olfactory tract*. The tract flattens posteriorly and separates into medial and lateral *olfactory striae*. Divergence of the two striae exposes a triangular area, the olfactory trigone. The striae distribute fibers to the *anterior olfactory nucleus, olfactory tubercule, septal nuclei*, subdivisions of the *amygdaloid complex*, parts of the *hypothalamus, pyriform cortex*, and *transitional entorhinal cortex*.

Central olfactory representation forms a rough triangle, proceeding posteriorly from the bulb at the apex, widening along the posterior medial orbital surface of the frontal lobe, and widening still more so laterally as the base of the triangle flows posteriorly and laterally over the anterior and medial surfaces of the tip of the temporal lobe. Many years ago (1941), Allen showed in dogs that the cortical projection system at the base of this triangle is necessary and sufficient (as compared with other cortical areas) for olfactory learning and olfactory conditional responses. His contribution was the more remarkable because at that time it was presumed that the whole "rhinencephalon" was involved in olfactory perception and behavior.

Heimer (1976) provides a useful critical review of olfactory projection systems and their wider relations. The distribution of tufted cells is limited to the more rostral of these olfactory projection areas: the anterior olfactory nucleus, medial part of the pyriform cortex, and lateral part of the olfactory tubercle. But mitral cells proceed farther and distribute terminals to all regions of olfactory representation (Price, 1977).

The *anterior olfactory nucleus*, located anterior to the division of the olfactory tract into medial and lateral striae, projects to the same structures

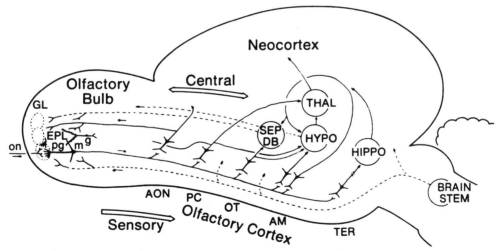

Figure 9.119. Central olfactory projections. Primary receptor-neurons enter the olfactory bulb glomerular layer (*GL*) where they are affected by periglomerular (*pg*) cells in the external plexiform layer (*EPL*) and by granule (*g*) cells deeper within the bulb. Abbreviations for olfactory projections: *AON*, anterior olfactory nucleus; *PC*, pyriform cortex; *OT*, olfactory tubercle; *AM*, amygdala; *TER*, transitional entorhinal cortex (mesocortex); *HIPPO*, hippocampus (archicortex). Abbreviations for subcortical regions: *SEP-DB*, septum and diagonal band; *HYPO*, hypothalamus; *THAL*, thalamus, which projects to precentral cortex in the frontal lobes (neocortex). Brain stem aminergic projections (serotonin and norepinephrine to the bulb and dopamine to the olfactory tubercle) are indicated by *dashed lines*. All destinations of mitral and tufted cell axons send back reciprocal projections which are schematically diagrammed by *upper dashed lines*. These modulate afferent throughput indirectly by influence on granule and periglomerular cells. [From Shepherd (1983).]

that also receive input directly from the olfactory bulb. Thus, the anterior olfactory nucleus represents an indirect relay station. Its physiology has been summarized by Daval and Leveteau (1974).

The *olfactory tubercle* corresponds to the anterior perforated space, gray matter lying immediately behind the olfactory trigone. It receives direct bulbar projections and projections that are relayed through the anterior olfactory nucleus. It also receives dense dopaminergic (DA) innervation from the substantia nigra and associated reticular formation in the brain stem (Björklund and Lindvall, 1976). This DA projection is of particular interest because it may be functionally implicated in schizoaffective disorders. The tubercle is considered important for its nonolfactory projections to temporal lobe, particularly to limbic structures. Olfactory association fibers to the limbic system presumably contribute to olfactory perceptions and memories during emotional experience and expression.

The *pyriform lobe* is the main region engaged in olfactory discrimination. It has three principal parts: (1) a large, heterogenous area of gray matter which extends along the medial posterior orbital surface of the frontal lobe, from pyriform cortex itself almost to the olfactory bulb—hence its designation as the *prepyriform area*; (2) *pyriform cortex* proper which covers the anterior and anteromedial surface of the temporal lobe; and (3) *entorhinal cortex* (Brodmann's area 28) which includes a large part of the anterior hippocampus.

The entorhinal destination of olfactory projections led neurologists for a long time to assume that what is now known as the *limbic system* would be entirely devoted to olfactory functions, a "smell brain." On this account it was called rhinencephalon.

The anterior and parts of the medial and posterior cortical nuclei of the *amygdaloid complex* also receive powerful projections from the bulb. The amygdaloid nuclei project to the *ventromedial nucleus of the hypothalamus*. By this means there is a relatively direct circuit between the olfactory system and that part of the hypothalamus which is involved in *initiating eating behavior*.

As described subsequently, the amygdala is involved in fundamental aspects of social behavior, and olfactory contributions to the amygdaloid complex undoubtedly participate instrumentally in this aspect of behavior.

Olfactory cortical projections include part of a triple ring of cortex which encircles the hylus of the cerebral hemispheres. It comprises one sector of the anterolateral zone of this encircling cortex, which is called *mesocortex* and *archicortex*. Mesocortex is

three-layered, intermediate in complexity between *archicortex* which is a one-layered cortex, and *neocortex*, which is six-layered. Neocortex, as its name implies, is the newest as well as the most elaborately layered cortex which makes its great evolutionary contributions to mammals.

Other sensory modalities: auditory-vestibular, visual, and somesthetic, all project to broad expanses of neocortex. Therefore, *olfaction not only has a primitive neuroepithelium but its central projections are limited to relatively primitive cortex*. Perhaps the most important distinction concerning olfactory projections in comparison with other sensory systems is that the others all reach cortex by relaying through the thalamus. *Olfaction does not pass through the thalamus* on its way to cortex, but reaches the thalamus secondarily by way of cortical efferent projections to the medial dorsal nucleus of the thalamus.

Organization of Olfactory Cortex

Olfactory cortex receives axons from mitral and tufted cells which enter along the cortical surface and terminate on the apical dendrites of a superficial layer of pyramidal cells. Again, this contrasts with other sensory cortices where the afferent entrance is from below and the main synaptic terminals synapse in the middle cortical layers. Cortical pyramidal cells are first excited by olfactory signals and then inhibited under the influence of intracortical circuits. Many of the pyramidal cells have extraordinarily long axon collaterals which arborize widely throughout olfactory cortex. Perhaps because of this excitatory feedforward circuitry, olfactory cortex is especially susceptible to seizure activity. *Seizure induction anywhere in the olfactory forebrain tends to spread throughout mesocortex* and into directly connected *subcortical structures*. It does not readily spread across the cytoarchitectural boundary into neocortical areas. When it does, it is likely to result in classical, grand mal seizure activity.

Mesocortex is nearly equally susceptible to seizure activity throughout its entire hilus-encircling ring structure. Archicortex is greatly expanded in the temporal lobe where it is known as the hippocampus. It, too, is extremely vulnerable to seizure activation. In addition, these circuits are especially vulnerable to the effects of a wide variety of psychotropic drugs, narcotic agents, and alcohol. The combination of *temporal cortical and subcortical seizure susceptibility* (including the amygdala) accounts for a major category of epilepsy, *temporal lobe epilepsy*. Temporal lobe seizures do not ordinarily spill over into neocortex to manifest classical grand mal seizures, with gross tonic and clonic rhythmic skeletal muscle contractions, breath holding, tongue biting, frothing at the mouth, urinary sphincter relaxation, etc. Rather, they result in *altered states of consciousness.*

Olfactory Associations with Other Brain Systems

Olfactory association fibers from the pyriform lobe project to the large *medial dorsal nucleus of the thalamus* and are reprojected onto the *neocortex of the frontal lobe* (*prefrontal cortex*) which is involved in anticipating and planning future behavior.

A massive fiber bundle, known as the olfactohypothalamic system connects the olfactory bulb and adjacent retrobulbar areas with the *septum* and the *anterolateral and posterior hypothalamus.* This powerful projection also contributes to the *stria medullaris* which courses posteriorly as a visible band on the surface of the thalamus, destined for the *medial dorsal nucleus of the thalamus*, and the *habenular complex* which lies in the posterior diencephalon, alongside the junction between third ventricle and aqueduct. Another part projects to the *anterior and lateral hypothalamus*, thereby *contributing to neuroendocrine functions and behaviors relating to reproduction* (Scott and Leonard, 1971; Powell et al, 1965).

Olfaction and Social Behavior

The amygdaloid complex receives highly complex derivations from activities in sensory association fields: auditory, visual, somesthetic, and olfactory. Gloor (1976) has proposed an interesting interpretation of the relationship of the amygdaloid complex to central mechanisms in perceptual experience. The amygdala not only sits astride these perceptual confluences, it has outflow access to appropriate behavioral systems through both the hypothalamus and the limbic system. Gloor suggests that the amygdaloid complex contributes an "affective bias," what we call "mood," in relation to familiar as well as unfamiliar sensory stimuli. It can accordingly influence hypothalamic appetitive, neuroendocrine, and behavioral channels. It can likewise accordingly influence limbic generation of emotional experience and emotional expression.

The amygdaloid complex, according to Gloor's interpretation, contributes to "decision-making processes" on which approach and avoidance behaviors depend, determining what is suitable and what is unsuitable (e.g., for eating and for reproductive behavior). It does this, as well, concerning who is friend and who is foe, evoking social cooperation with its associated feelings of warmth, well-being, and security, and alternatively, fleeing or

fighting, with the cautions, fears, and defensive/ aggressive behaviors associated with social conflict. The "*moods*," which represent superordinate integrations of combined perceptions, *set the stage for consequential behaviors* in a variety of contexts (Gloor, 1976).

Ethological studies show that, especially in lower animals, olfactory signals have an overriding importance in determining affective attitudes and affective behaviors (Schultze-Westrum, 1969; Ralls, 1971). It is also important to remember that olfactory memories are extremely powerful in respect to commanding attention and that they are also extremely resistant to decay (Engen et al., 1973a, 1973b). Gloor bases his interpretations on the fact that the amygdaloid complex is recipient of information relating to all complex sensory association systems, auditory, somesthetic, visual, and olfactory.

The combination of higher perceptions and perceptual memories available to the amygdaloid complex occurring in a decision-making context presumably results in decisions being executed through hypothalamic and limbic projections to the midbrain for appropriate approach and avoidance behaviors. Animals in which both amygdalae have been destroyed are seriously handicapped in their behavioral interactions within their social environment. They are evidently unable to link exteroceptive signals that *should* guide social behavior with appropriate affective states. Kling (1972, 1975) showed that wild monkeys in whom a bilateral amygdalectomy has been performed shun conspecifics, even the troop with which they have been reared, and fail to respond to various socially significant signals emitted by other monkeys. After release to the wild, the amygdalectomized monkeys become social isolates, usually hiding in inaccessible places, although they manage their nonsocial existence quite well.

Temporal Lobe Seizures

An instructive variety of seizure activity is *psychomotor epilepsy*, which has a primary focus in the anterior part of the temporal lobe. The best description was the first, published by Jackson in 1888. The patient, who was a physician, was able to provide an account of his subjective state at the outset of an attack. He had what Jackson called an "intellectual aura," feelings of *altered consciousness.* He might then go into a fugue state, with amnesia, lasting a few minutes to an hour. During such episodic fugue states, without later memory for what he had been doing, his behavior included examining a patient and writing admission orders for hospitalization, complete with a diagnosis of pneumonia. He was relieved to learn afterward that he had made a correct diagnosis and given appropriate orders. On another occasion, he experienced an onset of altered consciousness while on a train. He got off the train at the right station, giving up his ticket appropriately, and proceeded correctly, as he later reconstructed events, to find an unfamiliar address.

Psychomotor attacks are initiated by an abnormal focal epileptic discharge in the olfactory or limbic cortex or amygdala in the anterior part of the temporal lobe. There may be periods of obtunded consciousness, mental confusion, and fugue states with amnesia which may last up to several hours or days. Although rare, periods of prolonged abnormal function of temporal lobe structures are of medicolegal importance, since patients may commit some misdemeanor or even serious crime while they are in such amnestic states.

More commonly, psychomotor attacks are characterized by a briefer loss of consciousness which may be accompanied by lip smacking, incoherent speech, mental cloudiness, fugue states, and periods of amnesia lasting from 30 s to a few minutes, with partial clouding of consciousness after the attack has apparently ended.

Kindling

Direct electrical excitation of different cortical regions reveals large differences in susceptibility to seizure induction (French et al., 1956). Some regions of cortex have low thresholds while others are so refractory that any seizures induced through their stimulation may depend on spread of activity into the more susceptible regions. Olfactory and limbic cortex—and the underlying amygdala—are among cortical regions that are relatively more susceptible to direct seizure induction. Stimulation of this region at intensities well below threshold for seizure induction elsewhere has provided important new insights into the dynamics of brain circuits.

Kindling is a term borrowed from the expression for using small pieces of wood to ignite a larger fire; by analogy, repeated weak excitations of certain parts of the brain will, over time, generate full-blown seizures. Cortical and subcortical *kindling is observed when electrical stimulation* (or other exciting agent) *is repeatedly applied at low intensity or dosage* (well below seizure threshold), *and the response to that agent progressively increases until it evokes a major clinical convulsion* (Goddard, 1967; Goddard et al., 1969).

THE OLFACTORY BULB AND ALL OLFACTORY PROJECTIONS ARE READILY KINDLED

Temporal olfactory and limbic cortex and the amygdala are particularly suitable for kindling experiments. Other primary sensory pathways are also subject to kindling, although less facilely, providing only that the kindling stimulation is repeated long enough (Cain, 1979). With initial electrical stimulation, it is not necessary to elicit a local electrical afterdischarge because with repeated mild stimulation, the threshold for afterdischarge is progressively lowered (Racine, 1972a). Following initial appearance of an afterdischarge, it progressively increases in intensity and duration and spreads to remote areas of the brain, including particularly the mirror-image area in the contralateral hemisphere.

A typical protocol for kindling involves the application of a short train of electrical pulses (e.g., one or a few seconds in train of 25 Hz at 50 μA) through an electrode implanted chronically in the amygdala. The stimulus train is repeated once or a few times per day (see review by Goddard and Douglas, 1976). Sensitivity to the convulsive response typically continues to increase until, after many repetitions, convulsions occur and recur spontaneously (Wada et al., 1976).

Kindling has been observed in frogs as well as a wide range of mammals including primates and has been obtained with a wide variety of stimulants in addition to mild, local electrical excitation: focal chemical (cholinergic) stimulation as well as stimulation by a variety of pharmacological agents—pentylene tetrazol, flurothyl, cocaine, repeated alcohol withdrawal, repeated transcranial electroshock, and repeated auditory stimulation (Goddard et al., 1976). Kindling has not been elicitable from stimulation of certain areas of neocortex, much of the thalamus, midbrain reticular formation, red nucleus, substantia nigra, and cerebellum (Racine, 1972b). Control experiments demonstrate that the kindling effect does not result from tissue damage, edema, gliosis, or other effects of electrode implantation.

Until spontaneous seizures develop, the animal's behavior appears normal between kindling stimulus trials. Although the brain does not seem to be damaged by kindling, its circuits are altered, perhaps permanently, and the effects are widespread. Electrolytic destruction of all neurons which were directly activated during the kindling process does not prevent seizure induction by activating neurons on the periphery of the lesion, or by activating neurons in the mirror-focus in the contralateral hemisphere. Repeated stimulation of a second site results in more rapid kindling even though the primary focus has been eliminated (Goddard et al., 1969; Burnham, 1976).

Repetition of the kindling stimulus somehow increases the stability and fixes enhancement of responses evoked in the corresponding circuits. This enhancement of circuits is specific for the particular synapses affected during the kindling process, and may be pre- or postsynaptic, but is localized to the terminal axon and dendrite in the vicinity of those synapses that relay the kindling spike trains. Greater enhancement is obtained when two convergent pathways are stimulated independently. Since similar enhancements are predicted by a number of theories of memory, it is reasonable to conjecture that the underlying mechanisms of synaptic enhancement in kindling and the mechanisms of learning may be one and the same (Goddard et al., 1976).

Accessory Olfactory Bulb—Pheromone Circuits

The accessory olfactory bulb receives olfactory afferents carrying olfactory and particularly pheromone signals from the vomeronasal organ. It projects exclusively to parts of the medial and posterior cortical amygdaloid complex, which parts do not receive terminals from the principal olfactory bulb. It also projects exclusively to prepyriform and pyriform cortex. Thus, the accessory olfactory bulb projects *to different parts of the amygdaloid complex and to different fields of cortex, as compared with projections of the main olfactory bulb* (Scalia and Winans, 1975; Broadwell, 1975). This independence persists in next-order projections from the amygdala to the hypothalamus (Raisman, 1972). Like the olfactory bulb, the accessory olfactory bulb receives efferent projections from the same regions to which it projects. This means that *there is an apparently entirely separate central as well as peripheral representation of the two principal olfactory neuroepithelia.* Altogether these comprise dual independent olfactory systems, both capable of detecting a wide variety of odorants, *the vomeronasal-accessory olfactory system specializing in reception and response to sexual pheromones.*

Olfactory Cortical Associations

Each of the cortical and subcortical regions which receives projections from the olfactory bulb and from the accessory olfactory bulb sends fibers to one another. What has previously been maintained independently can now interact. The cortical ter-

minals of these association fibers end in the most superficial layer of the three-layered mesocortex, directly beneath the incoming bulbar afferents, and in the deepest third layer. Axons from pyramidal cells in the second and third layers go to the phylogenetically ancient hippocampus. Projections from olfactory cortex to the medial dorsal nucleus of the thalamus, the ventral putamen (basal ganglia), and the lateral hypothalamus, all phylogenetically more recent systems, however, arise from an additional, novel, deeper layer exclusive to olfactory cortex. It is as if this part of mesocortex were in the process of trying to add more layers, perhaps something to be functionally on a par with layers V and VI of neocortex. Olfactory corticocortical fibers also extend to neocortical areas adjoining the olfactory cortex, to infralimbic areas on the medial side of the hemisphere and to ventral posterior cortex of the insula.

It is clear that central systems for analysis of olfaction are both complex and orderly. Furthermore, *the olfactory tubercle, which lies more or less in the geometric center of olfactory representation, receives fibers from the brain stem which contain dopamine (DA)* (*Ungerstedt, 1974*). Because of close associations of the olfactory tubercle with the basal ganglia, as well as with the limbic system, the olfactory tubercle is probably involved in important dopamine-related psychological and behavioral disorders (Ungerstedt, 1974). It is also noteworthy that olfactory cortical association fibers project to the hippocampus where they are integrated with other sensory modalities in relation to emotional aspects of experience and behavior (MacLean, 1975).

EFFERENT PROJECTIONS TO THE OLFACTORY BULB

Kerr and Hagbarth (1955) were first to show that centrifugal input to the bulb can have an overall quieting influence or an excitatory or disinhibiting effect on resting and induced bulbar activity. Efferent fibers to the bulb arise from the anterior olfactory nucleus, anterior commissure, and nucleus of the diagonal band. They all terminate on granule and periglomerular cells. Central efferent control by way of these projections are therefore indirect in their effects on bulbar output by mitral and tufted cells. The anterior olfactory nucleus (AON) on each side projects to the ipsilateral and contralateral bulbs and to the contralateral AON.

Whether differences in concentration of odorants in the two nostrils receive early central comparison as a guide for odor tracking behavior is unknown, but watching how fast a good hunting dog can follow a warm trail, even when fresh snow has obliterated the tracks, suggests that internostral differences may contribute.

Olfactory-Neuroendocrine Influences

A typical sequence of reproductive behavior in mammals involves several sensory modalities which play their respective individual and interrelated roles. They may come into action successively, as when olfaction and vision are followed by acoustic interchange, touch, and taste. Such sensory modalities may operate alone, simultaneously, and even vicariously. Mammals characteristically exhibit frequent sniffing, lip smacking, and licking in the course of exploring the abundant chemical information in their environment. Animals are alert for food, conspecifics, and predators. Only a small fraction of the continuing central nervous system processing of olfactory and related taste information and information from other sensory modalities results in overt behavior.

Primer pheromones bring about slow, long-term responses of a developmental nature or a change of physiological state. *Releaser pheromones* bring about more immediate behavioral responses. *Alarm pheromones* cause prompt general excitement, defense, or escape behavior in individuals and may result in imitative release by conspecifics. Pheromones not only activate but they also habituate; heavy exposure may be followed by habituation lasting weeks. (For a general review of these mechanisms, see Mueller-Schwarze and Mozell, 1976).

A variety of neuroendocrine responses are elicited by pheromonal olfactory activation, including prolonged effects from release of growth hormone (GH), leuteinizing hormone (LH), and follicle-stimulating hormone (FSH), and more immediate effects from release of adrenal corticotrophic hormone (ACTH), and other factors, each with particular tissue targets, cellular and tissue effects, and operating time scales.

Studies in mice and hamsters indicate that sex-related olfactory cues will activate the hypothalamo-pituitary-gonadal axis in conspecifics (Macrides et al., 1977). Odorants emitted by females lead to a rapid increase of plasma testosterone levels in adult male mice. Prolonged exposure accelerates the growth of accessory sex organs in young males. In the adult male hamster, exposure to vaginal odors can be as effective in raising plasma testosterone levels as is pairing with a receptive female. Adult male mice respond similarly to conspecific female urine. Even urine from diestrous, proestrous, or ovariectomized female mice produce marked elevations of LH in male mice.

Chemosensory influences on gonadotropin secretions have emphasized olfactory control of LH and FSH. In male rats exposed to females, there is also a rise of plasma prolactin. In male hamsters, prolactin plays a major role in reactivating the testes on return of longer daylengths and with the onset of the breeding season. Moreover, prolactin enhances the response of sebaceous glands to testosterone and thereby contributes to dissemination of odorants. Thus, prolactin is involved in olfactorily related neuroendocrine readiness for sex and also in announcing that readiness.

It can be readily appreciated that chemosensory and social influences on gonadotropin and androgens are important for courtship activities as well as sexual motivation and fertilizing capacities of males. The dynamics of such complex system responses are illustrated by the fact that male mice paired with an intact female for 1 wk do not have higher plasma testosterone levels than do males in all-male control groups. The experimental group had all mated, and the females were pregnant and nonresponsive. Nevertheless, urine from these females would elevate hormone levels in other males.

In similar fashion, male odors affect neuroendocrine functions in females. Male odors can accelerate the sexual maturation of female mice. The estrous cycle in female mice can also be accelerated by an androgen factor in male mouse urine. Male odors of a newly introduced mouse can even interrupt pregnancy in female mice. Altogether, this means that olfactory stimulation and consequences of sexual behaviors combine to modulate olfactory-neuroendocrine response systems. *Hypothalamic mechanisms function not only on behalf of homeostatic mechanisms regulating the individual during development and throughout its adaptation to a dynamic environment, but specifically they function also in respect to the individual in social contexts.* Olfaction contributes importantly to the initiation of important neuroendocrine activities that relate to survival and reproduction.

TASTE

Sensations of Taste

Taste is chemical discrimination by touch. Taste sensations are generated by transactions among chemicals and receptors in vertebrate taste buds together with the consequent activities taking effect along their nerves and central pathways. Of course there is much sensory processing, convergence, and integration among related systems which altogether result in gustatory perceptions.

Olfaction lures us toward or propels us away from

airborne chemical sources. In contrast, chemicals important in taste are rarely airborne. Even when we are within tasting (contact) range, olfaction continues to present strong cues as to what may or may not be appropriate for tasting and ingesting. We may need to "hold our noses" in order to taste something offensive as an odorant even though it may not be distasteful.

Taste provides portal perceptual evaluations of what may be appropriate for ingestion. Foul or repulsive tastes provide a nearly compelling barrier to ingestion, even against force of will. By impulse, stuff may be ejected from the mouth and throat, and for what may have already been swallowed, by vomiting, from the stomach itself. It may be useful to sugar-coat or provide a tasteful vehicle to enable a patient to swallow something distasteful because it is important for nutrition and health. This subterfuge lets distasteful materials surmount the taste barrier. Taste thresholds are remarkably low as shown in Table 9.5.

THE ORGANIZATION OF TASTE BUDS AND PAPILLAE

Taste is a contact chemical sense, innervated by sensory components of the VIIth, IXth, and Xth cranial nerves (Fig. 9.113). The sensory epithelium consists of a scattered host of *taste buds*. Taste buds are found widely distributed over the tongue, pharynx, and larynx. They are aggregated in relation to three different kinds of papillae: fungiform, foliate, and circumvallate. Taste buds appear to be generally similar among these various papillary structures.

Fungiform papillae, generally distributed over the top and sides of the tongue, look like blunt pegs which stick up from the surface of the tongue with one to five taste buds mounted on top. They are innervated by the chorda tympani nerves (branches of VII). *Foliate papillae* resemble such pegs which have been lowered into a surrounding moat that is filled with serous fluid, so that the top of each peg is about even with the surrounding tongue surface.

Table 9.5
Absolute thresholds

Substance	% Concentration (approx.)	Molar Concentration
Sucrose	7×10^{-1}	2×10^{-2}
Sodium chloride	2×10^{-1}	3.5×10^{-2}
Hydrochloric acid	7×10^{-3}	2×10^{-3}
Crystallose (sodium salt of saccharin)	5×10^{-4}	2×10^{-5}
Quinine sulphate	3×10^{-5}	4×10^{-7}

From Stevens (1951).

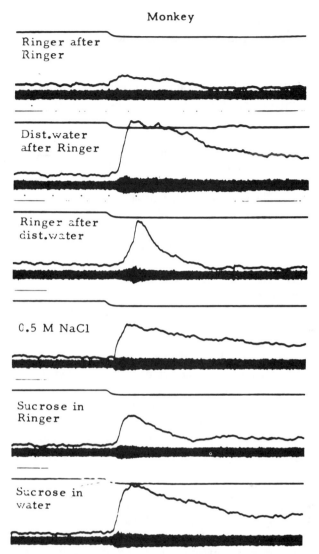

Monkey

Ringer after Ringer

Dist. water after Ringer

Ringer after dist. water

0.5 M NaCl

Sucrose in Ringer

Sucrose in water

Figure 9.120. Records from the undivided chorda tympani nerve in monkey. In each film strip, signal marks moment when test solution is applied on the tongue; below is the integrated response; lowest record is direct electrical response from the nerve. The response to sucrose in Ringer is smaller than that of sucrose in water, as Ringer inhibits the response to water. [From Zotterman (1967).]

Thousands of taste buds reside on the sides instead of the top of the foliate pegs and are facing into the serous fluid. Foliate papillae are found in the posterior and lateral surfaces of the tongue, in the region characterized by bitter taste reception. They are innervated by both the chorda tympani (VII) and glossopharyngeal (IX) nerves. Figure 9.120 shows electrical recordings, raw and integrated, from the whole chorda tympani nerve during application of solutions to fungiform papillae on the anterior two-thirds of the monkey tongue. *Circumvallate papillae*, innervated by the glossopharyngeal nerves (IXth), are very much larger organs, containing thousands of taste buds which face out from the sides of stout central stalks into large-scale circumferential serous-filled moats.

Each taste bud consists of 40–50 modified epithelial cells, grouped in a barrel-shaped aggregate beneath a small pore opening in the epithelial surface. *Receptor cells* crowd their narrow, specialized apical surfaces together beneath the side walls of the pore, forming the bottom of a cylindrical pit. Their exposed surfaces, dense with *microvilli*, lie in contact with the acidophilic contents of the pit. It is within this pit that chemicals exert their effects on receptor molecules embedded in the microvillous membranes of the receptor cells.

Each taste nerve arborizes before it innervates several taste buds and many cells within each taste bud. Each afferent thus reflects activity of many receptor cells. The effect of chemoactivation of the receptor cell is to generate a simultaneous receptor and generator potential which in turn activates the nerve endings that are attached to the base of the receptor cell. Receptor cells are surrounded by supporting cells. Replacement basal cells are found at the bottom of the taste bud. *Individual cells in each taste bud differentiate then degenerate and* are replaced on about a weekly basis from the stock of basal cells in the floor of the taste bud. This means that the taste nerves are continually remodeling synapses on newly generated receptor cells.

Trophic Influences of Gustatory Nerves

Neural innervation is essential for maintenance and regeneration of taste buds. Recordings can be made from the gustatory nerves, but if their ends are severed centrally, recording of chemically activated nerve impulses dies out within minutes to hours (Berland et al., 1977). Length of survival of physiological activity depends on the relative length of nerve remaining attached to the taste bud. The closer to the taste bud the nerve is sectioned, the shorter is the time of survival of recordable activity. Efferent fibers in the IXth nerve are not necessary for maintaining such responses. It is presumed that chemical interactions are required on a continuing basis between gustatory nerves and the taste buds and that this depends on continuing transport of chemical agents manufactured in the cell body. It takes about 10–20 hours for tritiated leucine to travel from the sensory ganglion of the rat vagus (Xth) nerve to the corresponding taste buds. This constitutes an example of *trophic function* of neurons.

Taste Bud Innervation

The tongue, along with other contents of the oral cavity, is innervated by the trigeminal (Vth) nerve.

Taste buds in the anterior two-thirds of the tongue are innervated by terminals of the intermediate nerve, a branch of the facial (VIIth) nerve (Fig. 9.113). The glossopharyngeal (XIIth) nerve innervates taste buds on the posterior third of the tongue. Branches of the vagus (Xth) nerve innervate taste buds found on the soft palate and palatal arches, the extreme posterior part of the dorsum of the tongue, and taste buds on the superior surface of the epiglottis and larynx. Cell bodies of these afferent nerves are located in the geniculate, superior petrosal, and nodosal ganglia, respectively (Fig. 9.113*B*).

The number of taste buds on the anterior two-thirds of the tongue is not great, and they tend to disappear progressively with age. Those on the larynx and epiglottis are abundant at birth but begin to disappear already during infancy. This leaves the tongue as the major organ of taste, with regionally separate peripheral channels of innervation. Nevertheless, the taste components of all three nerves, VII, IX, and X, converge in the bulb and pons to innervate cells in the *nucleus of the solitary tract (NST)*. This nucleus provides a familiar longitudinal landmark which courses through the entire length of the medulla oblongata. The NST consists of a long compact cylinder of cell bodies which are embraced by bundles of incoming afferent and ascending second order axons. The NST lies laterally along either side of the brain stem. In cross-sections they loosely speaking resemble "eyes," with crowded gray matter of cell bodies making a "pupil" within dense white matter.

The perception of "taste" is influenced by trigeminal nerve responses of tactile, thermal, and pain receptors in mouth regions. Moreover, taste is influenced by a substance's temperature and physical consistency as well as by its chemical characteristics. By tacit agreement, gustatory mechanisms omit consideration of associated sensations (mainly trigeminal) described as peppery, piquant, burning, etc. As noted earlier, *olfaction plays a very important role in what we assign to perceptions of "taste."* Onion and apple, for example, cannot be distinguished when the nasal passages are blocked.

The process of moving food about while it is in the mouth, by motion of lips, cheeks, jaws, tongue, and gullet, aided by secretion of salivary juices, serves to distribute food over the taste neuroepithelia, in much the same way that the eye moves over visual objects to improve visual perception. Chewing alters physical bulk and consistency and can mechanically release different tastes. The enzymatic actions of salivary secretions break down some compounds into tastier or less tasteful components. Liquification of food, from salivation and from drinking, allows material to find readier access to taste buds.

Combinations of taste may enhance one another, e.g., sweets without salt tend to be far less attractive. Taste is also influenced by the recent past history (adaptation) of taste receptors and by all longer standing past history (conditioning) which may associate special appreciation or revulsion to a particular taste. Taste in this sense is acquired.

The presence of certain substances in the bloodstream may lead to sensations of taste. As is true of material taken into the mouth, such tastes tend to adapt out. Alterations in some chemical constituents in the blood not only influence taste but may alter the sensitivity of the individual to the tastes of certain substances.

Tastes play a decisive role in maintaining appropriate physiological and nutritional balances (Pfaffmann, 1970). For example, animals with a salt deficiency show a definite preference for salt, and carbohydrate deficiency may be associated with an increase in appetite for sugar. Some deficiencies may not create a specific taste preference but may increase searching activity and preference for novel foods (Nachman and Cole, 1971).

SENSATIONS OF TASTE

The simplest categories of taste are sweet, salt, sour or acid, and bitter. Taste buds exhibit preferences along these lines in different regions of the human tongue (Fig. 9.113*A*). The four fundamental qualities of taste, combined with tactile and olfactory cues in various combinations, account for much of taste experience.

Taste in natural circumstances, of course, involves a wide spectrum of mixed chemical stimuli. Many different kinds of receptors are activated simultaneously. Each receptor presumably responds best at one part of the taste spectrum and is also sensitive to others. There is likely a broad overlap of sensitivities. Interactions undoubtedly include important peripheral and central inhibitory influences. Therefore, the discriminable information available from the whole system is vastly more than that of the sum of capabilities of individual chemosensory receptors.

Sugar molecules are among the simplest entities in which basic tastes can be correlated with stereochemical features. It is suggestive that hydrogen bonding is involved in sugar polarization with respect to taste receptors. Sugars and their derivatives are almost always sweet, bitter, or a combination of bitter-sweet in their perceptual effects. It is supposed that there may be two separate but closely adjoining receptors for sweet and bitter responses and that some molecules can span the two sites (Birch and Mylvaganam, 1976). Intramolecular hydrogen bonds in sugar molecules can render hydroxyl groups unavailable for complexing with receptors and hence can diminish sweetness. In amino acids, which possess only one asymmetric carbon atom, one enantiomer is usually sweet and the other is bitter.

The presence of hydrogen ions is fundamental to the identification of sour-tasting substances, although other factors contribute overall. Salty taste depends on ionization; the cation is more effective than the anion, and "saltiness" is most characteristic with NaCl. Other salts generate mixed tastes.

Central Gustatory Pathways

The principal components of central representation of taste are illustrated in the rat brain in Figure 9.121 which is by Pfaffmann, from Norgren, 1977. The first relay of taste impulses occurs in the cephalic half of the nucleus of the solitary tract (NST). Second order neurons residing in the anterior half of the NST receive afferents from the bipolar cells which extend their peripheral branches out to taste buds distributed throughout the oral cavity. Taste impulses enter the NST via terminal axons of these cells which belong to the VIIth, IXth, and Xth nerves. The cell bodies are located in the corresponding geniculate, superior petrosal, and nodosal ganglia.

NST neurons project directly to the medialmost part of the ventral posterior thalamus (VPM, or TTA), adjacent to other trigeminal oral fibers in juxtaposition to facial, oral, and particularly tongue representations. Third order axons project from thalamus to the somatosensory face (taste) area of cortex (CTA) in the lower postcentral gyrus, near the sylvian fissure. Another thalamocortical projection goes to anterior limbic cortex—entorhinal and anterior insular.

Brodal (1981, p. 503) cited a number of clinical observations which confirm much of this basic physiological and anatomical evidence: contralateral impairment of taste following tumors involving VPM of the thalamus; subjective sensations of taste on stimulation of the lowermost portion of the postcentral gyrus, where it approaches the insula; gustatory auras preceding epileptic convulsions originating in this region of cortex; and impairment of contralateral taste following cortical lesions in the same location. Altogether, the NST gives off axons or collaterals to many regions of the brain stem and diencephalon, including reticular formation, raphe nuclei, and other diffusely projecting

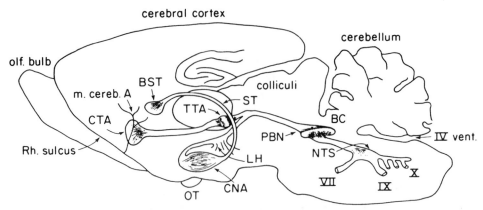

Figure 9.121. Basic pathways of the central gustatory system superimposed on a parasagittal section of the rat brain. *BC*, brachium conjunctivum; *BST*, bed nucleus of stria terminalis; *CNA*, central nucleus of the amygdala; *CTA*, cortical taste area; *IV vent.*, fourth ventricle; *IX*, glossopharyngeal nerve; *LH*, lateral hypothalamus; *m. cereb A*, middle cerebral artery; *NTS*, nucleus of the solitary tract; *olf. bulb*, olfactory bulb; *OT*, optic tract; *PBN*, parabrachial nuclei; *Rh. sulcus*, rhinal sulcus; *ST*, stria terminalis; *TTA*, thalamic taste area; *VII*, intermediate (facial) nerve; *X*, vagus nerve. [From Pfaffman; from Norgren (1977).]

systems (Richardo and Koh, 1978). The caudal (nontaste) part of the nucleus projects also to the spinal cord. The NST thus plays an important role in many visceral as well as gustatory reflexes.

Norgren has located a "parabrachial area" which projects bilaterally to the thalamus, and to the lateral hypothalamus and basal forebrain, for relays relating to taste sensation (thalamocortical), feeding responses (lateral hypothalamus), and visceral homeostasis. It projects also to hypothalamic nuclei (paraventricular, dorsomedial, and arcuate), the medial preoptic area, and the central nucleus of the amygdala (Norgren, 1980). By this second, bilateral, lower route, taste impulses can reach the amygdala and other temporal lobe structures juxtaposed to olfactory representation. Both limbic and basal forebrain structures probably contribute to the affective qualities associated with taste, and hypothalamic projections may contribute to the regulation of appetite and satiety.

CHAPTER 66

Touch, Pain, and Temperature

NEURAL MECHANISMS OF SOMESTHESIS

Somesthesis and somatic sensibility are synonyms, terms which include *touch-pressure, pain, temperature*, and *proprioception*. Proprioception includes sensibility of joint position and motion, and tension of muscle and tendon. Somesthesis includes all sensed events referable to body surface—except taste, and all sensed events referable to body wall—except vestibular.

Somesthesis is treated separately yet cannot entirely exclude inputs relating to autonomic mechanisms, central chemoreceptors, sensors associated with blood vessels and viscera—bodily activities of which we are generally unaware—nor sensed functions, partly visceral, relating to swallowing, coughing, defecating, and urinating. These latter functions, guarding as they do portals of entry and exit of visceral organs, are obviously advantageously linked to conscious perceptions and voluntary actions.

Somesthesis excludes sensory activities relating to remote events—auditory, visual, and olfactory. This exclusion must seem paradoxical because, except for olfaction, the physical events actually responsible for sensory transduction occur deep within body tissues, photoreception at the back of the retina, and mechanoreception along the inner spiral of the cochlea. These are considered distance sensory systems because the initiating events that are of perceptual interest take place at a distance. The corresponding sensory processes culminate in perceptions that are projected outside the body. This outward projection of internal activities is reinforced by continuing somesthetic experiences. We measure our way to near-distant events by our own movements, and we extrapolate from such experiences to remote perceptual projections.

The Body Image

Our "inmost self" consists of feeling-states: moods, appetites, and motivations which seek internal satisfactions. This subjective "core awareness"—of existence and state of being—is added to by visceral, vestibular, and somesthetic information. Altogether this combination constitutes a coherent *body image or body schema*.

Somesthesis organizes information concerning body wall and cutaneous conditions and composes these into an integrated perceptual whole. Somesthesis analyzes fitness and readiness for action: body position, tension, motion, and boundary conditions. With information from distance receptors, somesthesis contributes integrated information necessary for modeling transactions between the individual and his or her immediate extrapersonal space. *Somesthesis provides information about body machinery so that the individual can act in accordance with on-going motivations and with due regard for memories of how behaviors under similar circumstances have been evaluated in the past.*

Vestibular and distance sense data orient the body image with respect to three-dimensional space and dynamic environmental events. This image constitutes a model of the self to which all sensations are referred. Perception of things remote, through hearing, vision, and olfaction, are projected outward from this image. We experience ourself as a bundle of appetites occupying a generally capable but vulnerable body operating within a generally habitable physical and social environment. This body image is the subject of all touch, pain, and temperature experiences.

Touch

If any sensation among all of somesthesis is most important, that is touch. It is mainly through touch that we experience our body, and discover and measure the environment in which we find ourself. This probably is true life long—from earliest experiences of fetal life through our last moments of consciousness. A newborn deprived of varied, gentle touching fails to thrive and may die of inanition. It is interesting in this connection that hearing, which is probably the next most important sensation—

particularly for social transactions—is an evolutionary derivative of touch.

THE SENSE OF TOUCH

The sense of touch is primarily a cutaneous sense, although it is usually considered to include the touch receptors within the mouth, particularly on the tip of the tongue, and the receptors located in dermal tissue, tendons, the periosteum, and other deep-seated structures which respond to pressure.

The Stimulus

The sensory endings for touch are known to respond to uneven or nonuniform pressure which results in deformation of the skin or movement of the hairs which grow out of the skin. Sensitivity to touch varies in different regions of the body, but may require only very slight deformation or barely perceptible deflection of a single hair on the surface of the body. A wisp of cotton drawn over a hairy surface or even over bare skin is sufficient for stimulating the sense of touch.

Threshold for touch may be determined in terms of the pressure required to bend a hair. Surfaces of the body which are sparsely covered with hair tend to be more sensitive to direct contact. Shaving the hair from a hairy surface results in reduction of sensitivity to touch.

Minimum pressures required to elicit the sensation of touch vary by a factor of more than 20 to 1, depending on the region of the body considered. The nose, the lips, and the tips of the fingers require only 2 or 3 g/mm^2. Higher pressures are required on the backs of the fingers, the upper arm, and the outer surface of the thigh. Progressively higher pressures are required on the back of the hand, the calf, the shoulder, the abdomen, the front of the leg, the sole of the foot, and the back of the forearm. A pressure of nearly 50 g/mm^2 is required on the loin.

Touch Receptors

All cutaneous receptors are embedded in skin and hence are subject to the physical conditions of their surroundings: the deformability and hysteresis, elasticity, thermal and electrical conductile properties, etc., the density of their distribution, and the depths in which the receptors are located. Some cutaneous receptors arising from a single nerve fiber may be confined to a small region of skin, as in the finger tip, measuring slightly less than 1 mm^2. Others may be relatively widespread, measuring several cm^2. Some fibers which give rise to widely distributed endings may show spotty repre-

sentations or gaps between broad areas of distribution of their receptors.

There is usually considerable overlap of distribution of cutaneous receptors among adjacent neurons and likewise overlap between fibers that arise from neighboring peripheral nerve bundles. Thus the boundaries of anesthetized or denervated areas are broad and fuzzy, measured by a few millimeters width rather than being sharply bordered. From studies of regeneration following peripheral nerve injury, it is evident that there is a continuing competition for peripheral fields of representation between neighboring individual nerve fibers and between fibers emerging from adjacent different nerve bundles. As skin is abraded, superficial nerve endings are replaced, and there is a dynamic turnover of the various sensory end organs even in areas not subjected to wear and tear. The scene is one of dynamic morphological as well as functional activity.

There is a general distinction between the kinds of receptors and the sensations with which they are associated that depends on their location, whether in skin that supports hairs or in areas of hairless, glabrous, skin. The simplest kinds of receptors consist of bare nerve terminals which arborize among cells in the superficial epidermis and corneal surface. Other "free" nerve endings are found in deeper layers of cutaneous and subcutaneous tissues. Pain sensations are thought to originate from such bare nerve endings. Both pain and discriminative touch can be elicited from stimulation of free nerve endings, e.g., in the cornea where these are the only kinds of nerve terminals present. So at least some aspects of two different modalities, tactile and nociceptive, are subserved by similar looking undifferentiated free nerve endings. Temperature sensibility, in contrast, which might be expected to represent a continuum, is coded by two distinctive specific end organs, warm and cold thermoreceptors.

Touch reception, evoked by mechanical stimulation of the skin, varies from the faintest most evanescent tickle to firm pressure sensations. Changes in location of stimuli on the skin, two-point sensibility, textures, the rate at which stimuli are imposed and released, the depth of indentation of the skin, movements of single hairs and patterns of hair disturbance that don't depend on touch of the skin, appreciation of flexure and stretch of skin, and the spatial and temporal extent of moving and nonmoving stimuli are all relatively accurately appreciated. *Altogether, tactile sensibility depends on combinations of discriminations of place, intensity, and temporal order.*

Specialized receptors include Meissner's corpuscles which are found in abundance in the hand, the foot, the nipple, the lips, and the tip of the tongue; Merkel's disks found in the fingertips, the lips, and mouth; and basket-like terminations which surround hair follicles. Structures which look like Merkel's disks have been found in epidermal domes (Fig. 9.122) which appear to be specialized receptor regions in the skin (Iggo, 1966). Similar structures have been identified in other tissues (Cauna, 1966). The Pacinian corpuscle is an elliptical encapsulated ending which responds to rapid mechanical displacement of the skin. It is a layered structure (Figs. 9.123 and 9.124), the mechanical characteristics of which limit the stimulus energy transmitted to the nerve ending to relatively high frequencies. There is a transient response to sustained mechanical displacement. Response is apparently limited to the axonal membrane of the first node of Ranvier. Only one or two action potentials are seen when the generator potential is maintained at a steady level. Thus, repetitive discharge does not occur in response to any steady component of the generator

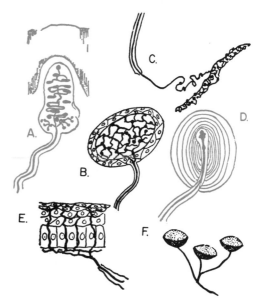

Figure 9.123. Various elaborations of cutaneous receptor endings. *A*, Meissner's corpuscle. *B*, Krause's end bulb. *C*, Ruffini end organ. *D*, Pacinian corpuscle. *E*, Bare nerve endings in the cornea (pain). *F*, Merkel's disks (touch). [From Bainbridge and Menzie (1925).]

potential which may be produced by temporal summation in response to repetitive stimulation. The receptor therefore produces impulses which follow the driving stimulus in the range of effective frequencies. At about 250 Hz, stimuli of 0.1 μm peak-to-peak amplitude elicit impulses in phase with the stimulus (Loewenstein and Mendelson, 1965). Some somatosensory receptor endings are illustrated schematically in Figure 9.123. There is evidence that the cells which comprise nerve endings change relatively frequently during the life of the individual (Cauna, 1966). This continuing renewal of receptor tissue is also characteristic of the gustatory and olfactory systems.

The innervation of the skin of various regions of the body differs with respect to the sense of touch. Even in a relatively homogeneous region, such as the back, there are differences. Localized regions, or spots, may be found which are more sensitive to touch or to temperature or to pain. Responsiveness to these qualitatively different dimensions of stimulation is not homogeneously distributed.

PHYSIOLOGICAL CLASSIFICATION OF CUTANEOUS MECHANORECEPTORS

Somatic sensation is served by *sensory units* which include sensory receptors together with their corresponding afferent neurons and central relay-destinations. Sensory units can be usefully classified according to their peripheral conduction velocities which range from about 2 m/s (equivalent to

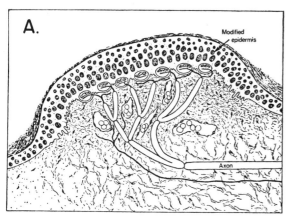

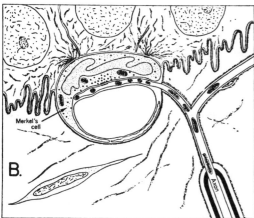

Figure 9.122. *A*, Cross-section of an epidermal dome with layer of nerve endings just inside the basement membrane. Endings are derived from a single myelinated axon. *B*, Detail of Merkel's cell at one of the nerve endings. [From Iggo (1966).]

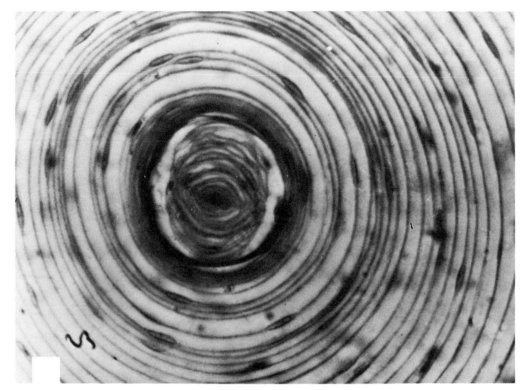

Figure 9.124. Unstained osmium-fixed transverse section of a cat Pacinian corpuscle as viewed by interference microscopy (×750). The axial nerve terminal is oval in cross-section. [From Quilliam (1966).]

a slow waking pace) to 75 m/s (equivalent to the speed of propeller-driven aircraft).

Other generalizable characteristics accompany this classification according to conduction velocity: the higher velocity sensory units tend to have increasingly elaborate sensory end organs, conducting impulses (at their high-velocity rates) to higher relay stations in the central nervous system. By virtue of their end organs, they have lower thresholds, and both respond and adapt more rapidly. Hence they can improve interpretations that depend on temporal order, e.g., flutter and vibration, and more accurately monitor movement.

In general, afferents of any given classification which innervate proximal regions of the body have somewhat slower conduction velocities than afferents that innervate distal body regions. This reduces the spread of central arrival times of signals from simultaneous events affecting different parts of trunk and limbs.

Slowly Conducting Fibers—Pain Sensory Units

C-fibers are the slowest conducting, at 2 m/s; they cannot follow repetitive stimuli above about 2/s. A-delta conduct an average of 10-fold more rapidly than C-fibers. Both are supposed to have arborizing free nerve endings in skin and subcuta-

neous tissues. A-δ fibers also terminate in the vicinity of hair follicles. They are abundant in tooth pulp, which is innervated by the trigeminal nerve. Both of these slowly conducting sensory units are capable of responding to touch but mainly subserve the perception of pain. They have relatively high thresholds and do not respond to brief stimuli. With more lasting stimulation they discharge after a slight delay, continue to respond throughout stimulation, and may exhibit an "afterdischarge" when stimulation is discontinued. The central destiny of C-fibers is limited to gray matter of the spinal segment of entry, and adjacent segments. A-δ fibers serve the same segments but also ascend the dorsal columns a few segments before terminating in gray matter of the dorsal horn.

Hair Innervation—Mechanoreceptor Sensory Units

Three types of afferents innervate tissues associated with hairs. They have more elaborate end organs which are spear-like, fence-like, and circular, all distributed in the vicinity of hair follicles. They vary in conduction velocities from 50–75 m/s and, beside distributing terminals to local spinal segments, ascend the dorsal columns most or all of the way to the bulb before terminating in high spinal dorsal horn or bulbar dorsal column nuclei.

They respond promptly with stronger "on" than "off" responses. They generate impulses during slow or rapid hair movement. They may respond to movement of hairs away from the rest position and be insensitive to return movements.

Cutaneous Field Sensory Units

All three types have specialized receptor end organs which manifest a variety of forms, including lamination (like Meissner's corpuscles) and encapsulation (like Krause's end-bulb) (refer to Fig. 9.123). Conduction velocities range from 50–65 m/s. They are prompt in onset and follow repetitive stimuli at moderately high rates. They respond best to deformation of skin and may continue to discharge for many seconds after movement has stopped.

Rapidly Responding Sensory Units

These are sensory units with even more elaborate receptors, consisting of Ruffini end organs, Pacinian corpuscles, and Merkel's disks (Fig. 9.123). Sensory units with Ruffini end organs may or may not show a resting discharge. They characteristically respond promptly with a regular firing rate to sustained light stimulation. They are classical tonic sensory receptors. Pacinian corpuscles, however, adapt so rapidly that their axons will follow high-frequency rates of stimulation (best at 250–300 Hz). They are classical phasic sensory units.

Pacinian corpuscles are rare in hairy skin but common in glabrous skin and otherwise prevalent throughout subcutaneous and intramuscular connective tissues and the periosteum. They are also found in the mesentery where they can be readily isolated for investigation. They respond to extremely slight perturbations generated by the pulse, respiratory movements, finger tremor, even a breath of air across an exposed Pacinian corpuscle in the mesentery.

A myelinated fiber enters the Pacinian corpuscle and a half-node of Ranvier gives rise to a bare nerve terminal which lies along the linear mechanical focus of the multilayered capsule (Figs. 9.123 and 9.124). Because the corpuscle is turgid, mechanical compression of its surface is transmitted to the primary transducer element, the bare nerve terminal. Slight pressure on the capsule leads to a generator potential in the nerve terminal which in turn leads to a regenerative depolarization at the node of Ranvier and an action potential that is conducted along the parent axon. *The Pacinian sensory unit thus illustrates both the local, gradable, summable,* *electrotonic generator potential and the all or none, regenerative, conductile action potential.*

Physiological Responses of Cutaneous Sensory Units

UNITS FROM GLABROUS SKIN

Glabrous skin on the palmar surfaces of the hand and foot are rich in innervation which is closely matched to special features such as dermal ridges which contribute to mechanoreception as well as provide finger and palm prints (Fig. 9.123). Anatomical and physiological properties of sensory units in this region are closely similar in humans, monkeys, and a few laboratory animals such as the raccoon.

C-fibers are sparse, and there are no hair receptors. Ruffini end organs and Merkel's disks give rise to regular and irregular tonic discharges which contribute to reception of velocity and acceleration as well as displacement components of cutaneous stimuli. Four types of sensory units innervate glabrous skin. Two of these are rapidly adapting (RA) and two are slowly adapting (SA). The RA sensory units respond only during skin movement, and not during static displacement, whereas the SA units continue to respond during sustained skin displacement (Fig. 9.125).

Rapidly Adapting (RA) Sensory Units

Fibers associated with *Meissner's corpuscles* discharge a brief burst of impulses when skin is indented, adapt rapidly to steady pressure, and deliver a brief burst on removal of the stimulus. Their rate of adaptation makes them selectively sensitive to stimulus frequencies in the range of 30–40 Hz. They therefore contribute to perception of what is called *flutter*.

Another type of RA unit is associated with *Pacinian corpuscles*. These, as noted previously, respond best in a higher frequency range (250–300 Hz) and subserve sensations of *vibration*. Both Meissner and Pacinian afferents are accurate in detecting acceleration, velocity, and displacement.

Slowly Adapting (SA) Sensory Units

Units associated with *Merkel's disks* are slowly adapting and very persistent, *SA type I*. A mechanical stimulation of 1 or 2 ms may evoke a burst of impulses lasting 10 ms during slow physical restoration from deformation of the surrounding tissue. Thus, "adaptation" on the part of these units is largely due to hysteresis occurring in the skin itself. The frequency of discharge of these units varies

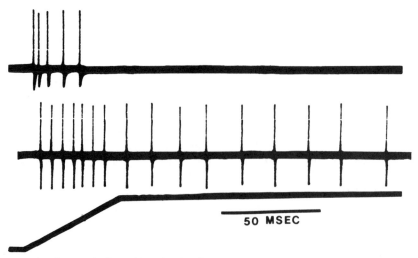

Figure 9.125. Touch reception from a single rapidly adapting fiber (*upper trace*) and a slowly adapting type I fiber (*middle trace*) to the same mechanical displacement of glabrous skin in the hand of a raccoon. Displacement ramp velocity 10 μm/s, final displacement, 480 μm/s (*lower trace*). [From Pubols and Pubols (1963).]

approximately with the amount of displacement of skin. When a smooth probe is drawn across the skin surface, Merkel disk units may respond at rates in excess of 1,000/s (Iggo, 1966). They are *normally quiescent in the absence of mechanical stimulation.*

Another sort of SA sensory unit, *SA type II,* is associated with *Ruffini end organs.* This type of sensory unit *may have a regular resting discharge* in the absence of any mechanical stimulation. Onset of response to stimulation is boosted briefly to an elevated rate which then settles down to a level somewhat above the resting rate and closely related to the degree of static skin displacement.

Referring again to Figure 9.125, we see responses of two cutaneous afferents, one RA and the other SA, in the same area of glabrous skin, responding to the same mechanical stimulation delivered to the paw of a raccoon. The RA unit responds very promptly and at a higher rate. It may cease to discharge even though mechanical displacement continues. The SA sensory unit fires at a relatively rapid but steady rate during the continuing ramp onset of mechanical displacement and then settles down to a slower rate of discharge which continues as long as skin indentation is maintained.

Adaptation in an SA glabrous skin mechanoreceptor is illustrated in Figure 9.126. A comparison is made between adaptation to reactive force (*A.*) and applied force (*B.*). The course of this unit's adaptation over time is depicted for both of these mechanical contexts (*C.*). Evidence suggests that mechanical properties of the skin contribute to, but do not completely account for, skin receptor discharge characteristics. Related studies indicate that what has been considered receptor fatigue is due in

part to mechanical characteristics of the skin. The generator potential, observed in other afferent systems, is also characteristically of lower amplitude (i.e., generates fewer impulses) when force is applied than when the tissue is displaced.

Figure 9.127 shows schematic maps of 45 glabrous skin receptive fields. On the left (*A*) are outlined fields associated with rapidly adapting units and on the right (*B*) are fields associated with SA units in the same hand of the same subject. To the right are action potentials recorded from a single ulnar nerve afferent of the same individual. This fiber innervated a receptive field on the volar surface of the middle phalanx of the fifth finger. The unit adapted slowly to each of three stimulating forces depicted below each tracing of responses (Knibestoel and Vallbo, 1970).

SENSORY UNITS IN HAIRY SKIN

In addition to endings similar to those found in glabrous skin, hairy skin has a variety of endings associated with hairs and structures allied with hairs. In hairy skin where innervation is less dense (i.e., on the back of the forearm), one can identify localized "spots" which may have a dominant single modality characteristic for sensory experience evoked by its stimulation, no matter how the tempo and intensity of the stimulus is modified. Thus, individual spots may be identified as "cold," "warm," "pain," and "touch" spots. This implies that, at least in relatively sparsely innervated areas, the firing of one or another specialized sensory unit can dominate the perceptual outcome.

Hair receptors may be usefully divided into those associated with short, fine downy hairs and those

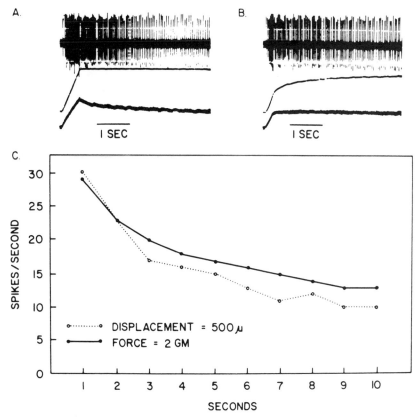

Figure 9.126.   Adaptation of cutaneous mechanoreceptors of raccoon to (*A.*) constant displacement of 500 μm and (*B.*) constant force of 2 g. *Upper traces,* spike discharges; *middle traces,* stimulus displacement; *lower traces,* reactive force *(solid line)* and applied force *(dashed line). C.,* Course of adaptation over a 10-s period. [From Pubols and Pubols (1963).]

associated with long, thick guard hairs. The former are innervated, characteristically, by thinly myelinated axons with conduction velocities in the A-δ range. They tend not to be sensitive to the direction of hair bending. Figure 9.128 illustrates the receptive field properties of a mechanosensitive cutaneous unit innervating "down" hair. A gentle air jet was employed as the stimulus, and the field was swept in one direction, left to right, in straight lines at half millimeter intervals. Responses mapped on the skin and correlated with the linear sweeping motion of the jet are seen on the left. The receptive field of the same unit was mapped by sweeping the air jet in the opposite direction, right to left. The receptive fields are seen to overlap one another to a large extent even though hair motion and slight skin deformation follows sequentially in opposite directions.

Guard hair receptors have a wider range of conduction velocities and tend to be specifically sensitive to the direction and velocity with which hairs are deflected. Figure 9.129 illustrates the transducer function of a unit bearing Merkel disk receptors which responds with directional sensitivity to bending of a single hair. It responds briskly during

displacement of the hair in one direction and emits a steady train of impulses while the hair is statically displaced. It ceases firing the instant the hair is released from displacement. When, the same hair is displaced in the opposite direction there is continuing silence until the hair is released from displacement in *that* direction, whereupon the unit manifests a rapid train of impulses during restoration of the hair to its normal position. This same unit also shows remarkable fidelity for dynamic vibration, following reliably to 1200 Hz (Gottschaldt and Vahle-Hinz, 1981). *Thus, in addition to signaling direction of hair displacement, these units respond to a range of vibratory stimuli which overlaps that of Pacinian cutaneous mechanoreceptors.*

Structure-Function Relations in Finely Myelinated Pain Fibers

A-δ nociceptive fibers are supposed to have "free nerve endings." Kruger et al. (1981) confirmed that the receptive fields of such small myelinated nociceptive endings may manifest numerous, large gaps between areas of local innervation. This observation is illustrated in Figure 9.130 which shows a discontinuity in the receptive field of a single no-

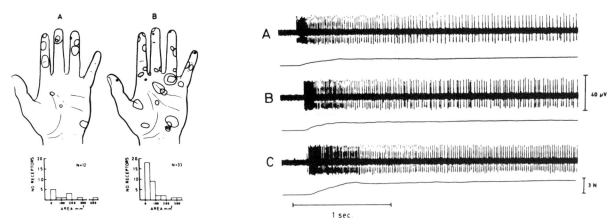

Figure 9.127. Schematic drawing of 45 cutaneous receptor fields of human glabrous skin *(above)* with histograms showing the distribution of receptive field sizes *(below). A,* Rapidly adapting units. *B,* Slowly adapting units. *Right,* A, B, C, Responses of a slowly adapting unit which showed distinctive receptive field borders to stimuli of different intensities. The unit was located on the volar aspect of the 5th finger on the middle phalanx. The approximate stimulating force, indicated by the lower traces, and the mean impulse frequency in the steady state were in *A,* 1.2 N and 28 impulses/s; *B,* 1.6 N and 37 impulses/s; and in *C,* 2.4 N and 48 impulses/s. [From Knibestol and Vallbo (1970).]

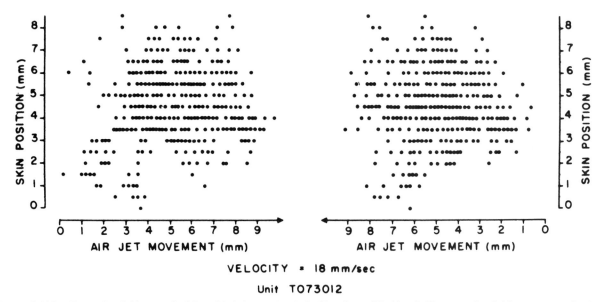

Figure 9.128. Receptive field map of a "down" hair (as contrasted with a "guard" hair) unit. The receptive field was scanned using an air jet, in two passes in opposite directions, in 0.5-mm steps with the same orientation. The density of spikes reflects the high sensitivity of this class of afferents to moving air stimulation. [From Ray and Kruger (1983).]

ciceptive fiber in hairy skin of a cat. Stimulation of areas of gap between spots responsive to nociceptive stimuli results in no response, although vigorous responses are elicited when stimuli are applied to adjacent innervated spots. The single sensory unit illustrated here served 17 separate nociceptive spots in the skin.

The authors mapped nociceptive spots by physiological means and then examined the corresponding innervation patterns with light and electron microscopy. Beneath each such nociceptive spot they found a thinly myelinated axon which, when

it entered the epidermal layer, lost its myelin sheath but continued to be associated with thin Schwann cell processes throughout its terminal extent. The unmyelinated terminal axons contained abundant, clear round vesicles, and sparse, large dense-core vesicles. The coapted Schwann cell processes exhibited many pinocytotic vesicles. The vesicles and pinocytosis imply communication between the neuron and glia.

The combination of unmyelinated axon terminal and companion Schwann cell processes are presumed to constitute the receptive apparatus for

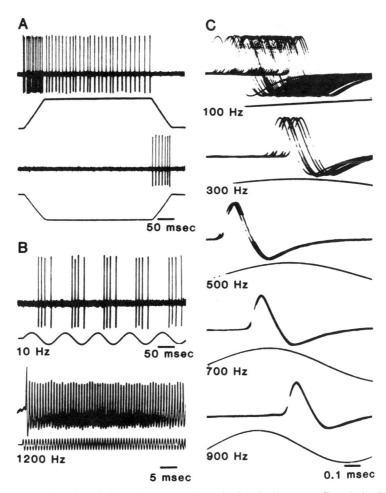

Figure 9.129.   Merkel cell transducer function. *A,* Directionally sensitive, slowly adapting nerve fiber during bending of a sinus hair in opposite directions. The stimulus signal is displayed below the spike records. *B,* Responses to hair movement at 10 and 1200 Hz. *C,* Phase jitter of at least 10 superimposed impulses in response to the same cycle of repeated vibratory stimuli at different frequencies. The electronic sine wave signal driving the stimulator is pictured below the spike records. The mechanoelectric transduction process is faster than that for any other mechanoreceptor, responding to vibration as well as displacement. It is the nerve endings themselves and not the Merkel cells that are the critical mechanoelectric transducer elements in these receptor complexes. [From Gottschaldt and Vahle-Hinz (1981).]

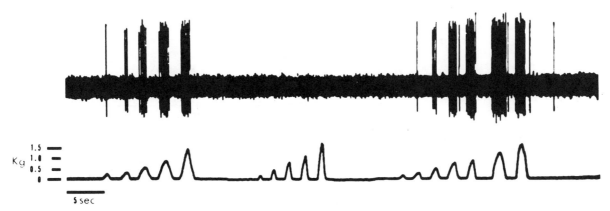

Figure 9.130.   Discontinuity of receptive field of a single nociception fiber. This unit had a 17-spot receptive field. *Upper trace,* Recording from two adjacent responsive points. *Lower trace,* Force applied with a 1-mm probe to first spot (*left*), an unresponsive intermediate point, and the second spot 5 mm from the first. [From Kruger et al. (1981).]

such nociceptive endings. As indicated by electron microscopy, A-δ pain terminals do not have truly "free nerve endings," and perhaps the term should be abandoned. The nerve endings are simply enshrouded by Schwann cell processes, not wrapped round and round by them as in the usual context of myelination.

Corpuscular end organs associated with sensory units are generally thought to be of Schwann cell origin. The intimate boundary between such elaborate end organs and the axon terminals they surround closely resembles that which is found between A-δ terminals and their accompanying Schwann cell processes. The relationship of terminal axons to Schwann cell processes may thus constitute a fundamental synergistic junction for both simple and elaborate end organs. The rest of the apparatus in elaborate end organs would accordingly be relegated to primarily mechanical roles: to impart selective mechanical advantages to the "real" receptor junctions—lying between Schwann and axonal membranes.

Temperature Receptor Mechanisms

Specific end organs have not been identified with thermoreceptors, but warm and cold sensations are mediated by two different systems, one for warmth, the other for cold. It is likely that both types of endings belong to the category of "free nerve endings," unencapsulated except perhaps for a shroud of Schwann cell membrane.

Both types of thermoreceptor are responsive to rate of change of temperature, linear at least with changes of intensity of cold pulses. Reception of warmth is not dependent on inhibition of cold receptors but requires an independent receptor population.

Figure 9.131 shows diagrammatically the effects of static and dynamic applications of warmth and cold in terms of impulse frequency with respect to temperature change over time, according to Hensel (1976). Warm fibers have been recorded from nerve bundles in human subjects using a Peltier thermode. These units were not excited by tactile or

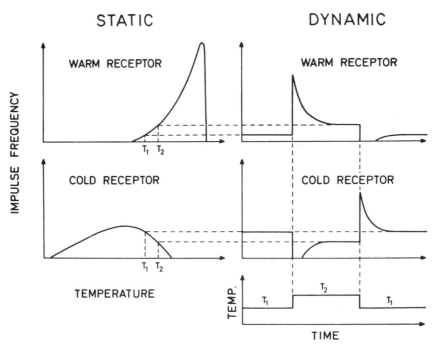

Figure 9.131. Diagram correlating neuronal activity with relative rate of thermal stimulation and time course for application of warmth and cold to cutaneous thermoreceptors. Impulse frequency is plotted against temperature and time course for each kind of receptor. Distinctive patterns of response reinforce evidence that there were two kinds of receptors, one for warmth, another for cold. When warmth is applied to a warm receptor there is a steady rise in impulse frequency to the point of discontinuation at 45°C skin temperature. At onset, with a steady-state background of unit activity, there is a prompt burst of activity which falls off to a steady-state level that is higher than previously, corresponding to the increased temperature. When the warmth is abruptly discontinued, the unit is silent but returns to its ambient level of activity. When warmth is applied to the cold receptor, its spontaneous activity is abruptly discontinued, recovering to a steady-state which is below its previous level of activity. When the warmth is discontinued, the cold fiber fires rapidly and slowly returns to its ambient level of activity. [From Hensel (1976).]

vibratory stimulation. Their conduction velocities were less than 1 m/s, indicating that they are certainly unmyelinated. The warm units were spontaneously active at skin temperatures between 33 and 35°C, with a frequency which rose with increasing skin temperatures up to 45°. Higher temperatures were not used. Dynamic cooling was followed by transient decreases in frequency.

Thermosensitive terminals tend to be sparsely dispersed, identifiable warm and cold spots which are persistent over time. Warming a cold spot is ineffective or yields paradoxical cold sensations.

PHYSIOLOGICAL RESPONSES OF CUTANEOUS SENSORY UNITS IN HUMAN SUBJECTS

Hagbarth and Vallbo (1967) pioneered by inserting microelectrodes through the skin into nerve trunks in humans and recording from single sensory and motor axons. The peripheral fields of sensory units could thereby be explored for distribution, modality specificity, response characteristics, and conduction velocity while at the same time, the authors could obtain the benefits of subjective testimony while the axon or its receptive field was being stimulated.

By stimulating as well as recording through such percutaneous microelectrodes in human subjects, Torebjöerk and Ochoa (1980) succeeded in establishing correspondences between single median nerve axons, their peripheral receptive fields as established by referred sensory localization, and characteristics of their responses to mechanical stimulation applied to the peripheral field.

Figure 9.132. illustrates microelectrode stimulation and recording from single myelinated cutaneous sensory units in man. Microstimulation of the unit at the elbow created a sensation referred to the thumb. Trains of stimuli applied to that unit were interpreted as flutter or vibration, depending on the stimulus frequency. Of course, the subject was deprived of audiovisual cues as to the timing of stimulus applications and the parameters of stimulation.

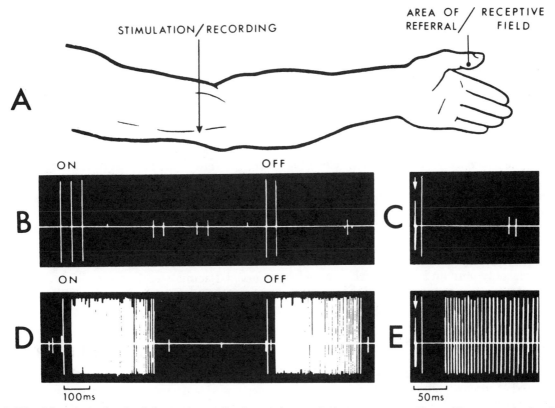

Figure 9.132. Microelectrode stimulation and recording in man from a single sensory nerve fiber, with perceptual referral to distal receptive field. *A,* Intraneural microstimulation in the median nerve at the elbow is referred as flutter-vibration to a focal area of referral in the thumb. *B,* Intraneural recording detects "on" and "off" responses to a single rapidly adapting unit responding to skin indentation in the area of referral. *C,* Electrical stimulation in the receptive field of unit excites a fiber with a conduction velocity of 45 m/s (*white arrow* indicates stimulus artifact). *D–E,* Prolonged intraneural microstimulation at high frequency induces posttetanic hyperexcitability, whereby mechanical (*D*) and electrical (*E*) stimulation elicits stereotyped bursts of impulses. Time bases given for *B–D* and *C–E,* respectively. [From Torebjork and Ochoa (1980).]

The authors also routinely recorded from single axons while exploring the periphery in order to detect units that might not give rise to conscious sensations. *Trains of impulses in single RA units innervating glabrous skin of the human hand usually generated conscious percepts. SA type II units which respond briskly to peripheral stimuli nevertheless only rarely reached consciousness. Excitation of C-fibers and A-δ fibers usually did arouse conscious percepts.* However, because of the small diameters of these latter fibers, it was uncertain whether the percept was due to stimulation of only a single or perhaps several axons by the microelectrode implanted in the nerve trunk.

Trained subjects find it easy to localize single unit activity in respect to its peripheral (referred) site. This does not require explorations using peripheral stimulation. When physiological stimuli are delivered to the referred site, subjects can provide additional useful information, beside accurate localization, concerning the stimulus modality and the sensory qualities elicited. Electrical responses of that same unit, recorded via the same electrode, can be related to the most effective peripheral stimulus parameters.

The subject's description "is concrete to begin with (painless-painful, intermittent-continuous, dull-bright) and then qualified (pressure, vibration, stinging pain, dull pain, itch…)." (a) *Activation of RA units yields painless "tapping" sensations which become "flutter" or "buzzing" at successively higher stimulus frequencies.* (b) *Painless continuous pressure is elicited by repetitive impulses delivered to SA type I units which have fast conduction velocities.* (c) *Sharp, stinging pain sensation is elicited from high threshold units with conduction velocities in the A-δ range.* (d) *Dull pain sensation is elicited by excitation of C-fibers, which are the slowest conducting.*

Perceptions evoked by single unit microstimulation is remarkably pure and specific, whereas subjective effects initiated by natural peripheral stimulation in the same subjects in the same area of referred sensation seem more ambiguous as if from activation of a mixture of otherwise pure and specific sensory units.

Firing frequency of single cutaneous sensory units is interpreted as *intensity*, in relation to continuous percepts (pressure, pain), and as *frequency*, in relation to intermittent percepts (flutter and vibration).

Trophic Influences of Afferent Neurons on Innervated Tissues

It has long been recognized that many nerve cells contribute trophic influences that affect the tissues they innervate, including other nerve cells, receptors, end organs, etc. Most afferent nerves generate such trophic effects. The effects are not triggered by synaptic transactions, but seem to depend on protein synthesis in the cell body and axonal transport to the site of innervation. The effects may be reciprocated, that is, some tissues and end organs can modify structural and functional characteristics of the nerves that innervate them.

Denervation (particularly partial denervation) can upset normal tissue functions by inducing hypertrophy, hyperactivity, and hypersensitivity. Partial denervation of nervous tissue may result in "supersensitivity" of the target populations of neurons. These may show exaggerated effects in response to both excitatory (greater and more prolonged excitation) and inhibitory (greater and more prolonged inhibition) synaptic influences from other sources of innervation, and to both activating and inactivating drugs (Stavraky, 1961).

Small diameter afferents associated with blood vessels may release substance P which contributes to neurogenic aspects of inflammatory responses (Kruger, 1983). Activation of one branch of an afferent nerve can generate an *axon reflex* in adjoining branches of the same neuron. Signals ascend the local afferent fiber to the nearest junction and then turn in the direction opposite to ordinary afferent conduction, and travel back to the periphery in adjoining branches of the same parent fiber. Axon reflexes contribute to vasodilation and inflammatory responses.

Taste buds depend on such trophic influences. If the afferent nerve is sectioned, the taste bud degenerates. It regenerates when regenerating taste afferents reapproach the mucosa. Merkel's disks similarly disappear and reappear with reinnervation. Pacinian corpuscles, however, do not degenerate and may be reinnervated by another afferent neuron, whether or not that neuron had previously served Pacinian end organs (Schiff and Loewenstein, 1972). In this case, the Pacinian corpuscle contributes reciprocal trophic effects on the reinnervating nerve fiber.

PERIPHERAL PAIN RECEPTION

Pain is ordinarily aroused with stimuli that damage and risk destruction of innervated tissues. Pain can be elicited by stimulation of any tissue except the brain itself. It consists of both *sensation* of pain, that which is experienced as hurting, and *reaction* to pain, somatic and autonomic reflex responses, voluntary efforts to remove the stimulus and escape, and emotional reactions which may be encumbered by past experiences associated with pain.

There are two principal types of cutaneous afferents which are categorized as including *nociceptive units*: a population of A-δ, small diameter, finely myelinated nociceptive fibers, and still smaller diameter C-fibers. Both of these types of cutaneous afferents include sensory units capable of responding discriminatively to light touch, but nociceptive units predominate and are most responsive to high threshold, tissue damaging stimuli which give rise to preponderantly painful sensations.

Activation of A-δ nociceptive sensory units yields sharp, pricking pain such as that elicited by abrupt sticking with a sharp pin. It is this kind of unit that is likely to be activated by introducing a not absolutely sharp hypodermic needle through the skin for drawing blood: an immediate pricking pain, accurately localized, which promptly goes away.

Activation of C-fiber nociceptive sensory units gives rise to a slower onset, "second" pain, more intense and longer lasting. Both types of nociceptive sensory units are most readily and effectively activated by relatively high intensity stimuli which introduce a high rate of destruction of innervated tissue, by mechanical, chemical, electrical, or thermal means. Threshold for thermal excitation of a pricking heat sensation is at a skin temperature of 45°. If such heat is prolonged, it can lead to tissue damage.

Collins et al. (1966) stimulated C-fibers in conscious human subjects after compression block of myelinated fibers. They found that a single C-fiber volley did not reach consciousness, whereas if stimuli were repeated at 3/s, subjects invariably felt pain. Higher frequency stimulation quickly rose to unbearable pain. Temporal summation therefore contributes importantly to pain perception.

Characteristically, C-fiber pain has a burning quality. Strong activation of C-fiber sensory units gives rise to a subjective feeling of *intense hurt*. An aching, paralyzing hurt can expand inexorably to affect a whole limb and more, after which it may slowly subside. While it is expanding, the pain can engulf all other sensations and absolutely transfix attention. C-fiber pain may be long persisting and, if so, it is likely to be even more burning in quality.

Other nerve endings, if sensitized, may also contribute to pain perception. If both peripheral and central circuits become exaggeratedly sensitized, a condition of intractable searing, burning pain may ensue, which is called *causalgia*, Greek for burning pain. Patients with causalgia describe the feeling "as if a red hot poker were continuously probing" the affected parts. In this context, any kind of stimulation, even the most gentle, a breath across the skin, may be perceived as excruciatingly painful (Livingston, 1943).

In chronic pain syndromes, like causalgia, there may be marked trophic changes in the distribution of the affected peripheral nerves. The skin may show marked loss of subcutaneous fat, elimination of normal creases in glabrous skin, reddened, mottled parchment-like appearance of the skin, overgrowth of the nails—to the appearance of claws, increased pigmentation and coarseness of hairs, and excessive activity of local, and, to a lesser extent, regional sweat glands. There may be rapid dripping of sweat from the affected parts. There may also be local osteoporosis, to an extent greater than expected from disuse atrophy. Segmental reflexes are likely to be affected: heightened withdrawal (flexion) reflexes, clonic reflex contractions of skeletal muscles, and spontaneous spasms or tonic-clonic contractions.

It is important to remember that although causalgia is rare, lesser symptoms of burning pain, and less obvious trophic changes may accompany "minor causalgias." Severe, refractory pain is more likely with partial than with complete severance of a peripheral nerve. The degree of pathophysiological changes, including trophic changes, and disabling pain, does not depend on the severity of the initial injury. A relatively minor trauma may give rise to full-blown causalgia (Livingston, 1943). Why it is that almost all instances of acute painful trauma fade away without residual effects, whereas occasionally a seemingly similar acute painful trauma may give rise to enduring, overwhelming signs and symptoms of intractable burning pain, is not known.

SPINAL MECHANISMS IN SOMESTHESIS

INVASION OF SPINAL CORD CENTERS BY SOMESTHETIC INFORMATION

Afferent Neuron Inputs

Afferent fibers from skin, body wall, and viscera enter the spinal cord by way of the dorsal roots. Their cell bodies lie in the dorsal root ganglia. They originate as bipolar cells. The two processes fuse into a single axon which bifurcates, sending one branch to a sensory receptive field, visceral or somatic, and the other to a central destination. There is no conduction delay at the point of bifurcation; peripheral impulses pass directly through the point of axonal bifurcation without hesitation. The cell body may be invaded by an impulse, secondarily.

Afferents innervating the face, the cranial analogues of spinal afferents, have cell bodies in the Gasserian ganglion (Vth cranial nerve) and enter the brain stem by way of the trigeminal nerve root. Visceral afferents arriving by way of the vagus (Xth cranial nerve) have cell bodies which lie in the nodosal ganglia, likewise homologous to dorsal root ganglia.

Functional Significance of Fiber Diameter Differences

According to a standard classification which takes into account fiber diameter, physiological role, and peripheral end organ, afferent nerves are grouped into four categories, I–IV. The largest diameter afferents, I-A, and many in the next category, II, arise from muscles and are described in Chapter 70, in relation to segmental motor controls. Group I afferents, A-α, 12–22 μm in diameter, arise solely from Golgi tendon organs. Group II includes those A-β units which arise from muscle spindles. Group II also includes the largest cutaneous afferents, likewise A-β, which relate to touch, pressure, flutter, and vibration receptors.

These largest *cutaneous* afferents all belong to group II and range in diameter from 5–12 μm. They include units with *slowly adapting* Merkel's disks; *quickly adapting* Meissner's corpuscles, Pacinian corpuscles, hair follicle and bare nerve endings; and *slowly adapting* Merkel's (type I) and Ruffini (type II) end organs.

Group III is composed exclusively of A-δ sensory units, 1–5 μm in diameter. These include *quickly adapting* mechanoreceptors which sense touch contact, *slowly adapting* thermoreceptors, and pain receptors.

Group IV consists solely of C-fiber sensory units, 0.2–1.5 μm in diameter. These include only *slowly adapting* receptors concerned with thermosensitivity or pain, probably also possessing some mechanoreceptor capabilities.

Temporal Distribution of Afferent Signals

Cutaneous sensory units belonging to all three groups, II, III, and IV, have representation in both glabrous and hairy skin areas. Considering the differences in conduction *distances*, in conduction velocity (ranging between 75 m/s and 0.5 m/s), and response *latency*, plus the differences in *adaptation* times, *there is a very broad distribution of arrival times and a prolonged spread of responses beyond the termination of stimuli.* There is a spread of several to many milliseconds between onset of stimuli and arrival of signals at spinal (dorsal root) and brain stem (trigeminal root) terminals. The spread will widen in proportion to the conduction distances. There are even greater time differences relating to the settling down of activity during steady-state conditions and following discontinuation of stimuli.

As nerve impulses approach their termination, they markedly slow in velocity by virtue of branching (accompanied by reductions in fiber diameters) and by further narrowing of the terminal branches. A considerable portion of the approximately 0.5 ms of "synaptic delay" is taken by the slowing of impulses in presynaptic terminals.

Information entering the spinal cord and brain stem moves on through subsequent orders of neurons; the range of diameters in the afferent fibers is often roughly repeated in the cell and fiber diameters of next order neurons. This means that what has been conveyed relatively slowly in the periphery is relayed and transmitted relatively slowly along subsequent central paths. This contributes to a further temporal spread of signals.

Biological Advantages of Rapidly Conducting Systems

Some of the biological advantages of rapid conduction relate to expedition of service for skeletal muscle reflexes. As will be detailed in Chapter 70, Golgi tendon organ and muscle spindle afferents serve to maintain, and, when necessary, in emergencies, to inhibit muscle "tone." This is accomplished by efficient *monosynaptic and paucisynaptic reflexes.* As noted, the largest diameter afferent fibers (groups I and II) originate from skeletal muscles and their tendons. These afferents communicate monosynaptically with large diameter motoneurons which serve skeletal muscles. Terminating afferents concentrate their synaptic connections on efferent neurons that serve the same muscles from which the afferents originate. They thereby contribute to the swift and measured control of stretch reflexes, contraction of antigravity muscles, and general muscular readiness for action.

A second important advantage of rapidly conducting systems relates to control of equilibrium, an urgent problem for all land dwelling animals. For vestibular, related brain stem and cerebellar control of balance, both afferent and efferent arcs of the peripheral apparatus are rapidly conducting. The central ascending and descending paths are likewise composed of large diameter, rapidly conducting neurons in spinal cord and brain stem.

Rapidly conducting sensory systems are thus involved in motor preparedness and maintenance of equilibrium to forefend the risks of gravity. This is additional confirmation of the fact that sensory mechanisms are primarily devoted to behavioral success. Information destined for perception is less pressing than is the subconscious information which keeps us from falling down.

Contributions of Slower Conducting Afferent Systems

Other body wall afferents, cutaneous and visceral afferents, do not possess monosynaptic access to motoneurons. Rather, they serve *pleurisynaptic, multineuronal spinal reflexes.* Their contributions are slower, centrally as well as peripherally. They induce ipsilateral flexor reflexes which involve *excitation of flexor muscles, inhibition of extensor muscles* on the ipsilateral side, with *excitation of extensor muscles (the crossed extensor reflex)* and *inhibition of flexor muscles* on the contralateral side of the body. This is accomplished by segmental crossing of second order sensory axons.

Consequently, strong excitation of any of these smaller diameter afferents contributes to ipsilateral withdrawal reflexes and leads to protective, guarding reflexes which may engage all flexor muscle groups in the corresponding limb, and even extend to contraction of flexor groups throughout the whole ipsilateral side of the body. If there is, for example, a steady visceral pain input from the right lower quadrant of the abdomen, as may accompany an inflamed appendix, skeletal muscles in the abdominal wall relating to that quadrant will reflexly contract sufficiently to serve as an abdominal splint. Muscle contraction resulting from such reflex guarding may be sufficiently strong to generate sources of tension pain in the corresponding muscles.

SPINAL AND BULBAR CONNECTIONS OF VISCERAL AFFERENTS

Impulses from the viscera are transmitted by afferent fibers which arise in visceral organs, mesentery, omentum, and pleural and peritoneal surfaces. Adequate stimuli of visceral organs for arousing conscious sensations include shearing forces, tension and pressure, as by pulling on mesentery or dilating a hollow viscus. The tension and pressure may arise in the event of vigorous peristaltic waves, particularly if forced against an obstruction, as by passing a stone through the common bile duct, a bolus that is too large through the gut, or tension arising from inflammation in visceral and associated tissues.

Visceral afferents are not distinguishable from afferents serving body wall and cutaneous surfaces, except for the distribution of their peripheral (visceral) end organs and their central terminations. They pass through various autonomic nerve trunks and plexuses and reach the central nervous system via vagus, splanchnic, pelvic, and other nerve roots (spinal dorsal roots and brain stem analogues).

Their centrally relayed projections constitute an *ascending visceral projection system* which serves visceral centers strung almost continuously throughout spinal, brain stem, and diencephalic levels. *This represents the innermost gray sensorimotor network and it serves primary visceral functions.*

Fibers arriving via the vagus and pelvic nerves have peripheral distributions which roughly correspond to parasympathetic efferent innervations. Afferent fibers of the *thoracolumbar division* are peripheral processes of ganglion cells located in the dorsal root ganglia from 1st thoracic to 3rd lumbar segments. No visceral afferents arise from sympathetic ganglia. No visceral afferents relay until they reach spinal cord or brain stem gray matter. This is in contrast to autonomic efferents all of which relay, outbound, in sympathetic, parasympathetic, and other peripheral autonomic ganglia.

Visceral afferents reach the thoracolumbar spinal cord from their transit through sympathetic ganglia via the white rami and are distributed to spinal gray matter at the level of entry and one or two segments above and below. They terminate on dendrites and cell bodies in layers I–IV of the dorsal horn. They also *send relayed signals which reach the intermediolateral gray matter* (lateral horn), a collection of small spinal visceral motor units whose axons pass out the ventral root as *preganglionic autonomic efferents.* This indirect communication provides for a pleurisynaptic autonomic reflex arc.

Visceral afferents also provide strong contributions to flexor reflexes and can increase tonic contractions of skeletal muscles to establish quadrantic or whole thoracic or abdominal wall splinting associated with acute visceral, pleural, or peritoneal irritation. Other collaterals activate small neurons scattered throughout the spinal central gray and white reticular formation.

Visceral afferents also contribute to a pleurisynaptic succession of relays which ascend the spinal cord and brain stem as the *ascending visceral projection system. This system contributes pleurisynaptically to gray matter all along the central core of the neuraxis including the floor of the IVth ventricle and midbrain periaqueductal gray matter.* It joins the medial forebrain bundle and therewith projects to numerous *diencephalic and basal forebrain regions.* Other nth order relays probably contribute to *spinothalamic projections and, via thalamic relays, to cerebral cortex.*

Because the ascending visceral projection system arises from relatively small cells that have thinly myelinated axons which ascend only a few segments and are repeatedly relayed, their signals are widely diffused over time. Because they make abundant collateral contributions to spinal, brain stem, and diencephalic gray matter, en route, their signals are likewise widely spatially diffused. Moreover, they are not dynamically highly responsive. *Their cortical representation is sparse and their normal activity does not reach consciousness.* It is only when there is massive visceral afferent activation that these normally subconscious signals "spill over" into paths to consciousness.

Visceral sensations, when evoked, tend to be poorly localized, loaded with feeling tone, and attention commanding. Ascending visceral activation, spreading as it does, gives rise to *referred sensations*, which are ordinarily bizarre or uncomfortable, and which may make normal body wall and cutaneous areas to which the sensation is referred feel acutely sensitive or spontaneously painful to local benign stimulation. Often the referred sensations are painful, *referred pain or hyperalgesia.* Signals coursing along the ascending visceral projection system may give rise to vague to severe chest wall or epigastric discomfort which may be the first sign of an acute thoracic or abdominal (visceral) emergency. Referred sensory distortion or pain, referred to the left shoulder or upper arm may be the first sign of myocardial hypoxia.

Because of their slow and diffuse central connections, this system is *functionally distinguished from spinothalamic projections to roughly the same degree (and in the same characteristics) as spinothalamic projections are distinguished from lemniscal.* That is, it is relatively even slower, less reliable in localization, timing, singularity of modality, etc.

SPINAL AND BULBAR CONNECTIONS OF CUTANEOUS SENSORY UNITS

Incoming afferent axons characteristically send collateral branches ascending and descending for one or two segments above and below the level corresponding to their dorsal root entry. These longitudinal, first order neuronal contributions are important for intersegmental integration.

Many incoming fibers send collaterals into the spinal gray matter while they send their main axons via dorsal column white matter to first order relays in the caudal brain stem, in the dorsal column nuclei. *These primary afferents that ascend the whole length of the spinal cord convey light discriminative touch, two-point sensitivity, and position sense.* When syphilis was more prevalent (before the introduction of penicillin), there were many cases of tertiary syphilis affecting the spinal cord. Persons afflicted with this disorder had a root radiculitis which led to lancinating, excruciating but brief "lightning" pains referred to the limb and body wall dermatomes of the affected roots. Their pathology revealed an almost total loss of fibers in the dorsal columns. As a consequence, they did not know the position of their limbs. They therefore favored the use of two canes and had to look carefully where they were walking. They were almost totally unable to get about in the dark, particularly if they had to go across rough terrain or walk up and down stairs.

Primary afferent axon extensions the length of the dorsal columns make these the longest nerve fibers in the body, the maximal spanning from toes to above the foramen magnum.

After synaptic relay in the dorsal column nuclei, second order neurons cross the midline becoming contralateral to the peripheral sensory fields represented. They assemble into a flat ribbon-like projection (hence *lemniscus*, meaning ribbon). The lemniscus proceeds to the ventral posterior thalamus. The lemniscal system is a phylogenetically new, rapidly conducting three neuron (two synaptic) path to sites of primary somesthetic representation in cerebral cortex. Just before these lemniscal fibers enter the thalamus, there is a mingling of fibers which have ascended in both ventrolateral and lateral spinothalamic pathways. These latter have been crossed from spinal levels, so the ventral posterior thalamus receives an all-crossed, multimodality sensory input from cutaneous, body wall, and visceral sources.

Spinal uncrossed dorsal columns and predominantly crossed spinothalamic tract give rise, on hemisection of the spinal cord, to loss of discriminative touch, two-point sensibility, position and vibratory sense, on the side of the body ipsilateral to the hemisection, and contralateral loss of pain and temperature sense.

Event-Related Potentials Recorded over Spinal Pathways in Humans

Considerable interest has been generated by the detection of far field potentials elicited by sensory stimulation (Jewett et al., 1970). Because of feebleness of the potentials, many signals must be repeated and averaged to obtain a reasonable signal. This is most thoroughly investigated in relation to the auditory pathway, as discussed in Chapter 62.

Figure 9.133 shows *far field potentials* recorded from electrodes placed at equal distances along the skin overlying the dorsal spinous processes. The "event" consists of repeated strong stimuli to the peroneal nerve. Although the amplitude of averaged event-related potentials gradually decrements and disperses as signals ascend the spinal cord, none-

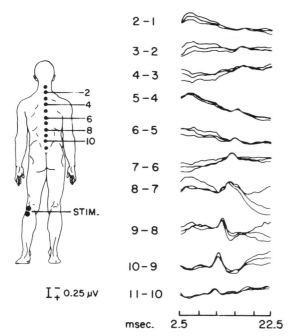

Figure 9.133. Somatosensory evoked potentials, recordings of spinal potential elicited by left peroneal nerve stimulation. The potential progressively increases in latency at more rostral recording locations. Over 8,000 samples were averaged, and 3 averaged potentials were superimposed at each level. [From Cracco (1973); Cracco and Celesia (1983).]

theless, the ascending time displacement of the potential provides a reasonable approximation of preponderant velocities of conduction along the ascending spinal pathways. This analysis can be used to detect interference with conduction systems as in multiple sclerosis affecting the spinal cord.

ORGANIZATION OF NEURONS IN SPINAL GRAY MATTER

From a functional point of view, neurons in spinal gray matter can be conveniently divided into four categories:

1. Somatic Motoneurons. These constitute cell groups which occupy most of the large ventral horns. They send axons via the ventral roots to innervate skeletal muscles.

2. Visceral Motoneurons. These neurons cluster in midlateral spinal gray matter ("lateral horns") throughout *thoracic and upper lumbar levels*. They send axons via the corresponding ventral roots to autonomic ganglia where they relay to units innervating visceral effectors.

Another kind of visceral motoneuron is seen in *parasympathetic craniosacral outflow*, encountered in efferent nuclei of the IIIrd, VIIth, IXth, and Xth cranial nerves in the brain stem, and in the lateral horn nuclear clusters in the 3rd and 4th sacral nerves (occasionally also S_2 and S_5).

3. Propriospinal Interneurons. The third type of neuron occupies the broad base of the dorsal horn and sends long ascending and descending axons which communicate with gray matter throughout many other spinal segments and in the caudal brain stem.

These propriospinal neurons stitch the neuraxis together longitudinally. Most of their axons enter the neighboring white matter to ascend or descend for varying distances, lying close to the gray matter, before sending collateral branches to reenter gray matter at higher or lower segmental levels. These neurons have elaborate dendrites which penetrate spinal gray matter for several segments up and down the cord.

Propriospinal neurons derive information from several neighboring spinal segments and contribute integrated signals to most spinal segments and caudal brain stem. They *contribute to propriospinal reflex mechanisms which control interlimb spinal and brain stem reflexes* (Shimamura and Livingston, 1963). They also integrate spinal afferent signals with intrinsic spinal activities and relay such information toward the forebrain.

Propriospinal neurons are *interneurons with a far wider ranging influence than classical interneurons, Golgi type II cells*. Golgi type II cells have only local dendrites and short axons and hence their influence is strictly limited to the vicinity of their cell bodies. There are relatively few Golgi type II neurons anywhere in the spinal cord; their place is presumably taken over by farther reaching, longitudinally integrating, propriospinal neurons.

4. First Order Sensory Relay and Sensory Control Neurons. A fourth type of intrinsic spinal neuron occupies the tapering dorsal half of the dorsal horn. They are located close to afferent input. They constitute sensory relay and sensory control nuclei and are organized in a series of layers. The dorsal horn neurons in the cat have been divided into six horizontal laminae on the basis of cell size, distribution of dendrites, and organization of synaptic relations (Rexed, 1952). Dorsal horn organization appears similar in primates. The layering of dorsal horn neurons (resembling sensory cortex) implies that the primary integrative role here is functional segregation of signals for *analysis*. The ventral horn is organized more like reticular formation, indicating a great deal of convergence and implying a dominant integrative role for *synthesis*.

Organization of Second Order Sensory Neurons

The most superficial lamina, layer I, contains large marginal neurons and horizontal fibers. The most spectacular part of the dorsal horn lies directly beneath, in layer II. This aggregation of neuropil, which has a gelatinous appearance, runs without interruption throughout the length of the neuraxis from caudal medulla to coccygeal spinal cord. It is referred to as the *substantia gelatinosa (SG)*. Apparently no cells in the substantia gelatinosa contribute to the lateral spinothalamic tract (STT). It provides a spinal relay and control station relating to sensory input.

Layers IV to VI constitute the *nucleus proprius* of the dorsal horn and contain large relay neurons. Cells that contribute to the (STT) are found in layers I and in IV–VI, even encroaching ventrally into territory that is dominated by propriospinal interneurons.

AFFERENT NERVE TERMINATIONS IN SPINAL GRAY MATTER

Second order sensory neurons in layer I are activated mainly by A-δ primary afferent fibers. These convey signals from cold thermoreception, nociceptive mechanoreceptors, and low threshold, long latency mechanoreceptors that appear to be particularly responsive to slowly moving stimuli, contributing to sensations of tickle and itch.

It is thought that *substance P* is the principal neurotransmitter among primary nociceptive afferents, A-δ and C-fibers alike. *Cells of the SG, layer II, are mainly activated by afferent impulses delivered by C-fibers.* These neurons send axon collaterals which terminate on dendrites of neurons in layers I, III, and IV.

Some of these interneurons are enkephalinergic and inhibitory, contributing to sensory relay control. Others contribute ascending fibers which relay upwards toward thalamic stations. Lateral spinothalamic projections, which convey mainly pain and temperature modalities, do not deliver their messages directly to the thalamus. Pain modality projections are relayed one or more times during their ascent to the thalamus.

Some *descending* projections which terminate in the SG bear *enkephalin terminals which end on* the cell body and proximal dendrites of *neurons that receive direct signals from cutaneous nociceptive units* (Ruda, 1982). Other descending projections directed to this part of the dorsal horn from the *nucleus raphé magnus* in the bulb contain *serotonin. Both the enkephalin and serotonin projections are considered inhibitory.* They contribute to modulation and control of pain sensibility at the first synaptic relay of impulses generated by noxious stimuli. *Thus, intrinsic SG neurons and descending projections of at least two kinds (enkephalinergic and serotonergic) exercise local control over A-δ and*

C-fiber pain and temperature signals before these are relayed to the spinothalamic tract.

Figure 9.134, from Bennett et al. (1980), shows two neurons in layer II of a cat, stained by horse-radish peroxidase after being tested by electrical and natural stimuli in their receptive fields. These cells did not respond to gentle mechanical stimulation and responded specifically to tissue-damaging stimuli. Responses to electrical stimulation and to sustained pinch with toothed forceps are depicted. Specific nociceptive neurons comprised 40% of units sampled in layer II. Similar second order nociceptive neurons have also been recorded in layers I, IV, and V.

Another 40% of neurons encountered in layer II responded to both gentle mechanical stimulation *and* to tissue-damaging stimuli. For this reason, they are called wide dynamic range neurons. Fifteen percent of the SG neurons respond to low threshold touch stimuli conveyed by unmyelinated fibers and relayed to local and ascending units.

PAIN MECHANISMS AT SPINAL AND BRAIN STEM LEVELS

THE SEARCH FOR MECHANISMS CONTROLLING PAIN

The use of morphine and morphine-like substances for the relief of pain has been known since antiquity. Recently it was found that cells in certain regions of the brain bind opiates stereospecifically and that the analgesic potency of a drug correlates directly with the affinity with which it binds to cell receptors (see Snyder and Matthysse, 1974).

This led to a search for naturally occurring endogenous morphine-like compounds. Two naturally occurring groups of compounds were discovered, *leucine and methionine enkephalin* (Hughes et al., 1975) and the *endorphins* (Loh et al., 1976), including β-endorphin, a part of pituitary β-lipoprotein.

A few years earlier, it had been discovered that focal electrical stimulation of the central nervous system in the rat produces analgesia without general behavioral depression (Reynolds, 1969). The same regions most effective in electrical induction of analgesia coincide with the most powerful opiate binding sites. Moreover, locally applied opiates contribute as effectively as electrical stimulation. *An opiate antagonist, naloxone, reverses analgesia produced by electrical stimulation* (Adams, 1976; Hosobuchi et al., 1977).

Attention focused on midbrain *periaqueductal gray matter* because cells in this region have great affinity for opiate and opiate agonist binding. The same cells produce endogenous morphine-like *enkephalins*. This area is also prepotent among sites

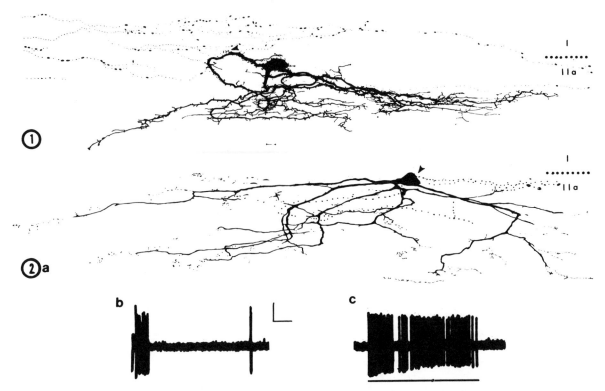

Figure 9.134. Second-order neurons in dorsal horn of spinal cord of cat display widely ramifying dendrites and axons throughout superficial layers where nociceptive primary neurons enter. These neurons did not respond to gentle mechanical stimulation but responded to tissue-damaging pricking and pinching within their receptive fields. Neuron 1 shows spiny cell body and extensive spiny dendritic ramifications in layer II, with faintly traced axonal ramifications in layer I. Axon arises from dendrite at *arrowhead*; axonal varicosities are shown as *larger dots*. Neuron 2 shows extensive axonal and dendritic ramifications in layer II. Fine extensions of dendrites are traced as *faint dashes*; axon which arises from cell body is shown by *dashed lines*. Below (*b*) are seven superimposed responses of neuron 2 to single electric shock showing early and late responses from myelinated (35 m/s) and unmyelinated (1.1 m/s) afferents. In *c*, this neuron's response to sustained pinch (underline shows duration) with small toothed forceps. Vertical scale bar = 10 μm. Time bar = 50 m/s (*b*); (*c*) 0.5 s. [From Bennett et al. (1980).]

for *electrically induced analgesia.* Since electrically induced analgesia suppresses flexion reflexes, the search was on to discover the descending pathway.

DESCENDING PAIN CONTROL MECHANISMS

There is a remarkable correspondence between cells that are rich in opiate-binding receptors, cells which produce endogenous opiates, and sites most effective for electrically induced analgesia. *Opiate administration activates pain suppression mechanisms which are organized at three levels of the neuraxis: midbrain, bulb (medulla oblongata), and spinal cord.* Activation of neurons in the midbrain periaqueductal gray matter is accomplished by direct electrical stimulation, by local introduction of small quantities of opiates, by larger doses of opiates administered systemically, and probably also by limbic activation associated with strong emotions. The limbic system has a widespread distribution of opiate receptors.

The result of activating neurons in the periaqueductal gray is direct and indirect activation of *serotonergic neurons in the rostral bulb, in the nucleus raphé magnus* (see Basbaum and Fields, 1978). These serotonergic cells project to the spinal root relay of the trigeminal and, via the *dorsolateral funiculus of the spinal cord,* to homologous nuclear (SG) relays in the dorsal horn. Interruption of this pathway produces loss of analgesia dependent on centrally introduced opiates or electrical stimulation. The interruption affects only ipsilateral segments below the level of cutting the dorsolateral fibers (Basbaum, 1976).

The effect of this descending pain control mechanism is to interfere with transmission between nociceptive afferents and the next order neurons. These are the relays which are presumably ordinarily activated by the release of substance P by nociceptive afferents. Next order neurons, if activated by pain stimuli, and insufficiently inhibited by descending pain control mechanisms, send signals up the lateral spinothalamic tract. Collaterals from this tract activate periaqueductal gray matter and the bulbar raphé nucleus. Thus is activated a feedback loop which augments activity in the descending pain control system.

The consequence of activating the periaqueductal neurons (which are rich in enkephalins) is to fire up the entire chain of neurons involved in the endogenous control of pain.

SUMMING UP SPINAL AND BRAIN STEM SENSORY MECHANISMS

A wide variety of sensory signals, spread out over space and time, enter the central nervous system by way of spinal dorsal roots and the brain stem trigeminal root. During the course of sensory relay, these afferent inputs are distributed mainly throughout the dorsal horn and nucleus proprius in the spinal cord and their trigeminal analogues in the brain stem.

The more penetrating afferent invasions of spinal gray matter concern reflexes: segmental, intersegmental, interlimb, ipsilateral, and contralateral, analyzed in Chapter 70. *The fastest, muscle and tendon afferents contribute mainly to skeletal muscle stretch reflexes and skeletal muscle motor "tone." The next fastest, cutaneous and body wall afferents, contribute mainly to flexion reflexes.* These latter are accompanied by ipsilateral inhibition of stretch reflexes and contralateral mirror image effects (i.e., crossed extensor reflex excitation and flexor reflex inhibition). The slowest, visceral afferents, contribute mainly to visceral and flexion reflexes.

Incoming sensory information contributes to four main ascending sensory systems:

1. Dorsal column fibers to dorsal column nuclei in the caudal brain stem (the "lemniscal" system) represent an ipsilateral continuation upward of axons of afferent, first order neurons, the longest nerve cells in the body. This is a phylogenetically recent (mammalian) contribution which represents the shortest—minimum two synaptic path—to cortex. The dorsal columns convey light discriminative touch, two-point sensibility, position, and vibratory sense.

2. Lateral spinothalamic tract serving crude touch, pain, and temperature sensations (the "spinothalamic" system). This is relayed by cells in the dorsal horn whose axons largely cross the midline and ascend in lateral white matter. There are ordinarily one to several additional synapses along this route to thalamus. The spinothalamic tract is not so dynamically accurate, modality specific, nor synaptically reliable as is the lemniscal system.

This pathway communicates in the bulb with the large raphé nucleus and in the midbrain with the periaqueductal gray matter. These serve as central stations responsible for descending pain control. The periaqueductal cells are opiate-receptive and produce endogenous opioids. The raphé nucleus, to which the periaqueductal cells project, is serotonergic. This latter nucleus projects to the region of substantia gelatinosa in brain stem (for the trigeminal nerve root) and spinal levels, where it contributes to inhibiting the relay of nociceptive signals. Lateral spinothalamic tract collaterals contribute to this central pain control feedback loop, thus augmenting descending traffic that controls pain relays.

3. Ascending visceral projections are phylogenetically the oldest, slowest, and most pleurisynaptic pathways, more diffused in space and time, less specific in modality characteristics, perhaps more than half crossed (the "archispinothalamic" tract). This undergoes repeated synaptic relays at successively higher spinal and brain stem levels on its way to diencephalon, primarily to the hypothalamus.

Each of these three systems gives off axon collaterals into the spinal cord and brain stem gray matter, thereby contributing to general state conditions of feeling, readiness for action, and arousal.

4. Propriospinal projections, far-reaching spinal interneurons which vertically stitch spinal segments together and contribute to intersegmental and interlimb reflexes. Afferent and intrinsic spinal signals contribute to vertically oriented spinal interneurons which have dendrites that reach upward and downward several segments, and axons that go up and down to reach most spinal segments, even caudal brain stem.

RELAY OF SOMESTHETIC INFORMATION THROUGH THALAMUS

Ventral Posterior Nuclei of the Thalamus

The ventral posterolateral nucleus (VPL) of the thalamus receives ascending fibers from the *medial lemniscus.* Information concerning peripheral stimulus location, form, modality, and dynamic characteristics are transmitted in an orderly fashion along this pathway. Signals pass through two reliable synaptic relays (dorsal column nuclei and VPL) to reach cortex. Precision in sensory processing along the lemniscal pathway derives from afferent neurons that have limited receptive fields, prefer single modality expressions, show dynamic responses, and transmit through conservative synaptic junctions to reliable third order (thalamocortical) neurons. Lateral inhibition increases the sharpness of spatial and temporal patterns at each synaptic relay. *Lemniscal information concerns light discriminative touch, two-point sensibility, flutter and vibratory sense, and position sense. This information relates to events in the contralateral side of the body.*

Lemniscal relay in VPL is not without corollary influences, however. VPL receives substantial inputs from the same somatosensory cortical areas to which it projects. It also receives projections from the median forebrain bundle (relating to frontal lobe and hypothalamic functions) and from midbrain reticular formation and nuclei in the thalamus (midline, intralaminar nuclei, and reticular nucleus) which are concerned with arousal. Ralston (1969) estimated that only about 8% of the boutons in VPL belong to lemniscal afferents.

The neighboring ventral posteromedial nucleus (VPM) receives axons from the main sensory nucleus of the trigeminal. Second order fibers ascend the brain stem as the *trigeminal lemniscus*. This projection is also somatotopically organized and consists of similarly precise representations of stimulus qualities and dynamics. The combination of VPL and VPM therefore comprises a comprehensive three-dimensional neurological model of the contralateral side of the body. This is organized in the ventral posterior part of the thalamus at its boundary with midbrain. Face representation is lowermost and medial, with the rest of the body positioned progressively more laterally and tilted foot upward. There is also a sparse lemniscal contribution to the medial part of the posterior nucleus of the thalamus, bordering the medial geniculate (auditory relay) nucleus.

Somatotopic representation in VPL and VPM contains partitioned distributions of neurons reflecting different sensory modalities. Cutaneous afferents are represented posteriorly, whereas deep fascial, periosteal, joint afferents, and muscle afferents are represented successively more anteriorly (Donaldson, 1973). This three-dimensional somatotopy is conserved two dimensionally in primary somatosensory cortex.

Comparison between Lemniscal and Spinothalamic Sensory Processing

Spinothalamic representation in the thalamus is relatively widely distributed, with an emphasis on activation of diffusely projecting thalamic nuclei (midline and intralaminar) which participate in forebrain arousal. Spinothalamic cortical representation is less abundant and less specific than that of lemniscal systems.

Spinothalamic projections contribute general features of sensations and qualitative experience relating to pain and temperature. Their information is much less precise concerning place, spatial organization, and dynamics. Their projections to spinal and bulbar reticular formation contribute importantly to readiness and comportment of visceral mechanisms. Spinothalamic tracts have a greater diversity of afferent fibers: A-δ and C-fibers as well as larger myelinated fibers. Their projections to midbrain reticular formation contribute importantly to mechanisms of arousal.

Evolution, in creating the lemniscal systems, not only shortened the pathways to higher centers but delivered vastly refined information concerning the body and environmental events impinging on the body. Information transmitted by the lemniscal systems travels in first order neurons by way of ipsilateral dorsal columns most or all the way to dorsal column nuclei. From those nuclei at the bulbar levels, all second order neurons cross the midline to aggregate in forming the *medial lemniscus* on the contralateral side. *Trigeminal* second order fibers, from the *nucleus proprius*, similarly all cross the midline to form the *trigeminal lemniscus*. Second order trigeminal neurons relayed through the spinal root of the trigeminal, comparable to spinothalamic projections, ascend as a mainly crossed, partly uncrossed, *trigeminothalamic projection*.

Characteristics of lemniscal precision stand in contract to the relative crudeness of information conveyed by the spinothalamic tracts. Spinothalamic projections synapse within a few spinal segments neighboring their dorsal root of entry. Second order spinothalamic neurons, located among several laminae of the dorsal horn, ascend in the lateralmost white matter of the spinal cord. The majority has crossed in the anterior commissure to the contralateral side, but some spinothalamic fibers ascend ipsilaterally.

The lateral spinothalamic tract, serving mainly crude touch, pain, and temperature, ascends the spinal cord and brain stem. This is closely followed by the trigeminothalamic tract. Trigeminothalamic and spinothalamic units present a greater diversity of receptive field sizes and distributions, a greater range of modality expressions, and a wider variety of dynamic responses than do the trigeminal and medial lemniscal paths. Spinothalamic·paths have at least one and ordinarily several additional synapses, and they have less refined mechanisms for representation of pattern and timing. They also show less precision and reliability in synaptic relays. Spinothalamic projections correspondingly contribute more to feeling states and the general sense of well or ill being, pain, temperature, and accompanying aspects of crude touch.

Brain Stem Pathways Activated by Tooth Stimulation

Popular consensus attributes nociceptive receptive capacities to tooth pulp stimulation. The tooth pulp is served by A-δ and C-fiber crude touch, pain, and temperature receptors. Stimulation of tooth pulp activates trigeminal lemniscal and trigeminothalamic pathways. It also activates certain additional ascending channels in the brain stem (Kerr et al., 1955). These include pathways in the ventral part of periaqueductal gray matter and adjoining gray and white reticular formation. Altogether there are five distinguishable ascending projection systems activated by tooth pulp stimulation.

These different sensory paths can be distinguished according to differences in latency, velocity, and other dynamic response characteristics, and to differential influences of anesthetic agents. Tooth pulp stimulation elicits short latency cortical responses in S I which are resistant to analgesic levels of anesthetic agents. It elicits in Sm II longer latency, bilateral responses that are susceptible to analgesic levels of anesthetic agents (Melzack and Haugen, 1957).

Furthermore, discrete brain stem lesions in cats affect the capacity to respond to noxious levels of heat and pin prick (Melzack et al., 1957). Bilateral lesions in the spinothalamic tract and in the central gray produce a markedly reduced capacity to respond to both noxious heat and pin prick, as compared with control groups of animals with lesions in the lemniscal systems. The effects of lesions in the central gray matter, however, only temporarily reduced responses to noxious heat stimuli.

Cats with bilateral lesions of a descending tract, the central tegmental fasciculus (CTT), responded as though they had become hypersensitive to pin prick. They exhibited responses suggestive of "spontaneous pain" in the absence of external stimulation. It is presumed that the spinothalamic tract and central gray are both involved in central transmission of pain signals. Central gray, however, is also responsible for part of the endogenous descending pain control process. The CTT is also presumed to exert control over the relay of pain signals. It is logical to suppose that defective central gray and CTT systems might play a role in chronic intractable pain syndromes, like causalgia.

As noted earlier, when the lemniscal system and spinothalamic tracts approach the thalamus, their fibers converge. Nevertheless, *the two systems maintain separate channels* in their thalamic relay

relations. Spinothalamic fibers project largely to the margins of VPL and overlap with terminals of the lemniscal system only in the medial posterior thalamic region where there is only sparse lemniscal representation. There appears to be a great deal of independence and segregation of information transmitted through the thalamus.

Powers of discrimination relating to the somatosensory system are remarkable. As in other sensory systems, stimulation of neighboring cutaneous fields results in powerful *lateral inhibition*. This pattern is reinforced from peripheral nerves through dorsal column and thalamic relays. Thus, edges and points tend to be sharpened in their localization. When these mechanisms of analysis of simple patterns are applied to more complex patterns, complex textures are similarly made more precise.

As with vision, skin detection systems can be moved about to enhance the combination of spatial and temporal discrimination. Thus, an individual who is adept at reading braille can discriminate separations of braille points, by applying fingertip movements, that approximate the distances between adjacent cutaneous stem afferents. In this respect, cutaneous searching defines discriminable touch-contact differences in the same way that visual searching detects differences in vernier displacements.

SOMATIC REPRESENTATION IN CEREBRAL CORTEX

Thalamic relay of somesthetic projections is predominantly to the postcentral gyrus. Somatotopy is preserved. Body representation projected onto the primate hemisphere is contralateral, with the foot represented at the top of the postcentral gyrus, the leg, pelvis, trunk, arm, hand, and face, all arranged in sequence laterally, ventrally, and inclining anteriorly, following the gyral pattern. Following face projections onto the superior bank of the Sylvian fissure are represented successively, mouth, tongue, throat, with visceral representation still more deeply infolded into the Sylvian fissure, refolding back onto the hidden surface of the insula.

This extended representation of the interior of the body approaches closure of a loop because, superiorly, on the medial surface of the hemisphere, over the crest of the postcentral gyrus, there is continuing representation of anus, perineum, buttocks, and, in tailed primates, the tail. It is as though a ring of somatic and visceral structures were linked in a nearly completed contiguous circle, interrupted medially only by the hylus of the hemisphere. The somatovisceral map lies like waistband not quite clasped around the middle of the hemisphere. The representational discontinuity is approximately at pelvic viscera.

Just as there are differences in density of distribution of cutaneous afferents in the periphery, there are even more spectacular differences in scale of representation of body parts in cortex. *Hand and fingertip representation, lip, and other specially discriminative parts have greatly enlarged representations.* The somatotopic representation of the contralateral cutaneous surface of the monkey in Figure 9.135, from the work of Nelson et al. (1980), shows relatively expanded cortical displays of foot and hand. These two parts occupy about half of the total primary cortical representation of the body. Note that the smallest areal representations relate

to dorsal hairy skin areas. Further, *cortical representation is banded, horizontally.* This involves a distinctive anteroposterior representation of *innervation relating to the same part of the body but organized in relation to different receptor modalities.*

In the depth of the central sulcus, that is, rising up along the posterior bank of the central sulcus (area 3a according to the nomenclature of Brodmann) is representation of muscle stretch afferents. This lies immediately adjacent, anteriorly, to motor cortex. Posteriorly are found (*3b* in Fig. 9.135) slowly adapting cutaneous representation, followed by (area 1) rapidly adapting cutaneous representation, followed in turn by (area 2) periosteal, fascial, and joint sensibility (Kaas et al., 1981). These separate zones can be traced back to distinctive cellular

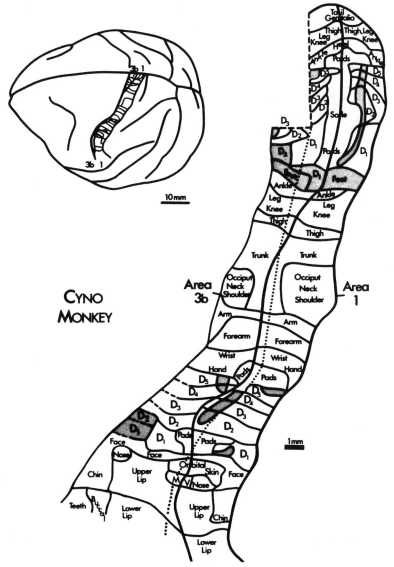

Figure 9.135. Unfolding of bodily representations in areas *3b* and *1* on the medial and superior surface of the right hemisphere. The *dotted line* indicates the position of the top of the central sulcus. *Shaded areas* indicate representations of hairy dorsum of foot and digits. [From Nelson et al. (1980).]

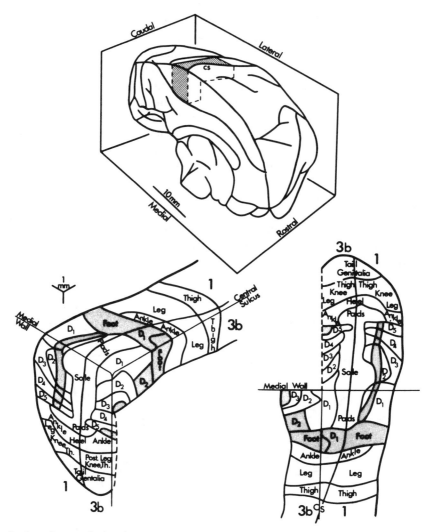

Figure 9.136. Organization of somesthetic primary cortex in monkey (SI), areas *3b* and *1*, shown on a dorsolateral view of brain (*left*) and as they appear "unfolded" from the central sulcus and medial wall of the hemisphere (*right*). The representations of individual digits of the foot and hand are outlined and numbered (*D1–D5*); *shaded areas* correspond to representation of hairy dorsum of digits. The *dashed line* indicates the region along the medial wall of the hemisphere where portions of the representation in areas *3b* and *1* are continued. [From Nelson et al. (1980).]

arrangements in the thalamus where these modalities are already segregated. It is thus clear that *sensory signals are projected along a series of parallel processing channels.*

Figure 9.136 shows a dorsolateral view of primary somatosensory cortex in monkey. Representations are "unfolded" from their enclosed fields of functional distribution along the medial wall of the hemisphere and in the depths of the central sulcus. Further details of digit representation in foot and hand are provided.

Dynamics of Cortical Organization

Effects of Deafferentation

These cortical representations are really dynamic. Apparent stability is the result of balancing among dynamic influences. When the median nerve

is sectioned, for example, there is only temporary inactivation of parts of the cortical map for innervation of the now denervated glabrous skin of the palm. Neurons in this "deafferented" cortical field, however, after a short interval, can be activated by stimulating parts of the hand outside the peripheral distribution of the median nerve. The new receptive fields are not random but are replaced by new maps of the neighboring dorsum of the hand. Glabrous skin representation in the cortex has therefore been succeeded by hairy skin representation (Kaas et al., 1981). The result of interruptions of input from part of the periphery is to expand cortical representation of other parts.

The authors conclude that *there is constant competition for cortical sites by adjacent skin areas.* Cortical maps change over time and, consequently, organization revealed by mapping experiments in normal animals reflects only part of the potential for discriminative organization that exists within the total nexus of anatomical connections.

Effects of Removal of Vibrissae

Many mammals have vibrissae (whiskers) which constitute marvelous sweeping probes (like a blind person's extended wand) which are used to explore the environment of the face and snout. These long whiskers are sensitive to touch, and like guard hairs, they are innervated by a number of directionally sensitive receptors. Vibrissae can be moved about, even somewhat independently as well as en masse. Vibrissae are provided with elaborate sensory mechanisms which lie at the base of each motile exploratory probe.

Woolsey and van der Loos (1970) found that each vibrissa has a large central cortical representation in the form of a nest of cells that constitutes a cortical "barrel" for each vibrissa. Barrels can be seen with the unaided eye when a slide of sensori-motor cortex is held up to light. Loosely speaking, these can be considered greatly expanded cortical columns.

The most interesting feature of vibrissae and their corresponding barrels is that if vibrissae are destroyed in a young animal, the corresponding cortical barrels die back and disappear. The consequences of destroying a row of such vibrissae are shown in Figure 9.137. What is astonishing is the anterograde retrogressive process which affects the entire sensory column. Following destruction of a row of vibrissae in the rat, changes are seen in primary trigeminal neurons, second order trigeminal lemniscal neurons, third order thalamic neurons projecting to cortex, and even in the cortical barrel itself.

Barrels demonstrate rudimentary yet adequate central circuitry at birth. This neonatal circuitry is tentative, however, and must be put to use during early life experience or it will retrogress. *If central nervous system parts are not activated from the periphery, central representations retrogress.* Thus, the central nervous system in somesthesis, as in other sensory systems, models the periphery and adjusts itself correspondingly.

Can the Nervous System Develop Better Than Usual Circuits?

There is evidence that the nervous system may be able to expand central representation as a result of enriched experiences, perhaps to develop better

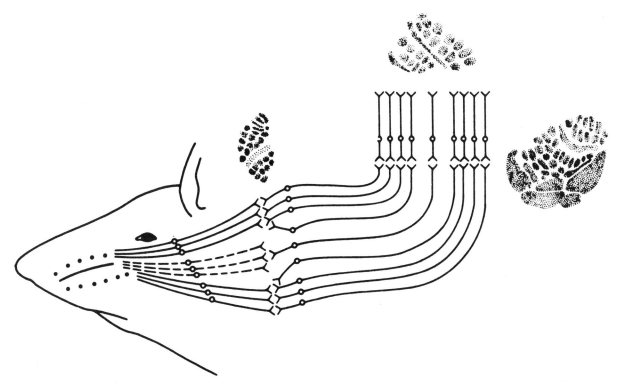

Figure 9.137. Summary of effects of vibrissae damage on pattern formation in the trigeminal distribution to somatosensory cortex in the rat. Following removal of the middle row (only three of the five rows are illustrated in this schematic), changes are detected in the termination of the primary neurons in the brain stem, secondary neurons in the thalamus, and tertiary neurons in the cortex. At all three levels, the effect is a band rather than discrete clusters. [From Killackey (1982).]

than usual circuitry (Bennett, 1976). Moreover, it is evident that the nervous system does not waste central processing capabilities on destroyed or neglected peripheral motor and sensory apparatus.

Relatively brief sensory deprivation may have significant physiological consequences. For instance, astronauts, after only a few days in a weightless environment, develop diminished cardiovascular reflexes needed for adequate cerebral circulation in a 1 *g* gravitation environment. They recover adequate reflex "tone" within a few days. They need to take prophylactic exercises to maintain integrity of these autonomic circuits during prolonged weightlessness. Space stations may need to be designed to obviate this difficulty by using centrifugal spin to supply an adequate equivalent of gravitation. This observation has obvious clinical implications, especially since retrogressive effects are relatively rapid.

Dynamic adaptive characteristics are also important when there has been injury to the nervous system because as a result of damage, many parts of the nervous system may be abruptly and lastingly deprived of normal activation.

Additional Major Somatosensory Representation

It is a matter of style how somatosensory cortex is defined. There are many areas from which potentials can be elicited following stimulation of somatic receptors. On this basis, a very large proportion of cortex might be claimed to have somatosensory representation. A more restricted interpretation suggests that the postcentral gyrus deserves priority as the *primary somatosensory cortex.* On that account it has been designated *S I.* Two additional areas also warrant special consideration. One is known as *Sensory II* and the other as *Supplementary Motor Cortex.* Because these represent motor as well as sensory functions, they are appropriately referred to as *sensorimotor cortex.* The first is predominantly sensory and is designated *Sm II,* emphasizing the sensory preponderance of that representation. The lower case m suggests a subordinate motor role. The second additional somatosensory field is predominantly motor and is therefore designated *Ms II* to indicate motoric preponderance over sensory.

Sm II is located laterally and posteriorly to face representation in S I. Its somatotopic organization suggests a figurine about one-tenth the length of S I, lying on its back, with its face viz a viz face representation in S I. The feet are extended back along the hidden upper bank of the Sylvian fissure.

Cortical areas in Sm II include portions of area 7b, retroinsular cortex (Ri), postauditory field (PA), and granular insula (Ig). These fields receive projections from the primary somatosensory cortex (S I) and as well from VPL and VPM somatosensory rely nuclei in the thalamus. They are all largely buried within the Sylvian fissure and lie mostly posterior to the caudal end of the insula (Robinson and Burton, 1980). Receptive fields are generally large (>10 cm^2) and responsive to stimulation on both sides of the body. Some units have variable receptive fields, depending on the level of wakefulness. The exception is Ri which has smaller receptive fields (generally <10 cm^2) contralateral representation, with stable field boundaries. Figure 9.138 illustrates Sm II (in this figure, designated S II) and surrounding cortical fields in the monkey. The cortical units in Sm II are less securely linked to peripheral unit activity than are cells in S I. Sm II has another distinguishing feature; it enjoys considerable bilateral representation. *Bilateral representation in Sm II may contribute to bilaterally coordinated body movements.*

Supplemental motor cortex (Ms II) is located in about a mirror image relationship with Sm II. It lies in front of motor cortex on the medial surface of the hemisphere. While Sm II lies face to face with S I on the lateral surface of the hemisphere, Ms II lies toe to toe with motor cortex on the medial surface of the hemisphere. In both secondary somatosensory representations, the face points anteriorly.

Columnar Organization of Somatosensory Cortex

The basic functional structure of neocortex lies in columns of cells which are vertically oriented throughout the thickness of cortex and perpendicular to the cortical surface. *A single cortical column constitutes a modular unit for cortical operations.* Cortical columns were first discovered by Mountcastle in S I and have subsequently been found in all areas of cortex, sensory, motor, and association.

COLUMNAR INPUT

Columns are powerfully connected vertically and sparsely so horizontally. They represent a chain of neurons, vertically disposed, which accomplish discrete tasks between input and output. Input arrives from thalamocortical projections, with terminals ending among pyramidal cells in layers III and V–VI. The incoming signals thus have rapid access to columnar output via cells in layers V and VI. They contribute and acquire modulatory influences through transactions with cells in more superficial parts (in layers I, II, and III) of the same column. These more superficial layers of cortical columns

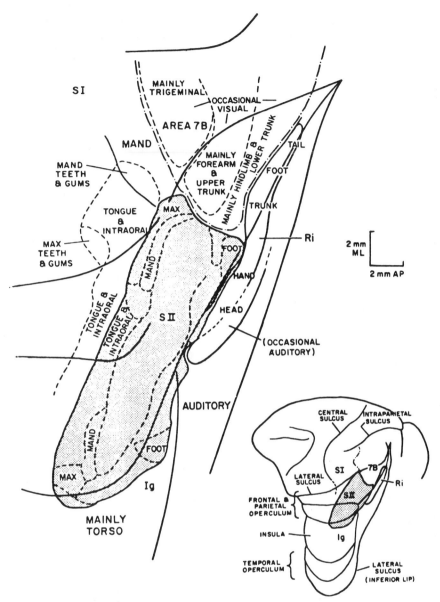

Figure 9.138. Somatosensory areas in monkey showing functional organization of secondary sensory area (*SII*), including exposed view of cortical surfaces normally buried within the lateral (Sylvian) fissure. *Lower right* of figure shows SII in relation to the left hemisphere and in more detail, *left center*. Note somatotopic arrangements, including parallel-processing representations and representations of additional sensory modalities, auditory and visual at the margins. [From Robinson and Burton (1980).]

receive input from other cortical areas via *collateral and commissural connections*. They also receive more general messages through projections from the *locus coeruleus* in the brain stem, and from *midline, intralaminar, and reticular nuclei* in the thalamus. This traffic is concerned with features of general state conditions such as *attention, sleep, and arousal*.

Cortical columns receive primary sensory input from thalamocortical fibers (VPL and VPM) which have closely overlapping receptive fields and which convey the same sensory modality. The same cortical columns fire back reciprocal impulses via cor-

ticothalamic projections which go to the same nuclear sites that provided the cortical input. *Cortical outputs* contribute impulses to descending projections destined to *influence sensory relay nuclei in a downward cascade which provides central control of sensory transmission*. Other cortical outputs are directed to *motor control systems, somatic and visceral*, in basal ganglia, diencephalon, brain stem, cerebellum, and spinal cord. Still other cortical outputs go to particular *motor and sensory fields in cortex*, thereby contributing to higher order motor command and to larger scale processes of sensory integration.

Vertical processing of incoming thalamic signals is ensured by intrinsic columnar recipient cells which possess vertically directed axons which synapse on dendrites of columnar pyramidal cells which have cell bodies in all cortical layers. This ensures that input distributed within the column will contribute to output.

COLUMNAR OUTPUT

Output from cortical columns consists of two major components: One is directed cortically, presumably contributing to integrative aspects of perceptual processes, culminating at the highest levels of perception and creative thought relating to body imagery, etc. The other component is directed subcortically, contributing directly and indirectly to influences on sensory receptors and ascending sensory signals, and directly and indirectly affecting motor control systems at various levels throughout the neuraxis.

Figure 9.139 is a diagrammatic representation of central control of sensory receptors and central sensory transmission. *Some 10% of efferent neurons leaving brain stem and spinal cord are destined to sensory receptors and sense organs in the periphery.* The best known of these are efferents to intrafusal fibers which control muscle spindle afferents. Many of the remainder were originally thought to be purely trophic in function. *They may have trophic functions but they also set the bias for excitability of many receptors and sensory organs* like the olfactory bulb, cochlea, taste buds, retina, etc.

This same system operates to control the level of excitability and bias for transmission at central sensory relays. In short, the nervous system has abundant mechanisms to control its own input. Centrifugal sensory control pathways follow two general anatomical patterns: (1) Some go from defined fields such as sensory cortex to the thalamic source of input to that same region, e.g., somatosensory cortex to VPL and VPM; primary visual cortex to LGB; pyramidal cells projecting from primary motor cortex to spinal motor nuclei send off collaterals which innervate, and exert control over dorsal column nuclear relay of the lemniscal system. (2) Other centrifugal fibers converge on brain stem and spinal reticular formation from a wide variety of cortical and subcortical regions. These seem to be more diffuse in the exercise of their control over sensory input.

Generally speaking, pyramidal cells in different cortical layers have axons that are directed to different destinations. *Pyramidal cells in upper layers of the column project mainly to other regions of cortex. Pyramidal cells in lower layers project mainly to subcortical regions.* Intrinsic short (Golgi type II) interneurons distribute lateral inhibitory influences close by, while basket cells distribute inhibitory influences more widely within the same and adjacent columns. Collaterals of axons which leave the column to enter white matter may reenter the cortex over a wider region, measured in millimeters, and contribute still more widespread lateral transactions.

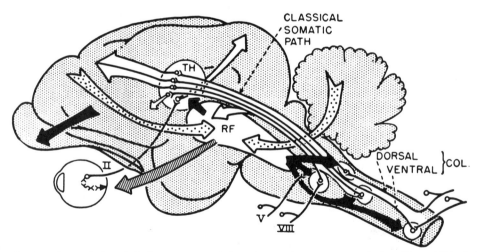

Figure 9.139. Systems for central control of sensory receptors and central sensory transmission. About 10% of efferent neurons are directed to sense receptors and to sensory organs like olfactory bulb, retina, cochlea, taste buds, etc. Sensory relay nuclei characteristically send feedback signals to lower stations, e.g., dorsal column nuclei (second order neurons) to spinal sensory relay throughout the spinal cord, and cortical sensory areas project downstream onto the thalamic (TH) sources of their input, e.g., primary visual cortex onto lateral geniculate neurons. Moreover, projections from cerebrum and cerebellum to brain stem reticular formation (*RF*) generate diffusely projecting influences which alter sensory input, presumably in more general, nonspecific ways. These sensory control systems are conditionable; hence the brain controls its own input in accordance with past experiences. The full biological significance with respect to sensory signal processing is only beginning to be understood. [From Hernández-Péon and Livingston (1978).]

Lateral Inhibition

We have seen that lateral transactions contribute to sensory analysis in each of the other sensory systems. The principle holds equally well for somesthesis. For example, lateral inhibition and lateral facilitation refine cutaneous directional sensitivity. Directionally sensitive neurons in S I were analyzed in response to moving edges applied to the skin of waking monkeys (Gardner and Costanzo, 1980). These units were either spontaneously active or were induced to fire tonically by a variety of cutaneous stimuli. Such cells showed depression of tonic activity only to off-directional stimuli, suggesting an underlying *active inhibitory process.* Inhibition was also elicited from within the receptive fields of directionally sensitive units, indicating that *directional sensitivity is produced by an asymmetric distribution of lateral inhibition and lateral facilitation. Evidently, lateral transactions play a significant role in analysis of spatial and temporal patterns in all sensory systems.*

SENSORY ASSOCIATION CORTEX

Evidence from the study of parietal cortex in waking monkeys indicates that this region is neither dominated by afferent input, as are primary sensory areas, nor dominating to peripheral effectors, as are primary motor areas. They are, instead, relatively remote from both input and output. The specific activities to which columnar units in parietal cortex, areas 5 and 7 of Brodmann, respond, share one common characteristic. They are active in relation to the animal's behavior in such a way that they *link pertinent sensory information to particular motives and particular opportunities for action.* They are active in relation to the self, and to the immediately surrounding extrapersonal space. They are concerned with spatial relations of body parts with one another in respect to the gravitational field and the near environment (Mountcastle, 1975, 1976).

Within area 7 there are sets of columns which are active during movement of the arm toward an object of interest. Other columns are active during manipulation of objects and others during the direction of visual attention to objects. Some columns are activated only when eye movements are elicited by sight of objects in the environment, during slow pursuit movements of the eyes, and during visual searching for an object. There are columns which are activated preferentially with the appearance of objects of interest in peripheral vision.

Electrode penetration along the vertical axis of columns shows preponderance of cells of only one class. Penetrations that slant across a series of columns encounter cells which are activated by systematically related attributes rather than by intermingled attributes. *These association areas appear to be organized so as to pool sensory information from a wide variety of sources, olfactory, visual, auditory, tactile, body wall, etc. and utilize the information to specify actions in accordance with specific motivations.* Units in a given column may be activated, for example, only in relation to the appearance of an object (food), and then *only* when the animal is hungry, *and only* when the object is within reach.

Entities of the nervous system known as dorsal horn, dorsal column nuclei, reticular formation, thalamic relays, neocortex, basal ganglia, cerebellum, brain stem, and spinal motor nuclei, etc. are composed of local circuits which make them functional modules. These modules are in turn grouped into larger functional entities according to their extrinsic connections, fields of topographic representation, and kinds of transactional connections with other functional entities (Mountcastle, 1978). These systems preserve topological order: connections run in parallel bundles which preserve topological relations with other bundles, thus dictating topographic representations at different levels of the neuroaxis as affected by requirements for sorting, changes in magnification, competition for space, and requirements imposed by vertebral column and skull design. Closely linked subsets of larger entities are engaged in precisely connected, distributed systems. This explains the parallel processing and distribution of functions that seem so baffling when they are examined at only one level of central nervous system organization. An article that is helpful in considering such global aspects of brain organization is Thinking about the Brain (Crick, 1979).

Parietal Lobe Deficits in Humans

Humans with parietal lobe lesions reveal fascinating disturbances of behavior. They show alterations in perception of body form and its relations with surrounding space. Unilateral lesions produce changes in behavior that are purely contralateral in some contexts, while other unilateral lesions can produce disorders affecting both sides of the body. Pure contralateral effects are commonplace following acute vascular lesions that affect one side. They may be manifested by *neglect or denial of one side of the body* (contralateral) *and of the corresponding side of objects in the environment.*

Mesulam (1981) has carefully considered clinical disturbances of unilateral neglect which he divides into four distinguishable functional categories relating to four different central nervous regions. Each anatomical location has a unique functional role that reflects its profile of anatomical connections and gives rise to different kinds of clinical neglect behavior.

Unilateral neglect follows disruption of mechanisms relating to spatial direction of attention. Clinical unilateral neglect syndromes are more common and severe after lesions in the right hemisphere. Right hemisphere mechanisms appear to be involved in the execution of most attentional tasks. The right hemisphere appears capable of operating the neural apparatus for attention on both sides of the body, whereas the left hemisphere seems able to direct attention only contralaterally. It is therefore assumed that the right hemisphere, at least in dextrals, is somehow specialized for direction of attention throughout extrapersonal space.

The parietal cortex has multiple systems representations such as "sensory association," "limbic," "reticular," and "motor." This provides for convergences between sensations, motivations, states of arousal, and motoric potentialities in a given environmental context. Sensory signals do not have access to parietal cortex until they have been extensively processed from unimodal to polymodal signals. Dorsolateral parietal cortex manifests the most elaborately processed sensory information from all sensory systems and sense modalities. *The pulvinar nucleus of the thalamus is in closest reciprocal relations with this cortical area.* Intralaminar and reticular nuclei of the thalamus contribute information relating to arousal. Limbic information enters from the subtantia innominata and lateral hypothalamic area and closely associated basal forebrain (orbital) cortex. Parietal cortex also receives projections from the claustrum. Besides reciprocal connections with each of these resource areas, posterior parietal cortex projects to frontal eye fields (area 8 of Brodmann) and the superior colliculus for control of direction of gaze. It sends relatively few projections to motor or premotor cortex but sends abundant connections to the striatum and subthalamic nucleus among basal ganglia.

CAUSES OF UNILATERAL NEGLECT SYNDROMES

Parietal Neglect

Parietal cortex provides a location for afferent integration whereby dynamic sensory representations of the body can operate integratively with dynamic sensory representations of extrapersonal space. Unilateral damage to the template of extrapersonal space biases central representations in favor of the environment ipsilateral to the damaged

hemisphere (Mesulam, 1981). As a consequence, the patient is likely to engage only or predominantly with that environment, being attracted to it, interacting with it, and neglecting the contralateral half of extrapersonal space.

The patient may respond to stimuli from the relatively neglected hemispace, but if *simultaneous double stimulation* is presented to both sides, the patient will be dominated by the stimulus that is favorably biased and will neglect stimuli on the other side.

Frontal Neglect

Another sort of neglect is associated with lesions in frontal cortex, particularly in the frontal eye fields. Just as parietal neglect is due to a shift in sensory bias, so the frontal eye field lesion leads not to a paresis of eye motor control, but to a disinclination (bias against), exerting eye movements toward the contralateral hemispace. Lesions that affect the premotor cortex can likewise yield bias against moving limbs on the affected side. With the frontal neglect syndrome, the patient may even be biased against moving unaffected limbs into the neglected hemispace.

Limbic Neglect

Still another form of neglect is *associated with lesions in the anterior part of the cingulate cortex.* This is an important part of the mesocortical ring of limbic cortex. Cingulate cortex is closely connected with the hippocampal formation, presubiculum, and amygdala, and with polymodal sensory association areas, including posterior parietal cortex. The cingulate cortex appears to constitute a parallel *representation of extrapersonal space,* one that is *primarily related to motivation.* Unilateral cingulate lesions in the monkey result in contralateral neglect, apparently because of an absence of adequate motivation.

Neglect Relating to Central State Conditions

A fourth systematic source of neglect points to central state conditions: *interference with attention or arousal.* This follows lesions affecting (1) the ascending noradrenergic projections that arise from the locus coeruleus, a rapheal nucleus that lies at the junction between posterior midbrain and anterior pons, at the efflux of the aqueduct into the IVth ventricle. Lesions that interrupt this bundle, which is projected extremely widely throughout the cerebellar as well as cerebral hemispheres, *interfere with global attentional processes* and hence result in general neglect. (2) Unilateral neglect may follow unilateral lesions in mesencephalic reticular formation and unilateral lesions in intralaminar and reticular nuclei of the thalamus.

CLINICAL SYNDROMES OF UNILATERAL NEGLECT

Severe unilateral neglect in humans is usually attributed to inferior parietal lobe involvement, particularly areas 39 and 40. (The closest match in monkey cortex is area 7.) Patients who are presented geometrical shapes to be viewed through a centrally placed slit behind which the objects are moved at a constant rate, one at a time, have difficulty detecting differences on the left side of the objects, whether moving left to right or right to left under the slit. Since the objects can be seen only centrally, under the slit, the outcome must be attributed to *neglect of the internal representation per se.* Further evidence for this interpretation stems from the fact that patients with unilateral neglect may not perceive hallucinated objects on the neglected side (Mesulam, 1981).

Directed attention to extrapersonal space constitutes a compound function that is based on transactions among at least three major cortical areas and depends also on mechanisms essential for maintaining attention and arousal. Other functional disorders, anosognosia (denial of one's own body parts), dressing apraxia (neglect to wash, dry, groom, and dress parts of the body), and construction apraxia (inability to assemble simple components or to draw all the numbers on a clock face) are additional clinical manifestations of parietal damage.

Altogether, evidence suggests that *components of a single complex function are represented within distinctive individual but interconnected central nervous locations which collectively constitute a network for that function.* Individual cortical areas apparently can support activities that are essential for several complex functions and may belong to several different functional complexes. The same general function may be interfered with by a lesion in one of several cortical areas, which represent component parts of such an integrated network.

According to this interpretation, lesions confined to one cortical site are likely to manifest multiple, perhaps ambiguous, deficits. Such an organizational arrangement has its protective advantages, because to induce a severe, lasting deficit of major functional importance requires simultaneous interference with several widely distributed components in that particular network. This organizational safeguard has been tested on the hard anvil of evolution. At least there is manifest tolerance for widely distributed functional representations which contribute to complex behavioral tasks.

Problems of Interpreting Somesthetic Experiences

What is perceived as a consequence of stimuli applied to skin and tissues at depth in various parts of the body depends on the (a) general density of innervation; (b) difference in distribution of specific types of sensory units; (c) physical characteristics of surrounding tissues and individual end organs; and (d) spatial and temporal patterns of activation and adaptation among given sensory units. If, for example, the skin is under tension, being stretched, deformed from outside pressure or pressure from within, tumescence, hemorrhage, inflammation, irritated from scarification, blistering, histamine release, etc., sensations will be correspondingly modified, heightened, numbed, etc. An example of an intervening variable is that single touch receptor activity in isolated skin of the frog is facilitated by stimulation of the sympathetic nerve supply to that region (Lowenstein, 1956). Moreover, circulating epinephrine modifies the excitability state of many end organs and may affect central reflexes and sensory relays.

Larger biological perspectives on perception suggest that both evolutionary and developmental trends operate to enhance the speed and accuracy of sensory processing by higher centers, and the equivalent motor responses. The payoff is increased survival and improved central satisfactions as a consequence of more appropriate behavior. Improved perception contributes to more successful behavior and to greater satisfactions that result from the more successful behavior. The phylogenetic addition of fast conducting, paucisynaptic pathways from receptors to higher centers shortens the time between events and percepts, reduces the number of potentially intervening variables, and thereby improves the quality of perception. Parallel fast conducting, paucisynaptic motor response pathways similarly improve behavioral responses to environmental events. Thus, *recent evolutionary contributions to sensory and motor pathways, and actual experiences encountered during development, contribute to reality testing and improvement of sensory processing mechanisms.*

Physiological consequences of learning early during development undoubtedly improve the analytic capability and reliability of sensory circuits. The experiences of musicians, artists, and athletes while sharpening their performance skills through disciplined practice suggest that this is the case. *It is not practice that makes perfect. It is practice the results of which are known that contributes to skill perfection.* Evolution and development contribute actively to a more advantageous biological system for perception, as needed to guide behavior more appropriately for survival and internal satisfaction.

Obviously, what takes place during perception is the resultant not only of receptor transduction and impulse signalling in first order neurons, but it includes events taking place throughout central sensory pathways and areas of cerebral representation where multiple inputs—including inputs from remote sources—converge and interact with one another and with intrinsic neurons. The central nervous system modulates incoming sensory signals via *efferent* projections to sensory receptors, sense organs, and all sensory relay nuclei. Corticocortical activities modify input into primary receiving cortex and other cortical association areas. The combination of these cerebral controls modifies sensory information and everything dependent on sensory information including all motor behavior, from reflexes to voluntary activities.

From Cutaneous Sensory Units to Perception

The evidence cited here strongly reinforces the notion that cutaneous sensory experience ordinarily originates with signals that are generated in a variety of quite specialized, relatively independent, peripheral sensory units. Although sensation ordinarily originates from signals generated by receptors, our consciousness probably cannot reach out so far as directly to tap events occurring in the receptors. The original sensory impulses, which already represent a notable compromise between the physical parameters of the initiating stimulus and the transduction processes of the receptor, give rise to information which may reach perception. If that perception were to be veridical with respect to activities in afferent neurons, there would need to be completely reliable and unmodified communications from all the sensory units in the periphery to those forebrain loci where perception takes place.

Completely veridical transmission simply cannot occur: there is a chain of transmissions from stimulus to receptor to neuron to neuron along which each signal transferred encounters convergence, divergence, plasticity, remote control, all the processes involved in integration. *Perception results from a complex sequence of integrative processes which directly and indirectly modify information coming in from peripheral receptors. Remote controls include influences from brain mechanisms relating to motivation, emotion, memory stores, and competing contemporary events, all of which can substantially modify any given sensory experience.*

Clearly, genetic as well as experiential factors (past conditioning and perceptual skill training, history of abuse to any part of the involved circuits, state of alertness, focus of attention, immediately antecedent stimulation, drug effects, etc.), all contribute and may even dominate perceptual experiences.

Laboratory Conditions Are Exceptional

In the laboratory, the aim is to minimize contributions from intervening variables. Conditions, including orienting and motivating instructions, are expressly arranged to stabilize many potential intervening variables. In the laboratory setting it is possible to obtain close correlations between physical stimuli and sensation. In these protected and somewhat artificial circumstances, the relationship between stimulus and percept can be tested, experienced, and stored in the subject's memory which contributes to training effects in the sensory test situation. The nervous system gains experience and skill in organizing itself for this special purpose. It is not simply a matter of will, but a correspondence between subject mindset and the purposeful environmental context. Nonetheless, even in the laboratory, reliability depends on the creation and maintenance of bonds of confidence and trust between subject and investigator.

When experimental animals are used, similarly strict conditions are arranged to smooth and regulate the paths between stimulus and response. For example, animals may be moderately food- or water-deprived in order to ensure their motivation and devotion of strict attention to given signals, and the carrying out of appropriate responses, which will be reinforced by food or liquids. We sometimes forget that such precautions provide a kind of experiential cocoon around the sensory pathways being investigated. This in itself marshalls extraordinary controls affecting perceptual conditions. Laudable scientific efforts define, as they specifically aim to do, optimal correspondence between physical events and percepts. *Everyday experiences in uncontrolled environments are not equivalent.*

In commonplace medical contexts, therefore, "adventitious intrusions" into sensorimotor events are impossible to exclude and may be difficult to interpret. Sometimes they trigger signs or symptoms that may be of crucial importance to a physician's insight. For example, bowel sounds picked up by a stethoscope on the abdominal wall may not be appropriate. What is going on? Evoking that question may start an entirely new train of thought. A patient recovering nicely from a medical or surgical treatment in hospital may say, "I don't feel quite right," but without being able to be more explicit. Perhaps this patient is experiencing small pulmonary emboli preliminary to massive pulmonary emboli that might be prevented.

Subtle signs and symptoms may be the only prodromata of serious, possible preventable catastrophes. Hence, the wise physician is sensitive and develops an "inner ear" for subjective testimony concerning a patient's internal experiences.

A few less subtle examples: (a) It is a commonplace observation that during intense concentration and emotional surfeit, injuries which would ordinarily be painful and disabling may go entirely unnoticed until substantially after the event. (b) Following convincing suggestion to a blindfolded subject, light tough on the skin with a cube of ice may feel "blistering hot." (c) Simple clinical sensory testing in cutaneous areas of "referred pain" which originates from disorder in a visceral organ, reveals radically altered skin sensibility, *hyperesthesia*, whereby all sensations are intensified and are usually perceived as disagreeable. (d) You may recall personally how the symptoms associated with the onset of a common cold may include severe generalized aching of joints and excessive cutaneous sensitivity to temperature changes and air movements.

Patients' Testimony Relating to Subjective Sensory Experiences

Aware of sensory physiology, you will readily retain the following practical principles: First, many processes intervene between stimulation of peripheral receptors and higher perceptual centers. Second, your patient's testimony is inevitably conditioned by his or her conception of what is physiologically reasonable, and by his or her comprehension of what you, as physician, seek and need to understand from whatever testimony he/she can provide. Third, there are far more patients accused of malingering than there are malingerers.

It is easy for a physician to disregard or discredit sensory experiences that seem improbable in light of present physiological knowledge. Mistakes are made, and unnecessary harm may be inflicted, when a physician assumes that testimony concerning sensory experiences are exaggerated or false. Thoughtful questioning and mutual educational efforts usually solve this problem. Therefore, it is important for the clinician to give the maximum benefit of doubt to the veracity of a patient's subjective analysis, at least pending the exclusion of important potential pathophysiological intrusions. Even then,

there needs to be a leavening of self-doubt in the physician's mind as to the possibility of as yet unrecognized and as yet undiscovered sensory intrusions that may also be contributing influentially to the patient's perceptual experiences. *Nowhere is the physician more vulnerable to mistaken interpretations than in considering testimony relating to pain. There is not yet enough knowledge as to what loops within loops may be disturbed in the pain pathway itself and in endogenous pain control pathways.*

Note with clinical judiciousness, therefore, that controlled observations which reveal fairly strict correspondence between stimulus, receptor, and perception, depend on idealized conditions. Most clinical transactions depend on remote recollections in history and relatively crude stimuli administered in a context of apprehension. These conditions can yield responses that may not conform nicely to textbook physiology.

The Physician Listens and Learns from Patient Testimony

Do not be surprised or disappointed if your patient's testimony relating to his or her sensory experiences doesn't dovetail with your basic science expectations. *Your prime responsibility is to discover how a particular nervous system responds to particular events in particular sensory fields and anywhere between those fields and central perceptual processing, including the particular central perceptual processing system itself.* Most of medical diagnosis has to do with visceral, touch, pain, and temperature sensibility. Begin by assuming that the nervous system under consideration, as a consequence of its evolution and development, is making a close approximation to delivering veridical information from tissue events to consciousness.

Listen Carefully

The patient's testimony regarding personal somesthetic perceptions may convey the most important, valuable, and conceivably the *only* clues to an underlying pathophysiology. If the testimony seems beyond reasonable bounds, perhaps you are on the track of a major discovery for that individual, perhaps of historical significance.

CHAPTER 67

Nervous System Communication and Control

CONTROL

The following chapters concern motor control: *How all effector systems of the body come under integrated, biologically purposeful control.* We have seen that sensory mechanisms contribute not only to ascending signals but to reflex mechanisms and to successively higher levels of sensorimotor integration, processes that become very elaborate at cortical levels. At lower levels, sensory mechanisms contribute quite automatically (without conscious control) to the organization of behavior. At successively higher levels, greater awareness is manifest and contributes increasingly to purposeful organization of ongoing behavior. In ways that are not yet understood, higher levels of perception obviously contribute to the organization of sensorimotor experiences in memory where they are stored and retrieved for imaginative recollection. By such means, sensory mechanisms can also contribute to forward planning of behavior.

SENSORY AND MOTOR MECHANISMS COMPARED

In the present chapter, we examine how motor activities are integrated and how communications are effected among neuronal units. Secondly, we examine some of the techniques by which central control mechanisms can be investigated.

All motor controls focus down on final executive neurons which innervate effector tissues. In the case of spinal motor neurons which innervate skeletal muscles, the convergence of controls comes from a wide variety of sources throughout all levels of spinal cord, brain stem, and forebrain. In effect, *the whole nervous system has access to motor neurons,* which represent the "final common path" to behavior.

In humans, motor cortex represented in the precentral gyrus has some pyramidal cells whose axons descend all the way down the contralateral side of the spinal cord to innervate skeletal motoneurons directly. This provides a *monosynaptic relationship between motor cortex and motor performance.* Although most motor controls act through interneurons, at least some cortical units can communicate directly with the final common path. Whereas the shortest sensory path to cortex involves two synapses, the shortest path from cortex to movement is even more direct. The most direct cortical motor control (pyramidal tract) is, like the most direct sensory pathway (medial lemniscus), a phylogenetically recent contribution.

Motor systems benefit from, but don't entirely depend on, sensory and perceptual information. Whereas sensory systems operate generally in an *ascending* direction, with abundant horizontal and descending feedback controls, motor systems operate generally in a *descending* direction, with abundant horizontal and ascending feedback controls.

Sensory systems involve three distinctive yet interrelated ascending systems, *visceral, spinothalamic,* and *lemniscal.* In approximately parallel fashion, motor systems involve three distinctive yet interrelated descending systems which govern *visceral, expressive,* and *manipulative behaviors.*

AFFERENTS, EFFERENTS, AND INTERMEDIATE NEURONS

Fibers entering by way of the dorsal roots seem to be strictly *afferent neurons.* Fibers leaving by way of the ventral roots seem to be strictly *efferent neurons.* Central neurons are more questionable as to whether they are sensory or motor. *A spinal interneuron which lies directly between an afferent and an efferent neuron is obviously functionally intermediate.* In a strict sense, *all central neurons,* even those belonging to lemniscal and pyramidal tracts, are also functionally intermediate. They *are likewise mediating between input and output, between sensory receptors and motor effectors.* By convention, we label those neurons contributing to specified ascending pathways *sensory* and those neurons contributing to specified descending pathways *motor*.

It is clear that *evolution "pays off" only for adequate performance*, regardless of what takes place along any of the sensory, perceptual, and motor control circuits, that is, right up until final action is taken.

BIDIRECTIONALITY OF INFORMATION FLOW

One would like to suppose that there are strictly *afferent neurons* and strictly *efferent neurons*. However, we must acknowledge certain compromising details such as bidirectional chemical messages conveyed by axoplasmic flow; bidirectional trophic influences; axon reflexes whereby afferent neuron branches carry efferent signals to contribute to inflammatory reactions; central control of sensory receptors by efferent neurons, and central control of ascending relays by descending projections, and many others. It is clear that *all the way from molecular transactions to systems transactions, bidirectional transmission of information is the rule.*

NERVOUS SYSTEMS ARE BUILT FOR BEHAVIORAL DECISION MAKING

The above information helps to clarify the intimate functional relationship that exists between "sensory" and "motor" systems. The two are laced together in their combined motoric aim. The nervous system is a paramount instrument for *decision making*. This function begins already with single cell organisms which have sensory and motor apparatus wherewith biologically determinative decisions are made (Koshland, 1982). Nervous systems greatly elaborate and improve decision making mechanisms. At the very least they provide greatly expanded freedom which in higher forms is recognized as consciously purposeful will.

Evolutionary advantages of nervous systems of higher organisms depend on more elaborate (better analyzed) sensory representations, more expanded memory storage and access systems, and improved final selection of processes for motor control. This involves improved convergent selection among control signals which originate from distributed sources. The whole system—sensory, transactional, and motor—is crowned by an enormous sheet of cortex which in humans provides decision-making operations in networks woven among some 600 million modular columns of sensory, motor, and association cortex.

Some decisions must be promptly forthcoming, while others can afford time and may even be consciously preconsidered. Hence there are swiftly acting spinal and brain stem reflex systems, to be compared with slower ascending, central routing, descending control systems, and longer distance yet nonetheless shortcutting lemniscal and pyramidal tract pathways.

Functional Parallels Between Sensory and Motor Systems

Sensory systems culminate in multiple, widely distributed, operationally distinctive representations and rerepresentations. There exists no single comprehensive map of a whole person or of a given environment. Nor is there any single neuronal field programmed for dynamic transactions between an individual and an environment. Sensory representations are systematically widely distributed. Perception is a moving feast.

Likewise, motor systems function with widely distributed systems of control, having motor patterns many times represented and rerepresented. There is no single arena for representation of any scheme of action. There is no one location for programming effectors in accordance with such a behavioral scheme. Behavior is a moving drama.

The more sensory systems become organized into distinctive functional networks, the more they gain *sensorimotor* attributes and responsibilities. Sensory and motor mechanisms are stitched intimately together all along the neuraxis. *At many brain loci sensory and motor mechanisms appear indistinguishable.*

Like sensory mechanisms, motor mechanisms are functionally organized from lower levels upward. In the execution of prompt behaviors, motor mechanisms don't wait for sensory signals to be processed into conscious percepts. They seize upon sensory information as it arrives and continue to utilize sensory information all the while sensory signals are ascending the neuraxis. Moreover, as disclosed in Chapter 66, highest order cortical defects involved in combined sensory integration result in sensory *and* motor types of neglect.

MECHANISMS OF BEHAVIOR

SEARCH FOR AN UNDERSTANDING OF BEHAVIOR

We need to inquire: How is behavior generated, how governed? How does the nervous system exercise control over glands, smooth muscles, cardiac muscles, and skeletal muscles in orderly and coherent ways as necessary for the pursuit of distinctive behavioral goals? Among different options available, how does the nervous system focus its motivational aims, organize, carry out, and refine different behaviors?

How does the nervous system manage visceral systems so as to transform and make available appropriate states of energy needed for biological existence and reproductive fitness? How do organisms control brain structures involved in viscerosomatic and linguistic expressions which are so important for social signalling? How do higher organisms exercise control over effectors that can be brought under conscious dominion?

We have a penchant for thinking of behavior as "a thing in itself," and of environment as an abstraction, a three-dimensional space containing a variety of physical and organic things among which individual behaviors are acted out. We think of time as existing within this environmental context—making a combination space-time abstraction.

As Yakovlev (1948) pointed out, there are two biological measures of time, one, *evolutionary time*, a chronometry of evolution, a calendar of natural history, and *time as experienced* by living organisms, in the case of humans, as a conscious accompaniment of experiencing living.

Previous Internal Adaptation of Environmental Features

We tend also to overlook the fact that a great deal of the environment is automatically incorporated into biological existence. For example, all of life (including that in the sea) has grown up in a field of modest gravitational force; life has incorporated gravity-relatedness, just as it has incorporated ("biologized") many other features of the environment: biologically utilized parts of the electromagnetic spectrum; cyclic phenomena which put demands on or set biological clocks; biological actions that take into account the "spring and weight of the air," etc.

On top of this, we ordinarily neglect consideration of the *values of behavior*, such as cooperation, faith, mutual trust, altruism, territoriality, combativeness, cruelty, etc., which along with the more obvious observables of behavior, are equally valid accompaniments and products of evolution and development, including social development. These are important aspects in consideration of motor control, for the thoughtful student as well as for the physician who must address "real life problems" in their entirety.

TWO CLASSIC TYPES OF CONTROLLED BEHAVIOR

In order to secure a conceptual handle on problems of integration of motor command, we present two classical types of controlled behavior. These represent extremes of motor patterns under voluntary control. One is a "set performance" in which specific motor gestures are known and practiced well in advance. The other is an "improvised performance" in which specific motor demands cannot be known in advance, no matter how long they may be anticipated and rehearsed.

Set Performance

A young woman concert pianist is playing Tchaikovsky's First Piano Concerto, accompanied by an established symphony orchestra. We are concerned solely with how she supports her right wrist with her skeletal muscle extensor digitorum at the exact moment when she strikes a particular key of the piano on contraction of a specific digital flexor. She must tighten the extensor to stabilize her wrist at the critical moment of the climactic final chord concluding the first movement. The stabilizing action, quite as important as the key-striking contraction, must counterpoise gravity, acceleration, and the torque on the wrist exerted by a specified combination of flexor muscles. All her professional life comes to focus during this experience in which we capture a few milliseconds of integrated motor performance.

We cannot altogether dismiss visceral preparations which have bowel and bladder quiescent, blood sugar and oxygen adequate (even after the taxing anticipations and beginning of this performance). We fix our attention on certain clusters of pyramidal neurons in the contralateral (left) motor cortex, representing this specific forearm muscle, indeed, individual fascicles of the muscle.

Auditory cortex, somatosensory cortex, basal ganglia, limbic system, brain stem, cerebellum, and other circuits are engaged and busy organizing dynamic postural and gestural movements. This involves activating muscles which control individual fingers of each hand in the execution of disparate and distinctive hand and finger movements in rapid fire sequences, and control both feet pedalling their proper tempi all at the same time.

At the appropriate instant, there comes a shower of impulses *originating from distributed sources*, including suitable emotional contributions to secure desired nuances of phrasing and expression, according to her carefully considered musical experience. These *converge on motor cortex, basal ganglia, cerebellum, inferior olive and the rest, finally to activate, at the appropriate instant, specific spinal motoneurons* which induce contraction of the wrist extensor exactly in phase to stabilize the wrist to play the climactic chord.

Of course, it has taken a great deal of practice to get this whole neurological apparatus under technical and finally under musicological control.

She is neurologically consciously responsible for this highly integrated performance. She has been refining precise individual motor actions, suitable phrases, and longer sequences so that she eventually makes few errors, in space, time, force, rate of attack, duration, etc. throughout her considerable musical repertoire. She has schooled herself to carry on even when she makes errors or does not deliver the delicate musical expressions to which she aspires. Her concentration locks onto the conducter and orchestra whereby she controls her playing to reciprocate the precise phrasing that unfolds during this particular concert occasion, while out of the corner of her mind, she senses audience rapport and support.

By and large, with a well-trained pianist, accompanied by a first-rate conductor and orchestra, with an appreciative audience, it all goes well. *Such a set performance establishes precise requirements for distributed motor control. The controls converge to activate relatively prespecified actions on the part of prespecifiable motor systems.*

Improvised Performance

We witness a windsurfer in Hawaii. He is likewise highly skilled and takes pleasure in doing a complete 360° summersault—windsurf, sail, and all. He does this by sailing his board some 25–30 mph at about a 50° angle off the tradewind and a corresponding angle to an incoming wave.

The combination of windsurf speed, momentum, and curve of wave is sufficient to accomplish the summersault most of the time. The trick is to complete the 360° and hit the water with the windsurfer tipped slightly upward, and all parts moving forward together so as to continue sailing without pause. The situation includes variable blanketing of the bottom of the sail during the near approach of the oncoming wave, the requirement to utilize the sail appropriately during the entire 360° turn when the tradewind catches the sail again, full and turning, and then to level off just in time to take the shock in the knees, adjust the wishbone, and sail on looking suitably "cool."

Nothing Can Be Precisely Set in Advance

The "sailingsault" has to be largely improvised. The duration of the 360° turn is approximately 150 ms. This requires practice that focuses attention on various complex, dynamic sensory cues which will identify the quickest intentional cooperation

needed to meld with powerful vestibular, brain stem, and spinal reflexes. Prompt precise adjustments of axial and limb musculature are required in accord with generally anticipated instructions. Immediate, specific unprespecifiable responses to wave and wind conditions are mandated during the course of a brief dynamic episode which demands whole body axial and limb muscle responses together with preservation of balance. There is no possibility of performing this 360° flip with a windsurf rig the same way another time.

Improvised performance requires a wide variety of inputs closely linked to a wide variety of quickly executable and reciprocally reversible skeletal muscle actions. There needs to be anticipatory control as well as multiple rapid adjustments during the brief 360° flight. Many of the same motor control circuits as for the pianist are likely to be involved, but they cannot be prescheduled and routinized to the same degree. The essence is *improvization involving multiple distributed controls.*

Three Fundamental Behavioral Control Systems

VISCERAL BEHAVIOR

This concerns nervous system governance of overall metabolism, respiration, circulation, ingestion, secretion, excretion, reproduction, and similar activities. The organism controls an exchange of energy between itself and the environment and differentially regulates states of available energy sources within itself. Yakovlev (1948) calls this behavior visceral motility or *visceration.*

EXPRESSIVE BEHAVIOR

This involves the outward expression of the internal state. It consists of motion outward: literally, *ex-motion, e-motion, emotion* which accompanies hunger, thirst, pleasure, fear, rage, grief, pain, and the like—such as blushing, facial mimicry, vocalization, bodily attitudes and postures—which may have little effect on the outside world except in the context of social signalling. This is identified as *expression.*

MANIPULATIVE BEHAVIOR

This imposes changes in the world of matter by moving, shaping, manipulating, and altering things. All human artifacts are the result, directly or indirectly, of manipulative behavior. Hence Yakovlev calls this category of behavior—having a deliberate effect, *effectuation.* Relatively little manipulative behavior fundamentally changes the world, except effectuation spawned by humankind, and that mostly by the pyramidal tract. It would perhaps have been more appropriate to have named our species *Homo faciens* rather than *Homo sapiens*.

The whole physical structure of civilization—buildings, statuary, art, factories, ships, vehicles, roads, aircraft, computers, the structures and instruments of education, this paper, this printed word, science and technology in their entirety—are all evidence of human effectuation. Anxiety attends the present potentiality, novel in the history of the earth, that mankind, using its manipulative skills, might destroy irrevocably the cornucopia of creative evolution.

Integration of the Three Types of Behavior

These three types of behavior are not independent but merge into one another: *behavior for any given organism must be all of a piece, a concerted whole.* Every heart beat, peristaltic wave, contraction of skeletal muscle, is part of a continuous integrative process of living, in which the nervous system plays a superordinate and controlling role.

The nervous system provides command and control over each of these three motoric domains. It integrates them in accordance with values attached to behavior which are internally generated. These values are derived from evolution and prioritized during individual development, which includes important social conditioning. Only some of these values surface to consciousness. Most of them are biologically automatic and subconsciously ordained.

The pianist engaged in a fairly set performance and the windsurfer mainly improvising for balance are each obviously utilizing all three motoric domains simultaneously. The amount of enduring effect—effectuation per se—is probably only slight unless the event is documented (reviews, photographs) and unless inspired memory leads to diaries, more enduring literature, expressive forms of art, etc. Nonetheless, at present, because of the human knack for effectuation, nearly everyone in the world *could*, in principle, witness such a documented event simultaneously.

All organisms are integrating visceral behavior continually. Often, emotional expression—expressive behavior—becomes an obvious manifestation. Expressive behavior has evolved to some degree into a signalling instrument for shaping the behavior of other organisms, especially conspecifics. *Only rarely do we lastingly alter the nonliving world around us, have an effect on the world, leave a trace.* Manipulative behavior can become effective, involve *effectuation*.

Nervous System Mechanisms Relating to Visceral, Expressive, and Manipulative Behaviors

VISCERAL CONTROL MECHANISMS

The phylogenetically oldest, most central component—common to all vertebrate nervous systems—is concerned mainly with visceral control. Phylogenetically more recent systems are built around, outside the central core (visceral) controls. Visceral control mechanisms are the first to derive from cells that line the walls of the primitive neural tube. They don't migrate very far, but establish themselves as an inner column of nucleated gray matter distributed along the neuraxis. The two later developing motor systems migrate through their midst to establish phylogenetically more recent (successively more external) systems of motor control.

Central visceral controls are found all the way from the lamina terminalis—the legal front end of the neural tube—to its coccygeal closure. The primordial nature of visceration is characterized by its being built up mostly of relatively small neurons which have a great range of lengths of dendritic, and particularly, of axonal processes. Some reach all the way from hypothalamus to lower spinal centers (Swanson et al., 1983), but most of them are much shorter. Generally, central visceral control units exhibit a lack of myelin which has made it difficult to trace these pathways until the recent emergence of immunohistochemical methods.

Conduction throughout visceral control systems tends to be pleurisynaptic and relatively diffusely projecting. This means that visceral organ systems—cardiovascular, gastrointestinal, respiratory, genitourinary—tend to be activated on a massive scale compared with the finesse of activation of individual skeletal muscle fibers during manipulative behavior.

Peripheral visceral control components include distributed chains and plexuses of autonomic ganglia and ganglionic networks which lie within the walls of the viscera. These peripheral controls somewhat resemble invertebrate nervous systems. It is tempting to suppose that they may even constitute evolutionary baggage left over in our midst from prevertebrate ancestry. Of course, any neurological apparatus of this sort has continued to evolve all this time. But evolution is conservative, especially with respect to a system that serves biological fitness suitably. Perhaps this is the most tried and tested part of our nervous system.

These peripheral ganglionic and nerve net systems exert control locally and with considerable autonomy. It is for this reason that peripheral visceral control mechansims are given a special designation, the *autonomic nervous system.* Even the neurotransmitters used in their diffuse channels of communication and distributed among central and peripheral controls suggest evolutionary antiquity.

EXPRESSIVE CONTROL MECHANISMS

Neural mechanisms involved in expressive behavior are organized all along the neuraxis, lying on top of (outside) visceral systems of control. The behavior controlled through this system manifests as an outward expression of the internal state. We are not surprised, therefore, that the *limbic system,* which is involved in emotional experience and expression, and the *basal ganglia,* which are involved in bodily attitudes and postural expressions, *are principal components of this system.* They cooperate closely with one another structurally and functionally and with visceral control mechanisms. They communicate reciprocally and abundantly with the hypothalamus, brain stem, and spinal gray and white reticular formation.

MANIPULATIVE CONTROL MECHANISMS

Effectuation can be claimed to have originated with the earliest cells which left their imprint in the oldest sedimentary rocks. Earlier fossils have been recycled in metamorphic rocks and have thus been lost to paleobiology. Aside from the greatest architectural structures designed by living systems, enormous, laboriously constructed coral monuments, and the conversion of earth's atmosphere to oxygen, a less tangible but nontheless instrumental monument of phytoplankton, there are relatively few obvious traces of animal life on earth. It is interesting that these two dominating features of biological effectuation, which have had such an impact on the earth and its atmosphere, were created through tremendous cooperative enterprise.

Effectuation by individuals awaited mammals, and its products are almost entirely the result of human activity. It is produced by means of instruments of consciousness, and expressed mainly by corticifugal motor controls affecting skeletal muscles. Effectual behavior relies upon stratified stabilities provided by visceration and expression. *Visceration provides the essential control of energy, and expression provides the motivational drive necessary to orient and sustain effectual behavior* as a consequence of visceral and emotional experience. It is quintessentially by means of the pyramidal tract that humans have been so effective in leaving their tracks and monuments on the face of the earth, more recently on the face of the moon, and with exported artifacts, on the surface of other planets

in the solar system, and in the form of voyagers which are sailing beyond the solar winds.

COMPONENTS GOVERNING BEHAVIOR

The Physiology of Motoneurons

Peripheral neural control systems are diverse and complex in their organization. Yet the controlling signals which leave the neuraxis are conveyed by only a few types of neurons. Moreover, the controls they exert affect only a few types of tissues, namely: glands, smooth muscles, cardiac muscles, and skeletal muscles. As a consequence of its controlling signals, the nervous system changes rates of secretion in glands and changes lengths and/or tensions in muscles. *The sum total of this activity is behavior.*

Neurons exerting such controls, whose axons leave brain stem and spinal cord, are known as *efferent neurons* inasmuch as they are conveying signals that travel away from the central nervous system. They are also called *motor neurons or motoneurons,* although the term is hardly appropriate for neurons which control glands.

Efferent neurons present three characteristic patterns of organization as they project axons from the central nervous system toward glands and muscles (Fig. 9.140, *bottom right*):

(1) *For cardiac and smooth muscles, exocrine glands, and certain endocrine glands, the efferent axons, after leaving the central nervous system, terminate in autonomic ganglia where their endings activate other neurons.* Second order neurons then relay efferent control signals to the effector tissues which they control by release of appropriate neurotransmitters.

Cardiac muscle is rhythmically spontaneously active and has its own specialized conduction system. Therefore, cardiac innervation influences ongoing activity of cardiac muscle fibers and triggers or blocks spontaneous initiatives generated at nodal points along the cardiac conduction system. Smooth muscle also has some degree of autonomy of contractile functions and is additionally controlled by neurotransmitters released by peripheral (second order) autonomic innervation.

(2) *For some endocrine glands there is no direct innervation. In this category, control is neurohumoral.* A neurotransmitter, neurohormone, or neurohumor is conveyed to the gland by the bloodstream. This blood-borne neuroendocrine signal is delivered to the bloodstream by nerve endings or glands affected by neural control signals originating elsewhere. Cells of the neurohypophysis are controlled in this way. Neurohumors are released at nerve endings in the hypothalamus into veins that

flow into the portal circulation which in turn perfuses the adenohypophysis. Other peripheral tissues may also be affected by local or general release of neurotransmitters and neurohumors, through tissue diffusion and widespread vascular distribution.

(3) *Skeletal muscles receive direct innervation from neurons whose cell bodies lie in motor nuclei in the brain stem and spinal cord.* Their axons, in contrast to those for glands, smooth and cardiac muscles, course directly to the muscle itself which they activate by release of a locally acting neurotransmitter—acetylcholine. This is a system of explicitly specified, focused control. The neuromuscular junction is considered highly reliable. The efferent neuron and the muscle fibers it innervates, taken together, are considered to constitute a single *motor unit.*

CONVERGENCE ONTO EFFERENT NEURONS

Important aspects of motor integration result from the confluence of multiple control messages which originate from a wide variety of sources, local and distributed. Motor systems in general and efferent neurons in particular are characterized by having extensive, highly branched dendritic arborizations which allow a great deal of signal convergence.

Any neuron that receives multiple convergent signals from distributed sources engages in an important decision-making role during the course of generating its output. The locus of origin of the axon, the axon hillock, represents an important cellular locus of decision making. In motoneurons, impulses which are initiated or held in check represent biologically important (go/no go) motor decisions. Unit decision making is affected by several factors: (1) the tendency for that neuron to exhibit spontaneous activity, (2) the recent past history of that neuron, (3) the level of excitation and inhibition taking place in its local environment, including specifically within local circuits, and (4) the precise geometry (spatial summation) and timing (temporal summation) of signals (excitatory, inhibitory, and modulatory) which are being delivered to various specific parts of that cell's dendritic tree.

Signals terminating on remote dendritic branches are least likely to exercise control over next order neuronal activity. Such signals are likely to decrement as they pass from dendrite to cell body or to be overwhelmed by signals which are arriving closer to the cell body. On the other hand, *signals reaching the major dendritic trunks and cell body are more likely to exert control over the decision-making process.* Timing of arrival as well as location of arrival of incoming signals are of importance in this process. It is instructive to contrast this convergent decision-making strategy with the

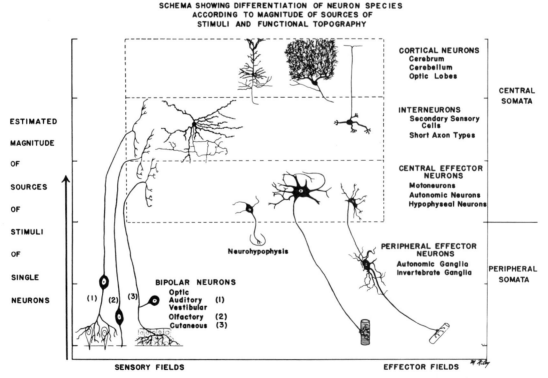

Figure 9.140. Major neuron types in mammalian central nervous system, classified according to general function hierarchical level, probable diversity, and magnitude of sources of synaptic connections. [From Bodian (1967).]

nearly one-to-one reliability, the minimal modification of throughput which is seen at relays along lemniscal sensory paths. There, speedy and reliable information is needed at higher centers. Here, in final motor control, decisions need to be postponed until the last possible instant, while information from local, remote, and distributed sources of control are taken into account.

Sherrington (1906) called the motoneuron the *"final common path."* This means that all neural paths lead, ultimately, to the efferent neuron: what happens among prior circuits may be relatively unimportant, at least until action is taken. *Cells engaged in the integration of convergent controls are where biologically important decision making counts.*

The Physiology of Interneurons

In brain stem and spinal motor nuclei there are many *interneurons*, or internuncial cells (see Fig. 9.140). Most centrifugal controls communicate only indirectly with efferent nerves; they ordinarily terminate on interneurons which communicate secondarily with efferent neurons. Descending control messages then can rouse either excitatory or inhibitory responses at their junctions with interneurons and can both directly and indirectly influence the junction between interneurons and efferent motoneurons. The interneurons themselves may be excitatory or inhibitory. This organization of descending controls provides a variety of different regulatory combinations in which spontaneous activity exercises an important influence.

In general, efferents to skeletal muscles are not spontaneously active, but efferents to other effectors are likely to be spontaneously active. Higher level centrifugal projections descending along motor pathways are often spontaneously active. This enables smoothing of motor control such as is made possible for subtle sensory detection where afferent units are likewise spontaneously active. To effect a change in rate of ongoing control activity does not require surpassing a "threshold."

Interneurons introduce great flexibility into lower motor control processes (Fig. 9.141). As has been emphasized, *propriospinal interneurons* have exceptionally long dendritic and axonal processes. Their dendrites reach down and up several spinal segments, and their axons may contribute impulses to all spinal segments and to caudal brain stem levels as well. The functional role of these long-ranging spinal interneurons is probably unique because they coordinate an extended chain of motor neurons which needs to be functionally integrated throughout the length of the spinal cord, including, e.g., interlimb reflexes for various swimming, walk-

ing, and running gaits. We need also to consider the more characteristic short process interneurons seen in Figure 9.141 as follows: *A*, A, F, K; *D*, C, D, E; *E*, A, B, C; and *F*, B. A classic example is the "Golgi type II neuron," but there are a number of other types, some of which are called "short axon neurons." But this excludes axonless interneurons like the amacrine cells in the retina. For this reason, the term "local circuit neurons" (LCNs) has recently been adopted (see Rakic (1975) for perspectives on LCNs).

LOCAL CIRCUIT NEURONS (LCNs)

Mammalian brains undoubtedly exhibit a greater variety of cells than all other organs and tissues of the body combined. Bullock (1982) estimates that there are probably more than 50 million distinctly different *types* of neurons in the human brain. Yet it is conceptually legitimate and useful to divide all neurons into two major classes: (1) cells with long axons which project from a given brain structure to one or more other such structures. These are considered *extrinsically projecting neurons.* (2) cells with short dendrites and short axons (or no axons) with connections confined to a given structure. These are considered *intrinsic, local circuit neurons* (*LCNs*). LCNs may be very densely packed as well as numerous. In the caudate nucleus, for example, one type of LCN, the common spiny neuron, dominates the cell population (constituting 95% of all caudate cells).

Of special interest is the fact that LCNs are involved in intricate local circuits and that the synaptic organization of local circuits appears to be highly plastic (Akert, 1973). For example, as illustrated in Figure 9.142, there are distinctive morphological differences, borne out statistically, between synaptic junctions in the spinal cord of anesthetized and unanesthetized rats. The synaptic cleft in the anesthetized condition is flatter, and in freeze fracture electron micrographs the presynaptic membrane manifests few if any pitted indentations. In the waking state, in contrast, the synaptic cleft shows (statistically) greater curvature and increased density of pitted indentations which are thought to be morphologic indicators of fusion of synaptic vesicles to the presynaptic membrane and therefore of active release of neurotransmitter. "Omega" figures, seen in thin section electron micrographs in spinal cords of waking rats, confirm this impression and are consistently more prevalent in the waking state. Two are shown diagrammatically to the left in Figure 9.142.

If such dynamic plastic changes are encountered in a transition between waking and anesthetized

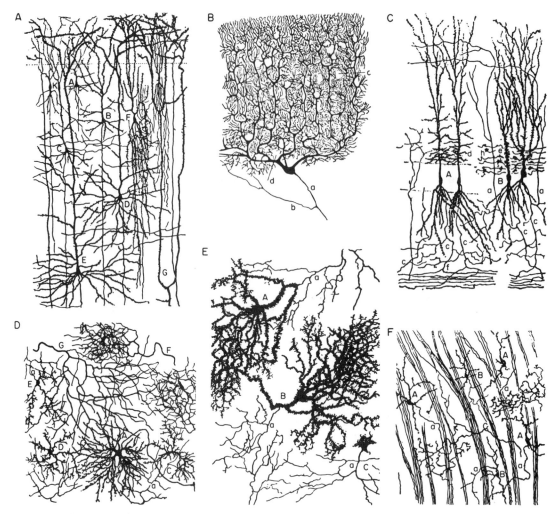

Figure 9.141. Examples of dendritic patterns of neurons which constitute the major cell types of elements in different structures of the mammalian brain. *A,* Outermost three layer of frontal cortex of a 1-month-old human infant. *Small (A, B, C,)* and *large (D and E)* medium pyramidal neurons as well as the dendritic process of a large pyramidal cell *(G)* of the fourth layer are shown. Cells with double dendritic bouquets *(F)* and fusiform appearance *(K)* are also identified. *B,* Purkinje cell of the cerebellum from adult human. Axon *(a)* and axon-collateral *(b),* capillary spaces *(c),* and spaces occupied by basket cells *(d)* are indicated. *C,* Pyramidal cells of hippocampus from a 1-month-old rabbit. *A,* small pyramidal cells of the superior region. *B,* large pyramidal cells of the inferior region. Large ascending collaterals *(a),* axons *(c),* and sites of contact of mossy fibers *(h)* are to be noted. *D,* Frontal section of the thalamic somatic sensory nucleus of the cat a few days old. *A,* cell with a long axon; *C, D,* and *E,* cells with short axons; *F,* sensory fibers; *G,* axons of cortical origin terminating in the sensory nucleus. *E,* Cells of the pons from a human infant, a few days old. *A* and *B,* Cells with axons arising from dendrites; *C,* cell with a bifurcated axon; *F,* Saggital section of the caudate nucleus from a newborn rat. *A,* Cells with a long axon; *B,* Cells with short axon; *C,* ascending afferent fibers. [From Ramón y Cajal, cited by Purpura (1967).]

states, it is not unreasonable to expect that other morphological and functional changes might be induced by other altered functional states such as attention, positive or negative reinforcement.

Bodian (1972) categorized various synaptic junctions according to schemata presented in Figure 9.143. It depicts only axonal endings and does not exhaust the number of structurally distinguished *varieties* of synapses. The figure illustrates several structural varieties which have been found to possess distinctive functional characteristics. In local circuits, such "microcircuits" are organized into systematically different "tools for management of information," morphologically and functionally characteristic of a given brain locus. *Local circuits contain detailed microcircuitry which provides microcontrol of information processing by that structure.*

There are terminal axons which end on dendrites and on cell bodies, as long inferred from light microscopy. There are also *axoaxonal* endings, including endings on the bare axonal intervals in nodes of Ranvier (Peters et al., 1970). In addition, there are *dendroaxonal, dendrodendritic, dendrosomatic, somatosomatic,* and *somatodendritic synaptic junctions.* These junctions can face either direction

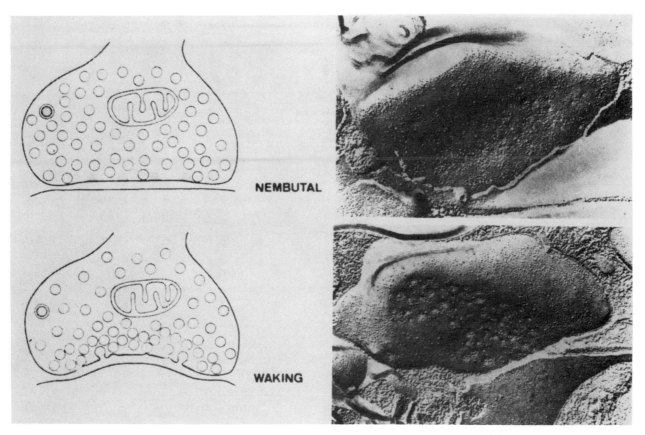

Figure 9.142. Distinctions revealed by freeze-etching techniques between anesthetized and unanesthetized junctions in the rat. Diagram illustrates *left* (*above*) a synaptic junction: presynaptic terminal containing numerous synaptic vesicles, synaptic cleft, and postsynaptic membrane. Effects of general nembutal anesthesia include flattening of synaptic cleft, reduction of crowding of synaptic vesicles in the vicinity of presynaptic membrane, and reduction or elimination of numbers of vesicles fused to presynaptic membrane having open access of vesicle contents to synaptic cleft. In waking state, *left* (*below*), vesicles crowd toward presynaptic membrane and "omega" figures are seen resulting from vesicle fusion and opening of vesicle contents to synaptic cleft. On *right* are representative electron micrographs of an anesthetized (*above*) and unanesthetized presynaptic terminal (*below*). The former shows only rare pitted indentations associated with vesicle fusion to presynaptic membrane, the latter shows numerous such indentations. [From Akert and Livingston (1973).]

across a synaptic cleft in respect to the side on which neurotransmitter is apparently released. Junctions adjacent to one other can send reciprocal signals in opposite directions from one neural process to the other (push me-pull you synapses). *All logically possible combinations between neuronal processes in terms of direction of apparent information flow have been observed in mammalian brains.* This suggests that each combination is likely to be functionally interesting and useful.

Serial synapses (Fig. 9.143) occur with three to five or more terminals stacked in serial succession, all linked together in avalanching interdependence. Other configurations allow LCN terminals to interfere with the final terminal release of neurotransmitter by neurons coming from disparate parts of the brain. *Such presynaptic terminals can, for example, interfere with axonal signals at the last possible instant before those signals would otherwise have been synaptically effective.*

Anatomical evidence that such local circuits are widespread is instructive in light of results from bioelectric studies that demonstrate special response properties of larger-scale, functionally comparable circuits: electrotonic coupling, effects of weak electric fields on local ionic flux in nervous tissue, etc. (Adey, 1982). Through such external influences, intraneuronal and interneuronal transport of substances that contribute to trophic effects of neurons on other neurons and on nonneural tissues are interwoven with intrinsic biochemical and bioelectric phenomena. The obvious implication, as yet unverifiable, is that local circuits may be similarly susceptible to similar sorts of outside electrical field forces.

DIMENSIONS OF NEUROPIL

There is probably no structure in the mammalian central nervous system that lacks LCNs. The relative proportion of LCNs to extrinsically projecting

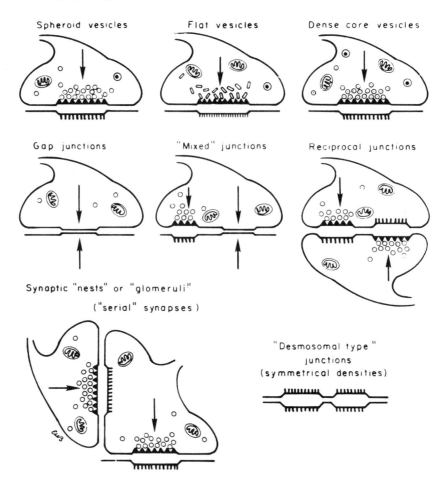

Figure 9.143. Schematic representation of some major variations of synaptic structures and topography, presumably associated with functional diversity. *Upper row,* "Asymmetric" contacts characteristic of most central synapses, with uniform cleft of about 16 nm width, dense pre- and postsynaptic materials and in the cleft (not shown). The presynaptic bulbs contain mitochondria and various kinds of microvesicles of about 20–40 nm. Spherical vesicles are ordinarily associated with known excitatory synapses and flattened vesicles with inhibitory synapses. Microvesicles with electron-dense cores are characteristic of adrenergic nerve terminals; those in the CNS have larger, 60–80 nm vesicles. *Middle row,* Neuron clefts of about 2 nm (gap junctions) characteristic of electrotonic transmission, devoid of junctional densities; they may occur on the same synaptic interface with chemical synapses. Reciprocal junctions are facing in opposite directions in the same interface. *Lower row,* Specialized interneuronal junctions may be arranged in configurations such as nest, glomeruli, and serial synapses which may be multiple. Desmosomal junctions are symmetric, with a cleft of 20 nm. [From Bodian (1972).]

neurons is probably greater than 3:1. *Both the proportion of LCNs and the complexity of local circuits increases with phylogeny.* Another phyletic sign of importance is the ratio of volume of gray matter to volume of nerve cell bodies, the *G/C* (*gray to cell volume*) *coefficient.* This gives a rough measure of the *relative amount of neuropil available for information processing, the local transactional space. Neuropil* is a term that implies a feltwork of interacting neuronal terminals, as in the "pile" of a rug. *Transaction* implies multiple, mutually interdependent components engaged in simultaneous action.

For primary visual cortex, Haug (1958) estimated G/C coefficients in a variety of species: man, 47; chimpanzee, 31; macaque, 19; rabbit, 15.4; guinea pig, 11.5; and mouse, 8.5. Cell dimensions in the

different species were not substantially different, but the space available for local neuropil increased dramatically going up the phylogenetic scale. *Expanding local transactional space undoubtedly provides increasing degrees of freedom for local decision-making processes.*

LOCAL CIRCUITS

Integrative processes in local circuits are enriched by the presence of numerous somatosomatic and dendrodendritic synapses and "gap junctions" among neurons (Peters et al., 1970; Sotelo and Llinas, 1972). This suggests that an enormous amount of neuronal crosstalk may be taking place at the receiving regions (dendrites and somata) of neurons, decision-making regions, prior to the arrival of signals at the critical zone where action potentials are generated. Gap junctions link cells electrotonically (by rapid exchange of ions) and biochemically (by exchange of chemicals below about 1000 MW). *Thus both synaptic and nonsynaptic processes are contributing to integration in local circuits.* Elsewhere, gap junctions are rare between neurons.

Neuroscientists speculate, without proof being accessible, that LCNs may play a significant role in higher nervous processes. Electron microscopy has revealed many other aspects of local circuit complexity which imply additional functional precision and flexibility (see Peters et al., 1970, for exemplary thin section electron microscopy, and Sandri et al., 1974, for exemplary freeze fracture electron microscopy). Local circuits exhibit an astonishing flow and counterflow of information among neighboring junctions facing upstream and downstream along assumed directions of general information flow (Schmitt et al., 1976). It is suggestive that there are many local feedback and feedforward "microcontrol" circuits which, in terms of circuit theory, can provide extraordinary functional plasticity on top of the kinds of morphological plasticity previously noted. *It is therefore reasonable to assume that distinctive properties of local circuits can contribute instrumentally at all levels of nervous system integration, including making contributions to higher neural processes (perceptual, ideational, linguistic, and the like).*

The Physiology of Neuroglia

Most nerve cells are surrounded by satellite cells, *neuroglia,* or *glia.* They probably outnumber neurons 10:1 and make up about half the volume of the nervous system (Fig. 9.144). Glia are divided according to location and cell type into (1) astrocytes, microglia, and oligodendrocytes in the central nervous system, and (2) Schwann cells in the periphery. Astrocytes are further divided into *fibrous astrocytes,* which contain many filaments and are more abundant among nerve fibers in white matter of the brain, and *protoplasmic astrocytes,* which have fewer filaments and are more commonplace among nerve cell bodies and neuropil.

Both types of astrocytes have extended processes which are applied to the outside of brain capillaries. These astrocytic foot processes may provide a route for bailing excess potassium out of the brain. They may even contribute to the "blood-brain barrier," although convincing evidence of an astrocytic role in this respect is lacking. Astrocytes respond expressly to injury of the brain by closing gaps and repairing wounds—providing scar formation in the brain. They also contribute to phagocytosis to some

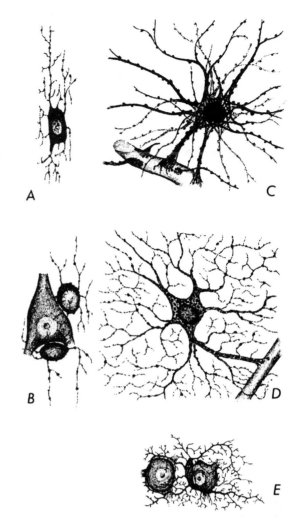

Figure 9.144. Drawings of typical glial cells. *A,* Oligodendrocyte. *B,* Two oligodendroglia lying against a neuron. *C,* Fibrous astrocyte with processes on blood vessel. *D,* Protoplasmic astrocyte. *E,* Microglial cell in vicinity of two neurons. [From Penfield, as reproduced by Copenhaver et al. (1972).]

extent. Microglia, on the other hand, are primarily phagocytotic.

Electrophysiology of Neuroglia

Glial membranes are distinctly different from those of neurons. Glia respond passively to electric current and do not conduct impulses. The glial membrane potential (about 90 mv) is higher than that of neurons. Moreover, the glial potential depends primarily on the distribution of potassium. Potassium is the principal cation inside glia, and glia may play a role as a source and sink for this ion. During impulse conduction, neurons release potassium into extracellular spaces. This provides a *nonsynaptic signal to glia,* causing them to depolarize. The signal is the same whether the neuron is delivering excitatory or inhibitory impulses: it is a straightforward sign of degree of neuron activity

(Fig. 9.145). The neuronal release of potassium is widespread. If neuron activity is further increased, glial depolarization is correspondingly augmented. The glial potential shift can affect the electroencephalogram and likewise the electroretinogram.

Myelin Sheathing of Axons by Neuroglia

In the central nervous system, myelin is wrapped around axons by *oligodendrocytes* (oligo-few, dendro-branches or simply *oligos*). In peripheral nerves, Schwann cells perform the same function. Schwann cells are distinguished chemically and also in terms of their pattern of axonal wrapping. In the central nervous system single oligos may be responsible for myelinating several segments of myelin in as many different axons (Bunge, 1968). In the periphery, single Schwann cells wrap only a single segment of myelin around a single axon. Both types of myelinating glia form glial-axonal junctions at the mar-

gins of the nodes of Ranvier, along the paranodal fringes of the spirally wound glial wrappings (Livingston et al., 1973). *Myelin sheaths can improve the reliability of conduction by a factor of about 2 and increase the velocity of conduction by a factor of about 6.* It is not known whether these septate-like structures at glial-axonal junctions are engaged in the exchange of ions as they appear to be elsewhere. If so, myelin may be participating in dynamic transactions between glial and axons, doing more than simply providing insulation.

Gap Junctions and Neuroglia

Glial cells are linked by numerous low-resistance gap junctions which permit the direct passage of ions and both neutral and negatively charged molecules up to a molecular weight of about 1,000 (Bennett and Goodenough, 1978). The permeability and electrical resistivity limits are consistent with a junctional structure involving a 1–2 nm hydrophilic channel. Neurons share gap junctions with other neurons much less frequently and then primarily in relation to local circuits. There appear to be no gap junctions between glia and neurons.

Glia probably do not provide mechanical support for neurons. Hydén finds that during microdissection in animal tissue or freshly removed human tissue, glial cells appear softer than neurons. It is more likely that the brain is held together by innumerable adhesions and junctions between individual cells than by cellular skeletal structures.

There has been a long dispute between electron microscopists and physiologists as to the dimension of the extracellular space in the brain. According to electron microscopy, the space appears small, about 5%. According to physiological observations (impedance, ionic concentrations, and inulin space), extracellular volume is in the range of 25–30%. The electron microscopic measurements are, however, artifactual because of acute swelling of the astrocytes during the process of securing and fixing brain tissue. When the brain tissue is rapidly frozen, the electron microscopically defined space is found to be equivalent to the space defined physiologically (Van Harreveld et al., 1965).

Once a drug enters the extracellular spaces of the brain, there are remarkably free movements among the cells and relatively free exchange with cerebrospinal fluid which can quite freely enter cerebral extracellular space in the brain.

Brain extracellular space constitutes a mixture of materials diffusing from and toward capillaries, substances selectively withdrawn from the extracellular pool and added to it by neural and glial metabolic activities, cerebrospinal fluid which en-

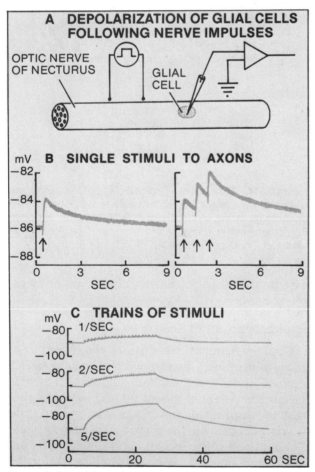

Figure 9.145. Effect of neural activity on glial cells in the optic nerve of the mudpuppy. Synchronous impulses in nerve fibers cause glial cells to become depolarized. Each volley of impulses leads to a depolarization that takes seconds to decline. The amplitude of the potential depends on the number of axons activated and the frequency of stimulation as shown in *B* and *C*. [From Kuffler and Nicholls (1967).]

ters from the walls of the ventricular system, subarachnoid space, and spinal canal, and any foreign substances (drugs, toxins, infectious agents, and the like) which may get across the blood-brain permeability barriers. There are some exceptional zones where blood-brain permeability barriers are reduced or absent, and in these areas extracellular fluid in the brain more closely resembles that in other tissues.

While distances between capillaries vary according to metabolic requirements, different brain tissues are also more or less remote from ventricular surfaces and therefore more or less freely bathed by cerebrospinal fluid.

DEVELOPMENT OF THE NERVOUS SYSTEM

Neurons are said to have their birthday at the time of their final cell division. Then they migrate from the vicinity of the inner wall of the neural tube to some intermediate or ultimate station. Peripheral sensory and autonomic neurons and cells of the adrenal medulla stem from the neural crest. *From the time of their birth, neurons seem to be committed to proceed to certain locations (in relation to other neurons) and to make certain kinds of connections on certain regions of target neurons, near and remote* (see Edelman, 1984).

When clusters of embryonic cells destined to form orderly arrangements such as the hippocampal or cerebellar cortex are removed and dissociated prior to such assembly and allowed to interact with one another in tissue culture, they sort themselves out and make a cooperative organization that resembles the structure for which they were originally ordained.

Many more cells are formed than ultimately survive. The dying back of nerve cells and their processes, including synapses, seems to represent a "pruning" process that depends on functional requirements, a survival of *components adequately utilized.*

The sequence of cell migration and formation of connections is an orderly process with specific spatial and temporal patterns (Schmitt et al., 1970, 1981). Levi-Montalcini (1964) found a factor that selectively influences the growth of sympathetic and sensory neurons. Presumably there are other such growth factors.

DYNAMIC FEATURES OF STRUCTURAL ORGANIZATION

PRENATAL DEVELOPMENT

During embryogenesis, neuroblasts divide a number of times in rapid succession. This provides for an enormous numerical increase. The human brain must produce, on average, on the order of one third of a million cells per minute throughout the 9 months of gestation. This proliferation of course accelerates as more and more cell divisions proceed, and it continues at a significantly high rate during the first months of postnatal life, tapering off gradually to a roughly zero base about the end of the first year of life. *The brain doubles in size during the first 6 months of life and doubles again by about the fourth birthday of the child, mainly by increasing the size and branching of neural processes, by increasing the number of glia, and by fabricating myelin.*

Growth and development of the brain is affected by a number of genetic, nutritional, and other environmental factors (Brazier, 1975). The *brain* is the first organ to differentiate, the fastest growing, and therefore the largest throughout gestation and early childhood. *It has a metabolic rate per unit volume that is double that of an adult.* Glial guides assure appropriate radial migration and contribute to the establishment of orderly stations for neurons in subcortical and cortical regions. Cells arising from a single location on the ventricular surface end up in the same radially oriented cortical column, with those generated later taking up positions external to the station occupied by their predecessors (Rakic, 1981).

POSTNATAL DEVELOPMENT

Much of the ultimate distribution of processes, particularly those to remote structures, and the dying back of synapses, processes, and cell bodies takes place after birth. In this way, the environment, and neural transactions with the environment, have a role in the determination of neuronal circuits. Figure 9.146 illustrates the orderliness of such early developmental processes in normal corticocortical projections compared with corticocortical projections following removal of the region of normal cortical destiny. Cortical projections deprived of their normal target seek out comparable targets and retain their orderly pattern of interhemispheric innervation, even though they are now directed to an abnormal target region of cortex.

Since human brain doubles in volume by 6 months of age and doubles again by the fourth birthday, it is essential that adequate nutrition be provided throughout gestation and early childhood in order to permit the child to fulfill its genetic potential. Lacking this opportunity, the brain will establish its organization but likely without parts which could not be manufactured at critical times or which died back for want of appropriate activa-

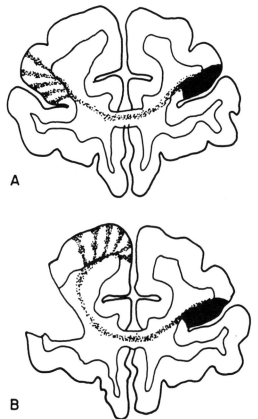

Figure 9.146. Plasticity of association cortex in primates. *A*, Drawing of prefrontal cortex in the left and right hemispheres of normal rhesus monkey whose superior bank of the principal sulcus in the right hemisphere was injected with a mixture of radioactive leucine and proline 3 weeks prior to sacrifice. The drawing illustrates the pattern of alternating callosal fiber bundles in the homotopic cortex of the left hemisphere. *B*, A monkey in which the dorsolateral prefrontal cortex was resected in the left hemisphere at 8 weeks of age. Two months later, intracortical injections of the radioactive amino acids were placed in the superior bank of the principal sulcus of the right hemisphere 3 weeks prior to sacrifice. Callosal fibers, deprived of their normal target in the cortex, project to areas dorsal and medial to the resected area, but the anomalous pathway retains its pattern of spatial periodicity. [From Goldman-Rakic (1981); Rakic and Goldman-Rakic (1982).]

tion by other missing parts or because of insufficient or inappropriate interactions with the environment.

The skeleton of this postnatal dynamic can be appreciated by experiencing Figure 9.147 from the work of Conel (1959) in which comparisons have been made in Golgi-stained neurons as seen in a human newborn (dead from nonneurological causes), a 3-month-old infant, and a 2-year-old child. These sections were taken from the same region of primary visual cortex. Other cortical regions have been shown by Conel to undergo similar progressive elaboration, although according to somewhat different age-correlated time schedules.

It is evident that during early postnatal life, neurons are elaborating additional processes and that the whole neuropil is becoming enormously expanded. Moreover, as can be determined by other anatomical methods, additional neurons are migrating into the cortex, especially into the more superficial layers. In monkeys, it appears highly suggestive that these late arriving neurons are establishing synaptic connections and contributing to cortical circuitry in accordance with ongoing experiences (Altman, 1966). When so much arborization and elaboration of processes of existing neurons and new circuit formation by additional postnatal neurons (not shown in Fig. 9.148) are combined, it indicates that even at the light microscopic level of magnification, the brain is exceedingly dynamic structurally. The architecture may be being affected according to ongoing life experiences during that period. If this assessment is correct, *the brain is being designed to fit the environment into which that infant has been born.* Additional dynamic refinements probably continue the rest of that individual's life, in terms of alterations of local synaptic and probably local circuit organization. The structural foundations at a light microscopic level of the complexity for such additional "custom-made" fine structural refinements may have already been fairly well established during early infancy.

Another way of demonstrating plasticity of neuronal organization during development of the nervous system is shown in Figure 9.148. Kittens were reared in the dark and then given a period of experience behaving in a lighted environment while one eye was allowed detailed visual experience and the other eye was deprived of detailed visual experience by having its eyelids sutured closed. The physiological responses of primary cortical units in terms of ocular dominance are show in bar graphs according to the number of units encountered which responsded to the deprived eye, equally to both eyes, or to the eye allowed detailed visual experience. On the left are shown counts of striate units in kittens dark-reared from birth to 4 months, followed by monocular vision for 4 to 8 weeks. It is obvious that many more cortical units responded solely to the nondeprived eye (about 5.5×) as compared to the deprived eye, and relatively few units responded equally to both eyes. On the right are shown similar counts of cortical neurons in kittens which had been allowed normal visual experience for 4 months followed by a substantially longer time (about double) of monocular deprivation by eyelid suture as compared with the previous group of kittens. In this group, there were fewer units responding solely to stimuli received in the deprived

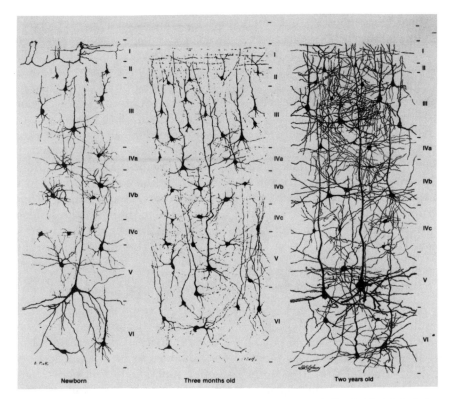

Figure 9.147. Development of dendritic arborizations in human visual cortex, as revealed by Golgi stain techniques, showing continuing development during early life, from newborn, 3-months, and 2-year-old infants. Depth of cortex has been arbitrarily expanded for newborn and 3-month-old (almost double for the newborn) in order to match layers as seen in the older infant. [From Conel (1959).]

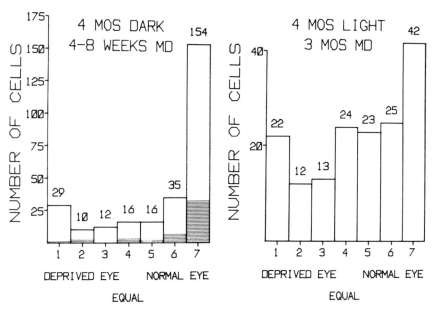

Figure 9.148. Influence of monocular visual experience on ocular dominance in kittens reared in dark as compared with kittens reared in light. *Left*, Numbers of cells encountered in striate cortex contralateral to the deprived eye which were dominated by the deprived eye or by the nondeprived eye. The incidence of units encountered that are driven strongly through the deprived eye is much less than that of units driven strongly through the exposed eye. Incidence of units with strong binocular convergence is reduced relative to that of monocularly driven cells. These kittens had been dark-reared from birth to 4 months, followed by 4–8 weeks of normally lighted environment with monocular vision (nondeprived eye, open; deprived eye, lids sutured closed). *Right*, Distribution of ocular dominance for units in cortex of kittens allowed normal visual experience for 4 months followed by 3 months of monocular visual experience (deprived eye, lids sutured closed). The number of cells driven through the exposed eye is approximately twice that of cells driven through the deprived eye. [From Cynader and Mitchell (1980).]

eye than in the normal eye (about ½) but many more units responding equally to both eyes. Relatively fewer compared to the other group (about one-third) of units responded solely to the nondeprived eye.

It must be emphasized that only through behavior can the nervous system obtain error signals relating to visual representation. Two features of this analysis illuminate problems of communication and control. *First, circuit usefulness (behavioral payoff) apparently plays a role in shaping circuit integrity between retina and visual cortex. Early life experience not only contributes to shaping circuits but it can stabilize circuits sufficiently so that later deprivation is not so devastating.*

ARE NEURONS ADDED DURING ADULT LIFE?

By means of thymidine autoradiography, it was discovered that dentate granule cells in the hippocampus of the rat continue to be produced during adult life (Altman, 1963; Altman and Das, 1965; Bayer, 1976; Bayer and Altman, 1975). It might be contended that the new cells are simply replacing cells that die during adult life. Such a replacement alternative has not been verified, but recent quantitative measurements by Altman's group demonstrated that there is an absolute numerical increase in dentate neuron population during adult life of the rat (Bayer et al., 1982). Therefore, in addition to genetic and early environment effects on brain architecture, and the presumption of continuing modeling of local circuit architecture associated with ongoing life experiences, there is evidence that at least in the hippocampus, new neurons may be added to the lot. This adds another dimension to brain plasticity.

NEURONAL SPROUTING

It has long been known that following damage to the central nervous system, traumatic or vascular, initially severe functional deficits may be reduced and considerable restoration of functions may take place. It is possible that undamaged parts of the brain take over the functional role of damaged parts, that *residual parts may function vicariously.* In addition and/or alternatively, *there may be functional reorganization.*

One mechanism known to be responsible for some restitution of function is *collateral sprouting* of neurons. Sprouting of axons in the peripheral nervous system had been known for some decades, but positive identification of such sprouting in the central nervous system was made initially by Liu and Chamber (1958) in the spinal cord of the cat.

They transected adjacent dorsal roots, leaving one isolated dorsal root intact. After an extended interval to allow restitution to stabilize, they examined the acute consequences of degeneration of the remaining dorsal root, and—as a control—of its contralateral counterpart. The side on which the neighboring dorsal roots had been transected during the first procedure showed that the remaining dorsal root had given rise to many additional branches which invaded areas that previously had been served by the adjoining but now completely degenerated roots. Such collateral sprouting has since been demonstrated in the hippocampus (Lynch et al., 1973; Zimmer, 1973) in the superior colliculus (Lund and Lund, 1971) and in the spinal cord following transection of descending spinal tracts (Liu and Chambers, 1958; Murray and Golberger, 1974).

Various responses of axons to damage to nervous and other tissues they innervate are illustrated diagrammatically in Figure 9.149. Typically, modest damage to skeletal muscle or to the neuromuscular junction is followed by sprouting from the nerve terminal itself and even from nodes of Ranvier farther back along the axons (A). If a myelinated nerve is transected, there is likely to be sprouting from the cut end of the fiber and from

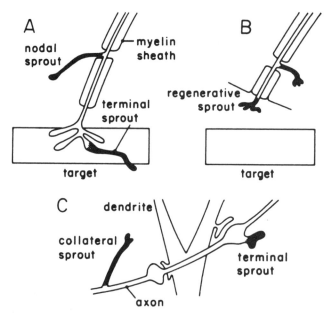

Figure 9.149. Axonal sprouting with consequences for regeneration, reinnervation, and synaptic restoration. Types of sprouting illustrated: *A*, Intact myelinated axon showing a collateral or nodal sprout arising from a node of Ranvier between myelin segments, and a terminal sprout. Most typical cases occur in the peripheral nervous system, e.g., neuromuscular junctions. *B*, Damaged myelinated axon showing regenerative sprouting. *C*, Axons with boutons *en passant* showing collateral and terminal sprouting, a situation arising frequently in the central nervous system. Growing sprouts are shaded. [From Cotman et al. (1981).]

proximal nodes of Ranvier (*B*). If the cells innervated by a neuron are damaged, the innervating neuron may exhibit collateral sprouting from more proximal regions of the axon, including where the neuron has synaptic junctions *en passant* (*C*). Exactly how the signal of damage is communicated to the sprouting neurons is unclear, but it probably involves trophic factors. The consequences for reestablishing old connections and the creating of additional new connections are potentially far reaching. It is presumable that after exuberant sprouting in response to injury, as in sprouting during early development, there may be later pruning back of redundant connections based on functional utility.

Central sprouting can establish new synaptic connections at vacated synaptic sites, as demonstrated by Raisman (1969a and 1969b) and Raisman and Field (1973). *Sprouting collaterals appear to be selective in the areas they will or will not invade and the kinds of connections they will establish.* They begin early and may be completed within a few days, and they may be long lasting (Zimmer, 1974). It is also evident that previously unrecognized pathways can become functional (see review by Wall, 1980).

Inasmuch as primary olfactory neurons are replenished from the basal cells in the olfactory neuroepithelium, Wright and Harding (1982) extirpated the olfactory bulbs in mice that had been previously trained to discriminate between scented and unscented air. Following bulbar extirpation, the simple olfactory discrimination was lost. But it returned again, without further training, when new primary olfactory neurons had regenerated as far as olfactory cortex. These results suggest that at least simple olfactory decisions can be made, even though the olfactory bulb has been bypassed.

GRAFTING OF FETAL BRAIN TISSUES INTO ADULT BRAINS

As already implied, the hippocampus may be a region of the nervous system which possesses a high potential for plastic morphological and functional alteration. Moore (1975) found that the monoamine neurons innervating the hippocampal formation and septum undergo extensive reorganization in response to injury. Björklund et al. (1976) successfully transplanted monoaminergic neurons from rat embryos into the hippocampus of adult rats. They found that adrenergic, serotonergic, and dopaminergic neurons from the embryonic tissue each had distinguishable patterns of innervation of the adult tissue and that the transplants developed different specifiable responses to damaged areas of the host brain. *This means that mechanisms for reestablish-*

ment of fiber pathways and connections must be retained by the adult brain (Björklund and Steveni, 1978).

Early fetal tissues survive better than do fetal transplants removed around the time of birth. Transplants into young brains are more successful than transplants into older brains. Thus both donor and host tissues have time-bound propensities for reorganization. It has been found that tissue from different parts of the fetal brain, implanted into the cerebellum, retains developmental patterns resembling those they would have exhibited if they had been left to develop in the donor brain. Embryonic retinas retained retina-like organization when transplanted into adult brain. Septal and striatal cholinergic tissues transplanted into the hippocampus both form normal patterns of cholinergic innervation there, despite the fact that the septal region normally does and the striatal region normally does not innervate the hippocampus (Cotman and Lewis, 1983).

The brain is an immunological protected site and for this reason may be less likely to reject grafts than are other tissues. There has even been some success with cross-species graft implantation into the adult brain. Experimental grafting in animals has involved grafts, explicitly selected to correct damage resulting in movement disorders, memory loss, hyperactivity, and overresponsiveness to stimuli. This work has succeeded in obtaining grafts that secrete dopamine, acetylcholine, norepinephrine, serotonin, vasopressin, and gonadotropin-releasing hormone.

Recently, Stenevi and Björklund have been injecting 1–2 μl of *brain cell suspensions* into adult rat brains rather than transplanting fetal tissue. The consequences of hippocampal damage involve short-term memory defects and an overreaction to stimuli. Björklund and Stenevi have been grafting into areas of hippocampal damage. By implanting grafts from the septum, they find that short-term memory is ameliorated, but the animals remain hyperactive and startle-prone. When they transplanted locus coeruleus cells, which secrete norepinephrine, the rats were no longer hyperactive, but they retained their short-term memory defects and exaggerated startle responses. The same team has found that suspensions containing dopamine-secreting cells introduced into the striatum of aging rats resulted in significant improvement in their motor coordination (Gage et al., 1983).

COMMUNICATION

Communication in the nervous system requires maintaining appropriate input and output relations with the environment and with the internal control of visceral mechanisms such as respiration, circulation, temperature regulation, appetite and digestion, excretion, hormonal and reproductive performance, and the like. This is accompanied by systems of appropriate receptors and effectors—external and internal—between which are stretched vast communications and decision-making networks. Specialized elements in the nervous system are particularly sensitive to various mechanical and physical perturbations, to electrical, chemical, gaseous, and ionic shifts, but most of the nervous system consists of neurons doing nothing more alluring than exciting and inhibiting one another. The bulk of the "business" of the nervous system consists of communicating *information*, all to the purpose of controlling behavior so as to optimize internal satisfactions. *Most nervous system communication involves synaptic transmission.* This process of communication was well advanced in evolution before axonal conduction was developed (Bishop, 1956). Axons evolved out of necessity to provide for decrementless delivery of information between nuclear constellations that were becoming very widely separated in large animals. It also permitted the longer circuiting of information that allows for elaborate perception, memory, learning, and forward planning.

Synaptic Transmission

Synaptic transmission differs from axonal transmission in several respects. Impulses travel only in one direction across synaptic junctions. In axons, impulses can travel in either orthodromic or antidromic directions, although, because of the directionality of junctions, axons ordinarily only conduct orthodromically. There is a junctional delay in synaptic conduction, whereas there is no similar delay in axonal conduction. Axonal conduction can follow high stimulation rates (above 1,000/s), but postsynaptic units can usually only follow slower transmission rates (below about 100/s). Action potentials are all-or-nothing in axons, whereas synaptic potentials summate algebraically, adding excitatory and subracting inhibitory potentials.

POSTSYNAPTIC POTENTIALS

Synaptic sites are not directly excitable; neurotransmitter is released across the junction and induces more or less prolonged postsynaptic potentials, depending on rates of diffusion, enzyme action to inactivate the neurotransmitter, and reuptake by the presynaptic terminal. *Almost all neurons which release a neurotransmitter actively reaccumulate the transmitter they relese.* The exception is acetylcho-

line which is rapidly dissociated, and only the choline is recovered. Peptides are apparently not involved in reuptake. Reuptake mechanisms are energy- and sodium-dependent.

Excitatory neurotransmitters released by the presynaptic terminal activate the postsynaptic membrane by increasing ionic conduction. This forces the membrane to shift toward the sodium equilibrium potential which results in *depolarization*. If the shift is sufficiently great, an all-or-nothing action potential is generated. Inhibitory neurotransmitters, similarly released, selectively activate chloride ion conductance which results in an inward flow of that ion and increased *polarization*. This further stabilizes the membrane and makes it less likely that a spike impulse will be discharged, thus counterbalancing the influence of excitatory neurotransmitter effects.

Most postsynaptic potentials are of relatively short duration (less than about 20 ms), but some postsynaptic potentials may last for many seconds. Long duration potentials often occur as a result of neurotransmitter inactivation of a tonically active postsynaptic membrane, one that has a slow sodium or potassium resting conductance. They can occur when tonically active excitatory or inhibitory terminals are themselves inactivated by presynaptic inhibition. Slow postsynaptic potentials may also occur as a result of neurotransmitter activation of cyclic nucleotides which in turn bring about the phosphorylation of membrane proteins which alter ionic permeability. In this way, a set point or bias toward activity or inactivity of the postsynaptic neuron is effected.

NEUROPHARMACOLOGY

Thousands of drugs and agents affect the nervous system. Aside from local anesthetics which can block transmission in axons, most drug effects seem to be involved primarily with synaptic events. The array of drugs specifically employed to modify nervous system actions include analgesics, narcotics, hypnotics, sedatives, anesthetics, anticonvulsants, psychotropic drugs, and a number of drugs that particularly affect the autonomic nervous system. Most drugs, insofar as they have been tested in suitable circumstances, appear to act pre- or postsynaptically. They do this in respect to the mechanisms of action of neurotransmitters. *Drugs may interfere with metabolism of the neurotransmitter or with its synthesis, storage, release, or reuptake mechanisms.* Alternatively, they may act *postsynaptically by activating or blocking receptors.*

Another aspect of neuropharmacology relates to blood-brain permeability barriers. Some of the barriers are physical: (1) the lack of fenestra in brain capillary endothelium, (2) the tight junctions which provide a double seal joining cerebral endothelial cells, and, to unknown extent, perhaps also (3) the foot processes of protoplasmic astrocytes. There are also chemical barriers including charges borne by transport channels across solvent partitions. Variations in kinds and rates of metabolic activity of local neuropil, differences in remoteness from sources of cerebrospinal fluid, and differences in diffusion barriers result in the fact that there may be significant local differences in neuronal environment and significant differences in drug availability, concentration, and access to synaptic junctions.

SECRETION OF NEUROTRANSMITTERS

Transmitter secretion can and usually does result from the occurrence of an action potential. Yet transmitters are also released from short axon neurons and axonless neurons, and from dendrites and cell bodies where graded potentials rather than action potentials occur. *Spikeless release occurs especially often with interneurons involved in local circuits.* Evidently, *calcium plays an instrumental role in all forms of neurotransmitter release.* Methods are now available to establish the intracellular concentration of ionized calcium. One of these measures the luminescence of aequorin, a protein isolated from jellyfish which reacts with Ca^{++} by emitting light: another involves x-ray or proton probe spectroscopy, and still another utilizes metallochromic dyes such as arsenazo III which changes its absorption spectrum when it complexes with Ca^{++}.

EXCITATION-SECRETION COUPLING

Axonal excitation is coupled to neurotransmitter secretion at *synaptic junctions.* There is a shift from sodium ion control of axonal membrane to calcium ion control of the secretory process. These two mechanisms which are ordinarily linked together can be separated by using the puffer fish poison *tetrodotoxin* which selectively blocks sodium channels. Then the terminal can be caused to initiate secretory actions by imposing electrical depolarization. If calcium ions are present in sufficient concentration, excitation-secretion coupling is enjoined.

It is observed that *most of the calcium in neurons is in bound form:* only a small proportion is free and ionized in cytoplasm. Because the level of ionized calcium is normally very low (about 10^{-7}) in the cytoplasm, the cell can make use of very small shifts in local Ca^{++} levels to obtain large physiological effects. *The invasion of the nerve terminal by an action potential is associated with an influx of Ca^{++}*

from extracellular space. Neurons then utilize calcium shifts from free to bound states to control secretion of neurotransmitters.

Similar Ca^{++} control evidently accounts for axoplasmic flow, many enzymatic functions, and membrane permeability. Endogenous phospholipase enzyme activity is enhanced by the presence of Ca^{++} and may play an important role in presynaptic neural events such as the aggregation of vesicles (Moskowitz et al., 1982). Ca^{++} activates material associated with the presynaptic grid of dense projections which have actin-like filaments that apparently pull synaptic vesicles toward the presynaptic membrane. These vesicle and cell membranes fuse, forming an "omega" figure which allows vesicle contents access to the synaptic cleft (see Fig. 9.142).

When these processes have completed exocytosis, the Ca^{++} must be cleared from the cytoplasm. This is accomplished by its binding to calmodulin and to endoplasmic reticulum and other organelles, by uptake in mitochondria, and by the slower pumping of the ion to the exterior of the cell. Mechanisms involved in the control of calcium availability are also coupled to pathways for neurotransmitter synthesis and the setting of regional metabolic activity by mitochondria. It is thought that by participating in such longer lasting processes, the control of calcium may contribute to synaptic plasticity in relation to neuronal development and possibly also to mechanisms of memory and learning.

RECEPTORS AND REGULATION

The process of synaptic transmission involves not only secretion of neurotransmitters but also their special affinity with receptors. *A specific transmitter substance is not limited to having only one specific postsynaptic effect.* The same transmitter can be used for different effects, both excitation and inhibition, and both excitation and inhibition can be elicited by different transmitters. Acetylcholine, for example, released by the motoneuron, is excitatory at the neuromuscular junction, whereas the same substance released by the vagus is inhibitory at the cardiac muscle junction. Similarly, dopamine and adrenaline can be both excitatory and inhibitory, depending on local receptors.

Kandel (1976) has summarized this evidence as follows:

1. The sign of synaptic action is not determined by the transmitter but by the properties of the receptors on the postsynaptic cell.
2. The receptors in the follower (postsynaptic) cells of a single presynaptic neuron can be pharmacologi-

cally distinct and can control different ionic channels.

3. A single follower cell may have more than one kind of receptor for a given transmitter, with each receptor controlling a different ionic conductance mechanism.

As a result of these three different features, cells can mediate opposite synaptic actions to different follower cells or to a single follower cell.

Presynaptic terminal uptake of norepinephrine (NE) from the synaptic cleft and neighboring extracellular space is evidently affected by the amount of NE present (Lee et al., 1983). The concentration of NE recognition sites on the presynaptic terminal varies according to the local concentration of NE. Depletion of NE by administration of reserpine reduces the number of reuptake sites; increase of the concentration of NE by administration of monoamine oxidase inhibitors increases the number of binding sites. This has the useful effect of *homeostatic regulation of NE availability at the synapse*. When NE is scarce, reuptake is retarded; when NE is abundant, reuptake is accelerated. This is another aspect of communications control using neurotransmitter actions.

NEUROTRANSMITTERS

The main categories of neurotransmitters include acetylcholine, biogenic amines, a few amino acids, and a large number of peptides. All transmitters have low molecular weight and are water-soluble compounds which are synthesized from precursor compounds. The amino acids and peptides are formed ultimately from circulating glucose. It is evident that these transmitters have great utility because the same categories are found in invertebrates as well as vertebrates despite great differences in organization of nervous systems. An excellent review of the role of peptides in neural function is presented by Baker and Smith (1980).

Individual Neurons May Contain More Than One Neurotransmitter

Evidence for this phenomenon to the extent of at least four transmitters is available in invertebrates and of at least three transmitters: 5-hydroxytryptamine (5-HT), thyrotropin-releasing hormone (TRH), and substance P, in the medulla oblongata and in the ventral horn of the spinal cord of rats (Hökfelt et al., 1980; Johansson et al., 1981). The significance of coexistence of more than one neurotransmitter in single central neurons is not entirely understood. In some autonomic paths in the cat, exocrine glands are innervated by two populations of axons. One population contains acetylcholine and the vasoactive intestinal polypeptide (VIP). The other population contains norepinephrine and a polypeptide which blocks vasodilation. Each of the peptides potentiates the transmitter with which it is associated. The two autonomic neuronal populations provide physiological responses of opposing nature. The VIP affects smooth muscle cells in the precapillaries to enhance secretion by increasing blood flow, increasing the effects of acetylcholine on the same exocrine gland. Norepinephrine and the peptide released with it from that type of neuron act synergistically to induce vasoconstriction. It remains to be seen how neurons with multiple transmitters may contribute to central nervous system mechanisms of control.

VESICLE HYPOTHESIS FOR NEUROTRANSMITTER RELEASE

Freeze-fracture electron microscopy, combined with thin section analysis, has provided essential proof that synaptic vesicles fuse with the axolemma when secretion of transmitter is elicited (see Fig. 9.142). Rapid freezing shows that fusion of the vesicle and the quantal release of acetylcholine occur within milliseconds, closer in time than the processes can be distinguished. Extracellular markers indicate that stimulated nerve terminals contain many vesicles that have formed from axolemma, suggesting that vesicle membrane may be recycled after fusion with the terminal membrane (Heuser and Reese, 1973). Additional vesicles are presumably formed from smooth endoplasmic reticulum and from membranous components, including preformed vesicles, transported into the terminal by axoplasmic flow.

In summarizing the evidence relating to the release of quanta of acetylcholine, Ceccarelli and Hurlbut (1980) state that although it cannot be absolutely excluded that quantal release of ACh may come from an extravesicular pool of the transmitter exiting from the nerve terminal, the preponderance of evidence indicates that quantal release comes from vesicles. Indeed, the sites for vesicular fusion, in the grid of dense projections arrayed on the presynaptic membrane, appear to be aligned opposite counterpart sites on the postsynaptic membrane where the highest densities of ACh receptors are aggregated. It is suggested that not only is the presynaptic grid of dense projections dynamically involved in vesicle attachment to the presynaptic membrane, but the organization of postsynaptic receptors is dynamically mobile in terms of aggregation along the postsynaptic membrane.

NONVESICULAR RELEASE OF NEUROTRANSMITTERS?

Because the dimensions of nerve terminals and synaptic junctions are so small, because the actual transmission processes take place in an even more confined region and at a pace of fractions of milliseconds, questions relating to alternative mechanisms for release of neurotransmitters are still alive, whether exclusive alternatives or additional release mechanisms (Tauc, 1982). It may eventuate that there are several mechanisms for transmitter release. Considerable ACh is found to leave the terminals in a nonquantal manner. At least part of this nonquantal release is also Ca^{++}-dependent, and the coupling between excitation and secretion appears to be similar to that for quantal release. Because of a lack of information, it is not known whether nonquantal release of neurotransmitters can be extended to nonacetylcholine neurotransmitter systems.

THE PROBLEM OF IDENTIFICATION OF NEUROTRANSMITTERS

Several different investigative methods need to be applied to establish whether a putative neurotransmitter can be identified as a bone fide neurotransmitter. Useful evidence includes: (1) presence and activity of pertinent enzymes necessary for synthesis of the substance in the presynaptic terminals and (2) presence and activity of enzymes for inactivation or removal of the substance at the synapse; (3) release of the candidate substance by stimulation of the presynaptic terminal; (4) reproduction of physiological responses by iontophoresis of the candidate substance at the synaptic site; (5) appropriate influence of drugs that affect the enzymes synthesizing and inactivating the candidate substance; (6) appropriate high-affinity receptors, release, uptake action, inactivation as agonists or antagonists of the substance, its analogues, and other pharmacologically active drugs.

In respect to the identification of *neuropeptides* which serve as neurotransmitters, advantage is taken of the use of *immunological methods*. The candidate oligopeptide is first injected into animals as an antigen to provide an antigenic response, producing a specific antibody to the peptide. It may be necessary to resort to monoclonal antibodies to achieve the necessary specificity. The antibody can then be applied to histological sections and a suitable *antibody to the antibody* is then applied. This second-echelon antibody carries a fluorescent molecule, metallic label for identification in electron microscopy, or horseradish peroxidase which can

be seen in both light and electron microscopic images.

Neuroactive peptides consitute an expanding list which includes: carnosine, thyrotropin-releasing hormone (TRH), met-enkephalin, leu-enkephalin, angiotensin II, cholecystokinin-like peptide (CCK), oxytocin, vasopressin, leuteinizing-hormone-releasing hormone (LHRN), substance P, neurotensin, bombesin, somatostatin, vasoactive intestinal polypeptide (VIP), β-endorphin, adrenocorticotropin hormone (ACTH), and others (Iverson, 1979). The list ranges from single amino acids to lengthy hormone molecules. The list is expanding very rapidly and may eventuate in scores and possibly hundreds of neuroactive peptides that may be identified as contributing to synaptic functions.

Some Methods of Analyzing Neurological Defects

Within the large and important category of *genetic defects*, methods of neurological analysis include securing detailed history and lineage of carriers as well as of individuals manifestating the disorder. It is to be remembered that genetic expression can vary considerably and that long-range interactions among chromosomes and genes can bring about additional complications even in persons having identical point gene defects (Rosenberg, 1980). *Congenital defects* may be traced to infections, nutritional deficiencies, immune disorders and induced sensitivities, and so forth.

Disease and trauma affecting the nervous system provide "natural experiments" that are difficult to analyze without adding to the damage, and are virtually impossible to standardize. *Lesions* result from nutritional deprivation, hemorrhage, adverse immune response, and indirect as well as direct effects of trauma. The consequences include *symptoms* recognized by the patient and *signs* identified and characterized with the help of the patient and employment of various diagnostic procedures. The overall diagnostic result is a measure of existing medical insight into physiological and pathophysiological mechanisms.

Neurological assessment depends on a combinatorial evaluation of many dynamic processes. In light of evidence for axonal collateral sprouting and synapse formation described previously, *the nervous system seems to exist in dynamic equilibrium both in health and disease, with restitutional changes occurring promptly after insult to any part of the nervous system and lasting indefinitely, presumably being stabilized on some utilitarian basis.*

Figure 9.150 illustrates a stereotactic apparatus

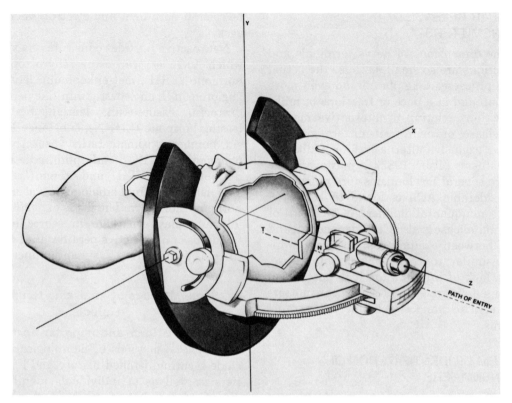

Figure 9.150. Stereotactic system for surgical procedures assisted by computed tomography (CT) provides fixation to hold the patient's skull firmly in place during operative procedures, a geometric frame to locate CT scan images within the frame's coordinates, and a mechanism for precise placement of surgical instruments. This system is designed for stereotactic operations conducted entirely in a CT scanner suite. The frame made of gas-sterilizable rigid plastic attaches to the CT scanner's patient table, causes no image artifact, and controls insertion to any desired 3-space within the cranium. This device is useful for tumor biopsy, freezing or electrocautery of localized tissue, implantation of radioactive pellets, and for X-ray therapy beam guidance. [From Rhodes et al. (1982).]

that is designed to permit precise local intervention within the 3-space of a patient's brain. This is useful in several respects: for intervention in order to localize and treat abnormalities that might otherwise lie beyond reasonable surgical or intravascular access. The apparatus is constructed of radiolucent material so that it can be used during computed axial tomographic (CAT) scanning. Serial images can thus be reconstructed three-dimensionally from scans made while the apparatus is attached to the patient's head. The apparatus can be used to direct a probe (electrode for recording, stimulation or making localized electrolytic lesion; a thermode for cooling—thus reducing local neuronal activity—or freezing, for permanent destruction, implantation of a source of ionizing radiation for local tumor destruction, etc.). Local events can be monitored, interfered with, or eliminated, depending on physiological information locally obtained as well as on the basis of gross morphological evidence. The objective is to establish the functional role of that locus in the manifestation of signs or symptoms.

Figure 9.151 demonstrates the identification of a conduction defect in the optic path between retina

and cortex on the right side in a young woman with normal perimetry and fundascopic findings. The altered evoked potentials helped to localize and objectify a defect suspected to be due to multiple sclerosis. Such clinical documentations can be compared over time in a disorder that has a fluctuating course of disability and a highly variable distribution of lesions.

A great variety of diagnostic opportunities is presently available involving physical (e.g., nuclear magnetic resonance imaging; positron emission tomography; computed axial tomography; radioisotope tracings of tumors; electroretinograph; electroencephalograph; evoked potentials; echoencephalograph, etc.) biochemical analyses of blood, urine, cerebrospinal fluid for particular drugs, and metabolic products. Of course, the patient is the best witness, having an attachment of multitudes of direct and indirect connections relating to the disorder. In parallel, neuropsychologists have been developing a powerful array of tests useful for neurological and psychological evaluation of patients with nervous system disorders. These are becoming increasingly specific in their neuroanatomical and

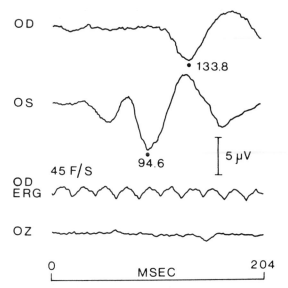

Figure 9.151. Visual evoked responses in a 36-year-old woman with diagnosis of suspected multiple sclerosis. Visual acuity was 20/30 for both eyes. Perimetry and fundoscopic examinations were normal. Note the greatly delayed visual evoked potential in response to monocular stimulation of the right eye (*OD*). The normal steady-state visual evoked potentials at the retinal level (*ERG*) to flash stimulation (45 flahses/s) contrast with absence of cortical responses (OZ-to reference) indicating dysfunction of the right optic nerve or optic tract and adequate retinal function. [From Cracco and Celesia (1983).]

physiological implications. It is now evident that *active pathological processes may have as much effect on the patient's mental status as the absence of parts.*

It is evident that pathological processes do not limit themselves to some particular anatomical part or physiological system. Moreover, with processes that are slow in onset, a great deal of compensation takes place without any signs or symptoms becoming manifest. For example, a child with glial tumor of the cerebellum may lose almost all of the cerebellum and yet the first signs may appear only when mass effects cause elevated intracranial pressure. Additionally, an infiltrative blood discrasia can block cerebrospinal fluid channels to the extent that the first presenting evidence may be headache associated with increased systemic blood pressure. Further, a focal lesion may given rise to remote symptoms, e.g., a frontal lobe tumor masking its presence by presenting cerebellar signs.

Destruction of part of the nervous system does not reveal the function of that part, even in a subtractive sense. It *reveals* instead only *what the rest of the nervous system can do without that part.*

Structure and function are two sides of the same coin. If examination of one does not reveal something about the other, it has not been studied adequately. Undertake the examination of any patient who presents a neurological problem with a sacred respect for the complexities of the nervous system puzzle. Do not rely upon knowledge frameworks that categorize normal and pathological physiology because such reliance may cause you to overlook some important unexpected clues. Neurological science is both too primitive and too fragmentary for such reliance; careful examination of your patient is likely to reveal information that will help fill in some gaps in the frontier of neurosciences. If death eventuates, by all means obtain permission for a thorough postmortem examination of the nervous system to disabuse yourself of fancy and to contribute your clinical observations together with autopsy findings to fundamental neurological understandings. Communication and control is what the nervous system is all about.

CHAPTER 68

Central Control Mechanisms

INTRODUCTION

Information is brought to the nervous system by afferent nerves, visceral and somatic, and by the bloodstream. There are on the order of about 10 million afferent nerve fibers. The number of efferent neurons conveying all the commands that operate visceral and somatic effector organs, glands, smooth muscle, cardiac muscle, and skeletal muscle is smaller. There are perhaps a half a million efferent fibers leaving the brain stem and spinal cord and an unknown number of second order autonomic neurons in scattered peripheral ganglia which innervate visceral organs. So there is already a considerable convergence from number of input channels to number of output channels available. If there are, roughly speaking, about 20 billion nerve cells entirely contained within the central nervous system, then there are on the order of 2,000 central neurons to each input and output channel.

Differences in behavior among the several phyla, orders, and species relate mainly to differences of interconnections between input and output. These differences are largely organized by distinctive connections among neurons, particularly interneurons and local circuits, and in pylogenetic differences in the amount of neuropil available for central decision making (see Bullock and Horridge, 1965).

Nomenclature

Names given to major structures in the central nervous system often describe how the parts looked to Greek anatomists (cortex = bark, claustrum = cloak, amygdala = almond, hippocampus = sea horse, mamillary body = nipple, striatum = furrowed, caudate = tailed, putamen = shell, globus pallidus = pale globule, substantia nigra = black substance, thalamus = chamber or curtained bed, medulla = marrow, spinal cord = spine string or bowstring, referring to vertebral spines, pyramidal tract = tract associated with a compact triangular bundle of fibers seen in the upper part of the medulla). Names are arbitrary in any event, and

these ancient names are generally unencumbered by implications of function. They have the advantage of traditional acceptance, and after a while they become friendly.

Other designations come from the five embryonic brain vesicles. These tie in with the concept of levels of functions as follows:

Telencephalon (*tele* = end, *encephalon* = brain)
Olfactory bulb and tract
Basal ganglia: claustrum, caudate, putamen, globus pallidus, substantia innominata (unnamed substance)
Limbic system: mesocortex, archicortex (including hippocampus), fornix (fornices = arches), septum (=partition), amygdala
Neocortex
Central white matter of the cerebrum, corpus callosum, anterior commissure, internal and external capsules
Diencephalon (*dia* = between hence 'tween)
Thalamus
Optic tract
Hypothalamus
Epithalamus
Subthalamus
Infundibular stalk (infundibulum = funnel), pituitary (pituita = phlegm, under the assumption that it produces nasal phlegm)
Mesencephalon (midbrain)
Superior and inferior colliculi (colliculus = hill)
Periaqueductal gray (around the water channel or duct)
Midbrain tegmentum (tegmentum = covering, more especially floor)
Red nucleus
Substantia nigra
Cerebral peduncles (peduncle = stem-like)
Metencephalon (afterbrain)
Cerebellum (diminutive of brain, hence little brain)
Pons (pons = bridge)
Pontine nuclei
Myelencephalon (myelos = marrow)

Medulla oblongata (oblong marrow, also called the bulb for its bulbous cone shape).

Some of these structures can be identified in the medial saggital view of the human brain as presented in Figure 9.152.

INTERNAL CAPSULE

The central white substance of each cerebral hemisphere contains a large number of fibers running from one hemisphere to the other, and also the afferent fibers ascending to cortex from thalamus and other lower structures, plus descending fibers from cortex to thalamus, striatum, pons, brain stem, and spinal cord. At the level of striatum and thalamus the ascending and descending fibers together constitute the *internal capsule* that separates caudate nucleus from putamen and lentiform nucleus from thalamus. Its descending fibers then flow downward to become the peduncle of the mesencephalon. The corticofugal fibers of this complex include both pyramidal and nonpyramidal axons. The internal capsule, however, is not exclusively motor in its function, since it incorporates both corticothalamic and thalamocortical fibers, including the principal ascending pathways for conscious sensation. The following are the most important tracts it transmits.

A. Descending
1. Corticospinal tracts.
2. Corticobulbar tracts.
3. Corticopontine tracts: four bundles of fibers, the frontopontine, parietopontine, temporopontine, and occipitopontine, from corresponding areas of the cerebral cortex to nuclei of the pons.
4. Corticorubral tract, from the frontal lobe to the red nucleus.
5. Corticothalamic fibers from almost all areas of the cortex to various nuclei of the thalamus.
6. Corticostriatal fibers from the precentral area of the cortex to the caudate nucleus and the putamen.
7. Occipitotectal fibers from the para- and peristriate cortex to the tectum of the midbrain.

B. Ascending
1. Thalamic radiations, four bundles ascending from the thalamus, anterior, superior, posterior, and inferior, which connect the thalamus with various parts of the cerebral cortex. The anterior group terminates in prefrontal areas, and the superior group in the somesthetic area (postcentral gyrus).
2. Auditory radiation fibers ascending from the

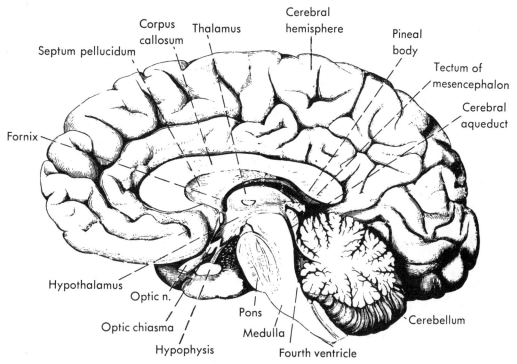

Figure 9.152. Medial sagittal section of the human brain. Note corpus callosum, the main interhemispheric connection. Its anterior end curves ventrally to form the thin line of lamina terminalis within which can be seen the anterior commissure which interconnects the two temporal lobes. The medial part of the orbital surface, subcallosal gyrus, cingulate gyrus, and retrosplenial cortex form a ring visible on the medial surface. The parahippocampal cortex lies largely hidden behind the brain stem, and the tip of the temporal lobe is visible. Together, these make a complete circle of primitive (limbic) cortex surrounding the hylus (neck) of the hemisphere. Several other important structures are labeled. [From Henle (1868); reproduced from Hollinshead (1967).]

medial geniculate body to the temporal cortex (geniculotemporal fibers).

3. Optic radiation fibers ascending from the lateral geniculate body to the visual area in the occipital cortex (geniculocalcarine) fibers.

The presence of the lentiform nucleus on its lateral aspect, together with caudate nucleus and thalamus medially, gives the internal capsule a right-angular profile when it is examined in a horizontal section of the cerebrum (Fig. 9.153). The region of the angle is called the genu (knee); the portions in front of and behind this are known, respectively, as the anterior and posterior limbs. The extension backwards of the posterior limb is known as the *retrolenticular part.*

The anterior limb is occupied by the frontopontine fibers and the anterior thalamic radiation (thalamofrontal fibers), the frontothalamic tract and corticostriatal fibers. In the anterior three-fifths of the posterior limb are transmitted the corticobulbar and corticospinal fibers. The fibers of these two tracts are organized in accordance with the portions of the body which they innervate, those carrying impulses for the eye muscles, jaw, the muscles of the tongue, and upper and lower extremities arranged in this order from before backwards; and those transmitting impulses destined for the proximal part of an extremity are placed in front of those for the more distal muscles. The remaining part of the posterior limb and the retrolenticular part carry in anteroposterior order the superior and posterior thalamic radiations, the auditory (geniculotemporal) and the inferior thalamic radiations, corticothalamic, and parietopontine fibers, and the optic (geniculooccipital) radiations.

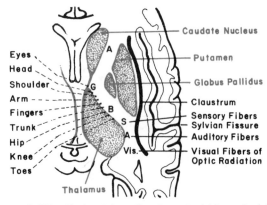

Figure 9.153. Horizontal section through right cerebral hemisphere showing relationships of thalamus, internal capsule, and basal ganglia including caudate, putamen and globus pallidus nuclei, and claustrum. Localization of fibers in internal capsule is also shown.

The internal capsule is supplied with blood through the lateral striate branches of the middle cerebral artery, by the recurrent artery of Hubner from the anterior cerebral artery, and the anterior choroidal artery from the internal carotid artery. The internal capsule is a frequent site of injury to the nervous system through hemorrhage from one of the striate branches. Both sensory and motor deficits may result on the side of the body opposite the lesion.

THALAMUS

Thalamus and Thalamocortical Projections

The thalamus is a functional way station for the entire forebrain. It contributes relay functions for all sensory systems except olfaction. It provides the relay mechanisms for major forebrain systems to communicate with one another. It relays back and forth with all areas of cerebral cortex in the course of contributing to association relay processing of higher neural activities. It is centrally concerned with diffusely projecting systems relating to sleep, wakefulness, and other state conditions. In effect, because it is so richly connected throughout the forebrain, its organization is like a brain in miniature, even though it represents only about 2% of the total brain volume.

At the level of the thalamus are third order somatosensory neurons which relay somesthetic signals from the lemniscal system to primary sensory cortex. Visual pathways relay through the thalamus on their way to cortex via the lateral geniculate body and the pulvinar nucleus. Auditory pathways project to cortex via the medial geniculate body. Olfactory signals do not relay through the thalamus but bypass it to reach primary olfactory cortex. Only secondarily does the thalamus become involved in higher order olfactory information relating to the control of behavior, relaying through the dorsal medial nucleus of the thalamus en route to frontal lobes.

The thalamus is directly engaged in relay of signals within the limbic system, between the basal ganglia and cerebral cortex, between cerebellar cortex and cerebral cortex, involving all the major systems in their mutual relations. At a more fundamental level, the thalamus distributes signals that have to do with generalized levels of brain activity. The thalamus is characteristically related to the other structures with which it communicates by reciprocal connections. Thus each region of cortex to which a given nucleus of the thalamus projects has neurons which project back onto that same thalamic nucleus.

The thalamus consists of two globular masses with stubby ventrolateral horns which curve in a blunt hook facing forward. These two masses are separated by the third ventricle. In about 20% of human brains, the thalamus is connected across the midline, producing a thin intermediary connection, the *massa intermedia.*

Roughly speaking, the *ventral half* of the thalamus is engaged in *corticocortical associations.* The posterior part of this region connects with the posterior regions of the cerebral hemisphere, the lateral part with the lateral regions of the hemisphere, medial part with the frontal lobe, so that there is a systematic relational organization. The dorsalmost and anterior parts interconnect the limbic elements of cortex which surround the hylus of the hemisphere. See Table 9.6 for a schema of functional organization of the thalamus (see also Figs. 9.154 and 9.155).

BASAL GANGLIA

The basal ganglia deserve the name because, during embryogenesis, the cells arise from the basement wall of the outpouching lateral ventricles. They end up as a massive organ in the base and lateral wall of each hemisphere. The *caudate* nucleus is the dorsalmost of this group of nuclei (Fig. 9.156). The head of the caudate forms the lateral border of the lateral ventricle. The body and tail of the caudate follow the inside curve of the lateral ventricle from its frontal horn to the tip of the temporal horn where the nucleus therefore lies in the roof of the ventricle. The internal capsule lies lateral to the caudate and ventricle. It divides the head of the caudate from its closest companion nucleus among the basal ganglia, the *putamen.* The caudate and putamen are joined anteriorly where bundles of fibers branching from the anterior limb of the internal capsule form striations as they pass through this nuclear union.

The anterior and posterior limbs of the internal capsule form a shallow V which is open laterally. The putamen and its medial neighbor, the *globus pallidus,* lie wedged into this V (Fig. 9.153). On the medial side of the internal capsule lies the thalamus. Outflow from the basal ganglia pass beneath and penetrate through the internal capsule as fibers leave the globus pallidus on their way to the thalamus and brain stem. Anteriorly, the globus pallidus is continuous with another nuclear mass that belongs to the basal ganglia, the *substantia innominata.* This is nestled between the amygdala laterally, the hypothalamus medially, and the anterior perforated space ventrally. The nucleus of the substantia innominata degenerates in presenile dementia known as Alzheimer's disease. The putamen is bounded laterally by the external capsule, the claustrum, and the extreme capsule and forms the bulge in the lateral wall of the hemisphere over which lies the insular cortex.

Other components of the basal ganglia include the *subthalamic nucleus,* a well encapsulated flattened nuclear mass located beneath the thalamus at the border between diencephalon and midbrain, and medial to the internal capsule. It is continuous caudally with the *substantia nigra.* The pars com-

Table 9.6
Functional organization of the thalamus

Sensory relay nuclei	Lateral geniculate body (LGB)	Visual relay to cortex from retina
	Pulvinar (P)	Visual relay to cortex from superior colliculus
	Medial geniculate body (MGB)	Auditory relay to cortex from inferior colliculus
	Ventral posterior lateral (VPL)	Somesthetic relay to cortex from lateral lemniscus (body)
	Ventral posterior medial (VPM)	Somesthetic relay to cortex from trigeminal lemniscus (face)
Relay-association nuclei	Anterior nuclear group (Ant)	Limbic relay association from hypothalamus to limbic (meso-) cortex
	Ventral lateral (VL) and ventral anterior (VA)	Relay association to motor cortex from basal ganglia and from cerebellum
	Reticularis (Ret)	Relay association to cortex from thalamus and midbrain (arousal)
Association nuclei	Medial dorsal nucleus (MD)	Frontal association, motor and premotor cortex; limbic, olfactory association
	Pulvinar (P)	Parietal, occipital, temporal association
Diffusely projecting nuclei	Lateral dorsal (LD), lateral posterior (LP) nuclei	Parietal, occipital, temporal arousal and state control via reticularis
	Centrum medianum (CM), intralaminar nuclei, midline nuclei	Information from hypothalamus and brain stem relating to mood, feeling tone, arousal, possibly attention, projecting to cortex via reticularis
	Parafasicularis (Pf)	Possible special thalamic representation for pain

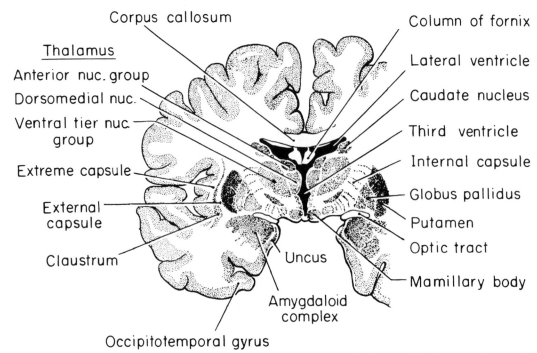

Figure 9.154. Coronal section of the human brain at the level of the interventricular foramen. The thalamus lies on either side of the third ventricle, bounded by the internal capsule. Beneath is the subthalamic nucleus, unlabeled. A band of cingulate cortex lies above the corpus callosum. The fornix is suspended beneath the corpus callosum. It will travel forward and curve ventrally to join the mamillary bodies, seen adjacent to the floor of the third ventricle. The putamen and globus pallidus lie external to the internal capsule. The amygdala is surrounded by parahippocampal cortex of which the uncus is a prominent part. [From Truex and Carpenter (1969).]

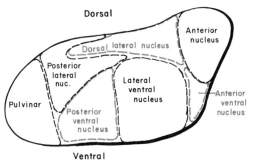

Figure 9.155. A schematic drawing of a longitudinal section through the thalamus from anterior (to the *right*) to posterior (to the *left*). Some of the major nuclei can be seen in this cut. Nuclei located medially in the thalamus are not included in the diagram.

pacta of this nucleus accumulates dark pigment due to the presence of melanin, a polymerized form of metabolites of dopamine and norepinephrine. This region of the nucleus and neighboring regions in midbrain tegmentum project dopaminergic axons to the *caudate* and *putamen*. These also receive information from all regions of cerebral cortex and from the intralaminar nuclei of the thalamus. These constitute the major inputs to the basal ganglia. The caudate and putamen project to the globus pallidus which in turn projects to motor cortex (Brodmann areas 4 and 6) via VA and VL of the thalamus (Fig. 9.156). With this circuit, the basal

ganglia can integrate information from many sources for the benefit of motor cortex. The same thalamic nuclear relays and projections to areas 4 and 6 are utilized by cerebellar output which partly relays through the red nucleus. The substantia nigra receives feedback projections from caudate and putamen and also projects to thalamus, brain stem reticular formation, and the tectum (Fig. 9.157). *The subthalamic nucleus is reciprocally related to the globus pallidus* (Fig. 9.156).

Functional Considerations

Vascular insufficiency in the subthalamic nucleus results in *hemiballismus*, a motor disorder in which spontaneous involuntary gestures of the face, extremities, and trunk are violently thrust here and there with ballistic force. Each movement is equivalent to an exaggerated normal gesture of expression. These involuntary gestures are extremely disturbing and fatiguing, but they tend to disappear after 3 to 4 weeks even though the subthalamic nucleus remains infarcted.

Parkinson's disease is clearly the most common degenerative disorder of the brain. It begins insidiously in older age groups and progresses slowly. Movements are slowed, the face becomes masked, limbs and trunks become increasingly rigid, there is often a characteristic "pill rolling" tremor at rest

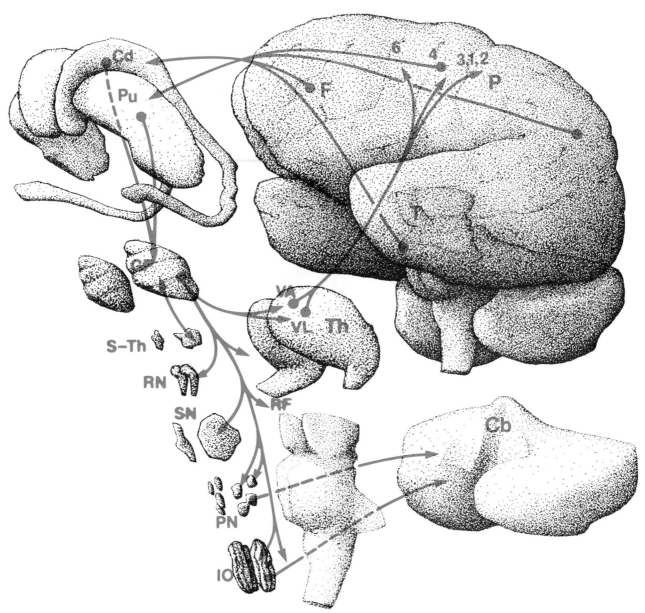

Figure 9.156. This and subsequent figures depict the whole brain in 3-D (*upper right*) as seen from a left frontal view (brain is rotated 30° from a straight frontal projection and tilted to be seen from slightly below a horizontal projection). Frontal, parietal, occipital, and temporal lobes can be seen, with the brain stem (midline) and cerebellum beneath. Structures to the left and beneath represent subcortical cerebral and brain stem nuclei, brain stem and cerebellum, from the same projection, approximately 2× expanded. From *top down*: caudate and putamen, collectively the striatum, the layered globus pallidus, subthalamic nucleus, red nucleus, and to the *right* of these, the thalamus. Beneath on the *left* are substantia nigra (flat disk shaped), pontine nuclei, and furrowed inferior olivary nuclei. Progressively to the *right* are the brain stem and cerebellum.

Cortical transaction with basal ganglia, brain stem, and cerebellum. All major regions of the cerebral hemisphere, frontal (*F*), parietal (*P*), temporal (*T*), and occipital (not labeled) project to caudate (*Cd*) and putamen (*Pu*). These project to globus pallidus (*GP*) which is reciprocally related to the subthalamic nucleus (*S-Th*) and which projects to VA and VL of the thalamus (which also receives and relays outflow from cerebellum) projecting to sensorimotor cortex (areas *6,4,3,1,2*). Globus pallidus also projects downstream to reticular formation (*arrows* to nonnucleated space), red nucleus (*RN*), substantia nigra (*SN*), pontine nuclei (*PN*), and inferior olivary nucleus (*IO*). The latter two project to contralateral cerebellar cortex as mossy and climbing fibers, respectively. [Drawing by Dana Livingston (1983).]

(interrupted by intentional movements), micrographia, disability in initiating movements such as getting up out of a chair, and difficulty in starting to walk. Walking consists of a shuffling gait as though the person were trying to keep up with the rate of forward falling. If suitably aroused emotionally, however, Parkinson patients can move briskly.

The discovery by Hornykiewicz that the brains of patients suffering from Parkinson's disease have markedly reduced concentrations of dopamine in

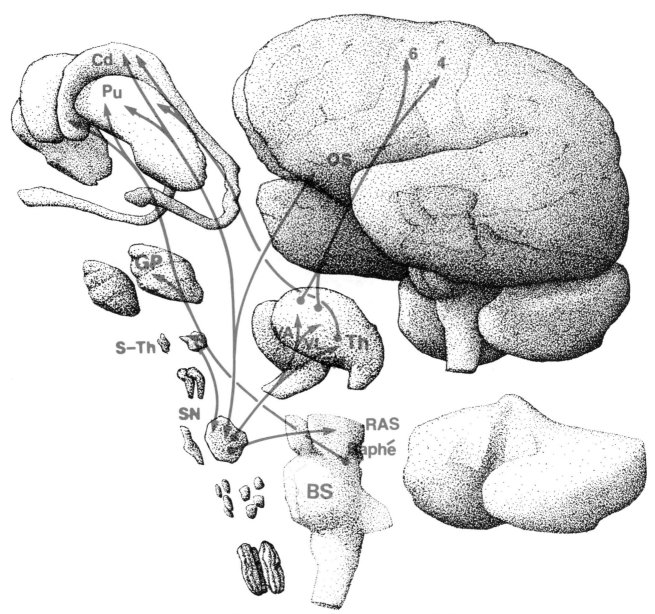

Figure 9.157. Brain stem mechanisms contributing to basal ganglia and motor cortex. Outflow from substantia nigra (*SN*) includes reciprocal projections to subthalamic nucleus (*S-Th*), caudate (*Cd*), and putamen (*Pu*), and to orbital surface of frontal lobe (*OS*). *SN* also projects to *VA* and *VL* of thalamus (*Th*), relay association nuclei in the thalamus, and to motor and premotor areas of cortex (4, 6). In addition, *SN* projects to brain stem reticular activating system (*RAS*) as well as to a diencephalic part of the same arousal (diffusely projecting) system, centrum medianum (*CM*) in thalamus. This activates *Cd* and *Pu*. Raphé nuclei in the brain stem also project to *Cd* and *Pu*, presumably contributing lasting changes in local circuitry. [Drawing from D. Livingston (1983).]

the caudate and putamen was the first instance of a disorder of the nervous system being attributed to a deficiency of a specific neurotransmitter. Since dopamine does not cross the blood-brain diffusion barrier, a precursor of dopamine, levodopa (L-dopa) has been used with palliative success. The patients' symptoms can be greatly reduced and crippling disabilities forestalled, but the basic disorder remains and continues to progress. The Parkinson disorder is complex, involving other monoaminergic

brain stem neurons and possibly also other neurotransmitters such as γ-aminobutyric acid (GABA).

Antipsychotic drugs, chlorpromazine and the like, which have made a substantial contribution by removing or ameliorating symptoms of schizophrenia, have unfortunate side effects, including a neurological disorder, *tardive dyskinesia*, associated with abnormal choreiform movements of face, mouth, and tongue particularly. If pronounced, the symptoms cannot be distinguished from Parkinson's disease. This drug-induced syndrome is thought to result from dopamine receptors becoming hypersensitive. The underlying suggestion, of course, is that there may be some kind of reciprocal relationship between schizophrenia and Parkinson's disease.

CEREBELLUM

Examined in respect to either its connections or its functions, the cerebellum has three types of interrelations. These three, however, are themselves interrelated, in that ultimately they control skeletal muscle and hence cannot be independent of one another. The three types are as follows.

1. Position of the body in space is conveyed via the vestibulocerebellar tract to the *archipallium* or archicerebellum; projections back to the vestibular nuclei and to other nuclei of the brain stem contribute to vestibular reflexes involving muscles of neck, trunk, and extremities.

2. Tension in skeletal muscles, joints, and tendons, as well as exteroceptive information, is relayed to the *paleocerebellum* via the dorsal and ventral spinocerebellar tracts and the cuneocerebellar fibers. This part of the cerebellar cortex also receives fibers from the reticular formation, and projects back to the same. It contributes to the central control of muscle tone, and like the reticular formation is in part responsible for the antigravity posture that typifies the decerebrate state.

3. Patterns of movements of muscles and the tonus adjustments that accompany movement are brought about through the *neocerebellum*. It is found only in higher animals, appearing phylogenetically with elaboration of the cerebral cortex and the pathways connecting cerebrum and cerebellum.

In addition to these three principal sources of input, viz., (a) vestibulocerebellar, (b) spino- and cuneocerebellar, and (c) corticocerebellar, the cerebellum has extensive connections with brain stem mechanisms. Two-way connections with the reticular formation probably include all portions of cerebellar cortex, as do those with the inferior olivary complex. Through these pathways the brain stem, cerebellar cortex, and cerebellar nuclei participate in all reactions of skeletal muscle, including tone, posture, involuntary movement, and volitional movements of the most refined variety (Fig. 9.158).

CEREBROCEREBELLAR RELATIONS

The many fibers that bring to the cerebellum neural activity that originates in the cerebral cortex run as corticopontine tracts to nuclei in the substance of the pons where they probably terminate somatotopically. Pontine neurons then send their fibers across the midline, through the middle peduncles, and into the cerebellar cortex as mossy fibers. Much of the cerebrocerebellar input comes from the precentral motor cortex, but there are also connections from the postcentral gyrus and from the auditory and visual cortices. In participating in the control of "voluntary" movement, therefore, the cerebellum receives information about activity in progress in all other major parts of the brain, including cerebral cortex and basal ganglia.

Similarly, the cerebellum relays information about its own activity back to those regions from which it receives input. The major pathway to the cerebrum is via the brachium conjunctivum, the red nucleus of the opposite side, anterior and lateral ventral nucleus (VA and VL) of the thalamus, and thalamocortical projections. These reciprocal connections provide a basis for control circuits; they are not, however, closed circuits exclusively, because the cerebellum has reciprocal relationships also with sensory and motor mechanisms of brain stem and spinal cord. These latter include spinal and vestibular projections to the cerebellum, and the fibers making linkage with reticular formation, inferior olivary complex and the red nucleus (Fig. 9.159).

It will be noted that the ascending fibers connect a given cerebellar hemisphere contralaterally with the red nucleus, thalamus, and cerebral cortex. But, as a result of the crossing of rubrospinal, corticospinal and corticobulbar tracts, the influence of the cerebellum on motor mechanisms tends to be ipsilateral.

External Landmarks

The cerebellum ("small brain") is the dorsal portion of the metencephalon. Lying in the posterior fossa of the cranium, it is connected with the tegmentum of the mesencephalon, metencephalon, and medulla oblongata by the superior, middle, and inferior peduncles, respectively (Fig. 9.160). The cerebellar surface has a characteristic patterning of parallel and curved furrows that separate the cortex into numerous laminae or *folia* ("leaves"). They are said to have a total area of about 100,000 mm^2, which is less than half the area of the cerebral cortex. Each portion of the cerebellum has an anatomical name (Fig. 9.161), only a few of which will be noted here.

The narrow, central part of the cortex is called the *vermis* ("worm"); the two larger, lateral masses are the right and left *hemispheres*. Demarcation between hemispheres and vermis is more definite

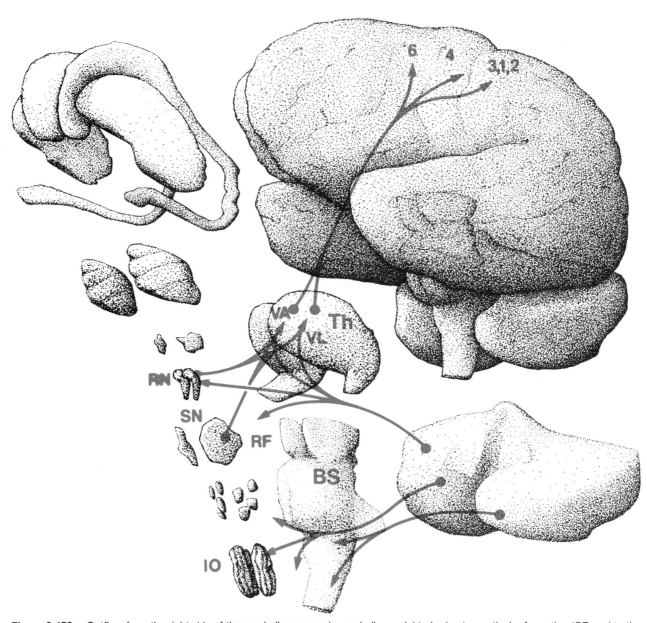

Figure 9.158. Outflow from the right side of the cerebellum goes via cerebellar nuclei to brain stem reticular formation (*RF*) and to the contralateral (*left*) red nucleus which relays to *VA* and *VL* of the thalamus (*Th*). These latter nuclei (which also receive information from the basal ganglia) relay combined integrated messages to sensorimotor cortex (areas *6,4,3,1,* and *2*). The flocculonodular lobe projects ipsilaterally to inferior olive (IO), reticular formation (RF), and vestibular nuclei in the brain stem (*BS*). [Drawing from D. Livingston (1983).]

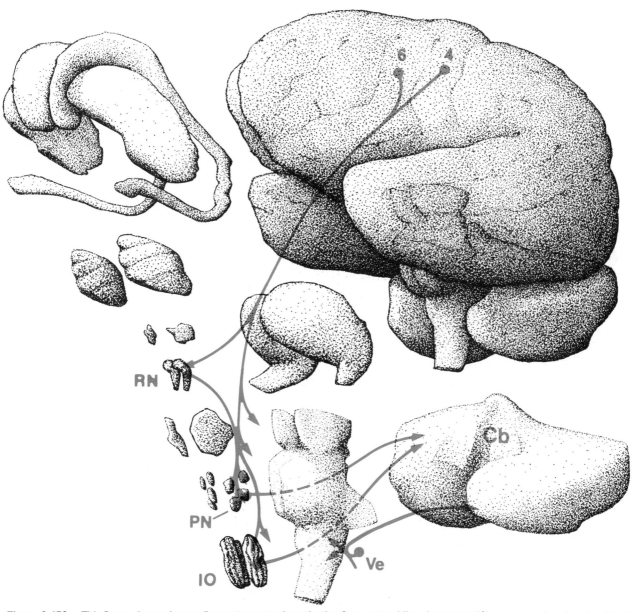

Figure 9.159.   This figure shows descending motor controls projecting from motor (*4*) and premotor (*6*) areas to red nucleus in midbrain and to brain stem reticular formation (*arrows* projecting to nonnucleated space). Red nucleus relays to pontine nuclei and inferior olivary nucleus. Pontine nuclei give rise to *crossed* mossy fibers which innervate the entire cerebellar cortex. Inferior olivary nucleus axons also *cross* and project to all cerebellar cortex as climbing fibers. First order vestibular afferents project to vestibular nuclei in brain stem and also directly to the posterior flocculonodular lobe of the cerebellum ipsilaterally. [Drawing from D. Livingston) 1983).]

on the inferior than on the superior surface of the cerebellum. The several lobes and principal fissures of the cortex are identified as follows, in a rostro-caudal direction:

> Anterior lobe (lingula, etc.)
> > *Fissura prima*
> Middle lobe
> > *Sulcus prepyramidalis*
> Posterior lobe (pyramis, uvula of the vermis)
> > *Fissura posterolateralis*
> Flocculonodular lobe

The anterior lobe is known as the *paleocerebel-*

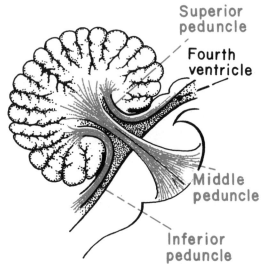

Figure 9.160. Relation of cerebellum to the brain stem and to its three peduncles.

lum, the middle and posterior lobes are the *neo-cerebellum,* and the flocculonodular lobe is the *archicerebellum.* Taken together the anterior, middle, and posterior lobes (but not the flocculonodular lobe) are known as the *corpus cerebelli.* The hemispheric portions of the middle and posterior lobes are called, together, the *ansiform lobe;* its relations are with the cerebropontile-cerebellar systems.

Neurons of the Cerebellar Cortex

Unlike the cerebral cortex, *the cerebellar cortex is histologically uniform.* Large Purkinje neurons dominate the middle layer of cerebellar cortex; the somata appear like onion bulbs arranged in a sheet that undulates up and down with the folia. Purkinje axons provide the only output of cerebellar cortex, and their signals are exclusively inhibitory. Rising into the upper layer of cerebellar cortex are the richly branching, fan-like dendrites of Purkinje cells. These dendrites are strictly confined to a flat plane perpendicular to the long axis of the folia (Fig. 9.162).

Nestled in the spaces between these planar Purkinje dendrites are the cell bodies and dendrites of *three distinctive interneurons, Golgi, stellate, and basket cells* which operate in that order, on the input, dendritic integration, and output of Purkinje cells. A fourth type of cerebellar interneuron, *the granule cell, lies in the deepest layer of cerebellar cortex.* Its axon rises through the layer of Purkinje cells. In the outer layer, granule cell axons bifurcate, sending straight branches in each direction along the long axis of the folia, at right angles and pene-

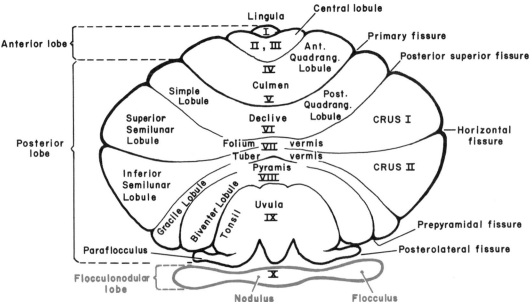

Figure 9.161. Subdivisions of the cerebellar cortex represented on an outline drawing. [From Carpenter (1976).]

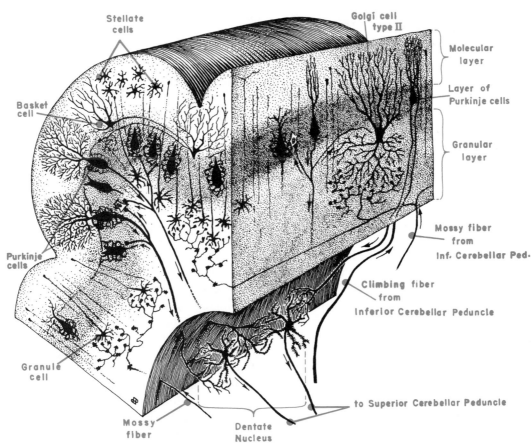

Figure 9.162. Cerebellar cortex drawn in three dimensions to show geometry of location of Purkinje cells. Golgi cells, granule, basket and stellate cells, with terminals of climbing and mossy fibers.

trating through the planar distributions of Purkinje dendrites. These *parallel fibers, arising from the granule cells, innervate the Purkinje dendrites and those of the other three interneurons, Golgi, stellate, and basket cells.*

CEREBELLAR INPUT

There are three types of fibers which enter cerebellar cortex. Each of these cerebellar inputs contributes to cerebellar nuclei as well as to cerebellar cortex. Each climbing fiber attaches itself closely and follows branching of dendrites of a few Purkinje cells. These fibers come from neurons in the *contralateral inferior olivary nucleus* which dominates the cephalic part of the medulla bordering the pyramidal tract. The more numerous mossy fibers, which come from pontine, reticular, vestibular, and spinal neurons, branch repeatedly in the white matter and branch further as they enter cerebellar cortex where they terminate on granule cells. The connections of mossy fibers to granule cells are organized in what is called a *glomerulus, a circumscribed specialized local circuit.* This involves a bul-

bous mossy terminal which has numerous synaptic connections with granule cell dendrites. Numerous axonal terminals of Golgi cells also present junctions for neurotransmitter release to granule cell dendrites. Evidently, the cerebellar cortex can exert a powerful feedback control over its own input. This Golgi axonal input is inhibitory and puts a check on granule dendritic activation whereby granule cell excitation of cerebellar cortex is sharpened. The granule cell dendrites, excited in one or several glomeruli, can initiate impulses in the granule cell axon which provides the long parallel fibers that run through the tracery of Purkinje, Golgi, stellate, and basket cell dendrites (Fig. 9.162).

Thus, *whereas climbing fibers provide powerful excitatory impulses to a few Purkinje cells, mossy fibers influence an immense number of granule cells which gain synaptic contact with a few hundred thousand Purkinje cells and cerebellar interneurons.* The Golgi interneurons, excited by parallel fibers, introduce a feedback inhibition of glomerular influence on granule cells, hence on parallel fibers. Golgis are therefore exerting control on cerebellar cortical inputs. Stellate cells, also excited by parallel fibers, have an inhibitory influence on Purkinje dendrites. They are thereby introducing direct inhibition. Basket cells, also activated by parallel fibers, send their axons to the bases of the Purkinje cell bodies and the points of origin of Purkinje axons where the basket cells exert a powerful inhibitory influence. The Purkinje cell is receiving direct (climbing fiber) and indirect (mossy-granule-parallel fiber) excitations, an indirect inhibition of granule input, and two direct inhibitory inputs—on dendrites (by stellates) and on the axon hillock (by basket cells).

Purkinje axons direct their inhibitory output onto cerebellar nuclei, and, from the vestibular (flocculonodular) portion of the cerebellum, direct inhibitory influence on neurons in the vestibular nuclei.

The picture is one of powerful Purkinje inhibition acting on spontaneously active cerebellar and vestibular nuclei, and inhibition of these Purkinje inhibitory effects by the contribution of interneurons in cerebellar cortex.

A third type of input to cerebellar cortex arises from *aminergic neurons*: serotoninergic neurons in the *raphé nuclei* of the brain stem, and norepinephrinergic neurons in the *locus coeruleus*. The terminals are partially climbing fibers but extremely widespread. These aminergic inputs are excitatory and very long-lasting in their effects, seconds as compared with millisecond influences on the part of climbing fibers from the inferior olivary nucleus. The aminergic neurons are modest in number, about 10,000 in the locus coeruleus in humans, but the axons are distributed extremely widely, throughout the cerebral and cerebellar cortex and subcortical nuclei (Bloom et al., 1970, 1971); this means that these aminergic neurons ramify extraordinarily and make generalized state-related functional contributions.

"LIMBIC CONTROLS"

For a long time, the double rings of *archicortex* and *mesocortex* which surround the hylus of the hemisphere on each side, representing the medialmost and adjacent circular bands of cortex, were thought to be olfactory in function. Thus, they were called the *rhinencephalon* (rhino = nose). This idea arose from the fact that the olfactory tract leads in to this cortex on the orbital surface, basal forebrain, and medial and anterior temporal cortex, all part of this double cortical ring. Papez (1937) made an extensive review of the comparative and functional anatomy of this cortex and associated subcortical structures and proposed that their function, beside

olfaction, included an array of sensory and motor activities which subserve emotion. MacLean (1949) linked this set of ideas with clinical evidence relating to visceral syndromes and psychosomatic diseases. MacLean and others have extended anatomical, physiological, biochemical, and evolutionary evidence relating to this brain region, calling it the *"limbic system"* (limbus = border) because it constitutes the medial border of the hemisphere, cortical and subcortical (see Livingston and Hornykiewicz, 1978).

Structures of the limbic system include: amygdala, hippocampus, fornix, septum, mamillary bodies, anterior nucleus of the thalamus (a relay association nucleus), and a ring of cortex which includes the cingulate gyrus, retrosplenial cortex, parahippocampal cortex, insula, orbital surface of the frontal lobe, and subcallosal cortex which joins up with the cingulate cortex. This forms a nearly complete loop, broken only between temporal insular and orbital cortex which are joined by a stout band of corticocortical connections, the uncinate fasciculus. Closely neighboring substantia innominata, part of the basal ganglia, and accumbens, part of the basal septal nuclei, link the limbic system intimately with the basal ganglia. This is understandable, inasmuch as *the limbic system is functionally related to emotional experience and expression* and *the basal ganglia to posture, attitude, and motor expressive behavior*. It is suggestive that they link up in the frontobasal region involving amygdala, hypothalamus, and olfactory representation (Fig. 9.163).

Klüver-Bucy Syndrome

Bilateral destruction of the anterior two-thirds of the temporal lobe is followed, in monkeys and humans, with a syndrome identified with Klüver and Bucy (1937). This comprises symptoms of: *visual agnosia*, loss of ability to recognize and relate significance of objects on the basis of visual criteria alone, and also a tendency to pay attention to every visual stimulus; *docility*, reduction of affectional and fear responses; *hyperorality*, putting the mouth to objects and putting objects into the mouth, eating inedible objects such as lighted matches, nuts, and bolts; *hypersexuality*, markedly increased and diversified solitary, paired, and group sexual activities; and inability to remember recent experiences, although past memories may be retained essentially intact. More about the limbic system and associated syndromes is presented in later chapters.

LEVELS OF CONTROL

A typical segment of the spinal cord seems to be relatively simple in its organization (Fig. 9.164). It

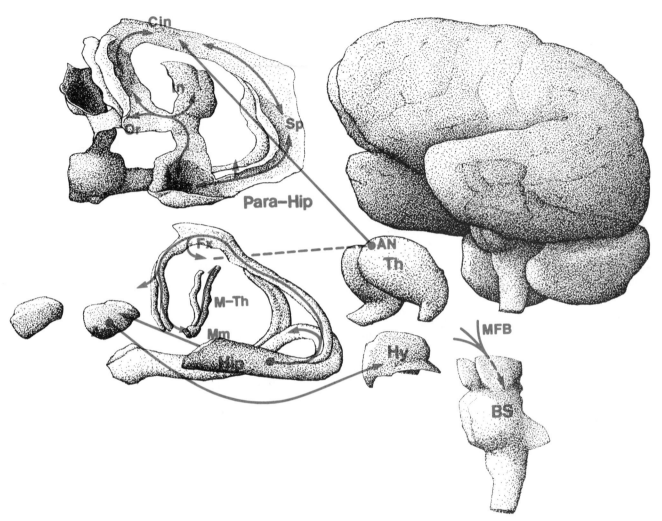

Figure 9.163. Organization of limbic system includes a ring of phylogenetically old cortex that lies on the medial and inferior surfaces of each hemisphere. This ring is richly interconnected by corticocortical projections going in both directions around the ring: cingulate gyrus (*Cin*), orbital surface (*Or*), insula (*In*), uncus (*Un*), parahippocampal cortex (*Para-Hip*), and splenium (*Sp*). *Vertical arrow* connects parahippocampal gyrus with hippocampus (*Hip*). The two arms of the horseshoe-shaped hippocampus (*Hip*) depicted below are interconnected in both directions by the hippocampal commissure. The fornix (*Fx*) arises from *Hip* on each side, arches high over thalamus (*Th*), and descends nearly vertically anteriorly, with *arrow* projecting forward into septum (not depicted), and continuing to enter hypothalamus. The mamillary bodies (*Mm*) in posterior hypothalamus project upwards via mamillothalamic tract (*M-Th*) to anterior nucleus of thalamus (*AN*). Hippocampus also project to amygdala (*Am*) which in turn projects to hypothalamus (*Hy*). The limbic system provides a circuit of cortical and subcortical structures which have mainly to do with emotional experience and expression. Combined with outflow from the frontal lobes having to do with behavioral planning, the limbic system outflow goes mainly by medial forebrain bundle (*MFB*) to midbrain central gray and adjacent brain stem reticular formation (*BS*) which is thereby known as the "frontal-limbic midbrain area." [Drawing from D. Livingston (1983).]

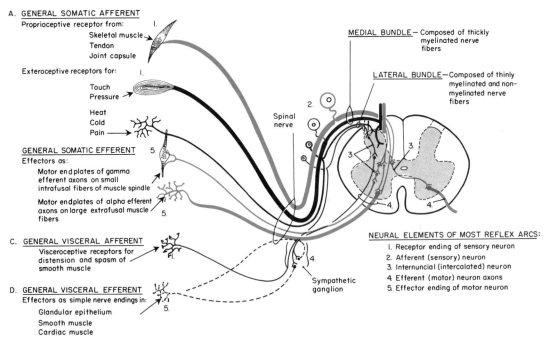

A. GENERAL SOMATIC AFFERENT
Proprioceptive receptor from:
 Skeletal muscle
 Tendon
 Joint capsule

Exteroceptive receptors for:
 Touch
 Pressure
 Heat
 Cold
 Pain

GENERAL SOMATIC EFFERENT
Effectors as:
 Motor endplates of gamma efferent axons on small intrafusal fibers of muscle spindle
 Motor endplates of alpha efferent axons on large extrafusal muscle fibers

C. GENERAL VISCERAL AFFERENT
Visceroceptive receptors for distension and spasm of smooth muscle

D. GENERAL VISCERAL EFFERENT
Effectors as simple nerve endings in:
 Glandular epithelium
 Smooth muscle
 Cardiac muscle

Spinal nerve

Sympathetic ganglion

MEDIAL BUNDLE— Composed of thickly myelinated nerve fibers

LATERAL BUNDLE— Composed of thinly myelinated and non-myelinated nerve fibers

NEURAL ELEMENTS OF MOST REFLEX ARCS:
1. Receptor ending of sensory neuron
2. Afferent (sensory) neuron
3. Internuncial (intercalated) neuron
4. Efferent (motor) neuron axons
5. Effector ending of motor neuron

Figure 9.164. Diagram of functional components of a thoracic spinal nerve showing distribution of dorsal root fibers as they enter the spinal cord. Skeletal muscle afferent and efferent fibers are shown in color. Numbers identify elements as listed at *lower right*. [From Carpenter (1976).]

receives fibers of sensory nerves through its posterior roots and gives off motor fibers to muscles (and glands) in its anterior roots. The central portion of the cord, the gray substance, contains terminations of the afferent fibers and the cell bodies of motor cells. It also contains internuncial neurons, including some that send their processes up or down the cord to other segments. Around the central gray there are longitudinally coursing bundles of fibers, many of them with myelin sheaths, that connect any given segment with others above and below, and with the higher centers of the neural axis. More on spinal controls is given in Chapter 70.

The brain stem contains the cranial nerve nuclei, sensory and motor. It also acts as the conduit and to some degree a controlling influence affecting all ascending and descending channels of communication between body and forebrain. The only exceptions are pathways for *neuroendocrine control* which are diencephalic (infundibular stalk and pituitary gland), the *visual pathway* which embraces both brain stem (superior colliculus) and diencephalon (thalamus and hypothalamus), and the *olfactory system* which is located so far anterior that it engages the telencephalon frontally and the diencephalon only secondarily.

The *cerebellum* stands astride the brain stem on three stout peduncular attachments on each side (Fig. 9.160). It provides a superordinate system of controls affecting the controls of the brain stem

itself. *The cerebellum receives from and projects mainly to brain stem nuclei. In addition it receives directly from the vestibular apparatus and from dorsal and lateral spinocerebellar tracts.* It projects to the thalamus, with some red nucleus relays in the brain stem. The cerebellum can be thought of (in a crude approximation) as a *servomechanism which accelerates and brakes cortical controls aimed at skeletal motor performance.* The cerebellum helps to accelerate and to brake smooth movements of body and limb musculature. With its intimate vestibular relations, the cerebellum contributes especially to muscular control of the body in three-dimensional space.

Somatotopic Organization

One of the most important features of organization of the central nervous system is called *somatotopy.* This term identifies the discovery that in sensory systems, for example, a virtual map of the region where the receptors are located is reproduced in the part of the brain where sensory channels have their terminal destination. As the sensory fibers leave a region such as the skin, the retina, or cochlea, they have certain positional relationships. These relationships are preserved through all stages of their ascent through the central nervous system and ultimately are displayed in the pattern of termination in the respective sensory cortex. Similarly, the motor cortex is topographically related to muscles of the body, and this relationship is reproduced also in motor connections in basal ganglia, cerebellum, and motor nuclei of brain stem and spinal cord.

Segmental Distribution of the Spinal Nerves

In the young mammalian embryo and in certain adult lower forms, e.g., fishes, the body is demarcated into a regular series of transverse segments or *metameres*. The muscles (*myotomes*), skin (*dermatomes*), and viscera of each of these embryonic blocks receive innervation from the nerve roots of a corresponding spinal segment. The anterior root of each spinal nerve supplies motor fibers (somatic efferent) to the respective myotome and autonomic fibers (visceral efferent) to the viscera and skin; the posterior root supplies sensory fibers (somatic and visceral afferent) to the corresponding dermatome as well as to the muscles and viscera (Fig. 9.165).

In mammals, as a result of the outgrowth of the embryonic limbs, the orderly arrangement of the metameres from before backwards becomes altered. In the body of the adult, the primitive metameric disposition is observed only in the trunk (Fig. 9.165). Although an apparent segmentation is present in all of the spinal nerves because they exit between pairs of vertebrae, this segmentation is not preserved in peripheral nerves because the spinal nerves supplying an extremity first combine to form the brachial or lumbosacral plexus where fibers intermingle before becoming reconstituted as the several nerves to arm or leg. A given nerve, therefore, may contain fibers derived from two or more spinal segments, while fibers from a given segment may pass to the periphery via several different nerves. One result of this redistribution is that muscles supplied by a given spinal segment do not necessarily lie adjacent to one another. For example, the coracobrachialis is innervated by the same spinal segments as those which supply the muscles of the thumb. Another result is that a single muscle may derive its nerve supply from more than one spinal segment. Finally, because the skin and muscles of a limb tend also to move away from the viscera with which they were originally associated, in the adult it often turns out that structures innervated by a common spinal segment may be widely separated. Thus, the diaphragm is innervated by the phrenic nerve from the third, fourth, and fifth cervical segments, which also supply skin and muscle in the region of neck and shoulder. Similarly, the heart receives sensory and autonomic fibers from the upper thoracic segments, which also supply sensory fibers to the skin over the inner

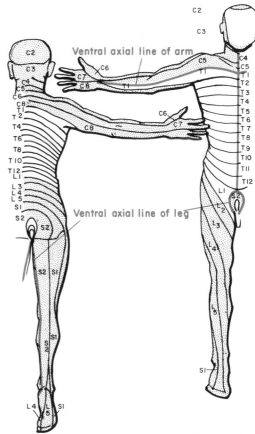

Figure 9.165. Levels of the spinal cord from which various areas of the surface of the body receive sensory innervation. Figure represents the data of Keegan and Garrett (1948).

aspect of the arm and hand and the upper part of the thorax. This topic has been considered in its relation to the phenomenon known as "referred pain."

Cranial Nerves

Being distributed at characteristic levels along the brain stem from rostral mesencephalon to caudal medulla, the sensory and motor components of the cranial nerves are convenient landmarks in studies of segmental localization. In clinical neurology the fact that a given nerve's function is or is not impaired may be the datum that permits localization of a lesion. Consequently, a working knowledge of the principal components of each nerve, their peripheral distribution, and their central connections is a worthwhile objective for any physician. The nerves are numbered from above down and will be reviewed briefly in that order, even though this is contrary to the outline of other material in this chapter (see Figs. 9.166 and 9.167). The nuclei of origin of motor components of cranial nerves are classified as follows:

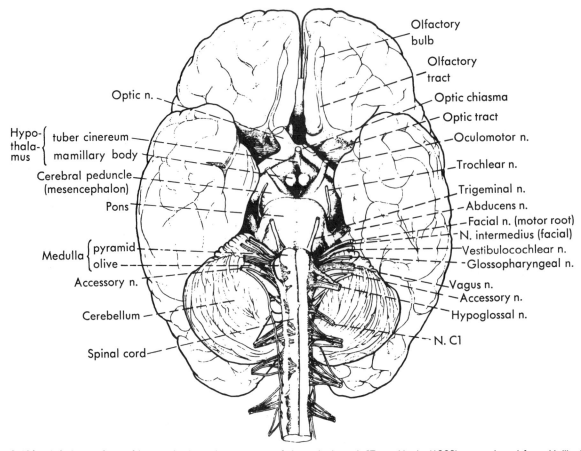

Figure 9.166. Inferior surface of human brain and upper part of the spinal cord. [From Henle (1968); reproduced from Hollinshead (1967).]

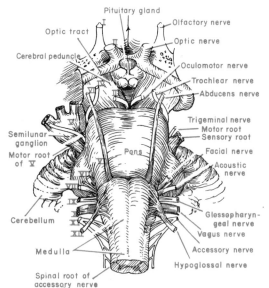

Figure 9.167. Inferior view of brain stem showing levels of attachment of cranial nerves. [From Chusid (1973).]

1. *Somatic efferent*—to striated muscles of orbit and tongue, via nerves III, IV, VI, and XII.
2. *Special visceral efferent*—to striated muscles derived from branchial arches, innervated by fibers in nerves V, VII, IX, and X (and XI).
3. *General visceral efferent*—to smooth muscle and glands, via nerves III, VII, IX, and X, the preganglionic neurons of the craniosacral autonomic system.

The nuclei of termination of afferent fibers—the nuclei that contain the cell bodies of second order sensory neurons—have the following classification:

4. *General somatic afferent*—sensory nucleus and nucleus of the spinal tract of nerve V, with a small contribution from nerve X, subserving proprioception, touch, temperature sense, and pain.
5. *Special somatic afferent*—vestibular and cochlear nuclei of nerve VIII.
6. *Visceral afferent*—site of terminations of taste and other visceral afferent fibers brought via nerves VII, IX, and X.

The general location of all of these nuclei is represented in relation to surface anatomy of the brain stem in Figure 9.168.

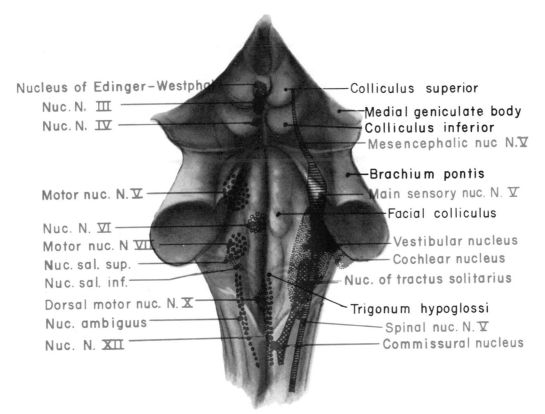

Figure 9.168. Dorsal view of human brain stem, with positions of cranial nerve nuclei projected upon the surface. Motor nuclei are on the *left*, sensory nuclei on the *right*. [From Herrick; from Ransom and Clark (1959).]

OLFACTORY AND OPTIC NERVES (I AND II)

These are not nerves like the other 10, in that they develop differently and have different types of connections. The afferent fibers they contain do not arise from cells located in ganglia outside the central nervous system. Rather, these two nerves should be regarded as fiber tracts of the brain. Their function is considered in Chapters 64 and 65. The olfactory nerve seems to contain only afferent fibers, whereas the optic nerve includes also fibers passing from the central nervous system to the retina.

OCULOMOTOR, TROCHLEAR, AND ABDUCENS NERVES (III, IV, AND VI)

The oculomotor nerve innervates all extrinsic muscles of the eye except the superior oblique muscle (innervated by trochlear nerve) and lateral rectus (by abducens nerve). Autonomic fibers to intrinsic muscles of the eye also run with the oculomotor nerve from their origin in the Edinger-Westphal nucleus in the mesencephalon—the most rostral of the four general visceral motor nuclei. Afferent fibers from extrinsic muscles of the eye enter the brain stem by way of the trigeminal nerve (V).

The oculomotor nerves originate in the mesen-

cephalon and leave the brain stem inferiorly and medially on each side between the midline and the medial border of the cerebral peduncle. Fibers of the trochlear nerve run dorsally around the central gray substance of the mesencephalon, cross in the region superior to and between the inferior colliculi, leave the brain stem, and then course around it en route to the orbit. By its crossing this nerve becomes unique among nerves supplying skeletal muscles. The abducens nerve makes its exit from the brain stem at the caudal border of the pons not far from the midline.

TRIGEMINAL PATHWAYS

The trigeminal nerve appears on the lateral surface of the pons. The nerve has three roots: (1) a large *sensory*, (2) a *mesencephalic* that is also sensory, and (3) a small *motor* root (Fig. 9.169).

The *main sensory nucleus* of the trigeminal corresponds to the nuclei of the posterior columns of the spinal cord (nuclei gracilis and cuneatus), and like them sends fibers to the cerebellum through the inferior cerebellar peduncle. The *spinal tract* of the trigeminal nerve corresponds to the tract of Lissauer, and the *spinal nucleus* is homologous with the substantia gelatinosa. A number of reflexes are mediated through the afferent fibers of the trigeminal nerve, e.g., corneal, blinking, sneezing, and the oculocardiac reflexes.

The *mesencephalic root* consists of a small bundle of afferent fibers which run in company with fibers of the motor root. Entering the pons they ascend to the *mesencephalic nucleus*, which is made up of

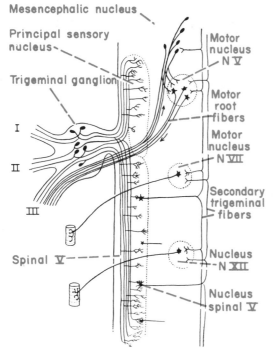

Figure 9.169. Diagram of peripheral and central distribution of sensory and motor fibers in trigeminal nerve. *I*, ophthalmic; *II*, maxillary; and *III*, mandibular divisions. [From Carpenter (1967).]

the first order cell bodies of this particular pathway. These fibers transmit proprioceptive impulses from the muscles of mastication, and extraocular muscles.

The *motor root* arises from a nucleus in the upper part of the pons underlying the lateral part of the floor of the fourth ventricle. The nucleus receives fibers from the corticobulbar tract of the opposite side, and also probably from that of the same side. The motor root after its emergence from the brain stem travels peripherally with the sensory root and joins the mandibular (third) division of the trigeminal nerve to supply the muscles of mastication, including the temporal, masseter, and pterygoid muscles.

FACIAL NERVE

This nerve consists of a large *motor* and a small *sensory* portion that appears at the lower border of the pons and enters the internal auditory meatus in company with the auditory nerve (Fig. 9.170).

The *sensory root* of the facial, which is also known as the *nervus intermedius of Wrisberg*, contains not only afferent fibers, but also secretory and vasodilator parasympathetic fibers as well. The sensory fibers arise from cells in the geniculate ganglion; the peripheral processes of most of these cells are distributed through the *chorda tympani* branch of the facial nerve to taste buds and the mucous membrane of the anterior two-thirds of the tongue (Fig. 9.170). A few additional sensory fibers pass by way

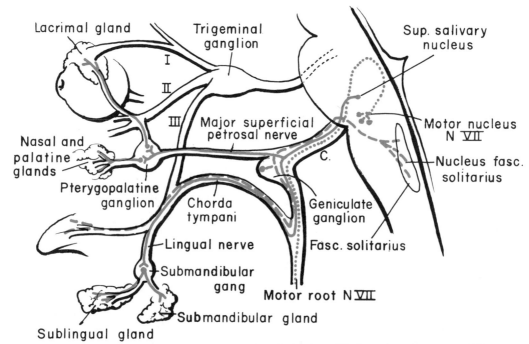

Figure 9.170. Components and distribution of the facial nerve. [Redrawn from Carpenter (1976).]

of other nerves to the lacrimal gland and to the mucosa of the soft palate and the posterior part of the nasal cavity. All of these sensory neurons send their central processes into the brain stem to end in the upper part of the *nucleus of the tractus solitarius*. Second order fibers arise in this nucleus and ascend in the trigeminal lemniscus of the opposite side to reach the thalamus.

The *parasympathetic* or general visceral motor fibers arise from the *superior salivatory nucleus* (which lies in close relation to the motor nucleus of the facial nerve) and run to the *sphenopalatine ganglion*; here the postganglionic fibers arise to pass to the lacrymal gland and to vessels and glands of the palate and posterior part of the nasal cavity.

The *motor* (special visceral) components of the facial nerve are distributed to muscles of the face, auricle, and forehead. These fibers arise from a nucleus in the lower part of the pontile tegmentum and pass backwards to the lower end of the nucleus of the abducens nerve. Then ascending to the upper end of the latter they bend and sweep forward and then anteriorly to their exit from the brain. As the facial fibers take this arched course they form, together with the abducens nucleus, a prominence in the floor of the fourth ventricle known as the *facial colliculus*.

VESTIBULOCOCHLEAR NERVE

The VIIIth cranial nerve is a true nerve, different from the olfactory and optic nerves, even though in function the three nerves seem comparable because they innervate organs of special sensation. One of the major parts of this nerve contains sensory fibers from neurons located in the vestibular ganglion, with peripheral processes innervating semicircular canals, utricle, and saccule. The other major part contains sensory fibers from cells of the spiral ganglion of the cochlea; their short, peripheral processes supply the hair cells of the organ of Corti. A third part of the nerve consists of efferent fibers to the organ of Corti, where they apparently modulate the sensitivity of the hair cells (Galambos, 1956). This group of fibers is known as the *efferent cochlear bundle*. Efferents also innervate vestibular hair cells.

More complete information about the functions of this nerve is given in Chapters 62 and 63.

GLOSSOPHARYNGEAL NERVE

The glossopharyngeal nerve contains motor, secretory, vasodilator, and sensory fibers. The nerve emerges from the side of the upper part of the medulla oblongata in the groove between the olive and the restiform body. Its special visceral motor fibers arise from the upper part of the *nucleus ambiguus* situated in the reticular formation of the medulla oblongata. They are distributed almost entirely to the stylopharyngeus muscle; a few terminate in the circular and longitudinal muscles of the upper part of the pharynx. Division of the glossopharyngeal nerve, however, as done for glossopharyngeal neuralgia, produces no apparent motor disability.

Secretory and vasodilator fibers are the axons of cells of the *inferior salivatory* nucleus, which lies below the superior nucleus of the same name. Via the tympanic branch of the nerve they innervate the parotid gland. The sensory fibers of this nerve arise from cells in the *superior* (jugular) and *inferior* (petrous) ganglia. Peripheral processes from cells of the superior ganglion mediate ordinary sensations (touch, heat, and cold, etc.) from the posterior third of the tongue and the mucosa of the pharynx and posterior part of the mouth. Their central processes terminate in the *dorsal sensory nucleus* of the *vagus nerve*. Fibers from cells in the inferior ganglion supply taste buds of the posterior third of the tongue and send their central processes to the lower part of the nucleus of the tractus solitarius, where second order sensory neurons give rise to fibers that are distributed as described earlier.

The *carotid sinus nerve* is another important afferent branch of the glossopharyngeal nerve. It conducts impulses from the carotid body and the carotid sinus to the medulla oblongata and participates in control of the circulation and of pulmonary ventilation.

VAGUS NERVE

The vagus nerve contains motor, secretory, vasodilator, and sensory fibers. The secretory and vasodilator fibers and the fibers to the smooth muscle of the bronchi, heart, esophagus, stomach, small intestine, gall-bladder, etc., are parasympathetic fibers; they arise from cells in the *dorsal motor nucleus of the vagus*, the principal autonomic motor mechanism in the brain stem. It belongs to the general visceral efferent group of nuclei and extends upwards from the lower, closed part of the medulla oblongata to beneath the floor of the 4th ventricle. The special visceral motor fibers arise in close relationship with the motor fibers of the glossopharyngeal nerve, namely, from the cells of the *nucleus ambiguus* lying below the glossopharyngeal neurons. They supply (through the superior laryngeal branch) the cricothyroid and arytenoid muscles of the larynx, and the inferior constrictor of the pharynx.

The pharyngeal and recurrent laryngeal branches of the vagus also convey motor fibers, many of which are derived from the bulbar nucleus of the accessory nerve (see below). The cell bodies of the *sensory fibers* lie in the inferior ganglion of the vagus (*ganglion nodosum*). The peripheral processes of these cells convey impulses from the lungs, heart, larynx, pharynx, esophagus, stomach, small intestine, and gallbladder. They also, through the anterior laryngeal branch, innervate the taste buds of the epiglottis and valleculae (the depressions lying at the sides of the fold running from the epiglottis to the base of the tongue). The taste fibers end centrally by synapsing with cells in the *gustatory nucleus* lying in the upper and medial part of the *nucleus of the tractus solitarius*. These impulses are relayed upwards along the same paths as those conveying other taste impulses. Other afferent vagal fibers from visceral structures terminate in the dorsal nucleus. This latter is, therefore, both motor and sensory in function and constitutes an important visceral reflex center.

Vagal afferent filaments travelling in Arnold's nerve mediate the general sensations of the skin lining the external auditory meatus and a small area behind the auricle. Irritation of these fibers may cause reflex coughing. These sensory fibers have their cell stations in the *jugular* or superior ganglion. Medullary reflexes involving the cardiovascular system with identification of central reflex representation are presented by Kostreva (1983).

SPINAL ACCESSORY NERVE

The spinal accessory nerve is entirely motor and is made up of a bulbar and a spinal root. The *bulbar* or special visceral motor root arises from the lower (caudal) end of the nucleus ambiguus from cells situated below those which give origin to the motor fibers of the vagus. The bulbar fibers join the vagus within and below the jugular foramen and are distributed, as already mentioned, in the pharyngeal and recurrent branches of the latter nerve; these fibers of the spinal accessory nerve innervate the muscles of the larynx (with the exception of the cricothyroid), the muscles of the pharynx (with the exception of stylopharyngeus), and those of the soft palate (with the exception of the tensor palati, which is supplied by the Vth cranial nerve). The *spinal* root is somatic motor and is composed of the axons of a group of cells in the anterior gray column of the spinal cord extending from the first to the fourth or fifth cervical segment inclusive. These fibers supply the sternocleidomastoid and trapezius muscles. The spinal part of the nerve exchanges fibers with the bulbar part in the jugular foramen.

HYPOGLOSSAL NERVE

The hypoglossal nerve is also said to be purely motor, although it may carry proprioceptive fibers from the tongue. Its fibers are derived from a somatic motor nucleus situated near the midline in the floor of the caudal part of the 4th ventricle and medial to the nucleus ambiguus. It supplies the thyrohyoid, styloglossus, hyoglossus, and genioglossus muscles, and the intrinsic muscles of the tongue.

Disabilities Following Lesions

SENSORY LOSS FOLLOWING LESIONS OF SPINAL CORD

In a *transverse spinal lesion* interrupting solely the posterior fasciculi, vibratory and position senses are lost, but the sensation of touch is retained, since the fibers which have entered the gray matter and crossed below the level of the lesion to ascend in the anterior spinothalamic tract have escaped injury. Likewise, although a *hemisection of the spinal cord* destroys vibratory and position senses on the same side and pain and temperature sensibilities on the opposite side, it does not abolish tactile sensibility, as this is conveyed in both crossed and uncrossed pathways (see Nordenbos and Wall, 1976).

A lesion in the *lower part of the brain stem* may involve one of the sensory pathways exclusive of the others. Thus, an injury localized to the outer part of the lower pons or of the medulla oblongata by injuring spinothalamic fibers may cause loss of sensation to pain, heat, and cold that is notable in that it involves the opposite side of the body, and leaves muscle sense and tactile discrimination intact because these fibers are protected from injury by their less superficial position. Sensory loss of this nature occurs as a result of occlusion of the posterior inferior cerebellar artery. The sensory changes are accompanied by signs of injury to motor neurons on the same side of the body as the lesion, as well as by cerebellar symptoms from ischemia of the nearby inferior cerebellar penduncle. In this syndrome it is usual to find a loss of pain and temperature sensibility of skin of the face on the same side as the lesion as a result of damage to first or second order neurons of the spinal nucleus of the trigeminal nerve. Tactile sensation from the face, however, may be preserved because it is transmitted via the main sensory nucleus that may be unaffected by the vascular occlusion.

A lesion more centrally placed may, by implicating the medial lemniscus alone, cause the converse type of dissociated sensory deficit, namely, loss of the sense of position of the limbs and of spatial discrimination, with retention of sensibility to pain, heat, and cold. In lesions at levels approaching the thalamus all forms of sensation are likely to be involved more or less equally.

LESIONS INVOLVING THE TRIGEMINAL PATHWAYS

Either the peripheral portions or the central connections of the nerve may be injured by a disease process. A lesion of the nerve peripheral to the ganglion is more likely to involve only one of its three divisions. Pain or loss of sensibility over the distribution of the division affected may result. If the gasserian ganglion is involved the disturbance of sensation usually involves the area of distribution of more than one division, often of the entire nerve. If as a result the eye is rendered anesthetic the patient may develop an ulceration of the cornea and an inflammation of the cornea and conjunctiva.

Herpetic eruptions may result from the involvement of the gasserian ganglion. Spontaneous herpes zoster usually involves the area of distribution of the ophthalmic division and is often serious because the lesions on the cornea may lead to corneal opacity and loss of vision. Such herpes is also commonly followed by a very distressing postherpetic neuralgia which is very difficult to treat satisfactorily.

The trigeminal nerve, or one of its divisions, is sometimes the seat of a severe and intractable type of pain which recurs in paroxysms (*trigeminal neuralgia; tic douloureux*). The cause of the affliction is unknown.

Paralysis of the muscles of mastication with reduction in the strength of the bite on the affected side and deviation of the mandible toward the paralyzed side when it is opened occurs when the motor division is involved.

Sensory Innervation

Details of sensory input to the nervous system, and of the functions of sensory receptors, are given in preceding chapters. In summary, afferent fibers to spinal cord and brain stem are classified according to the sensation they subserve and are both somatic and visceral in their origin (Fig. 9.164). In both their courses and their terminations the afferent fibers are somatotopically organized. When they reach the central nervous system they are distributed so as to initiate reflex actions at the same level. In the case of receptors from muscle spindles, the reflex connections are with motor cells innervating that part of a muscle from which the afferent messages originated (Cohen, 1953). Branches of afferent neurons also run up or down the cord to make connections in other segments, and certain of the processes run up to the medulla

oblongata before they terminate on neurons of the second order.

VISCERAL AFFERENT FIBERS

In the spinal cord the *general visceral* afferent fibers, small and myelinated, are believed to be distributed with other small fibers that subserve pain and temperature sense from somatic as well as from visceral structures. Their central courses are described briefly below. In the brain stem a similar distribution is exhibited by small fibers (presumably for light touch, pain, and temperature sense) that arrive in cranial nerve V. But the *special visceral* afferent fibers that innervate taste receptors and that run in nerves VII, IX, and X terminate in one of the landmarks of the brain stem, the *nucleus of the tractus solitarius*, where they make connection with second order neurons that ascend to the thalamus.

PROPRIOCEPTION AND TOUCH

Large myelinated fibers of the dorsal roots subserve the functions of position and movement sense from skeletal muscles, and of vibration, pressure, touch, and tactile and spatial localization. They enter the spinal cord as components of a medial division of dorsal root fibers (Fig. 9.164). Those from muscle and tendon receptors probably have local, segmental connections with both α- and γ-motor neurons of muscles from which the afferent fibers have arisen.

Afferent fibers that arise in muscle receptors and that terminate in the spinal cord also make connections with second order neurons that constitute the *posterior* and *anterior spinocerebellar tracts* (Fig. 9.171). The posterior tract is uncrossed; it arises from large multipolar neurons in the nucleus of Clarke (dorsal nucleus) of the thoracic or upper lumbar portion of the central gray substance. Connections between first and second order neurons are not made immediately upon entrance to the cord. Rather, the primary fibers may descend a few segments before entering the gray column for termination, or they may ascend similarly. Afferent fibers from sacral segments ascend to terminate in Clarke's nucleus of the lumbar cord. Those that enter via cervical nerves ascend in the fasciculus cuneatus to reach second order neurons in the accessory cuneate nucleus. The anterior spinocerebellar tract has a less well defined origin, but is made up largely of crossed fibers from cells in the posterior gray columns. Eventually both of these tracts reach the cerebellum. The posterior tract and fibers from the accessory cuneate nucleus arrive via the inferior cerebellar peduncle, those of the anterior tract, through the superior peduncle.

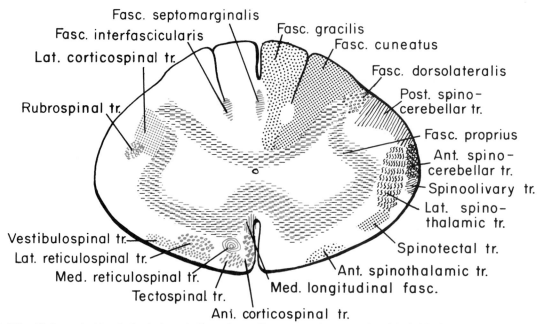

Figure 9.171. High cervical level of spinal cord. Ascending pathways are shown on the *right* in black, descending paths are on the *left* in color. Nuclear distributions are contained within the central open region, sensory dorsal to the central canal, motor ventral. Surrounding fasciculus proprius contains spinal interneuron fibers passing up and down the spinal cord. The dorsal columns (*Fasc. gracilis and cuneatus*) are stacked with lowermost entering fibers arranged most medially and upper limb fibers more laterally, as fibers enter successively higher dorsal roots. The main corticospinal projection (*Lat. corticospinal tr.*), which is the crossed pyramidal tract, has upper cervical fibers arranged most medially, so fibers can peal off and enter the spinal gray matter at successively lower segmental levels in an orderly fashion. [Redrawn from Carpenter (1967b).]

In addition to these intraspinal connections and pathways to cerebellum, second order neurons also relay input from muscle, tendon, joint, and touch receptors to the thalamus. These second order neurons, however, arise not in the spinal cord, but in the medulla oblongata. Long branches of primary sensory neurons ascend in the posterior white columns, the *fasciculi gracilis* and *cuneatus* (Fig. 9.171). The fibers are somatotopically arranged, and they are uncrossed. In the lower part of the medulla they terminate in the *nuclei gracilis* and *cuneatus* of the same side, where second order neurons arise and send their processes as internal arcuate fibers across the tegmentum. Having crossed the brain stem, these axons become the *medial lemniscus* that ascends through the lateral tegmentum to reach the thalamus.

LIGHT TOUCH, TEMPERATURE, AND PAIN

These sensory functions are subserved by smaller fibers. They enter the spinal cord with the lateral divisions of the dorsal roots, into the fasciculus dorsolateralis, and then terminate upon second order neurons in the substantia gelatinosa (Réthelyi, 1977). Processes of the latter cross the midline to become the spinothalamic tract serving light touch, pain, and temperature sense (Fig. 9.172). The cross-

ing in the anterior white commissure occurs at about the same level as that of the entering dorsal root fibers. The crossing, however, is not accomplished all at once. Consequently, section of the lateral spinothalamic tract for relief of intractable pain leaves a patient with a complete anesthesia on the opposite side of the body beginning about one segment below the level of the operation. The second order fibers ascend to the upper midbrain, where those of the spinothalamic tract become associated with the medial lemniscus—the second order neurons of posterior column nuclei previously mentioned.

Control from Higher Levels

Higher levels of the nervous system control motor elements of brain stem and spinal cord by way of descending tracts having the following characteristics: (1) They are somatotopically organized. (2) In mammals the main source of higher control is corticobulbar and corticospinal fibers that run without synapse between sensorimotor cortex and the level of motor neurons, where they end either on these latter cells or on interneurons. The most important descending tracts are shown on Figure 9.171.

In spite of the rich innervation of the cerebellum

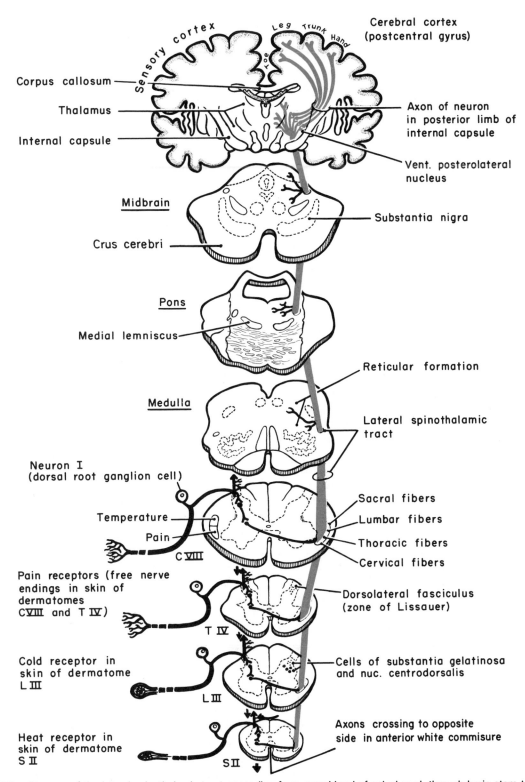

Figure 9.172. Diagram of the lateral spinothalamic tract ascending from sacral level of spinal cord, through brain stem to thalamus, and thence to postcentral gyrus of sensory cortex. [From Carpenter (1967).]

from spinal mechanisms, transmitted both directly and indirectly via the inferior olive, there is no direct communication between cerebellum and the motor nuclei of cranial or spinal nerves. Moreover,

the basal ganglia are not connected directly with the spinal cord motor systems, nor do these ganglia receive direct sensory input.

The most direct corticospinal motor system is

known as the "pyramidal" tract. It contains many large axons originating from large pyramid-shaped neurons of Betz in the motor cortex. The Betz cell axons are joined by other long axons which arise from motor cortex and from other frontal and parietal regions of the hemisphere.

In addition to this direct, corticospinal connection, there are other pathways by which the cortex may control motor neurons. These other pathways are multisynaptic in character, may be regarded as extrapyramidal or nonpyramidal, and in general are composed of connections between cortex and structures such as the red nucleus, reticular formation, and perhaps the inferior olive, all of which receive fibers from the cortex and (after synapse) project through recognized tracts to brain stem and spinal cord. In making a distinction between these subsystems one should not imply that they are independent of one another. Rather, the multisynaptic, indirect, and phylogenetically older connections provide the background or the foundation patterns of muscle contraction upon which the newer and more specific activity of the direct corticospinal system is superimposed. The indirect paths are concerned with posture, with automatic types of movement, and with unconscious adjustments of muscle tone and contraction that accompany movement. Moreover, they are believed to be involved in recovery of function that takes place following interruption of the direct corticospinal pathway.

CORTICOSPINAL AND CORTICOBULBAR PATHWAYS

For many years it was erroneously assumed that all of the fibers in the medullary pyramids arose from the giant pyramidal neurons of Betz in the precentral gyrus, and that they were responsible for all voluntary activity of skeletal muscles. In fact, only a small percentage of the corticospinal fibers arise from these particular cells, approximately 3–4% (Lassek, 1940). The remainder of the corticospinal fibers arise from other cells in the precentral gyrus, from neurons in the parietal cortex, and from other cortical areas. There is a question, too, as to whether all of the fibers in the medullary pyramids are of cortical origin, and whether all of them are descending fibers. Moreover, voluntary muscular activity is not achieved exclusively by corticospinal fibers nor by fibers contained in the medullary pyramids. When these fibers have been sectioned the multisynaptic or indirect corticospinal pathways are capable of producing voluntary movement, at least under some circumstances (Tower, 1940).

From their origins in the cerebral cortex, the corticospinal and corticobulbar tracts descend through the white matter of the hemisphere, aggregate in the internal capsule on each side, enter the peduncles in the midbrain, are dispersed in the pons by pontine nuclei, and form the dense pyramids in the cephalic end of the medulla or bulb (Fig. 9.173).

When the pyramidal tract passes through the pyramids and reaches the lower border of the medulla, each tract divides into two bundles of unequal size, the larger of which crosses to the opposite side and descends in the posterior part of the lateral funiculus of the cord as the *lateral corticospinal tract*. The remaining fibers descend uncrossed in the anterior funiculus as the *anterior corticospinal tract*. In most instances the direct or anterior corticospinal tract is but a small part of the total number of corticospinal fibers (usually not over 10%), but the number of fibers in the anterior tract varies. These fibers cross to the opposite side before terminating on the ganglion cells of the anterior gray column. There are also direct corticospinal fibers lying in the lateral columns of the spinal cord which do not cross to the opposite side. Through these the precentral motor cortex sends impulses to ipsilateral skeletal muscles.

All corticospinal fibers whether crossed or uncrossed connect with the large motor cells of the ventral gray columns (anterior horn cells). The connections are of two types, (1) direct synapses (20–30%) with the motor neurons; (2) indirect connections with these cells through an internuncial neuron whose cell body is also in communication with posterior root fibers on each side, and with other internuncial, intraspinal neurons. It has been estimated that from 75 to 90% of the corticospinal fibers terminate in the cervical (55%) and thoracic (20%) regions of the spinal cord, and about 25% in the lumbar and sacral regions.

The corticobulbar component of these paths includes all processes that originate from cells in the cerebral cortex and that terminate in the brain stem, especially in the tegmentum of midbrain, pons, and medulla. Those that come from the inferior part of the precentral motor area bear the same relationship to motor nuclei of cranial nerves as do the longer corticospinal tracts to anterior horn motor neurons. Other corticobulbar fibers originate in premotor cortex and postcentral gyrus. Leaving the corticospinal pathways at various levels of the brain stem, they cross and run dorsad into the tegmentum. Many of them apparently end on cells of the reticular formation, which serve as internuncials in this type of motor control. Others innervate structures such as the red nucleus and the olivary complex. A third category projects to sensory relay nuclei, including nuclei gracilis and

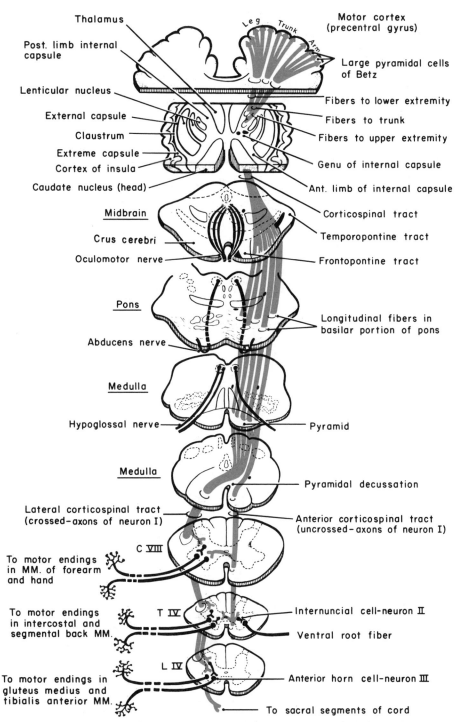

Figure 9.173. Diagram showing course of corticospinal fibers in lateral and anterior corticospinal tracts, beginning in pyramidal cells of cortex, running through internal capsule, basis pedunculi, pyramid, decussating and entering the spinal cord. Location of corticobulbar tracts is shown also (*right side*). [From Carpenter (1976).]

cuneatus, where they probably serve to modulate sensory afferent input from the periphery to the brain.

OTHER DESCENDING PATHWAYS

As the corticospinal and corticobulbar pathways are particularly, but not exclusively, concerned with

voluntary movement, so other descending pathways are particularly, but not exclusively, concerned with nonvoluntary movement. Certain of these other pathways modulate the tone of skeletal muscle through their influence on both α- and γ-motor neurons (e.g., the reticulospinal tracts). Others bring to spinal cord and brain stem patterns of activity, such as facilitation of extensor and anti-gravity contraction (vestibulospinal), or of flexor patterns (rubrospinal), or synchronization of movements of extrinsic eye muscles with those of head and neck (medial longitudinal fasciculus; tectospinal tracts). The general relationships of structures from which they originate are shown in Figure 9.173.

Reticular Formation

One of the main sites of higher control of brain stem tegmental and spinal motor mechanisms is the brain stem reticular formation. This has been appropriately defined as those neurons and processes "left over" after all well-defined, named nuclei and pathways have been accounted for. The retic-

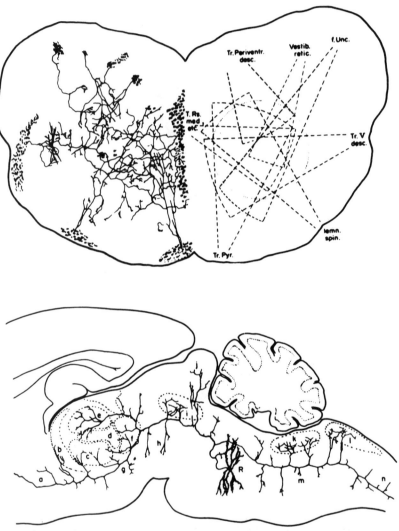

Figure 9.174. Illustrations of some of the morphological characteristics of diffusely projecting neurons in two different species. *Upper:* A transverse section through the upper third of the medulla oblongata of a 10-day-old kitten. The overlapping convergence of terminating and collateral fibers which enter the core of the reticular formation to form synapses are shown. *Left*, fibers traced from microscopic sections; *right*, some of the overlapping sectors which include, clockwise, fibers descending from the periventricular system (possibly relating to feeling states), projections from vestibular nuclei (relating to spatial orientation and movement), roof nuclei of the cerebellum (probably concerned with sensorimotor coordination), the descending root of the trigeminal (representing facial sensation), the spinothalamic (conveying touch, light-touch, and spinothalamic impulses for pain and temperature sensibility), the pyramidal tract (relating to cortical motor control), and medial reticular projections (diffusely projecting systems projecting onto diffusely projecting systems). *Lower:* A sagittal section through the brain of a young rat. The axonal trajectory of a single neuron with its cell body located in the nucleus reticularis gigantocellularis, *R*, is depicted. The rostrally coursing axonal component supplies collaterals to inferior colliculus, the region of the IIIrd and IVth cranial nerve nuclei, midbrain reticular formation, posterior nuclear groups in the thalamus (governing somatosensory relays), diffusely projecting nuclei of the thalamus, ventral thalamic nuclei (relaying basal ganglliar and cerebellar projections to motor cortex), hypothalamus, and basal forebrain region. The posteriorly directed axonal components send collaterals into the pontine and bulbar reticular formation, the hypoglossal nucleus (XIII), the dorsal column nuclear relay (n. gracilis), and the intermediate gray reticular formation of the spinal cord. [From Scheibel and Scheibel (1969).]

ular formation has developmental priority and provides the mother or *matrix* tissue, the groundwork of the nervous system. Named nuclear masses are formed by late arriving cells which climb through the reticular scaffolding to aggregate in more orderly, denser nuclear groups, often with well defined boundaries all of which makes them more suitable for naming.

Exemplary *reticular neurons* are shown in Figure 9.174. The upper figure illustrates how dendritic arborizations of a single neuron located to the left come into contact with collaterals of passing axons from a wide variety of ascending and descending sensory and motor pathways. Dendritic arborizations which reach out in all directions are characteristic of reticular units and so dominate brain stem organization as to give a "sunburst" effect to histological sections when held up to a bright light source. Such units *are functional "busybodies,"* responding to many different kinds of information coming from a wide variety of sources. They *engage in integrative processing of information from all traffic through the brain stem*; hence they are apprised of all that is important in the way of information going both ways.

The lower part of Figure 9.174 shows a single reticular neuron in the pontine brain stem. Its axon branches up and down the neuraxis and sends out *collaterals to reticular neurons and to each of the major sensory and motor nuclei throughout the brain stem, also* extending down (not shown) well into *the spinal cord*. The "busybody" dendrites and cell body then send integrated "gossip" to other reticular neurons and to all major nuclear constellations throughout the brain stem. The unit is thus *both diffusely receiving and diffusely projecting*.

This arrangement provides reticular formation with the *capacity to monitor virtually all that is going on in the body and in the environment and to contribute to priority decision making with respect to ongoing behavior. This is the essence of integration, and the reticular formation may represent high levels of such integration.* At the very least, large reticular units in the bulb which receive collaterals from the pyramidal tract can relay signals into the spinal cord by way of large diameter axons so that the information can reach motor neurons at about the same time or even ahead of information arriving by way of the original pyramidal axons (Gernandt and Gilman, 1960). This relayed, integrated message can presumably bias the outcome of the more direct motor command according to information gathered from many sources.

Not all access of reticular relays to spinal levels carry the same functional priority. When signals are introduced into the pyramidal tract, the brachial plexus, and vestibular afferent system, all of which relay by way of the reticular formation to somatic motor neurons at spinal levels, the signals generated by *vestibular afferents have a clear priority in both speed of access and duration of influence. Interlimb reflexes appear to have the next order of priority and the pyramidal least* (Gernandt and Gilman, 1960; Gernandt et al., 1959). Hierarchical *dominance in descending order for* relay through reticular formation to *visceral controls* to the vagus (Xth cranial nerve) motor outflow is: *bulbar respiratory center* (itself a functionally integrated constellation of reticular neurons), *vestibular afferents, limbic system* projections, *vagal afferents*, and *trigeminal afferents* (Akert and Gernandt, 1962).

There are also multitudes of intermediate and *small interneurons* in the reticular formation. These *serve as interneurons for interneurons*. Cajal concluded long ago that the sensory and motor tracts and nuclei, apart from the gray and white reticular formation of the brain stem, "appear to suffice, *a priori*, for all the exigencies of reflexes and combinations of movements," and that the addition of such intricately connected reticular neurons "is more complicated than it appears, the effect being more than additive, for it demands an organization so inextricable that one hesitates to consider it." (Cajal, 1909–1911).

Reticulospinal Tracts

The reticular formation gives rise to collections of fibers which descend ipsilaterally throughout all spinal cord levels. Midbrain and pontine units project into the anterior white column, while medullary reticular units descend progressively more laterally. They *end directly on both α and γ-motoneurons and on spinal central gray interneurons*.

Vestibulospinal Tract

A principal source of control of spinal mechanisms by the vestibular apparatus is the vestibulospinal tract. It arises from large neurons in the *lateral vestibular nucleus* (Deiter's), in the pontine tegmentum. This nucleus receives major input from fibers in the vestibular portion of the VIIIth cranial nerve and from the cerebellum. Thus, the vestibulospinal tract conveys vestibular information that is relayed directly through the vestibular nucleus and indirectly through the cerebellum. Moreover, the direct projections of vestibular afferents to cerebellar cortex and to the bulbar reticular formation provide abundant vestibulocerebellar and vestibuloreticular information that is all conveyed to the spinal cord. The vestibular projections from Deiter's nucleus are predominantly excitatory to the whole spinal motor column, especially strongly exciting at cervical and lumbar levels, whereas vestibular relay through the reticular formation results in both excitatory and inhibitory spinal influences (Gernandt et al., 1957).

The vestibular system is known to exert very powerful controls over brain stem and spinal motor mechanisms. Descending impulses to the spinal cord following direct stimulation of vestibular afferents result in a two-peaked response in all ventral roots on both sides of the spinal cord. The first peak is relayed through Deiter's nucleus, the second through the reticular formation. There is insufficient time for cerebellar relays to contribute to these earliest vestibulospinal reflex responses.

CEREBROSPINAL FLUID (CSF)

Within the central nervous system is one system, and around it is another, filled with cerebrospinal fluid (CSF) (Fig. 9.175). The *internal system* in- cludes all of the ventricles, their connections, and their projections. Thus, inside each cerebral hemisphere is one of the pair of *lateral ventricles*, communicating through the foramen of Monro with the unpaired *3rd ventricle* of the diencephalon. The latter, in turn, by way of the *aqueduct of Sylvius* of the mesencephalon joins the *4th ventricle*, which overlies the brain stem between pons, medulla, and cerebellum, and which is continuous caudally with the prolonged but relatively small central canal of the spinal cord.

The fourth ventricle possesses also three openings through which the internal and external systems are joined. These foramina are located laterally (two) and medially (one). Through them CSF can pass from the 4th ventricle into the *external system* that, in general, is defined by all the discrepancies in contour between (1) the internal surfaces of cranium and spinal column, and (2) the external surfaces of brain and spinal cord. That is, where the bone is not perfectly tailored to fit the

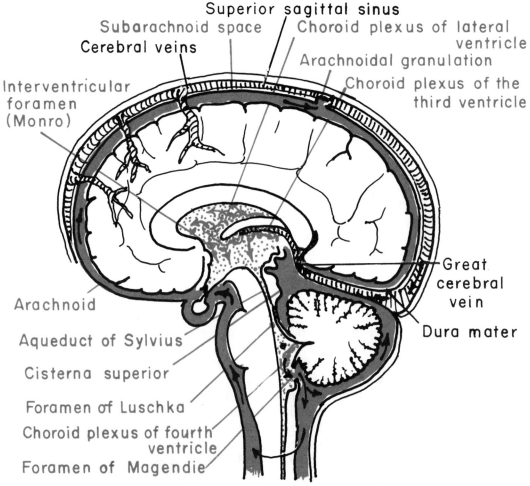

Figure 9.175.   Sagittal section of calvarium and brain to show relationship of internal (ventricular) and external (subarachnoid) systems, and distribution of cerebrospinal fluid.

nervous system it covers, the potential space is filled with cerebrospinal fluid. Moreover, between the dense membrane, the *dura mater*, that lines the cranium and spinal column, and the thin membrane, the *pia mater*, that covers the nervous system, is a potential space festooned by a net-like membrane, the *arachnoid*, the interstices of which are likewise filled with cerebrospinal fluid. In any region where the contour of the brain is concave the *subarachnoid space* is larger and is known as a *cistern*. For example, where the cerebellum overhangs the mesencephalon is the superior cistern; others are the cerebellomedullary, the pontine, and the interpeduncular cisterns, in addition to others less prominent.

Volume and Composition

All of the cerebrospinal fluid together has a volume of about 150 ml. It is believed to be produced at a rate of some 500–600 ml/day, which means that it undergoes a rather rapid turnover. A principal site of production is the choroid plexus of the lateral, third and fourth ventricles. A choroid plexus is a highly vascular structure, formed by elements from pia mater and ependyma where they come together along the margin of a ventricle. It projects into the ventricle as a convoluted, cauliflower-like ribbon, and produces a combination of active transport and filtration that yields fluid having much less protein than plasma. In Table 9.7 the two columns compare the ratio of several constituents in cerebrospinal fluid vs. plasma, and also in a protein-free dialysate of plasma vs. plasma, all concentrations being expressed with reference to water. The table shows that concentrations of potassium, calcium, and chloride are different from values to be expected in simple dialysate. Concentration of glucose, too, is less than if CSF were a plasma dialysate.

Another possible route of movement of this fluid is across the ependyma from brain into the ventricles. There is no question that materials, for example, drugs, injected into the ventricles find their way promptly across the ependyma into the substance of the brain. Consequently, a reverse movement probably takes place under appropriate conditions (see Bering, 1974).

Cerebrospinal fluid is returned to the circulation at a rate equal to its rate of production. The return is via *arachnoid villi*, invaginations of arachnoid-covered blood vessels into the lumen of a venous sinus. Transport at this site probably is passive, in response to hydrostatic pressure and diffusion. Some cerebrospinal fluid is taken up by lymphatics along the meningeal sleeves which enclose spinal nerve roots.

ACTIVE SECRETION OF CSF

Several types of evidence support the conclusion that cerebrospinal fluid is not merely a protein-free transudate (or dialysate) of plasma, as follows: (1) When a choroid plexus has been exposed at a neurosurgical operation, droplets of fluid can be seen forming on its outer surface. (2) Differences in concentration of cations and anions cannot be explained by any purely physicochemical mechanism (Table 9.8). (3) The work required to produce these separations of ions is judged to exceed perhaps 10-fold the work that might be accomplished by hydrostatic pressure in capillaries. (4) There is an electrical potential difference of some 4 to 5 mv between plasma (neg.) and cerebrospinal fluid (pos.), in keeping with the concept that electrochemical "work" has been accomplished in the formation of the CSF. (5) A high concentration of carbonic anhydrase in choroid plexus suggests that this structure is like gastric mucosa and renal tubules, where this enzyme is known to be essential for secretion. (6) Rate of production of CSF is inhibited by substances such as ouabain, which affects active transport of sodium, or acetazolamide, which inhibits carbonic anhydrase. In this latter instance the anionic composition of CSF is also changed (Maren and Broder, 1970).

Table 9.7
Comparison of cerebrospinal fluid with protein-free dialysate of plasma

Constituent	Ratio of Concentration in Water	
	CSF/Plasma[a]	Dial. Plasma/Plasma[b]
Na$^+$	1.00	0.95
K$^+$	0.66	0.96
Mg^{++}	1.24	0.80
Ca^{++}	0.46	0.65
Cl$^-$	0.81	1.04
Glucose	0.64	0.97

Data from Pollay (1974) and other sources. [a] Human. [b] Cat.

Table 9.8
Established and putative neurotransmitters

Acetylcholine	L-Glutamate	Taurine
Dopamine	γ-Aminobutyrate	5-Hydroxytryptamine
Norepinephrine	L-Aspartate	Carnosine
Epinephrine	Glycine	ATP and other purines
Octapamine	Histamine	Substance P and other
	L-Proline	peptides

From Agranoff (1975).

Pressure of CSF

Cerebrospinal fluid pressure is usually measured by inserting into the subarachnoid space a needle connected to a manometer. If the subject is lying down the pressure will be the same in the ventricles and the lumbar region. Since the brain is in a closed box, the total fluid content must remain constant in spite of a change in any of the individual compartments—that is, in arteries, veins, or ventricles and subarachnoid space. Thus, if the arterial system expands through a generalized vasodilatation, the venous system must diminish in volume. For this reason an increase in arterial pressure normally has little effect on pressure in ventricles, since compression of veins serves to relieve the influence of the arterial pressure change. The reverse, however, is not true. Changes in venous pressure have a rapid and profound effect on the cerebrospinal fluid pressure because the arteries cannot be compressed by changes in venous pressure. For example, if the external jugular veins are compressed, the pressure in veins inside the calvarium rises, and this pressure in turn is communicated to brain and to cerebrospinal fluid. This forms the basis for a clinical test for obstruction in the subarachnoid spaces, the Queckenstedt test. In this procedure the cerebrospinal fluid pressure of a lateral recumbent subject is measured via a lumbar puncture. When the jugular veins are compressed the cerebrospinal fluid pressure will rise quickly in a normal subject; when the jugular compression is released the pressure immediately falls. But if there is obstruction in any part of the path between the external jugular veins and the lumbar region, these pressure changes will be transmitted less freely or not at all.

CHAPTER 69

Visceral Control Mechanisms

INTRODUCTION

Multicellular existence depends on a continuing equitable distribution of nutrients and simultaneous disposal of wastes for all cells of the body. This must be achieved in a context of widely differing physiological demands and environmental stresses. It requires differential control of smooth muscles, cardiac muscle, and glands among several different interdependently controlled visceral systems. It also mandates functional integration of these systems with appropriate skeletal muscle activities. A general contrast is drawn between *somatic nervous system* mechanisms, generally governing skeletal musculature, and *visceral nervous system* mechanisms, generally governing smooth muscles, cardiac muscle, and glands. Visceral nervous controls maintain a dynamic internal environment that is necessary for proper functioning of cells, tissues, and organs. As with somatic nervous system organization, visceral organization includes reflexes and successively higher order integrative processes. Within the central nervous system, an interdependent coordination is maintained between somatic and visceral controls, and this viscerosomatic integration is achieved among *longitudinal* as well as *segmental* circuits.

Organization of Visceral Nervous Mechanisms

THE AUTONOMIC NERVOUS SYSTEM

Smooth muscles, cardiac muscle, and glands have a considerable capacity to be active on their own *initiative*, in contrast with skeletal muscles which in mammals are activated strictly by nervous impulses. Nervous system components that govern smooth and cardiac muscles and glands are largely controlled subconsciously and ordinarily function quite independently of volition. Nevertheless, some visceral mechanisms that have been considered entirely inaccessible to conscious control may be brought under considerable voluntary influence (Miller, 1969; Miller and Dworkin, 1980).

Peripheral nervous mechanisms that control smooth muscles, cardiac muscle, and glands are referred to as being *autonomic*, and the *efferent* (*motor*) components are collectively called the *autonomic nervous system*. Visceral afferent fibers which send signals to the central nervous system from the viscera are *not* considered part of the autonomic nervous system. The autonomic nervous system is divided into two major divisions: the *sympathetic* or thoracolumbar division, and the *parasympathetic*, or craniosacral division (Fig. 9.176). The autonomic nervous system is distinguished by having relays between the central nervous system and the viscera in numerous *peripheral ganglia*, complete with interneurons. This provides for extraordinary functional autonomy.

Organization of Peripheral Visceral Controls

The last neurons in the central nervous system to direct efferent impulses to influence smooth muscles, cardiac muscle, and glands do not represent a "final common path" in the same sense that somatic motoneurons do. The latter provide final integration for an unlimited variety of converging peripheral and central nervous system influences from remote as well as neighboring sources. These convergent controls contribute signals to the final neuronal outflow and can alter orders to skeletal muscles up until the last possible instant. Moreover, motoneurons have a precise and narrow motoric influence on one muscle or even a single fascicle of muscle fibers within a single muscle. Ventral horn motoneurons thus provide the final locus for nervous system decisions among channels for command and control of somatic behavior. In contrast, the last central neurons to issue visceral commands are only indirectly influential in respect to their effectors. Any such command is widely

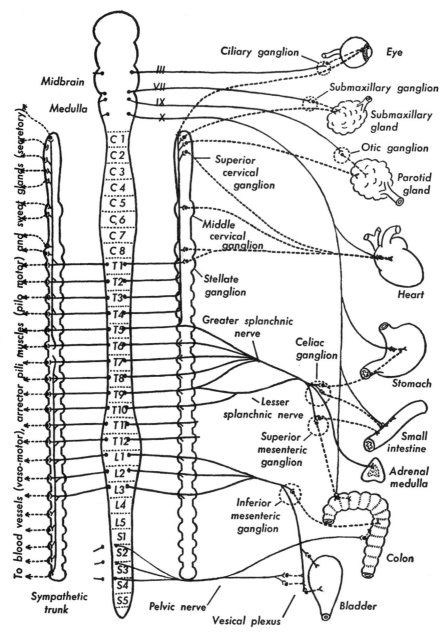

Figure 9.176. Diagram of connections of autonomic or visceromotor system. Craniosacral components are in black; thoracolumbar connections are in color. Course of postganglionic fibers is shown by broken lines. [From Carpenter (1976).]

distributed and also must pass through peripheral ganglionic relays which are engaged in integration of widespread and locally dominated visceral functions. This means that the central nervous system has much less control over visceral behavior.

Visceral systems govern *respiratory exchange* in the lungs and tissues, with pulmonary, renal, and cerebral controls; *cardiovascular performance* throughout the body with local need-regulated flow differences; *alimentary contributions*, ingestion, digestion involving gastrointestinal, liver, gallbladder, and pancreatic functions; differential local *metabolism*, plus elimination; *water balance*, drinking

renal, and bladder functions; and *reproduction* from start to finish. Visceral government is not only continuous, waking and sleeping, but is so highly complex and locally discriminative that it is obviously advantageous that we are are not obliged to depend on conscious perceptions and considered judgments in order to maintain a dynamically equilibrated internal milieu.

The last neurons in the central nervous system which contribute efferent signals to the viscera have cell bodies located along an interrupted string of gray matter in brain stem and spinal cord. Interruptions occur between midbrain and bulb at the levels of the pons, throughout cervical segments, and between lumbar 4 and sacral 1. Cranial, i.e., midbrain and bulbar visceral efferents, contribute to the IIIrd, VIIth, IXth, and Xth cranial nerves. These and three sacral segmental efferents (S 2–4) constitute the entire *parasympathetic outflow* from the central nervous system. The thoracocolumbar visceral motor outflow, from T1 to L3, constitutes the entire *sympathetic* outflow (Fig. 9.176).

Sympathetic and parasympathetic divisions of the autonomic nervous system differ in several morphological, pharmacological, and functional respects. Although they often operate antagonistically, it is appropriate to recognize their capacity for *integrated cooperation*. More generally, Hess, a winner of the Nobel Prize in Physiology and Medicine, emphasized that sympathetic activation allows the organism to mobilize and expend energy. Cardiac acceleration is accompanied by relaxation of coronary artery walls and contraction of peripheral arterioles with elevation of perfusion blood pressures; increased ventilation is fostered by relaxation of bronchial musculature; increased blood for skeletal muscles is provided by contraction of splanchnic veins; widening of the pupil admits more light into the eye. All such sympathetic contributions enable peak performance of all needed effectors on behalf of the organism during emergencies. Hess called these functions *ergotropic* (ergo = energy; tropic = turning or releasing). Parasympathetic functions, including narrowing of the pupil, gastrointestinal secretion and peristalsis with relaxation of intestinal sphincters, slowing of the heart and reduction of blood pressure, all contribute to conservative, restorative vegetative functions on behalf of the organism, hence are best characterized as *trophotropic* functions (trophos = relating to nutrition).

The morphology and pharmacology of these systems reflect these functional generalizations. The distribution of neurons and fibers in the *sympathetic* division ensure *widely distributed commands*. Sympathetic ganglia are remote from visceral effectors, preganglionic fibers often project to several ganglia, and postganglionic fibers often project to more than one visceral organ. Moreover, the predominant neurotransmitter (norepinephrine) is one that has prolonged and generalized effects. Further, norepinephrine is released in quantities that are taken up by the bloodstream whereby norepinephrine contributes to global energy mobilizing effects. Sympathetic preganglionic fibers also activate the medulla of the adrenal gland and cause release into the adrenal veins of a mixture of epinephrine and

norepinephrine, which thereby contributes to generalized energy-utilizing responses.

In contrast, *parasympathetic preganglionic* fibers ordinarily go directly to the region of the effector tissues in a specific visceral organ, where the peripheral parasympathetic ganglia reside. The *postganglionic cells serve discretely, locally,* and the predominant neurotransmitter (acetylcholine) is rapidly dissociated locally (Fig. 9.177).

AFFERENT VISCERAL INNERVATION

Impulses are transmitted from the viscera by *afferent fibers* whose cell bodies lie in the *dorsal root ganglia* (Fig. 9.178). Most visceral afferents are larger in diameter than visceral efferents and more myelinated, accounting for the whiteness of the white rami. Visceral afferents *closely resemble somatic afferents.* Afferents follow the same general course as visceral efferents, but unlike the latter, they do not relay during their transit through peripheral ganglia. They reach the central nervous system via the vagus, pelvic, splanchnic, and other autonomic nerves.

Afferent fibers, conveyed along the *thoracolumbar*

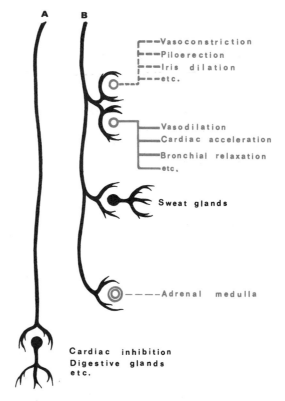

Figure 9.177. Distribution of cholinergic and adrenergic fibers, in parasympathetic (*left*) and sympathetic divisions of the autonomic system. Cholinergic mechanisms are in black; adrenergic neurons are in color.

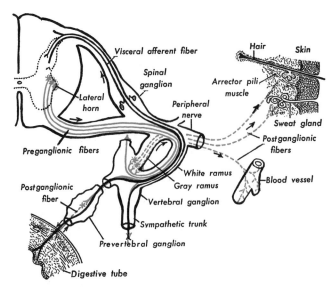

Figure 9.178. Diagram showing course of visceral afferent and efferent fibers in dorsal and ventral roots, respectively, and their distribution through the thoracolumbar system. [From Copenhaver, cited from Carpenter (1976).]

(*sympathetic*) division of the autonomic nervous system, pass from the viscera through one or more peripheral ganglia. They then *join somatic afferents by way of the white rami.* Their cell bodies reside in the corresponding dorsal root ganglia. They enter the spinal cord by way of dorsal roots from T1 to L3. Their central terminals pass deeply into the dorsal horn, ending on interneurons in the vicinity of the lateral horn where visceral efferent neurons take origin (Petras and Cummings, 1972).

Afferent fibers of the *cranial division* are axons whose cell bodies reside in the equivalent of the dorsal root ganglia of the VIIth, IXth, and Xth cranial nerves. The central processes of all these neurons end in the *nucleus of the tractus solitarius* (NTS). The IXth nerve projects to the intermediate zone of the NTS, parahypoglossal area, and nucleus amiguus in neighboring bulbar brain stem. These centers appear to be involved, along with supramedullary areas in the pons and hypothalamus, in modulating the important *carotid sinus reflex* (Kostreva, 1983). *Carotid body chemoreceptors* project to the medial part of the NTS. *Aortic depressor afferents* project to the caudal part of the NTS, a reflex mediated by the nucleus ambiguus, whereas the carotid sinus-induced decrease in heart rate is mediated by the external cuneate nucleus.

Altogether, the NTS receives *gustatory* and *taste afferents* from Vth, VIIth, IXth, Xth cranial nerves, *respiratory afferents, aortic baroreceptors, carotid body chemoreceptors, carotid sinus baroreceptors, and cardiopulmonary baroreceptors* from various sources. It also receives from the vestibular nuclei, fatigial nucleus (cerebellar roof nucleus), medullary

reticular formation, amygdala, and cerebral cortex. The *NTS projects to spinal sympathetic outflow,* including that to the adrenal medulla, inhibitory sympathetic loci in bulbar reticular formation, hypothalamic medial preoptic area (endocrine effects), olfactory gray matter in the basal forebrain, amygdala, diffusely projecting nuclei of the thalamus, inferior olive, vermis, and flocculus of cerebellum and bulbar efferents, including the dorsal motor nucleus of the vagus, visceral efferents to heart and lungs, somatic efferents to larynx and pharynx in the nucleus ambiguus, and other bulbar controls.

Afferent fibers coming from the visceral distributions of the *pelvic nerves,* serving bladder, urethra, and secondary sexual organs, stem from cell bodies in the dorsal root ganglia of *sacral segments 2, 3,* and *4.* Their reflexes are mediated locally, with ascending projections to unknown central representations.

Sympathetic Visceral Controls

Cells of origin of the *sympathetic division* of the autonomic nervous system are located in the *lateral horn* of spinal gray matter from the 8th cervical and 1st thoracic through the 3rd lumbar segment. The axons which are almost uniformly myelinated exit the spinal cord by way of local ventral nerve roots and pass through the *white rami* to the *sympathetic ganglia* (Fig. 9.178). The sympathetic ganglia are arranged in three groups: (1) *paravertebral;* (2) *collateral;* and (3) *peripheral.* In these ganglia, sympathetic preganglionics synapse with postganglionic nerve cells. Almost all *postganglionic axons* are unmyelinated. By returning *to peripheral nerves by way of the gray rami*, these axons proceed to innervate visceral structures in the body wall and skin. Other preganglionic sympathetic nerves pass up or down the *sympathetic chain* in order to gain ganglionic access closest to the destined visceral organs. Postganglionic sympathetic neurons serve smooth muscles and glands. Many investigators have shown that the sympathetic ganglia can mediate visceral reflexes without connection to the spinal cord (see Bosnjak and Kampine, 1982).

SYMPATHETIC GANGLIA AND THE SYMPATHETIC CHAIN

Sympathetic ganglia which are closest to spinal sympathetic outflow are located closely alongside the vertebral bodies on each side, underlying pleura in the thoracic region and peritoneum in the abdominal region, interconnected by fibers in transit and interganglionic communicating interneurons. They are spread out from their segmental origins from upper cervical to coccygeal levels. The lower four (sacral ganglia) are served by sacral parasympathetic outflow (Fig. 9.176).

Cervical Ganglia of the Sympathetic Chain

The cervical portion of the sympathetic chain consists of three ganglia—the *superior, middle,* and *inferior* cervical sympathetic ganglia. They are relatively large and represent the fusion of two or more smaller ganglia.

The *superior cervical ganglion*, located close to the base of the skull, is largest of the three. It receives preganglionic fibers from upper thoracic segments of the spinal cord. Its cells send postganglionic axons to upper cervical peripheral nerves to serve the blood vessels, glands, and smooth muscles in the head and upper neck, and to the *superior cardiac nerves*. The middle cervical ganglion innervates smooth muscles and glands in the distribution of the 5th and 6th cervical nerves, thyroid and parathyroid glands, and the *middle cardiac nerve*. The *inferior cervical ganglion* is usually fused with the first thoracic ganglion and occasionaly with the second as well, forming the large *stellate ganglion*. It serves visceral control needs in the distribution of cervical cranial nerves 7 and 8, the first two thoracic nerves, and the *inferior cardiac nerve*. Branches from the stellate ganglion form plexuses on the subclavian artery and its branches whereby sympathetic innervation is supplied to the vertebral, axillary, and brachial arteries.

Thoracic, Lumbar, and Sacral Ganglia

There are usually 10–12 sympathetic ganglia in chains on each side corresponding to each spinal segment. There are four lumbar and usually four or five sacral ganglia which latter may be fused with counterpart ganglia across the midline.

Collateral Sympathetic Ganglia

These lie in the thorax and pelvis and are closely related to the aorta and its branches (Fig. 9.179). The largest of these is the *celiac* (solar or semilunar) ganglion, related to the celiac artery. The *superior* and *inferior mesenteric* ganglia lie just caudal to the superior and inferior mesenteric arteries, respectively. *Terminal sympathetic ganglia* lie even closer to the visceral organs and especially close to the rectum and bladder in the pelvis.

DISTRIBUTION OF PREGANGLIONIC SYMPATHETIC FIBERS

A preganglionic fiber which arrives at a sympathetic ganglion may pursue one of three courses: it may (1) form synapses with cells in that ganglion; (2) traverse the ganglionated cord without interruption to find synaptic relay in a ganglion at a level higher or lower than the segment from which it originated; or (3) obtain a synaptic relay in a collateral sympathetic ganglion. Preganglionic fibers typically contribute synaptic junctions within several (five to nine) ganglia. Within each ganglion, each preganglionic axon usually serves several cells each with a large number of synaptic terminals. This obviously contributes to the widespread effects of sympathetic excitation.

DISTRIBUTION OF POSTGANGLIONIC SYMPATHETIC FIBERS

Ganglia of the sympathetic chain contribute postganglionic sympathetic fibers by way of gray rami to each of the peripheral nerves innervating body wall and skin. These supply blood vessels, sweat glands, and smooth muscles throughout the distribution of the corresponding peripheral nerves. Preganglionic fibers serving the upper limb arise from thoracic 2 to 7, reach the corresponding ganglia by way of white rami, and pass up and down the sympathetic chain. Postganglionic fibers exit via the gray rami which lie closest to the corresponding peripheral nerves. Lumbar preganglionics similarly travel downward along the sympathetic chain, and postganglionic fibers exit via gray rami to join the corresponding peripheral nerves emerging from lower lumbar, sacral, and coccygeal segments. Excision of the stellate and upper four thoracic ganglia, for example, deprives the upper limb, head, and neck of sympathetic innervation and disables most of the postganglionic sympathetic

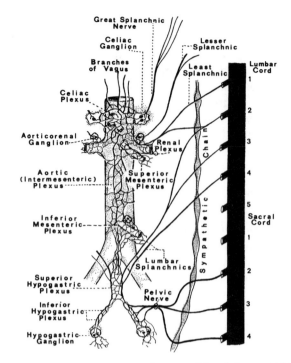

Figure 9.179. Diagram of nerve plexuses of the abdomen and pelvis.

innervation of the heart. The middle and superior cervical sympathetic preganglionics are of course severed when the stellate ganglion is removed.

Parasympathetic Craniosacral Visceral Controls

Cells giving rise to central parasympathetic outflow originate from *midbrain*, *bulbar* (medulla oblongata), and *midsacral* segments of the spinal cord. Parasympathetic ganglia, as previously noted, lie within or closely related to the visceral organ being innervated. Axons leaving the central nervous system, as in sympathetic outflow, are *preganglionic* fibers. These are myelinated, postganglionics unmyelinated. Autonomic postganglionic endings, sympathetic and parasympathetic, do not have elaborate junctions with smooth muscles and glands, but end with vesicle-filled sacculations. Sympathetic sacculations are shown in Figure 9.180.

MIDBRAIN OUTFLOW

Cells included in the Edinger-Westphal nucleus of the oculomotor system reside in the midbrain equivalent of the lateral motor column. Preganglionic axons pass through the short ciliary nerves to innervate the ciliary muscle for lens accommodation and sphincter of the iris for contraction of the pupil.

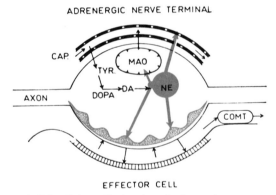

Figure 9.180. Schematic drawing of adrenergic nerve saccule. Pathways of synthesis of norepinephrine (*NE*) from tyrosine (*TYR*) is shown, together with presence of monoamine oxidase (*MAO*). Norepinephrine is released from storage sites, and also reaccumulated into these sites; some are oxidized by catechol-*O*-methyltransferase (COMT). [From von Euler (1971).]

BULBAR OUTFLOW

Efferent preganglionic parasympathetic fibers leave the brain stem via the VIIth (facial), IXth (glossopharyngeal), and Xth (vagal) nerves.

1. Parasympathetic vasodilatory and secretory fibers arise from the *superior salivatory* (*salivary*) nucleus. They emerge from the brain stem along with other fibers of VIIth cranial nerve, as the *intermediate nerve*, and pass through the facial canal of the temporal lobe. Thereafter they follow

either of two paths: (1) to the submaxillary (sub-mandibular) ganglion to innervate the *submaxillary and submandibular salivary glands* and the *mucous membranes of the mouth*, and (2) to the spheno-palatine ganglion to innervate the *lachrymal gland, mucous membrane of the soft palate, nasopharynx, and pharynx*, and send *vasodilator fibers into the cranium* by way of the middle meningeal artery and its branches.

2. Secretory and vasodilatory fibers of the IXth cranial nerve arise from the *inferior salivatory (salivary) nucleus*. They leave the brain stem with the glossopharyngeal nerve but separate from it to join fibers from the facial nerve and filaments from the internal carotid plexus which pass as the superficial petrosal nerve to the *otic ganglion*. Postganglionic fibers pass to the *parotid gland*. Some presumably preganglionic fibers bypass the otic ganglion and innervate the *mucous membranes of the tympanic cavity, the mastoid air cells, Eustachian tube*, and *internal ear*.

3. *The dorsal motor nucleus of the vagus* sends large number of preganglionic fibers through several branches of thoracic and abdominal viscera. The vagus innervates all viscera from throat to transverse colon. *All vagal ganglia are situated within the innervated organs.* Vagal impulses to the heart have synaptic relay through *ganglionic cells* in the *cardiac* walls; to the bronchi through intrinsic bronchial ganglia; to the esophagus, stomach, and intestine through ganglion cells of the *myenteric plexus of Auerbach* and the *submucous plexus of Meissner*. The preganglionic fibers are of course very long. The postganglionic fibers are very short. Cells of origin which provide most of the cardiac parasympathetic innervation constitute a discrete cardioinhibitory group of neurons which lie alongside the dorsal motor nucleus of the vagus.

SACRAL OUTFLOW

Cells of origin lie in the lateral columns of *sacral 2, 3, and 4*. They exit the spinal cord by way of the corresponding ventral roots. After separating from axons destined to accompany nerves to the body wall and surface, those destined for pelvic viscera form the *pelvic nerve* and *pelvic plexus*. They terminate with synaptic contacts on ganglion cells distributed in the wals of the *descending colon, rectum*, and *bladder*. They carry inhibitory impulses to the *internal anal, vesical*, and *uterine sphincters*, and *dilator impulses to blood vessels of the bladder, rectum*, and *genitalia*. By relaxing vasoconstrictor tone, these postganglionics control tumescence of labia and clitoris and penile erection.

Development of the Autonomic Nervous System

The entire autonomic nervous system (ANS), including the medulla of the adrenal gland, derives from the neural crest. The neural crest is an unusually pleomorphic embryonic tissue. It gives rise to all dorsal root ganglia including the analogous cranial sensory nerve nuclei, olfactory and taste neuroepithelia, parts of the pharynx and jaw contributing to food detection, engulfing, chewing, and swallowing, the squamous bones that cover the forebrain, and perhaps the forebrain itself (Gans and Northcutt, 1983). Neurons destined for the ANS undergo a number of binary "decisions" on their way from neuroblast to mature neuron: commitment to a neural rather than a glial lineage, commitment to an autonomic rather than a sensory lineage, and finally, commitment to a specific type of neurotransmitter.

There appear to be two stages of development of the ANS, first the establishment of principal neurons in the positions they will occupy, and second, differentiation to a specific transmitter economy based on trophic influences of the peripheral environment (Bunge et al., 1978). Enteric neuron diversity includes acetylcholine (ACh), norepinephrine (NE), 5-hydroxytryptamine (5-HT), dopamine (DA), and probably larger molecules as well: neuropeptides such as vasoactive intestinal polypeptide (VIP), substance P, somatostatin, enkephalin, cholecystokinin (CCK), neurotensin, and bombesin. ACh has been shown to be synthesized and released from the myenteric plexus. The peristaltic reflex can be interrupted by muscarinic antagonists such as atropine which interfere with transmission at the neuro-smooth muscle junction, and by nicotinic antagonists, such as hexamethonium, which block ganglionic transmission.

Neural crest derivatives, in contrast with cells stemming from the neural tube, seem to retain a capacity to continue to divide even after neuronal phenotypic expression is evident, and to switch neurotransmitter commitments, depending on environmental conditions (see Fed. Proc. Symposium, 1983).

The clinical importance of development of the ANS is obvious. There are a number of disorders due to abnormalities of ANS development: Hirschsprung's disease (congenital absence of ganglion cells in the distal segments of the large bowel, leading to megacolon); achalasia (failure to relax the esophagogastric sphincter on swallowing, leading to megaesophagus); Riley-Day syndrome (congenital dysautonomia, i.e., defective lacrymation, skin blotching, emotional instability); and certain disorders of late onset involving hypertension and other cardiovascular disorders.

The Enteric Nervous System

That part of the ANS which is most remote from control by the central nervous system, the peripheral myenteric plexuses, has been called the "enteric nervous system." There is a great disproportion of peripheral enteric neurons (over 10^8) compared to efferent vagal fibers to the gut (2×10^3) which further emphasizes peripheral autonomy.

Myenteric units, derived from neural crest, colonize the bowel before sympathetic and parasympathetic neurons migrate into their domains. *The microenvironment of the intestines influences phenotypic expression by cells in these enteric plexuses,* e.g., the proximodistal gradient of peristaltic reflex organization from esophagus to rectum. Moreover, cell division continues even after phenotypic expression is manifest (Gershon et al., 1983). Enteric reflexes appear to involve adrenergic-cholinergic axoaxonal synapses in the wall of the gut. Such synapses are widely distributed in the ANS. It is probable that the transmitter is 5-HT (Gershon, 1981) and that pressure in the gut releases 5-HT. Somatostatin is a good candidate for the interneuron neurotransmission that mediates the descending inhibition that is necessary for peristalsis.

Clear evidence for the existence of nonadrenergic, noncholinergic neurotransmitters in the autonomic nervous system emerged in the early 1960s (Burnstock et al., 1963; Martinson and Muren, 1963). This evidence now indicates that gastrointestinal, lung, bladder, and some blood vessels may use 5-hydroxytryptamine (5-HT), dopamine (DA), γ-aminobutyric acid (GABA), and adenosine triphosphate (ATP) as neurotransmitters. Some of these nerves store and release more than one transmitter. Some transmitters and cotransmitters act on presynaptic receptors to modulate nerve-mediated release of transmitter (Burnstock and Hökfelt, 1979).

Similarly, sympathetic ganglionic neurons show pleomorphic and pleochemic plasticity (Potter et al., 1983). Sweat glands and their innervation by sympathetic postganglionic fibers develop postnatally. By the end of 1 week postnatal, in the rat, axons innervating sweat glands exhibit catecholamine histofluorescence, by 2 weeks, combined cholinergic and vasoactive intestinal peptide (VIP) immunoreactivity is evident, and by 3 weeks, catecholaminergic signs are lacking. Neonatal treatment with a toxic norepinephrine congener results in a loss of cholinergic as well as catecholaminergic innervation (Landis, 1983). This implies that if NE

is throttled, ACh conversion is likewise blocked. Innervation of the levator palpebrae smooth muscle which widens the palpebral fissure is established by postganglionic sympathetics in the cervical chain, and functional activation by asphyxia and hypoglycemia is present even before ganglionic transmission from the central nervous system commences (Mills and Smith, 1983).

Functions of the Autonomic Nervous System

Autonomic activity creates an operating milieu for cells and tissues throughout the body, adjusting this milieu differentially depending on global as well as local demands, and specifically in full support of somatic behavior. It operates unremittingly and continuously, adjusting its functions in accordance with somatic as well as visceral requirements. Autonomic actions are not necessarily secondary to somatic functions, and as everyone knows from personal experience they occasionally absolutely dominate. Through actions associated with emotional behavior, the autonomic nervous system prepares the body for anticipated behavioral somatic and visceral requirements. For example, it increases cardiac output, channels blood flow to safeguard emergency metabolic requirements of heart, brain, and skeletal muscles, make sources of energy available, and in other ways prepares the organism for urgent vigorous activity. Thus, visceral behavior both supports and anticipates bodily needs.

Most of the preganglionic and postganglionic nerves are tonically active, and the relatively autonomous peripheral ganglia, e.g., those in the wall of the gut, are likewise tonically active. Tonic activity enables greater precision in visceral control. Tonic activity depends on the combination of: dynamics originating directly from the visceral organs, from peripheral and central reflex activities that engage larger dynamic constellations, and central nervous system mechanisms involved in patterns of visceral drives, appetite and satiety, emotions, and anticipations. Visceral control mechanisms thus constitute a system as a whole which intimately correlates and cooperates with somatic motor activities.

Cholinergic Transmission

Acetylcholine is a principal transmitter for all autonomic ganglia, where its actions are like those of *nicotine*. It is also a transmitter at many postganglionic endings of the parasympathetic craniosacral division, where its actions are like those of muscarine, a drug isolated from certain species of mushroom. Acetylcholine is therefore said to have both muscarinic and nicotinic actions. Acetylcho-

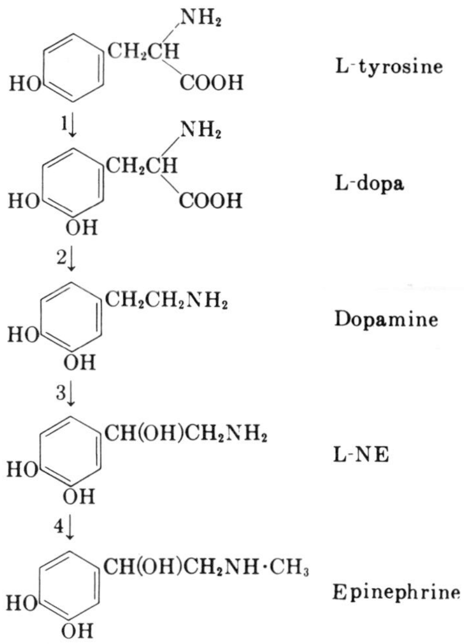

Figure 9.181. Biosynthesis of catecholamines from L-tyrosine (i.e., hydroxyphenylalanine). The enzyme for the fourth step, phenylethanolamine N-methyltransferase, is present in adrenal medulla and chromaffin tissues, but not in typical postganglionic adrenergic endings. [From Iversen (1967).]

line is unstable and readily hydrolyzed to choline and acetate ion, by *acetylcholinesterase*, present at cholinergic nerve terminals. Cholinesterases can be inhibited by substances such as physostigmine (eserine), from the calabar bean, and synthetic inhibitors such as neostigmine, diisopropylfluorophosphate (DFP) and tetraethyl pyrophosphate (TEPP). The latter two combine irreversibly with the enzyme; when they are used, cholinesterase must be newly synthesized to restore its activity.

Catecholamine Transmission

Epinephrine and norepinephrine are classical examples of compounds known as catecholamines, all synthesized in the body from tyrosine (Fig. 9.181). Besides epinephrine and norepinephrine, the list includes L-tyrosine, L-dopa, dopamine, and isoproterenol. These several compounds are synthesized by a series of enzymes having the following names and catalyzing the reactions numbered in Figure 9.181:

1. Tyrosine hydroxylase
2. L-dopa decarboxylase
3. Dopamine β-hydroxylase
4. Phenylethanolamine *n*-methyltransferase.

Catecholamines have been studied intensively by many investigators for some years, with the result that the pathways through which they are synthe-

sized and metabolized are well known (Iversen, 1967; Blaschko, 1973; Sharman 1973).

Vesicles in adrenergic nerve endings which store epinephrine also contain the enzyme β-hydroxylase, which converts dopamine to norepinephrine. Much of the norepinephrine release is taken back into the nerve and into the vesicles, and this is the primary inactivating mechanism. Primary amines can be oxidized by the enzyme *monoamine oxidase*. Epinephrine can be oxidized to form a quinone which in turn forms the cyclic structure *adrenochrome*. Circulating epinephrine and norepinephrine are metabolized by the methylating enzyme, O-methyltransferase (Fig. 9.182).

Different actions by the same compound on different tissues may be accounted for by different types of adrenergic receptor systems, a simple classification of which is presented in Table 9.9.

Examples of Sympathetic Actions

When the sympathetic system *as a whole* is activated, the combined sympathoadrenal responses resemble responses to severe stress and anger. The individual looks angry or frightened and exhibits acceleration of the heart, rise in blood pressure, elevation of blood glucose, piloerection, sweating, dilatation of the pupil, and other responses which prepare the individual for "fight or flight." Nevertheless, higher centers of visceral representation have considerable discretionary control which can allow quite subtle reflex actions and precise local control of tonic sympathetic activity.

Stimulation of the superior *sympathetic ganglion* results in dilatation of the pupil, exophthalmos (protrusion of the globe), and sweating on that side of the head. Interruption of that ganglion causes miosis (pupillary constriction), enophthalmos (the globe sinks slightly into orbit), and unilateral absence of sweating. This combination of signs is known as Horner's syndrome and can be induced by lesions in the cervical sympathetic chain, in the lateral column at T1 and T2 or higher in the spinal cord, and in the medial pons from which levels this sympathetic outflow originates.

Stimulation of the *stellate (inferior cervical)* ganglion results in acceleration of pacemaker activity in the heart, speeding up of cardiac conduction, and increasing force of cardiac muscle contraction. This effects an increase in stroke volume and cardiac output. There is also bronchial dilatation and moderate pulmonary vascular constriction. Denervation or central lesions can eliminate tonic sympathetic activity to these organs. Although innervation of abdominal viscera is orderly, elimination of one or a few ganglia from the sympathetic chain does not

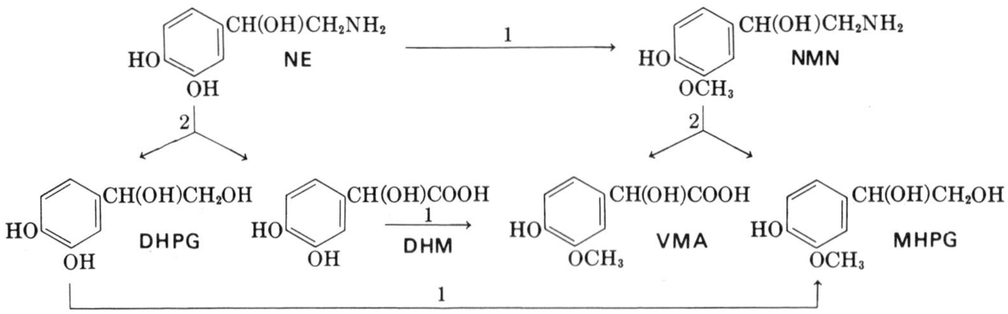

Figure 9.182. Pathways by which catecholamines are inactivated by catechol-*O*-methyltransferase (*1*) or by monoamine oxidase (*2*). *NE* is norepinephrine; *NMN* is normetanephrine; *DHPG* is 3,4-dihydroxyphenylethylglycol; *DHM* is 3,4-dihydroxymandelic acid; *VMA* is vanillymandelic acid; *MHPG* is 3-methoxy-4-hydroxymandelic acid. [From Iversen (1967).]

Table 9.9
Functions of α and β adrenergic receptors

α Receptors
  Vasoconstriction, especially in vessels of skin, mucosa, gut, salivary glands, lungs, cerebral circulation
  Stimulation of contraction of radial muscle of iris (pupillodilation), of nictating membrane, of pilomotor muscles, of muscle of uterus, and of ureter
  Stimulation of glycogenolysis
  Inhibition of insulin secretion
  Inhibition of smooth muscle of intestine

β₁ Receptors
  Stimulation of lipolysis
  Stimulation of heart, with cardioacceleration, increase in contractility of muscle of atria and ventricles

β₂ Receptors
  Vasodilatation in coronary circulation and in skeletal muscle, and possibly in the liver
  Relaxation of muscle of bronchi, and of uterus
  Stimulation of insulin secretion

ordinarily yield clinical signs. This is because several ganglia are served from each spinal level.

Activation of sympathetic fibers innervating the liver induces vasoconstriction and inhibits contraction of the gallbladder. *Thoracic sympathetic release of epinephrine and norepinephrine from the medulla of the adrenal gland* produces glycogenolysis and liberation of glucose, increases metabolic rate, inhibits insulin secretion, augments glucagon output, and reduces blood clotting time. Activation of sympathetic fibers to the spleen causes splenic contraction with a discharge of red and white blood cells into the general vascular bed. *Mechanical and secretory activities of the stomach and intestines are inhibited, and intestinal sphincters are constricted.*

Generalized venous and arteriolar constriction reduces the pool of blood in the veins and raises arterial blood prepssure. *Blood vessels to skeletal muscles may be dilated by sympathetic cholinergic neurons.* Sympathetic *activation of sweat glands* induces secretion through the local release of ace-

tylcholine. Sweating takes place in small bursts of activity at about 6–7/min during moderate sweating activity. Since different skin areas show synchronous sweating rhythms, there must be widespread rhythmic discharges of central origin. *Aprocrine sweat glands in the axilla are especially activated by mental stress. Piloerection associated with cold, fear, and anger is adrenergic.* Sympathetic innervation of the bladder causes *relaxation of the bladder wall and contraction of the internal sphincter.* Most pelvic structures undergo vasoconstriction, but some vasodilation occurs in external sex organs.

Examples of Parasympathetic Actions

Third nerve parasympathetic innervation provides tonic and phasic pupillary constriction. Denervation leaves a dilated pupil which dilates somewhat further with activation of the sympathetics. Stimulation of the lachrymal branches of the VIIth nerve causes tear production. VIIth and IXth nerve activation yields salivation. Parasympathetic innervation of the heart has an *inhibitory effect on the pacemaker* by decreasing the rate of depolarization. It *slows cardiac conduction* and *interferes with transmission at the atrioventricular node.* This results in reduction of blood pressure and decrease of cardiac output. *Bronchiolar constriction* and *increased secretion,* perhaps increased viscosity of bronchial secretion, affects ventilation.

The gastrointestinal tract is induced to *increased peristaltic activity, relaxed intestinal sphincters, shortened gastric emptying time,* and *increased secretion of gastrin* and *other digestive juices.* Parasympathetic activation elicits *contraction of the gallbladder, secretion of pancreatic digestive juices,* and *release of insulin.* Pelvic innervation from midsacral levels causes relaxation of sphincter muscles during urination and defecation and contributes to external genital tumescence and penile erection.

Effects of Spinal Cord Transection on Autonomic Functions

Spinal cord injuries in humans results in over-reactivity of visceral reflexes below the level of cord transection. Various stimuli, principally bladder and rectal in origin, and most marked in cases of high spinal lesion, may start paroxysmal bouts of arterial hypertension. Whereas orthostatic hypotension is exaggerated in such cases, cold immersion of the foot leads to temporary reflex hypertension. Bladder distention usually leads to excessive sweating, including above the level of the lesion. Similar mass action responses may yield headache, facial flushing, and pilomotor hyperactivity. These signs can be recognized by the patient and taken as an indication that the bladder or rectum needs evacuation or other attention.

The absence of central connections for temperature regulation results in *poikilothermia below the transection.* Spinalized individuals may be capable of reproduction, even though many of the pleasures are denied through sacrifice of sensory channels. Spinalized man may remain sexually potent and fertile if sympathetic outflow from T6 to L3 is preserved. Even if potency or ejaculation is precluded by nerve or cord mutilation, fertility may remain.

Higher Levels of Visceral Control

There are few data on the detailed functions of neurons in the lateral horn of the spinal cord, but there are abundant data as to visceral activities that can be elicited by stimulation of parts of the brain stem, hypothalamus, and limbic system. It was discovered by Claude Bernard in the last century, that glycosuria could be induced by surgical injury to the medulla oblongata. Subsequently, it was found that electrical stimulation of the floor of the 4th ventricle yields a rise in blood pressure, acceleration of the heart, peripheral vasoconstriction, and secretion of epinephrine. It soon became evident that *autonomic controls are a major function of the brain stem.* Bulbar loci are still classified as cardiovascular vasopressor and vasodepressor "centers," centers for respiratory inspiration and expiration, a pneumotactic center, etc. The terminology is misleading, however, for these are only nodal points within an extended system.

Systematic stimulation of the *hypothalamus* by Ranson and his associates (1936, 1937) and by Hess (1932, 1969) revealed functional representation of a number of *autonomic responses* relating to eating, *drinking, reproductive functions, temperature control,* and *sleep.* Still later, it became evident that

cortex, predominantly the phylogenetically old, limbic regions of cortex, have elaborate and various representations of autonomic functions.

The generalization is justified that the higher one goes in exploration of central representation of autonomic functions, from peripheral, to spinal, brain stem, hypothalamic, and limbic systems, the more distributed and probably also the more uniquely combined are the patterns of representation. Whereas *brain stem* representation appears to emphasize *reflexes* and controls that have primarily *vital responsibilities,* the *hypothalamus* has representations that relate to more *global and interdependent combinations,* such as control of thermal and nutritive economies, neuroendocrine controls, neuroendocrine and behavioral activities relating to reproduction, and the like. The *limbic system* appears to represent and rerepresent in various combinations a still wider variety of signs and symptoms. These combinations we associate with *emotional experience and expression.*

MacLean (1958) has made the appealing generalization that the anteromedial region of the limbic system, especially the septum, represents a cluster of functions relating to *survival of the species.* The anterolateral part of the same system, especially the amygdala, represents a cluster of functions relating to *survival of the individual.* Stimulation of each of these regions will elicit grooming, searching for food, protection, but in the septum, these functions relate to finding and grooming a mate, breeding and protecting the mate and offspring. In the amygdala, these same responses are directed to serving the needs of the individual. No one knows enough about the organization of exceedingly complicated representations to be able to evoke or mimic elaborate emotions in their full complexity by stimulation, but careful consideration of physiological responses and a variety of clinical disorders associated with limbic dysfunctions suggest that biologically useful behaviors of this kind may be organized by the limbic system.

Primitive Integrative Visceral Mechanisms

There are three great questions in biology: *How did it come about?* (evolution and development); *What does it do for the organism?* (how and in what measure is it adaptive); and *How does it function?* (mechanism). The autonomic nervous system operates with evolutionary primitive integrative processes, as though some early, probably invertebrate, evolutionary solutions to problems of survival were conserved through the vertebrate line. Organizational simplicity, unmyelinated and finely myelinated axons, primitive neuroeffector junctions, control of tissues that tend to be spontaneously active, all reflect primitive evolutionary mechanisms adapted to higher organisms. Of course, these mechanisms have been shaped by evolution throughout all this time. The autonomic nervous system establishes considerable functional competence early in fetal life. Many of the peripheral mechanisms initiate visceral controls even prior to the establishment of connections from the central nervous system.

Visceral mechanisms relating to endocrine controls illustrate the *primordial origins of chemical signalling and control systems.* For example, the release of thyroid and sex hormones involves a chain of four successive protein messages, leaving aside the corresponding coevolved receptors on membranes. First, hypothalamic *releasing factors* pass via the portal circulation to control the release by the anterior pituitary cells of corresponding stimulating hormones (e.g., TSH, LH, FSH). Second, these *pituitary hormones*, travelling via the general circulation, induce target *endocrine organs* to secrete thyroid and gonadal *hormones* which affect a number of tissues including the *hypothalamus* and *pituitary* as negative feedback controls. Analysis of protein sequences indicate that these and other neuroendocrine molecular controls are rigorously conserved during evolution. *Protein superfamilies* are engaged in signalling via this same chain of four specific protein sequences, and the chain can be traced all the way back through *early vertebrates* Dayhoff, 1976). Even though some of the hormones ultimately expressed may play distinctive roles in different animals, the controlling protein superfamilies persist. Some ubiquitous protein controls found in mammalian nervous systems can be traced back to prokaryotes, close to the origin of life itself.

REGULATION OF VISCERAL SYSTEMS

We have considered control mechanisms, but *regulation* is used to mean something more. It signifies the maintenance of relative equilibrium—close to constancies—in respect to variables that are essential for cellular well-being. This can be exceedingly complex, because of the demands of a free existence in a sometimes hostile environment. Since throughout the whole of evolution, none of our direct ancestors died before reaching reproductive age, and all succeeded in reproducing something possessing surviving and reproducing characteristics, our survivorship goes all the way back to some fateful catalysts in a primeval sea.

Multicellular organization must be supported by a large number of goal-seeking mechanisms which are successfully integrated. Integrative mechanisms represent physiological inventions to ensure survival which must be safeguarded at molecular, cellular, tissue, organ system, individual behavior, and societal levels of organization. Each level provides a stable foundation for successively more complex levels of organization, providing what Bronowski called *stratified stabilities* (1973, pp. 344–349).

For any physiological system, three questions are pertinent: (1) What is happening? (2) What is the rate of its happening? And, (3) What determines (regulates) events and sets the rates? These questions can be asked about the functions of cells or molecular systems within cells, or about systems like digestion, circulation, respiration, endocrine functions, and the central nervous system itself. The physiologist addresses a given level of stratified stability in terms of *its* regulation, leaving other levels to fend for themselves. All such systems are open systems, i.e., they depend on energy input and heat and waste disposal.

Regulation depends on a *detector* (exteroceptor or interoceptor) to gauge the variable or some derivative of it, and negative feedback of information from the detector to regulate the mechanism. Figure 9.183 illustrates schematically how feedback closes the loop, thus providing for regulation. The feedback may lead to *adaptation* which is more complicated than regulation but serves the same purpose. An example of adaptation is given by the vestibuloocular reflex (see Chapters 63 and 64). In that case, the "error message" is detected in a number of parts of the visual pathways. It feeds back with a complex pattern for adaptation which is then applied to one or both of the two central synaptic relays in the path from vestibular afferents to motoneurons governing the extraocular muscles.

Typical Regulations

Figure 9.184 illustrates how the several regulations of the body have features in common and can

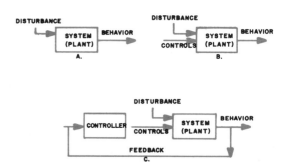

Figure 9.183. Development of feedback control patterns. *A.,* Uncontrolled system. *B.,* Open-loop control; *C.,* Closed-loop control with feedback. [From Grodins, in Yamamoto et al. (1969).]

be diagrammed in a similar manner. Other examples can be added to those given here. On these diagrams each noun or phrase is a variable that can be measured either physically or chemically, or it is a sequence of nerve impulses having a frequency, duration, and pattern that can be determined. The arrows do not have a mathematical significance; they merely denote a logical relationship in time, e.g., "is accompanied by," "leads to," "causes," or "brings about." Thus, a decrease in cardiac output is accompanied by (or causes) a decrease in stretching of aortic walls; this in turn leads to (is accompanied by) a decrease in the output of stretch detectors in the aortic wall and carotid sinus. In these diagrams the nature of the central computational process is not specified. It is represented by Sherrington's term, *integration.*

In bringing about these regulations the endocrine system may play a role, as well as the nervous system. The hormones of the endocrine system not only have specific actions upon their control systems as feedback loops, but they also have metabolic or other actions upon neurons in general. This means that in the absence of a hormone, thyroxine, for example, many portions of the neuraxis may show abnormal function. Similarly, there is reason to believe that sex steroids have specific actions as negative feedback upon control neurons in the hypothalamus, but also more generalized effects among the neural mechanisms responsible for behavior. These possibilities of interaction and control can be summarized by two statements as follows: 1. The central nervous system can exert a quick control of effector systems via neural messages, and also a slower and more prolonged control of certain organs by way of the endocrine glands. 2. Similarly, the nervous system may gather information quickly from neural detectors, or in a slower and more prolonged fashion from the action of hormones or the influence of other generalized changes directly affecting neuronal membranes and/or synaptic mechanisms.

CIRCULATION

The quantities regulated are (1) arterial blood pressure, which is monitored as stretching of elements in walls of major arteries, and (2) adequacy of blood supply in peripheral tissues, monitored as CO_2 tension (and pH) in, for example, skeletal muscle. The latter, the constancy of local blood supply, is maintained almost entirely through local, presumably nonneural mechanisms that alter smooth muscle tone of small blood vessels in response to accumulation of carbon dioxide. The general arterial blood pressure, by contrast, is preserved through a variety of mechanisms that control the following: (1) the force and frequency of the heart beat, and thus the cardiac output; (2) the tone of vessel walls, both in arteries and veins; (3) the degree of opening of arterioles through which blood flows out of the arterial system (this is the "peripheral resistance"); and (4) mechanical forces such as the action of gravity, the contraction of skeletal muscles, and the magnitude of intrathoracic pressure, that affect the return of blood to the heart. These mechanisms are controlled in large part by the autonomic nervous system, with participation of the somatic system and skeletal muscle, as in exercise, or in changes of posture. There is interaction of these controls with those of other systems. Thus, blood gas partial pressures affect the circulation as well as pulmonary ventilation by way of detectors in the carotid and aortic bodies.

These regulatory mechanisms can be influenced artificially in the medulla oblongata. They are integrated in the reticular formation. Similar stimulation evokes circulatory changes from hypothalamus and from the limbic system. It is hypothalamic regulation that is believed to be responsible for circulatory changes during exposure to heat or cold, and for those associated with emotional expression in decorticate animals. The hypothalamus is also regarded as having an important role in the changes that occur in circulation during exercise (Folkow and Rubinstein, 1965). The higher levels including the limbic cortex perhaps interact with the hypothalamus during emotional experience and expression. They may well serve also as components of feedback loops when exercise is initiated by cortical mechanisms. One sees, therefore, that the circulation is controlled by a complex of mechanisms distributed almost the full length of the neural axis (Korner, 1971).

RESPIRATION

Control of pulmonary ventilation depends upon the sensitivity of peripheral and central detectors to partial pressures of carbon dioxide and oxygen and is effected through changes in respiratory minute volume—that is, changes in respiratory frequency and depth (von Euler et al., 1970). Any illusion that the system is a simple one begins to disappear, however, when one moves from consideration of peripheral to central mechanisms. The nature of the central interaction of chemodetectors and the pulmonary stretch receptors is not fully understood, although it is known to take place in the medullary reticular formation, presumably upon the membranes of inspiratory neurons. Moreover, the problem of how the respiratory rhythm is generated in the first place is unsolved. As already noted, one possibility is that it represents a signal

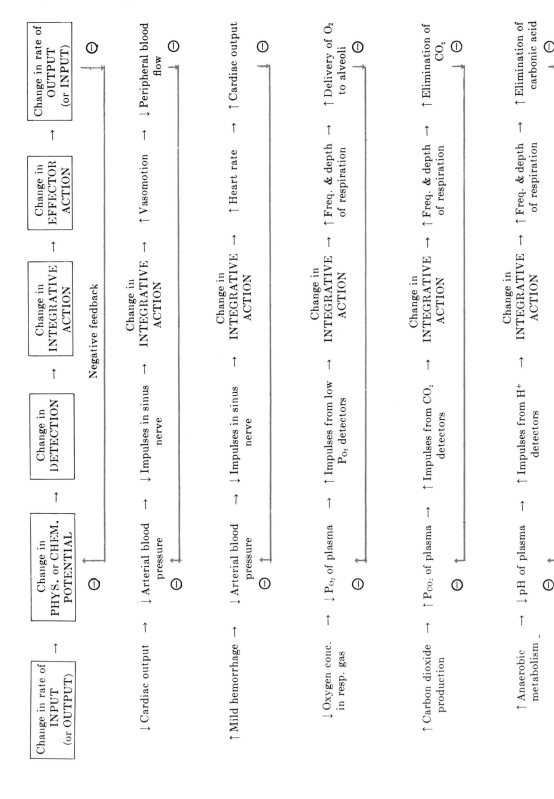

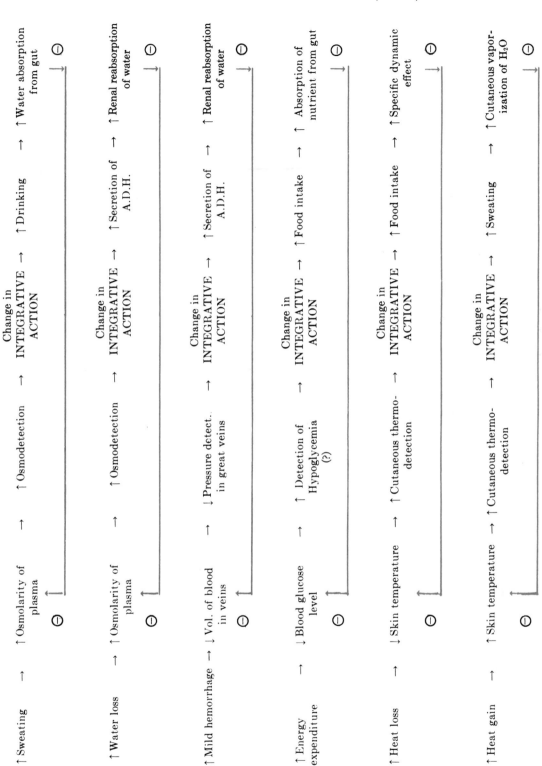

Figure 9.184. Diagrams that illustrate how a change in the magnitude of a physiological variable becomes a change in "potential" that is detected by specialized cells. They send signals to the nervous system as a result of which an effector response is created to alter the magnitude of the variable in an opposite direction. This in turn alters the potential, and hence the signal from detectors. A simple feedback loop of this type will usually oscillate.

generator function of central neurons, in which a train of impulses is modulated so as to create a second order rhythm by way of feedback loops. Control of pulmonary ventilation also interacts with other controls; for example, panting is a respiratory response to thermal stress. Other interrelations are apparent during speaking, singing, or playing of a wind instrument, as well as with exercise and in sleep.

Another characteristic of the control of pulmonary ventilation is the powerful behavioral reactions that can be evoked as part of it. Consider, for example, a subject with his nostrils and mouth forcibly covered; or a subject held involuntarily under water; or in the presence of a noxious gas. The behavioral component of the regulation may supersede all other kinds of behavior. The significant point is that regulations are not only automatic, involuntary, and reflex. They may command the highest levels of the nervous system and include a great variety of patterns of behavior. Respiration must be controlled for gesture vocalization, speech, and singing, for truncal activities such as lifting, shoving, defecation, and childbirth, and other occasions when the respiratory muscles serve auxiliary, nonrespiratory functions.

WATER BALANCE

Hypothalamic mechanisms control both the rate of water loss through the kidneys and the rate of water ingestion in drinking. Water loss was the first of these to be analyzed experimentally. A disorder of this control has been known for centuries, the condition named *diabetes insipidus*, where a patient excretes a large volume of dilute (i.e., "insipid") urine, and in compensation drinks a large volume of water daily. Early experiments did not disclose the nature of the diabetes, because it was assumed that the effective lesion would prove to lie either in the hypothalamus or in the pituitary gland. Later experiments showed that both structures are involved (Fisher et al., 1938) and that secretion of the antidiuretic hormone in the posterior lobe is controlled by neurons of the supraoptic nucleus. By perfusing the base of the brain with blood having an osmotic pressure either higher or lower than normal, Verney (1947) and others demonstrated that this region contains a mechanism sensitive to water concentration in the extracellular fluid. Brooks and his associates (Koizumi et al., 1964), have recorded changes in single unit activity from this nucleus in correlation with changes in osmolarity. Moreover, cats with the hypothalamus removed except for an "island" that includes the supraopticohypophyseal system do not exhibit diabetes insipidus (Woods and Bard, 1960). In any event, for control of water loss via the kidney this system is one that has two stages of transduction. First, there is transduction of water concentration ("activity") into nerve impulses; second, there is transduction of this neural activity into secretion of the antidiuretic hormone (Dreifuss et al., 1971).

Another kind of transduction evidently occurs within the major blood vessels, especially the left atrium, where changes in hydrostatic pressure bring about changes in secretion of antidiuretic hormone (Henry et al., 1956; Share, 1968). By these two systems the rate of water loss is made to vary in response to changes in (1) the concentration of water in the extracellular fluid and (2) the volume of this fluid as reflected in changes in central venous pressure.

Water Intake

For control of water intake the system contains additional detector mechanisms and reflexes. The first of these begins with perception of dryness in membranes of mouth and throat, as suggested by Cannon (1918) and later more fully described by Epstein and his associates (1964). The latter have found in rats that interruption of salivary ducts and extirpation of salivary glands lead to a condition where the animals must necessarily drink when they eat dry food, even though their general bodily state may not require water. A second mechanism is the water concentration in fluids of the body, and a third originates in the pressure-sensitive neurons of the great veins, as in the control of secretion of the antidiuretic hormone. And fourth, the kidney is placing in the circulation something that facilitates drinking. The material that does this is renin, the enzyme that produces angiotension II. Experiments have shown that angiotensin II induces drinking when it is perfused directly into the hypothalamus.

Lesions in the lateral hypothalamus abolish drinking (Montemurro and Stevenson, 1955 and 1956), except that arising from dryness of the mouth (Epstein and Teitelbaum, 1964). The latter can be interrupted by section of appropriate cranial nerves. The intensity of the behavior associated with water intake ranges all the way from simple, almost passive drinking to a really heroic effort when water supplies are scarce (Wolf, 1958). At some level in the neural or neuroendocrine circuits some completely unknown mechanism creates an awareness of need for water. At any instant, for example, any adult subject can state whether or not he feels thirsty. This awareness is presumed to require cortical function, but there is no evidence as to the nature or locus of this highest level component. In the hierarchy of mechanisms for regulation, water balance is somewhere near the middle. The body cannot sacrifice either circulation or respiration as a device for preserving water balance; yet water is more vital to the body than temperature regulation, energy balance, or reproduction.

ENERGY BALANCE

Still more complicated than the regulations heretofore described is the preservation of energy balance. This is assumed to be a regulation because of the constancy of its end product—body substance as measured by the body weight of an adult subject.

Mechanisms of gain of energy are known; in higher animals they are solely by way of food intake. Mechanisms of energy loss are more diverse. Energy can be lost as heat or as work in a truly physical sense—i.e., in raising a mass or giving it a momentum. Or energy can be stored within the body in the form of protein during growth or recovery from starvation; as carbohydrate (glycogen) during recovery from a brief fast; or as fat in almost unlimited quantities.

The possibility that there is some overall system for this regulation is suggested by the following convincing observations. (1) There is the common experience of adult men who find that their body weight is almost constant in spite of variations in activity, and often in the face of efforts to gain or lose. (2) Cohn and Joseph (1962) found that animals made obese by overfeeding would not eat a normal quantity of food until their body weight had returned to the control level (Fig. 9.185). (3) Lesions in the hypothalamus may cause a failure of feeding (Anand and Brobeck, 1951), or hyperphagia and obesity (Tepperman et al., 1941). (4) Injury to the hypothalamus will profoundly alter regulation of body temperature, and thus the energy converted into heat and lost from the body. (5) There is evidence that injury to the hypothalamus or the brain just ahead of it will increase motor output as locomotion.

What has been said about the results of hypothalamic injury can be said in most instances in reverse about hypothalamic stimulation. For example, if a lateral hypothalamic lesion abolishes feeding, stimulation of the lateral hypothalamus will induce feeding (Leibowitz, 1971).

The temperature of many warm-blooded (*homoiothermic*) animals remains practically unchanged, although the surrounding temperature may vary between 0° and 50°C or upward. On the contrary, the body temperature of a cold-blooded (*poikilothermic*) animal such as a frog, turtle, etc., is prac-

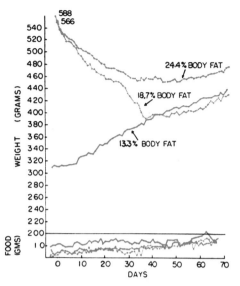

Figure 9.185. Two rats (566, 588) that had been made obese by forced feeding ate a reduced amount of food when they were fed ad libitum, until their body weight and estimated body fat concentration returned to normal levels. [From Cohn and Joseph (1962).]

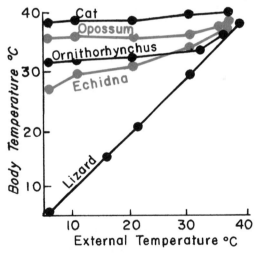

Figure 9.186. Body temperature of five species of animals as a function of environmental temperature. Duration of each exposure was 2 h. [From Martin (1930).]

tically that of its environment during laboratory tests (Fig. 9.186), even though under natural conditions these animals preserve an almost constant temperature by behavioral responses that include migration and hibernation. (See Roberts, 1974.)

The normal human body temperature recorded from the mouth is usually given as 37°C, with the rectal temperature one degree higher. The figure varies between individuals, ranging from 35.8 to 37.8°C orally. Variations also occur in any one individual throughout the day—a difference of 1.0°

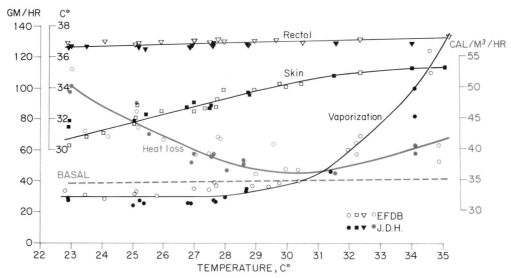

Figure 9.187. Graphs showing the influence of environmental temperature upon rectal and skin temperature (with mean body temperature derived arithmetically from them). Metabolic rate rises during cold exposure, mainly as a result of shivering. Vaporization rises in the heat mainly because of sweating. [From Crosbie et al. (1961), replotted.]

or even 2.0°C occurring between the maximum in the late afternoon or early evening, and the minimum between 3 and 5 o'clock in the morning. Temperature of the several regions of the body depends upon whether they are major sources of metabolic heat (e.g., liver, heart) or locations where heat is given off (e.g., skin, upper respiratory tract) (Fig. 9.187). Strenuous muscular exercise causes a temporary rise in body temperature that is proportional to the severity of the exercise; the level may go as high as 40.0°C.

Stability of body temperature varies with the size of the body. In small animals or in human infants where metabolic rate is relatively more intense and surface area relatively greater, the oral and even the rectal temperature may vary through a range perhaps twice that of larger animals or human adults. This reflects certain principles of the physics of heat storage and loss, rather than any inherent difference in the regulating system. In a small body the amount of heat "stored" is a smaller fraction of the metabolic rate than in larger individuals. In the latter, the stored heat provides an "inertia" that tends to counteract thermal changes. Heat storage is a function of thermal capacity, which varies with body mass; heat loss, by contrast, varies generally with surface area.

Under basal conditions when no external work is being done, all the metabolic energy ultimately appears as heat, and amounts to approximately 1 kcal/kg of body weight/h. Since the body is composed chiefly of water and has a specific heat of a little less than one, the body temperature would

rise about 1°C/h if no heat were lost externally. With strenuous physical exertion more than three-quarters of the increased metabolism appears as heat within the body; the remainder is converted either to work or to heat in the external system. Fever increases the metabolic rate by some 13% for each degree rise in mean body temperature (C). This change in metabolism corresponds to a temperature coefficient (Q_{10}) of between 2.0 and 3.0 overall.

The body's heat-regulating mechanisms are supposed to be in abeyance below an internal temperature of about 23°C, because the underlying neural mechanisms are suppressed. The body then gains or loses heat like an inanimate object. The lower lethal temperature for the human is about 26°C, although lower temperatures have been induced with survival during therapeutic hypothermia. Death is due to cardiac failure. The upper lethal limit is not clearly defined; a few individuals have recovered after attaining a rectal temperature of 43.5°C. Probably the average upper limit is about 43°C.

HEAT LOSS

Heat is lost from the body through (1) radiation, conduction, and convection from the skin; (2) warming and humidifying of inspired air; (3) evaporation of sweat and insensible perspiration; and (4) via urine and feces. Of these several routes only (1) and (3) are directly under physiological control. Radiation is responsible for about 50% of the total heat loss and convection for about 15% (Table

Table 9.10
Partitioning of heat loss of a subject whose daily energy exchange is 3,000 kcal, equivalent to a daily schedule that includes light work

Mechanism of Loss	Kcal	%
1 Radiation, convection, and conduction	1950	65
2 Evaporation of water from skin and lungs, and liberation of CO_2	900	30
3 Warming inspired air	90	3
4 Urine and feces (i.e., heat of these excreta over that of the food and water)	60	2
Total daily heat loss	3000	100

9.10). Urine and feces account for only about 2% or less of the total heat loss. Most of the remainder (about 30%) is lost in the evaporation of water. These percentages are given for an average man doing light work with an overall energy exchange of about 3,000 kcal/day.

Radiation

The rate of cooling of any object varies with the thermal gradient between its surface and objects in its environment. This gradient can be altered by physiological mechanisms that change skin temperature. These mechanisms include the following adjustments in the blood-vascular system:

Redistribution of Blood. By redistribution of blood flow the temperature of the skin may be adjusted to any temperature from about 15°C (during cold exposure) up to the central body temperature. At ambient temperature of 34°C the quantity of blood circulating through the skin may be as great as 12% of the cardiac output. Fingers, serve as highly efficient radiators of heat.

Variations in Blood Volume. An unacclimatized man is capable of increasing total circulating blood volume by about 10% in 2–4 h on exposure to heat of severity sufficient to cause a diffuse cutaneous vasodilatation. The blood is diluted by fluid drawn into the circulation from the tissues, chiefly the skin, muscles, and liver; cells are added from the spleen. Hematocrit drops and concentration of plasma protein decreases. The reverse changes occur during exposure to cold. A sudden transition from a warm to a cool environment causes a rapid decrease of blood volume associated with a marked diuresis.

Increased Circulation Rate. An increased cardiac output ensures a rapid blood flow through the dilated cutaneous vessels. This increase and the enhanced blood volume promote the transportation of heat from the interior to the surfaces of the body and so enhance the loss of extra heat.

Convection

The most important factor influencing heat loss by convection is air movement. For any given air temperature the loss increases with the square of the wind velocity up to 60 mph; beyond this velocity there is little further increase.

Evaporation of Water

It is obvious that the nearer the temperature of the environment comes to that of the blood the smaller will be the amount of heat lost by radiation and convection. Above an ambient temperature of 38°C the body gains heat by radiation from the environment. This heat is then dissipated, together with heat produced within the body, by processes in which water is converted from a liquid to a vapor state at surfaces of the body. The heat absorbed in evaporation of 1 ml of water amounts to 0.58 kcal. Even at ordinary room temperatures when there is no obvious perspiration, the heat lost through evaporation from the lungs and skin amounts to about 17 kcal/h. This is approximately one-fourth of the basal heat production. About two-thirds of this is lost from the skin as insensible perspiration, the remainder from respiratory passages. At higher temperatures the proportion of heat lost by vaporization of water, mainly from the skin, increases dramatically, so that at ambient temperatures above 35°C it accounts for nearly all of the heat lost (Fig. 9.187).

In animals that pant a comparable increase in vaporization occurs via the respiratory system. Rats enhance their heat loss by spreading saliva upon their fur (Hainsworth, 1967). The rate of evaporation of this water is influenced inversely by the degree to which the atmosphere is already saturated with water—by its relative humidity. This is why sweat drips instead of evaporating when the ambient air is humid. A man can maintain a normal temperature in an atmosphere of over 100°C provided the air is perfectly dry. On the other hand, in a damp atmosphere a temperature of 50°C causes body temperature to rise rapidly.

Sweating. Sweat is a weak solution of sodium chloride in water, together with urea and small quantities of potassium ion, other electrolytes, and lactic acid. It has a specific gravity of from 1.002 to 1.003, and a pH varying from 4.2 to 7.5. The concentration of NaCl varies from 50 to 100 meq/l. When sweating is profuse the concentrations of electrolytes rise towards the higher of these values, whereas with acclimatization the percentage is said to decline. Nevertheless, even after acclimatization the performance of strenuous exercise, e.g., marching, may lead to electrolyte and water depletion. If the water is replaced without salt, cramps occur in muscles of limbs and abdominal wall. They can be relieved by administration of NaCl tablets, or by assimilation of the electrolyte that is found naturally in most foods.

Control of Sweat Secretion. The sweat glands, which number over 2,500,000 in a man living in a temperate climate, are cholinergic, although their innervation is by means of postganglionic fibers of the thoracolumbar division of the visceral motor system. Sweating is controlled by mechanisms that are present in almost all levels of the nervous system. For example, in the initial stages of muscular exercise sweating is apparently initiated by the discharge of impulses from limbic cortex, and occurs before there is any significant increase in heat load. Spinal centers exist for segmental control of sweating; this can be demonstrated in quadriplegic patients who sweat reflexly in parts of the body innervated below the level of the transverse lesion of the spinal cord.

The few observations that have been made upon the secretion pressure of sweat indicate that it is high, 250 mm of Hg or more. Sweat is therefore a true secretion and not simply a filtrate. The rate of sweating may be really surprising; it may be as high as 1.6 l an hour. If it were all evaporated this would remove over 900 kcal of heat from the body in each hour.

HEAT PRODUCTION

When the environmental temperature is in the range between 28° and 31°C, the basal heat production can be dissipated to the environment by radiation, convection, and insensible vaporization of water by an unclothed male subject. If the environmental temperature is lowered below these levels, loss by radiation and convection becomes progressively greater, and the mean body temperature falls because the periphery becomes cooler. At the same time there is a gradual rise in heat production (Fig. 9.187). This rise in metabolic rate occurs principally in skeletal muscles as a result of increased tension, even before shivering is initiated. At a level of about 23°C (sometimes called the "critical temperature" for a nude subject) shivering begins (Fig. 9.188). When shivering is intense the overall heat production may be as much as three times greater than the basal rate. It is evident that this is a mechanism that is controlled by the somatic division of the nervous system that subserves an unconscious and automatic function in regulation of body temperature. After the skeletal muscles are paralyzed by a drug such as curare, an animal loses the power to maintain a normal temperature in a cold environment. Its ability to dissipate extra heat is not impaired, however, so that it can still resist high temperatures.

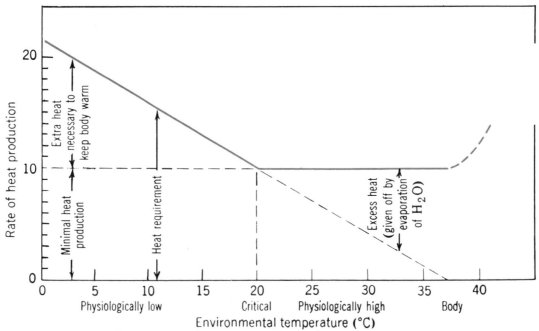

Figure 9.188. Diagram that illustrates the influence of environment temperature upon metabolic rate. At 20°C on this graph the subject can lose all metabolic heat without extra evaporation of water; similarly, body temperature can be maintained without shivering. [From Kleiber (1961).]

CENTRAL MECHANISMS FOR TEMPERATURE REGULATION

Integration of the autonomic mechanisms that control heat loss and the somatic mechanisms that govern heat production, posture, and behavior is accomplished in laboratory animals in the hypothalamus. In man the behavioral adjustments to thermal stress presumably involve higher levels of the nervous system. The significance of the hypothalamus can be illustrated by review of three types of evidence, as follows:

1. Injury to the hypothalamus almost invariably leads to changes in body temperature. If experimental lesions are placed in the rostral portion, the animal may exhibit a severe fever that may progress to death (Clark et al., 1939). This fever is regarded as a release phenomenon analogous to decerebrate rigidity. The rostral hypothalamus apparently contains an inhibitory mechanism that normally acts to suppress the mechanisms of heat production and that also facilitates the stimulation of mechanisms of heat loss. These particular results would follow the interruption of the loop by which information is returned to the controller with a negative sign. The usual therapy for fever of this type, particularly when it occurs in human patients after neurosurgical operations, is forced cooling by means of ice packs and cold water enemas. A more specific therapy is administration of a barbiturate; Ranson and his associates (Beaton et al., 1943) discovered that such a drug selectively decreases the activity of fever-producing mechanisms and so permits body temperature to return to normal. If the hypothalamic lesion is more caudal, and especially if it extends into the lateral hypothalamic areas, the patient or experimental animal will have a low temperature, particularly if the environment is cool. This hypothermia can be corrected by artificial heating, but if it is not so treated it, too, may terminate fatally.

2. The second type of evidence came from experiments such as those of Karplus and Kreidl (1910) and of Ranson (1936 to 1937) and his associates, who discovered that most of the reactions needed for temperature regulation can be brought into play by electrical stimulation of the hypothalamus.

3. The most explicit type of evidence comes from experiments in which the hypothalamus has been artificially warmed or cooled. If the temperature of the hypothalamus is changed by high frequency warming through electrodes, or by conduction via thermodes, an artificial feedback loop is created that is open or independent of what is taking place in the control system. Experiments of this variety were first done by Barbour (1912), who passed warmed or cooled water through fine tubes implanted in the brain of rabbits. Later the technique was refined by Magoun et al. (1939), who oriented electrodes using a stereotaxic method, for passage of a "diathermy" type of high frequency current through the hypothalamus. The warming of a sensitive region that lies over the chiasm in cats, and ahead of it in monkeys, led to polypnea, panting, and sweating on foot pads. Other investigators have extended the observations by use of implanted thermodes in unanesthetized animals and have shown that this region is sensitive to cooling as well as heating. It apparently serves as a thermostat for the rest of the body—as a thermal detector that originates a negative feedback (Benzinger, 1969). In a sense, it utilizes the rest of the body to stabilize its own temperature, cooling the body when it becomes warm, and retaining heat in the body and stimulating heat production when it is chilled (Fig. 9.189). Investigators using microelectrodes have studied the function of single units in this region. Although some neurons seem to be insensitive to temperature and have been proposed as generators of a constant or set point signal, others respond to heating (Figs. 9.190 and 9.191), whereas still others are stimulated by cold.

In addition to this central detector and feedback loop, there is another similar loop that arises from thermal detectors in the skin. Its function is illustrated in Figure 9.192 where the response of an

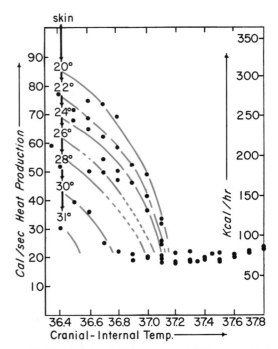

Figure 9.189. Relationship between central (tympanic) temperature and rate of heat production, with skin temperature as another variable (family of curves). [From Benzinger (1969).]

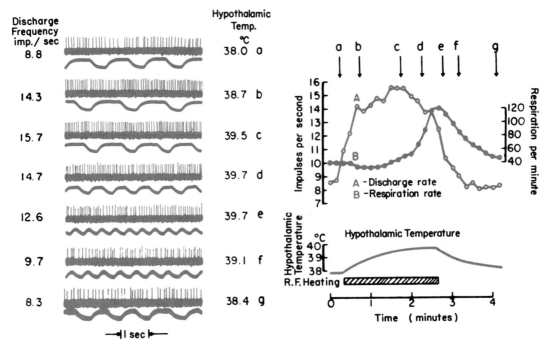

Figure 9.190. Action potentials recorded from single units in rostral hypothalamus during experiment in which the hypothalamus was warmed artificially through an implanted electrode. The graphs indicate the course of the experiment and the frequency of firing of the unit. [From Nakayama et al. (1961).]

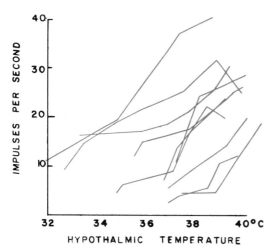

Figure 9.191. Frequency of firing of warm-sensitive units in rostral hypothalamus as a function of hypothalamic temperature. [From Hardy (1965).]

unanesthetized dog to heating of the hypothalamus is shown to depend upon ambient temperature. Stolwijk and Hardy (1966) have devised a mathematical formula that combines these two kinds of signals. According to their concept, the body does not respond to a change in skin temperature when there is no central change, and likewise does not respond to central change in the absence of peripheral deviation. They have concluded, therefore, that the central "integration" is some function of a mathematical product of the deviations in temperature in the two locations.

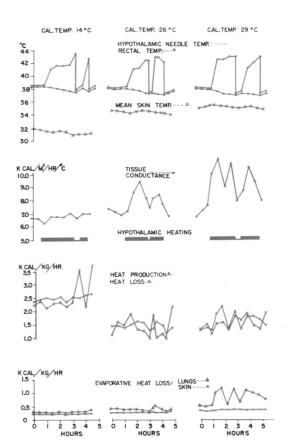

Figure 9.192. Records of experiments in which hypothalamus of a dog was warmed when the animal was exposed to three different environmental temperatures, 14°, 26°, and 29°C. "Tissue conductance" is a measure of vasodilatation in skin vessels. [From Fusco et al. (1961).]

The nature of the set point of this system has been considered by both Hardy (1965) and Hammell (1965); the latter suggested that the elevated temperature of exercise represents a change in set point that originates in neural adjustments inducing or accompanying the exercise. Likewise, there may be changes in the apparent set point in association with the several stages of sleep (Glotzbach and Heller, 1976). A similar explanation may account for the elevated temperature of infectious fevers, but on a biochemical basis at the neuronal level (Eisenman, 1969). Toxins known as *pyrogens* are liberated by bacteria and can be isolated also from white blood cells (Wood; see Moore et al., 1970; also Dinarello and Wolff, 1978). In effect, they stabilize the membrane of the central thermodetector neurons so that a higher temperature is required to induce the negative feedback. The possibility that the chemical intermediary of the action of pyrogens is the class of compounds known as prostaglandins has been suggested by Vane (1971). His suggestion is based upon the discovery that aspirin, a drug with pronounced antipyretic action, inhibits the synthesis of prostaglandins, and upon reports that prostaglandins injected directly into the hypothalamus induce fever. Prostaglandins are widely distributed compounds whose biological significance is not entirely clear; they are synthesized in the body from polyunsaturated fatty acids and are believed to be related to the cyclic adenosine monophosphate system (Horton, 1969).

ENDOCRINE GLANDS IN THERMOREGULATION

The thyroid and adrenal glands play significant roles in the regulation of body temperature. The calorigenic effects of the secretions of these glands are well known. Cannon (1932) observed that exposure to cold causes an increase in the rate of the denervated heart of a cat, as a result of epinephrine secretion. This hormone exerts a calorigenic effect which is immediate and of short duration. A less immediate and much more prolonged increase in heat production is brought about by stimulation of the thyroid gland. Rats exposed over a period of 3 weeks to low temperatures (7–12°C) show thyroid hyperplasia and a rise in metabolic rate of as much as 16%. Thyroidectomized rats, on the contrary, show little rise in metabolic rate under the same conditions. Uotila (1939) found that this stimulation of the thyroid did not occur if the pituitary stalk had been previously sectioned. He thus provided the earliest evidence that secretion of the thyrotrophic hormone by the anterior pituitary gland is under the control of hypothalamic mecha-

nisms. This control is now known to be exerted by way of thyrotrophin-releasing factor.

Some years ago Ring (1942) demonstrated another interesting mechanism of protection against cold exposure when he found that thyroid hormone potentiates the calorigenic action of epinephrine. This means that thyroidectomized animals not only are deficient in the action of thyroid hormone on metabolic rate, but also that the epinephrine they secrete in the cold has a reduced potency in elevating heat production even in a transient way.

EVIDENCE RELATING TO HYPOTHALAMIC INTEGRATION

The hypothalamus plays a critical role in regulation, and there is some knowledge concerning integrative mechanisms, particularly relating to the *supraoptic* and *paraventricular nuclei* (Swanson and Sawchenko, 1983). These nuclei contain cells that synthesize either oxytocin or vasopressin and release them into the bloodstream from their terminals in the posterior pituitary, the *neurohypophysis* (Fig. 9.193).

Oxytocin is liberated reflexly following distension of the uterus and vagina as in childbirth and to some degree during intercourse, and during stimulation of the nipples as in suckling. The effects are

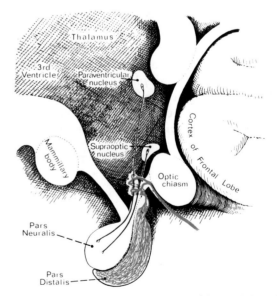

Figure 9.193. Schematic representation of key relations between hypothalamus and hypophysis (pituitary). The hypthalamohypophyseal tract carries neurosecretory granules between the paraventricular and supraoptic nuclei and the neurohypophysis (posterior lobe). The figure also depicts the hypothalamic nuclei, mostly in the region of the median eminence, which send axon terminals to discharge releasing factors into the blood vessels of the portal circulatory path to the anterior hypophysis. [From Noback (1967).]

on the uterus, which is induced both to secrete and to contract, and on the secretory epithelium and contractile ducts of the breast for "let down" of milk. Late in pregnancy, the uterus is particularly sensitive to oxytocin. The effects ensure strong uterine contractions, easier parturition followed by postpartum lactation, and further, suckling-induced uterine contractions.

Vasopressin, or antidiuretic hormone, is released as a consequence of an increase in plasma osmotic pressure from 290 mosmol/l by as little as 2%. This can be induced by injecting hypertonic saline solutions into the carotid artery which is followed by increase in firing rate of supraoptic neurons and an increase in plasma vasopressin (Fig. 9.194, Koisumi et al., 1964; Koisumi and Yamashita, 1978).

The *supraoptic nucleus* straddles the lateral border of the optic chiasma, and in the rat it contains roughly 5,000 cells. The *paraventricular nucleus* lies ventromedial to the descending column of the fornix as it leaves the septum to enter the hypothalamus (Fig. 9.193). This nucleus is parceled into several groups of cells with a total cell count of about 10,000 neurons. The dendrites of cells in each of these nuclei are largely confined to the limits of the respective nuclei, and *most of the axons of the large neurons pass to the neurohypophysis.* There are many soma-somatic gap junctions in both nuclei with conspicious *electrotonic coupling* of the cells. Interneurons support probably one-third of the synapses in the nuclei. Nonetheless, the two nonapeptides are found in separate neurons. The hormones and their respective neurophysin "carrier" proteins are localized exclusively in these two hypothalamic nuclei. *Enkephalin* as well as *vasopressin* have been detected in individual axon terminals in the posterior lobe. *Essentially all of the central neurotransmitters and neuropeptides thought to be involved in neurotransmission have been identified in cells or fibers in these two nuclei.* The propagation

of impulses in their predominantly unmyelinated fibers is less than 1 m/s. The *release of hormones at the nerve terminals is calcium-dependent and the consequence of electrical activity in the axons* (Lincoln and Wakerly, 1975).

In contrast to the supraoptic nucleus which apparently projects only to the neurohypophysis, neurons in the *paraventricular nucleus* also project to cells that contribute to the portal system in the *median eminence* and also directly to cells in the *brain stem* and *spinal cord.* Thus neurons in this hypothalamic nucleus *can influence hormonal secretion from the anterior as well as the posterior hypophysis.* Moreover, it can activate preganglionic *sympathetic* and *parasympathetic* neurons throughout the autonomic nervous system (Swanson and Swachenko, 1983). Individual neurons contribute to each of these separate destinations, and they are segregated within the nucleus, suggesting that there may be both coordinated autonomic and neuroendocrine controls, as implied by the gap junctions, and also some degree of independent activation.

Both nuclei are involved in the *release* of *oxytocin* from the neurohypophysis in response to dilation of the birth canal and suckling. The *paraventricular nucleus* is also responsive to baroreceptor activity from aortic, carotid sinus, the renal nerves and thereby *contributes to cardiovascular regulation.* It appears that several afferent systems contribute strictly segregated input to this nucleus.

MECHANISMS OF REINFORCEMENT

Visceral control mechanisms are organized to provide a satisfactory internal milieu, but organisms, especially higher forms, are engaged in additional pursuits. As the philosopher Santayana argued, all behavior is in pursuit of internal satisfactions. For higher animals, and especially humans, there are many activities which depend on satisfactions being met in relation to visceral housekeeping and yet which cannot be directly or simply ascribed to appetites and visceral drives. Some of these behaviors relate to satisfactions that may be only partially satisfied and largely postponed for years, like studying medicine. How can brain circuits be adapted or shaped to relatively attenuated as contrasted to immediate satisfactions. In this context we need to consider *mechanisms that underlie learning* and *memory* and *motivations* that may be relatively *remote from immediate visceral controls* (Livingston, 1967a and b).

Reinforcement subsumes processes involved in what are popularly called *reward* and *punishment,* *experiences liked* and *disliked,* and *behaviors impelled toward* (*approach*) and *away from objects* and

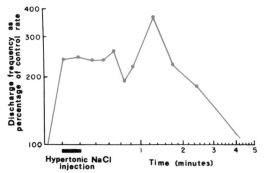

Figure 9.194. Increased frequency of discharge of neurons of supraoptic nucleus (ordinate) during and after carotid injection of hypertonic sodium chloride solution (0.5 ml of 1 M NaCl solution). [Koizumi et al. (1964).]

experiences (avoidance). The term reinforcement has been adapted from more elementary neurophysiological phenomena that are concerned with the strengthening of a unitary response to one stimulus by the concurrent advent of another stimulus. In the present context, *reinforcement relates to large scale strengthening of interactions affecting whole circuits as a consequence of stimuli,* as in the learning process.

Pavlovian (type I) conditioning involves somatic or visceral stimuli that are paired with somatic or visceral responses, in such combinations that a yielded response is more likely to recur to a stimulus that had not initially been effective. *Reinforcement,* e.g., food for a hungry animal, *is the means by which the conditioning* (the new or strengthened connections) *is established. Type II, operant conditioning,* usually associated with Konorski (1948) and more recently Skinner (1966), *involves reinforcement which is applied after* a desired or approaching desirable *behavior has been emitted* by the animal. In the Pavlovian case, the experimenter controls when the conditional stimulus (CS) will be administered; in the Konorskian case, the animal ventures a behavior which is then reinforced.

Reinforcement both positive (rewarding) and negative (punishment) has been a principal means of shaping behavior since time immemorial, but until 1954, it was known only in terms of external applications. In that year, independent discoveries by Delgado, Roberts and Miller, and by Olds and Milner revealed central *nervous loci are capable of positive or negative reinforcement.* In general, negative reinforcement can be elicited from *lateral midbrain and posterolateral diencephalon* extending into ventral hypothalamus and ventral thalamus. *Positive reinforcement is more widespread,* including the whole of the *limbic system* and most of the *median forebrain bundle,* sweeping from septum through medial hypothalamus and down into medial midbrain reticular formation and *periaqueductal gray.* There is no question that stimulation to certain areas yields negative reinforcement and that similar stimulation to other areas yields positive reinforcement. The volumetric extent of the latter is, fortunately, much greater, roughly 10-fold.

Mechanisms for central reinforcement relate closely to systems involved in the experience of *feeling states and moods;* indeed, they may consist of the same or overlapping circuits. They are similarly closely related to mechanisms concerned with motivation. As Pavlov wrote, "with a dog which has not long been fed, the unconditioned stimulus has only a small effect, and alimentary conditioned reflexes either are not formed at all or are estab-

lished very slowly." Moreover, it is evident that central stimulation in areas of positive or negative reinforcement can be used to modify appetitive and motivated states. The close relationship—anatomical, physiological, and subjective—between drive and reinforcement has been firmly established.

To survive, an organism must be able to modify its behavior in such ways that its basic needs are met despite changes in circumstances. The nervous system is organized so that behavior is not only goal-directed; it is guided by the *consequences* of both spontaneous and responsive behavior. Actions are initiated by internal feeling states and drives. They are maintained or altered or discontinued according to success or failure in fulfilling the satisfactions sought. Central reinforcement and drive mechanisms, affected by the consequences of behavior, modify subsequent behavior. *The nervous system is built for action; the actions are goal-directed; the goals are internal satisfactions; systems for satisfaction and dissatisfaction are built into the fabric of the brain stem, diencephalon and limbic system. Perceptual processes, memory, and learning, and the processes involved in consciousness are all directed toward biologically improved internal satisfactions.*

In the centralmost portions of the forebrain, neighboring the ventricular passages and forming a ring of cortical and subcortical structures around the hylus of the hemispheres, are *organized systems involved in feeling tone, approach and avoidance, appetite and satiety.* Lesions in alimentary appetitive loci lead to reduction of drive for food; lesions in food aversion loci lead to inadequacy of restraint on the drive that is normally exerted by satiety. The surrounding limbic system is less imperious, less compelling, and more modifiable through experience than are the underlying core systems. In the *limbic system,* there seems to be more *flexibility for exercise of judgments* as to whether the ongoing event is biologically significant and whether it is to be approached or avoided. *The limbic system can modulate or control the core systems.*

Out beyond the limbic system is phylogenetically newer *neocortex.* Neocortex appears to be even less imperious and perhaps even neutral in regard to approach or avoidance. It *is apparently not essential for exercise of judgment regarding biological significance.* Stimulation of neocortex does *not* give rise to bar-pressing for the satisfaction derived from local activation.

Positive and negative reinforcement mechanisms exist without obvious biological advantage beyond the consequences of activation of particular brain regions. Reinforcement is involved directly when there is actual biological gain, such as food for a hungry animal. But *it is not essential that there be an obvious biological gain for reinforcement to be manifested* through approach or avoidance behavior, including central self-stimulation. Behavior associated with reinforcement can also exist without any obvious drive reduction.

There is an interface between innate and reflexive behavioral patterns and learned patterns, between nonconscious and conscious patterns, which cannot be defined in static terms; it appears to involve an adaptive, dynamic boundary. *For behavior to be learned, for* improvement in biological directedness and goal satisfaction, the intervening steps do not need to be consciously experienced. And, as William James emphasized, even a consciously learned response can become so habitual that it submerges to a nonconscious level and thereafter is carried out quite automatically by the nervous system, more or less on the same plane as genetically endowed behavior.

CHAPTER 70

Segmental Control of Skeletal Muscle

ORGANIZATION OF SPINAL SEGMENTS

The immediate spinal control of skeletal muscles is governed by an uninterrupted column of anterior motor horn cells whose axons leave the ventral surface of the spinal cord on each side via a continuous longitudinal stream of *ventral roots*. These roots, and the corresponding dorsal roots which enter the dorsolateral aspect of the spinal cord, gather laterally to form a succession of *spinal nerves*.

The soft and vulnerable central nervous system is surrounded by skeletal armor. It is necessary for the corresponding cranial and spinal nerves to penetrate this armor to gain access to sensory and motor mechanisms in the body wall. Cranial nerves find their way through the skull via special *cranial foramina*. Spinal nerves exit the vertebral column by way of *intervertebral foramina*. According to the organization of spinal nerves and the intervals between intervertebral foramina, the spinal cord is divided into a series of *spinal segments*.

Since the intervals at which nerve roots are gathered into successive pairs of spinal nerves vary according to the intervals between intervertebral foramina, it follows that spinal segments vary in length. In the cervical region they average about 13 mm, in midthoracic about 26 mm, and in the lumbosacral levels they diminish from 16 mm to 4 mm.

Spinal segments innervate specific peripheral derivatives of embryonic dermatomes and myotomes. By careful sensory and motor examination, you can distinguish among peripheral, intraspinal, and central cord lesions, and localize the level of spinal cord defects. Based on functional organization within the spinal cord, some approximation can be made as to which zones of the spinal cord may be most compromised. Peripheral nerve defects can be determined with considerable precision because the patterning of the original spinal roots changes during convergence and rebranching within the brachial and lumbosacral plexuses.

Protection of the Spinal Cord

The bony vertebral column provides powerful vertical axial support. It permits a remarkable degree of skeletal flexion in all directions, while allowing considerable axial rotation. Flexibility is provided by the cushioned stack of vertebral bodies to which are attached, posteriorly, pedicles and arches which enclose the spinal canal. Between the pedicles, facing directly across the spinal canal, are the intervertebral foramina through which spinal nerves and vessels pass into and out of the spinal canal.

The spinal canal is lined with a smooth, continuous, tough, elastic, flexible ligamentous sheath. This is open at the foramen magnum, which provides entry into the skull, and at each of the paired intervertebral foramina.

The *dura mater* forms a slender tubular sheath suspended within a much larger spinal canal from its attachments along the base of the skull and around the foramen magnum. The spinal dura hangs freely suspended, a cul-de-sac, from the foramen magnum to its attachment at the lower level of the second sacral vertebra. The epidural space is filled with loose, fatty areolar tissue and a plexus of veins which cushion the spinal cord and spinal nerves within the canal.

The vertebral column grows at a faster rate than does the spinal cord. Spinal nerves, which in the embryo pass horizontally through their respective intervertebral foramina, become longer and more oblique from above downward, so that lumbar and sacral nerves descend almost vertically for several segments to reach their foramina. As they extend beyond the terminal end of the cord, the abundant spinal nerves resemble a horse's tail, hence the name, *cauda equina*. In the adult, the terminal end of the spinal cord lies at the first lumbar vertebra. A needle can be relatively safely inserted below that interspace to obtain samples of cerebrospinal fluid.

Spinal nerves, which are enclosed by their own

meningeal sheaths, are more rugged, pliant, and displaceable than is the spinal cord; hence, they contribute to freedom of the spinal cord to accommodate bending and twisting movements of head on neck and so forth along the spine, and especially in the lower back. The long trailing spinal nerves provide additional mechanical cushioning to lower parts of the spinal cord.

INJURY TO THE SPINAL CORD

Despite such relative freedom and protection, spinal cord injury occurs frequently, and the effects are usually devastating. Apart from intrinsic spinal cord malformations, tumors, cysts, and infectious processes, and mechanical problems due to arthritis and herniated disks (protrusion of the elastic nucleus pulposus), which mainly affect spinal nerves, *there are more than 200,000 people in the United States who are paralyzed as a consequence of spinal cord injury.* They are mostly young people, most of them victims of automobile and motor cycle accidents.

If spinal injury occurs below the second lumbar vertebra, damage is apt to affect spinal nerves rather than the spinal cord. This makes it likely that losses of sensory and motor function are less comprehensive. Peripheral nerve recovery is also more promising than is recovery from central nervous system damage. Injury to the spinal cord at a higher level is apt to result in a complete transection with a well-defined level of interruption of sensory and motor functions. If good care is provided, there is likely to be healthy survival of the detached portion of the spinal cord. Its reflex capabilities may be of some practical value for turning in bed and for triggering reflex bladder and bowel emptying, and at the same time they may be a conspicuous nuisance.

Many neuroscientists are investigating ways to provide a functional bridge across the gap in the severed spinal cord to establish recommunication of sensory and motor functions. Perhaps the most promising potential bridge is by using grafts of embryonic spinal tissue or cell suspensions of embryonic spinal cord which seem to be capable of establishing functional synaptic connections in adult central nervous systems of laboratory animals. It is not yet known to what extent such connections will result in useful functional restitution.

Paraplegia and Quadriplegia

A convenient way to remember spinal levels is that spinal transection at the level of the tenth thoracic spinal roots results in motor and sensory disconnection at the level of the umbilicus. This obviously includes loss of voluntary use of both legs and of bowel and bladder. The patient is said to be *paraplegic.* Lesions at cervical levels leave a person *quadriplegic.* People with lesions as high as the second cervical segment can survive if they are promptly given adequate artificial respiration. Inasmuch as both intercostal and phrenic nerves are cut off from the respiratory center, permanent artificial respiration must be maintained. A great deal of other sustained caregiving is also obviously mandatory. Nonetheless, such individuals may acquire mobility using a motorized wheelchair controlled by a joystick manipulated with the lips.

Important insight into what spinal cord injury entails from a physiological, psychological, and social point of view should be comprehended by all physicians: I recommend *Sitting on a Basketball: How it Feels to be a Paraplegic,* by the distinguished neurosurgeon, Paul C. Bucy (1973).

SPINAL CONTROL OF SKELETAL MUSCLE

Development of Spinal Motor Organization

Specific motoneurons are connected peripherally in precise and reproducible patterns to certain muscles. Early specification of motoneurons appears to be based on their position in the neural tube. Medially located motoneurons project to ventral muscles, and lateral motoneurons project to dorsal muscles. A specific recognition process takes place between the motoneuron and the muscle fibers once axons have reached the muscle. Studies of the development of chick motoneurons apparently exclude the hypothesis that outgrowth is random, followed by a dying back of cells that fail to connect appropriately (Landmesser, 1980).

Centrally, motoneurons appear to respond precisely to certain afferent neurons and interneurons. It is suspected but not yet proven that this central organization depends in part on trophic factors taken up by the motoneuron from the muscle, a myotrophic-neurotrophic-neurotrophic specification which can operate progressively transsynaptically in the retrograde direction, affecting functionally related afferents and spinal interneurons.

Specialization of Spinal Interneurons

Although spinal sensory and motor organization is extremely extended longitudinally, it functions as a well-integrated unit. To operate as a whole throughout its great length, the spinal cord is integratively bound together by a system of remarkable interneurons. *Almost all spinal interneurons have extraordinarily long dendrites and even longer ax-ons.* Their dendrites reach upward and downward along the neuraxis for several segments, tapping into many converging segmental afferent inputs and numerous channels that convey higher command signals. Their axons often course the whole length of the spinal cord and send collateral signals into all spinal segments and caudal brain stem as well.

FUNCTIONAL ORGANIZATION OF SPINAL INTERNEURONS

Spinal interneurons are divided into functional groups, each of which participates in organization of a particular type of spinal activity. Their functional role is determined by impulses arriving from supraspinal systems involved in motoneuron command and control and from particular peripheral receptors involved in reflexes (Kostyuk and Vasilenko, 1979). As noted above, interneuron specification may also derive from the muscles with which they are functionally related.

The various types of interneurons influence one another in a vast interchange of excitatory and inhibitory impulses. For example, low-threshold group Ia afferents from muscle spindles in several muscles converge on a single type Ia-interneuron where their excitatory impulses summate. Ia afferents from antagonistic muscles inhibit these Ia-interneurons. These same Ia-interneurons are activated by γ-motoneuron innervation of the muscle spindle in parallel with the motoneurons of the same muscle and send inhibitory signals to several antagonists of that muscle. Low-threshold cutaneous and high-threshold muscle afferents exert a facilitatory effect on the same Ia-interneurons through pleurisynaptic connections. Excitatory influences descending in the medial longitudinal fasciculus (MLF) and from the lateral vestibular (Deiters') nucleus converge on the same Ia-interneurons. The same Ia-interneuron-motoneuron combination receives inhibitory feedback from Renshaw interneurons which are excited by collaterals of the axons of the same motoneuron. It can be appreciated from these details that there is a great deal of functional specificity in the organization of spinal interneurons.

Shorter and Longer Propriospinal Interneurons

Relatively short propriospinal interneurons constitute a population of interneurons in the deeper laminae of the spinal gray matter (laminae V–VIII). Their axons pass into the lateral and ventral funiculi for distances of five or so segments to form short propriospinal paths, with conduction velocities that range from 35–40 m/s. More numerous longer interneurons in dorsolateral and in more ventral cord quadrants ascend and descend longer distances and have higher conduction velocities (110–120 m/s). These latter provide interconnections between lumbar and cervical levels, and between spinal and brain stem motor control systems. *The relatively longer distance fibers which have relatively more rapid conduction velocities contribute to the swiftest integrative operations of spinal and brain stem motor mechanisms as a whole.*

Shorter propriospinal interneurons receive corticospinal, rubrospinal, tectospinal, reticulospinal, and vestibulospinal collaterals. They receive descending monosynaptic excitation followed by subsequent pleurisynaptic excitation from these same sources. This means that such descending activation can be relatively sustained. Segmental afferents influencing these interneurons generate excitation to flexors and inhibition to extensors and thereby contribute to flexion reflexes. The interneurons in this descending motor control pathway are influenced by afferents only pleurisynaptically and weakly. Very strong afferent stimulation can, however, override descending command and result in flexion (protective, avoidance) responses.

Altogether, this suggests that *higher command obtains priority access to relatively shorter interneurons which contribute a penultimate integrative role prior to higher command's direct as well as indirect control of motoneurons.*

Glycine as an Inhibitory Neurotransmitter

Another feature peculiar to spinal integration is that much of the inhibition is conveyed by an inhibitory neurotransmitter that is characteristic of the spinal cord. *Inhibition,* which is so important in all integrative processes, is provided in the spinal cord predominantly by means of the chemically simplest neurotransmitter, *glycine. Although all other known neurotransmitters, including the most exotic, have been found in the spinal cord, glycine, as an inhibitory transmitter, is rarely found elsewhere.*

Spinal Interneurons Contrasted with Interneurons Elsewhere

Spinal interneuronal specialization, whereby single interneurons communicate throughout practically the entire motor column, stands in marked contrast to other central nervous system nuclei where interneurons have much more circumscribed dendritic and axonal processes. Generally speaking, most interneurons belong to the class of Golgi type II neurons, which by definition have very short processes. Many such interneurons have no axons (e.g., amacrine cells in the retina). Therefore, an important generalization is that *the nervous system provides local morphological and biochemical strategies to satisfy specific local integrative requirements.*

Spinal interneurons gather information from many segments and many sources and provide integrated motoneuron control influences, excitatory and inhibitory, directly and indirectly, to motoneurons over the full extent of the spinal cord and into the brain stem. This specialized organization helps to integrate segmental and interlimb reflexes and the important systems of reflexes that operate between head and body.

Specialization of Motoneurons

MOTONEURON DENDRITES AND VARIOUS INPUTS

Another notable contribution to spinal and cranial integration stems from the fact that *spinal and cranial motoneurons have great shaggy dendrites which branch three-dimensionally and reach out into neighboring segments, engaging with interneurons, reticular formation, motor, and sensory nuclei.* A comparable dendritic outreach is characteristic of neurons in brain stem and spinal reticular formation with which brain stem and spinal motoneurons and interneurons are functionally intimately related. In fact, neurons in the reticular formation are effectively interneurons for interneurons. This enables cranial and spinal motoneurons to take advantage of much convergent information that has already been synthesized by reticular neurons, making available two-tiered synthesis, synthesis *squared.*

The dendrites of spinal and cranial motoneurons thus present a very expanded arborization which makes available abundant synaptic surfaces for specialized interneurons, reticular neurons, incoming sensory signals, and suprasegmental motor commands. Spinal and cranial motoneurons are thereby engaged in condensed comprehensive convergence and synthesis of multiple sources of information. By this means, *almost everything that generates and shapes somatic behavior can obtain relatively direct access to the final common path to behavior—brain stem and spinal skeletal motoneurons.*

Integrative Organization of Brain Stem Motor Controls

As we have seen, spinal motor neurons governing skeletal muscle are somatotopically organized in a longitudinal column in the ventral horn. *The most ventral cranial motor column in the brain stem innervates somatic skeletal muscles derived embryologically from myotomes.* This cranial skeletal motor column is homologous to the spinal skeletal motor column in the ventral horn. It includes motor nuclei of the oculomotor (IIIrd), trochlear (IVth), abducens (VIth), and hypoglossal (XIIth) nerves.

In the brain stem, a unique motor column governing another kind of striated muscle is added: *A more lateral cranial motor column innervates special visceral striated muscles which derive from embryonic branchial arches, ultimately from the neural crest.* These special visceral striated muscles are innervated by motor nuclei of the trigeminal (Vth) and facial (VIIth) cranial nerves, in addition to the nucleus ambiguus. The nucleus ambiguus contributes motor fibers to the glossopharyngeal (IXth), vagus (Xth), and accessory (XIth) cranial nerves which together innervate striated muscles of the pharynx and larynx. Since this effector system is unique to the head, there is no spinal homology. (More dorsally in the brain stem is a column of cells which controls smooth muscle and glands, e.g., the dorsal motor nucleus of the vagus (Xth) cranial nerve. This is homologous to visceral preganglionic neurons in the lateral horn of the spinal cord.)

Brain stem somatic skeletal muscles and visceral striated muscles function in close harmony with one another and with autonomic systems which control smooth muscle and glands, e.g., pupillary widening (Edinger-Westphal nucleus of the IIIrd) and salivation (superior salivatory nucleus, VIIth, and inferior salivatory nucleus, IXth cranial nerves).

Now, imagine the coordinated activity that is involved in visually targeting a morsel of food and appreciating it with pupillary dilation and salivation. Then comes ingestion, chewing, and swallowing, and vocal gestures of satisfaction. This biologically fundamental and often practiced performance involves elements in all three of these cranial motor columns: somatic skeletal, visceral striated, and autonomic.

Integrative Organization of Brain Stem and Spinal Reflexes

Swallowing, coughing, and vomiting are extremely important reflexes. Stimulation of the pharynx and back of the tongue initiates *swallowing.* This is a linked chain of reflexes that delivers food and fluids safely across the glottal lid over the trachea to the stomach. As commonplace experience dictates, breathing and swallowing are incompatible: they must be reciprocally controlled when eating and drinking. Swallowing begins with signals carried in the glossopharyngeal and vagus nerves to the nucleus of the solitary tract and neighboring reticular formation. This region coordinates swallowing. Boluses of food and drink are virtually hurled into the esophagus by coordinated and exquisitely timed strong thrusting movements of the tongue (hypoglossal XIIth), palate, and pharynx (glossopharyngeal IXth, and vagus Xth). It is instructive to watch this swift and vigorously thrusting reflex in a slow motion film of a fluoroscopic examination of swallowing. The action needs to be slowed down in order to appreciate the deftness and accuracy of this vital action.

When swallowing falters and food, fluids, or other foreign objects get into the windpipe, irritation of the tracheal and bronchial mucosa elicits *coughing*. Afferent impulses travel in the glossopharyngeal (IXth) and vagus (Xth) nerves to the nucleus of the solitary tract. Motor impulses are mediated through the IXth, Xth, and XIIth nerves to the pharynx, larynx, and tongue. Other essential impulses control the coordinated expulsive movements executed by the diaphragm, intercostals, and abdominal muscles.

Vomiting is another propulsive gesture. In this case, there is a rapid ejection of gastric contents effected by coordinated contraction of abdominal muscles and diaphragm at the same time that the cardiac sphincter, throat, and esophagus are opened widely. Concomitant closure of the glottis prevents aspiration of vomitus. Vomiting is preceded and accompanied by copious salivation which provides lubrication and some dilution of acidity of the vomitus.

Vomiting can be initiated by the gag reflex but more generally is caused by irritation of the mucosa of the upper digestive tract. This initiates impulses that reach the solitary tract by way of the glossopharyngeal (IXth) and vagus (Xth) nerves. The coordinated response is controlled by a "vomiting center" in the medulla oblongata.

Vomiting can be triggered by centrally acting agents (e.g., apomorphine). These initiate responses from a chemosensitive zone in or near the *area postrema*, an area in the brain stem that is outside the blood-brain permeability barrier and is therefore accessible to many substances carried by the bloodstream. The area postrema then activates the vomiting center.

These reflexes, swallowing, coughing, and vomiting, are counted among vital reflexes because if any of them fails to function in nearly perfect form, the individual is bound to succumb, sooner or later, from aspiration pneumonia. Temporary bypass of the respiratory channels protects individuals from the consequences of temporary failure of these reflexes. Permanent tracheostomy is difficult to manage.

MOTOR UNITS AND THE CONTROL OF SKELETAL MUSCLE

Functional Plasticity of Motor Units

Chapter 4 depicts *excitation and contraction of muscle* along with the *physiology of the neuromuscular junction*. Recall that the combination of motoneuron and the muscle fibers under its command constitutes a motor unit. Functional unity is bolstered by contributions of trophic factors which are exchanged both ways between muscle and neuron. Note: *there is no spontaneous activity in normal mammalian skeletal muscles; the effects of a nerve impulse are always excitatory; and the neurotransmitter is invariably acetylcholine.*

RED, INTERMEDIATE, AND WHITE MUSCLE FIBERS

Skeletal muscle fibers are divided functionally into three types: "red," "intermediate," and "white." Most muscles consist of a mixture of the three, but proportions vary widely in different muscles. *Red fibers* are smaller, more granular, and darker. They are adapted for sustained posture-maintaining, antigravity contractions. Capillary density and mitochondrial and myoglobin contents in red fibers are greater, suggesting that these fibers are *suited for relatively higher levels of continuous metabolic activity.* They are activated by relatively low-frequency, steady trains of action potentials.

White muscle fibers are adapted to swift, phasic contractions. Their motoneurons generate intermittent bursts of relatively high-frequency impulses. Intermediate type muscle fibers are, not surprisingly, intermediate in respect to these characteristics.

Relatively speaking, larger motoneurons have larger diameter fibers which subdivide into more numerous branches and terminals, forming larger aggregations of muscle fibers. They serve white or white and mixed white and intermediate type muscles. Thus, they tend to discharge more phasically, with infrequent bursts, at higher discharge rates, than do smaller motoneurons.

Twitch and Tetanic Contractions

A refractory period is necessary for repolarization of the muscle membrane. Repetition of nerve impulses reestablishes or sustains tension. During the briefest contractions of individual muscles, *twitch contractions*, extraocular muscles show fastest onset and briefest duration of contraction, as compared with gastrocnemius, intermediate, and soleus, relatively the slowest muscle of the three. *The rate at which twitches overlap and summate (developing tetanus) depends on the combined speeds of contrac-*

tion and relaxation of the muscle fibers. Inasmuch as these rates vary greatly from muscle to muscle, it follows that *the frequency of nerve impulse discharges necessary to achieve smooth maximum contractions varies enormously.*

Fast and Slow Muscles

Fast and slow muscles are functionally adaptive. Their adaptation is dictated by the source of their innervation, and specifically by the pattern of nerve impulses by which they are driven. Reversing innervation between fast and slow muscles converts the characteristics of the muscle to the type of reinnervation: slow becomes fast and vice versa (Buller et al., 1960). Moreover, if implanted electrodes are applied to a fast motor unit axon and a consistent low-frequency stimulus is applied, the fast muscle assumes the characteristics of a slow muscle. A slow muscle given intermittent high-frequency stimulation assumes the characteristics of a fast muscle.

Much of what is achieved by athletic conditioning and training in dexterity, e.g., playing a musical instrument, relates to mutual functional improvements that can occur in motoneurons and muscle fibers, i.e. conditioning of motor units. Other conditioning effects relate to peripheral sensory contributions, especially the γ-motoneuron-muscle spindle system, and also functionally related circuits in the spinal cord and brain stem. *In sum, a great deal of motor conditioning and adaptation can be accomplished among components involved with immediate effector controls.*

THE SKELETAL NEUROMUSCULAR JUNCTION

Each motoneuron terminal ends at a specialized region called the *neuromuscular junction,* a localized, combined specialization of nerve terminal and sarcolemma. Transmission of the nerve impulse to the muscle takes place here, resulting in the generation of a muscle action current which activates contraction.

A nerve impulse induces a prolonged negative potential at the neuromuscular junction. This is not propagated but generates a muscle spike potential by depolarizing muscle membrane around the junction. *The neuromuscular junction is clinically important because it is the site of important neuromuscular disorders seen clinically, such as myasthenia gravis.*

Myasthenia Gravis

Myasthenia gravis is a neuromuscular disorder characterized by weakness and fatigability of skel-

etal muscles. Symptoms include weakening of the levator palpebrae which leads to lowering of the eyelid, weakness of the extraocular muscles, impairment of the voice, speech, and the ability to chew and swallow. Generalized weakness, especially of the proximal muscles and neck occur, and respiratory weakness may be life-threatening.

The disorder involves failure of neuromuscular transmission and resembles the neuromuscular block due to curare. Myasthenia symptoms respond to treatment with anticholinesterases whereby acetylcholine effects are enhanced and prolonged. Symptoms are worsened by repeated voluntary contractions or by repeated nerve stimulation. Symptoms are also worsened by one-tenth of the dose of *d*-tubocurarine that would yield minimal effects in normal subjects. Electron micrographs of the neuromuscular junction reveal a normal density of ACh vesicles in the presynaptic (nerve) terminals and a sparsity and shallowing of postsynaptic (muscle membrane) folds.

Isolation and characterization of snake venom neurotoxins by Lee (1972) led to isolation of acetylcholine receptors (AChR). Almost immediately, it was demonstrated that myasthenia gravis is associated with a shortage of AChR (Fambrough et al., 1973). It was soon shown that myasthenia gravis is the result of an antibody-mediated autoimmune attack (Lindstrom et al., 1976). IgG and AChR cross-link. This is followed by endocytosis and degradation of AChR within the muscle cell. Neural functions remain normal.

Treatment involves suppression of IgG with steroids, removal of the thymus gland, depletion of the antibody by plasmopheresis, and increasingly effectively, by specific immunotherapy. Solving the puzzle of myasthenia gravis is clearly one of the best examples of combined basic and clinical neurological research (Drachman, 1981).

Neuromuscular Changes Following Denervation

If the axon of a motor unit is severed, three dynamic effects soon follow: the neuromuscular junction becomes simplified, the receptors spread out on the sarcolemma, and the proximal portion of the nerve begins to sprout new processes. Nerve sprouting takes place at the point of severance of the axon and at more proximal nodes of Ranvier (see Fig. 9.149). Nodal sprouting is in response to a chemical stimulus stemming from processes associated with degeneration of the distal nerve. Terminal sprouting is in response to degenerative changes in the muscle (Brown et al., 1981).

It is not known whether nerve sprouting occurs because nerves have an intrinsic tendency to grow but are held in check by an active diffusible inhibitor, or whether they have an intrinsic tendency to withdraw and require stimulation by a diffusible growth substance emanating from their substrate. In any case, when sprouts reinnervate muscle fibers, the neuromuscular junction is reconstituted, excess sprouts are pruned back, and normal function may be restored.

THE MOTOR UNIT UNDER CENTRAL CONTROL

Motor units are affected primarily by what Sherrington called the *"central excitatory state."* This refers to the variable level of centrally generated excitement that is maintained by a combination of central sources of motor control. This may originate from spontaneously active neurons in brain stem and spinal cord, from brain stem reticular formation and raphé nuclei, and to a lesser extent from influences of afferent impulses on motoneurons. The central excitatory state remains, or is reconstituted, after complete deafferentation in the monkey (Taub, 1982). Some combination of spontaneous central activities contributes the state of readiness for motor unit action, voluntary or reflexive.

Suprasegmental motor control mechanisms ordinarily have priority for controlled motoneuron activation through interneurons. This, as we have seen, can be overridden by local afferents. Motor unit output can be greatly augmented by significant excitatory input by afferent neurons which are in turn excited by mechanical stretch of the same muscle they command, a positive feedforward mechanism involving the muscle spindle (see below). They also receive signals that further shape motor unit behavior from many other afferents, both muscular and cutaneous.

Motor units are also affected by nerve impulses that come from the opposite side of the spinal cord and brain stem and by impulses from lower and especially from higher level spinal and brain stem sources of reflex motor activity. The orderliness of this convergence onto motor units is demonstrated by analysis of mechanisms responsible for *spinal and brain stem reflexes.*

Characteristics of Spinal Reflex Activity

The anatomical basis for reflex activity is the *reflex arc.* Essential elements include afferents, interneurons (internuncials), and efferent neurons (Fig. 9.195). At each segmental level, some of the reflex connections between Ia muscle afferents and motoneurons are direct, that is, *monosynaptic,* with no intermediary neuron in the circuit. The vast

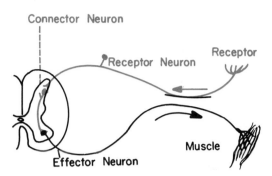

Figure 9.195.   Diagram of a simple reflex arc.

majority of reflex connections, however, are *pleurisynaptic,* involving a few to several intermediary neurons. Characteristically, afferent neurons contribute to reflexes through numerous collaterals and via numerous internuncial neurons whereby they can, in an organized way, influence a very large number of motoneurons.

Superficial reflexes are elicited by stimulation of mucous membranes and skin. They are all pleurisynaptic, usually involving a moving afferent stimulus, multiple interneurons, and a coordinated ensemble of effectors. Superficial reflexes include corneal, snout, rooting, suckling, abdominal, plantar, cremasteric, sphincter, and other reflexes. When the cornea is lightly stroked with a wisp of cotton, the eyelids blink reflexly, unilaterally. Stronger stimulation results in bilateral blinking accompanied by strong eye closure and secretion of tears. Any painful (nociceptive) stimuli can engage both superficial and deep reflexes and may result in predominantly ipsilateral withdrawal and crossed extension.

Snout, rooting, and suckling reflexes are functionally important in infants to ensure good feeding responses. They are pleasurable in their own right and reinforce thumb sucking, for example. They disappear in childhood but may reappear in senile patients. During childhood, infant feeding reflexes are put under some kind of central control, presumably inhibitory, and are subject to subsequent "release," as a consequence of frontal lobe damage or degeneration.

If one quadrant of the relaxed abdomen is slowly, gently stroked, the abdominal muscles reflexly and transiently pull in that direction. If the abdominal wall near the groin or skin inside the thigh is stroked, the cremaster muscle (which is a transposed abdominal muscle) draws the testicle up on that side. These superficial reflexes may be weakened or absent after damage to the internal capsule (a sign of "upper (suprasegmental) motor neuron lesion").

When the sole of the foot is slowly, firmly

stroked, the foot and toes flex, exhibiting the normal *plantar reflex*. This reflex becomes converted to the pathological "*sign of Babinski*" when there is an "upper motor neuron lesion." The question as to whether the loss of abdominal reflexes and pathological conversion of the plantar response constitute certain evidence of damage to the pyramidal tract is moot (for a careful analysis, see Brodal, 1981).

Deep reflexes are stretch reflexes in which abrupt short extension of a muscle is obtained by striking the tendon of that muscle. Stretch reflexes consist of combined monosynaptic and pleurisynaptic reflex actions which lead to a brisk concerted contraction of the muscle attached to the tendon that was struck. These include the jaw jerk, biceps, triceps, knee and ankle jerks.

Visceral reflexes include the pupillary reflex, evoked by shining light onto the retina, and the carotid sinus reflex, in which pressure over the sinus results in a fall in blood pressure. More on visceral reflexes is provided in Chapter 69.

Pathological reflexes include notably the classical sign of Babinski just mentioned. Instead of the normal plantar foot and toe flexion in response to appropriate stroking of the sole of the foot, the big toe turns up, the other toes fan outward, and there is a tendency to withdraw the foot. *The "sign of Babinski" is considered a sign of functional block at spinal or higher levels of suprasegmental motor control,* presumably affecting more than one descending motor control system.

The "sign of Babinski" is seen in patients with functional damage to contralateral sensorimotor cortex, the internal capsule (as commonly occurs in cases of stroke), or interference with any of several descending motor control paths in the brain stem. Interestingly, the sign is normally present in infants until about the age when they begin to walk and is thought to be related to the normal developmental lack of tonic descending suprasegmental inhibition affecting the segmental reflex arc.

REFLEX MECHANISMS DIFFERENTIATED

Two classical animal preparations have been useful for analyzing central nervous control of reflexes. The *spinal preparation* permits examination of reflex responses of the spinal cord in the absence of contributions from higher segmental and suprasegmental levels. The spinal cord is ordinarily transected a few segments above the level to be analyzed.

If the section is made higher than the midthoracic level, intercostal breathing is affected, but dia-

phragmatic breathing can compensate. If the section is made above middle cervical levels, artificial respiration is necessary because the motor neurons serving the phrenic and intercostals are separated from indispensable medullary and pontine respiratory controls.

The *decerebrate preparation* is obtained by transecting the brain stem at the mesencephalic level, usually between the inferior and superior colliculi. Both spinal and decerebrate transections are performed under deep surgical anesthesia. Usually a volatile anesthetic is employed so that the effects of the anesthetic will be dissipated shortly after decerebration. Then, the upper mesencephalon and entire forebrain can be removed so that any (possibly olfactory? or other?) higher level perceptual experiences can be excluded.

Ordinarily, decerebration is done prior to spinal transection. Otherwise the anesthetic must be continued, and anesthetics strongly influence spinal as well as brain stem reflexes. Indeed, *reflexes are useful for estimating the level of anesthesia* and the degree of coma which may result from any of a variety of cerebral and metabolic disorders.

In the spinal preparation there is flaccid paralysis and little or no spontaneous skeletal muscle activity. The limbs cannot support weight, but a number of important cutaneous and skeletal muscle reflexes can be elicited.

Classification of Reflexes

Reflexes may be classified as segmental, intersegmental, interlimb, and suprasegmental on the basis of their connections in spinal cord and brain stem. A typical *segmental reflex* is the deep tendon or stretch reflex such as the patellar reflex or knee jerk.

Interlimb reflexes engage multisegmental reflex patterns and involve ascending and descending pathways; e.g., pinching the hind limb of a cat yields reflex flexion of the contralateral forelimb. *Suprasegmental reflexes* involve reflex interconnections extending upward and downward between spinal and cranial levels. These are involved in important reflex controls relating head to body and vice versa and are important in semiautomatic somatovisceral skeletal and autonomic reflexes relating to swallowing, coughing, vomiting, etc., as detailed above. An example of suprasegmental reflexes is seen in decerebrate rigidity (seen in patients with functional transections at the midbrain level): turning of the head to one side activates vestibulospinal responses which alter muscle tone in all four extremities. Limbs ipsilateral to the side to which the face is turned extend and the contralateral limbs flex. When the head position is reversed, the limb reflex posture is reversed.

SPINAL AND BRAIN STEM REFLEXES

The Central Excitatory State

All sensory input has access to behavior by way of reflexes. In the chronic "spinal" cat, with no interference from higher centers, every type of sensory stimulus yields some kind of somatic or visceral motor or glandular response. The remarkable responsiveness of brain stem as well as spinal motoneurons to afferent influences can be attributed to the *central excitatory state* (CES). The existence of central excitation reflects the readiness of the nervous system to be active and responsive: *the nervous system is built for action.*

SPONTANEOUS ACTIVITY GENERATES THE CENTRAL EXCITATORY STATE

The CES represents convergent activity that is generated by a variety of spontaneously active neurons situated in different parts of the neuraxis, especially along its central core. Several motor control loci in the telencephalon show spontaneous activity as do appetitive regions of the hypothalamus. *Mainly, the CES can be attributed to spontaneous activity in the brain stem and spinal reticular formation and raphé nuclei.* These generate widespread excitatory and inhibitory patterns which influence the CES along ascending as well as descending trajectories (Magoun, 1950). Widespread inhibition of motoneurons, for example, occurs during rapid eye movement (REM) sleep as a consequence of widespread descending brain stem influences (Jouvet, 1973).

OTHER CONTRIBUTIONS TO THE CENTRAL EXCITATORY STATE

Of course, afferent activity also makes its contributions to the CES. Many afferent neurons are spontaneously active. But the CES does not depend upon afferent input; it persists even when essentially all sensory nerves have been severed. Certain drugs can have effects, heightening or reducing the CES. In the waking state, consciousness can presumably play a role, from barely sustaining wakefulness to the agitation, palpitation, exophthalmia, tremor, and insomnia characteristic of extreme anxiety. Naturally, all of these potential influences can interact with one another to augment or diminish the CES.

In terms of reflexes, the CES contributes to the amplitude and duration of single muscle twitches,

the rate at which individual muscle contractions reach sustained (tetanic) levels, and the thresholds at which afferent stimuli may be effective in initiating reflex responses. Any given motoneuron may therefore discharge more impulses at a higher frequency, and neurons not otherwise responsive may be recruited to be active.

Theoretical Organization of Reflex Mechanisms

Figure 9.196 suggests a useful conceptual framework for consideration of afferent input and motoneuron response. In schematic form it presents two excitatory afferent fibers, each of which has access to a separate "pool" of motoneurons. Depending on the CES and the degree of spatial overlap of sensory inputs and the timing of incoming signals, there may be several different patterns of reflex response. With due regard for the extreme simplification of this representation, think about the variety of potential responses implied in the organization depicted in Figure 9.196. Include consideration of the important fact that CES is going to contribute to ascending and descending sensory and motor pathways as well as to segmental activities.

Figure 9.197 illustrates schematically some of the interconnections that occur at a typical segment of the spinal cord. Flexor and extensor motoneurons are shown in respect to monosynaptic and pleurisynaptic connections relating to segmental and suprasegmental signals. Only one type of sensory

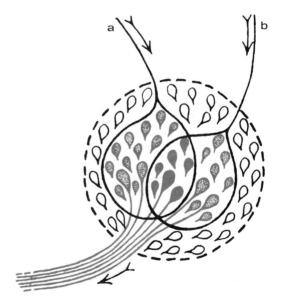

Figure 9.196. Diagram that illustrates hypothetical relationship between excitatory afferent fibers (*a, b*) as they end on motoneurons in two "pools" (enclosed by schematic afferent axons). Motor neurons in the portion common to the two terminal collections can be excited via either afferent. Subthreshold excitation from each afferent can summate to achieve contraction. [From Creed et al. (1932).]

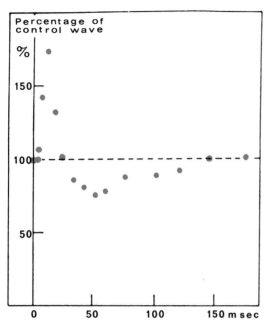

Figure 9.197. Time course for development of inhibition of the Renshaw type, when antidromic stimulation of a motor neuron's axon precedes excitation via a dorsal root or sensory nerve. [From Renshaw (1940).]

input has direct, monosynaptic access to motoneurons. That is type Ia afferents from muscle spindles which are involved in the most rapid of all reflexes, the stretch reflex.

Reflexes of Cutaneous Origin

FLEXOR REFLEXES

The most thoroughly studied of cutaneous reflexes are the flexor reflexes which *consist of contraction of functional flexors and relaxation of functional extensors.* Light stimulation of the skin elicits weak contraction of one or more flexors, but without necessarily having much effect on the skeleton. Painful stimulation of the skin elicits widespread contraction of flexors throughout the limb and effects brisk withdrawal. With nociceptive shocks of increasing intensity, the withdrawal response may become expanded regionally and globally and may be longer sustained. The pattern of reflex flexion of course varies with the nerve stimulated.

CROSSED EXTENSOR REFLEXES

When the "spinal" subject withdraws a limb in response to local noxious stimulation, the contralateral limb is simultaneously extended. *Collaterals of interneurons that excite ipsilateral flexors cross to the contralateral side of the spinal cord and excite extensor and inhibit flexor motor units at the same segmental level.* The function of this response is to support the weight of the body when flexion is

withdrawing another limb from making its weight-bearing contribution.

Reflexes of Muscular Origin

MUSCLE SPINDLES AND STRETCH REFLEXES

Stretching a muscle causes it to contract reflexly. This has the effect of recovering any desired posture or movement from any displacement in any direction. All skeletal muscles exhibit stretch reflexes. Appropriate stretch of flexor muscles yields a brief but unsustained flexor reflex contraction. The stretch reflex is more powerful and longer lasting in extensor, especially antigravity, muscles.

Muscle Spindle Afferents

The stretch reflex is the most rapid of all reflexes. Ia afferents are the largest diameter (fastest conducting) among all afferent nerves; they respond readily to stretch (less than 1 mm displacement may be sufficient); they have monosynaptic connections with motoneurons; motoneurons respond readily to Ia afferent signals; and the motoneurons discharge with an abrupt contraction. The reflex response is specific: it causes contraction in the same muscle from which the Ia afferent originated.

Ia fibers constitute the primary afferent component of the segmental stretch reflex control system for skeletal muscles (Fig. 9.198). Each of these fibers serves as a specialized detector element that has its own intrinsic motor innervation which is different from that of the surrounding bulk of the muscle. The γ-motoneurons induce contraction of striated muscle fibers contained within the muscle spindles. Their contraction excites activity of Ia afferents. The receptors involved are the annulospiral endings which coil around the equatorial region of the intrafusal (intraspindle) muscle fibers (Fig. 9.198). These primary afferents have to do mainly with phasic aspects of muscle contraction. Group II afferents, secondary endings, have to do more with measuring static forces relating to slow or sustained muscular contractions.

Gamma Efferents to Muscle Spindles

The *muscle spindle* contains motor as well as sensory elements. The *motor* parts consist of two types of skeletal muscle fibers which are contained within the spindle, that is, they are *intrafusal* in location. The multiple nuclei in the muscle fibers are collected centrally in the case of the nuclear "bag" fibers and are longitudinally distributed in the case of the nuclear "chain" fibers. Both muscle fiber types are innervated by γ-motoneurons.

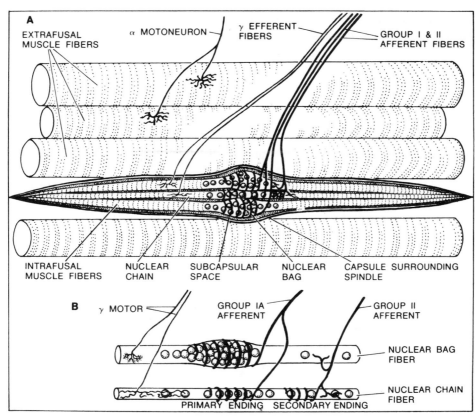

Figure 9.198. Mammalian muscle spindle. *A,* Schema of mammalian spindle innervation. The spindle is embedded in the bulk of the muscle made up of large extrafusal muscle fibers and supplied by α-motoneurons. *B,* Diagram of the two intrafusal muscle fiber types and their afferent innervation. [B from Matthews (1964), from Kuffler and Nicholls (1976).]

Spindle Afferent Activation and Control

The mechanically sensitive primary afferents arise as annulospiral endings within the equatorial zone of the nuclear bag cells. They are large diameter Ia fibers, greater than 12 μm in diameter, with conduction velocities greater than 90 m/s. The secondary afferents originate outside the nuclear bag region, in the myotubular region. These endings are flower spray in type. Their axons belong to group II afferents and have somewhat slower conduction velocities, less than 80 m/s. For each spindle, there is usually one type Ia afferent and from 0 to 5 type II afferents. Both type Ia and II spindle afferents are excitatory to the muscle which contains the spindle. *Type Ia spindle afferents are especially sensitive to abrupt, short stretch; hence they are responsive to vibratory stimuli applied to the muscle tendon. Type II, secondary afferents, are more sensitive to static stretch phenomena* (Matthews, 1982; Harris and Henneman, 1980).

Spindle Servocontrol of Stretch Reflexes

Muscle spindles are attached at each end to extrafusal muscle fibers, to other spindles, or to tendons. *Muscle spindles therefore are anatomically "in parallel" to extrafusal* muscle fibers. Contraction (hence shortening) of the latter relieves tension on the muscle spindles. If it were not for the γ-motoneuron control, the spindle afferents would always automatically decrease their rate of firing when the spindle is slackened, and in some instances, there is a corresponding "pause" in Ia afferent activity at the beginning of a muscle contraction. But then the γ-motoneurons which are fired by the same signals that fire the α-motoneuron, accelerate the firing of the Ia afferents so as to keep up or to exceed the slackening effects of contraction of the overall muscle. Of course, either passive or active stretching of a muscle lengthens the extrafusal fibers and exerts tension on the spindles. The γ-motoneurons adjust their firing rate to correspond to the aim of the movement that is under control. *The servocontrol provided by the γ-motoneuron and spindle afferents has the net effect of allowing the muscle spindle system to benefit from the full range of its effectiveness regardless of the momentary length of the parent muscle.* The net effect of γ-motoneuron muscle spindle feedback control at the segmental level is to increase the excitation—hence to increase the power of contraction—of the α-motoneurons, *in accordance with whatever load may be imposed on the muscle.* The Golgi tendon organ afferents, on the other hand, operate to prevent the muscle and tendon from overloading.

GOLGI TENDON ORGAN INHIBITORY CONTROL

Golgi tendon organs arise from tendons attached to skeletal muscles. These sensory organs are anatomically "in series" with the muscle. When the muscle is stretched, the Golgi tendon organ is stretched, and thereby its afferent fiber is activated. When the muscle contracts, the Golgi tendon organ is likewise stretch-activated. Golgi apparatus afferents are classified as Ib, having a modal conduction velocity of about 80 m/s. The length of stretch necessary to elicit a response from the Golgi tendon organ is greater than that to activate the annulospiral endings. *The reflex effects of the Golgi tendon organ afferents are inhibitory. Therefore, their servocontrol operates to prevent too powerful passive or active muscular stretch.*

GAMMA EFFERENT EXCITATORY CONTROL

Muscle spindles provide information about the state of muscles, their length, and the rate at which their length is changing. The skeletal muscles that haul the skeleton about are directly innervated by the large α-motoneurons in the ventral horn. The ventral horns also contain small γ-motoneurons, sometimes called fusimotor neurons, which are concerned with regulation of overall muscular contraction rather than with overall muscular contraction per se. The γ-motoneurons constitute one-third of all motoneurons, and their axons contribute an equal fraction to the ventral roots. The converging patterns of motoneuron activation that bear down on the α-motoneurons also converge on the γ-motoneurons. *The two types of motor units generally cocontract.*

Gamma Efferent Control of Spindle Afferents

Figure 9.199 illustrates the feedforward sensory control and feedback excitatory muscular control that is generated by the motoneuron. A single γ-

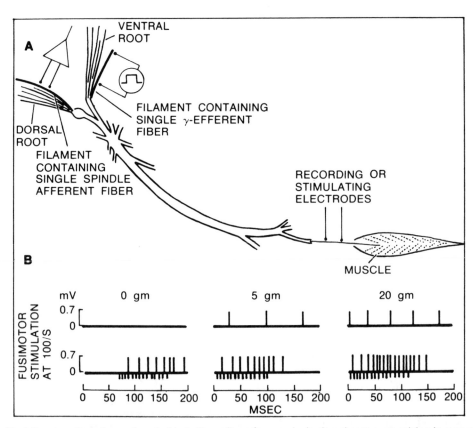

Figure 9.199. Centrifugal control of muscle spindle. *A,* Recordings from a single dorsal root axon arising in a muscle spindle; single fusimotor (γ) axon in a ventral root innervating the same spindle is stimulated. *B,* Upper records show sensory discharges when the muscle is slack (0-g tension) or lightly stretched (5- and 20-g tension). A brief train of 14 or 15 stimuli at 100/s either initiates sensory discharges (*lower left*) or accelerates them. Extrafusal muscle fibers remain inactive. [Adapted from Kuffler et al.; redrawn from Kuffler and Nicholls (1976).]

motoneuron is electrically stimulated to exert a standard control over a single muscle spindle. The primary spindle afferent (Ia) from that spindle is recorded. Extrafusal muscle efferents are eliminated. A brief train of stimuli to the γ-motoneuron axon yields a different response depending on the load (0–20 g) that is imposed on that muscle.

The finer the movements that are to be controlled, the greater the concentration of muscle spindles, and hence of γ-motoneurons and of groups Ia and II afferents. Intrinsic muscles of the hand, for example, are far more densely supplied with this regulatory apparatus than are the antigravity muscles.

Muscle Spindle Control in Human Subjects

Responses of muscle spindle afferents during voluntary contraction of extensor muscles indicate that γ-motoneurons are recruited in an orderly manner as are α-motoneurons, so that individual spindle endings become activated at specific contraction strengths. This provides evidence that in voluntary contractions, as in reflex contractions, muscle spindles are activated in parallel with α-motoneurons controlling the overall muscle.

Figure 9.200 shows the linear relationship between spindle activity and strength of contraction during voluntary isometric contractions. Different spindle afferents have widely different thresholds. Recruitment threshold for a given spindle afferent appears to depend on its fusimotor innervation and on the recruitment threshold for that γ-motoneuron. This depends on the desired muscle force rather than on the muscle length (Burke et al., 1978a). Even more important is the observation that muscle spindle afferents respond promptly and accurately to changes in load during voluntary accurate position maintenance (Burke et al., 1978b).

Muscle spindle afferent discharge appears to contribute to the body imagery. When vibrations of 100 Hz are applied to the tendon of the biceps or triceps muscles, subjects make systematic misjudgments as to the angle at the elbow. The subject supposes that the elbow is in a position that it would have assumed if the vibrated muscle had been stretched (Fig. 9.201).

REGULATION OF STIFFNESS BY SKELETOMOTOR REFLEXES

Skeletomotor reflexes tend to regulate length and force variables simultaneously in such a way that the ratio of force change to length change contributes the regulation of what is called "stiffness."

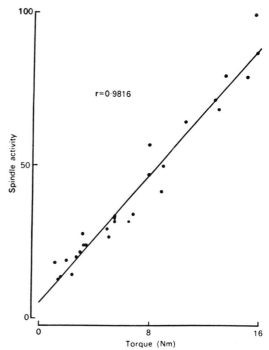

Figure 9.200. Comparison between human spindle afferent activities and strength of isometric voluntary contractions in the same muscle. Threshold for activation of different spindle endings varies but is evidently determined by its fusimotor neuron firing which has a recruitment order like that of the skeletal muscle motoneurons. The results indicate that fusimotor drive to spindles in a muscle is proportional to the skeletomotor drive to that muscle and is subject to similar if not identical descending command signals. [From Burke et al. (1978).]

Spindle feedback operates to maintain stable muscle length, and tendon organ feedback operates to keep muscle tension within bounds. These two feedback loops tend to oppose one another. The physiological requirement is to control the balance of these variables while the body or limb experiences variable external loads. These may be imposed by changes of the position of the body and limbs in the gravitational field, as by changing posture, or during centrifugation or submersion in water or more dense media or during weightlessness.

It appears that motoservo actions combine the regulation of length feedback from muscle spindles with the regulation of force feedback from tendon organs largely at the segmental reflex level. This amounts to *stiffness regulation* (see review by Houk, 1979). This suggests that higher motor command does not need to cope individually with either length or force but with the regulation of their ratio, to regulate stiffness. This simplifies higher command and provides a spring-like interface between the body and its mechanical environment. Compliance of this interface serves to absorb the impact of abrupt changes of load.

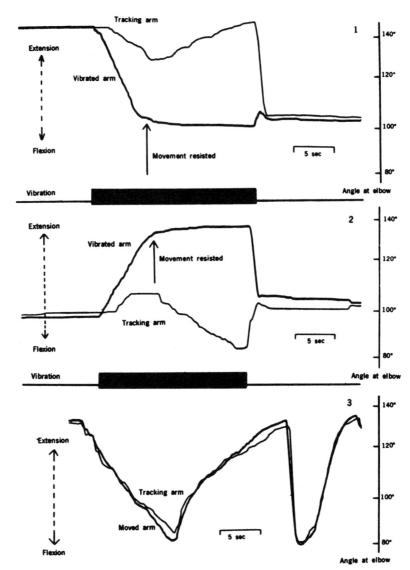

Figure 9.201. Muscle spindle afferent discharges apparently contribute to perception of position sense. *1,* The effect of vibrating the tendon of the right biceps muscle so as to produce tonic vibration reflex which optimally stimulates muscle spindles in vibrated muscle and which moves the arm into flexion is tested in blindfolded subject who uses left arm to track his perception of the position of the vibrated right arm. From arrow on, any further flexion of the vibrated arm was prevented by fixed string attached to splint on arm. *2,* Similar vibration applied to triceps muscle which induces arm extension. Subject's left arm expresses subjective tracking, movement similarly limited. *3,* Accuracy of tracking passively imposed movements applied by experimenter to splint and replicated by blindfolded subject's left arm. [From Goodwin et al. (1972).]

INTERSEGMENTAL AND SUPRASEGMENTAL REFLEXES

Spinal and Cranial Intersegmental Reflexes

PROPRIOSPINAL AND SPINOBULBOSPINAL REFLEXES

Two physiologically distinctive reflex systems interconnect the long sequence of spinal segments. One system, *propriospinal* (PS), is *carried by spinal interneurons* and is sufficient entirely within the spinal cord. That is, PS reflexes remain intact even when the spinal cord is separated from the brain stem. The other system, *spinobulbospinal* (SBS), involves a more circuitous *path through the brain stem reticular formation and reentry into the spinal cord from above downwards* (Shimamura and Livingston, 1963). The brain stem relay takes place in the caudal part of the reticular formation in the medulla oblongata (bulb) near the midline.

When electrical stimuli are delivered to the dorsal roots of any spinal segment, motor responses occur in ventral roots bilaterally all along the spinal cord. An initial direct response is followed by a delayed response that is relayed through the medial bulbar reticular formation. Latencies to the direct response vary with segmental distances from the point of stimulation. The ascending projection to the bulb, however, does not yield motor responses until after it has initiated impulses in the descending bulbospinal relay. Consequently, latencies to the bulbar relayed responses vary with segmental distances from the bulb (Fig. 9.202).

SBS reflexes involve flexor muscle responses when excited by cutaneous afferents, and extensor muscles when excited by extensor muscle afferents. Conversely, PS reflexes are more readily elicited by muscular afferents. They elicit contraction of both flexor and extensor muscles of the trunk and extremities and hence produce greater limb and truncal stability.

There is a "scissors crossing" relationship between the two reflex systems that depends on differences in spinal conduction velocities (Fig. 9.202). Although the bulbar pathway is longer, it is enough faster that it catches up with the time of arrival of the direct interlimb reflex. Thus, two reflex responses, direct and bulbar relayed, following stimulation of lower or upper limb afferents, reach the upper or lower limbs simultaneously. Therefore, *interlimb reflexes have bilateral simultaneous direct (propriospinal) and indirect (bulbar relayed) access to the other pair of limbs.*

Because the predominant and shortest latency responses to the SBS reflex system arise from cutaneous sources, it is evident that cutaneous sensory signals are the first to be integrated into motor control patterns by this suprasegmental control mechanism (Shimamura and Akert, 1965). Shimamura et al. (1976) have shown that the ascending path of the SBS reflex lies in the superficial layers of the dorsolateral quadrant of the spinal cord, and that the SBS reflex can be profoundly and prolongedly inhibited by stimulation of the contralateral sensorimotor cortex (Shimamura et al., 1967).

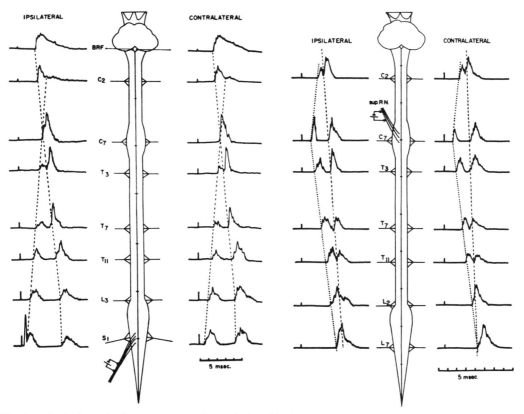

Figure 9.202. Longitudinal conduction systems serving spinal and brain stem coordination in the cat. Intersegmental and interlimb reflexes are illustrated, showing differing slopes of conduction velocities for direct propriospinal reflexes (intrinsic to the spinal cord), and indirect spinobulbospinal reflexes (which yield no response until following bulbar relay). *Left,* dorsal root S_1 is stimulated. *Right,* superficial branch of radial nerve (C_7) is stimulated. Recordings are made from ventral roots at a succession of lumbar, thoracic, and cervical ventral roots and from medial bulbar reticular formation (*BRF*). Sloping *dashed lines* identify direct propriospinal reflexes, showing latencies which increase with distance from the level of both sacral and cervical afferent inputs, and spinobulbospinal reflexes which yield only indirect responses that are relayed in the bulb. The latter therefore show increasing latencies from bulbar level downward. Small lesions in the medial bulbar relay eliminate the bulbar relayed spinobulbospinal reflexes but do not interfere with propriospinal reflexes. [From Shimamura and Livingston (1963).]

PROPRIOCRANIAL AND BULBAR RELAYED CRANIAL REFLEXES

In ways that parallel PS and SBS reflexes, when electrical stimuli are applied to cranial nerves, *direct (propriocranial) and indirect (craniobulbocranial) reflex responses occur at all cranial motor nerves* (Shimamura, 1963). Similar cutaneous and muscular afferent and motor response patterns obtain with these cranial reflexes. *The same medial bulbar reticular formation that relays the SBS interlimb reflexes relays cranial, craniobulbocranial (CBC) reflex responses in both ascending and descending directions* (Shimamura, 1963).

Figure 9.203 illustrates (A) how spinal input generates a bulbar relay that elicits cranial motor responses, and (B) how cranial input generates a bulbar relay that elicits spinal motor responses, both being relayed through the same bulbar reticular neurons that relay the SBS interlimb reflexes. *This represents an important reflex linkage between spinal and cranial sensorimotor controls.* As in the case of spinal projections to the bulbar relay, cranial impulses do not elicit motor responses until after their relay in the medial bulbar reticular formation where they generate an ascending (bulbocranial) and a descending (bulbospinal) response that is exhibited along the entire neuraxis.

Reflexes that are relayed in the bulbar reticular formation (spinobulbospinal, spinobulbocranial, craniobulbocranial, and craniobulbospinal) are relatively more susceptible to asphyxia and to the effects of anesthetic and stimulating drugs (barbiturates, strychnine) than are the more direct (propriospinal and propriocranial) and the even more direct cervicocranial reflexes, none of which depends on the reticular formation.

Convergence of Suprasegmental Control Signals

Inasmuch as these bulbar relay neurons receive converging signals from many different systems and regions of the central and peripheral nervous system, you can appreciate that this relay contributes important guidance to reflex behavior. *The most rudimentary choice seems to be between global*

SPINO-SPINAL & SPINO-CRANIAL COORDINATION

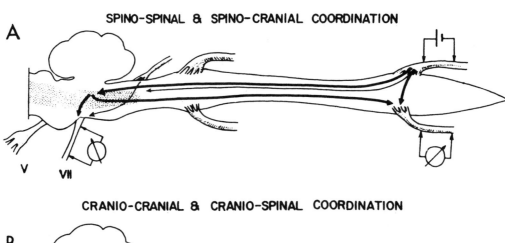

CRANIO-CRANIAL & CRANIO-SPINAL COORDINATION

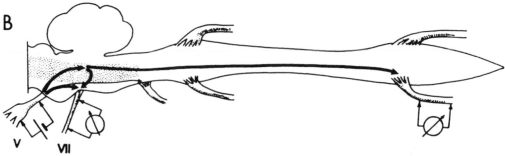

Figure 9.203. Longitudinal coordination systems of the neuraxis. *Above,* spinospinal and spinocranial coordination. *Below,* craniocranial and craniospinal coordination. Each dorsal root volley elicits motor dicharges along ascending (and descending) propriospinal pathway continuous with propriocranial coordination pathway to all spinal and cranial motor nuclei. Additional influences, relayed in the bulbar reticular formation course both upward to cranial motor nuclei (spinobulbocranial relays) and upward then downward to spinal motor nuclei (spinobulbospinal relays). In *A,* stimuli are introduced into a lumbar dorsal root and recorded from the ipsilateral ventral root (which reveals the segmental and appropriately delayed spinobulbospinal discharge) and from the facial nerve (VIIth) (which reveals a propriospinal-cranial discharge as well as a relayed spinobulbocranial discharge). In *B,* stimuli are introduced into the trigeminal nerve and recorded as propriocranial and relayed craniobulbocranial discharge recorded from the facial nerve, and a craniobulbospinal discharge recorded from the lumbar ventral root. [From Shimamura (1963).]

lexion withdrawal and pillar-like limb and trunk stability, depending on the priority of afferent (cutaneous and muscular) signals as interpreted at this bulbar relay. It amounts to an important emergency decision, "To stand or flee!"

Vestibular and Local Limb Supporting Reflexes

It is well known that proprioceptive and vestibular systems are both active in posture and locomotion. Impulses arising from receptors in each of these systems converge as they influence the final common path. The effects of their mutual interaction can be analyzed by recording ventral root responses following stimulation of the vestibular nerve and appropriate proprioceptive sources.

Vestibular stimulation alone elicits a rapidly conducted two-peaked response which can be recorded bilaterally from ventral roots all along the neuraxis, including the cranial counterparts of the ventral roots. *The first vestibular peak is facilitatory to skeletal motoneuron activity.* It appears to result from direct vestibulospinal impulses. *The second peak* results from relay through reticular neurons in the medulla oblongata (bulb) and *conveys both facilitatory and inhibitory influences.*

Even single shock vestibular stimulation elicits both facilitatory and inhibitory effects at the spinal level. If repetitive stimulation is introduced, the inhibitory influence becomes increasingly domi-

nant. If the vestibular nerve is stimulated at 30 impulses/s for a few seconds, as might be supposed to occur during strong natural vestibular stimulation, there is a long lasting suppression of all spinal reflexes (Gernandt et al., 1957). It is well known that loud sounds induce vestibular as well as cochlear responses. This may account for the Tulio response which occurs in people exposed to extremely loud noise (as with jets taking off from a flight deck). They may collapse from abrupt generalized failure of skeletal muscle tone and supportive postural reflexes.

The effects of such a train of strong vestibular stimulation represent an *active* descending inhibitory influence which can be interrupted by rapid cooling of the spinal cord, thereby promptly releasing the full amplitude of segmental reflexes below the level of cooling.

The only kind of peripheral stimulation found to facilitate the vestibular ventral root response is that of manipulating the foot by moving it into a weight-bearing position (Fig. 9.204). Upward displacement, or working of the tarsal-metatarsal joints in the foot, markedly enhances the facilitatory vestibular effects. These are strongly facilitatory to both flexors and extensors and therefore act to create pillar-like stability in that extremity.

It is clear that vestibular ventral root enhancement by this weight-bearing posture occurs only in

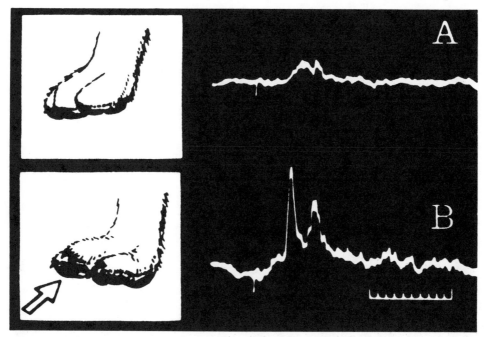

Figure 9.204. Effect of weight-bearing displacement of foot of cat on local ventral root responses to vestibular stimulation. In *A*, vestibular ventral root (L$_7$) responses are seen beginning about 4 ms following stimulus artefact and lasting for 4–5 ms with the foot in a non-weight-supporting position. In *B*, the cat's paw is pushed gently upwards as in light weight-bearing. The local vestibular responses are greatly augmented. The effect is to increase contraction of both flexors and extensors and hence to increase weight-bearing capability of the limb. [From Gernandt et al. (1963).]

the ventral roots which belong to the limb that is in a position to bear weight. Therefore, standing, stepping, springing, or landing will displace the foot in this way and will provide an adequate stimulus to initiate this local vestibular ventral root response. Given background vestibular excitation, there will be a correspondingly effective increase in the weight-bearing capabilities of that same limb. This should be an advantage to an individual proceeding in the dark, on uneven ground, to benefit from a reflex that provides automatic limb support whenever there is available a weight-bearing surface. This will also account for the positive supporting reactions—the so-called "magnet reaction" of Magnus (1924).

We sometimes think of monosynaptic reflexes as being private paths, but it is well to remember that *vestibular effects in the spinal cord facilitate for long intervals subsequent monosynaptic (Ia-induced) stretch reflexes.* The convergence of vestibular and proprioceptive impulses on spinal motoneurons that control skeletal muscles contributes specifically and efficiently to weight-bearing requirements for both posture and locomotion. It is an efficient reflex, inasmuch as it occurs only in those limbs that are manifestly in a position to bear weight. This contribution is to an automatic increase in the strength and stability of any potentially weight-bearing limb during any kind of standing, walking, jogging, dancing, and jumping behavior.

CHAPTER 71

Cortical and Subcortical Integrative Mechanisms

INTRODUCTION

Sensory and motor systems depicted in previous chapters account for all nervous system inputs and outputs. These systems provide central representations (and rerepresentations) for all sense data, plus diverse controls governing reflexes and all visceral and skeletal motor activity. Two questions arise which can be answered briefly immediately and later expanded.

1. *How can such diverse nervous mechanisms provide coherent subjective experiences and behavior?* Diverse mechanisms are made coherent by integrative actions, by the achievement of *nervous system integration.* Every living system—beginning with the most primitive—is highly integrated. Integration is a given requirement for the organization of life. Multicellular organisms require superordinate mechanisms in order to achieve integration as a whole. Every cell, tissue, and organ system has its own characteristic, limited integrative functions, while nervous and endocrine systems contribute superordinate specialized mechanisms for integration of the organism as a whole. We shall proceed to demonstrate some of what this entails.

2. *What else is there in addition to sensory and motor functions?* Many parts of the brain are devoted to making higher order associations, making cross-comparisons among a very wide and shifting variety of sources and complexities of information. The human brain in particular deals extensively with such higher order nervous system processing. This gives humankind the capacity for speech, language, and increasingly inclusive social and cultural integration. These activities require higher orders of whole nervous system integration and provide humankind astonishing capacities for creating and sustaining higher order individual and collective goals. Limits are by no means set.

PHYSIOLOGICAL ADAPTATIONS ARE REQUIRED TO SERVE CHANGING BIOLOGICAL PURPOSES

The principal function of the nervous system is to detect and express biological needs representing the organism as a whole and to goal direct behavior in timely ways so as to satisfy those needs. The nervous system strives to achieve harmony between the inside milieu of visceral needs and the outside world of opportunities and risks. The nervous system conducts a flexible and adaptive program to match biological requirements with environmental options. The brain uses means that are, as we have seen, adroit, flexible, and generally constructively adaptive.

Both survival and reproduction require continuing flexibility in goal pursuits. They require goal shifting over time-spans that range from seconds to decades and involve a whole panoply of different visceral and somatic activities. They also require adaptations that establish lasting changes within the nervous system to suit altered capabilities of the organism, as during development, and to acquire an almost limitless repertoire of novel behaviors.

Long-term changes in the nervous system are attributed to "plasticity" of the nervous system. Integrative mechanisms ensure that the nervous system as a whole operates so as to perfect a wide variety of goal-seeking activities. Adaptive plasticity of the nervous system has unknown limits. We know that limbs cannot be regenerated, for example, but we do not know to what extent existing adaptive capabilities might be stimulated for clinical advantage. Chronic neurological disorders are thought to become manifest only after the patient has reached his/her nervous system limits for reconstructive adaptation.

A LARGE PROPORTION OF THE HUMAN BRAIN IS AVAILABLE FOR HIGHER NERVOUS INTEGRATION

Sensory and motor mechanisms already considered constitute only a modest fraction of the total information processing capacities of the nervous system. Measured on 3-D computer graphics representation of a normal adult human brain, the total combined cortical representations of all sensory and motor functions described in previous chapters probably involve only about 20% of the total cortex. What does the remaining four-fifths of the cortex (and brain as a whole) do? How does the brain coordinate so many disparate systems?

The physiology of the remaining 80% of the volume of the brain undoubtedly represents additional decision-making capacity. How does this great amount of "extra" nervous tissue appear to function and how can everything be integrated? It is not a secret that nobody knows. Some "nonsensorimotor systems" were considered in Chapter 67 which dealt with mechanisms that link sensory and motor systems. Now, all of these mechanisms and the bulky "extra" need to be recognized as rearrangeable ensembles that must be organized into the dynamically integrated nervous system as a whole. Our lives depend on it. Successful perception, judgment, and behavior all depend on this overall integration being successful.

Our first objective, then, is to consider some of the simpler, better understood, neurophysiological mechanisms that contribute to the *"integrative action of the nervous system"* (Sherrington, 1906, 1947). This launches our consideration of cortical and subcortical integration, which is followed by a description of "plasticity of the nervous system."

Toward a Definition of Integration

ORGANISMS ARE INTERDEPENDENT WITH THEIR ENVIRONMENT AND WITH OTHER ORGANISMS

The behavior of any organism obviously reflects the organization and integrative capacities of its parts. Both external and internal activities of an organism are integrated according to *purposes established within that individual.* Survival requires dealing constructively with what is available in the environment. Integrative mechanisms are therefore obliged to integrate signals informative of the internal state with signals informative of the environment. In the course of this process, *the nervous system projects the values of that individual at that time onto all recognizable physical and social components of that environment.*

To identify the goals toward which integrative mechanisms must be oriented, we can examine "motivational mechanisms" as manifested by cells, tissues, organ systems, and whole organisms. *Survival depends on the successful satisfaction of biological needs as defined at each of these levels of integrative organization.* This is not simply true in the abstract, for animals and people in general; it is explicitly true for each individual patient. An anoxic cardiac muscle, a hyperactive reflex, an epileptic focus, a memory lapse, all represent defective operations of integrative processes at different levels of biological organization. There must at some point be a failure of integrative performance relating internal demands to resources required from the internal milieu or from the outside environment.

All organisms constitute "open" systems which means that they depend on throughput from the environment. They are engaged in dynamic interplay within a changing environment which includes other organisms. Their survival depends on their behavior being sufficiently individually integrated, and in a larger sense, integrated with respect to the required interplay with their environment and with other organisms on which they depend.

We are coming lately to realize the vital interdependence of living systems and the benefits that come from cultivating diversity of systems within the environment of a pond, a valley, a bioregion, and indeed, the whole planet. The idea of "survival of the fittest" has come to identify the fittest among organisms as those which contribute most constructively to the entire life support system in which they live, by contributing to their environment and to the support of adequate biological diversity within their environment. Humankind are the first creatures to be responsible stewards for the globe: much depends on how this insight is received and acted upon.

Considering the number of components, and the complexity of the interdependent physiological regulatory mechanisms involved, we are up against a baffling question: How can an organism keep so many dynamic control processes coordinated while guiding behavior to obtain satisfaction for interminably changing sequences of biological needs? Since biological needs and purposes shift, sometimes swiftly and radically, *no static rules for coordination and guidance will suffice.* Mechanisms exist which decide the direction and pattern for a very wide variety of goal-seeking activities. These mechanisms, simple and complex, are integrative, and the overall process requires overall integration. We can perhaps best define integration by contrast:

Cancer is disintegrative. It escapes the bounds of integrative restraint and thus can destroy itself as well as its host.

EXTEROCEPTIVE-INTEROCEPTIVE INTEGRATION: REQUIREMENTS FOR TRANSACTIONS WITH THE ENVIRONMENT

Organisms are continually in the process of regulating their activities to satisfy fluctuating internal needs and to accommodate changes that are occurring in their environment. As we have seen in previous chapters, distributed controls are used to influence motor and sensory fluctuations. These must be integrated in a coherent way to provide for continuing purposeful behavior of the organism as a whole. How is this possible? It is accomplished in part by higher levels of control acting through and upon lower level controls.

We have seen how different peripheral and central priorities for central command are imposed. On the sensory side, originating information is considerably safeguarded along its way upward toward higher level cortical and subcortical stations. On the motor side, final command is reserved until the last possible moment of convergent action impinging on the final common path.

Interlacing feedforward and feedback channels of information monitor and modulate activity at all levels. Integrative actions, as will be developed further, depend on systems of *"stratified stabilities."* Continuing survival, reproduction, development, and evolution depend on continuing successful satisfaction of a wide variety of internally established biological needs. *It is these internal needs that steer all integrative mechanisms and govern the dynamic course of all goal-seeking activities.*

THE OVERARCHING INTEGRATIVE INFLUENCE OF NERVOUS AND ENDOCRINE SYSTEMS

Nervous and endocrine systems contribute most of the integrative regulations that are required for multicellular existence. Nervous and endocrine systems evolved early and expressly satisfied requirements for increasingly large scale integration. This is confirmed by the fact that neurotransmitters, and proteins that are employed for nervous, endocrine, and neuroendocrine signalling from cell to cell and tissue to tissue, are among the oldest families of compounds encountered in evolutionary chemistry (Dayhoff, 1976).

Clinical Significance of Integrative Mechanisms

Knowledge concerning integration is abundant, but embarrassingly incomplete. Nonetheless, un-

derstanding integrative mechanisms constitutes the basis for physiological thinking and provides a principal source of insight for solving difficult clinical problems. These can be sensibly addressed by asking: *What are the principal biological requirements for integration facing this individual in any given context?* You can then apply insights that have been accumulated from a wide variety of physiological, pathophysiological, and clinical studies.

FUNDAMENTAL INTEGRATIVE ROLES ARE PLAYED BY INDIVIDUAL NEURONS

Integrative capabilities of single celled organisms are not abandoned in multicellular organisms; rather, cellular integrative mechanisms are utilized by neighboring cells (tissue integration) and built upon by systems expressly specialized for integrative actions (e.g., nervous and endocrine systems). Neurons also make use of intrinsic integrative capacities of neuroglia and other functionally associated cells, such as receptors and effectors, cells that may stem from other germ layers and reside in other tissues. Thus, we have seen in previous chapters how trophic influences are frequently expressed among neurons and between neurons and other cell types.

THE PRACTICAL UNIQUENESS OF INDIVIDUAL NEURONS

Many integrative mechanisms are intrinsic to individual neurons. Neurons have evolved elaborate integrative capabilities, and during development they may acquire elaborate individual structural and functional distinctions, including distinctive neuronal connections. According to Bullock's estimate (1982), the human brain is composed of *a likely fifty million varieties of neurons*, that is, structurally and functionally distinctive neuronal types. This implies that distinctive integrative characteristics are likely to exist for each type of neuron. By genetic endowment, by specified and experientially modified connections, and by inductive and trophic "experiences," any given neuron is likely to assume quite specific individual integrative roles. Once the nervous system is organized, each neuron can meet its individual integrative responsibilities throughout its neighborhood, which can include influencing and being influenced by very remote regions, to wherever its far reaching protoplasmic arms may extend, and from whenever its inputs may arise.

Individual Neurons Are Decision Makers

As a generalization, neurons act as receivers, decision makers, and distributors of thus modified information. Figure 9.205 provides a schematic rep-

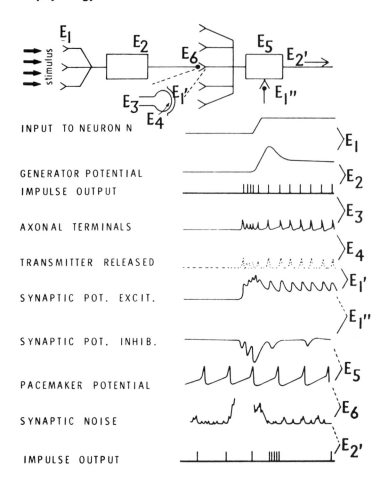

INPUT TO NEURON N

GENERATOR POTENTIAL

IMPULSE OUTPUT

AXONAL TERMINALS

TRANSMITTER RELEASED

SYNAPTIC POT. EXCIT.

SYNAPTIC POT. INHIB.

PACEMAKER POTENTIAL

SYNAPTIC NOISE

IMPULSE OUTPUT

Figure 9.205. Schematic diagram and summary of events plotted against time (duration of tracings about 0.1 s). This illustrates several successive stages in the transfer of information from one nerve cell to the next. *E*s represent independent excitabilities. Boxes in the schematic represent two nerve cells of order N and N + 1; the synaptic contact between them is enlarged to show the location of the *E*s involved. Release of neurotransmitter is shown by a broken line to indicate that the recording trace is hypothetical. *Dashed lines to Es in the bottom four records indicate that the events immediately above those Es are not the input for that particular E.* [From Bullock, 1968, 1977.]

resentation of successive stages of transfer of signals from one neuron to another. Different parts of the two neurons involved in this schema display different excitabilities, as represented in the circuit diagram and as depicted physiologically in the accompanying tracings (Bullock, 1968).

Each neuron, of course, receives and relays information from one or a variety of sources and relays to a single type of recipient cell or to a multitude of different types. *No neuron behaves like an all or nothing unit.* Its membrane consists of a mosaic of active and reactive parts. Its interior is capable of adjusting the bias to favor spontaneity and different sorts of responsivity. Localized functional individuation is reflected in the different membrane potentials (E1s-E6s, Fig. 9.205) for both receiving and sending contexts of neural activity. Local membrane processes may be continuously graded, while areas remote from decision making may be all or nothing. Input and output signals may be inhibitory

or excitatory, or inhibitory at one type of junction and excitatory at another, with the same neurotransmitter, or with more than one type of neurotransmitter. Sources and recipients of neuronal information may be near or remote. We must add to these conceptions the *important integrative contributions of local neuron circuits* as described in Chapter 67.

All neurons are influenced by their blood-borne and extracellular milieu as well as by neighboring neurons and glia, including cells neighboring their remote protoplasmic extensions. Pacemaker neurons characteristically generate information on their own and may be influenced by blood-borne molecules and by signals from other neurons.

Genetically endowed individuating characteristics of neurons are expressed on their surfaces as well as internally. This leads us to clues as to how structural organization of nervous systems may be initially established.

REGULATION OF NEURONAL MORPHOGENESIS AND ADHESION

Perhaps the most puzzling unsolved problem in all of biology is: *How does a one-dimensional genetic code specify the developmental organization of a three-dimensional adult nervous system?* Distinctive integrative characteristics of neurons must guide neuronal structure and function in respect to establishing and maintaining appropriate functional input/output relations. Edelman (1984) has advanced a hypothesis that is backed by evidence relating to cell adhesion molecules (CAMs). CAMs are thought to regulate morphogenetic movements responsible for the organization of neuronal circuits and overall nervous system structure.

According to Edelman, CAMs play a central role in morphogenesis by acting through stereospecific recognition molecules which are attached to the cell surface. These are attached by long sialic acid threads that can change length radically with slight changes in ionic balance and thus may account for specific distance recognition, movement, and cell to cell adhesion. Genes for CAMs may be expressed in schedules that relate to fate maps and sequential histories necessary for circuit building. They are considered responsible for bringing cells of different origins together in the course of various embryonic inductions. Small changes in genetic regulation of CAMs could lead to large changes of neuronal circuitry in relatively short evolutionary time.

HOW CAN OVERALL INTEGRATION BE ACCOMPLISHED?

Stratified Stabilities

Nervous system controls are distributed rather than centralized. Higher level, more elaborate controls are established by the organization of what may be called "*stratified stabilities.*" Organization at any given level of the nervous system, e.g., motoneuron and muscle, spinal segment, spinal cord, decerebrate preparation, etc., are each functionally relatively stable. Each may be controlled as a modifiable ensemble by a large number of other neural circuits. These in turn constitute larger scale modifiable ensembles that have correspondingly more elaborate, selective, and variable functional capabilities. Such large distributed ensembles are themselves functionally relatively stable.

Functional ensembles in the nervous system constitute stratified stabilities in the sense that this term was applied by Bronowski (1970, 1973) to the increasing functional complexity that is seen throughout inorganic and organic evolution. The stable units that compose one level of complexity establish the basis for construction of higher order configurations.

Each successive level of organization, from single cells to the nervous system as a whole, constitutes a relatively stable stratum which can be built upon by still more exalted, superordinate systems, thereby permitting successively more abstract and elaborate information processing and control. This is the great advantage of stratified stabilities. It is by linking together various combinations of stratified stabilities that the brain can serve the individual as a whole.

An important point is that stratified stabilities are not necessarily organized horizontally or segmentally; those are simply the easiest ways to examine them. Cutting across the neuraxis is obviously easier than splitting it longitudinally or trying to carve out a peculiarly structured network of neurons. Functionally stable ensembles may be horizontally or vertically organized, and they can be of any neurogeometric configuration that satisfies the functional requirements.

MOTIVATIONAL SOURCES PROPEL INTEGRATION

The basic driving force for integrative mechanisms is evidently generated by spontaneously active neurons associated with gray matter in medial diencephalon and brain stem, central gray matter, and floor of the fourth ventricle. Neurons lining ventricular channels, constituting the core of the neuraxis, provide respiratory, cardiovascular, appetitive, neuroendocrine, and arousal mechanisms. Altogether these constitute rhythmically active and tidally fluctuating visceral and somatic state control systems. Apparently, spontaneously active state control systems radiate their influences outward from the central core of the neuraxis to orchestrate the entire panoply of relatively stable neural organizations that represent sensory and motor mechanisms and higher neural processes as well.

It is undoubtedly significant that these central motivational systems are closely connected with central mechanisms for reward and punishment. A feedback loop can inform whether the goals of internal (visceral) needs are being met with corresponding visceral satisfactions.

More complex, higher level, modifiable ensembles are erected upon functionally stable strata of lower level modifiable ensembles. The whole nervous system is thereby governed and steered toward such behaviors as have in evolutionary and individual past history secured survival, reproduction, and other visceral satisfactions. The rhythms, appetites, and tides of visceral needs not only provide mood and feeling states, they literally *motivate* (drive, orient, bias, focus, entice, induce, propel, shape, spur, goad, and guide) activities throughout the remainder of the nervous system. Visceral needs are literally motivating and restless until satisfied. This is a feedforward program of activity that "seeks" visceral satisfactions. Motivations may be genetically endowed or acquired, and they may be short lasting in terms of seconds or fractions of a second, or enduring, measured in decades.

Hippocampal Pyramidal Cells Can Monitor Interoceptive and Exteroceptive States

Figure 9.206 provides an example of how information relating to the internal milieu can be compared with information relating to the external milieu. A cell in this context could play an important role deciding what the outside world has to offer to meet this individual's needs at this time. There may be many such decisive convergences in various parts of the nervous system. There are substantial reasons to believe that the limbic system, and specifically the hippocampus, can play such a role in biological and psychological integration.

The exteroceptive input stems from olfactory sources and is distributed to apical dendrites of hippocampal pyramidal cells. Activation of this input elicits excitatory postsynaptic potentials (EPSPs) but without neuronal discharge.

Interoceptive information stems from the hypothalamus and is relayed to the hippocampus by way of the septum. Septal input is distributed to the basal dendrites of the same pyramidal cells. This projection informs the hippocampus of conditions relating to the internal state. The effect of stimulating this input also excites EPSPs which are usually followed by pyramidal axonal impulses. The olfactory input, although excitatory, is more *conditional*.

Presumably this circuit could play a role in helping to decide whether approach or avoidance behavior is appropriate for food or objects of potential sexual interest. The outcome would depend on the internal state of food or sexual deprivation and the distinctive qualities of the olfactory signals (MacLean, 1970).

Activation of pathways from the septum which generate hippocampal discharge may be determinative of behavior and, simultaneously, memory storage. Through the septum, the hippocampus is informed of hypothalamic priorities relating to aversive, appetitive, visceral, and endocrine reactions of an unconditional nature. Internal signals

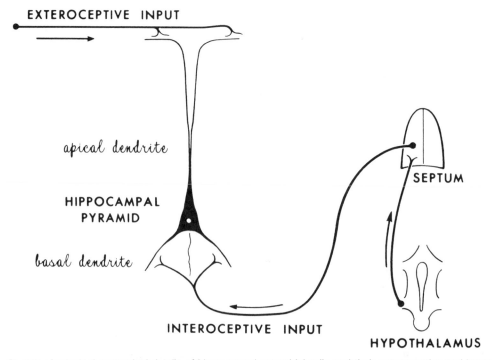

Figure 9.206. Sketch of principal anatomical details of hippocampal pyramidal cells and their exteroceptive and interoceptive inputs. Note that the exteroceptive input (olfactory) goes to the outermost apical dendrites. The interoceptive input, from hypothalamus via septum, goes to the basal dendrites. Pyramidal cell discharge will be conditional by comparing information between internal state and external opportunity. [From MacLean, 1969, 1970.]

have the overriding influence on responses by the decision-making neuron. The olfactory system carries information that provides guidance for ingestive and sexual behaviors. It is by such means that limbic integration can contribute to appropriate behavior and memory storage, the better to meet future contingencies.

LARGER SCALE INTEGRATION OF STRATIFIED STABILITIES

Intimate feedforward and feedback information, and longer looping circuitry that can bypass intermediate stable strata and deliver messages to and from otherwise remote strata, stitch the nervous system together in functionally adroit ways. Stratified stabilities represent variously distributed ensembles which may be combined and built upon during the course of evolution, development, and learning. They contribute an important creative ingredient to nervous system evolution. It is presumed that with relatively few genetic additions (some already tried out elsewhere) whole ensembles (stratified stabilities) could be added to brain organization. It is perhaps in this manner that certain cortical areas in the middle of the whole map of human neocortex (bordered by parietal, occipital, temporal, insular, and frontal cortex, lying considerably hidden within the Sylvian fissure) may have been added during the rapid emergence of speech and language functions during the transformation from common primate ancestors to Homo sapiens.

Each neuron, and each functional ensemble of neurons, has limited purview and capabilities. By operating integratively, these ensembles can contribute to the advantage of a given individual's survival and reproductive capabilities and thereby, in a more general sense, to survival of the species.

BREAST FEEDING AS AN ESSENTIAL MAMMALIAN NEUROENDOCRINE ENSEMBLE

Nervous and endocrine systems provide many familiar examples of interdependent mechanisms engaged in integrative activities. Breast feeding, as one example, involves the infant's *suckling and rooting* activities which stimulate reflex suffusion and *erection of the mother's areolae and nipples.* Simultaneously, this activity elicits impulses that ascend the spinal cord and brain stem to activate the *supraoptic nuclei of the hypothalamus.* This generates signals to the neurohypophysis which release *oxytocin* that induces contraction of collecting ducts in the breast and active *ejection of milk.*

Each period of suckling, moreover, results in periodic *increases in prolactin concentration,* by release of activity of lactotrophs in the anterior pituitary. This is accomplished by inhibition of the normal hypothalamic dopamine restraint of lactotroph growth and prolactin synthesis. The result of this inhibition of an inhibitory influence maintains milk production for subsequent feeding periods.

STRESS REACTIONS THAT CONTRIBUTE TO SUSCEPTIBILITY TO NEOPLASM

Studies demonstrates that anxiety-stress can be quantitatively induced and its consequences quantitatively measured according to biochemical and cellular effects (Riley, 1981). This demonstration requires that quiescent baselines of these parameters are obtained in experimental animals in low stress conditions. Emotional, psychosocial, and anxiety-evoked stress produces increases in plasma concentrations of adrenal corticoids and other hormones through well-known neuroendocrine pathways.

A direct consequence of abnormally increased corticoid concentrations is injury to elements of the immunological apparatus, which may leave the subject more vulnerable to the action of latent oncogenic viruses, newly transformed cancer cells, and other incipient pathological processes that are normally held in check by an intact immunological apparatus. There are adverse effects of increased plasma concentrations of adrenal corticoids on the thymus and thymus-dependent T cells, which elements constitute a major defense against various neoplastic processes and other pathologies.

EXAMPLES OF PERIPHERAL, SPINAL, AND BRAIN STEM INTEGRATIVE MECHANISMS

Prior to considering cortical and subcortical integrative mechanisms, it is worthwhile to have in mind a few examples of integrative mechanisms encountered in peripheral, spinal, and brain stem circuits. These are among numerous stratified stabilities which can be influenced and called upon by higher integrative circuits to participate in more elaborate sensory and behavioral repertoires. With the benefit of such stratified stabilities, higher command can greatly simplify its command creations.

A decerebrate cat, after removal of its entire forebrain, displays skin and hair responses to grooming and characteristic feline lashing of the tip of its tail. Milk put on its nose or tongue is licked and swallowed, but acid is rejected. Light stimulation of guard hairs on its pinna yields a brisk ear flick. Probing of hairs in its external auditory meatus results in a vigorous shaking of the head. An electrical "flea" stimulating the skin results in scratch reflexes that are directed accurately according to the location and direction of motion of the stimulus.

If a decerebrate cat is inverted and let fall, it alights on its feet. If its feet are placed, the decerebrate cat reflexly supports its own weight against gravity. If a treadmill moves backward under its feet, it walks, runs, or gallops according to the speed of the treadmill. If it is placed in the stream of a millrace, it swims accordingly. These are examples of mechanims that are patently useful to the behavior of the individual as a whole and all integrated at spinal and brain stem levels.

This does not mean that such circuits are unaffected in the intact state by forebrain mechanisms, but that even in the absence of higher circuitry, these lower circuits manifest remarkably elaborate and integrated activities. Still, the decerebrate animal exhibits no social reactions and does not learn in any conventional sense of that term. It behaves like a complex but "mindless" automaton.

ELABORATE CENTRIFUGAL CORRECTION OF A PROMPT BRAIN STEM REFLEX

As we demonstrated in Chapter 63, in an intact human, the vestibuloocular reflex (VOR), which is executed along a brain stem path, adapts, after a few days of error accumulation, to almost any systematic displacement or distortion of the retinal image. This goal-seeking reflex involves bilateral integration between sensory inputs that arise from five vestibular neuroepithelia and motor outputs that go to six extraocular muscles on each side of the brain stem. It involves a single interneuron and only two synapses (see Figs. 9.48 and 9.49 and accompanying test). These latter become elaborately reorganized in accordance with arbitrary requirements for direction and velocity of eye movements, the goal of which is to ensure stability of the retinal image during active and passive movements of the head.

The two synapses of the VOR lie in the brain stem, between upper pons and upper midbrain. Thus, they remain intact in a high decerebrate animal. Yet appropriate adaptation could not occur in such a high decerebrate preparation for the simple reason that the necessary error messages which depend on the visual system would not be available.

Adaptation in the VOR is instructive because it demonstrates the convergence of exceedingly complex information from more than one source onto a reflex arc that is very abbreviated (two synapses), brisk (15 ms), and machine-like (it works even in the dark, and following adaptation as well as in the original state). According to Miles and Lisberger, 1981, this adaptation takes place not through learning but depends on functional (and presumably morphological) reorganization affecting the two brain stem synaptic relays—in accordance with a cumulative record of error messages. *The goal-seeking nature of this adaptation uses visual cues to alter a reflex pathway involving a single brain stem interneuron in a very precise way.*

A notable feature of this adaptation is that nothing seems to happen while error messages are accumulating, until shortly before the adaptation is put into effect. Then, close approximations to the correct adaptation are manifested, and fine adjustments soon follow. This indicates that goal satisfaction (visual stability) can be automated and yet ultimately remotely controlled.

The VOR adaptation is particularly dramatic because the sensory control originates entirely outside the reflex path. *Goal-seeking mechanisms utilize remote and elaborate information to shape local reflex controls.* The VOR is also instructive because it affects such an abbreviated path so close to peripheral input-output relations.

Goal-seeking mechanisms can operate way "downstream." Decision making can be inserted into the most peripheral and mediate possible junctions to achieve desired *motor control for sensory purposes.* VOR control is exerted long before any message could be discriminated and controlled by conscious mediation. The implications of all this for complex behaviors such as playing a musical instrument are obvious but as yet unanalyzed. In a situation that demands great precision, the critical control may be given over to a reconstructed, automaton-like, built-in, anatomicophysiological system. Because minor error messages will still be received by the visual system, cumulative errors can be applied to occasionally modify the otherwise stable reflex circuit. Satisfaction may be usefully automated and remotely controlled.

Example of Large Scale Integration

CEREBROCEREBELLAR-NEURAXIAL-SKELETAL-ENVIRONMENTAL INTEGRATION

Understanding how a few small to medium scale ensembles produce goal-seeking integration enables us to visualize still larger scale integrative mechanisms. How does the cerebellum contribute to cerebral motor command of skeletal muscle? Obviously, the three major structures, cerebral hemisphere, cerebellar hemisphere, and the neuraxis—brain stem and spinal cord, muscles, joints, skin, visual witness, must be communicating with one another. There must be information as to what is the *goal* of a given movement, the motor *command*, as funnelled through the sensorimotor cortex—to brain stem and spinal levels, and *feedback information* as to the body configuration at the outset of the movement and during the movement. All this must be properly framed in accordance with bodily posture in three-dimensional space, together with coordinates relating to the organization of the environment within which the movement will be executed, and all abetted by visual contributions to the performance.

Figure 9.207 provides a schematic representation of cerebral, brain stem, spinal, and cerebellar functional relations with respect to skeletal muscle motor command. It is presumed that a command initiative originates in the cerebral hemisphere, prob-

ably by some motivational interplay between limbic, diencephalic, and frontal cortical zones relating to forward planning of behavior (not shown). The command would be executed according to orders expressed through the "cerebral motor cortex" via pyramidal and extrapyramidal pathways (heavy descending lines) to "spinal motor centers" and "muscles." Signals that monitor this command would be delivered to "pontine nuclei," "reticular nucleus," and "inferior olive." Relayed command monitoring from these three imtermediate stations would reach the "cerebellar anterior lobe" which provides a complete running account of body representation.

The cerebellum provides corrective feedback information to the cerebral motor cortex (*heavy dashed line at top*). This feedback has already bene-

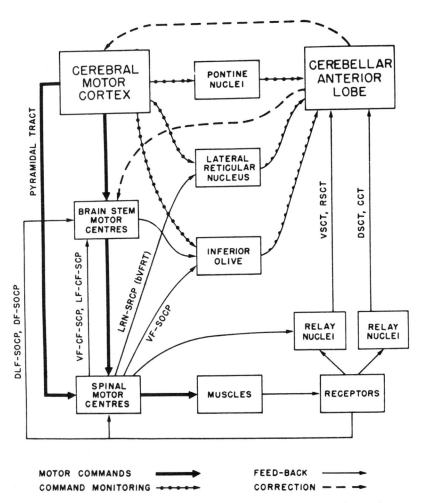

Figure 9.207. Cerebrocerebellar servomechanism to monitor and to correct motor command. Some of the pathways between cerebral motor cortex, anterior lobe of cerebellum, and lower motor centers. Added are indications of the functional role of each of these pathways. The anterior lobe is assumed to correct errors in motor activity elicited from cerebral cortex and carried out by command signals through pyramidal and extrapyramidal paths. Command signals are assumed to be monitored by the anterior lobe through paths relayed in the inferior olive, pontine, and reticular nuclei. Spinocerebellar paths are assumed to serve feedback channels which monitor activity in lower motor centers concerning the evolving movement. *DLF-SOCP*, dorsal spinoolivocerebellar path; *VF-CF-SCP*, ventral climbing fiber spinocerebellar path; *LF-CF-SCP*, lateral climbing fiber spinocerebellar path; *VSCT*, ventral spinocerebellar tract; *RSCT*, rostral spinocerebellar tract; *DSCT*, dorsal spinocerebellar tract; *CCT*, cuneocerebellar tract; *LRN-SRCP*, lateral reticular nucleus relay of spinoreticulocerebellar path; *VF-SOCP*, ventral spinoolivocerebellar path. [From Oscarsson, 1973.]

fitted from a comparison between command and execution of the movement as represented by direct and indirect feedback from "spinal motor centers" and muscle and other receptors. Within such a recoursive circuit, there is ample opportunity for the operation of ensembles of stratified stabilities, related to one another, and feedback as well as command circuits that can monitor the discrepancy between goal, command, and progress toward the goal. The cerebellum provides corrective signals during the course of the movement that will persist until the goal is properly satisfied. Deficits in functioning of any of the gross parts of this circuit can be analyzed in light of a patient's performance.

Synaptic Remodeling along Path between Cerebellum and Cerebral Cortex

Figure 9.208 illustrates synaptic remodeling, reactive synaptogenesis, in the central nervous system. Following a cerebellar lesion that eliminates one input to the red nucleus (BC), there is a change in response to cortical activation of that nucleus.

This is due to reorganization of cortical axonal terminals within the red nucleus. This is but one example of synaptic and circuit reorganization that takes place when part of a complex system is modified.

It is evident that reactive synaptogenesis can occur when there is deprivation of activity, or hyperactivity, along a given circuit. When an input is removed, the recipient cells undergo "sensitization of denervation." They become more sensitive to both excitatory and inhibitory influences, chemical (anesthetics and drugs) as well as synaptic. Perhaps this sensitization has something to do with evoking the reactive synaptogenesis. Whenever there is reduced competition for synaptic junctions, neighboring axons may sprout and establish additional junctions. The process is orderly, however, with certain replacement options having priority over others. *Evidently, the nervous system is continually remodeling itself, depending on ongoing experiences, the traffic flow of information—or absence thereof—and is continually adjusting circuits accordingly.*

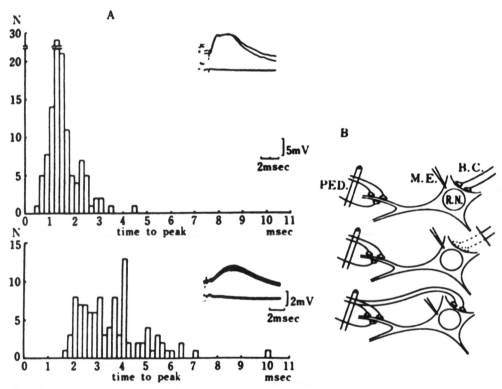

Figure 9.208.   Reactive synaptogenesis in the brain stem (red nucleus). *A*, Following a lesion in nucleus interpositus of the cerebellum which projects to the red nucleus there is a shortening of rise time of neocortically induced postsynaptic potentials as recorded from neurons in the red nucleus (*upper histogram*) as compared with normal rise time (*lower histogram*). Records of intracellular and corresponding extracellular responses are illustrated by inserts in each histogram. *B*, Diagram of experimental arrangement: *R.N.*, red nucleus neurons; *B.C.*, brachium conjunctivum input from interpositus nucleus; *PED.*, input from cerebrum obtained by stimulation of cerebral peduncles; *M.E.*, recording microelectrode. [From Tsukahara, 1978; from Cotman et al., 1981.]

A Miscellany of Important Integrative Mechanisms

THE CONTROL OF CEREBRAL BLOOD FLOW

A large share of the body's economy is utilized by the brain. At rest, about 20% of cardiac output and O_2 consumption is devoted to supporting brain activity. The products of metabolism, diminished O_2, elevated CO_2 and H^+, contribute to controlling cerebral blood flow locally in accordance with local metabolic requirements. In hyperthyroidism, however, the brain does not participate in the generalized increase in metabolism. The main action of thyroxin is on protein rather than carbohydrate metabolism. In deep anesthesia and coma, brain oxygen consumption may be reduced by 50%. During sleep, nevertheless, the utilization of oxygen is normal. Cerebral circulation is maintained normally in hypertension, despite perfusion pressures twice normal. This is accomplished by a combination of sympathetic vasoconstriction, humoral vasoconstriction, and homeostatic autoregulation. Hypertension may lead to overstretching of endothelial cells and result in leakage that leads to hypertensive encephalopathy.

Ingvar and Philipson (1977) have investigated cerebral blood flow patterns in neurologically normal subjects during motor ideation and motor performance. At rest, the normal hyperfrontal distribution of cerebral blood flow is seen, with highest flows in premotor and frontal regions, with very low flows temporally. During "hand movements conceived," without actual movements, mean flow was augmented about 15%. The most notable rise in flow was seen in the frontal region and over sylvian and temporal areas. During actual hand movements, rhythmical squeezing of a small rubber balloon, there was a definite peak in flow over the sylvian region, while flow in frontal and temporal areas was lower than in the "conceived" state, although higher than "at rest." This suggests that cerebral blood flow must be dynamically regulated during everyday shifts in mental and physical activity and indicates that control of motor ideation may occur in different cerebral loci than those that control actual movement.

BLOOD-BRAIN DIFFUSION BARRIERS

The brain enjoys a remarkably protected environment. A useful review of diffusion barrier, carrier, and active transport is provided by a symposium organized by Heistad, 1984. Cerebral capillary endothelium is specialized: bonded by tight junctions, lacking fenestra that are common in other capillary beds. It is moot whether the glial sheath of astrocytic foot processes serves special transport except to bail out excess K^+ ions. But the glial contribution may (by trophic effects on the endothelium) determine specific characteristics of the cerebral endothelium.

Extracerebral release of neurotransmitters does not interfere with synaptic transmission in the brain because the passage of neurotransmitters from blood to brain is greatly restricted. The concentration of ions that affect neuronal transmission is regulated by a barrier to their passage from blood to brain and by a powerful pump that actively transports them from brain to blood.

The blood-brain diffusion barrier also has carrier functions. It is almost impermeable to Na^+, but Na^+ is pumped into the cerebrospinal fluid by the choroid plexus.

The barrier function maintains the protected environment, while the carrier function enhances nourishment for the brain. Specific receptors for insulin are located on the endothelium. Facilitated transport is described for glucose, purines, nucleosides, ketone bodies, choline, and several amino acids. Certain vitamins and nucleosides do not enter the brain in appreciable amounts across the barrier but are transported by specific carrier-mediated mechanisms across the choroid plexus. From the cerebrospinal fluid these substances obtain access to the brain. The choroid plexus is important for the transport of ascorbic acid, folates, and nucleosides into the brain by way of cerebrospinal fluid. While steroid and thyroid hormones circulate bound to plasma proteins, the brain endothelia succeed in stripping off the circulating carrier proteins, enabling the hormones to cross the barrier.

RELATIONSHIPS AMONG LOCAL FUNCTIONAL ACTIVITY, ENERGY METABOLISM, AND BLOOD FLOW

Activity varies greatly in different parts of the brain, including innumerable subunits which are integrated into specialized functional networks. An interesting method of Sokoloff et al. (1977) provides means to measure quantitatively local rates of glucose utilization simultaneously in all light microscopically visible structures of the brain.

Glucose is almost the only substrate for cerebral oxidative metabolism, and its utilization is stoichiometrically related to oxygen consumption. Radioactive glucose does not have a sufficient biological half-life to stay in cerebral tissues long enough for these purposes. But the labeled analogue of glucose, 2-deoxy-D-[^{14}C]glucose, serves as a competitive substrate with glucose for both blood-brain transport and for hexokinase-catalyzed phosphorylation.

Unlike glucose-6-phosphate, deoxyglucose-6-phosphate cannot be converted to fructose-6-phosphate. 2-Deoxy-D-[^{14}C]glucose therefore gets trapped in cells long enough for purposes of radiographic measurement. The density of radioactivity emitted from this substance then serves as an accurate expression of the rate of glucose uptake, and this can be measured by detailed autoradiographic techniques.

The quantity of this trapped molecule following its introduction into the circulation is directly related to the amount of glucose that has been taken up during the same interval. This is the method utilized to provide the localized functional activity differences illustrated in Figures 9.86 and 9.89–9.91 in Chapter 64. From those, you can appreciate that a great deal can be quantitated by this means, down to the level of single cortical columns.

Similar autoradiographic methods permit quantitative measurements of local blood flow. *This combination of evidence establishes that rates of blood flow in various structural components of the brain parallel the local rates of glucose consumption, and both rates change in response to local changes in functional activity.*

Introduction of a local seizure-producing agent (potassium benzyl penicillin) into motor cortex of monkeys results in recurrent focal seizures in the corresponding skeletal muscles of the contralateral side (Caveness, 1969). Such focal seizure activity causes selective increases in glucose consumption in areas of motor cortex adjacent to the penicillin injection, and in correspondingly discrete loci in the putamen, globus pallidus, caudate nucleus, thalamus, and substantia nigra of the same side (Kennedy et al., 1975).

Analogously, decrements in functional activity result in reduced rates of glucose utilization. In the alert rat studied 24 h following unilateral enucleation of one eye, there are marked decrements in glucose uptake in the contralateral lateral geniculate body, visual cortex, and superior colliculus. Thus, functional ensembles can be picked out by quantitative methods in intact behaving animals. Such methods have some advantages over electrical recording, inasmuch as only a finite number of electrodes can be monitored, whereas the entire brain can be surveyed for localized differences in blood flow and metabolic rate.

Developments in computerized emission tomography permit analysis of local concentrations of labeled compounds in vivo in humans. Emission tomography using a positron-emitting derivative of radiodeoxyglucose, 2-[^{18}F]fluoro-2-deoxy-D-glucose (Reivich et al., 1979; Phelps et al., 1981) is proving useful in studies of the human visual system and clinical problems such as focal epilepsy (Kuhl et al., 1980). These methods hold considerable promise for understanding functionally localized disorders in neurological and psychiatric patients.

STRUCTURAL AND FUNCTIONAL SEX DIFFERENCES INTEGRATED BY STEROIDS

One of the integrative strategems of evolution has been to modify nerve cells in terms of their survival and function by giving them relatively brief exposure to hormones. Particular brain regions contain cells with cytoplasmic receptors specific for androgens, estrogens, or progestins. The bound complex of receptor-steroid moves into the nucleus where it presumably interacts with DNA to regulate specific RNA and protein output that alters cell function (McEwen, 1976).

The mammalian brain is inherently female. The absence of testosterone in the female at a critical perinatal period results in development of female sexual behavior patterns and cyclical release of luteinizing hormone necessary for ovulation.

Gonadal steroids can exert powerful effects on neuromorphogenesis and survival of special neurons in the hypothalamus and spinal cord. The timing of hormone action is critical. The "critical period" for sex steroid sensitivity begins a few days before birth and lasts into the early postnatal period. Gorski et al., 1978, found that natural release of testosterone at birth enduringly commits a sexually dimorphic nucleus in the medial preoptic area of the rat hypothalamus (SDN-POA). The median preoptic nucleus is important in the control of masculine sexual behavior and the cyclic release of gonadotropin. This nucleus grows three to seven times larger in the male than in the female. Castration of the newborn male rat produces a reduction to about half the normal adult volume of this nucleus; this effect can be prevented by a single administration of exogenous androgen one day postnatal (Jacobson et al., 1981).

Testosterone administered to a female newborn significantly increases the volume of the median preoptic nucleus in the adult. Although steroids given at birth can determine the adult size of the nucleus, permanent effects from a single injection disappear after about the 10th postnatal day. Hormones administered to males and females gonadectomized as adults will restore copulatory behavior but do not influence SDN-POA size. Sexual dimorphism depends on the release of testicular testosterone which is metabolized in the brain to form estradiol. This both masculinizes and defeminizes (Goy and McEwen, 1980). In fact, both estrogens and androgens contribute independently to masculinization of the nervous system.

Arendash and Gorski (1983) developed a "punch technique" for transplanting the SDN-POA of neonate males into the same region of littermate females. The technique involves use of a cannula to "punch out" the appropriate tissue from a fresh slice of donor brain and to extrude this transplant tissue stereotactically into the appropriate area of a recipient's brain. Histology verified that neural connectivity takes place between male transplant tissue and recipient female brain.

Steroids have a neurotrophic effect: both testosterone and estrogen enhance outgrowth of neural processes in explants of newborn median preoptic tissue. While they do not much change the adult secretory patterns of hormones in recipient neonate females, male transplants do have conspicuous

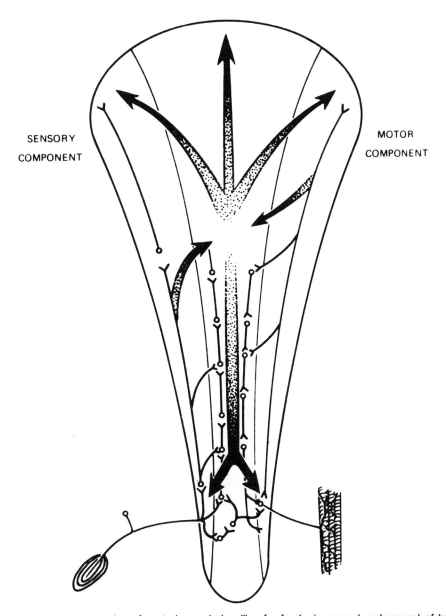

Figure 9.209. Diagrammatic representation of central neural signalling for forebrain arousal and arousal of brain stem and spinal sensory and motor background activity. The left side represents an ascending sensory column characterizing all sensory systems as having sequences of first-order, second-order, etc. neurons that may be directly and indirectly involved in sensorimotor reflexes and ascending relays to cortex. All sensory trajectories except olfactory send relatively direct signals into the brain stem reticular formation; olfactory arousal is powerful but less direct. The consequences of activation of brain stem reticular formation is ascending arousal affecting the whole forebrain and descending impulses which increase the central excitatory state all along the brain stem and spinal cord. The centrifugal effects influence all sensory as well as motor relays and also efferents such as muscle spindles that directly affect sensory input. The right side depicts a descending motor column representing all motor pathways from cortex to final motor neurons and effector organs. The motor as well as sensory columns contribute directly and by way of collaterals to brain stem and spinal reticular formation interneurons and associated diffusely projecting systems. [From Livingston, 1974.]

masculinizing effects on female behavior. Thus, *steroid actions are involved in alterations of CNS structure, hormonal secretory patterns, and sexual distinctions in behavior.*

Breedlove and Arnold (1980) found similar steroid effects in a special cluster of motor neurons in the spinal cord which innervate perineal muscles found only in males. Male and female rats exhibit sex differences in binding by serotonin receptors in discrete areas of the brain that have to do with ovulation and gonadotropin release. The number of binding sites are sex-specifically affected by steroids. The status of serotonin receptors may affect reproductive capacity of an organism and have an influence on CNS sensitivity to drugs and hormones (Fischette et al., 1983).

CORTICAL AND SUBCORTICAL CONTRIBUTIONS TO INTEGRATION

Harman published an important study in 1957, showing that *expansion of cortical volume* during primate evolution, from lemurs, monkeys, and apes,

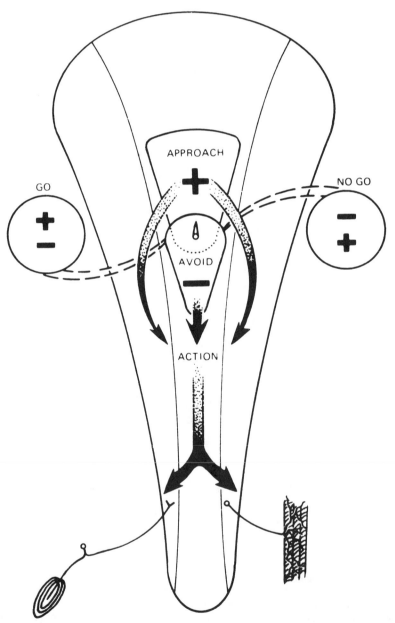

Figure 9.210. Diagrammatic representation of theoretical motivation-driven decision-making systems in cephalic diencephalon and brain stem. Neural activities relating to the status of appetite and satiety mechanisms and any perceived opportunities or threats in the environment relating to biological needs result in increased arousal followed by approach or avoidance behavior, "go," or freeze "no go." Posterior diencephalon and cephalic brain stem exhibit changes in electrical activity associated with switching on and off of both approach and avoidance behavior. Ambiguous stimuli lead to oscillatory electrical activities in this region which closely correlates with the animal's oscillatory behavior. [From Livingston, 1978.]

to man, *is in direct proportion to expansion of volume of the whole brain.* This indicates that primate cortical and subcortical structures evolved and expanded together, expansion of one being accompanied by parallel expansion of the other. This does not mean, however, that subcortical gray matter expanded exactly in parallel with expansion of gray matter in the cortex. Rather, additional cortico-cortical and corticosubcortical white matter make up a substantial part of the increase in subcortical mass. Szentagothai (1984) estimated that 80% of the connections of any given part of the cortex represent corticocortical connections. Thus expansion of cortex is accompanied by a considerable but not yet quantified expansion of subcortical gray matter, and the rest is made up by considerable increases in the white matter necessary to permit corticocortical integration and an expansion of corticosubcortical white matter that also must accompany the expansion of cortex.

This means that higher neural processes, and accompanying higher neural integrative mechanisms, have been greatly favored during recent evolution. It accounts for the fact that primate evolution is accompanied by a palpably appreciable increase in bulk of white matter. This would be necessary to support the additional structures and empower the increased degrees of freedom reflected in human thought, imagination, language, and social integration.

Figure 9.209 is a schematic functional diagram to suggest integration of the nervous system as a whole. Although this illustration is entirely theoretical, it intends to represent a reasonable neurophysiological frame of reference for overall integration.

A receptor, resembling a Pacinian corpuscle, is intended to represent all sensory neuroepithelia and all central nervous system input. A skeletal muscle fascicle is intended to represent all effectors, secretory as well as contractile. On the left are depicted phylogenetically old (pleurisynaptic) and new (paucisynaptic) sensory pathways, intended to represent all sensory contributions to spinal, brain stem (together with cerebellar), diencephalic, and forebrain levels of integration. On the right side of the figure are depicted phylogenetically old (pleurisynaptic) and new (paucisynaptic, in this case, monosynaptic) projections from cerebral cortex to brain stem, spinal cord, and final common path projections to effectors.

Broad ascending and descending arrows are intended to represent projections to spinal, brain stem, and diencephalic central core neurons which contribute cephalically and caudally to distributed readiness for action. These are spinal, brain stem,

and diencephalic reticular formations which control central state conditions all along the neuraxis. This control is evidenced by signs of behavioral and electroencepalographic arousal and alterations in brain stem and spinal cord central excitatory (and inhibitory) states.

Note that the central state control system projects upward and downward onto *sensory* as well as motor mechanisms and onto intermediate association areas of the brain as well as onto brain representations of all sensory and motor functions. Beside contributing to generalized arousal, these projections can be much more discriminating than simply influencing gross arousal and central readiness (Brazier and Hobson, 1980). They can also contribute to focusing of excitability and readiness for action according to circuits that are engaged by motivational imperatives, peripheral and central, visceral, sensory, motor, and associational. This, of course, includes integration of ongoing circumstances with remembered experiences. By observing overt behavior, we can appreciate that such an overarching integrative process must operate in a rich context of biological need, capabilities, and opportunities.

Figure 9.210 represents the functional convergence of central motivational and decison making processes on mechanisms that release or hold the final expressions of behavior in check. We have already seen that during sleep there is a check on the release of behavior that is controlled by a localized brain stem mechanism in the region between midbrain and pons (Jouvet, 1973).

Evidence from electrical recordings in animals during choice discrimination indicates that "go" and "no go" decisions evidently take place at the junction between posterior diencephalon and cephalic brain stem (Grastyan, 1966; Adey, 1974). At that level there is a convergence of motivational information represented in the "frontal-limbic midbrain area" whereby two alternatives, approach and avoidance, may be linked with two other alternatives to release behavior ("go") or to hold behavior in check ("no go"). The consequences of decision making at this level can of course affect sensory processing as well as action programs (Livingston, 1978).

Figure 9.211 represents a functional postulate for printing memory, based on suggestions by Livingston (1966, 1967a and b) and made explicit in terms of the likely role of biogenic amines by Kety (1967, 1974). Previous theories of memory presumed synaptic enhancement that was limited to synapses along circuits specifically engaged in signalling between unconditional and conditional stimulus-response events. The "now print" order implies that

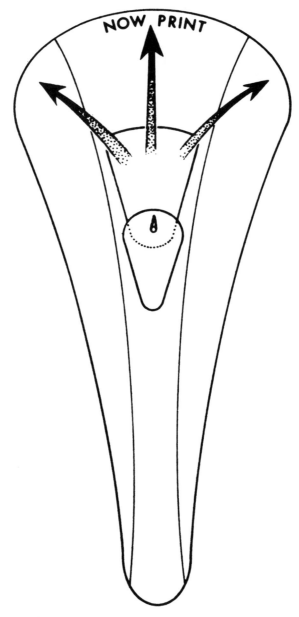

Figure 9.211. Diagrammatic representation of a theoretical mechanism for imprinting of memory. A schematic CNS shows brain stem reticular mechanism for arousal and theoretical "now print" order for consolidating memory. Theory assumes that when a "biologically meaningful event" occurs, there is activation of cerebral arousal (by known reticular activating mechanisms) and also activation of an additional widely distributed message (presumably norepinephrine from the pontine raphé nucleus, locus coeruleus). It is presumed that widely distributed norepinephrine might effect synaptic consolidation at all recently active synaptic junctions (Kety, 1969). Such a sequence would provide that any significant positive or negative reinforcement, relating to any peripheral or central signalling system or combination of systems, would be followed by strengthening of those (and other) recently active neural circuits. Repeated pavlovian or operant conditioning would continue consolidation of only those circuits which are instrumental to the reinforcement process. Other adventitious brain events would not be equivalently synaptically reinforced. [From Livingston, 1969.]

all recently active circuits may be "printed" whenever there is some biologically important event. According to this thesis, *all* recently active circuits would become stronger. This would account for stimulus and response generalization. Those central signals which uniquely and regularly preceded biologically meaningful events would naturally become relatively even stronger during the course of reiterated experiences, and noncorresponding events would accordingly become relatively weaker.

Figure 9.212 illustrates the upward paths from brain stem levels that may induce arousal and possibly memory storage throughout the forebrain. Nuclei such as the locus coeruleus, located at the junction of midbrain and pons, project directly to all areas of the cerebellum and forebrain and deliver norepinephrine which has been postulated to affect consolidation throughout all recently active synaptic contacts. The release of norepinephrine would have such an effect by stimulating protein synthesis through the medium of adenylcyclase, an effect that is potentiated by magnesium and potassium ions and inhibited by calcium, indicating that the adrenergic effect would be prepotent for recently active, as opposed to inactive, synapses (Kety, 1974).

More general arousal mechanisms, arising from adjacent brain stem reticular formation, project directly to cerebellum, hypothalamus, hippocampus and amygdala, and to the diffusely projecting systems of the thalamus. These latter involve principally the centrum medianum, midline, and intralaminar nuclei, and relay through the reticularis nucleus to all limbic areas and neocortex. Altogether these pathways are thought to relate to mechanisms for alerting, arousal, attention, and possibly also to mechanisms involved in learning and memory.

COLUMNAR ORGANIZATION OF CORTEX

A cortical column is about 300 μm in diameter and consists of about 110 neurons. This number is consistent in all mammals studied (Rockel et al., 1974). A notable exception is primate striate cortex where cortical columns characteristically include about 260 neurons. The cortical column, therefore, seems to be a functional module, a prepackageable design for some critical functions of analysis and decision making. It is impressive that this kind of processing module is apparently relatively uniform across the cortex and phylogenetically uniform as well.

There are about 600 million columns in human cortex, and the total cortical cell count is thought to approach 50 billion cells (Edelman and Mountcastle, 1978).

Each column is activated by inputs, specified genetically and ontogenetically. For instance, some columns in somatosensory cortex are activated by touch, others by motion of a joint, etc. Since all neurons within a single column respond to the same place and modality-specific source, this is the kind of elementary functional analysis and decision making they represent in all information processing systems. Their outputs project to specified cortical or subcortical destinations. There appear to be additional, secondary, access sources and projection targets, by means of which associative recognition is possible (Mountcastle and Edelman, 1978). It is presumed that, as a result of experience and the formation of secondary repertoires, columnar modules are "circuited" so as to discriminate new configurations of information.

The manner in which cortical columns process and distribute information appears to be qualitatively similar in all areas of neocortex. Cortical columns thereby constitute functional elements which are differentiated according to their detailed input and output connections. Cortical columns are embodied in functional systems that are widely and reciprocally interconnected, cortically and subcortically.

Larger entities we know as functional areas of cortex are themselves composed of replicated columns and local neuronal circuits which vary in their connections and processing modes from one area to another but are basically similar within a given area (Szentagothai, 1975).

Subcortical structures are organized in systematic ways with cortical areas, having a common or dominating system of connections. As we have seen previously, there is functional partitioning within subcortical structures (e.g., the thalamus) that links them not only to particular areas of cortex but to particular cortical (modality specific) representations within those areas. *The processes of representation, re-representation, and re-re-representation provide precisely connected but distributed information systems.*

THE BASAL GANGLIA

The basal ganglia contribute to the control of movement and posture. The aim here is to add insights about how the basal ganglia and related systems interact in the course of voluntary behavior.

Groves (1984) suggested a division of neostriatal efferents into two functional cell systems, a dominant lateral inhibitory network comprised of the common spiny cells and the far less numerous aspiny cells which are excitatory. Both types are long axon neurons which provide the major outflow of the neostriatum. Collateral inhibition of spiny neurons on aspiny neurons allows an interplay of action and braking action that provides a beautifully controlled dynamic system. Each spiny neuron is excited by coincidence of firing from a wide variety of cortical and brain stem sources which are specific and topographically orderly in their origins and destinations. These circuits are presumed to be active during anticipation of movement. Preparation for movement involves mostly interactions among cortex, neostriatum, other basal ganglia, and brain stem.

The circuits within the basal ganglia involve sequences of different neurotransmitters which may account for disabilities faced by Parkinson patients and patients with Huntington's chorea and other disorders, where degenerative processes may predominately affect one or another neuron species. This notion also holds promise of being able to intervene pharmacologically, as can be done by the administration of L-dopa in the case of Parkinsonism, and perhaps by genetic engineering to overcome the problem in the case of Huntington's patients.

Neostriatal contributions to the actual execution of movement is thought to be fulfilled by impulses which are routed via thalamus to sensorimotor cortex, where impulses provide time-space coordinated patterns of action controlled by that region.

The major efferent outflow from the basal ganglia originates in the globus pallidus (GP) and projects to the ventroanterior (VA) and ventrolateral (VL) nuclei in the thalamus. This pair of thalamic nuclei also receives the major ascending projections from cerebellar outflow, and they project to sensorimotor cortex. Many neurons in the globus pallidus of the monkey discharge phasically in relation to "voluntary" or instrumental limb movements (DeLong, 1971). The majority of these units fire in relation to contralateral limb movements. Clinical evidence indicates that lesions of the pallidothalamocortical pathway usually result in contralateral hemiballistic movements. Such movements are characteristic of damage to the subthalamic nucleus which receives from and projects back to the globus pallidus. The substantia innominata (ventral to the pallidum) projects to the hypothalamus and plays a role in feeding behavior.

Substantia Nigra

Branches from individual nigral axons innervate the ventromedial thalamus and the superior colliculus and may play a role at both cortical and collicular sites in controlling head and eye movements.

LIMBIC CONNECTIONS

Kuypers and colleagues introduced new retrogradely transported markers to neuroanatomy, mostly fluorescent dyes that can be distinguished for identification of doubly or multiply labeled cells (Kuypers et al., 1977; Van der Kooy et al., 1978). Cells in the dentate gyrus in the hippocampus project to the contralateral dentate and also send collaterals into the ipsilateral dentate gyrus. Individual pyramidal neurons in the hippocampus project by way of collaterals to both the septum and entorhinal cortex. Commissural, ipsilateral association, septal and subicular projections from the hippocampus are all similarly due to collaterals of individual axons. At least 80% of the cells in deep layers of the dentate gyrus give rise to both ipsilateral

associational and commissural crossed projections to dentate granule cells. In this respect, cortical and corticosubcortical projections of the hippocampus are quite different from those of neocortex. Neocortex characteristically projects axons that have single cortical or subcortical destinations.

FRONTAL CORTEX: COLUMNAR ORGANIZATION

By double-labeling, Goldman-Rakic and Schwartz found that in the prefrontal association cortex of monkeys, associational projections from the parietal lobe of one hemisphere interdigitate with callosal projections from the opposite frontal lobe, forming bands of adjoining columns 300 to 750 μm wide.

One of the most important concepts for understanding the structure and function of the neocortex has been that of the vertical compartmentalization of its cells and connections. Such organization was discovered during electrophysiological analysis of somatosensory cortex by Mountcastle; later functional and morphological research extended this principle to receptive field properties and connectivity of the primary visual and auditory as well as somatosensory areas of cortex. In these systems, vertical "columns" or "bands" relate to input serving one class of sensory receptor and alternate with input from another group of receptors within the same modality. Vertical organization of inputs is not solely a property of sensory systems but applies to association cortex as well. Callosal (contralateral) terminals alternate with associational (ipsilateral) terminals in selected cytoarchitectonic areas of primate association cortex (Goldman-Rakic and Schwartz, 1982). Side by side registration of inputs from the two hemispheres is probably relevant to the cerebral mechanisms underlying interhemispheric integration.

Analysis of autoradiographic data revealed that labeled fibers from the opposite prefrontal cortex are distributed in bands that extend across all layers of cortex, interspersed by blank areas found to be related to ipsilateral association projections. This means that callosal and associational projections reach exclusive columnar territories, in areas of close convergence. This is reminiscent of alternating ocular dominance bands in striate cortex (see Figs. 9.89 and 9.90).

HRP-labeled fibers originating from the inferior parietal cortex were also distributed in roughly the same area of the prefrontal cortex. They also formed discontinuous bands alternating with spaces of variable width. There are evidently many variations on the theme of having information that can be usefully related projecting to neighboring bands. Similar banding of spinal, pontine, and olivary projections onto cerebellar cortex suggests that this structural contribution to integrative processes is widespread.

In the striate cortex, adjacent bands from opposite eyes relate to the corresponding loci in the two retinae. The advantage this has for discriminating such things as motion and distance parallax have been emphasized in Chapter 64 (Vision). Banding of interhemispheric and intrahemispheric association projections may make important contributions to correlating functions between left brain and right brain. Undoubtedly additional functional advantages from juxtaposition (beside shortening corticocortical interconnections) will emerge as more is known in detail about the ubiquitous banding and columnar organization patterns in cortex. For example, Goldman-Rakic has identified cells in the prefrontal cortex of monkeys that fire only when the animals *plan* what response they will make in certain circumstances.

Occasionally segregation of callosal and associational inputs is a rule that can be broken. Some columnar spaces adjacent to callosal and associational columns may remain unlabeled. It is likely that these unaccounted-for territories are filled by terminals from one or more other sources. It is possible that some cortical projections interdigitate, as described, while other may overlap (Goldman-Rakic and Schwartz, 1982).

FRONTAL LOBE FUNCTIONS

Following removal of part of the frontal lobes to reduce intractable epileptic seizures, patients were unable to shift criteria for card-sorting tasks (Brenda Milner, 1974). Working with Milner, Michael Petrides developed a test to distinguish between left and right frontal lobe lesions. He gave patients 12 cards, each containing a list of the same 12 words arranged at random. The patient touches a different word on each card in such a way that each word is touched only once. Only those frontal lobe patients who have had surgery on the left hemisphere have serious difficulty with this task. Patients with right frontal lobe surgery have trouble with a similar task that uses 12 different drawings rather than words. To the researchers' surprise, however, patients with surgery in the left frontal lobe also have difficulty with the drawing test. They suppose that "the left hemisphere as a whole may be dominant for the initiation, programming, and monitoring of sequential tasks." Perhaps all strategies are aided to some degree by verbal formulation. Recently Petrides developed a test for monkeys to learn either to grip a stick or touch a button, depending on whether they were shown a green bottle top or a blue and yellow toy truck. A small lesion in the periarcuate region (the frontal eye field) interferes with this learning task.

If a monkey is trained to move his eyes in a certain direction, specific cells in the frontal eye-fields will fire just before that movement. Yet these cells would not fire if the same eye movement were merely incidental to what the monkey was doing. If brain cells drive an eye movement in a remembered direction even in the dark, without benefit of visual cues, Goldberg asks, "Is that activity the memory itself?" "The motor actvity is divorced from a sensory stimulus because of a previous experience; is that what happens during learning?"

"A human being whose frontal eye field has been taken out will be able to move his eyes perfectly well, but the frontal eye field is necessary for eye movements done according to complicated criteria," explains Goldberg. "The frontal eye fields do not send down a motor command; they send down a program for a desired eye movement—a plan."

PARIETAL LOBE FUNCTION

Neurons in the inferior parietal association cortex (area 7) have been studied in alert, behaving rhesus monkeys, trained to fixate and follow visual targets (Yin and Mountcastle, 1978). Visual tasks engage four classes of neurons. Three of the four classes are related to visuomotor functions: visual fixation, tracking, and saccades. They are activated only when the animal is visually tracking or fixing gaze and prior to visually elicited saccades. One is sensitive to visual stimuli and has large contralateral receptive fields with maximal sensitivity in the far temporal quadrants. None of the four classes is sensory or motor in the usual sense, and they are not activated during spontaneous saccades and fixations that the monkey makes while casually exploring the environment. It is presumed that the light-sensitive neurons provide visual input to the visuomotor cells which, in turn, control command signals for direction of visual attention and shift of focus of attention from one target to another. Neurons in area 7 integrate motivational and motor commands for direction of visual attention.

Motivation and Integration

Olds and Milner (1954) discovered localized regions in the brain stem, diencephalon, and limbic system that are positively reinforcing. They consist of pathways and gray matter with quite well defined boundaries. *The most powerful positive reinforcements are found in the median forebrain bundle, a massive circuit between frontal lobes, limbic system, hypothalamus, and midbrain.* The limbic system as a whole appears to participate closely with central pleasurable experiences and positive reinforcement activation. The termination of frontal lobe and limbic outflow is to the "frontal-limbic midbrain area" which is itself powerfully rewarding.

This important discovery was first appreciated when rats with implanted electrodes were allowed to roam freely, tethered only by long, flexible leads to their electrodes, in an open field environment. Whenever threshold stimulation was introduced, the animals stopped whatever they were doing. It dawned on Olds and Milner that the animals might "like" the experience of receiving such stimulation. Previously, all assumptions about rewarding behavior were based on the assumption that certain peripheral stimuli were pleasant and other stimuli indifferent as regards pleasure. Disagreeable peripheral stimuli were reckoned as being the origination of negative reinforcement.

Activation of positive reinforcement mechanisms fosters repetition of behavior occurring just prior to the stimulation. Olds and Milner found that rats

would cross an open field in any direction desired by the experimenter if a brief positive reinforcement is given whenever the animal moves in the right direction. Rats will learn a complex maze for no other reward than brief central reinforcement. They will cross an electrified grid to obtain such stimulation. Hungry animals will neglect feeding in favor of positive reinforcement. Sex-deprived animals will similarly neglect access to receptive sexual partners. It appears that at least the strongest of central reinforcement systems are preeminently powerful (Olds, 1958).

Central negative reinforcement was discovered at the same time as postive reinforcement by Delgado, Roberts, and Miller (1954). *Negative reinforcement constitutes a system which closely corresponds at the midbrain level to pain pathway projections and continues into ventral posterior hypothalamus and dorsally between hypothalamus and thalamus, forming a vertically forked upward pathway, like a serpent's tongue.* This pathway seems to disappear abruptly into central positive reinforcing areas and to go no higher. Delgado, Roberts, and Miller found that

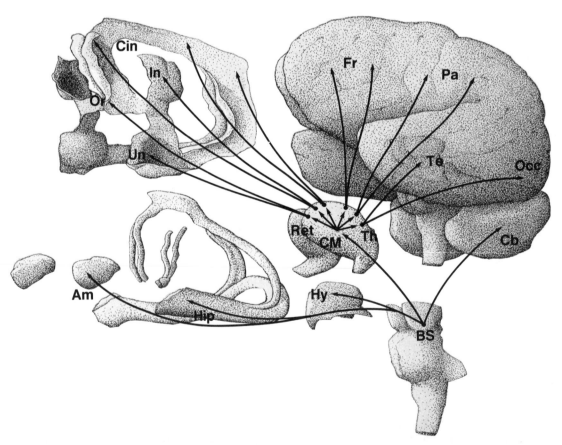

Figure 9.212.   Sketch by Dana R. Livingston from 3-D graphics display of normal human brain with selected brain stem, diencephalic and limbic structures expanded below and to the left. Activation of the reticular activating system in the tegmentum of the midbrain is followed by arousal of interneurons and other state-related neurons in the hypothalamus, hippocampus, and amygdala. This is by way of the midbrain, centrum medianum, intralaminar and reticularis nuclei in the thalamus. Impulses radiate to all areas of the forebrain, including the limbic system indicated by the double ring of archicortex and mesocortex in *upper left*.

rats will perform work and learn complicated behaviors in order to avoid or to escape such stimulation.

Fortunately, positive reinforcement systems are nearly 10-fold larger in volume than are negative reinforcement systems, and they reach up into more cephalic brain circuits. Both systems closely relate to central mechanisms that govern feeling states, appetites, and emotions.

Stimulation of other vast regions of the brain, such as the neocortex, results in neither positive nor negative reinforcement of behavior. Those regions seem to be quite indifferent with respect to motivations of any kind. They can be linked to behavior as signalling (cueing) events, but they are ineffectual for direct reinforcement purposes.

Before such lines of evidence were available, it was generally supposed that behavior is shaped on the basis of *sensations* originating from the outside. It is now clear that what makes certain sensations pleasant and others unpleasant depends not only on the sensory receptive field but on the central projections activated, directly and indirectly, by such stimuli. Further, it is now appreciated that motivations stemming from feelings as well as appetites depend largely on central nervous system activities, including events occurring in central reward and punishment systems. They themselves contribute to approach and avoidance behavior, motivate learning, and modulate many other activities, among sensory and motor, association, and higher nervous processes.

The primitive attitude of an organism toward its environment lies within the organism itself. This can be conditioned by sensory input. Behavior is goal-oriented and directed, and the goals are defined by central state conditions. Satisfactions need not be consciously experienced, nor the behavior consciously directed. Much visceral control is of this category, but can participate in shaping and emotionally coloring conscious experiences. The interface between unconscious and conscious experiences is difficult to localize: it may in fact be a dynamically movable complicated neurogeometric boundary, as close as columnar columns (viz., alternating strabismus involving "seeing" alternately by use of information from one or the other ocular dominance map). All behavior is organized in this elaborate way. Improvements in biological directedness and goal satisfaction do not need to be experienced consciously (viz., vestibuloocular reflex), and consciously experienced learning may become so habituated that it submerges beneath the level of consciousness and thereafter is pursued on more or less the same plane as genetically endowed behavior (Livingston, 1967).

CHAPTER 72

Higher Neural Functions

SLEEP

Although sleep is known to everyone from personal experience, it has a special significance in neurophysiology. Consider, for example, the results of a neurological examination performed upon a sleeping subject. If the subject remains asleep throughout the examination, the examiner will note what appear to be catastrophic deficiencies of central neural function—as contrasted with the waking state. These include loss of muscle tone and changes in both cutaneous and deep reflexes; elevated threshold for all types of sensory stimulation; almost complete dissociation from the environment; visceral changes including fall in metabolic rate, lowering of central body temperature, perhaps an elevation of skin temperature, and a lowering of blood pressure and heart rate; and a respiratory pattern different from that of the waking state to a degree that in infants may include periodic breathing. Taken together these neurological signs would signify the presence of some profound neurological injury, were it not that they disappear upon waking.

Mechanisms that control the sleep-waking cycle apparently lie in the tegmentum of the brain stem, especially the pons, mesencephalon, and lateral hypothalamus. Data now available suggest that, like many other control systems of brain stem and hypothalamus, two distinct portions can be identified. One neural complex is responsible for waking the body and keeping it awake; the other seems to initiate sleep and determine its depth or stage. To a degree these subsystems are characterized by different transmitter substances. Thus, dopamine is said to be the transmitter for the waking system, serotonin (5-hydroxytryptamine) for the sleep system, and norepinephrine for induction of the stage that is known by the terms rapid eye movement (REM), paradoxical (PS), or desynchronized sleep (see Jouvet, 1973).

Biological clocks refer to innate, endogenous, ubiquitous rhythms which can be observed to affect physiological processes in all organisms. When the period of oscillation approximates the rate of rotation of the earth, the rhythm is called *circadian* (circa = about; dies = day).

Circadian rhythms affect all human activities. The cycle is remarkably exact, measuring within 1 or 2 min of 24 h, but it may be reset stepwise about an hour at a time by light entrainment. The circadian rhythm has a conspicuous effect on the sleep cycle of air travellers going east or west, travelling east being noticeably more troublesome to accommodate. It takes nearly as many days to complete recovery as time zones traversed.

The circadian rhythm is not much affected by temperature, and it resists chemical perturbations by alcohol, anesthesia, anoxia, convulsions, hormones, and autonomic drug effects. The mammalian clock depends on the hypothalamus which when diseased or damaged may leave shorter, residual random oscillations of activity and rest.

It is moot whether fish sleep, but sharks and reptiles certainly do. Sleep is highly developed and complex in birds and mammals. Lower mammals generally sleep more and have a lesser proportion of REM sleep to total sleep than do humans. Sleep is really an altered state of consciousness. Perceptual mechanisms, although generally subtunded, are nonetheless capable of eliciting arousal on the basis of specific, quite subtle stimuli. A child's cry will often waken a parent even though the cry may be less loud than other sounds. The skipper of an Alaskan coastal steamer slept soundly in a curtained alcove off the wheelhouse: however, if anyone mentioned "iceberg," he would be instantly awake without being conscious of what had wakened him.

Normal sleep can be divided into a series of sleep stages I–IV, as depicted in Figure 9.213. An electroencephalogram (EEG) of a normal adult in the waking state shows a random pattern, over all scalp regions, of low voltage fast electrical activity (20–30 Hz). This is thought to derive from asynchronous activity in cortical and corticothalamic circuits. When she closes her eyes, the *alpha rhythm* (8–13

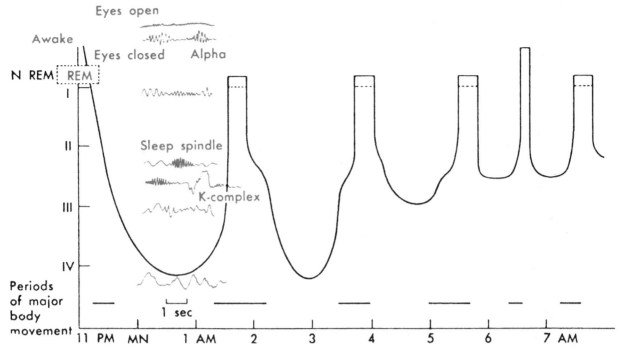

Figure 9.213. Progression of states of sleep during one night's sleep as recorded in a young adult; awake, eyes open; eyes closed with prominent alpha waves, sleep spindles, K-complex and deep slow-wave sleep, characterizing sleep stages I–IV of non-rapid-eye-movement (non-REM) sleep. REM sleep occurred four times during the night. One awakening took place shortly before 7 A.M. [From Willis and Grossman (1981).]

Hz), which is always present, becomes more obvious. Sleep spindles (14–16 Hz) may appear and are correlated with brief lapses of consciousness. The early period of well-defined sleep usually involves a rapid transit to stage IV—a fully relaxed, slow wave sleep (SWS). The high voltage slow waves correspond to highly synchronized cortical and corticothalamic activity.

As sleep progresses, after a relatively long period of SWS, her EEG shows low voltage fast (20–30 Hz), asynchronous activity, resembling that when she was alert. Muscular relaxation is profound, limbs are flaccid, and stretch reflexes are absent. There may be occasional muscular twitching and limb jerking. A hound asleep at the fireside may show feeble running movements and may vocalize softly, as if dreaming. At this time, even though her eyes remain closed, her eyelids reveal that she is having rapid horizontal eye movements. This stage of sleep is referred to as paradoxical sleep (PS) or rapid eye movement (REM) sleep.

Although the EEG tracing resembles the waking state, the individual is actually more difficult to arouse than when the record shows slower waves. When a person is awakened during or immediately following REM sleep, they can usually recall an interrupted dream or testify that they had just been

dreaming. Nightmares occur during REM sleep. Dreaming occurs predominantly during REM sleep periods, but it can also occur in non-REM sleep periods. Waking subjects during other sleep stages rarely correlates with dreaming. It is suggestive that the eye movements relate to the contents of the dreams.

During REM sleep, a pontine reticular locus is responsible for generalized spinal inhibition—a physiological measure that precludes acting out dreams. This inhibition may be responsible for the "paralyzed" feeling one may have of being unable to move during a dream.

There may be a succession of excursions to stages IV, III, or II, with REM stages occurring at approximately 90-min intervals, as sleep continues (Fig. 9.213). Before fully awakening, she may reach full consciousness but again slip back into light sleep. As we describe below, REM sleep is important especially because it may have to do with consolidation of recent learning experiences.

Figure 9.214, from Jouvet, shows EEG recordings from a cat with leads recording from sensorimotor cortex, ectosylvian cortex, hippocampus, medullary and pontine reticular formation, neck muscle EMG, extraocular eye muscle EMG, electrocardiogram, plethysmogram, and respiration. Records from

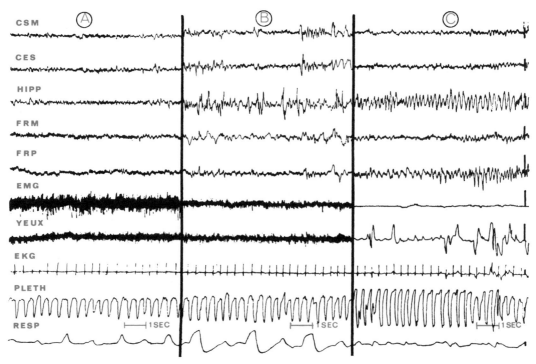

Figure 9.214.   Electrical recordings that show differences in waking state (*A*), slow wave sleep (*B*), and paradoxical (REM) sleep (*C*). [From Jouvet (1967).]

three distinctive states are depicted: in *A*, waking; *B*, slow wave sleep; and *C*, paradoxical or REM sleep.

The Brain Stem Reticular Activating System

In the 1950s, Magoun, Lindsley, and their colleagues discovered that mechanisms other than sensory systems are necessary for arousal and for sustaining the wakeful state. They made massive electrolytic lesions of upper brain stem and caudal diencephalon in cats and monkeys. Some of the lesions interrupted the major sensory pathways (all except olfaction and vision) but spared the central core of the neuraxis. Other lesions interrupted the core but spared the major sensory paths.

They found that if the brain stem reticular formation in the midbrain and its projection pathways into the diencephalon were spared, the animals went through regular wake-sleep cycles, sought and consumed food, and roughly cared for themselves, although they behaved like "zombies," exhibiting obviously inappropriate relations with their environment.

Animals with lesions in the midbrain reticular formation, sparing the main ascending sensory paths, were lastingly comatose (Fig. 9.215). The EEG tracing from the cat with the massive lesion just illustrated showed continuous high amplitude slow waves, interrupted only briefly by a loud buzzer, shrill whistle, or painful toe pinch (Fig. 9.216). There was no evidence of spontaneous behavioral or electrographic arousal. Such animals had to be tube-fed and nursed conscientiously. They closely resembled patients in coma. Patients in such refractory coma are likely to have similar core brain stem lesions, in some cases also sparing the major sensory paths so that they may have event-related potentials (ERPs), or they may have massive, generalized lesions of the cortex.

Moruzzi and Magoun (1949) demonstrated that electrical stimulation of the medial reticular formation transforms a high voltage, slow wave EEG (typical of slow wave sleep) into a low voltage, fast wave EEG (typical of the waking pattern). This capacity to activate the forebrain and to effect behavioral arousal is attributed to the *ascending reticular activating system* (Fig. 9.217). Access to this system for sensory arousal occurs by way of collaterals from the primary sensory pathways (lemniscal, spinothalamic, trigeminal lemniscus, trigeminothalamic, lateral lemniscus, etc.). There is also considerable pleurisynaptic ascending traffic that does not belong to any of the named pathways. Influences of the ascending reticular activating system are transmitted over multisynaptic pathways via midline, centrum medianum, intralaminar, and reticularis nuclei of the thalamus, and thence to the entire cortical mantle. Brain stem projections like-

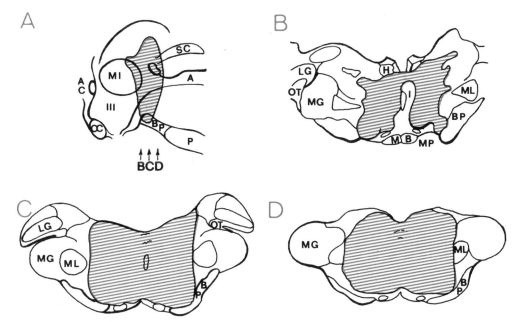

Figure 9.215. Midsagittal (*A*) and cross-section (*B–D*) through brain stem of cat with destruction of central portions of reticular formation. [From Lindsley et al. (1950).]

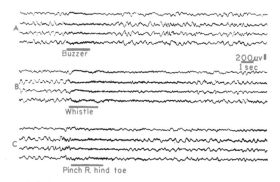

Figure 9.216. Electroencephalogram of cat with lesion shown in Figure 9.215. Record shows persistent sleep pattern, with brief transition to waking pattern during and just after stimulation by sound or pain. Four different cortical regions show similar wave forms. [From Lindsley et al. (1950).]

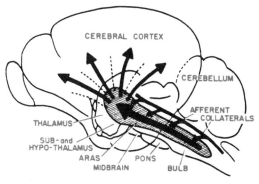

Figure 9.217. Outline of the brain of the cat, showing general location of the activating mechanism in the tegmentum of the brain stem, and its relation to principal afferent systems and projections via thalamus to cerebral cortex. [From Starzl et al., cited from Magoun 1952).]

wise arouse the cerebellum and contribute central excitation to brain stem and spinal gray matter.

Magoun provided a seminal review of caudal and cephalic influences of the brain stem reticular formation which had a strong influence on bridging the gap between neurophysiology and psychology (1950).

It was soon discovered that certain particular areas of the cerebral cortex influence the brain stem reticular formation and are capable of inducing widespread functional changes throughout the neuraxis and contributing to the state of arousal and the sustaining of wakefulness (French et al., 1955; Adey et al., 1957). Some among these centrifugal projections from cortex were likewise shown to be especially susceptible to seizure activation (French et al., 1956).

Initiation of Sleep

The Swiss neurophysiologist, W. R. Hess (1932), who originated techniques for implanting electrodes for stimulation in the waking state after an animal has fully recovered from anesthesia, discovered that slow frequency intense stimulation of the midline thalamus motivates an animal to sleep. He showed that it is not adynamia, catalepsy, or astasia abasia but rather an inclination of the animal to sleep. The animal carries out suitable individualistic rituals before curling up and going placidly to sleep. The sleep is normal in other respects, and the animal can be easily aroused again. This important work was reinforced by Akert et al. (1952, and R. Hess, Jr., 1953), including establishing the correlations between state of the animal and the EEG.

Nauta (1946), making lesions in the suprachiasmatic and supraoptic area, produced locomotor hyperactivity and failure of sleeping in rats. Locomotor hyperactivity was found by Ruch and Shenkin (1943) and analyzed further by Livingston et al. (1947a and b), in monkeys, following small bilateral ablations in the orbital surface of the frontal lobes. Hyperactivity follows the normal diurnal cycle of activity and may be 8- to 10-fold higher than normal. It is unaffected by environmental temperatures over a wide range.

In 1973, Jouvet observed prolonged wakefulness in cats following destruction of midline structures in the pons and bulb. It is presumed that this lesion causes hypersensitivity among catecholaminergic neurons in the cephalic brain stem.

HUMORAL MECHANISMS OF SLEEP AND WAKEFULNESS

Jouvet (1967) summarized previous evidence favoring a humoral influence on sleep mechanisms and concluded that serotonin is a neural transmitter that induces sleep. In the median raphé system where lesions lead to prolonged insomnia, serotonin is normally found in high concentrations (Ungerstedt, 1971). The existence of an activating substance was demonstrated by Dell (1952) and extended by Purpura (1956). Purpura connected each femoral artery of a cat to a femoral vein of a second cat, and vice versa, to obtain intermixing of the total blood volumes of the two animals. Stimulation of the brain stem reticular formation produced immediate activation of the EEG in the stimulated cat. After about 60 s an activation pattern could be seen in the EEG of the recipient cat. The reticular formation involved contains both norepinephrine and dopaminergic neurons (Dahlstrom and Fuxe, 1964). It is likely that NE activates cortex and DA produces behavioral alertness and readiness for action through its influence on the nigrostriatal system.

Experiments by Jouvet in which animals were differentially deprived of paradoxical (REM) sleep show that the animals build up an unusual need for REM sleep and will devote more sleep time to REM sleep if given an oportunity. Cats that were given monoamine oxidase inhibitors during the deprivation process did not show the excessive REM sleep during recovery.

Significance of Sleep

There has been a great deal of speculation about why sleep is needed. Yet almost nothing is known scientifically about sleep function. In the absence of sleep, higher brain functions are most susceptible to deterioration. Later, subcortical—more nearly automatic functions (lower stratified stabilities)—suffer. Tremors, nystagmus, ptosis, dysarthria, and affective disorders appear.

As Hobson (1969) has pointed out, certain forms of illness are most likely to occur when the patient is asleep; nocturnal angina, asthma, emphysematous anoxia, Cheyne-Stokes breathing, and some kinds of epilepsy occur only during sleep.

DISORDERS OF SLEEP

Somnolence, prolonged sleep patterns, and coma are encountered in medical practice. Coma due to a lesion in the cephalic brain stem may leave pulmonary ventilation relatively intact. Persistent coma of this kind is a serious ethical as well as medical problem. Many of the uncertainties of diagnosis of "brain death" beyond which it is inappropriate to continue life support actions hinge on the issue of level and extent of lesions affecting the cephalic brain stem. Animals with coma-producing lesions of the cephalic brain stem, if nursed carefully and long enough over a period of several months, may recover some capacities for spontaneous arousal. Additional lesions at the margins of the old lesion will reestablish the full coma. One is left with the supposition that the neighboring cells either shared responsibility for arousal or developed such capacities over time.

Narcolepsy

This is the term applied to a disturbance in the sleep mechanism in which sudden attacks of an irresistible desire for sleep occur during the daytime. The duration of the attacks, which resemble normal sleep, is quite brief—from a few seconds to 20 min or so. The condition may be a sequel to influenza or to epidemic encephalitis involving the hypothalamus or may result from a tumor or injury in this region. In other instances the condition appears without known cause (idiopathic narcolepsy). Although evidence is not conclusive, it is likely that in this latter condition a disordered hypothalamic function is responsible, for other features, e.g., obesity, polyuria, or impairment of sexual functions, pointing to an abnormality of this region, are frequently present.

*Cataplexy** is the term given to a condition allied to and very frequently associated with narcolepsy in which the patient as a result of some emotion—amusement, anger, fear, embarrassment, or surprise—is seized with complete muscular relaxation

---

* This should not be confused with catalepsy, an entirely different condition.

and weakness. The attack is brief, lasting for a few seconds, or for a minute or two at the most. Consciousness is not lost, but the muscles are completely toneless and powerless for the time, and if the attack supervenes while the subject is standing his knees fail him and he sinks to the ground. The deep reflexes are lost. Mirth is especially likely to precipitate a cataplectic attack. Though narcolepsy occurs without cataplexy, the converse is extremely rare. This association of the two conditions at once suggests a common pathogenesis, but the muscular atonicity characterizing the cataplectic attack cannot be explained upon any known physiological basis.

EMOTION

The Physiological Basis of Emotion

Self-awareness in relation to our moods and feeling states is so intimately bound to their underlying nervous causes that we are at least partially aware of our central state condition even as we sleep. Moods and feelings obtrude on our consciousness as soon as we waken, attend all our waking hours, and will color our last flicker of consciousness.

Emotional Experience

Moods and feelings are subjective aspects of our existence, purely private. They can be shared in part, verbally and otherwise, by bodily expressions, as by displaying what we call *emotion*, the outward expression of the internal state. *Moods and feelings are completely private affective experiences. Emotion is their partially publicly revealed behavioral expression.*

Emotional Expression

Emotional expression is to a considerable extent involuntary, although from childhood we acquire an encyclopedic lore of recipes for emotional constraints relating to what is, and what is not, socially appropriate, for intrafamily as well as public expression of emotions, and on which particular occasions. Learning the game of social-emotional finesse can be seen as one of life's most challenging and elaborate ritualistic acquisitions, entailing control by higher neural functions. Socially exacted measures of maturity and of mental health are manifested by almost universal compliance (within a given society) with respect to what is tolerable in public expressions of emotion. The fact that these measures differ across cultures emphasizes the learned nature of such viscerosomatic skills.

Social Duplicity in Emotional Expression

For social purposes, we may "tell lies" by emotional constraint, and sometimes by emotional ex-

aggeration. Actors, politicians, and to some extent, teachers, lawyers, physicians, and perhaps all professionals learn to do this as an explicit contribution to their professional communications.

CATEGORIES OF AFFECT

MacLean (1970) emphasizes that both affect and emotion are revealed to us only as internal information. He says, "There are no neutral affects because, emotionally speaking, it is impossible to feel unemotionally." MacLean divides affects into three categories: basic, specific, and general. *Basic affects* are subjectively identified as hunger, thirst, sexual need, urge to defecate, urinate, and on a shorter time leash, the urge to breathe. *Specific affects* are those introduced by activation of specific sensory systems (e.g., enthusiasm aroused by smelling odor of cooking food, or pain following noxious stimulation). *General affects* include traditionally regarded emotions such as love, anger, disappointment, happiness, and so forth. General affects are easily distinguished from either the basic or specific affects, inasmuch as the former may pertain to situations, individuals, and institutions. Moreover, general affects can persist or return after the inciting circumstances have occurred.

MacLean (1958) relates all general affects to two basic biological principles which he locates in different parts of the limbic system: preservation of the species (located by paths flowing into and from the septal region) and preservation of the self (located by paths flowing into and from the amygdala) (see Chapters 67–69 for further related neurophysiological insights).

Information that indicates threat to self or species is disagreeable. Information revealing the removal of such threats, or the gratification of individual, group, or species' needs, is satisfying to affect.

COMPELLING AFFECTS RELATING TO PERCEPTIONS AND JUDGMENTS

Temporal Lobe Contributions to Vividness of Experience

MacLean calls attention to limbic mechanisms underlying affective feelings of "individuality" and "what is important." He reminds us of the type of temporal lobe aura experience described by Dostoyevsky in *The Idiot*, as the "sensation of existence in the most intense degree." The temporal lobe is able to generate, totally out of context, what is subjectively real and important to us. It establishes feelings of *deja vu* ("already seen"—an illusion in which a new situation is incorrectly viewed as a repetition of a previous experience) and *jamais vu* ("never seen"—although previously familiar) which may be experienced by normal persons on occasion but may dominate experiences for someone with temporal lobe epilepsy.

INTELLECTUAL AURA ASSOCIATED WITH TEMPORAL LOBE EPILEPSY

In 1888 John Hughlings Jackson reported several interesting cases of "intellectual aura" associated with temporal lobe epilepsy. A description in the words of one of his patients, himself a medical doctor, shows that although an affected individual may be able to pursue quite complicated behavior, he/she may nonetheless have no memory of the occasion.

> I was attending a young patient. . . with some history of lung symptoms. . . Whilst he was undressing, I felt the. . . [onset of a temporal lobe seizure]. I remember taking out my stethoscope and turning away a little to avoid conversation. The next thing I recollect is that I was sitting. . . in the same room, speaking to another person, and as my consciousness became more complete, recollected my patient. . . I was interested to ascertain. . .[that] I had made a physical examination, written. . . [a diagnosis] of "pneumonia of the left base". . . and advised him to take to bed at once. I re-examined him with some curiosity, and found that my conscious diagnosis was the same as my unconscious,—or perhaps I should say, unremembered diagnosis had been.

The young doctor was demonstrated to have a cyst in the temporal lobe, beneath the uncus, by autopsy 10 yr later (Jackson and Colman, 1898).

In other cases, seizures can establish feelings of depersonalization, or objects may be seen as abnormally large or small, close or remote; one's body parts may be distorted, time as well as space warped, and so forth. Very commonly, there may be interference with experiences being remembered. Some of these experiences occur with organic and toxic psychoses and may be familiar to users of psychedelic drugs.

Human patients with lesions in these areas are said to undergo a form of epilepsy (psychomotor) where the symptoms are autonomic discharge and emotional experience; stimulation during neurosurgical operations by way of electrodes applied to these structures evokes responses comparable to psychomotor epilepsy (Penfield and Jasper, 1954). In a few persons, recordings have been made and stimulation has been applied through implanted electrodes, while they are living a more or less conventional life as hospital patients. The data so obtained are in general agreement with results obtained from experimental animals. For example, "spontaneous" bursts of anger were evoked by stim-

ulation of the right amygdala. Other kinds of experience elicited by the stimulation included pleasant sensations, elation, deep thoughtful concentrations, "odd feelings," relaxation, and colored visions (Delgado, 1970).

SPEECH AND ITS DISORDERS

Speech and its relations, writing and reading, are the most elaborate forms of communication and are limited to the human species. Cortical function is necessary for speech; according to the observations of Penfield and Roberts (1966) it is the cortex of the left cerebral hemisphere that is usually responsible for speech, even in left-handed persons where the right hemisphere is said to be "dominant" for motions such as writing and other skilled performance.

It is supposed that the first stage in the development of speech is the association of certain sounds (words) with visual, tactile, and other sensations aroused by objects in the external world. After definite meanings have been attached to certain words, pathways between the auditory area of the cortex and the motor area for the muscles of articulation become established, and the child attempts to formulate and pronounce the words which he has heard. This act of verbal expression involves the coordinated movements of respiratory, laryngeal, lingual, pharyngeal, and labial muscles. Later, as the child learns to read, auditory speech is associated with the visual symbols of speech; and finally, through an association between these and the motor area for the hand, the child learns to express his auditory and visual impressions by the written word (see Lenneberg, 1970).

Aphasia

Aphasia is the term applied to disorders of expression in speech, writing, or signs and symbols, as well as to disabilities in comprehension of spoken or written language. Lesions causing aphasia are usually in the cerebral cortex, but they do not necessarily involve the motor mechanisms of area 4, the corticobulbar fibers, cranial nerve nuclei, or peripheral nerves. Rather, the lesions typically lie in cortical "association" areas.

The faculty of speech can be studied in two ways, by analysis of defects in patients with naturally occurring lesions of the cortex or following operations or by stimulation of the cortex in a patient given only local anesthesia for a surgical operation. That this electrical stimulation is categorically different from natural neural activity is dramatically shown by the observation that in an area where a lesion causes aphasia, electrical stimulation when a patient is speaking will arrest the speech. After the stimulus is ended the patient may report that he does not know why he stopped talking, but he was unable to think of the words he wished to use. There are four such areas in the left hemisphere, the dominant hemisphere in a right-handed person. They include the following: (1) *lower prefrontal* cortex of Brodmann's area 44 (known as Broca's area); (2) *upper frontal* on the mesial surface of the hemisphere anterior to the foot area; (3) *parietal*, posterior to the lower part of the postcentral gyrus; and (4) *temporal*, in the posterior part of the temporal lobe (Wernicke's area). A lasting injury to any one of these areas causes persistent aphasia, with the possible exception of the one in the upper frontal region.

CLASSIFICATION OF THE APHASIAS

There is no one system for classifying aphasias, the several neurologists who have studied this phenomenon having usually proposed categories that seem consistent with their own findings. In this account the views of Sir Henry Head (1926), an English neurologist, are followed.

Head divided aphasia into four types, as follows:

1. Verbal Defects. The outstanding feature is that the power to express an idea in words is practically lost. The patient, however, is not entirely speechless but can usually utter a few monosyllables, "yes" or "no," etc., or ejaculations and emotional expressions, such as "damn," or "oh dear me." When the disorder is less severe, the words are mispronounced, but sentences are correctly constructed. For such patients reading is difficult, and writing is defective or impossible. They usually understand printed or oral commands.

2. Syntactical Defects (Agrammatism, Jargon Dysphasia). The patient is voluble but speaks a jargon in which, though the individual words may be fairly accurately pronounced, they are strung into short phrases or badly constructed sentences without articles, prepositions, or conjunctions. The ability to read aloud is impaired, and the understanding or ordinary conversation is defective.

3. Nominal (Naming) Defects. In this form of speech disorder the patient has difficulty in finding the right word to express his meaning or for naming a well-known object. He will often employ a descriptive phrase in substitution for the word which he cannot recall. These patients can draw from a model either directly or from memory, after it has been shown and then removed, but are usually unable to draw from imagination. They write a coherent letter with difficulty, usually fail to carry out simple arithmetical exercises, and confuse the values of coins.

4. Semantic Defects. A patient suffering from this type experiences little difficulty in articulate speech, can name objects, understands individual words and some sentences, but the general meaning of what he hears escapes him. He often fails to follow his own utterances to an intelligent conclusion, his sentences trailing off as though he had forgotten what he had started out to say. When shown a picture he picks out the details but fails to grasp the meaning which it conveys to others. Such a patient therefore misses the point of a joke whether this is printed, told to him, or is in pictorial form. He fails to comprehend the significance of much that he sees and hears. There is no impairment in the pronunciation of words and, though speech tends to be in short jerky sentences, syntax and intonation are not disturbed.

Other Disabilities of Speech and Perception

Anarthria or *dysarthria* is loss or difficulty of speech due to paresis or ataxis of the muscles concerned in articulation. There is no impairment of the psychical aspects of speech, i.e., "internal speech" is unaffected; there is no difficulty in the comprehension of spoken or written speech. Other functions, e.g., swallowing, which are dependent upon the same groups of muscles as those used in speech, are also frequently affected.

Agnosia (not knowing) is a defect higher than the mere inability to perceive tactile, visual, auditory or other forms of sensation; it results rather from the failure to interpret sensory impressions which enable an object, sound, or symbol to be recognized and have meaning. "Word blindness" and "word deafness" are forms of visual and auditory agnosia, respectively.

Agraphia or *dysgraphia*, the inability to write or difficulty in writing, may occur with visual agnosia because of a handicap in the recognition of written words, i.e., a word blindness.

Astereognosis is a disorder in which, though sensations of touch and muscle sense are retained, the patient cannot recognize an object placed in his hand if his eyes are closed. Visual agnosia is seen in lesions of the occipital lobe of the dominant hemisphere, auditory agnosia in injury to the temporal cortex, and astereognosis in lesions of the parietal lobe posterior to the postcentral gyrus.

Apraxia (unable to act) is the inability to perform purposeful movements at will, either at command or in imitation, though the muscles normally engaged in the act are not paralyzed. It is allied to aphasia, which might be called apraxia of the speech faculty. Apraxia may be sensory or motor. In the former, the patient does not recognize the significance of an object and therefore cannot put it to its

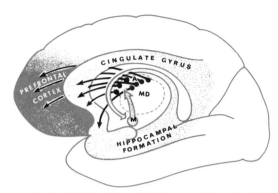

Figure 9.218. Relationships of thalamus with cerebral cortex, including cingulate gyrus and prefrontal cortex. *A*, anterior nucleus of thalamus; *MD*, dorsomedial nucleus of thalamus; *M*, mamillary body. [From MacLean (1967).]

proper use; this is *visual agnosia*. In motor apraxia the patient has no conception, or a very defective one, of the pattern of muscular movement required to perform a purposeful act. For example, apraxia of the tongue is frequently seen in hemiplegic patients. The tongue cannot be protruded upon request, but a moment later the patient may without thought lick his lips.

Important Convergence on Cortex of the Frontal Lobes

Figure 9.218 illustrates the important projections of brain stem and limbic information into the frontal lobe. Converging paths from brain stem and diecephalon, from olfactory and the other sensory systems, meet in various combinations in association cortex and in limbic cortex. New combinations are thereby enriched by information relating to moods, feelings, emotions, by information as to what is "me-myself," what is "real and important," and, generally, by information concerning ongoing motives.

There are several major outflows from amygdala and other temporal lobe and limbic structures: (1) direct projections from and to the diencephalon and brain stem; (2) projections by way of the stria terminalis and fornix to the hypothalamus; (3) impulses recurrent to the limbic system which pass from the mamillary bodies to the anterior nucleus of the thalamus and thence to the cingulate gyrus; (4) information combined from brain stem, limbic system, and olfactory inputs pass to the medial dorsal nucleus of the thalamus. This is one of the largest thalamic nuclei, matching the pulvinar. Between the two, the thalamus contributes to all cortical areas anterior to and posterior to a belt around midhemisphere; (5) medial dorsal thalamic projections onto the frontal lobe allow conjunction of motivational and emotional (value-attaching) information to be presented to cortex of the frontal

lobe which is occupied with behavioral planning into the future; (6) projections from the olfactory system, septum, amygdala, and limbic system, generally, also pass caudally directly into diencephalon and thence into midbrain, and, by way of the stria medullaris, less directly, also to the midbrain; and finally, (7) the frontal lobes join this downflow through abundant projections into the median forebrain bundle to midbrain.

The ultimate bounty of all of these projections is information about what are current biological feelings, needs, internal options, external opportunities, and behavioral plans. All of this biologically significant information ends up in the gray and white reticular formation in the core of the midbrain—the *frontal-limbic midbrain area*. It is from this locus that major overall decision making for behavior of the individual as a whole takes place (see Figure 9.210).

CONFLUENCE ONTO CORTEX OF THE FRONTAL LOBE

Basic stratified stabilities for aggression, drinking, and feeding responses can be identified at different levels of brain stem and spinal cord. Figure 9.219 schematizes some of these functional relations. A convergence of internal and external signals at the level of the hypothalamus and the limbic system leads to functional coupling of endocrine, autonomic, and somatic functions at the hypothalamic level. Neocortex is relatively unaffected by strongly motivating stimuli such as pain, and visceral sensory representation there is only slight. The rich and extensive exteroceptive influences on neocortex, however, make important contributions to limbic and hypothalamic decision making with respect to the goal direction of the behaviors in question. The upshot of these transactions yields well-organized behaviors that we call drinking, feeding, and defense reactions. In the case of humans, these behaviors may be extremely sophisticated.

Once more, this scheme of thinking emphasizes the first principle of biological importance to design behavior so that the needs of the whole animal will be appropriately satisfied in due time. The internal controls of circadian rhythms, metabolic requirements, temperature regulation, rest and recuperation, and other physiological phenomena have to be integrated (the needs have to be "prioritized") so that freedom for additional or more far-sighted pursuits can be undertaken. Reptiles lack neocortex, a "thinking cap" that liberates mammals from having to depend so greatly on rituals and stereotyped habits. Neocortex is sufficiently removed from the assessment of needs that it can (by de-

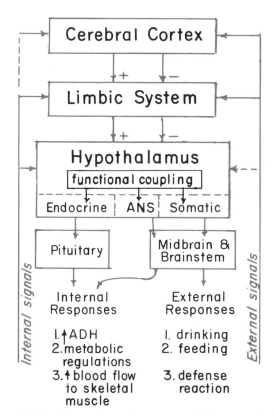

Figure 9.219. Basic reflexes for defense, drinking, and feeding operate through several levels of brain stem and spinal cord. Superimposed are facilitatory and inhibitory controls by limbic system and cerebral cortex. The hypothalamus integrates internal and external signals, as well as input from other neural structures, and thus provides a "functional coupling" of the endocrine, autonomic (*ANS*), and somatic efferent systems. [From Mogenson and Huang (1973).]

pending on lower stratified stabilities to take care of those matters) make longer range decisions and devise more elaborate strategies to obtain a biologically and socially more fulfilling future. The frontal lobe is more forward looking than simply planning future behaviors.

Sham Rage

About the turn of the century, it was found that an otherwise docile dog is readily provoked to anger or rage if it is deprived of its cerebral cortex. Decorticate cats, immediately following recovery from anesthesia, demonstrated *sham rage*: lashing of the tail, piloerection on the tail and back, protrusion of the claws, dilatation of the pupils, salivation, struggling and biting, tachypnea, and rise in blood pressure. Epinephrine was liberated as indicated by an increase in rate of the denervated heart. Mild stimuli elicited paroxysms of rage.

Hess found that similar organized rage responses can be elicited by stimulation of the hypothalamus (1949). It is thought that the cortex may play an inhibitory role, the removal of which allows

"release" phenomena such as can be evoked by direct hypothalamic stimulation.

In 1948, Bard and Mountcastle showed that removal limited to *neocortex* leads to extraordinary placidity in cats. The animals can scarcely be provoked to anger; noxious stimuli instead lead to purring and other gratulant behavior. Subsequent removal of parts of the limbic system (pyriform region, amygdala, and anterior hippocampal formation) transformed the animals into "tigers." The slightest provocation yielded all the signs of directed rage.

Klüver and Bucy (1938; 1939) removed the anterior two-thirds of the temporal lobes in monkeys. These animals exhibited an interesting constellation of behaviors: (1) *Visual agnosia.* The monkeys seem to lose their ability to recognize objects by vision (or acoustic, or somesthetic) criteria alone. (2) *Oral tendencies.* There is a strong tendency to mouth objects, to put their mouths to objects, even if it requires an awkward posture, and to put all sorts of objects into their mouths: bolts, nuts, inedible husks, even lighted matches, and to do so repeatedly. (3) *"Hypermetamorphosis."* A marked tendency to take notice of and to attend to every visual stimulus. (4) *Tameness.* The animals seem to lose their sense of fear of snakes or other objects that ordinarily are threatening and anxiety-evoking in monkeys. (5) *Hypersexuality.* There is a striking increase in a wide range of sexual activities and amount of time and effort expended. (6) *Changes in dietary habits.* Macaque monkeys do not ordinarily eat meat, but after removal of both temporal lobe segments, they do so without hesitation, even monkey meat which is normally abhorrent to them. They increase consumption of food as well.

These behaviors characterize what is known as the *Klüver-Bucy syndrome.* This syndrome is conspicuous in patients with massive bilateral temporal lobe injury, and in a few patients in whom bilateral temporal lobectomies were done for intractable bilateral temporal lobe seizures. The most striking consequence for temporal lobectomized humans is their failure to lay down new memories from ongoing experiences. The anterior two-thirds of the hippocampus has proven to be the critical feature of these ablations, with respect to the disabling memory deficit. Old memories are not disturbed, for storage and recall; hence they are not stored in the hippocampus nor dependent on hippocampal functions.

Attention and Selective Attention

Thus far we have concentrated on what comes into higher level transactions that must be discriminated in order to match information from the outside world. This provides the kind of decision making that is exemplified by the hippocampal pyramidal cell depicted in Figure 9.206. In that situation, a single cell receives on its basal dendrites information relating to internal needs (from hypothalamus via the septum) and olfactory information from the outside world on its apical dendrites. The input for internal needs is largely compelling; signals relating to "external options" are conditional. The pyramidal cell is thereby in a good position to reach a decision that will be reflected downstream, presumably in the frontal-limbic midbrain area, where "go" and "no go" decisions may be made in the context of the animal as a whole (see Fig. 9.210). Hippocampal pyramidal cells thereby make a modest contribution to this overall transaction.

HOW IS INCOMING INFORMATION SELECTIVELY DISCRIMINATED?

We have observed just above (Fig. 9.219) how large stratified stabilities can process information on a grand scale to account for eating, drinking, and defense reactions. A challenging question naturally follows: How can "higher neural functions" influence such multiprocess systems in order to have decisive influence? How does the neocortical "thinking cap" get hold of any parts within this stack of stratified stabilites?

Perhaps the place to begin is to inquire how higher neural mechanisms might select for attention among incoming signals. For example, how can an individual, given a specific task, make effective selection of information coming in along a designated sensory pathway?

In previous chapters, we considered how information gets into the nervous system, how it may be affected in various ways, but we have neglected considering how the conscious subject might selectively attend particular incoming information. In Chapter 62 (Hearing) we showed brain stem and cortical evoked responses (Fig. 9.26) and the responses to expected but omitted clicks (Fig. 9.29). This was expanded in the example of Figure 9.30, showing effects when "oddball" stimuli are introduced, and, particularly, when a completely novel sequence is introduced at random. These are examples of a subject responding to variations in cognitive input.

Figure 9.31 showed that the cognitive content of such an input can be selectively affected by the subject. But heretofore we have not tried to inquire how the attentive individual can discriminatively select information according to specific instructions.

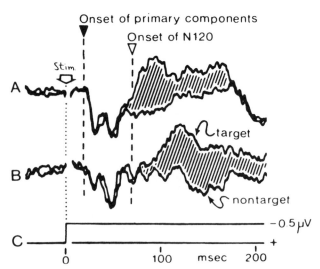

Figure 9.220. Early and middle range event-related potentials (ERPs) in selective attention. Experiment involves a bisensory paradigm with random sequences of acoustic clicks and brief electrical square pulses to one finger. All stimuli are near threshold and difficult to detect. Intervals between stimuli varied randomly between 1 and 15 s. *A* and *B*, ERPs to contralateral finger stimuli recorded from parietal scalp focus with earlobe reference. When the stimuli are targets to be mentally counted, they result in the superimposed traces showing a large N140. *A* and *B* correspond to two experiments on different subjects. The onset of primary components occurs at about 20 ms (*first vertical dashed line*). The primary components are not changed by the task. The N140 is elicited when the finger stimuli are selectively attended. The tracings diverge from control at about 70 ms (*second vertical dashed line*). [From Desmedt (1981).]

Figure 9.220, from Desmedt (1981) shows early and middle range event-related potentials (ERPs) as recorded during a selective attention task. The task is taxing because it involves bisensory stimulation (acoustic and somesthetic) with intensities near threshold—hence difficult to detect. Moreover, the intervals between stimuli are varied randomly over several seconds. When the finger is receiving signals that are to be counted, as compared with such signals not to be counted, there is a large amplitude negative far field potential that is maximal about 140 ms after stimulus onset. It is noteworthy that selective attention to the channel in question makes a detectable difference in averaged response amplitude. The subject is able, by selective attention, to pick out and thereby literally "signify" those events. It is also important that selective attention does not seem to have an influence on the far field potentials earlier than about 70 ms after stimulus onset.

Behavior: Higher Command Authority of Parietal Cortex

Figure 9.221 (from Mountcastle et al., 1975) illustrates the remarkable experimental control that

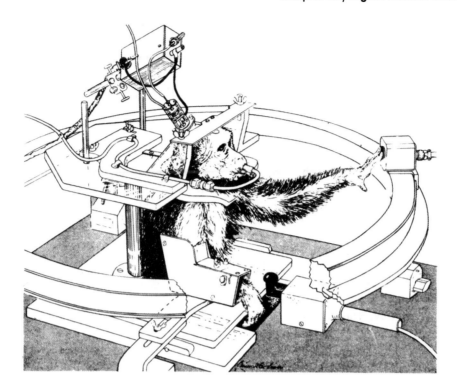

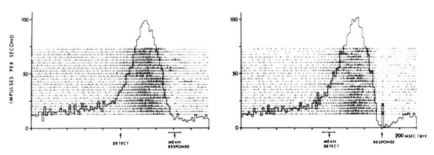

Figure 9.221. *Upper,* Drawing of monkey working in test apparatus used for study of posterior parietal cortex. The head fixation, implanted microelectrode drive, cathode follower, and reward tube are shown *upper left*. The signal key is shown through a cutaway of the circular race. The animal has just released the key with his left hand and projected that arm and hand forward and to his left to contact the lighted switch mounted on the moving carriage, shown *upper right*. The carriage can be moved from any present position in either direction at speed of 12 or 21°/s for preset distances. *Lower,* Replicas of original records, pre-, intra-, and postresponse histograms, made during study of an active projection neuron. The cell never responded to any passively delivered mechanical stimulus to the arm, to visual or auditory stimuli, or during aversive-aggressive movements of the contralateral arm. Each horizontal line is the time course of a single trial; each upstroke is the instant at which a nerve impulse occurred. *Left,* records and histograms aligned to the instant of detection *(arrow);* the bar indicates mean response time ± 1 sd. *Right,* the same records and histograms, now aligned by the instant closure of target switch by projected contralateral hand; the bar indicates mean instant detection ± 1 sd. Neuronal activity began to accelerate before release of detect key, reached a peak as arm moved toward target switch, and declined virtually to zero before the hand contacted the switch. [From Mountcastle et al., 1975.]

can be obtained by using well-trained, intelligent animals in circumstances in which they draw some positive reinforcement through individual excitement and challenge as well as visceral satisfactions designed into the ongoing experience. The choice of an item of behavior in respect to a given cortical area is made after preliminary microelectrode exploration of the region in waking but untrained

animals and from defects in behavior observed in animals in which the area has been removed. Here the posterior parietal cortex lent itself to tasks relating to how an animal explores and coordinates hand-eye movements within immediate extrapersonal space.

This particular experiment involves establishing the temporal pattern of response of a parietal cortical neuron that was shown to be involved in projecting the contralateral hand to touch a moving target. Since the carriage can be moved in either direction and with different velocities, the space-time coordination of movement is demanding. The animal is trained visually to fixate a standing target and to track the target when it moves, detect a change in a light on the target, and successfully project arm and hand to the target in order to receive a reward presented through the drinking tube.

Neurons in area 5 have consistent distinctive functional properties, cutaneous, muscle, other deep tissues, joint rotation, visual, including projection and hand manipulation. One of the latter type units is documented in Figure 9.221.

The alignment of histograms shows that neuronal activity began to accelerate, on average, before release of the detect key, reached a peak during the motion of the arm through the air towards the target, and declined to below resting rate before the hand touched the target switch. Parietal cortical neurons, in contrast to those in the precentral motor cortex, do not show much difference in discharge patterns when the projected movements differ greatly (by as much as 60°), or when the arm is forced to go under or over an obstacle to reach the target, or whether the arm is permitted to project to a visually fixated target only after vision is occluded. The discharge patterns of parietal neurons is independent of the sensory channel used for cueing signals, somesthetic or visual.

Mountcastle concludes that "the projection and hand manipulation cells ... compose a command apparatus for manual exploration of extrapersonal space. These command signals do not contain the detailed specification for the movements commanded, matters left ... to the precentral motor field. The parietal command for projection is holistic in nature; *it is a gestalt.*

Having explored conscious differential control of an incoming sensory signal, by human subjects, and "holistic" command by a neuron interested in exploration of immediate extrapersonal space, we now turn to consider plasticity of units which respond to combined sensory inputs.

CONDITIONING OF SINGLE CORTICAL NEURONS

Figure 9.222, from Morrell (1967), shows that paired stimuli, each of which yielded responses in this particular cortical neuron, will modify the response pattern of the neuron to more closely resemble the combined response even when only one stimulus channel is activated. Thus unit, in visual

cortex, showed plasticity in response to a particularly located, particularly oriented visual bar when that stimulus was repeatedly combined with electrical stimulus to the contralateral leg. Morrell found that units showing this kind of plasticity are relatively rare (10%), but that similarly plastic units can be found in like proportions in subcortical as well as cortical gray matter.

The scale of what is manifested as conditionable can be magnified from single neurons to include such a number as can influence an electrode across meninges, skull, and scalp. Figure 9.223 is representative of EEG changes associated with thinking and learning. When subject starts thinking, the alpha rhythm (8–13 Hz)—more prominent because subject is resting with eyes closed—abruptly shifts to low voltage fast activity characteristic of an aroused individual. This arousal outlasts the answer and then reverts to alpha rhythm again. Apparently the mistake sinks in, and a correction is made during a rearoused EEG.

Below is a classical paradigm of conditional stimulus preceeding an unconditional stimulus by a few seconds. After a few pairings, the conditional stimulus elicits a response similar to the unconditional response.

It is easy to overlook some very important principles underlying learning models. Gross brain events (as shown by EEG changes) presumably reflect large scale "brain circuit strategies." Microelectrode unit analyses obviously show one element in what might be interpreted as "tissue tactics" in the process. Beneath the level of individual neurons there is evidence for microscopic, ultrastructural, and biochemical changes at still finer resolution. It is likely that tissue tactics are virtually the same across all kinds of learners, invertebrate as well as vertebrate.

Additional common denominators are striking. Conditional stimulus (CS) must precede unconditional stimulus (US) or there will be little or no conditioning. Backward conditioning may involve looping in time from long previous stimuli. It does not appear to matter for purposes of creating conditional circuits in the brain whether the experimentor uses his/her discretion to trigger a signal to which the animal must respond appropriately (pavlovian, type I conditioning), or whether the animal emits a response which is then reinforced at the discretion of the experimentor (konorskian, skinnerian, type II conditioning).

What appears to be of complementary importance in conditioning is sometimes lost sight of because it is so routine. It is that some kind of motivation must be incorporated. Generally this is

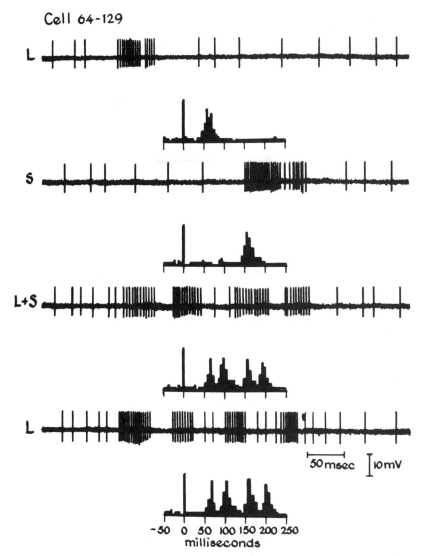

Figure 9.222. Electrical signs of sensory coding are demonstrated by experimentally induced modification of neuronal response pattern. Cell in visual cortex of cat responded to a dark, horizontal bar at 3:00 (*L*) and also to electric shock to the contralateral hind limb (*S*). Combining these two stimuli (*L* + *S*) resulted in a histogram very different from that which might occur from simple linear addition of the two separate responses. After 40 trials of such paired stimuli, the original visual stimulus (*L*) was presented alone. It elicited a pattern much more like that elaborated by paired stimulation than like that which it produced prior to pairing. Cells that are capable of transient modification of their response patterns in this way are relatively infrequently encountered, representing about 10% of the total cortical cells which respond to more than one modality of stimulation. In other experiments, it has become evident that "plasticity" of this kind may be seen in small proportion of neurons in various nuclei all along the neuraxis and not simply at cortical levels. The single traces are representative of pattern in each group of 20 which are summed in histograms. [From Morrell, 1967.]

taken into account by reinforcement regimes. But also, animals are kept slightly underweight to improve their performance on alimentary conditioning and relatively dehydrated (at least at the time of trials) to do likewise for liquid reinforcement. Human subjects invariably get something out of the experiment or it doesn't work in the expected way. Moreover, humans receive carefully standardized instructions.

It is obviously desirable to remember that these conditions have very widespread effects on the nervous system. They make it particularly receptive in predictable ways.

It is also true that animals and humans useful for conditioning experiments have a degree of liberty among their visceral priorities, for otherwise they would not attend and could not be conditioned. Central state conditions make a radical difference in outcome of learning experiences. Ask any student.

MEMORY

One of the early speculations concerning the nature of the memory trace is generally attributed to Müller and Pilzecker (1900). In order to account

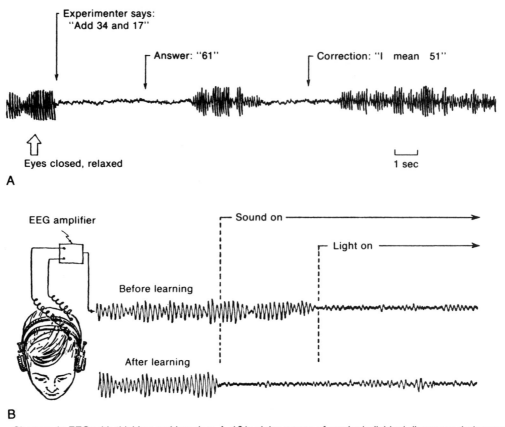

Figure 9.223. Changes in EEG with thinking and learning. *A,* 10/s alpha waves of awake individual disappear during mental effort on an arithmetic problem. *B,* A weak sound usually does not flatten the alpha waves, but a bright light always does. When sound is regularly paired with and precedes light by a fixed interval, the alpha rhythm flattens in response to the sound. [From Galambos, in Bullock (1977).]

for the phenomenon of decreased retention of recently learned material when another activity intervenes between initial learning and retention testing, Müller and Pilzecker proposed the existence of a neural perseverative process that was presumed to be easily interfered with by external influences and was necessary to the "consolidation" of the memory trace for recently learned material. Most of the more recent investigators have accepted this concept, namely, that the establishment of a memory trace takes place in two stages: an initial stage in which the memory trace is evanescent and easily disrupted, and a later, more stable state which is the "permanent" memory trace.

For many years the only evidence in support of such a proposal was the retrograde amnesia seen in many patients after cerebral trauma. Head injuries are often followed by loss of memory for events occurring during the period preceding the injury. With the advent of electroshock therapy for psychoses in the late 1930s, further evidence for this two-stage nature became available. Shock is administered through electrodes applied to the skin of the head. It leads to unconsciousness and presumably disrupts by electrical energy many if not all of the

neuronal circuits of the brain—at least temporarily. Many physicians and investigators who used or studied electroconvulsive shock (ECS) found that it too produces a differential loss of memory only for events immediately preceding the shock treatment (e.g., Williams, 1950).

Other procedures that have been successfully utilized to disrupt the consolidation phase include hypothermia, insulin coma, Metrazol convulsions, anoxia, some anesthetic agents, and application of chemicals or electrical stimulation to local regions of the brain.

In 1970, John summarized information relating to memory consolidation. His interpretation is depicted in Figure 9.224. Evidence indicates that electroconvulsive shock does not erase the memory trace. A temporary holding phase is capable of mediating retrieval decays within about a few hours to a day. This holding phase is not disrupted by blockage of RNA or protein synthesis, but strychnine given during that interval can forestall amnesia that would otherwise follow. Protein synthesis appears to be needed for long-term storage to occur. Various agents that interfere with protein synthesis have different effects on long-term storage. Cyclo-

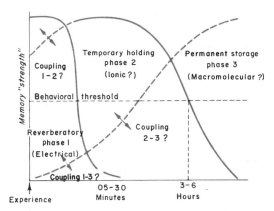

Figure 9.224. Relation of time to "strength" of memory, with *curves* showing hypothetical duration of three phases in memory process. [From John (1970).]

heximide or acetoxycycloheximide do not block retention, although protein synthesis is as effectively impaired as it is with doses of puromycin which does block retention. Cycloheximide plus puromycin actually improves retention. Puromycin has been found to cause abnormal hippocampal activity, while the other agents do not. It appears that learning does not depend on protein synthesis, but that retention involves a protein which probably requires less than an hour to complete synthesis.

HABITUATION AS NEGATIVE LEARNING

If a specific stimulus is given several times and then test stimuli are presented, the response is successively depressed, although other new signals can elicit full amplitude responses. The nervous system evidently elaborates a succession of adjusting filters that are particularly tuned to the particular stimulus under particular circumstances. The filters relate to each of the parameters of the habituated stimulus, its duration, intensity, color, frequency, etc. The application of such a multidimensional system of filters blocks only those signals arising from the particular stimulus.

Apparently, no habituation occurs in the afferent neuron. In hippocampal neurons which show convergence of signals from all modalities of the body's receptive surface, there is habituation to orientation toward the stimulus as well as to the characteristics of the stimulus itself. Different feature extractions project independently onto hippocampal neurons. The hippocampal neurons modify portions of their membrane-receptive surfaces in respect to specific feature detector signals.

Neocortical neurons are very stable. After 6 h of continuous stimulation little or no habituation could be found. The filters appear to be located either in the interneurons or in the hippocampal

neurons themselves (Sokolov, 1976). Evidence is accumulating that learning depends on plasticity, that the hippocampus is engaged in perhaps the most plastic of circuits. It appears that information about biologically important events can be generated or synthesized in the hippocampus. The outcome is a discharge to the frontal-limb midbrain area, from which something comparable to a "now print" order initiates the process of circuit consolidation among recently active neuron. According to Vinogradova (1970), the hippocampus plays a double role: comparing crucial signals and blocking or unblocking the reticular activating mechanisms which are necessary for registration of information on the basis of such comparisons, e.g., comparing novel vs. familiar signals.

It has been argued by John (1977) that the essential event taking place during learning is the elaboration of spatiotemporal configurations of neuronal activity pervading large populations of neurons. Thus, the installation of a memory trace for a particular experience involves a complex process that has some critical time periods. Immediately after the learning experience, there is a phase of consolidation. This phase continues during the first episode of paradoxical (PS) or rapid eye movement (REM) sleep. Experimental enhancement of memory consolidation suppresses REM large scale neuronal augmentation during subsequent periods of sleep. This does not mean that REM sleep is a part of the memory mechanism but does imply that it may play a role in enhancing the *conditions necessary for memory consolidation*. These conditions seem to depend upon whole brain activation, "now print," either by the reticular arousal mechanism in wakefulness or by REM activation during sleep (Bloch, 1976).

HIGHER CONSCIOUSNESS

The uncertainty principle in physics is associated in our ideas with the spatial smallness of units of matter. Physics has achieved its great intellectual and technical gains by focusing on the "few body" problem, and on the "large n problem," *provided* that the problem is limited to experimentation on pure gases, pure crystals, etc., where there is the homogeneity necessary to allow predictability. Living systems invariably involve a very high n, coupled with extremely great heterogeneity. Living systems therefore have freedom within the principle of unity of total pattern of the organism as a whole. Without integration, life fails. While integration is sufficient, there is freedom.

The earliest stages of consciousness in phylogeny probably arise, as proposed by Herrick and Coghill, when the customary automatisms of animals fail to provide satisfaction. An emergent affect or feeling, perhaps a feeling of effort associated with movement, may be the most elemental experience (Herrick, 1949). The basic principle of subjective experience is self-awareness, feeling, and mood, e.g., feeling of "self in action."

Preliminary stages of human type mentation are found in lower animals (Griffin, 1981). What we are concerned with is not the search for a liaison between brain as a physical existence and some other entity, which we call mind. Rather, we need to understand the properties of the brain as a functioning tissue which enable it to compose the feeling, mood, and awareness that we call mind (Herrick, 1956).

Predictable behavior which might be explained on a mechanistic basis, depending on stratified stabilities, prevails only briefly in vertebrate embryonic life. The individual as a whole soon expresses unpredictable behavior. This is a reflection of internal choice, the exercise of freedom. As the individual develops, choices multiply, degrees of freedom expand. In the case of humans, where self-consciousness is so obvious and where we have access through our own experience, freedom becomes moral freedom when the social implications of actions are taken into account. As Herrick wrote: "We have at our command all the physical and mental resources that are needed for further advance [toward more efficient techniques for mentation, for widening the range of experiences and rational interpretations of life] . . . if only we do not squander them in senseless and suicidal rivalry and conflict" (1956, pp. 220–221).

Higher Perceptual Processes

An illusion well described by Yellott (1981) relates to the possibility of seeing an inside out as a normal face seen in Figure 9.225 and explained in 9.226. This and other illusions (the rotating trapezoid of Ames, which appears as a window oscillating Ames, 1955; Cantril, 1949; 1960) reveal how facile, arbitrary, and capricious central perceptual mechanisms are.

Figure 9.227, from Bach-y-Rita (1972) illustrates a method for sensory substitution, in this case, skin stimulation as a substitute for retinal stimulation. The portable TV camera scans the environment and the circuit applies stimuli to the skin of the abdomen a corresponding pattern. In this way, patterns of skin stimulation corresponding to images formed via the video system are employed as a somesthetic substitute in place of visual loss experienced by blind persons. Patterns projected onto the skin have no "visual meaning" unless the indi-

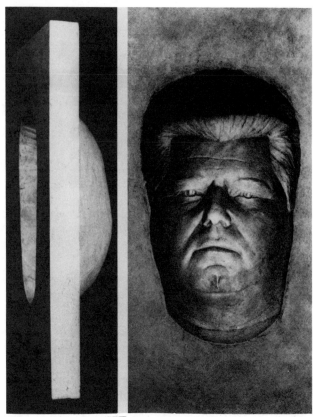

Figure 9.225. Inside-out face, made as the mold of a bust, is shown in side and front views. Looked at from the front, it is more easily seen as a normal face because the brain overrides the depth cues that suggest an object as improbable as an inside-out face. (The reversal is made easier when, as in the front view, the lighting eliminates shadows that might aid the brain in making the correct interpretation.) A three-dimensional inside-out face seen in reversed perspective seems to rotate and to follow an observer who is moving laterally past it. [From Yellott (1981).]

vidual is behaviorally active in directing the video camera. This can be done as well using head, hand, or body movements to direct the camera. After only a few hours of active direction of the camera, the subject no longer feels the impulses as messages relating to the skin but rather as form projected into the space that is being explored by the directed "gaze."

Lateral inhibitory mechanisms that operate along channels between skin and perceptual access (cortex?) serve to establish a perceived image that is sufficiently well defined that it carries prognostic value for behavior. After about 50 h of active "seeing" with the skin, the individual will make automatic averting movements when the "visual field" as projected dictates a threatening motion, by the subject or from some part of the environment.

Bach-y-Rita learned from a girl, blind from birth, who undertook such sensory substitution that, "You sighted people live in a strange world!" When he inquired how so, she replied "All your right

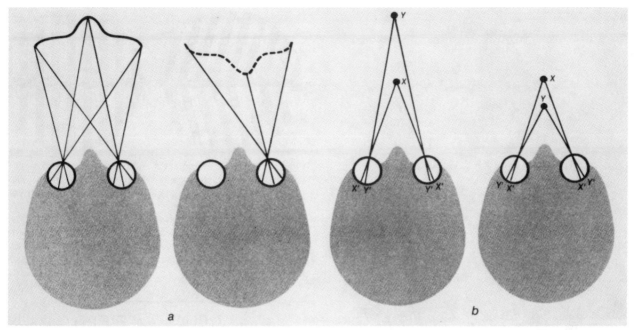

Figure 9.226. Two hypotheses on binocular depth inversion are monocular suppression (a) and disparity reversal (b). In each case, what is actually seen is portrayed as the left and the inversion the hypothesis would explain is shown on the right. In monocular suppression, the left eye's view is depicted as being suppressed, so that the inside-out face is seen depth-inverted as it would appear in a monocular inversion when it is viewed by the right eye alone. In disparity reversal point X is really nearer the observer than point Y, but the brain treats the retinal image as though it came from the left eye and vice versa, with the result that point Y appears to be closer than point X. [From Yellot (1981).]

angles are not right angles when perceived by vision." Her experience with tactile explorations had always confirmed rectilinearity. The truth of this insight helps us realize how much of what we have experienced in the past dictates what we perceive. Our experiential knowledge of rectilinear architecture overrides perception of images as actually projected.

Figures 9.228 illustrates the tilt after effect, an illusion that reveals the operation of hidden visual processes along the visual pathway. First, look at the pattern of stripes at the right. The top half of the pattern will appear to be aligned with the bottom half. Next, let your eyes scan back and forth across the chevron pattern to the left. When you turn to the pattern on the right, you should find that the bars will no longer seem to be aligned.

Experiments of this sort by Wolfe (1983) indicate that visual experience is made up by a series of specialized subsystems that act more or less independently. It suggests further that some of the visual subsystems are responsible for processes of which the viewer is normally unaware.

SPECIALIZATION OF THE CEREBRAL HEMISPHERES

Until relatively recently, asymmetric brain damage served as the primary source of knowledge concerning functional differences in the two hemi-

spheres. However, in the last two decades a great deal of valuable evidence has been harvested through examination of patients who have had complete or partial section of the corpus callosum. These operative procedures are designed to help control epilepsy that is otherwise refractory to treatment.

The *corpus callosum* contains more than 200 million fibers which mainly interconnect neocortical systems. The corpus callosum transfers information from one hemisphere to be integrated with information in the other. However, it limits the consolidation of memory traces to the hemisphere that originated the learning activity (Doty and Negrao, 1973). The corpus callosum, thus, deals with unilateral memory stores. This is an advantageous way to enable more information to be stored in the two hemispheres and to increase potentialities for hemispheric specialization.

The *anterior commissure* interconnects temporal lobe structures. In contrast to the corpus callosum, the anterior commissure sends information which contributes memory stores to the other hemisphere. Doty et al., 1977, consider that the anterior commissure contributes an essential link in mnemonic processes. When they tetanize the anterior commissure, neither storage nor retrieval of memories is possible. Recovery from tetanization is prompt. Tetanization during the delay period in delayed

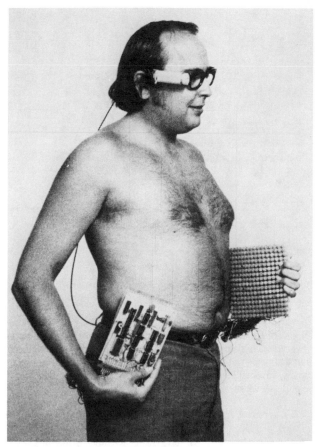

Figure 9.227. A blind subject with a 16-line portable electrical system which includes a TV camera attached to a pair of spectacle frames. Wires lead to an electrical stimulus drive circuitry (held in the right hand) which drives the matrix of 256 concentric silver electrodes which provide skin stimulation in patterns related to video scan density differences. When the blind subject moves the camera across a field or an object, he obtains an image that moves across receptors in his skin. Mechanisms similar to those in the retina, such as lateral inhibition, integrate skin receptor activity to produce edge enhancement. After some score or so hours of active experience while *moving* head with camera attached, blind subject begins to "project" objects perceived through the skin as being "out there" and related spatially to the subject in accordance with the position and direction of the head. [From Bach-y-Rita (1972).]

match to sample tasks does not interfere with learning.

In both animal and human studies it has been established that different aspects of experience can be represented in the two hemispheres at the same time, even involving stable contrary conditioning to the same stimulus (Gazzaniga, 1970; Sperry, 1974). Sperry and colleagues, studying humans with section of the corpus callosum, found two distinctive modes of perception which involve complementary specializations in the two sides of the brain (Fig. 9.229).

The right and left halves of the visual field are divided down the midline and present information to the contralateral hemispheres. The same is true

Figure 9.228. Tilt aftereffect is an illusion that serves investigation of hidden visual processes in pathways leading to visual perception. To experience the illusion, let your eyes scan back and forth along horizontal black bar crossing chevron to left. You should then find that two halves of the pattern at right are briefly not colinear. Tilt aftereffect is measured by making the pattern at right adjustable and asking subject to adjust it so that it seems to be colinear immediately after they have stared at the chevron. [From Wolfe (1983).]

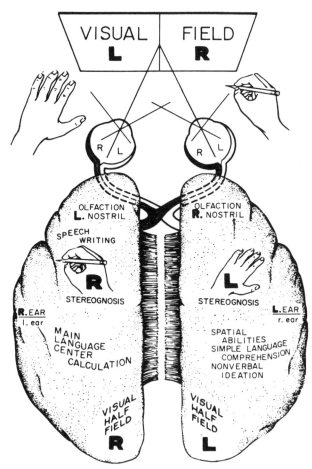

Figure 9.229. Section of the corpus callosum in humans, as a means to prevent spread of epileptic seizure activity, enabled Sperry and collaborators to demonstrate differences between the two hemispheres. This summarizing diagram shows functions known from neuroanatomy, cortical lesion data, and postoperative testing that are separated by the surgery. [From Sperry (1974).]

for cerebral representation of somesthetic information. Auditory functions are projected more bilaterally, yet, as we shall see, the auditory contributions to the two hemispheres are distinctive. The auditory system, by establishing special features for the extraction of acoustic signals for purposes of communicating language, may have originated the process of hemispheric specialization. It is noteworthy that hemispheric specialization is greatest in humans and birds, both of which depend greatly on acoustic analysis and do so with the contralateral hemisphere.

Surgical sectioning of the corpus callosum effectively separates the neocortex of the two hemispheres and makes much of what goes on in them independent. The first observation of interest is that individuals with total section of the corpus callosum, 2 yr after the surgery, show so few effects that these would not likely be recognized in a routine medical examination. All cerebral functions, speech and language, verbal intelligence, calculation, motor coordination, reasoning and recall, including personality and temperament, remain essentially intact.

There is also a surprising lack of symptoms in persons born with agenesis of the corpus callosum. Some of these individuals are essentially without any functional stigmata; they may be able to perform as well as do normal subjects when tested in apparatus that brings out defects in the performance of commissurotomized patients. Intracarotid anesthesia of one hemisphere at a time in one of these subjects indicated that speech had developed on both sides. In such individuals, the anterior commissure may be enlarged, and it might carry more intrahemispheric traffic than it does in the normal situation. It is likely that the congenital abnormality, which is acquired as a result of a developmental defect that occurs about the fourth month of gestation provides a suitable opportunity for early plasticity of the brain to compensate for this deficit.

THE BISECTED BRAIN

Sperry and others have demonstrated that despite the apparent normalcy of people with whole callosal section, appropriate tests reveal many distinctive impairments. Even though there remain brain stem and anterior commissural interconnections between the hemispheres, left and right sides of the brain function independently in most conscious activities. Each hemisphere is autonomous with respect to sensations, perceptions, and ideas, isolated from the experience in the contralateral hemisphere. Stored memories, at least those in neocortex, are unavailable to recall by the other hemisphere. Normal transfer of perceptual information across the midline is missing. Objects identified by touch with one hand cannot be recognized

by the other. Odors identified through one nostril are not recognized when exposed to the other nostril (Gordon and Sperry, 1969). There is a general inability to name objects in the left field of vision, or felt by the left hand or foot, or smelled through the left nostril.

These deficits are not very obvious in everyday existence because the individuals are usually exploring with eye movements, feeling things with both hands, and using language cues to bilateralize sensory information. Since the retinae are intact and the eyes move conjugately, there is less separation of visual experiences in the two hemispheres than would otherwise be the case. Mechanisms of orientation, arousal, and reinforcement all contribute to integration between the partially separated hemispheres. Sperry reports that emotional effects tend to spread easily into the opposite hemisphere, presumably through intact brain stem projections and through the anterior and hippocampal commissures which interconnect the two halves of the limbic system.

Although possible under experimental conditions, there is little separate simultaneous performance of distinctively different motor skill tasks. There may be brain stem and thalamic mechanisms that switch behaviors from one hemisphere to the other, activating and focusing in one hemisphere and repressing competitive activity in the other.

Figure 9.230, also from Sperry, shows that the two hemispheres, in the absence of the corpus callosum, cannot transfer information necessary to recognize hand and finger positions either when gestured by the subject or imposed by the examiner.

Figure 9.231 shows that the two hemispheres deal with composite figures which have different and conflicting images joined in the middle. As a result, the two hemispheres apparently see two different things occupying the same space. They are each unaware of the discordance between the two sides of the composite pictures.

The experimenters tried verbal and manual matching options and found that if linguistic processing is required, the response is dominated by the left hemisphere. If matching is required, the right hemisphere is dominant. An additional observation by Preilowski and Sperry, 1972, is that either hemisphere in patients with section of the corpus callosum can occasionally capture motor control for either side of the body, depending on which hemisphere is dominant for the task. Nevertheless, commissurotomy patients are impaired in performing new motor acts that require coordination between the right and left hands. Habitual bimanual tasks such as tying shoelaces and neckties are not noticeably affected (Preilowski, 1972).

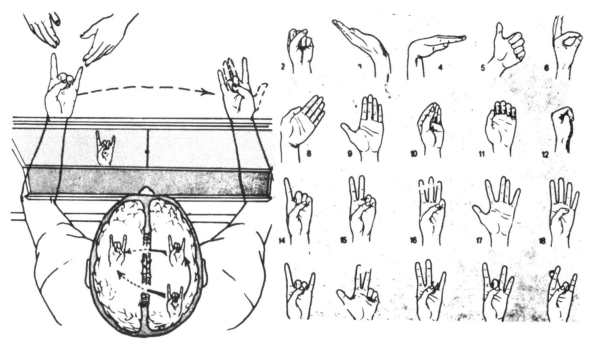

Figure 9.230. Tests used to detect lack of interhemispheric integration following section of the corpus callosum. Subject attepts to replicate complex hand and finger postures flashed to left and right visual half-fields or imposed directly on one hand by examiner. Interhemispheric combinations are performed successfully, but crossed combinations fail. [From Sperry (1974).]

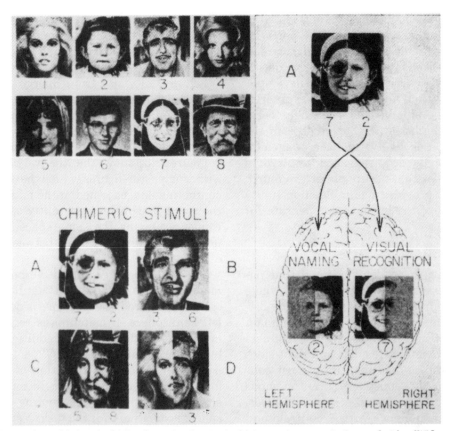

Figure 9.231. Perception with hemispheric disconnection to tachistoscopic presentations of chimeric faces viewed on center apparently involves two separate percepts for which the two test subjects could make correct verbal and manual responses. Manual responses were directed to a match for the left half of the stimulus and the verbal responses described the right half. With such double responses, a conflict between the verbal and the manual responses became evident to the subjects and resulted in considerable perplexity and confusion, lessened by brevity of the exposure. The perplexity evidently arises "as a result of responses by one hemisphere that were recognized to be incorrect by the other hemisphere." [From Levy, Trevarthen, and Sperry (1972).]

It has long been recognized that the left cerebral hemisphere is dominant for speech in most individuals. To a slight degree, a few percent only, this is reversed among right-handed individuals (Milner, 1974). All senses except possibly olfaction appear to be differently affected in terms of the kinds of information extracted by the two hemispheres.

The right ear, which projects predominantly into the left hemisphere, is employed more advantageously for speech. The left ear, which projects to its contralateral hemisphere, is better at discriminating music. Milner (1974) has shown that the right hemisphere is superior to the left in discrimination and remembrance of spatial patterns. Right frontal lobe deficits lead to faulty performance in the temporal ordering of nonverbal events.

Children appear to utilize both hemispheres in language functions during early development, prior to commitment of language to its characteristic main location, in the left hemisphere. Early damage to protospeech areas in the child's left cortex may induce a representation of speech in the *right* hemisphere. The results of such "cognitive crowding" are manifestations of cerebral plasticity, but such individuals are likely to be below average in general intelligence.

Berlucchi has found asymmetries by examining *normal* subjects with respect to visual discriminations, using very brief exposures to the two visual half-fields of noncorresponding, conflicting images (1974). Broadbent has conducted similar studies with normal subjects in relation to conflicting or competing information introduced separately into the two ears (1974).

Downer (1961) has made use of an interesting preparation in monkeys: the optic chiasma is split medially and the amygdala in one hemisphere is destroyed. This permits activation of either hemisphere by limiting visual stimuli to the eye on the same side. The hemisphere in which the amygdala is destroyed fails to integrate the normal emotional response to sight of a human. Interhemispheric transfer of information for appropriate reaction to that sight depends on integrity of the splenium of the corpus callosum. These animals register normal visual fear of humans as long as the posterior (splenial) portion of the corpus callosum remains intact. They fail to register such fear after the splenium is divided. With such methods, extremes of hemispheric specialization can be created surgically, and integration of their specialized functions can take place via callosal paths.

And men should know that from nothing else
but from the brain come joys,
delights, laughter, jests, and sorrows,
griefs, despondency and lamentations.
And by this, in an especial manner
we acquire wisdom and knowledge,
and see and hear and know what are foul
and what are fair, what sweet and unsavory.
And by the same organ we become mad and
delerious, and fears and terrors assail us,
some by night and some by day,
and dreams and untimely wanderings,
and cares that are not unsuitable,
and ignorance of present circumstances,
disuetude and unskillfulness.
All these things we endure from the brain,
when it is not healthy . . . or when it suffers
any other preternatural and unusual affliction.

Hippocrates
On the Sacred Disease

BIBLIOGRAPHY

ADAMS, J. E. Naloxone reversal of analgesia produced by brain stimulation in the human. *Pain* 2: 161–166, 1976.

ADEY, W. R. Spontaneous electrical brain rhythms accompanying learned responses. In: *The Neurosciences: Second Study Program*, edited by F. O. Schmitt. New York: The Rockefeller University Press, 1974, p. 224–243.

ADEY, W. R. Tissue interaction with nonionizing electromagnetic fields. *Physiol. Rev.* 61: 435–514, 1981.

ADEY, W. R., J. P. SEGUNDO, AND R. B. LIVINGSTON. Corticifugal influences on intrinsic brain-stem conduction in cat and monkey. *J. Neurophysiol.* 20: 1–16, 1957.

ADLER, F. H. *Physiology of the Eye*, 4th ed. St. Louis: Mosby, 1965.

ADRIAN, E. D. Discharges from vestibular receptors in the cat. *J. Physiol. (Lond.)* 101: 389–407, 1943.

ADRIAN, E. D. The electrical activity of the mammalian olfactory bulb. *Electroencephal. Clin. Neurophysiol.* 2: 377–388, 1950.

AKERT, K., AND B. E. GERNANDT. Neurophysiological study of vestibular and limbic influences upon vagal outflow. *Electroencephal. Clin. Neurophysiol.* 14: 904–914, 1962.

AKERT, K., W. P. KOELLA, AND R. HESS, JR. Sleep produced by electrical stimulation of the thalamus. *Am. J. Physiol.* 168: 260–267, 1952.

AKERT, K., H. MOOR, K. PFENNINGER, AND C. SANDRI. Contributions of new impregnation methods and freeze etching to the problems of synaptic fine structure. *Prog. Brain Res.* 31: 223–240, 1969.

AKERT, K., AND R. B. LIVINGSTON. Morphological plasticity of the synapse. In: *Surgical Approaches in Psychiatry*, edited by L. V. Laitinen and K. E. Livingston. Lancaster, UK: Medical and Technical Publishing Co. Ltd, 1973, p. 315–330.

ALLMAN, J. M. Evolution of the visual system in the early primates. In: *Progress in Psychobiology and Physiological Psychology*, edited by J. Sprague and A. Epstein. New York, Academic Press, 1977, vol. 7, p. 1–53.

ALLMAN, J. M., J. H. KAAS, R. H. LANE, AND F. M. MIEZEN. A representation of the visual field in the inferior nucleus of the pulvinar in the owl monkey (Aotus trivirgatus). *Brain Res.* 40: 291–302, 1972.

ALLEN, W. F. Effect of ablating the pyriform-amygdaloid areas and hippocampi on positive and negative olfactory conditioned reflexes and on conditioned olfactory differentiation. *Am. J. Physiol.* 132: 81–91, 1941.

ALLISON, A. C., AND R. T. T. WARWICK. Quantitative observation on the olfactory system of the rabbit. *Brain* 72: 186–197, 1949.

ALTMAN, J. Autoradiographic and histological studies of postnatal neurogenesis. II. A longitudinal investigation of the kinetics, migration and transformation of cells incorporating triated thymidine in infant rats, with special references to postnatal neurogenesis in some brain regions. *J. Comp. Neurol.* 128: 431–473, 1963.

ALTMAN, J. *Organic Foundations of Animal Behavior*. New York: Holt, Rinehart and Winston, 1966.

ALTMAN, J. Postnatal growth and differentiation of the mammalian brain, with implications for a morphological theory of memory. In: *The Neurosciences. A Study Program*, edited by G. C. Quarton, T. Melnechuk, and F. O. Schmitt. New York: Rockefeller Univ. Press, 1967, p. 723–743.

ALTMAN, J., AND G. D. DAS. Autoradiographic and histological studies of postnatal neurogenesis. *J. Comp. Neurol.* 126: 337–390, 1966.

ALTMAN, J., AND G. D. DAS. Post-natal origin of microneurons in the rat brain. *Nature (London)* 207: 953–965, 1965.

AMES, A., JR. *An Interpretive Manual for the Demonstrations in the Psychology Research Center, Princeton University: The Nature of Our Perceptions, Prehensions and Behavior.* Princeton, NJ: Princeton Univ. Press, 1955.

AMOORE, J. E., J. W. JOHNSTON, AND M. RUBIN. The stereochemical theory of odor. *Sci. Am.* 210 (Suppl 2): 42–49, 1964.

ANAND, B. K., AND J. R. BROBECK. Hypothalamic control of food intake in rats and cats. *Yale J. Bio. Med.* 24: 123–140, 1951.

ANDERSEN, R. A., AND V. B. MOUNTCASTLE. The influence of the angle of gaze upon the excitability of the light-sensitive neurons of the posterior parietal cortex. *J. Neurosci.* 3: 532–548, 1982.

ARENDASH, G., AND R. GORSKI. Enhancement of sexual behavior in female rats by neonatal transplantation of brain tissue from males. *Science* 217: 1276–1278, 1982.

ARENDASH, G., AND R. GORSKI. Effects of discrete lesions of sexually dimorphic nucleus of the preoptic area or other medial preoptic regions on the sexual behavior of male rats. *Brain Res. Bull.* 10: 147–154, 1983.

ARNOLD, A. P., AND R. A. GORSKI. Gonadal steriod induction of structural sex differences in the central nervous system. *Ann. Rev. Neurosci.* 7: 413–442, 1984.

BACH-Y-RITA, P. *Brain Mechanisms in Sensory Substitution.* New York: Academic Press, 1972.

BAINBRIDGE, F. A., AND J. A. MENZIE. *Essentials of Physiology*, 5th ed., edited and revised by C. L. Evans. New York: Longmans, 1925.

BARBOUR, H. G. Die Wirkung unmittelbarer Erwarmung und Abkuhlung del Warmezentra auf die Korpertemperatur. *Arch. Exp. Pathol. Pharmakol.* 70: 1–26, 1912.

BARD, P. Delimitation of central nervous mechanisms involved in motion sickness. *Fed. Proc.* 6: 72, 1947.

BARKER, J. L., AND T. G. SMITH, JR. (eds.) *The Role of Peptides in Neuronal Function.* New York: Dekker, 1980, 768 pp.

BARD, P., AND V. B. MOUNTCASTLE. Some forebrain mechanisms involved in expression of rage with special reference to suppression of angry behavior. *Res. Publ. Ass. Res. Nerv. Ment. Dis.* 27: 362–404, 1948.

BARIBEAU-BRAUN, J., T. W. PICTON, AND J. Y. GOSSELIN. Schizophrenia: a neurophysiological evaluation of abnormal information processing. *Science* 219: 874–876, 1983.

BARLOW, H. B., C. BLAKEMORE, AND J. D. PETTIGREW. The neural mechanisms of binocular depth discrimination. *J. Physiol. (London)* 193: 327–342, 1967.

BARTLETT, N. R. Dark adaptation and light adaptation. In: *Visual Perception*, edited by C. H. Graham. New York: Wiley, 1965b, ch. 8, p. 185–207.

BASBAUM, A. I., C. H. CLANTON, AND H. L. FIELDS. Ascending projections of nucleus raphe magnus in the cat. An autoradiographic study. *Anat. Rec.* 184: 354, 1976.

BASBAUM, A. I., AND H. L. FIELDS. Endogenous pain control mechanisms: review and hypothesis. *Ann. Neurol.* 4: 451–462, 1978.

BAUST, W., AND G. BERLUCCHI. Reflex responses to clicks of cat's tensor tympani during sleep and wakefulness and the influence thereon of the auditory cortex. *Arch. Ital. Biol.* 102: 686–712, 1964.

BAYER, S. A. The development of the septal region in the rat. I. Neurogenesis examined with 3H-thymidine autoradiography. *J. Comp. Neurol.* 183: 89–106, 1979; II. Morphogenesis in normal and x-irradiated embryos. *J. Comp. Neurol.* 183: 107–120, 1979.

BAYER, S. A., AND J. ALTMAN. The effects of x-irradiation in the postnatally-forming granule cell population in the olfactory bulb, hippocampus, and cerebellum of the rat. *Exp. Neurol.* 48: 167–174, 1975.

BAYER, S. A., J. W. YACKEL, AND P. S. PURI. Neurons in the rat dentate gyrus granular layer substantially increase during juvenile and adult life. *Science* 216: 890–892, 1982.

BEATON, L. E., C. LEININGER, W. A. McKINLEY, H. W. MAGOUN, AND S. W. RANSON. Neurogenic hyperthermia and its treatment with soluable pentobarbitol in the monkey. *Arch. Neurol. Psychiat. (Chicago)* 49: 518–536, 1943.

BECKSTEAD, R. M., J. R. MORSE, AND R. NORGREN. The nucleus of the solitary tract in the monkey; projection to the thalamus and brain stem nuclei. *J. Comp. Neurol.* 190: 259–282, 1982.

BELLUGI, U., AND E. S. KLIMA. Language: perspectives from another modality. Ciba Foundation Symposium, 69: 99–117, 1979 (See also KLIMA AND BELLUGI, 1979).

BENNETT, E. L. Cerebral effects of differential experience and training. In: *Neural Mechanisms of Learning and Memory*, edited by M. R. Rosenzweig and E. L. Bennett. Cambridge, MA: MIT Press, 1976.

BENNETT, G. J., M. ABDELMOUMENE, H. HAYASHI, AND R. DUBNER. Physiology and morphology of substantia gelatinosa neurons intracellularly stained with horseradish peroxide. *J. Comp. Neurol.* 194: 809–827, 1980.

BENNETT, M. V. L., AND D. A. GOODENOUGH. Gap junctions, electrotonic coupling, and intercellular communication. *Neurosci. Prog. Bull.* 16: 373–486, 1978.

BENSON, A. J. Modification of the response to angular accelerations by linear acceleration. In: *Handbook of Sensory Physiology*, vol. VI/2: Vestibular System, edited by H. H. Kornhuber. Heidelberg, FRG: Springer-Verlag, 1974, p. 281–320.

BENSON, A. J., AND M. A. BODIN. Accelerations on vestibular receptors in man. *Aerosp. Med.* 37: 144–154, 1966.

BENZINGER, T. H. Heat regulation: homeostasis of central temperature in man. *Physiol. Rev.* 49: 671–759, 1969.

BERGLAND, D. W., J. S. CHU, M. A. HOLSEY, L. B. JONES, J. M. KALISZEWSKI, AND B. OAKLEY. New approaches to the problem of the trophic function of neurons. In: *Proceedings of the Sixth International Symposium on Olfaction and Taste*, edited by J. Le Magnen and P. MacLeod. London: Information Retrieval Ltd., 1977, p. 217–224.

BERING, E. A., JR. The cerebrospinal fluid and the extracellular fluid of the brain. *Fed. Proc.* 33: 2061–2063, 1974.

BERLUCCHI, G. Cerebral dominance and interhemispheric communication in normal man. In: *The Neurosciences Third Study Program*, edited by F. O. Schmitt and W. G. Worden. Cambridge, MA: The MIT Press, 1974, p. 65–69.

BERRY, C. Recent advances in aerospace medicine. *Proc. 18th Int. Cong. Aviation. and Space Med.* Dordrecht, Netherlands: Reidel, 1970.

BERSON, D. M., AND A. M. GRAYBIEL. Some cortical and subcortical fiber projections to the accessory optic nuclei in the cat. *Neurosci.* 5: 2203–2217, 1980.

BERSON, D. M., AND A. M. GRAYBIEL. Organization of the striate-recipient zone of the cat's lateralis posterior-pulvinar complex and its relations with the geniculo-striate system. *Neuroscience* 9: 337–372, 1983.

BIRCH, G. G., AND A. R. MYLVAGANAM. Evidence for the proximity of sweet and bitter receptor sites. *Nature* 260: 632–634, 1976.

BIZZI, E. The coordination of eye-head movements. *Sci. Am.* 231: 100–106, 1974.

BIZZI, E., AND P. H. SHILLER. Single unit activity in the frontal eye fields of unanesthetized monkeys during eye and head movement. *Exp. Brain Res.* 10: 150–158, 1970.

BJÖERKLUND, A., AND O. LINDEVALL. The meso-telencephalic dopamine neuron system: a review of its anatomy. In: *Limbic Mechanisms: The Continuing Evolution of the Limbic System Concept*, edited by K. E. Livingston and O. Hornykiewicz. New York: Plenum Press, 1976, p. 307–331.

BJÖERKLUND, A., S. DUNNETT, U. STENEVI, M. LEWIS, AND S. IVERSEN. Reinnervation of the denervated striatum by substantia nigra transplants: functional consequences as revealed by pharmacological and sensorimotor testing. *Brain Res.* 199: 307–333, 1980.

BJÖRKLUND, A., U. STENEVI, AND S. DUNNETT. Functional reactivation of the deafferented neostriatum by nigral transplants. *Nature* 289: 497–499, 1981.

BJÖERKLUND, A., U. STENEVI, AND N. A. SVENDGAARD. Growth of transplanted monoaminergic neurones into the adult hippocampus along the perforant path. *Nature* 262: 787–790, 1976.

BJÖERKLUND, A., AND U. STENEVI. Intracerebral neural implants: neuronal replacement and reconstruction of damaged circuits. *Ann. Rev. Neurosci.* 7: 279–308, 1984.

BLAKEMORE, C. Developmental factors in the formation of feature extracting neurons. In: *The Neurosciences Third Study Program*, edited by F. O. Schmitt and F. G. Worden. Cambridge, MA: The MIT Press, 1974, p. 105–113.

BLAKEMORE, C., AND J. D. PETTIGREW. Eye dominance in the visual cortex. *Nature (London)* 225: 426–429, 1970.

BLASCHKO, H. Catecholamine biosynthesis. *Brain Med. Bull.* 29: 105–109, 1973.

BLEULER, E. *Textbook of Psychiatry*, New York: Macmillan, 1924; and *Dementia Praecox or the Group of Schizophrenias*. New York: International University Press, translated from German edition (1911) by J. Zinkin, 1950.

BLINKOV, S. M., AND I. I. GLEZER. *The Human Brain in Figures and Tables*. New York: Plenum Press, 1968.

BLOCH, V. Brain activation and memory consolidation. In: *Neural Mechanisms of Learning and Memory*, edited by M. R. Rosenzweig and E. L. Bennett. Cambridge, MA: The MIT Press, 1976, p. 583–590.

BLOOM, F. E., B. J. HOFFER, AND G. R. SIGGINS. Studies on norepinephrine-containing afferents to Purkinje cells of rat cerebellum. I. Localization of the fibers and their synapses. *Brain Res.* 25: 501–521, 1971.

BLUM, P. S., L. D. ABRAHAM, AND S. GILMAN. Vestibular, auditory and somatic input to the posterior thalamus of the cat. *Exp. Brain Res.* 34: 1–9, 1979.

BODIAN, D. Neuron junctions: a revolutionary decade. *Anat. Rec.* 174: 73–82, 1972.

BODIAN, D. Neurons, circuits, and neuroglia. In: *The Neurosciences. A Study Program*, edited by G. C. Quarton, R. Melnechuk, and F. O. Schmitt. New York: Rockefeller Univ. Press, 1967, p. 6–24.

BOGEN, J. E. The other side of the brain. I: Dysgraphia and dyscopia following cerebral commissurotomy. *Bull. Los Angeles Neurol. Soc.* 34: 73–105, 1969.

BORG, E., AND J. E. JACKRISSON. Stapedius reflex and speech features. *J. Acoust. Soc. Am.* 54: 525–527, 1973.

BOSNJAK, Z. B., AND J. P. KAMPINE. Intracellular recordings from the stellate ganglion of the cat. *J. Physiol. (London)* 324: 273–283, 1982.

BOULDING, K. E. *The Image*. Ann Arbor, MI: University of Michigan Press, 1956.

BOYNTON, R. M., AND D. N. WHITTEN. Visual adaptation in monkey cones: recordings of late receptor potentials. *Science* 170: 1423–1426, 1970.

BRAZIER, M. A. B. (ed.). *Growth and Development of the Brain: Nutritional, Genetic, and Environmental Factors*. International Brain Research Organization Series, Vol. I. New York: Raven Press, 1975.

BRAZIER, M. A. B., AND J. A. HOBSON. *The Reticular Formation Revisited: Specifying Functions for a Nonspecific System*, New York: Raven Press, 1980.

BREEDLOVE, S., AND A. ARNOLD. Hormone accumulation in a sexually dimorphic motor nucleus of the rat spinal cord. *Science* 210: 564–566, 1980.

BROADBENT, D. E. Division of function and integration of behavior. In: *The Neurosciences Third Study Program*, edited by F. O. Schmitt and F. G. Worden. Cambridge, MA: The MIT Press, 1974, p. 31–41.

BROADWELL, R. D. Olfactory relationships of the telencephalon and diencephalon in the rabbit. I: An autoradiographic study of the efferent connections of the main and accessory olfactory bulbs. *J. Comp. Neurol.* 163: 329–345, 1975.

BRODAL, A. *The Cranial Nerves: Anatomy and Anatomicoclinical Correlations.* Springfield, IL: Charles C Thomas, 1959.

BRODAL, A. *Neurological Anatomy in Relation to Clinical Medicine,* 3rd ed. New York: Oxford Univ. Press, 1981.

BRODAL, A. Some features in the anatomical organization of the vestibular nuclear complex in the cat. Basic aspects of central vestibular mechanisms. *Prog. Brain Res.* 37: 31–53, 1972.

BRONOWSKI. J. New concepts in the evolution of complexity. *Synthese* 21: 228–246, 1970.

BRONOWSKI, J. *The Ascent of Man.* Boston: Little, Brown, 1973.

BROWN, J. L. Flicker and intermittent stimulation. In: *Vision and Visual Perception,* edited by C. H. Graham. New York: Wiley, 1965.

BROWN, J. L., C. H. GRAHAM, H. LEIBOWITZ, AND H. B. RANKEN. Luminance thresholds for the resolution of visual detail during dark adaptation. *J. Opt. Soc. Am.* 43: 197–202, 1953.

BROWN, K. T. The electroretinogram: its components and their origin. *Vision Res.* 8: 633–677, 1968.

BROWN, M. C., R. L. HOLLAND, AND W. G. HOPKINS. Motor nerve sprouting. *Ann. Rev. Neurosci.* 4: 17–42, 1981.

BRUGGE, J. F., AND G. D. GEISLER. Auditory mechanisms of the lower brainstem. *Ann. Rev. Neurosci.* 1: 363–394, 1978.

BUCHWALD, J. S. Generators. In: *Basis of Auditory Brain-Stem Evoked Responses,* edited by E. J. Moore. New York: Grune & Stratton, 1983, p. 157–195.

BUCY, P. C. Sitting on a basketball: how it feels to be a paraplegic. *Perspect. Biol. Med.* 17: 151–163, 1974.

BULLER, A., J. C. ECCLES, AND R. M. ECCLES. Differentiation of fast and slow muscles in the cat hind limb. *J. Physiol.* 150: 399–416, 1960.

BULLOCK, T. H. Representation of information in neurons and sites for molecular participation. *Proc. Natl. Acad. Sci.* 60: 1058–1068, 1968.

BULLOCK, T. H., AND G. A. HORRIDGE. *Structure and Function in the Nervous Systems of Invertebrates.* San Francisco: Freeman and Company, 1965, 2 vols.

BULLOCK, T. H., R. ORKLAND, AND A. GRINNELL. *Introduction to Nervous Systems.* San Francisco: W. H. Freeman and Company, 1977.

BUNGE, R. P. Glial cells and central myelin sheath. *Physiol. Rev.* 48: 197–251, 1968.

BUNGE, R., M. JOHNSON, AND C. D. ROSS. Nature and nurture in development of the autonomic neuron. *Science* 199: 1409–1416, 1978.

BURKE, D., K-E. HAGBARTH, AND L. LÖFSTEDT. Muscle spindle responses in man to change in load during accurate position maintenance. *J. Physiol.* 276: 159–164, 1978a.

BURKE, D., K-E. HAGBARTH, AND N. F. SKUSE. Recruitment order of human spindle endings in isometric voluntary contractions. *J. Physiol.* 285: 101–112, 1978b.

BURNHAM, W. M. Primary and "transfer" seizure development in the kindled rat. In: *Kindling,* edited by J. A. Wada. New York: Raven Press, 1976, p. 61–84.

BURNSTOCK, G., G. CAMPBELL, M. BENNETT, AND M. E. HOLMAN. Inhibition of the smooth muscle of the taenia coli. *Nature (London)* 200: 581–582, 1963.

BURNSTOCK, G., AND T. HOEKFELT. Non-adrenergic, non-cholinergic autonomic neurotransmission mechanisms. *Neurosci. Res. Prog. Bull.* 17: 379–519, 1979.

BUTLER, R. A. Frequency spectrum of the auditory evoked response to simultaneously and successively presented stimuli. *Electroencephal. Clin. Neurophysiol.* 33: 277–282, 1972; see also BUTLER, R. A. The influence of spatial separation of sound sources on the auditory evoked response. *Neuropsychologia* 10: 219–225, 1972.

CAIN, D. P. Sensory kindling: implications for development of sensory prostheses. *Neurol.* 29: 1595–1599, 1979.

CAJAL, S. R. *Textura del Sistema Nerviosa del Hombre y de los Vertebrados.* Madrid: Moya, 1904.

CAJAL, S. R. *Histologie du Systeme Nerveux de l'Homme et des Vertebres.* (French edition revised and updated by author. Translated from Spanish by L. Azoulay, 2 vols. Paris: Maloine, 1909–1911. Republished, Madrid: Consejo Superior de Investigaciones Cientificas, 1950.

CALLAWAY, E., P. TEUTING, AND S. H. KOSLOW (eds.). *Event-related Brain Potentials in Man.* New York: Academic Press, 1978.

CANNON, W. B. The physiological basis of thirst. *Proc. Roy. Soc. (London) [Biol.] Ser. B.* 90: 283–301, 1918.

CANNON, W. B. *The Wisdom of the Body.* New York: Norton, 1932.

CANTRIL, H. (ed.) *The Morning Notes of Adelbert Ames, Jr.* New Brunswick, NJ: Rutgers University Press, 1960.

CANTRIL, H., A. A. AMES, JR., A. H. HASTORF, AND W. H. ITTELSON. Psychology and scientific research. *Science* 110: 461–464, 1949.

CARMEL, P. W., AND A. STARR. Acoustic and nonacoustic factors modifying middle-ear muscle activity in waking cats. *J. Neurophysiol.* 26: 598–616, 1963.

CARPENTER, M. B. Anatomical organization of the corpus striatum and related nuclei. *Res. Publ. Assoc. Res. Nerv. Ment. Dis.* 55: 1–36, 1958.

CARPENTER, M. B. *Human Neuroanatomy,* 7th ed. Baltimore: Williams & Wilkins, 1976, 741 p.

CASSEDAY, J. H., AND W. D. NEFF. Auditory localization: role of auditory pathways in brain stem of the cat. *J. Neurophysiol.* 38: 842–858, 1975.

CAUNA, N. Fine structure of the receptor organs and its probable functional significance. In: *Touch, Heat and Pain,* edited by A. V. S. deRenck and J. Knight. Boston: Little, Brown, 1966, p. 117–127.

CAVENESS, W. F. Ontogeny of focal seizures. In; *Basic Mechanisms of the Epilepsies,* edited by H. H. Jasper, A. A. Ward, and A. Pope. Boston: Little, Brown, 1969, p. 517–534.

CECCARELLI, B., AND W. P. HURLBUT. Vesicle hypothesis of the release of quanta of acetylcholine. *Physiol. Rev.* 60: 396–434, 1980.

CELESIA, G. G. Organization of auditory cortical areas in man. *Brain* 99: 403–414, 1976.

CHAPANIS, A., AND R. A. McCLEARY. Interposition as a cue for the perception of relative distance. *J. Gen. Psychol.* 48: 113–132, 1953.

CHANGEUX, J. P., AND S. G. DENNIS. Signal transduction across cellular membranes. *Neurosci. Res. Prog. Bull.* 20: 267–426, 1982.

CHUSID, J. G. *Correlative Neuroanatomy and Functional Neurology,* 15th ed. Los Altos, CA: Lange, 1973, 429 pp.

CLARK, B. Thresholds for the perception of angular acceleration in man. *Aerosp. Med.* 38: 443–450, 1967.

CLARK, B., AND J. D. STEWART. Comparison of sensitivity for the perception of bodily rotation and the oculogyral illusion. *Percept. Psychophys.* 3: 253–256, 1968.

CLARK, G., H. W. MAGOUN, AND S. W. RANSON. Hypothalamic regulation of body temperature. *J. Neurophysiol.* 2: 61–80, 1939.

COHN, D., AND D. JOSEPH. Influence of body weight and body fat on appetite of "normal" lean and obese rats. *Yale J. Biol. Med.* 34: 598–607, 1962.

COHN, M. L., M. COHN, AND F. H. TAYLOR. Guanosine 3′,5′-monophosphate: a central nervous system regulator of analgesia. *Science* 319–322, 1978.

COLLINS, W. E. Vestibular responses from figure skaters. *Aerosp. Med.* 37: 1098–1104, 1966.

COLLINS, W. F., F. E. NULSEN, AND C. N. SHEALY. Electrophysiological studies of peripheral and central pathways conducting pain. In: *Pain,* edited by R. S. Knighton and P. R. Dumke. Boston: Little, Brown, and Company, 1966.

CONE, R. A., AND W. H. COBBS. Rhodopsin cycle in the living eye of the rat. *Nature (London)* 221: 820–822, 1969.

CONEL, J. R. *The Postnatal Development of the Human Cerebral Cortex,* 6 vols. Cambridge, MA: Harvard University Press, 1939–1959.

CONSTANTINE-PATON, M., AND M. LAW. The development of maps and stripes in the brain. *Sci. Am.* 247: 62–70, 1982.

COPENHAVER, W. M., R. P. BUNGE, AND M. B. BUNGE. *Bailey's Textbook of Histology,* 16th ed. Baltimore: Williams & Wilkins, 1971, 745 pp.

COTMAN, C. W., M. NIETO-SAMPEDRO, AND E. W. HARRIS. Synapse replacement in the nervous system of adult vertebrates. *Physiol. Rev.* 61: 684–784, 1981.

COWEY, A. Sensory and non-sensory visual disorders in man and monkey. *Philos. Trans. Roy. Soc. Lond. (Biol.)* 25: 3–13, 1982.

CRACCO, R. Q. Spinal evoked responses: peripheral nerve stimulation in man. *Electroencephal. Clin. Neurophysiol.* 35: 379–386, 1973.

CRACCO, R. Q., AND G. C. CELESIA. Somatosensory and visual evoke potentials. In: *Bases of Auditory Brain-Stem Evoked Responses,* edited by E. J. Moore. New York: Grune & Stratton, 1983, p. 363–390.

CRICK, F. H. C. Thinking about the brain. *Sci. Am.* 241: 219–232, 1979.

CRICK, F. H. C., D. C. MARR, AND T. POGGIO. An information-processing approach to understanding the visual cortex. In: *The Organization of the Cerebral Cortex,* edited by F. O. Schmitt, F. G. Worden, G. Adelman, and S. G. Dennis. Cambridge, MA: The MIT Press, 1981, p. 505–533.

CROSBIE, R. J., J. D. HARDY, AND E. FESSENDEN. Electrical analog simulation of temperature regulation in man. *IRE Trans. Bio-Med. Elect.* 8: 245–252, 1961.

CROSBY, E. C., T. HUMPHREY, AND E. W. LAVER. *Correlative Anatomy of the Nervous System.* New York: Macmillan, 1962.

CYNADER, M., AND D. E. MITCHELL. Prolonged sensitivity to monocular deprivation in dark-reared cats. *J. Neurophysiol.* 43: 1026–1040, 1980.

DAHLSTROM, A., AND K. FUXE. Evidence for the existence of monoamine-containing neurons in the central nervous system. *Acta Physiol. Scand.* 62. (Suppl. 232): 55 pp., 1964.

DALLOS, P., J. SANTOS-SACCHI, AND A. FLOCK. Intracellular recordings from cochlear outer hair cells. *Science* 218: 582–584, 1982.

DAVAL, G., AND J. LEVETEAU. Electrophysiological studies of centrifugal and centripetal connections of the anterior olfactory nucleus. *Brain Res.* 78: 395–410, 1974.

DAVIS, H. Psychophysiology of hearing and deafness. In: *Handbook of Experimental Psychology,* edited by S. S. Stevens. New York: Wiley, 1951, p. 1116–1142.

DAVIS, H. A mechano-electrical theory of cochlear action. *Ann. Otol. Rhin. Laryngol.* 67: 789–801, 1958.

DAVIS, H. Guide for the classification and evaluation of hearing handicaps in relation to the international audiometric zero. *Trans. Am. Acad. Ophthal. Otolaryngol.* 69: 740–751, 1965b.

DAVIS, H., R. BENSON, W. P. COVELL, C. FERNANDEZ, R. GOLDSTEIN, Y. KATSUKI, J.-P. LEGOUIX, D. R. McAULIFFE, AND I. KATSUKI. Acoustic trauma in the guinea pig. *J. Acoust. Soc. Am.* 25: 1180–1189, 1953.

DAVIS, H., C. T. MORGAN, J. E. HAWKINS, JR., R. GALAMBOS, AND F. W. SMITH. Temporary deafness following exposure to loud tones and noise. *Acta Otolaryngol. Stockh.* (Suppl. 88), 57 pp., 1950.

DAVSON, H. Formation of the intra-ocular fluid. *Ophthal. Lit. (Denver)* 4: 3–13, 1950.

DAVSON, H. The intra-ocular fluids. In: *The Eye,* 2nd ed. New York: Academic Press, 1969, vol. 1.

DAYHOFF, M. O. The origin of protein superfamilies. *Fed. Proc.* 35: 2132–2138, 1976.

DeCASPER, A. J., AND W. P. FIFER. Of human bonding: newborns prefer their mother's voices. *Science* 208: 1174–1176, 1980.

DELGADO, J. M. R. Modulation of emotions by cerebral radio stimulation. In: *Physiological Correlates of Emotion,* edited by P. Black. New York: Academic Press, 1970, p. 189–202.

DELGADO, J. M., W. M. ROBERTS, AND N. E. MILLER. Learning motivated by electrical stimulation of the brain. *Am. J. Physiol.* 179: 587–593, 1954.

DeLONG, M. R. Activity of pallidal neurons during movements. *J. Neurophysiol.* 34: 414–427, 1971.

DeLORENZO, A. J. Studies on the ultrastructure and histophysiology of cell membranes, nerve fibers and synaptic junctions in chemoreceptors. In: *Olfaction and Taste,* edited by Y. Zotterman. New York: Macmillan, 1963, p. 5–17.

DELL, P. Correlations entre le systeme nerveux vegetatif et le systeme de la vie de relation: mesencephale, diencephale et cortex cerebrale. *J. Physiol. (Paris)* 44: 471, 1952.

DEMSKI, L. S., AND R. G. NORTHCUTT. The terminal nerve: a new chemosensory system in vertebrates. *Science* 220: 435–437, 1983.

DENSERT, O. Adrenergic innervation in the rabbit cochlea. *Acta. Ototlaryngol. (Stockholm)* 78: 345–356, 1974.

DESMEDT, J. E. Auditory-evoked potentials from cochlear to cortex as influenced by activation of the efferent olivocochlear bundle. *J. Acoust. Soc. Am.* 34: 1478–1496, 1962.

DESMEDT, J. E. Scalp-recorded cerebral event-related potentials in man as point of entry into the analysis of cognitive processing. In: *The Organization of the Cerebral Cortex,* edited by F. O. Schmitt. Cambridge, MA: The MIT Press, 1981, p. 441–473.

DeVALOIS, R. L. Color vision mechanisms in the monkey. *J. Gen. Physiol.* 43 (Suppl. 6): 115–128, 1960.

DeVALOIS, R. L. Central mechanisms of color vision. In: *Handbook of Sensory Physiology,* vol. VII/3A: Central Visual Information, edited by R. Jung. Heidelberg, FRG: Springer, 1973, p. 209–253.

DeVRIES, H., AND S. STUIVER. The absolute sensitivity of the human sense of smell. In: *Sensory Communication,* edited by W. A. Rosenblith. Cambridge, MA: M.I.T. Press, 1961, p. 159–167.

DEWSON, J. H., III. Efferent olivocochlear bundle: some relationships to noise masking and to stimulus attenuation. *J. Neurophysiol.* 30: 817–832, 1967.

DEWSON, J. H., III. Efferent olivocochlear bundle: some relationships to stimulus discrimination in noise. *J. Neurophysiol.* 31: 122–130, 1968.

DINARELLO, C. A., AND S. M. WOLFF. Pathogenesis of fever in man. *N. Engl. J. Med.* 298: 607–612, 1978.

DONALDSON, I. M. L. The properties of some human thalamic units. Some new observations and a critical review of the localization of thalamic nuclei. *Brain* 96: 419, 1973.

DOTY, R. W., AND N. NEGRÃO. Forebrain commissures and vision. *Handbook of Sensory Physiology,* edited by R. Jung. Berlin: Springer Verlag, 1973, vol. 7, p. 543–582.

DOTY, R. W., AND W. H. OVERMAN, JR. Mnemonic role of forebrain commissures in macaques. In: *Lateralization in the Nervous System,* edited by S. Harnand, R. W. Doty, J. Jaynes, L. Goldstein, and G. Krauthamer. New York: Academic Press, 1977, p. 75–88.

DØVING, K. B., AND A. J. PINCHING. Selective degeneration of neurons in the olfactory bulb following prolonged odour exposure. *Brain Res.* 52: 115–129, 1973.

DØVING, K. B., AND R. SELSET. Behavior patterns in cod released by electrical stimulation of olfactory tract bundles. *Science* 207: 559–560, 1980.

DOWNER, J. L., DeC. Changes in visualgnostic functions and emotional behavior following unilateral temporal pole damage in the "split-brain" monkey. *Nature (London)* 191: 50–51, 1961.

DRACHMAN, D. B. The biology of myasthenia gravis. *Ann. Rev. Neurosci.* 4: 195–225, 1981.

DREIFUSS, J. J., I. KALNINS, J. S. KELLY, AND K. B. RUF. Action potentials and release of neurohypophysial hormones in vitro. *J. Physiol (London)* 215: 805–817, 1971.

DUKE-ELDER, S. *Textbook of Ophthalmology,* 2nd ed. St. Louis: Mosby, 1937, vol. 1.

EASON, R. G. Visual evoked potential correlates of early neural filtering during selective attention. *Bull. Psychonomic. Soc.* 18: 203–206, 1981.

EASON, R. G., M. OAKLEY, AND L. FLOWERS. Central neural influences on the human retina during selective attention. *Physiol. Psychol.* 11: 18–28, 1983.

ECONOMO, C., AND L. HORN. Ueber Windungsrelief, Masse und Rindenarchitektonik der Supratemporalflaeche, ihre individuellen und ihre Seil Unterschiede. *Zeitschr. Ges. Neurol. Psychiat.* 130: 678–757, 1930.

EDELMAN, G. M. Cell adhesion and morphogenesis: the regulator hypothesis. *Proc. Natl. Acad. Sci.* 81: 1460–1464, 1984.

EDELMAN, G. M., AND V. B. MOUNTCASTLE. The Mindful Brain: Cortical Organization and Group-selective Theory of Higher Brain Function. Cambridge, MA: The MIT Press, 1978.

EISENMAN, J. S. Pyrogen-induced changes in the thermosensitivity of septal and preoptic neurons. *Am. J. Physiol.* 216: 330–334, 1969.

ENGEN, T., J. E. KUISMA, AND P. E. EIMAS. Short term memory for odors. *J. Exp. Psychol.* 99: 222–225, 1973a.

ENGEN, T., AND B. M. ROSS. Long term memory for odors with and without verbal descriptions. *J. Exp. Psychol.* 100: 221–227, 1973b.

ENGSTRÖM, H. The first-order vestibular neuron. In: *Fourth Symposium on the Role of Vestibular Organs in Space Exploration, NASA SP-187*. Pensacola, FL: National Aeronautics Space Administration, 1968, p. 123–134.

ENGSTRÖM, H., H. W. ADES, AND J. E. HAWKINS. The vestibular sensory cells and their innervation. (In: *Modern Trends in Neuromorphology*, edited by J. Szentagothai.) *Symp. Biol. Hungary* 5: 21–41, 1965b.

EPSTEIN, A. N., D. SPECTOR, A. SAMMAN, AND C. GOLDBLUM. Exaggerated prandial drinking in the rat without salivary glands. *Nature (London)* 29: 1342–1343, 1964.

EPSTEIN, A. N., AND P. TEITELBAUM. Severe and persistent deficits in thirst produced by lateral hypothalamic damage. In: *Thirst*, edited by M. J. Wayner. Oxford, UK: Symposium Publications Division, Permagon Press, 1964, p. 395–410.

ERULKAR, S. D. Comparative aspects of spatial localization of sound. *Phyiol. Rev.* 52: 237–360, 1972.

EVANS, E. F. Effects of hypoxia on the tuning of single cochlear nerve fibers. *J. Physiol.* 238: 65–67, 1974.

EVANS, E. F. The dynamic range problem: place and time coding at the level of cochlear nerve and nucleus. In: *Neuronal Mechanisms of Hearing*, edited by J. Syka and L. Aitkin. New York: Plenum Press, 1981, p. 69–85.

FAMBROUGH, D. M., D. B. DRACHMAN, AND S. SATYAMURTI. Neuromuscular junction in myasthenia gravis: decreased acetylcholine receptors. *Science* 182: 293–295, 1973.

FEDERATION PROCEEDINGS SYMPOSIUM. Development of the autonomic nervous system, chaired by P. M. Gootman. *Fed. Proc.* 42: 1619–1655, 1983.

FEX, J. The olivocochlear feedback system. In: *Sensorineural Hearing Processes and Disorders*, edited by A. B. Graham. London: Churchill, 1968, p. 77–86.

FISCHER, B., AND G. F. POGGIO. Depth sensitivity of binocular cortical neurons of behaving monkeys. *Proc. Roy. Soc. (Biol.)* 204: 409–414, 1979.

FISCHETTE, C. T., A. BIEGON, AND B. S. McEWEN. Sex differences in serotonin 1 receptor binding in rat brain. *Science* 222: 333–335, 1983.

FISCHER, C., W. R. INGRAM, AND S. W. RANSON. *Diabetes Insipidus and the Neuro-hormonal Control of Water Balance: A Contribution to the Structure and Function of the Hypothalamico-Hypophyseal System*. Ann Arbor, MI: Edwards Brothers, 1983.

FISHER, S. K., AND R. H. STEINBERG. Origin and organization of pigment epithelial apical projections to cones in cat retina. *J. Comp. Neurol.* 206: 131–145, 1982.

FLOCK, A. Transducing mechanisms in the lateral line canal organ receptors. *Cold Spring. Harb. Symp. Quant. Biol.* 30: 133–144, 1965.

FLOCK, A. Sensory transduction in hair cells. In: *Handbook of Sensory Physiology*, vol. 1: Principles of Receptor Physiology, edited by W. R. Lowenstein. Berlin: Springer-Verlag, 1971, p. 396–441.

FOERSTER, O. The dermatomes in man. *Brain* 56: 1–39, 1933.

FOLKOW, F., AND E. RUBINSTEIN. Behavioural and autonomic patterns evoked by stimulation of the lateral hypothalamic area in the cat. *Acta Physiol. Scand.* 65: 292–299, 1965.

FRENCH, J. D., R. HERNANDEZ-PEON, AND R. B. LIVINGSTON. Projections from cortex to cephalic brain stem (reticular formation) in monkey. *J. Neurophysiol.* 18: 74–95, 1955.

FRENCH, J. D., B. E. GERNANDT, AND R. B. LIVINGSTON. Regional differences in seizure susceptibility in monkey cortex. *Arch. Neurol. Psychiat. (Chicago)* 75: 260–274, 1956.

FRENCH, N. R., AND J. C. STEINBERG. Factors governing the intelligibility of speech sounds. *J. Acoust. Soc. Am.* 19: 90–119, 1947.

FROMME, F. S., T. P. PIANTANIDA, AND D. H. KELLY. Psychophysical evidence for more than two kinds of cone in dichromatic color blindness. *Science* 215: 417–418, 1982.

FURUKAWA, T., AND Y. ISHII. Neurophysiological studies on hearing in gold fish. *J. Neurophysiol.* 30: 1377–1403, 1967a.

FURUKAWA, T., AND Y. ISHII. Effects of static bending of sensory hair cells on sound reception in the goldfish. *Jpn. J. Physiol.* 17: 572–588, 1967b.

FUSCO, M. M., J. D. HARDY, AND H. T. HAMMEL. Interaction of central and peripheral factors in physiological temperature regulation. *Am. J. Physiol.* 200: 572–580, 1961.

FUSTER, J. M., AND J. P. JERVEY. Neuronal firing in the inferotemporal cortex of monkey in a visual memory task. *J. Neurosci.* 2: 361–375, 1982.

GACEK, R. R. Anatomical evidence for an efferent vestibular pathway. In: *Third Symposium on the Role of the Vestibular Organs in Space Exploration*. National Aeronautics Space Administration, NASA, SP-152: 203–211, 1967.

GACEK, R. R., AND M. LYON. The localization of vestibular efferent neurons in the kitten with horseradish peroxidase. *Acta Otolaryngol. (Stockholm)* 77: 92–101, 1974.

GAGE, F. H., S. B. DUNNETT, U. STENEVI, AND A. BJOERKLUND. Aged rats: recovery of motor impairments by intrastriatal nigral grafts. *Science* 221: 966–968, 1983.

GALAMBOS, R. *Nerves and Muscles*. Garden City, NJ: Doubleday, 1962.

GALAMBOS, R. Processing of auditory information. In: *Brain and Behavior*, edited by M. A. B. Brazier. Washington, D.C.: Amer. Instit. Biol. Sci., 1961, vol. 1, p. 171–203.

GALAMBOS, R. Suppression of auditory nerve activity by stimulation of efferent fibers to cochlea. *J. Neurophysiol.* 19: 424–437, 1956.

GALAMBOS, R. The human auditory evoked response. In: *Sensation and Movement*, edited by H. R. Moskowitz. Dordrecht, Netherlands: Reidel Publishing Co., 1974, p. 215–221.

GALAMBOS, R., AND H. DAVIS. The response of single auditory nerve fibers to acoustic stimulation. *J. Neurophysiol.* 6: 39–57, 1943.

GALAMBOS, R., AND S. A. HILLYARD. Electrophysiological approaches to human cognitive processing. *Neurosci. Res. Prog. Bull.* 20: 141–265, 1981.

GALAMBOS, R., J. SCHWARTZKOPFF, AND A. RUPERT. A microelectrode study of superior olivary nuclei. *Am. J. Physiol.* 197: 527–536, 1959.

GALABURDA, A. M., F. SANIDES, AND N. GESCHWIND. Human brain, cytoarchitectonic, left-right asymmetries in the temporal speech regions. *Arch. Neurol.* 35: 812–817, 1978.

GANS, C., AND R. G. NORTHCUTT. Neural crest and the origin of vertebrates: a new head. *Science* 220: 268–273, 1983.

GARDNER, E. P., AND R. M. COSTANZO. Neuronal mechanisms underlying direction sensitivity of somatosensory cortical neurons in awake monkeys. *J. Neurophysiol.* 43: 1342–1354, 1980.

GASSER, H. S. Effect of method of leading on the recording of the nerve fiber spectrum. *J. Gen. Physiol.* 43: 927–940, 1960.

GAZZANIGA, M. S. *The Bisected Brain*, New York: Appleton-Century-Crofts, 1970.

GEISERT, E. E., P. D. SPEAR, S. R. ZETLAND, AND A. LANGSETMO. Recovery of Y-cells in the lateral geniculate nucleus of the monocularly deprived cat. *J. Neurosci.* 2: 577–588, 1982.

GERNANDT, B. E., M. IRANYI, AND R. B. LIVINGSTON.

Vestibular influences on spinal mechanisms. *Exp. Neurol.* 1: 248–273, 1959.

GERNANDT, B. E., Y. KATSUKI, AND R. B. LIVINGSTON. Functional organization of descending vestibular influences. *J. Neurophysiol.* 20: 453–469, 1957.

GERNANDT, B. E., AND S. GILMAN. Interactions between vestibular, pyramidal, and cortically evoked extrapyramidal activities. *J. Neurophysiol.* 23: 516–533, 1960a.

GERNANDT, B. E., AND S. GILMAN. Vestibular and propriospinal interactions and protracted spinal inhibition by brain stem activation. *J. Neurophysiol.* 23: 269–287, 1960b.

GERSHON, M. D. The enteric nervous system. *Ann. Rev. Neurosci.* 4: 227–272, 1981.

GERSHON, M. D., R. F. PAYETTE, AND R. P. ROTHMAN. Development of the enteric nervous system. *Fed. Proc.* 42: 1620–1625, 1983.

GESTELAND, R. C., J. Y. LETTVIN, AND W. H. PITTS. Chemical transmission in the nose of the frog. *J. Physiol.* 181: 525–559, 1965.

GIBSON, J. J. *The Senses Considered as Perceptual Systems.* Boston: Houghton Mifflin, 1966.

GLOOR, P. Inputs and outputs of the amygdala: what the amygdala is trying to tell the rest of the brain. In: *Limbic Mechanisms: The Continuing Evolution of the Limbic System Concept,* edited by K. E. Livingston and O. Hornykiewicz. New York: Plenum Press, 1976, p. 189–210.

GLOTZBACH, S. F., AND H. C. HELLER. Central nervous regulation of body temperature during sleep. *Science* 194: 537–539, 1976.

GODDARD, G. V. Development of epileptic seizures through brain stimulation at low intensity. *Nature (London)* 214: 1020–1021, 1967.

GODDARD, G. V., AND R. M. DOUGLAS. Does the engram of kindling model the engram of normal long term memory? In: *Kindling,* edited by J. A. Wada. New York: Raven Press, 1976, p. 1–18.

GODDARD, G. V., D. C. McINTYRE, AND C. K. LEECH. A permanent change in brain function from daily electrical stimulation. *Exp. Neurol.* 25: 295–330, 1969.

GODDARD, G. V., B. L. McNAUGHTON, R. M. DOUGLAS, AND C. A. BARNES. Synaptic change in the limbic system: evidence from studies using electrical stimulation with and without seizure activity. In: *Limbic Mechanisms: The Continuing Evolution of the Limbic System Concept,* edited by K. E. Livingston and O. Hornykiewicz. New York: Plenum Press, 1976, p. 355–368.

GOGEL, W. C., B. O. HARTMAN, AND G. S. HARKER. The retinal size of a familiar object as a determiner of apparent distance. *Psychol. Monogr.* 70: 1–16, 1957.

GOLDBERG, J. M., AND P. B. BROWN. Responses of binaural neurons of dog superior olivary complex to dichotic tonal stimuli: some physiological mechanisms of sound localization. *J. Neurophysiol.* 32: 613–636, 1969.

GOLDMAN-RAKIC, P. S. Neuronal development and plasticity of association cortex in primates. *Neurosci. Res. Prog. Bull.* 20: 471–479, 1982.

GOLDMAN-RAKIC, P. S., AND M. L. SCHWARTZ. Interdigitation of contralateral and ipsilateral columnar projections to frontal association cortex in primates. *Science* 216: 755–757, 1982.

GOLDSTEIN, M. H., JR., AND M. ABELES. Note on tonotopic organization of primary auditory cortex in the cat. *Brain Res.* 100: 188–191, 1975.

GOODHILL, V. *Beethoven: Triumph over Silence.* Composite 37 min, film. Los Angeles: UCLA Film Library, 1983.

GORDON, H. W., AND R. W. SPERRY. Lateralization of olfactory perception in the surgically separated hemispheres in man. *Neuropsychologia* 7: 111–120, 1969.

GORSKI, R. A., J. H. GORDON, J. E. SHRYNE, AND A. M. SOUTHAM. Evidence for a morphological sex difference within the medial preoptic area of the rat brain. *Brain Res.* 148: 333–346, 1978.

GORSKI, R., R. HARLAN, C. JACOBSON, J. SHRYNE, AND

A. SOUTHAM. Evidence for the existence of a sexually dimorphic nucleus in the preoptic area of the rat. *J. Comp. Neurol.* 193: 529–539, 1980.

GOTTSCHALDT, K-M., AND C. VAHLE-HINZ. Merkel cell receptors: structure and transducer function. *Science* 214: 183–186, 1981.

GOY, R. W., AND B. S. McEWEN. *Sexual Differentiation of the Brain.* Cambridge, MA: M.I.T. Press, 1980.

GRAHAM, A. B. (ed.). *Sensorineural Hearing Processes and Disorders.* Boston: Little, Brown, 1966.

GRAHAM, C. H. Visual space perception. In: *Vision and Visual Perceptions,* edited by C. H. Graham. New York: Wiley, 1965b, p. 504–507.

GRAHAM, C. H., K. E. BAKER, M. HECHT, AND V. V. LLOYD. Factors influencing thresholds for monocular movements parallax. *J. Exp. Psychol.* 38: 205–223, 1948.

GRAHAM, C. H., AND J. L. BROWN. Color contrast and color appearances: brightness constancy and color constancy. In: *Vision and Visual Perception,* edited by C. H. Graham. New York: Wiley, 1965, p. 452–478.

GRAHAM, C. H., H. G. SPERLING, Y. HSIA, AND A. H. COULSON. The determination of some visual functions of a unilaterally color-blind subject: method and results. *J. Psychol.* 51: 3–32, 1961.

GRASTYAN, E., G. KARMOS, L. VERECZKEY, AND L. KELLENYI. The hippocampal electrical correlates of the homeostatic regulation of motivation. *Electroenceph. Clin. Neurophysiol.* 21: 34–53, 1966.

GRAYBIEL, A., B. CLARK, AND J. J. ZARIELLO. Observation on human subjects living in a "slow rotation room" for periods of two days. *Arch. Neurol.* 3: 55–73, 1960.

GRAZIADEI, P. P. C. Functional anatomy of the mammalian chemoreceptor system. In: *Chemical Signals in Vertebrates,* edited by D. Mueller-Schwarze and M. M. Mozell. New York: Plenum Press, 1976.

GREER, C. A., W. B. STEWART, M. T. TEICHER, AND G. M. SHEPHERD. Functional development of the olfactory bulb and a unique glomerular complex in the neonatal rat. *J. Neurosci.* 2: 1744–1759, 1982.

GRIFFIN, D. R. *The Question of Animal Awareness: Evolutionary Continuity of Mental Experience.* New York: Rockefeller University Press, 1981.

GRODIN, F. S. *Control Theory and Biological Systems.* New York: Columbia University Press, 1963.

GROVES, P. M. A theory of functional organization of the neostriatum and the neostriatal control of voluntary movement. *Brain Res. Rev.* 5: 109–132, 1983.

GUEDRY, F. E. Psychophysiological studies of vestibular function. In: *Contributions to Sensory Physiology,* edited by W. D. Neff, New York: Academic Press, 1965.

GUTH, P. S., AND R. P. BOBBIN. The pharmacology of peripheral auditory processes; cochlear pharmacology. *Adv. Pharmacol. Chemotherap.* 9: 93–130, 1971.

HABER, R. N., AND L. S. NATHANSON. Processing of sequentially presented letters. *Percept. Psychophys.* 5: 359–361, 1969.

HABER, R. N., AND L. G. STANDING. Direct measures of short-term visual storage. *Q. J. Exp. Psychol.* 21: 43–54, 1969.

HAGBARTH, K.-E., AND A. B. VALLBO. Mechanoreceptor activity recorded percutaneously with semimicroelectrodes in human peripheral nerves. *Acta Physiol. Scand.* 69: 121–122, 1967.

HAINSWORTH, F. R. Saliva spreading, activity and body temperature regulation in the rat. *Am. J. Physiol.* 212: 1288–1292, 1967.

HALPERN, M. Nasal chemical senses in snakes. *Advances in Vertebrate Neuroethology,* edited by J. P. Evarts, R. R. Capranica, and D. J. Ingle. New York: Plenum Press, 1983, p. 141–176.

HAMMEL, H. T. Neurons and temperature regulation. In: *Physiological Controls and Regulations,* edited by W. S. Yamamoto and J. R. Brobeck. Philadelphia: Saunders, 1965, p. 71–97.

HAMMES, G. G. Receptor biophysics and biochemistry: enzymes. *Neurosci. Res. Prog. Bull.* 11: 164–175, 1973.

HARDIN, W. B., AND V. F. CASTELLUCI. Analysis of somatosensory, auditory and visual averaged transcortical and scalp responses in the monkey. *Electroencephal. Clin. Neurophysiol.* 28: 488–498, 1970.

HARDY, J. D. The "set-point" concept in physiological temperature regulation. In: *Physiological Controls and Regulations*, edited by W. S. Yamamoto and J. R. Brobeck. Philadelphia: Saunders, 1965, p. 98–116.

HARMAN, P. J. *Paleoneurologic, Neoneurologic, and Ontogenetic Aspects of Brain Phylogeny. James Arthur Lecture on the Evolution of the Human Brain.* New York: American Museum of Natural History, 1957, 24 pp.

HARRIS, D. A., AND E. HENNEMAN. Feedback signals from muscle and their efferent control. In: *Medical Physiology*, 14th ed., edited by V. B. Mountcastle. St. Louis: C. V. Mosby, 1980, vol. 1, p. 703–717.

HARTLINE, H. K. The response of single optic nerve fibers of the vertebrate eye to illumination of the retina. *Am. J. Physiol.* 121: 400–415, 1938.

HARTLINE, H. K., AND C. H. GRAHAM. Nerve impulses from single receptors in the eye. *J. Cell Physiol.* 1: 277–295, 1932.

HAUG, H. Remarks on the determination and significance of the gray cell coefficient. *J. Comp. Neurol.* 104: 473–492, 1956.

HAUG, H. *Quantitative Untersuchungen an der Sehrinde; die individuelle Schwankungsbreite beim Menschen, verbunden mit einigen Bemerkungen uber die Schizophrenie. Die Entwicklung der menschlichen Sehrinde; die Volumenverhaltnisse bei einigen Mammalia.* Stuggart, FRG: G. Thieme, 1958.

HEAD, H. On disturbances of sensation with especial reference to the pain of visceral disease. *Brain* 16: 1–133, 1893.

HEAD, H. *Aphasia and Kindred Disorders of Speech.* Cambridge, UK: Cambridge Univ. Press, 1962, 2 vols.

HECHT, S., S. SHLAER, AND M. H. PIRENNE. Energy, quanta and vision. *J. Gen. Physiol.* 25: 819–840, 1942.

HECHT, S., AND C. D. VERRIJP. Intermittent stimulation by light. III. The relation between intensity and critical fusion frequency for different retinal locations. *J. Gen. Physiol.* 17: 251–265, 1933.

HEIMER, L. The olfactory cortex and the ventral striatum. In: *Limbic Mechanisms: The Continuing Evolution of the Limbic System Concept*, edited by K. E. Livingston and O. Hornykiewicz. New York: Plenum Press, 1976, p. 95–187.

HEISTAD, D. D. The blood-brain barrier. *Fed. Proc. (Symp.)* 43: 185–219, 1984.

HELD, R., AND J. A. BAUER, JR.. Visually guided reaching in infant monkeys after restricted rearing. *Science* 155: 718–720, 1967.

HELMHOLTZ, H. VON. *Treatise on Physiological Optics.* Translated from the third German edition by James P. C. Southard. Rochester, NY: Optical Society of America, 1924, 3 vols.

HENRY, J. P., O. H. GAUER, AND J. L. REEVES. Evidence of the atrial location of receptors influencing urine flow. *Circ. Res.* 4: 85–90, 1956.

HENSEL, H. Correlations of neural activity and thermal sensation in man. In: *Sensory Functions of the Skin*, edited by Y. Zotterman. Oxford, UK: Permagon Press, 1976, p. 331–353.

HERING, E. *Beitraege zur Physiologie.* Leipzig, E. Germany: Engelmann, 1861.

HERNANDEZ-PEON, R. Neurophysiological correlates of habituation and other manifestations of plastic inhibition (internal inhibition). *Electroencephal. Clin. Neurophysiol.* 13 (Suppl): 101–114, 1960.

HERNANDEZ-PEON, R., H. SCHERRER, AND M. JOUVET. Modification of electric activity in cochlear nucleus during "attention" in unanesthetized cats. *Science* 123: 331–332, 1956.

HERRICK, C. J. *George Ellett Coghill, Naturalist and Philosopher.* Chicago: Univ. of Chicago Press, 1949.

HERRICK, C. J. *The Evolution of Human Nature.* Austin, TX: Univ. of Texas Press, 1956.

HESS, R., JR., W. P. KOELLA, AND K. AKERT. Cortical and subcortical recordings in natural and artificially induced sleep in cats. *Electroenceph. Clin. Neurophysiol.* 5: 75–90, 1953.

HESS, W. R. *Die Methodik der lokalisierten Reizung und Ausschaltung subkortikaler Hirnabschnitte.* Leipzig, E. Germany: Georg Thieme, 1932.

HESS, W. R. *Das Zwischenhirn, Syndrome, Lokalisationen, Funktionen.* Basel, Switzerland: Schwabe, 1949, p. 187.

HESS, W. R. *Hypothalamus and Thalamus, Experimental Documentation.* Stuttgart, FRG: Georg Thieme, 1956.

HEUSER, J. E., AND T. S. REESE. Evidence for recycling of synaptic vesicle membrane during transmitter release at the frog neuromuscular junction. *J. Cell Biol.* 57: 315–344, 1973.

HILGARD, E. R. *Divided Consciousness: Multiple Controls in Human Thought and Action.* New York: Raven Press, 1977.

HOBSON, J. A. Sleep: Physiologic aspects. *N. Engl. J. Med.* 281: 1343–1345, 1969.

HÖKFELT, T., O. JOHANSSON, A. LJUNGDAHL, J. M. LUNDBERG, AND M. SCHULTZERG. Peptidergic neurones. *Nature (London)* 284: 515–521, 1980.

HOLLEY, A., AND K. B. DØVING. Receptor sensitivity, acceptor distribution, convergence and neural coding in the olfactory system. In: *Olfaction and Taste*, edited by J. LeMagnen and P. MacLeod. London: Information Retrieval Ltd., 1977, vol. 6, p. 113–123.

HOLLINSHEAD, W. H. *Textbook of Anatomy*, 2nd ed. New York: Hoeber, Harper & Row, 1967, 994 pp.

HONRUBIA, V. Temporal and spatial distribution of the CM and SP of the cochlea. In: *Frequency Analysis and Periodicity Detection in Hearing*, edited by R. Plomp and G. F. Smoorenburg, Leiden, Netherlands: Sijthoff, 1970, p. 94–106.

HONRUBIA, V., AND M. A. B. BRAZIER. *Nystagmus and Vertigo: Clinical Approaches to the Patient with Dizziness.* New York: Academic Press, 1982.

HORTON, E. W. Hypotheses on physiological roles of prostaglandins. *Physiol. Rev.* 39: 122–161, 1969.

HOSOBUCHI, Y., J. E. ADAMS, AND R. LIPSHITZ. Pain relief by electrical stimulation of the central gray matter in humans and its reversal by naloxone. *Science* 197: 183–186, 1977.

HOUK, J. D. Regulation of stiffness by skeletomotor reflexes. *Ann. Rev. Physiol.* 41: 99–114, 1979..

HSIA, Y., AND C. H. GRAHAM. Color blindness. In: *Vision and Visual Perception*, edited by C. H. Graham. New York: Wiley, 1965.

HUBBELL, W. L., AND M. D. BOWNDS. Visual transduction in vertebrate photoreceptors. *Ann. Rev. Neurosci.* 2: 17–34, 1979.

HUBEL, D. H., AND T. N. WIESEL. Cells sensitive to binocular depth in area 18 of the macaque monkey cortex. *Nature (London)* 225: 41–42, 1970.

HUBEL, D. H., AND T. N. WIESEL. Brain mechanisms of vision. *Sci. Am.* 241: 150–162, 1979.

HUDSPETH, A. J. The hair cells of the inner ear. *Sci. Am.* 248: 54–56, 1983.

HUDSPETH, A. J., AND R. JACOBS. Stereocilia mediate transduction in vertebrate hair cells. *Proc. Natl. Acad. Sci. U.S.A.* 76: 1506–1509, 1979.

HUGHES, J., T. W. SMITH, H. W. KOSTERLITZ, L. A. FOTHERGILL, B. A. MORGAN, AND H. R. MORRIS. Leuenkephalin and Met-enkephalin-endogenous ligands for opiate receptors. *Nature (London)* 258: 577–579, 1975.

IGGO, A. Gastric mucosal chemoreceptors with vagal afferent fibers in the cat. *Q. J. Exp. Physiol.* 42: 398–409, 1957.

IGGO, A. Cutaneous receptors with a high sensitivity to mechanical displacement. In: *Touch, Heat and Pain*. Ciba Foundation Symposium. Boston: Little, Brown, 1966.

IMIG, T. J., AND H. D. ADRIAN. Binaural columns in the primary field (AI) of cat auditory cortex. *Brain Res.* 138: 241–257, 1977.

IMIG, T. J., AND R. A. REALE. Ipsilateral corticocortical projections related to binaural columns in cat primary auditory cortex. *J. Comp. Neurol.* 203: 1–14, 1981.

INGVAR, D. H., AND L. PHILIPSON. Distribution of cerebral blood flow in the dominant hemisphere during motor ideation and motor performance. *Ann. Neurol.* 2: 230–237, 1977.

INTERNATIONAL ORGANIZATION FOR STANDARDIZATION (ISO Standards Publication, 1964).

ITO, M. The control mechanisms of cerebellar motor systems. In: *The Neurosciences Third Study Program*, edited by F. O. Schmitt and F. G. Worden. Cambridge, MA: The MIT Press, 1974, p. 293–303.

IURATO, S. Efferent innervation of the cochlea. In: *The Handbook of Sensory Physiology*, edited by W. D. Keidel. New York, Springer Verlag, 1974, vol. 5, p. 259–282.

INVERSON, L. L. The catecholamines. *Nature (London)* 214: 8–14, 1967.

IVERSEN, L. L., R. A. NICOLL, AND W. W. VALE. Neurobiology of peptides. *Neurosci. Res. Prog. Bull.* 16: 209–330, 1978.

JACKSON, J. H. On a particular variety of epilepsy ("intellectual aura"), one case with symptoms of organic brain disease. *Brain* 11: 179–207, 1888.

JACKSON, J. H., AND W. S. COLMAN. Case of epilepsy with tasting movements and "dreamy state"—very small patch of softening in the left uncinate gyrus. *Brain* 21: 580–590, 1898.

JACOBSON, C. D., V. J. CSERNUS, J. E. SHRYNE, AND R. A. GORSKI. The influence of gonadectomy, androgen exposure, or a gonadal graft in the neonatal rat on the volume of the sexually dimorphic nucleus of the preoptic area. *J. Neurosci.* 1: 1142–1147, 1981.

JANTSCH, E. *The Self-Organizing Universe: Scientific and Human Implications of the Emerging Paradigm of Evolution*, edited by E. Laszlo. New York: Permagon Press, 1980.

JEWETT, D. L. Volume-conducted potentials in response to auditory stimuli as detected by averaging in the cat. *Electroencephalogr. Clin. Neurophysiol.* 28: 609–618, 1970.

JEWETT, D. L., AND J. S. WILLISTON. Auditory-evoked far fields averaged from the scalp of humans. *Brain* 94: 681–696, 1971.

JOHN, E. R. *Mechanisms of Memory*. New York: Academic Press, 1967; See also THATCHER, R. W., AND E. R. JOHN: *Foundations of Cognitive Processes*. Hillsdale, NJ: Erlbaum; New York: distributed by Halsted Press, 1977.

JOHN, E. R. Summary: symposium on memory transfer, American Association for the Advancement of Science, New York, December 1976. In: *Molecular Approaches to Learning and Memory*, edited by W. L. Byrne. New York: Academic Press, 1970, pp 335–342.

JOHNSON, W. H., R. A. STUBBS, G. F. KELK, AND W. R. FRANKS. Stimulus required to produce motion sickness. *J. Aviat. Med.* 22: 363–374, 1951.

JONGKEES, L. B. W. Pathology of vestibular sensation. In: *Handbook of Sensory Physiology*, vol. 6/2; edited by H. H. Kornhuber. Berlin: Springer Verlag, 1974, p 413–450.

JOUVET, M. Neurophysiology of the states of sleep. *Physiol. Rev.* 47: 117–177, 1967.

JOUVET, M. A possible role of catecholamine-containing neurons of the brain-stem of the cat in the sleep-waking cycle. *Acta Physiol. Pol.* 24: 5–19, 1973.

JUDD, D. B. Color perceptions of deuteranopic and protanopic observers. *J. Opt. Soc. Am.* 39: 252–256, 1948a.

JUDD, D. B. Color perceptions of deuteranopic and protanopic observers. *J. Res. Natl. Bur. Stand.* 4: 247–271, 1948b.

JULESZ, B. Binocular depth perception without familiarity cues. *Science* 145: 356–362, 1964.

JULESZ, B. Texture and visual perception. *Sci. Am.* 212: 38–48, 1965.

JULESZ, B., AND R. A. SCHUMER. Early visual perception. *Ann. Rev. Psychol.* 32: 575–627, 1981.

JULESZ, B., AND G. J. SPIVACK. Stereopsis based on vernier acuity cues alone. *Science* 157: 563–565, 1967.

KAAS, J. H., R. J. NELSON, M. SUR, AND M. M. MEZENICH. Multiple representations of the body in the postcentral somatosensory cortex of the primates. In: *Cortical Sensory Organization*, vol. 1: *Multiple Somatic Areas*, edited by N. Woolsey. Clifton, NJ: Humana, 1981, p. 29–45.

KANEKO, A. Physiology of the retina. *Ann. Rev. Neurosci.* 2: 169–191, 1979.

KARPLUS, J. P., AND A. KREIDL. Gehirn und sympathicus: II. Ein Sympathicuszetntrum im Zwischenhirn. *Pfluegers Arch.* 135: 401–416, 1971.

KATSUKI, Y. Neural mechanisms of auditory sensation in cats. In: *Sensory Communication*, edited by W. A. Rosenblith. Cambridge, MA: M.I.T. Press, 1961, p. 561–583.

KATZ, B. *Nerve, Muscle, and Synapse*. New York: McGraw-Hill Book Co., 1966.

KEEGAN, J. J., AND F. D. GARRETT. The segmental distribution of the cutaneous nerves in the limbs of man. *Anat. Rec.* 102: 409–437, 1948.

KENNEDY, C., M. DES ROSIERS, J. W. JEHLE, M. REIVICH, F. SHARP, AND L. SOKOLOFF. Mapping of functional neural pathways by autoradiographic survey of local metabolic rates with ^{14}C. *Science* 187: 850–853, 1975.

KERR, D. I. B., AND K-E. HAGBARTH. An investigation of olfactory centrifugal fiber system. *J. Neurophysiol.* 18: 362–374, 1955.

KETY, S. S. The central physiological and pharmacological effects of the biogenic amines and their correlations with behavior. In: *The Neurosciences: A Study Program*, edited by G. C. Quarton, T. Melnechuk, and F. O. Schmitt. New York: The Rockefeller University Press, 1967, p. 444–451.

KETY, S. S. The biogenic amines in the central nervous system: Their possible roles in arousal, emotion, and learning. In: *The Neurosciences: Second Study Program*, edited by F. O. Schmitt. New York: The Rockefeller University Press, 1974, p. 324–336.

KHANNA, S. M., AND D. G. B. LEONARD. Basilar membrane tuning in the cat cochlea. *Science* 215: 305–306, 1982.

KIANG, N. Y-S., T. WATANABE, E. C. THOMAS, AND L. F. CLARK. *Discharge Patterns of Single Fibers in the Cat's Auditory Nerve*. Cambridge, MA: The MIT Press, 1965.

KILLACKEY, H. P. Pattern formation in the trigeminal system of the rat. *Trends Neurosci.* 3: 303–306, 1980a.

KILLACKEY, H. P. Development and plasticity of somatosensory cortex. *Neurosci. Res. Prog. Bull.* 20: 507–513, 1982.

KIMURA, D. Some effects of temporal lobe damage on auditory perception. *Can. J. Psychol.* 15: 156–171, 1961a.

KIMURA, D. Cerebral dominance and the perception of verbal stimuli. *Can. J. Physiol.* 15: 166–171, 1961b.

KIMURA, D. Left-right differences in the perception of melodies. *Q. J. Exp. Psychol.* 16: 355–368, 1964.

KITAHATA, L. M., Y. AMAKATA, AND R. GALAMBOS. Effects of halothane upon auditory recovery function in cats. *J. Pharmacol. Exp. Therapeut.* 167: 14–25, 1969.

KLEIBER, M. *The Fire of Life*. New York: Wiley, 1961, 454 pp.

KLIMA, E. S., AND U. BELLUGI. *The Signs of Language*, Cambridge, MA: Harvard University Press, 1979.

KLING, A. Effects of amygdalectomy on social-affective behaviour in non-human primates. In: *Neurobiology of the Amygdala*, edited by B. E. Eleftheriou. New York: Plenum Press, 1972, p. 511–536.

KLING, A. Brain lesions and aggressive behavior of monkeys in free living groups. In: *Neural Bases of Violence and Aggression*, edited by W. S. Fields and W. H. Sweet. St. Louis: Warren H. Green, Inc., 1975.

KLINKE, R., AND N. GALLEY. Efferent innervation of vestibular and auditory receptors. *Physiol. Rev.* 54: 316–357, 1974.

KLUVER, H., AND P. C. BUCY. An analysis of certain effects of bilateral temporal lobectomy in the rhesus monkey, with special reference to "psychic blindness." *J. Psychol.* 5: 33–54, 1938.

KLUVER, H., AND P. C. BUCY. Preliminary analysis of functions of the temporal lobes in monkeys. *Arch. Neurol. Psychiat. (Chicago)* 42: 979–1000, 1939.

KNIBESTOL, M., AND A. B. VALLBO. Single unit analysis of mechanoreceptors activity from the human glabrous skin. *Acta Physiol. Scand.* 80: 178–195, 1970.

KOIZUMI, K., T. ISIKAWA, AND C. McC. BROOKS. Control of activity of neurons in the supraoptic nucleus. *J. Neurophysiol.* 27: 878–892, 1964.

KOIZUMI, K., AND H. YAMASHITA. Influence of atrial stretch

receptors on hypothalamic neurosecretory neurons. *J. Physiol. (Lond.)* 285: 341–358, 1978.

KONORSKI, J. *Conditioned Reflexes and Neuron Organization.* Cambridge, UK: Cambridge University Press, 1948; See also: KONORSKI, J. *Integrative Activity of the Brain. An Interdisciplinary Approach.* Chicago: University of Chicago Press, 1967.

KORNER, P. I. Integrative neural cardiovascular control. *Physiol. Rev.* 51: 312–367, 1971.

KOSHLAND, D. E., JR. Bacterial chemotaxis in relation to neurobiology. *Ann. Rev. Neurosci.* 3: 43–75, 1980.

KOSHLAND, D. E., JR., A. GOLDBETER, AND J. B. STOCK. Amplification and adaptation in regulatory and sensory system. *Science* 217: 220–225, 1982.

KOSTREVA, D. R. Functional mappings of cardiovascular reflexes and the heart using 2-D deoxyglucose. *Physiologist* 26: 333–350, 1983.

KOSTYUK, P. G., AND D. A. VASILENKO. Spinal interneurons. *Ann. Rev. Physiol.* 41: 115–126, 1979.

KRUGER, L. Information processing in cutaneous mechanoreceptors: feature extraction at the periphery. Introduction: an overview. *Fed. Proc.* 42: 2519–2520, 1983; whole Symposium, *Fed. Proc.* 42: 2519–2552, 1983.

KRUGER, L., E. R. PERL, AND M. J. SEDIVEC. Fine structure of myelinated mechanical nociceptor endings in cat hairy skin. *J. Comp. Neurol.* 198: 137–154, 1981.

KUENZLE, H., AND A. AKERT. Efferent connections of cortical area 8 (frontal eye field) in *Macaca fascicularis*. A reinvestigation using the autoradiographic technique. *J. Comp. Neurol.* 17: 147–164, 1972.

KUFFLER, S. W., AND J. G. NICHOLLS. *From Neuron to Brain: A Cellular Approach to the Function of the Nervous System.* Sunderland. MA: Sinauer Assoc., 1976.

KUHL, D. E., J. ENGEL, JR., M. E. PHELPS, AND C. SELIN. Epileptic patterns of local cerebral metabolism and perfusion in humans determined by emission computed tomography of 18 FDG and 13 NH3. *Ann. Neurol.* 8: 348–360, 1980.

KUPPERMAN, R. The dynamic DC potentials in the cochlea of the guinea pig. *Acta Otolaryngol.* 62: 465–480, 1966.

KUPPERMAN, R. The SP connection with movements of the basilar membrane. In: *Frequency Analysis and Periodicity Detection in Hearing*, edited R. Plomp and G. F. Smoorenburg. Leiden, Netherlands: Sijthoff, 1970, p. 126–133.

KUTAS, M., AND S. A. HILLYARD. Reading senseless sentences: brain potentials reflect semantic incongruity. *Science* 207: 203–205, 1980.

KUYPERS, H. G., C. E. CATSMAN-BERREVOETS, AND R. R. PADT. Retrograde axonal transport of fluorescent substances in the rat's forebrain. *Neurosci. Lett.* 6: 127–135. 1977.

LAND, L. J. Localized projection of olfactory nerves to rabbit olfactory bulb. *Brain Res.* 63: 153–166, 1973.

LAND, L. J., R. P. EAGER, AND G. M. SHEPHERD. Olfactory nerve projections to the olfactory bulb in rabbit: demonstration by means of a simple ammoniacal silver degeneration method. *Brain Res.* 23: 250–254, 1971.

LANDIS, S. C. Developmental neurobiology. *Science* 219: 489–490, 1983.

LANDMESSER, L. T. The generation of neuromuscular specificity. *Ann. Rev. Neurosci.* 3: 179–302, 1980.

LASHLEY, K. S. In search of the engram. *Symp. Soc. Exp. Biol.* 4: 454–482, 1950.

LEE, C. Y. *Chemistry and Pharmacology of Polypeptide Toxins in Snake Venoms.* Pharmacological Institute, College of Medicine, National Taiwan University, Taipei, Taiwan, China, 1972, p. 265–286.

LEE, C-M., J. A. JAVITCH, AND S. H. SNYDER. Recognition sites for norepinephrine uptake: regulation by neurotransmitter. *Science* 220: 626–629, 1983.

LEIBOWITZ, S. F. Hypothalamic alpha- and beta-adrenergic systems regulate both thirst and hunger in the rat. *Proc. Natl. Acad. Sci. USA.* 68: 332–334, 1971.

LENNEBERG, E. H. Brain correlates of language. In: *The Neurosciences Second Study Program*, edited by F. O. Schmitt. New York: Rockefeller Univ. Press, 1970, p. 361–371.

LeVAY, S., T. N. WIESEL, AND D. H. HUBEL. The development of ocular dominance columns in normal and visually deprived monkeys. *J. Comp. Neurol.* 191: 1–51, 1980.

LeVAY, S., AND H. SHERK. The visual claustrum of the cat. I. Structure and connections. *J. Neurosci.* 1: 956–980, 1981.

LEVI-MONTALCINI, R. The nerve growth factor: its mode of action on sensory and sympathetic cells. *Harvey Lectures Ser.* 60: 217–259, 1966.

LEVY, J., C. TREVARTHEN, AND R. W. SPERRY. Perception of bilateral chimeric figures following deconnection. *Brain* 95: 61–78, 1972.

LEWIS, E. R., AND C. W. COTMAN. Neurotransmitter characteristics of brain grafts: striatal and septal tissues form the same laminated input to the hippocampus. *Neurosci.* 8: 57–66, 1983.

LIBERMAN, M. C. Auditory nerve responses from cats raised in a low-noise chamber. *J. Acoust. Soc. Amer.* 63: 442–455, 1978.

LICKLIDER, J. C. R. Basic correlates of the auditory stimulus. In: *Handbook of Experimental Psychology*, edited by S. S. Stevens. New York: Wiley, 1951, p. 985–1039.

LINCOLN, D. W., AND J. B. WAKERLY. Factors governing the periodic activation of supraoptic and paraventricular neurosecretory cells during suckling in the rat. *J. Physiol. (Lond.)* 250: 443–461, 1975.

LINDSLEY, D. B., J. W. BOWDEN, AND H. W. MAGOUN. Effect upon the EEG of acute injury to the brain stem activating system. *Electroencephal. Clin. Neurophysiol.* 1: 475–486, 1949.

LINDSLEY, D. B., L. H. SCHREINER, W. B. KNOWLES, AND H. W. MAGOUN. Behavioral and EEG changes following brain stem lesions in the cat. *Electroencephal. Clin. Neurophysiol.* 2: 483–498, 1950.

LINDSTROM, J. M., M. E. SEYBOLD, V. A. LENNON, S. WHITTINGHAM, AND D. D. DUANE. Antibody to acetylcholine receptor in myasthenia gravis: prevalence, clinical correlates, and diagnostic value. *Neurology* 26: 1054–1059, 1976.

LISBERGER, S. G., F. A. MILES, AND L. M. OPTICAN. Frequency-selective adaptations: evidence for channels in the vestibulo-ocular reflex? *J. Neurosci.* 3: 1234–1244, 1983.

LIU, C.-N., AND W. W. CHAMBERS. Intraspinal sprouting of dorsal root axons. *Arch. Neurol. Psychiat.* 79: 46–61, 1958.

LIVINGSTON, K. E., AND O. HORNYKIEWICZ (eds.). *Limbic Mechanisms: The Continuing Evolution of the Limbic Concept.* New York: Plenum Press, 1978, 542 pp.

LIVINGSTON, R. B. Reinforcement. In: *The Neurosciences Third Study Program*, edited by G. C. Quarton, T. Melnechuk, and F. O. Schmitt. New York: Rockefeller University Press, 1967, p. 568–577.

LIVINGSTON, R. B. Brain circuitry relating to complex behavior. In: *The Neurosciences Third Study Program*, edited by G. C. Quarton, T. Melnechuk, and F. O. Schmitt. New York: Rockefeller Univ. Press, 1967, p. 449–515.

LIVINGSTON, R. B. Neural integration. In: *Pathophysiology. Altered Regulatory Mechanisms in Disease*, edited by E. D. Frohlich. Philadelphia: J. B. Lippincott, 1972, p. 569–598.

LIVINGSTON, R. B. *Sensory Processing, Perception, and Behavior.* New York: Raven Press, 1978.

LIVINGSTON, R. B., J. F. FULTON, J. M. R. DELGADO, E. SACHS, JR., S. J. BRENDLER, AND G. D. DAVIS. Stimulation and regional ablation of orbital surface of frontal lobe. *Res. Publ. Assoc. Res. Nerv. Ment. Dis.* 27: 405–420, 1947.

LIVINGSTON, R. B., W. P. CHAPMAN, K. E. LIVINGSTON, AND L. KRAINTZ. Stimulation of orbital surface of man prior to frontal lobotomy. *Res. Publ. Assoc. Res. Nerv. Ment. Dis.* 27: 421–432, 1947.

LIVINGSTON, R. B., K. PFENNINGER, H. MOOR, AND K. AKERT. Specialized paranodal and interperinodal glial-axonal junctions in the peripheral and central nervous system: a freeze-etching study. *Brain Res.* 58: 1–24, 1973.

LIVINGSTON, W. K. *Pain Mechanisms.* New York: MacMillan, 1943; Republished, with a Forward by R. Melzack. New York: Plenum Press, 1976.

LIVINGSTONE, M. S., AND D. H. HUBEL. Anatomy and physiology of a color system in the primate visual cortex. *J. Neurosci.* 4: 309–356, 1984.

LOH, H. H., L. F. TSENG, E. WEI, AND C. H. LI. Beta-endorphin is a potent analgesic agent. *Proc. Natl. Acad. Sci. USA* 83: 2895–2898, 1976.

LOWENSTEIN, W. R. Modulation of cutaneous mechanoreceptors by sympathetic stimulation. *J. Physiol.* 132: 40–60, 1956.

LOWENSTEIN, W. R., AND M. MENDELSON. Components of receptor adaptation in a Pacinian corpuscle. *J. Physiol.* 177: 377–397, 1965.

LUND, R. D., AND J. S. LUND. Synaptic adjustment after deafferentation of the superior colliculus of the rat. *Science* 171: 804–807, 1971.

LYNCH, G., S. DEADWYLER, AND C. COTMAN. Post lesion axonal growth produces permanent functional connections. *Science* 180: 1364–1366, 1973.

MacLEAN, P. D. Psychosomatic disease and the "visceral brain." Recent developments bearing on the Papez theory of emotion. *Psychosom. Med.* 11: 338–353, 1949.

MacLEAN, P. D. The limbic system with respect to self-preservation and the preservation of the species. *J. Nerv. Ment. Dis.* 127: 1–11, 1958.

MacLEAN, P. D. The brain in relation to empathy and medical education. *J. Nerv. Ment. Dis.* 144: 374–382, 1967.

MacLEAN, P. D. The co-evolution of the brain and family. In: *Anthroquest,* The L. S. B. Leakey Foundation News, No. 24, p. 1, 14–15. Pasadena, CA: The L. S. B. Leakey Foundation, Winter 1982.

MacLEAN, P. D. The internal-external bonds of the memory process. *J. Nerv. Ment. Dis.* 149: 40–47, 1969a.

MacLEAN, P. D. The limbic brain in relation to the psychoses. In: *Physiological Correlates of Emotion,* edited by P. Black. New York: Academic Press, 1970, p. 130–146.

MacLEAN, P. D. An ongoing analysis of hippocampal inputs and outputs: microelectrode and anatomic findings in the squirrel monkey. In: *The Hippocampus,* vol. I: *Structure and Development,* edited by R. L. Isaacson and K. H. Pribram. New York: Plenum Press, 1975, p. 177–211.

MacNICHOL, E. F., JR., AND G. SVAETICHIN. Electric responses from the isolated retinas of fishes. *Am. J. Ophthal.* 46: 26–40, 1958.

MACRIDES, F. Dynamic aspects of central olfactory processing. In: *Chemical Signals in Vertebrates,* edited by D. Mueller-Schwarze and M. M. Mozell. New York: Plenum Press, 1976, p. 499–514.

MACRIDES, F., A. BARTKE, AND B. SVARE. Interactions of olfactory stimuli and gonadal hormones in the regulation of rodent social behavior. In: *Olfaction and Taste IV,* edited by J. Le Magnen and P. MacLeod. London: Information Retrieval Ltd, 1977, p. 143–147.

MAGNUS, R. *Koerperstellung,* Berlin: Springer, 1924.

MAGOUN, H. W. Caudal and cephalic influences of the brain stem reticular formation. *Physiol. Rev.* 30: 459–474, 1950.

MAGOUN, H. W. The ascending reticular activating system. *Organization in the Central Nervous System* 30: 480–492, 1952.

MAGOUN, H. W. An ascending reticular activating system in the brain stem. *Arch. Neurol. Psychiat.* 67: 145–154, 1952.

MAGOUN, H. W. *The Waking Brain.* Springfield, IL: Charles C Thomas, 1958.

MAGOUN, H. W., F. HARRISON, J. R. BROBECK, AND S. W. RANSON. Activation of heat loss mechanisms by local heating of the brain. *J. Neurophysiol.* 1: 101–114, 1938.

MAREN, T. H., AND L. E. BRODER. The role of carbonic anhydrase in anion secretion into cerebrospinal fluid. *J. Pharmacol. Exp. Ther.* 172: 197–202, 1970.

MARR, D. *Vision.* San Francisco: W. H. Freeman, 1982.

MARTIN, C. Thermal adjustment of man and animals to external conditions. *Lancet* 2: 561–567; 617–620; 673–678, 1930.

MARTINSON, J., AND A. MUREN. Excitatory and inhibitory effects of vagus stimulation on gastric motility in the cat. *Acta Physiol. Scand.* 57: 309–316, 1963.

MASERA, R. Sul'esistenza di un particolare organo olfatico nel setto nasale della cavia e di altri roditori. *Arch. Ital. Anat.*

Embriol. 48: 157–212, 1943.

MATHEWS, D. F. Response patterns of single neurons in the tortoise olfactory epithelium and olfactory bulb. *J. Gen. Physiol.* 60: 166–180, 1972.

MATTHEWS, P. B. C. Where does Sherrington's "muscular sense" originate? Muscles, joints, corollary discharges? *Ann. Rev. Neurosci.* 5: 189–218, 1982.

MATURNA, H. R., J. Y. LETTVIN, W. S. McCULLOCH, AND W. H. PITTS. Anatomy and physiology of vision in the frog (Rana pipens). *J. Gen. Physiol.* 43: 129–175, 1960.

MAYNE, R. A systems concept of the vestibular organs. In: *Handbook of Sensory Physiology VI/2,* edited by H. H. Kornhuber. Berlin: Springer Verlag, 1974, p. 494–580.

McEWEN, B. S. Steroid receptors in neuroendocrine tissues: topography, subcellular distribution and functional implications. In: *International Symposium on Subcellular Mechanisms in Reproductive Neuroendocrinology,* edited by F. Naftolin, K. J. Ryan, and J. Davies. Amsterdam: Elsevier, 1976, p. 277–304.

McKINLEY, R. W. (ed.). *IES Lighting Handbook,* New York: Illuminating Engineering Society, 1947.

MEIRY, J. L. *The Vestibular System and Human Dynamic Space Orientation.* National Aeronautics Space Administration, NASA Cr-628 (Doctoral thesis, M.I.T. June 1965), 1966.

MELLONI, B. *The Internal Ear, An Atlas of Some Pathological Conditions of the Eye, Ear and Throat.* Chicago: Abbott Laboratories, 1957, p. 26–31.

MELVILL JONES, G. Plasticity in the adult vestibulo-ocular reflex arc. *Phil. Trans. Roy. Soc. (Biol)* 278: 319–334, 1977.

MELZACK, R., AND F. P. HAUGEN. Responses evoked at the cortex by tooth stimulation. *Am. J. Physiol.* 190: 570–574, 1957.

MELZACK, R., W. A. STOTLER, AND W. K. LIVINGSTON. Effects of discrete brainstem lesions in cats on perception of noxious stimulation. *J. Neurophysiol.* 21: 353–367, 1958.

MEREDITH, M., D. M. MARQUES, R. D. O'CONNELL, AND F. L. STERN. Vomeronasal pump: significance for male hamster sexual behavior. *Science* 207: 1224–1226, 1980.

MERZENICH, M. M., AND J. H. KAAS. Principles of organization of sensory-perceptual systems in mammals. In: *Progress in Psychobiology and Physiological Psychology,* edited by J. M. Sprague and A. N. Epstein. New York: Academic Press, 1975, p. 2–43.

MESULAM, M-M. A cortical network for directed attention and unilateral neglect. *Ann. Neurol.* 10: 309–325, 1981.

MEYER, J. S., L. A. HAYMAN, T. AMANO, S. NAKAJIMA, T. SHAW, P. LAUZON, S. DERMAN, E. KARACAN, AND Y. HARATI. Mapping local blood flow of human brain by CT scanning during stable xenon inhalation. *Stroke* 12: 426–435, 1981.

MILES, F. A. Centrifugal control of the avian retina. IV. Effects of reversible cold block of the isthmooptic tract on the receptive field properties of cells in the retina and isthmooptic nucleus. *Brain Res.* 48: 131–145, 1972.

MILES, F. A., AND S. G. LISBERGER. Plasticity in the vestibulo-ocular reflex: a new hypothesis. *Ann. Rev. Neurosci.* 4: 273–299, 1981.

MILLER, E. F., AND A. GRAYBIEL. Otolith function as measured by ocular counterrolling. In: *The Role of the Vestibular Organs in Space Exploration.* National Aeronautics Space Administration, NASA SP-77: 121–130, 1965.

MILLER, E. F., JR., A. GRAYBIEL, AND R. S. KELLOG. Otolith organ activity within earth standard, one-half standard and zero gravity environments. *Aerosp. Med.* 37: 399–403, 1966.

MILLER, N. E. Learning of visceral and glandular responses. *Science* 163: 434–445, 1969.

MILLER, N. E., AND B. R. DWORKIN. Different ways in which learning is involved in homeostasis. In: *Neural Mechanisms of Goal-Directed Behavior and Learning,* New York: Academic Press, Inc., 1980.

MILLS, E., AND P. G. SMITH. Functional development of the cervical sympathetic pathway in the neonatal rat. *Fed. Proc.* 42: 1639–1642, 1983.

MILNER, B. Hemispheric specialization: scope and limits. In: *The Neuroscience Third Study Program*, edited by F. O. Schmitt and F. G. Worden. Cambridge, MA: The MIT Press, 1974, p. 75–89.

MITCHISON, G., AND F. CRICK. Long axons within the striate cortex: their distribution, orientation, and patterns of connection. *Proc. Natl. Acad. Sci. U.S.A.* 79: 3661–3665, 1982.

MOGENSON, G. J., AND Y. H. HUANG. The neurobiology of motivated behavior. In: *Progress in Neurobiology*, 1973, vol. I, p. 53–83.

MONTEMURRO, D. G., AND J. A. F. STEVENSON. The localization of hypothalamic structures in the rat influencing water consumption. *Yale J. Biol. Med.* 28: 396–403, 1955–1956.

MOORE, D. M., S. F. CHEUK, J. D. MORTON, R. D. BERLIN, AND W. B. WOOD, JR. Studies on the pathogenesis of fever. XVII. Activation of leucocytes for pyrogen production. *J. Exp. Med.* 131: 179–188, 1970.

MOORE, E. J. (Ed.). *Bases of Auditory Brain-Stem Evoked Responses.* New York: Grune & Stratton, 1983.

MOORE, M. J., AND D. M. CASPAERY. Strychnine blocks binaural inhibition in lateral superior olivary neurons. *J. Neurosci.* 3: 237–242, 1983.

MOORE, R. Y. Visual pathways and the central neural control of diurnal rhythms. In: *The Neurosciences Third Study Program*, edited by F. O. Schmitt and F. G. Worden. Cambridge, MA: The MIT Press, 1974, p. 537–542.

MOORE, R. Y. Monoamine neurons innervating the hippocampal formation and septum: organization and response to injury. In: *Hippocampus*, edited by R. L. Issacson and K. H. Pribram. New York: Plenum Press, 5: 215–237, 1975.

MOREST, D. K. Auditory neurons of the brain stem. *Adv. Oto Rhino Laryngol.* 20: 337–356, 1973.

MORRELL, F. Electrical signs of sensory coding. In: *The Neurosciences, A Study Program*, edited by G. C. Quarton, T. Melnechuk, and F. O. Schmitt. New York: Rockefeller University Press, 1967, p. 452–469.

MORUZZI, G., AND H. W. MAGOUN. Brain stem reticular formation and activation of the E. E. G. *Electroencephal. Clin. Neurophysiol.* 1: 455–473, 1949.

MOSKOWITZ, N., W. SCHOOK, AND S. PUSZKIN. Interaction of brain synaptic vesicles induce by endogenous CA^{2-} dependent phospholipase A_2. *Science* 216: 305–307, 1982.

MOULTON, D. G. Electrical activity in the olfactory system of rabbits with indwelling electrodes. In: *Olfaction and Taste.* New York: Macmillan, 1963.

MOULTON, G. Minimum odorant concentrations detectable by the dog and their implication for olfactory receptor sensitivity. In: *Chemical Signals in Vertebrates*, edited by D. Mueller-Schwarze and M. M. Mozell. New York: Plenum Press, 1976, p. 455–464.

MOUNTCASTLE, V. B. The view from within: pathways to the study of perception. *John Hopkins Med. J.* 136: 109–131, 1975.

MOUNTCASTLE, V. B. The world around us: neural command functions for selective attention. The F. O. Schmitt Lecture in Neuroscience for 1975. *Neurosci. Res. Prog. Bull.* 14: 1–47, 1976.

MOUNTCASTLE, V. B., J. C. LYNCH, A. GEORGOPOULOS, H. SAKATA, AND C. ACUNA. Posterior parietal association cortex of the monkey: command functions for operations within extrapersonal space. *J. Neurophysiol.* 38: 871–908, 1975.

MOVSHON, J. A., AND R. C. VAN SLYTERS. Visual neural development. *Ann. Rev. Psychol.* 32: 477–522, 1981.

MOZELL, M. M. Processing of olfactory stimuli at peripheral levels. In: *Chemical Signals in Vertebrates*, edited by D. Mueller-Schwarze and M. M. Mozell. New York: Plenum Press, 1976, p. 465–482.

MUELLER-SCHWARZE, D. Complex mammalian behavior and pheromone bioassay in the field. In: *Chemical Signals In Vertebrates*, edited by D. Mueller-Schwarze and M. M. Mozell. New York: Plenum Press, 1976, p. 413–433.

MUELLER-SCHWARZE, D., AND M. M. MOZELL (eds.). *Chemical Signals in Vertebrates.* New York: Plenum Press, 1976.

MÜLLER, G. E., AND A. PILZECKER. Experimentelle Beitraege zur Lehre vom Gedaechtnis. *Z. Psychol. Physiol. Sinnesorg.* (Suppl. 1): 1–300, 1900.

MÜLLER, J. *Beitraege zur vergleichenden Physiologie des Gesichtsinnes.* Leipzig, E. Germany: Knoblock, 1826.

MURRAY, M., AND M. E. GOLDBERGER. Restitution of function and collateral sprouting in the cat spinal cord: the partially hemisected animal. *J. Comp. Neurol.* 155: 19–36, 1974.

NACHMAN, M., AND L. P. COLE. Role of taste in specific hungers. *Handbook of Sensory Physiology*, vol. IV/2: *Chemical Senses*, edited by L. M. Beidler. Heidelberg, FRG: Springer-Verlag, 1971, p. 337–362.

NAKAYAMA, T., J. S. EISENMAN, AND J. D. HARDY. Single unit activity of anterior hypothalamus during local heating. *Science* 134: 560–561, 1961.

NAUTA, W. J. H. Hypothalamic regulation of sleep in rats. An experimental study. *J. Neurophysiol.* 9: 285–316, 1946.

NELSON, R. J., M. SUR, D. J. FELDMAN, AND J. H. KAAS. Representations of the body surface in postcentral parietal cortex of Macaca fascicularis. *J. Comp. Neurol.* 192: 611–643, 1980.

NEVILLE, H. J., AND S. L. FOOTE. Auditory event-related potentials in the squirrel monkey: parallel to human late wave responses. *Brain Res.* 298: 107–116, 1984; see also: GALAMBOS, R., AND S. A. HILLYARD. Electrophysiological approaches to human cognitive processes. *Neurosci. Res. Prog. Bull.* 20: 240–246, 1981.

NEWSOM, B. D., J. F. BRADY, W. A. SHAFER, AND S. FRENCH. Adaptation to prolonged exposures in the revolving space station simulator. *Aerosp. Med.* 37: 778–783, 1966.

NIEMYER, W., AND I. STARLINGER. Do the blind hear better? *Audiology* 20: 510–515, 1981.

NILSSON, S. E. G. Receptor cell outer segment development and ultrastructure of the disk membranes in the retina of the tadpole (Rana pipens). *J. Ultrastruct. Rev.* 11: 581–620, 1964.

NISHIHARA, H. K., AND T. POGGIO. Hidden cues in random-line stereograms. *Nature (London)* 300: 347–349, 1982.

NORGREN, R. A synopsis of gustatory neuroanatomy. In: *Olfaction and Taste VI*, edited by J. LeMagnen and P. MacLeod. London: Information Retrieval, Ltd., 1977, p. 225–232.

NORGREN, R. Projections from the nucleus of the solitary tract in the rat. *Neuroscience* 3: 207–218, 1978.

NORGREN, R. (1980; See BECKSTEAD, et al., 1982). OGLE, K. N. *Researches in Binocular Vision.* Philadelphia: Saunders, 1950.

OLDS, J. Self-stimulation of the brain. *Science* 127: 315–324, 1958.

OLDS, J., AND P. MILNER. Positive reinforcement produced by electrical stimulation of septal area and other regions of rat brain. *J. Comp. Physiol. Psychol.* 47: 419–427, 1954.

OLIVER, D. L., AND W. C. HALL. The medial geniculate body of the tree shrew, Tupuaia glis. I. Cytoarchitecture and midbrain connections. *J. Comp. Neurol.* 182: 423–458, 1978.

OLIVER, D. L., AND W. C. HALL. The medial geniculate body of the tree shrew, Tupuaia glis. II. Connections with the neocortex. *J. Comp. Neurol.* 182: 459–493, 1978.

OPTICAL SOCIETY of AMERICA. *The Science of Color.* New York: Crowell, 1953.

OSCARSSON, O. Functional organization of spinocerebellar paths. In: *Handbook of Sensory Physiology.* Vol. 2: *Somatosensory System*, edited by A. Iggo. Berlin: Springer, 1973, p. 339–380.

OSTERBERG, G. A. Topography of the layer of rods and cones in the human retina. *Acta Ophthalmol.* 13 (Suppl. VI): 103 pp., 1935.

OTTOSON, D. Analysis of electrical activity of the olfactory epithelium. *Acta Physiol. Scan.* 35: (Suppl. 122): 1–83, 1956.

OTTOSON, D. The electro-olfactogram. In: *Handbook of Sensory Physiology: Olfaction*, edited by L. M. Beidler. New York: Springer Verlag, 1971, p. 205–215.

OTTOSON, D., AND G. M. SHEPHERD. Experiments and concepts in olfactory physiology. In: *Sensory Mechanisms*, edited by Y. Zotterman. Amsterdam: Elsevier, 1967, p. 83–138.

PAPEZ, J. W. A proposed mechanism of emotion. *Arch. Neurol. Psychiat.* 38: 725–743, 1937.

PENFIELD, W., AND H. H. JASPER. *Epilepsy and the Functional Anatomy of the Human Brain.* Boston: Little, Brown, 1954, 896 pp.

PENFIELD, W., AND L. ROBERTS. *Speech and Brain-Mechanisms.* Princeton, NJ: Princeton University Press, 1959; reprinted 1966.

PERKEL, D. H., AND T. H. BULLOCK. Neural coding. *Neurosci. Res. Prog. Bull.* 6: 221–348, 1968.

PETERS, A., S. L. PALAY, AND H. DE F. WEBSTER. *The Fine Structure of the Nervous System.* New York: Harper & Row, 1970.

PETRAS, J. M., AND J. F. CUMMINGS. Autonomic neurons in the spinal cord of the rhesus monkey: a correlation of the findings of cytoarchitectonics and sympathectomy with fiber degeneration following dorsal rhizotomy. *J. Comp. Neurol.* 146: 189–218, 1972.

PFAFFMAN, C. Physiological and behavioural process of the sense of taste. In: *Taste and Smell in Vertebrates*, Ciba Foundation. London: J. & A. Churchill, 1970, p. 31–50.

PHELPS, M. E., J. C. MAZZIOTA, AND D. E. KUHL. Metabolic mapping of the brain's response to visual stimulation: studies in man. *Science* 211: 1445–1448, 1981.

PICTON, T. W., S. A. HILLYARD, H. I. KRAUS, AND R. GALAMBOS. Human auditory evoked potentials. I: Evaluation of components. *Electroencephal. Clin. Neurophysiol.* 36: 170–190, 1974.

POGGIO, G. F., F. H. BAKER, Y. LAMARRE, AND E. RIVA SANSEVERINO. Afferent inhibition at input to visual cortex of the cat. *J. Neurophysiol.* 32: 892–915, 1969.

POGGIO, G. F., AND B. FISHER. Binocular interaction and depth sensitivity of striate and prestriate cortical neurons of the behaving rhesus monkey. *J. Neurophysiol.* 40: 1392–1405, 1977.

POLICH, J. M., AND A. STARR. Middle-, late-, and long-latency auditory evoked potentials. In: *Bases of Auditory Brain-Stem Evoked Responses*, edited by E. J. Moore. New York: Grune & Stratton, 1983, p. 345–361.

POLYAK, S. *The Retina.* Chicago: University of Chicago Press, 1941.

POLYAK, S. *The Vertebrate Visual System.* Chicago: University of Chicago Press, 1957.

POTTER, D. D., E. J. FURSHPAN, AND S. C. LANDIS. Transmitter status in cultured rat sympathetic neurons; plasticity and multiple function. *Fed. Proc.* 42: 1626–1632, 1983.

POWELL, T. P. S., W. M. COWAN, AND G. RAISMAN. The central olfactory connexions. *J. Anat. (London)* 99: 791–813, 1965.

POWERS, J. B., R. B. FIELDS, AND S. S. WILLIAMS. Olfactory and vomeronasal system participates in male hamsters' attraction to female vaginal secretions. *Physiol. Behav.* 22: 77–84, 1979.

PRECHT, W. Vestibular mechanisms. *Ann. Rev. Neurosci.* 2: 265–189, 1979.

PREILOWSKI, U., AND R. W. SPERRY. Minor hemisphere dominance in a bilateral competitive tactual word recognition task. *Biol. Ann. Rep. (California Institute of Technology)* p. 83–84, 1972.

PRICE, J. L. Structural organization of the olfactory pathways. In: *Olfaction and Taste VI*, edited by J. LeMagnen and P. MacLeod. London: Information Retrieval, Ltd., 1977, p. 87–96.

PRIGOGINE, I. *From Being to Becoming: Time and Complexity in the Physical Sciences.* San Francisco: W. H. Freeman, 1980.

PUBOLS, B. H., JR., AND L. M. PUBOLS. Tactile receptor discharge and mechanical properties of glabrous skin. *Fed. Proc.* 42: 2528–2535, 1983.

PURPURA, D. P. A neurohumoral mechanism of reticulocorti-

cal activation. *Am. J. Physiol.* 186: 250–254, 1956.

PURPURA, D. P. Comparative physiology of dendrites. In: *The Neurosciences. A Study Program*, edited by G. C. Quarton, T. Melnechuk, and F. O. Schmitt. New York: Rockefeller University Press, 1967, p. 372–393.

QUILLIAM, T. A. Unit design and array patterns in receptor organs. In: *Touch, Heat and Pain*, edited by A. V. S. de Reuck and J. Knight. Boston: Little, Brown, 1966, p. 86–112.

RACINE, R. J. Modification of seizure activity by electrical stimulation: I. After-discharge threshold. *Electroencephal. Clin. Neurol.* 32: 269–279, 1972a.

RACINE, R. J. Modification of seizure activity by electrical stimulation: II. Motor seizure. *Electroencephal. Clin. Neurol.* 32: 281–294, 1972b.

RAISMAN, G. An experimental study of the projections of the amygdala to the accessory olfactory bulb and its relationship to the concept of a dual olfactory system. *Exp. Brain.* 14: 395–408, 1972.

RAISMAN, G. Neural plasticity in the septal nuclei of the adult rat. *Brain Res.* 14: 25–48, 1969.

RAISMAN, G., AND P. M. FIELD. A quantitative investigation of the development of collateral reinnervation after partial denervation of the septal nuclei. *Brain Res.* 50: 241–264, 1973.

RAKIC, P. Developmental events leading to laminar and areal organization of the neocortex. In: *The Organization of the Cerebral Cortex*, edited by F. O. Schmitt, F. G. Worden, G. Adelman, and S. G. Dennis. Cambridge, MA: The MIT Press, 1981.

RAKIC, P. *Local Circuit Neurons.* Cambridge, MA: The MIT Press, 1975.

RALLS, K. Mammalian scent marking. *Science* 171: 443–449, 1971.

RALSTON, H. J., III. The synaptic organization of lemniscal projections to the ventrobasal thalamus of the cat. *Brain Res.* 14: 99–115, 1969.

RANSON, S. W. Some functions of the hypothalamus. *Harvey Lectures Ser.* 32: 92–121, 1936.

RANSON, S. W. Some functions of the hypothalamus. *Bull. N.Y. Acad. Med.* 13: 241–271, 1937.

RANSON, S. W., AND S. L. CLARK. *The Anatomy of the Nervous System*, 10th ed. Philadelphia: W. B. Saunders, 1959.

RASMUSSEN, G. L. Efferent fibers in the cochlear nerve and cochlear nucleus. In: *Neural Mechanisms of the Auditory and Vestibular System*, edited by G. L. Rasmussen and W. Windle. Springfield, IL: Charles C Thomas, 1960, p. 105–115.

RATOOSH, P. On interposition as a cue for the perception of distance. *Proc. Natl. Acad. Sci. U.S.A.* 35: 357–359, 1949.

RAY, R. H., AND L. KRUGER. Spatial properties of receptive field of mechano-sensitive primary afferent nerve fibers. *Fed. Proc.* 42: 2536–2541, 1983.

REALE, R. A., AND T. J. IMIG. Tonotopic organization in auditory cortex of the cat. *J. Comp. Neurol.* 192: 265–291, 1980.

REGER, S. N. Effect of middle ear muscle action on certain psycho-physical measurements. *Ann. Otol. Rhinol. Laryngol.* 69: 1179–1198, 1960.

REIVICH, M., D. KUHL, A. WOLF, J. GREENBERG, M. PHELPS, T. IDO, V. CASSELLA, J. FOWLER, E. HOFFMAN, A. ALAVI, P. SOM, AND L. SOKOLOFF. The [18F] fluoro-deoxyglucose method for the measurement of local cerebral glucose utilization in man. *Circ. Res.* 44: 127–137, 1979.

RETHELYI, M. Preterminal and terminal axon arborizations in the substantia gelatinosa of cat's spinal cord. *J. Comp. Neurol.* 172: 511–528, 1977.

REYNOLDS, D. V. Surgery in the rat during electrical analgesia induced by focal brain stimulation. *Science* 164: 444–445, 1969.

REXED, B. The cytoarchitectonic organization of the spinal cord in the cat. *J. Comp. Neurol.* 96: 415–495, 1952.

RHODES, M. L., AND W. V. GLENN, JR. An improved system for CT aided neurosurgery. *Proceedings International Symposium IEEE Computer Soc.*, p. 589–597, 1982.

RICARDO, J. A., AND E. T. KOH. Anatomical evidence of direct projections from the nucleus of the solitary tract to the hypothalamus, amygdala, and other forebrain structures in rat. *Brain Res.* 153: 1–26, 1978.

RIGGS, L. A. Light as a stimulus for vision. In: *Vision and Visual Perception*, edited by C. H. Graham. New York: Wiley, 1965a.

RIGGS, L. A. Visual acuity. In: *Vision and Visual Perception*, edited by C. H. Graham. New York: Wiley, 1965b.

RILEY, V. Psychoneuroendocrine influence on immunocompetence and neoplasia. *Science* 212: 1100–1109, 1981.

RING, G. C. The importance of the thyroid in maintaining an adequate production of heat during exposure to cold. *Am. J. Physiol.* 137: 582–588, 1942.

ROBERTS, J. L. Temperature acclimation and behavioral thermoregulation in cold-blooded animals. *Fed. Proc.* 33: 2155–2161, 1974.

ROBINSON, C. J., AND H. BURTON. Somatotopographic organization in the second somatosensory area of M. fascicularis. *J. Comp. Neurol.* 192: 43–67, 1980.

ROBINSON, C. J., AND H. BURTON. Organization of somatosensory receptive fields in cortical areas, 7b, retroinsula, postauditory and granular insula of M. fascicularis. *J. Comp. Neurol.* 192: 69–92, 1980.

ROBINSON, D. A. Eye movement control in primates. *Science* 161: 1219–1224, 1968.

ROBINSON, K., AND P. RUDGE. Abnormalities of the auditory evoked potentials in patients with multiple sclerosis. *Brain* 100: 19–40, 1977.

ROBINSON, K., AND P. RUDGE. The stability of the auditory evoked potentials in normal man and patients with multiple sclerosis. *J. Neurol. Sci.* 35: 147–156, 1978.

ROCKEL, A. J., R. W. HIORNS, AND T. P. S. POWELL. Numbers of neurons through full depth of neocortex. *J. Anat.* 118: 371, 1974.

ROSENBERG, R. N. Genetic variation in neurological disease. *Trends Neurosci.* 3: 144–148, 1980.

RUCH, T. C., AND H. A. SHENKIN. The relation of area 13 on the orbital surface of the frontal lobes to hyperactivity and hyperphagia in monkeys. *J. Neurophysiol.* 6: 349–360, 1943.

RUDA, M. A. Opiates and pain pathways: demonstration of enkephalin synapses and dorsal horn projection neurons. *Science* 215: 1523–1524, 1982.

RUSHTON, W. A. H. *Visual Pigments in Man. Sherrington Lectures.* Springfield, IL: Charles C Thomas, 1962, 38 pp.

RUSHTON, W. A. H. Visual pigments and color blindness. *Sci. Am.* 232: 64–74, 1975.

RUSSELL, I. J., AND P. M. SELLICK. Tuning properties of cochlear hair cells. *Nature (London)* 267: 858–860, 1977.

RUSSO, A. F., AND D. E. KOSHLAND, JR. Separation of signal transduction and adaptation functions of the aspartate receptor in bacterial sensing. *Science* 220: 1016–1020, 1983.

SANDRI, C., J. VAN BUREN, AND K. AKERT. *Membrane Morphology of the Vertebrate Nervous System: A Study with Freeze-Etch Techniques.* Amsterdam: Elsevier, 1977.

SALOMON, G., AND A. STARR. Electromyography of middle ear muscles in man during motor activities. *Acta Neurol. Scand.* 39: 161–168, 1963.

SCALIA, F., AND S. S. WINANS. The differential projections of the olfactory bulb and accessory olfactory bulb in mammals. *J. Comp. Neurol.* 161: 31–56, 1975.

SCHIFF, J., AND W. R. LOEWENSTEIN. Development of a receptor on a foreign nerve fiber in a pacinian corpuscle. *Science* 177: 712–715, 1972.

SCHMIDT, R. F., ed. *Fundamentals of Neurophysiology.* New York: Springer, 1978.

SCHMITT, F. O., P. DEV, AND B. H. SMITH. Electrotonic processing of information by brain cells. *Science* 193: 114–120, 1976.

SCHMITT, F. O., G. C. QUARTON, T. MELNECHUK, AND G. ADELMAN. *The Neurosciences: Second Study Program.* New York: The Rockefeller University Press, 1970.

SCHMITT, F. O., F. G. WORDEN, G. ADELMAN, AND S. G. DENNIS. *The Organization of the Cerebral Cortex.* Cambridge, MA: The MIT Press, 1981.

SCHNEIDER, D. Insect olfaction: deciphering system for chemical messages. *Science* 163: 1031–1037, 1969.

SCHULMAN-GALAMBOS, C., AND R. GALAMBOS. Brainstem auditory-evoked responses in premature infants. *J. Speech Hearing Res.* 18: 456–465, 1975.

SCHULTZE-WESTRUM, T. G. Social communication by chemical signals in flying phalanger. In: *Olfaction and Taste III: Proceedings of the Third International Symposium*, edited by C. Pfaffman. New York: Rockefeller University Press, 1969, p. 269–277.

SCHWARZ, W. Electron microscopic observation of the human vitreous body. In: *The Structure of the Eye*, edited by G. K. Smelser. New York: Academic Press, 1961.

SCOTT, J. W., AND C. M. LEONARD. The olfactory connections of the lateral hypothalamus in the rat, mouse and hamster. *J. Comp. Neurol.* 141: 331–344, 1971.

SECHZER, J. A. Successful interocular transfer of pattern discrimination in "split-brain" cats with shock-avoidance motivation. *J. Comp. Physiol. Psychol.* 58: 76–83, 1964.

SEMPLE, M. N., AND L. M. AITKIN. Representation of sound frequency and laterality by units in central nucleus of cat inferior colliculus. *J. Neurophysiol.* 42: 1626–1638, 1979.

SHARE, L. Control of plasma ADH titer in hemorrhage: role of atrial and arterial receptors. *Am. J. Physiol.* 215: 1384–1389, 1968.

SHARMAN, D. F. The catabolism of catecholamines: recent studies. *Brain Med. Bull.* 29: 110–115, 1973.

SHARP, F., J. KAUER, AND G. M. SHEPHERD. Local sites of activity related glucose metabolism in the rat olfactory bulb during olfactory stimulation. *Brain Res.* 98: 596–600, 1975.

SHEPHERD, G. M. The olfactory bulb as a simple cortical system: experimental analysis and functional implications. In: *The Neurosciences Second Study Program*, edited by F. O. Schmitt, G. C. Quarton, T. Melnechuk, and G. Adelman. New York: Rockefeller University Press, 1970, p. 539–552.

SHEPHERD, G. M. *Neurobiology.* Oxford, UK: Oxford Univ. Press, 1983.

SHERMAN, S. M., AND P. D. SPEAR. Organization of visual pathways in normal and visually deprived cats. *Physiol. Rev.* 62: 738–855, 1982.

SHERRINGTON, C. S. *The Integrative Action of the Nervous System.* London: Constable, 1906; Reprinted, with a new Forward, 1947.

SHIMAMURA, M. Longitudinal coordination between spinal and cranial reflex systems. *Exp. Neurol.* 8: 505–521, 1963.

SHIMAMURA, M., AND K. AKERT. Peripheral nervous relations of propriospinal and spino-bulbo-spinal reflex systems. *Jpn. J. Physiol.* 15: 638–647, 1965.

SHIMAMURA, M., AND R. B. LIVINGSTON. Longitudinal conduction systems serving spinal and brain-stem coordination. *J. Neurophysiol.* 26: 258–272, 1963.

SHIMAMURA, M., S. MORI, AND T. YAMAUCHI. Interactions of spino-bulbo-spinal reflexes with cortically evoked pyramidal and extrapyramidal activities. *Brain Res.* 4: 93–102, 1967.

SHIMAMURA, M., I. KOGURE, AND Y. IGUSA. Ascending spinal tracts of the spino-bulbo-spinal reflex in cats. *Jpn. J. Physiol.* 26: 577–589, 1976.

SIMMONS, F. B., R. GALAMBOS, AND A. RUPERT. Conditioned response of middle ear muscles. *Am. J. Physiol.* 197: 537–538, 1959.

SKINNER, B. F. The phylogeny and ontogeny of behavior. *Science* 153: 1205–1213, 1966.

SLOAN, L. L. Congenital achromatopsia: a report of 19 cases. *J. Opt. Soc. Am.* 44: 117–128, 1954.

SMITH, C. A., AND G. L. RASMUSSEN. Recent observations on the olivocochlear bundle. *Ann. Otol. Rhinol. Laryngol.* 72: 489–506, 1963.

SMITH, C. A., AND G. L. RASMUSSEN. Nerve endings in the maculae and cristae of the chinchilla vestibule, with a special reference to the efferents. In: *Third Symposium on the Role of the Vestibular Organs in Space Exploration.* National Aeronautic Space Administration, NASA SP-152: 183–200, 1967.

SMITH, C. A., O. H. LOWRY, AND M-L WU. The electrolytes of the labyrinthine fluids. *Laryngoscope* 64: 141–153, 1954.

SMITH, H. D. Audiometric effects of voluntary contraction of the tensor typmani muscle. *Arch. Otolaryngol.* 38: 369–372, 1943.

SNYDER, S. H., AND S. MATTHYSSE. Opiate receptor mechanisms. *Neurosci. Res. Prog. Bull.* 13: 3–166, 1975.

SOKOLOFF, L. Local cerebral circulation at rest and during altered cerebral activity induced by anesthesia or visual stimulation. In: *Regional Neurochemistry*, edited by S. S. Kety. Oxford, UK: Pergamon Press, 1961, p. 107–117.

SOKOLOFF, L. The relationship between function and energy metabolism: its use in the localization of functional activity in the nervous system. *Neurosci. Res. Prog. Bull.* 19: 159–210, 1981.

SOKOLOFF, L., M. REIVICH, C. KENNEDY, M. H. DES ROSIERS, C. S. PATLAK, K. D. PETTIGREW, O. SAKURADA, AND M. SHINOHARA. The [^{14}C]deoxyglucose method for the measurement of local cerebral glucose utilization: theory, procedure, and normal values in the conscious and anesthetized albino rat. *J. Neurochem.* 28: 897–916, 1977.

SOKOLOV, E. N. Learning and memory: habituation as negative learning. In: *Neural Mechanisms of Learning*, edited by M. R. Rosenzweig and E. L. Bennett. Cambridge, MA: The MIT Press, 1976, p. 475–479.

SOTELO, C., AND R. LLINAS. Specialized membrane junctions between neurons in the vertebrate cerebellar cortex. *J. Cell. Biol.* 53: 271–289, 1972.

SPEAR, P. D., R. E. KALIL, AND T. TONG. Functional compensation in lateral suprasylvian visual area following neonatal visual cortex removal in cats. *J. Neurophysiol.* 43: 851–869, 1980.

SPERRY, R. W. Cerebral organization and behavior. *Science* 133: 1749–1757, 1961.

SPERRY, R. W. Changing priorities. *Ann. Rev. Neurosci.* 4: 1–15, 1981.

SPERRY, R. W. Lateral specialization in the surgically separated hemispheres. In: *The Neurosciences Third Study Program*, edited by F. O. Schmitt and F. G. Worden. Cambridge, MA: The MIT Press, 1974, p. 5–19.

SPERRY, R. W., AND M. S. GAZZANIGA. Language following surgical disconnection of the hemispheres. In: *Brain Mechanisms Underlying Speech and Language*, edited by C. H. Millikan and F. L. Darley. New York: Grune & Stratton, 1967, p. 177–184.

SPOENDLIN, H. H. Ultrastructural studies of the labyrinth in squirrel monkeys. In: *The Role of Vestibular Organs in the Exploration of Space*. National Aeronautics Space Administration, NASA SP-77, 7–22, 1965.

SPRAGUE, J. M., H. C. HUGHES, AND G. BERLUCCHI. Cortical mechanisms in pattern and form perception. In: *Brain Mechanisms and Perceptual Awareness*, edited by O. Pompeiano and C. Ajmone Marsan. New York: Raven Press, 1981, p. 107–132.

SPRINGER, A. D. Retinopetal cells in the goldfish olfactory bulb. *Investig. Ophthal. Vis. Sci.* 22: 246, 1982 (See also DEMSKI AND NORTHCUTT, 1983).

STARK, L. *Neurological Control Systems: Studies in Bioengineering*. New York: Plenum Press, 1968.

STARR, A. Auditory brainstem responses in brain death. *Brain* 99: 543–554, 1976.

STARR, A. Influence of motor activity on clicked-evoked responses in the auditory pathway of waking cats. *Exp. Neurol.* 10: 191–204, 1964.

STARR, A. Regulatory mechanisms of the auditory pathway. In: *Modern Neurology*. Boston: Little Brown, 1969, p. 101–114.

STEINHAUSEN, W. Ueber die Beobachtung der Arpula in den Bogengangsampullen de Labyrinthes de lebenden Hechts. *Pfluegers Arch.* 232: 500–512, 1933.

STEVENS, S. S. *Handbook of Experimental Psychology*. New York: Wiley, 1951, ch. 29, p. 1143–1171.

STEVENS, S. S. Some similarities between hearing and seeing. *Laryngoscope* 68: 508–527, 1958.

STOLWIJK, J. A. J., AND J. D. HARDY. Temperature regulation in man—a theoretical study. *Pfluegers Arch.* 291: 129–162, 1966.

SWANSON, L. W., P. E. SAWCHENKO, AND W. M. COWAN. Evidence for collateral projections by neurons in ammon's horn, the dentate gyrus, and the subiculum: a multiple retrograde labeling study in the rat. *J. Neurosci.* 1: 548–559, 1981.

SWANSON, L. W., AND P. E. SAWCHENKO. Hypothalamic integration: organization of the paraventricular and supraoptic nuclei. *Ann. Rev. Neurosci.* 6: 269–324, 1983.

SZENTAGOTHAI, J. Downward causation? *Ann. Rev. Neurosci.* 7: 1–11, 1984.

SZENTAGOTHAI, J., AND M. A. ARBIB. *Conceptual Models of Neural Organization*. Cambridge, MA: The MIT Press, 1975.

TASAKI, I. Nerve impulses in individual auditory nerve fibers of guinea pig. *J. Neurophysiol.* 17: 97–122, 1954.

TASAKI, I., H. DAVIS, AND J. P. LEGOUIX. Space-time pattern of the cochlear microphonic as recorded by differential electrodes. *J. Acoust. Soc. Am.* 24: 502–519, 1952.

TAUB, E., M. HARGER, H. C. GRIER, AND W. HODOS. Some anatomical observations following chronic dorsal rhizotomy in monkeys. *Neuroscience* 5: 389–401, 1980.

TAUC, L. Nonvesicular release of neurotransmitter. *Physiol. Rev.* 62: 857–887, 1982.

TEPPERMAN, J., J. R. BROBECK, AND C. N. H. LONG. A study of experimental hypothalamic obesity in the rat. *Am. J. Physiol.* 133: P468–P469, 1941.

THATCHER, R. W., AND E. R. JOHN. *Foundations of Cognitive Processes*. Hillsdale, NJ: Lawrence W. Erlbaum Associates. New York: distributed by Halsted Press, 1977.

TIMM, C. Physikalische Vorgaenge bei der Labyrinthreizung. *Z. Laryngol. Rhinol. Otol.* 32: 237–251, 1953.

TOMITA, T. Electrical response of single photoreceptors. *Proc. IEEE* 56: 1015–1023, 1968.

TOREBJOERK, H. E., AND J. L. OCHOA. Specific sensations evoked by activity in single identified sensory units in man. *Acta Physiol. Scand.* 110: 445–447, 1980.

TOWER, S. Pyramidal lesion in the monkey. *Brain* 63: 36–90, 1940.

TRUEX, R. C., AND M. B. CARPENTER. *Human Neuroanatomy*. Baltimore: Williams & Wilkins, 1969, 673 pp.

TSCHERMAK-SEYSENEGG, A. Ueber Parallaktoskopie. *Pfluegers. Arch. Ges. Physiol.* 251: 454–469, 1939.

TSUKAHARA, N. Synaptic plasticity in the red nucleus. *J. Physiol. (Paris)* 74: 339–345, 1978.

TYLER, D. B., AND P. BARD. Motion sickness. *Physiol. Rev.* 29: 311–369, 1949.

UNGERSTEDT, U. Stereotaxic mapping of the monoamine pathways in the rat brain. *Acta Physiol. Scand. (Suppl. 367)* 82: 1–48, 1971.

UNGERSTEDT, U. Brain dopamine neurons and behavior. In: *The Neurosciences Third Study Program*, edited by F. O. Schmitt and F. G. Worden. Cambridge, MA: The MIT Press, 1974, p. 695–703.

UOTILA, U. U. On the role of the pituitary stalk in the regulation of the anterior pituitary, with special reference to the thyrotropic hormone. *Endocrinology* 25: 605–612, 1939.

VAN DER KOOY, D., H. G. J. M. KUYPERS, AND C. E. CATSMAN-BERREVOETS. Single mammillary cells with divergent axon collaterals. Demonstration by a simple fluorescent retrograde double-labeling technique in the rat. *Brain Res.* 158: 189–196, 1978.

VAN ESSEN, D. C. Visual areas of the mammalian cerebral cortex. *Ann. Rev. Neurosci.* 2: 227–263, 1979.

VAN HERREVELD, A., J. CROWELL, AND S. K. MALHOTRA. A study of extracellular space in central nervous tissue by freeze-substitution. *J. Cell. Biol.* 25: 117–137, 1965.

VANE, J. R. Inhibition of prostaglandin synthesis as a mechanism of action for aspirin-like drugs. *Nature New Biol.* 231: 232–235, 1971.

VERNEY, E. G. Antidiuretic hormone and the factors which determine its release. *Proc. R. Soc. (London) [Biol] Ser. B* 135: 25–106, 1947.

VINGRADOVA, O. S., E. S. BRAZHNIK, A. M. KARANOV, AND S. D. ZHADINA. Neuronal activity of the septum following various types of deafferentation. *Brain Res.* 187: 353–368,

1980; See also, VINGRADOVA, O. S. *Hippocampus and Memory.* Moscow: Nauka, 1975 (in Russian).

VON BEKESY, G. The variation of phase along the basilar membrane with sinusoidal vibrations. *J. Acoust. Soc. Am.* 19: 452–460, 1947.

VON BEKESY, G. *Experiments in Hearing.* New York: McGraw-Hill, 1960.

VON EULER, C., F. HERRERO, AND I. WEXLER. Control mechanisms determining rate and depth of respiratory movements. *Respir. Physiol.* 10: 93–108, 1970.

VON EULER, U. S. Adrenergic neurotransmitter functions. *Science* 173: 202–206, 1971.

WADA, J. A., T. OSAWA, AND T. MIZOGUCHI. Recurrent spontaneous seizure state induced by prefrontal kindling in Senegalese baboons, Papio papio. In: *Kindling,* edited by J. A. Wada. New York: Raven Press, 1976, p. 173–202.

WALD, G. Visual pigments and photoreceptors—review and outlook. *Exp. Eye Res.* 18: 333–343, 1974.

WALD, G., AND P. K. BROWN. Human color vision and color blindness. *Cold Spring Harbor Symp. Quant. Biol.* 30: 345–359, 1965.

WALD, G., P. K. BROWN, AND I. R. GIBBONS. The problem of visual excitation. *J. Opt. Soc. Am.* 53: 20–35, 1963.

WALD, G., AND A. B. CLARK. Visual adaptation and the chemistry of the rods. *J. Gen. Physiol.* 21: 93–105, 1937.

WALL, P. D. Mechanisms of plasticity of connection following damage in adult mammalian nervous system. In: *Recovery of Function. Theoretical Considerations for Brain Injury Rehabilitation,* edited by P. Bach-y-Rita. Bern, Switzerland, Huber, 1980, p. 91–105.

WALLS, G. L. *The Vertebrate Eye.* Bloomfield Hills, MI: Cranbrook Institute of Science, 1942.

WARR, W. B. Olivocochlear and vestibular efferent neurons of the feline brain stem: their location, morphology and number determined by retrograde axonal transport and acetylcholinesterase histochemistry. *J. Comp. Neurol.* 161: 159–182, 1975.

WEBSTER, W. R. The effects of repetitive stimulation on auditory evoked potentials. *Electroencephal. Clin. Neurophysiol.* 30: 318–330, 1971.

WEISKRANTZ, L. The interaction between occipital and temporal cortex in vision: an overview. In: *The Neurosciences Third Study Program,* edited by F. O. Schmitt and F. G. Worden. Cambridge, MA: The MIT Press, 1974, p. 189–204.

WERBLIN, F. S. The control of sensitivity in the retina. *Sci. Am.* 228: 70–79, 1973.

WERNICK, J. S., AND A. STARR. Binaural interaction in the superior olivary complex of the cat: an analysis of field potentials evoked by binaural-beat stimuli. *J. Neurophysiol.* 31: 428–441, 1968.

WERSÄLL, J. S. Epithelium of the cristae ampullares of the guinea pig. *Acta Otolaryngol. (Stockh.) (Suppl. 126)*: 1–85, 1956.

WERTHEIM, T. Uber die indirekte Sehscharfe. *A. Psychol.* 7: 172–187, 1965.

WEYMOUTH, F. W. The eye as an optical instrument. In: *A Textbook of Physiology,* 17th ed., edited by J. F. Fulton. Philadelphia: Saunders, 1955, ch. 23.

WIESEL, T. N., D. H. HUBEL, AND D. M. K. LAM. Autoradiographic demonstration of ocular-dominance columns in the monkey striate cortex by means of transneuronal transport. *Brain Res.* 79: 273–279, 1974.

WILLIAMS, M. Memory studies in electric convulsion therapy. *J. Neurol. Neurosurg. Psychiat.* 13: 30–35, 1950.

WILLIS, W. D., AND R. G. GROSSMAN. *Medical Neurobiology.* St. Louis: C. V. Mosby Co., 1981.

WILLOT, J. F., AND S-M. LU. Noise-induced hearing loss can alter neural coding and increase excitability in the central nervous system. *Science* 216: 1331–1332, 1982.

WITELSON, S. F., AND W. PALLIE. Left hemisphere specialization for language in the newborn. Neuroanatomical evidence of asymmetry. *Brain* 96: 641–646, 1973.

WOLF, A. V. *Thirst: Physiology of the Urge to Drink and Problems of Water Lack.* Springfield, IL: Charles C Thomas, 1958, 536 pp.

WOLFE, J. M. Hidden visual processes. *Sci. Am.* 248: 94–103, 1983.

WOLTER, J. R. The centrifugal nerves in human optic tract, chiasm, optic nerve, and retina. *Trans. Am. Ophthalmol. Soc.* 63: 678–707, 1965.

WOLTER, J. R., AND O. E. LUND. Reaction of centrifugal nerves in the human retina. *Am. J. Ophthalmol.* 66: 221–232, 1968.

WOODS, J. W., AND P. BARD. Antidiuretic hormone secretion in the cat with a chronically denervated hypothalamus. Copenhagen: Proc. 1st Int. Congr. Endocr., 1960, p. 113.

WOODSON, W. E. *Human Engineering Guide for Equipment Designers.* Los Angeles: University of California Press, 1954.

WOOLSEY, T. A., AND H. VAN DER LOOS. The structural organization of layer IV in the somatosensory region (SI) of mouse cerebral cortex. The description of a cortical field composed of discrete cytoarchitectonic units. *Brain Res.* 17: 205–242, 1970.

WRIGHT, J. W., AND J. W. HARDING. Recovery of olfactory function after bilateral bulbectomy. *Science* 216: 322–324, 1982.

YAKOVLEV, P. I. Motility, behavior and the brain. *J. Nerv. Ment. Dis.* 107: 313–335, 1948.

YAMAMOTO, W. S., F. S. GRODINS, L. STARK, L. D. PARTRIDGE, F. E. YATES, R. D. BRENNAN, J. URQUHART, AND H. T. MILHORN, JR. Application of control systems theory to physiology. Physiology Society Symposium. *Fed. Proc.* 28: 46–88, 1969.

YAMASHITA, H., AND K. KOIZUMI. Influence of carotid and aortic baroreceptors on neurosecretory neurons in supraoptic nuclei. *Brain Res.* 170: 259–277, 1979.

YELLOTT, J. I., JR. Binocular depth inversion. *Sci. Am.* 245: 148–152, 1981.

YIN, T. C., AND V. B. MOUNTCASTLE. Mechanisms of neural integration in the parietal lobe for visual attention. *Fed. Proc.* 37: 2251–2257, 1978.

YIN, T. C., AND V. B. MOUNTCASTLE. Visual input to the visuomotor mechanisms of the monkey's parietal lobe. *Science* 197: 1381–1383, 1977.

ZEKI, S. M. Colour coding in rhesus monkey prestriate cortex. *Brain Res.* 53: 422–427, 1973.

ZEKI, S. M. Cells responding to changing image size and disparity in the cortex of the rhesus monkey. *J. Physiol.* 242: 827–841, 1974.

ZEKI, S. M. Colour coding in the superior temporal sulcus of the rhesus monkey visual cortex. *Proc. Roy. Soc. (Biol)* 197: 195–223, 1977.

ZIMMER, J. Extended commissural and ipsilateral projections in postnatally deentorhinated hippocampus and fascia dentata demonstrated in rats by silver impregnation. *Brain Res.* 64: 293–311, 1973.

ZIMMER, J. Proximity as a factor in the regulation of aberrant aonal growth in postnatally deafferented fascia dentata. *Brain Res.* 72: 137–142, 1974.

ZOTTERMAN, Y. The neural mechanisms of taste. In: *Sensory Mechanisms,* edited by Y. Zotterman. Amsterdam: Elsevier, 1967, p. 139–154.

ZWISLOCKI, A. Analysis of some auditory characteristics. In: *Handbook of Mathematical Psychology,* edited by R. D. Luce, R. R. Bush, and E. Galanter. New York: Wiley, 1965, vol. 3.

Index

Page numbers in italics denote figures; those followed by "t" or "f" denote tables or footnotes, respectively.